Study Guide and Workbook: An Interactive Approach
With sections in this guide numbered to match the concept spreads in the text, you can focus on smaller amounts of material. Understanding and retention are increased as you respond to questions by writing in the spaces provided.
(0-534-53010-9)

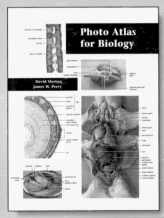

Study Skills for Science Students
Written specifically for science students, this book discusses how to develop good study habits, sharpen memory, learn more quickly, get the most out of lectures, prepare for tests, produce excellent term papers, and improve critical thinking skills.
(0-314-03983-X)

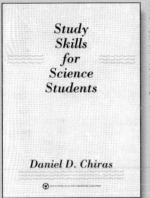

Photo Atlas for Biology
This invaluable aid not only helps you grasp the material more quickly in a lab setting, but also enables you to recall later what you've seen in the lab. By covering up the labels associated with the photos, you can effectively use the Atlas to study and review for exams.
(0-534-23556-5)

Electronic Study Guide
This study guide has an average of 40 multiple choice questions per chapter. It lets you respond to questions, and then an on-screen prompt allows you to review your answers and learn why they are correct or incorrect.
(Mac: 0-534-53017-6; IBM DOS: 0-534-53021-4; IBM Windows: 0-534-53016-8)

Answers to End-of-Chapter Review Questions
Contains detailed answers to the book's review questions. The self-quiz questions and genetics problems are answered in the book's appendix.
(0-534-53011-7)

Flash Cards for 1000 Glossary Terms
(0-534-16578-8)

Building Your Life Science Vocabulary
(0-534-26214-7)

Learning just got easier with your new CD-ROM and InfoTrac College Edition from Wadsworth Publishing Company. Please register your product today.

We want to hear from you! If you would like to be a student reviewer of these products for Wadsworth Publishing, please fill out this postcard and mail it to us so we can contact you.

Want to improve your grades?
Want to do well on quizzes and exams?

Wadsworth Publishing has developed a wide range of study tools to help you overcome the challenges of introductory biology. These tools facilitate your learning and understanding and will help you suceed.

To order any of these titles, talk with your instructor or ask your bookstore manager. Or you may order directly:

1 **Call us**   1-800-487-5510

2 **Fax us**   1-606-525-1543

3 **E-mail us**
   http://www.thomson.com/cgi-bin/plweb/search.cgi

Please indicate the quantity, title, and ISBN.

Name_____

School_____

Phone_____   E-mail_____

Did you use

Interactive Concepts in Biology CD-ROM?____ InfoTrac?____
Neither?____

How useful were these products? Please answer in the space below and continue your comments on the back of this postcard if necessary.

_____

_____

_____

_____

May we quote your remarks? Yes____ No____

PLACE
STAMP
HERE

Halee Dinsey
Wadsworth Publishing Company
10 Davis Drive
Belmont, CA 94002-3098

BIOLOGY

The Unity and Diversity of Life

EIGHTH EDITION

CECIE STARR / RALPH TAGGART

WADSWORTH PUBLISHING COMPANY

I(T)P® AN INTERNATIONAL THOMSON PUBLISHING COMPANY

Belmont, CA • Albany, NY • Bonn • Boston
Cincinnati • Detroit • Johannesburg • London • Madrid
Melbourne • Mexico City • New York
Paris • San Francisco • Singapore
Tokyo • Toronto • Washington

BIOLOGY PUBLISHER: Jack C. Carey

ASSISTANT EDITOR: Kristin Milotich

MEDIA PROJECT MANAGER: Pat Waldo

MARKETING MANAGER: Halee Dinsey

DEVELOPMENTAL EDITOR: Mary Arbogast

PROJECT EDITOR: Sandra Craig

EDITORIAL ASSISTANT: Michael Burgreen

PRINT BUYER: Karen Hunt

PRODUCTION: Mary Douglas, Rogue Valley Publications

TEXT AND COVER DESIGN, ART DIRECTION: Gary Head,
Gary Head Design

ART COORDINATOR: Myrna Engler-Forkner

EDITORIAL PRODUCTION: Mary Roybal, Karen Stough,
Susan Gall

PRIMARY ARTISTS: Raychel Ciemma; Precision Graphics
(Jan Troutt, J.C. Morgan)

ADDITIONAL ARTISTS: Robert Demarest, Darwen Hennings,
Vally Hennings, Betsy Palay, Nadine Sokol, Kevin Somerville,
Lloyd Townsend

PHOTO RESEARCH, PERMISSIONS: Stephen Forsling, Roberta Broyer

COVER PHOTOGRAPH: *Minnehaha Falls*, © Richard Hamilton Smith

COMPOSITION: Precision Graphics (Jim Gallagher, Kirsten
Dennison)

COLOR PROCESSING: H&S Graphics (Tom Anderson,
Nancy Dean, John Deady, Rich Stanislawski)

PRINTING AND BINDING: World Color, Versailles

BOOKS IN THE WADSWORTH BIOLOGY SERIES

Biology: Concepts and Applications, Third, Starr
Biology: The Unity and Diversity of Life, Eighth, Starr/Taggart
Human Biology, Second, Starr/McMillan
Laboratory Manual for Biology, Perry and Morton
General Botany, Rost et al.
Introduction to Biotechnology, Barnum
General Ecology, Krohne
Introduction to Microbiology, Ingraham/Ingraham
Living in the Environment, Tenth, Miller
Environmental Science, Seventh, Miller
Sustaining the Earth, Third, Miller
Environment: Problems and Solutions, Miller
Introduction to Cell and Molecular Biology, Wolfe
Molecular and Cellular Biology, Wolfe
Cell Ultrastructure, Wolfe
Marine Life and the Sea, Milne
Essentials of Oceanography, Garrison
Oceanography: An Invitation to Marine Science, Second, Garrison
Oceanography: An Introduction, Fifth, Ingmanson/Wallace
Plant Physiology, Fourth, Salisbury/Ross
Plant Physiology Laboratory Manual, Ross
Plants: An Evolutionary Survey, Second, Scagel et al.
Psychobiology: The Neuron and Behavior, Hoyenga/Hoyenga
Sex, Evolution, and Behavior, Second, Daly/Wilson
Dimensions of Cancer, Kupchella
Evolution: Process and Product, Third, Dodson/Dodson

Library of Congress Cataloging-in-Publication Data
Starr, Cecie.
    Biology : the unity and diversity of life / Cecie Starr, Ralph
Taggart.—8th ed.
        p.   cm.—(The Wadsworth biology series)
    Includes bibliographical references and index.
    ISBN 0-534-53001-X
    1. Biology.   I. Taggart, Ralph.   II. Title.   III. Series.
QH308.2.S72   1998
570—dc21   97-45639

For more information, contact Wadsworth Publishing Company,
10 Davis Drive, Belmont, California 94002, or electronically at
http://www.thomson.com/wadsworth.html

International Thomson Publishing Europe
Berkshire House 168-173, High Holborn
London, WC1V7AA, England

Thomas Nelson Australia
102 Dodds Street
South Melbourne 3205, Victoria, Australia

Nelson Canada
1120 Birchmount Road
Scarborough, Ontario, Canada M1K 5G4

International Thomson Editores
Campos Eliseos 385, Piso 7
Col. Polanco, 11560 México D.F. México

International Thomson Publishing GmbH
Königswinterer Strasse 418
53227 Bonn, Germany

International Thomson Publishing Asia
221 Henderson Road, #05-10 Henderson Building
Singapore 0315

International Thomson Publishing Japan
Hirakawacho Kyowa Building, 3F
2-2-1 Hirakawacho, Chiyoda-ku, Tokyo 102, Japan

International Thomson Publishing Southern Africa
Building 18, Constantia Park
240 Old Pretoria Road
Halfway House, 1685 South Africa

CONTENTS IN BRIEF

INTRODUCTION

1 Concepts and Methods in Biology 2

I PRINCIPLES OF CELLULAR LIFE

2 Chemical Foundations for Cells 20
3 Carbon Compounds in Cells 36
4 Cell Structure and Function 54
5 A Closer Look at Cell Membranes 80
6 Ground Rules of Metabolism 96
7 Energy-Acquiring Pathways 112
8 Energy-Releasing Pathways 130

II PRINCIPLES OF INHERITANCE

9 Cell Division and Mitosis 148
10 Meiosis 160
11 Observable Patterns of Inheritance 174
12 Chromosomes and Human Genetics 192
13 DNA Structure and Function 216
14 From DNA to Proteins 226
15 Controls Over Genes 242
16 Recombinant DNA and Genetic Engineering 256

III PRINCIPLES OF EVOLUTION

17 Emergence of Evolutionary Thought 270
18 Microevolution 280
19 Speciation 296
20 The Macroevolutionary Puzzle 310

IV EVOLUTION AND DIVERSITY

21 The Origin and Evolution of Life 332
22 Bacteria and Viruses 352
23 Protistans 370
24 Fungi 388
25 Plants 398
26 Animals: The Invertebrates 416
27 Animals: The Vertebrates 446
28 Human Evolution: A Case Study 468

V PLANT STRUCTURE AND FUNCTION

29 Plant Tissues 480
30 Plant Nutrition and Transport 500
31 Plant Reproduction 514
32 Plant Growth and Development 528

VI ANIMAL STRUCTURE AND FUNCTION

33 Tissues, Organ Systems, and Homeostasis 544
34 Information Flow and the Neuron 558
35 Integration and Control: Nervous Systems 570
36 Sensory Reception 588
37 Endocrine Control 608
38 Protection, Support, and Movement 626
39 Circulation 648
40 Immunity 670
41 Respiration 690
42 Digestion and Human Nutrition 710
43 The Internal Environment 730
44 Principles of Reproduction and Development 744
45 Human Reproduction and Development 762

VII ECOLOGY AND BEHAVIOR

46 Population Ecology 792
47 Community Interactions 812
48 Ecosystems 834
49 The Biosphere 854
50 Human Impact on the Biosphere 882
51 An Evolutionary View of Behavior 900

DETAILED CONTENTS

INTRODUCTION

1 **CONCEPTS AND METHODS IN BIOLOGY**

BIOLOGY REVISITED 2

1.1 **DNA, Energy, and Life** 4
Nothing Lives Without DNA 4
DNA and the Molecules of Life 4
The Heritability of DNA 4
Nothing Lives Without Energy 4
Energy Defined 4
Metabolism Defined 5
Sensing and Responding to Energy 5

1.2 **Energy and Life's Organization** 6
Levels of Biological Organization 6
Interdependencies Among Organisms 7

1.3 **So Much Unity, Yet So Many Species** 8

1.4 **An Evolutionary View of Diversity** 10
Mutation—Original Source of Variation 10
Evolution Defined 10
Natural Selection Defined 10

1.5 **The Nature of Biological Inquiry** 12
Observations, Hypotheses, and Tests 12
About the Word "Theory" 13

1.6 FOCUS ON SCIENCE: THE POWER AND PITFALLS OF EXPERIMENTAL TESTS 14

An Assumption of Cause and Effect 14
Experimental Design 14
Identifying Important Variables 15
Sampling Error 15
Bias in Reporting the Results 15

1.7 **The Limits of Science** 16

I PRINCIPLES OF CELLULAR LIFE

2 **CHEMICAL FOUNDATIONS FOR CELLS**

LEAFY CLEAN-UP CREWS 20

2.1 **Regarding the Atoms** 22
Structure of Atoms 22
Isotopes—Variant Forms of Atoms 23
When Atom Bonds With Atom 23

2.2 FOCUS ON SCIENCE: USING RADIOISOTOPES TO DATE THE PAST, TRACK CHEMICALS, AND SAVE LIVES 24
Radiometric Dating 24
Tracking Tracers 25
Saving Lives 25

2.3 **The Nature of Chemical Bonds** 26
Electrons and Energy Levels 26
Electrons and the Bonding Behavior of Atoms 27
From Atoms to Molecules 27

2.4 **Important Bonds in Biological Molecules** 28
Ion Formation and Ionic Bonding 28
Covalent Bonding 28
Hydrogen Bonding 29

2.5 **Properties of Water** 30
Polarity of the Water Molecule 30
Water's Temperature-Stabilizing Effects 30
Water's Cohesion 31
Water's Solvent Properties 31

2.6 **Acids, Bases, and Buffers** 32
The pH Scale 32
Acids and Bases 32
Buffers Against Shifts in pH 33
Salts 33

3 **CARBON COMPOUNDS IN CELLS**

CARBON, CARBON, IN THE SKY—ARE YOU SWINGING LOW AND HIGH? 36

3.1 **Properties of Organic Compounds** 38
Effects of Carbon's Bonding Behavior 38
Hydrocarbons and Functional Groups 39

3.2 **How Cells Use Organic Compounds** 40
Five Classes of Reactions 40
The Molecules of Life 40

3.3 FOCUS ON THE ENVIRONMENT: FOOD PRODUCTION AND A CHEMICAL ARMS RACE 41

3.4 **Carbohydrates** 42
The Simple Sugars 42
Short-Chain Carbohydrates 42
Complex Carbohydrates 42

3.5 **Lipids** 44
Fatty Acids 44
Triglycerides (Neutral Fats) 44
Phospholipids 45
Sterols and Their Derivatives 45
Waxes 45

3.6 **Amino Acids and the Primary Structure of Proteins** 46
Structure of Amino Acids 46
Primary Structure of Proteins 46

3.7   Emergence of the Three-Dimensional Structure of Proteins *48*
Second Level of Protein Structure *48*
Third Level of Protein Structure *48*
Fourth Level of Protein Structure *48*
Glycoproteins and Lipoproteins *49*
Structural Changes by Denaturation *49*

3.8   Nucleotides and Nucleic Acids *50*
Nucleotides With Roles in Metabolism *50*
Nucleic Acids—DNA and RNA *50*

4   CELL STRUCTURE AND FUNCTION

ANIMALCULES AND CELLS FILL'D WITH JUICES *54*

4.1   Basic Aspects of Cell Structure and Function *56*
Structural Organization of Cells *56*
The Lipid Bilayer of Cell Membranes *56*
Cell Size and Cell Shape *57*

4.2   FOCUS ON SCIENCE: MICROSCOPES—GATEWAYS TO THE CELL *58*
Light Microscopes 58
Electron Microscopes 58

4.3   The Defining Features of Eukaryotic Cells *60*
Major Cellular Components *60*
Typical Organelles in Plant Cells *60*
Typical Organelles in Animal Cells *60*

4.4   The Nucleus *64*
Nuclear Envelope *64*
Nucleolus *64*
Chromosomes *65*
What Happens to the Proteins Specified by DNA? *65*

4.5   The Cytomembrane System *66*
Endoplasmic Reticulum *66*
Golgi Bodies *66*
A Variety of Vesicles *67*

4.6   Mitochondria *68*

4.7   Specialized Plant Organelles *69*
Chloroplasts and Other Plastids *69*
Central Vacuole *69*

4.8   Components of the Cytoskeleton *70*
Microtubules *70*
Microfilaments *71*
Myosin and Other Accessory Proteins *71*
Intermediate Filaments *71*

4.9   The Structural Basis of Cell Motility *72*
Mechanisms of Cell Movements *72*
Case Study: Flagella and Cilia *72*

4.10 Cell Surface Specializations *74*
Eukaryotic Cell Walls *74*
Matrixes Between Animal Cells *74*
Cell-to-Cell Junctions *75*

4.11 Prokaryotic Cells—The Bacteria *76*

5   A CLOSER LOOK AT CELL MEMBRANES

IT ISN'T EASY BEING SINGLE *80*

5.1   Membrane Structure and Function *82*
The Lipid Bilayer of Cell Membranes *82*
Fluid Mosaic Model of Membrane Structure *82*
Overview of Membrane Proteins *83*

5.2   FOCUS ON SCIENCE: TESTING IDEAS ABOUT CELL MEMBRANES *84*
Testing Membrane Models 84
Observing Membrane Fluidity 85

5.3   How Substances Cross Cell Membranes *86*
Concentration Gradients and Diffusion *86*
Factors Influencing the Rate and Direction of Diffusion *86*
Mechanisms by Which Solutes Cross Cell Membranes *87*

5.4   The Directional Movement of Water Across Membranes *88*
Water Movement by Osmosis *88*
Effects of Tonicity *88*
Effects of Fluid Pressure *89*

5.5   Protein-Mediated Transport *90*
Passive Transport *90*
Active Transport *91*

5.6   Exocytosis and Endocytosis *92*
Transport To the Plasma Membrane *92*
Transport From the Plasma Membrane *92*
Membrane Cycling *93*

6   GROUND RULES OF METABOLISM

GROWING OLD WITH MOLECULAR MAYHEM *96*

6.1   Energy and the Underlying Organization of Life *98*
Defining Energy *98*
How Much Energy Is Available? *98*
The One-Way Flow of Energy *98*

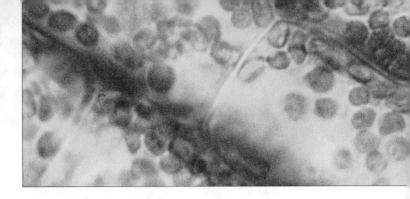

6.2   The Directional Nature of Metabolism *100*
Which Way Will a Reaction Run? *100*
No Vanishing Atoms at the End of the Run *101*
Energy Inputs Coupled With Outputs *101*

6.3   Energy Transfers and Cellular Work *102*
The Structure of ATP *102*
Phosphate-Group Transfers *102*
ATP Output and Metabolic Pathways *102*

6.4   Enzyme Structure and Function *104*
Four Features of Enzymes *104*
Enzyme-Substrate Interactions *104*

6.5   Factors Influencing Enzyme Activity *106*
Enzymes and Environmental Conditions *106*
Control of Enzyme Function *106*
Enzyme Helpers *107*

**6.6   Electron Transfers Through
Transport Systems** *108*

6.7   FOCUS ON SCIENCE: YOU LIGHT UP MY LIFE—
VISIBLE SIGNS OF METABOLIC ACTIVITY *109*
Bioluminescence *109*
Putting Bioluminescence To Use *109*

7   ENERGY-ACQUIRING PATHWAYS

SUN, RAIN, AND SURVIVAL *112*

7.1   Photosynthesis—An Overview *114*
Energy and Materials for the Reactions *114*
Where the Reactions Take Place *115*

7.2   Sunlight as an Energy Source *116*
Properties of Light *116*
Pigments—The Molecular Bridge From Sunlight
to Photosynthesis *117*

7.3   The Rainbow Catchers *118*
The Chemical Basis of Color *118*
On the Variety of Photosynthetic Pigments *118*
What Happens to the Absorbed Energy? *119*
About Those Roving Pigments *119*

7.4   The Light-Dependent Reactions *120*
The ATP-Producing Machinery *120*
Cyclic Pathway of ATP Formation *120*
Noncyclic Pathway of ATP Formation *120*
The Legacy—A New Atmosphere *121*

**7.5   A Closer Look at ATP Formation
in Chloroplasts** *122*

7.6   Light-Independent Reactions *123*
Capturing Carbon *123*
Building the Glucose Subunits *123*

7.7   Fixing Carbon—So Near, Yet So Far *124*
C4 Plants *124*
CAM Plants *125*

7.8   FOCUS ON THE ENVIRONMENT: AUTOTROPHS,
HUMANS, AND THE BIOSPHERE *126*

8   ENERGY-RELEASING PATHWAYS

THE KILLERS ARE COMING! THE KILLERS ARE COMING! *130*

8.1   How Cells Make ATP *132*
Comparison of the Main Types of Energy-Releasing
Pathways *132*
Overview of Aerobic Respiration *132*

**8.2   Glycolysis: First Stage of the
Energy-Releasing Pathways** *134*

8.3   Second Stage of the Aerobic Pathway *136*
Preparatory Steps and the Krebs Cycle *136*
Functions of the Second Stage *136*

8.4   Third Stage of the Aerobic Pathway *138*
Electron Transport Phosphorylation *138*
Summary of the Energy Harvest *138*

8.5   Anaerobic Routes of ATP Formation *140*
Fermentation Pathways *140*
Lactate Fermentation *140*
Alcoholic Fermentation *140*
Anaerobic Electron Transport *141*

**8.6   Alternative Energy Sources in the
Human Body** *142*
Carbohydrate Breakdown in Perspective *142*
The Fate of Glucose at Mealtime *142*
The Fate of Glucose Between Meals *142*
Energy From Fats *142*
Energy From Proteins *142*

8.7   COMMENTARY: PERSPECTIVE ON LIFE *144*

II   PRINCIPLES OF INHERITANCE

9   CELL DIVISION AND MITOSIS

SILVER IN THE STREAM OF TIME *148*

**9.1   Dividing Cells: The Bridge Between
Generations** *150*
Overview of Division Mechanisms *150*
Some Key Points About Chromosomes *150*
Mitosis and the Chromosome Number *150*

9.2   The Cell Cycle *151*

9.3   The Stages of Mitosis—An Overview *152*
Prophase: Mitosis Begins *152*
Transition to Metaphase *152*
From Anaphase Through Telophase *153*

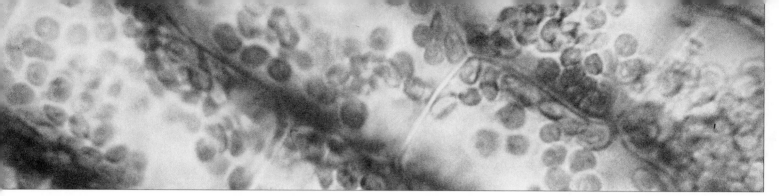

9.4   A Closer Look at the Cell Cycle *154*

The Wonder of Interphase *154*

Chromosomes, Microtubules, and the
Precision of Mitosis *154*

> *Organization of Metaphase Chromosomes 154*

> *Spindles Come, Spindles Go 154*

9.5   Division of the Cytoplasm *156*

Cell Plate Formation in Plants *156*

Cleavage of Animal Cells *157*

9.6   FOCUS ON SCIENCE: HENRIETTA'S
IMMORTAL CELLS *158*

10   MEIOSIS

OCTOPUS SEX AND OTHER STORIES *160*

**10.1   Comparison of Asexual and Sexual
Reproduction** *162*

**10.2   How Meiosis Halves the
Chromosome Number** *162*

Think "Homologues" *162*

Two Divisions, Not One *162*

10.3   A Visual Tour of the Stages of Meiosis *164*

10.4   Key Events of Meiosis I *166*

Prophase I Activities *166*

Metaphase I Alignments *167*

10.5   From Gametes to Offspring *168*

Gamete Formation in Plants *168*

Gamete Formation in Animals *168*

More Shufflings at Fertilization *168*

10.6   Meiosis and Mitosis Compared *170*

11   OBSERVABLE PATTERNS OF INHERITANCE

A SMORGASBORD OF EARS AND OTHER TRAITS *174*

**11.1   Mendel's Insight Into Patterns
of Inheritance** *176*

Mendel's Experimental Approach *176*

Some Terms Used in Genetics *177*

11.2   Mendel's Theory of Segregation *178*

Predicting the Outcome of Monohybrid Crosses *178*

Testcrosses *179*

11.3   Independent Assortment *180*

Predicting the Outcome of Dihybrid Crosses *180*

The Theory in Modern Form *181*

11.4   Dominance Relations *182*

Incomplete Dominance *182*

ABO Blood Types: A Case of Codominance *182*

11.5   Multiple Effects of Single Genes *183*

11.6   Interactions Between Gene Pairs *184*

Hair Color in Mammals *184*

Comb Shape in Poultry *185*

11.7   Less Predictable Variations in Traits *186*

Regarding the Unexpected Phenotype *186*

Continuous Variation in Populations *186*

**11.8   Examples of Environmental Effects
on Phenotype** *188*

12   CHROMOSOMES AND HUMAN GENETICS

TOO YOUNG TO BE OLD *192*

**12.1   The Chromosomal Basis of Inheritance—
An Overview** *194*

Genes and Their Chromosome Locations *194*

Autosomes and Sex Chromosomes *194*

Karyotype Analysis *194*

12.2   FOCUS ON SCIENCE: PREPARING A
KARYOTYPE DIAGRAM *195*

12.3   Sex Determination in Humans *196*

**12.4   Early Questions About
Gene Locations** *198*

Linked Genes—Clues to Inheritance
Patterns *198*

Crossing Over and Genetic Recombination *199*

**12.5   Recombination Patterns and
Chromosome Mapping** *200*

How Close Is Close? A Question of
Recombination Frequencies *200*

Linkage Mapping *201*

12.6   Human Genetic Analysis *202*

Constructing Pedigrees *202*

Regarding Human Genetic Disorders *202*

12.7   Patterns of Autosomal Inheritance *204*

Autosomal Recessive Inheritance *204*

Autosomal Dominant Inheritance *204*

12.8   Patterns of X-Linked Inheritance *206*

X-Linked Recessive Inheritance *206*

X-Linked Dominant Inheritance *207*

A Few Qualifications *207*

12.9   Changes in Chromosome Number *208*
Categories and Mechanisms of Change *208*
Change in the Number of Autosomes *208*
Change in the Number of Sex Chromosomes *209*
   Turner Syndrome 209
   Klinefelter Syndrome 209
   XYY Condition 209

12.10   Changes in Chromosome Structure *210*
Deletion *210*
Duplications *210*
Inversion *211*
Translocation *211*

12.11   FOCUS ON SCIENCE: PROSPECTS
IN HUMAN GENETICS *212*

---

13   DNA STRUCTURE AND FUNCTION

CARDBOARD ATOMS AND BENT-WIRE BONDS *216*

13.1   Discovery of DNA Function *218*
Early and Puzzling Clues *218*
Confirmation of DNA Function *218*

13.2   DNA Structure *220*
Components of DNA *220*
Patterns of Base Pairing *221*

13.3   DNA Replication and Repair *222*
How a DNA Molecule Gets Duplicated *222*
Monitoring and Fixing the DNA *222*
Regarding the Chromosomal Proteins *223*

13.4   FOCUS ON HEALTH: WHEN DNA CAN'T BE FIXED *224*

---

14   FROM DNA TO PROTEINS

BEYOND BYSSUS *226*

14.1   FOCUS ON SCIENCE: DISCOVERING THE CONNECTION
BETWEEN GENES AND PROTEINS *228*

14.2   Transcription of DNA Into RNA *230*
The Three Classes of RNA *230*
How RNA Is Assembled *230*
Finishing Touches on the mRNA Transcripts *230*

14.3   Deciphering the mRNA Transcripts *232*
The Genetic Code *232*
Roles of tRNA and rRNA *232*

14.4   Stages of Translation *234*

14.5   How Mutations Affect Protein Synthesis *236*
Common Types of Gene Mutations *236*
Causes of Gene Mutations *236*
The Proof Is in the Protein *237*

---

15   CONTROLS OVER GENES

HERE'S TO SUICIDAL CELLS! *242*

15.1   Overview of Gene Controls *244*

15.2   Controls in Bacterial Cells *244*
Negative Control of Transcription *244*
Positive Control of Transcription *245*

15.3   Controls in Eukaryotic Cells *246*
A Case of Cell Differentiation *246*
Selective Gene Expression at Many Levels *246*
   Controls Related to Transcription 246
   Post-Transcriptional Controls 246

15.4   Evidence of Gene Control *248*
Transcription in Lampbrush Chromosomes *248*
X Chromosome Inactivation *249*

15.5   Examples of Signaling Mechanisms *250*
Hormonal Signals *250*
Sunlight as a Signal *250*

15.6   FOCUS ON SCIENCE: CANCER AND THE CELLS
THAT FORGET TO DIE *252*
   The Cell Cycle Revisited 252
   Characteristics of Cancer 252

**16   RECOMBINANT DNA AND
GENETIC ENGINEERING**

MOM, DAD, AND CLOGGED ARTERIES *256*

**16.1   Recombination in Nature—
And in the Laboratory** *258*
Plasmids, Restriction Enzymes,
and the New Technology *258*
Producing Restriction Fragments *259*

16.2   Working With DNA Fragments *260*
Amplification Procedures *260*
Sorting Out Fragments of DNA *260*
DNA Sequencing *261*

16.3   FOCUS ON SCIENCE: RIFF-LIPS AND
DNA FINGERPRINTS *262*

16.4   Modified Host Cells *263*
Use of DNA Probes *263*
Use of cDNA *263*

**16.5   Bacteria, Plants, and the New
Technology** *264*
Genetically Engineered Bacteria *264*
Genetically Engineered Plants *264*

16.6   Genetic Engineering of Animals *266*
Supermice and Biotech Barnyards *266*
Applying the New Technology
to Humans *266*

16.7   FOCUS ON BIOETHICS: REGARDING
GENE THERAPY *267*

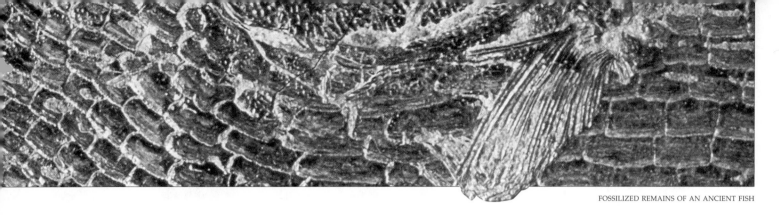

FOSSILIZED REMAINS OF AN ANCIENT FISH

III   PRINCIPLES OF EVOLUTION

17   EMERGENCE OF EVOLUTIONARY THOUGHT

FIRE, BRIMSTONE, AND HUMAN HISTORY 270

17.1   Early Beliefs, Confounding Discoveries 272
The Great Chain of Being 272
Questions From Biogeography 272
Questions From Comparative Anatomy 272
Questions About Fossils 273

17.2   A Flurry of New Theories 274
Squeezing New Evidence Into Old Beliefs 274
Voyage of the *Beagle* 274

17.3   Darwin's Theory Takes Form 276
Old Bones and Armadillos 276
A Key Insight—Variation in Traits 276

17.4   FOCUS ON SCIENCE: REGARDING THE FIRST
OF THE "MISSING LINKS" 278

18   MICROEVOLUTION

DESIGNER DOGS 280

18.1   Individuals Don't Evolve—Populations Do 282
Examples of Variation in Populations 282
The "Gene Pool" 282
Stability and Change in Allele Frequencies 283
Mutations Revisited 283

18.2   FOCUS ON SCIENCE: WHEN IS A POPULATION
NOT EVOLVING? 284

18.3   Natural Selection Revisited 285

**18.4   Directional Change in the Range
of Variation** 286
Directional Selection 286
The Case of the Peppered Moths 286
Pesticide Resistance 286
Antibiotic Resistance 287

**18.5   Selection Against or In Favor of
Extreme Phenotypes** 288
Stabilizing Selection 288
Disruptive Selection 289

18.6   Special Types of Selection 290
Sexual Selection 290
Balancing Selection 290
Sickle-Cell Anemia—Lesser of Two Evils? 290

18.7   Gene Flow 291

18.8   Genetic Drift 292
Chance Events and Population Size 292
Bottlenecks and the Founder Effect 292
Genetic Drift and Inbred Populations 293

19   SPECIATION

THE CASE OF THE ROAD-KILLED SNAILS 296

19.1   On the Road to Speciation 298
What Is a Species? 298
  Morphological Species Concept 298
  Biological Species Concept 298
Genetic Change and Speciation 299

19.2   Reproductive Isolating Mechanisms 300
Prezygotic Isolation 300
Postzygotic Isolation 301

**19.3   Speciation in Geographically Isolated
Populations** 302
Allopatric Speciation Defined 302
The Pace of Geographic Isolation 302
Allopatric Speciation on Archipelagos 303

19.4   Models for Other Speciation Routes 304
Sympatric Speciation 304
  Evidence From Cichlids in Africa 304
  Speciation by Way of Polyploidy 304
Parapatric Speciation 305

19.5   Patterns of Speciation 306
Branching and Unbranched Evolution 306
Evolutionary Trees and Rates of Change 306
Adaptive Radiations 307
Extinctions—End of the Line 307

20   THE MACROEVOLUTIONARY PUZZLE

OF FLOODS AND FOSSILS 310

20.1   Fossils—Evidence of Ancient Life 312
Fossilization 312
Interpreting the Geologic Tombs 313
Interpreting the Fossil Record 313

20.2   Evidence of a Changing Earth 314
An Outrageous Hypothesis 314
Drifting Continents and Changing Seas 315

20.3 **Evidence From Comparative Embryology** *316*
Developmental Program of Larkspurs *316*
Developmental Program of Vertebrates *316*

20.4 **Evidence of Morphological Divergence** *318*
Homologous Structures *318*
Potential Confusion From Analogous Structures *318*

20.5 **Evidence From Comparative Biochemistry** *320*
Molecular Clocks *320*
Protein Comparisons *321*
Nucleic Acid Comparisons *321*

20.6 **Identifying Species, Past and Present** *322*
Assigning Names to Species *322*
Groupings of Species—The Higher Taxa *323*
Putting Classification Schemes to Work *323*

20.7 **Finding Evolutionary Relationships Among Species** *324*
The Systematics Approach *324*
Actually, There Is No Single Systematics Approach *324*

20.8 FOCUS ON SCIENCE: CONSTRUCTING A CLADOGRAM *326*

20.9 **How Many Kingdoms?** *328*
Whittaker's Five-Kingdom Scheme *328*
Along Came the Archaebacteria *328*

IV EVOLUTION AND DIVERSITY

21 THE ORIGIN AND EVOLUTION OF LIFE

IN THE BEGINNING . . . *332*

21.1 **Conditions on the Early Earth** *334*
Origin of the Earth *334*
The First Atmosphere *334*
Synthesis of Organic Compounds *334*

21.2 **Emergence of the First Living Cells** *336*
Origin of Agents of Metabolism *336*
Origin of Self-Replicating Systems *336*
Origin of the First Plasma Membranes *337*

21.3 **Origin of Prokaryotic and Eukaryotic Cells** *338*

21.4 FOCUS ON SCIENCE: WHERE DID ORGANELLES COME FROM? *340*
Origin of the Nucleus and ER *340*
A Theory of Endosymbiosis *340*
Evidence of Endosymbiosis *341*

21.5 **Life in the Paleozoic Era** *342*

21.6 **Life in the Mesozoic Era** *344*
Speciation on a Grand Scale *344*
Rise of the Ruling Reptiles *344*

21.7 FOCUS ON SCIENCE: HORRENDOUS END TO DOMINANCE *346*

21.8 **Life in the Cenozoic Era** *347*

21.9 **Summary of Earth and Life History** *348*

22 BACTERIA AND VIRUSES

THE UNSEEN MULTITUDES *352*

22.1 **Characteristics of Bacteria** *354*
Splendid Metabolic Diversity *354*
Bacterial Sizes and Shapes *354*
Structural Features *355*

22.2 **Bacterial Growth and Reproduction** *356*
The Nature of Bacterial Growth *356*
Prokaryotic Fission *356*
Bacterial Conjugation *357*

22.3 **Bacterial Classification** *357*

22.4 **Eubacteria** *358*
A Sampling of Diversity *358*
Regarding the "Simple" Bacteria *359*

22.5 **Archaebacteria** *360*

22.6 **Summary of Major Bacterial Groups** *361*

22.7 **The Viruses** *362*
Defining Characteristics *362*
Examples of Viruses *362*
Infectious Agents Tinier Than Viruses *363*

22.8 **Viral Multiplication Cycles** *364*

22.9 FOCUS ON HEALTH: THE NATURE OF INFECTIOUS DISEASES *366*
Categories of Diseases *366*
An Evolutionary View *366*
Ebola and Other Emerging Pathogens *366*
On the Drug-Resistant Strains *367*
Pathogens Lurking in the Kitchen *367*
Prions Lurking in Sheep and Cows *367*

23 PROTISTANS

KINGDOM AT THE CROSSROADS *370*

23.1 **Parasitic and Predatory Molds** *372*
Chytrids *372*
Water Molds *372*
Slime Molds *372*

23.2 **The Animal-like Protistans** *374*

23.3 **Amoeboid Protozoans** *374*
Rhizopods *374*
Actinopods *375*

LICHENS AND PLANTS ON A FOREST FLOOR

23.4 **Major Parasites—Animal-like Flagellates and Sporozoans** *376*
Animal-Like Flagellates *376*
Representative Sporozoans *376*

23.5 FOCUS ON HEALTH: MALARIA AND THE NIGHT-FEEDING MOSQUITOES *377*

23.6 **A Sampling of Ciliated Protozoans** *378*

23.7 **A Sampling of the (Mostly) Single-Celled Algae** *380*
Euglenoids *380*
Chrysophytes *380*
Dinoflagellates *381*

23.8 **Red Algae** *382*

23.9 **Brown Algae** *383*

23.10 **Green Algae** *384*

24 FUNGI

DRAGON RUN *388*

24.1 **Characteristics of Fungi** *390*
Major Groups of Fungi *390*
Nutritional Modes *390*
Key Features of Fungal Life Cycles *390*

24.2 **Consider the Club Fungi** *390*
A Sampling of Spectacular Diversity *390*
Example of a Fungal Life Cycle *391*

24.3 **Spores and More Spores** *392*
Producers of Zygosporangia *392*
Producers of Ascospores *392*
Elusive Spores of the Imperfect Fungi *393*

24.4 **Beneficial Associations Between Fungi and Plants** *394*
Lichens *394*
Mycorrhizae *395*
As Fungi Go, So Go the Forests *395*

24.5 FOCUS ON SCIENCE: A LOOK AT THE UNLOVED FEW *396*

25 PLANTS

PIONEERS IN A NEW WORLD *398*

25.1 **Evolutionary Trends Among Plants** *400*
Overview of the Plant Kingdom *400*
Evolution of Roots, Stems, and Leaves *400*
From Haploid to Diploid Dominance *400*
Evolution of Pollen and Seeds *401*

25.2 **Bryophytes** *402*

25.3 **Existing Seedless Vascular Plants** *404*
Whisk Ferns *404*
Lycophytes *404*
Horsetails *404*
Ferns *405*

25.4 FOCUS ON SCIENCE: ANCIENT CARBON TREASURES *406*

25.5 **The Rise of the Seed-Bearing Plants** *407*

25.6 **Gymnosperms—Plants With "Naked" Seeds** *408*
Conifers *408*
Lesser-Known Gymnosperms *408*
Cycads *408*
Ginkgos *409*
Gnetophytes *409*

25.7 **A Closer Look at the Conifers** *410*

25.8 **Angiosperms—The Flowering, Seed-Bearing Plants** *412*
Dicots and Monocots *412*
Key Aspects of the Life Cycles *413*

26 ANIMALS: THE INVERTEBRATES

MADELEINE'S LIMBS *416*

26.1 **Overview of the Animal Kingdom** *418*
General Characteristics of Animals *418*
Diversity in Body Plans *418*
Body Symmetry and Cephalization *418*
Type of Gut *419*
Body Cavities *419*
Segmentation *419*

26.2 **Puzzles About Origins** *420*

26.3 **Sponges—Success in Simplicity** *420*

26.4 **Cnidarians—Tissues Emerge** *422*
Regarding the Nematocysts *422*
Cnidarian Body Plans *422*

26.5 **Variations on the Cnidarian Body Plan** *424*
Various Stages in Cnidarian Life Cycles *424*
A Sampling of Colonial Types *424*

26.6 **Comb Jellies** *425*

26.7   **Acoelomate Animals—And the Simplest Organ Systems** *426*

Flatworms *426*

   Turbellarians 426

   Flukes 426

   Tapeworms 427

Ribbon Worms *427*

26.8   **Roundworms** *428*

26.9   FOCUS ON SCIENCE: A ROGUE'S GALLERY OF WORMS *428*

26.10   **Rotifers** *430*

26.11   **Two Major Divergences** *430*

26.12   **A Sampling of Molluscan Diversity** *431*

26.13   **Evolutionary Experiments With Molluscan Body Plans** *432*

Twisting and Detwisting of Soft Bodies *432*

Hiding Out, One Way or Another *432*

On the Cephalopod Need for Speed *433*

26.14   **Annelids—Segments Galore** *434*

Advantages of Segmentation *434*

Annelid Adaptations—A Case Study *435*

26.15   **Arthropods—The Most Successful Organisms on Earth** *436*

Arthropod Diversity *436*

Adaptations of Insects and Other Arthropods *436*

26.16   **A Look at Spiders and Their Kin** *437*

26.17   **A Look at the Crustaceans** *438*

26.18   *How* **Many Legs?** *439*

26.19   **A Look at Insect Diversity** *440*

26.20   **The Puzzling Echinoderms** *442*

27   **ANIMALS: THE VERTEBRATES**

   MAKING DO (RATHER WELL) WITH WHAT YOU'VE GOT *446*

27.1   **The Chordate Heritage** *448*

Characteristics of Chordates *448*

Chordate Classification *448*

27.2   **Invertebrate Chordates** *448*

Tunicates *448*

Lancelets *448*

27.3   **Evolutionary Trends Among the Vertebrates** *450*

Puzzling Origins, Portentous Trends *450*

The First Vertebrates *451*

27.4   **Existing Jawless Fishes** *451*

27.5   **Existing Jawed Fishes** *452*

Cartilaginous Fishes *452*

Bony Fishes *452*

27.6   **Amphibians** *454*

Origin of Amphibians *454*

Salamanders *455*

Frogs and Toads *455*

Caecilians *455*

27.7   **The Rise of Reptiles** *456*

27.8   **Major Groups of Existing Reptiles** *458*

Crocodilians *458*

Turtles *458*

Lizards and Snakes *458*

Tuataras *459*

27.9   **Birds** *460*

27.10   **The Rise of Mammals** *462*

Distinctly Mammalian Traits *462*

Mammalian Origins and Radiations *463*

27.11   **A Portfolio of Existing Mammals** *464*

Cases of Convergent Evolution *464*

Human Impact on Mammalian Diversity *464*

28   **HUMAN EVOLUTION: A CASE STUDY**

   THE CAVE AT LASCAUX AND THE HANDS OF GARGAS *468*

28.1   **Evolutionary Trends Among the Primates** *470*

Primate Classification *470*

Key Evolutionary Trends *470*

   Enhanced Daytime Vision 470

   Upright Walking 471

   Power Grip and Precision Grip 471

   Teeth for All Occasions 471

28.2   **From Primates to Hominids** *472*

Origins and Early Divergences *472*

The First Hominids *472*

28.3   **Emergence of Early Humans** *474*

Defining "Human" *474*

Homo habilis and the First Stone Tools *474*

From *Homo erectus* to *Homo sapiens* *475*

28.4   FOCUS ON SCIENCE: OUT OF AFRICA—ONCE, TWICE, OR . . . *476*

V   **PLANT STRUCTURE AND FUNCTION**

29   **PLANT TISSUES**

   PLANTS VERSUS THE VOLCANO *480*

29.1   **Overview of the Plant Body** *482*

Shoots and Roots *482*

Three Plant Tissue Systems *482*

Meristems—Where Tissues Originate *483*

FLOWERING PLUM BUSILY ENTICING POLLINATORS

29.2 Types of Plant Tissues *484*

Simple Tissues *484*

Complex Tissues *484*

   Vascular Tissues 484

   Dermal Tissues 485

Dicots and Monocots—Same Tissues,
Different Features *485*

29.3 Primary Structure of Shoots *486*

How Stems and Leaves Form *486*

Internal Structure of Stems *486*

29.4 A Closer Look at Leaves *488*

Similarities and Differences
Among Leaves *488*

The Fine Structure of Leaves *488*

   Leaf Epidermis 488

   Mesophyll—Photosynthetic Ground Tissue 488

   Veins—The Leaf's Vascular Bundles 489

29.5 COMMENTARY: USES AND ABUSES OF SHOOTS *490*

29.6 Primary Structure of Roots *492*

Taproot and Fibrous Root Systems *492*

Internal Structure of Roots *492*

Regarding the Sidewalk-Buckling,
Record-Breaking Root Systems *493*

**29.7 Accumulated Secondary Growth—
The Woody Plants** *494*

Woody and Nonwoody Plants Compared *494*

Activity at the Vascular Cambium *494*

29.8 A Look at Wood and Bark *496*

Formation of Bark *496*

Heartwood and Sapwood *496*

Early Wood, Late Wood, and Tree Rings *496*

Limits to Secondary Growth *497*

30 PLANT NUTRITION AND TRANSPORT

FLIES FOR DINNER *500*

30.1 Soil and Its Nutrients *502*

Properties of Soil *502*

Nutrients Essential for Plant Growth *502*

Leaching and Erosion *503*

**30.2 Absorption of Water and Mineral Ions
Into Roots** *504*

Absorption Routes *504*

Specialized Absorptive Structures *504*

   Root Hairs 504

   Mycorrhizae 505

30.3 A Theory of Water Transport Through Plants *506*

Transpiration Defined *506*

Cohesion-Tension Theory of
Water Transport *506*

30.4 Conservation of Water in Stems and Leaves *508*

The Water-Conserving Cuticle *508*

Controlled Water Loss at Stomata *508*

**30.5 Distribution of Organic Compounds
Through the Plant** *510*

Translocation *510*

Pressure Flow Theory *510*

31 PLANT REPRODUCTION

A COEVOLUTIONARY TALE *514*

**31.1 Reproductive Structures of
Flowering Plants** *516*

Think Sporophyte and Gametophyte *516*

Components of Flowers *516*

Where Pollen and Eggs Develop *517*

31.2 FOCUS ON THE ENVIRONMENT: POLLEN
SETS ME SNEEZING *517*

31.3 A New Generation Begins *518*

From Microspores to Pollen Grains *518*

From Megaspores to Eggs *518*

From Pollination to Fertilization *518*

31.4 From Zygote to Seeds and Fruits *520*

Formation of the Embryo Sporophyte *520*

Formation of Seeds and Fruits *520*

31.5 Dispersal of Fruits and Seeds *522*

31.6 FOCUS ON SCIENCE: WHY SO MANY FLOWERS
AND SO FEW FRUITS? *523*

**31.7 Asexual Reproduction of
Flowering Plants** *524*

Asexual Reproduction in Nature *524*

Induced Propagation *524*

32 PLANT GROWTH AND DEVELOPMENT

FOOLISH SEEDLINGS AND GORGEOUS GRAPES *528*

**32.1 Patterns of Early Growth and
Development—An Overview** *530*

Seed Germination *530*

Genetic Programs, Environmental Cues *530*

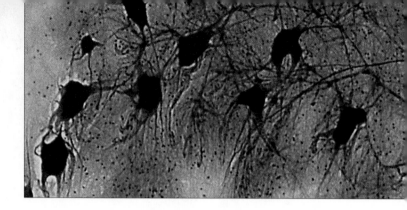

32.2 Hormonal Effects on Plant Growth and Development *532*

Growth Versus Development *532*

Categories of Plant Hormones *532*

    Gibberellins 532

    Auxins 532

    Cytokinins 533

    Ethylene 533

32.3 Adjustments in the Rate and Direction of Growth *534*

The Tropisms *534*

    Gravitropism 534

    Phototropism 534

    Thigmotropism 535

Responses to Mechanical Stress *535*

32.4 Biological Clocks and Their Effects *536*

Rhythmic Leaf Movements *536*

Flowering—A Case of Photoperiodism *536*

32.5 Life Cycles End, and Turn Again *538*

Senescence *538*

Entering Dormancy *538*

Vernalization *539*

Breaking Dormancy *539*

32.6 FOCUS ON SCIENCE: THE RISE AND FALL OF A GIANT *540*

VI   ANIMAL STRUCTURE AND FUNCTION

33   TISSUES, ORGAN SYSTEMS, AND HOMEOSTASIS

MEERKATS, HUMANS, IT'S ALL THE SAME *544*

33.1 Epithelial Tissue *546*

General Characteristics *546*

Cell-to-Cell Contacts *546*

Glandular Epithelium *547*

33.2 Connective Tissue *548*

Connective Tissue Proper *548*

Specialized Connective Tissue *548*

33.3 Muscle Tissue *550*

33.4 Nervous Tissue *551*

33.5 FOCUS ON SCIENCE: FRONTIERS IN TISSUE RESEARCH *551*

33.6 Organ Systems *552*

Overview of the Major Organ Systems *552*

Tissue and Organ Formation *553*

33.7 Homeostasis and Systems Control *554*

Concerning the Internal Environment *554*

Mechanisms of Homeostasis *554*

    Negative Feedback 554

    Positive Feedback 555

34   INFORMATION FLOW AND THE NEURON

TORNADO! 558

34.1 Neurons—The Communication Specialists *560*

Functional Zones of a Neuron *560*

A Neuron at Rest, Then Moved to Action *560*

Restoring and Maintaining Readiness *560*

34.2 A Closer Look at Action Potentials *562*

Approaching Threshold *562*

An All-or-Nothing Spike *562*

Propagation of Action Potentials *563*

34.3 Chemical Synapses *564*

A Smorgasbord of Signals *564*

Synaptic Integration *565*

Removing Neurotransmitter From the Synaptic Cleft *565*

34.4 Paths of Information Flow *566*

Blocks and Cables of Neurons *566*

Reflex Arcs *566*

34.5 FOCUS ON HEALTH: SKEWED INFORMATION FLOW *568*

35   INTEGRATION AND CONTROL: NERVOUS SYSTEMS

WHY CRACK THE SYSTEM? *570*

35.1 Invertebrate Nervous Systems *572*

Life-styles and Nervous Systems *572*

Regarding the Nerve Net *572*

On the Importance of Having a Head *572*

35.2 Vertebrate Nervous Systems— An Overview *574*

Evolutionary High Points *574*

Functional Divisions of the Vertebrate Nervous System *574*

35.3 The Major Expressways *576*

Peripheral Nervous System *576*

    Somatic and Autonomic Subdivisions 576

    Sympathetic and Parasympathetic Nerves 576

The Spinal Cord *577*

35.4 Functional Divisions of the Vertebrate Brain *578*

Hindbrain *578*

Midbrain *578*

Forebrain *579*

The Reticular Formation *579*

Brain Cavities and Canals *579*

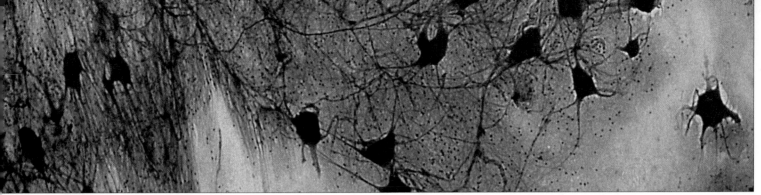

35.5 **A Closer Look at the Human Cerebrum** *580*

Organization of the Cerebral Hemispheres *580*

Connections With the Limbic System *581*

35.6 FOCUS ON SCIENCE: SPERRY'S SPLIT-BRAIN EXPERIMENTS *582*

35.7 **Memory** *583*

35.8 **States of Consciousness** *584*

35.9 FOCUS ON HEALTH: DRUGGING THE BRAIN *584*

Stimulants *584*

Depressants, Hypnotics *585*

Analgesics *585*

Psychedelics, Hallucinogens *585*

---

36 **SENSORY RECEPTION**

DIFFERENT STOKES FOR DIFFERENT FOLKS *588*

36.1 **Sensory Receptors and Pathways— An Overview** *590*

Types of Sensory Receptors *590*

Sensory Pathways *590*

36.2 **Somatic Sensations** *592*

Receptors Near the Body Surface *592*

Muscle Sense *592*

Regarding the Sensation of Pain *592*

36.3 **Senses of Taste and Smell** *594*

36.4 **Sense of Balance** *595*

36.5 **Sense of Hearing** *596*

Properties of Sound *596*

Evolution of the Vertebrate Ear *596*

Sensory Reception in the Human Ear *596*

Echolocation *597*

36.6 **Sense of Vision** *598*

Requirements for Vision *598*

A Sampling of Invertebrate Eyes *598*

Simple Eyes *598*

Complex Eyes *598*

36.7 **Structure and Function of Vertebrate Eyes** *600*

The Eye's Structure *600*

Visual Accommodation *601*

36.8 **Case Study: From Signaling to Visual Perception** *602*

Organization of the Retina *602*

Neuronal Responses to Light *602*

Rod Cells *602*

Cone Cells *603*

Responses at Receptive Fields *603*

On to the Visual Cortex *603*

36.9 FOCUS ON SCIENCE: DISORDERS OF THE HUMAN EYE *604*

---

37 **ENDOCRINE CONTROL**

HORMONE JAMBOREE *608*

37.1 **The Endocrine System** *610*

Hormones and Other Signaling Molecules *610*

Discovery of Hormones and Their Sources *610*

37.2 **Signaling Mechanisms** *612*

The Nature of Hormonal Action *612*

Characteristics of Steroid Hormones *612*

Characteristics of Peptide Hormones *613*

37.3 **The Hypothalamus and Pituitary Gland** *614*

Posterior Lobe Secretions *614*

Anterior Lobe Secretions *614*

Anterior Pituitary Hormones *614*

Regarding the Hypothalamic Triggers *615*

37.4 **Examples of Abnormal Pituitary Output** *616*

37.5 **Sources and Effects of Other Hormones** *617*

37.6 **Feedback Control of Hormonal Secretions** *618*

Negative Feedback From the Adrenal Cortex *618*

Local Feedback in the Adrenal Medulla *618*

Skewed Feedback From the Thyroid *618*

Feedback Control of the Gonads *619*

37.7 **Responses to Local Chemical Changes** *620*

Secretions From Parathyroid Glands *620*

Effects of Local Signaling Molecules *620*

Secretions From Pancreatic Islets *620*

37.8 **Hormonal Responses to Environmental Cues** *622*

Daylength and the Pineal Gland *622*

Comparative Look at a Few Invertebrates *623*

38   PROTECTION, SUPPORT, AND MOVEMENT

OF MEN, WOMEN, AND POLAR HUSKIES 626

38.1   Integumentary System 628
Vertebrate Skin and Its Derivatives 628
Functions of Skin 629

38.2   A Look at Human Skin 630
Structure of the Epidermis and Dermis 630
Sweat Glands, Oil Glands, and Hairs 630

38.3   FOCUS ON HEALTH: SUNLIGHT AND SKIN 631
    The Vitamin D Connection 631
    Suntans and Shoe-Leather Skin 631
    Sunlight and the Front Line of Defense 631

38.4   Types of Skeletons 632
Operating Principles for Skeletons 632
Examples From the Invertebrates 632
Examples From the Vertebrates 633

38.5   Characteristics of Bone 634
Functions of Bone 634
Bone Structure 634
    How Bones Develop 635
    Bone Tissue Turnover 635

38.6   Human Skeletal System 636
Appendicular and Axial Portions 636
Skeletal Joints 636

38.7   Skeletal-Muscular Systems 638
How Muscles and Bones Interact 638
Human Skeletal-Muscular System 639

38.8   Muscle Structure and Function 640
Functional Organization of Skeletal Muscle 640
Sliding-Filament Model of Contraction 640

38.9   Control of Muscle Contraction 642
The Control Pathway 642
The Control Mechanism 642
Sources of Energy for Contraction 643

38.10   Properties of Whole Muscles 644
Muscle Tension and Muscle Fatigue 644
Effects of Exercise and Aging 644

38.11   FOCUS ON SCIENCE: PEBBLES, FEATHERS, AND MUSCLE MANIA 645

39   CIRCULATION

HEARTWORKS 648

39.1   Circulatory Systems—An Overview 650
General Characteristics 650
Evolution of Vertebrate Circulatory Systems 651
Links With the Lymphatic System 651

39.2   Characteristics of Blood 652
Functions of Blood 652
Blood Volume and Composition 652
    Plasma 652
    Red Blood Cells 652
    White Blood Cells 653

39.3   FOCUS ON HEALTH: BLOOD DISORDERS 654

39.4   Blood Transfusion and Typing 654
Concerning Agglutination 654
ABO Blood Typing 655
Rh Blood Typing 655

39.5   Human Cardiovascular System 656

39.6   The Heart Is a Lonely Pumper 658
Heart Structure 658
Cardiac Cycle 658
Mechanisms of Contraction 659

39.7   Blood Pressure in the Cardiovascular System 660
Arterial Blood Pressure 660
Resistance to Flow at Arterioles 661
Controlling Mean Arterial Blood Pressure 661

39.8   From Capillary Beds Back to the Heart 662
Capillary Function 662
Venous Pressure 662

39.9   FOCUS ON HEALTH: CARDIOVASCULAR DISORDERS 664
    Risk Factors 664
    Hypertension 664
    Atherosclerosis 664
    Arrhythmias 665

39.10   Hemostasis 666

39.11   Lymphatic System 666
Lymph Vascular System 666
Lymphoid Organs and Tissues 667

40   IMMUNITY

RUSSIAN ROULETTE, IMMUNOLOGICAL STYLE 670

40.1   Three Lines of Defense 672
Surface Barriers to Invasion 672
Nonspecific and Specific Responses 672

40.2   Complement Proteins 673

40.3   Inflammation 674
The Roles of Phagocytes and Their Kin 674
The Inflammatory Response 674

40.4   The Immune System 676
Defining Features 676
Antigen-Presenting Cells—The Triggers for Immune Responses 676

SOLDIER TERMITES DEFENDING THEIR COLONY

Key Players in Immune Responses *676*

Control of Immune Responses *677*

40.5 Lymphocyte Battlegrounds *678*

40.6 Cell-Mediated Responses *678*

T Cell Formation and Activation *678*

Functions of Effector T Cells *678*

Regarding the Natural Killer Cells *679*

40.7 Antibody-Mediated Responses *680*

B Cells and the Targets of Antibodies *680*

The Immunoglobulins *681*

40.8 FOCUS ON HEALTH: CANCER AND IMMUNOTHERAPY *681*

Monoclonal Antibodies 681

Multiplying the Tumor Cells 681

Therapeutic Vaccines 681

40.9 Immune Specificity and Memory *682*

Formation of Antigen-Specific Receptors *682*

Immunological Memory *683*

40.10 Defenses Enhanced, Misdirected, or Compromised *684*

Immunization *684*

Allergies *684*

Autoimmune Disorders *685*

Deficient Immune Responses *685*

40.11 FOCUS ON HEALTH—AIDS—THE IMMUNE SYSTEM COMPROMISED *686*

Characteristics of AIDS 686

A Titanic Struggle Begins 686

How HIV Is Transmitted 687

Regarding Prevention and Treatment 687

---

41 **RESPIRATION**

CONQUERING CHOMOLUNGMA *690*

41.1 The Nature of Respiration *692*

The Basis of Gas Exchange *692*

Factors That Influence Gas Exchange *692*

Surface-to-Volume Ratio 692

Ventilation 692

Transport Pigments 692

41.2 Invertebrate Respiration *693*

41.3 Vertebrate Respiration *694*

Gills of Fishes and Amphibians *694*

Lungs *694*

41.4 Human Respiratory System *696*

Functions of the Respiratory System *696*

From Airways to the Lungs *697*

Sites of Gas Exchange in the Lungs *697*

41.5 Breathing—Cyclic Reversals in Air Pressure Gradients *698*

The Respiratory Cycle *698*

Lung Volumes *699*

Breathing and Sound Production *699*

41.6 Gas Exchange and Transport *700*

Gas Exchange *700*

Oxygen Transport *700*

Carbon Dioxide Transport *700*

Matching Air Flow With Blood Flow *701*

41.7 FOCUS ON HEALTH: WHEN THE LUNGS BREAK DOWN *702*

Bronchitis 702

Emphysema 702

Effects of Cigarette Smoke 702

Effects of Marijuana Smoke 703

41.8 Respiration in Unusual Environments *704*

Breathing at High Altitudes *704*

Carbon Monoxide Poisoning *704*

Respiration in Diving Mammals *705*

The Physiology of Diving 705

Human Deep-Sea Divers 705

41.9 FOCUS ON SCIENCE: RESPIRATION IN LEATHERBACK SEA TURTLES *706*

---

42 **DIGESTION AND HUMAN NUTRITION**

LOSE IT—AND IT FINDS ITS WAY BACK *710*

42.1 The Nature of Digestive Systems *712*

Incomplete and Complete Systems *712*

Correlations With Feeding Behavior *712*

42.2 Overview of the Human Digestive System *714*

42.3 Into the Mouth, Down the Tube *715*

42.4 Digestion in the Stomach and Small Intestine *716*

The Stomach *716*

The Small Intestine *716*

The Role of Bile in Fat Digestion *717*

Controls Over Digestion *717*

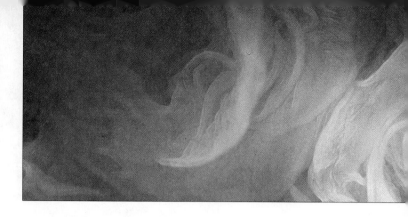

42.5 **Absorption in the Small Intestine** *718*
Structure Speaks Volumes About Function *718*
Mechanisms of Absorption *718*

42.6 **Disposition of Absorbed Organic Compounds** *720*

42.7 **The Large Intestine** *721*
Colon Functioning *721*
Colon Malfunctioning *721*

42.8 **Human Nutritional Requirements** *722*
New, Improved Food Pyramids *722*
Carbohydrates *722*
Lipids *723*
Proteins *723*

42.9 **Vitamins and Minerals** *724*

FOCUS ON SCIENCE—TANTALIZING ANSWERS
TO WEIGHTY QUESTIONS *726*

In Pursuit of the "Ideal" Weight 726
My Genes Made Me Do It 726

43 **THE INTERNAL ENVIRONMENT**

TALE OF THE DESERT RAT *730*

43.1 **Urinary System of Mammals** *732*
The Challenge—Shifts in Extracellular Fluid *732*
Water Gains and Losses 732
Solute Gains and Losses 732
Components of the Urinary System *733*
Nephrons—Functional Units of Kidneys *733*

43.2 **Urine Formation** *734*
Factors Influencing Filtration *734*
Reabsorption of Water and Sodium *735*
Reabsorption Mechanisms 735
Hormone-Induced Adjustments 735
Thirst Behavior 735

43.3 FOCUS ON HEALTH: WHEN KIDNEYS BREAK DOWN *736*

43.4 **The Acid-Base Balance** *736*

43.5 **On Fish, Frogs, and Kangaroo Rats** *737*

43.6 **Maintaining the Body's Core Temperature** *738*
Temperatures Suitable for Life *738*
Heat Gains and Heat Losses *738*
Ectotherms, Endotherms, and In-Betweens *739*

43.7 **Temperature Regulation in Mammals** *740*
Responses to Cold Stress *740*
Responses to Heat Stress *741*
Fever *741*

44 **PRINCIPLES OF REPRODUCTION AND DEVELOPMENT**

FROM FROG TO FROG AND OTHER MYSTERIES *744*

44.1 **The Beginning: Reproductive Modes** *746*
Sexual Versus Asexual Reproduction *746*
Costs and Benefits of Sexual Reproduction *746*

44.2 **Stages of Development—An Overview** *748*

44.3 **Early Marching Orders** *750*
Information in the Egg Cytoplasm *750*
Cleavage—The Start of Multicellularity *750*
Cleavage Patterns *751*

44.4 **How Specialized Tissues and Organs Form** *752*
Cell Differentiation *752*
Morphogenesis *753*

44.5 FOCUS ON SCIENCE: TO KNOW A FLY *754*
Superficial Cleavage in Drosophila 754
Genes as Master Organizers 755

44.6 **Pattern Formation** *756*
A Theory of Pattern Formation *756*
Embryonic Induction *756*

44.7 **From the Embryo Onward** *758*
Post-Embryonic Development *758*
Aging and Death *758*

44.8 COMMENTARY: DEATH IN THE OPEN *759*

45 **HUMAN REPRODUCTION AND DEVELOPMENT**

THE JOURNEY BEGINS *762*

45.1 **Reproductive System of Human Males** *764*
Where Sperm Form *764*
Where Semen Forms *764*
Cancers of the Prostate and Testis *765*

45.2 **Male Reproductive Function** *766*
Sperm Formation *766*
Hormonal Controls *766*

45.3 **Reproductive System of Human Females** *768*
The Reproductive Organs *768*
Overview of the Menstrual Cycle *768*

45.4 **Female Reproductive Function** *770*
Cyclic Changes in the Ovary *770*
Cyclic Changes in the Uterus *771*

45.5 **Visual Summary of the Menstrual Cycle** *772*

45.6 **Pregnancy Happens** *773*
Sexual Intercourse *773*
Fertilization *773*

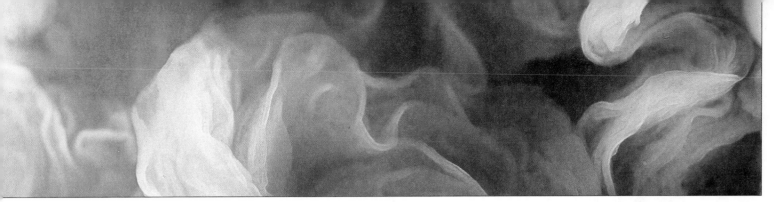

45.7 Formation of the Early Embryo *774*
Cleavage and Implantation *774*
Extraembryonic Membranes *775*

45.8 Emergence of the Vertebrate Body Plan *776*

45.9 On the Importance of the Placenta *777*

45.10 Emergence of Distinctly Human Features *778*

45.11 FOCUS ON HEALTH: MOTHER AS PROVIDER, PROTECTOR, POTENTIAL THREAT *780*
Some Nutritional Considerations 780
Risk of Infection 780
Effects of Prescription Drugs 781
Effects of Alcohol 781
Effects of Cocaine 781
Effects of Cigarette Smoke 781

45.12 From Birth Onward *782*
Giving Birth *782*
Nourishing the Newborn *782*
Regarding Breast Cancer *783*
Postnatal Development, Aging, and Death *783*

45.13 Control of Human Fertility *784*
Some Ethical Considerations *784*
Birth Control Options *784*

45.14 FOCUS ON HEALTH: SEXUALLY TRANSMITTED DISEASES *786*
AIDS 786
Gonorrhea 786
Syphilis 787
Pelvic Inflammatory Disease 787
Genital Herpes 787
Genital Warts 787
Chlamydial Infection 787

45.15 FOCUS ON SCIENCE: TO SEEK OR END PREGNANCY *788*

VII ECOLOGY AND BEHAVIOR

46 POPULATION ECOLOGY

TALES OF NIGHTMARE NUMBERS *792*

46.1 Characteristics of Populations *794*
Regarding Population Density *794*
Patterns of Dispersion *794*
Some Qualifiers *795*

46.2 Population Size and Exponential Growth *796*
How Population Size Changes *796*
From Ground Zero to Exponential Growth *796*
Regarding the Biotic Potential *797*

46.3 Limits on the Growth of Populations *798*
Limiting Factors *798*
Carrying Capacity and Logistic Growth *798*
Density-Dependent Controls *799*
Density-Independent Factors *799*

46.4 Life History Patterns *800*
Life Tables *800*
Patterns of Survivorship and Reproduction *800*

46.5 FOCUS ON SCIENCE: NATURAL SELECTION AND THE GUPPIES OF TRINIDAD *802*

46.6 Human Population Growth *804*
How We Began Sidestepping Controls *804*
Present and Future Growth *805*

46.7 Control Through Family Planning *806*

46.8 Population Growth and Economic Development *808*
Demographic Transition Model *808*
A Question of Resource Consumption *809*

46.9 Social Impact of No Growth *809*

47 COMMUNITY INTERACTIONS

NO PIGEON IS AN ISLAND *812*

47.1 Factors That Shape Community Structure *814*
The Niche *814*
Categories of Species Interactions *814*

47.2 Mutualism *815*

47.3 Competitive Interactions *816*
Categories of Competition *816*
Competitive Exclusion *816*
Resource Partitioning *817*

47.4 Predation *818*
Predation Versus Parasitism *818*
Dynamics of Predator-Prey Interactions *818*

47.5 FOCUS ON THE ENVIRONMENT: THE COEVOLUTIONARY ARMS RACE *820*
Camouflage 820
Warning Coloration 820
Mimicry 820
Moment-of-Truth Defenses 820
Adaptive Responses to Prey 821

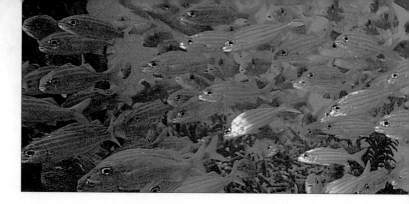

47.6 **Parasitic Interactions** *822*
Kinds of Parasites *822*
Evolution of Parasitism *823*
Regarding the Parasitoids *823*
Parasites as Biological Control Agents *823*

47.7 **Forces Contributing to Community Stability** *824*
The Successional Model *824*
The Climax-Pattern Model *824*
Cyclic, Nondirectional Changes *824*
Restoration Ecology *825*

47.8 **Community Instability** *826*
How Keystone Species Tip the Balance *826*
How Species Introductions Tip the Balance *827*

47.9 FOCUS ON THE ENVIRONMENT: NILE PERCH AND RABBITS AND KUDZU, OH MY! *828*
Hello Victoria, Good-bye Cichlids *828*
The Rabbits That Ate Australia *828*
The Plant That Ate Georgia *829*

47.10 **Patterns of Biodiversity** *830*
Mainland and Marine Patterns *830*
Island Patterns *830*

48 **ECOSYSTEMS**

CRÊPES FOR BREAKFAST, PANCAKE ICE FOR DESSERT *834*

48.1 **The Nature of Ecosystems** *836*
Overview of the Participants *836*
Structure of Ecosystems *836*
Trophic Levels *836*
Food Webs *837*

48.2 **Energy Flow Through Ecosystems** *838*
Primary Productivity *838*
Major Pathways of Energy Flow *838*
Ecological Pyramids *838*

48.3 FOCUS ON SCIENCE: ENERGY FLOW AT SILVER SPRINGS, FLORIDA *840*

48.4 **Biogeochemical Cycles—An Overview** *841*

48.5 **Hydrologic Cycle** *842*

48.6 **Carbon Cycle** *844*

48.7 FOCUS ON THE ENVIRONMENT: FROM GREENHOUSE GASES TO A WARMER PLANET? *846*
Greenhouse Effect *846*
Global Warming Defined *846*
Evidence of an Intensified Greenhouse Effect *846*

48.8 **Nitrogen Cycle** *848*
Cycling Processes *848*
Nitrogen Scarcity *849*
Human Intervention in the Nitrogen Cycle *849*

48.9 **Sedimentary Cycles** *850*
A Look at the Phosphorus Cycle *850*
Eutrophication *850*

48.10 **Predicting the Impact of Change in Ecosystems** *851*
Ecosystem Modeling *851*
Biological Magnification—A Case Study *851*

49 **THE BIOSPHERE**

DOES A CACTUS GROW IN BROOKLYN? *854*

49.1 **Air Circulation Patterns and Regional Climates** *856*

49.2 **The Ocean, Land Forms, and Regional Climates** *858*
Ocean Currents and Their Effects *858*
Regarding Rain Shadows and Monsoons *858*

49.3 **The World's Biomes** *860*
Biogeographic Distribution *860*
Biogeographic Realms and Biomes *860*

49.4 **Soils of Major Biomes** *862*

49.5 **Deserts** *863*

49.6 **Dry Shrublands, Dry Woodlands, and Grasslands** *864*

49.7 **Tropical Rain Forests and Other Broadleaf Forests** *866*

49.8 **Coniferous Forests** *868*

49.9 **Tundra** *869*

49.10 **Freshwater Provinces** *870*
Lake Ecosystems *870*
Seasonal Changes in Lakes *870*
Trophic Nature of Lakes *871*
Stream Ecosystems *871*

49.11 **The Ocean Provinces** *872*
Primary Productivity in the Ocean *872*
Hydrothermal Vents *872*

49.12 **Coral Reefs and Banks** *874*
The Master Builders *874*
The Once and Future Reefs *874*

49.13 **Life Along the Coasts** *876*
Estuaries *876*
The Intertidal Zone *876*
Upwelling Along Coasts *877*

49.14 FOCUS ON SCIENCE: EL NIÑO AND SEESAWS IN THE WORLD'S CLIMATES *878*

A GLIMPSE OF LIFE AT A TROPICAL REEF

50   **HUMAN IMPACT ON THE BIOSPHERE**

AN INDIFFERENCE OF MYTHIC PROPORTIONS *882*

50.1   **Air Pollution—Prime Examples** *884*
Smog *884*
Acid Deposition *884*

50.2   **Ozone Thinning—Global Legacy of Air Pollution** *886*

50.3   **Where to Put Solid Wastes, Where to Produce Food** *887*
Oh Bury Me Not, in Our Own Garbage *887*
Converting Marginal Lands for Agriculture *887*

50.4   **Deforestation—Concerted Assaults on Finite Resources** *888*

50.5   FOCUS ON BIOETHICS: YOU AND THE TROPICAL RAIN FOREST *890*

50.6   **Trading Grasslands for Deserts** *891*

50.7   **A Global Water Crisis** *892*
Consequences of Heavy Irrigation *892*
Water Pollution *892*
The Coming Water Wars *893*

50.8   **A Question of Energy Inputs** *894*
Fossil Fuels *894*
Nuclear Energy *894*
    Nuclear Reactors 894
    Nuclear Waste Disposal 895

50.9   **Alternative Energy Sources** *896*
Solar-Hydrogen Energy *896*
Wind Energy *896*
Fusion Power *896*

50.10   FOCUS ON BIOETHICS: BIOLOGICAL PRINCIPLES AND THE HUMAN IMPERATIVE *897*

51   **AN EVOLUTIONARY VIEW OF BEHAVIOR**

DECK THE NEST WITH SPRIGS OF GREEN STUFF *900*

51.1   **The Heritable Basis of Behavior** *902*
Genes and Behavior *902*
Hormones and Behavior *902*
Instinctive Behavior Defined *903*

51.2   **Learned Behavior** *904*

51.3   **The Adaptive Value of Behavior** *905*

51.4   **Communication Signals** *906*
The Nature of Communication Signals *906*
Examples of Communication Displays *906*
Illegitimate Signalers and Receivers *907*

51.5   **Mating, Parenting, and Individual Reproductive Success** *908*
Sexual Selection Theory and Mating Behavior *908*
Costs and Benefits of Parenting *909*

51.6   **Benefits of Living in Social Groups** *910*
Cooperative Predator Avoidance *910*
The Selfish Herd *910*
Dominance Hierarchies *911*

51.7   **Costs of Living in Social Groups** *912*

51.8   **Evolution of Altruism** *913*

51.9   FOCUS ON THE SOCIAL ENVIRONMENT: ABOUT THOSE SELF-SACRIFICING INSECTS *914*

51.10   FOCUS ON SCIENCE: ABOUT THOSE NAKED MOLE-RATS *916*

51.11   **An Evolutionary View of Human Social Behavior** *917*

APPENDIX I   BRIEF CLASSIFICATION SCHEME

APPENDIX II   UNITS OF MEASURE

APPENDIX III   ANSWERS TO GENETICS PROBLEMS

APPENDIX IV   ANSWERS TO SELF-QUIZZES

APPENDIX V   A CLOSER LOOK AT SOME MAJOR METABOLIC PATHWAYS

APPENDIX VI   PERIODIC TABLE OF THE ELEMENTS

A GLOSSARY OF BIOLOGICAL TERMS

PREFACE

Not too long from now we will cross the threshold of a new millennium, a rite of passage that invites reflection on where biology has been and where it might be heading. About 500 years ago, during an age of global exploration, naturalists first started to systematically catalog and think about the staggering diversity of organisms all around the world. Less than 150 years ago, just before the start of a civil war that would shred the fabric of a new nation, the naturalist Charles Darwin shredded preconceived notions about life's diversity. It was only about 50 years ago that biologists caught their first glimpse of life's unity at the molecular level. Until that happened a biologist could still hope to be a generalist—someone who viewed life as Darwin did, without detailed knowledge of mechanisms that created it, that perpetuate it, that change it.

No more. Biology grew to encompass hundreds of specialized fields, each focused on one narrow aspect of life and yielding volumes of information about it. Twenty years ago I wondered whether introductory textbooks could possibly keep up with the rapid and divergent splintering of biological inquiry. James Bonner, a teacher and researcher at the California Institute of Technology, turned my thinking around on this. He foresaw that authors and instructors for introductory courses must become the new generalists, the ones who give each generation of students broad perspective on what we know about life and what we have yet to learn.

And we must do this, for the biological perspective remains one of the most powerful of education's gifts. With it, students who travel down specialized roads can sense intuitively that their research and its applications may have repercussions in unexpected places in the world of life. With that perspective, students in general might cut their own intellectual paths through social, medical, and environmental thickets. And they might come to understand the past and to predict possible futures for ourselves and all other organisms.

CONCERNING THE EIGHTH EDITION

Like earlier editions, this book starts with an overview of the basic concepts and scientific methods. Three units on the principles of biochemistry, inheritance, and evolution follow. The principles provide the conceptual background necessary for deeper probes into life's unity and diversity, starting with a richly illustrated evolutionary survey of each kingdom. Units on the comparative anatomy and physiology of plants, then animals, follow. The last unit focuses on the patterns and consequences of organisms interacting with one another and with their environment. Thus the organization parallels the levels of biological organization, from cells through the biosphere. We adhere to this traditional approach for good reason: it works.

As before, we identify and highlight the key concepts, current understandings, and research trends for the major fields of inquiry. Through examples of problem solving and experiments, we give ample evidence of "how we know what we know" and thus demonstrate the power of critical thinking. We explain the structure and functioning of a broad sampling of organisms in enough detail so that students can develop a working vocabulary about life's parts and processes. We also updated the glossary.

CONCEPT SPREADS

In the first chapter, an overview of the levels of biological organization kicks off a story that continues through the rest of the book. Telling such a big, complex story might be daunting unless you remind yourself of the question *"How do you eat an elephant?"* and its answer, *"One bite at a time."* We who have told the story again and again know how the parts fit together, but many students need help to keep the story line in focus within and between chapters. And they need to chew on concepts one at a time.

In every chapter we present each concept on its own table, so to speak. That is, we organize the descriptions, art, and supporting evidence for it on two facing pages, at most. Think of this as a concept spread, as in Figure A. Each starts with a numbered tab and ends with boldface statements to summarize the key points. Students can use these cues as reminders to digest one topic before starting on another. Well-crafted transitions between spreads help students focus on where topics fit in the larger story and gently discourage memorization for its own sake. The clear demarcation also gives instructors greater flexibility in assigning or skipping topics within a chapter.

By restricting the space available for each concept, we force ourselves to clear away the clutter of superfluous detail. Within each concept spread, we block out headings and subheadings to rank the importance of its various parts. Any good story has such a hierarchy of information, with background settings, major and minor characters, and high points and an ending where everything comes together. Without a hierarchy, a story has all the excitement, flow, and drama of an encyclopedia. Where details are useful as expansions of concepts, we integrate them into suitable illustrations to keep them from disrupting the text flow.

Not all students are biology majors, and many of them approach biology textbooks with apprehension. If the words don't engage them, they sometimes end up hating the book, and the subject. It comes down to line-by-line judgment calls. During twenty-two years of authorship, we developed a sense of when to leave core material alone and when to loosen it up to give students breathing room. Interrupting, say, an account of mitotic cell division with a distracting anecdote does no good. Plunking a humorous aside into a chapter that ties together the evolution of the Earth and life trivializes a magnificent story. Including an entertaining story is fine, provided that doing so reinforces a key concept. Thus, for example, we include the story of a misguided species introduction that resulted in wild European rabbits running amok through Australia.

BALANCING CONCEPTS WITH APPLICATIONS

Each chapter starts with a lively or sobering application that leads into an adjoining list of key concepts. The list is an advance organizer for the chapter as a whole. At strategic points, examples of applications parallel the core material—not so many as to be distracting, but enough to keep minds perking along with the conceptual development. Many brief applications are integrated in the text. Others are in *Focus* essays, which give more depth on medical, environmental, and social issues for interested students but do not interrupt the text flow. A separate index on the book's last few pages (tinted *green*) lists all applications for quick reference.

FOUNDATIONS FOR CRITICAL THINKING

To help students develop a capacity for critical thinking, we walk them through experiments that yielded evidence in favor of or against hypotheses being discussed. The main index for the book will give you a sense of the number and types of experiments used (see the entry *Experiments*).

We use certain chapter introductions as well as entire chapters to show students some of the productive results of critical thinking. Among these are the introductions to the chapters on Mendelian genetics (11), DNA structure and function (14), speciation (19), immunology (40), and animal behavior (51).

Many *Focus on Science* essays provide more detailed, optional examples of how biologists apply critical thinking to problem solving. For example, one of these describes RFLP analysis (Section 16.3) and a few of its more jarring applications. Another essay (Section 21.4) helps convey to students that biology is not a closed book. Even when new research brings a sweeping story into sharp focus—in this case, the origin of the great prokaryotic and eukaryotic lineages—it also opens up new roads of inquiry.

This edition has *Critical Thinking* questions at the end of chapters. Katherine Denniston of Towson State University developed these thought-provoking questions. Chapters 11 and 12 also include a large selection of *Genetics Problems* that help students grasp the principles of inheritance.

To keep readers focused, we cover each concept on one or two facing pages, starting with a numbered tab . . .

5.1 MEMBRANE STRUCTURE AND FUNCTION

Earlier chapters provided you with a brief look at the structure of cell membranes and the general functions of their component parts. Here, we incorporate some of the background information in a more detailed picture.

The Lipid Bilayer of Cell Membranes

Fluid bathes the two surfaces of a cell membrane and is vital for its functioning. The membrane, too, has a fluid quality; it is not a solid, static wall between cytoplasmic and extracellular fluids. For instance, puncture a cell with a fine needle, and its cytoplasm will not ooze out. The membrane will flow over the puncture site and seal it!

How does a fluid membrane remain distinct from its fluid surroundings? To arrive at the answer, start by reviewing what we have already learned about its most abundant components, the phospholipids. Recall that a **phospholipid** has a phosphate-containing head and two fatty acid tails attached to a glycerol backbone (Figure 5.2a). The head is hydrophilic; it easily dissolves in water. Its tails are hydrophobic; water repels them. Immerse a number of phospholipid molecules in water, and they will interact with water molecules and with one another until they spontaneously cluster in a sheet or film at the water's surface. Their jostlings may even force them to become organized in two layers, with all fatty acid tails sandwiched between all hydrophilic heads. This **lipid bilayer** arrangement, remember, is the structural basis of cell membranes (Section 4.1 and Figure 5.2c).

The organization of each lipid bilayer minimizes the total number of hydrophobic

groups exposed to water, so the fatty acid tails do not have to spend a lot of energy fighting water molecules, so to speak. A "punctured" membrane exhibits sealing behavior precisely because a puncture is energetically unfavorable. It leaves far too many hydrophobic groups exposed to the surrounding fluid.

Ordinarily, few cells get jabbed by fine needles. But the self-sealing behavior of membrane phospholipids is good for more than damage control. Among other things, it functions in vesicle formation. For example, as vesicles bud away from ER or Golgi membranes, phospholipids interact hydrophobically with cytoplasmic water. They get pushed together, and the rupture seals. You will read more about vesicle formation later in the chapter.

Fluid Mosaic Model of Membrane Structure

Figure 5.3 shows a bit of membrane that corresponds to the **fluid mosaic model**. By this model, cell membranes are a mixed composition—a "mosaic"—of phospholipids, glycolipids, sterols, and proteins. The phospholipid heads as well as the length and saturation of the tails are not all the same. (Recall that unsaturated fatty acids have one or more double bonds in their backbone and fully saturated ones have none.) The glycolipids are structurally similar to phospholipids, but their head incorporates one or more sugar monomers. In animal cell membranes, cholesterol is the most abundant sterol (Figure 5.2b). Phytosterols are their equivalent in plant cell membranes.

Also by this model, the membrane is "fluid" owing to the motions and interactions of its component parts.

Figure 5.2
(a) Structural formula of phosphatidylcholine, a phospholipid that is one of the most common components of the membranes of animal cells. *Orange* indicates its hydrophilic head; *yellow* indicates its hydrophobic tails.

(b) Structural formula of cholesterol, the major sterol in animal tissues.

(c) Diagram showing how lipids that are placed in liquid water may spontaneously organize themselves into a bilayer structure.

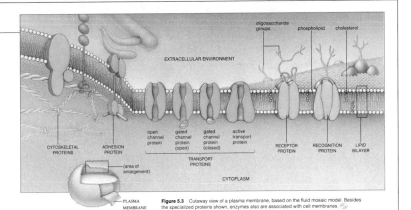

Figure 5.3 Cutaway view of a plasma membrane, based on the fluid mosaic model. Besides the specialized proteins shown, enzymes also are associated with cell membranes.

The hydrophobic interactions that give rise to most of a membrane's structure are weaker than covalent bonds. This means most phospholipids and some proteins are free to drift sideways. Also, the phospholipids can spin about their long axis and flex their tails, which keeps neighboring molecules from packing together in a solid layer. Short or kinked (unsaturated) fatty acid tails also contribute to membrane fluidity.

The fluid mosaic model is a good starting point for exploring cell membranes. But bear in mind, membranes differ in the details of their molecular composition and arrangements, and they are not even the same on both surfaces of their bilayer. For example, oligosaccharides and other carbohydrates are covalently bonded to protein and lipid components of a plasma membrane, but only on its outward-facing surface (Figure 5.3). Moreover, they differ in number and kind from one species to the next, even among the different cells of the same individual.

Overview of Membrane Proteins

The proteins embedded in a lipid bilayer or attached to one of its surfaces carry out most membrane functions. Many are enzyme components of metabolic machinery. Others are **transport proteins** that allow water-soluble substances to move through their interior, which spans the bilayer. They bind molecules or ions on one side of the membrane, then release them on the other side.

The **receptor proteins** bind extracellular substances, such as hormones, that trigger changes in cell activities.

For example, certain enzymes that crank up machinery for cell growth and division become switched on when somatotropin, a hormone, binds with receptors for it. Different cells have different combinations of receptors.

Diverse **recognition proteins** at the cell surface are like molecular fingerprints; their oligosaccharide chains identify a cell as being of a specific type. For example, "self" proteins pepper the plasma membrane of your cells. Certain white blood cells chemically recognize the proteins and leave your own cells alone, but they attack invading bacterial cells having "nonself" proteins at their surface. Finally, **adhesion proteins** of multicelled organisms help cells of the same type locate and stick to one another and stay positioned in the proper tissues. They are glycoproteins with oligosaccharides attached. After tissues form, the sites of adhesion may become a type of cell junction, as described earlier in Section 4.10.

A cell membrane has two layers composed mainly of lipids, phospholipids especially. This lipid bilayer is the structural foundation for the membrane and also serves as a barrier to water-soluble substances.

Hydrophilic heads of the phospholipids are dissolved in fluids that bathe the two outer surfaces of the bilayer. Their hydrophobic tails are sandwiched between the heads.

Proteins associated with the bilayer carry out most membrane functions. Many are enzymes, transporters of substances across the bilayer, or receptors for extracellular substances. Other types function in cell-to-cell recognition or adhesion.

82   Principles of Cellular Life

Chapter 5   A Closer Look at Cell Membranes   83

. . . and ending with one or more summary statements.

FIGURE A   *A concept spread from this edition.*

VISUAL OVERVIEWS OF MAJOR CONCEPTS

While writing the text, we simultaneously develop the illustrations as inseparable parts of the same story. This integrative approach appeals to students who are visual learners. When they can first work their way through a visual overview of some process, then reading through the corresponding text becomes less intimidating. Over the years, students have repeatedly thanked us for our hundreds of overview illustrations, which contain step-by-step, written descriptions of biological parts and processes. We break down the information into a series of illustrated steps that are more inviting than a complex, "wordless" diagram. Figure *B* is a sample. Notice how simple descriptions, integrated with the art, take students through the stages by which mRNA transcripts become translated into polypeptide chains, one step at a time.

Similarly, we continue to create visual overviews for anatomical drawings. The illustrations integrate structure and function. Students need not jump back and forth from the text, to tables, to illustrations, and back again in order to comprehend how an organ system is put together and what its parts do. Even individual descriptions of parts are hierarchically arranged to reflect the structural and functional organization of that system.

COLOR CODING

In line illustrations, we consistently use the same colors for the same types of molecules and cell structures. Visual consistency makes it easier for students to track complex parts and processes. Figure C is the color coding chart.

ZOOM SEQUENCES

Many illustrations in the book progress from macroscopic to microscopic views of the same subject. Figure 7.2 is an example; this zoom sequence shows where the reactions of photosynthesis proceed, starting with a plant growing by a roadside. As another example, Figures 38.19 and 38.20 move down through levels of skeletal muscle contraction, starting with a muscle in the human arm.

ICONS

Within the text, small diagrams next to an illustration help relate the topic to the big picture. For instance, in Figure *A*, a simple representation of a cell subtly reminds students of the location of the plasma membrane relative to the cytoplasm. Other icons serve as reminders of the location of reactions and processes in cells and how they

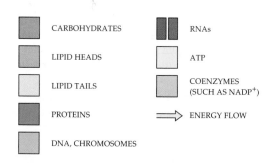

FIGURE C   *Color coding chart for the diagrams of biological molecules and cell structures.*

Step-by-step art with simple descriptions helps students visualize a process before reading text about it.

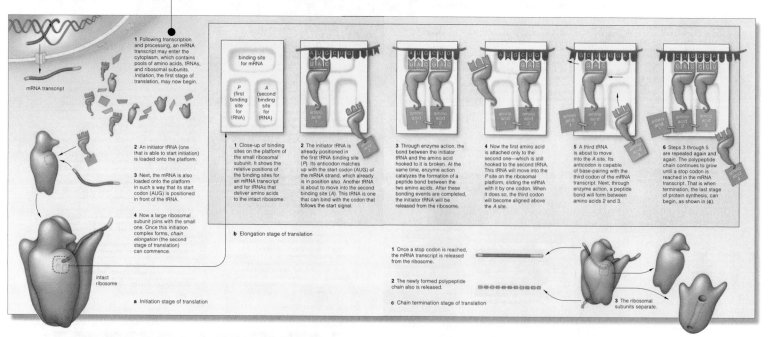

FIGURE B   *A visual overview from this edition.*

interrelate to one another. Still others remind students of the evolutionary relationships among groups of organisms, as in Chapter 26.

New to this edition are icons that invite students to use multimedia. One icon directs them to art in the CD-ROM enclosed with each student copy, another to supplemental material on the Web, and a third to InfoTrac:

CD-ROM ICON: WEB ICON: INFOTRAC ICON:

END-OF-CHAPTER STUDY AIDS

Figure *D* shows a sampling of our end-of-chapter study aids, which reinforce the key concepts. Each chapter ends with a summary in list form, review questions, a self-quiz, critical thinking questions, selected key terms, and a list of readings. Italicized page numbers tie the review questions and key terms to relevant text pages.

END-OF-BOOK STUDY AIDS

At the book's end, the detailed classification scheme in Appendix I is helpful for reference purposes. Appendix II includes metric-English conversion charts. Appendix III has detailed answers to the genetics problems in Chapters 11 and 12, and the fourth has answers to the self-quizzes at the end of each chapter. For those interested students and professors who prefer the added detail, Appendix V gives structural formulas for the major metabolic pathways. Appendix VI shows the periodic table of the elements.

A Glossary includes boldfaced terms from the text, with pronunciation guides and word origins to make the formidable words less so. The Appendixes as well as the Glossary are printed on paper that is tinted different colors to preclude frustrating searches for where one ends and the next begins. The Index is detailed enough to help readers find doors to the text more quickly. An Applications Index separately lists all of the book's examples of applications.

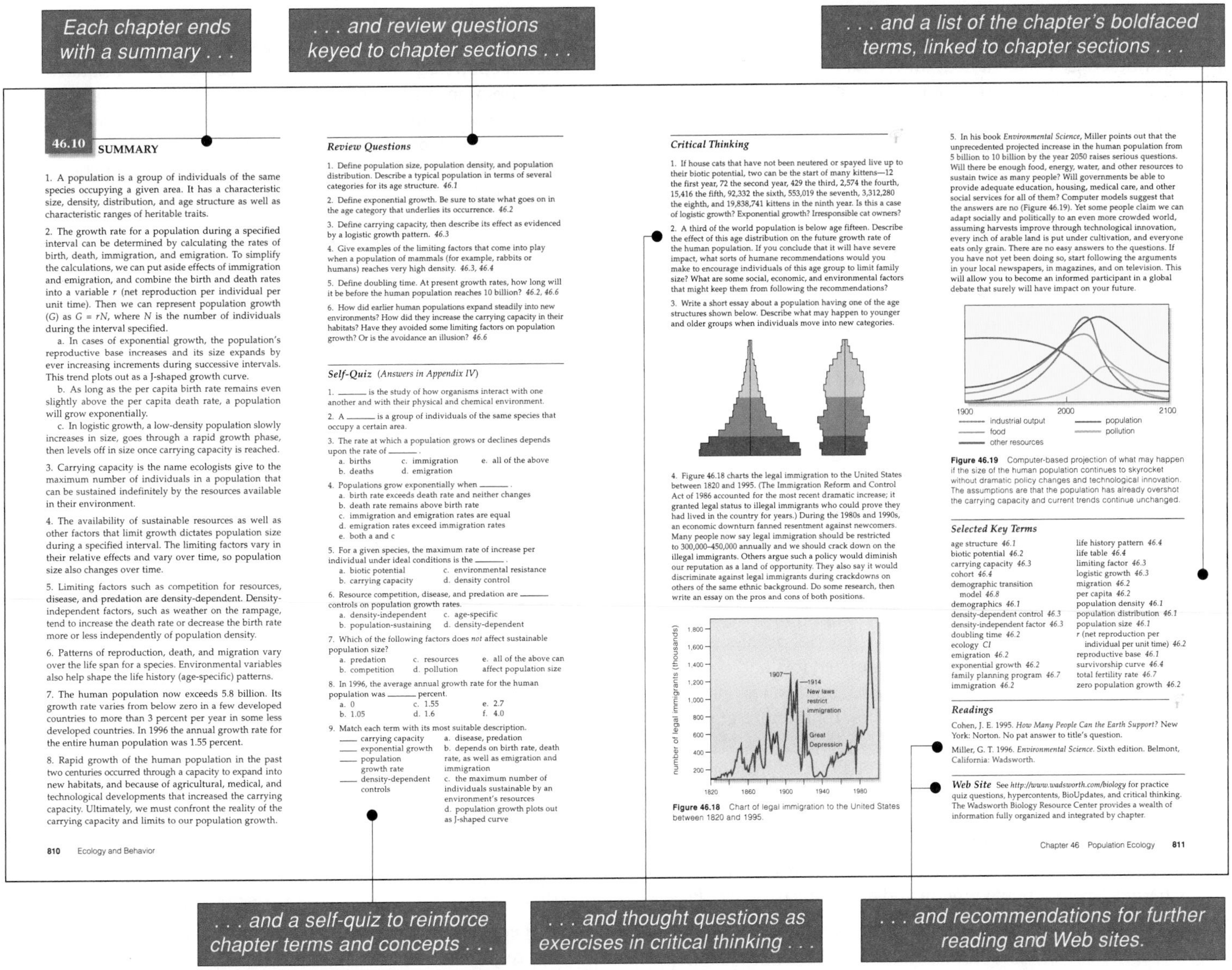

Each chapter ends with a summary . . .
. . . and review questions keyed to chapter sections . . .
. . . and a list of the chapter's boldfaced terms, linked to chapter sections . . .
. . . and a self-quiz to reinforce chapter terms and concepts . . .
. . . and thought questions as exercises in critical thinking . . .
. . . and recommendations for further reading and Web sites.
number of legal immigrants (thousands)
1,800
1,600
1,400
1,200
1,000
800
600
400
200
1820
1860
1900
1940
1980
1907
1914
New laws restrict immigration
Great Depression
1900
2000
2100
industrial output
food
other resources
population
pollution

46.10 SUMMARY

1. A population is a group of individuals of the same species occupying a given area. It has a characteristic size, density, distribution, and age structure as well as characteristic ranges of heritable traits.

2. The growth rate for a population during a specified interval can be determined by calculating the rates of birth, death, immigration, and emigration. To simplify the calculations, we can put aside effects of immigration and emigration, and combine the birth and death rates into a variable r (net reproduction per individual per unit time). Then we can represent population growth (G) as $G = rN$, where N is the number of individuals during the interval specified.

a. In cases of exponential growth, the population's reproductive base increases and its size expands by ever increasing increments during successive intervals. This trend plots out as a J-shaped growth curve.

b. As long as the per capita birth rate remains even slightly above the per capita death rate, a population will grow exponentially.

c. In logistic growth, a low-density population slowly increases in size, goes through a rapid growth phase, then levels off in size once carrying capacity is reached.

3. Carrying capacity is the name ecologists give to the maximum number of individuals in a population that can be sustained indefinitely by the resources available in their environment.

4. The availability of sustainable resources as well as other factors that limit growth dictates population size during a specified interval. The limiting factors vary in their relative effects and vary over time, so population size also changes over time.

5. Limiting factors such as competition for resources, disease, and predation are density-dependent. Density-independent factors, such as weather on the rampage, tend to increase the death rate or decrease the birth rate more or less independently of population density.

6. Patterns of reproduction, death, and migration vary over the life span for a species. Environmental variables also help shape the life history (age-specific) patterns.

7. The human population now exceeds 5.8 billion. Its growth rate varies from below zero in a few developed countries to more than 3 percent per year in some less developed countries. In 1996 the annual growth rate for the entire human population was 1.55 percent.

8. Rapid growth of the human population in the past two centuries occurred through a capacity to expand into new habitats, and because of agricultural, medical, and technological developments that increased the carrying capacity. Ultimately, we must confront the reality of the carrying capacity and limits to our population growth.

810 Ecology and Behavior

Review Questions

1. Define population size, population density, and population distribution. Describe a typical population in terms of several categories for its age structure. *46.1*

2. Define exponential growth. Be sure to state what goes on in the age category that underlies its occurrence. *46.2*

3. Define carrying capacity, then describe its effect as evidenced by a logistic growth pattern. *46.3*

4. Give examples of the limiting factors that come into play when a population of mammals (for example, rabbits or humans) reaches very high density. *46.3, 46.4*

5. Define doubling time. At present growth rates, how long will it be before the human population reaches 10 billion? *46.2, 46.6*

6. How did earlier human populations expand steadily into new environments? How did they increase the carrying capacity in their habitats? Have they avoided some limiting factors on population growth? Or is the avoidance an illusion? *46.6*

Self-Quiz (*Answers in Appendix IV*)

1. ______ is the study of how organisms interact with one another and with their physical and chemical environment.

2. A ______ is a group of individuals of the same species that occupy a certain area.

3. The rate at which a population grows or declines depends upon the rate of ______.
a. births c. immigration e. all of the above
b. deaths d. emigration

4. Populations grow exponentially when ______.
a. birth rate exceeds death rate and neither changes
b. death rate remains above birth rate
c. immigration and emigration rates are equal
d. emigration rates exceed immigration rates
e. both a and c

5. For a given species, the maximum rate of increase per individual under ideal conditions is the ______.
a. biotic potential c. environmental resistance
b. carrying capacity d. density control

6. Resource competition, disease, and predation are ______ controls on population growth rates.
a. density-independent c. age-specific
b. population-sustaining d. density-dependent

7. Which of the following factors does *not* affect sustainable population size?
a. predation c. resources e. all of the above can
b. competition d. pollution affect population size

8. In 1996, the average annual growth rate for the human population was ______ percent.
a. 0 c. 1.55 e. 2.7
b. 1.05 d. 1.6 f. 4.0

9. Match each term with its most suitable description.
___ carrying capacity a. disease, predation
___ exponential growth b. depends on birth rate, death rate, as well as emigration and immigration
___ population growth rate c. the maximum number of individuals sustainable by an environment's resources
___ density-dependent controls d. population growth plots out as J-shaped curve

Critical Thinking

1. If house cats that have not been neutered or spayed live up to their biotic potential, two can be the start of many kittens—12 the first year, 72 the second year, 429 the third, 2,574 the fourth, 15,416 the fifth, 92,332 the sixth, 553,019 the seventh, 3,312,280 the eighth, and 19,838,741 kittens in the ninth year. Is this a case of logistic growth? Exponential growth? Irresponsible cat owners?

2. A third of the world population is below age fifteen. Describe the effect of this age distribution on the future growth rate of the human population. If you conclude that it will have severe impact, what sorts of humane recommendations would you make to encourage individuals of this age group to limit family size? What are some social, economic, and environmental factors that might keep them from following the recommendations?

3. Write a short essay about a population having one of the age structures shown below. Describe what may happen to younger and older groups when individuals move into new categories.

4. Figure 46.18 charts the legal immigration to the United States between 1820 and 1995. (The Immigration Reform and Control Act of 1986 accounted for the most recent dramatic increase; it granted legal status to illegal immigrants who could prove they had lived in the country for years.) During the 1980s and 1990s, an economic downturn fanned resentment against newcomers. Many people now say legal immigration should be restricted to 300,000–450,000 annually and we should crack down on the illegal immigrants. Others argue such a policy would diminish our reputation as a land of opportunity. They also say it would discriminate against legal immigrants during crackdowns on others of the same ethnic background. Do some research, then write an essay on the pros and cons of both positions.

Figure 46.18 Chart of legal immigration to the United States between 1820 and 1995.

5. In his book *Environmental Science*, Miller points out that the unprecedented projected increase in the human population from 5 billion to 10 billion by the year 2050 raises serious questions. Will there be enough food, energy, water, and other resources to sustain twice as many people? Will governments be able to provide adequate education, housing, medical care, and other social services for all of them? Computer models suggest that the answers are no (Figure 46.19). Yet some people claim we can adapt socially and politically to an even more crowded world, assuming harvests improve through technological innovation, every inch of arable land is put under cultivation, and everyone eats only grain. There are no easy answers to the questions. If you have not yet been doing so, start following the arguments in your local newspapers, in magazines, and on television. This will allow you to become an informed participant in a global debate that surely will have impact on your future.

Figure 46.19 Computer-based projection of what may happen if the size of the human population continues to skyrocket without dramatic policy changes and technological innovation. The assumptions are that the population has already overshot the carrying capacity and current trends continue unchanged.

Selected Key Terms

age structure *46.1*
biotic potential *46.2*
carrying capacity *46.3*
cohort *46.4*
demographic transition model *46.8*
demographics *46.1*
density-dependent control *46.3*
density-independent factor *46.3*
doubling time *46.2*
ecology *CI*
emigration *46.2*
exponential growth *46.2*
family planning program *46.7*
immigration *46.2*
life history pattern *46.4*
life table *46.4*
limiting factor *46.3*
logistic growth *46.3*
migration *46.2*
per capita *46.2*
population density *46.1*
population distribution *46.1*
population size *46.1*
r (net reproduction per individual per unit time) *46.2*
reproductive base *46.1*
survivorship curve *46.4*
total fertility rate *46.7*
zero population growth *46.2*

Readings

Cohen, J. E. 1995. *How Many People Can the Earth Support?* New York: Norton. No pat answer to title's question.

Miller, G. T. 1996. *Environmental Science*. Sixth edition. Belmont, California: Wadsworth.

Web Site See *http://www.wadsworth.com/biology* for practice quiz questions, hypercontents, BioUpdates, and critical thinking. The Wadsworth Biology Resource Center provides a wealth of information fully organized and integrated by chapter.

Chapter 46 Population Ecology 811

FIGURE D *Study aids at the end of a chapter from this edition.*

CONTENT REVISIONS

Instructors who use *Biology: The Unity and Diversity of Life* may wish to evaluate this overview of key modifications. Overall, conceptual development is more integrated. The writing is still crisp but not too brief, because some topics can confuse students when presented in insufficient detail. New research called for some adjustments in the overall framework. For example, we subscribe to the six-kingdom classification scheme, which now has exceptional support from comparative biochemical studies. Even the end-of-chapter Critical Thinking questions incorporate current material, as on population growth models (Section 46.10).

INTRODUCTION This conceptual overview for the book is more focused, starting with introductions to the molecular trinity (DNA to RNA to protein), energy, and the levels of biological organization. An early glimpse of life's diversity reflects the six-kingdom model. We follow it with a simple explanation of evolution by natural selection, a dominant theme throughout the book. We treat scientific methods in more detail. To highlight the power of scientific inquiry, we have a new spread on experimental tests of an alternative to antibiotics. This detailed, timely example builds on an earlier description of selection for antibiotic resistance.

UNIT I. PRINCIPLES OF CELLULAR LIFE Chapter 1 opens with a new vignette on a chemical element and a current application (phytoremediation, as at Chernobyl). Sections on radioactive decay and on acids, bases, and buffers are rewritten. Chapter 3 opens with a "carbon story" about swings in atmospheric CO_2. It has an essay on pesticides. Chapter 4 has better treatment of cytoskeletal elements and accessory proteins, including myosin. Chapter 5 has improved explanations of membrane proteins, diffusion, osmosis, and membrane transport mechanisms. A rewrite of Chapter 6 makes the ground rules of metabolism more intelligible. The photosynthesis chapter has new icons, new concept spreads on properties of light and photosynthetic pigments, and a visual comparison of carbon-fixing routes. Chapter 8 has a new illustration of the energy harvest from aerobic respiration, and a more lively, updated treatment of alternative energy sources in humans and other mammals.

UNIT II. PRINCIPLES OF INHERITANCE Chapter 9 includes a new spread on the cell cycle and chromosome structure. Chapter 11 includes rewrites on pleiotropy, phenotypic variation, and some new, illustrated genetics problems. We strengthened Chapter 12 with reorganized text and new art on linkage, recombination patterns, linkage mapping, and changes in chromosome structure. It has more graphic examples of inheritance patterns. Chapter 13 has a visual overview as well as a detailed look at DNA replication and an essay on DNA mutations and cancer. We improved the Chapter 14 treatment of mutation. Chapter 15 opens with a new vignette on apoptosis, a prelude to an updated section on cancer at the chapter's end. Definitions are clarified. Chapter 16 has a new vignette and a new essay on RFLPs.

UNIT III. EVOLUTION Chapter 17 clarifies the influence of theories of catastrophism and uniformity on evolutionary thought. It has a new essay on "missing links." Chapter 18 has expanded treatment of phenotypic variation and of mutation. We now discuss natural selection first, with an overview followed by three concept spreads of examples. Distribution maps for malaria and sickle-cell anemia now accompany the text on balanced polymorphism. We have expanded, improved text and a new computer model for genetic drift. Chapter 19 on speciation is overhauled and has better art. Reorganized Chapter 20 integrates the geologic time scale with the fossil record, has new spreads on plate tectonics and on evolutionary systematics, and has a brief new section that compares the five-kingdom, three-domain, and six-kingdom schemes. Recently determined absolute dates for the Archean eon are incorporated in the chapter.

UNIT IV. EVOLUTION AND DIVERSITY Chapter 21 has a tree of life adapted to the six-kingdom scheme. Ediacaran and Cambrian forms are defined, and gymnosperm dominance is more clearly correlated with the Mesozoic. The global broiling theory is correlated with the K-T asteroid impact. Besides being a conceptual and chronological framework for the diversity chapters, this survey chapter may help students sense their place in nature. The greatly revised diversity chapters have crisp evolutionary story lines, a more balanced text-to-art ratio, and riveting applications (see Section 22.9 on infectious diseases). Chapter 23 states how opinions differ on which organisms are protistans. In keeping with current consensus, we treat chytrids, water molds, and algae in that chapter. We have a new red algal life cycle. Chapter 24 includes a better treatment of lichens. Chapter 25 includes more on the Carboniferous and new spreads on existing seed-bearing plants. In the Chapter 26 introduction, I shamelessly segue from my grandchild to the early invertebrates. The chapter has new sections on animal origins, molluscan evolution, and adaptations of arthropods. Chapter 27 is reorganized for clarity, as in the sections on vertebrate evolutionary trends, reptiles, and mammals. Chapter 28 integrates new fossil evidence.

UNIT V. PLANT STRUCTURE AND FUNCTION Chapter 29 has improved text and graphics, as in the section on wood and bark, and an applications essay (Section 29.5). Chapter 30 has a new section on soil and plant nutrients. Chapter 31 opens with a story about the coevolution of plants and pollinators. It includes revised sections on seed and fruit formation and dispersal. In Chapter 32, descriptions of the hormonal effects on plant growth and development are crisper.

UNIT VI. ANIMAL STRUCTURE AND FUNCTION The unit takes a more comparative approach, with selections from numerous invertebrate and vertebrate groups. Chapter 34 has a less abstract model of the sodium-potassium pump as well as reorganized treatment of chemical synapses and paths of information flow. In Chapter 35, sections on the vertebrate brain now adjoin. They describe the reticular formation, blood-brain barrier, and limbic system early on. The section on memory is updated. Chapter 36 has a better classification of sensory receptors, and new, updated sections on somatic sensations and special senses. Chapter 37 now organizes the description of endocrine functions in sections on feedback controls, responses to local chemical changes, and responses to environmental cues. Chapter 38 includes a new introduction and two revised essays. Chapter 39 has refined art, a new section on blood cell disorders, and new descriptions of arterial blood pressure as well as capillary function. Chapter 40 has more on inflammation, new art on the generation of antibody diversity, and updates on HIV replication. Chapter 41 has separated sections on the invertebrates and vertebrates. Its sections on breathing, gas exchange and transport, acclimatization, hypoxia, and the physiology of diving are rewritten and updated.

Chapter 42 opens with a new story on obsessions with dieting. A new essay traces how molecular detectives used obese mice to isolate the gene for leptin, a hormone involved in appetite suppression and higher rates of metabolism. Chapter 43 has better delineation between sections on kidney structure and function, and tightened sections on temperature regulation. Chapter 44 is overhauled to reflect extensive new research and to present a coherent picture of the vital topic of development. It has a new essay on *Drosophila* development, which is yielding astounding insights into development in general. We rewrote much of Chapter 45 and tightened definitions to improve clarity.

UNIT VII. ECOLOGY AND BEHAVIOR   The new Chapter 46 vignette extracts a lesson from Angel Island's fecund deer population, then raises questions concerning growth of the human population. Definitions are crisper. David Reznick provided some of his data for a rewrite of the essay on life history patterns for the guppies of Trinidad. Numbers and examples for the sections on human population growth are updated. We have new age structure diagrams for the baby boomers. Critical thinking questions (Section 46.10) invite reflection on immigration and population increases.

Chapter 47 includes Charles Krebs' long-term study of snowshoe hares as an update on predator/prey cycles. The descriptions of predator/prey adaptations are integrated in an essay on the coevolutionary arms race. The section on parasitic interactions is expanded. The climax-pattern model and restoration ecology are now included in the section on succession. Community instability is treated in more depth, as with new text on categories of geographic dispersal and a new essay on species introductions.

Chapter 48 has an updated essay on the greenhouse effect and the possibility of global warming. The section on the nitrogen cycle has better delineation between nitrogen cycling and nitrogen losses. The section on the phosphorus cycle has new material on eutrophication, including an experiment demonstrating lake eutrophication in Ontario, Canada. George Cox rewrote the section on biological magnification (Section 48.10). In Chapter 49, the section on soil profiles now builds on information presented earlier, in Section 30.1, and has revised art. It has more on coral reefs and banks and, for novice divers, a new photograph that clearly shows what a moray eel looks like. Material on ENSOs is updated and rewritten. Section 49.15 has a critical thinking question on wetlands, including mangrove swamps.

Chapter 50 opens with a new vignette and stunning art on the extent of human population growth in one part of the world (the San Francisco Bay area and the corridor between it and Sacramento). It has new maps of regional differences in atmospheric concentrations of fine particles. The text is updated, as on ozone thinning, the coming water wars, the aftermath of Chernobyl, and alternative energy sources. The chapter has a revised essay on deforestation in the tropics. Charts of energy consumption are updated. We eliminated redundancies in the treatment of animal behavior by combining two chapters (51). Definitions of instinctive behavior, its environmental cues, and response programs are clarified. The section on communication signals and displays is revised. There is crisper delineation between the costs and benefits of living in social groups.

MULTIMEDIA SUPPLEMENTS

1. *Interactive Concepts in Biology.* Packaged free with all student copies, this is the first CD-ROM to address the full sweep of biology. The cross-platform CD-ROM covers all concept spreads in the book. Because students can learn by doing, it encourages them to manipulate the book's art. A combination of text, graphics, photographs, animations, video, and audio enhances each book chapter. The revised CD-ROM has three times as many animations, many more interactive quizzes, and many more interactive exercises.

2. *InfoTrac College Edition.* This on-line library is available FREE with each copy of *Biology: The Unity and Diversity of Life.* It gives students access to full articles— not abstracts—from more than 600 scholarly and popular periodicals dating back as far as four years. The articles are available through InfoTrac's impressive database that has such periodicals as *Discover*, *Audubon*, and *Health*.

3. *Student Guide to InfoTrac College Edition.* This guide is on the Wadsworth Biology Resource Center site on the World Wide Web. It has an introduction to InfoTrac and a set of electronic readings for each chapter, updated frequently. It links some of each chapter's critical thinking questions to InfoTrac articles to invite deeper examination of issues.

4. *Biology Resource Center.* All information is arranged by the eighth edition's chapters. Every month it has new BioUpdates on relevant applications, and hyperlinks and an average of 40 practice quiz questions per chapter. It includes descriptions of degrees and careers in biology, a student feedback site, cool clip art, ideas for teaching on the Web, and a forum where instructors can share ideas on teaching courses. It also includes flashcards for all glossary terms, critical thinking exercises, news groups, surfing lessons including biology surfing, and a variety of search engines, biological games, a BioTutor, and a final Blitz set of practice questions. Perhaps most importantly, it also has Internet exercises for each chapter to guide students to doing more than randomly browse sites. A cool event of the quarter will have an ongoing experiment in which students and instructors can participate. The address for the Wadsworth Biology Resource Center is:

http://www.wadsworth.com/biology

5. *Internet Activities for General Biology.* This book guides students to more productive activities than browsing the Web. Students learn by interactive dissections, surveys, genetic crosses, lab experiments, notice postings, and other diverse activities. It has tear-out worksheets that may be handed in for evaluation.

6. *An Introduction to the Internet.* This 100-page booklet helps students learn how to get around the Internet when using a browser such as Netscape, search engines, e-mail, setting up home pages, and related topics. It lists useful biology sites on the Net that correspond to book chapters.

7. *The Biology Place (www.biology.com).* Wadsworth is an official distributor of this site, created by Peregrine Publishers, Inc., for instructors and students on the World Wide Web. A community of educators who developed and maintain it offer learning activities categorized by topic and type, including interactive study guides, lab investigations, study projects, and collaborative research. One can access Research News, Best of the Web, and *Scientific American Connection* for cutting-edge research. Each learning activity has interactive self-assessment worksheets and notebooks. Results can be printed or e-mailed directly to instructors.

8. *American Botanical Society.* This site will be maintained on the Wadsworth Web page at http://www.thomson.com.

9. *BioLink 2.* With this presentation tool, instructors can easily assemble art and database files with lecture notes to create a fluid lecture that may help stimulate even the least-engaged students. It includes all illustrations in the book, animations and films from the student CD, and art from other Wadsworth biology textbooks. BioLink 2 also has a Kudo Browser with an easy drag-and-drop feature that allows file export into such presentation tools as Power Point. Upon its creation, a file or lecture with BioLink 2 can be posted to the Web, where students can access it for reference or for studying needs.

10. *Overheads and 35mm Slides.* All the micrographs and diagrams in the book are available as overheads and slides that are reproduced in vivid color with large, bold-lettered labels. All of the diagrams are on *CD-ROM, Biolink 2.*

11. *Cycles of Life: Exploring Biology.* Twenty-six programs of this telecourse feature compelling footage from around the world, original microvideography, and spectacular 3D animation. A student study guide, faculty manual, and laboratory manual are included. For information on course licensing and pricing, send queries to Coast Telecourses by fax 714-241-6286 or telephone 1-800-547-4748.

12. *Animations and Films from Cycles of Life Telecourse.* Tape One has animations for cell structure and function, and for principles of inheritance and evolution. Tape Two has more on diversity, plant structure and function, animal structure and function, and ecology and behavior.

13. *CNN Videos.* Produced by Turner Learning, nine videos can stimulate and engage students. They cover general biology, anatomy-physiology, and environmental science. New tapes are offered every year for these topics.

14. *Protecting Endangered Species* and *Science in the Rain Forest Videos.* Current treatment of these major issues.

15. *Life Science Video Library.* Films for the Humanities and CNN created this library.

16. *Wadsworth Biology Videodisc.* Carolina Biological Supply and Bill Surver (Clemson University) collaborated on this videodisc with new animations and films. Line art has large, boldface labels and often step-by-step or full-motion animation. It is available with fill-in-the-blank labels for tests. There are 3,500 still photographs, a correlation directory, bar code guide, and hypercard and toolbook software. All items are organized by book chapter.

17. *West's Biology Videodisc.* This videodisc provides thousands of additional images.

18. *Liquid Assets—The Ecology of Water.* Two double-sided level-3 videodiscs compare the ecologies of the Everglades and San Francisco Bay. A study guide is included.

19. *STELLA II.* Adopters can get a version of this software tool to develop critical thinking skills, and a workbook. This modeling software has 23 simulations. The textbook's critical thinking questions and 150 additional ones are arranged by chapter in *Critical Thinking Exercises.*

20. *Electronic Study Guide.* This has an average of 40 multiple-choice questions per book chapter that differ from those in the test-item booklet. Students respond to each question, and then an on-screen prompt allows them to review their answers and learn why they are correct or incorrect.

21. *West Nutrition CD.* This interactive learning tool has animations, video, hands-on exercises, and a glossary with pronunciation guides. In-depth sections allow students to learn more about the biochemistry of particular topics.

ADDITIONAL SUPPLEMENTS

Seven respected test writers created the *Test Items* booklet. The booklet has more than 4,500 questions in electronic form for IBM and Macintosh in a test-generating data manager. Questions also are available in Microsoft Word for Windows, DOS, and Mac.

For each book chapter, an *Instructor's Resource Manual* has an outline, objectives, list of boldface or italic terms, and detailed lecture outline, ideas for lectures, classroom and lab demonstrations, discussion questions, research paper topics, and annotations for filmstrips and videos. For those who wish to modify material, this resource manual is available electronically on the Wadsworth Web page.

An interactive *Study Guide and Workbook* lets students write answers to questions, which are arranged by chapter section with references to specific text pages. For those who wish to modify or select parts, *chapter objectives* are available on disk in the testing file.

A 100-page *Answer Booklet* answers the textbook's review questions. (The answers to the self-quizzes and genetics problems are given in appendixes to the book itself.)

Flashcards show 1,000 glossary items. A booklet, *Building Your Life Science Vocabulary,* helps students learn biological terms by explaining root words and their applications.

Study Skills for Science Students. Daniel Chiras's guide explains how to develop good study habits, sharpen memory and learning, prepare for tests, and produce term papers.

Strategies for Success: Learning Skills Booklets. Individual learning skills chapters from Gardner/Jewler's best-selling college success text can be customized together or singly and bundled with the text. Some chapters cover managing time, test taking, writing and speaking, note taking, reading and memory, computers and the Internet, critical thinking, and campus resources including the library.

Jim Perry and David Morton's *Laboratory Manual* has 38 experiments and exercises, with 600+ full-color, labeled photographs and diagrams. Many experiments are divided in parts for individual assignment, depending on available time. Each consists of objectives, discussion (introduction, background, and relevance), list of materials for each part, procedural steps, prelab questions, and post-lab questions. An *Instructor's Manual* for the lab manual lists quantities, procedure for preparing reagents, time requirements for each part of an exercise, hints to make the lab a success, and vendors of materials with item numbers. It has more investigative exercises that can be copied for lab use.

Customized Laboratory Manuals by Phillip Shelp and its accompanying instructor's manual can be tailored for individual courses. A new *Photo Atlas* has 700 full-color, labeled photographs and micrographs of the cells and organisms that students typically deal with in the lab.

Eight additional readings supplements are available: • *Contemporary Readings in Biology* is a collection of articles on applications of interest. • *Current Perspectives in Biology* is another collection of articles. • *A Beginner's Guide to Scientific Method* is a supplement for those who wish to treat this topic in detail. • *The Game of Science* gives students a realistic view of what science is and what scientists do. • *Environment: Problems and Solutions* provides a brief 120-page introduction to environmental concerns. • *Green Lives, Green Campuses* is a hands-on workbook to help students evaluate the environmental impact of their own life-styles. • *Watersheds: Classic Cases in Environmental Ethics* is a collection of important case studies. • *Environmental Ethics: An Introduction to Environment Philosophy* surveys environmental ethics and recent philosophical positions.

A COMMUNITY EFFORT

One, two, or a smattering of authors can write accurately and often very well about their field of interest, but it takes more than this to deal with the full breadth of the biological sciences. For us, it takes an educational network that includes more than 2,000 teachers, researchers, and photographers in the United States, Canada, England, Germany, France, Australia, Sweden, and elsewhere. On the next two pages, we acknowledge reviewers whose contributions continue to shape our thinking. There simply is no way to describe the thoughtful effort these individuals and others before them gave to our books. We can only salute their commitment to quality in education.

In large part, *Biology: The Unity and Diversity of Life* is widely respected because it reflects the understandings of our general advisors and contributors and their abiding concern for students. Steve Wolfe, author of acclaimed books on cell biology, works out alternative phrasings with us, sometimes line by line over the phone. Daniel Fairbanks, scholar and gentleman, still finds the time to ferret out errors that creep into the genetics manuscripts. Katherine Denniston, another long-time advisor, wrote original critical thinking questions, itself no small feat. Aaron Bauer, Paul Hertz, Samuel Sweet, and Jerry Coyne helped chisel major parts of the evolution unit. Jerry also created our new computer simulation of genetic drift. The unit on plant structure and function is strong, thanks to the initial resource manuscripts from Cleon Ross.

And what would we do without Robert Lapen, who wrote the definitive manuscript on immunology and lived to tell about it? What would we do if our abiding friends Gene Kozloff, John Jackson, and Ron Hoham did not work diligently to stop us from inventing biology? Also for this edition, in collaboration with Linda Beidleman, Gene assembled the book's detailed Index.

We thank Rob Colwell, George Cox, and Tyler Miller for their contributions to the ecology unit. John Alcock, author of a respected book on behavior, provided resource manuscripts for our animal behavior chapter. Bruce Levin and Jim Bull provided prepublication data when we wrote a new essay on experimental testing. Lauralee Sherwood, author of a fine animal physiology text, helped refine the respiration chapter. As always, Jane Taylor and Larry Sellers were meticulous in their attention to detail.

Also over the years, Nancy Dengler, Bruce Holmes, David Morton, and Frank Salisbury have been guardians of accuracy and teachability. So has Tom Garrison, himself a seasoned author, who dispenses sympathy with wit better than just about anybody. This time, Tom and Don Collins generously opened their lecture hall and laboratory, and the revelations will shape our thinking for years to come.

Mary Douglas has now become the finest production manager in the business. We cannot imagine our complex revisions traveling through computerized production without guidance from talented, even-tempered, flexible Mary. She also falls in the category of author's shrink. Myrna Engler-Forkner, in charge of oversight and art coordination, has become as patient and superb as Mary. Gary Head, Quarkmeister, has the patience to teach even computer-challenged authors how to do page layouts. He designed the book and its cover. He went out on many photographic assignments and came back with such treats as Fred-and-Ginger, the streetwise snail in Figure 17.1.

Sandra Craig is a good sport and a fine shepherdess who keeps editorial production on its convoluted route. Kristin Milotich is in charge of a vast enterprise known as the book's supplements program. She is remarkable for her endurance, a big heart, and no attitude problem. Our developmental editor and good friend, Mary Arbogast, is in a class by herself. Pat Waldo, force of nature, outruns clocks and moves mountains to keep us at the forefront of multimedia for biology education. Thanks to Chris Evers, outstanding author, consultant, and Pat's secret weapon in digital publishing. Thanks also to the rest of the 0's and 1's Club, Stephen Rapley, Steve Bolinger, Jennie Redwitz, and Cooperative Media Group.

If authors could invent the most supportive president of a publishing house, they would come up with Susan Badger. If we could bottle Halee Dinsey's energy, we would be able to power all cities west of the Rockies. Gary Carlson, Michael Burgreen, Andrea Geanacopoulos, and John Walker add formidable talent to the biology team. Stephen Forsling signed on as photo researcher, and Roberta Broyer as permissions editor. Carol Lawson and Rebecca Linquist cheerfully kept us Quarking. Kathie Head, Pat Brewer, Peggy Meehan, and Stephen Rapley have made valuable contributions for many years.

Of the artists with whom we have worked, Raychel Ciemma remains the best. She works directly with us to turn rough sketches into works of art. We also are grateful for the dedication of Mary Roybal, Karen Stough, and the others listed on the copyright page.

Jim Gallagher, J.C. Morgan, and Jan Troutt at Precision Graphics are superb professionals. Tom Anderson, John Deady, Nancy Dean, and Rich Stanislawski at H&S Graphics cheerfully put up with our interminable pursuit of excellence.

Twenty-two years ago, Jack Carey convinced us to write this book. Ever since, he has remained close counselor and abiding friend. And nothing, in all that time, has shaken our shared belief in the intrinsic capacity of biology to enrich the lives of each new generation of students.

GENERAL ADVISORS AND CONTRIBUTORS

JOHN ALCOCK, *Arizona State University*
AARON BAUER, *University of Chicago*
ROBERT COLWELL, *University of Connecticut*
JERRY COYNE, *University of Chicago*
GEORGE COX, *San Diego State University*
KATHERINE DENNISTON, *Towson State University*
DANIEL FAIRBANKS, *Brigham Young University*
PAUL HERTZ, *Barnard College*
JOHN JACKSON, *North Hennepin Community College*
RONALD HOHAM, *Colgate University*
EUGENE KOZLOFF, *University of Washington*
ROBERT LAPEN, *Central Washington University*
WILLIAM PARSON, *University of Washington*
CLEON ROSS, *Colorado State University*
SAMUEL SWEET, *University of California, Santa Barbara*
STEPHEN WOLFE, *University of California, Davis*

REVIEWERS

ABBAS, ABDUL, *Brigham Womens Hospital*
ALDRIDGE, DAVID, *North Carolina State University*
AMOS, BONNIE, *Angelo State University*
ANDERSON, DEBRA, *Oregon Health Sciences University*
ARMSTRONG, PETER, *University of California, Davis*
ARNOLD, STEVAN, *University of Chicago*
ATCHISON, GARY, *Iowa State University*
BAJER, ANDREW, *University of Oregon*
BAKKEN, AIMEE, *University of Washington*
BAPTISTA, LUIS, *California Academy of Sciences*
BARKWORTH, MARY, *Utah State University*
BATZING, J. L., *State University of New York, Cortland*
BECK, CHARLES, *University of Michigan*
BEECHER, MICHAEL, *University of Washington*
BEEVERS, HARRY, *University of California, Santa Cruz*
BELL, MICHAEL, *Collin County Community College*
BELL, ROBERT, *University of Wisconsin, Stevens Point*
BENACERRAF, BARUJ, *Harvard School of Medicine*
BENDER, KRISTEN, *California State University, Long Beach*
BERGSTROM, BRADLEY, *Valdosta State College*
BIDLACK, JAMES, *University of Central Oklahoma*
BIRKEY, C. WILLIAM, *Ohio State University*
BISHOP, VERNON, *University of Texas, San Antonio*
BLYSTONE, ROBERT, *Trinity University*
BOHR, D. F., *University of Michigan Medical School*
BOOHAR, RICHARD, *University of Nebraska, Lincoln*
BOTTRELL, CLYDE, *Tarrant County Junior College South*
BRADBURY, JACK, *University of California, San Diego*
BRENGELMANN, GEORGE, *University of Washington*
BRINKLEY, B., *University of Alabama, Birmingham*
BROMAGE, TIM, *City University of New York*
BROOKS, VIRGINIA, *Oregon Health Sciences University*
BROWN, ARTHUR, *University of Arkansas*
BROWNING, MARK, *Purdue University*
BUCKNER, VIRGINIA, *Johnson County Community College*
BURNES, E. R., *University of Arkansas, Little Rock*
CABOT, JOHN, *State University of New York, Stony Brook*
CALVIN, CLYDE, *Portland State University*
CARLSON, ALBERT, *State University of New York, Stony Brook*
CASE, CHRISTINE, *Skyline College*
CASE, TED, *University of California, San Diego*
CASEY, KENNETH, *Veterans Administration Medical Center*
CENTANNI, RUSSELL, *Boise State University*
CHAMPION, REBECCA, *Kennesaw State College*
CHAPMAN, DAVID, *University of California, Los Angeles*
CHARLESWORTH, BRIAN, *University of Chicago*
CHISZAR, DAVID, *University of Colorado*
CHRISTENSEN, A. KENT, *University of Michigan Medical School*
CHRISTIAN, DONALD, *University of Minnesota*
CLAIRBORNE, JAMES, *Georgia Southern University*
COLAVITO, MARY, *Santa Monica College*
CONKEY, JIM, *Truckee Meadows Community College*
CONNELL, MARY, *Appalachian State University*
CROWCROFT, PETER, *University of Texas at Austin*
DANIELS, JUDY, *Washtenau Community College*
DAVIS, DAVID GALE, *University of Alabama*

DAVIS, JERRY, *University of Wisconsin*
DAVIS, ROGER, *University of Maryland, Baltimore County*
DELCOMYN, FRED, *University of Illinois, Urbana*
DEMMANS, DANA, *Finger Lakes Community College*
DENGLER, NANCY, *University of Toronto*
DENISON, WILLIAM, *Oregon State University*
DESAIX, JEAN, *University of North Carolina*
DEWALT, RICHARD, *Louisiana State University*
DIBARTOLOMEIS, SUSAN, *Millersville University of Pennsylvania*
DICKSON, KATHRYN, *California State University, Fullerton*
DIEHL, FRED, *University of Virginia*
DLUZEN, DEAN, *University of Illinois, Urbana*
DOYLE, PATRICK, *Middle Tennessee State University*
DRUGER, MARVIN, *Syracuse University*
DULING, BRIAN, *University of Virginia, Charlottesville*
DUOBINIS-GRAY, LEON, *Murray State University*
EDLIN, GORDON, *University of Hawaii, Manoa*
EICHINGER, DAVID, *Purdue University Main Campus*
ELMORE, HAROLD, *Marshall University*
ENDLER, JOHN, *University of California, Santa Barbara*
ENGLISH, DARREL, *Northern Arizona University*
ERICKSON, GINA, *Highline Community College*
ERWIN, CINDY, *City College of San Francisco*
ESCH, GERALD, *Wake Forest University*
EWALD, PAUL, *Amherst College*
FALK, RICHARD, *University of California, Davis*
FISHER, DAVID, *University of Hawaii, Manoa*
FLESCH, DAVID, *Mansfield University*
FLESSA, KARL, *University of Arizona*
FORTNEY, SUSAN, *Johnson Space Center*
FRAILEY, CARL, *Johnson County Community College*
FROEHLICH, JEFFREY, *University of New Mexico*
FROST-MASON, SALLY, *University of Kansas Main Campus*
FRYE, BERNARD, *University of Texas, Arlington*
FULCHER, THERESA, *Pellissippi State Technical Community College*
FULLER, STEPHEN, *Mary Washington College*
FUNK, FRED, *Northern Arizona University*
GAGLIARDI, GRACE, *Bucks County Community College*
GAINES, MICHAEL, *University of Miami*
GHOLZ, HENRY, *University of Florida*
GIBSON, THOMAS, *San Diego State University*
GOFF, CHRISTOPHER, *Haverford College*
GOODMAN, MAURICE, *University of Massachusetts Medical School*
GORDON, ALBERT, *University of Washington*
GOSZ, JAMES, *University of New Mexico*
GRAHAM, LINDA, *University of Wisconsin*
GRANT, BRUCE, *College of William and Mary*
GREENE, HARRY, *University of California, Berkeley*
GREGG, KATHERINE, *West Virginia Wesleyan College*
HALL, RICHARD, *Capitol University*
HAMM, MICHAEL, *Rutgers, the State University of New Jersey*
HANKEN, JAMES, *University of Colorado*
HANRATTY, PAMELA, *University of Southern Mississippi*
HARDIN, JOYCE, *Hendrix College*
HARLEY, JOHN, *Eastern Kentucky University*
HARRIS, JAMES, *Utah Valley Community College*
HARTNEY, KRISTINE BEHRENTS, *California State University, Fullerton*
HASSAN, ASLAM, *University of Illinois, Urbana*
HELGESON, JEAN, *Collin County Community College*
HELLER, LOIS JANE, *University of Minnesota*
HEPFER, CAROL, *Millersville University of Pennsylvania*
HESS, WILLIAM, *Brigham Young University*
HEWITSON, WALTER, *Bridgewater State College*
HILDEBRAND, MILTON, *University of California, Davis*
HILFER, S. ROBERT, *Temple University*
HINCK, LARRY, *Arkansas State University*
HODGSON, RONALD, *Central Michigan University*
HOLLINGER, TOM, *University of Florida, Gainesville*
HOLMES, BRUCE, *Western Illinois University*
HOLMES, KENNETH, *University of Illinois, Urbana*
HOOKE, ANNE, *Miami University of Ohio*
HOSICK, HOWARD, *Washington State University*
HUERTA, ALFREDO, *Miami University of Ohio*
JAKOBSON, ERIC, *University of Illinois, Urbana*
JENSEN, STEVEN, *Southwest Missouri State University*
JOHNSON, LEONARD, *University of Tennessee School of Medicine*
JOHNSON, TED, *St. Olaf College*
JONES, PATRICIA, *Stanford University*
JUILLERAT, FLORENCE, *Indiana University/Purdue University*
KAYE, GORDON, *Albany Medical School*
KAYNE, MARLENE, *Trenton State College*
KEIM, MARY, *Seminole College*

KELLY, DOUGLAS, *University of Southern California*
KELSEN, STEVEN, *Temple University Hospital*
KENDRICK, BRYCE, *University of Waterloo*
KIGER, JOHN, *University of California, Davis*
KILLIAN, JOELLA, *Mary Washington College*
KIMBALL, JOHN, *Tufts University*
KIRK, HELEN, *University of Western Ontario*
KIRKPATRICK, LEE, *Glendale Community College*
KNUTTGEN, HAROLD, *Boston University*
KREBS, CHARLES, *University of British Columbia*
KREBS, JULIA, *Francis Marion University*
KURIS, ARMAND, *University of California, Santa Barbara*
KUTCHAI, HOWARD, *University of Virginia Medical School*
LANZA, JANET, *University of Arkansas, Little Rock*
LASSITER, WILLIAM, *University of North Carolina, Chapel Hill*
LATIES, GEORGE, *University of California, Los Angeles*
LATTA, VIRGINIA, *Jefferson State Junior College*
LEVY, MATTHEW, *School of Medicine, City University of New York*
LEWIS, LARRY, *Bradford University*
LINDSEY, JERRI, *Tarrant County Junior College*
LITTLE, ROBERT, *Medical College of Georgia*
LOCKE, MICHAEL, *University of Western Ontario*
LOHMEIER, LYNNE, *Mississippi Gulf Coast Community College*
MACKLIN, MONICA, *Northeastern State University*
MADIGAN, MICHAEL, *Southern Illinois University, Carbondale*
MALLOCH, DAVID, *University of Toronto*
MANN, ALAN, *University of Pennsylvania*
MARGULIES, MAURICE, *Rockville, Maryland*
MARR, ELEANOR, *Dutchess Community College*
MARTIN, JAMES, *Reynolds Community College*
MATSON, RONALD, *Kennesaw State College*
MATTHAI, WILLIAM, *Tarrant County Junior College*
MAXSON, LINDA, *Pennsylvania State University*
MAXWELL, JOYCE, *California State University, Northridge*
McCLINTIC, J. ROBERT, *California State University, Fresno*
McCULLOCH, DAVID, *Collin County Community College*
McEDWARD, LARRY, *University of Florida*
McKEE, DOROTHY, *Auburn University, Montgomery*
McNABB, ANNE, *Virginia Polytechnic Institute and State University*
McREYNOLDS, JOHN, *University of Michigan Medical School*
MERTENS, THOMAS, *Ball State University*
MILLER, G. TYLER, *Pittsboro, North Carolina*
MITZNER, WAYNE, *Johns Hopkins University*
MOCK, DOUG, *University of Washington*
MOHRMAN, DAVID, *University of Minnesota*
MOISES, HYLAND, *University of Michigan Medical School*
MONTVILO, JEROME, *Bryant College*
MOORE-LANDECKER, ELIZABETH, *Glassboro State University*
MORBECK, MARY ELLEN, *University of Arizona*
MORRISON, WILLIAM, *Shippensburg University*
MORTON, DAVID, *Frostburg State University*
MOUNT, DAVID, *University of Arizona*
MUCH, DAVID, *Muhlenberg College*
MURPHY, RICHARD, *University of Virginia Medical School*
MURRISH, DAVID, *State University of New York, Binghamton*
MYERS, NORMAN, *Oxford University, England*
MYRES, BRIAN, *Cypress College*
NAGLE, JAMES, *Drew University*
NELSON, RILEY, *University of Texas at Austin*
NEMEROFSKY, ARNOLD, *State University of New York, New Paltz*
NICHOLS-KIRK, HELEN, *University of Western Ontario*
NORRIS, DAVID, *University of Colorado*
O'BRIEN, ELINOR, *Boston College*
OJANLATVA, ANSA, *Sacramento, California*
OLSON, MERLE, *University of Texas Health Science Center*
ORR, CLIFTON, *University of Arkansas*
ORR, ROBERT, *California Academy of Sciences*
PAI, ANNA, *Montclair State College*
PALMBLAD, IVAN, *Utah State University*
PAPPENFUSS, HERBERT, *Boise State University*
PARSONS, THOMAS, *University of Toronto*
PAULY, JOHN, *University of Arkansas for Medical Sciences*
PECHENIK, JAN, *Tufts University*
PECK, JAMES, *University of Arkansas, Little Rock*
PERRY, JAMES, *Frostburg State University*
PETERSON, GARY, *South Dakota State University*
PIERCE, CARL, *Washington University*
PIKE, CARL, *Franklin and Marshall College*
PIPERBERG, JOEL, *Millersville University*
PLEASANTS, BARBARA, *Iowa State University*
PLETT, HAROLD, *Fullerton College*
POWELL, FRANK, *University of California, San Diego*

RALPH, CHARLES, *Colorado State University*
RAWN, CARROLL, *Seton Hall University*
RICKETT, JOHN, *University of Arkansas, Little Rock*
RIEDER, CONLY, *Wadsworth Center for Laboratories and Research*
ROMANO, FRANK, *Jacksonville State University*
ROSE, GREIG, *West Valley College*
ROSEN, FRED, *Harvard University School of Medicine*
ROSS, GERALDINE, *Highline Community College*
ROSS, GORDON, *University of California Medical Center, Los Angeles*
ROSS, IAN, *University of California, Santa Barbara*
ROST, THOMAS, *University of California, Davis*
RUIBAL, RUDOLFO, *University of California, Riverside*
SACHS, GEORGE, *University of California, Los Angeles*
SALISBURY, FRANK, *Utah State University*
SCHAPIRO, HARRIET, *San Diego State University*
SCHECKLER, STEPHEN, *Virginia Polytechnic Institute and State University*
SCHIMEL, DAVID, *NASA Ames Research Center*
SCHLESINGER, WILLIAM, *Duke University*
SCHMID, RUDI, *University of California, Berkeley*
SCHMOYER, IRVIN, *Muhlenberg College*
SCHNERMANN, JURGEN, *University of Michigan School of Medicine*
SCHREIBER, DAN, *Mesa Community College*
SEARLES, RICHARD, *Duke University*
SEHGAL, PREM, *East Carolina University*
SHANHOLTZER, SHERYL, *DeKalb College*
SHARP, ROGER, *University of Nebraska, Omaha*
SHEPHERD, JOHN, *Mayo Medical University*
SHERMAN, PAUL, *Cornell University*
SHERWOOD, LAURALEE, *West Virginia University School of Medicine*
SHOEMAKER, DAVID, *Emory University School of Medicine*
SHONTZ, NANCY, *Grand Valley State University*
SHOPPER, MARILYN, *Johnson County Community College*
SILK, WENDY, *University of California, Davis*
SLATKIN, MONTGOMERY, *University of California, Berkeley*
SLOBODA, ROGER, *Dartmouth College*
SLONE, J. HENRY, *Francis Marion College*
SMILES, MICHAEL, *State University of New York, Farmingdale*
SMITH, MICHAEL, *Valdosta State University*
SOLOMON, NANCY, *Miami University*
SOLOMON, TRAVIS, *Kansas City V.A. Medical Center*
STEELE, KELLY, *Appalachian State University*
STEIN-TAYLOR, JANET, *University of Illinois, Chicago*
STEINERT, KATHLEEN, *Bellevue Community College*
STITT, JOHN, *John B. Pierce Foundation Laboratory*
SULLIVAN, LAWRENCE, *University of Kansas*
SUMMERS, GERALD, *University of Missouri*
SUNDBERG, MARSHALL, *Louisiana State University*
SWAIN, SARAH, *Middle Tennessee State University*
TERHUNE, JERRY, *Jefferson Community College/University of Kentucky*
THAMES, MARC, *Medical College of Virginia*
TIFFANY, LOIS, *Iowa State University*
TIZARD, IAN, *Texas A & M University*
TRAMMELL, JAMES, JR., *Arapahoe Community College*
TROUT, RICHARD, *Oklahoma City Community College*
TUTTLE, JEREMY, *University of Virginia, Charlottesville*
TYSER, ROBIN, *University of Wisconsin, La Crosse*
VALENTINE, JAMES, *University of California, Santa Barbara*
VALTIN, HEINZ, *Dartmouth Medical School*
WAALAND, ROBERT, *University of Washington*
WADE, MICHAEL, *University of Chicago*
WAHLERT, JOHN, *City University of New York, Baruch College*
WALDVOGEL, JERRY, *Clemson University*
WALSH, BRUCE, *University of Arizona*
WARING, RICHARD, *Oregon State University*
WARMBRODT, ROBERT, *University of Maryland*
WARNER, MARGARET, *Indiana University, Krennert Institute*
WEBB, JACQUELINE, *Villanova University*
WEIGL, ANN, *Winston-Salem State University*
WEISBRODT, NORMAN, *University of Texas Medical School, Houston*
WEISS, MARK, *Wayne State University*
WELKIE, GEORGE, *Utah State University*
WENDEROTH, MARY PAT, *University of Washington*
WHEELIS, MARK, *University of California, Davis*
WHIPP, BRIAN, *University of California, Los Angeles*
WHITENBERG, DAVID, *Southwest Texas State University*
WHITTOW, G. CAUSEY, *University of Hawaii School of Medicine*
WILSON, THOMAS, *Miami University*
WINICUR, SANDRA, *Indiana University, South Bend*
WISE, ROBERT, *Francis Scott Key Medical Center*
WOMBLE, MARK, *University of Michigan Medical School*
YONENAKA, SHANNA, *San Francisco State University*
ZIHLMAN, ADRIENNE, *University of California, Santa Cruz*

Current configurations of Earth's oceans and land masses —the geologic stage upon which life's drama continues to unfold. Thousands of separate images were pieced together to create this remarkable cloud-free view of our planet.

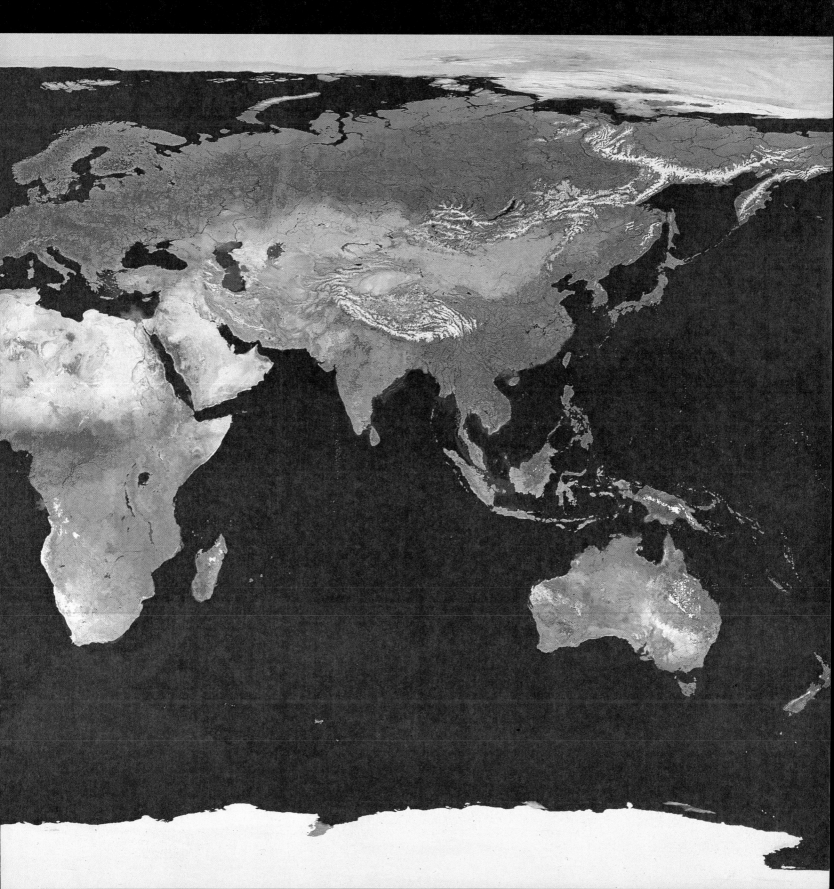

CONCEPTS AND METHODS IN BIOLOGY

Biology Revisited

Buried somewhere in that mass of tissue just above and behind your eyes are memories of first encounters with the living world. In that brain are early memories of discovering your hands and feet, your family, friends, the change of seasons, the smell of rain-drenched earth and grass. Still in residence are memories of early introductions to a great disorganized parade of spiders, flowers, frogs, and furred things, mostly living, sometimes dead. There, too, are memories of questions— "What is life?" and, inevitably, "What is death?" There are also memories of answers, some satisfying, others less so.

By making observations, asking questions, and gradually accumulating answers about the natural world, you have built up a store of knowledge about life. Education as well as experience has been refining your questions, and no doubt some answers are difficult to come by. Think of a young man, only twenty years old, whose motorcycle ran straight into a truck. Now he is in the hospital, with a brain that is functionally dead. If his breathing, heart rate, and other basic functions will proceed only as long as he stays hooked up to a respirator and other highly mechanized support systems, is he still "alive"? Or think of a recently fertilized egg

Figure 1.1 Think back on all you have ever known and seen. This is a foundation for your deeper probes into life.

inside a woman's body. At this early stage, a series of cell divisions has transformed the egg into a cluster of no more than a few dozen microscopically small cells. At what point in its development would you define the growing cell mass as a *human* life? If questions like this have ever crossed your mind, your thoughts about life obviously run deep.

The point is, the book you are starting to read is not your introduction to biology—the study of life—for you have been studying life ever since information started to penetrate your brain. The book simply is biology *revisited*, in ways that may help carry your thoughts to deeper, more organized levels of understanding.

Return to the question, *What is life?* Offhandedly, you might respond that you know it when you see it. To biologists, however, the question opens up a story that has been unfolding in countless directions for several billion years! "Life" is an outcome of ancient events by which nonliving materials became assembled into the first living cells. "Life" is a way of capturing and using energy and raw materials. "Life" is a way of sensing and responding to changes in the environment. "Life" is a capacity to reproduce, grow, and develop. And "life" evolves, meaning that details in the body plan and functions of organisms can change through successive generations.

Yet this short description only hints at the meaning of life. Deeper insight requires wide-ranging study of life's characteristics.

Throughout this book, you will come across many diverse examples of how organisms are constructed, how they function, where they live, and what they do. The examples support certain concepts which, taken together, will give you a sense of what "life" is. This chapter introduces you to the basic concepts. It also sets the stage for forthcoming descriptions of observations, experiments, and tests that help show how biologists develop, modify, and so refine their views of the world around them. As you continue reading the book, you may find it useful to return occasionally to this simple overview as a way of reinforcing your grasp of details.

KEY CONCEPTS

1. There is an underlying unity in the world of life, for all organisms are alike in key respects. They consist of the same kinds of substances, put together according to the same laws that govern matter and energy. Their activities depend on inputs of energy, which they must obtain from their environment. All organisms sense and respond to changing conditions in their environment. And they all grow and reproduce, based on instructions contained in their DNA.

2. There also is immense diversity in the world of life. Millions of different organisms inhabit Earth, and many millions more lived and became extinct over the past 3.8 billion years. And each kind of organism is unique in some of its traits—that is, in some aspects of its body plan, body functions, and behavior.

3. Theories of evolution, especially a theory of natural selection as first formulated by Charles Darwin, help explain the meaning of life's diversity.

4. Biology, like other branches of science, is based on systematic observations, hypotheses, predictions, and relentless observational and experimental tests. The external world, not internal conviction, is the testing ground for scientific theories.

Nothing Lives Without DNA

DNA AND THE MOLECULES OF LIFE   Picture a frog on a rock, busily croaking. Without even thinking about it, you know the frog is alive and the rock is not. But would you be able to explain why? At a fundamental level, both are no more than concentrations of the same units of matter, called protons, electrons, and neutrons. These units are the building blocks of atoms, which are building blocks of larger bits of matter called molecules. And it is at the molecular level that differences between living and nonliving things start to emerge.

You will never, ever find a rock made of nucleic acids, proteins, carbohydrates, and lipids. In the natural world, only cells build particular assortments of these complex molecules, in particular amounts. The **cell** is the smallest unit of matter having the capacity for life; all living things consist of one or more of them. And the cell's signature molecule is a nucleic acid popularly known as **DNA**. No chunk of granite or quartz has it.

Encoded in DNA's structure are the instructions for assembling a dazzling array of proteins from a limited number of smaller building blocks, the amino acids. By analogy, if you follow suitable instructions and invest some energy in the task, you can organize a heap of a few kinds of ceramic tiles (amino acids) into different patterns (diverse proteins), as shown in Figure 1.2.

Among the proteins are enzymes, a class of workers that build, juggle, and split *all* of the complex molecules of life when they get an energy boost. Some enzymes work as partners with another class of nucleic acids, the **RNAs**, in carrying out DNA's instructions. A simple way to think about this is to envision a flow from *DNA to RNA to protein*. As you will read in Chapter 14, this molecular trinity is central to our understanding of life.

THE HERITABILITY OF DNA   Although we humans tend to think we enter the world rather abruptly and leave it the same way, we are much more than this. *We and all other organisms are part of a journey that began almost 4 billion years ago, with the emergence of the first living cells.* Under present-day conditions on Earth, cells can arise only from cells that already exist. They do so by one of life's defining features, called **reproduction**. By this process, parents transmit DNA instructions for duplicating their traits to offspring. Why do baby storks look like storks and not pelicans? They inherited stork DNA, which is not exactly the same as pelican DNA.

The process typically starts with a single cell that contains the DNA of one or two parents. Think of the first cell that forms from the fusion of a sperm (a single cell) with an egg (another single cell). That fertilized egg would not even exist if the sperm and egg had not formed earlier, according to DNA instructions passed on from cell to cell through countless generations.

For frogs and humans and other large organisms, DNA also encodes a developmental program by which single cells divide again and again, with most of their descendants becoming specialized in ways that form different tissues and organs. Think of a moth. Is it only a winged insect? Then what of the fertilized egg that a female moth deposits on a leaf (Figure 1.3)? Inside are DNA instructions for becoming an adult. They guide the egg's development into a caterpillar: an immature, larval stage adapted for rapid feeding and growth. The caterpillar eats and grows until an internal alarm clock goes off. Then tissues undergo drastic remodeling into a different developmental stage, called a pupa. Many cells die; others multiply and become organized in different patterns. In time, an adult emerges that is adapted for reproduction. It contains organs that produce sperm or eggs. And it flutters its distinctly colored and patterned wings at a frequency suitable for attracting a mate.

None of these stages is "the insect." The insect is a series of organized stages, from one fertilized egg to the next. Each stage is vital for the ultimate production of new moths. The instructions for each stage were written into moth DNA long before each individual moment of reproduction—and so the ancient moth story continues.

Nothing Lives Without Energy

ENERGY DEFINED   Everything in the entire universe has some amount of **energy**, which is most simply defined as a capacity to do work. And nothing—absolutely nothing—happens in the universe without a complete or partial *transfer* of energy. For example, a solitary atom does nothing except vibrate incessantly with its

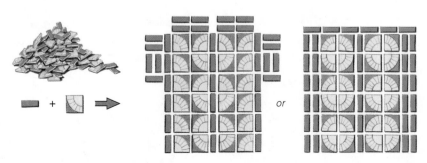

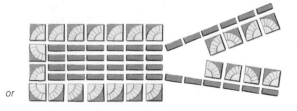

Figure 1.2   Examples of objects built from the same materials but with different assembly instructions.

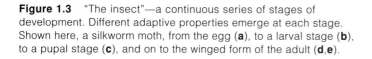

Figure 1.3 "The insect"—a continuous series of stages of development. Different adaptive properties emerge at each stage. Shown here, a silkworm moth, from the egg (**a**), to a larval stage (**b**), to a pupal stage (**c**), and on to the winged form of the adult (**d,e**).

own inherent energy. Suppose the atom absorbs extra energy from the sun's rays and starts vibrating faster. Now some energy is on the move, and it can do work by getting transferred elsewhere. By chance, the atom collides with a neighboring atom, and one may give up, grab, or share energy with the other. Molecules form, stay together, become rearranged, and may get split by such energy transfers. Without them, organisms cannot stay alive, grow, and reproduce.

METABOLISM DEFINED By now, you have an idea that a cell has the capacity to (1) obtain and convert energy from its surroundings and (2) use energy to maintain itself, grow, and make more cells. We call this capacity **metabolism**. Think of a food-producing cell in a leaf. By a process known as **photosynthesis**, it intercepts energy from the sun and uses it to form ATP, an energy carrier. ATP transfers energy to metabolic workers—in this case, enzymes that assemble sugars. Thus energy from the sun becomes stored as energy in sugar molecules. Later, by a process called **aerobic respiration**, some energy released from sugars and other molecules is intercepted to form more ATP, which helps drive hundreds of activities.

SENSING AND RESPONDING TO ENERGY It is often said that only organisms respond to the environment. Yet even a rock shows responsiveness, as when it yields to the force of gravity and tumbles down a hill or changes its shape slowly under the repeated battering of wind, rain, or tides. The difference is this: *Organisms sense changes in their surroundings, then they make controlled, compensatory responses to them.* How? Each organism has **receptors**, which are molecules and structures that can detect specific stimuli. A **stimulus** is any energy change in the environment, such as a variation in the amount of light or heat, that a receptor can detect.

Cells adjust metabolic activities in response to signals from receptors. Each cell (and organism) can withstand only so much heat or cold. It must rid itself of harmful substances. It requires certain foods, in certain amounts. Yet temperatures do shift, harmful substances might be encountered, and food is sometimes plentiful or scarce. Think of what happens after you finish a snack. Simple sugars leave the gut and enter your blood, which is part of your *internal* environment (the other part is tissue fluid that bathes cells). Over the long term, too much or too little blood sugar causes problems, such as diabetes. Normally, though, a glandular organ called the pancreas steps up its secretion of insulin as the sugar level rises. Most of your cells have receptors for this hormone, which stimulates cells to take up sugar. When they do, the sugar level in blood returns to normal.

What if you skip dinner and the sugar level declines? A different hormone stimulates liver cells to tap their stores of energy-rich molecules and degrade them to simple sugars. Sugars released from the cells enter the blood and help return the blood sugar level to normal.

Organisms respond so exquisitely to energy changes that their internal operating conditions remain within tolerable limits. We call this a state of **homeostasis**, and it is one of the key defining features of life.

All organisms consist of one or more cells, the smallest units of life. Under present-day conditions, new cells form only through the reproduction of cells that already exist.

DNA, the molecule of inheritance, encodes protein-building instructions, which RNAs help carry out. Many of the proteins are enzymes, and these metabolic workers are necessary to construct all of the complex molecules of life.

Cells live only as long as they engage in metabolism. They acquire and transfer energy about to assemble, break down, stockpile, and dispose of materials in ways that promote survival and reproduction.

Single cells and multicelled organisms sense and respond to specific energy changes in the environment in ways that help maintain their internal operating conditions.

1.2 ENERGY AND LIFE'S ORGANIZATION

Levels of Biological Organization

Taken as a whole, the metabolic activities of single cells and multicelled organisms maintain the great pattern of organization in nature, as sketched out in Figure 1.4. Consider this hierarchy. Life's properties emerge when DNA and other molecules become organized into cells. We can formally define the **cell** as the smallest unit of organization having a capacity to survive and reproduce on its own, given suitable conditions and building blocks, DNA instructions, and energy inputs. Amoebas, which are free-living single cells, clearly are like this. Does the definition still apply to **multicelled organisms**, which consist of specialized, interdependent cells that are typically organized in tissues and organs? Yes. You may find this a strange answer. After all, your own cells could never live alone in nature, because body fluids must continually bathe them. Yet even isolated human cells stay alive under controlled laboratory conditions. Investigators routinely maintain isolated human cells for use in important experiments, as in cancer studies.

BIOSPHERE
All those regions of Earth's waters, crust, and atmosphere in which organisms can exist

ECOSYSTEM
A community and its physical environment

COMMUNITY
The populations of all species occupying the same area

POPULATION
A group of individuals of the same kind (that is, the same species) occupying a given area

MULTICELLED ORGANISM
An individual composed of specialized, interdependent cells most often organized in tissues, organs, and organ systems

ORGAN SYSTEM
Two or more organs interacting chemically, physically, or both in ways that contribute to survival of the whole organism

ORGAN
A structural unit in which a number of tissues, combined in specific amounts and patterns, perform a common task

TISSUE
An organized group of cells and surrounding substances functioning together in a specialized activity

CELL
Smallest unit having the capacity to live and reproduce, independently or as part of a multicelled organism

ORGANELLE
Inside all cells except bacteria, a membrane-bound sac or compartment for a separate, specialized task

MOLECULE
A unit in which two or more atoms of the same element or different ones are bonded together

ATOM
Smallest unit of an element (a fundamental substance) that still retains the properties of that element

SUBATOMIC PARTICLE
An electron, proton, or neutron; one of the three major particles of which atoms are composed

Figure 1.4 Levels of organization in nature.

You typically find cells and multicelled organisms as part of a **population**, defined as a group of organisms of the same kind, such as a herd of zebras. The next level of organization is the **community**—all the populations of all species living in the same area (such as the African savanna's bacteria, grasses, zebras, lions, and so on). The next level, the **ecosystem**, is the community *and*

Producers trap, convert, and use or store some energy from the sun.

PRODUCERS

NUTRIENT CYCLING

CONSUMERS, DECOMPOSERS

ONE-WAY FLOW OF ENERGY

Energy is transferred from one organism to another; in time, all flows back to the environment.

Figure 1.5 An example of the one-way energy flow and the cycling of materials through the biosphere. (**a**) Plants of a warm, dry grassland called the African savanna capture energy from the sun and use it to build plant parts. Some of the energy ends up in plant-eating organisms, including this adult male elephant. He eats huge quantities of plants to maintain his eight-ton self and produces huge piles of solid wastes—dung—that still contain some unused nutrients. Thus, although most organisms would not recognize it as such, elephant dung is an exploitable food source.

(**b**) And so we next have little dung beetles rushing to the scene almost simultaneously with the uplifting of an elephant tail. Working rapidly, they carve fragments of moist dung into round balls, which they roll off and bury in burrows. In the balls the beetles lay eggs—a reproductive behavior that helps assure their forthcoming offspring (**c**) of a compact food supply. Thanks to beetles, dung does not pile up and dry out into rock-hard mounds in the intense heat of the day. Instead, the surface of the land is tidied up, beetle offspring get fed, and the leftover dung accumulates in beetle burrows—there to enrich the soil that nourishes the plants that sustain (among others) the elephants.

its physical and chemical environment. The **biosphere** encompasses all regions of Earth's atmosphere, waters, and crust in which organisms live. Astoundingly, *this globe-spanning organization begins with the convergence of energy, certain materials, and DNA in tiny, individual cells.*

Interdependencies Among Organisms

A great flow of energy into the world of life starts with the **producers**—plants and all other organisms that make their own food. Animals are **consumers**. Directly or indirectly, they depend on energy that became stored in the tissues of producers. For example, some energy is transferred to zebras after they browse on grasses. It is again transferred when lions devour a baby zebra that has wandered away from its herd. And it is transferred again when certain fungal and bacterial decomposers feed on tissues and remains of lions, elephants, or any other organism. **Decomposers** break down sugars and other biological molecules to simple materials—which

may be cycled back to producers. In time, all the energy that the producers initially captured from the sun's rays returns to the environment, but that's another story.

For now, keep in mind that organisms connect with one another by a one-way flow of energy through them and a cycling of materials among them, as in Figure 1.5. Their interconnectedness affects the structure, size, and composition of populations and communities. It affects ecosystems, even the biosphere. Understand the extent of their interactions and you will gain insight into acid rain, amplification of the greenhouse effect, and other modern-day problems.

Levels of organization exist in nature. The characteristics of life emerge at the level of single cells and extend through populations, communities, ecosystems, and the biosphere.

A one-way flow of energy through organisms and a cycling of materials among them organizes life in the biosphere. In nearly all cases, energy flow starts with energy from the sun.

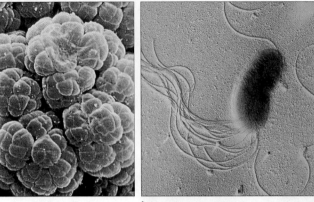
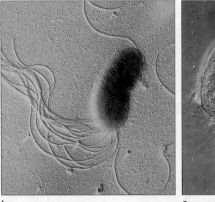
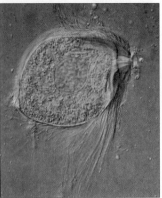

a

b

c

So far, we have focused on life's unity, on the characteristics that all living things have in common. Think of it! They are put together from the same "lifeless" materials. They remain alive by metabolism —by ongoing energy transfers at the cellular level. They interact in their requirements for energy and for raw materials. They have the capacity to sense and respond to the environment in highly specific ways. They all have a capacity to reproduce, based on instructions encoded in DNA. And they all inherited their molecules of DNA from individuals of a preceding generation.

Superimposed on the common heritage is immense diversity. You share the planet with what may be many millions of different kinds of organisms, or **species**. Many millions more preceded you at some time during the past 3.8 billion years, but their lineages vanished; they became extinct. A few centuries ago, attempts to make sense of the confounding diversity resulted in a classification scheme that assigns each newly identified species a two-part name. The first part designates the **genus** (plural, genera). Each genus encompasses all of the species that seem to be related by way of descent from a common ancestor. The second part of the name designates a particular species within that genus.

For example, *Quercus alba* is the scientific name for the white oak, and *Q. rubra* is the name for the red oak. As this example suggests, once you spell out the name of the genus in a document, you may abbreviate the name wherever else it appears in that document.

Biologists also classify life's diversity by assigning species to groups at more encompassing levels. Among other things, they group genera that apparently share a common ancestor into the same *family*, related families into the same *order*, related orders into the same *class*, and related classes into the same *phylum* (plural, phyla). At a higher level, they assign related phyla to the same *kingdom*. They are still refining the groups. For instance, ancient scholars recognized only two kingdoms (animals and plants). Modern biologists have recognized as many as five kingdoms, but new evidence strongly suggests there should be six—the **Archaebacteria**, **Eubacteria**, **Protista**, **Fungi**, **Plantae**, and **Animalia** (Figure 1.6). We use a six-kingdom scheme throughout this book. At this point, it is enough simply to become familiar with a few of the defining features of their members.

All of the world's bacteria (singular, bacterium) are single cells, of a type called *prokaryotic*. The word means they do not have a nucleus, a membranous sac that otherwise would keep their DNA separated from the

d

Figure 1.6 A few representatives of life's diversity.

KINGDOM ARCHAEBACTERIA. (**a**) From the muck of an anaerobic (oxygen-free) habitat, a colony of methanogenic cells.

KINGDOM EUBACTERIA. (**b**) A eubacterium sporting a number of bacterial flagella, a structure that has uses in motility.

KINGDOM PROTISTA. (**c**) A trichomonad that lives in a termite's gut. Compared to bacteria, most protistans are much larger and have greater internal complexity. They range from microscopically small single cells to giant seaweeds.

KINGDOM FUNGI. (**d**) A stinkhorn fungus. Some species of fungi are parasites and some cause diseases, but the vast majority are decomposers. Without decomposers, communities would gradually become buried in their own garbage.

KINGDOM PLANTAE. (**e**) A grove of redwoods near the coast of California. Like nearly all plants, redwoods produce their own food by photosynthesis. (**f**) Flower of a plant called a composite. Its colors and patterning guide bees to nectar. The bees get food, and the plant gets help reproducing. Like many other organisms, they interact in a mutually beneficial way.

KINGDOM ANIMALIA. (**g**,**h**) Male bighorn sheep competing for females and so displaying a characteristic of the kingdom— they move actively through their environment.

rest of their interior. Different bacterial species are producers, consumers, and decomposers. And of all organisms, bacteria are the ones that show the greatest metabolic diversity.

Archaebacteria live only in extreme habitats, such as the ones that prevailed when life originated. The eubacteria are more successful in distribution; different species live nearly everywhere, including in or on other organisms. Those living in your gut and on your skin outnumber the trillions of cells that make up your body.

Protistans include single-celled species as well as some multicelled forms. Most are larger than bacteria, and they have far greater internal complexity. All are *eukaryotic*, meaning that their DNA is enclosed within a nucleus. A dazzling variety of microscopically small, single-celled producers and consumers are protistans. So are multicelled brown algae and other "seaweeds," some of which are giants of the underwater world.

Most fungi, including the common field mushrooms sold in grocery stores, are multicelled. These eukaryotic decomposers and consumers feed in a distinctive way. They secrete substances that digest food outside of the fungal body, then their cells absorb the digested bits.

Plants and animals include many familiar species and an astounding number of obscure ones. Members of both kingdoms are eukaryotic. Plants are multicelled producers; nearly all are photosynthetic. Animals are multicelled consumers that include diverse plant eaters, parasites, and meat eaters. Unlike plants, animals move about during at least some stage of their life.

Pulling this all together, you start to get a sense of what it means when someone says life shows unity *and* diversity.

Unity threads through the world of life, for all organisms are alike in several ways. They are composed of the same substances, which are assembled in the same basic ways. They engage in metabolism, and they sense and respond to their environment. They all have a capacity to reproduce, based on heritable instructions encoded in their DNA.

Immense diversity also threads through the world of life. As an outcome of differences in their DNA instructions, organisms differ enormously in body form, in the functions of their body parts, and in their behavior.

To make the study of life's diversity more manageable, we group organisms into six great kingdoms—archaebacteria, eubacteria, protistans, fungi, plants, and animals.

AN EVOLUTIONARY VIEW OF DIVERSITY

Given that organisms are so much alike, what could account for their great diversity? One key explanation is called evolution by means of natural selection. A few simple examples will be enough to introduce you to its assumptions, which build on the simple observation that variation in traits exists in all populations.

Mutation—Original Source of Variation

DNA has two striking qualities. Its instructions work to assure that offspring will resemble their parents, yet they also permit variations in the details of most traits. As an example, having five fingers on each hand is a human trait. Yet some humans are born with six fingers on each hand instead of five. This is an outcome of a **mutation**, a molecular change in the DNA. Mutations are the original source of variations in heritable traits.

Many mutations are harmful. A change in even a bit of DNA may be enough to sabotage the body's growth, development, or functioning. One such mutation causes *hemophilia A*. If a person affected by this blood-clotting disorder gets even a small cut or bruise, an abnormally lengthy time passes before a clot forms and stops the bleeding. Yet some variations are harmless or beneficial. A classic case is a mutation in light-colored moths that results in dark offspring. Moths fly at night and rest during the day, when the birds that eat them are active. A light-colored moth resting on a light-colored tree trunk is camouflaged (it "hides in the open," as in Figure 1.7), so birds tend not to see it. Suppose people build coal-burning factories nearby. Over time, soot-laden smoke darkens the tree trunks. Now the dark moths are less conspicuous to bird predators, so they have a better chance of living long enough to reproduce. Under the sooty conditions, the variant (dark) form of the trait is more adaptive. An **adaptive trait** simply is any form of a trait that helps an organism survive and reproduce under a given set of environmental conditions.

Evolution Defined

Think about a population of light-colored moths in a sooty forest. At some point, a DNA mutation arose in the population, and it resulted in a moth of a different color. When the mutated individual reproduced, some of its offspring inherited the trait. Birds saw and ate many light-colored moths, but most of the dark ones escaped detection and lived long enough to reproduce. So did their dark offspring, and so did *their* offspring. Over the generations, the frequency of the dark form of the trait increased and that of the light form decreased. As more time passes by, the dark form of the trait might even become the more common, and people might end up referring to "the population of dark-colored moths."

Figure 1.7 Example of different forms of the same trait (surface coloration), adaptive to two different environmental conditions. (**a**) On a light-colored tree trunk, light moths (*Biston betularia*) are hidden from predators, but dark ones stand out. (**b**) The dark color is more adaptive in places where tree trunks are darkened with soot.

Evolution is under way. In biology, the word refers to change occurring in a line of descent over time. As is true of moths, the individuals of most populations typically show different forms of many (or most) of their traits, and the frequencies of those different forms relative to one another can change over successive generations.

Natural Selection Defined

Long ago, the naturalist Charles Darwin used pigeons to explain the conceptual connection between evolution and variation in traits. Domesticated pigeons display splendid variation in size, feather color, and other traits (Figure 1.8). As Darwin knew, pigeon breeders select certain forms of traits. For example, if breeders prefer tail feathers that are black with curly edges, they will allow only those individual pigeons having the most black and the most curl in their tail feathers to mate and produce offspring. Over time, "black" and "curly" will become the most common forms of tail feathers in the captive population, and other forms of the two traits will become less common or will be eliminated.

Pigeon breeding is a case of **artificial selection**, for the selection among different forms of a trait is taking place in an artificial environment, under contrived, manipulated conditions. Yet Darwin saw the practice as a simple model for *natural* selection, a favoring of some forms of traits over others in nature. Whereas breeders are "selective agents" that promote reproduction of some individuals over others in captive populations, a pigeon-eating peregrine falcon is one of many selective agents that operate across the range of variation among pigeons in the wild. Generation after generation, the swifter or more effectively camouflaged pigeons have a better chance of living long enough to reproduce than the not-so-swift or too-conspicuous ones among them.

a

Figure 1.8 A few of the 300+ varieties of domesticated pigeons produced by artificial selection practices. Breeders began with variant forms of traits in captive populations of wild rock doves (**a**).

What Darwin identified as **natural selection** is no more than the outcome of differences in the survival and reproduction of a population's individuals that differ in one or more traits.

Unless you happen to be a pigeon breeder, Darwin's pigeon example may not be firing rockets through your imagination. So think about an example closer to home. Certain bacteria and fungi make *antibiotics*, metabolic products that can kill their bacterial competitors for nutrients in soil. Starting in the 1940s, we learned to use antibiotics to treat and control a variety of diseases that result when bacteria invade the human body and use it as a source of nutrients. Doctors routinely prescribed these so-called wonder drugs for mild as well as serious infections. The one called penicillin was even added to toothpaste, mouthwash, and chewing gum.

It turned out that antibiotics are powerful agents of natural selection. Over time, owing to mutations, some bacterial neighbors of the antibiotic producers have developed antibiotic resistance. Consider streptomycin, which binds with some essential bacterial proteins and inhibits their activity. In some variant strains of bacteria, mutations slightly changed the form of the proteins, so streptomycin is unable to bind with them. Such bacteria can escape streptomycin's effects.

In infected patients, an antibiotic acts against bacteria that are susceptible to its action—but it actually favors variant strains that have resistance to it! Presently, such antibiotic-resistant strains are making it difficult to treat typhoid, tuberculosis, gonorrhea, staph (*Staphylococcus*) infections, and some other bacterial diseases. In a few patients, the "superbugs" responsible for tuberculosis cannot be successfully eliminated.

As antibiotic resistance evolves by natural selection, so must antibiotics evolve if they are to overcome the defenses we unwittingly "selected." For example, drug companies have modified portions of the streptomycin molecule. Such molecular changes in the laboratory have produced more effective antibiotics—at least until

future generations of more resistant superbugs enter the deadly evolutionary competition for nutrients.

Later in the book, we will consider the mechanisms by which populations of moths, pigeons, bacteria, and all other organisms evolve. Meanwhile, keep in mind the following points about natural selection. They are central to biological inquiry, for they have consistently proved useful in explaining a great deal about nature.

1. Individuals of a population vary in form, function, and behavior, and much of the variation is heritable.

2. Some forms of heritable traits are more adaptive to environmental conditions than others. They improve an individual's chance of surviving and reproducing, as by helping it secure food, a mate, hiding places, and so on.

3. Natural selection is the outcome of differences in the survival and reproduction of individuals that show variation in one or more traits.

4. In time, natural selection results in a better fit with prevailing environmental conditions. The adaptive forms of traits tend to become more common and other forms less so. The population changes in its defining characteristics; it evolves.

In short, in the evolutionary view, *life's diversity is the sum total of variations in traits that have accumulated in different lines of descent over time, as by natural selection and other processes of change.*

Mutations in DNA introduce variations in heritable traits.

Although many mutations are harmful, some give rise to variations in form, function, or behavior that are adaptive under prevailing environmental conditions.

Natural selection is the outcome of differences in survival and reproduction among individuals of a population that show variation in one or more traits. The process helps explain evolution—changes in lines of descent over time.

The preceding sections sketched out the major concepts in biology. Now consider approaching this or any other collection of "facts" with a critical attitude. *"Why should I accept that they have merit?"* The answer requires insight into how biologists make inferences about observations and then test the predictive power of their inferences against actual experiences in nature or the laboratory.

Observations, Hypotheses, and Tests

To get a sense of "how to do science," start by following some practices that are pervasive in scientific research:

1. Observe some aspect of nature, carefully check what others have found out about it, and then frame a question or identify a problem related to your observation.

2. Develop **hypotheses**, or educated guesses, about possible answers to questions or solutions to problems.

3. Using hypotheses as a guide, make a **prediction**— that is, a statement of what you should observe in the natural world if you were to go looking for it. This is often called the "if-then" process. (*If* gravity does not pull objects toward Earth, *then* it should be possible to observe apples falling up, not down, from a tree.)

4. Devise ways to **test** the accuracy of your predictions, as by making systematic observations, developing models, and conducting experiments.

5. If the tests do not turn out as you expected, check to see what might have gone wrong. For example, maybe you overlooked a factor that influenced the test results. Or maybe the hypothesis is not a good one.

6. Repeat the tests or devise new ones—the more the better, for hypotheses that withstand many tests are likely to have a higher probability of being useful.

7. Objectively analyze and report the test results and the conclusions you have drawn from them.

You might hear someone refer to these practices as "the scientific method," as if all scientists march to the drumbeat of an absolute, fixed procedure. They do not. Many observe, describe, and report on some subject, then leave it to others to hypothesize about it. Some are lucky; they stumble onto information they are not even looking for, although chance does favor the prepared mind. It is not one single method they have in common. It is a critical attitude about being shown rather than told, and taking a logical approach to problem solving.

Logic encompasses thought patterns by which an individual draws a conclusion that does not contradict the evidence used to support it. Lick a lemon and you notice it's extremely sour. Lick ten more and you notice the same thing each time, so you conclude all lemons are extremely sour. You have correlated one specific (lemon) with another (sour). By this pattern of thinking, called *inductive* logic, an individual derives a general statement from specific observations.

Express the generalization in "if-then" terms, and you have a hypothesis: "If you lick any lemon, then you will get an extremely sour taste in your mouth." By this pattern of thinking, called *deductive* logic, an individual makes inferences about specific consequences or specific predictions that must follow from a hypothesis.

You decide to test the hypothesis by tracking down and sampling all the varieties of lemons in the vicinity. One variety, the Meyer lemon, is actually mellow, for a lemon. You also discover that some people cannot taste anything. So you must modify the original hypothesis: "If most people lick any lemon *except* the Meyer lemon, they will get an extremely sour taste in their mouth." Suppose, after sampling all the known lemon varieties in the world, you conclude the modified hypothesis is a good one. You can never prove it beyond all shadow of a doubt, because there might be lemon trees growing in places people don't even know about. You *can* say the hypothesis has a high probability of not being wrong.

Comprehensive observations are a logical means to test the predictions that flow from hypotheses. So are **experiments**. These tests simplify observation in nature or the laboratory by manipulating and controlling the conditions under which observations are made. When suitably designed, observational and experimental tests allow you to predict that something will happen if a hypothesis isn't wrong (or won't happen if it *is* wrong).

For example, say you decided to use Darwin's view of natural selection to explain wing patterns of moths. By his reasoning, this trait or any other persists when it improves odds for reproductive success. You develop a hypothesis: *Moth wing patterning is a visual signal that helps males and females of a species identify each other and mate, and thereby produce offspring.* Next, you come up with a prediction based on it: *Moths mate at night, so if their wing pattern is a mating flag, then on moonless nights they will not see the pattern and I won't see moths mating.*

You test the prediction on a moonlit night, then on a moonless night. Moths mate both times. Do your direct observations mean the hypothesis is flawed? Maybe not; maybe your *prediction* is flawed. What if you overlooked a factor that affected the outcome? For example, what if moths, like cats, see better than you do in the dark?

You devise an experimental test to reveal whether wing pattern is not a mating flag. If moths having an *altered* wing pattern still attract mates, the prediction is probably flawed. So you capture moths, paint an altered pattern on their wings (Figure 1.9), and cage them with unaltered moths. You also set up a control group.

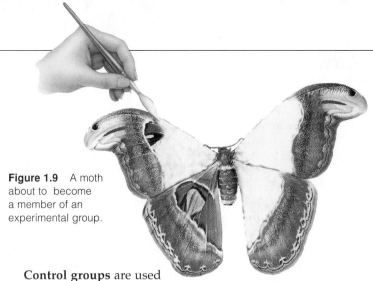

Figure 1.9 A moth about to become a member of an experimental group.

Control groups are used to identify side effects during a test that involves an experimental group. Ideally, a control group is identical to an experimental group in all respects *except* for the variable being studied. **Variables** are specific aspects of objects or events that may differ over time and among individuals. Experimenters directly manipulate a single variable that may support or disprove a prediction, and they also try to hold constant any other variables that may influence the results. In your case, a painted wing pattern is the key variable. But maybe paint fumes, not the painted pattern, will be repulsive to potential mates. Or maybe painting roughs up a moth in a way that will make it less attractive than moths of the control group. (For instance, wing fluttering attracts mates. What if the paint weighs enough to change the flutter frequency?)

To eliminate variables that are unrelated to testing your prediction, you paint moths of the control group with the same paint and brushes, and handle them the same way—but in their case you precisely *duplicate* the natural wing pattern. The point of the manipulations is to make the experimental and control groups identical *except* for the key variable. Now, if the moths having an altered wing pattern still turn out to be lucky in love, then the experiment will not support the prediction.

Have you been thinking that dabbing paint on moths is a fanciful idea of an experimental test? As reported in *Nature* in 1993, Karen Marchetti, a graduate student of the University of California at Davis, carefully used a paintbrush on birds in a forest in Kashmir, India. She found evidence for her hypothesis that bright feather color, not feather patterning, gives male yellow-browed leaf warblers a competitive edge in mating.

In early tests, Marchetti captured male birds and put a patternless dab of yellow paint on their head feathers. In later tests, she painted larger-than-normal yellow bands on the wings of an experimental group of male birds, then she painted out part of the bands of another group. She used transparent paint for a third group. (Can you guess why?) As Marchetti discovered, color-enhanced birds secured larger territories and produced more offspring, compared to the manipulatively toned-down birds in her experiments.

About the Word "Theory"

Suppose no one has disproved a hypothesis after years of rigorous tests. Suppose scientists use it to formulate *more* hypotheses that successfully explain a broad range of phenomena. When a hypothesis meets these criteria, it may become accepted as a **scientific theory**.

You may hear someone apply the word "theory" to a speculative idea, as in the expression "It's only a theory." However, a scientific theory differs from speculation for a simple reason: *Many researchers have tested its predictive power many times, in many different ways, and have found no evidence to disprove it.* That is why Darwin's view of natural selection is a highly regarded theory. Biologists use it successfully to explain such diverse issues as the origin of life, the relationship between plant toxins and plant-eating animals, the sexual advantages of brightly colored or patterned wings or feathers, the reason why certain cancers run in families, or why antibiotics that doctors often prescribe are no longer effective. By giving reasoned evidence that evolution occurred in the past, the theory even influenced views of Earth history.

An exhaustively tested theory might be as close to the truth as scientists can get with the evidence at hand. For example, Darwin's theory stands, with only minor modification, after more than a century of thousands of different tests. As biologists realize, we cannot show the theory holds under all possible conditions; an infinite number of tests would be required to do so. As for any theory, we can only say it has a *high or low probability* of being a good one. So far, biologists haven't found any evidence that calls Darwin's theory into question. Yet they still keep their eyes open for new information and new ways of testing that might disprove its premises.

And this point gets us back to the value of thinking critically. Scientists must keep asking themselves: *Will observations or experiments show that a hypothesis is false?* They expect one another to put aside pride or bias by testing ideas, even in ways that may prove them wrong. Even if an individual doesn't or won't do this, others will—for science proceeds as a community that is both cooperative and competitive. Ideally, its practitioners share their ideas, with the understanding that it is just as important to expose errors as it is to applaud insights. Individuals can and often do change their minds when presented with contradictory evidence. As you will see, this is a strength of science, not a weakness.

A scientific approach to studying nature is based on asking questions, formulating hypotheses, making predictions, devising tests, and objectively reporting the results.

A scientific theory is a testable explanation about the cause or causes of a broad range of related phenomena. It remains open to tests, revision, and tentative acceptance or rejection.

1.6

THE POWER AND PITFALLS OF EXPERIMENTAL TESTS

AN ASSUMPTION OF CAUSE AND EFFECT   Experiments start from a premise that any aspect of the natural world has one or more underlying causes, whether hidden or not. With this premise, science is distinct from faith in the supernatural (meaning "beyond nature"). Experiments must deal with *potentially falsifiable* hypotheses. Such hypotheses can be tested in the natural world in ways that might disprove them.

EXPERIMENTAL DESIGN   To get conclusive test results, experimenters rely on certain practices. They refine a test design by searching the literature for related information. They design their own experiments to test one prediction of a hypothesis at a time. Each time, they set up a control group as a standard for comparison with one or more experimental groups (Figure 1.10a).

Consider some experimental tests designed by Bruce Levin of Emory University and Jim Bull of the University of Texas. Both biologists were concerned with the ever increasing frequencies of antibiotic-resistant strains of pathogenic (disease-causing) bacteria. They had been studying literature on a possible alternative to antibiotics, a *biological therapy* that enlists the bacteriophages—the "bacteria-eaters." These viruses attack a narrow range of bacterial strains. When they contact a target, they typically inject a few enzymes and genetic material into it. What happens next is a hostile takeover of the cell's metabolic machinery. The cell itself builds many new viral particles, then it dies after its outer membrane ruptures. The virus particles released this way can infect more cells.

The biologists wondered, as others had decades ago, whether injections of bacteriophages could help people resist or fight off bacterial infections. The discovery of antibiotics in the 1940s had diverted attention away from the idea, at least in the West. Now, with so many lethal

pathogens breaching the antibiotic arsenal, bacteria eaters were starting to look good again. Levin and Bull focused on some promising phage therapy experiments conducted in 1982 by H. Williams Smith and Michael Huggins, two British researchers. They started their research by testing whether Smith and Huggins's results could be duplicated.

They selected 018:K1:H, a strain of *Escherichia coli* originally isolated from a human patient with meningitis. Like the normally harmless *E. coli* strain that lives in the intestines of humans and other mammals, this one departs from the body in feces. The bacteriophage typically used in laboratory work ignored 018:K1:H7, so Levin and Bull went looking for some *E. coli* killers in samples from an Atlanta sewage treatment plant. They successfully isolated two kinds of bacteriophages and named them *H* (for *Hero*, the most effective killer) and *W* (for *Wimp*).

With graduate student Terry DeRouin and laboratory technician Nina Moore Walker, the biologists grew a large population of 018:K1:H7 in a culture flask. They selected a specific strain of laboratory mice, all females of the same age. Then they ran the experiments outlined in Figure 1.10. Their results reinforced Smith and Huggins's conclusion that certain bacteriophages can be as good as or better than antibiotics at stopping specific bacterial infections.

Hypothesis: If bacteriophages specifically target and destroy cells of *E. coli* 018:K1:H7 in petri dishes, then they will do the same in laboratory mice that have been infected by that strain.

Prediction: Laboratory mice injected with a preparation that contains more than 10^7 particles of *H* bacteriophage will not die following an injection of *E. coli* strain 018:K1:H7.

Experimental test of the prediction:

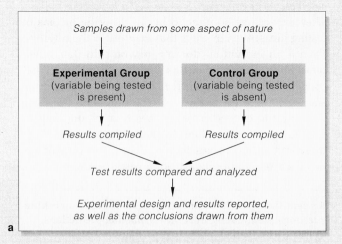

Researchers establish large populations of H bacteriophage and of E. coli 018:K1:H7 from which to draw their samples, and select a specific strain of laboratory mice.

EXPERIMENTAL GROUP	CONTROL GROUP
E. coli injected into right thigh of 15 of the mice; bacteriophage injected into their left thigh.	*E. coli injected into 15 other mice; <u>no</u> bacteriophage injected into this sampling.*

Test results:

All mice survive.	*All mice die within 32 hours.*

Figure 1.10 (**a**) Generalized sequence of steps involved in an experimental test of a prediction based on a hypothesis. (**b,c**) Two experiments comparing the effectiveness of bacteriophage injections and antibiotics in treating a bacterial infection.

Samples drawn from some aspect of nature

Experimental Group (variable being tested is present)	**Control Group** (variable being tested is absent)

Results compiled   *Results compiled*

Test results compared and analyzed

Experimental design and results reported, as well as the conclusions drawn from them

a

b

Each type of bacteriophage targets specific strains of one or at most a few bacterial species. At present, it takes too long for clinicians to discover which specific bacterium is infecting a patient, so the right bacteriophage might not be enlisted until it is too late to do the patient any good. On the bright side, procedures are now being developed that can dramatically accelerate the identification process.

IDENTIFYING IMPORTANT VARIABLES In nature, many factors can influence the outcome of a bacterial infection. For example, genetic differences among individuals lead to differences in how their immune system will respond to the invasion. So do the individual's age, nutrition, and health at the time of infection. Some strains of infectious bacteria are deadlier than others, and so on. That is why researchers try to simplify and control the variables in their experiments. Variables, recall, are specific aspects of objects or events that may differ over time and among individuals. All of the *E. coli* cells in Levin and Bull's experiments were descended from the same parent cell, raised on the same nutrients at the same temperature in the flask, and therefore could be expected to respond the same way to the bacteriophage attack. All of the mice were the same age and gender and were raised under identical

a Natalie, blindfolded, randomly plucks a single jellybean from a jar filled with 120 green and 280 black jumbo jellybeans. That is a ratio of 30 to 70 percent.

c Next, the still-blindfolded Natalie randomly plucks 50 jellybeans from the jar and ends up selecting 10 green and 40 black jellybeans.

b The jar is hidden before she removes her blindfold. She sees only a single green jellybean in her hand and assumes the jar contains green jellybeans only.

d Now, the larger sampling leads Natalie to assume that one-fifth of the jellybeans in the jar are green and four-fifths are black (a ratio of 20 to 80).

e Natalie's larger sampling more closely approximates the ratio of green to black jumbo jellybeans in the jar. The more times she repeats her sampling, the greater the likelihood that she will come close to determining the actual ratio.

Figure 1.11 A simple demonstration of sampling error.

laboratory conditions. Each mouse in every experimental group received the same amount of bacteriophages or streptomycin; every mouse in a given control group got the same injection of saline solution. Thus the focus was on *one variable*—a specific bacteriophage versus a specific antibiotic—in a simple, controlled, artificial situation.

SAMPLING ERROR As Levin and Bull did with their bacteriophage, *E. coli*, and mice, experimenters generally use samples (or subsets) of populations, events, and other aspects of nature. If they don't use large-enough samples, they run the risk of performing tests with groups that are not representative of the whole. In general, the larger the sample, the less likely it will be that differences among individuals will distort the results (Figure 1.11).

BIAS IN REPORTING THE RESULTS Whether intentional or not, experimenters run the risk of interpreting data in terms of what they want to prove or dismiss. A few have even been known to fake measurements or nudge findings in ways that reinforce their own bias. That is why science emphasizes presenting test results in *quantitative* terms—that is, with actual counts or some other precise form. Doing so allows other experimenters to check or test the results readily and systematically, as Levin and Bull did. At this writing, they are assembling a detailed report of their own experiments, for publication in a science journal.

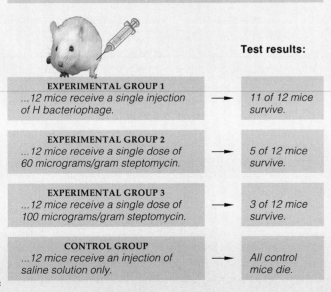

Another prediction derived from the same hypothesis:
H bacteriophage that target and destroy 018:K1:H7 will be more effective than single doses of streptomycin in treating mice infected with 018:K1:H7.

Experimental test of the new prediction:

Researchers inject 48 laboratory mice with 018:K1:H7, then divide them into four groups of 12 each. Eight hours later . . .

Test results:

EXPERIMENTAL GROUP 1 ...12 mice receive a single injection of H bacteriophage.	→ 11 of 12 mice survive.
EXPERIMENTAL GROUP 2 ...12 mice receive a single dose of 60 micrograms/gram steptomycin.	→ 5 of 12 mice survive.
EXPERIMENTAL GROUP 3 ...12 mice receive a single dose of 100 micrograms/gram steptomycin.	→ 3 of 12 mice survive.
CONTROL GROUP ...12 mice receive an injection of saline solution only.	→ All control mice die.

c

The call for objective testing strengthens the theories that emerge from scientific studies. It also puts limits on the kinds of studies that can be carried out. Beyond the realm of science, some events remain unexplained. Why do we exist, for what purpose? Why does any one of us have to die at a particular moment? Such questions lead to *subjective* answers, which come from within, as an outcome of all the experiences and mental connections shaping our consciousness. Because people differ vastly in this regard, subjective answers do not readily lend themselves to scientific analysis and experiments.

This is not to say subjective answers are without value. No human society can function for long unless its members share a commitment to certain standards for making judgments, even subjective ones. The moral, aesthetic, philosophical, and economic standards vary from one society to the next. But they all guide their members in deciding what is important and good, and what is not. All attempt to give meaning to what we do.

Every so often, scientists stir up controversy when they happen to explain some part of the world that was considered to be beyond natural explanation—that is, as belonging to the "supernatural." This is often true when a society's moral codes have become interwoven with religious narratives. Exploring some longstanding view of the world from a scientific perspective may be misinterpreted as questioning morality, even though the two are not the same thing.

For example, centuries ago in Europe, Nicolaus Copernicus studied the planets and concluded the Earth circles the sun. Today this seems obvious enough. Back then it was heresy. The prevailing belief was that the Creator made the Earth (and, by extension, humankind) the immovable center of the universe. Later a respected scholar, Galileo Galilei, studied the Copernican model of the solar system, thought it was a good one, and said so. He was forced to retract his statement publicly, on his knees, and put the Earth back as the fixed center of things. (Word has it that when he stood up he muttered, "Even so, it *does* move.") Later still, Darwin's theory of evolution ran up against the same prevailing belief.

Today, as then, society has sets of standards. Those standards might be questioned when a new, natural explanation runs counter to supernatural beliefs. This doesn't mean that the scientists who raise questions are less moral, less lawful, less sensitive, or less caring than anyone else. It simply means one more standard guides their work: *The external world, not internal conviction, must be the testing ground for scientific beliefs.*

Systematic observations, hypotheses, predictions, tests— in all these ways, science differs from systems of belief that are based on faith, force, or simple consensus.

1. There is unity in the living world, for all organisms have these characteristics in common:

   a. They are assembled from the same kinds of atoms and molecules, according to the same laws of energy.

   b. They survive by metabolism, and by sensing and responding to specific conditions in the environment.

   c. They have the capacity for growth, development, and reproduction, based on the heritable instructions encoded in the molecular structure of their DNA.

2. The characteristics of life extend from cells, through multicelled organisms, then populations, communities, ecosystems, and the biosphere.

3. Many millions of species (kinds of organisms) exist; many millions more lived in the past and became extinct. Classification schemes place species in ever more inclusive groupings, from genus on up through (for instance) family, order, class, phylum, and kingdom.

4. The diversity among organisms has arisen through mutations, which are changes in the structure of DNA molecules. The changes are the foundation for variation in heritable traits. These are traits that parents bestow on offspring, including most details of the body's form and functioning.

5. Darwin's theory of evolution by natural selection is a cornerstone of biological inquiry. Its key premises are:

   a. Individuals of a given population differ in their versions of the same heritable trait. Variant forms of traits may affect the ability to survive and reproduce.

   b. Natural selection is the outcome of differences in survival and reproduction among individuals that show variation in one or more traits. Adaptive forms of a given trait tend to become more common; less adaptive ones become less common or disappear. Thus the traits we use to define a population can change over time; the population can evolve.

6. There are many diverse methods of scientific inquiry. The following terms are important to all of them:

   a. Theory: An explanation of a broad range of related phenomena, supported by many tests. An example is Darwin's theory of evolution by natural selection.

   b. Hypothesis: A possible explanation of a specific phenomenon. Sometimes called an educated guess.

   c. Prediction: A claim about what can be expected in nature, based on the premises of a theory or hypothesis.

   d. Test: An attempt to produce actual observations that match predicted or expected observations.

   e. Conclusion: A statement about whether a theory or hypothesis should be accepted, modified, or rejected, based on tests of the predictions derived from it.

7. Systematic observations, hypotheses, predictions, and relentless tests are the foundation of scientific theories. The external world, rather than internal conviction, is the testing ground for those theories.

Review Questions

1. Why is it difficult to formulate a simple definition of life? *CI.* [For this and subsequent chapters, *italics* after the review questions identify section where you can find the answers to those questions. They include section numbers and *CI* (for *Chapter Introduction*).]

2. What is the molecule of inheritance? What is so important about the flow of events from DNA to RNA to protein? *1.1*

3. Write out simple definitions of the following terms: *1.1*
   a. cell   c. metabolism   e. ATP
   b. energy   d. photosynthesis   f. aerobic respiration

4. By what mechanisms do organisms sense changes in their surroundings? *1.1*

5. List the shared characteristics of life. *CI, 1.3*

6. Study Figure 1.4. Then, on your own, arrange and define the levels of biological organization. *1.2*

7. Study Figure 1.5. Then, on your own, make a sketch of the one-way flow of energy and the cycling of materials through the biosphere. To the side of the sketch, write out definitions of the producer, consumer, and decomposer organisms. *1.2*

8. Each kind of organism is called a separate species. What are the two components of its name? List the six kingdoms of species and name some of their general characteristics. *1.3*

9. Define mutation and adaptive trait. *1.4*

10. Explain the connection between mutations and the immense diversity of life. *1.4*

11. Write out simple definitions of evolution, artificial selection, and natural selection. *1.4*

12. Define and distinguish between: *1.5*
    a. hypothesis or speculation and scientific theory
    b. observational test and experimental test
    c. inductive and deductive logic

13. With respect to experimental tests, define variable, control group, and experimental group. What is meant by the statement that scientific experiments are based on an assumption of cause and effect? *1.5, 1.6*

14. What is a sampling error? *1.6*

Self-Quiz *(Answers in Appendix IV)*

1. The _____ is the smallest unit of life.

2. _____ may be defined as the capacity of cells to extract and transform energy from their environment and use it to maintain themselves, grow, and reproduce.

3. _____ is a state in which the internal environment is being maintained within tolerable limits.

4. If a form of a trait improves chances for surviving and reproducing in a given environment, it is a(n) _____ trait.

5. The capacity to evolve is based on variations in heritable traits, which originally arise through _____ .

6. You have some number of traits that also were present in your great-great-great-great-grandmothers and -grandfathers. This is an example of _____ .
   a. metabolism   c. a control group
   b. homeostasis   d. inheritance

7. DNA molecules _____ .
   a. contain instructions for traits
   b. undergo mutation
   c. are transmitted from parents to offspring
   d. all of the above

8. For many years in a row, a dairy farmer allowed his best milk-producing cows but not the poor producers to mate. Over many generations, milk production increased. This outcome is an example of _____ .
   a. natural selection
   b. artificial selection
   c. evolution
   d. both b and c

9. Match the terms with the most suitable descriptions.
   ____ adaptive trait
   ____ natural selection
   ____ theory
   ____ hypothesis
   ____ prediction

   a. statement of what you should observe in nature if you were to go looking for it
   b. educated guess
   c. improves chance of surviving and reproducing in environment
   d. related set of hypotheses that form a broad-reaching, testable explanation about nature
   e. outcome of differences in survival and reproduction that occurred among individuals that differ in one or more traits

Critical Thinking

1. A certain type of fishing spider (*Dolomedes*) that feeds on insects around ponds occasionally captures tadpoles and small fishes, as in Figure 1.12, and isn't that fun to think about. While they are immature, the female spiders confine themselves to a small area above the pond. When they are sexually mature, they mate and store sperm that fertilizes their eggs. Only then do they start to migrate through larger areas around the pond. Develop hypotheses to explain this observation, and then design an experiment to test each hypothesis.

Figure 1.12 A captured minnow, the unfortunate recipient of a paralyzing venom and digestive enzymes delivered by a fishing spider (*Dolomedes triton*), which is now sucking predigested juices from it.

2. Witnesses in a court of law are asked to "swear to tell the truth, the whole truth, and nothing but the truth." What are some of the problems inherent in the question? Can you think of a better alternative?

3. Design an experiment to support or refute the following hypothesis: Body fat appears yellow in certain rabbits—but only when those rabbits also eat leafy plants that contain an abundance of a yellow pigment molecule called xanthophyll.

4. A group of scientists devised the following experiments to shed light on whether different species of fishes of the same genus compete with one another in their natural habitat. They set up twelve ponds that were identical in chemical composition and physical characteristics. In each pond, they released the following:

Ponds 1,2,3:	species A	(300 individuals each pond)
Ponds 4,5,6:	species B	(300 individuals each pond)
Ponds 7,8,9:	species C	(300 individuals each pond)
Ponds 10,11,12:	species A,B,C	(300 of each in each pond)

Does this experimental design take into consideration all factors that can affect the outcome? If not, how would you change it?

5. Many popular magazines publish an astounding number of articles on diet, exercise, and other health-related topics. Often the authors recommend a specific diet or dietary supplement. What kinds of evidence do you think the articles should describe so that you can decide whether to accept their recommendations?

6. The males of many species of birds have brightly colored feathers, whereas the females are often rather drab. Why do you suppose drab coloration might be an adaptive trait for the females?

7. Surgical incisions in the skin of the African clawed frog (*Xenopus laevis*) do not become infected after the toad is placed in an aquarium with water that contains a variety of bacteria. Develop hypotheses to explain this observation. Then design an experiment to test each hypothesis.

8. An experimental pesticide is being sprayed in facilities where chickens are raised. The following year, an unusually large number of chicks with birth defects are hatched. Develop hypotheses to explain this observation. Then design an experiment to test each hypothesis.

9. The most recent manned mission to the planet Mars has returned. One of the astronauts recovered a gelatinous red substance from its surface. Terika, a NASA biologist, has been assigned the responsibility of determining whether the substance is alive or is a product of a living organism. What features should she look for?

Selected Key Terms

For this and subsequent chapters, these are the **boldface** terms that appear in the text, in the sections indicated here by *italic* numbers (or *CI*, short for Chapter Introduction). Make a list of the terms, write a definition next to each, and then check it against the one in the text. You will be using these terms later on. Becoming familiar with each one now will help give you a foundation for understanding the material in later chapters.

adaptive trait *1.4*	energy *1.1*	photosynthesis *1.1*
aerobic	Eubacteria *1.3*	Plantae *1.3*
respiration *1.1*	evolution *1.4*	population *1.2*
Animalia *1.3*	experiment *1.5*	prediction *1.5*
Archaebacteria *1.3*	Fungi *1.3*	producer *1.2*
artificial	genus *1.3*	Protista *1.3*
selection *1.4*	homeostasis *1.1*	receptor *1.1*
biosphere *1.2*	hypothesis *1.5*	reproduction *1.1*
cell *1.1*	logic *1.5*	RNA *1.1*
community *1.2*	metabolism *1.1*	species *1.3*
consumer *1.2*	multicelled	stimulus *1.1*
control group *1.5*	organism *1.2*	test, scientific *1.5*
decomposer *1.2*	mutation *1.4*	theory, scientific *1.5*
DNA *1.1*	natural selection *1.4*	variable *1.5*
ecosystem *1.2*		

Readings

Carey, S. 1994. *A Beginner's Guide to the Scientific Method*. Belmont, California: Wadsworth. Paperback.

Committee on the Conduct of Science. 1989. *On Being a Scientist*. Washington, D.C.: National Academy of Sciences. Paperback.

Dott, R., and R. Batten. 1988. *Evolution of the Earth*. Fourth edition. New York: McGraw-Hill.

Larkin, T. June 1985. "Evidence Versus Nonsense: A Guide to the Scientific Method." *FDA Consumer* 19: 26–29.

McCain, G., and E. Segal. 1988. *The Game of Science*. Fifth edition. Pacific Grove, California: Brooks/Cole. Paperback.

Moore, J. 1993. *Science as a Way of Knowing—The Foundations of Modern Biology*. Cambridge, Massachusetts: Harvard University Press.

Web Site See *http://www.wadsworth.com/biology* for practice quiz questions, hypercontents, BioUpdates, and critical thinking. The Wadsworth Biology Resource Center provides a wealth of information fully organized and integrated by chapter.

FACING PAGE: *Living cells of a green plant* (Elodea), *as seen with the aid of a microscope. Each rectangular cell contains efficient chemical factories called chloroplasts (the green spheres).*

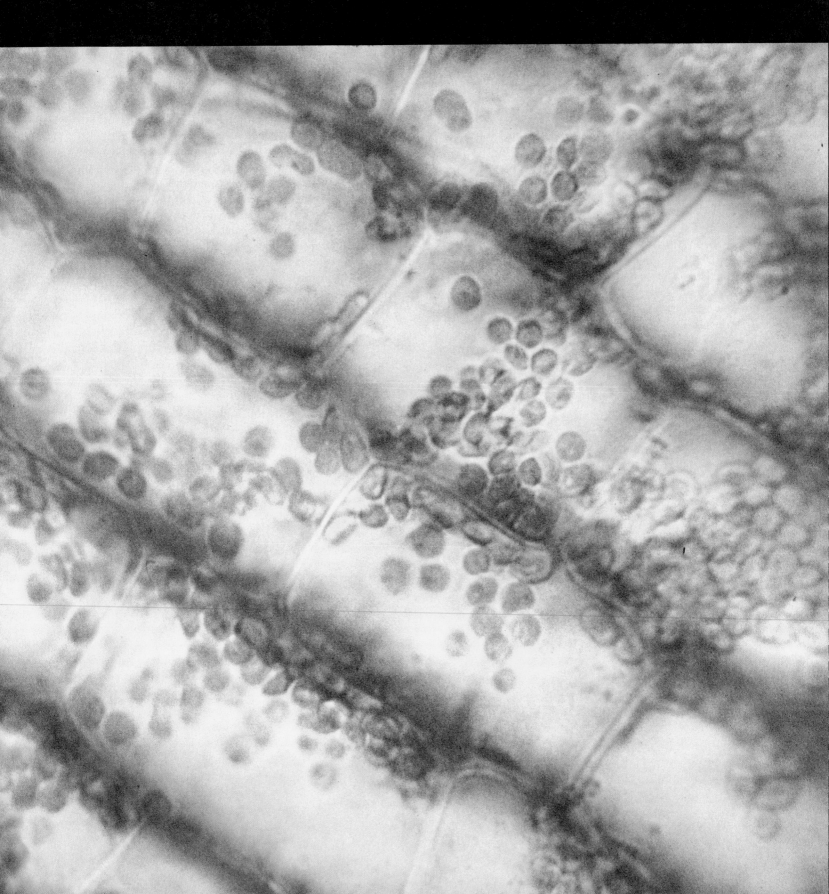

2 CHEMICAL FOUNDATIONS FOR CELLS

Leafy Clean-Up Crews

Right now you are breathing in oxygen. You would die without it. Two centuries ago, no one had a clue to what oxygen is, where it comes from, and how it helps keep people alive. Then researchers started unlocking the secrets of this chemical element and others. As their knowledge of chemistry deepened, they began to conjure up such amazing things as nuclear power, synthetic fertilizers, nylons, lipsticks, fabric cleaners, aspirin, antibiotics, and plastic parts of refrigerators, computers, television sets, jet planes, and cars.

Today, our chemical "magic" brings us benefits *and* problems. Think of the astounding scale of agriculture necessary to sustain the human population, which now surpasses 5.8 billion. If farmers did not use synthetic fertilizers to boost crop yields, more humans than you might ever imagine would starve to death. Fertilizers contain nitrogen, selenium, and other elements that serve as nutrients; they help plants grow. The nutrients dissolve in irrigation water, and a portion accumulates in runoff from the fields. Farmers commonly divert the runoff to ponds. There the water evaporates, and the leftover elements become even more concentrated. Especially troublesome is an ionized form of selenium

in irrigation water. At extremely high concentrations, it contaminates soil and kills wildlife.

In 1996, Norman Terry and his research colleagues at the University of California, Berkeley, started a large-scale experiment in Corcoran, California. They grew cattails, bulrushes, *Spartina*, and other grasses in ten experimental plots, each measuring a quarter of an acre, near some agricultural fields. As they knew, plants can incorporate selenium in their tissues and also have the metabolic means to convert some to dimethyl selenide —a gas about 600 times less toxic than selenium. Plants, they hypothesized, will be able to remove selenium from contaminated soil. They planned to measure how much selenium remains in mud of wetland plots, and how much gets incorporated in plant tissues and also escapes in gaseous form to the air. From earlier laboratory work, Terry knew that selenium levels can vary, depending on the soil type and plant species. He predicted that before runoff trickles on through to evaporation ponds,

Figure 2.1 At experimental wetlands in Corcoran, California, cattails, bulrushes, *Spartina*, and other grasses take up selenium from agricultural runoff and thereby help clean up the water.

a   **b**

Figure 2.2 (**a**) Young poplars taking up TCE-contaminated water. (**b**) Sunflowers, with their roots dangling in pondwater, take up radioactive elements near Chernobyl, in the Ukraine. As described in Section 50.8, the containment walls of Chernobyl's nuclear power plant exploded open in 1986, and dangerous radioactive elements entered the surrounding air, soil, and water.

the plants that he selected for his plots should be able to reduce selenium levels to less than 2 ppb. (That is, there should be less than two parts of selenium among every billion parts of other elements in the water.)

Terry got the idea for his experiment after gathering data on the selenium that became incorporated in plants growing in a wetland near an oil refinery in Richmond, California (Figure 2.1). As his measurements revealed, each day those plants remove 90 percent of the selenium present in 10 million liters of the refinery's wastewater!

Terry's research is an example of *phytoremediation*—the use of living plants to withdraw harmful substances from the environment. Similarly, Milton Gordon and his colleagues at the University of Washington utilize deep-rooted poplar plants to cleanse groundwater that became contaminated with TCE (Figure 2.2*a*). This chemical is now banned, but it was once common in dry cleaning and degreasing products. The poplars release TCE to the air, and sunlight breaks it down. As a final example, a phytoremediation company is using sunflowers to clean a pond that became contaminated with ^{90}strontium, ^{137}cesium, and other extremely nasty radioactive elements (Figure 2.2*b*).

As you might deduce from this story, safeguarding the environment, our food supplies, and our health depends on knowledge of chemistry. So do efforts to minimize side effects of its applications on biological systems. You owe it to yourself and others to gain insight into the structure and behavior of chemical substances. By demystifying chemistry's "magic," you will be better equipped to assess its benefits and risks.

KEY CONCEPTS

1. All substances consist of one or more elements, such as hydrogen, oxygen, and carbon. Each element is composed of atoms, which are the smallest units that still display the element's properties. In turn, its atoms are composed of protons, electrons, and (except for hydrogen) neutrons.

2. All of the atoms that make up each kind of element have the same number of protons and electrons, but they may vary slightly from one another in their number of neutrons. Variant forms of an element's atoms are called isotopes.

3. Atoms have no overall electric charge unless they become ionized—that is, unless they lose electrons or acquire more of them. An ion is an atom or molecule that has gained or lost one or more electrons and thereby has acquired an overall positive or negative charge.

4. Whether a given atom will interact with other atoms depends on how many electrons it has and how they are structurally arranged within the atom. When the electron structures of two or more atoms become united in some way, this is a chemical bond.

5. The molecular organization and the activities of living things arise largely from the interactions called ionic, covalent, and hydrogen bonds.

6. Life originated in water, and it is exquisitely adapted to its properties. Foremost among these properties are water's temperature-stabilizing effects, cohesiveness, and capacity to dissolve or repel a variety of substances.

7. Life depends on the controlled formation, use, and disposal of hydrogen ions (H^+). The pH scale is a measure of the concentration of these ions in different solutions.

People, pesticides, pumpkins, the solid earth beneath your feet, the very air you breathe—*everything in and around you is "chemistry."* Every solid, liquid, or gaseous substance you care to think about is a collection of one or more kinds of elements. Think of these **elements** as fundamental forms of matter that occupy space, have mass, and cannot be broken apart into something else, at least not by any ordinary means. If you could break a chunk of some element into ever smaller bits, the smallest bit you would end up with would be the same element. Ninety-two elements occur naturally on Earth, and many others have been produced under extreme conditions in laboratories (Appendix VI).

Each element has a unique symbol that is internationally recognized, regardless of its name in different countries. One such element is *nitrogen* in English, *azoto* in Italian, and *stickstoff* in German, but its symbol is always N. Similarly, sodium's symbol is always Na (from the Latin *natrium*). Table 2.1 lists the symbols for other elements.

Typically, more than 95 percent of the body weight of an organism is composed of only four elements: oxygen, carbon, hydrogen, and nitrogen. The remainder consists of calcium, phosphorus, potassium, sulfur, sodium, chlorine, magnesium, and more than a dozen trace elements. A **trace element** simply is one that represents less than 0.01 percent of body weight.

Figure 2.3 lists your own body's elements. The amounts may seem trivial, but normal functioning depends on them. For example, your daily meals usually include a trace of magnesium. Without it, muscles would weaken and your brain would not work properly. As another example, leaves droop and die without magnesium. Or think of a weed that absorbs 2,4D, a pesticide. It is stimulated to grow rapidly, but it cannot absorb trace elements fast enough to sustain the accelerated growth. It literally will grow itself to death.

Table 2.1	Atomic Number and Mass Number of Elements Common in Living Things		
Element	Symbol	Atomic Number	Most Common Mass Number
Hydrogen	H	1	1
Carbon	C	6	12
Nitrogen	N	7	14
Oxygen	O	8	16
Sodium	Na	11	23
Magnesium	Mg	12	24
Phosphorus	P	15	31
Sulfur	S	16	32
Chlorine	Cl	17	35
Potassium	K	19	39
Calcium	Ca	20	40
Iron	Fe	26	56
Iodine	I	53	127

Structure of Atoms

What are the smallest particles that retain the properties of an element? **Atoms**. A line of about a million of them would fit in the period ending this sentence. Small as atoms are, physicists have split them into more than a hundred kinds of smaller particles. The only subatomic particles you will need to consider in this book are the ones called **protons**, **electrons**, and **neutrons**.

All atoms have one or more protons, which carry a positive electric charge (p^+). Except for hydrogen atoms, they also have one or more neutrons, which carry no charge. Protons and neutrons make up the atom's core region, or atomic nucleus (Figure 2.4). Zipping about the nucleus and occupying most of the atom's volume are one or more electrons, which carry a negative charge (e^-). Each atom has just as many electrons as protons. This means atoms carry no net charge, overall.

Each element has a unique **atomic number**, which refers to the number of protons present in its atoms. To give examples, the atomic number is 1 for the hydrogen

EARTH'S CRUST	
Oxygen	46.6
Silicon	27.7
Aluminum	8.1
Iron	5.0
Calcium	3.6
Sodium	2.8
Potassium	2.6
Magnesium	2.1
Other elements:	1.5

HUMAN	
Oxygen	65
Carbon	18
Hydrogen	10
Nitrogen	3
Calcium	2
Phosphorus	1.1
Potassium	0.35
Sulfur	0.25
Sodium	0.15
Chlorine	0.15
Magnesium	0.05
Iron	0.004
Iodine	0.0004

PUMPKIN	
Oxygen	85
Hydrogen	10.7
Carbon	3.3
Potassium	0.34
Nitrogen	0.16
Phosphorus	0.05
Calcium	0.02
Magnesium	0.01
Iron	0.008
Sodium	0.001
Zinc	0.0002
Copper	0.0001
Other:	0.00005

Figure 2.3 Proportions of elements making up the human body and the fruit of pumpkin plants, compared to the proportions of elements in the materials of our planet's crust. In what respects are the proportions similar? In what respects do they differ?

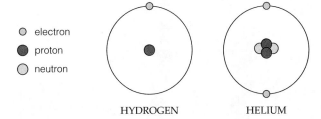

○ electron
● proton
○ neutron

HYDROGEN HELIUM

Figure 2.4 The hydrogen atom and helium atom according to one model of atomic structure. The model is highly simplified; the nucleus of these two representative atoms really would be an invisible speck at the scale employed here.

atom (which has a single proton) and 6 for the carbon atom (which has six protons).

Each element also has a **mass number**, which is the number of protons *and* neutrons in the atomic nucleus. A carbon atom with six protons and six neutrons has a mass number of 12. Some people call the relative mass of atoms "atomic weight." Mass is not quite the same thing as weight, but use of this imprecise term continues.

Why bother with atomic and mass numbers? They can give you an idea of whether and how substances will interact. *And this knowledge can help you predict how substances might behave in cells, in multicelled organisms, and in the environment, under a variety of conditions.*

Isotopes—Variant Forms of Atoms

All of the atoms of an element have the same number of protons and electrons, but they may not have the same number of neutrons. If an atom of a given element has more or fewer neutrons than those having the most common number, it is an **isotope**. As an example, "a carbon atom" might be carbon 12 (six protons and six neutrons, this being the most common form), carbon 13 (six protons and seven neutrons), or carbon 14 (with six protons and eight neutrons). Often you will see symbols for isotopes, as for the carbon isotopes ^{12}C, ^{13}C, and ^{14}C.

All isotopes of an element interact with other atoms in the same way. As you will see, this means that cells can use any isotope of an element during metabolism.

You have probably heard of radioactive isotopes, or **radioisotopes**. Henri Becquerel, a physicist, discovered them in 1896, after he put a heavily wrapped rock on an unexposed photographic plate in a desk drawer. The rock happened to contain uranium isotopes. A few days later, the plate bore a faint image of the rock, which must have been emitting energy. Becquerel's coworker, Marie Curie, named the phenomenon "radioactivity."

As we now know, radioisotopes are unstable atoms with dissimilar numbers of protons and neutrons, and they spontaneously lose subatomic particles and energy at a known rate. This process, called **radioactive decay**, transforms an unstable atom into an atom of a different

Figure 2.5 Chemical bookkeeping. We use symbols for elements when writing *formulas*, which identify the composition of compounds. For example, water has the formula H_2O. The subscript indicates two hydrogen (H) atoms are present for every oxygen (O) atom. We use such symbols and formulas when writing *chemical equations*, which are representations of the reactions among atoms and molecules. The substances entering a reaction (reactants) are to the left of the reaction arrow, and the products are to the right, as shown by the following chemical equation for photosynthesis:

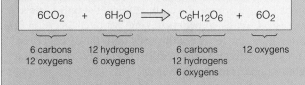

$$6CO_2 + 6H_2O \Longrightarrow C_6H_{12}O_6 + 6O_2$$

6 carbons	12 hydrogens	6 carbons	12 oxygens
12 oxygens	6 oxygens	12 hydrogens	
		6 oxygens	

You may read about reactions for which reactants and products are expressed in moles. A *mole* is a certain number of atoms or molecules of any substance, just as "a dozen" can refer to any twelve cats, roses, and so forth. Molar weight, in grams, equals the total atomic weight of all the atoms making up that substance.

For example, the atomic weight of carbon is 12, so one mole of carbon weighs 12 grams. A mole of oxygen (atomic weight 16) weighs 16 grams. Can you state why a mole of water (H_2O) weighs 18 grams, and why a mole of glucose ($C_6H_{12}O_6$) weighs 180 grams?

element. Carbon 14, for instance, decays to nitrogen 14. You will read about some uses of this radioisotope and others in the next section.

When Atom Bonds With Atom

Very shortly, we will consider **chemical bonds**, which are unions between the electron structures of atoms. Take a moment to review Figure 2.5, which summarizes a few conventions that are used to describe chemical reactions.

Elements are forms of matter that occupy space, have mass, and cannot be broken apart into something else by ordinary means.

Atoms, the smallest particles that are unique to each element, have one or more positively charged protons, negatively charged electrons, and (except for hydrogen) neutrons.

Isotopes are atoms of an element that differ in the number of neutrons. A radioisotope has uneven numbers of protons and neutrons. Such unstable atoms spontaneously emit particles and energy and so become transformed to a different atom.

USING RADIOISOTOPES TO DATE THE PAST, TRACK CHEMICALS, AND SAVE LIVES

RADIOMETRIC DATING   Probably for as long as they have been digging up the earth, people have been turning up fossilized leaves, shells, skeletons, and other stone-hard evidence of past life, entombed in rocks. Figure 2.6*a* and *b* shows examples. Some time ago, geologists hypothesized that if newly formed rocks slowly accumulate on top of older rocks, then fossils in older rock layers must be more ancient than those in more recently deposited ones. They used this perception to count backward through great numbers of rock layers and thereby construct a chronology of Earth history—a geologic time scale. As described in Section 20.1, they used sequences of fossils and other clues in the rocks to define the boundaries of the time scale's successive spans. However, no one was able to assign firm dates along the length of the scale until the discovery of radioactive decay, which made it possible to convert the relative ages into absolute ones.

By a method called **radiometric dating**, researchers now measure the proportions of (1) a radioisotope in a mineral that became trapped long ago in a brand-new rock and (2) a daughter isotope that formed from it in the same rock. Remember, the atomic nucleus of each radioactive element

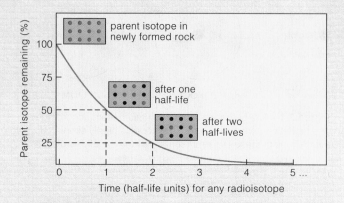

Figure 2.7   Generalized diagram of radioactive decay. Notice the proportion of the parent radioisotope in a rock sample over time, compared to the proportion of each decay product. During *each* half-life unit, the amount of the parent radioisotope will diminish by half.

consists of an *unstable* number of protons and neutrons. The nucleus spontaneously decays—it gives up energy and one or more particles of itself—until it reaches a more stable configuration. (Such nuclear emissions were the invisible culprits that exposed Becquerel's film.)

Half-life is the time it takes for half of a given quantity of any radioisotope to decay into a different, less unstable daughter isotope (Figure 2.7). The rate of decay is constant; changes in pressure, temperature, or chemical state cannot alter it. For example, ^{238}uranium occurs in zircon, which is a mineral in most volcanic rocks. Its nuclei decay into ^{234}thorium, a still-unstable daughter isotope that in turn decays into something else, and so on through a series of intermediate daughter isotopes to ^{206}lead—the final, most stable configuration for this series (Table 2.2). By using ^{238}uranium, with its half-life of 4.5 billion years, researchers realized the Earth formed more than 4.6 billion years ago.

Radiometric dating works for volcanic rocks or ashes. However, most fossil-containing rocks formed by the compaction of sand and other sediments. Thus the only way to date most fossil-containing rocks is to determine their position relative to volcanic rocks in the same area.

c

a   b

Figure 2.6   (**a**) Fossilized frond of a tree fern, one of many plant species that lived more than 250 million years ago. (**b**) Fossil of a worm that lived in the seas about 600 million years ago. (**c**) One of many glassy spherical particles, no bigger than sand grains, that are between 3.5 billion and 2.5 billion years old. The fact that it was buried in sediments indicates it was not volcanic in origin. It contains a large amount of iridium, an element that is rare on the Earth but abundant in meteorites. It may have formed early in Earth history, during the same bombardments from space that cratered the moon.

Table 2.2	Radioisotopes Commonly Used in Radiometric Dating		
Radioisotope (unstable)	More Stable Product	Half-Life (years)	Useful Range (years)
^{147}Samarium →	^{143}Neodymium	106 billion	>100 million*
^{87}Rubidium →	^{87}Strontium	48.8 billion	>100 million
^{232}Thorium →	^{208}Lead	14 billion	>200 million
^{238}Uranium →	^{206}Lead	4.5 billion	>100 million
^{40}Potassium →	^{40}Argon	1.25 billion	>100,000
^{235}Uranium →	^{207}Lead	700 million	>100 million
^{14}Carbon →	^{14}Nitrogen	5,730	0–60,000

* The symbol > means greater than.

You may have heard of dating methods based on "carbon 14." The greatest numbers of this radioisotope form continually in the upper atmosphere. There it combines with oxygen to form carbon dioxide. Along with the more stable isotopes of carbon, small amounts of ^{14}carbon enter the web of life by photosynthesis. Every organism incorporates it. When organisms die, the amount of ^{14}carbon starts to decrease as an outcome of radioactive decay. ^{14}Carbon has a half-life of 5,730 years, so amounts in bone, wood, shells, and other organic substances are too small to detect by direct analysis. Instead, researchers use scintillation counters to monitor emissions from a given sample. The older the sample, the fewer emissions (counts) will be recorded in a given period.

TRACKING TRACERS Radioisotopes also make fine tracers. A **tracer** is a substance with a radioisotope attached to it, rather like a shipping label, that researchers can track after they deliver it into a cell, a body, an ecosystem, or some other system. Scintillation counters and other devices can track a tracer's precise movement through some pathway or pinpoint its final destination.

Consider how Melvin Calvin and other botanists figured out the steps by which plants build carbohydrates during photosynthesis. They knew all isotopes of an element have the same number of electrons, so all must interact with other atoms the same way. Plant cells, they hypothesized, should be able to use any isotope of carbon when they build carbon compounds. By putting plant cells in a medium enriched with ^{14}carbon, these botanists were able to track the uptake of carbon through all the reaction steps leading to the formation of sugars and starches.

As another example, tracers are being used to identify how plants take up and use soil nutrients and synthetic fertilizers. The findings may help improve crop yields.

SAVING LIVES Radioisotopes have uses as diagnostic tools in medicine. For safety considerations, clinicians use only the kinds with extremely short half-lives. Consider the human thyroid, located in front of the windpipe. It is the only gland of ours that takes up iodine, a key building block for

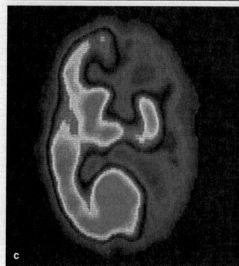

Figure 2.9 (**a**) Patient being moved into a PET scanner. Inside (**b**), a ring of detectors intercepts the radioactive emissions from labeled molecules that were injected into the patient. Computers analyze and color-code the number of emissions from each location in the scanned body region. (**c**) Brain scan of a child who has a neurological disorder. Different colors in a brain scan signify differences in metabolic activity. The right half of this brain shows very little activity. By comparison, cells of the left half absorbed and used the labeled molecule at expected rates.

the synthesis of thyroid hormones that affect growth and metabolism. If a patient's symptoms point to a thyroid hormone imbalance, clinicians may inject a trace amount of ^{123}iodine into his or her blood, then use a photographic imaging device to scan the gland. Figure 2.8 shows three examples of the resulting image.

Also consider *PET* (for *Positron-Emission Tomography*). It yields images of metabolically active and inactive tissues in a patient. Suppose clinicians attach a radioisotope to glucose (or some other molecule) that cells require for metabolism. They inject the radioactively labeled glucose into the patient, who is moved into a PET scanner of the sort shown in Figure 2.9*a*. Cells in the patient's tissues will absorb the glucose. Detectors intercept emissions from it, and these are used to form an image of any variations or abnormalities in metabolic activity, as in Figure 2.9*b,c*.

Finally, radioisotopes have uses in some therapies. For example, the energetic emissions from ^{238}plutonium drive artificial pacemakers, which help correct irregular heartbeats. In such devices, the radioisotope is sealed inside a case so that its dangerous emissions won't damage tissues. As another example, clinicians may use radioisotopes for *radiation therapy*, which destroys or impairs living cells. In some therapies, they bombard localized cancers with the energetic emissions from a source of ^{226}radium or ^{60}cobalt.

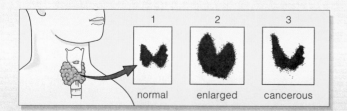

| normal | enlarged | cancerous |

Figure 2.8 Scans of the thyroid gland from three patients.

THE NATURE OF CHEMICAL BONDS

In a given chemical reaction, atoms may acquire extra electrons, share them, or donate them to another atom. The atoms of certain elements do this rather easily, but others do not. What determines whether one atom will interact with another in such ways? *The outcome depends on the number and arrangement of their electrons.*

Electrons and Energy Levels

If you've ever tinkered with magnets, you have a sense of the attractive force between unlike charges (+ −) and the repulsive force between like charges (+ + or − −). All electrons carry a negative charge. In an atom, they repel each other but are attracted to the positive charge of the protons. They spend as much time as possible near the protons and far away from each other by moving about in different orbitals. Think of **orbitals** as *volumes of space* around the atomic nucleus in which electrons are likely to be at any instant. (As a rough analogy, picture three preschoolers circling a cookie jar, not yet expert in the fine art of sharing. Each is drawn inexorably to the cookies but dreads being shoved away by the others. Two of them might duck and weave about on opposite sides of the jar in order to avoid a direct hit. But all three never, ever will occupy the same space at the same time.)

Just as the number of electrons differs among atoms, so does the number of occupied orbitals. One or at most two electrons can occupy any given orbital.

Consider hydrogen, the simplest atom of all. Its lone electron occupies a spherical orbital that is closest to the nucleus, which corresponds to the *lowest available energy level*. In all other atoms, two electrons fill this orbital.

electron orbital
nucleus

HYDROGEN ATOM

In other atoms, another two electrons occupy a second spherical orbital surrounding the first. Up to six more electrons may be distributed in three dumbbell-shaped orbitals, even more in other orbitals, and so on (Figure 2.10). These electrons spend more time farther away from the nucleus; they are at *higher energy levels.*

The **shell model** in Figure 2.11 is a simple though not quite accurate way to think about how electrons are distributed in a given atom. By this model, a series of shells encompasses all the orbitals that are available to electrons. The first shell encloses the first orbital, which is spherical. A second shell (at a higher energy level) encloses the first, and the next four available orbitals fit inside it. Additional orbitals fit inside a third shell, a fourth shell, and so on up to the large, complex atoms of heavier elements, as listed in Appendix VI.

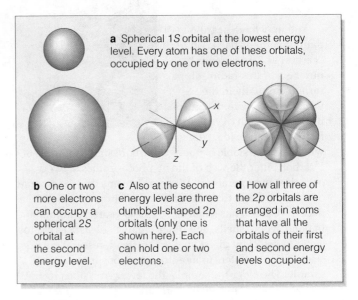

a Spherical 1*S* orbital at the lowest energy level. Every atom has one of these orbitals, occupied by one or two electrons.

b One or two more electrons can occupy a spherical 2*S* orbital at the second energy level.

c Also at the second energy level are three dumbbell-shaped 2*p* orbitals (only one is shown here). Each can hold one or two electrons.

d How all three of the 2*p* orbitals are arranged in atoms that have all the orbitals of their first and second energy levels occupied.

Figure 2.10 Arrangement of electrons in atoms. (**a**) One or at most two electrons occupy a ball-shaped volume of space—an orbital—close to the nucleus. When electrons occupy this orbital, they are at the atom's lowest energy level. (**b–d**) At the second energy level, there can be as many as eight more electrons; that is, up to two electrons in each of four more orbitals.

Orbital shapes get tricky, but for our purposes we can ignore them. Simply think of the total number of orbitals at a given energy level as being somewhere inside a "shell" that surrounds the nucleus of an atom. As you can deduce from Figure 2.11, higher energy levels correspond to shells farther from the nucleus.

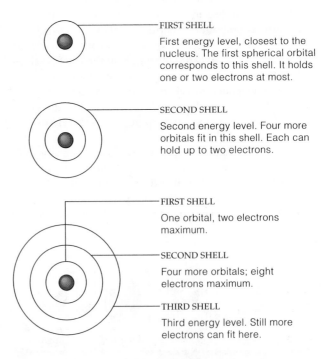

FIRST SHELL

First energy level, closest to the nucleus. The first spherical orbital corresponds to this shell. It holds one or two electrons at most.

SECOND SHELL

Second energy level. Four more orbitals fit in this shell. Each can hold up to two electrons.

FIRST SHELL

One orbital, two electrons maximum.

SECOND SHELL

Four more orbitals; eight electrons maximum.

THIRD SHELL

Third energy level. Still more electrons can fit here.

Figure 2.11 Shell model of electron distribution in atoms. Only three of the many possible energy levels (shells) are shown.

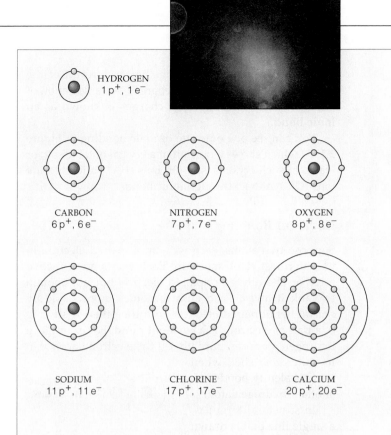

			Distribution of Electrons			
Element	Symbol	Atomic Number*	First Shell	Second Shell	Third Shell	Fourth Shell
Hydrogen	H	1	1	—	—	—
Helium	He	2	2	—	—	—
Carbon	C	6	2	4	—	—
Nitrogen	N	7	2	5	—	—
Oxygen	O	8	2	6	—	—
Neon	Ne	10	2	8	—	—
Sodium	Na	11	2	8	1	—
Magnesium	Mg	12	2	8	2	—
Phosphorus	P	15	2	8	5	—
Sulfur	S	16	2	8	6	—
Chlorine	Cl	17	2	8	7	—
Calcium	Ca	20	2	8	8	2

* The number of protons in the nucleus.

Figure 2.12 Examples of the shell model of the distribution of electrons in atoms. Hydrogen, carbon, and other atoms having electron vacancies (unfilled orbitals) in their outermost shell tend to give up, accept, or share electrons. Helium, neon, and other atoms with no electron vacancies in their outermost shell show no such tendency. Amazingly, physicists confined electrons in a bubble of liquid helium, then used fluctuating sound waves to pop the bubble. The photograph shows a white-centered red flash given off by an electron that escaped. It allows us to see how a single electron can make something happen.

Electrons and the Bonding Behavior of Atoms

How can you tell whether any two atoms will interact? Check for any electron vacancies in their outermost shell. Vacancies mean an atom might give up, gain, or share electrons under suitable conditions, which of course will change the distribution or number of its electrons. You can use the shell model to visualize what goes on here.

In Figure 2.12, which shows the electron distribution for several atoms, electrons are assigned to an energy level (a circle, or shell). Count the electron vacancies— that is, one or more unfilled orbitals in the outermost shell of each atom. As the table shows, helium and neon have no vacancies. They are among the *inert* atoms, which show little tendency to enter chemical reactions.

Now look again at Figure 2.3, which lists the most abundant of the elements making up a typical organism (a human). These elements include hydrogen, oxygen, carbon, and nitrogen. Atoms of all four elements have electron vacancies. Because of this, they tend to form *bonds* with other atoms—and thereby fill the vacancies.

From Atoms to Molecules

When two or more atoms bond together, the outcome is a **molecule**. Many molecules are composed of only one element. Molecular nitrogen (N_2), with its two nitrogen atoms, is like this.

The molecules of **compounds** are composed of two or more different elements in proportions that never vary. Water is a prime example of a compound. Each molecule of water has one oxygen atom chemically bonded to two hydrogen atoms. The molecules of water in rainclouds, the ocean, a Siberian lake, your bathtub, or anywhere else always have twice as many hydrogen as oxygen atoms.

By contrast, in a **mixture**, two or more elements are simply intermingling in proportions that can vary (and usually do). Consider the sugar sucrose, a compound of carbon, hydrogen, and oxygen. Swirl sucrose molecules and water molecules together, and you get a mixture.

In an atom, electrons occupy orbitals, which are volumes of space around the nucleus. By a simplified model, orbitals are arranged inside a series of shells that surround the nucleus. The shells correspond to levels of energy that become greater with distance from the nucleus.

One or two electrons at most occupy any orbital. Atoms with unfilled orbitals in their outermost shell tend to interact with other atoms; those with no vacancies do not.

In molecules of an element, all of the atoms are of the same kind. In molecules of a compound, atoms of two or more elements are bonded together, in unvarying proportions.

IMPORTANT BONDS IN BIOLOGICAL MOLECULES

Eat your peas! Drink your milk! Probably for longer than you care to remember, somebody has been telling you to eat foods that are rich in carbohydrates, proteins, and other "biological molecules." Only living organisms put together and use these molecules, which consist of a few kinds of atoms held together by only a few kinds of bonds. Foremost among the molecular interactions are the ionic, covalent, and hydrogen bonds.

Ion Formation and Ionic Bonding

An atom, recall, has just as many electrons as protons, so it carries no net charge. However, that balance can change for atoms with a vacancy—an unfilled orbital—in their outermost shell. For example, a chlorine atom can acquire an extra electron and thus fill the vacancy, as shown in Figure 2.13*a*. Similarly, the lone electron in a sodium atom's outermost shell can be knocked out of an orbital or pulled completely away from it.

Any atom that has either gained or lost one or more electrons is an **ion**. The balance between its protons and its electrons has shifted, so the atom is now ionized; it has become positively or negatively charged.

In living cells, neighboring atoms commonly accept or donate electrons among one another. When one atom loses an electron and one gains, both become ionized. Depending on cellular conditions, the two ions may not

separate; they may remain together as a result of the mutual attraction of opposite charges. An association of two ions that have opposing charges is known as an **ionic bond**.

You can see one outcome of ionic bonding in Figure 2.13*b*, which shows a portion of a crystal of table salt, or NaCl. In such crystals, sodium ions (Na^+) and chloride ions (Cl^-) interact through ionic bonds.

Covalent Bonding

Suppose two atoms, each with an unpaired electron in its outermost shell, meet up. Each exerts an attractive force on the other's unpaired electron but not enough to yank it away. Each atom becomes more stable by *sharing* its unpaired electron with the other. A sharing of a pair of electrons is a **covalent bond**. For example, a hydrogen atom can partially fill the electron vacancy in its outermost shell when it is covalently bonded to another hydrogen atom.

In structural formulas, a single line that is drawn

MOLECULAR HYDROGEN (H_2)

between two atoms represents a *single* covalent bond. Molecular hydrogen has such a bond: H—H. In a *double* covalent bond, two atoms share two pairs of electrons. This is the case with molecular oxygen (O_2), or O=O.

In a *triple* covalent bond, two atoms share three pairs of electrons. This is true of molecular nitrogen (N_2), which may be written as N≡N. These three examples happen to be gaseous molecules. Each time you breathe in some air, you draw a great number of H_2, O_2, and N_2 molecules into your nose.

Covalent bonds are nonpolar or polar. In a *nonpolar* covalent bond,

a

SODIUM ATOM
11 p$^+$
11 e$^-$

electron transfer

CHLORINE ATOM
17 p$^+$
17 e$^-$

SODIUM ION
11 p$^+$
10 e$^-$

CHLORIDE ION
17 p$^+$
18 e$^-$

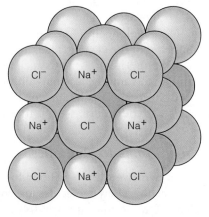

Figure 2.13 **(a)** Ionization by way of an electron transfer. In this case, a sodium atom donates the lone electron in its outermost shell to a chlorine atom, which has an unfilled orbital in *its* outermost shell. A sodium ion (Na^+) and a chloride ion (Cl^-) are the outcome of this interaction.

(b) In each crystal of table salt, or NaCl, many sodium and chloride ions remain together because of the mutual attraction of their opposite charges. Their interaction is an example of ionic bonding.

b A crystal of sodium chloride (NaCl)

participating atoms exert the same pull on the electrons and share them equally. The term "nonpolar" implies there is not any difference in charge at the ends of the bond (that is, at its two poles). Molecular hydrogen is a simple example of this. Its two H atoms, each with one proton, attract the shared electrons equally.

In a *polar* covalent bond, atoms of different elements (which have different numbers of protons) do not exert the same pull on shared electrons. The more attractive atom ends up with a slight negative charge; the atom is "electronegative." Its effect is balanced out by the other atom, which ends up with a slight positive charge. In other words, taken together, the atoms interacting in a polar covalent bond have no *net* charge—but the charge is distributed unevenly between the bond's two ends.

Consider the water molecule, which has two polar covalent bonds: H—O—H. In this molecule, electrons are less attracted to the hydrogens than to the oxygen, which has more protons. A water molecule carries no *net* charge, but you will see shortly that its polarity can weakly attract neighboring polar molecules and ions.

Hydrogen Bonding

The patterns of electron sharing in covalent bonds hold atoms together in specific arrangements in molecules. Some of the patterns also give rise to weak attractions and repulsions between charged functional groups of molecules, as well as between molecules and ions. Like interacting skydivers, such interactions break and form easily (Figure 2.14). Yet they have important roles in the structure and functioning of biological molecules.

For example, in a **hydrogen bond**, a small, highly electronegative atom of a molecule weakly interacts with a hydrogen atom that is already participating in a

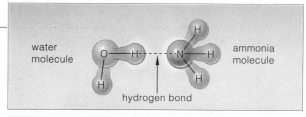

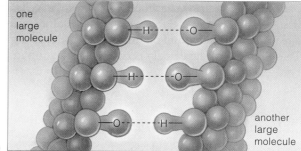

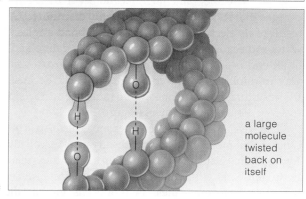

Figure 2.15 Examples of hydrogen bonds. Compared to covalent bonds, a hydrogen bond is easier to break. Collectively, however, such bonds are important in water and many other substances.

polar covalent bond. During this interaction, hydrogen shows a slight positive charge, and this attracts other atoms that carry a slight negative charge (Figure 2.15).

Hydrogen bonds may form between two or more molecules. They also may form in different parts of the *same* molecule, where it twists and folds back on itself. For example, many such bonds form between the two strands of a DNA molecule. Individually, the hydrogen bonds break easily. Collectively, they stabilize DNA's structure. Similarly, hydrogen bonds form between the molecules that make up water. As you will read next, they contribute to water's life-sustaining properties.

In an ionic bond, two ions of opposite charge attract each other and stay together. Ions form when atoms gain or lose electrons and so acquire a net positive or negative charge.

In a covalent bond, atoms share a pair of electrons. If the atoms share the electrons equally, the bond is nonpolar. If the sharing is not equal, the bond itself is polar—slightly positive at one end, slightly negative at the other.

In a hydrogen bond, a small, highly electronegative atom of a molecule interacts weakly with a neighboring hydrogen atom that is already taking part in a polar covalent bond.

Figure 2.14 Like skydivers who briefly clasp hands to form an orderly pattern, weak attractions within and between molecules, and between ions and molecules, can form and break easily.

PROPERTIES OF WATER

No sprint through basic chemistry is complete unless it leads us to the collection of molecules called water. Life originated in water. Many organisms still live in it. The ones that don't cart water around with them, in cells and tissue spaces. Many metabolic reactions require water as a reactant. Cell shape and internal structure depend on it. These topics will repeatedly occupy our attention in the book, so you may find it useful to become familiar with the following points about water's properties.

Polarity of the Water Molecule

A water molecule, remember, has no *net* charge, but the charges that it does carry are unevenly distributed. As a result of its electron arrangements and bond angles, the water molecule's oxygen "end" is a bit negative and its other end is a bit positive. Figure 2.16a shows a simple way to think about this charge distribution. Because of the resulting polarity, one water molecule attracts and hydrogen-bonds with others (Figure 2.16b).

The weak polarity of the water molecule also *attracts* other polar molecules, including sugars. We call these **hydrophilic** (water-loving) **substances**, for they readily hydrogen-bond with water. By contrast, water's polarity *repels* nonpolar molecules, including oils. We call these **hydrophobic** (water-dreading) **substances**. Observe this for yourself by shaking a bottle that contains water and salad oil, then setting it on a table. Soon, hydrogen bonds replace the ones that broke when you shook the bottle. As water molecules reunite, the oil molecules are pushed aside and forced to cluster as droplets or as a film at the water's surface. Such hydrophobic interactions are vital. For example, a thin, oily membrane is almost all that separates a cell's watery surroundings and its watery interior. The organization of that membrane starts with hydrophobic interactions, as described in Section 5.1.

Water's Temperature-Stabilizing Effects

Cells consist mostly of water, and they release a great deal of heat energy during their metabolic activities. If it were not for the hydrogen bonds in liquid water, cells might cook in their own juices. To see why this is so, start with these observations: Each molecule of water or any other substance is in constant motion, and when it absorbs heat, its motion increases. The **temperature** of a given substance is simply a measure of its molecular motion. Compared to most fluids, water can absorb more heat energy before its temperature increases measurably. Why? Much of the incoming energy disrupts hydrogen bonding *between* neighboring water molecules instead of directly causing the motion of individual molecules to increase. The stupendous number of hydrogen bonds in liquid water can buffer large swings in temperature

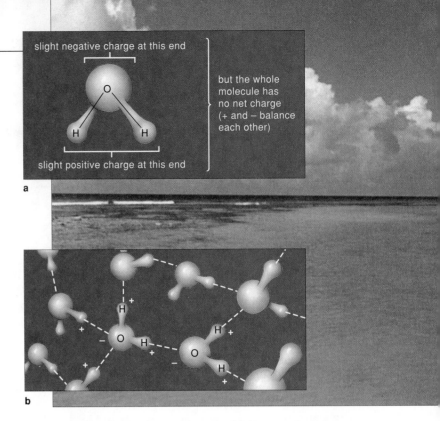

slight negative charge at this end

but the whole molecule has no net charge (+ and − balance each other)

slight positive charge at this end

a

b

Figure 2.16 Water—a substance vital for life. (**a**) Polarity of the water molecule. (**b**) Hydrogen bonding between water molecules in liquid water. Dashed lines are hydrogen bonds. (**c**) Hydrogen bonding between water molecules in ice, which in vast quantities blankets the habitat of choice of polar bears. Below 0°C, each water molecule becomes locked, by four hydrogen bonds, into a crystal lattice. In the bonding pattern of ice, molecules are spaced farther apart than in liquid water at room temperature. When water is liquid, constant molecular motion usually prevents the maximum number of hydrogen bonds from forming.

inside cells. Similarly, hydrogen bonds help stabilize the temperature of aquatic habitats.

Even when the temperature of liquid water is not shifting much, hydrogen bonds are constantly breaking, but they also are forming again just as fast. By contrast, a large energy input can increase molecular motion so much that hydrogen bonds *stay* broken, and individual molecules at the water's surface escape into the air. By this process, called **evaporation**, heat energy converts liquid water to the gaseous state. As large numbers of molecules break free and depart, they carry away some energy and lower the water's surface temperature.

Evaporative water loss can help cool you and some other mammals when you work up a sweat on hot, dry days. Under such conditions, sweat—which is about 99 percent water—evaporates from your skin.

Below 0°C, hydrogen bonds resist breaking and lock water molecules in the latticelike bonding pattern of ice (Figure 2.16c). Ice is less dense than water. During winter freezes, ice sheets may form near the surface of ponds, lakes, and streams. Like a blanket, ice "insulates" the liquid water beneath it and helps protect many fishes, frogs, and other aquatic organisms against freezing.

c

Figure 2.17 A water strider (*Gerris*). This long-legged bug feeds on lightweight insects that land or fall onto the surface of water. Because of its fine, water-resistant leg hairs (and water's high surface tension), the bug scoots easily across the surface.

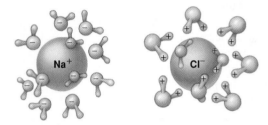

Figure 2.18 Spheres of hydration around two charged ions.

Water's Cohesion

Life also depends on water's cohesion. **Cohesion** means something has the capacity to resist rupturing when it is placed under tension—that is, stretched—as by the weight of a bug's legs (Figure 2.17).

Consider what goes on in a lake or swimming pool or some other large body of liquid water. Uncountable episodes of hydrogen bonding exert a continual inward pull on the water molecules at or near the surface. The ongoing, collective hydrogen bonding results in a high surface tension. It may be especially attention-grabbing if you swim in the lake or pool on a hot summer night, when far too many night-flying insects splat against the water and float on it.

Cohesion operates inside organisms as well as on the outside. For example, maple trees and other land plants require nutrient-laden water for their growth and metabolism. Largely because of the cohesion among water molecules, ribbonlike columns of liquid water move through vascular tissues that serve as pipelines from the roots to leaves and all other parts of the plant. On sunny days, water evaporates from the leaves. As individual molecules depart, hydrogen bonds "pull up" more water molecules into leaf cells as replacements, in ways that you will read about in Section 30.3.

Water's Solvent Properties

A final point to keep in mind is water's capacity as a solvent; ions and polar molecules readily dissolve in it. When dissolved in water, substances are called **solutes**. But what does "dissolved" mean? By way of example, pour some table salt (NaCl) into a glass of water. In time, the salt crystals separate into Na^+ and Cl^-. Each Na^+ attracts the negative end of water molecules; at the same time, each Cl^- attracts the positive end of others (Figure 2.18). Many water molecules cluster as "spheres of hydration" around ions and keep them dispersed in the fluid. In general, a substance is "dissolved" when such spheres have formed around its individual ions or molecules. This is what happens to solutes in the fluid portion of cells, in maple tree sap, in your own blood, and in all other fluids associated with life.

A water molecule has no net charge, yet it shows polarity. The polarity allows water molecules to hydrogen-bond with one another and with other polar (hydrophilic) substances. Water molecules tend to repel nonpolar (hydrophobic) substances.

Water has temperature-stabilizing effects, internal cohesion, and a capacity to dissolve many substances. These properties influence the structure and functioning of organisms.

A great variety of ions dissolved in the fluids inside and outside cells influence cell structure and functioning. Among the most influential are **hydrogen ions**, or H^+. Hydrogen ions are the same thing as free (unbound) protons. They have far-reaching effects largely because they are chemically active and there are so many of them.

The pH Scale

At any instant in liquid water, some water molecules break apart into hydrogen ions and **hydroxide ions** (OH^-). This ionization of water is the basis of the **pH scale**, as shown in Figure 2.19. Biologists use this scale when measuring the H^+ concentration of seawater, tree sap, blood, and other fluids.

Pure water always contains just as many H^+ as OH^- ions. This condition also may occur in other fluids, and it signifies neutrality. We assign it a value of 7 at the midpoint of the pH scale, which ranges from 0 (the highest H^+ concentration) to 14 (the lowest). *The greater the H^+ concentration, the lower the pH.*

Starting at neutrality, each change by one unit of the pH scale corresponds to a tenfold increase or decrease in the H^+ concentration. An easy way to sense the logarithmic difference in scale is to dissolve a bit of baking soda (pH 9) on your tongue, then egg white (pH 8). Next, sip pure water (pH 7), then lemon juice (2.3).

Acids and Bases

As certain substances dissolve in water, they donate protons (H^+) to some water molecules. For a fleeting instant, the water molecules become H_3O^+, but these just as instantaneously donate the protons elsewhere, so for our purposes it simply is easier to think of all the water molecules as being H_2O. All substances that *donate* H^+

when dissolved in water are called **acids**. Certain other substances are **bases**. They *accept* H^+ when dissolved in water, and OH^- forms directly or indirectly after they do so. All *acidic* solutions, such as lemon juice, gastric fluid, and coffee, contain more H^+ than OH^-; their pH is below 7. *Basic* solutions, such as baking soda, seawater, and egg white, contain more OH^- than H^+. Such solutions, which are also referred to as "alkaline" fluids, have a pH above 7.

The solution inside most of your body's cells is neutral; the cell interior is about 7 on the pH scale. Most fluids bathing those cells are slightly higher, with pH values that range between 7.3 and 7.5. The same is true of the fluid portion of blood. By contrast, seawater is more alkaline than body fluids of organisms that live in it.

You can think of most acids as being either weak or strong. Weak ones such as carbonic acid (H_2CO_3) are reluctant H^+ donors. Depending on the pH, they can just as easily accept H^+ after giving it up, so they alternate between acting as an acid and a base. By contrast, strong acids totally give up H^+ when they dissociate in water. Hydrochloric acid (HCl), nitric acid (HNO_3), and sulfuric acid (H_2SO_4) are examples.

Think about what happens after someone sniffs and eats fried chicken, then sends it on its way to gastric fluid inside the saclike stomach. The meal stimulates cells in the lining of the stomach to secrete a strong acid, HCl, that dissociates into H^+ and Cl^-. These ions make the gastric fluid more acidic. The increased acidity activates enzymes, which digest chicken proteins and help kill most of the bacteria that lurked in or on the chicken. When people eat too much fried chicken, they may get an *acid stomach* and reach for an antacid such

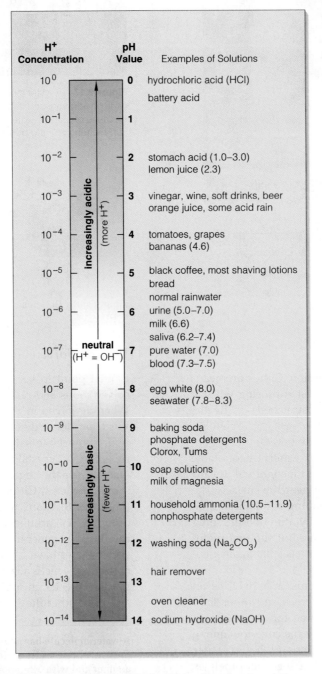

Figure 2.19 The pH scale, in which a liter of some solution is assigned a number according to the number of hydrogen ions that it contains. Also shown are approximate pH values for some common solutions. The pH scale ranges from 0 (most acidic) to 14 (most basic). A change of 1 on the scale means a tenfold change in H^+ concentration.

as milk of magnesia. When this strong base dissolves, it releases magnesium ions and OH⁻, which combines with some of the excess H⁺ in gastric fluid, raises the pH, and settles things down.

High concentrations of strong acids or bases can disrupt the environment and pose dangers to life. Read the labels on bottles of ammonia, drain cleaner, and other products that are often stored in homes. Many can cause severe *chemical burns*. So can the sulfuric acid in car batteries. Or think of strongly acidic industrial wastes. When airborne, they alter the pH of rain (Figure 2.20). In turn, the pH of acid rainwater can severely alter many freshwater and terrestrial habitats, as described in Section 50.1.

Figure 2.20 Sulfur dioxide emissions from a coal-burning power plant. Camera lens filters revealed the otherwise invisible emissions. Sulfur dioxide and other airborne pollutants dissolve in water vapor to form acidic solutions. They are a major component of acid rain.

Buffers Against Shifts in pH

Metabolic reactions are sensitive to even slight shifts in pH, for H⁺ and OH⁻ can combine with many different molecules and alter their functions. Normally, control mechanisms minimize unsuitable shifts in pH, as they do when HCl enters the stomach in response to a meal. Many of the controls involve buffer systems.

A **buffer system** is a partnership between a weak acid and the base that forms when it dissolves in water. The two work as a pair to counter slight shifts in pH. Remember, when a strong base enters a fluid, the OH⁻ level rises. But a weak acid neutralizes part of the added OH⁻ by combining with it. By this interaction, its partner forms. Later, if a strong acid floods in, the base will accept H⁺ and so become its partner in the system.

Bear in mind, the action of a buffer system cannot make or eliminate hydrogen ions. It can only keep them locked up, out of the way, until other metabolic events restore the balance.

In all complex multicelled organisms, diverse buffer systems operate in the internal environment—that is, in blood and tissue fluids. For example, later in the book you will see how reactions in lungs and kidneys help control the acid-base balance of this environment, at levels suitable for life. For now, simply think of what happens when blood loses H⁺ and is not as acidic as it should be. Carbonic acid happens to be dissolved in the blood. At such times, it releases H⁺ and so becomes the partner base, bicarbonate:

$$H_2CO_3 \longrightarrow HCO_3^- + H^+$$
CARBONIC ACID   BICARBONATE

When blood becomes more acidic, more H⁺ becomes bound to the base, thus forming the partner acid:

$$HCO_3^- + H^+ \longrightarrow H_2CO_3$$
BICARBONATE   CARBONIC ACID

Unless they are controlled, shifts in pH have drastic outcomes. If blood pH declines even to 7, an individual will enter into a *coma*, a sometimes irreversible state of unconsciousness. An increase to 7.8 can lead to *tetany*, a potentially lethal stage in which the body's skeletal muscles enter a state of uncontrollable contraction. In *acidosis*, carbon dioxide builds up in blood, too much carbonic acid forms, and blood pH severely decreases. *Alkalosis* is an uncorrected increase in blood pH. Both conditions weaken the body and can be lethal.

Salts

Salts are compounds that release ions *other than* H⁺ and OH⁻ in solutions. Salts and water often form when a strong acid and a strong base interact. Depending on a solution's pH value, salts can form and dissolve easily. Consider how sodium chloride forms, then dissolves:

$$\text{HCl (acid)} + \text{NaOH (base)} \longrightarrow \text{NaCl (salt)} + H_2O$$
HYDROCHLORIC   SODIUM   SODIUM CHLORIDE
ACID   HYDROXIDE

$$Na^+ \quad Cl^- \text{ (ionization)}$$

Many salts dissolve into ions that serve key functions in cells. For example, nerve cell activity depends on ions of sodium, potassium, and calcium; plant cells take up water with the aid of potassium ions; and blood clots and muscles contract with the help of calcium ions.

Hydrogen ions (H⁺) and other ions dissolved in the fluids inside and outside cells affect cell structure and function.

When dissolved in water, acidic substances release H⁺ and basic (alkaline) substances accept them. Certain acid-base interactions, as in buffer systems, help maintain the pH value of a fluid—that is, its H⁺ concentration.

A buffer system counters slight shifts in pH by releasing hydrogen ions when their concentration is too low or by combining with them when the concentration is too high.

Salts are compounds that release ions other than H⁺ and OH⁻, and many of those ions have key roles in cell functions.

SUMMARY

1. Chemistry helps us understand the nature of all of the substances that make up cells, organisms, the Earth, its waters, and the atmosphere. Each substance consists of one or more elements. Of the ninety-two naturally occurring elements, oxygen, carbon, hydrogen, and nitrogen are the most common in organisms. Organisms have lesser and varying amounts of other elements, such as calcium, phosphorus, potassium, and sulfur.

2. Table 2.3 summarizes some key chemical terms that you will encounter throughout this book.

3. Elements consist of atoms. An atom has one or more positively charged protons, an equal number of negatively charged electrons, and (except for hydrogen) one or more uncharged neutrons. The protons and neutrons occupy a core region, the atomic nucleus. Atoms of each element may vary in the number of neutrons. Those that deviate from the most common number of neutrons are isotopes.

4. An atom has no *net* charge. But it may gain or lose one or more electrons and so become an ion, which has an overall positive or negative charge.

5. By one simplified model, electrons can occupy orbitals (volumes of space) inside a series of shells around the atomic nucleus. Whether one atom interacts with others depends on the number and arrangement of its electrons. Atoms with one or more unfilled orbitals in the outermost shell tend to bond with other elements.

6. A chemical bond is a union between the electron structures of atoms.

   a. In an ionic bond, a positive ion and negative ion remain together because of a mutual attraction of their opposite charges.

   b. Various atoms can share one or more pairs of electrons, in single, double, or triple covalent bonds. Electron sharing is equal in a nonpolar covalent bond and unequal in a polar covalent bond. The interacting atoms have no net charge, but the bond is slightly negative at one end and slightly positive at the other.

   c. In a hydrogen bond, a highly electronegative, small atom weakly interacts with a hydrogen atom that is already taking part in a polar covalent bond in the same molecule or in a different one.

7. By the pH scale, a solution is assigned a number that reflects its H^+ concentration. This ranges from 0 (highest concentration) to 14 (lowest). At pH 7, the H^+ and OH^- concentrations are equal. Acids release H^+ in water; bases combine with them. Buffer systems help maintain the pH values of blood, tissue fluids, and the fluid inside cells.

8. Polar covalent bonds join together the atoms of water molecules. The polarity promotes extensive hydrogen bonding between water molecules. Such bonding is the basis of liquid water's ability to resist temperature changes more than other fluids do, to show internal cohesion, and to easily dissolve polar or ionic substances. These properties profoundly influence the metabolic activity, shape, and internal organization of cells.

Table 2.3 Summary of Key Players in the Chemical Basis of Life	
ELEMENT	Fundamental form of matter that occupies space, has mass, and cannot be broken apart into a different form of matter by ordinary physical or chemical means.
ATOM	Smallest unit of an element that still retains the characteristic properties of that element.
Proton (p^+)	Positively charged particle of the atomic nucleus. All atoms of an element have the same number of protons, which is the atomic number. A proton without an electron zipping around it is a hydrogen ion (H^+).
Electron (e^-)	Negatively charged particle that can occupy volumes of space (orbitals) around an atomic nucleus. All atoms of an element have the same number of electrons. Electrons can be shared or transferred among atoms.
Neutron	Uncharged particle of the nucleus of all atoms except hydrogen. For a given element, the mass number is the number of protons and neutrons in the nucleus.
MOLECULE	Unit of matter in which two or more atoms of the same element, or different ones, are bonded together.
Compound	Molecule composed of two or more different elements in unvarying proportions. Water is an example.
Mixture	Intermingling of two or more elements in proportions that can and usually do vary.
ISOTOPE	For a given element, an atom with more or fewer neutrons than atoms with the most common number.
Radioisotope	Unstable isotope, having an unbalanced number of protons and neutrons, that emits particles and energy.
Tracer	Molecule of a substance to which a radioisotope is attached. In conjunction with tracking devices, it can be used to follow the movement or destination of that substance in a metabolic pathway, the body, etc.
ION	Atom that has gained or lost one or more electrons, thus becoming positively or negatively charged.
SOLUTE	Any molecule or ion dissolved in some solvent.
Hydrophilic substance	Polar molecule or molecular region that can readily dissolve in water.
Hydrophobic substance	Nonpolar molecule or molecular region that strongly resists dissolving in water.
ACID	Substance that donates H^+ when dissolved in water.
BASE	Substance that accepts H^+ when dissolved in water; OH^- forms directly or indirectly afterward.
SALT	Compound that releases ions other than H^+ or OH^- when dissolved in water.

Review Questions

1. Name the four elements (and their symbols) that make up more than 95 percent of the body weight of all organisms. *2.1*

2. Define isotope, and describe how radioisotopes are used either in radiometric dating or as tracers. *2.1, 2.2*

3. How many electrons can occupy each orbital around an atomic nucleus? Using the shell model, explain how the orbitals available to electrons are distributed in an atom. *2.3*

4. Distinguish between:
   a. ionic and hydrogen bonds *2.4*
   b. polar and nonpolar covalent bonds *2.4*
   c. hydrophilic and hydrophobic interactions *2.5*

5. If a water molecule has no net charge, then why does it attract polar molecules and repel nonpolar ones? *2.5*

6. Define acid and base. Then describe the behavior of a weak acid in solutions having a high or low pH value. *2.6*

Self-Quiz *(Answers in Appendix IV)*

1. Electrons carry a(n) _____ charge.
   a. positive   b. negative   c. zero

2. Atoms share electrons unequally in a(n) _____ bond.
   a. ionic   c. polar covalent
   b. nonpolar covalent   d. hydrogen

3. An individual water molecule shows _____ .
   a. polarity   d. solvency
   b. hydrogen-bonding capacity   e. a and b
   c. heat resistance   f. all of the above

4. In liquid water, spheres of hydration form around _____ .
   a. nonpolar molecules   d. solvents
   b. polar molecules   e. b and c
   c. ions   f. all of the above

5. Hydrogen ions (H^+) are _____ .
   a. the basis of pH values   d. dissolved in blood
   b. unbound protons   e. both a and b
   c. targets of certain buffers f. all of the above

6. When dissolved in water, a(n) _____ donates H^+; however, a(n) _____ accepts H^+ and leads to OH^- formation.

7. Match the terms with their most suitable descriptions.
   ____ trace element   a. weak acid and its partner base work
   ____ buffer system   as a pair to counter pH shifts
   ____ chemical bond   b. union between electron structures
   ____ temperature   of two atoms
                         c. less than 0.01% of body weight
                         d. measure of molecular motion in
                            some defined region

Critical Thinking

1. A hospital patient is administered a 3.2-milligram dose of [131]iodine, which has a half-life of 8.1 days. How much of this radioisotope will remain in the patient's body after twenty-four days, assuming none is lost except by way of radioactive decay?

2. An ionic compound forms when calcium combines with chlorine. Referring to Figure 2.12, give the compound's formula. (Hint: Be sure the outermost shell of each atom is filled.)

3. The molecular weight of hydrochloric acid (HCl), a strong acid, is 36 grams/mole, and that of sodium hydroxide (NaOH), a strong base, is 40 grams/mole. A one molar (1M) solution of any compound can be made by dissolving one mole in a liter of water.

Tara has 1 liter of a 1M solution of HCl. How many grams of NaOH must she add to neutralize the solution (adjust pH to 7)?

4. David, an inquisitive three-year-old, touched the water in a metal pan on the stove and found it was warm. Then he touched the pan and got a nasty burn. Devise a hypothesis to explain why water in a metal pan heats up far more slowly than the pan itself.

5. When molecules absorb microwaves, a form of electromagnetic radiation, they move more rapidly. Explain why a microwave oven can heat foods.

6. From what you know about cohesion, devise a hypothesis to explain why water forms droplets.

7. Edward is trying to study a chemical reaction that an enzyme catalyzes (speeds up). H^+ forms during the reaction, but the enzyme is destroyed at low pH. What can he include in his reaction mix to protect the enzyme while he studies the reaction? Explain how your suggestion might solve the problem.

8. Many metabolic reactions proceed on molecular regions of enzymes and other proteins. Cells must have access to those regions. By interactions with water and ions, a soluble protein stays dispersed in cellular fluid rather than settling against some cell structure. An electrically charged cushion around the protein makes this happen. Using the diagram at right as a guide, explain the chemical interactions by which such a cushion forms, starting with major bonds within the protein itself.

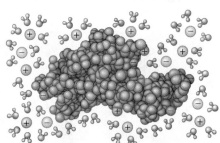

Selected Key Terms

acid *2.6*	hydrogen	orbital, atomic *2.3*
atom *2.1*	ion (H^+) *2.6*	pH scale *2.6*
atomic	hydrophilic	proton *2.1*
number *2.1*	substance *2.5*	radioactive
base *2.6*	hydrophobic	decay *2.1*
buffer system *2.6*	substance *2.5*	radioisotope *2.1*
chemical bond *2.1*	hydroxide	radiometric
cohesion *2.5*	ion (OH^-) *2.6*	dating *2.2*
compound *2.3*	ion *2.4*	salt *2.6*
covalent bond *2.4*	ionic bond *2.4*	shell model *2.3*
electron *2.1*	isotope *2.1*	solute *2.5*
element *2.1*	mass number *2.1*	temperature *2.5*
evaporation *2.5*	mixture *2.3*	trace element *2.1*
half-life *2.2*	molecule *2.3*	tracer *2.2*
hydrogen bond *2.4*	neutron *2.1*	

Readings

Pennisi, E. February 20, 1993. "Water, Water, Everywhere." *Science News.*

Ritter, P. 1996. *Biochemistry: A Foundation.* Pacific Grove, California: Brooks/Cole.

Web Site See *http://www.wadsworth.com/biology* for practice quiz questions, hypercontents, BioUpdates, and critical thinking. The Wadsworth Biology Resource Center provides a wealth of information fully organized and integrated by chapter.

3 CARBON COMPOUNDS IN CELLS

Carbon, Carbon, in the Sky—Are You Swinging Low and High?

High in the mountains of the Pacific Northwest, vast coniferous forests have endured one more murderously cold winter (Figure 3.1). As is true of other organisms, these evergreen, cone-bearing trees cannot grow or reproduce in the absence of liquid water. Yet through the winter months, water in the surroundings is locked away from them, in the form of snow and ice. The trees get by anyway. At the surface of their needle-shaped leaves, a layer of interconnected cells with thick, waxy walls forms a covering, an epidermis, that prevents water loss. The only way precious water can escape is through small gaps across the epidermis that close or open in response to changing conditions. During the cool, dry days of autumn, the trees enter dormancy. Metabolic activities idle and growth ceases, but water conserved inside them is enough to keep their cells alive.

With the arrival of spring, rising temperatures and water from melting snow stimulate renewed growth. Tree roots soak up mineral-laden water. And carbon dioxide, a gaseous molecule of one carbon atom and two oxygen atoms, moves in from the air, through gaps in the leaf epidermis. With their photosynthetic magic, the conifers turn these simple materials into sugars, starches, and other carbon-based compounds. They are premier producers of the northern forests. Producer organisms, recall, use the self-made compounds as structural materials and as packets of energy. So do other organisms. One way or another, every consumer and decomposer is nourished by the producer organisms of forests and all other ecosystems the world over.

Today, plants of the great prevailing forests and plains at northern latitudes

Figure 3.1 Conifers beneath the first snows of winter on Silver Star Mountain, Washington. As is true of all other organisms, the structure, activities, and very survival of these trees start with the carbon atom and its diverse molecular partners in organic compounds.

of the United States, Canada, and other parts of the Northern Hemisphere are breaking dormancy sooner than they did just two decades ago. Why? No one knows for sure, but the change in their life cycles might be one outcome of long-term change in the global climate.

Researchers in the Northern Hemisphere have been studying the concentration of carbon dioxide in the air around us since the early 1950s. Among other things, they found out that the concentration shifts with the seasons. It declines during spring and summer, when photosynthesizers take up stupendous amounts of the gas. It rises during other times of year, when huge populations of decomposers that release the gas as a metabolic by-product show rapid growth.

According to the researchers' measurements, the spring decline now starts a full week earlier than it did in the mid-1970s. Besides this, the seasonal swings are becoming more pronounced—by as much as 20 percent in Hawaii and a whopping 40 percent in Alaska.

Swings in the atmospheric concentration of carbon dioxide were greatest in 1981 and 1990. Intriguingly, temperatures of the lower atmosphere also have been rising—and in 1981 and 1990, they were uncommonly high. Are we in the midst of a long-term, worldwide rise in atmospheric temperature? And is this *global warming* promoting a longer growing season, hence the wider seasonal swings?

The picture gets more intricate. We humans burn great quantities of coal, gasoline, and other fossil fuels for energy. The fossil fuels are rich in carbon, which is released during the burning processes. The released carbon may be contributing to the global warming, in ways that you will read about in later chapters.

For now, the point to keep in mind is this: *Carbon permeates the entire world of life, from the energy-requiring activities and structural organization of individual cells, to physical and chemical conditions that span the globe and influence life everywhere.* With this chapter, we turn to the life-giving properties that emerge out of the molecular structure of carbon-rich compounds. Study the chapter well. It will serve as your foundation for understanding how different organisms put such compounds together, how they use them, and how the effects of these uses can ripple through the biosphere.

KEY CONCEPTS

1. Organic compounds have a backbone of one or more carbon atoms to which hydrogen, oxygen, nitrogen, and other atoms are attached. We define cells partly by their capacity to assemble the organic compounds known as carbohydrates, lipids, proteins, and nucleic acids.

2. Cells put together large biological molecules from their pools of smaller organic compounds, which include simple sugars, fatty acids, amino acids, and nucleotides.

3. Glucose and other simple sugars are carbohydrates. So are organic compounds composed of two or more sugar units, of one or more types, that are covalently bonded together. The most complex carbohydrates that cells assemble are polysaccharides, many of which consist of hundreds or thousands of sugar units.

4. Lipids are greasy or oily compounds that show little tendency to dissolve in water but that can dissolve in nonpolar compounds, including other lipids. They include neutral fats, phospholipids, waxes, and sterols.

5. Cells use carbohydrates and lipids as building blocks and as their major sources of energy.

6. Proteins have truly diverse roles. Many are structural materials. Many are enzymes, a type of molecule that enormously increases the rate of specific metabolic reactions. Other kinds transport cell substances, contribute to cell movements, trigger changes in cell activities, and defend the body against injury and disease.

7. For living organisms, ATP and other nucleotides are crucial players in metabolism. DNA and RNA, strandlike nucleic acids assembled from nucleotide subunits, are the basis of inheritance and reproduction.

PROPERTIES OF ORGANIC COMPOUNDS

The molecules of life are **organic compounds**, meaning they consist of one or more elements covalently bonded to carbon atoms. The term is a holdover from a time when chemists thought "organic" substances were the ones they got from animals and vegetables, as opposed to "inorganic" substances obtained from minerals. The term persists, even though researchers now synthesize organic compounds in laboratories. And it persists even though there are reasons to believe organic compounds were present on Earth *before* organisms were.

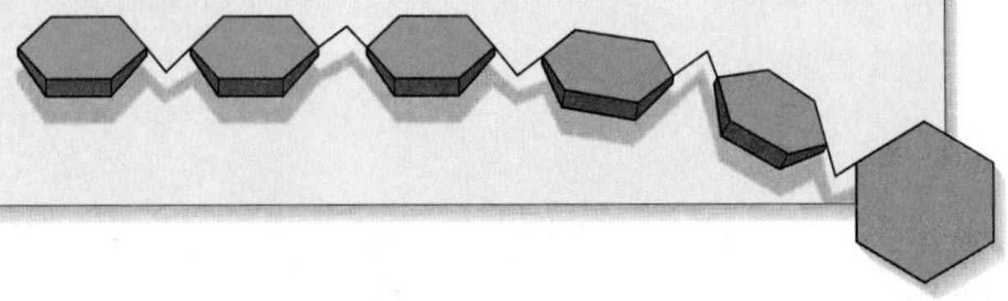

Effects of Carbon's Bonding Behavior

By far, organisms consist mainly of oxygen, hydrogen, and carbon (Figure 2.3). Much of the oxygen and the hydrogen is in the form of water. Remove the water, and carbon makes up more than half of what's left.

Carbon's importance in life arises from its versatile bonding behavior. *Each carbon atom can share pairs of electrons with as many as four other atoms.* Each covalent bond formed this way is quite stable. Such bonds link carbon atoms together in chains and rings. These serve as a backbone to which hydrogen, oxygen, and other elements become attached, as in Figure 3.2.

The flattened structural formulas shown in Figure 3.2 might lead you to believe that the molecules they represent are flat, like paper dolls. Yet the molecules have wonderful three-dimensional shapes, which begin with bonding arrangements in the carbon backbone.

When a carbon atom bonds with four other atoms, this three-dimensional shape emerges:

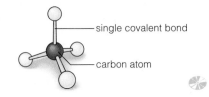

In a carbon backbone, some of the carbon atoms rotate freely around a *single* covalent bond:

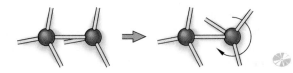

However, a *double* covalent bond uniting two carbon atoms restricts their rotation. Where such bonds occur in the backbone, the atoms rigidly hold their position in space:

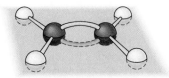

Now imagine a backbone with some atoms locked in position and others rotating at different angles in space. The resulting orientations of these atoms may promote

Figure 3.2 Carbon compounds. There is a Tinkertoy quality to carbon compounds, for a single atom can be the start of diverse molecules assembled from "straight-stick" covalent bonds. Start with the simplest hydrocarbon, methane (CH_4). Stripping one hydrogen from methane leaves a methyl group, which occurs in fats, oils, and waxes:

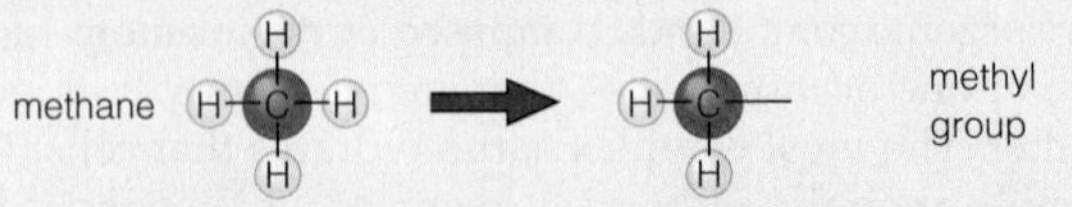

methane   →   methyl group

Imagine stripping a hydrogen atom from each of two methane molecules, then bonding the two molecules together. If the resulting structure were to lose a hydrogen atom, you would end up with an ethyl group:

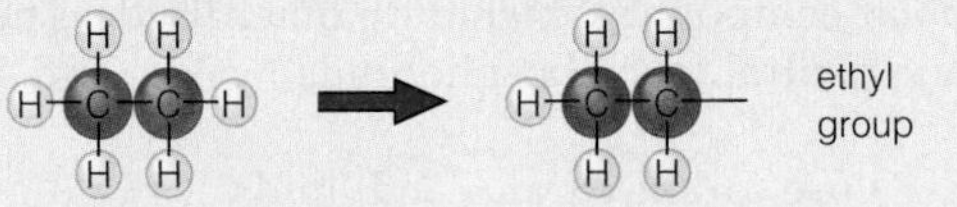

ethyl group

You could go on building a continuous chain:

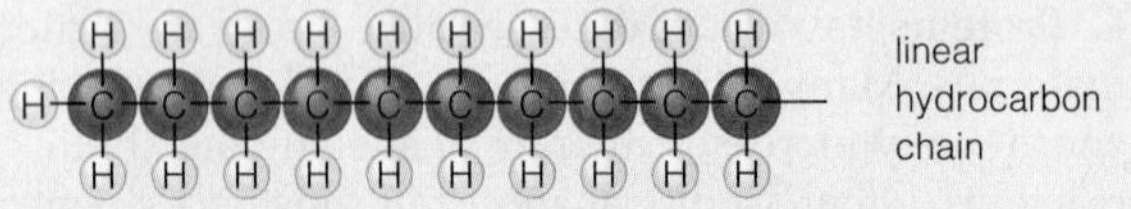

linear hydrocarbon chain

And you could add branches to the chain:

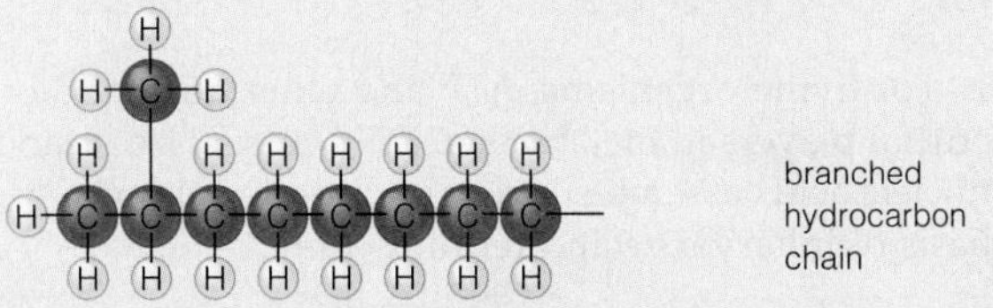

branched hydrocarbon chain

You might have the chain coil back on itself into a ring, which you could diagram in various ways:

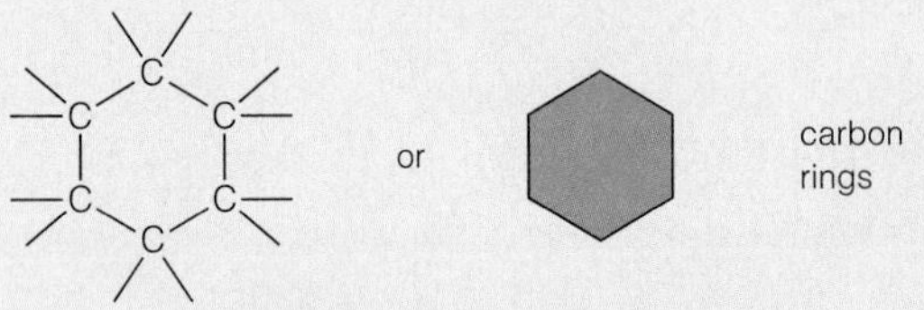

or   carbon rings

You might even join many carbon rings together and so produce larger molecules:

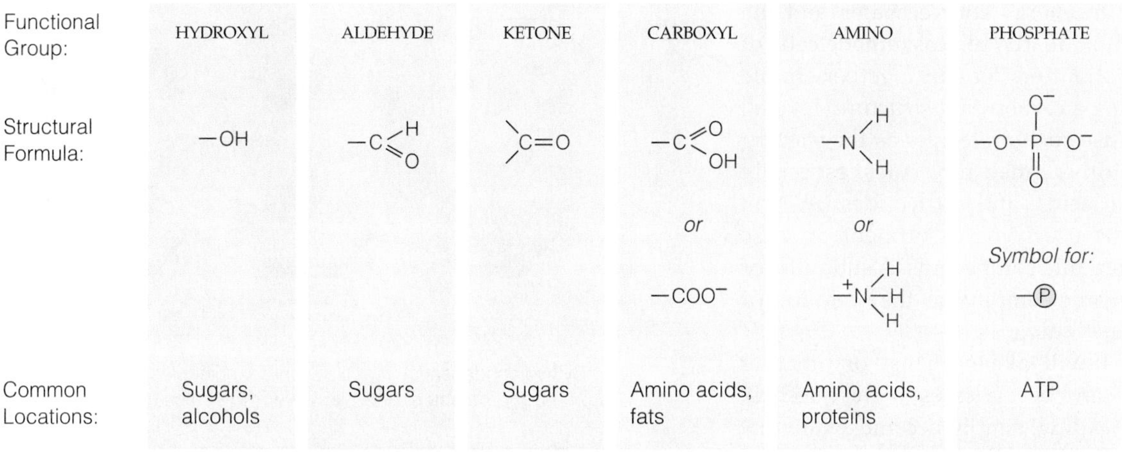

Functional Group:	HYDROXYL	ALDEHYDE	KETONE	CARBOXYL	AMINO	PHOSPHATE
Structural Formula:	—OH	$-C\stackrel{\nearrow H}{\diagdown O}$	$\diagup C{=}O$	$-C\stackrel{\nearrow O}{\diagdown OH}$	$-N\stackrel{\diagup H}{\diagdown H}$	$-O-P-O^-$ (with O^- above and O double-bonded below)
				or	or	Symbol for:
				—COO⁻	$^+_{}N{-}H$ (with H above and H below)	—Ⓟ
Common Locations:	Sugars, alcohols	Sugars	Sugars	Amino acids, fats	Amino acids, proteins	ATP

Figure 3.3 Examples of functional groups and some of their common locations in biological molecules. The carboxyl and amino groups also are shown in their ionized form, which is most common in cells and in the body fluids of multicelled organisms.

a female wood duck   male wood duck

b HO—⬡⬡⬡⬡=O AN ESTROGEN

c O=⬡⬡⬡⬡—OH TESTOSTERONE

Figure 3.4 Notable differences in traits between male and female wood ducks (*Aix sponsa*). Two different sex hormones have key roles in the development of feather color and other traits that help the two ducks recognize each other (and thereby influence reproductive success).

Both of the hormones have the same carbon ring structure. As you can see, however, the ring structures have different functional groups attached to them.

or discourage interactions with other substances in the vicinity. As you will see, *such interactions give rise to the shapes and functions of biological molecules.*

Hydrocarbons and Functional Groups

"Hydrocarbons" have only hydrogen atoms attached to a carbon backbone. They do not break apart very easily, and they form the stable portions of most biological molecules. Biological molecules also incorporate some number of **functional groups**, which are single atoms or clusters of atoms covalently bonded to the carbon backbone. To get a sense of their importance, consider a few of the groups shown in Figure 3.3. Sugars and other organic compounds classified as **alcohols** have one or more hydroxyl groups (—OH). Alcohols are quick to dissolve in water, for water molecules readily form

hydrogen bonds with —OH groups. Proteins have a backbone formed by reactions between amino groups and carboxyl groups. The unique three-dimensional structure of each protein starts with bonding patterns associated with this backbone. Also, amino groups can combine with H⁺ and act as buffers against decreases in pH. As a final example, Figure 3.4 shows two different functional groups that are a molecular starting point for differences between males and females of many species.

Organic compounds have diverse, three-dimensional shapes and functions. Their diversity begins with flexible and rigid bonding arrangements in their carbon backbones.

Functional groups covalently bonded to the carbon backbones of organic compounds add enormously to the structural and functional diversity.

HOW CELLS USE ORGANIC COMPOUNDS

By using sunlight as an energy source, water, and the carbon from carbon dioxide, the photosynthetic cells of those trees you read about earlier put together simple sugar molecules having a carbon ring structure. Like all living cells, they also use such molecules as the starting point for assembling other small molecules, especially the fatty acids, amino acids, and nucleotides. As you will read shortly, cells use some assortment of these four classes of small organic compounds as subunits for building all of the organic compounds they require for their structure and functioning.

How do they do it? It will take more than one chapter to sketch out answers (and best guesses) to the question. At this point in your reading, simply become aware that the reactions by which cells build organic compounds, and even rearrange them and break them apart, require more than an energy input. They also require the class of proteins called **enzymes**, which make specific metabolic reactions proceed faster than they would on their own.

Five Classes of Reactions

Different enzymes mediate different kinds of reactions, but most of the reactions fall into five categories:

1. **Functional-group transfer**. One molecule gives up a functional group, which another molecule accepts.

2. **Electron transfer**. One or more electrons stripped from one molecule are donated to another molecule.

3. **Rearrangement**. A juggling of internal bonds converts one type of organic compound into another.

4. **Condensation**. Through covalent bonding, two molecules combine to form a larger molecule.

5. **Cleavage**. A molecule splits into two smaller ones.

To get a sense of what goes on, consider two examples of these events. First, in many condensation reactions, enzymes remove a hydroxyl group from one molecule and an H atom from another, then speed the formation of a covalent bond between the two molecules at their exposed sites (Figure 3.5a). As a typical but incidental outcome of the reaction, the discarded atoms join to form a water molecule. A series of condensation reactions can produce starches and other polymers. A **polymer** is a large molecule with three to millions of subunits, which may or may not be identical. Often the subunits are called **monomers**, as in the sugar monomers of starch.

As the second example, a cleavage reaction called **hydrolysis** is a bit like condensation in reverse (Figure 3.5b). Enzymes that act on covalent bonds at functional groups split molecules into two or more parts, then attach —H and —OH derived from a water molecule

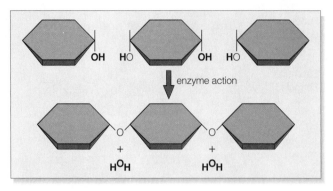

a By two condensation reactions, three molecules covalently bond into a larger molecule, and two water molecules typically form.

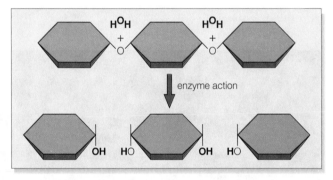

b Hydrolysis, a water-requiring cleavage reaction.

Figure 3.5 Examples of reactions by which most biological molecules are put together, rearranged, and broken apart.

(**a**) Two condensation reactions. (**b**) Two hydrolysis reactions.

to the exposed sites. Hydrolysis allows cells to cleave large polymers such as starch into smaller units when these are required for building blocks or energy.

The Molecules of Life

Under present-day conditions in nature, *only living cells synthesize carbohydrates, lipids, proteins, and nucleic acids.* These are the molecules characteristic of life. As you will see, different classes of biological molecules function as the cells' packets of instant energy, energy stores, structural materials, metabolic workers, and libraries of hereditary information.

Carbohydrates, lipids, proteins, and nucleic acids are the main biological molecules, the organic compounds that only living cells can assemble under conditions that now occur in nature.

Cells assemble, rearrange, and degrade organic compounds mainly through enzyme-mediated reactions involving the transfer of functional groups or electrons, rearrangement of internal bonds, and a combining or splitting of molecules.

3.3 FOOD PRODUCTION AND A CHEMICAL ARMS RACE

The next time you shop for groceries, reflect on what it takes to provide you with your daily supply of organic compounds. For example, those heads of lettuce typically grew in fertilized cropland. Possibly they competed with weeds, which didn't know the nutrients in the fertilizer were meant for the lettuce. Leaf-chewing insects didn't know the lettuce wasn't meant to be their salad bar. Each year, these food pirates and others ruin or gobble up nearly half of what people all over the world try to grow.

Most plants aren't entirely defenseless. They evolved under intense selection pressure from attacks by insects, fungi, and other organisms, and in natural settings they often can repel the attackers with toxins. A **toxin** is an organic compound, a normal metabolic product of one species, but its chemical effects can harm or kill individuals of a different species that come in contact with it. Humans, too, encounter traces of natural toxins in most of what they eat, even in such familiar edibles as hot peppers, potatoes, figs, celery, rhubarb, and alfalfa sprouts. Still, we do not die in droves from these natural toxins, so apparently our bodies have chemical defenses against them.

In 1945, we took a cue from plants and started using newly developed, synthetic toxins to protect crop yields, food stores, freshwater supplies, and even our health, pets, and ornamental plants. The *herbicides* kill weeds by disrupting metabolism and growth. Most *insecticides* clog a target insect's airways, jangle its nerve and muscle cells, or block its reproduction. *Fungicides* work against harmful fungi, including a mold that produces aflatoxin, one of the deadliest poisons known. By 1995, people in the United States alone were spraying or spreading more than 1.25 billion pounds of these toxins through fields, gardens, homes, offices, and industrial sites (Figure 3.6).

There are downsides to pesticide applications. Along with pests, some of the toxic organic compounds kill birds and other predators which, in natural settings, help control pest population sizes. Worse, the targeted pests have been developing resistance to the chemical arsenal, for reasons

Figure 3.6 One consumer of food crops, and a low-flying crop duster with its rain of pesticides.

that you read about in Chapter 1. Also, pesticides cannot be released haphazardly in the environment, for people might end up inhaling them, ingesting them with food, or absorbing them through the skin. Different types remain active for days, weeks, even years. Some trigger severe headaches, rashes, hives, asthma, and joint pain in millions of people. Some trigger life-threatening allergic reactions in abnormally sensitive individuals.

Presently, the long-lived pesticides are banned in the United States. Even rapidly degradable ones, including malathion and other organophosphates, are subjected to rigorous application standards and ongoing safety tests. However, with so many pests looming around us, the search for effective countermeasures goes on.

Consider first the carbohydrates—the most abundant of all biological molecules, which cells use as structural materials and transportable or storage forms of energy. Most **carbohydrates** consist of carbon, hydrogen, and oxygen in a 1:2:1 ratio, but there are exceptions. Three classes are the **monosaccharides**, **oligosaccharides**, and **polysaccharides**.

The Simple Sugars

"Saccharide" comes from a Greek word meaning sugar. A *mono*saccharide, meaning "one monomer of sugar," is the simplest carbohydrate. It has an aldehyde or a ketone group and at least two —OH groups joined to the carbon backbone. Most monosaccharides are sweet tasting and readily dissolve in water. The most common have a backbone of five or six carbon atoms that tends to form a ring structure when dissolved in cells or body fluids. Ribose and deoxyribose, the sugar components of RNA and DNA, respectively, have five carbon atoms. Glucose has six (Figure 3.7*a*). Besides being the main energy source for most organisms, glucose is a precursor (parent molecule) of many compounds and a building block for larger carbohydrates. Vitamin C (a sugar acid) and glycerol (an alcohol with three —OH groups) are other examples of compounds based on sugar monomers.

Short-Chain Carbohydrates

Unlike the simple sugars, an *oligo*saccharide is a short chain of two or more sugar monomers that are bonded covalently. (*Oligo-* means a few.) The simplest kinds, the *di*saccharides, have only two sugar units. Lactose, sucrose, and maltose are examples. Lactose (a glucose and a galactose unit) is present in milk. Sucrose, the most plentiful sugar in nature, consists of a glucose and a fructose unit (Figure 3.7*c*). Plants continually convert carbohydrate stores to sucrose, which can be transported through leaves, stems, and roots. Table sugar is sucrose crystallized from sugarcane and sugar beets. Proteins and other large molecules often have oligosaccharides attached as side chains to the carbon backbone. Some chains take part in the body's defenses against disease, others in cell membrane functions.

Complex Carbohydrates

The "complex" carbohydrates, or *poly*saccharides, are straight or branched chains of *many* sugar monomers, often hundreds or thousands of the same or different kinds. Starch, cellulose, and glycogen, the most common polysaccharides, consist only of glucose. In plants, cells store sugars in the form of large molecules of starch, which enzymes can readily hydrolyze to glucose units. Plant cells use cellulose as a structural material in their walls. Fibers made of cellulose molecules are tough and insoluble. Like the steel rods in reinforced concrete, the fibers withstand considerable weight and stress.

If both cellulose and starch consist of glucose, why are their properties so different? The answer lies with differences in covalent bonding patterns between the monomers, which in both cases are strung in chains. In starch, the bonds angle each monomer with respect to the next in line, and the chain coils like a spiral staircase (Figure 3.8*a,b*). Such coils are not especially stable, and in starches that have branched chains, they are even less so. Many —OH groups project outward from the coiled chains, so they are readily accessible to enzymes.

By contrast, in cellulose, many glucose chains stretch out side by side and hydrogen-bond to one another at —OH groups (Figure 3.8*c,d*). This bonding arrangement

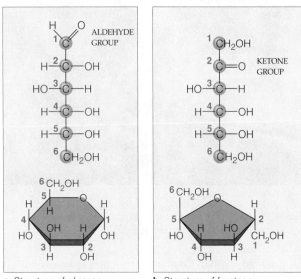

a Structure of glucose.   **b** Structure of fructose.

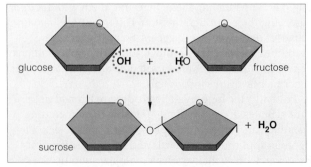

c Formation of sucrose by a condensation reaction.

Figure 3.7   Straight-chain and ring forms of (**a**) glucose and (**b**) fructose. For reference purposes, the carbon atoms of simple sugars are commonly numbered in sequence, starting at the end closest to the aldehyde or ketone group. (**c**) Condensation of two monosaccharides into a disaccharide.

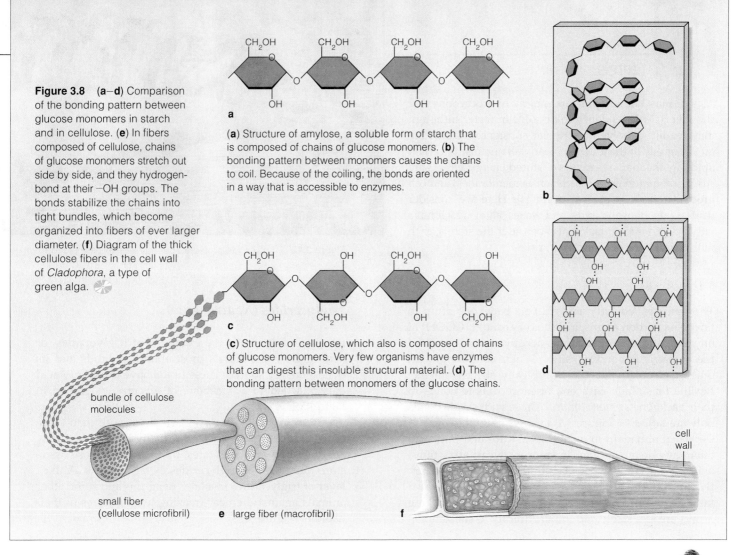

Figure 3.8 **(a–d)** Comparison of the bonding pattern between glucose monomers in starch and in cellulose. **(e)** In fibers composed of cellulose, chains of glucose monomers stretch out side by side, and they hydrogen-bond at their –OH groups. The bonds stabilize the chains into tight bundles, which become organized into fibers of ever larger diameter. **(f)** Diagram of the thick cellulose fibers in the cell wall of *Cladophora*, a type of green alga.

(a) Structure of amylose, a soluble form of starch that is composed of chains of glucose monomers. **(b)** The bonding pattern between monomers causes the chains to coil. Because of the coiling, the bonds are oriented in a way that is accessible to enzymes.

(c) Structure of cellulose, which also is composed of chains of glucose monomers. Very few organisms have enzymes that can digest this insoluble structural material. **(d)** The bonding pattern between monomers of the glucose chains.

bundle of cellulose molecules

small fiber (cellulose microfibril)

e large fiber (macrofibril)

f

cell wall

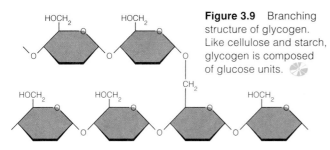

Figure 3.9 Branching structure of glycogen. Like cellulose and starch, glycogen is composed of glucose units.

Figure 3.10 Scanning electron micrograph of a tick. Its body covering is a protective, chitin-reinforced cuticle.

stabilizes the chains in a tightly bundled pattern that resists digestion, at least by most enzymes.

For long-term storage, plants squirrel away glucose in the form of amylose, the starch in Figure 3.8*a*. They also build amylopectin, a branched starch that invites rapid mobilization of glucose. In animals, especially in liver and muscle tissue, the sugar-storage equivalent of amylopectin is glycogen. When the sugar level in blood falls, liver cells degrade glycogen and release glucose to the blood. When you exercise briefly but intensely, muscle cells tap their glycogen stores for quick energy. Figure 3.9 shows many one of glycogen's branchings.

Chitin, another polysaccharide, has nitrogen atoms attached to the backbone. Chitin is the main structural

material in the external skeletons and other hard body parts of numerous animals, such as crabs, earthworms, insects, and ticks (Figure 3.10). Chitin also is the main structural material in the cell walls of many fungi.

Carbohydrates are either simple sugars (such as glucose) or molecules composed of many sugar units (such as starch). Every cell requires carbohydrates as structural materials, stored forms of energy, and transportable packets of energy.

3.5 LIPIDS

As you may know, **lipids** are mainly hydrocarbons that show very little tendency to dissolve in water, although they readily dissolve in nonpolar substances. Lipids are greasy or oily to the touch. In nearly all organisms, certain lipids are the main reservoirs of stored energy. Others are structural materials in cells (such as membranes) and cell products (such as surface coatings). Here we consider neutral fats, phospholipids, and waxes, all of which have fatty acid components. We also consider the sterols, each with a backbone of four carbon rings.

Figure 3.12 Triglyceride-protected penguins taking the plunge.

Fatty Acids

The lipids called **fatty acids** have a backbone of up to thirty-six carbon atoms, a carboxyl group (—COOH) at one end, and hydrogen atoms occupying most or all of the remaining bonding sites. When combined with other molecules, fatty acids typically stretch out like flexible tails. Tails with one or more double bonds in their backbone are *unsaturated*. Those with single bonds only are *saturated*. Figure 3.11*a* shows examples.

When abundant in a substance, saturated fatty acid tails snuggle in parallel by mutual weak attractions, and this gives the substance a rather solid consistency. By contrast, the double and triple bonds in unsaturated fatty acids put rigid kinks in the tails. These packing arrays are less stable and impart fluidity to substances.

Triglycerides (Neutral Fats)

Butter, lard, and oils are examples of **triglycerides**, or neutral fats—the body's most abundant lipids and its richest energy source. These lipids have three fatty acid tails attached to a backbone of glycerol, as shown in Figure 3.11*b*.

Gram for gram, triglycerides yield more than twice as much energy as complex carbohydrates when they are degraded. In vertebrates, the cells of adipose tissue store quantities of triglycerides as fat droplets. A thick layer of triglycerides insulates penguins and some other animals against the near-freezing temperatures of their habitats (Figure 3.12).

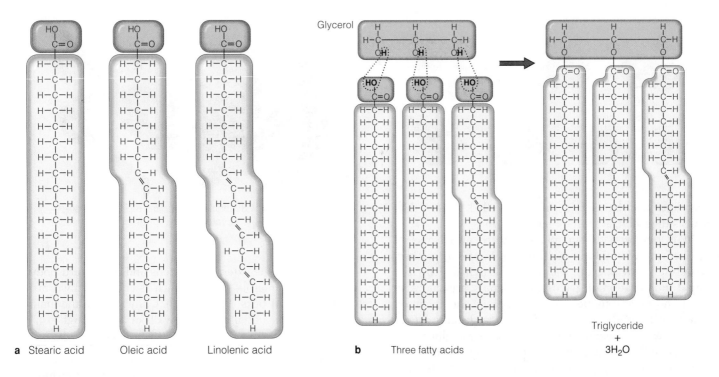

a Stearic acid   Oleic acid   Linolenic acid   **b** Three fatty acids   Triglyceride + 3H₂O

Figure 3.11 (**a**) Structural formulas for three fatty acids. Stearic acid's carbon backbone is fully saturated with hydrogen atoms. Oleic acid, with a double bond in its backbone, is unsaturated. Linolenic acid, with three double bonds, is a "polyunsaturated" fatty acid. (**b**) Condensation of fatty acids and glycerol into a triglyceride.

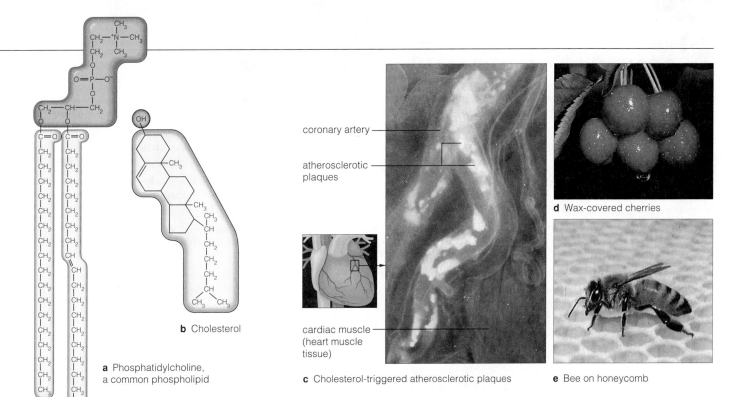

coronary artery

atherosclerotic plaques

cardiac muscle (heart muscle tissue)

c Cholesterol-triggered atherosclerotic plaques

d Wax-covered cherries

e Bee on honeycomb

a Phosphatidylcholine, a common phospholipid

b Cholesterol

Figure 3.13 (**a**) Structural formula of a typical phospholipid in animal and plant cell membranes. Its hydrophilic head is shaded *orange*. Are its hydrophobic tails (*yellow*) saturated or unsaturated? (**b**) Structural formula of cholesterol, the major sterol of animal tissues. (**c**) Your own liver synthesizes enough cholesterol for the body. A fat-rich diet may result in excessively high cholesterol levels in blood and the formation of abnormal masses of material in the blood vessels called arteries. Such *atherosclerotic plaques* may clog the arteries that deliver blood to the heart. (**d**) Demonstration of the water-repelling attribute of a cherry cuticle. (**e**) Honeycomb, a structural material constructed of a firm, water-repellent, waxy secretion called beeswax.

Phospholipids

A **phospholipid** has a glycerol backbone, two fatty acid tails, and a hydrophilic "head" with a phosphate group and another polar group (Figure 3.13*a*). Phospholipids are the main materials of cell membranes, which have two layers of lipids. Heads of one layer are dissolved in the cell's fluid interior, and heads of the other layer are dissolved in the surroundings. Sandwiched between the two are all the fatty acid tails, which are hydrophobic.

Sterols and Their Derivatives

Sterols are among the many lipids that have no fatty acid tails. Sterols differ in the number, position, and type of their functional groups, but they all have a rigid backbone of four fused-together carbon rings:

sterol backbone

Sterols occur in eukaryotic cell membranes. Cholesterol is the main type in animal tissues (Figure 3.13*b* and *c*). Cells also remodel cholesterol into compounds such as vitamin D (required for good bones and teeth), steroids (such as estrogen and testosterone, sex hormones that govern sexual traits and gamete formation), and bile salts (which assist fat digestion in the small intestine).

Waxes

The lipids called **waxes** have long-chain fatty acids tightly packed and linked to long-chain alcohols or to carbon rings. They have a firm consistency and repel water. Waxes and another lipid, cutin, make up most of the cuticles that cover aboveground plant parts. These coverings help plants conserve water and fend off some parasites. A waxy cherry cuticle is an example (Figure 3.13*d*). In many animals, waxy secretions from cells are incorporated into coatings that protect, lubricate, and impart pliability to skin or hair. Among waterfowl and other birds, wax secretions help keep feathers dry. As a final example, beeswax is the material of choice when bees construct their honeycombs (Figure 3.13*e*).

Being largely hydrocarbons, lipids can dissolve in polar substances but tend not to dissolve in water.

Neutral fats (triglycerides), which have a glycerol head and three fatty acid tails, are the body's main energy reservoirs. Phospholipids are the main components of cell membranes.

Sterols such as cholesterol are membrane components as well as precursors of steroid hormones and other compounds. Waxes are firm yet pliable components of water-repelling and lubricating substances.

AMINO ACIDS AND THE PRIMARY STRUCTURE OF PROTEINS

Of all the large biological molecules, **proteins** are the most diverse. The ones called enzymes make metabolic events proceed much faster than they otherwise would. Structural types are the stuff of feathers and webs, bone and cartilage, and a dizzying array of other body parts and products. Transport proteins move molecules and ions across cell membranes and through body fluids. Nutritious proteins abound in milk, eggs, and a variety of seeds. Protein hormones and other regulatory types are signals for change in cell activities. Many proteins are weapons against disease-causing bacteria and other invaders. Amazingly, cells build diverse proteins from their pools of only twenty kinds of amino acids.

Structure of Amino Acids

Every **amino acid** is a small organic compound that consists of an amino group, a carboxyl group (an acid), a hydrogen atom, and one or more atoms known as its R group. As you can see from the structural formula in Figure 3.14, these parts are covalently bonded to the same carbon atom. Figure 3.15 shows some specific amino acids that we will consider later in the book.

Primary Structure of Proteins

When a cell synthesizes a protein, amino acids become linked, one after the other, by peptide bonds. As Figure 3.16 shows, this is the type of covalent bond that forms between one amino acid's amino group (NH_3^+) and the carboxyl group ($-COO^-$) of the next amino acid.

When peptide bonds join two amino acids together, we have a **dipeptide**. When they join three or more, we have a **polypeptide chain**. In such chains, the carbon backbone has nitrogen atoms positioned in this regular pattern: $-N-C-C-N-C-C-$.

For each particular kind of protein, different amino acid units are selected one at a time from the twenty kinds available. Their orderly progression is prescribed by the cell's DNA. Overall, the resulting sequence of amino acids is unique for each kind of protein, and it represents the protein's *primary* structure (Figure 3.17).

Now consider this: different cells make thousands of different proteins. Many of the proteins are fibrous, with polypeptide chains organized as strands or sheets. Collectively, many such molecules contribute to the shape and internal organization of cells. Other kinds of proteins are globular, with one or more polypeptide

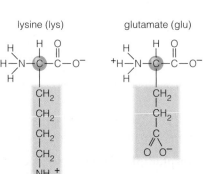

Figure 3.14 Generalized structural formula for the amino acids.

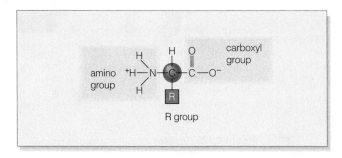

AN UNCHARGED, POLAR AMINO ACID

A POSITIVELY CHARGED, POLAR AMINO ACID

A NEGATIVELY CHARGED, POLAR AMINO ACID

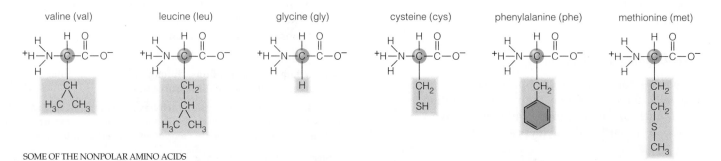

SOME OF THE NONPOLAR AMINO ACIDS

Figure 3.15 Structural formulas for nine of the twenty common amino acids. *Green* boxes highlight R groups (side chains that contain functional groups). The side chain gives each amino acid its distinctive properties.

Figure 3.16 Peptide bond formation during protein synthesis.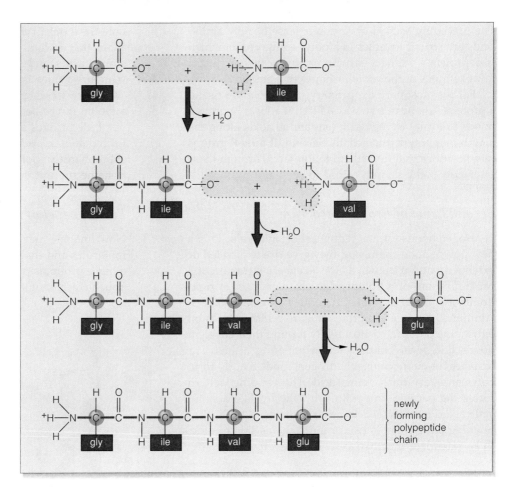

a The first two amino acids in this example are glycine and isoleucine. They are at the start of the sequence for one of two polypeptide chains that make up the protein insulin in cattle. Through a condensation reaction, the isoleucine becomes joined to glycine by a peptide bond. Water forms as a by-product of the reaction.

b A peptide bond forms between the isoleucine and valine, which is another amino acid, and water again forms.

c Remember, DNA specifies the order in which the different kinds of amino acids follow one another in a growing polypeptide chain. In this case, glutamate is the fourth amino acid specified. Chapter 14 provides more details on the steps that lead from DNA instructions to proteins.

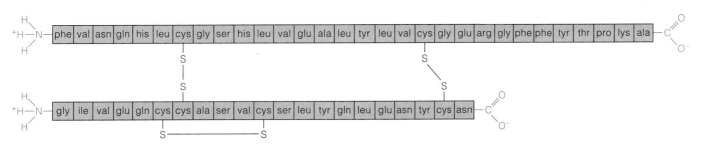

Figure 3.17 Simplified diagram of the amino acid sequence for the insulin molecule in cattle, as deduced by Frederick Sanger. This protein functions as a hormone, and it consists of two polypeptide chains. Disulfide bridges (—S—S—), each formed by a condensation reaction at two sulfhydryl groups, link the chains.

chains folded in compact, rather rounded shapes. Most enzymes are globular proteins. So are actin and most of the other proteins that contribute to cell movement.

Regardless of the type of protein, its shape and its function arise from the primary structure—that is, from information built into its amino acid sequence. As you will see next, that information dictates which parts of a polypeptide chain will coil, bend, or interact with other chains nearby. And the type and arrangement of atoms in the coiled, stretched-out, or folded regions determine whether that protein will function as, say, an enzyme, a transporter, a receptor, or sometimes as an inadvertent target for a bacterium or virus.

A protein consists of one or more polypeptide chains, in which amino acids are strung together. The amino acid sequence (which kind of amino acid follows another in the chain) is unique for each kind of protein and gives rise to its unique structure, chemical behavior, and function.

The preceding section gave you a sense of how amino acids are strung together in a polypeptide chain, which is a protein's primary structure. Now consider just a few examples of the protein shapes that emerge.

For the most part, the primary structure gives rise to a protein's shape in two ways. First, it allows hydrogen bonds to form between different amino acids along the length of a polypeptide chain. Second, it puts R groups into positions that force them to interact. Through these interactions, the chain is forced to bend and twist.

Second Level of Protein Structure

Hydrogen bonds form at regular, short intervals along a new polypeptide chain, and they give rise to a coiled or extended pattern known as the *secondary* structure of a protein. Think of a polypeptide chain as a set of rigid dominoes joined by links that can swivel a bit. Each "domino" is a peptide group (Figure 3.18a). Atoms on either side of it can rotate slightly around their covalent bonds and form bonds with neighboring atoms. For instance, in many chains, hydrogen bonds readily form between every third amino acid. The bonding pattern forces the peptide groups to coil helically, like a spiral staircase (Figure 3.18b). In other proteins, a hydrogen-bonding pattern holds two or more chains side by side, in an extended, sheetlike array (Figure 3.18c).

Third Level of Protein Structure

Most polypeptide chains that have a coiled secondary structure undergo more folding, owing to the number and the location of certain amino acids (such as proline) along their length. These amino acids bend a chain in certain directions and at certain angles. The chain loops out, and R groups far apart along its length interact to

hold the loops in characteristic positions. A polypeptide chain that is folded as a result of its bend-producing amino acids and R-group interactions has a *tertiary* structure, which is the third level of protein structure.

Figure 3.19a shows the compact, tertiary structure of a globin molecule, an outcome of hydrogen bonds and disulfide bridges at functional groups along its length. Each disulfide bridge links the sulfur atom of a cysteine subunit in the polypeptide chain with that of another cysteine subunit in the chain (refer also to Figure 3.17).

Fourth Level of Protein Structure

Now imagine that bonds also form between *four* globin molecules and that an iron-containing functional group, a heme group, is positioned near the center of each one. You have hemoglobin, an oxygen-transporting protein

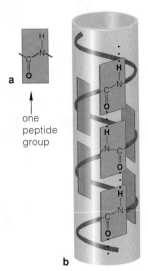
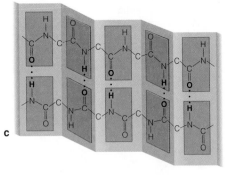

a

one peptide group

b

c

Figure 3.18 The major bonding patterns between the peptide groups (**a**) of polypeptide chains. Extensive hydrogen bonding (*dotted lines*) can bring about (**b**) coiling or (**c**) sheetlike formations.

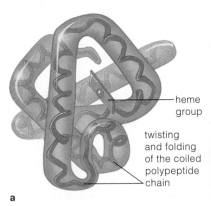

heme group

twisting and folding of the coiled polypeptide chain

a

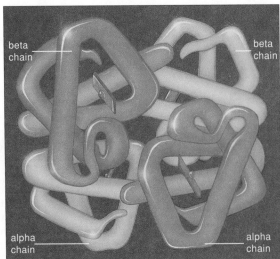

beta chain

beta chain

alpha chain

alpha chain

b

Figure 3.19 (**a**) A globin molecule. It is a single coiled polypeptide chain and a heme group associated with it. This iron-containing group is strongly attractive to oxygen. There is a whole family of globin molecules. They are identical in critical regions of their amino acid sequences, but they differ in other regions. In this diagram, the artist drew a transparent "green noodle" around the chain to help you visualize how it folds in three dimensions.

(**b**) Hemoglobin, a protein that transports oxygen in blood, is composed of four globin molecules. To keep this diagram from looking like a tangle of noodles, the two polypeptide chains in the foreground are tinted differently from the two in the background.

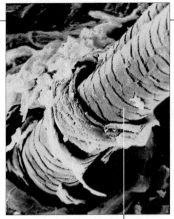

dead, flattened cells around a developing hair shaft

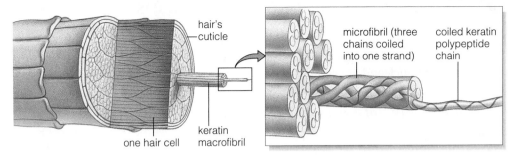

hair's cuticle

microfibril (three chains coiled into one strand)

coiled keratin polypeptide chain

keratin macrofibril

one hair cell

Figure 3.20 Structure of hair. Hair cells, which are derived from skin cells, synthesize polypeptide chains of keratin. Three chains become linked as fine fibers (microfibrils), which become bundled into larger, cablelike fibers (macrofibrils). These essentially fill the cells, which eventually die off. Dead, flattened cells form a tubelike cuticle around the developing hair shaft, shown in the photograph.

(Figure 3.19*b*). As you read this, each mature red blood cell in your body is transporting a billion or so oxygen molecules, bound to 250 million hemoglobin molecules.

Hemoglobin is a fine example of the fourth level of protein structure, or *quaternary* structure. In all proteins at this level of organization, two or more polypeptide chains have become joined together by numerous weak interactions (such as hydrogen bonds) and sometimes by covalent bonds between sulfur atoms of R groups.

Like most enzymes, hemoglobin is globular. Many other proteins having quaternary structure are fibrous. Keratin, a structural protein of hair and fur, is like this (Figure 3.20). So is collagen, the most common animal protein. Skin, bones, corneas, blood vessels, and other body parts depend on the strength inherent in collagen.

Glycoproteins and Lipoproteins

Some proteins have other organic compounds attached to their polypeptide chains. For example, **lipoproteins** form when certain proteins circulating in blood combine with cholesterol, triglycerides, and phospholipids that were absorbed from the gut after a meal. Lipoproteins differ in their particular protein and lipid components. Similarly, most **glycoproteins** have linear or branched oligosaccharides covalently bonded to them. Most form when new polypeptide chains are being modified into mature proteins. Nearly all the proteins at the surface of animal cells are glycoproteins. So are most protein secretions from cells and many proteins in blood.

Structural Changes by Denaturation

Breaking the weak bonds of a protein or any other large molecule disrupts its three-dimensional shape, an event called **denaturation**. For example, hydrogen bonds are weak, and they are sensitive to increases or decreases in temperature and pH. If the temperature or pH exceeds a protein's range of tolerance, its polypeptide chains will unwind or change shape, and the protein will no longer function.

Consider the protein albumin, concentrated in the "egg white" of uncooked chicken eggs. When you cook eggs, the heat doesn't disrupt the strong covalent bonds of albumin's primary structure. But it destroys weaker bonds contributing to the three-dimensional shape. For some proteins, denaturation might be reversed when normal conditions are restored—but albumin isn't one of them. There is no way to uncook a cooked egg.

We now leave this overview of protein structure, the levels of which are summarized in Table 3.1.

Proteins have a primary structure, which is the sequence of different kinds of amino acids along a polypeptide chain. An individual's DNA specifies that sequence.

Proteins have a secondary structure, a coiled or extended, sheetlike pattern that arises through hydrogen bonding at short, regular intervals along a polypeptide chain.

Many proteins have a folded, tertiary structure. It arises when certain bend-producing amino acids make a coiled chain loop at certain angles and in certain directions, and when interactions among certain R groups that are far apart in the chain hold the loops in characteristic positions.

Many proteins have a quaternary structure. They incorporate two or more polypeptide chains, joined together by numerous hydrogen bonds and other interactions.

Table 3.1 Levels of Protein Structure	
Primary structure	The sequence of amino acids in a polypeptide chain; unique for each type of protein
Secondary structure	Coiled or extended shape owing to hydrogen bonds at short intervals along the chain
Tertiary structure	Further folding of a coiled chain owing to bend-producing amino acids and interactions among R groups far apart on a looped-out chain
Quaternary structure	Two or more polypeptide chains linked tightly by many hydrogen bonds and other interactions

NUCLEOTIDES AND THE NUCLEIC ACIDS

We conclude this chapter with a brief introduction to the nucleotides and the larger nucleic acids, which are assembled from nucleotide monomers.

Nucleotides With Roles in Metabolism

Nucleotides are small organic compounds with a sugar, at least one phosphate group, and a base. The sugar is either ribose or deoxyribose. Both sugars have a five-carbon ring structure. The only difference is that ribose has an oxygen atom attached to carbon 2 in the ring and deoxyribose does not. The bases have a single or double carbon ring structure that incorporates nitrogen.

One nucleotide, **ATP** (adenosine triphosphate), has a string of three phosphate groups attached to its sugar component in the manner shown in Figure 3.21. An ATP molecule can readily transfer a phosphate group to many other molecules inside the cell, whereupon the acceptor molecules become energized enough to enter into a reaction. Although ATP acquires its phosphate groups at certain reaction sites, it can deliver them to nearly all other reaction sites in a cell, so in this respect ATP is absolutely central to metabolism.

Other nucleotides are subunits of **coenzymes**. These enzyme helpers accept hydrogen atoms and electrons stripped from molecules at a reaction site and transfer them elsewhere. Two important examples are NAD^+ (nicotinamide adenine dinucleotide) and FAD (flavin adenine dinucleotide). Still other nucleotides function as chemical messengers within and between cells. You will be reading about one of these (cyclic adenosine monophosphate, or cAMP) later in the book.

Nucleic Acids—DNA and RNA

Different kinds of nucleotides, including the ones shown in Figure 3.22, are building blocks for the large, single- or double-stranded molecules known as **nucleic acids**. In the backbone of such strands, the sugar component of each nucleotide is covalently bonded to the phosphate

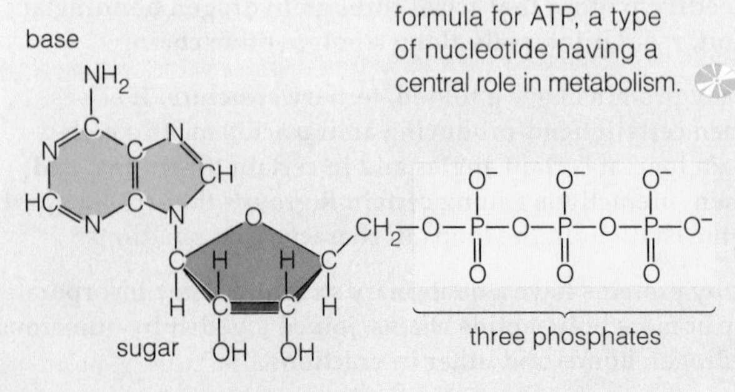

Figure 3.21 Structural formula for ATP, a type of nucleotide having a central role in metabolism.

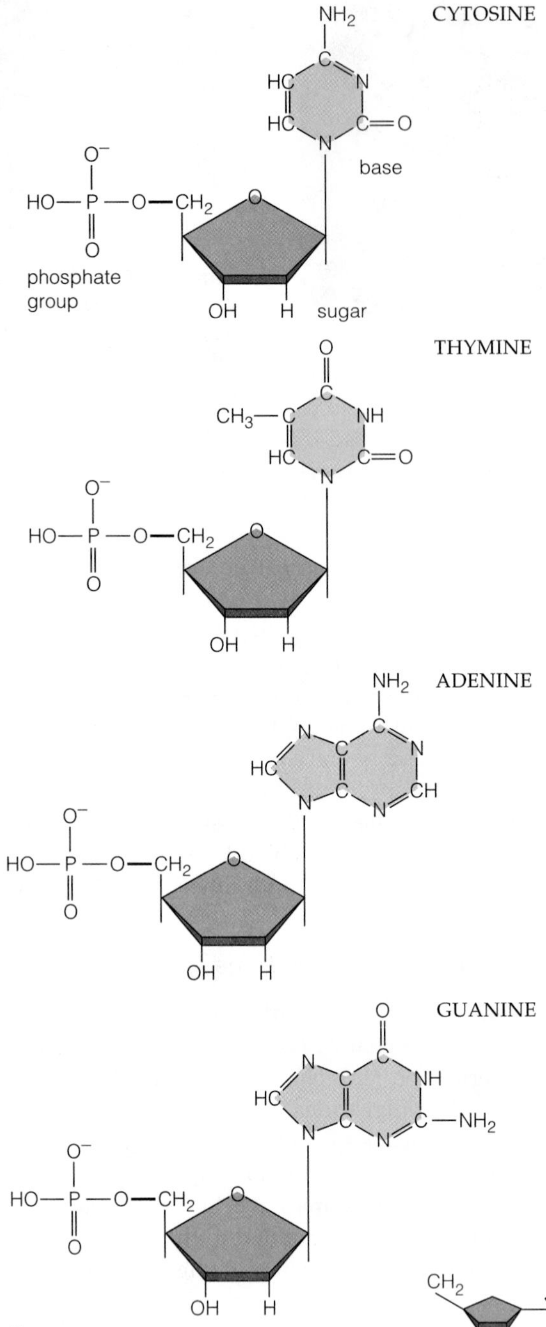

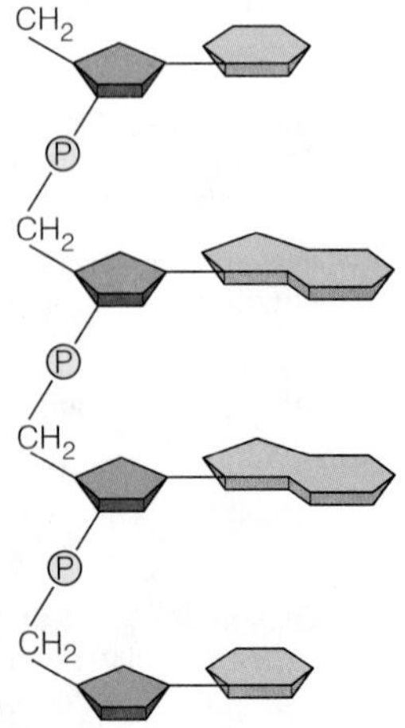

Figure 3.22 (**a**) The four kinds of nucleotides that serve as building blocks for the DNA molecule. Two of the bases, cytosine and thymine, have a single ring structure (*blue*). The other two, adenine and guanine, have a double ring structure. (**b**) In each strand of DNA or RNA, nucleotides are joined one after another in linear sequence. An individual's genetic instructions has particular nucleotides in a specified sequence.

Figure 3.23 (a) Bonding patterns within the double-stranded DNA molecule. Notice how hydrogen bonds connect the bases (*blue*) of one strand with bases of the other strand. The base pairs are like the rungs of a ladder, and the sugar-phosphate backbones are like the ladder's posts. The "ladder" twists in the shape of a double helix.

(**b**) A computer-generated, three-dimensional model of DNA structure. Each ball signifies a single atom. The two nucleotide strands are shown twisted together. Base pairs of the sugar-phosphate backbones are shaded *blue*. The sugar components of the backbone are shaded *light purple*, and the phosphate groups *red* and *yellow*.

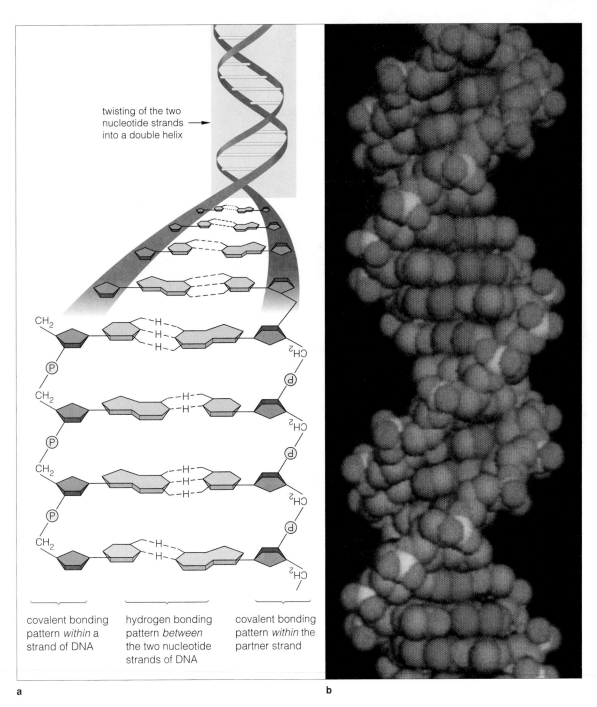

twisting of the two nucleotide strands into a double helix →

covalent bonding pattern *within* a strand of DNA

hydrogen bonding pattern *between* the two nucleotide strands of DNA

covalent bonding pattern *within* the partner strand

a

b

group of adjacent nucleotides. Most likely, you already know something about **DNA** (deoxyribonucleic acid). To refresh your memory, DNA is a long, double strand of nucleotides twisted helically, as shown in Figure 3.23. Hydrogen bonding between the nucleotide bases holds the two strands of a DNA molecule together. Heritable (genetic) information is encoded in the base sequences. And in at least some regions along DNA's length, the sequence of particular bases is unique to each species.

Unlike DNA, the **RNAs** (ribonucleic acids) usually are single nucleotide strands. Ever since living cells first appeared, RNAs have functioned in processes by which genetic information is used to build proteins.

Nucleotides are small, nitrogen- and phosphorus-containing organic compounds. They are building blocks for DNA and RNA. Some, including ATP, have central roles in metabolism.

DNA, a double-stranded nucleic acid, has genetic information encoded in its sequence of nucleotide bases. The RNAs are single-stranded nucleic acids with roles in the processes by which DNA's genetic information is used to build proteins.

SUMMARY

1. Organic compounds consist of one or more elements covalently bonded to carbon atoms that commonly form linear and ring-shaped backbones. Functional groups attached to the backbone impart diverse properties to organic compounds.

2. In nature, only living cells can assemble the large organic compounds called biological molecules—the complex carbohydrates and lipids, proteins, and nucleic acids. To do so, they draw from small pools of organic compounds, which include simple sugars, fatty acids, amino acids, and nucleotides. Table 3.2 summarizes the major categories of these compounds.

3. Cells use the simple sugars (such as glucose) and oligosaccharides (such as sucrose) as building blocks and transportable forms of energy.

4. Cells use complex carbohydrates (including cellulose and starch) as energy storage forms and structural materials. They use lipids (including triglycerides) as energy storage forms, as structural components of cell membranes, and as precursors of other compounds.

5. Proteins, built of amino acids, are the most diverse biological molecules. They function in enzyme activity, structural support, cell transport, cell movements, changes in cell activities, and defenses against disease. The nucleic acids DNA and RNA, built of nucleotides, are the basis of inheritance and reproduction.

Table 3.2 Summary of the Main Organic Compounds in Living Things

Category	Main Subcategories	Some Examples and Their Functions	
CARBOHYDRATES . . . contain an aldehyde or a ketone group, and one or more hydroxyl groups	**Monosaccharides** (simple sugars)	Glucose	Energy source
	Oligosaccharides	Sucrose (a disaccharide)	Form of sugar transported in plants
	Polysaccharides (complex carbohydrates)	Starch Cellulose	Energy storage Structural roles
LIPIDS . . . are largely hydrocarbon; generally do not dissolve in water but dissolve in nonpolar substances	**Lipids with fatty acids:** *Glycerides:* one, two, or three fatty acid tails attached to glycerol backbone	Fats (e.g., butter) Oils (e.g., corn oil)	Energy storage
	Phospholipids: phosphate group, one other polar group, and (often) two fatty acids attached to glycerol backbone	Phosphatidylcholine	Key component of cell membranes
	Waxes: long-chain fatty acid tails attached to alcohol	Waxes in cutin	Water retention by plants
	Lipids with no fatty acids: *Sterols:* four carbon rings; the number, position, and type of functional groups differ among various types	Cholesterol	Component of animal cell membranes; precursor of many steroids and of vitamin D
PROTEINS . . . are polypeptides (up to several thousand amino acids, covalently linked)	**Fibrous proteins:** Individual polypeptide chains, often linked into tough, water-insoluble molecules	Keratin Collagen	Structural element of hair, nails Structural element of bone, cartilage
	Globular proteins: One or more polypeptide chains folded and linked into globular shapes; many roles in cell activities	Enzymes Hemoglobin Insulin Antibodies	Increase in rates of reactions Oxygen transport Control of glucose metabolism Tissue defense
NUCLEIC ACIDS (AND NUCLEOTIDES) . . . are chains of units (or individual units) that each consist of a five-carbon sugar, phosphate, and a nitrogen-containing base	**Adenosine phosphates**	ATP	Energy carrier
	Nucleotide coenzymes	NAD^+, $NADP^+$, FAD	Transport of protons (H^+), electrons from one reaction site to another
	Nucleic acids: Chains of thousands to millions of nucleotides	DNA, RNAs	Storage, transmission, translation of genetic information

Review Questions

1. Define organic compound. Name the type of chemical bond that predominates in the backbone of such a compound. *3.1*

2. Name the "molecules of life." Do they break apart most easily at their hydrocarbon portion or at functional groups? *3.1, 3.2*

3. Select one of the carbohydrates, lipids, proteins, or nucleic acids described in this chapter. Speculate on how its functional groups as the bonds between carbon atoms in its backbone contribute to its final shape and function. *3.4, 3.5, 3.6, 3.7, or 3.8*

4. Which category includes all of the other items listed? *3.5*
   a. triglyceride   c. wax   e. lipid
   b. fatty acid   d. sterol   f. phospholipid

5. Explain how a hemoglobin molecule's three-dimensional shape arises, starting with its primary structure. *3.6, 3.7*

Self-Quiz   *(Answers in Appendix IV)*

1. Each carbon atom can share pairs of electrons with as many as _____ other atoms.
   a. one   b. two   c. three   d. four

2. Hydrolysis is a _____ reaction.
   a. functional group transfer   d. condensation
   b. electron transfer   e. cleavage
   c. rearrangement   f. both b and d

3. _____ are simple sugars (monosaccharides).
   a. glucose   c. ribose   e. both a and b
   b. sucrose   d. chitin   f. both a and c

4. A saturated fat has fatty acid tails with one or more _____ .
   a. single covalent bonds   b. double covalent bonds

5. _____ are to proteins as _____ are to nucleic acids.
   a. sugars; lipids   c. amino acids; hydrogen bonds
   b. sugars; proteins d. amino acids; nucleotides

6. Nucleotides include _____ .
   a. ATP   c. DNA and RNA subunits
   b. coenzyme subunits   d. all of the above

7. A denatured protein or DNA molecule has lost its _____ .
   a. hydrogen bonds   c. function
   b. shape   d. all of the above

8. Match each molecule with the most suitable description.
   ____ long sequence of amino acids   a. carbohydrate
   ____ the main energy carrier   b. phospholipid
   ____ glycerol, fatty acids, phosphate   c. protein
   ____ two strands of nucleotides   d. DNA
   ____ one or more sugar monomers   e. ATP

Critical Thinking

1. Jack decided to celebrate summer by making a crab salad and a peach pie for some friends. He had such a good time, he forgot to put the cap on the bottle of olive oil after making the salad. He also didn't tighten the lid on the can of shortening after making the pie crust. A few weeks later, he discovered the oil had turned rancid, but the shortening still smelled okay. Both substances are fats. Both were stored in a dark cupboard, at room temperature. Why did one spoil and the other stay fresh?

2. In the following list, identify which is the carbohydrate, fatty acid, amino acid, and polypeptide:

a. $^+NH_3$—CHR—COO$^-$ c. (glycine)$_{20}$

b. $C_6H_{12}O_6$ d. $CH_3(CH_2)_{16}COOH$

3. A clerk in a health-food store tells you that certain "natural" vitamin C tablets extracted from rose hips are better for you than synthetic vitamin C tablets. Given your understanding of the structure of organic compounds, what would be your response? Design an experiment to test whether these vitamins differ.

4. Rabbits that eat green, leafy vegetables containing xanthophyll accumulate this yellow pigment molecule in body fat but not in muscles. What chemical properties of the molecule might cause this selective accumulation?

5. Grasses can be digested in cows, but not in people. A cow's four-chambered stomach houses populations of certain microorganisms that are not normal inhabitants of the human stomach. What kind of metabolic reactions do you think the microorganisms carry out? What do you think might happen to a cow undergoing treatment with an antibiotic that killed the microorganisms?

6. A Gary Larsen cartoon states that sheep with steel wool have no natural enemies. You might laugh at the thought of such a preposterous animal—but what exactly *is* wool? It is the keratin-rich, soft undercoat of sheep, Angora goats, llamas, and other hairy mammals. Keratin, a protein, is a long polypeptide chain. Along its length, many hydrogen bonds hold it in a helical shape. Visualize three such chains coiled together. Many disulfide bridges in the end regions stabilize the trio in a tight, ropelike array. Each hair has linear aggregates of the ropelike fibers, which are highly resistant to stretching. Given keratin's structure, speculate why, when you wash a wool sweater or run it through the hot cycle of a dryer, it will shrink, pathetically and permanently.

Selected Key Terms

alcohol *3.1*	functional group *3.1*	phospholipid *3.5*
amino acid *3.6*	functional-group	polymer *3.2*
ATP *3.8*	transfer *3.2*	polypeptide chain
carbohydrate *3.4*	glycoprotein *3.7*	*3.6*
cleavage *3.2*	hydrolysis *3.2*	polysaccharide *3.4*
coenzyme *3.8*	lipid *3.5*	protein *3.6*
condensation *3.2*	lipoprotein *3.7*	rearrangement
denaturation *3.7*	monomer *3.2*	(of bonds) *3.2*
dipeptide *3.6*	monosaccharide *3.4*	RNAs *3.8*
DNA *3.8*	nucleic acid *3.8*	sterol *3.5*
electron transfer *3.2*	nucleotide *3.8*	toxin *3.3*
enzyme *3.2*	oligosaccharide *3.4*	triglyceride *3.5*
fatty acid *3.5*	organic compound *3.1*	wax *3.5*

Readings

Atkins, P. 1987. *Molecules.* New York: Scientific American Library. A molecular "glossary" includes the formula, three-dimensional stucture, and action of many organic compounds.

"The Molecules of Life." October 1985. *Scientific American* (253). The entire issue focuses on the structure and functioning of biological molecules.

Ritter, P. 1996. *Biochemistry: An Introduction.* Pacific Grove, California: Brooks/Cole. Chockful of human applications.

Wolfe, S. 1995. *Introduction to Molecular and Cellular Biology.* Belmont, California: Wadsworth.

Web Site See *http://www.wadsworth.com/biology* for practice quiz questions, hypercontents, BioUpdates, and critical thinking. The Wadsworth Biology Resource Center provides a wealth of information fully organized and integrated by chapter.

4

CELL STRUCTURE AND FUNCTION

Animalcules and Cells Fill'd With Juices

Early in the seventeenth century, a scholar by the name of Galileo Galilei arranged two glass lenses within a cylinder. With this instrument he happened to look at an insect, and later he described the stunning geometric patterns of its tiny eyes. Thus Galileo, who was not a biologist, was the first to record a biological observation made through a microscope. The study of the cellular basis of life was about to begin. First in Italy, then in France and England, scholars set out to explore a world whose existence had not even been suspected.

At midcentury Robert Hooke, Curator of Instruments for the Royal Society of England, was at the forefront of these studies. When Hooke first turned a microscope to thinly sliced cork from a mature tree, he observed tiny compartments (Figure 4.1c). He gave them the Latin name *cellulae*, meaning small rooms—hence the origin of the biological term "cell." They actually were the walls of dead cells, which is what cork is made of, but Hooke did not think of them as being dead because he did not know cells could be alive. In other plant tissues, he discovered cells "fill'd with juices" but could not imagine what they represented.

Given the simplicity of their instruments, it is amazing that the pioneers in microscopy observed as much as they did. Antony van Leeuwenhoek, a Dutch shopkeeper, had exceptional skill in constructing lenses and possibly the keenest vision (Figure 4.1a). By the late 1600s, he was observing wonders everywhere, including "many very small animalcules, the motions of which were very pleasing to behold," in scrapings of tartar from his own teeth. He discovered diverse protistans, sperm, even a bacterium—an organism so small that it would not be seen again for another two centuries!

In the 1820s, improvements in lenses brought cells into sharper focus. Robert Brown, a botanist, was noticing an opaque spot in

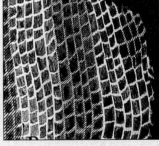

a

Figure 4.1 Early glimpses into the world of cells.
(a) Antony van Leeuwenhoek, with microscope in hand.
(b) Robert Hooke's compound microscope and (c) his drawing of cell walls from cork tissue. (d) One of van Leeuwenhoek's early sketches of sperm cells. (e) Some cartoon evidence of the startling impact of microscopic observations on nineteenth-century London.

b

c

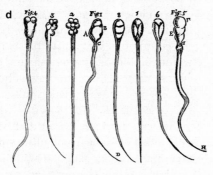

a variety of cells. He called it a nucleus. In 1838 still another botanist, Matthias Schleiden, wondered if the nucleus had something to do with a cell's development. As he hypothesized, each plant cell must develop as an independent unit even though it is an integral part of the plant.

By 1839, after years of studying animal tissues, the zoologist Theodor Schwann had this to say: Animals as well as plants consist of cells and cell products—and even though the cells are part of a whole organism, to some extent they have an individual life of their own.

A question remained: Where do cells come from? A decade later, Rudolf Virchow, a physiologist, completed his own studies of a cell's growth and reproduction—that is, its division into two daughter cells. Every cell, he decided, comes from a cell that already exists.

And so, by the middle of the nineteenth century, microscopic analysis had yielded three generalizations, which together constitute the **cell theory**. *First, every organism is composed of one or more cells. Second, the cell is the smallest unit having the properties of life. Third, the continuity of life arises directly from the growth and division of single cells.* All three insights still hold true.

This chapter provides an overview of our current understanding of the structure and function of cells. It also introduces some of the modern microscopes that transport us more deeply into the spectacular worlds of juice-fill'd cells and animalcules.

KEY CONCEPTS

1. All organisms are composed of one or more cells. The cell is the smallest unit that still retains the characteristics of life. And each new cell arises from a preexisting cell. These are the three generalizations of the cell theory.

2. All cells have an outermost, double-layered component, the plasma membrane, that helps their interior remain distinct from the surroundings. Cells contain cytoplasm, an internal region that is highly organized for energy conversions, protein synthesis, movements, and other activities necessary for survival. They also contain a nucleus or a comparable internal region in which their DNA is located.

3. The plasma membrane and internal cell membranes consist largely of phospholipid and protein molecules. The phospholipids form two adjacent layers that give the membrane its basic structure and prevent water-soluble substances from freely crossing it. Proteins embedded in those layers or positioned at its surfaces carry out most membrane functions.

4. The nucleus is one of many organelles. Organelles are membrane-bound compartments inside eukaryotic cells. They physically separate different metabolic reactions and allow them to proceed in orderly fashion. Prokaryotic cells (bacteria) do not have comparable organelles.

5. When cells grow, they increase faster in volume than in surface area. This physical constraint on increases in size influences cell size and shape.

6. Different microscopes modify light rays or accelerated beams of electrons in ways that allow us to form images of incredibly small specimens. They are the foundation for our current understanding of cell structure and function.

Inside your body and at its moist surfaces, trillions of cells live in interdependency. In scummy pondwater, a single-celled amoeba moves about freely, thriving on its own. For humans, amoebas, and all other organisms, the **cell** is the smallest entity that retains the properties of life. A cell either can survive on its own or has the potential to do so. Its structure is highly organized, and it engages in metabolism. A cell senses and responds to changes in the environment. And it has the potential to reproduce, based on inherited instructions in its DNA.

Structural Organization of Cells

Cells differ enormously in size, shape, and activities, as you might gather by comparing a tiny bacterium with one of your relatively giant liver cells. Yet they are alike in three respects. All cells start out life with a plasma membrane, a region of DNA, and a region of cytoplasm:

1. **Plasma membrane**. This thin, outermost membrane maintains the cell as a distinct entity. By doing so, it allows metabolic events to proceed apart from random events in the environment. A plasma membrane does not *isolate* the cell interior. Substances and signals continually move across it, in highly controlled ways.

2. **DNA-containing region**. DNA occupies part of the cell interior, along with molecules that can copy or read its hereditary instructions.

3. **Cytoplasm**. The cytoplasm is everything in the cell interior *except* for the region of DNA. It includes a fluid portion, **ribosomes** (two-unit structures upon which proteins are synthesized), and other components.

This chapter introduces two fundamentally different kinds of cells. The cytoplasm of **eukaryotic cells** includes organelles, which are tiny sacs and other compartments bounded by membranes. One, the nucleus, houses the DNA. It is the defining feature of eukaryotic cells:

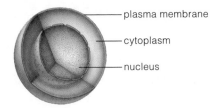

plasma membrane
cytoplasm
nucleus

By contrast, **prokaryotic cells** have no nucleus; they have no membranes intervening between their region of DNA and the cytoplasm that surrounds it. Bacteria are the only prokaryotic cells. Beyond the bacterial realm, all other organisms—from amoebas to peach trees and puffball mushrooms to zebras—are eukaryotes.

The Lipid Bilayer of Cell Membranes

All cell membranes have the same structural framework of two sheets of lipid molecules. Figure 4.2 shows this **lipid bilayer** arrangement for the plasma membrane, the continuous boundary that bars the free passage of water-soluble substances into and out of the cell. Within eukaryotic cells, membranes subdivide the cytoplasm into specific zones in which substances are synthesized, processed, stockpiled, or degraded.

Diverse proteins embedded in the lipid bilayer or positioned at or near one of its surfaces carry out most membrane functions. For example, some of the proteins serve as passive channels for water-soluble substances.

Figure 4.2   Organization of the lipid bilayer of cell membranes. (**a**) Symbol for phospholipid molecules, the most abundant membrane components. These and other lipids are arranged as two layers. (**b**) The hydrophobic tails of the molecules are sandwiched between their hydrophilic heads, which are dissolved in cytoplasm on one side of the bilayer and in extracellular fluid on the other (**c**).

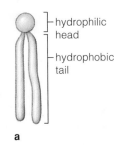

hydrophilic head
hydrophobic tail

a

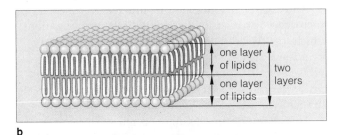

one layer of lipids
one layer of lipids
two layers

b

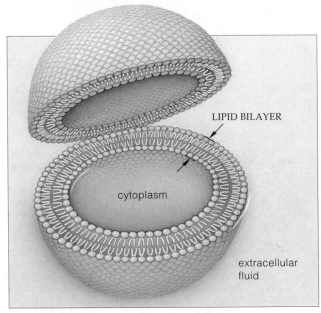

LIPID BILAYER
cytoplasm
extracellular fluid

c

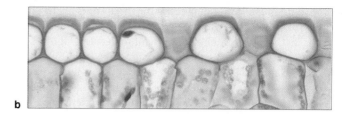

Diameter (cm):	0.5	1.0	1.5
Surface area (cm^2):	0.79	3.14	7.07
Volume (cm^3):	0.06	0.52	1.77
Surface-to-volume ratio:	13.17:1	6.04:1	3.99:1

a

Figure 4.3 (**a**) An example of the surface-to-volume ratio. This physical relationship between increases in volume and in surface area imposes restrictions on the size and shape of cells, including the ones near the surface of leaves (**b**) and in skeletal muscles (**c**).

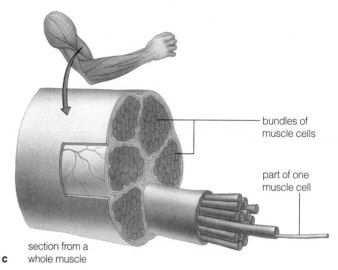

bundles of muscle cells

part of one muscle cell

section from a
c whole muscle

Others transport electrons or pump substances across the lipid bilayer. Still others are receptors, which latch onto hormones and other types of signaling molecules that trigger alterations in cell activities. You will read more about membrane proteins in the chapter to follow.

Cell Size and Cell Shape

You may be wondering how small cells really are. Can any be observed with the unaided human eye? There are a few, including the "yolks" of bird eggs, cells in the red part of watermelons, and the fish eggs we call caviar. However, most cells cannot be observed without microscopes. To give you a sense of cell sizes, red blood cells are about 8 millionths of a meter across. You could fit a string of 2,000 of them across your thumbnail!

Why are most cells so small? A physical relationship called the **surface-to-volume ratio** constrains increases in a cell's size. By this relationship, an object's volume increases with the cube of the diameter, but its surface area increases only with the square. Figure 4.3a gives an example of this. Simply put, *if a cell expands in diameter during growth, its volume will increase more rapidly than its surface area will.*

Suppose you figure out a way to make a round cell grow four times wider than it normally would. Its volume increases 64 times (4^3), but its surface area only increases 16 times (4^2). As a result, each unit of the cell's plasma membrane must now serve four times as much cytoplasm as it did previously! Moreover, past a certain point, the inward flow of nutrients and outward flow of wastes will not be fast enough, and the cell will die.

A large, round cell also would have trouble moving materials *through* its cytoplasm. In small cells, such as those shown in Figure 4.3b, the random, tiny motions of molecules can easily distribute materials. If a cell is not small, you usually can expect it to be long and thin or to have outfoldings and infoldings that increase its surface relative to its volume. *The smaller or narrower or more frilly-surfaced the cell, the more efficiently materials cross its surface and become distributed through the interior.*

Multicelled body plans show evidence of surface-to-volume constraints, also. For example, cells attach end to end in strandlike algae, and each one interacts directly with its surroundings. Your skeletal muscle cells are very thin, but each one is as long as a biceps or some other muscle of which it is part (Figure 4.3c). Your circulatory system delivers materials to muscle cells and trillions of others, and it also sweeps away their metabolic wastes. Its many "highways" cut through the volume of tissue and so shrink the distance to and from individual cells.

All cells have an outermost plasma membrane, an internal region of cytoplasm, and an internal region of DNA.

Besides the plasma membrane, eukaryotic cells have internal, membrane-bound compartments, including a nucleus.

Each membrane has a bilayer structure, consisting largely of phospholipids. The hydrophobic parts of its lipid molecules are sandwiched between the hydrophilic parts, which are dissolved in the fluid surroundings.

A lipid bilayer imparts structure to the cell membrane and serves as a barrier to water-soluble substances.

Proteins embedded in the bilayer or positioned at its surfaces carry out most membrane functions.

During their growth, cells increase faster in volume than they do in surface area. This physical constraint on increases in size influences the size and shape of cells. It also influences the body plans of multicelled organisms.

4.2 MICROSCOPES—GATEWAYS TO THE CELL

Modern microscopes, including the three kinds sketched in Figure 4.4, are gateways to astounding worlds. Some even afford glimpses into the structure of molecules. Figure 4.5 gives an idea of the range of magnifications possible. The micrographs in Figures 4.5 and 4.6 hint at the details now being observed. A **micrograph** simply is a photograph of an image that was formed with the aid of a microscope.

LIGHT MICROSCOPES Light microscopes work by bending (refracting) light rays. Any rays directly entering a glass lens, except at the very center, are bent. The farther they are from the center of the lens, the more they will bend:

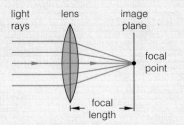

The angle at which rays of light enter a glass lens and the molecular structure of the glass dictate the extent to which the rays will bend.

In a *compound light microscope*, two or more sets of glass lenses bend incoming rays to form an enlarged image of a cell or some other specimen. A living cell must be small or thin enough for light to pass through. It would help if cell parts differed in color and density from the surroundings, but most are nearly colorless and appear uniformly dense. That is why microscopists stain cells (expose them to dyes that react with some parts but not others). Staining may alter and kill cells. Dead cells break down very fast, so they typically are pickled or preserved before being stained.

Suppose you are using the best glass lens system. When you magnify the diameter of a specimen by 2,000 times or more, you discover that cell parts appear larger but are not clearer. What limits the resolution of smaller details? The answer lies with a certain property of visible light.

Picture a train of waves moving across an ocean. Each **wavelength** is the distance from one wave's peak to the peak of the wave behind it. Light also travels as waves. In fact, different colors of light correspond to waves of certain, unvarying lengths. The wavelength is about 750 nanometers for red light and 400 nanometers for violet; all other colors fall in between. *And if a cell structure is less than one-half of a wavelength long, rays of light streaming by it will not be disturbed enough to make the object visible.*

ELECTRON MICROSCOPES Electrons are particles, but they, too, behave like waves. Electron microscopy is based on accelerating the flow of streams of electrons that have wavelengths of about 0.005 nanometer—about 100,000 times shorter than those of visible light. Electrons cannot pass through glass lenses, but they can be bent from their paths and focused by a magnetic field. In a *transmission electron microscope*, such a field is the "lens." Accelerated electrons are directed through a specimen, focused into an image, and magnified. With *scanning electron microscopes*, a narrow beam of electrons moves back and forth across a specimen thinly coated with metal. The metal responds by emitting some of its own electrons. A detector connected to electronic circuitry transforms the electron energy into an image of the specimen's surface on a television screen. Most of the images have fantastic depth.

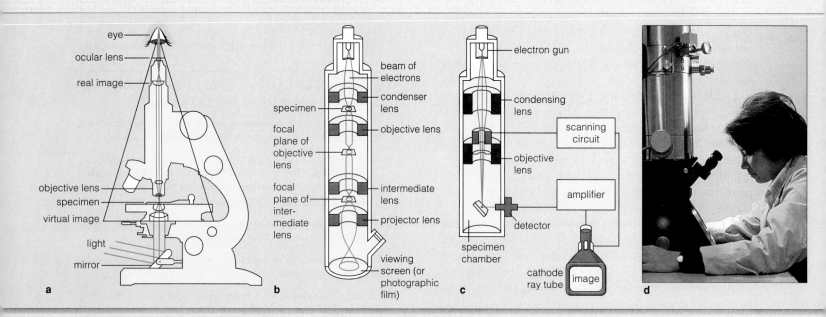

Figure 4.4 Diagrams of (**a**) a light microscope, (**b**) a transmission electron microscope, and (**c**) a scanning electron microscope. (**d**) Exterior view of an electron microscope.

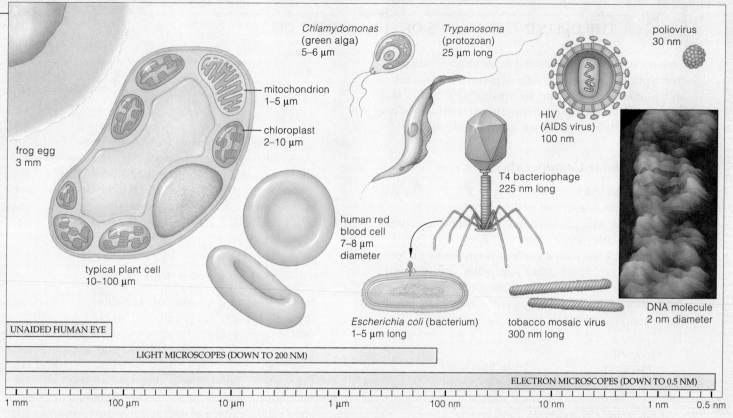

Figure 4.5 Units of measure used in microscopy. A *scanning tunneling microscope*, which can provide magnifications up to 100 million, gave us the photomicrograph of part of a DNA molecule. Its needlelike probe has a single atom at its tip. Voltage that is applied between the tip and an atom at a specimen's surface causes a detectable tunnel to form in the electron orbitals. As the tip moves over a specimen's contours, a computer analyzes the motion and creates a three-dimensional view of the surface atoms.

1 centimeter (cm)	= 1/100 meter, or 0.4 inch
1 millimeter (mm)	= 1/1,000 meter
1 micrometer (μm)	= 1/1,000,000 meter
1 nanometer (nm)	= 1/1,000,000,000 meter

1 meter = 10^2 cm = 10^3 mm = 10^6 μm = 10^9 nm

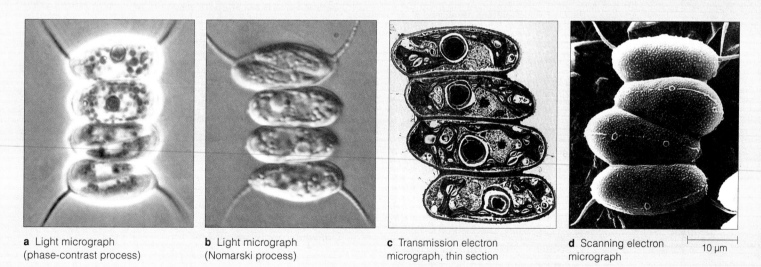

a Light micrograph (phase-contrast process)

b Light micrograph (Nomarski process)

c Transmission electron micrograph, thin section

d Scanning electron micrograph

10 μm

Figure 4.6 How different microscopes can reveal different aspects of the same organism—in this case, a green alga (*Scenedesmus*). The images of all four specimens are at the same magnification. The phase-contrast and Nomarski processes mentioned in (**a**) and (**b**) can create optical contrasts without staining the cells. Both processes enhance the usefulness of light micrographs.

As for other micrographs in the book, the short horizontal bar below the micrograph in (**d**) provides you with a visual reference for size. A micrometer (μm) is 1/1,000,000 of a meter.

THE DEFINING FEATURES OF EUKARYOTIC CELLS

We turn now to organelles and other structural features that typically occur in the cells of plants, animals, fungi, and protistans. We define an **organelle** as an internal, membrane-bound sac or compartment that serves one or more specialized functions inside these cells.

Major Cellular Components

Observe some micrographs of a typical eukaryotic cell, such as the ones described in the preceding section, and you probably will quickly notice that the nucleus is one of its most conspicuous features. Remember, any cell that starts out life with a nucleus is eukaryotic; it has a "true nucleus." Many other organelles and structures also are typical of these cells, although the numbers and kinds differ from one cell type to the next. Here is a list of the most common features:

Organelle or Structure	Main Function
Nucleus	*Localizing the cell's DNA*
Ribosomes	*Assembling polypeptide chains*
Endoplasmic reticulum	*Routing and modifying the newly formed polypeptide chains; also, synthesizing lipids*
Golgi body	*Modifying polypeptide chains into mature proteins; sorting and shipping proteins and lipids for secretion or for use inside cell*
Various vesicles	*Transporting or storing variety of substances; digesting substances and structures within the cell; other functions*
Mitochondria	*Producing many ATP molecules in highly efficient fashion*
Cytoskeleton	*Imparting shape and internal organization to cell; moving the cell and its internal structures*

Think about this list, and you might well find yourself asking: What is the advantage of partitioning the cell interior among many organelles? *Compartmentalization allows a large number of activities to proceed simultaneously in very limited space.* Consider a photosynthetic cell in a leaf. It can put together starch molecules by one set of reactions *and* break them apart by another set. Yet the cell would gain absolutely nothing if the synthesis and breakdown reactions proceeded at the same time on the same starch molecule. Without organelle membranes, the balance of diverse chemical activities that help keep eukaryotic cells alive would spiral out of control.

Figure 4.7 (*Facing page*) Generalized sketches showing some of the features that are typical of many plant cells (**a**) and animal cells (**b**).

Organelle membranes have another function besides physically separating incompatible reactions. They also allow reactions that are compatible and interconnected to proceed at different times. For instance, a plant's photosynthetic cells produce and store starch molecules in an organelle called a chloroplast, then later release starch for use in different reactions in the same organelle.

Typical Organelles in Plant Cells

Figure 4.7a can start you thinking about the location of organelles in a typical plant cell. Bear in mind, calling a cell "typical" is like calling a cactus or a water lily or an elm tree a "typical" plant. As is true of animal cells, variations on the basic plan are mind-boggling. With this qualification in mind, also take a close look at the micrograph in Figure 4.8, on the subsequent page. It shows the locations of organelles and structures you are likely to observe in many specialized plant cells.

Typical Organelles in Animal Cells

Next, start thinking about the organelles of a typical animal cell, such as the one shown in Figures 4.7b and 4.9. Like the plant cell, it contains a nucleus, numerous mitochondria, and the other components listed earlier. *These structural similarities point to basic functions that are necessary for survival, regardless of cell type.* We will return to this concept throughout the book.

Comparing Figures 4.7 through 4.9 also will give you an initial idea of how plant and animal cells differ in their structure. For example, you will never observe an animal cell surrounded by a cell wall. (You might see assorted fungal and protistan cells with one, however.) What other differences can you identify?

Eukaryotic cells contain a profusion of organelles, which are internal, membrane-bound sacs and compartments that serve specific metabolic functions.

Organelles physically separate chemical reactions, many of which are incompatible.

Organelles separate different reactions in time, as when certain molecules are put together, stored, and then used later in other reaction sequences.

All eukaryotic cells contain certain organelles (such as the nucleus) and structures (such as ribosomes) that perform functions essential for survival. Specialized cells also may incorporate additional kinds of organelles and structures.

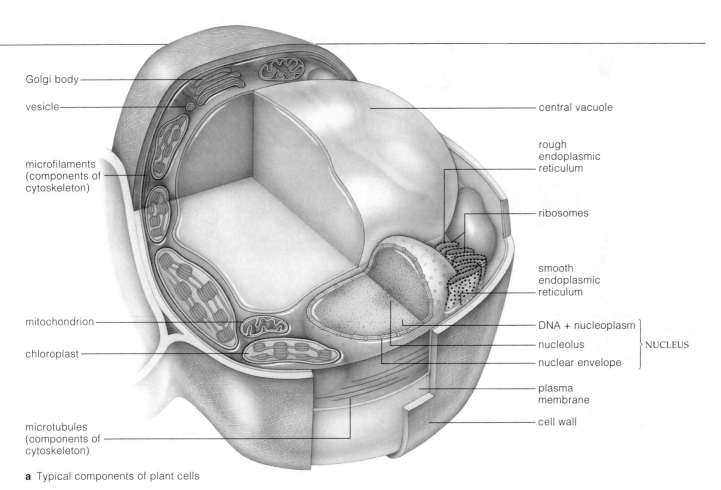

Golgi body
vesicle

microfilaments
(components of
cytoskeleton)

mitochondrion
chloroplast

microtubules
(components of
cytoskeleton)

central vacuole

rough
endoplasmic
reticulum

ribosomes

smooth
endoplasmic
reticulum

DNA + nucleoplasm
nucleolus   NUCLEUS
nuclear envelope

plasma
membrane

cell wall

a Typical components of plant cells

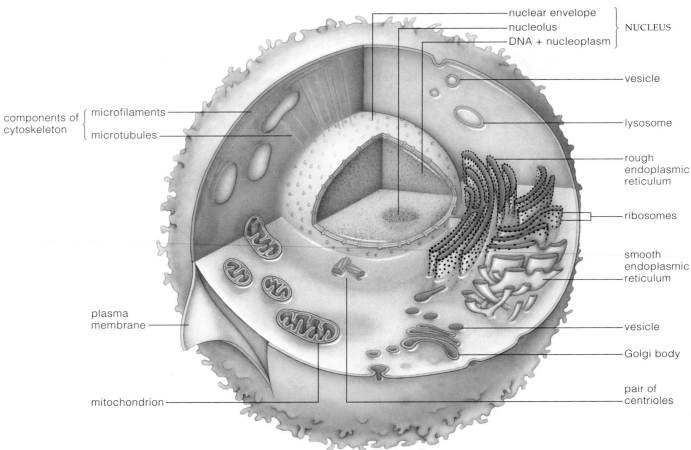

nuclear envelope
nucleolus   NUCLEUS
DNA + nucleoplasm

vesicle

components of { microfilaments
cytoskeleton   { microtubules

lysosome

rough
endoplasmic
reticulum

ribosomes

smooth
endoplasmic
reticulum

plasma
membrane

vesicle

Golgi body

pair of
centrioles

mitochondrion

b Typical components of animal cells

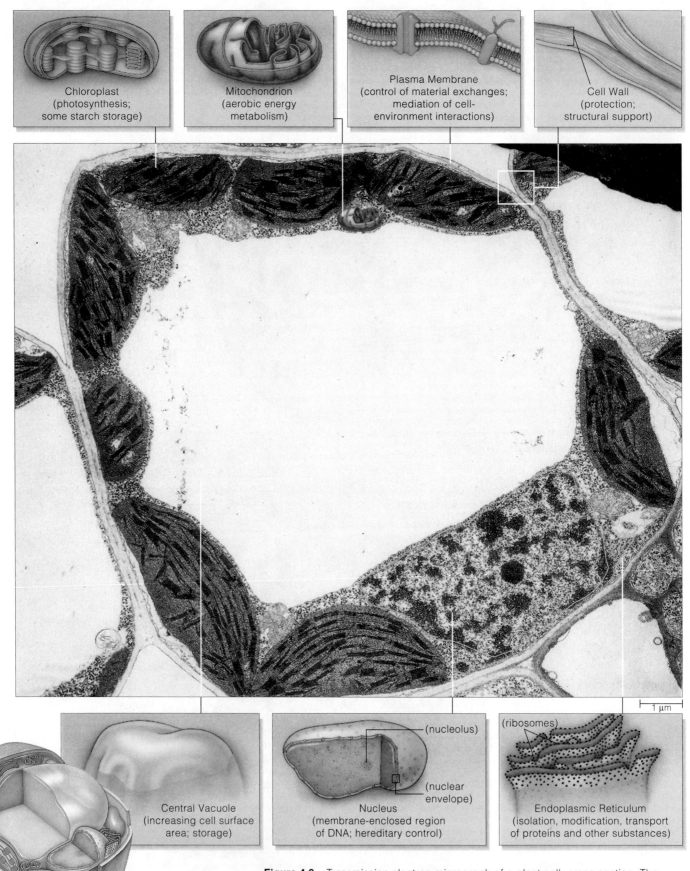

Chloroplast (photosynthesis; some starch storage)

Mitochondrion (aerobic energy metabolism)

Plasma Membrane (control of material exchanges; mediation of cell-environment interactions)

Cell Wall (protection; structural support)

1 μm

Central Vacuole (increasing cell surface area; storage)

Nucleus (membrane-enclosed region of DNA; hereditary control)

(nucleolus)

(nuclear envelope)

(ribosomes)

Endoplasmic Reticulum (isolation, modification, transport of proteins and other substances)

Figure 4.8 Transmission electron micrograph of a plant cell, cross-section. The sketches highlight key organelles. The specimen is a photosynthetic cell from a blade of Timothy grass (*Phleum pratense*). This plant is one of the important forage grasses that ranchers grow on lands that are suitable for grazing but not for agriculture.

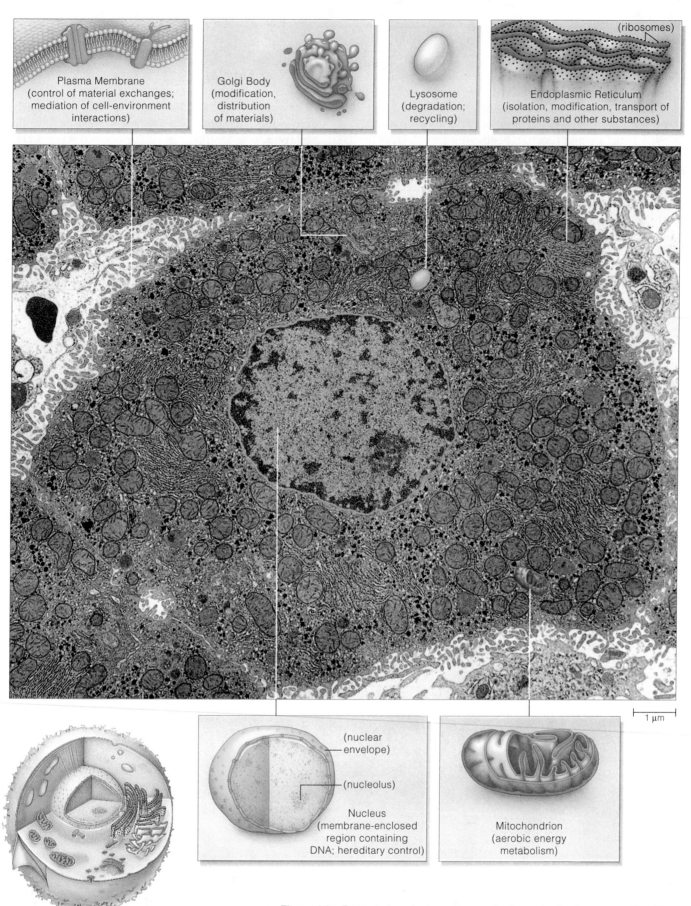

Plasma Membrane (control of material exchanges; mediation of cell-environment interactions)

Golgi Body (modification, distribution of materials)

Lysosome (degradation; recycling)

(ribosomes)

Endoplasmic Reticulum (isolation, modification, transport of proteins and other substances)

1 μm

(nuclear envelope)

(nucleolus)

Nucleus (membrane-enclosed region containing DNA; hereditary control)

Mitochondrion (aerobic energy metabolism)

Figure 4.9 Transmission electron micrograph of an animal cell, cross-section. The sketches highlight key organelles. The specimen is a cell from the liver of a rat.

Constructing, operating, and reproducing cells simply cannot be done without carbohydrates, lipids, proteins, and nucleic acids. It takes a class of proteins—enzymes—to build and use these molecules. Said another way, a cell's structure and function begin with proteins. *And instructions for building proteins are contained in DNA.*

Unlike bacteria, eukaryotic cells have their genetic instructions distributed through several to many DNA molecules of various lengths. For example, each of your body cells contains forty-six DNA molecules. Stretched end to end, they would be about one meter long. DNA in frog cells would be ten meters, end to end. Compared to the single molecule in bacteria, that's a lot of DNA!

Eukaryotic DNA resides in the **nucleus**. This type of organelle has a distinctive structure, as shown by the example in Figure 4.10, and it serves two key functions. *First*, the nucleus physically tucks away all of the DNA molecules, apart from the intricate metabolic machinery of the cytoplasm. This localization of DNA makes it easier to copy the genetic instructions before the time comes for a cell to divide. The DNA molecules can be assorted into parcels—one parcel for each new cell that is produced. *Second*, the membranous boundary of the nucleus helps control the exchange of substances and signals between the nucleus and the cytoplasm.

Nuclear Envelope

The outermost component of the nucleus is the **nuclear envelope**. This double-membrane system has *two* lipid bilayers, one wrapped over the other (Figure 4.11). It completely surrounds the fluid portion of the nucleus, or nucleoplasm. As is true of all other cell membranes, the lipid bilayers prevent water-soluble substances from moving across. However, pores composed of clusters of proteins span both bilayers. The nuclear pores allow ions and small, water-soluble molecules to move across the nuclear envelope. They also control the passage of ribosomal subunits, proteins, and other large molecules.

The innermost surface of the nuclear envelope has attachment sites for protein filaments. These anchor the DNA molecules to the membrane and help keep them organized. Peppering the outermost surface are many of the cell's ribosomes, which serve in protein synthesis.

Nucleolus

What is the dense, globular mass of material within the nucleus shown in Figure 4.10? As eukaryotic cells grow, one or more of the masses appear in the nucleus. Each is a **nucleolus** (plural, nucleoli). It is a site where the

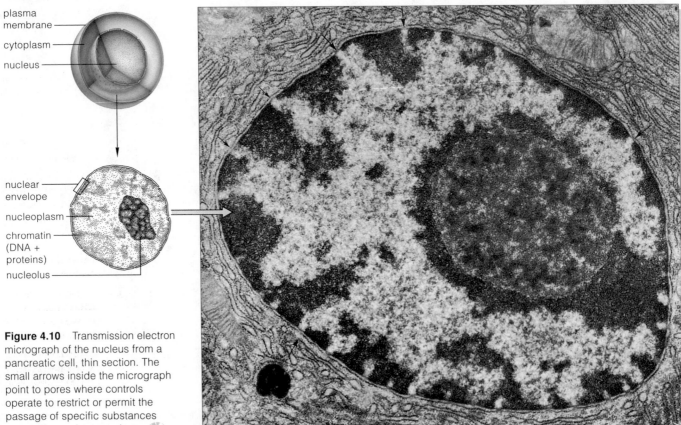

Figure 4.10 Transmission electron micrograph of the nucleus from a pancreatic cell, thin section. The small arrows inside the micrograph point to pores where controls operate to restrict or permit the passage of specific substances across the nuclear envelope.

plasma membrane
cytoplasm
nucleus

nuclear envelope
nucleoplasm
chromatin (DNA + proteins)
nucleolus

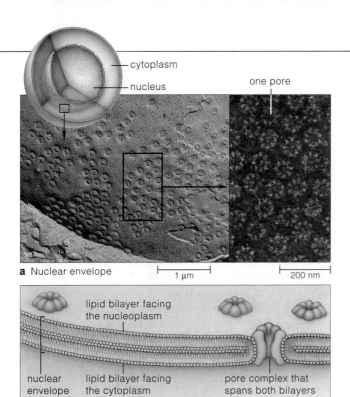

cytoplasm

nucleus

one pore

a Nuclear envelope

1 μm

200 nm

lipid bilayer facing
the nucleoplasm

nuclear
envelope

lipid bilayer facing
the cytoplasm

pore complex that
spans both bilayers

b

Figure 4.11 (**a**) Surface view of a nuclear envelope, which consists of two pore-studded lipid bilayers. *Left:* This specimen was deliberately fractured for microscopy, in the manner described in Section 5.2. *Right:* Closer view of the pores. Each pore across the envelope is an organized array of membrane proteins. It permits the selective transport of larger molecules into and out of the nucleus. (**b**) Sketch of the nuclear envelope.

Early microscopists bestowed the names *chromatin* on the seemingly grainy substance and *chromosomes* on the condensed structures. We now define **chromatin** as a cell's collection of DNA, together with all the proteins associated with it (Table 4.1). Each **chromosome** is one DNA molecule and its associated proteins, regardless of whether it is in threadlike or condensed form:

unduplicated,
not condensed
chromosome
(one DNA double
helix + proteins)

duplicated but
not condensed
chromosome
(two DNA double
helices + proteins)

duplicated and
now condensed
chromosome

In other words, the "chromosome" doesn't always look the same during the life of a eukaryotic cell. We will consider different aspects of chromosomes in chapters to come, so it will be helpful to keep that point in mind.

What Happens to the Proteins Specified by DNA?

Enzymes assemble polypeptide chains on ribosomes in the cytoplasm. What happens to the newly formed chains? Many are used or stockpiled in the cytoplasm. Many others pass through the cytomembrane system. As you will read in the next section, this system is a series of organelles, including endoplasmic reticulum, Golgi bodies, and vesicles.

Many proteins take on final form and get packaged in vesicles in the cytomembrane system. Also, lipids are assembled and packaged here. Some vesicles deliver proteins and lipids to regions of the cell where new membranes are to be built. Others store proteins or lipids for specific uses. Still others move to the plasma membrane and release their contents outside the cell.

protein and RNA subunits of ribosomes are assembled. Later, the subunits are shipped from the nucleus, into the cytoplasm. At the time when the polypeptide chains of proteins are synthesized, the subunits join together as intact, functional ribosomes.

Chromosomes

Between cell divisions, eukaryotic DNA is threadlike. Many enzymes and other proteins are attached to it, like beads on a string. Except at extreme magnification, the beaded threads look grainy, as in Figure 4.10. Before a cell divides, however, it duplicates its DNA molecules (so each new cell will get all of the required hereditary instructions). Besides this, the molecules get folded and twisted into condensed structures, proteins and all.

Table 4.1	Summary of the Components of the Nucleus
Nuclear envelope	Pore-riddled double-membrane system that selectively controls the passage of various substances into and out of the nucleus
Nucleolus	Dense cluster of RNA and proteins that will be assembled into subunits of ribosomes
Nucleoplasm	Fluid interior portion of the nucleus
Chromosome	One DNA molecule and many proteins that are intimately associated with it
Chromatin	Total collection of all DNA molecules and their associated proteins in the nucleus

The nucleus, a double-membrane organelle, sequesters the DNA molecules of eukaryotic cells away from the metabolic machinery of the cytoplasm.

The sequestration makes it easier to organize and copy the DNA, before a cell divides.

Pores across the nuclear envelope help control the passage of larger molecules between the nucleus and the cytoplasm.

The **cytomembrane system** is a series of organelles in which lipids are assembled and new polypeptide chains are modified into final proteins. Its products are sorted and shipped to different destinations. Figure 4.12 shows how its organelles—the ER, Golgi bodies, and various vesicles—are functionally related to one another.

Endoplasmic Reticulum

The functions of the cytomembrane system begin with **endoplasmic reticulum**, or **ER**. In animal cells, the ER membranes start at the nucleus and curve through the cytoplasm. Different regions of these membranes have a rough or smooth appearance. The difference arises largely from the presence or absence of ribosomes on the side of the membrane facing the cytoplasm.

Rough ER is often organized as stacked, flattened sacs that have many ribosomes attached (Figure 4.13a). All polypeptide chains are assembled on ribosomes. But only the newly forming chains having a built-in signal can enter the space within rough ER or get incorporated into ER membranes. (The signal is a string of fifteen to twenty specific amino acids.) Once the chains are inside rough ER, enzymes may attach oligosaccharides and other side chains to them. Many specialized cells secrete the final proteins. Rough ER is abundant in such cells. For example, ER-rich, glandular cells of the pancreas produce and secrete enzymes that end up in the small intestine, where they help digest your meals.

Smooth ER is free of ribosomes. It curves through the cytoplasm like connecting pipes (Figure 4.13b). In many cells, smooth ER is the main site of lipid synthesis. It is especially developed in seeds. In liver cells, it even inactivates certain drugs and harmful metabolic by-products. Sarcoplasmic reticulum, a type of smooth ER in skeletal muscle cells, functions in muscle contraction.

Golgi Bodies

In **Golgi bodies**, enzymes put the finishing touches on proteins and lipids, sort them out, and package them in vesicles for shipment to specific locations. For example, an enzyme in one Golgi region might attach a phosphate group to a new protein, thereby giving it a mailing tag to its proper destination.

Each Golgi body is composed of flattened membrane sacs that rather resemble a stack of pancakes (Figure 4.14). Vesicles form at the final region of a Golgi body (the uppermost pancakes) when parts of the membrane bulge, then break away. In animal cells, this is the region that is closest to the plasma membrane.

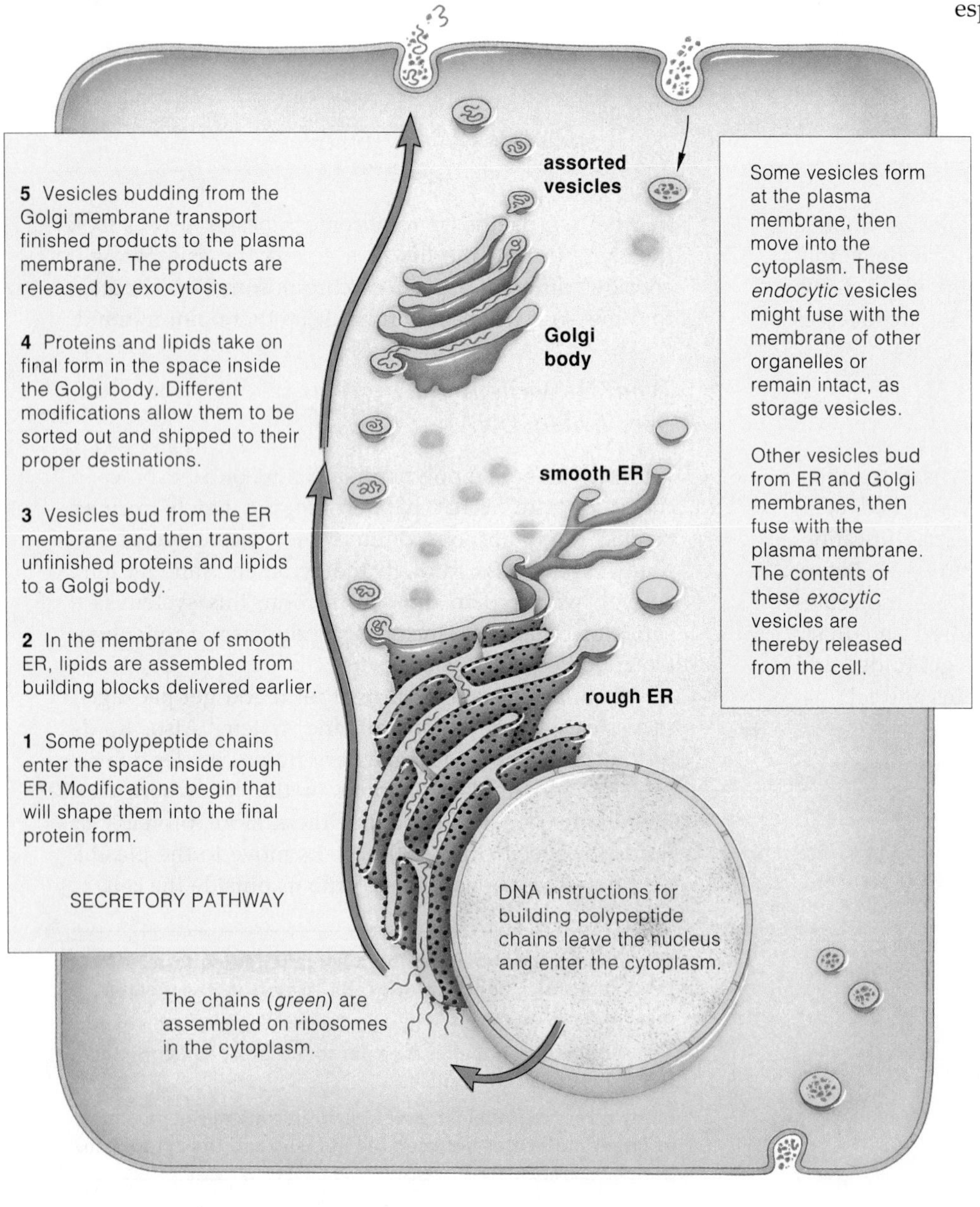

5 Vesicles budding from the Golgi membrane transport finished products to the plasma membrane. The products are released by exocytosis.

4 Proteins and lipids take on final form in the space inside the Golgi body. Different modifications allow them to be sorted out and shipped to their proper destinations.

3 Vesicles bud from the ER membrane and then transport unfinished proteins and lipids to a Golgi body.

2 In the membrane of smooth ER, lipids are assembled from building blocks delivered earlier.

1 Some polypeptide chains enter the space inside rough ER. Modifications begin that will shape them into the final protein form.

SECRETORY PATHWAY

assorted vesicles

Golgi body

smooth ER

rough ER

Some vesicles form at the plasma membrane, then move into the cytoplasm. These *endocytic* vesicles might fuse with the membrane of other organelles or remain intact, as storage vesicles.

Other vesicles bud from ER and Golgi membranes, then fuse with the plasma membrane. The contents of these *exocytic* vesicles are thereby released from the cell.

DNA instructions for building polypeptide chains leave the nucleus and enter the cytoplasm.

The chains (*green*) are assembled on ribosomes in the cytoplasm.

Figure 4.12 Cytomembrane system. This membrane system in the cytoplasm functions in the assembly, modification, packaging, and shipment of proteins and lipids. *Green* arrows highlight a secretory pathway by which certain proteins and lipids are packaged and then released from many types of cells. Among these are glandular cells that secrete mucus, sweat, and digestive enzymes.

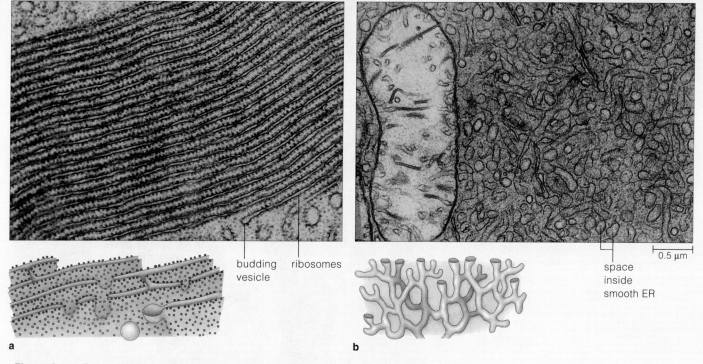

Figure 4.13 Transmission electron micrographs and sketches of endoplasmic reticulum. (**a**) Ribosomes pepper the flattened surfaces of rough ER that face the cytoplasm. (**b**) This section reveals the diameters of the many interconnected, pipelike regions of smooth ER. The large organelle at left is a mitochondrion.

budding vesicle   ribosomes

space inside smooth ER

0.5 µm

a   b

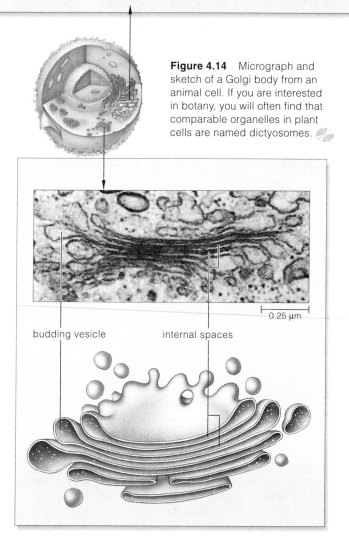

Figure 4.14 Micrograph and sketch of a Golgi body from an animal cell. If you are interested in botany, you will often find that comparable organelles in plant cells are named dictyosomes.

0.25 µm

budding vesicle   internal spaces

A Variety of Vesicles

Assorted vesicles move through the cytoplasm or take up positions inside it. Consider the **lysosome**, a type of vesicle that buds from the Golgi membranes of animal cells and some fungal cells. Lysosomes are organelles of intracellular digestion. They contain a potent brew, rich with enzymes that speed the breakdown of complex carbohydrates, proteins, nucleic acids, and some lipids. Often, lysosomes fuse with other vesicles that formed at the plasma membrane. As described in Section 5.6, such vesicles contain particles, bacteria, or other items that docked at the plasma membrane. Lysosomes may even digest whole cells or some of their parts. For example, when a tadpole is developing into an adult frog, its tail disappears. Lysosomal enzymes help destroy cells of the tail as part of a controlled program of development.

Or consider the **peroxisomes**. These sacs of enzymes break down fatty acids and amino acids. A potentially harmful product, hydrogen peroxide, forms during the reactions. But enzyme action converts it to water and oxygen or channels it into reactions that break down alcohol. After someone drinks alcohol, nearly half of it is degraded in peroxisomes of liver and kidney cells.

In the ER and Golgi bodies of the cytomembrane system, many proteins take on final form and lipids are synthesized.

Lipids, proteins (including enzymes), and other items become packaged in vesicles destined for export, storage, membrane building, intracellular digestion, and other cell activities.

MITOCHONDRIA

Recall, from Section 3.8, that ATP molecules are premier energy carriers. Energy associated with their phosphate groups can be delivered to nearly all reaction sites and drives nearly all cell activities. In eukaryotic cells, *many* ATP molecules can form when organic compounds are degraded in organelles called **mitochondria** (singular, mitochondrion). Figure 4.15 shows one of these organelles. The ATP-forming reactions require oxygen, and they can extract far more energy than can be done by any other means. When you breathe in, you are taking in oxygen primarily for all the mitochondria in your trillions of cells.

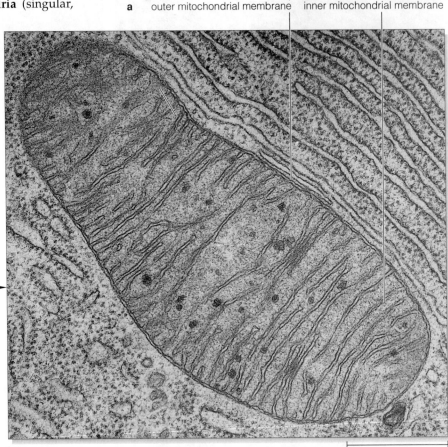

cristae

outer compartment

inner compartment

a   outer mitochondrial membrane   inner mitochondrial membrane

b   0.5 µm

Figure 4.15 (**a**) Sketch and (**b**) transmission electron micrograph of a thin slice through a typical mitochondrion. Reactions inside this organelle produce quantities of ATP, which is the major energy carrier between different reaction sites in cells. The mitochondrial reactions cannot proceed without oxygen.

A mitochondrion has a double-membrane system. As you can tell from Figure 4.7, the cytoplasm bathes the outermost membrane. Most often, the inner membrane repeatedly folds back on itself. Each inner fold is called a crista (plural, cristae). The double-membrane system creates two distinct compartments in the mitochondrion. Enzymes and other types of proteins embedded in the inner mitochondrial membrane or positioned near its surface function as operating equipment, used for the formation of ATP. Free oxygen helps keep the machinery running. Chapter 8 gives details of the reactions.

All eukaryotic cells have one or more mitochondria. Maybe you will find only one in a single-celled yeast. But you might find a thousand or more in energy-demanding cells, such as those of muscles. Take a look at the profusion of mitochondria in Figure 4.9, which is a micrograph of merely one thin slice from a liver cell. It alone tells you that the liver is an exceptionally active, energy-demanding organ.

In terms of size and biochemistry, mitochondria resemble bacteria. They even have their own DNA and some ribosomes, and they divide on their own. Possibly they evolved from ancient bacteria that were engulfed by a predatory, amoebalike cell yet managed to escape digestion. Perhaps they were able to reproduce inside the cell and its descendants. If they became permanent, protected residents, they may have lost structures and functions required for independent life while becoming mitochondria. We return to this topic later in the book, in Section 21.4.

The organelles called mitochondria are the ATP-producing powerhouses of all eukaryotic cells.

Energy-releasing reactions proceed at the compartmented, internal membrane system of mitochondria. The reactions, which require oxygen, produce far more ATP than can be made by any other cellular reactions.

SPECIALIZED PLANT ORGANELLES

Chloroplasts and Other Plastids

Many plant cells contain plastids, a general category of organelles that specialize in photosynthesis or function in storage. Three types are common in different parts of the plant. They are called chloroplasts, chromoplasts, and amyloplasts.

Chloroplasts occur only in photosynthetic, eukaryotic cells. Within these organelles, energy is harnessed from the sun's rays, ATP forms, and sugars and other organic compounds are synthesized from water and carbon dioxide. Chloroplasts commonly are oval or disk-shaped. They have two outer membrane layers, one wrapped over the other. The layers surround a semifluid interior, the stroma. Here, another membrane forms an inner system of interconnecting, disk-shaped compartments (Figure 4.16). The compartments are stacked together in many chloroplasts. Each stack is a granum (plural, grana).

The first stage of photosynthesis proceeds at light-trapping pigments, enzymes, and other proteins of the inner membrane system. That is where light energy is absorbed and ATP forms. The synthesis stage proceeds in the stroma. There, sugars and other organic compounds are put together. There also, starch grains (clusters of new starch molecules) may be temporarily stored.

Of the photosynthetic pigments, chlorophylls (green) are the most abundant, followed by carotenoids (yellow, orange, and red). As described in Section 7.3, the relative abundances of various pigments in plant parts influence their coloration. Some parts may be green, others golden brown, and so on.

In many ways, chloroplasts resemble photosynthetic bacteria. Like mitochondria, they might have evolved from bacteria that were engulfed by predatory cells, yet escaped digestion and became permanent residents in them. We return to this idea in Section 21.4.

Unlike chloroplasts, chromoplasts have an abundance of carotenoids but no chlorophylls. They are the source of red-to-yellow colors of many flowers, autumn leaves, ripening fruits, and carrots or other roots. The pigments also may visually attract animals that pollinate plants or that disperse seeds. Amyloplasts have no pigments. They often store starch grains and are abundant in cells of stems, potato tubers, and many seeds.

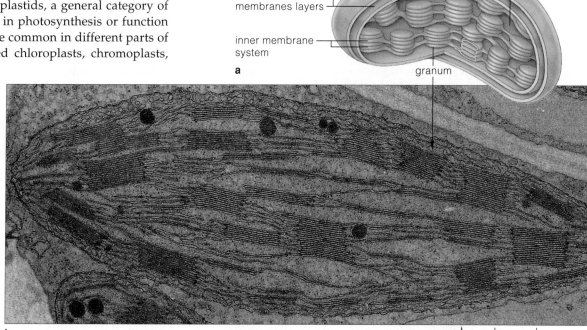

stroma (fluid interior)

two outermost membranes layers

inner membrane system

a

granum

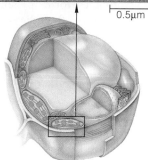

0.5μm

Figure 4.16 (**a**) Generalized sketch of the chloroplast, the key defining feature of every photosynthetic eukaryotic cell. (**b**) This transmission electron micrograph shows a chloroplast, thin section.

b

Central Vacuole

Many mature, living plant cells have a **central vacuole** (Figure 4.7). This fluid-filled organelle stores amino acids, sugars, ions, and toxic wastes. As it enlarges, it causes fluid pressure to build up inside the cell and so forces the cell's still-pliable cell wall to enlarge. The cell itself enlarges under the force, and its increased surface area enhances the rate at which substances can cross the plasma membrane. In most cases, the central vacuole increases so much in volume that it takes up 50 to 90 percent of the cell's interior. The cytoplasm ends up as a very narrow zone between the central vacuole and the plasma membrane.

Photosynthetic eukaryotic cells contain chloroplasts and other plastids that function in food production and storage.

Many plant cells have a central vacuole. When this storage vacuole enlarges during growth, cells are forced to enlarge, and this increases the surface area available for absorption.

An interconnected system of fibers, threads, and lattices extends between the nucleus and plasma membrane of eukaryotic cells. This system, the **cytoskeleton**, gives cells their internal organization, shape, and capacity to move. Some elements reinforce the plasma membrane and nuclear envelope; others form scaffolds that hold protein clusters in specific membranes or cytoplasmic regions. Still others are like railroad tracks upon which organelles are shipped from one site to another!

The micrograph in Figure 4.17 shows an animal cell, isolated from its home tissue, as it was stretching out from left to right across a glass slide. The internal mesh of filaments only hints at the extent and intricacy of this cell's cytoskeleton. In this case, researchers were studying its **microtubules** and **microfilaments**. These two classes of cytoskeletal elements, acting singly or collectively, underlie nearly all movements of eukaryotic cells. Some animal cells also have **intermediate filaments**, a class of ropelike cytoskeletal elements that impart mechanical strength to cells and tissues. Many of the elements are permanent. Others appear only at certain times in a cell's life. Before a cell divides, for instance, many new microtubules form a "spindle" structure that moves chromosomes, then disassemble when the task is done.

Microtubules

Figure 4.18*a* shows part of a microtubule. This hollow cylinder, the cytoskeleton's largest structural element, is twenty-five nanometers wide. Many tubulin subunits make up the cylinder. Tubulin is a protein consisting of two chemically distinct polypeptide chains, each folded into the shape of a ball. While a microtubule is being assembled, all the subunits become oriented in the same direction. The uniform orientation puts slightly different chemical and electrical properties at opposite ends of the growing microtubule. The difference causes one end (the *plus* end) to elongate much faster than the other, *minus* end, which tends to lose subunits if not stabilized. The minus end typically gets anchored in loose material in the cytoplasm or at an **MTOC** (microtubule organizing center). MTOCs are sites of dense material that give rise to quantities of microtubules. You will be reading about various types, which go by such names as centrioles, centrosomes, and kinetochores. Once a newly forming microtubule is anchored, its plus end rapidly grows in a prescribed direction. For example, this is how many microtubules form an organized spindle apparatus.

Microtubules govern the division of cells and some aspects of their shape as well as many cell movements. Cells can't do much without them. As you might well imagine, microtubules have become prime targets in the chemical warfare between vulnerable species and their attackers. For example, autumn crocus and other plants of the genus *Colchicum* synthesize colchicine. This poison inhibits the assembly and promotes the disassembly of microtubules, with dire effects on browsing animals. The plant cells themselves are insensitive to colchicine, which cannot bind well to their tubulin subunits.

Taxol, a poison produced by the western yew (*Taxus brevifolia*), stabilizes microtubules that already exist in a cell and so prevents formation of new ones that the cell may require for other tasks. For example, like colchicine and some other microtubule poisons, taxol interferes with the formation of microtubular spindles, so it can inhibit cell division. Doctors have used it to reduce the uncontrolled cell divisions that underlie the growth of some benign and malignant tumors.

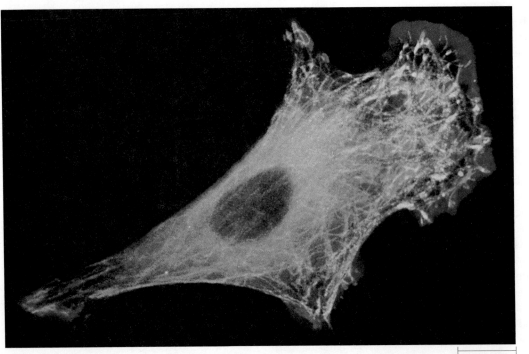

10 μm

Figure 4.17 A view of many of the structural elements that make up the cytoskeleton of a fibroblast, a type of animal cell that gives rise to certain animal tissues. This composite of three micrograph images reveals the presence of microtubules (tinted *green* in this image). Two kinds of microfilaments are tinted *blue* and *red*.

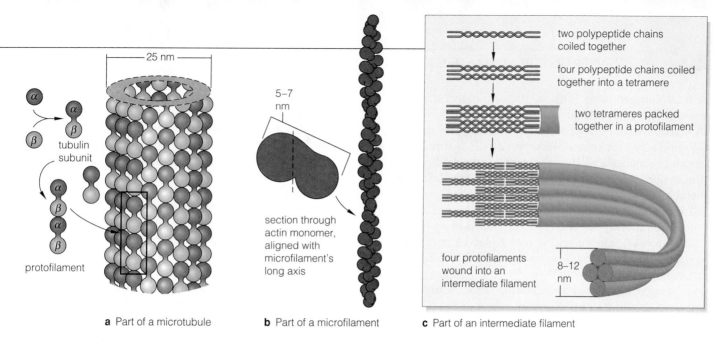

a Part of a microtubule   **b** Part of a microfilament   **c** Part of an intermediate filament

Figure 4.18 Structural organization of (**a**) microtubules, (**b**) microfilaments, and (**c**) an intermediate filament.

Microfilaments

Microfilaments, the thinnest of cytoskeletal elements, are only five to seven nanometers wide (Figure 4.18*b*). Each consists of two helically twisted, polypeptide chains that are assembled from U-shaped actin monomers. Like tubulins, actins are added preferentially to one end of a growing microfilament and removed from the other end.

As you will see in several chapters throughout this book, microfilaments take part in a great variety of movements, especially the kinds that proceed at the cell surface. They also contribute to the development and maintenance of animal cell shapes.

Myosin and Other Accessory Proteins

Both tubulin and actin have been highly conserved over evolutionary time; the microtubules and microfilaments of all eukaryotic species are assembled from them. In spite of the uniformity, monomers of a variety of other proteins can become attached to them. Thus embellished, microtubules and microfilaments that were assembled according to the same basic pattern can serve different functions in different regions of the cell.

For example, monomers of myosin, dynein, or some other "motor protein" typically extend from the surface of microtubules and microfilaments that have roles in cell movement. Myosin is abundant in muscle cells, and it is part of the machinery by which they contract. "Crosslinking proteins" splice adjacent microfilaments together. Certain kinds take part in the formation of the cell cortex, an extensive, three-dimensional network of microfilaments and other proteins just beneath the plasma membrane. The cortex reinforces the cell surface and facilitates movements as well as changes in shape. Other crosslinking proteins splice microfilaments in a

stable network having the properties of a gel. Spectrin, another accessory protein, attaches microfilaments to the plasma membrane. As a final example, the integrins span this outermost membrane, attach to microfilaments inside the cell, and connect them with proteins outside.

Intermediate Filaments

Intermediate filaments, the most stable elements of the cytoskeleton, are between eight and twelve nanometers wide (Figure 4.18*c*). The six known groups mechanically strengthen cells or cell parts and help maintain their shape. For example, the desmins and vimentins help support machinery by which muscle cells contract. The lamins help form a scaffold that reinforces the nucleus. Diverse cytokeratins structurally reinforce the cells that give rise to nails, claws, horns, and hairs.

Unlike the other two classes of cytoskeletal elements, the intermediate filaments occur only in animal cells of specific tissues. Moreover, because each cell usually has only one or sometimes two kinds, researchers can use intermediate filaments to identify cell type. Such typing has proved to be a useful tool in diagnosing the tissue origin of different forms of cancer.

Every eukaryotic cell has a cytoskeleton, the diverse elements of which are the basis of its shape, its internal structure, and its capacity for movements.

Microtubules are key organizers of the cytoskeleton and help move certain cell structures. Microfilaments take part in diverse movements and in the formation and maintenance of cell shape. Intermediate filaments structurally reinforce cells and internal cell structures.

THE STRUCTURAL BASIS OF CELL MOTILITY

If a cell lives, it moves. This is true of free-living single cells, such as the protistan and sperm cells in Figure 4.19. It is true also of cells with fixed positions in the tissues of multicelled organisms. Eukaryotic cells rearrange or shunt organelles and chromosomes, contract (shorten), thrust out long or fat lobes of cytoplasm, stir fluids, or bend a tail and propel the cell body forward. These and all other movements of cell structures, the cell itself, and entire multicelled organisms are forms of motility.

Section 4.8 introduced the major players in cellular motility—the microtubules, microfilaments, and other proteins that interact with them. Energy inputs, as from ATP, trigger motions at the molecular level. Many such motions combine to produce movement at the cellular level, as when a muscle cell contracts. The coordinated movements of many cells cause movements of tissues or organs, as when bundles of skeletal muscle cells interact with bones to make a thumb and finger turn a page.

Mechanisms of Cell Movements

Microfilaments, microtubules, or both take part in most aspects of motility. They do so by three mechanisms.

First, *the length of a microtubule or microfilament can grow or diminish by the controlled assembly or disassembly of its subunits*. When either one lengthens or shortens, a chromosome or some other structure attached to it is pushed or dragged through the cytoplasm.

Also, the rapid assembly of microfilaments produces dynamic surface extensions by which some cells crawl about. *Amoeba proteus*, a soft-bodied protistan, crawls on **pseudopods** ("false feet"; they actually are lobelike protrusions from the cell body). Rapid polymerization may distend the network of microfilaments beneath its plasma membrane. The animal cell in Figure 4.17 was migrating on sheetlike distensions of such a network.

Second, *parallel arrays of microfilaments or microtubules actively slide in controlled directions*. Consider a muscle cell. It contains a series of contractile units, each with parallel arrays of microfilaments and of motor proteins. Briefly, ATP activates the motor proteins (polymers of myosin), which have oarlike projections that bind and release the adjacent microfilaments in a series of power strokes. The repeated action causes the microfilaments to slide past them. Cumulatively, the directional sliding and shortening of all contractile units shortens the cell.

A similar sliding mechanism might be operating as cells crawl about. If the microfilament network beneath the plasma membrane at the cell's trailing end contracts, then some cytoplasmic gel would be squeezed into the leading end, which would bulge forward in response.

Third, *microtubules or microfilaments shunt organelles or portions of the cytoplasm from one location to another*. For example, chloroplasts in photosynthetic cells move to different light-intercepting positions in response to the changing angle of the sun's overhead position. Myosin monomers are "walking" over microfilaments and are carrying the attached chloroplasts with them. This shunt mechanism also causes cytoplasmic streaming: an active, directional, and often rapid flowing of components in cytoplasmic gel. Observe a living plant cell with a light microscope, and you may see pronounced streaming.

Case Study: Flagella and Cilia

To biologists, the **flagellum** (plural, flagella) and **cilium** (plural, cilia) are classic examples of cell motility. As Figure 4.20 shows, both motile structures contain a ring of nine pairs of microtubules around a central pair. A system of spokes and links holds the "9 + 2 array" together. A sliding mechanism operates in the outer ring of the array to bend the motile structure (Figure 4.21).

Each 9 + 2 array arises from a **centriole.** This barrel-shaped structure is one type of MTOC. The centriole remains at the base of the completed array, where it is often called a **basal body** (Figure 4.20).

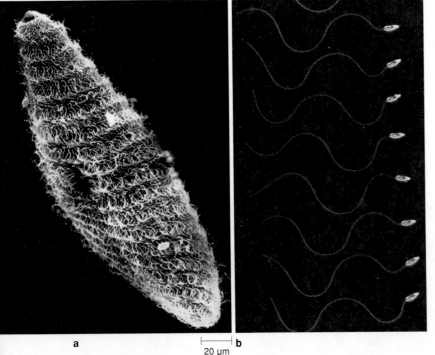

a   ⊢——⊣ b
20 µm

Figure 4.19 Two free-living, motile cells. (**a**) Scanning electron micrograph of *Paramecium*, a heterotrophic protistan. Multiple rows of cilia at its surface beat in a synchronized way. The beating propels the cell through its aquatic habitat. It also sweeps food-laden water into the gullet, visible at left as a depression in the cell surface. (**b**) Time-lapse micrographs arranged to show the beating of a sperm's flagellum. Wavelike motions start at the flagellum's base and move to its tip.

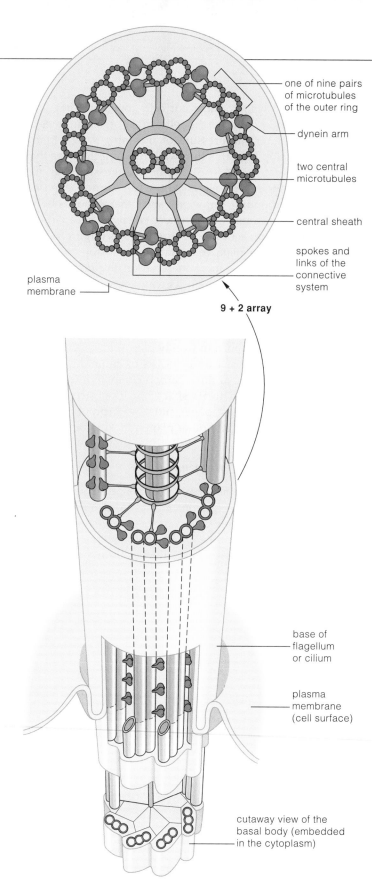

one of nine pairs
of microtubules
of the outer ring

dynein arm

two central
microtubules

central sheath

spokes and
links of the
connective
system

plasma
membrane

9 + 2 array

base of
flagellum
or cilium

plasma
membrane
(cell surface)

cutaway view of the
basal body (embedded
in the cytoplasm)

Figure 4.20 Internal organization of flagella and cilia. Both motile structures have a system of microtubules and a connective system of spokes and linking elements. Nine pairs (doublets) of the microtubules are arranged as an outer ring around two central microtubules. This is called a 9 + 2 array.

Figure 4.21 Model of a sliding mechanism responsible for the beating of flagella and cilia, as proposed by cell biologists K. Summers and R. Gibbons. Inside an unbent flagellum (or cilium), all microtubule doublets extend the same distance into the tip. When the flagellum bends, the doublets on the side that is bending the most are being displaced farthest from the tip, as shown here:

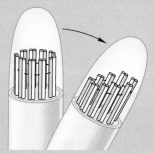

A sliding mechanism operates between the microtubule doublets of the outer ring. A series of motor proteins (dynein) form short arms that extend from each doublet toward the next doublet in the ring. Inputs of ATP energy cause the arms of one doublet to attach to the doublet in front of them, tilt in a short power stroke that pulls on the attached doublet, then release their hold. Repeated power strokes and cross-bridgings force the doublet to slide in a direction toward the base of the flagellum. As the attached doublet moves, *its* arms force the *next* doublet in line to slide down a bit, which forces the *next* doublet to slide down a bit also, and so on in sequence.

The system of interconnecting spokes and links extends through the length of the flagellum and prevents doublets from sliding totally out of the 9 + 2 array. This restriction forces the entire flagellum to bend to accommodate the internal displacement of the sliding doublets.

In short, in the 9 + 2 array of flagella and cilia, energized motor proteins make the nine doublets slide in sequence, and restrictions imposed by a system of spokes and links convert the doublet sliding into a bending motion.

The protistan cell in Figure 4.19*a* bears cilia, which typically are shorter and far more profuse than flagella. Sperm and many other free-living cells use flagella as whiplike tails to move in fluid environments (Figure 4.19*b*). In multicelled organisms, some ciliated epithelial cells stir their surroundings. For example, thousands line airways of your respiratory tract. Their beating cilia direct mucus-trapped particles away from the lungs.

Cell contractions and migrations, chromosome movements, and all other forms of cell motility arise at organized arrays of microtubules, microfilaments, and accessory proteins.

Different mechanisms cause these cytoskeletal elements to assemble or disassemble, slide past one another, and shunt structures to new locations.

This survey of eukaryotic cells concludes with a look at cell walls and some other specialized surface structures. Many of these architectural marvels are constructed of various secretions from the cells themselves. Others are cytoplasmic bridges or sets of membrane proteins that connect neighboring cells and allow them to interact.

Eukaryotic Cell Walls

Single-celled eukaryotic species are directly exposed to their surroundings. Many have a **cell wall**, an outermost structure that wraps continuously around the plasma membrane. The protistan shown in Figure 4.22 is an example. Cell walls help support and protect their owners. Because walls are porous, water and solutes can easily move to and from the plasma membrane. Walls also occur on various cells of plants and fungi.

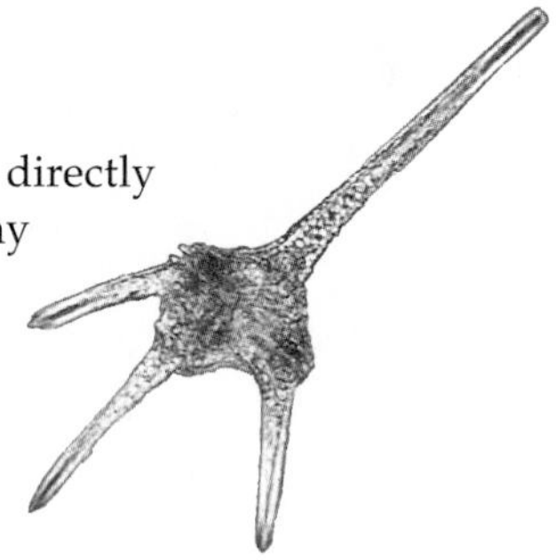

For example, young plant cells in actively growing regions secrete pectin and hemicellulose (two gluelike polysaccharides), glycoproteins, and cellulose. The cellulose molecules join into ropelike strands that become embedded in the gluey matrix. Together, the secretions form a **primary wall** (Figure 4.23*a*). Primary walls are sticky, and they

Figure 4.22 From a freshwater habitat, a single-celled, walled protistan (*Ceratium hirundinella*, one of the dinoflagellates).

cement the walls of adjacent cells together. They also are thin and pliable, so the cell surface area continues to enlarge under the pressure of incoming water. Many cells develop only this thin wall; they are the ones that retain the capacity to divide or change shape as the plant grows and develops. At the cell surfaces exposed to the surrounding air, waxes and other deposits build up as a semitransparent protective covering, a cuticle, that helps reduce evaporative water loss (Figure 4.24*a*).

When other plant cells mature, they stop enlarging and start secreting more material on the *inside* of the primary wall. These deposits combine to form a rigid, **secondary wall** that reinforces cell shape (Figure 4.23*c*). Whereas cellulose made up less than 25 percent of the primary wall, the added deposits now contribute more to structural support. In woody plants, up to 25 percent of the secondary cell wall consists of lignin. Lignin is a complex phenolic (a three-carbon chain and an oxygen atom are attached to its six-carbon ring structure). It makes plant parts stronger, more waterproof, and less inviting to insects and other plant-attacking organisms.

Matrixes Between Animal Cells

Although animal cells have no walls, diverse matrixes composed of cell secretions and even materials drawn from the surroundings intervene between many of them. Think of cartilage at the knobby ends of your leg bones.

Cartilage consists of scattered cells and collagen or elastin fibers of their own secretion, all embedded within a "ground substance" of secreted, modified polysaccharides. Similarly, mature bone is mostly an extensive intercellular matrix (Figure 4.24*b*).

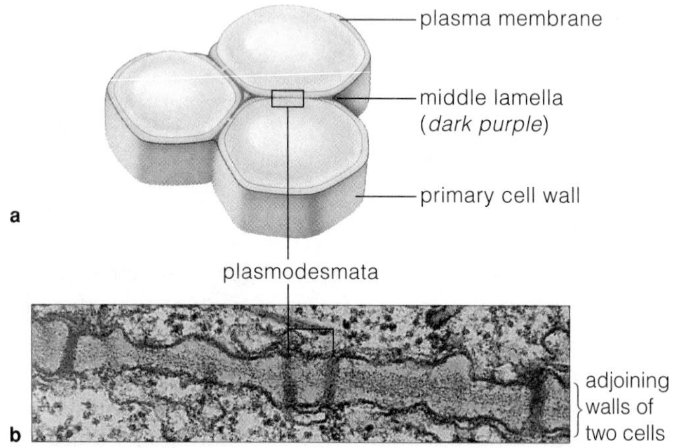

plasma membrane

middle lamella (*dark purple*)

primary cell wall

a

plasmodesmata

adjoining walls of two cells

b

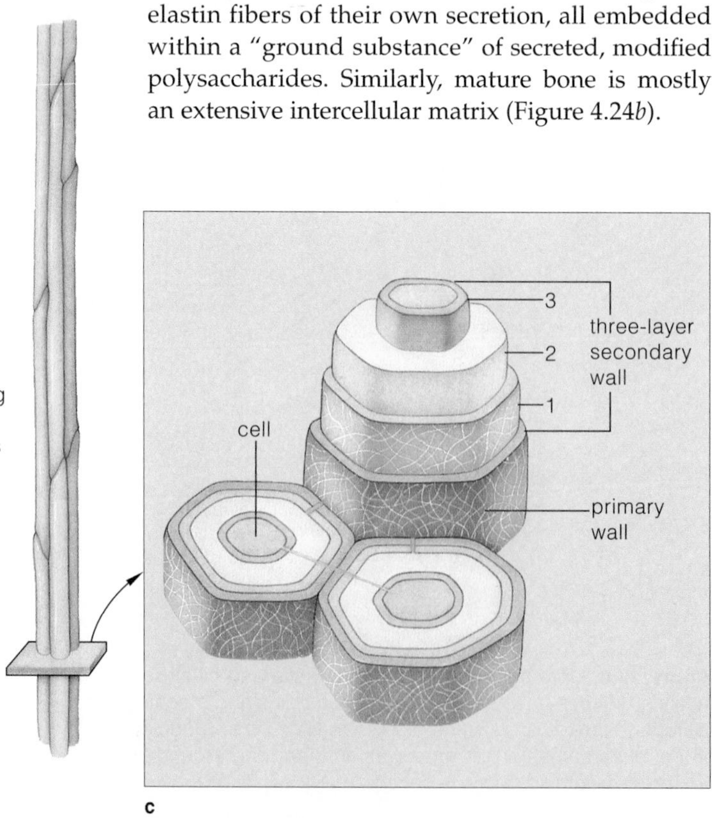

Figure 4.23 (**a**) Primary cell wall of young cells in plant tissues. Cell secretions form the middle lamella, a layer between the walls of adjoining cells. The layer is thickest in adjoining corners. (**b**) Plasmodesmata, which are membrane-lined channels across the adjacent walls, connect the cytoplasm of neighboring cells. (**c**) In many types of cells, such as the long fiber cells shown at right, more layers become deposited inside the primary wall. These stiffen the wall and help maintain its shape. Later, the cell may die, leaving the stiffened walls behind. This also is true of the water-conducting pipelines that thread through most plant tissues. They are "tubes" of many interconnected, stiffened cell walls.

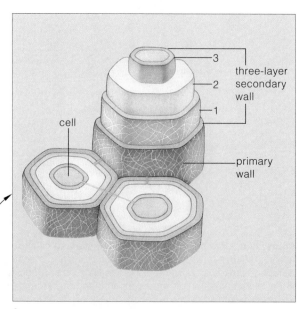

three-layer secondary wall

3

2

1

cell

primary wall

c

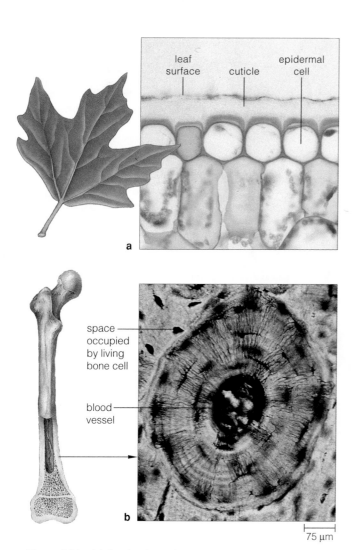

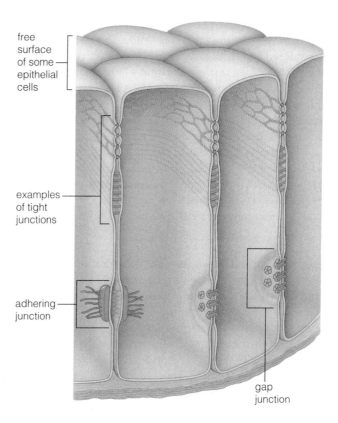

Figure 4.25 Composite of types of cell junctions in animals.

Figure 4.24 (**a**) Section through a plant cuticle, a surface layer composed of cell secretions. (**b**) Section through compact bone tissue, stained for microscopy.

Cell-to-Cell Junctions

Even when a wall or some other structure imprisons a cell in its own secretions, the only contact that cell has with the outside world is *through* its plasma membrane. Certain components of that membrane project outward, into adjacent cells as well as the surrounding medium. Among those components are junctions where the cell sends and receives diverse signals and materials, where it recognizes and cements itself to kindred cells.

In plants, for instance, numerous tiny channels cross the adjacent primary walls of living cells and connect their cytoplasm. Figure 4.23*b* shows a few. Each channel is a **plasmodesma** (plural, plasmodesmata). The plasma membranes of the adjoining cells have merged to fully line the channels, so there can be an uninterrupted flow of substances between cells. Thus, all living cells in the plant body have the potential to exchange substances.

In most animal tissues, three categories of cell-to-cell junctions are common (Figure 4.25). *Tight* junctions link the cells of epithelial tissues, which line the body's outer surface and internal cavities and organs. They seal adjoining cells together so water-soluble substances can't leak between them. Thus gastric fluid cannot leak across the stomach's lining and damage surrounding tissues. *Adhering* junctions join cells in tissues of the skin, heart, and other organs subject to stretching. *Gap* junctions, which link the cytoplasm of neighboring cells, are open channels for the rapid flow of signals and substances.

We will be returning to the cell walls, intercellular substances, and cell-to-cell interactions in later chapters. For now, these are the points to remember:

A variety of protistan, plant, and fungal cells have a porous but protective wall that surrounds the plasma membrane. The cells themselves secrete the wall-forming materials.

Secretions from the surface of certain cells form the cuticles of plants and some animals, the extracellular matrixes in animal tissues, and other specialized structures.

In multicelled organisms, coordinated cell activities depend on cell-to-cell junctions, which are protein complexes or cytoplasmic bridges that serve as physical links and forms of communication between cells.

PROKARYOTIC CELLS—THE BACTERIA

We turn now to the bacteria. Unlike the cells we have considered so far, all species of bacteria are prokaryotic; their DNA is *not* enclosed within a nucleus. *Prokaryotic* means "before the nucleus." Microbiologists selected this word as a reminder that bacteria existed on Earth before the nucleus appeared in the forerunners of all eukaryotic cells.

With only one recently discovered exception, bacteria are the smallest of all cells. They usually are not much more than one micrometer wide; even rod-shaped species are only about a few micrometers long. In structural terms, they

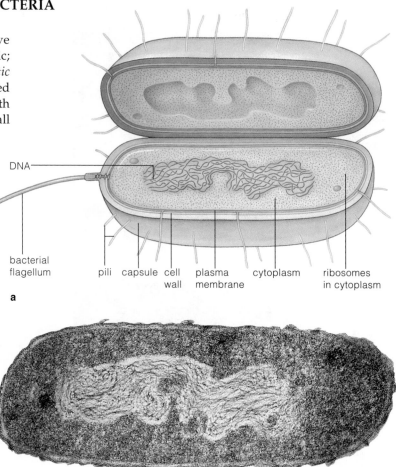

are the simplest kinds of cells to think about. Most species have a semirigid or rigid cell wall that surrounds the plasma membrane, physically supports the cell, and imparts shape to it (Figure 4.26a). Dissolved substances can still move freely to and from the plasma membrane, for the wall is porous. Sticky polysaccharides cover the cell wall of many species. They help a bacterium attach to rocks, teeth, the vagina, and many other interesting surfaces. In many disease-causing (pathogenic) species, the polysaccharides form a thick, jellylike capsule that surrounds and helps protect the wall.

As is true of eukaryotic cells, bacteria have a plasma membrane that controls the movements of substances to and from the cytoplasm. Their plasma membrane, too, contains proteins that function as channels, transporters, and receptors. It incorporates built-in machinery for metabolic reactions, such as the breakdown of energy-rich compounds. In the photosynthetic types, clusters of some membrane proteins harness light energy and convert it to the chemical energy of ATP.

Bacterial cells are too small to contain more than a small volume of cytoplasm, but they have many ribosomes upon which polypeptide chains are assembled. Apparently, these cells are small enough and so internally simple that they do not require a cytoskeleton.

The cytoplasm of a bacterial cell is distinct from that of eukaryotic cells in being continuous with an irregularly shaped region of DNA. Membranes do not surround this region, which is named a **nucleoid**. As Figure 4.26c indicates, a single, circular molecule of DNA occupies this region.

Extending from the surface of many bacterial cells are one or more threadlike motile structures known as **bacterial flagella** (singular, flagellum). These are not the same as eukaryotic flagella, for the 9 + 2 array of microtubules is absent. Bacterial flagella help a cell

DNA

bacterial flagellum | pili | capsule | cell wall | plasma membrane | cytoplasm | ribosomes in cytoplasm

a

b   0.5 µm

Figure 4.26 (**a**) Generalized prokaryotic body plan. (**b**) Micrograph of *Escherichia coli*.

Your own gut is home to a large population of a normally harmless strain of *E. coli*. In 1993, a harmful strain of contaminated meat that was sold to some fast-food restaurants. The same strain also contaminated hard apple cider sold at a few roadside stands. Cooking the meat thoroughly or boiling the cider would have killed the bacterial cells. Where this was not done, people who ate the meat or drank the cider became quite sick. Some died.

Facing page: (**c**) Researchers manipulated this *E. coli* cell to release its single, circular molecule of DNA. (**d**) Cells of different bacterial species are shaped like balls, rods, or corkscrews. The ball-shaped cells of *Nostoc*, a photosynthetic bacterium, stick together inside a thick, gelatinlike sheath of their own secretions. Chapter 22 gives other splendid examples. (**e**) Like this *Pseudomonas marginalis* cell, many species have one or more bacterial flagella, motile structures that propel the cell through fluid environments.

move rapidly through the fluid surroundings. Other surface projections include pili (singular, pilus), which are protein filaments that help many kinds of bacterial cells attach to various surfaces, even to one another.

Taken as a whole, the bacteria are certainly the most metabolically diverse organisms. They have managed to exploit energy and raw materials in just about every kind of environment. Besides this, ancient prokaryotes gave rise to all the protistans, plants, fungi, and animals

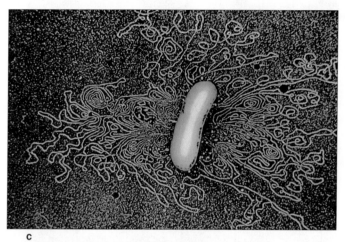

c

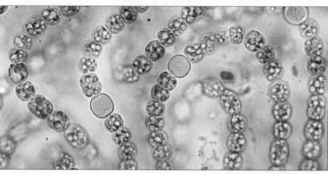

d

1 µm

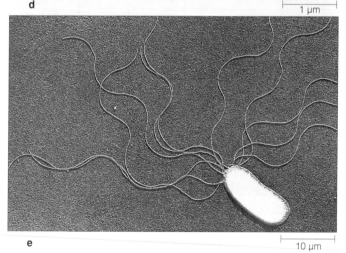

e

10 µm

ever to appear on Earth. The evolution, structure, and functioning of these remarkable cells are topics of later chapters.

Bacteria alone are prokaryotic cells; their DNA is not housed inside a nucleus. Most species have a cell wall around the plasma membrane. Generally, their cytoplasm does not have any organelles comparable to those of eukaryotic cells.

Bacteria are the simplest cells but as a group show the most metabolic diversity. Their metabolic activities proceed at the plasma membrane and at many ribosomes in the cytoplasm.

---

1. Three generalizations constitute the cell theory:
   a. All living things are composed of one or more cells.
   b. The cell is the smallest entity that retains the properties of life. That is, it either lives independently or has a built-in, genetic capacity to do so.
   c. New cells arise only from cells that already exist.

2. At the minimum, a newly formed cell has a plasma membrane, a region of cytoplasm, and a region of DNA.
   a. The plasma membrane (a thin, outer membrane) maintains the cell as a distinct, separate entity. It allows metabolic events to proceed apart from random events in the environment. Many substances and signals are continually moving across it, in highly controlled ways.
   b. The cytoplasm is all the fluids, ribosomes, structural elements, and (in eukaryotic cells) organelles between the plasma membrane and the region of DNA.

3. Membranes are crucial to cell structure and function. They consist of lipids (phospholipids, for the most part) and proteins. The lipids are arrayed as two layers, with all the hydrophobic tails of both layers sandwiched in between all the hydrophilic heads. This lipid bilayer imparts structure to the membrane and bars passage of water-soluble substances across it. Diverse proteins are embedded in the bilayer or attached to its surfaces.

4. Proteins carry out most cell membrane functions. For example, many serve as channels or pumps that allow or promote passage of water-soluble substances across the lipid bilayer. Others are receptors for extracellular substances that trigger changes in cell activities.

5. Cell membranes divide the cytoplasm of eukaryotic cells into functional compartments called organelles. Prokaryotic cells do not have comparable organelles.

6. Organelle membranes separate metabolic reactions in the space of the cytoplasm and allow different kinds to proceed in orderly fashion. (In bacteria, many similar reactions proceed at the plasma membrane.)
   a. The nuclear envelope functionally separates the DNA from the metabolic machinery of the cytoplasm.
   b. The cytomembrane system includes the ER, Golgi bodies, and vesicles. Many new proteins are modified into final form and lipids are assembled in this system. Finished products are packaged and then shipped off to destinations inside or outside the cell.
   c. Mitochondria are specialists in oxygen-requiring reactions that produce many ATP molecules.
   d. Chloroplasts trap sunlight energy and produce organic compounds in photosynthetic eukaryotic cells.

7. The cytoskeleton of a eukaryotic cell functions in cell shape, internal organization, and movements.

8. Table 4.2 on the next page summarizes the defining features of both prokaryotic and eukaryotic cells.

Table 4.2 Summary of Typical Components of Prokaryotic and Eukaryotic Cells

Cell Component	Function	PROKARYOTIC	EUKARYOTIC			
		Archaebacteria, Eubacteria	Protistans	Fungi	Plants	Animals
Cell wall	Protection, structural support	✔*	✔*	✔	✔	None
Plasma membrane	Control of substances moving into and out of cell	✔	✔	✔	✔	✔
Nucleus	Physical separation and organization of DNA	None	✔	✔	✔	✔
DNA	Encoding of hereditary information	✔	✔	✔	✔	✔
RNA	Transcription, translation of DNA messages into polypeptide chains of specific proteins	✔	✔	✔	✔	✔
Nucleolus	Assembly of subunits of ribosomes	None	✔	✔	✔	✔
Ribosome	Protein synthesis	✔	✔	✔	✔	✔
Endoplasmic reticulum (ER)	Initial modification of many of the newly forming polypeptide chains of proteins; lipid synthesis	None	✔	✔	✔	✔
Golgi body	Final modification of proteins, lipids; sorting and packaging them for use inside cell or for export	None	✔	✔	✔	✔
Lysosome	Intracellular digestion	None	✔	✔*	✔*	✔
Mitochondrion	ATP formation	**	✔	✔	✔	✔
Photosynthetic pigment	Light–energy conversion	✔*	✔*	None	✔	None
Chloroplast	Photosynthesis; some starch storage	None	✔*	None	✔	None
Central vacuole	Increasing cell surface area; storage	None	None	✔*	✔	None
Bacterial flagellum	Locomotion through fluid surroundings	✔*	None	None	None	None
Flagellum or cilium with 9 + 2 microtubular array	Locomotion through or motion within fluid surroundings	None	✔*	✔*	✔*	✔
Cytoskeleton	Cell shape; internal organization; basis of cell movement and, in many cells, locomotion	None	✔*	✔*	✔*	✔

* Known to occur in at least some groups.

** Oxygen-requiring (aerobic) pathways of ATP formation do occur in many groups, but mitochondria are not involved.

Review Questions

1. Label the organelles in this diagram of a plant cell. *4.3*

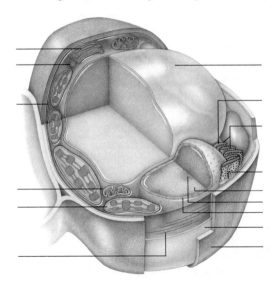

2. Label the organelles in this diagram of an animal cell. *4.3*

3. State the three key points of the cell theory. *CI*

4. Describe three features that all cells have in common. After reviewing Table 4.2, write a paragraph on the key differences between prokaryotic and eukaryotic cells. *4.1, 4.3, 4.11*

5. Suppose you want to observe the three-dimensional surface of an insect's eye. Would you benefit most by using a compound light microscope, transmission electron microscope, or scanning electron microscope? *4.2*

6. Briefly characterize the structure and function of the nucleus, the nuclear envelope, and the nucleolus. *4.4*

7. Define chromosome and chromatin. Do chromosomes always have the same appearance during a cell's life? *4.4*

8. Which organelles are part of the cytomembrane system? *4.5*

9. Is this statement true or false: Plant cells have chloroplasts, but not mitochondria. Explain your answer. *4.6, 4.7*

10. What are the functions of the central vacuole in mature, living plant cells? *4.7*

11. Define cytoskeleton. How does it aid in cell functioning? *4.8*

12. Are all components of the cytoskeleton permanent? *4.8, 4.9*

13. What gives rise to the 9 + 2 array of cilia and flagella? *4.9*

14. Cell walls are typical of which organisms: animals, plants, fungi, protistans, or bacteria? Are the walls solid or porous? *4.10*

15. In certain plant cells, is a secondary wall deposited inside or outside the surface of the primary wall? *4.10*

16. In multicelled organisms, coordinated interactions depend on linkages and communications between cells. What types of junctions occur between adjacent animal cells? Plant cells? *4.10*

Self-Quiz *(Answers in Appendix IV)*

1. Cell membranes consist mainly of a _____ .
   a. carbohydrate bilayer and proteins
   b. protein bilayer and phospholipids
   c. lipid bilayer and proteins
   d. none of the above

2. Organelles _____ .
   a. are membrane-bound compartments
   b. are typical of eukaryotic cells, not prokaryotic cells
   c. separate chemical reactions in time and space
   d. all of the above are features of the organelles

3. Cells of many protistans, plants, and fungi, but not animals, commonly have _____ .
   a. mitochondria   c. ribosomes
   b. a plasma membrane   d. a cell wall

4. Is this statement true or false: The plasma membrane is the outermost component of all cells. Explain your answer.

5. Unlike eukaryotic cells, prokaryotic cells _____ .
   a. lack a plasma membrane   c. do not have a nucleus
   b. have RNA, not DNA   d. all of the above

6. Match each cell component with its function.
   ____ mitochondrion   a. synthesis of polypeptide chains
   ____ chloroplast   b. initial modification of new
   ____ ribosome   polypeptide chains
   ____ rough ER   c. final modification of proteins; lipid
   ____ Golgi body   synthesis; sorting and shipping tasks
                        d. photosynthesis
                        e. site of oxygen-requiring pathway
                        of ATP formation

Critical Thinking

1. Why is it likely that you will never encounter a predatory two-ton living cell on the sidewalk?

2. Many compound light microscopes have blue filters. Think of the spectrum of visible light (as in Figure 7.3) and explain why blue light is efficient when viewing objects at high magnification.

3. Your biology professor shows you an electron micrograph of a cell that contains very large numbers of mitochondria and Golgi bodies. You notice that this particular cell also has a great deal of rough endoplasmic reticulum. What kinds of cellular activities would require such an abundance of the three kinds of organelles?

4. Do the events that proceed in the nucleus and cytoplasm of eukaryotic cells seem too remote from topics of human interest? Then try starting with the human body and moving down to the cellular level. For example, *cystic fibrosis* is a disabling and eventually fatal genetic disorder. Glands of an affected person secrete far more than they should, with far-reaching effects. In time, digestive enzymes clog a duct between the pancreas and small intestine, food cannot be digested properly, and even if food intake increases, malnutrition results. Cysts form in the pancreas, which degenerates and becomes fibrous (hence the name of the disorder). Thick mucus builds up in the respiratory tract, so affected individuals have difficulty expelling airborne bacteria and particles that enter the lungs. Also, too much sweat forms. Midwives used to lick the forehead of newborns, and if they tasted far too much salt in the sweat, they predicted the new individual would develop lung congestion.

   A defective form of a protein is implicated in the disorder. The protein is a plasma membrane component of cells of glands that secrete mucus, digestive enzymes, and sweat. After reviewing Sections 4.4 and 4.5, track the disorder back from the affected protein to its source at the cellular level.

Selected Key Terms

bacterial flagellum *4.11*	ER (endoplasmic reticulum) *4.5*	nucleolus *4.4*
basal body *4.9*	eukaryotic cell *4.1*	nucleus *4.4*
cell *4.1*	flagellum *4.10*	organelle *4.3*
cell theory *CI*	Golgi body *4.5*	peroxisome *4.5*
cell wall *4.10*	intermediate filament *4.8*	plasma membrane *4.1*
central vacuole *4.7*	lipid bilayer *4.1*	plasmodesma *4.10*
centriole *4.9*	lysosome *4.5*	primary wall *4.10*
chloroplast *4.7*	microfilament *4.8*	prokaryotic cell *4.1*
chromatin *4.4*	micrograph *4.2*	pseudopod *4.9*
chromosome *4.4*	microtubule *4.8*	ribosome *4.1*
cilium *4.9*	mitochondrion *4.6*	secondary wall *4.10*
cytomembrane system *4.5*	MTOC *4.8*	surface-to-volume ratio *4.1*
cytoplasm *4.1*	nuclear envelope *4.4*	wavelength, of light *4.2*
cytoskeleton *4.8*	nucleoid *4.11*	

Readings

deDuve, C. 1985. *A Guided Tour of the Living Cell.* New York: Freeman. Beautiful introduction to the cell; two short volumes.

Web Site See *http://www.wadsworth.com/biology* for practice quiz questions, hypercontents, BioUpdates, and critical thinking. The Wadsworth Biology Resource Center provides a wealth of information fully organized and integrated by chapter.

5

A CLOSER LOOK AT CELL MEMBRANES

It Isn't Easy Being Single

As small as it may be, a cell is a living thing engaged in the risky business of survival. Consider how something as ordinary as water can challenge its very existence. Water bathes cells inside and out, donates its individual molecules to many metabolic reactions, and dissolves ions that are necessary for cell functioning. If all goes well, the cell holds onto enough water and dissolved ions—not too little, not too much—to survive. But who is to say that life consistently goes well?

Think of a goose barnacle attached to the submerged side of a log that is drifting offshore, at the mercy of ocean currents. At feeding time, the barnacle opens its hinged shell and extends featherlike appendages, which trap bacteria and other bits of food suspended in the water (Figure 5.1a). The fluid bathing each living cell in the barnacle's body is salty, rather like the salt

composition of seawater. And seawater normally is in balance with the salty fluid inside the barnacle's cells.

However, suppose the log drifts into a part of the ocean where meltwater from a glacier has diluted the water (Figure 5.1b). For reasons you will explore in this chapter, salts inevitably move out of the barnacle's body—and out of its cells. The previously exquisite salt–water balance gradually spirals out of control, and the cells die. So, in time, does the barnacle.

The same thing happens to burrowing worms and other soft-bodied organisms that live between the high and low tide marks along a rocky shore. For instance, after an unpredictably fierce storm along the coast, seawater becomes highly dilute with runoff from the land. The dilution upsets the balance between body fluids that bathe cells of those organisms and the water

a

Figure 5.1 (a) Goose barnacles, which live attached to logs and other floating objects in the seas. As is true of every organism, drastic changes in salt concentration can threaten the cells of these marine animals. Such changes would occur if the barnacles were to accidentally end up in glacial meltwaters. (b) The dark-blue seawater in this photograph is quite salty. The lighter blue water is diluted by meltwater from the glacier in the background; it is much, much lower in salts than seawater.

moving in with the tides. At such times, the resulting death toll in the intertidal zone can be catastrophic.

With these examples we begin to see the cell for what it is: a tiny, organized bit of life in a world that is, by comparison, unorganized and sometimes harsh. How finely adapted a living cell must be to its environment! The cell must be built in such a way that it can bring in certain substances, release or keep out other substances, and conduct its internal activities with great precision.

For this bit of life, precision begins at the plasma membrane—a flimsy bilayer of lipids, dotted with diverse proteins, that surrounds the cytoplasm. Across the membrane, the cell exchanges substances with its surroundings in highly selective ways. For eukaryotic cells, precision continues at the membranes of internal compartments called organelles. Aerobic respiration, photosynthesis, and many other metabolic processes depend on the selective movement of substances across the membranes of organelles. Gain insight into the structure and function of cell membranes, and you will gain insight as well into survival at life's most fundamental level.

b

KEY CONCEPTS

1. A cell membrane consists largely of phospholipids and proteins. Its phospholipid molecules are organized as a double layer. This "bilayer" gives the cell membrane its basic structure and prevents the haphazard movement of water-soluble substances across it. Proteins embedded in the bilayer or associated with one of its surfaces carry out most membrane functions.

2. Many transport proteins span the bilayer of all cell membranes. Different kinds of transport proteins are open channels, gated channels, carriers, or pumps for specific water-soluble substances.

3. At the outer surface of a plasma membrane, diverse receptor proteins receive chemical signals that can trigger alterations in cell activities. Also present are recognition proteins, which are like molecular fingerprints; they identify a cell as being of a certain type.

4. Unless something prevents them from doing so, ions or molecules of a substance tend to move from one region into an adjacent region where they are less concentrated. This behavior is called diffusion.

5. In a form of molecular behavior known as osmosis, water diffuses across a membrane to a region where its concentration is lower. Osmosis only occurs at selectively permeable membranes that permit water to cross freely but that prevent the free passage of ions and large polar molecules.

6. Some membrane proteins function in passive transport. By this process, a solute enters or leaves a cell by diffusing through the protein's interior, in the direction that its concentration gradient takes it.

7. Other membrane proteins function in active transport. By this process, an energized transport protein actively pumps a solute across a cell membrane, against the concentration gradient.

MEMBRANE STRUCTURE AND FUNCTION

Earlier chapters provided you with a brief look at the structure of cell membranes and the general functions of their component parts. Here, we incorporate some of the background information in a more detailed picture.

The Lipid Bilayer of Cell Membranes

Fluid bathes the two surfaces of a cell membrane and is vital for its functioning. The membrane, too, has a fluid quality; it is not a solid, static wall between cytoplasmic and extracellular fluids. For instance, puncture a cell with a fine needle, and its cytoplasm will not ooze out. The membrane will flow over the puncture site and seal it!

How does a fluid membrane remain distinct from its fluid surroundings? To arrive at the answer, start by reviewing what we have already learned about its most abundant components, the phospholipids. Recall that a **phospholipid** has a phosphate-containing head and two fatty acid tails attached to a glycerol backbone (Figure 5.2*a*). The head is hydrophilic; it easily dissolves in water. Its tails are hydrophobic; water repels them. Immerse a number of phospholipid molecules in water, and they will interact with water molecules and with one another until they spontaneously cluster in a sheet or film at the water's surface. Their jostlings may even force them to become organized in two layers, with all fatty acid tails sandwiched between all hydrophilic heads. This **lipid bilayer** arrangement, remember, is the structural basis of cell membranes (Section 4.1 and Figure 5.2*c*).

The organization of each lipid bilayer minimizes the total number of hydrophobic

groups exposed to water, so the fatty acid tails do not have to spend a lot of energy fighting water molecules, so to speak. A "punctured" membrane exhibits sealing behavior precisely because a puncture is energetically unfavorable. It leaves far too many hydrophobic groups exposed to the surrounding fluid.

Ordinarily, few cells get jabbed by fine needles. But the self-sealing behavior of membrane phospholipids is good for more than damage control. Among other things, it functions in vesicle formation. For example, as vesicles bud away from ER or Golgi membranes, phospholipids interact hydrophobically with cytoplasmic water. They get pushed together, and the rupture seals. You will read more about vesicle formation later in the chapter.

Fluid Mosaic Model of Membrane Structure

Figure 5.3 shows a bit of membrane that corresponds to the **fluid mosaic model**. By this model, cell membranes are a mixed composition—a "mosaic"—of phospholipids, glycolipids, sterols, and proteins. The phospholipid heads as well as the length and saturation of the tails are not all the same. (Recall that unsaturated fatty acids have one or more double bonds in their backbone and fully saturated ones have none.) The glycolipids are structurally similar to phospholipids, but their head incorporates one or more sugar monomers. In animal cell membranes, cholesterol is the most abundant sterol (Figure 5.2*b*). Phytosterols are their equivalent in plant cell membranes.

Also by this model, the membrane is "fluid" owing to the motions and interactions of its component parts.

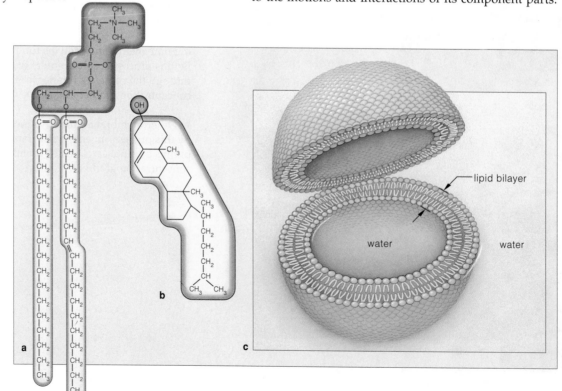

Figure 5.2

(**a**) Structural formula of phosphatidylcholine, a phospholipid that is one of the most common components of the membranes of animal cells. *Orange* indicates its hydrophilic head; *yellow* indicates its hydrophobic tails.

(**b**) Structural formula of cholesterol, the major sterol in animal tissues.

(**c**) Diagram showing how lipids that are placed in liquid water may spontaneously organize themselves into a bilayer structure.

lipid bilayer

water

water

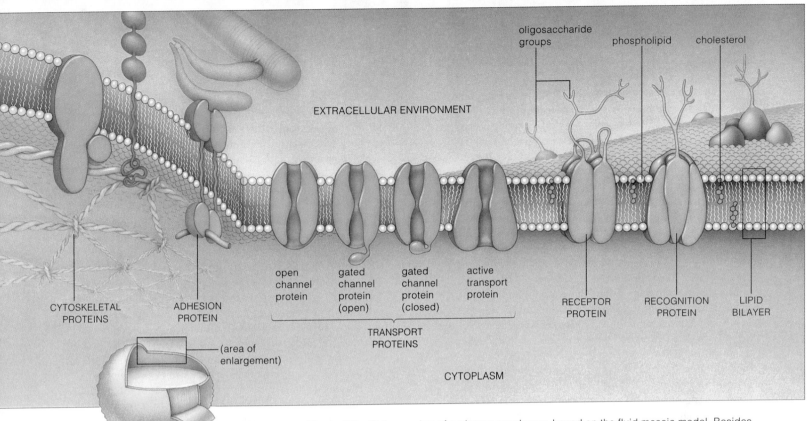

EXTRACELLULAR ENVIRONMENT

oligosaccharide groups   phospholipid   cholesterol

open channel protein

gated channel protein (open)

gated channel protein (closed)

active transport protein

CYTOSKELETAL PROTEINS

ADHESION PROTEIN

RECEPTOR PROTEIN

RECOGNITION PROTEIN

LIPID BILAYER

TRANSPORT PROTEINS

(area of enlargement)

CYTOPLASM

PLASMA MEMBRANE

Figure 5.3 Cutaway view of a plasma membrane, based on the fluid mosaic model. Besides the specialized proteins shown, enzymes also are associated with cell membranes.

The hydrophobic interactions that give rise to most of a membrane's structure are weaker than covalent bonds. This means most phospholipids and some proteins are free to drift sideways. Also, the phospholipids can spin about their long axis and flex their tails, which keeps neighboring molecules from packing together in a solid layer. Short or kinked (unsaturated) fatty acid tails also contribute to membrane fluidity.

The fluid mosaic model is a good starting point for exploring cell membranes. But bear in mind, membranes differ in the details of their molecular composition and arrangements, and they are not even the same on both surfaces of their bilayer. For example, oligosaccharides and other carbohydrates are covalently bonded to protein and lipid components of a plasma membrane, but only on its outward-facing surface (Figure 5.3). Moreover, they differ in number and kind from one species to the next, even among the different cells of the same individual.

Overview of Membrane Proteins

The proteins embedded in a lipid bilayer or attached to one of its surfaces carry out most membrane functions. Many are enzyme components of metabolic machinery. Others are **transport proteins** that allow water-soluble substances to move through their interior, which spans the bilayer. They bind molecules or ions on one side of the membrane, then release them on the other side.

The **receptor proteins** bind extracellular substances, such as hormones, that trigger changes in cell activities.

For example, certain enzymes that crank up machinery for cell growth and division become switched on when somatotropin, a hormone, binds with receptors for it. Different cells have different combinations of receptors.

Diverse **recognition proteins** at the cell surface are like molecular fingerprints; their oligosaccharide chains identify a cell as being of a specific type. For example, "self" proteins pepper the plasma membrane of your cells. Certain white blood cells chemically recognize the proteins and leave your own cells alone, but they attack invading bacterial cells having "nonself" proteins at their surface. Finally, **adhesion proteins** of multicelled organisms help cells of the same type locate and stick to one another and stay positioned in the proper tissues. They are glycoproteins with oligosaccharides attached. After tissues form, the sites of adhesion may become a type of cell junction, as described earlier in Section 4.10.

A cell membrane has two layers composed mainly of lipids, phospholipids especially. This lipid bilayer is the structural foundation for the membrane and also serves as a barrier to water-soluble substances.

Hydrophilic heads of the phospholipids are dissolved in fluids that bathe the two outer surfaces of the bilayer. Their hydrophobic tails are sandwiched between the heads.

Proteins associated with the bilayer carry out most membrane functions. Many are enzymes, transporters of substances across the bilayer, or receptors for extracellular substances. Other types function in cell-to-cell recognition or adhesion.

5.2   TESTING IDEAS ABOUT CELL MEMBRANES

TESTING MEMBRANE MODELS   Recall, from Chapter 1, that scientific methods are used to reveal nature's secrets. To gain insight into how researchers discovered some structural details of cell membranes, imagine yourself duplicating some of their test methods. You can start by attempting to identify the molecular components of the plasma membrane.

Your first challenge is to secure a membrane sample that is large enough to study and not contaminated with organelle membranes. Using red blood cells will simplify the task. These cells are abundant, easy to collect, and structurally simple. As they are maturing, their nucleus disintegrates. Ribosomes, hemoglobin, and just a few other components remain, but these are enough to keep the cells functioning for their brief, four-month life span.

By placing a sample of red blood cells in a test tube filled with distilled water, you can separate the plasma membranes from the other cell components. Cells have more solutes and fewer water molecules than distilled water does. For reasons that you will read about shortly, the difference in solute concentrations between the two regions causes water to move across a plasma membrane, into the cells. These particular cells have no mechanisms for actively expelling the excess water, so they swell. In time they burst, and their contents spill out.

Now you have a mixture of membranes, hemoglobin, and other cell components in water. How do you separate all the bits of membrane? You place a tube that holds a solution of the cell parts in a **centrifuge**, which is a motor-driven rotary device that can spin test tubes at very high speed (Figure 5.4). Structures and molecules move in response to the force of centrifugation. How far they move depends on their mass, density, and shape—as well as on the mass, density, and fluidity of the solution in the tubes. If the bits of membrane have greater mass and density than the solution, they will move downward in the tube. If they have lesser mass or density, they will stay near the top of the tube.

Each cell component has a molecular composition that gives it a characteristic density. As the centrifuge spins at an appropriate speed, the components having the greatest density move toward the bottom of the tube. Other cell components take up positions in layers above them, according to their relative densities.

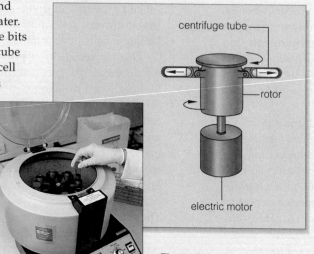

Figure 5.4   One example of a centrifuge. The diagram shows the arrangement of the tubes and the rotor of this high-speed spinning device.

You perform centrifugation properly, so one layer in the solution consists only of membrane shreds. You draw off this layer carefully and examine it with a microscope to check whether the membrane sample is contaminated.

Afterward, by using standard procedures of chemical analysis, you discover that the plasma membrane of red blood cells consists of lipids and proteins.

In the past, two competing structural models guided research into membranes. According to the "protein coat" model, a cell membrane consists of a lipid bilayer that has a layer of proteins coating both of its outward-facing surfaces. According to the "fluid mosaic" model, proteins are largely embedded within the bilayer.

You decide to test the protein coat model. First you calculate how much protein it would take to coat the inner and outer surfaces of a known number of red blood cells. Then you separate a membrane sample into its lipid and protein fractions and measure the amount of each. In this way, you can compare the *observed* ratio of proteins to lipids against the ratio *predicted* on the basis of the model.

Such calculations were performed with membrane samples. There are indeed enough lipids for a bilayer arrangement. And there *might* have been enough proteins to cover both of its surfaces *if* all the membrane proteins had a stretched-out (not globular) structure. However, that possibility was discarded for two reasons. First, it became clear that proteins stretched out in a thin layer over lipids would be a most unfavorable arrangement, in terms of the energy cost to maintain it. Second, biochemical analysis showed that the proteins are globular. Such evidence does not favor the protein coat model.

Now you do an observational test of the fluid mosaic model. You prepare cell samples for microscopy by *freeze-fracturing* and *freeze-etching* them. By these research methods, you immerse the cell sample in an extremely cold fluid: liquid nitrogen. The cells freeze instantly. Then you strike them with an exceedingly fine blade, as in Figure 5.5. A suitably directed blow can fracture cells in such a way that one lipid layer of the plasma membrane will separate from the other. Such preparations can be observed under extremely high magnifications.

If the protein coat model were correct, then you would observe a perfectly smooth, pure lipid layer. However, the freeze-fractured cell membranes you observe reveal many bumps and other irregularities in the separated layers of lipids. The bumps are proteins, incorporated directly in the bilayer—just as the fluid mosaic model predicts.

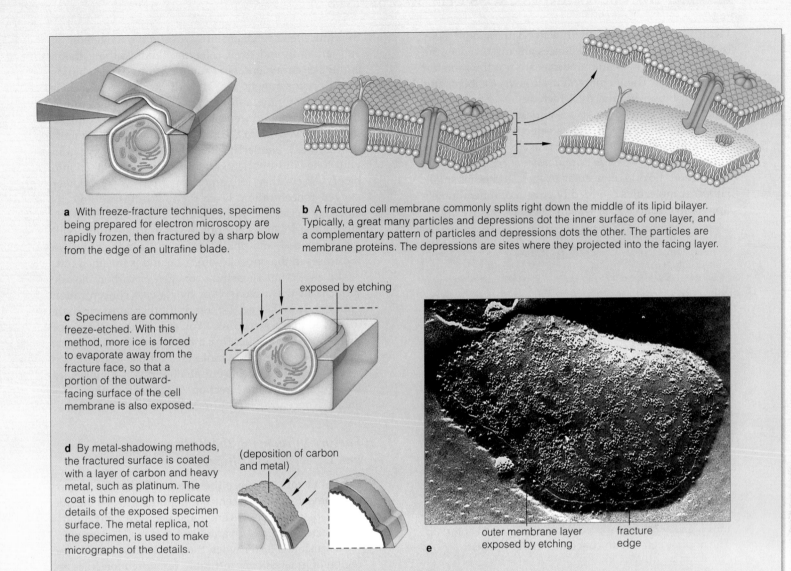

a With freeze-fracture techniques, specimens being prepared for electron microscopy are rapidly frozen, then fractured by a sharp blow from the edge of an ultrafine blade.

b A fractured cell membrane commonly splits right down the middle of its lipid bilayer. Typically, a great many particles and depressions dot the inner surface of one layer, and a complementary pattern of particles and depressions dots the other. The particles are membrane proteins. The depressions are sites where they projected into the facing layer.

exposed by etching

c Specimens are commonly freeze-etched. With this method, more ice is forced to evaporate away from the fracture face, so that a portion of the outward-facing surface of the cell membrane is also exposed.

d By metal-shadowing methods, the fractured surface is coated with a layer of carbon and heavy metal, such as platinum. The coat is thin enough to replicate details of the exposed specimen surface. The metal replica, not the specimen, is used to make micrographs of the details.

(deposition of carbon and metal)

outer membrane layer exposed by etching

fracture edge

e

Figure 5.5 Freeze-fracturing and freeze-etching methods. The micrograph in (**e**) shows part of a replica of a red blood cell that was prepared by the techniques described in the text.

OBSERVING MEMBRANE FLUIDITY Consider now an experiment that yielded evidence of the fluid motion of membrane proteins. Researchers induced an isolated human cell and an isolated mouse cell to fuse. The plasma membranes of the cells from the two species merged to form a single, continuous membrane in the new, hybrid cell. Analysis showed that, in less than an hour, most of the membrane proteins were mixed together; they were free to drift laterally through the hybrid membrane:

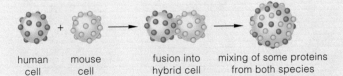

human cell   mouse cell   fusion into hybrid cell   mixing of some proteins from both species

Through observational tests, we also know that the fluidity of a membrane is subject to change, depending on environmental temperatures—and that at least some cells can respond to the danger.

For example, after some researchers decreased the temperature of a solution that contained live bacterial or yeast cells, the membranes of those cells started to stiffen. Such stiffening can disrupt the functioning of membrane proteins. However, the cells responded by rapidly synthesizing unsaturated fatty acids, and the infusion of those kinked lipids into the cell membranes countered the stiffening effect. However, past a certain low temperature, the membranes did solidify. The temperature at which this occurs varies, depending on the lipid composition of a given membrane.

HOW SUBSTANCES CROSS CELL MEMBRANES

Picture a cell membrane, with water bathing both sides of its bilayer. Plenty of substances are dissolved in the water, but the kinds and amounts are not the same on the two sides. The membrane itself helps establish and maintain those differences, which are essential for cell functioning. How does it accomplish this feat? Like every cell membrane, it shows **selective permeability**. *Because of its molecular structure, it allows some substances but not others to cross it in certain ways, at certain times.*

Carbon dioxide, molecular oxygen, and other small, nonpolar molecules can readily cross a cell membrane's lipid bilayer, and so can water molecules:

O_2, CO_2, other small, nonpolar molecules, as well as H_2O

Although water molecules do show polarity, they might move through transient gaps that open up in the bilayer when hydrocarbon chains of the lipids flex and bend. Glucose and other large, polar molecules almost never move freely across the bilayer. Neither do ions:

$C_6H_{12}O_6$, other large, polar, water-soluble molecules, ions (such as H^+, Na^+, K^+, Ca^{++}, Cl^-), along with H_2O

Such water-soluble substances must move across the bilayer through the interior of transport proteins. And the water they are dissolved in moves with them.

The question becomes this: Why does a substance move one way or another during any given interval? The answer starts with concentration gradients.

Concentration Gradients and Diffusion

"Concentration" refers to the number of molecules or ions of a substance in a specified region, as in volume of fluid or air. "Gradient" means the number in one region is not the same as it is in another. Thus, a **concentration gradient** is a difference in the number of molecules or ions of a given substance in two adjoining regions.

In the absence of other forces, a substance moves from a region where it is more concentrated to a region where it is less concentrated. *The energy inherent in its individual molecules, which keeps them in constant motion, drives the directional movement.* Although the molecules collide randomly and career back and forth, millions of times a second, the *net* movement is away from the place of greater concentration (and the most collisions).

Diffusion is the name for the net movement of like molecules or ions down a concentration gradient. It is a key factor in the movement of substances across cell membranes and through fluid parts of the cytoplasm.

In multicelled organisms, it figures in the movement of substances to and from cells, the fluids bathing them, and the environment. For example, when oxygen builds up in photosynthetic leaf cells, it diffuses across their plasma membrane, through water in leaves, and out to the air, where its concentration is lower. Carbon dioxide diffuses inward from the air, the region where it is most concentrated. Similarly, after you breathe in, oxygen in your lungs diffuses across a thin, moist lining, then into blood vessels, then into any tissue fluid having a lower oxygen concentration. From there, it diffuses into cells engaged in aerobic respiration. Carbon dioxide levels are highest in such cells but lower in tissue fluids, still lower in blood, and finally lowest in air in the lungs, so it diffuses in the opposite direction and is breathed out.

Like all other substances, oxygen or carbon dioxide diffuses in the direction set by its *own* concentration gradient. It makes no difference if other substances are dissolved in the same fluid. You can see the outcome of this tendency by squeezing a drop of dye into one part of a bowl of water. Dye molecules diffuse to the region where they are less concentrated, and water molecules move in the opposite direction, to the region where *they* are less concentrated (Figure 5.6).

Factors Influencing the Rate and Direction of Diffusion

Several factors influence the rate at which molecules or ions move down a concentration gradient. They include that gradient's steepness and molecular size, temperature, and electric or pressure gradients that may be present.

Diffusion is faster when the gradient is steep. In the region of greatest concentration, far more molecules are moving outward, compared to the number moving in. As the gradient decreases, the difference in the number of molecules moving one way or the other becomes less pronounced, and diffusion slows. When the gradient is gone, individual molecules are still in

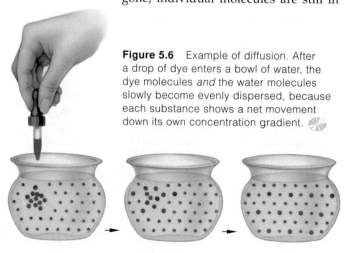

Figure 5.6 Example of diffusion. After a drop of dye enters a bowl of water, the dye molecules *and* the water molecules slowly become evenly dispersed, because each substance shows a net movement down its own concentration gradient.

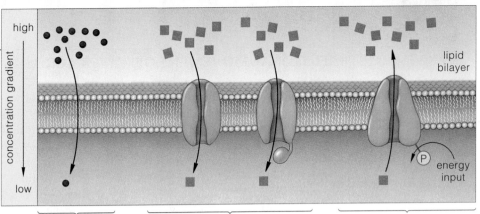

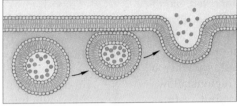

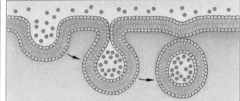

Figure 5.7 Overview of the major mechanisms by which solutes cross cell membranes. (Exocytosis and endocytosis proceed only at the plasma membrane.)

DIFFUSION ACROSS LIPID BILAYER

Lipid-soluble substances as well as water diffuse across.

PASSIVE TRANSPORT

Water-soluble substances, and water, diffuse through interior of transport proteins. No energy boost required. Also called facilitated diffusion.

ACTIVE TRANSPORT

Specific solutes are pumped through interior of transport proteins. Requires energy boost.

EXOCYTOSIS

Vesicle in cytoplasm moves to plasma membrane, fuses with it; contents released to the outside.

ENDOCYTOSIS

Vesicle forms from a patch of inward-sinking plasma membrane, enters cytoplasm.

motion. However, the total number moving one way or the other during a specified interval is approximately the same. When the net distribution of molecules becomes nearly uniform throughout two adjoining regions, we call this dynamic equilibrium.

Rates of diffusion are faster at higher temperatures. With more heat energy, molecules move faster and thus tend to collide more frequently. Diffusion rates also are influenced by molecular size. Small molecules tend to move faster than large ones.

In addition, the rate and direction of diffusion may be modified by an **electric gradient**: a difference in the electric charges of adjoining regions. For example, the fluid bathing each surface of a cell membrane has many kinds of dissolved ions, and each one contributes to its overall electric charge. Opposite charges attract, so the fluid with a more *negative* charge, overall, tends to exert the greatest pull on a *positively* charged substance—say, sodium ions. In the absence of other factors, the pull can enhance the flow of sodium across the membrane. Many processes, such as information flow in the nervous system, depend upon the combined force of electric and concentration gradients to attract substances across cell membranes.

Finally, as you will see shortly, the rate and direction of diffusion also may be modified by the presence of a **pressure gradient**, which is a difference in the pressure being exerted in two adjoining regions.

Mechanisms By Which Solutes Cross Cell Membranes

Figure 5.7 is an overview of the mechanisms that move solutes across all cell membranes. Through combinations of these mechanisms, cells and organelles are supplied with raw materials and are rid of wastes, at controlled rates. The mechanisms also help maintain the volume and pH of a cell or organelle within functional ranges.

Again, small nonpolar molecules (such as oxygen) and water diffuse across the lipid bilayer. By contrast, in **passive transport**, a polar substance diffuses down its concentration gradient through the interior of transport proteins spanning the bilayer. This directional movement

is also called *facilitated* diffusion. In **active transport**, a solute also cross the membrane through the interior of transport proteins, but in this case the net movement is against the concentration gradient. This mechanism can only operate when transport proteins are activated, as by an energy boost from ATP.

With exocytosis or endocytosis, substances move in bulk across the plasma membrane. As you will read in Section 5.6, **exocytosis** involves fusion of a vesicle (that formed earlier in the cytoplasm) with the cell's plasma membrane. **Endocytosis** involves an inward sinking of a small patch of plasma membrane, which then seals back on itself to form a cytoplasmic vesicle.

Molecules or ions of a substance constantly collide because of their inherent energy of motion. The collisions result in diffusion, a net outward movement of a substance from one region into an adjoining region where it is less concentrated.

A concentration gradient is a form of energy that can drive the directional movement of a substance across membranes.

The steepness of the concentration gradient affects diffusion rates. So do temperature, molecular size, and the presence of electric or pressure gradients.

Nonpolar substances and water can diffuse across the lipid bilayer of cell membranes. Polar substances and water cross by passive transport (facilitated diffusion) and by active transport. Substances also move in bulk across the plasma membrane by exocytosis and endocytosis.

THE DIRECTIONAL MOVEMENT OF WATER ACROSS MEMBRANES

By far, more water diffuses across cell membranes than any other substance, so the key factors that influence its directional movement deserve some attention before we get into the details of transport mechanisms.

Water Movement by Osmosis

Turn on a faucet or watch a waterfall, and the moving water provides a demonstration of bulk flow. **Bulk flow** is the mass movement of one or more substances in response to pressure, gravity, or some other external force. It accounts for some movement of water through complex plants and animals. With each beat, your heart creates fluid pressure that drives a volume of blood, which is mainly water, through interconnected blood vessels. Sap runs inside conducting tissues that thread through maple trees, and this, too, is a case of bulk flow.

For cells, the diffusion of water from one region to another is a different movement. It occurs *only* when a membrane that intervenes between two regions allows the small, polar water molecules to cross but restricts the passage of ions and large polar molecules. **Osmosis** is the name for the diffusion of water in response to a water concentration gradient between two regions that are separated by a selectively permeable membrane.

Osmotic movement depends on the concentrations of solutes in the water on both sides of a membrane. *The side with more solute particles has a lower concentration of water.* To see why this is so, dissolve a small amount of glucose in water. That glucose solution has fewer water molecules than an equivalent volume of water, for each glucose molecule now occupies some of the space formerly occupied by a water molecule. Also, some water molecules are no longer free to contribute to the water concentration. They have become part of spheres of hydration around polar groups at the surface of each glucose molecule (Section 2.5).

It is mainly the *total number* of molecules or ions, not the type of solute, that dictates the concentration of water. Dissolve one mole of an amino acid or urea in 1 liter of water, and the water concentration decreases about as much as it did in the glucose solution. Add one mole of sodium chloride (NaCl) to 1 liter of water, and it dissociates into equal numbers of sodium ions and chloride ions. There are now two moles of solute particles—twice as many as in the glucose solution—so the water concentration has decreased proportionately.

Effects of Tonicity

Given that water molecules tend to move osmotically to a region where water is less concentrated, the direction of movement will be toward a region where solutes are more concentrated. Figure 5.8 illustrates this tendency.

Suppose you decide to make a simple observational test of this statement. You construct three sacs out of a membrane that water but not sucrose can cross, and you fill each with a 2M sucrose solution. (*M* stands for Molarity, the number of moles of a specified solute in 1 liter of fluid.) Then you immerse one of the sacs in 1 liter of distilled water (which has no solutes), one in a 10M sucrose solution, and the third in a 2M sucrose solution. In each case, the extent and direction of water movement are dictated by tonicity (Figure 5.9*a*).

Tonicity refers to the relative solute concentrations of two fluids. When two fluids on opposing sides of a membrane differ in solute concentration, the one having fewer solutes is called the **hypotonic solution**, and the one with more is the **hypertonic solution**. Water tends to diffuse from hypotonic to hypertonic fluids. **Isotonic solutions** have the same solute concentrations, so water shows no net osmotic movement from one to the other.

Normally, the fluid inside your cells and the tissue fluid bathing them are isotonic. If a tissue fluid became drastically hypotonic, so much water would diffuse into your cells that they would burst. If it became far too hypertonic, the outward diffusion of water would make the cells shrivel. Most cells have built-in mechanisms for adjusting to shifts in tonicity. Red blood cells do not; Figure 5.9*b* shows what happened to such cells during a demonstration of the effects of tonicity differences. That is why patients who are severely dehydrated are given infusions of a solution that is isotonic with blood. Such solutions move by bulk flow from a bottle mounted above the patient, through a flexible tube, and directly into the blood of an incised vein.

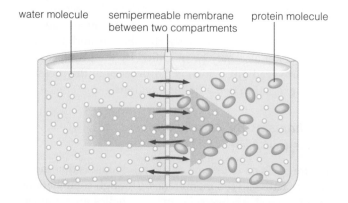

water molecule   semipermeable membrane   protein molecule
between two compartments

Figure 5.8 Effect of a solute concentration gradient on osmotic movement. Start with a container divided by a membrane that water but not proteins can cross. Pour water into the left half. Pour the same volume of a protein-rich solution into the right half. There, proteins occupy some of the space. And the polar groups of each protein are sites where spheres of hydration form and tie up some water molecules, which thereby cannot contribute to the concentration of free water. The net diffusion of water in this case is from left to right (*large blue arrow*).

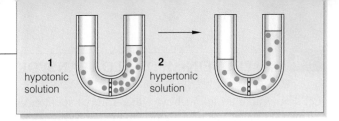

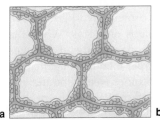

Figure 5.10 Increase in fluid volume as a result of osmosis. When the net diffusion across a membrane separating two compartments is equal, the fluid volume in compartment 2 is greater because the membrane is impermeable to solutes.

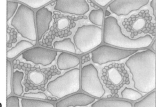

Figure 5.11 Plasmolysis, an osmotically induced loss of internal fluid pressure, in young plant cells. The cytoplasm and central vacuole shrink; the plasma membrane moves away from the wall.

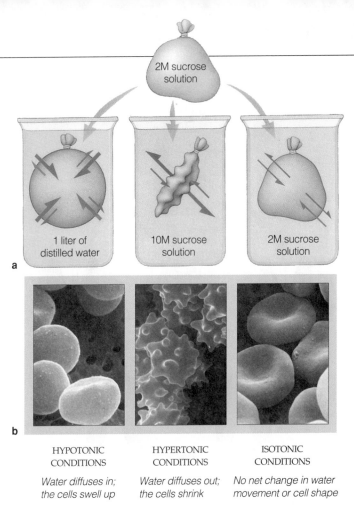

a

b

HYPOTONIC CONDITIONS	HYPERTONIC CONDITIONS	ISOTONIC CONDITIONS
Water diffuses in; the cells swell up	*Water diffuses out; the cells shrink*	*No net change in water movement or cell shape*

Figure 5.9 Effect of tonicity on water movement, as described in the text. (**a**) The direction and relative amounts of water movement are indicated by arrow widths. The micrographs (**b**) correspond to the sketches. They show the kinds of shapes that human red blood cells will assume when placed in fluids of higher, lower, and equal solute concentrations. Normally, the solutions inside and outside of red blood cells are in balance. This type of cell does not have built-in mechanisms for adjusting to drastic changes in solute levels in its fluid surroundings.

Effects of Fluid Pressure

Animal cells generally can avoid bursting by engaging in the ongoing selective transport of solutes across the plasma membrane. Cells of plants and many protistans, fungi, and bacteria also avoid that unpleasant prospect with the help of pressure exerted on their cell walls.

Pressure differences as well as solute concentrations influence the osmotic movement of water. Take a look at Figure 5.10. It shows how water continues to diffuse from a hypotonic solution to a hypertonic one until its concentration becomes the same on both sides of the membrane between them. As you can see, the *volume* of the formerly hypertonic solution has increased (because its solutes can't diffuse out). Any volume of fluid exerts **hydrostatic pressure**, a force directed against a wall, membrane, or some other structure that encloses the fluid. The greater the solute concentration of the fluid, the greater will be the hydrostatic pressure it exerts.

As you know, living cells cannot increase in volume indefinitely (Section 4.1). At some point, the hydrostatic pressure that develops inside a cell counters the inward diffusion of water. That point is the **osmotic pressure**, the amount of force which prevents any further increase in the volume of a solution.

Consider young plant cells, with a pliable, primary wall. While they grow, water diffuses in and hydrostatic pressure increases against the wall. The wall expands, and the cell volume increases. The thin walls are strong enough for the cell's internal fluid pressure to develop to the point where it counterbalances water uptake.

Plant cells are more vulnerable to water loss, which can happen when soil dries or becomes too salty. Water stops diffusing in, the cells lose water, and internal fluid pressure drops. Such osmotically induced shrinkage of cytoplasm is called **plasmolysis** (Figure 5.11). Plants can adjust somewhat to the loss of pressure, as when they actively take up potassium ions against a concentration gradient by mechanisms outlined in the next section.

As you will read in Chapters 39 and 43, hydrostatic and osmotic pressure also influence the distribution of water in the blood, tissue fluid, and cells of animals.

Osmosis is the net diffusion of water between two solutions that differ in water concentration and that are separated by a selectively permeable membrane. The greater the number of molecules and ions dissolved in a solution, the lower its water concentration will be.

Water tends to move osmotically to regions of greater solute concentration (from hypotonic to hypertonic solutions). There is no net diffusion between isotonic solutions.

The fluid pressure that a solution exerts against a membrane or wall also influences the osmotic movement of water.

PROTEIN-MEDIATED TRANSPORT

Let's now take a look at how water-soluble substances diffuse into and out of cells or organelles. Whether by passive or active transport mechanisms, a great variety of transport proteins have roles in moving molecules or ions of such substances across cell membranes. These proteins span the lipid bilayer, and their interior is able to open on both sides of it (Figure 5.12).

To understand how these proteins work, you have to know they are not rigid blobs of atoms. When the protein interacts with a particular solute, it changes from one shape to another shape, then back again. The changes start when a solute becomes weakly bound at a specific site on the protein surface. Part of the protein closes in behind the bound solute and part opens up to the other side of the membrane. The solute dissociates (separates) from the site on that side. Think of the solute as hopping onto the transport protein on one side of the membrane, then hopping off on the other side.

Passive Transport

Transport proteins permit solutes to move both ways across a cell membrane. In cases of passive transport, the *net* direction of movement during a given interval depends on how many molecules or ions of the solute are making random contact with vacant binding sites in the interior of the proteins (Figure 5.13). The binding and transport simply proceed more often on the side of the membrane where the solute is more concentrated. Because there are more molecules around, the random encounters with the binding sites are more frequent than they are on the other side of the membrane

By itself, the passive two-way transport of a solute would continue until its concentrations became equal on both sides of the membrane. In most cases, however, other processes affect the outcome. For example, the bloodstream delivers glucose molecules to all of your body's tissues. Nearly all cells use this simple sugar as an energy source and as a building block for larger

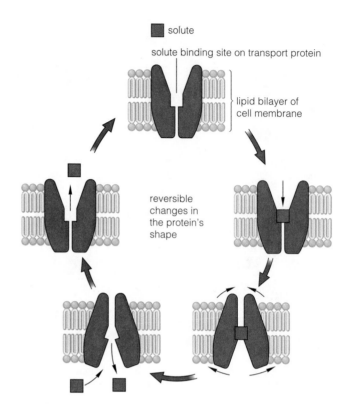

Figure 5.13 Passive transport across a cell membrane. A solute can move in both directions through transport proteins. In passive transport, however, the *net* movement will be down the solute's concentration gradient (from higher to lower concentration). This will continue until the concentrations become the same on both sides of the membrane.

compounds. When the glucose concentration in blood and tissues is high, cellular uptake is rapid. But as fast as some glucose molecules diffuse into a cell, others usually are leaving the cytoplasm and entering various metabolic reactions. Thus, when cells are rapidly using glucose, they are actually maintaining a concentration gradient that favors the uptake of *more* glucose.

Figure 5.12 Sketch of some transport proteins that span the lipid bilayer of a plasma membrane. Transport proteins passively allow or actively assist ions as well as glucose and other large polar molecules to move through their interior, from one side of the membrane to the other.

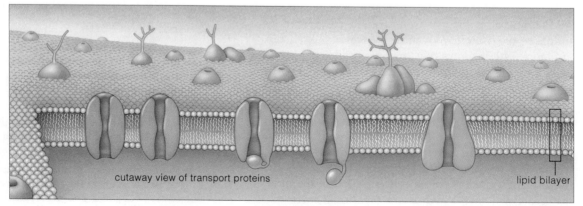

cutaway view of transport proteins

lipid bilayer

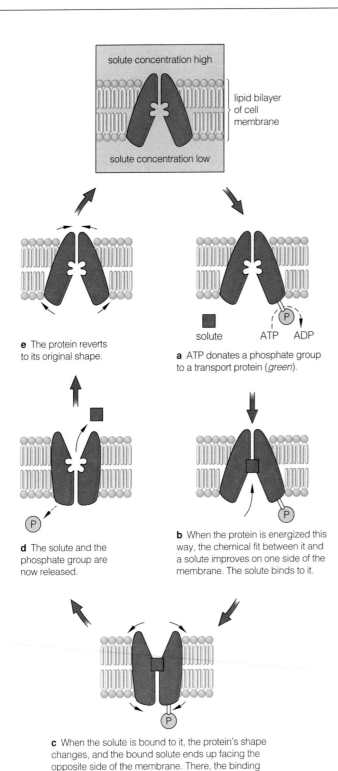

solute concentration high

lipid bilayer of cell membrane

solute concentration low

e The protein reverts to its original shape.

solute   ATP   ADP

a ATP donates a phosphate group to a transport protein (*green*).

d The solute and the phosphate group are now released.

b When the protein is energized this way, the chemical fit between it and a solute improves on one side of the membrane. The solute binds to it.

c When the solute is bound to it, the protein's shape changes, and the bound solute ends up facing the opposite side of the membrane. There, the binding site reverts to the less attractive configuration.

Figure 5.14 Active transport across a cell membrane. As with passive transport, a solute can diffuse in both directions through the interior of the transport protein. However, with active transport, ATP transfers a phosphate group to the protein. The transfer sets in motion reversible changes in the protein's shape that result in a greater net movement of solute particles against the concentration gradient.

Active Transport

Transport proteins with roles in active transport across cell membranes move a solute against its concentration gradient. Unlike passive transport, this mechanism can continue until the solute becomes *more* concentrated on the side of the membrane to which it is being pumped.

Active transport does not proceed spontaneously. It requires an energy boost, most often from ATP. Briefly, ATP donates a phosphate group to a transport protein. When it does, the chemical fit between the binding site and the solute improves on one side of the membrane (Figure 5.14*a* and *b*). After a solute particle binds at the site, the protein's folded shape changes in such a way that the bound solute becomes exposed to fluid bathing the opposite side of the membrane (Figure 5.14*c*). Now the binding site reverts to its less attractive state and the solute is released. A less attractive site means fewer molecules or ions of the solute make the return trip. So the *net* movement is to the side of the membrane where the solute is more concentrated. By analogy, picture hordes of skiers rapidly hopping onto the chairs of a ski lift that moves them up a mountain. At the top, the chairs tilt in a way that is not very inviting for skiers who want to hop on quickly for a downhill ride. Thus, more skiers will be transported uphill than downhill.

One type of active transport system, the **calcium pump**, helps keep the calcium concentration in a cell at least a thousand times lower than outside. Another, the **sodium-potassium pump**, moves potassium ions (K^+) across the plasma membrane. Activation of this protein facilitates binding of a sodium ion (Na^+) on one side of a cell membrane. After the Na^+ makes the crossing, it is released. Its release facilitates the binding of K^+ at a different binding site on the protein, which reverts to its original shape after K^+ is delivered to the other side.

Through operation of such active transport systems, concentration and electric gradients are maintained across membranes. The gradients are vital to many cell activities and physiological processes, including muscle contraction and information flow in nervous systems. We will return to the mechanisms of active transport in later chapters. For now, keep these points in mind:

Transport proteins span the lipid bilayer of cell membranes. When they bind a solute on one side of a membrane, they undergo a reversible change in shape that shunts the solute through their interior, to the other side.

With passive transport, a solute simply diffuses through the protein; its net movement is down its concentration gradient.

With active transport, the net diffusion of a solute is uphill, against its concentration gradient. The transporting protein must be activated, as by ATP energy, to counter the energy inherent in the gradient.

EXOCYTOSIS AND ENDOCYTOSIS

Transport proteins can move ions and small molecules into or out of cells. However, when it comes to taking in or expelling large molecules or particles, cells rely on vesicles that form through exocytosis and endocytosis.

Transport To the Plasma Membrane

By exocytosis, a cytoplasmic vesicle moves to the cell surface, and the protein-studded lipid bilayer of its own membrane fuses with the plasma membrane. While this exocytic vesicle is losing its identity, its contents are released to the surroundings (Figure 5.15*a*).

Transport From the Plasma Membrane

By three pathways of endocytosis, a cell internalizes substances next to its surface. In all three cases, a small indentation forms at the plasma membrane, balloons inward, and pinches off. The resulting endocytic vesicle transports its contents or stores them in the cytoplasm (Figure 5.15*b*). By *receptor-mediated* endocytosis, the first pathway, membrane receptors chemically recognize and bind specific substances, such as lipoproteins, vitamins, iron, peptide hormones, growth factors, and antibodies. The receptors become concentrated in tiny depressions, or pits, in the plasma membrane (Figure 5.16). Each pit looks like a loosely woven basket on its cytoplasmic side. The basket consists of protein filaments (clathrin), interlocked in stable, geometric patterns. When the pit sinks into the cytoplasm, the basket closes back on itself and becomes the structural framework for the vesicle.

Cells are less finicky with *bulk-phase* endocytosis, the second pathway. An endocytic vesicle forms around a

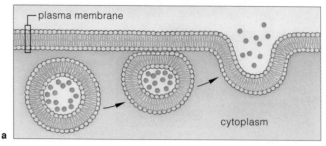

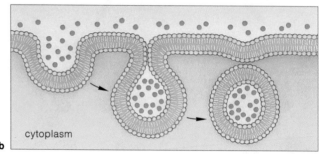

Figure 5.15 (**a**) Exocytosis. Substances are released outside the cell when an exocytic vesicle's membrane fuses with the plasma membrane. (**b**) Endocytosis. Substances are brought into the cell when a bit of plasma membrane balloons inward beneath them, self-seals, and so forms an endocytic vesicle.

small volume of extracellular fluid regardless of which substances happen to be dissolved in it. The bulk-phase pathway operates at a fairly constant rate in nearly all eukaryotic cells. By steadily sending patches of plasma membrane into the cytoplasm, the pathway compensates for membrane that steadily departs from the cytoplasm in the form of exocytic vesicles.

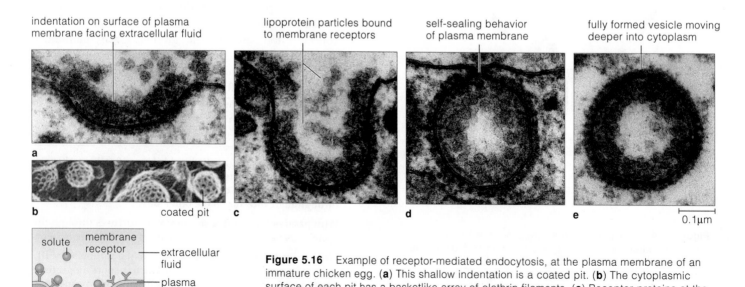

Figure 5.16 Example of receptor-mediated endocytosis, at the plasma membrane of an immature chicken egg. (**a**) This shallow indentation is a coated pit. (**b**) The cytoplasmic surface of each pit has a basketlike array of clathrin filaments. (**c**) Receptor proteins at the pit's outer surface preferentially bind lipoprotein particles. (**d**) The pit deepens and rounds out. (**e**) The formed endocytic vesicle encases lipoproteins that the cell will use or store.

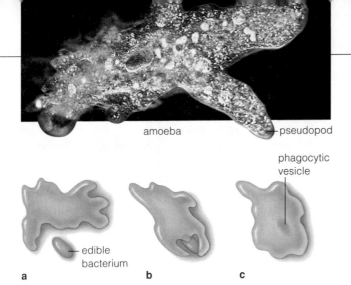

amoeba — pseudopod

phagocytic vesicle

edible bacterium

a   b   c

Figure 5.17 Phagocytosis in amoebas (including the *Amoeba proteus* cell in the photograph), certain white blood cells such as macrophages, and some other cells. Lobes of cytoplasm extend outward and surround the target. The plasma membrane of the extensions fuses together, thereby forming a phagocytic vesicle. Such vesicles move deeper into the cytoplasm and then fuse with lysosomes. Their contents are digested and, along with the vesicle's membrane components, are recycled elsewhere.

The third pathway, **phagocytosis**, is an active form of endocytosis by which the cell engulfs microorganisms, large edible particles, and cellular debris. (Phagocytosis literally means "cell eating.") Amoebas and some other protistans get food this way. In multicelled organisms, certain white blood cells engage in phagocytosis during responses that defend the body against harmful viruses, bacteria, and other threats to health.

As you read earlier in Section 4.9, a phagocytic cell shows amoeboid motion after a target binds to certain receptors that bristle from its plasma membrane. The binding sends signals into the cell. The signals trigger a directional assembly and crosslinking of microfilaments into a dynamic, ATP-requiring network just beneath the plasma membrane. The network contracts in ways that squeeze some cytoplasm toward the cell margins, thus forming the lobes called pseudopods (Figure 5.17). The pseudopods flow over the target and fuse at their tips. The result is a phagocytic vesicle, which sinks into the cytoplasm. There it fuses with lysosomes, the organelles of intracellular digestion in which trapped items are digested to fragments and smaller, reusable molecules.

Membrane Cycling

As long as a cell stays alive, exocytosis and endocytosis continually replace and withdraw patches of its plasma membrane. And they apparently do so at rates that can maintain the plasma membrane's total surface area.

As an example, consider the bursts of exocytosis by which neurons (nerve cells) release neurotransmitters, a type of signaling molecule that acts on neighboring cells. Two cell biologists, John Heuser and T. S. Reese,

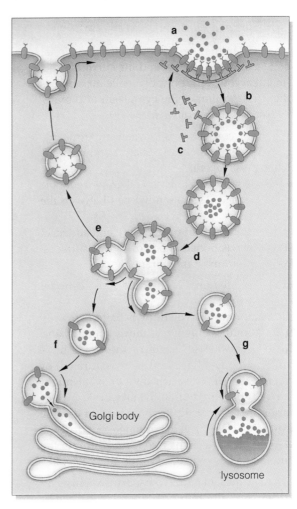

a Molecules get concentrated inside coated pits of plasma membrane.

b Endocytic vesicles form from the pits.

c Vesicles lose molecules of clathrin, which return to the plasma membrane.

d Enclosed molecules are sorted and often released from receptors.

e Many sorted molecules are cycled back to the plasma membrane.

f, g Many other sorted molecules travel to Golgi bodies and stay there. Others are routed to spaces inside nuclear envelope and ER membranes, and still others to lysosomes.

Golgi body

lysosome

Figure 5.18 The cycling of membrane lipids and proteins. This example starts with receptor-mediated endocytosis. The plasma membrane gives up small patches of itself to endocytic vesicles that form from coated pits. It gets membrane back from exocytic vesicles that budded from ER membranes and from Golgi bodies. The membrane that was initially appropriated by endocytic vesicles will cycle receptor proteins and lipids back to the plasma membrane.

documented a notably intense burst of endocytosis that immediately followed and counterbalanced an intense episode of exocytosis in neurons. Figure 5.18 shows some ways in which cells cycle their bits of membrane.

Whereas transport proteins in a cell membrane deal only with ions and small molecules, exocytosis and endocytosis move larger packets of material across the plasma membrane.

By exocytosis, a cytoplasmic vesicle fuses with the plasma membrane, so that its contents are released outside the cell. By endocytosis, a small patch of the plasma membrane sinks inward and seals back on itself, forming a vesicle inside the cytoplasm. Membrane receptors often mediate this process.

SUMMARY

Membrane Structure and Function

1. The plasma membrane is a structural and functional boundary between the cytoplasm and the surroundings of all cells. In eukaryotic cells only, organelle membranes subdivide the fluid portion of the cytoplasm into many functionally diverse compartments.

2. A cell membrane consists of two water-impermeable layers of lipids (phospholipids especially) and proteins associated with those layers.

   a. In the bilayer, fatty acid tails and other hydrophobic parts of the lipid molecules are sandwiched between the hydrophilic heads.

   b. Many different kinds of proteins are embedded in the lipid bilayer or positioned at one of its two surfaces. The proteins carry out most membrane functions.

3. These are the key features of the fluid mosaic model of membrane structure:

   a. A cell membrane shows fluid behavior, mainly because its lipid components twist, move laterally, and flex hydrocarbon tails. Also, some of the lipids have ring structures, and many have kinked (unsaturated) or short fatty acid tails, all of which disrupt what might otherwise be tight packing within the bilayer.

   b. A membrane is a mosaic, or composite, of diverse lipids and proteins. The proteins are embedded in the bilayer and are at its surface. Its two layers differ in the number, kind, and arrangement of lipids and proteins.

4. All cell membranes include transport proteins and proteins that structurally reinforce the membrane. The plasma membrane also has proteins that serve in signal reception, cell recognition, and adhesion. Differences in the number and types of proteins among cells affect metabolism, cell volume, pH, and responsiveness to substances that make contact with the membrane.

   a. Transport proteins allow water-soluble substances to pass through their interior, which opens to both sides of the membrane.

   b. Receptor proteins bind extracellular substances that trigger alterations in metabolic activities.

   c. Recognition proteins are molecular fingerprints at the surface of each cell type. Infection-fighting white blood cells chemically recognize self proteins (of the body's own cells) and nonself proteins of foreign cells.

   d. In multicelled organisms, adhesion proteins help cells adhere to one another and form cell junctions.

Movement of Substances Into and Out of Cells

1. Molecules or ions of a substance tend to move from a region of higher to lower concentration. Movement in response to a concentration gradient is called diffusion.

   a. Diffusion rates are influenced by the steepness of the concentration gradient, temperature, and molecular size, as well as by gradients in electrical charge and pressure that may occur between two regions.

   b. Cells have built-in mechanisms that work with or against gradients to move solutes across membranes.

2. Osmosis is the technical name for the diffusion of water across any selectively permeable membrane in response to a concentration gradient.

   a. The direction of osmotic movement between two solutions (such as cytoplasmic fluid and tissue fluid) can be predicted by comparing their tonicity. Tonicity is a measure of the solute concentration of one solution *relative to* the solute concentration of another solution.

   b. Water tends to move from a hypotonic solution (with the lower solute concentration) to a hypertonic solution (with the higher solute concentration).

   c. If fluids are isotonic (equal solute concentrations), water will show no net movement in either direction.

3. Oxygen, carbon dioxide, and other small nonpolar molecules diffuse across a membrane's lipid bilayer. Ions, glucose, and other large, polar molecules cross it with the passive or active help of transport proteins. Water can move through the bilayer and the proteins.

4. Transport proteins bind specific solutes on one side of a membrane, and they shunt solutes to the other side through reversible changes in their shape.

   a. By passive transport, the protein allows a solute simply to diffuse through its interior, in the direction of its concentration gradient. The protein's shape changes without an input of energy.

   b. By active transport, the protein pumps a solute across the membrane against its concentration gradient. The changes in protein shape that trigger the process require an energy boost, as from ATP.

5. By exocytosis, vesicles in the cytoplasm move to the plasma membrane. When their membrane fuses with it, their contents are automatically released to the outside.

6. By endocytosis, a small patch of plasma membrane sinks into the cytoplasm and seals back on itself to form a vesicle. The three pathways are receptor-mediated endocytosis (requires recognition of specific solutes), bulk-phase endocytosis (indiscriminate uptake of some extracellular fluid), and phagocytosis (active uptake of large particles, cell parts, or whole cells).

Review Questions

1. Describe the fluid mosaic model of cell membranes. What imparts fluidity to the membrane? What makes it a mosaic? *5.1*

2. Structurally, what do all cell membranes have in common? In what ways do the membranes of different cell types vary? *5.1*

3. Distinguish among transport proteins, receptor proteins, recognition proteins, and adhesion proteins. *5.1*

4. Define diffusion. Does diffusion occur in response to a solute concentration gradient, an electric gradient, a pressure gradient, or some combination of these? *5.3*

5. Define osmosis. How do the solute concentrations of two solutions on either side of a membrane influence the osmotic movement of water down the water concentration gradient? *5.4*

6. Define hypertonic, hypotonic, and isotonic solutions. Does each term refer to a property inherent in a given type of solution? Or are they used only when comparing one solution to another? *5.4*

7. If *all* transport proteins shunt substances across cell membranes by changing shape, then how do the passive transporters differ from the active transporters? *5.5*

8. Define exocytosis and endocytosis. Describe the main features of the three pathways of endocytosis. *5.3, 5.6*

Self-Quiz *(Answers in Appendix IV)*

1. Cell membranes consist mainly of a _____ .
   a. carbohydrate bilayer and proteins
   b. protein bilayer and phospholipids
   c. lipid bilayer and proteins

2. In a lipid bilayer, _____ of lipid molecules are sandwiched between _____ .
   a. hydrophilic tails; hydrophobic heads
   b. hydrophilic heads; hydrophilic tails
   c. hydrophobic tails; hydrophilic heads
   d. hydrophobic heads; hydrophilic tails

3. Most membrane functions are carried out by _____ .
   a. proteins   c. nucleic acids
   b. phospholipids   d. hormones

4. All cell membranes of multicelled organisms incorporate _____ .
   a. transport proteins   d. recognition proteins
   b. receptor proteins   e. all of the above
   c. adhesion proteins

5. Immerse a living cell in a hypotonic solution, and water will tend to _____ .
   a. move into the cell   c. show no net movement
   b. move out of the cell   d. move in by endocytosis

6. A _____ is a device that spins test tubes and separates cell components according to their relative densities.

7. _____ can readily diffuse across a lipid bilayer.
   a. Glucose   c. Carbon dioxide
   b. Oxygen   d. b and c

8. Sodium ions cross a membrane through transport proteins that receive an energy boost. This is an example of _____ .
   a. passive transport   c. facilitated diffusion
   b. active transport   d. a and c

Critical Thinking

1. Water moves osmotically into *Paramecium*, a single-celled protistan of aquatic habitats. If unchecked, the influx would bloat the cell and rupture its plasma membrane. An energy-requiring mechanism involving contractile vacuoles expels the excess (Figure 5.19). Water enters tubelike extensions of this organelle and collects in a central space in the vacuole. When full, the vacuole contracts and squirts excess water through a pore that opens to the surroundings. Are the fluid surroundings hypotonic, hypertonic, or isotonic relative to *Paramecium*'s cytoplasm?

2. The bacterium *Vibrio cholerae* causes the disease *cholera*. Infected people have severe diarrhea and may lose up to twenty liters of fluid in a day. The bacterium enters the body if someone drinks contaminated water, then it adheres to the intestinal lining. It secretes a metabolic product that is toxic to cells of the lining,

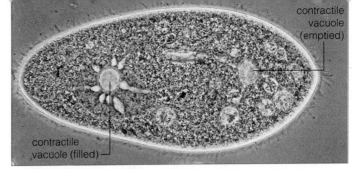

Figure 5.19 Contractile vacuoles of *Paramecium*, a protistan.

and they start secreting chloride ions (Cl^-). Sodium ions (Na^+) follow the chloride ions into the fluid in the intestines. Explain how this sequence of events causes the massive fluid loss.

3. Many cultivated fields in California require heavy irrigation. Over the years, most of the water has evaporated from the soil, leaving behind all of the irrigation water's solutes. What kinds of problems might the altered soil conditions cause for plants?

4. Certain species of bacteria thrive in environments where the temperatures approach the boiling point of water—for example, in the steam-venting fissures of volcanoes and in hot springs of Yellowstone National Park. Assume that the lipid bilayer of the bacterial cell membranes consists mainly of phospholipids. What features might the fatty acid tails of the phospholipids have that help stabilize the membranes at such extreme temperatures?

5. In hospitals, solutions of glucose having a concentration of 0.3M can be infused directly into the bloodstream of patients. The same is true of solutions of sodium chloride (NaCl), but these have a concentration of 0.15M. Only isotonic solutions can be infused safely into the blood. If that is so, then how can you explain the difference in the Molarity of the two solutions?

Selected Key Terms

active transport *5.3*	isotonic solution *5.4*
adhesion protein *5.1*	lipid bilayer *5.1*
bulk flow *5.4*	osmosis *5.4*
calcium pump *5.5*	osmotic pressure *5.4*
centrifuge *5.2*	passive transport *5.3*
concentration gradient *5.3*	phagocytosis *5.6*
diffusion *5.3*	phospholipid *5.1*
electric gradient *5.3*	plasmolysis *5.4*
endocytosis *5.3*	pressure gradient *5.3*
exocytosis *5.3*	receptor protein *5.1*
fluid mosaic model *5.1*	recognition protein *5.1*
hydrostatic pressure *5.4*	selective permeability *5.3*
hypertonic solution *5.4*	sodium-potassium pump *5.5*
hypotonic solution *5.4*	transport protein *5.1*

Readings

Bretscher, M. October 1985. "Molecules of the Cell Membrane." *Scientific American* 253(4): 100–108.

Dautry-Varsat, A., and H. Lodish. May 1984. "How Receptors Bring Proteins and Particles Into Cells." *Scientific American* 250(5): 52–58. Describes receptor-mediated endocytosis.

Singer, S., and G. Nicolson. 1972. "The Fluid Mosaic Model of the Structure of Cell Membranes." *Science* 175: 720–731.

Web Site See *http://www.wadsworth.com/biology* for practice quiz questions, hypercontents, BioUpdates, and critical thinking. The Wadsworth Biology Resource Center provides a wealth of information fully organized and integrated by chapter.

GROUND RULES OF METABOLISM

Growing Old With Molecular Mayhem

Somewhere in those slender strands of DNA in your cells are snippets of instructions for constructing two wonderful proteins. The proteins are called superoxide dismutase and catalase (Figure 6.1a). Both are enzymes, a type of molecule that makes metabolic reactions run much, much faster than they would spontaneously, on their own. And both kinds of enzymes help keep you from growing old before your time.

The two enzymes help your cells clean house, so to speak. Together, they produce and then disassemble hydrogen peroxide (H_2O_2), a normal but potentially toxic by-product of certain oxygen-requiring reactions. Oxygen (O_2) is supposed to pick up electrons from the

reactions. Sometimes it picks up only one—which is not enough to complete the reaction but is enough to give the oxygen a negative charge (O_2^-).

Like other unbound, molecular fragments with the wrong number of electrons, O_2^- is a **free radical**. Free radicals are *so* reactive, they even attach to molecules that usually will not take part in just any reaction—molecules such as DNA. An attack by free radicals can disrupt DNA's structure and destroy its function.

Enter superoxide dismutase. Under its chemical prodding, two of the rogue oxygen molecules combine with hydrogen ions. H_2O_2 and O_2 are the outcome.

Enter catalase. Under *its* prodding, two molecules of hydrogen peroxide react and split into ordinary water and ordinary oxygen: $2H_2O_2 \longrightarrow 2H_2O + O_2$.

As people age, their capacity to produce functional proteins, including enzymes, begins to falter. Among those enzymes are superoxide dismutase and catalase.

Figure 6.1 (**a**) Adult owner of skin with a spattering of age spots—visible evidence of free radicals on the loose. At one time he, like the boy shown in (**b**), had of a good supply of smoothly functioning superoxide dismutase (**c**) and (**d**) catalase molecules. Both of these enzymes help keep free radicals inside the body in check.

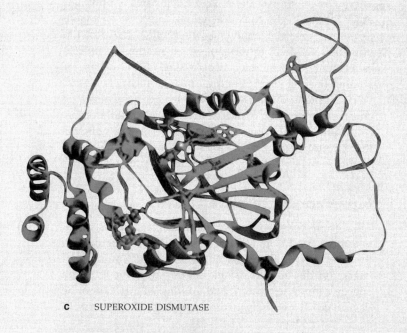

c   SUPEROXIDE DISMUTASE

d   CATALASE

Cells synthesize copies of the two enzymes in ever diminishing numbers, in crippled form, or both. When that happens, free radicals and hydrogen peroxide accumulate. Like loose cannons, they career through cells as tiny blasts at the structural integrity of proteins, DNA, membranes, and other vital components.

Cells under attack suffer or die outright. Those brown "age spots" you may have noticed on an older person's skin are evidence of assaults by free radicals. The irregular dark spots on the skin of the adult male in Figure 6.1a are examples. Each age spot is a mass of brownish-black pigment molecules that build up in cells when free radicals take over—all for the want of two enzymes.

With this chapter, we will start to examine the kinds of activities that keep all cells alive and functioning smoothly. At times the topics may seem remote from the world of your interests. But they help define who *you* are and who you will become, age spots and all.

KEY CONCEPTS

1. Cells engage in metabolism—that is, they use energy to build, stockpile, break apart, and eliminate substances in ways that help them survive and reproduce.

2. With each metabolic reaction, energy escapes into the environment. To stay alive, cells must balance their energy losses with energy gains. Yet they cannot create energy from scratch. They can only draw upon existing sources, such as light energy from the sun and chemical energy that has become tucked away in glucose and other substances.

3. Compared to their environment, cells maintain greater or lesser amounts of certain substances, as required for metabolism. They do so even though the molecules or ions of any substance have a natural tendency to diffuse into regions where they are less concentrated. Membrane transport proteins work with or against this tendency.

4. Metabolic pathways maintain, increase, or decrease the relative amounts of various substances in cells. Typically, the pathways couple reactions that release usable energy from substances to other reactions that require energy.

5. Chemical reactions proceed far too slowly on their own to sustain life. In living organisms, the action of specific enzymes greatly increases the rate of specific reactions.

6. ATP transports usable energy, in chemical form, from one reaction site to another. When it donates a phosphate group to glucose or some other substance, the substance becomes activated—that is, primed for chemical change. Metabolic pathways depend on such phosphate-group transfers.

Find a microscope and watch a living cell, suspended in a water droplet. The image might jar you: something that small pulsates with movement. Even as you watch, the cell takes in energy-rich solutes, builds membranes, stores things, replenishes enzymes, and checks out its DNA. It is alive; it is growing; it may divide in two. Multiply such activities by *trillions* of cells and you get an idea of what goes on in your own body, even when you do nothing more than sit quietly and watch a cell!

These dynamic activities are signs of **metabolism**— of a cell's capacity to acquire energy and use it to build, break apart, store, and release substances in controlled ways. Metabolism is the means by which cells survive and reproduce, and it all begins with energy.

Defining Energy

If you have ever watched a house cat stalking a mouse, you know it can "freeze" its position to avoid detection before springing at its unsuspecting prey. Like anything else in the whole universe that is stationary, the cat has a store of **potential energy**—a capacity to do work, simply owing to its position in space and the arrangement of its parts. When a cat springs, some of its potential energy is transformed into **kinetic energy**, the energy of motion. Energy on the move can do work by imparting motion to other things. In the cat's skeletal muscle cells, ATP gave up some potential energy to molecules of contractile units and set them in motion, which resulted in muscle movements. The transfer also resulted in the release of another form of kinetic energy—**heat**, or *thermal* energy.

The potential energy of molecules has its own name: **chemical energy**. It is measurable, as in kilocalories. A **kilocalorie** is the same thing as 1,000 calories, which is the amount of energy it takes to heat 1,000 grams of water from 14.5°C to 15.5°C at standard pressure.

How Much Energy Is Available?

Like single cells, you can't create energy from scratch; you must get it from someplace else. Why? According to the **first law of thermodynamics**, the total amount of energy in the universe remains constant. That is, more energy cannot be created, and existing energy cannot be destroyed. It can only be converted from one form to some other form.

Think about what the law means. The universe has only so much energy, distributed in a variety of forms. One form can be converted to another, as when corn plants absorb energy from the sun and convert it to the chemical energy of starch. After you eat corn, your cells extract energy from starch and convert it to other forms, such as kinetic energy for your movements. During each metabolic conversion, a bit of energy escapes to the surroundings, as heat. Even when you are "doing nothing," your body gives off about as much heat as a 100-watt light bulb because of conversions proceeding in your cells. The energy being released is transferred to atoms and molecules that make up the air and so "heats up" the surroundings (Figure 6.2). There, kinetic energy increases the number of ongoing, random collisions among molecules. And with each collision, a bit more energy is released as heat. However, none of the energy ever vanishes.

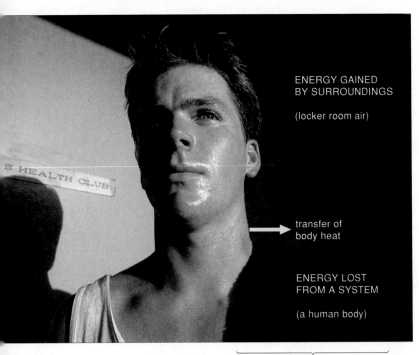

ENERGY GAINED
BY SURROUNDINGS

(locker room air)

transfer of
body heat

ENERGY LOST
FROM A SYSTEM

(a human body)

net energy change = 0

Figure 6.2 Example of how the total energy content of any system *together with its surroundings* remains constant.

"System" means all matter in a specific region, such as a human body, a plant, a DNA molecule, or a galaxy. "Surroundings" can be a small region in contact with the system or as vast as the entire universe. The system shown (a human male) is giving off heat to the surroundings (a locker room) by evaporative water loss from sweat. What one region loses, the other region gains, so the total energy content of both does not change.

The One-Way Flow of Energy

Most of the energy available for conversions in a cell resides in covalent bonds. Glucose, glycogen, starches, fatty acids, and other organic compounds have complex arrangements of many of these bonds and are said to have a high energy content. When the compounds enter metabolic reactions, certain bonds break or become rearranged. During the molecular commotion, some heat energy is lost to the surroundings. In general, cells cannot recapture energy lost as heat.

Figure 6.3 An example of the one-way flow of energy into the world of life that compensates for the one-way flow of energy out of it. The sun continuously loses energy, much of it in the form of wavelengths of light (Section 7.2). Living cells intercept some of the energy and convert it to useful forms of energy, stored in bonds of organic compounds. Each time a metabolic reaction proceeds in cells, stored energy is released—and some inevitably is lost to the surroundings, mostly as heat.

In the lower photograph, *green* "dots" are water-dwelling, photosynthetic cells (*Volvox*) living in tiny, spherical colonies. *Orange* dots are cells set aside for reproduction. They form new colonies inside the parent sphere.

ENERGY LOST
One-way flow of energy away from the sun

ENERGY GAINED
One-way flow of energy from the sun into organisms

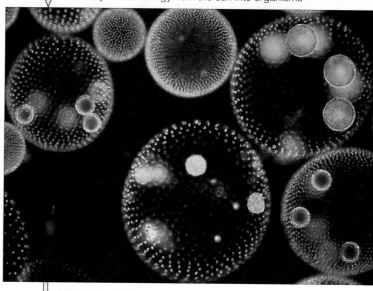

ENERGY LOST
One-way flow of energy away from organisms to the surroundings

Think of what happens when your cells break all of the covalent bonds in glucose that they can until, after many steps, six molecules of carbon dioxide and six of water remain. They do so to release usable energy, some of which is conserved in ATP. Compared to glucose, the leftovers have more stable arrangements of atoms, but the chemical energy in all of their bonds is much less than the total chemical energy of glucose. Why? *Some energy was lost at each step leading to their formation.* Said another way, as a source of energy that cells can use, carbon dioxide is lower in quality than glucose.

What about the small amount of heat energy that was transferred to the surroundings when the carbon dioxide formed? It is very low quality; it does not lend itself to conversions in cells and so cannot be used to do work.

Bad news for cells of the remote future: The amount of "low-quality" energy in the universe is increasing. Because no energy conversion can ever be 100 percent efficient, the total amount of energy in the universe is spontaneously flowing from forms of higher to lower quality. That, basically, is the point to remember about the **second law of thermodynamics**.

Without energy inputs to maintain it, any organized system tends to get disorganized over time. **Entropy** is a measure of the degree of a system's disorder. Think of the Egyptian pyramids—originally organized, presently crumbling, and many thousands of years from now, dust. The ultimate destination of those pyramids and everything else in the universe is a state of maximum entropy. Billions of years from now, all of the energy available for conversions will be dissipated.

Can it be that life is one glorious pocket of resistance to the depressing flow toward maximum entropy? After all, every time a new organism grows, new bonds form and hold atoms together in precise arrays, so molecules become more organized and have a far richer store of energy, not poorer! Yet a simple example will show that the second law does indeed apply to life on Earth.

The primary energy source for life on Earth is the sun, which has been losing energy since it first formed. Plants capture sunlight energy, convert it in various ways, then lose energy to other organisms that feed, directly or indirectly, on plants. At each energy transfer along the way, some energy is lost, usually as heat that joins the universal pool. *Overall, energy still flows in one direction.* The world of life maintains a high degree of organization only because it is being resupplied with energy that is being lost from someplace else (Figure 6.3).

The amount of energy in the universe remains constant. Energy can undergo conversions from one form to another, but it cannot be created out of nothing or destroyed.

The total amount of energy in the universe is spontaneously flowing from forms of higher to lower quality.

A steady flow of sunlight energy into the interconnected web of life compensates for the steady flow of energy leaving it.

Cells never stop building and tearing down molecules and moving them in specific directions until they die. What does the incessant juggling act accomplish? Think it through. Nearly all cells are microscopically small, they can take up and use only so many molecules at a given time, and they have only so much internal space. If they produce more of a substance than they are able to use, store, or secrete, they have big problems.

Consider *phenylketonuria*, or PKU. People affected by this heritable disorder produce a defective enzyme that prevents cells from using phenylalanine, which is one of the amino acids. When the amino acid accumulates, the excess enters reactions that produce phenylketones. An accumulation of *these* molecules damages the brain in less than a few months. In most developed countries, routine screening programs identify affected newborns, who can grow up symptom-free if they are placed on a phenylalanine-restricted diet.

Which Way Will a Reaction Run?

From the preceding chapter, you have an idea that cells control their internal concentrations of substances with respect to the surroundings, and that eukaryotic cells further control the concentrations of substances on both sides of organelle membranes. At any time, thousands of concentration gradients across cell membranes are helping to drive specific molecules and ions in specific directions. Even on their own, molecules or ions are in constant random motion, which puts them on collision courses. But the more concentrated they are, the more often they collide. Energy associated with the collisions might be enough to cause a **chemical reaction**—that is, to make a molecule combine with something else, split into smaller parts, or change its shape.

Nearly all chemical reactions in cells are reversible. In other words, they might start out in the "forward" direction, from starting substances to products. But they also can run in "reverse," with the products being converted back to starting substances. When you see a chemical equation with opposing arrows, this signifies the reaction is reversible. Each arrow means *yields*:

A + B ⇄ C
STARTING SUBSTANCES   PRODUCT

Which way such a reaction runs depends partly on the ratio of reactant to product. When the concentration of reactant molecules is high, this is an energetically favorable state, so the reaction runs spontaneously and strongly in the forward direction. When the product concentration is high enough, more molecules or ions of the product are available to revert spontaneously to reactants.

Any reversible reaction tends to run spontaneously toward **chemical equilibrium**, the time at which it will be running at about the same pace in both directions (Figure 6.4). Almost always, the *amounts* of the reactant and product molecules are not the same at that time. Picture a party with just as many people drifting in as drifting out of two rooms. The number in each room stays the same overall—say, thirty in one and ten in the other—even as the mix of people in each room changes.

Each type of reaction has a certain ratio of reactant to product molecules at chemical equilibrium. Consider the conversion of glucose-1-phosphate into glucose-6-phosphate (Figure 6.5). Both are glucose molecules with a phosphate group attached to one of their six carbon

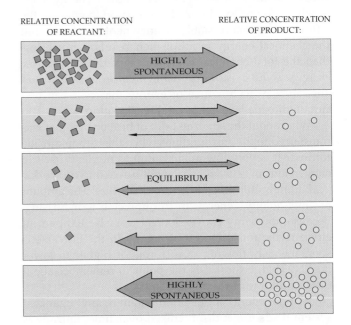

RELATIVE CONCENTRATION OF REACTANT: RELATIVE CONCENTRATION OF PRODUCT:

HIGHLY SPONTANEOUS

EQUILIBRIUM

HIGHLY SPONTANEOUS

Figure 6.4 Chemical equilibrium. When the concentration of reactant molecules is high, a reaction runs most strongly in the forward direction (to products). When the concentration of product molecules is high, it runs most strongly in reverse. At equilibrium, the rates of the forward and reverse reactions are the same.

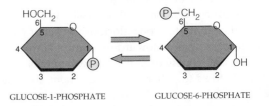

GLUCOSE-1-PHOSPHATE   GLUCOSE-6-PHOSPHATE

Figure 6.5 A reversible reaction. Glucose is primed to enter reactions when a phosphate group becomes attached to it. With a high concentration of glucose-1-phosphate, the reaction tends to run in the forward direction. With a high concentration of glucose-6-phosphate, it runs in reverse. (The 1 and 6 of these names simply identify which particular carbon atom of the glucose ring has a phosphate group attached to it.)

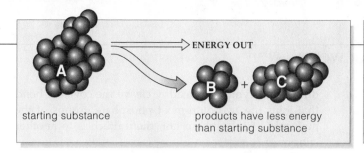

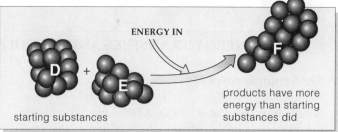

a In exergonic reactions, starting substances have more energy than the products. In some reactions, including some steps of aerobic respiration, usable energy is released. (A bit of energy also is lost, as heat.)

b During endergonic reactions, starting substances have less energy than the products. Energy inputs drive such reactions, as when energy from the sun is harnessed to drive photosynthesis.

atoms. Depending on the ratio of reactants to products at the outset, the reaction runs in the forward or reverse direction. In time the forward and reverse reactions will proceed at the same rate, but only when there are about nineteen glucose-6-phosphate molecules for every one of glucose-1-phosphate. For this particular reaction, the ratio at equilibrium is 19:1.

No Vanishing Atoms at the End of the Run

When metabolic reactions run in either the forward or reverse direction, they rearrange atoms, but they never destroy them. By the **law of conservation of mass**, the total mass of all substances entering a reaction equals the total mass of all the products. When you study any chemical equation, count up the individual atoms of each reactant and product molecule. There should be as many atoms of each element to the right of the arrow as there are to the left, even though they are combined in different forms. When you write out equations for metabolic reactions, they must balance this way.

Energy Inputs Coupled With Outputs

Many metabolic reactions can release usable energy, as when cells degrade glucose to carbon dioxide and water. The products have more stable structures; it takes more energy to break them apart. Their total bond energies are lower than they were in glucose, however, for the released energy is channeled elsewhere or lost as heat.

As Figure 6.6a suggests, an **exergonic reaction** is one that ends with a net *loss* in energy. (*Exergonic* means "energy outward.") Such reactions proceed continually in living cells. In a well-fed adult human, for example, the cumulative daily breakdown of glucose and other organic compounds by exergonic reactions releases an average of 1,200 to 2,800 kilocalories of usable energy.

An **endergonic reaction** ends with a net *increase* in energy. (*Endergonic* means "energy in.") Such reactions are energetically unfavorable. They tend not to run toward equilibrium without prodding, as with an input of energy from an outside source (Figure 6.6b).

Think of an exergonic reaction as a downhill run, with equilibrium being the bottom of the hill. Getting it

c Cells conserve energy by coupling exergonic with endergonic reactions. For instance, some usable energy released by glucose breakdown is sent elsewhere, as to the energy-requiring steps of biosynthesis.

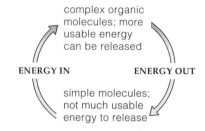

Figure 6.6 Energy changes during metabolic reactions.

to run in the reverse direction is an uphill battle. For example, that is what happens when photosynthetic cells assemble glucose molecules from carbon dioxide and water. The synthesis steps are a steep climb, and they depend on inputs of energy from the sun.

Are steep energy hills worth the battle? Absolutely. The farther endergonic reactions can be made to run uphill, the more potential energy can be stored in the products. The trick is to drive these energy-requiring reactions with dependable sources of energy.

In cells, usable energy that is released during certain reactions becomes channeled into the formation of ATP. From earlier chapters, you are already familiar with the idea that this molecule delivers energy to thousands of reaction sites in cells. Now you know its deliveries can make reactions run in directions that are energetically unfavorable, that would not happen on their own. This is how cells build starch, fatty acids, and other "energy-rich" molecules from glucose, which they build from still smaller molecules (carbon dioxide and water) of lower energy content. With this thought in mind, turn now to the remarkable role of ATP in metabolism.

A cell can simultaneously increase, decrease, and maintain the concentrations of thousands of different substances by coordinating thousands of metabolic reactions.

Certain metabolic reactions release energy when they run toward equilibrium; others require energy inputs that drive them away from equilibrium. The molecules participating in the reaction show a net loss or net gain in energy, but the total number of atoms does not change.

The coupling of energy-releasing reactions with energy-requiring reactions, as by ATP, is central to metabolism.

The Structure of ATP

Recall, from Section 3.8, that **ATP** is the abbreviation for adenosine triphosphate, which is one of the small organic compounds known as nucleotides. An ATP molecule is composed of a five-carbon sugar (ribose) to which adenine (a nucleotide base) and a string of three phosphate groups are attached, as in Figure 6.7*a* and *b*. As you will see, the triphosphate tail of the molecule is where the action is, so to speak.

Phosphate-Group Transfers

In itself, a covalent bond is a stable interaction between atoms, and this is the type of bond that holds all the component parts of an ATP molecule together. Even so, not all bonds within the molecule are equally stable. Many hundreds of

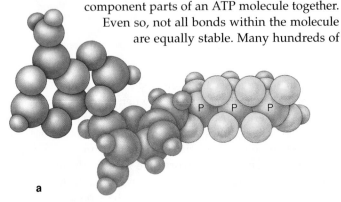

a

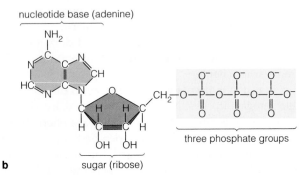

nucleotide base (adenine)

three phosphate groups

b   sugar (ribose)

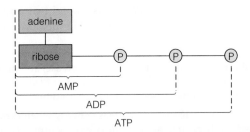

c

Figure 6.7 ATP—adenosine triphosphate, the main energy carrier in all cells. (**a**) Three-dimensional model showing ATP's component atoms. (**b**) Structural formula for ATP. (**c**) Adenosine diphosphate (ADP) forms when ATP gives up one phosphate group to another molecule. Two phosphate-group transfers leave adenosine monophosphate (AMP). 🪓

different enzymes can readily cleave the covalent bond that joins the two outermost phosphate groups of the molecule's triphosphate tail, then attach it to another substance.

The transfer of a phosphate group to a molecule of any sort is known as **phosphorylation**. A great deal of usable energy is transferred during this event. The energy is sufficient to activate hundreds of different molecules and drive hundreds of cellular activities. For example, phosphorylation is the energy transfer that drives the synthesis and breakdown of molecules, the active transport of substances across cell membranes against concentration gradients, and the contraction of muscle cells. It's been said that ATP molecules are like the coins of a nation; their phosphate groups are the common energy currency in all cells. That is why you often see a cartoon "coin" used to symbolize ATP, as shown by the sketch to the right.

Cells can renew their ATP supplies.  At certain steps of many metabolic processes, such as aerobic respiration, an unbound phosphate atom (P_i) or a phosphate group that an enzyme cleaves from some substance becomes attached to adenosine diphosphate, or **ADP**. The result is an ATP molecule. When ATP transfers a phosphate group elsewhere, it reverts to ADP, thereby completing the steps of the **ATP/ADP cycle**:

energy input → ATP → energy output (for diverse cellular reactions) → ADP + P_i

Another enzyme can remove and transfer the two outermost phosphate groups of an ATP molecule. The nucleotide that forms this way is cyclic adenosine monophosphate, or cAMP (Figure 6.7*c*). It is a player in pathways by which cells respond to outside signals.

In sum, phosphate-group transfers from ATP are a renewable, rapid, and near-universal mechanism for delivering energy where and when it is required to do work. Through such controlled transfers, energy drives transport proteins to pump solutes across membranes against concentration gradients. Transferred energy drives exergonic reactions uphill. It drives contractions and other movements of cell structures and organelles, of the cell itself, and of the multicelled organism. Figure 6.8 gives two representative examples.

ATP Output and Metabolic Pathways

At any instant, phosphate-group transfers are driving reactions that involve thousands of substances within the confines of a cell. All of the reactions don't proceed

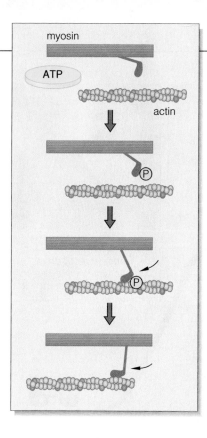

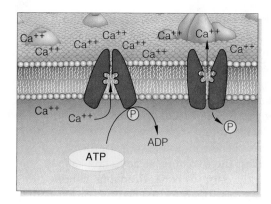

a In a contractile unit of a muscle cell, ATP transfers a phosphate group to a filament of myosin, which binds a neighboring actin filament to itself and drags it along during a short power stroke, in a direction that helps shorten the contractile unit.

b ATP transfers a phosphate group to an active transport protein, which changes shape and pumps calcium ions out of the cell, against calcium's concentration gradient.

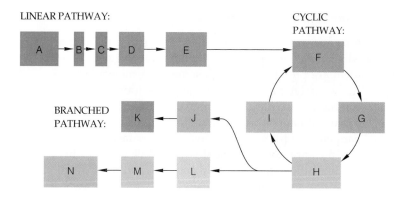

Figure 6.8 Two examples of the type of cellular work that can be triggered by phosphate-group transfers from ATP to enzymes and other participants in metabolic reactions.

willy-nilly. Most proceed in precisely ordered, enzyme-mediated sequences called **metabolic pathways**.

The main metabolic pathways are biosynthetic or degradative, overall. In a *biosynthetic* pathway, small molecules are assembled into complex carbohydrates, proteins, and other large molecules of higher energy content. In a *degradative* pathway, large molecules are broken down to products of lower energy content. Later in the book, you will consider examples of both kinds of pathways. You may find it useful to start thinking about their participants, as outlined here:

1. **Substrates** (also called reactants or precursors). The substances that enter a reaction.

2. **Intermediates**. Any of the substances that form between the start and end of a pathway.

3. **End products**. Substances remaining at the end of a metabolic reaction or pathway.

4. **Energy carriers**. ATP and a few other compounds that activate substances by transferring energy (via functional groups) to them.

5. **Enzymes**. Primarily proteins that speed specific reactions. (Some RNAs also show enzyme activity.)

6. **Cofactors**. Certain organic compounds or metal ions that assist enzymes or that transport electrons, atoms, or functional groups released during a reaction.

7. **Transport proteins**. Membrane-bound proteins by which concentration gradients are adjusted in ways that influence the direction of metabolic reactions.

Figure 6.9 Types of reaction sequences of metabolic pathways.

The steps of many metabolic pathways proceed in linear sequence, from substrate to end products. Many others proceed in a circle, with the final end product serving as the beginning reactant if its concentration becomes high enough. Intermediates or end products of one pathway very commonly become substrates that enter different metabolic pathways. In other words, two pathways (or more) often become coupled, in linear or branching fashion. Figure 6.9 shows some of the many possibilities.

ATP is the main, renewable energy carrier between sites of metabolic reactions in cells. Its deliveries couple energy-releasing reactions with energy-requiring reactions.

When an organic compound becomes phosphorylated, as by a phosphate-group transfer from ATP, its store of energy increases and it becomes primed to enter a reaction.

Metabolic reactions and pathways proceed from substrates to the end products, often with intermediates forming in between. The participants include enzymes, cofactors, and transport proteins as well as energy carriers.

ENZYME STRUCTURE AND FUNCTION

Even when you are about to fall asleep, when you think your body is shutting down for the night, its metabolic machinery is synthesizing and degrading uncountable numbers of molecules. Much of the glucose from your latest meal was dismantled earlier for a quick energy fix in cells of your brain, heart, and skeletal muscles. More glucose is being converted to fat or something else. Hemoglobin molecules built sixty days ago are being digested to bits even as new molecules replace them.

Without enzymes, the dynamic, steady state called "you" would quickly cease to exist. Reactions simply would not proceed fast enough for the body to process food, build and tear down hemoglobin and other vital molecules, send signals to and from brain cells, make muscles contract, and do everything else to stay alive.

Four Features of Enzymes

Enzymes are *catalytic* molecules; they speed the rate at which reactions approach equilibrium. Nearly all of them are proteins, although some RNA molecules also show catalytic activity. Enzymes have four features in common. *First*, they do not make anything happen that could not happen on its own. But they usually make it happen hundreds to millions of times faster. *Second*, enzymes are not permanently altered or used up in a reaction; the same one may act over and over again. *Third*, the same enzyme usually works for the forward and the reverse directions of a reaction. *Fourth*, each type of enzyme is highly selective about its substrates.

An enzyme's substrates are specific molecules that it can chemically recognize, bind, and modify in certain ways. For example, thrombin, an enzyme involved in blood clotting, recognizes a side-by-side arrangement of two specific amino acids (arginine and glycine) and cleaves a peptide bond between them:

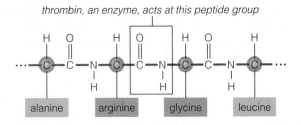

thrombin, an enzyme, acts at this peptide group

Enzyme-Substrate Interactions

A reaction occurs when participating molecules collide with some minimum amount of energy, known as the **activation energy**. This is true regardless of whether the reaction occurs spontaneously or with enzymes. Think of activation energy as an "energy hill" that must be surmounted before a reaction will proceed (Figure 6.10).

Enzymes make the energy hill smaller, so to speak.

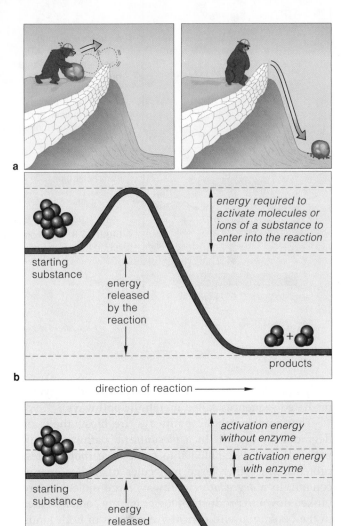

Figure 6.10 Activation energy. (**a,b**) Activation energy is like a barrier at the crest of a hill. It takes an input of energy to move a rock over the barrier before it can roll down spontaneously. (**c**) An enzyme enhances the rate at which a given number of molecules complete a reaction. It lowers the amount of energy required to boost reactants to the crest (transition state) of an energy hill.

How do they do this? Every enzyme has one or more **active sites**, or crevices in its surface where substrates interact with the enzyme and where a specific reaction is catalyzed. Figure 6.11a and b shows two views of the active site of an enzyme that phosphorylates glucose.

According to Daniel Koshland's **induced-fit model**, each substrate has a surface region that almost but not quite matches chemical groups in an active site. When substrates first settle in the site, the contact strains some of their bonds. Strained bonds are easier to break, which

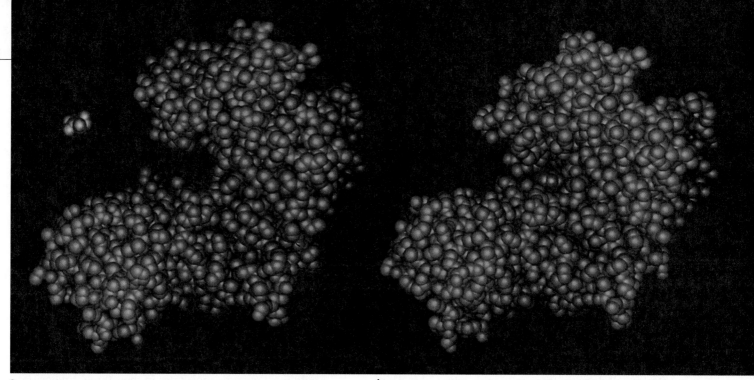

a

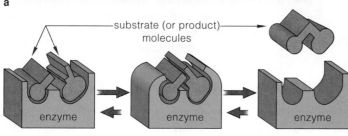

substrate (or product) molecules

enzyme   enzyme   enzyme

transition state
(tightest binding but least stable)

b

Figure 6.11 Model of an enzyme, hexokinase, at work. (**a**) The substrate is a glucose molecule (color-coded *red*). It is heading toward the active site, a cleft in the enzyme (*green*). (**b**) When it makes contact with the site, parts of the enzyme temporarily close in around it and prod the molecule to enter a reaction.

(**c**) Induced-fit model of enzyme-substrate interactions. Only when the substrate is bound in place is an enzyme's active site complementary to it. The fit is most precise during the transition state of a reaction. An enzyme-substrate complex is short-lived, for the attractive forces holding it together are usually weak.

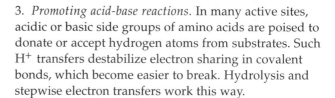

promotes formation of new bonds (in products). Also, interactions among charged or polar groups in the site often favor a redistribution of electric charge in substrates that primes them for conversion to an activated state.

When substrates fit most precisely in the active site of an enzyme, they reach an activated, *transition* state (Figure 6.11c). Substrates in the transition state react spontaneously, just as a rock that has been pushed up and over a barrier at the crest of a hill rolls down on its own (Figure 6.10a).

What induces the transition state or gets substrates over the energy barrier once the state is reached? Several mechanisms are involved, including the following:

1. *Helping substrates get together*. Substrate molecules rarely collide if their concentrations are low. By binding substrates, the active site effectively boosts their local concentrations, which may increase reaction rates by 10,000 to a billion times.

2. *Orienting substrates in positions favoring reaction*. On their own, substrates collide from random directions. By contrast, weak but extensive bonding at an active site puts their mutually attractive chemical groups on precise collision courses much more frequently.

3. *Promoting acid-base reactions*. In many active sites, acidic or basic side groups of amino acids are poised to donate or accept hydrogen atoms from substrates. Such H^+ transfers destabilize electron sharing in covalent bonds, which become easier to break. Hydrolysis and stepwise electron transfers work this way.

4. *Shutting out water*. Some active sites bind substrates so tightly that they exclude some or all water molecules. The nonpolar environment can reduce the activation energy for certain reactions, such as the attachment of a carboxyl group ($-COO^-$) to a molecule, by as much as 500,000 times.

Depending on the enzyme, such mechanisms work alone or in combination to bring about the straining and warping that convert substrates to the transition state.

Enzymes catalyze (speed) the rate at which specific reactions reach equilibrium. They do so by lowering the amount of activation energy necessary to make substrates react.

Enzymes change the rate, not the outcome, of a reaction. They only act on specific substrates. And they may catalyze the same reaction repeatedly, as long as substrates are available.

FACTORS INFLUENCING ENZYME ACTIVITY

You probably don't get much done when you feel too hot or cold, or out of sorts because you ate too many sour plums or salty potato chips. Probably when the cupboard is bare, you focus on food. Maybe you call a friend to go shopping with you, and if you drive too fast to the grocery store, police tend to slow you down. In such respects, you have a lot in common with enzymes. They, too, respond to shifts in temperature, pH, and salinity, and to the relative abundances of particular substances. Many even engage helpers for specific tasks. And all normal enzymes respond to metabolic police.

Enzymes and Environmental Conditions

Temperature, recall, is a measure of molecular motion. You may think increases in temperature must improve the rate of enzyme-mediated reactions by making the substrates collide more frequently with active sites. This is true, but only until some point on the temperature scale. Past that point—which differs among enzymes—the increased molecular motion disrupts weak bonds holding the enzyme in its three-dimensional shape. The substrates can no longer bind to the active site, and so the reaction rate declines sharply, as in Figure 6.12a.

Expose an organism to temperatures that are far higher than it normally faces and its enzymes will unravel, thereby throwing its metabolic activities into turmoil. This can happen to sick people who develop extremely high fevers. They usually will die if their internal temperature reaches 44°C (112°F). Yet enzymes of a certain bacterium (*Thermoanaeobacter*), which lives in hot springs of Yellowstone National Park, perk along until the temperature passes 78°C!

Similarly, pH values that rise or sink beyond each enzyme's range of tolerance disrupt enzyme structure and functioning (Figure 6.13). Nearly all enzymes work best in the pH range between 6 and 8. For example, trypsin is active in the mammalian intestine, where pH is around pH 8. Pepsin, a protein-digesting enzyme, is one of the exceptions. It functions in gastric fluid, even though this highly acidic fluid denatures most enzymes.

Enzyme activity also will suffer if the environment gets far saltier than is normally encountered. Extremely high ion concentrations disrupt interactions that help hold most enzyme molecules in their three-dimensional shape. Here again we find some exceptions. Enzymes of *Halobacterium* and some algae function in such salty places as Utah's Great Salt Lake.

Control of Enzyme Function

Built into each living cell are controls over its enzyme activity. By coordinating its control mechanisms, the cell maintains, increases, and decreases concentrations of substances. Certain controls adjust the rate at which some enzymes are synthesized. They can limit or beef up the number of enzyme molecules for a key step in a metabolic pathway. Other controls switch on or inhibit enzymes that are already synthesized. For example, by **allosteric control**, enzymes that have already been formed are activated or inhibited when a specific substance combines with them at a binding site *other than* the active site (*allo-* means

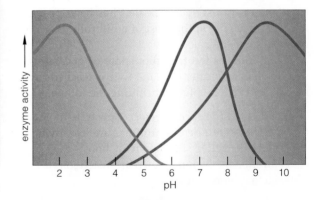

Figure 6.12 Example of how temperatures that fall outside the range of tolerance for an enzyme influence its activity. (**a**) Changes in the activity of one type of enzyme that had been subjected to changing temperatures. (**b**) Siamese cats show observable effects of such changes. Fur on the ears and paws contains more of a dark brown pigment, melanin, than the rest of the body does. A heat-sensitive enzyme controlling melanin production is less active in warmer body regions, and this results in lighter fur in those regions.

Figure 6.13 Diagram showing how the activity of three different enzymes is influenced by pH. One enzyme (the *brown* line) functions best in neutral solutions. Another enzyme (*red* line) functions best in basic solutions, and another (*purple* line) in acidic solutions.

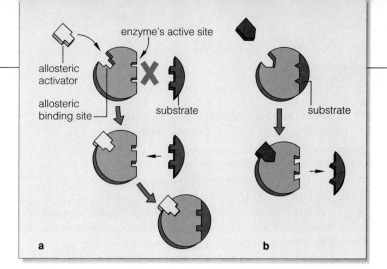

Figure 6.14 Allosteric control. (**a**) Activation of an allosteric enzyme. (**b**) Inhibition of an allosteric enzyme.

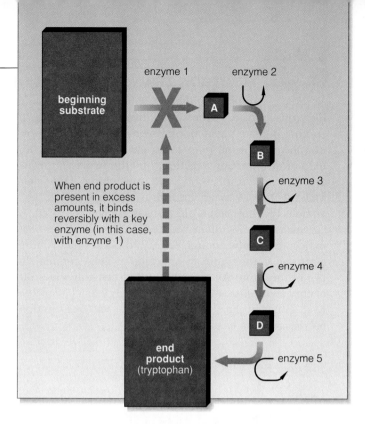

Figure 6.15 Example of feedback inhibition of a metabolic pathway. Five kinds of enzymes act in sequence to convert a substrate to an end product (tryptophan). When end product accumulates, some of the excess molecules bind to molecules of the first enzyme and thereby block the entire pathway.

different; *steric* means structure, or state). Figure 6.14 shows models of the binding, which is reversible.

Picture a bacterium, busily synthesizing tryptophan and other amino acids necessary to construct its proteins. After a bit, protein synthesis slows, so tryptophan is no longer required. But the tryptophan pathway is in full swing, so the cellular concentration of its end product continues to rise. Now **feedback inhibition** kicks in: A cellular change, caused by a specific activity, *shuts down the activity that brought it about*. In this case, the feedback loop starts and ends with a key allosteric enzyme in the pathway. Unused tryptophan molecules bind with the allosteric site. This inactivates the enzyme and blocks the pathway. By contrast, if few tryptophan molecules are around when the demand for them increases, the enzyme remains free of inhibition, and so tryptophan production rises. Such feedback loops can quickly adjust the concentrations of many substances (Figure 6.15).

In humans and other multicelled organisms, control of enzyme activity is just amazing. Cells not only work to keep themselves alive, they work with other cells in ways that benefit the whole body! Consider hormones, which are key signaling agents in this vast enterprise. Specialized cells release hormones into the blood. Any cell having receptors that are specific for a particular hormone will take it up, then its program for building a particular protein or some other activity will be altered. The hormone trips internal control agents into action—and the activities of specific enzymes change.

Enzyme Helpers

During many reactions, enzymes speed the transfer of one or more electrons, atoms, or functional groups from one substrate to another. Cofactors assist the reactions or briefly act as transfer agents. They include complex organic compounds called **coenzymes** as well as metal ions that associate with the enzyme molecule.

NAD$^+$ (for nicotinamide adenine dinucleotide) and FAD (for flavin adenine dinucleotide) are examples of coenzymes that are derived from vitamins. Both accept electrons and hydrogen atoms released during many reactions, such as glucose breakdown, and then transfer the electrons to other reaction sites. Being attracted to electrons, the unbound protons (H$^+$) go along for the ride. When loaded with electrons and hydrogen, NAD$^+$ and FAD are, respectively, abbreviated NADH and FADH$_2$. The coenzyme NADP$^+$ (nicotinamide adenine dinucleotide phosphate) functions in photosynthesis. Like NAD$^+$, it is a derivative of the vitamin niacin. When loaded down, it is abbreviated NADPH.

Ferrous iron (Fe^{++}), one of the metal ions that act as cofactors, is a component of cytochrome molecules. Cytochromes are proteins that show enzyme activity. They are embedded in cell membranes, including the membranes of chloroplasts and mitochondria.

Enzymes function best when the cellular environment stays within limited ranges of temperature, pH, and salinity. The actual ranges depend on the type of enzyme.

Control mechanisms govern the synthesis of new enzymes and stimulate or inhibit the activity of existing enzymes. By controlling enzymes, cells control the concentrations and kinds of substances available to them.

Some enzymes require cofactors (coenzymes and metal ions), which help catalyze reactions or transfer electrons, atoms, and functional groups from one substrate to another.

ELECTRON TRANSFERS THROUGH TRANSPORT SYSTEMS

Electrons, too, are transferred from one molecule to another in many metabolic pathways. Occasionally you may hear someone refer to an electron transfer as an **oxidation-reduction reaction**, but the name simply is a technical way of saying the same thing. The *donor* molecule that gives up electrons is said to be "oxidized." A molecule that *accepts* electrons is the one "reduced."

Such transfers often proceed after some atom of a molecule absorbs enough energy to boost one or more electrons farther from the nucleus or even completely out of the atom. An electron excited this way gives off energy as it returns to the lowest energy level available to it, either in the same atom or in a new one. Section 6.7 describes an example of this.

of atoms. Breaking the bonds in glucose puts electrons up for grabs, so to speak. In this metabolic pathway and many others in the cell, the released electrons are made to flow in ways that do useful work.

For example, consider an **electron transport system**, an organized array of enzymes and coenzymes that transfer electrons in sequence. One molecule donates electrons, and the next molecule in line accepts them. You find these systems in cell membranes, such as the ones inside mitochondria and chloroplasts (Figure 6.17).

An electron transport system operates by accepting electrons at higher energy levels, then releasing them at lower energy levels. Think of it as a staircase (Figure 6.16*b*). Excited electrons at the top step have the most

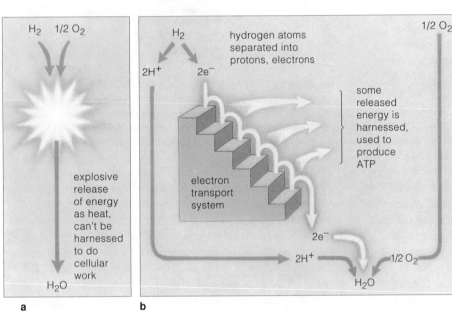

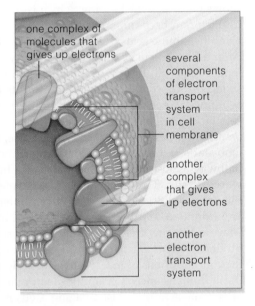

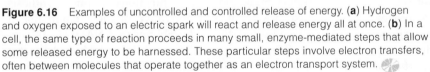

Figure 6.16 Examples of uncontrolled and controlled release of energy. (**a**) Hydrogen and oxygen exposed to an electric spark will react and release energy all at once. (**b**) In a cell, the same type of reaction proceeds in many small, enzyme-mediated steps that allow some released energy to be harnessed. These particular steps involve electron transfers, often between molecules that operate together as an electron transport system.

Figure 6.17 Example of electron transport systems embedded in the internal membrane system of chloroplasts. Each is positioned to accept electrons that sunlight energy knocks out of a neighboring complex of pigment molecules.

Cells conserve usable energy during certain electron transfers. Imagine tossing a bit of glucose into a wood fire. Its atoms would quickly let go of one another and combine with oxygen in the air, thereby forming CO_2 and H_2O. In that case, all of the released energy would be lost as heat to the surroundings (Figure 6.16*a*). By contrast, cells do not "burn" glucose all at once, so they do not waste much energy. Instead, enzymes cleave atoms from intermediates that form at different steps in a pathway by which glucose molecules are dismantled. Each step releases only some of the potential energy that has been stored in chemical bonds. Such bonds, remember, are unions between the electron structures

energy. They drop down, one step at a time, and release a bit of energy as they do. At certain steps, some energy is harnessed to do work. For instance, energy can move ions into a membrane-bound compartment and thereby set up concentration and electric gradients across the compartment's membrane. Such gradients are central to ATP formation. And that metabolic event, an absolutely central aspect of a cell's life, will occupy our attention in the next two chapters.

In many metabolic pathways, electrons as well as energy are transferred among molecules, as by electron transport systems.

6.7 YOU LIGHT UP MY LIFE—VISIBLE SIGNS OF METABOLIC ACTIVITY

BIOLUMINESCENCE Fireflies and some other organisms show **bioluminescence**: they give off flashes of light when enzymes called luciferases convert chemical energy to light energy. The reactions start when ATP phosphorylates luciferin, a protein that, when activated, emits fluorescent light. Luciferases, with the help of oxygen, convert the activated luciferin to a different chemical form. The action excites electrons in the luciferin to a higher energy level. The excited electrons quickly return to a lower energy level. As they do, they release energy in the form of light.

In Jamaica's tropical forests, click beetles known locally as kittyboos are renowned for their communal flashings (Figure 6.18*a* and *b*). Different varieties emit green, yellow-green, yellow, or orange flashes. Amazingly, biologists now make *bioluminescent gene transfers*. They insert copies of the genes that specify bioluminescence into bacteria, plants, and other organisms (Figure 6.18*c* and *d*).

PUTTING BIOLUMINESCENCE TO USE Besides being fun to think about, the transfers have practical applications. Each year, for example, 3 million people die from a lung disease caused by *Mycobacterium tuberculosis*. Different strains of this bacterium are resistant to different kinds of antibiotics, so effective treatment depends on identifying which strain is causing an infection. Bacterial cells taken from a patient can be cultured and exposed to luciferase genes. Some bacterial cells take up the genes, which, in a few cases, get inserted into their DNA. Clinicians identify the genetically modified cells, then expose colonies of their descendants to different antibiotics. If an antibiotic has no effect, the bacterial cells churn out gene products, including luciferase, and the colony glows. If it doesn't glow, this signifies the antibiotic destroyed the strain.

Christopher Contag and Pamela Contag, postdoctoral students at Stanford University, lit up bacteria that cause *Salmonella* infections in live laboratory mice! Researchers of viral or bacterial diseases infect dozens to hundreds of laboratory mice for their experiments. Before the Contags' discovery, the only option was to kill infected mice and examine their tissues to find out whether infection had occurred—a costly, tediously painstaking practice that also happens to offend animal rights activists.

The Contags approached David Benaron, a medical imaging researcher at Stanford, with a novel idea: If the infecting bacteria were bioluminescent, would the light

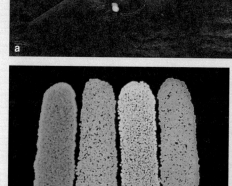

Figure 6.18 (**a**) Two kittyboos (*Pyrophorus noctilucus*). (**b**) Time-lapse photograph that reveals random paths of kittyboos on the wing. These beetles light up the night sky with bioluminescent flashes, which help potential mates find each other in the dark. (**c**) Micrograph of four colonies of bacterial cells. Each colony started with a parent bacterial cell that had taken up a kittyboo gene for a different glowing color. (**d**) A genetically modified tobacco plant that glows in the dark.

shine through tissues of *living* animals? In an experimental test of this idea, the researchers put glowing *Salmonella* cells in a thawed chicken breast. The glow showed through.

And so, in a series of experimental tests, the researchers transferred genes for bioluminescence into three strains of *Salmonella* and inoculated mice in three experimental groups. They used a digital imaging camera to track the course of infection in each group. The first strain was weakest; mice in the first group successfully fought off the infection within six days and did not glow. The second remained localized; it did not spread through the body. The third strain was dangerous; it spread rapidly through the mouse gut, and the entire gut glowed.

Bioluminescent gene transfer, in combination with the imaging of luciferase activity, has now proved its potential for verifying the effectiveness of new drug treatments (Figure 6.19). It may have application as well in *gene therapy*, whereby clinicians insert copies of functional genes into patients as replacements for defective or cancer-causing genes of their own. This is a fine story in itself, one to which we will return in Chapter 16.

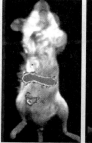
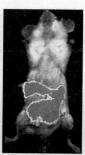

Figure 6.19 Using bioluminescent bacterial cells to chart the location and spread of infectious bacteria inside living laboratory mice. (**a**) False-color images in this pair of photographs show how infection spread in a control group that had not been dosed with antibiotics. (**b**) This pair shows how antibiotics had killed most of the infectious bacterial cells.

SUMMARY

1. Cells store, break down, and dispose of substances by acquiring and using energy from outside sources. The sum total of these energy-driven activities, called metabolism, underlies the survival of each living thing.

2. Two laws of thermodynamics affect life. First, energy undergoes conversion from one form to another, but its total amount never increases or decreases as a result of any conversion. Thus the total amount of energy in the universe holds constant. Second, energy spontaneously flows in one direction, from forms of higher to lower quality.

 a. All matter has some amount of potential energy (as measured by the capacity to do work) by virtue of its position in space and the arrangement of its parts.

 b. Potential energy may be transformed into kinetic energy, the energy of motion. Mechanical movements and heat, which corresponds to the degree of molecular motion, are two common forms of kinetic energy.

 c. Chemical energy (the potential energy inherent in molecular bonds) is often measured in kilocalories.

3. Like all organized systems, the cell tends to become disorganized without energy. It inevitably loses some of its chemical potential energy during every metabolic reaction, mainly in the form of heat. It stays organized and alive as long as it counters its energy expenditures (outputs) with energy replacements (inputs).

4. The sun is life's primary energy source. In plants and other photosynthesizers, cells trap energy from the sun's rays and convert it to the chemical energy of organic compounds. The plants, and then organisms that feed on plants and one another, use some of the energy that was stored in the organic compounds to do cellular work.

5. Most metabolic reactions proceed strongly in the forward direction (to products) when concentrations of starting substances are high. They also run strongly in reverse (to starting substances) when the concentrations of products are high.

 a. When left to themselves, reversible reactions run toward chemical equilibrium, at which time they are proceeding at about the same rate in both directions.

 b. But cellular mechanisms work with and against this tendency. At any time, reversible reactions help to increase, decrease, and maintain the concentrations of thousands of different substances required for specific reactions.

 c. Regardless of the direction of a metabolic reaction or the extent of the molecular cleavages, combinations, and rearrangements, the total number of atoms that end up in all the products will equal the total number of atoms that were present in the starting substances.

6. Metabolic reactions may release energy when they run toward equilibrium, or they may require energy inputs to drive them away from equilibrium.

7. An exergonic reaction (energy out) ends with a net loss in energy. An endergonic reaction (energy in) ends with a net gain in energy. Cells conserve a great deal of energy by coupling energy-releasing reactions with energy-requiring ones.

8. ATP is the main energy carrier in cells. It is the key coupling agent between energy-releasing and energy-requiring pathways.

9. ATP forms when energy released during specific reactions drives the phosphorylation (attachment of a phosphate molecule) to ADP. ATP later donates energy to other reactions by phosphorylating either starting substances or intermediates, which thereby become primed to enter into specific reactions.

10. Metabolic pathways are orderly, stepwise sequences of enzyme-mediated reactions.

 a. A substrate (or reactant) is a substance that enters a reaction or pathway. Intermediates are the substances that form between the reactants and end products.

 b. By the end of a biosynthetic pathway, energy-rich organic compounds have been assembled from smaller molecules of lower energy content.

 c. By the end of a degradative pathway, energy-rich molecules have been broken down to smaller ones of lower energy content.

11. Enzymes are catalysts; they greatly enhance the rate of a reaction involving specific substrates but do not alter the outcome. Nearly all enzymes are proteins, although some RNAs also show catalytic activity.

 a. Enzymes lower the activation energy required to start a reaction. They bind substrates at an active site, strain its bonds, and make the bonds easier to break.

 b. Each kind of enzyme acts best within limited ranges of temperature, pH, and salinity.

 c. Cofactors assist enzymes in speeding a reaction, or they carry electrons, hydrogen, or functional groups stripped from substrates to other sites. They include coenzymes (such as NAD^+) and metal ions.

 d. Controls stimulate or inhibit enzyme activity at key steps in metabolic pathways. They help coordinate the kinds and amounts of substances available.

12. Many energy conversions in cells involve a flow of electrons, as through electron transport systems.

Review Questions

1. Define free radical. How does it relate to aging? *CI*

2. State the first and second laws of thermodynamics. Does life violate the second law? *6.1*

3. Define and give examples of potential and kinetic energy. *6.1*

4. What can a chemical reaction do to a substrate molecule? *6.2*

5. Make a simple diagram of the ATP molecule; then highlight which part of it can be transferred to another molecule and later replaced. Why is such a transfer possible at so many different sites in the cell, and what can it accomplish? *6.3*

6. Define activation energy, then briefly describe four ways in which enzymes may lower it. 6.4

7. Define feedback inhibition as it relates to the activity of an allosteric enzyme. 6.5

8. What is an oxidation-reduction reaction? Explain how such reactions proceed in an electron transport system. 6.6

Self-Quiz (Answers in Appendix IV)

1. _____ is the primary source of energy for life on Earth.
   a. Food   c. The sun
   b. Water   d. ATP

2. Which is *not* true of chemical equilibrium?
   a. Product and reactant concentrations are always equal.
   b. The rates of the forward and reverse reactions are the same.
   c. There is no further net change in product and reactant concentrations.

3. If we liken chemical equilibrium to the bottom of an energy hill, then an _____ reaction is an uphill run.
   a. endergonic   c. ATP-assisted
   b. exergonic   d. both a and c

4. Phosphate-group transfers from ATP to another molecule are a _____ mechanism for delivering energy.
   a. rapid   c. near-universal
   b. renewable   d. all of the above

5. Which statement is *not* correct? A metabolic pathway _____ .
   a. has an orderly sequence of reaction steps
   b. is mediated by one enzyme, which initiates the reactions
   c. may be biosynthetic or degradative, overall
   d. all of the above

6. Enzymes are _____ .
   a. enhancers of reaction rates   d. not influenced by salinity
   b. influenced by temperature   e. a through c
   c. influenced by pH   f. all of the above

7. All enzymes incorporate a(n) _____ .
   a. active site   c. metal ion
   b. coenzyme   d. all of the above

8. The coenzymes NAD^+, FAD, and $NADP^+$ are _____ .
   a. cofactors   c. metal ions
   b. allosteric enzymes   d. both a and b

9. Electron transport systems involve _____ .
   a. enzymes and cofactors   c. cell membranes
   b. electron transfers   d. all of the above

10. Match each substance with the most suitable description.
    ____ coenzyme or metal ion   a. reactant or substrate
    ____ adjusts gradients at membrane b. enzyme
    ____ substance entering a reaction c. cofactor
    ____ substance formed while   d. intermediate
         a reaction is proceeding   e. product
    ____ substance at end of reaction f. energy carrier
    ____ enhances reaction rate   g. transport protein
    ____ mainly ATP

Critical Thinking

1. When cyanide, a toxic organic compound, binds to an enzyme that is a component of electron transport systems, the outcome is *cyanide poisoning*. Binding prevents the enzyme from donating electrons to a nearby acceptor molecule in the system. What effect will this have on ATP production? From what you know of ATP function, what effect will this have on a person's health?

2. AZT, or azidothymidine, is a drug used to alleviate symptoms of *AIDS* (acquired immunodeficiency syndrome). AZT is very similar in molecular structure to thymidine, one of the DNA nucleotides. AZT also can fit into the active site of an enzyme produced by the virus that causes AIDS. Infection puts the viral enzyme and viral genetic material (RNA) into a cell. There, the RNA is used as a template (structural pattern) for joining nucleotides to form a strand of DNA. The viral enzyme takes part in the assembly reactions. Propose a model to explain how AZT might inhibit replication of the virus inside cells.

3. The bacterium *Clostridium botulinum* is an obligate anaerobe, meaning it dies quickly in the presence of oxygen. It lives in oxygen-free pockets in soil, and it can enter a metabolically inactive, resting state by forming spores. The spores may end up on the surfaces of garden vegetables. When the picked vegetables are being canned, the spores must be destroyed. If they are not destroyed, *C. botulinum* may grow and produce botulinum, a toxin. When tainted canned foods are not cooked properly, the toxin remains active and causes a type of food poisoning called *botulism*. This bacterium is not able to produce either superoxide dismutase or catalase. Develop a hypothesis to explain how the inability to make the enzymes is related to its anaerobic life-style.

4. *Pyrococcus furiosus* thrives at 100°C, the boiling point of water. This species of bacterium was discovered growing in a volcanic vent in Italy. Enzymes isolated from *P. furiosus* cells do not function well below 100°C. What is it about the structure of these enzymes that allows them to remain stable and active at such high temperatures? (Hint: Review Section 3.7, which summarizes the interactions that maintain protein structure.)

Selected Key Terms

activation energy 6.4	exergonic reaction 6.2
active site 6.4	feedback inhibition 6.5
ADP 6.3	first law of thermodynamics 6.1
allosteric control 6.5	free radical CI
ATP 6.3	heat 6.1
ATP/ADP cycle 6.3	induced-fit model 6.4
bioluminescence 6.7	intermediate 6.3
chemical energy 6.1	kilocalorie 6.1
chemical equilibrium 6.2	kinetic energy 6.1
chemical reaction 6.2	law of conservation of mass 6.2
coenzyme 6.5	metabolic pathway 6.3
cofactor 6.3	metabolism 6.1
electron transport system 6.6	oxidation-reduction reaction 6.6
end product 6.3	phosphorylation 6.3
endergonic reaction 6.2	potential energy 6.1
energy carrier 6.3	second law of thermodynamics 6.1
entropy 6.1	substrate (reactant) 6.3
enzyme 6.3	transport protein 6.3

Readings

Fenn, J. *Engines, Energy, and Entropy*. 1982. New York: Freeman. Deceptively simple paperback.

Rusting, R. December 1992. "Why Do We Age?" *Scientific American* 267(6): 131–141.

Web Site See *http://www.wadsworth.com/biology* for practice quiz questions, hypercontents, BioUpdates, and critical thinking. The Wadsworth Biology Resource Center provides a wealth of information fully organized and integrated by chapter.

ENERGY-ACQUIRING PATHWAYS

Sun, Rain, and Survival

Just before dawn in the American Midwest, the air is dry and motionless. Heat has scorched the land for weeks; it still rises from the earth and hangs in the air of a new day. There are no clouds in sight. There is no promise of rain. For hundreds of miles, crops stretch out, withered and nearly dead. All the marvels of modern agriculture can't save them. In the absence of one vital resource—water—life in each cell of those hundreds of thousands of plants has ceased.

In Los Angeles, a student wonders if the Midwest drought will bump up food prices. In Washington, D.C., economists analyze crop failures in terms of the tonnage available for domestic consumption and for export.

Thousands of kilometers away, in the vast Sahel Desert of Africa, grasses and cattle are dying after a similarly unrelenting drought. Children with bloated bellies and spindly legs wait passively for death. Deprived of nourishment for too long, living cells of their bodies will never function normally again.

You are about to explore key pathways by which cells secure and use energy. At first,

Inputs of energy from sunlight drive the reactions

PHOTOSYNTHESIS
The main energy-acquiring pathway

1. Reactions convert energy from the sun to chemical bond energy of ATP.

2. ATP energy drives reactions that produce glucose, other energy-rich organic compounds.

Carbon dioxide and water are required

Oxygen is released

Carbon dioxide and water are released

AEROBIC RESPIRATION
The main energy-releasing pathway

1. Usable energy is released as reactions break down glucose, other compounds.

2. The released energy is coupled to electron transfers that bring about the formation of many ATP molecules.

Oxygen is required

ATP energy available to drive cellular tasks

Figure 7.1 Links between photosynthesis and aerobic respiration, the main energy-acquiring and energy-releasing pathways in the world of life.

these pathways may seem far removed from your daily interests. *Yet the food that nourishes you and nearly all other organisms cannot be produced or used without them.*

We will return to this point in later chapters, when we address the important issues of human population growth, nutrition, limits on agriculture, and genetic engineering, as well as the impact of pollution on crops. Here, our point of departure is the *source* of our food—which isn't a farm or a supermarket or a refrigerator. What we call "food" was put together somewhere in the world by living cells from glucose and other organic compounds. Such compounds are built on a framework of carbon atoms, so the questions become these:

1. *Where does the carbon come from in the first place?*

2. *Where does the energy come from to drive the synthesis of carbon-based compounds?*

Answers to these questions depend on an organism's mode of nutrition.

Nutritionally, many organisms are **autotrophs**. They are "self-nourishing," which is what autotroph means. They latch onto carbon dioxide (CO_2) as their source of carbon. The air all around us has an abundance of this gaseous compound, which also is dissolved in aquatic habitats. What about energy? Plants, many protistans, and some bacteria are *photo*autotrophs; they trap energy from the sun. A few bacteria are *chemo*autotrophs; they use energy that is released when they break apart a variety of inorganic substances, such as sulfur.

Many other organisms are **heterotrophs**. They can't nourish themselves with inorganic substances. To secure carbon and energy, they feed on autotrophs, each other, and organic wastes. (*Hetero·* means other, as in "being nourished by other organisms.") That is how animals, fungi, many protistans, and most bacteria stay alive.

When you think about this grand pattern of who feeds on whom, one thing becomes clear. *Survival of nearly all organisms ultimately depends on photosynthesis, the main pathway by which carbon and energy enter the web of life.* Once glucose and other organic compounds are assembled, cells put them in storage or use them as building blocks. When cells require energy, they break such compounds apart by several pathways. *Of all the energy-releasing pathways, the most common is aerobic respiration.* Figure 7.1 provides you with a preview of the chemical links between photosynthesis and aerobic respiration, the focus of this chapter and the next.

KEY CONCEPTS

1. Organic compounds, with their backbone of carbon atoms, are the structural materials and energy stores of life. Plants, like other photoautotrophs, produce their own organic compounds by photosynthesis.

2. In the first stage of photosynthesis, sunlight energy is trapped and converted to chemical bond energy in ATP molecules. In the second stage, ATP transfers energy to reaction sites where glucose is synthesized from carbon dioxide and water. The glucose monomers become the building blocks for starch and other molecules.

3. Photosynthesis is the biosynthetic pathway by which most carbon and energy enter the web of life.

4. In plant cells, photosynthesis proceeds in organelles called chloroplasts. The pathway starts at a membrane system inside the chloroplast. The required metabolic machinery includes organized arrays of light-absorbing pigments, enzymes, and the coenzyme $NADP^+$, which delivers hydrogen and electrons to the synthesis reactions.

5. Photosynthesis is often summarized this way:

$$12H_2O + 6CO_2 \xrightarrow{\text{LIGHT ENERGY}} 6O_2 + C_6H_{12}O_6 + 6H_2O$$

PHOTOSYNTHESIS—AN OVERVIEW

Photosynthesis is an ancient pathway, and it evolved in distinct ways in different photoautotrophic organisms. To keep things simple, we can focus on what goes on in weeds, lettuces, and other leafy plants (Figure 7.2).

Energy and Materials for the Reactions

The pathway has two stages, each with its own set of reactions. In the *light-dependent* reactions, energy from the sun is absorbed and converted to ATP energy. Water molecules are split and a coenzyme, $NADP^+$, picks up the liberated hydrogen and electrons to form NADPH. In the *light-independent* reactions, ATP donates energy to sites where glucose ($C_6H_{12}O_6$) is put together from carbon, hydrogen, and oxygen atoms. NADPH delivers the hydrogen, and carbon dioxide (CO_2) that diffuses

Figure 7.2 Diagrams and photomicrographs showing where photosynthesis takes place inside the leaves of sow thistle (*Sonchus*), a common weed in many parts of the world. This plant is growing next to a country road in Germany.

a

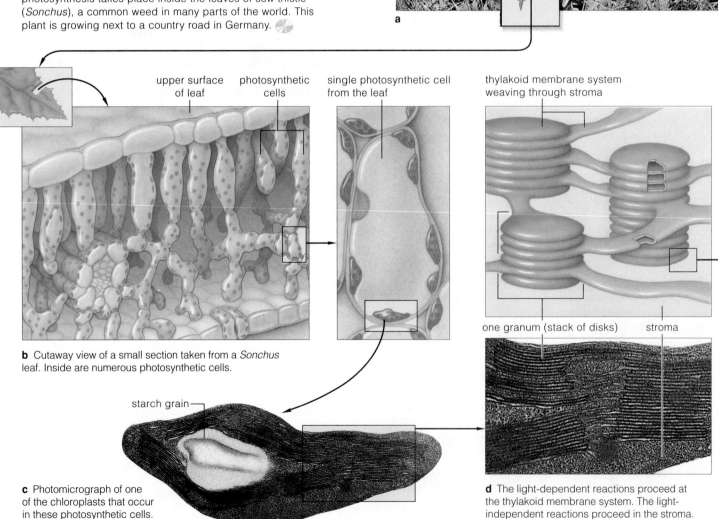

upper surface of leaf   photosynthetic cells

single photosynthetic cell from the leaf

thylakoid membrane system weaving through stroma

one granum (stack of disks)   stroma

b Cutaway view of a small section taken from a *Sonchus* leaf. Inside are numerous photosynthetic cells.

starch grain

c Photomicrograph of one of the chloroplasts that occur in these photosynthetic cells.

d The light-dependent reactions proceed at the thylakoid membrane system. The light-independent reactions proceed in the stroma.

into the leaf donates the carbon and oxygen. Overall, the reactions are often summarized in this manner:

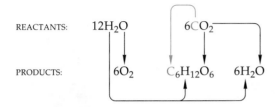

$$12H_2O + 6CO_2 \xrightarrow{\text{LIGHT ENERGY}} 6O_2 + C_6H_{12}O_6 + 6H_2O$$

If you were to track the destinations of all the atoms of the reactants, you might end up with this picture:

REACTANTS: $12H_2O$ $6CO_2$

PRODUCTS: $6O_2$ $C_6H_{12}O_6$ $6H_2O$

Notice how the summary equation shows glucose as the pathway's carbon-rich end product. Doing so does keep the chemical bookkeeping simple. However, the pathway does not really end with glucose molecules. Glucose and other simple sugars combine at once to form sucrose, starch, and other carbohydrates—the true end products of photosynthesis.

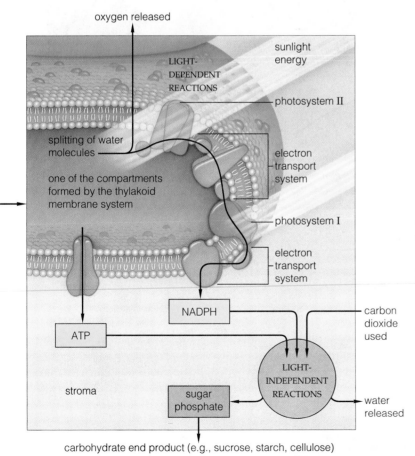

e Overview of the sites where the key steps of both stages of reactions proceed. The sections to follow will fill in the details.

Where the Reactions Take Place

The two stages of photosynthesis proceed at different sites inside each **chloroplast**. Only the photosynthetic cells of plants and some protistans contain this type of organelle (Section 4.7). Each chloroplast, recall, has two outer membranes that are wrapped around a largely fluid interior, called the **stroma**. Another membrane extends through the stroma. Often this membrane is repeatedly folded into numerous channels and stacked disks. Each stack of disks is a granum (plural, grana).

The first stage of photosynthesis proceeds at this much-folded inner membrane, which is the **thylakoid membrane system** (Figure 7.2*e*). The spaces within its channels and disks interconnect, as a compartment that collects hydrogen ions. The flow of these ions across the membrane is required for ATP production. The second stage of photosynthesis—the set of reactions by which sugars are assembled—takes place in the stroma.

The diagram below is a simplified version of Figure 7.2. Where you see it repeated in sections to follow, use it as a reminder to refer back to this figure to reinforce your grasp of how the details fit into the big picture:

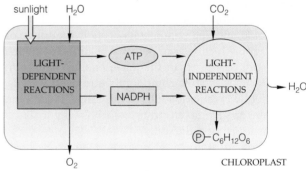

Before turning to the details, think about this: Two thousand chloroplasts, lined up single file, would be no wider than a dime. Imagine all the chloroplasts in just one weed or one lettuce leaf—each a tiny factory for producing sugars and starch—and you get an idea of the magnitude of metabolic events required to feed you and all other organisms on this planet.

Chloroplasts are organelles that are present only in plants and some protistans. They specialize in photosynthesis.

In the first stage of photosynthesis, energy from the sun's rays is absorbed and converted to ATP energy. This stage takes place at a membrane system inside the chloroplast.

In the second stage, hydrogen, carbon, and oxygen combine to form carbohydrates. This second stage takes place in the stroma, which is located between the chloroplast's outer and inner membrane systems.

SUNLIGHT AS AN ENERGY SOURCE

Each second, more than 2 million metric tons of the sun's mass enter thermonuclear reactions that release stupendous amounts of energy. It takes a mere eight minutes for a fraction of the released energy to reach the Earth's atmosphere, 160 million kilometers away. Each day, that fraction averages 7,000 kilocalories per square meter of the Earth as a whole. Almost a third is reflected back into space. Of the amount that does reach the Earth's surface, only about 1 percent is intercepted by photoautotrophs, the entry point for a one-way flow of energy through nearly all webs of life.

Properties of Light

Energy released from the sun radiates through space in an undulating motion, a bit like waves crossing the ocean. These are different forms of energy in motion. In both cases, however, the horizontal distance between the crests of every two successive waves is known as a **wavelength** (Figure 7.3*a*).

Wavelengths of radiant energy range from gamma rays, which are less than 10^{-5} nanometers long, to radio waves, which are more than 10 kilometers long. The entire range of all of these wavelengths represents the **electromagnetic spectrum** (Figure 7.3*b*).

Photoautotrophs can only harness the wavelengths between 400 and 750 nanometers. This is the range of **visible light**, which we can perceive as different colors. Wavelengths shorter than this have enough energy to break bonds of organic compounds and destroy cells. Ultraviolet (UV) radiation falls in this range. It can even force atoms to give up electrons, which is why it also is known as ionizing radiation.

Hundreds of millions of years ago, an ozone layer formed in the atmosphere and shielded the Earth from incoming ultraviolet radiation. Before this happened, the lethal bombardment kept photosynthetic bacteria

and other organisms confined below the surface of the seas. We return to this topic in Section 21.3.

Besides having wavelike properties, visible light also behaves as if its energy is contained in distinct *packets*. We call such a packet of light energy a **photon**. Different photons have different but fixed amounts of energy. The most energetic photons travel as very short wavelengths that correspond to blue-violet light. The least energetic photons travel as long wavelengths that correspond to red light (Figure 7.3*b*).

a

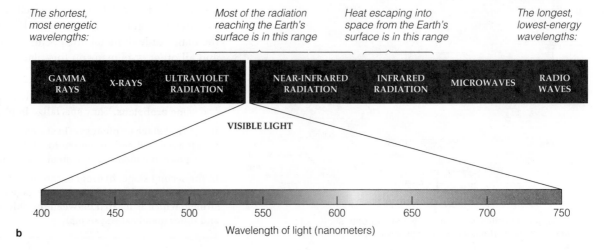

Figure 7.3
(**a**) Examples of wavelengths, the horizontal distance between crests of successive waves.
(**b**) The entire electromagnetic spectrum. Visible light is only one of many forms of electromagnetic radiation in the spectrum.

b

The shortest, most energetic wavelengths:

Most of the radiation reaching the Earth's surface is in this range

Heat escaping into space from the Earth's surface is in this range

The longest, lowest-energy wavelengths:

| GAMMA RAYS | X-RAYS | ULTRAVIOLET RADIATION | NEAR-INFRARED RADIATION | INFRARED RADIATION | MICROWAVES | RADIO WAVES |

VISIBLE LIGHT

400   450   500   550   600   650   700   750

Wavelength of light (nanometers)

Figure 7.4 T. Englemann's 1882 observational test that correlated photosynthetic activity of *Spirogyra*, a strandlike green alga, with certain wavelengths of light. Oxygen is a by-product of photosynthesis. Like many organisms, certain free-living bacterial cells use oxygen for aerobic respiration. These cells move toward places where conditions favor their activities and away from less favorable conditions.

As Englemann reasoned, oxygen-requiring bacteria living in the same habitats as the green alga would congregate where oxygen was being produced. He put a strand of the alga in a drop of water containing the bacteria and then mounted it on a microscope slide. He positioned a crystal prism to cast a spectrum of colors across the slide. The bacterial cells congregated mostly where violet and red wavelengths crossed the algal strand. Englemann concluded these wavelengths are most effective for photosynthesis (and oxygen production).

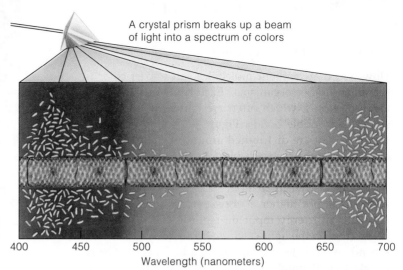

A crystal prism breaks up a beam of light into a spectrum of colors

Wavelength (nanometers)

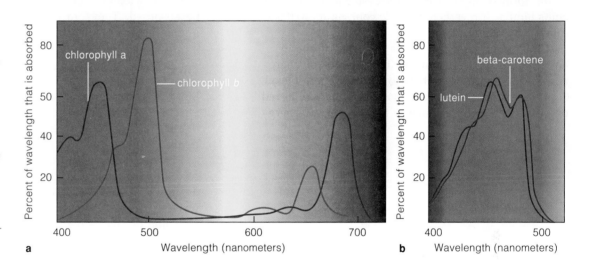

Figure 7.5 (**a**) Two absorption spectra showing the efficiency with which two kinds of chlorophylls absorb certain wavelengths of visible light. (**b**) Similar absorption spectra for two kinds of accessory pigments. Peaks and valleys in the ranges of absorption correspond to measured amounts of energy absorbed from wavelengths and used in photosynthesis.

a   Wavelength (nanometers)

b   Wavelength (nanometers)

Pigments—The Molecular Bridge From Sunlight to Photosynthesis

The sun's rays contain all wavelengths of light, which collectively look white to us. A pyramid-shaped crystal prism that intercepts white light can sort it out into its component colors by bending different wavelengths by different degrees. As early as the 1800s, T. Englemann, a botanist, used prisms to find out which wavelengths were driving photosynthesis in a green alga (Figure 7.4). However, molecular biology was far into the future, so he could not identify the bridge between sunlight and photosynthetic activity.

Pigments are the molecular bridge. These molecules absorb wavelengths of light, and organisms put them to many uses. Most absorb only some wavelengths and transmit the rest. A few, such as the melanins in animals, absorb so many wavelengths they appear dark or black.

An **absorption spectrum** is a diagram that shows how effectively a given pigment molecule absorbs different wavelengths in the spectrum of visible light. Consider

chlorophylls, the main pigments of photosynthesis. As Figure 7.5a shows, chlorophylls absorb all wavelengths except very little of the green and yellow-green ones, which they mainly transmit. (That is why plant parts that are chockful of chlorophylls appear green to us.) Other, *accessory* pigments that impart different colors also are present in different photoautotrophs. Figure 7.5b and the next section show how their assistance extends the range of wavelengths that can drive photosynthesis.

Radiation from the sun travels in waves that differ in length and energy content. Short wavelengths pack the most energy.

We perceive wavelengths of visible light as different colors and measure their energy content in packets called photons.

Chlorophylls and certain other pigments absorb specific wavelengths of visible light. They are the molecular bridge between the sun's energy and photosynthetic activity.

THE RAINBOW CATCHERS

The Chemical Basis of Color

How is it that pigment molecules can impart color to plants and other organisms? Each has a light-catching array of atoms, which often are joined by alternating single and double bonds. Figure 7.6 shows examples. Electrons that are distributed around the atomic nuclei of such arrays can absorb photons of specific energies, which correspond to specific colors of light.

Remember, when an atom's electrons absorb energy, they move to a higher energy level (Section 2.3). In a pigment, an energy input destabilizes the distribution of electrons in the light-catching array. Within 10^{-15} of a second, excited electrons return to a lower energy level, the electron distribution stabilizes, and energy is emitted in the form of light. When any destabilized molecule emits light as it reverts to a stable configuration, this is called a **fluorescence**.

Excitation occurs *only* if the quantity of energy of an incoming photon matches the amount of energy required to boost an electron to a higher energy level. Suppose a sunbeam strikes a pigment. Photons of its red or orange wavelengths match the amount of energy

required for the boost, so they are absorbed. Photons of its blue or violet wavelengths are a mismatch; they can't be absorbed. They are transmitted, reflected, and so impart a blue or violet color to the molecule.

On the Variety of Photosynthetic Pigments

Most pigments respond to only part of the rainbow of visible light. If acquiring energy is so vital for life, why doesn't each photosynthetic pigment go after the whole rainbow? In other words, why isn't each one black?

If the first photoautotrophs evolved in the seas, then so did their pigments. Ultraviolet and red wavelengths do not penetrate water as deeply as green and blue wavelengths do. Possibly natural selection favored the evolution of diverse pigments at different depths. Many existing red algae *are* nearly black, and they live in deep water. Green algae live in shallow water, and their main pigments respond to red wavelengths. Their accessory pigments help harvest light, and some even function as shields against ultraviolet radiation.

Today, the **chlorophylls** are the main pigments in all but one marginal group of photoautotrophs. Remember, photosynthetic pigments respond best to red and blue-to-violet light. The grass-green chlorophyll *a* is the main pigment inside chloroplasts (Figure 7.6). Chlorophyll *b*, a bluish-green pigment, occurs in plants, green algae, as well as a few photoautotrophic bacteria.

All photoautotrophs also produce **carotenoids**. These accessory pigments absorb blue-violet wavelengths that chlorophylls miss. Usually their backbone has a carbon ring at each end. Many types, including beta-carotene, are pure hydrocarbons (Figure 7.6). The xanthophylls also incorporate oxygen. Carotenoids impart yellow, orange, and red colors to many flowers, fruits, and vegetables. Often they have proteins attached, in which case they appear purple, violet, blue, green, brown, or black.

Carotenoids are less abundant than the chlorophylls in green leaves, but in many plants they become visible in autumn (Figure 7.7). Three weeks a year, tourists

CHLOROPHYLL *a*

BETA-CAROTENE

Figure 7.6 Structural formulas for two pigments, chlorophyll *a* and beta-carotene. The light-catching array is shaded in the color of light it transmits. The backbone is pure hydrocarbon, which readily dissolves in the lipid bilayer of cell membranes.

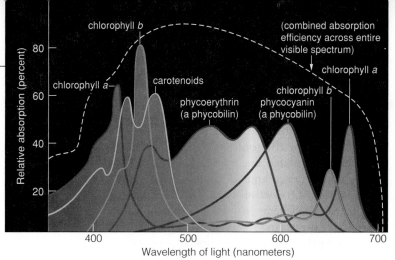

Figure 7.7 Changes in leaf color in autumn. In intensely green leaves, the abundance of chlorophyll masks the presence of carotenoids and other pigments. Chlorophyll molecules are synthesized and degraded on a regular basis, but in many species, the gradual decline of daylength in autumn and other seasonal factors cause their synthesis to lag behind. Accessory pigments are unmasked, and more colors show through.

Also in autumn, water-soluble anthocyanins accumulate in central vacuoles of leaf cells. These pigments appear red if fluids being transported through plants are slightly acidic, blue if the fluids are basic (alkaline), or colors in between if they are at intermediate pH levels. Soil conditions contribute to the pH differences. Birch, aspen, and other deciduous trees always have the same color in autumn. The leaf color of maple, ash, sumac, and other species varies around the country. It even varies from one leaf to the next, depending on the pigment combinations.

Figure 7.8 Combined energy absorption efficiency, indicated by the dashed line, of certain chlorophylls (*green lines*), carotenoids (*yellow-orange lines*), and phycobilins (*purple and blue lines*).

spend about a billion dollars just to watch the demise of chlorophyll in the deciduous trees of New England.

Other accessory pigments include the red and blue **anthocyanins** and **phycobilins**. The anthocyanins are pigments in many flowers, and the phycobilins are the signature pigments of red algae and cyanobacteria.

Collectively, the chlorophylls and accessory pigments can absorb nearly all wavelengths of visible light, as the absorption spectra in Figure 7.8 suggest.

What Happens to the Absorbed Energy?

Let's now consider what happens after electrons of a pigment's light-catching array have absorbed photons. Again, when electrons return to a lower energy level, a pigment molecule quickly gives up energy. If nothing else were around to intercept it, all the released energy would escape as fluorescent light and heat. However, pigments are not isolated. They have neighbors.

For example, peppering the thylakoid membrane of chloroplasts are **photosystems**, which are clusters of

proteins and 200 to 300 pigments. Most of the pigments only harvest photon energy. Then, instead of releasing energy as a fluorescent afterglow, a harvester directly transfers energy to a neighbor. As it reverts to a stable state, its neighbor enters a state of excitation, and so on. In this way, excitation energy randomly "walks" among the light-harvesting pigment molecules (Figure 7.9).

With each energy transfer, a bit of energy is lost as heat. Within 10^{-15} of a second, though, the energy still left corresponds to a wavelength that only a specialized chlorophyll a molecule can trap. By virtue of its position, that molecule is at the photosystem's **reaction center**. It accepts excitation energy but doesn't pass it on. Rather, the reaction center *gives up electrons* for photosynthesis. As you will see, a primary acceptor molecule, poised at the start of a transport system, accepts the gift.

About Those Roving Pigments

There are only about eight classes of pigments, but this limited group gets around in the world. For example, animals synthesize some pigments (such as melanin) but not carotenoids. These originate with photoautotrophs and move up through food webs, as when tiny aquatic snails feed on green algae and flamingos eat the snails. Flamingos modify the ingested carotenoids in diverse ways. For instance, beta-carotene molecules are split to form two molecules of vitamin A, the precursor of a visual pigment (retinol) that transduces light to electric signals in the flamingo's eyes. Beta-carotene molecules also are dissolved in fat reservoirs under the flamingo's skin, where they are taken up by skin cells that differentiate and give rise to glorious pink feathers.

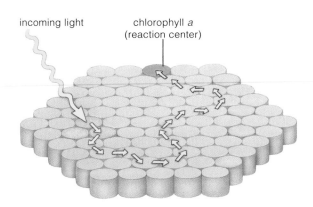

Figure 7.9 Random walk of energy from photons among the pigments of a photosystem. Excitation energy quickly reaches a specialized chlorophyll a molecule, at the system's reaction center. Energy that reaches it cannot escape. When sufficiently activated, a reaction center gives up electrons to a nearby transport system.

Chlorophylls and accessory pigments interact in light-harvesting photosystems. Photon absorption initiates a random walk of excitation energy among them and ends at a specialized chlorophyll a at the system's reaction center, which gives up electrons to ATP-producing machinery.

THE LIGHT-DEPENDENT REACTIONS

We are now ready to look more closely at the first stage of photosynthesis, the **light-dependent reactions**, as it proceeds in chloroplasts. Three events unfold during this stage. *First*, the pigments of photosystems absorb photon energy and give up excited electrons. *Second*, transfers of electrons and hydrogen through electron transport systems lead to ATP and NADPH formation. *Third*, the pigments that gave up electrons in the first place get electron replacements.

The ATP-Producing Machinery

The chloroplast's thylakoid membrane incorporates the light-harvesting photosystems, many thousands of them in some plants (Figure 7.10). Excited electrons from the reaction center of each photosystem are picked up by a primary acceptor molecule, which transfers them to an adjacent electron transport system in the membrane.

Electron transport systems, remember, are organized arrays of enzymes, coenzymes, and other proteins in certain cell membranes. As described earlier in Section 6.6, one component transfers electrons to another in orderly sequence. Some energy escapes at each transfer, but much of it is harnessed to drive specific reactions. We will focus on some of these reactions in Section 7.5. For now, simply note that photosystems and electron transport systems work together as the machinery that produces ATP, NADPH, or both. Chloroplasts happen to contain two different models of this machinery. They allow plants to produce ATP by two different pathways —one cyclic and the other noncyclic.

Cyclic Pathway of ATP Formation

The simplest pathway starts when a primary acceptor molecule picks up excited electrons from the reaction center, designated *P700*, of a *type I* photosystem. In the **cyclic pathway of ATP formation**, the electrons "cycle"

from the reaction center, through an adjacent electron transport system, then back to P700. Energy associated with the electron flow drives the formation of ATP from the chloroplast's pool of ADP and unbound phosphate:

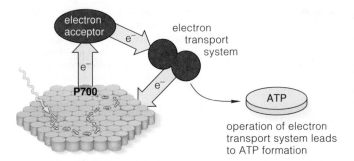

operation of electron transport system leads to ATP formation

The cyclic pathway is probably the oldest means of ATP production. The first cells to use it were as tiny as existing bacteria, so their body-building programs were scarcely enormous. ATP alone would have provided enough energy to build enough organic compounds. Building larger organisms requires far greater amounts of organic compounds—and vast amounts of hydrogen atoms and electrons. Long ago, in the forerunners of plants, the machinery of the cyclic pathway apparently underwent expansion and became the basis of a more efficient ATP-forming pathway. Amazingly, this bit of modified machinery changed the course of evolution.

Noncyclic Pathway of ATP Formation

The cyclic pathway still operates in the chloroplasts of trees, weeds, lettuces, and other members of the plant kingdom. But the **noncyclic pathway of ATP formation** now dominates. Electrons are not cycled through this pathway. They depart with NADPH, *and electrons from water molecules replace them.*

The pathway goes into operation when rays from the sun bombard a *type II* photosystem. Photon energy

Figure 7.10 Location of photosystems and electron transport systems relative to the lipid bilayer of the chloroplast's inner, thylakoid membrane system. All the components of electron transport systems are dark green. (The components of one transport system extend from P680 of photosystem II to P700 of photosystem I. The components of the second transport system extend from P700 and deliver electrons and hydrogen to NADP+.)

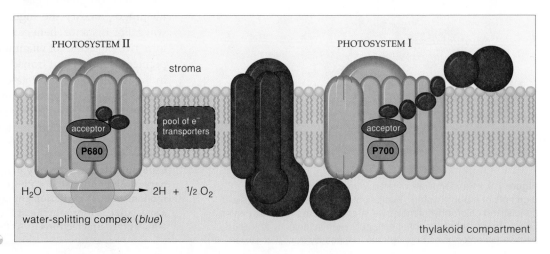

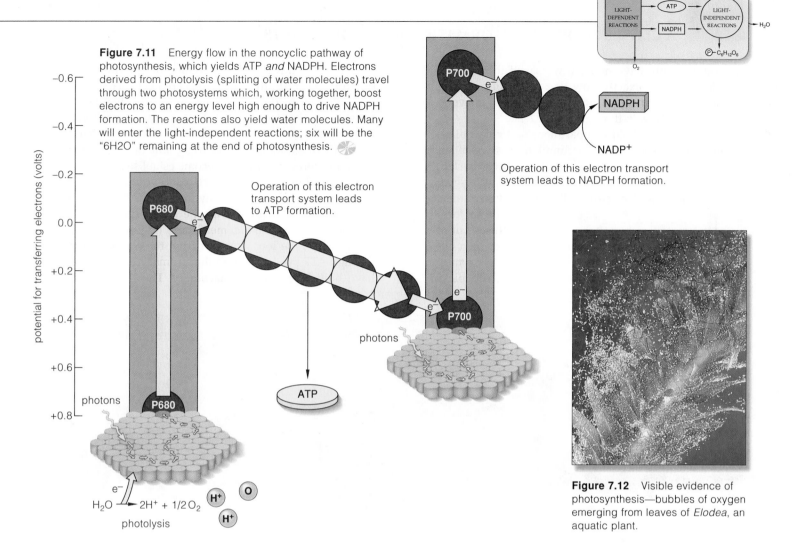

Figure 7.11 Energy flow in the noncyclic pathway of photosynthesis, which yields ATP *and* NADPH. Electrons derived from photolysis (splitting of water molecules) travel through two photosystems which, working together, boost electrons to an energy level high enough to drive NADPH formation. The reactions also yield water molecules. Many will enter the light-independent reactions; six will be the "6H2O" remaining at the end of photosynthesis.

Operation of this electron transport system leads to ATP formation.

Operation of this electron transport system leads to NADPH formation.

NADPH

NADP$^+$

photons

ATP

potential for transferring electrons (volts)

photons

e$^-$

$H_2O \longrightarrow 2H^+ + 1/2 O_2$

photolysis

Figure 7.12 Visible evidence of photosynthesis—bubbles of oxygen emerging from leaves of *Elodea*, an aquatic plant.

makes this photosystem's reaction center (chlorophyll P680) give up electrons. The energy input also triggers **photolysis**, a reaction sequence in which molecules of water split into oxygen, hydrogen ions, and electrons. P680 attracts the less excited electrons as replacements for the excited ones that got away (Figure 7.11).

Meanwhile, excited electrons are moving through a transport system, then to P700 of *photosystem I*. They still have some extra energy left, and they get even more upon their arrival—because photons also happen to be bombarding P700. The energy boost in photosystem I puts electrons at a higher energy level that allows them to enter a second transport system. There a coenzyme, NADP$^+$, assists one of the enzyme components of the second electron transport system. It becomes NADPH when it accepts two of the electrons and a hydrogen ion from the enzyme, *then transfers them to the sites where organic compounds are built.*

The Legacy—A New Atmosphere

On sunny days, on the surfaces of aquatic plants, you can see bubbles of oxygen (Figure 7.12). The oxygen is a by-product of the noncyclic pathway of photosynthesis. This pathway may have evolved more than 2 billion years ago. At first, the oxygen simply dissolved in seas, lakes, wet mud, and other bacterial habitats. By about 1.5 billion years ago, significant amounts of dissolved oxygen were escaping into what had been an oxygen-free atmosphere. The accumulation of oxygen changed the atmosphere forever. And it made possible aerobic respiration—the most efficient pathway for releasing usable energy stored in organic compounds. Ultimately, the emergence of the noncyclic pathway allowed you and all other animals to be around today, breathing the oxygen that helps keep your cells alive.

In the light-dependent reactions, electrons of photosystem reaction centers are raised to higher energy levels, released, then transferred through nearby electron transport systems. Their energy drives ATP and NADPH formation. ATP can deliver energy and NADPH can deliver hydrogen and electrons to sites where organic compounds are built.

Oxygen, a by-product of photosynthesis, profoundly changed the early atmosphere and made aerobic respiration possible.

A CLOSER LOOK AT ATP FORMATION IN CHLOROPLASTS

As you probably have noticed, we saved the trickiest question for last. Exactly how does ATP form during the noncyclic pathway of photosynthesis? To arrive at the answer, let's walk through Figure 7.13.

Inside the thylakoid compartment of a chloroplast, many hydrogen ions (H⁺) accumulate as a result of photolysis, the first step of the pathway. At this step, enzyme action splits water molecules into oxygen, hydrogen ions, and electrons. The two oxygen atoms combine at once to form O_2, which diffuses out of the chloroplast, then out of the cell. A primary acceptor molecule picks

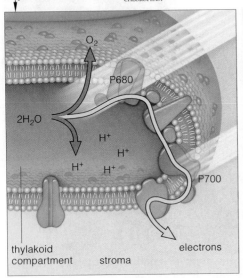

Therefore, through a combination of photolysis and electron transport, hydrogen ions are ever increasingly concentrated inside the compartment than they are in the stroma. The unequal distribution of these positively charged ions also creates a difference in electric charge across the membrane. An electric gradient, as well as a concentration gradient, has become established.

The combined force of the H⁺ concentration gradient and electric gradient propels hydrogen ions through the interior of ATP synthases, a type of transport protein that spans the thylakoid membrane (Figure 7.13c). In other words, the ions flow out from the compartment, through ATP synthases, into the stroma. ATP synthases have built-in enzymatic machinery. The flow of ions

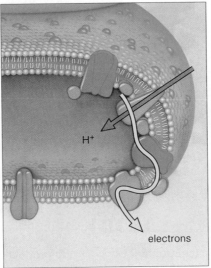
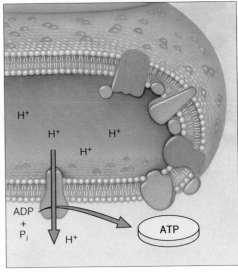

a Hydrogen ions derived from the splitting of water molecules (photolysis) accumulate inside the thylakoid compartment. (The two oxygen atoms combine to form O_2 and simply diffuse out of the chloroplast, then out of the photosynthetic cell.)

b More hydrogen ions accumulate in the compartment as certain components of electron transport systems accept excited electrons. They pick up hydrogen ions in the stroma and shunt them across the membrane at the same time.

c The ion accumulation sets up concentration and electric gradients across the thylakoid membrane. The ions follow the gradient and flow into the stroma, through the interior of ATP synthases embedded in the membrane. The flow drives ATP formation from ADP and P_i.

Figure 7.13 How ATP forms in chloroplasts during the noncyclic pathway of photosynthesis.

up the excited electrons and transfers them to a nearby electron transport system that is also positioned in the thylakoid membrane. However, as indicated in Figure 7.13a, the hydrogen ions derived from water remain in the thylakoid compartment.

Still more hydrogen ions enter the compartment when the electron transport systems are operating. This is true of both the cyclic and noncyclic pathways. As Figure 7.13b indicates, at the same time that certain components of the transport system accept electrons, they also pick up hydrogen ions from the stroma. They immediately shunt the ions across the membrane.

through them drives the machinery, which catalyzes the attachment of unbound phosphate to ADP. In this way, an ATP molecule forms.

The sequence of events just described is often called the *chemiosmotic theory* of ATP formation. As you will see in the next chapter, the same theory applies to ATP formation in mitochondria as well as in chloroplasts.

In chloroplasts, H⁺ concentration and electric gradients form across the thylakoid membrane. The flow of ions from the thylakoid compartment to the stroma drives ATP formation.

LIGHT-INDEPENDENT REACTIONS

Light-independent reactions are the "synthesis" part of photosynthesis. ATP delivers the required energy for the reactions. NADPH delivers the required hydrogen and electrons. Carbon dioxide (CO_2) in the air around photosynthetic cells provides the carbon and oxygen.

We say the reactions are light-independent because they do not depend directly on sunlight. They proceed even in the dark, as long as ATP and NADPH hold out.

Capturing Carbon

Let's track a CO_2 molecule that diffuses into air spaces inside a leaf and ends up next to a photosynthetic cell. It diffuses into the cell, then into a chloroplast's stroma. An enzyme attaches the carbon atom of CO_2 to **RuBP** (ribulose bisphosphate), a compound with a backbone of five carbon atoms. The enzyme is called "rubisco" (for RuBP carboxylase). Its action produces an unstable six-carbon intermediate that splits into two molecules of **PGA** (phosphoglycerate). PGA is a stable molecule with a three-carbon backbone. The incorporation of a carbon atom from CO_2 into a stable organic compound is called **carbon fixation**.

Carbon fixation is the first step of a cyclic pathway, the **Calvin-Benson cycle**, that yields a sugar phosphate molecule and regenerates RuBP. For our purposes, we can simply focus on the carbon atoms of the substrates, intermediates, and end products, as Figure 7.14 shows.

Building the Glucose Subunits

Each PGA receives a phosphate group from ATP, then hydrogen and electrons from NADPH. The resulting intermediate is called **PGAL** (phosphoglyceraldehyde). To build *one* six-carbon sugar phosphate, the carbon from six CO_2 molecules must be fixed and twelve PGAL must form. Most of the PGAL is rearranged to form new RuBP, which can be used to fix more carbon. But two of the PGAL combine, thus forming glucose (six carbons) with an attached phosphate group.

When phosphorylated this way, glucose is primed to enter other reactions. Plants use it as a building block for their main carbohydrates—sucrose, cellulose, and starch. *The synthesis of these organic compounds by other pathways marks the end of the light-independent reactions.*

With six turns of the cycle, enough RuBP molecules form to replace the ones used in carbon fixation. The ADP, NADP+, and phosphate leftovers diffuse through the stroma, to the light-dependent reaction sites. There they are converted back to NADPH and ATP.

Photosynthetic cells convert newly formed sugar phosphates to sucrose or starch during daylight hours. Of all plant carbohydrates, sucrose is the most easily transportable. Starch is the most common storage form.

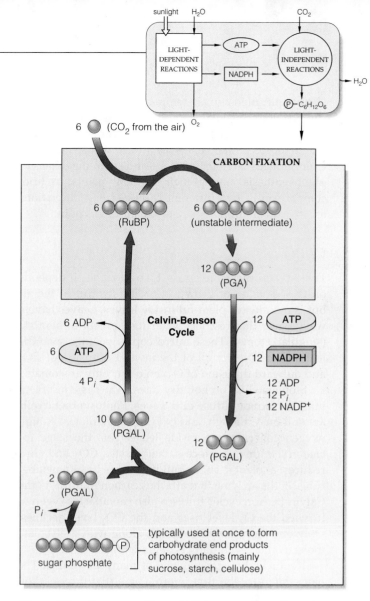

Figure 7.14 Summary of the light-independent reactions of photosynthesis. *Red* circles signify the carbon atoms of the key molecules. All of the intermediates have one or two phosphate groups attached. For simplicity, this diagram shows only the phosphate group on the resulting sugar phosphate. Remember also, many water molecules that formed in the light-dependent reactions enter this pathway, and six remain at its conclusion.

These cells also convert excess PGAL to starch. They briefly store starch as grains in the stroma. (You can see one of these grains in Figure 7.2c.) After the sun goes down, cells convert starch to sucrose, for export to other living cells in leaves, stems, and roots. Ultimately, the products and intermediates of photosynthesis end up as energy sources and building blocks for all of the lipids, amino acids, and other organic compounds that are required for growth, survival, and reproduction.

During the Calvin-Benson cycle, carbon is "captured" from carbon dioxide, a sugar phosphate forms during reactions that require ATP and NADPH, and RuBP (necessary to capture the carbon) is regenerated.

FIXING CARBON—SO NEAR, YET SO FAR

If sunlight intensity, air temperature, rainfall, and the composition of soil were uniform everywhere in the world, all year long, maybe photosynthetic reactions would proceed the same way in all plants. However, environments obviously differ—and so do the details of photosynthesis among their resident plants. A brief comparison of two different carbon-fixing adaptations to stressful environments will illustrate this point.

C4 Plants

Think of Kentucky bluegrass. Like all plants, it depends on CO_2 uptake for growth. CO_2 is plentiful in the air but is not always plentiful inside leaves. Leaves have a waxy cover that restricts water loss, except at **stomata** (singular, stoma). These microscopic openings span the leaf surface. Nearly all of the inward diffusion of CO_2 and outward diffusion of O_2 can occur only at stomata.

Stomata close on hot, dry days. Water is conserved, but CO_2 cannot diffuse into leaves. Photosynthetic cells are still busy, though, and oxygen accumulates. A high oxygen concentration inside leaves sets the stage for *photorespiration*, a process that wastes CO_2 and thus reduces a plant's sugar-building capacity. Remember rubisco, the enzyme that attaches carbon to RuBP in the Calvin-Benson cycle? Rubisco also can attach oxygen to it when the O_2 level rises and the CO_2 level declines. Only one (not two) PGA forms, along with a glycolate molecule that eventually is degraded back to CO_2.

Kentucky bluegrass is one of many **C3 plants**. The name refers to the first intermediate (the three-carbon PGA) of its carbon-fixing pathway. By contrast, the first intermediate formed when crabgrass, corn, and many other plants fix carbon is the four-carbon oxaloacetate. Hence *their* name, **C4 plants** (Figures 7.15 and 7.16*b*).

C4 plants maintain adequate amounts of CO_2 inside leaves even though they, too, close stomata on hot, dry days. *They fix carbon not once but twice*, in two types of photosynthetic cells. First mesophyll cells use the CO_2 to produce oxaloacetate, which is transferred to bundle-sheath cells around leaf veins. There CO_2 is released and fixed again—in the Calvin-Benson cycle. With this pathway, C4 plants get by with tinier stomata and lose less water, and they can produce more sugars than C3 plants can when conditions are hot, dry, and bright.

For example, 80 percent of the plants that evolved in Florida's heat are C4 species, compared to 0 percent in Manitoba, Canada. Kentucky bluegrass and other C3 species have greater advantage where temperatures drop below 25°C (they are not as sensitive to cold). Mix C3 and C4 species from different regions in a garden, and one or the other will do better during at least part of the year. That's why a bluegrass lawn that does well during cool spring weather in San Diego is overwhelmed during a hot summer by a C4 plant—crabgrass.

Especially in hot weather, photorespiration greatly lowers the photosynthetic efficiency of many important C3 crop plants, including tomatoes, rice, barley, wheat, soybeans, and potatoes. In some experiments, tomatoes were grown in hothouses at CO_2 concentrations high enough to eliminate photorespiration. The growth rate increased, in some cases by as much as five times. If photorespiration is so significantly wasteful, then why hasn't natural selection eliminated it? The answer may lie with RuBP carboxylase, the switch-hitting enzyme that sets the process in motion. The enzyme evolved in

Figure 7.15 (**a**) The internal structure of a leaf from corn (*Zea mays*), a typical C4 plant. Its photosynthetic mesophyll cells (*light green*) surround bundle-sheath cells (*light blue*) that surround tubes in leaf veins. The tubes transport water into leaves and photosynthetic products from them. (**b**) The carbon-fixing system that precedes the Calvin-Benson cycle in C4 plants.

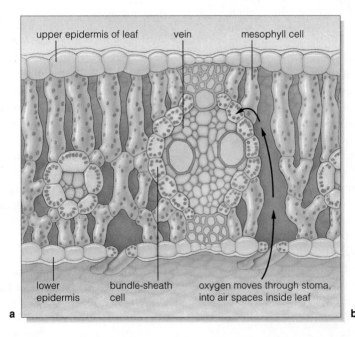

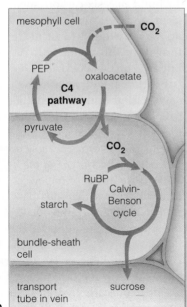

ancient times, when atmospheric levels of oxygen were still low and carbon dioxide levels high. Maybe its structure cannot undergo any mutation without adversely affecting its essential carbon-fixing activity. Or maybe the pathway that breaks down glycolate to carbon dioxide has proved so adaptive that it cannot be eliminated. Glycolate can be toxic at high concentrations.

The C4 pathway evolved separately in many lineages of flowering plants over the past 50 to 60 million years. By then, atmospheric CO_2 had declined to levels that gave C4 plants a selective advantage over C3 plants.

It will be interesting to see which pathway will be the most adaptive in years to come. The atmospheric levels of CO_2 have been rising for decades, and some ecologists are predicting that they will double during the next fifty years. If this happens, photorespiration will once again decline—possibly with beneficial effects on many of our vital crops.

CAM Plants

We see another splendid carbon-fixing adaptation in deserts and other dry environments. Think of a cactus plant, a succulent with juicy, water-storing tissues and thick surface layers that restrict water loss. It cannot open its stomata on blistering-hot days without losing precious water. Instead, it opens them and fixes CO_2 *at night*. Its cells store the resulting intermediate in their central vacuoles and then use it for photosynthesis the next day, when the stomata close.

Many plants are adapted this way. They are known as **CAM plants** (short for *Crassulacean Acid Metabolism*). Unlike C4 species, CAM plants do not fix carbon twice, in different types of cells. They fix it in the same cells, but at different times (Figure 7.16c).

During prolonged droughts, when many plants die, some CAM plants survive by keeping their stomata closed even at night. They repeatedly fix the CO_2 that forms during aerobic respiration. Not that much forms, but it is enough to allow these plants to maintain very low rates of metabolism. As you may have deduced,

a With low CO_2/high O_2, photorespiration predominates in C3 plants.

b With low CO_2/high O_2, Calvin-Benson cycle predominates in C4 plants.

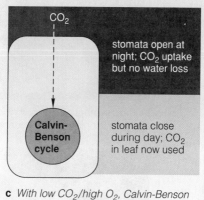

c With low CO_2/high O_2, Calvin-Benson cycle predominates in CAM plants.

Figure 7.16 Three ways of fixing carbon in hot, dry weather, when the CO_2 level is low and the O_2 level is high in leaves. (**a**) The C3 pathway is common among evergreen trees and shrubs as well as many nonwoody plants of temperate zones, such as sunflowers. (**b**) The C4 pathway is common among grasses and other plants that evolved in the tropics. Examples are corn, sorghum, crabgrass, and Bermuda grass, and the sugarcane shown here. (**c**) CAM plants open their stomata and fix carbon at night. Examples are pineapple, cacti and many other succulents (plants having a low surface-to-volume ratio), orchids, and Spanish moss.

CAM plants grow slowly. Try growing a cactus plant in Seattle or some other place with a mild climate and it will compete poorly with C3 and C4 plants.

C4 plants and CAM plants both have modified ways of fixing carbon for photosynthesis. The modifications counter the stress imposed by hot, dry conditions in their environments.

AUTOTROPHS, HUMANS, AND THE BIOSPHERE

As we conclude this chapter, remind yourself of where photosynthesis "fits" in the world of living things. Think about the mind-boggling numbers of single-celled and multicelled photoautotrophic organisms in which the light-trapping and sugar-building reactions are proceeding at this very moment, on land and in the sunlit waters of the Earth. The sheer volume of reactant molecules and product molecules that the photosynthesizers deal with might surprise you.

For example, drifting through the surface waters of the oceans and seas are uncountable numbers of single cells. You can't see them without a microscope—a row of 7 million cells of one aquatic species would be less than a quarter-inch long. Yet are they abundant! In some parts of the world, a cup of seawater may hold 24 million cells of one species, and that doesn't even include all the cells of *other* aquatic species.

Nearly all of the drifters are photoautotrophic bacteria, plants, and protistans. Together they are the pastures of the seas, the food base for diverse consumers. The pastures "bloom" in the spring, when seawater becomes warmer and enriched with nutrients that winter currents churn up from the deep. Then, populations burgeon, by rapid cell divisions. Until NASA gathered data from satellites in space, biologists had no idea that the number of cells and their distribution were so stupendous. For example, the satellite image in Figure 7.17*b* shows a springtime bloom in the surface waters of the North Atlantic Ocean. It stretches from North Carolina all the way past Spain!

Collectively, these single cells have enormous impact on the global climate. When they engage in photosynthesis, they sponge up nearly half the carbon dioxide we humans release each year, as when we burn fossil fuels or burn forests. Without the aquatic photoautotrophs, carbon dioxide in the air would accumulate more rapidly and possibly accelerate global warming, as described in Section 49.7. If our planet warms too much, all lowlands along the coasts of islands and continents may become submerged, and nations that are now the greatest food producers may be hit hard.

Amazingly, every day, humans dump tons of industrial wastes, fertilizers, and raw sewage into the ocean and thereby alter the living conditions for those drifting cells. How much of the noxious chemical brew will they continue to tolerate?

One final point: Photosynthesis is so central to our understanding of nature, it is sometimes easy to overlook other, less common energy-acquiring routes. Near hydrothermal vents on the floor of deep oceans, in hot springs, even in waste heaps of coal mines, we find diverse bacteria that are classified as **chemoautotrophs**. They obtain energy not from the sun's rays but rather from a variety of inorganic compounds in their environment. For example, certain species strip hydrogen and electrons from ammonium ions, and other species obtain them from iron and sulfur compounds. These organisms, too, exist in monumental numbers. They influence the global cycling of nitrogen and other vital elements through the biosphere. We will return later to their environmental effects. In this unit, we turn next to pathways by which cells release energy from glucose and other biological molecules, the chemical legacy of autotrophs everywhere.

a Photosynthetic activity in winter. In this color-enhanced image, *red-orange* shows where chlorophyll concentrations are greatest.

b Photosynthetic activity in spring.

Figure 7.17 Two satellite images that help convey the astounding magnitude of photosynthetic activity during the spring in the North Atlantic portion of the world ocean.

Pigments and Radiation From the Sun:

1. Plants and other photoautotrophs use wavelengths of visible light as an energy source and carbon dioxide as the carbon source for building organic compounds. Animals and other heterotrophs get carbon and energy from organic compounds synthesized by plants and other autotrophs.

2. Photosynthesis is the main biosynthetic pathway by which carbon and energy enter the web of life. Figure 7.18 summarizes its two sets of reactions. The reactions start by trapping light energy ("photo") and end with assembly reactions ("synthesis").

   a. In plants, the light-dependent reactions take place at the thylakoid membrane system inside chloroplasts. These reactions produce ATP and NADPH.

   b. The light-independent reactions proceed outside the membrane system, in the chloroplast's stroma. They produce the sugar phosphates used in building sucrose, starch, and other end products of photosynthesis.

3. The sun continually radiates energy, which travels as waves through space. Of the entire electromagnetic spectrum, photoautotrophs harness the wavelengths of visible light, which consist of distinct packets of energy called photons.

   a. The shorter the wavelength (that is, the horizontal distance between crests of two successive waves), the more energetic its photons.

   b. We perceive wavelengths of visible light as colors that range from violet and blue (the most energetic) to red (the least energetic).

4. All but one group of photoautotrophs have chlorophyll *a* and various accessory pigments that collectively can absorb all the wavelengths of visible light.

   a. Chlorophylls absorb all wavelengths of visible light except green and yellow-green ones, which they transmit.

   b. Accessory pigments, including the carotenoids, anthocyanins, and phycobilins, absorb wavelengths that the chlorophylls cannot absorb.

5. Each pigment has a light-catching array of atoms that absorbs photons of specific energies. If photons of a given wavelength match the amount of energy required to boost electrons in that array to a higher energy level, they will be absorbed. If not, the wavelength will be transmitted and will impart color to the pigment.

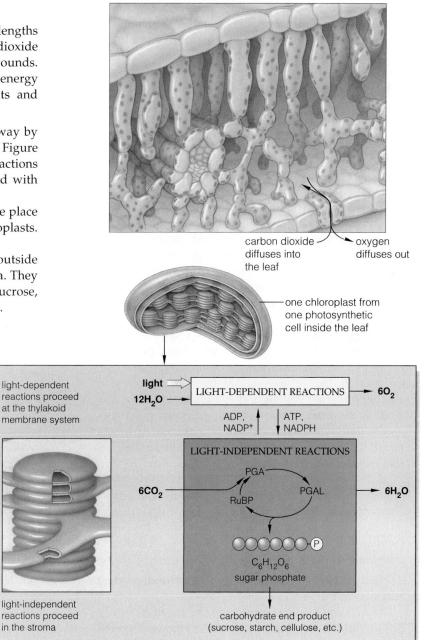

carbon dioxide diffuses into the leaf — oxygen diffuses out

one chloroplast from one photosynthetic cell inside the leaf

light-dependent reactions proceed at the thylakoid membrane system

light ⟹ LIGHT-DEPENDENT REACTIONS → **6O$_2$**

12H$_2$O →

ADP, NADP$^+$ ↑ | ATP, NADPH ↓

LIGHT-INDEPENDENT REACTIONS

PGA

6CO$_2$ → RuBP ⟲ PGAL → **6H$_2$O**

○○○○○○—Ⓟ

C$_6$H$_{12}$O$_6$
sugar phosphate

light-independent reactions proceed in the stroma

carbohydrate end product (sucrose, starch, cellulose, etc.)

Figure 7.18 Summary of the main reactants, intermediates, and products of photosynthesis, corresponding to the equation:

$$12H_2O + 6CO_2 \xrightarrow{\text{LIGHT ENERGY}} 6O_2 + C_6H_{12}O_6 + 6H_2O$$

WATER CARBON DIOXIDE — OXYGEN GLUCOSE WATER

Starting with the light-dependent reactions, six water molecules are split, yielding twelve electrons that are necessary to form a sugar phosphate (glucose with a phosphate group attached). Other water molecules enter and leave reactions; a net 6H$_2$O remain at the end of the remaining pathway. For every four electrons, three ATP and two NADPH form. During the light-independent reactions, *each turn* of the Calvin-Benson cycle requires one CO$_2$, three ATP, and two NADPH. Each sugar phosphate that forms has a backbone of six carbon atoms, so its formation requires six turns of the cycle.

The Reactions of Photosynthesis:

1. Photosystems are clusters of pigments and proteins in thylakoid membranes (or in the plasma membrane of single-celled photoautotrophs). Thylakoid membranes are peppered with two types: photosystems I and II.

a. Photosystems and neighboring electron transport systems work together in pathways of ATP formation.

b. Photosystem I operates during a cyclic pathway of ATP formation.

c. Photosystems I and II operate together during the noncyclic pathway of ATP formation.

2. Here are key points concerning the light-dependent reactions at the thylakoid membrane of chloroplasts:

a. When photons are absorbed, a photosystem's reaction center (a special molecule of chlorophyll *a*) gives up excited electrons. A primary acceptor molecule transfers these to an electron transport system.

b. In the cyclic pathway, electrons travel from the reaction center of photosystem I (P700 chlorophyll), then give up their extra energy in a transport system, and then return to photosystem I.

c. In the noncyclic pathway, electrons travel from the reaction center of photosystem II (P680 chlorophyll), give up extra energy in a transport system, then enter photosystem I. Photon energy absorbed here excites the electrons further. They give up their extra energy in another transport system, and end up in NADPH.

d. At the start of the noncyclic pathway, photolysis occurs: water molecules are split into oxygen, hydrogen ions (H^+), and electrons. The electrons replace the ones that P680 initially gives up.

e. Photolysis and operation of the electron transport systems dump H^+ into the thylakoid membrane's inner compartment. Their action sets up concentration and electric gradients across the membrane that drive ATP formation at nearby transport proteins (ATP synthases) in the membrane.

3. Here are key points concerning the light-independent reactions in the stroma of chloroplasts:

a. ATP and NADPH formed during the first stage of photosynthesis are required for the reactions. The ATP delivers chemical energy and the NADPH delivers hydrogen and electrons to the stroma, where sugar phosphates form and then are combined into starch, cellulose, and other end products.

b. The sugar phosphates form in the Calvin-Benson cycle. This cyclic pathway begins when carbon from CO_2 in the air is affixed to RuBP, making an unstable intermediate that splits into two PGA. ATP transfers a phosphate group to each PGA. The resulting molecule receives H^+ and electrons from NADPH to form PGAL.

c. For every six carbon atoms that enter the cycle by way of carbon fixation, twelve PGAL form. Two PGAL are used to produce a six-carbon sugar phosphate. The remainder are used to regenerate the RuBP.

Carbon-Fixing Adaptations:

1. Rubisco, the carbon-fixing enzyme of the Calvin-Benson cycle, evolved when the atmosphere had far more carbon dioxide and far less oxygen. Today, when the O_2 level is higher than the CO_2 level in leaves, the enzyme attaches oxygen rather than carbon to RuBP. This results in formation of only one PGA (not two) and glycolate, a compound that can't be used to form sugars but instead is degraded to carbon dioxide and water. This wasteful process is called photorespiration.

2. Photorespiration predominates in C3 plants such as sunflowers under hot, dry conditions. Then, stomata close and O_2 from photosynthesis builds up in leaves to levels higher than CO_2 levels. Sugarcane and other C4 plants raise the CO_2 level by fixing carbon twice, in two cell types. CAM plants such as cacti raise it by fixing carbon at night, when stomata are open.

Review Questions

1. A cat eats a bird, which earlier speared and ate a caterpillar that had been chewing on a weed. Which of these organisms are the autotrophs? the heterotrophs? *CI*

2. Summarize the photosynthesis reactions as an equation. State the key events of both stages of reactions, then fill in the blanks on the diagram on the facing page. *7.1*

3. Which of the following pigments are most visible in a maple leaf in summer? Which become the most visible in autumn? *7.3*
   a. chlorophylls
   b. carotenoids
   c. anthocyanins
   d. phycobilins

4. Identify which of the following substances accumulates inside the thylakoid compartment of chloroplasts: glucose, chlorophyll, carotenoids, hydrogen ions, or fatty acids. *7.5*

5. Which is *not* used in the light-independent reactions: ATP, NADPH, RuBP, carotenoids, free oxygen, CO_2, or enzymes? *7.6*

6. How many carbon atoms from CO_2 must enter the Calvin-Benson cycle to produce one sugar phosphate? Why? *7.6*

7. A busily photosynthesizing plant takes up molecules of CO_2 that have incorporated radioactively labeled carbon atoms ($^{14}CO_2$). Identify the compound in which the labeled carbon will appear first: NADPH, PGAL, pyruvate, or PGA. *7.6*

Self-Quiz (Answers in Appendix IV)

1. Photosynthetic autotrophs use _____ from the air as a carbon source and _____ as their energy source.

2. In plants, light-*dependent* reactions proceed at the _____ .
   a. cytoplasm   c. stroma
   b. plasma membrane   d. thylakoid membrane

3. The light-*independent* reactions proceed in the _____ .
   a. cytoplasm   c. stroma
   b. plasma membrane   d. grana

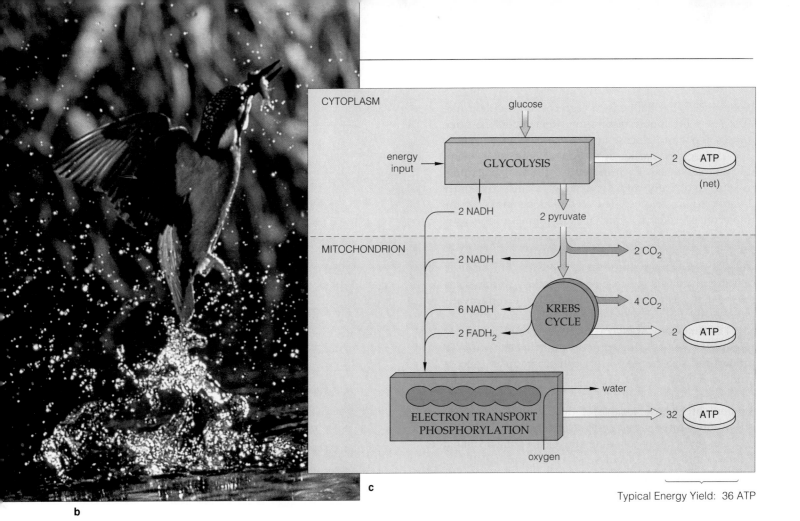

CYTOPLASM

glucose

energy input → GLYCOLYSIS → 2 ATP (net)

2 NADH   2 pyruvate

MITOCHONDRION

2 NADH ← 2 CO_2

6 NADH ← KREBS CYCLE → 4 CO_2

2 $FADH_2$ ← → 2 ATP

ELECTRON TRANSPORT PHOSPHORYLATION → water → 32 ATP

oxygen

c

Typical Energy Yield: 36 ATP

b

However, as you can see, the summary equation only tells us what the substances are at the start and finish of the pathway. In between are three reaction stages.

Let's use Figure 8.3 and the following descriptions as a brief overview of these reactions. The initial stage, again, is glycolysis. During the second stage, which is mainly a cyclic pathway of reactions called the **Krebs cycle**, pyruvate molecules are broken down to carbon dioxide and water by a series of enzyme-mediated steps. During both glycolysis and the Krebs cycle, coenzymes assist the enzymes by picking up electrons and hydrogen that are being stripped from the intermediates. Not much ATP forms in either stage. The large energy harvest comes *after* coenzymes deliver their cargo to an electron transport system.

In the third stage, the transport system functions as machinery for **electron transport phosphorylation**. It sets up hydrogen ion (H^+) concentration and electric gradients that drive ATP formation at nearby transport proteins. It is during this final stage that so many ATP molecules are produced. As it ends, oxygen inside the mitochondrion accepts the "spent" electrons from the

Figure 8.3 Overview of aerobic respiration, which proceeds through three stages. (**a**, **b**) Only this pathway delivers enough ATP to construct and maintain giant redwoods and other large, multicelled organisms. Only great amounts of ATP can sustain birds, bees, humans, and other highly active animals.

(**c**) Glucose partially breaks down to pyruvate in the first stage of reactions (glycolysis). In the second stage, which is mainly the Krebs cycle, pyruvate breaks down completely to carbon dioxide. Coenzymes (NAD$^+$ and FAD) pick up electrons and hydrogen stripped from intermediates at both stages.

In the final stage (electron transport phosphorylation), the loaded-down coenzymes (NADH and $FADH_2$) give up the electrons and hydrogen to a transport system. Energy released during the flow of electrons through the system drives ATP formation. Oxygen accepts the electrons at the end of the third stage.

From start (glycolysis) to finish, the typical net energy yield from each glucose molecule is thirty-six ATP.

last component of the transport system. Oxygen picks up hydrogen at the same time and thereby forms water.

Cells drive nearly all metabolic activities by releasing energy from glucose and other organic compounds and converting it to the chemical bond energy of ATP.

All of the main energy-releasing pathways start inside the cytoplasm with glycolysis, a stage of reactions that break down glucose to pyruvate.

The most common anaerobic pathways, which include the fermentation routes, end in the cytoplasm. Each has a net energy yield of two ATP.

Aerobic respiration, an oxygen-dependent pathway, runs to completion in the mitochondrion. From start (glycolysis) to finish, it commonly has a net energy yield of thirty-six ATP.

GLYCOLYSIS: FIRST STAGE OF THE ENERGY-RELEASING PATHWAYS

Let's track what happens to a glucose molecule in the first stage of aerobic respiration. Remember, the same things happen to glucose in the anaerobic routes.

As described earlier in Section 3.4, glucose is one of the simple sugars. Each molecule of it has six carbon, twelve hydrogen, and six oxygen atoms covalently bonded to one another. The carbon atoms make up the molecule's backbone, which we may represent this way:

During glycolysis, glucose or some other carbohydrate in the cytoplasm is partially broken down to **pyruvate**, a molecule with a backbone of three carbon atoms.

The first steps of glycolysis are *energy-requiring*. As Figure 8.4 shows, they proceed only when two ATP molecules each transfer a phosphate group to glucose and so donate energy to it. Such transfers, recall, are "phosphorylations." In this case, they raise the energy content of glucose to a level that is high enough to allow entry into the *energy-releasing* steps of glycolysis.

The first energy-releasing step cleaves the activated glucose into two molecules, which we can call PGAL (phosphoglyceraldehyde). Each PGAL is converted to an unstable intermediate, each of which allows ATP to form by giving up a phosphate group to ADP. The next intermediate in the sequence does the same thing.

Thus, a total of four ATP form by **substrate-level phosphorylation**. This metabolic event is defined as the direct transfer of a phosphate group from a substrate of a reaction to some other molecule, such as ADP. Remember, though, two ATP were invested to start the reactions. So the *net* energy yield is only two ATP.

Meanwhile, the coenzyme **NAD$^+$** (nicotinamide adenine dinucleotide) picks up electrons and hydrogen liberated from each PGAL, thereby becoming NADH. When the NADH gives up its cargo at another reaction site, it becomes NAD$^+$ once more. As is true of other enzyme helpers, then, NAD$^+$ is reusable (Section 6.5).

In sum, glycolysis converts a bit of the energy stored in glucose to a transportable form of energy, in ATP. A coenzyme picks up electrons and hydrogen stripped from glucose. These have key roles in the next stage of reactions. So do the end products of glycolysis—the two molecules of pyruvate.

Glycolysis is an energy-releasing stage of reactions in which glucose or some other carbohydrate is partially broken down to two molecules of pyruvate.

A total of two NADH and four ATP molecules form at certain steps in the reaction sequence. However, subtracting the two ATP required to start the reactions puts the *net* energy yield of glycolysis at two ATP.

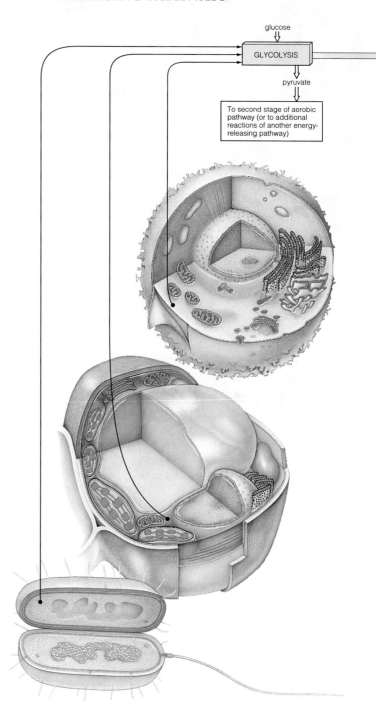

Figure 8.4 Glycolysis, first stage of the main energy-releasing pathways. The reaction steps proceed inside the cytoplasm of every living prokaryotic and eukaryotic cell.

In this example, glucose is the starting material. By the time the reactions end, two pyruvate, two NADH, and four ATP have been produced. Cells invest two ATP to start glycolysis, however, so the *net* energy yield of glycolysis is two ATP.

Depending on the type of cell and on environmental conditions, the pyruvate may be used in the second set of reactions of the aerobic pathway, which includes the Krebs cycle. Or it may be used in other reactions, such as those of fermentation pathways.

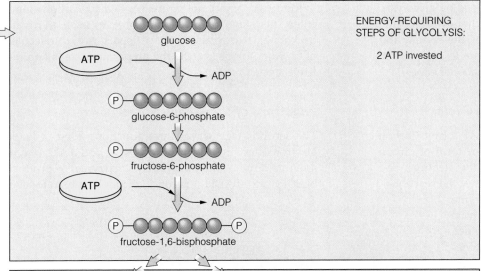

ENERGY-REQUIRING
STEPS OF GLYCOLYSIS:

2 ATP invested

a *Glycolysis starts with an energy investment of two ATP*. First, enzyme action promotes the transfer of a phosphate group from ATP to glucose, which has a backbone of six carbon atoms. With this transfer, the glucose molecule becomes slightly rearranged.

b Enzyme action promotes the transfer of a phosphate group from another ATP to the rearranged molecule.

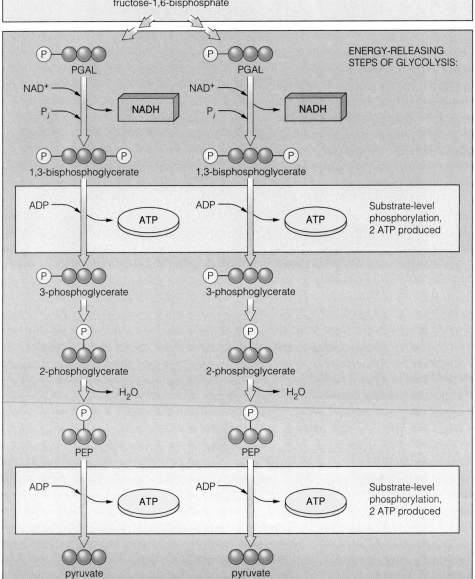

ENERGY-RELEASING
STEPS OF GLYCOLYSIS:

c The resulting fructose-1,6-bisphosphate molecule splits at once into two molecules, each with a three-carbon backbone. We can call these two PGAL.

d During enzyme-mediated reactions, two NADH form after each PGAL gives up two electrons and a hydrogen atom to NAD$^+$. Each PGAL also combines with inorganic phosphate (P$_i$) present in the cytoplasm, then donates a phosphate group to ADP.

e *Thus two ATP have formed by the direct transfer of phosphate from two intermediate molecules that serve as substrates in the reactions*. With this formation of two ATP molecules, the original energy investment of two ATP is paid off.

f In the next two enzyme-mediated reactions, each of the two intermediate molecules releases a hydrogen atom and an —OH group, which combine to form water.

g The resulting intermediates (two molecules of 3-phosphoenolpyruvate, or PEP) are rather unstable. Each gives up a phosphate group to ADP. *Once again, two ATP have formed by substrate-level phosphorylation.*

h Thus the net energy yield from glycolysis is two ATP for each glucose molecule entering the reactions. The end products of glycolysis are two molecules of pyruvate, each with a three-carbon backbone.

NET ENERGY YIELD: 2 ATP

SECOND STAGE OF THE AEROBIC PATHWAY

Suppose two pyruvate molecules, formed by glycolysis, leave the cytoplasm and enter a **mitochondrion** (plural, mitochondria). In this organelle alone, the second and third stages of the aerobic pathway run to completion. Figure 8.5 shows its structure and functional zones.

Preparatory Steps and the Krebs Cycle

During the second stage, a bit more ATP forms. Carbon atoms depart from the pyruvate, in the form of carbon dioxide. And coenzymes latch onto the electrons and hydrogen stripped from intermediates of the reactions.

In a few preparatory steps, an enzyme removes a carbon atom from each pyruvate molecule. Coenzyme A, an enzyme helper, becomes **acetyl-CoA** when it combines with the remaining two-carbon fragment. It transfers this fragment to **oxaloacetate**, the entry point of the Krebs cycle. The name of this cyclic pathway honors Hans Krebs, who began working out its details

in the 1930s. Notice, in Figure 8.6, that *six* carbon atoms enter this stage of reactions (three in each pyruvate backbone). Notice also that *six* depart, in six molecules of carbon dioxide, during the preparatory steps and the Krebs cycle proper.

Functions of the Second Stage

The second stage serves three functions. First, it loads electrons and hydrogen onto NAD$^+$ and **FAD** (flavin adenine dinucleotide, a different coenzyme) to produce NADH and FADH$_2$. Second, through substrate-level

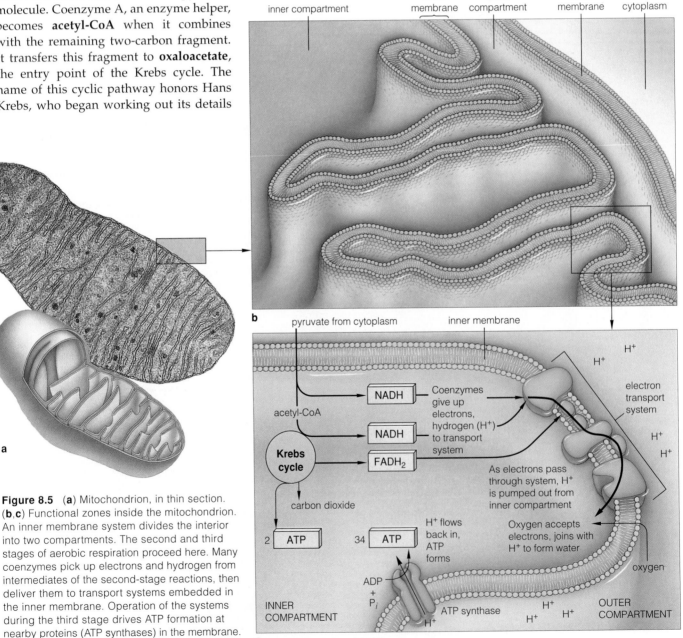

Figure 8.5 (a) Mitochondrion, in thin section. (b,c) Functional zones inside the mitochondrion. An inner membrane system divides the interior into two compartments. The second and third stages of aerobic respiration proceed here. Many coenzymes pick up electrons and hydrogen from intermediates of the second-stage reactions, then deliver them to transport systems embedded in the inner membrane. Operation of the systems during the third stage drives ATP formation at nearby proteins (ATP synthases) in the membrane.

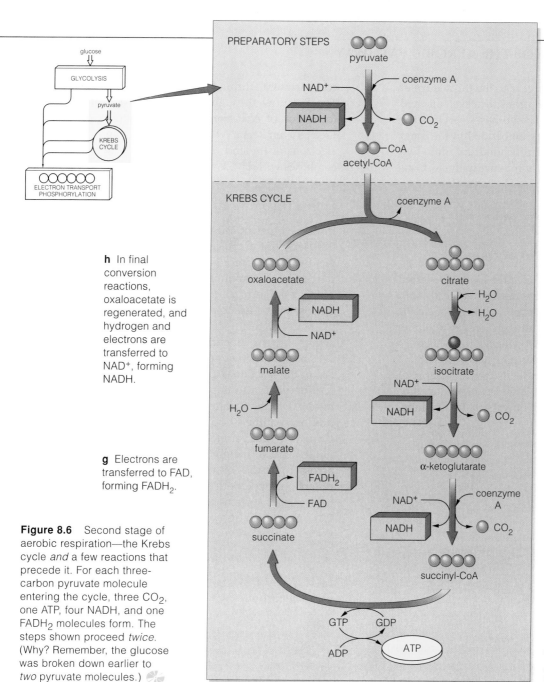

Figure 8.6 Second stage of aerobic respiration—the Krebs cycle *and* a few reactions that precede it. For each three-carbon pyruvate molecule entering the cycle, three CO_2, one ATP, four NADH, and one $FADH_2$ molecules form. The steps shown proceed *twice*. (Why? Remember, the glucose was broken down earlier to *two* pyruvate molecules.)

PREPARATORY STEPS

pyruvate

NAD^+ — coenzyme A

NADH

CO_2

acetyl-CoA —CoA

KREBS CYCLE

coenzyme A

oxaloacetate

citrate

H_2O
H_2O

h In final conversion reactions, oxaloacetate is regenerated, and hydrogen and electrons are transferred to NAD^+, forming NADH.

NADH
NAD^+

malate

isocitrate

NAD^+

NADH

CO_2

H_2O

α-ketoglutarate

fumarate

g Electrons are transferred to FAD, forming $FADH_2$.

$FADH_2$

FAD

NAD^+ — coenzyme A

NADH

CO_2

succinate

succinyl-CoA

GTP — GDP

ADP — ATP

a A pyruvate molecule enters a mitochondrion. It undergoes preparatory conversions before entering cyclic reactions (Krebs cycle).

b First, the pyruvate is stripped of a functional group (COO^-), which departs as CO_2. Next, it gives up hydrogen and electrons to NAD^+, forming NADH. A coenzyme joins with the remaining two-carbon fragment, forming acetyl-CoA.

c The acetyl-CoA is transferred to oxaloacetate, a four-carbon compound that is the point of entry into the Krebs cycle. The result is citrate, with a six-carbon backbone.

d Citrate enters conversion reactions in which a COO^- group departs (as CO_2). Also, hydrogen and electrons are transferred to NAD^+, forming NADH.

e Another COO^- group departs (as CO_2) and another NADH forms. *At this point, three carbon atoms have been released, balancing out the three that entered the mitochondrion (in pyruvate).*

f After further reactions, ATP forms by substrate-level phosphorylation.

glucose

GLYCOLYSIS

pyruvate

KREBS CYCLE

ELECTRON TRANSPORT PHOSPHORYLATION

phosphorylations, it produces two ATP. And third, it rearranges Krebs cycle intermediates into oxaloacetate. Cells have only so much oxaloacetate, and it must be regenerated to keep the cyclic reactions going.

The two ATP that form don't add much to the small yield from glycolysis. However, *many* coenzymes pick up electrons and hydrogen for transport to the sites of the third and final stage of the aerobic pathway:

Glycolysis:	2 NADH
Pyruvate conversion before Krebs cycle:	2 NADH
Krebs cycle:	2 $FADH_2$ + 6 NADH
Coenzymes sent to third stage:	2 $FADH_2$ + 10 NADH

Overall, these are the key points to remember about the second stage of aerobic respiration:

During the second stage of aerobic respiration, two pyruvate molecules from glycolysis enter a mitochondrion.

Each pyruvate molecule gives up a carbon atom, then its remnant enters the Krebs cycle. All of the carbon atoms of pyruvate eventually end up in carbon dioxide.

The preparatory steps and the cycle proper yield only two ATP. However, the reactions regenerate oxaloacetate, the entry point for the cycle. And many coenzymes pick up electrons and hydrogen that were stripped from substrates, for delivery to the final stage of the pathway.

ATP production goes into high gear in the third stage of the aerobic pathway. Electron transport systems and neighboring proteins called ATP synthases serve as the production machinery. They are embedded in the inner membrane that divides the mitochondrion into two compartments (Figure 8.7). They interact with electrons and unbound hydrogen—that is, H^+ ions. Remember, coenzymes deliver this bounty from reaction sites of the first two stages of the aerobic pathway.

Electron Transport Phosphorylation

Briefly, during the final stage, electrons get transferred from one molecule of each transport system to the next in line. When certain molecules accept and then donate the electrons, they also pick up hydrogen ions in the inner compartment. Quickly afterward, they release them to the outer compartment. Their shuttling action sets up H^+ concentration and electric gradients across the inner mitochondrial membrane. Nearby in the membrane, the ions follow the gradients and flow back to the inner

compartment, through the interior of ATP synthases. The H^+ flow through these transport proteins drives formation of ATP from ADP and unbound phosphate. Free oxygen keeps ATP production going. It withdraws electrons at the end of the transport systems and then combines with H^+. Water is the result.

Summary of the Energy Harvest

In many types of cells, thirty-two ATP form during the third stage of aerobic respiration. Add these to the net yield from the preceding stages, and the total harvest is thirty-six ATP from one glucose molecule (Figure 8.8). That's a lot! An anaerobic pathway may use eighteen glucose molecules to produce the same amount of ATP.

Think of thirty-six ATP as a typical yield only. The actual amount depends on cellular conditions, as when cells require a given intermediate elsewhere and pull it out of the reaction sequence.

The yield also depends on how particular cells use the NADH that formed during glycolysis. Any NADH produced in the cytoplasm can't enter a mitochondrion. It can only deliver electrons and hydrogen *to* the outer mitochondrial membrane, where proteins shuttle them across to NAD^+ or to FAD molecules already in the mitochondrion. Both coenzymes deliver the electrons to transport systems of the inner membrane. However, FAD puts them at a *lower* entry point in the transport system, so *its* deliveries produce less ATP (Figure 8.8).

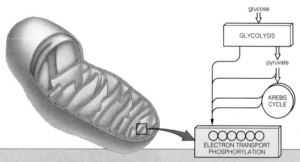

Figure 8.7   Electron transport phosphorylation, the third and final stage of aerobic respiration. The reactions proceed at electron transport systems and at ATP synthases, a type of transport protein, in the inner mitochondrial membrane. Each electron transport system consists of specific enzymes, cytochromes, and other proteins that act in sequence.

The inner membrane functionally divides the mitochondrion into two compartments. The third-stage reactions start in the inner compartment, when NADH and $FADH_2$ give up electrons and hydrogen to transport systems. Electrons are transferred *through* the system, but electron transport proteins drive unbound hydrogen (H^+) *to* the outer compartment:

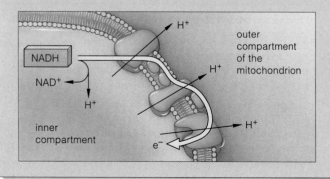

In very short order, there is a higher concentration of H^+ in the outer compartment compared to the inner one. Concentration and electric gradients now exist across the membrane. The ions follow the gradients and flow across the membrane, through the interior of the ATP synthases. Energy associated with the flow drives the formation of ATP from ADP and unbound phosphate (P_i). Hence the name, electron transport *phosphorylation*:

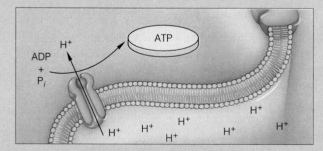

Do these metabolic events sound familiar? They should. As you may recall from Section 7.5, ATP forms in much the same way inside chloroplasts. According to the *chemiosmotic* theory, H^+ concentration and electric gradients across a cell membrane drive ATP formation. The theory applies also to mitochondria, although the ions flow in the opposite direction compared to chloroplasts.

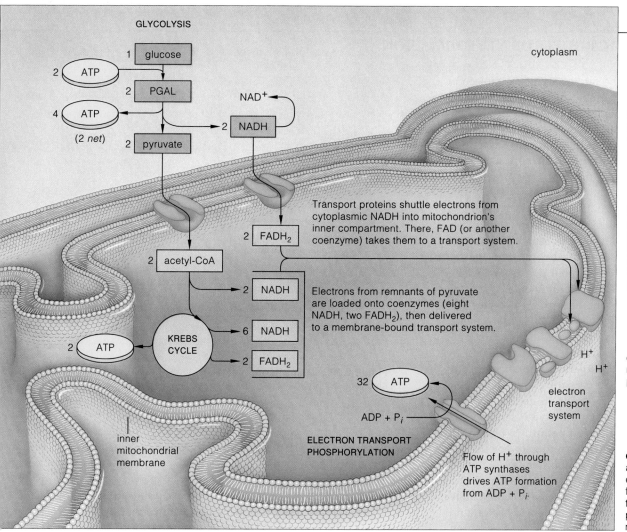

a Two ATP formed in first stage in cytoplasm (during glycolysis, by *substrate-level* phosphorylations).

b NADH that formed in cytoplasm during first stage delivers electrons and hydrogen that help drive the formation of four ATP during third stage at the inner mitochondrial membrane (by *electron transport* phosphorylations).

c Two ATP form at second stage in mitochondrion (by *substrate-level* phosphorylations of Krebs cycle).

28 ATP

d Coenzymes from Krebs cycle and its preparatory steps deliver electrons and hydrogen that drive formation of twenty-eight ATP at third stage (by *electron transport* phosphorylations at the inner mitochondrial membrane).

36 ATP

TYPICAL NET
ENERGY YIELD

Figure 8.8 Summary of the harvest from the energy-releasing pathway of aerobic respiration. Commonly, thirty-six ATP form for each glucose molecule that enters the pathway. However, the net yield varies according to shifting concentrations of reactants, intermediates, and end products of the reactions. It also varies among different types of cells.

For example, cells differ in how they use the NADH from glycolysis. These NADH cannot enter mitochondria. They only deliver their cargo of electrons and hydrogen *to* certain transport proteins of the outer mitochondrial membrane. The proteins shuttle the electrons and hydrogen across the membrane, to NAD^+ or FAD already inside the mitochondrion, to form NADH or $FADH_2$.

Any NADH inside the mitochondrion delivers electrons to the highest possible point of entry into a transport system. When it does, enough H^+ can be pumped across the inner membrane to produce *three* ATP. By contrast, any $FADH_2$ delivers them to a lower entry point in the transport system. Fewer hydrogen ions can be pumped, so only *two* ATP can be produced.

In liver, heart, and kidney cells, for example, electrons and hydrogen enter the highest entry point of transport systems, so the overall energy harvest is thirty-eight ATP. More commonly, as in skeletal muscle and brain cells, they are transferred to FAD—so the overall harvest is thirty-six ATP.

One final point. Glucose, recall, has more energy (stored in more covalent bonds) than carbon dioxide or water does. When it breaks down to those more stable end products, about 686 kilocalories are released. Much of this energy escapes (as heat), but 7.5 kilocalories or so are conserved in each ATP molecule. Therefore, when 36 ATP form through the breakdown of a glucose molecule, the energy-conserving efficiency of aerobic respiration is $(36)(7.5)/(686) \times 100$, or 39 percent.

In the final stage of the aerobic pathway, coenzymes deliver electrons to transport systems of the inner mitochondrial membrane. As electrons move through the system, they set up H^+ gradients that drive ATP formation at nearby proteins in the membrane. Oxygen is the final electron acceptor.

Again, from start (glycolysis in the cytoplasm) to finish (in mitochondria), the pathway commonly has a net yield of thirty-six ATP for every glucose molecule metabolized.

So far, we have tracked the fate of a glucose molecule through the pathway of aerobic respiration. We turn now to its use as a substrate for fermentation pathways. Remember, these are anaerobic pathways; they do *not* use oxygen as the final acceptor of the electrons that ultimately drive the ATP-forming machinery.

Fermentation Pathways

Diverse kinds of organisms use fermentation pathways. Many are bacteria and protistans that make their homes in marshes, bogs, mud, deep-sea sediments, the animal gut, canned foods, sewage treatment ponds, and other oxygen-free settings. Some kinds of fermenters actually die if exposed to oxygen. The bacteria responsible for many diseases, including botulism and tetanus, are like this. Other kinds of fermenters, including the bacterial "employees" of yogurt manufacturers, are indifferent to the presence of oxygen. Still others can use oxygen, but they also can use a fermentation pathway when oxygen becomes scarce. Even your muscle cells do this.

As is true of aerobic respiration, glycolysis serves as the first stage of the fermentation pathways. Here also, enzymes split glucose and rearrange the fragments into two pyruvate molecules. Here again, two NADH form, and the net energy yield is two ATP. However, as you can see from Figure 8.9, the reactions do not completely break down glucose to carbon dioxide and water, and they produce no more ATP beyond the tiny yield from glycolysis. *The final steps serve only to regenerate NAD$^+$, a coenzyme with central roles in the breakdown reactions.*

Fermentation yields enough energy to sustain many single-celled anaerobic organisms. It even helps carry some aerobic cells through times of stress. But it is not enough to sustain large, active, multicelled organisms,

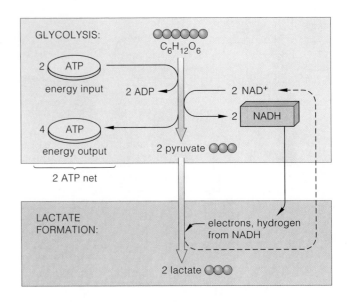

Figure 8.10 Lactate fermentation. In this anaerobic pathway, electrons end up in lactate, the reaction product.

this being one reason why you never will meet up with an anaerobic elephant.

LACTATE FERMENTATION With these points in mind, take a look at Figure 8.10, which tracks the main steps of **lactate fermentation**. During this anaerobic pathway, the *pyruvate* molecules from the first stage of reactions (glycolysis) accept the hydrogen and electrons from NADH. The transfer regenerates the NAD$^+$ and, at the same time, converts each pyruvate to a three-carbon compound called lactate. You may hear people refer to this compound as "lactic acid." However, its ionized form (lactate) is far more common in cellular fluids.

Some bacteria, such as *Lactobacillus*, rely exclusively on this anaerobic pathway. Left to their own devices, their fermentation activities often spoil food. Yet certain fermenters have commercial uses, as when they break down glucose in huge vats where cheeses, yogurt, and sauerkraut are produced.

In humans, rabbits, and many other animals, some types of cells also can switch to lactate fermentation for a quick fix of ATP. When your own demands for energy are intense but brief—say, during a short race—muscle cells use this pathway. They cannot do so for long; they would throw away too much of glucose's stored energy for too little ATP. When the glucose stores are depleted, muscles fatigue and lose their ability to contract.

ALCOHOLIC FERMENTATION In the anaerobic route of **alcoholic fermentation**, each pyruvate molecule that forms during glycolysis is converted to an intermediate

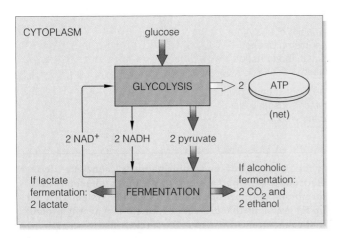

Figure 8.9 Overview of two fermentation routes. In this type of energy-releasing pathway, only the initial stage (glycolysis) has a net energy harvest, in the form of two ATP molecules. The remaining reactions serve to regenerate NAD$^+$.

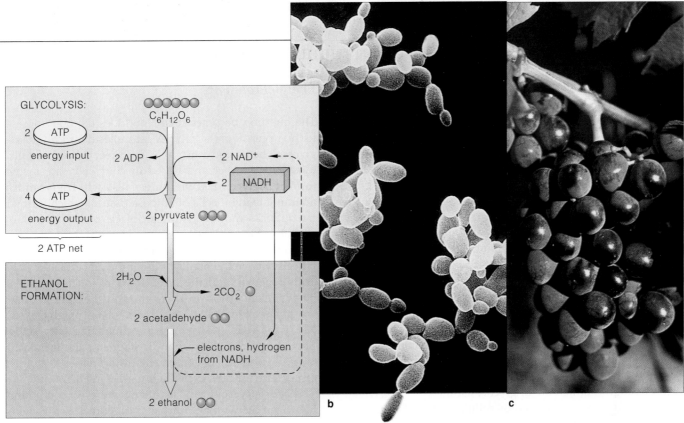

GLYCOLYSIS:

$C_6H_{12}O_6$

2 ATP
energy input

2 ADP

2 NAD^+

2 NADH

4 ATP
energy output

2 pyruvate

2 ATP net

ETHANOL
FORMATION:

$2H_2O$

$2CO_2$

2 acetaldehyde

electrons, hydrogen
from NADH

2 ethanol

a

b

c

Figure 8.11 (**a**) Alcoholic fermentation. In this anaerobic pathway, acetaldehyde, an intermediate of the reactions, is the final acceptor of electrons. Ethanol is the end product. Yeasts, single-celled organisms, use this pathway. (**b**) One species of *Saccharomyces* makes bread dough rise. Another (**c**) lives on sugar-rich tissues of ripened grapes.

form, acetaldehyde. The NADH transfers electrons and hydrogen to this form and so converts it to an alcoholic end product—ethanol (Figure 8.11).

Certain species of single-celled fungi called yeasts are renowned for their use of this pathway. One type, *Saccharomyces cerevisiae,* makes bread dough rise. Bakers mix the yeast with sugar, then blend both into dough. When yeast cells degrade the sugar, they release carbon dioxide. Bubbles of the gas expand the dough (make it rise). Oven heat forces the gas out of the dough, and a porous product remains.

Beer and wine producers use yeasts on a large scale. Vintners use wild yeasts living on grapes and cultivated strains of *S. ellipsoideus,* which remain active until the alcohol concentration in wine vats exceeds 14 percent. Wild yeasts die when the concentration passes 4 percent. Then, birds get drunk on naturally fermenting berries. That is why landscapers don't plant prodigious berry-producing shrubs near highways; drunk birds doodle into windshields. Wild turkeys get similarly tipsy when they gobble fermenting apples in untended orchards.

Anaerobic Electron Transport

Especially among the bacteria, we find less common energy-releasing pathways, some of which are topics of later chapters in the book. For example, many bacterial species have key roles in the global cycling of sulfur, nitrogen, and other crucial elements. Collectively, their metabolic activities influence nutrient availability for organisms everywhere.

For example, certain bacteria use **anaerobic electron transport**. Electrons stripped from some type of organic compound move on through transport systems of their plasma membrane. Commonly, an inorganic compound in the environment serves as the final electron acceptor. The net energy yield varies, but it is always small.

Even as you read this, some anaerobic bacteria that live in waterlogged soil are stripping electrons from a variety of compounds. They dump electrons on sulfate. Hydrogen sulfide, a putrid-smelling gas, is the result. The sulfate-reducing bacteria also live in many aquatic habitats that are enriched with decomposed organic material. They even live on the deep ocean floor, near hydrothermal vents. As described in Section 49.11, they form the food production base for unique communities.

In fermentation pathways, an organic substance that forms during the reactions serves as the final acceptor of electrons from glycolysis. The reactions regenerate NAD^+, which is required to keep the pathway operational.

In anaerobic electron transport, an inorganic substance (but not oxygen) usually serves as the final electron acceptor.

For each glucose molecule metabolized, anaerobic pathways typically have a net energy yield of two ATP, which only form during glycolysis.

So far, you have looked at what happens after a lone glucose molecule enters an energy-releasing pathway. Now you can start thinking about what cells do when they have too many or too few molecules of glucose.

Carbohydrate Breakdown in Perspective

THE FATE OF GLUCOSE AT MEALTIME Consider what happens to you or any other mammal during a meal. Glucose and certain other small organic molecules are being absorbed across the gut lining, then the blood transports them through the body. A rise in the glucose level in blood prompts the pancreas to release insulin, a hormone that stimulates cells to take up glucose at a faster rate. The cells convert the windfall of glucose to glucose-6-phosphate and so "trap" it in the cytoplasm. (When phosphorylated, glucose cannot be transported back out, across the plasma membrane.) Look again at Figure 8.4, and you see that glucose-6-phosphate is the first activated intermediate of glycolysis.

If your glucose intake exceeds cellular demands for energy, ATP-producing machinery goes into high gear. Unless a cell is rapidly using ATP, its concentration of ATP can rise to a high level. Then, glucose-6-phosphate is diverted into a biosynthesis pathway that assembles glucose units into glycogen, a storage polysaccharide (Section 3.4). This is especially true of liver cells and muscle cells, which maintain the largest glycogen stores.

THE FATE OF GLUCOSE BETWEEN MEALS When you are not eating, glucose is not entering your bloodstream and its level in the blood declines. If the decline were not countered, that would be bad news for the brain, your body's glucose hog. The brain constantly takes up more than two-thirds of the freely circulating glucose because its many hundreds of millions of cells simply use this sugar alone as their preferred energy source.

The pancreas responds to the decline by secreting glucagon, a hormone that prompts liver cells to convert glucose-6-phosphate back to glucose and send it back to the blood. Only liver cells do this; muscle cells won't give it up. The glucose level rises, and brain cells keep on trucking. Thus, *hormones control whether the body's cells use free glucose as an energy source or tuck it away.*

A word of caution: Don't let the preceding examples lead you to believe cells squirrel away large amounts of glycogen. In adult humans, glycogen makes up merely 1 percent or so of the body's total energy reserves, the energy equivalent of two cups of cooked pasta. Unless you eat on a regular basis, you will deplete the liver's small glycogen stores in less than twelve hours. Of the total energy reserves in, say, a typical adult American, 78 percent (about 10,000 kilocalories) is concentrated in body fat and 21 percent in proteins.

Energy From Fats

The question becomes this: How does the body access its huge reservoir of fats? A fat molecule, recall, has a glycerol head and one, two, or three fatty acid tails. Most fats that become stored in your body are in the form of triglycerides, with three tails each. Triglycerides accumulate in fat cells of adipose tissues, which form at the buttocks and other strategic places beneath the skin.

When blood glucose levels decline, triglycerides can be tapped as an energy alternative. Then, enzymes in fat cells cleave the bonds holding the glycerol and fatty acids together, and the breakdown products enter the bloodstream. Afterward, enzymes in the liver convert the glycerol to PGAL—an intermediate of glycolysis. Nearly all cells can take up the circulating fatty acids. Enzymes cleave the carbon backbone of the fatty acid tails and convert the fragments to acetyl-CoA—which can enter the Krebs cycle (Figures 8.6 and 8.12).

Each fatty acid tail has many more carbon-bound hydrogen atoms than glucose, so its breakdown yields much more ATP. In between meals or during sustained exercise, fatty acid conversions supply about half of the ATP that muscle, liver, and kidney cells require.

What happens if you eat too many carbohydrates? Exceed the glycogen-storing capacity of your liver and muscle cells, and the excess gets converted to fats. *Too much glucose ends up as excess fat.* Worse yet, 25 percent of the people in the United States are blessed with a combination of genes that allows them to eat as much as they like without gaining weight—but a diet far too rich in carbohydrates keeps the other 75 percent fat. For them, insulin levels remain elevated, which "tells" the body to store fat rather than use it for energy. We will return to this topic in Section 42.10.

Energy From Proteins

Eat more proteins than your body requires to grow and maintain itself, and its cells won't store them. Enzymes split these proteins into amino acid units. Then they remove the amino group ($-NH_3^+$) from each unit, and ammonia (NH_3) forms. What happens to the leftover carbon backbones? Depending on conditions in the cell, the outcome varies. These backbones can be converted to carbohydrates or fats. Or they may enter the Krebs cycle, as in Figure 8.12, where coenzymes can pick up hydrogen and electrons stripped away from the carbon atoms. The ammonia that forms undergoes conversions to become urea. This nitrogen-containing waste product would be toxic if it accumulated to high concentrations. Normally your body excretes ammonia, in urine.

As this brief discussion makes clear, maintaining and accessing the body's energy reserves is complicated

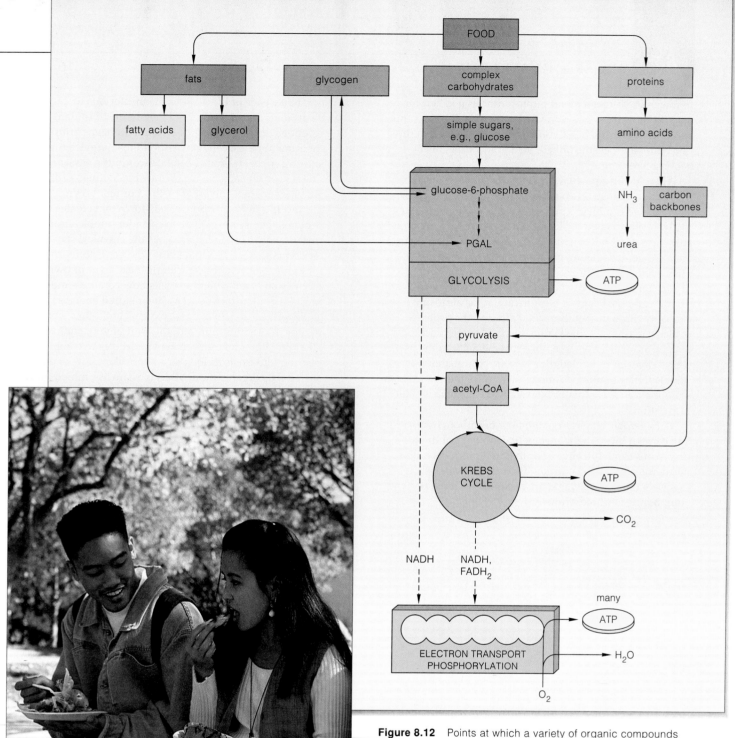

Figure 8.12 Points at which a variety of organic compounds can enter the reaction stages of aerobic respiration. Complex carbohydrates, fats, and proteins cannot enter the aerobic pathway directly. In humans and other mammals, the digestive system as well as individual cells must first break apart these large molecules into simpler, degradable subunits.

business. Hormonal controls over the disposition of glucose are special only because glucose is the fuel of choice for the all-important brain. However, as you will see in later chapters, providing all of your cells, organs, and organ systems with energy starts with the kinds and proportions of food you put in your mouth.

This concludes our look at aerobic respiration and other energy-releasing pathways. The section to follow may help you get a sense of how they fit into the larger picture of life's evolution and interconnectedness.

In humans and other mammals, the entrance of glucose or other organic compounds into an energy-releasing pathway depends on the kinds and proportions of carbohydrates, fats, and proteins in the diet as well as on the type of cell.

PERSPECTIVE ON LIFE

In this unit, you read about photosynthesis and aerobic respiration—the main pathways by which cells trap, store, and release energy. What you might not know is that the two pathways became linked, on a grand scale, over evolutionary time.

When life originated more than 3.8 billion years ago, the Earth's atmosphere had little free oxygen. The earliest single-celled organisms probably used reactions similar to glycolysis to make ATP. Without oxygen, fermentation pathways must have dominated. About 1.5 billion years later, oxygen-producing photosynthetic cells had emerged. They irrevocably changed the course of evolution.

Oxygen, a by-product of the noncyclic pathway of photosynthesis, began to accumulate in the atmosphere. Probably through mutations that affected the proteins of electron transport systems, some cells started using oxygen as an electron acceptor. At some point in the past, descendants of those fledgling aerobic cells abandoned photosynthesis entirely. Among them were the forerunners of animals and all other organisms that engage in aerobic respiration.

With aerobic respiration, the flow of carbon, hydrogen, and oxygen through the metabolic pathways of living organisms came full circle. For the final products of this key aerobic pathway—carbon dioxide and water—are precisely the materials that are necessary to build organic compounds in photosynthesis:

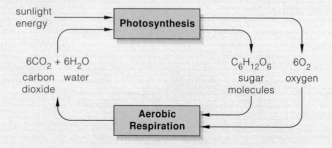

sunlight energy

Photosynthesis

$6CO_2 + 6H_2O$
carbon dioxide water

$C_6H_{12}O_6$
sugar molecules

$6O_2$
oxygen

Aerobic Respiration

Perhaps you have difficulty fathoming the connection between yourself—an intelligent being—and such remote-sounding events as energy flow and the cycling of carbon, hydrogen, and oxygen. Is this really the stuff of humanity?

Think back, for a moment, on the structure of a water molecule. Two hydrogen atoms sharing electrons with an oxygen atom may not seem close to your daily life. And yet, through that sharing, water molecules show polarity—and they hydrogen-bond with one another. Their chemical behavior is a beginning for the organization of lifeless matter that leads to the organization of all living things.

For now you can imagine other molecules interspersed through water. The nonpolar kinds resist interaction with water; the polar kinds dissolve in it. On their own, the phospholipids among them assemble into a two-layered film. Such lipid bilayers, remember, serve as the very framework of all cell membranes, hence all cells. From the beginning, the cell has been the fundamental *living* unit.

The essence of life is not some mysterious force. It is metabolic control. With a cell membrane to contain them, reactions *can* be controlled. With mechanisms built into their membranes, cells can respond to energy changes and shifting concentrations of substances in the environment. The response mechanisms operate by "telling" proteins—enzymes—when and what to build or tear down.

And it is not some mysterious force that creates the proteins themselves. DNA, the slender double-stranded treasurehouse of inheritance, has the chemical structure—*the chemical message*—that allows molecule to reproduce molecule, one generation after the next. In your own body, DNA strands tell trillions of cells how countless molecules must be built or torn apart for their stored energy.

So yes, carbon, hydrogen, oxygen, and other atoms of organic molecules represent the stuff of you, and us, and all of life. But it takes more than molecules to complete the picture. Life exists as long as an unbroken flow of energy sustains its organization. Molecules are assembled into cells, cells into organisms, organisms into communities, and so on up through the biosphere. It takes energy inputs from the sun to maintain these levels of organization. And energy flows through time in one direction—from organized to less organized forms. Only as long as energy continues to flow into the web of life can life continue in all its rich diversity.

In short, life is no more *and no less* than a marvelously complex system of prolonging order. Sustained by energy transfusions from the sun, life continues onward, through its capacity for self-reproduction. For with the hereditary instructions contained in DNA, energy and materials can be organized, generation after generation. Even with the death of individuals, life elsewhere is prolonged. With each death, molecules are released and may be cycled once more, as raw materials for new generations.

In this flow of energy and cycling of material through time, each birth is affirmation of our ongoing capacity for organization, each death a renewal.

SUMMARY

1. Metabolic reactions run on the energy, inherent in phosphate groups, that ATP molecules deliver to them. Aerobic respiration, fermentation, and other pathways that release chemical energy from organic compounds, such as glucose, produce ATP.

2. After glucose enters these pathways, enzymes strip electrons and hydrogen from intermediates that form along the way. Coenzymes pick these up and deliver them to other reaction sites at which the pathway is completed. NAD^+ is the main coenzyme; the aerobic route also uses FAD. When loaded with electrons and hydrogen, they are designated NADH and $FADH_2$.

3. The main energy-releasing pathways all start with glycolysis, a stage of reactions that begin and end in the cytoplasm. The glycolytic reactions can be completed either in the presence of oxygen or in its absence.

   a. During glycolysis, enzymes break down a glucose molecule to two pyruvate molecules. Two NADH and four ATP form during the reactions.

   b. The *net* energy yield is two ATP (because two ATP had to be invested up front to get the reactions going).

4. Aerobic respiration continues on through two more stages: (1) the Krebs cycle and a few steps preceding it, and (2) electron transport phosphorylation. These stages proceed only inside the organelles called mitochondria, which occur only in eukaryotic cells.

5. The second stage of the aerobic pathway starts when an enzyme strips a carbon atom from each pyruvate. Coenzyme A binds the remaining two-carbon fragment (to form acetyl-CoA), then transfers it to oxaloacetate, the entry point of the Krebs cycle. The cyclic reactions, along with the steps immediately preceding them, load up ten coenzymes with electrons and hydrogen (eight NADH and two $FADH_2$). Two ATP form. Three carbon dioxide molecules are released for each pyruvate that entered this second stage.

6. The third stage of the aerobic pathway proceeds at a membrane that divides the interior of a mitochondrion into two compartments. Electron transport systems and ATP synthases are embedded in this inner membrane.

   a. Coenzymes deliver electrons from the first two stages to transport systems. In the outer compartment, hydrogen ions accumulate, so concentration and electric gradients form across the membrane.

   b. Hydrogen ions follow the gradients and flow from the outer to the inner compartment, through the interior of ATP synthases. Energy released during the ion flow drives the formation of ATP from ADP and unbound phosphate.

   c. Oxygen withdraws electrons from the transport system and at the same time combines with hydrogen ions to form water molecules. The oxygen is the final acceptor of electrons that initially resided in glucose.

7. Aerobic respiration has a typical net energy yield of thirty-six ATP for each glucose molecule metabolized. Yields vary, according to cell type and cell conditions.

8. Fermentation pathways as well as anaerobic electron transport also start with glycolysis, but they do not use oxygen; they are anaerobic, start to finish.

   a. Lactate fermentation has a net energy yield of two ATP, which form in glycolysis. The remaining reactions regenerate the NAD^+. The two NADH from glycolysis transfer electrons and hydrogen to two pyruvate from glycolysis. Two lactate molecules are the end products.

   b. Similarly, alcoholic fermentation has a net energy yield of two ATP from glycolysis, and its remaining reactions regenerate NAD^+. Enzymes convert pyruvate from glycolysis to acetaldehyde, and carbon dioxide is released. The NADH from glycolysis transfer electrons and hydrogen to the two acetaldehyde molecules, thus forming two ethanol molecules, the end products.

   c. Certain bacteria use anaerobic electron transport. Electrons are stripped from various organic compounds and travel through transport systems in the bacterial cell's plasma membrane. An inorganic compound in the environment often serves as the final electron acceptor.

9. In humans and other mammals, simple sugars such as glucose from carbohydrates, glycerol and fatty acids from fats, and carbon backbones of amino acids from proteins can enter ATP-producing pathways.

Review Questions

1. Is this true or false: Aerobic respiration occurs in animals but not plants, which make ATP only by photosynthesis. *8.1*

2. For this diagram of the aerobic pathway, fill in all the blanks and write in the number of molecules of pyruvate, coenzymes, and end products. Also write in the net ATP formed in each stage, and the net ATP formed from start (glycolysis) to finish. *8.1*

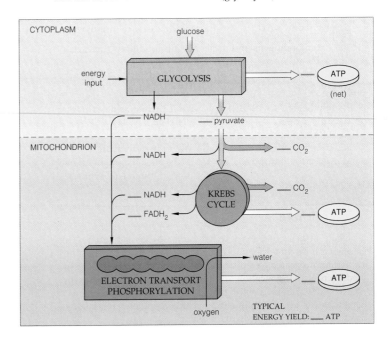

3. Is glycolysis energy-*requiring* or energy-*releasing*? Or do both kinds of reactions occur during glycolysis? *8.2*

4. In what respect does *electron transport* phosphorylation differ from *substrate-level* phosphorylation? *8.2, Figure 8.7*

5. Sketch the double-membrane system of the mitochondrion and show where transport systems and ATP synthases are located. *8.3*

6. Name the compound that is the entry point for the Krebs cycle and state whether it directly accepts the pyruvate from glycolysis. For each glucose molecule, how many carbon atoms enter the Krebs cycle? How many depart from it, and in what form? *8.3*

7. Is this statement true or false: Muscle cells cannot contract at all when deprived of oxygen. If true, explain why. If false, name the alternative(s) available to them. *8.5*

Self-Quiz *(Answers in Appendix IV)*

1. Glycolysis starts and ends in the _____ .
   a. nucleus   c. plasma membrane
   b. mitochondrion   d. cytoplasm

2. Which of the following does *not* form during glycolysis?
   a. NADH   c. $FADH_2$
   b. pyruvate   d. ATP

3. The pathway of aerobic respiration is completed in the _____ .
   a. nucleus   c. plasma membrane
   b. mitochondrion   d. cytoplasm

4. In the last stage of aerobic respiration, _____ is the final acceptor of electrons that originally resided in glucose.
   a. water   c. oxygen
   b. hydrogen   d. NADH

5. _____ engage in lactate fermentation.
   a. *Lactobacillus* cells c. Sulfate-reducing bacteria
   b. Muscle cells   d. a and b

6. In alcoholic fermentation, _____ is the final acceptor of the electrons stripped from glucose.
   a. oxygen   c. acetaldehyde
   b. pyruvate   d. sulfate

7. The fermentation pathways produce no more ATP beyond the small yield from glycolysis, but the remaining reactions _____ .
   a. regenerate ADP   c. dump electrons on an
   b. regenerate NAD^+ inorganic substance (not oxygen)

8. In certain organisms and under certain conditions, _____ can be used as an energy alternative to glucose.
   a. fatty acids   c. amino acids
   b. glycerol   d. all of the above

9. Match the event with its most suitable metabolic description.
   ____ glycolysis   a. ATP, NADH, $FADH_2$, CO_2,
   ____ fermentation   and water form
   ____ Krebs cycle   b. glucose to two pyruvate
   ____ electron transport c. NAD^+ regenerated, two ATP net
        phosphorylation   d. H^+ flows through ATP synthases

Critical Thinking

1. Diana suspects that a visit to her family doctor is in order. After eating carbohydrate-rich food, she always experiences sensations of being intoxicated and becomes nearly incapacitated—as if she had been drinking alcohol. She even wakes up with a hangover the next day. Having completed a course in freshman biology, Diana has an idea that something is affecting the way her body is metabolizing glucose. Explain why.

2. The cells of your body absolutely do not use nucleic acids as alternative energy sources. Suggest why.

3. The body's energy needs and its programs for growth depend on balancing the levels of amino acids in blood with proteins in cells. Cells of the liver, kidneys, and intestinal lining are especially important in this balancing act. When the levels of amino acids in blood decline, lysosomes in cells can rapidly digest some of their proteins (structural and contractile proteins are spared, except in cases of malnutrition). The amino acids released this way enter the blood and thereby help maintain the required levels.

Suppose you embark on a body-building program. You already eat plenty of carbohydrates, but a nutritionist advises a protein-rich diet that includes protein supplements. Speculate on how the extra dietary proteins will be put to use, and in which tissues.

4. Each year, Canada geese lift off in precise formation from their northern breeding grounds. They head south to spend the winter months in warmer climates, and then they make the return trip in spring. As is true of other migratory birds, their flight muscle cells are efficient at using fatty acids as an energy source. (Remember, the carbon backbone of fatty acids can be cleaved into fragments that can be converted to acetyl-CoA for entry into the Krebs cycle.)

Suppose a lesser Canada goose from Alaska's Point Barrow has been steadily flapping along for three thousand kilometers and is nearing Klamath Falls, Oregon. It looks down and notices a rabbit sprinting like the wind from a coyote with a taste for rabbit. With a stunning burst of speed, the rabbit reaches the safety of its burrow.

Which energy-releasing pathway predominated in the rabbit's leg muscle cells? Why was the Canada goose relying on a different pathway for most of its journey? And why wouldn't the pathway of choice in goose flight muscle cells be much good for a rabbit making a mad dash from its enemy?

5. Reflect on this chapter's Introduction, then on Question 4. Now speculate on which energy-releasing pathway is predominating in an agitated Africanized bee chasing a farmer across a cornfield.

Selected Key Terms

acetyl-CoA *8.3*   Krebs cycle *8.1*
aerobic respiration *8.1*   lactate fermentation *8.5*
alcoholic fermentation *8.5*   mitochondrion *8.3*
anaerobic electron transport *8.5* NAD^+ *8.2*
electron transport   oxaloacetate *8.3*
   phosphorylation *8.1*   pyruvate *8.2*
FAD *8.3*   substrate-level
glycolysis *8.1*   phosphorylation *8.2*

Readings

Levi, P. October 1984. "Travels with C." *The Sciences.* Journey of a carbon atom through the world of life.

Wolfe, S. 1995. *An Introduction to Molecular and Cellular Biology.* Belmont, California: Wadsworth. Exceptional reference text.

Web Site See *http://www.wadsworth.com/biology* for practice quiz questions, hypercontents, BioUpdates, and critical thinking. The Wadsworth Biology Resource Center provides a wealth of information fully organized and integrated by chapter.

FACING PAGE: *Human sperm, one of which will penetrate this mature egg and so set the stage for the development of a new individual in the image of its parents.*

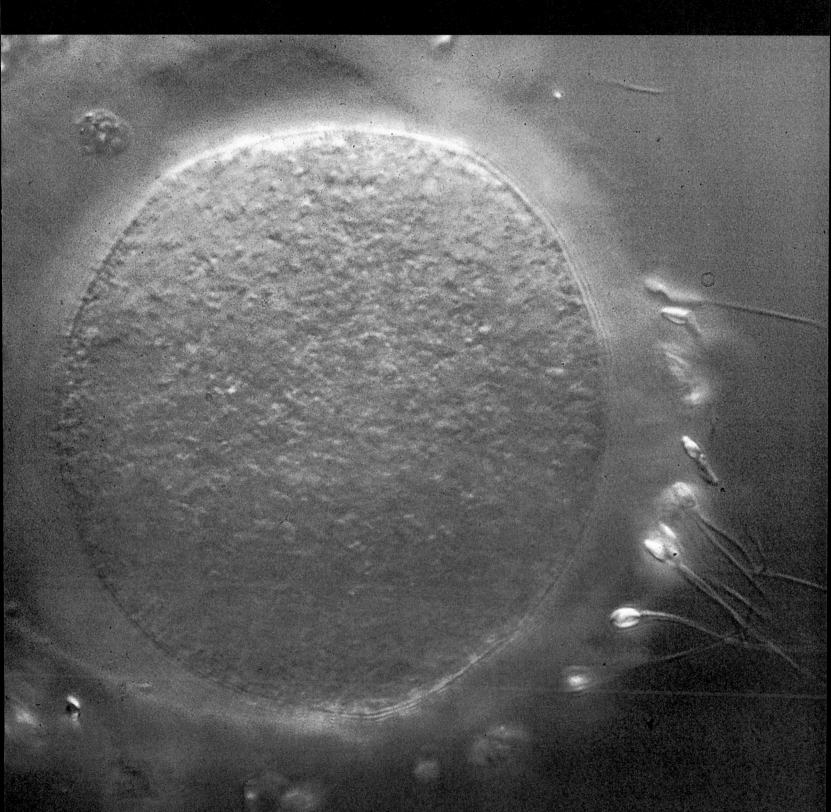

9

CELL DIVISION AND MITOSIS

Silver In the Stream of Time

Five o'clock, and the first rays from the sun dance over the wild Alagnak River of the Alaskan tundra. It is September, and life is ending and beginning in the clear, frigid waters. By the thousands, mature silver salmon have returned from the open ocean to spawn in their shallow native home. The females are tinged with red, the color of spawners, and they are dying.

This morning, a female salmon releases translucent pink eggs into a shallow "nest," hollowed out by her fins in the gravel riverbed (Figure 9.1). Moments later a male sheds a cloud of sperm, and fertilization follows. Trout and other predators eat most of the eggs, but some eggs survive and give rise to a new generation.

Within three years, the pea-size eggs have become streamlined salmon, fashioned from billions of cells. A few of their cells will develop into eggs or sperm. In time, on some September morning, they will take part in an ongoing story of birth, growth, death, and rebirth.

For you, as for salmon and all other multicelled species, growth as well as reproduction depends on *cell division*. Inside your mother, a fertilized egg divided in two, then the two into four, and so on until billions of cells were growing, developing in specialized ways, and dividing at different times to produce all of your genetically prescribed body parts. Your body now has roughly 65 trillion living cells—and many of them are still dividing. Every five days, for instance, divisions replace the tissue that lines your small intestine.

Understanding cell division—and, ultimately, how new individuals are put together in the image of their parents—begins with answers to three questions. *First*, what instructions are necessary for inheritance? *Second*, how are those instructions duplicated for distribution into daughter cells? *Third*, by what mechanisms are those instructions divided into daughter cells? We will require more than one chapter to consider the nature of

Figure 9.1   The last of one generation and the first of the next in Alaska's Alagnak River.

cell reproduction and other mechanisms of inheritance. Even so, the points made early in this chapter can help you keep the overall picture in focus.

Begin with the word **reproduction**. In biology, this means that parents produce a new generation of cells or multicelled individuals like themselves. The process starts in cells that are programmed to divide. And the ground rule for division is this: *Parent cells must provide their daughter cells with hereditary instructions, encoded in DNA, and enough metabolic machinery to start up their own operation.*

DNA, recall, contains instructions for synthesizing proteins. Some proteins are structural materials. Many are enzymes that speed the assembly of specific organic compounds, such as the lipids that cells use as building blocks and sources of energy. Unless a daughter cell receives the necessary instructions for making proteins, it simply will not be able to grow or function properly.

Also, the cytoplasm of a parent cell already contains enzymes, organelles, and other operating machinery. When a daughter cell inherits what looks like a blob of cytoplasm, it really is getting start-up machinery, which will keep it operating until it can use its inherited DNA for growing and developing on its own.

KEY CONCEPTS

1. The continuity of life depends on reproduction, by which parents produce a new generation of cells or multicelled individuals like themselves. Cell division is the bridge between generations.

2. When a cell divides, its two daughter cells must each receive a required number of DNA molecules as well as cytoplasm. For eukaryotic cells, a division mechanism called mitosis sorts out the DNA into two new nuclei. A separate mechanism divides the cytoplasm.

3. Mitosis is one part of the cell cycle. The other part is interphase, an interval when each new cell formed by mitosis and cytoplasmic division increases in mass, doubles the number of its components, then duplicates its DNA. The cycle ends when that cell divides.

4. In eukaryotic cells, many proteins with structural and functional roles are attached to the DNA. Each DNA molecule, with its attached proteins, is a chromosome.

5. Members of the same species have a characteristic number of chromosomes in their cells. The chromosomes differ from one another in length and shape, and they carry different portions of the hereditary instructions.

6. The body cells of humans and many other organisms have a diploid chromosome number; they contain two of each type of chromosome characteristic of the species.

7. Mitosis keeps the chromosome number constant, from one cell generation to the next. Therefore, if a parent cell is diploid, so will be the daughter cells.

8. Mitotic cell division is the basis of growth and tissue repair in multicelled eukaryotes. It also is the means by which single-celled eukaryotes and many multicelled eukaryotes reproduce asexually.

DIVIDING CELLS: THE BRIDGE BETWEEN GENERATIONS

Overview of Division Mechanisms

In plants, animals, and all other eukaryotic organisms, hereditary instructions are distributed among a number of DNA molecules. Before the cells of such organisms reproduce, they must undergo *nuclear* division. **Mitosis** and **meiosis** are two nuclear division mechanisms. Both sort out and package a parent cell's DNA into new nuclei, for the forthcoming daughter cells. Some other mechanism splits the cytoplasm into daughter cells.

Multicelled organisms grow by way of mitosis and the cytoplasmic division of body cells, which are called **somatic cells**. They also repair tissues that way. Nick yourself peeling a potato, and mitotic cell divisions will replace the cells that the knife sliced away. Besides this, many protistans, fungi, plants, and even some animals reproduce asexually by mitotic cell division.

By contrast, meiosis occurs only in **germ cells**, a cell lineage set aside for the formation of gametes (such as sperm and eggs) and sexual reproduction. As you will read in the next chapter, meiosis has much in common with mitosis, but the end result is different.

Prokaryotic cells, or bacteria, reproduce asexually by a different mechanism, called prokaryotic fission. We will consider the bacteria later, in Section 22.2.

Some Key Points About Chromosomes

In a nondividing cell, the DNA molecules are stretched out like thin threads, with many proteins attached to them. Each DNA molecule, with its attached proteins, is a **chromosome**. When a cell prepares for mitosis, each

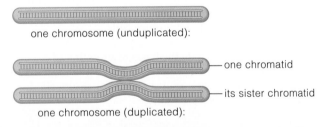

one chromosome (unduplicated):

— one chromatid

— its sister chromatid

one chromosome (duplicated):

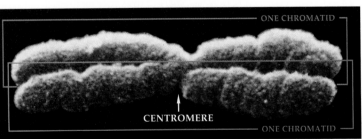

ONE CHROMATID

CENTROMERE

ONE CHROMATID

Figure 9.2 Sketch of a chromosome in the unduplicated and duplicated states. The scanning electron micrograph shows a human chromosome in the duplicated state; it consists of two sister chromatids, attached at the centromere.

threadlike chromosome is duplicated. It now consists of two DNA molecules, which will stay together until late in mitosis. As long as they remain attached, the two are called **sister chromatids** of the chromosome.

Figure 9.2 illustrates a eukaryotic chromosome in the unduplicated and duplicated states. Notice how the duplicated chromosome narrows down in one region along its length. This is the **centromere**, a small region with attachment sites for the microtubules that move the chromosome during nuclear division. Bear in mind, the sketch in Figure 9.2 is simplified. For example, the location of the centromere differs among chromosomes. And the two strands of a DNA molecule do not look like a ladder; they twist rather like a spiral staircase and are much longer than can be shown here.

Mitosis and the Chromosome Number

Each species has a characteristic **chromosome number**, which refers to the sum total of chromosomes in cells of a given type. Human somatic cells have 46, those of gorillas have 48, and those of pea plants have 14.

Actually, your 46 chromosomes are like volumes of two sets of books. Each set is numbered 1 to 23. For example, you have two "volumes" of chromosome 22—that is, *a pair of them*. Generally, both members of each pair have the same length and shape, and they carry the same portion of hereditary instructions for the same traits. Think of them as two sets of books on how to build a house. Your father gave you one set. Your mother had her own ideas about storage, plumbing, and so on, so she gave you a revised edition. Her set covers the same topics but says slightly different things about many of them.

We say the chromosome number is **diploid**, or 2*n*, if a cell has two of each type of chromosome characteristic of the species. The body cells of humans, gorillas, pea plants, and a great many other organisms are like this. (By contrast, as explained in Chapter 10, their eggs and sperm have a *haploid* chromosome number, or *n*. This means they have only one of each type of chromosome characteristic of the species.)

With mitosis, a diploid parent cell can produce two diploid daughter cells. This doesn't mean each merely gets forty-six or forty-eight or fourteen chromosomes. If only the total mattered, one cell might get, say, two pairs of chromosome 22 and no pairs whatsoever of chromosome 9. However, neither cell would function properly *without two of each type of chromosome*.

Mitosis keeps the chromosome number constant, division after division, from one cell generation to the next. Thus, if a parent cell is diploid, its daughter cells will be diploid also.

THE CELL CYCLE

Mitosis is only one phase of the **cell cycle**. Such cycles start each time new cells are produced and end when those cells complete their own division. The cycle starts again for each new daughter cell (Figure 9.3). Usually, the longest phase of the cell cycle is **interphase**, which has three parts. During interphase, a cell increases its mass, roughly doubles the number of its cytoplasmic components, and then duplicates its DNA. The different parts of the cycle are often abbreviated this way:

G1 Of interphase, a "*Gap*" (interval) of cell growth before the onset of DNA replication

S   Of interphase, a time of "*Synthesis*" (DNA replication)

G2 Of interphase, a second "*Gap*" (interval) following DNA replication, when the cell prepares for division

M   *Mitosis*; nuclear division only, usually followed by cytoplasmic division

The cell cycle lasts about the same length of time for cells of a given type. Its duration differs among cells of different types. For example, all of the neurons (nerve cells) in your brain are arrested at interphase, and they usually will not divide again. By contrast, cells of a new sea urchin may double in number every two hours.

Adverse conditions may disrupt a cell cycle. When deprived of a vital nutrient, for instance, the free-living cells called amoebas do not leave interphase. Even so, if any cell proceeds past a certain point in interphase, the cycle normally will continue regardless of outside conditions, owing to built-in controls over its duration.

Figure 9.3 Eukaryotic cell cycle, generalized. The length of each part differs among different cell types.

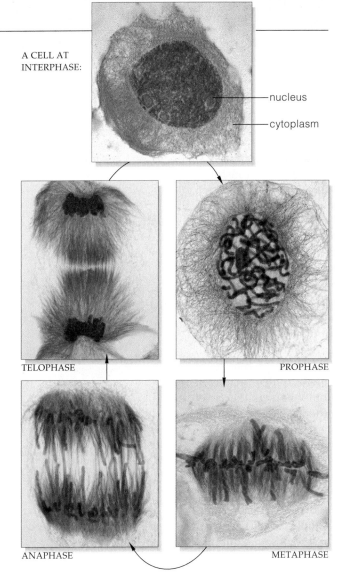

A CELL AT INTERPHASE:

nucleus

cytoplasm

TELOPHASE

PROPHASE

ANAPHASE

METAPHASE

Figure 9.4 Mitosis in a cell from the African blood lily, *Haemanthus*. The chromosomes are stained *blue*, and the many microtubules are stained *red*. Before reading further, take a moment to become familiar with the labels on the micrographs.

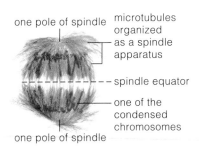

one pole of spindle

microtubules organized as a spindle apparatus

spindle equator

one of the condensed chromosomes

one pole of spindle

We turn now to mitosis and to how it maintains the chromosome number through turn after turn of the cell cycle. Figure 9.4 only hints at the divisional ballet that begins as a cell leaves interphase.

A cell cycle begins at interphase, when a new cell (formed by mitosis and cytoplasmic division) increases its mass, roughly doubles the number of its cytoplasmic components, then duplicates its chromosomes. The cycle ends when the cell divides.

THE STAGES OF MITOSIS—AN OVERVIEW

When a cell makes the transition from interphase to mitosis, it has stopped constructing new cell parts, and its DNA has been replicated. Within that cell, profound changes will now proceed smoothly, one after the other, through four stages. The sequential stages of mitosis are **prophase**, **metaphase**, **anaphase**, and **telophase**.

Figure 9.5 illustrates mitosis in an animal cell. By comparing the series of photographs against those of the plant cell in Figure 9.4, it becomes clear that chromosomes in both cells are moving about. They don't do so on their own. A **spindle apparatus** harnesses and moves them.

A spindle consists of microtubules arranged into two sets. Each set extends from one of the two poles (end points) of the spindle. The two sets overlap each other a bit at the spindle equator, or midway between the two poles. The formation of this bipolar, microtubular spindle establishes what will be the ultimate destinations of chromosomes during mitosis, as you will see shortly.

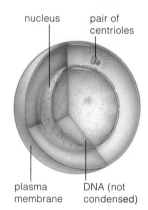

nucleus — pair of centrioles

plasma membrane — DNA (not condensed)

CELL AT INTERPHASE
The cell duplicates its DNA; then it prepares for nuclear division.

Figure 9.5 Mitosis. This nuclear division mechanism ensures that every daughter cell will have the same chromosome number as the parent cell. For clarity, the diagram shows only two pairs of chromosomes from a diploid (2n) animal cell. With only rare exceptions, the picture is more involved than this, as indicated by the micrographs of mitosis in a whitefish cell (*facing page*).

MITOSIS

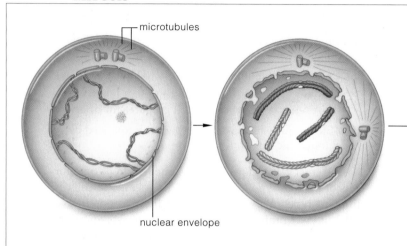

microtubules

nuclear envelope

EARLY PROPHASE
The DNA and its associated proteins have started to condense. The two chromosomes colored *purple* were inherited from the male parent. The other two (*blue*) are their counterparts, inherited from the female parent.

LATE PROPHASE
Chromosomes continue to condense. New microtubules are assembled, they move one of the two centriole pairs to the opposite end of the cell. The nuclear envelope starts to break up.

Prophase: Mitosis Begins

We know a cell is in prophase when its chromosomes become visible in the light microscope as threadlike forms. ("Mitosis" comes from the Greek *mitos*, meaning thread.) Each chromosome was duplicated earlier, in interphase. In other words, each now consists of two sister chromatids, joined at the centromere. Early on, the two sister chromatids twist and fold into a more compact form. By late prophase, all the chromosomes will be condensed into thicker, rod-shaped forms.

Meanwhile, in the cytoplasm, most microtubules of the cytoskeleton are breaking apart into their tubulin subunits (Section 4.8). The subunits reassemble near the nucleus, as *new* microtubules of the spindle. Many of these microtubules will extend from one spindle pole or the other to the centromere of a chromosome. The remainder will not interact at all with the chromosomes. They will extend from the poles and overlap each other.

While new microtubules are assembling, the nuclear envelope physically prevents them from interacting with the chromosomes inside the nucleus. However, the nuclear envelope starts breaking up as prophase ends.

Many cells have two barrel-shaped **centrioles**. Each centriole started duplicating itself during interphase, so there are two pairs of them when prophase is under way. Microtubules start moving one pair to the opposite pole of the newly forming spindle. Centrioles, recall, give rise to flagella or cilia. If you observe them in cells of an organism, you can bet that flagellated or ciliated cells (such as sperm cells) develop during its life cycle.

Transition to Metaphase

So much happens between prophase and metaphase that researchers give this transitional period its own name, "prometaphase." The nuclear envelope breaks up completely, into numerous tiny, flattened vesicles. Now the chromosomes are free to interact with microtubules that are extending toward them, from the poles of the forming spindle. Microtubules from both poles harness each chromosome and start pulling on it. The two-way pulling orients the chromosome's two sister chromatids toward opposite poles. Meanwhile, overlapping spindle

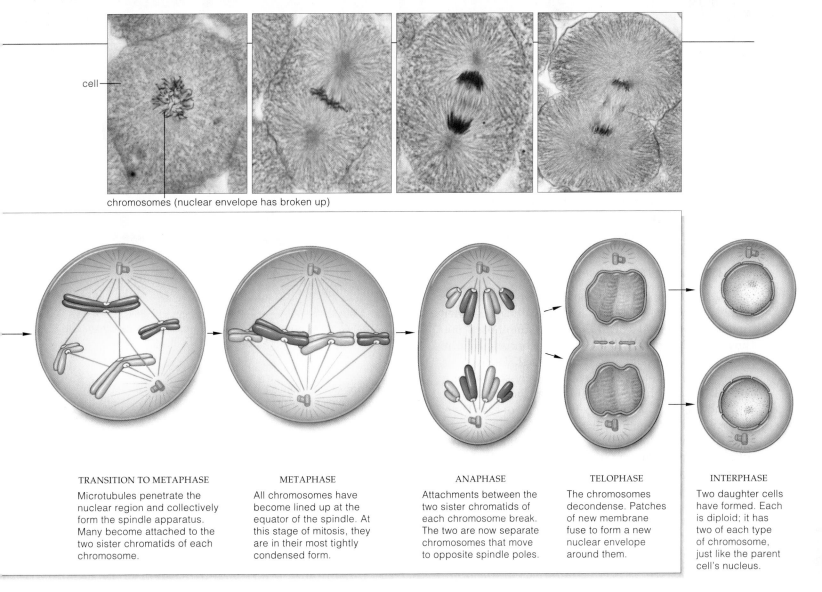

cell —

chromosomes (nuclear envelope has broken up)

TRANSITION TO METAPHASE

Microtubules penetrate the nuclear region and collectively form the spindle apparatus. Many become attached to the two sister chromatids of each chromosome.

METAPHASE

All chromosomes have become lined up at the equator of the spindle. At this stage of mitosis, they are in their most tightly condensed form.

ANAPHASE

Attachments between the two sister chromatids of each chromosome break. The two are now separate chromosomes that move to opposite spindle poles.

TELOPHASE

The chromosomes decondense. Patches of new membrane fuse to form a new nuclear envelope around them.

INTERPHASE

Two daughter cells have formed. Each is diploid; it has two of each type of chromosome, just like the parent cell's nucleus.

microtubules ratchet past each other and push the poles of the spindle apart. The push–pull forces are balanced when the chromosomes reach the spindle's midpoint.

When all of the duplicated chromosomes are aligned midway between the poles of a completed spindle, we call this metaphase (*meta-* means "midway between"). The alignment is crucial for the next stage of mitosis.

From Anaphase Through Telophase

At anaphase, the sister chromatids of each chromosome separate from each other and move to opposite poles by two mechanisms. First, the microtubules attached to the centromere regions shorten and *pull* the chromosomes to the poles. Second, the spindle elongates as overlapping microtubules continue to ratchet past each other and *push* the two spindle poles even farther apart. Once each chromatid is separated from its sister, it has a new name. It is now a separate chromosome in its own right.

Telophase gets under way as soon as each of two clusters of chromosomes arrives at a spindle pole. The

chromosomes are no longer harnessed to microtubules, and they return to threadlike form. Vesicles of the old nuclear envelope fuse and form patches of membrane around the chromosomes. Patch joins with patch, and soon a new nuclear envelope separates each cluster of chromosomes from the cytoplasm. If the parent cell was diploid, each cluster contains two chromosomes of each type. With mitosis, remember, each new nucleus has the same chromosome number as the parent nucleus. Once two nuclei form, telophase is over—and so is mitosis.

Prior to mitosis, each chromosome in a cell's nucleus is duplicated, so that it consists of two sister chromatids.

Mitosis proceeds through four consecutive stages, called prophase, metaphase, anaphase, and telophase.

A microtubular spindle moves sister chromatids of every chromosome apart, to opposite spindle poles. Around each of two clusters of chromosomes, new nuclear envelope forms. Both of the daughter nuclei formed this way have the same chromosome number as the parent cell's nucleus.

Close this book for a moment and reflect on what you have just learned from the two preceding sections. Be sure you have a clear picture of the flow of events in interphase and mitosis, up to the time of cytoplasmic division and the formation of new daughter cells. How easily you will get through many subsequent chapters depends on how well you understand the cell division story. If parts of the picture are still not clear, it may be worth your time to read Sections 9.2 and 9.3 once again, before continuing with the details presented here.

The Wonder of Interphase

If you could coax the DNA molecules from just one of your somatic cells to stretch in a single line, one after another, they would extend from your armpit past your fingertips. Salamander DNA is even more amazing. A single line of it would extend ten meters! The wonder is, enzymes and other proteins selectively scan all of a cell's DNA, switch protein-building instructions on and off, and even produce base-by-base copies of each DNA molecule—all during interphase.

G1, S, and G2 of interphase have distinct patterns of biosynthesis. During G1, most of the carbohydrates, lipids, and proteins for a cell's own use and for export are assembled. During S, the cell copies its DNA and synthesizes the proteins that will become organized into structural scaffolding for the condensed versions of chromosomes. Finally, during G2, the proteins that will drive mitosis to completion are produced.

Once S begins, events normally proceed at about the same rate in all cells of a species and continue all the way through mitosis. Given this observation, you might well assume the cycle has built-in, molecular brakes. Apply the brakes that operate in G1, and the cycle stalls in G1. Lift the brakes and the cycle runs to completion. Said another way, *control mechanisms govern the rate of cell division*.

Imagine a car losing its brakes just as it starts down a steep mountain road. As you will read later in the book, that is how cancer starts. Controls over division are lost, and the cell cycle simply cannot stop turning.

Chromosomes, Microtubules, and the Precision of Mitosis

Equally impressive is the precision with which mitosis parcels out DNA for forthcoming daughter cells. That precision depends upon chromosome organization and interactions among microtubules and motor proteins.

ORGANIZATION OF METAPHASE CHROMOSOMES Even during interphase, eukaryotic DNA has many proteins bound tightly to it. **Histones** are among them. Many

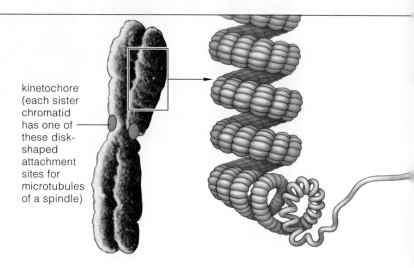

kinetochore (each sister chromatid has one of these disk-shaped attachment sites for microtubules of a spindle)

a A duplicated human chromosome during metaphase, when it is in its most condensed form. Interactions among certain chromosomal proteins keep its loops of DNA tightly packed in a "supercoiled" array.

Figure 9.6 One model of the levels of organization in a human metaphase chromosome.

histones are like spools for winding up small stretches of DNA. Each histone-DNA spool is a single structural unit called a **nucleosome**. Another histone stabilizes the spools. During mitosis (and meiosis also), interactions between histones and DNA make the chromosome coil back on itself repeatedly. The coiling greatly increases the chromosome's diameter. Other proteins besides the histones form a structural scaffold when the DNA folds further, into a series of loops (Figure 9.6).

A chromosome acquires its distinct shape and size late in prophase, when condensation is nearly complete. By then, each of its sister chromatids has at least one constricted region, the most prominent of which is the centromere. At the surface of the centromere is a disk-shaped structure, a **kinetochore**, which will serve as a docking site for spindle microtubules (Figure 9.6a).

When the threadlike DNA molecules are condensing into such compact chromosome structures, why don't they get tangled up? Actually, it appears that they do, but an enzyme called DNA topoisomerase chemically recognizes the tangling and puts things right. When researchers deliberately interfered with this enzyme's activity, the tangles persisted. Later, sister chromatids failed to separate at anaphase.

SPINDLES COME, SPINDLES GO Different species have between ten and tens of thousands of microtubules in the mitotic spindle. The plant cell shown in Figure 9.4 probably has 10,000 of them. In each case, the spindle is in exquisite balance with the cell's pool of tubulin subunits. Its microtubules are continually assembling and disassembling, so subunits are being withdrawn

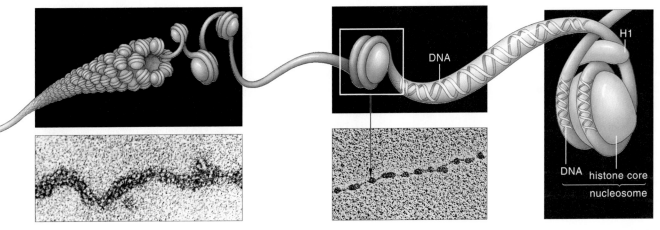

b At a deeper level of organization, the chromosomal proteins and the DNA are arranged as a cylindrical fiber (solenoid) thirty nanometers in diameter.

c Immerse a chromosome in saltwater and it loosens up to a beads-on-a-string organization. The "string" is one DNA molecule. Each "bead" is a nucleosome.

d A nucleosome consists of a double loop of DNA around a core of eight histones. Other histones stabilize the structural array.

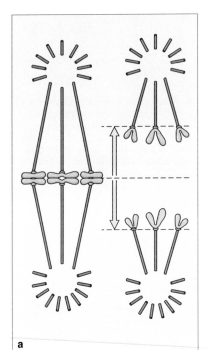

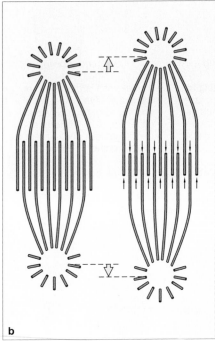

Figure 9.7 Models of two of the mechanisms that separate sister chromatids of a chromosome from each other at anaphase. In (**a**), the microtubules attached to the chromatids *shorten* and thereby decrease the distance between the kinetochores and spindle poles. In (**b**), overlapping microtubules ratchet past each other, thereby moving the spindle poles apart and increasing the distance between sister chromatids of each chromosome.

kinetochore is like a train chugging along a railroad track—except the track disassembles after it. Two motor proteins called dynein and kinesin have been isolated from kinetochores and may drive the sliding motion. As you read in Sections 4.8 and 4.9, **motor proteins** are a class of monomers that project from the surface of microtubules and some other cytoskeletal elements with roles in cell movements.

What about the microtubules that do not attach to kinetochores? Those extending from one pole actively slide past those extending from the other pole at the zone where they overlap (Figure 9.7b). Dynein and kinesin have been identified in this zone of overlap, also.

from and returned to the pool all the time. The balance can be tipped experimentally, as by exposing cells to the microtubule poison colchicine (Section 4.8). The spindle disassembles almost at once, but often it reassembles in seconds or minutes if the disrupting agent is removed.

At anaphase, some movements reduce the distance between kinetochores and the poles; others make the whole spindle lengthen (Figure 9.7). The kinetochores slide over the microtubules. Experimental observations show that the position of the microtubules stays put all through anaphase. The microtubules *do* shorten, in the manner shown in Figure 9.7a, so they must disassemble where a kinetochore passes over them. By analogy, the

Once the S stage of interphase begins, the cell cycle proceeds at about the same rate in all cells of a given type, all the way through mitosis. Molecular mechanisms control whether a cell enters S and thereby control the rate of cell division.

The condensed form of metaphase chromosomes arises by interactions among DNA and structural proteins, including histones, that associate with DNA throughout the cell cycle.

During mitosis, motor proteins act on microtubules of the spindle to produce chromosomal movements.

DIVISION OF THE CYTOPLASM

The cytoplasm usually divides at some time between late anaphase and the end of telophase. As you might gather from Figures 9.8 and 9.9, the mechanism of this **cytoplasmic division** (or cytokinesis, as it is commonly called) differs among organisms.

Cell Plate Formation in Plants

As described in Section 4.10, most plant cells are walled, which means their cytoplasm cannot be pinched in two. Cytoplasmic division of such cells involves **cell plate formation**, as shown in Figure 9.8. By this mechanism, vesicles chockful of wall-building materials fuse with

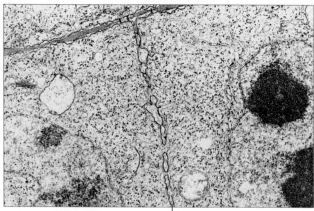

part of a newly forming cell plate ⌐

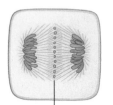

former spindle equator

a As mitosis ends, vesicles converge at the spindle equator. They contain cementing materials and structural materials for a new primary cell wall.

vesicles converging

b A cell plate starts forming as membranes of the vesicles fuse. Materials inside the vesicles get sandwiched between two new membranes that elongate along the plane of the cell plate.

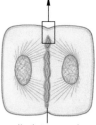

cell plate growing

c Cellulose is deposited on the inside of the "sandwich" and will form two cell walls. Other deposits will form a middle lamella and cement the walls together (Section 4.10).

two new primary walls

d A cell plate grows at its margins until it fuses with the parent cell's plasma membrane. During plant growth, when cells expand and their walls are still thin, new material gets deposited on the old primary wall.

Figure 9.8 Cytoplasmic division of a plant cell, as brought about by cell plate formation.

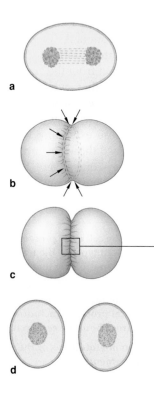

a

b

c

d

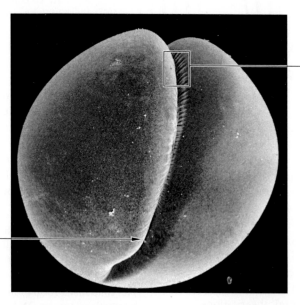

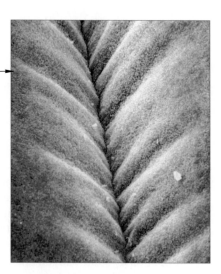

Figure 9.9 Cytoplasmic division of an animal egg. (**a**) Mitosis is completed, and the spindle is now disassembling. (**b**) Just beneath the plasma membrane of the parent cell, rings of microfilaments at the former spindle equator contract and close around the cell, rather like a purse string being tightened. (**c,d**) Contractions continue and in time will divide the cell in two. The micrographs show how the surface of the plasma membrane sinks inward. The depression defines the cleavage plane.

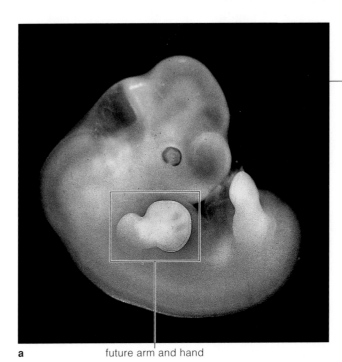

a   future arm and hand

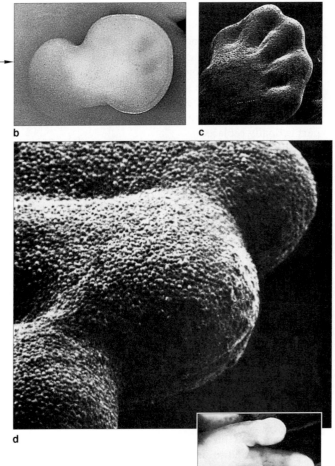

b   c

d

e

Figure 9.10 Development of the human hand by way of mitosis, cytoplasmic divisions, and other processes. Many individual cells resulting from mitotic cell divisions are visible in (**d**).

one another and with remnants from the microtubular spindle. Together, they form a disklike structure—a cell plate. At this location, deposits of cellulose accumulate and form a crosswall that divides the parent cell into two daughter cells.

Cleavage of Animal Cells

Unlike plant cells, an animal cell is not confined within a cell wall, and its cytoplasm typically "pinches in two." Consider a newly fertilized egg. Through a mechanism called **cleavage**, a shallow, ringlike depression forms at the cell surface, above the cell's midsection (Figure 9.9). The depression is known as the cleavage furrow. It is visible evidence of parallel arrays of microfilaments in the cytoplasm that connect with the plasma membrane. Microfilaments, remember, are threadlike cytoskeletal elements. These particular ones are organized so that they slide past one another (Section 4.9). When they do so, they pull the plasma membrane inward and thereby cut the cell in two.

This concludes our picture of mitotic cell division. Look now at your hands and try to envision all the cells in your palms, thumbs, and fingers. Imagine the divisions that produced all the generations of cells that preceded them when you were developing, early on, inside your mother (Figure 9.10). And be grateful for the astonishing precision of the mechanisms that led to their formation, for the alternatives can be terrible indeed. Why? Good health, and survival itself, depends absolutely on the proper timing and completion of cell cycle events, including mitosis. Genetic disorders often arise from mistakes in the duplication or distribution of just one chromosome. As you will read in Section 15.6, cancer may rapidly destroy a mature tissue if controls are lost that otherwise prevent cells from dividing. The section that concludes this chapter describes a landmark case of such unchecked cell divisions.

Following mitotic cell division, a separate mechanism cuts the cytoplasm into two daughter cells, each with a daughter nucleus.

In plants, cytoplasmic division often involves the formation of a cell plate and a crosswall in between the adjoining, new plasma membranes of daughter cells.

Cytoplasmic division in animals may involve cleavage. Rings of microfilaments around a parent cell's midsection slide past one another in a way that pinches the cytoplasm in two.

Focus on Science

HENRIETTA'S IMMORTAL CELLS

Each human starts out as a single fertilized egg. By the time of birth, mitotic cell divisions and other processes have resulted in a human body of about a trillion cells. Even in an adult, billions of cells are still dividing. For example, cells of the stomach's lining divide every day. Liver cells usually do not divide, but if part of the liver becomes injured or diseased, repeated cell divisions will keep on producing more new cells until the damaged part is finally replaced.

In 1951, George and Margaret Gey of Johns Hopkins University were trying to develop a way to keep human cells dividing *outside* the body. With such isolated cells, these researchers and others could investigate basic life processes. They also could conduct studies of cancer and other diseases without having to experiment directly on patients and thereby gamble with human lives.

The Geys had samples of normal and diseased human cells, which local physicians had sent to them. But they just couldn't stop the descendants of those precious cells from dying out within a few weeks.

Mary Kubicek, a laboratory assistant, worked with the Geys in their efforts to start a self-perpetuating lineage of cultured human cells. She was about to give up after dozens of failed attempts. Even so, in 1951 she decided to prepare one more sample of cancer cells for culture. She gave the sample the code name **HeLa cells**, for the first two letters of the patient's first and last names.

The HeLa cells began to divide. And divide. And divide again! By the fourth day there were so many cells that Kubicek had to subdivide them into more culture tubes. As months passed, the culture continued to thrive.

Unfortunately, the tumor cells inside the patient's body were just as vigorous. Six months after the patient was first diagnosed as having cancer, tumor cells had spread to tissues throughout her body. Only eight months after the diagnosis, Henrietta Lacks, a young woman from Baltimore, was dead.

Although Henrietta passed away, some of her cells lived on in the Geys' laboratory as the first successful human cell culture. HeLa cells were soon shipped to other researchers, who passed cells on to others, and so on. HeLa cells came to live in laboratories all over the world. Some of the cultured cells even journeyed far into space, as components of experiments to be carried out aboard the *Discoverer XVII* satellite. Every year, hundreds of scientific papers describe research that is based on work with HeLa cells.

Henrietta was only thirty-one years old when runaway cell divisions quickly killed her. Yet now, many decades later, her legacy is still benefiting humans everywhere—in cellular descendants that are still alive and dividing, day after day after day.

SUMMARY

1. Through specific division mechanisms, summarized in Table 9.1, a parent cell provides each daughter cell with hereditary instructions (DNA) and cytoplasmic machinery necessary to start up its own operation.

  a. In eukaryotic cells, the nucleus divides by mitosis or meiosis. Cytoplasmic division typically follows.

  b. Prokaryotic cells divide by prokaryotic fission.

2. Each eukaryotic chromosome is one DNA molecule with numerous proteins attached. The chromosomes in a given cell differ in length, shape, and which part of the hereditary instructions they carry.

  a. "Chromosome number" refers to the sum total of chromosomes in the cells of a given type. Cells with a diploid chromosome number ($2n$) contain two of each kind of chromosome.

  b. Mitosis divides the nucleus into two equivalent nuclei, each having the same chromosome number as the parent cell. Therefore, it maintains the chromosome number from one cell generation to the next.

  c. Mitosis is the basis of growth and tissue repair among multicelled eukaryotes, and often of asexual reproduction among single-celled eukaryotes. (Meiosis occurs only in germ cells used in sexual reproduction.)

3. A cell cycle starts when a new cell forms. It proceeds through interphase and ends when the cell reproduces by mitosis and cytoplasmic division. In interphase, a cell increases in its mass and cytoplasmic components, then duplicates its chromosomes.

4. Each duplicated chromosome has two molecules of DNA, attached at the centromere. For as long as the two are attached, they are called sister chromatids.

5. Mitosis proceeds through four continuous stages:

  a. Prophase. Duplicated chromosomes, in threadlike form, start to condense. New microtubules start to assemble near the nucleus into a spindle apparatus. The nuclear envelope starts to break up.

  b. Metaphase. During the *transition* to metaphase, the nuclear envelope breaks up totally into vesicles. Microtubules of opposite poles of the forming spindle

Table 9.1	Summary of Cell Division Mechanisms	
Mechanisms	**Functions**
MITOSIS, CYTOPLASMIC DIVISION	In multicelled eukaryotes, the basis of bodily growth. In single-celled and many multicelled eukaryotes, the basis of asexual reproduction.
MEIOSIS, CYTOPLASMIC DIVISION	In single-celled and multicelled eukaryotes, the basis of gamete formation and of sexual reproduction.
PROKARYOTIC FISSION	In bacterial cells, the basis of asexual reproduction.

attach to only one of the two sister chromatids of each chromosome. *At* metaphase, all of the chromosomes are aligned at the spindle equator.

   c. Anaphase. Microtubules pull sister chromatids of each chromosome away from each other, to opposite spindle poles. Each type of parental chromosome is represented by a daughter chromosome at both poles:

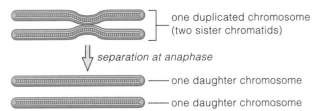

one duplicated chromosome
(two sister chromatids)

separation at anaphase

— one daughter chromosome

— one daughter chromosome

   d. Telophase. The chromosomes decondense to the threadlike form. A new nuclear envelope forms around them. Each nucleus has the same chromosome number as the parent cell. Mitosis is completed.

6. Different mechanisms divide the cytoplasm near the end of nuclear division or afterward. In plant cells, a cell plate forms, then a crosswall and plasma membranes form. Animal cells pinch in two, as by cleavage.

Review Questions

1. Define the two types of division mechanisms that operate in eukaryotic cells. Does either one divide the cytoplasm? *9.1*

2. Define somatic cell and germ cell. *9.1*

3. What is a chromosome called when it is in the unduplicated state? In the duplicated state (with two sister chromatids)? *9.1*

4. Describe the microtubular spindle and its functions, then name and briefly describe the stages of mitosis. *9.3*

5. Briefly explain how cytoplasmic division differs in typical plant and animal cells. *9.5*

Self-Quiz *(Answers in Appendix IV)*

1. A somatic cell with two of each type of chromosome characteristic of the species has a(n) _____ chromosome number.
   a. diploid   c. abnormal
   b. haploid   d. a and c

2. A duplicated chromosome has _____ chromatid(s).
   a. one   c. three
   b. two   d. four

3. In a chromosome, a _____ is a constricted region with attachment sites for microtubules.
   a. chromatid   c. cell plate
   b. centromere   d. cleavage furrow

4. Interphase is the part of the cell cycle when _____ .
   a. a cell ceases to function
   b. a germ cell forms its spindle apparatus
   c. a cell grows and duplicates its DNA
   d. mitosis proceeds

5. After mitosis, the chromosome number of a daughter cell is _____ the parent cell's.
   a. the same as   c. rearranged compared to
   b. one-half   d. doubled compared to

6. Mitosis and cytoplasmic division function in _____ .
   a. asexual reproduction of single-celled eukaryotes
   b. growth, tissue repair, and sometimes asexual reproduction in many multicelled eukaryotes
   c. gamete formation in prokaryotes
   d. both a and b

7. Only _____ is not a stage of mitosis.
   a. prophase   b. interphase   c. metaphase   d. anaphase

8. Match each stage with the events listed.
   ____ metaphase   a. sister chromatids move apart
   ____ prophase   b. chromosomes start to condense
   ____ telophase   c. chromosomes decondense and daughter nuclei form
   ____ anaphase
                  d. all duplicated chromosomes are aligned at spindle equator

Critical Thinking

1. Suppose you have a means of measuring the amount of DNA in a single cell during the cell cycle. You first measure the amount during the G1 phase. At what points during the remainder of the cycle would you predict changes in the amount of DNA per cell?

2. A cell from a tissue culture has 38 chromosomes. After mitosis and cytoplasmic division, one daughter cell has 39 chromosomes, and the other has 37. Speculate on what might have occurred to cause the abnormal chromosome numbers. Generally speaking, how might the abnormality affect cell structure, function, or both?

3. Pacific yews (*Taxus brevifolius*) face extinction. People started stripping its bark and killing the trees when they heard that *taxol*, a chemical extract from the bark, may be useful for treating breast and ovarian cancer. (Synthesizing taxol in the laboratory may save the species.) Taxol prevents the disassembly of microtubules. What does this tell you about its potential as an anticancer drug?

4. X-rays and gamma rays emitted from some radioisotopes cause chemical damage to DNA, especially in cells engaged in DNA replication. High-level exposure can result in *radiation poisoning*. Hair loss and damage to the lining of the stomach and intestines are two early symptoms. Speculate why. Also speculate on why highly focused radiation therapy is used against some cancers.

Selected Key Terms

anaphase *9.3*	cytoplasmic	metaphase *9.3*
cell cycle *9.2*	division *9.5*	mitosis *9.1*
cell plate	diploid (chromosome	motor protein *9.4*
formation *9.5*	number) *9.1*	nucleosome *9.4*
centriole *9.3*	germ cell *9.1*	prophase *9.3*
centromere *9.1*	HeLa cell *9.6*	reproduction *CI*
chromosome *9.1*	histone *9.4*	sister chromatid *9.1*
chromosome	interphase *9.2*	somatic cell *9.1*
number *9.1*	kinetochore *9.4*	spindle apparatus *9.3*
cleavage *9.5*	meiosis *9.1*	telophase *9.3*

Reading

Murray, A., and M. Kirschner. March 1991. "What Controls the Cell Cycle?" *Scientific American* 264(3): 56–63.

Web Site See *http://www.wadsworth.com/biology* for practice quiz questions, hypercontents, BioUpdates, and critical thinking. The Wadsworth Biology Resource Center provides a wealth of information fully organized and integrated by chapter.

Octopus Sex and Other Stories

The couple clearly are interested in each other. First he caresses her with one tentacle, then another—and then another and another. She reciprocates with a hug here, a squeeze there. This goes on for hours. Finally the male reaches under his mantle, a fold of tissue that drapes around most of his body. He removes a packet of sperm from a reproductive organ and inserts it into an egg chamber underneath the female's mantle. For every sperm that fertilizes an egg, a new octopus may develop.

Unlike the one-to-one coupling between a male and female octopus, sex for the slipper limpet is a group activity. Slipper limpets are marine animals, relatives of those familiar snails on land. Before becoming transformed into a sexually mature adult, a slipper limpet must pass through a free-living stage of development called a larva. When a limpet larva is about to undergo transformation, it settles down on a pebble or shell or rock. If it settles down all by itself, it will become a female. Then, if another larva settles and develops on the

Figure 10.1 Variations in the reproductive modes among eukaryotic organisms.

(**a**) Some slipper limpets, busily perpetuating the species through group participation in sexual reproduction. The tiny crab in the foreground is merely a passerby, not a voyeur. (**b**) Live birth of an aphid, a type of insect that reproduces sexually in autumn but can switch to an asexual mode in summer.

With this chapter, we turn to the kinds of cells that serve as the bridge between generations of organisms. For many eukaryotic species, specialized phases of reproduction and development, including asexual episodes, loop out from the basic life cycle. Regardless of their specialized details, all of these life cycles turn on two basic events: *gamete formation* and *fertilization*.

first one, it will function right off as a male. However, if *another* male develops on top of it, that first male will gradually become a female. That third male *also* will become a female if a fourth male develops on top of it, and so on amongst ten or more limpets.

Slipper limpets typically live in such piles, with the bottom one always being the oldest female and the uppermost one being the youngest male (Figure 10.1*a*). Until they make the gender switch, the males release sperm; these fertilize a female's eggs, which then grow to become males and then, most likely, females—and so it goes, from one limpet generation to the next.

Even within a single species, we often come across variations in the mode of reproduction. For example, the life cycle of many sexually reproducing organisms also includes asexual episodes that are based on mitotic cell divisions. Consider the orchids, dandelions, and many other species of plants that are able to reproduce without engaging in sex. Consider the flatworms, a group of aquatic animals. They can split their small body into two roughly equivalent parts that each grow into a new flatworm.

Or consider the aphids. During the summer, nearly all aphids are females, which produce more females from *unfertilized* egg cells (Figure 10.1*b*). Only when autumn approaches do male aphids finally develop and do their part in the sexual phase of the life cycle. Even then, the females that manage to survive through the winter can do without males. Come summer, the females begin another round of producing offspring all by themselves.

These examples only hint at the immense variation in reproductive modes among eukaryotic organisms. And yet, despite the variation, *sexual* reproduction dominates their life cycles, and it inevitably involves certain events. Briefly, before the time of cell division, chromosomes are duplicated in reproductive cells. In animals, for instance, immature reproductive cells are called **germ cells**. In both males and females, the germ cells undergo meiosis and, later, cytoplasmic division. In time, their cellular descendants mature and may function as **gametes**, or sex cells. When gametes join at fertilization, they form the first cell of a new individual.

Meiosis, the formation of gametes, and fertilization are the hallmarks of sexual reproduction. As you will see in this chapter, these interconnected events contribute to the splendid diversity of life.

KEY CONCEPTS

1. Sexual reproduction proceeds through three events: meiosis, formation of gametes, and fertilization. Sperm and eggs are familiar gametes.

2. Meiosis, a nuclear division mechanism, occurs only in immature cells that are set aside for sexual reproduction. The germ cells of male and female animals are examples. Meiosis sorts out a germ cell's chromosomes into four new nuclei. After meiosis, gametes form by way of cytoplasmic division and other events.

3. Cells with a diploid chromosome number contain two of each type of chromosome characteristic of the species. The two function as a pair during meiosis. Typically, one chromosome of the pair is maternal, with hereditary instructions from a female parent. The other is paternal, with the same categories of hereditary instructions from a male parent.

4. Meiosis divides the chromosome number by half for each forthcoming gamete. Thus, if both parents are diploid ($2n$), the gametes that form are haploid (n). Later, the union of two gametes at fertilization restores the diploid number in the new individual ($n + n = 2n$).

5. During meiosis, each pair of chromosomes may swap segments. Each time they do, they exchange hereditary instructions about certain traits. Also, meiosis assigns one of every pair of chromosomes to a forthcoming gamete—but *which* gamete is its destination is a matter of chance. Hereditary instructions are further shuffled at fertilization. All three of these reproductive events lead to variations in traits among offspring.

6. In most species of plants, spore formation and other events intervene between meiosis and gamete formation.

10.1 COMPARISON OF ASEXUAL AND SEXUAL REPRODUCTION

When an orchid, flatworm, or aphid reproduces all by itself, what sort of offspring does it get? By the process of **asexual reproduction**, one parent alone produces offspring, and each offspring inherits the same number and kinds of genes as its parent. **Genes** are specific stretches of chromosomes—that is, of DNA molecules. Taken together, the genes for every species contain all of the heritable bits of information that are necessary to produce new individuals. Rare mutations aside, this means asexually produced offspring can only be clones, or genetically identical copies of the parent.

Inheritance gets much more interesting with **sexual reproduction**. This process involves meiosis, gamete formation, and fertilization (the union of two gametes). In humans and other species that use this process, the first cell of a new individual ends up with *pairs of genes*, on pairs of homologous chromosomes. Typically, one of each pair is maternal and the other paternal in origin.

If instructions encoded in every pair of genes were identical down to the last detail, sexual reproduction would produce clones, also. Just imagine—you, every single person you know, the entire human population might be a clone, with everybody looking alike. But the two genes of a pair may *not* be identical. Why not? The molecular structure of genes can change; this is what we mean by mutation. Depending on their structure, two genes that happen to be paired in a person's cells may "say" slightly different things about a trait. Each unique molecular form of the same gene is called an **allele**.

Such tiny differences affect thousands of traits. For example, whether your chin has a dimple depends on which pair of alleles you inherited at one chromosome location. One kind of allele at that location says "put a dimple in the chin," another kind says "no dimple." This leads us to a key reason why members of sexually reproducing species don't all look alike. *Through sexual reproduction, offspring inherit new combinations of alleles, which lead to variations in physical and behavioral traits.*

This chapter describes the cellular basis of sexual reproduction. More importantly, it starts us thinking about far-reaching effects of gene shufflings at different stages of the process. The process introduces variations in traits among offspring that may be acted upon by agents of natural selection. Thus, *variation in traits is a foundation for evolutionary change.*

Asexual reproduction produces genetically identical copies of the parent. Sexual reproduction introduces variations in the details of traits among offspring.

Sexual reproduction dominates the life cycle of eukaryotic species. Meiosis, formation of gametes, and fertilization are the basic events of this process.

10.2 HOW MEIOSIS HALVES THE CHROMOSOME NUMBER

Think "Homologues"

Think back on the preceding chapter and its focus on mitotic cell division. Unlike mitosis, **meiosis** divides chromosomes into separate parcels not once *but twice* prior to cell division. Unlike mitosis, it is the first step leading to the formation of gametes.

Gametes, recall, are sex cells such as sperm or eggs. In nearly all multicelled eukaryotic organisms, gametes arise from immature reproductive cells, such as germ cells, that form inside specialized structures and organs. Figure 10.2 shows examples of where gametes form.

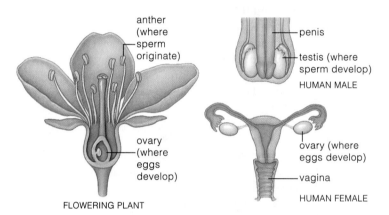

Figure 10.2 Examples of gamete-producing structures.

From Chapter 9, you know that the "chromosome number" is the sum total of chromosomes in cells of a given type. Germ cells start with the same chromosome number as somatic cells (the rest of the body's cells). If a cell has a **diploid number** (2n), it has *a pair* of each type of chromosome, often from two parents. In general, the chromosomes of each pair have the same length and shape. Their genes deal with the same traits. And they line up with each other during meiosis. We call them **homologous chromosomes** (*hom-* means alike).

As you can probably deduce from Figure 10.3, your own germ cells have 23 + 23 homologous chromosomes. After meiosis, 23 chromosomes—one of each type—end up in gametes. Thus meiosis halves the chromosome number, so that gametes have a **haploid number** (n).

Two Divisions, Not One

Meiosis resembles mitosis in some respects, even though the outcome is different. Before interphase gives way to meiosis, a germ cell duplicates its DNA. Each duplicated chromosome now consists of two DNA molecules. These

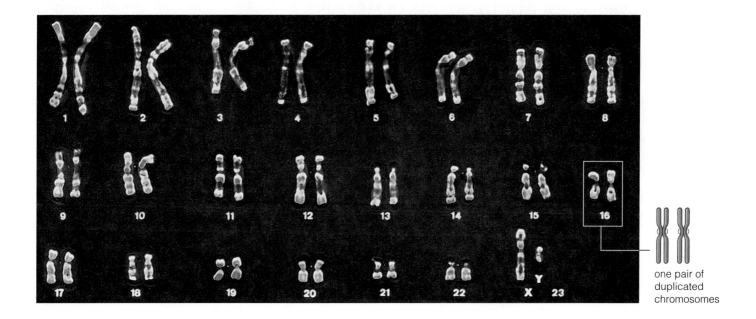

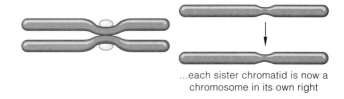

one pair of duplicated chromosomes

remain attached at a narrowed-down region, called the centromere. For as long as the two remain attached, they are known as **sister chromatids** of the chromosome:

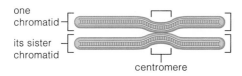

As in mitosis, the microtubules of a spindle apparatus move the chromosomes in prescribed directions.

With meiosis alone, however, *chromosomes proceed through two consecutive divisions, which end with the formation of four haploid nuclei.* The two divisions are called meiosis I and II:

	MEIOSIS I		MEIOSIS II
DNA is replicated during interphase	PROPHASE I	DNA is *not* replicated between divisions	PROPHASE II
	METAPHASE I		METAPHASE II
	ANAPHASE I		ANAPHASE II
	TELOPHASE I		TELOPHASE II

During meiosis I, each duplicated chromosome lines up with its partner, *homologue to homologue*; and then the partners are moved apart from each other:

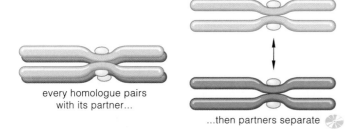

Figure 10.3 From a diploid cell of a human male, twenty-three pairs of homologous chromosomes. The sketch, corresponding to the two chromosomes inside the white box, indicates that the chromosomes are in the duplicated state. That is, each consists of two sister chromatids.

The cytoplasm typically starts to divide at some point after homologues have been separated from each other. The cytoplasmic division results in two daughter cells. Each daughter cell is haploid; it has only one of each type of chromosome. But remember, the chromosomes are still in the duplicated state.

Next, during meiosis II, *the two sister chromatids of each chromosome are separated from each other:*

...each sister chromatid is now a chromosome in its own right

Four nuclei form, then the cytoplasm often divides once more. The final outcome is four haploid cells.

Figure 10.4 on the next two pages illustrates the key events of meiosis I and II.

Meiosis is a type of nuclear division mechanism that reduces the parental chromosome number by half, to the haploid number (*n*).

Meiosis proceeds only in immature cells, such as germ cells, that are set aside for sexual reproduction. It is the first step leading to the formation of gametes.

A VISUAL TOUR OF THE STAGES OF MEIOSIS

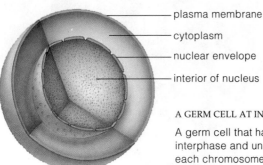

- plasma membrane
- cytoplasm
- nuclear envelope
- interior of nucleus

A GERM CELL AT INTERPHASE

Figure 10.4 Meiosis: the nuclear division mechanism by which a parental number of chromosomes in an immature reproductive cell is reduced by half (to the haploid number) for forthcoming gametes. In this case, only two of the pairs of homologous chromosomes of a germ cell are shown. The maternal chromosomes are shaded *purple*. The paternal chromosomes are shaded *blue*.

A germ cell that happens to have a diploid chromosome number (2n) is about to leave interphase and undergo the first division of meiosis. Its DNA is already duplicated, so each chromosome is in the duplicated state: it consists of two sister chromatids.

MEIOSIS I

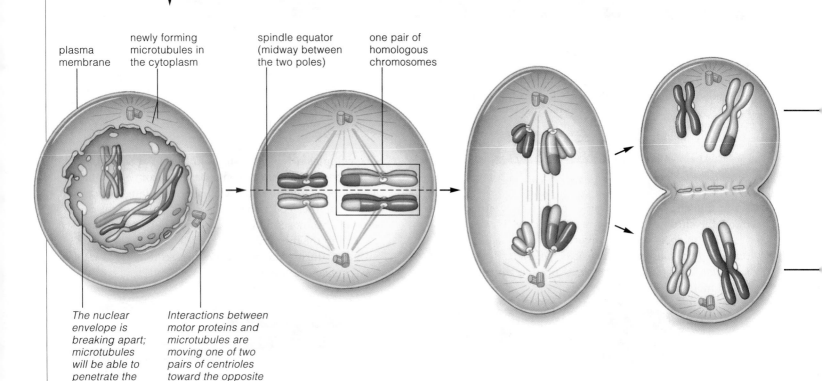

plasma membrane

newly forming microtubules in the cytoplasm

spindle equator (midway between the two poles)

one pair of homologous chromosomes

The nuclear envelope is breaking apart; microtubules will be able to penetrate the nuclear region.

Interactions between motor proteins and microtubules are moving one of two pairs of centrioles toward the opposite spindle pole.

PROPHASE I

Each duplicated chromosome is in threadlike form, but now it starts to twist and fold into more condensed form. It pairs with its homologue, and the two typically swap segments. The swapping, called crossing over, is indicated by the break in color on the pair of larger chromosomes. Each chromosome becomes attached to some microtubules of a newly forming spindle.

METAPHASE I

Motor proteins have been driving the movement of microtubules that became attached to the kinetochores of chromosomes. As a result, the chromosomes have been pushed and pulled into position, midway between the spindle poles. Now the spindle is fully formed, owing to dynamic interactions of motor proteins, microtubules, and the chromosomes themselves.

ANAPHASE I

Microtubules extending from the poles and overlapping at the spindle equator *lengthen* and push the poles apart. At the same time, the microtubules extending from the poles to the kinetochores of chromosomes *shorten*, and each chromosome is thereby pulled away from its homologous partner. These motions move the homologous partners to opposite poles.

TELOPHASE I

At some point, the cytoplasm of the germ cell divides. Two cells, each with a haploid chromosome number (n), result. That is, the cells have one of each type of chromosome that was present in the parent (2n) cell. All chromosomes are still in the duplicated state.

Of the four haploid cells that form by way of meiosis and cytoplasmic divisions, one or all may proceed to develop into gametes and function in sexual reproduction.

(In plants, the cells that form after meiosis has been completed may develop into spores, which take part in a stage of the life cycle that precedes the formation of gametes.)

MEIOSIS II

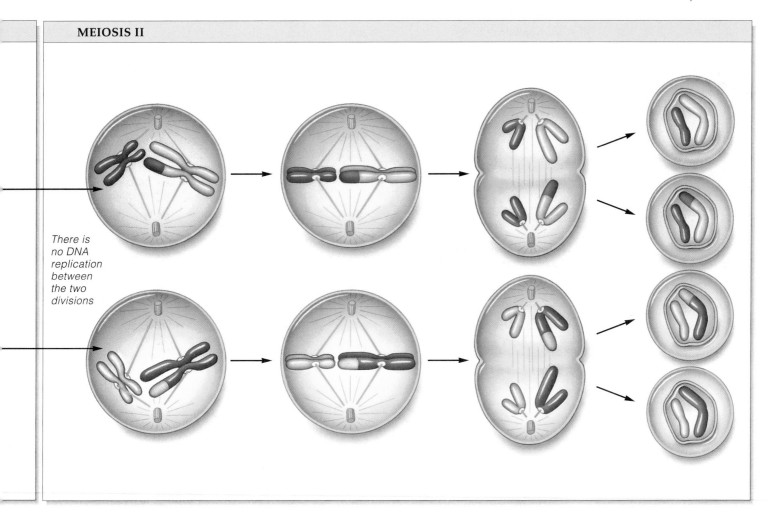

There is no DNA replication between the two divisions

PROPHASE II

In each of the two daughter cells, microtubules have already moved one member of the centriole pair to the opposite pole of the spindle during the transition to prophase II. Now, at prophase II, microtubules attach to the kinetochores of chromosomes, and motor proteins drive the movement of chromosomes toward the spindle's equator.

METAPHASE II

Now, in each daughter cell, interactions among motor proteins, spindle microtubules, and each duplicated chromosome have moved all of the chromosomes so that they are positioned at the spindle equator, midway between the two poles.

ANAPHASE II

The attachment between the two chromatids of each chromosome breaks. The former "sister chromatids" are now chromosomes in their own right. Motor proteins interact with kinetochore microtubules to move the separated chromosomes to opposite poles of the spindle.

TELOPHASE II

By the time telophase II is completed, there will be four daughter nuclei. At the time when the division of the cytoplasm is completed, each daughter cell will have a haploid chromosome number (*n*). All of those chromosomes will be in the unduplicated state.

The preceding overview, in Sections 10.2 and 10.3, is enough to convey the overriding function of meiosis—that is, *the reduction of the chromosome number by half for forthcoming gametes.* However, as you will now read, two other events that take place during prophase and metaphase of meiosis I contribute greatly to the adaptive advantage of sexual reproduction.

That advantage, recall, is the production of offspring with new combinations of alleles. Those combinations translate into a new generation of individuals that display variation in the details of some number of traits.

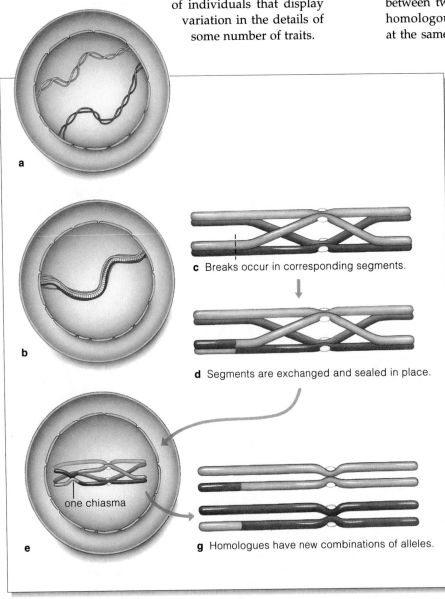

a

b

c Breaks occur in corresponding segments.

d Segments are exchanged and sealed in place.

one chiasma

e

g Homologues have new combinations of alleles.

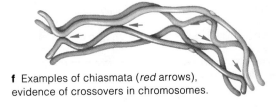

f Examples of chiasmata (*red* arrows), evidence of crossovers in chromosomes.

Prophase I Activities

Prophase I of meiosis is a time of major gene shufflings. Consider Figure 10.5a, which shows two chromosomes condensed to threadlike form. All chromosomes in a germ cell condense this way. When they do, each is drawn close to its homologue by a process called synapsis. It is as if homologues become stitched together point by point along their length, with little space between. The intimate, parallel array favors **crossing over**, a molecular interaction between two of the *non*sister chromatids of a pair of homologous chromosomes. Nonsister chromatids break at the same places along their length and then exchange corresponding segments—that is, genes—at the break points. Figure 10.5c and d is a simplified picture of this interaction.

Figure 10.5 Key events of prophase I, the first stage of meiosis. For clarity, the diagram shows only one of the pairs of homologous chromosomes in the cell and one crossover event. *Blue* signifies the paternal chromosome; *purple* signifies its maternal homologue.

(**a**) Both chromosomes have been duplicated previously, at interphase. Early in prophase I, each duplicated chromosome is in threadlike form, attached at both ends to the nuclear envelope. Its two sister chromatids are so close together, they look like a single thread.

(**b**) Homologous chromosomes become zippered together, so all four chromatids are intimately aligned. A pair of sex chromosomes having different forms (such as X paired with Y) also become zippered together in a small region.

(**c,d**) One or more crossovers typically occur at intervals along the chromosomes. Each time, two nonsister chromatids break at identical sites. Then they swap segments at the breaks, and enzymes seal the broken ends. For clarity, these chromosomes are shown in condensed form and pulled apart, except at their ends. (Think of the blue chromosome as a spider and the purple one as its reflection in a mirror.) But remember, crossing over only proceeds while all four chromatids are extended like threads and tightly aligned together.

(**e,f**) As prophase I ends, the chromosomes continue to condense and become thicker, rodlike forms. They detach from the nuclear envelope and from each other except at contact points known as chiasmata. Each chiasma is indirect evidence that a crossover occurred at some place in the chromosomes. They don't indicate *which* point, because they slip down to the chromosome tips and vanish.

(**g**) Crossing over breaks up old combinations of alleles and puts new ones together in pairs of homologous chromosomes.

Gene swapping would be rather pointless if each type of gene never varied from one chromosome to the next. But remember, a gene can have slightly different forms—alleles. You can safely bet that some alleles on one chromosome will *not* be identical to those on the homologue. Therefore, every crossover represents a chance to swap a *slightly different version* of hereditary instructions for a particular trait.

We will look at the mechanism of crossing over in later chapters. For now, it is enough to remember this: *Crossing over leads to genetic recombination, which in turn leads to variation in the traits of offspring.*

Metaphase I Alignments

Major shufflings of whole chromosomes begin during the transition from prophase I to metaphase I, which is the second stage of meiosis. Suppose the shufflings are proceeding at this moment in one of your germ cells. By now, crossovers have made genetic mosaics out of the chromosomes, but put this aside in order to simplify tracking. Just call the twenty-three chromosomes you inherited from your mother the *maternal* chromosomes and their twenty-three homologues from your father the *paternal* chromosomes.

Kinetochore microtubules have already oriented one chromosome of each pair toward one spindle pole and its homologue toward the other pole (compare Section 9.4). Now they are moving all the chromosomes, which soon will become positioned at the spindle's equator.

Are all the maternal chromosomes attached to one pole and all the paternal ones attached to the other? Maybe, but probably not. Remember, the first contacts between kinetochore microtubules and chromosomes are random. Because of the random grabs, *the eventual positioning of a maternal or paternal chromosome at the spindle equator at metaphase I follows no particular pattern.* Carrying this one step further, either one of each pair of homologous chromosomes might end up at either pole of the spindle after they move apart at anaphase I.

Think about the possibilities when you are tracking merely three pairs of homologues. As you can see from Figure 10.6, by metaphase I, the homologues may be arranged in any one of four possible positions. In this case, 2^3 or eight combinations of maternal and paternal chromosomes are possible for forthcoming gametes.

Of course, a human germ cell has twenty-three pairs of homologous chromosomes, not just three. So a grand total of 2^{23} or *8,388,608 combinations* of maternal and paternal chromosomes is possible every time a germ cell gives rise to sperm or eggs!

Moreover, in each sperm or egg, many hundreds of alleles inherited from the mother might not "say" the exact same thing about hundreds of different traits as

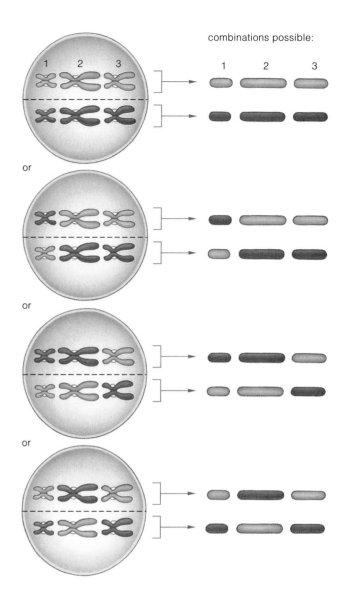

combinations possible:

Figure 10.6 The possible outcomes of the random alignment of three pairs of homologous chromosomes at metaphase I of meiosis. Three types of chromosomes are labeled 1, 2, and 3. Maternal chromosomes are *purple;* paternal ones are *blue.* With merely four possible alignments, eight combinations of maternal and paternal chromosomes are possible in forthcoming gametes.

the alleles inherited from the father. Are you beginning to get an idea of why such splendid mixes of traits show up even in the same family?

Crossing over is an interaction between a pair of homologous chromosomes. It breaks up old combinations of alleles and puts new ones together during prophase I of meiosis.

The random attachment and subsequent positioning of each pair of maternal and paternal chromosomes at metaphase I lead to different combinations of maternal and paternal traits in each generation of offspring.

The gametes that form after meiosis are not all the same in their details. For example, human sperm have one tail, opossum sperm have two, and roundworm sperm have none. Crayfish sperm look like pinwheels. Most eggs are microscopic, yet ostrich eggs tucked inside a shell are as large as a baseball. From appearance alone, you might not believe a plant gamete is even remotely like an animal's.

Later chapters contain details of how gametes form in the life cycles of representative organisms, including humans. Figure 10.7 and the following points may help you keep the details in perspective.

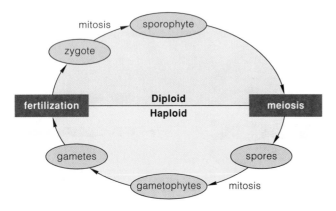

a Generalized life cycle for most kinds of plants

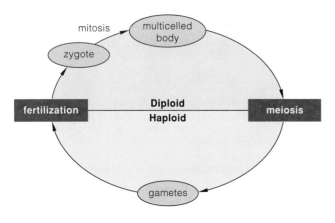

b Generalized life cycle for animals

Figure 10.7 Generalized life cycles for (**a**) most plants and (**b**) animals. The zygote is the first cell that forms when the nuclei of two gametes fuse together at fertilization.

For plants, a sporophyte (spore-producing body) develops, by way of mitotic cell divisions, from the zygote. Following meiosis, gametophytes (gamete-producing bodies) form. A pine tree is a sporophyte. Gametophytes develop in its cones.

Chapters 25, 26, and 45 include many illustrated examples of the life cycles of representative plants and animals.

Gamete Formation in Plants

For pine trees, apple trees, roses, dandelions, corn, and nearly all other familiar plants, certain events intervene between the times of meiosis and gamete formation. Among other things, spores form.

Spores are haploid cells, often with walls that help them resist episodes of drought, cold, or other adverse environmental conditions. When favorable conditions return, spores germinate and develop into a haploid body or structure that will produce gametes. Thus, *gamete*-producing bodies and *spore*-producing bodies develop during the life cycle of most kinds of plants. Figure 10.7*a* is a generalized diagram of these events.

Gamete Formation in Animals

In male animals, gametes form by a process that is known as spermatogenesis. Inside the male reproductive system, a diploid germ cell increases in size. It becomes a large, immature cell (a primary spermatocyte) that undergoes meiosis and cytoplasmic divisions. The outcome is four haploid daughter cells, which develop into immature cells called spermatids (Figure 10.8). The spermatids change in form, and each develops a tail. In this way, each becomes a **sperm**, a common type of mature male gamete.

In female animals, gametes form by a process that is known as oogenesis. In human females, for example, a diploid germ cell develops into an **oocyte**, or immature egg. Unlike a sperm cell, an oocyte accumulates many cytoplasmic components. As Figure 10.9 indicates, its daughter cells also differ in size and function. When the oocyte undergoes cytoplasmic division after meiosis I, one cell (the secondary oocyte) gets nearly all of the cytoplasm. The other cell, called the first polar body, is quite small. At some time afterward, both cells undergo meiosis II, then cytoplasmic division. One daughter cell of the secondary oocyte develops into a second polar body. The other receives most of the cytoplasm and develops into the gamete. A mature female gamete is called an ovum (plural, ova) or, more commonly, an **egg**.

Thus, after the first polar body divides, there are three polar bodies. These do not function as gametes. They serve as dumping grounds for chromosomes, so the egg will end up with a suitable (haploid) number of chromosomes at fertilization. Polar bodies do not have much in the way of nutrients or metabolic machinery. In time they degenerate.

More Shufflings at Fertilization

The chromosome number characteristic of the parents is restored at **fertilization**, the time when a male gamete penetrates a female gamete and their haploid nuclei

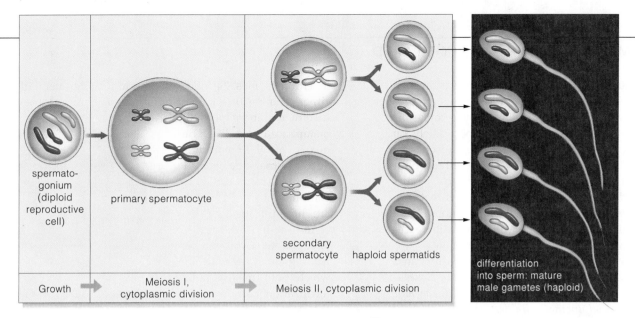

Figure 10.8 Generalized sketch of sperm formation in male animals.

spermato-
gonium
(diploid
reproductive
cell)

primary spermatocyte

secondary
spermatocyte

haploid spermatids

differentiation
into sperm: mature
male gametes (haploid)

Growth → Meiosis I,
cytoplasmic division → Meiosis II, cytoplasmic division

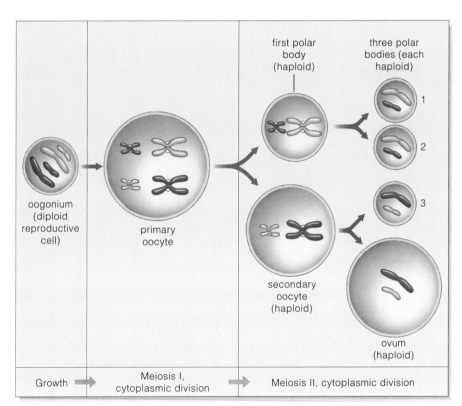

oogonium
(diploid
reproductive
cell)

primary
oocyte

first polar
body
(haploid)

three polar
bodies (each
haploid)

1

2

3

secondary
oocyte
(haploid)

ovum
(haploid)

Growth → Meiosis I,
cytoplasmic division → Meiosis II, cytoplasmic division

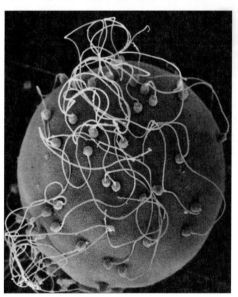

Figure 10.9 Egg formation in female animals. The generalized sketch is not at the same scale as Figure 10.8. Eggs are *far* larger than sperm, as the micrograph of sea urchin gametes shows. Also, the three polar bodies that form in meiosis are extremely small, compared to the egg.

fuse. If meiosis did not precede it, fertilization would double the chromosome number each new generation. Such changes in chromosome number would disrupt the hereditary instructions encoded in chromosomes, usually for the worse, for those instructions operate as a complex, fine-tuned package in each individual.

Fertilization contributes to variation in offspring. Reflect on the possibilities for humans alone. During prophase I, an average of two or three crossovers takes place in each human chromosome. Even without the crossovers, the random positioning of pairs of paternal and maternal chromosomes at metaphase I results in one of millions of possible chromosome combinations in each gamete. And of all the male and female gametes that are produced, which two actually get together is a matter of chance. The sheer number of combinations that can exist at fertilization is staggering!

Cumulatively, crossing over, the distribution of random mixes of homologous chromosomes into gametes, and fertilization contribute to variation in the traits of offspring.

MEIOSIS AND MITOSIS COMPARED

In this unit, our focus has been on two nuclear division mechanisms. Single-celled eukaryotic species reproduce asexually by way of mitosis, followed by cytoplasmic division. Many multicelled eukaryotic species rely on mitosis and cytoplasmic division during episodes of asexual reproduction in their life cycle, and all rely on it for growth and tissue repair. By contrast, meiosis occurs only in reproductive cells, such as germ cells that give rise to the gametes used in sexual reproduction. Figure 10.10 summarizes the basic similarities and differences between the two nuclear division mechanisms.

The end results of the two mechanisms differ in a crucial way. *Mitotic cell division only produces clones— genetically identical copies of a parent cell. But meiotic cell division, in conjunction with fertilization, promotes variation in traits among offspring.* First, crossing over at prophase I of meiosis puts new combinations of alleles in chromosomes. Second, the random assignment of either member of a pair of homologous chromosomes to either spindle pole at metaphase I puts different mixes of the maternal and paternal alleles into gametes. Third, different combinations of alleles are brought together by chance at fertilization. In later chapters, you will be reading about the ways in which both meiosis and fertilization contribute to the staggering diversity and evolution of sexually reproducing organisms.

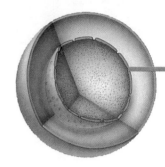

A *somatic cell* with a diploid chromosome number (2n) is at interphase. Before mitotic division begins, its DNA is replicated (all chromosomes are duplicated).

Figure 10.10 Summary of mitosis and meiosis. Both diagrams use a diploid (2n) animal cell as the example. They are arranged to help you compare similarities and differences between the division mechanisms. Maternal chromosomes are coded *purple*, and paternal chromosomes are coded *blue*.

MEIOSIS I

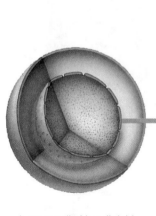

A *germ cell* with a diploid chromosome number (2n) is at interphase. Before mitotic division begins, its DNA is replicated (all chromosomes are duplicated).

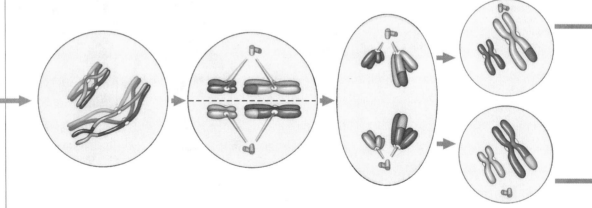

PROPHASE I	**METAPHASE I**	**ANAPHASE I**	**TELOPHASE I**
Each duplicated chromosome (consisting of two sister chromatids) condenses to threadlike form, then rodlike form. *Crossing over* occurs. Each chromosome unzips from its homologue. Each gets attached to the spindle in transition to metaphase.	All chromosomes are now positioned at the spindle's equator.	*Each chromosome is separated from its homologue.* They are moved to opposite poles of the spindle.	When the cytoplasm divides, there are two cells. Each has a haploid (n) number of chromosomes, but these are still in the duplicated state.

MITOSIS

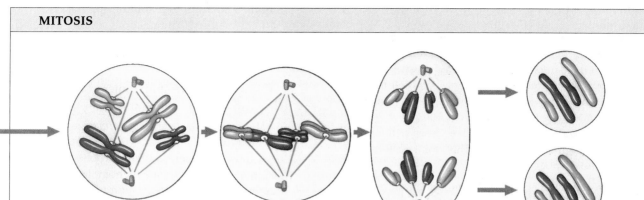

PROPHASE

Each duplicated chromosome (consisting of two sister chromatids) condenses from threadlike form to rodlike form. Each gets attached to the spindle during the transition to metaphase.

METAPHASE

All chromosomes are now positioned at the spindle's equator.

ANAPHASE

Sister chromatids of each chromosome are separated from each other. These new, daughter chromosomes are moved to opposite poles of the spindle.

TELOPHASE

When the cytoplasm divides, there are two cells. Each is diploid (2n)—*it has the same chromosome number as the parent cell.*

MEIOSIS II

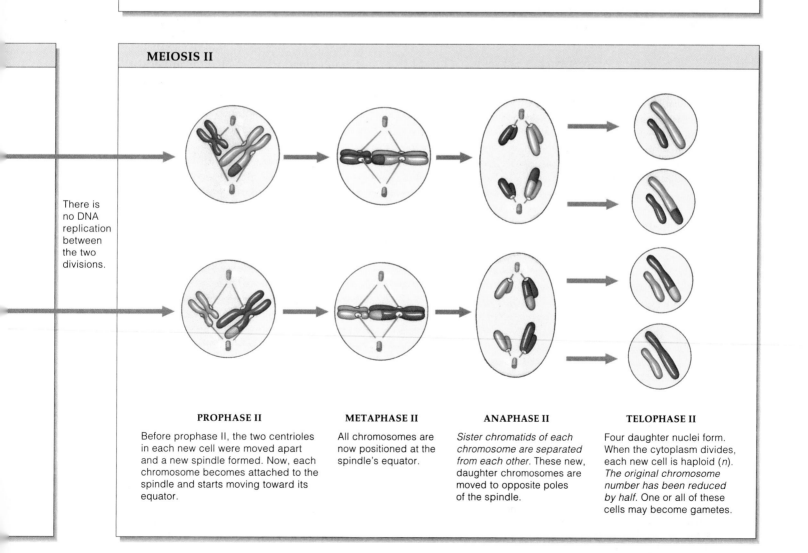

There is no DNA replication between the two divisions.

PROPHASE II

Before prophase II, the two centrioles in each new cell were moved apart and a new spindle formed. Now, each chromosome becomes attached to the spindle and starts moving toward its equator.

METAPHASE II

All chromosomes are now positioned at the spindle's equator.

ANAPHASE II

Sister chromatids of each chromosome are separated from each other. These new, daughter chromosomes are moved to opposite poles of the spindle.

TELOPHASE II

Four daughter nuclei form. When the cytoplasm divides, each new cell is haploid (n). *The original chromosome number has been reduced by half.* One or all of these cells may become gametes.

SUMMARY

1. The life cycle of each sexually reproducing species includes meiosis, gamete formation, and fertilization.

   a. Meiosis, a nuclear division mechanism, reduces the chromosome number of a parent germ cell by half. It *precedes* the formation of haploid gametes (typically, sperm in males, eggs in females).

   b. At fertilization, a sperm nucleus fuses with an egg nucleus. As Figure 10.11 shows, this event restores the chromosome number.

2. A germ cell with a diploid chromosome number ($2n$) has *two* of each type of chromosome characteristic of its species. Commonly, one of each pair of chromosomes is maternal (inherited from a female parent), and the other is paternal (from a male parent).

3. Each pair of maternal and paternal chromosomes shows homology, meaning the two are alike. In general, the two have the same length, same shape, and same sequence of genes. They interact during meiosis.

4. Chromosomes are duplicated during interphase. So before meiosis even begins, each consists of two DNA molecules that remain attached, as sister chromatids.

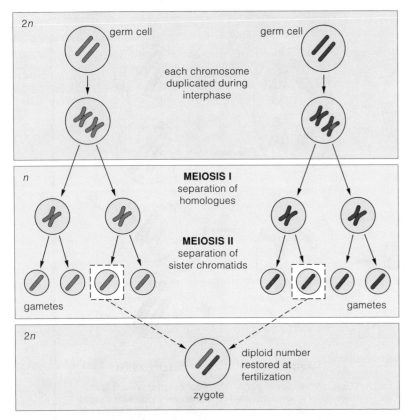

Figure 10.11 Summary of changes in the chromosome number at different stages of sexual reproduction, using diploid ($2n$) germ cells as the example. Meiosis reduces the chromosome number by half (n). Then the union of haploid nuclei of two gametes at fertilization restores the diploid number.

5. Meiosis consists of two consecutive divisions that require a microtubular spindle apparatus.

   a. In meiosis I, interactions among motor proteins, microtubules, and kinetochores of chromosomes move each type of chromosome away from its homologous partner.

   b. In meiosis II, similar interactions move the sister chromatids of each chromosome away from each other.

6. The following events characterize the first nuclear division (meiosis I):

   a. Crossing over occurs in prophase I. Two nonsister chromatids of each pair of homologous chromosomes commonly break at corresponding sites and exchange segments, which puts new combinations of alleles together. Alleles (slightly different molecular forms of the same gene) code for different forms of the same trait. Different combinations of alleles lead to variation in the details of a given trait among offspring.

   b. Also in prophase I, a microtubular spindle starts to form outside the nucleus, and the nuclear envelope starts to break up. If the cell has a duplicated pair of centrioles, one pair starts moving to the opposite pole of the newly forming spindle.

   c. At metaphase I, all of the pairs of homologous chromosomes have become positioned at the spindle equator. For each pair, either the maternal chromosome or its homologue can be oriented toward either pole.

   d. In anaphase I, interactions among motor proteins, microtubules, and the chromosomal kinetochores move each duplicated chromosome away from its homologue, toward opposite spindle poles.

7. The following events characterize the second nuclear division (meiosis II):

   a. At metaphase II, all the duplicated chromosomes are positioned at the spindle equator.

   b. Sister chromatids are moved apart in anaphase II. Each is now a separate, unduplicated chromosome.

   c. By the end of telophase II, four nuclei that have a haploid chromosome number (n) have been formed.

8. When the cytoplasm divides, there are four haploid cells. One or all of these may function as gametes (or as spores, in flowering plants).

9. Crossing over, the chance allocation of different mixes of pairs of maternal and paternal chromosomes to different gametes, and the chance of any two gametes meeting at fertilization all contribute to the immense variation in the details of traits among offspring.

Review Questions

1. Genetically speaking, what is the key difference between the outcomes of sexual and asexual reproduction? *10.1, 10.6*

2. Suppose a diploid germ cell contains four pairs of homologous chromosomes, designated AA, BB, CC, and DD. How would the chromosomes of the gametes be designated? *10.2*

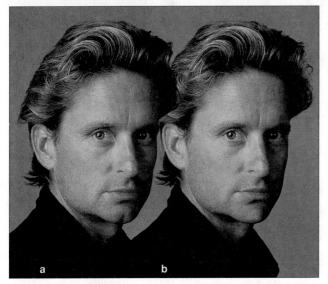

Figure 10.12
Example of the chin dimple trait (actually a fissure in the chin surface).

3. From each of his parents, actor Michael Douglas (Figure 10.12*a*) inherited a gene that influences the chin dimple trait. One form of the gene called for a dimple and the other didn't, but one is all it takes in this case. Figure 10.12*b* shows what his chin might have looked like if he had inherited two ordinary forms of the gene instead. What are alternative forms of the same gene called? *10.1*

4. To the right, the diploid chromosome numbers in somatic cells of a few kinds of organisms are listed. Figure out how many chromosomes would end up in gametes of each organism. *10.2*

Fruit fly, *Drosophila melanogaster*	8
Garden pea, *Pisum sativum*	14
Corn, *Zea mays*	20
Frog, *Rana pipiens*	26
Earthworm, *Lumbricus terrestris*	36
Human, *Homo sapiens*	46
Chimpanzee, *Pan troglodytes*	48
Amoeba, *Amoeba*	50
Horsetail, *Equisetum*	216

5. Define meiosis and then describe its main stages. In what key respects is meiosis *not* like mitosis? *10.3, 10.6*

6. Take a look at all of the chromosomes in the germ cell in the diagram at the right. Is this cell at anaphase I or anaphase II? *10.3, 10.6*

The cell is at anaphase ____ rather than anaphase ____ . I know this because:

_____
_____
_____
_____ .

7. Outline the steps by which sperm and eggs form in animals. *10.5*

Self-Quiz *(Answers in Appendix IV)*

1. Sexual reproduction requires _____ .
   a. meiosis
   b. gamete formation
   c. fertilization
   d. all of the above

2. Meiosis is a division mechanism that produces _____ .
   a. two cells
   b. two nuclei
   c. four cells
   d. four nuclei

3. An animal cell having two rather than one of each type of chromosome has a _____ chromosome number.
   a. diploid
   b. haploid
   c. normal gamete
   d. both b and c

4. Meiosis _____ the parental chromosome number.
   a. doubles
   b. reduces
   c. maintains
   d. corrupts

5. Generally, a pair of homologous chromosomes _____ .
   a. carry the same genes
   b. are the same length, shape
   c. interact at meiosis
   d. all of the above

6. Before the onset of meiosis, all chromosomes are _____ .
   a. condensed
   b. released from protein
   c. duplicated
   d. both b and c

7. Each chromosome moves away from its homologue and ends up at the opposite spindle pole during _____ .
   a. prophase I
   b. prophase II
   c. anaphase I
   d. anaphase II

8. Sister chromatids of each chromosome move apart and end up at opposite spindle poles during _____ .
   a. prophase I
   b. prophase II
   c. anaphase I
   d. anaphase II

9. Match each term and its description.
   ____ chromosome number
   ____ alleles
   ____ metaphase I
   ____ interphase
   a. different molecular forms of the same gene
   b. none between meiosis I and II
   c. pairs of homologous chromosomes are aligned at spindle equator
   d. sum total of chromosomes in all cells of a given type

Critical Thinking

1. Assume you can measure the amount of DNA in a primary oocyte, then in a primary spermatocyte, which gives you a mass *m*. What mass of DNA would you expect to find in each mature gamete (egg and sperm) that forms after meiosis? What mass of DNA would you expect to find (1) in an egg fertilized by one of the sperm and (2) in that egg after the first DNA duplication?

2. Adam has a pair of alleles that influence whether a person is right- or left-handed. One allele says "left," and its partner says "right." Visualize one of his germ cells, in which chromosomes are being duplicated prior to meiosis. Visualize what happens to the chromosomes during anaphase I and II. (It may help to use index cards as models of the sister chromatids of each chromosome.) What fraction of Adam's sperm will carry the gene for right-handedness? For left-handedness?

3. Adam also has one allele for long eyelashes, and a partner allele (on the homologous chromosome) for short eyelashes. What fraction of his sperm will have these gene combinations:
   right-handed, long eyelashes   *left-handed, long eyelashes*
   right-handed, short eyelashes   *left-handed, short eyelashes*

Selected Key Terms

allele *10.1*
asexual reproduction *10.1*
crossing over *10.4*
diploid *10.2*
egg *10.5*
fertilization *10.5*
gamete *CI*
gene *10.1*
germ cell *CI*
haploid *10.2*
homologous chromosome *10.2*
meiosis *10.2*
oocyte *10.5*
sexual reproduction *10.1*
sister chromatid *10.2*
sperm *10.5*
spore *10.5*

Readings

Klug, W., and M. Cummings. 1994. *Concepts of Genetics.* Fourth edition. New York: Macmillan.

Wolfe, S. 1995. *Introduction to Molecular and Cellular Biology.* Belmont, California: Wadsworth.

Web Site See *http://www.wadsworth.com/biology* for practice quiz questions, hypercontents, BioUpdates, and critical thinking. The Wadsworth Biology Resource Center provides a wealth of information fully organized and integrated by chapter.

11

OBSERVABLE PATTERNS OF INHERITANCE

A Smorgasbord of Ears and Other Traits

Basketball ace Charles Barkley has them. So does actor Tom Cruise. Actress Joan Chen doesn't, and neither did a monk named Gregor Mendel. To see how you fit in with these folks, use a mirror to check out your ears. Is the fleshy lobe at the base of each ear attached to the side of your head? If so, you and Barkley and Cruise have something in common. Or is the fleshy lobe not attached, so that you can flap it back and forth? If so, you are like Chen and Mendel (Figure 11.1).

Whether a person is born with attached or detached earlobes depends on a single kind of gene. That gene

Charles Barkley

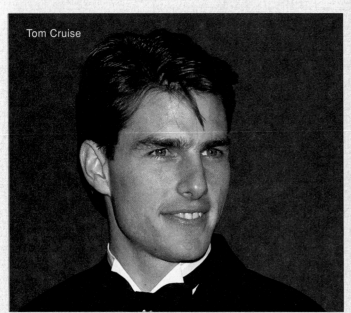

Tom Cruise

comes in slightly different molecular forms—alleles. Only one form has information about detached lobes. The information is put to use while a human body is developing inside the mother. It calls for a death signal, which is sent to all the cells positioned between the newly forming lobes and the head. Without the signal, the cells don't die, and earlobes don't detach.

We all have genes for thousands of traits, including earlobes, cheeks, lashes, and eyeballs. Most traits vary in their details from one person to the next. Remember, we inherit pairs of genes, on pairs of chromosomes. In some pairs, one allele has strong effects and overwhelms the other allele's contribution. The outgunned allele is said to be recessive to the dominant one. If you have detached earlobes, dimpled cheeks, long lashes, or large eyeballs, you have at least one and quite possibly two dominant alleles that affect the trait in a particular way.

Figure 11.1 Attached and detached earlobes of representative humans. This sampling provides observable evidence of a trait governed by a certain gene, which exists in different molecular forms in the human population. Do you have one or the other version of the trait? It depends on which molecular forms of the gene you inherited from your mother and father. As Gregor Mendel perceived, such easily observable traits can be used to identify patterns of inheritance that exist from one generation to the next.

When both alleles of a pair are recessive, nothing masks their effect on a trait. You get *attached* earlobes with one pair of recessive alleles (and *flat* feet with another, a *straight* nose with another, and so on).

How did we discover such remarkable things about our genes? It started with Gregor Mendel. By analyzing pea plants generation after generation, Mendel found indirect but *observable* evidence of how parents bestow units of hereditary information—genes—on offspring.

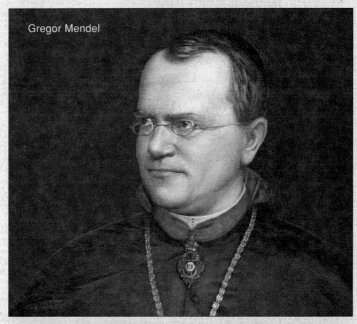
Gregor Mendel

Joan Chen

This chapter focuses on both the methods and the results of Mendel's experiments. They remain a classic example of how a scientific approach can pry open important secrets about the natural world. And to this day, they serve as the foundation for modern genetics.

KEY CONCEPTS

1. Genes are units of information about heritable traits. Alleles, which are slightly different molecular forms of a gene, specify different versions of the same trait.

2. Each gene has a particular location on a particular chromosome of a species. Humans, pea plants, and other organisms with a diploid chromosome number inherit *pairs* of genes, at equivalent locations on pairs of homologous chromosomes.

3. When each pair of homologous chromosomes is moved apart at meiosis, their paired genes are moved apart also, and they end up in different gametes. Gregor Mendel found indirect evidence of this gene segregation when he crossbred plants showing different versions of the same trait, such as purple or white flowers.

4. Each pair of homologous chromosomes (and the genes they carry) is sorted out for distribution into one gamete or another independently of how the other pairs of homologous chromosomes are assorted. Mendel found indirect evidence of this when he tracked many plants having observable differences in two traits, such as flower color and height.

5. The contrasting traits that Mendel happened to study were specified by nonidentical alleles. One allele was dominant, in that its effect on a trait masked the effect of a recessive allele paired with it.

6. Not all traits have such clearly dominant or recessive forms. One allele of a pair may be fully or partially dominant over its partner or codominant with it. Also, two or more gene pairs often influence the same trait, and some single genes influence many traits. Besides this, environmental factors induce further variation in traits.

More than a century ago, people wondered about the basis of inheritance. It was common knowledge that sperm and eggs both transmit information about traits to offspring. But few suspected that the information is organized in units (genes). Instead, the idea was that a father's blob of information "blended" with a mother's blob at fertilization, like cream into coffee.

However, carried to its logical conclusion, blending would slowly dilute a population's shared pool of hereditary information until there was only a single version left of each trait. If that were so, why did, say, freckles keep showing up among the children of nonfreckled parents over the generations? Why weren't all the descendants of a herd of white stallions and black mares gray? The theory of blending scarcely explained the obvious variation in traits that people could observe with their own eyes. Nevertheless, few disputed the theory.

Charles Darwin was among the scholarly dissidents. According to a key premise of his theory of natural selection, individuals of a population show variation in heritable traits. Through the generations, variations that improve chances of surviving and reproducing occur with greater frequency than those that do not. The less advantageous variations may persist among fewer individuals or may even disappear. It is not that separate versions of a trait are "blended out" of the population. Rather, *each version of a trait may persist in a population, at frequencies that may change over time*.

Even before Darwin presented his theory, someone was gathering evidence that eventually would support his key premise. A monk, Gregor Mendel, had already guessed that sperm and eggs carry distinct "units" of information about heritable traits. By carefully analyzing traits of pea plants generation after generation, Mendel found indirect but *observable* evidence of how parents transmit genes to offspring.

Mendel's Experimental Approach

Mendel spent most of his adult life in a monastery in Brno, a city near Vienna that has since become part of the Czech Republic. The monastery of St. Thomas was somewhat removed from the European capitals, which were then the centers of scientific inquiry. Yet Mendel was not a man of narrow interests who accidentally stumbled onto principles of great import.

Having been raised on a farm, Mendel was aware of agricultural principles and their applications. He kept abreast of breeding experiments and developments described in the available literature. He was a member of the regional agricultural society. He also won several awards for developing improved varieties of vegetables and fruits. Shortly after entering the monastery, he spent two years studying mathematics at the University of Vienna. Few scholars of his time had such combined talents in plant breeding and mathematics.

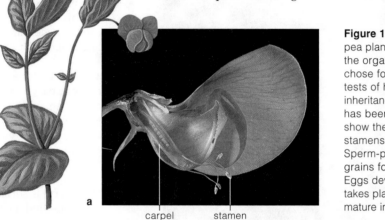

a

carpel   stamen

Figure 11.2 The garden pea plant (*Pisum sativum*), the organism Mendel chose for experimental tests of his ideas about inheritance. (**a**) This flower has been sectioned to show the location of its stamens and carpel. Sperm-producing pollen grains form in stamens. Eggs develop, fertilization takes place, and seeds mature inside the carpel.

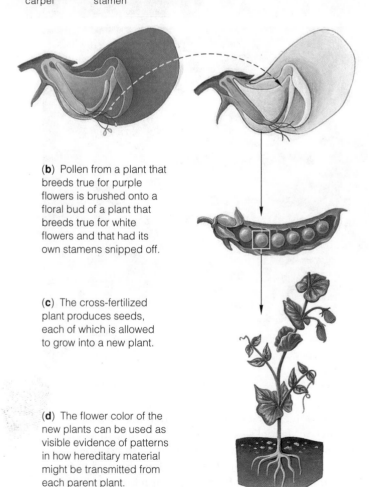

(**b**) Pollen from a plant that breeds true for purple flowers is brushed onto a floral bud of a plant that breeds true for white flowers and that had its own stamens snipped off.

(**c**) The cross-fertilized plant produces seeds, each of which is allowed to grow into a new plant.

(**d**) The flower color of the new plants can be used as visible evidence of patterns in how hereditary material might be transmitted from each parent plant.

Shortly after his university training, Mendel began experimenting with the garden pea plant, *Pisum sativum* (Figure 11.2). This plant is self-fertilizing. Male as well as female gametes (call them sperm and eggs) develop in different parts of the same flower, where fertilization takes place. Nearly all the plants breed true for certain traits. In other words, successive generations are just like their parents in one or more traits, as when all of the offspring grown from seeds of self-fertilized, white-flowered parent plants have white flowers.

As Mendel knew, pea plants also will cross-fertilize when sperm and eggs from different plants are mixed together under controlled conditions. For his studies, he could open flower buds of a plant that bred true for a trait—say, white flowers—and snip out the stamens. (Stamens bear pollen grains in which sperm develop.) Then he could brush the "castrated" buds with pollen from a plant that bred true for a *different* version of the same trait—purple flowers. As Mendel hypothesized, he could use such clearly observable differences to track a given trait through many generations. If there were patterns to the trait's inheritance, *those patterns might tell him something about the hereditary material itself.*

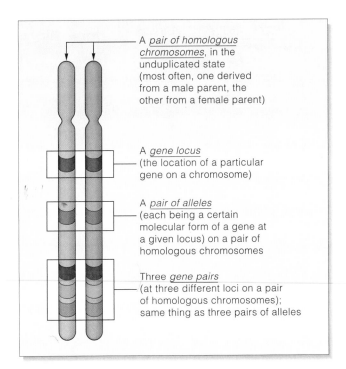

A *pair of homologous chromosomes*, in the unduplicated state (most often, one derived from a male parent, the other from a female parent)

A *gene locus* (the location of a particular gene on a chromosome)

A *pair of alleles* (each being a certain molecular form of a gene at a given locus) on a pair of homologous chromosomes

Three *gene pairs* (at three different loci on a pair of homologous chromosomes); same thing as three pairs of alleles

Figure 11.3 A few genetic terms illustrated. Diploid organisms have pairs of genes, on pairs of homologous chromosomes. For example, you inherited one chromosome of each pair from your mother, and the other, homologous chromosome from your father.

Most genes can have slightly different molecular forms, called alleles. Different alleles specify different versions of the same trait. An allele at one location on a chromosome may or may not be identical to its partner on the homologous chromosome.

Some Terms Used in Genetics

Having read the chapter on meiosis, you already have insight into the mechanisms of sexual reproduction, which is more than Mendel had. Neither he nor anyone else of his era knew about chromosomes. So he could not have known that a chromosome number is reduced by half in gametes, then restored when gametes meet at fertilization. Even so, Mendel sensed what was going on. As we follow his thinking, let's simplify the story by substituting a few of the modern terms used in studies of inheritance (see also Figure 11.3):

1. **Genes** are units of information about specific traits, and they are passed from parents to offspring. Each gene has a specific location (locus) on a chromosome.

2. Cells with a diploid chromosome number ($2n$) have pairs of genes, on pairs of homologous chromosomes.

3. Mutation can change a gene's molecular structure and thus its information about a trait (as when the gene for flower color specifies purple and a mutated version specifies white). All the different molecular forms of the same gene are called **alleles**.

4. When offspring of genetic crosses inherit a pair of *identical* alleles for a trait, generation after generation, they are a **true-breeding lineage**. By contrast, when offspring of a genetic cross inherit a pair of *nonidentical* alleles for a trait, they are **hybrid offspring**.

5. When both alleles of a pair are identical, this is a *homozygous* condition. When the two are not identical, this is a *heterozygous* condition.

6. An allele is said to be *dominant* when its effect on a trait masks that of any *recessive* allele paired with it. We use capital letters for dominant alleles and lowercase letters for recessive ones; for instance, *A* and *a*.

7. Putting this all together, a **homozygous dominant** individual has a pair of dominant alleles (*AA*) for a trait under study. A **homozygous recessive** individual has a pair of recessive alleles (*aa*). And a **heterozygous** individual has a pair of nonidentical alleles (*Aa*).

8. Two terms help keep the distinction clear between genes and the traits they specify. **Genotype** refers to the particular genes an individual carries. **Phenotype** refers to an individual's observable traits.

9. When tracking the inheritance of traits through generations of offspring, these abbreviations apply:

P	parental generation
F_1	first-generation offspring
F_2	second-generation offspring

Mendel had an idea that in every generation, a plant inherits two "units" (genes) of information for a trait, one from each parent. To test this idea, he performed what we now call **monohybrid crosses**. The offspring of such a cross are heterozygous for the one trait being studied (which is what monohybrid means). The two parents breed true for different versions of the trait, so their offspring inherit a pair of nonidentical alleles.

Predicting the Outcome of Monohybrid Crosses

Mendel tracked many individual traits through two generations. For instance, in one series of experiments, he crossed true-breeding purple-flowered plants and true-breeding white-flowered ones. All plants grown from the seeds that resulted from this cross had purple flowers. Mendel allowed these plants to self-fertilize. Some plants grown from the seeds had white flowers!

If Mendel's hypothesis were correct—if each plant had inherited two units of information about flower color—then the unit for "purple" had to be dominant, because it had masked the unit for "white" in F_1 plants.

Let's rephrase his thinking. Germ cells of pea plants are diploid, with pairs of homologous chromosomes. Assume one parent is homozygous dominant (*AA*) and the other is homozygous recessive (*aa*) for flower color. After meiosis, a sperm or egg carries one allele for flower color (Figure 11.4). Thus, when a sperm fertilizes an egg, only one outcome is possible: $A + a = Aa$.

Before continuing, you should know Mendel crossed hundreds of plants and tracked thousands of offspring. Besides this, he counted and recorded the number of plants showing dominance or recessiveness. As you can see from Figure 11.5, an intriguing ratio emerged. On average, three of every four F_2 plants had the dominant phenotype, and one had the recessive phenotype.

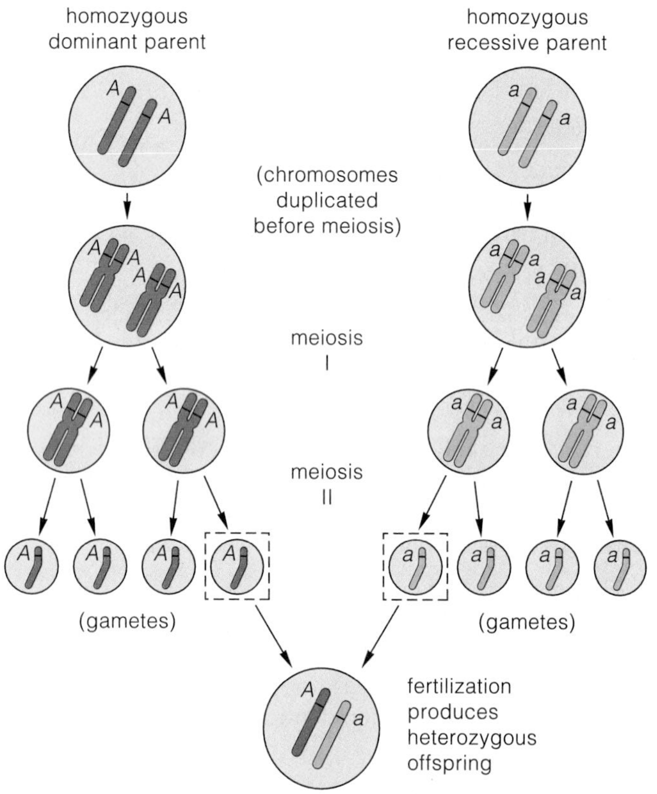

Figure 11.4 A monohybrid cross, showing how one gene of a pair segregates from the other. Two parents that breed true for two different versions of a trait can produce only heterozygous offspring.

Figure 11.5 (*Right*) Results from Mendel's monohybrid cross experiments with the garden pea plant (*P. sativum*). The numbers given are his counts of the F_2 plants that carried dominant or recessive hereditary "units" (alleles) for the trait. On average, the dominant-to-recessive ratio was 3:1.

Trait Studied	Dominant Form	Recessive Form	F_2 Dominant-to-Recessive Ratios:
seed shape	5,474 round	1,850 wrinkled	2.96:1
seed color	6,022 yellow	2,001 green	3.01:1
pod shape	882 inflated	299 wrinkled	2.95:1
pod color	428 green	152 yellow	2.82:1
flower color	705 purple	224 white	3.15:1
flower position	651 along stem	207 at tip	3.14:1
stem length	787 tall	277 dwarf	2.84:1

Average ratio for all traits studied: 3:1

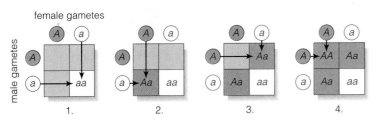

female gametes

male gametes

1. 2. 3. 4.

Figure 11.6 Punnett-square method of predicting the probable outcome of a genetic cross. Circles represent gametes. *Italic* letters on the gametes represent dominant or recessive alleles. The different squares show the different genotypes possible among offspring. In this case, gametes are from a self-fertilizing heterozygous (*Aa*) plant.

Figure 11.7 (*Right*) Results from one of Mendel's monohybrid crosses. On average, the dominant-to-recessive ratio among the second-generation (*F*$_2$) plants was 3:1.

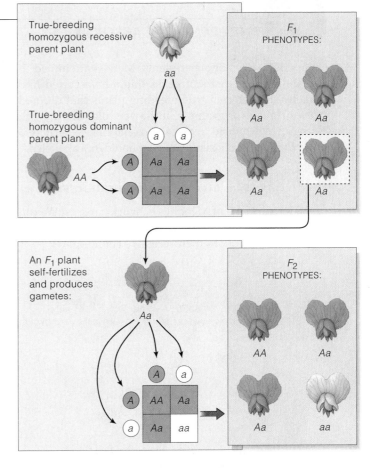

To Mendel, the ratio suggested that fertilization is a chance event, with a number of possible outcomes. And he had an understanding of probability, which applies to chance events *and therefore could help him predict the possible outcomes of crosses*. **Probability** simply means this: The chance that each outcome of a given event will occur is proportional to the number of ways it can be reached.

The **Punnett-square method**, explained in Figure 11.6 and applied in Figure 11.7, may help you visualize the possibilities. As you can see, if half of a plant's sperm or eggs were *a* and half were *A*, then four outcomes were possible each time a sperm fertilized an egg:

POSSIBLE EVENT:	PROBABLE OUTCOME:
sperm *A* meets egg *A*	1/4 *AA* offspring
sperm *A* meets egg *a*	1/4 *Aa*
sperm *a* meets egg *A*	1/4 *Aa*
sperm *a* meets egg *a*	1/4 *aa*

By this prediction, an *F*$_2$ plant had three chances in four of getting at least one dominant allele (purple flowers). It had one chance in four of getting two recessive alleles (white flowers). That is a probable phenotypic ratio of three purple to one white, or 3:1.

Mendel's observed ratios weren't *exactly* 3:1. You can see this for yourself by looking at the numerical results listed in Figure 11.5. Why did Mendel put aside the deviations? To understand why, flip a coin a couple of times. As we all know, a coin is just as likely to end up heads as tails. But often it ends up heads, or tails, several times in a row. So if you flip the coin only a few times, the observed ratio may differ greatly from the predicted ratio of 1:1. Flip the coin many, many times, and you are more likely to come close to the predicted ratio. Mendel understood the rules of probability—and he performed a large number of crosses. Almost certainly, this kept him from being confused by minor deviations from the predicted results of the experimental crosses.

Testcrosses

By running **testcrosses**, Mendel gained support for his prediction. In this type of experimental test, an organism shows dominance for a specified trait but its genotype is unknown, so it is crossed to a known homozygous recessive individual. Test results may reveal whether the organism is homozygous dominant or heterozygous.

Regarding the monohybrid crosses just described, Mendel tested his prediction that the purple-flowered *F*$_1$ offspring were heterozygous by crossing them with true-breeding, white-flowered plants. If they were all homozygous dominant, then all the *F*$_2$ offspring would show the dominant form of the trait. If heterozygous, there would be about as many dominant as recessive plants. Sure enough, about half of the *F*$_2$ plants had purple flowers (*Aa*) and half had white (*aa*). Can you construct two Punnett squares that show the possible outcomes of this testcross?

The results from Mendel's monohybrid crosses and testcrosses became the basis of a theory of **segregation**, which we state here in modern terms:

MENDEL'S THEORY OF SEGREGATION. **Diploid cells have pairs of genes, on pairs of homologous chromosomes. During meiosis, the two genes of each pair are segregated from each other. As a result, they end up in different gametes.**

By another series of experiments, Mendel attempted to explain how *two* pairs of genes might be assorted into gametes. He selected true-breeding plants that differed in two traits—for example, in flower color and height. With such **dihybrid crosses**, F_1 offspring inherit two gene pairs, each consisting of two nonidentical alleles.

Predicting the Outcome of Dihybrid Crosses

Let's diagram one of Mendel's dihybrid crosses. We can use A for flower color and B for height as the dominant alleles, and a and b as their recessive counterparts:

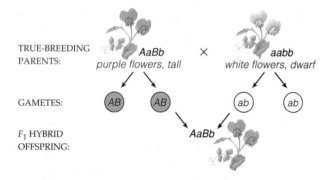

TRUE-BREEDING PARENTS: *AaBb*   ×   *aabb*
purple flowers, tall   *white flowers, dwarf*

GAMETES: AB   AB   ab   ab

F_1 HYBRID OFFSPRING: *AaBb*

As Mendel would have predicted, the F_1 offspring from this cross are all purple-flowered and tall ($AaBb$). When the F_1 plants mature and reproduce, how will the two gene pairs be assorted into gametes? The answer partly depends on the chromosomal locations of the gene pairs. Assume one pair of homologous chromosomes carries the Aa alleles and a *different* pair carries the Bb alleles. Next, think of how all chromosomes become positioned at the spindle equator during metaphase I of meiosis (Figures 8.4 and 11.8). The chromosome with the A allele might be positioned to move to either spindle pole (and then into one of four gametes). The same is true of its homologue. And the same is true of the chromosomes with the B and b alleles. Thus, after meiosis and gamete formation, four combinations of alleles are possible in the sperm or eggs: $1/4\ AB$, $1/4\ Ab$, $1/4\ aB$, and $1/4\ ab$.

Given the alternative alignments of chromosomes at metaphase I, several allelic combinations are possible at fertilization. Simple multiplication (four kinds of sperm times four kinds of eggs) tells us sixteen combinations of gametes are possible in the F_2 offspring of a dihybrid cross. Use the Punnett-square method to diagram the probabilities (Figure 11.9). Now add up all the possible phenotypes and you get 9/16 tall purple-flowered, 3/16 dwarf purple-flowered, 3/16 tall white-flowered, and 1/16 dwarf white-flowered plants. That is a probable phenotypic ratio of 9:3:3:1. Results from one dihybrid cross that Mendel described were close to this ratio.

Figure 11.8 Independent assortment. This example tracks two pairs of homologous chromosomes. An allele at one locus on a chromosome may or may not be identical with its partner allele on the homologous chromosome. At meiosis, either chromosome of a pair may become attached to either pole of the spindle. Thus, in this case, two different lineups are possible at metaphase I.

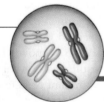

Nucleus of a diploid (2n) reproductive cell with only two pairs of homologous chromosomes

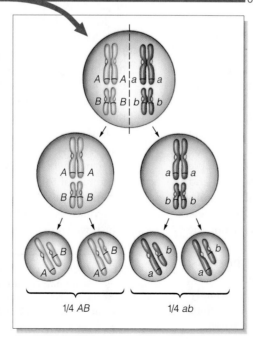

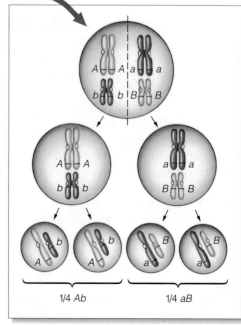

Possible alignments of the homologous chromosomes at metaphase I of meiosis, as shown by two diagrams:

The resulting alignments of chromosomes at metaphase II:

The combinations of alleles possible in the forthcoming gametes:

1/4 AB   1/4 ab   1/4 Ab   1/4 aB

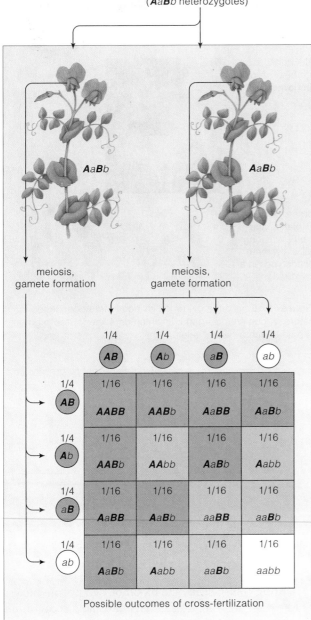

AABB
purple-flowered
tall parent
(homozygous
dominant)

AB ☓ (ab)

aabb
white-flowered
dwarf parent
(homozygous
recessive)

F₁ OUTCOME: All F₁ plants purple-flowered, tall
(**A**a**B**b heterozygotes)

AaBb

AaBb

meiosis,
gamete formation

meiosis,
gamete formation

	1/4 **AB**	1/4 **A**b	1/4 a**B**	1/4 (ab)
1/4 **AB**	1/16 **AABB**	1/16 **AAB**b	1/16 **A**a**BB**	1/16 **A**a**B**b
1/4 **A**b	1/16 **AAB**b	1/16 **AA**bb	1/16 **A**a**B**b	1/16 **A**abb
1/4 a**B**	1/16 **A**a**BB**	1/16 **A**a**B**b	1/16 aa**BB**	1/16 aa**B**b
1/4 (ab)	1/16 **A**a**B**b	1/16 **A**abb	1/16 aa**B**b	1/16 aabb

Possible outcomes of cross-fertilization

ADDING UP THE F₂ COMBINATIONS POSSIBLE:

☐ 9/16 or 9 purple-flowered, tall

☐ 3/16 or 3 purple-flowered, dwarf

☐ 3/16 or 3 white-flowered, tall

☐ 1/16 or 1 white-flowered, dwarf

Figure 11.9 Results from Mendel's dihybrid cross between parent plants that bred true for different versions of two traits: flower color and plant height. *A* and *a* represent dominant and recessive alleles for flower color. *B* and *b* represent dominant and recessive alleles for plant height. As the Punnett square indicates, the probabilities of certain combinations of phenotypes among F₂ offspring occur in a 9:3:3:1 ratio, on the average.

The Theory in Modern Form

Mendel could do no more than analyze the numerical results from his dihybrid crosses, because he didn't know that seven pairs of homologous chromosomes carry the pea plant's "units" of inheritance. It just seemed to him that the two units for the first trait he was tracking had been assorted into gametes independently of the two units for the other trait. In time, his interpretation became known as the theory of **independent assortment**, which we state here in modern terms: By the end of meiosis, each pair of homologous chromosomes—and the genes they carry—have been sorted for shipment into gametes independently of how the other pairs were sorted out.

Independent assortment and hybrid crossing lead to staggering variety among offspring. In a monohybrid cross involving only a single gene pair, three genotypes are possible: *AA*, *Aa*, and *aa*. We can represent this as 3^n, where *n* is the number of gene pairs. When we consider more gene pairs, the number of possible combinations increases dramatically. Even if parents differ in merely ten pairs of genes, nearly 60,000 genotypes are possible among their offspring. If they differ in twenty pairs of genes, the number approaches 3.5 billion!

In 1865 Mendel presented his ideas to the Brünn Natural History Society. His ideas had little impact. The next year his paper was published, and apparently it was read by few and understood by no one. In 1871 he became an abbot of the monastery, and his experiments gradually gave way to administrative tasks. He died in 1884, never to know that his work would be the starting point for the development of modern genetics.

Today, Mendel's theory of segregation still stands. Hereditary material is indeed organized in units (genes) that retain their identity and are segregated from each other for distribution into different gametes. However, the theory of independent assortment has undergone some modification, as you will see in the next chapter.

MENDEL'S THEORY OF INDEPENDENT ASSORTMENT. **By the end of meiosis, the genes on pairs of homologous chromosomes have been sorted out for distribution into one gamete or another independently of gene pairs of other chromosomes.**

DOMINANCE RELATIONS

For the most part, Mendel studied traits having clearly dominant or recessive forms. As the remaining sections of this chapter will make clear, however, the expression of other traits is not as straightforward.

Incomplete Dominance

In **incomplete dominance**, one allele of a pair isn't fully dominant over its partner, so a heterozygous phenotype *somewhere in between* the two homozygous phenotypes emerges. Cross a true-breeding red snapdragon and a true-breeding white one. All F_1 offspring will have pink flowers. Cross two F_1 plants and expect red, pink, and white snapdragons in a predictable ratio (Figure 11.10). What causes this inheritance pattern? Red snapdragons have two alleles that allow them to make an abundance of red pigment molecules. White snapdragons have two mutated alleles that render them pigment-free. The pink ones are heterozygous. Their one red allele can specify enough pigment to make flowers pink but not red.

ABO Blood Types: A Case of Codominance

In **codominance**, a pair of nonidentical alleles specify two phenotypes, which are both expressed at the same time in heterozygotes. As an example, think about one of the recognition proteins at the surface of your red blood cells. This type of protein helps give red blood cells their unique identity, although it comes in slightly different molecular forms. A method of analysis known as *ABO blood typing* reveals which form a person has.

In humans, the gene that determines the protein's final structure has three alleles. Two alleles, I^A and I^B, are codominant when paired. The third, i, is recessive; a pairing with I^A or I^B masks its effects. Altogether, they represent a **multiple allele system**, which we define as the presence of three or more alleles of a gene among individuals of a population.

Before each protein molecule becomes positioned at the cell surface, it gets modified in the cytomembrane system (Section 4.5). There, an oligosaccharide chain is attached to it, then an enzyme specified by the gene of interest attaches a sugar monomer to the end of the chain. Alleles I^A and I^B code for two slightly different versions of the enzyme. The two attach different sugars, which give the protein its identity—either A or B.

Which alleles do you have? With either I^AI^A or I^Ai, you have type A blood. With I^BI^B or I^Bi, your blood is type B. With codominant alleles I^AI^B, it is AB—meaning you have both versions of the sugar-attaching enzyme. But if you are homozygous recessive (ii), the molecules never did get a sugar monomer attached to them. Your blood type is neither A nor B; that is what type "O" means. Figure 11.11 summarizes the possibilities.

homozygous parent ✕ homozygous parent

All F_1 offspring are heterozygous for flower color:

Cross two of the F_1 plants, and the F_2 offspring will show three phenotypes in a 1:2:1 ratio:

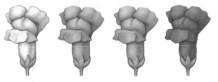

Figure 11.10 Visible evidence of incomplete dominance in heterozygous snapdragons, in which an allele for red pigment is paired with a "white" allele.

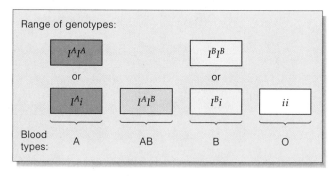

Range of genotypes:

I^AI^A			I^BI^B	
or			or	
I^Ai	I^AI^B	I^Bi		ii

Blood types: A AB B O

Figure 11.11 Allelic combinations related to ABO blood typing.

During *transfusions*, the blood of two people mixes. Unless their red blood cells have the same recognition proteins, the recipient's immune system perceives the red blood cells from the donor as "nonself." It will act against those cells and may cause death (Chapter 40).

One allele may be fully dominant, incompletely dominant, or codominant with its partner on the homologous chromosome.

MULTIPLE EFFECTS OF SINGLE GENES

Expression of the alleles at just a single location on a chromosome may have positive or negative effects on two or more traits. This phenotypic outcome of a single gene's activity is known as **pleiotropy** (after the Greek *pleio-*, meaning more, and *-tropic*, meaning to change).

The genetic disorder *sickle-cell anemia* is a classic example of how the alleles at a single locus can have pleiotropic effects. The disorder arises from a mutated gene for beta-globin, one of two kinds of polypeptide chains in the hemoglobin molecule. Hemoglobin, recall, is the oxygen-transporting protein in red blood cells. We designate the mutated allele as Hb^S instead of Hb^A. Heterozygotes (Hb^A/Hb^S) usually show few symptoms of the disorder. Their red blood cells are able to produce enough normal hemoglobin molecules to compensate for the abnormal ones. In homozygotes (Hb^S/Hb^S), the red blood cells can only produce abnormal hemoglobin. This one abnormality may have drastic repercussions throughout the body, for it disrupts the concentration of oxygen in the bloodstream.

Humans, like most organisms, depend on the intake of oxygen for aerobic respiration. Oxygen in air flows into the lungs, then diffuses into the blood. There it binds to hemoglobin, which transports it through arteries, arterioles, and then small-diameter, thin-walled capillaries threading past all living cells in the body. A steep oxygen concentration gradient exists between blood and the fluid within the surrounding tissues, so most of the oxygen diffuses into the tissues, and on into cells. With this extensive cellular uptake, the concentration of oxygen in blood declines. The decline is most pronounced at high altitudes and during strenuous activity.

In the red blood cells of people who bear the mutant gene, abnormal hemoglobin molecules stick together into rodlike arrangements. The rods distort the cells into a sickle shape, as in Figure 11.12*b*. (A sickle is a farm tool having a crescent-shaped blade.) The distorted cells rupture easily, and their remnants clog and then rupture capillaries. When these oxygen transporters are swiftly destroyed, the cells in affected tissues become starved for oxygen. Also, their clumping effect leads to local failures in the capacity of the circulatory system to deliver oxygen and carry away carbon dioxide and other metabolic wastes.

Over time, ongoing expression of the mutant gene can cause widespread damage. Figure 11.12*c* tracks how the successive changes in phenotype may proceed.

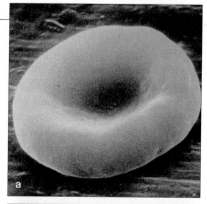
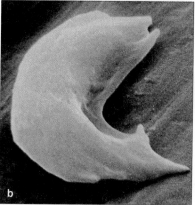

Figure 11.12 (**a**) A red blood cell from a person affected by sickle-cell anemia, which is a genetic disorder. This scanning electron micrograph shows the surface appearance of the affected individual's red blood cells when the blood is adequately oxygenated. (**b**) This scanning electron micrograph shows the sickle shape that red blood cells assume when the concentration of oxygen in blood is low. (**c**) This diagram summarizes the range of symptoms that are characteristic of an individual who is homozygous recessive for the disorder.

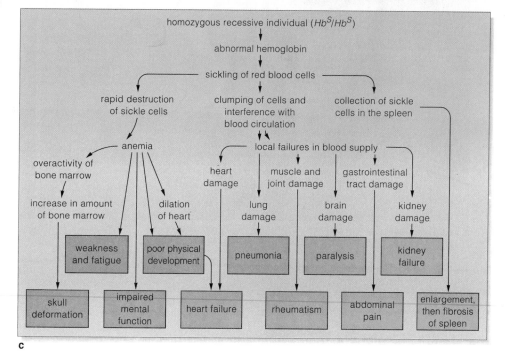

In chapters to come, you will read about other aspects of this genetic disorder.

The alleles at a single gene location may have positive or negative effects on two or more traits.

The effects may not be simultaneous. Rather, they may have repercussions over time. The gene may lead to an alteration in one trait, that change may alter another trait, and so on.

INTERACTIONS BETWEEN GENE PAIRS

Often a trait results from interactions among products of two or more gene pairs. For example, two alleles of one gene may mask expression of another gene's alleles, so some expected phenotypes may not appear at all. Such interactions between pairs of genes are called **epistasis** (meaning the act of stopping).

Hair Color in Mammals

Epistasis is common among the gene pairs responsible for the coloration of fur or skin in mammals. Consider the black, brown, or yellow fur of Labrador retrievers (Figure 11.13). The different colors arise from variations in the amount and distribution of melanin, a brownish black pigment. Enzymes and other products of many gene pairs influence different steps in the production of melanin and its deposition in certain body regions.

The alleles of one gene specify an enzyme required to produce melanin. Expression of allele B (black) has a more pronounced effect and is dominant to b (brown). Alleles of a different gene control the extent to which molecules of melanin will be deposited in a retriever's hairs. Allele E permits full deposition. Two recessive alleles (ee) reduce deposition, and the fur will be yellow.

Interactions between these two gene pairs may not even be possible if a certain allelic combination exists at still another gene location. There, a gene (C) calls for tyrosinase, the first of several enzymes in the metabolic pathway that produces melanin. An individual having one or two dominant alleles (CC or Cc) can produce the

a BLACK LABRADOR   **b** YELLOW LABRADOR   **c** CHOCOLATE LABRADOR

Figure 11.13 The heritable basis of coat color among Labrador retrievers. The trait arises through interactions among the alleles of two pairs of genes.

One kind of gene is involved in melanin production. Allele B (black) of this gene is dominant to allele b (brown). A different kind of gene influences the deposition of melanin pigment in individual hairs. Allele E of this gene promotes deposition, but a pairing of recessive alleles (ee) of the gene blocks deposition, and a yellow coat results.

F_1 offspring of a dihybrid cross produce F_2 offspring in a 9:3:4 ratio, as the Punnett-square diagram to the right indicates.

The yellow Labrador in photograph (**b**) probably has genotype BBee, because it can produce melanin but cannot deposit pigment in hairs. After studying the photograph, can you say why?

HOMOZYGOUS PARENTS: BBEE × bbee

F_1 PUPPIES: BbEe

ALLELIC COMBINATIONS POSSIBLE AMONG F_2 PUPPIES:

	BE	Be	bE	be
BE	BBEE	BBEe	BbEE	BbEe
Be	BBEe	BBee	BbEe	Bbee
bE	BbEE	BbEe	bbEE	bbEe
be	BbEe	Bbee	bbEe	bbee

RESULTING PHENOTYPES:

- 9/16 or 9 black
- 3/16 or 3 brown
- 4/16 or 4 yellow

Figure 11.14 A rare albino rattlesnake. Like other animals that cannot produce melanin, it has pink eyes and its body is white, overall. (Eyes look pink because the absence of melanin from a tissue layer in the eyeball allows red light to be reflected from blood vessels in the eyes.) In birds and mammals, surface coloration results from pigments in feathers, fur, or skin. In fishes, amphibians, and reptiles, color-bearing cells give skin its surface coloration. Some of the cells contain melanin pigments or yellow-to-red pigments. Others contain crystals that reflect light and alter the effect of other pigments present.

The mutation affecting melanin production in the snake shown here had no effect on the production of yellow-to-red pigments and of light-reflecting crystals. Therefore, the snake's skin appears to be iridescent yellow as well as white.

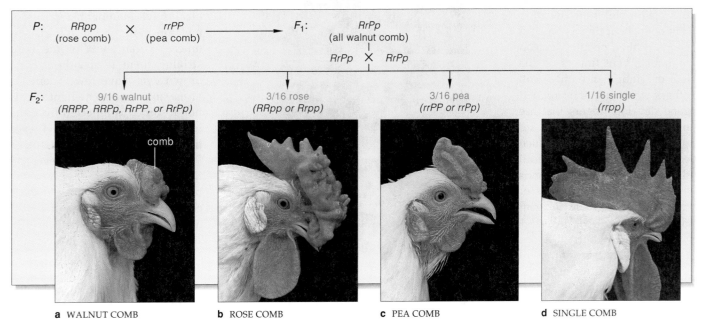

P: RRpp   ×   rrPP   ⟶   F₁: RrPp
      (rose comb)   (pea comb)   (all walnut comb)

                                        RrPp × RrPp

F₂: 9/16 walnut   3/16 rose   3/16 pea   1/16 single
      (RRPP, RRPp, RrPP, or RrPp)   (RRpp or Rrpp)   (rrPP or rrPp)   (rrpp)

comb

a WALNUT COMB   **b** ROSE COMB   **c** PEA COMB   **d** SINGLE COMB

Figure 11.15 Interaction between two genes that affect the same trait in domestic breeds of chickens. The initial cross is between a Wyandotte (with a rose comb, **b**, on the crest of its head) and a brahma (pea comb, **c**). With complete dominance at the locus for pea comb and at the locus for rose comb, products of the two gene pairs interact and give rise to a walnut comb (**a**). With complete recessiveness at both gene loci, products interact and give rise to a single comb (**d**).

functional enzyme. An individual bearing two recessive alleles (*cc*) cannot. When the biosynthetic pathway for melanin production is blocked, *albinism*—the absence of melanin—is the resulting phenotype (Figure 11.14).

Comb Shape in Poultry

In some cases, interaction between two gene pairs results in a phenotype that neither pair can produce alone. The geneticists W. Bateson and R. Punnett identified two interacting gene pairs (*R* and *P*) that affect comb shape in chickens. Allelic combinations of *rr* at one gene locus and *pp* at the other locus result in the least common phenotype, the single comb. The presence of dominant alleles (*R*, *P*, or both) results in varied phenotypes. The Figure 11.15 diagram shows the combinations of alleles that lead to rose, pea, or walnut combs.

Genes often interact, as when alleles of one gene mask the expression of another gene, and some expected phenotypes may not appear at all.

Regarding the Unexpected Phenotype

As Mendel demonstrated, the phenotypic effects of one or two pairs of certain genes show up in predictable ratios from one generation to the next. Besides this, certain interactions among two or more gene pairs also may produce phenotypes in predictable ratios, as the example of Labrador coat color in Section 11.6 clearly demonstrated. However, even when you are tracking a single gene through the generations, you may discover that the resulting phenotypes are not quite what you had expected.

Consider *camptodactyly*, a rare genetic abnormality that affects both the shape and the movement of fingers. Certain people who carry the gene for this heritable trait develop immobile, bent fingers on both hands. Other people develop immobile, bent fingers on the left or right hand only. Still others who carry the mutated gene develop fingers that are not affected either way.

What is the source of such confounding variation? Recall that most organic compounds are synthesized by a series of metabolic steps, *and different enzymes—each a gene product—regulate different steps.* Maybe one gene is mutated in one of a number of different ways. Maybe the product of another gene blocks the pathway or causes it to run nonstop, or not long enough. Or maybe poor nutrition or another factor that can vary in the individual's environment affects a crucial enzyme in the pathway. These are the sorts of variable factors that commonly introduce far less predictable variations in the phenotypes resulting from gene expression.

Continuous Variation in Populations

Generally, the individuals of a population display a range of small differences in most traits.

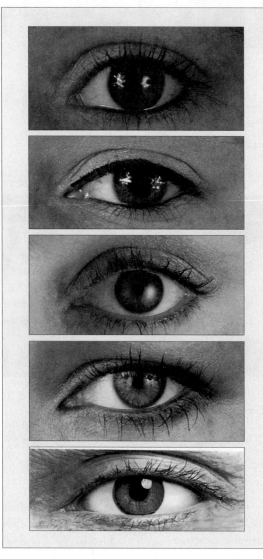

Figure 11.16 Samples from the range of continuous variation in human eye color. Different pairs of genes interact to produce and deposit melanin. Among other things, this pigment helps color the eye's iris. Different combinations of alleles result in small differences in eye color. Thus the frequency distribution for the eye-color trait appears to be continuous over a range from black to light blue.

This characteristic of populations, which is known as **continuous variation**, is largely an outcome of the number of genes affecting a trait and the number of environmental factors that influence their expression. In most cases, the greater the number of genes and environmental factors, the more continuous will be the expected distribution of all the versions of that trait.

Think about your own eye color. The colored part is the iris, a doughnut-shaped, pigmented structure beneath the cornea. As is true of all humans, the color of your iris is the cumulative outcome of a number of gene products. Those products take part in the stepwise production and distribution of melanin, the same light-absorbing pigment that influences coat color in mammals. Dark eyes that appear almost black have abundant deposits of melanin molecules in the iris—so much so that most of the light striking them is absorbed. Dark brown eyes have fewer deposits of melanin, and some light that is not absorbed is reflected from the iris. Light brown or hazel eyes have even fewer (Figure 11.16).

Green, gray, or blue eyes do not contain green, gray, or blue pigments. In these cases, the iris contains different quantities of melanin, but not very much of it. As a result, many or most of the blue wavelengths of light that do enter the eye are reflected out.

How might you describe the continuous variation of some trait within a group, such as the college students in Figure 11.17? They range from very short to very tall, with average heights much more common than either extreme. You can start out by dividing the full range of the different phenotypes into measurable categories. Next, count all the individual students in each category. This will give you the relative frequencies of phenotypes distributed across the range of measurable values.

The bar chart in Figure 11.17c plots the proportion of students in each category against the range of the measured phenotypes. The vertical bars that are the shortest

Number of individuals	1	4	8	10	16	16	15	15	14	13	13	11	9	8	8	5	1	2
Height (inches)	60	61	62	63	64	65	66	67	68	69	70	71	72	73	74	75	76	77

a Students at Brigham Young University, organized according to height, as a splendid example of continuous variation

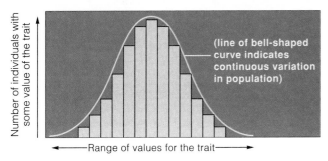

(line of bell-shaped curve indicates continuous variation in population)

◄——Range of values for the trait——►

b Idealized bell-shaped curve for a population that displays continuous variation in some trait

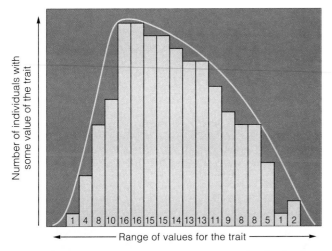

◄—— Range of values for the trait ——►

c Specific bell-shaped curve corresponding to the distribution of the trait (height) illustrated by the photograph in (**a**)

Figure 11.17 Continuous variation in body height, a trait that is one of the characteristics of the human population.

(**a**) Suppose you want to find the frequency distribution for height in a group of 169 biology students at Brigham Young University. You decide on how finely the range of possible heights should be divided. Then you measure each student and assign her or him to the appropriate category. Finally, you divide the number in each category by the total number of all students in all categories.

(**b**) Often a bar graph is used to depict continuous variation in a population. In such graphs, the proportion of individuals in each category is plotted against the range of measured phenotypes. Notice the curved line above the bars. It is an idealized example of the kind of "bell-shaped" curve that emerges for populations showing continuous variation in a trait. The bell-shaped curve in (**c**) is a specific example of this type of diagram.

represent categories with the least number of students. The bar that is tallest represents the category with the greatest number of students. Finally, draw a graph line around all of the bars and you end up with a "bell-shaped" curve. Such curves are typical of populations that show continuous variation in a trait.

Enzymes and other products of genes regulate each step of most metabolic pathways. Mutations, gene interactions, and environmental conditions may affect one or more of the steps, and this leads to variations in phenotypes.

For most traits, the individuals of a population display continuous variation—that is, a range of small differences.

The greater the number of genes and environmental factors that can influence a trait, the more continuous will be the expected distribution of all the versions of that trait.

EXAMPLES OF ENVIRONMENTAL EFFECTS ON PHENOTYPE

We have mentioned, in passing, that environmental conditions often contribute to variable gene expression among individuals of a population. Before leaving this chapter, consider just two examples of environmentally induced variations in phenotype.

Possibly you have observed a Himalayan rabbit and probably a Siamese cat. Both of these furry mammals carry an allele that specifies a heat-sensitive version of one of the enzymes necessary for melanin production. At the surface of warm body regions, the enzyme is less active. Fur growing there is lighter in color than fur in cooler regions, which include the ears and other body parts that project away from the main body mass. The experiment shown in Figure 11.18 provided observable evidence of an environmental effect on gene expression.

The environment also influences genes that govern phenotypes of plants. You may have observed the color variation in the floral clusters of *Hydrangea macrophylla*, a species widely favored in home gardens (Figure 11.19). In this type of plant, the action of genes responsible for floral color can produce different phenotypes, depending on the acidity of the soil in which the plant is growing.

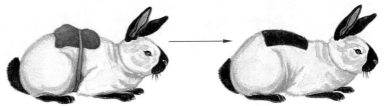

Figure 11.18 Example of the observable effect of different environmental conditions on gene expression in animals. A Himalayan rabbit normally has black hair only on its long ears, nose, tail, and lower leg limbs. For one experiment, a patch of a rabbit's white fur was plucked clean, and then an icepack was secured over the hairless patch. Where the colder temperature had been maintained, the hairs that grew back were black.

Himalayan rabbits are homozygous for the *ch* allele of a gene that codes for tyrosinase, an enzyme required to produce melanin. The allele specifies a heat-sensitive version of the enzyme, which is able to function only when the surrounding temperature is below about 33°C. When cells that give rise to hairs grow under warmer conditions, they cannot produce melanin and the hairs appear light. This happens in body regions that are massive enough to conserve a fair amount of metabolic heat. Ears and other slender extremities are cooler because they tend to lose metabolic heat more rapidly.

Figure 11.19 Effect of environmental conditions on gene expression in a favorite garden plant (*Hydrangea macrophylla*). Even plants that carry the same alleles have floral colors ranging from pink to blue, depending on the acidity of the soil in which they happen to be growing.

And so we conclude this chapter, which has dealt with heritable and environmental factors that give rise to variations in phenotype. What is the take-home lesson? Simply this: An individual's phenotype is an outcome of complex interactions among genes, enzymes and other gene products, and environmental factors.

Owing to gene mutations, cumulative gene interactions, and environmental effects on genes, individuals of a population show degrees of variation for many traits.

SUMMARY

1. A gene is a unit of information about a heritable trait. Alleles of a gene are different molecular versions of that information. Through experimental crosses with pea plants, Mendel gathered evidence that diploid organisms have two genes for each trait and that genes retain their identity when transmitted to offspring.

2. Mendel performed monohybrid crosses between two true-breeding plants that displayed different versions of a single trait, such as flower color. The crosses provided indirect evidence that some forms of a gene may be dominant over other, recessive forms.

3. A homozygous dominant individual has inherited two dominant alleles (AA) for the trait being studied. A homozygous recessive has two recessive alleles (aa), and a heterozygote has two nonidentical alleles (Aa).

4. In Mendel's monohybrid crosses ($AA \times aa$), all F_1 offspring were Aa. Crosses between F_1 plants resulted in these combinations of alleles in F_2 offspring:

	A	a	
A	AA	Aa	AA (dominant)
			Aa (dominant)
a	Aa	aa	Aa (dominant)
			aa (recessive)

This produced the expected phenotypic ratio of 3:1.

5. Results from Mendel's monohybrid crosses led to the formulation of a theory of segregation. In modern terms, diploid organisms have pairs of genes, on pairs of homologous chromosomes. The two genes of each pair segregate from each other at meiosis, such that each gamete formed ends up with one or the other gene.

6. Mendel performed dihybrid crosses between two true-breeding plants that displayed different versions of two traits. Results were close to a 9:3:3:1 phenotypic ratio:

    9 dominant for both traits
    3 dominant for A, recessive for b
    3 dominant for B, recessive for a
    1 recessive for both traits

7. Mendel's dihybrid crosses led to the formulation of a theory of independent assortment. In modern terms, by the end of meiosis, the gene pairs of two homologous chromosomes have been sorted out for distribution into one gamete or another, independently of how the gene pairs of other chromosomes were sorted out.

8. Four factors commonly influence gene expression:
    a. Degrees of dominance may occur between some pairs of genes.
    b. The products of pairs of genes may interact to influence the same trait.

c. One gene may have positive or negative effects on two or more traits, a condition called pleiotropy.
    d. Environmental conditions to which an individual is subjected may affect gene expression.

Review Questions

1. Define the difference between these terms: *11.1*
    a. gene and allele
    b. dominant allele and recessive allele
    c. homozygote and heterozygote
    d. genotype and phenotype

2. Define a true-breeding lineage. What is a hybrid? *11.1*

3. Distinguish between monohybrid and dihybrid crosses. What is a testcross, and why is it useful in genetic analysis? *11.2, 11.3*

4. Do segregation and independent assortment proceed during mitosis, meiosis, or both? *11.2, 11.3*

5. What do the vertical and horizontal arrows of this diagram represent? What do the bars and the curved line represent? *11.7*

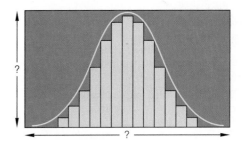

Self-Quiz *(Answers in Appendix IV)*

1. Alleles are _____ .
    a. different molecular forms of a gene
    b. different molecular forms of a chromosome
    c. self-fertilizing, true-breeding homozygotes

2. A heterozygote has a _____ for the trait being studied.
    a. pair of identical alleles
    b. pair of nonidentical alleles
    c. haploid condition, in genetic terms
    d. a and c

3. The observable traits of an organism are its _____ .
    a. phenotype   c. genotype
    b. sociobiology   d. pedigree

4. F_1 offspring of the monohybrid cross $AA \times aa$ are _____ .
    a. all AA c. all Aa
    b. all aa d. 1/2 AA and 1/2 aa

5. Second-generation offspring from a cross are the _____ .
    a. F_1 generation   c. hybrid generation
    b. F_2 generation   d. none of the above

6. Assuming complete dominance will occur, the offspring of the cross $Aa \times Aa$ will show a phenotypic ratio of _____ .
    a. 3:1   b. 9:1   c. 1:2:1   d. 9:3:3:1

7. Crosses between F_1 individuals resulting from the cross $AABB \times aabb$ lead to F_2 phenotypic ratios close to _____ .
    a. 1:2:1   b. 3:1   c. 1:1:1:1   d. 9:3:3:1

8. Match each example with the most suitable description.
    ____ dihybrid cross   a. *bb*
    ____ monohybrid cross   b. *Aa Bb × Aa Bb*
    ____ homozygous condition   c. *Aa*
    ____ heterozygous condition   d. *Aa × Aa*

Critical Thinking—Genetics Problems

(Answers in Appendix III)

1. One gene has alleles *A* and *a*. Another has alleles *B* and *b*. For each genotype listed, what type(s) of gametes will be produced? (Assume independent assortment occurs before gametes form.)

    a. *AABB*   c. *Aabb*

    b. *AaBB*   d. *AaBb*

2. Still referring to Problem 1, what will be the genotypes of the offspring from the following matings? Indicate the frequencies of each genotype among them.

    a. *AABB* × *aaBB*   c. *AaBb* × *aabb*

    b. *AaBB* × *AABb*   d. *AaBb* × *AaBb*

3. In one experiment, Mendel crossed a pea plant that bred true for green pods with one that bred true for yellow pods. All the F_1 plants had green pods. Which form of the trait (green or yellow pods) is recessive? Explain how you arrived at your conclusion.

4. Return to Problem 1, and assume you now study a third gene having alleles *C* and *c*. For each genotype listed, what type(s) of gametes will be produced?

    a. *AABBCC*   c. *AaBBCc*

    b. *AaBBcc*   d. *AaBbCc*

5. Mendel crossed a true-breeding tall, purple-flowered pea plant with a true-breeding dwarf, white-flowered plant. All F_1 plants were tall and had purple flowers. If an F_1 plant self-fertilizes, then what is the probability that a randomly selected F_2 offspring will be heterozygous for the genes specifying height and flower color?

6. At a certain gene location on a human chromosome, a dominant allele controls *tongue rolling*, an ability to curl up the sides of the tongue (Figure 11.20). People who are homozygous for a recessive allele at that locus cannot roll the tongue. At a different gene locus, a dominant allele controls whether the earlobes will be attached or detached (refer to Figure 11.1). These two pairs of genes assort independently. Suppose a tongue-rolling, detached-earlobed woman marries a man who has attached earlobes and cannot roll his tongue. Their first child has the father's phenotype. Given this outcome;

    a. What are the genotypes of the mother, father, and child?

    b. What is the probability that a second child of theirs will have detached earlobes and won't be a tongue roller?

7. Bill and his wife, Marie, hope to have children. Both have notably flat feet and long eyelashes, and they tend to sneeze a lot (hence the name of the *achoo syndrome*). A dominant allele gives rise to each of these traits: *A* (foot arch), *E* (eyelash length), and *S* (chronic sneezing). Bill is heterozygous and Marie is homozygous for all three traits.

    a. What is Bill's genotype? What is Marie's genotype?

    b. Marie becomes pregnant four times. What is the probability that each child will show all three of Bill's traits? Of Marie's traits?

    c. What is the probability that each child will have short lashes, high arches, and no chronic tendency to sneeze?

8. *DNA fingerprinting* is a method of identifying individuals based on locating unique base sequences in their DNA (Section 16.3). Before researchers refined the method, attorneys often relied on the ABO blood-typing system to settle disputes over paternity. Suppose, as a geneticist, you were called upon to testify during a paternity case in which the mother has type A blood, the child has type O blood, and the alleged father has type B blood. How would you respond to the following statements?

    a. Attorney of the alleged father: "The mother's blood is type A, so the child's type O blood must have come from the father. Because my client has type B blood, he simply could not be the father."

    b. Mother's attorney: "Because further tests prove this man is heterozygous, he must be the father."

9. Suppose you identify a new gene in mice. One of its alleles specifies white fur color. A second allele specifies brown fur color. You are asked to determine whether the relationship between the two alleles is one of simple dominance or incomplete dominance. What sorts of genetic crosses would give you the answer? On what types of observations would you base your conclusions?

10. Your sister moves away and gives you her purebred Labrador retriever, a female named Dandelion. Suppose you decide to breed Dandelion and sell puppies to help pay for your college tuition. Then you discover that two of her four brothers and sisters have a heritable hip disorder. If Dandelion mates with a male Labrador known to be free of the harmful allele, can you guarantee to a buyer that puppies will not carry the allele? Explain your answer.

11. A dominant allele *W* confers black fur on guinea pigs. If a guinea pig is homozygous recessive (*ww*), it has white fur. Fred would like to know whether his pet black-furred guinea pig is homozygous dominant (*WW*) or heterozygous (*Ww*). How might he determine his pet's genotype?

12. Red-flowering snapdragons are homozygous for the allele R^1. White-flowering snapdragons are homozygous for a different allele (R^2). Heterozygous plants (R^1R^2) bear pink flowers. What phenotypes should appear among F_1 offspring of the crosses listed, and what are the expected proportions for each phenotype?

    a. $R^1R^1 \times R^1R^2$ c. $R^1R^2 \times R^1R^2$

    b. $R^1R^1 \times R^2R^2$ d. $R^1R^2 \times R^2R^2$

Notice, in Problem 12, that in cases of incomplete dominance it is inappropriate to refer to either allele of a pair as dominant or recessive. When the phenotype of a heterozygous individual is halfway between those of the two homozygotes, then there is no dominance. Such alleles

Figure 11.20 A student at San Diego State University exhibiting the tongue-rolling trait for the benefit of a tongue-roll-challenged student.

are usually designated by superscript numerals, as shown here, rather than by uppercase letters for dominance and lowercase letters for recessiveness.

13. In chickens, two pairs of genes affect the comb type (Figure 11.15). When both genes are recessive, a chicken will have a single comb. A dominant allele of one gene, *P*, gives rise to a pea comb. A dominant allele of the other (*R*) gives rise to a rose comb. An epistatic interaction occurs when a chicken has at least one of both dominants, *P— R —* , which gives rise to a walnut comb. Predict the F_1 ratios resulting from a cross between two walnut-combed chickens that are heterozygous for both genes (*PpRr*).

14. As described in Section 11.5, a mutated allele gives rise to an abnormal form of hemoglobin (Hb^S instead of Hb^A). Homozygotes (Hb^SHb^S) develop sickle-cell anemia. But heterozygotes (Hb^AHb^S) show few outward symptoms.

   Suppose a female's mother is unaffected by this genetic disorder, but her father is homozygous for the Hb^S allele. She marries a male who is heterozygous for the allele, and they plan to have children. For *each* pregnancy, state the probability that this couple will have a child who is:

   a. homozygous for the Hb^S allele
   b. homozygous for the Hb^A allele
   c. heterozygous Hb^AHb^S

15. Certain dominant alleles are so vital for normal development that an individual who is homozygous recessive for a mutant recessive form of the allele cannot survive. Such recessive, *lethal alleles* can be perpetuated by heterozygotes. Consider the Manx allele (M^L) in cats. Homozygous cats (M^LM^L) die when they are still embryos inside the mother cat. In heterozygotes (M^LM), the spine develops abnormally, and the cats end up with no tail whatsoever (Figure 11.21).

   Suppose two M^LM cats mate. Among their *surviving* progeny, what is the probability that any one kitten will be heterozygous?

16. A recessive allele *a* is responsible for albinism, an inability to produce or deposit melanin in tissues. Humans and some other organisms can have this phenotype (Figure 11.22). In each of the following cases, what are the possible genotypes of the father, of the mother, and of their children?

   a. Both parents have normal phenotypes; some of their children are albino and others are unaffected.

Figure 11.21 Manx cat.

Figure 11.22 An albino male in India.

   b. Both parents are albino and have only albino children.

   c. The woman is unaffected, the man is albino, and they have one albino child and three unaffected children. (What gives rise to this 3:1 ratio?)

17. Kernel color in wheat plants is determined by two pairs of genes. Alleles of one pair show incomplete dominance over alleles of the other pair. For the gene pair at one locus on the chromosome, allele A^1 imparts one dose of red color to the kernel, whereas allele A^2 does not. At the second locus, allele B^1 gives one dose of red color to the kernel, whereas allele B^2 does not. A kernel with genotype $A^1A^1B^1B^1$ is dark red. A kernel with genotype $A^2A^2B^2B^2$ is white. All other genotypes have kernel colors in between these two extremes.

   a. If you cross a plant grown from a dark-red kernel with one grown from a white kernel, what genotypes and phenotypes would be expected among the offspring? In what proportions?

   b. If a plant with genotype $A^1A^2B^1B^2$ self-fertilizes, what genotypes and what phenotypes would be expected among the offspring? In what proportions?

Selected Key Terms

allele *11.1*	hybrid offspring *11.1*
codominance *11.4*	incomplete dominance *11.4*
continuous variation *11.7*	independent assortment *11.3*
dihybrid cross *11.3*	monohybrid cross *11.2*
epistasis *11.6*	multiple allele system *11.4*
F_1 *11.1*	phenotype *11.1*
F_2 *11.1*	pleiotropy *11.5*
gene *11.1*	probability *11.2*
genotype *11.1*	Punnett-square method *11.2*
heterozygous *11.1*	segregation *11.2*
homozygous dominant *11.1*	testcross *11.2*
homozygous recessive *11.1*	true-breeding lineage *11.1*

Readings

Cummings, M. 1994. *Human Heredity.* Third edition. St. Paul: West.

Orel, V. 1984. *Mendel.* New York: Oxford University Press.

Web Site See *http://www.wadsworth.com/biology* for practice quiz questions, hypercontents, BioUpdates, and critical thinking. The Wadsworth Biology Resource Center provides a wealth of information fully organized and integrated by chapter.

Too Young to Be Old

Imagine being ten years old, trapped in a body that is rapidly getting a bit more shriveled, more frail—*old*—each day. You are only tall enough to peer over the top of the kitchen counter, and you weigh less than thirty-five pounds. Already you are bald and have a crinkled nose. Possibly you have a few more years to live. Yet, like Mickey Hayes and Fransie Geringer, you can still play and laugh with your friends (Figure 12.1).

Of every 8 million newborn humans, one is destined to grow old far too soon. That rare individual possesses a mutated gene on one of the forty-six chromosomes inherited from its mother or father. Through hundreds, thousands, then many billions of DNA replications and mitotic cell divisions, terrible information encoded in that gene was systematically distributed to every cell in the growing embryo, and later in the newborn. Its legacy will be accelerated aging and a greatly reduced life expectancy. These are the defining features of *Hutchinson-Gilford progeria syndrome*. There is no cure.

A **syndrome** is a set of symptoms that characterize a disorder. In this case, the mutation is the start of severe disruptions in interactions among the genes that bring about bodily growth and development. Outwardly apparent symptoms begin to emerge when the child is not even two years old. Skin that should be plump and resilient starts to thin. Skeletal muscles are weakened. Limb bones

that should lengthen and grow stronger start to soften. Hair loss becomes pronounced; extremely premature baldness is inevitable.

Most progeriacs die in their early teens as a result of strokes or heart attacks. These final insults are brought on by a hardening of the walls of arteries, a condition that is typical of advanced age.

There are no documented cases of progeria running in families, which suggests that the gene spontaneously mutates at random. It apparently is dominant over its normal partner on the homologous chromosome.

We began this unit of the book by looking at cell division, the starting point of inheritance. Then we

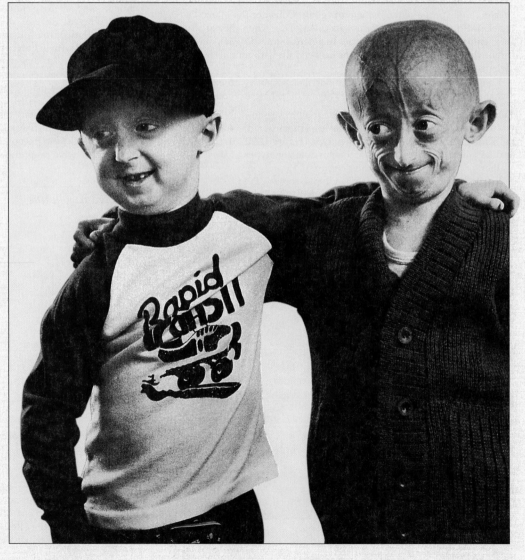

Figure 12.1 Two boys who met during a gathering of progeriacs at Disneyland, California, when they were not yet ten years old. Progeria, a heritable disorder, arises through a mutation in a dominant allele. Its defining characteristics are accelerated aging and a greatly reduced life expectancy.

started thinking about how chromosomes—and the genes they carry—are shuffled during meiosis and at fertilization. In this chapter, we delve more deeply into patterns of chromosomal inheritance. If at times the described methods of analysis seem remote from the world of your interests, remember this: For better or worse, *an inherited collection of specific bits of information gives rise to traits that help define each organism*—and this includes yourself and other human individuals. When Mickey Hayes turned all of eighteen, he was the oldest living progeriac. Fransie was only seventeen years old when he died.

So in that sense, our story of Mickey and Fransie takes us back to the 1880s, and to the rediscovery of Gregor Mendel's studies of inheritance. In 1884 Mendel himself had just passed away, and his paper on pea plants had been gathering dust in a hundred libraries for nearly two decades. However, improvements in the resolving power of microscopes had rekindled efforts to locate the hereditary material within cells. By 1882, Walther Flemming had observed threadlike bodies—chromosomes—inside the nuclear region of dividing cells. By 1884, a question was taking shape: Could chromosomes be the hereditary material?

Then researchers realized each gamete has half the number of chromosomes of a fertilized egg. In 1887, August Weismann hypothesized that a special division process must reduce the chromosome number by half before gametes form. Sure enough, in that same year meiosis was discovered. Weismann began to promote this theory of heredity: The chromosome number is halved during meiosis, then restored at fertilization; therefore, a cell's hereditary material is half paternal and half maternal in origin. His theory was hotly debated among biologists, and it prompted a flurry of experimental crosses—just like the ones Mendel had carried out.

Finally, in 1900, researchers came across Mendel's paper while checking literature related to their own genetic crosses. To their surprise, their experimental results confirmed what Mendel's results had already suggested: Diploid cells have two units (genes) for each heritable trait, and the two units are segregated from each other before gametes form.

During the decades that followed, many researchers learned a great deal more about chromosomes. Let's turn now to a few high points of their work, which will serve as background for our understanding of human inheritance.

KEY CONCEPTS

1. One gene follows another in sequence along the length of a chromosome. Each gene has its own position in that sequence.

2. The combination of alleles in a chromosome does not necessarily remain intact through meiosis and gamete formation. During an event called crossing over, some alleles in the sequence swap places with their partners on the homologous chromosome. The alleles that are swapped may or may not be identical.

3. Allelic recombinations contribute to variations in the phenotypes of offspring.

4. The structure of a chromosome may change, as when a chromosome segment is deleted, duplicated, inverted, or moved to a new location. Also, the chromosome number may change as a result of an improper separation of duplicated chromosomes during meiosis or mitosis.

5. Changes in chromosome structure and in chromosome number are rare events. When they do occur, they often give rise to genetic abnormalities or disorders.

Genes and Their Chromosome Locations

Earlier chapters provided you with a general sense of the structure of chromosomes and their behavior in meiosis. To refresh your memory and get a glimpse of where you are going from here with your reading, take a moment to study the following list:

1. **Genes** are units of information about heritable traits. The genes of eukaryotic species are distributed among a number of chromosomes, and each has its own location (locus) in one type of chromosome.

2. A cell with a diploid chromosome number ($2n$) has inherited pairs of **homologous chromosomes**. All but one pair are identical in their length, shape, and gene sequence. The exception is a pairing of nonidentical sex chromosomes, such as X with Y. In all cases, the two members of a pair can interact during meiosis.

3. A pair of homologous chromosomes can carry identical or nonidentical alleles at a given locus. **Alleles**, which arise through mutation, are different molecular forms of the same gene. Although many different forms may have arisen in a population, a diploid cell can only have a pair of them.

4. A *wild-type* allele is the most common form of a gene, either in a natural population or in standard, laboratory-bred strains of a species. Any other form of the gene is called a mutated allele.

5. Think of genes on the same chromosome as being linked. The farther apart two linked genes are, the more vulnerable they are to **crossing over**, an event by which homologous chromosomes break and exchange corresponding segments. Crossing over results in **genetic recombination**, or nonparental combinations of linked alleles in gametes, then in offspring.

6. The random alignment of each pair of homologous chromosomes at metaphase I results in nonparental combinations of alleles in gametes, then in offspring.

7. Abnormal occurrences during meiosis or mitosis occasionally change the structure of chromosomes and the parental chromosome number.

As you will see, these are the points that will help you make sense of the basic patterns of human inheritance.

Autosomes and Sex Chromosomes

In all but one case, a pair of homologous chromosomes are exactly like each other in their length, shape, and gene sequence. Microscopists discovered the exception in the late 1800s. For many organisms, a distinctive chromosome is present in female *or* male individuals, but not in both. For example, a diploid cell of a human male has one **X chromosome** and one **Y chromosome** (written as XY). That of a human female has two X chromosomes (XX). This inheritance pattern is common among many organisms, including all mammals and fruit flies. By contrast, in butterflies, moths, birds, and certain fishes, inheriting two identical sex chromosomes results in a male, and inheriting two nonidentical sex chromosomes results in a female.

Figure 12.2 shows examples of the human X and Y chromosomes. Notice that they are physically different; one is much shorter than the other. Besides this, they do not carry the same genes. Despite the differences, the two are able to synapse in a small region along their length, and this allows them to function as homologues during meiosis.

The human X and Y chromosomes fall in the more general category of **sex chromosomes**. The term refers to distinctive types of chromosomes which, in certain combinations, govern gender—that is, whether a new individual will develop into a male or a female. All other chromosomes in an individual's cells are the same in both sexes; they are called **autosomes**.

Karyotype Analysis

Today, microscopists can routinely analyze the physical appearance of a cell's sex chromosomes and autosomes. Chromosomes, recall, are in the most highly condensed state at metaphase of mitosis. At that time, their size, length, and centromere location are easiest to identify. In addition, after microscopists stain them in certain ways, the chromosomes of many species show distinct banding patterns. The most condensed chromosome regions often take up more stain and therefore appear as darker bands. Figure 10.3 gives an example.

A **karyotype** for an individual (or for a species) is a preparation of metaphase chromosomes, sorted out by their defining features. The next section explains how to construct a karyotype diagram: a cut-up, rearranged photograph in which all autosomes are lined up, largest to smallest, and sex chromosomes are positioned last.

Diploid cells have pairs of genes, on pairs of homologous chromosomes. At each gene locus, the alleles (alternative forms of a gene) may be identical or nonidentical.

As a result of crossing over and other events during meiosis, new combinations of alleles and parental chromosomes end up in offspring. Abnormal events during meiosis or mitosis also can change the structure and number of chromosomes.

Autosomes are the pairs of chromosomes that are the same in males and females of a species. One other pair, the sex chromosomes, govern a new individual's gender.

12.2 PREPARING A KARYOTYPE DIAGRAM

Karyotype diagrams can help answer questions about an individual's chromosomes. Chromosomes are in their most condensed form, and easiest to identify, in cells that are proceeding through metaphase of mitosis. Technicians don't count on finding a cell that happens to be dividing in the body when they go looking for it. Instead, they culture cells **in vitro** (literally, "in glass"). They put a small sample of blood, skin, bone marrow, or some other tissue in a glass container. In that container is a solution that stimulates cell growth and mitotic cell division for many generations.

Dividing cells can be arrested at metaphase by adding colchicine to a culture medium. As Section 4.8 describes, colchicine is an extract of the autumn crocus (*Colchicum autumnale*). Technicians and researchers use it to block spindle formation and so prevent duplicated chromosomes from separating during nuclear division. With suitable colchicine concentrations and exposure times, many metaphase cells can accumulate, and this increases the chances of finding candidates for karyotype diagrams.

Following colchicine treatment, the culture medium is transferred to the tubes of a centrifuge, a motor-driven spinning device (Section 5.2 and Figure 12.2*a*). Cells have greater mass and density than the surrounding solution, so the spinning force moves them farthest from the center of rotation, to the bottom of the attached tubes. Separation in response to a spinning force is called **centrifugation**.

Afterward, the cells are transferred to a saline solution. When immersed in this hypotonic fluid, they swell (by osmosis) and move apart. And so do the metaphase chromosomes. The cells are ready to be dropped onto a microscope slide, fixed (as by air-drying), and stained.

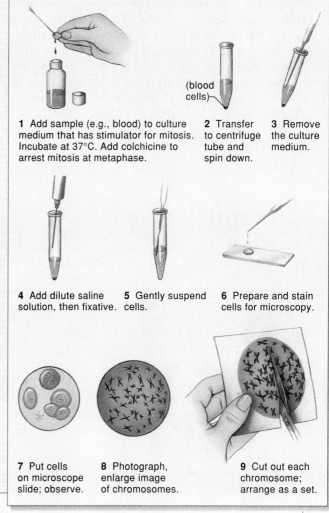

1 Add sample (e.g., blood) to culture medium that has stimulator for mitosis. Incubate at 37°C. Add colchicine to arrest mitosis at metaphase.

2 Transfer to centrifuge tube and spin down.

3 Remove the culture medium.

(blood cells)

4 Add dilute saline solution, then fixative.

5 Gently suspend cells.

6 Prepare and stain cells for microscopy.

7 Put cells on microscope slide; observe.

8 Photograph, enlarge image of chromosomes.

9 Cut out each chromosome; arrange as a set.

a

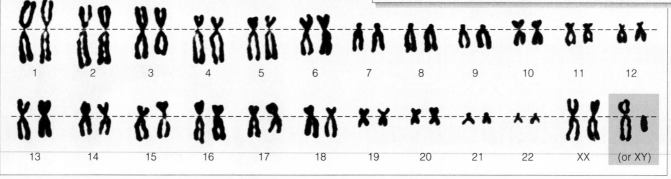

1 2 3 4 5 6 7 8 9 10 11 12

13 14 15 16 17 18 19 20 21 22 XX (or XY)

b

Chromosomes take up some stains uniformly along their length, and this allows identification of chromosome size and shape. With other staining procedures, horizontal bands show up along the length of the chromosomes of certain species. If researchers direct a ray of ultraviolet light at these chromosomes, the bands will fluoresce. You can see an example of this in Figure 10.3, which is another human karyotype diagram.

In the last steps of karyotype preparation, metaphase chromosomes are photographed, then the microscope

Figure 12.2 (**a**) Karyotype preparation. (**b**) Human karyotype. Human somatic cells have 22 pairs of autosomes and 1 pair of sex chromosomes (XX or XY). That is a diploid number of 46. These are metaphase chromosomes; each is in the duplicated state.

image is enlarged. The photographed chromosomes are cut apart, one at a time, then arranged according to their size, shape, and length of arms. Finally, all the pairs of homologous chromosomes are horizontally aligned by their centromeres, as shown in Figure 12.2*b*.

SEX DETERMINATION IN HUMANS

Analyzing human cells, as by karyotyping, has yielded evidence that each egg produced by a female carries one X chromosome. Half the sperm cells produced by a male carry an X chromosome and half carry a Y.

If an X-bearing sperm fertilizes an X-bearing egg, the new individual will develop into a female. If the sperm happens to carry a Y chromosome, the individual will develop into a male (Figure 12.3).

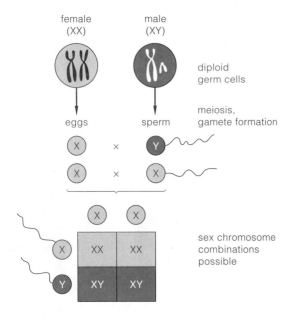

Figure 12.3   Pattern of sex determination in humans.

Among the very few genes on the Y chromosome is a "male-determining gene." As Figure 12.4 indicates, expression of this particular gene leads to the formation of testes, which are primary male reproductive organs. In the gene's absence, ovaries will form automatically. Ovaries are primary female reproductive organs. Testes and ovaries both produce sex hormones that influence the development of particular sexual traits.

A human X chromosome carries more than 2,300 genes. Like other chromosomes, it carries some genes associated with sexual traits, such as the distribution of body fat and hair. However, most of its genes deal with *nonsexual* traits, such as blood-clotting functions. These genes can be expressed in males as well as in females (males, remember, also carry one X chromosome).

A certain gene on the human Y chromosome dictates that a new individual will develop into a male. In the absence of the Y chromosome (and the gene), a female develops.

Figure 12.4   Boys, girls, and the Y chromosome.

For about the first four weeks of its existence, a human embryo has neither male nor female traits, even though it normally carries XY or XX chromosomes. However, ducts and other structures start forming that can go either way.

(**a–c**) In an XX embryo, the primary female reproductive organs (ovaries) start to form automatically—*in the absence of a Y chromosome*. By contrast, in an XY embryo, the primary male reproductive organs (testes) start to form during the next four to six weeks. Apparently, a gene region on the Y chromosome governs a fork in the developmental road that can lead to maleness.

The newly forming testes start to produce testosterone and other sex hormones. These hormones are crucial for the development of a male reproductive system. By contrast, in the XX embryo, the newly forming ovaries start to produce different kinds of sex hormones. These are crucial for the development of a female reproductive system.

The master gene for male sex determination is named SRY (short for *S*ex-determining *R*egion of the *Y* chromosome). So far, the same gene has been identified in DNA from male humans, chimpanzees, rabbits, pigs, horses, cattle, and tigers. None of the females tested had the gene. Tests with mice indicate that the gene region becomes active about the time that testes are starting to develop.

The SRY gene resembles regions of DNA that are known to specify regulatory proteins. As described in Section 15.1, such proteins bind with certain parts of DNA and thereby turn genes on and off. It seems that the SRY gene product regulates a cascade of reactions that are necessary for male sex determination.

umbilical cord (lifeline between embryo and maternal tissues)   amnion (a protective, fluid-filled sac surrounding the embryo)

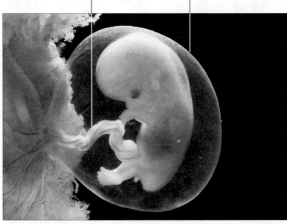

a A human embryo, eight weeks old. Male reproductive organs have already started to develop.

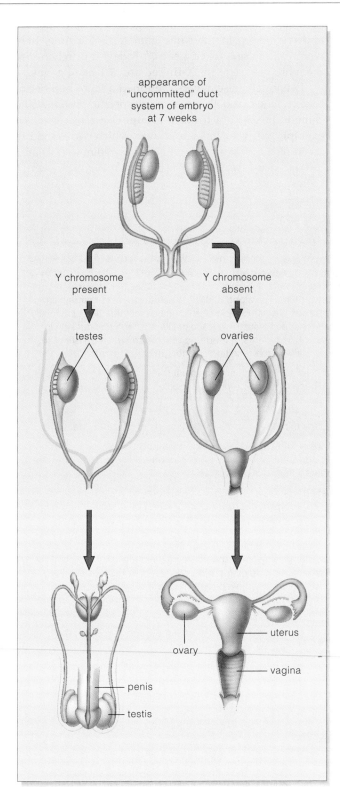

appearance of
"uncommitted" duct
system of embryo
at 7 weeks

Y chromosome
present

Y chromosome
absent

testes

ovaries

ovary

uterus

penis

vagina

testis

b The duct system in the early human embryo that may develop into the primary male reproductive organs *or* into the primary female reproductive organs.

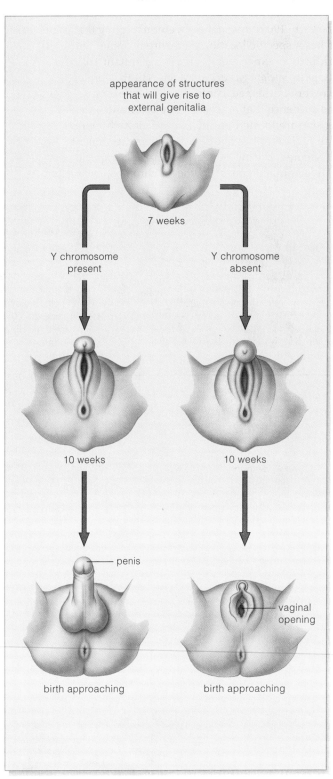

appearance of structures
that will give rise to
external genitalia

7 weeks

Y chromosome
present

Y chromosome
absent

10 weeks

10 weeks

penis

birth approaching

vaginal
opening

birth approaching

c External appearance of the newly forming reproductive organs in human embryos.

EARLY QUESTIONS ABOUT GENE LOCATIONS

Linked Genes—Clues to Inheritance Patterns

By the early 1900s, researchers were suspecting that each gene has a specific location on a chromosome. Through hybridization experiments involving mutant fruit flies (*Drosophila melanogaster*), Thomas Hunt Morgan and his coworkers helped confirm this. For example, they found evidence that a gene for eye color and a gene for wing size are located on the *Drosophila* X chromosome. Figure 12.5 describes one series of their experiments.

It seemed, during the early *Drosophila* experiments, that two mutant genes on the X chromosome (*w* for white eyes and *m* for miniature wings) were "linked." That is, they were traveling together during meiosis and ending up in the same gamete. For a time, they were called "sex-linked genes." Today, researchers use the more precise terms **X-linked** and **Y-linked genes**.

Eventually, researchers identified a large number of linked genes, and the ones on a specific chromosome came to be called a **linkage group**. *D. melanogaster*, for example, has four linkage groups, which correspond to its four pairs of homologous chromosomes. Similarly, Indian corn (*Zea mays*) has ten linkage groups, which correspond to ten pairs of homologous chromosomes.

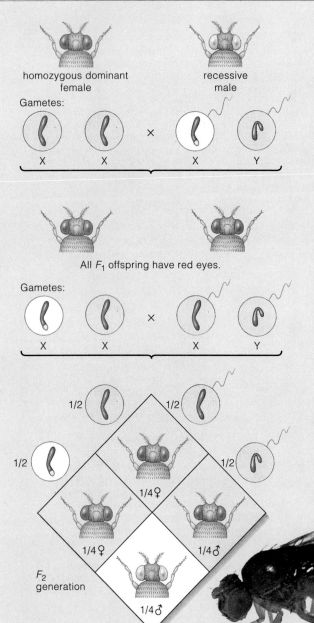

Figure 12.5 X-linked genes as clues to inheritance patterns.

In the early 1900s, the embryologist Thomas Morgan was studying inheritance patterns. He and his coworkers discovered an apparent genetic basis for the connection between gender and certain nonsexual traits. For example, human males and females both have blood-clotting mechanisms. Yet hemophilia, a blood-clotting disorder, shows up most often in males, not females, of a family lineage. This gender-specific outcome was not like anything Mendel saw in his hybrid crosses of pea plants. It made no difference which parent plant carried a recessive allele; the resulting phenotype was the same.

Morgan studied eye color and other nonsexual traits of fruit flies (*Drosophila melanogaster*). These flies can be raised in bottles on cornmeal, molasses, and agar. A female lays hundreds of eggs in a few days, and offspring can themselves reproduce in less than two weeks. Morgan could track hereditary traits through nearly thirty generations of thousands of flies in a year's time.

At first, all flies were wild type for eye color; they had brick-red eyes. Then, as a result of an apparent mutation in a gene controlling eye color, a white-eyed male appeared in one of the bottles.

Morgan established true-breeding strains of white-eyed males and females. Then he did paired, **reciprocal crosses**. (In the first of such paired crosses, one parent displays the trait of interest. In the second, the other parent displays it.) For the first cross, Morgan allowed white-eyed males to mate with homozygous red-eyed females. All F_1 offspring had red eyes. Of the F_2 offspring, however, only some of the males had white eyes. In the second cross, white-eyed females were mated with true-breeding red-eyed males. Half of the F_1 offspring were red-eyed females and half were white-eyed males. Of the F_2 offspring, 1/4 were red-eyed females, 1/4 white-eyed females, 1/4 red-eyed males, and 1/4 white-eyed males!

The seemingly odd results implied a relationship between an eye-color gene and gender. Probably the gene was located on a sex chromosome. But which one? Because females (XX) could be white-eyed, the recessive allele would have to be on one of their X chromosomes. Suppose white-eyed males (XY) also carry the recessive allele on their X chromosome—and suppose there is no corresponding eye-color allele on the Y chromosome. If that were so, then the males would have white eyes, for they have no dominant allele to mask the effect of the recessive one.

The diagram at left illustrates the expected results when Morgan's idea of an X-linked gene is combined with Mendel's concept of segregation. By proposing that one particular gene is located on an X chromosome but not on the Y, Morgan was able to explain the outcome of his reciprocal crosses. His experimental results matched the predicted outcomes.

Inside the figure (left diagram labels):

homozygous dominant female · recessive male
Gametes: X X × X Y
All F_1 offspring have red eyes.
Gametes: X X × X Y
1/2 · 1/2 · 1/2 · 1/2
1/4 ♀ · 1/4 ♀ · 1/4 ♂ · 1/4 ♂
F_2 generation

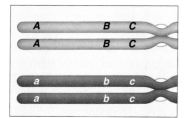

a One pair of homologous chromosomes in the duplicated state (each consists of two sister chromatids). One is shaded blue, the other purple. Three different genes are shown. The alleles at all three gene loci are nonidentical (A with a, B with b, and C with c).

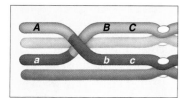

b In prophase I of meiosis, two nonsister chromatids break while tightly aligned. (All four chromatids are shown pulled apart for clarity, as in Figure 8.5.) They swap segments, then enzymes seal the broken ends. The breakage and exchange represent one crossover event.

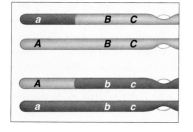

c The outcome of the crossover is genetic recombination between two of four chromatids. (They are shown after meiosis, as unduplicated, separate chromosomes.)

Figure 12.6 Simplified diagram of crossing over. This event occurs in prophase I of meiosis (Figure 8.4).

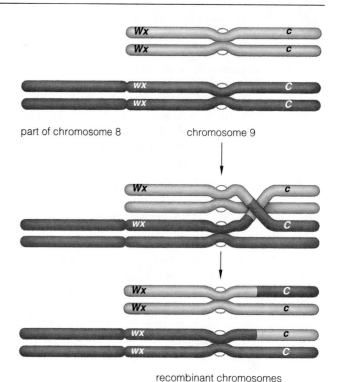

part of chromosome 8   chromosome 9

recombinant chromosomes

Figure 12.7 Harriet Creighton and Barbara McClintock's correlation of a cytological difference with a genetic difference in chromosome 9 from a strain of Indian corn (*Zea mays*).

Crossing Over and Genetic Recombination

If linked genes always stayed together through meiosis, then a dihybrid cross between true-breeding parents should always have a predictable outcome. Specifically, the most frequent phenotypes among the F_2 offspring should be those of the original parents. (Here you may wish to review Section 11.3 and Figure 12.6.) However, a number of puzzling results from the early *Drosophila* experiments did not match this expectation.

For example, Morgan crossed a *Drosophila* female that was recessive for white eyes and miniature wings with a wild-type male (red eyes and long wings). As expected, all F_1 males had white eyes and miniature wings, and all females were wild type. After crossing F_1 flies, Morgan analyzed 2,441 of the F_2 offspring. A significant number had white eyes and long wings or red eyes and miniature wings! As you will see in the next section, 900 (or 36.9 percent) were recombinants. As Morgan hypothesized, physical exchanges must have occurred between X chromosomes during meiosis.

It was not until 1931 that two genetic researchers, Harriet Creighton and Barbara McClintock, discovered evidence of such exchange. They were experimenting with two Z. *mays* chromosomes that differed physically in a way that could be distinguished with a microscope. Such distinguishable features are **cytological markers** for the genes being studied.

The researchers used corn that was heterozygous for two genes on chromosome 9. Two alleles of one gene specify colored (C) or colorless (c) seeds. One allele of the other gene specifies the synthesis of two forms of starch (Wx), and the other allele specifies only one (wx). One chromosome 9 that was used in an experiment had genotype *cWx* and a normal appearance (Figure 12.7). Another chromosome 9 had genotype *Cwx* and was longer. An abnormal event caused a piece of a different chromosome to become attached to one of its ends.

During meiosis, crossing over sometimes occurred between the two gene loci on a pair of the cytologically different chromosomes. The outcome was two kinds of genetic recombinants: *CWx* and *cwx*. As Creighton and McClintock realized, when genetic recombination had occurred, cytological features of the chromosomes also had changed. When *wx* ended up with *c* instead of *C*, so did the extra length. They observed no such correlation in F_1 generations that retained the parental genotype.

Genes on the same chromosome belong to the same linkage group and do not assort independently during meiosis. Crossing over between homologous chromosomes disrupts gene linkages and results in the production of nonparental combinations of genes in chromosomes.

Correlations between specific genes and cytological markers provide evidence of genetic recombination.

As we now know, crossing over is not a rare event. In fact, for humans and most other eukaryotic species, meiosis cannot even be completed properly unless each pair of homologous chromosomes takes part in at least one crossover. The preceding section briefly described experimental evidence of this remarkable event. Let us now consider a few specific examples of the ways in which crossing over disrupts gene linkages and what researchers do with this information.

How Close Is Close? A Question of Recombination Frequencies

Figure 12.8 shows Morgan's cross between a wild-type *Drosophila* male and a female that was recessive for white eyes and for miniature wings. Again, the two parental genotypes showed up equally among the F_1 offspring (50 percent white-eyed, miniature-winged males and 50 percent wild-type females). Of the 2,441 F_2 offspring analyzed, 36.9 percent were recombinants.

Morgan's group extended their investigations to other X-linked genes. In one series of experiments, a true-breeding white-eyed, yellow-bodied mutant female was crossed with a wild-type male (with red eyes and a gray body). As expected, 50 percent of the F_1 offspring showed one or the other parental phenotype. In this case, however, only 129 of 2,205 of the F_2 offspring that were analyzed—or 1.3 percent—were recombinants.

From the results of many such experimental crosses, it appeared that the genes controlling eye color and wing size were not as "tightly linked" as those for eye color and body color. Also, the two parental genotypes were represented in approximately equal numbers of the F_2 offspring, and the same was true of the two kinds of recombinant genotypes. What was going on here? As Morgan hypothesized, certain alleles tend to remain together during meiosis more often than others *because they are positioned closer together on the same chromosome.*

Imagine any two genes at two different locations on the same chromosome. *The probability that a crossover*

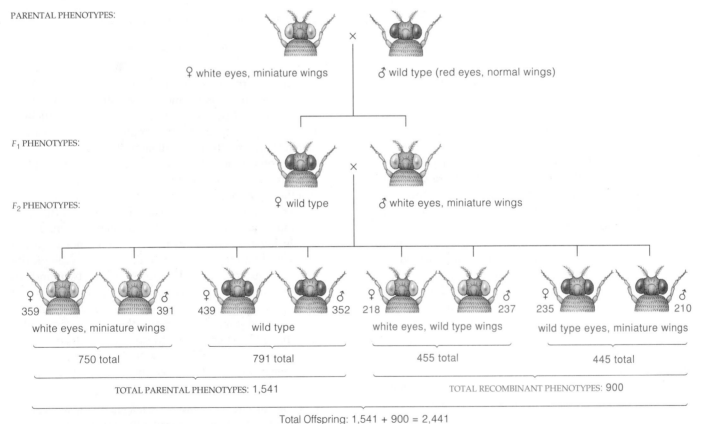

PARENTAL PHENOTYPES:

♀ white eyes, miniature wings   ×   ♂ wild type (red eyes, normal wings)

F_1 PHENOTYPES:

F_2 PHENOTYPES:

♀ wild type   ×   ♂ white eyes, miniature wings

♀ 359   391 ♂ — white eyes, miniature wings — 750 total

♀ 439   352 ♂ — wild type — 791 total

♀ 218   237 ♂ — white eyes, wild type wings — 455 total

♀ 235   210 ♂ — wild type eyes, miniature wings — 445 total

TOTAL PARENTAL PHENOTYPES: 1,541   TOTAL RECOMBINANT PHENOTYPES: 900

Total Offspring: 1,541 + 900 = 2,441

PERCENT RECOMBINANTS: 900/2,441 × 100 = 36.9%

Figure 12.8 Experimental crosses with *Drosophila melanogaster* that provided indirect evidence of how crossing over disrupts linkage groups. In this example, Morgan's group tracked a mutant gene for white eyes and a different mutant gene for miniature wings. As you probably know, the symbol ♀ signifies female and the symbol ♂ signifies male.

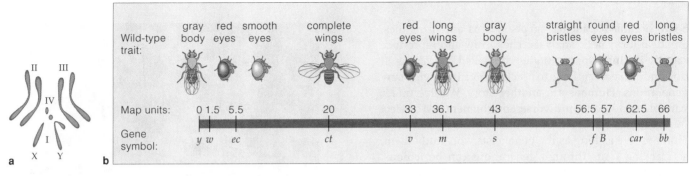

Figure 12.9 (**a**) *Drosophila melanogaster* chromosomes. (**b**) A linkage map for a number of genes on chromosome I (here, the X chromosome). As you can see, more than one gene can influence the same trait, such as eye color.

will disrupt their linkage is proportional to the distance that separates them. Suppose genes *A* and *B* are twice as far apart as two other genes, *C* and *D*:

We would expect crossing over to disrupt the linkage between *A* and *B* much more often.

Two genes are very closely linked when the distance between them is small; their allelic combinations nearly always end up in the same gamete. Gene linkage is more vulnerable to crossover if the distance between genes is greater. When two genes are very far apart, crossing over disrupts linkage so often that those genes assort independently of each other into gametes.

Linkage Mapping

After using testcrosses to analyze crossover patterns, Alfred Sturtevant, one of Morgan's students, proposed that the percentage of recombinants in gametes might be a quantitative measure of the relative positions of genes along a chromosome. This investigative approach is now called **linkage mapping**.

The relative, linear distance between any two linked genes on a genetic map is expressed in map units. One map unit corresponds to a crossover frequency of 1 percent, twenty map units correspond to a crossover frequency of 20 percent, and so on. For instance, take a look at Figure 12.9, which is a linkage map for one of the *D. melanogaster* chromosomes. The amount of gene recombination to be expected between "smooth eyes" and "long wings" would be 30.6 percent (5.5 map units subtracted from 36.1 map units).

Linkage maps do not show *actual* physical distances between genes. The most accurate approximations have been calculated only for closely linked genes. Why? As the map distance increases, multiple crossovers occur and skew recombination frequencies. Even so, the *map distance* between two genes as estimated by genetic crosses generally correlates with the *physical distance*, or length of DNA between them, as calculated by other methods. Exceptions to this generalization are known to occur in the centromere region.

Of the several thousand known genes in the four types of *Drosophila* chromosomes, the positions of about a thousand have been mapped. What about the 50,000 to 100,000 genes on the twenty-three types of human chromosomes? Unlike fruit flies, humans do not lend themselves to experimental crosses. Nevertheless, some tight gene linkages have been identified by tracking the resulting phenotypes, one generation after another, in certain families.

For example, *color blindness* and a blood-clotting disorder, *hemophilia*, are disorders that are caused by recessive alleles at two gene loci on the X chromosome. One female carried both alleles, although she herself was symptom-free. So was her father, so she must have inherited a normal X chromosome from him (remember, males have only one X and one Y). The X chromosome she inherited from her mother must have carried both mutant alleles. The female gave birth to six sons. Three sons developed color blindness and hemophilia, and two were unaffected. Here was phenotypic evidence that recombination had not occurred. But her sixth son started life as a fertilized, recombinant egg; he was color-blind only. Thus, for the two mutant alleles in this family, the recombination frequency is 1/6 (or 0.167 percent). Many such affected families would have to be examined to get a good estimate of the map distance between the two genes. Generally, these genes do not have high recombination frequencies because they are very closely linked at one end of the X chromosome.

The farther apart two genes are on a chromosome, the greater will be the frequency of crossing over and therefore of genetic recombination between them.

Linkage mapping is a method of measuring the relative, linear distances between genes on the same chromosome. Such maps roughly correspond to actual physical distances, or lengths of DNA, as calculated by other methods.

HUMAN GENETIC ANALYSIS

Some organisms, including pea plants and fruit flies, are ideal for genetic analysis. They grow and reproduce rapidly in small spaces, under controlled conditions. It does not take very long to track a trait through many generations. Humans are another story. We live under variable conditions in diverse environments. We select our own mates and reproduce if and when we want to. Humans live as long as the geneticists who study them, so tracking traits through generations can be tedious. Most human families are so small, there are not enough offspring for easy inferences about inheritance.

Constructing Pedigrees

To get around the problems associated with analyzing human inheritance, geneticists put together pedigrees. A **pedigree** is a chart that shows genetic connections among individuals. Genetic researchers construct them by using standardized methods and definitions, as well as standardized symbols to represent individuals, as shown in Figure 12.10.

An affected child with six digits on each hand

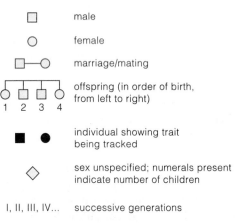

a Standardized symbols in pedigrees

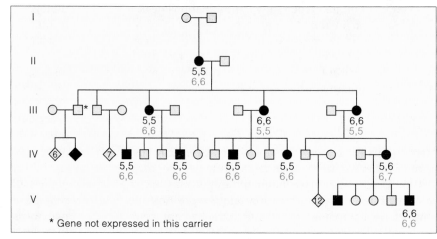

b Pedigree for a family in which polydactyly recurs

Figure 12.10 (a) Some of the symbols used in constructing pedigrees. (b) One pedigree for *polydactyly*, a condition in which a person has extra fingers, extra toes, or both. Expression of the gene for this trait can vary from one individual to the next. *Black* numerals signify the number of fingers on each hand where data were available; *blue* numerals signify the number of toes on each foot.

When analyzing pedigrees, geneticists rely on their knowledge of probability and of Mendelian inheritance patterns, which might yield clues to the genetic basis for a trait. For instance, they might determine that the responsible allele is dominant or recessive, or that it is located on an autosome or a sex chromosome.

Gathering a great many family pedigrees increases the numerical base for analysis. When a trait shows a simple Mendelian inheritance pattern, a geneticist may have confidence in predicting the probability of its occurrence among the children of prospective parents. We will return to this topic later in the chapter.

Regarding Human Genetic Disorders

Table 12.1 lists some of the heritable traits that have been studied in detail. A few of these are abnormalities, or deviations from the average condition. Said another way, a **genetic abnormality** is nothing more than a rare, uncommon version of a trait, as when a person is born

Table 12.1   Examples of Human Genetic Disorders and Genetic Abnormalities

Disorder or Abnormality*	Main Consequences	Disorder or Abnormality*	Main Consequences
AUTOSOMAL RECESSIVE INHERITANCE		**X-LINKED DOMINANT INHERITANCE**	
Albinism 11.6; CT 11.9	Absence of pigmentation	Faulty enamel trait 12.8	Problems with teeth
Blue offspring CT 18.9	Bright blue skin coloration	**X-LINKED RECESSIVE INHERITANCE**	
Cystic fibrosis CT 4.12	Excessive glandular secretions leading to tissue, organ damage	Color blindness 12.5, 36.6	Inability to distinguish among all or some colors of the spectrum of visible light
Ellis-van Creveld Syndrome 18.8	Extra fingers, toes, short limbs		
Galactosemia 12.7	Brain, liver, eye damage	Duchenne muscular dystrophy 12.8 CT 15.7	Muscles waste away
Phenylketonuria (PKU) 6.2; 12.11	Mental retardation	Hemophilia, 11.5, 12.8	Impaired blood-clotting ability
Sickle-cell anemia 11.5; 14.1; 18.6	Adverse plleiotropic effects on organs throughout body	Testicular feminization syndrome 37.2	XY individual but having some female traits, sterility
AUTOSOMAL DOMINANT INHERITANCE		X-linked anhidrotic ectodermal dysplasia 15.4	In human females, mosaic patches of skin with or without sweat glands
Achondroplasia 12.7	One form of dwarfism	**CHANGES IN CHROMOSOME NUMBER**	
Achoo syndrome CT 11.9	Chronic sneezing	Down syndrome 12.9	Mental retardation, heart defects
Amyotrophic lateral sclerosis (ALS) 12.6	Loss of all muscle function	Klinefelter syndrome 12.9	Sterility, retardation
Camptodactyly 12.8	Rigid, bent little fingers	Turner syndrome 12.9	Sterility; abnormal ovaries, abnormal sexual traits
Familial hypercholesterolemia CI 16	High cholesterol levels in blood; eventually clogged arteries	XYY condition 12.9	Mild retardation or free of symptoms
Huntington disorder 12.6; 12.7	Nervous system degenerates progressively, irreversibly	**CHANGES IN CHROMOSOME STRUCTURE**	
Polydactyly 12.6	Extra fingers, toes, or both	Cri-du-chat syndrome 12.10	Mental retardation, abnormally formed larynx
Progeria CI12; 12.7	Drastic premature aging	Fragile X syndrome 12.10	Mental retardation
Tay-Sachs disorder 12.7	Progressive deterioration of the nervous system		

*Italic numbers indicate sections in which a disorder is described. CI signifies Chapter Introduction. CT signifies an end-of-chapter Critical Thinking question.

with six toes on each foot instead of five. Whether an individual or society at large views an abnormal trait as disfiguring or merely interesting is subjective. As the classic novel *The Hunchback of Notre Dame* so clearly emphasized, there is nothing inherently life-threatening or even ugly about it.

By comparison, a **genetic disorder** is an inherited condition that sooner or later causes mild to severe medical problems. The alleles underlying severe genetic disorders do not abound in populations, for they put individuals at great risk. Why do they not disappear entirely? There are two reasons. First, rare mutations introduce new copies of the alleles into the population. Second, in heterozygotes, the harmful allele is paired with a normal one that might cover its functions, so it still can be passed on to offspring.

You may hear someone refer to a genetic disorder as a disease, but the terms are not always interchangeable. A disease results from infection by bacteria, viruses, or some other environmental agent that invades the body and multiplies inside its tissues. Illness follows only if

the infection leads to tissue damage that interferes with normal body functions. When a person's genes increase susceptibility to infection or weaken the response to it, the resulting illness might be called a **genetic disease**.

With these qualifications in mind, we will turn next to examples of inheritance in the human population. As you will see, genetic analyses of family pedigrees have often revealed simple Mendelian inheritance patterns for certain traits. Researchers have traced many of the traits to dominant or recessive alleles on an autosome or X chromosome. They have traced others to changes in the structure or number of chromosomes.

For many genes, pedigree analysis might reveal simple Mendelian inheritance patterns that will allow inferences about the probability of their transmission to children.

A genetic abnormality simply is a rare or less common version of an inherited trait. A genetic disorder is an inherited condition that results in mild to severe medical problems.

Autosomal Recessive Inheritance

For some traits, inheritance patterns reveal two clues that point to a recessive allele on an autosome. *First*, if both parents are heterozygous, any child of theirs will have a 50 percent chance of being heterozygous and a 25 percent chance of being homozygous recessive, as Figure 12.11 indicates. *Second*, if the parents are both homozygous recessive, any child of theirs will be, also.

About 1 in 100,000 newborns are homozygous for a recessive allele that causes *galactosemia*. They cannot produce functional molecules of an enzyme that stops a product of lactose breakdown from accumulating to toxic levels. Lactose normally is converted to glucose and galactose, then to glucose-1-phosphate (which is broken down by glycolysis or converted to glycogen). In affected persons, the full conversion is blocked:

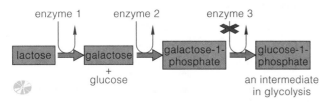

A high blood level of galactose can damage the eyes, liver, and brain. Malnutrition, diarrhea, and vomiting are early symptoms. A high galactose level, the telling symptom, can be detected in urine samples. Untreated galactosemics often die in childhood. But if affected individuals are placed on a restricted diet that excludes dairy products, they can grow up symptom-free.

People who are homozygous for another autosomal recessive allele develop *Tay-Sachs disorder*. They appear normal at birth, but their brain and spinal cord start to deteriorate before they reach their first birthday. Mental retardation, blindness, and loss of neural and muscle function follow. Affected children usually die between ages three and four. The mutant allele responsible for the disorder causes a deficiency in an enzyme that is necessary for the metabolism of sphingolipids, a type of lipid that is an especially abundant component of the plasma membrane of cells in nerves and the brain.

Autosomal Dominant Inheritance

Two clues of a different sort indicate that an autosomal dominant allele is responsible for a trait. *First,* the trait typically appears in each generation, for the allele is usually expressed even in heterozygotes. *Second,* if one parent is heterozygous and the other is homozygous recessive, there is a 50 percent chance that any child of theirs will be heterozygous (Figure 12.12).

A few dominant alleles cause severe disorders, yet they persist in populations. Some are perpetuated by

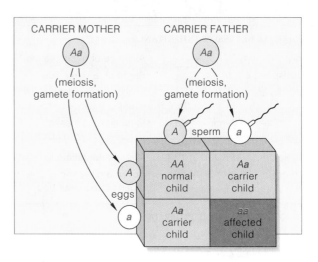

Figure 12.11 One pattern for autosomal recessive inheritance. In this example, both parents are heterozygous carriers of the recessive allele (shown in *red*).

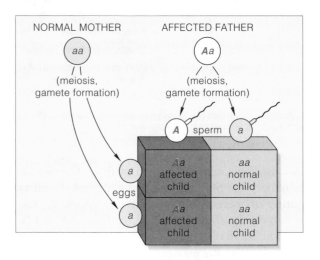

Figure 12.12 One pattern for autosomal dominant inheritance. In this example, the dominant allele (*red*) is fully expressed in carriers.

spontaneous mutations. This is true of progeria, the rare aging disorder described in the chapter introduction. In other cases, expression of the dominant allele may not interfere with reproduction, or the affected individuals have children before symptoms become severe.

For example, Huntington disorder is a condition characterized by a gradual increase in involuntary movements, a progressive deterioration of the nervous system, and eventual death. The symptoms may not

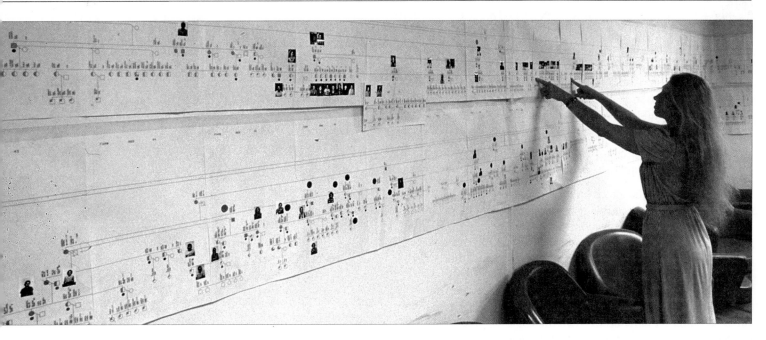

Figure 12.13 From the human genetic researcher Nancy Wexler, a pedigree for Huntington disorder, by which the nervous system progressively degenerates. Wexler's team constructed an extended family tree for nearly 10,000 people in Venezuela. Wexler has a special interest in Huntington disorder; she herself has a 50 percent chance of developing it.

Figure 12.14 A painting by Velázquez of Infanta Margarita Teresa of the Spanish court and her maids, including the achondroplasic woman at the far right.

Analysis of affected and unaffected individuals in one notably extended family in Venezuela revealed that a dominant allele on human chromosome 4 is the genetic culprit responsible for Huntington disorder. Part of the pedigree for this family is shown in Figure 12.13.

As another example, *achondroplasia* affects about 1 in 10,000 people. The homozygous dominant condition commonly leads to stillbirth. Heterozygotes are able to reproduce. However, while they are young and their limb bones are forming, the cartilage components of those bones cannot form properly. As an outcome, at maturity, affected individuals have abnormally short arms and legs, relative to their other body parts. Figure 12.14 shows a famous example. Adult achondroplasics are less than 4 feet, 4 inches tall. The dominant allele often has no other phenotypic effects than this.

Most individuals affected by an autosomal recessive disorder have heterozygous, symptom-free parents. If both parents are homozygous for the allele, all of their children will be, also.

Most often, an autosomal dominant disorder does not skip generations, because the responsible allele is expressed even in heterozygotes. If one parent is heterozygous and the other is homozygous recessive, each child they may produce has a 50 percent chance of being heterozygous.

Genetic disorders arising from a pair of autosomal dominant alleles are rare in populations. They persist because of rare, spontaneous mutations or because they do not interfere with reproduction.

even start to show up until the affected individual is past age thirty. For most people, reproduction already has occurred by then. Affected individuals usually die in their forties or fifties, before they may realize that they have passed on the mutant allele to their children.

An X- or Y-linked trait, recall, arises from expression of one or more genes on the X or Y chromosome. In most cases, evidence for Y-linked inheritance is questionable or nonexistent. Better documentation exists for X-linked inheritance patterns, a few of which will be described in the paragraphs to follow.

X-Linked Recessive Inheritance

Distinctive clues often show up when a recessive allele on an X chromosome causes a trait. First, males show the recessive phenotype far more often than females do. A recessive allele can be masked in females, who may inherit a dominant allele on their other X chromosome. The allele cannot be masked in males, who have only one X chromosome (Figure 12.15). Second, a son cannot inherit the recessive allele from his father. A daughter can. If a daughter is heterozygous, there is a 50 percent chance each son of hers will inherit the allele.

Hemophilia A, a blood-clotting disorder, is a case of X-linked recessive inheritance. In most people, a blood-clotting mechanism quickly stops bleeding from minor injuries. Reactions that result in clot formation require the products of several genes. If any of the genes is mutated, its defective product will allow bleeding to proceed for an abnormally long time. About 1 in 7,000 males inherits the gene for hemophilia A. Clotting time is close to normal in heterozygous females.

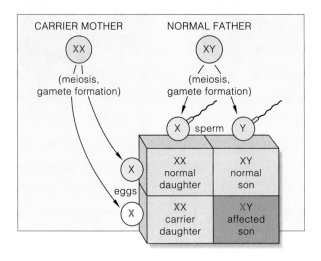

Figure 12.15 One pattern for X-linked inheritance. Assume the mother carries a recessive allele on one X chromosome (*red*).

The frequency of hemophilia A was unusually high in royal families of nineteenth-century Europe. Queen Victoria of England was a carrier (Figure 12.16). At one time, the recessive allele was present in eighteen of her sixty-nine descendants. A hemophilic great-grandchild, Crown Prince Alexis, was a focus of political intrigue that helped trigger the Russian Revolution of 1917.

Another, rare case of X-linked recessive inheritance is *Duchenne muscular dystrophy.* Muscles slowly enlarge

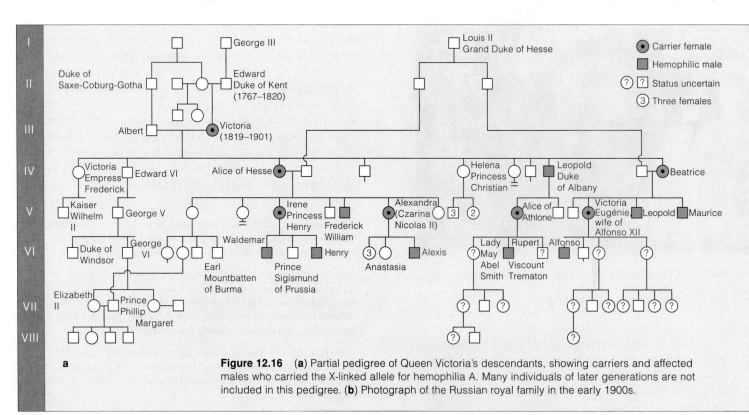

a

Figure 12.16 **(a)** Partial pedigree of Queen Victoria's descendants, showing carriers and affected males who carried the X-linked allele for hemophilia A. Many individuals of later generations are not included in this pedigree. **(b)** Photograph of the Russian royal family in the early 1900s.

with fat and connective tissue deposits, but the muscle cells atrophy (waste away). At first, children with this form of dystrophy appear normal. Between the second and tenth birthdays, muscle weakening makes them progressively more imbalanced and clumsier. After they are twelve years old, affected persons no longer can walk. In time, the chest and head muscles deteriorate. Typically they die of respiratory failure during their early twenties. At present, there is no cure.

X-Linked Dominant Inheritance

The *faulty enamel trait* is one of the few known cases of X-linked dominant inheritance. With this disorder, the hard, thick enamel coating that normally protects teeth fails to develop properly (Figure 12.17). The inheritance pattern is like that for X-linked recessive alleles, except the allele is also expressed in heterozygous females. The phenotype tends to be less pronounced in females than in males. A son cannot inherit the dominant allele for the trait from an affected father, but all daughters will. If a woman is heterozygous, she will transmit the allele to half of her offspring, regardless of their sex.

A Few Qualifications

Don't take the preceding examples of autosomal and X-linked traits too seriously. We include them not to turn

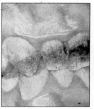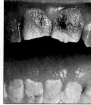

Figure 12.17 The discolored, abnormal tooth enamel of two individuals affected by the faulty enamel trait.

you into an armchair geneticist, but rather to give you a general idea of the kinds of clues that hold meaning for trained geneticists. Before diagnosing a case, geneticists often find it necessary to pool together many pedigrees. Typically they make detailed analyses of clinical data and keep abreast of research. Why? More than one type of gene might be responsible for a given phenotype. Geneticists already know of dozens of conditions that arise from a mutated gene on an autosome *or* a mutated gene on the X chromosome. They know of autosomal genes that show dominance in males and recessiveness in females, so the traits appear to be due to X-linked recessive inheritance even though they are not.

Also, don't assume that errant genes automatically relegate a person to life's sidelines. More than twenty years ago, Stephen Hawking noticed his muscles were weakening. In time he had trouble swallowing and speaking. Motor neurons in his brain and spinal cord were deteriorating; scar tissue was forming along his spinal cord. His muscles could not get proper signals from the nervous system and they began to waste away. Such are the symptoms of *amyotrophic lateral sclerosis*, or ALS. The disorder, which killed baseball's Lou Gehrig, affects about 1 in 1,000 people. Possibly a virus issues the death warrant, but a mutated allele of an autosomal dominant gene, EAAT2, may be the executioner. The gene specifies a membrane transport protein that is supposed to sponge up excess glutamate in the brain and spinal cord. Glutamate serves as a communication signal between nerve cells (neurons) that control the body's muscles. In excess amounts, it kills the cells.

Hawking's affliction has not immobilized his mind, which continues to dance freely around some of the most challenging questions we humans have dared to ask. Through his acclaimed research in astrophysics and his eloquent yet accessible writings, Hawking has changed the way we view time and the universe.

b The Russian royal family members. All are believed to have been executed near the end of the Russian Revolution. They were recently exhumed from their hidden graves, but DNA fingerprinting indicates the remains of Alexis and one daughter, Anastasia, are not among them.

Czarina Alexandra (a carrier; descendant of Queen Victoria)

Czar Nikolas II (free of allele for hemophilia A)

Alexis (hemophilic son)

A recessive X-linked allele may be masked in heterozygous females but not in males. A father can pass on the allele to daughters but not sons. A heterozygous daughter has a 50 percent chance of passing on the allele to any son of hers.

An X-linked dominant allele is expressed each generation, even in heterozygotes. If one parent is heterozygous and the other is homozygous recessive, there is a 50 percent chance that any child of theirs will be heterozygous.

Occasionally, abnormal events occur before or during cell division, then gametes and new individuals end up with the wrong chromosome number. The consequences range from minor to lethal physical changes.

Categories and Mechanisms of Change

With **aneuploidy**, individuals have one extra or one less chromosome. This condition is a major cause of human reproductive failure. Quite possibly it affects half of all fertilized eggs. Autopsies show that most of the human embryos that were miscarried (spontaneously aborted before pregnancy reached full term) were aneuploids.

With **polyploidy**, individuals have three or more of each type of chromosome. About one-half of all species of flowering plants are polyploid (Section 19.4). So are some insects, fishes, and other animals. Polyploidy is lethal for humans. It may disrupt interactions between the genes of autosomes and sex chromosomes at key steps in the complex pathways of development and reproduction. All but 1 percent of human polyploids die before birth. The rare newborns die within a month.

Chromosome numbers can change during mitotic or meiotic cell divisions. Suppose a cell cycle goes through DNA duplication and mitosis but is arrested before the cytoplasm divides. The cell is *tetra*ploid, with four of each type of chromosome. Suppose one or more pairs of chromosomes fail to separate in mitosis or meiosis, an event called **nondisjunction**. Some or all forthcoming cells will have too many or too few chromosomes, as in the example in Figure 12.18. For certain experiments, developmental biologists commonly expose living cells to colchicine to induce nondisjunction (Section 12.2).

The chromosome number also may become changed at fertilization. Visualize a normal gamete uniting by chance with an $n + 1$ gamete (one extra chromosome). The new individual will be "trisomic" ($2n + 1$); it will have three of one type of chromosome and two of every other type. What if the other gamete is $n - 1$? In that case, the new individual will be "monosomic" ($2n - 1$).

Change in the Number of Autosomes

Most changes in the number of autosomes arise through nondisjunction during gamete formation. Let's consider one of the most common of the resulting disorders. A trisomic 21 newborn, with three chromosomes 21, will show the effects of *Down syndrome*. The symptoms vary greatly, but most affected individuals show moderate to severe mental impairment. About 40 percent develop heart defects. Abnormal development of the skeleton means older children have shortened body parts, loose joints, and poorly aligned bones of the hips, fingers, and toes. Muscles and muscle reflexes are weaker than normal, and speech and other motor skills develop slowly. With special training, trisomic 21 individuals often take part in normal activities (Figure 12.19). As a group, they are cheerful, affectionate people who derive great pleasure from socializing.

Down syndrome is one of many disorders that can be detected by prenatal diagnosis (Section 12.11). Before detection procedures were widespread, about 1 in 700 newborns of all ethnic groups were trisomic 21. Now the number is closer to 1 in 1,100. The risk is greater if pregnant women are more than thirty-five years old, as indicated by the graph in Figure 12.19b.

Figure 12.18 Nondisjunction. In this example, a pair of chromosomes fails to separate at anaphase I of meiosis. The chromosome number changes in gametes. (Make a sketch of nondisjunction at anaphase II. What will the chromosome numbers be in gametes?)

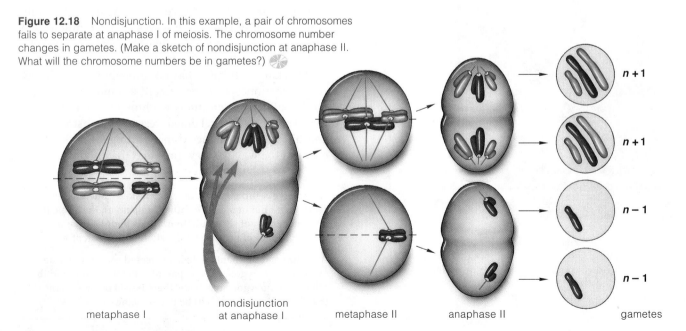

metaphase I   nondisjunction at anaphase I   metaphase II   anaphase II   gametes

$n + 1$
$n + 1$
$n - 1$
$n - 1$

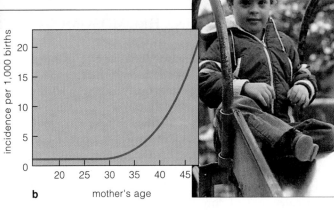

Figure 12.19 Down syndrome. (**a**) Karyotype revealing a trisomic 21 condition (*red* arrows). (**b**) Relation between the frequency of Down syndrome and mother's age at the time of childbirth. Results are from a study of 1,119 affected children who were born in Victoria, Australia, between 1942 and 1957. The young lady above was a lively participant in the Special Olympics, held annually in San Mateo, California.

Change in the Number of Sex Chromosomes

Most sex chromosome abnormalities arise as a result of nondisjunction during meiosis and gamete formation. Let's look at a few phenotypic outcomes.

TURNER SYNDROME   Inheriting one X chromosome without a partner X or Y chromosome gives rise to *Turner syndrome*, which affects 1 in 2,500 to 10,000 or so newborn girls. Nondisjunction affecting sperm accounts for 75 percent of the cases. We see fewer people with Turner syndrome compared to people having other sex chromosome abnormalities. The likely reason is that at least 98 percent of all X0 zygotes spontaneously abort early in pregnancy. Approximately 20 percent of all spontaneously aborted embryos in which chromosome abnormalities were detected have been X0.

Despite the near-lethality, X0 survivors are not as disadvantaged as other aneuploids. They grow up well proportioned, albeit short—4 feet, 8 inches tall, on the average. Generally, their behavior is normal during childhood. But most Turner females are infertile. They do not have functional ovaries and so cannot produce eggs or sex hormones. Without sex hormones, breast enlargement and the development of other secondary sexual traits are reduced. Possibly as a result of their arrested sexual development and small size, females often become passive and are easily intimidated by peers during their teens. Some patients have benefited from hormone therapy and corrective surgery.

KLINEFELTER SYNDROME   One out of every 500 to 2,000 liveborn males has two X chromosomes and one Y chromosome. This XXY condition results mainly from nondisjunction in the mother (about 67 percent of the time, compared to 33 percent in the father). Symptoms of the resulting *Klinefelter syndrome* develop after the onset of puberty.

XXY males are taller than average and are sterile or nearly so. Their testes usually are much smaller than average; the penis and scrotum are not. Facial hair is often sparse, and breasts may be somewhat enlarged. Injections of the hormone testosterone can reverse the feminized traits but not the low fertility. Some XXY males show mild mental impairment, although many fall within the normal range of intelligence. Except for low fertility, many show no outward symptoms at all.

XYY CONDITION   About 1 in 1,000 males has one X and two Y chromosomes. *XYY males* tend to be taller than average. Some may be mildly retarded, but most are phenotypically normal. At one time, XYY males were thought to be genetically predisposed to become criminals. The erroneous conclusion was based on small numbers of cases in highly selected groups, such as prison inmates. Investigators often knew who the XYY males were, and this may have biased their evaluations. There were no **double-blind studies**, by which different investigators gather data independently of one another and then match them up only after both sets of data are completed. In this case, the same investigators gathered the karyotypes and the personal histories. Fanning the stereotype was a sensationalized report in 1968 that a mass-murderer of young nurses was XYY. He wasn't.

In 1976, a Danish geneticist issued a report on a large-scale study based on records of 4,139 tall males, twenty-six years old, who had reported to their draft board. Besides giving results of physical examinations and intelligence tests, the records provided clues to socioeconomic status, educational history, and any criminal convictions. Only twelve of the males were XYY, which left more than 4,000 for the control group. The only significant finding was that tall, mentally impaired males who engage in criminal activity are more likely to get caught—irrespective of karyotype.

Most changes in chromosome number arise as a result of nondisjunction during meiosis and gamete formation.

Rare genetic disorders or abnormalities also arise when the physical structure of a chromosome changes. Such aberrations often occur spontaneously in nature. They also are induced in the laboratory through exposure to chemical substances or irradiation. Either way, they often can be detected through microscopic examination and karyotype analysis of cells undergoing mitosis or meiosis.

For our purposes, we can limit our review to four major categories of chromosome aberrations that may have serious or even lethal consequences for the human individual:

1. With a **deletion**, a segment of a chromosome is lost.

2. With **duplication**, the same linear stretch of DNA within a chromosome is repeated, often several to many times in the same chromosome or a different one.

3. With an **inversion**, a linear stretch of DNA within a chromosome becomes oriented in the reverse direction, with no molecular loss.

4. With **translocation**, a stretch of one chromosome's DNA physically moves to another location in the same chromosome or a different one, with no molecular loss.

Deletion

Viral attack, irradiation (especially ionizing radiation), chemical assaults, or some other environmental factor can trigger a deletion, or the loss of some segment of a chromosome (Figure 12.20).

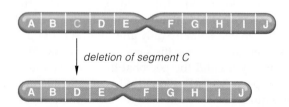

deletion of segment C

Figure 12.20 Simple diagram of a chromosome deletion.

Any chromosome region is vulnerable to deletion. The loss may even occur at one end, which makes the chromosome unstable. Wherever it happens, a deletion permanently excises one or more of the chromosome's genes. The loss of genetic information usually causes problems for the individual.

For example, one deletion from human chromosome 5 results in mental retardation and the development of an abnormally shaped larynx. When affected infants cry, the sounds they produce are like a cat meowing. Hence the name of the disorder, *cri-du-chat* (cat-cry). Figure 12.21 shows an affected child.

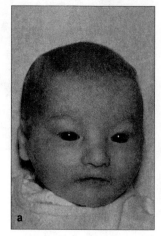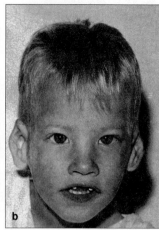

Figure 12.21 Cri-du-chat syndrome. (**a**) A male infant who will develop the cri-du-chat syndrome as a result of a deletion on the short arm of chromosome 5. The photograph was taken just after birth. Notice the ears positioned low on the side of the head relative to the eyes. (**b**) The same boy, photographed four years later. By this age, affected individuals no longer make the mewing sounds typical of the syndrome.

You might be thinking that species with diploid cells have an advantage when deletions do occur. After all, the presence of the same segment on the homologous chromosome may mediate the effect of the loss. But the sword of chance cuts both ways. If a missing gene's partner on the homologous chromosome is normal, all well and good. If it is a harmful recessive allele, nothing will mask or compensate for *its* effects.

Duplications

Even normal chromosomes contain duplications, which are gene sequences that are repeated several to many times. Figure 12.22 shows different forms that such a duplication can take. Often the same gene sequence has been repeated thousands of times.

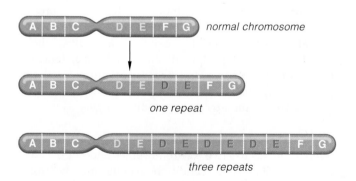

normal chromosome

one repeat

three repeats

Figure 12.22 Simple diagrams of chromosome duplications.

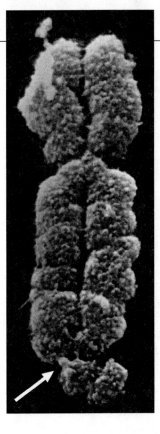

Consider the *fragile X syndrome*, which affects 1 of every 1,500 males and 1 of every 2,500 females. The disorder correlates with multiple copies of a nucleotide sequence in one gene (FMR-1) on the X chromosome. Nearly all people have about 29 of the repeats; however, affected people have 700 or even more. When the number of repeats exceeds a certain threshold, gene expression apparently gets reduced. In cultured human cells, most X chromosomes with the unstable array of repeats have an odd constricted region (Figure 12.23).

Duplications that have harmful or lethal effects tend not to be conserved over the course of evolution. An individual's growth, development, and maintenance activities all depend on intricate interactions among a number of gene products, so mutations that drastically alter genes are usually selected against. Even so, many other duplications apparently have done their bearers no harm. Although they are relatively rare events, they have accumulated over millions, even billions of years and are now built into the DNA of all species.

Biologists speculate that duplicates of some gene sequences with neutral effects could have an adaptive advantage. In effect, a copy could free up a gene for chance mutations that might turn out to be useful, for the normal gene would still issue the required product. One or more duplicated gene sequences could become slightly modified, and then the products of those genes could function in slightly different or new ways.

Several kinds of duplicated, modified genes seem to have had pivotal roles in evolution. Consider the gene regions for the polypeptide chains of the hemoglobin molecule, as shown in Section 3.7. In humans and other primates, these regions contain copies of remarkably similar gene sequences. They produce whole families of polypeptide chains with slight structural differences. Each of the resulting hemoglobin molecules performs transport functions with slightly different efficiencies, depending on cellular conditions.

Inversion

During an inversion, a segment that loops out from a chromosome becomes reinserted at the same place, but in the reverse order. The reversal alters the position and order of the chromosome's genes (Figure 12.24).

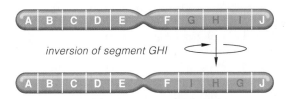

inversion of segment GHI

Figure 12.24 Simple diagram of a chromosome inversion.

Together with duplications, certain inversions and translocations probably helped put primate ancestors of humans on a unique evolutionary road. Of the twenty-three pairs of human chromosomes, eighteen are nearly identical to their counterparts in the chimpanzee and in the gorilla. The other five pairs differ at inverted and translocated regions.

Translocation

In most translocations, part of a chromosome exchanges places with a corresponding part of a *non*homologous chromosome, as in Figure 12.25.

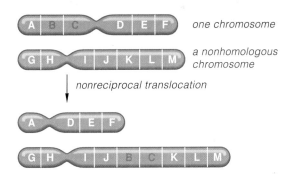

one chromosome

a nonhomologous chromosome

nonreciprocal translocation

Figure 12.25 Simple diagram of a chromosome translocation.

Chromosome 14, for instance, may end up with part of chromosome 8 (and chromosome 8 with a part of 14). In this case, normal controls over the segment's genes are lost at the new location, and a form of cancer results.

On rare occasions, a segment of a chromosome may get lost, inverted, moved to a new location, or duplicated.

Most chromosome aberrations are harmful or lethal. But over evolutionary time, many have been conserved; they confer adaptive advantages or have done their bearers no harm.

12.11 PROSPECTS IN HUMAN GENETICS

With the arrival of their newborn, parents typically ask, "Is our baby normal?" Quite naturally, they want their baby to be free of genetic disorders, and most of the time it is. But what are the options when it is not?

We do not approach diseases and heritable disorders the same way. Diseases follow infection by bacteria and other agents from the outside, and we attack them with antibiotics, surgery, and other weapons. But how do we attack an "enemy" within our genes? Do we institute regional, national, or global programs to identify people who might be carrying harmful alleles? Do we tell them they are "defective" and run a risk of bestowing their disorder on their children? Who decides which alleles are "harmful"? Should society bear the cost of treating genetic disorders before and after birth? If so, should society also have a say in whether affected embryos will be born at all, or whether they should be aborted?

Questions such as these are only the tip of an ethical iceberg. And we do not have answers that are universally acceptable throughout our society.

PHENOTYPIC TREATMENTS   Often, the symptoms of genetic disorders can be either minimized or suppressed by exerting dietary controls, making adjustments to specific environmental conditions, and even intervening surgically or by way of hormone replacement therapy.

For example, dietary control works in PKU, or phenylketonuria. A certain gene specifies an enzyme that converts one amino acid to another —phenylalanine to tyrosine. If an individual is homozygous recessive for a mutated form of the gene, the first of these amino acids accumulates inside the body. If excess amounts are diverted into other pathways, then phenylpyruvate and other compounds may form. High levels of phenylpyruvate in the blood can impair the functioning of the brain. When they restrict their intake of phenylalanine, affected persons are not required to dispose of excess amounts, and they can therefore lead normal lives. Among other things, they can avoid soft drinks and other food products that are sweetened with aspartame, a compound that contains phenylalanine.

Environmental adjustments help counter or minimize the symptoms of some disorders, as when albinos avoid exposure to direct sunlight. Surgical reconstructions also can minimize many problems. For example, surgeons can close up a form of *cleft lip* in which a vertical fissure cuts through the lip and extends into the roof of the mouth.

GENETIC SCREENING   Through large-scale screening programs in the general population, affected persons or carriers of a harmful allele often can be detected early enough to start preventive measures before symptoms develop. For example, most hospitals in the United States routinely screen newborns for PKU, so today it is less common to see people with symptoms of this disorder.

GENETIC COUNSELING   If a first child or close relative has a severe heritable problem, prospective parents may worry about their next child. They may request help in evaluating their options from a qualified professional counseler. *Genetic counseling* often includes diagnosis of parental genotypes, detailed pedigrees, and biochemical testing for hundreds of known metabolic disorders. Geneticists, too, may be contacted to help predict risks for disorders that follow simple Mendelian inheritance, although not all disorders do this. During the genetic counseling, prospective parents also must be reminded that the same risk applies to each pregnancy.

PRENATAL DIAGNOSIS   Methods of *prenatal diagnosis* can be used to determine the sex as well as more than a hundred genetic conditions of fetuses ("prenatal" means "before birth").

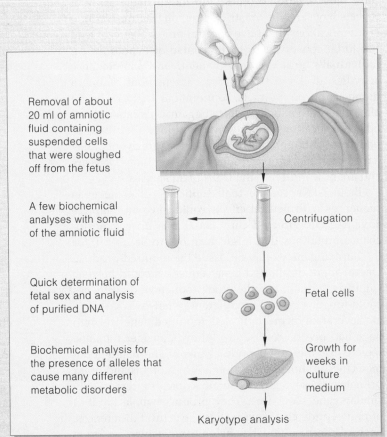

Removal of about 20 ml of amniotic fluid containing suspended cells that were sloughed off from the fetus

A few biochemical analyses with some of the amniotic fluid

Centrifugation

Quick determination of fetal sex and analysis of purified DNA

Fetal cells

Biochemical analysis for the presence of alleles that cause many different metabolic disorders

Growth for weeks in culture medium

Karyotype analysis

Figure 12.26   Steps in amniocentesis.

For example, suppose a woman who is forty-five years old and pregnant worries that her forthcoming child may develop Down syndrome. With **amniocentesis**, a clinician withdraws a tiny sample of the fluid inside the amnion, a membranous sac around the fetus. Some cells that the fetus has sloughed off are suspended in the sample. These are cultured and then analyzed, as by karyotyping. Figure 12.26 is a simplified diagram of the procedure.

By contrast, with **chorionic villi sampling** (CVS), a clinician withdraws some cells from the chorion, which is a fluid-filled, membranous sac that surrounds the amnion. It is possible to perform CVS weeks before amniocentesis. It can yield results as early as the ninth week of pregnancy.

Both procedures carry risks. Either one may accidentally cause infection or puncture the fetus. Also, a 1994 report from the Centers for Disease Control pointed out that mothers who request CVS might run a 0.3 percent risk that their future child will have missing or underdeveloped fingers and toes. Mothers-to-be probably should be counseled to balance that finding against (1) the overall risk of 3 percent that any child will have some kind of birth defect, (2) the severity of genetic disorder that her forthcoming child might be at risk of developing, and (3) how old she is at the time of her pregnancy.

REGARDING ABORTION   What happens if prenatal diagnosis does reveal a serious problem? Do prospective parents opt for induced **abortion**—that is, the expulsion of the embryo from the uterus? We can only say here that they must weigh their awareness of the crushing severity of the genetic disorder against ethical and religious beliefs. Worse still, they must play out their personal tragedy on a larger stage, dominated now by a nationwide battle between fiercely vocal "pro-life" and "pro-choice" factions. We return to this volatile issue in Sections 43.13 and 45.15.

PREIMPLANTATION DIAGNOSIS   Another procedure, *preimplantation diagnosis*, relies on **in-vitro fertilization**. "In vitro," recall, literally means "in glass." Sperm and eggs that have been taken from prospective parents are quickly transferred to an enriched medium in a glass

petri dish. One or more eggs may become fertilized. In two days, mitotic cell divisions may convert a fertilized egg into a ball of eight cells, such as the one shown in Figure 12.27a.

According to one view, the tiny, free-floating ball is a *pre*-pregnancy stage. Like the unfertilized eggs discarded monthly from a woman, the ball is not attached to the

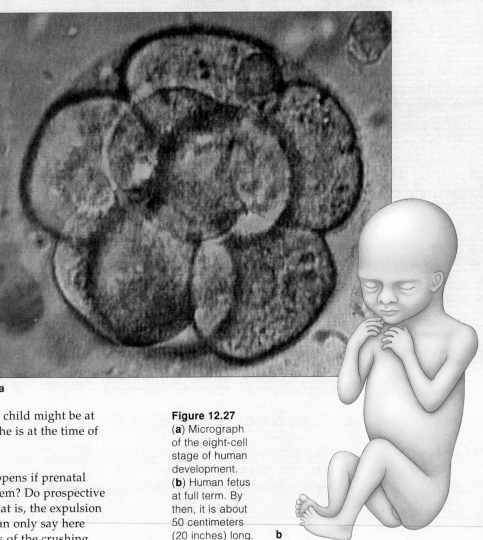

Figure 12.27
(**a**) Micrograph of the eight-cell stage of human development.
(**b**) Human fetus at full term. By then, it is about 50 centimeters (20 inches) long.

uterus. All cells in the ball have the same genes and are not yet committed to giving rise to specialized cells of a heart, lungs, and other organs. Doctors take one of the undifferentiated cells and analyze its genes for suspected disorders. If the cell has no detectable genetic defects, the ball is inserted into the uterus.

Several couples who are at risk of passing on muscular dystrophy, cystic fibrosis, and other disorders have opted for the procedure. Several *"test-tube"* babies have been born in good health and are free of the harmful genes.

SUMMARY

1. Genes, the units of instruction for heritable traits, are arranged one after the other along chromosomes.

2. Human cells are diploid ($2n$), with twenty-three pairs of homologous chromosomes that interact during meiosis. Each pair has the same length, shape, and gene sequence (except for an XY pairing).

3. Human females have two X chromosomes. Males have one X paired with one Y. All other chromosomes are autosomes (the same in both females and males). A gene on the Y chromosome determines gender.

4. Pedigrees, or charts of genetic connections through lines of descent, often provide clues to inheritance of a trait. Certain patterns are characteristic of dominant or recessive alleles on autosomes or on the X chromosome.

5. Genes on the same chromosome represent a linkage group. However, crossing over (breakage and exchange of segments between homologues) disrupts linkages, as summarized in Figure 12.28. The farther apart two gene loci are along the length of a chromosome, the greater will be the frequency of crossovers between them.

6. A chromosome's structure may become altered on rare occasions. A segment might be deleted, inverted, moved to a new location (translocated), or duplicated.

7. On rare occasions, the parental chromosome number may change. Gametes and new individuals may end up with one more or one less chromosome than the parents (aneuploidy) or with three or more of each type of chromosome (polyploidy). Nondisjunction at meiosis (prior to gamete formation) accounts for most of these chromosome abnormalities.

8. Crossing over adds to potentially adaptive variation in traits among members of a population. By contrast, most changes in chromosome number or structure are harmful or lethal. Over evolutionary time, such changes occasionally have proved to be neutral or have offered adaptive advantage, and so have accumulated in DNA.

Review Questions

1. What is a gene? What are alleles? *12.1*

2. Distinguish between: *12.1, 12.2*
   a. homologous and nonhomologous chromosomes
   b. sex chromosomes and autosomes
   c. karyotype and karyotype diagram

3. Define genetic recombination, and describe how crossing over can bring it about. Also give an example of cytological evidence that a crossover has occurred in a cell. *12.1, 12.4*

4. Define pedigree. Also explain why a genetic abnormality is not the same as a genetic disorder or genetic disease. *12.6*

5. Contrast a typical pattern of autosomal recessive inheritance with that of autosomal dominant inheritance. *12.7*

6. Contrast a typical pattern of X-linked recessive inheritance with that of X-linked dominant inheritance. *12.8*

7. Define aneuploidy and polyploidy. Make a sketch of an example of nondisjunction. *12.9*

8. Distinguish among a chromosomal deletion, duplication, inversion, and translocation. *12.10*

Self-Quiz *(Answers in Appendix IV)*

1. _____ segregate during _____ .
   a. Homologues; mitosis
   b. Genes on one chromosome; meiosis
   c. Homologues; meiosis
   d. Genes on one chromosome; mitosis

2. The probability of a crossover occurring between two genes on the same chromosome is _____ .
   a. unrelated to the distance between them
   b. increased if they are closer together on the chromosome
   c. increased if they are farther apart on the chromosome
   d. impossible

3. Chromosome structure can be altered by _____ .
   a. deletions   d. translocations
   b. duplications   e. all of the above
   c. inversions

4. Nondisjunction can be caused by _____ .
   a. crossing over in mitosis
   b. segregation in meiosis
   c. failure of chromosomes to separate during meiosis
   d. multiple independent assortments

Figure 12.28
Summary of the results of hybrid crosses when gene linkage is an example of complete and when crossing over affects the outcome.

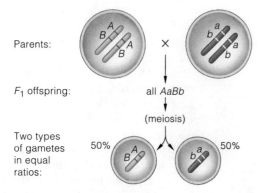

Parents: ×

F_1 offspring: all *AaBb*

(meiosis)

Two types of gametes in equal ratios: 50% 50%

Complete gene linkage (no crossovers); half the gametes have on parental genotype and half have the other.

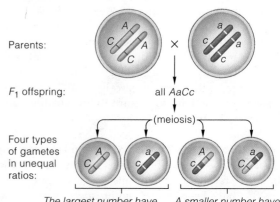

Parents: ×

F_1 offspring: all *AaCc*

(meiosis)

Four types of gametes in unequal ratios:

The largest number have the parental genotype

A smaller number have recombinant genotypes

5. A gamete affected by nondisjunction would have _____ .
   a. a change from the normal chromosome number
   b. one extra or one missing chromosome
   c. the potential for a genetic disorder
   d. all of the above

6. Genetic disorders can be caused by _____ .
   a. altered chromosome number   c. mutation
   b. altered chromosome structure d. all of the above

7. Which of the following events contributes to phenotypic variation in a population?
   a. independent assortment
   b. crossing over
   c. changes in chromosome structure and number
   d. all of the above

8. Match the chromosome terms appropriately.
   ___c___ crossing over
   ___e___ deletion
   ___d___ nondisjunction
   ___b___ translocation
   ___a___ karyotype

   a. number and defining features of individual's metaphase chromosomes
   b. chromosome segment moves to a nonhomologous chromosome
   c. disrupts gene linkages at meiosis
   d. causes gametes to have abnormal chromosome numbers
   e. loss of a chromosome segment

Critical Thinking—Genetics Problems
(Answers in Appendix III)

1. Human females are XX and males are XY.
   a. Does a male inherit the X from his mother or father?
   b. With respect to X-linked genes, how many different types of gametes can a male produce?
   c. If a female is homozygous for an X-linked gene, how many types of gametes can she produce with respect to that gene?
   d. If a female is heterozygous for an X-linked gene, how many types of gametes can she produce with respect to that gene?

2. Suppose one allele of a Y-linked gene results in nonhairy ears in males. Another allele results in rather long hairs, a condition called *hairy pinnae* (Figure 12.29).
   a. Why would you *not* expect females to have hairy pinnae?
   b. Any son of a hairy-eared male will also be hairy-eared, but no daughter will be. Explain why.

3. Suppose you carry two linked genes with alleles *Aa* and *Bb*, respectively, as in Figure 12.30. If the crossover frequency between these two genes is zero, what genotypes would be expected among the gametes you produce, and with what frequencies?

4. In *D. melanogaster*, a gene governs wing length. A dominant allele at this gene locus results in the formation of long wings. A recessive allele results in the formation of vestigial (short) wings, as shown in Figure 12.31. You cross a homozygous dominant, long-winged fly with a homozygous recessive,

Figure 12.31 Vestigial-winged *D. melanogaster.*

vestigial-winged fly. You ask a technician to expose the fertilized eggs to a level of x-rays known to induce mutation and deletions. Later, when irradiated fertilized eggs develop into adults, most of the flies are heterozygous and have long wings. A few have vestigial wings. What might explain these results?

5. Say you have alleles for lefthandedness and straight hair on one chromosome and alleles for righthandedness and curly hair on the homologous chromosome. If the two loci for the alleles are very close together along the chromosome, how likely is it that crossing over will occur between them? If they are distant from each other, is a crossover more or less likely to occur?

6. Individuals affected by *Down syndrome* typically have an extra chromosome 21. In other words, their body cells contain a total of 47 chromosomes.
   a. At which stages of meiosis I and II could a mistake occur that could result in the altered chromosome number?
   b. In a few cases, 46 chromosomes are present. Included in this total are two normal-appearing chromosomes 21 and a longer-than-normal chromosome 14. Explain how these few individuals can have a normal chromosome number.

7. For the mugwump, a tree-dwelling mammal, the male is XX and the female is XY. However, sex-linked genes are found to have the same effect as in humans. For instance, a recessive X-linked allele *c* produces *red-green color blindness*. If a normal female mugwump mates with a phenotypically normal male mugwump whose mother was color blind, what is the probability that a son from that mating will be color blind? A daughter?

8. One type of childhood *muscular dystrophy* is a recessive, X-linked trait. A slowly progressing loss of muscle function leads to death, usually by age twenty or so. Unlike color blindness, this disorder is restricted to males. Suggest why.

Selected Key Terms

abortion *12.11*	homologous chromosome *12.1*
allele *12.1*	inversion *12.10*
amniocentesis *12.11*	in-vitro *12.2*
aneuploidy *12.9*	in-vitro fertilization *12.11*
autosome *12.1*	karyotype *12.1*
centrifugation *12.2*	linkage group *12.4*
chorionic villi sampling *12.11*	linkage mapping *12.5*
crossing over *12.1*	nondisjunction *12.9*
cytological marker *12.4*	pedigree *12.6*
deletion *12.10*	polyploidy *12.9*
double-blind study *12.9*	reciprocal cross *12.4*
duplication *12.10*	sex chromosome *12.1*
gene *12.1*	syndrome *CI*
genetic abnormality *12.6*	translocation *12.10*
genetic disease *12.6*	X chromosome *12.1*
genetic disorder *12.6*	X- or Y-linked gene *12.4*
genetic recombination *12.1*	Y chromosome *12.1*

Readings

Russell, P. 1992. *Genetics.* Third edition. New York: Harper Collins.

Travis, J. 30 November 1996. "Clue to Lou Gehrig's Disease Emerges." *Science News* 150: 340.

Web Site See *http://www.wadsworth.com/biology* for practice quiz questions, hypercontents, BioUpdates, and critical thinking. The Wadsworth Biology Resource Center provides a wealth of information fully organized and integrated by chapter.

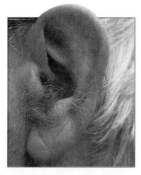

Figure 12.29 Hairy pinnae.

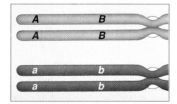

Figure 12.30 Two of your linked genes, on a pair of homologous chromosomes.

13 DNA STRUCTURE AND FUNCTION

Cardboard Atoms and Bent-Wire Bonds

One might have wondered, in the spring of 1868, why Johann Friedrich Miescher was collecting cells from the pus of open wounds and, later, from the sperm of a fish. Miescher, a physician, wanted to identify the chemical composition of the nucleus. These particular cells have very little cytoplasm, which makes it easier to isolate the nuclear material for analysis.

Miescher finally succeeded in isolating an organic compound with the properties of an acid. Unlike other substances in cells, it incorporated a notable amount of phosphorus. Miescher called the substance nuclein. He had discovered what came to be known many years later as **deoxyribonucleic acid**, or **DNA**.

The discovery did not cause even a ripple through the scientific community. At the time, no one really knew much about the physical basis of inheritance—that is, *which chemical substance encodes the instructions for reproducing parental traits in offspring*. Few even suspected that the cell nucleus might hold the answer. For a time, researchers generally believed that hereditary instructions had to be encoded in the structure of some unknown class of proteins. After all, heritable traits are spectacularly diverse. Surely the molecules encoding information about those traits were structurally diverse also. Proteins are put together from potentially limitless combinations of twenty different amino acid subunits, so they almost certainly could function as the sentences (genes) in each cell's book of inheritance.

By the early 1950s, however, the results of many ingenious experiments clearly indicated that DNA was the substance of inheritance. Moreover, in 1951, Linus Pauling did something that no one had done before. Through his training in biochemistry, a talent for

Figure 13.1 James Watson and Francis Crick posing in 1953 by their newly unveiled structural model of DNA.

model building, and a few great educated guesses, he deduced the three-dimensional structure of the protein collagen. Pauling's discovery was truly electrifying. If someone could pry open the secrets of proteins, then why not assume the same might be done for DNA? And once the structural details of the DNA molecule were understood, wouldn't they provide clues to its biological functions? *Who would go down in history as having discovered the very secrets of inheritance?*

Scientists around the world started scrambling after that ultimate prize. Among them were James Watson, a young post-doctoral student from Indiana University, and Francis Crick, an exuberant researcher working at Cambridge University. Exactly how could DNA, a molecule consisting of only four kinds of subunits, hold genetic information? Watson and Crick spent long hours arguing over everything they had read about the size, shape, and bonding requirements of the subunits of DNA. They fiddled with cardboard cutouts of the subunits. They even badgered chemists to help them identify any potential bonds they might have overlooked. Then they assembled models from bits of metal, held together with wire "bonds" bent at seemingly suitable angles.

In 1953, they put together a model that fit all of the pertinent biochemical rules and all of the facts about DNA they had gleaned from other sources (Figures 13.1 and 13.2). They had discovered the structure of DNA. And the breathtaking simplicity of that structure enabled them to solve another long-standing riddle—*how life can show unity at the molecular level and yet give rise to such spectacular diversity at the level of whole organisms.*

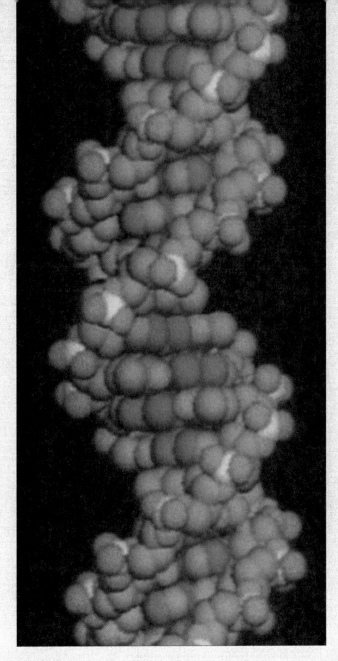

Figure 13.2 A more recent, computer-generated model of DNA. Although more sophisticated in appearance, it is much the same as the prototype Watson and Crick put together decades ago.

KEY CONCEPTS

1. In all living cells, DNA molecules are the storehouses of information about heritable traits.

2. In a DNA molecule, two strands of nucleotides twist together, like a spiral stairway. Each strand consists of four kinds of nucleotides, which are the same except for one component—a nitrogen-containing base. The four bases are adenine, guanine, thymine, and cytosine.

3. Great numbers of nucleotides are arranged one after another in each strand of the DNA molecule. In at least some regions, the order in which one kind of nucleotide follows another is unique for each species. Hereditary information is encoded in that particular sequence.

4. Hydrogen bonds connect the bases of one strand of the DNA molecule to bases of the other strand. As a rule, adenine pairs (hydrogen-bonds) with thymine, and guanine with cytosine.

5. Before a cell divides, its DNA is replicated with the assistance of enzymes and other proteins. Each double-stranded DNA molecule starts unwinding. A new, complementary strand is assembled bit by bit on the exposed bases of each parent strand, according to the base-pairing rule.

With this chapter, we turn to investigations that led to our current understanding of DNA. The story is more than a march through details of its structure and function. *It also is revealing of how ideas are generated in science.* On the one hand, having a shot at fame and fortune quickens the pulse of men and women in any profession, and scientists are no exception. On the other hand, science proceeds as a community effort, with individuals sharing not only what they can explain but also what they do not understand. Even when an experiment fails to produce the anticipated results, it might turn up information that others can use or lead to questions that others can answer. Unexpected results, too, might be clues to something important about the natural world.

DISCOVERY OF DNA FUNCTION

Early and Puzzling Clues

The year was 1928. Frederick Griffith, an army medical officer, was attempting to develop a vaccine against *Streptococcus pneumoniae*, a bacterium that causes a form of pneumonia. (When introduced into a person's body, vaccines can mobilize internal defenses against a real attack. Many are preparations of either killed or weakened bacterial cells.) Griffith never did develop a vaccine. But his work unexpectedly opened a door to the molecular world of heredity.

Griffith isolated and cultured two different strains of the bacterium. He noticed that colonies of one strain had a rough surface appearance, but those of the other strain appeared smooth. He designated the two strains *R* and *S* and used them in a series of four experiments:

1. Laboratory mice were injected with live R cells. As Figure 13.3 indicates, they did not develop pneumonia. *The R strain was harmless.*

2. Mice were injected with live S cells. The mice died. Blood samples taken from them teemed with live S cells. *The S strain was pathogenic* (disease-causing).

3. S cells were killed by exposure to high temperature. Mice injected with these cells did not die.

4. Live R cells were mixed with heat-killed S cells and injected into mice. The mice died—and blood samples from them teemed with *live* S cells!

What was going on in the fourth experiment? Maybe heat-killed S cells in the mixture weren't really dead. But if that were true, then mice injected with heat-killed S cells alone (experiment 3) would have died. Maybe harmless R cells in the mixture had mutated into a killer form. But if that were true, then mice injected with the R cells alone (experiment 1) would have died.

The simplest explanation was as follows: *Heat killed the S cells but did not destroy their hereditary material— including the part that specified "how to cause infection."* Somehow, that material had been transferred from dead S cells to living R cells, which put it to use.

Further experiments showed the harmless cells had indeed picked up information on causing infections and were permanently transformed into pathogens. After hundreds of generations, descendants of transformed bacterial cells were still infectious!

The unexpected results of Griffith's experiments intrigued Oswald Avery and his fellow biochemists. In time, they were even able to transform harmless bacterial cells with *extracts* of killed pathogens. Finally in 1944, after rigorous chemical analyses, they were able to report that the hereditary substance in their extracts probably was DNA—not proteins, as was then widely believed. To give experimental evidence of this conclusion, they reported that they had added certain protein-digesting enzymes to some extracts, but cells exposed to those extracts were transformed anyway. To other extracts they had added an enzyme that digests DNA but not proteins. Doing so blocked hereditary transformation.

Despite these impressive experimental results, most biochemists refused to give up on the proteins. Avery's findings, they said, probably applied only to bacteria.

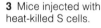
genetic material
viral coat
sheath
baseplate
tail fiber

Confirmation of DNA Function

By the early 1950s molecular detectives, including Max Delbrück, Alfred Hershey, Martha Chase, and Salvador Luria, were using viruses as experimental subjects. The viruses they had selected, called **bacteriophages**, infect *Escherichia coli* and other bacteria.

Viruses are as biochemically simple as you can get. Although they are not alive, they do contain hereditary information about building more new virus particles. At some point after they infect a host cell, viral enzymes take over a portion of the cell's metabolic machinery— which starts churning out substances that are necessary to construct new virus particles.

1 Mice injected with live cells of harmless strain R.

2 Mice injected with live cells of killer strain S.

3 Mice injected with heat-killed S cells.

4 Mice injected with live R cells *plus* heat-killed S cells.

Figure 13.3 Results of Griffith's experiments with a harmless and a pathogenic strain of *Streptococcus pneumoniae*, as described in the text above.

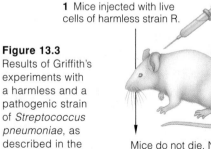

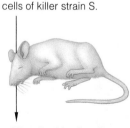

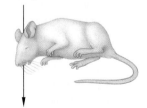

Mice do not die. No live R cells in their blood.

Mice die. Live S cells in their blood.

Mice do not die. No live S cells in their blood.

Mice die. Live S cells and live R cells in their blood.

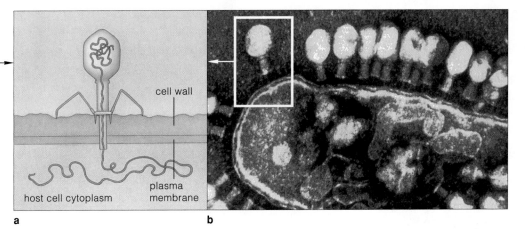

Figure 13.4 (**a**) *Far left*: Structural organization of a T4 bacteriophage. The diagram shows the genetic material of this type of virus being injected into the cytoplasm of a host cell. In this case, it is DNA (the *blue,* threadlike strand). (**b**) Electron micrograph of T4 virus particles infecting a bacterium (*Escherichia coli*) that has just become an unwilling host.

a   b

cell wall

host cell cytoplasm

plasma membrane

Figure 13.5 Two examples of the landmark experiments that pointed to DNA as the substance of heredity. In the 1940s, Alfred Hershey and his colleague, Martha Chase, were studying the biochemical basis of inheritance. They were aware that certain bacteriophages consisted of proteins and DNA. *Did the proteins, DNA, or both contain the viral genetic information?*

To find a possible answer, Hershey and Chase designed two experiments, based on two known biochemical facts. First, the proteins of bacteriophages incorporate sulfur (S) but not phosphorus (P). Second, their DNA incorporates phosphorus but not sulfur.

(**a**) In one experiment, some bacterial cells were grown on a culture medium that incorporated the radioisotope ^{35}S. Therefore, when the bacterial cells synthesized proteins, they would take up that radioisotope of sulfur—which would serve as a tracer. (Here you may wish to review Section 2.2.) After the cells became labeled with the tracer, bacteriophages were allowed to infect them. As the infection ran its course, viral proteins were synthesized inside the host cells. These proteins also became labeled with ^{35}S. And so did the new generation of virus particles.

Next, the labeled bacteriophages were allowed to infect a new batch of unlabeled bacteria that were suspended in a fluid culture medium. Afterward, Hershey and Chase whirred the fluid in a kitchen blender. Whirring dislodged the viral protein coats from the cells. The particles became suspended in the fluid medium. Analysis revealed the presence of labeled protein in the fluid—but no protein *inside* the bacterial cells.

(**b**) For the second experiment, Hershey and Chase cultured more bacterial cells. The phosphorus that was available to them for synthesizing DNA included the radioisotope ^{32}P. Later, bacteriophages were allowed to infect the cells.

As predicted, viral DNA synthesized inside the infected cells became labeled, as did the new generation of virus particles. Next, the labeled particles were allowed to infect bacteria that were suspended in a fluid medium. Then they were dislodged from the host cells. Analysis revealed that the labeled viral DNA was not in the fluid. DNA remained *inside* the host cells, where its hereditary information had to be put to use to make more virus particles. Here was strong evidence that DNA is the genetic material of this type of virus.

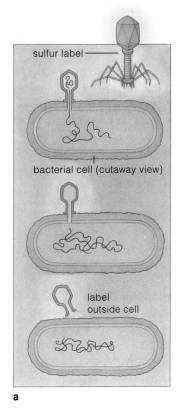

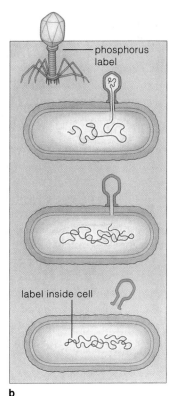

sulfur label

bacterial cell (cutaway view)

label outside cell

phosphorus label

label inside cell

a   b

By 1952, researchers knew that some bacteriophages consist only of DNA and a protein coat. Also, electron micrographs revealed that the main part of the viruses remains *outside* the cells they are infecting (Figure 13.4). Possibly, such viruses were injecting genetic material alone *into* host cells. If that were true, was the material DNA, protein, or both? Through many experiments, researchers accumulated strong evidence that DNA, not proteins, serves as the molecule of inheritance. Figure 13.5 describes two of these landmark experiments.

Information for producing the heritable traits of single-celled and multicelled organisms is encoded in DNA.

Components of DNA

Long before the bacteriophage studies were under way, biochemists knew that DNA contains only four types of nucleotides that are the building blocks of nucleic acids. A **nucleotide** consists of a five-carbon sugar (which is deoxyribose in DNA), a phosphate group, and one of the following nitrogen-containing bases:

adenine	guanine	thymine	cytosine
(A)	(G)	(T)	(C)

As you can see from Figure 13.6, all four types of nucleotides in DNA have their component parts joined together in much the same way. However, T and C are pyrimidines, which are single-ring structures. A and G are purines, which are larger, bulkier molecules; they have double-ring structures.

By 1949, Erwin Chargaff, a biochemist, had shared two crucial insights into the composition of DNA with the scientific community. First, the amount of adenine relative to guanine differs from one species to the next. Second, the amount of adenine in DNA always equals that of thymine, and the amount of guanine always equals that of cytosine. We may show this as:

$$A = T \quad \text{and} \quad G = C$$

The relative proportions of the four kinds of nucleotides were tantalizing clues. In some way, those proportions almost certainly were related to the arrangement of the two chains of nucleotides in a DNA molecule.

The first convincing evidence of that arrangement came from Maurice Wilkins's laboratory in England. One of Wilkins's colleagues, Rosalind Franklin, had obtained especially good **x-ray diffraction images** of DNA fibers. By this process, a beam of x-rays is directed at a molecule, which scatters the beam in patterns that can be captured on film. The pattern itself consists only of dots and streaks; in itself, it doesn't reveal molecular structure. However, the images of those patterns can be used to calculate the positions of the molecule's atoms.

ADENINE (A) — double-ring structure

GUANINE (G) — double-ring structure

THYMINE (T) — single-ring structure

CYTOSINE (C) — single-ring structure

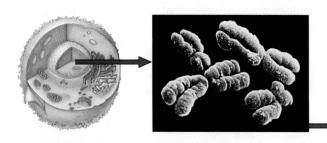

All chromosomes contain DNA. What does DNA contain? Four kinds of nucleotides. A nucleotide has a five-carbon sugar (shaded *red*), which has a phosphate group attached to the fifth carbon of its carbon ring structure. It also has one of four kinds of nitrogen-containing bases (*blue*) attached to its first carbon atom. The four nucleotides differ only in *which* base is attached to that carbon atom.

Figure 13.6 The four kinds of nucleotide subunits of DNA. Small numerals on the structural formulas identify the carbon atoms to which other parts of the molecule are attached.

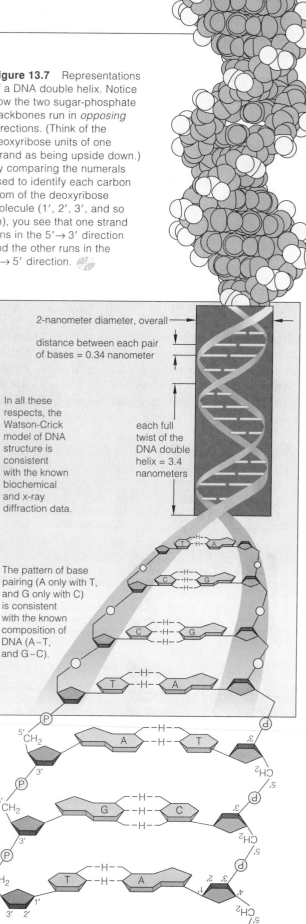

Figure 13.7 Representations of a DNA double helix. Notice how the two sugar-phosphate backbones run in *opposing* directions. (Think of the deoxyribose units of one strand as being upside down.) By comparing the numerals used to identify each carbon atom of the deoxyribose molecule (1', 2', 3', and so on), you see that one strand runs in the 5'→3' direction and the other runs in the 3'→5' direction.

2-nanometer diameter, overall

distance between each pair of bases = 0.34 nanometer

In all these respects, the Watson-Crick model of DNA structure is consistent with the known biochemical and x-ray diffraction data.

each full twist of the DNA double helix = 3.4 nanometers

The pattern of base pairing (A only with T, and G only with C) is consistent with the known composition of DNA (A–T, and G–C).

DNA does not readily lend itself to x-ray diffraction. However, researchers could rapidly spin a suspension of DNA molecules, spool them onto a rod, and gently pull them into gossamer fibers, like cotton candy. If the atoms in DNA were arranged in a regular order, x-rays directed at a fiber should scatter in a regular pattern that could be captured on film.

Calculations based on Franklin's images strongly indicated that the DNA molecule had to be long and thin, with a 2-nanometer diameter. Some molecular configuration was being repeated every 0.34 nanometer along its length, and another one, every 3.4 nanometers.

Could the sequence of nucleotide bases be twisting, like a circular stairway? Certainly Pauling thought so. After all, he discovered the helical shape of collagen. He and everybody else—including Wilkins, Watson, and Crick—were thinking "helix." Watson later wrote, "We thought, why not try it on DNA? We were worried that *Pauling* would say, why not try it on DNA? Certainly he was a very clever man. He was a hero of mine. But we beat him at his own game. I still can't figure out why."

Pauling, it turned out, made a big chemical mistake. His model had hydrogen bonds at phosphate groups holding DNA's structure together. That does happen in highly acidic solutions. It doesn't happen in cells.

Patterns of Base Pairing

As Watson and Crick perceived, DNA consists of *two* strands of nucleotides, held together at their bases by hydrogen bonds. The bonds form when the two strands run in opposing directions and twist together into a double helix (Figure 13.7). Two kinds of base pairings form along the length of the molecule: A—T and G—C. This bonding pattern permits variation in the order of bases in any given strand. For example, in even a tiny stretch of DNA from a rose, gorilla, human, or any other organism, the sequence might be:

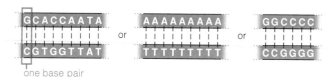

one base pair

In fact, even though all DNA molecules show the same bonding pattern, each species has unique base sequences in its DNA. *This molecular constancy and variation among species is the foundation for the unity and diversity of life.*

The pattern of base pairing between the two strands in DNA is constant for all species—A with T, and G with C. However, the DNA molecules of each species show unique differences in the sequence of base pairs along their length.

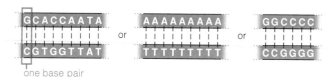

DNA REPLICATION AND REPAIR

How a DNA Molecule Gets Duplicated

The discovery of DNA structure was a turning point in studies of inheritance. Until then, no one could explain **DNA replication**, or how the molecule of inheritance is duplicated before the cell divides. The Watson-Crick model suggested at once how this might be done.

Enzymes can easily break hydrogen bonds between the two nucleotide strands of a DNA molecule. When they act on the molecule, one strand unwinds from the other, thereby exposing some nucleotide bases. Cells have stockpiles of free nucleotides, and these pair with exposed bases. Each parent strand remains intact, and a companion strand is assembled on each one according to this base-pairing rule: A to T, and G to C. As soon as a stretch of a new, partner strand forms on a stretch of the parent strand, the two twist into a double helix.

Because the parent strand is conserved, each "new" DNA molecule is really half old, half new (Figure 13.8). That is why biologists sometimes refer to this process as *semiconservative* replication.

DNA replication requires a large team of molecular workers. For instance, one kind of enzyme unwinds the two nucleotide strands, as in Figure 13.9. At the same time, many other proteins bind to unwound portions and hold them apart while replication proceeds. **DNA polymerases** are key players. These enzymes attach free nucleotides to a growing strand. Other enzymes, the **DNA ligases**, seal new short stretches of nucleotides into a continuous strand. Some of these enzymes have uses in recombinant DNA technology (Section 16.1).

Monitoring and Fixing the DNA

DNA polymerases, DNA ligases, and other enzymes also engage in **DNA repair**. By this process, enzymes excise and repair altered parts of the base sequence in one strand of a double helix. DNA polymerases "read" the complementary sequence on the other strand. With the aid of other repair enzymes, they restore the original sequence. Section 13.4 gives you an idea of what can happen if the excision-repair function is impaired.

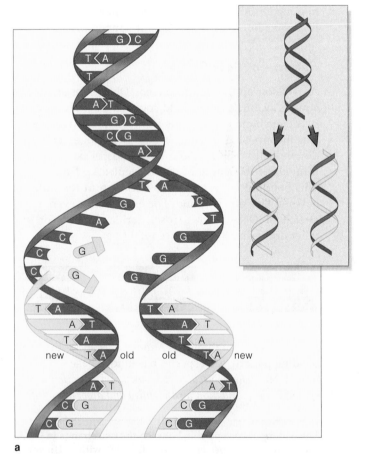

a

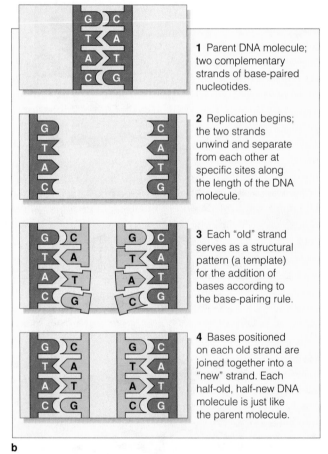

1 Parent DNA molecule; two complementary strands of base-paired nucleotides.

2 Replication begins; the two strands unwind and separate from each other at specific sites along the length of the DNA molecule.

3 Each "old" strand serves as a structural pattern (a template) for the addition of bases according to the base-pairing rule.

4 Bases positioned on each old strand are joined together into a "new" strand. Each half-old, half-new DNA molecule is just like the parent molecule.

b

Figure 13.8 (**a**) Overview of the semiconservative nature of DNA replication. The original two-stranded DNA molecule is shown in *blue*. Each parent strand remains intact, and a new strand (*yellow*) is assembled on each one. (**b**) A closer look at the base additions.

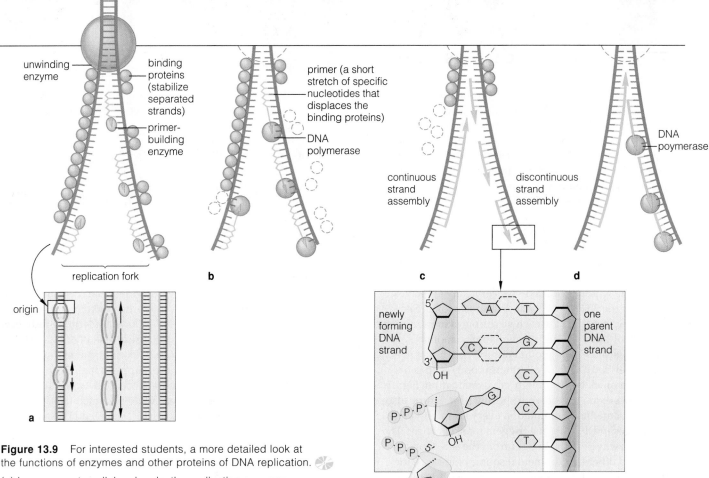

unwinding enzyme

binding proteins (stabilize separated strands)

primer-building enzyme

primer (a short stretch of specific nucleotides that displaces the binding proteins)

DNA polymerase

b

continuous strand assembly

discontinuous strand assembly

DNA poymerase

c

d

replication fork

origin

a

5'

3'
OH

A — T

C --- G

newly forming DNA strand

one parent DNA strand

C

C

T

P·P·P·

G

P·P·P· 5'

3'
OH

G
OH

Figure 13.9 For interested students, a more detailed look at the functions of enzymes and other proteins of DNA replication.

(**a**) In response to cellular signals, the replication enzymes become active at many short, specific base sequences called *origins*. Organized complexes of these enzymes as well as other proteins unwind the two strands of the parent DNA molecule, prevent them from rewinding, and assemble a new strand on each one. They do this only in *replication forks.* These are limited, V-shaped regions that advance in both directions away from an origin. Enzymes rewind the half-old, half-new molecules while they are being completed, behind each advancing replication fork.

(**b**) At each fork, some binding proteins are displaced when enzymes synthesize primers at intervals along both of the parent templates. DNA polymerases recognize the primers as "start" tags. They join nucleotide monomers together behind the primers. Exposed bases of a parent template dictate which kind of nucleotide will be added in sequence according to the base-pairing rule: A with T, and G with C.

(**c**) As discovered by Reiji Okazaki, strand assembly is a *continuous* process on one parent template. It is *discontinuous* on the other strand, where nucleotides must be assembled in short stretches only. Why? They can only be joined together in the 5'→3' direction, because this leaves one of the —OH groups of the growing sugar-phosphate backbone exposed. This exposed group is the only kind of site where nucleotide units can be joined together.

(**d**) Primers are removed. DNA polymerases fill in the gaps but cannot make the final connection between the stretches of nucleotides. DNA ligases must seal the tiny nicks that remain. The result is continuous strands on both parent templates.

Free nucleotides brought up for strand assembly also provide the energy that drives the replication process. Each has three phosphate groups attached. The action of DNA polymerase splits away two of these groups, and some of the released energy is used to attach the nucleotide to a growing strand.

Regarding the Chromosomal Proteins

Earlier, in Section 9.4, you read that eukaryotic DNA has a tremendous number of histones and other protein molecules bound tightly to it. Many of these proteins interact with the DNA and form the scaffolding for the condensed organization of the metaphase chromosome. You may well wonder: Wouldn't the proteins interfere with DNA replication and other functions? Apparently, some parts of the protein scaffold intervene *between* genes, not in regions with protein-building information. Do proteins organize the DNA into "domains" having distinct functions? Does the organization make it easier for enzymes to replicate the DNA, even to "read" the genes as the first step in protein synthesis? These are just two of the possibilities being probed by the new generation of molecular detectives.

DNA is replicated prior to cell division. Enzymes unwind its two strands. Each strand remains intact throughout the process—it is conserved—and enzymes assemble a new, complementary strand on each one.

Enzymes involved in replication also repair the DNA where base-pairing errors have crept into the nucleotide sequence.

WHEN DNA CAN'T BE FIXED

1992 was an unforgettable year for Laurie Campbell. She finally turned eighteen. And in that same year she just happened to notice a peculiar mole on her skin. This one was suspiciously black. It had an odd lumpiness about it, a ragged border, and an encrusted surface. Laurie quickly made an appointment with her family doctor, who just as quickly ordered a biopsy. The mole turned out to be a *malignant melanoma*—the deadliest form of skin cancer.

Laurie was lucky. She detected the cancer in its earliest stage, before it could spread through her body. Ever since, she consistently checks out the appearance of other moles that pepper her skin. She is painfully aware of having become a statistic—one of 500,000 people in the United States alone who develop skin cancer in any given year, and one of the 23,000 with malignant melanoma. She knows now that 7,500 die each year from skin cancer, and that 5,600 of them die from malignant melanoma.

Laurie is smart. She plotted out the position of every mole on her body. Once a month, this body map is her guide for a quick but thorough self-examination. Figure 13.10 shows examples of what she looks for. Laurie also schedules a medical examination every six months.

Changes in DNA are the triggers for skin cancer. The ultraviolet wavelengths in light from the sun, tanning lamps, and other environmental sources can cause the abnormal molecular changes. Among other things, the wavelengths can promote covalent bonding between two adjacent thymine bases in a nucleotide strand. In this way, the two nucleotides to which the bases belong are combined into an abnormal, bulky structure called a thymine dimer within the DNA.

Normally, a DNA repair mechanism gets rid of such bulky lesions. The mechanism requires at least seven gene products. Mutation in one or more of the required genes can skew the repair machinery. Thymine dimers can accumulate in skin cells of individuals with this type of mutation. The accumulation may trigger development of lesions, including skin cancer.

The risk of skin cancer is greater for some than others. At one extreme, a newborn destined to develop *xeroderma pigmentosum* literally faces a dim future. The individuals affected by this genetic disorder cannot be exposed to sunlight, even briefly, without risking disfiguring skin tumors and possible early death from cancer.

You are at risk if you have moles that are chronically irritated, as by shaving or abrasive clothing. You are at risk if your skin, including your lip surface, is chronically chapped, cracked, or sore. You are at risk if your family has a history of cancer or if you have undergone radiation therapy. And, like Laurie, you are at risk if you have pale skin and burn easily in the sun. Even on slightly overcast days, fair-skinned people must apply a high SPF (Sun Protection Factor) sunblock, especially between 10 A.M. and 2 P.M. They should wear wide-brimmed hats, long sleeves, and long pants or skirts. Damaged DNA and skin cancer are the reality, in spite of the ill-advised, socially promoted allure of a golden tan.

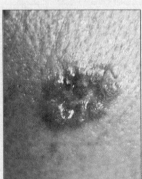

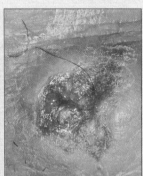

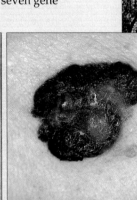

a Basal cell carcinoma **b** Squamous cell carcinoma **c** Malignant melanoma

Figure 13.10 Some examples of what can happen when repair enzymes cannot fix changes in the nucleotide sequence of DNA. (**a**) *Basal cell carcinoma*, the most common form of skin cancer. This slow-growing, raised lump may be uncolored, reddish-brown, or black. (**b**) *Squamous cell carcinoma*, the second most common skin cancer. These pink growths, firm to the touch, grow rapidly under the surface of skin exposed to the sun. (**c**) *Malignant melanoma*, which spreads most rapidly. The malignant cells form very dark, encrusted lumps. They may itch like an insect bite or bleed easily. (**d**) Laurie Campbell, avoiding the sun—and melanoma.

d

SUMMARY

1. The hereditary information of cells and multicelled organisms is encoded in DNA (deoxyribonucleic acid).

2. DNA consists of nucleotide subunits. Each of these has a five-carbon sugar (deoxyribose), one phosphate group, and one of four kinds of nitrogen-containing bases (adenine, thymine, guanine, or cytosine).

3. A DNA molecule consists of two nucleotide strands twisted together into a double helix. The bases of one strand pair (hydrogen-bond) with bases of the other.

4. The bases of the two strands in a DNA double helix pair in constant fashion. Adenine pairs with thymine (A to T), and guanine with cytosine (G to C). *Which* base pair follows the next (A—T, T—A, G—C, or C—G) varies along the length of the strands.

5. Overall, the DNA of one species includes a number of unique stretches of base pairs that set it apart from the DNA of all other species.

6. During DNA replication, enzymes unwind the two strands of a double helix and assemble a new strand of complementary sequence on each parent strand. Two double-stranded molecules result. One strand of each molecule is "old" (it is conserved); the other is "new."

7. Some of the enzymes involved in DNA replication also repair DNA where base-pairing errors have been introduced into the nucleotide sequence.

Review Questions

1. Name the three molecular parts of a nucleotide in DNA. Name the four different bases that occur in these nucleotides. *13.2*

2. What kind of bond joins two DNA strands in a double helix? Which nucleotide base-pairs with adenine? With guanine? *13.2*

3. Explain how DNA molecules can show constancy and variation from one species to the next. *13.2*

Self-Quiz (Answers in Appendix IV)

1. Which is *not* a nucleotide base in DNA?
   a. adenine   c. uracil   e. guanine
   b. thymine   d. cytosine

2. What are the base-pairing rules for DNA?
   a. A–G, T–C   b. A–C, T–G   c. A–U, C–G   d. A–T, G–C

3. A DNA strand having the sequence C–G–A–T–T–G would be complementary to the sequence _____ .
   a. C–G–A–T–T–G   c. T–A–G–C–C–T
   b. G–C–T–A–A–G   d. G–C–T–A–A–C

4. One species' DNA differs from others in its _____ .
   a. sugars   c. base sequence
   b. phosphate groups   d. all of the above

5. When DNA replication begins, _____ .
   a. the two DNA strands unwind from each other
   b. the two DNA strands condense for base transfers
   c. two DNA molecules bond
   d. old strands move to find new strands

6. DNA replication requires _____ .
   a. free nucleotides   c. many enzymes
   b. new hydrogen bonds   d. all of the above

7. Match the DNA terms appropriately.
   ____ DNA polymerase   a. two nucleotide strands
        and DNA ligase   twisted together
   ____ constancy in   b. A with T, G with C
        base pairing   c. hereditary material
   ____ replication   duplicated
   ____ DNA double helix   d. replication enzymes

Critical Thinking

1. Chargaff's data suggested that adenine pairs with thymine, and guanine pairs with cytosine. What other data available to Watson and Crick suggested that adenine-guanine and cytosine-thymine pairs normally do not form?

2. One of Matthew Meselson and Frank Stahl's experiments supported the semiconservative model of DNA replication. The researchers made "heavy" DNA by growing *Escherichia coli* in a medium enriched with ^{15}N, a heavy isotope of nitrogen. They prepared "light" DNA by growing *E. coli* in the presence of ^{14}N, the more common isotope. An available technique helped them identify which replicated molecules were heavy, light, or hybrid (one heavy strand, one light). Use two pencils of two different colors, one for heavy strands and one for light strands. Starting with a DNA molecule having two heavy strands, sketch the daughter molecules that would form after one replication in an ^{14}N-containing medium. Now sketch the four DNA molecules that would result if these daughter molecules were replicated a second time in the ^{14}N medium.

3. Mutations (permanent changes in base sequences of genes) are the original source of genetic variation. This variation is the raw material of evolution. Yet how can both statements be true, given that cells have efficient mechanisms to repair DNA before mutations can become established?

4. As indicated in Section 4.11, a pathogenic strain of *E. coli* has acquired an ability to produce a dangerous toxin that has caused medical problems and fatalities. This is especially true of young children who have ingested undercooked, contaminated beef. Develop a hypothesis to explain how a normally harmless bacterium such as *E. coli* can become a pathogen.

Selected Key Terms

adenine (A) *13.2*   DNA repair *13.3*
bacteriophage *13.1*   DNA replication *13.3*
cytosine (C) *13.2*   guanine (G) *13.2*
deoxyribonucleic acid (DNA) *CI*   nucleotide *13.2*
DNA ligase *13.3*   thymine (T) *13.2*
DNA polymerase *13.3*   x-ray diffraction image *13.2*

Readings

Watson, J. 1978. *The Double Helix.* New York: Atheneum. Highly personal view of scientists and their methods, interwoven into an account of how DNA structure was discovered.

Wolfe, S. 1995. *Introduction to Molecular and Cellular Biology.* Belmont, California: Wadsworth. Comprehensive, current, and accessible.

Web Site See *http://www.wadsworth.com/biology* for practice quiz questions, hypercontents, BioUpdates, and critical thinking. The Wadsworth Biology Resource Center provides a wealth of information fully organized and integrated by chapter.

14

FROM DNA TO PROTEINS

Beyond Byssus

Picture a mussel, of the sort shown in Figure 14.1. Hard-shelled but soft of body, it is using its muscular foot to probe a wave-scoured rock. At any moment, pounding waves can whack the mussel into the water, hurl it repeatedly against the rock with shell-shattering force, and so offer up a gooey lunch for gulls.

By chance, the mussel's foot comes across a crevice in the rock. The foot moves, broomlike, and sweeps the crevice clean. It presses down, forcing air out from underneath it, then arches up. The result is a vacuum-sealed chamber, rather like the one that forms when a plumber's rubber plunger is being squished down and up to unclog a drain. Into this vacuum chamber the mussel spews a fluid, consisting of keratin and other proteins, which bubbles into a sticky foam. Now, by curling its foot into a small tubular shape and pumping the foam through it, the mussel forms sticky threads about as wide as a human whisker. As a final touch, it varnishes the threads with another type of protein and thereby ends up with an adhesive called byssus, which anchors the mussel to the rock.

Byssus is the world's premier underwater adhesive. Nothing that humans have manufactured even comes close. (Sooner or later, water chemically degrades or deforms synthetic adhesives.) Byssus truly fascinates biochemists, dentists, and surgeons looking for better ways to

Figure 14.1 Mussels busily demonstrating the importance of proteins for survival. When mussels come across a suitable anchoring site, they use their muscular foot rather like a plumber's plunger to create a vacuum chamber. In this chamber they manufacture the world's best underwater adhesive from a mix of proteins. The adhesive anchors them to substrates in their wave-swept habitat.

do tissue grafts and to rejoin severed nerves. Genetic engineers insert mussel DNA into yeast cells, which reproduce in large numbers and serve as "factories" for translating mussel genes into useful quantities of proteins. This exciting work, like the mussel's own byssus building, starts with one of life's universal concepts: *Every protein is synthesized in accordance with instructions contained in DNA.*

You are about to trace the steps leading from DNA to protein. Many enzymes are players in this pathway, and so is another kind of nucleic acid besides DNA. The same steps produce *all* proteins, from mussel-inspired adhesives to the keratin in your hair and fingernails to the insect-digesting enzymes of a Venus flytrap.

Start out by thinking of each cell's DNA as a book of protein-building instructions. The alphabet used to create the book is simple enough: A, T, G, and C (for the nucleotide bases adenine, thymine, guanine, and cytosine). How do you get from this alphabet to a protein? The answer starts with DNA's structure.

DNA, recall, is a double-stranded molecule. Which kind of nucleotide base follows the next along the length of a strand—that is, the **base sequence**—differs from one kind of organism to the next. As you read in the preceding chapter, before a cell divides, its DNA is replicated and the two strands unwind entirely from each other. However, at other times in a cell's life, the two strands unwind only in certain regions to expose particular base sequences—genes. Most of those genes contain instructions for building proteins.

It takes two steps, **transcription** and **translation**, to carry out a gene's protein-building instructions. In eukaryotic cells, transcription proceeds in the nucleus. In this step, a selected base sequence in DNA serves as a structural pattern—as a template—for assembling a strand of **ribonucleic acid** (**RNA**) from the cell's pool of free nucleotides. Afterward, the RNA moves into the cytoplasm, where translation proceeds. In this second step, RNA directs the assembly of amino acids into polypeptide chains. The newly formed chains become folded into the three-dimensional shapes of proteins.

In short, DNA guides the synthesis of RNA, then RNA guides the synthesis of proteins:

$$\text{DNA} \xrightarrow{\textit{transcription}} \text{RNA} \xrightarrow{\textit{translation}} \text{PROTEIN}$$

The newly synthesized proteins will have structural and functional roles in cells. Some even will have roles in building more DNA, RNA, and proteins.

KEY CONCEPTS

1. Life cannot exist without enzymes and other proteins. Proteins consist of polypeptide chains, which consist of amino acids. The sequence of amino acids corresponds to a gene, which is a sequence of nucleotide bases in a DNA molecule.

2. The path leading from genes to proteins has two steps, called transcription and translation.

3. In transcription, the double-stranded DNA molecule is unwound at a gene region, then an RNA molecule is assembled on the exposed bases of one of the strands.

4. In translation, a certain type of RNA directs the linkage of one amino acid after another, in the sequence required to produce a specific kind of polypeptide chain.

5. With few exceptions, the genetic "code words" by which DNA instructions are translated into proteins are the same in all species of organisms.

6. A mutation is a permanent change in a gene's base sequence. Such changes are the original source of genetic variation in populations.

7. Mutations give rise to alterations in protein structure, protein function, or both. The alterations may lead to small or large differences in traits among the individuals of a population.

14.1   DISCOVERING THE CONNECTION BETWEEN GENES AND PROTEINS

GARROD'S HYPOTHESIS   During the early 1900s a physician, Archibald Garrod, was puzzling over certain illnesses. They appeared to be heritable, for they kept recurring in the same families. They also appeared to be metabolic disorders, for blood or urine samples from the affected patients contained abnormally high levels of a substance that was known to be produced at a certain step in a metabolic pathway.

Most likely, the enzyme that operated at the *next* step in the metabolic pathway was defective. Something was preventing it from chemically recognizing or interacting properly with the substance that is supposed to be its substrate. If Garrod's reasoning were correct, then the metabolic pathway would be blocked from that step onward, as Figure 14.2 indicates.

Garrod's hypothesis could explain why the molecules of a particular substance were accumulating in excess amounts in the body fluids of affected individuals. He suspected that the metabolic activities of his patients differed from unaffected individuals in one important

respect: They had inherited a single metabolic defect. Thus, Garrod concluded, specific "units" of inheritance (genes) must function through the synthesis of specific enzymes.

BEADLE, TATUM, AND A BREAD MOLD   Thirty-three years later two researchers, George Beadle and Edward Tatum, were experimenting with the red bread mold (*Neurospora crassa*). This fungus is a common spoiler of baked goods (Figure 14.3). But it also lends itself to genetic experiments and has become an organism of choice in the laboratory. *N. crassa* can be grown easily on an inexpensive culture medium that contains only sucrose, mineral salts, and biotin, which is one of the B vitamins. The fungal cells can synthesize all of the other nutrients they require, including other vitamins.

Suppose a fungal enzyme that takes part in a synthesis pathway is defective as a result of a gene mutation. The researchers suspected that this had happened in some of the strains of *N. crassa* that they were studying. One strain grew only when it was supplied with vitamin B_6, another with vitamin B_1, and so on.

Chemical analysis of cell extracts revealed a different defective enzyme in each mutant strain. In other words, *each inherited mutation corresponded to a defective enzyme.* Here was evidence favoring the "one-gene, one-enzyme" hypothesis.

CLUES FROM GEL ELECTROPHORESIS   The one-gene, one-enzyme hypothesis was refined as a result of investigations into the genetic basis of *sickle-cell anemia.* This heritable disorder arises from the presence of an abnormal version of a protein, hemoglobin, in the red blood cells of affected individuals. The abnormal hemoglobin is designated HbS instead of HbA (Section 11.5).

In 1949, the biochemists Linus Pauling and Harvey Itano subjected molecules of HbS and HbA to **gel electrophoresis**. This laboratory procedure uses an electric field to move molecules through a viscous gel and separate them according to their size, shape, and net surface charge. Often the gel is sandwiched between glass or plastic plates to form a viscous slab. The two ends of the slab are suspended in two salt solutions that are connected by electrodes to a power source (Figure 14.4). When voltage is applied to the apparatus, the molecules present in the gel migrate through the electric field according to their individual charge, and they move away from one another in the gel. Later on, the molecules can be pinpointed by staining the gel after a predetermined period of electrophoresis.

Pauling and Itano carefully layered a mixture of HbS and HbA molecules on the gel at the top of the slab. As the molecules gradually migrated down through the gel, they separated into distinct bands. The band that moved fastest carried the greatest

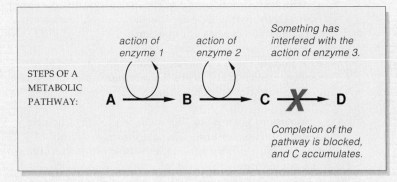

STEPS OF A METABOLIC PATHWAY:

action of enzyme 1   action of enzyme 2   Something has interfered with the action of enzyme 3.

A → B → C ✗ D

Completion of the pathway is blocked, and C accumulates.

Figure 14.2   How a defective enzyme can block completion of a metabolic pathway.

fungal colony on a tortilla, and isn't *that* appetizing

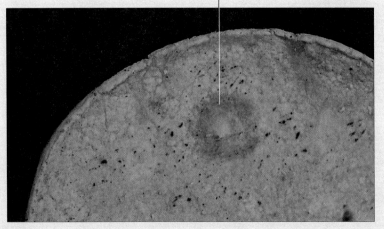

Figure 14.3   Colony of the red bread mold (*Neurospora crassa*) and other fungal species on a stale tortilla.

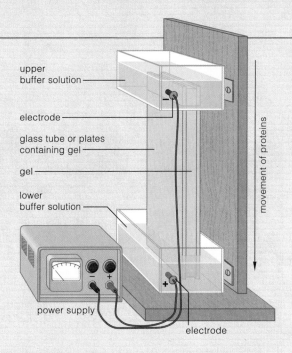

upper buffer solution

electrode

glass tube or plates containing gel

gel

lower buffer solution

power supply

electrode

movement of proteins

Figure 14.4 One type of apparatus that is used for gel electrophoresis studies.

beta chain

beta chain

alpha chain

alpha chain

a Arrangement of the four polypeptide chains of a hemoglobin molecule, and a closer view of one of the beta chains.

Figure 14.5 A single amino acid substitution that starts with a gene mutation and ends with the symptoms of sickle-cell anemia, which are described in Section 11.5.

surface charge, and this turned out to be composed of HbA molecules. HbS molecules moved more slowly.

ONE GENE, ONE POLYPEPTIDE Later, Vernon Ingram pinpointed the difference between HbS and HbA. Recall that hemoglobin contains four polypeptide chains (Figure 14.5). Two are designated alpha and the other two beta. An abnormal HbS chain arises from a gene mutation that affects protein synthesis. The mutation causes valine instead of glutamate to be added as the sixth amino acid of the beta chain (Figure 14.5c). Whereas glutamate carries an overall negative charge, valine has no net charge—and so HbS behaved differently in the electrophoresis studies.

As a result of the one mutation, HbS hemoglobin has a "sticky" (hydrophobic) patch. In blood capillaries, where oxygen concentrations are at their lowest, hemoglobin molecules interact at the sticky patches. They aggregate into rods and distort red blood cells, and the consequences adversely affect organs throughout the body (Section 11.5).

The discovery of the genetic difference between alpha and beta chains of hemoglobin suggested that *two* genes code for hemoglobin—one for each kind of polypeptide chain. More importantly, it further suggested that genes must code for proteins in general, not just for enzymes.

And so a more precise hypothesis emerged: *The amino acid sequences of polypeptide chains—the structural units of proteins—are encoded in genes.*

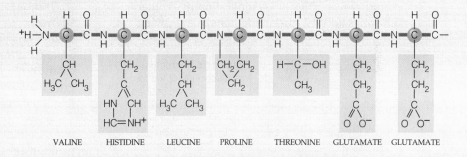

VALINE HISTIDINE LEUCINE PROLINE THREONINE GLUTAMATE GLUTAMATE

b The normal sequence of amino acids at the start of the beta chain of the HbA molecule

VALINE HISTIDINE LEUCINE PROLINE THREONINE VALINE GLUTAMATE

c The single amino acid substitution (coded *yellow*) that gives rise to the abnormal beta chain of the HbS molecule

TRANSCRIPTION OF DNA INTO RNA

The Three Classes of RNA

Before turning to the details of protein synthesis, let's clarify one point. The introduction to this chapter might have left you with the impression that protein synthesis requires only one class of RNA molecules. Actually, it requires three. Transcription of most genes produces **messenger RNA**, or **mRNA**—the only class of RNA that carries *protein-building* instructions. Transcription of some other genes produces **ribosomal RNA**, or **rRNA**, a major component of ribosomes. Ribosomes, recall, are structural units upon which polypeptide chains can be assembled. Transcription of still other genes produces **transfer RNA**, or **tRNA**. This delivers amino acids one by one to a ribosome in the order specified by mRNA.

How RNA Is Assembled

An RNA molecule is almost but not quite like a single strand of DNA. RNA, too, consists of only four types of nucleotides. Each nucleotide has a five-carbon sugar ribose (not DNA's deoxyribose), a phosphate group, and a base. Three of RNA's bases—adenine, cytosine, and guanine—are the same as in DNA. However, the fourth type of base is **uracil**, not thymine (Figure 14.6). Like thymine, uracil can pair with adenine. This means a new RNA strand can be put together on a DNA region according to base-pairing rules (Figure 14.7).

Transcription resembles DNA replication in another respect. Enzymes add the nucleotide units to a growing

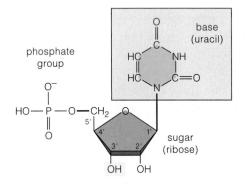

Figure 14.6 Structural formula for one of the four types of RNA nucleotides. The three others have a different base (adenine, guanine, or cytosine instead of uracil, shown here). Compare Figure 13.6, which shows DNA's four nucleotides. Notice that the sugars of DNA and RNA differ at one group only (*yellow*).

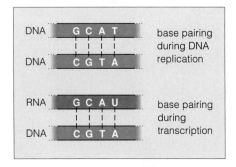

Figure 14.7 Comparison of how nucleotide bases pair with one another in DNA and in RNA.

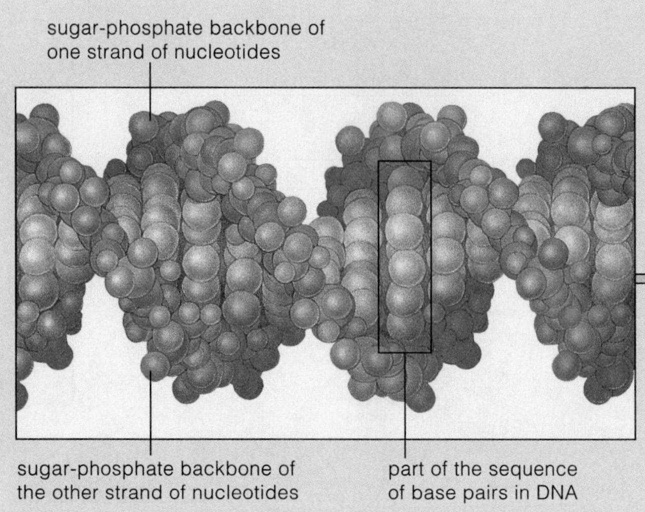

sugar-phosphate backbone of one strand of nucleotides

sugar-phosphate backbone of the other strand of nucleotides

part of the sequence of base pairs in DNA

a This sketch shows a gene region in part of a DNA double helix. In this region, the base sequence of one of the two nucleotide strands (not both) is about to be transcribed into an RNA molecule.

Figure 14.8 The process of gene transcription, by which an RNA molecule is assembled on a DNA template.

RNA strand one at a time, in the 5'→ 3' direction. (Here you may wish to refer to Figure 13.9c.)

Transcription *differs* from DNA replication in three key respects. First, only a selected stretch of one DNA strand, not the whole molecule, serves as the template. Second, different enzymes, called **RNA polymerases**, catalyze the addition of nucleotides to the 3' end of a growing strand. Third, transcription results in a single, free strand of RNA nucleotides.

Transcription starts at a **promoter**, a base sequence in DNA that signals the start of a gene. Proteins help position an RNA polymerase on the DNA so it binds to the promoter. The enzyme then moves along the DNA strand, joining nucleotides one after another (Figure 14.8). When it reaches a base sequence that serves as a stop signal, the RNA is released as a free transcript.

Finishing Touches on the mRNA Transcripts

In eukaryotic cells alone, a newly formed molecule of mRNA is unfinished; it must undergo modifications before its protein-building instructions can be put to use. Just as a dressmaker might snip off some threads or add some bows on a dress before it leaves the shop, so do eukaryotic cells tailor their pre-mRNA.

Enzymes immediately attach a cap to the 5' end of a pre-mRNA molecule. The cap is a nucleotide that has a methyl group and phosphate groups bonded with it.

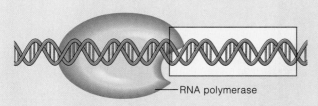

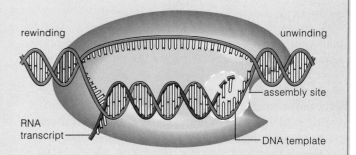

b A molecule of RNA polymerase binds with a promoter region in the DNA. It will recognize the base sequence positioned "downstream" from that site as a template for linking together the nucleotides adenine, cytosine, guanine, and uracil into a strand of RNA.

d All through transcription, the DNA double helix becomes unwound just in front of the RNA polymerase. Short lengths of the newly forming RNA strand temporarily wind up with the DNA template strand. Then they unwind from it, and the two strands of DNA wind up together again.

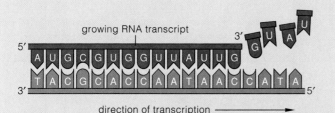

c During transcription, RNA nucleotides are base-paired, one after another, with exposed bases on the DNA template.

new RNA transcript

e At the end of the gene region, the last stretch of the RNA is unwound and released from the DNA template.

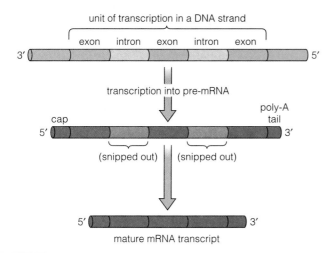

Figure 14.9 Transcription and modification of newly formed mRNA in the nucleus of eukaryotic cells. The cap simply is a nucleotide with functional groups attached. The poly-A tail is a string of adenine nucleotides.

Also, enzymes attach a tail of about 100 to 200 adenine-containing nucleotides to the 3' end of most pre-mRNA transcripts. Hence the name, "poly-A tail" (for multiple adenine units). The tail gets wound up with proteins. Later, in the cytoplasm, the cap will assist the binding of mRNA to a ribosome. Also, enzymes will gradually destroy the wound-up tail from the tip on back, then the mRNA. Such tails "pace" enzyme access to mRNA.

Apparently they help keep protein-building messages intact for as long as the cell requires them.

Besides these alterations, the mRNA message itself is modified. Most eukaryotic genes have one or more **introns**, base sequences that do not get translated into an amino acid sequence. The introns intervene between **exons**, the only parts of mRNA that get translated into protein. As Figure 14.9 shows, introns are transcribed right along with exons, but enzymes snip them out before the mRNA leaves the nucleus in mature form.

It could be that some introns are evolutionary junk, the leftovers of past mutations that led nowhere. Yet other introns are sites where instructions for building a particular protein can be snipped apart and spliced back together in various ways. The alternative splicing allows different cells in your body to use the same gene to make different versions of a pre-mRNA transcript, and therefore different versions of the resulting protein. We will return to this topic in the next chapter.

During gene transcription, a sequence of exposed bases in one of the two strands of a DNA molecule serves as the template for assembling a single strand of RNA. The assembly follows base-pairing rules (adenine only with uracil, cytosine only with guanine).

Before leaving the nucleus, each new mRNA transcript, or pre-mRNA, undergoes modification into final form.

The Genetic Code

Like a strand of DNA, an mRNA molecule is a linear sequence of nucleotides. What are the protein-building "words" encoded in that sequence? Gobind Khorana, Marshall Nirenberg, and other investigators came up with the answer. They deduced that RNA polymerases "read" nucleotide bases *three at a time*, as triplets. In an mRNA strand, such base triplets are called **codons**.

Figure 14.10 will give you an idea of how the order of different codons in an mRNA strand dictates the order in which particular amino acids will be added to a growing polypeptide chain.

Count the codons listed in Figure 14.11, and you see there are sixty-four kinds. Notice how most of the twenty kinds of amino acids correspond to more than one codon. Glutamate corresponds to the code words GAA *or* GAG, for example. Also notice how AUG has dual functions. It codes for the amino acid methionine, and it also is an initiation codon, the START signal for translating the mRNA transcript at a ribosome. That is,

a Base sequence of a gene region in DNA:

G	C	A	C	C	A	A	T	A	A	C	C	A	T	A

b Part of an mRNA strand, transcribed from the DNA:

C	G	U	G	G	U	U	A	U			U	A	U

c What the amino acid sequence will be when the mRNA is translated into a polypeptide chain:

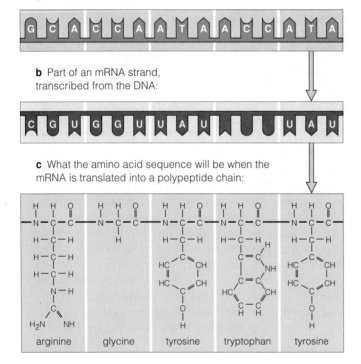

arginine   glycine   tyrosine   tryptophan   tyrosine

Figure 14.10 The steps from genes to proteins. (**a**) This diagram represents a region of a DNA double helix that was unwound during transcription. (**b**) The exposed bases on one DNA strand served as a template for assembling an mRNA strand. In the new mRNA transcript, every three nucleotide bases equaled one codon. Each codon calls for one amino acid in a polypeptide chain. (**c**) Referring to Figure 14.7, can you fill in the blank codon for tryptophan in the chain?

First Letter	Second Letter				Third Letter
	U	C	A	G	
U	phenylalanine	serine	tyrosine	cysteine	U
U	phenylalanine	serine	tyrosine	cysteine	C
U	leucine	serine	STOP	STOP	A
U	leucine	serine	STOP	tryptophan	G
C	leucine	proline	histidine	arginine	U
C	leucine	proline	histidine	arginine	C
C	leucine	proline	glutamine	arginine	A
C	leucine	proline	glutamine	arginine	G
A	isoleucine	threonine	asparagine	serine	U
A	isoleucine	threonine	asparagine	serine	C
A	isoleucine	threonine	lysine	arginine	A
A	methionine (or START)	threonine	lysine	arginine	G
G	valine	alanine	aspartate	glycine	U
G	valine	alanine	aspartate	glycine	C
G	valine	alanine	glutamate	glycine	A
G	valine	alanine	glutamate	glycine	G

Figure 14.11 The genetic code. The codons in mRNA are nucleotide bases, "read" in blocks of three. Sixty-one of the base triplets correspond to specific amino acids. Three others serve as signals that stop translation. The left column of the diagram shows the first of the three nucleotides in each codon in mRNA. The middle columns show the second nucleotide. The right column shows the third. Reading from left to right, for instance, the triplet U G G corresponds to tryptophan. Both U U U and U U C correspond to phenylalanine.

the "three-bases-at-a-time" selections start at the first AUG in the nucleotide sequence of the transcript. The codons UAA, UAG, UGA do not correspond to amino acids. They are STOP signals that prevent the further addition of amino acids to a new polypeptide chain.

The set of sixty-four different codons is the **genetic code**. It is the basis of protein synthesis in all organisms.

Roles of tRNA and rRNA

In a cell's cytoplasm are pools of free amino acids and free tRNA molecules. The tRNAs each have a molecular "hook," an attachment site for amino acids. They also have an **anticodon**, a nucleotide triplet that is able to

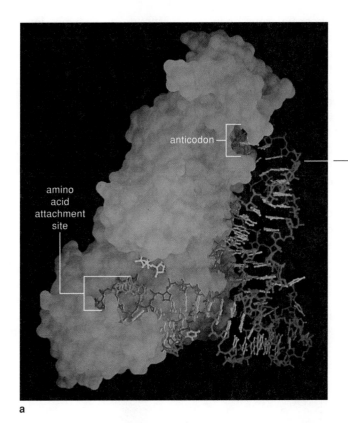

a

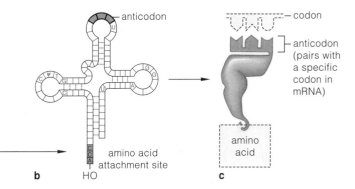

b   HO

c

anticodon

amino acid
attachment site

codon

anticodon
(pairs with
a specific
codon in
mRNA)

amino
acid

amino
acid
attachment
site

anticodon

Figure 14.12 (**a**) Computer-generated, three-dimensional model of one type of tRNA molecule. The tRNA (*reddish brown*) is shown attached to a bacterial enzyme (*green*), along with an ATP molecule (*gold*). This particular enzyme catalyzes the attachment of amino acids to tRNAs.

(**b**) Structural features common to all tRNAs. (**c**) Simplified model of tRNA that you will come across in illustrations to follow. The "hook" that is sketched at one end is the site to which a specific amino acid can become attached.

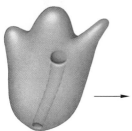

One small ribosomal subunit, with a platform beneath the surface of the site shown by the *red* arrow.

+

One large ribosomal subunit, with a tunnel through part of its interior.

In an intact ribosome, the platform and tunnel are aligned, as indicated by this side view.

Figure 14.13 Model of eukaryotic ribosomes. Polypeptide chains are assembled on the platform of the small ribosomal subunit. Newly forming chains may move through the tunnel of the large ribosomal subunit.

base-pair with codons (Figure 14.12). When tRNAs bind to the codons, they position their attached amino acids automatically, in the order specified by mRNA.

A cell has sixty-four kinds of codons, but it is able to utilize fewer kinds of tRNAs. How do they match up? According to base-pairing rules, adenine must pair with uracil, and cytosine with guanine. However, for codon-anticodon interactions, the rules loosen up at the third base. As an example, CCU, CCC, CCA, and CCG all specify proline. All three can pair with the tRNA that carries proline. Such freedom in the codon-anticodon pairing at a third base is known as the "wobble effect."

Even before anticodons interact with the codons of an mRNA strand, that strand must bind to specific sites on the surface of ribosomes. As shown in Figure 14.13, each ribosome has two subunits. These are assembled inside the nucleus from rRNA and protein components, some of which show enzyme activity. At some point,

the subunits are shipped separately to the cytoplasm. There they will combine as functional units only during translation of the message encoded in mRNA.

The nucleotide sequence of both DNA and mRNA encodes protein-building instructions. The genetic code is a set of sixty-four base triplets (nucleotide bases, read in blocks of three). A codon is a base triplet in mRNA.

Different combinations of codons specify the amino acid sequence of different polypeptide chains, start to finish.

mRNAs are the only molecules that carry protein-building instructions from DNA to the cytoplasm.

tRNAs deliver amino acids to ribosomes, where they base-pair with codons in the order specified by mRNA. Their action translates mRNA into a sequence of amino acids.

rRNAs are components of ribosomes, the structures upon which amino acids are assembled into polypeptide chains.

Translation of the mRNA proceeds in the cytoplasm. It has three stages: initiation, elongation, and termination.

In *initiation*, a tRNA that can start transcription and an mRNA transcript are both loaded onto a ribosome. First, the initiator tRNA binds with the small ribosomal subunit. Its anticodon can base-pair with AUG, the start codon in the transcript. The AUG also binds with the small subunit. Second, a large ribosomal subunit binds with the small subunit and thereby forms the initiation complex (Figure 14.14*a*). The next stage can begin.

In *elongation*, a new polypeptide chain forms as the mRNA passes between the ribosomal subunits, like a thread being moved through the eye of a needle. Again,

ribosomal components that function as enzymes join amino acids in the sequence dictated by the codons of mRNA. As you can see from Figure 14.14*b*, they catalyze the formation of a peptide bond between the growing polypeptide chain and each new amino acid delivered to the intact ribosome. (Here you may wish to refer to the diagram of peptide bond formation in Section 3.6.)

In *termination*, a stop codon is reached and there is no tRNA with a corresponding anticodon. Now release factors bind to the A site and trigger enzyme action that detaches the mRNA and the chain from the ribosome (Figure 14.14*c*). As Section 4.5 indicates, the detached chain may simply join the pool of free proteins in the cytoplasm. Or even before it is finished, the chain may enter the cytomembrane system, starting with the inner

Figure 14.14   Translation, the second step of protein synthesis.

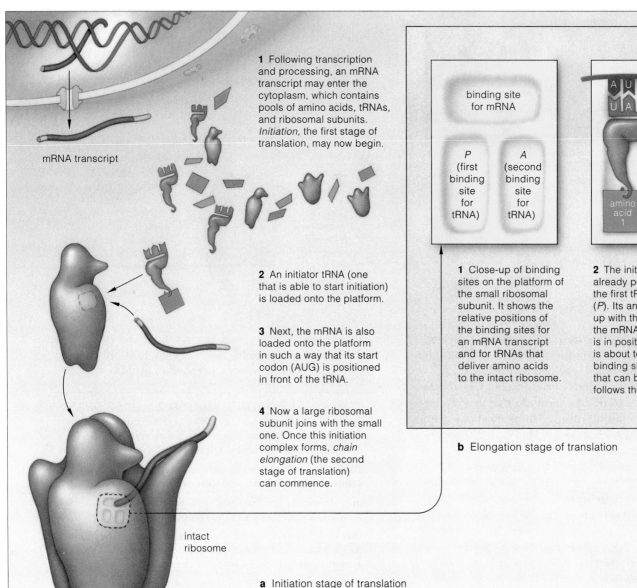

mRNA transcript

1 Following transcription and processing, an mRNA transcript may enter the cytoplasm, which contains pools of amino acids, tRNAs, and ribosomal subunits. *Initiation*, the first stage of translation, may now begin.

2 An initiator tRNA (one that is able to start initiation) is loaded onto the platform.

3 Next, the mRNA is also loaded onto the platform in such a way that its start codon (AUG) is positioned in front of the tRNA.

4 Now a large ribosomal subunit joins with the small one. Once this initiation complex forms, *chain elongation* (the second stage of translation) can commence.

intact ribosome

a   Initiation stage of translation

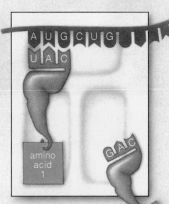

binding site for mRNA

P (first binding site for tRNA)   *A* (second binding site for tRNA)

1 Close-up of binding sites on the platform of the small ribosomal subunit. It shows the relative positions of the binding sites for an mRNA transcript and for tRNAs that deliver amino acids to the intact ribosome.

amino acid 1

2 The initiator tRNA is already positioned in the first tRNA binding site (*P*). Its anticodon matches up with the start codon (AUG) of the mRNA strand, which already is in position also. Another tRNA is about to move into the second binding site (*A*). This tRNA is one that can bind with the codon that follows the start signal.

b   Elongation stage of translation

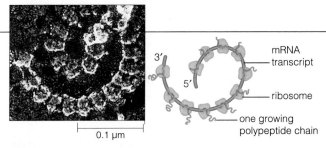

mRNA
transcript

3′
5′

ribosome

one growing
polypeptide chain

0.1 µm

Figure 14.15 From a eukaryotic cell, a polysome, or numerous ribosomes simultaneously translating the same mRNA molecule.

spaces of rough ER. Many of the newly formed chains take on final form in the system before they are shipped to their ultimate destinations inside or outside the cell.

Often, numerous ribosomes and tRNAs translate the same mRNA transcript simultaneously. The transcript threads through all the ribosomes, which are arranged one after another in assembly-line fashion, as in Figure

14.15. Such close arrays of ribosomes on the transcript are often called **polysomes**. Their presence indicates that a cell is rapidly producing a number of copies of a polypeptide chain from the same mRNA transcript.

Translation is initiated through the convergence of a small ribosomal subunit, an initiator tRNA, an mRNA transcript, and then a large ribosomal subunit.

More tRNAs deliver amino acids to the ribosome in the order dictated by the sequence of mRNA codons, to which the tRNA anticodons base-pair. A polypeptide chain grows as peptide bonds form between every two amino acids.

Translation is over when a stop codon triggers events that cause the chain and the mRNA to detach from the ribosome.

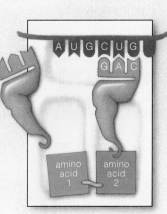
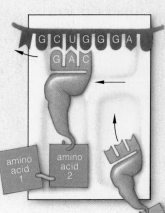

3 Through enzyme action, the bond between the initiator tRNA and the amino acid hooked to it is broken. At the same time, enzyme action catalyzes the formation of a peptide bond between the two amino acids. After these bonding events are completed, the initiator tRNA will be released from the ribosome.

4 Now the first amino acid is attached only to the second one—which is still hooked to the second tRNA. This tRNA will move into the *P* site on the ribosomal platform, sliding the mRNA with it by one codon. When it does so, the third codon will become aligned above the *A* site.

5 A third tRNA is about to move into the *A* site. Its anticodon is capable of base-pairing with the third codon of the mRNA transcript. Next, through enzyme action, a peptide bond will form between amino acids 2 and 3.

6 Steps 3 through 5 are repeated again and again. The polypeptide chain continues to grow until a stop codon is reached in the mRNA transcript. That is when termination, the last stage of protein synthesis, can begin, as shown in (**c**).

---

1 Once a stop codon is reached, the mRNA transcript is released from the ribosome.

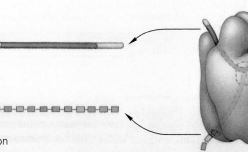

2 The newly formed polypeptide chain also is released.

c Chain termination stage of translation

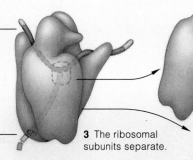
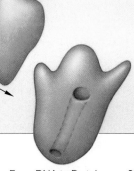

3 The ribosomal subunits separate.

Whenever a cell puts its genetic code into action, it is making precisely those proteins that it requires for its structure and functions. If something changes a gene's code words, the resulting protein may also change. If the protein is central to cell architecture or metabolism, we can expect the outcome to be an abnormal cell, such as those sickled red blood cells that make HbS instead of HbA hemoglobin.

Gene sequences do change. Sometimes one base gets substituted for another in the nucleotide sequence. At other times, an extra base is inserted into it or a base is lost. Such small-scale changes in the nucleotide sequence of a DNA molecule are **gene mutations**. There is some leeway here, for more than one codon often specifies the same amino acid. For instance, if a mutation changes UCU to UCC, it probably would not have dire effects, because both codons specify serine (Figure 14.11). More often, though, gene mutations give rise to proteins with skewed or blocked functions.

Common Types of Gene Mutations

Figures 14.16 and 14.17 show examples of two common types of gene mutations. In the first, adenine is wrongly paired with a cytosine unit in the DNA strand. Repair enzymes probably will detect the error, then remove and replace one of the bases. However, they may "fix" the mismatch by substituting the wrong base for that pair. The outcome of such a **base-pair substitution** may be the replacement of one amino acid with a different one during protein synthesis. This is what happened, recall, in individuals who carry the mutated gene that gives rise to sickle-cell anemia (Section 14.1).

In the second example, an extra base is inserted into a gene region of DNA. Remember, polymerases read a nucleotide sequence in blocks of three. This insertion shifts the three-at-a-time reading frame; hence the name "frameshift mutation." The gene now has a different message, and an abnormal protein will be synthesized.

Frameshift mutations fall within broader categories of gene mutation, called **insertions** and **deletions**. In such cases, one to several base pairs are inserted into a DNA molecule or deleted from it.

As a final example, Barbara McClintock discovered that mutations can result when **transposable elements** are on the move. These are DNA regions that move spontaneously from one location to another in the same DNA molecule or to a different one. Often they will inactivate the genes into which they become inserted. As Figure 14.18 indicates, the unpredictability of such jumps can cause interesting variations in phenotype.

Causes of Gene Mutations

Many gene mutations arise spontaneously while DNA is being replicated. This should not come as a surprise, given the rapid pace of replication and the huge pools of free nucleotides concentrated around the growing

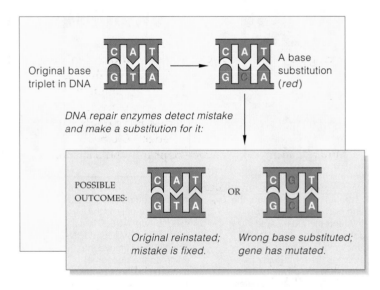

Figure 14.16 Example of a base-pair substitution.

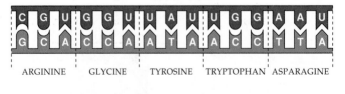

Resulting mRNA transcript of DNA:

Part of parental DNA template:

Resulting amino acid sequence: ARGININE   GLYCINE   TYROSINE   TRYPTOPHAN   ASPARAGINE

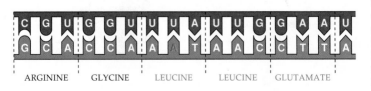

Now mRNA has altered message:

Parental template has one base insertion (red):

Resulting amino acid sequence: ARGININE   GLYCINE   LEUCINE   LEUCINE   GLUTAMATE

Figure 14.17 Example of an insertion, a type of mutation in which an extra base gets inserted into a gene region of DNA. This insertion has caused a *frameshift*; it has changed the reading frame for base triplets in the DNA and in the mRNA transcript of that region. As a result, the wrong amino acids will be called up when the mRNA transcript is translated into protein.

Figure 14.18 Barbara McClintock, who won a Nobel Prize for her insight that some genes can move from one site to another in DNA molecules, as transposable elements. In her hands is an ear of Indian corn (*Zea mays*). Its kernels sent her on the road to discovery.

In the kernels, all cells have the same pigment-coding genes. But some kernels are colorless or spottily colored. In the ancestor of the plant from which this ear of corn was plucked, a gene in a germ cell left its position in a DNA molecule, invaded another DNA molecule, and shut down a pigment gene. The plant inherited the mutation. As cell divisions proceeded in the growing plant, none of the mutated cell's descendants was able to synthesize pigment molecules. Each gave rise to colorless kernel tissue. Later, in some cells, the movable gene slipped out of the gene specifying pigment. And all the descendants of *those* cells produced pigment—and colored kernel tissue.

strands. Proofreading and repair enzymes detect most of them but, like most people, they are good but not perfect. A low number of mistakes do slip past these enzymes with predictable frequency.

Each gene has a characteristic **mutation rate**, which is the probability it will mutate spontaneously during a specified interval, such as each DNA replication cycle. (This is not the same as mutation *frequency*, the number of times a gene mutation has occurred in a population, as in 1 million gametes that produced 500,000 people.)

Mutation rates vary. For bacteriophages and bacteria, it is about 10^{-5} to 10^{-7} per generation, or once every 100,000 to 100 million replication cycles. For eukaryotes in general, it ranges between 10^{-4} and 10^{-6} per gene per generation. Consider the gene for hemoglobin. The rate at which it mutates to H^a, the recessive allele that causes hemophilia, is 3×10^{-5}. The rate at which the red-eye gene in *Drosophila melanogaster* mutates to the white-eye allele is 4×10^{-5}. It might be that differences in proofreading functions give rise to such variations in rates. Or maybe some mutations are silent, meaning we just can't detect them by ordinary genetic tests.

Not all mutations are spontaneous. Many result after exposure to mutagens, or mutation-causing agents in the environment. Ultraviolet radiation, especially the 260-nanometer wavelength in sunlight, is mutagenic. This is the wavelength DNA absorbs most strongly, and it often induces crosslinks to form between pyrimidine neighbors on the same DNA strand. Skin cancers are one outcome (Section 13.4). Other mutagens are gamma rays and x-rays. They ionize water and other molecules around the DNA, so free radicals form. These molecular fragments, which have an unpaired electron, can attack the structure of DNA. Such **ionizing radiation** causes base substitutions or breaks in one or both strands.

Natural and synthetic chemicals in the environment can accelerate the rate of spontaneous mutations. For example, **alkylating agents** might transfer methyl or ethyl groups to reactive sites on the bases or phosphate groups of DNA. At an alkylated site, the DNA becomes more susceptible to breaks and base-pair disruptions that invite mutation. Many cancer-causing agents, or **carcinogens**, operate by alkylating DNA.

The Proof Is in the Protein

When you think about the examples of mutation rates, you can deduce that spontaneous gene mutations are rare in terms of a human lifetime. If one arises in a somatic cell, any good or bad consequences will not endure, for it cannot be passed on to offspring. If the gene mutation arises in a germ cell or gamete, however, it may enter the evolutionary arena. The same is true of a mutation in an asexually reproducing organism or cell. In all such cases, the test is this: *A protein specified by a heritable mutation may have harmful, neutral, or beneficial effects on the ability of an individual to function in the prevailing environment.* As you will read in the next unit of the book, the outcomes of gene mutations can have powerful evolutionary consequences.

A gene mutation is an alteration in one to several bases in the nucleotide sequence of DNA.

Each gene has a spontaneous and characteristic mutation rate, which may be accelerated by exposure to harmful radiation and certain chemicals in the environment.

A protein specified by a mutated gene may have harmful, neutral, or beneficial effects on the ability of an individual to function in the prevailing environment.

SUMMARY

1. Cells, and multicelled organisms, cannot stay alive without enzymes and other proteins. A protein consists of one or more polypeptide chains, each of which is composed of a linear sequence of amino acids.

a. The amino acid sequence of a polypeptide chain corresponds to a gene region in a double-stranded DNA molecule. Each gene is a sequence of nucleotide bases in one of the two strands. The bases are adenine, thymine, guanine, and cytosine (A, T, G, and C).

b. For most genes, that sequence corresponds to a linear sequence of specific amino acids for a particular polypeptide chain. (Some genes specify tRNA or rRNA, not the mRNA that is translated into proteins.)

2. The path from genes to proteins has two steps, called transcription and translation:

$$\text{DNA} \xrightarrow{\textit{transcription}} \text{RNA} \xrightarrow{\textit{translation}} \text{PROTEIN}$$

a. During transcription, the double-stranded DNA is unwound at a gene region. Enzymes use its exposed bases as a template, or a structural pattern, to assemble a strand of ribonucleic acid (RNA) from the cell's pool of free nucleotides.

b. During translation, three different classes of RNAs interact in the synthesis of polypeptide chains, which later twist, fold, and often become modified into the final, three-dimensional shape of the protein.

c. Figure 14.19 is a visual summary of this flow of genetic information from DNA to proteins, as it occurs in eukaryotic cells. The DNA is transcribed into RNA in the nucleus, but RNA is translated in the cytoplasm. Prokaryotic cells (bacteria) lack a nucleus; transcription *and* translation proceed in their cytoplasm.

d. Understanding of the connection between genes and proteins started with studies of mutations that affected particular enzymes known to catalyze steps in metabolic pathways. Comparisons between normal and abnormal proteins, hemoglobin especially, led to the hypothesis that the amino acid sequence of polypeptide chains is encoded in genes.

3. Here are the key points concerning transcription:

a. When RNA is transcribed from exposed bases of DNA, base-pairing rules govern its assembly. Guanine pairs with cytosine, as in DNA replication, but uracil (not thymine) pairs with adenine in RNA:

DNA:	thymine	adenine	guanine	cytosine
RNA:	adenine	**uracil**	cytosine	guanine

b. Different gene regions in DNA serve as templates for assembling different RNA molecules.

c. Messenger RNA (mRNA) is the only class of RNA that has protein-building instructions.

d. Ribosomal RNA (rRNA) becomes a component of ribosomes, the physical units upon which polypeptide chains will be assembled.

e. Transfer RNA (tRNA) is the vehicle of translation; it will latch onto free amino acids in the cytoplasm and delivers them to ribosomes. It will do so in a sequence that corresponds to the sequential message of mRNA.

f. RNA transcripts of eukaryotic cells are processed into final form before being shipped from the nucleus. For example, mRNA's noncoding portions (introns) are excised, and its coding portions (exons) are spliced together. Only mature mRNA transcripts get translated.

4. Here are the key points concerning translation:

a. mRNA interacts with tRNAs and ribosomes so that amino acids get linked in the sequence required to produce a specific kind of polypeptide chain.

b. Translation is based on the genetic code, a set of sixty-four base triplets. A triplet is a series of nucleotide bases that ribosomal proteins "read" in blocks of three.

c. In mRNA, a base triplet is a codon. An anticodon is a complementary triplet in tRNA. Some combination of codons specifies what the amino acid sequence of a polypeptide chain will be, start to finish.

5. Translation proceeds through three stages:

a. Initiation. A small ribosomal subunit, an initiator tRNA, and an mRNA transcript converge. The small subunit then binds with a large ribosomal subunit.

b. Chain elongation. tRNAs deliver amino acids to the ribosome. Their anticodons base-pair with codons in mRNA. The amino acids become linked by peptide bonds to form a new polypeptide chain.

c. Chain termination. A stop codon in mRNA causes the chain and the mRNA to detach from the ribosome.

6. Gene mutations are potentially heritable, small-scale alterations in the nucleotide sequence of DNA.

a. Many gene mutations arise spontaneously during DNA replication. Others arise after the DNA is exposed to mutagens, such as ultraviolet radiation and other mutation-causing agents in the environment.

b. A base-pair substitution (replacement of one base pair by a different one) affects a single codon, but one amino acid substitution may alter protein function. Insertions of one or more bases into a gene or deletions from it can shift the reading frame to specify different amino acids. Transposable (movable) elements typically inactivate genes into which they become inserted.

c. Each gene has a characteristic mutation rate: the probability that it will spontaneously mutate in some specified time interval, such as a DNA replication cycle.

7. A protein specified by a mutated gene might have harmful, neutral, or beneficial effects on the individual. The outcome depends on prevailing conditions in the internal and external environments. Somatic mutations affect individuals only. Mutations in reproductive cells are heritable and can enter the evolutionary arena.

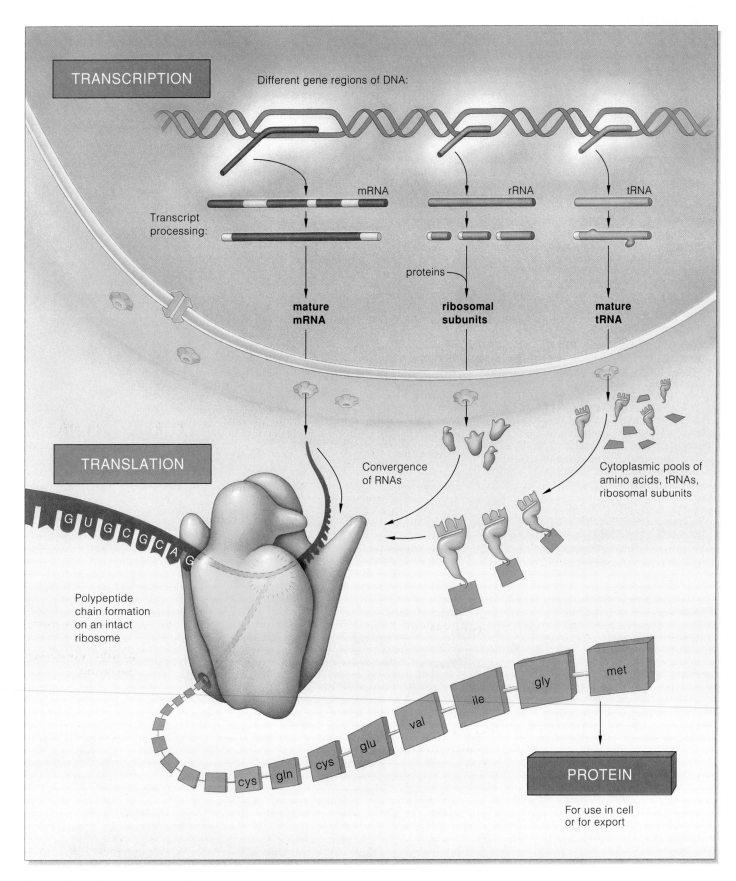

Figure 14.19 Visual summary of the steps of protein synthesis—transcription and translation—as it proceeds in all eukaryotic cells.

Review Questions

1. Are the polypeptide chains of proteins assembled on DNA? If so, state how. If not, state how they are assembled, and on which molecules. *CI, 14.2*

2. Define gene transcription and translation, the two stages of events by which proteins are synthesized. Both stages proceed in the cytoplasm of prokaryotic cells. Where does each stage proceed in eukaryotic cells? *CI*

3. Briefly state how gel electrophoresis, a common laboratory procedure, works. How did it yield a clue that small differences in normal and abnormal versions of the same protein may lead to big differences in how the proteins function? *14.1*

4. Name the three classes of RNA and briefly describe their functions. *14.2, 14.3*

5. In what key respect does the sequence of nucleotide bases in RNA differ from those in DNA? *14.2*

6. How does the process of gene transcription resemble DNA replication? How does it differ? *14.2*

7. The pre-mRNA transcripts of eukaryotic cells contain introns and exons. Are the introns or exons snipped out before the transcript leaves the nucleus? *14.2*

8. Distinguish between codon and anticodon. *14.3*

9. Cells use the set of sixty-four codons in the genetic code to build polypeptide chains from twenty kinds of amino acids. Do different codons specify the same amino acid? If so, in what respect do they differ? *14.3*

10. Name the three stages of translation and briefly describe the key events of each one. *14.4*

11. Review Figure 14.19. Then, on your own, fill in the blanks of the diagram at the right.

12. Define gene mutation. Do all mutations arise spontaneously? Do environmental agents serve as the trigger for change in each case? *14.5*

13. Define and state the possible outcomes of the following types of mutation: a base-pair substitution, a base insertion, and an insertion of a transposable element at a new location in the DNA. *14.5*

14. Define and explain the difference between mutation rate and mutation frequency. What determines whether an altered product of a mutation will have helpful, neutral, or harmful effects? *14.5*

Self-Quiz (*Answers in Appendix IV*)

1. Garrod, then Beadle and Tatum, deduced that the individual, heritable mutations they investigated corresponded to _____ .
   a. metabolic disorders
   b. defective enzymes
   c. defective tRNAs
   d. both a and b

2. DNA contains many different genes that are transcribed into different _____ .
   a. proteins
   b. mRNAs
   c. tRNAs and rRNAs
   d. all are correct

3. An RNA molecule is _____ .
   a. a double helix
   b. usually single-stranded
   c. always double-stranded
   d. usually double-stranded

4. An mRNA molecule is produced by _____ .
   a. replication
   b. duplication
   c. transcription
   d. translation

5. Each codon calls for a specific _____ .
   a. protein
   b. polypeptide
   c. amino acid
   d. carbohydrate

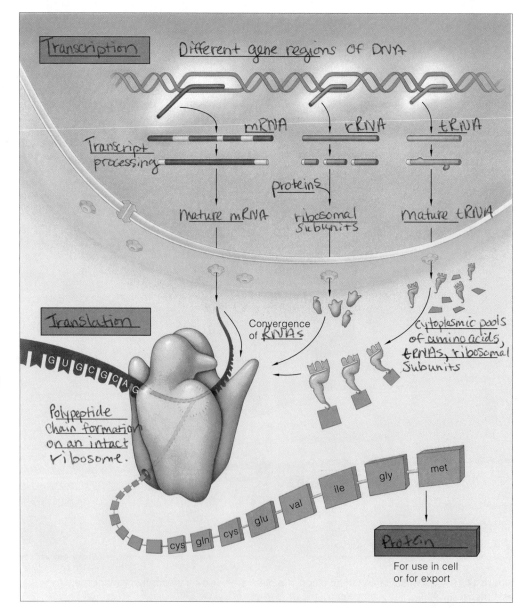

Transcription

Different gene regions of DNA

Transcript processing

mRNA

rRNA

tRNA

proteins

Mature mRNA

ribosomal subunits

mature tRNA

Convergence of RNAs

Cytoplasmic pools of amino acids, tRNAs, ribosomal subunits

Translation

G U G C G C A G

Polypeptide chain formation on an intact ribosome.

cys gln cys glu val ile gly met

Protein

For use in cell or for export

6. Referring to Figure 14.11, use the genetic code to translate the mRNA sequence UAUCGCACCUCAGGAGACUAG. Notice that the first codon in the frame is UAU. Which amino acid sequence is being specified?

    a. TYR — ARG — THR — SER — GLY — ASP — STOP

    b. TYR — ARG — THR — SER — GLY

    c. TYR — ARG — TYR — SER — GLY — ASP — STOP

    d. none of the above

7. Anticodons pair with _____ .
    a. mRNA codons   c. tRNA anticodons
    b. DNA codons   d. amino acids

8. In translation, an initiation complex consists of _____ .
    a. an initiator tRNA
    b. the start codon of mRNA
    c. a small ribosomal subunit
    d. a large ribosomal subunit
    e. all of the above

9. A polysome is _____ .
    a. the initiation site for translation
    b. a mutated ribosome
    c. a number of ribosomes on the same transcript
    d. a chromosomal duplication during transcription

10. Match the terms with the suitable description.

    ___ alkylating agent   a. coding part of mRNA transcript
    ___ chain   b. base triplet coding for amino acid
          elongation   c. second stage of translation
    ___ exon   d. base triplet that pairs with codon
    ___ genetic code   e. one kind of environmental agent
    ___ anticodon   that can induce mutation in DNA
    ___ intron   f. set of sixty-four codons for mRNA
    ___ codon   g. noncoding part of mRNA
                          transcript

Critical Thinking

1. Noah has isolated a tRNA with a mutation in the anticodon 3'-AAU instead of 3'-AUU. What effect will the mutation have on protein synthesis in cells having the mutated tRNA?

2. Carefully examine the diagram of the genetic code in Figure 14.11. Identify the similarities and differences among all of the codons that specify the same amino acid. Can you discern any patterns in the similarities and differences? Researchers did, and on the basis of their insight, they formulated a hypothesis that the original genetic code in the first living cells on Earth may have been a simpler, *two-letter* code that specified fewer amino acids. In what respects do the patterns in Figure 14.11 seem to support their hypothesis?

3. A DNA polymerase made an error during the replication of an important gene region of DNA. None of the DNA repair enzymes detected or repaired the damage. A portion of the DNA strand with the error is shown here:

After the DNA molecule is replicated and two daughter cells have formed, one cell is carrying a mutation and the other cell is normal. Develop a hypothesis to explain this observation.

4. This diagram shows the nucleotide sequence for a gene:

5' ATGCCCGCCTTTGCTACTTGGTAG 3'
3' TACGGGCGGAAACGATGAACCATC 5'

When the gene is transcribed, this mRNA is produced:

5' AUGCCCGCCUUUGCUACUUGGUAG 3'

An insertion, indicated in *red*, occurs in the gene:

5' ATGCCCGCCTAATTGCTACTTGGTAG 3'
3' TACGGGCGGATTAACGATGAACCATC 5'

What effect will this particular mutation have on the structure of the protein product?

5. In the bacterium *E. coli*, the end product (E) of the following metabolic pathway is absolutely essential for life:

Ginetta, a geneticist, is attempting to isolate mutations in the genes for the four enzymes of this pathway. She has been able to isolate mutations in the genes for enzymes 1 and 2. In each case, the mutant *E. coli* cells synthesize a reduced amount of the pathway's end product E. However, Ginetta has not been able to isolate *E. coli* cells that have mutations in the genes for enzymes 3 and 4. Develop a hypothesis to explain why.

Selected Key Terms

alkylating agent *14.5*
anticodon *14.3*
base-pair substitution *14.5*
base sequence *CI*
carcinogen *14.5*
codon *14.3*
deletion (of base) *14.5*
exon *14.2*
gel electrophoresis *14.1*
gene mutation *14.5*
genetic code *14.3*
insertion (of base) *14.5*
intron *14.2*

ionizing radiation *14.5*
mRNA (messenger RNA) *14.2*
mutation rate *14.5*
polysome *14.4*
promoter (RNA) *14.2*
ribonucleic acid (RNA) *CI*
RNA polymerase *14.2*
rRNA (ribosomal RNA) *14.2*
transcription *CI*
translation *CI*
transposable element *14.5*
tRNA (transfer RNA) *14.2*
uracil *14.2*

Readings

Crick, F. October 1966. "The Genetic Code: III." *Scientific American.*

Nowak, R. 4 February 1994. "Mining Treasures From Junk DNA." *Science* 263: 608–610.

Russell, P. 1992. *Genetics.* Third edition. New York: Harper Collins. Chapter 18 has a good introduction to gene mutation.

Web Site See *http://www.wadsworth.com/biology* for practice quiz questions, hypercontents, BioUpdates, and critical thinking. The Wadsworth Biology Resource Center provides a wealth of information fully organized and integrated by chapter.

15 CONTROLS OVER GENES

Here's to Suicidal Cells!

Every day, a portion of you disappears as millions of body cells of your skin, intestines, thymus gland, and elsewhere commit suicide. Fortunately, millions of freshly dividing cells are just as rapidly replacing them and are thereby contributing to the survival of your various parts and, ultimately, you.

For all multicelled organisms, the very first cell of a new individual contains marching orders that will take its descendants through a program of growth, development, and reproduction, then on to eventual death. As part of that program, many cells heed calls to self-destruct when they finish a prescribed function. They also can execute themselves when they become altered in ways that might pose a threat to the body as a whole, as by infection or cancerous transformation. **Apoptosis** (pronounced APP-oh-TOE-sis) is the name for this form of cell death. As Figure 15.1a shows, it starts with molecular signals that activate and thus unleash lethal weapons of self-destruction that were stockpiled earlier within the cell itself. Apoptosis is not the same as necrosis, which is the passive death of many cells that results from severe tissue damage.

Figure 15.1b shows a cell in the act of suicide. First it shrank away from its neighbors. Now its cytoplasm seems to be roiling, and its surface repeatedly bubbles outward and inward. No longer are the chromosomes extended through the nucleoplasm; they have bunched together near the nuclear envelope. The nucleus, then the whole cell, breaks apart. Phagocytic white blood cells patrol the body's tissues, and when they encounter suicidal cells or remnants of them, the patrolling cells swiftly engulf them.

The timing of cell death is predictable in some cells, such as the pigment-packed keratinocytes that form the densely packed sheets of dead cells that are continually sloughed off and replaced at the surface of your skin. They have a three-week life span, more or less. And yet keratinocytes and other body cells, even the kinds that are supposed to last a lifetime, can be induced to die ahead of schedule.

All it takes is sensitivity to specific signals that can activate weapons of death. Those weapons are protein-cleaving enzymes, the **ICE-like proteases**. They are named after the first to be discovered (*Interleukin-1 Converting Enzyme*). Think of them as folded pocket-knives or lethal Ninja weapons. When popped open, they can chop apart structural proteins, including the building blocks of cytoskeletal elements and

signal to die

a

Figure 15.1 (a) Computer model of the suicidal response of a cell to an external triggering signal. In this case, a membrane receptor directs the message to ICE-like proteases. When activated, these enzymes chop up the cytoskeleton and activate other enzymes that slice up the DNA. (b) A cell caught in the act of suicide, as evidenced by its surface blebs.

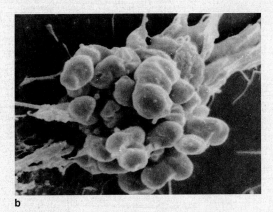
b

nucleosomes that organize the DNA (Figure 15.1*a*). They will do so, for example, if a cell is deprived of the signalling molecules called growth factors, if it loses contact with its neighbors, or if it receives altered signals about when to grow, divide, or cease dividing. The weapons also will be unleashed in the presence of certain regulatory proteins that can induce apoptosis.

The knives remain sheathed in cancer cells, which are supposed to—but don't—kill themselves on cue. As described at the end of this chapter, cancer arises as a result of mutations in several genes that govern cell growth and division. Researchers already know that the gene coding for p53 has been tampered with or inactivated in many masses of transformed cells. The normal form of the protein induces apoptosis when a cell's DNA is damaged. Intriguingly, p53 is absent or malfunctioning in more than half of the cancer patients that the researchers have investigated to date.

If cancer cells live long enough, they may end up with a number of mutations that do more than allow their descendants to divide uncontrollably. The cells may also end up ignoring the death signal if they manage to break away from their home tissue. Then they will be free to establish new colonies of cancer cells in distant tissues.

This fleeting glimpse of Ninja knives and cancerous transformations invites us to reflect on how lucky we are when proper gene controls are in place and our cells are operating as they should. As you will see in this chapter, controls govern whether any particular gene will be transcribed and translated in a particular cell. They also govern which products of genes will be activated or inhibited at any given time. The extent to which you and all other organisms on Earth depend on controls over genes and gene products is just astounding.

KEY CONCEPTS

1. In cells, a variety of controls govern when, how, and to what extent genes are expressed. The control elements operate in response to changing chemical conditions and to signals from the outside environment.

2. Control is exerted by way of regulatory proteins and other molecules that operate before, during, or after transcription. Different kinds interact with DNA, with RNA that has been transcribed from the DNA, or with gene products—that is, with the resulting polypeptide chains or final proteins.

3. Prokaryotic cells depend on rapid control over short-term shifts in nutrient availability and other aspects of their surrounding environment. Commonly, they rely on a small number of regulatory proteins that exert rapid, on-off control of transcription.

4. All eukaryotic cells depend on controls over short-term shifts in diet and levels of activity. In multicelled species, they also depend on controls over an intricate, long-term program of growth and development.

5. Controls over eukaryotic cells come into play when new cells contact one another in developing tissues, and as cells start interacting with their neighbors by way of hormones and other signaling molecules.

6. Although all cells of a multicelled organism inherit the same genes, different cell types activate or suppress many of those genes in different ways. The controlled, selective use of genes leads to synthesis of the proteins that give each type of cell its distinctive structure, function, and products.

At this very moment, bacteria are feeding on nutrients in your gut. Red blood cells are binding, transporting, or giving up oxygen, and great numbers of epithelial cells in your skin are busily synthesizing the protein keratin. Like cells everywhere, they are functioning by virtue of the protein products of genes.

Cells don't express all of their genes all of the time. Instead, cells use some genes and their products only once. They use other genes at certain times, all of the time, or not at all. *Which genes are being expressed depends on the type of cell, its moment-by-moment adjustments to changing chemical conditions, which signals from the outside it happens to be receiving—and its built-in control systems.*

For example, availability of nutrients shifts rapidly and often for the enteric bacteria, which inhabit animal intestines. Like other prokaryotic cells, they can rapidly transcribe certain genes and synthesize many nutrient-digesting enzyme molecules when nutrients happen to be moving past, and they can restrict synthesis when nutrients are scarce. By contrast, the composition and solute concentrations of the fluid bathing your cells do not shift drastically. And few of your cells exhibit rapid shifts in transcription.

The control systems consist of certain molecules, such as **regulatory proteins**, that come into play during transcription or translation, or after translation. Various components of the systems have a capacity to interact with DNA, RNA, or gene products (either newly minted polypeptide chains or final proteins, such as enzymes). Some operate in response to extracellular signals such as hormones. Others operate in response to changing concentrations of substances within the cell.

Negative control systems block the activity of their target molecule, and **positive control systems** promote it. Thus, one regulatory protein inhibits transcription of a particular gene when it binds with DNA, but the action of a different regulatory protein enhances that gene's transcription. Bear in mind, regulatory proteins do not always act alone. For instance, as you will read shortly, some types release their grip on a target when a different type comes along and interacts with them.

Summing up, these are the points to keep in mind as you read through the rest of this chapter:

Cells exert control over when, how, and to what extent each of their genes is expressed.

The expression of a given gene depends on the type of cell and its functions, on chemical conditions, and on signals from the outside environment.

Many regulatory proteins and other molecules exert control over gene expression through their interactions with DNA or RNA. Others exert control through their interaction with gene products, either polypeptide chains or final proteins.

Let's first consider some examples of gene control in prokaryotic cells—that is, bacteria. When nutrients are plentiful and other environmental conditions also favor growth, bacteria tend to grow and divide indefinitely. Gene controls promote the rapid synthesis of enzymes that have roles in nutrient digestion and other growth-related activities. Transcription is rapid, and translation is initiated even before mRNA transcripts are finished. Remember, bacteria have no nucleus; nothing separates their DNA from ribosomes in the cytoplasm.

When a nutrient-degrading pathway utilizes several enzymes, all genes for those enzymes are transcribed, often into one continuous mRNA molecule. The genes are not transcribed when conditions turn unfavorable. The rest of this section provides two cases of this all-or-nothing control of transcription.

Negative Control of Transcription

All mammals house the enteric bacterium *Escherichia coli*. It lives on glucose, lactose (a sugar in milk), and other ingested nutrients. Like other adult mammals, you probably don't drink milk around the clock. When you do, *E. coli* cells in your gut rapidly transcribe three genes for enzymes with roles in breakdown reactions that begin with lactose.

A promoter precedes the three genes, which are next to one another. A **promoter**, recall, is a base sequence that signals the start of a gene. Another sequence, an **operator**, intervenes between a promoter and bacterial genes. It is a binding site for a **repressor**, a regulatory protein that can block transcription (Figure 15.2). An arrangement in which a promoter and operator service more than one gene is an **operon**. Elsewhere in *E. coli* DNA, a different gene codes for this repressor, which can bind with the operator *or* with a lactose molecule.

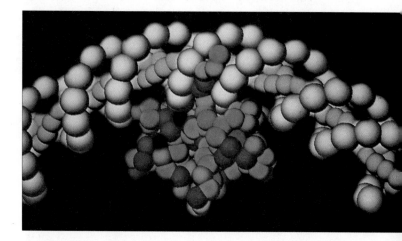

Figure 15.2 Model of a repressor protein (*green*) binding to an operator at a site in a bacterial DNA molecule (*blue*).

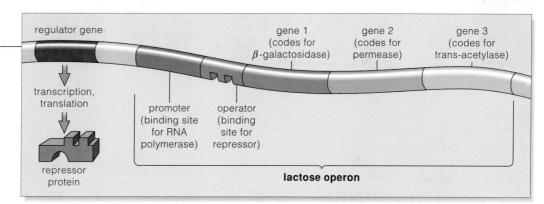

a A repressor protein exerts negative control over three genes of the lactose operon by binding to the operator and inhibiting transcription.

regulator gene

gene 1 (codes for β-galactosidase)

gene 2 (codes for permease)

gene 3 (codes for trans-acetylase)

transcription, translation

promoter (binding site for RNA polymerase)

operator (binding site for repressor)

repressor protein

lactose operon

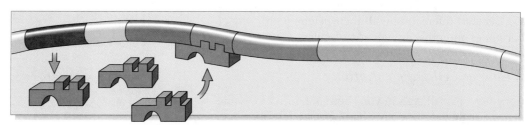

b When the concentration of lactose is low, the repressor is free to block transcription. Being bulky, it overlaps the promoter and prevents binding by RNA polymerase. The enzymes (not needed) are not produced.

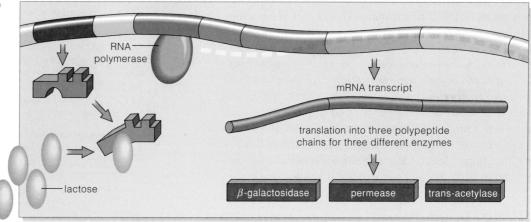

c At high concentration, lactose is an inducer of transcription. It binds to and distorts the shape of the repressor—which now cannot bind to the operator. The promoter is exposed and the genes can be transcribed.

RNA polymerase

mRNA transcript

translation into three polypeptide chains for three different enzymes

lactose

β-galactosidase   permease   trans-acetylase

Figure 15.3 Negative control of the lactose operon. The first gene of this operon codes for an enzyme that splits lactose, a disaccharide, into two subunits (glucose and galactose). The second codes for an enzyme that transports lactose into cells. The third enzyme functions in metabolizing certain sugars.

When the lactose concentration is low, a repressor binds with the operator, as in Figure 15.3. Being a large molecule, it overlaps the promoter, so transcription is blocked. Thus, *lactose-degrading enzymes are not built if they are not required.* When the concentration of lactose is high, odds are greater that a lactose molecule will bind with the repressor. Binding alters the repressor's shape, so it cannot bind with the operator—and RNA polymerase can now transcribe the genes. Thus, *lactose-degrading enzymes are synthesized only when required.*

Positive Control of Transcription

E. coli cells pay far more attention to glucose than to lactose. They transcribe genes for glucose breakdown continually, at faster rates. Even if lactose is present, the lactose operon isn't used much—*unless glucose is absent.*

At such times, a regulatory protein called CAP acts on the operon. CAP is an **activator protein**, a key player in a positive control system. To understand how it works, you have to know that the lactose operon's promoter is not good at binding RNA polymerase. It does a better job when CAP adheres to it first. But CAP won't do this unless it is activated by a small molecule called cAMP.

Among other things, cAMP is produced from ATP, which *E. coli* can produce by glucose breakdown (that is, by way of glycolysis). Not much cAMP is available in the cell when glucose is plentiful and glycolysis is proceeding full bore. When these conditions prevail, the activator protein does not become primed to adhere to the promoter, and transcription of the lactose operon genes slows almost to a standstill. Now suppose that glucose is scarce and lactose becomes available. cAMP can accumulate, CAP-cAMP complexes can form—and the lactose operon genes can be transcribed.

Prokaryotic cells, which must respond rapidly to changing conditions, commonly rely on a small number of regulatory proteins that exert rapid, on-off control of transcription.

CONTROLS IN EUKARYOTIC CELLS

Like bacteria, all eukaryotic cells depend on controls over short-term shifts in diet and level of activity. If the cells are merely one of hundreds, millions, or even trillions of cells in the body of a multicelled organism, they also require more intricate controls. For them, gene activity changes as a program of development unfolds, as new cells contact one another in the developing tissues, and as cells start interacting with their neighbors by way of hormones and other signaling molecules.

A Case of Cell Differentiation

Consider this: All cells in your body inherited the same genes, for they descended from the same fertilized egg. Many of the genes specify proteins that are basic to any cell's structure and functioning. That is why the protein subunits of ribosomes are the same from one cell to the next, as are many enzymes. *Yet nearly all of your cells became specialized in composition, structure, and function*. This process of **cell differentiation** proceeds while all multicelled species develop. It arises as embryonic cells and their descendants activate and suppress some of the total number of genes in unique, selective ways. Thus, only immature red blood cells use the genes for making hemoglobin. Only certain white blood cells use the genes for making weapons called antibodies.

Selective Gene Expression at Many Levels

Eukaryotic cells exert highly selective control of gene expression at many levels. Figure 15.4 and the next few paragraphs will give you a sense of what goes on at these levels. You will come across specific examples in the sections that follow.

CONTROLS RELATED TO TRANSCRIPTION Many genes concerned with housekeeping tasks in eukaryotic cells are under positive controls that promote continuous, low levels of transcription. For many other genes, we see less of the rapid, all-or-nothing control over gene transcription that is common among bacteria. Why? In multicelled organisms, the internal environment (that is, tissue fluids and blood) affords much more stable operating conditions for individual cells. Most often, transcription rates rise or fall by degrees in response to slight shifts in concentrations of signaling molecules, substrates, and products in that environment.

As the examples in Figure 15.4a indicate, some gene sequences are repeatedly duplicated or rearranged in genetically programmed ways prior to transcription. Besides this, programmed chemical modifications often shut down many genes. So does the orderly packaging of DNA by histones and other chromosomal proteins,

a CONTROLS RELATED TO TRANSCRIPTION. At any given time, most genes of a multicelled organism are shut down, either permanently or temporarily. The genes necessary for a cell's everyday tasks are under positive controls that promote ongoing, low levels of transcription. With the help of these controls, a cell is assured of having enough enzymes and other proteins to carry out its most basic functions. Transcription of many other genes often shifts only slightly. In this case, controls work to assure chemical responsiveness even when concentrations of specific substances rise or fall only slightly.

Also, even before some genes are transcribed, a portion of their base sequences may be amplified, rearranged, or chemically modified in temporary or reversible ways. These are not mutations; they are heritable, programmed events that affect the manner in which particular genes will be expressed (if at all).

1. *Gene amplification.* Some cells that require enormous numbers of certain molecules temporarily increase the number of the required genes. In response to a molecular signal, multiple rounds of DNA replication sometimes produce hundreds or thousands of gene copies prior to transcription! This happens in immature amphibian eggs and glandular cells of some insect larvae (Section 15.4).

2. *DNA rearrangements.* In a few cell types, many base sequences are alternative "choices" for parts of a gene. Prior to transcription, they are snipped out of the DNA. Combinations of the snippets are spliced together as the gene's final base sequence. For instance, this happens when B lymphocytes, a special class of white blood cells, are forming (Section 40.9). The cells transcribe and then translate their uniquely rearranged DNA into staggeringly diverse versions of protein weapons called antibodies, which act specifically against one of staggeringly diverse kinds of pathogens and other foreign agents in the body.

3. *Chemical modification.* Numerous histones and other proteins interact with eukaryotic DNA in highly organized fashion. The DNA-protein packaging, as well as chemical modifications to particular sequences, influences gene activity. Usually, only a small fraction of a cell's genes are available for transcription. A more dramatic shutdown is called X chromosome inactivation in female mammals, as described in Section 15.4.

Figure 15.4 Examples of the levels of control over gene expression in eukaryotes.

as you may realize after reflecting on the organization of eukaryotic chromosomes (Section 9.4).

POST-TRANSCRIPTIONAL CONTROLS Control also occurs after gene transcription (Figure 15.4b–d). Many controls govern transcript processing, transport of mature RNAs from the nucleus, and translation rates. Others deal with modification of the new polypeptide chains. Still others deal with activating, inhibiting, and degrading

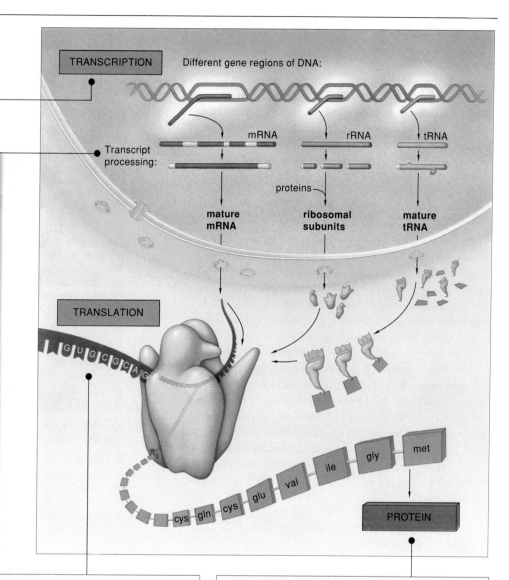

TRANSCRIPTION

Different gene regions of DNA:

mRNA rRNA tRNA

Transcript processing:

proteins

mature mRNA **ribosomal subunits** **mature tRNA**

TRANSLATION

G U G C G C A G

cys gln cys glu val ile gly met

PROTEIN

b TRANSCRIPT PROCESSING CONTROLS. As you read earlier in Section 14.2, pre-mRNA transcripts undergo modifications before they leave the nucleus.

For example, introns are removed and exons are spliced together in more than one way. In other words, a transcript from a single gene may undergo *alternative splicing*.

Pre-mRNA transcribed from a gene that specifies a contractile protein, troponin-1, is like this. Enzymes excise different portions of the pre-mRNA transcript in different cells. When the exons are spliced together, the final protein-building message is slightly different from one cell to the next. The resulting proteins are all very similar, but each is unique in a certain region of its amino acid sequence. The proteins function in slightly different ways, which may account for the subtle variations we observe in the functioning of different types of muscles in the body.

c CONTROLS OVER TRANSLATION. Diverse controls govern when, how rapidly, and how often a given mRNA transcript will be translated. Sections 37.2, 44.3, and other parts of the book provide elegant examples.

The stability of a transcript influences the number of protein molecules that can be produced from it. Enzymes destroy transcripts from the poly-A tail on up (Section 14.2). The tail's length and its attached proteins affect the pace of degradation. Also, after leaving the nucleus, some transcripts are inactivated, temporarily or permanently. For example, in unfertilized eggs, many transcripts are inactivated and stored in the cytoplasm. These "masked messengers" will not be available for translation until after fertilization, when great numbers of protein molecules will be required for the early cell divisions of the new individual.

d CONTROLS FOLLOWING TRANSLATION. Before they can become fully functional, many polypeptide chains must pass through the cytomembrane system (Section 4.5). There they undergo modification, as by having specific oligosaccharides or phosphate groups attached to them.

Also, a variety of control mechanisms govern the activation, inhibition, and stability of the enzymes and other molecules that are involved in protein synthesis. Allosteric control of tryptophan synthesis, as described earlier in Section 6.5, is an example.

the final proteins. Consider enzymes alone, the proteins that catalyze nearly all metabolic reactions. In addition to selectively transcribing and translating genes for those enzymes, control systems activate and inhibit enzyme molecules that already exist. Just imagine the coordination that governs which of thousands of types of enzymes become stockpiled, deployed, or degraded in a given interval. *That coordination governs all short-term and long-term aspects of cell structure and function.*

In each multicelled organism, gene controls underlie basic, short-term housekeeping tasks in cells and more intricate, long-term patterns of bodily growth and development.

All of those cells inherit the same genes, yet most become specialized in composition, structure, and function. This process of cell differentiation arises as different populations of cells activate and suppress some of their genes in highly selective, unique ways.

15.4 EVIDENCE OF GENE CONTROL

By some estimates, the cells of a multicelled organism rarely use more than 5 to 10 percent of their genes at a given time. One way or another, controls are keeping most of the genes repressed. In addition, which genes are being expressed varies, depending on the stage of growth and development that the organism is passing through. The following examples hint at the kinds of control mechanisms that operate at different stages.

Transcription in Lampbrush Chromosomes

While the unfertilized eggs of amphibians mature, they grow extremely fast. Growth requires numerous copies of enzymes and other proteins. The germ cells that give rise to the eggs stockpile large numbers of RNAs and ribosomes, all of which will have roles in rapid protein synthesis. During prophase I of meiosis, chromosomes in the germ cells decondense in such a way that the DNA loops out profusely from their protein scaffold.

The chromosomes look so bristly at this time, they are said to have a "lampbrush" configuration (Figure 15.5). At such times, histones and other proteins that structurally organize the DNA have loosened their grip, so that the genes specifying RNA molecules are now accessible. Histones, recall, are part of nucleosomes, the basic unit of organization in eukaryotic chromosomes (Section 9.4). Each nucleosome consists of a stretch of DNA looped around a core of histone molecules. The diagram in Figure 15.6b is based on photomicrographs that show the nucleosome packaging being loosened during transcription.

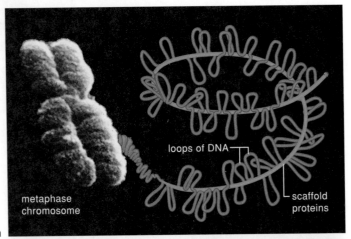

loops of DNA

scaffold proteins

metaphase chromosome

a

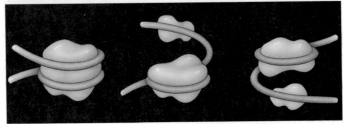

b

one nucleosome (a bit of DNA looped around a histone core)

one nucleosome loosened up

Figure 15.6 Changes in DNA organization that may promote transcription. (**a**) At certain times, DNA decondenses into loops that extend from attachment sites on scaffold proteins. (**b**) The tight DNA-histone packing in nucleosomes also may loosen up. Transcription proceeds mainly in regions where chromosome packaging has become most relaxed.

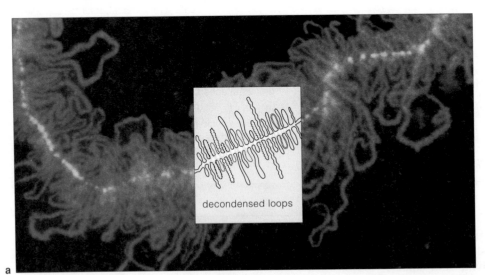

decondensed loops

a

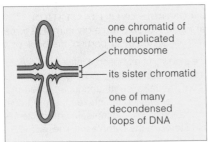

one chromatid of the duplicated chromosome

its sister chromatid

one of many decondensed loops of DNA

b

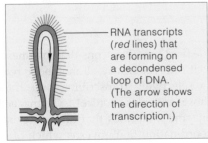

RNA transcripts (*red* lines) that are forming on a decondensed loop of DNA. (The arrow shows the direction of transcription.)

c

Figure 15.5 Lampbrush chromosome from a germ cell of a newt (*Notophthalmus viridiscens*). During prophase I, gene regions of this duplicated chromosome decondensed into thousands of loops. A *red* fluorescent dye labeled one of the components of ribonucleoproteins. The presence of ribonucleoproteins here is evidence of gene activity. (A *white* fluorescent dye labeled the proteins making up the axis of the duplicated chromosome's structural framework.)

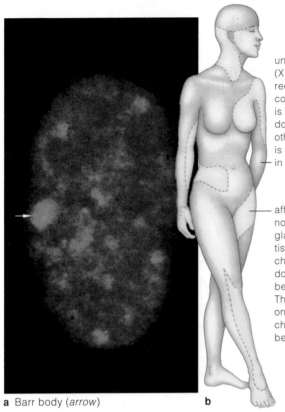

a Barr body (*arrow*)

b

unaffected skin (X chromosome with recessive allele was condensed; its allele is inactivated. The dominant allele on other X chromosome is being expressed in this tissue.)

affected skin with no normal sweat glands (In this tissue, the X chromosome with dominant allele has been condensed. The recessive allele on the other X chromosome is being transcribed.)

Figure 15.7 (**a**) Micrograph of an inactivated X chromosome, called a Barr body, as it appears in a human female's somatic cell during interphase. The X chromosome is not condensed this way in a human male's cells. (**b**) Anhidrotic ectodermal dysplasia, a mosaic pattern of gene expression. The condition arises as a result of random X chromosome inactivation.

Figure 15.8 Why is this female calico cat "calico"? In her cells, one X chromosome carries a dominant allele for the brownish-black pigment melanin. The allele on her other X chromosome specifies yellow fur. At an early stage of the cat's embryonic development, one of the two X chromosomes was inactivated at random in each cell that had formed by then. In all descendants of those cells, the same chromosome also became inactivated, leaving only one functional allele for the coat-color trait. We see patches of different colors, depending on which allele was inactivated in cells that formed a given tissue region. (The white patches result from a gene interaction involving the "spotting gene," which blocks melanin synthesis entirely.)

X Chromosome Inactivation

A mammalian zygote destined to become a female has two X chromosomes (one from the mother, one from the father). As it develops, *one* of the two condenses in each cell. The condensation is a programmed event, but the outcome is random. *One or the other chromosome may become inactivated.* You may observe such a condensed X chromosome in the interphase nucleus; it shows up as a dark spot (Figure 15.7*a*). It is called a **Barr body** after its discoverer, Murray Barr.

When the maternal (or paternal) X chromosome is inactivated in a cell, it also becomes inactivated in all of the cell's descendants. By adulthood, each adult female is "mosaic" for the X chromosomes. *She has patches of tissues where maternal genes of the X chromosome are being expressed—and patches of tissue in which paternal genes of the X chromosome are being expressed.* Because any pair of alleles on her two X chromosomes may or may not be identical, the tissue patches may or may not have the same characteristics.

Mary Lyon discovered the mosaic tissue effect that arises from random X chromosome inactivation. We see this effect in human females who are heterozygous for a recessive allele on the X chromosome that prevents sweat glands from forming. The lack of sweat glands is one symptom of *anhidrotic ectodermal dysplasia.* In the affected females, the X chromosome with the dominant

allele has condensed, and genes on the one bearing the mutant allele are being transcribed in the patches of skin with no sweat glands. Figure 15.7*b* is a diagram of the mosaic tissue effect. The same effect is apparent in female calico cats. Such cats are heterozygous for black and yellow coat-color alleles on their X chromosomes. The coat color in a given body region depends on which X chromosome's genes are transcribed (Figure 15.8).

In a typical cell of multicelled organisms, controls repress all but about 5 to 10 percent of the genes at a given time.

The remaining fraction of genes is selectively expressed at different stages of growth and development. Lampbrush chromosome formation, X chromosome inactivation, and other events provide evidence of this.

In the chapter introduction, we introduced the idea that a great variety of signals influence gene activity. The following examples from animals and plants will give you an idea of their effects at the molecular level.

Hormonal Signals

Hormones, a major category of signaling molecules, can stimulate or inhibit gene activity in target cells. Any cell with receptors for a given hormone is a target. Animal cells secrete hormones into tissue fluid. Most hormone molecules are picked up by the bloodstream, which distributes them to cells some distance away. In plants, hormones do not travel far from cells that secrete them.

Certain hormones bind to membrane receptors at the surface of target cells. Others enter target cells, bind to regulatory proteins, and so help initiate transcription. In many cases, a hormone molecule must first touch bases, so to speak, with an enhancer. **Enhancers** are base sequences that serve as binding sites for suitable activator proteins. They may or may not be adjacent to a promoter in the same DNA molecule. When some distance separates the two, a loop forms in the DNA and brings them together. RNA polymerase binds avidly to the bound complex at the base of the loop.

Consider the effect of **ecdysone**, a hormone with key roles in the life cycle of many insects. Immature stages of the insects, called larvae, grow rapidly. They feed continually on organic matter, such as decaying leaves, and require copious amounts of saliva to prepare food for digestion. DNA in their salivary gland cells has been replicated repeatedly. The copies of DNA molecules have remained together in parallel array, forming what is known as a *polytene* chromosome. When ecdysone binds to a receptor on the gland cells, its signal triggers very rapid transcription of multiple copies of genes in the DNA. The gene regions affected by this hormonal signal puff out during transcription, as in Figure 15.9. Afterward, translation of the mRNA transcripts from the genes produces the protein components of saliva.

In vertebrates, certain hormones have widespread effects on gene expression because many types of cells have receptors for them. As one example, the pituitary gland secretes somatotropin, or growth hormone. This hormonal signal stimulates synthesis of all the proteins required for cell division and, ultimately, the body's growth. Most cells have receptors for somatotropin.

Other vertebrate hormones signal only certain cells at certain times. Prolactin, a secretion from the pituitary gland, is like this. Beginning a few days after a female mammal gives birth, researchers can detect prolactin in the blood. This hormone activates genes in mammary gland cells that have receptors for it. Those genes have responsibility for milk production. Liver cells and heart cells have the same genes but do not have the receptors necessary to respond to signals from prolactin.

Explaining hormonal control of gene activity is like explaining a full symphony orchestra to someone who has never seen one or heard it perform. Many separate parts must be defined before their interactions can be understood! We will return to this topic, starting with Chapter 37 on the endocrine system. As you will see in Chapters 44 and 45, some of the most elegant examples of hormonal controls are drawn from studies of animal reproduction and development.

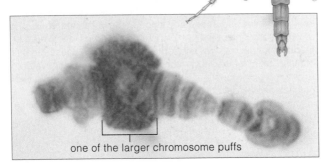

one of the larger chromosome puffs

Figure 15.9 Visible evidence of transcription in a polytene chromosome from a larva of a midge (*Chironomus*). Midges are flies, one to ten millimeters long. The short-lived, winged adults often congregate in great swarms, which help them find mates in a hurry. They might seem to be pests merely because of their large numbers. They also look somewhat like mosquitoes, but they don't bite.

Most midge larvae develop in aquatic habitats. To sustain their rapid growth, they feed continuously, as on decaying organic material. They require a lot of saliva, and they must continuously transcribe genes for saliva's protein components. Those genes have undergone amplification. Ecdysone, a hormone, serves as a regulatory protein that helps promote their transcription. Midge chromosomes loosen and puff out in regions where the genes are being transcribed in response to the hormonal signal. The puffs become large and appear quite diffuse when transcription is most intense. Staining techniques reveal banding patterns in the chromosomes, as evident in the micrograph.

Sunlight as a Signal

Plant a few seeds from a corn or bean plant in a pot that contains moist, nutrient-rich soil. Next, let the seeds germinate, but keep them in total darkness. After eight days have passed, they will develop into seedlings that are spindly and notably pale, owing to the absence of chlorophyll (Figure 15.10). Now expose the seedlings to a single burst of dim light from a flashlight. Within ten minutes, they will start converting stockpiles of certain molecules to the activated form of chlorophylls, the

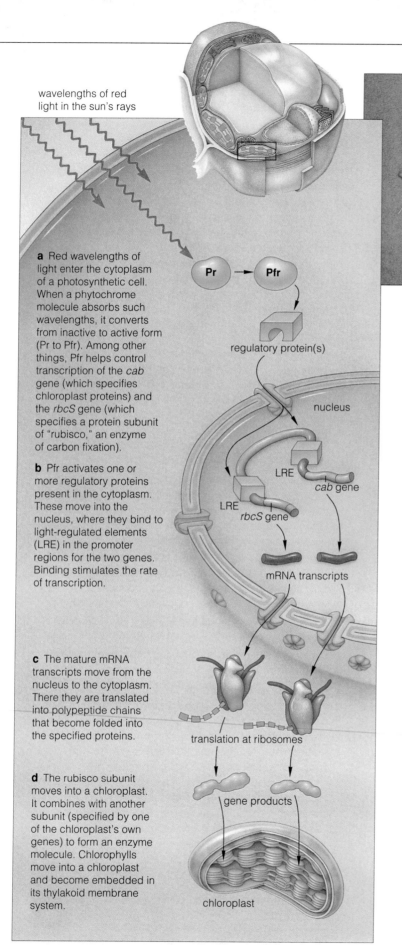

wavelengths of red light in the sun's rays

a Red wavelengths of light enter the cytoplasm of a photosynthetic cell. When a phytochrome molecule absorbs such wavelengths, it converts from inactive to active form (Pr to Pfr). Among other things, Pfr helps control transcription of the *cab* gene (which specifies chloroplast proteins) and the *rbcS* gene (which specifies a protein subunit of "rubisco," an enzyme of carbon fixation).

b Pfr activates one or more regulatory proteins present in the cytoplasm. These move into the nucleus, where they bind to light-regulated elements (LRE) in the promoter regions for the two genes. Binding stimulates the rate of transcription.

c The mature mRNA transcripts move from the nucleus to the cytoplasm. There they are translated into polypeptide chains that become folded into the specified proteins.

d The rubisco subunit moves into a chloroplast. It combines with another subunit (specified by one of the chloroplast's own genes) to form an enzyme molecule. Chlorophylls move into a chloroplast and become embedded in its thylakoid membrane system.

Pr → Pfr

regulatory protein(s)

nucleus

LRE

cab gene

LRE

rbcS gene

mRNA transcripts

translation at ribosomes

gene products

chloroplast

Figure 15.10 Sunlight as a signal for gene activity. The photograph shows the effect of the absence of light on corn seedlings. The two seedlings to the left were the control group; they were grown in sunlight, in a greenhouse. The two seedlings positioned next to them were grown in total darkness for eight days. The dark-grown plants were not able to convert their stockpiled precursors of chlorophyll molecules to active form, and they never did green up. The diagram illustrates one model for the mechanism by which phytochrome helps control gene transcription in plants. Red wavelengths can convert the phytochrome molecule from inactive form (here designated Pr) to active form (Pfr). In this form, the phytochrome can serve as a regulator of transcription.

light-trapping pigment molecules that are absolutely central to photosynthesis.

Phytochrome, a blue-green pigment, is a signaling molecule that helps plants adapt over the short term to changes in light conditions. Section 32.4 takes a close look at this molecule. For now, simply be aware that it alternates between active and inactive forms. At sunset, during the night, or in the shade, far-red wavelengths predominate, and the phytochrome in cells is inactive. It becomes activated at sunrise, when red wavelengths dominate the sky. In addition, the quantity of incoming red or far-red wavelengths varies as days alternate with nights, and as days grow shorter and then longer with the changing seasons. Such variations serve as signals that help regulate phytochrome activity, which in turn influences transcription of certain genes at certain times of day and year. The genes specify enzymes and other proteins with key roles in germination, stem elongation, branching, leaf expansion, and the formation of flowers, fruits, and seeds.

Figure 15.10 shows a model of phytochrome control of transcription. Experiments by Elaine Tobin and her coworkers at the University of California, Los Angeles, provide evidence in favor of such a model. Working with dark-grown seedlings of duckweed (*Lemna*), they discovered a marked increase in the number of certain mRNA transcripts after a one-minute exposure to red light. Exposure enhanced transcription of the genes for proteins that bind chlorophylls and for an enzyme that mediates carbon fixation. Both proteins are required for the development and greening of chloroplasts.

Hormones and other signaling molecules, including diverse kinds that respond to environmental stimuli, have profound influence on gene expression.

CANCER AND THE CELLS THAT FORGET TO DIE

THE CELL CYCLE REVISITED   Every second, millions of cells in your skin, gut lining, liver, and other body parts divide and replace their worn-out, dead, and dying predecessors. They do not divide willy-nilly. Controls govern the expression of genes that specify enzymes and other proteins required for cell growth, DNA replication, spindle formation, chromosome movements, and the division of the cytoplasm. Regulatory proteins control the synthesis and use of these gene products, and they control when the division machinery is put to rest.

Controls over the cell cycle are extensive. At various interrelated points, gene activity can be stepped up or slowed down, and gene products can advance, delay, or block the cycle. Some genes directly regulate passage through the cycle and thereby are its primary controllers. Among the most crucial are **protein kinases**, a class of enzymes that attach phosphate groups to proteins. You will find molecules of them operating at the boundary between G1 and S of the cycle, and at the transition from G2 to mitosis. (Here you might refer to Section 9.2.)

Other genes encode proteins that modify the activity of the primary control genes or their products. Among them are genes that specify **growth factors**, which are transcriptional signals sent by one cell to trigger growth in other cells. The products include CSF (short for *Colony Stimulating Factor*), which normally stimulates growth of certain white blood cells. EFG (for *Epidermal Growth Factor*) stimulates skin cells to grow and divide; NFG stimulates growth of neurons. Still other genes specify cell receptors for particular growth factors.

Also, certain genes specify enzymes and other factors required to maintain DNA replication, stockpile the ICE-like proteases, and perform other vital activities. They do not act directly as controls. But if they become altered, as by mutations, a cell may be deprived of important enzymes or proteins, and some controls over the cycle may be lost.

CHARACTERISTICS OF CANCER   When controls are lost, cell division will not stop as long as conditions for growth stay favorable. When cells are not responding to normal controls over growth and division, they form a tissue mass called a **tumor** (Figure 15.11). The cells of common skin warts and other *benign* tumors grow slowly, in an

unprogrammed way. They still have surface recognition proteins that can hold them together in their home tissue. Surgically remove a benign tumor, and you remove its potential threat to the surrounding tissues.

In a *malignant* tumor, abnormal cells grow and divide more rapidly, with destructive physical and metabolic effects on surrounding tissues. These cells are disfigured, grossly so. They cannot construct a normal cytoskeleton or plasma membrane, and they cannot synthesize normal versions of recognition proteins. Such cells can break loose from their home tissue, enter lymph or blood vessels, travel through the body, and become lodged where they do not belong. This process of abnormal cell migration and invasion is called **metastasis** (Figure 15.12).

The photographs in Section 13.4 show merely 3 of the more than 200 types of malignant tumors that have been identified so far. **Cancer** is the general category in which all malignant tumors are grouped. Each year in developed countries alone, 15 to 20 percent of all deaths result from cancer. And it is not just a human affliction. Cancer has also been observed in most of the animals that have been studied. Similar abnormalities have even been observed in many kinds of plants.

At the minimum, all cancer cells show the following characteristics:

1. *Profound changes in the plasma membrane and cytoplasm.* Membrane permeability increases. Membrane proteins are lost or altered, and different ones form. The cytoskeleton becomes disorganized, shrinks, or both. Enzyme activity shifts, as in amplified reliance on glycolysis.

2. *Abnormal growth and division.* Controls that prevent overcrowding in tissues are lost. Cell populations reach high densities. New proteins trigger abnormal increases in small blood vessels that service the growing cell mass.

3. *Weakened capacity for adhesion.* Recognition proteins are altered or lost; cells can't stay anchored in proper tissues.

4. *Lethality.* Unless cancer cells are eradicated, they will kill the individual.

Any gene having the potential to induce a cancerous transformation is called an **oncogene**. Oncogenes were first identified in retroviruses, which are a class of RNA viruses. They are altered forms of normal genes, now called **proto-oncogenes**, that specify certain proteins required for normal cell functioning. In other words, normal expression of proto-oncogenes is vital, which helps explain why their *abnormal* expression is lethal.

Transformation may start with mutations in control elements that govern proto-oncogenes or with the

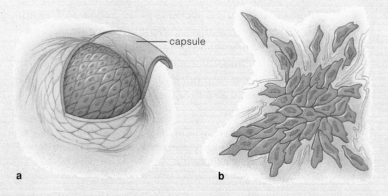

capsule

a   b

Figure 15.11   Evidence of loss of gene control: (**a**) A benign tumor, a mass of cells with an outwardly normal appearance that are enclosed in a capsule of connective tissue. (**b**) A malignant tumor, a disorganized collection of cancer cells.

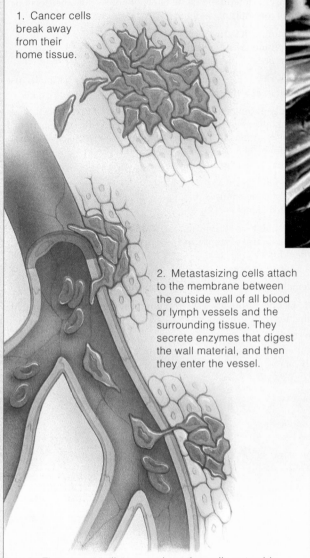

1. Cancer cells break away from their home tissue.

2. Metastasizing cells attach to the membrane between the outside wall of all blood or lymph vessels and the surrounding tissue. They secrete enzymes that digest the wall material, and then they enter the vessel.

3. The cancer cells creep along the walls or tumble through blood or lymph. In time, possibly in response to molecular cues at particular sites, they leave the blood or lymph vessels the same way they got in and are free to found new tumor masses in the new tissue.

a

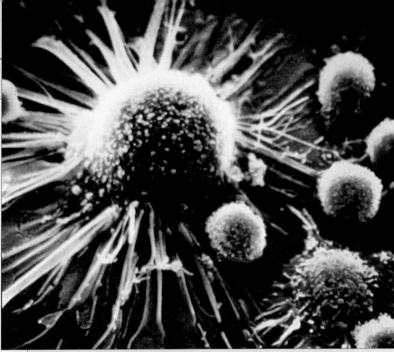

b

Figure 15.12 (**a**) Metastasis, a migration of cancer cells from the tissue in which they originated and their invasion of other tissues in the body. (**b**) The scanning electron micrograph shows a few of the body's defenders—white blood cells—flanking a cancer cell. The defenders may or may not be able to destroy it. Cancer arises through loss of controls that govern specific genes, many of which influence cell growth and division.

genes themselves. For instance, this happens with certain insertions of viral DNA into cellular DNA. It can happen when carcinogens (cancer-inducing agents) cause changes in the DNA. Ultraviolet radiation and ionizing radiation (x-rays and gamma rays) are common carcinogens. So are many natural and synthetic compounds, including asbestos and certain components of tobacco smoke.

Some cancers arise if base substitutions or deletions alter a proto-oncogene or one of the controls over its transcription. Others arise if a gene becomes abnormally amplified. And still others arise if a whole gene or part of it moves to a new location in the same chromosome or to an entirely different chromosome.

Consider the *myc* gene, which is located in human chromosome 8. It specifies a regulatory protein that helps control the cell cycle. Sometimes a break occurs in this gene region and in a region of chromosome 14. The two chromosomes swap segments. The *myc* gene ends up in a new location, where it may escape normal controls over its transcription and come under the influence of new ones. Whatever the case, a transformation begins that converts *myc* from a proto-oncogene to an oncogene— and Burkitt's lymphoma may be the outcome.

Burkitt's lymphoma is a malignant tumor of the white blood cells called B lymphocytes. This is the only type of cell that produces antibody molecules, which are protein weapons against specific agents of disease (Section 40.7). In activated B cells, a gene region that specifies antibody production is transcribed at an extremely high rate. The translocation between chromosomes 8 and 14 puts the *myc* gene next to this region and makes it abnormally active.

Yet cancer is a multistep process, involving more than one oncogene. Researchers have already identified three such genes with roles in nearly all colon cancers. In 1995, they identified a key gene in ovarian and breast cancers. Through such discoveries, it may be possible to diagnose carriers early and frequently. A cancer detected early enough may still be curable by surgery.

Perhaps more than any other example, cancerous transformations bring home the extent to which you and all other organisms depend on controls over transcription, translation, and when and where gene products will be used. This example alone reinforces the vital nature of the topics introduced in this chapter.

15.7 SUMMARY

1. Cells control gene expression; that is, which gene products appear, when, and in what amounts. When the control mechanisms come into play depends on the type of cell, prevailing chemical conditions, and outside signals that can change the cell's activities.

2. Regulatory proteins, enzymes, hormones, and other control elements interact with one another, with DNA and RNA, and with the gene products. They operate before, during, and after transcription and translation.

3. In all cells, two of the most common types of control systems operate to block or enhance transcription. Their effects are reversible.

   a. In negative control systems, a regulatory protein binds at a specific DNA sequence and prevents one or more genes from being transcribed.

   b. In positive control systems, a regulatory protein binds to DNA and promotes transcription.

4. Most prokaryotic cells (bacteria) do not require many genes to grow and reproduce. Most of their control systems affect transcription rates. Control of operons (groupings of related genes and control elements) are examples.

5. Promoters are DNA binding sites that signal the start of a gene. In bacteria alone, operators are binding sites for regulatory proteins that can inhibit transcription. Enhancers are binding sites for activator proteins that promote transcription.

6. Eukaryotic cells, particularly those of multicelled organisms, require more complex gene controls. Gene activity must change rapidly in response to short-term shifts in the surroundings, as in prokaryotic cells. But it also must be adjusted in intricate ways during long-term growth and development, when great numbers of cells multiply, make physical contact with one another in tissues, and interact chemically.

7. Being descended from the same cell, all of the cells in a multicelled organism inherit the same assortment of genes. However, different types of cells activate and suppress some fraction of the genes in different ways. This behavior is called selective gene expression.

8. One outcome of selective gene expression is called cell differentiation. By this process, different lineages of cells in the developing multicelled organism become specialized in appearance, composition, and function.

9. Cancer results when controls over the cell cycle and the cell's death machinery are lost. Some genes encode proteins that are the primary regulators of the cycle. Others encode proteins that modify the activity of the primary controllers or their products. Others encode enzymes and other proteins that can, when altered, indirectly affect the cycle.

Review Questions

1. In what fundamental way do negative and positive control of transcription differ? Is the effect of one or the other form of control (or both) reversible? *15.1*

2. Distinguish between: *15.2*
   a. promoter and operator
   b. repressor protein and activator protein

3. Describe one type of control over transcription of the lactose operon in *E. coli*, a prokaryotic cell. *15.2*

4. A plant, fungus, or animal is composed of diverse cell types. How might this diversity arise, given that the body cells in each organism inherit the same set of genetic instructions? As part of your answer, define cell differentiation and explain how selective gene expression brings it about. *15.3*

5. Review Figure 15.4. Using the diagram below, define five types of gene controls in eukaryotic cells and indicate where they take effect. *15.3*

6. If a polytene chromosome and a lampbrush chromosome are both evidence of transcriptional activity, in what respect do they differ? *15.4, 15.5*

7. What is a Barr body? Does it appear in the cells of males, females, or both? Explain your answer. *15.4*

8. Define hormones. Why do hungry midge larvae depend on the hormone ecdysone? *15.5*

9. What are the characteristics of cancer cells? Explain the difference between a benign tumor and one that is malignant. *15.6*

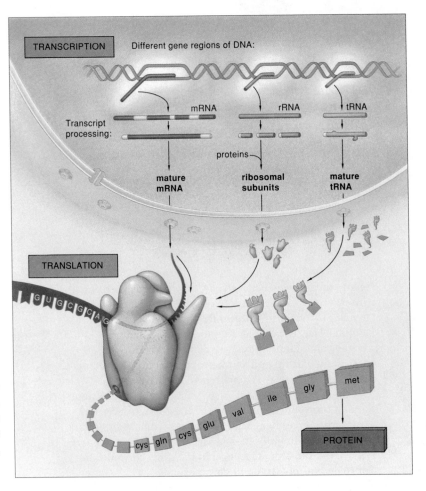

Self-Quiz (Answers in Appendix IV)

1. Apoptosis is _____ .
   a. cell death by severe tissue damage
   b. cell death by suicide
   c. a popping sound in mutated toes

2. ICE-like proteases are _____ .
   a. regulatory proteins   c. environmental signals
   b. lethal weapons   d. low-temperature enzymes

3. The expression of a given gene depends on _____ .
   a. cell type and functions   c. environmental signals
   b. chemical conditions   d. all of the above

4. Regulatory proteins interact with _____ .
   a. DNA   c. gene products
   b. RNA   d. all of the above

5. In prokaryotic but not eukaryotic cells, a(n) _____ precedes the genes of an operon.
   a. lactose molecule   c. operator
   b. promoter   d. both b and c

6. A base sequence signaling the start of a gene is a(n) _____ .
   a. promoter   c. enhancer
   b. operator   d. activator protein

7. An operon most typically governs _____ .
   a. bacterial genes   c. genes of all types
   b. a eukaryotic gene d. DNA replication

8. Prokaryotic cells rely most heavily on _____ controls over transcription.
   a. slow, continuous   c. slow, on-off
   b. rapid, continuous   d. rapid, on-off

9. Cell differentiation _____ .
   a. occurs in all multicelled organisms
   b. requires different genes in different cells
   c. involves selective gene expression
   d. both a and c
   e. all of the above

10. Some gene controls in eukaryotic cells operate by _____ .
    a. amplifying genes   d. rearranging DNA
    b. processing RNA transcripts e. all of the above
    c. chemically modifying DNA

11. X chromosome inactivation in mammalian females may result in a _____ for some traits.
    a. male phenotype   c. rise in transcription rates
    b. mosaic tissue effect d. rise in translation rates

12. Hormones interact with _____ .
    a. membrane receptors   c. enhancers
    b. regulatory proteins d. all of the above

13. Match the terms with their most suitable descriptions.
    ____ phytochrome   a. inhibits gene transcription
    ____ Barr body   b. gene with the potential to induce
    ____ oncogene   cancerous transformation
    ____ repressor   c. normally required, lethal when
    ____ proto-oncogene   mutated
                          d. helps plants adapt to changes
                             in light
                          e. inactivated X chromosome

Critical Thinking

1. Kristen and Dan have isolated a strain of *E. coli* in which a mutation has affected the capacity of CAP to bind to a region of the lactose operon, as it would do normally. State how the mutation affects transcription of the lactose operon when cells of this bacterial strain are subjected to the following conditions:
   a. Lactose and glucose are available.
   b. Lactose is available but glucose is not.
   c. Both lactose and glucose are absent.

2. *Duchenne muscular dystrophy*, a genetic disorder, affects boys almost exclusively. Early in childhood, muscles begin to atrophy (waste away) in affected individuals, who typically die in their teens or early twenties as a result of respiratory failure. Muscle biopsies of women who carry a gene associated with the disorder reveal some regions of atrophied muscle tissue. Yet muscle tissue adjacent to these regions was normal or even larger and more chemically active, as if to compensate for the weakness of the adjoining region. How can you explain these observations?

3. Unlike most rodents, guinea pigs are already well developed at the time of birth. Within a few days, they are able to eat grass, vegetables, and other plant material. Suppose a breeder decides to separate the baby guinea pigs from their mothers after three weeks. He wants to keep the males and females in different cages, but it is quite difficult to determine the gender of guinea pigs when they are so young. Suggest a simple test that the breeder can perform to determine gender.

4. Individuals affected by *pituitary dwarfism* cannot synthesize somatotropin (growth hormone). Children with this genetic abnormality will not grow to normal height unless they receive injections of somatotropin. Develop a hypothesis to explain why this hormone therapy is effective.

5. In this chapter, you have read that transcription is generally controlled by the binding of a protein to a DNA sequence that is "upstream" from a gene, rather than by modification of RNA polymerase. Develop a hypothesis to explain why this is so.

Selected Key Terms

activator protein *15.2*	oncogene *15.6*
apoptosis *CI*	operator *15.2*
Barr body *15.4*	operon *15.2*
cancer *15.6*	phytochrome *15.5*
cell differentiation *15.3*	positive control system *15.1*
ecdysone *15.5*	promoter *15.2*
enhancer *15.5*	protein kinases *15.6*
growth factor *15.6*	proto-oncogene *15.6*
hormone *15.5*	regulatory protein *15.1*
ICE-like proteases *CI*	repressor *15.2*
metastasis *15.6*	tumor *15.6*
negative control system *15.1*	

Readings

Duke, R. D. Ojcius, and J. Ding-E Young. December 1996. "Cell Suicide in Health and Disease." *Scientific American* 80–87.

Feldman, M., and L. Eisenbach. November 1988. "What Makes a Tumor Cell Metastatic?" *Scientific American* 259(5): 60–85.

Murray, A., and M. Kirschner. March 1991. "What Controls the Cell Cycle?" *Scientific American* 264(3): 56–63.

Tijan, R. February 1995. "Molecular Machines That Control Genes." *Scientific American* 54–61.

Web Site See *http://www.wadsworth.com/biology* for practice quiz questions, hypercontents, BioUpdates, and critical thinking. The Wadsworth Biology Resource Center provides a wealth of information fully organized and integrated by chapter.

16 RECOMBINANT DNA AND GENETIC ENGINEERING

Mom, Dad, and Clogged Arteries

Butter! Bacon! Eggs! Ice cream! Cheesecake! Possibly you think of such foods as enticing, off-limits, or both. After all, who among us doesn't know about animal fats and the dreaded cholesterol?

Soon after you feast on these fatty foods, cholesterol enters the bloodstream. Cholesterol is important. It is a structural component of animal cell membranes, and without membranes, there would be no cells. Cells also remodel cholesterol into various molecules, including the vitamin D that is necessary for the development of good bones and teeth. Normally, however, your liver synthesizes enough cholesterol for your cells.

Some proteins circulating in the blood combine with cholesterol and other substances to form lipoprotein particles. The *HDLs* (high-density lipoproteins) collect cholesterol and transport it to the liver, where it can be metabolized. *LDLs* (low-density lipoproteins) normally end up in cells that store or use cholesterol.

Sometimes too many LDLs form, and the excess infiltrates the elastic walls of arteries. There they promote formation of abnormal masses called atherosclerotic plaques (Figure 16.1). These interfere with blood flow and narrow the arterial diameter. If the plaques clog one of the tiny coronary arteries that deliver blood to the heart, the resulting symptoms can range from mild chest pains to a heart attack.

How your body handles dietary cholesterol depends on what you inherited from your parents. Consider the gene for a protein that serves as the cell's receptor for LDLs. Inherit two "good" alleles of that gene, and your blood level of cholesterol will tend to remain so low that your arteries will never get clogged, even with a high-fat diet. Inherit two copies of a certain mutated allele, however, and you are destined to develop a rare genetic disorder called *familial cholesterolemia*. With this disorder, cholesterol builds up to abnormally high levels. Many affected individuals die of heart attacks during childhood or their teens.

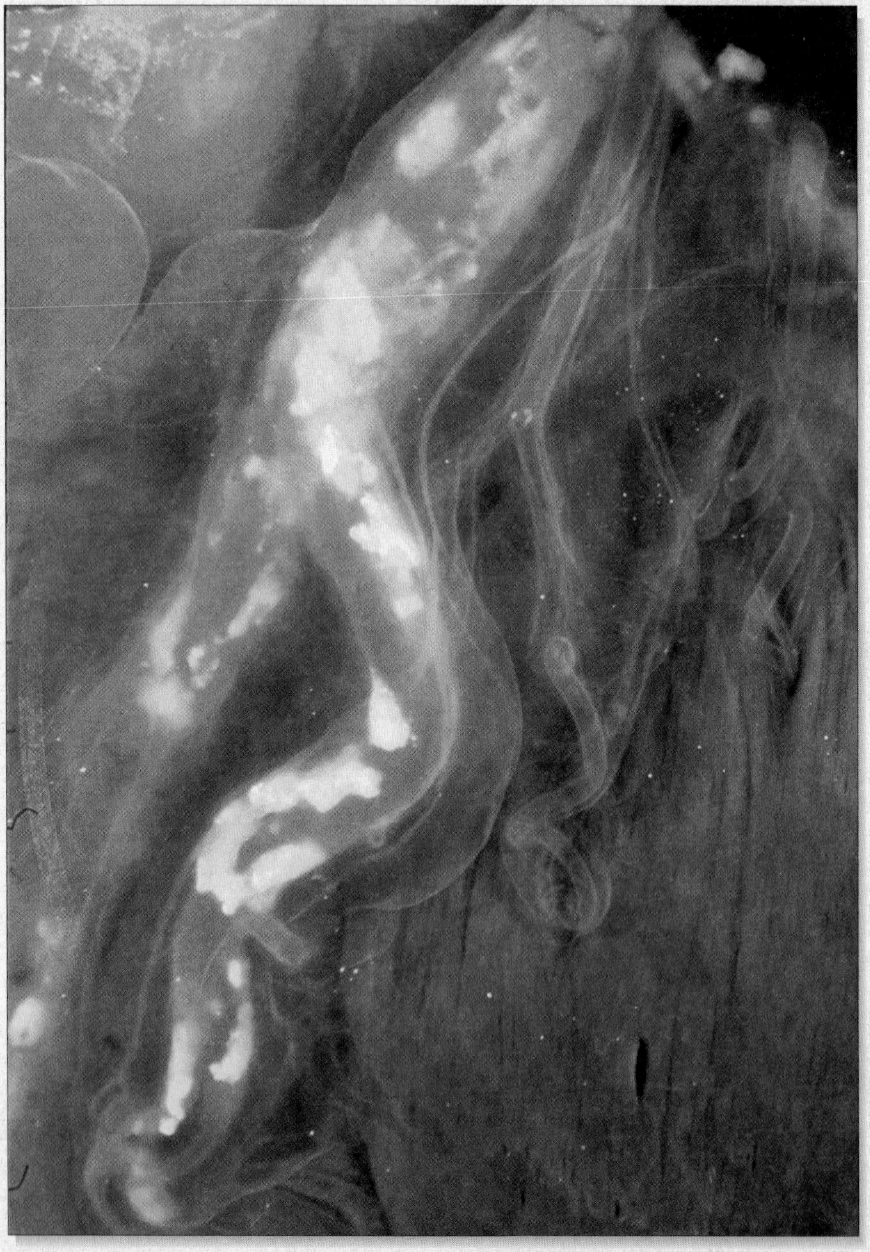

Figure 16.1 Potentially life-threatening plaques (*bright yellow*) inside one of the coronary arteries, which are small vessels that deliver blood to the heart. The plaques are the legacy of abnormally high levels of cholesterol.

In 1992 a woman from Quebec, Canada, became a milestone in the history of genetics. She was thirty years old. Like two of her younger brothers who had died from heart attacks in their early twenties, she inherited the defective gene for the LDL receptor. She herself survived a heart attack when she was sixteen. At twenty-six, she had coronary bypass surgery.

At the time, people were hotly debating the risks and promise of **gene therapy**—the transfer of one or more normal or modified genes into an individual's body cells to correct a genetic defect or boost resistance to disease. Even so, the woman consented to undergo an untried, physically wrenching procedure designed to give her body working copies of the good gene.

Medical researchers removed about 15 percent of the woman's liver. They placed liver cells in a nutrient-rich medium that promoted growth and division. *And they spliced the good gene into the genetic material of a harmless virus.* That modified virus served roughly the same function as a hypodermic needle. The researchers allowed it to infect the cultured liver cells and thereby insert copies of the good gene into them.

Later, the researchers infused about a billion of the modified cells into the woman's portal vein, a major blood vessel that leads directly to the liver. There, at least some cells took up residence, and they started to produce the missing cholesterol receptor. Two years after this, between 3 and 5 percent of the woman's liver cells were behaving normally and sponging up cholesterol from the blood. Her blood levels of LDLs had declined nearly 20 percent. Scans of her arteries showed no evidence at all of the progressive clogging that had nearly killed her. At a recent press conference, the woman announced she is active and doing well.

Her cholesterol levels do remain more than twice as high as normal, and it is too soon to know whether the gene therapy will prolong her life. Yet the intervention provides solid proof that the concept of gene therapy is sound, and hopes are high.

As you might gather from this pioneering clinical application, recombinant DNA technology has truly staggering potential for medicine. It also has great potential for agriculture and industry. The technology does not come without risks. With this chapter, we consider some basic aspects of the new technology. At the chapter's end, we also address some ecological, social, and ethical questions related to its application.

KEY CONCEPTS

1. Genetic experiments have been proceeding in nature for billions of years, through gene mutations, crossing over and recombination, and other natural events.

2. Humans are now purposefully bringing about genetic changes by way of recombinant DNA technology. Such enterprises are called genetic engineering.

3. With this technology, researchers isolate, cut, and splice together gene regions from different species, then greatly amplify the number of copies of the genes that interest them. The genes, and in some cases the proteins they specify are produced in quantities that are large enough to use for research and for practical applications.

4. Three activities are at the heart of recombinant DNA technology. First, procedures based on specific types of enzymes are used to cut DNA molecules into fragments. Second, the fragments are inserted into cloning tools, such as plasmids. Third, the fragments containing the genes of interest are identified, then copied rapidly and repeatedly.

5. Genetic engineering involves isolating, modifying, and inserting genes back into the same organism or into a different one. The goal is to beneficially modify traits that the genes influence. Human gene therapy, which focuses on controlling or curing genetic disorders, is an example.

6. The new technology raises social, legal, ecological, and ethical questions regarding its benefits and risks.

RECOMBINATION IN NATURE—AND IN THE LABORATORY

For at least 3 billion years, nature has been conducting uncountable numbers of genetic experiments, through mutation, crossing over, and other events that introduce changes in genetic messages. This is the source of life's diversity.

For many thousands of years, we humans have been changing numerous genetically based traits of species. By artificial selection practices, we produced new crop plants and breeds of cattle, birds, dogs, and cats from wild ancestral stocks. We developed meatier turkeys and sweeter oranges, larger corn, seedless watermelons, flamboyant ornamental roses, and other useful plants. We produced splendid hybrids, including the tangelo (tangerine × grapefruit) and mule (horse × donkey).

Researchers now use **recombinant DNA technology** to analyze genetic changes. With this technology, they cut and splice DNA from different species, then insert the modified molecules into bacteria or other types of cells that engage in rapid replication and cell division. The cells copy the foreign DNA right along with their own. In short order, huge populations produce useful quantities of recombinant DNA molecules. The new technology also is the basis of **genetic engineering**, by which genes are isolated, modified, and inserted back into the same organism or into a different one.

Plasmids, Restriction Enzymes, and the New Technology

Believe it or not, this astonishing technology originated with the innards of bacteria. Bacterial cells have a single chromosome, a circular DNA molecule that has all the genes they require to grow and reproduce. But many species also have **plasmids**, or small, circular molecules of "extra" DNA that contain a few genes (Figure 16.2).

Usually, plasmids are not essential for survival, but some of the genes they carry may benefit the bacterium. For instance, some plasmid genes confer resistance to antibiotics. (*Antibiotics*, remember, are toxic metabolic products of microorganisms that can kill or inhibit the growth of competing microorganisms.) The bacterium's replication enzymes copy and reproduce plasmid DNA, just as they copy chromosomal DNA.

In nature, many bacteria are able to transfer plasmid genes to a bacterial neighbor of the same species or a different one (Section 22.2). Replication enzymes may even integrate a transferred plasmid into the bacterial chromosome of a recipient cell. A recombinant DNA molecule is the result.

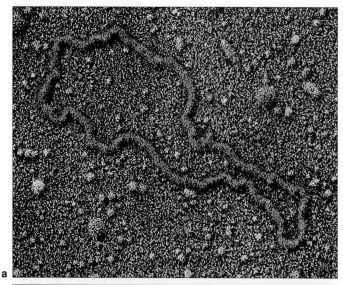

a

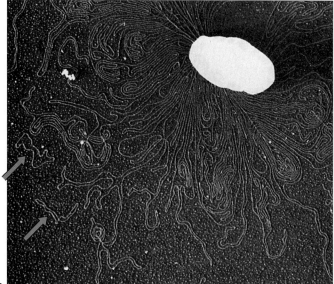

b

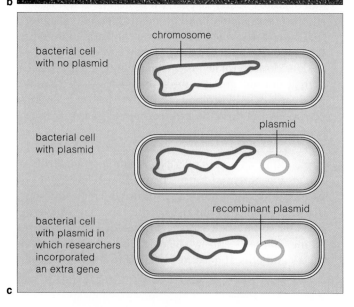

bacterial cell with no plasmid

chromosome

bacterial cell with plasmid

plasmid

bacterial cell with plasmid in which researchers incorporated an extra gene

recombinant plasmid

c

Figure 16.2 (**a**) A plasmid at high magnification. (**b**) Plasmids (*blue* arrows) released from a ruptured *Escherichia coli* cell. (**c**) Naturally occurring and modified plasmids.

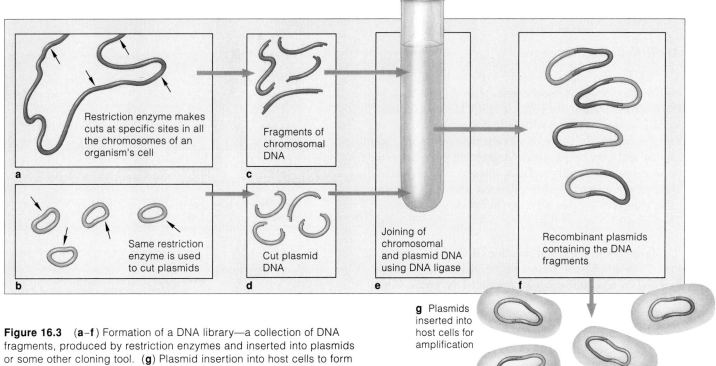

Figure 16.3 (**a–f**) Formation of a DNA library—a collection of DNA fragments, produced by restriction enzymes and inserted into plasmids or some other cloning tool. (**g**) Plasmid insertion into host cells to form cloned DNA, or multiple, identical copies of the DNA fragments.

Infectious particles called viruses as well as bacteria dabble in gene transfers and recombinations. And so do most eukaryotic species. As you might imagine, viral infection does a bacterium no good. Over evolutionary time, bacteria developed an arsenal against invasion by harmful genes. They became equipped with many types of **restriction enzymes**, which are able to recognize and cut apart foreign DNA that may enter a cell. Eventually, researchers learned how to use plasmids *and* restriction enzymes for genetic recombination in the laboratory.

Producing Restriction Fragments

Each type of restriction enzyme makes a cut wherever it recognizes a specific, very short nucleotide sequence in the DNA. Cuts at two identical sequences in the same DNA molecule produce a fragment. Because some types of enzymes make *staggered* cuts, some fragments have single-stranded portions at both ends. Sometimes these are referred to as sticky ends:

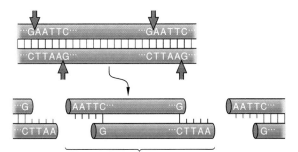

liberated DNA fragment with sticky ends

By "sticky," we mean the short, single-stranded ends of a DNA fragment will have the chemical capacity to base-pair with any other DNA molecule that also has been cut by the same restriction enzyme.

For example, suppose you use the same restriction enzyme to cut plasmids *and* DNA molecules that you have isolated from a human cell. When you mix the cut molecules together, they base-pair at the cut sites. After this, you add **DNA ligase** to the mixture. DNA ligase is an enzyme that seals DNA's sugar-phosphate backbone at the cut sites, just as it does during DNA replication. In this way, you create "recombinant plasmids," which have pieces of DNA from another organism inserted into them (Figure 16.2c).

You now have a **DNA library**. It is a collection of DNA fragments, produced by restriction enzymes, that have been incorporated into plasmids, as illustrated in Figure 16.3.

With recombinant DNA technology, DNA from different species is cut and spliced together; then the recombinant molecules are amplified, by way of copying mechanisms, to produce useful quantities of the genes of interest.

Recombination is made possible by the use of restriction enzymes that make specific cuts in DNA molecules and of DNA ligases that seal the cut ends.

Recombinant DNA technology has uses in basic research.

The new technology also has uses in genetic engineering—the deliberate modification of genes, followed by their insertion into the same individual or a different one.

Amplification Procedures

A DNA library is almost vanishingly small. To obtain useful amounts of it, biochemists resort to methods of **DNA amplification**, by which a DNA library is copied again and again. One such method uses "factories" of bacteria, yeasts, or some other cells that can reproduce rapidly and take up plasmids. A growing population of these cells can amplify a DNA library in short order. Their repeated cycles of replication and cell division yield cloned DNA that has been inserted into plasmids. The "cloned" part of this name refers to the multiple, identical copies of DNA fragments.

The **polymerase chain reaction**, or **PCR**, is a newer method of amplifying fragments of DNA. The reactions proceed in test tubes, not in microbial factories. First, researchers identify short nucleotide sequences located just before and just after a region of DNA from a cell of the organism that interests them. Then they synthesize **primers**. These short nucleotide sequences, recall, will base-pair with any complementary sequences in DNA. And the replication enzymes called **DNA polymerases** recognize them as START tags (Section 13.3).

For PCR, a DNA polymerase from a bacterium that lives in hot springs, even water heaters, is the enzyme of choice, because it remains functional at the elevated temperatures necessary to unwind DNA and also at the lower temperatures necessary for base pairing. Researchers mix together the primers, the polymerases, all the DNA from one of the organism's cells, and free nucleotides. Next, they expose the mixture to precise temperature cycles. During each cycle, the two strands of all the DNA molecules unwind from each other. And primers become positioned on exposed nucleotides at the targeted sites according to base-pairing rules (Figure 16.4). With each round of reactions, the number of DNA molecules doubles. For example, if there are 10 such molecules in the test tube, there soon will be 20, then 40, 80, 160, 320, 640, 1,280, and so on. Very quickly, a target region from a single DNA molecule can be amplified to *billions* of molecules.

In short, *PCR amplifies samples that contain even tiny amounts of DNA.* As you will see in the next section, such samples can be obtained from fossils—even from a single hair or drop of blood left at the scene of a crime.

Sorting Out Fragments of DNA

When restriction enzymes cut DNA, the fragments they produce are not all the same length. Researchers can use **gel electrophoresis** to separate fragments from one another according to length. This laboratory procedure employs an electric field to force molecules through a viscous gel and to separate them according to physical

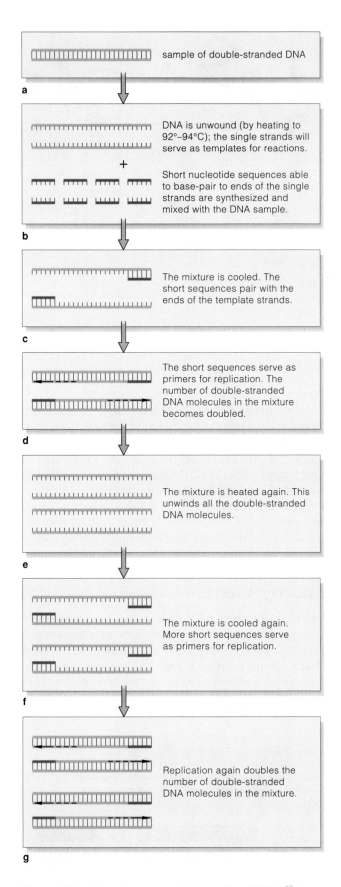

a sample of double-stranded DNA

b DNA is unwound (by heating to 92°–94°C); the single strands will serve as templates for reactions.

+

Short nucleotide sequences able to base-pair to ends of the single strands are synthesized and mixed with the DNA sample.

c The mixture is cooled. The short sequences pair with the ends of the template strands.

d The short sequences serve as primers for replication. The number of double-stranded DNA molecules in the mixture becomes doubled.

e The mixture is heated again. This unwinds all the double-stranded DNA molecules.

f The mixture is cooled again. More short sequences serve as primers for replication.

g Replication again doubles the number of double-stranded DNA molecules in the mixture.

Figure 16.4 The polymerase chain reaction (PCR).

Figure 16.5 One of the methods used for sequencing DNA, as first developed by Frederick Sanger. The method is employed to determine the nucleotide sequence of specific DNA fragments, as this example illustrates.

a Single-stranded DNA fragments are added to a solution in four different test tubes:

All four tubes contain DNA polymerases, short nucleotide sequences that can serve as primers for replication, and the nucleotide subunits of DNA (A, T, C, and G). Each tube contains a modified, labeled version of only one of the four kinds of nucleotides. We can show these as A*, T*, C*, and G*. The labeled form is present in low concentration, along with a generous supply of the unmodified form of the same nucleotide. Let's follow what happens in the tube with the A* subunits.

b As expected, the DNA polymerase recognizes a primer that has become attached to a fragment, which it uses as a template strand. The enzyme assembles a complementary strand according to base-pairing rules (A only to T, and C only to G). Sooner or later, the enzyme picks up an A* subunit for pairing with a T on the template strand. The modified nucleotide is a chemical roadblock—it prevents the enzyme from adding more nucleotides to the growing complementary strand. In time, the tube contains labeled strands of different lengths, as dictated by the location of each A* in the sequence:

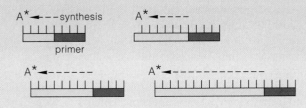

The same thing happens in the other three tubes, with strand lengths dictated by the location of T*, C*, and G*.

c DNA from each of the four tubes is placed in four parallel lanes in the same gel. Then the DNA can be subjected to electrophoresis. The resulting nucleotide sequence can be read off the resulting bands in the gel. Look at the numbers running down the side of this diagram:

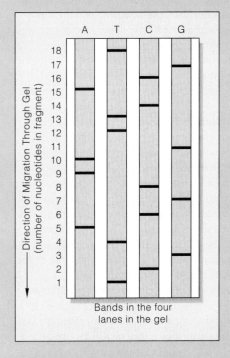

Start with "1" and read across the four lanes (A, T, C, G). As you can see, T is the closest to the start of the nucleotide sequence; it has migrated farthest through the gel. At "2," the next nucleotide is C, and so on. The entire sequence, read from the first nucleotide to the last, is

T C G T A C G C A A G T T C A C G T

And now, by applying the rules of base pairing, you can deduce the sequence of the DNA fragment that served as the template.

and chemical properties. For DNA, size alone affects how far the molecules move. (For proteins, remember, a molecule's size, shape, and net surface charge are the determining factors.)

A gel that contains DNA fragments is immersed in a buffered solution, and electrodes connect the solution to a power source. Apply voltage, and negatively charged phosphate groups of the fragments respond to it. The fragments move toward the positively charged pole at different rates and thereby become separated into bands according to their lengths. Again, how far they migrate depends only on their size; the larger fragments cannot move as fast through it. After a predetermined period of electrophoresis, fragments of different length can be identified by staining the gel.

DNA Sequencing

Once DNA fragments from a sample have been sorted out according to length, researchers can work out the nucleotide sequence of each type. The Sanger method of DNA sequencing, as detailed in Figure 16.5, will give you a sense of one of the ways this can be done.

Restriction fragments can be rapidly amplified by PCR and by large populations of bacterial cells or yeast cells.

Restriction fragments can be sorted out according to length, as by gel electrophoresis. Then sequencing methods can be employed to determine the nucleotide sequences of the DNA fragments.

16.3 RIFF-LIPS AND DNA FINGERPRINTS

Suppose a researcher has isolated a good-size sample of your DNA molecules. She cuts them up with restriction enzymes, then subjects the restriction fragments to gel electrophoresis. Large ones cannot move as fast as small ones, so the fragments separate from one another in the gel. Now the researcher uses the *Southern blot method*. She transfers the electrophoretically separated fragments from the gel to a membrane filter and then immerses the filter in a solution containing a labeled probe, which can bind to the fragment that interests her. Afterward, she applies the same method to DNA samples from several other people.

When she compares the banding patterns, she finds small variations among them. Why? Although the DNA molecules of any two people are alike in some regions, they differ significantly in others. *The slight molecular differences shift the number and location of sites where restriction enzymes can make their cuts.* As a result, some fragments from corresponding regions of DNA from two or more individuals differ in length. The differences in DNA electrophoresis patterns among individuals are called **RFLPs** (pronounced RIFF-lips). The term stands for **restriction fragment length polymorphisms**.

As you know, the set of fingerprints of each human is unique, a marker of his or her identity. As is true of other sexually reproducing species, each human also has a **DNA fingerprint**—a unique array of RFLPs, inherited from each parent in a Mendelian pattern.

RFLP analysis is a wonderful new procedure for basic research. It also has startling applications in society at large. Consider the human genome project. As described in Section 16.6, this is an ambitious effort to decipher the nucleotide sequence of human DNA, with its 3.2 billion base pairs. Or consider how evolutionary biologists are analyzing RFLPs to decipher DNA from mummies, from fossilized insects and plants, and from mammoths and humans that were preserved many thousands of years ago in glacial ice. Investigators even used RFLP analysis to confirm that bones exhumed from a shallow pit in Siberia belonged to five members of the Russian imperial family, all shot to death in secrecy in 1918.

RFLP analysis has spectacular potential for medicine. Researchers have already pinpointed unique restriction sites in several mutated genes, including the ones that cause sickle-cell anemia, cystic fibrosis, and other genetic disorders. They can use the sites to find out whether an individual carries a mutated gene, as clinicians are now doing in prenatal diagnosis (Section 12.11).

What about social issues? Think of all the paternity and maternity cases against celebrities and ordinary folks alike. They now can be resolved by comparing the child's DNA fingerprints with those of the disputed parent. Or think about forensic medicine. A few drops of blood, semen, or even cells from a hair follicle at a crime scene or on a suspect's clothing often yield enough DNA to identify the perpetrator (Figure 16.6).

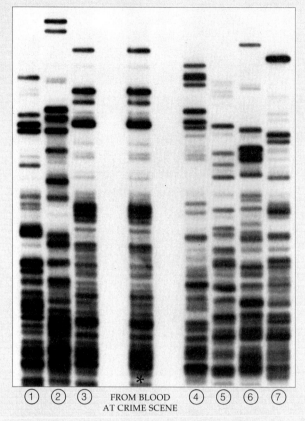

① ② ③ FROM BLOOD AT CRIME SCENE ④ ⑤ ⑥ ⑦

Figure 16.6 DNA fingerprint from a bloodstain discovered at a crime scene, with DNA fingerprints from blood samples of seven suspects (*circled numbers*). Which one of the seven was a match?

Britain, which started the first DNA database in the world, is at the forefront of solving crimes with DNA fingerprinting. Also, police in Britain can lawfully collect samples of hair or saliva from crime suspects, even if by force, for RFLP analysis. At this writing, detectives in Cardiff, Wales, are engaged in a massive "blood drive" to locate the man who raped and killed a local teenager. They expect to contact as many as 5,000 "donors." The detectives have let it be known that anyone who declines to donate blood voluntarily risks calling attention to himself—and inviting exceedingly fine scrutiny. Mass DNA screening has already helped solve other murders in Britain and Germany.

Recently in the United States, defense lawyers in a high-profile and seemingly interminable criminal case hammered away at DNA evidence. Someone nearly cut off the head of a celebrity's wife, slashed her friend to death, and apparently left a trail of blood leading to the celebrity's house. RFLP analysis seems to implicate him. His lawyers pounded on the competence and character of just about everybody who found blood at the crime scene and the house, collected samples, ran the PCRs, and did the RFLP analyses. And one can only wonder if sensationalized efforts of defense attorneys alone will send DNA research back to the twelfth century.

Use of DNA Probes

Recombinant plasmids are not much use in themselves. They must be mixed with living cells that can take them up. So how do you find out which cells do this? You can utilize **DNA probes**: short DNA sequences synthesized from radioactively labeled nucleotides. Part of a probe must be designed to base-pair with some portion of the DNA of interest. Any base pairing between nucleotide sequences (that is, RNA as well as DNA) from different sources is called **nucleic acid hybridization**.

The challenge is to "select" the cells that have taken up recombinant plasmids and the gene of interest. One way to do this is to use a plasmid containing a gene that confers resistance to a particular antibiotic. You put prospective host cells on a culture medium that has the antibiotic added to it. The antibiotic prevents growth of all cells *except* the ones that house plasmids with the antibiotic-resistance gene.

As the cells divide, they form colonies. Assume the colonies are growing on agar, a gel-like substance, in a petri dish. You blot the agar against a nylon filter. Some cells stick to the filter at sites that mirror the locations of the original colonies (Figure 16.7). You use solutions to rupture the cells, fix the released DNA onto the filter, and make the double-stranded DNA unwind. Then you add DNA probes, which hybridize only with the gene region having a complementary base sequence. Probe-hybridized DNA emits radioactivity and allows you to tag the colonies that harbor the gene of interest.

Use of cDNA

Researchers study genes to find out about the protein products and how they are put to use. However, even if a host cell takes up a gene, the cell may not be able to synthesize the protein. For example, recall that new mRNA transcripts of human genes contain noncoding sequences (introns). They cannot be translated until the introns are snipped out and coding regions (exons) are spliced together into mature form. Bacterial enzymes do not recognize the splice signals, so bacterial host cells cannot always properly translate human DNA.

Sometimes researchers get around the problem by using **cDNA**, which is a DNA strand "copied" from a *mature* mRNA transcript for the gene. By a backwards process called **reverse transcription**, a complementary DNA strand is assembled on mRNA. The outcome is a hybrid mRNA-cDNA molecule (Figure 16.8). Enzymes remove the RNA and then synthesize a complementary DNA strand, thus making double-stranded cDNA.

Double-stranded cDNA can be further modified, as by attaching signals for transcription and translation. The modified cDNA can be inserted into a plasmid for

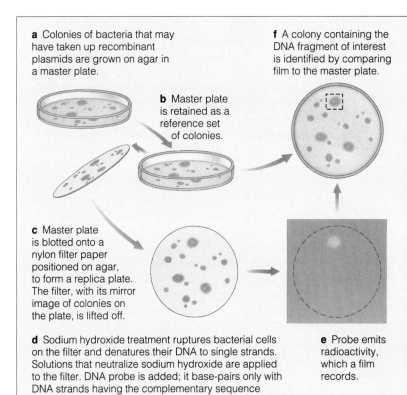

a Colonies of bacteria that may have taken up recombinant plasmids are grown on agar in a master plate.

b Master plate is retained as a reference set of colonies.

c Master plate is blotted onto a nylon filter paper positioned on agar, to form a replica plate. The filter, with its mirror image of colonies on the plate, is lifted off.

d Sodium hydroxide treatment ruptures bacterial cells on the filter and denatures their DNA to single strands. Solutions that neutralize sodium hydroxide are applied to the filter. DNA probe is added; it base-pairs only with DNA strands having the complementary sequence

e Probe emits radioactivity, which a film records.

f A colony containing the DNA fragment of interest is identified by comparing film to the master plate.

Figure 16.7 Use of a DNA probe to identify a colony of bacterial cells that have taken up plasmids with a specific DNA fragment.

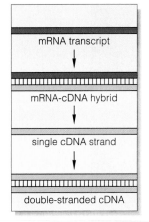

a A mature mRNA transcript of a gene of interest is used as a template to assemble a single strand of DNA. The enzyme reverse transcriptase catalyzes the assembly. An nRNA-cDNA hybrid molecule is the result.

b By enzyme action, the mRNA is removed and a second strand of DNA is assembled on the DNA strand remaining. The outcome is a double-stranded cDNA, as "copied" from an mRNA template.

mRNA transcript

mRNA-cDNA hybrid

single cDNA strand

double-stranded cDNA

Figure 16.8 Formation of cDNA from an mRNA transcript.

amplification. Recombinant plasmids can be inserted into bacterial cells, which may use the cDNA directions for synthesizing the protein of interest.

Host cells that take up modified genes may be identified by procedures involving DNA probes.

Modified eukaryotic genes may not be expressed in host bacterial cells unless first introduced into the cells as cDNA.

Many years have passed since foreign DNA was first transferred into a plasmid, yet that transfer started a debate that will continue into the next century. The issue is this: *Do the benefits of gene modifications and gene transfers outweigh the potential dangers?* Before you personally come to any conclusions, reflect upon a few examples of the work with bacteria, then with plants.

Genetically Engineered Bacteria

Imagine a miniaturized factory that churns out insulin or another protein having medical value. This is an apt description of the huge stainless steel vats of genetically engineered, protein-producing bacteria.

Think of *diabetics*, who need insulin injections for as long as they live. The pancreas produces insulin, a protein hormone. Medical supplies of insulin once were obtained from pigs and cattle, although some diabetics developed allergic reactions to foreign insulin. Eventually, synthetic genes for human insulin were transferred into *E. coli* cells. (Can you say why the genes had to be synthesized?) This was the start of huge bacterial factories that manufacture human insulin, hemoglobin, somatotropin, interferon, albumin, and other valuable proteins.

In addition, in research laboratories around the world, certain strains of bacteria are now being engineered to degrade oil

Figure 16.9 Spraying an experimental strawberry patch in California with "ice-minus" bacteria. The sprayer used elaborate protective gear to meet government regulations in effect then.

spills from tankers, to manufacture alcohol and other chemicals, to process minerals, or to leave crop plants alone. The strains are harmless to begin with. The ones meant to be confined to the laboratory are "designed" to prevent escape. As added precautions, the foreign DNA usually includes "fail-safe" genes. Such genes are silent *unless* the engineered bacteria somehow are exposed to conditions that occur in the environment. Exposure will activate the genes, with lethal results for the bacteria.

For example, foreign DNA may include a *hok* gene next to a promoter of the lactose operon (Section 15.2). If an engineered bacterium does manage to escape to the environment, where sugars are common, the gene will trip into action. The protein the gene specifies will destroy membrane function and so destroy the cell.

Even so, some people worry about the possible risks of introducing genetically engineered bacteria into the environment. Consider how Steven Lindow engineered a bacterium that can make many crop plants resist frost. A protein at the bacterial surface promotes formation of ice crystals. Lindow excised the ice-forming gene from some cells. As he hypothesized, spraying his "ice-minus bacteria" on strawberry plants in an isolated field prior to a frost would make plants less vulnerable to freezing. Even though Lindow deleted a *harmful* gene from an organism, it triggered a bitter legal battle over releasing engineered bacteria into the environment. In time, the courts did rule in favor of allowing such a release, and researchers sprayed a small strawberry patch (Figure 16.9). Nothing bad ever happened. Since then, the rules governing the release of genetically engineered species have become less restrictive.

Genetically Engineered Plants

A HUNT FOR BENEFICIAL GENES   Currently, botanists are combing the world for seeds and other living tissues from wild ancestors of potatoes, corn, and other plants. They send their prizes—genes from a plant's lineage— to a safe storage laboratory in Colorado. Why? *Farmers now rely on only a few strains of high-yield crop plants that feed most of the human population.* The near-absence of genetic diversity means our food base is dangerously vulnerable to many disease-causing viruses, bacteria, and fungi. It is a race against time. Our population has reached an astounding size. People, bulldozers, and power saws are encroaching on previously uninhabited areas; hundreds of wild plant species disappear weekly.

The danger is real. Recently a new fungal strain that rots potatoes entered the United States from Mexico. In 1994, crop losses from the resulting *potato blight* topped $100 million. Pennsylvania farmers had little to harvest. Possibly laboratory-stored, fungus-resistant strains of potato plants will save crops in the future.

PLANT REGENERATION Botanists also search for good genes in the laboratory. Years ago, Frederick Steward and his coworkers cultured cells of carrot plants and induced them to grow into small embryos, some of which grew into whole plants (Section 31.7). Other species, including many crop plants, are regenerated today from cultured cells. The methods raise mutation rates, so cultures are a source of genetic modifications.

Researchers can pinpoint a useful mutation among millions of cells. Suppose a culture medium contains a toxic product of a disease agent. If a few cells have a mutated gene that confers resistance to the toxin, only they will live in the culture. If cells regenerate whole plants that can be hybridized with other varieties, the hybrids may get the good gene.

METHODS OF GENE TRANSFER Now genetic engineers insert genes into cultured plant cells. For example, they insert DNA fragments into the "Ti" plasmid from the bacterium *Agrobacterium tumefaciens*, which can infect many species of flowering plants. Some plasmid genes invade a plant's DNA and induce formation of abnormal tissue masses called crown gall tumors (Figure 16.10*a*). Before introducing the Ti plasmid into plants, researchers first remove the tumor-inducing genes and insert desired genes into it. They grow the modified bacterial cells with cultured plant cells, then regenerate plants from the cells that take up the genes. In some instances, the foreign genes have been expressed in plant tissues, with observable effects (Figure 16.10*b*).

A. tumefaciens can only infect beans, peas, potatoes, and other dicots. However, wheat, corn, rice, and many major food crops are monocots, which the bacterium cannot infect. In some cases, genetic engineers can use electric shocks or chemicals to deliver modified DNA into protoplasts. (Protoplasts are plant cells stripped of their walls.) At this writing, regenerating some species of plants from protoplast cultures is not yet possible. But researchers have had some success in delivering genes into cultured plant cells by actually shooting them with pistols. Instead of bullets, they use blanks to drive DNA-coated, microscopic particles into the cells.

ON THE HORIZON Despite many obstacles, improved varieties of crop plants have been developed or are in the works. For example, genetically engineered cotton

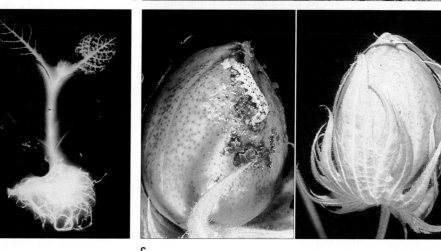

Figure 16.10 Gene transfers in plants. (**a**) Genes in a plasmid from a common bacterium (*Agrobacterium tumefaciens*) cause crown gall tumors on woody plants. Scientists use this plasmid to move desirable genes into cultured plant cells. (**b**) Evidence of a successful gene transfer: a modified tobacco plant that glows in the dark. A firefly gene became incorporated into the plant's DNA and is being translated. The protein product is luciferase, an enzyme required for bioluminescence (Section 6.7). (**c**) Example of a valuable plant that benefits from genetic engineering. *Left:* Buds of cotton plants are vulnerable to worm attack. *Right:* Buds of a genetically modified cotton plant resist attack.

plants now resist worm attacks without any help from pesticides, which have the disadvantage of also killing off beneficial pest-eating insects (Figure 16.10*c*).

Also on the horizon are engineered plants that can serve as factories for pharmaceuticals. A few years ago, genetically engineered tobacco plants that make human hemoglobin, melanin, and other proteins were planted in a field in North Carolina. Ecologists later found no trace of foreign genes or proteins in the soil or in other plants or animals in the vicinity. Recently, mustard plant cells made plastic beads in a Stanford University laboratory. (Plastics are long-chain polymers of identical subunits.) The plant DNA incorporates three bacterial enzymes that are able to convert carbon dioxide, water, and nutrients into inexpensive, biodegradable plastic.

Fail-safe measures must be built into genetically engineered organisms that may end up where they shouldn't be.

Genetically engineered plants are one of the few options available for protecting our vulnerable food supply.

GENETIC ENGINEERING OF ANIMALS

Supermice and Biotech Barnyards

The first mammals enlisted for experiments in genetic engineering were laboratory mice. Consider an example of this work. R. Hammer, R. Palmiter, and R. Brinster corrected a hormone deficiency that leads to dwarfism in mice. Insufficient levels of somatotropin give rise to the abnormality. The researchers used a microneedle to inject the gene for rat somatotropin into fertilized mouse eggs. After the eggs were implanted in an adult female, the gene was successfully integrated into the mouse DNA. Later, the baby mice in which the foreign gene was expressed were 1-1/2 times larger than their dwarf littermates. In other experiments, researchers transferred the gene for human somatotropin into a mouse embryo, where it became integrated into the DNA. The modified embryo grew up to be "supermouse" (Figure 16.11).

Today, as part of research into the molecular basis of Alzheimer's disease and other genetic disorders, a few human genes are being inserted into mouse embryos. Besides microneedles, microscopic laser beams are used to open up temporary holes in the plasma membrane of cultured cells, although such methods have varying degrees of success. Retroviruses also are used to insert genes into cultured cells, which may incorporate the foreign genes into their DNA. But the genetic material of retroviruses can undergo rearrangements, deletions, and other alterations that shut down the introduced genes. And if the virus particles escape from laboratory isolation, they may infect organisms.

Animals of "biotech barnyards" are competing with bacterial factories as genetically engineered sources of proteins. For example, goats produce CFTR protein (for treating cystic fibrosis) and TPA (which diminishes the severity of heart attacks). Cattle may soon produce human collagen for repairing cartilage, bone, and skin. Also, in 1996 researchers made a genetic duplicate of an adult ewe. They reprogrammed one of her mammary gland cells so all of its genes could be expressed. They fused that cell with an egg (from another ewe) from which the nucleus had been removed. Signals from the egg cytoplasm triggered the development of a cluster of embryonic cells, which was implanted into a surrogate mother. The result was a clone, a lamb named *Dolly*. At this writing, the clone is thriving. The experiment opens up the possibility of producing genetically engineered clones of sheep, cattle, and other farm animals that can supply consistently uniform quantities of proteins, even tissues and organs, for medical uses and for research.

Applying the New Technology to Humans

Researchers around the world are working their way through the 3.2 billion base pairs of the twenty-three

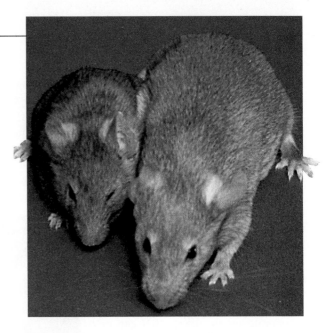

Figure 16.11 Ten-week-old mouse littermates. The mouse to the left weighs 29 grams and the one to the right, 44 grams. The larger mouse grew from a fertilized egg into which the gene for human somatotropin (growth hormone) had been inserted.

pairs of human chromosomes. This ambitious effort is the **human genome project**. Some researchers focus on specific chromosomes. Others are deciphering certain gene regions only. For example, rather than studying the noncoding sequences (introns), J. Venter and Sidney Brenner isolate mRNAs from brain cells and use them to make cDNA. By sequencing only cDNAs, they have identified hundreds of previously unknown genes.

About 99.9 percent of the nucleotide sequence is the same in all humans on Earth. The remaining 0.1 percent —about 3,200,000 base pairs—are mutations and other sequence variations sprinkled throughout the genome. They account for *all* genetic differences in the human population. Therefore, completion of the project should give us the ultimate reference book of human biology and genetic disorders. What will we do with all of that information? Certainly we will use it in the search for treatments and cures for genetic disorders. Of the 2,000 or so genes studied so far, 400 have already been linked to genetic disorders. And the knowledge opens doors to gene therapy, the transfer of one or more normal or modified genes into the body cells of an individual to correct a genetic defect or boost resistance to disease.

But what about forms of human gene expression that are neither disabling nor life-threatening? Will we tinker with them, also? We now leave the chapter with a *Focus* essay that looks at these questions.

As with any new technology, the potential benefits of recombinant DNA technology and genetic engineering should be weighed carefully against the potential risks.

This chapter opened with a historic, inspiring case, the first proof that gene therapy can help save human lives. It closes with questions that invite you to consider some social and ethical issues related to the application of recombinant DNA technology to our rapidly advancing knowledge of the human genome.

To most of us, human gene therapy to correct genetic abnormalities seems like a socially acceptable goal. Let's take this idea one step further. Is it also socially desirable or acceptable to change certain genes of a normal human individual (or sperm or egg) to alter or enhance traits?

The idea of selecting desirable human traits is called *eugenic engineering.* Yet who decides which forms of a trait are most "desirable"? For example, what happens if prospective parents start picking the sex of a child by way of genetic engineering? In one survey group, three-fourths of the people who were asked this question said they would choose a boy. What would be the long-term social implications of a drastic shortage of girls?

As other examples, would it be okay to engineer taller or blue-eyed or fair-skinned boys and girls? Would it be okay to engineer "superhumans" with amazing strength or intelligence? Suppose a person of average intelligence moved into a town of 800 Einsteins. Would the response go beyond mutterings of "There goes the neighborhood"? Are there people narcissistic enough to commission a clone of themselves, one that has just a few genetically engineered "improvements"? Researchers can alter genes now, and they have already cloned a sheep from a fully differentiated cell.

There are those who say that the DNA of any organism must never be altered. Put aside the fact that nature itself alters DNA much of the time, and has done so for nearly all of life's history. The concern is that *we* do not have the wisdom to bring about beneficial changes without causing great harm to ourselves or to the environment.

When it comes to manipulating human genes, one is reminded of our human tendency to leap before we look. When it comes to restricting genetic modifications of any sort, one also is reminded of the old saying, "If God had wanted us to fly, he would have given us wings." Yet something about the human experience has given us a capacity to imagine wings of our own making—and that capacity has carried us to the frontiers of space.

Where are we going from here with recombinant DNA technology, this new product of our imagination? To gain perspective on the question, spend some time reading the history of the human species. It is a history of survival in the face of all manner of new challenges, threats, bumblings, and sometimes disasters on a grand scale. It is also a story of our connectedness with the environment and with one another.

The basic questions confronting you today are these: Should we be more cautious, believing that one day the risk takers may go too far? And what do we as a species stand to lose if the risks are not taken?

1. Uncountable gene mutations and other forms of genetic "experiments" have been proceeding in nature for at least 3 billion years.

2. Through artificial selection practices, humans have been manipulating the genetic character of species for many thousands of years. Today, recombinant DNA technology has enormously expanded our capacity to genetically modify organisms.

3. With recombinant DNA technology, researchers can isolate, cut, and then splice together gene regions from different species. They can greatly amplify the number of copies of those regions into usefully large quantities. Certain enzymes make the cuts and do the splicing.

4. Researchers employ a variety of restriction enzymes and DNA ligase to cut and insert DNA into plasmids from bacterial cells.

   a. Plasmids are small, circular DNA molecules that contain extra genes—that is, genes in addition to those of the circular bacterial chromosome.

   b. A DNA library is a collection of DNA fragments produced by restriction enzymes and later incorporated into plasmids.

5. It is possible to amplify the quantity of restriction fragments by enlisting a population of rapidly dividing cells, such as bacteria or yeasts. Short DNA fragments can be amplified hugely and far more rapidly in a test tube by the polymerase chain reaction, or PCR.

6. Following amplification, the DNA fragments can be sorted out, according to length, and their nucleotide sequences can be determined.

7. Each individual of a species has a DNA fingerprint, which is a unique array of restriction fragment length polymorphisms (RFLPs). Small molecular variations in the base sequence of the DNA of different individuals lead to small, identifiable differences in the restriction fragments cut from the DNA.

8. Cells that take up foreign genes can be identified by DNA probes, which base-pair with the genes of interest. The base pairing of nucleotide sequences from different sources is called nucleic acid hybridization.

9. In genetic engineering, genes are isolated, modified, and inserted into the same organism or a different one. In gene therapy, copies of normal or modified genes are inserted into an individual in order to correct a genetic defect or boost resistance to disease.

10. Both recombinant DNA technology and genetic engineering have enormous potential for research and applications in medicine, agriculture, the home, and industry. As with any new technology, the potential benefits must be weighed against the potential risks, including ecological and social repercussions.

Review Questions

1. Distinguish these terms from one another:
   a. recombinant DNA technology and genetic engineering *16.1*
   b. restriction enzyme and DNA ligase *16.1*
   c. cloned DNA and cDNA *16.2, 16.4*
   d. PCR and DNA sequencing *16.2*

2. Define RFLPs. Briefly explain one method by which they are produced, then name some applications for RFLP analysis. *16.3*

3. Explain how DNA probes and nucleic acid hybridization can be used to identify genetically modified host cells. *16.4*

Self-Quiz *(Answers in Appendix IV)*

1. _____ are small circles of bacterial DNA that are separate from the circular bacterial chromosome.

2. DNA fragments result when _____ cut DNA molecules at specific sites.
   a. DNA polymerases   c. restriction enzymes
   b. DNA probes   d. RFLPs

3. PCR stands for _____ .
   a. polymerase chain reaction
   b. polyploid chromosome restrictions
   c. polygraphed criminal rating
   d. politically correct research

4. A _____ is a collection of DNA fragments, produced by restriction enzymes and incorporated into plasmids.
   a. DNA clone   c. DNA probe
   b. DNA library   d. gene map

5. A _____ is multiple, identical copies of a collection of DNA fragments inserted into plasmids.
   a. DNA clone   c. DNA probe
   b. DNA library   d. gene map

6. Gel electrophoresis, a standard laboratory procedure, can be used to separate DNA restriction fragments according to _____ .
   a. shape   c. net surface charge
   b. size   d. all of the above

7. In reverse transcription, _____ is assembled on _____ .
   a. mRNA; DNA   c. DNA; enzymes
   b. cDNA; mRNA   d. DNA; agar

8. _____ is the transfer of normal genes into body cells to correct a genetic defect.
   a. Reverse transcription   c. Gene mutation
   b. Nucleic acid hybridization d. Gene therapy

9. Match the terms with the most suitable description.
   ____ DNA fingerprint   a. selecting "desirable" traits
   ____ Ti plasmid   b. deciphering 3.2 billion base
   ____ nature's genetic   pairs of 23 human chromosomes
        experiments   c. used in some gene transfers
   ____ nucleic acid   d. unique array of RFLPs
        hybridization   e. base pairing of nucleotide
   ____ human genome   sequences from different
        project   DNA or RNA sources
   ____ eugenic engineering   f. mutations, crossovers

Critical Thinking

1. Besides this chapter's examples, list what you believe might be some potential benefits and risks of genetic engineering.

2. Ryan, a forensic scientist, obtained a very small DNA sample from a crime scene. In order to examine the sample by DNA fingerprinting, he must first amplify the sample by the polymerase chain reaction. He estimates there are 50,000 copies of the DNA in his sample. Derive a simple formula and calculate the number of copies he will have after fifteen cycles of PCR.

3. A game warden in Africa confiscated eight ivory tusks from elephants. Some tissue is still attached to the tusks. Now he must determine whether the tusks were taken illegally from northern populations of endangered elephants, or from other elephants from populations to the south that can be hunted legally. How can he use DNA fingerprinting to find the answer?

4. A restriction enzyme dubbed EcoRI is specified by a gene on an R plasmid in *E. coli*. In nature, it cleaves the circular DNA of a small virus that infects mammals. Lisa, a biotechnologist, has cloned an EcoRI fragment into a plasmid. The cloned EcoRI fragment is 850 base pairs (bp) long. And 300 bp from one end of the fragment is a site where a different restriction enzyme (BamHI) can cleave it. The plasmid is 1,900 bp long and has no BamHI sites.
   a. When EcoRI and the plasmid are mixed together in solution, what length does Lisa expect the resulting fragments to be? What can she expect when she uses BamHI alone to cut it? When she uses both EcoRI and BamHI?
   b. Draw a circular map of the recombination plasmid, showing the relative distance between the two restriction enzyme sites.

5. Last winter, Lunardi's Market put out a bin of tomatoes having splendid vine-ripened redness, flavor, and texture. The sign posted above the bin identified them as genetically engineered produce. Most shoppers selected unmodified tomatoes in the adjacent bin, even though those tomatoes were pale pink, mealy-textured, and tasteless. Which ones would you pick? Why?

Selected Key Terms

cDNA *16.4*   human genome project *16.6*
DNA amplification *16.2*   nucleic acid hybridization *16.4*
DNA fingerprint *16.3*   plasmid *16.1*
DNA library *16.1*   polymerase chain reaction (PCR) *16.2*
DNA ligase *16.1*   primer *16.2*
DNA polymerase *16.2*   recombinant DNA technology *16.1*
DNA probe *16.4*   restriction enzyme *16.1*
gel electrophoresis *16.2*   restriction fragment length
gene therapy *CI*   polymorphism (RFLP) *16.3*
genetic engineering *16.1*   reverse transcription *16.4*

Readings

Anderson, W. F. 1992. "Human Gene Therapy." *Science* 256: 808–813.

Gasser, C. S., and R. T. Fraley. 1989. "Genetically Engineering Plants for Crop Improvement." *Science* 244: 1293–1299.

Joyce, G. December 1992. "Directed Molecular Evolution." *Scientific American* 267(6): 90–97.

10 March 1997. "Send in the Clones." *Newsweek*.

Watson, J. D. 1990. "The Human Genome Project: Past, Present, and Future." *Science* 248: 44–49.

Web Site See *http://www.wadsworth.com/biology* for practice quiz questions, hypercontents, BioUpdates, and critical thinking. The Wadsworth Biology Resource Center provides a wealth of information fully organized and integrated by chapter.

FACING PAGE: *Millions of years ago, a bony fish died, and sediments gradually buried it. Today its fossilized remains are studied as one more piece of the evolutionary puzzle.*

17   EMERGENCE OF EVOLUTIONARY THOUGHT

Fire, Brimstone, and Human History

With this unit, we turn to evolutionary theories and to the ways in which they can be used to interpret the past and present, even to predict possible futures for the natural world. Today the theories are routinely used to guide scientific investigations in many fields, and they are widely accepted in society at large. But this was not the case when some individuals in Europe first started working out the details. The early evolutionists rankled more than a few people with their astounding theories, which opened a window on the interrelatedness of the geologic record, the fossil record, and the existing sweep of biological diversity.

As convincing evidence accumulated in support of evolutionary theories, many zoologists, geologists, and other scholars of the time came to accept them as useful guides for further research. Others were unable to reconcile them with the Bible, the premier book of the Western world. For generation after generation, that book had been helping people cope with the uncertainties of life, death, and unpredictable forces of nature. Presumably eyewitness accounts of volcanism, great floods, and other catastrophes were interwoven into its narrative. Modern scientists have indeed found physical evidence of many ancient catastrophic events, so we know bad things did happen. However, *those accounts had been assigned meaning through the prism of prevailing cultural beliefs and a limited understanding of geologic processes.*

For example, the Jordan Valley is a long depression in the Earth's crust, between the Red Sea to the south and the Dead Sea to the north. About 4,000 years ago, the notorious cities of Sodom and Gomorrah apparently flourished in this valley, at the south end of the Dead Sea (Figure 17.1). By biblical account, "brimstone and fire rained upon the cities . . . and the smoke of the land

Figure 17.1 Map showing the location of Thera, a volcanic island in the eastern Mediterranean Sea, and the area most devastated by its violent eruption in prehistoric times. The photograph shows an incandescent lava flow from a modern-day volcanic eruption.

went up like the smoke of a furnace." Both cities were destroyed, and stories of the horrified survivors found their way into the Biblical narrative.

Scientists now have satellite images of the straight walls of the valley. They see physical evidence of hot springs past and present, great lava flows, and violent tremors—all signs of deep cracks (faults) in the Earth's crust. Severe earthquakes must have tilted a portion of the crust along the fault at the southern end of the Dead Sea. As the ground heaved, lava poured from the depths, hot springs spewed forth a sulfurous brew, and the violently displaced waters of the sea undoubtedly flooded cities along its shores.

As another example, during the heydey of ancient Egyptian civilization some 3,500 years ago, a volcanic island near Greece blew apart and generated seismic sea waves that may have been 100 meters high. Twenty minutes later, the giant waves slammed into the island of Crete to the south (Figure 17.1). Apparently the abrupt collapse of the Minoan civilization coincided with this catastrophe; the Minoan capital on Crete is thought to have been leveled at the same time as the eruption. Those who survived the deluge would have been further terrified, for the skies must have been darkened for days as prevailing winds blew volcanic ash across the island. This account comes from modern-day studies of Thera, a small, crescent-shaped island between Greece and Crete. The island is merely the exposed uppermost part of a great submerged crater, 400 meters deep.

Quite possibly, that violent episode in Earth history was transformed into the biblical account of a Great Deluge. If, as European scholars believed, the very first humans were expelled from paradise as punishment for their sins, and if their descendants had continued to indulge in sinful ways, then the Great Deluge must have been divinely invoked to punish their offspring, too. Those scholars also believed that the Earth was only about 6,000 years old. If so many catastrophes had been inflicted on humans in so short a time, then all changes to the Earth's surface must have occurred by sudden, violent geologic processes. These were the core hypotheses of what came to be known as the theories of **catastrophism**.

Biological science emerged within the framework of such prevailing beliefs, and so did awareness of change in the Earth and its creatures. Understand this fact, and you will gain insight into why acceptance of the very idea of biological evolution was so long in coming.

KEY CONCEPTS

1. In the past, devastating floods, volcanic eruptions, and other catastrophic events in the natural world were interpreted through the prism of prevailing cultural beliefs and with limited knowledge of geologic forces. Evolutionary theories gave early scholars a new way of interpreting and investigating the natural world.

2. Today we define biological evolution as heritable changes in lineages, or lines of descent. Evidence of evolution comes from investigations that began nearly two centuries ago.

3. Through the early investigations, certain relationships among major groups of animals were discerned on the basis of comparisons of body structure and patterning. At the same time, explorers discovered differences in the world distribution of species that could not be explained unless those species had evolved in separate regions. Also, geologists found evidence of the evolution of life in fossil sequences, as recorded in the cakelike layers of sedimentary rock.

4. As Charles Darwin and Alfred Wallace perceived, evolution can occur by way of natural selection. In any population, this process leads to differences in survival and reproduction among individuals that differ in one or more heritable traits.

5. Here are the key premises of the theory: Populations of each species are characterized by the heritable traits that its individuals share. Those traits vary in their details from one individual to the next. Over time, the forms of traits that prove to be most adaptive in a given environment increase in a population, and other forms do not. Thus the traits that characterize a population can change over time; the population can evolve.

EARLY BELIEFS, CONFOUNDING DISCOVERIES

The Great Chain of Being

Our story begins more than two thousand years ago, when the seeds of biological inquiry were beginning to take hold among the ancient Greeks. At the time, popular belief held that supernatural beings intervened directly in human affairs. For example, angry gods were said to inflict a common ailment known as the sacred disease. And yet, from one physician of the school of Hippocrates, these thoughts come down to us:

It seems to me that the disease called sacred . . . has a natural cause, as other diseases have. Men think it divine merely because they do not understand it. But if they called everything divine that they did not understand, there would be no end of divine things! . . . If you watch these fellows treating the disease, you see them use all kinds of incantations and magic— but they are also very careful in regulating diet. Now if food makes the disease better or worse, how can they say it is the gods who do this? . . . It does not really matter whether you call such things divine or not. In Nature, all things are alike in this, in that they can be traced to preceding causes.

—On the Sacred Disease (400 B.C.)

Passages such as this reflect the early attempts to find natural explanations for observable events.

Aristotle was foremost among the early naturalists, and he described the world around him in excellent detail. He had no reference books or instruments to guide him, for biological science in the Western world began with the great thinkers of this age. Yet here was a man who was no mere collector of random tidbits of information. In his descriptions, we find evidence of a mind perceiving connections between observations and attempting to explain the order of things.

Aristotle believed (as did others) that each kind of organism was distinct from all the rest. Nevertheless, he wondered about organisms that seemed to have rather blurred positions in nature. For example, even though some sponges look very much like plants, they do not make their own food, as plants do. They capture and digest it, as animals do. In time, Aristotle came to view nature as a continuum of organization, from lifeless matter through complex forms of plant and animal life.

By the fourteenth century, Aristotle's idea had been transformed into a rigid view of life. A great Chain of Being was seen to extend from the lowest forms of life to humans and on up to spiritual beings. Each kind of being, or "species" as it was called, was a separate link in the chain. All the links were designed and forged at the same time, at the same center of creation, and had not changed since. Scholars thought that once they had discovered, named, and described all of the links, the meaning of life would be revealed to them.

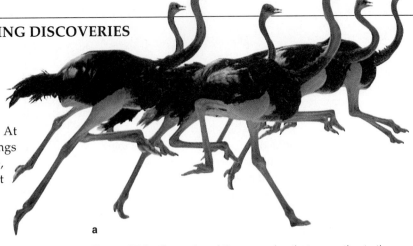

a

Figure 17.2 Examples of three species that are native to three geographically separate parts of the world. (**a**) *Above:* Ostrich of Africa. *Right:* (**b**) Rhea of South America and (**c**) emu of Australia. All three species are like one another and unlike most birds in several ways, including their inability to get airborne.

Questions From Biogeography

Until the fifteenth century, naturalists were not aware that the world is a great deal bigger than Europe, so the task of locating and describing all species seemed manageable. Then global explorations began, and the known world expanded enormously. Naturalists were soon overwhelmed by descriptions of tens of thousands of plants and animals that explorers were discovering in Asia, Africa, the Pacific Islands, and the New World.

In 1590 the naturalist Thomas Moufet attempted to sort through the bewildering array. He simply gave up and wrote such gems as this description of locusts and grasshoppers: "Some are green, some black, some blue. Some fly with one pair of wings; others with more; those that have no wings they leap; those that cannot fly or leap they walk; some have long shanks, some shorter. Some there are that sing, others are silent." It was not exactly a work of subtle distinctions.

Even so, a few scholars began to examine the world distribution of organisms, a discipline now known as **biogeography**. It became clear that many unique plants and animals live in isolated places, including remote oceanic islands. It also became clear that some species have very similar traits yet live in areas isolated by great oceans or impenetrable mountains (Figure 17.2).

How did so many species ever get from one center of creation to oceanic islands and other isolated locations?

Questions From Comparative Anatomy

By the eighteenth century, many scholars were engaged in **comparative anatomy**, the systematic study of the similarities and differences in body plans of different mammals, reptiles, and other major groups. Think of the bones of a human arm, whale flipper, and bat wing.

b

c

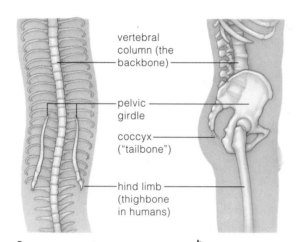

vertebral
column (the
backbone)

pelvic
girdle

coccyx
("tailbone")

hind limb
(thighbone
in humans)

a   b

Figure 17.3 (**a**) Python bones corresponding to the pelvic girdle of other vertebrates, including humans (**b**). Small "hind limbs" protrude through the skin on the underside of the snake.

These body parts differ in size, shape, and function. Yet all have similar locations in the body. They consist of the same tissues, arranged in the same patterns. And they develop in similar ways in embryos. Comparative anatomists who discovered all of this wondered: Why are certain animals that are so different in some features so much alike in others?

By one hypothesis, basic body plans were so perfect that there was no need to come up with a totally new one for each organism at the time of their creation. Yet if that were true, then how could there be parts with no function? For instance, some snakes have bones that correspond to a pelvic girdle—a set of bones to which hind legs attach (Figure 17.3). Snakes don't have legs. Why the bones? Humans have bones that correspond to a few of the tailbones of many other mammals, but they don't have a tail. Why do they have parts of one?

Questions About Fossils

From the late 1600s on, geologists added to the growing confusion. They began mapping layers of rocks at sites where erosion or quarrying had cut deep into the earth. They found similar layers around the world. (Such beds consist of distinct, multiple layers of sedimentary rock. Figure 20.4 shows an example.) Most agreed that sand and other sediments had been deposited at different times and had slowly compacted into rock layers. They realized that certain layers contained distinctive **fossils**, which were known to be evidence of life in the past.

For example, deep layers contained fossils of simple marine organisms. Some fossils in the overlying layers were similar but structurally more complex. And fossils in layers above these closely resembled living marine organisms. Were these *sequences* of fossils, separated in time? If so, what did increasing structural complexity among fossils of a given type of organism represent?

Taken as a whole, the findings from biogeography, comparative anatomy, and geology did not fit well with prevailing beliefs. Georges-Louis Leclerc de Buffon and a few others started to formulate novel hypotheses. If dispersal of all species from a center of creation was not possible, given the vast oceans and other barriers, *then perhaps species had originated in more than one place.* And if they were not created in a perfect state (and fossil sequences and the occurrence of "useless" body parts in certain organisms suggested they were not), *then perhaps species had been modified over time.* Awareness of change in lines of descent—**evolution**—was in the wind.

Awareness of biological evolution emerged over centuries, through the cumulative observations of many naturalists, biogeographers, comparative anatomists, and geologists.

A FLURRY OF NEW THEORIES

Squeezing New Evidence Into Old Beliefs

In the nineteenth century, naturalists tried to reconcile the growing evidence of change in lines of descent with a traditional conceptual framework that did not allow for change. Foremost among them was Georges Cuvier, a respected anatomist. For years he had compared body plans of fossils and living organisms. He acknowledged the abrupt changes in the fossil record, corresponding to discontinuities between certain layers of sedimentary beds. Was the record evidence of changing populations of organisms that lived in those ancient environments? Cuvier thought so. And he was right, as you will see from the evolutionary story in Chapter 21.

Based on this inference, Cuvier constructed his own theory of catastrophism. There was one time of creation, he said, that populated the world with all species. A global catastrophe destroyed many of them. Survivors repopulated the world. It wasn't that survivors were *new* species; naturalists simply hadn't yet found earlier fossils that would date to the time of creation. Later catastrophes wiped out more species, and repopulation by the survivors followed, as recorded by fossils.

Many scholars accepted his theory, but others kept at the puzzle. Consider Jean-Baptiste Lamarck's **theory of inheritance of acquired characteristics**. During the life of each individual, said Lamarck, environmental pressures and internal "needs" bring about permanent changes in the body, and offspring inherit the needed changes. And so life, created long ago in a simple state, gradually improved. The force for change was a drive for perfection, up the Chain of Being. The drive was centered in nerves that directed an unknown "fluida" to body parts in need of change.

For instance, suppose modern giraffes had a short-necked ancestor. Pressed by the need to find food, it kept stretching its neck to browse on leaves beyond the reach of other animals. Stretching directed fluida to its neck, which lengthened permanently. The longer neck was bestowed on offspring, which stretched their necks also. And so generations of animals desiring to reach higher leaves led to the modern giraffe.

As Lamarck correctly inferred, the environment *is* a factor in evolution. But his overall theory, like others proposed at the time, has not been supported by any investigation carried out since then. No one has found evidence that environmental modifications of the traits of an existing individual can be passed to offspring.

Voyage of the Beagle

In 1831, in the midst of the confusion, Charles Darwin was twenty-two years old and wondering what to do with his life. Ever since he was eight, he had wanted to

a

Figure 17.4 (**a**) Charles Darwin and a blue-footed booby, one of the species he encountered during his five-year voyage around the world on H.M.S. *Beagle*. A replica of this ship is shown in (**b**), sailing off the coast of South America. During stops along Argentina's coast, Darwin explored parts of the Andes. He saw fossils of marine organisms embedded in rocks 3.6 kilometers above sea level. (**c,d**) The Galápagos Islands, isolated bits of land in the ocean, far to the west of Ecuador. We now know they arose through volcanic action about 5 million years ago, so organisms could not have originated there. Winds or ocean currents must have carried them to the new islands.

hunt, fish, collect shells, or simply watch insects and birds—anything but sit in school. Later, at his father's insistence, he did try to study medicine in college. The crude, painful procedures used on patients at that time sickened him. Then his father strongly urged him to become a clergyman, so Darwin packed for Cambridge. His grades were good enough to earn him a degree in theology. But he spent most of his time among faculty members with leanings toward natural history.

John Henslow, a botanist, perceived Darwin's real interests. He arranged for Darwin to function as ship's naturalist aboard H.M.S. *Beagle*. The *Beagle* was about to embark on a five-year voyage that would take Darwin around the world (Figure 17.4). The young man who hated schoolwork and who had no formal training as a naturalist suddenly began to work enthusiastically.

The *Beagle* sailed first to South America, to complete work on mapping the coastline. During the Atlantic crossing, Darwin collected and examined marine life. He also read Henslow's parting gift, the first volume of Charles Lyell's *Principles of Geology*. During stops along the coast and at various islands, he observed diverse species in environments ranging from sandy shores to high mountains. And he started circling the question of

b

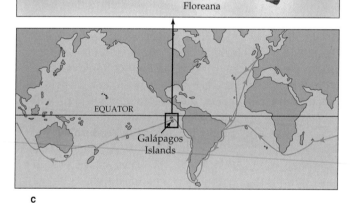

Darwin
Wolf

Pinta

Marchena   Genovesa

EQUATOR

Santiago   Bartolomé
                    Seymour
        Rábida   Baltra
Fernandina   Pinzón
                    Santa Cruz

                Santa Fe
                    San Cristóbal
    Tortuga
Isabela
                    Española
    Floreana

EQUATOR

Galápagos
Islands

c

d

evolving life, which was now on the minds of many respected individuals—including his own grandfather.

Darwin started mulling over a rather radical theory that Lyell was advancing in his book. Lyell and other geologists were arguing that catastrophes had no more and possibly less effect on Earth history than did subtle processes of change. They had thought about how long it takes for rains, the pounding surf, and other forces of nature to sculpt the landscape. For years geologists had chipped away at layers of sandstones, limestones, and

other rocks, which form after sediments erode from the land and accumulate in the beds of rivers and seas. And they thought about how many stacked layers occurred in many sedimentary beds. If, they hypothesized, the deposition had occurred as gradually in the past as it did in their own era, then surely it must have taken many millions of years—not a few thousand—for such thick stacks to form. They even managed to incorporate earthquakes and other infrequent events into their view of Earth history. After all, numerous major floods, about 150 great earthquakes, and about 20 volcanic eruptions occur every year, so catastrophes are not unusual.

Their view of gradual, uniformly repetitive change became the **theory of uniformity**. It challenged the prevailing views of the age of the Earth.

The theory bothered scholars who firmly believed the Earth was less than 6,000 years old. They thought people had recorded everything that happened during those thousands of years, and in all that time no one ever mentioned seeing a species evolve. Yet by Lyell's calculations, it must have taken millions of years to mold the present landscape. *Wasn't that enough time for species to evolve in diverse ways?* Later, Darwin thought so. But exactly *how* did they evolve? He would end up devoting the rest of his life to that burning question.

Prevailing beliefs can influence how we interpret clues to natural processes and their observable outcomes.

Darwin's observations during a global voyage helped him think about these processes in a novel way.

Old Bones and Armadillos

After he returned to England in 1836, Darwin talked with other naturalists about possible evidence that life evolves. By carefully studying all of the notes from his journey, he came up with some possibilities.

For example, he had observed fossils of glyptodonts in Argentina. Of all the animals on Earth, only living armadillos are like glyptodonts, which are now extinct (Figure 17.5). Of all places on Earth, armadillos live only in the same regions where glyptodonts once lived. If the two kinds of animals had been created at the same time, lived in the same place, and were so much alike, why is only one still alive? Wouldn't it be reasonable to assume glyptodonts were early relatives of armadillos? Many of their shared traits might have been retained through countless generations. Other traits might have been modified in the armadillo branch of a family tree. Descent with modification—it seemed possible. What, then, was the driving force for evolution?

A Key Insight—Variation in Traits

While Darwin assessed his notes, an influential essay by Thomas Malthus, a clergyman and economist, made him take a closer look at a topic of social interest. Malthus had correlated famine, disease, and war with population size. Humans, he claimed, tend to reproduce faster than food supplies, living space, and other resources can be sustained. The larger a population gets, the more people there are to reproduce in each generation. As population size burgeons, resources dwindle—and so the struggle for existence intensifies. Many people starve, get sick, and engage in war and other forms of competition for the remaining resources.

After thinking about his personal observations, Darwin suspected that any population has a capacity to produce more individuals than the environment can support. To give an example, a single sea star can release 2,500,000 eggs every year, but the seas obviously are not filled with

sea stars. In each generation, nearly all eggs and larvae end up in the bellies of predators. Still others starve or succumb to disease or some other environmental insult.

Assume that the environment keeps the number of reproducing individuals in check. *Which* individuals are the winners and losers? What might affect the outcome?

Darwin thought about the populations of animals and plants he had observed during his voyage. As he recalled, individuals in those populations weren't alike down to the last detail. They showed variations in size, coloration, and other traits. *And it dawned on him that variations in traits might affect an individual's ability to secure resources—and therefore to survive and reproduce in particular environments.*

Did the Galápagos Islands show evidence of this? Between these volcanic islands and the South American coast are 900 kilometers of open ocean. The islands offer a variety of habitats along rocky shorelines, in deserts, and on mountain flanks. Most of their inhabitants live nowhere else—although they do resemble species on the mainland. Were those remote points of land in the ocean colonized by species that flew, floated, or were blown over from the mainland? If so, then in different island habitats, island-hopping descendants of the colonizers may have become adapted over time to local conditions.

Darwin learned, through conversations with other naturalists, that populations of thirteen finch species were distributed among the various Galápagos Islands. He himself had collected specimens of the birds, and he attempted to correlate the variations in their traits with environmental challenges. Imagine yourself in his place. For one population, you notice the finches have a large, strong bill, suitable for cracking seeds (Figure 17.6). Yet the bill of a few birds is a bit stronger, and they can crack open seeds that other birds find too tough to deal with. If most of the seeds being produced in a particular habitat have hard coats, then a strong-billed bird will have a competitive edge that ultimately will affect success in producing offspring. Assuming the trait has a heritable basis, the same will be true of that bird's strong-billed descendants.

Now take this thought one step further. Over time, if factors in the environment continue to "select" the most adaptive version of the trait, the population will become one of mostly strong-billed birds. *And a population is evolving when the forms of its heritable traits are changing over successive generations.*

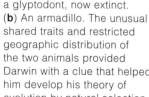

Figure 17.5 (**a**) Painting of a glyptodont, now extinct. (**b**) An armadillo. The unusual shared traits and restricted geographic distribution of the two animals provided Darwin with a clue that helped him develop his theory of evolution by natural selection.

a

b

a b c

d

Figure 17.6 Four species of finches from the Galápagos Islands. (**a**) *Geospiza conirostris* and (**b**) *G. scandens*, both with a bill adapted for eating cactus flowers and fruits. Other finches have thick, strong bills that can crush cactus seeds. (**c**) *Certhidea olivacea*, a tree-dwelling finch, resembles warblers in song and behavior. It uses its slender beak to probe for insects. (**d**) *Camarhynchus pallidus* feeds on wood-boring insects such as termites. It does not have the woodpecker's long, probing tongue. It has learned to break cactus spines and twigs to suitable lengths, then hold the "tools" and use them to probe bark for insects hidden from its view.

Recall, from Chapter 1, that Darwin used pigeon breeding and other *artificial* selection practices as a way to explain *natural* selection of adaptive traits in the wild. Thus, when breeders favor pigeons with black tail feathers, they encourage every new black-tailed pigeon to mate—but they might put white-tailed ones in the stewpot. It was an easy way to show how selection could cause an increase in the frequency of one version of a trait in a captive population, and how it might eliminate the other version.

Anticipating that his view of evolution by **natural selection** would generate controversy, Darwin waited to announce it and searched for flaws in his reasoning. He waited too long. More than a decade after he wrote but never published an essay on his theory, he received a paper from Alfred Wallace, another naturalist (Figure 17.7). On his own, Wallace had developed the same theory, then quickly wrote up and circulated his ideas. Darwin's colleagues prevailed on him to formally present a paper at the same time as Wallace (who also believed that Darwin deserved most of the credit). The following year, in 1859, Darwin's impressively detailed evidence in support of the theory was published under the title, *On the Origin of Species.*

Although you may have heard that Darwin's book fanned an intellectual firestorm, the idea that diversity is the product of evolution was accepted almost at once by most naturalists. Darwin's specific explanation, of gradual evolution by natural selection, was fiercely debated. Nearly seventy years passed before advances

Figure 17.7 Alfred Wallace. Darwin worked out a theory of natural selection long before Wallace did, but Wallace was first to circulate a letter outlining his view of the process.

in a new field, genetics, led to widespread acceptance of that remarkable explanation. Meanwhile, his name was associated primarily with the idea that life evolves—something others had suggested before him.

DARWIN'S THEORY OF EVOLUTION BY NATURAL SELECTION. **Any population can evolve (change over time) when individuals differ in one or more heritable traits that are responsible for differences in the ability to survive and reproduce.**

REGARDING THE FIRST OF THE "MISSING LINKS"

At the time when Darwin formally presented his theory of evolution by natural selection, it faced a serious challenge. If scholars subscribed to the theory, then they would be tentatively accepting that all species are related, by way of descent, to ever more ancient species. Darwin saw with his own eyes that artificial selection practices by pigeon breeders and others could mold the traits of a population in no time at all. So he argued it was entirely possible for one species to evolve gradually into a separate species, with one or more traits that were uniquely its own, over hundreds or thousands of generations. But if that were so, then where were the "missing links"? That is, if each species evolved from others, then where were the fossils of all the *transitional forms*? Where were the species with traits that bridged two major groups of organisms?

More than a century would pass before more fossil finds as well as the application of molecular biology and genetic analysis would yield plausible answers to that question. In Darwin's time, though, the presumed absence of transitional forms was a major point of contention in discussions of the theory. Ironically, a fossil of just such a transitional form had already been unearthed at the Solnhofen limestones of Bavaria, in southern Germany.

The fossilized skeleton initially was labeled a small theropod (meat-eating) dinosaur, about the size of a pigeon. The specimen's shape said "dinosaur." And like the theropods, it had a long, bony tail, clawed fingers, and a heavy jaw with short, spiky teeth (Figure 17.8).

Later, another fossil of the same type was unearthed. And later still, someone noticed the feathers. If those fossils were of dinosaurs, what were they doing with *feathers*? Close examination revealed that the feathers were like those of modern birds, and the specimen type was named *Archaeopteryx* (meaning "ancient winged one").

Between 1860 and 1988, six specimens of *Archaeopteryx* and a fossilized feather were found. Anti-evolutionists tried to dismiss them as forgeries. Someone, they asserted, had pressed bones and feathers of existing birds against wet plaster. After the imprinted plaster casts dried, they merely looked like fossils. But microscopic examination confirmed that the fossils are real. Further confirmation of their antiquity is found in the remains of clearly ancient species of worms, jellyfishes, and many other kinds of organisms preserved in the same limestone layers.

Through radiometric dating methods, we now know that *Archaeopteryx* lived 150 million years ago. When you examine the photograph in Figure 17.8, you might wonder: How could the remains of those winged creatures be so well preserved after so much time? *Archaeopteryx* happened to live in tropical forests next to a large lagoon. The lagoon was warm, still, and stagnant, because coral reefs barred inputs of oxygenated water from the sea. Thus it was not a favorable habitat for scavenging animals that might otherwise have feasted on—and obliterated the remains of—*Archaeopteryx* and other organisms that fell from the skies or drifted offshore. With each tropical storm surge, though, fine sediments were swept over the reefs, and they gently buried the carcasses littered at the bottom of the lagoon. Over time, the soft mud compacted and hardened to become the exceptionally fine limestone tombs of more than 600 species, including *Archaeopteryx*.

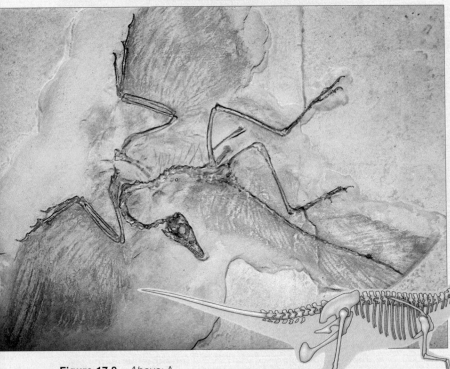

Figure 17.8 *Above:* A photograph of one of the *Archaeopteryx* fossils. Right: Comparison of the skeletons of a dinosaur (a type of reptile) and of *Archaeopteryx*, which was on or very near the evolutionary road from reptiles to birds.

Dromaeosaurus, a small, two-legged dinosaur that had the most traits in common with birds. Like birds, it had long, slender hind legs, three toes forward and one toe pointing in reverse, and long forelimbs. Unlike birds, it had no feathers.

Archaeopteryx, which had feathers. Like dinosaurs, though, it also had a long, slender bony tail, a heavy jaw with serrated teeth, and three long fingers.

17.5 SUMMARY

1. Awareness of evolution—changes in lines of descent over time—emerged through comparisons of the body structure and patterning for major groups of animals (comparative anatomy), questions concerning the world distribution of plants and animals (biogeography), and observations of fossils of different types present in different layers of sedimentary rocks.

2. Here are the key points of the theory of evolution by natural selection, as proposed by Darwin and Wallace:

   a. Individuals of all populations have the capacity to produce more offspring than the environment is able to support, so individuals must compete for resources.

   b. Individuals of a population vary in size, form, and other traits. The variant forms of a trait may be more or less adaptive under prevailing conditions.

   c. When some form of a trait is adaptive under prevailing conditions, and when it has a heritable basis, its bearers tend to survive and reproduce more frequently than individuals with less adaptive forms of the trait. Over the generations, the adaptive version becomes more common in the population.

   d. Natural selection is a difference in survival and reproduction among individuals of a population that differ from one another in one or more traits.

   e. Natural selection results in modifications of traits within a line of descent. Over time, it may bring about the evolution of a new species, with an array of traits that is uniquely its own relative to other species.

3. One "test" of Darwin's theory would be evidence of one major kind of organism changing into another kind. *Archaeopteryx*, an extinct, transitional form between reptiles and birds, provided early evidence. Not until many decades later did studies at many levels of biological organization offer a large body of evidence in support of his theory of evolution by natural selection.

Review Questions

1. Define and contrast the theories of catastrophism and of uniformity. *CI, 17.2*

2. Define biogeography and comparative anatomy. How did studies in both disciplines contradict the idea that species have remained unchanged since the time of creation? *17.1*

3. Cuvier and Lamarck interpreted the fossil record differently. Briefly state how their interpretations differed. *17.2*

4. Define evolution. Define evolution by natural selection. Can an individual evolve? *17.1, 17.3*

Self-Quiz *(Answers in Appendix IV)*

1. Individuals in a population with the _____ tend to make up more of the reproductive base through the generations.
   a. most ancestors in the fossil record
   b. widest world distribution of their species
   c. most adaptive traits

2. Natural selection may occur when there are _____ .
   a. heritable traits
   b. differences in the adaptiveness of forms of traits to prevailing environmental conditions
   c. differences in survival and reproduction among individuals that differ in one or more traits
   d. all of the above

3. Most biologists strongly support the theory that species evolve because _____ .
   a. it was first suggested by Darwin and Wallace
   b. studies at many levels of biological organization provide a large body of evidence that the theory is not wrong
   c. another, better explanation of the great body of observations and test results has yet to appear
   d. both b and c

4. Match the following individuals with their ideas.
   _____ Cuvier   a. theory of natural selection
   _____ Lamarck   b. populations outgrow resources
   _____ Lyell   c. catastrophism
   _____ Malthus   d. inheritance of traits acquired
   _____ Darwin   through environmental pressures
         and Wallace   and internal desires for change
                            e. geologic evidence that the Earth is
                               extremely ancient

Critical Thinking

1. Darwin was not aware of Gregor Mendel's investigations into patterns of inheritance, as described in Chapter 11. How might knowledge of Mendel's results have helped him when he was developing his theory of natural selection?

2. In 1996, after reflecting on evidence of evolution that has been accumulating for more than a hundred years, Pope John Paul II acknowledged that the theory of evolution "... has been progressively accepted by researchers, following a series of discoveries in various fields of knowledge. The convergence, neither sought nor fabricated, of the results of work that was conducted independently is in itself a significant argument in favor of this theory." His words infuriated individuals who believe in a strict, literal interpretation of the biblical account of creation. Some of these individuals have been demanding that biology instructors treat evolution as "just a theory" and give equal time to the biblical interpretation.

   Refer to Section 1.5 on the nature of scientific inquiry. Do you equate scientific and religious explanations of life's origin and history as alternative theories? Why or why not?

Selected Key Terms

biogeography *17.1*
catastrophism (theory of) *CI*
comparative anatomy *17.1*
evolution *17.1*
fossil *17.1*
inheritance of acquired characteristics (theory of) *17.2*
natural selection *17.2*
uniformity (theory of) *17.2*

Reading

Darwin, C. 1957. *Voyage of the Beagle.* New York: Dutton. An account of what Darwin saw and thought about during his voyage, in his own words.

Web Site See *http://www.wadsworth.com/biology* for practice quiz questions, hypercontents, BioUpdates, and critical thinking. The Wadsworth Biology Resource Center provides a wealth of information fully organized and integrated by chapter.

18 MICROEVOLUTION

Designer Dogs

We humans have tinkered rather ruthlessly with the modern descendants of a long and distinguished lineage. That lineage originated some 40 million years ago with the appearance of small, weasel-shaped, tree-dwelling carnivores. They were the forerunners of bears, raccoons, pandas, badgers—and dogs. About 10,000 years ago, we began domesticating wild dogs.

No doubt the advantages of doing so were obvious. Times were tough in the days before supermarkets and police protection. Dogs could guard people and their possessions. They could corner, kill, eat, and thereby dispose of rats and other vermin that were not at all welcome in the home.

It didn't take us long to develop different varieties (breeds) through artificial selection. Individual dogs having the desired forms of traits were selected from each new litter and, later, encouraged to breed. Those having undesired forms of traits were passed over.

After favoring the pick of the litter over hundreds or thousands of generations, we ended up with sheep-herding collies, badger-hunting dachshunds, wily retrievers, and snow-traversing, sled-pulling huskies. And at some point we began to delight in the odd, extraordinary dog. Imagine! In practically no time at all, evolutionarily speaking, we picked our way through the pool of variant dog alleles and came up with such extremes as Great Danes and chihuahuas (Figure 18.1).

Of course, canine designs can exceed the limits of biological common sense. For example, how long would a tiny, finicky-eating, nearly hairless, nearly defenseless chihuahua last in the wild? Not very long. Or what about the English bulldog, bred for a very short snout and a compressed face? Long ago, breeders thought these particular traits would allow the dogs to get a better grip on the nose of bulls. (Why they wanted dogs to bite bulls is a story in itself.) So now the roof of the bulldog mouth is ridiculously wide and it is often flabby, so bulldogs have trouble breathing. They sometimes get so short of air, they pass out.

Through our centuries-old fascination with artificial selection, we produced thousands of varieties of crop plants, cats, cattle, and birds as well as dogs. With the currently available technologies of genetic engineering, we are now mixing the genes of different species and producing incredible new varieties, including tobacco plants that produce hemoglobin and mustard plants that produce plastic.

So, when you hear someone wonder about whether "evolution" takes place, remind yourself that evolution simply means *genetic change through time*. Selective breeding practices provide abundant, tangible evidence that heritable changes do, indeed, occur. The actual mechanisms by which those changes are brought about in a given line of descent are the subject of this chapter.

KEY CONCEPTS

1. In general, all individuals of a population have the same number and kinds of genes, which give rise to the same array of traits.

2. In the population as a whole, each gene may exist in two or more slightly different molecular forms, called alleles. Individuals may or may not inherit the same combinations of alleles, so they may not be exactly alike in the details of their traits.

3. Any allele at a given gene locus may become more or less common in the population, relative to the other kinds, or it may disappear. *Microevolution* means that changes have occurred in a population's allele frequencies over time.

4. Allele frequencies can change through mutation, gene flow, genetic drift, and natural selection. Mutation alone produces new alleles. Gene flow, genetic drift, and natural selection shuffle existing alleles into, through, or out of populations.

5. Natural selection is not an "agent," something that is purposefully searching for the "best" individuals in a population. Natural selection simply is the difference in survival and reproduction among individuals that differ in one or more traits. The difference, however, leads to a continuing adaptation of a species to its environment.

Figure 18.1 Two designer dogs. About 10,000 years ago, humans began domesticating wild dogs. From that ancestral stock, artificial selection produced diverse yet rather closely related breeds, such as the Great Dane (*legs, left*) and the chihuahua (*possibly fearful of being stepped on, right*).

Examples of Variation in Populations

As Charles Darwin perceived, the individual doesn't evolve; populations do. By definition, a **population** is a group of individuals of the same species occupying a given area. To understand how a population evolves, start with the variation in traits among its individuals.

Certain features characterize a population. All of its members have the same body plan, as when jays have wings, feathers, feet with three toes forward and one toe back, and so on. These are *morphological* traits (*morpho-* means form). The cells and body parts of all individuals in the population operate much the same way during short-term metabolic tasks, growth, and reproduction. These are *physiological* traits, which relate to the body's functioning. Also, individuals respond the same way to basic stimuli, as when babies instinctively imitate adult facial expressions. These are *behavioral* traits.

Most traits differ in their details from one individual to the next, especially in sexually reproducing species. Pigeon feathers and snail shells differ in patterning or coloration in a given population (Figures 1.7 and 18.2). Some individuals of a frog population might be more sensitive to winter cold or better at attracting a mate than others. Humans differ greatly in the distribution, color, texture, and amount of hair. And these examples can only hint at the staggering variation in populations; almost every trait of every species is variable.

In addition, many traits, such as those studied by Gregor Mendel, show *qualitatively different* variation. They come in two or more distinct forms (morphs) in a population, as when bird feathers are yellow or white. This feature is called **polymorphism**. But other traits, such as human eye color, show *continuous* variation. That is, individuals of the population show small, incremental differences that can be quantified, as described in Section 11.7.

The "Gene Pool"

The information about heritable traits resides in genes, which are specific regions of DNA molecules. In general, all individuals of a population have the same number and kinds of genes. We say "in general" because males and females of sexually reproducing populations differ in some genes on the sex chromosomes.

Think of all the genes in the entire population as a **gene pool**—a pool of genetic resources that, in theory at least, is shared by all members of a population and passed on to the next generation. Each kind of gene in the pool usually exists in two or more slightly different molecular forms, called **alleles**.

Individuals inherit different combinations of alleles. This leads to variations in phenotype (to differences in the details of traits). For example, whether your hair is black, brown, red, or blond depends on which alleles of certain genes you inherited from your two parents. When studying the basis of evolution, remember this: *Offspring inherit genes, not phenotypes.* What may appear to be a heritable trait may actually be the outcome of environmental effects on gene expression (Section 11.8).

Which alleles end up in a given gamete and, later, in a new individual? The outcome depends on five events, as described in earlier chapters and summarized here:

1. Gene mutation (produces new alleles)

2. Crossing over at meiosis I (puts novel combinations of alleles in chromosomes)

3. Independent assortment at meiosis I (puts mixes of maternal and paternal chromosomes in gametes)

4. Fertilization (combines alleles from two parents)

5. Change in chromosome number or structure (leads to the loss, duplication, or repositioning of genes)

Of the five events just listed, mutation alone *creates* new alleles. The other four only shuffle *existing* alleles into different combinations. But what a shuffle! Consider

Figure 18.2 Variation in shell color and banding patterns in populations of a single species of snail that lives on islands of the Caribbean. For most traits, different individuals carry different alleles for the genes that specify these traits. Members of a population vary in traits because they carry different combinations of alleles at particular gene locations along their chromosomes.

this: A human gamete will end up with one of 10^{600} possible combinations of alleles. Not even 10^{10} humans are alive today. So unless you have an identical twin, it is extremely unlikely that another person with your exact genetic makeup has ever lived, or ever will.

Stability and Change in Allele Frequencies

Imagine yourself in a large flower garden in summer. Gradually you notice the butterflies flitting about. They appear to be all the same, except in wing coloration. A few wings are white, but most are blue. Most likely, you muse, the "blue" allele must be more common than the "white" allele. With genetic analysis, you could identify the **allele frequencies**, or the abundance of each kind of allele in the population as a whole. By doing so, you could track the rate of genetic change over time.

You can start with the "Hardy-Weinberg rule," as given in Section 18.2, to set up a theoretical reference point for measuring patterns of change. At this point, called **genetic equilibrium**, the frequencies of alleles at a given gene locus remain stable, one generation after the next. The population is *not* evolving with respect to that gene, for five conditions are being met. First, there have been no gene mutations. Second, the population is very large. Third, it is isolated from other populations of the species. Fourth, the gene has no effect at all on survival or reproduction. Finally, all mating is random.

Rarely, if ever, do all five conditions prevail at the same time in any natural population. Gene mutation is an infrequent but inevitable occurrence. Besides this, three processes—*natural selection, gene flow,* and *genetic drift*—may drive the population away from genetic equilibrium, even in the span of a few generations. The term **microevolution** refers to the small-scale changes in allele frequencies brought about by mutation, natural selection, gene flow, and genetic drift.

Mutations Revisited

Gene mutations, recall, are heritable changes in DNA that typically give rise to altered gene products. They are the only source of new alleles. We cannot predict exactly when or in which individual they will appear. Yet each gene has a characteristic **mutation rate**, which is the probability of its mutating between or during DNA replications. On average, the rate is between 10^{-5} and 10^{-6} per gene locus per gamete, each generation. In a single reproductive season, only 1 gamete in 100,000 to 1,000,000 has a new mutation at any given locus.

Mutations often cause changes in structure, function, or behavior that decrease the individual's chances of surviving and reproducing. Even a single biochemical change sometimes has devastating effects throughout the body. For example, the growth and development of vertebrate skin, bones, tendons, and many other organs depend on the protein collagen. If the gene specifying the molecular form of this one protein is mutated, then pronounced changes in the skeleton, arteries, lungs, liver, and other body parts may follow. Remember, the genes of every complex organism are expressed in the context of a truly intricate developmental program. A mutation that has drastic effects on phenotype usually causes the individual's death; it is a **lethal mutation**.

By contrast, a **neutral mutation** neither helps nor harms the individual. Natural selection cannot increase or decrease the frequency of neutral mutations in the population, for these do not influence an individual's chance of surviving or reproducing. For example, if you have the mutated gene for attached earlobes instead of detached earlobes, this alone should not stop you from surviving and reproducing just as well as anybody else.

Finally, every so often, a new mutation bestows an advantage on the individual. For example, a mutated growth-regulating gene might make a corn plant grow larger or faster and so give it the best access to sunlight and nutrients. Or maybe a previously neutral mutation proves advantageous when environmental conditions change. Even if the advantage is small, chance events or natural selection may preserve the mutated gene and assure its representation in the next generation.

Mutations are so rare, they usually have little or no immediate effect on allele frequencies of a population. But beneficial mutations, and neutral ones, have been accumulating in different lineages for billions of years. Through all that time, they have been the raw material for evolutionary change—the basis for the staggering range of biological diversity, past and present.

In the evolutionary view, then, the reason you don't look like a bacterium or a grass plant or an earthworm or even your neighbors down the street began with different mutations that originated at different times in the past, in different lines of descent.

Certain morphological, physiological, and behavioral traits characterize a population. The traits differ in their details from one individual to the next.

Differences in the combinations of alleles that individuals of the population carry give rise to variations in phenotype. By phenotypic variation, we mean differences in the details of structural, functional, and behavioral traits shared by those individuals.

For sexually reproducing species, individuals of a population represent a pool of genetic resources—that is, a gene pool.

Mutation alone creates *new* alleles. Natural selection, gene flow, and genetic drift change the *frequencies* of alleles in the gene pool. The evolutionary story begins with these changes.

WHEN IS A POPULATION *NOT* EVOLVING?

Look again at Figure 11.1, which shows the earlobes of Tom Cruise, Joan Chen, and other celebrated individuals. Is the number of individuals with *detached* earlobes staying the same in the human population, one generation after the next? What about the number of individuals who have *attached* earlobes? Stated more broadly, how do we know whether a population is evolving with respect to earlobes or any other trait?

Almost a hundred years ago, a mathematician and a doctor came up with the answer. Working independently, they thought about what it would take to maintain the frequencies of alleles of an idealized population. They came up with what is now called the **Hardy-Weinberg rule**, in their honor. They started out with this formula: In a population at genetic equilibrium, the proportions of genotypes at one gene locus, for which there are two kinds of alleles, are

$$p^2 \; AA + 2pq \; Aa + q^2 \; aa = 1$$

where *p* is the frequency of allele *A*, and *q* is the frequency of allele *a*. Their rule is this: *The allele frequencies will stay the same through the generations if there is no mutation, if the population is infinitely large and is isolated from other populations of the same species, if mating is random, and if all individuals survive and reproduce equally.*

To test whether allele frequencies of a given population will remain constant from one generation to the next in the absence of evolutionary forces, you decide to track a pair of alleles through a population of butterflies. These sexually reproducing organisms have pairs of genes, on pairs of homologous chromosomes. You start by assuming that the population is currently at genetic equilibrium. The pair of alleles you are interested in govern wing color. Allele *A* specifies dark-blue wings. Allele *a* is associated with white wings. The heterozygous (*Aa*) condition results in medium-blue wings.

In the population as a whole, the frequencies of *A* and *a* must add up to 1. For example, if *A* occupies half of all the loci for this gene in the population, then 0.5 + 0.5 = 1. If *A* occupies 90 percent of all the loci, then *a* must occupy the remaining 10 percent (0.9 + 0.1 = 1). No matter what the proportions of the two kinds of alleles,

$$p + q = 1$$

During meiosis in germ cells, each allele segregates from its partner, and the two end up in separate gametes. Therefore, *p* is also the proportion of gametes with the *A* allele, and *q* is the proportion with the *a* allele.

To find the expected frequencies of the three possible genotypes (*AA*, *Aa*, and *aa*) in the next generation, you construct a Punnett square:

	p A	*q* a
p A	AA (*p²*)	Aa (*pq*)
q a	Aa (*pq*)	aa (*q²*)

The frequencies of the genotypes add up to 1:

$$p^2 + 2pq + q^2 = 1$$

To see whether the allele frequencies and genotypic frequencies will remain the same through the generations, you work through an example. You decide the population has 1,000 butterflies, each of which produces two gametes:

490 *AA* individuals produce 980 *A* gametes
420 *Aa* individuals produce 420 *A* and 420 *a* gametes
90 *aa* individuals produce 180 *a* gametes

You notice the frequency of *A* among the 2,000 gametes is

$$(980 + 420)/2{,}000 = 0.7$$

Also,

$$q = (420 + 180)/2{,}000 = 0.3$$

At fertilization, the gametes combine at random and give rise to the next generation, as given in the Punnett square. And so, assuming the population remains constant at 1,000 individuals, you now have

$p^2 \; AA = 0.7 \times 0.7 = 0.49$		490 *AA* individuals
$2pq \; Aa = 2 \times 0.7 \times 0.3 = 0.42$	*or*	420 *Aa* individuals
$q^2 \; aa = 0.3 \times 0.3 = 0.09$		90 *aa* individuals

and

$$p^2 + 2pq + q^2 = 0.49 + 0.42 + 0.09 = 1$$

The allele frequencies have not changed:

$$A = \frac{2 \times 490 + 420}{2{,}000 \text{ alleles}} = \frac{1{,}400}{2{,}000} = 0.7 = p$$

$$a = \frac{2 \times 90 + 420}{2{,}000 \text{ alleles}} = \frac{600}{2{,}000} = 0.3 = q$$

You notice that the genotype frequencies have not changed, either. And as long as the five assumptions of the Hardy-Weinberg rule hold true, frequencies should stay the same through the generations. To test this, you calculate allele frequencies in the gametes of the *next* generation:

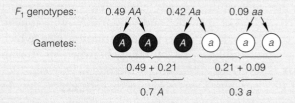

which is back where you started from. Because the allele frequencies for dark-blue, medium-blue, and white wings are the same as they were in the original gametes, they will yield the same phenotypic frequencies as you saw in the second generation.

You could do similar calculations for additional wing colors. You could go on with your calculations until you ran out of paper (or patience). However, as long as the five assumptions hold true, allele frequencies and the range of values for the wing-color trait will not change, as you can see from Figure 18.3.

Therefore, when genotypes and phenotypes do *not* show up in the proportions that you predicted on the basis of the Hardy-Weinberg rule, this tells you that one or more conditions of the rule are being violated. And the hunt can begin for the specific evolutionary force, or forces, driving the change.

Starting generation:

490 *AA* butterflies
(dark-blue wings)

420 *Aa* butterflies
(medium-blue wings)

90 *aa* butterflies
(white wings)

The next generation:

490 *AA* butterflies

420 *Aa* butterflies

90 *aa* butterflies

(NO CHANGE)

The next generation:

490 *AA* butterflies

420 *Aa* butterflies

90 *aa* butterflies

(NO CHANGE)

Figure 18.3 A hypothetical population of butterflies at genetic equilibrium.

We now turn from our idealized population that never changes to the real-world processes of change. Of these processes, natural selection probably accounts for most of the morphological and physiological changes that have occurred throughout the history of life.

Darwin, recall, was able to explain natural selection after correlating his understanding of inheritance with certain features of populations and the environment. Before we consider the modes of natural selection, let's restate his correlations in modern terms:

1. *Observation:* All populations in nature have the reproductive capacity to increase in numbers over the generations.

2. *Observation:* No population is able to increase indefinitely, for its individuals will run out of food, living space, and other resources that sustain it.

3. *Inference:* Sooner or later, the individuals of a population will end up competing for resources.

4. *Observation:* All of the individuals have the same genes, which specify the same assortment of traits. Collectively, their genes represent a pool of heritable information.

5. *Observation:* Most, if not all, kinds of genes occur in different molecular forms (alleles), which give rise to differences in phenotypic details.

6. *Inference:* Some phenotypes are better than others at helping the individual compete for resources, and therefore to survive and reproduce. Thus, the alleles for those phenotypes increase in the population, and other alleles decrease. Over time, the genetic change leads to increased **fitness**—that is, to an increase in adaptation to the environment.

7. *Conclusion:* **Natural selection** is the outcome of differences in survival and reproduction among individuals that vary in heritable traits. Adaptation is one outcome of this microevolutionary process.

Evolutionary biologists have been documenting the results of natural selection in thousands of field studies of populations of all kinds of organisms. They find that this microevolutionary process has different results. As you will see in sections to follow, sometimes the result is a shift in the range of values for a given trait in some direction. At other times, the result may be stabilization or disruption of an existing range of values.

As Darwin perceived, the process of natural selection is the outcome of differences in survival and reproduction among individuals that show variation in heritable traits. Over the generations, natural selection can lead to increased fitness, or an increase in adaptation to the environment.

Directional Selection

In cases of **directional selection**, allele frequencies that are responsible for a range of phenotypic variation shift in a consistent direction. Such shifts are a response to a directional change in the environment or to one or more new environmental conditions. They also occur when a mutation appears and proves to be adaptive. As Figure 18.4 indicates, the forms of a trait at one end of the range become more common than the midrange forms. Consider now a few documented cases of this outcome.

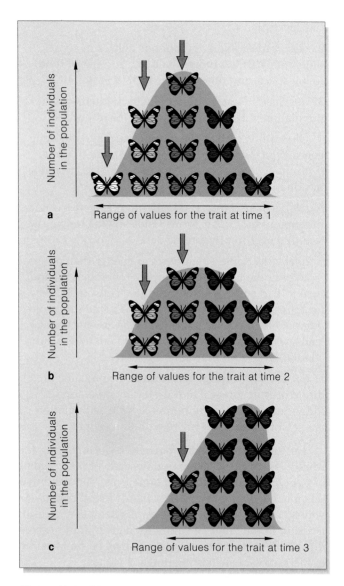

Figure 18.4 Directional selection, using phenotypic variation within a population of butterflies as the example. A bell-shaped curve (*brown*) represents the range of continuous variation in wing color. The most common forms (powder blue) are between extreme forms of the trait (white at one end of the curve, deep purple at the other). The *orange* arrows signify which forms are being selected against over time.

The Case of the Peppered Moths

In England, biologists tracked directional selection in about a hundred moth species, including the peppered moth (*Biston betularia*). Peppered moths feed and mate at night. During the day, they rest motionless on birches and other trees. Their wings and body have a mottled pattern, in shades that range from light gray to nearly black. Apparently, their behavior, coloration, and wing patterns help camouflage them from moth-eating birds, which actively hunt for food during the day.

An industrial revolution began in England in the 1850s. Outpourings of sooty smoke altered conditions in many parts of the surrounding countryside. Before then, the light moths were the most common form, and a dark form was rare. Also before conditions changed, light-gray speckled lichens grew quite profusely on tree trunks. Lichens can camouflage light moths that rest on them, but not dark ones (Figure 18.5a).

Lichens are sensitive to air pollution. Between 1848 and 1898, the soot and other pollutants from factories started killing the lichens and darkening tree trunks. Now the rare form of the moth was better camouflaged, as Figure 18.5b suggests. Moth collectors hypothesized that conditions had previously favored light moths—but that the changed conditions would favor dark ones.

In the 1950s, H. B. Kettlewell used a *mark-release-recapture* method to test the prediction. He bred both forms of the moth in captivity, then marked hundreds so that they could be identified. He released moths near heavily industrialized areas around Birmingham *and* in Dorset, an unpolluted area. After a time, he recaptured as many moths as he could. More dark moths were recaptured in the polluted area, and more light moths in the pollution-free area (Table 18.1). By stationing observers in blinds near groups of moths tethered to tree trunks, Kettlewell gathered direct evidence of birds capturing more light moths around Birmingham and more dark moths around Dorset. Directional selection was operating.

In 1952, strict pollution controls went into effect. Lichens made comebacks. Tree trunks became free of soot, for the most part. As you might have predicted, directional selection started to operate in the reverse direction. Where levels of pollution have declined, the frequency of dark moths is declining, also.

Pesticide Resistance

Widespread use of chemical pesticides in agriculture has resulted in directional selection. Initial applications kill most of the insects, worms, or other pests, but some individuals usually manage to survive. Some aspect of their structure, physiology, or behavior allows them to

a

b

Figure 18.5 Individuals of a population of peppered moths (*Biston betularia*) that has undergone directional selection in response to changes in the environment. Light-winged and dark-winged individuals are resting on a lichen-covered tree trunk in (**a**) and on a soot-darkened tree trunk in (**b**).

Table 18.1 Marked Moths (*Biston betularia*) Recaptured in a Polluted Area and a Nonpolluted Area

	Near Birmingham (pollution high)		Near Dorset (pollution low)	
LIGHT-GRAY MOTHS:				
Released	64		393	
Recaptured	16	(25%)	54	(13.7%)
DARK-GRAY MOTHS:				
Released	154		406	
Recaptured	82	(53%)	19	(4.7%)

Data after H. B. Kettlewell.

resist the chemical effects. When their resistance has a heritable basis, it becomes more common in the next generation, and the next, and the next. The chemicals are the agents of selection; they favor the most resistant forms! Today, 450 different species are resistant to one or more pesticides. Worse, the pesticides also kill the natural predators of the pests. When freed from natural constraints, the populations of resistant pests burgeon,

and crop damage is greater than ever. This outcome of directional selection is called **pest resurgence**.

Genetic engineering may reduce pesticide use. Even though plants that are engineered to resist pests won't escape the coevolutionary arms race, they might help keep our food supplies one step ahead of the pests. But many consumers are leery of genetically engineered food. What about *biological control*? By this practice, natural enemies of pests, including parasitic wasps and predatory beetles, are raised in commercial insectaries in great numbers and then released at selected sites. The practice has an advantage, in that a control species can coevolve with the pests. But farmers must replace the ones that migrate from the fields or are destroyed at harvest time, and the cost is passed on to consumers.

Antibiotic Resistance

When your grandparents were young, up to one-fourth of the annual deaths in the United States alone were caused by bacterial agents of tuberculosis, pneumonia, and scarlet fever. In the 1940s, we started treating such bacterial-induced diseases with antibiotics. **Antibiotics**, recall, are metabolic products of certain microorganisms that can kill their bacterial competitors for nutrients in soil (Section 1.4). The streptomycins, for example, block protein synthesis in target cells. The penicillins disrupt the formation of covalent bonds that hold bacterial cell walls together. Penicillin derivatives cause the wall to weaken until it ruptures.

Antibiotics should be prescribed with restraint and care. Besides performing their intended function, some disrupt populations of bacteria that normally live in the intestines and of yeast cells in the vaginal canal. Such disruptions can lead to secondary infections.

Worse yet, antibiotics have been overprescribed in the human population. Too frequently they have been used for simple infections that many individuals could have overcome successfully on their own. Disturbingly, antibiotics have lost their punch. Over time, they did destroy the most susceptible cells of target populations. But they also favored their replacement by much more resistant cells. Millions of people around the world are now dying each year of tuberculosis, cholera, and other bacterial infections. Even vancomycin, held in reserve as the antibiotic of last resort, is no longer effective against certain pathogenic strains of enterobacteria. In 1996 the World Health Organization announced that, in the race for supremacy, pathogens are sprinting ahead.

With directional selection, allele frequencies underlying a range of variation tend to shift in a consistent direction in response to directional change in the environment.

As you have seen, natural selection can bring about a directional shift in a population's range of phenotypic variation. Depending on the prevailing environmental conditions, the process also may favor either the most common or the most extreme phenotypes in that range.

Stabilizing Selection

In **stabilizing selection**, intermediate forms of a trait are favored and alleles that specify extreme forms are eliminated from a population (Figure 18.6). This mode of selection tends to counter the effects of mutation, gene flow, and genetic drift and to preserve the most common phenotypes.

Consider the selection pressures on a gallmaking fly, *Eurosta solidaginis*. After a fly larva emerges from an egg, it bores into a stem of a tall goldenrod (*Solidago altissima*). In response, plant cells multiply rapidly and surround the invader with a tumorous mass known as a gall. The larva feeds on juicy plant tissues and becomes an adult. As genetic analysis reveals, flies of different phenotypes induce formation of galls of different sizes.

The wasp *Eurytoma gigantea* can only parasitize the larvae inside small galls (Figure 18.7), so it promotes selection in favor of individual flies that are associated with the formation of *large* galls. But some insect-eating birds, including the downy woodpecker, preferentially assault large-diameter galls. As researchers discovered through a number of studies in Pennsylvania, flies that induced the formation of *intermediate-size* galls had the highest survival rate and fitness. Extreme phenotypes of the fly—which actually invited natural enemies with small galls and large galls of their own doing—were selected against. And so were the alleles for traits to which the goldenrod plants responded.

Thus, wasps work against one extreme in the range of phenotypes, birds work against the other—and the net result is an increase in intermediate phenotypes.

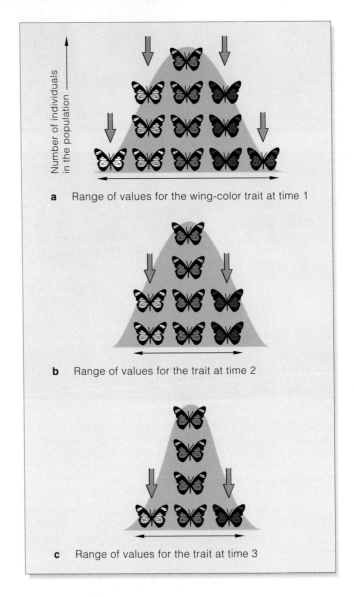

a   Range of values for the wing-color trait at time 1

b   Range of values for the trait at time 2

c   Range of values for the trait at time 3

Figure 18.6   Stabilizing selection, using phenotypic variation within a population of butterflies as the example.

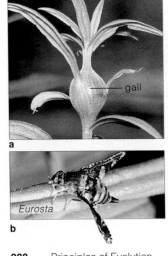

gall

Eurosta

a

b

c

d

Figure 18.7   Example of stabilizing selection. (**a**,**b**) The larvae of the fly *Eurosta solidaginis* induce formation of tumors (galls) on goldenrod stems. (**c**) Downy woodpeckers (*Dendrocopus pubescens*) and other birds prefer to chisel into large-size galls and eat the larvae. (**d**) The egg-laying device of the wasp *Eurytoma gigantea* can only penetrate the thin wall of small galls. Its eggs develop into larvae, then *its* larvae eat fly larvae. Warren Abrahamson and his coworkers monitored twenty populations of *Eurosta* in Pennsylvania. They found that the larvae in small and large galls have low relative fitnesses, and larvae in intermediate-size galls have relatively high fitnesses. Thus, there is a stabilizing component to selection pressures created by the fly's natural enemies.

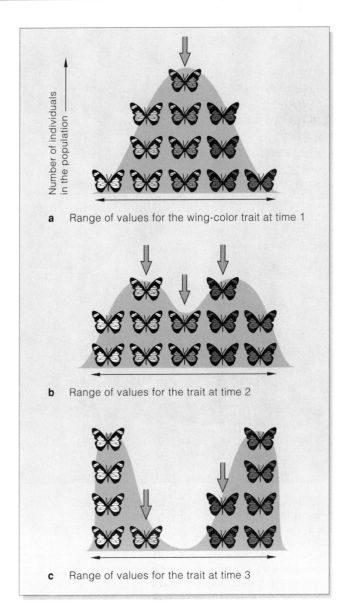

a   Range of values for the wing-color trait at time 1

b   Range of values for the trait at time 2

c   Range of values for the trait at time 3

Figure 18.8 Disruptive selection, using phenotypic variation within a population of butterflies as the example.

Disruptive Selection

In **disruptive selection**, forms at both ends of the range of variation are favored and intermediate forms are selected against (Figure 18.8). Thomas Smith discovered a splendid example of this in a remote rain forest in Cameroon, West Africa. Smith had read about unusual variation in bill size in populations of the black-bellied seedcracker (*Pyrenestes ostrinus*). These African finches have large or small bills, but no sizes in between. The pattern holds for females and males, throughout their geographic range. (This is simply remarkable; imagine every person in Texas being 4 feet *or* 6 feet tall, with no intermediates.) If the pattern is unrelated to gender or geography, what causes it?

a

Figure 18.9 Disruptive selection among African finches. In feeding trials conducted by biologist Thomas Smith, large-billed birds were more efficient at using hard seeds. Small-billed birds were most efficient at using soft seeds, not hard ones. (**a**) Two specimens displaying small and large bill sizes.

(**b**) Survival of juvenile birds during the dry season, when competition for resources is most intense. Individuals with *very* small, *very* large, or intermediate-size bills cannot feed efficiently on either type of seed; they survive poorly. For this graph, *tan* bars show the number of nestlings; *brown* bars show the only survivors among them. The findings are based on measurements of 2,700 netted individuals.

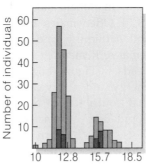

b   Measured width of bill's lower portion (millimeters)

If, as Smith hypothesized, the persistence of only two bill sizes in the populations relates to seed-cracking ability (which directly affects survival), then disruptive selection may be eliminating birds with intermediate-size bills. But what factors cause disruptive selection pressure on feeding performance? Cameroon's swamp forests flood in the wet season. Lightning-sparked fires burn in the dry season. Two species of sedges (grasslike, fire-resistant plants) dominate the region. One has hard seeds; the other has soft seeds. When finches reproduce, both hard and soft seeds are abundant. All birds prefer soft seeds for as long as they can get them. But food dwindles as the dry season peaks. Then, young birds especially are at a competitive disadvantage; many do not survive (Figure 18.9).

Smith also performed experimental crosses between large- and small-billed birds. All offspring had large *or* small bills. Along with other data, this suggests that bill size and feeding performance may be controlled mainly by a single autosomal gene with two alleles.

With stabilizing selection, intermediate phenotypes are favored, and extremes at both ends of the range of variation are eliminated.

With disruptive selection, intermediate forms are selected against; extreme forms in the range of variation are favored.

SPECIAL TYPES OF SELECTION

Sexual Selection

As you may have noticed, individuals of most sexually reproducing species have a distinctively male or female phenotype. We call this **sexual dimorphism** (*dimorphos* means "having two forms"). How does this condition come about, and what maintains it? Here again, natural selection is at work. In this case, it is **sexual selection**, for it favors traits with no advantage for survival and reproduction, other than the fact that males or females prefer them. Through nonrandom mating, the alleles for preferred traits prevail over the generations.

Sexual dimorphism is especially striking among many mammals and birds, including the pair in Figure 18.10. Often males are larger, have flashier coloration and patterning, and are far more aggressive than the females. Remember those male bighorn sheep butting heads in Figure 1.6g? Fighting wastes time, it wastes energy, and it may cause injuries. Why, then, do alleles that contribute to aggressive behavior persist in the population? The increased chance of mating offsets the costs. Male bighorn sheep fight only to control areas where receptive females gather during a winter rutting season. Winners mate often, with a number of females; and the losers will not challenge a stronger, larger male.

Females are the main agents of selection. They exert direct control over reproductive success by choosing their mates. We return to this topic in Chapter 51.

Balancing Selection

Balancing selection includes all forms of selection that are maintaining two or more alleles for a given trait in a population. When this type of genetic variation persists, we call it **balanced polymorphism** (after *polymorphos*, "having many forms"). A population is in a state of balanced polymorphism when nonidentical alleles for a trait are being maintained at frequencies greater than 1 percent. Frequencies may shift slightly, but over time they often will bounce back to the same values. Smith's work with African finches is a fine example of the effect of balancing selection through the generations.

Sickle-Cell Anemia—Lesser of Two Evils?

We sometimes observe cases of balanced polymorphism where environmental conditions favor heterozygotes (which carry nonidentical alleles for a specified trait), not the homozygotes (which carry identical alleles for the trait). Said another way, *the heterozygote has a higher fitness than either homozygote*.

Let's look at the environmental pressures that favor the pairing of an Hb^A and an Hb^S allele in humans. Hb^S is responsible for a mutated form of hemoglobin, an

Figure 18.10 One ▼ outcome of sexual selection. This male bird of paradise (*Paradisaea raggiana*) is engaged in a flashy courtship display. He caught the eye (and, perhaps, the sexual interest) of the smaller, less colorful female. Males of this species compete fiercely for females, which serve as selective agents. (Why do you suppose drab-colored females have been favored?)

oxygen-transporting protein in the blood. Homozygotes (Hb^S/Hb^S) develop *sickle-cell anemia*, a genetic disorder with potentially severe phenotypic outcomes (Section 11.5). The Hb^S allele frequency is high in tropical and subtropical regions of Africa and Asia. Often, Hb^S/Hb^S homozygotes die in their early teens or early twenties. But heterozygotes (Hb^A/Hb^S) make up nearly a third of the human populations in these regions! Why is this allelic combination maintained at such high frequency?

The balancing act, an outcome of natural selection, is most pronounced in areas with the highest incidence of *malaria* (Figure 18.11). There a mosquito transmits *Plasmodium*, the parasite responsible for the disease, to humans. The parasite multiplies in the liver, then in red blood cells. The cells rupture and release new parasites during severe, recurring bouts of infection. People are far more likely to survive the awful recurrences *if* they are Hb^A/Hb^S heterozygotes. The allelic combination gives

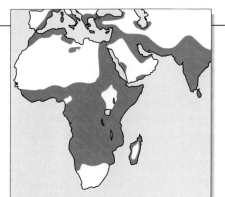

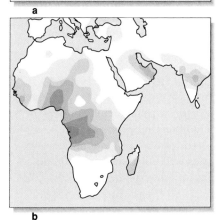

Figure 18.11
(a) Distribution of cases of malaria in parts of Africa, the Middle East, and Asia in the 1920s, before extensive mosquito control programs had been instituted. (b) From the same regions, the distribution and the frequency of individuals with the sickle-cell trait. Notice the close correlation between the color-tinted areas of the two maps.

- ☐ less than 1 in 1,600
- ☐ 1 in 400–1,600
- ☐ 1 in 180–400
- ☐ 1 in 100–180
- ■ 1 in 64–100
- ■ more than 1 in 64

them two forms of hemoglobin, with interesting results. Each has enough nonmutated molecules to maintain body functions, even if at reduced levels. But the *altered* molecules distort the shape of red blood cells, and this interferes with normal blood circulation. The slowdown in blood flow is a factor. It hampers the parasite's ability to rapidly infect new cells during an infective cycle.

Thus the persistence of the "harmful" Hb^S allele is a matter of relative evils. Natural selection has favored one allelic combination, Hb^A/Hb^S, because its bearers show greatest fitness in environments where malaria is most prevalent. Hb^A/Hb^S has a higher fitness than either Hb^S/Hb^S or Hb^A/Hb^A in these regions.

In tropical and subtropical parts of Asia and the Middle East, malaria has been a selective force for more than 2,000 years. Although sickle-cell anemia occurs at high frequencies in these regions, its symptoms are much less severe than they are in Central Africa, where the Hb^S allele became established much later in time. Most likely, other gene products might be modifying some of the widespread effects of the Hb^S allele in ways that can minimize the symptoms of the disorder.

With sexual selection, a trait gives an individual an advantage in reproductive success. Sexual dimorphism is one outcome of sexual selection.

Balanced polymorphism is a state in which natural selection is maintaining two or more alleles over the generations at frequencies greater than 1 percent.

With emigration, individuals leave a population. With immigration, new individuals enter it. Either way, the frequencies of alleles can change. This physical flow of alleles, or **gene flow**, helps keep separate populations genetically similar. Over time, the flow of alleles tends to counter genetic differences between populations that are brought about through mutation, genetic drift, and natural selection.

Think of the acorns that blue jays disperse when they store nuts for the winter. Each fall the jays may make hundreds of round trips from acorn-bearing oak trees to bury acorns in the soil of their home territories, which may be up to a mile away (Figure 18.12). The alleles flowing with "immigrant acorns" help reduce genetic differences that might otherwise arise among neighboring stands of oaks.

Figure 18.12 Gene flow among oak populations, courtesy of feathered travel agents. Blue jays hoard acorns in their home territory, but they might shop at nut-bearing trees up to a mile away. Some acorns contribute to the allele pool of an oak population some distance away from the parent tree.

Or think of the millions of people from economically bankrupt, politically explosive countries who seek more stable homes. The sheer scale of their movement is unprecedented, but hardly unique. Throughout human history, immigrations may have minimized many of the genetic differences that otherwise would have built up among geographically separated groups of people.

Gene flow is the physical movement of alleles into and out of a population, through immigration and emigration.

Chance Events and Population Size

Genetic drift is a random change in allele frequencies over the generations, as brought about by chance alone. The magnitude of its effect on genetic diversity and on the range of phenotypes relates to population size. Its impact tends to be minor or insignificant in very large populations but significant in small ones.

Sampling error, a rule of probability, helps explain the difference. By this rule, the fewer times a chance event occurs, the greater will be the variance from the expected outcome of that occurrence. Think back on the coin-flipping example given earlier, in Section 11.2. Each time you flip a coin, there is a 50 percent chance it will turn up heads or tails. With, say, only ten flips, the odds are great that the coin will turn up heads 7 times and tails 3 times; with a thousand flips, the odds that the coin will turn up heads 700 times and tails 300 times are virtually nil. Similarly, sampling error applies each time random mating and fertilization take place in a population.

Bear in mind, genetic drift has nothing to do with how a population got small in the first place. *It simply increases the chance of any given allele becoming more or less prevalent when the number of individuals in a population is small.*

Figure 18.13 is a computer simulation of the effect of genetic drift in two populations: one large, the other small. The outcomes parallel the findings of an actual experiment with beetles (*Tribolium*) that mate with one another at random. Researchers grouped 1,320 beetles into twelve populations of 10 beetles and twelve of 100. Which beetles ended up in which group was a matter of chance. At the start, the frequency of a wild-type allele (call it *A*) was 0.5. The researchers tracked allele *A* for twenty generations. Each time, they randomly removed some offspring to maintain each population's original size. At the experiment's end, *A* wasn't the only allele left in the *large* groups, but it had become fixed in seven of the *small* groups. **Fixation** means only one kind of allele remains at a specified locus in a population; all individuals are homozygous for it.

In sum, *over time, and in the absence of other forces, random change in allele frequencies leads to the homozygous condition and a loss of genetic diversity.* This happens in all populations; it just happens more rapidly in small ones. Once alleles inherited from an original population are fixed, their frequencies will not change again unless new alleles appear by mutation or gene flow.

Bottlenecks and the Founder Effect

Genetic drift is pronounced when very few individuals rebuild a population or found a new one. A **bottleneck** is a severe reduction in population size, as brought about by intense selection pressure or natural calamity. Suppose contagious disease, loss of habitat, hunting, or a massive volcanic blast destroys much of a population.

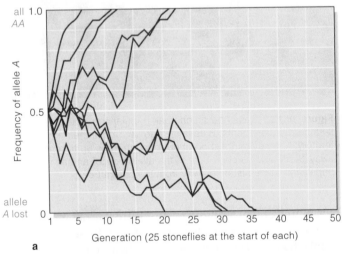

Generation (25 stoneflies at the start of each)

a

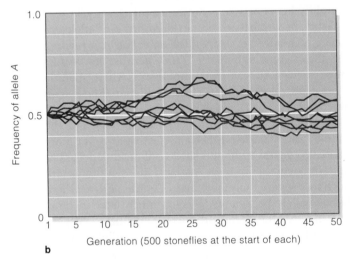

Generation (500 stoneflies at the start of each)

b

Figure 18.13 Computer simulation of the effect of genetic drift on one allele's frequency in small and large populations. (**a**) The size of nine populations of stoneflies was maintained at 25 breeding individuals each generation, through fifty generations. (**b**) The size of nine other populations was maintained at 500 individuals each generation, through fifty generations.

The lines reaching the top of the graph in (**a**) tell you that allele *A* became fixed in five of the small populations. The lines plummeting off the bottom of the graph tell you it was lost from four of them. As you can see, *alleles can be fixed or lost even in the absence of selection.* As you can see from (**b**), allele *A* did not become fixed in any of the large populations. The magnitude of drift was much less in every generation than in the small populations.

Equal fitnesses assumed for these three simulations:

AA = 1
Aa = 1
aa = 1

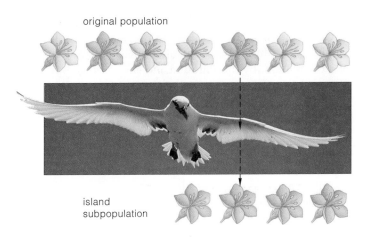

original population

island
subpopulation

Figure 18.14 Example of the founder effect. A seabird carries a few seeds, stuck to its feathers, to a remote oceanic island. By chance, most of the seeds carry an allele for orange flowers that was not common in the original population. In the absence of further gene flow or selection for flower color, genetic drift will fix the allele in the island population.

Figure 18.15 A few of the remaining cheetahs, all of which have some very bad alleles that made it through a severe bottleneck.

Even if a moderate number do survive the bottleneck, allele frequencies will have been altered at random.

In the 1890s, hunters killed all but twenty of a large population of northern elephant seals. The population recovered. When its size passed 30,000, electrophoretic analysis of a sample of twenty-four genes revealed no variation in the population. A number of alleles had been lost in the bottleneck.

The genetic outcome can be similarly dicey after a few individuals leave a population and establish a new one elsewhere. This form of bottlenecking is called a **founder effect**. By chance, the allele frequencies of the founders may not be the same as those of the original population. In the absence of further gene flow, natural selection will influence those frequencies in drastically different ways, because of its interaction with genetic drift. As you might deduce, the founder effect is quite pronounced on isolated islands (Figure 18.14).

Genetic Drift and Inbred Populations

Inbreeding refers to nonrandom mating among closely related individuals, which have many identical alleles in common. Inbreeding can be viewed as a form of drift in a small population; that is, in the group of relatives that are preferentially interbreeding. Like genetic drift, inbreeding leads to the homozygous condition. It also can lower fitness when the alleles that are increasing in frequency are recessive and have harmful effects.

Most human societies forbid or discourage incest (inbreeding between parents and children or between siblings). But inbreeding among other close relatives is common in small communities that are geographically or culturally isolated from a larger population. The Old

Order Amish of Pennsylvania, for instance, is a group with distinct genotypes. One outcome of inbreeding is the high frequency of a recessive allele that causes *Ellis-van Creveld syndrome*. Affected individuals have extra fingers or toes and short limbs (compare Section 12.6). The allele probably was rare when the small number of founders immigrated to Pennsylvania. Now 1 in 8 are heterozygous and 1 in 200 are homozygous for it.

Bottlenecks and inbreeding are an especially awful combination for **endangered species**, the populations of which are very small and vulnerable to extinction. Consider the cheetah (Figure 18.15), which apparently went through a drastic bottleneck during the nineteenth century. The surviving parents mated with their own offspring when no other options were available.

Inbreeding among survivors and their descendants left the 20,000 existing cheetahs with strikingly similar alleles. One of these is a mutated allele with bad effects on fertility. Typically, a male cheetah has a low sperm count, and 70 percent of the sperm are abnormal. Other shared alleles result in much lower resistance to disease. Infections that are seldom life-threatening to other cat species can be devastating to cheetahs. In one outbreak of *feline infectious peritonitis* in a wild animal park, the pathogen (a coronavirus) had little effect on the captive lions but killed a great many cheetahs. This infection triggers an uncontrollable inflammatory response. Fluid fills the body cavity housing the heart and other internal organs, and the cat dies in agony. There is no vaccine.

Pressure from hunting and urban sprawl has also put the highly inbred Florida panther on the endangered species list. Only fifty of these cats remain.

Genetic drift is the random change in allele frequencies over the generations, brought about by chance alone. The magnitude of its effect is greatest in small populations, such as the ones that make it through a bottleneck.

Barring mutation, selection, and gene flow, the chance losses and increases of the various alleles at a given locus lead to the homozygous condition and a loss of genetic diversity.

SUMMARY

1. Individuals of a population generally have the same number and kind of genes. But genes come in different allelic forms, and this leads to variations in their traits.

2. A population is evolving when some forms of a trait (and the alleles specifying them) are becoming more or less common with respect to the other kinds, over the generations. This happens as a result of mutation, gene flow, genetic drift, and natural selection (Table 18.2).

3. Genetic equilibrium, a state in which a population is not evolving, is used as a baseline to measure change. It occurs only if there is no mutation, if the population is very large and isolated from other populations of the same species, and if there is no selection (all members survive and reproduce equally by random mating).

4. Gene mutations, heritable changes in DNA, are the only source of *new* alleles. New combinations of *existing* alleles occur by crossing over, independent assortment at meiosis, and mixing of alleles at fertilization.

5. Natural selection is an outcome of differences in survival and reproduction among the individuals of a population that differ in one or more heritable traits. The process affects the relative abundances of alleles responsible for adaptive and maladaptive versions of traits over the generations.

   a. Selection pressure may shift the range of variation for a trait in one direction (*directional* selection), favor extremes (*disruptive* selection), or favor intermediate forms and eliminate the extremes from a population (*stabilizing* selection). Most species are well adapted to fairly stable habitats, so extremes are often lost. Thus stabilizing selection may be the most common mode.

   b. Selection can result in balanced polymorphism. A population is in this state when nonidentical alleles for a given trait are being maintained over the generations at frequencies greater than 1 percent.

   c. Sexual selection, by one sex or the other, leads to forms of traits that offer an advantage in reproductive success. Sexual dimorphism, which is the persistence of phenotypic differences between males and females of a species, is one outcome of sexual selection.

6. Gene flow is a change in allele frequencies brought about by the physical movement of alleles into and out of a population (by immigration and emigration).

7. Genetic drift is a change in allele frequencies over the generations due to chance events alone.

   a. Because of sampling error, the magnitude of its effect is greater in small populations than in large ones. In the absence of mutation, selection, and gene flow, sooner or later genetic drift can result in a homozygous condition and loss of genetic diversity.

   b. Genetic drift has great impact after a bottleneck, a severe reduction in population size from which the population recovers. The alleles that do make it through a bottleneck may give rise to a different range of phenotypes, compared to the original population. In one type of bottlenecking, called the founder effect, a few emigrants from one population successfully establish a small subpopulation in a new environment.

Review Questions

1. Define genetic equilibrium. Explain how these processes can send allele frequencies out of equilibrium: *18.1, 18.2*
   a. mutation
   b. natural selection
   c. gene flow
   d. genetic drift

2. Define fitness, as biologists use the term. *18.3*

3. Define bottleneck and the founder effect. Are these cases of genetic drift, or do they set the stage for it? *18.8*

4. Consider the brilliantly hued male sugarbird and the subdued-hued female (Figure 18.16). In the evolutionary view, what is maintaining the morphological difference over time? *18.6*

Figure 18.16 A male and a female sugarbird.

Table 18.2	Summary of Microevolutionary Processes
MUTATION	A heritable change in DNA
NATURAL SELECTION	Change or stabilization of allele frequencies as a result of differences in survival and reproduction among variant individuals of a population
GENETIC DRIFT	Random fluctuation in allele frequencies over time, due to chance occurrences alone
GENE FLOW	Change in allele frequencies as individuals leave or enter a population

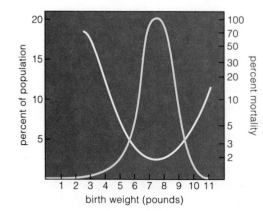

Figure 18.17 Weight distribution for 13,730 human newborns (*yellow* curve) correlated with mortality rate (*white* curve).

Figure 18.18 Bell-shaped curves characteristic of three modes of natural selection.

5. As long-term studies indicate, the prospects for human newborns of very high or very low birth weight are not good. Being born too small increases the risk of stillbirth and early infant death (Figure 18.17). Similarly, pre-term rather than full-term pregnancies also increase the risks. Which of the microevolutionary processes is at work here? *18.5*

6. In Figure 18.18, identify the modes of selection (stabilizing, directional, and disruptive) in each diagram. *18.4, 18.5*

Self-Quiz (*Answers in Appendix IV*)

1. Individuals don't evolve; _____ do.

2. Genetic variation gives rise to variation in _____ traits.
   a. morphological   c. behavioral
   b. physiological   d. all of the above

3. Sickle-cell anemia first appeared in Asia, the Middle East, and Africa. The causative allele entered the population of the United States when people were forcibly brought over from Africa prior to the Civil War. In microevolutionary terms, this is a case of _____ .
   a. mutation   c. gene flow
   b. genetic drift   d. natural selection

4. Directional selection _____ .
   a. eliminates uncommon forms of alleles
   b. shifts allele frequencies in a steady, consistent direction
   c. favors intermediate forms of a trait
   d. works against adaptive traits

5. Disruptive selection _____ .
   a. eliminates uncommon forms of alleles
   b. shifts allele frequencies in a steady, consistent direction
   c. doesn't favor intermediate forms of a trait
   d. both b and c

6. Match the evolution concepts.
   ____ gene flow   a. source of new alleles
   ____ natural   b. changes in a population's allele
        selection   frequencies due to chance alone
   ____ mutation   c. immigration, emigration change
   ____ genetic   allele frequencies
        drift   d. differences in survival and
                            reproduction among variant individuals

Critical Thinking

1. For centuries, *tuberculosis* (TB) has caused many deaths around the world. A bacterium (*Mycobacterium tuberculosis*) causes this contagious lung disease. TB declined steadily in the United States, where public health officials even predicted there would be no more cases by the year 2000. But caseloads are increasing. The AIDS epidemic, the influx of immigrants from regions where TB is still common, and severe overcrowding in tenements and homeless shelters contribute to the increase. At one time, antibiotics could cure TB within six to nine months. But now antibiotic-resistant strains of *M. tuberculosis* have evolved. Currently, two or even three different antibiotics must be used simultaneously to block different metabolic pathways of this bacterium. How would this approach help avoid the problem of antibiotic resistance?

2. A few families that live in a remote region of Kentucky show a high frequency of *blue offspring*, an autosomal recessive disorder. The skin of affected individuals has a bright blue appearance. Homozygous recessives lack the enzyme diaphorase. The enzyme catalyzes reactions that help maintain hemoglobin (the oxygen-carrying red pigment in blood) in its normal molecular form. Without the enzyme, a blue form of hemoglobin accumulates in the blood. Skin and the blood capillaries associated with it are transparent, so the color of pigments in blood shows through and contributes to skin coloration. This gives the skin of nonmutated individuals a pinkish cast—and blue skin its color. Formulate a hypothesis to explain why this trait is rather common among a cluster of families but rare in the human population at large.

Selected Key Terms

allele *18.1*	genetic equilibrium *18.1*
allele frequency *18.1*	Hardy-Weinberg rule *18.2*
antibiotic *18.4*	inbreeding *18.8*
balanced polymorphism *18.6*	lethal mutation *18.1*
balancing selection *18.6*	microevolution *18.1*
bottleneck *18.8*	mutation rate *18.1*
directional selection *18.4*	natural selection *18.3*
disruptive selection *18.5*	neutral mutation *18.1*
endangered species *18.8*	pest resurgence *18.4*
fitness *18.3*	polymorphism *18.1*
fixation *18.8*	population *18.1*
founder effect *18.8*	sampling error *18.8*
gene flow *18.7*	sexual dimorphism *18.6*
gene pool *18.1*	sexual selection *18.6*
genetic drift *18.8*	stabilizing selection *18.5*

Readings

Gould, S. J. 1977. *Ever Since Darwin*. New York: Norton.

Smith, T. B. January 1991. "A Double-Billed Dilemma." *Natural History*: 14–21.

Weiner, J. 1994. *The Beak of the Finch: Evolution in Real Time*. New York: Alfred A. Knopf. An account of continuing investigations of evolution among the Galápagos finches.

Web Site See *http://www.wadsworth.com/biology* for practice quiz questions, hypercontents, BioUpdates, and critical thinking. The Wadsworth Biology Resource Center provides a wealth of information fully organized and integrated by chapter.

19 SPECIATION

The Case of the Road-Killed Snails

If you happen to be a snail living in a garden in Bryan, Texas, it doesn't take much to keep your genes away from snails in a backyard across the street. By day, the sunbaked asphalt would be about as inviting to a snail as a desert would be to a catfish. Besides, day or night, a street-traversing snail is vulnerable to cars, trucks, skateboards, and bicycles (Figure 19.1*a*). Whatever else it might be, that strip of asphalt is a formidable barrier to gene flow between populations. For snails, that is.

Whether any physical barrier deters gene flow depends in large part on the organism's mode of locomotion or dispersal. It also depends on how fast and how long an organism *can* move in response to environmental factors or its own hormonal signals.

A snail is not swift, and it does not roam far from its home population. Compare it to the wild black duck, banded in Virginia in 1969, that turned up eight years later in Korea. Compare it to the wandering albatross, one of the supreme barrier busters. After lifting off from Kerguelen Island in the Indian Ocean, one of these birds soared westward past the southern tip of Africa, across the Atlantic, and around South America's Cape Horn. After it had traveled thirteen thousand kilometers, it landed in Chile. With its lightweight body and its wingspan of 3.65 meters (12 feet across), that albatross was able to exploit the great prevailing winds of the Southern Hemisphere.

And yet, in 1859, some snails did cross an ocean. Humans had transported garden-variety snails (*Helix aspersa*) all the way from France to California, then turned them loose. The idea

Figure 19.1 (**a**) One snail (*Helix aspersa*) encountering a major barrier to gene flow. It only appears to be reading the warning label. (**b**) Results from a study of neighboring populations of snails, all descended from founders that ended up in a small town in Texas in the 1930s. For each population, a circle represents relative abundances of three alleles (coded *pink*, *orange*, or *brown*) for an enzyme, leucine aminopeptidase. The genetic variation is greater between populations living on opposite sides of Twenty-Second Street. For example, notice the higher frequencies of the allele color-coded *orange* in the block to the west of the street.

a

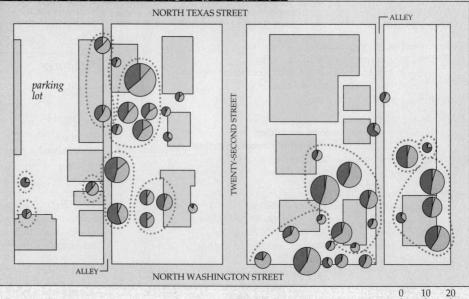

b

was that the snails would mate, multiply, and so meet the demand for that French delicacy, *escargots aux fines herbes*. It was bad enough that the importers mistakenly brought over a less tasty species. Worse yet, the snails exceeded expectations, and they became an absolute nuisance in gardens and nurseries throughout much of the southwestern United States.

By the 1930s, *H. aspersa* had hitched rides, possibly as eggs in the soil of plant containers, to Bryan, Texas. They founded small colonies in the local vegetation. Forty years later, Robert Selander, now of Pennsylvania State University, was down on his hands and knees with a few graduate students, scouring patches of vegetation on two adjacent city blocks. Why? Selander wanted to determine the effect of genetic drift on the introduced populations. He and his students collected every single snail—2,218 of them—from fourteen local populations. For each population, they determined the allele frequencies for five different genes.

For each of the five genes, the results from their analysis pointed to some genetic variation among the colonies on the same block—and to major differences in allele frequencies *between* different blocks. Figure 19.1b shows the results for one of the genes studied.

Assume the genetic differences between populations will continue to increase, as through natural selection or genetic drift. Will the time come when snails from opposite sides of the street can no longer interbreed successfully, even if they do manage to get together? In other words, *will they become members of separate species?* Or would stabilizing selection work against significant genetic divergence by eliminating extreme phenotypes from the neighboring populations? After all, how much can a workable package of *H. aspersa* genes evolve in adjacent patches of vegetation—and under very similar environmental pressures—in a small town in Texas?

And with these questions, we find ourselves arriving at the topic of **speciation**—that is, to changes in allele frequencies that are significant enough to mark the formation of daughter species from a parent species. Obviously, no one was around to watch the formation of species during the past. No one lives long enough to know whether many existing populations are at some intermediate stage leading to speciation. Evolutionary biologists are still working out their theories about speciation processes, and what you are about to read may change in the near or distant future.

KEY CONCEPTS

1. A species consists of one or more populations of individuals that can interbreed under natural conditions and produce fertile offspring, and that are reproductively isolated from other such populations. This definition is restricted to sexually reproducing species.

2. The populations of a species have a shared genetic history, they are maintaining genetic contact over time, and they are evolving independently of other species.

3. Speciation is the process by which daughter species evolve from a parent species.

4. By one model, speciation starts when a geographic barrier arises between populations or subpopulations of a species. Thereafter, mutation, natural selection, and genetic drift operate independently in each population and lead to irreversible genetic divergence of one from the other.

5. Such populations may come to differ in certain alleles that affect morphological, physiological, or behavioral traits associated with reproduction. When the genetic differences affecting reproduction become great enough, the populations can no longer interbreed even if they later coexist in the same area. Speciation is completed.

6. The model of speciation just described, which applies to geographically separated populations, is known as allopatric speciation. This might be the way most species originate in nature. Sympatric speciation and parapatric speciation might be less prevalent routes.

7. With sympatric speciation, species form within the home range of the parent species. With parapatric speciation, adjacent populations become distinct species while still maintaining contact along a common border between their home ranges.

8. The timing, rate, and direction of speciation vary within and between lineages. The extinction of some number of species is inevitable for all lineages.

ON THE ROAD TO SPECIATION

What Is a Species?

MORPHOLOGICAL SPECIES CONCEPT "If it looks like a duck, walks like a duck, and quacks like a duck, then probably it's a duck." Let's use this familiar saying as a starting point for defining what is and what is not a species. **Species** is a Latin word. Generally, it simply means "kind," as in "a particular kind of duck."

Based on the appearance of the body alone—that is, on morphological traits—few of us would have trouble distinguishing a duck from, say, a chicken. At the very least, chickens don't have a flattened bill and webbed feet. And a chicken's body is not adapted to tipping bottoms up in a pond, head down, to strain food from the mud or water. But think about this: There are forty known species of ducks around the world. Especially in the interior of the Northern Hemisphere's continents, ducks that show great morphological diversity gather in large numbers to nest in marshlands. How do we know a particular duck belongs to one species and not another? More to the point, how does the *duck* know it?

Even when early naturalists were defining species on the basis of morphological traits, common sense told them other factors must also be considered. Sometimes individuals of the same species have strikingly different appearances simply because they have responded to different environmental conditions. Those arrowhead plants in Figure 19.2 are one example. Sometimes pronounced differences in form and coloration emerge as part of the life history of a species, so that the young do not look at all like the adults. You have only to think about the wriggly larval beginnings of adult butterflies to know this is so. And sometimes sexual dimorphism can be extreme. Thus, at first glance, you might well dismiss a busily mating male anglerfish as a puny extension of the female's body (Figure 19.3).

Variation in morphology is not the only source of potential confusion. Individuals that outwardly appear to be almost identical may belong to genetically distinct species. Consider the gray treefrogs *Hyla versicolor* and *H. chrysoscelis*. We find both species in portions of the central and eastern United States. Both have warty skin and prominent sticky pads on their toes. Both have a white belly, yellow or yellow-orange coloration on the inside of the hind legs, and a large white spot beneath each eye. Both also live in the same forest habitats and prefer the same kind of prey. It took careful behavioral and cytological studies to identify *H. versicolor* and *H. chrysoscelis* as separate species. During the breeding season, the distinctive call of each type of male frog attracts only females of the same species (Figure 19.4). Also, the two species show a chromosomal difference; one is diploid and the other is tetraploid.

Figure 19.2 Pronounced morphological differences between individuals of the same species. Mature leaves of arrowheads (*Sagittaria sagittifolia*) growing on land (**a**) or in the water (**b**) differ. This is attributable to adaptive responses to different environmental conditions, not to genetic differences.

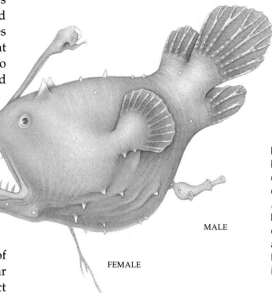

Figure 19.3
Pronounced sexual dimorphism in the deep-sea angler fish *Linophryne*, also known as the net devil. The female is about ten centimeters long. The tiny male is attached to her.

MALE

FEMALE

BIOLOGICAL SPECIES CONCEPT Morphological details can vary so enormously, perhaps we should extend our consideration past the details, down to a basic function that unites members of a species and isolates them from all other species.

For example, using *reproduction* as a basic, defining function is central to the **biological species concept**. The evolutionary biologist Ernst Mayr has phrased the concept this way: "Species are groups of interbreeding natural populations that are reproductively isolated from other such groups." No matter how extensive the phenotypic variation, individuals will remain members of the same species as long as their form, physiology,

a *Hyla versicolor*   **b** *H. chrysoscelis*

Figure 19.4 Which frog is shown here—*Hyla versicolor* or *H. chrysoscelis*? You cannot tell the two species apart on the basis of external appearance. Identification depends on other clues, such as the female-attracting calls made by the male frogs during the breeding season. (**a**, **b**) Sound spectrograms (visual records of each note's frequency, or pitch) reveal a behavioral difference between the two species.

and behavior permit them to interbreed and produce fertile offspring. Their capacity to contribute to a shared gene pool is their qualification for membership. Mayr's concept cannot be applied to organisms that reproduce asexually, such as bacteria. However, it has proved to be a useful guide for research into factors that define sexually reproducing species—and the vast majority of species reproduce sexually.

Genetic Change and Speciation

If we subscribe to Mayr's concept, then speciation is the attainment of reproductive isolation. Bear in mind, reproductive isolation does not evolve purposefully to promote the formation of a species or to maintain its identity. Rather, *any structural, functional, and behavioral difference that brings about reproductive isolation is simply a by-product of genetic change.*

Recall that genetic changes between populations of the same species can only be countered by **gene flow**, the movement of alleles into and out of populations by immigration and emigration. Gene flow can exert its homogenizing effect even when two populations are separated geographically, as long as some gene exchange continues between them. This microevolutionary process helps maintain their common reservoir of alleles.

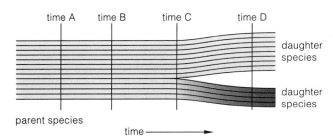

Figure 19.5 Simplified diagram of genetic divergence. Each horizontal line represents a different population of a species.

What if some geographic barrier arises and prevents the intermingling of genes between populations or even subpopulations of a species? Then, each will embark on its own evolutionary road. Through **genetic divergence**, differences will build up between the allele pools of the genetically separated populations, for mutation, natural selection, and genetic drift will now be free to operate independently in each one. Figure 19.5 is a simple way to think about genetic divergence. Each horizontal line of the diagram represents a different population of a species. At times A and B, gene flow is helping to keep them genetically united. At time C, a geographic barrier has arisen and divergences begin.

As you will see shortly, speciation usually proceeds gradually, through genetic divergence. For some groups of organisms, it also has proceeded rapidly, following changes in chromosome number. Said another way, *the process of speciation can vary in its duration.*

Depending on how the affected populations interact and on their patterns of distribution, *speciation also may vary in its details.* The preceding paragraphs started you thinking about what is probably the main route, which is called allopatric speciation. Later in the chapter you will be taking a closer look at this route and two others, called sympatric and parapatric speciation.

THE BIOLOGICAL SPECIES CONCEPT. A species is one or more populations of individuals that (1) are interbreeding under natural conditions and producing fertile offspring, and (2) are reproductively isolated from other such populations.

Speciation is the process by which a daughter species forms from a population or subpopulation of the parent species. The process can vary in its details and in the length of time it takes before reproductive isolation is complete.

By what is thought to be the most common speciation route, a geographic barrier arises and separates populations of a species. Differences build up in their gene pools as mutation, natural selection, and genetic drift operate independently in each one. This outcome is called genetic divergence.

Reproductive isolating mechanisms may evolve simply as by-products of the genetic changes. They are heritable traits that, one way or another, prevent interbreeding.

REPRODUCTIVE ISOLATING MECHANISMS

Let's now define **reproductive isolating mechanisms** as any heritable feature of body form, functioning, or behavior that prevents interbreeding between one or more genetically divergent populations. Some prevent successful mating or pollination between individuals of the divergent populations so that hybrid zygotes cannot form. Others take effect *after* zygotes have formed.

Prezygotic Isolation

The mechanisms of *prezygotic* isolation come into play before or during fertilization.

TEMPORAL ISOLATION   Suppose that the potential for interbreeding still exists between individuals of two divergent populations. If they reproduce at different times, this makes no difference. Most animals mate and most plants get pollinated quickly, sometimes in less than a day, so there is not much chance of overlap with others. We find extreme temporal isolation among the periodical cicadas (*Magicicada*) of the eastern United States. These insects mature underground, where they feed on juices of tree roots. Three species that differ in size, color, and song emerge and reproduce every 17 years (Figure 19.6). Each has a *sibling* species, meaning their morphologies are indistinguishable. But its sibling species emerges every 13 years. Thus, only once every 221 years do the two release gametes at the same time!

Figure 19.6 Adult *Magicicada septendecim*, a periodical cicada that matures and then emerges from underground to reproduce every 17 years. Often its populations overlap the habitats of its sibling species (*M. tredecim*), which reproduces every 13 years. Adults live only a few weeks. They are a taste thrill to birds, which often gorge themselves so frenziedly that they vomit; which makes one suspect why periodical cicadas stay underground for so long.

BEHAVIORAL ISOLATION   Behavioral isolation is a major barrier to gene flow among related species living in the same territory. For example, before male and female birds copulate, they often engage in intricate courtship rituals (Figure 19.7). A female is genetically equipped to recognize singing, head bobbing, wing spreading, or prancing of a male of the same species as an overture to sex. Females of another species usually are not.

MECHANICAL ISOLATION   An incompatibility between body parts of potential mates or pollinators is a form of mechanical isolation. Consider two sage species, *Salvia apiana* and *S. mellifera*. Their nectar is cupped inside a cluster of petals. It attracts pollinators, which use other petals as a landing platform (Figure 19.8). *S. apiana*'s pollen-bearing stamens extend some distance from the nectar cup, right above a platform of petals that is large enough to accommodate large pollinators. *S. mellifera* has a small landing platform, more suitable for small pollinators. When small bees happen to visit *S. apiana*, stamens of this larger flower often do not brush against them, so the pollen is not transported to flowers of the other species. Similarly, the large pollinators of *S. apiana* cannot land on and cross-pollinate *S. mellifera*.

ECOLOGICAL ISOLATION   Populations that are adapted to different microenvironments in the same habitat are a case of ecological isolation. In the seasonally dry foothills of the Sierra Nevada are populations of two species of woody, treelike shrubs called manzanita (Figure 19.9). *Arctostaphylos patula* grows in open forests of conifers at elevations ranging from 600 to 1,850 meters. *A. viscida* grows at elevations ranging from 750 to 3,350 meters. Where their ranges overlap, the two rarely hybridize. Differences in traits confine them to different parts of the same forests. For example, like all manzanitas, both species have physiological mechanisms that enhance water conservation during the dry season. But *A. patula* is adapted to more sheltered sites, where water stress is not as intense as it is on more exposed (and drier) rocky hillsides, which *A. viscida* prefers.

Figure 19.7   A sampling of courtship displays that precede copulation (sexual union) between a male and female albatross. Members of the same species recognize the visual, acoustical, and tactile components of such displays.

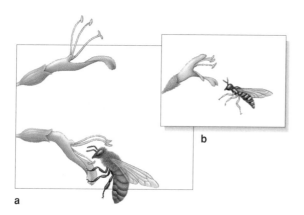

Figure 19.8 Mechanical isolation between two species of sage. (**a**) Some petals of *Salvia apiana* flowers are arranged as a large landing platform for large pollinators. A tight cluster of other petals forms a cup for nectar. Pollen-bearing stamens of those flowers extend some distance away from the nectar cup. Large pollinators brush against the stamens; small ones usually don't. (**b**) By contrast, some petals of *S. mellifera* form a small landing platform for small or medium-size pollinators. Hence the small pollinators of *S. mellifera* are mostly incapable of spreading pollen to flowers of *S. apiana*. The large pollinators of *S. apiana* cannot land on and cross-pollinate *S. mellifera*.

a   b

Figure 19.9 Two species of manzanita: (**a**) *Arctostaphylos patula* and (**b**) *A. viscida*. The two species are ecologically isolated in the same forests in the Sierra Nevada.

GAMETIC MORTALITY In cases of gametic mortality, gametes of different species have evolved in ways that render them incompatible at the molecular level. For example, certain molecular parts of flowers are signals to pollen grains. They can trigger their growth down through the plant's tissues, to the egg. If a pollen grain of one species happens to touch down on a flower of

Figure 19.10 A mixed herd of zebroids and horses. Zebroids are interspecies hybrids, from crosses between horses and zebras.

a different species, it will not chemically recognize the signals in most cases, so hybridization between the two will not occur.

Postzygotic Isolation

The mechanisms of *postzygotic* isolation take effect after fertilization, while an embryo is still developing. Unsuitable interactions among some of the individual's genes or gene products lead to early death, sterility, or hybrids of low fitness. Hybrid offspring commonly are weak, and their survival rates are not good. A few types are sturdy but sterile. Mules, which result from a cross between a female horse and a male donkey, are like this.

Before leaving this section on reproductive isolating mechanisms, think about this final point: Attaining reproductive isolation in nature does not necessarily require a large number of genetic changes. Consider the zebroids, which are hybrid offspring of wild zebras and domesticated horses that were confined to the same pasture (Figure 19.10). The lineages that gave rise to horses and zebras diverged more than 3 million years ago. Existing members of this family of mammals range from true horses (*Equus*), to zebras (*Hippotrigris* and *Dolichohippus*), to the donkeys and asses (*Asinus*). The unnatural confinement of zebras with horses promoted a breach of the reproductive barriers that isolate them in nature. The production of zebroids in captivity is evidence that considerable genetic compatibility still exists between individuals of the divergent lineages.

Mechanisms of reproductive isolation that evolve between genetically divergent populations take effect before, during, or after fertilization.

The prezygotic mechanisms prevent mating or pollination between individuals of the two populations, so that hybrid zygotes cannot form. The postzygotic mechanisms take effect after hybrid zygotes have formed.

Allopatric Speciation Defined

If we assume physical separation between populations promotes the genetic changes that are required for speciation, then allopatry may be the main speciation route. By the model for **allopatric speciation**, some type of physical barrier arises and prevents gene flow between populations or subpopulations of a species. (*Allo-* means different, and *patria* can be taken to mean homeland.) Reproductive isolating mechanisms evolve in the genetically diverging populations. The process of speciation is completed when individuals of the two populations no longer will interbreed even if changing circumstances put them back together in the same area.

Whether a geographic barrier proves to be effective at blocking gene flow between populations depends on an organism's means of travel, how fast it can travel, and whether it is inclined or compelled to disperse. You have only to think back on those garden snails and the wandering albatross described at the start of this chapter to conclude that this is so.

The Pace of Geographic Isolation

Some measurable distance separates the populations of most species, so that gene flow among them is more of an intermittent trickle than a steady stream. It would not take much of a barrier to shut off the trickles. For example, a wide barrier of water formed between some insect populations when a major earthquake changed the course of the Mississippi River in the 1800s. The shift abruptly separated those populations of insects, which could not swim or fly.

Geographic isolation also can proceed slowly, over great spans of time. We find evidence of such extended events in the fossil record, which affords glimpses into the breakup of formerly continuous environments.

For example, vast glaciers advanced down through North America and Europe many times in the past, and they slowly cut off populations from one another. When the glaciers retreated, plants, animals, and other groups descended from those populations came in contact. In some cases, individuals of the populations no longer were reproductively compatible; they had evolved into separate species. In other cases, divergences did not proceed far enough and the descendant populations can still interbreed. For them, reproductive isolation was not completed; speciation did not occur.

As another example, as you may already know, the Earth's crust is fractured into gigantic plates, rather like a cracked eggshell. In the past, imperceptibly slow but colossal movements of the plates caused land masses to collide and to break up. Such geologic movements uplifted part of the seafloor in the region now called the Isthmus of Panama. The uplifting divided an ancient ocean basin, and it set the stage for allopatric speciation among populations of fishes and other marine species.

In the 1980s, John Graves compared four enzymes from the muscle cells of related Isthmus fishes (Figure 19.11). All of the fishes he studied are strong swimmers. As Graves knew, temperature affects the activity of different enzymes in different ways. He also knew that seawater on the Pacific side of the Isthmus is cooler by about 2°–3°C than it is on the Atlantic side, and that it varies more with the changing seasons.

Graves found that all four "Pacific" enzymes function better at lower temperatures than the four "Atlantic" enzymes can. Electrophoretic analysis

ISTHMUS
OF
PANAMA

Figure 19.11 (**a**) Blue-headed wrasse (*Thalassoma bifasciatum*) from the Atlantic side of the Isthmus of Panama and (**b**) Cortez rainbow wrasse (*T. lucasanum*) from the Pacific side. The fishes apparently are related by descent from a common ancestral population that split when geologic forces created the Isthmus. As is common among reef fishes, these individuals of the same species differ in body coloration and patterning.

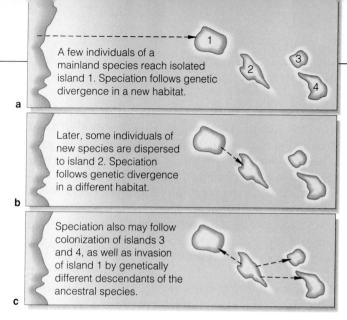

a A few individuals of a mainland species reach isolated island 1. Speciation follows genetic divergence in a new habitat.

b Later, some individuals of new species are dispersed to island 2. Speciation follows genetic divergence in a different habitat.

c Speciation also may follow colonization of islands 3 and 4, as well as invasion of island 1 by genetically different descendants of the ancestral species.

Figure 19.12 Allopatric speciation on an isolated archipelago.

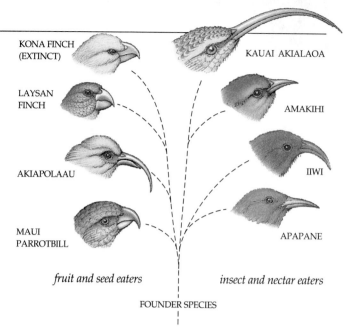

KONA FINCH (EXTINCT)

KAUAI AKIALAOA

LAYSAN FINCH

AMAKIHI

AKIAPOLAAU

IIWI

MAUI PARROTBILL

APAPANE

fruit and seed eaters   *insect and nectar eaters*

FOUNDER SPECIES

Figure 19.13 A few Hawaiian honeycreepers, an example of how a new arrival in species-poor habitats on an isolated archipelago can be the start of a flurry of allopatric speciation.

revealed slight differences in electric charge between two of the four pairs of enzyme molecules, owing to slight differences in their amino acid sequences.

Graves drew these tentative conclusions: First, the selection pressure imposed by even small differences in environmental conditions may bring about divergences in the structure of enzymes. Second, closely related populations of fishes on the two sides of the Isthmus are apparently starting to diverge from each other. The alternative molecular forms of the enzymes he studied are already showing detectable differences in activity.

How might we interpret the results from his study of geographically isolated populations? For these fishes, a gradual accumulation of mutations that are neutral or possibly adaptive might signify an evolutionary "foot in the door." In other words, the genetic divergences may be evidence of speciation in progress.

Allopatric Speciation on Archipelagos

An **archipelago** is an island chain some distance away from a continent. Some, including the Florida Keys, are so near to the mainland that gene flow remains more or less unimpeded, so there is little, if any, speciation. Other archipelagos are so isolated that they are like laboratories for the study of evolution (Figure 19.12). Foremost among these are the Galápagos Islands, about 900 kilometers from the South American coast, and the Hawaiian Archipelago, nearly 4,000 kilometers from the California coast. The islands of both chains are the tips of great volcanoes, some still active, that rise from the seafloor. When each first broke the surface of the sea, it was devoid of life.

Remember those Galápagos finches (Section 17.3)? A few finches from the mainland apparently evolved in isolation on one Galápagos island. Later, some of their descendants reached other islands in the chain. In those new and unoccupied habitats, even in different parts of

the same habitats, conditions varied, so that the island hoppers were subject to a variety of selection pressures. In time, genetic divergences within and between the islands paved the way for new episodes of allopatric speciation. Later on, some island-hopping new species even invaded the island of their ancestors. The distances between these islands are enough to foster divergence, but not enough to stop occasional invasions.

Bursts of speciation in the Hawaiian Archipelago have been so dramatic, they are a premier example of adaptive radiation in a species-poor environment, as described in Section 19.5. The youngest island, Hawaii, is less than a million years old. Here alone we find a great diversity of habitats, ranging from rain forests to alpine grasslands below high, snow-capped volcanoes. When the ancestors of Hawaiian honeycreepers arrived, they found a veritable buffet of fruits, seeds, nectars, and tasty insects—and not many competitors for them. The near-absence of competition from other species fanned allopatric speciations. Figure 19.13 only hints at the resulting variation that arose among the different species of Hawaiian honeycreepers. Today, these and thousands of other species of animals as well as plants that evolved in the archipelago are found nowhere else in the world. As a final example of the potential for speciation, the Hawaiian Islands represent less than 2 percent of the world's land mass, yet they are home to 40 percent of all species of *Drosophila*.

ALLOPATRIC SPECIATION. Some type of physical barrier becomes interposed between populations or subpopulations of a species and prevents gene flow among them, thereby favoring genetic divergence and speciation.

Geographic isolation may not always be required for reproductive isolation. By the models of sympatric and parapatric speciation, genetically divergent populations that are *not* separated by geographic barriers and are still in contact can become reproductively isolated, also.

Sympatric Speciation

By the model for **sympatric speciation**, species may form *within* the home range of an existing species, in the absence of a physical barrier. (*Sym-* means together with, as in "together with others in the homeland.")

EVIDENCE FROM CICHLIDS IN AFRICA In 1994, Ulrich Schliewen and his coworkers found evidence in favor of the model of sympatric speciation. They had studied certain fishes (cichlids) that were living together in two lakes in Cameroon, West Africa. The lake basins are the collapsed cones of volcanoes. Each lake is quite small (Figure 19.14), yet eleven different kinds of cichlids coexist in one, and nine kinds coexist in the other.

The researchers analyzed mitochondrial DNA from all the species in one lake. They did the same for all the species in the other lake, as a basis for comparison. In their mitochondrial DNA sequences, the species of each lake are like each other and *not* like related species in nearby lakes and rivers. For example, the nine species in one lake are the only ones that carry an unusual base-pair substitution in the gene that specifies the protein cytochrome *b*.

Physical and chemical conditions are too uniform in the lakes to promote geographic separation. For instance, their uniform shorelines do not present even small-scale topographic barriers, so allopatric populations could not have formed even on a small scale. The crater rim isolates these lakes from all but tiny creeks trickling in from higher elevations, so gene flow from outside is now restricted. These lakes must have been colonized before their past connection with a river system on the outside was severed.

Cichlids are mobile, not sluggish, fishes. Also, it is safe to assume individuals of different species often encounter one another in such small lakes. In other words, they all must live in sympatry.

Inside each lake, species do show a small degree of ecological separation that is based on feeding preferences. Some feed in open waters and others at the lake bottom. Even so, they all *breed* close to the bottom, in sympatry. Small-scale ecological separation may have been enough to influence sexual selection among potential mates and, over time, the reproductive isolation that can lead to speciation.

SPECIATION BY WAY OF POLYPLOIDY Quite possibly, sympatric speciation has been a prevalent evolutionary route among flowering plants. Consider that about half of all known flowering plant species are polyploid. **Polyploidy**, remember, is a change in the chromosome number; offspring inherit three or more of each type of chromosome characteristic of the parental stock. Such changes arise when chromosomes separate improperly during meiosis or mitosis. They also arise when a **germ** cell replicates its DNA but fails to divide, then goes on to function as a gamete (Section 12.9).

Speciation may have been rapid for many flowering plants that engage in self-fertilization or in some asexual reproductive mode. If the offspring inherited a novel chromosome number, maybe extra chromosomes paired with each other at meiosis, and maybe the extra set of genes did no harm. Some species, such as common bread wheat, may have formed when polyploidy was followed by cross-fertilization, as suggested by Figure 19.15.

Polyploid animals are rare. Allen Orr has evidence that this is a result of a failure of "dosage compensation," whereby genes on sex chromosomes are expressed at the same levels in females *and* males. Among mammals, for instance, inactivation of one of the X chromosomes in females means the cells of both males and females have a single active X chromosome (Section 15.4). Polyploidy skews dosage compensation, with bad or fatal effects. This mechanism is absent in plants, so chromosome doublings are not as problematic as they are in animals.

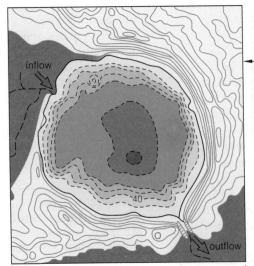

Figure 19.14 Topographical map of Lake Barombi Mbo in Cameroon, West Africa. Apparently, nine kinds of cichlids evolved by sympatric speciation in this small, isolated crater lake, from which even microgeographic separation is absent. There is separation by feeding preferences, but all species breed near the lake bottom, in sympatry.

1 kilometer

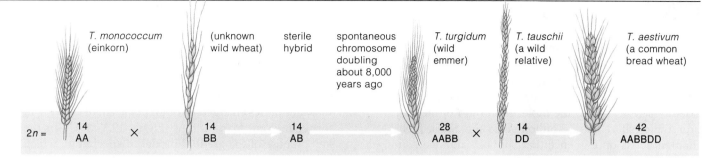

| 2n = | T. monococcum (einkorn) 14 AA | × | (unknown wild wheat) 14 BB | → | sterile hybrid 14 AB | spontaneous chromosome doubling about 8,000 years ago → | T. turgidum (wild emmer) 28 AABB | × | T. tauschii (a wild relative) 14 DD | → | T. aestivum (a common bread wheat) 42 AABBDD |

a By 11,000 years ago, humans had already started to cultivate wild wheats. Einkorn wheat (*Triticum monococcum*) is still around, and we know it has a diploid chromosome number of 14 (two sets of 7 chromosomes, shown above as 14AA). Long ago, einkorn wheat probably hybridized with another species with the same chromosome number.

b If the AB hybrid offspring were sterile but self-fertilizing, an interbreeding population of AB plants could have arisen by asexual reproduction. About 8,000 years ago, we know that polyploidy arose in such a population. Wild emmer (*T. turgidum*) plants are tetraploid (AABB), with a chromosome number of 28 (two sets of 14). They are fertile; at meiosis, the A chromosomes pair with each other, and B chromosomes pair with each other.

c Later, an AABB plant probably hybridized with *T. tauschii*, a wild relative of the wild emmer. Its diploid chromosome number must have been 14 (two sets of 7 DD). Populations of the hybrid descendants now include common bread wheats, such as *T. aestivum*, which have a chromosome number of 42 (six sets of 7 AABBDD).

Figure 19.15 Presumed sympatric speciation in wheat by polyploidy and hybridizations over time. Wheat grains 11,000 years old have been found in the Near East. Diploid wild wheats still grow there.

BULLOCK'S ORIOLE

BALTIMORE ORIOLE

Parapatric Speciation

By the model for **parapatric speciation**, neighboring populations become distinct species while maintaining contact along a common border. (*Para-* means near, as in "near another homeland.") Interbreeding individuals produce hybrid offspring in this region, which is called a **hybrid zone**. Evidence of parapatric speciation is sketchy. Why? Whether or not geographically separated populations are merely "subspecies," or geographically distinct populations of a single species, is often difficult to determine.

For example, Bullock's orioles differ from Baltimore orioles in the color patterning of most of their feathers. The males have different territorial songs and generally mate with their own kind. As K. Corbin determined, these birds also differ in the frequencies of alleles for several enzymes. They have different home ranges, but these overlap in the American Midwest (Figure 19.16). Interbreeding was once common in the hybrid zone, although it is now becoming less frequent. This may be an example of parapatric speciation in progress. But it

Figure 19.16 A possible setting for parapatric speciation? Hybrids arise along the common border of the ranges of two species of orioles. To the east are Baltimore orioles; to the west are Bullock's orioles. Hybridization was once so common, they were considered subspecies. In 1997, the American Ornithologist Union decided they are separate species.

may also be an example of "secondary contact." That is, previously isolated subspecies, which had diverged in the past from a common ancestor, have been getting together again.

SYMPATRIC SPECIATION. **Daughter species arise from a group of individuals within an existing population. This may have been a common speciation route among polyploid flowering plants.**

PARAPATRIC SPECIATION. **Adjacent populations evolve into distinct species even while maintaining contact along their common border.**

Branching and Unbranched Evolution

All species, past and present, are related by descent. They are genetically connected through lineages that extend back in time to the molecular origin of the first prototypic cells, some 3.8 billion years ago. Subsequent chapters focus on evidence that supports this view. In anticipation of those chapters, let's start thinking about ways to interpret the large-scale histories of species.

The fossil record yields evidence of two patterns of evolutionary change in lineages—one branching, and the other unbranched. The first is called **cladogenesis** (from the Greek *klados*, meaning branch, and *genesis*, meaning origin). This is the pattern by which a lineage splits, with populations becoming genetically isolated and then diverging in different evolutionary directions. *It is the pattern of speciation*, as described earlier.

By the second pattern, called **anagenesis**, changes in allele frequencies and morphology accumulate within an unbranched line of descent. (In this context, *ana-* means renewed.) The directional changes are confined to a single lineage. Gene flow does not cease among the populations of that lineage. They continue to maintain reproductive cohesion over time, and daughter species do not form. Such changes are occurring in populations of those peppered moths described earlier.

Evolutionary Trees and Rates of Change

Evolutionary trees summarize information about the continuity of relationship among species. Figure 19.17 is a simple way to start thinking about how tree diagrams are constructed. Each *branch* represents a single line of descent from a common ancestor. Each *branch point* represents a time of genetic divergence and speciation, as brought about by microevolutionary processes.

Figure 19.17 also shows how an evolutionary tree diagram can be used to convey rates of change—that is, the approximate length of time between two speciation events. The branches that have slight angles indicate that the species emerged through many small changes in form over long spans of time (Figure 19.17*a*). This is the key premise of the **gradual model of speciation**, which actually shows a good fit with many fossil sequences. For example, in many layers of sedimentary rock we find sequences of intricately perforated shells of single-celled foraminiferans, a type of protistan. The sequences provide evidence of slow change.

Alternatively, evolutionary tree diagrams for some lineages are constructed with short, horizontal branches that abruptly make a 90-degree turn (Figure 19.17*b*). These diagrams are consistent with the **punctuation model of speciation**. According to this model, most morphological changes are compressed within a brief period when populations first start to diverge—say, within merely hundreds or thousands of years. The idea is that bottlenecks, founder effects, strong directional selection, or some combination of these brings about rapid speciation. Daughter species recover quickly from the adaptive wrenching, then change little over the next 2 million to 6 million years or so. Said another way, reproductive cohesion seems to have prevailed for about 99 percent of the history of most lineages. This pattern seems to pervade many parts of the fossil record.

Apparently, changes in evolutionary trees have been gradual, abrupt, or both. Species originated at different times, and they differed in how long they persisted on the evolutionary stage. Remember, some lineages have endured without much change, producing a species here, losing a species there, often over many millions of years. Other lineages have branched bushily and sometimes spectacularly, during times of adaptive radiation.

Figure 19.17 How to read evolutionary tree diagrams.

present

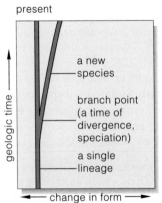

geologic time

← change in form →

a Softly angled branching means speciation occurred through gradual changes in traits over geologic time.

a new species

branch point (a time of divergence, speciation)

a single lineage

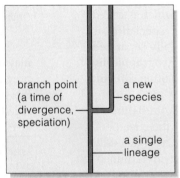

branch point (a time of divergence, speciation)

a new species

a single lineage

b Horizontal branching means traits changed rapidly around the time of speciation. Vertical continuation of a branch means traits of the new species did not change much thereafter.

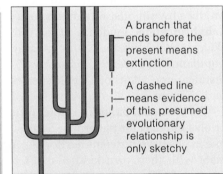

A branch that ends before the present means extinction

A dashed line means evidence of this presumed evolutionary relationship is only sketchy

c Many branchings of the same lineage at or near the same point in geologic time means that an adaptive radiation occurred.

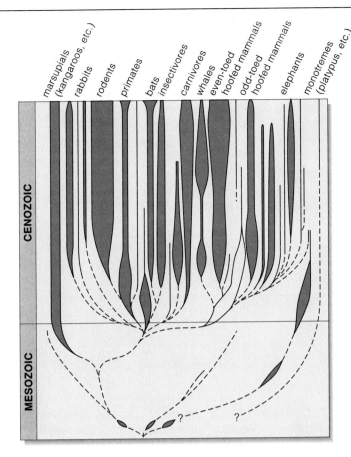

marsupials (kangaroos, etc.)
rabbits
rodents
primates
bats
insectivores
carnivores
whales
even-toed hoofed mammals
odd-toed hoofed mammals
elephants
monotremes (platypus, etc.)

CENOZOIC

MESOZOIC

? ?

Figure 19.18 Evolutionary tree diagram of the great adaptive radiation of mammals, which began about 65 million years ago at the start of a geologic era called the Cenozoic. Variations in the width of a given branch correspond to the range of species diversity (that is, to the number of groups) within the lineage as represented at different points in time. The wider the branch, the greater the diversity.

Adaptive Radiations

An **adaptive radiation** is a burst of divergences from a single lineage that give rise to many new species, each adapted to a new or unoccupied habitat or to using a novel resource. Figure 19.18 shows an example. Often, lineages have diversified this way when their member species were presented with unfilled **adaptive zones**. Think of such zones as ways of life, such as "burrowing in seafloor sediments" or "catching insects in the air at night." A lineage may radiate into such zones when it has physical, evolutionary, or ecological access to them.

Physical access means a lineage happens to be there when adaptive zones open up. For example, mammals were once distributed through uniform tropical regions of a single continent. That continent split into several land masses. Habitats and resources changed in many different ways on those land masses and set the stage for independent radiations.

Evolutionary access means that modification of some structure or function will permit a lineage to exploit the environment in more efficient or novel ways. We call such modifications **key innovations**. For example, when the forelimbs of certain five-toed vertebrates evolved into wings, this innovation opened up new adaptive zones to the ancestors of birds and bats.

Ecological access means that the lineage can enter an unoccupied adaptive zone or displace resident species. We will say more about this in Chapter 47.

Extinctions—End of the Line

Some number of species within a lineage inevitably disappear as local conditions change. This expected rate of disappearance over time is called **background extinction**. Table 19.1 lists examples. By contrast, an abrupt rise in extinction rates above the background level is called **mass extinction**. It is a catastrophic, global event in which entire families and other major groups are wiped out simultaneously.

Table 19.1	Examples of the Estimated Average Durations of Species	
	Group	Duration (millions of years)
Protistans:	Foraminiferans	20–30
	Diatoms	25
Plants:	Bryophytes	20+
	Higher plants	8–20+
Animals:	Gastropods	10–13.5
	Ammonites	1–2, 6–15
	Trilobites	1+
	Beetles	2+
	Freshwater fishes	3
	Snakes	2+
	Mammals	1–2+

Lineages that are not widely dispersed tend to be hit hard in times of mass extinctions. This is especially true of the tropics. Global temperatures drop at such times. Species that are adapted to cool climates survive, but those in the tropics have nowhere else to go.

Finally, luck has a lot to do with it. In the past, for instance, asteroids hit the Earth and leveled the playing field. Some survivors of one such impact radiated into adaptive zones occupied earlier by diverse dinosaurs. Among them were the mammals that had previously remained, wisely, hidden in the shrubbery.

Taken together, the persistence, branchings, and extinctions of species account for the full range of biological diversity at any point in geologic time.

SUMMARY

1. A species is a single kind of organism, recognized partly in terms of its morphology.

2. By the biological species concept, a species consists of one or more populations of individuals that are, under natural conditions, interbreeding and producing fertile offspring, and that are reproductively isolated from other such populations. The concept identifies a species mainly in terms of the portion of alleles that promote or maintain reproductive isolation. It applies only to sexually reproducing organisms.

3. The populations of a species have a shared genetic history, they are maintaining genetic contact over time, and they are evolving independently of other species.

4. Speciation is the process by which daughter species form from a population or subpopulation of a parent species. It is the attainment of reproductive isolation.

a. As an example, if a geographic barrier prevents gene flow between such populations, they will undergo genetic divergence; for mutation, natural selection, and genetic drift can operate independently in each one. Genetic divergence is a buildup of differences in allele frequencies between populations or subpopulations of a species.

b. Among their varied effects, the microevolutionary processes may accidentally give rise to reproductive isolating mechanisms that prevent interbreeding. By doing so, they may bring about irreversible genetic differences between the populations.

c. Speciation occurs gradually by genetic divergence. It also occurs instantaneously, as by polyploidy or other changes in chromosome number.

5. Prezygotic isolating mechanisms prevent mating or pollination between populations (see Table 19.2). They include differences in reproductive timing or behavior, incompatibilities in reproductive structures or gametes, and occupation of different microenvironments in the same area. Postzygotic mechanisms lead to early death, sterility, or unfit hybrid offspring. They take effect after fertilization, while the embryo develops.

6. There are three current models of speciation:

a. Allopatric speciation. Geographic barriers prevent gene flow between populations of a species, which then genetically diverge in such ways that interbreeding will not occur in nature even if their individuals later make contact. Allopatric speciation may be the most prevalent speciation route.

b. Sympatric speciation. Species form after there is reproductive isolation of individuals that share the same home range. Speciation by polyploidy is an example.

c. Parapatric speciation. Adjacent populations give rise to a new species while maintaining contact along a border between their home ranges.

Table 19.2 Summary of Isolating Mechanisms	
PREZYGOTIC ISOLATION (*Mating or zygote formation is blocked*)	
Temporal isolation	Potential mates occupy overlapping ranges but reproduce at different times
Behavioral isolation	Potential mates meet but cannot figure out what to do about it
Mechanical isolation	Potential mates attempt engagement, but sperm cannot be successfully transferred
Gametic mortality	Sperm is transferred, but egg is not fertilized (gametes die or gametes are incompatible)
Ecological isolation	Potential mates occupy different local habitats within the same area
POSTZYGOTIC ISOLATION (*Hybrids don't work*)	
Zygotic mortality	Egg is fertilized, but zygote or embryo dies
Hybrid inviability	First-generation hybrid forms but shows very low fitness
Hybrid infertility	Hybrid is sterile or partially so

After Alan Templeton, Jerry Coyne, and others.

7. Lineages differ in the time, rate, and direction of speciation. Speciation may proceed gradually, rapidly, or both. Extensive branching of a lineage (divergence of one or more of its populations) in the same geologic time span is an adaptive radiation. Often this occurs as member species of a lineage become adapted to new or unoccupied habitats or to using novel resources.

8. Evolutionary tree diagrams show the relationship among groups of species. Each branch of such trees represents a line of descent (lineage). Branch points are speciation events, as brought about by natural selection, genetic drift, and other microevolutionary processes.

9. All lineages inevitably lose some number of species over time. This loss is called background extinction. By contrast, in a mass extinction, major groups of species perish abruptly in a catastrophic, global event.

10. The persistences, branchings, and extinctions of species account for the full range of biological diversity through geologic time (Table 19.3).

Review Questions

1. Describe how the biological species concept differs from a definition of species based on morphological traits alone. *19.1*

2. Define speciation. Define and give examples of reproductive isolating mechanisms. *CI, 19.2*

3. Define each of the three speciation models. *19.3, 19.4*

4. Interpret these features of evolutionary tree diagrams: a single line, softly angled branching, horizontal branching, vertical continuation of a branch, many branchings of the same line, a dashed line, a branch that ends before the present. *19.5*

Table 19.3 Summary of the Processes and Patterns of Evolution

MICROEVOLUTIONARY PROCESSES

Mutation	Original source of alleles	Stability or change in a species is the outcome of balances or imbalances among all of these processes, the effects of which are influenced by population size and by prevailing environmental conditions.
Gene flow	Preserves species cohesion	
Genetic drift	Erodes species cohesion	
Natural selection	Preserves or erodes species cohesion, depending on environmental pressures	

MACROEVOLUTIONARY PATTERNS

Genetic persistence	The basis of the unity of life. The biochemical and molecular basis of inheritance extends from the origin of the first cells through all subsequent lines of descent.
Genetic divergence	The basis of the diversity of life, brought about by adaptive shifts, branchings, and radiations. The rates and times of change have varied within and between lineages.
Genetic disconnect	The steady loss (background extinction) or abrupt, catastrophic loss (mass extinction) of lineages.

Self-Quiz (Answers in Appendix IV)

1. Sexually reproducing individuals of the same species _____ .
   a. can interbreed under natural conditions
   b. can produce fertile offspring
   c. have a shared genetic history
   d. all of the above

2. Reproductive isolating mechanisms _____ .
   a. prevent interbreeding   c. reinforce genetic divergence
   b. prevent gene flow   d. all of the above

3. When potential mates occupy overlapping ranges but reproduce at different times, this is a case of _____ isolation.
   a. postzygotic   c. temporal
   b. mechanical   d. gametic

4. In an evolutionary tree diagram, a branch point represents _____ , and a branch that ends represents _____ .
   a. a single species; incomplete data on lineage
   b. a single species; a time of extinction
   c. a time of divergence; extinction
   d. a time of divergence; speciation complete

5. An evolutionary tree diagram with horizontal branches that abruptly become vertical is consistent with _____ .
   a. the gradual model of speciation
   b. the punctuation model of speciation
   c. the idea of small changes in form over long spans of time
   d. both a and c

6. Match each term with its most suitable description.
   ____ cladogenesis   a. burst of microevolutionary
   ____ anagenesis   activity within a lineage
   ____ adaptive radiation   b. catastrophic disappearance
   ____ background   of major groups of organisms
        extinction   c. branching lineages
   ____ mass extinction   d. expected rate of species loss
                                 within a lineage
                              e. genetic and morphological
                                 change in unbranched lineage

Critical Thinking

1. You notice several duck species in the same lake habitat. The females of different species look very similar to one another, but the males of each particular species have feathers with distinctive patterns and colors. Speculate on which forms of reproductive isolation may be keeping each species distinct. How does the appearance of the male ducks provide a clue to the answer?

2. Suppose a small population founded by a few individuals is cut off from gene exchange with the main population of their species. By Mayr's view of the *founder effect*, the frequencies of alleles for a few gene products change owing to genetic drift, and the change triggers a number of changes in other genes that are affected by those alleles. If that is so, then genetic changes in newly founded populations may be so significant that speciation occurs rapidly—so rapidly that we will seldom be able to document it in the fossil record. Does this view correspond to the gradual or punctuated model of speciation?

3. A key innovation, recall, is some modification in structure or function that permits a species to exploit the environment in a more efficient or novel way, compared to the ancestral species. Identify a key innovation of an existing species, such as humans. Then describe how that innovation might be the basis of an adaptive radiation in environments of the distant future.

Selected Key Terms

adaptive radiation *19.5*
adaptive zone *19.5*
allopatric speciation *19.3*
anagenesis *19.5*
archipelago *19.3*
background extinction *19.5*
biological species concept *19.1*
cladogenesis *19.5*
evolutionary tree *19.5*
gene flow *19.1*
genetic divergence *19.1*
gradual model, speciation *19.5*
hybrid zone *19.4*
key innovation *19.5*
mass extinction *19.5*
parapatric speciation *19.4*
polyploidy *19.4*
punctuation model, speciation *19.5*
reproductive isolating mechanism *19.2*
speciation *CI*
species *19.1*
sympatric speciation *19.4*

Readings

Futuyma, D. 1986. *Evolutionary Biology*. Second edition. Sunderland, Massachusetts: Sinauer.

Mayr, E. 1976. *Evolution and the Diversity of Life*. Cambridge, Massachusetts: Belknap Press of Harvard University Press.

Otte, D., and J. Endler (editors). 1989. *Speciation and Its Consequences*. Sunderland, Massachusetts: Sinauer.

Web Site See *http://www.wadsworth.com/biology* for practice quiz questions, hypercontents, BioUpdates, and critical thinking. The Wadsworth Biology Resource Center provides a wealth of information fully organized and integrated by chapter.

20

THE MACROEVOLUTIONARY PUZZLE

Of Floods and Fossils

About 500 years ago, Leonardo da Vinci was brooding about seashells entombed in layered rocks of northern Italy's high mountains, hundreds of kilometers from the sea. How did they get there? If he accepted the traditional explanation, he would have to agree that stupendous floodwaters deposited those shells in the mountains during the Great Deluge (Figure 20.1). But many shells were thin and fragile—yet intact. Surely they would have been battered to bits if they had been swept across such distances, then up the mountains.

Da Vinci also brooded about the rock layers. They were stacked like cake layers, and some had shells but others had none. Then he remembered how large rivers deposit silt with each spring flood. *Did the layers slowly accumulate long ago, as a series of silt deposits where rivers emptied into the sea?* If so, then shells in the mountains would be evidence of a progression of communities of organisms that once lived in the seas! Da Vinci did not announce his novel idea, perhaps knowing it would be met with deafening silence, imprisonment, or worse.

By the 1700s, fossils were accepted as evidence of past life. They were still being interpreted through the prism of prevailing cultural beliefs, as when a Swiss naturalist excitedly unveiled the remains of a giant salamander and announced they were the skeleton of a man who had drowned in the Deluge.

By midcentury, however, scholars began to question such interpretations. Extensive mining, quarrying, and canal excavations were under way. The diggers were finding similar rock layers and similar fossil sequences in distant places, such as the cliffs on both sides of the English Channel. More than a few scholars began to view the findings as evidence of connections between Earth history and the history of life.

Ever since, fossils have been analyzed in increasingly refined ways. Together with biochemical studies and other modern sources of information, they yield good evidence of evolution through vast spans of time.

Only populations that already exist can evolve. As a result of mutations, natural selection, and genetic drift, each existing species is a mosaic of ancestral and novel traits that emerged earlier, along unbranched or branching evolutionary roads that began with the first populations of the very first species on Earth. Thus, *all*

species that have ever evolved are related to one another, by way of descent. This principle of evolution guides efforts to make sense of scattered and often puzzling scraps of evidence of past life. It guides the task of identifying and sorting out the many lines of descent, or **lineages**, that connect all species, past and present.

Ultimately, then, life is a story of *species*—of how and when each originated, whether its defining traits persisted or were modified, and whether it vanished or endured. Life also is a story of **macroevolution**—of large-scale patterns, trends, and rates of change among families and other more inclusive groups of species. We turn to this larger picture in the next unit. Here we begin with the nature of the evidence that supports it.

Figure 20.1 Two interpretations of the past: (*Left*) From the Sistine Chapel, Michelangelo's painting of the onset of the Great Deluge, a catastrophic flood that became integrated into biblical thought. (*Below*) A modern-day photographer captures the pattern of fossilized shells of ammonites. About 65 million years ago, all of these marine animals abruptly perished, along with many other groups of organisms. That mass extinction is but one intriguing piece of the evolutionary puzzle.

KEY CONCEPTS

1. In the evolutionary view, all species that have ever lived are related, some closely, others remotely so. This relatedness exists because each new species had to evolve from variant individuals of species that already existed, starting with the first living cells to appear on Earth.

2. The term macroevolution refers to the patterns, trends, and rates of change among lineages over geologic time.

3. The fossil record, the geologic record, and radiometric dating of rocks yield evidence of macroevolution.

4. Anatomical comparisons between major lineages help us understand and reconstruct patterns of change through time. Among the most revealing aspects of anatomy are homologous structures in the embryos or adult forms of different lineages. They signify descent from a common ancestor.

5. Biochemical comparisons within and between major lineages also provide evidence of macroevolution.

6. Stunning diversity characterizes the distribution of species through time and through the global environment. Biological systematics attempts to discern patterns in life's diversity through taxonomy, phylogenetic reconstruction, and classification.

7. Taxonomy is concerned with identifying and naming new species. Phylogenetic reconstruction is concerned with working out evolutionary connections. Classification is an attempt to organize information about species into retrieval systems.

Fossilization

"Fossil" comes from a Latin word for something that has been "dug up." In general, **fossils** are recognizable, physical evidence of ancient life, as in Figure 20.2.

Most of the fossils discovered so far are bones, teeth, shells, seeds, capsules of spores, and other hard parts. (Soft parts usually are the first to decompose when an organism dies.) Imprints of leaves, stems, tracks, trails, burrows, and other *trace* fossils yield indirect evidence of past life. Even fossilized feces, or coprolites, contain residual evidence of which species were being eaten—and therefore were present—in ancient environments.

Fossilization is a very slow process that starts when an organism, or traces of it, becomes buried in volcanic ash or in sediments at the bottom of some lake, lagoon, or sea. Sooner or later, water infiltrates the organic remains, which become infused with dissolved metal ions and other inorganic compounds. More and more sediments gradually accumulate above the burial site and exert ever increasing pressure on the remains. Over great spans of time, the pressure and chemical changes transform the remains to stony hardness.

Preservation is favored when organisms are buried rapidly, in the absence of oxygen. Gentle entombment by volcanic ash or anaerobic mud is best. Preservation also is favored when a burial site stays undisturbed. Most often, however, erosion and other geologic insults crush, deform, break, or scatter the fossils.

Figure 20.2 (a) A fossil hunter's dream: a complete skeleton of a bat that lived 50 million years ago. Erosion and other forces of nature have left few ancient burial sites undisturbed, so intact fossils are rare. Even jumbled parts of the ducklike birds in (b) are a good find. It will take hours of preparation and analysis to identify the species. (c) Fossilized parts of the oldest known land plant (*Cooksonia*). Its stems were not even seven centimeters tall. (d) A complete fossil—skeletal remains of an ichthyosaur, a dolphin-like marine reptile that lived 200 million years ago.

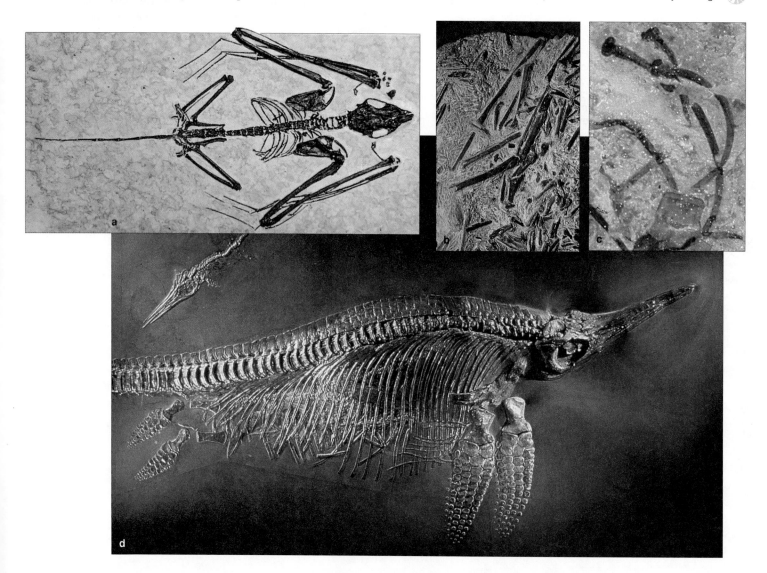

Figure 20.3 The geologic time scale. The major boundaries mark times of mass extinctions. Radiometric dating methods (Section 2.2) allowed researchers to assign absolute dates. The "Proterozoic" turned out to be so vast, it is subdivided into eons, each of which lasted for hundreds of millions of years. One of these, the Archean, was the time of life's origin.

The time spans are not to scale. If they were, the Archean and Proterozoic portions would run off the page. Think of the spans as minutes on a clock that runs from midnight to noon. If we say that life originated at midnight, the Paleozoic began at 10:04 A.M., the Mesozoic at 11:09 A.M., and the Cenozoic at 11:47 A.M. The recent epoch of the Cenozoic started during the last 0.1 second before noon.

	Period	Epoch	Millions of Years Ago (mya)
CENOZOIC ERA	QUATERNARY	Recent	0.01–
		Pleistocene	1.55
	TERTIARY	Pliocene	5
		Miocene	25
		Oligocene	38
		Eocene	54
		Paleocene	65
MESOZOIC ERA	CRETACEOUS	Late	100
		Early	138
	JURASSIC		205
	TRIASSIC		240
PALEOZOIC ERA	PERMIAN		290
	CARBONIFEROUS		360
	DEVONIAN		410
	SILURIAN		435
	ORDOVICIAN		505
	CAMBRIAN		550
PROTEROZOIC EON			2,500
ARCHEAN EON AND EARLIER			4,600

Interpreting the Geologic Tombs

We find similar fossil-containing layers of sedimentary rock over vast areas, even on different continents. Such layers formed long ago by the gradual deposition of volcanic ash, silt, and other materials, one above the other. This layering of sedimentary deposits is called **stratification**. Generally, the deepest layers were the first to form. Layers closest to the surface formed last. Because particles tend to settle in response to gravity, we can assume that most sedimentary layers must have formed horizontally. Where they are tilted or ruptured, this is evidence of subsequent geologic disturbance.

Understand how rock layers form, and you realize that fossils within a particular layer are from a similar age in Earth history. *And the older the layer, the older the fossils.* Given that rock layers formed in sequence, then the fossil assemblages that they contain were unique to sequential ages. Therefore, fossils can be used to assign relative dates to the record of the rocks.

Early geologists identified four abrupt transitions in the sequence of fossil assemblages. They found them in the fossil record from regions around the world. They used these transitions as the boundaries between four great intervals in time. The oldest fossils were from the first interval, the Proterozoic. These were followed in time by fossils from the Paleozoic era, the Mesozoic era, and finally the "modern" era, the Cenozoic.

We still use the boundaries between these intervals in a chronological chart of Earth history. Figure 20.3 is a modern version of this **geologic time scale**.

Interpreting the Fossil Record

At present, we have fossils for about 250,000 species, which give insights into evolutionary history. However, judging from the current range of diversity, there must have been many, many millions of ancient, now-extinct species. We never will be able to recover fossils for most of them, so our "record" of past life is incomplete, with built-in biases. Why is this so?

Most important, colossal movements in the Earth's crust obliterated evidence from crucial intervals in the past. Besides this, most species of ancient communities simply weren't preserved. For example, bony fishes and hard-shelled mollusks are well represented in the fossil record. Soft-bodied worms and jellyfishes are not, even though they may have been just as common or more so. Population density and body size also skew the record. For example, a population of plants may have released millions of spores in a single growing season, whereas the earliest humans lived in small groups and produced few offspring. What are the chances of finding even one fossilized skeleton of an early human, as compared to finding spores of plants that lived at the same time?

The fossil record also is heavily biased toward some environments. Most species we know about lived either on land or in shallow seas which, as a result of geologic uplifting, became part of continents. We have recovered very few fossils from the sediments beneath the ocean—which covers nearly three-fourths of the Earth's surface! Finally, most fossils have been found in the Northern Hemisphere simply because that's where most fossil hunters have lived.

Fossils, the stone-hard physical evidence of ancient life, are present in layers of sedimentary rock. The deeper the layers, the older the fossils. The geologic time scale is based on sequences of fossils in sedimentary rocks.

The completeness of the fossil record varies as a function of the kinds of organisms represented, where they lived, and the stability of their burial sites.

EVIDENCE OF A CHANGING EARTH

While early geologists were busily discovering fossils and mapping the record of the rocks, Charles Darwin was so taken with one of their theories of Earth history that it helped shape his view of life's evolution. By the **theory of uniformity**, recall, mountain building and erosion had repeatedly worked over the surface of the Earth, in exactly the same ways, through time (Section 17.2). Great stacks of sedimentary layers, as in Figure 20.4, were evidence of this. Yet the more geologists clinked their hammers against the rocks, the more they realized repetitive change was only part of the picture. Like life, the Earth had changed irreversibly; it evolved.

An Outrageous Hypothesis

For example, with their refined maps, geologists could not help but notice a striking parallelism between the southern continents on opposite sides of the Atlantic. If the huge continents were once joined, like pieces of a

Figure 20.4 A splendid slice through time: the Grand Canyon of the American Southwest, once part of an ocean basin. Its layers of sedimentary rock formed gradually over hundreds of millions of years. Geologic forces lifted them above sea level. Later, the erosive force of rivers carved the deep canyon walls and so exposed the ancient stacked layers.

Figure 20.5 Baker's wishful reconstruction of the ancient supercontinent Pangea.

jigsaw puzzle, then they must have slowly drifted apart! This idea came to be known as **continental drift**. In 1908, F. Taylor argued that the existing great mountain ranges formed as drifting land masses collided. One of his fellow Americans, H. Baker, quickly sketched the continents as a single land mass, although he conveniently squished together and rotated islands and peninsulas to improve the fit (Figure 20.5). A few years later, Alfred Wegener proposed a refined model of the supercontinent, which he named **Pangea**, by aligning the continental shelves rather than eroded coastlines. He used clues from glacial deposits and other rocks (to infer ancient climatic zones) and clues from the world distribution of fossils and existing species.

Most scholars could not accept the hypothesis. The continents were known to be thinner and less dense than the underlying mantle. Continents drifting under their own power seemed about as plausible as Kleenex tissues plowing on their own across a bathtub. Clearly, the theory of uniformity was so powerfully entrenched, it was shutting out alternative ideas.

Then scientists made an amazing discovery in the 1950s. When iron-rich rocks first form, their internal structure takes on a north-south orientation in response to the Earth's magnetism. Even so, the orientations in ancient rocks were *not* aligned with the magnetic poles. Either the poles moved—or the continents did. Scientists tried to rotate North American and European rocks that formed 200 million years ago to their inferred, original north-south alignment. The alignments dovetailed only when they joined the two continents together!

Besides this, deep-sea probes revealed evidence of **seafloor spreading**. Molten rock erupts from great, continuous ridges on the ocean floor, flows laterally in both directions, and hardens to form new crust. As the seafloor spreads out, it forces the older crust down into great trenches elsewhere in the seafloor (Figure 20.6). It turned out the ridges and trenches are edges of huge but relatively thin plates, rather like a cracked eggshell. When the crustal plates move, they raft continents to new positions. The drift hypothesis could now become incorporated in a broader theory of crustal movements.

In the 1960s researchers demonstrated the predictive power of the new theory, which is now known as **plate tectonics**. Earlier, in South America, Africa, India, and Australia, fossils of the *same* plants (*Glossopteris*, Figure 20.7) and a mammal-like reptile had been found in the *same* succession of glacial deposits, coal seams, and a top layer of basalt. The plant seeds were far too heavy and the animals too small for dispersal across oceans, so they

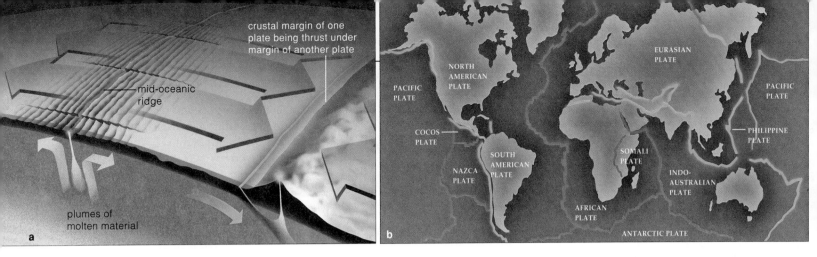

a — mid-oceanic ridge

plumes of molten material

crustal margin of one plate being thrust under margin of another plate

b — EURASIAN PLATE / NORTH AMERICAN PLATE / PACIFIC PLATE / PACIFIC PLATE / COCOS PLATE / PHILIPPINE PLATE / NAZCA PLATE / SOUTH AMERICAN PLATE / SOMALI PLATE / INDO-AUSTRALIAN PLATE / AFRICAN PLATE / ANTARCTIC PLATE

Figure 20.6 Some forces of geologic change. (**a**) The Earth's crust is fractured into rigid plates that split apart, drift, and collide. Driving the movements are great plumes that well up from the Earth's interior and spread laterally beneath the crust. In the past, such plumes ruptured the crust at ridges on the ocean floor. At the mid-oceanic ridges, molten material seeps out, cools, and forces the seafloor on either side of them away from the rupture. Seafloor spreading displaces the plates away from the ridges. Often the displacement forces other parts of a plate under an adjoining plate, which is slowly uplifted as a result. Major mountain ranges paralleling the coasts of continents formed this way. Plumes also cause deep rifting and splitting within continents. Such rifting is proceeding in Missouri, at Lake Baikal in Russia, and in eastern Africa. Long ago, *superplumes* violently ruptured the crust. Volcanoes still form at these "hot spots," as they do today in the Hawaiian Archipelago. (**b**) Current configuration of the crustal plates.

Glossopteris fossil

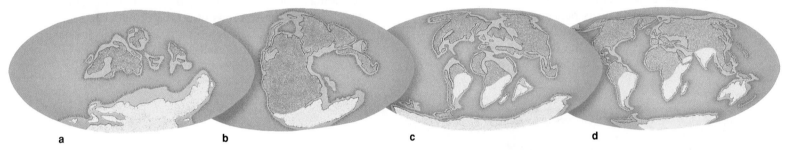

a   b   c   d

Figure 20.7 Reconstructions of drifting continents. (**a**) Gondwana (*yellow*), 420 million years ago. (**b**) All land masses collided to form Pangea. About 260 million years ago, certain seed ferns (*Glossopteris*) and other plants lived on the Gondwana portion of this supercontinent but nowhere else. (**c**) Position of fragments from Pangea's breakup 65 million years ago, when the last dinosaurs became extinct. (**d**) The fragments 10 million years ago, closing in on their present positions.

must have evolved on the same land mass, which was originally part of the early supercontinent **Gondwana**. Explorers also discovered *Glossopteris* fossils near the South Pole, but these were not identified correctly until the 1930s. In that year, A. du Toit predicted *Glossopteris* fossils, the mammal-like reptile, and the succession of glacial deposits, coal, and basalt on the other southern continents would also be discovered on Antarctica. In the 1960s, evidence gathered during Antarctic expeditions supported the prediction—and the plate tectonics theory.

Drifting Continents and Changing Seas

Take a good look at Figure 20.7. In the remote past, plate movements put huge land masses on collision courses. In time the land masses converged and formed supercontinents, which later split open at deep rifts that became new ocean basins. One of the early continents,

Gondwana, drifted south from the tropics, across the south polar region, then north until it crunched into other land masses to form Pangea. That single world continent extended from pole to pole, and an immense ocean spanned the rest of the globe! All the while, the slow erosive forces of water and wind were resculpting the surface of the land through time. And as if this were not enough, asteroids and meteorites bombarded the crust, and the impacts have had long-term effects on global temperature and climate.

These changes in the land, oceans, and atmosphere profoundly influenced the evolution of life. Think of early life flourishing in the warm, shallow waters along the shorelines of continents. When continents collided, the shorelines vanished. Such changes were devastating for many lineages. Yet, even as old habitats vanished, new ones opened for the survivors—and evolution took off in new directions. As you will see in the unit to follow, the histories of the Earth and life are thus inseparable.

Over the past 3.8 billion years, changes in the Earth's crust, the atmosphere, and the oceans profoundly affected the evolution of life. Certain boundaries of the geologic time scale mark abrupt transitions in the evolutionary story.

Strong evidence of evolution comes from anatomical comparisons of major lineages. This field of inquiry is called **comparative morphology**. A few examples from comparative embryology (the ways in which embryos of different plant and animal lineages develop) will give insight into one of its guiding principles. Namely, *when it comes to making changes in the body's development, evolution tends to follow the line of least resistance.*

Developmental Program of Larkspurs

Most constraints on evolution take effect as embryos are developing. The sequence of a developmental program is so finely tuned, mutations that alter especially the early steps in the sequence tend to be selected against. Every so often, though, a mutation allows evolution to proceed—but only in particular directions.

Flowering plants give evidence of evolution by way of mutations that changed the developmental program. Consider the flowers of *Delphinium decorum*, a larkspur in California. A ring of its petals guides honeybees to the nectar-storing tube (Figure 20.8). Outward-bulging reproductive structures at the flower's center give the bees something to hang onto while they draw nectar—

and pollinate the flowers. *D. nudicaule*, a larkspur of more recent origin, has compact flowers that discourage bees. Its pollinators are hummingbirds, which hover in front of a flower as they draw nectar from it.

As you can see from Figure 20.8b–d, the mature flowers of *D. nudicaule* strongly resemble the buds of *D. decorum*. The difference in morphology is an outcome of a gene mutation that slowed down the rate of floral development in *D. nudicaule*.

Developmental Program of Vertebrates

Now consider the vertebrates, which range from fishes to amphibians, reptiles, birds, and mammals. These are diverse animals. Yet comparisons of the ways in which their embryos develop provide compelling evidence of their evolutionary connection with one another.

Every vertebrate starts out life as a fertilized egg. It proceeds through specific stages of development before becoming an adult. Intriguingly, early in the program of development, the embryos of all the different lineages proceed through strikingly similar stages. Take a look at Figure 20.9. Without the labels, would you know which embryo is from a fish, lizard, chicken, or human?

c Bud and mature flower (side and front views) of *D. decorum*

d Mature flower of *D. nudicaule*

a *D. decorum*

b *D. nudicaule*

Figure 20.8 From comparative embryology, evidence of evolutionary relationship among two species of larkspurs (*Delphinium*). Although the time required for flowers to develop and mature is much the same in both species, the *rate* of development is much slower for most of the floral structures of *D. nudicaule*, so the changes in petal shape are not nearly as great as they are in the other species, *D. decorum*.

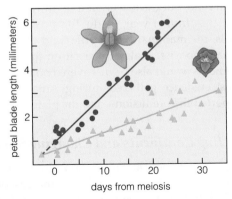

e Lengthening of petals of *D. decorum* (*dark blue* graph line) and of *D. nudicaule* (*light blue*), plotted against development of pollen and eggs in floral structures, starting with meiosis.

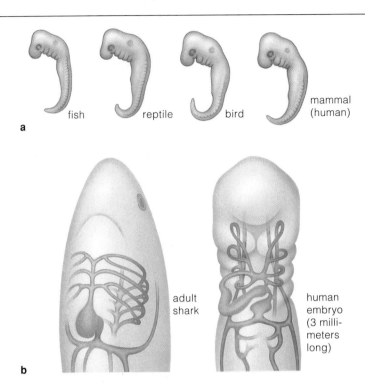

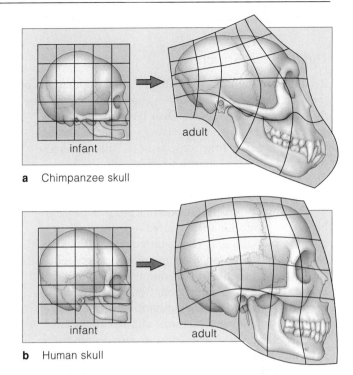

a   Chimpanzee skull

b   Human skull

Figure 20.9 From comparative embryology, some evidence of evolutionary relationship among vertebrates.

(**a**) Adult vertebrates show great diversity, yet the very early embryos retain striking similarities. This is evidence of change in a common program of development. (**b**) Fishlike structures still form in early embryos of reptiles, birds, and mammals. For example, a two-chambered heart (*orange*), certain veins (*blue*), and portions of arteries called aortic arches (*red*) develop in fish embryos—and they persist in adult fishes. The same structures form in an early human embryo.

Figure 20.10 Pronounced morphological differences between two lines of descent, thought to be an outcome of changes in the timing of developmental steps. This example compares proportional changes in the skull bones of a chimpanzee and a human. Both skulls are similar in infants.

Think of the representations of infant skulls as paintings on a blue rubber sheet divided into a grid. Stretching the sheet deforms the grid's squares. For the adult skulls, differences in size and shape within corresponding grid sections reflect differences in growth patterns.

The early embryos of vertebrates strongly resemble one another because they have inherited the same ancient plan for development. According to this plan, tissues can form only when cells divide in certain patterns and interact in prescribed ways. Later on, the heart, bones, skeletal muscles, gut, and other body parts grow in prescribed ways. But they do so only if each developmental step is properly completed before the next step begins.

During vertebrate evolution, most mutations that disrupted early stages of development must have had lethal effects on the organized interactions required for later stages. The embryos of different groups remained similar because mutations that altered steps in early developmental stages were strongly selected against.

How, then, did adults of different groups get to be so different? At least some differences probably came about by mutations that altered the onset, rate, or time of completion of particular developmental steps. Such mutations would bring about changes in shape through increases or decreases in the relative size of body parts. (These are called allometric changes.) They also could lead to adult forms that retain some juvenile features.

For example, Figure 20.10 illustrates how changes in the growth rate at key developmental steps may have caused differences in the proportions of chimpanzee and human skull bones. Consider that the skull bones of both kinds of primates are alike at the time of birth. Thereafter, they change dramatically for chimps but only slightly for humans. As several lines of evidence indicate, both chimps and humans arose from the same ancestral stock. *Their genes are nearly identical.* However, somewhere along the separate evolutionary road that led to humans, one or more genes that affected growth rates apparently mutated in ways that proved adaptive. Ever since that time, instead of promoting the rapid growth required for dramatic changes in skull bones (as in chimps), mutated human genes have blocked it.

Similar patterns of embryonic development may be a clue to evolutionary relationship among lineages.

Mutations that affect certain steps in a shared program of early development may be enough to bring about major differences in the adult forms of related lineages.

Homologous Structures

Recall, from the preceding chapter, that populations of the same species start to diverge genetically after gene flow ceases between them. Over large spans of time, the populations also may diverge considerably in their appearance, functions, or both. This pattern of macroevolution—that is, change from the form of a common ancestor—is called **morphological divergence**. (*Morpho-* means body form.)

Even with great morphological divergence, however, evolutionarily related species remain alike in many ways. They started out with the same body plan, and their evolution proceeded through conservative modifications to the shared body plan. Look carefully enough, and you can discover their underlying similarities.

For example, reptiles, birds, and mammals are land-dwelling vertebrates, presumably descended from amphibians that started to venture onto land. Our knowledge of the earliest land vertebrates is based on the fossilized, five-toed limb bones of a rather squat reptile (Figure 20.11*a*). Descendants of the reptile expanded into new habitats on land. A few of the later descendants even retreated from the land and ended up living in the seas.

Thus, for the pterosaur, bird, and bat lineages, the five-toed limb was a key innovation. It served as evolutionary clay that was gradually molded into different kinds of wings (Figure 20.11*b–d*). In the lineages that led to porpoises and penguins, the five-toed limb evolved into flippers (Figure 20.11*e,f*). In still other lineages, it slowly became modified into the long, one-toed limbs of modern horses, the stubby limbs of burrowing mammals such as moles, and the strong, pillarlike limbs of elephants. And it was modified into the human arm and five-fingered hand (Figure 20.11*g*).

What we have been describing is an example of **homology**, a similarity in one or more body parts in different organisms that we can attribute to descent from a common ancestor. (*Homo-* means the same.) Such similarities are probably the most important clues for researchers who are looking for connections among the pieces of the macroevolutionary puzzle.

Potential Confusion From Analogous Structures

Body parts that have similar form and functions in different lineages aren't *always* homologous. If the last shared connection between two lineages was in the remote past, major differences in form

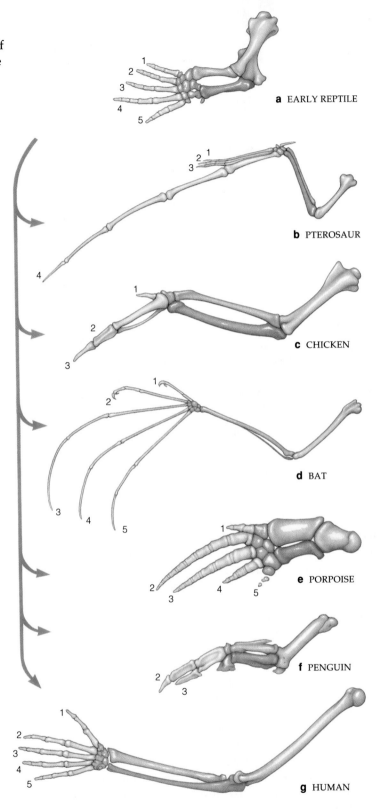

Figure 20.11 Morphological divergence in the vertebrate forelimb, starting with a generalized form of ancestral early reptiles. Diverse forms evolved even as similarities in the number and position of bones were preserved. The drawings are not to the same scale.

Figure 20.12 A case of morphological convergence among three lineages that diverged considerably over evolutionary time. In the lineages that led to the modern sharks, penguins, and porpoises, natural selection apparently favored adaptations that promoted rapid swimming. As one outcome, sharks, penguins, and porpoises have body forms that are all quite similar, even though these vertebrates are related only remotely.

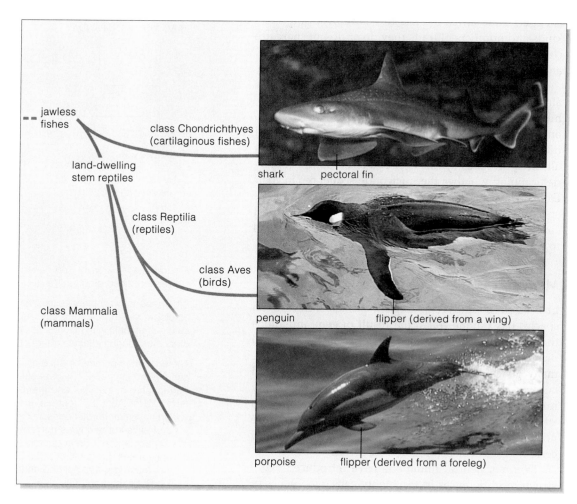

jawless fishes

class Chondrichthyes (cartilaginous fishes)

land-dwelling stem reptiles

class Reptilia (reptiles)

class Aves (birds)

class Mammalia (mammals)

shark   pectoral fin

penguin   flipper (derived from a wing)

porpoise   flipper (derived from a foreleg)

may have emerged during their separate evolutionary journeys. Suppose, by chance, some of their descendants encountered similar kinds of environmental pressures. Suppose those only remotely related organisms started using comparable body parts in much the same way. Most likely, those body parts would have undergone modification, as an outcome of natural selection, and would have ended up resembling one another in their structure and function.

This macroevolutionary pattern, in which lineages that are only remotely related slowly evolve in similar directions, is called **morphological convergence**.

For example, sharks, penguins, and porpoises are all fast-swimming predators of the seas. Their lineages diverged very early in vertebrate evolution. As you can see from Figure 20.12, penguin and porpoise flippers are similar in shape to the shark pectoral fin. All three structures stabilize the body in water. However, they are *not* homologous structures.

Sharks never left the water, and shark fins haven't changed much from the ancestral form. By contrast, penguins are descended from four-legged vertebrates that evolved into birds, which later became adapted to

life in water instead of in the air. So penguin flippers are modified wings, which are modified forelimbs. And porpoises descended from four-legged vertebrates that evolved into mammals, which later became adapted to life in the seas. Their flippers are modified front legs. Thus, the penguin and porpoise flippers converged on the pectoral fins of sharks.

The body parts just mentioned are an example of **analogy**. The word refers to body parts that were once quite different in evolutionarily distant lineages, but that converged in structure and function because those lineages responded to similar environmental pressures. (*Analogos* means similar to one another.)

Homologous structures provide very strong evidence of morphological divergence. These are the same body parts that became modified in different ways in different lines of descent from a common ancestor.

Analogous structures indicate morphological convergence. At one time these body parts, now similar in form and function, differed in evolutionarily remote lineages, but then were used in comparable ways in similar environments.

20.5 EVIDENCE FROM COMPARATIVE BIOCHEMISTRY

All species are a mix of ancestral and novel traits. The kinds and numbers of traits they do or do not share are clues to how closely they are related. As Figures 20.13 and 20.14 suggest, this is also true of biochemical traits.

For example, on the basis of morphological studies alone, you might already have an idea that monkeys, humans, chimpanzees, and other primates are related to one another. You could test this idea by studying small differences in the amino acid sequences of proteins that all three primates synthesize. You also could test it by determining whether the nucleotide sequences in their DNA match up closely or not much at all. Logically, the species that are most closely related share the greatest similarities in their biochemistry. This premise is central to comparative analysis at the molecular level.

Molecular Clocks

Like its close relatives among the primates, the human species has unique traits, the result of gene mutations that accumulated after a divergence from the ancestral primate stock. Yet humans and other species still have many genes in common. For example, the last shared ancestor of certain yeast cells and humans lived many hundreds of millions of years ago. However, more than 90 percent of their known gene products (proteins) have not changed. In other words, the overall structure of certain genes has been highly conserved.

Even so, neutral mutations have introduced slight structural differences in the highly conserved genes of different lineages. **Neutral mutations**, recall, have little or no effect on survival or reproduction. They can slip past agents of selection and accumulate in the DNA.

By some calculations, neutral mutations in highly conserved genes accumulated at a regular rate. Think of the accumulation of neutral mutations in a lineage as a series of predictable ticks of a **molecular clock**. Then turn the clock back, so that the total number of ticks "unwinds" down through the great geologic intervals of the past. Where the last tick stops, that is roughly the time of origin for the lineage.

RACCOON
RED PANDA
GIANT PANDA
DIVERGENCE
15–20 million years ago
DIVERGENCE
approximately
40 million years ago
COMMON ANCESTOR
(meat eater)

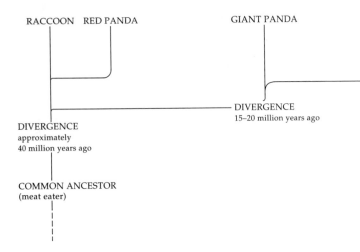

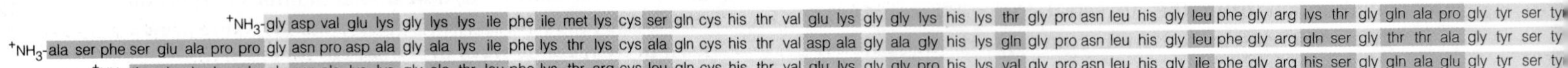

Figure 20.14 How the red panda, the giant panda, and bears are related, according to DNA-DNA hybridization studies.

A giant panda is a plant-eating mammal with paws. Unlike antelope, deer, cattle, and other ungulates, which are plant-eating mammals with hooves, the panda cannot efficiently digest the tough cellulose fibers of plants.

The ungulates pummel plant material inside their elaborately chambered stomach. They benefit from bacteria that reside permanently in their gut and that produce cellulose-digesting enzymes for them. By contrast, the panda has the gut of a *meat-eating* mammal. Its diet consists only of bamboo which, pound for pound, is not nearly as nutrient-packed as meat. A panda must grasp and eat great quantities of bamboo to get enough carbohydrates, proteins, and fats. Its rounded, short-toed paws are not much good at holding and stripping leaves from bamboo stalks. But each front paw also has a thumblike bony digit. When a panda uses a "thumb" in opposition to the five toes on one of its paws, it gets a better grip on a bamboo meal.

The giant panda certainly looks like a bear. It also resembles the red panda, which lives in the same general area in China and eats bamboo. Then again, neither bears nor the red panda has thumbs. Is the giant panda more closely related to red pandas or bears?

Researchers studied the extent to which single-stranded DNA from the giant panda would hybridize with DNA from the red panda and from bears. The more mismatched nucleotide bases, the greater the evolutionary distance between these mammals.

Results from DNA-DNA hybridization studies supported an earlier hypothesis that these three kinds of mammals are related by descent from a common ancestor that apparently lived more than 40 million years ago. The results indicate a divergence took place among the descendants. One branch eventually led to modern raccoons and to the red panda. Another branch led to bears. Much later, between 20 million and 15 million years ago, a split from the bear lineage put the ancestors of the giant panda on a unique evolutionary road. Therefore, the giant panda seems to be much more closely related to bears than it is to the red panda.

Figure 20.13 (*Below*) Primary structure of three versions of cytochrome *c* molecules. The amino acid sequence of this protein has not changed much, even in evolutionarily distant lineages. The three sequences shown are from a representative yeast (*top row*), wheat (*middle row*), and primate (*bottom row*). The *gold* portions highlight the amino acids that are identical in all three species. The probability that this striking molecular resemblance resulted from chance alone is extremely low.

$^{+}NH_3$-gly asp val glu lys gly lys lys ile phe ile met lys cys ser gln cys his thr val glu lys gly gly lys his lys thr gly pro asn leu his gly leu phe gly arg lys thr gly gln ala pro gly tyr ser ty
$^{+}NH_3$-ala ser phe ser glu ala pro pro gly asn pro asp ala gly ala lys ile phe lys thr lys cys ala gln cys his thr val asp ala gly ala gly his lys gln gly pro asn leu his gly leu phe gly arg gln ser gly thr thr ala gly tyr ser ty
$^{+}NH_3$-thr glu phe lys ala gly ser ala lys lys gly ala thr leu phe lys thr arg cys leu gln cys his thr val glu lys gly gly pro his lys val gly pro asn leu his gly ile phe gly arg his ser gly gln ala glu gly tyr ser ty

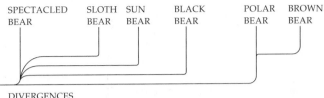

SPECTACLED BEAR | SLOTH BEAR | SUN BEAR | BLACK BEAR | POLAR BEAR | BROWN BEAR

DIVERGENCES
2 million years ago

RED PANDA

GIANT PANDA

BROWN BEAR

Protein Comparisons

Suppose two species have the same conserved gene. The amino acid sequences of the gene's product are the same or nearly so. The absence of mutation implies that the species are closely related. What if the sequences differ quite a bit? Then many neutral mutations must have accumulated in them. A very long time must have passed since the species shared a common ancestor.

Consider the highly conserved gene that specifies cytochrome *c*, a protein component of electron transport chains. Species ranging from aerobic bacteria to humans synthesize it. In humans it has a primary structure of 104 amino acids. Figure 20.13 shows how the amino acid sequences for cytochrome *c* from a fungus, plant, and animal are strikingly similar. Now think about this: Compared to human cytochrome *c*, the *entire* sequence is identical in chimpanzees. It differs by 1 amino acid in rhesus monkeys, 18 in chickens, 19 in turtles, and 56 in yeasts. On the basis of this biochemical information, would you assume humans are more closely related to a chimpanzee or a rhesus monkey? A chicken or a turtle?

Nucleic Acid Comparisons

Typically, structural alterations that resulted from gene mutations pepper the nucleotide sequences of DNA and RNA molecules. Therefore, *the extent to which a strand of DNA or RNA that has been isolated from one species will base-pair with a comparable strand from a different species is a rough measure of the evolutionary distance between them.*

As an example, some nucleic acid comparisons are based on **DNA-DNA hybridization**. By this method, recall, two strands of a double-stranded DNA molecule are induced to unwind from each other, in the manner described in Section 16.4. A molecule of DNA from another species is induced to unwind, also. Then the single strands are allowed to recombine into "hybrid" molecules. Next, the two double-stranded, hybridized molecules are subjected to heating, which breaks the hydrogen bonds holding them together.

The amount of heat energy required to pull apart a hybrid DNA molecule is a measure of the similarity between its two strands. It takes more heat energy to disrupt the hybrid DNA of closely related species. Figure 20.14 describes one application of this method.

Biochemical similarities are greatest among the most closely related species and weakest among the most distantly related.

ala ala asn lys asn lys gly ile ile trp gly glu asp thr leu met glu tyr leu glu asn pro lys lys tyr ile pro gly thr lys met ile phe val gly ile lys lys lys glu glu arg ala asp leu ile ala tyr leu lys lys ala thr asn glu-COO⁻
ala ala asn lys asn lys ala val glu trp glu glu asn thr leu tyr asp tyr leu leu asn pro lys lys tyr ile pro gly thr lys met val phe pro gly leu lys lys pro gln asp arg ala asp leu ile ala tyr leu lys lys ala thr ser ser-COO⁻
asp ala asn ile lys lys asn val leu trp asp glu asn asn met ser glu tyr leu thr asn pro lys lys tyr ile pro gly thr lys met ala phe gly gly leu lys lys glu lys asp arg asn asp leu ile thr tyr leu lys lys ala cys glu-COO⁻

So far in this book, we have mentioned a few of the kinds of organisms that originated and later vanished over the past 3.8 billion years. Many millions of species traveled on that long evolutionary road, and at present the surviving, descendant species also number in the millions.

Back in the 1500s, European naturalists were content simply to identify and describe all the new specimens that were being discovered during an exciting period of global exploration (Section 17.1). Previously unknown species have been turning up ever since. For example, almost on a weekly basis, biologists in tropical forests of Brazil and elsewhere find new examples of diversity, especially among insects. Recently, dozens of unknown species were discovered living in total darkness in a cave 24 meters (80 feet) underground in southwestern Romania. Their ancestors had been trapped inside 5.5 million years ago and had been nourished, directly or indirectly, by chemoautotrophic bacteria that lived in warm water, rich in hydrogen sulfide, on the cave floor.

Even this brief look at the magnitude of diversity is enough to convey the challenges facing **taxonomy**. This field of biology is concerned with identifying, naming, and classifying species. As you will see, the taxonomists attempt to be objective when they make their judgment calls, but available information about species can be interpreted differently and arranged in different ways.

Assigning Names to Species

Let's start with something easy: the system of naming species, which all biologists agree upon. Corn plants, vanilla orchids, houseflies, humans—these and all other known species have a scientific name that is recognized immediately, all over the world (Figure 20.15). Each name was assigned in accordance with a system that Carl von Linné devised in the eighteenth century. You may have heard this Swedish naturalist referred to by his more prestigious, latinized name, Linnaeus.

Linnaeus was a naturalist whose enthusiasm knew no bounds. He sent ill-prepared students around the world to gather specimens of plants and animals, and is said to have lost a third of his collectors to the rigors of their expeditions. Although perhaps not commendable as a student adviser, Linnaeus did go on to develop a **binomial system** by which species are assigned a two-part Latin name.

The first part of the species name is generic; it is descriptive of similar species that are thought to be the same type of organism and so are grouped together. Such a grouping is called a **genus** (plural, genera). The second part is the **specific name** which, in combination with the generic name, refers to one kind of organism only. Such scientific names provide an unambiguous way for people anywhere to discuss the same organism.

KINGDOM	Plantae	Plantae	Animalia	Animalia
PHYLUM	Anthophyta (flowering plants)	Anthophyta	Arthropoda	Chordata
CLASS	Monocotyledonae (monocots)	Monocotyledonae	Insecta	Mammalia
ORDER	Commelinales	Orchidales	Diptera	Primates
FAMILY	Poaceae	Orchidaceae	Muscidae	Hominidae
GENUS	Zea	Vanilla	Musca	Homo
SPECIES	Z. mays	V. planifolia	M. domestica	H. sapiens
COMMON NAME:	corn	vanilla orchid	housefly	human

Figure 20.15 Taxonomic classification of four representative organisms. Each has been assigned to seven ever more inclusive categories, or taxa: species, genus, family, order, class, phylum, kingdom.

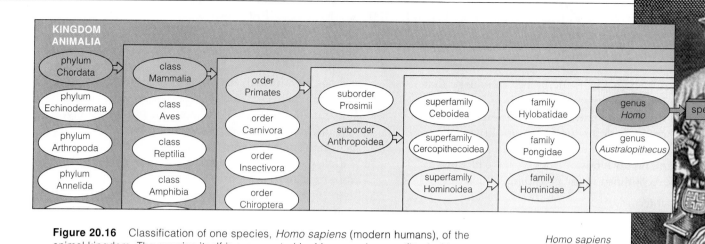

Figure 20.16 Classification of one species, *Homo sapiens* (modern humans), of the animal kingdom. The species itself is represented by Linnaeus, here outfitted with equipment suitable at the time for exploration. Appendix I includes additional taxa.

Homo sapiens
(only living species of this genus)

For example, there is only one *Ursus maritimus*, or polar bear. Other bears include *Ursus arctos* (the brown bear) and *Ursus americanus* (the black bear). Notice how the first letter of the generic name is capitalized and the second (specific) name is not. The specific name is never used without the full or abbreviated generic name preceding it, for it also can be the second name of a species in a different group. Thus, *U. americanus* does indeed mean black bear, but *Homarus americanus* means Atlantic lobster, and *Bufo americanus* means American toad. (Hence one would not order *americanus* for dinner unless one is willing to take what one gets.)

Groupings of Species—The Higher Taxa

Shortly after Linnaeus formulated the binomial system, it became the core organizing unit for a classification scheme. **Classification schemes** are organized ways of retrieving information about particular species. The Linnaean scheme was based on perceived similarities or differences in morphological traits, such as the number of legs, body size, wings or no wings, coloration, and so forth. Later, biologists invented ever more inclusive groupings that are meant to reflect relationships among species. Such groupings of species are known as **higher taxa** (singular, taxon). Some familiar higher taxa include family, order, class, phylum, and kingdom.

As you can deduce from Figure 20.15, ranking by taxa alone does not yield information about patterns of relationships. In themselves, the higher taxa reflect only relative degrees of morphological similarity. A corn plant and vanilla orchid are closely related (both are monocots) but are grouped in separate orders. A human and a housefly are both animals but are only distantly related; they are not even in the same phylum. Humans belong to a superfamily designated Hominoidea, not to superfamily Ceboidea. And they are closer to Ceboidea than to suborder Prosimii, as Figure 20.16 indicates.

Today, taxonomists attempt to assign species to higher taxa in terms of **phylogeny**. This word refers to the evolutionary relationships among species, starting with what are perceived to be the most ancestral forms and including all the branches that lead to all of their descendants.

Putting Classification Schemes To Work

Classifying organisms has its own rewards for anyone who burns with curiosity about life's diversity. But it has its practical applications as well. For example, how might you go about deciding which type of habitat should be protected to help conserve a rare, poorly understood species? You might start with species that are classified as being closely related to it. If those are better understood species, then you can make inferences on the basis of *their* needs for particular habitats. As another example, how would you go about testing the effectiveness of a new antibiotic against a pathogenic bacterium, such as the one that causes Legionnaires' disease in humans? You might conduct tests on a close primate relative of humans, the assumption being that its physiological responses to the infection and to the antibiotic will be similar, if not the same. In short, the information that becomes organized in classification schemes can touch our lives in many different ways, at many different levels.

Each species is assigned a two-part scientific name, which consists of its genus and specific name. It also is assigned to ever more inclusive groupings of species, the higher taxa.

Classification schemes organize information about species and simplify its retrieval. The newer, phylogenetic schemes attempt to reflect evolutionary relationships among species.

The Systematics Approach

Evolutionary systematics is the branch of biology that applies evolutionary theory to the task of identifying patterns of diversity over time and in the environment. It approaches the patterns through work in taxonomy, classification, and phylogenetic reconstruction. Reflect, for a moment, on the kinds of clues we have to patterns of diversity, as outlined in the preceding sections:

> The fossil record
> The geologic record
> Comparative embryology
> Homologous structures
> Comparative biochemistry

Such clues are interpreted through an understanding of microevolutionary processes. For example, suppose you start with the assumption that the forerunners of living reptiles evolved by gene mutation, natural selection, and genetic drift. If this is true, then lizards, dinosaurs, and other reptilian groups arose from their ancestral stock by way of genetic divergences and speciation events. To determine the similarities and differences in morphology, life-styles, and habitat pressures of such groups, you study reptiles in a Costa Rican rain forest, then in the Gila Desert. You also search the literature on

biochemical studies for clues to how closely or distantly the different groups are related. You compare all the findings against both the fossil record and information on environments to which earlier reptiles were adapted.

After analyzing all of the available information, you conclude that turtles are at one end of the evolutionary spectrum, and dinosaurs and crocodiles are at the other. You also suspect birds share a common ancestor with dinosaurs. To portray your perceptions of the pattern of diversity among the reptiles, you decide to construct an evolutionary tree diagram. How will you do it?

Actually, There Is No Single Systematics Approach

Within systematics there are different schools of thought on how to represent patterns of diversity. With *classical* taxonomy, classification schemes and evolutionary tree diagrams are constructed to reflect the perceived degree of morphological divergences among major lineages. In Figure 20.17a, for example, birds are not grouped with dinosaurs and other reptiles. Feathers alone are seen to be a key innovation that allowed ancestors of birds to diverge so far from the dinosaurs that their descendants are a distinct group.

With *cladistic* taxonomy, groups are arranged by branch points in an evolutionary tree diagram (*clad-* is from a Greek word for branch). Only species that share derived traits are grouped past a given branch point, which represents the last shared common ancestor. A **derived trait** is a novel feature that evolved only once and is shared only among descendants of the ancestral species in which it evolved.

Figure 20.17 Reconstructions of the evolutionary history of a few groups of vertebrates, according to two schools of systematics. (**a**) By classical systematics, each group is defined by a number of traditionally accepted morphological traits, such as the presence or absence of scales, feathers, and fur. (**b**) In a cladistic scheme, they are grouped on the basis of presumed shared ancestries.

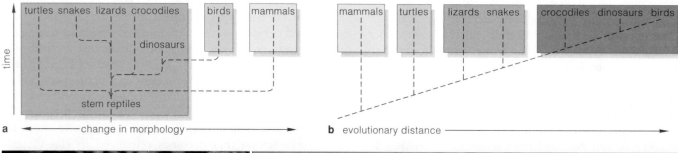

a ← change in morphology →   b evolutionary distance →

GREEN PYTHON   NILE CROCODILE

OWL

DINOSAURS

SEA TURTLE

CHAMELEON

As you can see by comparing Figure 20.17*a* and *b*, these two approaches to assessing patterns of diversity among organisms can translate into large differences in how taxonomists end up grouping them. For example, by using branchings as the cut-off points, so to speak, crocodiles and dinosaurs have far more in common with birds than they do with lizards and snakes.

The evolutionary tree diagrams based on a cladistic approach are known as **cladograms**. Section 20.8 shows how they are constructed. Bear in mind, cladograms do not directly convey "who came from whom." They only portray *relative* relationships among groups. The groups that are closer together on a cladogram simply share a more recent common ancestor than those farther apart. As such, cladograms may be more objective (hence less treacherous) ways to reconstruct evolutionary history. When classical taxonomists assign more importance to one morphological divergence over another, they make judgment calls, and some of the calls may be subjective.

Why bother thinking about the difference between these two approaches? Which approach you subscribe to may influence more than how you look at crocodiles and birds. It may influence how you decide to assign relative importance to the similarities and differences among, say, regional human populations. It may affect how you interpret the kinds and distributions of fossils that are presumably on or near the evolutionary road to modern humans. As you will see shortly, it may also profoundly shape your view of the entire sweep of life's evolution. Especially with the recent developments in comparative biochemistry, we have startling insights into the first major divergences from the first organisms ever to appear on Earth.

Reconstructing the evolutionary history of a given lineage must be based on detailed understanding of the morphology, life-styles, and habitats of the groups represented, and on biochemical comparisons of its existing members.

With classical taxonomy, taxa are grouped by assigning relative importance to morphological divergences.

With cladistic taxonomy, taxa are grouped by derived traits that reflect how recently they shared a common ancestor.

CONSTRUCTING A CLADOGRAM

To see how phylogenetic reconstruction works in practice, consider how you might go about constructing and then interpreting a cladogram. Briefly, you will be analyzing an organism from different groups in terms of a number of morphological, physiological, and behavioral traits (also called characters) that clearly differ among them. Your choice of organisms to study represents the *ingroup*, and it is based on at least some understanding that they are likely to be related. You also select a different organism, one not too distantly related, that can serve as your point of reference for evaluating the evolutionary distances within the ingroup.

Let's suppose you are interested in the evolution of vertebrates, which are all animals with a backbone. You select seven of them: a lamprey, trout, lungfish, turtle, cat, gorilla, and human. For each vertebrate, you know investigators have recorded the six different traits shown in Figure 20.18a. To keep things simple, you focus only

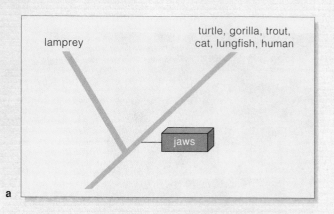

Figure 20.19 Step-by-step construction of a cladogram, an evolutionary tree diagram that groups taxa by derived traits.

on the presence (+) or absence (−) of these traits. From information that is presented in Chapter 27, you decide that the lamprey is only distantly related to the other vertebrates listed. For example, it lacks jaws and paired appendages. This means you can use the lamprey as the outgroup. Compared to the ingroup, an *outgroup* is the organism exhibiting the fewest of the shared traits. The idea is to interpret any deviation from an outgroup as being a derived trait, which suggests a morphological divergence and a branching in the evolutionary tree.

In Figure 20.18b, each zero (0) across the columns of traits for each vertebrate indicates an outgroup condition. Each one (1) means the vertebrate shows the derived trait.

As the next step, you look for the derived traits that your selected groups of vertebrates do or do not share. For example, you see that the human and the gorilla share five derived traits, whereas the human and the cat share only four. For so few traits, you can now sketch out a simple cladogram. In practice, systematists may resort to computers. They often use a great many traits to examine many taxa, and it takes a computer to find the pattern that is best supported by so much information.

Figure 20.19a shows how you would sketch information based on the jaw trait. All of the vertebrates *other than the outgroup* have jaws. The jaw is a derived trait. Figure 20.19b shows what happens to the cladogram when you add information based on lungs. All the vertebrates except the lamprey and the trout have lungs. The lungfish, human, gorilla, turtle, and cat appear to be a *monophyletic* group, meaning they have a common evolutionary heritage. (They share two derived traits with respect to the lamprey and one with respect to the trout.)

Figure 20.19c through e shows the resulting changes in the cladogram as you keep adding information about other derived traits to it.

LAMPREY

Figure 20.18 A selection of seven vertebrates about to be cladogramed.

TURTLE

CAT

GORILLA

Taxon	Traits (Characters)					
	Jaws	Limbs	Hair	Lungs	Tail	Shell
Lamprey	−	−	−	−	+	−
Turtle	+	+	−	+	+	+
Cat	+	+	+	+	+	−
Gorilla	+	+	+	+	−	−
Lungfish	+	−	−	+	+	−
Trout	+	−	−	−	+	−
Human	+	+	+	+	−	−

a Tabulation of traits present or absent

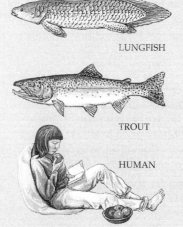

LUNGFISH

TROUT

HUMAN

Taxon	Traits (Characters)					
	Jaws	Limbs	Hair	Lungs	Tail	Shell
Lamprey	0	0	0	0	0	0
Turtle	1	1	0	1	0	1
Cat	1	1	1	1	0	0
Gorilla	1	1	1	1	1	0
Lungfish	1	0	0	1	0	0
Trout	1	0	0	0	0	0
Human	1	1	1	1	1	0

b Tabulation of outgroup and ingroup conditions

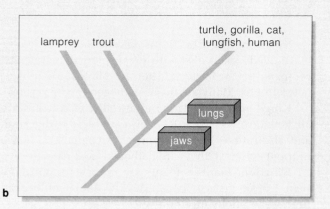

b

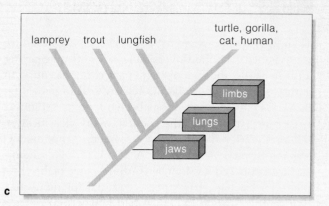

c

Wisely, you did not use the presence of a hardened shell as a trait. Only the turtle has a shell. Although the shell is a derived trait relative to the outgroup, it is not shared with any of the other animals under consideration. Therefore, it is not useful in reconstructing this phylogeny.

Every species has many such characters. The traits may be useful in identifying the species but not in determining relationships. In this example, none of the traits provides information that is contradictory to other traits. In reality, such contradictions often show up. Why? All organisms are a mixture of evolutionarily ancient traits and derived traits. Also, not all of the traits you initially regarded as homologous may actually be homologous. Morphological convergences may have occurred (Section 20.4).

How will you "read" the final cladogram? Remember, it indicates relative relatedness, not ancestors. As you can see, a human is more closely related to a gorilla than to a cat. The gorilla isn't the *ancestor* of humans (it, too, is a modern organism), but both organisms shared a common ancestor more recently than either did with the cat. The cat, human, and gorilla as a group are more closely related to one another than they are to the turtle. These are easy relationships to perceive. Perhaps less obviously, your cladogram also tells you a lungfish is more closely related to a human than to a trout. Follow the lines or branches of the cladogram from the human and lungfish back to the intersection (or node) where they meet (Figure 20.19*f*, node 1). Next, trace back the branches of the lungfish and trout to their intersection (Figure 20.19*f*, node 2). As you can see, the trout-lungfish intersection is nearer to the base of the tree, compared with the lungfish-human intersection. The higher up the evolutionary tree diagram, the more derived traits are shared by the taxa. The lower they are on the tree, the fewer traits are shared.

Another way to think about this is to regard the axis of the cladogram as a time bar, but one without absolute dates. *The lower the position of the branch point between two groups on a cladogram, the more distant is the most recent common ancestor of the taxa being compared.*

Finally, even though the axis of the cladogram can be viewed as a time line, bear in mind that the ordering of taxa from left to right is not important. No matter how the taxa are arranged across the top of the cladogram, follow the branches back to the lungfish-human and lungfish-

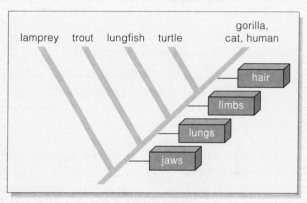

d

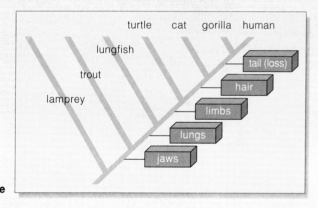

e

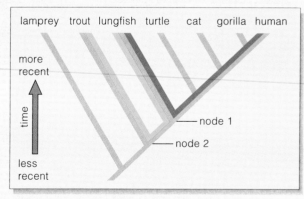

f

trout intersections and you will get the same result. And remember this: All species at the top of the cladogram are modern, *existing* species; they are neither ancestors nor descendants of one another.

HOW MANY KINGDOMS?

There was a time when just about everything in nature was viewed as animal, vegetable, or mineral. Starting with Aristotle, individuals who paid closer attention to nature started to perceive its hierarchical organization in more detailed ways. It has been a long investigative road from Aristotle to the present. Many classification schemes were put forth, then discarded or modified as more information emerged. A two-kingdom scheme, as first proposed by Linnaeus, divided all of life into plants or animals. His scheme endured for more than two centuries, even though it had nowhere to put the tens of thousands of known species of bacteria.

Whittaker's Five-Kingdom Scheme

In 1969, the ecologist Robert Whittaker proposed a *five-kingdom* classification scheme. Entry into one kingdom or another was based on similarities and differences in cell structure, morphology, developmental features, and modes of nutrition, as sketched out here:

Kingdom Monera. In Whittaker's scheme, all the bacteria are grouped in a separate kingdom. They are single-celled organisms that show very little internal complexity but far more diversity in their metabolism than eukaryotic cells. Different species are producers (photoautotrophs and chemoautotrophs) and decomposers (heterotrophs).

Kingdom Protista. Protistans are single-celled or multicelled eukaryotes with great internal complexity, compared to the bacteria. They are photoautotrophs or heterotrophs. Many are major pathogens and parasites.

Kingdom Fungi. These multicelled eukaryotes are all heterotrophs. They feed by extracellular digestion and absorption. Many are major decomposers, hence nutrient cyclers. Others are pathogens and parasites.

Kingdom Plantae. Plants are eukaryotes. Of more than 295,000 known species, nearly all are multicelled producers (photoautotrophs).

Kingdom Animalia. All of these eukaryotes are multicelled, and they range from microscopic size to giants. Of more than a million known species, all are heterotrophs, such as predators and parasites.

Figure 20.20 is a modified version of Whittaker's scheme, which enjoyed wide acceptance until recent findings from comparative biochemistry called parts of it into question.

Along Came the Archaebacteria

For years, biologists have known about archaebacteria, those single-celled prokaryotes that thrive in unusually harsh places, including scalding water near volcanic vents on the seafloor. Compared to *Escherichia coli* or some other typical bacterium, archaebacteria differ in chemical composition, cell wall, and membrane characteristics. Even so, the differences did not seem to be significant enough to pull them out of the kingdom Monera.

Then, in the 1970s, the microbiologist Carl Woese and his colleagues at the University of Illinois took advantage of the new nucleic acid hybridization and sequencing methods. After comparing the base sequences of rRNAs from a variety of organisms, they proposed that bacteria should be divided into two major taxa. Archaebacteria, they said, are as different from eubacteria as they are from eukaryotes.

Strong evidence favoring their conclusion came in 1996. Carol Bult and her colleagues at the Institute for Genomic Research sequenced all of the 1.7 million base pairs of DNA from the archaebacterium *Methanococcus jannaschii*. Woese collaborated with Bult on deciphering the sequence. More than half of the genes had never been seen before. And some are closer to the genes of humans and other eukaryotes than to the ones in eubacteria.

A consensus is growing among biologists to replace the five-kingdom classification scheme with one that is a better reflection of the evolutionary history of life. The first great divergence appears to have been the one dividing **Eubacteria** and **Archaebacteria**. Shortly after

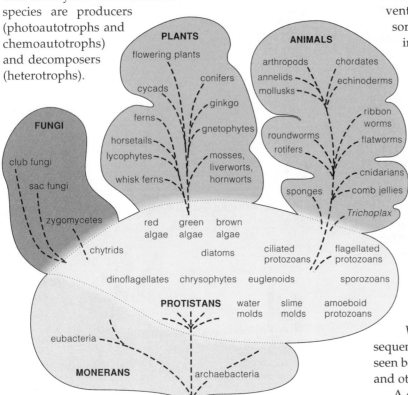

Figure 20.20 Modified version of Robert Whittaker's five-kingdom classification scheme.

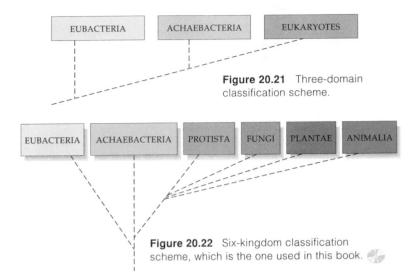

Figure 20.21 Three-domain classification scheme.

Figure 20.22 Six-kingdom classification scheme, which is the one used in this book.

this, the next major branching occurred when certain archaebacteria gave rise to the ancestors of single-celled eukaryotic species—some of which later gave rise to all multicelled forms of life. Evidence of this primordial branching has given rise to a *three-domain* scheme, as in Figure 20.21. As you might well suspect, its fiercest advocates are microbiologists, who know the most about bacteria. Those showing the most reluctance to adopt it argue that it does not give suitable weight to the far greater sweep of diversity among eukaryotes. For our introductory book, it is probably enough to recognize a **six-kingdom classification scheme**. By this scheme, the archaebacteria and eubacteria are each accorded the rank of kingdom, together with the protistans, fungi, plants, and animals (Figure 20.22).

Our understanding of evolutionary relationships is itself evolving, but don't lose sleep over this. Simply know that *species* are the real entities in such schemes. Higher taxa—the groupings of "twigs, branches, and limbs" of evolutionary trees—are not real. As you read through the next unit, focus on the underlying genetic continuity that threads through the history of life. The boundaries of kingdoms and other taxa cut across lines of descent that have continued unbroken through time. The boundaries have not been assigned arbitrarily, but neither are they absolute. All classification schemes are subject to ongoing modification as more pieces of the macroevolutionary puzzle come together.

A widely accepted classification scheme groups organisms into five kingdoms, called Monera, Protista, Fungi, Plantae, and Animalia. The groupings are based on cell structure, morphology, developmental traits, and modes of nutrition.

Recent evidence from comparative biochemistry strongly favors a six-kingdom scheme, which divides the Monera into two kingdoms: the Archaebacteria and Eubacteria.

20.10 SUMMARY

1. There is an underlying continuity of relationship among all species, past and present. The patterns, trends, and rates of change among groups of species over long spans of time are called macroevolution.

2. The fossil record, the geologic record, comparative morphology, and comparative biochemistry all yield extensive evidence of evolution. The evidence is based on similarities and differences in body form, functions, behavior, and biochemistry.

3. The completeness of the fossil record is variable in terms of the species represented, where they lived, and the stability of their tombs since fossilization occurred.

a. Fossilization starts when an organism or traces of it are buried in volcanic ash or in sediments. Dissolved inorganic compounds infuse the remains. Sediments gradually accumulate above them, and pressure and chemical changes transform the remains to stone.

b. Just as life evolved irreversibly from the first cells, so has the Earth evolved in irreversible ways.

c. By the plate tectonics theory, the Earth's crust is fractured into thin plates, like a cracked eggshell, and the movements of these plates slowly raft continents to new positions. Ocean and atmospheric currents shift accordingly. The fossil record and geologic record show that such changes have influenced life's evolution.

4. Comparative morphology has revealed similarities in embryonic development that indicate evolutionary relationship among major groups. It also has identified homologous structures, which are shared as a result of descent from a common ancestor, among groups.

5. Comparative biochemistry identifies similarities and differences among species at the molecular level. All species have an accumulation of neutral mutations in highly conserved genes. Such mutations are like ticks of a molecular clock; they help date divergences from a common ancestor. Other comparisons use nucleic acid hybridization, gene sequencing, and other methods to discern relative evolutionary distances among species.

6. Taxonomists identify, name, and classify species. By the Linnaean binomial system, each kind of organism is assigned a two-part Latin name. The first part (genus) identifies species with shared similarities and presumed derivation from common ancestors. In conjuction with the second part, it is the name of the particular species.

7. Classification systems use higher taxa, or ever more inclusive groupings, from species to genera, families, orders, classes, phyla, and kingdoms. Such groupings reflect phylogeny (presumed evolutionary relatedness).

8. This book uses a six-kingdom phylogenetic scheme: the Eubacteria, Archaebacteria, Protista, Fungi, Plantae, and Animalia.

Review Questions

1. Will the fossil record ever be complete? Why or why not? *20.1*

2. Explain the difference between:
   a. microevolution and macroevolution *CI*
   b. homologous and analogous structures *20.4*
   c. morphological divergence and convergence *20.4*

3. Give reasons why two organisms that are quite different in outward appearances may belong to the same lineage. *20.4*

4. What mutations are the basis of a molecular clock, and why? *20.5*

5. Name a protein specified by a gene that has been highly conserved in organisms ranging from bacteria to humans. *20.5*

6. Why do evolutionary biologists apply heat energy to hybrid molecules of DNA from two species? *20.5*

7. Define taxonomy. In what basic respect does cladistic taxonomy differ from classical taxonomy? *20.6, 20.7*

8. On the basis of recent evidence about a certain group of organisms, consensus has grown to change a five-kingdom classification scheme to a six-kingdom scheme. Which group of organisms is this, and what type of evidence favors putting it in its own kingdom? *20.9*

Self-Quiz *(Answers in Appendix IV)*

1. Morphological convergences may lead to _____ .
   a. analogous structures   c. divergent structures
   b. homologous structures   d. both a and c

2. A classification system that is _____ is based on presumed evolutionary relationship.
   a. epigenetic   c. credited to Linnaeus
   b. phylogenetic   d. both b and c

3. *Pinus banksiana, Pinus strobus,* and *Pinus radiata* are _____ .
   a. three families of pine trees
   b. three different names for the same organism
   c. several species belonging to the same genus
   d. both a and c

4. Increasingly inclusive taxa range from _____ to _____ .
   a. kingdom; species   c. genera; kingdom
   b. kingdom; genera   d. species; kingdom

5. Match these terms suitably.
   ____ phylogeny   a. accumulation of neutral mutations
   ____ fossil   b. stone-hard evidence of life in past
   ____ stratification   c. similar body parts in different
   ____ homology   lineages owing to common descent
   ____ molecular clock d. e.g., shark fins, penguin flippers
   ____ analogy   e. evolutionary relationship among
                            species, ancestors to descendants
                         f. layers of sedimentary rock

Critical Thinking

1. You probably have heard about "missing links," undiscovered transitional forms between major groups of organisms. What are some possible reasons for apparent gaps in the fossil record?

2. Protein comparisons and nucleic acid hybridization studies help us estimate evolutionary relationship and approximate times for divergences from ancestral stocks. DNA sequence comparisons yield even more accurate estimates. Reflect on the genetic code (Section 14.3), then suggest why this may be a more accurate measure of mutations, mutation rates, and biochemical relatedness.

3. At the end of your backbone are several small, fused bones, called the coccyx. Is the coccyx a vestigial structure—all that is left of a tail that was a feature of the evolutionarily distant vertebrate (and primate) ancestors of humans? Or is it the start of a newly evolving structure with an as-yet undetermined function? Take an educated guess, then describe some ways in which you might test whether your answer is plausible.

Selected Key Terms

analogy *20.4*	homology *20.4*
Animalia *20.9*	lineage *CI*
Archaebacteria *20.9*	macroevolution *CI*
binomial system *20.6*	molecular clock *20.5*
cladogram *20.7*	Monera *20.9*
classification scheme *20.6*	morphological
comparative	convergence *20.4*
morphology *20.3*	morphological
continental drift *20.2*	divergence *20.4*
derived trait *20.7*	neutral mutation *20.5*
DNA-DNA	Pangea *20.2*
hybridization *20.5*	phylogeny *20.6*
Eubacteria *20.9*	Plantae *20.9*
evolutionary	plate tectonics *20.2*
systematics *20.7*	Protista *20.9*
fossil *20.1*	seafloor spreading *20.2*
fossilization *20.1*	six-kingdom classification
Fungi *20.9*	scheme *20.9*
genus *20.6*	specific name *20.6*
geologic time scale *20.1*	stratification *20.1*
Gondwana *20.2*	taxonomy *20.6*
higher taxon (taxa) *20.6*	theory of uniformity *20.2*

Readings

Brooks, D. R., and D. A. McLennan. 1991. *Phylogeny, Ecology, and Behavior*. Chicago: University of Chicago Press.

Bult, C., et al. 23 August 1996. "Complete Genome Sequence of the Methanogenic Archaeon *Methanococcus jannaschii*." *Science* 273: 1058–1073.

Morell, V. April 1997. "On the Origin of (Amazonian) Species." *Discover* 18(4): 56–64. An account of how evolutionary biologists Jim Patton and Maria da Silva are attempting to correlate the geologic history of the Amazon basin, which is astoundingly rich in species diversity, with patterns of speciation.

Web Site See *http://www.wadsworth.com/biology* for practice quiz questions, hypercontents, BioUpdates, and critical thinking. The Wadsworth Biology Resource Center provides a wealth of information fully organized and integrated by chapter.

FACING PAGE: *Patterns of diversity in nature, here represented by different species of plants and fungi.*

THE ORIGIN AND EVOLUTION OF LIFE

In The Beginning . . .

Some clear evening, watch the moon as it rises from the horizon and think of the 380,000 kilometers between it and you. *Five billion trillion times* the distance between the moon and you are galaxies—systems of stars—at the boundary of the known universe. Wavelengths traveling through space move faster than anything else—millions of meters per second. Yet the long wavelengths that originated from faraway galaxies many billions of years ago are only now reaching the Earth.

By every known measure, all the near and distant galaxies suspended in the vast space of the universe are moving away from one another, which means that the universe is expanding. And the prevailing view of how the colossal expansion came about accounts for every bit of matter in the universe, in every living thing.

Think about how you rewind a videotape on a VCR, then imagine "rewinding" the universe. As you do this, the galaxies start moving back together. After 12 to 15

Figure 21.1 Part of the great Eagle nebula, a hotbed of star formation 7,000 light-years from Earth, in the constellation Serpens. (The Latin *nebula* means mist.) Not shown in this image are a few huge, young stars above the pillars. For the past few million years, intense ultraviolet radiation from the stars has been eroding the less dense surface of the pillars. (By analogy, visualize a strong prevailing wind blowing away sand in a desert and exposing rocks.) Globules of denser gases and dust that have resisted erosion are visible at the surface. Each one is wider than our solar system—more than 10 billion miles across! New stars are hatching from the protruding globules; some are shining brightly on the tips of gaseous streamers.

billion years of rewinding, all galaxies, all matter, and all of space are compressed into a hot, dense volume about the size of the sun. You have arrived at time zero.

That incredibly hot, dense state lasted only for an instant. What happened next is known as the **big bang**, a stupendous, nearly instantaneous distribution of all matter and energy everywhere, through all of the known universe. About a minute later, temperatures dropped to a billion degrees. Fusion reactions produced most of the light elements, including helium, which are still the most abundant elements throughout the universe. Radio telescopes have detected relics of the big bang in the form of cooled and diluted background radiation, left over from the beginning of time.

Over the next billion years, stars started forming as gaseous material contracted in response to the force of gravity. When stars became massive enough, nuclear reactions were ignited in their central region, and they gave off tremendous light and heat. As massive stars continued to contract, many became dense enough to promote the formation of heavier elements.

All stars have a life history, from birth to an often spectacularly explosive death. In what might be called the original stardust memories, the heavier elements released during the explosions became swept up during the gravitational contraction of new stars, and they became raw materials for the formation of even heavier elements. Even as you are reading this page, the Hubble space telescope is providing astounding glimpses of star-forming activity, as in the dust clouds of Orion, Serpens, and other constellations (Figure 21.1).

Now imagine a time long ago, when explosions of dying stars ripped through our galaxy and left behind a dense cloud of dust and gas that extended trillions of kilometers in space. As the cloud cooled, countless bits of matter gravitated toward one another. By 4.6 billion years ago, the cloud had flattened out into a slowly rotating disk. At the dense, hot center of that disk, the shining star of our solar system—the sun—was born.

The remainder of this chapter is a sweeping slice through time, one that cuts back to the formation of the Earth and the chemical origins of life. It is the starting point for the next four chapters, which will take us along lines of descent that led to the present range of species diversity. The story is not complete. Even so, all the available evidence, from many avenues of research, points to a principle that can help us organize separate bits of information about the past: *Life is a magnificent continuation of the physical and chemical evolution of the universe, of galaxies and stars, and of the planet Earth.*

KEY CONCEPTS

1. A great body of evidence suggests that life originated more than 3.8 billion years ago. Its origin and subsequent evolution have been linked to the physical and chemical evolution of the universe, the stars, and the planet Earth.

2. All of the inorganic and organic compounds necessary for self-replication, membrane assembly, and metabolism —that is, for the structure and functioning of living cells— could have formed spontaneously under conditions that existed on the early Earth.

3. The history of life, from its chemical beginnings to the present, spans five intervals of geologic time. It extends through two great eons—the Archean and the Proterozoic— and the Paleozoic, Mesozoic, and Cenozoic eras.

4. Not long after life originated, divergences led to two great prokaryotic lineages, called the archaebacteria and eubacteria. Shortly afterward, the ancestors of eukaryotes diverged from the archaebacterial lineage.

5. Archaebacteria and eubacteria dominated the Archean and Proterozoic eons. Eukaryotic cells originated late in the Proterozoic era and became spectacularly diverse. A theory of endosymbiosis helps explain the profusion of specialized organelles that arose in eukaryotic cells.

6. All six kingdoms of organisms are characterized by the persistences, extinctions, and radiations of many different lineages over time.

7. Throughout the history of life, asteroid impacts, drifting and colliding continents, and other environmental insults have had profound impact on the direction of evolution.

Origin of the Earth

Cloudlike regions of the universe, as shown in Figure 21.1, are mostly hydrogen gas. These clouds also contain water, iron, silicates, hydrogen cyanide, ammonia, methane, formaldehyde, and some other simple organic and inorganic substances. The contracting cloud from which our solar system evolved probably was similar in composition. Between 4.6 and 4.5 billion years ago, the periphery of the cloud cooled. Mineral grains and ice orbiting the sun started to clump together as a result of electrostatic attraction and the pull of gravity (Figure 21.2). In time, larger, faster clumps started colliding and shattering. Some became more massive by sweeping up asteroids, meteorites, and the other rocky remnants of collisions, and gradually they evolved into planets.

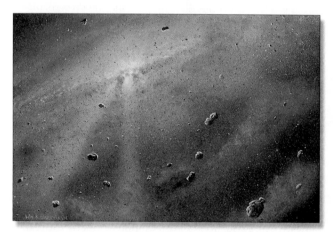

Figure 21.2 Representation of the cloud of dust, gases, and clumps of rock and ice around the early sun.

As the Earth was forming, much of its inner rocky material melted. Asteroid impacts as well as its own internal compression and radioactive decay of minerals could have generated the heat necessary to do this. As the rocks melted, nickel, iron, and other heavy materials moved to the Earth's interior; lighter ones floated to the surface. This process produced a crust, mantle, and core. The **crust** is an outer region of basalt, granite, and other low-density rocks. It envelops the intermediate-density rocks of the **mantle**, which wraps around a core of very high-density, partially molten nickel and iron.

Four billion years ago, the Earth was a thin-crusted inferno (Figure 21.3a). In less than 200 million years, life had originated on its surface! We have no record of the event. As far as we know, movements in the mantle and crust, volcanic activity, and erosion obliterated all traces of it. Still, we can put together a plausible explanation of how life originated by considering three questions:

First, *when life originated, what physical and chemical conditions prevailed on Earth?*

Second, *based on physical, chemical, and evolutionary principles, could large organic molecules have spontaneously formed and then evolved into molecular systems displaying the fundamental properties of life?*

Third, *can we devise experiments to test whether living systems could have emerged by chemical evolution?*

The First Atmosphere

When the first patches of crust were forming, hot gases blanketed the Earth. This first atmosphere probably was a mix of gaseous hydrogen (H_2), nitrogen (N_2), and carbon monoxide (CO), as well as carbon dioxide (CO_2). Was gaseous oxygen (O_2) also present? Probably not. Rocks release oxygen when they are subjected to the intense heat of volcanic eruptions, but not much. Besides, free oxygen would have reacted at once with other elements. At first, water at the Earth's molten surface must have evaporated into the atmosphere. After the crust cooled and solidified, rains from the clouds blanketing the Earth drenched the parched rocks. For millions of years the runoff eroded mineral salts and other compounds from the rocks. And salt-laden waters gradually collected in depressions in the crust and formed the early seas.

If the early atmosphere and seas had not been free of oxygen, then the organic compounds that became organized into the first cells might not have formed on their own. Oxygen would have attacked them, in the manner described in the introduction to Chapter 6. If liquid water had not accumulated, then cell membranes could not have formed. Cells are the basic units of life. *Each has a capacity to survive and reproduce on its own.*

Synthesis of Organic Compounds

Reduce a cell to its lowest common denominator and all that remains are proteins, complex carbohydrates and lipids, and nucleic acids. Existing cells assemble these molecules from smaller organic compounds: the simple sugars, fatty acids, amino acids, and nucleotides. Energy from the environment drives the synthesis reactions. Were small organic compounds also present on the early Earth? Were there sources of energy that spontaneously drove their assembly into the large molecules of life?

Mars, meteorites, the Earth's moon, and the Earth all formed at the same time, from the same cosmic cloud. Rocks collected from Mars, meteorites, and the moon contain precursors of biological molecules, so the same precursors must have been present on the Earth. If this were indeed the case, *then energy from sunlight, lightning, or even heat escaping from the crust could have been enough to drive their combination into organic molecules.*

Stanley Miller conducted the first experimental test of that prediction. First he mixed methane, hydrogen,

Figure 21.3 (a) Representation of the Earth during its formation, when the moon's orbit was much closer than it is today. If the Earth had condensed into a planet of smaller diameter, its gravitational mass would not have been great enough to hold onto an atmosphere. If it had settled into an orbit closer to the sun, water would have evaporated from its hot surface. If the Earth's orbit had been more distant from the sun, its surface would have been far colder, and water would have been locked up as ice. Without liquid water, life as we know it never would have originated on Earth.

(b) Stanley Miller's experimental apparatus, used to study the synthesis of organic compounds under conditions that presumably existed on the early Earth. The condenser cools circulating steam so that water droplets form.

ammonia, and water inside a reaction chamber of the sort depicted in Figure 21.3b. Then he kept the mixture circulating and bombarded it with a spark discharge to simulate lightning. In less than a week, amino acids and other small organic compounds had formed.

In other experiments that simulated conditions on the early Earth, glucose, ribose, deoxyribose, and other sugars formed spontaneously from formaldehyde, and adenine from hydrogen cyanide. Ribose and adenine are found in ATP, NAD, and other nucleotides vital to cells.

However, if *complex* organic compounds had formed directly in the seas, they would not have lasted long. The spontaneous direction of the necessary reactions would have been toward hydrolysis, not condensation, in water. How did more lasting bonds form?

By one hypothesis, clay in the rhythmically drained muck of tidal flats and estuaries served as templates (structural patterns) for the spontaneous assembly of proteins and other complex organic compounds. Clay consists of thin, stacked layers of aluminosilicates with metal ions at their surfaces, which attract amino acids. Expose amino acids to some clay, warm the clay with rays from the sun, then alternately moisten and dry it. Condensation reactions will proceed at its surfaces and yield proteins and other complex organic compounds.

By another hypothesis, complex organic compounds formed spontaneously near hydrothermal vents on the seafloor, where species of archaebacteria are thriving

today. As experimental tests by Sidney Fox and others show, when amino acids are placed in water and then heated, they spontaneously order themselves into small protein-like molecules, which Fox calls "protenoids."

However the first proteins formed, their molecular structure dictated how they would interact with other compounds. Suppose some proteins had the structure to function as weak enzymes. That is, they were able to hasten bond formation between amino acids. Enzyme-directed synthesis would have a selective advantage. In the chemical competition for available amino acids, the protein configurations that promoted reactions would win. Also, proteins have the capacity to bind metal ions and other carbon-based compounds. As you will see next, such chemical modification was the foundation for the evolution of metabolism. For now, simply reflect on the possibility that selection was at work before the origin of living cells, favoring the chemical evolution of enzymes and other complex organic compounds.

Many diverse experiments yield indirect evidence that the complex organic molecules characteristic of life could have formed under conditions that existed on the early Earth.

Origin of Agents of Metabolism

A defining characteristic of life is metabolism. The word refers to all the reactions by which cells harness energy and use it to drive their activities, such as biosynthesis. During the first 600 million years or so of Earth history, enzymes, ATP, and other organic compounds may have assembled spontaneously, and they may have done so in the same physical locations. If so, their close association would have naturally promoted chemical interactions and the beginning of metabolic pathways.

Imagine an ancient estuary, rich in clay deposits. It is a coastal region where seawater mixes with mineral-rich water being drained from the land. There, beneath the sun's rays, countless aggregations of organic molecules stick to the clay. At first there are quantities of an amino acid; call it *D*. Throughout the estuary, *D* molecules get incorporated into new proteins—until the supply of *D* dwindles. However, suppose an enzyme-like protein molecule also is present in the estuary. It can promote the formation of *D* by acting on an abundant, simpler substance—call it *C*. By chance, some aggregations of organic molecules include that particular enzyme-like protein, and so they have an advantage in the chemical competition for starting materials.

In time, *C* molecules become scarce. At that point, the advantage tilts to aggregations that can promote formation of *C* from even simpler organic substances *B* and *A*—say, from carbon dioxide and water. As you know, carbon dioxide and water occur in essentially unlimited amounts in the atmosphere and in the seas. Chemical selection has favored a synthetic pathway:

$$A + B \longrightarrow C \longrightarrow D$$

Finally, suppose some aggregations are better than others at absorbing and using energy. Which molecules could bestow such an advantage? Think of the energy-trapping pathway that now dominates the world of life: photosynthesis. It starts at pigments called chlorophylls. The light-absorbing and electron-donating portion of a chlorophyll molecule is called a porphyrin ring structure. Porphyrins are also present in cytochromes, which are part of electron transport systems in all photosynthetic and aerobically respiring cells. Porphyrin molecules can spontaneously assemble from formaldehyde—one of the molecular legacies of cosmic clouds (Figure 21.4). Was porphyrin an electron transporter of some of the early metabolic pathways? Perhaps.

Origin of Self-Replicating Systems

Another defining characteristic of life is a capacity for reproduction, which now starts with protein-building instructions in DNA. The DNA molecule is fairly stable,

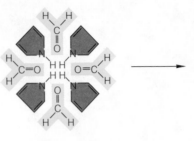

four pyrrole rings plus four
formaldehyde molecules

Figure 21.4   One hypothetical sequence by which formaldehyde, an organic compound, underwent chemical evolution into porphyrin. Formaldehyde was present when the Earth formed. Porphyrin is the light-trapping and electron-donating component of all existing chlorophyll molecules. It also is a component of cytochrome, which is a protein component of electron transport systems that are part of many metabolic pathways.

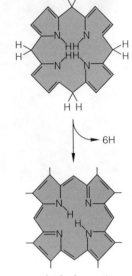

porphyrin ring system

chlorophyll *a*, one of the
light-trapping pigments of
photosynthetic plant cells

and it is easily replicated before each cell division. As you know from earlier chapters, arrays of enzymes and RNA molecules operate together to carry out DNA's encoded instructions.

Most existing enzymes are assisted by the small organic molecules or metal ions that are called coenzymes. Intriguingly, some types of coenzymes have a structure that is identical to that of the RNA nucleotides. Another clue: Heat nucleotide precursors and short chains of phosphate group together, and they will self-assemble into strands of RNA. On the early Earth, energy from the sun or from geothermal events would have been sufficient to drive the spontaneous formation of RNA molecules.

Very simple self-replicating systems of RNA, enzymes, and coenzymes have been created in some laboratories. Did RNA later become the information-storing templates for protein

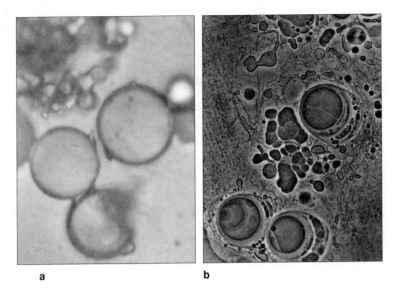

a   b

Figure 21.5 Microscopic spheres of (**a**) proteins and (**b**) lipids that self-assembled under abiotic conditions.

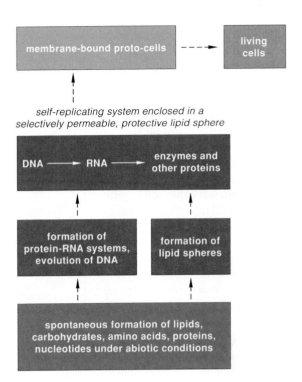

Figure 21.6 One possible sequence of events that led to the first self-replicating systems, then to the first living cells.

synthesis? Perhaps it initially did so, although existing RNA molecules are just too chemically fragile for such a role. Yet the use of RNA templates may have set up an **RNA world** that preceded DNA's dominance as the main informational molecule. Whatever the case, DNA eventually assumed this function, probably because it can form long nucleotide chains in more stable fashion.

We still don't know how DNA entered the picture. Until we identify the likely chemical ancestors of RNA and DNA, the story of life's origin will be incomplete. Filling in the details will require imaginative sleuthing. For instance, a few researchers ran a program on a rapid, advanced computer. The program included information about simple compounds, of the sort thought to have been present on the early Earth. The researchers asked the computer to subject the compounds that they had chosen to random chemical competition and to natural selection. They ran the program repeatedly. And the outcome was always the same: The simple precursors inevitably evolved to form interacting systems of large, complex molecules.

Origin of the First Plasma Membranes

Experimental tests are more revealing of the origin of the plasma membrane, the outermost component of every living cell. A plasma membrane consists of a lipid bilayer, studded with proteins that carry out diverse functions. Its main role is to control which substances move into and out of the cell. Without this control, cells cannot exist. Before cells, then, there must have been **proto-cells**. Maybe they were little more than membrane sacs that protected information-storing templates and

metabolic agents from the environment. We know that simple membrane sacs can form spontaneously.

For one experiment, Fox heated amino acids until they formed protein-like chains, which he then placed in hot water. After cooling, the chains assembled into small, stable spheres of the sort shown in Figure 21.5a. Like the membranes of cells, these proteinoid spheres were selectively permeable to a variety of substances. They picked up free lipids from the surroundings, and a lipid-protein film formed at their surface.

In still other experiments by David Deamer and his coworkers, fatty acids and glycerol combined to form long-tail lipid molecules under laboratory conditions that simulated evaporating tidepools. The lipids self-assembled into small, water-filled sacs. Many were like cell membranes (Figure 21.5b).

In short, there are major gaps in the story of life's origins. But there also is strong experimental evidence that chemical evolution probably led to the molecules and structures that are characteristic of life. Figure 21.6 summarizes the milestones in that chemical evolution, which preceded the first cells.

Although the story is not yet complete, many laboratory experiments and computer simulations indirectly show that chemical and molecular evolution gave rise to proto-cells.

The first living cells originated in the **Archean** eon, which lasted from 3.9 billion to 2.5 billion years ago. Those cells emerged as molecular extensions of the evolving universe, of our solar system and Earth. They may have appeared in tidal flats or in muddy sediments of an ancient seafloor. Fossils indicate they were like existing bacteria. Specifically, they were **prokaryotic cells**, with no nucleus. At first they probably were no more than self-replicating, membrane-bound sacs of DNA and other complex organic molecules. Given the absence of free oxygen, they must have secured energy by anaerobic pathways—fermentation, most likely. Energy was plentiful. Geologic processes had enriched the seas with organic compounds. So "food" was available, predators were absent, and cellular structures were free from oxygen attacks.

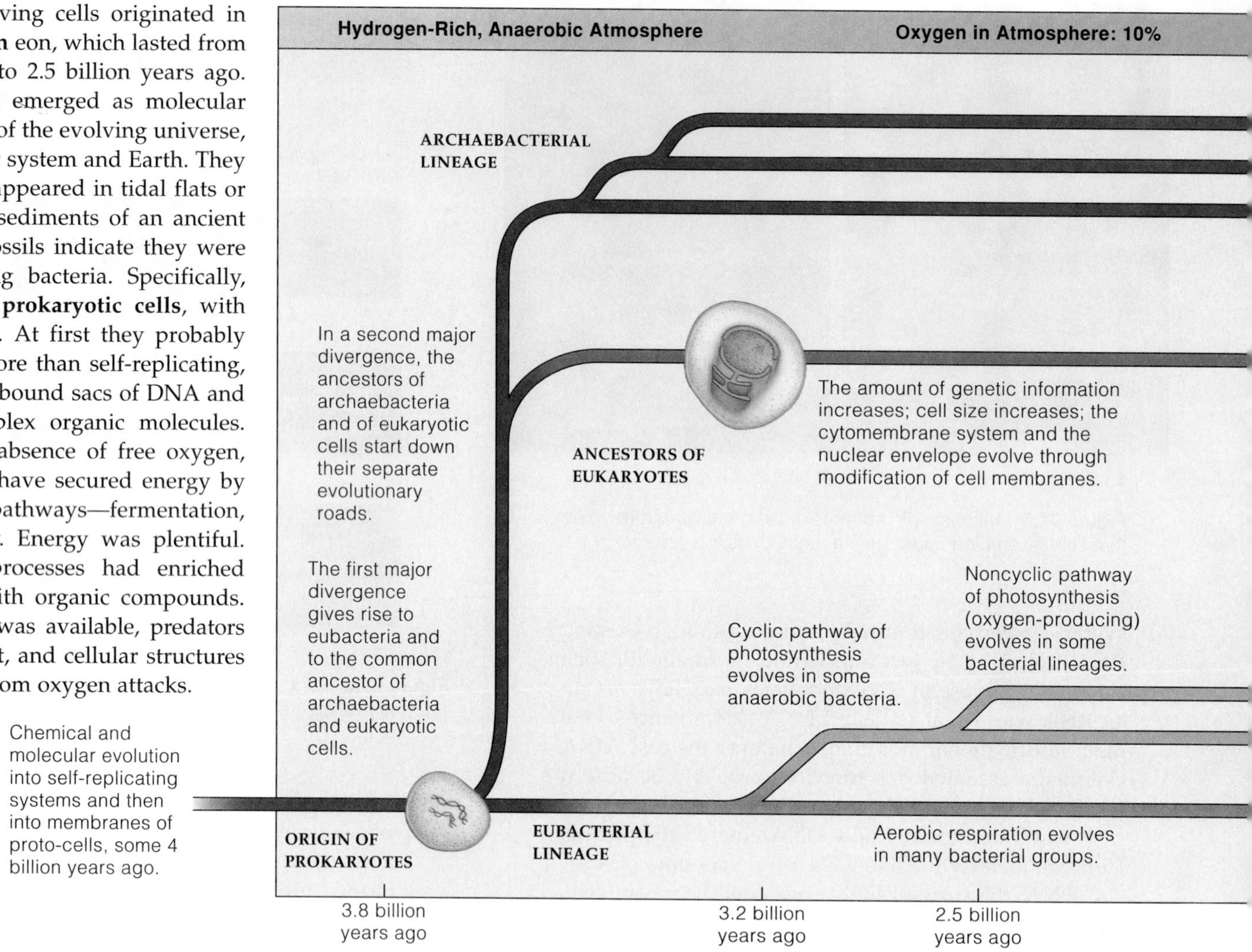

Hydrogen-Rich, Anaerobic Atmosphere | **Oxygen in Atmosphere: 10%**

ARCHAEBACTERIAL LINEAGE

In a second major divergence, the ancestors of archaebacteria and of eukaryotic cells start down their separate evolutionary roads.

The first major divergence gives rise to eubacteria and to the common ancestor of archaebacteria and eukaryotic cells.

Chemical and molecular evolution into self-replicating systems and then into membranes of proto-cells, some 4 billion years ago.

ANCESTORS OF EUKARYOTES

The amount of genetic information increases; cell size increases; the cytomembrane system and the nuclear envelope evolve through modification of cell membranes.

Noncyclic pathway of photosynthesis (oxygen-producing) evolves in some bacterial lineages.

Cyclic pathway of photosynthesis evolves in some anaerobic bacteria.

ORIGIN OF PROKARYOTES

EUBACTERIAL LINEAGE

Aerobic respiration evolves in many bacterial groups.

3.8 billion years ago | 3.2 billion years ago | 2.5 billion years ago

Figure 21.7 An evolutionary tree of life that reflects mainstream thinking about the connections among major lineages. The diagram incorporates ideas about the origins of some eukaryotic organelles.

Some populations of those first prokaryotic cells apparently diverged in two major directions shortly after the time of origin. One evolutionary road led to the **eubacteria**. The other gave rise to the shared ancestors of the **archaebacteria** and **eukaryotic cells** (Figure 21.7).

Between 3.5 and 3.2 billion years ago, light-trapping pigments, electron transport systems, and other metabolic machinery evolved in some of the anaerobic eubacteria, which thereby became the first practitioners of the cyclic pathway of photosynthesis. (You read about this ATP-forming pathway in Section 7.4.) An unlimited source of energy—sunlight—had been tapped. For nearly 2 billion years, the photosynthetic descendants of those bacterial cells dominated the world of life. Their tiny but numerous populations formed very large mats in which sediments collected. The mats accumulated, one atop the other. Calcium deposits hardened and preserved the mats, which came to be called **stromatolites** (Figure 21.8).

Figure 21.8 Stromatolites exposed at low tide in Western Australia's Shark Bay. These mounds started forming 2,000 years ago. They are identical in structure to stromatolites that formed more than 3 billion years ago.

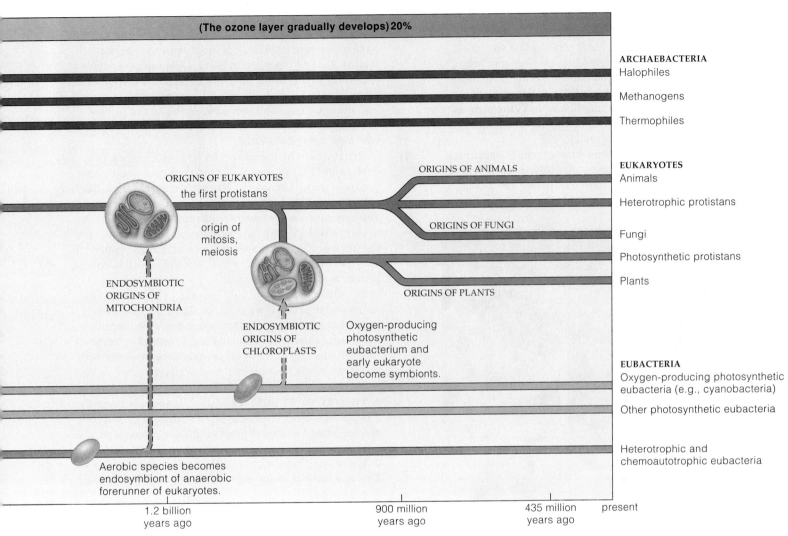

(The ozone layer gradually develops) 20%

ARCHAEBACTERIA
Halophiles

Methanogens

Thermophiles

ORIGINS OF EUKARYOTES
the first protistans

ORIGINS OF ANIMALS

EUKARYOTES
Animals

Heterotrophic protistans

origin of
mitosis,
meiosis

ORIGINS OF FUNGI

Fungi

Photosynthetic protistans

Plants

ENDOSYMBIOTIC
ORIGINS OF
MITOCHONDRIA

ENDOSYMBIOTIC
ORIGINS OF
CHLOROPLASTS

Oxygen-producing
photosynthetic
eubacterium and
early eukaryote
become symbionts.

ORIGINS OF PLANTS

EUBACTERIA
Oxygen-producing photosynthetic
eubacteria (e.g., cyanobacteria)

Other photosynthetic eubacteria

Heterotrophic and
chemoautotrophic eubacteria

Aerobic species becomes
endosymbiont of anaerobic
forerunner of eukaryotes.

1.2 billion
years ago

900 million
years ago

435 million
years ago

present

By the dawn of the **Proterozoic** eon, 2.5 billion years ago, the photosynthetic machinery had become altered in some eubacterial species, and the noncyclic pathway of photosynthesis emerged. Oxygen, one of the pathway's by-products, started to accumulate. In time, this had two irreversible effects. First, *an oxygen-rich atmosphere stopped the further chemical origin of living cells.* Except in restricted anaerobic habitats, complex organic compounds could no longer form spontaneously and resist attack. Second, *aerobic respiration became the dominant energy-releasing pathway.* In many prokaryotic lineages, selection favored metabolic equipment that "neutralized" oxygen by using it as an electron acceptor. This innovation contributed to the rise of multicelled eukaryotes and their invasion of far-flung environments (Section 8.7).

Eukaryotic cells evolved in the Proterozoic, possibly before 1.2 billion years ago. We have fossils, 900 million years old, of well-developed red algae, green algae, fungi, and plant spores. Organelles, remember, are the hallmark of eukaryotic cells. *Where did they come from?* The next section describes a few plausible hypotheses.

About 800 million years ago, stromatolites began to decline dramatically. They had become a concentrated source of food for newly evolved, tiny bacteria-eating animals. About the same time, tiny soft-bodied animals were making tracks and digging burrows in seafloor sediments. They lived near the shores of Laurentia, an early supercontinent. About 570 million years ago, in "Precambrian" times, some of their descendants started the first adaptive radiation of animals.

The first cells evolved by about 3.8 billion years ago, during the Archean. All were prokaryotic cells, and most, if not all, probably made ATP by fermentation routes.

In the first major divergence, the ancestors of archaebacteria and of eukaryotic cells branched away from the road that would lead to modern eubacteria.

Oxygen-releasing photosynthetic bacteria evolved. In time, the oxygen-enriched atmosphere put an end to the further spontaneous chemical evolution of life. That atmosphere was a key selection pressure in the evolution of eukaryotic cells.

21.4

WHERE DID ORGANELLES COME FROM?

Thanks to Andrew Knoll, William Schopf, Jr., and other globe-hopping microfossil hunters, we have tantalizing evidence of early life. One fossil treasure is a strand of bacterial cells that lived 3.5 billion years ago, not long after the time that life originated. Others are from the Proterozoic. They were eukaryotic cells that contained a few membrane-bound organelles in the cytoplasm, as shown in Figure 21.9. Their living descendants have a profusion of organelles (Figure 21.10).

Where did eukaryotic organelles come from? Speculations abound. Some organelles probably evolved through gene mutations and natural selection. For others, researchers make a good case for evolution by way of endosymbiosis.

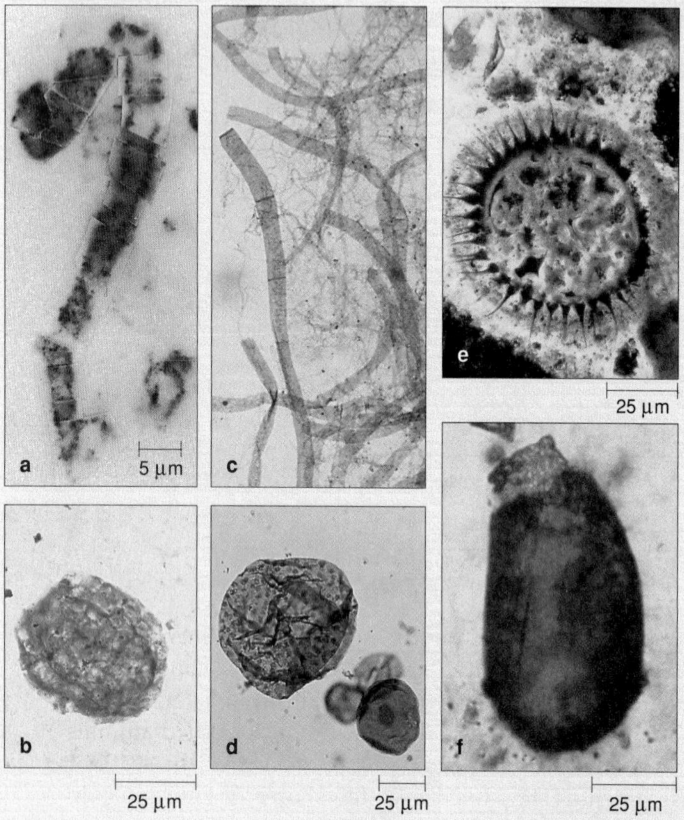

ORIGIN OF THE NUCLEUS AND ER Prokaryotic cells do not have an abundance of organelles, but some species have interesting infoldings of the plasma membrane (Figure 21.11). Embedded in that membrane are enzymes and other agents of metabolism. In the early forerunners of eukaryotic cells, similar infoldings may have extended into the cytoplasm and served as routes to the surface. They may have evolved into ER channels and into an envelope around the DNA.

What would be the advantage of such membranous enclosures? Maybe they protected the genes and protein products from "foreigners." Remember, bacterial species often transfer plasmid DNA among themselves. So do yeasts, which are very simple eukaryotic cells. At first, a nuclear envelope may have been favored because it got the cell's genes, replication enzymes, and transcription enzymes out of the cytoplasm. It would have allowed vital genetic messages to be copied and read, free of metabolic competition from what could become an unmanageable hodgepodge of foreign genes. Similarly, ER channels might have kept important proteins and other organic compounds away from metabolically hungry "guests"— foreign cells that one way or another became permanent residents inside the host cell, as described next.

A THEORY OF ENDOSYMBIOSIS It appears likely that accidental partnerships between a variety of prokaryotic species formed countless times on the evolutionary road to eukaryotic cells. Some partnerships possibly resulted in the origin of mitochondria, chloroplasts, and other organelles. This is a story of endosymbiosis, as developed in greatest

Figure 21.9 From Australia, (**a**) a strand of walled prokaryotic cells 3.5 billion years old and (**b**) one of the oldest known eukaryotes— a protistan 1.4 billion years old. From Siberia, (**c**) a multicelled alga 900 million to 1 billion years old and (**d**) eukaryotic microplankton 850 million years old. (**e**) From China, a splendid eukaryotic cell that lived 560 million years ago. (**f**) From Spitsbergen, Norway, a protistan that was alive 50 million years before the dawn of the Cambrian.

Figure 21.10
A fine example of the profusion of diverse organelles that are hallmarks of eukaryotic cells: *Euglena*, a single-celled protistan, sliced lengthwise for this micrograph. It also has a long flagellum, which could not fit in the image area at this magnification.

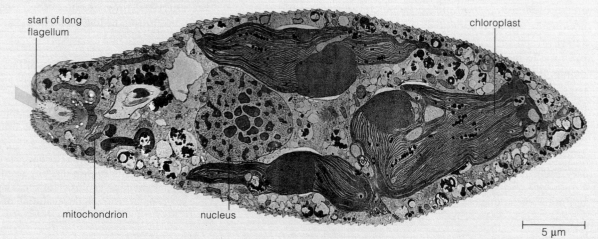

start of long flagellum

chloroplast

mitochondrion

nucleus

5 µm

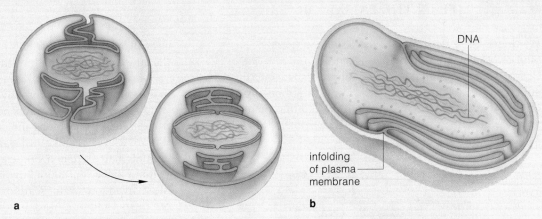

Figure 21.11 (**a**) One idea concerning the origin of endoplasmic reticulum and the nuclear envelope. In the prokaryotic ancestors of eukaryotic cells, infoldings of the plasma membrane may have given rise to these structures. (**b**) Such infoldings are present in the cytoplasm of many kinds of existing bacteria, including *Nitrobacter*, sketched here in cutaway view.

DNA

infolding of plasma membrane

a

b

detail by Lynn Margulis. *Endo-* means within; *symbiosis* means living together. In cases of **endosymbiosis**, one species (a guest) lives permanently inside another species (the host), and the interaction benefits both.

According to one theory, eukaryotic cells evolved by way of endosymbiosis long after the noncyclic pathway of photosynthesis had emerged and oxygen had accumulated to significant levels in the atmosphere. In some bacterial groups, certain electron transport systems in the plasma membrane had already expanded and now included "extra" cytochromes. Those cytochromes were able to donate electrons to oxygen. The bacteria could extract energy from organic compounds by aerobic respiration. By 1.2 billion years ago, and possibly much earlier, the forerunners of eukaryotes were engulfing aerobic bacteria. Maybe they were like existing soft-bodied, amoebalike cells that weakly tolerate free oxygen. If so, they would have trapped food by sending out cytoplasmic extensions from the cell body. Endocytic vesicles could form around food and deliver it to the cytoplasm for digestion.

A key point of the theory is that some aerobic bacteria resisted digestion. They actually thrived in the protected, nutrient-rich environment. In time, they were releasing extra ATP, which the host cells came to depend upon for growth, increased activity, and the assembly of hard body parts and other new structures. The guests were no longer duplicating metabolic functions that the hosts performed for them. The anaerobic and aerobic cells were incapable of independent life. The guests had become mitochondria, supreme suppliers of ATP.

EVIDENCE OF ENDOSYMBIOSIS Strong evidence favors Margulis's theory. There are plenty of examples of nature continuing to tinker with endosymbionts, including the cell in Figure 21.12. Its mitochondria are like bacteria in size and structure. The inner mitochondrial membrane is like a bacterial plasma membrane. Each mitochondrion replicates its own DNA and divides independently of the host cell's division. A few genetic code words in its DNA and mRNA have unique meanings. They are translated into a few proteins required for specialized mitochondrial tasks. Thus, compared to the genetic code of cells, the "mitochondrial code" has a few distinct differences.

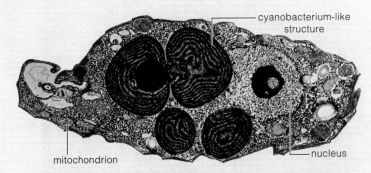

cyanobacterium-like structure

mitochondrion

nucleus

Figure 21.12 *Cyanophora paradoxa*, a protistan. Its mitochondria resemble bacteria. Its photosynthetic structures look like spherical cyanobacteria (which are photosynthetic) without the cell wall.

Also consider chloroplasts, which may be the stripped-down descendants of oxygen-evolving, photosynthetic bacteria. Perhaps predatory aerobic bacteria engulfed such photosynthetic cells, which escaped digestion, absorbed nutrients from their host's cytoplasm, and continued to function. By providing their respiring host with oxygen, their endosymbiotic existence would have been favored.

In their metabolism and overall nucleic acid sequence, chloroplasts resemble some eubacteria. Their DNA is self-replicating, and they divide independently of the cell's division. Chloroplasts vary in shape and in their array of light-absorbing pigments, just as various photosynthetic eubacteria do. They may have originated a number of times, in a number of different lineages. Adding to the intrigue are species of ciliated protistans, even marine slugs, that "enslave" chloroplasts! The slugs eat algae but retain the algal chloroplasts in their gut. The chloroplasts draw nutrients from the host tissues, and they continue to photosynthesize and release oxygen for weeks.

However they arose, new kinds of cells did appear on the evolutionary stage. They had become equipped with a nucleus, cytomembranes, and mitochondria, chloroplasts, or both. They were eukaryotic cells, the first **protistans**. With their efficient metabolic strategies, early protistans underwent rapid divergences and adaptive radiations. In no time at all, evolutionarily speaking, some of their descendants gave rise to the great kingdoms of plants, fungi, and animals, as sketched out earlier in Figure 21.7.

We divide the **Paleozoic** into the Cambrian, Ordovician, Silurian, Devonian, Carboniferous, and Permian periods. Before the dawn of that era, gradual rifting had split the supercontinent Laurentia apart. From Cambrian times on into the Silurian, its fragments straddled the equator, and warm, shallow seas lapped their margins:

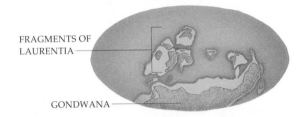

FRAGMENTS OF LAURENTIA

GONDWANA

The global conditions restricted pronounced seasonal changes in the prevailing winds, ocean currents, and the churning of nutrients upward from deep waters. As a result, supplies of nutrients along shorelines at or near the equator were stable but limited.

Most of the major animal phyla had evolved earlier, in Precambrian seas. Possibly some of their ancestors were among the **Ediacarans**, odd organisms shaped like fronds, disks, and blobs that nearly defy classification (Figure 21.13a,b). Like Ediacarans, the early Cambrian animals had flattened bodies, with a good surface-to-volume ratio for taking up nutrients (Figure 21.13c). Most lived on or in seafloor sediments, where dead organisms and organic debris settled. They ranged from sponges to simple vertebrates, and they were diverse.

How could so much diversity arise? Possibly genes governing early growth and development were far less intertwined than they are today, so there may not have been as much selection against mutant alleles and novel traits. Also, warm waters near vast new coasts afforded vacant adaptive zones, with splendid opportunities for new ways to secure food.

Entombed in sedimentary beds from the Cambrian are fossilized organisms that have punctures, missing chunks, and healed wounds. These are not artifacts of fossilization; the animals were injured while they were alive. Diverse predators and prey with armorlike shells, spines, mouths, and novel feeding structures evolved in short order. Things were starting to get lively!

Later in the Cambrian, temperatures in the shallow seas changed drastically. Trilobites (Figure 21.13d), one of the most common animals, almost vanished. In the Ordovician, the supercontinent Gondwana had been drifting south, and parts became submerged in shallow seas. Vast new marine environments opened up and promoted adaptive radiations. Many new reef organisms appeared. Among them were swift, shelled predators called nautiloids. Their surviving descendants include the chambered nautiluses, as shown in Section 49.12.

Later on, Gondwana straddled the South Pole, and vast glaciers formed across its surface. When enormous volumes of water were locked up as ice, shallow seas throughout the world were drained. This was the first ice age that we know about. It may have triggered the first global mass extinction. At the Ordovician-Silurian boundary, reef life everywhere collapsed.

Gondwana drifted north during the Silurian and on into the Devonian. This was a pivotal time of evolution. Reef communities recovered. Armor-plated fishes with massive jaws diversified. In the wet lowlands, small

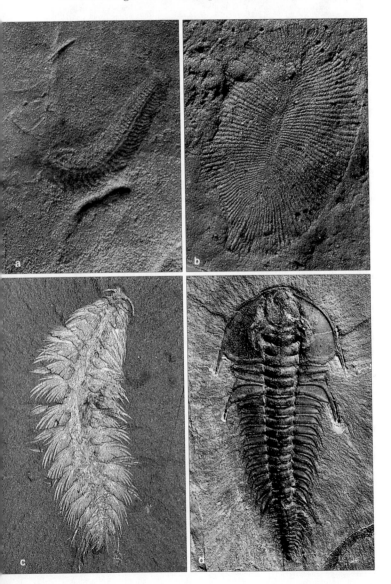

Figure 21.13 Representatives from the Precambrian and Cambrian seas. Two Ediacarans, about 600 million years old: (**a**) *Spriggina* and (**b**) *Dickensonia*. The oldest known Ediacarans lived 610 million years ago and the most recent in Cambrian times, 510 million years ago. (**c**) From the Burgess Shale of British Columbia, a fossilized marine worm. (**d**) A beautifully preserved fossil of one of the earliest trilobites.

a

b

c

d

Figure 21.14 (**a**) Life in the Silurian seas. Some of these shelled animals (nautiloids) were twelve feet long. (**b**) A Silurian swamp, dominated by the forerunners of modern ferns and club mosses. (**c**) Fossils of a Devonian plant (*Psilophyton*), possibly one of the earliest ancestors of seed-bearing plants. (**d**) Reptiles (*Dimetrodon*) of a largely hotter and drier time, the Permian. Fossils of these carnivores were found in Texas. Giant club mosses and horsetails had declined, and conifers, cycads, and other gymnosperms replaced them.

stalked plants appeared (Figure 21.14*b,c*). So did fungi and many invertebrates, such as segmented worms. In Devonian times, the fishes that would become ancestors to amphibians invaded the land. Those fishes had lobed fins, the forerunners of legs and other limbs. And they had simple lungs. As described in Section 27.6, lobed fins and lungs were key innovations—they would prove most advantageous for life out of water, in dry land habitats.

Then, as they say, something bad happened. At the Devonian-Carboniferous boundary, sea levels swung catastrophically, for unknown reasons, and triggered another mass extinction. Afterward, plants and insects embarked on adaptive radiations on land.

Throughout the Carboniferous, land masses were gradually submerged and drained many times. Organic debris piled up, then it was compacted and converted to coal, in the manner described in Section 25.4.

Insects, amphibians, and early reptiles flourished in vast swamp forests of Permian times (Figure 21.14*d*). Ancestors of the seed-producing plants called cycads, ginkgos, and conifers dominated those forests.

As the Permian drew to a close, the greatest of all mass extinctions took place. Nearly all known species perished. Pangea was forming at that time; all land masses were colliding together. The vast supercontinent

eventually extended from pole to pole, and a single world ocean lapped its margins:

PANGEA

TETHYS SEA

As you will see, the new distribution of oceans, land masses, and land elevations had catastrophic effects on global climates—and on the course of life's evolution.

Early in the Paleozoic era, organisms of all six kingdoms were flourishing in the seas. By the end of the era, many lineages had successfully invaded the land, including the wet lowlands of the supercontinent Pangea.

LIFE IN THE MESOZOIC ERA

Speciation on a Grand Scale

We divide the **Mesozoic** into the Triassic, Jurassic, and Cretaceous periods. It lasted about 175 million years. Early on in the Cretaceous, the supercontinent Pangea started to break up. Its huge fragments slowly drifted apart, and we can assume that the resulting geographic isolation favored divergences and speciation:

This was an era of spectacular expansion in the range of global diversity. In the seas, invertebrates and fishes underwent adaptive radiations. On land, conifers and other seed-bearing **gymnosperms** as well as insects and reptiles became the most visibly dominant lineages. Flowering plants, or **angiosperms**, originated before the end of the era. Within a mere 30 to 40 million years, they would displace the conifers and related plants in nearly all environments (Figure 21.15 and Section 25.7).

Rise of the Ruling Reptiles

Early in the Triassic, the first **dinosaurs** evolved from a reptilian lineage. They were not much larger than a turkey. Possibly most species had high metabolic rates, and maybe they were warm-blooded. Many sprinted about on two legs. The dinosaurs weren't the dominant land animals then. Center stage belonged to *Lystrosaurus* and other plant-eating, mammal-like reptiles that were too large to be bothered by most predators.

In time, adaptive zones did open up for dinosaurs, possibly when an asteroid struck the Earth. In central Quebec, a crater about the size of Rhode Island shows how huge such an impact could have been. The blast wave and global firestorm, earthquakes, and lava flows would have been stupendous. Most of the animals that survived this time of mass extinction (and later ones) were smaller, equipped with higher metabolic rates, and less vulnerable than others to drastic changes in outside temperatures. Descendants of the surviving dinosaurs became the ruling reptiles; they endured for 140 million years. Some species reached monstrous proportions. Among them were the ultrasaurs, fifteen meters tall.

Many dinosaurs perished in another mass extinction at the end of the Jurassic, then in a pulse of extinctions in the Cretaceous. Perhaps plumes of molten material ruptured the crust and triggered changes in the global

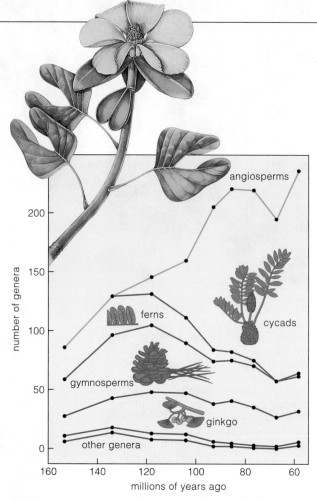

Figure 21.15 Range of diversity among vascular plants during the Jurassic and Cretaceous. Conifers and other gymnosperms were dominant before then. They were already declining when flowering plants began a spectacular radiation that continued into the Cenozoic. Also shown is a floral shoot of *Archaeanthus linnenbergeri*, a flowering plant of Cretaceous times. In many of its traits, this now-extinct species resembled living magnolias.

climate. Or perhaps a swarm of asteroids inflicted the blows. Whatever the causes, conditions changed for the dinosaurs and not all of them lived through it. Yet some lineages recovered and new ones evolved. Duckbilled dinosaurs appeared in forests and swamps. Tanklike *Triceratops* and other plant eaters flourished in open regions. They were prey for the fearsomely toothed, agile, and swift *Velociraptor* of motion picture fame.

About 120 million years ago, global temperatures shot up by 25 degrees. By one theory, plumes spread out beneath the crust and "greased" the crustal plates into moving twice as fast. A superplume or a rash of them broke through the crust. The crust in what is now the South Atlantic opened like a zipper. Basalt and lava poured from the fissures; volcanoes spewed nutrient-rich ashes. Simultaneously, the plumes released great amounts of carbon dioxide, one of the "greenhouse" gases that absorb some of the heat radiating from the Earth before it escapes into space. The nutrient-enriched

Figure 21.16 If dinosaurs of this sort had not disappeared at the end of the Cretaceous, would the then-tiny mammals ever have ventured out from under the shrubbery? Would *you* even be here today?

planet warmed up—and it stayed warm for 20 million years. On land and in the shallow seas, photosynthetic organisms flourished. Their remains were slowly buried and converted into the world's oil reserves.

About 65 million years ago, the last dinosaurs and many marine organisms vanished in a mass extinction (Figure 21.16). As described in the next section, their disappearance apparently coincided when an asteroid

the size of Mount Everest slammed into the Earth. Over time, the impact site drifted into a position that we now call the northern Yucatán peninsula.

The Mesozoic was a time of major adaptive radiations and of a mass extinction in which the last dinosaurs and many marine organisms disappeared.

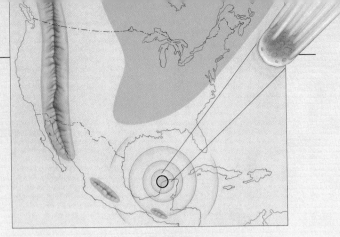

HORRENDOUS END TO DOMINANCE

It has only been about 50,000 years since the first fully modern humans walked the Earth. Since then, how many times have people puffed up with self-importance and set out to conquer neighbors, the land, and the seas? Think about it—then think about the dinosaurs. Were they good at reigning supreme? No question about it; their lineage dominated the land for 140 *million* years. In the end, did it matter? Not a bit. Sixty-five million years ago, at the Cretaceous-Tertiary (K-T) boundary, nearly all remaining members of their most excellent lineage perished. Why? Bad luck.

A thin layer of iridium-rich rock distributed around the world dates precisely to the K-T boundary. Iridium is rare on the Earth's surface but common in asteroids. **Asteroids** are rocky, metallic bodies, 1,000 kilometers to a few meters in diameter, that are hurtling through space. When the planets were forming, their gravitational pull swept most asteroids from the sky. At least 6,000 still orbit the sun in a belt between Mars and Jupiter (Figure 21.17). The orbits of dozens of others take them across Earth's orbit, like Russian roulette on a cosmic scale.

By analyzing iridium levels in soils, gravity maps, and other evidence, Walter Alvarez and Luis Alvarez hypothesized that an asteroid impact caused the K-T mass extinction. Later, researchers identified the impact site. Massive movements in the crust transported the site to what is now the northern Yucatán peninsula of Mexico (Figure 21.18). The impact crater is 9.6 kilometers deep and 300 kilometers across—wider than Connecticut. This crater as well as other evidence strongly supports what has become known as the **K-T asteroid impact theory**.

To make a crater that large, the asteroid had to hit the Earth at 160,000 kilometers (100,000 miles) per hour. At least 200,000 cubic kilometers of debris and dense gases were blasted skyward. The crust itself heaved violently. Monstrous waves, 120 meters high, raced across the ocean, obliterating life on islands and then slamming

Figure 21.18 Artist's interpretation of what might have happened in the last few minutes of the Cretaceous.

into the coasts of continents. Researchers long thought that atmospheric debris blocked out sunlight for months. If so, plants and other producers that sustained the web of life would have withered and died; animals and other consumers would have starved to death. The theory has problems. By some calculations, the volume of debris that was blasted aloft would not have been enough to have such significant consequences.

Then, in 1994, the comet Shoemaker-Levy 9 slammed into Jupiter. Particles blasted into the Jovian atmosphere caused intense heating over an area larger than the Earth. This outcome favors a **global broiling theory**, proposed first by H. J. Melosh and his colleagues. Energy released at the K-T impact site was equivalent to detonating 100 million nuclear bombs. Trillions of tons of vaporized debris rose in a colossal fireball, then rapidly condensed into particles the size of sand grains. Seconds later, a cooler fireball of steam, carbon dioxide, and unmelted rock formed. When the debris fell to the Earth, it raised the atmospheric temperature by thousands of degrees. The sky above the whole planet must have glowed with heat ten times more intense than the noonday sun above Death Valley in summer. In one horrific hour, nearly all plants erupted in flames and every animal out in the open was broiled alive.

Things haven't settled down much since the demise of the dinosaurs. For example, about 2.3 million years ago, a huge chunk of matter from space apparently hit the Pacific Ocean. At about the same time, vast ice sheets started forming abruptly in the Northern Hemisphere. Long-term shifts in climate may have been ushering in this most recent ice age, but the global impact would have accelerated it. Water vaporized by the impact could have contributed to a global cloud cover that limited the amount of sunlight reaching the surface. The ancestors of humans were around when this happened. The extreme climate shift surely tested their adaptability.

Almost certainly, severe environmental tests await all existing lineages. When we become too smitten with our importance in the world of life, we would do well to step back, from time to time, and reflect on what is going on above and beneath the Earth's surface. Asteroids and superplumes do have a way of leveling the playing field, as they did for the tiny mammals and gigantic dinosaurs.

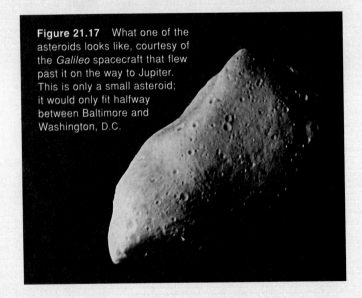

Figure 21.17 What one of the asteroids looks like, courtesy of the *Galileo* spacecraft that flew past it on the way to Jupiter. This is only a small asteroid; it would only fit halfway between Baltimore and Washington, D.C.

LIFE IN THE CENOZOIC ERA

The breakup of Pangea triggered events that continued into the present era, the **Cenozoic**. At the dawn of the Cenozoic, land masses were on collision courses:

Coastlines fractured. The Cascades, Andes, Himalayas, and Alps rose through volcanic activity, uplifting, and other events at crustal rifts and plate boundaries. These geologic changes brought about major shifts in climate that influenced the further evolution of life.

During the Paleocene epoch, climates were warmer and wetter. Tropical and subtropical forests extended farther north and south than they do today. Woodlands and forests spread even into polar regions. Although their key traits developed before the dinosaurs left the scene, mammals now began their major radiation.

The global climate warmed even more in the Eocene epoch, and subtropical forests extended north into the polar regions. A variety of mammals, including primates, bats, rhinos, hippos, elephants, horses, and assorted carnivores, emerged. By the late Eocene, climates became cooler and drier, and seasonal changes became pronounced. Woodlands and semiarid grasslands now dominated vast tracts of land. Patterns of vegetational growth changed, and this drove many mammals to extinction.

From the Oligocene through the Pliocene, an abundance of grazing and browsing animals thrived in the woodlands and grasslands. Among them were camels and the "giraffe rhinoceros," along with some fearsome carnivores that stalked them (Figure 21.19).

Today the distribution of land masses favors species diversity. The richest ecosystems are the tropical forests of South America, Madagascar, and Southeast Asia, as well as the marine ecosystems of archipelagos in the tropical Pacific. Yet we are in the midst of what may be one of the greatest mass extinctions. About 50,000 years ago, nomadic humans followed migrating herds of wild animals around the Northern Hemisphere. Within a few

Figure 21.19 Representatives from Cenozoic times. (**a**) In the early Paleocene, in what is now Wyoming, diverse mammals lived in dense forests of sequoia trees, laurel, and other plants. On the ground is the raccoonlike *Chriacus*, facing a tree-climbing rodent (*Ptilodus*). Higher in the tree is *Peradectes*, a marsupial. (**b**) From the late Eocene to the early Miocene, *Indricotherium* browsed on woodland trees. This "giraffe rhinoceros," the largest land mammal known, weighed 15 tons and was 5.5 meters (18 feet) at the shoulder. (**c**) From the Pleistocene, a small horse and the saber-tooth cat (*Smilodon*). Both mammals are extinct. Fossils of them have been found in a pitch pool at Rancho LaBrea, California.

thousand years, major groups of mammals were extinct. The pace of extinction has since accelerated as humans hunt for food, fur, feathers, or fun, and as they destroy habitats to clear land for farm animals or crops. Chapter 50 focuses on the global repercussions.

Major geologic changes during the Cenozoic triggered shifts in climate. The great adaptive radiation of mammals began, first in tropical forests, then in woodlands and grasslands.

SUMMARY OF EARTH AND LIFE HISTORY

We conclude this overview chapter with an illustration that correlates milestones in the evolution of life with the evolution of the Earth. As you study Figure 21.20, keep in mind that it is only a generalized summary. For example, it charts five of the greatest mass extinctions, but there were many others in between. The diagram of the range of global diversity conveys the shrinking and expansion of species over time. However, the range represents *all* of the major groups on land and in the seas combined. Remember, each major group has its own history of persisting lineages, of radiations, and of extinctions.

And now, with these qualifications in mind, we are ready to turn to the next chapters in this unit. They will provide you with richer detail of the history and the current range of diversity for all six kingdoms of life.

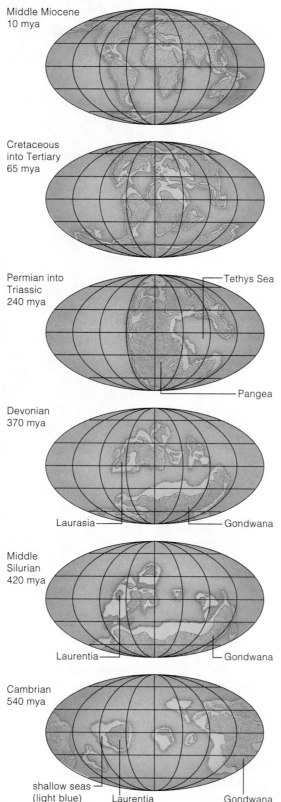

Figure 21.20 Summary of major events in the evolution of the Earth and of life. As you read through the chapters to follow, you may wish to return to this illustration now and then to remind yourself of how the details fit into the greater evolutionary picture.

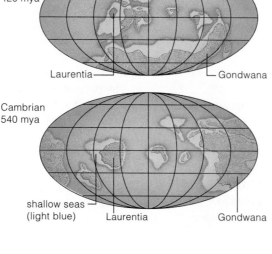

	Period	Epoch
CENOZOIC ERA	Quaternary	Recent
		Pleistocene
	Tertiary	Pliocene
		Miocene
		Oligocene
		Eocene
		Paleocene
MESOZOIC ERA	Cretaceous	Late
		Early
	Jurassic	
	Triassic	
PALEOZOIC ERA	Permian	
	Carboniferous	
	Devonian	
	Silurian	
	Ordovician	
	Cambrian	
PROTEROZOIC EON		
ARCHEAN EON AND EARLIER		

Millions of Years Ago (mya)	Range of Global Diversity (marine and terrestrial)	Times of Major Geological and Biological Events

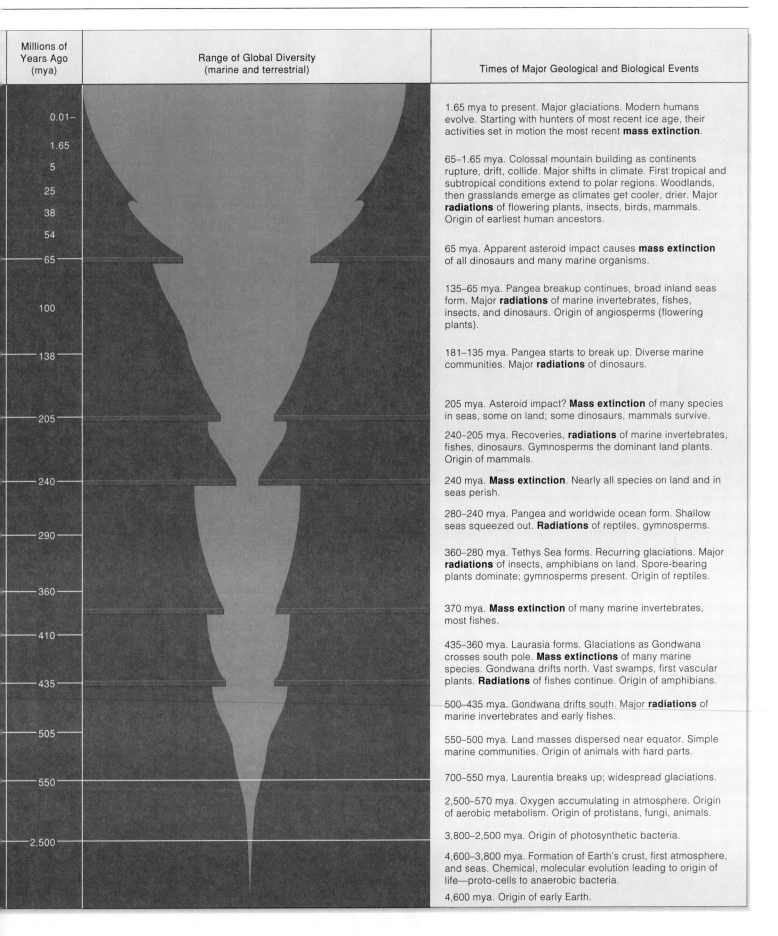

0.01–
1.65
5
25
38
54
65
100
138
205
240
290
360
410
435
505
550
2,500

1.65 mya to present. Major glaciations. Modern humans evolve. Starting with hunters of most recent ice age, their activities set in motion the most recent **mass extinction**.

65–1.65 mya. Colossal mountain building as continents rupture, drift, collide. Major shifts in climate. First tropical and subtropical conditions extend to polar regions. Woodlands, then grasslands emerge as climates get cooler, drier. Major **radiations** of flowering plants, insects, birds, mammals. Origin of earliest human ancestors.

65 mya. Apparent asteroid impact causes **mass extinction** of all dinosaurs and many marine organisms.

135–65 mya. Pangea breakup continues, broad inland seas form. Major **radiations** of marine invertebrates, fishes, insects, and dinosaurs. Origin of angiosperms (flowering plants).

181–135 mya. Pangea starts to break up. Diverse marine communities. Major **radiations** of dinosaurs.

205 mya. Asteroid impact? **Mass extinction** of many species in seas, some on land; some dinosaurs, mammals survive.

240–205 mya. Recoveries, **radiations** of marine invertebrates, fishes, dinosaurs. Gymnosperms the dominant land plants. Origin of mammals.

240 mya. **Mass extinction**. Nearly all species on land and in seas perish.

280–240 mya. Pangea and worldwide ocean form. Shallow seas squeezed out. **Radiations** of reptiles, gymnosperms.

360–280 mya. Tethys Sea forms. Recurring glaciations. Major **radiations** of insects, amphibians on land. Spore-bearing plants dominate; gymnosperms present. Origin of reptiles.

370 mya. **Mass extinction** of many marine invertebrates, most fishes.

435–360 mya. Laurasia forms. Glaciations as Gondwana crosses south pole. **Mass extinctions** of many marine species. Gondwana drifts north. Vast swamps, first vascular plants. **Radiations** of fishes continue. Origin of amphibians.

500–435 mya. Gondwana drifts south. Major **radiations** of marine invertebrates and early fishes.

550–500 mya. Land masses dispersed near equator. Simple marine communities. Origin of animals with hard parts.

700–550 mya. Laurentia breaks up; widespread glaciations.

2,500–570 mya. Oxygen accumulating in atmosphere. Origin of aerobic metabolism. Origin of protistans, fungi, animals.

3,800–2,500 mya. Origin of photosynthetic bacteria.

4,600–3,800 mya. Formation of Earth's crust, first atmosphere, and seas. Chemical, molecular evolution leading to origin of life—proto-cells to anaerobic bacteria.

4,600 mya. Origin of early Earth.

SUMMARY

1. The story of life begins with the "big bang," a model of the origin of the universe.

   a. By this model, all matter and all of space were once compressed in a fleeting state of enormous heat and density. Time began with the near-instantaneous distribution of all matter and energy throughout the known universe, which has been expanding ever since.

   b. Nearly all helium and other light elements, the most abundant elements of the universe, formed right after the big bang. Heavier elements originated during the formation, evolution, and death of stars.

   c. Every element of the solar system, the Earth, and life itself is a product of the physical and chemical evolution of the universe and its stars.

2. Four billion years ago, the Earth had a high-density core, a mantle of intermediate density, and a thin, extremely unstable crust of low-density rocks. Probably gaseous hydrogen, nitrogen, and carbon monoxide, as well as carbon dioxide, made up the first atmosphere. Free oxygen and water could not have accumulated at the surface under the prevailing conditions.

3. Water accumulated after the Earth's crust cooled. Runoff from rains carried dissolved mineral salts and other compounds to crustal depressions, where early seas formed. Life could not have originated without this salty liquid water.

4. Many diverse studies and experiments have yielded indirect evidence that life originated under conditions that presumably existed on the early Earth.

   a. Comparative investigations of the composition of cosmic clouds, rocks from other planets, and rocks from the Earth's moon suggests that precursors of complex molecules associated with life were available.

   b. In laboratory tests that simulated the primordial conditions, including the absence of free oxygen, the precursors spontaneously assembled into sugars (such as glucose), amino acids, and other organic compounds.

   c. Known chemical principles as well as advanced computer simulations indicate that metabolic pathways could have evolved through chemical competition for the limited supplies of organic molecules (which had accumulated by natural geologic processes in the seas).

   d. Self-replicating systems of RNA, enzymes, and coenzymes have been synthesized in the laboratory. How DNA entered the picture is not yet understood.

   e. In laboratory simulations of conditions thought to have existed on the early Earth, lipids as well as lipid-protein membranes having some of the properties of cell membranes have formed spontaneously.

5. Life originated by 3.8 billion years ago. Since then, it has been influenced by major changes in the Earth's crust, atmosphere, and oceans. Forces of change have included plate movements, asteroid impacts, and the activities of organisms (including oxygen-producing photosynthesizers and, currently, the human species).

6. Abrupt discontinuities in the fossil record mark the times of global mass extinctions. We use them as boundary markers for five great intervals in a geologic time scale. Radiometric dating has allowed us to assign absolute dates to this time scale:

   a. Archean: 3.9 billion to 2.5 billion years ago
   b. Proterozoic: 2.5 billion to 550 million years ago
   c. Paleozoic: 550 to 240 million years ago
   d. Mesozoic: 240 to 65 million years ago
   e. Cenozoic: 65 million years ago to the present

7. The first living cells were prokaryotic (bacteria). Not long after they appeared, the first divergence began that led to eubacteria, and to a common prokaryotic ancestor of archaebacteria and eukaryotes. Some eubacteria used a cyclic pathway of photosynthesis.

8. During the Proterozoic, the noncyclic pathway of photosynthesis evolved in some lineages of eubacteria. Oxygen, a by-product of the pathway, gradually started to accumulate in the atmosphere.

   a. Eventually, the atmospheric concentration of free oxygen prevented the further spontaneous formation of organic molecules. From that time on, the spontaneous origin of life was no longer possible on Earth.

   b. The abundance of atmospheric oxygen was a selective pressure that brought about the evolution of aerobic respiration. Aerobic respiration was a key step toward the origin of the first eukaryotic cells.

   c. Mitochondria and chloroplasts, two important eukaryotic organelles, probably evolved as an outcome of endosymbiosis between certain aerobic bacteria and the anaerobic bacterial forerunners of eukaryotes.

   d. The oxygen-rich atmosphere promoted formation of a layer of ozone (O_3). In time, that shield against destructive ultraviolet radiation allowed some lineages to move out of the seas, into low wetlands.

9. Early in the Paleozoic, diverse organisms of all six lineages had become established in the seas. By its end, the invasion of land was under way. From that time on, there have been pulses of mass extinctions and adaptive radiations. Asteroid impacts and other catastrophes triggered many of these events. So did plate movements that changed the distribution of oceans and land as well as the prevailing global and regional climates.

Review Questions

1. Compare the presumed chemical and physical conditions that are thought to have prevailed on the Earth 4 billion years ago with conditions that exist today. *21.1*

2. Describe examples of the kinds of experimental evidence for the spontaneous origin of (1) large organic molecules, (2) the self-assembly of proteins, and (3) the formation of organic membranes and spheres, under laboratory conditions similar to those of the early Earth. *21.1, 21.2*

3. Summarize the key points of the theory of endosymbiotic origins for mitochondria and chloroplasts. Cite evidence that favors this theory. *21.4*

4. Describe the prevailing conditions that probably favored the Cambrian "explosion" of diversity among marine animals, as evidenced by the fossil record. *21.5*

5. During which geologic time spans did plants, fungi, and insects invade the land? What kind of vertebrates first invaded the land, and when? *21.5*

6. What were global conditions like when gymnosperms and dinosaurs originated? *21.6*

7. Briefly explain how an asteroid impact and "global broiling" may have caused the mass extinctions at the K-T boundary. *21.7*

8. Would you expect the Paleozoic, Mesozoic, or Cenozoic to be called "the age of mammals"? As part of your answer, explain the differences between global conditions in each era. *21.8*

Self-Quiz (*Answers in Appendix IV*)

1. Life originated by _____ .
   a. 4.6 billion years ago   c. 3.8 billion years ago
   b. 2.8 million years ago   d. 3.8 million years ago

2. Through study of the geologic record, we know that the evolution of life has been profoundly influenced by _____ .
   a. tectonic movements of the Earth's crust
   b. bombardment of the Earth by celestial objects
   c. profound shifts in land masses, shorelines, and oceans
   d. physical and chemical evolution of the Earth
   e. all of the above

3. _____ was the first to obtain indirect evidence that organic molecules could have been formed on the early Earth.
   a. Darwin   c. Fox
   b. Miller   d. Margulis

4. An abundance of _____ was conspicuously absent from the Earth's atmosphere 4 billion years ago.
   a. hydrogen   c. carbon monoxide
   b. nitrogen   d. free oxygen

5. Which of the following statements is false?
   a. The first living cells were prokaryotes.
   b. The cyclic pathway of photosynthesis first appeared in some eubacterial species.
   c. Oxygen began accumulating in the atmosphere after the noncyclic pathway of photosynthesis evolved.
   d. In the Proterozoic, increasing levels of atmospheric oxygen enhanced the spontaneous formation of organic molecules.
   e. All are correct.

6. The first eukaryotic cells emerged during the _____ .
   a. Paleozoic   c. Archean   e. Cenozoic
   b. Mesozoic   d. Proterozoic

7. Match the geologic time interval with the events listed.
   ____ Archean   a. major radiations of dinosaurs, origin
   ____ Proterozoic   of flowering plants and mammals
   ____ Paleozoic   b. chemical evolution, origin of life
   ____ Mesozoic   c. major radiations of flowering plants,
   ____ Cenozoic   insects, birds, mammals; emergence of
                       human forms
                       d. oxygen present; origin of aerobic
                       metabolism, protistans, fungi, animals
                       e. rise of early plants, origin of
                       amphibians, origin of reptiles

Critical Thinking

1. Explain, in terms of hydrophilic and hydrophobic interactions, how proto-cells might have formed in water from aggregations of lipids, proteins, and nucleic acids.

2. The Atlantic Ocean is gradually widening, and the Pacific Ocean and Indian Ocean are closing. Many millions of years from now, the continents will collide and form a second Pangea. Write a short essay on what environmental conditions might be like on that future supercontinent and on what types of species might survive there.

3. By some estimates, there is a chance that about 10^{20} planets have formed in the universe that are capable of sustaining life—but only one chance at intelligent life per planet. Given your knowledge of molecular biology and evolutionary processes, speculate on why the odds are so low.

Selected Key Terms

angiosperm *21.6*	global broiling theory *21.7*
archaebacterium *21.3*	gymnosperm *21.6*
Archean *21.3*	K-T asteroid impact theory *21.7*
asteroid *21.7*	mantle, of Earth *21.1*
big bang *CI*	Mesozoic *21.6*
Cenozoic *21.8*	Paleozoic *21.5*
crust, of Earth *21.1*	prokaryotic cell *21.3*
dinosaur *21.6*	Proterozoic *21.3*
Ediacaran *21.5*	protistan *21.4*
endosymbiosis (theory of) *21.4*	proto-cell *21.2*
eubacterium *21.3*	RNA world *21.2*
eukaryotic cell *21.3*	stromatolite *21.3*

Readings

Bambach, R., C. Scotese, and A. Ziegler. 1980. "Before Pangea: The Geographies of the Paleozoic World." *American Scientist* 68(1): 26–38.

de Duve, C. September–October 1995. "The Beginnings of Life on Earth." *American Scientist* 83: 428–437.

———. April 1996. "The Birth of Complex Cells." *Scientific American* 274(4): 50–57.

Dobb, E. February 1992. "Hot Times in the Cretaceous." *Discover* 13: 11–13.

Dott, R., Jr., and R. Batten. 1988. *Evolution of the Earth.* Fourth edition. New York: McGraw-Hill.

Hartman, W., and Chris Impey. 1994. *Astronomy: The Cosmic Journey.* Fifth edition. Belmont, California: Wadsworth.

Horgan, J. February 1991. "Trends in Evolution: In the Beginning . . . ." *Scientific American* 264(2): 116–125.

Wright, K. March 1997. "When Life Was Odd." *Discover* 18(3): 52–61. Update on the bizarre Ediacarans.

Web Site See *http://www.wadsworth.com/biology* for practice quiz questions, hypercontents, BioUpdates, and critical thinking. The Wadsworth Biology Resource Center provides a wealth of information fully organized and integrated by chapter.

22 BACTERIA AND VIRUSES

The Unseen Multitudes

Did a friend ever mention that you are nearly 1/1,000 of a mile tall? Probably not. What would be the point of measuring people in units as big as miles? Even so, we think this way, in reverse, whenever we measure microorganisms. For the most part, **microorganisms** are single-celled organisms that are too small to be seen without the aid of a microscope.

The bacterial cells shown in Figure 22.1 are a case in point. To measure them, you would have to divide one meter into a thousand units, or millimeters. Next, you would have to divide one of the millimeters into a thousand smaller units, or micrometers. To give you a sense of how small that is, a single millimeter would be about as small as the dot of this "i." And a *thousand* bacteria would fit side by side on top of the dot!

To be sure, viruses are smaller still, as you might deduce after looking at Figure 22.2. We measure them in nanometers (billionths of a meter). But viruses are not alive. We consider them in this chapter because one kind or another infects just about every organism.

Bacteria generally are the smallest microorganisms, but they vastly outnumber the individuals in all other kingdoms combined. Their reproductive potential is staggering. Under ideal conditions, some divide about every twenty minutes. If that rate of reproduction were to hold constant, a single bacterium would have nearly a billion descendants in ten hours!

So why don't the unseen multitudes take over the world? Sooner or later, their burgeoning populations use up available nutrients and pollute the surroundings with their own metabolic wastes. In other words, they alter the very conditions that initially favored their reproduction. Besides this, certain kinds of viruses attack just about every species of bacterium and help keep their population sizes in check. So do seasonal changes in living conditions.

Of course, you probably do not find much comfort in this when you serve as host for one of the pathogenic types. **Pathogens** are infectious, disease-causing agents that invade target organisms and multiply inside or on

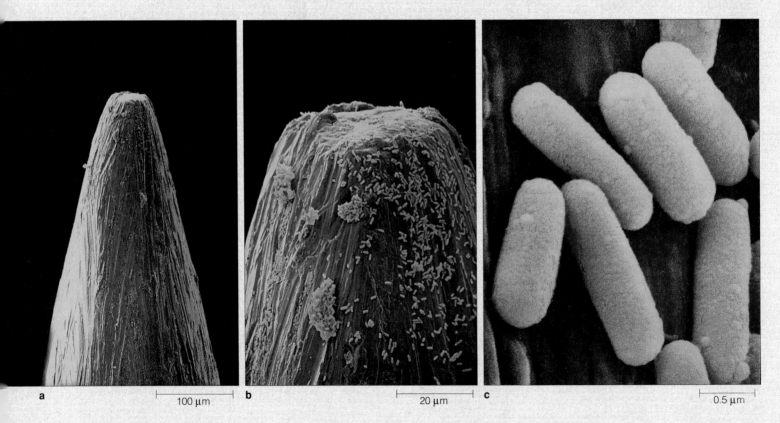

a   ⊢——— 100 μm ———⊣   b   ⊢——— 20 μm ———⊣   c   ⊢——— 0.5 μm ———⊣

Figure 22.1 (**a–c**) How small are bacteria? Shown here, *Bacillus* cells peppering the tip of a pin. The cells in (**c**) are magnified more than 16,600 times.

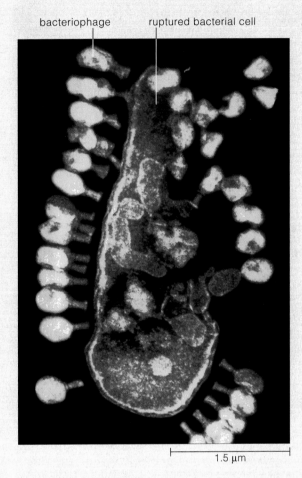

bacteriophage   ruptured bacterial cell

1.5 µm

Figure 22.2 How small are the viruses? Shown here, bacteriophage particles, each about 225 nanometers tall, that infected a bacterial cell. The cell has ruptured open.

them. Disease follows when the metabolic activities of pathogenic cells damage the tissues and interfere with the body's normal functioning.

Certain pathogens can indeed make you suffer, but they should not give every microorganism a bad name. For example, think back on the uncountable numbers of photosynthetic bacterial cells in the seas (Section 7.8). Collectively, they help provide food and oxygen for entire communities and have major roles in the global cycling of carbon. Or think about the kinds of bacterial species that feed on organic debris. Together with other decomposers, they help cycle nutrients that sustain entire communities.

From the human perspective, microorganisms are good or bad, even dangerous. Basically, however, they are simply surviving and reproducing like the rest of us, in ways that are the topics of this chapter.

KEY CONCEPTS

1. In structural terms, we find the simplest forms of life among the bacteria. Most of these are microscopically small. Smaller still are the viruses.

2. Bacteria alone are prokaryotic cells. They do not have a profusion of internal, membrane-bound organelles in their cytoplasm, as eukaryotic cells do. However, as a group, the bacteria show great metabolic diversity, and many species show complex behavior.

3. Most bacteria reproduce by prokaryotic fission. This cell division mechanism follows DNA replication and divides the parent cell into two genetically equivalent daughter cells.

4. Bacteria were the first living organisms on Earth. Not long after they originated, they diverged into two lineages. One lineage gave rise to eubacteria, and the other to the common ancestors of archaebacteria and eukaryotic cells.

5. A virus is a noncellular infectious particle. It consists of nucleic acid (either DNA or RNA), a protein coat, and sometimes an outer envelope. Viruses cannot be replicated without pirating the metabolic machinery of a specific type of host cell.

6. Nearly all viral multiplication cycles proceed through five steps: attachment to a host cell, penetration of its plasma membrane, replication of viral DNA or RNA and synthesis of viral proteins, then assembly of new viral particles, and finally release from the infected cell.

7. Most of us tend to judge microorganisms through the prism of human interests. Yet their lineages are the most ancient, their adaptations are astoundingly diverse, and they are simply surviving and reproducing like the rest of us.

Of all organisms, bacteria are the most abundant and far-flung. Many thousands of species live in places that range from deserts to hot springs, glaciers, and seas. The hardiest have been carrying on for millions of years 2,780 meters (9,600 feet) below the surface of the Earth! Billions of bacterial cells may occupy a handful of rich soil. The ones in your gut and on your skin outnumber your body cells. (Your cells are larger, so you are only "a few percent bacterial" by weight.) Bacteria also have the longest evolutionary history. Trace any lineage back far enough and you will find bacterial ancestors. From *Escherichia coli* to amoebas, elephants, clams, and coast redwoods, all living organisms interconnect, regardless of their differences in size, numbers, and evolutionary distance. Table 22.1 and Figure 22.3 introduce features that help characterize the remarkable bacteria.

Splendid Metabolic Diversity

All organisms take in energy and carbon to meet their nutritional requirements. Compared with other species, however, the bacteria show the greatest diversity in the means of securing these resources.

Like plants, *photoautotrophic* bacteria build organic compounds by photosynthesis; they are "self-feeders." They tap sunlight for energy and use carbon dioxide as their carbon source. Their plasma membrane houses the photosynthetic machinery. Some photosynthesizers use a noncyclic pathway, with electrons and hydrogen from water molecules being used for the synthesis reactions and oxygen being released as a by-product. Others are anaerobic; they cannot use oxygen or die in its presence. Those species use a cyclic pathway in which electrons and hydrogen are stripped from inorganic compounds, such as gaseous hydrogen and hydrogen sulfide.

Chemoautotrophic bacteria also are self-feeders. Most get carbon from carbon dioxide. For energy, some types use organic compounds as a source of electrons (and hydrogen). Others (the chemolithotrophs) use inorganic substances, such as gaseous hydrogen, sulfur, nitrogen compounds, and a form of iron.

By contrast, *photoheterotrophic* bacteria are not self-feeders. They can use sunlight for photosynthesis but cannot get carbon from carbon dioxide. Instead, their carbon sources are fatty acids, complex carbohydrates, and other compounds that other organisms produce.

Chemoheterotrophic bacteria are either parasites or saprobes, not self-feeders. The parasitic types live on or in a living host and draw glucose and other nutrients from it. One way or another, the saprobic types obtain nutrients from the organic products, wastes, or remains of other organisms.

Bacterial Sizes and Shapes

So far, you have a general sense of the microscopically small sizes of bacteria. Typically, the width or length of these cells falls between 1 and 10 micrometers.

Three basic shapes are common among bacteria. A spherical shape is a **coccus** (plural, cocci; from a word that means berries). A rod shape is a **bacillus** (plural, bacilli, meaning small staffs). A cell body with one or more twists is a **spirillum** (plural, spirilla):

coccus   bacillus   spirillum

Don't let these simple categories fool you. Cocci may also be oval or flattened. Bacilli may be skinny (like straws) or tapered (like cigars). Surface extensions give some bacteria a starlike shape. Even square bacteria live in salt ponds in Egypt. Besides this, after daughter cells divide, they may stay stuck together in chains, sheets,

---

Table 22.1   Characteristics of Bacterial Cells

1. All bacterial cells are prokaryotic; they do not have a membrane-bound nucleus.

2. Bacterial cells in general have a single chromosome (a circular DNA molecule); many species also have plasmids.

3. Most bacteria have a cell wall that is composed of peptidoglycan.

4. Most bacteria reproduce by prokaryotic fission.

5. Collectively, bacteria show great diversity in their modes of metabolism.

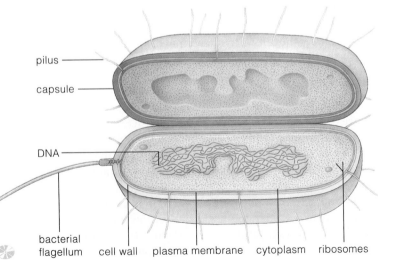

pilus

capsule

DNA

bacterial flagellum   cell wall   plasma membrane   cytoplasm   ribosomes

Figure 22.3   (*Right*) Generalized body plan of a bacterium.

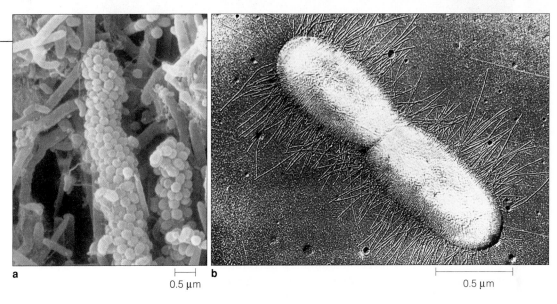

Figure 22.4 (a) Surface view of numerous bacteria that have become attached, by means of a sticky mesh, to the surface of a human tooth; and doesn't this micrograph make you want to grab a toothbrush?
(b) One example of pili, filamentous structures that project from the surface of many bacterial cells. This particular *Escherichia coli* cell is dividing in two.

a

0.5 µm

b

0.5 µm

and other aggregations, as in Figure 22.4. Some spiral species are curved, like a comma, and others are flexible or like stiff corkscrews.

Structural Features

Bacteria alone are **prokaryotic cells**, meaning they were around before the evolution of nucleated cells (*pro-*, before; and *karyon*, nucleus). Few have membrane-bound compartments of any sort for isolating metabolic events. Reactions take place in the cytoplasm or at the plasma membrane. For example, protein synthesis proceeds at ribosomes that are distributed through the cytoplasm or attached to the plasma membrane. This does not mean bacteria are in some way inferior to the eukaryotic cells. Being tiny, fast reproducers, they do not require great internal complexity.

A wall usually surrounds the plasma membrane. A **cell wall** is a semirigid, permeable structure that helps the cell maintain its shape and resist rupturing when internal fluid pressure increases (Section 5.4). Cell walls of eubacteria are composed of peptidoglycan molecules. In such molecules, peptide groups crosslink numerous polysaccharide strands to one another.

Clinicians can identify many bacterial species on the basis of their wall structure and composition. Consider the staining reaction called the **Gram stain**. A sample of unknown bacterial cells is exposed to a purple dye, then to iodine, an alcohol wash, and a counterstain. The cell wall of *Gram-positive* species remain purple. The wall of *Gram-negative* species loses color after the wash, but the counterstain turns it pink (Figure 22.5).

A sticky mesh, or **glycocalyx**, often surrounds the cell wall. It consists of polysaccharides, polypeptides, or both. When highly organized and attached firmly to the wall, we call the mesh a capsule. When less organized and loosely attached to the wall, we call it a slime layer. The mesh helps a bacterial cell attach to teeth, mucous membranes, rocks in streambeds, and other interesting surfaces (Figure 22.4a). It even helps some encapsulated species avoid being engulfed by phagocytic, infection-fighting cells of their host organism.

Some bacteria have one or more **bacterial flagella** (singular, flagellum), which are used in motility. These don't have the same structure as a eukaryotic flagellum, and they don't operate the same way. They move the cell by rotating like a propeller. Many species also have **pili** (singular, pilus). These short, filamentous proteins project above the cell wall, as in Figure 22.4b. Some pili help cells adhere to surfaces. Others help them attach to one another as a prelude to conjugation, an interaction that is described in the next section.

Figure 22.5 Gram staining. Cocci and rods smeared on a slide are stained with a purple dye (such as crystal violet), washed off, and then stained with iodine. All the cells are now purple. The slide is washed with alcohol, which renders the Gram-negative cells colorless. Now the slide is counterstained (with safranin), washed, and dried. Gram-positive cells (in this case, *Staphylococcus aureus*) remain purple. But the counterstain colors Gram-negative cells (in this case, *Escherichia coli*) pink.

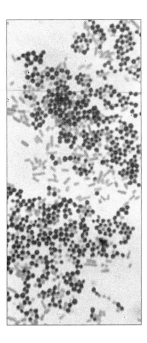

- ■ stain with purple dye
- ☐ stain with iodine
- ☐ wash with alcohol
- ▨ counterstain with safranin

Bacteria are microscopic, prokaryotic cells. They generally have one circular bacterial chromosome, and often they have extra DNA in the form of plasmids.

Nearly all bacteria have a wall around the plasma membrane, and many have a capsule or slime layer around the cell wall.

The Nature of Bacterial Growth

Between cell divisions, bacteria grow through increases in their component parts. We measure the growth of a large, multicelled organism in terms of increases in size, but doing so for a microscopically small bacterium would be a bit pointless. Instead, we measure bacterial growth as an increase in the number of cells in a given population. Under ideal conditions, each cell divides in two, the division of two cells results in four, four result in eight, and so forth. Many types can divide every half hour; a few can do so every ten minutes. Such rates of increase lead to large population sizes in short order.

In nature, conditions that promote the growth of some bacterial populations might not sustain the growth of others. Most species could not establish themselves in the most extreme environments—say, Antarctica, the Negev Desert, or deep inside the Earth. There, the few species that do endure, inside rocks, rarely reproduce. And few can live in the highly acidic wastewater from mining operations. K-12, a strain of *E. coli* originally isolated from the human gut, has been cultivated for such a long time in the laboratory that it no longer is able to grow when reintroduced into the gut. Through microevolutionary processes, it has become adapted to the conditions in its artificial environment.

Prokaryotic Fission

When a bacterium has nearly doubled in size, it divides in two. Each daughter cell inherits a single **bacterial chromosome**, which is a circularized, double-stranded DNA molecule that has a few proteins attached to it. In some species, the daughter cell merely buds from the parent cell. Most often, however, bacteria reproduce by a division mechanism called **prokaryotic fission**.

As prokaryotic fission begins, a parent cell replicates its DNA (Figure 22.6). The two DNA molecules that result are attached to the plasma membrane at adjacent sites. The cell synthesizes lipid and protein molecules, which become incorporated in the membrane between the attachment sites. Membrane growth moves the two DNA molecules apart. New wall material is deposited above the membrane. The membrane and wall grow through the cell midsection and divide the cytoplasm. The result is two genetically equivalent daughter cells.

(Especially in microbiology, you may hear someone refer to this division mechanism as *binary* fission, but such usage can cause confusion. The same term applies to a form of asexual reproduction among flatworms and some other multicelled animals. It refers to growth by way of mitotic cell divisions, then division of the whole body into two parts of the same or different sizes.)

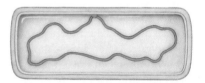 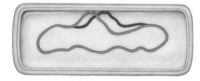 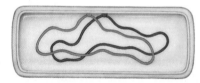

a Cutaway view of a bacterial cell before its DNA is replicated. The DNA molecule is attached to the plasma membrane.

b Replication starts and proceeds in two directions, away from some point in the bacterial DNA molecule.

c The DNA copy is attached at a site on the plasma membrane, near the attachment site of the parent DNA molecule.

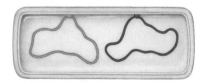  

d New lipids and proteins are added to the plasma membrane between the two attachment sites; the membrane growth moves the two DNA molecules apart.

e New wall material is deposited on the outer surface of new membrane, and both start growing through the cell's midsection.

f The membrane and wall material that were deposited at the cell midsection divide the cytoplasm in two. When the two new walls part company, two separate daughter cells result.

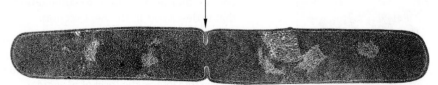

Figure 22.6 Bacterial reproduction by way of prokaryotic fission, a cell division mechanism. The micrograph shows the division of the cytoplasm of *Bacillus cereus*, as brought about by the formation of new membrane and wall material. This cell has been magnified 13,000 times its actual size.

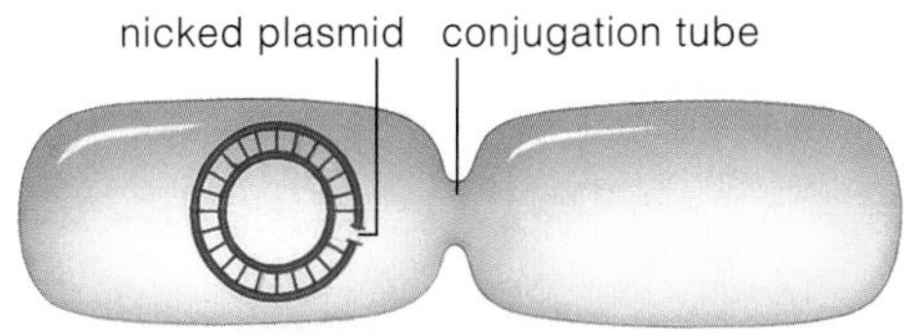

a A conjugation tube has already formed between a donor and a recipient cell. An enzyme has nicked the donor's plasmid.

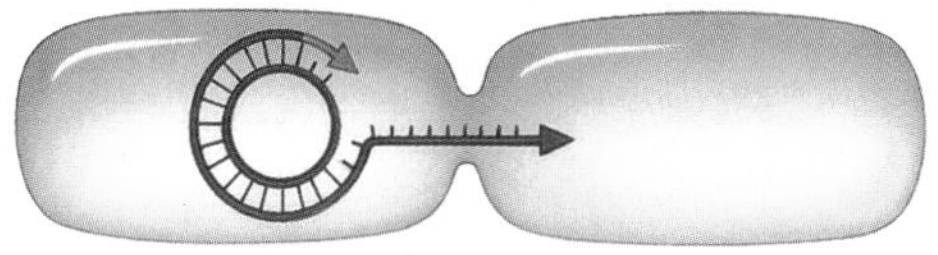

b DNA replication starts on the nicked plasmid. The displaced DNA strand moves through the tube and enters the recipient cell.

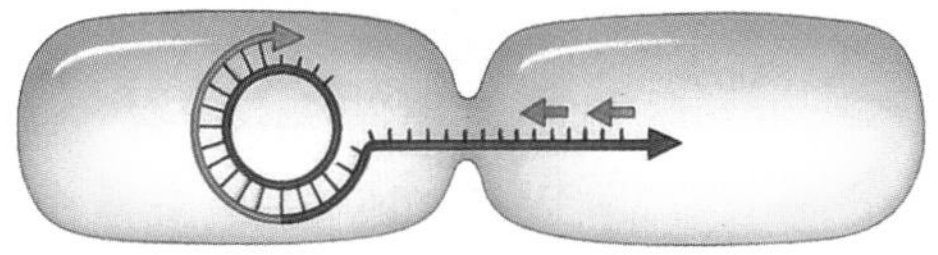

c In the recipient cell, replication starts on the transferred DNA.

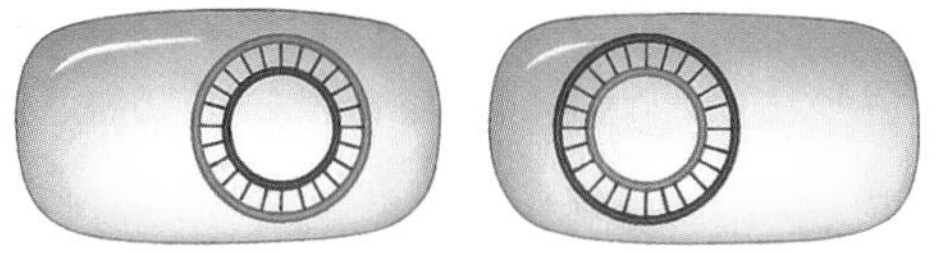

d The cells separate from each other; the plasmids circularize.

Figure 22.7 A simplified sketch of bacterial conjugation. For the sake of clarity, the bacterial chromosome is not included in this sketch, and the size of the plasmid is greatly enlarged. (Compare Section 16.1.)

Bacterial Conjugation

Daughter bacterial cells also may inherit one or more plasmids. A **plasmid**, recall, is a small, self-replicating circle of extra DNA that has a few genes (Section 16.1). Some of the genes on the F (Fertility) plasmid confer the means to engage in **bacterial conjugation**. By this mechanism, a donor cell transfers plasmid DNA to a recipient cell. The transfers are known to occur among many types of bacteria, such as *Salmonella*, *Streptococcus*, and *E. coli*—even between *E. coli* and yeast cells, in the laboratory. The F plasmid carries genetic instructions for synthesizing a structure called a sex pilus. Sex pili at the surface of a donor cell can hook onto a recipient cell and pull it right next to the donor. Shortly after the two cells make contact, a conjugation tube develops between them. After this, plasmid DNA is transferred through the tube, in the manner shown in Figure 22.7.

Most bacteria reproduce by way of prokaryotic fission, a cell division mechanism that follows DNA replication. Each daughter cell inherits a single bacterial chromosome (one DNA molecule). Many species also transfer plasmid DNA.

22.3 BACTERIAL CLASSIFICATION

Just a few decades ago, reconstructing the evolutionary history of bacteria appeared to be an impossible task. Except for stromatolites (Section 21.3), the most ancient groups of bacteria are not well represented in the fossil record. Most groups are not represented at all.

Given their elusive histories, the many thousands of known species of prokaryotic cells traditionally have been classified by **numerical taxonomy**. By this practice, traits of an unidentified cell are compared with those of a known bacterial group. Such traits typically include cell shape, motility, staining attributes of the cell wall, nutritional requirements, metabolic patterns, and the presence or absence of endospores. The greater the total number of traits that the cell has in common with the known group, the closer is their inferred relatedness.

Since the 1970s, nucleic acid hybridization studies, gene sequencing, and other methods of comparative biochemistry have been revealing compelling evidence of bacterial phylogenies (Section 20.5). Comparisons of ribosomal RNAs are notably useful. Remember, rRNAs are indispensable to protein synthesis, and they cannot undergo drastic changes in their overall base sequence without loss of function. The numerous small changes that accumulated in the rRNAs of different lineages can be directly measured. Surprisingly, the measurements are uniting some groups that did not even appear to be related on the basis of other tests. For example, it now seems a key divergence began shortly after prokaryotic cells first appeared on Earth (Section 20.9). One branch led to the **eubacteria**, the most common prokaryotic cells. (Here, the *eu-* is meant to signify "typical.") The other branch led to the common ancestor of both the **archaebacteria** and the first eukaryotic cells. Figure 22.8 is a tree diagram of these evolutionary relationships.

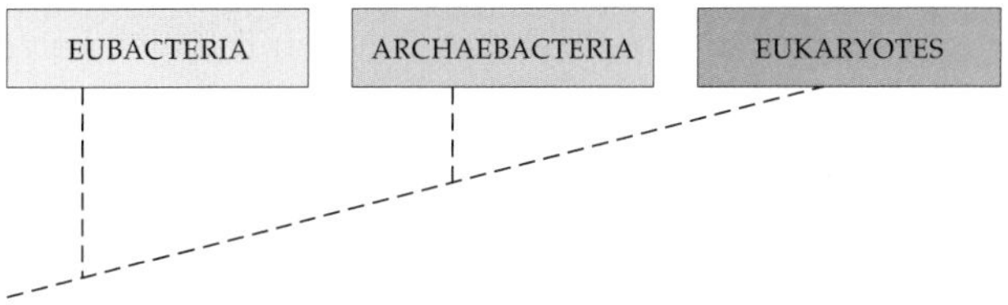

Figure 22.8 Evolutionary tree diagram showing the relationship of archaebacteria to eubacteria and eukaryotes, as suggested by evidence from comparative biochemistry.

Prokaryotic cells are classified by numerical taxonomy (the total percentage of observable traits they have in common with a known bacterial group). They also are classified more directly by comparisons at the biochemical level.

Prokaryotic cells are now assigned to one of two lineages, called the eubacteria and archaebacteria.

EUBACTERIA

More than 400 genera of prokaryotes are recognized, and by far, most are eubacteria. Unlike other bacterial cells, eubacteria have fatty acids incorporated into their plasma membrane. And when a species has a cell wall, as nearly all do, the wall incorporates peptidoglycan. We still do not know enough about the evolutionary histories of eubacteria to move much beyond taxonomic classification. Here we focus on modes of nutrition as a way to give you a sense of the diversity in this kingdom.

A Sampling of Diversity

PHOTOAUTOTROPHIC EUBACTERIA The cyanobacteria, once called blue-green algae, are a good example of the photoautotrophic eubacteria. They are also among the most common photoautotrophs on Earth. Cyanobacteria are aerobic cells, and they themselves release oxygen when photosynthesizing. Most live in ponds and other freshwater habitats. They may grow as mucus-sheathed chains of cells, which can form dense, slimy mats near the surface of nutrient-enriched water.

Anabaena and other types also convert nitrogen gas (N_2) to ammonia, which they use in biosynthesis. When nitrogen compounds are scarce, some cells develop into **heterocysts**. These modified cells make a nitrogen-fixing enzyme (Figure 22.9). They produce and share nitrogen compounds with photosynthetic cells and they receive carbohydrates in return. They freely share substances by way of cytoplasmic junctions between cells.

Anaerobic photoautotrophs, such as green bacteria, get electrons from hydrogen sulfide or hydrogen gas, not from water. They may resemble anaerobic bacteria in which the cyclic pathway of photosynthesis evolved.

CHEMOAUTOTROPHIC EUBACTERIA Many eubacteria in this category influence the global cycling of nitrogen, sulfur, and other nutrients. For example, as you know, nitrogen is a key building block for amino acids and proteins. Without it, there would be no life. Nitrifying bacteria in soil strip electrons from ammonia. Plants use the end product, nitrate, as a nitrogen source.

CHEMOHETEROTROPHIC EUBACTERIA Most bacteria fall in this category. Many, including the pseudomonads, are decomposers; their enzymes break down organic compounds and even pesticides in soil. Other "good" species, at least by human standards, are *Lactobacillus* (employed in the manufacture of pickles, sauerkraut, buttermilk, and yogurt) and actinomycetes (sources of antibiotics). *E. coli* produces vitamin K and compounds that help humans digest fat, it helps newborns digest milk, and its metabolic activities help keep many foodborne pathogens from colonizing the gut. Sugarcane and corn benefit from a symbiont, the nitrogen-fixing

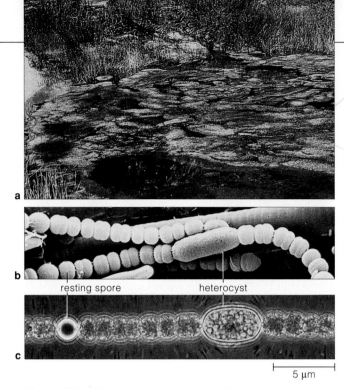

resting spore   heterocyst

Figure 22.9 Cyanobacteria—common photoautotrophs. (**a**) A cyanobacterial population near the surface of a nutrient-enriched pond. (**b,c**) Resting spores form when conditions do not favor growth. A nitrogen-fixing heterocyst is also shown.

spirochete *Azospirillum*. The plants take up some of the nitrogen initially fixed by the bacterium, and they give up some sugars to it. Beans, peas, and other legumes are mutualists with *Rhizobium*, a symbiont in their roots.

Also in this category are most pathogenic bacteria. We admire pseudomonads as decomposers in soil, not when they grow on "our" soaps, antiseptics, and other carbon-rich goods. They are especially bad because they can transfer plasmids with antibiotic-resistance genes.

Some *E. coli* strains cause a form of diarrhea that is the main cause of infant death in developing countries. *Clostridium botulinum* can taint fermented grain as well as food in improperly sterilized or sealed cans and jars. Its toxins cause *botulism*, a form of poisoning that can disrupt breathing and lead to death. *C. tetani*, one of its relatives, causes the disease *tetanus* (Section 34.5).

Like many other bacteria that commonly live in soil, *C. tetani* can form an **endospore**. This resting structure forms inside the cell, around a copy of the bacterial chromosome and part of the cytoplasm (Figure 22.10). Endospores form when the depletion of nitrogen or some other nutrient stops cell growth. They are released

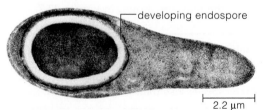

developing endospore

Figure 22.10
An endospore developing in *Clostridium tetani*, one of the dangerous pathogens.

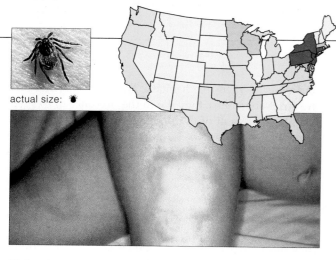

actual size: ●

Figure 22.11 An extreme reaction to an infection by *Borrelia burgdorferi*, a spirochete. The rash is a symptom of what is now the most common tick-borne disease in the United States: Lyme disease. Tick bites deliver the spirochete to new hosts. The map shows statewide cases reported in 1996 (*tan*, fewer than 10; *yellow*, 11 to 99; *gold*, 100 to 600; *red*, more than 2,000).

as free spores when the plasma membrane ruptures. They are exceptionally resistant to heat, and they also resist drying out, irradiation, acids, disinfectants, and boiling. Endospores can remain dormant, sometimes for many decades. When favorable conditions return, they become metabolically active and then develop into a single bacterial cell.

Many chemoautotrophic eubacteria taxi from host to host inside the gut of insects. For example, bites from blood-sucking ticks can transmit *Borrelia burgdorferi* from deer and some other wild animals to humans, who develop *Lyme disease*. A "bull's-eye" rash often develops around the bite (Figure 22.11). Severe headaches, backaches, chills, and fatigue follow. Without prompt treatment, the condition worsens.

Regarding the "Simple" Bacteria

Bacteria are small. Their insides are not elaborate. *But bacteria are not simple.* A brief look at their behavior will reinforce this point. Bacteria move toward nutrient-rich regions. Aerobes move toward oxygen; anaerobes avoid it. Photosynthetic types move into light and away from light that is too intense. Many species can tumble away from toxins. Such behaviors often start with membrane receptors that are stimulated by changes in chemical conditions or in the intensity of light coming in from a particular direction. Such a change triggers a shift in metabolic activities inside the cell, which might then lead to an adjustment in the direction of movement.

Magnetotactic bacteria contain a chain of magnetite particles that serves as a tiny compass (Figure 22.12*a*). The compass helps them sense which way is north and also down. These bacteria swim toward the bottom of a body of water, where oxygen concentrations are lower and therefore more suitable for their growth.

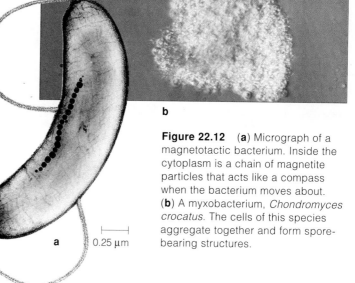

b

Figure 22.12 (**a**) Micrograph of a magnetotactic bacterium. Inside the cytoplasm is a chain of magnetite particles that acts like a compass when the bacterium moves about. (**b**) A myxobacterium, *Chondromyces crocatus*. The cells of this species aggregate together and form spore-bearing structures.

a ⊢———⊣ 0.25 µm

Some species even show *collective* behavior, as when millions of *Myxococcus xanthus* cells form a "predatory" colony. The cells secrete enzymes that digest "prey," such as cyanobacteria, that get stuck to the colony. Then they absorb the breakdown products. What's more, the cells migrate, change direction, and move as a single unit toward what may be food!

Many myxobacteria colonies form **fruiting bodies** (spore-bearing structures). Under suitable conditions, some cells in the colony differentiate and form a slime stalk, others form branchings, and others form clusters of spores (Figure 22.12*b*). Spores disperse when a cluster bursts open; each may give rise to a new colony. As you will see in the next chapter, certain eukaryotic species also form spore-bearing structures.

Eubacteria are the most common and diverse prokaryotic cells. They are adapted to nearly all environments.

We divide the kingdom of archaebacteria into three major groups, called the methanogens, halophiles, and extreme thermophiles. In many respects, archaebacteria are unique in their composition, structure, metabolism, and nucleic acid sequences. They differ as much from other bacteria as they do from eukaryotic cells. Many investigators believe that the existing archaebacteria, which can withstand conditions as hostile as those on the early Earth, resemble the first living cells. Hence the name of the kingdom (*archae-* means beginning).

The **methanogens**, or "methane makers," are quite at home in swamps, sewage, stockyards, animal guts,

and other oxygen-free habitats. These prokaryotic cells are strict anaerobes; they die in the presence of oxygen. Mainly as a result of the findings from ribosomal RNA sequencing studies, the seventeen known genera have been assigned to seven categories. Figure 22.13 shows representatives. Although evolutionarily distant from humans, the methanogens still are closer relatives to you than are the eubacteria.

The methanogens make ATP by anaerobic electron transport, a pathway described in Section 8.5. Usually, they use hydrogen gas (H_2) as their source of electrons, although some groups get them from ethanol and other alcohols. Nearly all use carbon dioxide as their carbon source and as a final electron acceptor for the reactions, which end with the formation of methane (CH_4). As a group, the methanogens produce about 2 billion tons of methane every year. Collectively, their activities affect the levels of carbon dioxide in the atmosphere and the cycling of carbon through ecosystems.

Long ago, vast methane deposits accumulated on the seafloor. For example, geologists recently found 35 billion tons of it 400 kilometers off the South Carolina coast. If the ocean circulation were to change, as it has in the past, then all of that gas could move abruptly and explosively to the surface. The rapid release of so much methane would drastically change the atmosphere and therefore the global climate (Section 48.7).

The "salt lovers," or **halophiles**, live in brackish ponds, salt lakes, near volcanic fissures on the seafloor (hydrothermal vents), and other high-salinity settings (Figure 22.13c). Halophiles can spoil salted fish, animal hides, and commercially produced sea salt. Most make ATP by aerobic pathways. When oxygen levels are low, some also produce ATP by photosynthesis. A unique photosynthetic pigment (bacteriorhodopsin) forms in the plasma membrane and absorbs light energy. Absorption starts a metabolic process that leads to an increase in an H^+ gradient across the membrane. ATP forms when these ions flow down the gradient, through the interior of membrane proteins (compare Section 7.5).

The "heat lovers," or **extreme thermophiles**, live in such places as highly acidic soils, hot springs, even coal mine wastes. Some species are the basis of remarkable food webs in sediments around hydrothermal vents. Sometimes the water above the sediments is as hot as 110°C. Extreme thermophiles use the hydrogen sulfide spewing from the vents as a source of electrons for ATP formation. Their existence at vents is cited as evidence that life could have originated deep in the oceans.

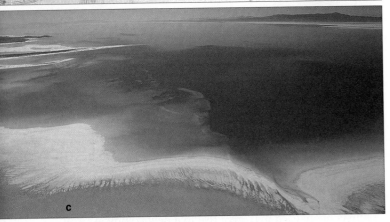

Figure 22.13 Archaebacteria. (**a**) Scanning electron micrograph of a dense population of *Methanosarcina* cells. (**b**) Transmission electron micrograph of *Methanococcus jannaschii*, which had its DNA fully sequenced (Section 20.9). (**c**) Pinkish, saline water in Great Salt Lake, Utah—a sign of colonies of halophilic bacteria and certain algae that contain pinkish to red-orange carotenoid pigments.

Like the first cells on Earth, archaebacteria live in extremely inhospitable habitats. In their properties, they differ as much from eubacteria as from eukaryotic cells.

SUMMARY OF MAJOR BACTERIAL GROUPS

By now, you probably have sensed that the *species* is the basic unit in bacterial classification schemes. However, the definition of species that fits sexually reproducing organisms does not fit bacteria, which do not form reproductively isolated populations of interbreeding individuals. Each bacterial cell generally does its own thing; genetic recombination is infrequent. Besides this, bacteria do not show spectacular variation in traits, and variations that do occur are dictated by relatively few genes. If two types show only minor differences, one may be classified as a **strain**, not a new species.

These are just a few of the problems we face when attempting to bring order to our understanding of the bacterial kingdoms. Until evolutionary relationships are sorted out, bacteria will continue to be grouped mainly according to numerical taxonomy, as described earlier. Table 22.2 summarizes some of the major groupings that are touched upon in this book.

Table 22.2 Summary of Representative Eubacteria and Archaebacteria

Some Major Groups	Main Habitats	Characteristics	Representatives
EUBACTERIA			
Photoautotrophs:			
Cyanobacteria, green sulfur bacteria, and purple sulfur bacteria	Mostly lakes, ponds; some marine, terrestrial habitats	Photosynthetic; use sunlight energy, carbon dioxide; cyanobacteria use oxygen-producing noncyclic pathway; some also use cyclic route	*Anabaena, Nostoc, Rhodopseudomonas, Chloroflexus*
Photoheterotrophs:			
Purple nonsulfur and green nonsulfur bacteria	Anaerobic, organically rich muddy soils, and sediments of aquatic habitats	Use sunlight energy; organic compounds as electron donors; some purple nonsulfur may also grow chemotrophically	*Rhodospirillum, Chlorobium*
Chemoautotrophs:			
Nitrifying, sulfur-oxidizing, and iron-oxidizing bacteria	Soil; freshwater, marine habitats	Use carbon dioxide, inorganic compounds as electron donors; influence crop yields, cycling of nutrients in ecosystems	*Nitrosomonas, Nitrobacter, Thiobacillus*
Chemoheterotrophs:			
Spirochetes	Aquatic habitats; parasites of animals	Helically coiled, motile; free-living and parasitic species; some major pathogens	*Spirochaeta, Treponema*
Gram-negative aerobic rods and cocci	Soil, aquatic habitats; parasites of animals, plants	Some major pathogens; some fix nitrogen (e.g., *Rhizobium*)	*Pseudomonas, Neisseria, Rhizobium, Agrobacterium*
Gram-negative facultative anaerobic rods	Soil, plants, animal gut	Many major pathogens; one bioluminescent (*Photobacterium*)	*Salmonella, Escherichia, Proteus, Photobacterium*
Rickettsias and chlamydias	Host cells of animals	Intracellular parasites; many pathogens	*Rickettsia, Chlamydia*
Myxobacteria	Decaying organic material; bark of living trees	Gliding, rod-shaped; aggregation and collective migration of cells	*Myxococcus*
Gram-positive cocci	Soil; skin and mucous membranes of animals	Some major pathogens	*Staphylococcus, Streptococcus*
Endospore-forming rods and cocci	Soil; animal gut	Some major pathogens	*Bacillus, Clostridium*
Gram-positive nonsporulating rods	Fermenting plant, animal material; gut, vaginal tract	Some important in dairy industry, others major contaminators of milk, cheese	*Lactobacillus, Listeria*
Actinomycetes	Soil; some aquatic habitats	Include anaerobes and strict aerobes; major producers of antibiotics	*Actinomyces, Streptomyces*
ARCHAEBACTERIA			
Methanogens	Anaerobic sediments of lakes, swamps; animal gut	Chemosynthetic; methane producers; used in sewage treatment facilities	*Methanobacterium*
Halophiles	Brines (extremely salty water)	Heterotrophic; also, unique photosynthetic pigments (bacteriorhodopsin) form in some	*Halobacterium*
Extreme thermophiles	Acidic soil, hot springs, hydrothermal vents	Heterotrophic or chemosynthetic; use inorganic substances as electron donors	*Sulfolobus, Thermoplasma*

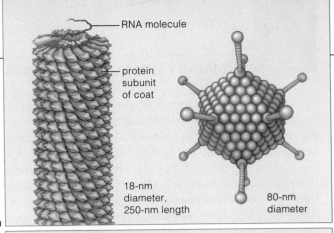
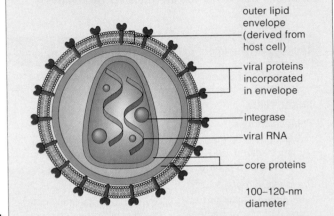
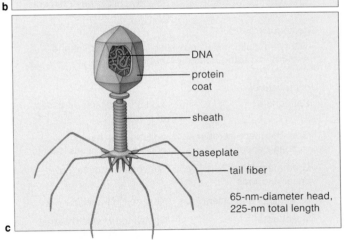

22.7 THE VIRUSES

Defining Characteristics

In ancient Rome, *virus* meant "poison" or "venomous secretion." In the late 1800s, this rather nasty word was bestowed on newly discovered pathogens, smaller than the bacteria being studied by Louis Pasteur and others. Many viruses deserve the name. They attack humans, cats, cattle, birds, insects, plants, fungi, protistans, and bacteria. You name it, there are viruses that can infect it.

Today we define a **virus** as a noncellular infectious agent that has two characteristics. First, a viral particle consists of a protein coat surrounding a nucleic acid core—that is, around its genetic material. Second, a virus cannot reproduce itself. It can be reproduced only after its genetic material and a few enzymes enter a host cell and subvert the cell's biosynthetic machinery.

The genetic material of a virus is DNA *or* RNA. The coat consists of one or more types of protein subunits organized into a rodlike or many-sided shape (Figure 22.14). The coat protects the genetic material during the journey from one host cell to the next. It also contains proteins that can bind with specific receptors on host cells. Some coats are enclosed in an envelope of mostly membrane remnants from the previously infected cell. These envelopes bristle with glycoprotein spikes. Coats of complex viruses have sheaths, tail fibers, and other accessory structures attached.

The immune system of vertebrates can detect certain viral proteins. The problem is, the genes for many viral proteins mutate so frequently that a virus may elude the immune fighters. For example, people susceptible to lung infections get new "flu shots" each year because envelope spikes on influenza viruses keep changing.

Examples of Viruses

Each kind of virus can multiply only in certain hosts. It cannot be studied easily unless investigators culture living host cells. This is why much of our knowledge of viruses comes from **bacteriophages**, a group of viruses that infect bacterial cells. Unlike cells of humans and other complex, multicelled species, bacterial cells can be cultured easily and rapidly. This is also why bacteria and bacteriophages were used in early experiments to determine DNA function (Section 13.1). They are still used as research tools in genetic engineering.

Table 22.3 lists some major groups of animal viruses. They contain double- or single-stranded DNA or RNA, which is replicated in various ways. Animal viruses range in size from parvoviruses (18 nanometers) to the brick-shaped poxviruses (350 nanometers). Many cause diseases, including the common cold, certain cancers, warts, herpes, and influenza (Figure 22.15). One, HIV, is the trigger for *AIDS*. By attacking certain white blood

cells, it weakens the immune system's ability to fight infections that might not otherwise be life threatening. Researchers attempting to develop drugs against HIV and other diseases use HeLa cells and other immortal cell lines for early experiments (Section 9.6). Later they must use laboratory animals, then human volunteers, to test a new drug for toxicity and effectiveness. Why? It takes a functioning immune system to test responses.

Figure 22.14 Body plans of viruses. (**a**) To the left, *helical* viruses have a rod-shaped coat of protein subunits, coiled helically around the nucleic acid. The upper subunits have been removed from this portion of a tobacco mosaic virus to reveal the RNA. To the right, *polyhedral* viruses, such as this adenovirus, have a many-sided coat. (**b**) *Enveloped* viruses, such as HIV, have an envelope around a helical or polyhedral coat. (**c**) *Complex* viruses, such as T-even bacteriophages, have additional structures attached to the coat.

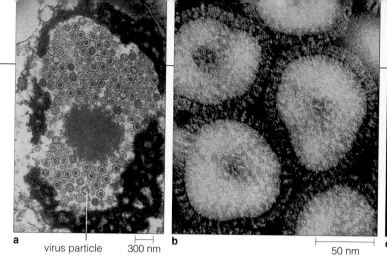

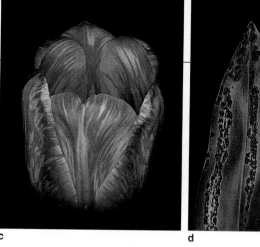

a virus particle |—— 300 nm b |—— 50 nm c d

Table 22.3	Classification of Some Major Animal Viruses
DNA Viruses	**Some Diseases**
Parvoviruses	Gastroenteritis; roseola (fever, rash) in small children; aggravation of symptoms of sickle-cell anemia
Adenoviruses	Respiratory infections; some cause tumors in animals
Papovaviruses	Benign and malignant warts
Poxviruses	Smallpox, cowpox
Herpesviruses:	
H. simplex type I	Oral herpes, cold sores
H. simplex type II	Genital herpes
Varicella-zoster	Chicken pox, shingles
Epstein-Barr	Infectious mononucleosis, some forms of cancer
Hepadnavirus	Hepatitis B, severe liver damage

RNA Viruses	**Some Diseases**
Picornaviruses:	
Enteroviruses	Polio, hemorrhagic eye disease, hepatitis A (infectious hepatitis)
Rhinoviruses	Common cold
Hepatitis A virus	Inflammation of liver, kidneys, spleen
Togaviruses	Forms of encephalitis, rubella
Flaviviruses	Yellow fever (fever, chills, jaundice), dengue (fever, severe muscle pain)
Coronaviruses	Upper respiratory infections, colds
Rhabdoviruses	Rabies, other animal diseases
Filoviruses	Hemorrhagic fevers, as by *Ebola* virus
Paramyxoviruses	Measles, mumps
Orthomyxoviruses	Influenza
Bunyaviruses	Hemorrhagic fevers, as by hantaviruses
Arenaviruses	Hemorrhagic fevers
Retroviruses:	
HTLV I, II	Some cancers (leukemia)
HIV	AIDS
Reoviruses	Respiratory and intestinal infections

Figure 22.15 Some viruses and their effects. (**a**) Particles of a DNA virus that causes a herpes infection in humans. (**b**) Particles of an enveloped RNA virus that causes influenza in humans. Spikes project from the lipid envelope. (**c**) Streaking of a tulip blossom. A harmless virus infected pigment-forming cells in the colorless parts. (**d**) An orchid leaf infected by a rhabdovirus.

Plant viruses must breach plant cell walls to cause diseases. They typically hitch rides on the piercing or sucking devices of insects that feed on plant juices. Some RNA viruses infect tobacco plants (the tobacco mosaic virus), barley, potatoes, and other major crop plants. Certain DNA viruses infect such valuable crops as cauliflower and corn. Figure 22.15*c,d* shows visible effects of two viral infections.

Infectious Agents Tinier Than Viruses

Some infectious agents are more stripped down than viruses. **Prions** are small proteins linked to eight rare, fatal degenerative diseases of the nervous system. They are altered products of a gene that is present in normal as well as infected individuals. Unaltered and altered forms of the protein are found at the surface of neurons (communication cells of the nervous system). You may have heard of *kuru* and *Creutzfeldt-Jakob* diseases. They slowly destroy muscle coordination and brain function in humans (Section 22.9). *Scrapie*, a disease of sheep, is so named because infected animals rub against trees or posts until they scrape off most of their wool.

Viroids are tightly folded strands or circles of RNA, smaller than anything in viruses. They resemble introns (noncoding portions of eukaryotic DNA) and may have evolved from them. Viroids have no protein coat, but their tight folding may help protect them from a host's enzymes. These bits of "naked" RNA are known to cause plant diseases. Each year, they destroy millions of dollars' worth of potatoes, citrus, and other crop plants.

A virus is a nonliving infectious particle that consists of nucleic acid enclosed in a protein coat and sometimes an outer envelope. It cannot be replicated without pirating the metabolic machinery of a specific type of host cell.

Viruses multiply in a variety of ways. Even so, nearly all of their multiplication cycles proceed through five basic steps, as outlined here:

1. *Attachment.* The virus attaches to a host cell. Any cell is a suitable host if a virus can chemically recognize and lock onto specific molecular groups at the cell surface.

2. *Penetration.* Either the whole virus or its genetic material alone penetrates the cell's cytoplasm.

3. *Replication and Synthesis.* In an act of molecular piracy, the viral DNA or RNA directs the host cell into producing many copies of viral nucleic acids and proteins, including enzymes.

4. *Assembly.* The viral nucleic acids and viral proteins are put together to form new infectious particles.

5. *Release.* New virus particles are released from the cell.

Consider this list with respect to some bacteriophages. Lytic and lysogenic pathways are common among their replication cycles. In a **lytic pathway**, steps 1 through 4 proceed rapidly, and new particles are released when the host cell undergoes lysis (Figure 22.16). **Lysis** means the plasma membrane, cell wall, or both are damaged, so that cytoplasm leaks out and the cell dies. Late into most lytic pathways, a viral enzyme is synthesized that causes swift destruction of the bacterial cell wall.

In a **lysogenic pathway**, a latent period extends the cycle. The virus does not kill its host outright. Instead, a viral enzyme cuts a host chromosome, then integrates viral genes into it. When the infected cell prepares to divide, the recombinant molecule gets replicated. As a result of that single instance of genetic recombination, miniature time bombs get passed on to all of that cell's descendants. Later on, a molecular signal or some other stimulus may reactivate the cycle.

Latency occurs in the multiplication cycles of many viruses, not just among bacteriophages. Type I *Herpes simplex*, which causes *cold sores*, is an example. Nearly everybody harbors this virus. It remains latent inside a ganglion (a cluster of cell bodies of neurons) in the face. Stress factors, such as sunburn, can reactivate the virus. Then, virus particles move down the neurons to their tips near the skin. There they infect epithelial cells and cause painful skin eruptions.

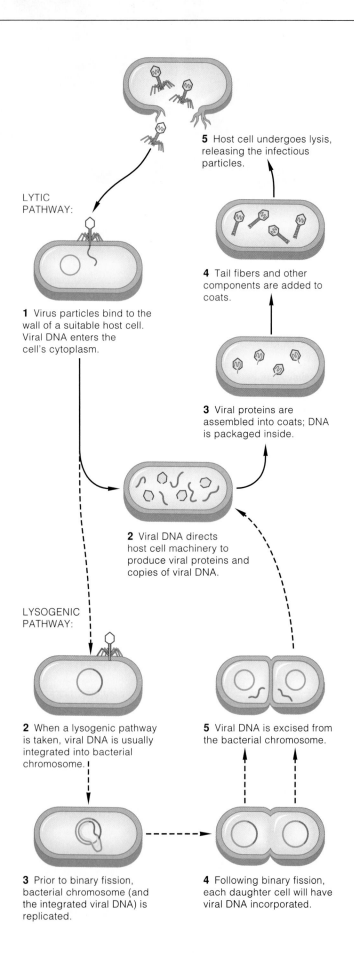

LYTIC PATHWAY:

1 Virus particles bind to the wall of a suitable host cell. Viral DNA enters the cell's cytoplasm.

5 Host cell undergoes lysis, releasing the infectious particles.

4 Tail fibers and other components are added to coats.

3 Viral proteins are assembled into coats; DNA is packaged inside.

2 Viral DNA directs host cell machinery to produce viral proteins and copies of viral DNA.

LYSOGENIC PATHWAY:

2 When a lysogenic pathway is taken, viral DNA is usually integrated into bacterial chromosome.

5 Viral DNA is excised from the bacterial chromosome.

3 Prior to binary fission, bacterial chromosome (and the integrated viral DNA) is replicated.

4 Following binary fission, each daughter cell will have viral DNA incorporated.

Figure 22.16 Generalized multiplication cycle for some of the bacteriophages. New viral particles may be produced and then released by way of a lytic pathway alone. For certain viruses, the lytic pathway may be expanded upon to include a lysogenic pathway.

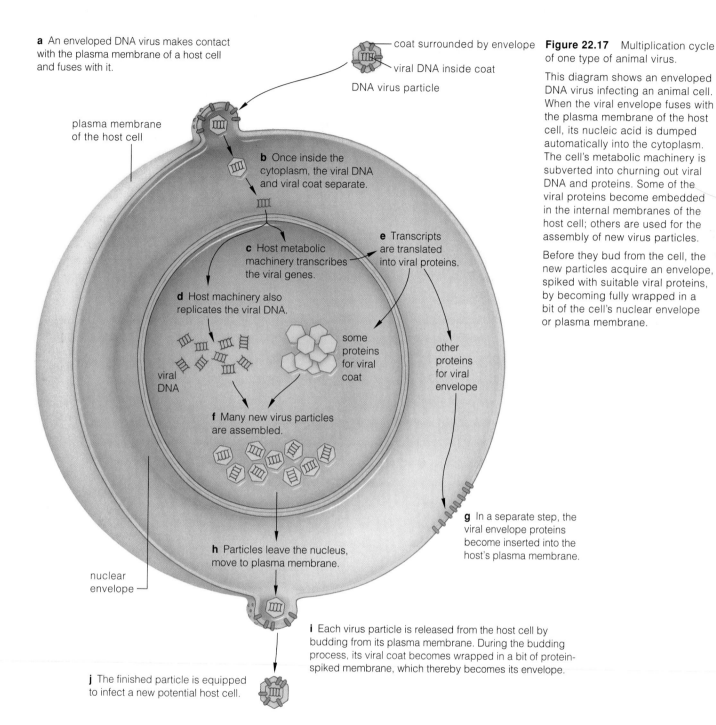

a An enveloped DNA virus makes contact with the plasma membrane of a host cell and fuses with it.

coat surrounded by envelope

viral DNA inside coat

DNA virus particle

plasma membrane of the host cell

b Once inside the cytoplasm, the viral DNA and viral coat separate.

c Host metabolic machinery transcribes the viral genes.

e Transcripts are translated into viral proteins.

d Host machinery also replicates the viral DNA.

viral DNA

some proteins for viral coat

other proteins for viral envelope

f Many new virus particles are assembled.

h Particles leave the nucleus, move to plasma membrane.

nuclear envelope

g In a separate step, the viral envelope proteins become inserted into the host's plasma membrane.

i Each virus particle is released from the host cell by budding from its plasma membrane. During the budding process, its viral coat becomes wrapped in a bit of protein-spiked membrane, which thereby becomes its envelope.

j The finished particle is equipped to infect a new potential host cell.

Figure 22.17 Multiplication cycle of one type of animal virus.

This diagram shows an enveloped DNA virus infecting an animal cell. When the viral envelope fuses with the plasma membrane of the host cell, its nucleic acid is dumped automatically into the cytoplasm. The cell's metabolic machinery is subverted into churning out viral DNA and proteins. Some of the viral proteins become embedded in the internal membranes of the host cell; others are used for the assembly of new virus particles.

Before they bud from the cell, the new particles acquire an envelope, spiked with suitable viral proteins, by becoming fully wrapped in a bit of the cell's nuclear envelope or plasma membrane.

Like other enveloped viruses, the herpesviruses can enter a host cell by their own version of endocytosis, then leave it by budding from the plasma membrane. Figure 22.17 shows how they accomplish this.

The multiplication cycle of the RNA viruses has an interesting twist to it. In the host cell's cytoplasm, their RNA serves as a template for synthesizing either DNA or mRNA. For example, HIV, a retrovirus, carts its own enzymes into cells. These assemble DNA on viral RNA by reverse transcription (Sections 16.4 and 40.11).

The multiplication cycles of nearly all viruses include five basic steps: attachment to a suitable host cell, penetration into the cell, viral DNA or RNA replication and synthesis of viral proteins, assembly of new viral particles, and then release from the infected cell.

Bacteriophage replication cycles commonly follow a rapid, lytic pathway and an extended, lysogenic pathway.

Replication cycles of RNA viruses involve the use of viral RNA as a template for synthesizing DNA or mRNA.

22.9 THE NATURE OF INFECTIOUS DISEASES

CATEGORIES OF DISEASES   Just by being human, you are a potential host for a great many pathogenic bacteria, viruses, fungi, protozoans, and parasitic worms. When a pathogen has invaded your body and is multiplying in host cells and tissues, this is an **infection**. Its outcome, **disease**, results if the body's defenses cannot be mobilized fast enough to prevent the pathogen's activities from interfering with normal body functions. You may have heard of *contagious* diseases. This simply means that the pathogenic agents can be transmitted by direct contact with body fluids secreted or otherwise released (as by explosive wet sneezes) from infected individuals.

During an **epidemic**, a disease spreads rapidly through part of a population for a limited time, then the outbreak subsides. What happens when an epidemic breaks out in several countries around the world at the same time? We call this a **pandemic**. AIDS, an incurable disease, is a prime example. Millions of people around the world are infected by the causative agent, HIV (short for the *Human Immunodeficiency Virus*).

Sporadic diseases such as whooping cough break out irregularly and affect few people. *Endemic* diseases pop up more or less continuously, but they don't spread far in large populations. Tuberculosis is like this. So is impetigo, a highly contagious bacterial infection that often spreads no further than, say, a single day-care center.

AN EVOLUTIONARY VIEW   Now look at disease in terms of the pathogen's prospects for survival. Just like you, *a pathogen stays around only for as long as it has access to outside sources of energy and raw materials*. To a microscopic organism or viral particle, a human host is a veritable treasurehouse of both. With such a bountiful supply of resources, it can multiply or replicate itself to amazing population sizes. And, evolutionarily speaking, the ones that leave the most descendants win.

There are two significant barriers to complete world dominance by pathogens. First, any species with a history of being attacked by a particular pathogen has coevolved with it and has built-in defenses against it. The premier example is the vertebrate immune system, as described in Chapter 40. Second, if a pathogen is too good at rapidly killing an individual host, it may vanish before having time to infect a new one. This is one reason why most pathogens have less-than-fatal effects on a host. After all, an infected individual that lives longer can spread more of the next generation of a pathogen and so contribute to its reproductive or replicative success.

Usually, a host dies only if a pathogen enters its body in overwhelming numbers, if it is a novel host (one with no coevolved defenses against that pathogen)—or if a mutant strain of the pathogen emerges and can surmount the host's current array of defenses.

Being equipped with the evolutionary perspective, you probably can work out the numbers on your own. To wit: *The greater the population density of host individuals, the greater will be the kinds and frequencies of infectious diseases transmitted among them.* And this brings us to the bad news.

EBOLA AND OTHER EMERGING PATHOGENS   Thanks to planes, trains, and automobiles, people now travel often and in droves all around the world. Among their more exotic destinations are virgin tropical forests and other remote environments where the human body was an infrequent (or nonexistent) opportunity for pathogens.

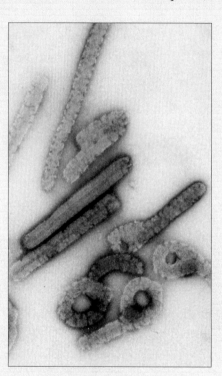

Figure 22.18   Transmission electron micrograph of particles of *Ebola* virus.

Today, strange and often dangerous pathogens are opportunistic about the novel, two-legged treasurehouses of metabolic machinery and nutrients that are walking smack into their habitats. Human travelers can become infected within a matter of hours and taxi such pathogens far away, and eventually back home.

We know little about the deadly, **emerging pathogens**. Maybe they have been around for a long time and only now are taking great advantage of the increased presence of the novel human hosts. Maybe they are newly mutated strains of existing species.

Consider *Ebola*, one of the deadliest of the viruses that cause hemorrhagic fever. *Ebola*, an RNA virus, appears either filamentous or circular in electron micrographs (Figure 22.18). It may have coevolved with monkeys in tropical forests in Africa. By 1976 and possibly earlier, it was infecting humans. It kills between 70 and 90 percent of its victims. There is no vaccine against it. There is no treatment for the disease, and what eventually happens is not pretty. High fever and flu-like aches mark the onset of the disease. Within a few days, nausea, vomiting, and diarrhea begin. Cells making up the lining of blood vessels are destroyed. Blood is free to seep out from the circulatory system, into the surrounding tissues and out through all of the body's orifices. The liver and the kidneys may rapidly turn to mush. Patients often become deranged, and soon after that they die of circulatory shock. There have been four *Ebola* epidemics since 1976. Each time, government agencies around the world were mobilized; quarantine procedures were implemented to limit the spread of the disease.

ON THE DRUG-RESISTANT STRAINS   There is an old saying that, when you attack nature, it comes back at you with a pitchfork. We have already considered antibiotic resistance in several contexts, in Sections 1.4, 1.6, and 18.4. Here we reinforce the point that infectious diseases are on the upsurge.

A mere twenty-five years ago, antibiotics and vaccines were considered invincible, and the Surgeon General of the United States announced we could "close the book on infectious diseases." There are now more than 5.8 billion people. Most live in crowded cities, and at any time, as many as 50 million are on the move within and between countries in search of a better life (Section 46.6). Even putting aside the emerging pathogens, is it any wonder that pathogens responsible for cholera, tuberculosis, and other familiar diseases are striking with a vengeance?

Consider how one of these pathogens is having a field day. Because of economic pressures over the past several decades, the number of working mothers has skyrocketed in the United States. So has the number of preschoolers enrolled in day-care centers. In effect, each of these centers is a population of host individuals whose immune systems are still developing and vulnerable to contagious diseases. Now think of *Streptococcus pneumoniae*. This bacterium causes pneumonia, meningitis, and middle-ear infections in people of all ages; about 40,000 to 50,000 cases end in death every year. The risk of infection is 36 times greater in large day-care centers than it is for children cared for at home. As recently as 1988, drug-resistant strains of *S. pneumoniae* were practically unheard of in the United States. They are becoming the rule, not the exception.

PATHOGENS LURKING IN THE KITCHEN   Each year, food poisoning hits more than 80 million people in the United States alone. It kills about 9,000 of them, mainly the very young, the very old, or those with compromised immune systems. Yet the cases reported to health agencies number only in the thousands; people often dismiss their misery as "the 24-hour flu." *Salmonella enteriditis* as well as certain strains of *E. coli* are prominent among the culprits. They typically are ingested with undercooked beef or poultry, contaminated water, unpasteurized milk or cider, and vegetables grown in fields fertilized with manure.

A few examples: In 1966, *S. enteriditis* slipped into ingredients used to make a popular ice cream. In a single outbreak, possibly as many as 224,000 people suffered from stomach cramps, diarrhea, and other symptoms of *Salmonella*. In 1993 in Washington State, more than 500 people became sick and three children died after eating restaurant hamburgers tainted with a pathogenic strain of *E. coli*. A recent outbreak of food poisoning was traced to unpasteurized apple juice.

Outbreaks such as these certainly grab our attention, yet pathogens lurking in the kitchen probably account for at least half of all cases of food poisoning. Examine wood or plastic cutting boards or sleek, stainless steel

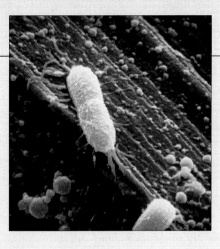

Figure 22.19 Among microscopic bits of food, *Pseudomonas* cells that are attaching to stainless steel by means of pili at their surface—and think about *that* the next time you use an unwashed kitchen knife previously used on raw poultry, beef, or seafood.

knives with electron microscopes and you see nooks and crannies where microbes can lurk (Figure 22.19). Carlos Enriquez and his colleagues at the University of Arizona sampled 75 dishrags and 325 sponges in several homes. Most harbored colonies of *Salmonella* and *E. coli*, as well as *Pseudomonas* and *Staphylococcus*. Bacteria can live two weeks in a wet sponge. Use the sponge to wipe down the kitchen, and you spread them about. Antibacterial soaps, detergents, and weak solutions of household bleach get rid of them. You can sanitize sponges simply by putting them through a dishwasher cycle.

Do you think food poisoning is of small concern? The annual cost of treating the infections caused by food-borne pathogens ranges between 5 and 22 *billion* dollars.

PRIONS LURKING IN SHEEP AND COWS   Also weighing in as pathogenic heavyweights are the prions (pronounced PREE-ons), those proteinaceous infectious particles you read about in Section 22.7. In 1996, 150,000 cattle in Great Britain were staggering about, drooling, and showing other symptoms of *BSE* (bovine spongiform encephalopathy), or *mad cow disease*. Prions had triggered conversion of normal proteins in their neurons, and spongy holes and amyloid deposits had formed in the cattle brains. BSE is fatal.

How did it happen? Ground-up tissues of sheep that had died of scrapie—another prion-caused disease—had been used as a supplement in cattle feed. The practice was banned in 1988, but prions were already in the food chain.

Medical researchers already knew that, each year, 1 in 1 million people around the world develops Creutzfeldt-Jakob disease (CJD). Prions are linked to this fatal disease, as well. The symptoms, which include loss of vision and speech, rapid mental deterioration, and spastic paralysis, may take decades to develop. However, at the same time as the outbreak of BSE in Great Britain, ten people were diagnosed with a variant of CJD; four had already died. Genetic analysis and interviews linked all of the cases to exposure to BSE. Coincidentally, a variant of CJD may have caused the death of a woman in Stockton, California, in 1996, also.

Ten other countries had reported cases of BSE, possibly as a result of importing cattle feed from Great Britain. The incidence of BSE in Great Britain is now declining since strict precautionary measures have been instituted. In the United States, importation of live cattle or meat products from BSE-affected countries has been banned since 1989.

SUMMARY

1. A major divergence occurred soon after the origin of life. One lineage gave rise to the eubacteria, and the other to the common ancestors of archaebacteria and eukaryotic cells.

   a. By far, the most common existing prokaryotic cells are eubacteria.

   b. Archaebacteria (the methanogens, halophiles, and extreme thermophiles) live in extreme environments, like those in which life probably originated. They differ from eubacteria in wall structure and other features, and they share some features with eukaryotic cells.

2. All bacteria are prokaryotic cells; they have no nucleus. Some have membrane infoldings and other structures in the cytoplasm, but none has the profusion of organelles that is typical of eukaryotic cells.

   a. Three basic bacterial shapes are common: cocci (spheres), bacilli (rods), and spirilla (spirals).

   b. Nearly all eubacteria have a cell wall composed of peptidoglycan. It protects the plasma membrane and helps it resist rupturing. The composition and structure of the cell wall help identify particular bacterial species.

   c. A sticky mesh of polysaccharides (glycocalyx) may surround the wall as a capsule or slime layer. It helps a bacterium attach to substrates and sometimes to resist a host organism's infection-fighting mechanisms.

   d. Some species have bacterial flagella, structures that rotate a bit like a propeller and function in motility. Many species have pili, filamentous proteins that help cells adhere to a surface or facilitate conjugation.

3. Prokaryotic fission is a cell division mechanism used only by bacteria. It involves replication of the single bacterial chromosome and division of a parent cell into two genetically equivalent daughter cells.

4. Many species have plasmids, small circles of DNA that are replicated independently of the single, circular bacterial chromosome. Certain genes on some plasmids confer antibiotic resistance. Certain genes on one type confer the ability to engage in bacterial conjugation. Plasmids are transmitted to daughter cells and may be transferred, by way of bacterial conjugation, to cells of the same species or a different species.

5. Bacteria as a group show great metabolic diversity, as in their modes of acquiring energy and carbon.

   a. *Photoautotrophs* use sunlight and carbon dioxide during photosynthesis. They include cyanobacteria and other oxygen-releasing species of eubacteria. They also include green nonsulfur and purple nonsulfur bacteria that do not produce oxygen; these obtain electrons from sulfur and other inorganic compounds, not from water.

   b. *Photoheterotrophs* use sunlight energy and organic compounds as sources of carbon. They include certain archaebacteria and eubacteria.

   c. *Chemoautotrophs*, including the nitrifying bacteria, use carbon dioxide but not sunlight. Different kinds get energy by stripping electrons from various organic or inorganic substances.

   d. The *chemoheterotrophs* are either parasites (which draw carbon and energy from living hosts) or saprobes (which feed on the organic products, wastes, or remains of other organisms). Most bacteria are in this category. They include major decomposers and pathogens.

6. Bacteria can make behavioral responses to stimuli, as when they move toward regions with more nutrients.

7. Viruses are nonliving, noncellular agents that infect particular species of nearly all organisms.

   a. Each virus particle consists of a core of DNA or RNA and a protein coat that sometimes is enclosed in a lipid envelope. Many glycoprotein spikes project from the envelopes. Coats of complex viruses have sheaths, tail fibers, and other accessory structures.

   b. A virus particle cannot reproduce on its own. Its genetic material must enter a host cell and direct the cellular machinery to synthesize the materials necessary to produce new virus particles.

8. Nearly all viral multiplication cycles include five steps: attachment to a suitable host cell, penetration of it, viral DNA or RNA replication and protein synthesis, assembly of new viral particles, and release.

9. Two pathways are common in the multiplication cycle of bacteriophages (bacteria-infecting viruses). In a lytic pathway, multiplication is rapid and new viral particles are released by lysis. In a lysogenic pathway, the infection enters a latent period. The host cell is not killed outright, and the viral nucleic acid may undergo genetic recombination with a host cell chromosome.

10. Multiplication cycles of viruses are diverse. They proceed rapidly or enter a latent phase. Penetration and release of most enveloped types occur by endocytosis and exocytosis. The DNA viruses spend part of the cycle in the nucleus of a host cell. The RNA viruses complete the cycle in the cytoplasm. The viral RNA is the template for mRNA synthesis and for protein synthesis.

Review Questions

1. Label the structures on this generalized bacterial cell: *22.1*

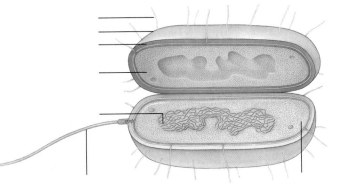

2. Describe the key metabolic and structural features of bacteria. Make sketches of the three basic shapes of bacterial cells. *22.1*

3. Compared to your own bodily growth, how is bacterial growth measured? *22.2*

4. With respect to bacterial classification, what are some of the pitfalls of numerical taxonomy? How are comparisons of rRNA sequences from different bacterial groups assisting classification efforts? *22.3, 22.6*

5. Name a few of the photoautotrophic, chemoautotrophic, and chemoheterotrophic eubacteria. Describe some that are likely to give humans the most trouble, medically speaking. *22.4*

6. What is a virus? Why is a virus considered to be no more alive than a chromosome? *22.7*

7. Distinguish between:
   a. microorganism and pathogen *CI*
   b. infection and disease *22.9*
   c. epidemic and pandemic *22.9*

Self-Quiz *(Answers in Appendix IV)*

1. Nondividing bacteria have _____ chromosome(s) and may have extra circles of _____ called plasmids.
   a. one; RNA   c. one; DNA
   b. two; RNA   d. two; DNA

2. _____ live in habitats much like those of the early Earth.
   a. Cyanobacteria   c. Archaebacteria
   b. Eubacteria   d. Protozoans

3. Which of the following is not grouped with archaebacteria?
   a. halophiles   c. thermophiles
   b. cyanobacteria   d. methanogens

4. Bacteria reproduce by _____ .
   a. mitosis   c. prokaryotic fission
   b. meiosis   d. longitudinal fission

5. Eubacterial cell walls are composed of _____ ; and a sticky mesh of polysaccharides, a _____ , often surrounds the wall.
   a. peptidoglycan; plasma membrane
   b. cellulose; glycocalyx
   c. cellulose; plasma membrane
   d. peptidoglycan; glycocalyx

6. Most bacteria are _____ and include major decomposers and pathogens.
   a. photoautotrophs   c. chemoautotrophs
   b. photoheterotrophs d. chemoheterotrophs

7. Viruses are _____ .
   a. the simplest living organisms   d. both a and b
   b. infectious particles   e. both b and c
   c. nonliving

8. Viruses have a _____ and a _____ .
   a. DNA core; carbohydrate coat
   b. DNA or RNA core; plasma membrane
   c. DNA-containing nucleus; lipid envelope
   d. DNA or RNA core; protein coat

9. Match the organisms with the most suitable descriptions.
   ____ archaebacteria   a. bacterial cell division mechanism
   ____ eubacteria   b. nonliving infectious particle with
   ____ virus   nucleic acid core and protein coat
   ____ plasmid   c. small circle of bacterial DNA
   ____ prokaryotic   d. methanogens, halophiles,
        fission   extreme thermophiles
                          e. most common prokaryotic cells

Critical Thinking

1. *Salmonella* bacteria, which cause a form of food poisoning, often live in poultry and eggs, but not in newly hatched chicks until chicks eat bacteria-laden feces of healthy adult chickens. Harmless bacteria ingested this way colonize the surface of intestinal cells, leaving no place for *Salmonella* to take hold. Some farmers raise thousands of chicks in confined quarters with no adult chickens. Should they consider feeding the chicks a known mixture of bacteria from a lab or a mixture of unknown bacteria from healthy adult chickens? Devise an experiment to test which approach may be more effective.

2. In 1852 in the Coorong district of Australia, a rubbery substance was discovered near a lake that had formed after a period of heavy rainfall. The substance, later named coorongite, consisted of the same kinds of hydrocarbons as crude oil. The discovery sparked sixty years of feverish oil drilling, even though geologists knew the place wasn't likely to yield oil. Later, biologists wondered if microorganisms had formed the substance. Drilling continued. Then someone found a green, scummy mat of bacteria (*Botyrococcus braunii*) in a natural lagoon. After water around the mat evaporated, the mat congealed into coorongite. Chemical analysis suggests it may be an early stage in the formation of shale oil. Formulate a hypothesis to explain how the Earth's great oil reserves formed.

3. Reflect on the description of mad cow disease in Section 22.9. The FDA is moving to prohibit all protein supplements for sheep, cattle, and other ruminants. Investigate the extent of this practice in the United States and how it has been monitored.

Selected Key Terms

archaebacterium *22.3*	infection *22.9*
bacillus (bacilli) *22.1*	lysis *22.8*
bacterial chromosome *22.2*	lysogenic pathway *22.8*
bacterial conjugation *22.2*	lytic pathway *22.8*
bacterial flagellum *22.1*	methanogen *22.5*
bacteriophage *22.7*	microorganism *CI*
cell wall *22.1*	numerical taxonomy *22.3*
coccus (cocci) *22.1*	pandemic *22.9*
disease *22.9*	pathogen *CI*
emerging pathogen *22.9*	pilus (pili) *22.1*
endospore *22.4*	plasmid *22.2*
epidemic *22.9*	prion *22.7*
eubacterium *22.3*	prokaryotic cell *22.1*
extreme thermophile *22.5*	prokaryotic fission *22.2*
fruiting body *22.4*	spirillum (spirilla) *22.1*
glycocalyx *22.1*	strain (bacterial) *22.6*
Gram stain *22.1*	viroid *22.7*
halophile *22.5*	virus *22.7*
heterocyst *22.4*	

Readings

Brock, T., et al. 1994. *Biology of Microorganisms*. Seventh edition. Englewood Cliffs, New Jersey: Prentice-Hall.

Geunno, B. October 1995. "Emerging Viruses." *Scientific American* 273(4): 56–62.

Hively, W. May 1997. "Looking for Life in All the Wrong Places." *Discover*: 76–85. Account of microbes that live in the most extreme environments on and in the Earth.

Web Site See *http://www.wadsworth.com/biology* for practice quiz questions, hypercontents, BioUpdates, and critical thinking. The Wadsworth Biology Resource Center provides a wealth of information fully organized and integrated by chapter.

23

PROTISTANS

Kingdom at the Crossroads

More than 2 billion years ago, in tidal flats and soils, in estuaries, streams, lakes, and lagoons, prokaryotic cells were inconspicuously changing the world. Ever since the time of life's origin, the Earth's atmosphere had been free of oxygen, and anaerobic bacteria had reigned supreme. Their realm was about to shrink, drastically. An oxygen-producing pathway of photosynthesis had been operating in vast populations of cells. Gaseous oxygen had been dissolving in aquatic habitats and escaping into the air, and its atmospheric concentration was beginning to approach its current level.

The oxygen-enriched atmosphere was a selection pressure of global dimensions. Thereafter, anaerobic species that could not neutralize the highly reactive, potentially lethal gas were restricted to black muds, stagnant waters, and other anaerobic habitats. Others developed oxygen tolerance. It must have been a short evolutionary step from having a capacity to neutralize oxygen to using it in metabolic reactions; some aerobic species emerged in nearly all bacterial groups.

This metabolic innovation was the start of rampant competition for resources. In the presence of so much free oxygen, energy-rich organic compounds could no longer accumulate by geochemical processes. Organic compounds formed by living cells became the premier source of carbon and energy. Through evolutionary experimentation, novel ways of acquiring and using the compounds developed. Many diverse bacteria engaged in new kinds of partnerships, predations, and parasitic interactions, as outlined in Section 21.4. In less than 500 million years, some of the experiments led to the first eukaryotic cells. And in short order, eukaryotic lineages were branching in different directions (Figure 23.1).

Of all existing organisms, **protistans** are most like the earliest, structurally simple eukaryotes. Although comparative biochemistry is now yielding insights into their phylogenies, it is still said that protistans are most easily classified by what they are *not*. In later chapters, you will read about structural and functional traits that distinguish protistans from plants, fungi, and animals. By a process of elimination, "protistans" are organisms that do not show these traits and that are not bacteria.

Like all other eukaryotes, protistans differ from the bacteria in several respects. At the least, they have a nucleus, large ribosomes, mitochondria, endoplasmic reticulum, and Golgi bodies. Their chromosomes are DNA molecules with a great many histones and other proteins. Protistans assemble microtubules for use in a cytoskeleton, in spindles that move chromosomes, and in the 9 + 2 core of flagella or cilia. Many species have chloroplasts. And depending on the species, protistans reproduce by way of mitosis, meiosis, or both.

a

b

Figure 23.1 A sampling of the confounding diversity among protistans. (**a**) Thriving in the surf pounding against a rocky shoreline, a stand of *Postelsia palmaeformis*, one of the multicelled brown algae. (**b**) *Micrasterias*, a single-celled freshwater green alga, here dividing in two. (**c**) *Physarum*, a plasmodial slime mold. This aggregation of yellow-gold cells is migrating along a rotting log. (**d**) Mealtime for *Didinium*, a ciliated protozoan with a big mouth. *Paramecium*, a different ciliated protozoan, is poised at the mouth (*left*) and swallowed (*right*).

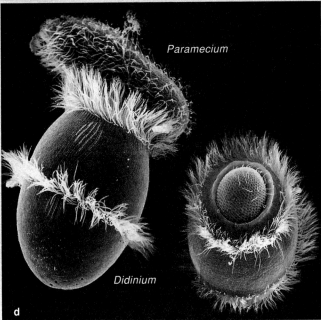

Paramecium

Didinium

50 µm

The photosynthetic types range from microscopic single cells to giant seaweeds. The saprobes resemble some bacteria and fungi. Some of the predators and parasites resemble animals. The vast majority are single-celled, yet nearly every lineage has multicelled forms, and some have very close evolutionary ties to other kingdoms. This chapter introduces the major lineages. As you will see, the chytrids, water molds, slime molds, protozoans, and sporozoans are heterotrophs. Among the euglenoids, chrysophytes, and dinoflagellates are species that are photosynthetic, heterotrophic, or both. Most red, brown, and green algae are evolutionarily committed to photosynthesis.

Opinions differ on how to classify many of these confoundingly diverse organisms. For example, many biologists view the multicelled algae as protistans, and about as many view them as plants. Don't worry about memorizing where the lines of classification schemes are being drawn. Instead, simply become familiar with the groups themselves. In time, biochemical studies will help bring the evolutionary picture into sharper focus.

KEY CONCEPTS

1. The kingdom Protista includes all of the free-living eukaryotic cells and a number of the structurally simple, multicelled species. Some protistan groups have close evolutionary ties to other kingdoms.

2. Protistans are easily distinguishable from prokaryotes (the bacteria) but are difficult to classify with respect to other eukaryotes. They also differ enormously from one another in their morphology and life-styles.

3. Among the fungus-like protistans are parasitic and predatory molds that produce spores. The majority of the chytrids and water molds are single-celled decomposers in aquatic habitats. The phagocytic slime molds live as single amoeboid cells and as aggregations of cells that migrate together and form spore-producing structures.

4. The animal-like protistans include nonphotosynthetic flagellated protozoans, sporozoans, and ciliates. Among their ranks are single-celled predators, grazers, and parasites. In any given year, hundreds of millions of people are affected by infections caused by a few dozen species of flagellated protozoans and sporozoans.

5. Among the plant-like protistans are single-celled, photosynthetic flagellates. Astounding numbers of them live freely or as colonies in the open ocean and other bodies of water. These protistans are important members of phytoplankton, the food producers that are the start of nearly all food webs of aquatic habitats.

6. The protistans most like plants are the red, brown, and green algae; indeed, many botanists prefer placing them in the plant kingdom. Most species are photosynthetic. They range from single-celled to multicelled forms that show great diversity in size, morphology, life-styles, reproductive modes, and habitats.

PARASITIC AND PREDATORY MOLDS

Three groups of protistans have members that resemble fungi in some respects and were once classified with them. These groups are the **chytrids**, **water molds**, and **slime molds**. Like the fungi, all produce spore-bearing structures, and all are heterotrophs. Like fungi, many of their species are saprobic decomposers or parasites. **Saprobes** secrete digestive enzymes that break down organic compounds produced by other organisms, then they absorb the breakdown products. Unlike fungi, the slime molds are phagocytic predators. Also, members of all three groups differ from the fungi in producing motile cells during the life cycle.

Chytrids

The chytrids (Chytridiomycota) are common in marine and freshwater habitats, where they absorb nutrients from plant debris or from necrotic plant tissues. Most of the 575 species are saprobic decomposers, but some are parasites.

When chytrids grow in damp environments, their populations become fuzzy, pale masses that resemble fungal molds. Chytrids are actually similar to fungi in some biochemical respects. Also, chitin reinforces the cell wall of some species, as it does in fungi.

Single-celled species of chytrids produce flagellated asexual spores. When their spores are released into the surroundings, they settle on a host cell, germinate, then develop into globular cells having rootlike absorptive

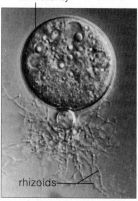

globe-shaped cell that will be a spore-producing structure at maturity

rhizoids

25 µm

Figure 23.2 One of the common chytrids, *Chytridium confervaie*.

structures called rhizoids (Figure 23.2). At maturity, the globe-shaped cells become spore-producing structures.

As in many fungi, complex multicelled species have a mesh of fine absorptive filaments called a **mycelium** (plural, mycelia). Walls of individual cells usually do not cut across the filaments, and cytoplasmic streaming throughout the mycelium distributes enzymes and the nutrients from digested food (compare Section 4.9).

Water Molds

Water molds (Oomycota) may be distantly related to the yellow-green and brown algae. Most of the 580 known types are key saprobic decomposers in aquatic habitats. Of these, many are free-living species that get nutrients from plant debris in ponds, lakes, and streams. Others live inside necrotic tissues of living plants. Some water molds parasitize aquatic organisms. The pale, cottony growths you may have seen on some aquarium fish are mycelia of a parasitic type, *Saprolegnia* (Figure 23.3a).

Most species of water molds produce an extensive mycelium, and some of its filaments differentiate into gamete-producing structures. At fertilization, a male and female gamete fuse and thus form a diploid zygote, which develops into a thick-walled resting spore. After germination, a new mycelium develops from the spore. Water molds also produce flagellated, asexual spores.

Some water molds are major plant pathogens. For example, grapes with *downy mildew* have been attacked by *Plasmopara viticola* (Figure 23.3b). Another pathogen seriously influenced human affairs. Over a century ago, Irish peasants grew potatoes as their main food crop. Between 1845 and 1860, the growing seasons were cool and damp, year after year, and conditions encouraged the rapid spread of *Phytophthora infestans*. This water mold causes *late blight*, a rotting of potato (and tomato) plants. Its abundant spores were dispersed unimpeded through the watery film on the plants. Destruction was rampant. During a fifteen-year period, one-third of the population in Ireland starved to death, died during the outbreak of typhoid fever that followed as a secondary effect, or fled to the United States and other countries.

Slime Molds

All slime molds produce free-living, amoebalike cells during part of the life cycle. The group includes *cellular* slime molds (Acrasiomycota) and *plasmodial* slime molds (Myxomycota). Their cells crawl on rotting plant parts, such as decaying leaves and bark. Like true amoebas, they are phagocytic predators. They engulf bacteria and yeasts, spores, and various organic compounds. When nutrients are scarce and cells are starving, many of the cells aggregate and form a slimy mass that may migrate

Figure 23.3 Effects of two water molds. (**a**) The parasitic water mold *Saprolegnia* has destroyed tissues of this aquarium fish tail. (**b**) *Plasmopara viticola* causes downy mildew in grapes. At times it has threatened huge vineyards in Europe and North America.

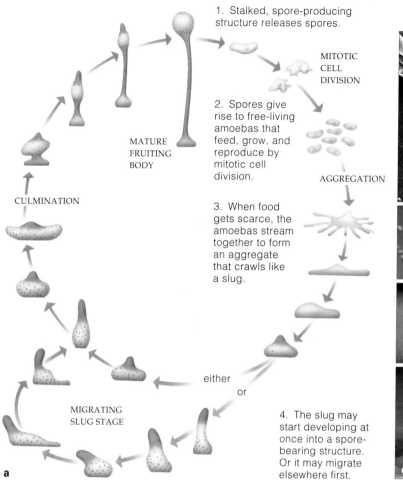

1. Stalked, spore-producing structure releases spores.

MITOTIC CELL DIVISION

MATURE FRUITING BODY

2. Spores give rise to free-living amoebas that feed, grow, and reproduce by mitotic cell division.

CULMINATION

AGGREGATION

3. When food gets scarce, the amoebas stream together to form an aggregate that crawls like a slug.

either

or

MIGRATING SLUG STAGE

4. The slug may start developing at once into a spore-bearing structure. Or it may migrate elsewhere first.

a

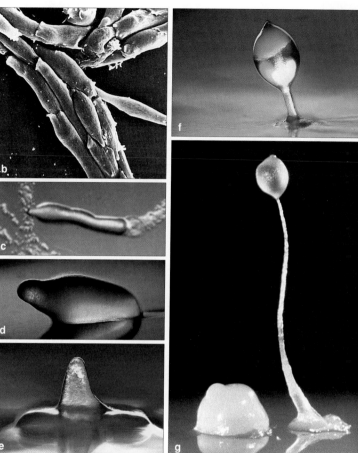

b

c

d

e

f

g

Figure 23.4 *Dictyostelium discoideum*, a cellular slime mold. (**a**) The life cycle includes a spore-producing stage. The spores give rise to free-living amoebas, which grow and divide until food (soil bacteria) dwindles. In response to cyclic AMP, a chemical signal that they secrete, amoebas stream toward one another. As the amoebas aggregate, their plasma membranes become sticky and they adhere to one another. A cellulose sheath forms around the aggregation, which starts crawling like a slug (**b–d**). Some slugs may incorporate 100,000 or more amoebas.

As a slug migrates, the amoebas develop into different types of cells: prestalk (*red*), prespore (*white*), and anteriorlike cells (*brown dots*). Prestalk cells secrete ammonia in amounts that vary with temperature and light intensity. The slug moves most rapidly in response to intermediate amounts of ammonia (not too little, not too much), which correspond to warm, moist conditions (not too cold or hot, not soppy or dry). Where such conditions exist at the surface of a substrate, prestalk cells and prespore cells differentiate and form a stalked, spore-bearing structure (**e–g**). Anteriorlike cells, which sort into two groups, may function in elevating the nonmotile spores for dispersal from the top of the spore-bearing structure.

to a new location where conditions are more favorable for growth. Collectively, the contractions of single cells move the mass. Later in the life cycle, the amoebalike cells develop into a few cell types that differentiate and

form a spore-bearing structure. After the spores are released, they germinate on warm, damp surfaces. Each gives rise to one amoebalike cell. Sexual reproduction by way of gametes also is common among slime molds.

Dictyostelium discoideum, one of 70 species of cellular slime molds, is commonly used for laboratory studies of development. Figure 23.4 shows its life cycle. Figure 23.1c shows one of the 500 species of plasmodial slime molds. When amoebalike cells of the members of that phylum aggregate, each cell's plasma membrane breaks down. Cytoplasm flows unimpeded and so distributes nutrients and oxygen through the mass. The streaming mass (the plasmodium) often will occupy several square meters and migrate if food runs out. You may have seen one crossing a lawn or a road, or even climbing a tree.

Most chytrids and water molds are saprobic decomposers of aquatic habitats. Some are single cells. Like slime molds, many are free-living predators some of the time and simple experiments in multicellularity at other times.

During a slime mold life cycle, amoeboid cells aggregate to form a migrating mass. Cells in the mass differentiate, then they form reproductive structures and spores or gametes.

About 65,000 named species of single-celled protistans are informally called the **protozoans** ("first animals"), because they are thought to resemble the single-celled, heterotrophic protistans that gave rise to animals. In the next four sections we limit our sampling to amoebas, animal-like flagellates, sporozoans, and ciliates. Among their ranks are predators, parasites, and grazers that actively move about to secure a meal (Table 23.1).

Table 23.1   Major Groups of Animal-Like Protistans

SARCODINA

Amoeboid protozoans. Soft or shelled bodies; locomotion by pseudopods. Free-living or endosymbiotic heterotrophs.

   RHIZOPODS.   Naked amoebas, foraminiferans.

   ACTINOPODS. Radiolarians, heliozoans

MASTIGOPHORA

Animal-like flagellates. Includes free-living and parasitic heterotrophs. Locomotion by one or more flagella.

APICOMPLEXA

Parasitic heterotrophs with a complex of structures, such as rings, tubules, and cones, at the head end. The most familiar members of a group commonly called sporozoans.

CILIOPHORA

Diverse free-living, sessile, and motile heterotrophs with numerous cilia. Predators or symbionts, some parasitic.

Free-living species live in damp soil, in ponds, lakes, and other freshwater habitats, and in marine habitats. The symbionts and parasites live in or on moist tissues of other organisms. Some species are major pathogens.

Protozoans reproduce either asexually or sexually, although many species alternate between reproductive modes in response to environmental conditions. Most commonly, asexual reproduction is by **binary fission**, a process by which the individual's body divides in two. The division plane is random for amoebas, longitudinal for flagellates, and transverse for ciliates. Budding from a parent organism also occurs. Some species undergo **multiple fission**. More than two nuclei form, then each nucleus and a bit of cytoplasm separate as a daughter cell. Many parasitic types go through an encysted stage. The **cyst**, a resistant body covering made of their own secretions, helps them wait out stressful conditions.

Protozoans are single-celled protistans that are animal-like in some respects, as in the way they actively move about to secure their food. They include predators, parasites, and grazers. Some species are pathogens that cause serious diseases in humans and other animals.

Let's first look at **amoeboid protozoans** (Sarcodina), which are naked amoebas, foraminiferans, heliozoans, and radiolarians. None of them has permanent motile structures. Each moves by a combination of cytoplasmic streaming and formation of **pseudopods** ("false feet"). As described in Section 4.9, pseudopods are temporary cytoplasmic extensions of the cell body. Figure 23.5 shows the rather thick pseudopods of a naked amoeba.

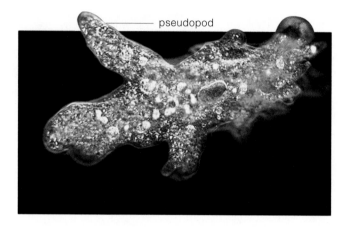

— pseudopod

Figure 23.5 *Amoeba proteus*, one of the naked amoebas. Compared with most other kinds of amoeboid protozoans, the pseudopods projecting from its body are stubbier. This species is a favorite for laboratory experiments in biology classes.

Rhizopods

The naked amoebas and foraminiferans are members of a group known as **rhizopods**. The constantly changing cytoskeleton alone lends structural support to a naked amoeba's soft body, which never shows any symmetry. You find these protistans in damp soil, freshwater, and saltwater. Most species are free-living phagocytes that engulf other protozoans and bacteria. Some live in the gut of certain invertebrates or vertebrates and normally do no harm. But a few are opportunistic parasites, and so are certain free-living species that happen to enter the gut with food or water. When conditions in the gut fan their population growth, these types cause intestinal problems.

Entamoeba histolytica is a pathogenic type that must complete its life cycle in human intestines. The infective stage becomes encysted and leaves the body in feces. The encysted stage is inactive in water or soil. After it enters a new host, the trophozoite—the active, motile feeding stage—emerges from it. Trophozoites multiply rapidly by binary fission and release lysozymes, which destroy the cells making up the lining of the colon and rectum. Painful abdominal cramps and diarrhea are symptoms of this infectious disease, called *amoebic dysentery*.

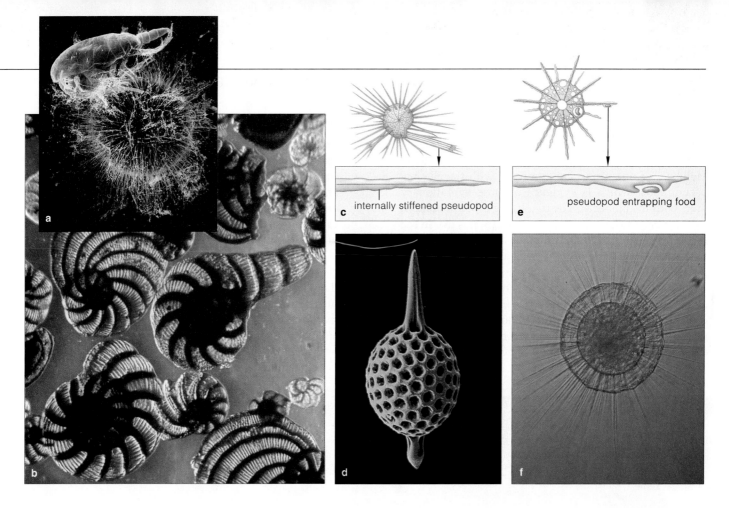

Figure 23.6 (**a**) Foraminiferan with prey. (**b,c**) Foraminiferan skeletons and body plan. (**d**) Radiolarian skeleton. (**e,f**) Body plan and micrograph of a living heliozoan. As in radiolarians, a core of microtubules reinforces the many fine pseudopods.

Labels within figure: internally stiffened pseudopod (**c**); pseudopod entrapping food (**e**)

Amoebic dysentery is most prevalent in subtropical and tropical regions with contaminated water supplies and poor sanitation. Where cyst-harboring water, fruits, or vegetables abound, between 10 and 50 percent of the human population may be infected. The severity of an infection depends on the virulence of the *E. histolytica* strain and the state of an individual's immune system. People who recover show full or partial resistance to subsequent infection. Worldwide, amoebic dysentery is a leading cause of death among infants and toddlers, whose immune systems are not yet well developed.

Unlike naked amoebas, the foraminiferans have a highly perforated external skeleton of secreted organic substances that are hardened with calcium carbonate (Figure 23.6 *a–c*). Thin pseudopods extend through the perforations, and prey becomes trapped in mucus that covers them. Most foraminiferans live on the seafloor. All but 1 percent of the named species are extinct. The fossilized, compacted shells of countless foraminiferans make up distinctive sedimentary rock layers, including the famed white cliffs of Dover, England. We now use their calcified remains in cement and blackboard chalk.

Actinopods

Actinopods ("ray feet") include the radiolarians and heliozoans. Their name refers to the numerous slender, reinforced pseudopods that radiate from the body. Like foraminiferans, the radiolarians are well represented in the fossil record, for they have ornate, silica-hardened parts that resist degradation (Figure 23.6 *d*). Many of these skeletal parts are thin and project outward with the stiffened pseudopods. Nearly all radiolarians drift with ocean currents, as part of **plankton**. The word refers to aquatic communities of passively drifting or motile organisms, mostly microscopic in size. A few species of radiolarians form colonies in which numerous cells are cemented together.

Heliozoans ("sun animals") have many pseudopods radiating like the sun's rays (Figure 23.6 *e,f*). Most are free-floating or bottom-dwellers in freshwater habitats. As in radiolarians, a membrane or capsule divides the single-celled body into two biochemically distinct zones. The inner zone contains the nucleus. The outer zone has so many vacuoles in which digestion occurs that it appears almost frothy. The vacuoles impart buoyancy to the cell, which can remain suspended in the water.

Amoeboid protozoans are soft-bodied single cells, some with hardened skeletal elements. All form pseudopods for use in motility and prey capture. Free-living and gut-dwelling species, some parasitic, belong to this group.

Animal-Like Flagellates

Animal-like flagellates (Mastigophora) are free-living predators and parasites bearing one to several flagella. Members of one group are equipped with flagella *and* pseudopods, which is evidence of close evolutionary links with the amoebas. The free-living types abound in freshwater or marine habitats. Parasitic types live in the moist tissues of plants and animals, including humans.

Giardia lamblia (Figure 23.7*a*) is a notorious internal parasite. In any given year, it may infect as many as 10 percent of the human population in the United States alone. It usually causes mild intestinal upsets, such as diarrhea, but it has severe effects on a few susceptible people. Wild animals and foraging cattle often contain the parasite. It leaves the body encysted in feces, then

Representative Sporozoans

Sporozoan is an informal name for parasitic protistans that complete part of the life cycle *inside specific cells of host organisms*. These parasites form sporozoites, a type of motile infective stage; some have encysted stages. At one end of the body is a complex of structures that function in penetrating host cells. The sporozoans are often grouped as phylum Apicomplexa.

Many sporozoans cause human diseases. *Plasmodium* causes malaria (Section 23.5). *Cryptosporidium* strains that cause serious intestinal disorders now contaminate most water supplies. One outbreak in 1993 hospitalized thousands of people in Milwaukee. Symptoms normally subside after ten days. However, people with weakened immune systems can suffer for months or years.

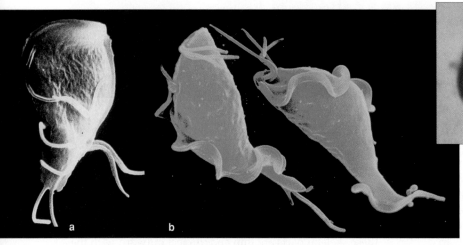

trypanosome   red blood cell

Figure 23.7 Three notorious animal-like flagellates. (**a**) *Giardia lamblia* causes severe intestinal ailments. (**b**) *Trichomonas vaginalis* causes trichomoniasis, one of the sexually transmitted diseases. (**c**) *Trypanosoma brucei* causes African sleeping sickness.

infects people who drink water or ingest food tainted with the feces. Because even remote streams may harbor cysts, streamwater should be boiled before being drunk.

Many trichomonads are among the parasitic members of the group. *Trichomonas vaginalis*, a worldwide nuisance, can infect new human hosts during sexual intercourse (Figure 23.7*b*). Unless trichomonad infections are treated, they can damage the urinary and reproductive tracts.

Trypanosomes also are parasites. One, *Trypanosoma brucei*, causes *African sleeping sickness* (Figure 23.7*c*). Bites of the tsetse fly transmit it to hosts, where it may damage the nervous system. *T. cruzi*, prevalent in South America and Mexico, causes *Chagas disease*. Bugs pick up the parasite when they feed on infected humans and other animals. The parasite multiplies in the insect gut, then it may be excreted onto a host. Scratches in skin invite infection. The liver and spleen enlarge, eyelids and the face swell up, then the brain and heart become severely damaged. There is no treatment or cure.

Toxoplasma completes the sexual phase of its life cycle in cats and asexual phases in humans, cattle, and other animals. Flies and cockroaches can spread cysts from cat feces to the table. Humans also can get infected if they eat undercooked or raw meat that contains cysts. The resulting disease, *toxoplasmosis*, is not prevalent, but an infection can cause birth defects. Pregnant women who become infected may transmit the parasite to the fetus, which may suffer brain damage or die. They should never empty litterboxes or clean up after cats.

Fewer than two dozen species of the animal-like flagellates and sporozoans cause serious diseases in humans. But they infect hundreds of millions each year. There are no effective vaccines against them.

Animal-like flagellates are motile heterotrophs, free-living or parasitic. Sporozoans are internal parasites that produce infective, motile stages. Members of both phyla often form encysted stages.

MALARIA AND THE NIGHT-FEEDING MOSQUITOES

Each year in Africa alone, about a million people die of *malaria*, a long-lasting disease that four different parasitic species of *Plasmodium* can cause. This type of sporozoan currently has infected more than 100 million people.

Only female mosquitoes of the genus *Anopheles* can transmit the parasites to human hosts. Their bite delivers sporozoites (the infective, motile stage) to the host's blood. The bloodstream transports sporozoites to the liver, where they undergo multiple fission. Some of the resulting cells, called merozoites, penetrate and asexually reproduce in red blood cells, which they destroy. Others develop into male and female gametocytes, which will mature into gametes (Figure 23.8).

Symptoms of malaria begin after infected cells abruptly rupture and release merozoites, metabolic wastes, and cellular debris into the individual's bloodstream. Shaking, chills, a burning fever, and drenching sweats are classic symptoms. After one episode, symptoms subside for a few weeks or months. Infected individuals might even feel normal, but relapses inevitably recur. In time, anemia and gross enlargement of the liver and the spleen may result.

Amazingly, the *Plasmodium* life cycle is sensitive to the body temperature and oxygen levels inside humans and mosquitoes. The gametocytes cannot mature in humans, who are warm-bodied and have little free oxygen in their bloodstream (most is bound to hemoglobin). They are induced to mature inside the mosquito, which has a lower body temperature and which incidentally slurps in oxygen from the air along with the gametocyte-containing blood.

Mature gametes fuse to form zygotes. These repeatedly divide and form many sporozoites, which migrate to the female mosquito's salivary glands and await the next bite.

Malaria has been around for a long time. People were describing its symptoms more than 2,000 years ago. It received its name in the seventeenth century, when Italians made a connection between the disease and noxious gases from swamps near Rome, where mosquitoes flourished (*mal*, bad; *aria*, air). By severely incapacitating so many people, malaria contributed to the decline of the ancient empires of Greece and Rome. Much later in history, it also incapacitated soldiers during the United States Civil War, World War II, and the Korean and Vietnam conflicts.

Throughout human history, malaria has been most prevalent in tropical and subtropical parts of Africa. Today, however, the numbers of cases reported in North America and elsewhere are increasing dramatically, owing to the hordes of globe-hopping travelers and unprecedented levels of human immigration.

Travelers who intend to visit countries with high rates of malaria are advised to use antimalarial drugs such as chloroquine. However, certain strains of *Plasmodium* are now resistant to the drugs, and a vaccine has been difficult to develop. *Vaccines* are preparations that induce the body to build up resistance to a specific pathogen. Experimental vaccines for malaria are not equally effective against all the different stages that develop during sporozoan life cycles. This is generally the case for vaccines that researchers hope to develop against most parasites with complex life cycles.

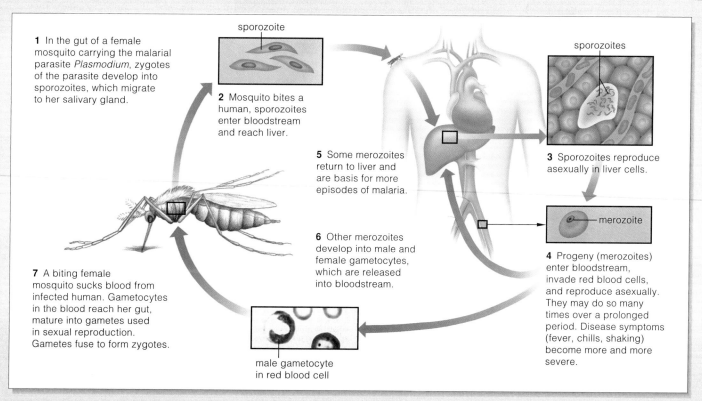

1 In the gut of a female mosquito carrying the malarial parasite *Plasmodium*, zygotes of the parasite develop into sporozoites, which migrate to her salivary gland.

sporozoite

2 Mosquito bites a human, sporozoites enter bloodstream and reach liver.

5 Some merozoites return to liver and are basis for more episodes of malaria.

6 Other merozoites develop into male and female gametocytes, which are released into bloodstream.

7 A biting female mosquito sucks blood from infected human. Gametocytes in the blood reach her gut, mature into gametes used in sexual reproduction. Gametes fuse to form zygotes.

male gametocyte in red blood cell

sporozoites

3 Sporozoites reproduce asexually in liver cells.

merozoite

4 Progeny (merozoites) enter bloodstream, invade red blood cells, and reproduce asexually. They may do so many times over a prolonged period. Disease symptoms (fever, chills, shaking) become more and more severe.

Figure 23.8   Life cycle of one of the sporozoans (*Plasmodium*) that causes malaria.

A SAMPLING OF CILIATED PROTOZOANS

Ciliated protozoans (Ciliophora) typically have profuse arrays of cilia at their surface. You read about these fine motile structures in Section 4.9. Most ciliates use them for swimming through freshwater and marine habitats, where they prey on bacteria, tiny algae, and one another, as Figure 23.1*d* so aptly suggested. The cilia beat in such a synchronized pattern over the body surface, they call to mind a soft wind through a field of tall grasses.

Paramecium is typical of the group (Figure 23.9). A fully grown cell is about 150 to 200 micrometers long. Outer membranes form a **pellicle**, a body covering that may be rigid or quite flexible, depending on how those membranes are organized. Inside the gullet, which starts at an oral depression at the body surface, some cilia sweep food-laden water into the cell body. There, the food becomes enclosed in enzyme-filled vesicles and is digested. Like the amoeboid protozoans, *Paramecium's* internal solute concentrations are greater than those of its surroundings. Therefore, it must continually counter

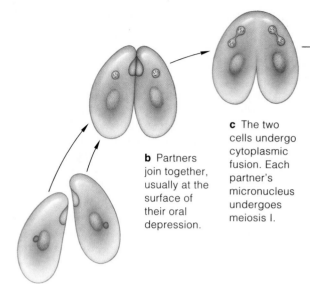

c The two cells undergo cytoplasmic fusion. Each partner's micronucleus undergoes meiosis I.

b Partners join together, usually at the surface of their oral depression.

a Prospective partners meet up.

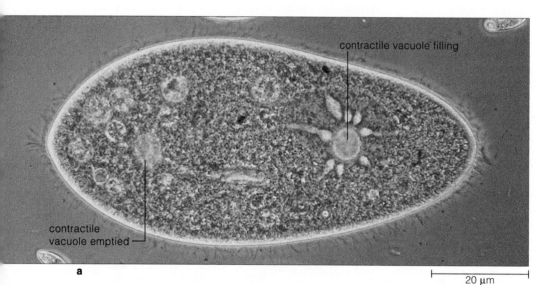

contractile vacuole filling

contractile vacuole emptied

a

20 μm

Figure 23.9 Representatives of two groups of ciliated protozoans.

Left and below: Paramecia. (**a**) Light micrograph of a living paramecium. (**b**) Generalized diagram of the body plan for the genus *Paramecium*. (**c**) Components of the pellicle. Trichocysts, which span the pellicle, are organelles that discharge threads when the cell is irritated. They may be a defense against predators.

Facing page: (**d**) Scanning electron micrograph of the arrays of cilia at the surface of these single-celled predators. Hypotrichs. (**e**) This free-living species lives in the Bahamas. The hypotrichs are the most animal-like of the ciliates. They run around on leglike tufts of cilia. Some have a "head" end with modified sensory cilia.

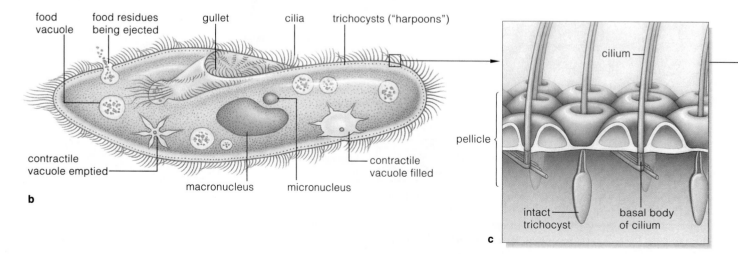

food vacuole

food residues being ejected

gullet

cilia

trichocysts ("harpoons")

cilium

contractile vacuole emptied

macronucleus

micronucleus

contractile vacuole filled

pellicle

intact trichocyst

basal body of cilium

b

c

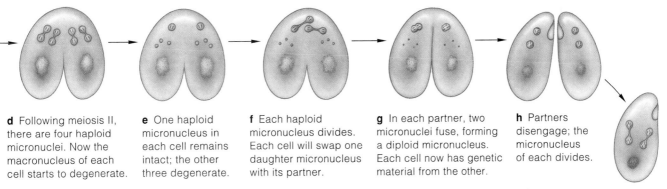

d Following meiosis II, there are four haploid micronuclei. Now the macronucleus of each cell starts to degenerate.

e One haploid micronucleus in each cell remains intact; the other three degenerate.

f Each haploid micronucleus divides. Each cell will swap one daughter micronucleus with its partner.

g In each partner, two micronuclei fuse, forming a diploid micronucleus. Each cell now has genetic material from the other.

h Partners disengage; the micronucleus of each divides.

i Micronuclei divide again in each cell; the original macronucleus degenerates.

Figure 23.10 A generalized diagram of protozoan conjugation, a form of sexual reproduction, as demonstrated by the two ciliates that first encounter each other in (**a**).

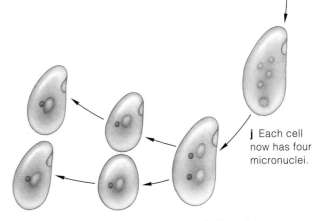

j Each cell now has four micronuclei.

k Two of the micronuclei develop into macronuclei.

l Cytoplasmic division now begins.

m Two daughter cells result. Each has a micronucleus and a macronucleus.

water's tendency to diffuse inward by osmosis. (Here you may wish to refer to Section 5.4.) Like the amoebas, it depends on **contractile vacuoles**. Tiny tubes extend from the center of these organelles and collect excess water that moves osmotically into the cell body. A filled vacuole contracts and forces water through a small pore to the outside.

As is true of other protistan groups, the ciliates show diversity in life-styles. Like *Paramecium*, about 65 percent of the known species are free-living and motile. Others attach permanently or temporarily to substrates, often by stalks. Some form colonies. About 30 percent live in or on other organisms, as symbionts. *Balantidium coli* is the largest protozoan parasite of humans, with effects like those of *Entamoeba histolytica* infections.

Ciliates can reproduce sexually and asexually, and things get interesting because each cell commonly has

two types of nuclei. When a cell reproduces asexually by binary fission, a small, diploid *micro*nucleus divides by mitosis. And the large *macro*nucleus lengthens and splits in two a bit sloppily; some of the DNA may spill out. Most often, sexual reproduction is by a unique form of **conjugation**, as shown in Figure 23.10. In brief, partner ciliates repeatedly divide their micronuclei, swap two daughter micronuclei, and allow two others to fuse and form a diploid macronucleus to replace the one that disappears. And you probably thought sex among the single-celled critters was simple!

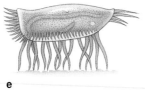

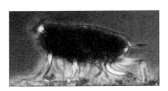

e

d

Ciliated protozoans bear a profusion of cilia. These motile structures are used for swimming and for beating food into an oral cavity. Each ciliate usually has two types of nuclei, which are distributed to daughter cells in unique ways during asexual and sexual reproduction.

Like other protistans, the ciliates show great diversity in life-styles and diverse variations on the basic body plan.

A SAMPLING OF THE (MOSTLY) SINGLE-CELLED ALGAE

The rest of this chapter surveys the protistans, largely photosynthetic, that are informally called "the algae." Let's first consider the euglenoids, chrysophytes, and dinoflagellates. Single-celled species dominate all three groups, which is not the case for the red, brown, and green algae (Table 23.2). The majority are members of **phytoplankton**: communities of aquatic photosynthetic species, mainly microscopic, that drift or swim weakly through water. As food producers, they are the basis for nearly all food webs in aquatic habitats. Diverse species occur in great numbers in inland seas, rivers, streams, lakes and ponds, and the open ocean (Section 7.8).

Euglenoids

The **euglenoids** (Euglenophyta) are a classic example of evolutionary experimentation. Euglenoids are free-living, flagellated cells that abound in freshwater or stagnant ponds and lakes. Most of the 1,000 known species are *photosynthetic*. The rest are *heterotrophs*; they subsist on organic compounds dissolved in the water.

Euglena has a profusion of organelles (Figure 23.11). Among these are chloroplasts with chlorophylls *a* and *b* and carotenoids, just like plant chloroplasts. *Euglena* has flagella (one long, one short) and a contractile vacuole, as animal-like protozoans do. Like some protozoans, it has a pellicle, this one being a flexible cover with spiral strips of a translucent, protein-rich material. Pigments form an "eyespot" that partly shields a light-sensitive receptor. By moving its long flagellum, the single cell keeps the receptor exposed to light and so stays where the light is most suitable for its activities.

Generally, "self-feeders" make their own vitamins, which are required for growth. But all photosynthetic euglenoids must get vitamin B_{12} from their surroundings (most cannot make vitamin B_1, either). Probably their chloroplasts originated from endosymbiotic green algae, in the manner described in Section 21.4. Given their array of traits, could euglenoids be very close relatives of some existing protozoans? Maybe, but probably not. Researchers believe their similarities are more likely an outcome of convergent evolution.

Chrysophytes

Chrysophytes (Chrysophyta) are mainly photosynthetic, free-living cells with chlorophylls *a*, c_1, and c_2. They include diatoms, golden algae, and yellow-green algae.

Most **diatoms** are photosynthetic. The silica shells of the 5,600 existing species are two perforated structures that overlap like a pillbox (Figure 23.12*a*). Substances move to and from the cell's plasma membrane through numerous perforations in the shell. For 100 million years, the shells of at least 35,000 species of now-extinct diatoms accumulated on the bottom of lakes and seas, and many sediments now contain their finely crumbled shells. We use the deposits in insulation, abrasives, and filters. More than 270,000 metric tons are quarried annually near Lompoc, California.

Among the 500 species of **golden algae** are many species with no obvious cell wall; they have silica scales

Table 23.2	The Mostly Photosynthetic Protistans
EUGLENOPHYTA	Euglenoids. Free-living cells. Autotrophic *and* heterotrophic.
CHRYSOPHYTA	Free-living cells, mostly autotrophic.
	CHRYSOPHYTES (golden algae, yellow-green algae). Mostly photosynthetic; some are heterotrophs.
	DIATOMS. Mostly photosynthetic; some symbionts with foraminiferans.
PYRRHOPHYTA	Dinoflagellates. Free-living cells. Autotrophic *and* heterotrophic.
RHODOPHYTA	Red algae. Mostly multicelled and photosynthetic.
PHAEOPHYTA	Brown algae. Multicelled; all photosynthetic.
CHLOROPHYTA	Green algae. Mostly multicelled and photosynthetic.

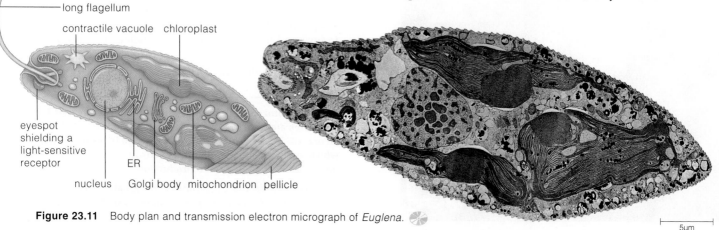

long flagellum

contractile vacuole · chloroplast

eyespot shielding a light-sensitive receptor

nucleus · Golgi body · mitochondrion · pellicle

ER

Figure 23.11 Body plan and transmission electron micrograph of *Euglena*.

5µm

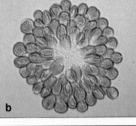
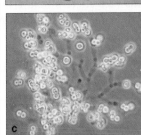

Figure 23.12 Some of the chrysophytes. (**a**) Shells of diatoms. (**b**) *Synura*, a golden alga. Its colonies in phytoplankton have a fishy odor. (**c**) A colonial type of yellow-green alga, *Mischococcus*, which occurs in phytoplankton. (**d**) *Vaucheria*, a yellow-green alga, growing on red sandstone in Arizona.

a

b

c

Figure 23.13 (**a**) Red tide in a central California bay. (**b**) Part of a fish kill that resulted from a dinoflagellate "bloom." (**c**) Scanning electron micrograph of one dinoflagellate, *Gymnodinium breve*, that causes red tides along Florida's coast.

or skeletal elements. As Figure 23.12*b* suggests, their chlorophylls are masked by a golden-brown carotenoid, fucoxanthin. Except for their chloroplasts, some of the amoeboid species closely resemble the amoebas.

Fucoxanthin is absent from the 600 known species of **yellow-green algae** (Figure 23.12*c*,*d*). Most species are nonmotile, but they have flagellated gametes. Yellow-green algae are common in many aquatic habitats.

Dinoflagellates

More than 1,200 species of **dinoflagellates** live in marine phytoplankton. Most are photosynthetic but many are heterotrophs. Some have flagella and cellulose plates. Dinoflagellates are yellow-green, green, brown, blue, or red, depending on their pigments and endosymbiotic

history. Some undergo population explosions and color the seas red or brown. A few produce a toxin, so the resulting **red tides** can be devastating (Figure 23.13). The toxin accumulates in tissues of mollusks and fishes, and it can poison humans who eat them. Such "algal blooms" may be causing massive die-offs of fish-eating migratory birds in California's Salton Sea, including 150,000 eared grebes in 1992–1993 and nearly 5,000 brown pelicans, an endangered species, in 1996. Raw sewage and fertilizer-enriched runoff from croplands fan such algal blooms, which are worldwide occurrences.

Nearly all single-celled photosynthetic protistans, including most euglenoids, chrysophytes, and dinoflagellates, belong to phytoplankton—the food-producing base of aquatic habitats.

Of 4,100 known species of **red algae** (Rhodophyta), nearly all are marine; only 200 live in freshwater. Red algae are most abundant in warm currents and tropical seas, often at surprising depths (265 meters below the surface) in clear water. A few occur in phytoplankton. Some encrusting types contribute to the formation of coral reefs and banks (Section 49.12).

Cells of some red algae have walls hardened with calcium carbonate. Different species appear red, green, purple, or nearly black, depending on which accessory pigments (phycobilins, mostly) mask their chlorophyll *a*. Phycobilins are good at absorbing the green and blue-green wavelengths, which can penetrate deep waters. (Chlorophylls are more efficient at absorbing red and blue wavelengths, which may not reach as far below the water's surface.) The chloroplasts of red algae resemble cyanobacteria, which suggests endosymbiotic origins.

The life cycle of most species includes multicelled stages, but these lack tissues or organs (Figure 23.14). Reproductive modes are diverse, with complex asexual and sexual phases. Figure 23.15 shows one example.

Red algae have a flexible, slippery texture owing to mucous material in their cell walls. Agar is made from extracts of wall material of several species. This inert, gelatinous substance is used as a moisture-preserving agent in baked goods and cosmetics, as a setting agent

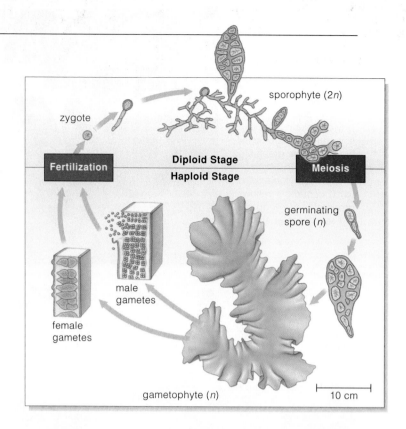

Figure 23.15 Life cycle of *Porphyra*. From 1623 to the 1950s, Japanese fishermen cultivated and harvested a species of this red alga only in the early fall. The rest of the year, it seemed to disappear. Then Kathleen Drew-Baker studied the sheetlike form of *P. umbilicus* in the laboratory. She observed gametes forming in packets interspersed between vegetative cells near the sheet margins; the sheet was a gametophyte (gamete-producing body). She also observed the gametes in a petri dish. The zygotes that formed after fertilization grew on pieces of shell in the dish—and each developed into a tiny branching, filamentous form. This was how *Porphyra* spent most of the year! People already knew about the filamentous form as common pinkish growths on shells. But it was so different from the sheetlike form that it had been classified as a separate species. It is the alga's diploid sporophyte (spore-producing body). Drew-Baker's discovery of these alternating stages in the alga's life history, and the realization that the diploid stage could be grown on shells or other calcium-rich surfaces, revolutionized the nori industry. Within a few years, researchers had worked out the life cycle of *P. tenera*, the harvested species. By 1960, *nori cultivation* was a billion-dollar industry.

for jellies, and as culture gels. We also shape it into soft capsules for delivery of drugs and food supplements. Carrageenan, extracted from *Eucheuma*, is a stabilizer in paints, dairy products, and many other emulsions.

Humans find different species of *Porphyra* tasty as well as nutritious. You may know it from sushi bars as *nori*, a wrapping for rice and fish (Figure 23.15).

Figure 23.14 (**a**) A red alga (*Bonnemaisonia hamifera*). Such a growth pattern (branching, filamentous) is the most common. (**b**) From a tropical reef, a red alga showing sheetlike growth.

Red algae are photosynthetic protistans with phycobilins and other accessory pigments that mask their chlorophyll *a*. Usually, multicelled stages develop during the life cycles. Most species are aquatic. They show great diversity in size, morphology, life-styles, reproductive modes, and habitats.

23.9   BROWN ALGAE

Walk along a rocky coast at low tide and you may come across brownish seaweeds (Figure 23.16a). They are among the 1,500 species of **brown algae** (Phaeophyta). Nearly all thrive in cool or temperate marine waters, from the intertidal zone to the open ocean. Extensive masses of a floating brown alga, *Sargassum*, are major ecosystems in the Sargasso Sea, which lies between the Azores and the Bahamas.

a

Different species appear olive-green, golden, or dark brown, depending on their array of pigments. Like the chrysophytes, to which they may be related, brown algae contain fucoxanthin and other carotenoids as well as chlorophylls a, c_1, and c_2. They range from microscopic, filamentous *Ectocarpus* to the giant kelps, twenty to thirty meters long. Like the red algae, they have diverse and complex life cycles, with sexual and asexual phases. In their life cycles, too, gametophytes alternate with sporophytes.

Macrocystis, *Laminaria*, and the other giant kelps are the largest, most complex protistans. Their multicelled sporophytes have stipes (stemlike parts), blades (leaflike parts), and holdfasts (anchoring structures). Hollow, gas-filled bladders impart buoyancy to the stipes and blades and keep them upright in the water. Within the stipes are tubelike arrangements of elongated cells. The tubes rapidly carry dissolved sugars and other products of photosynthesis to living cells throughout the body. Flowering plants also transport sugars through similar kinds of tubes. This is a case of convergent evolution in two unrelated groups of large, multicelled organisms.

Giant kelp beds function as productive ecosystems. Think of them as underwater forests within which great numbers of diverse bacteria and protistans, as well as fishes and other animals, carry out their lives. The sizes of the kelp forests are variable, depending on changes in ocean currents and in the abundances of sea urchins (which feed on kelp debris) and sea otters (which feed on sea urchins). For example, in the late 1950s, a warm current displaced the cooler currents off the California coast. *Macrocystis* did not fare well with the temperature shift, and extensive beds off La Jolla and Palos Verdes almost disappeared. Without organic debris for the sea urchins to feed upon, these invertebrates fed directly on the kelps instead. Quantities of quicklime were dumped in the water to reduce the number of sea urchins, and the kelps made a comeback. So did fishes, lobster, abalone,

— bladder

— blade

— stipe

— holdfast

b

Figure 23.16   (a) Closer view of *Postelsia*, the brown alga shown in Figure 23.1a. You will find it thriving along coasts exposed to heavy surf from Vancouver Island on down to central California. More than a hundred deeply grooved blades top a highly resilient stipe, which is fastened to rocky substrates by a mound of anchoring structures. Its spores never disperse far from the parent sporophyte. At low tide, they simply drip onto rocks from the grooved blades. (b) Diagram of *Macrocystis* and an underwater view of a kelp forest. A few species live in the coastal waters of North and South America, New Zealand, Tasmania, and most of the islands in subantarctic waters.

and other species that make the kelp beds their home.

Macrocystis and certain other brown algae are commercially harvested. Extracts from them are used in ice cream, pudding, jellybeans, salad dressings, beer, canned and frozen foods, cough syrup, toothpaste, cosmetics, floor polish, and paper. Alginic acid in the cell walls of some species is used to make algins, which are added to various products as thickening, emulsifying, and suspension agents. In the Far East especially, people harvest kelps as sources of food and mineral salts, and as a fertilizer for crops.

The brown algae range in size from the microscopic to the largest of all of the multicelled protistans. Nearly all species live in marine habitats, ranging from the intertidal zone to the surface waters of the open ocean.

Of all protistans, the **green algae** (Chlorophyta) are structurally and biochemically most like the plants and may be their closest relatives. All are photosynthetic. Like plants, their chlorophylls are the molecular types designated *a* and *b*, and they, too, store carbohydrates as starch grains inside their chloroplasts. The cell walls of some species are composed of cellulose, pectins, and other polysaccharides typical of plants.

Figure 23.17 is a sampling of the more than 7,000 known species of green algae. They include single-celled, colonial, filamentous, sheetlike, and tubular forms. You won't be able to see many species without the aid of a microscope. Most, including the *Micrasterias* cell shown in Figure 23.1*b*, live in freshwater. *Micrasterias* is one of thousands of species of desmids, which are important food producers in nutrient-poor ponds and peat bogs. Green algae also grow at the ocean surface, in marine sediments, just below the surface of soil, and on rocks, tree bark, other organisms, and snow. Some types are symbionts with fungi, protozoans, and a few marine animals. A colonial form (*Volvox*) is a hollow whirling sphere of 500 to 60,000 flagellated cells. Those beautiful white, powdery beaches in tropical regions are largely the work of uncountable numbers of green algal cells

Figure 23.17 Representative green algae. (**a**) One of the marine species (*Codium*) with a pronounced branching form. Although many green algae are microscopic, one member of this genus (*C. magnum*) is taller than you are. In 1956 still another species of *Codium* from Puget Sound, Washington, was accidentally introduced into the Connecticut River where it empties into the Atlantic. In Puget Sound, populations of this alga were kept in check by herbivores that evolved with them. In its new habitat, there were no native *Codium* species or herbivores that were adapted to grazing on it. The introduced alga spread along the coast as far north as Maine and as far south as North Carolina in a little more than two decades.

(**b**) Also from a marine habitat, *Acetabularia*, fancifully called the mermaid's wineglass. Each individual in the cluster is a multinucleate cell mass with a rootlike structure, stalk, and cap in which gametes form.

(**c**) Sea lettuce (*Ulva*) grows in estuaries and attaches to kelps in the seas. Reproductive cells form within and are released from the margins of the sporophyte and gametophyte, both of which have the same form shown in the diagram.

(**d**) *Volvox*, a colony of interdependent green algal cells that bear resemblances to free-living, flagellated cells of the genus *Chlamydomonas*.

(**e**) Above the summer timberline in Utah, phycologist Ron Hoham's footprint in "red snow" reveals the presence of dormant cells of *Chlamydomonas nivalis*, one of the snow algae. Many red accessory pigments protect the chlorophylls of these cells.

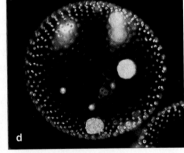

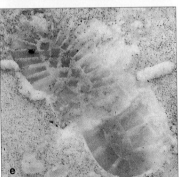

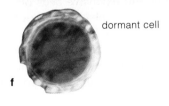

dormant cell

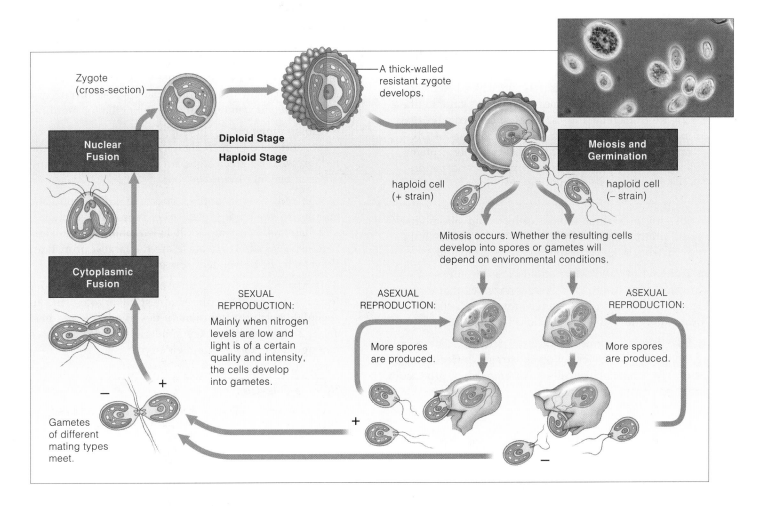

Zygote
(cross-section)

A thick-walled
resistant zygote
develops.

Nuclear Fusion

Diploid Stage

Haploid Stage

Meiosis and Germination

haploid cell
(+ strain)

haploid cell
(− strain)

Cytoplasmic Fusion

Mitosis occurs. Whether the resulting cells
develop into spores or gametes will
depend on environmental conditions.

SEXUAL REPRODUCTION:
Mainly when nitrogen levels are low and light is of a certain quality and intensity, the cells develop into gametes.

ASEXUAL REPRODUCTION:

ASEXUAL REPRODUCTION:

More spores are produced.

More spores are produced.

+

+

−

−

Gametes of different mating types meet.

Figure 23.18 Life cycle of a species of *Chlamydomonas*, one of the most common green algae of freshwater habitats. This single-celled species reproduces asexually most of the time and sexually under certain environmental conditions.

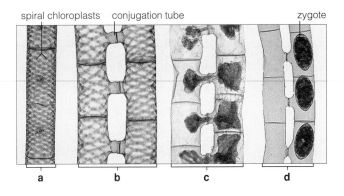

spiral chloroplasts   conjugation tube   zygote

a   b   c   d

Figure 23.19 One mode of sexual reproduction in *Spirogyra*, or watersilk. (**a**) This green alga has spiral, ribbonlike chloroplasts. (**b**) A conjugation tube forms between cells of adjacent haploid filaments of different mating strains. (**c,d**) The cellular contents of one strain pass through the tubes into cells of the other strain, where zygotes form. Zygotes develop thick walls. They undergo meiosis when they germinate and give rise to haploid filaments.

(*Halimeda*) that formed calcified cell walls, then died and disintegrated. Someday, green algae might even accompany astronauts in space. They can grow in very small spaces, give off vital oxygen, and take up carbon dioxide exhaled by the aerobically respiring crew.

Green algae employ diverse modes of reproduction. *Chlamydomonas* provides a classic example. This single-celled alga is twenty-five or so micrometers wide. It can reproduce sexually, but most of the time it engages in asexual reproduction, with up to sixteen daughter cells forming by mitotic cell division within the confines of the parent cell wall. Daughter cells may live at home for a while, but sooner or later they leave by secreting enzymes that digest what's left of their parent. Figure 23.18 shows the life cycle of one species. To give a final example, Figure 23.19 shows a filamentous green alga, *Spirogyra*, reproducing sexually.

The green algae show great diversity in size, morphology, life-styles, and habitats. The structure and biochemistry of some groups indicate they have close evolutionary ties to the plant kingdom.

SUMMARY

1. Recall that protistans and other eukaryotes differ from bacteria in key respects. At the least, eukaryotic cells have a double-membraned nucleus, mitochondria, endoplasmic reticulum, and large ribosomes. They have microtubules, which have uses as cytoskeletal elements, as spindles for chromosome movements, and in the $9 + 2$ core of flagella and cilia. These cells contain two or more chromosomes, each of which consists of DNA complexed with numerous histones and other proteins. Eukaryotic cells alone engage in mitosis and meiosis. Many have chloroplasts and other plastids.

2. Many protistans are heterotrophs. They include water molds, chytrids, slime molds (cellular and plasmodial types), protozoans (the amoebas, animal-like flagellates, and ciliated species), and sporozoans. See Table 23.3.

3. The chytrids and water molds are decomposers or parasites. Like fungi, they secrete enzymes that digest organic matter, then absorb breakdown products. Some form a mycelium (a mesh of absorptive filaments).

4. Like animals, slime molds are predatory. For part of the life cycle, they crawl on substrates as free-living, phagocytic, amoebalike cells. Like fungi, they produce spore-producing structures.

5. Like animals, protozoans are predators, grazers, or parasites. The amoebas, foraminiferans, heliozoans, and radiolarians are amoeboid protozoans with or without skeletal elements. Animal-like protozoans are internal parasites or live freely in aquatic habitats. Many cause human diseases. Sporozoans are internal parasites that may form cysts during the life cycle. An example is *Plasmodium*, which causes malaria. Ciliated protozoans such as *Paramecium* typically use their cilia for motility.

6. Most euglenoids, chrysophytes (diatoms, golden algae, and yellow-green algae), dinoflagellates, and the red, brown, and green algae are photoautotrophs. Many are members of phytoplankton.

7. The red, brown, and green algae differ from one another in their body plan, size, reproductive modes, and habitats. Most are multicelled photosynthesizers. Some species of brown algae are the largest protistans. Plants may be descended from early green algae.

8. It has taken us more than one chapter to consider features of protistans, the simplest eukaryotes, and to speculate on their evolutionary ties to other kingdoms. Table 23.4 summarizes key similarities and differences among the prokaryotes and eukaryotes.

Table 23.3	Summary of Major Groups of Protistans
Group	Common Name
HETEROTROPHIC (decomposers, predators, grazers, parasites):	
Chytridiomycota	Chytrids
Oomycota	Water molds
Acrasiomycota	Cellular slime molds
Myxomycota	Plasmodial slime molds
Sarcodina	Amoeboid protozoans:
	Rhizopods (naked amoebas, foraminiferans)
	Actinopods (heliozoans, radiolarians)
Mastigophora	Animal-like flagellated protozoans
Apicomplexa	Major group of sporozoans
Ciliophora	Ciliated protozoans
AUTOTROPHIC AND HETEROTROPHIC:	
Euglenophyta	Euglenoids
Pyrrhophyta	Dinoflagellates
MOSTLY AUTOTROPHIC (photosynthesizers; some parasites):	
Chrysophyta	Chrysophytes (golden algae, yellow-green algae, diatoms)
Rhodophyta	Red algae
Phaeophyta	Brown algae
Chlorophyta	Green algae

Review Questions

1. Outline some of the general characteristics of protistans. Then explain what is meant by this statement: Protistans are often classified by what they are *not*. CI

2. Review Table 23.2, then cover it with a sheet of paper. Now list the common names for the major categories of protistans. CI, *Table 23.3*

3. Correlate some structural features of a protistan from each of the groups listed with conditions in their environments:
   a. chytrids and water molds *23.1*
   b. slime molds *23.1*
   c. amoeboid protozoans *23.3*
   d. animal-like flagellates and sporozoans *23.4, 23.5*
   e. ciliated protozoans *23.6*
   f. euglenoids, chrysophytes, and dinoflagellates *23.7*
   g. red, brown, and green algae *23.8–23.10*

4. Select three different protistan species, then briefly explain how they adversely affect our affairs, such as crop yields and human health. *23.1, 23.3, 23.4–23.7*

5. Select and briefly explain how the activities of photosynthetic protistan species have positive benefits for some communities of organisms, including human communities. *23.7–23.10*

Self-Quiz (*Answers in Appendix IV*)

1. Most chytrids and water molds are _____ decomposers in aquatic habitats.
   a. parasitic   c. autotrophic
   b. saprobic   d. chemosynthetic

2. Free-living, amoebalike cells crawl on rotting plant parts as they engulf bacteria, spores, and organic compounds. This description best fits the _____ .
   a. water molds   c. sporozoans
   b. amoeboid protozoans   d. slime molds

Table 23.4 Comparison of Prokaryotes With Eukaryotes

	Prokaryotes	Eukaryotes
Organisms represented:	Bacteria only	Protistans, fungi, plants, and animals
Ancestry:	Two major lineages (archaebacteria and eubacteria) that evolved more than 3.5 billion years ago	Equally ancient prokaryotic ancestors gave rise to forerunners of eukaryotes, which evolved more than 1.2 billion years ago.
Level of organization:	Single-celled	Protistans, single-celled or multicelled. Nearly all others are multicelled, with a division of labor among differentiated cells, tissues, and often organs.
Typical cell size:	Small (1–10 micrometers)	Large (10–100 micrometers)
Cell wall:	Most have distinctive walls.	Cellulose or chitin; none in animal cells
Membrane-bound organelles:	Very rarely	Typically profuse
Modes of metabolism:	Both anaerobic and aerobic	Aerobic modes predominate.
Genetic material:	Bacterial chromosome (and sometimes plasmids)	Complex chromosomes (DNA, many associated proteins) within a nucleus
Mode of cell division:	Prokaryotic fission, mostly; also budding	Nuclear division (mitosis, meiosis, or both), associated with one of various modes of cytoplasmic division

3. During a _____ life cycle, amoeboid cells aggregate and form a migrating mass. Cells in the mass then differentiate, forming reproductive structures and spores or gametes.
    a. slime mold   c. protozoan
    b. water mold   d. chytrid

4. Amoebas, foraminiferans, radiolarians, and heliozoans are all classified as _____ .
    a. ciliated protozoans   c. amoeboid protozoans
    b. animal-like protozoans d. sporozoans

5. Parasitic, flagellated protozoans of tropical regions known as trypanosomes are associated with which disease(s)?
    a. African sleeping sickness and Chagas disease
    b. toxoplasmosis
    c. malaria
    d. amoebic dysentery

6. Euglenoids and chrysophytes are mostly _____ .
    a. photoautotrophic   c. heterotrophic
    b. chemoautotrophic   d. omnivorous

7. Single-celled photosynthetic protistans, including most of the euglenoids, chrysophytes, and dinoflagellates, are members of the _____ , the "pastures" of most aquatic habitats.
    a. zooplankton   c. brown algae
    b. red algae   d. phytoplankton

8. Algin is used in ice cream, pudding, salad dressing, jelly beans, beer, cough syrup, toothpaste, cosmetics, and other products. Certain _____ are sources of algin.
    a. green algae   c. red algae
    b. brown algae   d. dinoflagellates

Critical Thinking

1. Suppose you decide to vacation in a developing country where sanitation practices and standards of personal hygiene are poor. Having read about some of the parasitic protozoans that lurk in water and damp soil, what would you consider safe to drink, once you arrive there? Which kinds of foods might be best to avoid and what kinds of food preparations might make them safe to eat?

2. As you read in this chapter, red tides are associated with "algal blooms." Such blooms may follow enrichment of aquatic habitats with water that drains into them from heavily fertilized croplands or from concentrated sources of raw sewage. After thinking about it, do you accept the resulting destruction of aquatic species, birds, and other forms of wildlife as an unfortunate but necessary side effect of human activities? If you do not accept it, how would you stop the water pollution? And if you could stop it, what sorts of measures would you take to feed the enormous human population, which is now extremely dependent on high-yield (and very heavily fertilized) crops? How would you propose to dispose of or prevent the accumulation of fecal matter and other wastes of 7.5 billion people?

Selected Key Terms

actinopod 23.3	contractile	plankton 23.3
amoeboid	vacuole 23.6	protistan CI
protozoan 23.3	cyst 23.2	protozoan 23.2
animal-like	diatom 23.7	pseudopod 23.3
flagellate 23.4	dinoflagellate 23.7	red alga 23.8
binary fission 23.2	euglenoid 23.7	red tide 23.7
brown alga 23.9	golden alga 23.7	rhizopod 23.3
chrysophyte 23.7	green alga 23.10	saprobe 23.1
chytrid 23.1	multiple	slime mold 23.1
ciliated	fission 23.2	sporozoan 23.4
protozoan 23.6	mycelium 23.1	water mold 23.1
conjugation	pellicle 23.6	yellow-green
(protozoan) 23.6	phytoplankton 23.7	alga 23.7

Readings

Bold, H., and M. Wynne. 1985. *Introduction to the Algae*. Second edition. Englewood Cliffs, New Jersey: Prentice-Hall. Includes descriptions of the economic importance of major algal groups.

Margulis, L. 1993. *Symbiosis in Cell Evolution*. Second edition. New York: Freeman. Paperback.

Margulis, L., and K. Schwartz. 1992. *Five Kingdoms*. Second edition. New York: Freeman. Paperback.

Satchell, M. 28 July 1997. "The Cell From Hell." *U.S. News and World Report*. 26–28. Among other cases, correlates hundreds of millions of gallons of raw sewage from North Carolina's factory-like log farms with a bloom of the dinoflagellate *Pfiesteriia* that killed 14 million fish.

Web Site See *http://www.wadsworth.com/biology* for practice quiz questions, hypercontents, BioUpdates, and critical thinking. The Wadsworth Biology Resource Center provides a wealth of information fully organized and integrated by chapter.

24 FUNGI

Dragon Run

The first winter storm is blasting through southeastern Virginia, and Dragon Run—part swamp, part marsh, part old-growth woodland—pays tribute to the wind. Oaks, maples, gums, and beeches release dead leaves by the millions and these shower to the earth, where they pile up as crisp, ankle-deep mounds. Branches snap off trees; sheaves of dead grasses sink into marsh

b Big laughing mushroom (*Gymnophilus*)

c Purple coral fungus (*Clavaria*)　　d Rubber cup fungus (*Sarcosoma*)

Figure 24.1 Fungal species from southeastern Virginia. This small sampling merely hints at the rich diversity within the kingdom Fungi.

muds, and other plants buckle into the shallow waters of the bottomlands. The storm kills uncounted numbers of insects, some birds, a few squirrels. By late December, Dragon Run is partially buried in organic debris.

Dig down through the debris and you will discover the accumulated litter of many past seasons—moist cushions of decayed leaves, bits of spiders and insects, mouldering branches, and carcasses of small mammals.

A new growing season begins with the warm rains of February. Buds open on shrubs and trees, including a massive silver beech that has been leafing out every spring for three centuries. Woodland violets are the first to sprout in the thawing soil. By April, the resurgence of growth has imparted such a sharp, green freshness to Dragon Run that you might overlook the organisms on which the resurgence largely depends. On logs, under leaves, and in soil, fungi are commandeering resources and engaging in vegetative growth (Figure 24.1).

Fungi are premier decomposers. Like every other heterotroph, they feast on organic compounds that other organisms produce. But few organisms besides fungi digest their dinner out on the table, so to speak. As fungi grow in or on organic matter, they secrete enzymes that digest it into bits that individual cells can

a Sulfur shelf fungus (*Polyporus*) on a living tree

e Scarlet hood (*Hygrophorus*)

f Frost's bolete (*Boletus*)

g Yellow coral fungus (*Clavaria*)

h Trumpet chanterelle (*Craterellus*)

absorb. Such a mode of nutrition is called **extracellular digestion and absorption**. Some of the carbon and other nutrients from organic compounds also can be absorbed by plants, the primary producers of Dragon Run and nearly all other ecosystems.

Keep the global perspective in mind. Why? As you will see, the metabolic activities of some fungi do cause diseases in humans, pets and farm animals, ornamental plants, and important crop plants. Some species are notorious spoilers of food supplies. Others have uses in the commercial production of substances ranging from antibiotics to excellent cheeses. We tend to assign "value" to fungi and other organisms in terms of their direct effect on our lives. There is nothing wrong with battling dangerous species and admiring beneficial ones—as long as we do not lose sight of the greater roles of fungi or any other kind of organism in nature.

KEY CONCEPTS

1. Fungi are heterotrophs. Together with heterotrophic bacteria, they are decomposers of the biosphere. Saprobic types obtain nutrients from nonliving organic matter. Parasitic types obtain them from tissues of living hosts.

2. Fungi secrete enzymes that digest food outside their body, then fungal cells absorb breakdown products. Their metabolic activities release carbon dioxide to the atmosphere and return many nutrients to the soil, where they become available to producer organisms.

3. Most fungi are multicelled. A mycelium, which is the food-absorbing portion of a fungal body, develops during the fungal life cycle. Each mycelium is a mesh of hyphae, which are elongated filaments that grow and develop by repeated mitotic cell divisions.

4. Commonly, a number of fungal hyphae become modified and weave together to form a reproductive structure in or upon which fungal spores develop. A "mushroom" is such a structure. Germinating spores grow and develop into a new mycelium.

5. Many fungi are symbionts with other organisms. Some are partners with algae and other organisms, in lichens. Many are part of mycorrhizae; they are locked in mutually beneficial relationships with young roots of land plants. Metabolic activities of cells making up the fungal hyphae provide the plants with nutrients, and the plants provide the fungi with carbohydrates.

6. We tend to assign value to plants and fungi in terms of their direct effect on our lives. Our battles with "bad" ones and reliance on the "good" ones should start from a solid understanding of their long-established roles in nature.

Major Groups of Fungi

For many of us, "fungi" are drab mushrooms sold in grocery stores. These are simply fungal body parts, and they are produced by a fungus with stunningly diverse relatives. The few species in Figure 24.1 don't do justice to the 56,000 fungal species we know about—and there may be at least a million more we don't know about!

We know, from the fossil record, that fungi evolved before 900 million years ago. Some accompanied plants onto the land 430 million years ago. About 100 million years later, three major lineages were well established. We call them the **zygomycetes** (Zygomycota), **sac fungi** (Ascomycota), and **club fungi** (Basidiomycota). Other, puzzling kinds known as "imperfect fungi" are lumped together but are not a formal taxonomic group. The vast majority of species in all these groups are multicelled.

Nutritional Modes

Fungi are heterotrophs, meaning they require organic compounds that other organisms synthesize. Most are **saprobes**; they obtain nutrients from nonliving organic matter and so cause its decay. Others are **parasites**; they extract nutrients from tissues of a living host. When the cells of any species grow in or on organic matter, they secrete digestive enzymes and then absorb breakdown products. Again, their extracellular digestion benefits plants, which absorb some of the released nutrients as well as the carbon dioxide by-products. Without the fungi and heterotrophic bacteria, communities would slowly become buried in their own garbage, nutrients would not be cycled, and life could not go on.

Key Features of Fungal Life Cycles

Fungi reproduce asexually most often, but given the opportunity, they also reproduce sexually. They form great numbers of nonmotile spores. As in plants, their **spores** are reproductive cells or multicelled structures, often walled, that germinate after dispersal from the parent. In multicelled species, spores give rise to a mesh of branched filaments. The mesh, a **mycelium** (plural, mycelia), rapidly grows over or into organic matter and has a good surface-to-volume ratio for food absorption. Each filament in a mycelium is called a **hypha** (plural, hyphae). Hyphal cells commonly have chitin-reinforced walls. Their cytoplasm interconnects, so that nutrients flow unimpeded throughout the mycelium.

Fungi are major decomposers that engage in extracellular digestion and absorption of organic matter. Multicelled types form absorptive mycelia and spore-producing structures.

A Sampling of Spectacular Diversity

Fungal life cycles and life-styles show dizzying variety. The most we can do here is to sample a few species, starting with the club fungi. The 25,000 or so club fungi include mushrooms, shelf fungi, coral fungi, puffballs, and stinkhorns. Figures 24.1, 24.2, and 1.6d show fine examples. Some of the saprobic species are important decomposers of plant debris. As you will see, other species are symbionts that live in close association with the young roots of forest trees. The fungal rusts and smuts can destroy fields of wheat, corn, and other crop plants. Cultivation of the common mushroom (*Agaricus brunnescens*) is a multimillion-dollar business. It is the mushroom of grocery-store and pizza-topping fame. Yet some of its relatives produce toxins that can kill you or any other organism that nibbles on them.

Have you ever wondered which organisms are the oldest and the largest? *Armillaria bulbosa* is one of them. The mycelium of one individual, discovered in a forest in northern Michigan, extends through fifteen hectares of soil. (Each hectare is the equivalent of 10,000 square meters.) By some estimates, this fungus weighs more than 10,000 kilograms and has been spreading beneath the forest floor for more than 1,500 years!

a

b

Figure 24.2 Two club fungi. (**a**) The light-red coral fungus *Ramaria*. (**b**) Closer view of the shelf fungus *Polyporus* shown in Figure 24.1a. With the exception of the rubber cup fungus, all of the species shown in Figure 24.1 are club fungi.

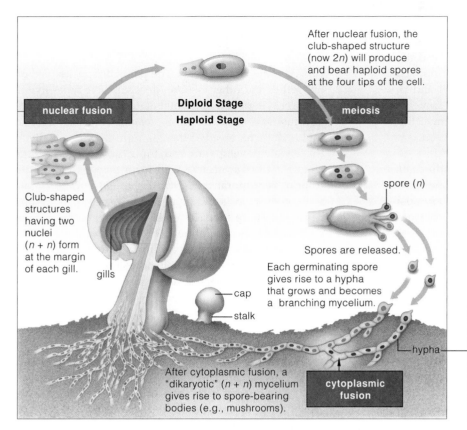

After nuclear fusion, the club-shaped structure (now 2n) will produce and bear haploid spores at the four tips of the cell.

nuclear fusion | **Diploid Stage** | meiosis
Haploid Stage

Club-shaped structures having two nuclei (n + n) form at the margin of each gill.

gills

cap

stalk

spore (n)

Spores are released.

Each germinating spore gives rise to a hypha that grows and becomes a branching mycelium.

After cytoplasmic fusion, a "dikaryotic" (n + n) mycelium gives rise to spore-bearing bodies (e.g., mushrooms).

cytoplasmic fusion

hypha

Figure 24.3 Generalized life cycle for many club fungi. When two compatible mating strains grow together, the cytoplasm (but not nuclei) of two hyphal cells may fuse. Cell divisions result in a dikaryotic mycelium (its cells each have two nuclei). Mushrooms form, with club-shaped spore-bearing structures on the surface of the gills. The two nuclei in each structure fuse to form a diploid zygote, which starts the cycle anew.

When you see mushrooms or any other fungus growing outdoors, think twice before devouring them. Unlike cultivated species, such as the common mushroom, many are toxic, including the species shown in Figure 24.4b. *No one should eat any fungi that were gathered in the wild unless they have been accurately identified as edible.* As the saying goes, there are old mushroom hunters and bold mushroom hunters, but no old bold mushroom hunters.

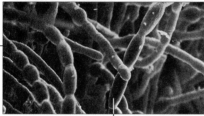

hypha in mycelium

a   b

Figure 24.4 (**a**) Fly agaric mushroom (*Amanita muscaria*) causes hallucinations when it is eaten. It was used to induce trances in ancient rituals in Central America, Russia, and India. (**b**) From California, *A. ocreata*. It can be fatal if ingested. A close relative, the death cap mushroom (*A. phalloides*), lives up to its name. Nibble even as little as five milligrams of its toxin and vomiting and diarrhea will begin eight to twenty-four hours later. Your liver and kidneys will degenerate; death can follow within a few days.

Example of a Fungal Life Cycle

Probably you already know what a common mushroom (*A. brunnescens*) looks like, so let's use it as our example of a fungal life cycle. Like most of the other club fungi, this species produces short-lived reproductive bodies—**mushrooms**—that are merely its aboveground parts; its living mycelium is buried in soil or decaying wood. A mushroom has a stalk and a cap. Fine tissue sheets, or gills, line the cap's inner surface. On these gills, club-shaped, spore-bearing structures form. The spores that form here are **basidiospores**. When a spore dispersed from a particular strain of mushroom lands on a suitable site, it germinates and gives rise to a haploid mycelium.

Suppose hyphae of two compatible mating strains make contact. They may undergo cytoplasmic fusion, but their nuclei do not fuse at once. The fused part may be the start of a *dikaryotic* mycelium, in which hyphal cells have one nucleus of each mating type (Figure 24.3). An extensive mycelium forms. And when conditions are favorable, mushrooms will form. Each spore-producing structure of the mushroom is initially dikaryotic, but then its two nuclei fuse to form a short-lived zygote. The zygote undergoes meiosis, haploid spores develop, and then air currents disperse them.

Club fungi, the fungal group with the greatest diversity, produce spores inside distinctive club-shaped structures.

A fungus has a thing about spores. It produces sexual spores, asexual spores, or both, depending on contact with a suitable hypha, food availability, and how cool or damp conditions are. Its spores are small and dry, and air currents easily disperse them. Each spore that germinates can be the start of a hypha and a mycelium. Stalked reproductive structures may develop on many of the hyphae and produce asexual spores. After these spores germinate, each may be the start of still *another* extensive mycelium. In no time at all, that one fungus and staggering numbers of its descendants are busily decomposing organic stuff or pirating nutrients from a host! Look at what can happen to a slice of stale bread:

Each fungal class produces unique sexual spores. Club fungi form basidiospores, zygomycetes form spores by way of zygosporangia, and sac fungi form ascospores.

Figure 24.5 Life cycle of the black bread mold *Rhizopus stolonifer*. Asexual phases are common. Different mating strains (+ and −) also reproduce sexually. Either way, haploid spores form and give rise to mycelia. Chemical attraction between a + hypha and a − hypha causes them to fuse. Two gametangia form, each with several haploid nuclei. Later their nuclei fuse to form a zygote. The zygote develops a thick wall, so becoming a zygosporangium, and may remain dormant for several months. Meiosis occurs as the zygosporangium germinates, and sexual spores form.

Producers of Zygosporangia

Consider the zygomycetes. The parasitic species feed on houseflies and other insects. Most saprobic types live in soil, decaying plant or animal material, and stored food. You just saw what *Rhizopus stolonifer*, the black bread mold, does to bread. When it reproduces sexually, a thick wall develops around the zygote. Together, the zygote *and* its protective wall form a **zygosporangium** (plural, zygosporangia), as in Figure 24.5a. The zygote undergoes meiosis and gives rise to a specialized hypha bearing a spore sac. The entire contents of the sac are converted to some number of spores, each of which may give rise to a new mycelium. Stalked hyphae grow out of such mycelia. Asexual spores form in a spore sac perched on top of each stalk (Figure 24.5b).

Producers of Ascospores

We know of more than 30,000 kinds of sac fungi. Most form sexual spores called **ascospores** inside sac-shaped cells. They alone form these cells, called asci (singular, ascus). The asci of multicelled species are enclosed in reproductive structures, which consist of tightly interwoven hyphae. The structures resemble diverse flasks, globes, and shallow cups, as in Figure 24.6.

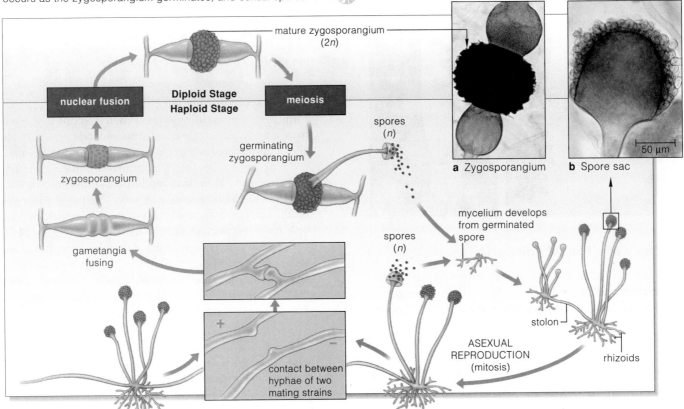

a Zygosporangium   b Spore sac

50 μm

mature zygosporangium (2n)

Diploid Stage
Haploid Stage

nuclear fusion   meiosis

spores (n)

germinating zygosporangium

zygosporangium

gametangia fusing

spores (n)

mycelium develops from germinated spore

+
−

contact between hyphae of two mating strains

stolon

ASEXUAL REPRODUCTION (mitosis)

rhizoids

ascospore (sexual spore)

spore sac

spore-bearing hypha of this ascocarp

a

b ascocarp   **c** ascocarp   **d** conidia (chains of asexual spores)   **e** budding yeast cell

Figure 24.6 Sac fungi. (**a**) Diagram and (**b**) photograph of *Sarcoscypha coccinia,* the scarlet cup fungus. Saclike structures on the cup's inner surface produce sexual spores (ascospores) by meiosis. (**c**) One of the morels (*Morchella esculenta*). This edible species has a poisonous relative. (**d**) From *Eupenicillium,* chains of asexual spores of a type called conidiospores. These drift away from the chains, like dust, even after the slightest jiggling. "Conidia" means dust. (**e**) Cells of *Candida albicans,* agent of "yeast" infections of the vagina, mouth, intestines, and skin.

The vast majority of sac fungi are multicelled. They include highly prized truffles and morels (Figure 24.6*c*). The truffles are underground symbionts with oak and hazelnut tree roots. Trained pigs and dogs snuffle out truffles in the wild. In France, truffles are now being cultivated on the roots of inoculated seedlings.

Other multicelled sac fungi include certain species of *Penicillium* that "flavor" Camembert and Roquefort cheeses and species that make penicillins, widely used as antibiotics. We use *Aspergillus* to make citric acid for candies and soft drinks, and to ferment soybeans for soy sauce. Most of the red, bluish-green, and brown fungal molds that spoil stored food also are multicelled. One species, the salmon-colored *Neurospora sitophila,* wreaks havoc in bakeries and in research laboratories. It produces so many spores that it is extremely difficult to eradicate. One of its relatives, *N. crassa,* is an important organism in genetic research.

Sac fungi also include about 500 species of single-celled yeasts (although still other yeasts are classified as club fungi). Yeasts reproduce sexually when two cells fuse and become a spore-producing sac. Some types live in the nectar of flowers and on fruits and leaves. Bakers and vintners put fermenting by-products of vast populations of yeasts to use. For example, the carbon dioxide by-products of *Saccharomyces cerevisiae* leaven bread. The commercial production of wine and beer depends on its ethanol end product. Many yeast strains with desirable properties have been developed through artificial selection and genetic engineering. Then again, *Candida albicans,* a notorious relative of "good" yeasts, causes vexing infections in humans (Figure 24.6*e*).

roundworm   noose formed by hypha

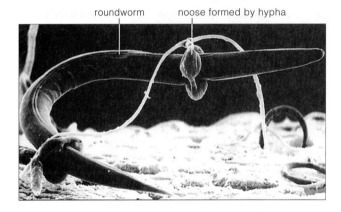

Figure 24.7 *Arthrobtrys dactyloides,* an imperfect fungus. This predatory species forms a nooselike ring that swells rapidly with turgor pressure when stimulated. The "hole" in the noose shrinks and captures a worm, into which hyphae will grow.

Elusive Spores of the Imperfect Fungi

Imperfect fungi are set aside in a taxonomic holding station not because they are somehow defective but mainly because no one has yet discovered what kind of sexual spores they produce (if any). Figure 24.7 shows one of the species, a puzzling predatory fungus, that is awaiting formal classification. Investigators recently reunited the previously orphaned *Aspergillus, Candida,* and *Penicillium* with their kin—other sac fungi.

Through their exuberant and rapid production of asexual and sexual spores, fungi take quick advantage of available organic matter, whether it has been discarded or is part of a living or dead organism. Their penchant for spore production is central to their success as decomposers and parasites.

Recall, from earlier chapters, that **symbiosis** refers to species that live in close association. The word literally means "living together." In many cases, one species is a victim, not a partner, of a parasite. In other cases, called **mutualism**, the interaction benefits both partners or at least does one of them no harm. We find classic examples of mutualists among fungi that enter into partnerships with algae and cyanobacteria, as well as with tree roots.

Lichens

A **lichen** is a single vegetative body in which a fungus has become intimately intertwined with one or more photosynthetic organisms. The fungal part of a lichen is the *myco*biont (after the Greek *myes*, meaning fungus). The photosynthetic component is the *photo*biont. About 13,500 types of lichens have been identified. Nearly half of the mycobionts are sac fungi. However, only 100 or so species function as photobionts, and most commonly they are green algae (Chlorophyta) and cyanobacteria.

Lichens typically colonize sites that most organisms would find hostile. We find one type or another growing slowly in habitats ranging from the Antarctic to the Arctic. They can colonize almost any substrate, including sunbaked or frozen rocks, recently hardened lava, fence posts, gravestones, and the bark or leaves of a variety of plants. The long-lived Galápagos tortoises have lichens clinging to their shells; so do the tops of giant Douglas firs in the Pacific Northwest. Certain lichens even withstand periodic submergence along seashores and riverbanks.

In almost every instance, the fungus is the largest part of the lichen. Cyanobacterial cells typically reside in a separate structure inside or outside the main body. The fungus benefits by having a long-term source of nutrients that it absorbs from cells of the photobiont. By giving up some nutrients, the photobiont suffers a bit in terms of its own growth, although it might benefit a bit from the lichen's sheltering effect. In cases where more than one fungus is present, the added species might be a mycobiont. Then again, it might be parasitizing the lichen or merely using the lichen as a substrate.

A lichen forms after the tip of a fungal hypha binds with a suitable host cell. Either both lose their wall and undergo cytoplasmic fusion or the hypha induces its host cell to cup around it. Now the mycobiont and the photobiont grow and multiply together. The structure of the lichen depends on how cells of the photobiont become distributed among fungal cells. Sometimes they are distributed more or less uniformly throughout the lichen. In most cases, however, the lichen has distinct layers, as in Figure 24.8a. The overall growth pattern

may be leaflike, flattened, pendulous, or erect. Figure 24.8*b–e* provides examples of these growth patterns.

Lichens absorb mineral ions from their substrates and nitrogen from the air. Among the products of their metabolic activities are antibiotics against bacteria that might decompose the lichen body and toxins that inhibit the larval development of invertebrates that might graze on them. Coincidentally, the products also contribute to

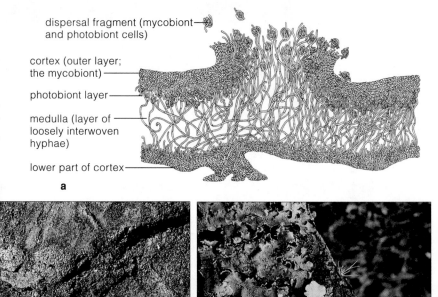

dispersal fragment (mycobiont and photobiont cells)

cortex (outer layer; the mycobiont)

photobiont layer

medulla (layer of loosely interwoven hyphae)

lower part of cortex

a

b   **c**

Figure 24.8 Lichens. (**a**) Diagram of a stratified lichen, cross-section. Notice the distinct layers. (**b**) Leaflike lichen on birch tree bark. (**c**) Encrusting lichens. (**d**) *Usnea* (old man's beard), a pendant lichen. (**e**) An erect, branching lichen, *Cladonia rangiferina*, sometimes called reindeer moss. Reindeer and caribou feed mainly on this species during winter.

the formation or enrichment of soils. When lichens are pioneers in new habitats, such as soil newly exposed by a retreating glacier, their metabolic activities can set the stage for colonization by different species. Actually, lichens may have been among the first invaders of the land. Also, some ecosystems depend on cyanobacteria-containing lichens as nitrogen sources. For example, up to 20 percent of the nitrogen entering old-growth forests of the Pacific Northwest starts with the lichen *Lobaria*.

Lichens also serve as early warnings of deteriorating environmental conditions. They absorb but cannot rid themselves of toxins. From extensive studies in New York City and in England, we know that when lichens die around cities, air pollution is getting bad.

d

e

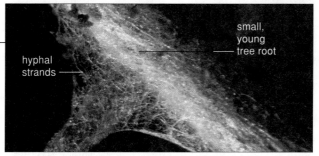

a

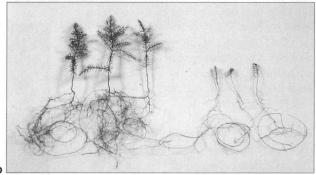

b

Figure 24.9 (**a**) Mycorrhiza from a hemlock tree. (**b**) Effects of the presence or absence of mycorrhizae on plant growth. The juniper seedlings at the left are six months old. They were grown in sterilized, phosphorus-poor soil with a mycorrhizal fungus. The seedlings at the right were grown without the fungus.

labels: hyphal strands; small, young tree root

Mycorrhizae

Fungi also enter into symbiotic interactions with young tree roots in which benefits flow both ways. This form of mutualism is a **mycorrhiza** (plural, mycorrhizae), which means "fungus-root." The fungus benefits by absorbing carbohydrates from the plant, which benefits by absorbing minerals from the fungus. Collectively, the fungal hyphae have a huge surface area for absorbing water and dissolved mineral ions. The fungus can take up ions of phosphorus and other minerals when they are abundant in soil and release them to the plant when ions are scarce. Many plants do not grow as efficiently in the absence of mycorrhizae.

The example in Figure 24.9*a* is an *exo*mycorrhiza, in which hyphae form a dense net around living cells in roots but don't penetrate them. Other hyphae form a velvety wrapping around the root, and the mycelium radiates outward from it. Exomycorrhizae are common in temperate forests, and they help the trees withstand adverse seasonal changes in temperature and rainfall.

About 5,000 fungal species enter the associations. Most are club fungi, including those truffles described earlier.

The more common *endo*mycorrhizae form in about 80 percent of all vascular plants. These fungal hyphae penetrate plant cells, as they do in lichens. Fewer than 200 species of zygomycetes serve as the fungal partner. Their hyphae branch extensively, forming tree-shaped absorptive structures within cells. They also extend for several centimeters into the surrounding soil. Section 30.2 provides a closer look at these beneficial species.

As Fungi Go, So Go the Forests

Since the early 1900s, collectors have been recording data on the wild mushroom populations in the forests of Europe. A look at the records tells us that the number and diversity of fungi are now declining at alarming rates. Certainly mushroom gatherers aren't to blame; toxic as well as edible species are vanishing. But the decline does correlate with rising air pollution. Vehicle exhaust, smoke from coal burning, and emissions from nitrogen fertilizers pump ozone, nitrogen oxides, and sulfur oxides into the air. Normally, as a tree ages, one species of mycorrhizal fungus gives way to another in predictable patterns. When fungi die, the tree loses its support system and becomes vulnerable to severe frost and drought. Are the North American forests at risk? Conditions there are deteriorating in comparable ways.

Lichens and mycorrhizae are both symbiotic associations between fungi and other organisms, with mutual benefits.

A LOOK AT THE UNLOVED FEW

You know you are a serious student of biology when you view organisms objectively in terms of their place in nature, not in terms of their impact on humans generally and you in particular. As a student you salute saprobic fungi as vital decomposers and praise parasitic fungi that help keep populations of harmful insects and weeds in check. The true test is when you open the fridge to get a bowl of high-priced raspberries and discover that a fungus beat you to them. The true test is when a fungus starts feeding on warm, damp tissues between your toes and turns skin scaly, reddened, and cracked (Figure 24.10*a*).

Which home gardeners wax poetic about black spot or powdery mildew on their roses? Which farmers happily give up millions of dollars each year to sac fungi that attack corn, wheat, peaches, and apples (Figure 24.10*b*)? Who cares that the sac fungus *Cryphonectria parasitica* blitzed most of the chestnut trees in eastern North America, leaving them to sprout as stubby versions of their former selves?

And who willingly would inhale airborne spores of *Ajellomyces capsulatus*? These are dimorphic beasties. When they alight on soil, they form mycelia. When they alight in moist lung tissues, however,

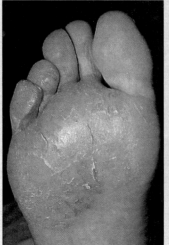
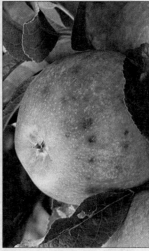

a   b

they form populations of yeastlike cells that cause the respiratory disease *histoplasmosis*. Often, aggregations of macrophages that converge on and defend the infected tissues become packed with the invading cells. Usually the defenders eliminate the invaders, and their aggregations

Figure 24.10
Love those fungi! (**a**) Athlete's foot, courtesy of *Epidermophyton floccosum*. (**b**) Apple scab, trademark of *Venturia inaequalis*.

become calcified as the tissue heals. With heavy exposure to the spores, however, pneumonia develops. In rare cases, the calcified masses form in the lymph nodes, liver, and other organs. Progressive histoplasmosis usually ends in death. Where nitrogen-rich droppings of many chickens, starlings, pigeons, and other birds accumulate, the fungus thrives. For example, kick up dust from spore-containing regions drained by the Ohio and Mississippi rivers, and you put yourself at risk. Go spelunking in caves where roosting bats busily pile up droppings on the cave floor, and you similarly invite the spores into your lungs.

Certain fungi even have had impact on human history. Consider one notorious species, *Claviceps purpurea*, which parasitizes rye and other cereal grains. Give it credit; we use some of its by-products (alkaloids) to treat migraine headaches and, following childbirth, to shrink the uterus to prevent hemorrhaging. However, the alkaloids are toxic when ingested in large amounts. Eat a lot of bread made with tainted rye flour and you end up with *ergotism*. The symptoms include vomiting, diarrhea, hallucinations, hysteria, and convulsions. Untreated, the disease can turn limbs gangrenous, and it can end in death.

Ergotism epidemics were common in Europe during the Middle Ages, when rye was a major crop. Ergotism also thwarted Peter the Great, the Russian czar who was obsessed with conquering ports along the Black Sea for his nearly landlocked empire. Soldiers laying siege to the ports ate mostly rye bread and fed rye to their horses. The former went into convulsions and the latter into "blind staggers." Possibly, outbreaks of ergotism were used as an excuse to launch witch-hunts in colonial Massachusetts and elsewhere. 🔷

Table 24.1 Some Pathogenic and Toxic Fungi*

ZYGOMYCETES	
Rhizopus	Food spoilage
ASCOMYCETES	
Aspergillus (certain species)	Aspergilloses (allergic reactions, sinus, ear, and lung infections; *A. flavus* toxin linked to cancers)
Ajellomyces capsulatus	Histoplasmosis
Candida albicans	Infection of mucous membranes
Claviceps purpurea	Ergot of rye, ergotism
Coccidioides immitis	Valley fever
Cryphonectria parasitica	Chestnut blight
Ophiostoma ulmi	Dutch elm disease
Microsporum, Trichophyton, Epidermophyton	Various species cause ringworms, of scalp, body, nails, beard, athlete's foot (Figure 24.10*a*)
Monilinia fructicola	Brown rot of peaches and other stone fruits
Venturia inaequalis	Apple scab (Figure 24.10*b*)
Verticillium	Plant wilt
BASIDIOMYCETES	
Amanita (some species)	Severe mushroom poisoning
Pneumocystis carinii	A virulent form of pneumonia in individuals with weakened immune systems, as with AIDS
Puccinia graminis	Black stem wheat rust
Tilletia indica	Smut of cereal grains
Ustilago maydis	Smut of corn

* After C. Alexopoulos, C. Mims, and M. Blackwell. 1996. *Introductory Mycology*, fourth edition. Wiley.

SUMMARY

1. Fungi are heterotrophs and major decomposers. The saprobes feed on nonliving organic matter; the parasites feed on the tissues of living organisms. Some species of fungi are symbiotic partners with other organisms. The cells of all species secrete digestive enzymes that break down food into small molecules, which the cells absorb.

2. Nearly all fungi are multicelled. The food-absorbing portion, the mycelium, consists of a mesh of filaments (hyphae). Aboveground reproductive structures, such as mushrooms, form from tightly interwoven hyphae.

3. The major groups of fungi are the zygomycetes, the ascomycetes (sac fungi), and the basidiomycetes (club fungi). Each is characterized by distinctive sexual and asexual spores. When a sexual phase cannot be detected or is absent from the life cycle, a fungus is assigned to an informal category called the imperfect fungi.

4. Lichens are mutualistic associations of fungi with photosynthetic partners (green algae and cyanobacteria, mostly). Mycorrhizae are mutualistic associations of a fungus and the young roots of plants. Fungal hyphae provide nutrients for their symbiont, which provides the fungus with carbohydrates.

Review Questions

1. Describe the fungal mode of nutrition, and explain how the structure of mycelia facilitates this mode. *24.1*

2. How does a lichen differ from a mycorrhiza? *24.4*

Self-Quiz *(Answers in Appendix IV)*

1. New mycelia form after _____ germinate.
   a. hyphae
   b. spores
   c. mycelia
   d. mushrooms

2. A "mushroom" is _____ .
   a. the food-absorbing part of a fungal body
   b. the part of the fungal body not constructed of hyphae
   c. a reproductive structure
   d. a nonessential part of the fungus

3. A mycorrhiza is a _____ .
   a. fungal disease of the foot
   b. fungus-plant relationship
   c. parasitic water mold
   d. fungus endemic to barnyards

4. Parasitic fungi obtain nutrients from _____ .
   a. tissues of living hosts
   b. nonliving organic matter
   c. only living animals
   d. none of the above

5. Saprobic fungi derive nutrients from _____ .
   a. nonliving organic matter
   b. living organisms
   c. root hairs
   d. both b and c

6. Match the terms appropriately.
   ____ zygomycetes
   ____ conidia
   ____ hypha
   ____ basidiomycetes
   ____ asci
   ____ ascomycetes

   a. mushrooms, shelf fungi
   b. some yeasts and morels
   c. chains of asexual spores
   d. each filament in a mycelium
   e. black bread mold
   f. sac-shaped cells in which asexual spores are produced

Figure 24.11 Reproductive structures of *Pilobolus*, a Greek word meaning "hat-thrower." The "hats" are spore sacs.

Critical Thinking

1. *Pilobolus* is a type of fungus that commonly dines on horse dung. Each morning, stalked reproductive hyphae emerge from irregularly spaced piles of dung. By early afternoon, they have completed the task of dispersing spores to sunlit grasses where horses feed. The spores pass through the horse gut unharmed and exit with their very own pile of dung. At the tip of each stalked hypha is a dark-walled, spore-containing sac (Figure 24.11). Just below the sac, the stalk is differentiated into a vesicle, swollen with a fluid-filled central vacuole. At the base of the vesicle is a ring of light-sensitive, pigmented cytoplasm. The stalk bends as it grows until its wall is parallel with the sun's rays and light strikes all of the ring. When that happens, turgor pressure builds up inside the central vacuole until the vesicle ruptures. The forceful blast can propel spore sacs two meters away—which is amazing, considering that the stalk is less than ten millimeters tall. Reflect on the examples of fungi discussed in this chapter. Would you say that *Pilobolus* is a zygomycete, club fungus, or sac fungus?

2. Diana discovers in the laboratory that a fungus (*Trichoderma*) can grow well in distilled water. It continues to do so even after she rigorously treats the water and glassware to remove all traces of organic carbon. This fungus is not a photoautotroph. Suggest a metabolic life-style that would allow it to grow under these conditions.

3. Some strains of *Trichoderma* are being tested as a natural pest control agent. Laboratory experiments demonstrated that the strains of this sac fungus combat other fungi that cause plant diseases. Some even promote seed germination and plant growth. In one set of twenty trials, workers increased lettuce yields by 54 percent. What concerns must be addressed before *Trichoderma* can be released into the environment for commercial applications?

Selected Key Terms

ascospore *24.3*	hypha *24.1*	sac fungus
basidiospore *24.2*	lichen *24.4*	(ascomycetes) *24.1*
club fungus	mushroom *24.2*	saprobe *24.1*
(basidiomycetes) *24.1*	mutualism *24.4*	spore (fungal) *24.1*
extracellular digestion	mycelium *24.1*	symbiosis *24.4*
and absorption *C1*	mycorrhiza *24.4*	zygomycetes *24.1*
fungus *CI*	parasite *24.1*	zygosporangium *24.3*

Reading

Moore-Landecker, E. 1990. *Fundamentals of the Fungi.* Third edition. Englewood Cliffs, New Jersey: Prentice-Hall.

Web Site See *http://www.wadsworth.com/biology* for practice quiz questions, hypercontents, BioUpdates, and critical thinking. The Wadsworth Biology Resource Center provides a wealth of information fully organized and integrated by chapter.

25 PLANTS

Pioneers In a New World

Seven hundred million years ago, no shorebirds stirred and noisily announced the dawn of a new day. There were no crabs to clack their claws together and skitter off to burrows. The only sounds were the rhythmic muffled thuds of waves in the distance, at the outer limits of another low tide. More than 3 billion years before, life had its beginning somewhere in the waters of the Earth—and now, quietly, the invasion of the land was under way.

Astronomical numbers of photosynthetic cells had come and gone, and the oxygen-producing types had slowly changed the atmosphere. High above the Earth, the sun's energy had converted much of the oxygen into a dense ozone layer. That layer became a shield against lethal doses of ultraviolet radiation, which had kept early organisms beneath the water's surface.

Were cyanobacteria the first to adapt to intertidal zones, where mud dried out with each retreating tide? Were they the first to spread into shallow, freshwater streams flowing down to the coasts? Probably so. From fossil evidence, we know that later in time, green algae and fungi made the same journey together.

Every plant around you is a descendant of ancient species of green algae that lived near the water's edge or made it onto the land. Diverse fungi still associate with nearly all of them. Together, the plants and fungi became the basis of communities in coastal lowlands, near the snow line of high mountains, and just about all places in between (Figure 25.1).

We have a few tantalizing fossils of the first pioneers. We also are learning about them through comparative biochemistry and studies of existing species. Today, as in the late Precambrian, cyanobacteria and green algae grow in mats in nearshore waters and on the banks of freshwater streams (Figure 25.1a). After a volcano erupts or a glacier retreats, cyanobacteria are the first to colonize the barren rocks. Then, symbiotic associations between green algae and fungi follow. Gradually their organic products and remains accumulate and create pockets of soil. Mosses and other plant species can take hold in the newly forming soil and further enrich it.

With this chapter we turn to the plant kingdom. With only a few exceptions, plants are multicelled photoautotrophs. They absorb energy from the sun, carbon dioxide from the air, and some minerals dissolved in water to synthesize organic compounds. These metabolic wizards also can split water molecules. In doing so, they obtain the stupendous numbers of the electrons and hydrogen atoms required for growth into multicellular

a

Figure 25.1 (a) Filaments of a green alga, massed in a shallow stream. More than 400 million years ago, green algal species that may have been ancestral to all plants, past and present, lived in similar streams that meandered down to the shores of early continents. (b) One land-dwelling descendant of those ancestral forms—a Ponderosa pine high above the floor of Yosemite Valley in the Sierra Nevada of California. (c) Flowers of one of the most highly prized flowering plants—orchids—growing on a branch of a living tree in a tropical rain forest. With this chapter, we turn to the beginning—and ends of the line—of some ancient lineages.

b

c

KEY CONCEPTS

1. With very few exceptions, the plant kingdom consists of multicelled photoautotrophs. From earlier chapters, you know that plants use chlorophylls *a* and *b* as their main photosynthetic pigments. In this respect they are like green algae, which are their closest relatives.

2. Unlike their algal ancestors, which were adapted to aquatic habitats, nearly all existing plants live on land.

3. In general, plants are structurally adapted to intercept sunlight, absorb water and mineral ions, and conserve water. Their lignin-reinforced tissues permit upright growth. Root systems mine the soil for water and ions, and internal tissues conduct water and solutes to all living cells in belowground and aboveground parts.

4. Land plants are reproductively adapted to withstand dry periods. During the life cycle, a sporophyte develops roots, stems, and leaves. It holds onto its developing gametes and supplies them with water and food resources. And it disperses the new generation in ways that are responsive to the conditions in specific habitats.

5. Early divergences gave rise to the bryophytes, then seedless vascular plants, and then seed-bearing vascular plants. Of these categories, the seed producers were the most successful in radiating into drier environments.

6. The seed-bearing vascular plants called gymnosperms include the cycads, ginkgo, gnetophytes, and conifers. The angiosperms, another group of vascular plants, bear flowers as well as seeds. There are two classes of flowering plants, informally called the dicots and monocots.

forms as tall as the giant redwoods, as extensive as an aspen forest that is one continuous clone.

We know of at least 295,000 kinds of existing plants. Be glad their ancient ancestors left the water. Without them, we humans and other land-dwelling animals never would have made it onto the evolutionary stage.

EVOLUTIONARY TRENDS AMONG PLANTS

Overview of the Plant Kingdom

The plant kingdom includes at least 295,000 species of photoautotrophs and a few heterotrophs. Most kinds are **vascular plants**, with internal tissues that conduct water and solutes through the plant body. Such plants have roots, stems, and leaves, defined in part by the presence of vascular tissues. Fewer than 19,000 species are *non*vascular plants called **bryophytes**. Plants, like photoautotrophic bacteria and protistans, are producers of organic compounds for entire communities.

Liverworts, hornworts, and mosses are bryophytes. The whisk ferns, lycophytes, horsetails, and ferns are *seedless* vascular plants. Cycads, ginkgo, gnetophytes, and conifers belong to a group of *seed-bearing* vascular plants called **gymnosperms**. The **angiosperms**, another group of vascular plants, bear flowers as well as seeds. Dicots and monocots are two classes of flowering plants.

The ancestors of plants evolved in the seas by 700 million years ago. About 265 million more years passed before simple stalked plants appeared along coasts and streams. Evolutionarily speaking, the pace picked up after that. Within a mere 60 million years, plants had radiated through much of the land. Some long-term changes in their structure and reproductive events help explain how the diversity came about.

Evolution of Roots, Stems, and Leaves

Simple underground structures started to evolve when plants first colonized the land. In lineages that led to vascular plants, they developed into **root systems**. Most root systems have a number of underground absorptive structures that collectively afford a large surface area. These rapidly take up soil water and dissolved mineral ions; often they anchor the plant. Aboveground, **shoot systems** evolved. Shoot systems have stems and leaves, which efficiently absorb energy from the sun's rays and carbon dioxide from the air. Stems grew and branched extensively only after plants developed the biochemical capacity to synthesize and deposit **lignin**, an organic compound, in cell walls. Cells with lignified walls can structurally support stems, which grow in patterns that increase the total light-intercepting surface of leaves.

Many plants became equipped with cellular pipelines for water and solutes. The pipelines were key factors in the evolution of roots, stems, and leaves. They evolved as components of two vascular tissues called xylem and phloem. **Xylem** distributes water and dissolved ions to all living cells throughout the plant. **Phloem** distributes dissolved sugars and other photosynthetic products.

Life on land also depended on water conservation, which wasn't a problem in most aquatic habitats. Shoots became protected by a **cuticle**, a waxy coat that helps conserve water on hot, dry days. **Stomata** (singular, stoma), tiny openings across the surfaces of leaves and some stems, became strategic in controlling absorption of carbon dioxide and restricting evaporative water loss. Later chapters describe these tissue specializations.

From Haploid to Diploid Dominance

As early plants radiated into higher, drier places, their life cycles changed. Think about the gametes of algae, which cannot get together without the presence of liquid water. As shown in earlier chapters, the *haploid* (n) phase, in the form of **gametophytes** (gamete-producing bodies), dominates their life cycle. The diploid ($2n$) phase is the zygote, which forms when gametes fuse at fertilization.

Now look at Figure 25.2. *In most plant life cycles, the diploid phase dominates.* After a diploid zygote forms at fertilization, mitotic cell divisions and cell enlargements transform it into a multicelled diploid body, of a type called a **sporophyte**. Pine trees are an example. In time, some cells of the sporophyte undergo meiosis and give rise to haploid cells of a type called **spores** (sporophyte means spore-producing body). Later, the spores divide by way of mitosis and give rise to the gametophytes.

The shift to diploid dominance was an adaptation to habitats on land, most of which show seasonal changes in the availability of free water and dissolved nutrients. Long ago in those challenging habitats, natural selection must have favored sporophytes with well-developed root systems. Young roots of such systems interact with fungal symbionts (Section 24.4). The association, called a mycorrhiza, enhances the plant's uptake of water and scarce minerals, even during dry seasons.

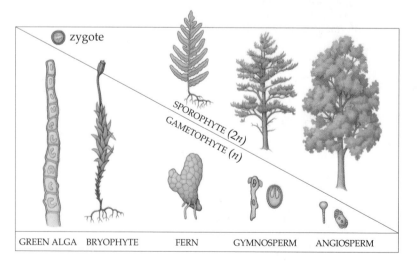

zygote

SPOROPHYTE (2n)

GAMETOPHYTE (n)

GREEN ALGA   BRYOPHYTE   FERN   GYMNOSPERM   ANGIOSPERM

Figure 25.2 Evolutionary trend from gametophyte (haploid) dominance to sporophyte (diploid) dominance, represented here by existing species ranging from a green alga (*Ulothrix*) to a flowering plant. This trend occurred when early plants were colonizing habitats on land. See also Section 10.5.

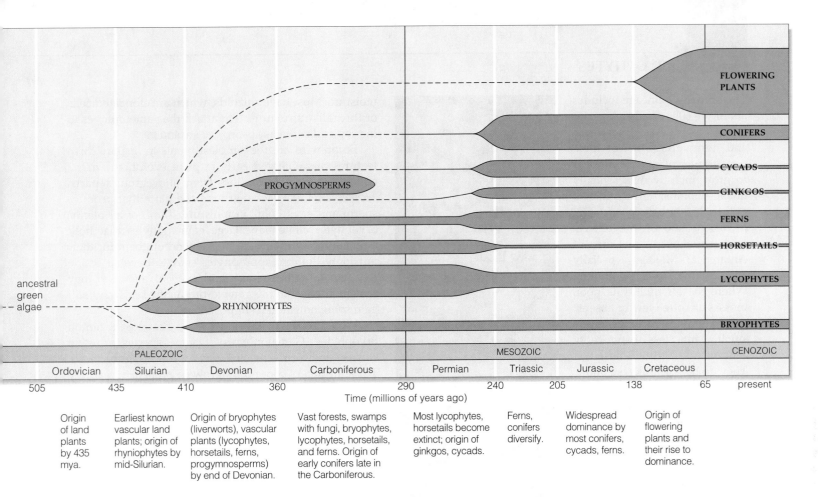

							FLOWERING PLANTS
							CONIFERS
							CYCADS
	PROGYMNOSPERMS						GINKGOS
							FERNS
							HORSETAILS
							LYCOPHYTES
ancestral green algae	RHYNIOPHYTES						BRYOPHYTES

PALEOZOIC				MESOZOIC			CENOZOIC
Ordovician	Silurian	Devonian	Carboniferous	Permian	Triassic	Jurassic	Cretaceous

505   435   410   360   290   240   205   138   65   present

Time (millions of years ago)

| Origin of land plants by 435 mya. | Earliest known vascular land plants; origin of rhyniophytes by mid-Silurian. | Origin of bryophytes (liverworts), vascular plants (lycophytes, horsetails, ferns, progymnosperms) by end of Devonian. | Vast forests, swamps with fungi, bryophytes, lycophytes, horsetails, and ferns. Origin of early conifers late in the Carboniferous. | Most lycophytes, horsetails become extinct; origin of ginkgos, cycads. | Ferns, conifers diversify. | Widespread dominance by most conifers, cycads, ferns. | Origin of flowering plants and their rise to dominance. |

Figure 25.3 Milestones in plant evolution.

Unlike algae and bryophytes, vascular plants have a sporophyte that is larger and structurally more complex than the gametophyte. As you will see, the gametophytes of seedless vascular plants develop independently of the sporophyte that produces them. But they protect their gametes, and, after fertilization, they nourish the embryo sporophytes. The sporophyte became most dominant among the gymnosperms and, later, the angiosperms. It retains, nourishes, and protects developing gametophytes as well as the young sporophytes. *And it does so until environmental conditions are suitable for fertilization and for dispersal of the new generation.*

Evolution of Pollen and Seeds

Like some seedless species, seed-bearing plants produce not one but two kinds of spores. This condition is called *hetero*spory, as opposed to *homo*spory. In gymnosperms and angiosperms, one type of spore gives rise to **pollen grains**, which will develop into mature, sperm-bearing male gametophytes. The other type of spore develops into female gametophytes, where eggs form and later become fertilized. The pollen grains hitch rides on air currents, insects, birds, and so on; they do not require free-standing water to reach the eggs. In this respect, they differ enormously from the algae. The evolution of pollen grains is one reason for the successful radiation of seed-bearing plants into high and dry habitats.

Seed production also was adaptive in drier habitats. Female gametophytes (and eggs) of seed-bearing plants form inside nutritive tissues and a jacket of cell layers. Each **seed** consists of an embryo sporophyte, nutritive tissues, and a protective coat (which develops from the jacket of cell layers). Seeds withstand hostile conditions. Not coincidentally, the seed plants rose to dominance in Permian times, when shifts in climate were extreme.

Before turning to the spectrum of diversity among plants, take a look at Figure 25.3. You can use it as a map for tracking the branching evolutionary roads.

The plant kingdom includes multicelled, photosynthetic species called bryophytes, seedless vascular plants, and seed-bearing vascular plants. Most of these live on land.

In most lineages, structural adaptations to life on land included root and shoot systems, waxy cuticles, stomata, vascular tissues, and lignin-reinforced tissues.

Sporophytes with well-developed roots, stems, and leaves came to dominate the life cycle of most land plants. Parts of these complex sporophytes nourish and protect fertilized eggs and embryos through unfavorable conditions.

Some plants started producing two types of spores, not one. This led to the evolution of male gametes adapted for dispersal without liquid water and to the evolution of seeds.

The bryophyte lineage includes about 18,600 existing species known as **mosses**, **liverworts**, and **hornworts**. Most of these nonvascular plants are adapted to grow in fully or seasonally moist habitats, although you will find some mosses growing in deserts and on the bitterly cold, windswept plateaus of Antarctica. Mosses especially are sensitive to air pollution. Where the air quality is poor, mosses are often few or absent.

Bryophytes are generally small plants, less than twenty centimeters (eight inches) tall. They do have leaflike, stemlike, and rootlike parts, but these don't contain xylem or phloem. Bryophytes, like lichens and some algae, can dry out, then revive after they absorb some moisture. Most have rhizoids, which are elongated cells or threadlike structures that attach the gametophytes to the soil and serve as absorptive structures.

Bryophytes are the simplest plants to display three features that emerged early in plant evolution. *First*, a cuticle prevents water loss from aboveground parts. *Second*, a cellular jacket around the parts that produce sperm and eggs holds in moisture. *Third*, of all plants, bryophytes alone have large gametophytes that hold onto sporophytes and do not depend on them for their nutrition. Embryo sporophytes start to develop inside the gametophyte tissues—and even at maturity, they remain *attached to* the gamete-producing body and still gain some nutritional support from it.

With 10,000 species, mosses are the most common bryophytes. Gametophytes of some species grow in clusters and form low, cushiony mounds (Figure 25.4*a*). Those of others commonly grow in branched, feathery

Figure 25.4 (**a**) Moss-covered rocks near a small stream. (**b**) Photograph and life cycle of a representative bryophyte, a moss (*Polytrichum*). The moss sporophyte remains attached to the gametophyte and depends upon it for water and nutrients.

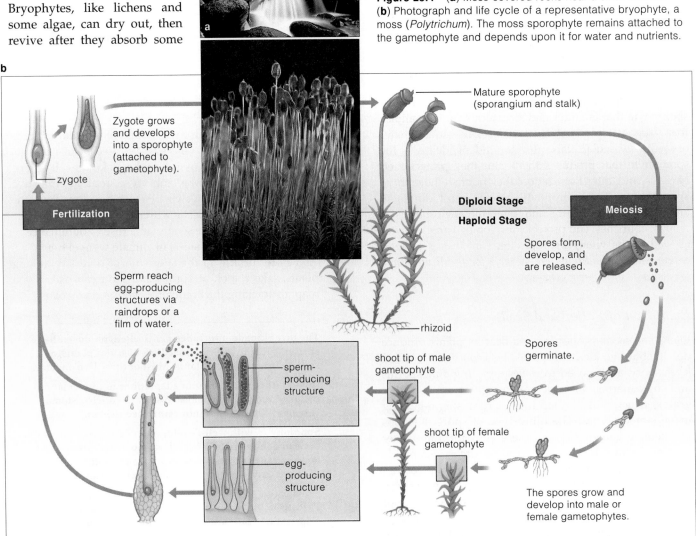

Mature sporophyte
(sporangium and stalk)

Zygote grows and develops into a sporophyte (attached to gametophyte).

zygote

Fertilization

Sperm reach egg-producing structures via raindrops or a film of water.

Diploid Stage

Haploid Stage

Meiosis

Spores form, develop, and are released.

Spores germinate.

sperm-producing structure

shoot tip of male gametophyte

egg-producing structure

shoot tip of female gametophyte

rhizoid

The spores grow and develop into male or female gametophytes.

Figure 25.5 (a) A peat bog in Ireland. A family is cutting blocks of peat, and stacking them to dry, as a fuel source for their home. Peat also is harvested on a scale large enough to generate electricity in peat-burning power plants. (b) The gametophyte of a peat moss (*Sphagnum*) with a few sporophytes attached. The sporophytes are the brown, jacketed structures on the white stalks. Because of its good antiseptic properties and high absorbency, peat moss was used as an emergency poultice on the wounds of soldiers during World War I.

a

b

male gametophyte   female gametophyte   gemmae

a   b   c

Figure 25.6 *Marchantia*, one of the liverworts. Like other liverworts, this nonvascular plant can reproduce sexually. Unlike other types, it forms (a) male reproductive structures and (b) female reproductive structures on different plants. (c) *Marchantia* also reproduces asexually by way of gemmae, multicelled vegetative bodies that develop in tiny cups on the plant body. Gemmae grow and develop into individual plants after splashing raindrops transport them to suitable sites.

patterns on tree trunks and branches in humid climates. Eggs and sperm develop in tiny, jacketed vessels at the shoot tips of gametophytes. Sperm reach the eggs by swimming through a film of water on plant parts. After fertilization, the zygotes give rise to sporophytes, each consisting of a stalk and a jacketed structure in which spores will develop.

Figure 25.5 shows one of 350 kinds of peat mosses (*Sphagnum*). Whereas most bryophytes grow slowly, the peat mosses can grow fast enough to yield twelve metric tons of organic matter per hectare, which is about twice as much as corn plants yield. They soak up five times as much water as cotton does, owing to large, dead cells in their leaflike parts. They also produce acids that inhibit the growth of bacterial and fungal decomposers. Their remains accumulate into compressed, excessively moist mats called **peat bogs**. In cold and temperate regions, peat bogs cover an area equal to one-half of the United States. Only the most acid-tolerant plants, such as larch,

cranberries, blueberries, and Venus flytraps, can grow in the bogs, which can be as acidic as vinegar.

Nearly all of the peat harvested and dried in Ireland and elsewhere is burned to generate electricity in power plants. Compared with coal burning, peat fires generate fewer pollutants. Every so often, peat harvesters come across the exceedingly well-preserved bodies of humans that lived 2,000 to 3,000 years ago. Some ancient bogs apparently were sites of ceremonial human sacrifices.

So as not to dwell on the macabre, let us leave this section with the liverworts and their interesting ways of reproducing, as described in Figure 25.6.

Bryophytes are nonvascular plants with flagellated sperm that require liquid water to reach and fertilize the eggs.

A sporophyte of these plants develops within gametophyte tissues. It remains attached to the gametophyte and receives some nutritional support from it.

Figure 25.7a shows one of the early seedless vascular plants. Descendants of certain lineages are still with us; we call them **whisk ferns**, **lycophytes**, **horsetails**, and **ferns**. Like their ancestors, they differ from bryophytes in three key respects. The sporophyte does not remain attached to a gametophyte; it has true vascular tissues; and it is the larger, longer lived phase of the life cycle.

Most seedless vascular plants live in wet, humid places, and their gametophytes lack vascular tissues. Water droplets clinging to the plants are the only means by which flagellated sperm can reach the eggs. The few species living in dry habitats reproduce sexually during brief, seasonal pulses of heavy rains. Thus, whisk ferns, lycophytes, horsetails, and ferns are the "amphibians" of the plant kingdom. They have not fully escaped the aquatic habitats of their ancestors.

Whisk Ferns

Whisk ferns (Psilophyta), which are not ferns, resemble whisk brooms. Florist suppliers commonly cultivate them in Hawaii, Texas, Louisiana, Florida, Puerto Rico, and other tropical or subtropical regions. One genus,

Psilotum, is a unique vascular plant, for its sporophytes have no roots or leaves. The photosynthetic, branched stems have xylem, phloem, and scalelike projections (Figure 25.7b). Belowground are **rhizomes**, branching, short, mostly horizontal stems that serve in absorption.

Lycophytes

About 350 million years ago, lycophytes (Lycophyta) included tree-sized members of swamp forests. About 1,100 far tinier species exist today. The most familiar are club mosses—members of communities in the Arctic, the tropics, and regions in between. Many form mats on forest floors. One type, called the resurrection plant, is common in Texas, New Mexico, and Mexico.

Sporophytes of most club mosses have leaves and a branching rhizome that gives rise to vascularized roots and stems. Some have nonphotosynthetic, cone-shaped clusters of leaves that bear spore sacs (Figure 25.7c). Each cluster is a **strobilus** (plural, strobili). After spores disperse, they germinate and then develop into small, free-living gametophytes. In one genus (*Selaginella*), two kinds of spores develop in the same strobilus.

Horsetails

Tree-sized sphenophytes (Sphenophyta) flourished in ancient swamp forests. Twenty-five or so smaller species of one genus (*Equisetum*) made it to the present. These are the horsetails. Of all existing plants, they may be the oldest, and their body plan changed little over the past 300 million years. They grow in streambank mud and vacant lots, roadsides, and other disrupted habitats. Figure 25.7d–f shows the fertile stems and vegetative, photosynthetic stems of one species. Its spores give rise

Figure 25.7 (a) *Cooksonia*, one of the earliest known vascular plants, no more than a few centimeters tall. It probably grew in mud flats. Its upright, branching stems had a cuticle. Its spores formed in sporangia at stem tips. Compare Figure 21.2c. (b) Sporophytes of a whisk fern (*Psilotum*), a seedless vascular plant. Pumpkin-shaped, spore-producing structures form at the ends of stubby branchlets. (c) Sporophyte of one of the lycophytes (*Lycopodium*). (d) Vegetative stem of *Equisetum*, which resembles a horsetail. (e) Fertile, nonphotosynthetic stems of *Equisetum*. (f) Closer look at the spore-bearing structure of a fertile stem.

strobilus, an aggregation of spore-producing structures, at tip of a vegetative shoot of a horsetail sporophyte

f Closer look at a strobilus. Inside each petal-shaped structure, many spores have been forming by way of meiotic cell divisions.

b   c   d   e

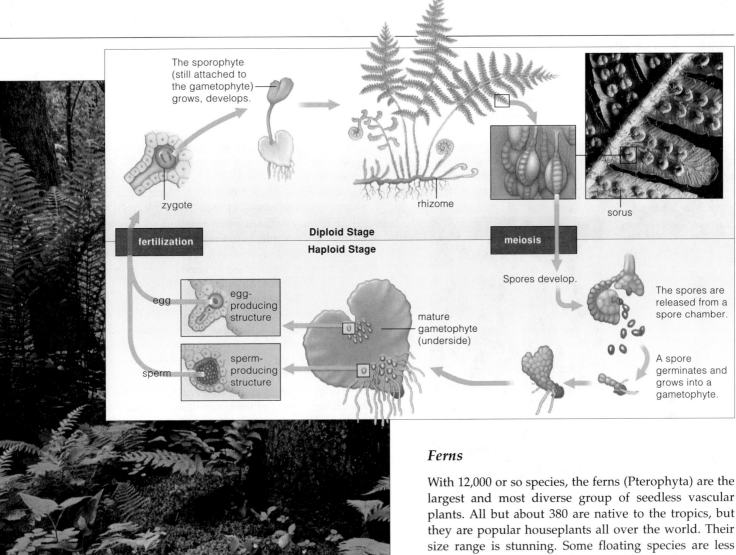

The sporophyte
(still attached to
the gametophyte)
grows, develops.

zygote

rhizome

sorus

fertilization

Diploid Stage

Haploid Stage

meiosis

Spores develop.

egg

egg-
producing
structure

sperm

sperm-
producing
structure

mature
gametophyte
(underside)

The spores are
released from a
spore chamber.

A spore
germinates and
grows into a
gametophyte.

Figure 25.8 Life cycle of a fern. The photograph shows ferns growing in a moist habitat in Indiana. Ferns with finely divided fronds are in the foreground.

to free-living gametophytes 1 millimeter to 1 centimeter across. Sporophytes of most horsetails have rhizomes, hollow photosynthetic stems, and scale-shaped leaves. Clusters of xylem and phloem are arrayed as a ring in the stems. Silica-reinforced ribs structurally support the stems and give them a texture like sandpaper. Pioneers of the American West, who had to travel light, gathered horsetails along the way to use as pot scrubbers.

Ferns

With 12,000 or so species, the ferns (Pterophyta) are the largest and most diverse group of seedless vascular plants. All but about 380 are native to the tropics, but they are popular houseplants all over the world. Their size range is stunning. Some floating species are less than 1 centimeter across. Some tropical tree ferns are 25 meters (82 feet) tall. One climbing fern has a modified leaf stalk about 30 meters long.

Most ferns have vascularized rhizomes that give rise to roots and leaves. Exceptions include tropical tree ferns and epiphytes. (*Epiphyte* refers to any aerial plant that grows attached to tree trunks or branches.) While they develop, young fern leaves are coiled, rather like a fiddlehead. At maturity, the leaves (fronds) commonly are divided into leaflets.

You may have noticed rust-colored patches on the lower surface of many fern fronds. Each patch, a cluster of spore chambers, is called a sorus (plural, sori). At dispersal time, the chambers snap open and cause the spores to catapult through the air. Each germinating spore develops into a small gametophyte, such as the green, heart-shaped type shown in Figure 25.8.

Seedless vascular plants (whisk ferns, lycophytes, horsetails, and ferns) have sporophytes adapted to conditions on land. Yet they have not entirely escaped their aquatic ancestry. When they reproduce sexually, their flagellated sperm cannot reach the eggs unless liquid water is clinging to the plant.

25.4 ANCIENT CARBON TREASURES

Three hundred million years ago, about halfway through the Carboniferous, mild climates prevailed and swamp forests carpeted the wet lowlands of the continents. The absence of pronounced seasonal swings in temperature favored plant growth through much of the year. The plants having lignin-reinforced tissues and well-developed root and shoot systems had the competitive edge under these growth conditions, and some of them evolved into giants. Massive-stemmed lycophyte trees—the giant club mosses—topped out at nearly forty meters (Figure 25.9). Each of their strobili produced as many as 8 billion microspores or several hundred megaspores. And being so high above the forest floor, dispersal of each new generation was a cinch. Giant horsetails, including species of *Calamites*, were close to twenty meters tall. Often their aboveground stems, which grew from rapidly spreading underground rhizomes, formed dense thickets.

As it happened, sea levels rose and fell fifty times during the Carboniferous. Each time the seas receded, the swamp forests flourished. When the seas moved back in, forest trees became submerged and buried in sediments that protected them from decay. Gradually the sediments compressed the saturated, undecayed remains into peat. Each time more sediments accumulated, they subjected the peat to increased heat and pressure that made it even more compact. In this way, compressed organic remains were transformed into great seams of **coal** (Figure 25.9).

With its high percentage of carbon, coal is energy rich and is one of our premier "fossil fuels." It took a fantastic amount of photosynthesis, burial, and compaction to form each major seam of coal in the Earth. It has taken us only a few centuries to deplete much of the world's known coal deposits. Often you will hear about annual "production rates" for coal or some other fossil fuel. But how much do we really produce each year? None. We simply *extract* it from the Earth. Coal is a nonrenewable source of energy.

Lepidodendron

stem of a giant lycophyte
(*Lepidodendron*)

seed fern (*Medullosa*); probably related to the progymnosperms, which may have been among the earliest seed-bearing plants

stem of a giant horsetail
(*Calamites*)

Figure 25.9   Reconstruction of a Carboniferous forest. The boxed inset shows part of a seam of coal.

25.5 THE RISE OF THE SEED-BEARING PLANTS

About 360 million years ago, as the Devonian gave way to the Carboniferous, the first seed-bearing plants arose. In terms of diversity, numbers, and distribution, they would become the most successful groups of the plant kingdom. Seed ferns, gymnosperms, and (much later) angiosperms were the dominant groups. They differed from seedless vascular plants in three crucial respects.

First, seed-bearing plants produce pollen grains, the sperm-bearing male gametophytes. Remember, these plants produce two types of spores. Their **microspores** give rise to pollen grains. Unlike the spores of seedless vascular plants, they do not have a "tetrad scar," which marks the cleavage planes between four spores that form during meiotic cell division (Figure 25.10).

Like a suitcase, a pollen grain is a means of getting its contents (the sperm) to the eggs even during times of prolonged drought. Seedless vascular plants do not have such an advantage; without predictable rains and moisture, their sperm simply cannot reach the eggs, and this has adverse effects on reproductive success. By contrast, pollen grains of gymnosperms simply drift with air currents. Those of angiosperms also are loaded onto insects, birds, bats, and other animals that truck them to the eggs. **Pollination** is the name for the arrival of pollen grains on the female reproductive structures. By this process, seed-bearing plants escaped dependence on free water for fertilization.

Second, besides microspores, seed-bearing plants also produce **megaspores**. These develop into **ovules,** the female reproductive structures which, at maturity, are seeds (Figure 25.11). Each ovule consists of a female gametophyte (with its egg cell), nutrient-rich tissue, and a jacket of cell layers which, recall, develops into the seed coat. A zygote will form inside the ovule when a sperm reaches and fertilizes the egg. An embryo sporophyte will develop, and when the time comes to say good-bye to the parent plant, the seed coat around it will afford protection for the journey.

Third, compared to seedless vascular plants, these plants had thicker cuticles, stomata recessed below the surface of the leaf, and other traits that gave them competitive advantages in drier and cooler environments. Such environments were ushered in when the Carboniferous gave way to the Permian (Section 21.5). **Seed ferns** of the type shown in Figure 25.9 rose to dominance and then prevailed for about 70 million years. These seed-

bearing plants apparently arose from **progymnosperms**. Progymnosperms may have been the first seed-bearing plants, although their seeds may only have been seed-like structures. When the global climate grew cooler and drier, swamplands disappeared, and so did the seed ferns. They were replaced by the cycads, conifers, and other gymnosperms.

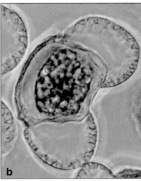 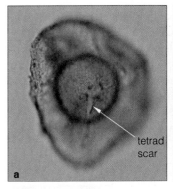

tetrad scar

Figure 25.10 (**a**) From the Devonian-Carboniferous boundary, a fossilized spore of a lycophyte. Its tetrad scar is typical of the spores of seedless plants. (**b**) A pine pollen grain (*Pinus*). LIke the pollen of other seed-bearing plants, it has no tetrad scar.

ovule, which may mature into a seed

seed (mature ovule); the surrounding ovary matured to form the fleshy fruit

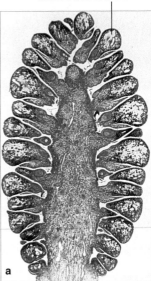 

Figure 25.11 (**a**) Longitudinal section through a seed-bearing cone of a pine. These seeds are not protected by sporophyte tissues; they are exposed on scales of a cone. (**b**) A mature ovule, or seed, of a peach (*Prunus*), one of the angiosperms. It is protected by a seed coat, which is enclosed within the fleshy, edible tissue of the fruit.

In time, gymnosperms were the primary producers that sustained the dinosaurs. What some folks call the Age of Dinosaurs, botanists call the Age of Cycads.

Seed-bearing plants rely on pollen grains, ovules that mature into seeds, and tissue changes adapted to dry conditions.

GYMNOSPERMS—PLANTS WITH "NAKED" SEEDS

With a bit of history behind us, we turn now to a survey of some of the existing gymnosperms. Unlike the seeds of flowering plants, which are enclosed in a chamber called an ovary, gymnosperm seeds are perched, in an exposed way, on a spore-producing structure. (*Gymnos* means naked; *sperma* is taken to mean seed.)

Conifers

Conifers (Coniferophyta) are woody trees and shrubs that have needlelike or scalelike leaves and that bear seeds exposed on cone scales. Conifer **cones** are clusters of modified leaves that surround the spore-producing structures. Figure 25.12 gives examples.

Most conifers shed some leaves all year long yet stay leafy, or *evergreen*. A few are *deciduous*, meaning they shed all of their leaves in the fall. Among the conifers are the most abundant trees of the Northern Hemisphere (pines), the tallest (coast redwoods), as well as the oldest (the bristlecone pine; one 4,725-year-old individual sprouted when Egyptians were building the Great Sphinx). The firs, yews, spruces, junipers, larches, cypresses, bald cypress, dawn redwood, and podocarps also belong to this group.

Lesser Known Gymnosperms

CYCADS About 100 species of **cycads** (Cycadophyta) made it to the present day. Figure 25.13 shows examples of their pollen-bearing and seed-bearing cones, which form on separate plants. Insects or air currents transfer pollen from "male" to "female" plants. At first glance, you might mistake cycad leaves for those of a palm tree; but palms are flowering plants.

These plants mainly inhabit tropical and subtropical areas. Two species (*Zamia*) grow wild in Florida and are planted as ornamentals. Elsewhere, their seeds and a flour made from the trunks are edible after their toxic alkaloids are removed. Many species are vulnerable to extinction.

Figure 25.12 (**a**) Bristlecone pine (*Pinus longaeva*) growing near the timberline in the Sierra Nevada. (**b**) Male pine cones releasing pollen. (**c**) Appearance of a female pine cone at the time of pollination. (**d**) From a juniper (*Juniperus*), cones with a berry-like appearance. These cones are made of fused-together, fleshy scales.

Figure 25.13
(**a**) Pollen-bearing cone of a "male" cycad (*Zamia*).
(**b**) Seed-bearing cone of a "female" cycad. Of all the existing species of gymnosperms, the cycads produce the largest seed-bearing cones. Some of these grow as long as one meter and weigh more than fifteen kilograms.

Figure 25.14 (**a**) Mature ginkgo. (**b**) Fossilized ginkgo leaf compared with a leaf from its existing descendant. The fossil formed at the Cretaceous-Tertiary boundary. Even though 65 million years have passed since then, the leaf structure has not changed much, if at all. (**c**) Pollen-bearing cones and (**d**) fleshy-coated seeds of this type of gymnosperm.

Figure 25.15 (**a**) Sporophyte of *Ephedra* and (**b**) its pollen-bearing cones and (**c**) a seed-bearing cone. (**d**) Sporophyte of *Welwitschia mirabilis* and (**e**) its seed-bearing cones.

GINKGOS   The **ginkgos** (Ginkgophyta) were a diverse group in dinosaur times. The only surviving species is the maidenhair tree, *Ginkgo biloba*. Like a few species of larch and some other gymnosperms, these plants are deciduous. Several thousand years ago, ginkgo trees were widely planted around temples in China. Then the natural populations nearly became extinct, even though ginkgos seem hardier than many other trees. Perhaps they became targets for firewood. Today, male ginkgo trees are again widely planted. They have attractive, fan-shaped leaves and are resistant to insects, disease, and air pollutants. Female trees are not favored. Their thick, fleshy seeds, which are the size of small plums, give off an awful stench (Figure 25.14).

GNETOPHYTES   At present, there are three genera of woody plants known as the **gnetophytes** (Gnetophyta). Trees and leathery leafed vines of *Gnetum* thrive in the humid tropics. The shrubby *Ephedra* lives in California

deserts and some other arid regions (Figure 25.15*a–c*). Photosynthesis proceeds in its green stems. *Welwitschia mirabilis* grows in hot deserts of south and west Africa. Its sporophyte is mainly a deep-reaching taproot. Its exposed part, a woody disk-shaped stem, has cones and one or two strap-shaped leaves that split lengthwise repeatedly as the plant ages (Figure 25.15*e*).

Conifers, cycads, ginkgos, and gnetophytes are groups of existing gymnosperms. Like their ancestors, they bear their seeds on the exposed surfaces of cones and other spore-producing structures.

A CLOSER LOOK AT THE CONIFERS

Before we leave the gymnosperms, let's use the conifers as an example of reproductive strategies. Depending on the species, a conifer's life cycle lasts a year or more.

Who among us hasn't noticed the woody, shelflike scales of "a pine cone"? The scales are actually parts of a mature female cone in which megaspores formed and developed into female gametophytes. Pine trees also produce male cones, in which microspores form and develop into pollen grains (Figure 25.16). Each spring, millions of pollen grains drift away from the male cones.

Pollination is completed when some land on ovules of female cones. After each germinates, a tubular structure forms from it. This germinating pollen grain, the sperm-bearing male gametophyte, grows toward the egg in the female gametophyte. For species of pines, fertilization occurs months or a year after pollination.

As in other gymnosperms, seed formation begins at the ovule (Figure 25.16). An embryo sporophyte starts developing from the fertilized egg. The outer layers of the jacket around the female gametophyte and embryo mature into a hard coat. The seed coat will protect the embryo sporophyte after it is dispersed from the parent plant. The nutrients will help it through the critical time

Figure 25.16 Life cycle of one conifer, the ponderosa pine.

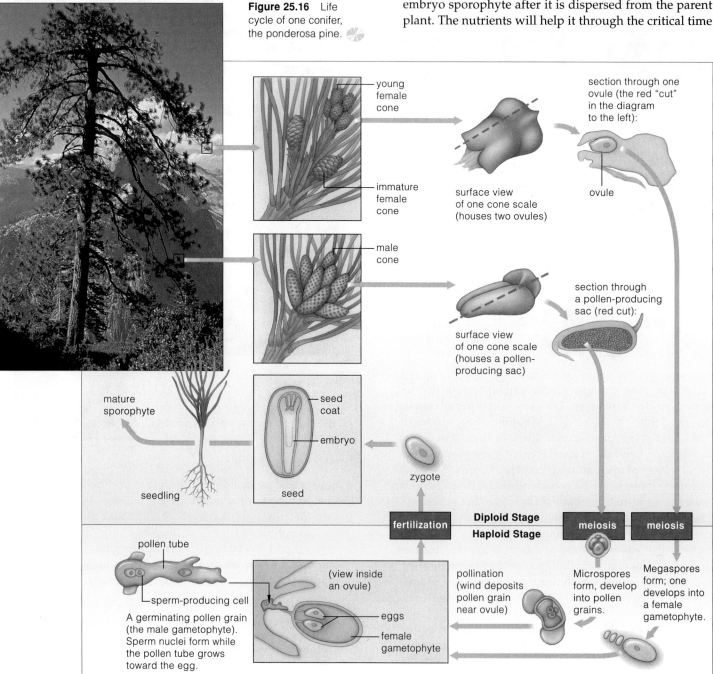

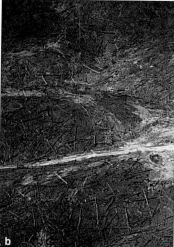

Figure 25.17 A sampling of the rampant deforestation under way around the world. Only deforested tracts in North America are shown; Section 50.4 focuses on the tropical rain forests.

From America's heartland, logged-over acreage in Arkansas (**a**). From the eastern seaboard, a denuded piece of North Carolina (**b**). Clear-cut peaks in Washington (**c**) and Alaska (**d**). These are not isolated examples. In the early 1980s, 400 million board feet of timber were being cut in Washington's Olympic Peninsula every year. In Arkansas, about one-third of the Ouachita National Forest was clear-cut. Its once-diverse forest communities have been replaced by "tree farms" of a single species of pine. Throughout the world, huge tracts of land that were deforested years ago still show no signs of recovery.

of germination, before its roots and shoots become fully functional.

Conifers dominated many land habitats during the Mesozoic, but their slow reproductive pace put them at a competitive disadvantage when the flowering plants began their great adaptive radiation (Section 21.6). Coniferous forests still predominate in the far north, at higher elevations, and in some parts of the Southern Hemisphere. However, existing conifers face more than competition with flowering plants for resources. Now

they are vulnerable to **deforestation**: the removal of all trees from large tracts, as by clear-cutting (Figure 25.17). Conifers just have the bad luck to be premier sources of lumber, paper, and other wood products required in human societies. We return to this topic in Chapter 50.

Where flowering plants flourish, conifers are at a competitive disadvantage, partly because they take so long to reproduce. Rampant deforestation isn't helping them one bit, either.

ANGIOSPERMS—THE FLOWERING, SEED-BEARING PLANTS

Only angiosperms produce specialized reproductive structures called **flowers** (Figure 25.18). *Angeion*, which means vessel, refers to the female reproductive parts at the center of a flower. The enlarged base of the "vessel" is the floral ovary, where ovules and seeds develop.

Most flowering plants coevolved with **pollinators**—insects, bats, birds, and other animals that withdraw nectar or pollen from a flower and, in so doing, transfer pollen to its female reproductive parts. The recruitment of animals as assistants in reproduction probably has contributed to the success of flowering plants, which have dominated the land for 100 million years.

At least 260,000 species now live in a great variety of habitats. They range in size from the tiny duckweeds (about a millimeter long) to towering *Eucalyptus* trees, some of which are more than 100 meters tall. A few species, including mistletoes and Indian pipe, are not even photosynthetic; they directly or indirectly feed on other plants.

Figure 25.18 The unique trademark of angiosperms —the flower, a reproductive structure that has roles in pollination and formation of seeds. (**a**) A hummingbird is sipping nectar from this passion flower. (**b**) A few angiosperms, including the water lilies (*Nymphaea*), live in water. (**c**) Dwarf mistletoe (*Arceuthobium*), a parasitic plant, limits the growth of forest trees in the western United States. (**d**) Indian pipe (*Monotropa uniflora*) is one of the few nonphotosynthetic species. It gets nutrients from fungal symbionts that form mycorrhizae with young roots of photosynthetic plants. (**e**) Parts of flowers and seeds.

Dicots and Monocots

There are two great classes of flowering plants, called the **dicots** and **monocots** (formally, the Dicotyledonae and Monocotyledonae). Among the 180,000 dicots are most herbaceous (nonwoody) plants, such as cabbages and daisies; most flowering shrubs and trees, such as oaks and apple trees; water lilies; and cacti. Among the 80,000 or so species of monocots are orchids, palms, lilies, and grasses, such as rye, sugarcane, corn, rice, wheat, and other highly valued crop plants.

In flowers (reproductive structures), seeds develop from ovules and some tissues of the parent sporophyte after fertilization.

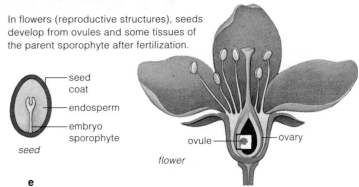

seed coat
endosperm
embryo sporophyte

seed

ovule — ovary

flower

e

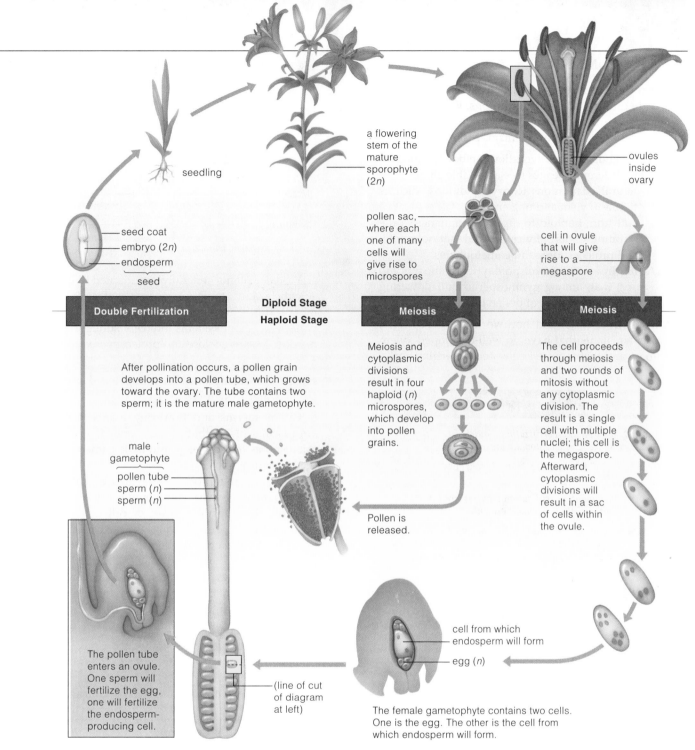

seedling

a flowering stem of the mature sporophyte (2n)

ovules inside ovary

seed coat
embryo (2n)
endosperm
seed

pollen sac, where each one of many cells will give rise to microspores

cell in ovule that will give rise to a megaspore

Double Fertilization	Diploid Stage	Meiosis	Meiosis
	Haploid Stage		

After pollination occurs, a pollen grain develops into a pollen tube, which grows toward the ovary. The tube contains two sperm; it is the mature male gametophyte.

Meiosis and cytoplasmic divisions result in four haploid (n) microspores, which develop into pollen grains.

The cell proceeds through meiosis and two rounds of mitosis without any cytoplasmic division. The result is a single cell with multiple nuclei; this cell is the megaspore. Afterward, cytoplasmic divisions will result in a sac of cells within the ovule.

male gametophyte
pollen tube
sperm (n)
sperm (n)

Pollen is released.

The pollen tube enters an ovule. One sperm will fertilize the egg, one will fertilize the endosperm-producing cell.

(line of cut of diagram at left)

cell from which endosperm will form
egg (n)

The female gametophyte contains two cells. One is the egg. The other is the cell from which endosperm will form.

Figure 25.19 Life cycle of a lily (*Lilium*), a monocot. "Double" fertilization is a distinctive feature of flowering plant life cycles. A male gametophyte delivers *two* sperm to an ovule. One sperm fertilizes the egg, and the other fertilizes a cell that gives rise to endosperm, a tissue that will nourish the forthcoming embryo. Figure 32.4 provides a closer look at flowering plant life cycles, using a dicot as the example.

Key Aspects of the Life Cycles

The next unit deals with the structure and function of flowering plants. For now, simply start thinking about the large sporophyte that dominates their life cycles (Figure 25.19). It retains and nourishes gametophytes; its sperm are dispersed within pollen grains. Inside flowering plant seeds, endosperm (a nutritive tissue) surrounds the embryo. Also, as seeds develop, tissues of most ovaries and some other structures mature into **fruits**, which protect and help disperse the embryos.

Angiosperms are the most successful plants in terms of their diversity, numbers, and distribution. They alone produce flowers. Most species coevolved with animal pollinators.

SUMMARY

1. Green algae probably gave rise to plants, nearly all of which are multicelled photoautotrophs. Plants invaded the land more than 430 million years ago. Table 25.1 summarizes and compares the major plant phyla.

2. Several trends in plant evolution can be identified by comparing different lineages (see also Table 25.2):

a. Structural adaptations to dry conditions, such as vascular tissues (xylem and phloem).

b. A shift from haploid to diploid dominance in the life cycle. Complex sporophytes evolved that hold onto, nourish, and protect spores and gametophytes.

c. A shift from one to two spore types (homospory to heterospory) that, among gymnosperms and flowering plants, led to the evolution of pollen grains and seeds.

3. Mosses, liverworts, and hornworts are bryophytes, nonvascular plants that have no well-developed xylem or phloem and that require free water for fertilization.

Table 25.1 Comparison of Major Plant Groups

Nonvascular land plants. Fertilization requires free water. Haploid dominance. Cuticle, stomata present in some.

BRYOPHYTES	18,600 species. Moist, humid habitats.

Seedless vascular plants. Fertilization requires free water. Diploid dominance. Cuticle, stomata present.

WHISK FERNS	7 species, sporophytes with no obvious roots or leaves. *Psilotum*.
LYCOPHYTES	1,100 species with simple leaves. Mostly wet or shady habitats.
HORSETAILS	25 species of single genus. Swamps, disturbed habitats.
FERNS	12,000 species. Wet, humid habitats in mostly tropical, temperate regions.

Gymnosperms—vascular plants with "naked seeds." Diploid dominance. Cuticle, stomata present.

CONIFERS	550 species, mostly evergreen, woody trees and shrubs having pollen- and seed-bearing cones. Widespread distribution.
CYCADS	185 slow-growing species. Tropics, subtropics.
GINKGO	1 species, a tree with fleshy-coated seeds.
GNETOPHYTES	70 species. Limited distribution in deserts and tropics.

Angiosperms—vascular plants with flowers and protected seeds. Diploid dominance. Cuticle, stomata present.

FLOWERING PLANTS

Monocots	80,000 species. Floral parts often arranged in threes or in multiples of three; one seed leaf; parallel leaf veins common.
Dicots	At least 180,000 species. Floral parts often arranged in fours, fives, or multiples of these; two seed leaves; net-veined leaves common.

Table 25.2 Evolutionary Trends Among Plants

Bryophytes	Ferns	Gymnosperms	Angiosperms

Nonvascular → Vascular ————————————————→

Haploid → Diploid ————————————————→
dominance   dominance

Spores of → Spores of ————————————→
one type   two types

Motile gametes ——————————————→ Nonmotile →
  gametes*

Seedless —————————→ Seeds ——————————→

* Require pollination by wind, insects, etc.

4. Vascular plants generally are adapted to life on land. A cuticle and stomata conserve water. Root systems mine soil for nutrients. Upright and branching growth patterns of shoot systems intercept sunlight and carbon dioxide. Tissues enclose and protect spores and gametes.

5. Whisk ferns, lycophytes, horsetails, and ferns are seedless vascular plants. Their flagellated sperm require ample water to swim to the eggs.

6. Gymnosperms and flowering plants (angiosperms) are vascular plants that produce pollen grains (matured microspores that develop into male gametophytes) and seeds (mature ovules).

a. Ovules are reproductive structures that contain the egg-producing female gametophytes, the precursor of nutritive tissue, and a jacket of cell layers, the outer portion of which develops into the seed coat.

b. The evolution of pollen grains freed these plants from dependence on water for fertilization. Their seeds are efficient means of dispersing new generations even during hostile conditions. Pollen grains and seeds were key adaptations in the move to high, dry habitats.

7. Only angiosperms produce flowers. Most coevolved with pollinators, which enhance the transfer of pollen grains to female reproductive parts. Their seeds have a unique nutritive tissue (endosperm) and are usually surrounded by fruits, which aid in dispersal.

Review Questions

1. Identify a few of the structural and reproductive modifications that helped plants invade and diversify in habitats on land. *25.1*

2. Does the haploid phase or diploid phase dominate the life cycles of most plants? *25.1*

3. Name representatives of the following groups of plants and compare their key characteristics: (*also refer to Table 25.1*)
   a. Bryophytes and seedless vascular plants *25.2, 25.3*
   b. Gymnosperms and angiosperms *25.6–25.8*

4. Distinguish between:
   a. Root system and shoot system *25.1*
   b. Xylem and phloem *25.1*
   c. Sporophyte and gametophyte *25.1*
   d. Ovule and seed *25.1, 25.5*
   e. Microspore and megaspore *25.5*

Figure 25.20 Where many conifers end up.

Self-Quiz *(Answers in Appendix IV)*

1. Which of the following statements is *not* true?
   a. Monocots and dicots are two classes of angiosperms.
   b. Bryophytes are nonvascular plants.
   c. Lycophytes and angiosperms are both vascular plants.
   d. Gymnosperms are the simplest vascular plants.

2. Of all land plants, bryophytes alone have independent _____ and attached, dependent _____.
   a. sporophytes; gametophytes   c. rhizoids; zygotes
   b. gametophytes; sporophytes   d. rhizoids; stalked sporangia

3. Psilophytes, lycophytes, horsetails, and ferns are _____ plants.
   a. multicelled aquatic   c. seedless vascular
   b. nonvascular seed   d. seed-bearing vascular

4. Which does *not* apply to gymnosperms and angiosperms?
   a. vascular tissues   c. single spore type
   b. diploid dominance   d. all of the above

5. A seed is _____ .
   a. a female gametophyte   c. a mature pollen tube
   b. a mature ovule   d. an immature embryo

6. Match the terms appropriately.
   ____ gymnosperm   a. produces haploid gametes
   ____ sporophyte   b. control water loss
   ____ seedless vascular   c. "naked" seeds
        plant   d. protects, disperses embryo
   ____ ovary   sporophyte
   ____ bryophyte   e. produces haploid spores
   ____ gametophyte   f. nonvascular land plant
   ____ stomata   g. lycophyte
   ____ angiosperm seed   h. usually a fruit at maturity

Critical Thinking

1. Elliot Meyerowitz of the California Institute of Technology has studied the genetic basis of flower formation in *Arabidopsis thaliana*. By inducing mutations in seeds of this small weed, he discovered three genes (call them *A*, *B*, and *C*) that interact in different parts of a developing flower. The gene interactions lead to the formation of different structures—sepals, petals, stamens, and carpels—from the same mass of undifferentiated tissue. From what you know of gene regulation, suggest ways in which the *A*, *B*, and *C* genes might be controlling flower development.

2. Genes nearly identical to the *A*, *B*, and *C* genes of *A. thaliana* also have been isolated from snapdragons and other flowering plants. About 150 million years ago, these plants originated and rose to dominance rather abruptly, in nearly all land habitats.

Speculate on how their rapid, spectacular radiation might have come about.

3. Figure 25.20 shows a forest in the Nahmint Valley of British Columbia, before and after logging. It also shows the wood frames of homes that are in the process of being built. Reflect on these photographs and the ones in Figure 25.17. As a way of stopping the loggers, would you chain yourself to a tree in an old-growth forest scheduled for clear-cutting? If your answer is yes, would you also give up the possibility of owning a wood-frame home (as most homes are in developed countries)? What about forest products, including wood for fireplaces and camp fires, newspapers, and toilet tissue?

4. With respect to question 3, multiply each of your answers by 5.8 billion (there are about that many people in the world) and describe what might happen when, inevitably, we run out of trees. Also describe what you might consider to be some of the pros and cons of tree farms of, say, a single species of pine.

Selected Key Terms

angiosperm 25.1	gymnosperm 25.1	pollinator 25.8
bryophyte 25.1	hornwort 25.2	progymnosperm 25.5
coal 25.4	horsetail 25.3	rhizome 25.3
cone 25.6	lignin 25.1	root system 25.1
conifer 25.6	liverwort 25.2	seed 25.1
cuticle 25.1	lycophyte 25.3	seed fern 25.5
cycad 25.6	megaspore 25.5	shoot system 25.1
deforestation 25.7	microspore 25.5	spore 25.1
dicot 25.8	monocot 25.8	sporophyte 25.1
fern 25.3	moss 25.2	stoma (plural,
flower 25.8	ovule 25.5	stomata) 25.1
fruit 25.8	peat bog 25.2	strobilus 25.3
gametophyte 25.1	phloem 25.1	vascular plant 25.1
ginkgo 25.6	pollen grain 25.1	whisk fern 25.3
gnetophyte 25.6	pollination 25.5	xylem 25.1

Readings

Gensel, P., and H. Andrews. 1987. "The Evolution of Early Land Plants." *American Scientist* 75: 478–489.

Moore, R., W. D. Clark, and K. Stern. 1995. *Botany*. Dubuque, Iowa: W. C. Brown.

Raven, P., R. Evert, and S. Eichhorn. 1986. *Biology of Plants*. Fourth edition. New York: Worth.

Web Site See *http://www.wadsworth.com/biology* for practice quiz questions, hypercontents, BioUpdates, and critical thinking. The Wadsworth Biology Resource Center provides a wealth of information fully organized and integrated by chapter.

26

ANIMALS: THE INVERTEBRATES

Madeleine's Limbs

In August of 1994, about 900 million years after the first animals appeared on Earth, Madeleine made *her* entrance. As they are wont to do, grandmothers and aunts made a quick count on the sly—arms, legs, ears, and eyes, two of each; fully formed mouth and nose—just to be sure these were present and accounted for.

One grandmother, having been too long in the company of biologists, experienced an epiphany as she witnessed Madeleine's birth. In that profound instant she sensed ancestral connections, emerging from the distant past and through her, into the future.

Madeleine's body plan did not emerge out of thin air. Thirty-five thousand years ago, people just like us were having children just like Madeleine. And if we are interpreting the fossil record correctly, then five million years ago the offspring of individuals on the road to modern humans resembled her in some respects but not others. Sixty million years ago, the primate ancestors of those individuals were giving birth precariously, up in the trees. Two hundred and fifty million years ago,

mammalian ancestors of those primates were giving birth—and so on back in time to the very first animals, which had no limbs or eyes or noses at all.

We have very few clues to what those first animals looked like, but one thing is clear. By the dawn of the Cambrian, they had given rise to all major groups of invertebrates—animals without backbones—even to Madeleine's backboned but limbless ancestors.

And what stories those Cambrian animals tell! One bunch flourished 530 million years ago, in a submerged basin that had formed between a reef and the coast of an early continent. Protected from ocean currents, the sediments had piled against the steep reef. About 500 feet below the surface, the water was oxygenated and clear. Tiny, well-developed animals lived in, on, and above the dimly lit muddy sediments (Figure 26.1*a*).

Like castles built from wet sand along a seashore, their living quarters were unstable. Part of the bank above the community slumped abruptly and obliterated it. The sediments from that underwater avalanche kept scavengers from reaching and removing all traces of the dead. Gradually through time, muddy silt rained down on the tomb. The increased pressure and

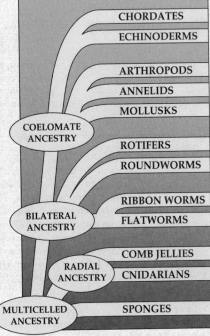

Figure 26.1 (a) Reconstruction of a few Cambrian animals, known from the Burgess Shale fossils. (b) Presumed evolutionary relationships among major groups of animals. Take a moment to study this evolutionary tree diagram. We will use it repeatedly as our road map through discussions of each group. (c) Madeleine.

chemical changes transformed the sediments into finely stratified shale; and soft parts of the flattened animals became shimmering mineralized films.

Sixty-five million years ago, part of the seafloor was plowing under the North American plate, and western Canada's mountain ranges were slowly rising. By 1909, the fossils were high in the eastern mountains of British Columbia. There, a fossil hunter tripped over a chunk of shale, which split apart into fine layers—and so the Burgess Shale story came to light.

In this chapter and the next, you will be comparing the body plans of different groups of animals. Such comparisons give insight into evolutionary relatedness and help us construct family trees, such as the one in Figure 26.1b. Don't assume that the structurally simple animals of the most ancient lineages are somehow primitive or evolutionarily stunted. As you will see, they, too, are exquisitely adapted to their environment.

As you poke through the branches of the animal family tree, keep the greater evolutionary story in mind. At each branch point, microevolutionary processes gave rise to workable changes in body plans. Madeleine's uniquely human traits, and yours, emerged through modification of certain traits that had evolved earlier in countless generations of vertebrates and, before them, in ancient invertebrate forms.

KEY CONCEPTS

1. All animals are multicelled, aerobic heterotrophs that ingest or parasitize other organisms. Nearly all kinds have tissues, organs, and organ systems, and most are motile during at least part of their life cycle. Animals reproduce sexually and often asexually, and their embryos develop in a series of continuous stages.

2. Animals originated approximately 900 million years ago. More than 2 million existing species have been identified. Of these, more than 1,950,000 are invertebrates (animals with no backbone). Fewer than 50,000 species are vertebrates (animals with a backbone).

3. Comparisons of the body plans of existing animals, in conjunction with the fossil record, reveal that there were several trends in the evolution of certain lineages. The most revealing aspects of an animal's body plan are its type of symmetry, gut, and cavity (if any) between the gut and body wall; whether it has a distinct head end; and whether it is divided into a series of segments.

4. The placozoans and sponges are structurally simple animals with no body symmetry. Both are at the cellular level of body construction. The cnidarians and the comb jellies have radial symmetry. They are at the tissue level of body construction.

5. Flatworms, roundworms, rotifers, and nearly all other animals that are more complex than the cnidarians show bilateral symmetry. They consist of tissues, organs, and organ systems.

6. Not long after the flatworms evolved, divergences gave rise to two major lineages. One evolutionary branching gave rise to the mollusks, annelids, and arthropods. The other gave rise to the echinoderms and chordates.

7. By biological measures, including diversity, sheer numbers, and distribution, the arthropods—especially insects—have been the most successful animal group.

General Characteristics of Animals

What, exactly, are **animals**? We can only define them by a list of general characteristics, not with just a sentence or two. *First*, animals are multicelled, and in most cases their body cells form tissues that are arranged as organs and organ systems. The body cells are diploid in nearly all species. *Second*, animals are heterotrophs that obtain carbon and energy by ingesting other organisms or by absorbing nutrients from them. *Third*, animals require oxygen, for use in aerobic respiration. *Fourth*, animals reproduce sexually and, in many cases, asexually. *Fifth*, most animals are motile during at least part of the life cycle. *Sixth*, their life cycles include a series of stages of embryonic development. In brief, mitotic cell divisions transform the animal zygote into a multicelled embryo. The embryonic cells are the forerunners of **ectoderm**, **endoderm**, and, in most species, **mesoderm**. These are the primary tissue layers which give rise to all tissues and organs of the adult, as described in Section 44.2.

Diversity in Body Plans

Mammals, birds, reptiles, amphibians, and fishes are the most familiar animals. All are **vertebrates**—the only animals with a "backbone." And yet, of probably more than 2 million species of animals, fewer than 50,000 are vertebrates! What we call **invertebrates** are animals with diverse features, but not a backbone.

We group animals into more than thirty phyla. Table 26.1 lists the ones that are described in this book. The characteristics they share with one another arose early in time, before divergences from a common ancestor gave rise to separate lineages. Later, as morphological differences accumulated among them, animal lineages took off in amazingly diverse directions. How might we get a conceptual handle on their modern descendants— on animals as different as flatworms, hummingbirds, platypuses, humans, and giraffes? We can compare their similarities and differences with respect to five basic features. These are body symmetry, cephalization, type of gut, type of body cavity, and segmentation.

BODY SYMMETRY AND CEPHALIZATION   With very few exceptions, animals are radial or bilateral. Those with **radial symmetry** have body parts arranged regularly around a central axis, like spokes of a bike wheel. Thus a cut down the center of a hydra (Figure 26.2*a*) divides it into equal halves; another cut at right angles to the first divides it into equal quarters. All radial animals live in water. Their body plan is adapted to intercepting food that is coming toward them from any direction.

Animals with **bilateral symmetry** have right and left halves that are mirror images of each other. Most

Table 26.1   Animal Phyla Described in This Book

Phylum	Some Representatives	Existing Species
PLACOZOA (*Trichoplax*)	Simplest animal; like a tiny plate, but with layers of cells	1
PORIFERA (poriferans)	Sponges	8,000
CNIDARIA (cnidarians)	Hydrozoans, jellyfishes, corals, sea anemones	11,000
CTENOPHORA (comb jellies)	Like jellyfishes with comblike structures of modified cilia	11,000
PLATYHELMINTHES (flatworms)	Turbellarians, flukes, tapeworms	15,000
NEMATODA (roundworms)	Pinworms, hookworms	20,000
NEMERTEA (ribbon worms)	Proboscis-equipped worms closely related to flatworms	800
ROTIFERA (rotifers)	Tiny body with crown of cilia, great internal complexity	2,000
MOLLUSCA (mollusks)	Snails, slugs, clams, squids, octopuses	110,000
ANNELIDA (segmented worms)	Leeches, earthworms, polychaetes	15,000
ARTHROPODA (arthropods)	Crustaceans, spiders, insects	1,000,000+
ECHINODERMATA (echinoderms)	Sea stars, sea urchins	6,000
CHORDATA (chordates)	Invertebrate chordates: Tunicates, lancelets	2,100
	Vertebrates: Fishes	21,000
	Amphibians	3,900
	Reptiles	7,000
	Birds	8,600
	Mammals	4,500

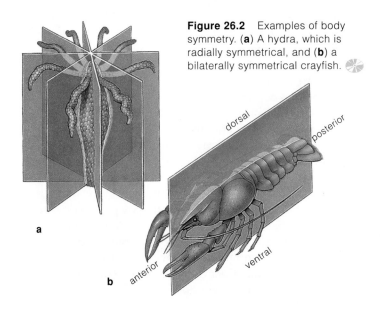

Figure 26.2 Examples of body symmetry. (**a**) A hydra, which is radially symmetrical, and (**b**) a bilaterally symmetrical crayfish.

a

dorsal

posterior

anterior

ventral

b

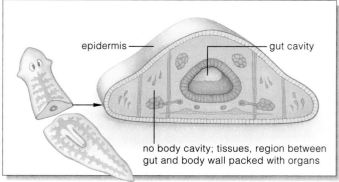

a No coelom (*acoelomate* animals)

epidermis — gut cavity

no body cavity; tissues, region between gut and body wall packed with organs

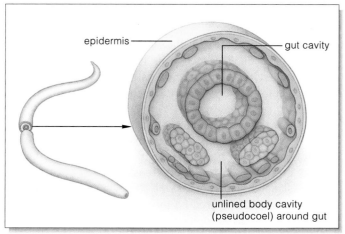

b Pseudocoel (*pseudocoelomate* animals)

epidermis — gut cavity

unlined body cavity (pseudocoel) around gut

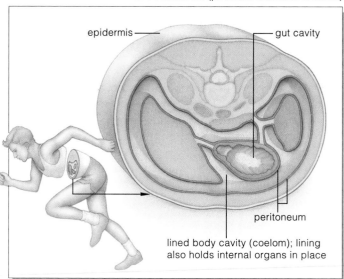

c Coelom (*coelomate* animals)

epidermis — gut cavity

peritoneum

lined body cavity (coelom); lining also holds internal organs in place

Figure 26.3 Type of body cavity (if any) in animals.

have an *anterior* end (head) and an opposite, *posterior* end. They have a *dorsal* surface (a back) and an opposite, *ventral* surface (Figure 26.2b). This body plan evolved among the first forward-creeping species. Mostly, their forward end must have been the first to encounter food and other stimuli. We can imagine there was selection for **cephalization**. By this evolutionary process, sensory

structures and nerve cells became concentrated in the head. The joint evolution of bilateral body plans and cephalization resulted in pairs of muscles and pairs of sensory structures, nerves, and brain regions.

TYPE OF GUT The **gut** is a region inside the body in which food is digested, then absorbed into the internal environment. Saclike guts have one opening (a mouth) for taking in food and expelling residues. Other guts are part of a tubelike system with a mouth and an anus— that is, openings at both ends. Such "complete" digestive systems are very efficient. Different parts of them have specialized functions, such as preparing, digesting, and storing material. The evolution of such systems helped pave the way for increases in body size and activity.

BODY CAVITIES A body cavity separates the gut and the body wall of most bilateral animals (Figure 26.3). One type of cavity, a **coelom**, has a unique tissue lining called a peritoneum. This lining also encloses organs in the coelom and helps hold them in place. For example, your body houses a coelom, which a sheetlike muscle divides into two smaller cavities. Your heart and lungs are positioned inside the upper (thoracic) cavity, and your stomach, intestines, and other organs occupy the lower (abdominal) cavity.

Some invertebrates don't have a body cavity; tissues fill the region between their gut and body wall. Others have a pseudocoel ("false coelom"), a body cavity with no peritoneum. The coelom was a key innovation in the evolution of animals that were larger and more complex than ancestral forms. It favored increases in size and activity by cushioning and protecting internal organs.

SEGMENTATION Segmented animals have a repeating series of body units that may or may not be similar to one another. The many segments of earthworms have a similar outward appearance. Insect segments are fused into three units (head, thorax, and abdomen) and differ greatly from one another. Among insects especially, diverse head parts, legs, wings, and other appendages evolved from less specialized segments.

Animals are multicelled, and most have tissues, organs, and organ systems. The body cells of most species are diploid.

Animals are aerobically respiring heterotrophs that ingest other organisms or absorb nutrients from them.

Animals reproduce sexually and, in many cases, asexually. They go through a period of embryonic development, and most are motile during at least part of the life cycle.

Body plans of animals differ with respect to five features: body symmetry, cephalization, type of gut, type of body cavity, and segmentation.

From telltale tracks and burrows they left behind in marine sediments, we suspect that multicelled animals probably evolved by 900 million years ago. *Where did the first animals come from?* They probably originated with protistan lineages, although we don't know which ones (Sections 21.3 and 21.4).

By one hypothesis, the forerunners of animals were ciliates, much like *Paramecium*, that had multiple nuclei in a single-celled body; over time, each nucleus became compartmentalized in one cell of a multicelled body. Yet no existing animal develops by compartmentalization.

By another hypothesis, multicelled animals arose from spherical colonies of a number of flagellated cells, maybe like the *Volvox* colonies shown in Figure 6.3. In time, as a result of mutations, some cells in the colony became modified in ways that enhanced reproduction and other specialized tasks. So began the division of labor that characterizes multicellularity.

Suppose such colonies became flattened and started creeping about on the seafloor. This may have led to the evolution of layers of cells, such as those of *Trichoplax adhaerens*. This soft-bodied marine animal, shaped a bit like pita bread, has no symmetry and no mouth. It is the only known **placozoan** (Placozoa, after *plax*, meaning plate, and *zoon*, meaning animal). Its several thousand cells are arrayed in two distinctly different layers. When it glides over food, its body briefly humps up (Figure 26.4). Glandular cells present in the lower layer secrete digestive enzymes onto the food, then individual cells absorb the breakdown products. Reproduction might be asexual (by budding or fission) or it might be sexual, by mechanisms not yet understood. In sum, structurally and functionally, *Trichoplax* is as simple as animals get.

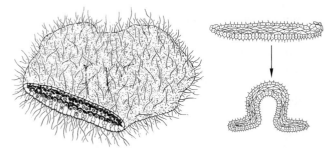

Figure 26.4 Cutaway view of *Trichoplax adhaerens*, an animal with a two-layer body measuring about three millimeters across.

Possibly the question of origins requires more than one answer. It may be that different lineages descended from more than one group of protistan-like ancestors.

Multicelled animals arose from protistan-like ancestors that may have resembled the ciliates, the colonial flagellates, or both.

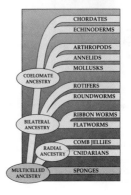

Sponges (Porifera) are animals with no symmetry, tissues, or organs, yet they are one of nature's success stories. They have been abundant in the seas ever since the Precambrian, especially in waters off coasts and along reefs. Of the 8,000 or so known species, only about 100 live in freshwater. Tiny worms, shrimps, and other animals make their home inside or on sponges.

Some sponges are large enough to sit in, and others are as small as a fingernail. Figures 26.5 and 26.6 show a few of the sprawling, flattened, lobed, compact, tubular, cuplike, and vaselike shapes. Regardless of its shape, the body is not symmetrical. Flattened cells line its outer surface and inner cavities, but the linings are not much more than the cell layers of *Trichoplax*, and they differ from the tissues of other animals. Amoeboid cells live in a gelatin-like substance between the two linings (Figure 26.6b). Glasslike spicules (of calcium carbonate or silica), tough fibers (of the protein spongin), or both stiffen the sponge body and impart structure to it.

The skeletal elements may be a reason why sponges as a group have endured so long. Cleveland Hickman put it this way: Most potential predators discover that sampling a sponge is about as pleasant as eating a mouthful of glass splinters embedded in fibrous gelatin. Besides, chemically speaking, many sponges stink.

Water flows into a sponge body through microscopic pores and chambers, and then out through one or more openings known as oscula (singular, osculum). It does so mainly when thousands or millions of **collar cells** are beating their flagella. These cells are components of the sponge body's inner lining. Besides having a flagellum, they have "collars" of food-trapping structures called microvilli (Figure 26.6c). Bacteria and other bits of food dissolved in the water become trapped in the collars, then are engulfed by way of phagocytosis. Some of the

Figure 26.5 A sprawling, red-orange sponge, one of many types that encrust underwater ledges in temperate seas.

Figure 26.6 (**a**,**b**) Body plan of a simple sponge. The outer lining consists of flattened cells. It also has some contractile cells, arranged as bands around the tiny openings across the body. These cells contract slowly, independently of the other cell types, and so influence water flow through the body. Amoeboid cells inside the gelatin-like matrix between the inner and outer linings secrete materials from which the body's spicules and fibers are constructed. Other amoeboid cells digest and transport food. They do not lose the capacity to divide, and their cellular descendants can differentiate into any other type of sponge cell. They also have roles in asexual processes, such as gemmule formation.

(**c**) A great many flagellated cells line inner canals and chambers of the sponge body. Each has a collar of food-trapping structures called microvilli. Fine filaments connect the microvilli to each other; they form a "sieve" that strains food particles from the water. At the base of the collar, the cell engulfs the trapped food, by phagocytosis.

(**d**) An example of skeletal elements—Venus's flower basket (*Euplectella*). This marine sponge has six-rayed spicules of silica fused in a rigid, elaborate network. A thin layer of cells stretches over all of the interconnecting spicules. At the base of the body is an anchoring tuft of spicules.

(**e**) A basket sponge of the Caribbean, releasing a cloud of sperm into the water.

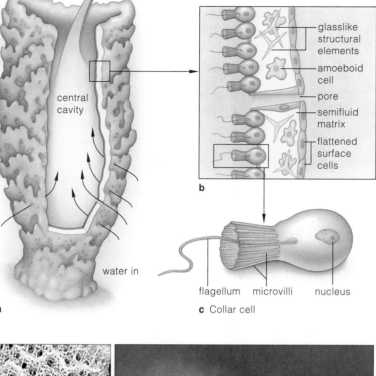

water out

central cavity

water in

a

glasslike structural elements

amoeboid cell

pore

semifluid matrix

flattened surface cells

b

flagellum   microvilli   nucleus

c Collar cell

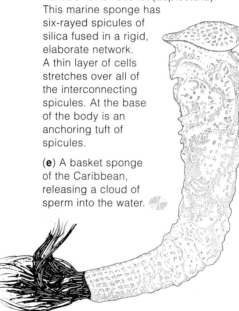

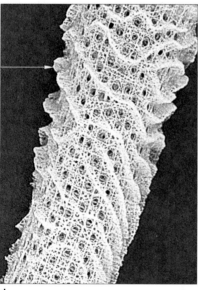

d

e

engulfed food also gets transferred to the amoebalike cells for further breakdown, storage, and distribution.

Sponges reproduce sexually when some cells give rise to eggs and to sperm. Most release sperm into the water (Figure 26.6*e*), but the eggs typically are retained until after they have been fertilized and embryos start their development. Young sponges proceed through a microscopic, swimming larval stage. A **larva** (plural, larvae) is a sexually immature stage that grows and develops into the sexually mature form of the species, the **adult**. Many animal life cycles include larval stages.

Some kinds of sponges also can reproduce asexually by fragmentation (small fragments break away from the

parent and grow into new sponges). Most freshwater species also reproduce asexually by way of gemmules. These are clusters of sponge cells, some of which form a hard covering around others. The clusters inside are protected from extreme cold or drying out. Later, when favorable conditions return, the gemmules germinate and establish a new colony of sponges.

Sponges have no symmetry, tissues, or organs; they are at the cellular level of construction. Yet they have successfully endured through time, possibly because most predators find their spicule-rich and often stinky bodies unappetizing.

CNIDARIANS—TISSUES EMERGE

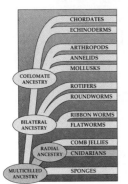

Think of a jellyfish, such as the one shown in Figure 26.7. It is one of the scyphozoans, which, together with anthozoans (such as sea anemones) and hydrozoans (including *Hydra*) are tentacled, radial animals of phylum **Cnidaria**. Most cnidarians live in the seas. Of 11,000 known species, fewer than 50 are adapted to freshwater habitats.

Regarding the Nematocysts

Of all animals, cnidarians alone produce **nematocysts**: capsules that house dischargeable, tubular threads. The threads of some types have prey-piercing barbs and an open tip that delivers toxins (Figure 26.8). Other nematocysts discharge threads that ooze a sticky substance from their tip or long threads that entangle prey. Many swimmers have learned that the toxin-tipped threads of some species can sting. Hence the name of the phylum, Cnidaria, after the Greek word for nettle.

Cnidarian Body Plans

The **medusa** (plural, medusae) and **polyp** are the most common cnidarian body forms. Both have a saclike gut (Figure 26.7*a,b*). Medusae float. Some look like bells and others like upside-down saucers. The mouth, centered under the bell, may have extensions that assist in prey capture and feeding. Polyps have a tubelike body with a tentacle-fringed mouth at one end. Usually the other end is attached to a substrate. When *Trichoplax* is draped over and digesting food, it is rather like a gut on the run. By contrast, the cnidarian gut is a permanent food-processing chamber. This gut has a gastrodermis, a sheetlike lining with many glandular cells that secrete digestive enzymes. An epidermis lines the rest of the body's surfaces (Figure 26.9).

operculum (capsule's lid at cell's free surface)

barbs

trigger (modified cilium)

barbed thread coiled inside capsule

nematocyst (capsule at free surface of epidermal cell)

Figure 26.8 One type of nematocyst before and after a prey organism (not shown) touched its trigger. The contact made the capsule more "leaky" to water. Water diffused inward, turgor pressure built up within the capsule, and the thread was forced to turn inside out. The thread's tip pierced the prey's body.

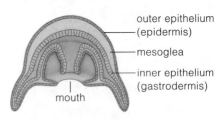

outer epithelium (epidermis)

mesoglea

inner epithelium (gastrodermis)

mouth

a Medusa, midsection

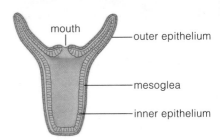

mouth

outer epithelium

mesoglea

inner epithelium

b Polyp, midsection

Figure 26.7 Cnidarian body plans. (**a**,**b**) Diagrams of a medusa and a polyp, sliced through the midsection. (**c**) The medusa of a sea nettle (*Chrysaora*), one of the jellyfishes. (**d**) A hydrozoan polyp (*Hydra*) attached to a substrate, as it captures and digests its prey.

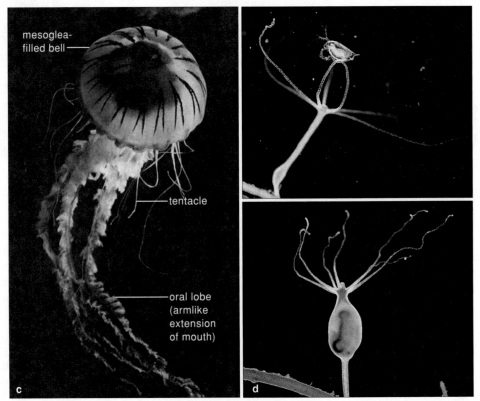

mesoglea-filled bell

tentacle

oral lobe (armlike extension of mouth)

c

d

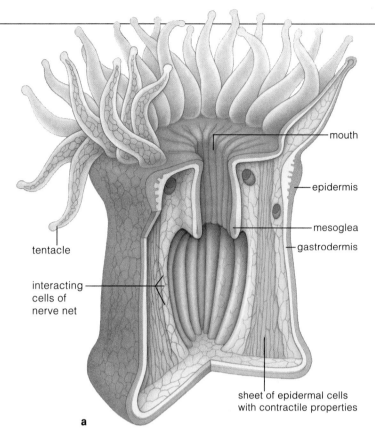

tentacle

mouth

epidermis

mesoglea

gastrodermis

interacting cells of nerve net

sheet of epidermal cells with contractile properties

a

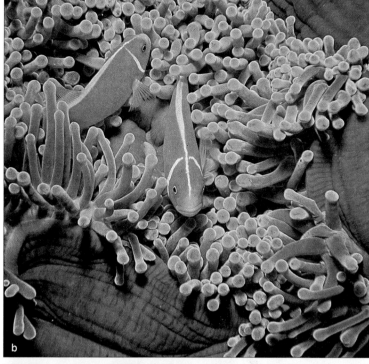

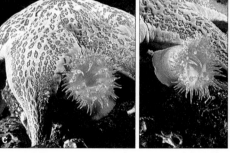

b

c

Figure 26.9 (**a**) Tissue organization of a sea anemone. (**b**) Tentacles fringing the mouth of a sea anemone, which eats many fishes but not clownfishes. A clownfish swims out, captures food, and returns to the tentacles, which protect it from predators. The sea anemone eats food scraps falling from the fish mouth. This is a case of mutualism, a two-way flow of benefits between species. (**c**) A sea anemone using its hydrostatic skeleton to escape from a sea star. It closes its mouth, and the force generated by contractile cells in epithelial tissues acts against water in the gut. The body changes shape and the anemone thrashes about, generally in a direction away from the predator.

Each lining is an **epithelium** (plural, epithelia), a tissue with a free surface that faces the environment or some type of fluid inside the body. All animals more complex than sponges have epithelia. Cnidarian epithelia house **nerve cells**, which receive signals from receptors that sense changes in the surroundings and send signals to **contractile cells** that can carry out suitable responses. (When stimulated, contractile cells *shorten*, then return to their original length when stimulation stops.) The nerve cells interact as a "nerve net," a simple nervous tissue, to control movement and changes in shape.

Between the epidermis and gastrodermis is a layer of gelatinous secreted material, the mesoglea ("middle jelly"). Jellyfishes contain enough mesoglea to impart buoyancy and serve as a firm yet deformable skeleton against which contractile cells act. Imagine many cells contracting in coordinated ways in a jellyfish bell. Their action narrows the bell and forces water to jet out from underneath it, and the jet propels the jellyfish forward. The bell returns to its original position, cells contract

again, and the animal is propelled forward. Although the swimming movements are not Olympian, they work well enough for an animal that secures food by dragging its tentacles through the water for prey moving past.

Any fluid-filled cavity or cell mass against which contractile cells can act is a **hydrostatic skeleton**. With coordinated contractions, the cavity's volume or mass does not change but rather is shunted about, so that the shape of the body changes. The contractile cells of most polyps, which have little mesoglea, act against water in their gut. As you will see, nearly all animals have some form of skeletal-muscular system of movement.

Cnidarians are radial animals with tentacles, a saclike gut, epithelia, a nerve net, and a hydrostatic skeleton. They alone produce nematocysts.

Cnidarians are at the tissue level of construction; compared with sponges, they have layers of cells interacting in more coordinated fashion in the performance of specific tasks.

Various Stages in Cnidarian Life Cycles

From the preceding section, you might have concluded that the body form of a particular cnidarian is a medusa or a polyp. This is indeed the case for many species. However, the life cycle of *Obelia, Physalia,* and some other cnidarians includes both body forms. By using the life cycle of *Obelia* as an example, you can get a general sense of how these forms grow and develop (Figure 26.10).

The medusa is the sexual stage of the cnidarian life cycle. It has simple **gonads**, which are primary (gamete-producing) reproductive organs. Either the epidermis or gastrodermis houses the gonads, which release gametes by rupturing. Most of the zygotes formed at fertilization develop into **planulas**—a kind of swimming or creeping larva, usually with ciliated epidermal cells. In time, a mouth opens at one end, the larva is transformed into a polyp or medusa, and the cycle begins anew.

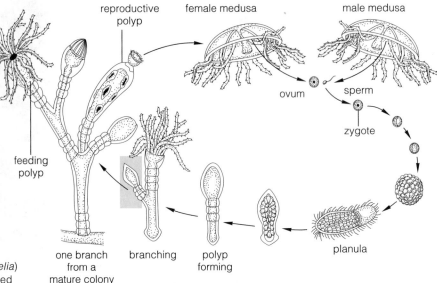

Figure 26.10 (*Right*) Life cycle of a hydrozoan (*Obelia*) that includes medusa and polyp stages. An established colony may contain thousands of feeding polyps.

A Sampling of Colonial Types

The reef-forming corals and other colonial anthozoans are a fine example of variations on the basic cnidarian body plan. The reef formers thrive in clear, warm water that is at least 20°C (68°F). As Figure 26.11 shows, the colonies consist of polyps that have secreted calcium-

Figure 26.11 (**a**) One of the reef-building, colonial corals. External skeletons of their polyps interconnect with one another. Dinoflagellate mutualists live in the polyp tissues. (**b**) Aerial view of a barrier reef, mainly an accumulation of the compacted skeletons of certain coral species. (**c**) Not all corals are reef builders; cup corals such as *Tubastrea* are solitary forms. With tentacles extended, their polyps look like tiny sea anemones.

reinforced external skeletons, which interconnect with one another. Over time, the skeletons accumulate and so become the main building material for reefs. Today the most impressive accumulation, the Great Barrier Reef, parallels the eastern coast of Australia for about 1,600 kilometers.

Reef-building corals receive nutrient inputs from the changing tides and from dinoflagellate mutualists living in their tissues. Masses of these photosynthetic protistans supply corals with oxygen, recycle their mineral wastes, and adjust the pH of the surrounding water in a way that enhances the rate of calcium deposition for their host's skeletons. The host corals reciprocate by giving the dinoflagellates a relatively safe, sunlit habitat that has plenty of dissolved carbon dioxide and mineral ions. In the sunlit parts of a reef, nutrients are cycled quickly, directly, and efficiently.

As a final example of cnidarian diversity, consider *Physalia*, informally called the Portuguese man-of-war. The toxin in the nematocysts of this infamous colonial hydrozoan poses a danger to bathers and fishermen as well as to prey organisms (fish). Although *Physalia* lives mainly in warm waters, currents sometimes move it up to the Atlantic coasts of North America and Europe. A blue, gas-filled float that develops from the planula keeps the colony near the water's surface, where winds move it about (Figure 26.12). Under the float, groups of

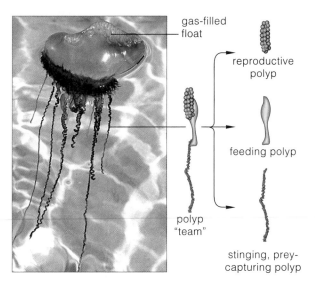

Figure 26.12 Portuguese man-of-war (*Physalia*).

polyps and medusae interact as "teams" in feeding, reproduction, defense, and other specialized tasks.

Cnidarians, including the colonial forms, show notable variation in their body plans and life-styles.

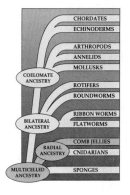

If you have ever been diving in coastal waters, anywhere from the tropics to the poles, you may have observed cnidarians that look a bit like jellyfishes with combs running down the sides. These are **comb jellies**, of phylum Ctenophora (meaning "comb-bearing"). All are weak-swimming predators in planktonic communities, and all show modified radial symmetry. Slice them in two equal halves, then slice them into quarters, and two of the quarters will be mirror images of the other two.

A comb jelly has eight rows of comblike structures made of thick, fused cilia (Figure 26.13). All the combs in a row beat in waves and propel the animal forward, usually (and handily) mouth first. Some species have two long, muscular tentacles with branches that are equipped with sticky cells. Comb jellies do not produce nematocysts, but sometimes they opportunistically save and use the ones from jellyfish they have eaten. Others use their sticky lips to capture prey.

Evolutionarily, comb jellies are interesting because they have cells with multiple cilia. This trait evolved in

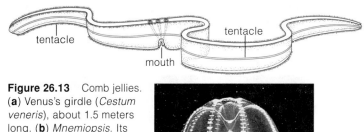

Figure 26.13 Comb jellies. (**a**) Venus's girdle (*Cestum veneris*), about 1.5 meters long. (**b**) *Mnemiopsis*. Its expanded and contractile mouth lobes waft prey into the mouth. It is about 15 centimeters (6 inches) long.

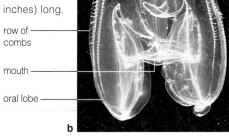

many complex animals. In addition, the comb jellies are the simplest animals with embryonic tissues that are like mesoderm. In complex animals, mesoderm is the embryonic source of muscles as well as organs of the circulatory, excretory, and reproductive systems.

Comb jellies have a modified radial body and comblike structures of modified cilia. They have an embryonic tissue that resembles the mesoderm of more complex animals.

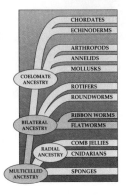

When we move beyond the cnidarians in our survey, we find animals that range from flatworms to humans. All of these animals have simple or complex organs. An **organ** is an association of one or more kinds of tissues, arranged in particular proportions and patterns. Most often, one organ interacts with others to carry out an activity that helps the body function. By definition, two or more organs that are interacting efficiently in the performance of some task represent an **organ system**. We turn now to the simplest animals at the *organ-system* level of construction. They are not what you would call breathtakingly complex but, as you will see shortly, a few can make our lives breathtakingly miserable.

Flatworms

Among the 15,000 or so known species of **flatworms** (phylum Platyhelminthes) are turbellarians, flukes, and tapeworms. Most of these bilateral, cephalized animals have a flattened body (hence their name) and simple organ systems (Figure 26.14). Their digestive system, for example, has a pharynx (a muscular tube, which is used for feeding) and a saclike and often branching gut. Their reproductive systems vary, but most flatworms are **hermaphrodites**. That is, an individual has female and male gonads. Two individuals reproduce sexually through the mutual transfer of sperm. Each has a penis (a sperm-delivery structure) and glands that produce a protective capsule around fertilized eggs.

TURBELLARIANS Most turbellarians (class Turbellaria) live in the seas; only planarians and a few others live in freshwater. Some eat tiny animals or suck tissues from dead or wounded ones. Commonly, a planarian will reproduce asexually by transverse fission. It divides in half at its midsection, then each half regenerates the missing parts. Like you, a planarian is able to adjust the composition and volume of its body fluids. Its water-regulating system has one or more tiny, branched tubes called protonephridia (singular, protonephridium). The tubes extend from pores at the body surface to bulb-shaped flame cells in body tissues. When excess water diffuses into the flame cells, a tuft of cilia "flickering" in the bulb drives the water through the tubes and on out to the surroundings (Figure 26.14*b*).

FLUKES Flukes (class Trematoda) are parasitic worms. A parasite, recall, lives in or on a living host and feeds on its tissues. Most do not kill the host, at least not until they have reproduced. Fluke life cycles have sexual and asexual phases and at least two kinds of hosts. As the

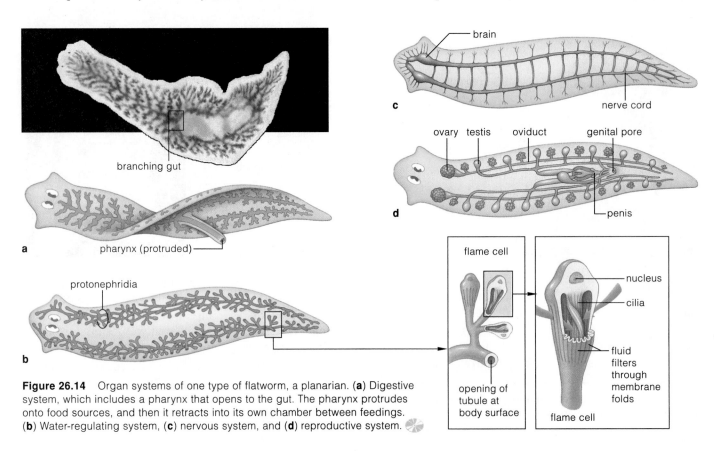

Figure 26.14 Organ systems of one type of flatworm, a planarian. (**a**) Digestive system, which includes a pharynx that opens to the gut. The pharynx protrudes onto food sources, and then it retracts into its own chamber between feedings. (**b**) Water-regulating system, (**c**) nervous system, and (**d**) reproductive system.

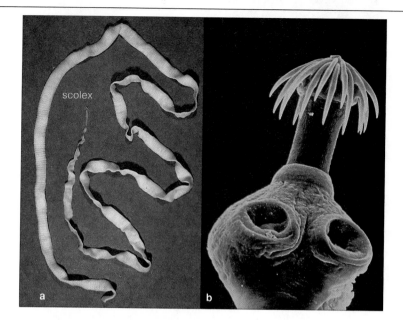

Figure 26.15 (**a**) Tapeworm scolex. This one attaches to the gut lining of a shorebird, its primary host. (**b**) Sheep tapeworm.

bilateral animals evolved from planula-like ancestors, through increased cephalization and the development of tissues derived from mesoderm.

Ribbon Worms

Ribbon worms (phylum Nemertea) are bilateral, soft-bodied, elongated predators that swallow or suck tissue fluids from small worms, mollusks, and crustaceans. Most crawl, burrow, or lurk in shallow marine habitats, as in Figure 26.16, but some live in freshwater or humid tropical habitats. Like flatworms, which may be their close relatives, ribbon worms have a ciliated surface, and they secrete mucus and then move by beating their cilia through it. Ribbon worms also resemble flatworms in their tissue organization. However, they differ from flatworms in having a circulatory system, a complete gut, and separation of sexes. Ribbon worms also have a proboscis, which in this case is a tubular, prey-piercing,

examples in Section 26.9 show, the flukes grow and reach sexual maturity in an animal that serves as their *primary* host. Larval stages use an *intermediate* host, in which they either develop or become encysted.

TAPEWORMS Tapeworms (class Cestoda) parasitize intestines of vertebrates. It seems probable that the ancestral tapeworms had a gut but later lost it during their evolution in animal intestines, which happen to be habitats that are rich in predigested food. Their existing descendants attach to the intestinal wall by a scolex, a structure equipped with suckers, hooks, or both (Figure 26.15). **Proglottids** are new units of the tapeworm body that bud just behind the scolex. They are hermaphroditic bodies; they can mate and transfer sperm with one another. The older proglottids (those farthest from the scolex) store fertilized eggs. They break off from younger ones, then leave the body in feces. Later, an intermediate host may meet up with their eggs. Section 26.9 includes a look at proglottid formation in one tapeworm life cycle.

In some respects, the simplest turbellarians, larval flukes, and larval tapeworms resemble the planulas of cnidarians. The resemblance inspires speculation that

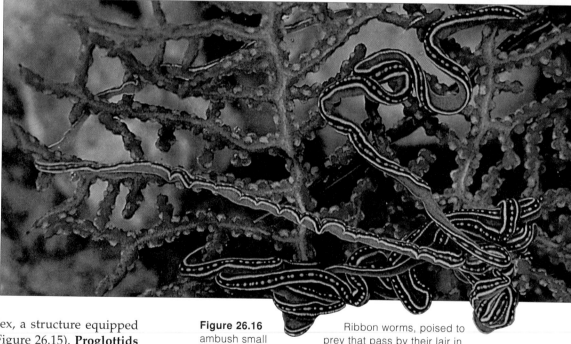

Figure 26.16 Ribbon worms, poised to ambush small prey that pass by their lair in an orange-colored gorgonian (horny coral) colony.

venom-delivering device tucked inside the head end. Muscle contractions force the proboscis inside out and into prey, which the venom paralyzes.

Flatworms are among the simplest bilateral, cephalized animals with organ systems. Ribbon worms may be related to them, but they have a complete gut, a circulatory system, and other traits that are notable departures from flatworms.

ROUNDWORMS

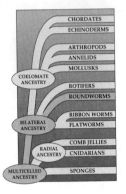

Roundworms (phylum Nematoda) are pseudocoelomate worms that live nearly everywhere and are probably the most abundant of all multicelled animals alive today. A million nematodes in each square meter of sediments is typical in shallow saltwater and freshwater. Thousands of them may occupy a handful of rich soil, in which scavenging types make fast work of dead earthworms or rotting vegetation. We know of 20,000 species, but a hundred times that many may yet be discovered. Each has a bilateral, cylindrical body, usually tapered at both ends and protected by a cuticle. In animals, a **cuticle** is a tough, often flexible body covering (Figure 26.17). The roundworms are the simplest animals outfitted with a complete digestive system. Between the gut and the body wall is a false coelom, typically jam-packed with reproductive organs. The cells in all tissues absorb nutrients from coelomic fluid and give up wastes to it.

Parasitic roundworms can do extensive damage to their hosts, which include humans, cats, dogs, cows, and sheep as well as soybeans, potatoes, and other valued crop plants. They are all thin but can grow long; some in female sperm whales are nine *meters* long. One of the most rapidly spreading diseases in the world, elephantiasis, is the handiwork of a few species of roundworms (Section 26.9).

Unfortunately, the parasitic types have given roundworms everywhere a bad name. Yet most are harmless, free-living types that do beneficial work, as when they help cycle nutrients in a variety of communities. One roundworm, *Caenorhabditis elegans*, is a key experimental organism for laboratory studies into inheritance, development, and aging.

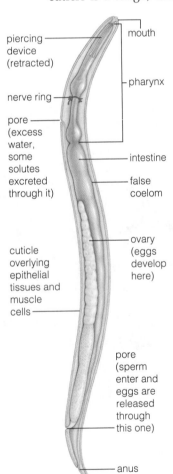

piercing device (retracted)

mouth

nerve ring

pharynx

pore (excess water, some solutes excreted through it)

intestine

false coelom

ovary (eggs develop here)

cuticle overlying epithelial tissues and muscle cells

pore (sperm enter and eggs are released through this one)

anus

Figure 26.17 Body plan of a female roundworm (*Paratylenchus*) that parasitizes plant roots.

Roundworms are cylindrical, bilateral, cephalized animals with a false coelom and a complete digestive system. Most cycle nutrients in communities; the parasites are notorious.

A ROGUE'S GALLERY OF WORMS

A number of parasitic flatworms and roundworms call the human body home. In any given year, for example, about 200 million people house blood flukes responsible for *schistosomiasis*. One of these, the Southeast Asian blood fluke *Schistosoma japonicum*, requires a human primary host, standing water in which larvae can swim, and an aquatic snail as an intermediate host. The flukes grow, become sexually mature, and mate inside a human host (Figure 26.18*a*). After being fertilized, the female's eggs leave the human body in feces and hatch into ciliated, swimming larvae (*b*) that burrow into a snail and multiply asexually (*c*). In time, many fork-tailed larvae develop (*d*). These leave the snail (*e*) and swim about until they contact human skin (*f*). They bore in and migrate to thin-walled intestinal veins, and the cycle begins anew. In infected humans, white blood cells that defend the body attack the masses of fluke eggs, and grainy masses form in tissues. In time, the liver, spleen, bladder, and kidneys deteriorate.

Some tapeworms parasitize humans. Different species use pigs, freshwater fish, or cattle as intermediate hosts. Humans become infected when they eat pork, fish, or beef that is raw, improperly pickled, or insufficiently cooked—and contaminated with tapeworm larvae (Figure 26.19).

Or consider a parasitic roundworm that causes thin, serpentlike ridges in human skin. For several thousand years, healers have been extracting the "serpents" by winding them out slowly, painfully, around a stick. The roundworms called pinworms and hookworms cause other problems. *Enterobius vermicularis*, a pinworm of temperate regions, parasitizes humans. It lives in the large intestine, but at night the centimeter-long females migrate to the anal region of the host and lay eggs. Their presence causes itching, and scratchings made in response transfer

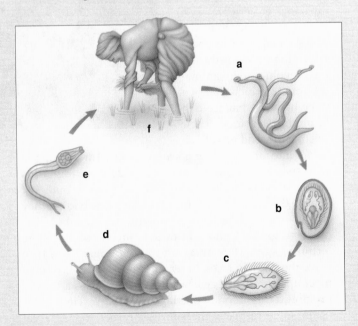

Figure 26.18 Life cycle of a dangerous blood fluke, *Schistosoma japonicum*.

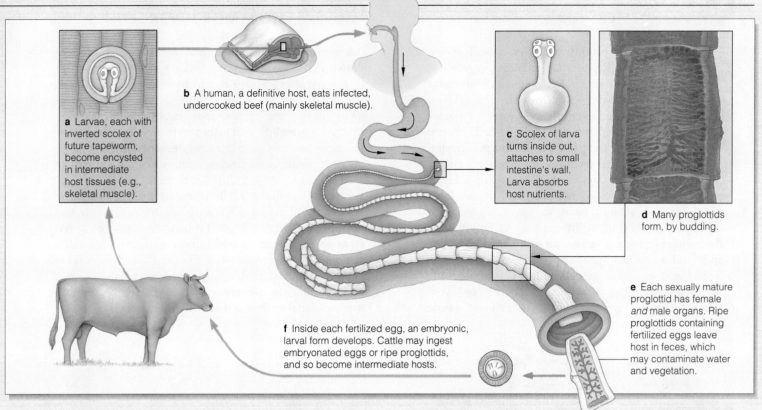

a Larvae, each with inverted scolex of future tapeworm, become encysted in intermediate host tissues (e.g., skeletal muscle).

b A human, a definitive host, eats infected, undercooked beef (mainly skeletal muscle).

c Scolex of larva turns inside out, attaches to small intestine's wall. Larva absorbs host nutrients.

d Many proglottids form, by budding.

e Each sexually mature proglottid has female *and* male organs. Ripe proglottids containing fertilized eggs leave host in feces, which may contaminate water and vegetation.

f Inside each fertilized egg, an embryonic, larval form develops. Cattle may ingest embryonated eggs or ripe proglottids, and so become intermediate hosts.

Figure 26.19 Life cycle of a beef tapeworm, *Taenia saginata*.

some eggs to hands, then to other objects. Newly laid eggs contain embryonic pinworms, but within a few hours they have developed into juveniles and are ready to hatch if another human inadvertently ingests them.

Hookworms are especially serious in impoverished areas of the tropics and subtropics. Adult hookworms live in the small intestine. After the toothlike devices or sharp ridges bordering their mouth cut into the intestinal wall, they feed on blood and other tissues and so compete with their host for nutrients. Adult females, about a centimeter long, can release a thousand eggs daily. These leave the body in feces, then hatch into juveniles. Walk barefoot, and a juvenile hookworm may penetrate the skin. Inside a host, the parasite travels the bloodstream to the lungs. There it works its way into the air spaces. After moving up the windpipe, the parasite moves into the gut when the host swallows. Soon it is in the small intestine, where it may mature and live for several years.

Another roundworm, *Trichinella spiralis*, causes painful, sometimes fatal symptoms. Adults live in the lining of the small intestine. Females release juveniles (Figure 26.20*a*), which work their way into blood vessels and travel to muscles. There they become encysted; they secrete a covering around themselves and enter a resting stage. Humans become infected mainly by eating insufficiently cooked meat from pigs or some game animals. It is not easy to detect the encysted juveniles when fresh meat is being examined, even in a slaughterhouse.

Figure 26.20*b* shows the results of prolonged, repeated infections by *Wuchereria bancrofti*, another roundworm.

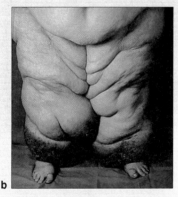

Figure 26.20 (**a**) Juveniles of a roundworm, *Trichinella spiralis*, living inside the muscle tissue of a host animal. (**b**) Legs of a woman infected by the roundworm *Wuchereria bancrofti*.

Adult worms become lodged in the lymph nodes, which filter lymph (excess tissue fluid) that normally flows into the bloodstream. In these nodes, the worms can obstruct the flow of lymph. When such an obstruction causes fluid to back up and accumulate in tissues, legs and other body regions undergo grotesque enlargement. This condition is called *elephantiasis*.

A mosquito is *Wuchereria*'s intermediate host. Females of this parasite produce active young that travel about at night, in the bloodstream. If a mosquito sucks blood from an infected person, the juveniles may enter the insect's tissues. In time they move near the insect's sucking device and enter a new host when it draws blood again.

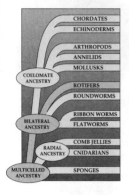

Like the roundworms just described, the **rotifers** (Rotifera) are bilateral, cephalized animals with a false coelom. All but about 5 percent live in freshwater, such as lakes, ponds, and even films of water on mosses and other plants. We typically find between 40 and 500 rotifers in a liter of pondwater; 5,000 were recorded on a few occasions. They eat bacteria and microscopic algae. Most types are not even a millimeter long, yet seldom have so many organs been packed in so little space. As Figure 26.21 shows, rotifers have a pharynx, an esophagus, digestive glands and a stomach, protonephridia, and usually an intestine and anus. Nerve cell bodies clustered in the head integrate body activities. Some rotifers have "eyes" (clusters of absorptive pigments). Two "toes" exude

Figure 26.21
A rotifer (*Philodina roseola*). Males are unknown in this species and many others. Females produce diploid eggs that become diploid females. The females of other species do the same, but they also can produce haploid eggs that develop into haploid males. If a haploid egg happens to be fertilized by a male, it develops into a female. The male rotifers appear only occasionally and are dwarfed and short-lived.

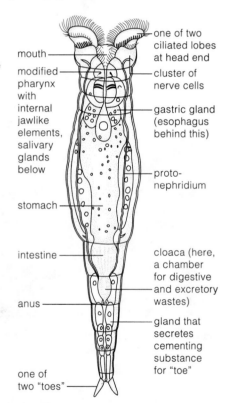

one of two ciliated lobes at head end

mouth

modified pharynx with internal jawlike elements, salivary glands below

cluster of nerve cells

gastric gland (esophagus behind this)

proto-nephridium

stomach

intestine

cloaca (here, a chamber for digestive and excretory wastes)

anus

gland that secretes cementing substance for "toe"

one of two "toes"

substances that attach free-living species to substrates at feeding time. A feature unique to rotifers is a crown of cilia at the head end that assists in swimming and in wafting food toward the mouth. Its rhythmic motions reminded early microscopists of a turning wheel; hence the name of the phylum ("rotifer" means wheel-bearer).

The rotifers are bilateral, cephalized animals with ciliated lobes at their head end and a false coelom that is packed with diverse organs.

Bilateral animals not much more complex than modern flatworms evolved in Cambrian times. Shortly after this, some species gave rise to two lineages of animals equipped with a coelom (Figure 26.1*b*). We call these great lineages the **protostomes** and the **deuterostomes**. Mollusks, annelids, and arthropods are all protostomes. Echinoderms and chordates are deuterostomes.

As a result of mutations, embryos of the animals in each lineage develop differently from fertilized eggs. For example, mitotic cell divisions cut the egg cytoplasm repeatedly, along prescribed planes, to form a tiny ball of cells (the early embryo). Protostomes undergo **spiral cleavage**, a pattern in which these early cuts are made at oblique angles relative to the genetically prescribed body axis. But deuterostomes undergo **radial cleavage**, a developmental pattern in which the early cuts are made parallel with and perpendicular to the axis:

An early protostome embryo, consisting only of four cells, as it undergoes cleavages that are *oblique to* the original body axis:

An early deuterostome embryo, also at the four-cell stage, undergoing cleavages that are *parallel with* and *perpendicular to* the original body axis:

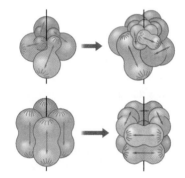

As another example, the very first opening to form at the surface of a protostome embryo becomes the mouth; an anus forms elsewhere. In a deuterostome embryo, the first opening becomes the anus; the second becomes the mouth. As a final example, a protostome coelom arises from spaces in the mesoderm, but a deuterostome coelom forms from outpouchings of the gut wall:

How the coelom forms in protostomes:

pouch that will form mesoderm, enclose coelom

embryonic gut

coelom

How the coelom forms in deuterostomes:

solid mass of mesoderm

embryonic gut

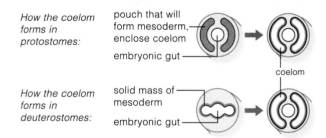

Such modifications to the embryonic stages of the two kinds of animals led to major differences in body plans.

Soon after the coelomate animals evolved in Cambrian times, two great lineages—the protostomes and deuterostomes—evolved through mutations that affected how their embryos develop, and this led to major differences in body plans.

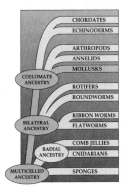

From children's books and explorations in gardens, few of us would have trouble recognizing a land snail when we see one (Figure 26.22*a*). Yet few of us know much about its 110,000 relatives in one of the largest of all animal groups, the phylum Mollusca. As their name implies, **mollusks** have fleshy soft bodies (*molluscus*, a Latin word, means soft). These bilateral animals have a small coelom. *Most* have a shell, or a reduced version of one, made of calcium carbonate and protein. These components are secreted from cells of a tissue that drapes like a skirt over the body mass. This tissue, the **mantle**, is unique to mollusks. The respiratory organs, gills of a type called ctenidia, contain thin-walled leaflets for gas exchange. *Most* mollusks have a fleshy foot. *Many* have a radula, a

their foot spreads out as they crawl. Many species have spirally coiled or conical shells. Coiling compacts the organs into a mass that can be balanced above the body, rather like a backpack. Other species have a reduced shell or none at all (Figure 26.22*a,b*).

Chitons are slow-moving or sedentary grazers with a dorsal shell divided into eight plates (Figure 26.22*c*). The bivalves, or animals with a "two-valved shell," include clams, scallops, oysters, and mussels (Figure 26.22*d*). Some bivalves are only a millimeter across. A few giant clams are over a meter across and weigh 225 kilograms (close to 500 pounds). Humans have been eating one type of bivalve or another since prehistoric times.

Cephalopods are highly active predators of the seas. They include the swiftest invertebrates (jet-propelled squids), the largest of all known invertebrates (the giant squid), and the smartest (octopuses, Figure 22.6*e*). For

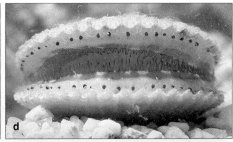
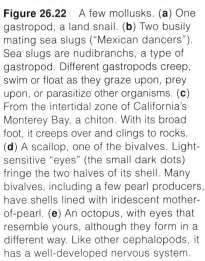

Figure 26.22   A few mollusks. (**a**) One gastropod, a land snail. (**b**) Two busily mating sea slugs ("Mexican dancers"). Sea slugs are nudibranchs, a type of gastropod. Different gastropods creep, swim or float as they graze upon, prey upon, or parasitize other organisms. (**c**) From the intertidal zone of California's Monterey Bay, a chiton. With its broad foot, it creeps over and clings to rocks. (**d**) A scallop, one of the bivalves. Light-sensitive "eyes" (the small dark dots) fringe the two halves of its shell. Many bivalves, including a few pearl producers, have shells lined with iridescent mother-of-pearl. (**e**) An octopus, with eyes that resemble yours, although they form in a different way. Like other cephalopods, it has a well-developed nervous system.

tonguelike, toothed organ that shreds food destined for the gut. Mollusks with a well-developed head have eyes and tentacles, but not all have a head. Beyond these generalizations, there are no "typical" mollusks. They range from tiny snails in treetops to huge predators of the seas. In this section and the next, we sample four classes: chitons, gastropods, bivalves, and cephalopods.

The largest class, with 90,000 species of snails and slugs, is gastropods ("belly foots"), so named because

example, show an octopus an object with a distinctive shape and then give it a mild electric shock, and it will thereafter avoid that particular object. With respect to memory and learning abilities, octopuses and certain squids are the world's most complex invertebrates.

Mollusks are bilateral, soft-bodied, coelomate animals that vary tremendously in body details, size, and life-styles.

Maybe it was their fleshy, soft bodies—so forgiving of chance evolutionary changes in morphology—that gave the ancestors of mollusks the potential to diversify in so many ways and to radiate into so many habitats. Let's explore this idea by using a few characteristics of our representative mollusks. As a point of departure, start by studying the body plan shown in Figure 26.23.

Twisting and Detwisting of Soft Bodies

Notice, in Figure 26.23*a,* how evolution put an unusual twist in the soft snail body. Its anus dumps wastes near the mouth! As a gastropod embryo develops, a cavity between its mantle and the shell gets twisted 180° counterclockwise, and so does nearly all of the visceral mass (the gut, heart, gills, and other internal organs). This process, called **torsion**, occurs only in gastropods:

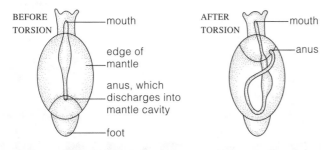

Such a drastic rearrangement of body parts could have come about through mutations that affected retractor muscles, which attach a gastropod embryo or larva to its shell. The muscles on the body's right side develop before those on the left. As they grow toward the front of the body, they drag the mantle cavity and organs with them. It takes them a few hours at most to do this.

By one hypothesis, torsion proved to be adaptive, for the head could withdraw into the mantle cavity in times of danger. However, in some 1985 experiments by J. Pennington and F. Chia, predators ate just as many torsioned larvae as "pre-torsioned" ones. Was torsion a bad evolutionary experiment? By putting the

gills, anus, and kidneys above the mouth, it certainly created a potentially awful sanitation problem. At the very least, the uptake of discharged wastes in ancestral torsioned species must have been distasteful.

In fact, we find evolutionary compensations for this state of affairs. Most gastropods now have enough cilia in this region to create currents that sweep the wastes away. Also, torsion is not as pronounced as it once was in some lineages. Nudibranchs have even undergone an apparent detorsion; the soft larval body twists, but then it untwists. At some point in their evolution, they also ended up losing most of their mantle cavity and all of their ctenidia. Most species have other outgrowths that function in gas exchange (Figures 26.22*b* and 26.23*c*).

Hiding Out, One Way or Another

If you were small, edible, and soft of body, an external shell would be a distinct advantage, as it is for chitons and clams. When a chiton is disturbed by predators or surf or exposed by a receding tide, it hunkers under its shell. Muscles in its foot pull the body mass down, and the mantle's edge around the shell's rim presses like a suction cup against a rock. That eight-plated shell is flexible. Pull a chiton from a rock, and it can roll up in a ball until it can unroll and become reattached elsewhere.

Besides having a shell, protection also can be had by hiding in sediments and other substances. A bivalve's head is not much to speak of, but its foot is usually large and specialized for burrowing. Bivalves burrowed

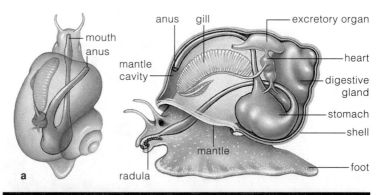

Figure 26.23 (**a**) Body plan of an aquatic snail, a gastropod. (**b**) Close-up of a radula, a feeding device of snails and some other mollusks. As the radula is rhythmically protracted and retracted, it rasps food and then draws it toward the gut on the retraction stroke. (**c**) The sea slug *Aplysia,* also called the sea hare. The two flaps above its dorsal surface are foot extensions that undulate and so help ventilate the mantle cavity. Like many other gastropods, *Aplysia* is hermaphroditic. It can function as a male, a female, or both.

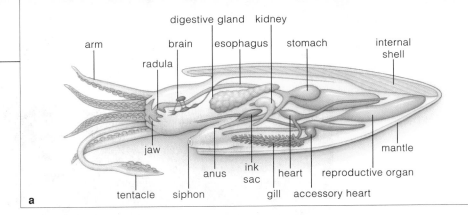

digestive gland kidney
arm brain esophagus stomach internal shell
radula
jaw anus ink sac heart reproductive organ mantle
tentacle siphon gill accessory heart

a

b

c

in sand or mud have a pair of siphons: extensions of the mantle edges, fused into tubes (Figure 26.24a). Water is drawn into the mantle cavity through one siphon and leaves through the other, carrying wastes. One wonders what preyed on the ancestors of geoducks of the Pacific Northwest, which have siphons more than a meter long.

On the Cephalopod Need for Speed

Some 500 million years ago the cephalopods, with their buoyant, chambered shells, were the supreme predators in Ordovician seas (Section 21.5). And yet, of a lineage having more than 7,000 ancestral species, the shell of all but one of the existing descendant species is reduced or gone (Figure 26.25). What happened? This evolutionary trend coincided with an adaptive radiation of the bony fishes—which preyed on cephalopods or were strong competitors for the same prey. During what may have been a long-term race for speed and wits, cephalopods lost their thick external shell and became streamlined and highly active. Of all mollusks, they now have the largest brain relative to body size and display the most

Figure 26.25 (**a**) Body plan (generalized) of a cuttlefish, a cephalopod. Its tentacles, thinner than the arms, are specialized for capturing prey. (**b**) A squid (*Dosidiscus*) and a diver inspecting each other. (**c**) A chambered nautilus, the only existing cephalopod that has an external shell. Being so active, cephalopods have great demands for oxygen. They are the only mollusks with a closed circulatory system. Their blood is pumped from a main heart to two gills, each with a booster (accessory) heart at its base that speeds blood flow, hence the uptake of oxygen (for muscle cells especially) and removal of carbon dioxide.

mouth left mantle retractor muscle
retractor muscle water flows out through exhalant siphon

water flows in through inhalant siphon

a foot palps left gill shell

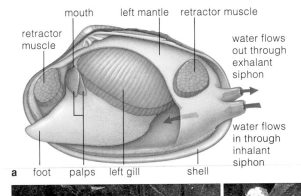

b

Figure 26.24 (**a**) Body plan of a clam, with half of its shell removed. In nearly all bivalves, gills serve in collecting food and in respiration. As water moves through the mantle cavity, mucus on the gills traps food. Cilia move the mucus and food to palps, where final sorting takes place before suitable bits are driven to the mouth. (**b**) A scallop escaping from a sea star by clapping its valves and producing a propulsive water jet.

complex behavior. Nerves connect the brain to muscles that can make quick responses to food or danger. Blood circulation and respiration are highly efficient (Figure 26.25). Except for the chambered nautilus, cephalopods can discharge dark fluid from an ink sac, maybe to confuse predators.

Jet propulsion became the name of the game. Cephalopods force a jet of water out of the mantle cavity and a funnel-shaped siphon. As mantle muscles relax, water is drawn into the cavity. As they contract, a water jet is squeezed out. When the mantle's free edge closes down on the head at the same time, a jet shoots out through the siphon. The brain controls the siphon's activity and so influences the direction of escape or pursuit.

Lively stories emerge when evolutionary theory is used to interpret the fossil record and the range of existing species diversity, as we have done for the mollusks.

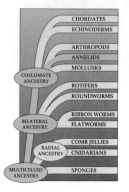

Maybe you've noticed earthworms after a downpour, when they wriggle out of their burrows to avoid drowning. Earthworms are among 15,000 or so species of bilateral, segmented animals known as the **annelids** (Annelida). Their relatives are a few kinds of leeches and the polychaetes, which are far more diverse but not nearly as well known (Figure 26.26). The phylum name means "ringed forms." But the "rings" are really a series of repeating body units, and this segmentation is pronounced. Also, except for leeches, nearly all segments have pairs or clusters of chitin-reinforced bristles on each side of the body. The bristles are also called *setae* or *chaetae*, but these are just formal names for "bristles." When pushed into soil, the bristles provide the traction required for crawling or burrowing. They have become broadened paddles in some swimming species. Earthworms, one of the oligochaetes, have few setae per body segment, and marine polychaete worms typically have many of them (*oligo-*, few; *poly-*, many).

Advantages of Segmentation

A segmented body has great evolutionary potential, for individual parts can undergo modification and become highly adapted for specialized tasks. Although most of an earthworm's segments are similar, the leeches have suckers at both ends, and polychaetes have an elaborate head and fleshy-lobed appendages known as parapods ("closely resembling feet"). By analyzing existing species, we catch glimpses of developments that led to increases in size and to more complex internal organs.

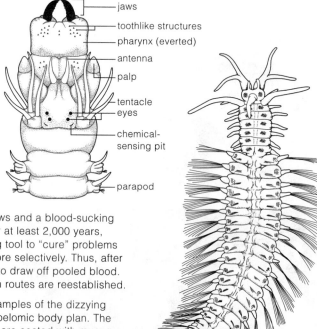

Figure 26.26 Representative annelids. (**a**) Most leeches have sharp jaws and a blood-sucking device. This one is shown before and after gorging on human blood. For at least 2,000 years, *Hirudo medicinalis*, a freshwater leech, has been used as a blood-letting tool to "cure" problems ranging from nosebleeds to obesity. Today leeches are still used, but more selectively. Thus, after surgeons reattach a severed ear, lip, or fingertip, leeches may be used to draw off pooled blood. A patient's body cannot do this on its own until severed blood circulation routes are reestablished.

(**b**) A less scary annelid—an earthworm. (**c,d**) From the polychaetes, examples of the dizzying variety of modifications that have evolved, starting from a segmented, coelomic body plan. The photograph shows a tube-dweller. Featherlike structures at its head end are coated with mucus. After the mucus has trapped bacteria and other bits of food, coordinated beating of cilia sweeps them to the mouth. Most polychaetes live in marine habitats. They actually are one of the most common types of animals along coasts. Many are predators or scavengers; others dine on algae.

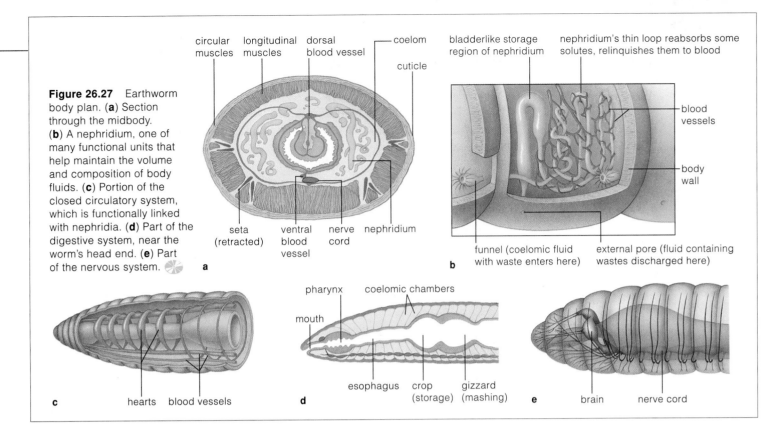

Figure 26.27 Earthworm body plan. (**a**) Section through the midbody. (**b**) A nephridium, one of many functional units that help maintain the volume and composition of body fluids. (**c**) Portion of the closed circulatory system, which is functionally linked with nephridia. (**d**) Part of the digestive system, near the worm's head end. (**e**) Part of the nervous system.

Labels for (a): circular muscles · longitudinal muscles · dorsal blood vessel · coelom · cuticle · seta (retracted) · ventral blood vessel · nerve cord · nephridium

Labels for (b): bladderlike storage region of nephridium · nephridium's thin loop reabsorbs some solutes, relinquishes them to blood · blood vessels · body wall · funnel (coelomic fluid with waste enters here) · external pore (fluid containing wastes discharged here)

Labels for (c): hearts · blood vessels

Labels for (d): pharynx · coelomic chambers · mouth · esophagus · crop (storage) · gizzard (mashing)

Labels for (e): brain · nerve cord

Annelid Adaptations—A Case Study

Earthworms are the usual textbook examples of annelids. As Figure 26.27 shows, partitions divide the body into a series of coelomic chambers. In most chambers we find repeats of muscles, blood vessels, branching nerves, and other organs. The gut extends through all the chambers, from mouth to anus. Like all annelids, an earthworm has a cuticle of secreted material that surrounds the body surface. It bends easily and is permeable to water as well as gases, which is one reason why annelids are restricted to aquatic habitats or moist habitats on land.

Earthworms are scavengers. They ingest moist soil and mud that contains decomposing plant material and other organic matter. Each worm ingests its own weight every twenty-four hours. Collectively, their burrowing and feeding activities aerate soil and lift nutrients to the surface, to the benefit of many plants.

As in other annelids, the fluid-cushioned coelomic chambers serve as a hydrostatic skeleton against which muscles act. Each segment's wall incorporates a layer of circular muscles (Figure 26.27*a*). When longitudinal muscles that span several segments are contracting, the circular ones relax, so that segments shorten and fatten. When the pattern reverses, the segments lengthen. While this is going on, bristles on different segments are protracting and retracting. When the first few segments lengthen, the body is extended forward. Bristles of the segments behind them plunge into the ground and hold the body in its extended position. As the first segments plunge *their* bristles into the ground, the ones behind them retract their bristles and are pulled forward. The alternating contractions and elongations proceed along the body's length and move the whole worm forward.

Figure 26.27*b* shows part of a system of **nephridia** (singular, nephridium), units that regulate the volume and composition of body fluids. In many annelids, cells of these units are similar to flame cells, which implies an evolutionary link between flatworms and annelids. More often, a nephridium starts as a funnel that collects excess fluid from one coelomic chamber. The funnel connects with a tubular part of the nephridium, which delivers fluid to a surface pore in the body wall of the next coelomic chamber in line.

The worm's head end has a rudimentary **brain**, an aggregation of nerve cell bodies that integrate sensory input and muscle responses for the whole body. Paired **nerve cords**, each a bundle of extensions of nerve cell bodies, lead away from the brain. They are pathways for rapid communication. In each segment, the paired nerve cords broaden into a **ganglion** (plural, ganglia), a cluster of nerve cell bodies that controls local activity.

Finally, as is true of most annelids, earthworms have a closed circulatory system, with blood confined in hearts and muscularized blood vessels. Contractions keep blood circulating in one direction. Smaller blood vessels service the gut, nerve cord, and body wall.

Annelids are bilateral, coelomate, segmented worms that have complex organ systems. Some species show the degree of specialization possible with a segmented body plan.

ARTHROPODS—THE MOST SUCCESSFUL ORGANISMS ON EARTH

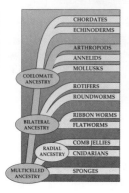

Arthropod Diversity

Evolutionarily speaking, "success" means having the greatest number of species, producing the most offspring, occupying the most habitats, effectively fending off predators and competitors, and having a capacity to exploit the greatest amounts and kinds of food. These are the features that come to mind when we attempt to characterize the **arthropods** (Arthropoda).

Over a million species (mostly insects) are known, and new ones are being discovered weekly. Of four major lineages, the trilobites are extinct (Section 21.5). The other three lineages are chelicerates (spiders and their kin), crustaceans (such as barnacles and crabs), and uniramians (centipedes, millipedes, and insects).

Adaptations of Insects and Other Arthropods

Six important adaptations contributed to the success of arthropods in general and insects in particular:

1. A hardened exoskeleton
2. Jointed appendages
3. Fused and modified segments
4. Specialized respiratory structures
5. Efficient nervous system and sensory organs
6. Division of labor in the life cycle

HARDENED EXOSKELETONS Arthropods have a cuticle of chitin, proteins, and surface waxes. It is sufficiently hardened, as by calcium carbonate deposits, to serve as a rigid, protective external skeleton—an exoskeleton. Such cuticles might have evolved as defenses against predation, but they took on added functions when arthropods invaded the land. They support a body deprived of water's buoyancy. Their waxy surface restricts evaporative water loss. Hard cuticles do restrict increases in size, but arthropods grow in spurts, by **molting**. At certain stages of their life cycle, they secrete a new, soft cuticle under their old one, which they shed (Figure 26.28). The body mass then increases by repeated, rapid cell division before the new cuticle hardens.

Figure 26.28 Molting, as demonstrated by a bright orange centipede backing out of its old exoskeleton.

JOINTED APPENDAGES If arthropods had a uniformly hardened cuticle, they wouldn't move much. However, the cuticle thins down at joints (places where different body parts abut). Muscles associated with the joints can make the thinned cuticle bend in specific directions and so move the attached body parts. A jointed exoskeleton was an evolutionary innovation that led to specialized appendages, including diverse wings and antennae as well as legs (arthropod means "jointed foot").

FUSED AND MODIFIED SEGMENTS The first arthropods were segmented, like the annelid stock that presumably gave rise to them. In most of their existing descendants, however, the serial repeats of the body wall and organs are masked. Segments in different parts of the body have become fused together and modified for far more specialized functions. For example, in the ancestors of insects, different segments fused to form three regions —head, thorax, and abdomen—which morphologically diverged from one another in astounding ways.

RESPIRATORY STRUCTURES Many aquatic arthropods depend on gills for gas exchange. Air-conducting tubes evolved among insects and other land-dwellers. Insect tracheas begin as pores on the body surface and branch into tubes that deliver oxygen directly to tissues. They support energy-consuming activities, such as flight.

SPECIALIZED SENSORY STRUCTURES Intricate eyes and other sensory organs contributed to arthropod success. Numerous species have a wide angle of vision and can process visual information from many directions.

DIVISION OF LABOR Moths, butterflies, beetles, flies, and many other species divide the job of surviving and reproducing among different stages of development. As their immature forms grow, body tissues are massively reorganized and body parts become remodeled. (Figure 1.4). Growth and major transformation of an immature form into the adult is called metamorphosis. Typically, immature stages such as caterpillars specialize in feeding and increasing in size, and the adult specializes mainly in dispersal and reproduction. A life cycle that turns on such a *division of labor* among different developmental stages is adaptive to environmental changes, including seasonal variation in food sources and water supplies.

As a group, the arthropods are exceptionally abundant and widespread, and they have enormously different life-styles.

Their success arises largely from their hardened, jointed exoskeletons; fused, modified body segments; specialized appendages; specialized respiratory, nervous, and sensory organs; and often a division of labor in the life cycle.

A LOOK AT SPIDERS AND THEIR KIN

The chelicerates originated in shallow seas early in the Paleozoic. The only surviving marine species are a few mites, sea spiders, and horseshoe crabs (Figure 26.29). The familiar chelicerates—scorpions, spiders, ticks, and chigger mites—are classified as arachnids. Most species of arachnids live on land, and we might say this about them: Never have so many been loved by so few.

Figure 26.29 Horseshoe crab. Its hard, shieldlike cover hides *five* pairs of legs, one of its defining features.

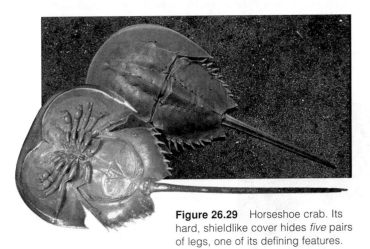

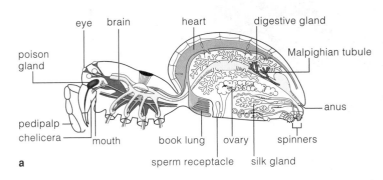

Scorpions and spiders are predators; they sting or bite and may subdue prey with venom. When they sting or bite us, the reaction might be painful but is rarely serious. Spiders especially are beneficial in that they prey on great numbers of pestiferous insects. Bites of the blood-sucking ticks that parasitize deer, mice, and other vertebrates can cause maddening itches and often serious diseases. For example, the bites of some ticks transmit the bacterial agents of Rocky Mountain spotted fever or Lyme disease to humans (Section 22.4). Most mites are free-living scavengers.

Arachnids have segments fused into a forebody and hindbody. The forebody's jointed appendages include four pairs of legs, a pair of pedipalps with primarily sensory functions, and a pair of chelicerae that inflict wounds and discharge venom. The appendages of the hindbody spin out silk threads for webs and for egg cases. Most webs are netlike. One type of spider spins a

Figure 26.30 (a) Internal organization of a spider body. (b) Wolf spider. Like most spiders, it helps keep insect populations in check and its bite is harmless to humans. (c) Female black widow. Its bite can be painful and sometimes dangerous. (d) Brown recluse, with a violin-shaped mark on its forebody. Its bite can be severe to fatal.

vertical thread with a ball of sticky material at the end. It uses a leg to swing the ball at insects passing by! Inside the body is an *open* circulatory system, with a heart that pumps blood into tissues, then receives blood through small openings in its wall. Some blood travels through moist folds of book lungs. These respiratory organs resemble book pages, and they greatly increase the surface area available for gas exchange with the air. Figure 26.30*a* shows the arrangement of book lungs and other major organs inside the spider body.

The spiders, scorpions, and their relatives have a variety of appendages specialized for predatory or parasitic life-styles.

A LOOK AT THE CRUSTACEANS

Shrimps, lobsters, crabs, barnacles, pillbugs, and other crustaceans got their name because they have a hard yet flexible "crust" (an external skeleton), but so do nearly all arthropods. Only some of the 35,000 species live in freshwater or on land. The vast majority live in marine habitats, where they are so abundant they have been dubbed the insects of the seas. Lobsters and crabs are the "giants" of this subphylum; most crustaceans are less than a few centimeters long. All have major roles in food webs, and humans harvest many edible types.

The simplest crustaceans have many pairs of similar appendages for most of their length and may resemble

their annelid ancestors. In other lineages, unspecialized appendages evolved into diverse structures of the sort shown in Figure 26.31. The strong claws of lobsters and crabs are used to collect food, intimidate other animals, and sometimes dig burrows. Feathery appendages of barnacles comb microscopic bits of food from the water.

Many crustaceans have sixteen to twenty segments; some have more than sixty. In crabs, lobsters, and some other crustaceans, the dorsal cuticle extends back from the head as a shield-like cover (carapace) over some or all of the segments. The head incorporates two pairs of antennae, a pair of mandibles (jawlike appendages),

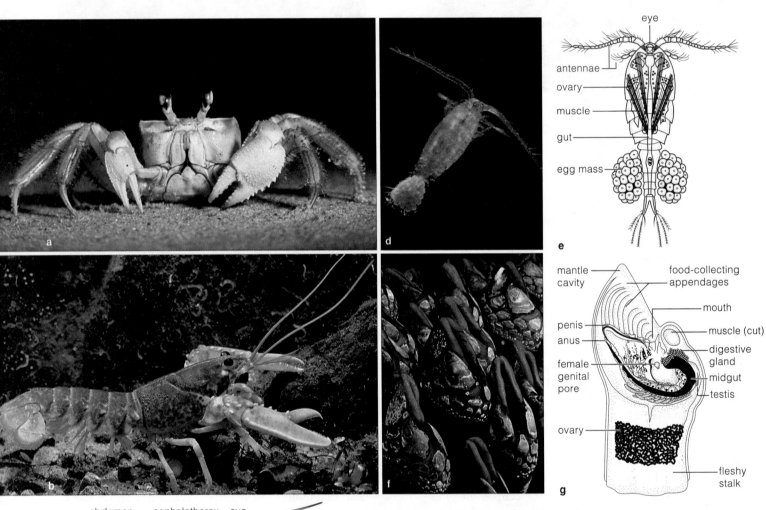

Figure 26.31 A sampling of crustaceans and their diverse life-styles. A crab (**a**). Photograph and body plan of (**b,c**) a lobster, (**d,e**) a copepod, and (**f,g**) stalked barnacles. Lobsters are secretive for most of their lives. Crabs skitter actively and openly across sand and rocks. The copepods are free-living filter feeders, predators, or parasites. This female has a pair of long antennae and is carrying her eggs around with her. As adults, barnacles cement themselves to one spot. The goose barnacles shown earlier in Figure 5.1 have no problem attaching to boat hulls, floating logs or bottles, and other available objects in the water. You might mistake barnacles for mollusks, but as soon as they open their hinged shell to filter feed, their jointed appendages—the hallmark of arthropods—are apparent.

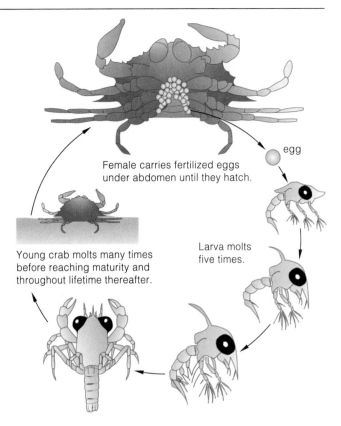

Figure 26.32 Life cycle of a crab. The larval and juvenile stages molt repeatedly during their growth.

Female carries fertilized eggs under abdomen until they hatch.

egg

Larva molts five times.

Young crab molts many times before reaching maturity and throughout lifetime thereafter.

and two pairs of maxillae (food-handling appendages). Crayfish, crabs, lobsters, shrimps, and their kin have five pairs of walking legs.

Figure 26.31d shows a copepod. Copepods are less than two millimeters long and are the most numerous animals in aquatic habitats, maybe even in the world. About 1,500 kinds parasitize various invertebrates and fishes. The majority—8,000 species—are consumers of phytoplankton, the "pastures" of aquatic habitats. Some also eat larval or small adult invertebrates, fish eggs, and fish larvae, which they grab with their pair of food-handling appendages. In turn, the copepods are food for different invertebrates, fishes, and baleen whales.

Of all the arthropods, only barnacles have a calcified "shell," a modified external skeleton that protects them from predators, drying out, and battering currents and surf. Adult barnacles cement themselves to rocks, wharf pilings, and similar surfaces (Figure 26.31f,g). A few kinds attach themselves only to the skin of whales.

Like other arthropods, crustaceans molt repeatedly to shed the exoskeleton during their life cycle. Figure 26.32 shows a crab's larval stages and increases in size.

Crustaceans differ greatly in the number and kind of their appendages. As for arthropods generally, they repeatedly replace their external skeleton by molting.

The **millipedes** and **centipedes** have a long, segmented body with many legs. Of course, millipedes don't have "a thousand," as their name implies. Most have about 100, although one exuberant individual grew 752. And centipedes have between 15 and 177 pairs of legs, not a nicely rounded number of "one hundred."

Millipedes have a rounded body (Figure 26.33a). As they develop, pairs of segments fuse, so what looks like a single segment in the adult has *two* pairs of legs. They typically bulldoze through soil and forest litter, mostly as scavengers of decaying vegetation. Centipedes have a flattened body, and all but two segments have a pair

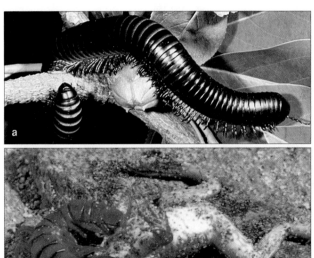

Figure 26.33 (a) Millipede. (b) A Southeast Asian centipede.

of walking legs. All species are fast-moving, aggressive predators, outfitted with fangs and venom glands. They prey on insects, earthworms, and snails. The one shown in Figure 26.33b subdues small lizards, toads, and frogs. The house centipede (*Scutigera coleoptrata*) often lurks in buildings, where it actively preys on cockroaches, flies, and other pests. Although helpful in this respect, its vaguely terrifying body keeps it from being welcomed.

The mild-mannered, scavenging millipedes and aggressive, predatory centipedes do not lend themselves to leg counts as they walk by.

A LOOK AT INSECT DIVERSITY

As a group, insects share the adaptations listed earlier in Section 26.15. Here we expand the list a bit. Insects have a head, a thorax, and an abdomen. The head has paired sensory antennae and paired mouthparts used in biting, chewing, sucking, or puncturing (Figure 26.34). The thorax has three pairs of legs and usually two pairs of wings. Most abdominal appendages are reproductive structures, such as egg-laying devices. An insect has a foregut, midgut (where most digestion proceeds), and hindgut (where water is reabsorbed). Insects get rid of waste material by **Malpighian tubules**, small tubes that connect with the midgut. Nitrogen-containing wastes from protein breakdown diffuse from blood into the tubules and are converted into harmless crystals of uric acid. The crystals are eliminated with feces. This system allows the land-dwelling insects to get rid of potentially toxic wastes without losing precious water.

We've already catalogued more than 800,000 species of insects. If we use sheer numbers and distribution as the yardstick, the most successful insects are small in size and have a staggering reproductive capacity. For example, you may find some of these species growing and reproducing in great numbers on a single plant that might be only an appetizer for another animal. By one estimate, if all the progeny of a single female fly were to survive and reproduce through six more generations, that fly would have more than 5 trillion descendants!

Besides this, the most successful insect species are winged. In fact, they are the *only* winged invertebrates. They can move among food sources that are too widely scattered to be exploited by other kinds of animals. The capacity for flight contributed to their success on land.

Finally, insect life cycles commonly proceed through stages that allow exploitation of different resources at different times. As an insect embryo develops, organs required for feeding and other vital activities form and become functional. Before an insect becomes an adult (the sexually mature form of the species), it proceeds through immature, post-embryonic stages. **Nymphs** and **pupae**, as well as larvae, are examples of the stages.

Like human infants, some insect nymphs are shaped like miniature adults; but unlike children, they undergo growth and molting (Figure 26.35a). Other insects go through post-embryonic stages of reactivated growth, tissue reorganization, and remodeling of body parts. Metamorphosis, recall, is the name for this resumption of growth and transformation into an adult form. Figure 26.35b,c shows how the transformation is more drastic in some species than in others.

The factors that contribute to insect success also make them our most aggressive competitors. Insects destroy crops, stored food, wool, paper, and timber. As they stealthily draw blood from us and from our pets, they often transmit pathogenic microorganisms. On the

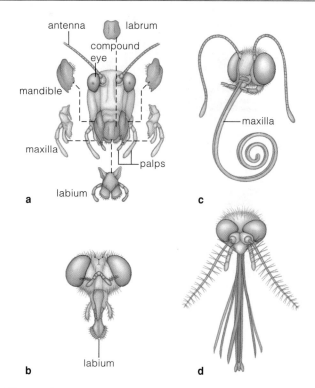

Figure 26.34 Examples of insect appendages. Headparts of (**a**) grasshoppers, a chewing insect; (**b**) flies, which sponge up nutrients with a specialized labium; (**c**) butterflies, which siphon up nectar with a specialized maxilla; and (**d**) mosquitoes, with piercing and sucking appendages.

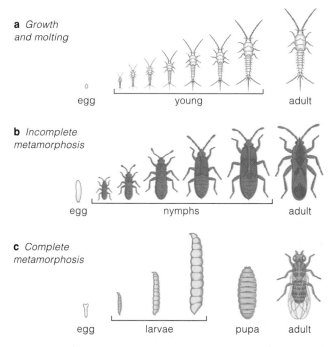

Figure 26.35 Examples of post-embryonic development. (**a**) Young silverfish are adults in miniature, changing little except in size and proportion as they mature into adults. (**b**) True bugs show *incomplete* metamorphosis, which involves gradual, partial change from the first immature form until the last molt. (**c**) Fruit flies show *complete* metamorphosis. Tissues of immature forms are destroyed and replaced before emergence of the adult.

Figure 26.36 Insects. (**a**) Stinkbugs (order Hemiptera), newly hatched. (**b**) A honeybee (order Hymenoptera) attracting its hive mates with a dance, as described in Section 51.4. (**c**) Ladybird beetles (order Coleoptera) swarming. These beetles are raised commercially and released as biological controls of aphids and other pests. Also in this order, the scarab beetle (**d**). With more than 300,000 species, Coleoptera is the largest order of the animal kingdom. (**e**) Luna moth (order Lepidoptera) of North America. Like most other moths and butterflies, its wings and body are covered with microscopic scales. (**f**) Dragonfly (order Odonata), which swiftly captures and eats insects in midflight.

(**g**) Duck louse (order Mallophaga). It eats bits of feathers and skin. (**h**) European earwig (order Dermaptera), a common household pest. (**i**) Flea (order Siphonaptera), with strong legs for jumping onto and off animal hosts. (**j**) Mediterranean fruit fly (order Diptera). Its larvae destroy citrus fruit and other crops.

bright side, many insects pollinate flowering plants in general and crop plants in particular. And many "good" insects attack or parasitize the ones we would rather do without. We now leave our survey of insects and other arthropods with Figure 26.36 and its representatives from some of the major orders of insects.

As a group, insects show immense variation on the basic arthropod body plan. Many have wings, and their life cycles have stages that allow exploitation of different and often widely scattered food sources. Many species produce great numbers of small individuals that pass through immature, post-embryonic stages before the adult form emerges.

THE PUZZLING ECHINODERMS

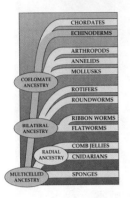

We turn, finally, to the second lineage of the coelomate animals, the deuterostomes. The major invertebrate members of this lineage are **echinoderms** (Echinodermata). The feather star, sea urchin, sea cucumber, and brittle stars shown in Figure 26.37 belong to this phylum. So do the sea lilies (crinoids), sand dollars, and sea biscuits. Sea lilies, which look a bit like stalked plants, flourished in Silurian times (Figure 21.14). About 13,000 echinoderm species are known from the fossil record, but most of these became extinct. Nearly all of the 6,000 or so existing species live in marine habitats.

An echinoderm body wall bears a number of spines, spicules, or plates made stiff with calcium carbonate. (Echinodermata means spiny skinned.) These structures serve defensive functions, as you might suspect if you have ever stepped barefoot on a sea urchin. Its spines trigger painful swelling if they break off under the skin. Most echinoderms also have a well-developed internal skeleton, which is composed of calcium carbonate and other substances secreted from specialized cells.

Oddly, the adult echinoderms are radial with some bilateral features. Most species even produce bilateral larvae during their life cycle. Did bilateral invertebrates give rise to the ancestors of echinoderms, which later picked up some radial features as they evolved? Maybe.

Adult echinoderms have no brain. However, their decentralized nervous system allows them to respond to information about food, predators, and so forth that is coming from different directions. For instance, any arm of a sea star that senses the shell of a tasty scallop

Figure 26.37 Representative echinoderms. (**a**) Feather star, with finely branched food-gathering appendages. (**b**) Sea urchin, which moves about on spines and tube feet. (**c**) Sea cucumber, with rows of tube feet along its body. (**d**) Brittle stars. Their arms (rays) make rapid, snakelike movements.

tube feet   spine

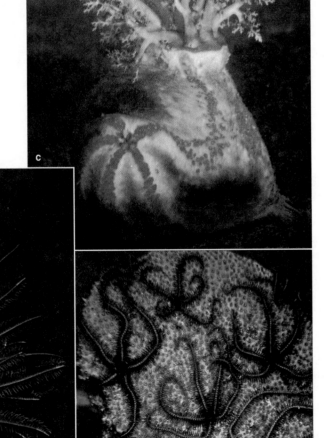

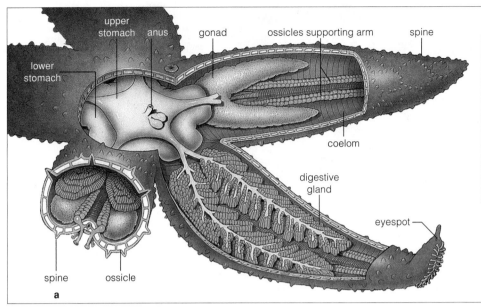

upper stomach · anus · gonad · ossicles supporting arm · spine
lower stomach
coelom
digestive gland
eyespot
spine · ossicle

a

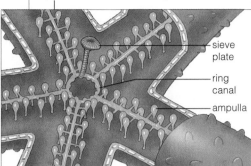

part of the water-vascular system
sieve plate
ring canal
ampulla

b

c

tube feet on the ventral surface of a sea star arm

Figure 26.38 (**a**) Some key aspects of the radial body plan of a sea star. The water-vascular system, in combination with great numbers of tube feet, is the basis of locomotion. (**b**,**c**) Five-armed sea star, with closer views of tube feet.

can become the leader, directing the rest of the body to move in a direction suitable for prey capture.

Figures 26.37 and 26.38 show examples of tube feet. These fluid-filled, muscular structures have suckerlike adhesive disks. Sea stars use their tube feet for walking, burrowing, clinging to a rock, or gripping a clam or a snail about to become a meal. Tube feet are part of a **water-vascular system** unique to echinoderms. In sea stars, the system includes a main canal in each arm. Short side canals extend from them and deliver water to the tube feet. Each tube foot has an ampulla, a fluid-filled, muscular structure shaped rather like the rubber bulb on a medicine dropper. As an ampulla contracts, it forces fluid into the foot and causes it to lengthen.

Tube feet change shape constantly as muscle action redistributes fluid through the water-vascular system. Hundreds of tube feet may move at a time. After being released, each one swings forward, reattaches to the substrate, then swings backward and is released before swinging forward again. Their motions are splendidly coordinated, so sea stars glide rather than lurch along.

On their ventral surface, sea stars have a formidable feeding apparatus, such as the one shown here:

Some eager types simply swallow their prey whole. Others push part of their stomach outside the mouth and around their prey, then start digesting their meal even before swallowing it. Sea stars get rid of coarse, undigested residues through the mouth. They do have a small anus, but this is of no help in getting rid of empty clam or snail shells.

With their curious traits, echinoderms are a suitable point of departure for this chapter. Even though we can identify broad trends in animal evolution, we should keep in mind that there are confounding exceptions to the perceived macroevolutionary patterns.

Echinoderms are coelomate animals with spines, spicules, or plates in the body wall. From the evolutionary perspective, they are a puzzling mix of bilateral and radial features.

SUMMARY

1. Animals are multicelled, aerobic heterotrophs that ingest or parasitize other organisms. Nearly all have diploid body cells, arranged as tissues, organs, and organ systems. Animals reproduce sexually and often asexually. They go through embryonic development, and most are motile during at least part of the life cycle.

2. Animals range from structurally simple placozoans and sponges to vertebrates. The species of each phylum share certain characteristics.

   a. By comparing major animal phyla and integrating the information with the fossil record, biologists have identified major evolutionary trends among them.

   b. The revealing aspects of body plans are the type of symmetry, gut, and cavity (if any) between the gut and body wall; whether there is a head end; and whether the body is divided into a series of segments (Figure 26.39).

3. *Trichoplax*, the only known placozoan, is the simplest animal. It consists of little more than two layers of cells, with a fluid matrix in between.

4. A sponge body is at the cellular level of construction and has no symmetry. A sponge has several kinds of cells, but these are not organized as epithelia and other tissues of complex animals.

5. Cnidarians (such as jellyfishes, sea anemones, and hydras) have radial symmetry and are at the tissue level of construction. Cnidarians alone produce nematocysts (capsules with dischargeable threads, used mainly in prey capture).

6. Nearly all animals more complex than cnidarians show bilateral symmetry, and they form tissues, organs, and organ systems. Their gut may be saclike, as it is in flatworms, but it usually is complete, with an anus and a mouth. Like most animals, flatworms have a coelom or false coelom, which are cavities between the gut and body wall. A coelom has a special lining (peritoneum).

7. Two major lineages diverged shortly after flatworms evolved. One (protostomes) gave rise to the mollusks, annelids, and arthropods. The other (deuterostomes) gave rise to echinoderms and chordates.

8. All mollusks have a fleshy soft body, a mantle, and a shell or remnant of one. They vary greatly in size, body details, and life-styles.

9. Annelids (the earthworms, polychaetes, and leeches) have a segmented body, complex organs, and a series of coelomic chambers.

10. Collectively, arthropods are the most successful of all groups in terms of diversity, numbers, distribution, defenses, and capacity to exploit food resources.

   a. The arthropods, including arachnids, crustaceans, insects, all have hardened, jointed exoskeletons, as well as modified segments, specialized appendages, highly

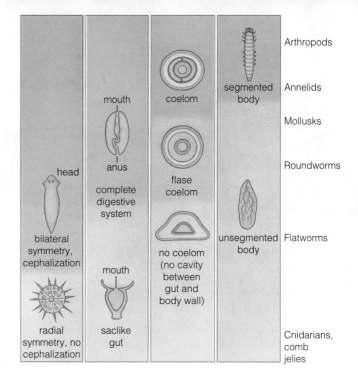

Figure 26.39 Summary of key trends in the evolution of animals, as identified by comparing body plans of major phyla. All of these features did not appear in every group.

specialized respiratory, nervous, and sensory organs, and (in insects only) wings.

   b. Arthropods develop by growth in size and molting or develop through a series of immature stages, such as larvae and nymphs. Many metamorphose; the tissues of immature forms undergo major reorganization and body parts are remodeled before the adult emerges.

11. Echinoderms have spines, spicules, or plates in their body wall. Evolutionarily, the phylum is puzzling. The larvae of most species form bilateral features, but they go on to develop into basically radial adults.

Review Questions

1. List the six main features that characterize animals. *26.1*

2. When attempting to discern evolutionary relationships among major groups of animals, which aspects of their body plans provide the most useful clues? *26.1*

3. What is a coelom? Why was it important in the evolution of certain animal lineages? *26.1*

4. Name some animals with a saclike gut. Evolutionarily, what advantages does a complete gut afford? *26.1, 26.4–26.6*

5. Choose a species of insect that lives in your neighborhood and describe some of the observable adaptations that underlie its success. *26.15, 26.19*

Self-Quiz *(Answers in Appendix IV)*

1. Which is *not* characteristic of the animal kingdom?
   a. multicellularity; cells form tissues, organs
   b. exclusive reliance on sexual reproduction
   c. motility at some stage of the life cycle
   d. embryonic development during the life cycle

Figure 26.40 *Telesto*, one of the soft branching corals.

2. Jellyfishes, sea anemones, and their relatives have _____ symmetry, and their cells form _____ .
   a. radial; mesoderm   c. radial; tissues
   b. bilateral; tissues   d. bilateral; mesoderm

3. In sheer numbers and distribution, _____ are the most successful animals.
   a. arthropods   c. snails and clams   e. vertebrates
   b. sponges   d. sea stars

4. Bilateral, segmented bodies and hardened exoskeletons occur among the _____ .
   a. arthropods   c. snails and clams   e. vertebrates
   b. sponges   d. sea stars

5. Which phylum contains members that are notorious for causing serious diseases in humans?
   a. cnidarians   c. segmented worms
   b. flatworms   d. chordates

6. More complex animals have a _____ between the gut and body wall.
   a. pharynx   c. coelom
   b. pseudocoelom   d. archenteron

7. Most animals that are more complex than cnidarians have _____ symmetry, and _____ forms in their embryos.
   a. radial; mesoderm   c. bilateral; mesoderm
   b. bilateral; endoderm   d. radial; endoderm

8. Match the terms with the appropriate groups.
   ____ sponges   a. spiny-skinned
   ____ cnidarians   b. vertebrates and kin
   ____ flatworms   c. flukes and tapeworms
   ____ roundworms   d. no tissue organization
   ____ rotifers   e. no males for some
   ____ mollusks   f. nematocysts, radial symmetry
   ____ annelids   g. hookworms, elephantiasis
   ____ arthropods   h. jointed exoskeleton
   ____ echinoderms   i. "belly-foots" and kin
   ____ chordates   j. segmented worms

Critical Thinking

1. A carnivorous sponge was discovered in an underwater cave in the Mediterranean Sea. Unlike other sponges, it has no pores or canals. Projecting from branching outgrowths of its body surface are hooklike spicules that act like Velcro to trap shrimp and other animals. The outgrowths envelop prey, which sponge cells then digest. Which components of the sponge body had to evolve for such a feeding strategy?

2. Tapeworms are hermaphroditic. What selective advantages might this feature offer, and in what kinds of environments?

3. People who eat raw oysters or clams harvested from sewage-polluted waters develop mild to severe gastrointestinal ailments.

Think about the feeding modes of these mollusks and develop a hypothesis to explain why mollusk eaters can get sick.

4. You are diving in calm, warm waters behind a tropical reef. You see something that looks like a bright-red and white plant, but it has a profusion of tiny tentacles (Figure 26.40). It is a branching soft coral with small individual polyps. You will never see such a coral growing on reef surfaces exposed to the open sea. Propose two possible explanations for this.

5. The animal shown in Figure 26.41a is packed with organs and has a crown of cilia at its head end. The marine worm in Figure 26.41b has a segmented body. Most of its segments are similar to one another, and each has bristles on the sides that were used to dig its burrow in sediments. To which groups do the two animals belong?

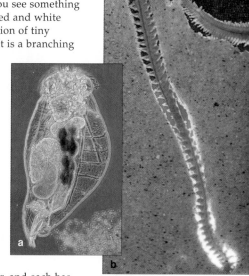

Figure 26.41 Go ahead, name the mystery animals.

Selected Key Terms

adult *26.3*	ganglion *26.14*	nymph *26.19*
animal *26.1*	gonad *26.5*	organ *26.7*
annelid *26.14*	gut *26.1*	organ system *26.7*
arthropod *26.15*	hermaphrodite *26.7*	placozoan *26.2*
bilateral	hydrostatic	planula *26.5*
symmetry *26.1*	skeleton *26.4*	polyp *26.4*
brain *26.14*	invertebrate *26.1*	proglottid *26.7*
centipede *26.18*	larva *26.3*	protostome *26.11*
cephalization *26.1*	Malpighian	pupa *26.19*
cnidarian *26.4*	tubule *26.19*	radial cleavage *26.11*
coelom *26.1*	mantle *26.12*	radial symmetry *26.1*
collar cell *26.3*	medusa *26.4*	ribbon worm *26.7*
comb jelly *26.6*	mesoderm *26.1*	rotifer *26.10*
contractile cell *26.4*	metamorphosis *26.15*	roundworm *26.8*
cuticle *26.8*	millipede *26.18*	spiral cleavage *26.11*
deuterostome *26.11*	mollusk *26.12*	sponge *26.3*
echinoderm *26.20*	molting *26.15*	torsion *26.13*
ectoderm *26.1*	nematocyst *26.4*	vertebrate *26.1*
endoderm *26.1*	nephridium *26.14*	water-vascular
epithelium *26.4*	nerve cell *26.4*	system *26.20*
flatworm *26.7*	nerve cord *26.14*	

Readings

Kozloff, E. 1990. *Invertebrates*. Philadelphia: Saunders.

Pearse, V., et al. 1987. *Living Invertebrates*. Palo Alto, California: Blackwell.

Pechenik, J. 1995. *Biology of Invertebrates*. Third edition. Dubuque: Wm. C. Brown.

Web Site See *http://www.wadsworth.com/biology* for practice quiz questions, hypercontents, BioUpdates, and critical thinking. The Wadsworth Biology Resource Center provides a wealth of information fully organized and integrated by chapter.

27

ANIMALS: THE VERTEBRATES

Making Do (Rather Well) With What You've Got

In 1798, a few naturalists were skeptically probing a specimen delivered to the British Museum in London, looking for signs that a prankster had slyly stitched the bill of an oversized duck onto the pelt of a small furry mammal. They didn't know it, but they were examining the remains of a platypus, a web-footed mammal about half the size of a housecat. Like the other mammals, the duck-billed platypus (*Ornithorhynchus anatinus*) has mammary glands and hair. And yet, like birds and reptiles, it has a cloaca, a single enlarged duct through which gametes, feces, and excretions from the kidneys pass. It lays shelled eggs, as birds and most reptiles do. Its young hatch pink and unfinished, as late embryonic stages too helpless to fend for themselves (Figure 27.1*a*). Its fleshy bill does look ducklike, and its broad, flat, furry tail does look like the one on a beaver (Figure 27.1*b*).

With its unusual traits, the platypus invites us to challenge preconceived notions of what constitutes "an animal." This particular combination of traits started coming together when the supercontinent Pangea was breaking up. As you will see, the platypus's ancestors happened to be stuck on a huge fragment that remained isolated from the other continents for 150 million years; eventually it became Australia. The geographic separation was a springboard for genetic divergence, and unique mutations in combination with selection pressures gave rise to the platypus body. Even though that body plan is much ridiculed, we scarcely can call it a failure. It has endured far longer than the one we humans inherited.

The platypus inhabits streams and lagoons in remote regions of Australia and Tasmania. During the day it hides in underground burrows. At night it slips into the water, where it hunts for small invertebrates. Even when it stays in water all night, its dense fur helps maintain body temperature, just as it does all day long when

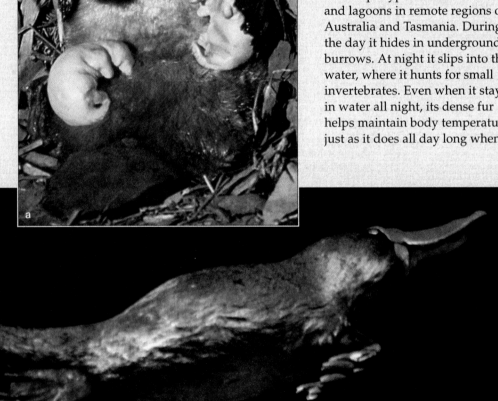

Figure 27.1 One of evolution's success stories—the platypus, (**a**) raising its incompletely formed offspring in a burrow and (**b**) underwater, eyes and ears shut, yet homing in on prey.

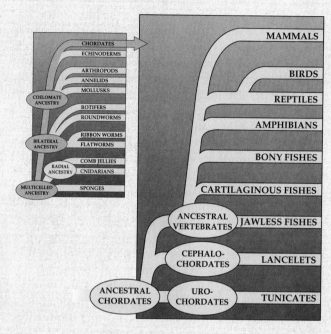

Figure 27.2 Family tree for vertebrates and other chordates.

The following labels appear in the figure:

CHORDATES
ECHINODERMS
ARTHROPODS
ANNELIDS
MOLLUSKS
ROTIFERS
ROUNDWORMS
RIBBON WORMS
FLATWORMS
COMB JELLIES
CNIDARIANS
SPONGES

COELOMATE ANCESTRY
BILATERAL ANCESTRY
RADIAL ANCESTRY
MULTICELLED ANCESTRY

MAMMALS
BIRDS
REPTILES
AMPHIBIANS
BONY FISHES
CARTILAGINOUS FISHES
JAWLESS FISHES
LANCELETS
TUNICATES

ANCESTRAL VERTEBRATES
CEPHALO-CHORDATES
ANCESTRAL CHORDATES
URO-CHORDATES

the platypus is resting in its cool burrow. The broad, thick tail is a most excellent rudder during underwater maneuvers and a storehouse for energy-rich fat reserves. The broad, flattened bill is well-shaped for scooping up aquatic snails, mussels, worms, shrimps, and insect larvae. Horny pads lining the platypus jaws grind up and make short work of shelled and hard-cuticled meals. Back at the burrow, the platypus uses strong claws on its hind feet for digging.

Like a submarine, a platypus closes the hatches, so to speak, when it dives into the water. Its nostrils as well as a fleshy groove around the eyes and ears snap shut. No need for eyes or ears to zero in on prey; many of the 800,000 sensory receptors in its bill can detect tiny oscillations in water pressure as prey swim past. Other receptors detect changes in electric fields as weak as those initiated by the flicks of a shrimp tail.

The platypus is one of 4,500 kinds of mammals that now occupy the vertebrate branch of the animal family tree (Figure 27.2). Its vertebrate relatives include fishes, amphibians, reptiles, and birds, which are described in this chapter. As you consider each group, keep a key concept in mind. *Each animal is a mosaic of traits, many conserved from remote ancestors and others unique to its branch on the animal family tree.* This concept sets the stage for the chapter to follow. There you will trace the evolutionary history of the human species, starting with its mammalian and primate stocks.

KEY CONCEPTS

1. The chordate branch of the family tree for animals includes invertebrate and vertebrate species. All are bilateral animals with a supporting rod for the body (notochord), a dorsal nerve cord, a pharynx, and gill slits in the pharynx wall. These features typically appear in embryos or larvae; some or all may persist in the adults.

2. Existing invertebrate chordates include the tunicates and lancelets.

3. Of eight classes of vertebrates, seven have living representatives. These are jawless fishes, cartilaginous fishes, bony fishes, amphibians, reptiles, birds, and mammals. The other class, the placoderms, became extinct early in vertebrate history.

4. We can identify four major trends in the evolution of certain vertebrate lineages. First, structural support and movement came to depend less on the notochord and more on a backbone. Second, after jaws evolved, the nerve cord evolved into a spinal cord and brain. Third, during the invasion of land, gills became less important than lungs for gas exchange, a trend enhanced by the evolution of more efficient circulatory systems. Fourth, among the pioneers on land, fleshy fins with skeletal supports evolved into limbs, which evolved further among amphibians, reptiles, birds, and mammals.

Characteristics of Chordates

The preceding chapter left off with echinoderms, one of the most ancient lineages of the deuterostome branch of the animal family tree. Dominating this branch are their more recently evolved relatives, the **chordates** (phylum Chordata). Of these bilateral animals, only about 2,100 are, like echinoderms, invertebrates. The vast majority—about 47,000 species—are vertebrates.

Vertebrates are chordates with a backbone, either of cartilage or bone, and a brain located inside a protective chamber of skull bones. The "invertebrate chordates" share certain features with these dominant members of the phylum, but a backbone isn't one of them.

Four features are evident in chordate embryos, and in many species these features persist into adulthood. *First*, a **notochord**, a long rod of stiffened tissue (not cartilage or bone), helps support the body. *Second*, the nervous system has its foundation in a tubular, dorsal **nerve cord** that runs parallel to the notochord and gut. As the embryo develops, the anterior end of the nerve cord increases in mass and becomes modified to form the brain. *Third*, a **pharynx**, which is a muscularized tube, functions in feeding, respiration, or both. The wall of the chordate pharynx has distinctive slits. *Fourth*, a tail forms in embryos and extends past the anus.

Chordate Classification

Biologists group nearly all of the chordates into three subphyla. These are the Urochordata (tunicates and their kin), Cephalochordata (lancelets), and Vertebrata (vertebrates). There are eight classes of vertebrates:

Agnatha	*Jawless fishes*
Placodermi	*Jawed, armored fishes (extinct)*
Chondrichthyes	*Cartilaginous fishes*
Osteichthyes	*Bony fishes*
Amphibia	*Amphibians*
Reptilia	*Reptiles*
Aves	*Birds*
Mammalia	*Mammals*

Appendix I has an expanded classification scheme for the vertebrates. Unit VI provides details of their body plans and functions. Here we become acquainted with major trends in their evolution. We find clues to those trends among the invertebrate chordates.

The embryos of chordates alone have this combination of features: a notochord, a tubular dorsal nerve cord, a pharynx with slits in its wall, and a tail extending past the anus.

Tunicates

The 2,000 species of existing urochordates are baglike animals one to several centimeters long. More often they are called the **tunicates**, after the gelatinous or leathery "tunic" that adults secrete around themselves. Adults of the most common species, the "sea squirts," squirt water through a siphon when something irritates them.

Tunicates live in marine habitats, from the intertidal zone to surprising depths. Most adults remain attached to rocks and other suitably hard substrates. Some live solitary lives; others are colonial. Figure 27.3a shows an adult sea squirt. It develops from a bilateral, swimming larva that resembles a tadpole (Figure 27.3b–d). A larva, recall, is an immature stage between the embryonic and adult stages of an animal life cycle. The firm, flexible notochord of a tunicate larva is a series of cells that acts like a torsion bar. As muscles on one side of the tail or the other contract, the "bar" bends, then it springs back when the muscles relax. The strong, side-to-side motion propels the animal forward. Most fishes use muscles and the backbone for the same kind of motion.

Like some other animals, tunicates are **filter feeders**: they filter food from a current of water that is directed through part of their body. Water flows in through one siphon and passes through **gill slits**, or openings in the thin pharynx wall. Water flows out through a different siphon. Their pharynx also acts as a respiratory organ. Dissolved oxygen is more concentrated in the water than in the blood in vessels adjoining the pharynx, so it diffuses from water into the blood. Carbon dioxide's concentration gradient is such that it diffuses from blood into water leaving the pharynx.

Sea squirts undergo metamorphosis. By this process of development, recall, remodeling and reorganization of body tissues transform an immature stage into the adult. As a sea squirt larva metamorphoses, its tail and notochord disappear, and a tunic forms. The pharynx enlarges, and perforations in its wall get subdivided into many slits. The nerve cord regresses, so that all that remains is a greatly simplified nervous system.

Lancelets

About twenty-five species of fish-shaped, translucent animals called cephalochordates live just offshore, on seafloors around the world. Most are shorter than your little finger. Much of the time they are buried, almost up to their mouth, in sand and sediments. Their common name, **lancelets**, refers to the sharp tapering of their body at both ends. The lancelet body plan, shown in Figure 27.4a, clearly shows the four chordate features. Notice the segmented pattern of muscles on both sides of the notochord. Like tunicate larvae, lancelets use the

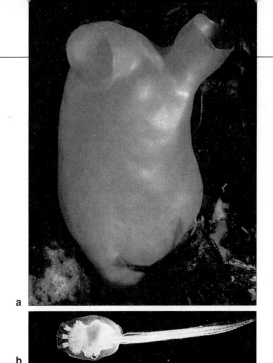

a

b

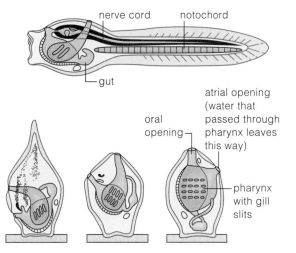

c Section through the tadpole-like larva.

nerve cord   notochord

gut

oral opening

atrial opening (water that passed through pharynx leaves this way)

pharynx with gill slits

d A new larva swims about for a brief period. Metamorphosis begins when its head attaches to a substrate. The notochord, tail, and most of the nervous system are resorbed (recycled to form new tissues). Slits in the pharynx wall multiply. Organs rotate until the openings through which water enters and leaves the pharynx are directed away from the substrate.

Figure 27.3 Example of a tunicate. (**a**) Adult form of a sea squirt. (**b**,**c**) Photograph and diagram of a sea squirt larva. (**d**) Metamorphosis of the larva into the adult.

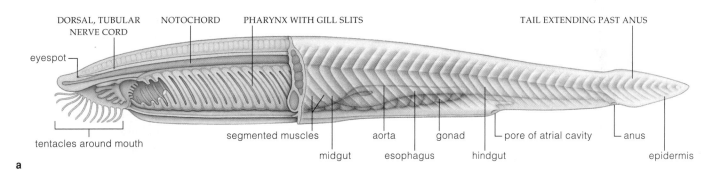

DORSAL, TUBULAR NERVE CORD   NOTOCHORD   PHARYNX WITH GILL SLITS   TAIL EXTENDING PAST ANUS

eyespot

tentacles around mouth

segmented muscles

midgut   esophagus   hindgut

aorta   gonad   pore of atrial cavity   anus   epidermis

a

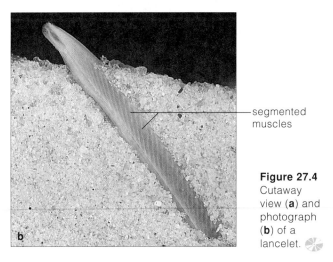

segmented muscles

Figure 27.4 Cutaway view (**a**) and photograph (**b**) of a lancelet.

b

notochord and muscles to produce swimming motions. Unlike tunicates, they have a closed circulatory system (but no red blood cells). A complex brain is nowhere in sight, but the anterior end of the dorsal nerve cord is expanded, and pairs of nerves extend into each muscle segment. The mode of respiration is simple yet effective for such a small body. Dissolved oxygen and carbon dioxide diffuse across the body's thin skin.

Like tunicates, lancelets are filter-feeding animals. Cilia line their mouth cavity and create a current that draws water through it. The water then moves through the pharynx, where food becomes trapped in mucus. The trapped food is delivered to the rest of the gut. By itself, the collective beating of tiny cilia cannot deliver sufficient food to a filter-feeding animal. Delivery also depends on having a large food-trapping surface area. In lancelets, the pharynx provides the area. It is large, relative to the overall body length (Figure 27.4). Besides this, as many as 200 ciliated, food-trapping gill slits perforate its wall. As you will read in sections to follow, the pharynx turned out to be an organ with interesting evolutionary possibilities for the vertebrates.

Tunicate larvae and the lancelets use their notochord and muscles for fishlike swimming movements. Their pharynx, a muscularized tube, has finely divided openings across its thin wall.

The tunicates and lancelets are filter feeders. They use their pharynx to filter microscopic food from a current of water that they draw in through the body. Tunicate larvae also use the pharynx as a respiratory organ, for gas exchange.

Puzzling Origins, Portentous Trends

Did vertebrates arise from tadpole-shaped chordates? It's possible, when you consider an obscure phylum called the hemichordates (*hemi-* meaning half, as in "halfway to chordates"). Evolutionarily, these marine invertebrates are midway between echinoderms and chordates. They don't have a notochord, but they do have a gill-slitted pharynx and a dorsal, tubular nerve cord. Their larvae resemble those of echinoderms *and* tunicates. What if mutations in an echinoderm lineage accelerated the rates at which sex organs developed? If those organs started functioning earlier, in tadpole-shaped *larvae*, then metamorphosis—and the original adult form—could be dispensed with. This idea is not far-fetched. A few existing tunicates look like larvae but have sex organs. Certain amphibians become sexually mature even though they are larvae in some respects, and they can reproduce, generation after generation.

Assuming tadpole-shaped chordates emerged, how did they evolve into vertebrates? The key trends started in fishes (Figures 27.5 and 27.6). One involved a shift from the notochord to reliance on a column of separate, hardened segments: **vertebrae** (singular, vertebra). This was the start of an endoskeleton (internal skeleton) that muscles could work against. *A vertebral column evolved in fast-moving predators, some of which were ancestral to all other vertebrates.*

In a related trend, part of the nerve cord expanded into a brain after **jaws** evolved. These skeletal elements arose through modification of the first in a series of structural elements that supported the gill slits. Jaws meant new feeding possibilities and greater competition

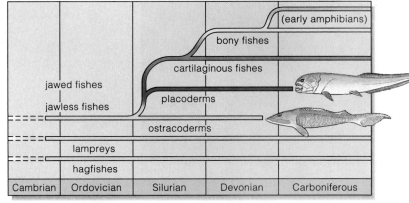

Figure 27.5 Evolutionary history of the fishes.

among predators. Fishes able to smell or see predators or food farther away would now have the advantage—provided that their brain also had evolved enough to process the new information. *A trend toward complex sensory organs and nervous systems did indeed begin in fishes, and it continued among land vertebrates.*

Another trend began when paired fins evolved. **Fins** are appendages that help propel, stabilize, and guide the body in water (compare Figure 27.9). In some lineages, ventral fins became fleshy and equipped with skeletal supports—the forerunners of limbs. *Paired, fleshy fins were the starting point for the legs, arms, and wings that evolved among amphibians, reptiles, birds, and mammals.*

Change in respiration was another trend. Think of the lancelet. Except when buried in sand, oxygen and carbon dioxide just diffuse across its body surface. In most vertebrate lineages, however, **gills** evolved. These respiratory organs have a moist, thin, and much-folded surface, they are richly endowed with blood vessels, and they offer a large surface area for gas exchange. For example, five to seven pairs of a shark's gill slits are continuous with gills extending from the pharynx to the body's surface. When a shark opens its mouth and closes its

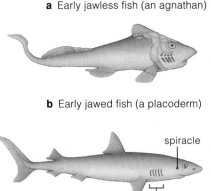

a Early jawless fish (an agnathan)

b Early jawed fish (a placoderm)

spiracle

other gill slits

c Modern jawed fish (a shark)

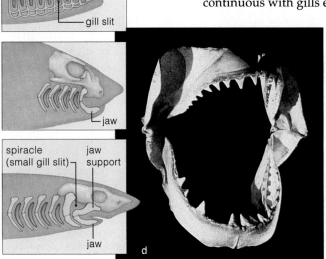

supporting structures

gill slit

jaw

spiracle (small gill slit) jaw support

jaw

d

Figure 27.6 Comparison of gill-supporting structures in jawless fishes and jawed fishes. In the placoderms and other early jawed vertebrates, cartilage supported the mouth's rim. In modern jawed fishes, a gill slit between the jaws and a nearby supporting element serves as a spiracle, an opening through which water is drawn. In (**a**), the gill supports are just under the skin. In (**b**) and (**c**), they are internal to the gill surface. (**d**) Jaws of a modern shark.

external gill openings, the pharynx expands. Oxygen diffuses *from* water in the mouth into the gills, as carbon dioxide diffuses *into* that water. Muscles constrict the pharynx and so force oxygen-depleted, carbon dioxide-enriched water out through the gill slits.

As some fishes became larger and more active, gills became more efficient. But gills can't work out of water; they stick together unless water flows through them and keeps them moist. In fishes ancestral to the land vertebrates, pouches developed from the gut wall. The pouches evolved into **lungs**—internally moistened sacs for gas exchange. In a related trend, modifications to the heart enhanced the pumping of oxygen and carbon dioxide through the body. *Ancestors of land vertebrates relied less on gills and more on lungs. And more efficient circulatory systems accompanied the evolution of lungs.*

The First Vertebrates

Figure 27.5 gives a time frame for vertebrate evolution. Free-swimming species originated in Cambrian times and gave rise to two kinds of fishes—those without and those with jaws. One or the other kind probably gave rise to all vertebrate lineages that followed. The earliest jawless fishes (Agnatha) included **ostracoderms**. Figure 27.6a shows one of those bottom-dwelling filter feeders. Their skeleton probably consisted of a notochord and a protective covering for the enlarged brain. Armorlike plates on the body consisted of bony tissue and dentin, a hardened tissue still present in vertebrate teeth. The armor might have been useful against the giant pincers of sea scorpions, but it apparently wasn't much good against jaws. Ostracoderms disappeared when jawed fishes began their adaptive radiations.

Placoderms were among the first fishes with jaws and paired fins (Figure 27.6b). These bottom-dwelling scavengers or predators had a notochord reinforced with bony elements. The first gill-supporting structures were enlarged and had bony projections something like teeth, and they functioned as jaws. Before placoderms, feeding strategies had been limited to filtering, sucking, or rasping away at food material. When the placoderms started biting and tearing off large chunks of prey, they set off an evolutionary race of offensive and defensive adaptations that has continued to the present.

Placoderms diversified, then became extinct during the Carboniferous period. New kinds of predators, the cartilaginous and bony fishes, replaced them in the seas. We will consider the living descendants of the new predators after a look at existing jawless fishes.

The emergence of a vertebral column, jaws, paired fins, and lungs was pivotal in the evolution of certain vertebrates.

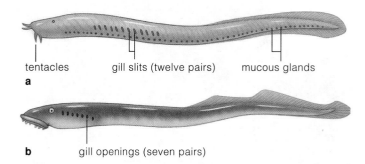

MAMMALS
BIRDS
REPTILES
AMPHIBIANS
BONY FISHES
CARTILAGINOUS FISHES
ANCESTRAL VERTEBRATES   JAWLESS FISHES

The **hagfishes** and the **lampreys** are living descendants of some early jawless fishes. All seventy-five or so of the known species have a cylindrical body, a cartilaginous skeleton, and no paired fins (Figure 27.7). Most are less than 1 meter (3.3 feet) long.

Hagfishes live together in groups on the sediments of continental shelves. They prey on polychaete worms or scavenge for weakened or dead organisms. Even though they lack jaws, hagfishes do eat well. They have sensory tentacles around the mouth and a tongue that rasps soft tissues from prey. They defend themselves by secreting copious amounts of sticky, smelly, slimy mucus from a series of glands along their body.

Lampreys are specialized predators that are almost parasites. Their suckerlike oral disk has horny, toothlike parts that rasp flesh from prey. Some types latch onto salmon, trout, and other commercially valuable fishes,

tentacles   gill slits (twelve pairs)   mucous glands
a

b   gill openings (seven pairs)

Figure 27.7 Body plan of (**a**) a hagfish and (**b**) a lamprey. (**c**) A lamprey's toothed oral disk, pressed against aquarium glass.

then suck out juices and tissues. Just before the turn of the century, lampreys began invading the Great Lakes of North America. Their introduction resulted in the collapse of populations of lake trout and other large, valued fishes, as it has done in other aquatic habitats.

Hagfishes and lampreys get along without jaws; they latch onto prey with their efficient, specialized mouthparts.

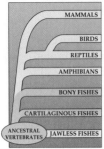

Few of us make a career of studying life beneath the surface of the seas and other bodies of water, so fishes are not widely recognized as being the world's dominant vertebrates. Their numbers exceed those of all other vertebrate groups combined. They also show far more diversity than the other groups; we have identified more than 21,000 species of bony fishes alone.

The form and behavior of a fish tell us something about the challenges it faces in water. For example, being about 800 times more dense than air, water resists fast movements. As an adaptation to this constraint, sharks and other predatory marine fishes are streamlined for pursuit (Figure 27.8a). Their long, trim body reduces friction; their tail muscles are organized for propulsive force and forward motion. By contrast, some bottom-dwelling fishes, such as the rays, have a flattened body (Figure 27.8b). Theirs is not a high-speed body plan; it is good for hiding from predators or prey.

A motionless trout, suspended in shallow water, is another example of adaptation to water's density. Like many fishes, it maintains neutral buoyancy with a **swim bladder**, an adjustable flotation device that exchanges gases with blood. When a trout gulps air at the water's surface, it is adjusting the volume of its swim bladder.

Cartilaginous Fishes

Cartilaginous fishes (Chondrichthyes) include about 850 species of skates, sharks, and chimaeras (Figure 27.8). These marine predators have pronounced fins, a skeleton of cartilage, and five to seven gill slits on both sides of the pharynx. Most species have a few scales or many rows of them. **Scales** are small, bony plates at the body surface. Fish scales often protect the body without weighing it down.

The skates and rays are mainly bottom dwellers with flattened teeth suitable for crushing hard-shelled prey. Both have enlarged fins that extend onto the side of the head. The largest species, the manta ray, measures up to six meters from fin tip to fin tip. A venom gland in the tail of sting rays probably helps deter predators. Other rays have electric organs in the tail or fins that can stun prey with as much as 200 volts of electricity.

At fifteen meters from head to tail, some sharks are among the largest living vertebrates. As Figure 27.6d shows, sharks have formidable jaws. They continually shed and replace their sharp, triangular teeth (modified scales), which they use to grab prey and rip off chunks of flesh. The relatively few shark attacks on humans give the whole group a bad reputation. Surfboards with legs dangling over the side are a recent temptation for some sharks, which have been dining on invertebrates, fishes, and marine mammals for many millions of years.

The thirty or so species of chimaeras feed mostly on mollusks. With their bulky body and slender tail, they look a bit like a rat; hence the common name, ratfishes. They have a venom gland in front of the dorsal fin.

Bony Fishes

Bony fishes (Osteichthyes) are the most numerous and diverse vertebrates. They make up all but 4 percent of the existing species of modern fishes. Their ancestors arose during the Silurian and soon gave rise to three lineages: the ray-finned fishes, lobe-finned fishes, and lungfishes. Descendants of the early forms radiated into nearly all aquatic habitats.

Body plans vary greatly. Marine predators typically have a torpedo shape, a flexible body, and strong tail

Figure 27.8 Cartilaginous fishes: (**a**) shark, (**b**) blue-spotted reef ray, and (**c**) chimaera, sometimes called a ratfish.

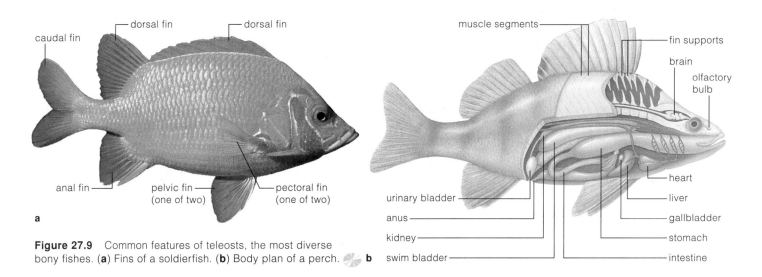

Figure 27.9 Common features of teleosts, the most diverse bony fishes. (**a**) Fins of a soldierfish. (**b**) Body plan of a perch.

Figure 27.10 A few variations on the body plan of bony fishes as shown in Figure 27.9. (**a**) Moray eel. (**b**) Sea horse. (**c**) Long-nose gar. (**d**) Coelacanth (*Latimeria*), a "living fossil" that shares many traits with early lobe-finned fishes.

fins that aid in swift pursuit. Many reef dwellers are small finned and box shaped; they move easily through narrow spaces. Those with an elongated, flexible body, including the moray eel, wriggle through mud and slip into concealing crevices. Sea horses and many bottom-dwelling species have a cryptic body shape that helps conceal them from prey or predators. Figures 27.9 and 27.10 show examples of these varied body plans.

In ray-finned fishes, rays derived from the dermis (one of the skin's layers) support paired fins. Most of these fishes have highly maneuverable fins and light, flexible scales, both of which contribute to a capacity for complex movements. Their efficient respiratory system rapidly delivers oxygen to metabolically active tissues. One group of ray-finned fishes resembles their early

ancestors. It includes sturgeons and paddlefishes of the Mississippi River basin. Teleosts, the most abundant group, include salmon, tuna, rockfish, catfish, perch, minnows, moray eels, flying fish, sculpins, blennies, scorpionfish, and pikes. All are thin scaled or scaleless.

By contrast, only one species of lobe-finned fishes and three genera of lungfishes made it to the present. As their name suggests, a **lobe-finned fish** is unique in having paired fins that incorporate fleshy extensions from the body. As you will see next, neither it nor the lungfishes have changed much from their ancient stock.

Of all existing vertebrates, the bony fishes are the most spectacularly diverse and the most abundant.

AMPHIBIANS

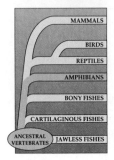

Origin of Amphibians

The lobe-finned fishes (Figure 27.10*d*) are living fossils, relics of a time when the first vertebrates moved onto land. Remember the earlier description of living conditions during the Devonian? Sea levels rose and fell repeatedly, and swamps fringing the coasts flooded and drained many times. Ancestors of lobe-finned fishes evolved in those trying times. They probably used their lobed fins to pull themselves from dried-up ponds to ones that were still habitable. What else made their pond-to-pond lurchings possible? They had sac-shaped outpouchings from the wall of the esophagus, which is a tube leading to the stomach. The sac began to be used as a way to supplement gas exchange. In other words, the ancestors of lobe-finned fishes had simple lungs.

Existing lungfishes provide more clues to how the ancestral forms might have made it through stressful times. They live in stagnant water but surface to gulp air. In dry seasons, when streams dwindle to mud, they encase themselves in a slathering of mud and slime that keeps them from drying out until the next rainy season.

Devonian lobe-finned fishes lurched over land only as a way of reaching more hospitable ponds. Yet the very act of traveling out of water favored the evolution of stronger fins and more efficient lungs. Among the evolving forms were the ancestors of amphibians. An **amphibian** (Amphibia) is a vertebrate with a body plan and reproductive mode somewhere between fishes and reptiles. Most kinds have a largely bony endoskeleton, and they have four legs (or four-legged ancestors).

Early amphibians were spending time on land by the close of the Devonian (Figure 27.11*a*). For them, life in those drier habitats was dangerous—and promising. Temperatures shifted more on the land than in water, air didn't support the body as well as water did, and water was not always plentiful. However, air has far more oxygen, and the lungs continued to be modified in ways that enhanced oxygen uptake. Also, circulatory systems became better at rapidly distributing oxygen to cells throughout the body. Both modifications increased the energy base for more active life-styles.

New sensory information also challenged the early amphibians. Swamp forests supported vast numbers of edible insects and other invertebrate prey. Animals with good vision, hearing, and balance—the senses that are most advantageous on land—were favored. And brain regions concerned with those senses expanded.

All of the salamanders, frogs, toads, and caecilians alive today are descended from those first amphibians. None has escaped the water entirely. Even when gills or lungs are present, amphibians can use their thin skin as a respiratory surface (that is, for gas exchange). But respiratory surfaces must be kept moist, and skin dries easily in air. Some species still live their entire lives in water; others lay their eggs in water or produce aquatic larvae. Even species that have adapted to habitats on land must lay their eggs in moist places.

Figure 27.11 (**a**) *Ichthyostega*, one of the first of the Devonian amphibians. Fossils of this species have been recovered in Greenland. The skull, deep tail, and fins were decidedly fishlike. Unlike fishes, this species had four limbs adapted for moving on land, and a short neck intervened between its head and the rest of the body. Its vertebral column and rib cage were adapted to support the body's weight out of water. (**b,c**) Proposed evolution of skeletal elements inside the lobed fins of certain fishes into the limb bones of early amphibians.

a

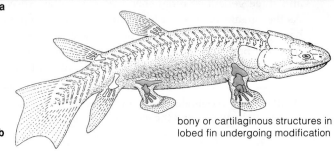

b   bony or cartilaginous structures in lobed fin undergoing modification

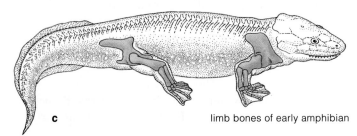

c   limb bones of early amphibian

Salamanders

Diverse salamanders and their kin, the newts, live in the world's north temperate regions and tropical parts of Central and South America. Most are less than fifteen centimeters long. And most have legs at right angles to the body, with forelimbs and hindlimbs about the same size (Figure 27.12*a*). Like fishes and early amphibians, salamanders bend from side to side when they walk:

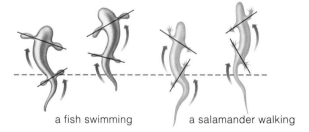

a fish swimming   a salamander walking

Probably the first four-legged vertebrates also walked this way. Both the larvae and adults are carnivores. Adults of some species retain several larval features. For example, the Mexican axolotl retains the external gills of the larval form, and the development of its teeth and bones is arrested at an early stage. As in some other species, its larvae are sexually precocious; they breed.

Frogs and Toads

With more than 3,000 species, frogs and toads are the most successful amphibians (Figure 27.12*b,c*). Their long

Figure 27.12 Amphibians. (**a**) Terrestrial stage in the life cycle of a red-spotted salamander. (**b**) A frog, splendidly jumping. (**c**) An American toad. (**d**) A caecilian.

hindlimbs and powerful muscles allow them to catapult into the air or barrel through the water. Most frogs flip a sticky-tipped, prey-capturing tongue from their mouth. An adult eats just about any animal it can catch; only its head size limits the size of prey. Skin glands of some species produce toxins, and poisonous types often have bright warning coloration. The skin of frogs, including the South African clawed frog (*Xenopus laevis*), harbors antibiotics against many pathogens in the water. In Unit VI, you will have opportunities to take a look at how frogs are put together and how they function.

Caecilians

The ancestors of caecilians lost their limbs and most of their scales. They gave rise to decidedly worm-shaped amphibians (Figure 27.12*d*). Nearly all of the 160 or so existing species burrow through soft, moist soil as they pursue insects and earthworms. A few live in shallow freshwater habitats. Adults of most species are blind.

Amphibians are halfway between fishes and reptiles in their body form and behavior. Regardless of how far they venture onto land, they have not fully escaped dependency on water.

THE RISE OF REPTILES

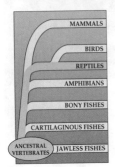

Reptiles (Reptilia) arose from amphibians during the late Carboniferous. Of all the vertebrates, they were the first to escape dependency on standing water. They did so mainly through four adaptations that clearly distinguish them from fishes and amphibians.

First, the reptiles have tough, dry, scaly skin that restricts loss of water from the body (Figure 27.13a). Second, fertilization is internal. A copulatory organ can deposit sperm into a female's body; sperm do not need free water to reach eggs. Third, reptilian kidneys are good at conserving water. Fourth, most species of reptiles produce **amniote eggs**, within which embryos develop to an advanced stage before being hatched or born into dry habitats. Amniote eggs contain membranes that retain water and protect or provide metabolic support to the embryo. Most of them have a leathery or calcified shell (Figure 27.13b,c). We will take a closer look at the structure and development of amniote eggs in Chapter 45.

Compared to amphibians, the early reptiles chased prey with far greater cunning and speed. With their well-muscled jawbones and formidable teeth, reptiles could seize and apply sustained, crushing force on prey. Their prey included other vertebrates as well as insects. Also, reptilian limbs generally were more efficient at supporting the trunk of the body on land. For many species, the nervous system increased in complexity. A reptile's brain is small compared to the rest of the body mass, but it governs complex forms of behavior that are unknown among amphibians. For example, the cerebral cortex is the most complex part of the forebrain. Here, information from diverse sensory organs is integrated and stored, and commands for responses are issued. The cerebral cortex evolved first among the reptiles.

b

c

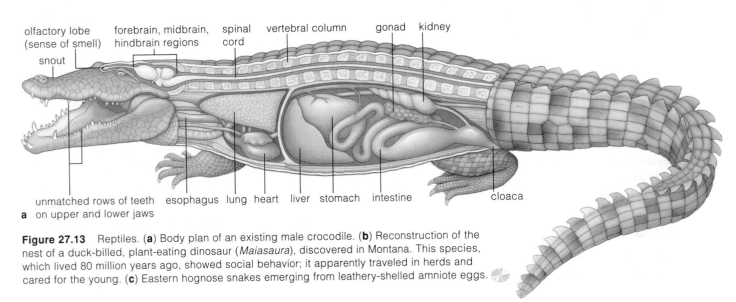

a

Figure 27.13 Reptiles. (**a**) Body plan of an existing male crocodile. (**b**) Reconstruction of the nest of a duck-billed, plant-eating dinosaur (*Maiasaura*), discovered in Montana. This species, which lived 80 million years ago, showed social behavior; it apparently traveled in herds and cared for the young. (**c**) Eastern hognose snakes emerging from leathery-shelled amniote eggs.

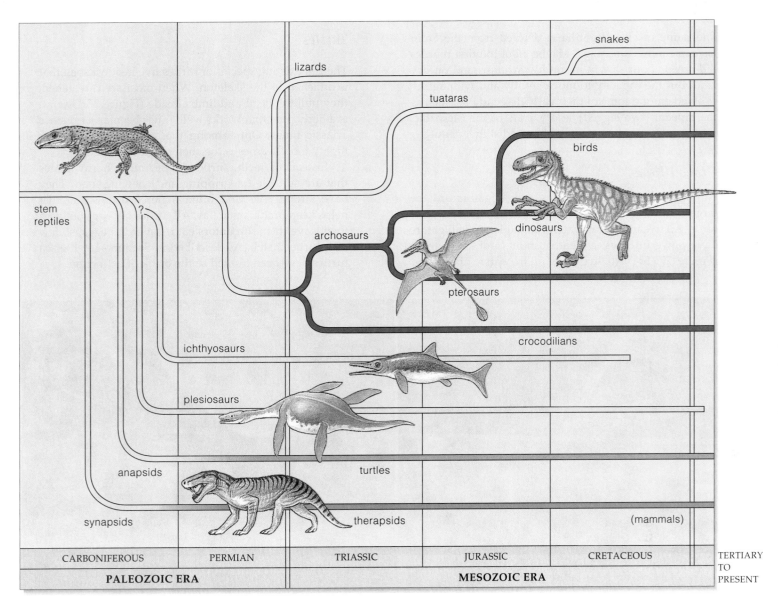

					TERTIARY TO PRESENT
CARBONIFEROUS	PERMIAN	TRIASSIC	JURASSIC	CRETACEOUS	
PALEOZOIC ERA		**MESOZOIC ERA**			

Figure 27.14 Evolutionary history of the reptiles.

The group of reptiles called crocodilians were the first vertebrates with a muscular, four-chambered heart fully separated into two halves. (Blood enters the first chamber of each half, and the second chamber pumps it out.) As Chapter 39 describes, such separation permits oxygen-rich blood to travel from the lungs to the rest of the body, and oxygen-poor blood from the body to the lungs, in two independent circuits. Also, gas exchange across the skin, so vital for amphibians, was abandoned by reptiles—nearly all of which depend on lungs.

The early reptiles underwent adaptive radiations that led to fabulously diverse forms. The group called dinosaurs evolved during the Triassic and for the next 125 million years were the dominant land vertebrates. The fossilized nests of one kind (*Maiasaura*) contain not only eggs but also juveniles a few months old, at most (Figure 27.13*b*). Such evidence suggests that at least some dinosaurs showed parental behavior; that is, they took care of their young during a period of dependency. (*Maiasaura* means "good mother lizard.")

When the Cretaceous ended abruptly, so did the last dinosaurs (Sections 21.6 and 21.7). As you can see from Figure 27.14, the reptilian groups that made it to the present are crocodilians, turtles, tuataras, snakes, and lizards.

Reptiles, with their tough, scaly skin, reliance on internal fertilization, water-conserving kidneys, and amniote eggs, were the first vertebrates to escape dependency on free-standing water in their habitats.

The move onto land also required major modifications in the nervous, circulatory, and respiratory systems.

The name of class Reptilia is derived from the Latin *repto*, meaning "to creep." Maybe all of the first reptiles did creep about slowly in muddy swamps and on dry land. But the capacity to move swiftly and with agility evolved among some of their early descendants, such as the bipedal (two-legged) *Velociraptors* of the Mesozoic. Existing descendants race, lumber, and slither about.

Crocodilians

Among the modern crocodiles and alligators are the largest living reptiles, and the closest living relatives of birds. All live in or near water. Crocodiles and alligators have powerful jaws, a slender snout, and sharp teeth (Figure 27.15*a*). All live near or in water. The feared

Turtles

The 250 existing species of turtles live inside a shell that is attached to the skeleton. When most are threatened, they pull the head and limbs inside (Figure 27.15*b*,*c*). It is a body plan that works well; it has been around since Triassic times. Only among the sea turtles and other highly mobile types is the shell reduced in size.

Instead of teeth, turtles have tough, horny plates that are suitable for gripping and chewing food. They have strong jaws and often a fierce disposition that helps keep predators at bay. All turtles lay eggs on land, then leave them. Predators eat most of the eggs, so few new turtles hatch. As described in Section 41.9, the sea turtles have been hunted to the brink of extinction.

a

b

Figure 27.15 Reptiles. (**a**) Spectacled caiman. As is true of other crocodilians, its peglike upper teeth do not match up with peglike lower ones. The body plan and life-styles of crocodilians haven't changed much for nearly 200 million years. Their future is not rosy. Housing developments are encroaching on many of their habitats, and their belly skin is in demand for wallets, shoes, and handbags. (**b**) A heavy-shelled Galápagos tortoise. (**c**) Section through the skeleton and shell of a turtle, with head withdrawn and extended. (**d**) Frilled lizard, flaring a large ruff of neck skin as defensive behavior. (**e**) A rattlesnake in mid-strike. (**f**) Tuatara (*Sphenodon*).

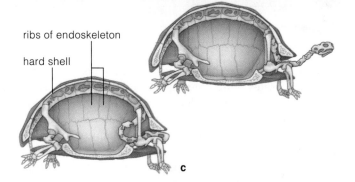

ribs of endoskeleton

hard shell

c

Lizards and Snakes

"man-eaters" of southern Asia and the Nile crocodiles weigh up to 1,000 kilograms. They drag a mammal or bird into the water, tear it apart by violently spinning about, then gulp down the torn chunks. Crocodilians are able to adjust body temperature by behavioral and physiological mechanisms, just as other reptiles and birds can do. Like birds, they also show complex social behavior, as when the parents guard nests and assist their hatchlings into the water.

About 95 percent of the existing reptiles, the lizards and snakes, are distant relatives of the dinosaurs. Most are small, but the Komodo monitor lizard is big enough to hunt young water buffalo. The longest modern snake would stretch across ten yards of a football field.

Most of the 3,750 kinds of lizards are insect eaters of deserts and tropical forests. They include aggressive, adhesive-toed geckos and iguanas; the photograph of the Unit VI introduction shows one of the more colorful types. Most lizards grab prey with small, peglike teeth. Chameleons capture prey with accurate flicks of their tongue, which is longer than their body (Section 33.4).

Being small themselves, most lizards are prey for many other animals. Some attempt to startle predators and intimidate rivals by flaring their throat fan (Figure 27.15d). Many give up their tail when a predator grabs them. The detached tail wriggles for a bit and may be distracting enough to permit a getaway.

During the early Cretaceous, some of the short-legged, long-bodied lizards gave rise to the elongated, limbless snakes. Most of the 2,300 existing species slither in S-shaped waves, much like salamanders. "Sidewinders"

f

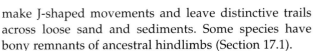

d   e

make J-shaped movements and leave distinctive trails across loose sand and sediments. Some species have bony remnants of ancestral hindlimbs (Section 17.1).

Snakes have highly movable jaws; some swallow animals wider than they are. The pythons and boas coil around their prey and suffocate it into submission before eating. Fanged types, including rattlesnakes and coral snakes, bite and subdue prey with venom (Figure 27.15e). Snakes usually do not act aggressively toward humans. However, each year in the United States alone, rattlesnakes (one of the pit vipers) and other poisonous types bite 8,000 or so people and kill about 12 of them. In India, king cobras and other types bite 200,000 or so people and kill about 9,000 of them.

Even the most feared snakes are vulnerable during their life cycle; birds and other predators relish snake eggs. The female snakes store sperm and lay several clutches of fertilized eggs at intervals after they mate, which improves the odds that at least some will hatch.

Tuataras

Only two species of tuataras are still around, on small, windswept islands near New Zealand (Figure 27.15f). Their body plan has not changed much since Mesozoic times. The tuataras resemble lizards, but their lineage is more recent. At the top of their head is a third "eye," with retina, lens, and nerves to the brain. Being covered with skin, it can only register changes in daylength and light intensity. Does it have roles in hormonal controls over reproduction? Maybe (compare Section 37.8). The tuataras, like turtles, may live longer than sixty years. They engage in sex only after they are twenty years old.

Existing crocodilians are the closest relatives of dinosaurs and birds. Turtles, like tuataras, have changed little in body plan for millions of years. About 95 percent of all living reptiles are lizards or snakes.

BIRDS

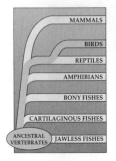

a

Of all organisms, only **birds** (Aves) grow feathers (Figure 27.16). **Feathers**, derived from skin, are lightweight structures that birds use in flight, as body insulation, or both. Judging from the fossil record, birds are descended from tiny reptiles that ran around on two legs 160 million years ago. *Archaeopteryx* was on or near the lineage leading to modern birds. Figure 17.7 shows how it had many reptilian traits as well as feathers and other avian traits.

In many respects, birds still resemble reptiles. For example, they have scales on their legs and a number of the same internal structures. They, too, lay their eggs and commonly engage in parental behavior, such as guarding the nest. One of Darwin's champions, Thomas Huxley, was the first to argue that birds are glorified reptiles. Today, many biologists do indeed classify birds as a branching from the reptilian lineage.

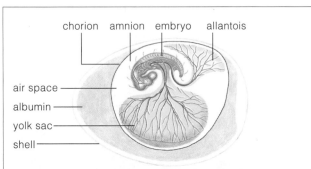

b

Figure 27.16 Characteristics of birds. (**a**) Flight. Of all living vertebrates, only birds and bats fly by flapping their wings. (**b**) All birds lay hard-shelled eggs. The photograph below this generalized sketch shows speckled eggs of a magpie.

(**c**) Feathers, the key defining characteristic of birds. The flamboyant plumage of this male pheasant is an outcome of sexual selection. This bird is a native of the Himalaya Mountains of India, and it is an endangered species. As is the case for many other kinds of birds, its jewel-colored feathers end up adorning humans—in this case, on the caps of native tribespeople.

(**d**) Many birds, including these Canada geese, show migratory behavior. *Animal migration* is a recurring pattern of movement between two or more locations in response to environmental rhythms. For example, seasonal change in daylength is a cue that acts on internal timing mechanisms (biological clocks), which in turn trigger physiological and behavioral changes. Such changes induce migratory birds to make round trips between distant regions that differ in climate. Canada geese spend the summer nesting in marshes and lakes in the northern United States and Canada. Their wintering grounds are in New Mexico and other parts of the southern United States.

c

d

There are almost 9,000 named species of birds. They show stunning variation in their body size, proportions, coloration, and capacity for flight. One of the smallest hummingbirds barely tips the scales at 2.25 grams (0.08 ounce). The largest existing bird, the ostrich, weighs about 150 kilograms (330 pounds). Ostriches cannot fly, but they are impressively long-legged sprinters (Figure 17.2). Many birds, such as warblers and other perching types, differ markedly in feather coloration and in their territorial songs. Bird songs and other complex social behaviors are topics of later chapters.

Flight demands high metabolic rates, which require a good flow of oxygen through the body. Just as you do,

birds have a large, durable, four-chambered heart. The heart pumps oxygen-enriched blood to the lungs and to the rest of the body along separate routes, as it probably did in their reptilian ancestors. Also, a bird's respiratory system incorporates a unique ventilating apparatus that enormously enhances oxygen uptake. Bird respiration is described in Section 41.3.

Flight also demands an airstream, low weight, and a powerful downstroke that can provide lift (a force at right angles to the airstream). The bird wing, a modified forelimb, consists of feathers and lightweight bones attached to powerful muscles. The bones weigh very little (because of profuse air cavities in the bone tissue).

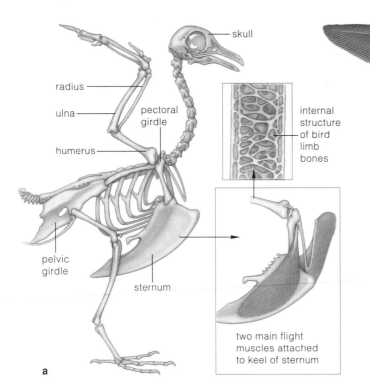

Figure 27.17 (**a**) Body plan of a typical bird. Flight muscles attach to the large, keeled breastbone (sternum). (**b**) A bird wing is a complex system of lightweight bones and feathers. Feathers gain strength from a hollow central shaft and from tiny barbules that are interlocked in a latticelike array.

For example, the skeleton of a frigate bird, which has a seven-foot wingspan, weighs only four ounces. That is less than the feathers weigh! A bird's flight muscles attach to an enlarged breastbone, or sternum, and to upper limb bones adjacent to it (Figure 27.17). Contraction of the muscles creates the powerful downstroke required for flight. The wings, with their long flight feathers, serve as the airfoils. Usually, a bird spreads out long feathers on a downstroke and thus increases the size of the surface pushing against air (Figure 27.16*a*). On the upstroke, it folds the feathers somewhat, so that each wing presents the least possible resistance to the air.

Of all animals, birds alone have feathers, which they use in flight, in heat conservation, and in socially significant communication displays.

THE RISE OF MAMMALS

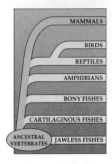

Distinctly Mammalian Traits

Mammals are vertebrates with hair and mammary glands (hence the name of class Mammalia, from the Latin *mamma*, which means breast). Of all organisms, they are the only ones having these traits. Females feed their young with milk, a nutritious fluid that mammary glands secrete into ducts that lead to openings at the body's ventral or anterior surface (Figure 27.18*a*).

Mammals differ greatly in the amount, distribution, and type of hair. A few aquatic mammals, including the whales, lost most of their hair after their land-dwelling ancestors returned to the seas. But if you look closely, you see that even a whale snout has some "whiskers." These modified hairs serve sensory functions, just as they do in dogs or cats. More typically, mammals have a furry coat of *underhair* (a dense, soft, insulative layer that can trap heat) and coarser, longer *guard hairs* that protect the insulative layer from wear and tear (Figure 27.18*b*). When wet, the guard hairs of platypuses, otters, and other aquatic mammals become matted down like a protective blanket, and the underhair stays dry.

Mammals also are distinctive in that they care for the young for an extended period, and they also serve as models for their behavior. Young mammals are born with a capacity to learn and to repeat behaviors that have survival value. It is among mammals that we find the most stunning shows of **behavioral flexibility**—a capacity to expand on basic activities with novel forms of behavior—although the trait is far more pronounced in some species than in others. Behavioral flexibility coevolved with expansion of the brain, especially the cerebral cortex. Remember, this outermost layer of the forebrain receives, processes, and stores information from sensory structures, and it issues commands for complex responses. We find the most highly developed cerebral cortex among the mammals called primates. Chapter 28 starts with their story.

Unlike reptiles, which generally swallow their prey whole, most mammals secure, cut, and sometimes chew food before swallowing. They also differ from reptiles in **dentition** (that is, in the type, number, and size of teeth). Mammals have four distinctive types of upper and lower teeth that match up and work together to crush, grind, or cut food (Figure 27.18*c*). Their incisors (flat chisels or cones) nip or cut food. Horses and other grazing mammals have large incisors. Canines, with piercing points, are longest in meat-eating mammals. Premolars and molars (cheek teeth) are a platform with surface bumps (cusps); these crush, grind, and shear food. If a mammal has large, flat-surfaced cheek teeth, you can safely bet its ancestors evolved in places where fibrous plants were abundant foods. As you will see from the chapter that follows, fossilized jaws and teeth from early primates and species on the evolutionary road to modern humans offer many clues to their life-styles.

a

b

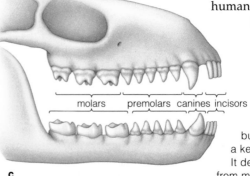

c

molars   premolars canines incisors

Figure 27.18 Three distinctly mammalian traits. (**a**) A human baby busily demonstrating a key defining feature: It derives nourishment from mammary glands. (**b**) A pair of juvenile raccoons displaying their fur coat. (**c**) Unlike the teeth of their reptilian ancestors, the upper and lower rows of mammalian teeth match up. (Compare Figure 27.15*a*.)

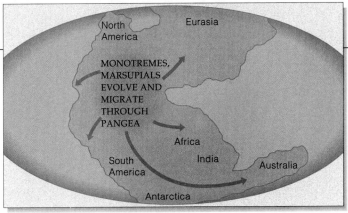

a About 150 million years ago (during the Jurassic)

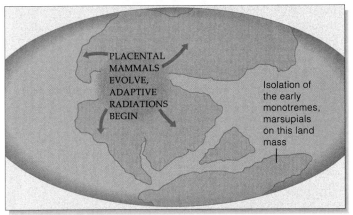

b Between 100 and 85 million years ago (Cretaceous times)

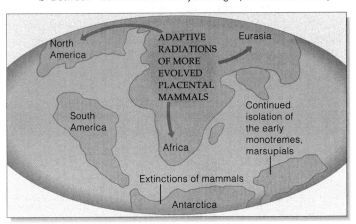

c About 20 million years ago (Miocene)

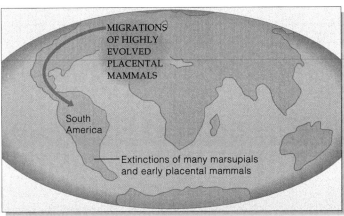

d About 5 million years ago (Pliocene)

Figure 27.19 Adaptive radiations of mammals.

Mammalian Origins and Radiations

More than 200 million years ago, during the Triassic, a genetic divergence from small, hairless reptiles called synapsids gave rise to the **therapsids**, the ancestors of mammals (Figure 27.14). And by Jurassic times, diverse plant-eating and meat-eating mammals called **therians** had evolved. Most were the size of a mouse, and they were endowed with hair and major changes in the jaws, teeth, and body form. For instance, their four limbs were positioned upright, under the body's trunk. This skeletal arrangement made it easier to walk erect, but a trunk higher from the ground was not as stable. At this time the cerebellum, a brain region dealing with the body's balance and spatial positioning, started expanding.

The therians coexisted with dinosaurs through the Cretaceous. When the last of the dinosaurs vanished, diverse adaptive zones opened up for three lineages of those previously inconspicuous mammals (Sections 19.5 and 21.8). New opportunities, differences in traits, and key geologic events put those lineages—the **monotremes** (egg-laying mammals), **marsupials** (pouched mammals), and **eutherians** (placental mammals)—on very different paths to the present. Compared with the monotremes and marsupials, which retained many archaic traits, the placental mammals had the competitive edge. They had higher metabolic rates, more precise ways of regulating body temperature, and a new way of nourishing their developing embryos. You will read about these traits in later chapters. For now, it is enough to know that the placental mammals radiated into many new adaptive zones, at the expense of their less competitive relatives.

As Figure 27.19a shows, the ancestors of monotremes and marsupials had entered the southern part of the supercontinent Pangea by the late Jurassic. Following the breakup of Pangea, the ones on the land mass that would become Australia were already isolated from the ancestors of placental mammals, which were evolving elsewhere (Figure 27.19b). In the future South America, monotremes were replaced by marsupials and early placental mammals, which evolved independently of their relatives on other continents. When a land bridge rejoined South and North America during the Pliocene, highly evolved placental mammals radiated southward and rapidly drove many of those previously isolated mammals to extinction. Only opossums and a few other species successfully invaded land in the other direction.

Mammals alone have hair and mammary glands. They have distinctive dentition, a highly developed nervous system, and a notable capacity for behavioral flexibility.

Much of mammalian history was a matter of luck, of species with particular traits being in the right or wrong places on a changing geologic stage at particular times.

Cases of Convergent Evolution

Evolutionarily distant, geographically isolated lineages have sometimes evolved in similar ways in similar habitats, so that they come to resemble each other in form and function. We call this **convergent evolution**, as defined in Section 20.4. With their interesting history, the three lineages of mammals offer classic examples.

The only living monotremes are two species of spiny anteaters and the duck-billed platypus described at the start of the chapter. One spiny anteater (*Tachyglossus*) is common in Australia; the other lives in the mountains of New Guinea. Both are small, burrowing mammals that feed mostly on ants. Like porcupines, they bristle with protective spines that are modified hairs (Figure 27.20*a*). The females lay eggs, like those of the platypus. Unlike their relative, they do not dig nests. A single egg is incubated, and the youngster suckles and completes development in a skin pouch that forms temporarily on the mother's ventral surface by muscle contractions.

Of the 260 existing species of marsupials, most are native to Australia and nearby islands; a few live in the Americas (Figure 27.20*b,c*). The tiny, blind, and hairless newborns suckle and complete development inside a permanent pouch on the female's ventral surface.

Every other existing descendant of the therians is a placental mammal. The **placenta** is a spongy tissue of maternal and fetal membranes. It forms in a pregnant

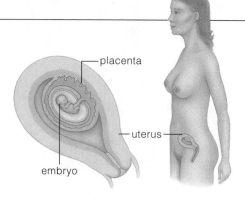

Figure 27.21
Location of the placenta in a human female.

female's uterus, the chamber in which embryos develop in relative freedom from harsh conditions in the outside world (Figure 27.21). Across the placenta, the female sends nutrients and oxygen to an embryo and removes its metabolic wastes. Placental mammals grow faster in the uterus than marsupials do in a pouch, and many are fully formed at birth. Figure 27.22 shows a few species; Appendix I lists the major groups. The body form and function, behavior, and ecology of the mammals will occupy us in later chapters. For now, simply reflect on Table 27.1, which lists some convergences among them.

Human Impact on Mammalian Diversity

In terms of species diversity, mammals don't even begin to approach, say, the mollusks or arthropods. Yet in terms of size, body form, functioning, and life-styles, the 4,500 known species are breathtaking. (In size alone they range from the 1.5-gram Kitti's hognosed bat to

Figure 27.20 (a) Spiny anteater (*Tachyglossus*), a monotreme. Marsupials: (b) From the southeastern United States, a female opossum and her young. (c) From Australia, a young kangaroo (joey) in its mother's pouch.

Table 27.1 Convergences Among Mammalian Groups

Life Style	Home	Mammalian Family
Aquatic invertebrate eater	North America Central America Australia	Water shrew (Soricidae) Water mouse (Cricedidae) Platypus (Ornithorrhynchidae)
Land-dwelling carnivore	North America Australia	Wolf (Canidae) Tasmanian wolf (Thylacinedae)
Land-dwelling ant eater	South America Africa Australia	Giant anteater (Myrmecophagidae) Aardvark (Orycteropodidae) Spiny anteater (Tachyglossidae)
Ground-dwelling leaf, tuber eater	North America South America Eurasia	Pocket gopher (Geomyidae) Tuco-tuco (Ctenomyidae) Mole rat (Spalacidae)
Tree-dwelling leaf eater	South America Africa Madagascar Australia	Howler monkey (Cebidae) Colobus monkey (Cercopithecidae) Woolly lemur (Indriidae) Koala (Phascolarctidae)
Tree-dwelling nut, seed eater	Southeast Asia Africa Australia	Flying squirrel (Sciuridae) Flying squirrel (Anomaluridae) Flying squirrel (Phalangeridae)

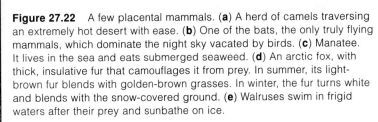

Figure 27.22 A few placental mammals. (**a**) A herd of camels traversing an extremely hot desert with ease. (**b**) One of the bats, the only truly flying mammals, which dominate the night sky vacated by birds. (**c**) Manatee. It lives in the sea and eats submerged seaweed. (**d**) An arctic fox, with thick, insulative fur that camouflages it from prey. In summer, its light-brown fur blends with golden-brown grasses. In winter, the fur turns white and blends with the snow-covered ground. (**e**) Walruses swim in frigid waters after their prey and sunbathe on ice.

whales, some of which weigh in at more than 100 tons.) They survived the dinosaurs and then one another, only to confront a modern challenge—a human population bent on hunting them or encroaching on their habitats, which commonly entails introducing novel species or favoring a few domesticated ones at the expense of wild stock. Think of it. In the 1800s, bison were nearly driven to extinction as part of a program to starve the Plains Indians into submission by killing off their major food source. Modern whaling industries based in Japan and elsewhere are exterminating the very mammals they hunt. About 300 mammalian species, including most whales, wild cats, otters, and primates other than humans, are currently on the endangered species list.

Convergent evolution has occurred among families of all three lineages of mammals that occupy similar habitats in different parts of the world. More recently, humans have introduced new challenges into the evolutionary equation.

SUMMARY

1. Nearly all chordate embryos have a notochord, a dorsal hollow nerve cord, a pharynx with gill slits (or hints of these), and a tail that extends past the anus. Some or all of these traits persist in the adult forms. Chordates with a backbone are known as vertebrates. Tunicates (including the sea squirts) and lancelets are invertebrate chordates.

2. The eight classes of vertebrates are jawless fishes, jawed armored fishes (now extinct), cartilaginous fishes, bony fishes, amphibians, reptiles, birds, and mammals.

3. The first vertebrates, jawless fishes that arose during the Cambrian era, included the ostracoderms. Their living descendants are lampreys and hagfishes. Jawed fishes also arose in the Cambrian. A dominant lineage, the placoderms, is extinct. Other lineages gave rise to the cartilaginous and bony fishes.

4. Four trends occurred during the evolution of certain vertebrate lineages:

a. A vertebral column supplanted the notochord as a structural element against which muscles act. This led to swift predatory animals.

b. Jaws evolved from gill-supporting elements. This led to increased predator-prey competition; it favored far more efficient nervous systems and sensory organs.

c. In one lineage of bony fishes, paired fins evolved into fleshy lobes with structural elements (forerunners of paired limbs).

d. In some fish lineages, respiration by gills came to be supplemented by lungs, which proved adaptive in the invasion of the land. In a related development, the circulatory system became more efficient at distributing oxygen through the body.

5. Amphibians were the first vertebrates to invade the land, but they never fully escaped the water. Their skin dries out, and aquatic stages persist in most life cycles.

6. Unlike amphibians, reptiles escaped dependence on standing free water. The key adaptations that allowed them to do so are their toughened, scaly skin, which restricts water loss from the body; a reliance on internal fertilization; more efficient, water-conserving kidneys; and amniote eggs, often leathery or shelled, that protect and metabolically support the embryos.

7. Reptiles, and the birds and mammals that descended from certain reptilian lineages, have efficient circulatory and respiratory systems and a highly developed nervous system and sensory organs.

8. Of all vertebrates, birds alone have feathers, which they use in flight, heat conservation, and social displays.

9. Of all animals, mammals alone have milk-producing mammary glands, and they have hair or thick skin that functions in insulation. They have distinctive dentition and a highly developed cerebral cortex, the brain's most complex region. Adults nurture their young through an extended period of dependency and learning.

Review Questions

1. List and describe features that distinguish chordates from other animals. 27.1

2. Name and describe the features of an organism from both groups of invertebrate chordates discussed in this chapter. 27.2

3. List four major trends that occurred during the evolution of at least some vertebrate lineages. 27.3

4. Which evolutionary modifications in fishes set the stage for the emergence of amphibians? 27.3, 27.6

5. List some of the characteristics that distinguish reptiles from amphibians. 27.7, 27.8

6. List some of the characteristics that distinguish birds from reptiles. 27.9

7. List some of the characteristics that distinguish each of the three mammalian lineages from the reptiles. 27.10, 27.11

Self-Quiz (Answers in Appendix IV)

1. Only _____ have a notochord, a tubular dorsal nerve cord, a pharynx with slits in the wall, and a tail extending past the anus.
   a. echinoderms   d. both b and c
   b. tunicates and lampreys   e. all are correct
   c. vertebrates

2. Gill slits function in _____ .
   a. respiration   d. water regulation
   b. circulation   e. both a and c
   c. food trapping

3. A shift from a reliance on _____ to reliance on _____ was pivotal in the evolution of all vertebrates.
   a. the notochord; a backbone   c. gills; lungs
   b. filter feeding; jaws   d. all are correct

4. The first vertebrates were _____ .
   a. bony fishes   c. jawed fishes
   b. jawless fishes   d. both a and b

5. Of all existing vertebrates, _____ are the most diverse.
   a. cartilaginous fishes   d. reptiles
   b. bony fishes   e. birds
   c. amphibians   f. mammals

6. The bony fishes include _____ .
   a. ray-finned fishes   c. lungfishes
   b. lobe-finned fishes   d. all are correct

7. The only amphibians to entirely escape dependence on aquatic habitats are _____ .
   a. salamanders   c. caecilians
   b. desert toads   d. none is correct

8. Reptiles moved fully onto land owing to _____ .
   a. tough skin   d. amniote eggs
   b. internal fertilization   e. both b and d
   c. good kidneys   f. all are correct

9. A four-chambered heart is characteristic of _____ .
   a. bony fishes   d. mammals
   b. amphibians   e. both b and c
   c. birds   f. both c and d

10. _____ have highly efficient circulatory and respiratory systems, and a complex nervous system and sensory organs.
   a. Reptiles
   b. Birds
   c. Mammals
   d. all are correct

11. Birds use feathers in _____ .
   a. flight
   b. heat conservation
   c. social functions
   d. all are correct

12. Various mammals _____ .
   a. hatch
   b. complete their embryonic development in pouches
   c. complete their embryonic development in the uterus
   d. both b and c
   e. all are correct

13. Match the organisms with the appropriate features.
   _____ jawless fishes
   _____ cartilaginous fishes
   _____ bony fishes
   _____ amphibians
   _____ reptiles
   _____ birds
   _____ mammals

   a. complex cerebral cortex, thick skin or hair
   b. respiration by skin and lungs
   c. include coelacanths
   d. include hagfishes
   e. include sharks and rays
   f. complex social behavior, feathers
   g. first with amniote eggs

Critical Thinking

1. Describe the factors that might contribute to the collapse of native fish populations in a lake following the introduction of a novel predator, such as the lamprey.

2. Think about the flight muscles of birds and their demands for oxygen and ATP energy. What type of organelle would you expect to be profuse in these muscles? Explain your reasoning.

3. Kathie and Gary, two amateur fossil hunters, have unearthed the complete fossilized remains of a mammal. How can they determine whether their find is a herbivore, carnivore, or omnivore?

4. In Australia, many species of marsupials are competing with recently introduced placental mammals (such as rabbits) for resources—but they are not winning. Explain how it is that placental mammals that did not even evolve in the Australian habitats show greater fitness than the native mammals.

5. About seventy-five species of coral snakes live in the rain forests, grasslands, mountains, and desertlike regions of North, Central, and South America. Coral snakes of family Colubridae are harmless; those of family Elapidae are poisonous. In both families, the most poisonous types bite mostly to kill prey, yet moderately poisonous types can be extremely vicious. All coral snakes, including the deadliest species (*Micrurus*) in Figure 27.23, have distinctive color banding. When they are gliding along, even snake experts (herpetologists) may have trouble knowing which is which. Yet their relatives in other parts of the world have no such banding; neither do bigger snakes.

Thus, color banding in the two families can't be explained in terms of environmental differences, phylogeny, or lethality. What brought it about? According to a theory proposed by herpetologist R. Mertens, the ancestors of *Micrurus* migrated across a land bridge that connected Asia with North America more than 60 million years ago. Because predators learned to avoid them by associating the color banding with "taste trials," the ancestral snakes must *not* have been as deadly as some of their descendants. (If the snakes were *too* deadly, all trials might have ended in death.) Therefore, the protective function of the color banding must depend on unpleasant trials with only *moderately* poisonous snakes.

Among the harmless snakes (Colubridae), identical warning coloration also evolved. This was a case of *mimicry*, in which one

Figure 27.23 One of the extremely poisonous coral snakes (*Micrurus*).

group (mimics) came to bear deceptive resemblance to another group (models) that enjoy a survival advantage. For this to work, of course, mimics can't outnumber models. That said, explain the following numbers of trapped snakes that were brought to the Butantan Institut in São Paulo for the years shown (trappers did not distinguish among poisonous and nonpoisonous types):

	1950	1951	1952	1953
Micrurus:	59	41	58	56
Others:	218	246	265	284

Selected Key Terms

amniote egg *27.7*
amphibian *27.6*
behavioral flexibility *27.10*
bird *27.9*
bony fish *27.5*
cartilaginous fish *27.5*
chordate *27.1*
convergent evolution *27.11*
dentition *27.10*
eutherian *27.10*
feather *27.9*
filter feeder *27.2*
fin *27.3*
gill *27.3*
gill slit *27.2*
hagfish *27.4*
jaw *27.3*
lamprey *27.4*
lancelet *27.2*
lobe-finned fish *27.5*
lung *27.3*
mammal *27.10*
marsupial *27.10*
monotreme *27.10*
nerve cord *27.1*
notochord *27.1*
ostracoderm *27.3*
pharynx *27.1*
placenta *27.11*
placoderm *27.3*
reptile *27.7*
scale (fish) *27.5*
swim bladder *27.5*
therapsid *27.10*
therian *27.10*
tunicate *27.2*
vertebra *27.3*
vertebrate *27.1*

Readings

Gould, S. J. (general editor). 1993. *The Book of Life.* New York: Norton. Splendid, easy-to-read essays, gorgeous illustrations.

Romer, A. S., and T. S. Parsons. 1986. *The Vertebrate Body.* Sixth edition. Philadelphia: Saunders.

Strickberger, M. 1996. *Evolution.* Second edition. Boston: Jones and Bartlett.

Wickler, W. 1968. *Mimicry in Plants and Animals.* New York: McGraw-Hill. Paperback.

Web Site See *http://www.wadsworth.com/biology* for practice quiz questions, hypercontents, BioUpdates, and critical thinking. The Wadsworth Biology Resource Center provides a wealth of information fully organized and integrated by chapter.

28 HUMAN EVOLUTION: A CASE STUDY

The Cave at Lascaux and the Hands of Gargas

Half a century ago, on a warm autumn day, four boys out for a romp stumbled onto a cave near Lascaux, a town in the Perigord region of France. What they discovered in that intricately tunneled cave stunned the world. Magnificent sketches, engravings, and paintings swept across the cave walls (Figure 28.1). The red, yellow, purple, and brown pigments were as vivid as if they had just been painted. Radiometric dating revealed that they were 17,000 to 20,000 years old. The prehistoric artists worked deep within the cave, where sunlight did not fade the images and neither winds nor water could wear them away. By the light of crude oil lamps, they captured the graceful, dynamic lines of bison, stags, stallions, ibexes, lions, a rhinoceros, and a heifer, since named the Great Black Cow.

Even earlier, art was committed to the walls of caves throughout southern France, northern Spain, and Africa, as when people made more than 150 imprints and outlines of hands in the cave of Gargas, in the Pyrenees, some 25,000 years ago.

Who were these artists? From their fossilized remains, we know they were people anatomically like us. From the way they planned and executed their art, we sense a level of abstract thinking that is unique to humans.

The quality of "humanness" did not emerge out of thin air. If we could go back 5 million years in time, to the place where the most recent ancestors of humans apparently originated, we might find ourselves in Africa, hiding in grasslands and dry woodlands of the sort shown in Figure 28.2. From there, our journey would take us back an additional 55 million years, to ancient tropical forests in which primates originated. We could not stop there. The mammalian ancestors of the primates evolved 250 million years ago, vertebrate ancestors of mammals evolved long before that, ancestors of vertebrates and all other animal groups evolved 900 million years ago—and so on back to the origin of the first living cells.

As we explore our own branch of the animal family tree, keep this greater evolutionary story in mind. *Our "uniquely" human traits emerged through modification of traits that evolved earlier, in ancestral forms.* From that perspective, the "ancient" cave paintings are the legacy of people who departed only yesterday, so to speak. The artists of Lascaux and Gargas are not remote from us. They *are* us.

Figure 28.1 Part of the human cultural heritage—prehistoric paintings from a cave in Lascaux, France, a unique outcome of a long history of biological evolution.

Figure 28.2 (*Below*) East Africa's Rift Valley, 3,200 kilometers long. Somewhere in this immense valley, sparsely wooded grasslands were the birthplace of the human species.

KEY CONCEPTS

1. In the primate branch of the mammalian lineage are the prosimians, tarsioids, and anthropoids (monkeys, apes, and humans). Apes and humans are hominoids. Only humans and some extinct species with a mosaic of apelike and humanlike traits are further classified as hominids.

2. When some early primates were adapting to life in the trees, their skull and eye sockets underwent modifications that led to increased reliance on daytime vision, compared to their night-foraging ancestors.

3. Hominoids evolved in Miocene times. In some of the apelike forms, certain skeletal modifications permitted an upright stance and bipedalism, or walking on two legs. Bipedalism freed the forelimbs for novel manipulative functions. Hands evolved, and among certain hominid descendants of Miocene apes, modifications in the bones and associated muscles led to a capacity to hold, carry, use, and make objects.

4. Starting in the Miocene, a long-term cooling trend led to seasonal changes in habitats and food sources. Forests gave way to grasslands with scattered stands of trees, and some hominids became scavengers of meat as well as foragers for fruits and nuts. The sizes and shapes of their jaws and teeth became adapted to processing new foods.

5. As food sources became scarcer, hominids underwent adaptive radiations through Africa and, later, into Europe and Asia. In some lineages, the brain increased in volume and became more complex. Eventually, the evolution of the brain became interlocked with cultural evolution.

6. Unlike earlier hominids, *Homo erectus* and *H. sapiens* displayed remarkable behavioral flexibility and creative experimentation with their environment, as when they started using fire. This characteristic allowed them to survive the challenges of dispersing into novel and often harsh environments around the world.

Having traversed the mammalian branch of the animal family tree in Chapter 27, we are now ready to follow it along the branchings leading to primates, then humans.

Primate Classification

The order **Primates** includes prosimians, tarsioids, and anthropoids (Table 28.1 and Appendix I). Figure 28.3 shows a few of these mammals. The prosimians are the oldest primate lineage (*pro*, before; *simian*, ape). They were **arboreal** (tree-dwelling), and they dominated the trees in North America, Europe, and Asia for millions of years before monkeys and apes evolved and almost displaced them. Tarsiers of Southeast Asia are the only living tarsioids. The traits of these small primates are somewhere between prosimians and anthropoids. All

monkeys, apes, and humans are **anthropoids**. In their biochemistry and structure, apes are closer to humans than to monkeys. Apes, humans, and extinct species of the human lineage are grouped together, as **hominoids**.

By 5 million years ago, a divergence from the last shared ancestor of apes and humans was under way. All species that evolved on that separate road leading to humans are further classified as **hominids**.

Key Evolutionary Trends

Most primates live in tropical or subtropical forests, woodlands, or **savannas**, which are open grasslands with a few stands of trees, as in Figure 28.2. Like their ancestors, the vast majority are tree dwellers. Yet no one feature sets the primates apart from other mammals. Each lineage evolved in a distinct way and has its own defining traits. Five trends define the lineage leading to humans, which is our focus. They were set in motion when primates started adapting to life in the trees, and they contributed to the emergence of modern humans. *First*, there was less reliance on the sense of smell and more on daytime vision. *Second*, skeletal changes led to upright walking, which freed the hands for novel tasks. *Third*, changes in bones and muscles led to refined hand movements. *Fourth*, teeth became less specialized. *Fifth*, elaboration of the brain and changes in the skull led to truly complex behaviors. These developments became interlocked with one another and with cultural evolution.

ENHANCED DAYTIME VISION Early primates had an eye on each side of their head. Later ones had forward-directed eyes, an arrangement that is much better for sampling shapes and movements in three dimensions. Other modifications allowed their eyes to respond to

Figure 28.3 Representative primates. (**a**) A gibbon, with body and limbs adapted for swinging arm over arm in the trees. (**b**) A spider monkey, which is a four-legged climber, leaper, and runner. (**c**) Tarsiers, which are vertical climbers and leapers.

Table 28.1	Primate Classification
Taxon	Representatives
PROSIMIANS	
Lemuroids	Lemurs, lorises
TARSIOIDS	
Tarsioids	Tarsiers
ANTHROPOIDS	
Ceboids (New World monkeys)	Spider monkeys
Cercopithecoids (Old World monkeys)	Baboons
Hominoids:	
Hylobatids	Gibbon, siamang
Pongids	Orangutan, gorilla, chimpanzee
Hominids	Humans, extinct humanlike forms

Figure 28.4 Comparison of the skeletal organization and stance of a monkey, ape, and human. Monkeys climb and leap. Gorillas and chimps use forelimbs to climb and help support their weight, but most of the time they walk on all fours on the ground. Humans are two-legged walkers. The different modes of locomotion arose through modifications of the basic skeletal plan of mammals.

variations in color and in light intensity (dim to bright). Being able to interpret and respond quickly to these stimuli is advantageous when living in the trees.

UPRIGHT WALKING A primate's mode of locomotion depends on arm length and the shape and positioning of its shoulder blades, pelvic girdle, and backbone. With arm and leg bones about the same length, a monkey can climb, leap, and run on four legs, but not two (Figure 28.4). A gorilla has long arms and walks on all fours, including its knuckles. Only humans show **bipedalism**: they walk freely on two legs. Compared with monkeys and apes, the human backbone is shorter, S-shaped, and flexible. Bipedalism was a key innovation that evolved in the ancestors of hominids.

POWER GRIP AND PRECISION GRIP How did we get such splendid hands? The first mammals spread their toes apart to help support the body as they walked or ran on four legs. Primates still spread fingers or toes. Many can cup their fingers, as when monkeys lift food to the mouth. Among ancient tree-dwelling primates, handbone modifications allowed them to wrap their fingers around objects (*prehensile* movements) and to touch their thumb to the tip of each finger (*opposable* movements). In time, their hands became freed from load-bearing functions. Later, when hominids evolved, refinements led to the power grip and precision grip:

These hand positions gave early humans a capacity to make and use tools. They were a foundation for unique technologies and cultural development.

TEETH FOR ALL OCCASIONS Before hominids evolved, modifications in primate jaws and teeth accompanied a shift from eating insects to fruits and leaves, and on to a mixed diet. Later, rectangular jaws and long canines

came to be additional defining features of monkeys and apes. Along the road leading to humans, a bow-shaped jaw and smaller teeth of about the same length evolved.

BETTER BRAINS, BODACIOUS BEHAVIOR Among some early primates, an arboreal life-style required shifts in reproductive and social behavior. In many lineages, parents put more effort in fewer offspring. They formed stronger bonds with offspring, maternal care grew more intense, and the learning period increased (Figure 28.5). Expansions of the brain became interlocked with selection for increasingly complex behavior. New behavior promoted new neural connections in the brain regions that dealt with sensory inputs, a brain with more intricate wiring favored more new behavior, and so on. We find evidence of such

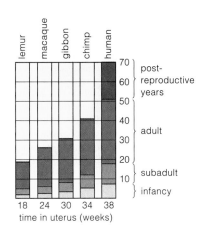

Figure 28.5 Trend toward longer life spans and longer dependency among primates.

interlocking in the parallel evolution of the human brain and culture. **Culture** is the sum total of behavior patterns of any social group, handed down from one generation to the next by learning and by symbolic behavior, especially language. A capacity for language arose among ancestral humans, through changes in the skull bones and brain.

Certain key adaptations evolved along the pathway from arboreal primates to modern humans: complex, forward-directed vision; bipedalism; refined hand movements; generalized dentition; and an interlocked elaboration of brain regions and cultural behavior.

Origins and Early Divergences

Primates evolved from mammals more than 60 million years ago, in tropical forests of the Paleocene. The first ones resembled small rodents or tree shrews (Figures 28.6 and 28.7). Like tree shrews, they probably had huge appetites and foraged at night for insects, seeds, buds, and eggs beneath the trees. They had a long snout and a good sense of smell, suitable for snuffling food and predators. They clawed their way up stems, although not with much speed or grace.

During the Eocene (between 54 and 38 million years ago), some primates were staying up in the trees. They had a shorter snout, enhanced daytime vision, a larger brain, and refined grasping movements. How did these traits evolve?

Consider the trees. An arboreal life-style had advantages: abundant food and safety from ground-dwelling predators. However, they also were habitats of uncompromising selection. Visualize an Eocene morning with dappled sunlight, boughs swaying in the breezes, colorful fruit hidden among the leaves, and perhaps predatory birds. A long, odor-sensitive snout would not have been of much use up in the trees, where air currents disperse odors. But a brain that could assess movement, depth, shape, and color would have been a definite plus. And so would a brain that could work fast when its owner was running and leaping (especially!) from branch to branch. The body's weight, the wind speed, and the distance and suitability of a destination had to be estimated all at the same time, and any adjustments for miscalculations had to be quick.

Figure 28.6 A nocturnal tree shrew of Indonesia.

By at least 36 million years ago, before the dawn of the Oligocene, tree-dwelling anthropoids had evolved in the forests. One form, the squirrel-sized *Catopithecus*, was on or very close to the evolutionary road that led to monkeys and apes. It had forward-directed eyes. Given its snoutless, flattened face and upper jaw with shovel-shaped front teeth, it must have used its hands to grab fruit and insects. Some of the early anthropoids lived in swamplands infested with formidable, predatory reptiles and rarely ventured to the ground. And maybe that is why it became imperative to think fast and grip strongly. Slip-ups were always possible; many primates still fall out of trees.

Between 25 and 5 million years ago (Miocene), apelike forms—*the first hominoids*—evolved, and they underwent an adaptive radiation into Africa, Europe, and southern Asia. Most forms became extinct. However, fossils as well as genetic analyses suggest that between 10 million and 5 million years ago, divergences from lineages of some Miocene apes led to gorillas and chimps, and to the first *hominids*.

The First Hominids

Most of the early hominids we know about lived in the East African Rift Valley. By 5 million years ago, at the dawn of the Pliocene, a long-term cooling trend was well under way as a result of drifting continents and changes in ocean circulation patterns. Figure 28.8 is one model of the geologic triggers for the trend. The once-

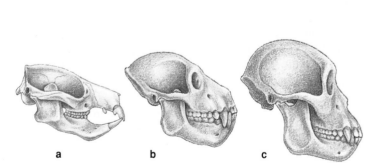

a   b   c

Figure 28.7 Comparison of skull shape and teeth of some early primates. (**a**) *Plesiadapis* (Paleocene) had rodentlike teeth. (**b**) *Aegyptopithecus* (Oligocene anthropoid) probably predates the divergence leading to Old World monkeys and apes. (**c**) Apelike dryopiths lived in the Miocene. *Plesiadapis* was as tiny as a tree shrew. *Aegyptopithecus* was monkey-sized, and some dryopiths, chimpanzee-sized.

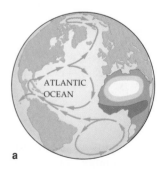
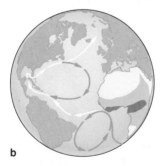

a   b

Figure 28.8 Long-term shift in an ancient climate. (**a**) Before the Isthmus of Panama formed, salinity levels of seawater were much the same around the world, and circulation patterns kept Arctic waters warmer. (**b**) After the Isthmus formed, North Atlantic waters became saltier and heavier; they sank before reaching the Arctic region. The Arctic ice cap formed and triggered a long-term trend toward a cooler, drier climate in Africa.

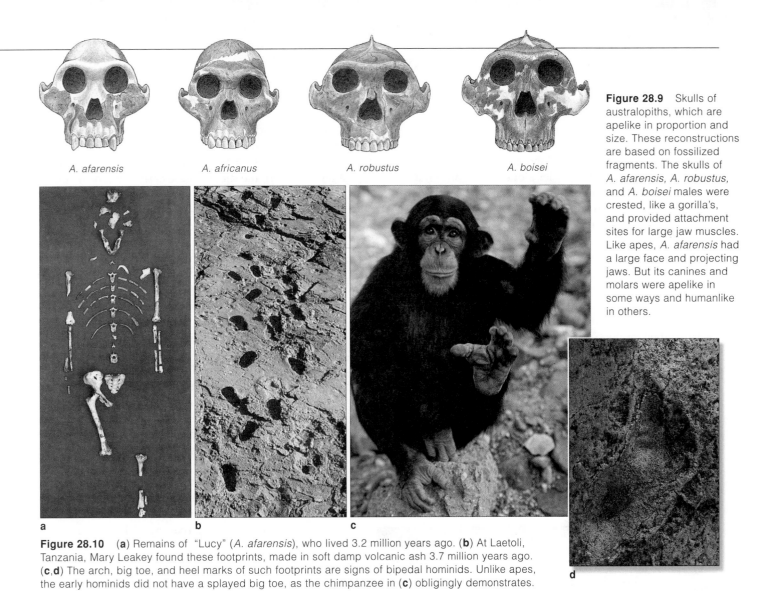

A. afarensis   A. africanus   A. robustus   A. boisei

Figure 28.9 Skulls of australopiths, which are apelike in proportion and size. These reconstructions are based on fossilized fragments. The skulls of *A. afarensis*, *A. robustus*, and *A. boisei* males were crested, like a gorilla's, and provided attachment sites for large jaw muscles. Like apes, *A. afarensis* had a large face and projecting jaws. But its canines and molars were apelike in some ways and humanlike in others.

a   b   c   d

Figure 28.10 (**a**) Remains of "Lucy" (*A. afarensis*), who lived 3.2 million years ago. (**b**) At Laetoli, Tanzania, Mary Leakey found these footprints, made in soft damp volcanic ash 3.7 million years ago. (**c,d**) The arch, big toe, and heel marks of such footprints are signs of bipedal hominids. Unlike apes, the early hominids did not have a splayed big toe, as the chimpanzee in (**c**) obligingly demonstrates.

vast tropical forests—with their bounty of edible soft fruits, leaves, and insects—gave way to dry woodlands and grasslands as the climate became more seasonal. The African savanna had emerged. The new foods were harder, and harder to find, and early hominids had two options: Move into the new adaptive zones or die out.

Fossil fragments 4 to 4.5 million years old indicate this was a "bushy" period of evolution; a confounding variety of hominids appeared (Figures 28.9 and 28.10). At present it is impossible to figure out their family tree, so they are simply grouped as **australopiths** (meaning southern apes). *Australopithecus anamensis* is the oldest of these. Like *A. afarensis* and *A. africanus,* it was gracile (slightly built). The ones called *A. boisei* and *A. robustus* were robust (muscular and heavily built).

With their large face, protruding jaws, and small skull (and brain) size, australopiths were apelike. Yet they differed from previous hominids. For example, their molars had thicker enamel and could grind harder foods. And they were good at walking upright. We know this from studies of fossilized hip and limb bones.

More telling, some australopiths left footprints. About 3.7 million years ago, *A. afarensis* individuals walked on newly fallen volcanic ash, which a light rain turned into quick-drying cement (Figure 28.10*b*). Bipedalism started to evolve earlier, when certain hominoids were forced to the ground late in the Miocene. They had mastered the power grip and precision grip in the trees. Rather than becoming specialized in running fast on all fours, they used their manipulative skills to advantage. They kept their hands free to carry offspring, and probably to carry precious food on their foraging expeditions.

Primates evolved from small, rodentlike mammals that moved into arboreal habitats 60 million years ago. During the Miocene, the first hominoids (apelike forms) evolved and radiated through Africa, Europe, and southern Asia.

Between 10 million and 5 million years ago, divergences from Miocene apes led to australopiths, the first hominids. They were apelike in many skeletal details, but they were humanlike in a crucial respect: They walked upright.

Defining "Human"

What can fossilized fragments of the early hominids tell us about our own origins? The fossil record is still too sketchy for us to know how all the diverse australopiths were related to one another, let alone which ones may have been ancestral to humans. Besides, what *are* the traits that distinguish **humans** (members of the genus *Homo*) from other groups? Well, there's always the brain. In modern humans, it is the basis of great analytical and verbal skills, complex social behavior, and technological innovation. It easily sets us apart from apes, which have a skull volume and brain size much smaller than ours (Figure 28.11). Yet this feature alone cannot tell us when certain hominids made the transition to being human, because their brain size probably fell within the range for apes. They were makers of simple tools, but so are chimps and certain parrots. And it goes without saying that their behavior did not lend itself to fossilization.

And so we are left to speculate about a continuum of physical traits among a number of fossils—a skeleton adapted for bipedalism, manual dexterity, and larger brain volume; a smaller face; and smaller, more thickly enameled teeth. These traits, which originated late in the Miocene, were evident in what many consider to be the earliest humans, *Homo habilis* (meaning "handy man").

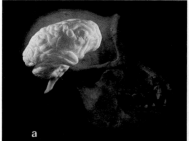
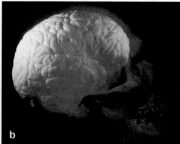

c

Homo habilis *and the First Stone Tools*

Between 2.5 and 1.6 million years ago, one or two forms of *H. habilis* lived in dry woodlands that punctuated the savannas of eastern and southern Africa. Judging from their dentition, the diet of these early humans included hard-shelled nuts and seeds as well as soft fruits, leaves, and insects. Supplies of these foods changed with the seasons. Most likely, *H. habilis* had to think ahead, to plan when to venture about to gather and possibly store foods that would help it survive the cold dry seasons.

H. habilis shared its habitat with saber-tooth cats and other formidable predators. The teeth of those cats could impale prey and shear off flesh, but they could not crush open the bones for marrow. Carcasses of their kills, with shreds of meat clinging to the marrow bones, were concentrated stores of nutrients in places that were nutrient-stingy. Although *H. habilis* was a forager, not a full-time carnivore, it opportunistically supplemented its diet by scavenging the carcasses (Figure 28.11c).

Fossil hunters have found numerous stone tools that date to the time of *H. habilis*. But they cannot say with certainty that *H. habilis* was the only species that made them. Possibly australopiths as well as *H. habilis* used sticks and other perishable tools before then, as modern apes do, but we have no way of knowing.

Maybe individuals on the road to modern humans started down a toolmaking road by picking up rocks to crack marrow bones. Maybe they started to scrape flesh from bones with small, sharp flakes that had fractured naturally from rocks. Eventually, early humans started *shaping* stone implements. Paleoanthropologist Mary Leakey was the first to discover evidence of toolmaking at Africa's Olduvai Gorge, which cuts through a great sequence of sedimentary rock layers. The most ancient tools at this site are crudely chipped pebbles that were buried in the deepest layers (Figure 28.12). They may have been used to smash marrow bones, dig for roots, and poke insects from tree bark. More recent layers have more complex tools. Large numbers of bones and tools have been found along the shores of ancient lakes that would have beckoned plenty of thirsty animals.

At such sites we find fossils of one form of *H. habilis* that was twice as brainy as the australopiths and that obviously ate well. There apparently was no selection pressure for more creativity in securing food resources; the stone tools of *H. habilis* did not change much for the next 500,000 years.

Figure 28.11 (a) Image of the brain of a modern chimpanzee, superimposed on a chimp skull. The brain of Lucy, one of the australopiths (*A. afarensis*), was only a bit larger than this and much smaller than (b) the modern human brain. (c) Artist's rendition of *Homo habilis* males in an East African woodland.

Figure 28.12 A few of the 37,000+ stone tools from Olduvai Gorge. (**a**) A crude chopper and (**b**) a more advanced form, with a joint and sharp edge. (**c**) Hand ax and (**d**) cleaver.

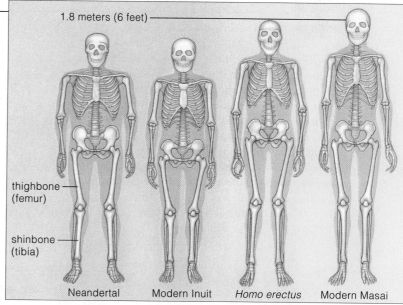

Figure 28.13 Climate and body build. Humans adapted to cold climates have a heat-conserving body (stockier, with shorter legs, compared with humans adapted to hot climates).

From Homo erectus *to* Homo sapiens

The ancestors of modern humans apparently stayed put in Africa until about 2 million years ago. At that time, genetic divergence from early members of *Homo* led to *Homo erectus*, a human species that the fossil record places on the evolutionary road to modern humans. Its name means "upright man." Although its forerunners also were upright, two-legged walkers, *H. erectus* did the name justice; its populations walked out of Africa, turned left into Europe, and right into Asia. It took some of them a long time to walk the 14,000 kilometers to China. *H. erectus* fossils from Southeast Asia and the former Soviet republic of Georgia are 1.8 million and 1.6 million years old. And the treks were strenuous. More than once, immense glaciers advanced down through northern Europe, southern Asia, and North America.

Whatever selection pressures triggered the adaptive radiations, this was a time of physical changes, as in skull size and leg length (Figure 28.13). It also was the time of cultural lift-off for the human lineage. *H. erectus* had a larger brain, it was a more creative toolmaker, and its social organization and communication skills must have been well developed to make such successful treks. From southern Africa on into England, different populations were using the same variety of hand axes and other tools designed to pound, scrape, shred, cut, and whittle. They withstood environmental challenges by building fires and using furs for clothing. Remains of fire use date from an ice age in the early Pleistocene.

Judging from fossils in Africa, **Homo sapiens** had evolved by 100,000 years ago. The origin and radiations of early *H. sapiens* are hotly debated topics (Section 28.4). Early *H. sapiens* had smaller teeth and jaws than *H. erectus*, and often it had a chin. Its facial bones were smaller, the skull was higher and rounder, and the brain was larger. Analysis of fossils indicates that these early forms may have had the capacity for complex language.

One group of early humans, the Neandertals, lived in Europe and the Near East from 200,000 to 35,000 years ago. Massively built and large brained, some of their populations were the first to adapt to the coldest regions (Figure 28.13). Their disappearance coincided with the appearance of anatomically modern humans in the same regions 40,000 to 30,000 years ago. There is no evidence that they warred or interbred with these later arrivals. We don't know yet what happened to them.

From 40,000 years ago to today, human evolution has been almost entirely cultural, not biological—and so we leave the story with these conclusions: Humans spread rapidly through the world by devising *cultural* means to deal with a broad range of environments. Compared with their predecessors, they developed rich and varied cultures. Even though hunters and gatherers persist in parts of the world, others moved from "stone-age" technology to the age of "high tech," attesting to the great plasticity and depth of human adaptations.

Cultural evolution has outpaced the biological evolution of the only remaining human species, *H. sapiens*. Today, humans everywhere rely on cultural innovation to adapt rapidly to a broad range of environmental challenges.

28.4   OUT OF AFRICA—ONCE, TWICE, OR . . .

If researchers are interpreting the fossil record of human evolution correctly, then it would seem that Africa was the cradle for us all. At this writing, at least, no one has found any fossils of humans that are older than 1 million years *except* in Africa. *H. erectus* coexisted for a time with earlier humans (*H. habilis*) before dispersing from the African savannas to the cooler grasslands, forests, and mountains of Europe and Asia. They apparently left Africa in waves between about 2 million and 500,000 years ago. Judging from recent examination of *H. erectus* fossils from Java, some populations may have survived, in relative isolation, until 27,000 to 53,000 years ago.

Where, on the larger geologic stage, do we place the origin of *H. sapiens*? *Here we find a good example of how the same body of evidence can be interpreted in different ways.* These interpretations are called the multiregional model and African emergence model for modern human origins. Both attempt to explain the world distribution of fossils of particular ages and the measured genetic distances among modern, existing human populations. For example, Figure 28.14 includes locations of *H. sapiens* fossils that have been dated to particular times. As another example, evidence from biochemical and immunological studies suggests the greatest genetic distance separates *H. sapiens* populations native to Africa from populations everywhere else; the next greatest distance separates Southeast Asia (and Australia) from everywhere else (Figure 28.15).

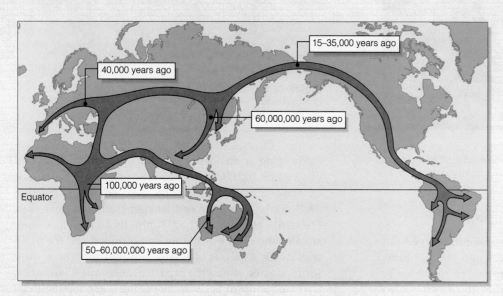

Figure with labels: 40,000 years ago; 15–35,000 years ago; 60,000,000 years ago; 100,000 years ago; Equator; 50–60,000,000 years ago

Figure 28.14 Dates when early *H. sapiens* populations were colonizing different parts of the world, based on fossil evidence. As shown here, the presumed dispersal routes (*brown* arrows) seem to support the African emergence model.

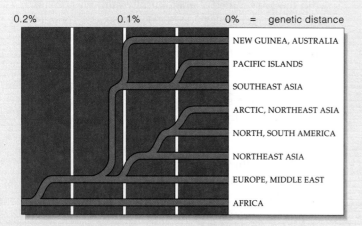

0.2%   0.1%   0%   =   genetic distance

NEW GUINEA, AUSTRALIA
PACIFIC ISLANDS
SOUTHEAST ASIA
ARCTIC, NORTHEAST ASIA
NORTH, SOUTH AMERICA
NORTHEAST ASIA
EUROPE, MIDDLE EAST
AFRICA

Figure 28.15 One proposed family tree for populations of modern humans (*Homo sapiens*) that are native to different regions of the world. Branch points indicate presumed genetic divergences. The tree is based on nucleic acid hybridization studies of many genes (including those for the ABO blood group) and immunological comparisons.

MULTIREGIONAL MODEL   By this model, *H. erectus* populations had spread through much of the world by 1 million years ago. Those geographically separated groups were subject to different selection pressures, so their traits evolved in regionally distinctive ways. Then subpopulations ("races") of *H. sapiens* evolved from them in different places. Although they differed phenotypically, they did not evolve into separate species because gene flow continued among them, even to the present day. (Thus, for example, while the armies of Alexander the Great were sweeping eastward, they also contributed blue-eye genes from the Greeks to the allele pool of generally brown-eyed subpopulations in Africa, the Near East, and Asia.)

AFRICAN EMERGENCE MODEL   This model does not dispute fossil evidence that populations of *H. erectus* evolved in distinctive ways in different regions. But it holds that *H. sapiens*—modern humans—originated in sub-Saharan Africa somewhere between 200,000 and 100,000 years ago. Only later did *H. sapiens* populations move out of Africa, then into other regions along the routes indicated by the Figure 28.14 map. In each region that *H. sapiens* populations settled, they replaced the archaic *H. erectus* populations that had preceded them. Only then did regional phenotypic differences become superimposed on the original *H. sapiens* body plan.

In support of this model, the oldest known *H. sapiens* fossils are indeed from Africa. Also, in Zaire, new finds of finely wrought barbed-bone harpoons and other tools suggest the African populations were as skilled at making tools as *Homo* populations known earlier from Europe.

SUMMARY

1. Like other mammals, humans and other primates have an internal skeleton, a complex brain and sensory organs fully enclosed in a skull, mammary glands (in females), and distinctive teeth. Adults nourish, protect, and serve as behavioral models for the young.

2. Primates include the prosimians (lemurs and related forms) tarsioids, and the anthropoids (which include monkeys, apes, and humans). Only apes and humans are hominoids. Only the anatomically modern humans (*H. sapiens*) and others of their lineage (from *A. afarensis* to *H. erectus*) are further classified as hominids.

3. The first primates were small, rodentlike mammals that evolved by 60 million years ago in tropical forests. Lineages evolving in the trees developed a power grip, a precision grip, and good daytime vision. Some lineages gave rise to anthropoids, including the ancestors of monkeys, apes, and humans, by 35 million years ago.

4. Apelike forms (the first hominoids) evolved between 25 and 13 million years ago, in the Miocene. In Africa, some had given rise to australopiths, the earliest known hominids, before 4 million years ago (Figure 28.16). The evolution of late Miocene apes and their hominid descendants has been correlated with strong selection pressures that emerged during a long-term trend from tropical climates to cooler, drier climates. The African savanna emerged, and the hominids that survived were the ones with physical changes that permitted upright walking (bipedalism), a more varied diet (changes in dentition), and more brain complexity to figure out how to gather scarcer and seasonally available food.

5. *H. habilis*, the earliest known members of the genus *Homo*, had evolved by 2.5 million years ago. It may have been the first stone toolmaker. It opportunistically supplemented its diet of nuts, fruits, and seeds with meat and marrow from carcasses.

6. *Homo erectus*, the presumed ancestor of modern human populations, evolved by 2 million years ago. *H. erectus* populations radiated out of Africa, into Asia and Europe. The oldest known fossils of early modern humans (*H. sapiens*) are from Africa; they are 100,000 years old. About 40,000 years ago, cultural evolution outstripped biological evolution of the human form.

7. Modern humans are adapted to a wide range of environments. This capacity resulted from evolutionary modifications in certain primate lineages. Starting with arboreal primate ancestors, there was less reliance on the sense of smell and more on enhanced daytime vision. Also among arboreal primates, manipulative skills increased as the hands began to be freed from load-bearing functions. Starting with Miocene apes, there was a shift from four-legged climbing to bipedalism, a shift from specialized to omnivorous eating habits, and increases in brain complexity and behavior.

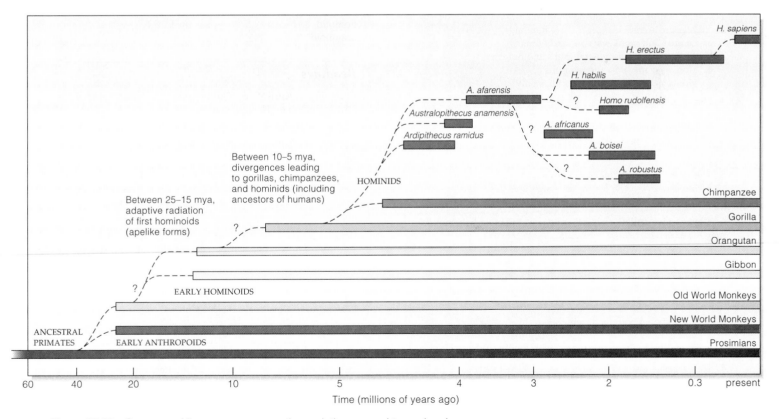

Figure 28.16 Summary of lineages on or near the evolutionary road to modern humans.

Review Questions

1. What is the difference between "hominoid" and "hominid"? Are we hominoids, hominids, or both? 28.1

2. List the presumed macroevolutionary trends among certain primate lineages that were ancestral to modern humans. 28.1

3. What environmental changes are correlated with the great adaptive radiation of apelike forms during the Miocene? 28.2

4. On the basis of the fossil record, where did the first humans (of genus *Homo*) originate? 28.3

5. Why is it difficult to determine how australopiths were related and which ones might have been ancestral to early humans? 28.2

6. Did *H. erectus* populations coexist for a time with *H. habilis*, which may have been ancestral to them? 28.3

7. Briefly describe some of the conserved physical traits that link anatomically modern humans with their mammalian ancestors, then with their primate ancestors. What are the characteristics that set modern humans apart from other primates? 28.3

8. Explain the difference between the multiregional model and African emergence model of modern human origins. 28.4

Self-Quiz *(Answers in Appendix IV)*

1. Primates include _____ .
   a. lemurs
   b. monkeys
   c. apes
   d. humans
   e. all of the above

2. Behavioral flexibility can be observed among _____ .
   a. mammals
   b. primates
   c. humans
   d. all of the above

3. _____ are hominoids; _____ are hominids.
   a. Lemurs and monkeys; apes and humans
   b. Apes and humans; australopiths and humans
   c. Monkeys; apes and australopiths
   d. Monkeys, apes, and humans; apes only

4. Hominids _____ .
   a. adapted to a wide range of environments
   b. adapted to a narrow range of environments
   c. had flexible bones that cracked easily
   d. were limber enough to swing through the trees

5. The oldest known primates date from the _____ .
   a. Miocene
   b. Paleocene
   c. Oligocene
   d. Pliocene

6. The adaptive radiations of Miocene apes and of early hominids may have been correlated with _____ .
   a. a long-term trend toward warmer global climates
   b. a long-term trend toward cooler global climates
   c. continental drift and changes in ocean circulation
   d. a and c are most likely
   e. b and c are most likely

7. The first known hominids were the _____ .
   a. *Homos*
   b. dryopiths
   c. cercopiths
   d. australopiths

8. The earliest known stone tools are _____ years old.
   a. 8 million
   b. 6 million
   c. 4.2 million
   d. 2.5 million

9. Match the primates with their descriptions.
   ____ prosimian
   ____ tarsioid
   ____ anthropoid
   ____ hominoid
   ____ hominid
   a. monkeys, apes, and humans
   b. lemurs
   c. humans and recent ancestors
   d. apes and humans
   e. tarsiers

Critical Thinking

1. Fossil evidence suggests that Neandertals and *H. erectus* populations coexisted in the same parts of the Middle East for 25,000 to 50,000 years. By one hypothesis, the Neandertals evolved in Europe, from *H. erectus* stock that had dispersed from Africa much earlier, then migrated to the Middle East. It appears that both groups used the same kinds of tools, and neither developed the techniques required for cave painting or for constructing decorative artifacts. Their extended coexistence implies that they did not compete with each other. However, the Neandertals became extinct and the other group gave rise to populations of anatomically modern humans. Does this scenario fit better with the multiregional model or African emergence model of modern human origins?

2. When it comes to human origins, different researchers "read" the fossil record in different parts of the world in different ways. Do you interpret this as an indication that the primate fossils they are citing might have nothing to do with human origins? Why or why not?

Selected Key Terms

anthropoid *28.1*	*Homo erectus 28.3*
arboreal (life-style) *28.1*	*Homo habilis 28.3*
australopith *28.2*	*Homo sapiens 28.3*
bipedalism *28.1*	human *28.3*
culture *28.1*	Primates (order) *28.1*
hominid *28.1*	savanna *28.1*
hominoid *28.1*	

Readings

Beardsley, T. April 1996. "Out of Food?" *Scientific American.* 20–22. Update on *H. erectus* migrations.

Larick, R., and R. Ciochon. November–December 1996. "The African Emergence and Early Asian Dispersal of the Genus *Homo.*" *American Scientist.*

Monastersky, R. 3 August 1996. "Out of Arid Africa." *Science News* (150): 74–75. Account of debates on whether climatic changes sparked pulses of evolutionary change among the hominids.

Waters, T. May 1990. "Almost Human." *Discover.* 43–53. Account of how noted sculptor John Gurche is painstakingly reconstructing the heads and facial features of seven forms of early hominids and of a modern human.

Weiss, M., and A. Mann. 1990. *Human Biology and Behavior.* Fifth edition. New York: Harper Collins.

Web Site See *http://www.wadsworth.com/biology* for practice quiz questions, hypercontents, BioUpdates, and critical thinking. The Wadsworth Biology Resource Center provides a wealth of information fully organized and integrated by chapter.

FACING PAGE: *A flowering plant* (Prunus) *busily doing what it does best: producing flowers for the fine art of reproduction.*

29 PLANT TISSUES

Plants Versus the Volcano

On a clear spring day in 1980, in a richly forested region of the Cascade Range of southwestern Washington, Mount Saint Helens exploded and 500 million metric tons of ash were blown skyward. Within minutes, shock waves from the blast flattened or incinerated hundreds of thousands of mature trees near the northern flanks of the mountain. Hot volcanic ash surged down the slopes at rates exceeding 44 meters per second, then turned into rivers of cementlike mud when the intense heat melted and released more than 75 billion liters of water from the mountain's snowfields and glacial ice.

In one mind-numbing moment, 40,500 hectares—100,000 acres—of magnificent forests dominated by hemlock and Douglas fir were transformed into barren sweeps of land (Figure 29.1*a,b*). In the aftermath of the violent eruption, more than a few observers gained stunning insight into what the world must have looked like long ago, before the first plants started to colonize habitats on land.

Yet it did not take long for existing plants to move back into habitats that their ancestors had claimed. In less than a year's time, seeds of a variety of flowering plants, including fireweed and blackberry, sprouted near the grayed trunks of fallen trees around Mount Saint Helens. Even before ten years passed, young willows and alders took hold near riverbanks, and low shrubs cloaked the land (Figure 29.1*c*). Their presence

a

b

c

Figure 29.1 (**a,b**) A grim reminder of what our world would be like without plants: the devastation following the violent eruption of Mount Saint Helens in 1980. Nothing remained of the extensive forest that had surrounded this Cascade volcano. (**c**) In less than a decade, however, seed-bearing vascular plants were making a comeback. (**d**) Merely twelve years after the volcanic eruption, seedlings of a dominant species (Douglas fir) were starting to reclaim the land.

afforded pockets of shade, which favored germination of the seeds of slower growing but ultimately dominant species, the hemlocks and Douglas fir (Figure 29.1*d*). In less than a century, the forest will be as it once was.

With this example, we open a unit dedicated to the seed-bearing vascular plants, with emphasis on the flowering types. In terms of their distribution and

d

diversity, they are the most successful plants on Earth. This first chapter provides an overview of their tissues and body plans. Chapter 30 explains how they absorb water and mineral ions, conserve water, and distribute organic substances through the plant body. Chapters 31 and 32 provide closer looks at their patterns of growth, development, and reproduction. As you will see, their structure and physiology (that is, the functioning of the plant body) help them survive sometimes hostile conditions on land—even momentary takeovers by volcanoes.

KEY CONCEPTS

1. Angiosperms (flowering plants) and, to a lesser extent, gymnosperms are the groups that now dominate the plant kingdom. These seed-bearing vascular plants have complex aboveground shoot systems, which consist of stems, leaves, flowers, and some other structures. They also have complex root systems that typically grow downward and outward through soil.

2. There are three major categories of tissue systems in these plants. A ground tissue system makes up the bulk of the plant body. A vascular tissue system distributes water, dissolved minerals, and photosynthetic products through roots and shoots. A dermal tissue system covers and protects plant surfaces exposed to the surroundings.

3. The simple plant tissues—parenchyma, collenchyma, and sclerenchyma—are each composed of no more than one type of cell.

4. Complex plant tissues incorporate two or more types of cells. Xylem and phloem, which are vascular tissues, are like this. So are the dermal tissues called epidermis and periderm.

5. Plants grow by way of mitotic cell divisions and cell enlargements at meristems, which are localized regions of self-perpetuating, embryonic cells.

6. Each growing season, shoots and roots lengthen. The lengthening, called primary growth, originates only at apical meristems, in the tips of shoots and roots.

7. Each growing season, the shoots and roots of many plants also thicken. Typically, lateral meristems inside shoots and roots give rise to an increase in diameter, which is called secondary growth. Wood is one outcome of secondary growth.

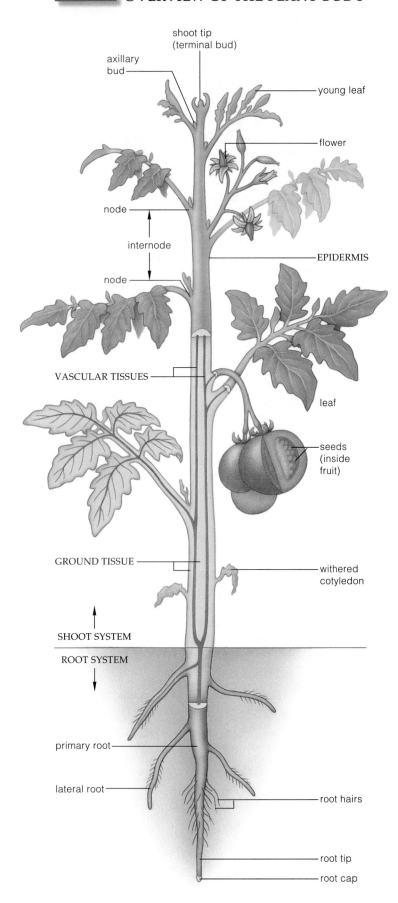

shoot tip
(terminal bud)

axillary
bud

young leaf

flower

node

internode

node

EPIDERMIS

VASCULAR TISSUES

leaf

seeds
(inside
fruit)

GROUND TISSUE

withered
cotyledon

SHOOT SYSTEM

ROOT SYSTEM

primary root

lateral root

root hairs

root tip

root cap

Earlier, in Chapter 25, we surveyed representatives of the 295,000 known species of plants. Even that sprint through diversity revealed why no one species can be used as a typical example of plant body plans. Even so, when we hear the word "plant," we usually think of well-known species of seed-bearing vascular plants—gymnosperms (such as pine trees) and angiosperms (flower-producing plants, such as roses, corn, cacti, and elms). With 260,000 species, angiosperms dominate the plant kingdom, and they will be our main focus.

Shoots and Roots

Many flowering plants have a body plan similar to that shown in Figure 29.2. Its aboveground parts, or **shoots**, are stems, leaves, flowers (reproductive shoots), and other structures. Its stems afford structural support for upright growth, and some parts of the stems conduct water and solutes. Upright growth gives photosynthetic cells in young stems and leaves favorable exposure to light. **Roots** are specialized structures that typically penetrate the soil and spread downward and outward through it. A root system absorbs water and dissolved minerals, and it commonly anchors aboveground parts. It also stores food, then releases it as required for root cells or for distribution to aboveground parts.

Three Plant Tissue Systems

The stems, branches, leaves, and roots of a flowering plant are similar in a key respect. Each has three major tissue systems. The **ground tissue system** is the most extensive; its tissues make up the bulk of the plant body. The **vascular tissue system** includes two kinds of conducting tissues that distribute water and solutes throughout the plant body. The **dermal tissue system** covers and protects the plant's surfaces. Figure 29.2 shows the general locations of these systems.

Some tissues in each system are simple, in that they contain one type of cell only. Parenchyma, collenchyma, and sclerenchyma fall in this category. Other tissues are complex, with organized arrays of two or more types of cells. Xylem, phloem, and epidermis are like this.

The next sections describe the tissue organization of shoots and roots. You may find it easier to interpret the photographs of this organization by studying Figure

Figure 29.2 Body plan for a representative angiosperm, a tomato plant (*Solanum esculentum*). Vascular tissues (*purple*) conduct water, dissolved minerals, and organic substances. They thread through ground tissue, which makes up the bulk of the plant body. A dermal tissue (in this case, epidermis) covers the surfaces of both the root system and shoot system.

Figure 29.3
Terms that identify how tissue specimens were cut from a plant. Cuts perpendicular to a stem or root's long axis give *transverse* sections (or cross-sections). Longitudinal cuts along the radius of a stem or root give *radial* sections. Cuts made at right angles to the radius of a stem or root give *tangential* sections.

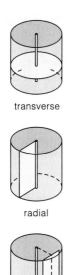

transverse

radial

tangential

Figure 29.4 (*Right*) Regions where growth generally occurs in vascular plants.

29.3. It can help you recognize the main ways in which tissue specimens are cut from plants.

Meristems—Where Tissues Originate

During the growing season, vascular plants do not grow everywhere at the same time. Plant growth is confined to **meristems**: localized regions of embryonic, self-perpetuating cells. In other regions, descendants of meristems are maturing or have reached maturity. Figure 29.4 shows the locations of meristems.

The lengthening of all shoots and roots originates at **apical meristems** located in their dome-shaped tip. Some of the cell populations that form here become *transitional* meristems known as protoderm, ground meristem, and procambium. These are the developmental bridges to the epidermis, ground tissue, and vascular tissues, respectively. Taken as a whole, the *lengthening* of stems and roots represents the plant's primary growth.

Also during the growing season, the older stems and roots of many plants thicken. The increases in girth start with lateral meristems inside the stems and roots. One of these, **vascular cambium**, produces secondary vascular tissues. The other lateral meristem, **cork cambium**, gives rise to a sturdier covering that replaces epidermis. Taken

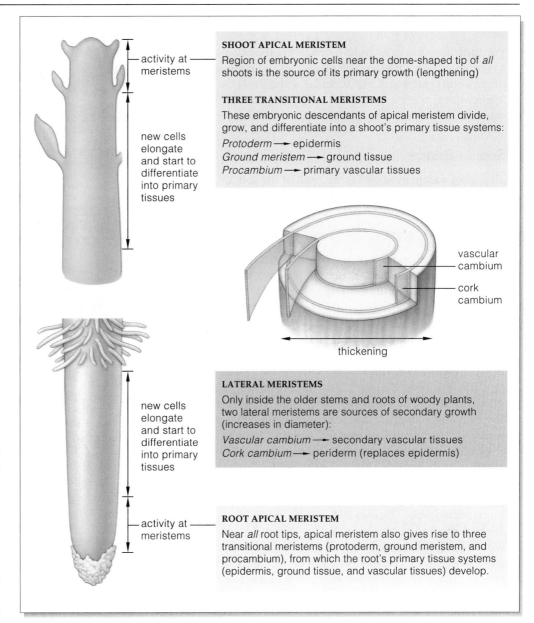

activity at meristems

new cells elongate and start to differentiate into primary tissues

new cells elongate and start to differentiate into primary tissues

activity at meristems

SHOOT APICAL MERISTEM
Region of embryonic cells near the dome-shaped tip of *all* shoots is the source of its primary growth (lengthening)

THREE TRANSITIONAL MERISTEMS
These embryonic descendants of apical meristem divide, grow, and differentiate into a shoot's primary tissue systems:

Protoderm → epidermis
Ground meristem → ground tissue
Procambium → primary vascular tissues

vascular cambium

cork cambium

thickening

LATERAL MERISTEMS
Only inside the older stems and roots of woody plants, two lateral meristems are sources of secondary growth (increases in diameter):

Vascular cambium → secondary vascular tissues
Cork cambium → periderm (replaces epidermis)

ROOT APICAL MERISTEM
Near *all* root tips, apical meristem also gives rise to three transitional meristems (protoderm, ground meristem, and procambium), from which the root's primary tissue systems (epidermis, ground tissue, and vascular tissues) develop.

as a whole, the *thickening* of stems and roots represents secondary growth.

Vascular plants have stems that support upright growth and conduct substances, leaves that function in photosynthesis, shoots specialized for reproduction, and other structures. They also have roots that absorb water and solutes, anchor aboveground parts, and often store food.

A ground tissue system makes up most of the plant body. A vascular tissue system distributes water, dissolved ions, and photosynthetic products through it. A dermal tissue system covers and protects plant surfaces.

Shoots and roots lengthen (put on primary growth) when their apical and transitional meristems are active. In many plants, older stems and roots also thicken (add secondary growth) when lateral meristems called vascular cambium and cork cambium are active.

TYPES OF PLANT TISSUES

We turn now to an overview of the organization and functions of plant tissues. Simple tissues are composed of only one type of cell. The vascular and dermal tissues are complex, with a variety of cell types. Figures 29.5 through 29.9 show examples from these tissue categories.

Simple Tissues

Tissues of **parenchyma** make up most of the soft, moist, primary growth of roots, stems, leaves, flowers, and fruits. Most parenchyma cells are pliable, thin-walled, and many-sided (Figure 29.6). When mature, they stay alive and retain the capacity to divide. Often their cell divisions heal wounded plant parts. **Mesophyll**, a type of parenchyma in leaves, is photosynthetic. Air spaces between the photosynthetic parenchyma cells enhance gas exchange. Other types of parenchyma have roles in storage, secretion, and other tasks. Parenchyma cells also thread through the vascular tissue systems.

Collenchyma provides flexible support for primary tissues (Figure 29.6). Its living cells often form patches or cylinders near the surface of a lengthening stem. They also form pliable ribs in many leaf stalks. Most of the cells are elongated and have unevenly thickened walls. The walls are impregnated with pectin, which is a gluelike polysaccharide that binds cellulose fibrils together and imparts pliability to the tissue.

Sclerenchyma supports mature plant parts and also protects many seeds. Most of its cells have thick, lignin-impregnated walls (Figures 29.6 and 29.7). Lignin, recall, strengthens and waterproofs walls. Without it, plants would not have evolved on land (Section 25.1).

Sclerenchyma consists of fibers or sclereids. *Fibers* are long, tapered cells in the vascular tissue systems of some stems and leaves. They can flex and twist without stretching. We use certain fibers to make cloth, rope, paper, and other commercial products (Figure 29.7a,b). *Sclereids* are stubbier cells. They impart a gritty texture to pears and certain other fleshy fruits (Figure 29.7c), and they form seed coats, coconut shells, and peach pits.

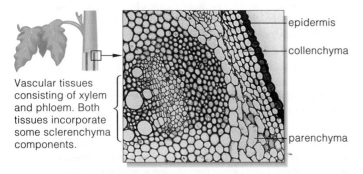

Vascular tissues consisting of xylem and phloem. Both tissues incorporate some sclerenchyma components.

epidermis
collenchyma
parenchyma

Figure 29.5 Locations of simple tissues and complex tissues in a portion of one kind of plant stem, transverse section.

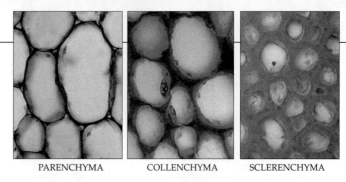

PARENCHYMA   COLLENCHYMA   SCLERENCHYMA

Figure 29.6 From the stem of a sunflower plant (*Helianthus*), examples of simple tissue, transverse section.

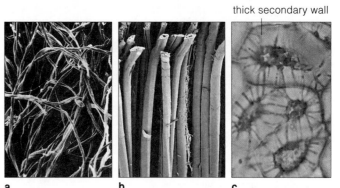

thick secondary wall

a   b   c

Figure 29.7 Different types of sclerenchyma. Fibers from (**a**) cotton plants and (**b**) flax plants. The loosely twisted cotton fibers interlock and form threads when spun. (**c**) From a pear, one type of sclereid: stone cells with thick, lignified walls.

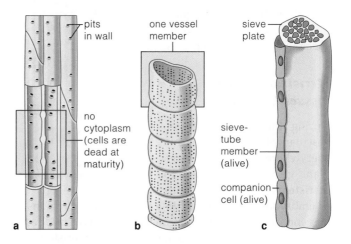

pits in wall
one vessel member
sieve plate
no cytoplasm (cells are dead at maturity)
sieve-tube member (alive)
companion cell (alive)

a   b   c

Figure 29.8 From xylem, portions of (**a**) three tracheids and (**b**) a vessel. Pipelines made of such cells conduct water and dissolved ions. (**c**) One type of phloem cell. Long tubes of many such cells conduct sugars and other organic compounds.

Complex Tissues

VASCULAR TISSUES Two vascular tissues, called xylem and phloem, function in the distribution of substances throughout the plant. Often their conducting cells are associated with a sheath of fibers and parenchyma cells.

Xylem conducts water and dissolved mineral ions. It also helps mechanically support a plant. Figure 29.8a,b shows examples of its conducting cells. The cells, called

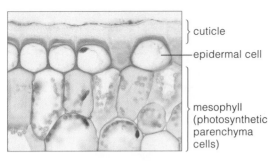

Figure 29.9 Light micrograph of a section through the upper surface of a kaffir lily leaf. The cuticle is made of secretions from epidermal cells. Inside the leaf are many photosynthetic parenchyma cells.

Figure 29.10 (*Right*) Comparison of the defining features of dicots and monocots. Both classes of flowering plants consist of the same simple and complex tissues, but their body parts show some differences in structural organization.

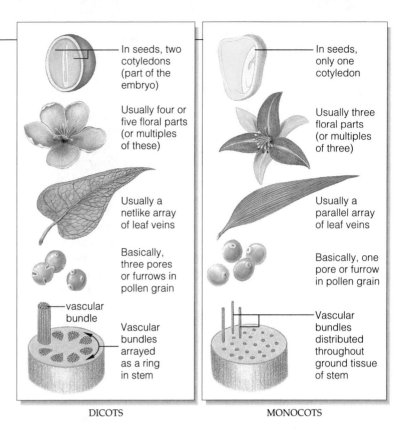

DICOTS   MONOCOTS

vessel members and *tracheids*, are dead at maturity, and their lignified walls interconnect. Collectively, the walls form water-conducting pipelines and also strengthen plant parts. Water flows into and out of the adjoining cells through numerous pits in the cell walls.

Phloem conducts sugars and other solutes through the plant. Its main conducting cells, called *sieve-tube members*, are alive at maturity (Figure 29.8c). Adjacent cells interconnect at openings in their side walls and at their open or perforated end walls. Sugars produced in leaves are loaded into sieve-tube members. Specialized, living parenchyma cells called *companion cells* typically assist the loading process. Sugars traveling through the pipelines of phloem are unloaded in any region where cells are growing or storing food. The chapter to follow describes this process.

DERMAL TISSUES All surfaces of primary plant parts are covered and protected by a dermal tissue system called **epidermis**. In most plants, epidermis is mainly a single layer of unspecialized cells. Waxes and cutin (a fatty substance) coat the outermost cell walls. We call the surface coating a **cuticle**. The plant cuticle functions in restricting water loss and often in resisting attacks by certain microorganisms (Figure 29.9).

Stem and leaf epidermis contains a large number of specialized cells, such as pairs of guard cells that change shape in response to changing conditions. A gap called a **stoma** (plural, stomata) can close or open up between them. The next chapter looks at how stomata are control points for the movement of water vapor, oxygen, and carbon dioxide across the epidermis. **Periderm** replaces epidermis in stems and roots with secondary growth. Dead cork cells in this protective covering have walls heavily impregnated with suberin, a fatty substance.

Dicots and Monocots—Same Tissues, Different Features

The **dicots** and **monocots**, recall, are the two classes of flowering plants (Section 25.8). Most trees and shrubs other than conifers—such as maples, elms, roses, cacti, peas, beans, lettuces, cotton, and carrots—are dicots. Palm trees, lilies, orchids, ryegrass, bamboos, wheat, corn, sugarcane, and pineapples are familiar monocots.

Dicots and monocots are similar in structure and function, but they differ in some distinctive ways. For example, dicot seeds have two cotyledons and monocot seeds have only one. Cotyledons are leaflike structures, commonly known as seed leaves. They form in seeds as part of a plant embryo, and they store or absorb food for it. After the seed germinates, the cotyledons wither and leaves start functioning. Figure 29.10 shows other differences between dicots and monocots.

Most of the plant body (ground tissue system) consists of parenchyma, collenchyma, and sclerenchyma. Each of these simple tissues is composed of only one type of cell.

Xylem and phloem are vascular tissues. In xylem, pipelines made of tracheids and vessel members conduct water and dissolved ions. In phloem, sieve tube members interact with companion cells to distribute organic compounds.

Of two dermal tissues, epidermis covers the surfaces of the primary plant body. Periderm replaces it on plant parts that have extensive secondary growth.

Monocots and dicots consist of the same tissues, but each has some of the tissues organized in distinctive ways.

PRIMARY STRUCTURE OF SHOOTS

How Stems and Leaves Form

Next time you or a friend eats a bundle of bean sprouts or alfalfa sprouts, carefully pull one aside and look at its structure. That seedling, which started forming while it was still inside a seed coat, already has a primary root and shoot. Inside the tip of the primary shoot, apical meristem and its derivative meristematic tissues are laying out an orderly framework for the stem's primary structure (Figure 29.11). Beneath the apical meristem, most tissue regions become progressively specialized as the cells divide at different rates, in different directions, and as they differentiate in size, shape, and function. All of this meristematic activity results in distinct stem regions, leaves, and axillary buds from which lateral shoots will develop. In turn, the lateral shoots give rise to side branches and reproductive structures.

Briefly, as a typical shoot lengthens, bulges of tissue develop along the flanks of the apical meristem. Each bulge is a leaf primordium (plural, primordia), or a rudimentary leaf (Figures 29.11 and 29.12). As growth continues, the stem lengthens between tier after tier of new leaves. Each part of the stem where one or more leaves are attached is a node. As shown in Figure 29.2, the stem region between two successive nodes is an internode. Buds develop in the leaf axils (that is, in the upper angle where leaves attach to the stem). A **bud** is an undeveloped shoot of mostly meristematic tissue, often protected by modified leaves called bud scales. As you will see in Chapter 31, buds give rise to new stems and to leaves, flowers, or both.

Internal Structure of Stems

While the primary plant body of a monocot or dicot is forming, the ground, vascular, and dermal tissues of the stem become organized in distinctive ways. Most often, primary xylem and phloem develop inside the same

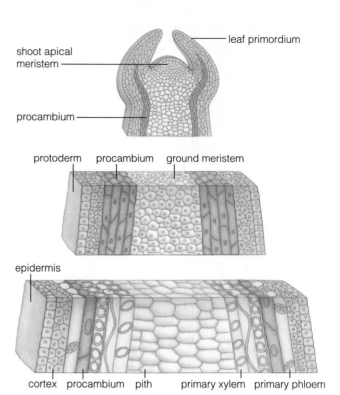

Figure 29.11 Successive stages in primary growth that started with activity at the shoot apical meristem of a typical dicot, then continued at transitional meristems derived from it. Notice the progressive differentiation of most of the tissue regions.

sheath of cells, as **vascular bundles**. Such bundles are multistranded cords that thread lengthwise through the ground tissue system of the primary and lateral shoots. They commonly develop according to two genetically dictated patterns. In most dicot stems, the long bundles are arranged as a ring that divides the ground tissue into a cortex and pith (Figure 29.13a). The stem's **cortex** is the region in between the vascular bundles and the epidermis. Its **pith** is the stem's center, inside the ring of vascular bundles. The ground tissue of the plant's roots

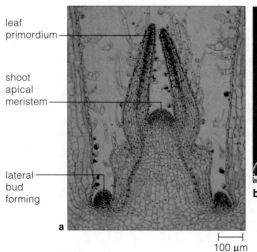

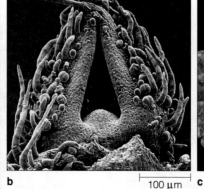

Figure 29.12 (a) Light micrograph of a *Coleus* shoot tip, cut longitudinally through its center. (b) Scanning electron micrograph of its surface. (c) New *Coleus* leaves.

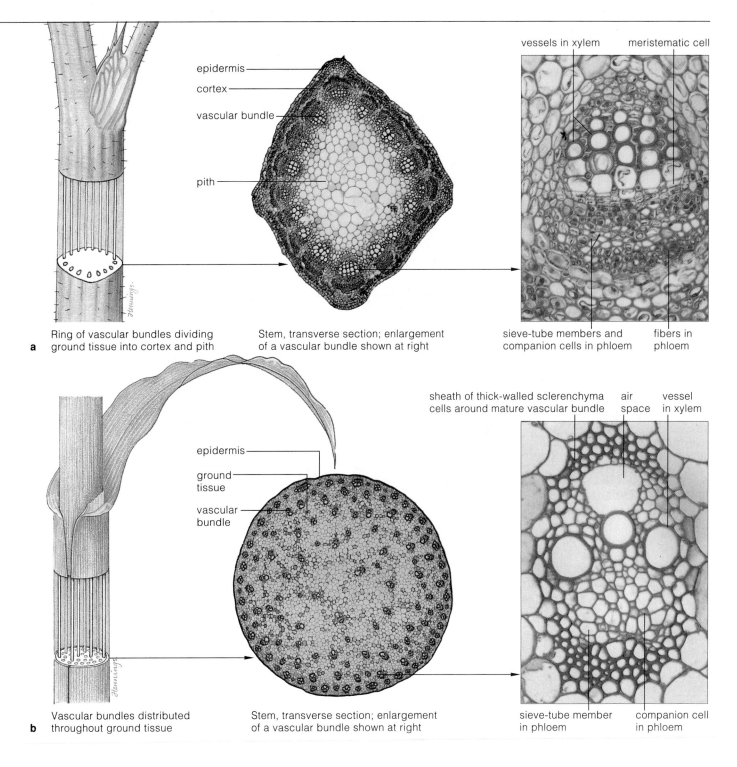

a Ring of vascular bundles dividing ground tissue into cortex and pith

epidermis
cortex
vascular bundle
pith

Stem, transverse section; enlargement of a vascular bundle shown at right

vessels in xylem   meristematic cell

sieve-tube members and companion cells in phloem   fibers in phloem

b Vascular bundles distributed throughout ground tissue

epidermis
ground tissue
vascular bundle

Stem, transverse section; enlargement of a vascular bundle shown at right

sheath of thick-walled sclerenchyma cells around mature vascular bundle   air space   vessel in xylem

sieve-tube member in phloem   companion cell in phloem

Figure 29.13 Internal organization of the stems from a dicot and a monocot.

(**a**) Part of a stem from alfalfa (*Medicago*), a dicot. In many species of dicots and conifers, the vascular bundles develop in a more or less ringlike array in the ground tissue system, as shown here. The portion of the ground tissue between the ring and the surface of the stem is the cortex; the portion enclosed within the ring is the pith.

(**b**) Part of a stem from corn (*Zea mays*), a monocot. In most monocots and some of the nonwoody dicots, vascular bundles are scattered through the ground tissue, as shown here.

becomes similarly divided, into root cortex and pith. A different pattern is common inside the stems of most monocots and some dicots. Their long vascular bundles are scattered through the ground tissue (Figure 29.13*b*). How different substances are conducted through such vascular systems is a topic of the next chapter.

The primary plant body has a distinctive internal structure, as in the pattern in which its vascular bundles are arranged.

Similarities and Differences Among Leaves

Every **leaf** that forms is a metabolic factory, equipped with many photosynthetic cells. Yet leaves vary greatly in size, shape, surface details, and internal structure. A duckweed leaf is no more than 1 millimeter (0.04 inch) across; the leaves of some water lilies are 2 meters (6.5 feet) wide. Various leaves resemble blades, spikes, cups, needles, feathers, tubes, and other structures. They also differ greatly in coloration, odor, and edibility; many are poisonous. The leaves of birches and other species of *deciduous* plants drop away from the stem as winter approaches. The leaves of camellias and other *evergreen* plants also drop, but not all at once.

A typical leaf has a flat blade and petiole, or stalk, that attaches it to the stem (Figure 29.14*a*). The *simple*

Figure 29.14 Common leaf forms of (**a**) dicots and (**b**) monocots. Examples of (**c**) simple leaves and (**d**) compound leaves.

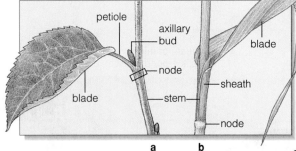

petiole
axillary
bud
blade
node
sheath
blade
stem
node

a   b

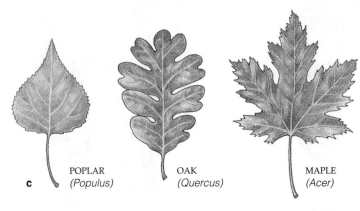

POPLAR
(*Populus*)

OAK
(*Quercus*)

MAPLE
(*Acer*)

c

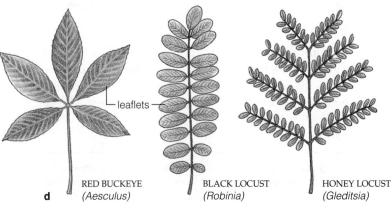

leaflets

RED BUCKEYE
(*Aesculus*)

BLACK LOCUST
(*Robinia*)

HONEY LOCUST
(*Gleditsia*)

d

leaves are not divided, although many are lobed (Figure 29.14*c*). The *compound* leaves have blades divided into leaflets, all arranged in the same plane (Figure 29.14*d*). Most monocot leaves, such as those of ryegrass and corn, are flat surfaced like a knife blade. The blade's base encircles and sheathes the stem, as in Figure 29.14*b*.

Most leaves are thin, with a high surface-to-volume ratio. Often their flat surface is oriented perpendicular to sunlight, and the plant's individual leaves commonly project from the stems in patterns that minimize the shading of their neighbors. For example, *Coleus* leaves form in pairs on opposite sides of a stem, and petioles of each pair attach to the stem at right angles to those of the pair above it (Figure 29.12*c*). Such leaf adaptations afford maximum interception of sunlight, which is the plant's energy source. They also promote the inward diffusion of carbon dioxide and outward diffusion of oxygen. When a leaf is thick, you can safely assume it belongs to a succulent or some other plant that lives in a dry habitat, and that it serves in water storage as well as photosynthesis. The leaves of many desert plants also orient themselves parallel with the sun's rays and thereby reduce heat absorption to tolerable levels.

The Fine Structure of Leaves

In its fine structure also, the leaf is adapted to intercept sunlight energy and promote gas exchange. In addition, many leaves have distinctive surface specializations.

LEAF EPIDERMIS Epidermis covers every leaf surface exposed to air. It may be smooth surfaced, sticky, or slimy, with hairs, scales, spikes, hooks, glands, and other surface specialties. The introduction to Chapter 30 gives examples. Covering the sheetlike, compact array of epidermal cells is a cuticle that minimizes the loss of precious water (Figures 29.9 and 29.15). Most plant leaves have far more stomata on the lower surface than on the upper surface. (Water lily leaves are one of the exceptions.) Stomata of plants growing in arid habitats commonly are located in depressions in the leaf surface, along with thick-coated epidermal hairs. Their location, and that of the hairs, helps restrict water loss.

MESOPHYLL—PHOTOSYNTHETIC GROUND TISSUE As you read earlier, mesophyll, a type of parenchyma with photosynthetic cells and a large volume of air spaces, extends throughout the interior of a leaf. Figure 29.15 shows an example. The air spaces, which connect with outside air through stomata, promote rapid diffusion of carbon dioxide to cells and oxygen away from them. Leaves that are oriented perpendicular to the sun's rays have two layers of mesophyll. Columnar parenchymal

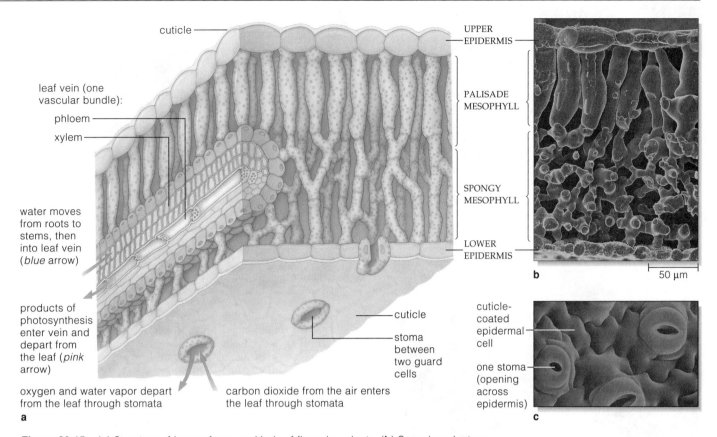

cuticle

leaf vein (one
vascular bundle):

phloem

xylem

water moves
from roots to
stems, then
into leaf vein
(*blue* arrow)

products of
photosynthesis
enter vein and
depart from
the leaf (*pink*
arrow)

oxygen and water vapor depart
from the leaf through stomata

a

carbon dioxide from the air enters
the leaf through stomata

UPPER
EPIDERMIS

PALISADE
MESOPHYLL

SPONGY
MESOPHYLL

LOWER
EPIDERMIS

b

50 µm

cuticle

stoma
between
two guard
cells

cuticle-
coated
epidermal
cell

one stoma
(opening
across
epidermis)

c

Figure 29.15 (**a**) Structure of leaves for many kinds of flowering plants. (**b**) Scanning electron
micrograph showing the tissue organization of a leaf from the kidney bean plant (*Phaseolus*).
Notice the compact array of epidermal cells. (**c**) Stomata, tiny openings across the epidermis,
appear when paired guard cells are in their plumped configuration.

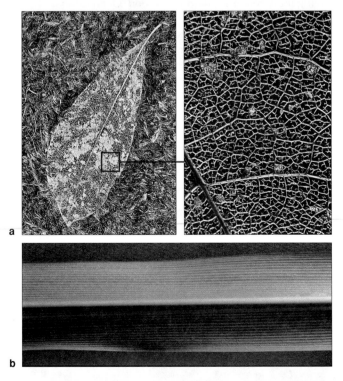

a

b

Figure 29.16 (**a**) Decaying dicot leaf, showing its netlike
veins. (**b**) Parallel veins of a monocot leaf (*Agapanthus*).

cells are attached to the upper epidermis. They seem
densely packed, but most of their cell walls are exposed
to the air. These cells of the *palisade* mesophyll have far
more chloroplasts and carry out more photosynthesis,
compared to the cells making up the layer of *spongy*
mesophyll below them (Figure 29.15). Monocot leaves
grow vertically and intercept light from all directions;
their mesophyll is not organized as two layers.

VEINS—THE LEAF'S VASCULAR BUNDLES   A leaf's **veins**
are vascular bundles, often strengthened with fibers.
Their continuous strands of xylem rapidly move water
and dissolved nutrients to all of the mesophyll cells,
and continuous strands of phloem carry photosynthetic
products, especially sugars, away from them. In most
dicots, veins branch lacily into a number of minor veins
that are embedded in the mesophyll. In most monocots,
veins are more or less similar in length and run parallel
with the leaf's long axis (Figure 29.16).

A leaf's structure is adapted for sunlight interception, gas
exchange, and distribution of water, dissolved nutrients,
and photosynthetic products. Each species has leaves of a
distinctive size and shape, and often surface specializations.

29.5 USES AND ABUSES OF SHOOTS

Imagine grocery carts deprived of stems and leaves. Out go the lettuces and cabbages, the celery and parsley, the spinach and mint. No thyme, basil, rosemary, and oregano for spaghetti sauce. No more green or black or oolong teas (Figure 29.17). Imagine parks and yards without sweeps of lawns and flowerbeds, or a cupboard without sugar and flour, cookies and bread (Figure 29.18). Or think of a bride deprived of her special bouquet of those splendidly modified shoots called flowers (Figure 29.19).

Now think about the earliest human species—the first hominids—as described in the preceding chapter. Through trial and error, they must have already started to acquire intimate knowledge of which plants provide taste thrills and which can kill. By 300,000 years ago, *H. erectus* clans living in China were already stashing pine nuts, walnuts,

Figure 29.19 A bride tossing her bouquet of floral shoots to all the single women who attended her wedding ceremony, a ritual that supposedly reveals who will be married next. (A sixty-eight-year-old optimist caught this one.)

hazelnuts, and rose hips in caves and roasting hackberry seeds. At least by then, adults were teaching their children which plants were edible and which were toxic. Through language, youngsters learned about the plants that had evolved in their parts of the world.

Starting about 11,000 years ago, wheat, barley, and other plants were domesticated, this being a way to count on having more reliable quantities of food. Of an estimated 3,000 species that various human populations recognized as food, only about 200 eventually became *the* major crops.

Plant lore still threads through our lives in many other ways. Landscapers and homeowners know to plant shrubs next to a building to help keep it from heating up in the summer and losing heat in the winter. We make twine and

Figure 29.17 Indonesian woman gathering tender shoots of tea plants, which are evergreen shrubs in the same genus as camellias. Plants grown on hillsides in moist, cool regions yield leaves with the most prized flavors. Only the terminal bud and two or three of the youngest leaves are picked for fine teas.

Figure 29.18 More of the highly prized shoots. (**a**) Flax (*Linum usitatissimum*). Lustrous fibers from its wiry stems are used to make linen. They are about three times stronger than cotton fibers. (**b**) From Hawaii, sugarcane (*Saccharum officinarum*). The wild stock of this cultivated species probably evolved in New Guinea. The sap extracted from its cut stems is boiled down to make sucrose crystals (table sugar) or syrups. (**c**) Triticale is a hybrid grain;the parental stocks are wheat (*Triticum*) and rye (*Secale*). This popular grain has wheat's high yield and rye's tolerance of harsh environmental conditions. (**d**) From the American Midwest, mechanized harvesting of a field of common bread wheat.

a

b

c

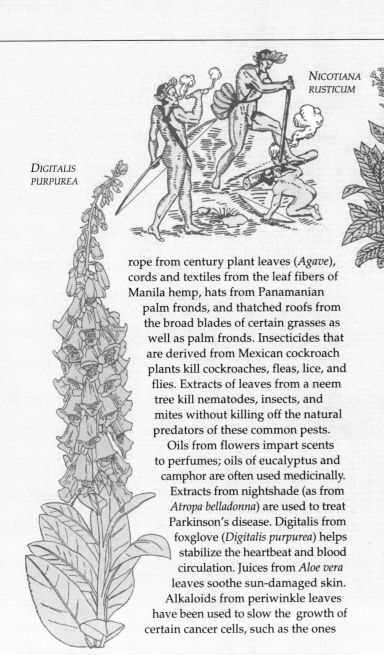

DIGITALIS PURPUREA

NICOTIANA RUSTICUM

CANNABIS SATIVA

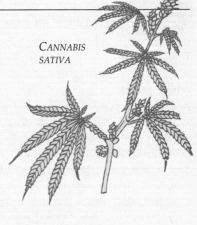

HYOSCYAMUS

rope from century plant leaves (*Agave*), cords and textiles from the leaf fibers of Manila hemp, hats from Panamanian palm fronds, and thatched roofs from the broad blades of certain grasses as well as palm fronds. Insecticides that are derived from Mexican cockroach plants kill cockroaches, fleas, lice, and flies. Extracts of leaves from a neem tree kill nematodes, insects, and mites without killing off the natural predators of these common pests.

Oils from flowers impart scents to perfumes; oils of eucalyptus and camphor are often used medicinally. Extracts from nightshade (as from *Atropa belladonna*) are used to treat Parkinson's disease. Digitalis from foxglove (*Digitalis purpurea*) helps stabilize the heartbeat and blood circulation. Juices from *Aloe vera* leaves soothe sun-damaged skin. Alkaloids from periwinkle leaves have been used to slow the growth of certain cancer cells, such as the ones

that cause malignant tumors of the lymphatic system (this cancer is known as Hodgkins disease).

Through much of recorded history, people also have figured out how to use leaves in harmful ways. Ancient Mayans cultivated tobacco plants (*Nicotiana tabacum* and *N. rusticum*) and introduced European explorers to tobacco smoking. Mayan priests thought that the smoke rising from pipes carried their priestly thoughts to the gods. People continue to smoke, chew, and tuck into their mouth the leaves of tobacco plants and so are candidates for the hundreds of thousands of annual deaths from lung, mouth, and throat cancers. Heavy smoking of *Cannabis sativa*, the source of marijuana and other mind-altering substances, is linked to abnormally low sperm counts. Cocaine, a substance derived from coca leaves, is used medicinally. It also is abused by millions of people who have have become addicted to its mind-altering properties, with devastating social and economic effects.

And what about that henbane (*Hyoscyamus niger*) and belladonna? Their alkaloids have been tapped for the occasional murder as well as for medicinal purposes. Remember Hamlet's father? As he slept, he was sneakily dispatched by someone who poured a solution of henbane into his ear. Remember Juliet's heartbroken, suicidal Romeo? After he sipped a potion of nightshade, he dropped dead. Those self-proclaimed witches of the Middle Ages used henbane, and atropine from nightshade, in some of their suspect rituals. They used sticks to apply atropine solutions to their body and induce sensations of weightlessness. During their atropine-induced hallucinatory sprees, they assumed they were flying off to meet with demons. Hence those Halloween cartoons of witches flying hither and yon on broomsticks.

ATROPA BELLADONNA

d

PRIMARY STRUCTURE OF ROOTS

Taproot and Fibrous Root Systems

When you think you can't bear one more gratuitously violent action-hero movie, give your mind a rest and watch a seed germinate. The first part to emerge from the seed coat is a **primary root**. In most dicot seedlings, the primary root increases in diameter while it grows downward. Later, **lateral roots** start forming in internal tissues at an angle perpendicular to the primary root's axis, then they erupt through epidermis. The youngest lateral roots are closest to the root tip. A primary root with lateral branchings is a **taproot system**. Dandelions, carrots, oak trees, and poppies are examples of plants with a taproot system (Figure 29.20a).

By contrast, the primary root of monocots, including rye plants and other grasses, is generally short-lived. In its place,

adventitious roots arise from the stem, then lateral roots branch from these. (*Adventitious* structures develop at an unusual location.) All of the lateral roots are more or less alike in diameter and length. Collectively, roots that form this way are a **fibrous root system** (Figure 29.20b).

Internal Structure of Roots

Figure 29.21 shows the meristems in the tip of one root of such systems. Many of the cellular descendants of these meristems divide, enlarge, elongate, and become cells of the primary tissue systems. Notice the root cap, a dome-shaped mass of cells at the tip. Apical meristem produces the cap and in turn is protected by it.

Protoderm gives rise to the root epidermis, a plant's absorptive interface with the soil. Some epidermal cells send out extensions called **root hairs** (Figure 29.21a). Collectively, such root hairs greatly increase the surface

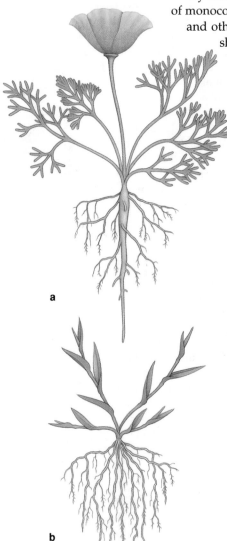

a

b

Figure 29.20 (**a**) Taproot system of a California poppy (*Eschscholzia californica*). (**b**) Fibrous root system of a grass plant.

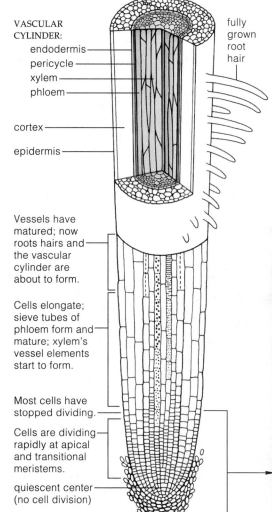

VASCULAR CYLINDER:
endodermis
pericycle
xylem
phloem

cortex

epidermis

Vessels have matured; now roots hairs and the vascular cylinder are about to form.

Cells elongate; sieve tubes of phloem form and mature; xylem's vessel elements start to form.

Most cells have stopped dividing.

Cells are dividing rapidly at apical and transitional meristems.

quiescent center (no cell division)

a   root cap

fully grown root hair

Figure 29.21 (**a**) Generalized diagram of a primary root, showing its zones of cell division, elongation, and differentiation. (**b**) Light micrograph of root tip of corn (*Z. mays*), longitudinal section. The oldest cells are farthest from the meristems, which the root cap protects. Root cap cells secrete mucigel, a polysaccharide-rich slime that lubricates the root as cell divisions push it through the soil. Mucigel ends up covering all of the root's epidermis. It may enhance the uptake of dissolved mineral ions and the formation of mycorrhizae.

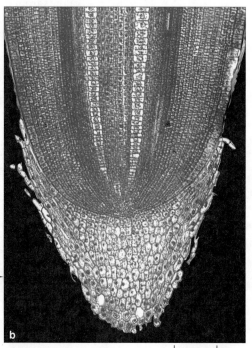

b

100 μm

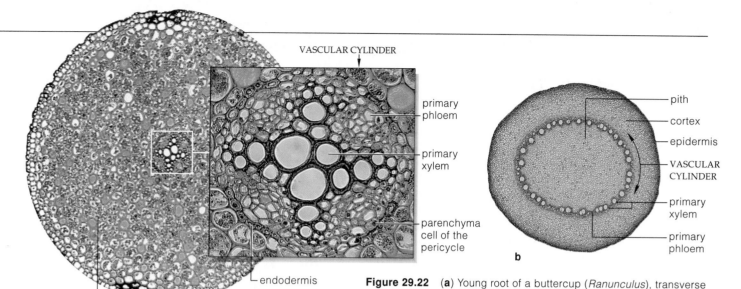

Figure 29.22 (**a**) Young root of a buttercup (*Ranunculus*), transverse section. The inset shows details of its vascular cylinder. (**b**) Root of a corn plant (*Z. mays*), transverse section. Its vascular cylinder divides the ground tissue into cortex and pith.

area available for taking up water and nutrients. Only first-time or foolish gardeners would yank a plant from the ground when transplanting it. They would tear off too much of the highly fragile absorptive surface.

Apical meristem also gives rise to the ground tissue system and to a **vascular cylinder**. A vascular cylinder consists of primary xylem and phloem and one or more layers of parenchyma cells called the pericycle. Figure 29.22a shows an example of a vascular cylinder at the center of the cortex of a dicot root. Figure 29.22b shows

how a monocot's vascular cylinder divides the ground tissue system into cortex and pith regions. Either way, there are plenty of air spaces in between cells of the ground tissue system, and oxygen can easily diffuse through them. Like other cells in the plant, living root cells depend on oxygen for aerobic respiration.

When water enters a root, it moves from cell to cell until it reaches the endodermis, a layer of cells around the vascular cylinder. Where endodermal cells abut, their walls are waterproof, so incoming water is forced to pass through their cytoplasm. As described in Chapter 30, this arrangement controls the movement of water and dissolved substances into the vascular cylinder.

The pericycle lies just inside the endodermis. Here, meristematic activity gives rise to lateral roots, which erupt through the cortex and epidermis (Figure 29.23).

Regarding the Sidewalk-Buckling, Record-Breaking Root Systems

Unless tree roots start to buckle a sidewalk or choke off a sewer line, most of us don't pay much attention to the root systems of flowering plants. Most roots manage to mine the soil to a depth of 2 to 5 meters. In hot deserts, where free water is scarce, one hardy mesquite shrub is known to have sent roots down 53.4 meters (175 feet) near a streambed. Some "simple" cacti have shallow roots radiating outward for 15 meters. Someone once measured the roots of a young rye plant that had been growing for four months in only 6 liters of soil water. If the surface area of that root system were laid out in one sheet, it would occupy more than 600 square meters!

Primary roots provide a plant with a tremendous surface area for absorbing water and solutes.

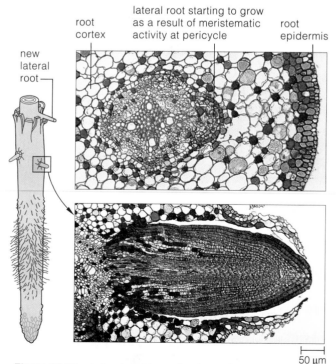

Figure 29.23 Lateral root formation in a primary root from a willow tree (*Salix*), transverse section.

Woody and Nonwoody Plants Compared

Flowering plant life cycles extend from germination to seed formation, then death. **Annuals** complete the life cycle in a single growing season, and they generally are *nonwoody*, or herbaceous, plants. Alfalfa and marigolds are like this. **Biennials**, such as carrots, live for two growing seasons. Their roots, stems, and leaves form the first season; flowers form, seeds form, and the plant dies in the next season. **Perennials** continue vegetative growth and seed formation year after year. Secondary tissues form in a number of them.

Like all gymnosperms, some of the monocots and many dicots add secondary growth during two or more growing seasons; they are *woody* plants. Early in life, their stems and roots are similar to those of nonwoody plants. Differences emerge after their lateral meristems become active and start producing large amounts of secondary vascular tissues, especially secondary xylem. The differences are especially pronounced in some of the perennial plants in which the vascular cambium has been reactivated every growing season for hundreds or thousands of years. Such ongoing meristematic activity has produced giants. For example, at last measure, the massive trunk of a coast redwood of the sort shown in Figure 29.24a was towering more than 110 meters above the forest floor. Its accumulated secondary growth has been estimated to weigh nearly 100 metric tons. The tree with greatest girth is a chestnut (*Castanea*) growing in Sicily. To walk completely around the base of it, you would have to pace off 58 meters.

Activity at the Vascular Cambium

Massive, woody stems and roots originate at the lateral meristem called vascular cambium. Take a look at the stem in Figure 29.24b. Every spring, primary growth resumes at its buds, and secondary growth proceeds inside it. When vascular cambium in the stem is fully developed, it is like a cylinder one cell or a few cells thick. Some cells (*fusiform initials*) give rise to secondary xylem and phloem, which extend longitudinally through

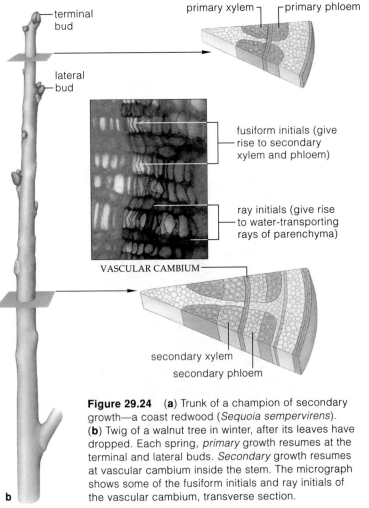

terminal bud

lateral bud

primary xylem — primary phloem

fusiform initials (give rise to secondary xylem and phloem)

ray initials (give rise to water-transporting rays of parenchyma)

VASCULAR CAMBIUM

secondary xylem
secondary phloem

Figure 29.24 (a) Trunk of a champion of secondary growth—a coast redwood (*Sequoia sempervirens*). (b) Twig of a walnut tree in winter, after its leaves have dropped. Each spring, *primary* growth resumes at the terminal and lateral buds. *Secondary* growth resumes at vascular cambium inside the stem. The micrograph shows some of the fusiform initials and ray initials of the vascular cambium, transverse section.

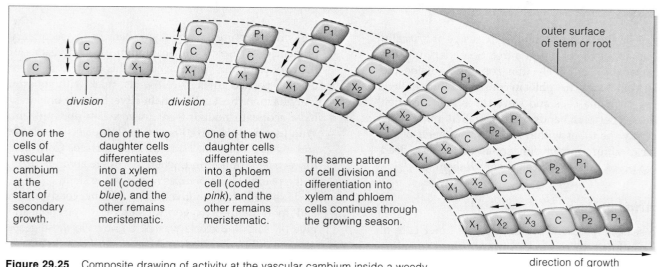

One of the cells of vascular cambium at the start of secondary growth.

One of the two daughter cells differentiates into a xylem cell (coded *blue*), and the other remains meristematic.

One of the two daughter cells differentiates into a phloem cell (coded *pink*), and the other remains meristematic.

The same pattern of cell division and differentiation into xylem and phloem cells continues through the growing season.

outer surface of stem or root

direction of growth

Figure 29.25 Composite drawing of activity at the vascular cambium inside a woody stem. Ongoing cell divisions enlarge the inner core of secondary xylem and displace the vascular cambium toward the surface of the stem or root.

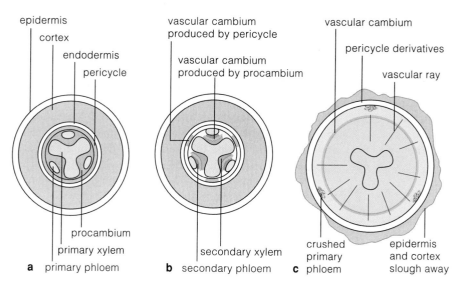

a
- epidermis
- cortex
- endodermis
- pericycle
- procambium
- primary xylem
- primary phloem

b
- vascular cambium produced by pericycle
- vascular cambium produced by procambium
- secondary xylem
- secondary phloem

c
- vascular cambium
- pericycle derivatives
- vascular ray
- crushed primary phloem
- epidermis and cortex slough away

Figure 29.26 Secondary growth in the root of one type of woody plant. (**a**) This is how tissues are organized as primary growth ends. (**b,c**) A thin cylinder of vascular cambium forms and gives rise to the secondary xylem and phloem. Cell divisions are parallel with the vascular cambium. The cortex ruptures as the root thickens.

the stem. Other cells of vascular cambium (*ray initials*) produce many rays of parenchyma cells. The rays deliver water "sideways" through the stem, in a radial pattern that is like a sliced pie. Without all of these longitudinal and radial vascular tissues, substances could not travel up, down, and sideways in enlarging woody stems.

Figure 29.25 shows a simple way to think about the pattern of secondary growth at vascular cambium. As you can see, secondary xylem forms on the *inner* face of this meristematic tissue, and secondary phloem forms on its *outer* face. Does an accumulation of secondary growth eventually squash the vascular cambium deep inside the stem? Not at all. As the inner core of xylem

gets thicker and thicker, it automatically and steadily displaces cells of the vascular cambium toward the stem surface. Those meristematic cells also maintain the ring of vascular cambium by dividing sideways, in an ever widening circle that is never far from the surface.

We have been focusing on how stems thicken, but bear in mind, secondary xylem and phloem form at vascular cambium in the plant's roots, also. Figure 29.26 shows one of the patterns of secondary growth at the vascular cambium in a typical root.

What are the selective advantages of having woody stems and roots? Remember, plants as well as other organisms compete for resources. In this case, the plants with taller stems or wider canopies that defy the pull of gravity can intercept more of the energy streaming in from the sun. With a greater energy supply for photosynthesis, they have the metabolic means to develop large root and shoot systems, hence to be more competitive in acquiring resources—and ultimately to be reproductively successful in particular habitats.

In woody plants, secondary vascular tissues form at a ring of vascular cambium inside older stems and roots. Wood is an accumulation of secondary xylem especially.

With their sturdier tissues, woody plants defy gravity and grow taller and broader. Where competition for sunlight is intense, the ones that intercept the most sunlight win. Other factors being equal, having more energy to drive photosynthesis provides advantages in terms of metabolic capabilities, growth, and reproductive success.

Where secondary xylem (wood) is extensive, it typically makes up about 90 percent of a tree. Secondary phloem is restricted to a relatively thin zone just outside the vascular cambium. This phloem consists of thin-walled, living parenchyma cells and sieve tubes that are often interspersed between bands of thick-walled, reinforcing fibers. Only the tubes within about a centimeter of the vascular cambium remain functional; the rest are dead and help protect the living cells beneath them.

Formation of Bark

As the seasons pass and a tree ages, its inner core of xylem continues to expand outward. The expanding core exerts pressure that is directed toward the stem or root surface. Eventually it ruptures the cortex and the outer portion of secondary phloem. Part of the cortex and epidermis splits away, and a new surface covering, the periderm, forms from cork cambium. Together, the periderm and secondary phloem constitute **bark**. In other words, bark is composed of all tissues external to the vascular cambium (Figure 29.27).

Periderm consists of cork, secondary cortex, and the cork cambium that produces these tissues. Soon after vascular cambium forms, cork cambium forms from the outermost parenchyma cells of the stem or root cortex. Such cells, recall, retain the capacity to divide. (When the cortex ruptures, parenchyma cells in the secondary phloem give rise to cork cambium.) Cell divisions at the cork cambium produce **cork**. This tissue consists of densely packed rows of cell walls, thick with suberin. Only its innermost cells remain alive, because only they have access to nourishment from xylem and phloem. With its many suberized layers, cork protects, insulates, and waterproofs the stem or root surface. Cork also forms over wounds and, when leaves are about to drop, at the places where petioles attach to the stem.

Like all living plant cells, the cells in woody stems and roots require oxygen for aerobic respiration and give off carbon dioxide wastes. So how do these gases get across the suberized, corky surface of bark? They do so through lenticels, which are localized areas where the packing of cork cells is loosened up a bit. Those dark spots you may have noticed on the cork of a wine bottle are all that is left of lenticels.

Heartwood and Sapwood

As a tree ages, changes also occur in the appearance and function of the wood itself. In the center of its older stems and roots is **heartwood**, which is dry tissue (it no longer transports water and solutes) and a dumping ground for some metabolic wastes. Resins, oils, gums, and tannins are among the metabolites. Eventually they clog and fill in the oldest xylem pipelines. Typically, they darken heartwood, strengthen it, and make it more aromatic.

Heartwood helps the tree defy gravity, but the big trees can get along without it. In the early 1900s, when lumbermen were active in California's groves of old-growth redwoods, someone with knowledge of heartwood cut a tunnel through a few of the biggest trees, the better to drive an automobile through them. Was this a cute idea, or was it analogous to cutting a tunnel through Grandpa?

By contrast, **sapwood** is secondary growth located between heartwood and the vascular cambium. Compared to heartwood, it is wet, usually pale, and not as strong. Maple trees provide a famous example. Each spring, from early March through early April, farmers in New England insert metal tubes into the sapwood of sugar maples (*Acer saccharum*). Sap, the sugar-rich fluid in secondary phloem, drips out into buckets.

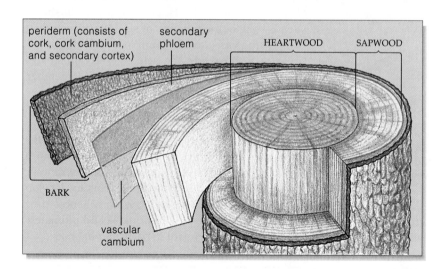

Figure 29.27 Structure of a stem with extensive secondary growth. Heartwood, the mature tree's core, has no living cells. Sapwood, a cylindrical zone of xylem between heartwood and vascular cambium, contains some living parenchyma cells among nonliving conducting cells of xylem. Everything outside the vascular cambium is bark.

Girdling is a deliberate stripping away of a band of secondary phloem around a trunk's circumference. With cuts through all of its vertical phloem, photosynthetically derived food cannot reach roots, which die—and so, in time, does the tree.

Early Wood, Late Wood, and Tree Rings

Vascular cambium becomes inactive during parts of the year in regions having cool winters or prolonged dry spells. The first xylem cells produced at the start of the

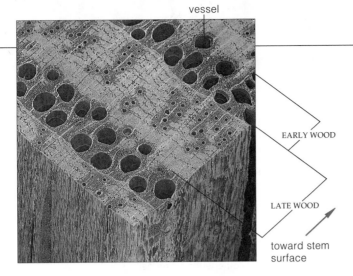

vessel

EARLY WOOD

LATE WOOD

toward stem surface

Figure 29.28 Scanning electron micrograph of early and late wood in a block cut from a red oak (*Quercus rubra*).

early and late wood. By contrast, the xylem of pines, spruces, redwoods, and other conifers have tracheids and parenchyma rays but no vessels or fibers. Conifers are **softwood** trees; without fibers, they are weaker and less dense than hardwoods (Figure 29.29*b,c*).

Limits to Secondary Growth

Unlike people, trees cannot run away from impending attacks. And if a pathogen penetrates their tissues, they cannot count on an immune system because they have none. Some trees live in habitats that are too harsh and too remote for most invaders, and their meristematic capacity for renewal has kept them going for many

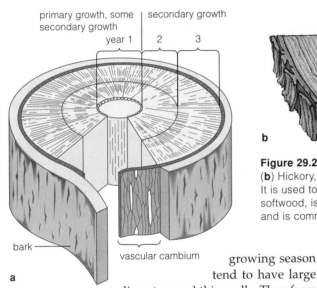

primary growth, some secondary growth | secondary growth

year 1 | 2 | 3

bark

vascular cambium

a

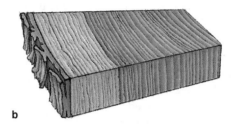

b

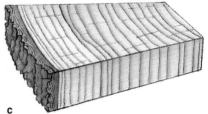

c

Figure 29.29 (**a**) Radial cut through a woody stem that has three annual rings. (**b**) Hickory, one of the hardwood dicots, is durable, strong, and yet resilient. It is used to make handles of hammers and other high-impact tools. (**c**) Pine, a softwood, is lightweight yet resists warping. Pine grows faster than hardwoods and is commercially "farmed" as a source of relatively cheap lumber.

growing season tend to have large diameters and thin walls. They form *early* wood. During the drier days of summer, vascular cambium gives rise to cells with smaller diameters and thicker walls. Such cells form *late* wood. When you come across "tree rings" in a transverse section from a trunk, these are alternating bands of early and late wood, which reflect light differently (Figures 29.28 and 29.29). The differences are called **growth rings**.

Seasonal change is predictable in temperate regions, and trees growing there usually add one growth ring per year. In deserts, thunderstorms rumble through at different times of year, and trees respond by adding more than one growth ring. Seasonal change is almost nonexistent in the tropics; as you might suspect, growth rings are not a feature of tropical trees.

Notice the large vessels in the block of oak shown in Figure 29.28. Oak is a **hardwood**. Like hickory, maple, and other dicot trees that evolved in temperate and tropical regions, it has vessels, tracheids, and fibers in its xylem. Depending on the species, the vessels may occur only in early wood or they may occur in both

thousands of years. (Remember the bristlecone pines?) Most trees growing elsewhere limit their vulnerability by walling off invaders, building a fortress of thickened cell walls around wounds, and deploying phenols and other toxic compounds. Taken as a whole, the responses to attack are called **compartmentalization**.

So potent are some tree toxins that they also kill cells of the tree itself. As compartments form around injured or infected or poisoned sites, the tree lays down new tissues over them. This works well when the tree is not under massive attack or when it can respond fast enough. Some don't make it; they die when too many compartments form and interfere with the flow of water and solutes through the vascular system.

Bark consists of all living and nonliving tissues outside the vascular cambium—that is, secondary phloem and periderm.

Periderm consists of cork (the outermost covering of woody stems and roots), cork cambium, and secondary cortex.

Wood may be classified by its location and functions (as in heartwood versus sapwood) and by the type of plant (many dicots produce hardwood, and conifers produce softwood).

---

1. Seed-bearing vascular plants include gymnosperms and angiosperms, which are flowering plants. Their shoots (stems, leaves, and other structures) and roots consist of dermal, ground, and vascular tissue systems.

2. Plant growth originates at meristems, which are localized regions of self-perpetuating, embryonic cells.

a. Primary growth, or lengthening of stems and roots, originates at apical meristems in root and shoot tips.

b. In many plants, secondary growth (or increases in diameter) originates inside stems and roots, at lateral meristems called vascular cambium and cork cambium.

3. Parenchyma, sclerenchyma, and collenchyma are the simple tissues, each with only one cell type (Table 29.1).

a. Parenchyma cells, alive and metabolically active at maturity, make up the bulk of ground tissue systems. They function in a variety of tasks; the ones making up mesophyll, for example, are photosynthetic.

b. Collenchyma helps strengthen some growing plant parts. Sclerenchyma supports mature plant parts.

4. Complex tissues include vascular tissues (xylem and phloem) and dermal tissues (epidermis and periderm). Each consists of two or more cell types (Table 29.1).

a. Vascular tissues distribute water and dissolved substances throughout the plant body. Vascular bundles, which are xylem and phloem inside a cellular sheath, thread through the ground tissue.

b. Water-conducting cells of xylem are not alive at maturity. Their lignified, pitted walls interconnect as pipelines for water and dissolved minerals.

c. Phloem's conducting cells are alive at maturity. The cytoplasm of the adjoining cells interconnects across perforated end walls and side walls. In leaves, sugars and other photosynthetic products are loaded into these cells, often with the help of companion cells. They are unloaded where cells are growing or storing food.

d. Epidermis covers and protects the outer surfaces of primary plant parts. Periderm replaces epidermis on plants showing extensive secondary growth.

5. Stems function in support of upright growth and in conducting substances through the plant body by way of vascular bundles. Most monocot stems have vascular bundles distributed throughout the ground tissue. Most dicot stems have a ring of bundles that divides the ground tissue into cortex and pith.

6. Leaves contain veins and mesophyll (photosynthetic parenchyma) between the upper and lower epidermis. Air spaces around the photosynthetic cells enhance gas exchange. Water vapor and gases cross the epidermis through numerous tiny openings called stomata.

7. Roots absorb water and mineral ions for distribution to aboveground parts. They also anchor the plant. Most store food, and some help support the shoot.

Table 29.1 Summary of Flowering Plant Tissues and Their Components

SIMPLE TISSUES

Parenchyma	Parenchyma cells
Collenchyma	Collenchyma cells
Sclerenchyma	Fibers or sclereids

COMPLEX TISSUES

Xylem	Conducting cells (tracheids, vessel members); parenchyma cells; sclerenchyma cells
Phloem	Conducting cells (sieve-tube members); parenchyma cells; sclerenchyma cells
Epidermis	Undifferentiated cells; also guard cells and other specialized cells
Periderm	Cork; cork cambium; secondary cortex

8. Wood (secondary xylem) is classified by location and function (as in heartwood or sapwood) and plant type (as in hardwood of many dicots, softwood of conifers). Bark consists of secondary phloem and periderm.

Review Questions

1. Choose a flowering plant and list some functions of its roots and shoots. *29.1*

2. Name and define the basic functions of a flowering plant's three main tissue systems. *29.1*

3. Describe the differences between:
   a. apical, transitional, and lateral meristems *29.1*
   b. parenchyma and sclerenchyma *29.2*
   c. xylem and phloem *29.2*
   d. epidermis and periderm *29.2, 29.8*

4. Study Figure 29.30. Is the plant that produced the yellow flower a dicot or monocot? Is the plant that produced the purple flower a dicot or monocot? *29.2*

Figure 29.30 Flower of (**a**) St. John's wort (*Hypericum*) and (**b**) a lily (*Lilium*).

5. Which of the following stem sections is typical of most dicots? Which is typical of most monocots? Label the main tissue regions of both sections. *29.2, 29.3*

6. Label the components of this three-year-old tree section, then label the growth rings in the stem section below it. *29.8*

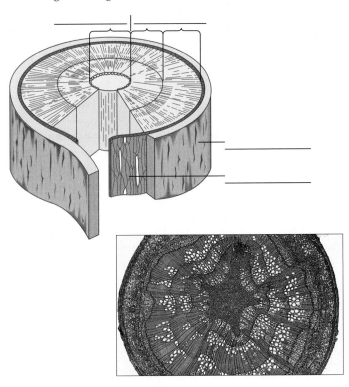

Self-Quiz (Answers in Appendix IV)

1. Fleshy plant parts consist mostly of _____ cells.
   a. parenchyma   c. collenchyma
   b. sclerenchyma   d. epidermal

2. Xylem and phloem are _____ tissues.
   a. ground   c. dermal
   b. vascular   d. both b and c

3. _____ conducts water and ions; _____ conducts food.
   a. Phloem; xylem   c. Xylem; phloem
   b. Cambium; phloem   d. Xylem; cambium

4. Roots and shoots lengthen through activity at _____ .
   a. apical meristems   c. vascular cambium
   b. lateral meristems   d. cork cambium

5. Older roots and stems thicken through activity at _____ .
   a. apical meristems   c. vascular cambium
   b. lateral meristems   d. both b and c

6. Buds give rise to _____ .
   a. leaves   c. stems
   b. flowers   d. all of the above

7. Mesophyll consists of _____ .
   a. waxes and cutin   c. photosynthetic cells
   b. lignified cell walls   d. cork but not bark

8. In early wood, cells have _____ diameters, _____ walls.
   a. small; thick   c. large; thick
   b. small; thin   d. large; thin

9. Match the plant parts with the most suitable description.
   ___ apical meristem   a. masses of xylem
   ___ lateral meristem   b. source of primary growth
   ___ xylem, phloem   c. corky surface covering
   ___ periderm   d. source of secondary growth
   ___ vascular cylinder   e. conduct water, food
        wood   f. central column in roots

Critical Thinking

1. Think about the kinds of conditions that might prevail in a hot desert in New Mexico or Arizona; on the floor of a shady, moist forest in Georgia or Oregon; and in the arctic tundra in Alaska. Also consider the kinds of animals living in each type of environment. Then "design" a type of flowering plant that would be suitably adapted to each place.

2. Fitzgerald, who is known for his underdeveloped sense of nature, sneaks into a forest preserve at night and maliciously girdles an old-growth redwood. Then he settles down and watches the tree, waiting for it to die. However, the tree will not die that night or any time soon. Explain why.

3. Sylvia lives in Santa Barbara, where droughts are common and a long-term abundance of water is not. She replaced most of the plants in her garden with drought-tolerant species and drastically cut back the size of the lawn. The lawn doesn't get a light sprinkling every day. Sylvia only waters it twice a week in the evening, after the sun goes down. Then the lawn gets a good soak, down to a depth of several inches. Why is her strategy good for lawn grasses?

Selected Key Terms

annual (plant) *29.7*
apical meristem *29.1*
bark *29.8*
biennial *29.7*
bud *29.3*
collenchyma *29.2*
compartmentalization *29.8*
cork *29.8*
cork cambium *29.1*
cortex *29.3*
cuticle (plant) *29.2*
dermal tissue system *29.1*
dicot *29.2*
epidermis *29.2*
fibrous root system *29.6*
ground tissue system *29.1*
growth ring *29.8*
hardwood *29.8*
heartwood *29.8*
lateral root *29.6*
leaf *29.4*
meristem *29.1*

mesophyll *29.2*
monocot *29.2*
parenchyma *29.2*
perennial *29.7*
periderm *29.2*
phloem *29.2*
pith *29.3*
primary root *29.6*
root *29.1*
root hair *29.6*
sapwood *29.8*
sclerenchyma *29.2*
shoot *29.1*
softwood *29.8*
stoma (stomata) *29.2*
taproot system *29.6*
vascular bundle *29.3*
vascular cambium *29.1*
vascular cylinder (root) *29.6*
vascular tissue system *29.1*
vein (leaf) *29.4*
xylem *29.2*

Readings

Esau, K. 1977. *Anatomy of Seed Plants.* Second edition. New York: Wiley. Well, sometimes the oldies are still goodies.

Moore, R. W., D. Clark, and K. Stern. 1995. *Botany.* Dubuque, Iowa: W. C. Brown. Stunning illustrations and interesting selection of experiments.

Simpson, B., and M. Conner-Ogorzaly. 1995. *Economic Botany.* Second edition. New York: McGraw-Hill. Fascinating book on uses and abuses of plants.

Web Site See *http://www.wadsworth.com/biology* for practice quiz questions, hypercontents, BioUpdates, and critical thinking. The Wadsworth Biology Resource Center provides a wealth of information fully organized and integrated by chapter.

30

PLANT NUTRITION AND TRANSPORT

Flies for Dinner

How often do we think that plants actually do anything impressive? Being mobile, intelligent, and emotional, we tend to be fascinated more with ourselves than with immobile, expressionless plants. Yet plants don't just stand around soaking up sunlight. Consider the Venus flytrap (*Dionaea muscipula*), a native flowering plant of bogs in North and South Carolina. Its two-lobed, spine-fringed leaves open and close much like a steel trap (Figure 30.1*a–d*). Like all other plants, it cannot grow properly without nitrogen and other nutrients, which happen to be scarce in the soil of bogs. But plenty of insects fly in from regions around the bogs.

Sticky sugars ooze from epidermal glands onto the surface of the flytrap's leaf. The sugars entice insects to land. As they do, they brush against hairlike structures that project from the leaf surface. These are triggers for the trap. When an insect touches two hairs at the same time or the same hair twice in rapid succession, the two lobes of the leaf snap shut. Now digestive juices pour out from cells of the leaf. They pool around the insect, dissolve it, and so release nutrients from it. In other words, the Venus flytrap makes its own nutrient-rich water, which it proceeds to absorb!

Plants with bizarre nutrient-acquiring habits also live in the shallow waters of many freshwater lakes and streams, which contain only dilute concentrations of dissolved minerals. For example, the bladderworts (*Utricularia*) are aquatic plants equipped with hollow, pear-shaped bladders, each with an opening guarded by a bristle-fringed door (Figure 30.2*a*). When a tiny protozoan or a mosquito larva brushes against the bristles, the door springs open and water rushes into the bladder, carrying the animal with it. When the door snaps shut, cells in the bladder wall secrete digestive enzymes. So do mutualistic species of bacteria that have taken up residence in these actively moving traps.

Like the Venus flytrap, bladderworts are one of several species of **carnivorous plants**. We call them this even though it takes a flying leap of the imagination to put their peculiar mode of nutrient acquisition (a form of extracellular digestion and absorption) in the same category as the chompings of lions, dogs, and similar meat eaters. Not all carnivorous plants have active traps; some lure in prey and then simply let them drown (Figure 30.2*b*). But they all evolved in habitats where nitrogen and other nutrients are hard to come by.

Given the variety and numbers of insects and other animals that attack plants, you can just imagine how endearing the carnivorous plants are to botanists. With their plucky modes of nutrition, these plants also are a fine way to start thinking about **plant physiology**—the study of adaptations by which plants function in their

a

VENUS FLYTRAP, OPEN FOR DINNER

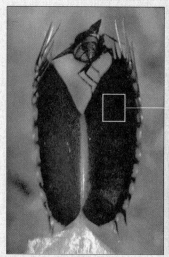

b

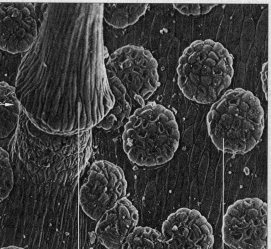

c

base of hairlike projection   secretory gland

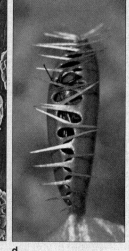

d

Figure 30.1 Do plants take nutrition seriously? You bet. (**a**) A Venus flytrap (*Dionaea muscipula*). This carnivorous plant makes up for scarce nutrients by turning the dinner table on animals that alight upon its leaves. (**b**) A fly stuck in sugary goo on a lobed leaf. (**c**) It brushes against hairlike triggers on the leaf surface; the base of one is shown here. (**d**) Activated, the leaf snaps shut.

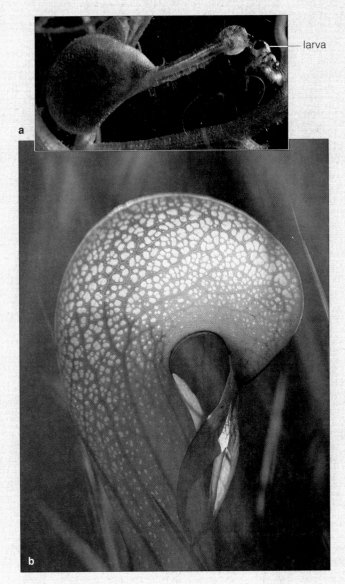

larva

a

b

Figure 30.2 (**a**) Bladderwort (*Utricularia*), magnified six times, soon to be nourished by its ensnared mosquito larva. (**b**) Cobra lily (*Darlingtonia californica*). Its leaves form a "pitcher" that is partly filled with digestive juices. Insects lured in by irresistible odors often cannot find the way back out; light shining through the pitcher's patterned dome confuses them. They just wander around and down, adhering to downward-pointing leaf hairs—which become slickened with wax above the potent vat.

environment. As you already know, nearly all plants are photoautotrophs that use energy from sunlight to drive the synthesis of organic compounds from water, carbon dioxide, and some minerals. Like people, they do not have unlimited supplies of the resources required to nourish themselves. Of every 1 million molecules of air, only 350 are carbon dioxide. Unlike the soggy habitats of Venus flytraps, most soils are frequently dry. And nowhere except in overfertilized gardens does soil water hold lavish amounts of dissolved minerals. As you will see shortly, *many aspects of plant structure and function are responses to low environmental concentrations of essential resources.*

KEY CONCEPTS

1. Many aspects of the structure and function of plants are adaptive responses to low concentrations of water, minerals, and other environmental resources.

2. A plant's root system spreads through soil and mines its nutrients. Mycorrhizae and bacterial symbionts typically assist in nutrient uptake. The properties of soil in a given habitat influence the availability of nutrients and water.

3. A plant's cuticle and its many stomata function in the conservation of water, a scarce resource in most habitats on land. Stomata are passageways across the epidermis of leaves and, to a lesser extent, stems. They help control water loss, yet they permit gas exchange.

4. During the day, stomata remain open so that carbon dioxide, required for photosynthesis, is able to diffuse into leaves. Water loss is rapid when stomata are open. Most plants conserve water by closing stomata at night.

5. In flowering plants and other vascular plants, the flow of water and solutes through xylem and phloem functionally connects all living cells of the roots, stems, and leaves. Xylem serves in the uptake and distribution of water and dissolved nutrients. Phloem serves in the distribution of photosynthetically produced sugars and other organic compounds through the plant body.

6. Water absorbed from soil moves on up through xylem and into leaves. By the process of transpiration, the dry air around leaves promotes evaporation. The force of evaporation is enough to pull continuous columns of water molecules, which are hydrogen-bonded to one another, all the way from roots to aboveground parts.

7. By the energy-requiring process of translocation, sucrose and other organic compounds are distributed throughout the plant. Organic compounds produced by photosynthetic cells in leaves are loaded into conducting cells of phloem. They are unloaded at actively growing regions or storage regions of the plant body.

Properties of Soil

At the surface of most habitats on land is a thin cloak of soil. **Soil** consists of particles of minerals mixed with variable amounts of decomposing organic material, or **humus**. Weathering of hard rocks produces the soil's minerals. Dead organisms and litter from them (leaves, feces, and so on) are the source of humus. Water and oxygen occupy the spaces in between the particles and organic bits.

In different regions, even in different parts of the same habitat, soils vary in the proportions of mineral particles and in the extent to which these are compacted together. Generally, these particles come in three sizes, known as sand, silt, and clay. If you have ever dribbled beach sand through your fingers, you already have an idea that sand particles are large (0.05 to 2 millimeters across). Rub silt from pond mud between your fingers and you won't see individual particles; they are only 0.002 to 0.05 millimeter across. Clay particles are the finest of all.

How suitable is a given soil for plant growth? Is it gummy when wet because it does not have enough air spaces? Does it form hard clods when dry? The answer depends partly on the relative proportions of sand, silt, and clay. The more clay, the finer the soil's texture.

Each clay particle consists of thin, stacked layers of aluminosilicates with negatively charged ions at their surfaces. Clay attracts and holds (adsorbs) positively charged mineral ions dissolved in the water trickling through soil, as well as water molecules themselves. Ions and water cling reversibly to the clay, and this is crucial for plants. With its high adsorption capacity, clay holds onto many nutrients for plants even as the water percolates on past and drains away.

Too much clay, however, is bad for the plants. Clay particles pack together so tightly, they do not leave enough space for the oxygen that root cells require for aerobic respiration. The packing also retards penetration of water into the soil. In heavy clay soils, runoff is pronounced, so water and its dissolved nutrients are not available for plant growth. Plants do best in **loams**, which are the soils having more or less equal proportions of sand, silt, and clay.

The amount of humus in a given soil also affects plant growth. Generally, humus has an abundance of negatively charged organic acids, so that it can weakly bind with and retain dissolved mineral ions of opposite charge. Humus also has a high capacity to absorb and swell with water, then shrink as the water is gradually released. Its alternating swelling and shrinking aerate the soil. As decomposers work it over, humus slowly releases nutrients, which become available for plants.

In general, soils that are 10 to 20 percent humus are most favorable for plant growth. The worst have less than 10 percent or more than 90 percent humus (which is characteristic of swamps and bogs).

Soils can be classified by *profile* properties. The term refers to the layered characteristics of soils, which are in different stages of development in different places. Figure 30.3 is an example. **Topsoil**, the uppermost part of soil, is called the A horizon. This is the most essential layer for plant growth, and its depth is highly variable.

Nutrients Essential for Plant Growth

We have mentioned nutrients in passing, but what does the term mean? **Nutrients** are elements that are essential for a given organism because, directly or indirectly, they have roles in metabolism (hence growth and survival) that no other element can fulfill. For plants, the essential elements include oxygen, hydrogen, and carbon, which the plants get from water and carbon dioxide during

O HORIZON
Fallen leaves and other organic material littering the surface of mineral soil

A HORIZON
Topsoil, which contains some percentage of decomposed organic material and which is variably deep; only a few centimeters deep in deserts, but elsewhere extending as much as thirty centimeters below the soil surface

B HORIZON
Compared with the A horizon, larger soil particles, not much organic material, but greater accumulation of minerals; extends thirty to sixty centimeters below soil surface

C HORIZON
No organic material, but partially weathered fragments and grains of rock from which soil forms; extends to underlying bedrock

BEDROCK

Figure 30.3 Some of the soil horizons that have developed in one habitat in Africa.

Table 30.1 Essential Elements and Plant Function		
MACRONUTRIENT	Some Functions	Some Deficiency Symptoms
Carbon Hydrogen Oxygen	Basic ingredients for photosynthesis	Available in abundance from water and from carbon dioxide in the air
Nitrogen	Component of proteins, nucleic acids, coenzymes, chlorophylls	Stunted growth; light-green older leaves; older leaves yellow and die (these symptoms define a condition called chlorosis)
Potassium	Activation of enzymes; key role in maintaining water-solute balance and so influences osmosis*	Reduced growth; curled, mottled, or spotted older leaves; burned leaf edges; weakened plant
Calcium	Regulation of many cell functions; cementing of cell walls	Leaves deformed; terminal buds die; poor root growth
Magnesium	Component of chlorophyll; activation of enzymes	Chlorosis; drooped leaves
Phosphorus	Component of nucleic acids, phospholipids, ATP	Purplish veins; stunted growth; fewer seeds, fruits
Sulfur	Component of most proteins, two vitamins	Light-green or yellowed leaves; reduced growth

MICRONUTRIENT	Some Functions	Some Deficiency Symptoms
Chlorine	Role in root and shoot growth; role in photolysis	Wilting; chlorosis; some leaves die
Iron	Roles in chlorophyll synthesis and in electron transport	Chlorosis; yellow and green striping in grasses
Boron	Roles in germination, flowering, fruiting, cell division, nitrogen metabolism	Terminal buds, lateral branches die; leaves thicken, curl, and become brittle
Manganese	Chlorophyll synthesis; coenzyme action	Dark veins, but leaves whiten and fall off
Zinc	Role in formation of auxin, chloroplasts, and starch; enzyme component	Chlorosis; mottled or bronzed leaves; abnormal roots
Copper	Component of several enzymes	Chlorosis; dead spots in leaves; stunted growth
Molybdenum	Part of enzyme used in nitrogen metabolism	Pale green, rolled or cupped leaves

* All mineral elements contribute to the water-solute balance, but potassium is notable because there is so much of it.

Figure 30.4 In agricultural fields, the erosive force of water can form gullies that channel runoff from the land. When gullies grow deeper and wider, erosion becomes more and more rapid. When topsoil is depleted, productivity declines and fertilizers usually must be trucked in to replace the lost nutrients.

plant's dry weight (weighed after all the water has been removed from it). The other elements are *micro*nutrients; they make up traces (a few parts per million, usually) of the dry weight. As Table 30.1 indicates, even those trace amounts are essential for normal growth.

Leaching and Erosion

Leaching refers to the removal of some of the nutrients in soil as water percolates through it. Leaching is most pronounced in sandy soils, which are not as good as clay at binding nutrients. It is not the same as **erosion**, which is the movement of land under the force of wind, running water, and ice (Figure 30.4). For example, each year, erosion from farmlands in the Mississippi River watershed puts about 25 billion metric tons of topsoil into the Gulf of Mexico. Whether by leaching or erosion, the loss of nutrients from soil is bad for plants and for all the organisms that depend on plants for survival.

The mineral component of soil includes particles ranging from large-grained sand to silt and fine-grained clay. These particles, clay especially, reversibly bind water molecules and dissolved mineral ions and thereby make them more accessible for uptake by plant roots.

Soil also contains humus—a reservoir of organic material, rich in organic acids, in different stages of decay.

Most plants grow best in soils having equal proportions of sand, silt, and clay and 10 to 20 percent humus.

Nutrients are essential elements; no other element can substitute for their direct or indirect roles in the metabolic activities that sustain growth and keep organisms alive.

photosynthesis. Besides these elements, plants depend on the uptake of at least thirteen others, listed in Table 30.1. Typically these elements are dissolved in soil water in ionic forms that can reversibly bind with clay. Ions of calcium (Ca^{++}) and potassium (K^+) are examples. Plants give up hydrogen ions to the clay in exchange for these weakly bound elements.

Nine of the essential elements are *macro*nutrients. They usually are required in amounts above 0.5 percent of the

Energetically speaking, mining the soil for mineral ions and water molecules that are clinging to clay particles is an expensive proposition. Plants spend considerable energy on building extensive root systems that can grow and branch out through parts of the soil around them. As the texture and composition of a given region of soil change, new roots must form to replace old ones and to branch in different regions. It is not that roots "explore" the soil for resources. Rather, the portions of the soil where concentrations of water and mineral ions are greater stimulate the outward growth.

Absorption Routes

Think back on the preceding chapter's discussion of a typical root's structure (Section 29.6). Water molecules in soil are only weakly bound to clay particles, so they can readily move across the root epidermis and continue to the **vascular cylinder**, the central column of vascular tissue in the root. There a cylindrical layer of cells, the **endodermis**, wraps around the column. A waxy band, the **Casparian strip**, runs through all abutting cell walls in this layer (Figure 30.5). Water molecules cannot penetrate this strip; they can only diffuse across the portions of the cell walls that *do not* have the strip. Therefore, water and solutes reach the vascular cylinder only by moving into, through, and out of endodermal cells. Like all cells, endodermal cells have many transport proteins embedded in the plasma membrane. These proteins allow some solutes but not others to cross it (Section 5.3). *Thus, transport proteins of endodermal cells are control points where a plant adjusts the quantity and types of solutes that it is absorbing from soil water.*

Many vascular plants also have an **exodermis**, a layer of cells just inside their roots (Figure 30.5a). These cells also have a Casparian strip that functions just like the one next to the root vascular cylinder.

Specialized Absorptive Structures

ROOT HAIRS Vascular plants require huge amounts of water. For example, the roots of a single, fully grown corn plant are absorbing more than three liters of water daily. They could not do this without their **root hairs**. Recall, from the preceding chapter, that root hairs are slender extensions of specialized epidermal cells, and that they greatly increase the surface area available for absorption (Section 29.6 and Figure 30.6). When a plant puts on primary growth, its root system may develop millions or billions of root hairs.

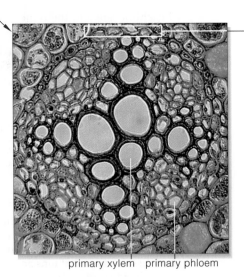

exodermis
root hair
epidermis
newly forming vascular cylinder
cortex
Casparian strip (*gold*) within all the abutting walls of cells of the endodermis

a

b Vascular cylinder, transverse section

primary xylem   primary phloem

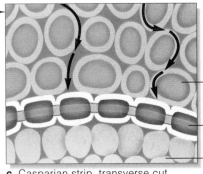

cells of root cortex; water molecules move through and between the walls of these cells

Casparian strip (*gold*)

inside vascular cylinder

c Casparian strip, transverse cut

d Cutaway view of two endodermal cell walls. Water and solutes move into a root vascular cylinder only by passing through the cytoplasm of endodermal cells.

Figure 30.5 Control of the uptake of water and dissolved nutrients in roots. (**a,b**) Roots of most flowering plants have an endodermis (a cell layer wrapped around the vascular cylinder) and an exodermis (a cell layer just beneath the epidermis). (**c**) Cells of both layers have a waxy Casparian strip embedded in their abutting walls. The strip keeps water from moving indiscriminately *around* the cells and into the vascular column. It makes water move *through* the cells, as shown in (**d**). In this way, transport proteins that span the plasma membrane of those cells can selectively control the uptake of water and nutrients.

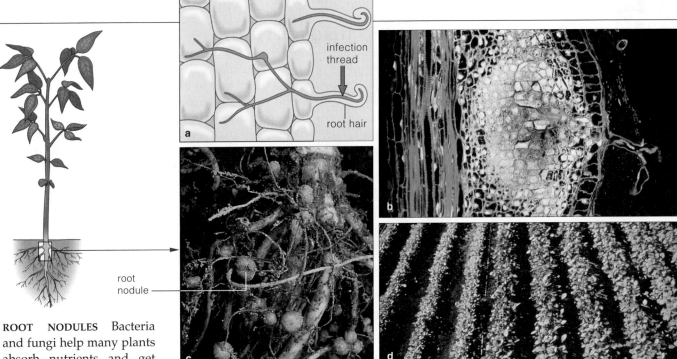

Figure 30.7 (a) Nutrient uptake at root nodules of legumes, which interact mutualistically with nitrogen-fixing bacteria (*Rhizobium* and *Bradyrhizobium*). When bacterial cells infect root hairs, the cells are induced to form an "infection thread" of cellulose deposits. The bacteria use the thread like a highway to invade cells inside the root cortex. (b,c) The infected plant cells and the bacterial cells within them divide repeatedly, forming a swollen mass that becomes a root nodule. The bacteria start fixing nitrogen when plant cell membranes surround them. The plant takes up some of the nitrogen, and the bacteria take up some photosynthetic compounds. (d) To the left, rows of soybean plants growing in nitrogen-poor soil. The plants in the rows to the right were inoculated with *Rhizobium* cells and developed root nodules.

ROOT NODULES Bacteria and fungi help many plants absorb nutrients and get something in return. Such a two-way flow of benefits between species, remember, is a symbiotic interaction called **mutualism** (Section 24.4). For example, think of how nitrogen deficiency limits plant growth. There is an abundance of gaseous nitrogen (N≡N) in the air. But plants do not have the metabolic means to engage in **nitrogen fixation**. By this process, certain enzymes break the three covalent bonds in gaseous nitrogen and attach the nitrogen atoms to organic compounds. To get high crop yields, farmers often provide crop plants with nitrogen compounds in fertilizers or they encourage the growth of nitrogen-fixing bacteria that are naturally present in soil. Such bacteria convert gaseous nitrogen to forms they—and the plants—use. String beans, peas, alfalfa, clover, and other legumes have an advantage in this respect. Nitrogen-fixing bacteria live in their roots, in local swellings called **root nodules** (Figure 30.7b). These symbiotic bacteria withdraw some of the organic compounds from the plant tissues, but they provide the plants with some of their fixed nitrogen in return.

MYCORRHIZAE Also think back on the **mycorrhizae** (singular, mycorrhiza). As described in Section 24.4, a mycorrhiza is a symbiotic interaction between a young root and a fungus (hence the name, meaning "fungus-root"). Fungal filaments (hyphae) either form a velvety covering around roots or they penetrate the root cells. Collectively, the hyphae have a large surface area that can absorb mineral ions from a larger volume of soil. The fungus absorbs sugars and nitrogen-containing compounds from root cells. The root cells obtain some scarce minerals that the fungus is better able to absorb.

Gymnosperms and flowering plant roots control the uptake of water and dissolved nutrients at the vascular cylinder's endodermis and at a similar layer near the root surface.

Root hairs, root nodules, and mycorrhizae enhance the uptake of water and scarce nutrients at the root.

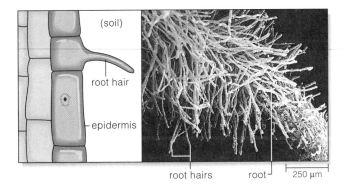

Figure 30.6 Sketch of a root hair (a slender extension of a specialized root epidermal cell), and a scanning electron micrograph of the root hairs of a portion of a young root.

Transpiration Defined

By now, you have a sense of how the distribution of water and dissolved mineral ions to all living cells is central to plant growth and functioning. Let's turn now to the prevailing model of how water actually moves from a plant's roots to its stems, then into leaves.

Recall that plants use only a fraction of the water that they absorb for growth and metabolism. Most of the water is lost, mainly through the numerous stomata in leaves. The evaporation of water from leaves as well as from stems and other plant parts is a process called **transpiration**.

Cohesion-Tension Theory of Water Transport

This brings up an interesting question. Assuming that plants lose most of the absorbed water from leaves, how does the water actually get *to* the leaves? What gets it to the top of leafy plants, including redwoods and other trees that may be more than 100 meters tall?

In a plant's vascular tissues, water moves through the pipelines called **xylem**. Recall, from Chapter 29, that water-conducting cells of xylem are called **tracheids** and **vessel members**. Figure 30.8 shows examples. These cells are dead at maturity, and only their lignin-reinforced walls remain. Therefore, xylem's conducting cells cannot be actively pulling the water "uphill."

Some time ago, the botanist Henry Dixon came up with a good way to explain water transport in plants. By his **cohesion-tension theory**, the water in xylem is being pulled upward by the drying power of air, which creates continuous negative pressures (tensions) that extend downward from leaves to roots. Figure 30.9 illustrates this theory. Think about the following points of Dixon's explanation as you review that illustration:

1. The drying power of air causes transpiration, or the evaporation of water from plant parts exposed to air. Transpiration puts the water confined in the conducting tubes of xylem in a state of tension. The tension extends from veins inside leaves, down through the stems, and on into the young roots where water is being absorbed.

2. Unbroken, fluid columns of water show *cohesion*; they resist rupturing as they are pulled upward under *tension* (compare Section 2.5). The collective strength of hydrogen bonds between water molecules, which are confined in the narrow, tubular xylem cells, imparts this cohesion.

3. As long as water molecules continue to escape from the plant, the continuous tension in the xylem permits more molecules to be pulled upward from the roots to replace them.

Hydrogen bonds are strong enough to hold water molecules together inside the xylem. But they are not

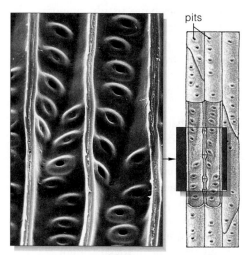

a Tracheids have tapered, unperforated end walls. In intact xylem, small pits in the walls of adjoining tracheids match up. Although the pits are too small to permit rapid flow, they do confine air bubbles— which obstruct water transport— to individual tracheids in the system.

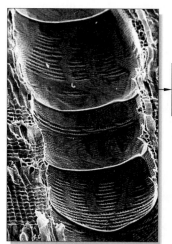

b Three individual vessel members. Together, the thick, perforated walls of these dead cells interconnect to form vessels, another type of water-conducting tube in xylem.

c Perforation plate at the end wall of one type of vessel member. The perforated ends permit water and air bubbles to flow unimpeded through the conducting tube. This may be why natural selection has favored retention of both tracheids and vessel members in the same plants.

Figure 30.8 Scanning electron micrographs and diagrams of typical tracheids and vessel members, which are conducting tubes in xylem. These cells are dead at maturity. Their walls remain interconnected and form the water-conducting tubes. Tracheids probably evolved before vessel members, but both are present in nearly all vascular plants.

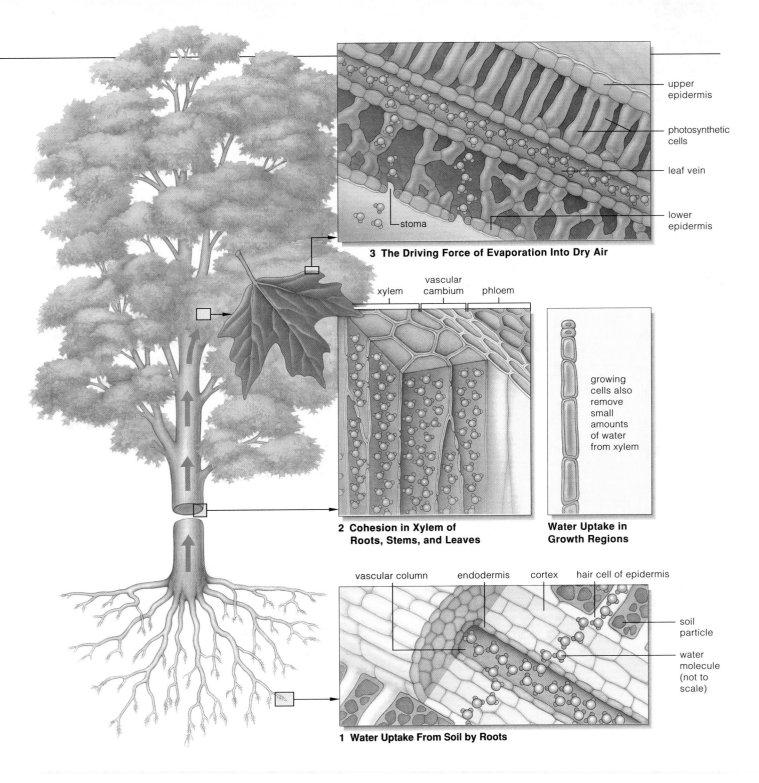

3 The Driving Force of Evaporation Into Dry Air

upper epidermis

photosynthetic cells

leaf vein

lower epidermis

stoma

vascular cambium

xylem   phloem

growing cells also remove small amounts of water from xylem

2 Cohesion in Xylem of Roots, Stems, and Leaves

Water Uptake in Growth Regions

vascular column   endodermis   cortex   hair cell of epidermis

soil particle

water molecule (not to scale)

1 Water Uptake From Soil by Roots

Figure 30.9 Diagram illustrating the cohesion-tension theory of water transport.

Point 1: Transpiration (evaporation of water from aboveground plant parts) puts the water in xylem in a state of tension that extends from roots to leaves.

Point 2: The collective strength of hydrogen bonds among water molecules, which are confined within the narrow water-conducting tubes in xylem, impart cohesion to the water. Hence the narrow columns of water in xylem resist rupturing under the continuous tension.

Point 3: As long as water molecules escape by transpiration, that tension drives the uptake of replacement water molecules.

strong enough to prevent the water molecules from breaking away from one another during transpiration and then escaping from leaves, through stomata.

Transpiration is the process of evaporation from plant parts.

By the cohesion-tension theory of water transport, this process is the key source of tensions in water in xylem. The tensions extend from leaves to roots, and they allow columns of water molecules that are hydrogen-bonded to one another to be pulled upward through the plant body.

CONSERVATION OF WATER IN STEMS AND LEAVES

At least 90 percent of the water that enters a leaf goes right on through it and evaporates into the surrounding air. Cells use only 2 percent of the water that stays in the leaf for photosynthesis, membrane functions, and other events. That tiny amount is vital. Plants wilt and

Figure 30.10 Osmosis and the wilting of leafy plants. When a plant with soft green leaves is growing well, you can safely bet that the soil water is dilute, or hypotonic, compared to fluids in the plant's living cells.

Recall, from Section 5.4, that when water responds to solute concentration gradients and moves osmotically into a plant cell, internal fluid pressure builds up against the cell wall. Botanists call this turgor pressure. Water also is squeezed out when the turgor pressure is enough to counter the attractive force of cytoplasmic fluid, which usually has more solutes than soil water does. Both forces have the potential to cause water to move directionally. The sum of these two opposing forces is the "water potential."

When soft plant parts are erect, as much water is moving into cells as is moving out, and the constant pressure keeps cell walls plump. Suppose the soil dries or gets too salty. Water's concentration gradient reverses and cells lose water. The osmotically induced shrinkage of cytoplasm in all the young cells of soft plant parts results in wilting; the parts droop with the loss of turgor pressure.

The experiment at left demonstrates the wilting effect. Place 10 grams of table salt (NaCl) in 60 milliliters of water. Pour the salty solution into the soil around a tomato plant. The plant starts to collapse after 5 minutes. In less than 30 minutes, wilting is severe.

water-dependent events are severely disrupted when water loss exceeds water uptake at roots for extended periods (Figure 30.10). Yet land plants are not entirely at the mercy of changes in the availability of soil water. They have a cuticle, and they have stomata.

The Water-Conserving Cuticle

Even mildly water-stressed plants would rapidly wilt and die without their **cuticle** (Figure 30.11). Epidermal cells secrete this translucent, water-impermeable layer, which coats their wall regions exposed to the air. At the cuticle surface are deposits of waxes: water-insoluble lipids with long fatty-acid tails. The cuticle proper is made of waxes embedded in **cutin**, an insoluble lipid polymer. Beneath the waxes, cellulose threads through the cutin matrix. A layer of polysaccharides (pectins) often helps bind the cuticle to cell walls.

A cuticle does not bar the passage of light rays into photosynthetic parts of a plant. It does restrict water loss. It also restricts the *inward* diffusion of the carbon dioxide required for photosynthesis and the *outward* diffusion of oxygen by-products. (Recall, from Section 7.7, that a buildup of oxygen in the air spaces inside a leaf has bad effects on the rate of photosynthesis.)

Controlled Water Loss at Stomata

How do carbon dioxide and oxygen get past the cuticle-covered, water-conserving epidermis? They do so at the openings called **stomata** (singular, stoma). Here, water evaporates from the plant and carbon dioxide moves in. A pair of highly specialized cells, the **guard cells**, define each opening (Figure 30.12). When the two cells swell

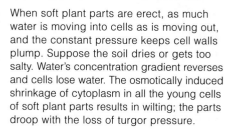

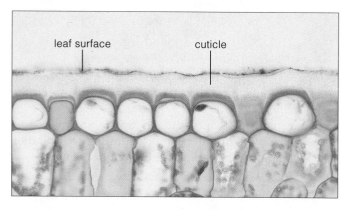

Figure 30.11 Waxy cuticle on the upper epidermis of a leaf. Water, carbon dioxide, and oxygen cannot cross the cuticle proper. They enter or depart from the leaf mainly at stomata. In general, plants in deserts and other seasonally dry habitats have thick cuticles, which provide greater protection against water loss. Aquatic plants have thin cuticles or none at all.

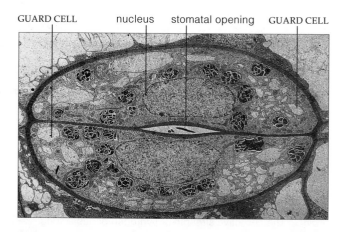

GUARD CELL   nucleus   stomatal opening   GUARD CELL

Figure 30.12 Guard cells. This pair spans the stem epidermis from a prickly pear (*Opuntia*), also called the beavertail cactus. Cacti have their stomata on stems and only tiny leaves, if any. They also bear spines—modified nonphotosynthetic shoots that protect succulent stems from browsing animals while reducing the surface area from which water can evaporate.

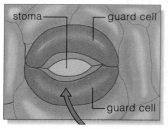

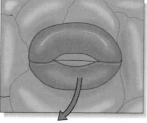

a Stomatal opening. Water enters collapsed guard cells, which swell under turgor pressure and move apart, forming the stoma.

b Stomatal closing. Water leaves swollen guard cells, which collapse against each other and close the stoma.

Figure 30.13 Stomatal action. Whether a stoma remains open or closed at a given time of day or night depends on changes in the shape of the paired guard cells, which together define one opening across the cuticle-covered epidermis.

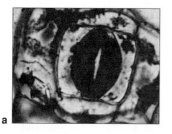

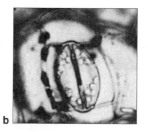

Figure 30.14 Experimental evidence that potassium ions are accumulating in guard cells that are expanding with incoming water. For this experiment, strips from the leaf epidermis of a dayflower (*Commelina communis*) were immersed in solutions containing dark-staining substances that bind preferentially with potassium ions. (**a**) In the leaf samples having opened stomata, most of the potassium ions were concentrated in the guard cells. (**b**) In the leaf samples having closed stomata, very little of the potassium was in guard cells; most of it was present in normal epidermal cells.

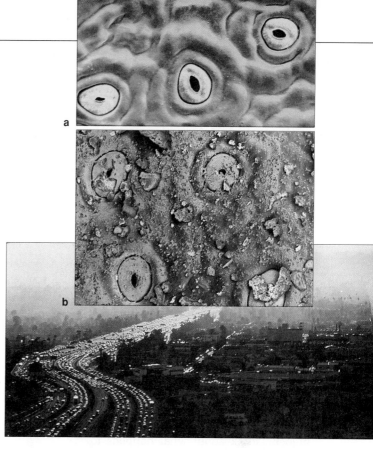

Figure 30.15 (**a**) Surface view of holly leaf stomata. (**b**) What the leaf looks like when a holly plant grows in industrialized regions. Gritty airborne pollutants clog its stomata and prevent much of the sun's rays from reaching photosynthetic cells inside the leaf.

up with water, the force of the internal fluid pressure distorts the cell walls. An osmotically induced, internal fluid pressure that builds up against cell walls is called **turgor pressure**. Under such pressure, guard cells bend in such a way that a gap (one stoma) forms between them. When the two guard cells lose water and internal fluid pressure drops inside them, they collapse against each other and so close the gap (Figure 30.13).

In most plants, stomata stay open during the day, when photosynthesis takes place. They do lose water—but they gain carbon dioxide for the reactions. Stomata stay closed at night. Then, plants conserve water, and precious carbon dioxide accumulates inside the leaf as the cells busily engage in aerobic respiration.

Whether a stoma opens or closes depends on how much water and carbon dioxide are in the two guard cells. Photosynthesis starts when the sun comes up. As the morning progresses, carbon dioxide levels drop in cells—including guard cells. The decrease helps trigger active transport of potassium ions into the guard cells (Figure 30.14). So do blue wavelengths of light, which penetrate the atmosphere better as the sun arcs higher in the sky. Water follows potassium ions into the cells, by osmosis. The inward movement of water provides the fluid pressure to open the stoma. Carbon dioxide levels rise when the sun goes down and photosynthesis stops. Potassium ions and then water move out of the guard cells, and the stoma closes.

CAM plants, including most cacti, conserve water differently. They open stomata at night, when they fix carbon dioxide by the metabolic C4 pathway described in Section 7.7. The next day, when their stomata close, CAM plants use carbon dioxide in photosynthesis.

As this section makes clear, plant survival depends on stomata. Think about this when you are out and about on smog-shrouded days (Figure 30.15).

Water-dependent events in plants are severely disrupted when water loss exceeds uptake at roots for extended periods. Wilting is one observable outcome.

Transpiration and gas exchange occur mainly at stomata. These numerous small openings span the waxy cuticle, which covers all plant epidermal surfaces exposed to air.

Plants open and close stomata at different times to control water loss, carbon dioxide uptake, and oxygen disposal, all of which affect rates of photosynthesis and plant growth.

Whereas xylem distributes water and minerals through the plant, the vascular tissue called **phloem** distributes organic products of photosynthesis. Like xylem, phloem consists of many conducting tubes, fibers, and strands of parenchyma cells, all of which extend through the plant body. Unlike xylem, phloem has conducting tubes of a type called **sieve tubes**. These tubes consist of *living* cells, positioned side by side and end to end in the vascular tissues. Figure 30.16 shows an example of the cells themselves, which are called sieve-tube members. Dissolved sugars and other organic compounds flow rapidly through large pores in the abutting end walls of sieve-tube members.

Adjoining the long sieve tubes are **companion cells**, which are a type of specialized parenchyma cell. As you will see shortly, companion cells help load organic compounds into sieve tubes.

To understand what goes on inside phloem, think about what happens to sucrose and other organic products of photosynthesis. Leaf cells use some of these products for their own activities. The remainder travels to roots, stems, buds, flowers, and fruits. In most cells, carbohydrates become stored as starch, in plastids. The proteins and fats, which cells synthesize from carbohydrates and the amino acids distributed to them, are stored in many seeds. Avocados and some other fruits also accumulate fats.

Starch molecules are too large for transport across the plasma membrane of cells that produce them. Also, they are too insoluble for transport to other plant parts. Proteins are simply too large and fats are too insoluble overall for transport from the storage sites. *But cells convert the storage forms of organic compounds to solutes of smaller size that are more easily transported through the phloem.* For instance, the cells degrade starch to glucose monomers. When one of these monomers then combines with fructose, the result is sucrose—which is an easily transportable sugar.

Experiments with the insects called aphids revealed that sucrose is the main carbohydrate transported in phloem. The experimenters exposed aphids feeding on the juices in the conducting tubes of phloem to high levels of carbon dioxide to anesthetize them. Then they detached the body of the aphids from the mouthparts, which were still embedded in the sieve tubes. For most of the plants studied, sucrose was the most abundant carbohydrate in the fluid being forced out of the tubes.

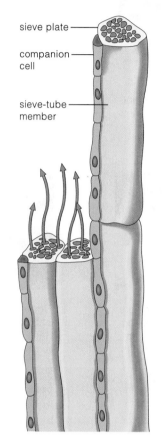

sieve plate

companion cell

sieve-tube member

Figure 30.16   Example of a few sieve-tube members, the living cells that interconnect to form a conducting tube in phloem. The companion cells expend ATP energy to load sugars and other organic compounds into sieve-tube members adjacent to them.

Figure 30.17   A honeydew droplet exuding from the end of an aphid gut. The insect's tubular mouthpart penetrated a conducting tube in phloem. The sugary fluid was under such high pressure in phloem, it was forced out through the gut's terminal opening.

Translocation

Translocation is the technical name for the transport of sucrose and all other organic compounds through the phloem of a vascular plant. High pressure drives this process. Often that pressure is five times as high as it is in automobile tires. Aphids demonstrate the magnitude of it when they force their mouthparts into sieve tubes and feed upon the dissolved sugars. The high pressure can force fluid through an aphid gut and out the other end, as "honeydew" (Figure 30.17). Park a car under some trees being attacked by aphids, and it might get spattered by sticky honeydew droplets, thanks to the high fluid pressure in phloem.

Pressure Flow Theory

Phloem translocates organic compounds along gradients of decreasing pressure and solute concentrations. The *source* of the flow is any region of the plant where any organic compounds are being loaded into the sieve tubes. Common sources are mesophylls, the photosynthetic tissues in leaves. The flow ends at a *sink*, which is any region of the plant where the organic compounds are being unloaded, used, or stored. As they are developing, flowers and fruits are common sink regions.

Why do the organic compounds flow from a source to a sink? According to the **pressure flow theory**, internal pressure builds up at the source end of the sieve-tube system and *pushes* the solute-rich

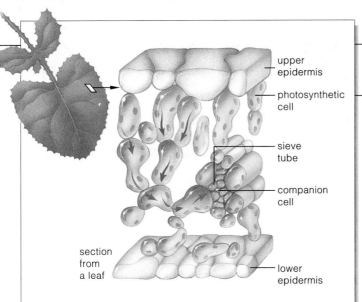

upper epidermis

photosynthetic cell

sieve tube

companion cell

section from a leaf

lower epidermis

a *Loading at a source.* Photosynthetic cells in leaves are a common source of organic compounds that must be distributed through a plant. Small, soluble forms of these compounds move from the cells into phloem (a leaf vein).

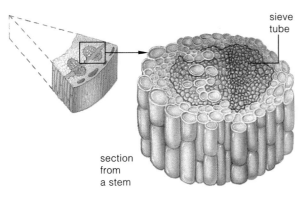

sieve tube

section from a stem

b *Translocation along a distribution path.* Fluid pressure is greatest inside sieve tubes at the source. It pushes the solute-rich fluid to a sink, which is any region where cells are growing or storing food. There, the pressure is lower because cells are withdrawing solutes from the tubes.

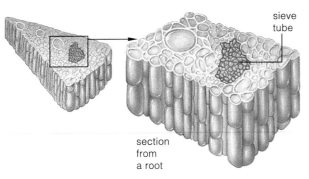

sieve tube

section from a root

b *Unloading at the sink.* Solutes are unloaded from sieve tubes into cells at the sink; water follows. Translocation continues as long as solute concentration gradients and a pressure gradient exist between the source and the sink.

Figure 30.18 Translocation of organic compounds in sow thistle (*Sonchus*). Compare Figure 7.2 to see how translocation relates to photosynthesis in the same plant.

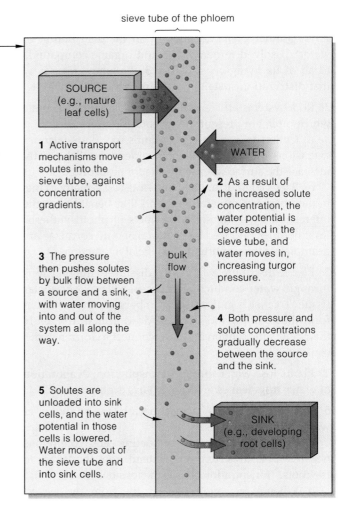

sieve tube of the phloem

SOURCE (e.g., mature leaf cells)

WATER

1 Active transport mechanisms move solutes into the sieve tube, against concentration gradients.

2 As a result of the increased solute concentration, the water potential is decreased in the sieve tube, and water moves in, increasing turgor pressure.

3 The pressure then pushes solutes by bulk flow between a source and a sink, with water moving into and out of the system all along the way.

bulk flow

4 Both pressure and solute concentrations gradually decrease between the source and the sink.

5 Solutes are unloaded into sink cells, and the water potential in those cells is lowered. Water moves out of the sieve tube and into sink cells.

SINK (e.g., developing root cells)

solution on toward any sink, where solutes are being removed. Experimental evidence for the pressure flow theory was obtained from studies with sow thistle (*Sonchus*). Use Figure 30.18 to track what happens after sucrose moves from photosynthetic cells into small veins in the leaves of this plant. Companion cells in the veins spend energy to load sucrose into adjacent sieve tube members. As the sucrose concentration increases in the tubes, water also moves in, by osmosis. More internal fluid pressure is exerted against the tube walls. As the pressure increases, it pushes the sucrose-laden fluid out of the leaf, into the stem, and on to the sink.

Plants store organic compounds in the form of starch, fats, and proteins. They convert the storage forms to sucrose and other small units that are soluble and easily translocated.

Translocation is the distribution of organic compounds to different plant regions. It depends on concentration and pressure gradients in the sieve tube system of phloem.

Gradients exist as long as companion cells load compounds into the sieve tubes at sources, such as mature leaves, and as long as compounds are removed at sinks, such as roots.

30.6 SUMMARY

1. A vascular plant depends upon the distribution of water, dissolved mineral ions, and organic compounds to all of its living cells. Figure 30.19 summarizes how that distribution sustains plant growth and survival.

2. Root systems efficiently take up water and nutrients, which often are present in low concentrations in soil.

    a. Collectively, a vascular plant's root hairs (slender extensions of specialized root epidermal cells) greatly increase the surface area available for absorption.

    b. Bacterial and fungal symbionts assist many plants in the uptake of mineral ions, and they benefit from the interaction by using some products of photosynthesis. Root nodules (induced by bacteria) and mycorrhizae are examples of these mutually beneficial interactions.

3. Plants distribute water and dissolved mineral ions through water-conducting tubes of xylem, a vascular tissue. Tracheids and vessel members are the cells that form these tubes, and they are dead at maturity. Only their cell walls remain and interconnect to form narrow, continuous pipelines.

4. Plants lose water through transpiration: evaporation of water from leaves and other parts exposed to air.

5. Here are the points of the cohesion-tension theory of water transport in xylem:

    a. Inside xylem's water-conducting cells, continuous negative pressures (tensions) extend from the leaves to the roots. Transpiration causes the tension.

    b. When water molecules escape from the leaves, replacements are pulled into the leaf under tension.

    c. The collective strength of numerous hydrogen bonds between water molecules imparts cohesion that allows water molecules to be pulled up as continuous fluid columns. The "cohesion" part of the theory refers to the capacity of the hydrogen bonds between water molecules to resist rupturing when put under tension.

6. Most plants have a waxy, water-impermeable cuticle covering their aboveground parts. They lose water and take up carbon dioxide at stomata. A pair of specialized parenchyma cells called guard cells define each of these small openings across leaf (and stem) epidermis.

7. Stomata open and close at different times. Control of their action balances water conservation with carbon dioxide uptake and the release of oxygen.

    a. In many kinds of plants, stomata open during the day. Such plants lose water but take in carbon dioxide for photosynthesis. Stomata close at night; then, water and the carbon dioxide released by aerobically respiring cells are conserved for the next day.

    b. CAM plants open stomata and fix carbon dioxide by the C4 pathway at night. During the day, they close their stomata and use the carbon fixed the night before for photosynthesis.

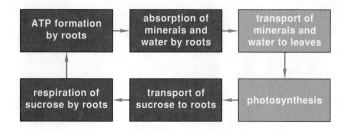

Figure 30.19 Summary of the interdependent processes that sustain the growth of vascular plants. All living cells in plants require oxygen, carbon, hydrogen, and at least thirteen mineral ions. They all produce ATP to drive metabolic activities.

8. Plants distribute organic compounds through sieve tubes of phloem, a vascular tissue. The distribution of organic compounds in phloem is called translocation.

9. According to the pressure flow theory, translocation is driven by differences in solute concentrations and pressure between source and sink regions. A source is any site where organic compounds are being loaded into sieve tubes, such as mature leaves. A sink is any site where compounds are being unloaded from sieve tubes. Roots are examples. Solute concentration gradients and pressure gradients exist for as long as companion cells expend energy to load solutes into the sieve tubes and for as long as solutes are removed at the sink.

Review Questions

1. Define soil in terms of its components. What is topsoil? *30.1*

2. Distinguish between: *30.1*
    a. humus and loam
    b. leaching and erosion
    c. macronutrient and micronutrient (for plants)

3. Refer to Table 30.1. What are some signs that a plant suffers a deficiency in one of the essential elements listed? *30.1*

4. What is the function of the Casparian strip in roots? *30.2*

5. Using Dixon's model, explain how water moves from soil upward through tall plants. *30.3*

6. Which type of ion influences stomatal action? *30.4*

7. Explain translocation according to the pressure flow theory described in this chapter. *30.5*

Self-Quiz *(Answers in Appendix IV)*

1. Water can be pulled up through a plant due to the cumulative strength of _____ between water molecules.

2. In plants, mineral ions _____ .
    a. have roles in metabolism
    b. help establish gradients across membranes
    c. influence water movement into cells
    d. maintain cell shape and growth
    e. all of the above

3. The nutrition of some plants depends on a root-fungus association known as a _____ .
    a. root nodule   c. root hair
    b. mycorrhiza   d. root hypha

4. The nutrition of some plants depends on a root-bacterium association known as a _____ .
   a. root nodule
   b. mycorrhiza
   c. root hair
   d. root hypha

5. Water evaporation from plant parts is called _____ .
   a. translocation
   b. expiration
   c. transpiration
   d. tension

6. Water transport from roots to leaves is explained by _____ .
   a. the pressure flow theory
   b. differences in source and sink solute concentrations
   c. the pumping force of xylem vessels
   d. the cohesion-tension theory

7. The _____ in endodermal cell walls forces water and solutes that entered roots to move through the cells, not around them.
   a. cutin strip
   b. lignin strip
   c. Casparian strip
   d. cellulose strip

8. During the day, most plants lose _____ and take up _____ .
   a. carbon dioxide; water
   b. water; oxygen
   c. oxygen; water
   d. water; carbon dioxide

9. At night, most plants conserve _____ ; _____ accumulates.
   a. carbon dioxide; oxygen
   b. water; oxygen
   c. oxygen; water
   d. water; carbon dioxide

10. In phloem, organic compounds flow through _____ .
    a. collenchyma cells
    b. sieve tubes
    c. vessels
    d. tracheids

11. Match the concepts of plant nutrition and transport.
    _____ stomata
    _____ sink
    _____ root system
    _____ hydrogen bonds
    _____ transpiration
    _____ translocation

    a. evaporation from plant parts
    b. balancing water loss with carbon dioxide requirements
    c. cohesion in water transport
    d. sugars unloaded from sieve tubes
    e. distribution of organic compounds through the plant
    f. response to scarce soil nutrients

Critical Thinking

1. Home gardeners, like farmers, must ensure that their plants have access to nitrogen from either nitrogen-fixing bacteria or fertilizer. Insufficient nitrogen stunts plant growth; leaves turn yellow and die. Which major classes of biological molecules incorporate nitrogen? How would a low nitrogen level in plants affect biosynthesis and cause symptoms of nitrogen deficiency?

2. When moving a plant from one place to another, it helps to include some native soil around the roots. Explain why, given what you know about mycorrhizae and root hairs.

3. Henry discovered a way to keep all of a plant's stomata open at all times. He also figured out how to keep all the stomata of another plant closed all the time. Both plants died. Explain why.

4. Allen is studying the rate of transpiration from tomato plant leaves. He notices that several environmental factors, including wind and relative humidity, affect the rate. Explain why.

5. You have just returned home from a three-day vacation. Your plants tell you, by their severe wilting, that you forgot to water them before you departed. Being aware of the cohesion-tension theory of water transport, explain what happened to them.

6. Not having a green thumb, your friend Stephanie decides to grow zucchini plants, which are most forgiving of poor soils and amateur gardeners. She plants too many seeds and ends up with far too many plants. Among them you happen to notice a stunted plant and decide to find out what happened to it. After

Figure 30.20 Middle fork of the Salmon River, Idaho.

many experiments, you decide that its leaves are producing a mutated, malfunctioning form of an enzyme that is necessary for the formation of sucrose. Knowing what you do about the pressure flow theory, explain why plant growth was hindered.

7. At the middle fork of the Salmon River in Idaho, clear water from a wilderness area (visible at the *right* in Figure 30.20) converges on brown-colored water (visible at *left* and *center*). The brown water is enriched with silt from a wilderness area that was disturbed by cattle ranching and some other human activities. Knowing what you do about the nature of topsoil, what is probably happening to plant growth in the disturbed habitat? Knowing how fishes acquire oxygen for aerobic respiration, how might the silt be affecting their survival?

Selected Key Terms

CAM plant *30.4*	leaching *30.1*	root nodule *30.2*
carnivorous plant *CI*	loam *30.1*	sieve tube *30.5*
Casparian strip *30.2*	mutualism *30.2*	soil *30.1*
cohesion-tension theory *30.3*	mycorrhiza *30.2*	stoma (stomata) *30.4*
companion cell *30.5*	nitrogen fixation *30.2*	topsoil *30.1*
cuticle *30.4*	nutrient *30.1*	tracheid *30.3*
cutin *30.4*	phloem *30.5*	translocation *30.5*
endodermis *30.2*	plant physiology *CI*	transpiration *30.3*
erosion *30.1*	pressure flow theory *30.5*	turgor pressure *30.4*
exodermis *30.2*		vascular cylinder *30.2*
guard cell *30.4*	root hair *30.2*	vessel member *30.3*
humus *30.1*		xylem *30.3*

Readings

Galston, A., P. Davies, and R. Satter. 1980. *The Life of a Green Plant.* Englewood Cliffs, New Jersey: Prentice-Hall.

Hausenbuiller, R. 1985. *Soil Science.* Third edition. Dubuque, Iowa: W. C. Brown.

Hewitt, E., and T. Smith. 1975. *Plant Mineral Nutrition.* New York: Wiley.

Salisbury, F., and C. Ross. 1992. *Plant Physiology.* Fourth edition. Belmont, California: Wadsworth.

Web Site See *http://www.wadsworth.com/biology* for practice quiz questions, hypercontents, BioUpdates, and critical thinking. The Wadsworth Biology Resource Center provides a wealth of information fully organized and integrated by chapter.

PLANT REPRODUCTION

A Coevolutionary Tale

Flowering plants thrive almost everywhere, from icy tundra to deserts to oceanic islands. What accounts for their distribution and diversity? Consider the **flower**, which is a specialized reproductive shoot. About 435 million years ago, when plants first invaded the land, insects that fed on decaying plants and spores probably were not far behind them. The smorgasbord of aboveground plant parts seems to have favored natural selection of winged insects with an amazing array of sucking, piercing, and chewing mouthparts.

By 390 million years ago, in humid coastal forests, seed-bearing plants were producing pollen grains—those sperm-bearing suitcases that can travel to eggs, which develop inside ovules in separate plant parts. Pollen is rich in nutrients. At first, it simply may have drifted on air currents to the ovules. Then insects made the connection between "plant parts with pollen" and "food." Plants lost some pollen to hungry insects but gained a reproductive advantage. Why? Compared to air currents, pollen-dusted insects clambering over the plants could make more accurate deliveries of pollen to ovules. The tastier the pollen, the more home deliveries, and the more seeds. *And the greater the number of seeds, the greater were the chances of reproductive success.*

What we are describing here is a case of **coevolution**. The word refers to two or more species jointly evolving as an outcome of close ecological interactions. When one of the species evolves, the change affects selection pressures operating between the two, and so the other species evolves also.

In our coevolutionary tale, new or modified plant structures that were more enticing to pollen-delivering insects were favored. Individual insects that were quicker to recognize and locate particular plants also enjoyed a competitive edge. Hence the evolution of plants with distinctive flowers, fragrances, and sugar-rich nectar, and of pollinators specifically adapted to take advantage of them. A **pollinator** is any agent that transfers pollen to a landing platform near an ovule in the same flower or in a different one. Besides insects, pollinating agents include air currents, water currents, bats, birds, and other animals (Figures 31.1 and 31.2).

Today you can correlate many floral features with specific pollinators. For example, the positioning of a flower's reproductive parts ensures that its pollinating agent will brush past them. Many parts are strategically located above nectar-filled floral tubes that are as long as the feeding devices of their preferred pollinators. Red and yellow flowers attract birds, which have excellent daytime vision but a poor sense of smell. As you might suspect, plants that birds visit do not divert metabolic

Figure 31.1 Examples of adaptations uniting flowering plants with their pollinators. (**a**) A giant saguaro of Arizona's Sonoran Desert produces large, showy white flowers at the tips of spiny arms. Insects visit the flowers by day, and bats visit by night. The plant offers the animals nectar, and the animals transport pollen grains that stick to their body from one cactus plant to another. (**b**) *Right:* Shine ultraviolet light upon a gold-petaled marsh marigold and you will reveal its bee-attracting pattern.

a

Figure 31.2 Bahama woodstar sipping nectar from a hibiscus blossom. Like other hummingbirds, it forages for nectar in midflight. Its long, narrow bill coevolved with long, narrow floral tubes.

resources to producing fragrances. By contrast, the red flowers do not attract beetles, and neither do flowers with large, deep nectar cups in which the beetles might drown. Flowers that are pollinated by beetles (and flies) smell like rotten meat, moist dung, or the litter on a forest floor, where beetles first evolved.

Daisies and other fragrant flowers with distinctive patterns, shapes, and red or orange components attract butterflies, which forage by day. Nectar-sipping bats and most moths forage by night. They pollinate intensely sweet-smelling flowers with white or pale petals, which are more visible in the dark than colored petals. Moths and butterflies have long, narrow mouthparts adapted to drinking from narrow floral tubes or floral spurs. The Madagascar hawkmoth uncoils a mouthpart twenty-two centimeters long—precisely as long as the floral tube of the orchid *Angraecum sesquipedale*!

how we see it

b

how bees see it

Flowers having sweet odors and yellow, blue, and purple parts attract bees. Some of the pigments absorb ultraviolet light and form patterns that say, "Nectar here!" Unlike us, bees see ultraviolet light, and they find the nectar guides alluring (Figure 31.1*b*). Unlike beetles, bees have long feeding devices that extend into floral tubes.

In short, far from being merely decorative, flowers are adaptations that contribute to a plant's reproductive success. This chapter and the next focus on the modes of reproduction, growth, and development of the plants that bear them.

KEY CONCEPTS

1. Sexual reproduction is the premier reproductive mode of flowering plant life cycles. Many species of flowering plants also can reproduce asexually.

2. Sexual reproduction of flowering plants involves the formation of spores and gametes, both of which develop inside specialized reproductive shoots called flowers.

3. Microspores form in a flower's male reproductive parts and develop into pollen grains, the sperm-bearing male gametophytes. Animals, air currents, and other pollinating agents transfer pollen grains to the female reproductive parts of flowers.

4. Megaspores form within ovules, which are tissue masses in the ovary of a flower's female reproductive parts. Egg-producing female gametophytes develop in ovules.

5. After sperm fertilize eggs, seeds develop. Each seed is a mature ovule. It consists of an embryo sporophyte and tissues that function in its nutrition and protection.

6. While seeds are developing, tissues of the ovary and sometimes neighboring tissues mature into fruits, which function in seed dispersal. Fruits differ considerably in structure, depending on the agents that disperse them. Air currents, water currents, and animals function as dispersal agents.

Think Sporophyte and Gametophyte

You probably don't think about this very often, but flowering plants engage in sex. Like humans, they have splendid reproductive systems that produce, nourish, and protect sperm and eggs. Like human females, they house embryos during early development. They issue floral invitations to third parties—pollinators that help sperm and egg get together. Long before we thought of it, flowering plants were using tantalizing fragrances and colors to improve the odds for sexual success.

When you hear the word "plant," you may think of something like a cherry tree (Figure 31.3a). The tree is an example of a typical **sporophyte**, a vegetative body that grows, by mitotic cell divisions, from a fertilized egg. At some point during the life cycle, the sporophyte bears flowers, the floral shoots that are specialized for reproduction. In those flowers, haploid spores form and develop into the haploid bodies called **gametophytes**. Sperm form inside the male gametophytes; eggs form inside the female gametophytes. As you know, the first cell that forms at fertilization has two sets of genetic instructions,   from two gametes.

Components of Flowers

Flowers grow at floral shoots of the primary plant body. As flowers form, they differentiate into nonfertile parts (the sepals and petals) and fertile parts (the stamens and carpels). Figures 31.3b and 31.4 show how floral components are arranged. Directly or indirectly, all the parts are attached to a receptacle, the modified base of the floral shoot. Peel open a rosebud, and you see that sepals, or the outermost whorl of leaflike parts, enclose leaflike petals. In turn, those petals enclose the flower's fertile components. This is a common arrangement. The sepals are the flower's calyx. The petals, which ring the male and female parts, are the flower's corolla.

Like leaves, sepals and petals have ground tissues, vascular tissues, and epidermis. Why are many flowers sweet smelling? Some epidermal cells of petals produce fragrant oils. What gives the petals their shimmer and color? Cells of the ground tissue contain pigments, such as carotenoids (yellow to red-orange) and anthocyanins (red to blue), and small, light-refracting crystals. What does a flowers' fragrance or its coloration, patterning, or arrangement of petals do? It attracts pollinators.

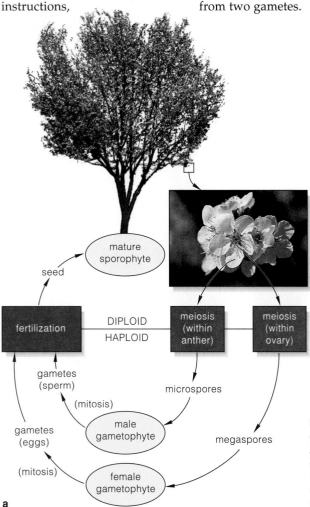

a

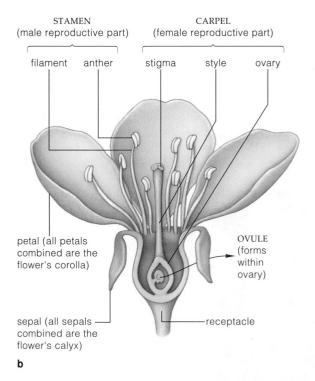

b

Figure 31.3   (**a**) Overview of flowering plant life cycles, using a cherry (*Prunus*) as the example. (**b**) Structure of one flower, a cherry blossom. As is true of the flowers of many plants, this blossom has a single carpel (a female reproductive part) and stamens (the male reproductive parts). Flowers of other plants have two or more carpels, often fused to form a single structure. Whether single or fused, carpels consist of an ovary and a stigma. Often the ovary extends upward as a slender column, the style.

corolla
(all petals
combined)
calyx
(all sepals
combined)
receptacle

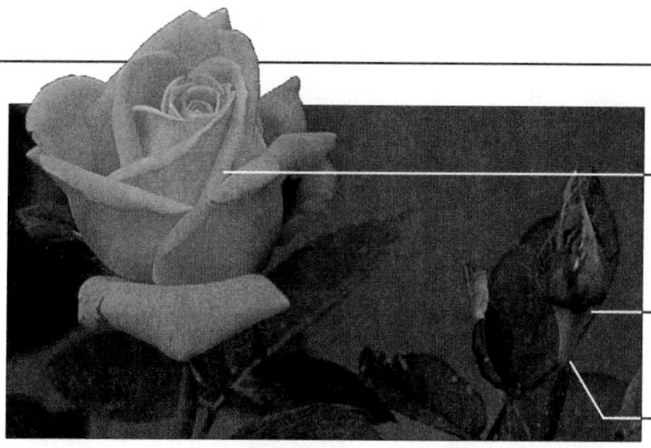

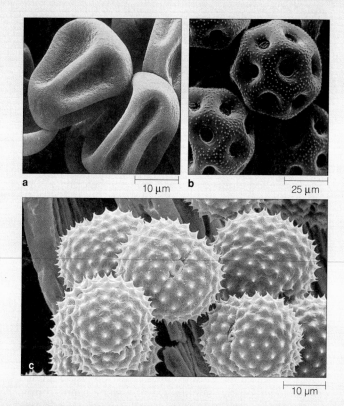

Figure 31.4 Location of a few floral parts in one of the most prized cultivated plants, the rose (*Rosa*).

Where Pollen and Eggs Develop

Just inside a flower's corolla are male reproductive parts, or **stamens**, nearly all of which consist of an anther and a single-veined stalk (filament). An anther is internally divided into pollen sacs, which are chambers in which walled, haploid spores form and develop into the male gametophytes called **pollen grains** (Figure 31.5).

A flower's female reproductive parts are located at its center. You may have heard someone call these parts pistils, but their more recent name is **carpels**.

Some flowers have one carpel. Others have more, often fused together to form a compound structure. The lower portion of either single or fused carpels is the **ovary**, where the eggs develop, fertilization takes place, and seeds mature. Angiosperm, a more formal name for flowering plants, refers to the carpel (from the Greek *angeion*, meaning vessel; and *sperma*, meaning seed). The upper portion of a carpel, the stigma, is a sticky or hairy surface tissue that captures pollen grains and favors their germination. Commonly the stigma is elevated on a style, a slender, upward extension of the ovary's wall.

Not every kind of flower has stamens and carpels. Some plant species produce *perfect* flowers, with both male and female parts. Other species produce *imperfect* flowers, which contain male or female parts. In certain species, such as oaks, the same individual plant bears male and female flowers. In willows and other species, they are on separate individual plants.

Sexual reproduction is the dominant reproductive mode of flowering plant life cycles. Such cycles alternate between the production of sporophytes (spore-producing bodies) and gametophytes (gamete-producing bodies).

Flowering plant sporophytes grow by mitotic cell divisions from a fertilized egg. The sporophyte is all of the plant body except for the male and female gametophytes, which form inside its flowers.

Gametophytes arise from haploid spores, which form in a flower's stamens (male reproductive parts) and ovaries (female reproductive parts). Male gametophytes develop into sperm-producing pollen grains. Female gametophytes are structures in which eggs develop, fertilization takes place, and seeds mature.

31.2 *Focus on the Environment*

POLLEN SETS ME SNEEZING

In the year 1835 Sidney Smith wrote, "I am suffering from my old complaint, the hay-fever . . . that sets me sneezing; and if I begin sneezing at twelve, I don't leave off till two o'clock, and am heard distinctly in Taunton [six miles distant] when the wind sets that way."

Sidney suffered what is now called *allergic rhinitis*. This is the name for a hypersensitivity to a normally harmless substance, and such hypersensitivity is common.

The pollen of ragweed, shown in Figure 31.5, brought on Sidney's dreaded hay fever, but it wasn't out to get him. Every spring and summer, flowering plants release pollen grains. In many millions of people, white blood cells respond by mounting an immune response against some of the proteins that project from the surface of pollen grain walls. The pollen grains from different plants have differences in their wall proteins. The immune system of a person who is hypersensitive might chemically respond to some of these but not others. The misdirected response results in a profusely runny nose, reddened and itchy eyelids, congestion, and bouts of sneezing.

Hay fever is a genetic abnormality; it runs in families. Some individuals simply are genetically predisposed to overreact to certain kinds of pollen. In addition, infections, emotional stress, and changes in temperature may trigger immune reactions to pollen.

a
10 μm
b
25 μm
c
10 μm

Figure 31.5 Pollen grains from (**a**) grass, (**b**) rose, and (**c**) ragweed plants. Pollen grains of most families of plants differ in size, wall sculpturing, and number of wall pores.

A NEW GENERATION BEGINS

From Microspores to Pollen Grains

We turn now to pollen grain formation. While anthers are growing, four masses of spore-producing cells form by mitotic cell divisions. Walls develop around them. Each anther now has four chambers, called pollen sacs (Figure 31.6a). Cells in the sacs undergo meiosis and cytoplasmic division. This results in **microspores**. The haploid spores go on to develop an elaborately sculpted wall. The walled microspores divide once or twice by way of mitosis and become pollen grains, which now enter a period of arrested growth. In time, they will be released from the anther. Their wall components will protect them from decomposers in their surroundings.

As soon as many types of pollen grains form, they produce sperm nuclei, which are the male gametes of flowering plants. Other types do not do this until after they arrive at a carpel and then start growing toward its ovule. In short, a pollen grain is a mature *or* immature male gametophyte, depending on the plant species.

From Megaspores to Eggs

Meanwhile, one or more masses of cells are forming on the inner wall of a flower's ovary. Each is the start of an **ovule**, a structure that contains a female gametophyte and that may become a seed. As each cell mass grows, a tissue forms within it, and one or two protective layers called integuments form around it. Inside the mass, a cell divides by meiosis, and four haploid spores form. The spores formed in flowering plant ovaries are generally larger than microspores and are called **megaspores**.

Commonly, all megaspores but one disintegrate. The one remaining undergoes mitosis three times without cytoplasmic division. At first it is a cell with eight nuclei (Figure 31.6b). Its cytoplasm divides after each nucleus migrates to a specific location. In this case, the result is a seven-celled embryo sac, or female gametophyte. One cell (the "endosperm mother cell") has two nuclei and will help form the **endosperm**, a nutritive tissue for the forthcoming embryo. Another cell is the egg.

From Pollination to Fertilization

Every spring, flowering plants release pollen. You are acutely aware of this reproductive event if you are one of the millions of people who suffer from hay fever, the allergic reaction to wall proteins of pollen grains from ragweed and many other plants (Section 31.2).

Specifically, **pollination** means the transfer of pollen grains to a receptive stigma. Air or water currents, birds, and other pollinating agents make the transfer. You already looked at the relationship between flowering plants and their pollinators, in the chapter introduction.

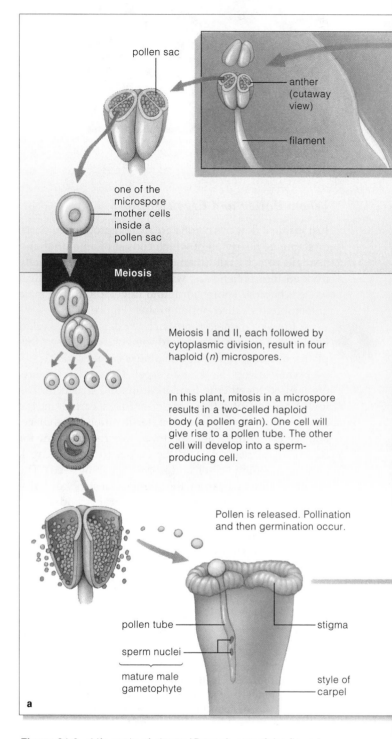

Meiosis I and II, each followed by cytoplasmic division, result in four haploid (*n*) microspores.

In this plant, mitosis in a microspore results in a two-celled haploid body (a pollen grain). One cell will give rise to a pollen tube. The other cell will develop into a sperm-producing cell.

Pollen is released. Pollination and then germination occur.

Figure 31.6 Life cycle of cherry (*Prunus*), one of the flowering plants classified as a dicot. (**a**) This part of the diagram shows how pollen grains (male gametophytes) develop and germinate. (**b**) This part shows what goes on inside the ovule in this ovary.

Once a pollen grain lands on a receptive stigma, it germinates. For this event, germination means a pollen grain resumes growth and then develops into a tubular structure. This "pollen tube" grows down through the tissues of the ovary and carries the sperm nuclei with it

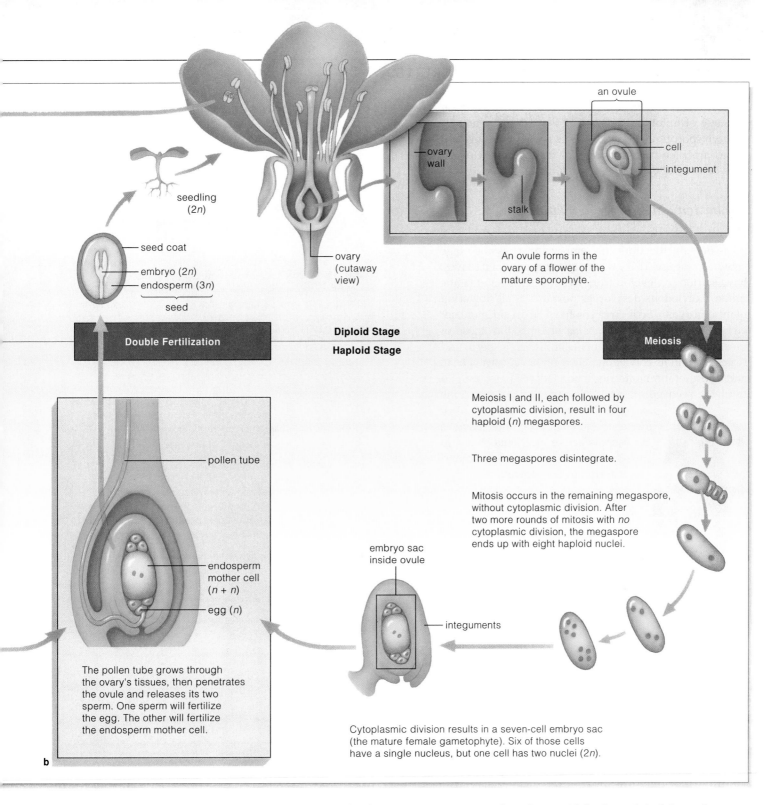

seedling
(2n)

seed coat

embryo (2n)

endosperm (3n)

seed

ovary
(cutaway
view)

an ovule

ovary
wall

stalk

cell

integument

An ovule forms in the
ovary of a flower of the
mature sporophyte.

Double Fertilization

Diploid Stage

Haploid Stage

Meiosis

Meiosis I and II, each followed by
cytoplasmic division, result in four
haploid (n) megaspores.

Three megaspores disintegrate.

Mitosis occurs in the remaining megaspore,
without cytoplasmic division. After
two more rounds of mitosis with *no*
cytoplasmic division, the megaspore
ends up with eight haploid nuclei.

pollen tube

endosperm
mother cell
(n + n)

egg (n)

embryo sac
inside ovule

integuments

The pollen tube grows through
the ovary's tissues, then penetrates
the ovule and releases its two
sperm. One sperm will fertilize
the egg. The other will fertilize
the endosperm mother cell.

b

Cytoplasmic division results in a seven-cell embryo sac
(the mature female gametophyte). Six of those cells
have a single nucleus, but one cell has two nuclei (2n).

(Figure 31.6b). Chemical and molecular cues guide the
pollen tube's growth through the ovary's tissues, toward
the egg chamber and sexual destiny. When the pollen
tube reaches an ovule, it penetrates the embryo sac, its
tip ruptures, and the two sperm are released.

"Fertilization" generally means fusion of a sperm
nucleus with an egg nucleus. But **double fertilization**
takes place in flowering plants. In the diploid species,
one sperm nucleus fuses with that of the egg, and so
produces a diploid (2n) zygote. Meanwhile, the other

sperm nucleus fuses with both nuclei of the endosperm
mother cell, forming a cell with a triploid (3n) nucleus.
That 3n cell gives rise to the nutritive tissue.

In flowering plants, sperm cells form within pollen grains,
the male gametophytes. Eggs develop inside ovules, which
contain the female gametophytes.

After pollination and double fertilization, an embryo and
nutritive tissue form in the ovule, which becomes a seed.

FROM ZYGOTE TO SEEDS AND FRUITS

After fertilization, the newly formed zygote embarks on a course of mitotic cell divisions that lead to a mature embryo sporophyte. It develops as part of an ovule and is accompanied by formation of a **fruit**: a mature ovary, with or without other, neighboring tissues (Table 31.1).

Formation of the Embryo Sporophyte

Consider how the embryo forms in shepherd's purse (*Capsella*). By the time this embryo reaches the stage shown in Figure 31.7*e*, two **cotyledons**, or seed leaves, have started to develop from two lobes of meristematic tissue. Cotyledons develop as portions of all flowering plant embryos. Dicot embryos have two, and monocot embryos have one. The *Capsella* embryo, like those of most dicots, absorbs nutrients from the endosperm and stores them in its cotyledons. By contrast, in corn, wheat, and most other monocots, endosperm is not tapped until the seed germinates. Digestive enzymes become stockpiled inside the thin cotyledons of monocot embryos. When the enzymes do become active, nutrients stored in the endosperm will be released and transferred to the growing seedling.

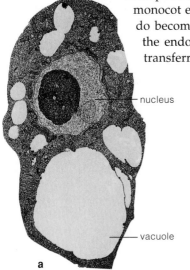

— nucleus

— vacuole

a

Figure 31.7 Stages in the development of a dicot seed— shepherd's purse (*Capsella*). The micrographs are not to the same scale. (**a**) Single-celled zygote. (**b–d**) The early embryo is identified by the *yellow* boxes. The row of cells below each box transfers nutrients to the embryo sporophyte from the parent plant. The embryo in (**e**) is well developed, and the one in (**f**) is mature. (**g**) *Capsella* fruit.

Formation of Seeds and Fruits

From the time a zygote forms until a mature embryo develops, a parent plant transfers nutrients to tissues of the ovule. Food reserves accumulate in endosperm or cotyledons. Eventually, the ovule separates from the ovary wall, and its integuments thicken and harden into a seed coat. The embryo, food reserves, and coat are a

Table 31.1 Some Major Categories of Fruits

Category	Characteristics	Examples
SIMPLE DRY FRUITS (from one simple or compound ovary)		
1. *Fruit wall dry, does not split when mature*:		
Nut	Hard pericarp, 1 seed	Hazelnut
Grain	Pericarp, ovary wall fused, 1 seed	Rice, corn, wheat
Achene	Thin pericarp; seed coat separate from ovary wall	Strawberry, sunflower
2. *Fruit wall dry, splits when mature to release seeds*:		
Legume	Opposite edges of pods split	Peas, beans
SIMPLE FLESHY FRUITS (from one simple or compound ovary)		
True berry	Mesocarp fleshy or slimy, endocarp soft; many seeds	Grape, tomato, papaya
Drupe	Mesocarp fleshy, endocarp usually hard; 1 or 2 seeds	Peach, plum, cherry, olive
Hesperidium	Berry with thick rind, oil glands	Orange, lemon
ACCESSORY FRUITS (from simple ovary plus other tissues)		
"Berries"	Superficially resemble berries	Banana
Pepo	Pericarp, other tissues form rind	Squashes, melons
Pome	Mostly enlarged receptacle	Apples, pears
AGGREGATE FRUITS		
From several separate ovaries of one flower		Raspberries
MULTIPLE FRUITS		
Combined from numerous separate flowers		Pineapples, figs

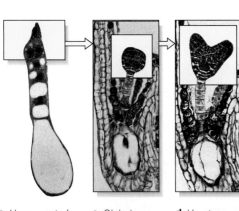

b Upper part of the zygote gives rise to embryo

c Globular stage of embryo

d Heart-shaped stage of embryo

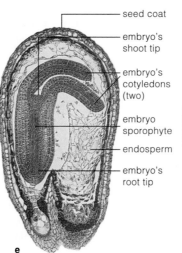

— seed coat

— embryo's shoot tip

— embryo's cotyledons (two)

— embryo sporophyte

— endosperm

— embryo's root tip

e

mature embryo within ovule

f

A fruit (a mature ovary) cut open to show seeds (the mature ovules). The embryo in each seed is at stage (**f**).

g

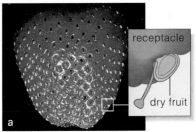

Figure 31.8 Fruits. (**a**) Strawberry (*Fragaria*), a fragrant, wildly popular, easily bruised fruit—actually a fleshy receptacle with many dry fruits at its surface. Genetic engineering has yielded strains that retain the strong fragrance yet are firm enough to resist packing damage. (**b**) How fruit forms on an apple (*Malus*) tree. Eggs usually are fertilized when the petals drop from the blossoms. The ovary and receptacle expand, and sepals and stamens remain on this immature fruit, which is a pome. (**c**) Pineapple (*Ananas comosus*), a multiple fruit. Besides running into the New World, Christopher Columbus also ran into cultivated pineapples, viewed by native Americans as a symbol of hospitality. To Columbus, the fruits resembled pine cones, hence the name. At the time, native peoples used pineapple juice as a base for an alcoholic beverage and for a poison in which to dip arrow tips.

b Expansion of each ovary and receptacle after petals drop from apple blossoms

individual fruit

remnants of petals and sepals

vascular bundles in outer portion of the ovary

seeds

enlarged receptacle

self-contained package—a **seed**, which is defined as a mature ovule.

While seeds are forming, changes in other parts of the flower are forming fruits. Different fruits are fleshy or dry. They may consist of one or more ovaries, and they may incorporate tissues besides those of the ovary, such as tissues of the receptacle. An acorn is a dry fruit that stays intact at maturity, whereas a pea pod is a dry fruit that unzips and releases its seeds. A strawberry also is a dry fruit, not a "berry." Its soft, red flesh is an expanded receptacle that has many seeds dotting its surface (Figure 31.8*a*). The bulk of an apple is an expanded receptacle, also (Figure 31.8*b*). A pineapple is a multiple fruit; it is derived from a number of individual flowers that were grouped close together on the parent plant (Figure 31.8*c*). Tomatoes, grapes, peaches, and cherries are classified as fleshy fruits with tissues derived only from the ovary.

Botanists often refer to three divisions of a fleshy fruit, although these are sometimes not clearly visible. The *endo*carp is the innermost portion around the seed or seeds. *Meso*carp is the fleshy portion, and *exo*carp is the skin. Together, these three regions of the fruit are called a **pericarp**. In walnuts and other dry fruits, the pericarp typically is very thin. In true berries, such as tomatoes and grapes, the pericarp is thin skinned and relatively soft at maturity.

A mature ovule, which encases an embryo sporophyte and food reserves inside a protective coat, is a seed.

A mature ovary, with or without additional floral parts that have become incorporated into it, is a fruit.

Fruits are classified according to whether they are dry or fleshy, whether they are derived from one or more ovaries, and whether they incorporate other tissues besides those of the ovary.

Fleshy fruits have three regions. Innermost is the endocarp, which surrounds the seeds. The mesocarp is the fleshy part, and the exocarp is the skin. Together, the three regions are known as the pericarp.

The fruits you read about in the preceding section have a common function: *seed dispersal*. To gain insight into their structure, think about how they coevolved with particular dispersing agents—currents of air or water, or animals passing by.

Consider the wind-dispersed fruit of maples (*Acer*), as shown in Figure 31.9*a*. The pericarp of a maple fruit extends out like wings, and when the fruit drops from the tree, it interacts with air currents and spins sideways. The wings can whirl the fruit far enough away that the embryo sporophytes inside the seeds will not have to compete with the parent plant for water, minerals, and sunlight. Brisk winds have been known to transport the

Besides digesting the fruit's flesh, these enzymes digest some of the seed coat. The digested portions make it easier for an embryo to break through its hard coat after the seeds are expelled from the animal body and start to germinate. We turn to this topic in the next chapter.

Some water-dispersed fruits have heavy wax coats, and others, including sedges, have sacs of air that help them float. With its thick, water-repelling pericarp, the fruit of coconut palms is adapted to traveling across the open ocean. If they don't beach soon enough, saltwater may penetrate the pericarp and kill their embryos, but coconuts can travel hundreds of kilometers before this happens.

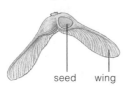

seed   wing

Figure 31.9 (**a**) Winged seeds of maple (*Acer*). (**b**) Students taking a chocolate ice cream break and indirectly contributing to the reproductive success of cacao (*Theobroma cacao*). (**c**) Cacao fruits, or pods, each with as many as forty seeds ("beans") that are processed into cocoa butter and essences of chocolate products. Interactions among the 1,000 or so compounds in chocolate exert compelling (some say addictive) effects on the human brain. Each year, the average American buys 8 to 10 pounds of chocolate, and the average Swiss citizen, a whopping 22 pounds. With that kind of demand, *T. cacao* seeds are cherished and "dispersed" in orderly arrays in plantations, where the trees that grow from them are carefully tended.

fruits as much as ten kilometers from a tree. Similarly, air currents easily lift the "parachute" pluming outward from each tiny fruit of dandelions and easily disperse orchid seeds, which are as fine as dust particles.

Other fruits taxi to new locations on or in animals, including many birds, mammals, and insects. Some can adhere to feathers, feet, or fur with hooks, spines, hairs, and sticky surfaces. The fruits of cocklebur, bur clover, and bedstraw are like this. Embryo sporophytes inside the seed coats of fleshy fruits survive being eaten and assaulted by the digestive enzymes in an animal's gut.

Humans are grand dispersing agents. In the past, explorers carried seeds all over the world, although most places now control imports. Cacao (Figure 31.9), oranges, corn, and other plants were domesticated and encouraged to reproduce in new habitats, in far greater numbers than they did on their own. And in evolutionary terms, reproductive success is what life is all about.

Seeds and fruits are structurally adapted for dispersal by air currents, water currents, and many kinds of animals.

31.6

WHY SO MANY FLOWERS AND SO FEW FRUITS?

In the Sonoran Desert of Arizona (Figure 31.10a), the giant saguaro produces as many as a hundred flowers at the tips of its huge, spiny arms. Remember its large, showy flowers as shown in Figure 31.1a? Each flower only blooms for twenty-four hours. Again, insects visit the flowers by day, and bats by night. The animals benefit by getting nectar, and the giant saguaro benefits because the animals deliver their dusting of pollen grains to other giant saguaros.

The flowers that do get pollinated, courtesy of insects or bats, now begin the task of producing seeds and fruits. Petals wilt, then they wither as the many egg cells inside the ovary become fertilized. Each egg-containing ovule expands and matures into a seed. During the same interval, the ovary develops into a fruit (Figure 31.10). Weeks later,

Common sense might lead you to conclude, "The more fruits, the better." And yet saguaros, *like many other plants*, commonly produce far more flowers than they do mature fruits. They seem to be producing "excess" flowers and giving up chances to make seeds and leave descendants. Doing so is not a reproductive adaptation that you would expect, based on Darwinian evolutionary theory.

Well, suppose that some plants simply did not receive enough pollen to have all their egg cells fertilized. After all, unfertilized flowers cannot produce seeds and fruits. This hypothesis has been experimentally tested for some plant species. Researchers carefully brushed pollen on every single flower. In many instances, the plants still failed to produce fruit for every flower!

a

Figure 31.10 (a) Giant saguaro growing in Arizona's Sonoran Desert. (b) Two fruits. The one above set, and the one below it failed to set. (c) A red, maturing fruit.

b

c

the plum-sized fruit splits open, exposing its bright red interior and dark seeds.

White-winged doves feast on the ripe fruits. When they fly away, each might carry hundreds of seeds in its gut. A few seeds may elude the gut's digestive enzymes and later end up on the ground. There, each seed will have a tiny chance of growing into a giant saguaro.

One of the puzzling aspects of this annual event is the frequency with which saguaro flowers *fail* to give rise to fruit (Figure 30.10b, for example). Why would a cactus invest so much energy constructing a hundred flowers if only thirty or so of them will set fruit? Compare one of these "inefficient" plants with a species that produces a hundred flowers, *all* of which develop into fruits. Which do you think will leave more descendants?

Alternatively, suppose the presumed "excess" flowers are formed strictly to produce pollen for export to other plants. Pollen grains are small. In energy terms, they are inexpensive to produce, compared to large, calorie-rich, seed-containing fruits. Thus, for a fairly small investment, a plant might reap a large reward in offspring that carry its genes. The idea that some plants actually set aside flowers exclusively for pollen export is only now being tested for saguaros. Perhaps you might like to design and carry out experiments that will help clear up the mystery of the "excess" flowers of saguaros and similar plants.

Asexual Reproduction in Nature

The features of sexual reproduction that we considered in the preceding sections dominate flowering plant life cycles. Bear in mind, many species also can reproduce asexually by various modes of **vegetative growth**. Table 31.2 lists some of these modes. In essence, new roots and shoots grow right out of extensions or fragments of a parent plant. As you know, such asexual reproduction proceeds by way of mitosis, so offspring are genetically identical to the parent; they are a clone.

One "forest" of quaking aspen (*Populus tremuloides*) provides us with an astounding example of vegetative reproduction. The leaves of this flowering plant species tremble, even in the slightest breeze; hence the name. Figure 31.11 is a panoramic view of the shoot systems of a single individual. The root system of the parent plant gave rise to adventitious shoots, which became separate shoot systems. Barring rare mutations, individuals at the north end of this vast clone are genetically identical to individuals at the south end. Water travels from roots near a lake all the way to shoot systems in much drier soil. Dissolved ions can travel in the opposite direction.

As long as environmental conditions favor growth and regeneration, such clones are as close as one can get to being immortal. No one knows how old aspen clones are. The oldest known clone, a ring of creosote bushes (*Larrea divaricata*) that is growing in the Mojave Desert, has been around for the past 11,700 years.

The possibilities are astounding. Raise strawberry plants and watch them send out runners (horizontal aboveground stems); then watch new roots and shoots develop at every other node. Eat an orange, and it may be descended from one tree in Southern California that reproduced by way of **parthenogenesis**. By this process, an embryo develops from an unfertilized egg.

Parthenogenesis may be stimulated when pollen has contacted a stigma even though a pollen tube has not grown down the style. Maybe certain hormones from the stigma or from pollen grains diffuse to the unfertilized egg and trigger formation of an embryo, which is 2*n* as a result of fusion of the products of mitotic cell division in the egg. A 2*n* cell outside the gametophyte also may be stimulated to develop into an embryo.

Induced Propagation

Most houseplants, woody ornamentals, and orchard trees are clones. They commonly are propagated from cuttings or fragments of shoot systems. For example, with suitable encouragement, a severed African violet leaf may form a callus from which adventitious roots develop. A callus is one type of meristem. Remember, a meristem is a localized region of still-embryonic cells that retain the potential for division.

Or consider a twig or bud from one plant, grafted onto a different variety of some closely related species. Vintners in France, for example, graft prize grapevines onto disease-resistant root stock from America.

Frederick Steward and his coworkers pioneered in **tissue culture propagation**. They cultured small bits of phloem from the differentiated roots of carrot plants (*Daucus carota*) in rotating flasks. They used a liquid growth medium that contained sucrose, mineral ions, and vitamins. It also contained coconut milk, which Steward knew was rich in then-unidentified growth-inducing substances. As the flasks rotated, individual cells that were torn away from the tissue bits divided and formed multicelled clumps, which sometimes gave rise to roots (Figure 31.12*a*). Steward's experiments were among the first to demonstrate that some cells of

Table 31.2	Asexual Reproductive Modes of Flowering Plants	
Mechanism	Representative	Characteristics
VEGETATIVE REPRODUCTION ON MODIFIED STEMS		
1. Runner	Strawberry	New plants arise at nodes along an aboveground, horizontal stem
2. Rhizome	Bermuda grass	New plants arise at nodes of underground horizontal stem
3. Corm	Gladiolus	New plant arises from axillary bud on short, thick, vertical underground stem
4. Tuber	Potato	New shoots arise from axillary buds on tubers, which are the enlarged tips of slender underground rhizomes
5. Bulb	Onion, lily	New bulb arises from an axillary bud on a short underground stem
PARTHENOGENESIS		
	Orange, rose	An embryo develops without nuclear or cellular fusion (for example, from an unfertilized haploid egg or by developing adventitiously, from tissue surrounding the embryo sac)
VEGETATIVE PROPAGATION		
	Jade plant, African violet	A new plant develops from tissue or structure (a leaf, for instance) that drops from the parent plant or is separated from it
TISSUE CULTURE PROPAGATION		
	Orchid, lily, wheat, rice, corn, tulip	A new plant is induced to arise from a parent plant cell that has not become irreversibly differentiated

Figure 31.11 A mere portion of Pando the Trembling Giant, so named by Michael Grant and his coworkers at the University of Colorado, who studied its genetic makeup. This "forest" in Utah is actually a single asexually reproducing male organism, of a type called quaking aspen (*Populus tremuloides*). Its root system extends through about 262 hectares (about 106 acres) of soil and functionally supports its 47,000 genetically identical shoots; that is, the trees. By one estimate, this clone weighs more than 5,915,000 metric tons.

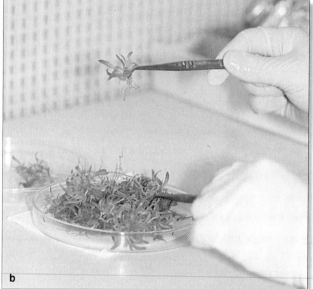

Figure 31.12 (a) Cultured cells from a carrot plant. At the time this photograph was taken, roots and shoots of embryonic plants were already forming. (b) Young orchid plants, developed from cultured meristem and young leaf primordia. Orchids are one of the most highly valued cultivated plants. Before meristems were cloned, they were difficult to hybridize. From the time seeds form, it can take seven years or more until a new plant flowers. In the natural habitat, seeds normally will not germinate unless they interact with a specific fungus.

specialized tissues still house the genetic instructions required to build new individuals.

Tissue culture propagations are now done with shoot tips or other parts of individual plants. The techniques are useful when an advantageous mutant arises. One such mutant might show resistance to a disease that is crippling to wild-type plants of the same species. Tissue culture propagation can result in hundreds and even thousands of identical plants from merely one mutant specimen. The techniques are already being employed in efforts to improve major food crops, such as corn, wheat, rice, and soybeans. They also are being used to increase the production of hybrid orchids, lilies, and other prized ornamental plants (Figure 31.12*b*).

Besides reproducing sexually, flowering plants engage in asexual reproduction, as by vegetative growth.

---

1. Sexual reproduction is the main reproductive mode of flowering plant life cycles. Diploid sporophytes, or spore-producing plant bodies, develop, as do haploid gametophytes (gamete-producing bodies).

a. The sporophyte is a multicelled vegetative body with roots, stems, leaves, and, at some point, flowers. Flowers of many species coevolved with animals that feed on nectar or pollen and also serve as pollinators. The gametophytes form in male and female floral parts.

b. Many flowering plants also reproduce asexually. They can do this naturally (as by runners, rhizomes, and bulbs) and artificially (as by cuttings and grafting).

2. Flowers typically have sepals, petals, and one or more stamens, which are male reproductive structures. They also have carpels, the female reproductive structures. Most or all of the reproductive structures are attached to a receptacle, or the modified end of a floral shoot.

a. Anthers of stamens contain pollen sacs in which cells divide by meiosis. A wall develops around each resulting haploid cell (microspore), which develops into a sperm-bearing pollen grain (male gametophyte).

b. A carpel (or two or more carpels fused together) has an ovary where eggs develop, fertilization takes place, and seeds mature. A stigma is a sticky or hairy surface tissue above the lower portion of the ovary. It captures pollen grains and promotes their germination.

3. Inside carpels, ovules form on the inner wall of the ovary. Each consists of a female gametophyte with an egg cell, an endosperm mother cell, a surrounding tissue, and one or two protective layers called integuments.

a. At maturity, each ovule is a seed; its integuments form the seed coat.

b. The ovary, and sometimes other tissues, matures into a fruit, which is a seed-containing structure.

4. Female gametophytes commonly form as follows:

a. Four haploid megaspores form after meiosis. All but one usually disintegrate.

b. The one remaining megaspore undergoes mitosis three times but not cytoplasmic division. Its multiple nuclei migrate to prescribed positions in the cytoplasm.

c. With cytoplasmic division, a female gametophyte (in this case, a seven-celled, eight-nuclei body) results. One cell is the egg. The cell with two nuclei (endosperm mother cell) will help give rise to endosperm, which is a nutritive tissue for the forthcoming embryo.

5. Pollination refers to the arrival of pollen grains on a suitable stigma. Once it has been transferred, a pollen grain germinates. It becomes a pollen tube that grows down through tissues of the ovary, carrying the two sperm nuclei with it.

6. At double fertilization, one sperm nucleus fuses with one egg nucleus, thereby forming a diploid ($2n$) zygote.

The other sperm nucleus fuses with both nuclei of the endosperm mother cell to form a cell that will give rise to nutritive tissue in the seed.

7. After double fertilization, the endosperm forms, the ovule expands, and the seed and fruit mature.

8. Fruits function to protect and disperse seeds, which germinate following dispersal from the parent plant. For example, fruits attract animals and other dispersal agents; lightweight fruits can be dispersed by winds. Some fruits have hooks and such that attach to animals.

9. Many flowering plants can reproduce asexually by vegetative growth. For example, new shoot systems may arise by mitotic divisions at nodes or buds along modified stems of the parent plant. New plants also may arise by vegetative propagation, from fragments or parts severed from a parent plant.

Review Questions

1. Define flower and define pollinator. What kinds of animals are attracted to flowers with red and orange components? What kinds respond to the flowers that smell like decaying organic material? What are some of the ways in which night-foraging animals find flowers in the dark? *CI*

2. Label the floral parts. Explain the role each part plays in the reproduction of flowering plants. *31.1*

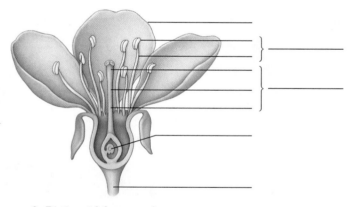

3. Distinguish between these terms:
   a. Sporophyte and gametophyte *31.1*
   b. Stamen and carpel *31.1*
   c. Ovule and ovary *31.1, 31.3*
   d. Microspore and megaspore *31.3*
   e. Pollination and fertilization *CI, 31.3*
   f. Pollen grain and pollen tube *31.1, 31.3*

4. Describe the steps by which an embryo sac, a type of female gametophyte, forms in a dicot such as a cherry (*Prunus*). *31.3*

5. Define the difference between a seed and a fruit. *31.4*

6. Do food reserves accumulate in endosperm, cotyledons, or both? If both, in what ways do these structures differ? *31.4*

7. Name an example of a simple dry fruit, simple fleshy fruit, accessory fruit, aggregate fruit, and multiple fruit. *31.4*

8. Name the three regions of a fleshy fruit. What is the name for all three regions combined? *31.4*

9. Define and give an example of an asexual reproductive mode used by a flowering plant. *31.7*

Self-Quiz (Answers in Appendix IV)

1. The _____ , which bears flowers, roots, stems, and leaves, dominates the life cycle of flowering plants.
   a. sporophyte   b. gametophyte

2. The flowers of many species coevolved with insects, birds, and other agents that function as _____ .

3. Male gametophytes of flowering plants produce _____ , and the female gametophytes produce _____ .
   a. megaspores; eggs   c. eggs; sperm
   b. sperm; microspores   d. sperm; eggs

4. Seeds are mature _____ ; fruits are mature _____ .
   a. ovaries; ovules   c. ovules; ovaries
   b. ovules; stamens   d. stamens; ovaries

5. A _____ is a closed vessel that contains an ovary in which eggs develop, fertilization occurs, and seeds mature.
   a. pollen sac   c. receptacle
   b. carpel   d. sepal

6. After meiosis within pollen sacs, haploid _____ form.
   a. megaspores   c. stamens
   b. microspores   d. sporophytes

7. Following meiosis in ovules, _____ megaspores form.
   a. two   c. six
   b. four   d. eight

8. The seed coat forms from which structure(s)?
   a. integuments   c. endosperm
   b. ovary   d. residues of sepals

9. Cotyledons develop as part of all flowering plant _____ .
   a. seeds   c. fruits
   b. embryos   d. ovaries

10. The outermost region of a fleshy fruit is the _____ .
    a. pericarp   c. mesocarp
    b. endocarp   d. exocarp

11. Development of a new plant from a tissue or structure that drops or is separated from the parent plant is called _____ .
    a. parthenogenesis   c. vegetative propagation
    b. exocytosis   d. nodal growth

12. Match the terms with the most suitable description.
    ____ double fertilization   a. formation of zygote and first cell of endosperm
    ____ ovule   b. outcome of two species interacting in close ecological fashion over geologic time
    ____ mature female gametophyte
    ____ asexual reproduction   c. contains a female gametophyte and has potential to be a seed
    ____ coevolution   d. an embryo sac, commonly with seven cells (one has two nuclei)
       e. mitotic cell division at bud or node produces a new plant

Critical Thinking

1. Wanting to impress her many friends with her sophisticated botanical knowledge, Dixie Bee is preparing a plate of tropical fruits for a party and cuts open a papaya (*Carica papaya*) for the first time. Inside are great numbers of slime-covered seeds, surrounded by soft flesh and soft skin (Figure 31.13). Knowing that her friends will ask what sort of fruit this is and which parts are the endocarp, mesocarp, and exocarp, she panics, runs to her biology book, and opens it to Section 31.4. What does she find out? What will she tell them when they ask what kinds of agents disperse the seeds of such a fruit from the parent plant?

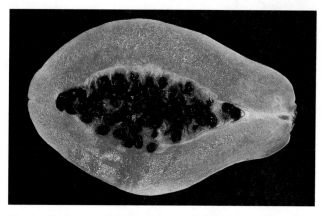

Figure 31.13 Seeds in the fruit of a papaya (*Carica papaya*).

2. Observe several kinds of flowers growing in the area where you live. Given the coevolutionary links between flowering plants and their pollinators, describe what sorts of pollination agents your floral neighbors might depend upon.

3. Elaine, a plant physiologist, succeeded in cloning genes for pest resistance into petunia cells. How can she utilize tissue culture propagation to produce many petunia plants with the good genes?

4. Before cherries, apples, peaches, and many other fruits ripen and the seeds inside them mature, their flesh is bitter or sour. Only later does it become tasty to animals that assist in seed dispersal. Develop a hypothesis of how this feature improves the odds for the plant's reproductive success.

Selected Key Terms

carpel *31.1*	ovule *31.3*
coevolution *CI*	parthenogenesis *31.7*
cotyledon *31.4*	pericarp *31.4*
double fertilization *31.3*	pollen grain *31.1*
endosperm *31.3*	pollination *31.3*
flower *CI*	pollinator *CI*
fruit *31.4*	seed *31.4*
gametophyte *31.1*	sporophyte *31.1*
megaspore *31.3*	stamen *31.1*
microspore *31.3*	tissue culture propagation *31.7*
ovary *31.1*	vegetative growth *31.7*

Readings

Grant, M. October 1993. "The Trembling Giant." *Discover* 14(10): 82–89.

Proctor, M., and P. Yeo. 1973. *The Pollination of Flowers.* New York: Taplinger.

Raven, P., R. Evert, and S. Eichhorn. 1992. *Biology of Plants.* Fifth edition. New York: Worth.

Salisbury, F., and C. Ross. 1991. *Plant Physiology.* Fourth edition. Belmont, California: Wadsworth.

Stern, K. 1997. *Introductory Plant Biology.* Seventh edition. Dubuque, Iowa: W. C. Brown. The photographs in this accessible little paperback are exceptional.

Web Site See *http://www.wadsworth.com/biology* for practice quiz questions, hypercontents, BioUpdates, and critical thinking. The Wadsworth Biology Resource Center provides a wealth of information fully organized and integrated by chapter.

32 PLANT GROWTH AND DEVELOPMENT

Foolish Seedlings and Gorgeous Grapes

A few years before the American stock market grew feverishly and then collapsed catastrophically in 1929, a researcher in Japan came across a substance that caused runaway growth and subsequent collapse of rice plants. Ewiti Kurosawa was studying what the Japanese call *bakane*, or the "foolish seedling" effect on rice plants. Stems of rice seedlings that had become infected with *Gibberella fujikuroi*, a fungus, grew twice as long as the stems of uninfected plants. The elongated stems were weak and spindly. Eventually they collapsed, and the infected plants died. Soon, Kurosawa discovered that he could trigger the disease by applying extracts of the fungus to plants. Many years later, other researchers purified the disease-causing substance from fungal extracts. The substance was named gibberellin.

Gibberellin, as we now know, is one of the premier plant hormones. **Hormones**, remember, are signaling molecules. After being produced and secreted by some cells, they travel to target cells and stimulate or inhibit gene activity (Section 15.5). Any cell with receptors for a given hormone is its target. Plant hormones never do travel far from the cells that secrete them.

Researchers have isolated more than eighty different forms of gibberellin from the seeds of flowering plants as well as from fungi. Possibly this same hormone is present in other groups of plants as well. The particular changes that gibberellins trigger make stems lengthen (Figure 32.1). In nature, gibberellins help seeds and buds break dormancy and resume growth in spring. They also influence the flowering process.

Figure 32.1 Demonstrations of the effects of hormones on plant growth and development. (**a**) Seedless grapes radiating market appeal. Gibberellin made their stems lengthen, which improved air circulation around the grapes and gave them more growing room. The grapes got larger and weighed more, to the delight of growers; grapes are sold by the pound. (**b**) This young California poppy (*Eschscholzia californica*) was left alone. (**c**) Gibberellin was applied to this young poppy plant.

two cabbages used as controls (untreated)   cabbages treated with gibberellins

Figure 32.2 Foolish cabbages!

Expose a radish plant to a suitable concentration of a gibberellin and it may grow as large as a beach ball. Do the same to a cabbage plant and it may well astound you (Figure 32.2). Applications of gibberellins can make celery stalks longer and crispier, and they can prevent the skin of navel oranges from maturing too quickly in orchards. Walk past plump seedless grapes in the produce bins of grocery stores and observe how the fleshy fruits of the grape plant (*Vitis*) grow in bunches along the stems. Gibberellin applications made the stems lengthen between internodes. With more physical space in between one another, the individual grapes grew larger. Also, air circulated more freely between them, which made it harder for disease-causing, grape-juice-loving fungi to take hold and do damage.

Gibberellin and other plant hormones orchestrate a plant's growth and development by responding to signals from one another. They also respond to cues afforded by the rhythmic changing of the seasons, as when days grow longer and warmer in spring, after the short, cold nights of winter. Those grapes, cabbage leaves, or celery stalks that find their way into your mouth are the culmination of a plant's own splendidly controlled program of growth and development.

Here we continue with a story that we started in the preceding chapter. That chapter traced events by which a zygote of a flowering plant forms and then develops into a mature embryo, housed inside a protective seed coat. At some point following its dispersal from the parent plant, the tiny embryo becomes transformed into a seedling, which in turn grows and develops into the mature sporophyte. In due time, the sporophyte forms flowers, then new seeds and fruits. Depending on the species, it drops old leaves throughout the year or all at once, in autumn.

This part of the story surveys the heritable, internal mechanisms that govern plant development, and the environmental cues that turn such mechanisms on or off at different times, and in different seasons.

KEY CONCEPTS

1. From the time a plant seed germinates, a number of hormones influence growth and development. This is true of seed-bearing plants as well as for other organisms.

2. Hormones are signaling molecules between cells. One cell type produces and then secretes a particular kind of hormone, which stimulates or inhibits gene activity in other cell types that take up molecules of the hormone. The changes in gene activity have predictable effects, as when they trigger the mitotic cell divisions and other processes necessary for stem elongation.

3. The known plant hormones are auxins, gibberellins, cytokinins, abscisic acid, and ethylene. We have indirect evidence of other kinds of hormones, which have not yet been identified.

4. Plant hormones help bring about predictable patterns of development, including the extent and direction of growth and cell differentiations in particular plant parts.

5. Plant hormones help adjust patterns of growth and development in response to environmental rhythms, including seasonal changes in daylength and temperature. Additionally, they help adjust the patterns in response to environmental circumstances in which an individual plant finds itself, such as the amount of sunlight or shade, moisture, and so on at a given site.

6. Commonly, two or more kinds of plant hormones must interact with one another to bring about specific effects on growth and development.

PATTERNS OF EARLY GROWTH AND DEVELOPMENT—AN OVERVIEW

Seed Germination

Let's first consider the overall patterns of growth and development for monocots and dicots, starting with the events that occur inside a seed. Figure 32.3 shows the embryo sporophyte inside a grain of corn. (Remember, a grain is a seed-containing dry fruit.) Before or after a seed is dispersed from a parent plant, the embryo's growth idles. Later, if all goes well, the seed germinates. **Germination** means an immature stage in the life cycle resumes growth after a period of arrested development.

Germination depends on environmental factors, such as the temperature, moisture, and oxygen level in soil and the number of daylight hours. The factors can vary seasonally. For example, mature seeds do not contain enough water for cell expansion or metabolism. In most habitats on land, ample water only becomes available on a seasonal basis, so that seed germination coincides with the return of spring rains. By a process known as **imbibition**, water molecules move into the seed, being attracted mainly to the hydrophilic groups of proteins stored inside endosperm or cotyledons. As more water moves in, the seed swells and its coat ruptures.

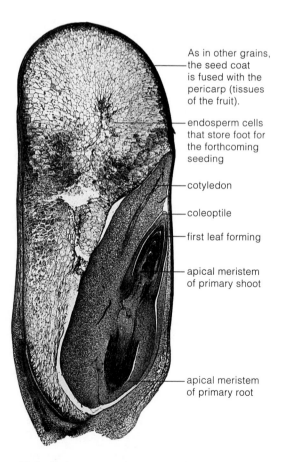

As in other grains, the seed coat is fused with the pericarp (tissues of the fruit).

endosperm cells that store foot for the forthcoming seeding

cotyledon

coleoptile

first leaf forming

apical meristem of primary shoot

apical meristem of primary root

Figure 32.3 Embryo sporophyte and its food reserves inside a grain of corn (*Zea mays*).

Figure 32.4 (**a,b**) The pattern of growth and development of a dicot, using the common bean plant (*Phaseolus vulgaris*) as the example. When a bean seed germinates, the embryo inside resumes growth. After the seed has germinated, the hypocotyl (a hook-shaped portion of the shoot beneath the coleoptile) forces a channel through soil. The food-storing cotyledons are pulled up through the channel without being torn apart. At the soil surface, exposure to sunlight causes the hook to straighten. Photosynthetic cells in this seedling's cotyledons produce food for several days, then the cotyledons wither and fall off. In time, foliage leaves (each divided into three leaflets) take over the task of photosynthesis. Flowers will develop in buds at the nodes. (**c**) This photograph shows the cotyledons breaking through the bean seed coat.

(**d,e**) The pattern of growth and development of a monocot, using corn (*Zea mays*) as the example. After germination of a corn grain, the coleoptile protects young leaves as the seedling grows through the soil. In corn seedlings, adventitious roots develop at the base of the coleoptile. On the facing page, two photographs show (**f**) the coleoptile and primary root of a corn seedling and (**g**) the coleoptile and first foliage leaf of two seedlings that have poked above the soil surface.

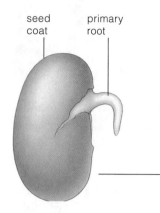

seed coat   primary root

a Bean seedling at the close of germination

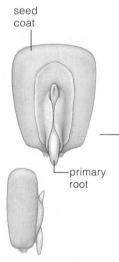

seed coat

primary root

b Corn seedling at the close of germination, front and side views

Once the seed coat splits, more oxygen reaches the embryo, and aerobic respiration moves into high gear. The embryo's meristematic cells start to divide rapidly. In general, the root meristem is the first to be activated. Its meristematic descendants divide, elongate, and give rise to the primary root, the first root of the seedling sporophyte. Germination is over when the seedling's primary root breaks through the seed coat.

Genetic Programs, Environmental Cues

Figure 32.4 shows two patterns of germination, growth, and development for monocots and dicots. The patterns have a heritable basis; they are dictated by the plant's genes. All cells in a new plant arise from the same cell (zygote), so they generally inherit the same genes. But unequal cytoplasmic divisions between daughter cells

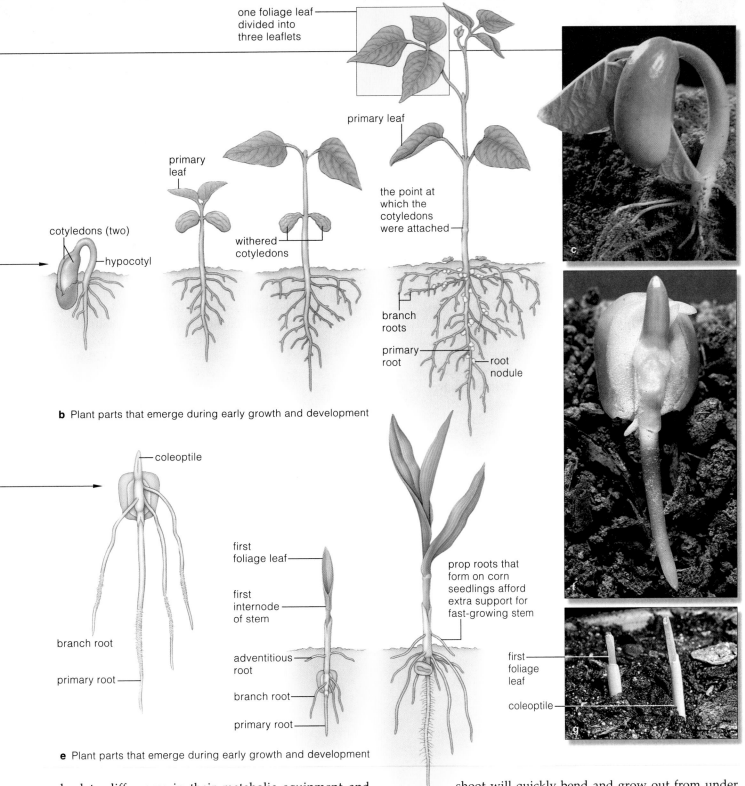

one foliage leaf divided into three leaflets

primary leaf

primary leaf

the point at which the cotyledons were attached

primary leaf

cotyledons (two)

hypocotyl

withered cotyledons

branch roots

primary root

root nodule

c

b Plant parts that emerge during early growth and development

coleoptile

first foliage leaf

first internode of stem

prop roots that form on corn seedlings afford extra support for fast-growing stem

branch root

primary root

adventitious root

branch root

primary root

first foliage leaf

coleoptile

g

e Plant parts that emerge during early growth and development

lead to differences in their metabolic equipment and output. Activities in daughter cells start to vary as a result of selective gene expression, as described in Chapter 15. For instance, genes governing the synthesis of growth-stimulating hormones are activated in some cells but not others. As you will see shortly, such events seal the developmental fate of the different cell lineages.

Bear in mind, the prescribed growth patterns are often adjusted in response to unusual environmental pressures. Suppose a seed germinates in a vacant lot and a heavy paper bag blows on top of it. The primary

shoot will quickly bend and grow out from under the bag, in the direction of sunlight impinging on it. Enzymes, hormones, and other gene products in its cells carry out this growth response.

Plant growth involves cell divisions and cell enlargements. Plant development requires cell differentiation, as brought about by selective gene expression.

Interactions among genes, hormones, and the environment govern how an individual plant grows and develops.

Growth Versus Development

Before getting into the effects of hormones on growth and development, make sure you understand the basic difference between these processes. In general, **growth** of a multicelled organism means the number, size, and volume of cells increase. **Development** is the emergence of specialized, morphologically different body parts. In other words, we measure plant growth in *quantitative* terms and plant development in *qualitative* terms.

As you know from Section 29.1, the cell divisions and enlargements underlying growth proceed at apical and transitional meristems. Approximately half of the resulting daughter cells do not increase in size but rather retain the capacity to divide. The rest enlarge, often by twenty times (Figure 32.5). Cellular descendants of the meristems divide in prescribed planes and expand in specific directions, as shown by the examples in Figure 32.6. The eventual outcome will be plant parts that have specific shapes and functions.

When young cells of an embryo sporophyte or a seedling take up water, turgor pressure (internal fluid pressure) increases against their primary wall and drives their enlargement. Imagine blowing up a balloon. When soft, the balloon inflates easily. Similarly, cells having a soft wall expand rapidly under turgor pressure. The wall thickens when polysaccharides are added to it, and more cytoplasm forms between it and a central vacuole.

Plant hormones have central roles in the selective gene expression underlying cell differentiation and the emergence of basic patterns of development. At least five different classes of hormones are known to have predictable and important effects on flowering plants, starting with the germination of the new seed. As with humans and other animals, genetic instructions in a plant's DNA govern the synthesis of enzymes and other proteins necessary for metabolism, hence for all of a cell's activities. *And how and when each type of enzyme functions depend on hormonal action.*

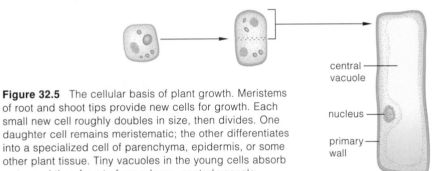

Figure 32.5 The cellular basis of plant growth. Meristems of root and shoot tips provide new cells for growth. Each small new cell roughly doubles in size, then divides. One daughter cell remains meristematic; the other differentiates into a specialized cell of parenchyma, epidermis, or some other plant tissue. Tiny vacuoles in the young cells absorb water and then fuse to form a large, central vacuole.

central
vacuole

nucleus

primary
wall

Categories of Plant Hormones

Table 32.1 lists five main categories of plant hormones. Besides gibberellins, they are **auxins**, **cytokinins**, **ethylene**, and **abscisic acid**. Here we define their known functions. The remainder of the chapter gives examples of their effects during different stages of the life cycle.

GIBBERELLINS As you saw earlier, the gibberellins promote stem lengthening (Figures 32.1 and 32.2). They also help buds as well as seeds break dormancy and resume growth in the spring. In at least some species, they influence the flowering process.

AUXINS Auxins affect the lengthening of stems and responses to gravity and light; they also make coleoptiles grow longer (Figure 32.7). The **coleoptile** is a sheath around the primary shoot of grass seedlings, such as corn plants. It helps keep the shoot from being torn apart during its growth through the soil. The most important natural auxin is indoleacetic acid (IAA). Orchardists spray IAA and other auxins on fruit trees to thin overcrowded seedlings in spring. They also use auxins to prevent premature fruit drop, which cuts farm labor costs; all the fruit can be picked at the same time.

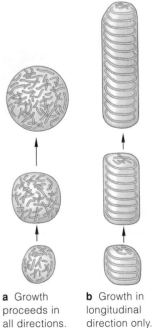

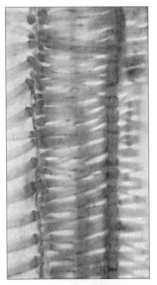

Figure 32.6 How different cell shapes can arise. In each cell, microtubules in the cytoplasm that it inherited from the parent cell are already oriented in prescribed patterns that govern how cellulose microfibrils will be oriented in the cell wall. With random orientation (**a**), the primary wall is elastic all over, so the cell can expand in all directions and assume a spherical shape. (**b,c**) When the cellulose strands become transversely oriented, the cell can only lengthen.

a Growth proceeds in all directions.

b Growth in longitudinal direction only.

c Cellulose rings thicken the secondary wall of this tracheid, a conducting tube of xylem.

Figure 32.7 Examples of the effects of auxin. (**a**) A cutting from a gardenia plant (*left*), four weeks after an auxin had been applied to its base. Another cutting, used as a control (*right*), was untreated.

(**b**) Experiments demonstrating how IAA (an auxin) in a coleoptile tip promotes elongation of cells below it. (*1*) Cut off the tip of an oat coleoptile. The cut stump does not elongate as much as a normal oat coleoptile (*2*) used as the control. (*3*) Next, place a tiny block of agar under the cut tip and leave it for several hours. During that interval, IAA diffuses into the agar. (*4*) Now place the agar on another de-tipped coleoptile. Its elongation will proceed about as rapidly as in an intact coleoptile growing next to it (*5*).

treated with auxin

untreated

a

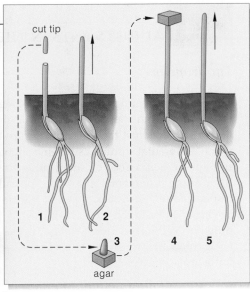

cut tip

agar

b

Table 32.1	Main Plant Hormones and Some of Their Known or Suspected Effects

GIBBERELLINS. Promote stem elongation; help end dormancy of seeds and buds; contribute to flowering. Notable amounts in apical meristems of buds, roots, and leaves and in embryos.

AUXINS. Promote cell elongation in coleoptiles and stems; roles in phototropism and gravitropism. Notable amounts in bud and leaf apical meristems and in embryos in seeds.

CYTOKININS. Promote cell division and leaf expansion; retard leaf aging. Synthesized in roots and travel elsewhere.

ABSCISIC ACID. Promotes stomatal closure; promotes bud and seed dormancy. Notable amounts in leaves, stems, and unripened fruit.

ETHYLENE. Promotes ripening of fruit and abscission of leaves, flowers, and fruits. Notable amounts in fruit, stems, leaves, roots, and seeds.

Certain synthetic auxins are used as **herbicides**. At suitable concentrations, these are compounds that kill some plants but not others. 2,4-Dichlorophenoxyacetic acid (2,4-D) can control broadleaf weeds that compete vigorously with cereal plants for nutrients. Apparently 2,4-D does not harm humans when properly handled. This was not true of 2,4,5-T, a related compound. When mixed in equal proportions, the two produce *Agent Orange*. During the Vietnam conflict, the United States Armed Forces used this herbicide to defoliate thickly forested, nearly impenetrable war zones. Later, results from laboratory experiments suggested that traces of dioxin, which is a contaminant of 2,4,5-T, may cause miscarriages, birth defects, leukemia, and disorders of the liver; it also is a carcinogen. The manufacture and use of 2,4,5-T is now banned in the United States.

CYTOKININS Cytokinins stimulate cell division (hence the name, which refers to cytoplasmic division). They are most abundant in root and shoot meristems and in the tissues of maturing fruits. Natural and synthetic versions are used to prolong the shelf life of cut flowers as well as lettuces, mushrooms, and other vegetables.

ABSCISIC ACID Abscisic acid (ABA) helps plants adapt to seasonal changes, as by inducing bud dormancy and inhibiting cell growth or premature seed germination. ABA also contributes to stomatal closure when a plant is water stressed. Often ABA is applied to nursery stock that is about to be shipped, because when plants are dormant, they are more resistant to damage.

ETHYLENE Ethylene induces various aging responses, including fruit ripening and leaf drop. Ancient Chinese burned incense to make fruit ripen faster, and growers in the 1900s put citrus fruits in sheds with kerosene stoves to ripen them; the smoke contained ethylene. Today, food distributors use ethylene to ripen green tomatoes and other fruit after they ship it. Fruit picked green doesn't bruise or deteriorate as fast. Oranges and other citrus fruits are exposed to ethylene to brighten their rind before being displayed in the market.

Besides the hormones just described, root and leaf cells make their own hormones. One or more unknown hormones might trigger flowering; and another kind, associated with shoot tips, blocks the growth of lateral buds. This form of growth inhibition is called **apical dominance**. By pinching off shoot tips, gardeners can prevent the hormone from diffusing through a plant's stems and exerting its inhibitory effect. Lateral buds are free to branch out, and gardeners get bushier plants.

Growth is a quantitative increase in the number, size, and volume of cells. Development, the emergence of qualitative differences in body parts, is basically a result of selective gene expression, hormone action, and cell differentiation.

The main categories of plant hormones are gibberellins, auxins, cytokinins, abscisic acid, and ethylene.

ADJUSTMENTS IN THE RATE AND DIRECTION OF GROWTH

The Tropisms

Generally, the young roots of land plants grow down through soil, and the shoots grow upright through air. Both also can adjust the direction of growth in response to environmental stimuli, as when a shoot turns toward sunlight. When a root or shoot turns toward or away from an environmental stimulus, we call this a **plant tropism** (after the Greek *trope*, for turning). Hormone-mediated shifts in the rates at which different cells grow and elongate bring about these responses, as the following examples illustrate.

GRAVITROPISM   After germination, the first root to emerge through a seed coat always curves downward and coleoptiles and stems curve up. Such growth responses to Earth's gravitational force are called forms of **gravitropism**.

Auxin, together with a growth-inhibiting hormone, may trigger this response in roots. Turn a young root on its side and remove its root cap, and it will *not* curve down. Put the cap back on, and the root will do so. Elongating root cells will not stop growing if you remove the root cap; if anything, they will grow faster.

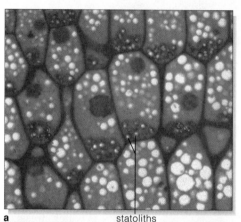
a   statoliths

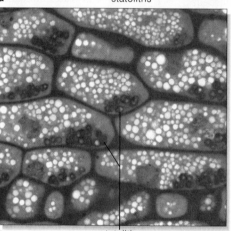
b   statoliths

Figure 32.8   Two observational tests to demonstrate gravitropic responses by young shoots and roots. (**a**) For one test, examine the root cap on a corn root. Normally, statoliths inside plastids of the root cap cells elongate vertically. This light micrograph clearly shows their normal orientation. (**b**) Now turn the root sideways. Within seconds after being reoriented, plastids will settle to the bottom of the root cap cells. A gravity-sensing mechanism based on statoliths may be sensitive to such redistribution of auxin through a root tip. A difference in auxin concentrations may cause cells on the "top" of a root turned sideways to elongate faster than cells on the bottom. Different elongation rates will make the root tip curve downward.

(**c**) For a different test, force a new sunflower seedling to grow in the dark for five days. Then turn it on its side and mark the young shoot at 0.5-centimeter intervals. A gravity-sensing mechanism will cause the stem to turn upright.

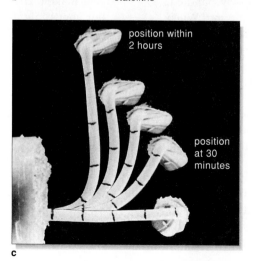
position within 2 hours

position at 30 minutes

c

Suppose a growth inhibitor that is present in root cap cells becomes *redistributed* in a root turned on its side. If gravity somehow causes the inhibitor to move out of the cap and accumulate in cells on the lower side of the root, those cells would not elongate as much as cells on the upper side—and the root would curve downward.

Generally, gravity-sensing mechanisms are based on **statoliths**, which are clusters of certain particles in some types of cells. In plants, statoliths are clusters of unbound starch grains in modified plastids. Figure 32.8*a* and *b* shows how they collect near the bottom of plastids in root cells according to the direction in which gravity is tugging on them. The plastids settle through the cytoplasm until they rest upon the lowest region of the cell. Possibly the redistribution of the statoliths inside them triggers a redistribution of auxin in the cells and thereby triggers this gravitropic response.

Similarly, Figure 32.8*c* shows how you can track what happens when you turn a potted seedling on its side in a dark room. The stem curves up, even in the absence of light. Why? In the horizontally positioned stem, cell elongations on the side facing upward slow down—markedly so—and cell elongations on the lower side rapidly increase. Different rates of elongation on opposing sides of the stem are enough to make it bend in an upward direction. Something must make the cells on the bottom of a sideways stem *more* sensitive to auxin or some other hormone, and those on top *less* so.

PHOTOTROPISM   Whenever stems or leaves adjust the rate and direction of their growth in response to light, they are showing a form of **phototropism**. The adaptive advantage is obvious. In the absence of certain wavelengths of light, the light-dependent reactions of photosynthesis stop; plants move and thereby maximize light interception.

Charles Darwin wondered about phototropism in plants after watching a coleoptile grow toward light that was striking one side of its tip. But it wasn't until the 1920s that Frits Went, a graduate student in Holland, made the connection between phototropism

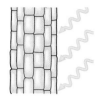
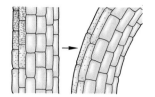

a Rays from sun strike one side of a coleoptile.

b The coleoptile bends when auxin diffuses down from tip to cells on its shaded side.

Figure 32.9 (**a,b**) Hormone-mediated differences in cell elongation rates that cause coleoptiles and stems to bend toward light. (**c**) Phototropism in seedlings. The plant physiologist Frank Salisbury grew these seedlings in darkness, then allowed light to strike their right side for a few hours before photographing them.

c

and a growth-promoting substance. He was the one who named the substance auxin, after the Greek *auxein,* meaning "to increase." Went showed that auxin moves from the tip of a coleoptile into cells less exposed to light and makes them elongate faster than cells on the illuminated side. The difference in growth rates brings about bending toward light (Figure 32.9a,b).

You can observe a phototropic response by putting bean sprouts or seedlings of another sun-loving plant in shade, next to a window through which sunlight is streaming in. They will start curving in the direction where the most light is available (Figure 32.9c).

As we now know, light of blue wavelengths induces plants to make the strongest phototropic response. Thus **flavoprotein**, a yellow pigment molecule that absorbs blue wavelengths, may be a central component of the phototropic bending mechanism.

THIGMOTROPISM Plants also shift their direction of growth when they contact solid objects. This response is called **thigmotropism** (after the Greek *thigma*, which means "touch"). Auxin and ethylene may have roles in the response. **Vines**, which are stems too slender or soft to grow upright without support, make this contact response. So do **tendrils**, which are modified leaves or stems that wrap around objects and thus help support the plant (Figure 32.10). You can observe what happens when they grow against a stem of another plant. Within minutes, cells on the contact side stop elongating and the vine or tendril starts curling around the stem, often more than once. Afterward, the cells on both sides will resume growth at the same rate.

Responses to Mechanical Stress

Mechanical stress, such as prevailing strong winds and grazing animals, can inhibit stem elongation and plant growth. You can see its effects on trees growing near the snowline of windswept mountains; they are stubby compared to trees of the same species that are growing

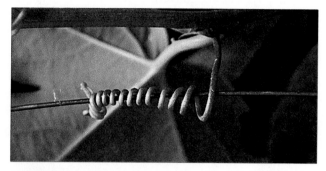

Figure 32.10 Passion flower (*Passiflora*) tendril twisting thigmotropically.

Figure 32.11 Effect of mechanical stress on tomato plants. (**a**) This plant, the control, grew in a greenhouse. (**b**) Each day for twenty-eight days, this plant was mechanically shaken for thirty seconds. (**c**) This one had two shakings each day.

at lower elevations. Similarly, plants grown outdoors commonly have shorter stems than plants grown in a greenhouse. You can observe this response to stress by shaking a plant daily for a brief period. Doing so will inhibit growth of the entire plant (Figure 32.11).

Plants adjust their direction and rate of growth in response to environmental stimuli.

Like other organisms, flowering plants have internal mechanisms that set the time for recurring changes in biochemical events. These internal timing mechanisms are called **biological clocks**. They trigger shifts in daily activities, and they also help bring about seasonal adjustments in the basic patterns of growth, development, and reproduction.

The alarm button for some biological clocks in plants is the blue-green pigment molecule **phytochrome**. This pigment can absorb both red and far-red wavelengths of light, with different results (Figure 32.12). At sunrise, red wavelengths dominate the sky, and phytochrome converts to its active molecular form (Pfr). At sunset, at night, or in shaded habitats, far-red wavelengths predominate. Then, phytochrome reverts to its inactive form (Pr). Its activation may induce cells to take up free calcium ions (Ca^{++}) or induce certain plant organelles to release them. Either way, when the ions combine with calcium-binding proteins in cells, they initiate rhythmic movements of leaves and certain other responses to light.

Rhythmic Leaf Movements

Each day, some plants position their leaves horizontally, then fold them closer to stems at night (Figure 32.13). Keep such a plant in full light or darkness for a few days and it continues to move its leaves into and out of the "sleep" position!

Rhythmic leaf movements are one type of **circadian rhythm**, which is a biological activity that recurs in cycles of about twenty-four hours. (*Circadian* means about a day.) Some experiments by Ruth Satter, Richard Crain, and their coworkers at the University of Connecticut showed that the phytochrome molecule plays a role in these

1 A.M.

6 A.M.

NOON

3 P.M.

10 P.M.

MIDNIGHT

leaf movements. Dr. Satter, who was a pioneer in the study of internal time-keeping mechanisms, was the first to correlate sleep movements of plants with "hands of a biological clock."

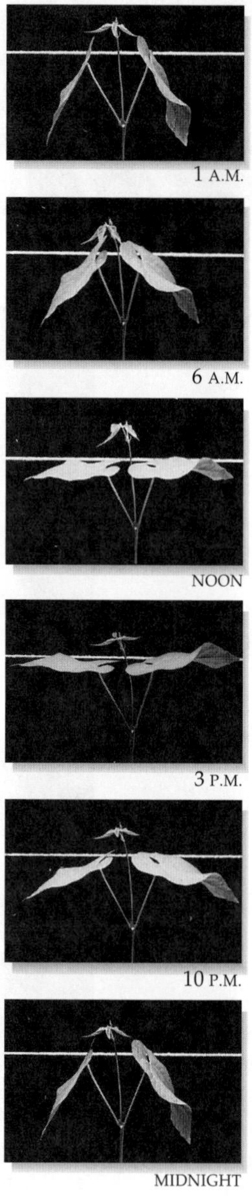

Figure 32.13 *Left:* Rhythmic movements of bean plant leaves. The investigator kept a bean plant in total darkness for twenty-four hours. Its leaves continued to move independently of sunrise (6 A.M.) and sunset (6 P.M.). Leaves folded close to stems may keep moonlight from activating phytochrome and interrupting the dark period that triggers flowering. Or folding may reduce heat loss from leaves exposed to cold night air.

Flowering—A Case of Photoperiodism

As one more example of internal timekeeping, consider that flowering plants reset biological clocks as seasons change. At a certain time of year, for example, a plant starts diverting more energy to the formation of flowers. Figure 32.14 gives you an idea of how this activity and others are predictable responses to rhythmic cues from the environment, such as more daylight hours and warmer temperatures in summer.

Photoperiodism is any biological response to change in the relative length of daylight and darkness in the cycle of twenty-four hours. Pfr, phytochrome's active form, may be a switching mechanism for the process; it may trigger the synthesis of specific enzymes in specific types of cells. Certainly different enzymes are required for germination of seeds, stem lengthening and branching, expansion of leaves, and formation of flowers, fruits, and seeds.

Plants do not all flower in response to the same cues. Spinach, irises, and other **long-day plants** produce their flowers in spring, when daylength exceeds a critical value (Figure 32.15a). Poinsettias, chrysanthemums, and cockleburs are **short-day plants**. They produce flowers in late summer or early fall, when daylength is shorter than a critical value (Figure 32.15b). **Day-neutral plants** flower when mature enough to do so.

Spinach plants will not flower and produce seeds unless they are exposed to ten hours of darkness (that is, fourteen hours of daylight) for two weeks. That is why you would not want to start, say, a spinach seed farm in the tropics, where this set of cues doesn't exist. By contrast, flower growers stall the flowering process by exposing chrysanthemum plants to a flash of light nightly, thereby breaking up one long night into two "short" nights. Thus you can buy chrysanthemums in spring even though they naturally bloom in autumn. Or think about cockleburs, which normally flower after a

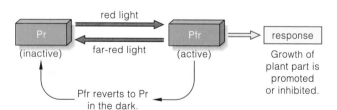

red light

Pr (inactive) — far-red light → Pfr (active) → response

Pfr reverts to Pr in the dark.

Growth of plant part is promoted or inhibited.

Figure 32.12 Interconversion of the phytochrome molecule from active form (Pfr) to inactive form (Pr). This blue-green pigment molecule is part of a switching mechanism that promotes or inhibits the growth of a variety of plant parts.

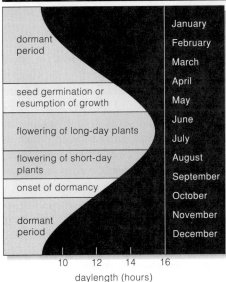

dormant period	January
	February
	March
seed germination or resumption of growth	April
	May
flowering of long-day plants	June
	July
flowering of short-day plants	August
	September
onset of dormancy	October
	November
dormant period	December

10   12   14   16
daylength (hours)

Figure 32.14
Plant growth and development as correlated with the number of hours of light available each day. The number changes with the passing of the seasons.

The data reflect the photoperiodic responses of plants growing in temperate regions of North America, where both rainfall and temperature shift seasonally.

— flowers

no flowering

flowers form

flash of light, no flowering

0   12   24
a   time (hours)

no flowering

flowers form

flash of light, no flowering

0   12   24
b   time (hours)

Figure 32.15 *Right:* Experimental results showing different flowering responses of (**a**) spinach, a long-day plant, and (**b**) chrysanthemum, a short-day plant. In each photograph, the plant positioned at the left grew under short-day conditions, and the one at the right grew under long-day conditions.

Each horizontal bar represents a span of twenty-four hours. The hours of daylight are coded *yellow*; hours of darkness are *black*. As you can deduce from the diagrams, the key factor is an uninterrupted period of *darkness* of some critical value, which differs among species of plants. Spinach will not flower unless exposed to ten hours of darkness (or fourteen hours of daylight) for two weeks. Even a brief flash of artificial light during the critical period of darkness (the third horizontal bar) will inhibit flowering. Chrysanthemums will not flower unless exposed to *uninterrupted* darkness for about ten hours each night.

single night longer than eight and one-half hours. If artificial light interrupts their dark period for even a minute, they refuse to flower. As an example of a poor decision by landscapers, why didn't the poinsettias that were planted along California's interstate highways put out flowers? Headlights from cars and trucks zipping by during the night blocked the flowering response.

Besides phytochrome, other kinds of hormones that are not yet identified probably influence flowering and other growth responses. Some of the hormones may be produced in leaves and transported to new buds. For example, if you trim all but one leaf from a cocklebur, then cover the remaining leaf with black paper for eight and one-half hours, the plant will flower. If you cut off

that one leaf right after the dark period is over, you will not see any cocklebur flowers.

Like other organisms, flowering plants have internal time-keeping mechanisms called biological clocks.

Phytochrome, a blue-green pigment, is part of a switching mechanism for phototrophic responses to light of red and far-red wavelengths. Its active form, Pfr, might interact with other hormones to control which kinds of enzymes are being produced in particular cells.

Different enzymes are necessary to complete flowering and other growth responses that are influenced by sunlight and other environmental cues.

Senescence

While leaves and fruits are growing, cells inside them produce auxin (IAA), which moves into stems. There it interacts with cytokinins and gibberellins to maintain growth. However, when autumn approaches and the length of daylight decreases, plants start to withdraw nutrients from leaves, stems, and roots, then distribute them to flowers, fruits, and seeds. Deciduous plants, which shed their leaves when the growing season ends, channel nutrients to storage sites in twigs, stems, and roots before the leaves die and drop. The dropping of flowers, leaves, fruits, and other plant parts is called **abscission**. The abscission zone is a narrow region of thin-walled parenchyma cells at the base of a petiole or some other part about to drop from the plant. Figure 32.16 shows one example.

Senescence is the sum total of processes that lead to death of a plant or some of its parts. The recurring cue for senescence is a decrease in daylength, but other environmental factors, such as drought, wounds, and nutrient deficiencies, can also bring it about. Either way, the cue triggers a decline in IAA production in leaves and fruits. A different kind of signal, possibly produced by the plant itself, stimulates cells in abscission zones to produce ethylene. There the presence of this hormone stimulates the cells to enlarge, deposit suberin in their walls, and produce enzymes that can digest cellulose and pectin in the middle lamella between the walls. (A middle lamella, recall, is the cementing layer between plant cells.) As the cells continue to enlarge and as their walls are digested, they separate from one another, and the leaf or other plant part above the abscission zone drops away.

Interrupt the diversion of nutrients into flowers, seeds, or fruits, and you can prevent senescence in a plant's leaves, stems, and roots. For example, if you remove each new flower or seed pod from a plant, its leaves and stems will remain vigorous and green much longer. Gardeners routinely remove flower buds from many kinds of plants to maintain vegetative growth. Figure 32.17 shows an example of delayed senescence.

Entering Dormancy

Many perennial and biennial plants start to shut down growth as autumn approaches and days grow shorter. They do so even when temperatures are still mild, the sky is bright, and water is plentiful. When a plant stops growing under conditions that seem (to us) suitable for growth, it has entered a state of **dormancy**, in which its metabolic activities idle. Ordinarily, the plant's buds will not resume growth until there is a convergence of precise environmental cues in early spring.

Short days and long, cold nights are strong cues for dormancy. (So is dry, nitrogen-deficient soil.) You can test this for yourself by interrupting the long dark period of, say, Douglas firs with a short period of red light. The plants will respond as if nights are shorter and days are longer. They will continue to grow taller (Figure 32.18). In this experiment, conversion of Pr to Pfr by red light during the dark period prevented dormancy. In nature, buds may enter dormancy because less Pfr can form when daylength shortens in late summer.

The requirement for multiple dormancy cues has adaptive value. For example, if temperature were the only cue, warm autumn weather might make plants flower and seeds germinate—and winter frost would kill them. By contrast, with artificial selection, growers have developed seeds that will germinate promptly in greenhouses any time of year.

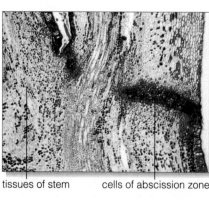

tissues of stem   cells of abscission zone

Figure 32.16 Light micrograph of the abscission zone in maple (*Acer*). This is a longitudinal section through the base of a leaf petiole.

Figure 32.17 Observable results from an experiment in which seed pods were removed from a soybean plant. Removal delayed its senescence.

control (pods not removed)   experimental plant (pods removed)

Figure 32.18 Experiment to test the effect of the relative length of day and night on Douglas firs. The plant at left was exposed to 12-hour light/12-hour dark cycles for a year. Its buds became dormant; daylength was too short. The plant at right was exposed to 20-hour light/4-hour dark cycles; growth continued. The middle plant was exposed to 12-hour light/11-hour dark cycles with 1 hour of light in the middle of the dark period. Light interruption prevented bud dormancy; it caused Pfr formation at a sensitive time in the normal day-night cycle.

Figure 32.19 Localized effect of cold temperature on dormant buds of a lilac plant. For this experiment, a branch of the plant was positioned so that it protruded from a greenhouse during a cold winter. The rest of the plant was kept inside and exposed only to warmer temperatures. Only the buds on the branch exposed to low outside temperatures resumed growth in spring.

Vernalization

Temperatures as well as daylength also change through the seasons in most parts of the world. The changes in temperature influence many plant responses, including flowering. For example, unless buds of some biennials and perennials are exposed to low winter temperatures, flowers will not form on their stems when spring rolls around. The low-temperature stimulation of flowering is called **vernalization** (from the Latin *vernalis*, which means "to make springlike"). Figure 32.19 shows how you can gather experimental evidence of this effect.

As long ago as 1915, the plant physiologist Gustav Gassner studied the flowering of some cereal plants after exposing their seeds to controlled temperatures.

Figure 32.20 A profusion of new leaves harvesting sunlight, a sure sign of spring and the resurgence of plant growth.

For example, he kept germinating seeds of winter rye (*Secale cereale*) at near-freezing temperatures. The seeds flowered the following summer even when they were planted very late in the spring. Vernalization is now a common agricultural practice.

Breaking Dormancy

A dormancy-breaking process is at work between fall and spring. The temperature may become milder, and rains and nutrients become available again. With the return of favorable conditions, the cycle of life begins anew. Seeds germinate; buds resume growth and give rise to new leaves, then to flowers (Figure 32.20).

Depending on the plant species, breaking dormancy probably involves gibberellins and abscisic acid, and it requires exposure to low winter temperatures at specific times. The temperature required to break dormancy varies quite a bit among plants. For example, Delicious apple trees grown in Utah require 1,230 hours near 43°F (6°C); whereas apricot trees grown there need only 720 hours. Generally, trees growing in the southern United States require less cold exposure than the trees growing in northern states and Canada. If you live in Colorado and order a young peach tree from a Georgia nursery, the tree might start spring growth too soon and be killed by late frost or heavy snow.

Multiple cues from the environment influence hormonal secretions that stimulate or inhibit processes of growth and development during the life cycle of plants. These cues include changes in daylength, temperature, moisture, and nutrient availability.

32.6 THE RISE AND FALL OF A GIANT

In this unit, you surveyed the spectrum of vascular plant tissues and the mechanisms and outcomes of plant growth, development, and reproduction. To pull all of this into a coherent picture, imagine walking in the steep, rounded, sandstone hills near the central California coast. Those hills started forming 65 million years ago as a jagged new coastal range. Over time, erosive rains and winds softened their stark contours and sent mineral-laden sediments into the canyons. Grasses took hold, and their remains slowly accumulated. More than 10 million years ago, in that enriched soil, the coast live oak (*Quercus agrifolia*) evolved.

As you walk, you come across a giant tree that is 300 years old. You rest briefly under trees thirty meters tall, with evergreen branches that spread even wider. It's early spring, so you see clusters of golden catkins among the leaves. A light wind is dispersing pollen from these male flowers to female flowers near the branch tips of the same tree or neighboring ones. Long after you pass by, sperm-bearing pollen tubes will still be growing toward ovules in the trees. In time, newly formed zygotes will undergo the first of millions of cell divisions. Seed coats will form, and ovarian walls will become hard shells. By early fall, the trees will shed their seeds in fruits called acorns.

Three centuries ago, long before Gaspar de Portola sent landing parties ashore to found colonies through upper California, oaks were shedding seeds of a new generation. By chance, one particular acorn escaped the attention of foraging squirrels and blue jays. It germinated the next spring and embarked on a journey of continued growth. Meristems in the primary root gave rise to the root cortex, epidermis, and a vascular cylinder through which water and ions flowed. Lateral roots emerged. As new roots grew longer, their absorptive surfaces increased. In the primary shoot, vascular bundles began forming.

The oak seedling's developing roots, stems, and leaves demanded more water and dissolved nutrients. As fungi formed velvety, mycorrhizal partnerships with the roots, the uptake of water and mineral ions increased. Stomatal action in the leaves conserved precious water. As seasons passed, woody stems developed and supported upright growth. By transpiration, water and dissolved nutrients kept on flowing from roots, through secondary xylem, to ever higher stems and leaves. Secondary phloem kept on shuttling sugars to cells actively growing or storing food everywhere in the growing plant.

By chance, the seed had sprouted in a well-drained, sunlit basin at the foot of a canyon. Each winter, rainwater accumulated and kept the soil moist enough to encourage growth through spring and the dry summers. The sun's red wavelengths activated phytochrome and so triggered hormonal events that encouraged stem branching and leaf expansion. Other intricate, hormone-mediated responses were made to the winds, the sun, the tug of gravity, and the changing seasons.

And so the oak flourished every season. Century after century, roots snaked through a huge volume of the moist soil. Branches continued to spread beneath the sun. Leaves proliferated, and mesophyll cells put together food with sunlight energy, water, carbon dioxide, and the few simple minerals being mined from the soil.

In 1849, prospectors on their way to California's gold fields rested in the shade of the oak's canopy (Figure 32.21). The great earthquake of 1906 scarcely disturbed the giant, anchored as it was by a root system extending twenty-four meters through the soil. The few small brush fires that swept through the canyon did not seriously damage the giant tree. Parasitic fungi that could have rotted its roots never took hold in the well-drained soil. Birds and the tree's own defensive chemicals kept insects in check.

In the 1960s, suburban housing started to encroach on the once-wild hills. A developer turned his tractors into the canyon but spared the giant oak. Death came later.

The new homeowners were ignorant of the ancient, fragile relationships between the giant tree and the land that sustained it. They graded the soil between the trunk and the drip line of the overhanging canopy, mounded flower beds against the trunk, and planted lawns beneath the branches, then kept sprinklers busy. Overwatering put standing water next to the trunk. Before, the tree had successfully resisted the oak root fungus (*Armillaria*). In the changed environment, the fungus took hold. With roots rotting away, the oak tree began to suffer massive disruptions to the feedback relations among its roots, stems, and leaves. Eventually it had to be cut down. In their fifth winter, in their small brick fireplace, the owners began burning three centuries of firewood.

Figure 32.21 Canopy of a coast live oak (*Quercus agrifolia*).

1. Plant growth involves increases in the number, size, and volume of cells, by way of mitotic cell divisions and cell enlargements. Plant development refers to the emergence of different body parts, as brought about by selective gene expression and cell differentiation. Thus, plant growth is measurable in quantitative terms, and plant development in qualitative terms.

2. How individual plants grow and develop depends on interactions among their genes, their hormones, and cues from the environment.

  a. A plant's genes govern the synthesis of enzymes and other proteins necessary for metabolism, hence for all cell activities. How and when each type of enzyme functions depend on hormonal action.

  b. Hormones are a category of signaling molecules. After being produced and secreted by some cells, they travel to target cells in different parts of the plant body and stimulate or inhibit gene activity. Any cell with receptors for a particular hormone is its target. Plant hormones do not travel far from cells that secrete them.

  c. The prescribed growth patterns are influenced by predictable environmental cues. Often they are adjusted in response to unusual environmental pressures.

3. Five types of plant hormones have been identified, and the existence of others is suspected.

  a. Gibberellins promote stem elongation, they help seeds and buds break dormancy in spring, and they may help induce the flowering process.

  b. Auxins promote coleoptile and stem elongation. They have roles in phototropism and gravitropism.

  c. Cytokinins stimulate cell division, promote leaf expansion, and retard leaf aging.

  d. Abscisic acid promotes bud and seed dormancy, and it limits water loss by promoting stomatal closure.

  e. Ethylene promotes fruit ripening and abscission.

4. These plant hormones and others interact with one another in ways that bring about patterns of growth and development. They help adjust growth patterns in response to environmental rhythms, such as the seasonal changes in daylength and temperature that occur in most regions outside the tropics. They also adjust patterns to the amount of sunlight, shade, and other environmental circumstances.

5. Following dispersal from the parent plant, seeds germinate. Metabolic activities have been idling in the embryo sporophyte inside a seed, but now the embryo absorbs water and resumes growth. Once its primary root breaks through the seed coat, germination is over.

6. Following germination, a plant increases in volume and mass. Tissues and organs of the seedling develop. Later, flowers, fruits, and new seeds form, and then older leaves drop away from the plant.

7. Plants make tropic responses to the environment.

  a. With gravitropism, roots grow downward and stems generally grow upright in response to the Earth's gravitational force. At least some of their gravity-sensing mechanisms are based on statoliths (particle clusters).

  b. With phototropism, stems and leaves adjust rates and directions of growth in response to light. A yellow pigment (flavoprotein) may be involved; it absorbs blue wavelengths that trigger the strongest response.

  c. With thigmotropism, plants adjust the direction of growth in response to contact with solid objects.

8. Plants respond to mechanical stress, as when strong winds inhibit stem elongation and plant growth.

9. Like other organisms, plants have biological clocks, or internal time-measuring mechanisms. They can reset the clocks and thereby make seasonal adjustments in patterns of growth, development, and reproduction. A prime example is photoperiodism.

  a. In photoperiodism, plants respond to a change in the relative length of daylight and darkness, as occurs seasonally. Phytochrome, a blue-green pigment, is the switching mechanism of a clock that promotes or inhibits seed germination, stem elongation and branching, leaf expansion, and flower, fruit, and seed formation.

  b. Long-day plants flower in spring or summer, when daylength is long relative to the night. Short-day plants flower when daylength is relatively short. Flowering of day-neutral plants is not regulated by light.

10. Senescence is the sum total of processes leading to the death of a plant or plant part, such as a leaf. The dropping of flowers, leaves, fruits, and other parts is called abscission.

11. Dormancy is a state in which a perennial or biennial stops growing even though conditions appear to be suitable for continued growth. A decrease in Pfr levels may trigger dormancy. Breaking dormancy may involve exposure to certain temperatures and hormonal action, including gibberellins and abscisic acid.

Review Questions

1. Distinguish between plant growth and development. *32.2*

2. List the five known types of plant hormones and describe the known functions of each. *32.2*

3. Define plant tropism and give a specific example. *32.3*

4. What is phytochrome, and what role does it play in the flowering process? *32.4*

Self-Quiz *(Answers in Appendix IV)*

1. Seed germination is over when the _____ .
  a. embryo sporophyte absorbs water
  b. embryo sporophyte resumes growth
  c. primary root pokes out of the seed coat
  d. cotyledons unfurl

2. Which of the following statements is false?
   a. Auxins and gibberellins promote stem elongation.
   b. Cytokinins promote cell division but retard leaf aging.
   c. Abscisic acid promotes water loss and dormancy.
   d. Ethylene promotes fruit ripening and abscission.

3. Plant hormones _____ .
   a. interact with one another
   b. are influenced by environmental cues
   c. are active in plant embryos within seeds
   d. are active in adult plants
   e. all of the above

4. Plant growth depends on _____ .
   a. cell division   c. hormones
   b. cell enlargement   d. all of the above

5. Light of _____ is the strongest stimulus for phototropism.
   a. red wavelengths   c. green wavelengths
   b. far-red wavelengths d. blue wavelengths

6. Light of _____ wavelengths causes phytochrome to switch from inactive to active form; light of _____ wavelengths has the opposite effect.
   a. red; far-red   c. far-red; red
   b. red; blue   d. far-red; blue

7. The flowering process is a _____ response.
   a. phototropic   c. photoperiodic
   b. gravitropic   d. thigmotropic

8. Abscission occurs during _____ .
   a. seed germination   c. senescence
   b. flowering   d. dormancy

9. Senescence involves a decrease in _____ in leaves and fruits and an increase in _____ at abscission zones.
   a. IAA; ethylene   c. Pfr; gibberellin
   b. ethylene; IAA   d. gibberellin; abscisic acid

10. Match the plant reproduction and development terms.
    ____ vernalization   a. water moves into seeds
    ____ senescence   b. unequal growth following
    ____ imbibition   contact with solid objects
    ____ thigmotropism   c. inhibition of formation of
    ____ apical   lateral buds
         dominance   d. low-temperature stimulation of
                                the flowering process
                             e. all processes leading to death
                                of plant or plant part

Critical Thinking

1. Given what you know about the primary growth of plants (Section 29.1), propose an explanation of why plant hormones need not travel very far from the cells that secrete them.

2. Plant growth depends on photosynthesis, which depends on inputs of energy from the sun. How, then, can seedlings that were germinated in a dark room grow taller than seedlings that germinated in the sun?

3. *Solar tracking* refers to the observation that many plants are able to maintain the flat blades of their leaves at right angles to the sun throughout the day. Sunflowers provide us with a good example (Figure 32.22). This tropic response maximizes light harvest by leaves. Suggest the name of one type of molecule that might be involved in the response.

4. Belgian scientists isolated a mutant of wall cress (*Arabidopsis thaliana*) that produces excess amounts of auxin. Predict what some of this mutant plant's phenotypic traits might be.

Selected Key Terms

abscisic acid *32.2*	development *32.2*	photoperiodism *32.4*
abscission *32.5*	dormancy *32.5*	phototropism *32.3*
apical	ethylene *32.2*	phytochrome *32.4*
dominance *32.2*	flavoprotein *32.3*	plant tropism *32.3*
auxin *32.2*	germination *32.1*	senescence *32.5*
biological	gibberellin *CI*	short-day
clock *32.4*	gravitropism *32.3*	plant *32.4*
circadian rhythm	growth *32.2*	statolith *32.3*
32.4	herbicide *32.2*	tendril *32.3*
coleoptile *32.2*	hormone *CI*	thigmotropism *32.3*
cytokinin *32.2*	imbibition *32.1*	vernalization *32.5*
day-neutral	long-day	vine *32.3*
plant *32.4*	plant *32.4*	

Readings

Bowley, J. D., and M. Black. 1985. *Seeds: Physiology of Development and Germination.* New York: Plenum.

Nickell, L. 1982. *Plant Growth Regulators: Agricultural Uses.* New York: Springer-Verlag. Concise explanations of agricultural practices that include use of growth regulators.

Salisbury, F., and C. Ross. 1991. *Plant Physiology.* Fourth edition. Belmont, California: Wadsworth.

Web Site See *http://www.wadsworth.com/biology* for practice quiz questions, hypercontents, BioUpdates, and critical thinking. The Wadsworth Biology Resource Center provides a wealth of information fully organized and integrated by chapter.

Figure 32.22 A field of sunflowers (*Helianthus*).

FACING PAGE: *How many and what kinds of body parts does it take to function as a lizard in a tropical forest? Make a list of what comes to mind as you start reading Unit VI, then see how resplendent the list can become at the unit's end.*

TISSUES, ORGAN SYSTEMS, AND HOMEOSTASIS

Meerkats, Humans, It's All the Same

After a cold night in Africa's Kalahari Desert, animals small enough to fit inside a coat pocket emerge stiffly from their burrows. These "meerkats" are a type of mongoose. They stand on their hind legs and face east, exposing their chilled bodies to the warm rays of the morning sun (Figure 33.1). Meerkats don't know it, but sunning behavior helps their enzymes. If the internal temperature of their body were to fall below a tolerable range, the activity of countless enzyme molecules in their cells would falter. With such a change in enzyme activity, metabolism would suffer.

Once meerkats warm up, they fan out from their burrows and look for food. Into the meerkat gut go insects and an occasional lizard. These are pummeled, dissolved, and then digested into glucose and other nutritious tidbits small enough to move across the gut wall, into the bloodstream, and on to cells throughout the body. In cells, aerobic machinery cracks molecules of glucose and other organic compounds apart and so releases vital energy. A respiratory system works with the bloodstream to supply the machinery with oxygen and take away its carbon dioxide leftovers.

All of this activity changes the composition and volume of the **internal environment**, which consists of interstitial fluid (tissue fluids) and blood that bathes

the living cells of any complex animal. Drastic changes in those fluids would kill the cells, but a urinary system works to keep this from happening. Governing this system and all others are the body's central command posts—a nervous system and an endocrine system. The two systems work together and mobilize the body as a whole for everything from simple housekeeping tasks to heart-thumping flights from predators.

And so meerkats start us thinking about this unit's central topics: how the animal body is structurally put together (its *anatomy*) and how the body functions (its *physiology*). This chapter is an overview of the animal tissues and organ systems that we will be considering. It also introduces the central concept of **homeostasis**. With respect to the animal body, the word refers to stable operating conditions in the internal environment, as brought about by the coordinated activities of cells, tissues, organs, and organ systems.

Amazingly, the body of all complex animals consists of only four basic types of tissues. These are epithelial, connective, muscle, and nervous tissues. A **tissue** is an interacting group of cells and intercellular substances that take part in one or more particular tasks. As one example, muscle tissue takes part in contraction. An **organ** consists of different tissues that are organized in

specific proportions and patterns. Thus every vertebrate heart has predictable proportions and arrangements of epithelial, connective, muscle, and nervous tissues. An **organ system** consists of two or more organs that are interacting physically, chemically, or both in a common task, as when interconnected arteries and other vessels transport blood through the body under the driving force of the beating heart.

Cells, tissues, organs, and organ systems split up the work, so to speak, in ways that contribute to the survival of the body as a whole. This is sometimes known as a **division of labor**. By the end of this unit, you may have an abiding appreciation of the sheer magnitude of the division of labor among the separate parts and of the extent to which their activities are integrated. As you will see, regardless of whether you look at a flatworm or salmon, a meerkat or human, the body of every complex animal is structurally and physiologically adapted to perform four overriding tasks:

1. *Maintain conditions in the internal environment within ranges that are most favorable for cell activities.*

2. *Acquire nutrients and other raw materials, distribute them through the body, and dispose of wastes.*

3. *Afford protection against injury or attack from viruses, bacteria, and other agents of disease.*

4. *Reproduce, then often help nourish and protect the new individuals during their early growth and development.*

Figure 33.1 In the Kalahari Desert, gray meerkats (*Suricata suricatta*) face the warming rays of the morning sun, just as they do every morning. This simple behavior helps maintain internal body temperature. How animals function in their environment is the subject of this unit.

KEY CONCEPTS

1. The cells of most animals interact at three levels of organization—in tissues, many of which are combined in organs, which are components of organ systems.

2. Most animals are constructed of only four types of tissues, which are called epithelial, connective, muscle, and nervous tissues.

3. Each animal cell engages in basic metabolic activities that assure its own survival. At the same time, animal cells of a given tissue perform one or more activities that contribute to the survival of the animal as a whole.

4. The body's internal environment consists of all fluids that are not inside cells—that is, blood and interstitial fluid.

5. The combined contributions of cells, tissues, organs, and organ systems help maintain stability in the internal environment, which is required for the survival of each individual cell. This concept helps us understand the functions of any organ or organ system.

6. Homeostasis is the formal name for stable operating conditions in the internal environment.

EPITHELIAL TISSUE

General Characteristics

We commonly refer to an epithelial tissue as **epithelium** (plural, epithelia). This tissue has a free surface, which faces either a body fluid or the outside environment. *Simple* epithelium, with a single layer of cells, functions as a lining for body cavities, ducts, and tubes. *Stratified* epithelium, which has two or more cell layers, typically functions in protection, as it does in skin. Figure 33.2 shows a few examples of this type of animal tissue.

All cells in epithelium are close together, with little intervening material. As is true of nearly every animal tissue, specialized junctions provide both structural and functional links between its individual cells, which absorb, synthesize, and secrete substances.

Cell-to-Cell Contacts

Figure 33.3 shows three kinds of cell junctions that occur in epithelium and other tissues. **Tight junctions** are strands of proteins that help stop substances from leaking across a tissue. **Adhering junctions** cement cells together. **Gap junctions** help cells communicate by promoting the rapid transfer of ions and small molecules among them.

Consider the lining of your stomach. If highly acidic, gastric fluid in the stomach were to leak across this epithelium, it would digest proteins of your own body instead of those brought in with meals. (As described in Section 42.4, this actually is an outcome of a peptic ulcer.) Tight junctions in this epithelial lining and others form a leakproof barrier between the cells near their free surface. Other junctions serve as spot welds and as open channels between cells.

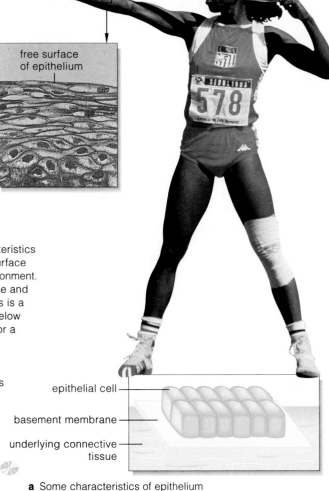

Figure 33.2 (a) Some basic characteristics of epithelium. Epithelia have a free surface exposed to a body fluid or to the environment. Between the epithelium's other surface and the connective tissue on which it rests is a basement membrane. The diagram below this athlete shows this arrangement for a simple epithelium, which has a single layer of cells. The light micrograph shows the upper portion of stratified epithelium. This type of epithelium has more than one layer of cells, which are flattened out near the surface.

(b) Light micrographs and sketches of three simple epithelia, which will give you an idea of the three basic shapes of cells in this type of tissue.

free surface of epithelium

epithelial cell

basement membrane

underlying connective tissue

a Some characteristics of epithelium

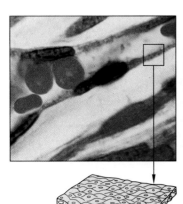

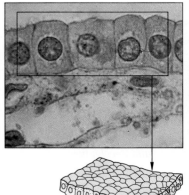

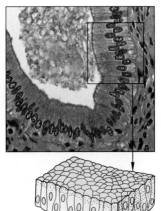

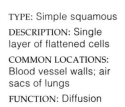

TYPE: Simple squamous
DESCRIPTION: Single layer of flattened cells
COMMON LOCATIONS: Blood vessel walls; air sacs of lungs
FUNCTION: Diffusion

TYPE: Simple cuboidal
DESCRIPTION: Single layer of cubelike cells; free surface may have microvilli (absorptive structures)
COMMON LOCATIONS: Glands and nephrons (slender tubes) in kidneys
FUNCTION: Secretion, absorption

TYPE: Simple columnar
DESCRIPTION: Single layer of tall, slender cells; free surface may have microvilli
COMMON LOCATIONS: Part of lining of gut and respiratory tract
FUNCTION: Secretion, absorption

b Examples of cell shape in simple epithelium

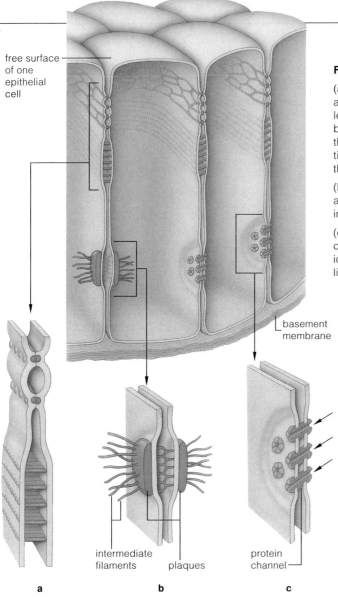

free surface
of one
epithelial
cell

basement
membrane

a

TIGHT JUNCTION

Rows of proteins, running parallel to the tissue's free surface, form strands that prevent leaks between adjacent cells.

intermediate
filaments   plaques

b

ADHERING JUNCTION

Adjoining cells adhere at a plaque (a mass of proteins) that is firmly anchored just inside the plasma membrane of each by intermediate filaments of the cytoskeleton.

protein
channel

c

GAP JUNCTION

Paired, aligned protein cylinders form channels that span the membranes of adjoining cells and interconnect their cytoplasm.

Figure 33.3 *Left:* Examples of cell junctions.

(**a**) In some epithelia, protein strands that ring each cell form tight seals and fuse it with its neighbors. The tight junctions prevent substances from leaking across the epithelium's free surface. Substances reach the tissues below only by passing *through* the epithelial cells. Built-in controls make the plasma membrane of these cells selectively permeable. At a given time, they are allowing some substances but not others to move across, through the interior of transport proteins (Section 5.3).

(**b**) Adhesion junctions are spot welds that hold the cells of epithelium (and all other tissues) together, so that they function as a unit. They are profuse in the skin's surface layer and in other tissues subjected to abrasion.

(**c**) Gap junctions are channels across the plasma membrane that allow the cytoplasm of adjoining cells to interconnect. They promote the diffusion of ions and small molecules from cell to cell. They are abundant in the heart, liver, and other organs in which cell activities must be rapidly coordinated.

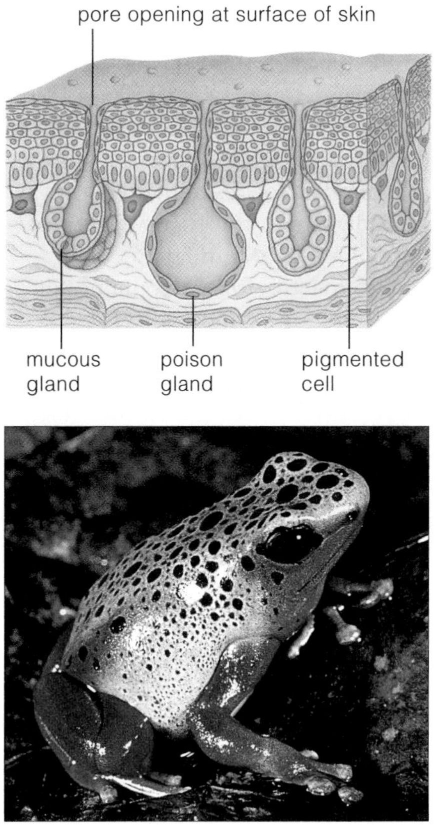

pore opening at surface of skin

mucous
gland   poison
gland   pigmented
cell

Figure 33.4
Section through a frog's glandular epithelium. The photograph shows a frog of the genus *Dendrobates*, which produces one of the most lethal glandular secretions known. Some natives of a tribe in Colombia use this exocrine gland secretion to poison dart tips, which they shoot through a blowgun. Pigment-rich cells that produce the surface coloration branch into lower epithelial layers of skin. The showy coloration of all poisonous frogs serves as a strong warning signal to potential predators.

Glandular Epithelium

Glands are secretory cells or multicelled structures that are derived from epithelium and often remain linked to it. **Exocrine glands** secrete mucus, saliva, earwax, milk, oil, digestive enzymes, and other cell products. Usually the products are released onto a free epithelial surface through ducts or tubes. Figure 33.4 shows an example.

By contrast, **endocrine glands** have no ducts. Their products are hormones, which are secreted directly into the fluid bathing the gland. Typically, the bloodstream picks up the hormone molecules and distributes them to target cells elsewhere in the body.

Epithelia are sheetlike tissues that have one free surface. Different types of epithelia line the body's surface and its cavities, ducts, and tubes.

Profuse cell-to-cell contacts bind the cells closely together, with little intercellular material between them.

Glands are secretory cells or multicelled structures derived from epithelium and often connected to it.

Of all tissues in complex animals, connective tissues are the most abundant and widely distributed. They range from connective tissue proper to specialized types, which include cartilage, bone, adipose tissue, and blood (Table 33.1). In all types except blood, fibroblasts and other kinds of cells secrete fibers of collagen or elastin, which are structural proteins. (This is the collagen that plastic surgeons use to plump wrinkled skin, sunken acne scars, and lips.) Fibroblasts also secrete modified polysaccharides. Secreted material accumulates between cells and fibers, as the tissue's "ground substance."

Connective Tissue Proper

All of these tissues have mostly the same components but in different proportions. **Loose connective tissue** has its fibers and cells loosely arranged in a semifluid ground substance (Figure 33.5a). Often it serves as a support framework for epithelium. Besides fibroblasts, it contains infection-fighting white blood cells. When small cuts or other wounds allow bacteria to breach the skin or lining of the digestive, respiratory, or urinary tracts, these cells mount an early counterattack.

Dense, irregular connective tissue contains fibers, mostly collagen-containing ones, and a few fibroblasts. This tissue forms protective capsules around organs that do not stretch much. It also is present in the deeper

Table 33.1   Types of Connective Tissue
CONNECTIVE TISSUE PROPER:
Loose connective tissue
Dense, irregular connective tissue
Dense, regular connective tissue (ligaments, tendons)
SPECIALIZED CONNECTIVE TISSUE:
Cartilage
Bone
Adipose tissue
Blood

part of skin (Figure 33.5b). **Dense, regular connective tissue**, which has parallel bundles of many collagen fibers, resists being torn apart. Rows of fibroblasts often intervene between the bundles. This is true of tendons, which attach skeletal muscle to bones (Figure 33.5c), and of elastic ligaments, which attach bones to each other.

Specialized Connective Tissue

Intercellular material of **cartilage** is solid yet pliable, like solid rubber, and resists compression. The material is produced by cells that in time become imprisoned in small cavities in their own secretions (Figure 33.5d). Cartilage elements in vertebrate embryos are structural models for bones that replace most of them. Cartilage

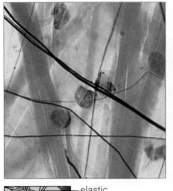
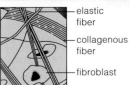

TYPE: Loose connective tissue

DESCRIPTION: Fibroblasts, other cells, plus fibers loosely arranged in semifluid ground substance

COMMON LOCATIONS: Under the skin and most epithelia

FUNCTION: Elasticity, diffusion

a

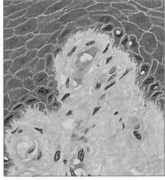

TYPE: Dense, irregular connective tissue

DESCRIPTION: Collagenous fibers, fibroblasts, less ground substance

COMMON LOCATIONS: In skin and capsules around some organs

FUNCTION: Support

b

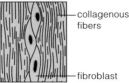

TYPE: Dense, regular connective tissue

DESCRIPTION: Collagen fibers in parallel bundles, long rows of fibroblasts, little ground substance

COMMON LOCATIONS: Tendons, ligaments

FUNCTION: Strength, elasticity

c

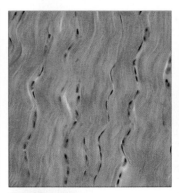
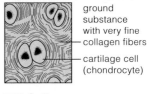

TYPE: Cartilage

DESCRIPTION: Cells embedded in pliable, solid ground substance

COMMON LOCATIONS: Ends of long bones, nose, parts of airways, skeleton of vertebrate embryos

FUNCTION: Support, flexibility, low-friction surface for joint movement

d

Figure 33.5 Examples of connective tissue proper and of specialized connective tissue.

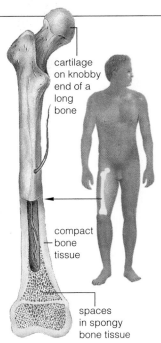

cartilage on knobby end of a long bone

compact bone tissue

spaces in spongy bone tissue

Figure 33.6 Cartilage and bone tissue. Spongy bone tissue has tiny, needlelike hard parts with spaces in between. Compact bone tissue is more dense. Bone is a load-bearing tissue that resists compression. Over time, it was the basis of increases in the body size of many land vertebrates, including giraffes. It gives large animals selective advantages. Among other things, they can ignore most predators with impunity, roam farther for food and water, and heat up and cool off more slowly than small animals (because of a lower surface-to-volume ratio and greater heat production).

compact bone tissue
space that contained living bone cell (osteocyte)

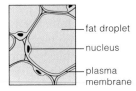

fat droplet
nucleus
plasma membrane

TYPE: Bone tissue

DESCRIPTION: Collagen fibers, ground substance hardened with calcium

COMMON LOCATIONS: Bones of vertebrate skeleton

FUNCTION: Movement, support, protection

e

TYPE: Adipose tissue

DESCRIPTION: Large, tightly packed fat cells occupying most of ground tissue

COMMON LOCATIONS: Under skin, around heart, kidneys

FUNCTION: Energy reserves, insulation, padding

f

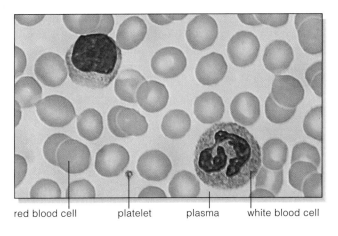

red blood cell   platelet   plasma   white blood cell

Figure 33.7 Some components of human blood, a vascular tissue. Its straw-colored, liquid matrix (plasma) is mostly water in which nutrients, diverse proteins, oxygen, carbon dioxide, ions, and other substances are dissolved.

maintains the shape of the nose, outer ear, and other body parts. It cushions joints between adjacent bones of the vertebral column, limbs, hands, and elsewhere.

Bone (Figures 33.5*e* and 33.6) is the weight-bearing tissue of vertebrate skeletons, which support or protect softer tissues and organs. Minerals harden this tissue; its collagen fibers and ground substance are loaded with calcium salts. Limb bones, such as the long bones of your legs, interact with the skeletal muscles that are attached to them to bring about movements. Tissues in some bones also are production sites for blood cells.

Adipose tissue is so chockful of large fat cells, it no longer looks like a connective tissue (Figure 33.5*f*). The body's excess carbohydrates and proteins are converted to storage fats and tucked away in this tissue. Adipose tissue is richly supplied with blood, which serves as an immediately accessible "highway" along which fats can move to and from the tissue's individual cells.

Blood, derived mainly from connective tissue, has transport functions. Circulating within *plasma*, its fluid medium, are a great many red blood cells, white blood cells, and platelets (Figure 33.7). Red blood cells deliver oxygen to metabolically active tissues, and carry carbon dioxide and other wastes away from them. Plasma is largely water, but it has a great number of different kinds of proteins, ions, and other substances dissolved in it. Section 39.2 describes this complex tissue.

Diverse types of connective tissues bind together, support, strengthen, protect, and insulate other tissues in the body.

Most connective tissues consist of protein fibers and a variety of cells in a ground substance. One type, blood, is a fluid tissue. Another type, adipose tissue, serves as a reservoir of stored energy.

In muscle tissue alone, cells *contract* (that is, shorten) in response to stimulation, then lengthen and so return to their uncontracted state. Many long, cylindrical cells are arranged in parallel in these tissues. Their coordinated contraction and relaxation help move the body through the environment and maintain or change the position of

Figure 33.9 The location and general arrangement of cells in a typical skeletal muscle.

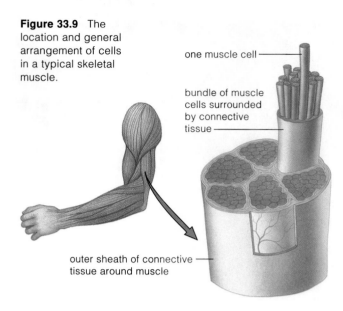

one muscle cell

bundle of muscle cells surrounded by connective tissue

outer sheath of connective tissue around muscle

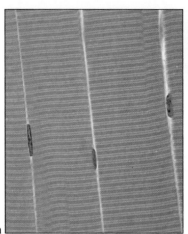

width of one muscle cell

cell nucleus

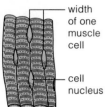

TYPE: Skeletal muscle

DESCRIPTION: Bundles of long, cylindrical, striated, contractile cells

LOCATION: Associated with skeleton

FUNCTION: Locomotion, movement of body parts

a

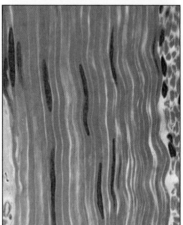

cells, teased apart for clarity

TYPE: Smooth muscle

DESCRIPTION: Contractile cells with tapered ends

LOCATION: Wall of internal organs, such as stomach

FUNCTION: Movement of internal organs

b

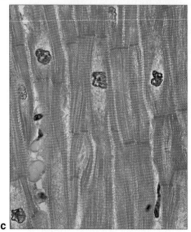

junction between adjacent cells

TYPE: Cardiac muscle

DESCRIPTION: Cylindrical, striated cells that have specialized end junctions

LOCATION: Wall of heart

FUNCTION: Pump blood within circulatory system

c

Figure 33.8 Characteristics and examples of skeletal muscle, smooth muscle, and cardiac muscle tissues.

its individual parts. The three types of muscle tissue are called skeletal, smooth, and cardiac muscle tissues.

The muscles that are connected to the bones of your skeleton consist of **skeletal muscle tissue** (Figure 33.8*a*). In a typical muscle, such as the biceps, striated skeletal muscle cells are bundled together in parallel. (*Striated* means striped.) A sheath of tough connective tissue encloses several bundles of the muscle cells, as Figure 33.9 indicates. The structure and function of skeletal muscle tissue are topics of Chapter 38.

The contractile cells of **smooth muscle tissue** taper at both ends (Figure 33.8*b*). Cell junctions hold them together, and a connective tissue sheath encloses them. The wall of internal organs, such as blood vessels, the stomach, and the intestines, contains this type of muscle tissue. Smooth muscle action is sometimes said to be "involuntary," because we usually are not able to make it contract merely by thinking about it (as we can do with skeletal muscle).

Cardiac muscle tissue is a contractile tissue that is present only in the heart (Figure 33.8*c*). Cell junctions fuse together the plasma membranes of cardiac muscle cells. Junctions at some fusion points allow the cells to contract as a unit. When one cell receives a signal to contract, its neighbors are stimulated to contract, also.

Muscle tissue, which can contract (shorten) in response to stimulation, helps move the body and specific body parts.

Skeletal muscle is the only muscle tissue attached to bones. Smooth muscle is a component of internal organs. Cardiac muscle alone makes up the contractile walls of the heart. Connective tissue sheaths all three types of tissues.

NERVOUS TISSUE

Of all tissues, **nervous tissue** exerts the greatest control over the body's responsiveness to changing conditions. In a human nervous system, much of the tissue consists of **neurons**, which are a type of excitable cell. More than half of it consists of **neuroglia**, diverse cells that protect and structurally and metabolically support the neurons.

When a neuron is suitably stimulated, an electrical disturbance swiftly travels along its plasma membrane. Arrival of the disturbance at the neuron's endings, or output zone, triggers events that may cause stimulation or inhibition of adjacent neurons and other cells.

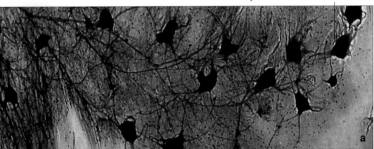

cell body of one motor neuron

Figure 33.10 (**a**) A few motor neurons that form part of the human nervous system. This type relays signals from the brain or spinal cord to muscles and glands. Collectively, different types of neurons detect and process a great number and variety of signals about environmental changes, and initiate suitable responses. (**b**) Without neurons, for instance, this chameleon could not detect an edible insect, calculate its distance, and command a long, sticky, prey-capturing tongue to uncoil with stunning speed.

For example, more than a hundred billion neurons are organized as communication lines throughout your body. Some detect specific changes in environmental conditions. Others coordinate immediate and long-term responses to change. The type shown in Figure 33.10*a* relays signals from the brain to muscles and glands. How these cells function is a topic of later chapters.

Neurons are the basic units of communication in nervous tissue. Different kinds detect specific stimuli, integrate information, and issue or relay commands for response.

FRONTIERS IN TISSUE RESEARCH

As you probably will gather after thinking about the many tissues that are used as examples throughout this unit, a tissue is more than a sum of its cells. As each new animal grows and develops, its cells interact and become organized in particular ways to give rise to the body's diverse tissues. And as each new tissue develops, its cells synthesize specific gene products that are vital for normal body functioning.

For many decades, medical researchers have attempted to find a way to construct artificial tissues in quantity in the laboratory. Currently, they can grow extensive sheets of epidermis from a patient and use it to regenerate skin that was destroyed through third-degree burns and other types of severe damage. In some laboratories, researchers are taking small sections of epidermis from the skin of patients and exposing them to a culture medium that contains nutrients and growth factors. The cells proliferate and form *laboratory-grown epidermis*. When surgeons place an epidermal sheet over a wound, the cells in the sheet interact biochemically and structurally with underlying cells. As an outcome of the interactions, the damaged or missing tissue is regenerated.

On the horizon are *designer organs*. These encapsulated, selected groups of living cells might synthesize specific hormones, enzymes, and other substances. The idea is to surgically snip a bit of epithelium from a patient and then use it to enclose the cells. Because the capsule is derived from epithelial cells of the patient's own body, it will not be chemically recognized as "foreign" and attacked by the patient's immune system. As you will see in Chapter 40, rejection of tissue and organ implants can have serious medical consequences.

Currently, biotechnologists are close to understanding how to synthesize molecular cues that will allow designer organs to stick to appropriate sites in the body. Once the synthetic organs stick, they may become integral parts of normal body functioning.

Ultimately, the goal of this research is to put together packages of cells capable of producing specific life-saving substances that are absent in patients who suffer from severe genetic disorders or chronic diseases. For example, imagine the potential for people who have type I *diabetes mellitus*. This metabolic disorder results in an elevated concentration of glucose in the blood. Affected people produce little if any insulin, which is the hormone that signals cells to take up glucose from the blood. Unless they receive insulin injections on a regular basis, they will die. However, if a customized, insulin-secreting organ can be successfully installed inside the body of such individuals, their daily injections of insulin might be a thing of the past.

33.6 ORGAN SYSTEMS

Overview of the Major Organ Systems

Figure 33.11 gives an overview of the organ systems of a typical vertebrate, the adult human. Figure 33.12 lists some terms that are used when describing the positions of the various organs. It also shows major body cavities in which they are located.

Each organ system contributes to the survival of all living cells in the animal body. You may think this is stretching things a bit. For example, how could muscles and bones be helping each microscopically small cell stay alive? And yet interactions between the skeletal and muscular systems allow us to move about—toward sources of nutrients and water, for example. Some parts of the two organ systems help keep blood circulating to cells, as when leg muscle contractions help move blood in veins back to the heart. Blood inside the circulatory system rapidly transports oxygen, nutrients, and other substances to cells, and transports products and wastes away from them. The respiratory system imports and exports the gases, skeletal muscles assist the respiratory system—and so it goes, throughout the entire body.

Figure 33.12 (**a**) The major cavities in the human body. (**b,c**) Directional terms and planes of symmetry for the vertebrate body. Notice how the *midsagittal* plane divides the body into right and left halves. Most vertebrates, such as fishes and rabbits, move with the main body axis parallel with Earth's surface. For them, *dorsal* pertains to their back or upper surface, and *ventral* pertains to the opposite, lower surface.

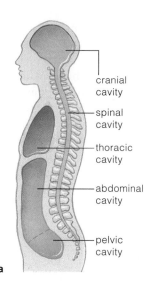

a

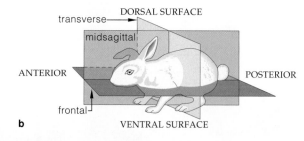

b

Figure 33.11 *Below*: Human organ systems and their functions.

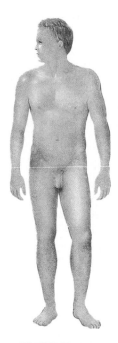

INTEGUMENTARY SYSTEM	MUSCULAR SYSTEM	SKELETAL SYSTEM	NERVOUS SYSTEM	ENDOCRINE SYSTEM	CIRCULATORY SYSTEM
Protect body from injury, dehydration, and some pathogens; control its temperature; excrete some wastes; receive some external stimuli	Move body and its internal parts; maintain posture; produce heat (by high metabolic activity)	Support and protect body parts; provide muscle attachment sites; produce red blood cells; store calcium, phosphorus	Detect both external and internal stimuli; control and coordinate responses to stimuli; integrate all organ system activities	Hormonally control body function; work with nervous system to integrate short-term and long-term activities	Rapidly transport many materials to and from cells; help stabilize internal pH and temperature

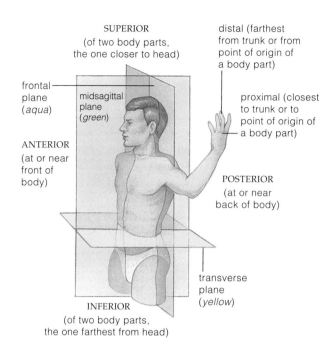

SUPERIOR
(of two body parts,
the one closer to head)

distal (farthest
from trunk or from
point of origin of
a body part)

frontal
plane
(*aqua*)

midsagittal
plane
(*green*)

proximal (closest
to trunk or to
point of origin of
a body part)

ANTERIOR
(at or near
front of
body)

POSTERIOR
(at or near
back of body)

transverse
plane
(*yellow*)

INFERIOR
(of two body parts,
the one farthest from head)

c Unlike quadrupedal animals, humans walk upright, with their main body axis perpendicular to the ground. *Anterior* refers to the front of the body; it corresponds to ventral, as shown in (**b**). *Posterior* refers to the back; it corresponds to dorsal.

Tissue and Organ Formation

Where do the tissues of organ systems come from? To get a sense of how they originate, start with a sperm and egg. Recall that sperm and eggs develop from germ cells, which are immature reproductive cells. (All other cells in the body are "somatic," after the Greek word for body.) After a zygote forms at fertilization, mitotic cell divisions form an early embryo. In vertebrates, the cells become arranged as three primary tissues—ectoderm, mesoderm, and endoderm. The three are the embryonic forerunners of all tissues in the adult. **Ectoderm** gives rise to the skin's outer layer and to tissues of a nervous system. **Mesoderm** gives rise to tissues of the muscles, bones, and most of the circulatory, reproductive, and urinary systems. **Endoderm** gives rise to the lining of the digestive tract and to organs derived from it.

In general, vertebrates have the same kinds of organ systems. Each organ system serves specialized functions, such as gas exchange, blood circulation, and locomotion.

Vertebrate tissues, organs, and organ systems arise from three primary tissues in the developing embryo. These primary tissues are ectoderm, mesoderm, and endoderm.

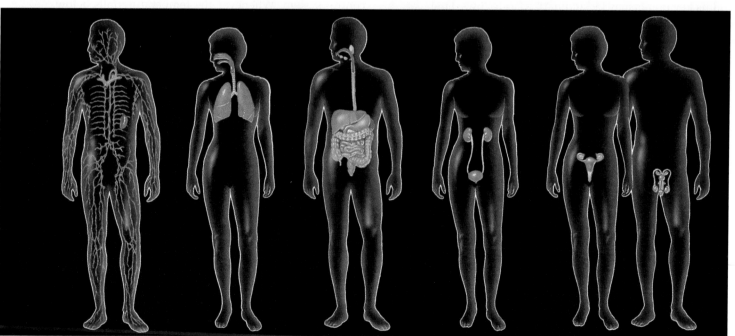

LYMPHATIC SYSTEM	RESPIRATORY SYSTEM	DIGESTIVE SYSTEM	URINARY SYSTEM	REPRODUCTIVE SYSTEM
Collect and return some tissue fluid to the bloodstream; defend the body against infection and tissue damage	Rapidly deliver oxygen to the tissue fluid that bathes all living cells; remove carbon dioxide wastes of cells; help regulate pH	Ingest food and water; mechanically, chemically break down food, and absorb small molecules into internal environment; eliminate food residues	Maintain the volume and composition of internal environment; excrete excess fluid and blood-borne wastes	*Female:* Produce eggs; after fertilization, afford a protected, nutritive environment for the development of new individual. *Male:* Produce and transfer sperm to the female. Hormones of both systems also influence other organ systems.

Concerning the Internal Environment

To stay alive, your cells must remain bathed in a fluid that offers nutrients and carries away metabolic wastes. In this they are no different from an amoeba or some other free-living, single-celled organism. The difference is, trillions of cells coexist in your body. They must draw nutrients from and dump wastes into the same fifteen liters of fluid, which is less than sixteen quarts.

The fluid *not* inside cells is **extracellular fluid**. Much of it is *interstitial*, meaning it occupies spaces between cells and tissues. The remainder is *plasma*, which is the fluid portion of blood. The interstitial fluid exchanges substances with the cells it bathes and with blood.

In functional terms, extracellular fluid is continuous with the fluid in cells. That is why drastic changes in the composition and volume of extracellular fluid have drastic effects on cell activities. The type and number of ions are especially crucial, for they must be maintained at concentrations that are compatible with metabolism. Otherwise, the animal itself cannot survive.

It makes no difference whether an animal is simple or complex. *The component parts of any animal work together to maintain the stable fluid environment required by all of its living cells.* This concept is absolutely central to understanding the structure and function of animals, and its key points may be summarized as follows. First, each cell of the animal body engages in basic metabolic activities that ensure its own survival. Second, the cells of a given tissue also perform one or more activities that contribute to the survival of the whole organism. Third, the combined contributions of individual cells, organs, and organ systems help maintain the stable internal environment—that is, the extracellular fluid—required for individual cell survival.

Mechanisms of Homeostasis

Homeostasis, recall, refers to stable operating conditions in the internal environment. Three components interact to maintain this state. They are called sensory receptors, integrators, and effectors. **Sensory receptors** are cells or cell parts that can detect a **stimulus**, which is a specific change in the environment. When someone kisses you, for example, there is a change in pressure on your lips. Receptors in the skin of your lips translate the stimulus into a signal that can be sent to the brain. Your brain is an **integrator**, a central command post where different bits of information are pulled together in the selection of a response. The brain sends signals to your muscles or glands (or both). Muscles and glands are **effectors**—they carry out the response. In this particular

case, the response might include flushing with pleasure and kissing the person back. Of course, you cannot engage in a kiss indefinitely, for this would prevent you from eating and carrying out other activities necessary to maintain operating conditions inside your body.

So how does your brain reverse the physiological changes induced by the kiss? Receptors only provide it with information about how things *are* operating. The brain also receives information about how things *should be* operating—that is, information from "set points." When physical or chemical conditions deviate sharply from a set point, the brain functions to bring them back to an effective operating range. It does this by way of signals that cause specific muscles and specific glands to increase or decrease their activity.

NEGATIVE FEEDBACK Feedback mechanisms are among the controls that operate to keep physical and chemical aspects of the body within tolerable ranges. As one example, with a **negative feedback mechanism**, some activity alters a condition in the internal environment, and this triggers a response that reverses the altered condition (Figure 33.13).

Think of a furnace with a thermostat. A thermostat senses the air temperature and "compares" it against a preset point on a thermometer built into the furnace's control system. Whenever the temperature falls below the preset point, the thermostat signals a switching mechanism that can turn on the furnace. When the air becomes heated enough to match the prescribed level, the thermostat signals the switching mechanism, which shuts off the furnace.

Similarly, feedback mechanisms help keep the body temperature of meerkats, humans, huskies, and many other animals near 37°C (98.6°F), even during hot or cold weather. Visualize a young husky running around on a hot summer day. Soon its body gets hot, and receptors trigger events that slow down the whole dog *and* its cells. The husky searches for shade and rests under a tree. Moisture from its respiratory system evaporates from the tongue and carries away some body heat with

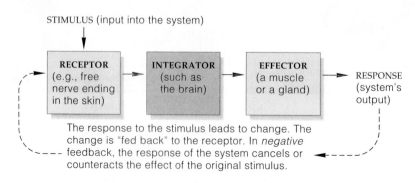

STIMULUS (input into the system)

| RECEPTOR (e.g., free nerve ending in the skin) | INTEGRATOR (such as the brain) | EFFECTOR (a muscle or a gland) | RESPONSE (system's output) |

The response to the stimulus leads to change. The change is "fed back" to the receptor. In *negative* feedback, the response of the system cancels or counteracts the effect of the original stimulus.

Figure 33.13 Components necessary for negative feedback at the organ level.

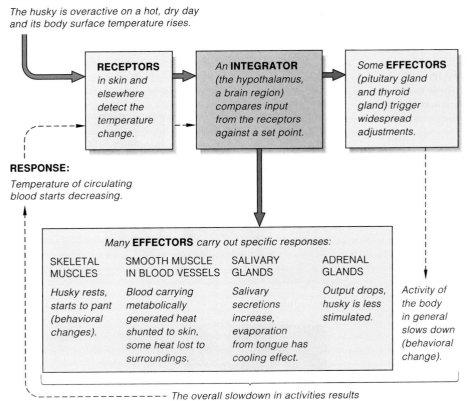

STIMULUS:

The husky is overactive on a hot, dry day and its body surface temperature rises.

RECEPTORS
in skin and elsewhere detect the temperature change.

An INTEGRATOR
(the hypothalamus, a brain region) compares input from the receptors against a set point.

Some EFFECTORS
(pituitary gland and thyroid gland) trigger widespread adjustments.

RESPONSE:

Temperature of circulating blood starts decreasing.

Many **EFFECTORS** *carry out specific responses:*

SKELETAL MUSCLES	SMOOTH MUSCLE IN BLOOD VESSELS	SALIVARY GLANDS	ADRENAL GLANDS
Husky rests, starts to pant (behavioral changes).	*Blood carrying metabolically generated heat shunted to skin, some heat lost to surroundings.*	*Salivary secretions increase, evaporation from tongue has cooling effect.*	*Output drops, husky is less stimulated.*

Activity of the body in general slows down (behavioral change).

The overall slowdown in activities results in less metabolically generated heat.

Figure 33.14 Homeostatic controls over the internal temperature of a husky's body. *Blue* arrows indicate the main control pathways. The *dashed line* shows how a feedback loop is completed.

it, as shown in Figure 33.14. These control mechanisms and others counter overheating by curbing activities that naturally generate metabolic heat and by giving up the body's excess heat to the surrounding air.

POSITIVE FEEDBACK In some cases, **positive feedback mechanisms** operate. These controls set in motion a chain of events that *intensify* a change from an original condition—and after a limited time, the intensification reverses the change. Positive feedback is associated with instability in a system. For example, during sexual intercourse, chemical signals from a female's nervous system can induce her to make intense physiological responses to her sexual partner. Her responses stimulate changes in her partner that stimulate the female even more—and so on until an explosive, climax level of excitation is reached. Normal conditions now return, and homeostasis prevails.

As another example, at childbirth, a fetus exerts pressure on the wall of its mother's uterus. Pressure stimulates the production and secretion of oxytocin, a hormone. Oxytocin causes wall muscles to contract and exert pressure on the fetus, which exerts more pressure on the wall, and so on until the fetus is expelled.

What we have been describing is a general pattern of monitoring and responding to a constant flow of information about an animal's internal and external environments. During all of this activity, organ systems operate together in astoundingly coordinated fashion. Throughout this unit, we will be asking the following questions about their operation:

1. *What physical or chemical aspects of the internal environment are organ systems working to maintain as conditions change?*

2. *By what means are organ systems kept informed of the various changes?*

3. *By what means do they process incoming information?*

4. *What mechanisms are set in motion in response?*

As you will see in later chapters, operation of all organ systems is under neural and endocrine control.

Each living cell of an animal body engages in basic metabolic activities that ensure its own survival. Concurrently, the cells of any given tissue are performing one or more activities that contribute to the survival of the whole animal.

The combined contributions of cells, organs, and organ systems help maintain the stable internal environment (the extracellular fluid) required for individual cell survival.

Homeostatic control mechanisms help maintain physical and chemical aspects of the body's internal environment within ranges that are most favorable for cell activities.

33.8 SUMMARY

1. A tissue is an aggregation of cells and intercellular substances that perform a common task. An organ is a structural unit of different tissues combined in definite proportions and patterns that allow them to perform a common task. An organ system has two or more organs interacting chemically, physically, or both in ways that contribute to the survival of the body as a whole.

2. Epithelial tissues cover external body surfaces and line internal cavities and tubes. These tissues have one free surface exposed to body fluids or the environment.

3. A great variety of connective tissues bind together, support, strengthen, protect, and insulate other tissues. Most types are composed of fibers of structural proteins (especially collagen), fibroblasts, and other cells within a ground substance.

   a. Loose connective tissue, with a semifluid ground substance, is present under skin and most epithelia.

   b. Dense, irregular connective tissue contains mostly collagen fibers and fibroblasts. It is present in skin, and it forms protective capsules around a number of organs.

   c. Dense, regular connective tissue, with its parallel bundles of collagen fibers, provides structural support for tendons, skin, and other organs.

   d. Cartilage, with its solid yet pliable intercellular material, has structural and cushioning roles. Bone, the weight-bearing tissue of vertebrate skeletons, interacts with skeletal muscle to bring about movement.

   e. Blood, a specialized connective tissue, consists of plasma, cellular components, and dissolved substances. Adipose tissue, another specialized connective tissue, is a reservoir of energy; it consists mainly of fat cells.

4. Muscle tissues contract (shorten), then return to the resting position. They help move the body or parts of it. The three types of muscle tissue are skeletal muscle, smooth muscle, and cardiac muscle tissue.

5. Nervous tissue intercepts and integrates information about internal and external conditions, and governs the body's responses to change. Neurons of this tissue are the basic units of communication in nervous systems.

6. Tissues, organs, and organ systems work together to maintain the stable internal environment (that is, the extracellular fluid) required for individual cell survival. At homeostasis, conditions in the internal environment are balanced at levels most favorable for cell activities.

7. Feedback controls help maintain internal operating conditions for the body's cells. With negative feedback, for example, a change in a particular condition triggers a response that results in a reversal of the change.

8. Homeostasis depends on receptors, integrators, and effectors. Receptors detect stimuli, or specific changes in the environment. Integrating centers, such as a brain, process the information and direct muscles and glands (the body's effectors) to carry out responses.

Review Questions

1. Describe the characteristics of epithelial tissue in general. Then describe the various types of epithelial tissues in terms of specific characteristics and functions. *33.1*

2. List the major types of connective tissues; add the names and characteristics of their specific types. *33.2*

3. Identify and describe the following tissues: *33.1–33.3*

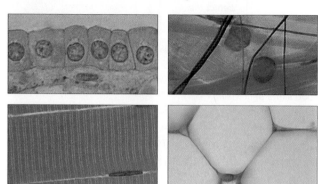

4. Identify this category of tissue and its characteristics. *33.4*

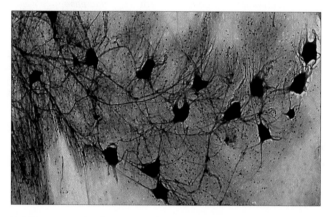

5. What type of cell serves as the basic unit of communication in nervous systems? *33.4*

6. Define animal tissue, organ, and organ system. List and define the functions of the eleven major organ systems of the human body. *CI, 33.6*

7. Define extracellular fluid, interstitial fluid, and plasma. *33.7*

8. Define homeostasis. *CI, 33.7*

9. Briefly describe two major categories of the homeostatic mechanisms operating in the human body. *33.7*

Self-Quiz *(Answers in Appendix IV)*

1. _____ tissues have closely linked cells and one free surface.
   a. Epithelial   b. Connective   c. Nervous   d. Muscle

2. In most _____ , cells secrete fibers of collagen and elastin.
   a. epithelial tissue   c. muscle tissue
   b. connective tissue   d. nervous tissue

3. _____ has a semifluid ground substance and occurs under skin and most epithelia.
   a. Dense, irregular connective tissue
   b. Loose connective tissue
   c. Dense, regular connective tissue
   d. Cartilage

4. _____ , a specialized connective tissue, is mostly plasma with cellular components and various dissolved substances.
   a. Irregular connective tissue   c. Cartilage
   b. Blood   d. Bone

5. After you eat too many carbohydrates and proteins, your body converts the excess to storage fats, which accumulate in _____ .
   a. connective tissue proper   c. adipose tissue
   b. dense connective tissue   d. both b and c

6. Components of _____ detect and coordinate information about changes and control responses to those changes.
   a. epithelial tissue   c. muscle tissue
   b. connective tissue   d. nervous tissue

7. In your own body, _____ can shorten (contract).
   a. epithelial tissue   c. muscle tissue
   b. connective tissue   d. nervous tissue

8. Cells of complex animals _____ .
   a. survive by their own metabolic activities
   b. contribute to the survival of the whole animal
   c. help maintain extracellular fluid
   d. all of the above

9. At _____ , physical and chemical conditions in the internal environment are being kept within tolerable ranges.
   a. positive feedback   c. homeostasis
   b. negative feedback   d. metastasis

10. With negative feedback mechanisms, _____ .
   a. a stimulus brings about a response that tends to return internal operating conditions to the original state
   b. a stimulus suppresses internal operating conditions to levels below a set point for the body
   c. a stimulus raises internal operating conditions to levels above a set point for the body
   d. fewer solutes are fed back to the affected cells

11. Of the components that exert feedback control of organ activity, _____ detect specific changes in the environment, an _____ pulls together different bits of information and selects a suitable response, and _____ carry out the response.

12. Match the terms with the suitable description.
   ____ epithelium   a. rather pliable, like rubber
   ____ cartilage   b. covers or lines body surfaces
   ____ homeostasis   c. stable internal environment
   ____ muscles and glands   d. integrating center
   ____ positive feedback   e. the most common type of
   ____ negative feedback   homeostatic control mechanism
   ____ brain   f. effectors
                    g. chain of events intensifies original condition in body

Critical Thinking

1. *Anhidrotic ectodermal dysplasia*, a genetic disorder described in Section 15.4, has been associated with a recessive allele on the mammalian X chromosome. Among other symptoms of this disorder, affected males and females have no sweat glands in the tissues where the recessive allele is expressed. What type of tissue are we talking about?

2. Adipose tissue and blood are often said to be "atypical" connective tissues. Compared with other connective tissues, which of their features are *not* typical?

3. After graduating from high school, Jeff and Ryan set out on a trip through the desert roads of California and Arizona. One hot, dry morning in Joshua Tree National Monument, they saw an unusual rock formation that didn't appear to be too far from the road. They left the car and started hiking toward it. Their destination turned out to be farther away than they thought and the sun's rays were more relentless than they had anticipated. They reached the shade of the rocks in the early afternoon. Their canteen was nearly empty, and the physiological meaning of "thirst" made itself known to them in a scary way. They knew they had to locate and drink water (or some other fluid), which is what the brain usually tells us to do when the body starts to get dehydrated.

From what you read in this chapter, would you suspect that Ryan and Jeff's *thirst behavior* is part of a positive or negative feedback control mechanism?

4. *Porphyria* is a genetic disorder that shows up in about 1 in every 25,000 individuals. Affected individuals lack certain enzymes that are part of a metabolic pathway leading to formation of heme, which is the iron-containing group of hemoglobin. An accumulation of porphyrins, which are intermediates of the pathway, causes awful symptoms, especially after exposure to sunlight. Lesions and scars form on the skin. Hair grows thickly on the face and hands. As gums retreat from teeth, the canines take on a fanglike appearance. Symptoms worsen upon exposure to a variety of substances, including garlic and alcohol. Affected individuals avoid sunlight and aggravating substances, and get injections of heme from normal red blood cells.

If you are familiar with vampire stories, which date from the Middle Ages or earlier, speculate on how they may have evolved among superstitious folk who did not have medical knowledge of porphyria.

Selected Key Terms

adhering junction *33.1*	homeostasis *CI*
adipose tissue *33.2*	integrator *33.7*
blood *33.2*	internal environment *CI*
bone *33.2*	loose connective tissue *33.2*
cardiac muscle tissue *33.3*	mesoderm *33.6*
cartilage *33.2*	negative feedback mechanism *33.7*
dense, irregular	nervous tissue *33.4*
connective tissue *33.2*	neuroglia *33.4*
dense, regular	neuron *33.4*
connective tissue *33.2*	organ *CI*
division of labor *CI*	organ system *CI*
ectoderm *33.6*	positive feedback mechanism *33.7*
effector *33.7*	sensory receptor *33.7*
endocrine gland *33.1*	skeletal muscle tissue *33.3*
endoderm *33.6*	smooth muscle tissue *33.3*
epithelium *33.1*	stimulus *33.7*
exocrine gland *33.1*	tight junction *33.1*
extracellular fluid *33.7*	tissue *CI*
gap junction *33.1*	

Readings

Bloom, W., and D. W. Fawcett. 1995. *A Textbook of Histology.* Twelfth edition. Philadelphia: Saunders.

Leeson, C. R., T. Leeson, and A. Paparo. 1988. *Textbook of Histology.* Philadelphia: Saunders.

Ross, M., L. Romrell, and G. Kaye. 1988. *Histology: A Text and Atlas.* Baltimore: Williams & Wilkins.

Web Site See *http://www.wadsworth.com/biology* for practice quiz questions, hypercontents, BioUpdates, and critical thinking. The Wadsworth Biology Resource Center provides a wealth of information fully organized and integrated by chapter.

TORNADO!

Figure 34.1 A terrifying view across the Kansas prairie—a tornado about to touch down. Imagine yourself alone in the prairie when you first see a tornado, knowing you have only minutes to remove yourself from harm's way. By what means do you conceive of plans of action? Can the plans be put together and evaluated quickly enough? The answers begin with the functioning of neurons in your nervous system.

It is spring in the American Midwest, and you are fully engrossed in photographing wildflowers in an expanse of shortgrass prairie. So intent are you on capturing all the species on film that you fail to notice the rapidly darkening sky. What's that rumbling you hear in the distance? It sounds something like a freight train. You wonder how can that be, when there is no train track, anywhere, in this part of the prairie. You turn to identify the source of the sound. Then you see it, but you don't want to believe your eyes. A dark funnel cloud is advancing across the prairie and heading right for you! *TORNADO!* The ominous image rivets your attention as nothing else has ever done (Figure 34.1). You know that you cannot remain where you are and survive. Suddenly you remember that hiding in a low area is better than standing out in the open. Commands rush from your brain to your four limbs: *GET MOVING OUT OF HERE!* With heart thumping, you run along a path, looking frantically for safety. You're in luck! Just ahead is a steep-banked creek. With a stunning burst of speed you reach the creek in less than a minute and scramble down the muddy bank. There you find a small ledge over the water. You quickly wedge yourself under it, wishing wildly to be inconspicuous, to be overlooked by a force of nature on the rampage.

It takes a few minutes before you realize the tornado has roared past the creek. You

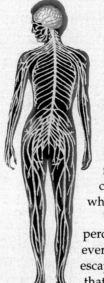

remain motionless, fingers clutching mud. Finally you do sit up and look around. Some distance away, you see a swath of twisted grass that marks the tornado's path. Your heart no longer feels as if it is slamming against your chest wall, but your legs feel like rubber when you stand up.

Thank your nervous system for every perception, every memory, and nearly every action that collectively helped you escape the tornado. Thank the information that traveled swiftly, in suitable directions, among interacting cells called **neurons**. These are the cells that work together to monitor conditions in and around the body and to issue commands for responsive actions that benefit the body as a whole. They represent the communication lines of your brain, spinal cord, and nerves.

This chapter begins with the structure and function of neurons. In chapters to follow, you will consider how these cells interact with one another in nervous systems and sensory systems. Through such interactions, you detect tornadoes and other events that have bearing on whether you survive from one day to the next.

There are three classes of neurons. Different types of **sensory neurons** are adapted to respond to specific stimuli and relay information about them to the spinal cord and brain. A **stimulus** is a specific form of energy, such as light and pressure. In the spinal cord and brain you find **interneurons**. These receive sensory input, integrate it with other incoming information and with stored information, then influence the activity of other neurons. **Motor neurons** relay information from the brain and spinal cord to effectors—muscles or glands—that carry out the specified responses:

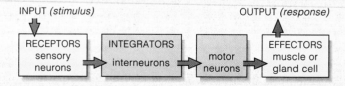

INPUT (*stimulus*)   OUTPUT (*response*)

Neurons are not the only cells in vertebrate nervous systems. As you will see in this chapter and the next, a variety of cells, collectively called neuroglia, make up more than half the volume of the vertebrate systems. Different kinds of neuroglial cells metabolically assist, protect, and structurally support the neurons.

Functional Zones of a Neuron

To understand how your own nervous system works, start with how its neurons function. Neurons consist of a nucleated cell body and cytoplasmic extensions that differ in number and length (Figure 34.2). Typically, the cell body and the slender extensions called **dendrites** are *input* zones, where the neuron receives information. Another slender but often longer extension, called an **axon**, is a *conducting* zone. Axons rapidly propagate signals that arise at a neuron's trigger zone. Except in sensory neurons, the *trigger* zone is a patch of plasma membrane at the junction between the cell body and an axon. The branched endings of an axon are *output* zones, where messages are sent to other cells.

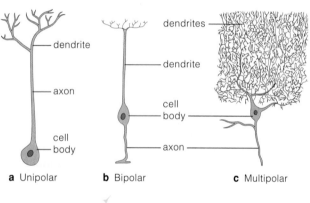

a Unipolar   **b** Bipolar   **c** Multipolar

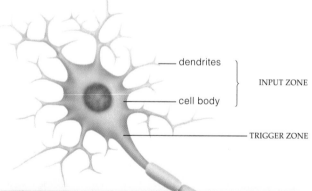

dendrites

cell body

INPUT ZONE

TRIGGER ZONE

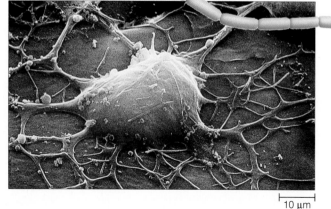

10 μm

A Neuron at Rest, Then Moved to Action

Different forms of signals arise in the nervous system. Let's start with one that begins and ends on the same neuron and is not transferred to another cell.

When a neuron is not being bothered, a difference in electric charge is being maintained across its plasma membrane. The cytoplasmic fluid next to the membrane remains negatively charged, compared to the interstitial fluid right outside. We measure these charges in units called millivolts. The amount of energy inherent in the steady voltage difference across the plasma membrane is the **resting membrane potential**. For many neurons, that amount is about −70 millivolts.

Suppose a weak signal reaches a patch of membrane in a neuron's input zone. The voltage difference across the patch changes only slightly, if at all. By contrast, a strong signal might trigger an **action potential**, which is an abrupt, short-lived reversal in the voltage difference across a plasma membrane. For a fraction of a second, the inside becomes positive with respect to the outside. The localized reversal invites an action potential at the adjoining membrane patch, which invites another at the next patch, and so on away from the initiation point. In short, *stimulation of a neuron disturbs the distribution of electric charge across its plasma membrane.*

Restoring and Maintaining Readiness

How does the neuron restore the voltage difference across each patch of plasma membrane and maintain it between action potentials? It relies on two membrane properties. First, a membrane has a lipid bilayer, which bars the passage of potassium ions (K^+), sodium ions (Na^+), and other charged substances. Thus, the neuron can build up differences in ion concentrations across the membrane. Second, ions can flow from one side to the other through the interior of transport proteins that span the bilayer, and that flow is controlled (Figure 34.3).

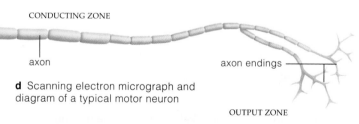

CONDUCTING ZONE

axon

axon endings

d Scanning electron micrograph and diagram of a typical motor neuron

OUTPUT ZONE

Figure 34.2 (**a**–**c**) Classification of neurons based on the number of cytoplasmic extensions of the cell body. *Unipolar* cells have a single axon with dendritic branchings. *Bipolar* cells have one dendrite and one axon. Many sensory neurons are like this. *Multipolar* cells, with one axon and many dendrites, predominate in vertebrate nervous systems. (**d**) Functional zones of a motor neuron, a multipolar cell.

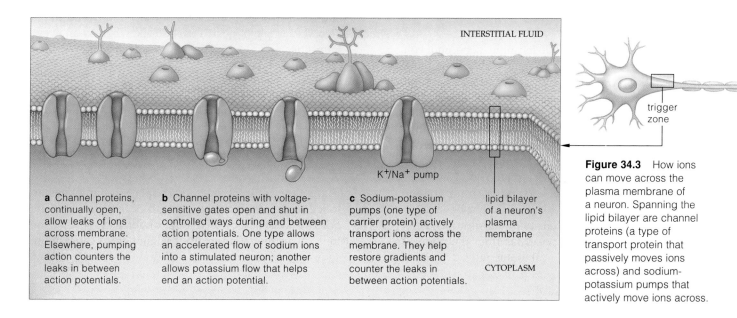

a Channel proteins, continually open, allow leaks of ions across membrane. Elsewhere, pumping action counters the leaks in between action potentials.

b Channel proteins with voltage-sensitive gates open and shut in controlled ways during and between action potentials. One type allows an accelerated flow of sodium ions into a stimulated neuron; another allows potassium flow that helps end an action potential.

c Sodium-potassium pumps (one type of carrier protein) actively transport ions across the membrane. They help restore gradients and counter the leaks in between action potentials.

lipid bilayer of a neuron's plasma membrane

CYTOPLASM

K⁺/Na⁺ pump

trigger zone

Figure 34.3 How ions can move across the plasma membrane of a neuron. Spanning the lipid bilayer are channel proteins (a type of transport protein that passively moves ions across) and sodium-potassium pumps that actively move ions across.

Suppose a motor neuron has 15 sodium ions inside the membrane for every 150 outside. Suppose it has 150 potassium ions inside for every 5 on the outside. You can depict each ion's concentration gradient in this way (from the large to the small letters):

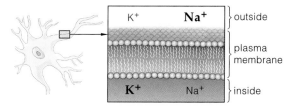

Gradients such as these determine the net direction in which sodium and potassium ions diffuse. The ions can diffuse through the interior of channel proteins, a type of transport protein in the membrane. Some of the channels never shut, so ions leak (diffuse) through them all of the time. Others have molecular gates, which can open only after the neuron is adequately stimulated.

Suppose that motor neuron is not being stimulated. Its sodium channels are shut, so sodium ions can't rush inside. Some potassium is leaking out through a few open channels and making the cytoplasm a bit more negative, so some potassium is attracted back in. When the inward pull of electric charge balances the outward force of diffusion, there is no more net movement of potassium. The concentration and electric gradients now existing across the plasma membrane will permit the neuron to respond to stimulation.

After the gradients have reversed during an action potential, **sodium-potassium pumps** restore them. These are transport proteins that span the plasma membrane (Section 5.5). When they get an energy boost from ATP,

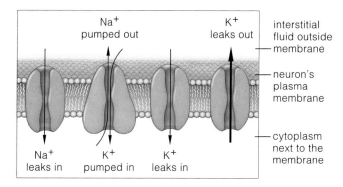

Figure 34.4 Pumping and leaking processes that influence the distribution of sodium and potassium ions across the plasma membrane of a neuron at rest. Arrow widths indicate the magnitude of the movements. Notice how the total inward and outward movements for each kind of ion are balanced.

the pumps actively transport potassium into the neuron and sodium out at the same time. Sodium-potassium pumps must maintain as well as restore the gradients. Why? Even in a resting neuron, a tiny fraction of the outward-leaking potassium isn't attracted back in. Also, a tiny fraction of sodium leaks in, through a few open channels (Figure 34.4). If the leaks went unattended, the crucial gradients would gradually disappear.

An undisturbed neuron is maintaining a resting membrane potential—a voltage difference across its plasma membrane. An action potential is an abrupt, short-lived reversal in that voltage difference in response to adequate stimulation.

After an action potential, sodium-potassium pumps restore and maintain the resting membrane potential.

Approaching Threshold

Weakly stimulate a neuron at its input zone and you disturb the ion balance across the membrane, but not much. Suppose your toes gently tap a cat and put a bit of pressure on its skin. Tissues beneath the skin surface have receptor endings—input zones of sensory neurons. Patches of plasma membrane at the receptor endings deform under the pressure. Some ions now flow across, so the voltage difference across the membrane changes slightly. The pressure produced a graded, local signal.

Graded means that signals arising at an input zone can vary in magnitude. Such signals might be large or small, depending on the intensity of the stimulus. *Local* means the signals do not spread far from the point of stimulation. It takes specialized types of

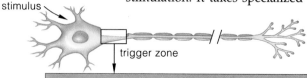

ion channels to propagate a signal along the membrane, and input zones simply don't have them.

However, when a stimulus is intense or long lasting, graded signals can spread out from the input zone and into an adjacent trigger zone. At this zone, a certain minimum amount of change in the voltage difference across the plasma membrane can trigger an action potential. That amount is known as the **threshold level**. *Threshold can be reached at any membrane patch that has voltage-sensitive, gated channels for sodium ions.*

As Figure 34.5 shows, the stimulus causes sodium ions to flow across the membrane, into the neuron. With the influx of positively charged ions, the cytoplasmic side of the plasma membrane becomes less negative. This causes more gates to open, more sodium to enter,

and so on. The ever increasing inward flow of sodium is a good example of **positive feedback**, whereby an event intensifies as a result of its own occurrence:

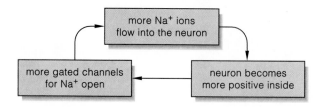

At threshold, the opening of more sodium gates no longer depends upon the strength of the stimulus. The positive-feedback cycle is now under way, so that the inward-rushing sodium itself is enough to cause more sodium gates to open.

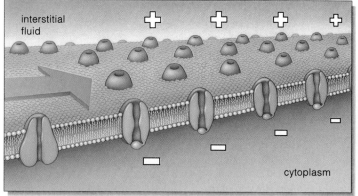

a Membrane at rest (inside negative with respect to the outside). An electrical disturbance (*red arrow*) spreads from an input zone to an adjacent trigger region of the membrane, which has a great number of gated sodium channels.

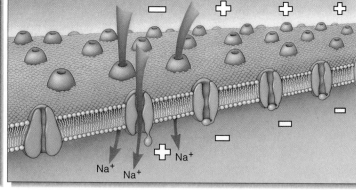

b A strong disturbance initiates an action potential. Sodium gates open. The sodium inflow decreases the negativity inside the neuron. The change causes more gates to open, and so on, until threshold is reached and the voltage difference across the membrane reverses.

Figure 34.5 Propagation of an action potential along the axon of a motor neuron.

An All-or-Nothing Spike

Figure 34.6 shows a recording of the voltage difference across the plasma membrane before, during, and after an action potential. Notice how the membrane potential spikes once threshold is reached. Every single action potential in the neuron spikes to the same level above threshold as an *all-or-nothing* event. That is, once the positive-feedback cycle starts, nothing will stop the full spiking. If the threshold is not reached, however, the disturbance to the plasma membrane will subside when the stimulus is removed.

Each spike lasts only for a millisecond or so. Why? At the membrane site of the charge reversal, the gated sodium channels closed and shut off the sodium inflow.

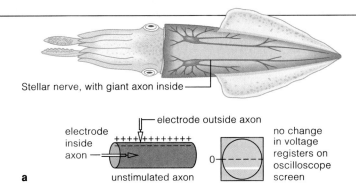

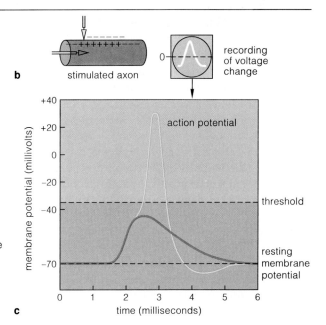

recording of voltage change

Figure 34.6 Action potentials. (**a**) When early researchers studied neural function, the squid *Loligo* provided them with evidence of action potential spiking. The squid's "giant" axons were large enough to slip electrodes inside. (**b**) When such an axon was stimulated, electrodes positioned inside and outside detected voltage changes, which showed up as deflections in a beam of light across the screen of an oscilloscope. (**c**) This is a typical waveform (*yellow* line) for an action potential on an oscilloscope screen. The *red* line represents a recording of a local signal that did not reach the threshold of an action potential; spiking did not occur.

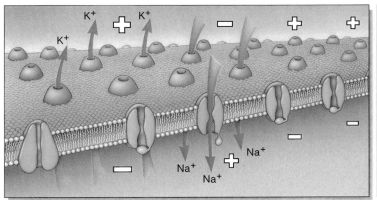

c With the reversal, sodium gates shut and potassium gates open (*pink arrows*). Potassium follows its gradient out of the neuron. Voltage is restored. The disturbance triggers an action potential at the adjacent site, and so on, away from the point of stimulation.

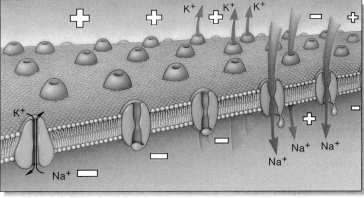

d Following each action potential, the inside of the plasma membrane becomes negative once again. However, the sodium and potassium concentration gradients are not yet fully restored. Active transport at sodium-potassium pumps restores them.

Also, about halfway through the reversal, potassium channels opened, so many more potassium ions flowed out and restored the original voltage difference across the membrane. And sodium-potassium pumps restored the ion gradients. Later on, after the resting membrane potential has been restored, most potassium gates close and sodium gates are in their initial state, ready to be opened with the arrival of a suitable disturbance.

Propagation of Action Potentials

The membrane disturbances leading up to an action potential are self-propagating, and they do not diminish in magnitude. As they spread to an adjacent membrane patch, an equivalent number of gated channels open. With this new disturbance, gated channels open in the *next* adjacent patch, then the next, and so on. For a brief period after each membrane patch has been excited, it is insensitive to stimulation. Sodium gates in the patch are inactivated and ions cannot move through them. This is the reason why action potentials do not spread back to the trigger zone (the site where they were initiated) but rather are self-propagating away from it.

The cytoplasm next to the plasma membrane of a neuron at rest is more negative than the interstitial fluid just outside the membrane.

During an action potential, the inside of a disturbed patch of membrane becomes more positive than the outside.

After an action potential, resting conditions are restored at the membrane patch.

When action potentials reach a neuron's output zone, they usually do not proceed farther than this. But their arrival may induce the neuron to release one or more **neurotransmitters**, which are signaling molecules that diffuse across chemical synapses. A **chemical synapse** is a narrow junction, or cleft, between the output zone of a neuron and an input zone of an adjacent cell (Figure 34.7). Some clefts intervene between two neurons, and others between a neuron and a muscle cell or gland cell.

At each chemical synapse, one of the two cells stores neurotransmitter molecules inside synaptic vesicles in its cytoplasm. Think of it as the *pre*synaptic cell. Here, gated channels for calcium ions span the membrane, and they open when an action potential arrives. There are more calcium ions outside the cell. When they flow into the cell (down their gradient), the synaptic vesicles are induced to fuse with the plasma membrane, so that neurotransmitter is released into the synaptic cleft.

The neurotransmitter molecules diffuse through the cleft. On the *post*synaptic cell membrane are protein receptors that bind specific neurotransmitters. Binding changes the receptor shape and creates a passageway through its interior. Ions cross the plasma membrane by diffusing through the passageway (Figure 34.8).

A postsynaptic cell's response depends on the type and concentration of neurotransmitter in the cleft, what kinds of receptors the cell bears, and the number and responsiveness of gated channels in its membrane. Such factors influence whether a neurotransmitter will have an *excitatory* effect and help drive the postsynaptic cell's membrane toward the threshold of an action potential. They also influence whether it will have an *inhibitory* effect and drive the membrane away from threshold.

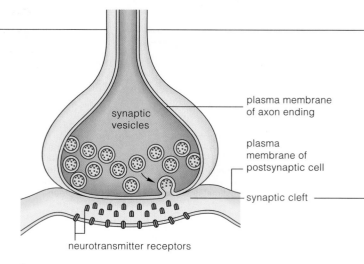

a

Figure 34.8 Closer look at a chemical synapse. (**a**,**b**) The plasma membranes of a motor neuron and a muscle cell face each other across a cleft between them. The arrival of an action potential at an axon ending triggers the release of molecules of neurotransmitter from the neuron. (**c**) These molecules diffuse through the cleft and then bind to receptors on gated channel proteins of the muscle cell membrane. The gates open; ions flow in and trigger a graded potential at the membrane site.

Consider **acetylcholine** (ACh), a neurotransmitter. It has excitatory and inhibitory effects on the brain, spinal cord, glands, and muscles. For example, it acts at the chemical synapse between a motor neuron and muscle cell, as in Figure 34.7. ACh released from the motor neuron diffuses across the cleft and binds to receptors on the muscle cell membrane. In this kind of cell it has excitatory effects; it can trigger action potentials, which in turn initiate muscle contraction (Section 38.9).

A Smorgasbord of Signals

Acetylcholine is only one of a veritable smorgasbord of signals that neurons deliver to target cells. For example, serotonin acts on neurons in brain regions that govern sleeping, sensory perception, temperature control, and emotions. Norepinephrine works in brain regions that control emotions, dreaming, and waking up. Dopamine also works in brain regions that deal with emotions. GABA (gamma aminobutyric acid) is the most common inhibitory signal in the brain. Antianxiety drugs, such as Valium, may exert their effects by enhancing GABA's effects. Except for ACh, these neurotransmitters and the other types are amino acids or are derived from them.

Signaling molecules known as **neuromodulators** can magnify or reduce the effects of a neurotransmitter on neighboring or distant neurons. They include substance P, which induces pain perception, and endorphins—the natural pain killers that inhibit the release of substance P from sensory nerves. Neuromodulators might also influence memory and learning, sexual activity, control of body temperature, and emotional states.

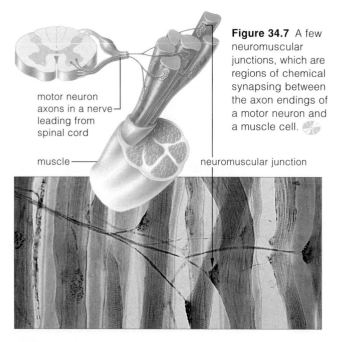

Figure 34.7 A few neuromuscular junctions, which are regions of chemical synapsing between the axon endings of a motor neuron and a muscle cell.

motor neuron axons in a nerve leading from spinal cord

muscle

neuromuscular junction

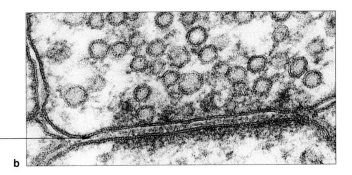

b

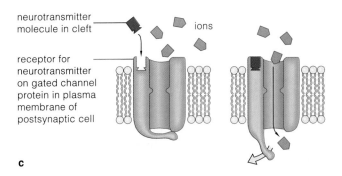

neurotransmitter molecule in cleft — ions

receptor for neurotransmitter on gated channel protein in plasma membrane of postsynaptic cell

c

Synaptic Integration

Anywhere from 1,000 to 10,000 communication lines form synapses with a typical neuron in your brain. And your brain contains at least 100 billion neurons. As long as you are alive, those neurons continually hum with messages about doing what it takes to be a human.

At any moment, a great number of excitatory and inhibitory signals may be washing over the input zones of a postsynaptic cell. Some signals drive its membrane closer to threshold; others maintain the resting level or drive it away from threshold. Said another way, signals compete for control of the neuron's membrane.

All synaptic signals are graded potentials. The ones we call **EPSPs** (for excitatory postsynaptic potentials) have a *depolarizing* effect. This simply means they bring the membrane closer to threshold. The **IPSPs** (inhibitory postsynaptic potentials) might have a *hyperpolarizing* effect (drive the membrane away from threshold) or might help maintain the membrane at its resting level.

With **synaptic integration**, competing signals that reach an input zone of a neuron at the same time are summed. By this process, two or more signals arriving

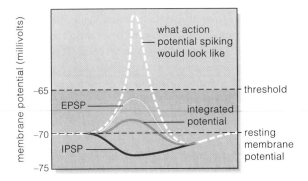

Figure 34.9 Example of synaptic integration. The *yellow* line shows how an EPSP of a certain magnitude would register on an oscilloscope screen if it were acting alone. The *purple* line shows the effect of an IPSP if *it* were acting alone. Suppose both signals arrive at a postsynaptic cell membrane at the same time. The *red* line shows their effect. In this case, when the two signals are integrated, threshold is not reached. Thus an action potential cannot be initiated in the target cell.

at a neuron may be dampened, suppressed, reinforced, or sent onward to other cells in the body.

Figure 34.9 shows what two separate recordings of an EPSP and an IPSP might look like, and it includes a recording of their summation. Such integration occurs when neurotransmitter from several presynaptic cells reaches the input zone of a neuron at the same time. It also occurs after neurotransmitter is released swiftly and repeatedly from a single presynaptic cell that has been whipped into a frenzy of excitability by a rapid series of action potentials.

Removing Neurotransmitter From the Synaptic Cleft

The flow of information through the nervous system depends on the prompt, precisely controlled removal of neurotransmitter molecules from synaptic clefts. Some amount of these molecules simply diffuses out of the cleft. Enzymes cleave others right in the cleft, as when acetylcholinesterase breaks apart ACh. Also, transport proteins actively pump the molecules back into the presynaptic cells or into neighboring neuroglial cells.

What happens if neurotransmitter accumulates in the cleft? As one example, cocaine blocks the uptake of dopamine. Molecules of this neurotransmitter linger in synaptic clefts and just keep on stimulating target cells. At first the stimulation produces euphoria (intense pleasure). Later on, it has disastrous effects, as you will read in the chapter to follow.

Neurotransmitters are signaling molecules that bridge a synaptic cleft, which is a tiny gap between two neurons or between a neuron and a muscle cell or gland cell.

Neurotransmitters have excitatory or inhibitory effects on different kinds of receiving cells. Synaptic integration is the moment-by-moment combining of excitatory and inhibitory signals acting on a postsynaptic cell.

With the summation process, messages traveling through the nervous system can be reinforced or downplayed, sent onward or suppressed. The process is essential for normal body functioning.

Blocks and Cables of Neurons

Through synaptic integration, messages arriving at any neuron in the body might be reinforced and sent on to neighboring neurons. What determines the direction in which a given message will travel? That depends on the organization of neurons in different body regions.

For example, your brain deals with its staggering numbers of neurons in a manner analogous to block parties. Regional blocks of hundreds or thousands of neurons receive excitatory and inhibitory signals. They integrate signals entering the block, then send out new

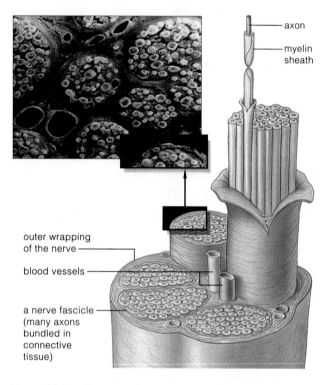

outer wrapping
of the nerve

blood vessels

a nerve fascicle
(many axons
bundled in
connective
tissue)

axon

myelin
sheath

Figure 34.10 Structure of a nerve. Axons in the nerve are bundled together inside wrappings of connective tissue.

ones in response. Some regions have neurons organized as *divergent* circuits, as when their processes fan out from one block and form connections with many others. Different regions have neurons arranged in *convergent* circuits, with signals from many sent on to just a few. In still other regions, neurons synapse back on themselves and repeat signals like gossip that just won't go away. Such neurons form *reverberating* circuits. They include the circuits that make your eye muscles rhythmically twitch while you sleep.

In the cablelike **nerves**, long axons of many sensory neurons, motor neurons, or both permit long-distance communication between the brain or spinal cord and the rest of the body. Connective tissue bundles most of the axons in parallel array (Figure 34.10). Each axon has a **myelin sheath** that enhances the rate of action potential propagation. The sheath is a series of **Schwann cells**, a type of neuroglial cell, wrapped like jellyrolls around the long axon. An exposed node, or gap, separates each cell from adjacent ones. There, voltage-sensitive, gated sodium channels pepper the plasma membrane (Figure 34.11). The sheathed regions in between nodes hamper ion movements across the membrane. Ion disturbances tend to flow along the membrane until the next node in line. At each node, ion flow can produce a new action potential. In large sheathed axons, action potentials are propagated at a remarkable 120 meters per second.

With *multiple sclerosis*, myelin sheaths around axons in the spinal cord degenerate slowly but inevitably. Gene mutation may predispose a person to the disease, but viral infection may be the actual trigger. Symptoms include weakening of muscles, fatigue, and numbness.

Reflex Arcs

Figure 34.12 provides a specific example of the direction of information flow through nervous systems. It shows how sensory and motor neurons of certain nerves take part in a path known as the stretch reflex. **Reflexes** are simple movements made in response to specific sensory stimuli. In the simplest reflex arcs, sensory neurons synapse directly on motor neurons.

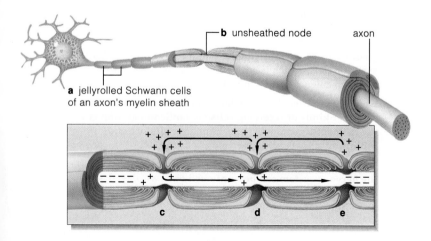

a jellyrolled Schwann cells
of an axon's myelin sheath

b unsheathed node   axon

c   d   e

Figure 34.11 High-speed travel in the nervous system—how an action potential is propagated along sheathed neurons. (**a**) A myelin sheath is a series of Schwann cells, each wrapped like a jellyroll around an axon. (**b**) Each jellyroll blocks ion movements across the membrane, but ions can cross at nodes in between. The unsheathed nodes have very dense arrays of gated sodium channels. (**c,d**) A disturbance caused by an action potential spreads down the axon. When it reaches a node, sodium gates open, sodium ions rush inward, and another action potential results. (**e**) The new disturbance spreads rapidly to the next node and triggers another action potential, and so on down the line.

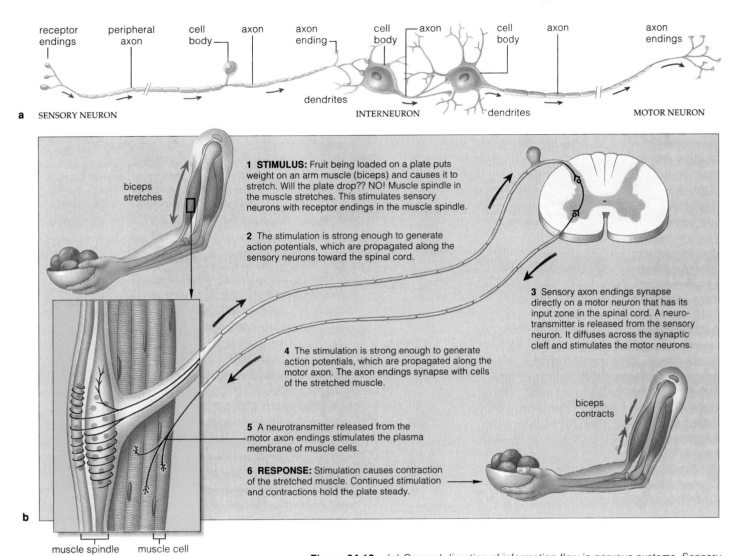

a SENSORY NEURON — receptor endings, peripheral axon, cell body, axon, axon ending, dendrites, INTERNEURON, cell body, axon, cell body, axon, axon endings, dendrites, MOTOR NEURON

1 STIMULUS: Fruit being loaded on a plate puts weight on an arm muscle (biceps) and causes it to stretch. Will the plate drop?? NO! Muscle spindle in the muscle stretches. This stimulates sensory neurons with receptor endings in the muscle spindle.

2 The stimulation is strong enough to generate action potentials, which are propagated along the sensory neurons toward the spinal cord.

biceps stretches

3 Sensory axon endings synapse directly on a motor neuron that has its input zone in the spinal cord. A neurotransmitter is released from the sensory neuron. It diffuses across the synaptic cleft and stimulates the motor neurons.

4 The stimulation is strong enough to generate action potentials, which are propagated along the motor axon. The axon endings synapse with cells of the stretched muscle.

5 A neurotransmitter released from the motor axon endings stimulates the plasma membrane of muscle cells.

6 RESPONSE: Stimulation causes contraction of the stretched muscle. Continued stimulation and contractions hold the plate steady.

biceps contracts

b

muscle spindle   muscle cell

The stretch reflex works to contract a muscle after gravity or some other load has caused the muscle to stretch. Suppose you hold out a large bowl and keep it stationary as someone puts several peaches into it. The peaches add weight to the bowl, and when your hand starts to drop, a muscle in your arm (the biceps) is stretched.

In the muscle, stretching activates receptor endings that are a part of muscle spindles—sensory organs in which specialized cells are enclosed in a sheath that runs parallel with the muscle. The receptor endings are the input zones of sensory neurons, the axons of which synapse with motor neurons within the spinal cord (Figure 34.12). Axons of the motor neurons lead back to the stretched muscle. Action potentials that reach the axon endings trigger the release of ACh, which initiates contraction. As long as receptor activity continues, the motor neurons are excited further, and this allows them to maintain your hand's position.

Figure 34.12 (**a**) General direction of information flow in nervous systems. Sensory neurons relay information *into* the spinal cord and brain, where they synapse with interneurons. Interneurons *within* the spinal cord and brain integrate signals. Many synapse with motor neurons, which carry signals *away* from the spinal cord and brain. (**b**) Organization of nerves in a reflex arc that deals with muscle stretching. In a skeletal muscle, stretch-sensitive receptors of a sensory neuron are located in muscle spindles. The stretching generates action potentials, which reach axon endings in the spinal cord. These synapse with a motor neuron that carries signals to contract, from the spinal cord back to the stretched muscle.

In the vast majority of reflexes, sensory neurons also make connections with a number of interneurons, which then activate or suppress all the motor neurons necessary for a coordinated response.

Vertebrates have interneurons organized in information-processing blocks. Cablelike nerves that have long axons of sensory neurons, motor neurons, or both connect the brain and spinal cord with the rest of the body.

Reflex arcs, in which sensory neurons synapse directly on motor neurons, are the simplest paths of information flow.

SKEWED INFORMATION FLOW

What happens when something disrupts information flow in the nervous system? To give one example, think about *Clostridium botulinum*. This anaerobic, endospore-forming bacterium normally lives in soil, but it can cause the disease *botulism*. When its endospores contaminate improperly stored, preserved, or canned food, they can germinate and produce a dangerous toxin, which people might ingest and absorb. The toxin can bind to neurons that synapse with muscle cells and block their release of acetylcholine (ACh). Once that happens, muscles cannot contract. The muscles become progressively flaccid and paralyzed. Death can follow within ten days, usually from respiratory and cardiac failure. Recovery is possible if antitoxins are quickly administered.

A related bacterium, *C. tetani*, lives in the gut of horses, cattle, and other grazing animals, even many people. Its endospores survive in soil, especially manure-rich ones. They persist for years if sunlight and oxygen do not reach them. They resist disinfectants, heat, and boiling water. When they enter the body through a deep puncture or a deep cut, they germinate in anaerobic, dead tissues. These bacteria do not spread from the dead tissue. They produce a toxin that the blood or nerves deliver to the spinal cord and brain. In the spinal cord, the toxin affects interneurons that help control motor neurons. It blocks the release of inhibitory neurotransmitters (GABA and glycine) and frees the motor neurons from normal inhibitory control. Symptoms of the disease *tetanus* are about to begin.

Muscles overstimulated for four to ten days stiffen and go into spasm. They cannot be released from contraction, and prolonged, spastic paralysis follows. Fists and jaws may remain clenched (the disease is also called lockjaw). The back may arch permanently. Spasms may be strong enough to break bones. When respiratory and cardiac muscles become paralyzed, death nearly always follows.

Vaccines were not available for soldiers of early wars, when dead cavalry horses and manure littered battlefields (Figure 34.13). Today, vaccines have all but eradicated tetanus in the United States.

Figure 34.13 Painting of a young victim of a contaminated battle wound, as he lay dying in a military hospital.

SUMMARY

1. The nervous system senses, interprets, and issues commands for responses to specific aspects of the environment. In nearly all animals, its communication lines consist of sensory neurons, interneurons, and motor neurons, which activate muscle and gland cells.

 a. Sensory neurons are receptors that detect specific stimuli (specific forms of energy, such as light).

 b. Interneurons are integrators in the brain and spinal cord. They receive signals and process signals from sensory receptors, and then issue commands for suitable responses.

 c. Motor neurons carry commands away from the brain and spinal cord to the body's effectors, which are muscle cells and gland cells that carry out responses.

2. A neuron's dendrites and cell body are input zones. If arriving signals spread to a trigger zone (such as the start of an axon), they may trigger an action potential that can be propagated to an output zone (axon endings).

 a. Transport proteins pepper the plasma membrane from the trigger zone to axon endings of a neuron. They serve as gated or open channels for the passage of ions, Na^+ and K^+ especially, across the membrane.

 b. The controlled flow of ions across the membrane is the basis of message propagation along a neuron.

3. For a neuron at rest, the ion distribution between the cytoplasm and extracellular fluid results in a difference in electric charge (voltage difference) across the plasma membrane. The cytoplasmic fluid is more negatively charged, by an amount called the resting membrane potential. An action potential is an abrupt, short-lived reversal in the voltage difference across the membrane in response to adequate stimulation.

 a. Stimuli give rise to local, graded potentials, which vary in their magnitude and do not spread very far. A disturbance caused by a number of graded potentials may spread to a trigger zone and drive the membrane of a neuron to threshold of an action potential.

 b. A strong disturbance makes many gated channels for sodium ions open in an ever accelerating, self-propagating way. As a result, the voltage difference across the membrane reverses abruptly. Accelerated openings proceed in patch after patch of membrane, to the neuron's output zone.

 c. In between action potentials, sodium-potassium pumps counter small ion leaks across the membrane and so maintain the gradients. After an action potential, they restore the gradients.

4. Action potentials arriving at an output zone trigger the release of neurotransmitters. These diffuse across chemical synapses, where the neuron forms a junction with another neuron, a muscle cell, or a gland cell. They may excite the cell's plasma membrane (drive it closer to threshold) or inhibit it (drive it away from threshold).

5. Integration is the moment-by-moment summation of excitatory and inhibitory signals reaching all of the synapses on a neuron. It is a means of playing down, suppressing, reinforcing, or sending on information to other neurons of the nervous system.

Review Questions

1. Describe sensory neuron, interneuron, and motor neuron in terms of their structure and functions. *CI, 34.1, 34.4*

2. Label the functional zones of this motor neuron. *34.1*

3. Define resting membrane potential, graded potential, and action potential. *34.1, 34.2*

4. A neuron at rest is controlling the ion distribution across its plasma membrane. Identify the two major kinds of ions. Do they leak across the membrane, are they pumped across, or both? As part of your answer, describe gated and ungated transport proteins embedded in the neural membrane. *34.1*

5. With respect to action potentials, explain what is meant by threshold level, by all-or-nothing spikes, and by self-propagation of an action potential. *34.2*

6. Define chemical synapse and neurotransmitter. Choose an example of a neurotransmitter and state where it acts. *34.3*

7. Define synaptic integration. Include definitions of EPSPs and IPSPs as part of your answer. *34.3*

8. What is a myelin sheath? Do all neurons have one? *34.4*

9. How do divergent, convergent, and reverberating circuits among blocks of neurons differ from one another? *34.4*

10. Define reflex, then give an example of a reflex arc. *34.4*

Self-Quiz *(Answers in Appendix IV)*

1. Action potentials occur when _____ .
   a. a neuron receives adequate stimulation
   b. sodium gates open in an ever accelerating way
   c. sodium-potassium pumps kick into action
   d. both a and b

2. Compared with interstitial fluid near the plasma membrane of a neuron at rest, the cytoplasm near the membrane's inner surface carries a slight _____ charge.
   a. positive   c. graded and local
   b. negative   d. both b and c

3. The resting membrane potential is maintained by _____ .
   a. ion leaks   c. neurotransmitters
   b. ion pumps   d. both a and b

4. Neurotransmitters diffuse across a _____ .
   a. chemical synapse   c. myelin sheath
   b. channel protein   d. both a and b

5. An action potential lasts only briefly because _____ at the membrane region where it occurred.
   a. gates for sodium open, gates for potassium close
   b. gates for sodium close, gates for potassium open
   c. sodium-potassium pumps restore gradients
   d. both b and c

6. A nerve may consist of bundled-together axons of _____ .
   a. sensory neurons   c. sensory and motor neurons
   b. motor neurons   d. all are correct

7. Integration is the _____ .
   a. self-propagation of ion flow along a neuron
   b. trigger that releases neurotransmitters
   c. summation of signals acting at all synapses on a neuron
   d. collection of all information about several stimuli

8. Match the terms with their most suitable description.
   ____ synaptic integration   a. occurs at neuron's input zone
   ____ muscle spindle   b. spatial, temporal summation
   ____ graded, local   of all signals arriving at neuron
        potential   c. arises at trigger zone
   ____ action potential   d. stretch-sensitive receptor

Critical Thinking

1. With *multiple sclerosis*, recall, myelin sheaths around axons in the spinal cord slowly degenerate. Affected individuals progressively lose the ability to control movements of their skeletal muscles. Reflect on the role of myelin sheaths in the nervous system, then formulate a hypothesis of how their degeneration results in a loss of control over skeletal muscles.

2. *Epilepsy* is a neurological disorder characterized by short but recurring episodes of sensory and motor malfunctioning. The episodes, or epileptic seizures, might arise when reverberating circuits involving millions of interneurons in the brain become abnormally activated. Affected individuals typically experience involuntary muscle contractions and often sense lights, sounds, and odors even if receptors in the eyes, ears, and nose have not been stimulated. The drug valproic acid can eliminate or lessen the severity of the seizures by stimulating the body's synthesis of GABA. Why would stepping up GABA production help?

3. A mud-covered nail punctured Evita's foot. To protect Evita from tetanus, her doctor administered an antitoxin. This one contained enough protein molecules that specifically bind with and neutralize the *C. tetanus* toxin molecules. Will Evita now have long-lasting protection against future attacks by this bacterium?

Selected Key Terms

acetylcholine (ACh) *34.3*	neuron *CI*
action potential *34.1*	neurotransmitter *34.3*
axon *34.1*	positive feedback *34.2*
chemical synapse *34.3*	reflex *34.4*
dendrite *34.1*	resting membrane potential *34.1*
EPSP *34.3*	Schwann cell *34.4*
interneuron *CI*	sensory neuron *CI*
IPSP *34.3*	sodium-potassium pump *34.1*
motor neuron *CI*	stimulus *CI*
myelin sheath *34.4*	synaptic integration *34.3*
nerve *34.4*	threshold level *34.2*
neuromodulator *34.3*	

Readings

Dunant, Y., and M. Israel. April 1985. "The Release of Acetylcholine." *Scientific American* 252(4): 58–83.

Nicholls, J., A. Martin, and B. Wallace. 1992. *From Neuron to Brain.* Third edition. Sunderland, Massachusetts: Sinauer.

Web Site See *http://www.wadsworth.com/biology* for practice quiz questions, hypercontents, BioUpdates, and critical thinking. The Wadsworth Biology Resource Center provides a wealth of information fully organized and integrated by chapter.

INTEGRATION AND CONTROL: NERVOUS SYSTEMS

Why Crack the System?

Suppose your biology instructor asks you to volunteer for an experiment. You will get a microchip implanted in your brain. It will make you feel really good. But it may mess up your health, lop ten years off your life, and destroy a good part of your brain. Your behavior will change for the worse, so you might have trouble completing school, getting or keeping a job, or even having a normal family life.

The longer the chip is implanted, the less you will want to give it up. You won't get paid. You will pay the experimenter—first at bargain rates, then a little more each week. The chip is illegal. If you get caught using it, you and the experimenter will go to jail.

Sometimes Jim Kalat, a professor at North Carolina State University, proposes this experiment, which of course is hypothetical. Hardly any students volunteer.

Figure 35.1 Owners of an evolutionary treasure—a complex brain that is the foundation for our memory and reasoning, and our future.

Then he substitutes *drug* for microchip and *dealer* for experimenter, and an amazing number of students come forward! Like 30 million other Americans, the "volunteers" seem ready to engage in self-destructive uses of drugs that alter emotional and behavioral states.

The destruction shows up in unexpected places. Each year, for instance, about 300,000 newborns are already addicted to crack, thanks to their addicted mothers. *Crack* is a cheap form of cocaine. It causes relentless stimulation of brain regions that govern the sense of pleasure. It dampens normal urges to eat and to sleep, and blood pressure rises. Elation and sexual desire intensify. In time, however, the brain cells that produce the stimulatory chemicals cannot keep up with the abnormal demand. The chemical vacuum makes crack users frantic and then profoundly depressed. Only crack makes them feel good again.

Addicted babies cannot know all of this. They can only quiver with "the shakes" and respond to the world

with chronic irritation. And they are abnormally small. While they were developing inside their mother, their body tissues simply were not provided with enough oxygen and nutrients. As one of its side effects, crack causes blood vessels to constrict, and maternal blood vessels are the only supply lines that the developing individual has.

Paradoxically, crack babies are exceptionally fussy, yet they cannot respond to rocking and other forms of stimulation that normally have soothing effects. It may be a year or more before they recognize even their own mother. Without treatment, they are likely to grow up as emotionally unstable children, prone to aggressive outbursts and stony silences. Why? Their mother's drug habit crippled their nervous system.

Think about it. The nervous system evolved as a way to sense and respond, with exquisite precision, to changing conditions inside and outside the body. Awareness of sounds and sights, of odors, of hunger and passion, fear and rage—all such things begin with the flow of information along the communication lines of the nervous system. Do those lines remain silent until they receive outside signals, much as telephone lines wait to carry calls from all over the country? Absolutely not. Even before you were born, excitable cells called neurons became organized into extensive gridworks in your newly forming tissues and started chattering among themselves. All through your life, in moments of danger or reflection, supreme excitement or sleep, their chattering has never ceased and will not cease until the time you die.

Each of us possesses a body of great complexity. Its architecture, its functioning are legacies of millions of years of evolution. Its nervous system is unparalleled in the living world. One of its most astonishing products is language, the encoding of shared experiences of groups of individuals in time and space. Through the evolution of our nervous system, the sense of history was born, and the sense of destiny. Through this system we can ask how we have come to be what we are, and where we are headed from here. Perhaps the sorriest consequence of drug abuse is its implicit denial of this legacy—the denial of self when we cease to ask, and cease to care.

This chapter can give you insight into the structure and function of the nervous system. Use these insights to think about what can happen when its operation falters, whether by disease or mutation, or by deliberate and abusive use of drugs.

KEY CONCEPTS

1. Nervous systems are composed of neurons and a variety of cells called neuroglia, which structurally and functionally support the neurons. The neurons interact in the detection and integration of information about external and internal conditions, and in commanding muscles and glands to carry out suitable responses.

2. The simplest nervous systems of all are the nerve nets of radial animals, such as sea anemones. The nervous systems of most animals show pronounced cephalization and bilateral symmetry.

3. Vertebrate nervous systems are functionally divided into central and peripheral regions. The brain and spinal cord make up the central nervous system. Paired nerves that thread through the rest of the body are the key components of the peripheral nervous system.

4. The somatic nerves of the peripheral nervous system deal with skeletal muscles. The autonomic nerves deal with the heart, lungs, and other internal organs.

5. The vertebrate brain is functionally divided into three regions, called the hindbrain, midbrain, and forebrain. Its most ancient parts deal with reflex control over breathing, blood circulation, and other basic functions that are essential for staying alive.

6. Long ago, in certain vertebrate lineages, the brain expanded in volume and complexity. Its newer regions appropriated more and more control over the ancient reflex functions.

7. In birds and mammals especially, the portion of the forebrain called the cerebrum contains the most complex centers for receiving, integrating, storing, and responding to sensory information.

8. The hypothalamus, another part of the forebrain, is the main homeostatic control center over the internal environment and the functioning of internal organs. Being part of the limbic system, the hypothalamus also influences emotional states and memory.

Life-styles and Nervous Systems

Even if you stare at it for hours on end, a sea urchin will never dazzle you with spectacular bursts of speed or precision acrobatics. Because it seems only a bit livelier than the pincushion it resembles, you might suspect this animal is brainless, and in fact it is. However, the sea urchin does have a nervous system, even if it is a decentralized one.

At this point in the book, you know that all animals except sponges have some type of a **nervous system** in which nerve cells, such as neurons, are oriented relative to one another in signal-conducting and information-processing pathways. At the minimum, communication lines of its component cells receive information about changing conditions outside and inside the body, then elicit suitable responses from muscle and gland cells.

To appreciate any animal's nervous system, you have to ask, *What does the animal do?* Many sea urchins spend most of their lives moving about slowly on little tube feet, scraping bits of algae off rocks with a toothy feeding apparatus (Section 26.20 and Figure 35.2). Their nervous system includes three sets of nerves. A **nerve**, recall, is like a cable in which sensory axons, motor

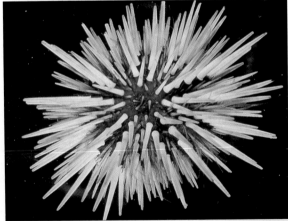

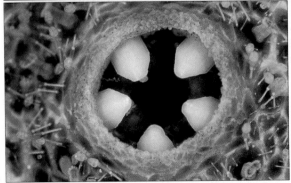

Figure 35.2 A spiny sea urchin, (**a**) right-side up and (**b**) flipped over to show its toothy, all-important mouth.

axons, or both are bundled together inside a sheath of connective tissue (Section 34.4). The ones in sea urchins deal mainly with sampling the environment, moving through it, and (you guessed it) getting food into the gut. They interconnect with one another by a **nerve net**, a loose mesh of nerve cells intimately associated with epithelial tissue. Information flow through the nerve net is not highly directional. Signals travel diffusely, in all directions, from a point of stimulation. You might not find this type of nervous system impressive, but if you did little more than slowly inch about eating algae, it would be quite sufficient.

Regarding the Nerve Net

Animals first evolved in the seas, and it is in the seas that we still find the animals with the simplest nervous systems. They are sea anemones, jellyfishes, and other cnidarians (Sections 26.4 and 26.5). These invertebrates show *radial* symmetry. Their body parts are arranged about a central axis, much like spokes of a bike wheel.

Like sea urchins, cnidarians have a nerve net (but no nerves). It extends throughout the body and is equally responsive to food or potential danger coming from any direction. Its nerve cells interact with sensory cells and contractile cells of the epithelium in **reflex pathways**. In such pathways, remember, sensory stimulation directly triggers simple, stereotyped movements (Section 34.4).

For example, in jellyfishes, reflex pathways permit slow swimming movements and keep the body right-side up. In all cnidarians, a reflex pathway concerned with feeding behavior extends from sensory receptors present in the tentacles, along nerve cells, to contractile cells around the mouth. Figure 35.3 illustrates the cells involved in such a pathway.

On the Importance of Having a Head

Flatworms are the simplest animals having a bilateral nervous system (Figure 35.4). *Bilateral* symmetry means having equivalent body parts on the left and right sides of the body's midsagittal plane. (Imagine sliding down a staircase banister that turns into a razor and you may never forget the location of the midsagittal plane.) Both sides have the same array of muscles that function in moving the body forward. Both have the same array of nerves to control the muscles, and so on.

The ladderlike nervous system of the flatworm has two cordlike nerves running longitudinally through the body. Besides the two nerve cords, some systems have **ganglia** (singular, ganglion), which are clusters of nerve cell bodies that serve as integrating centers. At the head end of this animal, ganglia form a two-part brainlike structure that coordinates signals from paired sensory

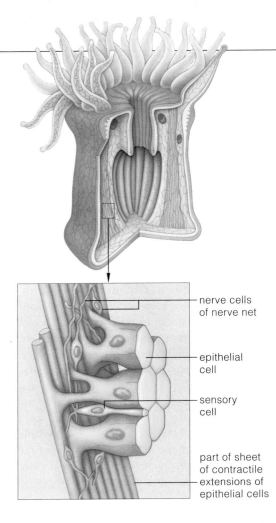

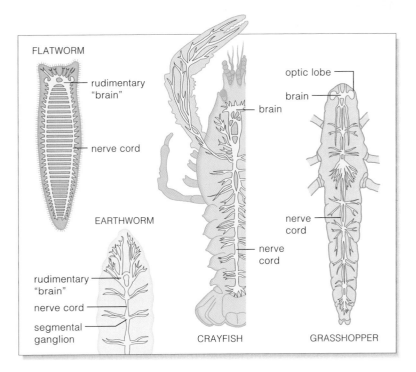

Figure 35.4 Bilateral nervous systems of a few invertebrates. The sketches are not to scale relative to one another.

Figure 35.3 Nerve net of a sea anemone, one of the cnidarians. The nerve cells interact with sensory cells and with contractile cells of a sheetlike epithelium between the outer epidermis and a jellylike middle layer (mesoglea) of the body wall.

organs, including two eyespots, and provides a degree of control over the nerve cords.

Did bilateral nervous systems evolve from nerve nets? Maybe. Consider that some cnidarian life cycles include a self-feeding larval stage, called a **planula**. Like flatworms, a planula has a flattened body and uses cilia to swim or crawl about:

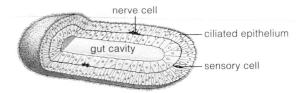

Imagine a few ancient planulas crawling about on the seafloor in Cambrian times. By chance, mutations of regulatory genes blocked their transformation into adults but had no effect on maturation of reproductive organs. (As you read in Section 27.6, this does happen in the larvae of certain salamanders and other animals.)

Such planulas kept on crawling, they reproduced, and so they passed on the mutated genes responsible for the capacity for forward mobility. The planula-like animals must have crawled into different, potentially dangerous or splendid situations. There, they would not have been helped much by sensory cells located at the trailing end of their body. But if such cells became concentrated at the leading end, rapid, effective responses to stimuli would have been possible. Probably natural selection did favor animals with concentrations of sensory cells at the leading end. Cephalization (formation of a head) and bilateral symmetry may have started this way. Both features do occur in most invertebrate lineages (Figure 35.4). And as you will see in the next section, *patterns of bilateral symmetry and cephalization also are evident in paired nerves and muscles, paired sensory structures, and paired brain centers of yourself and all other vertebrates.*

In nervous systems, nerve cells receive and process sensory input, then elicit responses from muscle and gland cells.

Radial animals have a nerve net, a diffuse mesh of nerve cells that take part in simple reflex pathways involving sensory cells and contractile cells of an epithelial tissue.

Nervous systems of cephalized, bilateral animals include a brain (or ganglia at the head end) as well as paired nerves and paired sensory structures.

Nerves are cablelike communication lines of sensory axons, motor axons, or both bundled in a connective tissue sheath.

Evolutionary High Points

Hundreds of millions of years ago, the earliest fishlike vertebrates were evolving (Section 27.3). A column of bony segments was taking over the function of the notochord, a long rod of stiffened tissue that worked with muscles to bring about movements. Above the notochord, a hollow, tubular nerve cord was evolving also. It was the forerunner of the spinal cord and brain. Remember, these developments were a foundation for new life-styles. Not long afterward, genetic divergences from lineages of filter-feeding, scavenging fishes gave rise to swift, jawed predators of the seas.

At first, simple reflex pathways predominated. The sensory neurons that were tripped into action synapsed on motor neurons, which directly stimulated muscles to contract. No other neurons altered the flow of signals from reception of a stimulus to the response. However, in the changing world of the fast-moving vertebrates, predators and prey that were better equipped to sense one another's presence had the competitive edge.

The senses of smell, hearing, and balance became even keener among vertebrates that invaded land. Bones and muscles evolved in ways that enhanced specialized motor activities. The brain became variably thickened with nervous tissue that could integrate rich sensory information and issue orders for complex responses.

The oldest parts of the vertebrate brain still deal with reflex coordination of breathing and other vital functions. Now, however, many interneurons synapse on sensory and motor neurons of ancient pathways and with one another in the newer brain regions. In the most complex vertebrates, interneurons receive, store, and compare information about experiences. They weigh possible responses. And they give our own species the capacity to reason, remember, and learn.

The nerve cord persists in all vertebrate embryos. We call it the **neural tube**. As the embryo grows and develops, the neural tube undergoes expansion and regional modifications into the brain and spinal cord (Figure 35.5a). The spinal cord becomes enclosed within the vertebral column. Adjacent tissues in the embryo give rise to nerves that thread through all body regions and connect with the spinal cord and brain.

FOREBRAIN. Receives, integrates sensory information from nose, eyes, and ears; in land-dwelling vertebrates, contains the highest integrating centers

MIDBRAIN. Coordinates reflex responses to sight, sounds

HINDBRAIN. Reflex control of respiration, blood circulation, other basic tasks; in complex vertebrates, coordination of sensory input, motor dexterity, and possibly mental dexterity

(start of spinal cord)

a Diagram of how the anterior end of the dorsal, hollow nerve cord of vertebrates expanded into functionally distinct regions, which also increased in complexity in certain lineages

Functional Divisions of the Vertebrate Nervous System

Figure 35.6 gives you a general sense of the expansions of nervous tissue in the human nervous system. This diagram shows the major paired nerves of this bilateral system. What it cannot possibly show are the 100 billion interneurons in the brain alone. To be sure, humans do have the most intricately wired nervous system in the animal world. Even so, you find similar patterns among other vertebrates.

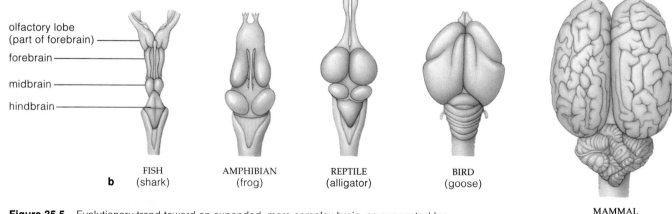

olfactory lobe
(part of forebrain)

forebrain

midbrain

hindbrain

b

FISH
(shark)

AMPHIBIAN
(frog)

REPTILE
(alligator)

BIRD
(goose)

MAMMAL
(horse)

Figure 35.5 Evolutionary trend toward an expanded, more complex brain, as suggested by comparing the brains of some existing vertebrates. These dorsal views are not to the same scale.

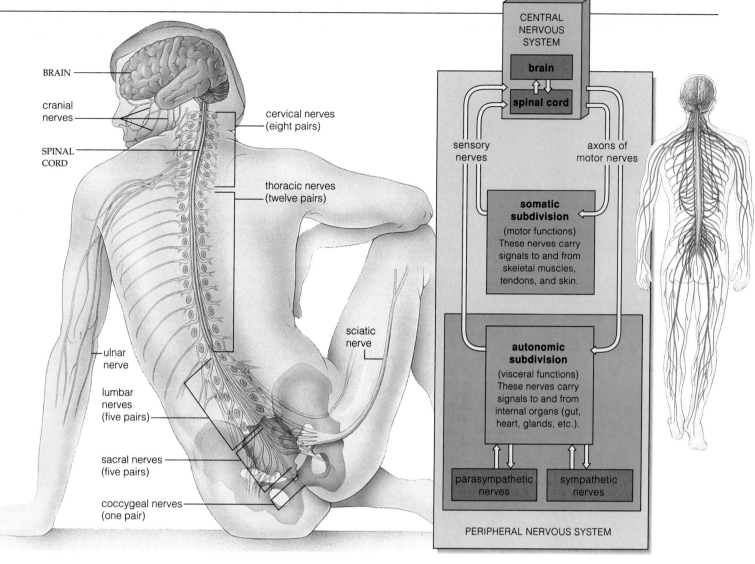

Figure 35.6 Sketch of the human nervous system, showing the brain, spinal cord, and some of the major peripheral nerves. This nervous system also includes twelve pairs of cranial nerves that originate from the brain. Other vertebrates have a similar system.

Figure 35.7 Functional divisions of the nervous system. The central nervous system is color-coded *blue*, the somatic nerves *green*, and the autonomic nerves *red*. Sometimes the nerves carrying sensory input to the central nervous system are said to be *afferent* (a word meaning "to bring to"). The ones carrying motor output away from the central nervous system to muscles and glands are *efferent* ("to carry outward").

Investigators typically approach the complexity of the vertebrate nervous system by functionally dividing it into central and peripheral regions (Figure 35.7). All of the interneurons are confined to the **central nervous system**, or the spinal cord and brain. The **peripheral nervous system** consists mainly of nerves that thread through the rest of the body and carry signals into and out of the central nervous system.

Inside the brain and spinal cord, the communication lines are called **tracts**, not nerves. The tracts making up **white matter** contain axons with glistening white myelin sheaths and specialize in rapid signal transmission. By contrast, **gray matter** consists of unmyelinated axons, dendrites, and nerve cell bodies and neuroglial cells. **Neuroglial cells**, remember, protect or structurally and functionally support neurons. They make up more than half the volume of vertebrate nervous systems.

The coevolution of nervous, sensory, and motor systems made possible more complex life-styles among vertebrates.

The vertebrate nervous system has become so intricately wired that the structure and functions of its central and peripheral regions are described separately.

The central nervous system consists of the brain and spinal cord. The peripheral nervous system consists of nerves that thread through the rest of the body and carry signals into and out of the central region.

Myelinated axons of tracts make up the white matter of the brain and spinal cord. Neuroglia as well as unmyelinated axons, dendrites, and cell bodies of neurons make up their gray matter.

Let's now take a look at the peripheral nervous system and the spinal cord, which interconnect as the major expressways for information flow through the body.

Peripheral Nervous System

SOMATIC AND AUTONOMIC SUBDIVISIONS   In humans, the peripheral nervous system includes thirty-one pairs of *spinal* nerves, which connect with the spinal cord. The system also includes twelve pairs of *cranial* nerves, which connect directly with the brain.

Cranial and spinal nerves can be further classified according to function. The ones that carry signals about moving your head, trunk, and limbs are the **somatic nerves**. The sensory axons inside these nerves deliver information from receptors in the skin, skeletal muscles, and tendons to the central nervous system. Their motor axons deliver commands from the brain and spinal cord to the body's skeletal muscles.

By contrast, spinal and cranial nerves dealing with smooth muscle, cardiac (heart) muscle, and glands are the **autonomic nerves**. They deal with the visceral parts of the body—in other words, with its internal organs and structures. Autonomic nerves carry signals to and from these organs and structures.

SYMPATHETIC AND PARASYMPATHETIC NERVES   Figure 35.8 shows the two categories of autonomic nerves. We call them parasympathetic and sympathetic. Normally they work antagonistically, with the signals from one opposing signals from the other. Both carry excitatory and inhibitory signals to internal organs. Often their signals arrive at the same time at muscle or gland cells and compete for control over them. In such instances, synaptic integration at the cellular level leads to minor adjustments in the organ's level of activity.

Parasympathetic nerves dominate when the body is not receiving much outside stimulation. They tend to

SYMPATHETIC OUTFLOW:

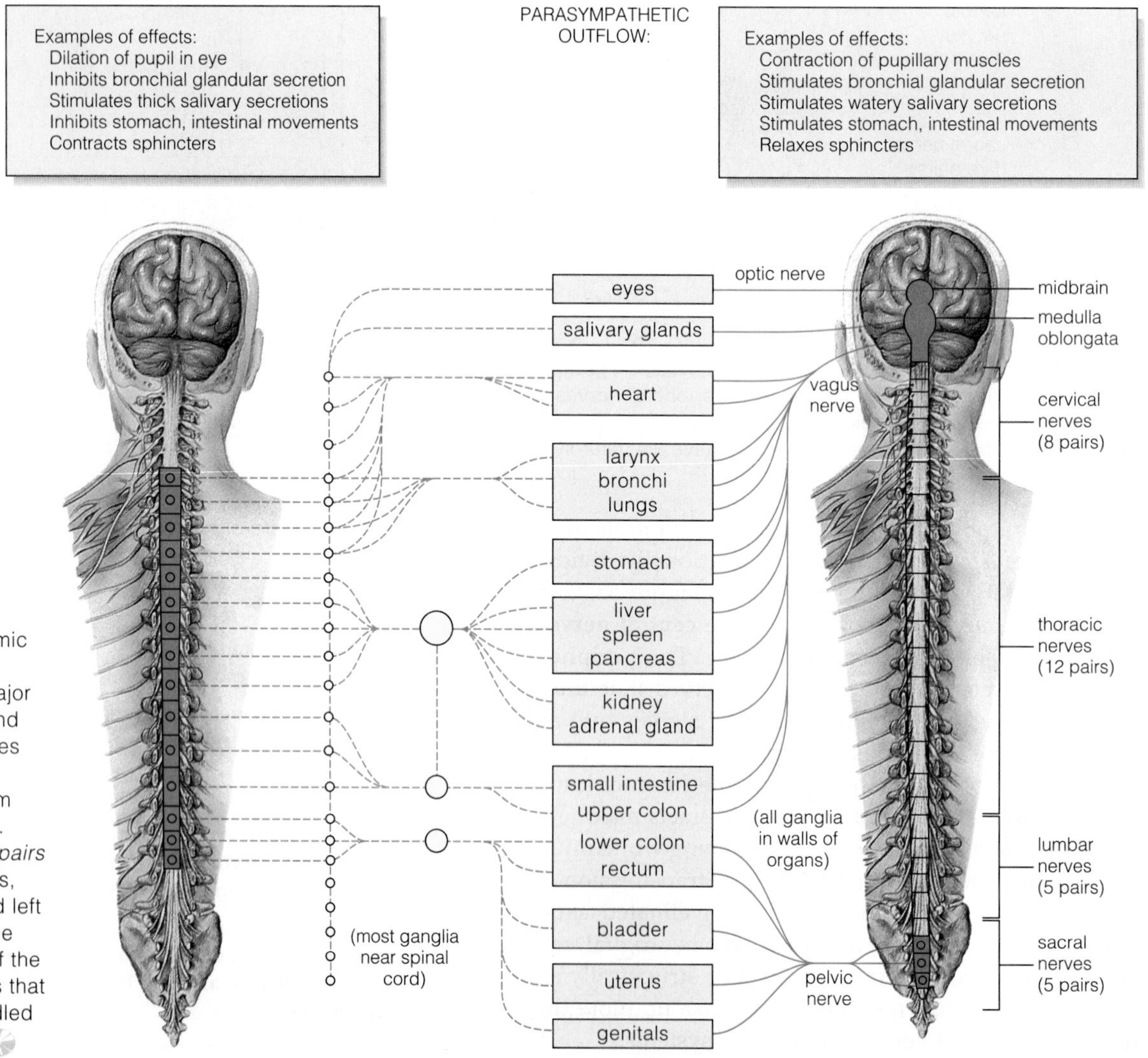

Examples of effects:
Dilation of pupil in eye
Inhibits bronchial glandular secretion
Stimulates thick salivary secretions
Inhibits stomach, intestinal movements
Contracts sphincters

PARASYMPATHETIC OUTFLOW:

Examples of effects:
Contraction of pupillary muscles
Stimulates bronchial glandular secretion
Stimulates watery salivary secretions
Stimulates stomach, intestinal movements
Relaxes sphincters

Figure 35.8   Autonomic nervous system. This diagram shows the major sympathetic nerves and parasympathetic nerves leading out from the central nervous system to some major organs. Remember, there are *pairs* of both kinds of nerves, servicing the right and left halves of the body. The ganglia are clusters of the cell bodies of neurons that have their axons bundled together in nerves.

eyes — optic nerve — midbrain
salivary glands — medulla oblongata
heart — vagus nerve — cervical nerves (8 pairs)
larynx bronchi lungs
stomach
liver spleen pancreas — thoracic nerves (12 pairs)
kidney adrenal gland
small intestine upper colon — (all ganglia in walls of organs)
lower colon rectum — lumbar nerves (5 pairs)
bladder
uterus — pelvic nerve — sacral nerves (5 pairs)
genitals

(most ganglia near spinal cord)

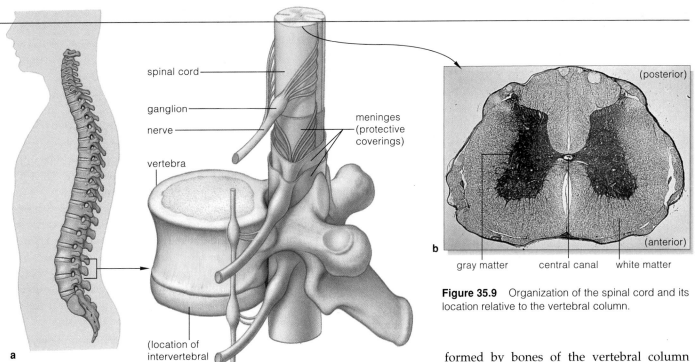

spinal cord

ganglion

nerve

vertebra

meninges (protective coverings)

(posterior)

(anterior)

b

gray matter   central canal   white matter

a

(location of intervertebral disk)

Figure 35.9 Organization of the spinal cord and its location relative to the vertebral column.

slow down the body overall and divert energy to basic "housekeeping" tasks, such as digestion.

Sympathetic nerves dominate in times of sharpened awareness, excitement, or danger. They tend to shelve housekeeping tasks. At the same time, they prepare the animal to fight or escape (when threatened) or to frolic (as in play or sexual behavior). Right now, sympathetic nerves are commanding your heart to beat a bit faster, and parasympathetic nerves are commanding it to beat a bit slower. Integration of the opposing signals adjusts the heart rate. If something scares or excites you, the parasympathetic input to the heart drops. Sympathetic signals cause the release of epinephrine, which makes your heart beat faster. It also makes you breathe faster and start to sweat. In this state of intense arousal, you are primed to fight (or play) hard or to get away fast. Hence the name **fight-flight response**.

Suppose the stimulus for the fight-flight response ends. Sympathetic activity may decrease abruptly, and parasympathetic activity may rise suddenly. You might observe this "rebound effect" after someone has been instantly mobilized to rush onto a highway to save a child from an oncoming car. The person may well faint as soon as the child has been swept out of danger.

The Spinal Cord

By definition, the **spinal cord** is a vital expressway for signals between the peripheral nervous system and the brain. Also, sensory and motor neurons make direct reflex connections in the spinal cord. (The stretch reflex, described in Section 34.4, arcs through the spinal cord this way.) The spinal cord threads through a canal

formed by bones of the vertebral column (Figure 35.9a). The bones and the ligaments attached to them protect the cord. So do the **meninges**, three tough, tubelike coverings around the spinal cord and brain. The coverings are not impervious to all attacks. *Meningitis*, an often-fatal disease, results from viral or bacterial infection. Disease symptoms include severe headaches, fever, a stiff neck, and nausea.

Signals travel rapidly up and down the spinal cord in bundles of axons with glistening myelin sheaths. The sheaths distinguish the cord's white matter from its gray matter (Figure 35.9b). Gray matter, recall, consists of dendrites and cell bodies of neurons, and neuroglial cells. It plays an important role in controlling reflexes for limb movements (as when you walk or wave your arms) and organ activity (such as bladder emptying).

Experiments with frogs provide evidence for these pathways. Between the frog's spinal cord and brain are neural circuits that deal with straightening legs after they have been bent. If these circuits are severed at the base of the brain, the legs become paralyzed—but only for about a minute. So-called extensor reflex pathways in the spinal cord recover quickly and have the frog hopping about in no time. Recovery is minimal after similar damage in humans and other primates—the vertebrates with the greatest cephalization.

The nerves of the peripheral nervous system connect the brain and spinal cord with the rest of the body.

Its somatic nerves deal with skeletal muscle movements. Its autonomic nerves deal with the functions of internal organs, such as the heart and glands.

The spinal cord is a vital expressway for signals between the brain and peripheral nerves. Some of its interneurons also exert direct control over certain reflex pathways.

The anterior end of the spinal cord merges with the **brain**, the body's master control center. The brain receives, integrates, stores, and retrieves information. It also coordinates responses to information by adjusting activities throughout the body. Like the spinal cord, the brain is protected by bones and membranes.

Reflect for a moment on Figures 35.5 and 35.10. The forebrain, midbrain, and hindbrain of vertebrates form from three successive portions of the neural tube. The nervous tissue that evolved first in all three regions is called the **brain stem**. The brain stem is still identifiable in an adult brain, and it still contains many simple, basic reflex centers. Over evolutionary time, expanded layers of gray matter developed from the brain stem. The more recent additions have been correlated with an increasing reliance on three major sensory organs: the nose, ears, and eyes. The forebrain and possibly parts of the cerebellum, which is part of the hindbrain region, contain the newest additions of gray matter. Figure 35.10 provides an overview of their main functions.

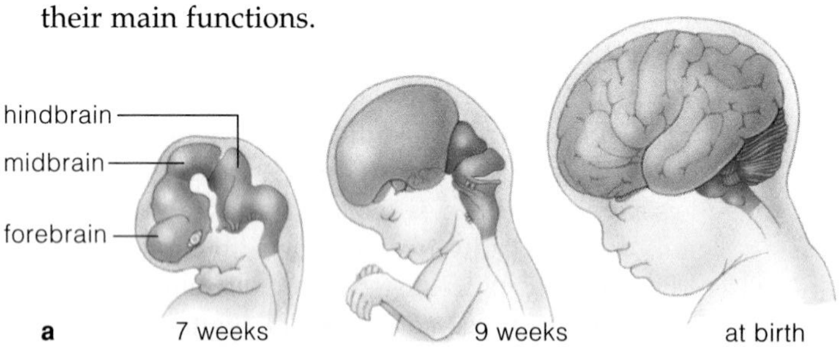

Hindbrain

The medulla oblongata, cerebellum, and pons are all components of the hindbrain. The **medulla oblongata** has reflex centers for vital tasks, such as respiration and blood circulation. It coordinates motor responses with complex reflexes, such as coughing. It influences other brain regions that help you sleep or wake up.

The **cerebellum** integrates sensory input from the eyes, ears, and muscle spindles with motor commands from the forebrain. It helps control motor dexterity. More recent expansions of the human cerebellum may be crucial in language and some other forms of mental dexterity.

Bands of many axons extend from both sides of the cerebellum to the **pons** (which means bridge). Although lodged in the brain stem, the pons is a major traffic center for information passing between the cerebellum and the higher integrating centers of the forebrain.

Midbrain

The midbrain coordinates reflex responses to sights and sounds. The **tectum** is a more recent layering of gray matter; it forms the roof of the midbrain. The fish and amphibian tectum is a center for coordinating nearly all sensory input and initiating motor responses. A frog surgically deprived of its highest brain center but with its tectum intact still does almost everything frogs do.

hindbrain
midbrain
forebrain

a   7 weeks   9 weeks   at birth

Division	Main Parts	Typical Functions
FOREBRAIN	Cerebrum	Localizes and processes sensory inputs; initiates and controls skeletal muscle activity; in the most complex vertebrates, roles in memory, mediating emotions, and abstract thought
	Olfactory lobes	Relaying of sensory input from the nose to olfactory centers of the cerebrum
	Thalamus	Relay stations for conducting sensory signals to and from cerebral cortex; role in memory
	Hypothalamus	Together with pituitary gland, homeostatic control center over the volume, composition, and temperature of the internal environment, influences behavior relating to organ functions (such as hunger and thirst) and physical expressions of emotions (such as sweating)
	Limbic system	A complex of structures that govern emotions and that have roles in memory
	Pituitary gland	With hypothalmus, endocrine control of metabolism, growth, and development
	Pineal gland	Control of some circadian rhythms; role in mammalian reproductive physiology
MIDBRAIN	Tectum	In fishes and amphibians, key coordinating center for sensory input and motor responses; in mammals, mainly reflex centers that rapidly relay sensory input to forebrain
HINDBRAIN	Pons	"Bridge" of tracts between cerebrum and cerebellum; its other tracts connect forebrain and spinal cord; works with medulla oblongata to control rate and depth of respiration
	Cerebellum	Coordinates motor activity for limb movements, maintaining posture, and spatial orientation
	Medulla oblongata	Contains tracts between pons and spinal cord; reflex centers involved in control of heart rate, blood vessel diameter, respiratory rate, vomiting, coughing, other vital functions

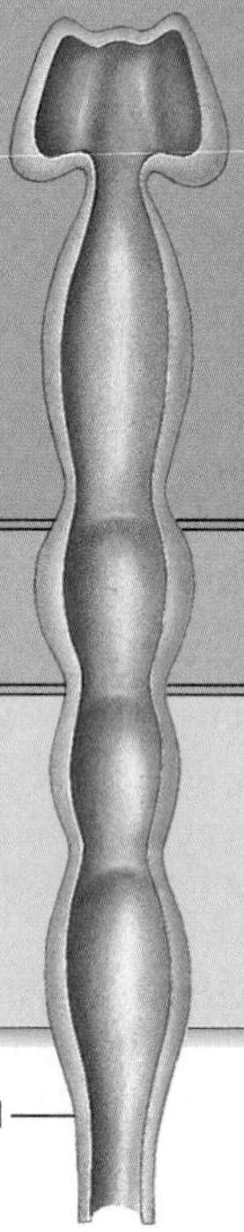

b

anterior end of spinal cord

Figure 35.10   (**a**) How the anterior end of a human embryo's neural tube develops into a brain. (**b**) Overview of the brain's subdivisions, correlated with the neural tube at an early developmental stage.

In most vertebrates (but not mammals), the midbrain includes a pair of brain centers called optic lobes, which deal with sensory input from eyes. Mammalian eyes deserted the tectum, so to speak; they formed important functional connections with integrating centers in the forebrain. The mammalian tectum ended up as a reflex center that swiftly relays sensory signals to those higher integrating centers.

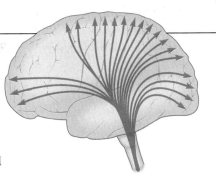

Figure 35.11 Communication pathways of the reticular formation, an evolutionarily ancient, diffuse network of neurons that extends from the anterior end of the spinal cord on up to the highest integrative centers of the cerebral cortex.

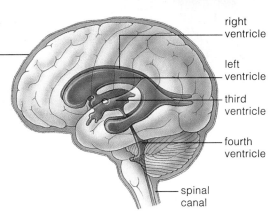

right ventricle
left ventricle
third ventricle
fourth ventricle
spinal canal

Figure 35.12 Cerebrospinal fluid (*blue*) in the human brain. This extracellular fluid surrounds and cushions the spinal cord and the brain. It also fills four interconnected cavities (cerebral ventricles) within the brain and the spinal cord's central canal.

Forebrain

For much of vertebrate history, chemical odors diffusing in aquatic environments were the major clues to survival. An olfactory lobe dealing mainly with odors from predators, prey, and potential mates was the main forebrain structure (Figure 35.5), along with paired outgrowths of the brain stem where olfactory input and responses to it were integrated. In time the outgrowths expanded greatly, especially during and after certain vertebrates invaded the land. We now call them the two hemispheres of the **cerebrum**. Another forebrain region, called the **thalamus**, evolved as a coordinating center for sensory input and as a relay station for signals to the cerebrum. Below this, the **hypothalamus** evolved as the main center for homeostatic control over the internal environment. It became central to behaviors related to internal organ activities, such as thirst, hunger, and sex, and to emotional expression, such as sweating with fear.

The Reticular Formation

By now, you may be thinking that the brain is tidily subdivided into three main regions. This is not the case. An evolutionarily ancient mesh of interneurons still extends from the uppermost part of the spinal cord, on through the brain stem, and on into higher integrative centers of the cerebral cortex (Figure 35.11). This major network of interneurons is the **reticular formation**. It persists as a low-level pathway to motor centers of the medulla oblongata and spinal cord. It also can activate centers in the cerebral cortex and thereby help govern the activities of the nervous system as a whole.

Brain Cavities and Canals

The hollow neural tube that first develops in vertebrate embryos persists in the adults, as a continuous system of fluid-filled cavities and canals. Within this system is the **cerebrospinal fluid**, a clear extracellular fluid that cushions the tissues of the brain and spinal cord from sudden, jarring movements (Figure 35.12).

A mechanism called the **blood-brain barrier** exerts some control over which solutes enter the cerebrospinal fluid and thereby helps to protect the brain and spinal cord. In no other portion of the extracellular fluid are solute concentrations maintained within such narrow limits. If the brain were exposed to the compositional shifts that typically accompany, say, digesting a meal or exercising, it would suffer. Why? Some of the hormones and amino acids in blood are signaling molecules that affect neurons. Also, shifting levels of certain ions (such as K^+) can skew the threshold for action potentials.

The barrier operates at the plasma membrane of cells making up the walls of capillaries that service the brain. In most brain regions, continuous tight junctions fuse together the abutting walls of these cells, so all water-soluble substances must move *through* the cells to reach the brain. Membrane transport proteins allow essential nutrients (such as glucose) and some ions to move into and out of the cells but bar urea and other metabolic wastes, certain toxins, and most drugs. The barrier does not bar fat-soluble substances, including oxygen, carbon dioxide, anesthetics, alcohol, caffeine, and nicotine. The barrier is nonexistent around the hypothalamus and the brain stem's vomiting center. (Can you guess why?)

The vertebrate brain develops from a hollow neural tube, which persists in adults as a system of cavities and canals filled with cerebrospinal fluid. The fluid cushions nervous tissue from sudden, jarring movements.

The vertebrate brain is functionally subdivided into three regions called the hindbrain, forebrain, and midbrain.

In all three regions, the nervous tissue that was the first to evolve is still recognizable and is called the brain stem. The brain stem affords reflex control over the most basic functions necessary for survival.

In the most complex vertebrates, the highest integrative centers reside in the forebrain, especially in its cerebral hemispheres.

A CLOSER LOOK AT THE HUMAN CEREBRUM

Organization of the Cerebral Hemispheres

Study any nervous system or even the best computers and you realize the human brain is the most complex integration center of all. It only weighs 1,300 grams (3 pounds) in a person of average size, and more than half of that mass consists of neuroglial cells. But remember, the human brain also has at least *100 billion* interneurons!

The human cerebrum, housed in a chamber of hard skull bones, resembles the much-folded nut in a walnut shell (Figure 35.13a). Interneurons here contribute most to your humanness. A longitudinal fissure divides the cerebrum into left and right **cerebral hemispheres**. Most of the brain centers dealing with analytical skills,

temporal, parietal, and frontal lobes. Figure 35.14 gives a few examples. (*EEG*, short for electroencephalogram, is simply a recording of summed electrical activity in a given brain region. Section 2.2 describes PET scans.)

The *occipital* lobe at the back of each hemisphere has centers for vision. Severe blows to it may destroy part of the visual field, the outside world that the individual sees. The *temporal* lobe near each temple has centers that deal with hearing and visual associations. These centers have become highly developed among humans and other primates. A blow here won't leave you blind, but it might impair your ability to recognize complex visual patterns, such as faces. Centers in the temporal lobe also influence emotional behavior.

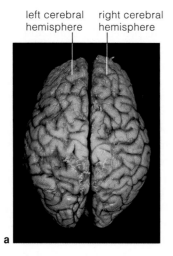

left cerebral hemisphere | right cerebral hemisphere

a

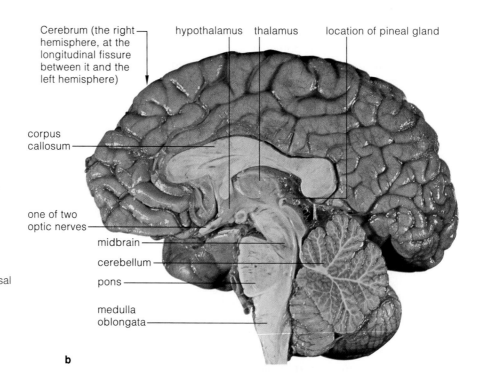

Cerebrum (the right hemisphere, at the longitudinal fissure between it and the left hemisphere)

hypothalamus | thalamus | location of pineal gland

corpus callosum

one of two optic nerves

midbrain

cerebellum

pons

medulla oblongata

b

Figure 35.13 (**a**) Human brain. This dorsal view shows how a longitudinal fissure separates the two cerebral hemispheres. (**b**) Sagittal view of the left cerebral hemisphere. Not visible is the reticular formation, which extends between the upper spinal cord and the cerebrum.

speech, and mathematics reside in the left hemisphere. The centers dealing with spatial relationships, music, and other nonverbal (abstract) skills reside mainly in the right hemisphere. Each half receives, processes, and coordinates responses to sensory input mostly from the opposite side of the body. For example, signals about pressure on the right arm reach the left hemisphere. A transverse band of nerve tracts, the corpus callosum, carries signals in both directions and coordinates the functioning of both hemispheres.

Above the brain's axons is the **cerebral cortex**, a thin layer of gray matter (the cell bodies of interneurons). Different parts receive and process different signals. EEGs as well as PET scans have localized the activities in each hemisphere to four tissue lobes: the occipital,

In the *parietal* lobe is the somatosensory cortex, the main receiving area for sensory input from the skin and joints. In the *frontal* lobe, the motor cortex deals with signals for motor responses. Thumb, finger, and tongue muscles get much of the brain's attention. This gives you an idea of how much control is required for hand movements and verbal expression (Figure 35.15).

The prefrontal cortex, a region in front of the motor cortex, is crucial in planning movements and inhibiting unsuitable behavior. As you will read later, it also has roles in some aspects of memory. PET scans suggest that this cortical region and part of the cerebellum (especially a structure called the dentate nucleus) probably interact to govern the motor abilities underlying language and some thought processes.

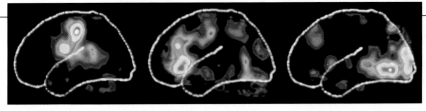

Motor cortex activity when speaking   Prefrontal cortex activity when generating words   Visual cortex activity when observing words

Figure 35.14 *Right:* Primary receiving and integrating centers for the human cerebral cortex. Primary cortical areas receive signals from receptors on the body's periphery. Association areas coordinate and process sensory input from different receptors. The three PET scans above identified which brain regions were active when a person performed three specific tasks: speaking, generating words, and observing words.

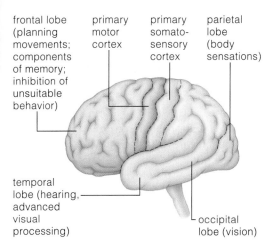

frontal lobe (planning movements; components of memory; inhibition of unsuitable behavior)

primary motor cortex

primary somato-sensory cortex

parietal lobe (body sensations)

temporal lobe (hearing, advanced visual processing)

occipital lobe (vision)

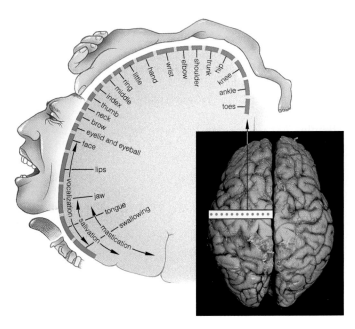

Figure 35.15 Body parts under control of the primary motor cortex. This diagram is a slice through the motor cortex of the left cerebral hemisphere. It controls muscles on the body's left side. The distortions to the human body that is draped over the diagram of the primary motor cortex indicate which body parts receive the most precise control.

Figure 35.16 The location of the limbic system, which deals with memories as well as emotional reactions to stimuli. It consists of a complex group of nerve tracts and gray matter at the middle of the cerebral hemispheres. See also Section 35.7.

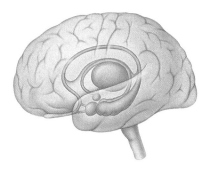

when your brain recalls the cologne of a special person who wore it. Connections from the cerebral cortex and other brain centers pass through this system, which includes the amygdala, hippocampus, hypothalamus, and parts of the thalamus. The connections enable us to correlate organ activities with self-gratifying behavior, such as eating and sex. They can make you feel that (say) your heart and stomach are on fire—with passion or with indigestion. Neural signals based on reasoning in the cerebral cortex often can override or dampen rage, hatred, and other "gut reactions."

The cerebrum of humans and other complex vertebrates is divided into hemispheres. Two-way signals along a nerve tract, the corpus callosum, afford communication between both hemispheres.

The cerebral cortex, the thin, outermost layer of gray matter of the hemispheres, contains the brain's primary receiving and integrating centers for sensory input.

The left hemisphere affords most of the control over speech, mathematics, and analytical skills. The right hemisphere affords most control over spatial abilities, music, and other abstract, nonverbal skills.

The cerebral cortex makes functional connections with the limbic system, which deals with emotions and memory.

Many ingenious experiments have provided us with information on how the cerebral hemispheres function. Section 35.6 describes a classic example. For a lyrical example (no pun intended), consider reading "Music of the Hemispheres," an article in the October 1996 issue of *Discover* magazine that explores the cerebral circuits dealing with melodies and language.

Connections With the Limbic System

The **limbic system**, which is located at the middle of the cerebral hemispheres, governs our emotions and has roles in memory (Figure 35.16). It is distantly related to olfactory lobes and still deals with the sense of smell. That's one reason why you may feel warm and fuzzy

35.6

SPERRY'S SPLIT-BRAIN EXPERIMENTS

Some time ago a neural surgeon, Roger Sperry, and his colleagues demonstrated some intriguing differences in perception between the two halves of the cerebrum of epileptics. Severe *epilepsy* is characterized by seizures, sometimes as often as every half hour. The seizures are analogous to an electrical storm in the brain. Sperry asked: Would *cutting* the corpus callosum of epileptics confine the electrical storm to one hemisphere, leaving at least the other hemisphere to function normally? Earlier studies of laboratory animals and humans whose corpus callosum had been damaged suggested this might be so.

Sperry performed the surgery on some patients. The electrical storms did subside in frequency and intensity. Cutting the neural bridge ended what must have been a positive feedback loop of ever intensifying electrical disturbances between the two hemispheres. The "split-brain" patients were able to lead what seemed, on the surface, entirely normal lives. But then Sperry devised some elegant experiments to test whether their conscious experience was indeed "normal." Given that the corpus callosum contains 200 million axons, surely *something* was different. Something was. "The surgery," Sperry later reported, "left these people with two separate minds, that is, two spheres of consciousness. What is experienced in the right hemisphere seems to be entirely outside the realm of awareness of the left."

Sperry presented the two hemispheres of split-brain patients with two different portions of the same visual stimulus. Researchers knew at the time that the visual connections to and from one hemisphere are mainly concerned with the opposite half of the visual field, as described in Figure 35.17. Sperry projected words—say, COWBOY—onto a screen so that COW fell in the left half of the visual field, and BOY fell in the right (Figure 35.18).

The subjects of this experiment reported *seeing* the word BOY. The left hemisphere, which controls language, perceived only the letters BOY. However, when asked to write the perceived word with the left hand—a hand that was deliberately blocked from a subject's view—the subject wrote COW. The right hemisphere "knew" the other half of the word (COW) and had directed the left hand's motor response. But it could not tell the left hemisphere what was going on because of the severed corpus callosum. The subject knew that a word was being written, but could not say what it was!

Thus Sperry showed that signals across the corpus callosum coordinate the functioning of the two cerebral hemispheres, each of which had responded to visual signals from the opposite side of the body.

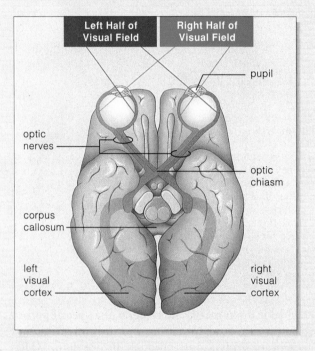

Figure 35.17 Pathway by which sensory input about visual stimuli reaches the visual cortex of the human brain.

Each eye gathers visual information at the retina, a layer of densely packed photoreceptors at the back of the eyeball (Section 36.8). Light from the *left* half of the visual field strikes receptors on the right side of both retinas. Parts of two optic nerves carry signals from the receptors to the right cerebral hemisphere. Light from the *right* half of the visual field strikes receptors on the left side of both retinas. Parts of the optic nerves carry signals from them to the left hemisphere.

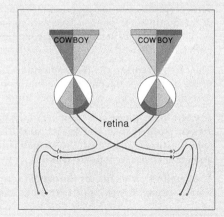
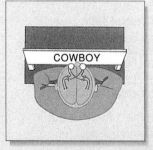

Figure 35.18 Response of a split-brain patient to different portions of a visual stimulus.

MEMORY

Memory is the capacity of an individual's brain to store and retrieve information about past sensory experience. Learning and adaptive modifications of our behavior would be impossible without it. Information is stored in stages. *Short-term* storage is a stage of neural excitation that lasts a few seconds to a few hours. It is limited to a few bits of sensory information—numbers, words of a sentence, and so on. In *long-term* storage, seemingly unlimited amounts of information get tucked away more or less permanently, as shown in Figure 35.19.

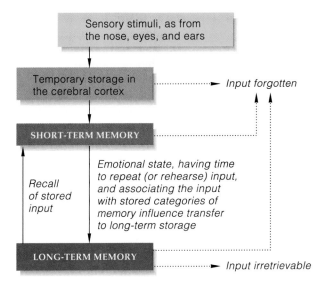

Figure 35.19 Stages of memory processing, starting with the temporary storage of sensory inputs in the cerebral cortex.

Not all of the sensory input bombarding the cerebral cortex ends up in memory storage. Only some is selected for transfer to brain structures involved in short-term memory. Information in these temporary holding bins is processed for relevance, so to speak. If irrelevant, it is forgotten; otherwise it is consolidated with the banks of information in long-term storage structures.

The human brain processes facts separately from skills. Dates, names, faces, words, odors, and other bits of explicit information are *facts*, soon forgotten or filed away in long-term storage, along with the circumstance in which they were learned. That is why you might associate, say, the smell of warm watermelon with a special picnic held long ago at the beach. By contrast, *skills* are gained by practicing specific motor activities. A skill such as slam-dunking a basketball or playing a piano concerto is best recalled by actually performing it, rather than by recalling the circumstances in which the skill was initially learned.

Separate memory circuits handle different kinds of input. A circuit leading to fact memory (Figure 35.20*a*) starts with inputs at the sensory cortex that flow to the

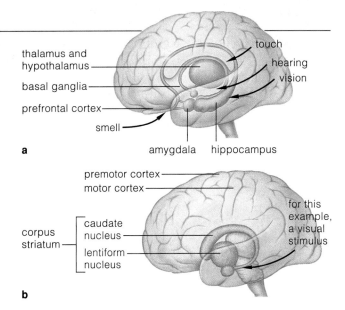

Figure 35.20 Possible circuits involved in (**a**) fact memory and (**b**) skill memory.

amygdala and hippocampus, which are structures in the limbic system. The amygdala is the gatekeeper; it connects the sensory cortex with parts of the thalamus and parts of the hypothalamus that govern emotional states. The hippocampus mediates learning and spatial relations. Information flows on to the prefrontal cortex, where multiple banks of fact memories are retrieved and used to stimulate or inhibit other parts of the brain. The new input also flows to basal ganglia, which send it back to the cortex in a feedback loop that reinforces the input until it can be consolidated in long-term storage.

Skill memory also starts at the sensory cortex, but this circuit routes sensory input to the corpus striatum, which promotes motor responses (Figure 35.20*b*). Motor skills involve muscle conditioning, and as you might suspect, the circuit extends to the cerebellum, the brain region that coordinates motor activity.

Amnesia is a loss of memory, the severity of which depends on whether the hippocampus, amygdala, or both are damaged, as by a severe head blow. Amnesia does not affect capacity to learn new skills. By contrast, basal ganglia are destroyed and learning ability is lost during *Parkinson's disease*, yet skill memory is retained. *Alzheimer's disease*, which usually has its onset later in life, is linked to structural changes in the cerebral cortex and hippocampus. Affected people often can remember long-standing information, such as their Social Security number, but they have difficulty remembering what has just happened to them. In time they become confused, depressed, and unable to complete a train of thought.

Memory, the storage and retrieval of sensory information, results from circuits between the cerebral cortex and parts of the limbic system, thalamus, and hypothalamus. Sensory input is processed through short-term and long-term storage.

35.8 STATES OF CONSCIOUSNESS

The spectrum of **consciousness** includes sleeping and aroused states, during which neural chattering shows up as wavelike patterns in EEGs. Specifically, EEGs are electrical recordings of the frequency and strength of membrane potentials at the surface of the brain (Figure 35.21). PET scans also can show the precise location of brain activity as it is proceeding.

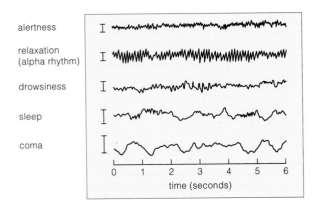

Figure 35.21 EEG patterns. Vertical bars indicate a fifty-range microvolt of electrical activity. The irregular horizontal graph lines indicate the electrical response over time.

The prominent wave pattern for someone who is relaxed, with eyes closed, is an *alpha rhythm*. The EEG waves are recorded in "trains" of one after the other, about ten per second. Alpha waves predominate during the state of meditation. During the transition to sleep, wave trains become larger, slower, and more erratic. This *slow-wave sleep* pattern dominates when sensory input is low and the mind is more or less idling. When subjects are awakened from slow-wave sleep, they usually report that they were not dreaming; rather, they often seemed to be mulling over recent, ordinary events.

Slow-wave sleep is punctuated by brief spells of *REM sleep*. *Rapid Eye Movements* accompany this pattern (the eyes jerk under closed lids), as do irregular breathing, faster heartbeat, and twitching fingers. Most people awakened from REM sleep report they were experiencing vivid dreams. The transition from sleep or deep relaxation into alert wakefulness is marked by a shift to low-amplitude, higher frequency wave trains. The transition, *EEG arousal*, occurs when conscious effort is made to focus on external stimuli or even on one's own thoughts.

Part of the reticular formation promotes chemical changes that influence whether you stay awake or fall asleep. Serotonin, a neurotransmitter released from one of its sleep centers, inhibits other neurons that arouse the brain and maintain wakefulness. At high levels, serotonin triggers drowsiness and sleep. Substances released from another brain center inhibit serotonin's effects and bring about wakefulness.

The spectrum of consciousness, which includes sleeping and states of arousal, is influenced by the reticular formation.

35.9 DRUGGING THE BRAIN

Broadly speaking, a **drug** is a substance introduced into the body to provoke a specific physiological response. Some drugs help a person cope with illness or emotional stress. Others act on the brain, artificially fanning pleasure associated with sex and other self-gratifying behaviors.

Many drugs are habit-forming. That is, even if the body is able to function well without them, a person continues to use drugs for the real or imagined relief they afford. Often the body develops tolerance of such drugs; it takes larger or more frequent doses to produce the same effect. Habituation and tolerance are signs of **drug addiction**, a chemical dependence on a drug (Table 35.1). *With an addiction, the drug has assumed an "essential" biochemical role in the body.* Abruptly deprive addicts of the drug, and they go through physical pain and mental anguish. The entire body goes through a period of biochemical upheavals. Stimulants, depressants, hypnotics, narcotic analgesics, hallucinogens, and psychedelics have such effects.

STIMULANTS Caffeine, nicotine, amphetamines, cocaine, and other stimulants increase alertness and body activity, then cause depression. Caffeine, a truly common stimulant, is present in coffee, tea, chocolate, and many soft drinks. Low doses arouse the cerebral cortex first, leading to increased alertness and restlessness. Higher doses acting at the medulla oblongata disrupt motor coordination and mental coherence. Nicotine, found in tobacco, has powerful effects on the nervous system. It mimics acetylcholine and directly stimulates a variety of sensory receptors. Its short-term effects include water retention, irritability, increased heart rate and high blood pressure, and gastric upsets.

Amphetamines, including "speed," mimic two of the neurotransmitters: dopamine and norepinephrine. In time, the brain produces less and less of these signaling molecules and depends more on artificial stimulation.

Table 35.1 Warning Signs of Drug Addiction*
1. Tolerance—it takes increasing amounts of the drug to produce the same effect.
2. Habituation—it takes continued drug use over time to maintain self-perception of functioning normally.
3. Inability to stop or curtail use of the drug, even if there is persistent desire to do so.
4. Concealment—not wanting others to know of the drug use.
5. Extreme or dangerous behavior to get and use the drug— as by stealing, asking more than one doctor to write drug prescriptions, or jeopardizing employment by drug use at work.
6. Deterioration of professional and personal relationships.
7. Anger and defensive behavior when someone suggests there may be a problem.
8. Preference of drug use over previously customary activities.

* Three or more of these signs may be cause for concern.

Possibly 2 million Americans are cocaine abusers. This drug gives a rush of pleasure by *blocking* the reabsorption of norepinephrine, dopamine, and some other neurotransmitters. Because these molecules accumulate in synaptic clefts, they incessantly stimulate postsynaptic cells for an extended period. Heart rate, blood pressure, and sexual appetite increase. In time the molecules diffuse away. However, neurons cannot synthesize replacements fast enough to counter the loss, and the sense of pleasure evaporates as the hypersensitized postsynaptic cells demand stimulation. After prolonged, heavy use of cocaine, "pleasure" can no longer be experienced. Addicts become anxious and depressed. They lose weight and cannot sleep properly. Their immune system weakens, and heart abnormalities set in.

Granular cocaine, which is inhaled (snorted), has been around for some time. Abusers burn crack cocaine and inhale the smoke, as in Figure 35.22. As suggested at the start of this chapter, crack is extremely addictive. Its highs are higher, the crashes are more devastating, and the social and economic tolls are extreme.

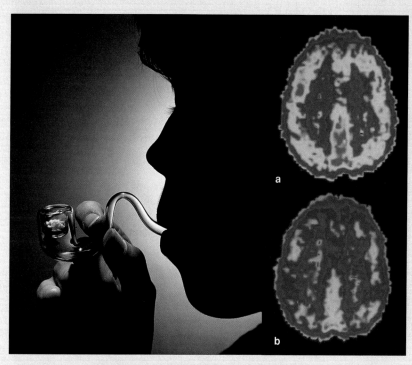

Figure 35.22 Smoking crack puts cocaine in the brain in less than eight seconds. (**a**) PET scan of a horizontal section of the brain, showing normal activity. (**b**) PET scan of a comparable section showing cocaine's effect. *Red* indicates greatest activity; *yellow, green,* and *blue* indicate successively inhibited activity.

DEPRESSANTS, HYPNOTICS   These drugs lower activity in nerves and parts of the brain. Some act at synapses in the reticular formation and thalamus. Responses depend on dosage and emotional and physiological states. They range from emotional relief through drowsiness, sleep, anesthesia, and coma to death. Low doses have the most effect on inhibitory synapses, so a person feels excited or euphoric at first. Increased doses suppress excitatory synapses as well and therefore lead to depression. Both the depressants and the hypnotics have additive effects. One amplifies another, as when alcohol plus barbiturates increases the depression.

Alcohol, or ethyl alcohol, acts directly on the plasma membrane to alter cell function. Like nicotine and cocaine, it is lipid soluble and easily crosses the blood-brain barrier to exert rapid effects. Some people mistakenly think of it as a harmless stimulant because of the initial "high" that it produces. However, alcohol is one of the most powerful psychoactive drugs and a major factor in many deaths. In the short term, small amounts cause disorientation and uncoordinated motor functions, and diminish judgment. Long-term addiction can lead to *cirrhosis*. Then, connective tissue permanently replaces damaged liver cells, so the self-regenerating capacity of liver tissue slowly diminishes. In time, the outcome is chronic liver failure and death.

ANALGESICS   When severe stress leads to physical or emotional pain, the brain produces natural analgesics, or pain relievers. Endorphins and enkephalins are two examples. They act on many parts of the nervous system, including brain centers that deal with emotions and pain. Narcotic analgesics, including codeine and heroin, sedate the body and relieve pain. They are among the most addictive substances known. Deprivation after massive doses of heroin results in hyperactivity and anxiety, fever, chills, violent vomiting, cramping, and diarrhea.

PSYCHEDELICS, HALLUCINOGENS   These drugs skew sensory perception by interfering with the activity of acetylcholine, norepinephrine, or serotonin. LSD (or lysergic acid diethylamide) alters serotonin's influence on inducing sleep, on temperature regulation, and on sensory perception. Even in small doses, LSD warps perceptions. For example, some users "perceived" they could fly and "flew" off buildings.

Marijuana, another hallucinogen, is made from crushed leaves, flowers, and stems of the plant *Cannabis*. In low doses it is like a depressant. It slows down but does not impair motor activity; it relaxes the body and elicits mild euphoria. It also causes disorientation, increased anxiety bordering on panic, delusions (including paranoia), and hallucinations.

Like alcohol, marijuana affects the performance of complex tasks, such as driving a car. In one study, pilots showed a marked deterioration in instrument-flying ability for more than two hours after smoking marijuana. Over time, marijuana smoking can suppress the immune system and impair mental functions.

1. Nervous systems detect, interpret, and issue calls for response to stimuli in the animal's external and internal environments. At their most basic operating level are reflexes: simple, stereotyped movements made directly in response to stimuli.

   a. The simplest nervous systems are nerve nets, such as those of cnidarians and other radial animals. These meshworks of nerve cells form reflex connections with contractile and sensory cells of the epithelium.

   b. Most animals have a bilateral, cephalized nervous system, with a brain or ganglia at the anterior end. Such animals also have cordlike nerves and tracts (bundled axons of sensory neurons, motor neurons, or both).

2. The vertebrate central nervous system consists of the brain and spinal cord. The peripheral nervous system consists mainly of nerves that carry signals between all body regions and the spinal cord and brain.

   a. Somatic nerves of the peripheral nervous system deal with skeletal muscles. Autonomic nerves deal with the heart, lungs, glands, and other internal organs.

   b. Autonomic nerves are either parasympathetic or sympathetic. Parasympathetic nerves dominate when outside stimulation is low and tend to divert energy to basic housekeeping tasks. Overall, sympathetic nerves step up body activities during heightened awareness or danger. They govern the fight-flight response.

3. Reflex connections for limb movements and internal organ activity are made in the spinal cord. Many tracts in the spinal cord also carry signals between the brain and the peripheral nervous system.

4. During development, the brain develops by regional expansions of a hollow neural tube in the embryo. The tube's anterior end becomes the brain stem, which is still discernible in an adult brain. The brain's most recent expansions of gray matter deal with storing, comparing, and using experiences to initiate novel action.

   a. A neural tube's interior becomes a system of canals and cavities filled with cerebrospinal fluid. A protective mechanism called the blood-brain barrier exerts some control over which water-soluble substances enter the cerebrospinal fluid, which must be maintained within very narrow limits for neurons to function properly.

   b. The reticular formation, a mesh of interneurons extending from the top of the spinal cord to the cerebral cortex, is a low-level route to centers of motor activity.

5. The vertebrate brain has three main divisions:

   a. The hindbrain includes the medulla oblongata, pons, and cerebellum. It contains reflex centers for vital functions and muscle coordination.

   b. The midbrain coordinates and relays visual and auditory information to other brain regions.

   c. The forebrain includes the cerebrum, thalamus, and hypothalamus. The cerebral hemispheres contain the highest centers for receiving, processing, storing, and responding to sensory information. The thalamus relays signals and helps coordinate motor responses. The hypothalamus, the main homeostatic control center over the internal environment and internal organs, also governs thirst, sexual activity, and other behaviors related to organ functions. It is part of the limbic system, which governs emotions and memory.

Review Questions

1. Generally describe the type of nervous systems found in radially symmetrical animals and in bilaterally symmetrical animals. *35.1*

2. What constitutes the central nervous system? The peripheral nervous system? *35.2*

3. Distinguish between the following:
   a. central and peripheral nervous system *35.2*
   b. cranial and spinal nerves *35.3*
   c. somatic and autonomic nerves *35.3*
   d. parasympathetic and sympathetic nerves *35.3*

4. Distinguish between:
   a. white matter and gray matter *35.2*
   b. nerve and tract *35.2*

5. Describe the structure and function of the spinal cord. *35.3*

6. Label the major parts of the human brain. *35.4, 35.5*

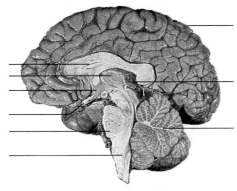

7. Explain the difference between the reticular formation and the limbic system in terms of their component parts and main functions. *35.4, 35.5*

8. Define cerebrospinal fluid and the blood-brain barrier. *35.4*

9. Briefly describe one of the brain centers of the cerebral cortex; mention the tissue lobe in which it is located. *35.5*

10. Distinguish between: *35.7*
    a. short-term memory and long-term memory
    b. fact memory and skill memory

11. Which part of the brain has major influence over states of consciousness, which include sleeping and arousal? *35.8*

12. What is a habit-forming drug? Briefly describe the effects of one such drug on the central nervous system. *35.9*

Self-Quiz (Answers in Appendix IV)

1. In sea anemones and jellyfishes, the _____ is a loose mesh of nerve cells that interact with sensory and contractile cells of an epithelium to bring about reflex responses to stimuli.

2. The oldest parts of the vertebrate brain provide _____ .
 a. reflex control of breathing, circulation, and other basic activities
 b. coordinating and relaying visual and auditory input
 c. storing, comparing, and retrieving sensory information
 d. both a and c

3. Is this statement true or false: White matter and gray matter are components of the spinal cord alone.

4. Is this statement true or false: The peripheral nervous system has nerves; the brain and spinal cord have tracts.

5. Overall, _____ nerves slow down the body and divert energy to digestion and other basic housekeeping tasks; and _____ nerves slow down housekeeping tasks and increase overall activity in times of heightened awareness or excitement.
 a. autonomic; somatic
 b. sympathetic; parasympathetic
 c. parasympathetic; sympathetic
 d. peripheral; central

6. Parasympathetic and sympathetic nerves _____ , to bring about minor adjustments in internal organ activity.
 a. service entirely different internal organs
 b. work antagonistically, most often
 c. come into play only during a fight-flight response
 d. none of the above

7. In tracts of the brain and spinal cord, _____ is associated with myelinated axons; _____ consists of neuroglia as well as unmyelinated axons, dendrites, and cell bodies of neurons.
 a. gray matter; white matter
 b. white matter; gray matter

8. In an adult brain, the brain stem is present in _____ .
 a. the hindbrain only
 b. the midbrain only
 c. the forebrain only
 d. all three divisions of the brain

9. The hindbrain, which includes the medulla oblongata, pons, and cerebellum, contains _____ .
 a. part of the reticular formation   c. the tectum
 b. the limbic system   d. all of the above

10. The _____ are located in the cerebrum.
 a. medulla oblongata, pons, and cerebellum
 b. thalamus, hypothalamus, and limbic system
 c. medulla oblongata, pons, and cerebral cortex
 d. cerebellum, medulla oblongata, pons, and limbic system
 e. hypothalamus, limbic system, pons, and cerebral cortex

11. The _____ is central to planning movements and blocking unsuitable behavior, and also has roles in memory.
 a. visual cortex
 b. motor cortex
 c. somatosensory cortex
 d. prefrontal cortex

12. Match the component with its main functions.
 _____ spinal cord   a. receives, integrates, issues commands for response to sensory information
 _____ medulla oblongata
 _____ hypothalamus   b. homeostatic control of internal environment, internal organs
 _____ limbic system   c. deals with emotions, memory
 _____ cerebral cortex   d. reflex control of respiration, blood circulation, other basic activities
                            e. expressway for signals between brain and peripheral nervous system; also, direct reflex connections

Critical Thinking

1. In human newborns and premature babies, the blood-brain barrier is not fully developed. Explain why this might be reason enough to pay careful attention to their diet.

2. When Jennifer was only six years old, a man lost control of his car and hit a tree in front of her house. She ran over to the car and screamed when she saw blood from a deep head wound dripping onto a huge bouquet of red roses. Thirty-five years later, someone gave Jennifer a bottle of *Tea Rose* perfume for a birthday present. When she sniffed the perfume, she became frightened and extremely anxious. A few minutes later she also had a vivid recollection of the accident. Explain this incident in terms of what you have learned about memory.

3. Eric typically drinks a cup of coffee nearly every hour, all day long. Today, by midafternoon, he is having trouble concentrating on his studies, and he feels tired and more than a little clumsy. Drinking still another cup of coffee didn't make Eric more alert. Explain how the caffeine in coffee might produce such symptoms.

Selected Key Terms

autonomic nerve 35.3
blood-brain barrier 35.4
brain 35.4
brain stem 35.4
central nervous system 35.2
cerebellum 35.4
cerebral cortex 35.5
cerebral hemisphere 35.5
cerebrospinal fluid 35.4
cerebrum 35.4
consciousness 35.8
drug 35.9
drug addiction 35.9
fight-flight response 35.3
ganglion (ganglia) 35.1
gray matter 35.2
hypothalamus 35.4
limbic system 35.5
medulla oblongata 35.4
memory 35.7

meninges (singular, meninx) 35.3
nerve 35.1
nerve net 35.1
nervous system 35.1
neural tube 35.2
neuroglia 35.2
parasympathetic nerve 35.3
peripheral nervous system 35.2
planula 35.1
pons 35.4
reflex pathway 35.1
reticular formation 35.4
somatic nerve 35.3
spinal cord 35.3
sympathetic nerve 35.3
tectum 35.4
thalamus 35.4
tract 35.2
white matter 35.2

Readings

Bloom, F., and A. Lazerson. 1988. *Brain, Mind, and Behavior.* Second edition. New York: Freeman.

Churchland, P., and P. Churchland. January 1990. "Could a Machine Think?" *Scientific American* 262(1): 32–37.

Fischbach, G. D. September 1992. "Mind and Brain." *Scientific American* 267(3): 48–57. This entire issue is devoted to the development, function, and certain disorders of the brain.

Romer, A., and T. Parsons. 1986. *The Vertebrate Body.* Sixth edition. Philadelphia: Saunders. Fine insights into the evolution of vertebrate nervous systems.

Shepherd, G. 1994. *Neurobiology.* Third edition. New York: Oxford. Paperback.

Web Site See *http://www.wadsworth.com/biology* for practice quiz questions, hypercontents, BioUpdates, and critical thinking. The Wadsworth Biology Resource Center provides a wealth of information fully organized and integrated by chapter.

36 SENSORY RECEPTION

Different Stokes for Different Folks

Even though you might be reluctant to pet a snake or scratch a bat behind the ears, you have to give them credit for being two of your vertebrate relatives with some special sensory traits.

Consider the python. Figure 36.1a shows one of these reptiles, which have thermoreceptors in rows of pits above and below the mouth. Thermoreceptors detect infrared energy—in this case, body heat of small, night-foraging mammals that are the snake's prey of choice. When stimulated, the receptors send messages to the brain, where several centers process the signals and issue commands to muscle cells. In no time at all, a strike is aimed and executed with stunning accuracy.

A motionless, edible frog would not have the same stimulatory effect on the receptors, so the snake would slither past it. The skin of a frog is not warm, and it blends with the colors of the frog's habitat. A python does not have the types of receptors that can detect a frog or a neural program for responding to it.

Or consider the mammals called bats. Nearly all species of bats sleep during the day and spread their webbed wings at dusk. Different kinds take to the air in search of nectar, fruit, frogs, or insects. Many sensory receptors that are located inside their eyes, nose, ears, mouth, and skin are not all that different from yours. However, other receptors provide bats with a sense of hearing that you cannot even begin to match. Even tiny-eyed, nearly blind species navigate and capture flying insects with precision in the dark!

Bats are masters of **echolocation**. They emit calls, as the bat in Figure 36.1b is doing. When sound waves of the calls bounce off insects, trees, and other objects, acoustical receptors inside the bat's ears detect the echoes and send signals about them to the bat brain.

As an echolocating bat flies, it emits a steady stream of about ten clicking sounds per second—sounds you cannot hear. The clicks are intense "ultrasounds," above the range of sound waves that receptors in human ears

a

Figure 36.1 Examples of sensory receptors. (**a**) Thermoreceptors in pits above and below this python's mouth detect body heat, or infrared energy, of nearby prey. (**b**) Some bats listen to echoes of their own high-frequency sounds. Their brain constructs a sound map, based on echoes bouncing back from objects in the surroundings. Such maps help this night-flying bat capture insects in midair, without the help of eyes.

are able to detect. When a bat hears a pattern of distant echoes from, say, an airborne mosquito or moth, it increases the rate of ultrasonic clicks to as many as 200 per second, which is faster than a machine gun fires bullets. In the few milliseconds of silence between the clicks, sensory receptors detect the echoes and deliver messages about them to the bat brain. The brain rapidly constructs a "map" of the sounds that the bat uses in its maneuvers through the night world.

b

With this chapter, we turn to the means by which animals receive signals from the external and internal environments, then decode signals in ways that give rise to awareness of sounds, sights, odors, pain, and other sensations. Sensory neurons, nerve pathways, and specific brain regions are required for these tasks. Together, they represent the portions of the nervous system called sensory systems.

KEY CONCEPTS

1. Sensory systems are portions of the nervous system. Each consists of specific types of sensory receptors, nerve pathways from receptors to the brain, and brain regions that receive and process the sensory information.

2. A stimulus is a form of energy that activates a specific type of sensory receptor, which is either a sensory neuron or a specialized cell adjacent to it. Photoreceptors detect light energy, thermoreceptors detect infrared energy, and so on.

3. A sensation is conscious awareness of change in some aspect of the external or internal environment. It begins when sensory receptors detect a specific stimulus. The stimulus energy becomes converted to a graded, local signal that may help initiate an action potential.

4. Information concerning the stimulus becomes encoded in the number and frequency of action potentials sent to the brain along particular nerve pathways. Then specific brain regions translate the information into a sensation.

5. The somatic sensations include touch, pressure, pain, temperature, and muscle sense.

6. The special senses include taste, smell, hearing, and vision.

Sensory systems, the front doors of the nervous system, receive information about specific changes inside and outside the body and notify the spinal cord and brain of what is going on. They consist of sensory receptors, nerve pathways leading from receptors to the brain, and brain regions where sensory information is translated into sensations. A **sensation** is conscious awareness of a stimulus. It is not the same as **perception** (understanding what a sensation means). *Compound* sensations arise when information about different stimuli is integrated at the same time. For example, "wetness" is not a single stimulus; our perception of it arises from simultaneous inputs concerning pressure, touch, and temperature.

Types of Sensory Receptors

Sensory receptors are the receptionists at the front door of the nervous system. As listed in Table 36.1, there are six major categories of sensory receptors, based on the type of stimulus energy that each type detects:

Mechanoreceptors detect forms of mechanical energy (changes in pressure, position, or acceleration).

Thermoreceptors detect infrared energy (heat).

Pain receptors (nociceptors) detect tissue damage.

Chemoreceptors detect chemical energy of specific substances dissolved in the fluid surrounding them.

Osmoreceptors detect changes in water volume (solute concentration) in the surrounding fluid.

Photoreceptors detect visible and ultraviolet light.

Depending upon the kinds and numbers of their sensory receptors, animals sample the environment in different ways and differ in their awareness of it. Unlike bees, for example, you have no receptors for ultraviolet light and do not see many flowers the way they do. The introduction to Chapter 31 provides an example of this. Unlike many bats, you and bees have no receptors for ultrasound. Unlike pythons, you, bees, and bats have no receptors for detecting warm-blooded prey in the dark.

Sensory Pathways

Recall, from Chapter 34, that sensory axons carry signals from receptors to the brain. Before they can do so, the stimulus energy must be converted to action potentials, the basis of neural messages. Briefly, when a stimulus disturbs the plasma membrane of a receptor's ending, certain ions flow across a local patch of the membrane. In other words, the disturbance triggers a local, graded potential. This type of signal does not spread far from the point of stimulation, and it can vary in magnitude.

When a stimulus is intense or when it is repeated fast enough for a summation of local signals, action potentials may be the result. Action potentials propagate themselves from the receptor to the axon endings of sensory neurons (Figure 36.2). There, neurotransmitter is released from the presynaptic cell, and it influences the electrical activity of the cell adjacent to it (another interneuron or a motor neuron). The disturbance may trigger action potentials in the postsynaptic cell, which is part of an information pathway leading to the brain.

Table 36.1	Major Categories of Sensory Receptors	
Category	Examples	Stimulus
MECHANORECEPTORS		
Touch, pressure	Certain free nerve endings and Pacinian corpuscles in skin	Mechanical pressure against body surface
Baroreceptor	Carotid sinus (Section 39.7)	Pressure changes in fluid that bathes them
Stretch	Muscle spindle in skeletal muscle	Stretching
Auditory	Hair cells in organ inside ear	Vibrations (sound or ultrasound waves)
Balance	Hair cells in organ inside ear	Fluid movement
THERMORECEPTORS	Certain free nerve endings	Change in temperature (heating, cooling)
PAIN RECEPTORS (NOCICEPTORS)*	Certain free nerve endings	Tissue damage (e.g., distortions, burns)
CHEMORECEPTORS		
Internal chemical sense	Carotid bodies in blood vessel wall	Substances (O_2, CO_2, etc.) dissolved in extracellular fluid
Taste	Taste receptors of tongue	Substances dissolved in saliva, etc.
Smell	Olfactory receptors of nose	Odors in air, water
OSMORECEPTORS	Hypothalamic osmoreceptors (Section 43.2)	Change in water volume (solute concentration) of fluid that bathes them
PHOTORECEPTORS		
Visual	Rods, cones of eye	Wavelengths of light

* Extremely intense stimulation of any sensory receptor also may be perceived as pain.

Action potentials that are traveling along a sensory neuron are not like a wailing ambulance siren. That is, *they do not vary in amplitude.* How, then, does the brain assess the nature of a given stimulus? The answer lies with (1) *which* nerve pathways happen to be carrying action potentials, (2) the *frequency* of action potentials traveling along each axon of the pathway, and (3) the *number* of axons that the stimulus has recruited.

First, the network of neurons in each animal's brain is genetically prescribed, and it can interpret action potentials only in certain ways. That is why you "see stars" when one of your eyes has been poked, even when it gets poked in a dark room. Photoreceptors inside the eye were mechanically disturbed enough to trigger messages that traveled along an optic nerve to your brain. And your brain always interprets any signals that are arriving from an optic nerve as "light."

Second, when a stimulus is strong, receptors fire action potentials more frequently than they do with a weak stimulus. The same receptor detects the sounds of a throaty whisper and a wild screech, and the brain senses the difference through frequency variations in signals that the receptor sends to it.

Third, a strong stimulus recruits more sensory receptors than a weaker stimulus does. Gently tap a spot of skin on one of your arms and you activate just a few receptors. Press hard on the same spot and you activate more receptors in a larger area. The increased disturbance translates into action potentials in many sensory axons at the same time. The brain interprets signals about the combined activity as

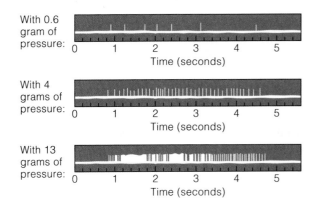

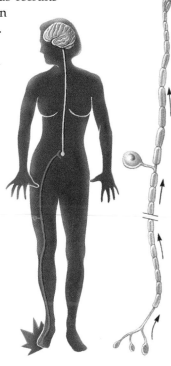

Figure 36.2 Example of a sensory nerve pathway that leads away from a sensory receptor to the brain. The sensory neuron is coded *red* and the interneurons are coded *yellow*.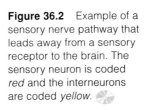

Receptor endings of a sensory neuron are stimulated when a bare foot lands on a tack.

Figure 36.3 Recordings of action potentials from a single pressure receptor with endings in the skin of a human hand. The recordings correspond to variations in stimulus strength. The investigator pressed a thin rod against the skin with the amount of pressure indicated on the left side of each diagram. The vertical bars above each thick horizontal line represent individual action potentials. In each case, the increases in frequency correspond to increases in the strength of the applied stimulus.

an increase in the stimulus intensity. Figure 36.3 shows the effect of increases in stimulus strength.

In some cases, the frequency of action potentials decreases or stops even when a stimulus is being maintained at constant strength. A decrease in the response to a stimulus is known as **sensory adaptation**. After you put on clothing, for example, awareness of its pressure against your skin ceases. Some of the mechanoreceptors inside your skin adapt rapidly to the sustained stimulation; they are of a type that only signals a change in a stimulus (its onset and its removal). By contrast, other receptors adapt slowly or not at all; they help the brain monitor certain stimuli all the time. The stretch receptors you read about in Section 34.4 are like this. Many continually inform the brain about the length of particular muscles that help maintain balance and posture.

We turn next to some specific examples of sensory receptors. The types that are present at more than one body location contribute to **somatic sensations**. Other types are restricted to particular locations, such as inside the eyes or ears. They contribute to the **special senses**.

A sensory system has sensory receptors for specific stimuli, nerve pathways that conduct information from receptors to the brain, and brain regions that receive the information.

The brain assesses a given stimulus based on which nerve pathways are carrying action potentials, the frequency of action potentials traveling along each axon of that pathway, and the number of axons that have been recruited to action.

Somatic sensations begin with receptors in the body's surface tissues, skeletal muscles, and walls of internal organs. The receptors are most highly developed in birds and mammals; amphibians have few, and apparently fishes have none. Inputs from these receptors travel into the spinal cord and on to the **somatosensory cortex**, part of the surface layer of gray matter of the cerebral hemispheres (Section 35.5). Interneurons of this part of the brain are organized like maps, which correspond to particular parts of the body's surface. Map regions with the largest areas correspond to body parts that have the most sensory acuity and that require the most intricate control. Such body parts include the fingers, thumbs, and lips (Figure 36.4).

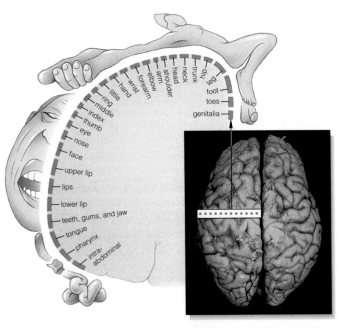

Figure 36.4 Differences in the representation of different body parts in the primary somatosensory cortex. This region is a strip of cerebral cortex, a little wider than 2.5 centimeters (about an inch), from the top of the head to above the ear.

Receptors Near the Body Surface

You and other mammals discern sensations of touch, pressure, cold, warmth, and pain near the body surface. Regions with the greatest number of sensory receptors, such as the fingertips and the tip of the tongue, are the most sensitive to stimulation. Other regions, such as the back of the hand and neck, do not have nearly as many receptors and are far less sensitive.

Free nerve endings are the simplest receptors. They are thinly myelinated or unmyelinated (naked) branched endings of sensory neurons in the epidermis or in the underlying dermis of skin. Different types function as mechanoreceptors, thermoreceptors, and pain receptors,

and all of them adapt very slowly to stimulation. One subpopulation gives rise to a sensation of prickling pain, as when you jab a finger with a pin. Another contributes to sensations of itching or warming that are elicited by chemicals, including histamine. Two thermoreceptive types have peak sensitivities that are higher and lower than the normal body temperature, respectively. One mechanoreceptive type coils around the hair follicle and detects movement of the hair inside (Figure 36.5).

Encapsulated receptors also are common near the body surface. A Meissner's corpuscle adapts slowly to vibrations of low frequencies. It is notably abundant in the lips, fingertips, eyelids, nipples, and genitals. The bulb of Krause is an encapsulated thermoreceptor that is activated by temperatures below 20°C (68°F). Below 10°C, it gives rise to painful freezing sensations. Ruffini endings are other slowly adapting encapsulated types. They are sensitive to steady touching and pressure, and to temperatures above 45°C (113°F).

The Pacinian corpuscle is an encapsulated receptor that is widely distributed in dermis, where it may have a role in perceptions of fine textures. It also is present deeper in the body, as in the membrane near freely movable joints and in some internal organs. Concentric, onionlike layers of membrane alternating with fluid-filled spaces surround this sensory ending (Figure 36.5). The construction enhances the detection of rapid pressure changes associated with touch and vibrations.

Muscle Sense

Sensing limb motions and the body's position in space requires mechanoreceptors in skeletal muscle, joints, tendons, ligaments, and skin. Examples include stretch receptors of muscle spindles. As described in Section 34.4, these sensory organs run parallel with skeletal muscle cells. Their responses to stimulation depend on how much and how fast the muscle stretches.

Regarding the Sensation of Pain

Pain is the perception of injury to some body region. The most important pain receptors are subpopulations of free nerve endings, several million of which are distributed throughout the skin and internal tissues. They trigger awareness of pain in the thalamus, but the type and intensity of pain are interpreted in the cerebral cortex.

Sensations of *somatic* pain start with pain receptors in skin, skeletal muscles, joints, and tendons. Sensations of *visceral* pain, which is associated with the internal organs, are related to excessive chemical stimulation, muscle spasms, muscle fatigue, inadequate blood flow to organs, and other abnormal conditions.

Figure 36.5 *Right:* A sampling of receptors in human skin. Subpopulations of free nerve endings function as mechanoreceptors (including those around hair follicles, which detect hair movements), thermoreceptors, and pain receptors. Pacinian corpuscles detect rapid pressure changes associated with touch and vibrations. Ruffini endings detect steady touching and pressure. The bulb of Krause is most sensitive to cold. Meissner's corpuscle detects vibrations of lower frequency than those detected by Pacinian corpuscles.

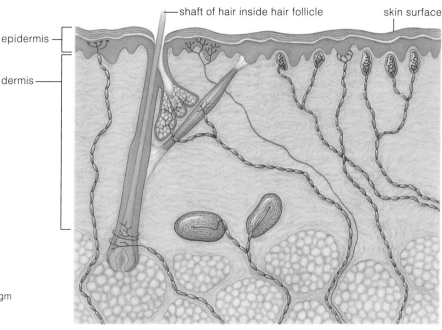

shaft of hair inside hair follicle · skin surface

epidermis

dermis

free nerve endings · Pacinian corpuscle · Ruffini endings · bulb of Krause · Meissner's corpuscle

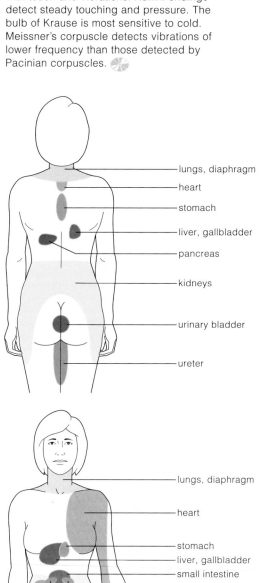

lungs, diaphragm
heart
stomach
liver, gallbladder
pancreas
kidneys
urinary bladder
ureter

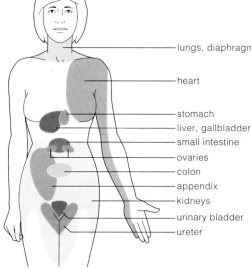

lungs, diaphragm
heart
stomach
liver, gallbladder
small intestine
ovaries
colon
appendix
kidneys
urinary bladder
ureter

Figure 36.6 Referred pain. Receptors in some internal organs detect painful stimuli. Instead of localizing the pain at the organs, the brain projects the sensation to the skin areas indicated.

Responses to pain depend on the ability of the brain to identify the affected tissue and project the sensation back to it. For example, get smacked in the face with a snowball and you "feel" the contact on facial skin. However, sensations of pain from certain internal organs may be wrongly projected to part of the skin surface. All people have a capacity to experience this response, which is called *referred pain*, so it is related to the way the nervous system is constructed. Possibly, sensory inputs from the skin and from certain internal organs enter the spinal cord along common nerve pathways, so that the brain cannot accurately identify their source. For example, the brain commonly projects the pain of a heart attack to the skin above the heart and along the left shoulder and arm (Figure 36.6).

As a final example, amputees often sense *phantom pain* from a missing body part, as if it were still there. In some undetermined way, the brain projects pain past the healed regions where sensory nerves were severed.

Diverse sensory receptors near the body surface and in internal organs detect touch, pressure, temperatures, pain, motion, and positional changes of body parts. Their input travels through the spinal cord to the somatosensory cortex and other brain regions where somatic sensations arise.

SENSES OF TASTE AND SMELL

We turn now to examples of the special senses, starting with those of taste and smell. Both are *chemical* senses; their sensory pathways start at chemoreceptors, which are activated when they bind a chemical substance that is dissolved in the fluid bathing them. The receptors themselves wear out and new ones replace them on an ongoing basis. In both pathways, sensory input travels from the receptors through the thalamus and on to the cerebral cortex, where perceptions of the stimulus take shape and undergo fine-tuning. The input also travels to the limbic system, which can integrate it with emotional states and stored memories (Sections 35.5 and 35.7).

Different animals taste substances with their mouth, antennae, legs, tentacles, or fins. It all depends on where the chemoreceptors called **taste receptors** are located. In your throat and mouth, especially the tongue's upper surface, they are located in about 10,000 sensory organs called taste buds (Figure 36.7). Each taste bud has a pore through which fluids in the mouth contact the surface of receptor cells. Of thousands of perceived tastes, all are combinations of four primary sensations: sweet (elicited by sucrose, glucose, and other simple sugars), sour (acids), salty (NaCl and other salts), and bitter (alkaloids and other potentially toxic plant substances).

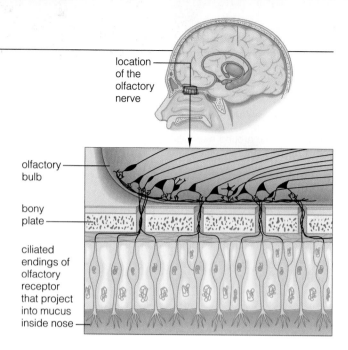

Figure 36.8 Sensory pathway leading from the sensory endings of olfactory receptors in the nose to the cerebral cortex and limbic system. Axons from these receptors pass through holes in a bony plate that separates the nasal lining from the brain. In the earliest vertebrates, an olfactory bulb and an olfactory lobe dominated the forebrain. In certain lineages, the olfactory bulb became reduced in importance, and the cerebrum expanded greatly in size and functions, as described in Sections 35.2 and 35.4.

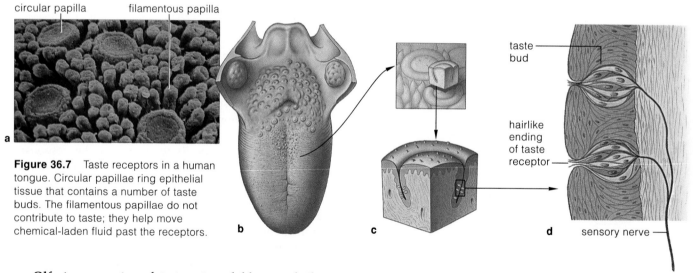

Figure 36.7 Taste receptors in a human tongue. Circular papillae ring epithelial tissue that contains a number of taste buds. The filamentous papillae do not contribute to taste; they help move chemical-laden fluid past the receptors.

Olfactory receptors detect water-soluble or volatile (easily vaporized) substances. A bloodhound nose has more than 200 million; a human nose has about 5 million. Axons of such receptors lead into one of two small brain structures called olfactory bulbs (Figure 36.8). There they synapse with groups of cells that sort out the components of a given scent and relay the information onward, by way of an olfactory tract, for further processing.

A vomeronasal organ, or "sexual nose," is common among animals, including humans. Its receptors detect **pheromones**, a type of signaling molecule with roles in social aspects of reproduction. These exocrine gland

secretions affect the behavior of other individuals of the same species, especially potential mates. Consider the effect of bombykol on the olfactory receptors of a male silk moth. Contact with merely one bombykol molecule per second sends action potentials to the moth brain. They help a male locate a female in the dark, even if she is more than a kilometer upwind from him.

The senses of taste and smell both start at chemoreceptors. Both involve sensory pathways that lead to processing regions in the cerebral cortex and in the limbic system.

36.4 SENSE OF BALANCE

Most animals assess and respond to displacements from their natural, *equilibrium* position, in which the body is balanced in relation to gravity, velocity, acceleration, and other forces that influence its positions and movement. For example, animals right themselves after tilting or turning upside-down. A variety of fishes, amphibians, and reptiles have paired **inner ears**, which evolved first as organs of equilibrium and had little, if anything, to do with hearing. These are systems of fluid-filled sacs and canals on both sides of the brain. Some parts detect rotational, accelerated motions of the head, as when you ride a looping roller coaster (Figure 36.9). Other parts detect linear motion of the head. In amphibians, birds, and mammals, the brain integrates sensory input from internal ears with input from receptors in the eyes, skin, and joints to arrive at sensations of balance.

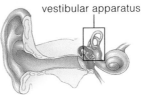
vestibular apparatus

Figure 36.9 Location of the internal ear in humans. Inside the vestibular apparatus (a system of fluid-filled sacs and canals), organs detect the head's linear, rotational, and accelerated motions.

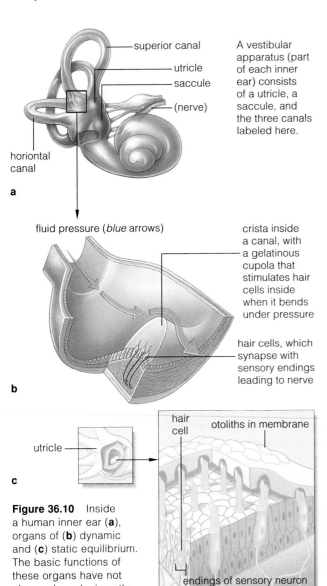

superior canal
utricle
saccule
(nerve)

horiontal canal

a

A vestibular apparatus (part of each inner ear) consists of a utricle, a saccule, and the three canals labeled here.

fluid pressure (*blue* arrows)

crista inside a canal, with a gelatinous cupola that stimulates hair cells inside when it bends under pressure

hair cells, which synapse with sensory endings leading to nerve

b

utricle

hair cell
otoliths in membrane

c

endings of sensory neuron that is part of a nerve

Figure 36.10 Inside a human inner ear (**a**), organs of (**b**) dynamic and (**c**) static equilibrium. The basic functions of these organs have not changed much since they first evolved in fishes.

In humans, organs of equilibrium are located in the part of the inner ear called the **vestibular apparatus** (Figures 36.9 and 36.10*a*). Cristae, the organs concerned with *dynamic* equilibrium, detect accelerated, rotational motions of the head. One is present at the swollen base of each of three semicircular canals that are oriented in three directions. When your head rotates horizontally, vertically, or diagonally, fluid in a canal corresponding to that direction is displaced in the opposite direction. The fluid presses against a cupola, a gelatinous mass into which hair cells project (Figure 36.10*b*). **Hair cells** are a type of mechanoreceptor. When they bend under pressure, the deformation to their plasma membrane may trigger action potentials. These hair cells synapse with the endings of nearby sensory neurons, the axons of which help form a vestibular nerve.

An organ of *static* equilibrium, which is activated when the animal starts and stops moving in a straight line, resides on the floor of each utricle and saccule of the vestibular apparatus (Figure 36.10*c*). A thickened membrane lying over the floor is weighted by deposits of calcium carbonate crystals. These crystalline masses are **otoliths**, or "ear stones." When the membrane slides in response to linear acceleration, hair cells projecting into it bend and are thereby activated.

Motion sickness may result when monotonous linear, angular, or vertical motion overstimulates hair cells in the canals of the organs of balance. It also may result from conflicting signals from the ears and eyes about motion or the head's position. People who are carsick, airsick, or seasick suffer nausea and may vomit; action potentials triggered by the sensory input reach a brain center that governs the vomiting reflex.

Organs of equilibrium help keep the body balanced in relation to gravity, velocity, acceleration, and other forces that influence its position and movement. In humans, such organs occur in the vestibular apparatus of the inner ear.

Properties of Sound

Many arthropods and most vertebrates have a sense of **hearing**, which is the perception of sounds. Sounds are traveling vibrations, or wavelike forms of mechanical energy. Think of what happens when you clap your hands and produce waves of compressed air. Each time hands clap together, molecules are forced outward, so a low-pressure state is created in the region they vacated. Such pressure variations are depicted as wave forms, in which *amplitude* corresponds to loudness (Figure 36.11). Typically, amplitude is measured in decibels. Every ten of these units signifies a tenfold increase in intensity above the faintest sound that humans can hear. The *frequency* of a sound is the number of wave cycles per second. Each cycle extends from the start of one wave peak to the start of the next peak. The more cycles per second, the higher are the frequency and perceived pitch.

Evolution of the Vertebrate Ear

After vertebrates invaded the land, the sense of hearing became far more important than it had been in aquatic habitats. Although the parts of the *internal* ear dealing with equilibrium did not change much, other parts slowly expanded into structures that could receive and process the faint sounds traveling through air. Also, the **middle ear** evolved, and structures inside it amplified and transmitted air waves to the inner ear. In reptiles, a shallow depression formed on each side of the head and led to an eardrum (a thin membrane that rapidly

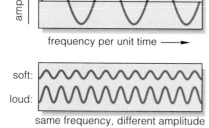

Sounds differ in *amplitude* (they vary in pressure, depicted here as wave forms) and in *frequency* (number of wave cycles/unit time).

A sound's *intensity* (loudness) depends on its amplitude.

A sound's *pitch* (tone) depends on its frequency.

Figure 36.11 Properties and examples of sound waves. Compared to the pure tone of a tuning fork, most sounds are combinations of sound waves of different frequencies. Combinations of overtones give the sounds their timbre, or quality, which is one of the properties that help you recognize, say, voices of people you know over the phone.

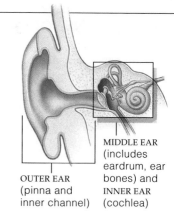

Figure 36.12 Sensory reception in the human ear, which collects, amplifies, and sorts sound waves (acoustical stimuli).

a *Right*: Each ear is divided into three regions, the outer, middle, and inner ear.

OUTER EAR (pinna and inner channel)

MIDDLE EAR (includes eardrum, ear bones) and INNER EAR (cochlea)

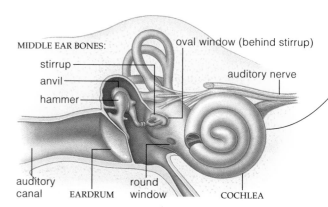

b Components of the human ear. External flaps of the outer ear collect sound waves, which move into and auditory canal and arrive at the eardrum (alson called the tympanic membrane).

vibrates in response to air waves). Behind the eardrum of all existing crocodiles, birds, and mammals is an air-filled cavity and small bones, which transmit vibrations to the inner ear. Long ago, among early fishes, the bones structurally supported gill pouches, then they became part of the jaw joint. Thus, among reptiles, birds, and mammals, bones that once functioned in gas exchange became specialized for feeding, then for hearing.

Among mammals, an **external ear** that functions in collecting sound waves also became well developed. The external ear of nearly all species has a pinna (sound-collecting, skin-covered flaps of cartilage, typically with elaborate folds, that project from both sides of the head) and a deep channel leading to the middle ear.

Sensory Reception in the Human Ear

Figure 36.12 illustrates the outer, middle, and inner ear of humans. Like other mammals and also birds, humans have a pea-size but highly developed **cochlea**, the portion of the inner ear that receives and sorts out sound waves. Here we find **acoustical receptors**, or vibration-sensitive mechanoreceptors. The ones in the human ear are hair cells, which bend back and forth in

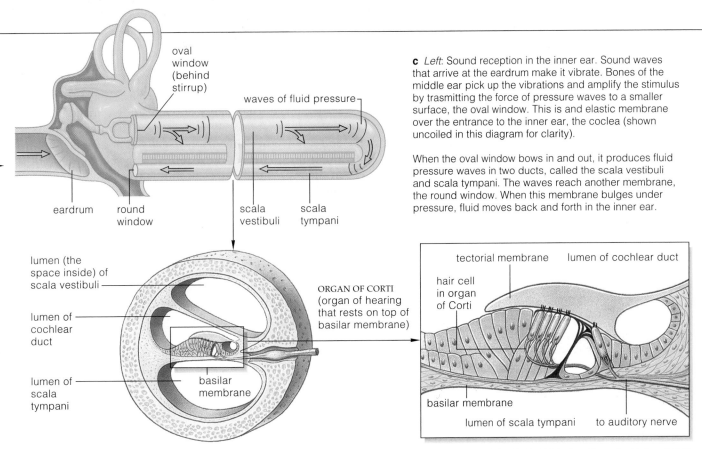

oval window (behind stirrup)

waves of fluid pressure

eardrum   round window

scala vestibuli   scala tympani

c *Left*: Sound reception in the inner ear. Sound waves that arrive at the eardrum make it vibrate. Bones of the middle ear pick up the vibrations and amplify the stimulus by trasmitting the force of pressure waves to a smaller surface, the oval window. This is and elastic membrane over the entrance to the inner ear, the coclea (shown uncoiled in this diagram for clarity).

When the oval window bows in and out, it produces fluid pressure waves in two ducts, called the scala vestibuli and scala tympani. The waves reach another membrane, the round window. When this membrane bulges under pressure, fluid moves back and forth in the inner ear.

lumen (the space inside) of scala vestibuli

lumen of cochlear duct

lumen of scala tympani

basilar membrane

ORGAN OF CORTI (organ of hearing that rests on top of basilar membrane)

tectorial membrane   lumen of cochlear duct

hair cell in organ of Corti

basilar membrane

lumen of scala tympani   to auditory nerve

d Pressure waves are sorted out at the cochlear duct, the third duct of the coiled inner ear. One of its membranes (basilar membrane) starts out narrow and stiff. It is broader and flexible deep in the coil. At different points, the membrane vibrates more strongly to sounds of different frequencies.

e At the organ of Corti, 16,000 hair cells project into a jellylike flap (techtorial membrane) over the basilar membrane. When the flap vibrates suitably, hair cells bend, action potentials arise, and messages travel along an auditory nerve to the brain.

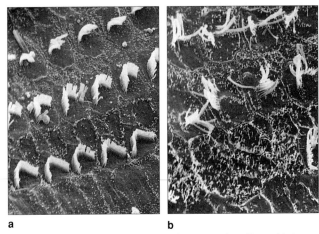

a   b

Figure 36.13 Results of an experiment on the effect of intense sound on the inner ear. (**a**) From a guinea pig's ear, three rows of hair cells that project into the tectorial membrane in the organ of Corti. (**b**) Hair cells in the same organ after twenty-four hours of exposure to noise levels comparable to extremely loud music.

To give you a sense of what "loud" is, a ticking watch measures 10 decibels (100 times louder than the hearing threshold for humans). A normal conversation is about 60 decibels (about a million times louder), a food blender operating at high speed is about 90 decibels (a billion times louder), and a raging rock concert, about 120 decibels (a trillion times louder).

response to pressure waves. This is the start of a flow of information along an auditory nerve that leads from the receptors to the brain. Hair cells can be permanently damaged by prolonged exposure to intense sounds. They are not adapted to amplified music, jet planes, and other recent developments (Figure 36.13).

Echolocation

Earlier, you read that a bat brain responds to signals generated by echolocation, or the use of echoes from ultrasounds that the bat itself produces. Ultrasounds are extremely high-frequency waves, in the 100-decibel range, which humans cannot even hear. Dolphins and whales also emit ultrasounds that travel through water. By perceiving frequency variations in the echoes, these mammals also can pinpoint the distance and direction of movement of one another and of predators or prey.

Hearing among land vertebrates involves structures that collect, amplify, and sort out sound waves traveling through air. In the inner ear, sound waves produce fluid pressure variations, which hair cells transduce into action potentials.

Requirements for Vision

All organisms are sensitive to light. Although they have no photoreceptor cells, sunflowers track the sun as it arcs across the sky. Even single-celled amoebas abruptly stop moving when you shine a light on them. We do not find photoreceptor cells until we poke about among the invertebrates. Even then, not all of the species that have photoreceptors "see" as you do. Many are able to detect a change in light intensity, as a mollusk on the seafloor might do when a fish swims above it. However, they are unable to discern the size or shape of such objects.

At the minimum, the sense of **vision** requires two components: eyes, and a capacity for image formation in brain centers that receive and interpret patterns of visual stimulation being sent to them. Such images are pieced together from information about the shape, brightness, position, and movement of visual stimuli. **Eyes** are sensory organs that contain a tissue with a dense array of photoreceptors. All photoreceptors have this in common: *They incorporate pigment molecules that can absorb photon energy, which can be converted into excitation energy in sensory neurons.*

Nearly all photoreceptor cells have some portion of their surface infolded to a small or large extent, which increases the surface area available for photochemical reactions. Their structure can be traced to the kind of epidermal cells that gave rise to them. The *rhabdomeric* photoreceptors are derived from cells having microvilli. They predominate in the flatworms, mollusks, annelids, arthropods, and echinoderms. By contrast, the *ciliary* photoreceptors have a photoreceptive surface derived from the membrane of a cilium. We find these among the cnidarians, some flatworms, and all vertebrates.

A Sampling of Invertebrate Eyes

SIMPLE EYES Earthworms and some other animals have photoreceptors dispersed through their integument, so they can respond to light even though they have no eyes. The simplest eye is an **ocellus** (plural, ocelli), a patch or cuplike depression of the integument in which photoreceptors and pigmented cells are concentrated. Figure 36.14*a* shows one of these. Such eyes allow the animal to use light for orienting the body, detecting a predator's shadow, or influencing biological clocks.

COMPLEX EYES Vision evolved among fast-moving predators that had to discriminate quickly among prey and other objects that crossed their rapidly changing visual field. Remember, a **visual field** is simply the part of the outside world that an animal sees. At the least, different groups of photoreceptors must sample the

light intensity of different parts of the visual field. The brain interprets the intensities as contrasting parts of a visual image. The quality of the resulting image varies among invertebrates. With the exception of cephalopods and a few other animals, it usually is not good. Why? Good image formation requires many photoreceptors in the eye, and the eyes of most invertebrates are too small to have a lot of them. For example, a planarian's pigment cup only has about 200, compared with the 70,000 per square millimeter in an octopus eye.

Image formation benefits from a **lens**, a transparent body that bends all light rays from a given point in the visual field so that they converge onto photoreceptors (compare Section 4.2). Abalones have a spherical lens that bends incoming light rays but not to the same points, so images that form are blurred (Figure 36.14*b*).

Things improve with a space between the back of the lens and the photoreceptors, as in a snail eye. They get better when the eye has a **cornea**, a transparent cover that directs light rays onto the lens, as happens in a snail or conch (Figure 36.14*c,d*).

Invertebrates on land are better adapted for visually discriminating objects. This is the case for spiders and

epidermis —
photoreceptor —
pigmented cell —

a

epidermis —
transparent body (lens) —
photoreceptor —
sensory cell —

b

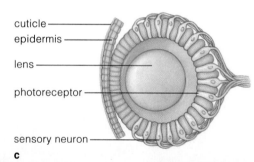

cuticle —
epidermis —
lens —
photoreceptor —
sensory neuron —

c

d

Figure 36.14 Organization of invertebrate eyes, longitudinal section. There are far more photoreceptors than can be shown in these diagrams. (**a**) A limpet's ocellus, a shallow depression in the epidermis that incorporates light-sensitive receptors. (**b**) An abalone eye, with its spherical, transparent lens. (**c**) Eye of a land snail. (**d**) Well-developed eyes of a conch, here peering into the waters of the Great Barrier Reef along the east coast of Australia.

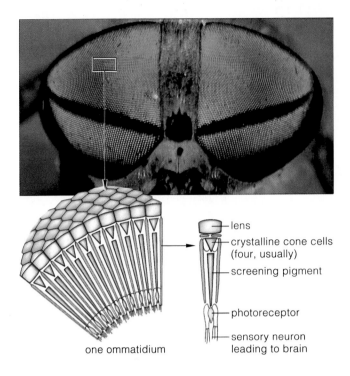

lens

crystalline cone cells
(four, usually)

screening pigment

photoreceptor

sensory neuron
leading to brain

one ommatidium

Figure 36.15 Compound eyes of a deerfly. The lens of each photosensitive unit (ommatidium) directs light onto a crystalline cone, which focuses light on photoreceptor cells below it.

Figure 36.16 Approximation of light reception in an insect eye. This image of a butterfly was formed when a photograph was taken through the outer surface of a compound eye that had been detached from an insect. It may not be what the insect "sees," because integration of signals sent to the brain from photoreceptors may produce a sharper image. This representation is useful insofar as it suggests how the overall visual field may be *sampled* by separate ommatidia.

most other arthropods having simple eyes, in which a single lens services all the photoreceptors. It also is the case with the complex, *compound* eyes of crustaceans and insects. A **compound eye** has many closely packed photosensitive units, each with a bundle of rhabdomeric photoreceptors (Figure 36.15). Some have thousands of units, called ommatidia (singular, ommatidium).

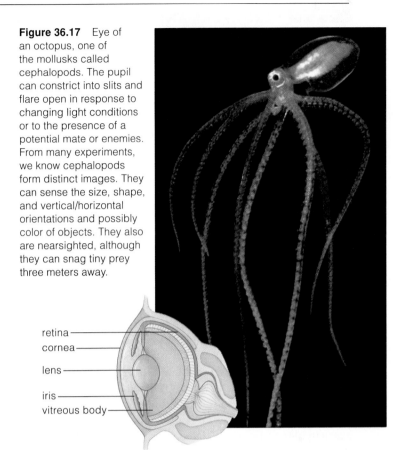

Figure 36.17 Eye of an octopus, one of the mollusks called cephalopods. The pupil can constrict into slits and flare open in response to changing light conditions or to the presence of a potential mate or enemies. From many experiments, we know cephalopods form distinct images. They can sense the size, shape, and vertical/horizontal orientations and possibly color of objects. They also are nearsighted, although they can snag tiny prey three meters away.

retina

cornea

lens

iris

vitreous body

According to the mosaic theory of image formation, each unit samples only a small part of the visual field. The brain builds images based on signals about differences in light intensities, with each unit contributing a small bit to its formation of a visual mosaic (Figure 36.16).

Of all invertebrates, octopuses and squids have the most complex eyes. Like vertebrates, they have **camera eyes**, so named because the eyeball is structured along the lines of a camera. Its interior is a darkened chamber. Light can enter it only through the pupil, an opening in a ring of contractile tissue called the iris (equivalent to a camera's diaphragm). Behind the pupil is a lens that focuses light onto the **retina**, a tissue with a densely packed array of photoreceptors (in a camera, onto light-sensitive film). Axons of sensory nerves converge as an optic tract that leads to the brain (Figure 36.17).

The cephalopods and vertebrates are only remotely related, so similarities in the structure and functions of their eyes may be a result of convergent evolution. By one theory, a group of genes that originally had roles in the development of the nervous system was coopted for the task of eye formation. Whether this happened more than once in different lineages is not yet known.

Vision requires eyes (sensory organs with a dense array of photoreceptors) and a capacity for image formation in the brain, based on patterns of visual stimulation.

The Eye's Structure

Reflect on some points made in the preceding section. *Vision* requires eyes (sensory organs that incorporate a tissue of densely packed photoreceptors) and image formation in brain centers that not only receive but also interpret patterns of stimulation from different parts of the eyeballs. Such images, again, require information about the shape, brightness, position, and movement of visual stimuli. With this in mind, take a look at Figure 36.18, which shows the human eye. Like most vertebrate eyes, the eyeball has a three-layered wall (Table 36.2).

The eyeball's outermost layer consists of the sclera and cornea. The sclera, the dense, fibrous "white" of the eye, protects most of the eyeball. The cornea, made of transparent collagen fibers, covers the rest of it.

The middle layer has a choroid, ciliary body, iris, and pupil. The choroid is a dark-pigmented, vascularized tissue. It absorbs light that photoreceptors have not absorbed and prevents it from scattering inside the eye. The doughnut-shaped, pigmented iris is suspended behind the cornea (the Latin *iris* refers to the rings of a rainbow). The dark "hole" in its center is the pupil, the entrance for light. When rays of bright light strike the eye, circular muscles inside the iris contract and thus shrink the size of the pupil. In dim light, radial muscles contract and so enlarge the pupil.

The inner layer, which includes the retina, is at the back of your eyeball. In birds of prey, it is more toward the eyeball's roof (Figure 36.19).

The eye's interior has a lens. A clear fluid called the aqueous humor bathes the lens, which consists of layers of transparent proteins. The vitreous body, a jellylike substance, fills the chamber behind the lens.

Figure 36.19 Here's looking at you! In owls and some other birds of prey, photoreceptors are concentrated more on top of the inner eyeball, not at back. Such birds look down more than up when they fly and scan the ground for a meal. When they are on the ground, they cannot see something overhead very easily unless they turn their head almost upside-down.

Eyes with a large pupil and iris tell us something about the life-style of their owners. Animals that move about actively at night or in dimly lit habitats are less likely to stumble, bump into objects, or fall off cliffs when they intercept as much of the available light as they can in a given interval. The greater the amount of incoming light, the better is the visual discrimination. A

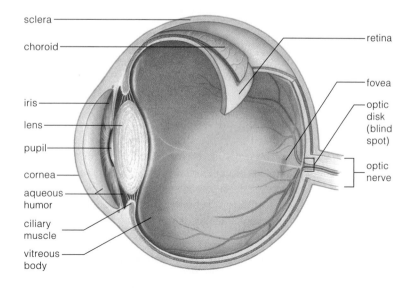

Figure 36.18 Structure of the human eye.

Table 36.2	Components of the Vertebrate Eye	
THREE LAYERS FORMING THE WALL OF EYEBALL:		
Fibrous Tunic:	Sclera. *Protects eyeball*	
	Cornea. *Focuses light*	
Vascular Tunic:	Choroid. *Blood vessels nutritionally support wall cells; pigments prevent light scattering*	
	Cilary body. *Its muscles control lens shape; its fine fibers hold lens in upright position*	
	Iris. *Adjustments here control incoming light*	
	Pupil. *Serves as entrance for light*	
Sensory Tunic:	Retina. *Absorbs and transduces light energy*	
	Fovea. *Increases visual acuity*	
	Start of optic nerve. *Carries signals to brain*	
INTERIOR OF EYEBALL:		
Lens	*Focuses light on photoreceptors*	
Aqueous humor	*Transmits light, maintains pressure*	
Vitreous body	*Transmits light, supports lens and eyeball*	

Figure 36.20 Pattern of retinal stimulation in the human eye. Being curved, the transparent cornea that is positioned in front of the pupil changes the trajectories of light rays as they enter the eye. The pattern is upside-down and reversed left to right, compared to the stimulus in the visual field.

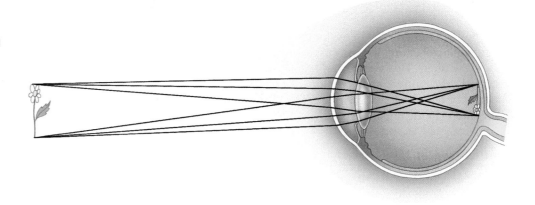

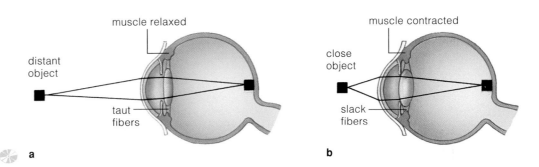

Figure 36.21 Two focusing mechanisms. A ciliary muscle encircles the lens and attaches to it. (**a**) When the muscle relaxes, the lens flattens; the focal point moves farther back and brings distant objects into focus. (**b**) When the muscle contracts, the lens bulges; the focal point moves closer and brings close objects into focus.

muscle relaxed

distant object

taut fibers

a

muscle contracted

close object

slack fibers

b

large pupil lets in large amounts of light. A pupil that is ringed by a large iris can be dilated or constricted to admit more or less light, which is useful for animals that are active at night as well as during the day.

When light rays do converge at the back of the eye, they stimulate the retina in a distinct pattern. Because the cornea has a curved surface, light rays coming from a given point in the visual field hit it at different angles, so their trajectories change. (As described in Section 4.2, rays of light bend at the boundaries between substances of different densities, and bending sends them in new directions.) Because of their newly angled trajectories, the light rays that converge at the back of the eyeball stimulate the retina in a pattern that is upside-down and reversed left to right, relative to the original source of the light rays. Figure 36.20 is a simplified diagram of this outcome.

Visual Accommodation

Light rays emanating from sources at varying distances from the eye strike the cornea at different angles, which means they could end up being focused at different distances behind it. This would do the brain no good. However, adjustments in the position or shape of the lens normally focus all incoming stimuli onto the retina. We call these lens adjustments **visual accommodation**. Without them, light rays from distant objects would be improperly focused in front of the retina, and rays from close objects would be focused behind it.

In fishes and reptiles, eye muscles move the lens forward or back, like a camera's focusing apparatus. Extending the distance between the lens and the retina moves the focal point forward; shrinking the distance moves it back. By contrast, in birds and mammals, the shape of the lens is adjusted. A ciliary muscle encircles the lens and attaches to it by fiberlike ligaments (Figure 36.18). When this muscle relaxes, the lens flattens and thereby moves the focal point farther back, as in Figure 36.21*a*. When the muscle contracts, the lens bulges and moves the focal point forward, as in Figure 36.21*b*.

In some cases, the lens cannot be adjusted enough to make the focal point match up precisely with the retina. For example, in some people, the eyeball is not shaped quite right, and the position of the lens is either too close or too far away from the retina. When this is the case, accommodation alone cannot bring about an exact match. As described in the next section, eyeglasses can correct both visual disorders, which are commonly called nearsightedness and farsightedness.

For most vertebrate eyes, the eyeball has three layers, and it encloses a lens, an aqueous humor, and a vitreous body.

A protective sclera and a light-focusing cornea make up the outer layer. The middle layer has a vascularized, pigmented choroid and other parts that admit and control incoming light. Photoreception occurs at the retina of the inner layer.

Adjustments in the positioning or shape of the lens focus incoming visual stimuli onto the retina.

CASE STUDY: FROM SIGNALING TO VISUAL PERCEPTION

We conclude this chapter with one of the best examples of neuronal architecture, the sensory pathway from the retina to the brain. This is how raw visual information is received, transmitted, and combined; and it leads to awareness of light and shadows, of colors, of near and distant objects in the outside world.

Organization of the Retina

Information flow begins as light strikes the retina. The retina's basement layer, a pigmented epithelium, covers the choroid. Resting on this layer are densely packed arrays of **rod cells** and **cone cells**, which are two classes of ciliary photoreceptors (Figure 36.22). Rod cells detect very dim light. During the night or in darkened places, they contribute to coarse perception of movement by detecting changes in light intensity across the visual field. Cone cells detect bright light. They contribute to sharp vision and color perception during the day.

Distinct layers of neurons are located above the rods and cones. The first to accept the visual information from the photoreceptors are bipolar sensory neurons, then ganglion cells. Axons of these ganglion cells form the two optic nerves to the brain (Figure 36.23).

Before signals depart from the retina, they converge dramatically. The input from 125 million photoreceptors converges on a mere 1 million ganglion cells. Signals also flow laterally among horizontal cells and amacrine cells. Both kinds of neurons act in concert to dampen or

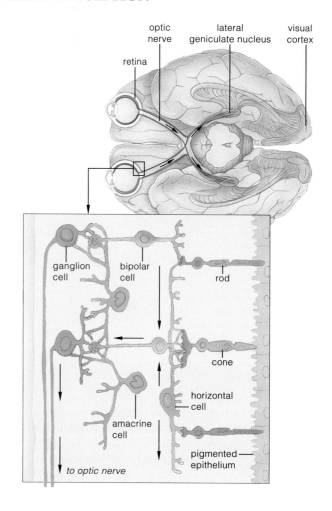

Figure 36.23 The sensory pathway from the retina to the brain.

enhance the signals before ganglion cells get them. Thus, *a great deal of synaptic integration and processing goes on even before visual information is sent to the brain.*

Neuronal Responses to Light

ROD CELLS A rod cell's outer segment contains several hundred membrane disks, each peppered with about 10^8 molecules of a visual pigment called rhodopsin. The membrane stacking and the extremely high density of pigments greatly increase the chances of intercepting packets of light energy—that is, photons of particular wavelengths. The action potentials that result from the absorption of even a few photons can lead to conscious awareness of objects in dimly lit surroundings, as in dark rooms or late at night.

Rhodopsin effectively absorbs wavelengths in the blue-to-green portion of the visible spectrum (Section 7.2). Absorption changes the shape of the pigment. This triggers a cascade of reactions that alter activity at ion

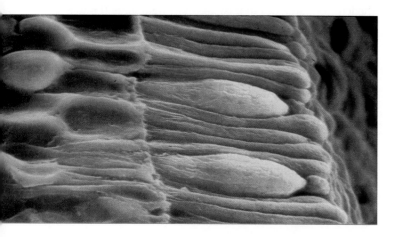

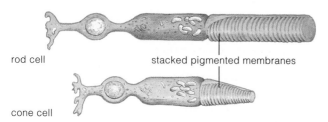

Figure 36.22 Mammalian photoreceptors: rods and cones.

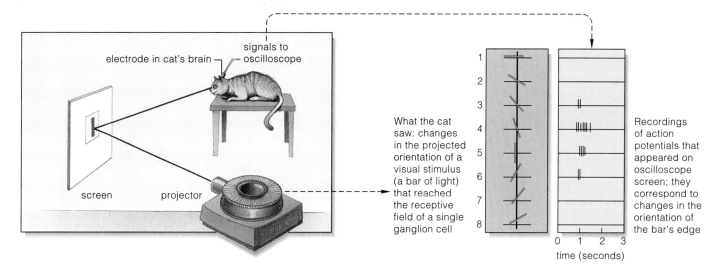

Figure 36.24 Example of experiments into the nature of the receptive fields for visual stimuli. David Hubel and Torsten Wiesel implanted an electrode in an anesthetized cat's brain. They placed the cat in front of a small screen upon which different patterns of light were projected—in this case, a hard-edged bar. Light or shadow falling on part of the screen excited or inhibited signals sent to a single neuron in the visual cortex.

Tilting the bar at different angles produced changes in the neuron's activity. A vertical bar image produced the strongest signal (*numbered 5 in the sketch*). When the bar image tilted slightly, signals were less frequent. When it tilted past a certain angle, signals stopped.

channels and ion pumps across the rod cell's plasma membrane. Gated sodium channels close, the voltage difference changes across the membrane, and a graded potential results. This potential reduces the ongoing release of a neurotransmitter with inhibitory effects on adjacent sensory neurons. Released from inhibition, the sensory neurons start sending signals about the visual stimulus on toward the brain.

CONE CELLS The sense of color and of daytime vision starts with photon absorption by red, green, and blue cone cells, each with a different kind of visual pigment. Here again, photon absorption reduces the release of a neurotransmitter that otherwise inhibits the neurons that are adjacent to the photoreceptors. Near the center of the retina is a funnel-shaped depression called the fovea. This is the retinal area with the greatest density of photoreceptors and the greatest visual acuity. Its cone cells are the ones that discriminate the most precisely between adjacent points in space.

RESPONSES AT RECEPTIVE FIELDS Collectively, the neurons in the eye respond to light in organized ways. The retinal surface is organized into receptive fields, restricted areas that influence the activity of individual sensory neurons. For example, for each ganglion cell, the field is a tiny circle. Some cells respond best to a tiny spot of light, ringed by dark, in the field's center. Others respond to a rapid change in light intensity, to a spot of one color, or to motion. During one experiment, sensory cells generated signals after they detected a suitably oriented bar (Figure 36.24). Such cells do not respond to diffuse, uniform illumination, which is just as well; doing so would produce a confusing array of signals to the brain.

ON TO THE VISUAL CORTEX The part of the outside world that an individual actually sees is its visual field. The right side of both retinas intercepts light from the *left* half of the visual field; the left side intercepts light from the *right* half. The optic nerve leading out of each eye delivers the signals about a stimulus from the *left* visual field to the right cerebral hemisphere, and signals from the *right* visual field to the left hemisphere (Figure 36.23).

Axons of the optic nerves end in a layered brain region, the lateral geniculate nucleus. Each layer has a map corresponding to receptive fields of the retina. Its interneurons deal with one aspect of the stimulus— form, movement, depth, color, texture, and so on. After initial processing, all the visual signals travel rapidly, at the same time, to different parts of the visual cortex. There, final integration produces organized electrical activity and the sensation of sight.

Each eye is an outpost of the brain, collecting and analyzing information about the distance, shape, brightness, position, and movement of visual stimuli. Its sensory pathway starts at the retina and proceeds along an optic nerve to the brain.

Organization of visual signals begins at receptive fields in the retina, it is further processed in the lateral geniculate nucleus, and it is finally integrated in the visual cortex.

36.9 DISORDERS OF THE HUMAN EYE

Of all the sensory receptors that the human brain requires to help keep you alive and functioning independently in the environment, fully two-thirds are located in your pair of eyes. They are photoreceptors, and they do more than detect light. They also allow you to see the world in a rainbow of colors. The eyes are the single most important source of information about the outside world.

Given their essential role, it is no surprise that so much attention is paid to eye disorders, as brought about by injuries, inherited abnormalities, diseases, and advancing age. The consequences range from relatively harmless conditions, such as nearsightedness, to total blindness. Each year, many millions of people must deal with such consequences.

FOCUSING PROBLEMS Other heritable abnormalities arise from misshapen features of the eye that affect the focusing of light. *Astigmatism*, for example, results from corneas with an uneven curvature; they cannot bend incoming light rays to the same focal point.

Nearsightedness, or myopia, commonly occurs when the horizontal axis of the eyeball is longer than the vertical axis. It also occurs when the ciliary muscle responsible for adjusting the lens contracts too strongly. The outcome is that images of distant objects are focused in front of the retina instead of on it (Figure 36.25).

Farsightedness, or hyperopia, is the opposite of myopia. Here, the vertical axis of the eyeball is longer than the horizontal axis (or the lens is "lazy"). As a result, close images are focused behind the retina (Figure 36.26).

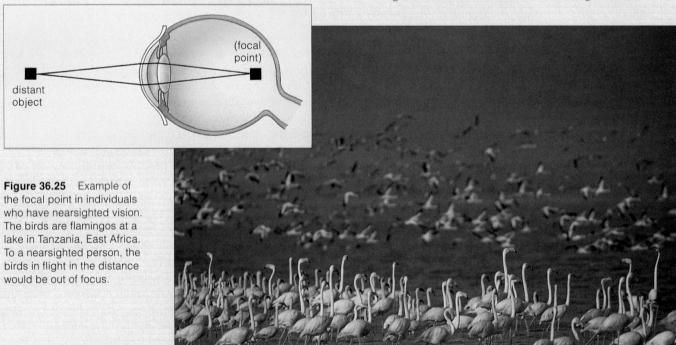

Figure 36.25 Example of the focal point in individuals who have nearsighted vision. The birds are flamingos at a lake in Tanzania, East Africa. To a nearsighted person, the birds in flight in the distance would be out of focus.

COLOR BLINDNESS Occasionally, some or all of the cone cells that selectively respond to light of red, green, or blue wavelengths fail to develop in individuals. The rare individuals who have only one of three kinds of cones are totally *color-blind*. They can perceive the world only in shades of gray.

Consider a common genetic abnormality, *red-green color blindness*. It is an X-linked, recessive trait that shows up most often in males. The retina of affected individuals lacks some or all of the cone cells with pigments that normally respond to light of red or green wavelengths. Most of the time, people who are affected by red-green color blindness merely have trouble distinguishing red from green in dim light. Some cannot distinguish between the two even in bright light.

Even a normal lens loses some of its natural flexibility as a person grows older. That is why people over forty years old often start wearing eyeglasses.

EYE DISEASES The structure and functioning of the eyes are vulnerable to infectious diseases. For example, *histoplasmosis*, a fungal infection of the lungs that is described in Section 24.5, can lead to retinal damage, which can cause partial or total loss of vision. As another example, *Herpes simplex*, a virus that causes skin sores, also can infect the cornea and cause it to ulcerate.

Trachoma is a highly contagious disease that has blinded millions, mostly in North Africa and the Middle East. The culprit is a bacterium that also is responsible for the sexually transmitted disease chlamydia (Section 45.14).

The eyeball and the lining of the eyelids (the conjunctiva) become damaged. The damaged tissues are entry points for bacteria that can cause secondary infections. In time the cornea can become so scarred that blindness follows.

AGE-RELATED PROBLEMS   You have probably heard of *cataracts*, a gradual clouding of the lens. It is a problem associated with aging, although it also may arise through eye injury or diabetes. Possibly cataracts form when the transparent proteins that make up the eye's lens undergo structural changes. The clouding may skew the trajectory of incoming light rays. If the lens becomes opaque, light cannot enter the eye at all.

Glaucoma, a different age-related problem, results when excess aqueous humor accumulates inside the eyeball. The

time it may peel away entirely, leaving its blood supply behind.

Early symptoms of a detached retina include blurred vision, flashes of light that occur even in the absence of outside visual stimulation, and loss of peripheral vision. Without medical intervention, the person may become totally blind in the damaged eye.

NEW TECHNOLOGIES   Today a variety of tools can be used to correct some eye disorders. In *corneal transplant surgery*, for example, a defective cornea can be removed, then an artificial cornea (made of clear plastic) or a natural cornea from a donor can be stitched in its place. Within a year, the patient can be fitted with eyeglasses

(focal point)

close object

Figure 36.26   Example of the focal point in farsighted vision. To a farsighted person, the birds standing in water in the foreground would be out of focus.

blood vessels that service the retina collapse under the increased fluid pressure. Vision deteriorates as sensory neurons of the retina and optic nerve die off.

Although chronic glaucoma often is associated with advanced age, the conditions that give rise to the disorder actually start to develop in middle age. If detected early enough, the fluid pressure that tends to build up in the eye can be relieved by drugs or surgery before damage becomes severe.

EYE INJURIES   *Retinal detachment* is the eye injury that we read about most often. It may follow a physical blow to the head or illness that causes a tear in the retina. As the jellylike vitreous body oozes through the torn region, the retina is lifted away from the underlying choroid. In

or contact lenses. Similarly, cataracts can be surgically corrected by removing the lens and replacing it with an artificial one, although the operation is difficult and not always successful.

Severely nearsighted people sometimes opt for *radial keratotomy*, a still-controversial surgical procedure in which tiny, spokelike incisions are made around the edge of the cornea to flatten it more. When all goes well, the adjustment eliminates the need for corrective lenses. Sometimes the result is overcorrected or undercorrected vision, and more surgery is required

As a final example, retinal detachment may be treated with *laser coagulation*, a painless technique in which a laser beam seals off leaky blood vessels and "spot welds" the retina to the underlying choroid.

SUMMARY

1. A stimulus is a specific form of energy that the body detects by means of sensory receptors. A sensation is an awareness that stimulation has occurred. Perception is understanding what the sensation means.

2. Sensory receptors are the endings of sensory neurons or specialized cells adjacent to them. They respond to stimuli, which are specific forms of energy, such as light and mechanical pressure. Animals respond to aspects of the internal or external environment when they have receptors that are sensitive to the energy of stimuli.

   a. Mechanoreceptors, such as free nerve endings, can detect mechanical energy related to touch, pressure, and motion and changes in position.

   b. Thermoreceptors detect the presence of or changes in radiant energy from heat sources.

   c. Pain receptors (nociceptors) detect tissue damage.

   d. Chemoreceptors, such as olfactory receptors and taste receptors, detect chemical substances dissolved in fluids that are bathing them.

   e. Osmoreceptors detect changes in water volume (hence solute concentrations) in the surrounding fluid.

   f. Photoreceptors, which include the rods and cones of the retina in the human eye, detect light.

3. A sensory system has sensory receptors for specific stimuli and nerve pathways from those receptors to receiving and processing centers in the brain. The brain assesses a particular stimulus based on which nerve pathway is delivering the signals, the frequency of signals traveling along each axon of that pathway, and the number of axons that were recruited into action.

4. Somatic sensations include touch, pressure, pain, temperature, and muscle sense. The receptors associated with these sensations are not localized in a single organ or tissue. The special senses include taste, smell, hearing, balance, and vision. The receptors associated with these senses typically reside in sensory organs, such as eyes, or some other particular body region.

5. The senses of taste and smell both involve sensory pathways from chemoreceptors to processing regions in the cerebral cortex and limbic system.

6. Organs of equilibrium, as in the vertebrate inner ear, detect gravity, velocity, acceleration, and other forces that affect the body's positions and movements. The vertebrate sense of hearing (sound perception) requires components of the outer, middle, and inner ear that respectively collect, amplify, and sort out sound waves.

7. Vision requires eyes (sensory organs with a dense array of photoreceptors, as in a retina) and a capacity for image formation in the brain, based upon incoming patterns of visual stimulation. The sense of vision and discrimination among objects evolved first among fast-moving, predatory animals.

Review Questions

1. What is a stimulus? When sensory receptors detect a specific stimulus, what happens to the stimulus energy? *36.1*

2. Name six categories of sensory receptors and the type of stimulus energy that each kind detects. *36.1*

3. How do somatic sensations differ from special senses? *36.1*

4. What is pain? Describe one type of pain receptor. *36.2*

5. Describe the properties of sound. Which evolved first, the sense of balance or sense of hearing? Do both senses require participation of the outer, middle, and inner ear? *36.4, 36.5*

6. How does vision differ from light sensitivity? What sensory organs and structures does vision require? *36.6, 36.7*

7. Label the component parts of the human eye: *36.7*

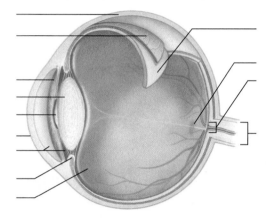

8. How does the vertebrate eye focus light rays from a visual stimulus? What do nearsighted and farsighted mean? *36.7, 36.9*

9. On the bell-shaped rim of the jellyfish shown in Figure 36.27 are tiny saclike structures that hold calcium particles next to a sensory cilium. When the bell tilts, the particles slide over the cilium. Would you assume that these structures contribute to the sense of taste, smell, balance, hearing, or vision? *36.4*

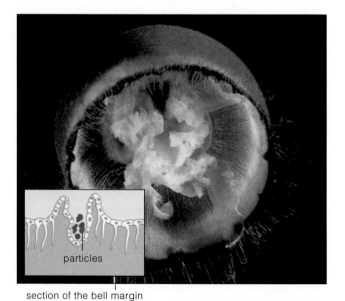

particles

section of the bell margin

Figure 36.27 Sensory structures located on the bell-shaped rim of a medusa (jellyfish).

Self-Quiz (Answers in Appendix IV)

1. A _____ is a specific form of energy that is detected by a sensory receptor.

2. Conscious awareness of a stimulus is called a _____ .

3. _____ is understanding what particular sensations mean.

4. Each sensory system consists of _____ .
   a. nerve pathways from specific receptors to the brain
   b. sensory receptors
   c. brain regions that deal with sensory information
   d. all of the above

5. _____ is (are) a decrease in the response to an ongoing stimulus.
   a. Perception   c. Visual accommodation
   b. Sensory adaptation d. both b and c

6. _____ detect mechanical energy associated with changes in pressure, position, or acceleration.
   a. Chemoreceptors   c. Photoreceptors
   b. Mechanoreceptors d. Thermoreceptors

7. _____ are chemoreceptors.
   a. Taste receptors   d. Photoreceptors
   b. Olfactory receptors e. both a and b
   c. Auditory receptors   f. both a and c

8. _____ detect infrared energy (heat).
   a. Chemoreceptors   c. Photoreceptors
   b. Mechanoreceptors d. Thermoreceptors

9. Detecting light energy is the function of _____ .
   a. chemoreceptors   c. photoreceptors
   b. mechanoreceptors d. thermoreceptors

10. Vision requires _____ .
    a. a tissue with dense arrays of photoreceptors
    b. eyes
    c. image-forming centers in brain
    d. all of the above

11. The outer layer of the human eyeball includes the _____ .
    a. lens and choroid   c. retina
    b. sclera and cornea d. both a and c

12. The inner layer of the human eyeball includes the _____ .
    a. lens and choroid   c. retina
    b. sclera and cornea d. both a and c

13. Match each term with the most suitable description.
    ___ somatic senses
    ___ stimulus
    ___ special senses
    ___ variations in stimulus intensity
    ___ sensory receptor

    a. sensory neuron endings or specialized cells next to them
    b. taste, smell, hearing, balance, and vision
    c. form of energy that a specific sensory receptor can detect
    d. encoded in the frequency and number of action potentials
    e. touch, pressure, temperature, pain, and muscle sense

Critical Thinking

1. Wayne, on standby for the last flight from San Francisco to New York, was assigned the last available seat on the plane. It was in the last row, where vibrations and noise from the engines are most pronounced. When Wayne got off the plane in New York, he was speaking very loudly and having trouble hearing what other people were saying. The next day, things were back to normal. Speculate on what happened to his sense of hearing during the flight.

2. Juanita made an appointment with her doctor because she was experiencing recurring episodes of dizziness. Her doctor immediately asked her whether "dizziness" meant she had sensations of lightheadedness, as if she were going to faint, or whether it meant she had sensations of *vertigo*—that is, a feeling that she herself or objects near her were spinning around. Why did her doctor consider this clarification important early in his evaluation of her condition?

3. Michael, now three years old, experiences chronic *middle-ear infection*, which is becoming quite common among youngsters enrolled in crowded day-care centers. This year, despite antibiotic treatment, an infection became so advanced that he had some trouble hearing. Then his left eardrum ruptured and a thickened, jellylike substance dribbled out. His mother was startled and upset, but the pediatrician told her not to worry, that if the eardrum had not ruptured on its own, he would have had to insert a small drainage tube into it. Speculate on why the pediatrician concluded that this would have been a necessary surgical intervention.

Selected Key Terms

acoustical receptor 36.5	otolith 36.4
camera eye 36.6	pain 36.2
chemoreceptor 36.1	pain receptor 36.1
cochlea 36.5	perception 36.1
compound eye 36.6	pheromone 36.3
cone cell 36.8	photoreceptor 36.1
cornea 36.6	retina 36.6
echolocation CI	rod cell 36.8
encapsulated receptor 36.2	sensation 36.1
external ear 36.5	sensory adaptation 36.1
eye 36.6	sensory system 36.1
free nerve ending 36.2	somatic sensation 36.1
hair cell 36.4	somatosensory cortex 36.2
hearing 36.5	special sense 36.1
inner ear 36.4	taste receptor 36.3
lens 36.6	thermoreceptor 36.1
mechanoreceptor 36.1	vestibular apparatus 36.4
middle ear 36.5	vision 36.6
ocellus 36.6	visual accommodation 36.7
olfactory receptor 36.3	visual field 36.6
osmoreceptor 36.1	

Readings

Kandel, E., and J. Schwartz. 1995. *Essentials of Neural Science and Behavior*. Norwalk, Connecticut: Appleton and Lange.

Nicholls, J., A. Martin, and B. Wallace. 1992. *From Neuron to Brain*. Third edition. Sunderland, Massachusetts: Sinauer.

Romer, A., and T. Parsons. 1986. *The Vertebrate Body*. Sixth edition. New York: Saunders.

Sherwood, L. 1997. *Human Physiology*. Third edition. Belmont, California: Wadsworth.

Wright, Karen. April 1994. "The Sniff of Legend." *Discover* 15(4): 60–67. Speculation on the existence of a sensory pathway activated by human pheromones.

Zeki, S. September 1992. "The Visual Image in Mind and Brain." *Scientific American* 267(3): 68–76.

Web Site See *http://www.wadsworth.com/biology* for practice quiz questions, hypercontents, BioUpdates, and critical thinking. The Wadsworth Biology Resource Center provides a wealth of information fully organized and integrated by chapter.

37 ENDOCRINE CONTROL

Hormone Jamboree

In the 1960s, at her camp in a forest by the shore of Lake Tanganyika in Tanzania, the primatologist Jane Goodall let it be known that bananas were available. One of the first chimpanzees attracted to the delicious food was a female—Flo, as she came to be called (Figure 37.1). Flo brought along her two offspring, an infant female and a juvenile male, and exhibited commendable parental behavior toward them.

Three years passed, and Goodall observed that Flo's preoccupation with motherhood gave way to a preoccupation with sex. She also observed that male chimpanzees followed Flo to the camp by the lake and stayed for more than the bananas.

Sex, as Goodall discovered, is *the* premier binding force in the social life of chimpanzees. These primates do not mate for life as eagles do, or wolves. Before the rainy season begins, mature females that are entering their fertile cycle

a

b

Figure 37.1 (**a**) Jane Goodall in Gombe National Park, near the shores of Lake Tanganyika, scouting for chimpanzees. (**b**) Flo and three of her offspring, all subjects of long-term field observations that clarified the central role of sex—and of the hormones that orchestrate it—in the social life of these primate relatives of humans.

become sexually active. As one of the more dramatic responses to changing concentrations of hormones in their bloodstream, the external sexual organs of the females become swollen and vivid pink. The swellings are strong visual signals to males. They are the flags of sexual jamborees, of great gatherings of highly stimulated chimps in which any males present may copulate in sequence with the same females.

The gathering of many flag-waving females draws together individuals that forage alone or in small family groups for most of the year. It reestablishes the bonds that unite their rather fluid community. As infants and juveniles play with one another and with the adults, their aggressive and submissive jostlings help map out future dominance hierarchies. Consider Flo, a high-ranking member of the social hierarchy. Through her sexual attractiveness and direct solicitations, she built alliances with many male chimps. Through her status and aggressive behavior, she helped her male offspring win confrontations with other young male chimps.

The hormone-induced swelling during a female's fertile period lasts somewhere between ten and sixteen days. Yet she is fertile for only one to five days. Sex hormones induce the swelling even after a female has become pregnant. Almost certainly, sexual selection has favored the prolonged swelling. Males groom a sexually attractive female more often, protect her, give her more food, and let her tag along to new foraging sites. The more that males accept a female, the higher she rises in the social hierarchy—and the more her individual offspring benefit.

Through their effects, hormones help orchestrate the growth, development, and reproductive cycles of nearly all animals, from the invertebrate worms to chimpanzees and humans. They influence minute-by-minute and day-to-day metabolic functions. Through interplays with one another and with the nervous system, hormones have profound influence over the physical appearance, the well-being, and the behavior of individuals. And even the behavior of individuals affects whether they will survive, either on their own or as part of social groups.

This chapter focuses mainly on hormones—on their sources, targets, and interactions as well as the mechanisms involved in their secretion. If the details start to seem remote, remember that this is the stuff of life. Hormones underwrote Flo's appearance, behavior, and rise through the chimpanzee social hierarchy—and just imagine what they have been doing for you.

KEY CONCEPTS

1. For nearly all animals, hormones and other signaling molecules have central roles in integrating the activities of individual cells in ways that benefit the whole body.

2. Only the cells with molecular receptors for a specific hormone are its targets. Hormones operate by serving as signals for change in the activities of their targets.

3. Many types of hormones influence gene transcription and protein synthesis in target cells. Other types call for alterations in existing molecules and structures in cells. Some exert their effects by binding with and altering membrane characteristics, such as the permeability of the plasma membrane to a particular solute.

4. Some hormones help the body adjust to short-term shifts in diet and in levels of activity. Others help induce long-term adjustments in cell activities that bring about bodily growth, development, and reproduction.

5. Among vertebrates, the hypothalamus and pituitary gland interact in ways that coordinate the activities of a number of endocrine glands. Together, they exert control over many of the body's functions.

6. Besides hormonal signals, neural signals, changes in local chemical conditions, and environmental cues serve as the triggers for hormonal secretion.

Hormones and Other Signaling Molecules

Throughout their lives, cells must respond to changing conditions by taking up and releasing various chemical substances. In vertebrates, the responses of millions to many billions of cells must be integrated in ways that benefit the whole body.

Integration of cell activities depends upon signaling molecules. These include hormones, neurotransmitters, local signaling molecules, and pheromones. Each type of signaling molecule acts on target cells, which are any cells that have receptors for the molecule and that may alter their activities in response to it. A target may or may not be adjacent to the cell that sends the signal.

By definition, **hormones** are the secretory products of endocrine glands, endocrine cells, and some neurons, and they travel the bloodstream to nonadjacent target cells. They are this chapter's focus.

By contrast, **neurotransmitters** are released from axon endings of neurons, then act swiftly on target cells by diffusing across the tiny gap that separates them. Section 34.3 describes their sources and their action. Also, **local signaling molecules,** released by many types of body cells, alter conditions within localized regions of tissues. Prostaglandins, for instance, target smooth muscle cells in bronchiole walls, which then constrict or dilate and so alter air flow in lungs.

Pheromones, nearly odorless secretions of particular exocrine glands, diffuse through water or air to targets outside the animal body. These hormone-like secretions act on cells of other individuals of the same species and help integrate social behavior, such as behaviors related to sexual reproduction. Pheromones are common signals among animals. For example, female silk moths secrete bombykol as a sex attractant (Section 36.3); and soldier termites secrete alarm signals when ants attack their colony (Section 51.9). Researchers have also discovered a pheromone detector, a *vomeronasal* organ, in humans. Do pheromones act at the subconscious level to trigger inexplicable impressions, such as spontaneous good or bad "feelings" about someone you just met? We do not know the answer, but studies are under way.

Discovery of Hormones and Their Sources

The word "hormone" dates back to the early 1900s. Then, W. Bayliss and E. Starling were attempting to find out what triggers the secretion of pancreatic juices when food is traveling through the canine gut. As they knew, acids are mixed with food inside the stomach, and the pancreas then secretes an alkaline solution after the acidic mixture has been propelled forward, into the small intestine. Was the nervous system or something else stimulating the pancreatic response?

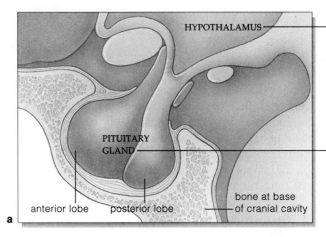

Figure 37.2   (**a**) A major neural-endocrine control center. The pituitary gland interacts intimately with the hypothalamus, a brain region that secretes some hormones. (**b**) *Facing page:* Overview of the key components of the human endocrine system and of the primary effects of their main hormonal secretions. The system also includes endocrine cells of many organs, including the liver, kidneys, heart, small intestine, and skin.

To find the answer, Bayliss and Starling blocked nerves—but not blood vessels—leading to a laboratory animal's upper small intestine. Later, when acidic food entered the small intestine, the pancreas still secreted the alkaline solution. Even more telling, extracts of cells from the intestinal lining—a glandular epithelium—also induced the response. Glandular cells had to be the source of the pancreas-stimulating substance.

The substance came to be called secretin. Proof of its existence and mode of action confirmed a centuries-old idea: *The bloodstream picks up internal secretions that can influence the activities of organs inside the body.* Starling coined the word "hormone" for such internal glandular secretions (after *hormon,* meaning to set in motion).

Later on, researchers identified many other kinds of hormones and their sources. Figure 37.2 is a simplified picture of the locations of the following major sources of hormones in the human body. Bear in mind, these sources are typical of most vertebrates:

Pituitary gland

Adrenal glands (*two*)

Pancreatic islets (*numerous cell clusters*)

Thyroid gland

Parathyroid glands (*in humans, four*)

Pineal gland

Thymus gland

Gonads (*two*)

Endocrine cells of the stomach, small intestine, liver, kidneys, heart, placenta, skin, adipose tissue, and other organs

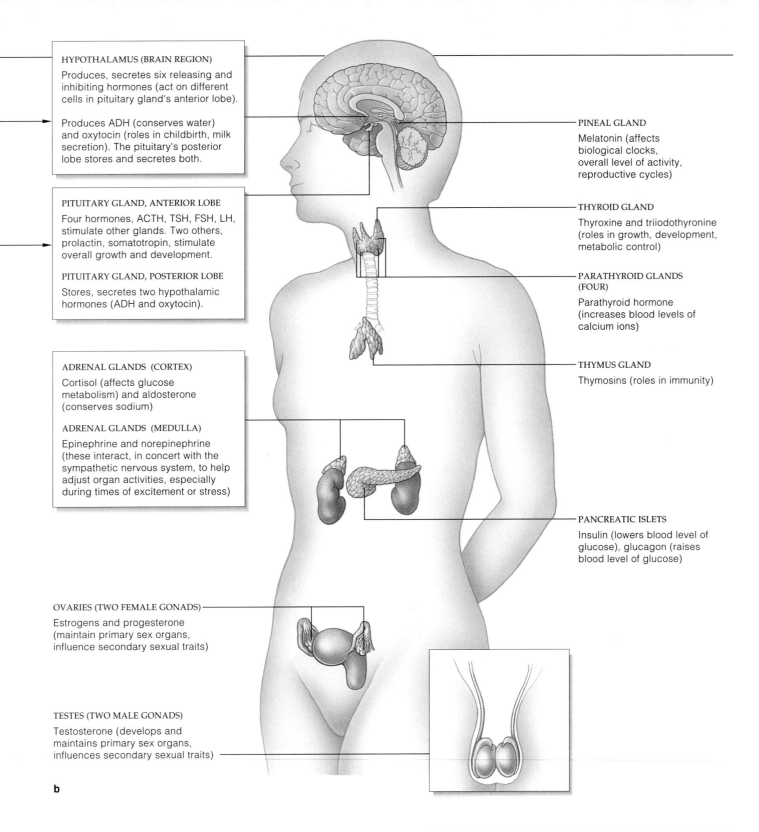

HYPOTHALAMUS (BRAIN REGION)

Produces, secretes six releasing and inhibiting hormones (act on different cells in pituitary gland's anterior lobe).

Produces ADH (conserves water) and oxytocin (roles in childbirth, milk secretion). The pituitary's posterior lobe stores and secretes both.

PITUITARY GLAND, ANTERIOR LOBE

Four hormones, ACTH, TSH, FSH, LH, stimulate other glands. Two others, prolactin, somatotropin, stimulate overall growth and development.

PITUITARY GLAND, POSTERIOR LOBE

Stores, secretes two hypothalamic hormones (ADH and oxytocin).

ADRENAL GLANDS (CORTEX)

Cortisol (affects glucose metabolism) and aldosterone (conserves sodium)

ADRENAL GLANDS (MEDULLA)

Epinephrine and norepinephrine (these interact, in concert with the sympathetic nervous system, to help adjust organ activities, especially during times of excitement or stress)

OVARIES (TWO FEMALE GONADS)

Estrogens and progesterone (maintain primary sex organs, influence secondary sexual traits)

TESTES (TWO MALE GONADS)

Testosterone (develops and maintains primary sex organs, influences secondary sexual traits)

PINEAL GLAND

Melatonin (affects biological clocks, overall level of activity, reproductive cycles)

THYROID GLAND

Thyroxine and triiodothyronine (roles in growth, development, metabolic control)

PARATHYROID GLANDS (FOUR)

Parathyroid hormone (increases blood levels of calcium ions)

THYMUS GLAND

Thymosins (roles in immunity)

PANCREATIC ISLETS

Insulin (lowers blood level of glucose), glucagon (raises blood level of glucose)

b

Collectively, the body's sources of hormones came to be called the **endocrine system**. The name implies that there is a separate control system for the body, apart from the nervous system. (*Endon* means within; *krinein* means separate.) Later, however, biochemical research and electron microscopy studies revealed that endocrine sources and the nervous system function in intricately connected ways, as you will see shortly.

Integration of cell activities depends on hormones and other signaling molecules. Each type of signaling molecule acts on target cells, which are any cells that have receptors for it and that may alter their activities in response to it.

Components of the endocrine system and certain neurons produce and secrete hormones, which the bloodstream takes up and distributes to nonadjacent target cells.

The Nature of Hormonal Action

Hormones and other signaling molecules interact with protein receptors of target cells, and their interactions have diverse effects on physiological processes. Some kinds of hormones induce target cells to increase their uptake of glucose, calcium, or some other substance from the surroundings. Other kinds stimulate or inhibit target cells into altering their rates of protein synthesis, modifying the structure of proteins or other elements in the cytoplasm, or changing cell shape.

Two factors exert considerable influence over the responses to hormonal signals. *First,* different hormones activate different cellular mechanisms. *Second,* not all cell types are equipped to respond to a given signal. For example, many cell types have receptors for cortisol, so this hormone has widespread effects on the body. By contrast, only a few cell types have the receptors for hormones that stimulate a highly directed response.

Let us now briefly consider the effects of two main categories of these signaling molecules: the steroid and peptide hormones (Table 37.1).

Characteristics of Steroid Hormones

Steroid hormones, recall, are lipid-soluble molecules derived from cholesterol (Section 3.5). The cholesterol remodeling proceeds at multiple-enzyme systems in the endoplasmic reticulum and the mitochondria of steroid-secreting cells. We find such cells in adrenal glands and primary reproductive organs. Testosterone, one of the sex hormones, is an example of the final products.

Consider testosterone's effects on the development of the secondary sexual traits associated with maleness. The developmental steps will proceed as normal only if target cells have working receptors for testosterone. In a genetic disorder called *testicular feminization syndrome,* these receptors are defective. Genetically, the affected person is male (XY); he has functional testes that are

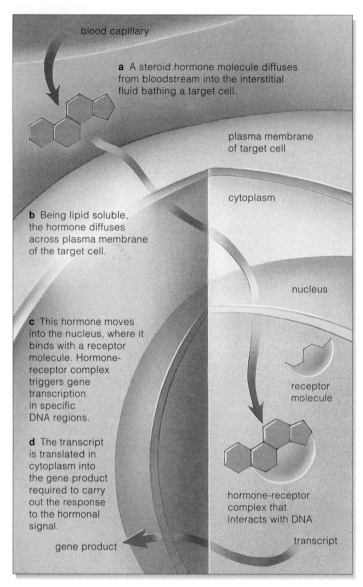

Figure 37.3 Example of a mechanism by which a steroid hormone initiates changes in a target cell's activities.

(image labels:)
blood capillary

a A steroid hormone molecule diffuses from bloodstream into the interstitial fluid bathing a target cell.

plasma membrane of target cell

cytoplasm

b Being lipid soluble, the hormone diffuses across plasma membrane of the target cell.

nucleus

c This hormone moves into the nucleus, where it binds with a receptor molecule. Hormone-receptor complex triggers gene transcription in specific DNA regions.

receptor molecule

d The transcript is translated in cytoplasm into the gene product required to carry out the response to the hormonal signal.

hormone-receptor complex that interacts with DNA

gene product

transcript

able to secrete testosterone. Target cells cannot respond to the hormone, however, so the secondary sexual traits that do develop are like those of females.

How does a steroid hormone such as testosterone exert effects on cells? Being lipid soluble, it may diffuse directly across the lipid bilayer of a target cell's plasma membrane (Figure 37.3). Once inside the cytoplasm, the hormone molecule usually moves into the nucleus and binds to some type of protein receptor. Or it binds to a receptor molecule in the cytoplasm in some cases, then the hormone-receptor complex enters the nucleus. The configuration of the complex allows it to interact with a specific region of the cell's DNA. Different complexes inhibit or stimulate transcription of certain gene regions

Table 37.1	Two Main Categories of Hormones
Type	Examples
Steroid and steroid-like hormones	Estrogens (feminizing effects), progestins (related to pregnancy), androgens (such as testosterone; masculinizing effects), cortisol, aldosterone. In addition, thyroid hormones and vitamin D act like steroids.
Peptide hormones:	
Peptides	Glucagon, ADH, oxytocin, TRH
Proteins	Insulin, somatotropin, prolactin
Glycoproteins	FSH, LH, TSH

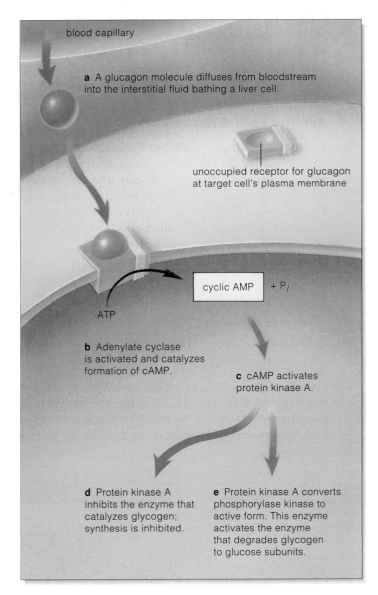

a A glucagon molecule diffuses from bloodstream into the interstitial fluid bathing a liver cell.

blood capillary

unoccupied receptor for glucagon at target cell's plasma membrane

cyclic AMP + P_i

ATP

b Adenylate cyclase is activated and catalyzes formation of cAMP.

c cAMP activates protein kinase A.

d Protein kinase A inhibits the enzyme that catalyzes glycogen; synthesis is inhibited.

e Protein kinase A converts phosphorylase kinase to active form. This enzyme activates the enzyme that degrades glycogen to glucose subunits.

Figure 37.4 Example of a mechanism by which a peptide hormone initiates changes in a target cell's activities. When the hormone glucagon binds at a receptor, it initiates reactions inside the cell. In this case cyclic AMP, which is one type of second messenger, relays the signal into the cell interior.

into mRNA. The translation of such mRNA transcripts results in enzymes and other proteins that can carry out a response to the hormonal signal.

Steroid hormones also may exert effects in another way. It now appears they can bind to cell membranes and alter the membrane properties in ways that modify the functions of the target cell.

One final point: Thyroid hormones and vitamin D are not steroid hormones, but they do behave like them. Also, the genes coding for their receptors are part of the same group that codes for steroid hormones.

Characteristics of Peptide Hormones

Traditionally, **peptide hormones** have been defined as water-soluble signaling molecules that may incorporate anywhere from 3 to 180 amino acids. Various peptides, polypeptides, and glycoproteins fall into this category. All peptide hormones issue signals right at a receptor of a target cell's plasma membrane. The signals activate specific membrane-bound enzyme systems that initiate reactions leading to the cellular response.

Consider a liver cell with receptors for glucagon, a peptide hormone. Every receptor spans the plasma membrane, and part of it extends into the cytoplasm. After a receptor binds glucagon, the response requires assistance from a **second messenger**. Such messengers are small molecules in the target cell's cytoplasm, and they relay a signal from a hormone-receptor complex at the plasma membrane into the cell interior. In this case, cAMP (cyclic adenosine monophosphate) is the second messenger. As Figure 37.4 indicates, glucagon binding triggers activity at a membrane-bound enzyme system. First an activated form of the enzyme adenylate cyclase converts ATP to cyclic AMP, which starts a cascade of reactions. Many molecules of cyclic AMP form and act as signals for the conversion of many molecules of an enzyme, a protein kinase, to active form. These act on other enzymes, and so on until a final reaction converts stored glycogen in the cell to glucose. In short order, the number of molecules involved in the final response to the glucagon-receptor complex is enormous.

Or consider a muscle cell with receptors for insulin, a protein hormone. Among other things, a signal from a hormone-receptor complex triggers the movement of molecules of glucose transporter proteins through the cytoplasm and insertion into the plasma membrane, so that the cell can take up glucose faster. In addition, it switches on enzymes of glucose metabolism.

Bear in mind, there are other hormone categories, including the catecholamines such as epinephrine. Like glucagon, epinephrine combines with specific surface receptors and triggers the release of cyclic AMP as a second messenger that assists in the cellular response.

Hormones interact with receptors located either at the plasma membrane or within the cytoplasm of target cells.

Steroid hormone-receptor complexes typically enter a target cell and interact with the cell's DNA. Also, some types interact with membranes and alter membrane functions.

Peptide hormones do not exert their effects by entering a cell. When they bind to a membrane receptor, the binding itself is a signal for enzyme-mediated, intracellular events. Often a second messenger inside the cytoplasm relays the signal into the cell interior.

Deep in the forebrain is the **hypothalamus**. This brain region monitors internal organs and activities related to their functioning, such as eating and sexual behavior. It also secretes some hormones. Suspended from its base by a slender stalk of tissue is a lobed gland, about the size of a pea. The hypothalamus and this **pituitary gland** interact as a major neural-endocrine control center.

The pituitary's *posterior* lobe secretes two hormones that are synthesized by the hypothalamus. Its *anterior* lobe produces and secretes its own hormones, most of which help control the release of hormones from other endocrine glands (Table 37.2). The pituitary of many vertebrates—*not* humans—has an intermediate lobe as well. In many cases, the third lobe secretes a hormone that governs reversible changes in skin or fur color.

Posterior Lobe Secretions

Figure 37.5*a* shows the cell bodies of certain neurons in the hypothalamus. Their axons extend down into the posterior lobe, then terminate next to a capillary bed. The neurons produce antidiuretic hormone (ADH) and oxytocin, then store them in the axon endings. After either hormone is released into the interstitial fluid, it diffuses into capillaries, then travels the bloodstream to its targets. ADH acts on cells of nephrons and collecting ducts in the kidneys, which filter the blood and rid the body of excess water and salts (in the form of urine). ADH promotes water reabsorption when the body must conserve water. Oxytocin has roles in reproduction. It triggers contractions of the uterus during labor, and it causes milk release when offspring are being nursed.

Anterior Lobe Secretions

ANTERIOR PITUITARY HORMONES Inside the pituitary stalk, a capillary bed picks up hormones secreted by the hypothalamus and delivers them into a *second* capillary bed in the anterior lobe. There, the hormones leave the bloodstream and then act on target cells. As Figure 37.6 shows, different cells of the anterior pituitary secrete six hormones that they themselves produce:

Corticotropin	ACTH
Thyrotropin	TSH
Follicle-stimulating hormone	FSH
Luteinizing hormone	LH
Prolactin	PRL
Somatotropin (or growth hormone)	STH (or GH)

The effects of these hormones are widespread through the body. ACTH and TSH orchestrate secretions from the adrenal glands and thyroid gland, respectively, as described shortly. FSH and LH act through the gonads

Table 37.2 Hormones Released From the Mammalian Pituitary Gland

Pituitary Lobe	Secretions	Designation	Main Targets	Primary Actions
POSTERIOR Nervous tissue (extension of hypothalamus)	Antidiuretic hormone	ADH	Kidneys	Induces water conservation as required during control of extracellular fluid volume and solute concentrations
	Oxytocin	OCT	Mammary glands	Induces milk movement into secretory ducts
			Uterus	Induces uterine contractions
ANTERIOR Glandular tissue, mostly	Corticotropin	ACTH	Adrenal cortex	Stimulates release of adrenal steroid hormones
	Thyrotropin	TSH	Thyroid gland	Stimulates release of thyroid hormones
	Follicle-stimulating hormone	FSH	Ovaries, testes	In females, stimulates estrogen secretion, egg maturation; in males, helps stimulate sperm formation
	Luteinizing hormone	LH	Ovaries, testes	In females, stimulates progesterone secretion, ovulation, corpus luteum formation; in males, stimulates testosterone secretion, sperm release
	Prolactin	PRL	Mammary glands	Stimulates and sustains milk production
	Somatotropin (or growth hormone)	STH (GH)	Most cells	Promotes growth in young; induces protein synthesis, cell division; roles in glucose, protein metabolism in adults
INTERMEDIATE* Glandular tissue, mostly	Melanocyte-stimulating hormone	MSH	Pigmented cells in skin and other integuments	Induces color changes in response to external stimuli; affects some behaviors

* Present in most vertebrates (not humans). MSH is associated with the anterior lobe in humans.

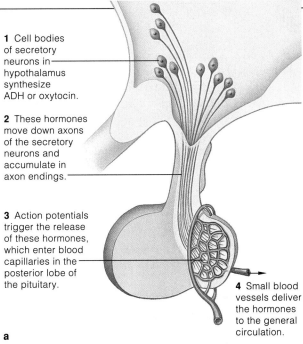

1 Cell bodies of secretory neurons in hypothalamus synthesize ADH or oxytocin.

2 These hormones move down axons of the secretory neurons and accumulate in axon endings.

3 Action potentials trigger the release of these hormones, which enter blood capillaries in the posterior lobe of the pituitary.

4 Small blood vessels deliver the hormones to the general circulation.

a

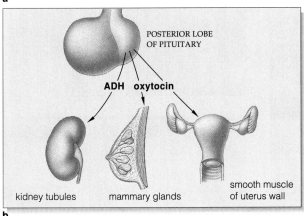

POSTERIOR LOBE OF PITUITARY

ADH oxytocin

kidney tubules   mammary glands   smooth muscle of uterus wall

b

Figure 37.5 (**a**) Functional links between the hypothalamus and the posterior lobe of the pituitary gland. (**b**) Main targets of the posterior lobe secretions.

1 Cell bodies of secretory neurons in hypothalamus secrete releasing and inhibiting hormones.

2 First capillary bed, in base of hypothalamus, picks up hormones.

3 Hormones are delivered into second capillary bed, in anterior lobe of pituitary.

4 Releasing or inhibiting hormones diffuse out of the capillaries, act on endocrine cells in the anterior lobe.

5 Hormones secreted from anterior lobe cells enter venules that lead to the general circulation.

a

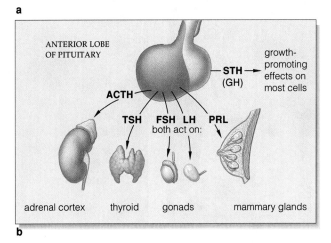

ANTERIOR LOBE OF PITUITARY

ACTH

TSH FSH LH PRL
both act on:

STH (GH) → growth-promoting effects on most cells

adrenal cortex   thyroid   gonads   mammary glands

b

Figure 37.6 (**a**) Functional links between the hypothalamus and the anterior lobe of the pituitary gland. (**b**) Main targets of the anterior lobe secretions.

to influence gamete formation and secretions of the sex hormones required in sexual reproduction, the central topic of Chapter 34. Somatotropin affects metabolism in many tissues and liver secretions that influence growth of bone and soft tissues (Table 33.2 and Figure 37.6*b*). Prolactin affects different cell types but it is best known for stimulating and then sustaining milk production in mammary glands, after other hormones have primed the tissues. In some species, prolactin affects hormone production in the ovaries as well.

REGARDING THE HYPOTHALAMIC TRIGGERS Most of the hypothalamic hormones that act in the anterior lobe of the pituitary are **releasers**, meaning they stimulate the secretion of hormones from target cells. For example, the one called GnRH (gonadotropin-releasing hormone) brings about the secretion of FSH and LH, which are

classified as gonadotropins. Similarly, TRH, a releasing hormone, stimulates the secretion of thyrotropin.

But some hypothalamic hormones are **inhibitors** of secretion from their targets in the anterior pituitary. For instance, the one known as somatostatin brings about a decrease in somatotropin and thyrotropin secretion.

The hypothalamus and pituitary gland produce eight kinds of hormones and interact to control their secretion.

The posterior lobe of the pituitary stores and secretes two hypothalamic hormones, ADH and oxytocin, both of which target specific cell types.

The anterior lobe of the pituitary produces and secretes six hormones, ACTH, TSH, FSH, LH, PRL, and STH. These all trigger the release of other hormones from other endocrine glands, with wide-ranging effects throughout the body.

EXAMPLES OF ABNORMAL PITUITARY OUTPUT

The body does not churn out enormous quantities of hormone molecules. (Two researchers, Roger Guilleman and Andrew Schally, realized this when they isolated the first known releasing hormone. In their four-year attempt to secure TRH, they dissected 500 metric tons of brains and 7 metric tons of hypothalamic tissue from sheep and ended up with only a single milligram of it.) Yet normal body function depends on the tiny amounts.

Endocrine glands release their tiny but significant secretions mainly in short bursts. Elegant controls over the frequency of those abrupt secretory events prevent underproduction or overproduction of a

hormone. When something interferes with the controls, disorders may be the outcome. For example, *gigantism* results when anterior pituitary cells produce too much somatotropin. Proportionally, affected adults are similar to a person of normal size but are much larger (Figure 37.7a). By contrast, *pituitary dwarfism* results when the body does not produce enough somatotropin. Affected adults are proportionally similar to an average person but much smaller (Figure 37.7b). What if somatotropin output becomes excessive in adulthood, when the long bones no longer can lengthen? Then, *acromegaly* results. Bone, cartilage, and other connective tissues in hands, feet, and jaws thicken abnormally. So do epithelia of the skin, nose, eyelids, lips, and tongue (Figure 37.7c).

Age nine

Sixteen

Thirty-three

Fifty-two

Figure 37.7 Examples of the outcome of abnormal secretion of a hormone.

(**a**) Manute Bol, an NBA center, is 7 feet 6-3/4 inches tall owing to excessive secretion of somatotropin (STH) during his childhood.

(**b**) More examples of the effect of STH on body growth. The male at the center of this photograph is affected by gigantism, which resulted from excessive STH production in childhood. The person at the right displays pituitary dwarfism, which resulted from underproduction of STH in childhood. The person at the left is of average height.

(**c**) Acromegaly, which resulted from excessive production of STH during adulthood. Before this female reached maturity, she was symptom-free.

As another example, ADH secretion can diminish or stop when the pituitary's posterior lobe is damaged, as by a blow to the head or by a brain tumor. This is one cause of *diabetes insipidus*. The symptoms include excretion of large volumes of dilute urine, which may cause life-threatening dehydration. Diabetes insipidus responds to hormone replacement therapy based on injections or nasal spray applications of synthetic ADH.

Generally, endocrine glands release very small amounts of hormones in short bursts, the frequency of which depends on control mechanisms. When controls fail, the resulting oversecretion or undersecretion may cause disorders.

SOURCES AND EFFECTS OF OTHER HORMONES

Table 37.3 lists hormones from endocrine sources other than the pituitary. The remainder of this chapter will provide you with a few examples of their effects and of the controls over their output. The examples will make more sense if you keep the following points in mind.

First, hormones often interact with one another. In other words, one or more hormones may oppose, add to, or prime target cells for another hormone's effects. *Second*, negative feedback mechanisms often control the secretions. When a hormone's concentration increases or decreases in some body region, the change triggers events that respectively dampen or stimulate further secretion. *Third*, a target cell may react differently to a hormone at different times. Its response depends on the hormone's concentration and on the functional state of the cell's receptors. *Fourth*, environmental cues may be important mediators of hormonal secretion.

The secretion of a hormone and its effects are influenced by hormone interactions, feedback mechanisms, variations in the state of target cells, and sometimes environmental cues.

Table 37.3	Hormone Sources Other Than the Mammalian Hypothalamus and Pituitary		
Source	**Secretion(s)**	**Main Targets**	**Primary Actions**
ADRENAL CORTEX	Glucocorticoids (including cortisol)	Most cells	Promote protein breakdown and conversion to glucose
	Mineralocorticoids (including aldosterone)	Kidney	Promote sodium reabsorption (sodium conservation); help control the body's salt–water balance
ADRENAL MEDULLA	Epinephrine (adrenaline)	Liver, muscle, adipose tissue	Raises blood level of sugar, fatty acids; increases heart rate and force of contraction
	Norepinephrine	Smooth muscle of blood vessels	Promotes constriction or dilation of blood vessel diameter
THYROID	Triiodothyronine, thyroxine	Most cells	Regulate metabolism; have roles in growth, development
	Calcitonin	Bone	Lowers calcium level in blood
PARATHYROIDS	Parathyroid hormone	Bone, kidney	Elevates calcium level in blood
GONADS:			
Testes (in males)	Androgens (including testosterone)	General	Required in sperm formation, development of genitals, maintenance of sexual traits; growth, development
Ovaries (in females)	Estrogens	General	Required for egg maturation and release; preparation of uterine lining for pregnancy and its maintenance in pregnancy; genital development; maintenance of sexual traits; growth, development
	Progesterone	Uterus, breasts	Prepares, maintains uterine lining for pregnancy; stimulates breast development
PANCREATIC ISLETS	Insulin	Liver, muscle, adipose tissue	Lowers sugar level in blood
	Glucagon	Liver	Raises sugar level in blood
	Somatostatin	Insulin-secreting cells	Influences carbohydrate metabolism
THYMUS	Thymosins, etc.	Lymphocytes	Have roles in immune responses
PINEAL	Melatonin	Gonads (indirectly)	Influences daily biorhythms, seasonal sexual activity
STOMACH, SMALL INTESTINE	Gastrin, secretin, etc.	Stomach, pancreas, gallbladder	Stimulate activities of stomach, pancreas, liver, gallbladder required for food digestion, absorption
LIVER	Somatomedins	Most cells	Stimulate cell growth and development
KIDNEYS	Erythropoietin	Bone marrow	Stimulates red blood cell production
	Angiotensin*	Adrenal cortex, arterioles	Helps control blood pressure, secretion of aldosterone (hence sodium reabsorption)
	1,25-hydroxyvitamin D_6* (calcitriol)	Bone, gut	Enhances calcium reabsorption from bone and calcium absorption from gut
HEART	Atrial natriuretic hormone	Kidney, blood vessels	Increases sodium excretion; lowers blood pressure

* Kidneys produce *enzymes* that modify precursors of this substance, which then enters the general circulation as an activated hormone.

By considering just a few of the endocrine glands listed in Table 37.3, you can sense how feedback mechanisms control hormonal secretions. Briefly, the hypothalamus, pituitary, or both signal these glands to alter secretory activity. The outcome is a change in the concentration of the secreted hormone in blood or elsewhere. With the shift in chemical information, a feedback mechanism trips into action and blocks or promotes further change.

With **negative feedback**, an increase or decrease in the concentration of a secreted hormone triggers events that *inhibit* further secretion. With **positive feedback**, an increase in the concentration of a secreted hormone triggers events that *stimulate* further secretion.

Negative Feedback From the Adrenal Cortex

Humans have a pair of adrenal glands, one above each kidney (Figure 37.2b). Some cells of the **adrenal cortex**, the outer portion of an adrenal gland, secrete hormones such as glucocorticoids. Glucocorticoids help maintain the concentration of glucose in blood and help suppress inflammatory responses. Cortisol, for instance, blocks the uptake and use of glucose by muscle cells. It also stimulates liver cells to form glucose from amino acids.

A negative feedback mechanism operates when the glucose level in blood declines below a set point. This chemical condition is known as *hypoglycemia*. Take a look at Figure 37.8. When the hypothalamus detects the decrease, it secretes CRH in response. This releasing hormone stimulates the anterior pituitary to secrete corticotropin (ACTH). In turn, ACTH stimulates cells of the adrenal cortex to secrete cortisol, which helps raise the glucose level in blood. How? Cortisol stops muscle cells from taking up glucose that blood is delivering through the body. These cells are major glucose users.

During severe stress, painful injury, or prolonged illness, the nervous system overrides feedback control of cortisol secretion. It initiates a *stress response* in which cortisol helps to suppress inflammation. If unchecked, prolonged inflammation damages tissues. That is why doctors often prescribe cortisol-like drugs for asthma and other chronic inflammatory disorders.

Local Feedback in the Adrenal Medulla

The **adrenal medulla** is the inner portion of the adrenal gland (Figure 37.8). It has hormone-secreting neurons that release epinephrine and norepinephrine. (These substances are neurotransmitters in some contexts and hormones in others.) Suppose the axons of sympathetic nerves carry hypothalamic signals that call for secretion of norepinephrine. Molecules of norepinephrine collect in the synaptic cleft between the axon endings and the target cells. In this case, a localized, negative feedback mechanism operates at receptors on the axon endings. The excess norepinephrine binds to the receptors and causes a shutdown of its further release.

In times of excitement or stress, epinephrine and norepinephrine help adjust blood circulation and fat and carbohydrate metabolism. They increase heart rate, trigger vasoconstriction and vasodilation of arterioles in different regions, and dilate airways to the lungs. The controlled activity directs more of the total volume of blood to heart and muscle cells, and more oxygen flows to energy-demanding cells through the body. These are features of the *fight-flight response* (Section 35.3).

Skewed Feedback From the Thyroid

The human **thyroid gland** is located at the base of the neck in front of the trachea, or windpipe (Figures 37.2b and 37.9a,b). Thyroxine and triiodothyronine, its main hormones, have widespread effects. In their absence, many tissues cannot develop normally. Also, the overall metabolic rates of warm-blooded animals, including humans, depend on them. The importance of feedback control over the secretion of these hormones is brought into sharp focus by cases of abnormal thyroid output.

Consider how thyroid hormone synthesis requires iodine, which we obtain from food. Iodine is converted to an iodized form, iodide, when absorbed from the gut. Without iodide, the blood levels of thyroid hormones decline. The anterior pituitary responds by secreting

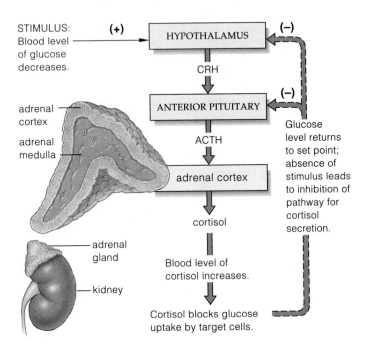

Figure 37.8 Location of the adrenal glands. One rests on top of each kidney. The diagram shows a negative feedback loop that governs secretion of cortisol from the adrenal cortex.

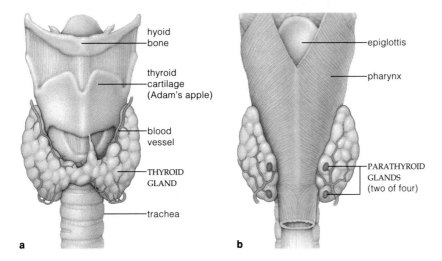

Figure 37.9 Human thyroid gland. (**a**) Anterior and (**b**) posterior views showing the location of four parathyroid glands. (**c**) A mild case of goiter, displayed by Maria de Medici in the year 1625. A rounded neck was considered to be a sign of great beauty during the late Renaissance. It occurred regularly in parts of the world where iodine supplies were insufficient for normal thyroid function.

thyroid-stimulating hormone (TSH). But without iodine, thyroid hormones cannot be made. The feedback signal continues, and so does TSH secretion. Excess TSH in blood overstimulates the thyroid gland, which enlarges in response. The enlargement is a form of *goiter*. Goiter resulting from iodine deficiency is no longer common in countries where people use iodized salt (Figure 37.9c).

When blood levels of thyroid hormones are too low, *hypothyroidism* results. Hypothyroid adults often are overweight, sluggish, dry-skinned, intolerant of cold, confused, and depressed. Affected women commonly show menstrual disturbances.

When blood levels of the thyroid hormones are too high, *hyperthyroidism* results. Affected adults show an increased heart rate, heat intolerance, elevated blood pressure, profuse sweating, and weight loss even when caloric intake increases. Affected individuals typically are nervous and agitated, and have trouble sleeping.

Feedback Control of the Gonads

Gonads are *primary* reproductive organs, which produce gametes and also synthesize and secrete sex hormones. Testes (singular, testis) in males and ovaries in females are examples. Testes secrete testosterone; ovaries secrete estrogens and progesterone. All of these sex hormones influence secondary sexual traits (as they did for the chimps described earlier), and feedback controls govern their secretion. Figure 37.10 is a preview of the feedback loops from ovaries to the hypothalamus and pituitary during the menstrual cycle, a key topic of Chapter 45.

Feedback mechanisms control secretions from endocrine glands. In many cases, feedback loops to the hypothalamus, pituitary, or both govern the secretory activity.

With negative feedback, further secretion of a hormone slows down. With positive feedback, further secretion is enhanced.

HYPOTHALAMUS

ANTERIOR PITUITARY

(+) (−) (−)

a GnRH, one of the releasing hormones, prods the pituitary's anterior lobe to secrete FSH and LH.

d LH surge *stimulates* egg release and formation of a glandular structure that produces progesterone and estrogen.

ovary

c Blood level of estrogen rises and will *stimulate* a surge in LH secretion.

b FSH and LH *stimulate* egg maturation, estrogen production, other events.

e The rising blood levels of progesterone and estrogen *inhibit* further LH and FSH secretion in last phase of cycle.

Figure 37.10 Feedback loops to the hypothalamus and pituitary gland from the ovaries during the menstrual cycle, a recurring reproductive event. Positive feedback triggers egg release from an ovary. Negative feedback after its release prevents release of another egg until the cycle is completed.

Some endocrine glands or cells don't respond primarily to signals from other hormones or nerves. They respond homeostatically to chemical change in the immediate surroundings, as the following examples illustrate.

Secretions From Parathyroid Glands

Humans have four **parathyroid glands** located on the posterior surface of the thyroid (Figure 37.9*b*). These glands secrete parathyroid hormone, or PTH, the main regulator of the calcium level in blood. Calcium ions, recall, have roles in muscle contraction, enzyme action, blood clot formation, and other tasks. The parathyroids secrete PTH in response to a low calcium level in blood. Their secretory activity slows when the calcium level rises. PTH acts on cells of the skeleton and kidneys.

PTH induces living bone cells to secrete enzymes that digest bone tissue and thereby release calcium and other minerals to interstitial fluid, then to the blood. It enhances calcium reabsorption from the filtrate flowing through the nephrons of kidneys. PTH also prods some kidney cells to secrete enzymes that act on blood-borne precursors of the active form of vitamin D_3, a hormone (Table 37.3). Vitamin D_3 stimulates intestinal cells to increase the absorption of calcium from the gut lumen. In children who have vitamin D deficiency, insufficient calcium and phosphorus are absorbed, so rapidly growing bones develop improperly. The resulting bone disorder, *rickets*, is characterized by bowed legs, a malformed pelvis, and in many cases a malformed skull and rib cage (Figure 37.11).

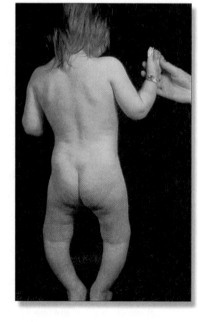

Figure 37.11 A child who is affected by rickets.

Effects of Local Signaling Molecules

Many cells detect changes in the surrounding chemical environment and alter their activity, often in ways that counteract or amplify the change. The cells secrete various local signaling molecules, the effects of which are confined to the immediate vicinity of change. Target cells take up most signaling molecules so rapidly that few enter the general circulation.

Prostaglandins are examples of signaling molecules. Cells in many tissues continually produce and release a variety of prostaglandins. But the rate of synthesis often increases in response to local chemical changes. Section 45.4 includes a fine example of this response.

Growth factors are other examples. They influence growth by regulating the rate at which cells divide. Thus an epidermal growth factor (EGF) discovered by Stanley Cohen influences growth of many cell types. A nerve growth factor (NGF) discovered by Rita Levi-Montalcini helps assure the survival of neurons and influences the direction of their growth in an embryo. One experiment demonstrated that certain immature neurons will survive indefinitely in tissue culture when NGF is present but will die within a few days if it is not.

Secretions From Pancreatic Islets

The pancreas is a gland with exocrine and endocrine functions. Its *exocrine* cells secrete digestive enzymes. Its *endocrine* cells are located in about 2 million clusters within the pancreas. Each small cluster, a **pancreatic islet**, contains three types of hormone-secreting cells:

1. *Alpha* cells in the pancreas secrete the hormone glucagon. In between meals, cells throughout the body take up and use glucose from the blood. The blood level of glucose decreases. At such times, glucagon secretion causes glycogen (a storage polysaccharide) and amino acids to be converted to glucose in the liver. In such ways, *glucagon raises the glucose level.*

2. *Beta* cells secrete the hormone insulin. After meals, when the blood glucose level is high, insulin stimulates glucose uptake by muscle and adipose cells especially. It promotes the synthesis of proteins and fats, and it inhibits protein conversion to glucose. Thus, *insulin lowers the glucose level.*

3. *Delta* cells secrete somatostatin, a hormone that helps control digestion. It also can block secretion of insulin and glucagon.

Figure 37.12 shows how pancreatic hormones interact to maintain the level of glucose in blood even though the times and amounts of food intake vary. Bear in mind, insulin is the only hormone that prods cells to take up and store glucose in forms that can be rapidly tapped when required. Its central role in carbohydrate, protein, and fat metabolism becomes clear when we observe people who cannot produce enough insulin or who lack body cells that can respond to it.

For example, insulin deficiency may lead to *diabetes mellitus*, a disorder in which excess glucose accumulates in blood, then in urine. Urination becomes excessive, so the body's water-solute balance is disrupted. Affected

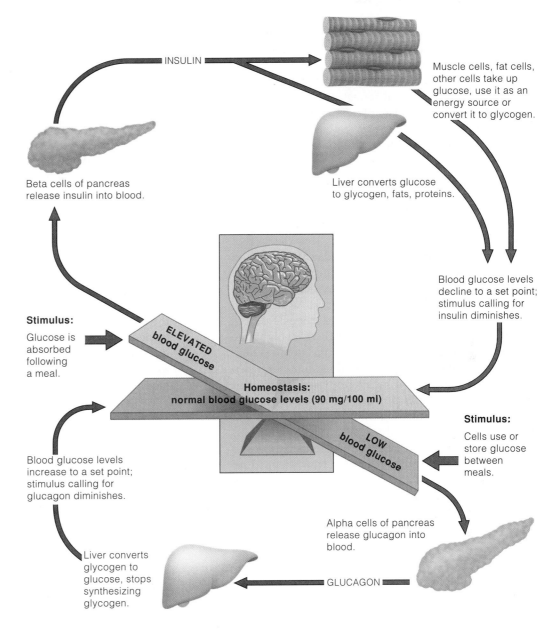

Figure 37.12 Some of the homeostatic controls over glucose metabolism.

Following a meal, glucose enters the bloodstream faster than cells can use it. The blood glucose level rises, and pancreatic beta cells are stimulated to secrete insulin. Insulin's main targets—liver, fat, and muscle cells—not only use glucose but store excess amounts of it in the form of glycogen.

Between meals, the blood glucose level decreases. Pancreatic alpha cells are stimulated to secrete glucagon. The target cells with receptors for this hormone convert glycogen back to glucose, which enters the blood.

Glucose metabolism also is subject to indirect controls. For example, the hypothalamus commands the adrenal medulla to secrete certain hormones. The hormones speed the conversion of glycogen to glucose in the liver and slow the reverse process, especially in cells of the liver, adipose tissue, and muscle tissue.

INSULIN

Muscle cells, fat cells, other cells take up glucose, use it as an energy source or convert it to glycogen.

Beta cells of pancreas release insulin into blood.

Liver converts glucose to glycogen, fats, proteins.

Blood glucose levels decline to a set point; stimulus calling for insulin diminishes.

Stimulus:
Glucose is absorbed following a meal.

ELEVATED blood glucose

Homeostasis: normal blood glucose levels (90 mg/100 ml)

LOW blood glucose

Stimulus:
Cells use or store glucose between meals.

Blood glucose levels increase to a set point; stimulus calling for glucagon diminishes.

Alpha cells of pancreas release glucagon into blood.

Liver converts glycogen to glucose, stops synthesizing glycogen.

GLUCAGON

people become dehydrated and thirsty—abnormally so. Without a steady supply of glucose, their body cells start depleting their own fats and proteins as sources of energy. Weight loss is one outcome. Another is that ketones accumulate in blood and urine. Ketones are normal acidic products of fat breakdown. When they accumulate, they contribute to water losses and alter the body's acid-base balance. Such imbalances disrupt brain function. In extreme cases, death may follow.

In "type 1 diabetes," the body mistakenly mounts an autoimmune response against its insulin-secreting beta cells. Certain lymphocytes identify the beta cells as foreign and destroy them. A combination of genetic susceptibility and environmental triggers produces the disorder, which is less common but more immediately dangerous than other forms of diabetes. Usually, the

symptoms first appear in childhood and adolescence (the disorder is also known as juvenile-onset diabetes). Type 1 diabetic patients survive with insulin injections.

In "type 2 diabetes," insulin levels are close to or above normal, but the target cells cannot respond to insulin. As affected persons grow older, their beta cells produce less and less insulin. Type 2 diabetes usually is manifested during middle age. Affected persons lead normal lives by controlling their diet and weight, and sometimes by taking drugs to enhance insulin action or secretion.

The secretions from some endocrine glands and cells are direct homeostatic responses to a change in the localized chemical environment.

This last section of the chapter invites you to reflect on a key point. An individual's growth, development, and reproduction begin with genes and hormones, and so does behavior. *But certain environmental factors commonly influence gene expression and hormonal secretion, and they do so in predictable ways.* Chapter 51 invites analysis of the environmental influence on animal behavior. For now, it is enough to consider the following examples.

Daylength and the Pineal Gland

Embedded in the brain is a photosensitive organ, the **pineal gland** (Section 35.5). In the absence of light, the gland secretes the hormone melatonin. Thus the level of melatonin in the blood varies from day to night, and with the seasons. The variations influence the growth and development of gonads, the primary reproductive organs. In a variety of species, they have important roles in reproductive cycles and reproductive behavior.

Think about a hamster. In winter, when nighttime darkness is longest, the blood level of melatonin is high and sexual activity is suppressed. In the summer, when daylength is longest, the melatonin level is low, and hamster sex reaches its peak. Or think about a male white-throated sparrow (Figure 37.13a). In the fall and winter, melatonin indirectly suppresses growth of its gonads by inhibiting gonadotropin secretion. It does so until days start to lengthen in spring. Now, stepped-up gonadal activity leads to production of hormones that influence singing behavior, as described in Section 51.1. With his distinctive song, the male sparrow defines his territory and may hold the interest of a mate.

Does melatonin also influence human behavior? Perhaps. Clinical observations and studies suggest that decreased melatonin secretion may trigger **puberty**, the age during which human reproductive organs and structures start to mature. For example, in cases where disease caused the destruction of an individual's pineal gland, puberty began prematurely.

Melatonin is known to act on certain neurons that lower your body's core temperature and that make you drowsy after sunset, when light is waning. At sunrise, when melatonin secretion slows, your core temperature increases and you wake up and become active.

An internal, biological clock governs the cycle of sleep and arousal. It seems to tick in synchrony with daylength. Think of night workers who try to sleep in the morning but end up staring groggily at sunbeams on the ceiling. Think of travelers from the United States to Paris who go through four days of "jet lag." Two or three hours past midnight they are sitting up in bed, wondering where the coffee and croissants are. Two hours past noon they are ready for bed. They will shift to a new routine when the melatonin signals arrive at their target neurons on Paris time.

In winter, some individuals experience *winter blues*. They get abnormally depressed, go on carbohydrate binges, and have an almost overwhelming desire to sleep (Figure 37.13b). Winter blues might arise when a biological clock gets out of sync with the seasonally shorter daylengths. Intriguingly, clinically administered doses of melatonin make the seasonal symptoms worse. And exposure to intense light, which can shut down pineal activity, may lead to dramatic improvement.

Figure 37.13 (a) A male white-throated sparrow, belting out a song that began, indirectly, with an environmentally induced decline in melatonin secretion. (b) Annie blanketing her winter blues.

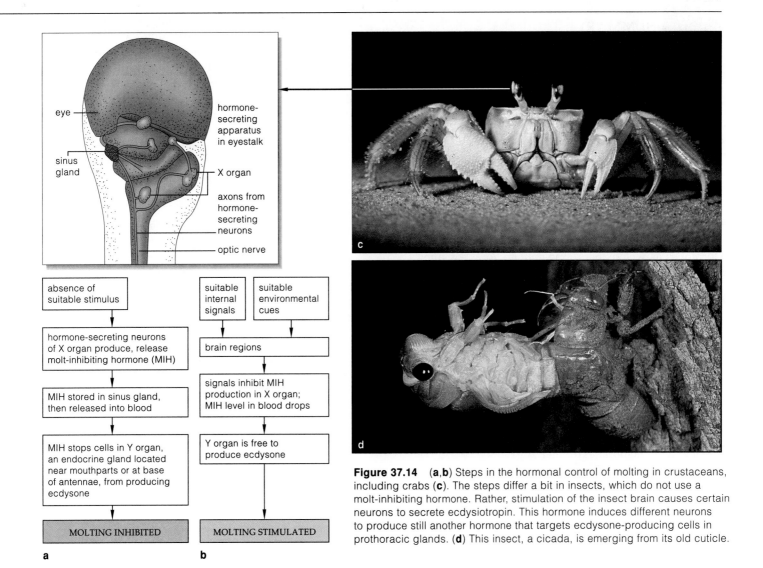

eye

hormone-
secreting
apparatus
in eyestalk

sinus
gland

X organ

axons from
hormone-
secreting
neurons

optic nerve

absence of suitable stimulus		suitable internal signals	suitable environmental cues

↓

hormone-secreting neurons of X organ produce, release molt-inhibiting hormone (MIH)

brain regions

↓

MIH stored in sinus gland, then released into blood

signals inhibit MIH production in X organ; MIH level in blood drops

↓

MIH stops cells in Y organ, an endocrine gland located near mouthparts or at base of antennae, from producing ecdysone

Y organ is free to produce ecdysone

↓

MOLTING INHIBITED

MOLTING STIMULATED

a   b

Figure 37.14 (**a,b**) Steps in the hormonal control of molting in crustaceans, including crabs (**c**). The steps differ a bit in insects, which do not use a molt-inhibiting hormone. Rather, stimulation of the insect brain causes certain neurons to secrete ecdysiotropin. This hormone induces different neurons to produce still another hormone that targets ecdysone-producing cells in prothoracic glands. (**d**) This insect, a cicada, is emerging from its old cuticle.

Comparative Look at a Few Invertebrates

Although this chapter's focus has been on vertebrates, do not lose sight of the fact that all organisms produce signaling molecules of one sort or another. Consider the hormonal control of **molting**, a periodic discarding and replacement of a hardened cuticle that otherwise would limit increases in body mass. As described in Section 26.15, molting occurs during the life cycle of all insects, crustaceans, and other invertebrates with thick cuticles.

Although details vary from group to group, molting is largely under the control of ecdysone. This steroid hormone is derived from cholesterol and is chemically related to many important vertebrate hormones. In insects and crustaceans, molting glands produce and store ecdysone, then release it for distribution through the body at molting time. Hormone-secreting neurons in the brain seem to regulate its release. The hormone-secreting neurons apparently respond to a combination of environmental cues, including light and temperature, as well as internal signals.

Figure 37.14 gives examples of the control steps, which differ in crustaceans and insects.

During premolt and molting periods, coordinated interactions among ecdysone and other hormones bring about major structural and physiological changes. The interactions cause the old cuticle to detach from the epidermis and muscles. They induce the dissolving and recycling of the cuticle's inner layers. The interactions also trigger shifts in metabolism and in the composition and volume of the internal environment. They promote cell divisions, secretions, and pigment formation, all of which go into producing a new cuticle. Simultaneously, hormonal interactions control heart rate, muscle action, color changes, and other physiological processes.

Environmental cues, such as changes in light intensity from day to night and seasonal changes in daylength, influence certain hormonal secretions.

SUMMARY

1. The cells of complex animals continually exchange substances with the body's internal environment. Their myriad withdrawals and secretions are integrated in ways that ensure cell survival through the whole body.

2. Integration of cell activities requires the stimulatory or inhibitory effects of signaling molecules.

    a. Signaling molecules are chemical secretions from a cell that adjust the behavior of other, target cells.

    b. Any cell with molecular receptors for a signaling molecule is a target. The target cells may or may not be next to the cell that sends the signal.

    c. There are different kinds of signaling molecules. Hormones as well as neurotransmitters, local signaling molecules, and pheromones are the main kinds.

    d. Certain steroids, steroidlike molecules, amines, peptides, proteins, and glycoproteins are hormones.

3. In target cells, hormones influence gene activation, protein synthesis, and alterations in existing enzymes, membranes, and other cellular components. Hormones exert their physiological effects through interactions with specific protein receptors at the plasma membrane or in the cytoplasm of target cells.

    a. Steroid hormones are lipid soluble. Complexes of steroid hormones and receptors interact with DNA and possibly with membranes of their targets.

    b. Protein hormones are water soluble. Some enter the cytoplasm complexed with receptors. The effect of others is exerted with the help of membrane transport proteins and second messengers in the cytoplasm, some of which trigger the actual response.

4. The posterior lobe of the pituitary stores and secretes two hypothalamic hormones, ADH and oxytocin. ADH targets cells in kidneys and affects extracellular fluid volume. Oxytocin acts on cells in mammary glands and the uterus to influence reproductive events.

5. The hypothalamic hormones called releasing and inhibiting hormones control secretions from different cells of the anterior lobe of the pituitary gland.

6. The anterior lobe makes and secretes six hormones, ACTH, TSH, FSH, LH, PRL, and STH. These trigger secretions from the adrenal cortex, thyroid, gonads, and mammary glands. By doing so, they exert wide-ranging responses throughout the body.

7. The vertebrate body has other sources of hormones, including the adrenal medulla, the parathyroid, thymus, and pineal glands, pancreatic islets, and endocrine cells in the stomach, small intestine, liver, and heart.

8. Hormone interactions, feedback mechanisms, the number and kind of receptors on target cells, variations in the state of target cells, and sometimes cues from the environment influence the secretion of a hormone and its effects.

9. Many cellular responses to hormones help the body adjust to short-term shifts in diet and levels of activity. Other kinds help bring about long-term adjustments for growth, development, and reproduction.

    a. In general, secretion of hormones such as insulin and parathyroid hormone can change rapidly when the extracellular concentration of some substance must be controlled homeostatically.

    b. Hormones such as somatotropin have prolonged, gradual, often irreversible effects, as on development.

Review Questions

1. Name the endocrine glands that occur in most vertebrates and state where each is located in the human body. 37.1

2. Distinguish among hormones, neurotransmitters, local signaling molecules, and pheromones. 37.1

3. A hormone molecule binds to a receptor on a cell membrane. It does not enter the cell, however. Rather, binding activates a second messenger inside the cell that triggers an amplified response to the hormonal signal. State whether the signaling molecule is a steroid hormone or a peptide hormone. 37.2

4. Which secretions of the posterior lobe of the pituitary gland have the targets indicated? (Fill in the blanks.) 37.3

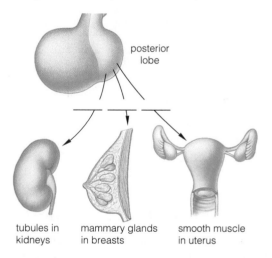

posterior lobe

tubules in kidneys   mammary glands in breasts   smooth muscle in uterus

5. Which secretions of the anterior lobe of the pituitary gland have the targets indicated? (Fill in the blanks.) 37.3

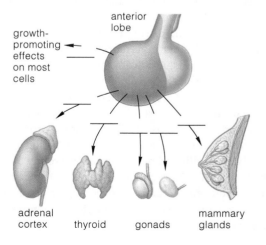

growth-promoting effects on most cells

anterior lobe

adrenal cortex   thyroid   gonads   mammary glands

Self-Quiz (Answers in Appendix IV)

1. _____ are molecules released from a signaling cell that have effects on target cells.
   a. Hormones
   b. Neurotransmitters
   c. Pheromones
   d. Local signaling molecules
   e. both a and b
   f. a through d

2. Hormones are products of _____ .
   a. endocrine glands or cells
   b. some neurons
   c. exocrine cells
   d. a and b
   e. a and c
   f. a, b, and c

3. ADH and oxytocin are hypothalamic hormones secreted from the _____ lobe of the pituitary gland.
   a. anterior
   b. posterior
   c. intermediate
   d. secondary

4. _____ has effects on body tissues in general.
   a. ADH
   b. Oxytocin
   c. Bombykol
   d. Somatotropin

5. Which do *not* stimulate hormone secretions?
   a. neural signals
   b. local chemical changes
   c. hormonal signals
   d. environmental cues
   e. All of the above can stimulate hormone secretion.

6. _____ lowers blood sugar levels; _____ raises it.
   a. Glucagon; insulin
   b. Insulin; glucagon
   c. Gastrin; insulin
   d. Gastrin; glucagon

7. The pituitary detects a rising hormone concentration in blood and inhibits the gland secreting the hormone. This is a _____ feedback loop.
   a. positive
   b. negative
   c. long-term
   d. b and c

8. Second messengers include _____ .
   a. steroid hormones
   b. protein hormones
   c. cyclic AMP
   d. both a and b

9. Match the hormone source with the closest description.
   ____ adrenal gland
   ____ thyroid gland
   ____ parathyroids
   ____ pancreatic islets
   ____ pineal gland
   ____ thymus gland
   a. affected by daylength
   b. key roles in immunity
   c. raise blood calcium level
   d. epinephrine source
   e. insulin, glucagon
   f. hormones require iodine

Critical Thinking

1. The zebra offspring being nursed in Figure 37.15 is too young to nourish itself by eating grasses. Its source of nutrients is its mother's milk. Explain how secretions from the hypothalamus and both lobes of the pituitary gland influence the production and secretion of milk.

2. In winter, with its far fewer daylight hours compared to the summer, Maxine became very depressed, craved carbohydrate-rich foods, and stopped exercising regularly. And she put on a great deal of weight. Her doctor diagnosed her condition as *seasonal affective disorder* (SAD), or the winter blues. Maxine was advised to purchase a cluster of intense, broad-spectrum lights and to sit near them at least an hour every day. The cloud of depression started to lift quickly. Use your understanding of the secretory activity of the pineal gland to explain why Maxine's symptoms appeared and why the prescribed therapy worked.

3. Marianne is affected by *type 1 insulin-dependent diabetes*. One day, after injecting herself with too much insulin, she starts to shake and feels confused. Her doctor recommends a glucagon injection. What caused her symptoms? How would an injection of glucagon help?

Figure 37.15 Female zebra nursing her offspring.

4. Through recombinant DNA technology, somatotropin (growth hormone) is now available commercially for treating pituitary dwarfism. Although it is illegal to do so, some athletes are using somatotropin instead of anabolic steroids. (Here you may wish to compare Section 38.11.) They do so because somatotropin cannot be detected by the drug test procedures employed in sports medicine. Explain how the athletes might believe this hormone can improve their performance.

5. *Osteoporosis* is a condition in which loss of calcium results in thin, brittle bones. Combined with other treatments, vitamin D_3 injections are sometimes recommended. Explain why.

Selected Key Terms

adrenal cortex *37.6*
adrenal medulla *37.6*
endocrine system *37.1*
gonad *37.6*
hormone *37.1*
hypothalamus *37.3*
inhibitor (hypothalamic) *37.3*
local signaling molecule *37.1*
molting *37.8*
negative feedback *37.6*
neurotransmitter *37.1*
pancreatic islet *37.7*
parathyroid gland *37.7*
peptide hormone *37.2*
pheromone *37.1*
pineal gland *37.8*
pituitary gland *37.3*
positive feedback *37.6*
puberty *37.8*
releaser (hypothalamic) *37.3*
second messenger *37.2*
steroid hormone *37.2*
thyroid gland *37.6*

Readings

Goodall, J. 1986. *The Chimpanzees of Gombe*. Cambridge, Massachusetts: Belknap Press of Harvard University Press.

Goodman, H. 1994. *Basic Medical Endocrinology*. Second edition. New York: Raven Press.

Hadley, M. 1995. *Endocrinology*. Fourth edition. Englewood Cliffs, New Jersey: Prentice-Hall.

Snyder, S. October 1985. "Molecular Basis of Communication Between Cells." *Scientific American* 253(4): 132–141.

Web Site See *http://www.wadsworth.com/biology* for practice quiz questions, hypercontents, BioUpdates, and critical thinking. The Wadsworth Biology Resource Center provides a wealth of information fully organized and integrated by chapter.

38

PROTECTION, SUPPORT, AND MOVEMENT

Of Men, Women, and Polar Huskies

In 1989 Will Steger and his dogsled team walked on ice for seven months, endured temperatures of −113°F, and lived through a blizzard that lasted for more than seven weeks. They crossed Antarctica—all 6,023 kilometers (3,741 miles) of it. In 1995 that legendary polar explorer set out with four men, two women, and thirty-three sled dogs to cross 3,220 kilometers of the Arctic Ocean in one season. Ice blankets this northernmost ocean in winter, but the ice becomes treacherously thin during the spring thaw. The sled dogs were with the team for two-thirds of the journey. They were flown out only when the team encountered too much melting ice and had to switch to using canoes.

To Steger's mind, the polar huskies were the heroes of the polar crossings, the members of the team that worked hardest and pulled all the weight (Figure 38.1). They are a mixed breed, the traits of which have been modified through years of artificial selection among

Canadian and Greenland huskies (bred for size and strength), Siberian huskies (bred for intelligence), and Alaskan racing dogs (bred for spirit and endurance). Steger's polar huskies show a combination of these traits as well as the loyalty of cared-for pets.

A husky's leg bones are sturdy yet lightweight. Its forelegs move freely, thanks to a rib cage that is deep but not too broad. Its hind legs have massive muscles. These are not the muscles of sprinting greyhounds or cheetahs. They are the muscles of a load-pulling, long-distance runner. The husky also has tough, calloused foot pads—cushions against sharp ice and frozen rock. Like many other mammals, it has a fur coat. The coat's underhair, a dense, soft insulative layer, traps heat. Its coarser, longer, and slightly oily guard hairs protect the insulative layer from wear and tear. On winter nights, the husky settles into a comfortable position and covers its nose with its furry tail, oblivious of drifting snow.

Figure 38.1   In Ely, Minnesota, Will Steger and his polar huskies warming up for their Arctic crossing.

Steger and his teammates, Victor Boyarksy, Julie Hanson, Martin Hignell, Paul Pregont, and Takako Takano, could not even approach the polar husky's stamina and built-in protection against the elements. Long before the polar crossings, they were adhering to a regimen of diet and exercise to put their arm and leg muscles in peak condition for the extraordinary effort that lay ahead. Human legs are not adapted for load-pulling motion, but rather for long-distance walking. Also, human skin cannot withstand bitter cold. Lacking the fur coat of mammals that evolved in polar climates, the team had to depend on special clothing that could insulate and protect them from cold without restricting body movements. From this perspective, it was human ingenuity that allowed humans to keep company with the huskies, which are supremely adapted for the challenges of life on ice.

With this chapter, we turn to the three systems that together are responsible for the superficial features, shape, and movements of most kinds of animals. Figure 38.2 serves as the starting point for our consideration of the structural organization and functions of these systems. Traveling from the outside in, it depicts the integumentary system (skin and its derived structures), muscle system, and skeletal system of one of the more familiar vertebrates.

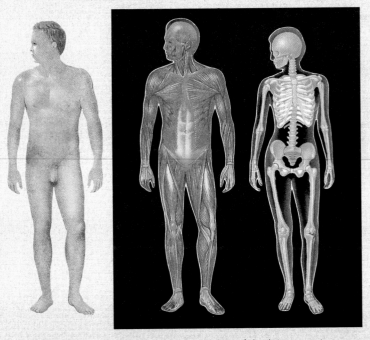

Figure 38.2 From left to right, overview of the integumentary, muscular, and skeletal systems of humans.

Animals ranging from invertebrate worms to humans have an outer covering, or **integument** (after the Latin *integere*, meaning to cover). Most coverings are tough, pliable barriers against many environmental insults.

The integument of roundworms and insects, crabs, and other arthropods is a protective **cuticle,** hardened with chitin. Chitin, remember, is a polysaccharide that incorporates nitrogen atoms. Sections 26.8 and 26.15 describe this type of covering.

Vertebrate Skin and Its Derivatives

For vertebrates, the integument consists of a covering called **skin** as well as a variety of structures derived from epidermal cells of the skin's outer tissue layers (Figures 38.3 and 38.4). Beyond the typical assortment of epithelial tissues and glands, great variation exists within vertebrate groups as well as between them. For example, birds have the unique epidermal derivatives

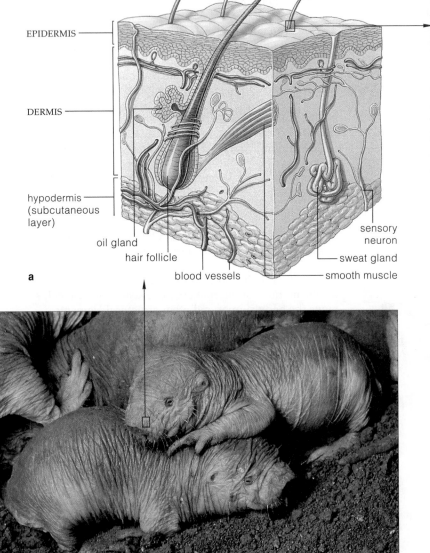

Dead, flattened epidermal cells around a shaft of hair that is projecting from the skin surface. Refer to Section 3.7, which includes a description of the molecular structure of hair.

Figure 38.3 (**a**) Structure of vertebrate skin. The uppermost portion is the epidermis; the lower portion is the dermis. (**b**) Scanning electron micrograph of a hair. (**c**) This display of skin is provided by naked mole-rats (*Heterocephalus glaber*), which live underground in burrows. Most of their hairs occur on the snout and are modified for sensory functions.

called feathers as well as bills and claws. Many herbivorous mammals have hooves and horns, porcupines have quills, you and your primate relatives have nails, and so on. As you read in Sections 27.4 and 27.5, the hagfishes have bare skin, which they coat with a lathering of slime; and epidermal cells of many fishes give rise to hardened scales of diverse thicknesses, shapes, and colors.

The skin itself has two regions: an outermost **epidermis** and an underlying **dermis**. Below this, a tissue region called the hypodermis anchors the skin to underlying structures yet still allows it to move a bit (Figure 38.3*a*). Fats that become stored in the hypodermis help insulate the body and cushion some of its parts.

Your skin weighs about four kilograms (nine pounds). Stretched out, its surface area would be fifteen to twenty square feet. For the most part, human skin is as thin as a paper towel. It thickens only on the soles of the feet and in other regions that are subjected to pounding or abrasion.

Figure 38.4 A few examples of skin and of structures that are derived from it.

(**a**) Richly pigmented skin of a tree frog. Its pigment-producing cells, called chromatophores, reside primarily in the dermis. The kind deepest in the dermis produces a silvery pigment. Wavelengths reflected and scattered from such pigment molecules appear blue. Other kinds of pigment cells higher in the dermis produce a yellow pigment, which filters the blue wavelengths to produce a rich green coloration. Section 33.1 has a detailed diagram of the skin of a poisonous frog.

(**b**) Richly pigmented skin and hairs of the mountain gorilla. An abundance of melanin-producing cells in the hair follicles supply the pigments. Each hair consists of a cuticle around a central shaft. The cuticle is composed of dead, flattened cells that are packed with the protein keratin.

Very little melanin is distributed among the epidermal layers of naked mole-rats (Figure 38.3c). Blood inside vessels that thread through the dermis contributes to the skin's reddish hue.

(**c**) Peacock feathers, one of the more spectacular examples of structures derived from skin. (**d–f**) This series of diagrams shows how each feather grows from its base, at a region of actively dividing epidermal cells.

a

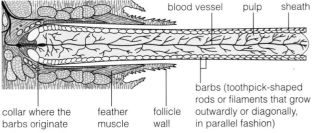
b

c

d One of many small buds that appear on a chick embryo's surface (in this case, on the sixth day of incubation).

e One of the buds growing into a cone-shaped structure. Its base sinks into the skin. The depression around it will become one feather follicle.

f A layer of horny cells at the cone's surface differentiates into a sheath. An epidermal layer just beneath the sheath will give rise to a feather. The dermis beneath this region of epidermis is richly supplied with blood vessels. It becomes the pulp, which nourishes the growing feather but does not contribute to its structure.

epidermis

dermis

blood vessel   pulp   sheath

collar where the barbs originate   feather muscle   follicle wall   barbs (toothpick-shaped rods or filaments that grow outwardly or diagonally, in parallel fashion)

Functions of Skin

No garment ever made comes close to the qualities of skin. What besides skin holds its shape after repeated stretchings and washings, blocks harmful rays from the sun, kills many bacteria on contact, holds in moisture, fixes small cuts and burns, *and* can last as long as you do? Skin also produces vitamin D, required for calcium metabolism. It plays a passive role in adjusting internal temperature; the nervous system can rapidly adjust the flow of blood (which transports metabolic heat) to and from skin's great numbers of tiny blood vessels. And signals from sensory receptor endings in skin help the brain assess what is going on in the outside world.

The body of most animals has an integument, a protective covering that usually is tough yet pliable. A number of invertebrate species have a cuticle; vertebrates have skin, which consists of epidermis and dermis.

A LOOK AT HUMAN SKIN

Structure of the Epidermis and Dermis

Like puff pastry, epidermis consists of sheetlike layers; it is a *stratified* epithelium. Its cells are structurally and functionally knit together by an abundance of junctions of the sort described in Section 33.1. Its inner sheets are composed of many living, rapidly dividing cells. The most abundant are **keratinocytes**, which produce the tough, water-insoluble protein keratin. One reason why skin is such a strong, cohesive integument is that the adhesion junctions between keratinocytes are anchored to numerous, crosslinked keratin fibers inside them.

Other cells, the **melanocytes**, produce and donate the brownish-black pigment melanin to keratinocytes. Melanin screens out harmful ultraviolet radiation from the sun. Humans generally have the same number of melanocytes, but skin color varies owing to differences in the distribution and metabolic activity of these cells. For example, melanocytes in albinos cannot produce all of the enzymes required for melanin production. Pale skin contains little melanin, so the pigment hemoglobin inside red blood cells is not masked. The skin appears pink because hemoglobin's red color shows through thin-walled blood vessels and the epidermis itself, both of which are transparent. Carotene, which is a yellow-orange pigment, also contributes to skin color.

Section 38.3 describes two less common cell types in epidermis, the Langerhans and Granstein cells.

The rapid, ongoing mitotic divisions push epidermal cells from deeper layers toward the skin's free surface. Because of pressure from the continually growing cell mass and from normal wear and tear at the surface, older cells are dead and flattened by the time they reach the outer layers (Figure 38.5). There, they are abraded off or flake away on an ongoing basis. Besides replacing the outermost, keratinized layer, the rapid cell divisions also help skin mend quickly after cuts or burns.

Beneath the epidermis is the dermis. This is mostly dense connective tissue with many elastin fibers (which resist daily stretching) and collagen fibers (which impart strength). Blood vessels, lymph vessels, and receptor endings of sensory nerves thread through the dermis. Nutrients from the bloodstream reach living epidermal cells by diffusing through the dermal ground tissue.

Sweat Glands, Oil Glands, and Hairs

Human skin typically has sweat glands, oil glands, and husklike cavities, or follicles, for hairs. These reside mostly in the dermis but epidermal cells give rise to them.

Fluid secreted by **sweat glands** is 99 percent water, with dissolved salts, traces of ammonia, vitamin C, and other substances. You have 2.5 million sweat glands, controlled by sympathetic nerves. One type abounds in

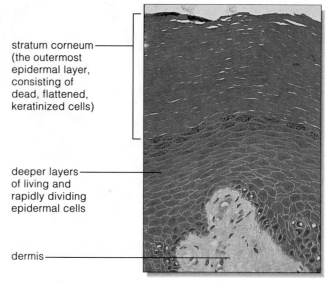

stratum corneum (the outermost epidermal layer, consisting of dead, flattened, keratinized cells)

deeper layers of living and rapidly dividing epidermal cells

dermis

Figure 38.5 Micrograph of a section through human skin.

your palms, soles, forehead, and armpits. They have roles in controlling body temperature and in *cold sweats*, a response to frightening, unsettling situations (Section 35.3). Secretions from a different type of sweat gland increase during stress, pain, and sexual foreplay and prior to menstruation.

Except on the palms of hands and the soles of feet, skin contains **oil glands** (also called sebaceous glands). Oil glands lubricate and soften hair and the skin, and their secretions kill many potentially harmful surface bacteria. *Acne* is an inflammation of the skin that occurs after bacteria have successfully infected oil gland ducts.

Each **hair**, a flexible structure of mostly keratinized cells, has a root embedded in skin and a shaft above its surface (Figure 38.3). Cells divide near the root's base, are pushed upward, then flatten and die. Flattened cells of the shaft's outer layer overlap like roof shingles. When mechanically abused, these frizz out as "split ends." An average human scalp has about 100,000 hairs, although genes, nutrition, and hormones influence hair growth and density. Protein deficiency causes hair to thin (amino acids are required for keratin synthesis). So do high fever, emotional stress, and excess vitamin A intake. When the body produces abnormal amounts of testosterone, *hirsutism*, or excessive hairiness, may be one result. This hormone influences patterns of hair growth and other secondary sexual traits.

The skin's multiple layers of keratinized, melanin-shielded epidermal cells help the body conserve water, avoid damage by ultraviolet radiation, and resist mechanical stress.

Sweat glands, oil glands, hairs, and other structures derived from epidermal cells are embedded largely in the dermis. Blood vessels, lymph vessels, and the receptor endings of sensory neurons also reside in the dermis.

38.3 SUNLIGHT AND SKIN

THE VITAMIN D CONNECTION   Even when you do little more than sit outside in the sun, you are giving some of the epidermal cells in your skin the opportunity to make vitamin D, or cholecalciferol. This steroid-like compound helps the body absorb calcium from food. When exposed to sunlight, some type of cell in the skin produces it from a precursor molecule that is related to cholesterol. The cells then release vitamin D to the bloodstream, which transports it to absorptive cells in the intestinal lining. This is a hormone-like action, which means the skin acts like an endocrine gland when exposed to sunlight.

Humans, remember, evolved beneath the intense sun of the African savanna, so skin alone would have provided our early ancestors with enough vitamin D. When humans started moving out of tropical environments and, later, into caves, animal skins, and layers of clothing, they also started to depend more on dietary sources of the essential vitamin D, as described in Section 42.9.

SUNTANS AND SHOE-LEATHER SKIN   Do you like to tan your body by rotating beneath the sun's rays or a tanning lamp, like a chicken in an oven broiler? If so, think about what a broiler does to the chicken. The sun's ultraviolet wavelengths stimulate melanin production in skin cells. Continued exposure increases melanin concentrations in light skin and visibly darkens it, thereby producing the "tan" that so many people covet (Figure 38.6). Tanning does protect the body against ultraviolet radiation. Even in naturally dark skin, however, prolonged exposure to sunlight causes the elastin fibers in connective tissue of the dermis to clump together, so skin loses its resiliency. In time it starts to look like old shoe leather.

In this respect, tanning accelerates the *aging* of skin. As any person grows older, epidermal cells divide less often.

Skin gets thinner and more susceptible to injury. Glandular secretions that once kept it soft and moistened dwindle. Collagen and elastin fibers in the dermis break down and become sparser, so skin loses elasticity and its wrinkles deepen. By themselves or in combination with excessive tanning, prolonged exposure to dry wind and tobacco smoke also accelerates the aging process.

SUNLIGHT AND THE FRONT LINE OF DEFENSE   Besides the cells that produce melanin and keratin, skin contains two other types of cells, which help defend the surface of the body against invasion by pathogenic cells and protect it against cancer. These components of the epidermis are called Langerhans cells and Granstein cells.

Langerhans cells are phagocytes that develop in bone marrow, then they take up stations in skin. After they engulf virus particles or bacterial cells, they pepper the surface of their plasma membrane with molecular alarm signals that mobilize the body's immune system.

Ultraviolet radiation can damage Langerhans cells. This might be why sunburns can trigger *cold sores*, the small, painful blisters that announce the recurrence of a *Herpes simplex* infection. Nearly everyone harbors the *H. simplex* virus. It remains hidden in the face, inside a ganglion (a cluster of neuron cell bodies). Sunburns and other stress factors can activate the virus. Virus particles move down the neurons to their endings in skin. There they infect epithelial cells and cause skin eruptions.

When ultraviolet radiation damages Langerhans cells, it weakens one of the body's first lines of defense against invasions. It also can activate proto-oncogenes and trigger cancerous transformation of skin cells. As described in Section 13.4, skin cancers grow rapidly and can spread to adjacent lymph nodes unless they are surgically removed.

In some as-yet-undetermined way, **Granstein cells** apparently interact with the white blood cells that can put the brakes on immune responses in the skin. By issuing suppressor signals, they help keep the responses from spiraling out of control. Although the functions of Granstein cells are not completely understood, these cells are known to be less vulnerable than the Langerhans cells to the damaging effects of ultraviolet radiation.

Figure 38.6   Demonstration of how shoe-leather skin forms.

TYPES OF SKELETONS

Operating Principles for Skeletons

Many of the responses an animal makes to external and internal conditions involve the movement of the whole body or parts of it. The activation, contraction, and relaxation of muscle cells bring about the movements. But muscle cells alone cannot produce them. *All muscles require the presence of some medium or structural element against which the force of contraction can be applied.* A skeletal system fulfills this requirement.

Three types of skeletons predominate in the animal world. With a **hydrostatic skeleton**, the muscles work against an internal body fluid and redistribute it within a confined space. Like a filled waterbed, the confined fluid resists compression. By contrast, an **exoskeleton** has rigid, *external* body parts, such as shells, that can receive the applied force of muscle contraction. An **endoskeleton** has rigid, *internal* body parts, such as bones, that receive the applied force of contraction.

Examples From the Invertebrates

Many invertebrates with a soft body have a hydrostatic skeleton. Reflect on the sea anemone, with its soft, vase-shaped body and saclike gut (Figure 38.7). Its body wall incorporates longitudinal and radial muscles. Between meals, when the longitudinal muscles are contracted (shortened) and radial ones are relaxed (lengthened), a sea anemone looks short and squat. When the body lengthens into an upright feeding position, its radial muscles are contracting (and forcing some fluid out of the gut cavity), and its longitudinal ones are relaxing.

Or reflect on the earthworm. This annelid, recall, has a series of coelomic compartments, each with its own muscles, nerves, and bristles. Its body moves forward by contracting and relaxing its fluid-filled segments one after another (Section 26.14). By coordinating contractions on one side or the other of the segments, the worm also thrashes sideways and moves forward and back.

As another example, jumping spiders have a hinged exoskeleton, as other arthropods do. In addition, they use body fluids to transmit force when

a Resting position   **b** Feeding position

Figure 38.7   Outcomes of contractile force applied against the sea anemone's hydrostatic skeleton. This invertebrate's body wall has contractile cells running longitudinally (parallel with the body axis) and radially around the gut cavity. (**a**) In this case, radial cells are relaxed and longitudinal ones are contracted. Anemones typically are in this resting position at low tide, when currents cannot bring food morsels to them. (**b**) Radial cells are contracted, longitudinal ones are relaxed, and the body is extended to its upright feeding position.

they leap at prey. Muscle contractions cause blood in body tissues to surge into the hind leg spines. It is like giving a water-filled rubber glove a quick squeeze to make the glove's skinny fingers rigidly erect. Figure 38.8 shows this resourceful use of hydraulic pressure (*hydraulic* means fluid pressure inside tubes).

The hinged arthropod exoskeleton has advantages. Some of its hard parts can be moved like levers by sets of muscles attached to them. Thus small contractions can bring about large movements of wings or some other body parts. This is especially true of the cuticle of a winged insect. It extends over all body segments and over gaps between segments (Figure 38.9). At these gaps, the cuticle remains pliable. It acts like a hinge when muscles alternately raise and lower either the wing or the body parts to which the wings are attached.

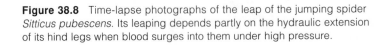

Figure 38.8   Time-lapse photographs of the leap of the jumping spider *Sitticus pubescens*. Its leaping depends partly on the hydraulic extension of its hind legs when blood surges into them under high pressure.

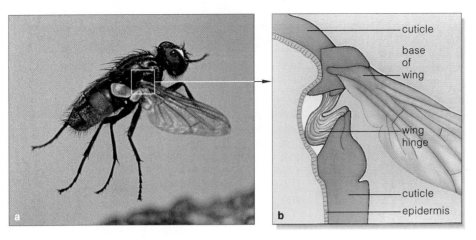

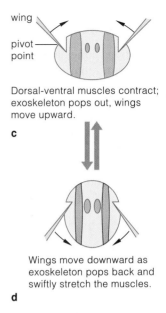

Dorsal-ventral muscles contract; exoskeleton pops out, wings move upward.

c

Wings move downward as exoskeleton pops back and swiftly stretch the muscles.

d

Figure 38.9 Housefly wing movement, an outcome of the contraction of sets of muscles that extend from dorsal to ventral regions of the exoskeleton. The contractile force works against the exoskeleton, near hinge points where wings are attached. When the muscles contract, the exoskeleton pops inward and wings move up. When the exoskeleton pops outward by elastic force, the wings move down. The quick-stretched muscles invite a fast repeat of the events.

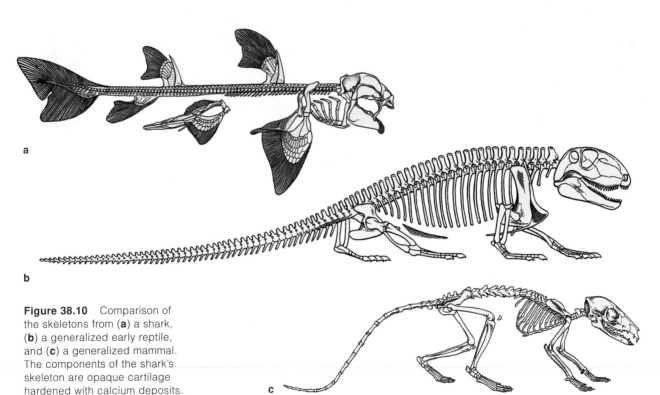

Figure 38.10 Comparison of the skeletons from (**a**) a shark, (**b**) a generalized early reptile, and (**c**) a generalized mammal. The components of the shark's skeleton are opaque cartilage hardened with calcium deposits.

Examples From the Vertebrates

Vertebrates have endoskeletons of one sort or another. For example, sharks have a skeleton of an opaque form of cartilage, hardened with calcium deposits (Figure 38.10*a*). Some other fishes have a flexible skeleton of an elastic, translucent form of cartilage that almost looks like glass. For most other vertebrates, however, the endoskeleton consists mainly of bone (Figure 38.10*b,c*).

We turn next to the functions and the characteristics of bones. Afterward, we will consider how different types of bones are arranged in the human skeletal system.

Animal skeletons have structural elements or body fluids against which the force of contraction can be applied.

Functions of Bone

By definition, **bones** are complex organs that function in movement, protection, support, mineral storage, and formation of blood cells, as listed in Table 38.1. Bones that interact with skeletal muscles maintain or change the positions of body parts. Bones also support and anchor muscles. Certain bones form hard compartments that enclose and protect the brain, the lungs, and other internal organs. Bones are reservoirs for mineral ions, the deposits and withdrawals of which help maintain body fluids and support metabolic activities. Only some bones are   sites of blood cell formation.

Table 38.1 Functions of Bone
1. *Movement.* Bones interact with skeletal muscles to maintain or change the position of body parts.
2. *Support.* Bones support and anchor muscles.
3. *Protection.* Many bones form hard compartments that enclose and protect soft internal organs.
4. *Mineral storage.* Bones are a reservoir for mineral ions, the deposits and withdrawals of which help maintain ion concentrations in body fluids.
5. *Blood cell formation.* Some bones contain regions where blood cells are produced.

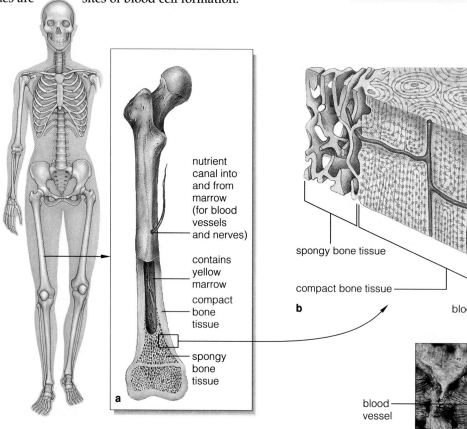

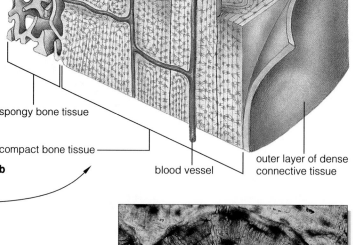

Bone Structure

Human bones range in size from tiny middle earbones to the clublike femurs, or thighbones. Bones are long, short (or cubelike), flat, and irregular. All of them have connective, epithelial, and bone tissues. The bone tissues are calcium hardened, with living cells and collagen fibers in a ground substance.

Consider the thighbone in Figure 38.11. *Compact* bone tissue in its shaft and at its ends resists mechanical shock. This type of bone tissue is deposited as many thin, dense, cylindrical layers around interconnected canals that contain blood vessels and nerves. Each cylindrical array is called a Haversian system. The blood

Figure 38.11 (**a**) Structure of a femur, one of the long bones of mammals. Femurs also are called thighbones. (**b**) Appearance of spongy bone tissue and compact bone tissue in a femur. The thin, dense layers of compact bone tissue form cylindrical arrays around interconnecting canals, which contain blood vessels and nerves. Each array is called a Haversian system. (**c**) Micrograph of one Haversian system. The blood vessel at its center services osteocytes, living bone cells in small spaces in bone tissue. Small tunnels connect neighboring spaces.

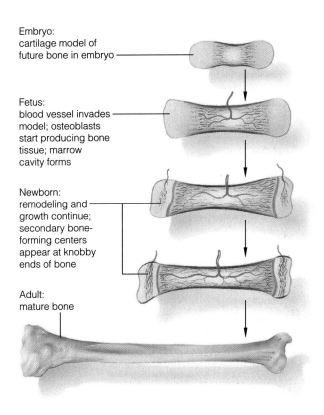

Embryo:
cartilage model of
future bone in embryo

Fetus:
blood vessel invades
model; osteoblasts
start producing bone
tissue; marrow
cavity forms

Newborn:
remodeling and
growth continue;
secondary bone-
forming centers
appear at knobby
ends of bone

Adult:
mature bone

Figure 38.12 Long bone formation, starting with osteoblast activity in a cartilage model (here, already formed in the embryo). Bone-forming cells are active first in the shaft region, then at the knobby ends. In time, the only cartilage left is at the ends.

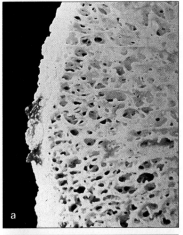

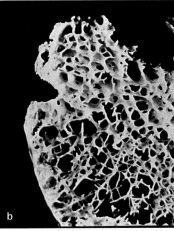

Figure 38.13 An example of bone tissue affected by osteoporosis.

(**a**) Section through normal bone tissue. In such tissue, the mineral deposits continually replace the withdrawals.

(**b**) After the onset of osteoporosis, replacements of mineral ions lag behind withdrawals. In time the tissue erodes, and bones become hollow and brittle.

vessels and nerves inside it service living bone cells. *Spongy* bone tissue in the bone ends and shaft imparts strength without adding much weight. Its abundant spaces make the tissue appear spongy, but its flattened parts are firm. **Red marrow**, a major site of blood cell formation, fills the spaces in some bones, such as the breastbone. The cavities in most mature bones contain **yellow marrow**. Yellow marrow consists largely of fat. It converts to red marrow and produces new red blood cells when blood loss from the body is severe.

HOW BONES DEVELOP The cartilage models for many bones form in the animal embryo (Figure 38.12). Bone-forming cells, or osteoblasts, secrete material inside the model's shaft and onto its surface. A marrow cavity opens up as cartilage breaks down inside the model. In time, osteoblasts become surrounded by their secretions and thereafter are called **osteocytes**, or living bone cells. Their metabolic activities maintain mature bones.

BONE TISSUE TURNOVER Minerals are continually deposited *and* withdrawn from bone tissue in ways that maintain the adult body's calcium levels. In this **bone tissue turnover**, bone cells secrete enzymes that digest bone tissue. The released ions of calcium and other minerals enter interstitial fluid, then the blood, which distributes them to metabolically active cells.

For instance, a thighbone becomes much thicker and stronger as its bone cells deposit minerals at the shaft's surface. At the same time, it becomes less heavy as other bone cells destroy bone tissue within its shaft.

Exercising stimulates calcium deposition and tends to increase bone density. Stress or injury triggers its withdrawal. Also, as a person ages, the backbone, hip bones, and other bones decrease in mass, especially in women. Figure 38.13 shows the effect of *osteoporosis*. A weakened backbone may collapse, curve abnormally, and lower the rib cage, which puts stress on internal organs. Decreasing osteoblast activity, calcium loss, sex hormone deficiencies, excessive intake of protein, and decreased physical activity contribute to the disorder.

Bones are collagen-rich, mineralized organs that function in movement, protection, support, storage of calcium and other minerals, and blood cell formation.

Appendicular and Axial Portions

The human skeleton has 206 bones, which anatomists have subdivided into appendicular and axial portions. Its *appendicular* portion has pectoral girdles (at the shoulders), pelvic girdle (at the hips), and paired arms, hands, legs, and feet. Its pectoral girdles have slender collarbones, and flat shoulder blades. Fall on an outstretched arm and you might dislocate a shoulder or fracture a collarbone, which is flimsily arranged and the bone most frequently broken.

The human skeleton's *axial* portion includes skull bones, twelve pairs of ribs, and the breastbone. It also has twenty-six **vertebrae** (singular, vertebra). These are bony segments of the vertebral column, or backbone. The curved backbone extends from the skull's base to the pelvic girdle. There the backbone transmits the weight of the torso to the lower limbs. A spinal cord threads through a series of bony canals at the rear of the column. Between are **intervertebral disks**—cartilaginous shock absorbers and flex points that permit movement.

Sometimes a severe or rapid shock forces a disk to slip out of place or rupture. Such painful, *herniated disks* are a less-than-advantageous outcome of bipedalism. Remember, the primate ancestors of the human lineage were quadrupedal. The first hominids started to walk upright about 4 million years ago, and a pronounced S-shaped curve in the backbone was a result. Today, the older we get, the longer we have been fighting gravity in a compromised way, and the more back pain we suffer.

Skeletal Joints

Joints are areas of contact or near-contact between bones, and each has a distinctive bridge of connective tissue. Very short connecting fibers join bones at *fibrous* joints. Straps of cartilage join them at *cartilaginous* joints. Long straps of dense connective tissue, or **ligaments,** bridge the gap between the bones at *synovial* joints.

Fibrous joints hold teeth in their sockets. They also connect the flat skull bones of a fetus. At childbirth, the loose connections allow the bones to slide over each other a bit and thereby prevent skull fractures. The skull of a newborn still has fibrous joints as well as membranous areas known as "soft spots," or fontanels. In childhood, the fibrous tissue hardens and skull bones are fused into a single unit.

Table 38.2 Components of the Human Skeleton
APPENDICULAR PORTION: Pectoral girdles: clavicle (collarbone) and scapula (shoulder blade) Arm bones: humerus, radius, ulna Hand bones: carpals, metacarpals, phalanges (of fingers) Pelvic girdle (six fused bones at the hip) Leg bones: femur (thighbone), patella, tibia, fibula Foot bones: tarsals, metatarsals, phalanges (of toes) AXIAL PORTION: Skull: cranial bones and facial bones Rib cage: sternum (breastbone) and ribs (12 pairs) Vertebral column: vertebrae (26) and intervertebral disks

Cartilaginous joints bridging vertebrae, ribs, and the breastbone permit slight movements. Synovial joints, such as knee joints, move freely. Ligaments stabilize the knee joints, as in Figure 38.14. Where one bone touches another, cartilage cushions them and absorbs shocks. A flexible capsule of dense connective tissue surrounds the region of contact. Cells of a membrane that lines the capsule's interior secrete a fluid that lubricates the joint.

Like many other joints, the knee joint is vulnerable to stress. This is the joint that lets you swing, bend, and turn the long bones below it. When you run, it absorbs the force of your weight each time the foot below it hits the ground. Stretch or twist a knee joint suddenly and too far, and you may *strain* it. Tear its ligaments or tendons and you *sprain* it. Move the wrong way and you may well dislocate the attached bones. During collision sports, such as football, blows to the knees frequently sever a ligament. The severed portion must be reattached surgically before ten days pass. Why? Phagocytes in a lubricating fluid within the joint normally clean up after everyday wear and tear. When presented with torn ligaments, they will indiscriminately turn the tissue to mush.

Joint inflammation as well as degenerative disorders are collectively called "arthritis." In *osteoarthritis*, cartilage at the knees and other freely movable joints wears off as a person ages. Joints in the fingers, knees, hips, and the vertebral column are affected most often. In *rheumatoid arthritis*, synovial membranes in joints become inflamed and thickened, cartilage degenerates, and bone deposits

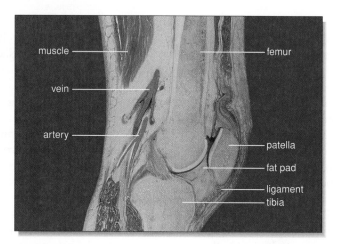

Figure 38.14 Human knee joint, longitudinal section.

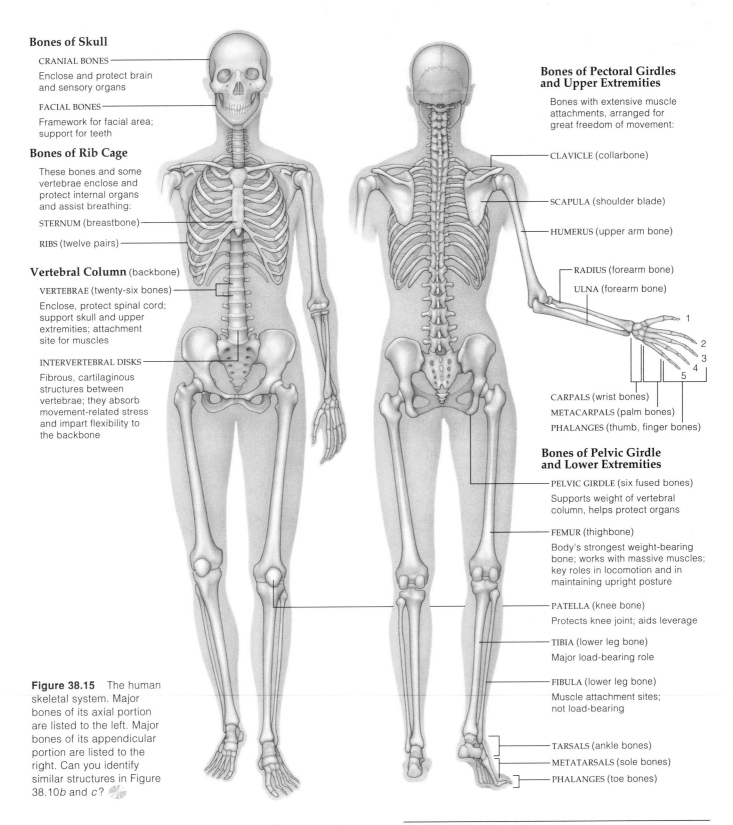

Bones of Skull

CRANIAL BONES

Enclose and protect brain
and sensory organs

FACIAL BONES

Framework for facial area;
support for teeth

Bones of Rib Cage

These bones and some
vertebrae enclose and
protect internal organs
and assist breathing:

STERNUM (breastbone)

RIBS (twelve pairs)

Vertebral Column (backbone)

VERTEBRAE (twenty-six bones)

Enclose, protect spinal cord;
support skull and upper
extremities; attachment
site for muscles

INTERVERTEBRAL DISKS

Fibrous, cartilaginous
structures between
vertebrae; they absorb
movement-related stress
and impart flexibility to
the backbone

**Bones of Pectoral Girdles
and Upper Extremities**

Bones with extensive muscle
attachments, arranged for
great freedom of movement:

CLAVICLE (collarbone)

SCAPULA (shoulder blade)

HUMERUS (upper arm bone)

RADIUS (forearm bone)

ULNA (forearm bone)

CARPALS (wrist bones)

METACARPALS (palm bones)

PHALANGES (thumb, finger bones)

**Bones of Pelvic Girdle
and Lower Extremities**

PELVIC GIRDLE (six fused bones)

Supports weight of vertebral
column, helps protect organs

FEMUR (thighbone)

Body's strongest weight-bearing
bone; works with massive muscles;
key roles in locomotion and in
maintaining upright posture

PATELLA (knee bone)
Protects knee joint; aids leverage

TIBIA (lower leg bone)
Major load-bearing role

FIBULA (lower leg bone)
Muscle attachment sites;
not load-bearing

TARSALS (ankle bones)

METATARSALS (sole bones)

PHALANGES (toe bones)

Figure 38.15 The human
skeletal system. Major
bones of its axial portion
are listed to the left. Major
bones of its appendicular
portion are listed to the
right. Can you identify
similar structures in Figure
38.10*b* and *c*?

build up. This degenerative disorder may be triggered
by a bacterial or viral infection, but it also appears to
have a genetic component. It can begin at any age, but
symptoms usually emerge before age fifty.

A human skeleton has an axial portion (a backbone, skull
bones, and rib cage) and an appendicular portion (pelvic
girdle, pectoral girdles, and arm, hand, leg, and foot bones).

How Muscles and Bones Interact

Skeletal muscles are the functional partners of bones. Each skeletal muscle contains bundles of hundreds to many thousands of muscle cells, which look like long, striped fibers. In muscle tissue, remember, muscle cells contract (shorten) in response to adequate stimulation. They lengthen in response to gravity and other loads. When you dance, breathe, scribble notes, or tilt your head, contracting muscle cells are helping to move your body or change the positions of some of its parts.

Connective tissue bundles muscle cells together and extends beyond them to form **tendons**. Each tendon, a cord or strap of dense connective tissue, attaches some muscle to bone (Figure 38.16). Most of the attachment sites are like a car's gearshift. *They form a lever system, in which a rigid rod is attached to a fixed point but able to move about at it.* Muscles connect to bones (rigid rods) near a joint (fixed point). As the muscles contract, they transmit force to bones and make them move.

Skeletal muscles interact with one another as well as with bones. Some are arranged in pairs or groups that work together to promote the same movement. Others work in opposition, with the action of one opposing or reversing the action of another. Figure 38.17a shows how opposing muscle groups work to move frog legs. Also look at Figure 38.17b, then extend your right arm forward. Now place your left hand over the biceps in the upper arm and slowly "bend your elbow." Feel the biceps contract? Even when your biceps contracts only a bit, it causes a large movement in the forearm bone that is connected to it. This is true of most leverlike arrangements.

Tendons can rub against bones, but sheaths help reduce the resulting friction.

Figure 38.16 Tendon sheath. This is not the same as a bursa, a small sac also filled with synovial fluid. Bursae are cushions interposed *between* bone and skin or bone and tendons.

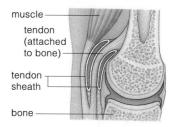

muscle

tendon (attached to bone)

tendon sheath

bone

Tendons slide within the sheaths, which are fluid-filled membranous sacs wrapped around them (Figure 38.16). Your knees, wrists, and finger joints have such sheaths.

Bear in mind, only *skeletal* muscle is the functional partner of bone. As mentioned earlier, smooth muscle is mainly a component of the walls of internal organs, such as the stomach (Section 33.3). Cardiac muscle is present only in the wall of the heart. We will consider the structure and functioning of smooth muscle and cardiac muscle in later chapters in this unit.

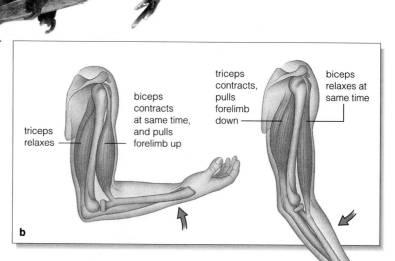

3 Now the first muscle group in the frog's upper hindlimb contracts again, drawing the limb back toward the body.

2 An opposing muscle group attached to the same limb contracts forcefully and pulls the limb back. The contractile force, directed against the ground, propels the frog forward.

1 A muscle group attached to each upper hindlimb contracts and pulls the leg forward a bit.

triceps relaxes

biceps contracts at same time, and pulls forelimb up

triceps contracts, pulls forelimb down

biceps relaxes at same time

a

b

Figure 38.17 (**a**) A frog demonstrating how a small decrease in the length of contracting muscles can produce a large movement. Its leap depends on opposing muscle groups attached to the upper limb bone of each hind leg. One muscle group pulls the limb forward and toward the body's midline. Another pulls it back and away from the body. (**b**) Two opposing muscle groups in a human arm. When a triceps contracts, the forearm extends straight out. When the triceps relaxes and its opposing partner (biceps) contracts, the elbow joint flexes and the forearm bends upward.

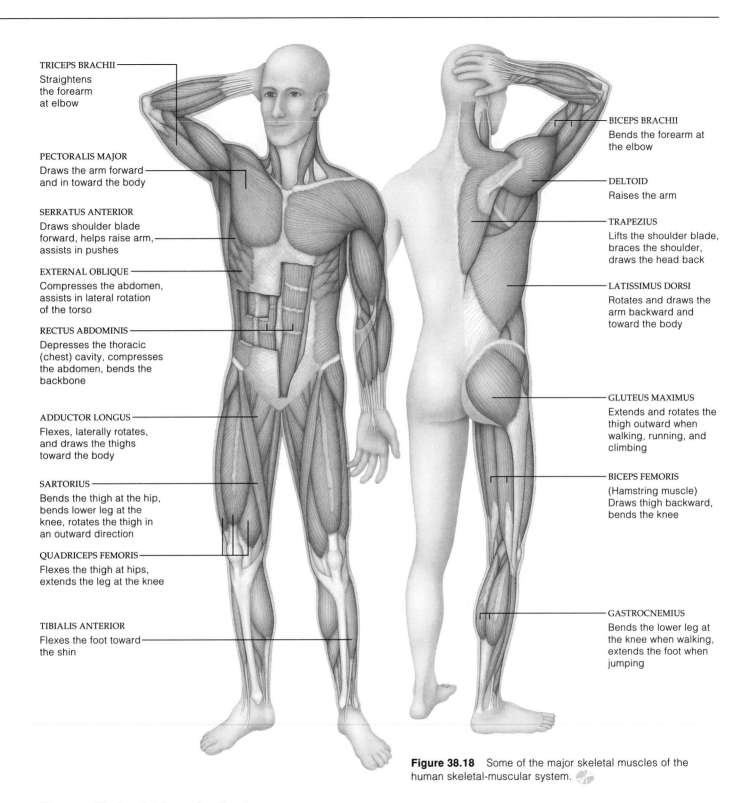

TRICEPS BRACHII
Straightens
the forearm
at elbow

PECTORALIS MAJOR
Draws the arm forward
and in toward the body

SERRATUS ANTERIOR
Draws shoulder blade
forward, helps raise arm,
assists in pushes

EXTERNAL OBLIQUE
Compresses the abdomen,
assists in lateral rotation
of the torso

RECTUS ABDOMINIS
Depresses the thoracic
(chest) cavity, compresses
the abdomen, bends the
backbone

ADDUCTOR LONGUS
Flexes, laterally rotates,
and draws the thighs
toward the body

SARTORIUS
Bends the thigh at the hip,
bends lower leg at the
knee, rotates the thigh in
an outward direction

QUADRICEPS FEMORIS
Flexes the thigh at hips,
extends the leg at the knee

TIBIALIS ANTERIOR
Flexes the foot toward
the shin

BICEPS BRACHII
Bends the forearm at
the elbow

DELTOID
Raises the arm

TRAPEZIUS
Lifts the shoulder blade,
braces the shoulder,
draws the head back

LATISSIMUS DORSI
Rotates and draws the
arm backward and
toward the body

GLUTEUS MAXIMUS
Extends and rotates the
thigh outward when
walking, running, and
climbing

BICEPS FEMORIS
(Hamstring muscle)
Draws thigh backward,
bends the knee

GASTROCNEMIUS
Bends the lower leg at
the knee when walking,
extends the foot when
jumping

Figure 38.18   Some of the major skeletal muscles of the human skeletal-muscular system.

Human Skeletal-Muscular System

The human body has more than 600 skeletal muscles, some superficial, others deep in the body wall. Some, such as facial muscles, attach to the skin. The trunk has muscles of the thorax, backbone, abdominal wall, and pelvic cavity. Other groups of muscles attach to upper and lower limb bones. Figure 38.18 shows a few of the main skeletal muscles and lists their functions. We turn next to the mechanisms underlying their contraction.

Skeletal muscles transmit contractile force to bones and make them move. Tendons strap skeletal muscles to bones.

Functional Organization of Skeletal Muscle

Bones move—they are pulled in some direction—when the skeletal muscles that are attached to them shorten. When a skeletal muscle shortens, its component muscle cells are shortening. When a muscle cell shortens, many units of contraction within that cell are shortening. The basic units of contraction are called **sarcomeres**.

Figure 38.19 gives you an idea of how bundles of cells in a skeletal muscle run parallel with the muscle itself. Inside each muscle cell are myofibrils, threadlike structures all bundled together in parallel array. Each myofibril is functionally divided into many sarcomeres, arranged one after another along its length. Dark bands called Z lines define the two ends of every sarcomere.

As Figure 38.19c shows, a sarcomere contains many filaments, oriented parallel with its long axis. Certain differences in their length and positioning give rise to the striped appearance of skeletal muscle (and cardiac muscle). Some of the filaments are thin; others are thick. Each *thin* filament is like two strands of pearls, twisted together. The "pearls" are molecules of **actin**, a globular protein with contractile functions:

Other proteins (coded *green*) are near actin's surface grooves. Each *thick* filament is made of molecules of **myosin**, another protein with contractile functions, in parallel array. A myosin molecule has a long tail and a double head that projects from the filament's surface:

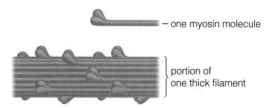

Thus the myofibrils, muscle cells, and muscle bundles of a skeletal muscle all run in the same direction. What function does this consistent, parallel orientation serve? It focuses the force of muscle contraction onto a bone in a particular direction.

Sliding-Filament Model of Contraction

How do sarcomeres shorten and bring about contraction of a skeletal muscle? The answer lies with sliding and pulling interactions among the sarcomere's filaments. A set of actin filaments extends from each Z line partway

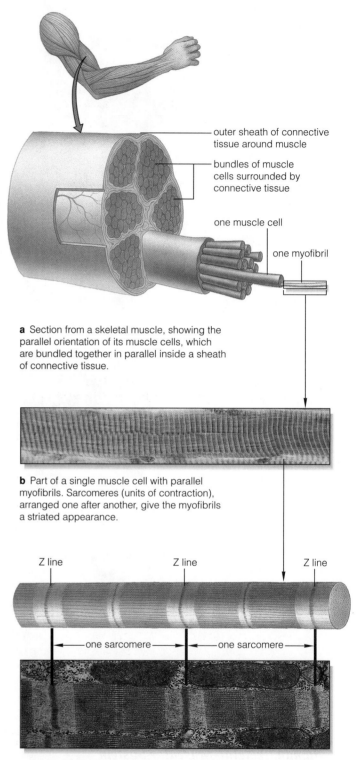

a Section from a skeletal muscle, showing the parallel orientation of its muscle cells, which are bundled together in parallel inside a sheath of connective tissue.

b Part of a single muscle cell with parallel myofibrils. Sarcomeres (units of contraction), arranged one after another, give the myofibrils a striated appearance.

c Diagram and transmission electron micrograph of two of the sarcomeres from a myofibril. This closer view shows how dark "bands," called Z lines, define the two ends of each sarcomere. Mitochondria (the oval-shaped organelles) adjacent to the sarcomeres provide ATP energy for muscle action.

Figure 38.19 Components of a skeletal muscle.

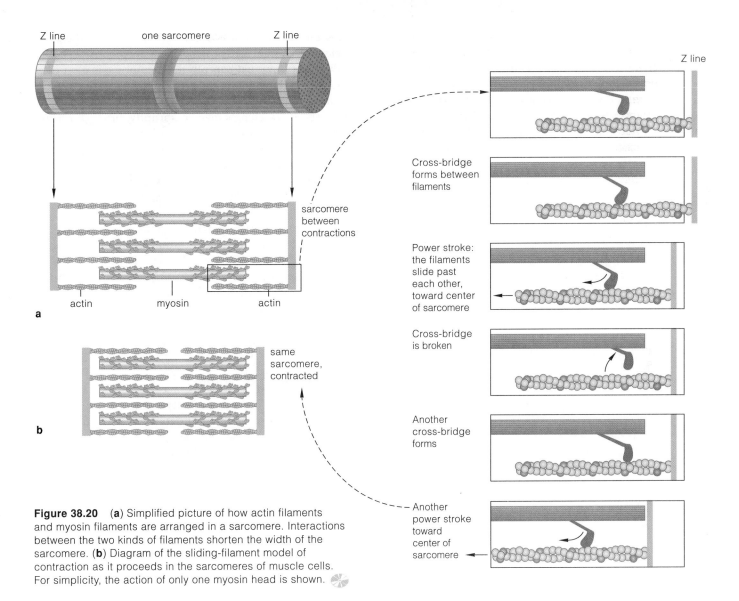

Figure 38.20 (**a**) Simplified picture of how actin filaments and myosin filaments are arranged in a sarcomere. Interactions between the two kinds of filaments shorten the width of the sarcomere. (**b**) Diagram of the sliding-filament model of contraction as it proceeds in the sarcomeres of muscle cells. For simplicity, the action of only one myosin head is shown.

to the center portion of each sarcomere. A set of myosin filaments partially overlaps the sets of actin, but it does not extend all the way to the Z lines. As Figure 38.20 indicates, when a muscle is contracting, these myosin filaments are physically sliding along and pulling each of the two sets of actin filaments toward the center of the sarcomere. The sarcomere shortens as a result. The physical interaction of actin and myosin is a key premise of the **sliding-filament model** of muscle contraction.

The myosin and actin filaments interact by way of **cross-bridge formation**. As indicated in Figure 38.20, these particular cross-bridges are attachments between a myosin head and a binding site on actin. The myosin heads are activated when messages from the nervous system stimulate the muscle cell. They physically attach to an adjacent actin filament and tilt in a short power stroke, driven by ATP energy, toward the sarcomere's

center. During the power stroke, the heads pull the actin filament along with them. Another energy input makes the heads let go, attach to another region of the filament, tilt in another power stroke, and so on down the line. A single contraction of a sarcomere requires a whole series of power strokes.

A skeletal muscle shortens through combined decreases in the length of its numerous sarcomeres. Sarcomeres are the basic units of contraction.

The parallel orientation of a muscle's component parts directs the force of contraction toward a bone that must be pulled in some direction.

By energy-driven interactions between the myosin and actin filaments, the many sarcomeres of a muscle cell shorten and collectively account for its contraction.

CONTROL OF MUSCLE CONTRACTION

The Control Pathway

When skeletal muscles contract, they help move the body and its assorted parts at certain times, in certain ways—and they do so in response to commands from the nervous system. The commands, which reach the muscles by way of motor neurons, can stimulate or inhibit contraction of the sarcomeres in muscle cells.

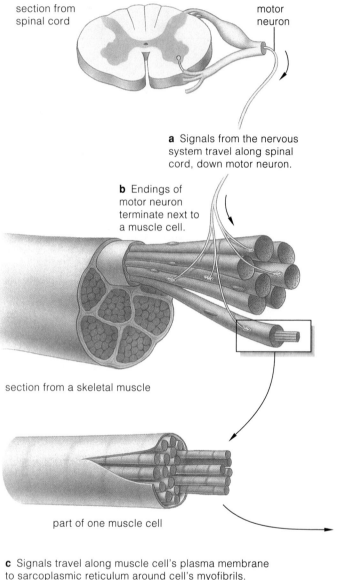

a Signals from the nervous system travel along spinal cord, down motor neuron.

b Endings of motor neuron terminate next to a muscle cell.

section from a skeletal muscle

part of one muscle cell

c Signals travel along muscle cell's plasma membrane to sarcoplasmic reticulum around cell's myofibrils.

Like all cells, a muscle cell shows a difference in electric charge across its plasma membrane. That is, the cytoplasm just beneath the membrane is a bit more negative than interstitial fluid outside it. But only in muscle cells, neurons, and other *excitable* cells does the difference in charge reverse abruptly, briefly, and in a predictable way in response to adequate stimulation.

The abrupt reversal in charge, an **action potential**, occurs as charged ions flow across the membrane in an accelerating way. The electrical commotion, remember, spreads without diminishing along the membrane, away from the point of stimulation (Section 34.1).

Suppose action potentials arise in a muscle cell. They spread rapidly away from the stimulation point, then along the small, tubelike extensions of the plasma membrane shown in Figure 38.21. The tubes connect with a system of membranous chambers that thread lacily around the muscle cell's myofibrils. That system, called the **sarcoplasmic reticulum**, takes up, stores, and releases calcium ions in controlled ways.

The Control Mechanism

The arrival of action potentials causes an outward flow of calcium ions from the sarcoplasmic reticulum. The released ions diffuse into the myofibrils and reach actin

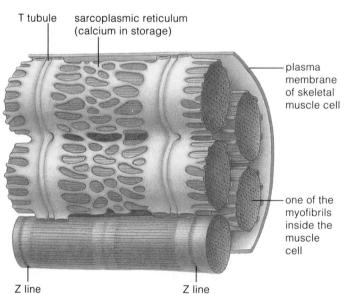

d Signals trigger the release of calcium ions from sarcoplasmic reticulum threading among the myofibrils. The arrival of calcium allows actin and myosin filaments in the myofibrils to interact and bring about contraction.

Figure 38.21 Pathway for signals from the nervous system that stimulate or inhibit contraction of skeletal muscle. The plasma membrane of each muscle cell surrounds the cell's myofibrils and connects with inward-threading T tubules. These membranous tubes are close to the sarcoplasmic reticulum, a calcium-storing system that functions in the control of contraction.

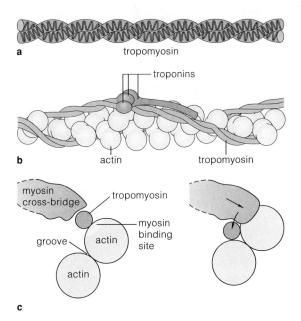

a tropomyosin

b actin / troponins / tropomyosin

c myosin cross-bridge / tropomyosin / groove / actin / actin / myosin binding site

Figure 38.22 Arrangement of troponins, tropomyosin, and actin filaments in skeletal muscle cells. When calcium binds with a troponin, tropomyosin moves away from the actin and thereby exposes the cross-bridge binding sites.

Figure 38.23 Three of the metabolic routes by which ATP forms in muscle cells in response to the demands of physical exercise.

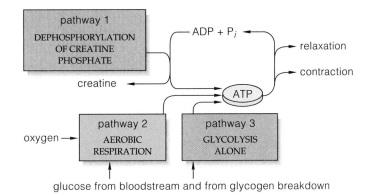

glucose from bloodstream and from glycogen breakdown

rises, however, calcium binds with the troponin and thereby alters its shape. When troponin is in its altered shape, it has a different molecular grip on tropomyosin, which is now free to move into the groove and expose the binding site (Figure 38.22c).

Sources of Energy for Contraction

All cells require ATP, but only in muscle cells does the demand skyrocket in so short a time. When a muscle cell at rest is called upon to contract, phosphate donations from ATP must proceed twenty to a hundred times faster. But the cell has only a small supply of ATP at the start of contractile activity. At such times, it forms ATP by a very fast reaction. An enzyme simply transfers phosphate from **creatine phosphate**, an organic compound, to ADP. The cell has about five times as much creatine phosphate as ATP, so this reaction is good for a few contractions. And it is enough to buy time for a relatively slower ATP-forming pathway to kick in (Figure 38.23).

During prolonged, moderate exercise, the oxygen-requiring reactions of aerobic respiration typically can provide most of the ATP required for contraction. For the first five to ten minutes, a muscle cell taps its store of glycogen for glucose, the starting substrate. For the next half hour or so of sustained activity, the muscle cell depends on glucose and fatty acid deliveries from the blood. For contractile activity longer than this, fatty acids are the main fuel source (Section 8.6).

What happens when exercise is so intensive that it exceeds the capacity of the respiratory system and circulatory system to deliver oxygen for the aerobic pathway? At such times, glycolysis alone will contribute more of the total ATP that is being produced. Remember, by this set of anaerobic reactions, a glucose molecule is only partly broken down, so the net ATP yield is small. But muscle cells can use this metabolic route as long as glycogen stores continue to provide glucose.

After intense exercise, deep, rapid breathing helps repay the body's **oxygen debt**, incurred when ATP use by muscles exceeded the aerobic pathway's deliveries.

filaments. Before this happened, the muscle was at rest (it was not contracting). Its actin binding sites were blocked and myosin could not form cross-bridges with them. However, the arrival of calcium ions clears the binding sites, so that contraction can proceed. After the contraction, the calcium ions are actively transported back into the membrane storage system.

What blocks cross-bridge binding sites in a muscle at rest? Two proteins, tropomyosin and troponin, are located in or near the surface grooves of actin filaments (Figure 38.22). At a low calcium level, the proteins are joined so tightly together that the tropomyosin is forced slightly outside the groove. In this position, it blocks the cross-bridge binding site. When the calcium level

It takes commands from the nervous system to initiate action potentials in muscle cells. The action potentials are signals for cross-bridge formation, hence for contraction.

During exercise, the availability of ATP inside muscle cells affects whether contraction will proceed, and for how long.

PROPERTIES OF WHOLE MUSCLES

Muscle Tension and Muscle Fatigue

Whether a muscle actually shortens during cross-bridge formation depends on the external forces acting upon it. Collectively, the cross-bridges exert **muscle tension**. By definition, this is a mechanical force that a contracting muscle exerts on an object, such as a bone. Opposing it is a load, either the weight of an object or gravity's pull on the muscle. Only when muscle tension exceeds the load does a stimulated muscle shorten.

An *isometrically* contracting muscle develops tension but does not shorten. It supports a load in a constant position, as when you hold a glass of lemonade in front of you. An *isotonically* contracting muscle shortens and moves a load. With *lengthening* contraction, though, an external load is greater than the muscle tension, so the muscle lengthens during the period of contraction. This happens to leg muscles when you walk down stairs.

A muscle's tension relates to the formation of cross-bridges in its cells and the number of cells recruited into action. Consider a **motor unit**: a motor neuron and all muscle cells that form junctions with its endings. By stimulating the motor unit with an electrical impulse, we can induce an action potential and make a recording of an isometric contraction. It takes a few milliseconds for tension to increase, then it peaks and declines. This response is a **muscle twitch** (Figure 38.24a). Its duration depends on the load and cell type. For example, fast-acting muscle cells rely on glycolysis (not efficient but fast) and use up ATP faster than slow-acting cells do.

If another stimulus is applied before the response is over, the muscle twitches again. **Tetanus** is a large contraction resulting from the repeated stimulation of a motor unit, so that twitches mechanically run together. (In a disease by the same name, toxins interfere with muscle relaxation.) Figure 38.24d shows a recording of tetanic contraction.

Continuous, high-frequency stimulation that keeps a muscle in a state of tetanic contraction leads to *muscle fatigue*, or a decline in tension. After a few minutes of rest, a fatigued muscle will contract again in response to stimulation. The extent of recovery depends largely on how long and how frequently it was stimulated before. Muscles associated with brief, intense exercise (such as weightlifting) fatigue fast but recover fast. The muscles associated with prolonged, moderate exercise fatigue slowly but take longer to recover, often up to twenty-four hours. The molecular mechanisms causing muscle fatigue are unknown, but glycogen depletion is a factor.

Effects of Exercise and Aging

A muscle's properties depend on how often, how long, and how intensely it is put to use. With regular **exercise**

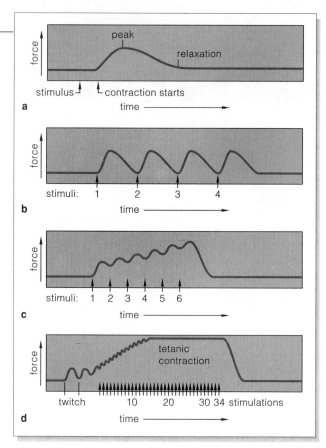

Figure 38.24 Recordings of twitches in muscles artificially stimulated in different ways. (**a**) A single twitch. (**b**) Two stimulations per second cause a series of twitches. (**c**) Six per second cause a summation of twitches, and (**d**) about twenty per second cause tetanic contraction.

(that is, increased levels of contractile activity), muscle cells do not increase in number. But they increase in size and metabolic activity, and they become more resistant to fatigue. Consider *aerobic exercise*, which is not intense but is long in duration. Aerobic exercise increases the number of mitochondria in both fast and slow muscle cells, and it increases the blood capillaries that service them. Such physiological changes improve endurance. By contrast, *strength training* (intense, short-duration exercise such as weightlifting) affects fast-acting muscle cells. These form more myofibrils and more enzymes of glycolysis. Strong, bulging muscles of the sort shown in Figure 38.25 may result, although they don't have much endurance. They fatigue rapidly.

Muscle tension decreases in adult humans thirty to forty years old. These people may exercise just as long and intensely as younger ones, but their muscles cannot adapt (change) in response to the same extent. Even so, some adaptation can be beneficial. Aerobic exercise can improve blood circulation. And, as it turns out, even modest strength training slows the loss of muscle tissue that is an inevitable part of the aging process.

Properties of muscles vary with age and levels of activity.

38.11 PEBBLES, FEATHERS, AND MUSCLE MANIA

A male penguin has a bit of a problem when scouting for a female penguin, because superficially they both look too much alike. The only way he can find one is to drop a pebble at the feet of a likely prospect. If those feet belong to a male, the pebble may be perceived as an insult, and it may start a fight. If female, his stony overture to courtship might be ignored or fancied, depending on her receptivity at that moment.

Peacocks and the males of many other species have no such problem. For them, sexual dimorphism is visible and quite pronounced (Figure 38.4c). Maybe the peacock's eye-stopping feathers evolved through sexual selection, with peahens serving as the deciders of their reproductive success. Charles Darwin certainly thought so. He viewed cases of extreme sexual dimorphism as the outcome of male competition for females and of females choosing among males.

What about extreme sexual dimorphism in body size, which involves increases in muscle mass? This may be one measure of how much mammalian males invest in fighting capacity. Possibly the cost of securing and using resources for growth and maintenance of a massive body is offset by great reproductive rewards.

All of this might make you wonder: Exactly what are the "rewards" for astoundingly muscled human athletes (Figure 38.25)? Is it a misdirected will to win, or a sign of modern athletic competition? Consider this: Each year in the United States alone, about a million athletes use anabolic steroids. And the vast majority of professional football players have used them to increase their "brute power." Even adolescent boys use them as a way to gain the winning edge in wrestling, football, and weightlifting tournaments.

Anabolic steroids are synthetic hormones. They mimic testosterone, a sex hormone that (among other things) governs secondary sexual traits. Testosterone makes boys get a deeper voice; more hair on their face, underarms, and pubic skin; and greater muscle mass in the arms, legs, shoulders, and elsewhere. Besides this, testosterone stimulates heightened aggressive behavior, which is often associated with maleness.

Anabolic steroids stimulate the synthesis of protein molecules, including proteins in muscle cells. Supposedly, they induce very rapid gains in muscle mass and muscle strength when taken during weight-training and exercise programs. This claim is disputed; results of most studies are based on too few subjects. Even so, not everyone believes that these drugs do enough damage to outweigh the edge they presumably give in athletic competition—or the wealth and hero status they accord the "winners."

Yet steroid-using athletes do suffer minor and major side effects. In men, acne, baldness, shrinking testes, and infertility are early signs of toxicity. The symptoms begin when a high level of anabolic steroids in blood triggers a sharp decline in the body's production of testosterone. Anabolic steroids also may trigger early heart disease. Even brief or occasional use may damage the kidneys or set the stage for cancer of the liver, testes, and prostate gland. Among women, anabolic steroids deepen the voice, and they produce pronounced facial hair. Menstrual cycles become irregular. Breasts may shrink, and the clitoris may become grossly enlarged.

Not every steroid user develops severe physical side effects. Far more common are mental difficulties, called *'roid rage* or *body-builder's psychosis*. In such cases, users become irritable and increasingly aggressive. Some men become wildly aggressive, uncontrollably manic, and delusional. For example, one steroid user accelerated his car to high speed and deliberately drove it into a tree; and doesn't that make you wonder just how superior some of us are to a pebble-toting penguin.

Figure 38.25 A human male with pumped-up biceps.

SUMMARY

1. Most animals have an integumentary system, which covers the body's surface. Examples include the cuticle of roundworms and arthropods, as well as vertebrate skin and the structures derived from it.

2. Skin protects against abrasion, ultraviolet radiation, dehydration, and many pathogenic bacteria. It also helps control internal temperature (blood flow to the skin can dissipate heat). Sensory receptors in skin detect stimuli in the external environment. When exposed to sunlight, skin serves an endocrine function; it produces vitamin D, a hormone-like substance required for absorption of calcium from food.

3. Skin consists of two regions: an outermost epidermis and an underlying dermis. The most abundant cells are keratinocytes (keratin producers). Also present are the melanocytes (melanin producers), and Langerhans cells and Granstein cells (which help defend the body against pathogens and cancer cells).

   a. Epidermis consists primarily of multiple layers of dead, keratinized, and melanin-shielded epithelial cells.

   b. The dermis is where rapid cell divisions produce replacements for cells shed on an ongoing basis. Hair, oil glands, sweat glands, and other structures derived from epidermal cells are embedded mainly in the dermis.

4. Movement of the animal body or parts of it requires contractile cells and some medium or structure against which contractile force can be applied.

   a. In hydrostatic skeletons, as in sea anemones, body fluids accept the contractile force and are redistributed inside a confined space.

   b. In exoskeletons, as in insects and other arthropods, rigid external body parts accept the contractile force.

   c. In endoskeletons, rigid internal body parts (mainly bones) receive the applied force of contraction.

5. Bones are complex organs that have osteocytes (living bone cells) in a mineralized, collagen-containing ground substance. Bones function in the movement, protection, and support of body parts; in mineral storage; and, in bones with red and yellow marrow, blood cell formation.

6. The human skeleton is divided into two portions.

   a. The appendicular region includes the pelvic girdle, pectoral girdles (collarbone and shoulderblade), arm and hand bones, and leg and foot bones.

   b. The axial region includes the skull bones, rib cage, and vertebral column, or backbone. The backbone is composed of vertebrae (bony segments) cushioned by cartilaginous, intervertebral disks.

7. Skeletal joints are areas of contact or near-contact where connective tissue bridges adjacent bones. Fibrous joints (short fibers), cartilaginous joints (of cartilage), or synovial joints (with straplike ligaments) bridge the gap between bones at different joints.

8. Cells of smooth, cardiac, and skeletal muscle contract (shorten) in response to adequate stimulation.

9. Tendons attach skeletal muscles to bones. The skeletal muscles and bones interact as a system of levers, with rigid rods (bones) moving at fixed points (joints). Many muscles work together or in opposition to bring about movement or positional changes in body parts.

10. Inside each skeletal muscle cell are many myofibrils arranged parallel with its long axis. These threadlike structures contain actin and myosin filaments organized in parallel. Each myofibril is functionally divided into sarcomeres, the basic units of contraction. The parallel orientation of the muscle's components directs the force of contraction at a bone to be pulled in some direction.

11. In response to stimulation, skeletal muscles shorten by decreases in the length of all of their sarcomeres. The nervous system stimulates muscle cells. Its commands are delivered by motor neuron endings that terminate on muscle cells, and they can trigger action potentials at the muscle cell's plasma membrane.

12. Here are the main points of the sliding-filament model of muscle contraction:

   a. Action potentials cause the release of calcium ions from a membrane system (sarcoplasmic reticulum) that threads around the cell's myofibrils. Calcium diffuses inside sarcomeres, binds to actin filaments, and makes the binding sites change shape. Then myosin filament heads can form cross-bridges with actin filaments.

   b. Each cross-bridge is a *brief* attachment between a myosin head and an actin binding site. Cross-bridges form during repeated, ATP-driven power strokes. The repeated strokes make actin filaments slide past myosin filaments, and collectively they shorten the sarcomere.

13. Muscle cells obtain the ATP required for contraction by three metabolic pathways:

   a. Dephosphorylation of creatine phosphate. This is direct, fast, and good for a few seconds of contraction.

   b. Aerobic respiration. This pathway predominates during prolonged, moderate exercise.

   c. Glycolysis. This pathway takes over when intense exercise exceeds the body's capacity to deliver oxygen to muscle cells.

14. Muscle tension refers to a mechanical force created by cross-bridge formation. The load (gravity or weight of objects) is an opposing force. A stimulated muscle shortens when tension exceeds the load and lengthens when it is less than the load. Levels of exercise and aging affect the properties of muscles.

   a. A motor unit is a motor neuron and all muscle cells that form junctions with its endings.

   b. A muscle twitch is a brief, weak contraction in response to a single action potential at a motor unit. Tetanus is a large contraction resulting from repeated stimulation at a motor unit.

Review Questions

1. List the functions of skin, then distinguish between its regions. Is the hypodermis part of skin? *38.1*

2. Name four cell types in skin and their functions. *38.2, 38.3*

3. Distinguish between:
   a. sweat gland and oil gland *38.2*
   b. hydrostatic skeleton, exoskeleton, and endoskeleton *38.4*
   c. red marrow and yellow marrow *38.5*
   d. ligament and tendon *38.6, 38.7*

4. What are the functions of bones? *38.5*

5. Name the three types of muscle, then state the function of each and where they are located in the body. *38.7*

6. Look at Figures 38.19 and 38.20. Then, on your own, sketch and label the fine structure of a muscle, down to one of its individual myofibrils. Identify the basic unit of contraction in the myofibrils. *38.8*

7. What role does calcium play in the control of contraction? What role does ATP play, and by what routes does it form? *38.9*

Self-Quiz *(Answers in Appendix IV)*

1. Nearly all animals have a(n) _____ system that protects the body from abrasion, ultraviolet radiation, bacterial attack, and other environmental stresses.

2. _____ and _____ systems work together to move the body and specific body parts.

3. The three categories of muscle tissue are _____ , _____ , and _____ .

4. Which is *not* a function of skin?
   a. resist abrasion   c. initiate movement
   b. restrict dehydration d. help control temperature

5. _____ are shock pads and flex points.
   a. Vertebrae   c. Marrow cavities
   b. Femurs   d. Intervertebral disks

6. Blood cells form in _____ .
   a. red marrow   c. certain bones only
   b. all bones   d. a and c

7. In a skeletal muscle cell, the _____ is the basic unit of contraction.
   a. myofibril   c. muscle fiber
   b. sarcomere   d. myosin filament

8. Muscle contraction requires _____ .
   a. calcium ions   c. action potential arrival
   b. ATP   d. all of the above

9. ATP for muscle contraction can be formed by _____ .
   a. aerobic respiration   d. a and b only
   b. glycolysis   e. a and c only
   c. creatine phosphate breakdown   f. a, b, and c

10. Match the M words with their defining feature.
   ____ muscle   a. actin's partner
   ____ muscle twitch   b. all in the hands
   ____ muscle tension   c. blood cell production
   ____ melanin   d. decline in tension
   ____ myosin   e. brownish-black pigment
   ____ marrow   f. motor unit response
   ____ metacarpals   g. force exerted by cross-bridges
   ____ myofibrils   h. muscle cells bundled in
   ____ muscle fatigue   connective tissue
                          i. threadlike parts in muscle cell

Critical Thinking

1. For Rizza and other young women, the recommended daily allowance (RDA) of calcium is 800 milligrams per day. During Rizza's pregnancy, the RDA is 1,200 milligrams per day. Why would a pregnant woman need the larger amount? What might happen to her bones without it?

2. Kate found a malnourished, stray cat that had just given birth to a litter of four kittens. The cat was trying to hunt, but her muscles were twitching so badly she could scarcely walk. Kate recalled what she had learned about muscle contraction. She offered milk as well as solid food to the cat—which, after several days, regained muscle control and was able to care for her kittens. How might milk help restore muscle function?

3. Compared to most people, Joe and other long-distance runners have a greater number of muscle fibers with more mitochondria. Sprinters have a greater number of muscle fibers that have more of the enzymes necessary for glycolysis but fewer mitochondria. Think about how these two forms of exercise differ and explain why the muscle fibers differ between the two kinds of runners.

Selected Key Terms

actin *38.8*	ligament *38.6*
action potential *38.9*	melanocyte *38.2*
anabolic steroid *38.11*	motor unit *38.10*
bone *38.5*	muscle tension *38.10*
bone tissue turnover *38.5*	muscle twitch *38.10*
creatine phosphate *38.9*	myosin *38.8*
cross-bridge formation *38.8*	oil gland *38.2*
cuticle *38.1*	osteocyte *38.5*
dermis *38.1*	oxygen debt *38.9*
endoskeleton *38.4*	red marrow *38.5*
epidermis *38.1*	sarcomere *38.8*
exercise *38.10*	sarcoplasmic reticulum *38.9*
exoskeleton *38.4*	skeletal muscle *38.7*
Granstein cell *38.3*	skin *38.1*
hair *38.2*	sliding-filament model *38.8*
hydrostatic skeleton *38.4*	sweat gland *38.2*
integument *38.1*	tendon *38.7*
intervertebral disk *38.6*	tetanus *38.10*
joint *38.6*	vertebra (vertebrae) *38.6*
keratinocyte *38.2*	yellow marrow *38.5*
Langerhans cell *38.3*	

Readings

Brusca, R., and G. Brusca. 1990. *Invertebrates*. Sunderland, Massachusetts: Sinauer.

Huxley, H. E. December 1965. "The Mechanism of Muscular Contraction." *Scientific American* 213(6): 18–27. Old article, great illustrations.

Sherwood, L. 1997. *Human Physiology*. Third edition. Belmont, California: Wadsworth.

Weeks, O. December 1989. "Vertebrate Skeletal Muscle: Power Source for Locomotion." *BioScience* 39(11): 791–798.

Web Site See *http://www.wadsworth.com/biology* for practice quiz questions, hypercontents, BioUpdates, and critical thinking. The Wadsworth Biology Resource Center provides a wealth of information fully organized and integrated by chapter.

CIRCULATION

Heartworks

For Dr. Augustus Waller, Jimmie the bulldog was no ordinary pooch. Connected to wires and soaked to his ankles in buckets of salty water, Jimmie was a four-footed explorer of the workings of the heart (Figure 39.1*a*). Press your fingers to your chest a few inches left of the center, between the fifth and sixth ribs, and feel the repetitive thumpings of your heart. The same rhythms intrigued Waller and other nineteenth-century physiologists. They wondered: Does each beat of the heart produce a pattern of electrical currents? Could they find out by devising a painless way to record such currents at the body surface?

That's where Jimmie and the buckets of salty water came in. Saltwater happens to be an efficient conductor of electricity. In Waller's experiment, it picked up faint signals from Jimmie's beating heart through the skin of his legs and conducted them to a crude monitoring device. With that device, Waller made one of the first recordings of heart activity (Figure 39.1). Today we call such recordings ECGs, an abbreviation for **electrocardiograms**.

A graph of the normal electrical activity of your own heart would look much the same. The pattern emerged a few weeks after you started growing, by way of mitotic cell divisions, from a single fertilized egg inside your mother. Early on, some of the embryonic cells differentiated into cardiac muscle cells, and these started to contract spontaneously. One small patch of the cells took the lead, and it has functioned as your heart's pacemaker ever since.

If all goes well, that patch of cardiac muscle cells will continue to contract as it should until the day you die. It is a natural pacemaker; it sets the baseline rate at which blood is pumped out of the heart, through blood vessels, then back to the heart. The rate is moderate, about seventy beats every minute; but commands from the nervous

a
BUCKET

b
JIMMIE   DR. WALLER

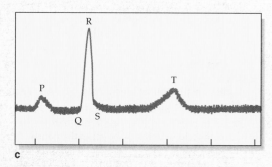

c

Figure 39.1 A bit of history in the making. (**a**) Jimmie the bulldog, taking part in a painless experiment. (**b**) Augustus Waller and his beloved pet bulldog sharing a quiet moment in Waller's study after the experiment, which yielded one of the world's first electrocardiograms (**c**).

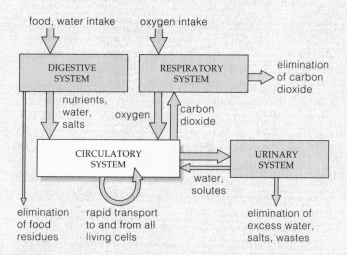

food, water intake   oxygen intake

DIGESTIVE SYSTEM   RESPIRATORY SYSTEM   elimination of carbon dioxide

nutrients, water, salts   oxygen   carbon dioxide

CIRCULATORY SYSTEM   URINARY SYSTEM

water, solutes

elimination of food residues   rapid transport to and from all living cells   elimination of excess water, salts, wastes

Figure 39.2 Diagram of the functional connections between the circulatory, respiratory, and digestive systems, which interact in transporting substances to and from all living cells in the animal body. Their integrated activities help maintain favorable operating conditions in the internal environment.

system and endocrine system continually adjust it. When you jog, for example, your skeletal muscle cells demand much more blood-borne oxygen and glucose than they do when you sleep. At such times, your heart starts pounding more than twice as fast, and this helps deliver sufficient blood to them.

We have come a long way from Waller and Jimmie in our monitorings of the heart. Sensors can now detect the faint signals characteristic of an impending heart attack. Internists now use computers to analyze a patient's beating heart, and they use ultrasound probes to build images of it on a video screen. Cardiologists routinely substitute battery-powered pacemakers for malfunctioning natural ones.

With this chapter, we turn to the circulatory system, the means by which substances move rapidly to and from the interstitial fluid that bathes living cells in nearly all animals. The circulatory system continually accepts the oxygen, nutrients, and other substances that the animal secures, by way of respiratory and digestive systems, from the outside environment (Figure 39.2). Simultaneously it picks up carbon dioxide and other wastes from cells and delivers them to the respiratory and urinary systems for disposal. Its smooth operation is absolutely central to maintaining operating conditions in the internal environment within a tolerable range—a state we call homeostasis.

KEY CONCEPTS

1. All cells survive by exchanging substances with their surroundings. In most animals, substances rapidly move to and from cells by way of a closed circulatory system.

2. Blood, a fluid connective tissue, is the transport medium of circulatory systems. It transports oxygen, carbon dioxide, plasma proteins, vitamins, hormones, lipids, and other solutes. It also transports metabolically generated heat.

3. In birds and mammals, a four-chambered, muscular heart pumps blood through two separate circuits of blood vessels, both of which lead back to the heart.

4. In the pulmonary circuit, the heart pumps oxygen-poor blood to the lungs, where it picks up oxygen; then blood flows back to the heart. In the systemic circuit, the oxygenated blood is pumped from the heart to all body regions, where it gives up oxygen and picks up carbon dioxide before flowing back to the heart.

5. Arteries are large-diameter, low-resistance vessels that rapidly transport oxygenated blood from the heart. Veins are large-diameter, low-resistance vessels that transport oxygen-poor blood to the heart and serve as blood volume reservoirs.

6. Arterioles are sites where the flow volume through each organ is controlled. In response to signals, their diameters widen in some regions and narrow in others. Their coordinated responses divert more of the flow volume to organs that are most active at a given time.

7. Numerous small-diameter, thin-walled capillaries interconnect in capillary beds, which are diffusion zones. As a volume of blood spreads out through a bed, it slows owing to the total cross-sectional area of the capillaries, which is greater than that of arterioles. The slowdown allows time for exchanges with interstitial fluid.

General Characteristics

Imagine that an earthquake has closed off the highways around your neighborhood. Grocery trucks can't enter and waste-disposal trucks can't leave, so food supplies dwindle and garbage piles up. Cells would face similar predicaments if your body's highways were disrupted. The highways are part of a **circulatory system**, which functions in the rapid internal transport of substances to and from cells. The system helps maintain favorable neighborhood conditions, so to speak, and this is a vital task. All of your differentiated cells perform specialized tasks and cannot fend for themselves. Different types interact in coordinated ways to maintain the volume, composition, and temperature of **interstitial fluid**, the tissue fluid that bathes them. A circulating connective tissue—**blood**—interacts with interstitial fluid. Blood makes continual deliveries and pickups that help keep conditions tolerable for enzymes and other molecules that carry out cell activities. Together, interstitial fluid and blood are the body's "internal environment."

Blood flows in blood vessels, or tubes that differ in wall thickness and in diameter. A muscular pump, the **heart**, generates pressure that keeps blood flowing. Like many animals, you have a *closed* circulatory system, in which blood is confined within continuously connected walls of the heart and blood vessels. This is not true of *open* circulatory systems of arthropods, such as insects, and most mollusks. In open systems, blood flows out through the vessels and into sinuses, which are small spaces in body tissues. There, blood mingles with tissue fluids, then moves back to the heart through openings in the blood vessels or the heart wall (Figure 39.3a).

Think about the overall "design" of a closed system. Because the heart pumps incessantly, the *volume* of blood flowing through blood vessels has to equal the heart's output at any given time. The flow's *velocity* (speed) is highest in large-diameter transport vessels. It decreases in specific body regions, where there must be enough time for blood to exchange substances with cells. The required slowdown proceeds at **capillary beds**, where blood spreads out through many small-diameter blood vessels called **capillaries**. During any interval, the same volume of blood is moving forward through the beds as elsewhere in the system. But it is doing so at a more leisurely pace, owing to the great total cross-sectional area of the blood capillaries. The simple analogy shown in Figure 39.3e may help you grasp this concept.

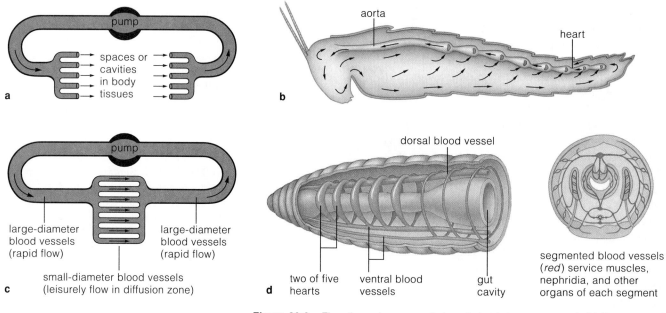

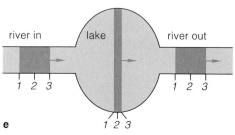

Figure 39.3 Flow through open and closed circulatory systems. (**a,b**) Open system of a grasshopper. A "heart" pumps blood through a vessel (aorta). Blood moves into tissue spaces and mingles with fluid bathing cells, then it reenters the heart through openings in the heart wall. (**c,d**) Closed system of an earthworm. Blood is confined within several pairs of muscular "hearts" near the head end and within blood vessels.

(**e**) Relation between flow velocity and total cross-sectional area in a closed circulatory system. Visualize two fast rivers flowing into and out of a lake. The flow *rate* is the same in all three places; an identical volume of water moves from point *1* to point *3* during the same interval. Flow *velocity* decreases in the lake, for the volume spreads through a larger cross-sectional area and moves forward a shorter distance during the same time.

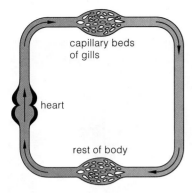 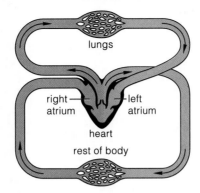 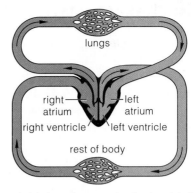

a In fishes, a two-chambered heart (atrium, ventricle) pumps blood in one circuit. Blood picks up oxygen in gills, delivers it to rest of body. Oxygen-poor blood flows back to heart.

b In amphibians, a heart pumps blood through two partially separate circuits. Blood flows to lungs, picks up oxygen, returns to heart. It mixes with oxygen-poor blood still in heart, flows to rest of body, returns to heart.

c In birds and mammals, the heart is fully partitioned into two halves. Blood circulates in two circuits: from the heart's right half to lungs and back, then from the heart's left half to oxygen-requiring tissues and back.

Figure 39.4 Comparison of the closed circulatory systems of vertebrate groups.

Evolution of Vertebrate Circulatory Systems

Humans and other existing vertebrates have a closed circulatory system, although the pump and plumbing of fishes, amphibians, birds, and mammals differ in their details. The differences evolved over hundreds of millions of years and corresponded to the move of some vertebrate lineages onto land. Recall, from Section 27.3, that fishes, the first vertebrates, had gills. Gills, like all respiratory structures, have a thin, moist surface that oxygen and carbon dioxide can diffuse across. Later, in the ancestors of land vertebrates, lungs evolved that supplemented gas exchange. Being *internally moistened* sacs, lungs had advantages for the move onto dry land. Also advantageous were concurrent modifications in circulatory systems, which pick up oxygen from lungs and deliver carbon dioxide wastes to them.

Consider this: In fishes, blood flows in *one* circuit (Figure 39.4a). Pressure generated by a two-chambered heart forces it through capillary beds of gills, the largest artery, capillary beds of organs, then back to the heart. The vast gill capillaries offer so much resistance to flow that the pressure drops considerably before the main artery. The blood delivery is fine for the activity level of most fishes. But it would not be sufficient for the more active life-styles of most land vertebrates.

When the amphibians evolved, their heart became partitioned into right and left halves—only partly so, but enough to pump blood through *two* partially separated circuits (Figure 39.4b). The separation of flow continued in crocodilians, which are reptiles. It became complete in birds and mammals; their heart acts like two side-by-side pumps (Figure 39.4c). The *right* half of their heart pumps oxygen-poor blood to the lungs, where blood picks up oxygen and gives up carbon dioxide. Then the freshly oxygenated blood flows to the heart's left half. This route is the **pulmonary circuit.**

In the **systemic circuit,** the heart's *left* half pumps the freshly oxygenated blood to every tissue and organ where oxygen is used and carbon dioxide forms. Then the oxygen-poor blood flows to the heart's right half.

The double circuit is a rapid and efficient mode of blood delivery. It supports the high levels of activity typical of vertebrates whose ancestors evolved on land.

Links With the Lymphatic System

The heart's pumping action puts pressure on blood flowing through the circulatory system. Partly because of the pressure, small amounts of water and a few of the proteins dissolved in blood are forced out of capillaries and become part of interstitial fluid. However, a rather elaborate network of drainage vessels picks up excess interstitial fluid and reclaimable solutes, then returns them to the circulatory system. This network is part of the **lymphatic system.** Later, you will see how other parts of the lymphatic system help cleanse bacteria and other pathogens from fluid being returned to the blood.

Closed circulatory systems confine blood within one or more hearts and a network of blood vessels. In the open systems, blood also intermingles with tissue fluids.

The closed system of vertebrates transports substances to and from interstitial fluid bathing the body's cells. It is functionally connected with the lymphatic system.

Blood flows rapidly in large-diameter vessels between the heart and capillary beds. There, the flow velocity slows and exchanges are made between blood and interstitial fluid.

In fishes, blood flows in one circuit from and back to the heart. In birds and mammals, blood flows in two circuits, through a heart partitioned as two side-by-side pumps. The double circuit supports the high levels of activity typical of vertebrates that evolved on land.

CHARACTERISTICS OF BLOOD

Functions of Blood

Blood is a connective tissue with multiple functions. It transports oxygen, nutrients, and other solutes to cells. It carries away their metabolic wastes and secretions, including hormones. Blood helps stabilize internal pH. Blood also serves as a highway for phagocytic cells that scavenge tissue debris and fight infections. In birds and mammals, blood helps equalize body temperature. It does this by carrying excess heat from skeletal muscles and other regions of high metabolic activity to the skin, where heat can be dissipated.

Blood Volume and Composition

The volume of blood depends on body size and on the concentrations of water and solutes. Blood volume for average-size adult humans is about 6 to 8 percent of the total body weight. That amounts to about four or five quarts. As is true of all vertebrates, the blood of humans is a sticky fluid, thicker than water and slower flowing. Its components are the **plasma**, **red blood cells**, **white blood cells**, and **platelets**. Plasma normally accounts for 50 to 60 percent of the total blood volume.

PLASMA   Prevent a blood sample in a test tube from clotting, and it separates into a layer of straw-colored liquid, the plasma, which floats over the red-colored cellular portion of blood (Figure 39.5). Plasma is mostly water, and it functions as a transport medium for blood cells and platelets. It also serves as a solvent for ions

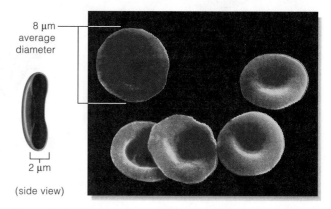

Figure 39.6 Size and shape of red blood cells.

8 μm average diameter

2 μm

(side view)

and molecules, including hundreds of different plasma proteins. Some of the plasma proteins transport lipids and fat-soluble vitamins through the body. Others have roles in blood clotting or in defense against pathogens. Collectively, the concentration of the plasma proteins affects the blood's fluid volume, for it influences the movement of water between blood and interstitial fluid. Glucose and other simple sugars, as well as lipids, amino acids, vitamins, and hormones, are dissolved in plasma. So are oxygen, carbon dioxide, and nitrogen.

RED BLOOD CELLS   Erythrocytes, or red blood cells, are biconcave disks, like doughnuts with a squashed-in center instead of a hole (Figure 39.6). They transport the oxygen used in aerobic respiration and they carry away some carbon dioxide wastes. When oxygen diffuses into

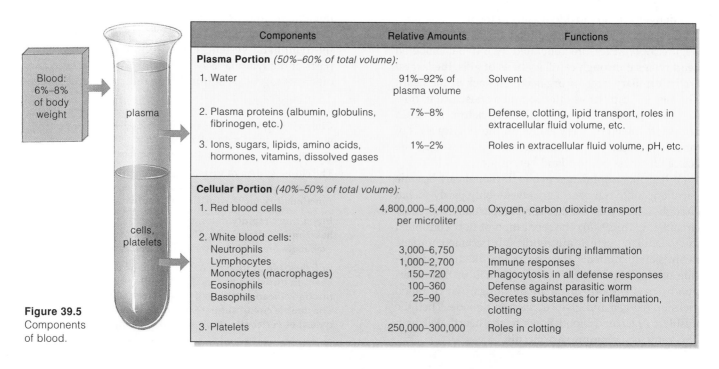

Figure 39.5 Components of blood.

Blood: 6%–8% of body weight

plasma

cells, platelets

Components	Relative Amounts	Functions
Plasma Portion *(50%–60% of total volume):*		
1. Water	91%–92% of plasma volume	Solvent
2. Plasma proteins (albumin, globulins, fibrinogen, etc.)	7%–8%	Defense, clotting, lipid transport, roles in extracellular fluid volume, etc.
3. Ions, sugars, lipids, amino acids, hormones, vitamins, dissolved gases	1%–2%	Roles in extracellular fluid volume, pH, etc.
Cellular Portion *(40%–50% of total volume):*		
1. Red blood cells	4,800,000–5,400,000 per microliter	Oxygen, carbon dioxide transport
2. White blood cells:		
Neutrophils	3,000–6,750	Phagocytosis during inflammation
Lymphocytes	1,000–2,700	Immune responses
Monocytes (macrophages)	150–720	Phagocytosis in all defense responses
Eosinophils	100–360	Defense against parasitic worm
Basophils	25–90	Secretes substances for inflammation, clotting
3. Platelets	250,000–300,000	Roles in clotting

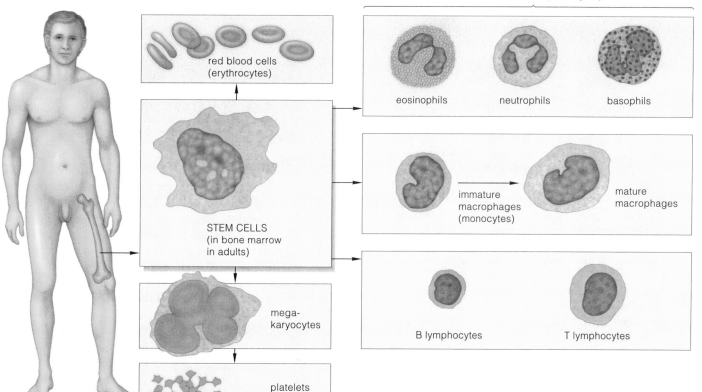

white blood cells (leukocytes)

red blood cells (erythrocytes)

eosinophils   neutrophils   basophils

STEM CELLS (in bone marrow in adults)

immature macrophages (monocytes)   mature macrophages

mega-karyocytes

B lymphocytes   T lymphocytes

platelets

Figure 39.7 The cellular components of blood.

blood, it binds with hemoglobin, the iron-containing pigment that gives red blood cells their color. (Here you may wish to review hemoglobin's molecular structure, as shown in Section 3.7.) Oxygenated blood is bright red. Poorly oxygenated blood is darker red but appears blue inside blood vessel walls near the body surface.

Red blood cells are derived from stem cells in bone marrow (Figure 39.7). Generally speaking, **stem cells** remain unspecialized and retain the capacity for mitotic cell division. Their daughter cells also divide, but only a portion go on to differentiate into specialized types.

Mature red blood cells no longer have their nucleus, nor do they require it. They have enough hemoglobin, enzymes, and other proteins to function for about 120 days. Phagocytes continually engulf the oldest red blood cells or those already dead, but ongoing replacements keep the cell count fairly stable. A **cell count** is a measure of the number of cells of a given type in a microliter of blood. For example, the average number of red blood cells is 5.4 million in males and 4.8 million in females.

WHITE BLOOD CELLS   Leukocytes, or white blood cells, arise from stem cells in bone marrow. They function in daily housekeeping and defense. Many patrol tissues, where they target or engulf damaged or dead cells and anything chemically recognized as foreign to the body. Many others are massed together in the lymph nodes

and spleen, which are part of the lymphatic system. There they divide to produce armies of cells that battle specific viruses, bacteria, and other invaders.

White blood cells differ in size, nuclear shape, and staining traits. There are five categories: neutrophils, eosinophils, basophils, monocytes, and lymphocytes (Figure 39.7). Their numbers can vary, depending on whether an individual is active, healthy, or under siege, as described in the next chapter. The neutrophils and monocytes are search-and-destroy cells. The monocytes follow chemical trails to inflamed tissues. There they develop into macrophages ("big eaters") that can engulf invaders and debris. Two classes of lymphocytes, B cells and T cells, make highly specific defense responses.

PLATELETS   Some stem cells in bone marrow give rise to giant cells (megakaryocytes). These shed fragments of cytoplasm enclosed in a bit of plasma membrane. The membrane-bound fragments are what we call the platelets. Each platelet only lasts five to nine days, but hundreds of thousands are always circulating in blood. They can release substances that initiate blood clotting.

Vertebrate blood has roles in transport, defense, clotting, and maintaining the volume, composition, and temperature of the internal environment.

The body continually replaces blood cells for good reason. Besides aging and dying off regularly, blood cells typically encounter a variety of pathogens that use them as places to complete their life cycle. Besides this, sometimes blood cells malfunction as a result of gene mutations.

RED BLOOD CELL DISORDERS   Consider the **anemias**, disorders that result from too few red blood cells or deformed ones. Oxygen levels in blood cannot be kept high enough to support normal metabolism. Shortness of breath, fatigue, and chills follow. *Hemorrhagic* anemias result from sudden blood loss, as from a severe wound; *chronic* anemias result from ongoing but slight blood loss, as from an undiagnosed bleeding ulcer, hemorrhoids, or the monthly blood loss of premenopausal women.

Certain infectious bacteria and parasites replicate inside blood cells, then escape by lysis; they cause some *hemolytic* anemias. Insufficient iron in the diet results in *iron deficiency* anemia, for red blood cells cannot produce enough normal hemoglobin without it. B_{12} *deficiency* anemia is a potential hazard for strict vegetarians and alcoholics. Red blood cells form but they cannot divide without vitamin B_{12}. Normally, meats, poultry, and fish in the diet provide enough of the vitamin.

As described elsewhere in the book, a gene mutation that gives rise to an abnormal form of hemoglobin can result in *sickle-cell* anemia. Another gene mutation, which blocks or lowers the synthesis of the globin chains that make up hemoglobin, causes *thalassemias*. Too few red blood cells can form; those that do are thin and fragile.

Or consider the *polycythemias*—symptoms of far too many red blood cells—which make blood flow sluggish. Some bone marrow cancers can result in this condition. So can "blood doping" by some athletes who compete in strenuous events. They withdraw their own red blood cells, store them, then reinject them a few days prior to competition. The withdrawal triggers red blood cell formation as the body attempts to replace the "lost" cells, so the cell count is bumped up when the withdrawn cells are put back in the body. The idea is to increase the body's oxygen-carrying capacity and endurance. Blood doping leads to temporary high blood pressure and lowers blood's viscosity. Some believe the practice works, but others call it unethical. It is banned from the Olympics.

WHITE BLOOD CELL DISORDERS   You may have heard of *infectious mononucleosis*. An Epstein-Barr virus causes this highly contagious disease, which results from too many monocytes and lymphocytes. Following a few weeks of fatigue, aches, low fever, and a chronic sore throat, the patient usually recovers.

Recovery is chancy for *leukemias*, a category of cancers that suppress or impair white blood cell formation in bone marrow. Radiation therapy and chemotherapy can kill the cancer cells. Remissions may last for months or years. (A remission is a symptom-free period of a chronic illness.)

Concerning Agglutination

Whenever blood volume or blood cell counts decline, countermeasures kick in automatically. However, if the volume were to decrease by more than 30 percent, then circulatory shock would follow and could lead to death.

Blood from donors can be transfused into patients affected by a blood disorder or substantial blood loss. Such **blood transfusions** cannot be decided willy-nilly. Why not? *A potential donor and the recipient may not have the same kinds of recognition proteins at the surface of their red blood cells.* Some of the proteins are "self" markers; they identify the cells as belonging to one's own body. During a blood transfusion, if cells from a donor have the "wrong" marker, the recipient's immune system will recognize them as foreign, with serious consequences.

Figure 39.8 shows what happens when blood from incompatible donors and recipients intermingle. In a defensive response called **agglutination**, proteins called

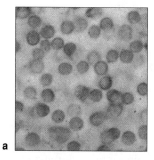 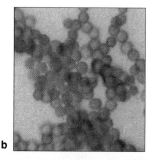

a   b

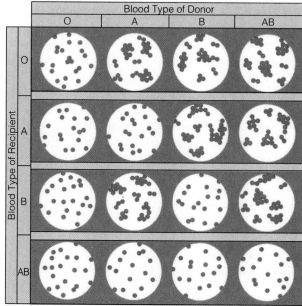

c

Figure 39.8   Light micrographs showing (**a**) the absence of agglutination in a mixture of two different but compatible blood types and (**b**) agglutination in a mixture of incompatible types. (**c**) Agglutination responses in blood types A, B, AB, and O when mixed with samples of the same and different types.

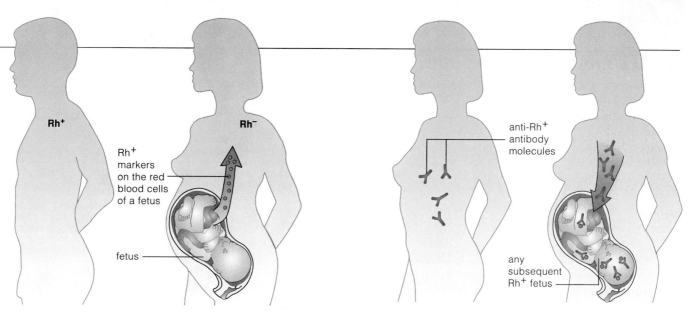

a A forthcoming child of an Rh⁻ woman and Rh⁺ man inherits the gene for the Rh⁺ marker. During pregnancy or childbirth, some of its cells bearing the marker may leak into the maternal bloodstream.

b The foreign marker stimulates her body to make antibodies. If she gets pregnant again and if this second fetus (or any other) inherits the gene for the marker, the circulating anti-Rh⁺ antibodies will act against it.

Figure 39.9 Antibody production in response to Rh⁺ markers on red blood cells of a fetus.

antibodies that are circulating in plasma act against the foreign cells and cause them to clump together. (As described in the next chapter, antibodies bind to specific foreign markers and so target the cell or particle bearing them for destruction by the immune system.) The sheer number of "foreign" cells transfused into the recipient translates into numerous clumps, which can clog small blood vessels and damage tissues. Without treatment, death may follow. The same thing can happen during certain pregnancies if antibodies diffuse from a mother's circulatory system into that of her unborn child.

Based on an understanding of cell surface markers and antibodies, scientists have devised ways to analyze the forms of self markers on a person's red blood cells. Blood typing is based on these analyses.

ABO Blood Typing

Molecular variations in one kind of self-marker on red blood cells are analyzed with **ABO blood typing**. The genetic basis of such variation is described in Section 11.4. People with one form of the marker are said to have type A blood; those with another form have type B blood. Those with both forms of the marker on their red blood cells have type AB blood. Others do not have either form of the marker; they have type O blood.

If you are type A, your antibodies ignore A markers but will act against B markers. If you are type B, your antibodies will ignore B markers but will act against A markers. If you are type AB, your antibodies ignore both forms of the marker, so you can tolerate donations of type A, B, AB, or O blood. If you are type O, you have antibodies against both forms of the marker, so your options are limited to type O donations.

Rh Blood Typing

Rh blood typing is based on the presence or absence of an Rh marker (so named because it was first identified in blood samples of *Rh*esus monkeys). If you are type Rh⁺, your blood cells bear this marker at their surface. If you are type Rh⁻, they don't. Ordinarily, people do not have antibodies against Rh markers. But a recipient of transfused Rh⁺ blood produces antibodies against them, and the antibodies remain in the blood.

If an Rh⁻ woman becomes impregnated by an Rh⁺ man, there is a chance the fetus will be Rh⁺. During pregnancy or childbirth, some of its red blood cells may leak into the mother's bloodstream. If they do, her body will produce antibodies against Rh (Figure 39.9). If she becomes pregnant *again*, Rh antibodies will enter the bloodstream of this new fetus. If its blood is type Rh⁺, then her antibodies will cause its red blood cells to swell, rupture, and release hemoglobin.

Erythroblastosis fetalis is a severe outcome of mixing Rh⁺ and Rh⁻ types. The fetus dies, for too many cells are destroyed. If diagnosed before birth or delivered alive, it can survive if its blood is slowly replaced with transfusions that are free of Rh antibodies. Currently, a known Rh⁻ woman can be treated right after her first pregnancy with an anti-Rh gamma globulin (RhoGam) that will protect her next fetus. The drug will inactivate any Rh⁺ fetal blood cells that are circulating through her bloodstream before she can become sensitized and begin producing potentially dangerous antibodies.

To avoid the symptoms of blood incompatibilities, red blood cells should be typed before transfusions or pregnancies.

HUMAN CARDIOVASCULAR SYSTEM

"Cardiovascular" comes from the Greek *kardia* (heart) and Latin *vasculum* (vessel). In a human cardiovascular system, blood is pumped by a muscular heart into large-diameter **arteries**. It then flows into small, muscular **arterioles**, which branch into the even smaller diameter capillaries introduced earlier. Blood flows continuously from capillaries into small **venules**, and then into large-diameter **veins** that return blood to the heart.

As in most vertebrates, a partition separates the heart into a double pump, which drives blood through two cardiovascular circuits (Figure 39.10). Each circuit has its own set of arteries, arterioles, capillaries, venules, and veins. The *pulmonary* circuit, a short loop, rapidly oxygenates blood. It leads from the heart's right half to capillary beds in both lungs, then returns to the heart's left half. The *systemic* circuit is a longer loop starting at the heart's left half. Its main artery, the **aorta**, accepts the oxygenated blood, which flows on through arterioles, capillary beds in all body regions, and then veins that deliver the oxygen-poor blood to the heart's right half. Figure 39.11 identifies the location of the major blood vessels of both circuits and defines their functions.

For most of the systemic branchings, a given volume of blood flows through one capillary bed. The branch to the intestines is one of the exceptions. First blood picks up glucose and other absorbed substances from one bed, then it moves through another capillary bed, in the liver—an organ with a key role in nutrition. The second bed gives the liver time to process absorbed substances.

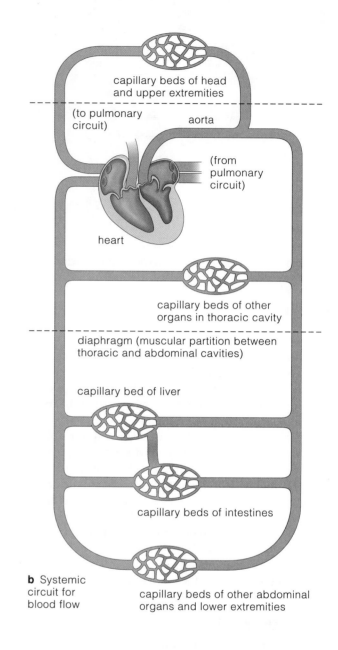

b Systemic circuit for blood flow

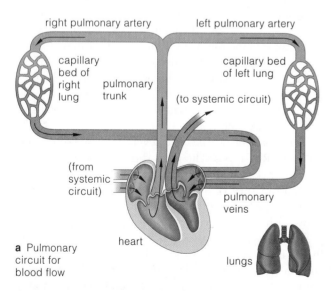

a Pulmonary circuit for blood flow

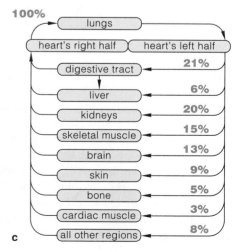

Figure 39.10 (**a**,**b**) Diagrams of the pulmonary and systemic circuits for blood flow through the human cardiovascular system. The blood vessels that are carrying oxygenated blood are color-coded *red*. Those carrying oxygen-poor blood are color-coded *blue*. (**c**) Distribution of the heart's output in a person at rest. The lungs receive all blood pumped out of the heart's right half. The organs serviced by the systemic circuit receive the portions indicated from the heart's left half. Control mechanisms adjust the flow distribution when necessary.

100%
lungs
heart's right half | heart's left half
digestive tract — 21%
liver — 6%
kidneys — 20%
skeletal muscle — 15%
brain — 13%
skin — 9%
bone — 5%
cardiac muscle — 3%
all other regions — 8%

c

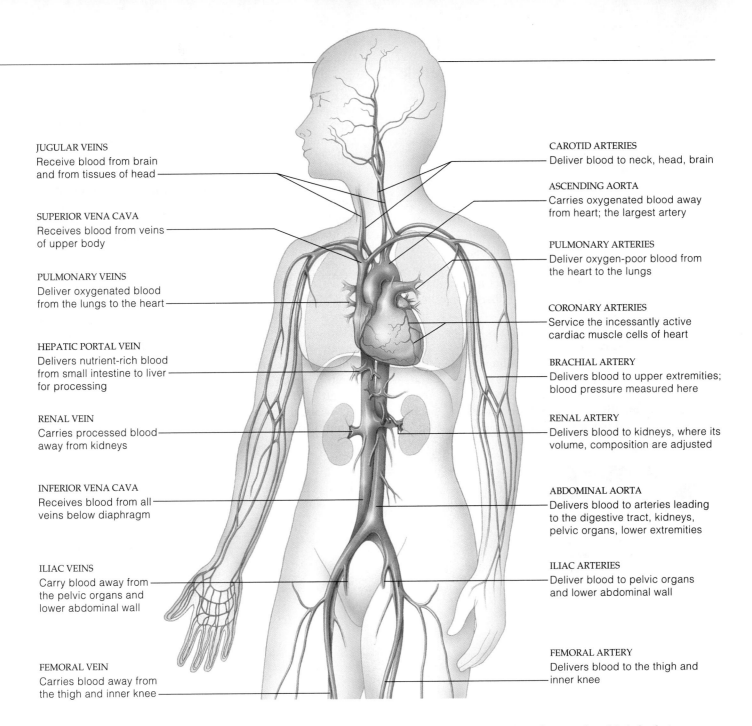

JUGULAR VEINS
Receive blood from brain and from tissues of head

SUPERIOR VENA CAVA
Receives blood from veins of upper body

PULMONARY VEINS
Deliver oxygenated blood from the lungs to the heart

HEPATIC PORTAL VEIN
Delivers nutrient-rich blood from small intestine to liver for processing

RENAL VEIN
Carries processed blood away from kidneys

INFERIOR VENA CAVA
Receives blood from all veins below diaphragm

ILIAC VEINS
Carry blood away from the pelvic organs and lower abdominal wall

FEMORAL VEIN
Carries blood away from the thigh and inner knee

CAROTID ARTERIES
Deliver blood to neck, head, brain

ASCENDING AORTA
Carries oxygenated blood away from heart; the largest artery

PULMONARY ARTERIES
Deliver oxygen-poor blood from the heart to the lungs

CORONARY ARTERIES
Service the incessantly active cardiac muscle cells of heart

BRACHIAL ARTERY
Delivers blood to upper extremities; blood pressure measured here

RENAL ARTERY
Delivers blood to kidneys, where its volume, composition are adjusted

ABDOMINAL AORTA
Delivers blood to arteries leading to the digestive tract, kidneys, pelvic organs, lower extremities

ILIAC ARTERIES
Deliver blood to pelvic organs and lower abdominal wall

FEMORAL ARTERY
Delivers blood to the thigh and inner knee

Figure 39.11 Location and functions of major blood vessels of the human cardiovascular system.

Figure 39.10c shows how the pulmonary and systemic circuits distribute the heart's output to different organs. The percentages listed are for a person at rest. As you will see, the flow distribution along the systemic route is adjusted in response to changes in levels of physical activity and to shifts in external and internal conditions. For instance, the flow to and from a skeletal muscle varies greatly, depending on what it is being called upon to do at a given time. The same is true of skin. When the body gets too cold, the flow of blood (and metabolic heat) is diverted away from skin's vast capillary beds.

When the body is hot, flow to the skin's beds increases and heat radiates from the skin's surface. Only the brain is exempt; it cannot tolerate variations in flow.

The human cardiovascular system consists of two separate circuits, pulmonary and systemic, for blood flow.

The pulmonary circuit is a short loop from the heart's right half, through both lungs, to the heart's left half. Oxygen-poor blood flowing into the circuit rapidly picks up oxygen and gives up carbon dioxide at capillary beds in the lungs.

The longer systemic circuit starts at the heart's left half and aorta. It delivers oxygen and accepts carbon dioxide at capillary beds of all metabolically active regions; then its veins deliver oxygen-poor blood to the heart's right half.

THE HEART IS A LONELY PUMPER

Heart Structure

Think about the fact that the human heart beats about 2.5 billion times during a seventy-year life span, and you know it must be a truly durable pump. Figure 39.12 shows its structure, which speaks of its durability. The pericardium, a double sac of tough connective tissue, protects the heart and anchors it to nearby structures (*peri*, around). A fluid between its layers lubricates the heart during its perpetual twisting motions. The inner layer serves as the outer part of the heart wall. The bulk of that wall, the myocardium, consists of cardiac muscle cells tethered to elastin and collagen fibers. The fibers are so densely crisscrossed, they serve as a "skeleton" against which the force of contraction is applied. The oxygen-demanding cardiac muscle cells have their own, *coronary* circulation; coronary arteries branching off the aorta lead to a capillary bed that services only them. The wall's glistening, innermost layer consists of connective tissue and **endothelium**, a one-cell-thick epithelial sheet. Only the heart and blood vessels have an endothelium.

Each half of the heart has two chambers: an **atrium** (plural, atria) and a **ventricle**. Blood flows into the atria, down into the ventricles, then out through great arteries (the aorta or pulmonary trunk). Between each atrium and ventricle is an AV valve (short for atrioventricular). Between each ventricle and the artery leading out from it is a semilunar valve. Both types are *one-way* valves, with membranes that rhythmically flap open, flap shut, and thereby help keep blood moving in one direction.

Cardiac Cycle

Each time the heart beats, its four chambers go through phases of contraction (systole) and relaxation (diastole). The sequence of contraction and relaxation is a **cardiac cycle**. When relaxed, the atria fill with blood. Increasing fluid pressure forces the AV valves open. Blood flows into the ventricles, which completely fill when the atria contract (Figure 39.13*a*). As the filled ventricles start to contract, the rising fluid pressure forces the AV valves shut. It rises so sharply above the pressure in the great

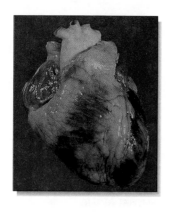

a

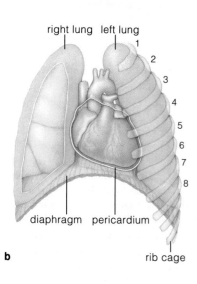

b

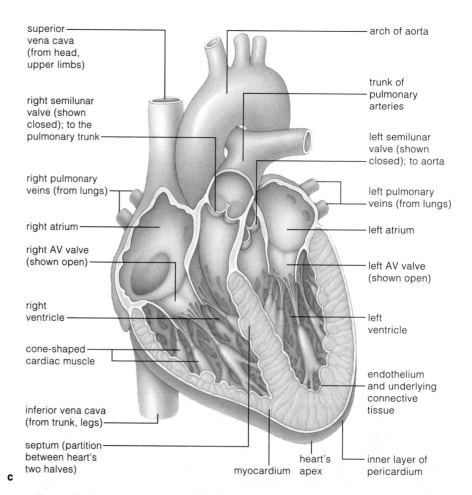

c

right lung   left lung
1
2
3
4
5
6
7
8

diaphragm   pericardium

rib cage

superior vena cava (from head, upper limbs)

right semilunar valve (shown closed); to the pulmonary trunk

right pulmonary veins (from lungs)

right atrium

right AV valve (shown open)

right ventricle

cone-shaped cardiac muscle

inferior vena cava (from trunk, legs)

septum (partition between heart's two halves)

arch of aorta

trunk of pulmonary arteries

left semilunar valve (shown closed); to aorta

left pulmonary veins (from lungs)

left atrium

left AV valve (shown open)

left ventricle

endothelium and underlying connective tissue

myocardium   heart's apex   inner layer of pericardium

Figure 39.12 (**a**) Photograph of the human heart and (**b**) its location in the thoracic cavity. (**c**) Cutaway view of the human heart, showing its wall and internal organization.

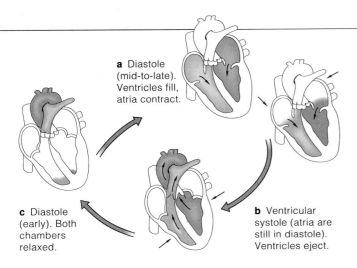

a Diastole (mid-to-late). Ventricles fill, atria contract.

b Ventricular systole (atria are still in diastole). Ventricles eject.

c Diastole (early). Both chambers relaxed.

Figure 39.13 Blood flow during part of a cardiac cycle. Blood and heart movements generate a "lub-dup" sound at the chest wall. At each "lub," AV valves are closing as ventricles contract. At each "dup," semilunar valves are closing as ventricles relax.

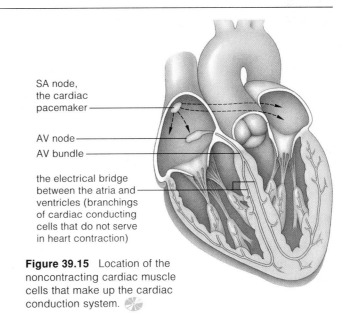

SA node, the cardiac pacemaker

AV node

AV bundle

the electrical bridge between the atria and ventricles (branchings of cardiac conducting cells that do not serve in heart contraction)

Figure 39.15 Location of the noncontracting cardiac muscle cells that make up the cardiac conduction system.

arteries, it forces the semilunar valves open—and blood leaves the heart (Figure 39.13*b*). The ventricles relax, the semilunar valves close, and the already filling atria are ready to repeat the cycle. Thus, in a cardiac cycle, atrial contraction simply helps fill the ventricles. *Contraction of the ventricles is the driving force for blood circulation.*

Mechanisms of Contraction

Like skeletal muscle, cardiac muscle is striated (striped). In response to action potentials, the sarcomeres of its cells also contract in the manner predicted by the sliding-filament model (Section 38.8). This tissue also requires energy for contraction; large numbers of mitochondria in the myocardium provide the ATP. But cardiac muscle cells are structurally unique. They are branching, short, and joined at the end regions. Communication junctions span the plasma membranes of abutting regions and let

action potentials spread swiftly among cells, in waves of excitation that wash over the heart (Figure 39.14).

In addition, about 1 percent of the cardiac muscle cells *do not* contract. They function instead as a **cardiac conduction system**. About seventy times each minute, these specialized cells initiate and propagate waves of excitation that travel in rhythmic, orderly sequence from the atria to ventricles. The synchronized excitation underlies the heart's truly efficient pumping. Each wave starts at the SA node, a cluster of cell bodies in the right atrium's wall (Figure 39.15). It passes through the wall to another cell body cluster, the AV node. This is the only electrical bridge between the atria and ventricles, which connective tissue insulates everywhere else. After the AV node, conducting cells are arranged as a bundle in the partition between the heart's two halves. These cells branch, and the branches deliver the excitatory wave up the ventricle walls. The ventricles contract in response with a twisting movement, upward from the heart's apex, that ejects blood into the great arteries.

The SA node fires action potentials faster than the rest of the system and serves as the **cardiac pacemaker**. Its spontaneous, rhythmic signals are the foundation for the normal rate of heartbeat. The nervous system can only *adjust* the rate and strength of contractions dictated by the pacemaker. Even if all nerves leading to a heart are severed, the heart will keep on beating!

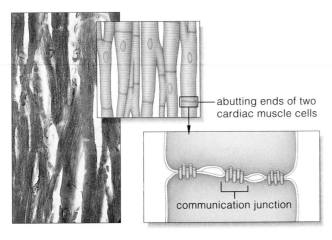

abutting ends of two cardiac muscle cells

communication junction

Figure 39.14 Light micrograph of cardiac muscle cells and sketches showing communication junctions between them.

The heart's construction reflects its role as a durable pump. Under the spontaneous, rhythmic signals from the cardiac pacemaker, its branching, abutting cardiac muscle cells contract in synchrony, almost as if they were a single unit.

Although the heart has four chambers (each half has one atrium and one ventricle), contraction of the ventricles is the driving force for blood circulation away from the heart.

As you have seen, blood flows to and from the human heart through arteries, arterioles, capillaries, venules, and finally veins. Figure 39.16 shows the structure of these vessels. The arteries are the major transporters of oxygenated blood throughout the body. Arterioles in each region are sites where the volume of blood flow through each organ can be controlled. The capillaries and, to a lesser extent, venules are diffusion zones, and veins are the major transporters of oxygen-poor blood to the heart.

Two key factors influence the rate of flow through each type of blood vessel. First, the flow rate is directly proportional to the pressure gradient between the start and end of the vessel. Second, the flow rate is inversely proportional to the vessel's resistance to flow.

Blood pressure is fluid pressure imparted to blood by heart contractions. The beating heart establishes the higher blood pressure at the beginning of a vessel. As flowing blood rubs against the vessel's inner wall, the friction causes some energy (in the form of pressure) to be lost. The lower pressure at the end of the vessel is due to the frictional losses. Even small changes in blood vessel diameter have major influence over the flow rate. In the pulmonary and systemic circuits, the diameters decrease and present more resistance to flow. And so, because of heart contractions and the subsequent events (mainly frictional losses), blood pressure is highest in the contracting ventricles, still high at the beginning of arteries, then continues to drop until reaching its lowest value—in the relaxed atria (Figure 39.17).

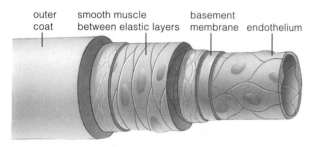

a ARTERY

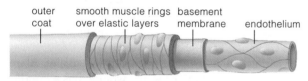

b ARTERIOLE

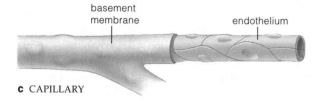

c CAPILLARY

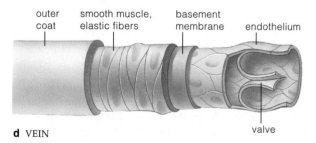

d VEIN

Figure 39.16 Structure of blood vessels. The basement membrane around the endothelium of each vessel is a noncellular layer, rich with proteins and polysaccharides.

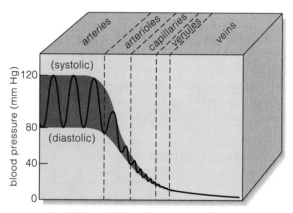

Figure 39.17 Blood pressure. This is a plot of measurements made of the drop in fluid pressure for a given volume of blood moving through the systemic circuit.

Arterial Blood Pressure

From Figure 39.16a and the preceding discussion, you see that arteries have a large diameter and present low resistance to flow. Thus they serve as rapid transporters of oxygenated blood. They also are pressure reservoirs that smooth out pulsations in pressure caused by each cardiac cycle. Their thick, muscular, elastic wall bulges as ventricular contraction forces a large volume of blood into them. Then the wall recoils and so forces the blood forward through the circuit while the heart is relaxing.

Suppose you decide to measure your blood pressure daily when you are resting. Typically, you would take a reading from the brachial artery of an upper arm, as in Figure 39.18. *Systolic* pressure is the peak pressure that the contracting ventricles exert against the artery's wall during a cardiac cycle. *Diastolic* pressure is the lowest pressure when blood is draining into the vessels after it. You can estimate the mean arterial pressure as *diastolic pressure + 1/3 pulse pressure* (which is the difference between the highest and lowest pressure readings).

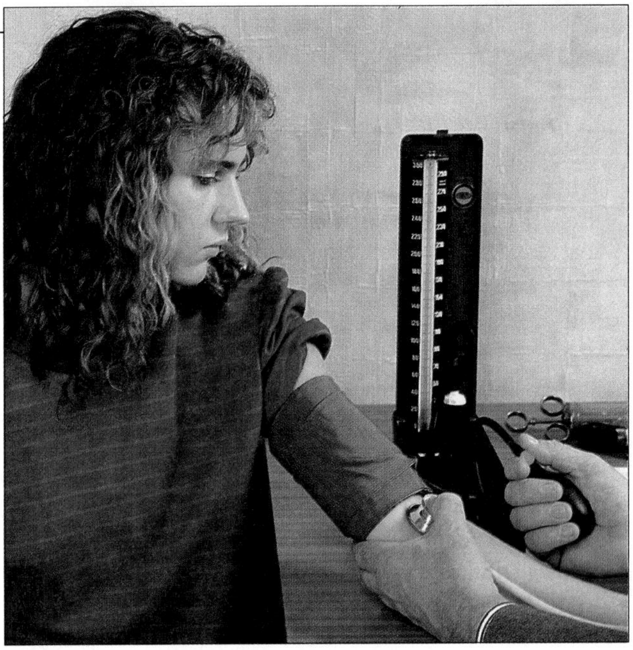

Figure 39.18 Measuring blood pressure.

A hollow cuff attached to a pressure gauge is wrapped around a person's upper arm. The cuff is inflated with air to a pressure above the highest pressure of the cardiac cycle (at systole, when the ventricles contract). Above this pressure, no sounds can be heard through a stethoscope positioned below the cuff and over the artery (because no blood is flowing through the vessel). Air in the cuff is slowly released, so that some blood flows into the artery. The turbulent flow causes soft tapping sounds, and when this first occurs, the value on the gauge is the systolic pressure, or about 120 mm mercury (Hg) in young adults at rest. This particular value means that the measured pressure would force mercury to move upward 120 millimeters in a narrow glass column.

More air is released from the cuff. Just after the sounds become dull and muffled, blood flows continuously. So the turbulence and tapping sounds stop. The silence corresponds to the diastolic pressure (at the end of a cardiac cycle, just before the heart pumps out blood). The reading is usually about 80 mm Hg. In this example, pulse pressure (the difference between the highest and lowest pressure readings) is 120 – 80, or 40 mm Hg. Assuming you are an adult in good health, this average resting value stays fairly constant over a few weeks or even months.

Resistance to Flow at Arterioles

Track a volume of blood through the systemic route and you find the greatest pressure drop at arterioles (Figure 39.17), for these offer the greatest resistance to blood flow. The slowdown at arterioles allows time for control mechanisms to divert greater or lesser portions of the total flow volume to different regions. Control is exerted at arteriole walls, which contain rings of smooth muscle. Some control signals cause the smooth muscle cells to relax, which results in **vasodilation**. The word means an enlargement (dilation) of the blood vessel diameter. Other signals cause contraction of the smooth muscle cells, which results in a decrease in the blood vessel diameter, or **vasoconstriction**.

The nervous and endocrine systems exert control over arteriole diameter. For example, sympathetic nerve endings terminate on the smooth muscle cells of many arterioles. Increased activity along the nerves triggers action potentials in the cells to bring about contraction, hence widespread vasoconstriction. Epinephrine and angiotensin are two of the signaling molecules that can trigger changes in arteriole diameter.

Arteriole diameter also is adjusted when changes in metabolic activity shift the localized concentrations of substances in a tissue. Such local chemical changes are "selfish," in that they invite or divert blood flow to meet the tissue's own metabolic needs. Think of what happens as you run. The oxygen level drops in skeletal muscle tissue, and levels of carbon dioxide, hydrogen and potassium ions, and other substances increase. The changed conditions trigger vasodilation of arterioles in the vicinity. Now more blood flows past active muscles, delivering more raw materials and carrying away cell products and metabolic wastes. When muscles relax and demand fewer blood deliveries, the oxygen level increases and triggers vasoconstriction of the arterioles.

Controlling Mean Arterial Blood Pressure

Mean arterial pressure depends on cardiac output and on total resistance through the vascular system. Cardiac output is influenced by controls over the rate and strength of heartbeats, and total resistance mainly by vasoconstriction at the arterioles. Given that organs and tissues vary in their demands for blood, maintaining blood pressure is not easy. The balance between cardiac output and total resistance must be continually juggled.

A **baroreceptor reflex** is the main short-term control over arterial pressure. It starts at baroreceptors, which detect fluctuations in arterial mean pressure and pulse pressure. The most critical of these sensory receptors are located in the carotid arteries (which supply blood to the brain) and aortic arch (which supplies blood to the rest of the body). They continually generate action potentials, and their signals reach a control center in a hindbrain region, the medulla oblongata (Section 35.4). In response, the center adjusts its commands flowing along sympathetic and parasympathetic nerves to the heart and blood vessels. For example, when the mean arterial pressure rises, the center commands the heart to beat more slowly and contract less forcefully.

Long-term control of blood pressure is exerted at kidneys, which adjust the volume and composition of the blood. And that is a topic of another chapter.

Mean arterial pressure is an outcome of controls over the heart's output and over the total resistance to blood flow through the vascular system, as exerted mainly at arterioles.

Capillary Function

THE NATURE OF THE EXCHANGES   Capillary beds, again, are diffusion zones for exchanges between blood and interstitial fluid. When the living cells in any part of the body are deprived of those exchanges, they die quickly; brain cells are dead within four minutes.

By some estimates, the human body has between 10 billion and 40 billion capillaries, which collectively represent an astounding surface area for the exchanges. These components of the circulatory system extend into nearly every tissue, and at least one is as close as 0.001 centimeter to every living cell. Their proximity to cells is crucial, for shorter distances mean faster diffusion rates. (Think about this: It would take years for oxygen to diffuse on its own from your lungs all the way down your legs. By that time your toes, and the rest of you, would long be dead.) Although capillaries are three to seven micrometers across, they deform amazingly; red blood cells (eight micrometers across) squeeze through them single file. Therefore, those oxygen-transporting cells, as well as the substances dissolved in plasma, are in direct contact with the exchange area—the capillary wall—or only a short distance away from it.

And remember, taken as a whole, capillaries present a greater cross-sectional area than the arterioles leading into them, so the flow velocity is not as great. With this slowdown, there is plenty of time for interstitial fluid to exchange substances with the 5 percent or so of the total blood volume that is moving forward through the capillary beds at a given time. Here you may wish to reflect on Figure 39.3e.

Each capillary is a tube of endothelial cells, a single layer thick (Figure 39.16c). The cells abut one another, but the "fit" is different in different parts of the body. In most cases, there are narrow, water-filled clefts between the endothelial cells. In capillaries that service the brain, tight junctions join the cells and clefts are nonexistent; substances cannot "leak" between cells but must move through them. Thus the junctions are a functional part of the blood-brain barrier, as described in Section 35.4.

Oxygen, carbon dioxide, and most other small lipid-soluble substances cross the capillary wall by diffusing through the lipid portion of an endothelial cell's plasma membrane and through its cytoplasm. Certain proteins enter and leave the cells by endocytosis or exocytosis. Small, water-soluble substances such as ions enter and leave at the clefts between endothelial cells. So do white blood cells, as you will see in the chapter to follow.

MECHANISMS OF EXCHANGE   Diffusion is not the only means by which substances are exchanged across the capillary wall. At any time, the fluid pressures acting on the wall may not be balanced, and there may be bulk flow one way or the other across it. Bulk flow, recall, is a movement of water and solutes in the same direction in response to fluid pressure.

As Figure 39.19 shows, the concentrations of water and solutes in blood and in interstitial fluid influence the direction in which the fluid flows. At the beginning of a capillary bed, the outward-directed force of blood pressure is greater than the inward-directed osmotic force of interstitial fluid. A small amount of protein-free plasma is pushed out in bulk through clefts in the wall. This process is called **ultrafiltration**. Farther on in the bed, the balance shifts. Whereas blood pressure has been continually declining, the osmotic force has stayed the same. When the inward-directed force exceeds the outward force of blood pressure, tissue fluid moves through the clefts between endothelial cells and into the capillary. This process is called **reabsorption**.

Normally the outcome is a very small, *net* outward movement of fluid from the capillary bed, which the lymphatic system returns to the blood. Such bulk flow helps maintain the fluid balance between blood and interstitial fluid. This is important, for blood pressure is maintained only when there is adequate blood volume. When the blood volume plummets, as happens during hemorrhage, interstitial fluid can be tapped by way of reabsorption to help counter the loss.

Sometimes blood pressure increases so much that it triggers excessive ultrafiltration. When excess fluid accumulates in interstitial spaces, this condition is called *edema*. It occurs to some extent during physical exercise, when arterioles dilate in many tissue regions. Edema also can result from an obstructed vein or from heart failure. It becomes extreme during *elephantiasis*. As described in Section 26.9, elephantiasis is a disease brought on by roundworm infection and a subsequent obstruction of lymphatic vessels.

Venous Pressure

What happens to the flow velocity after it continues on past the vast cross-sectional area of the capillary beds? Remember, the capillaries merge into venules, or "little veins." These in turn merge into large-diameter veins. The total cross-sectional area is once again reduced and flow velocity increases as blood returns to the heart.

In functional terms, venules are a bit like capillaries. Some solutes diffuse across their wall, which is only a bit thicker than that of a capillary. Some control over capillary pressure also is exerted at these vessels.

Veins are large-diameter, low-resistance transport tubes to the heart (Figure 39.16d). They have valves that prevent backflow. When gravity beckons, venous flow reverses direction and pushes the valves shut. Their wall can bulge considerably under pressure, more so

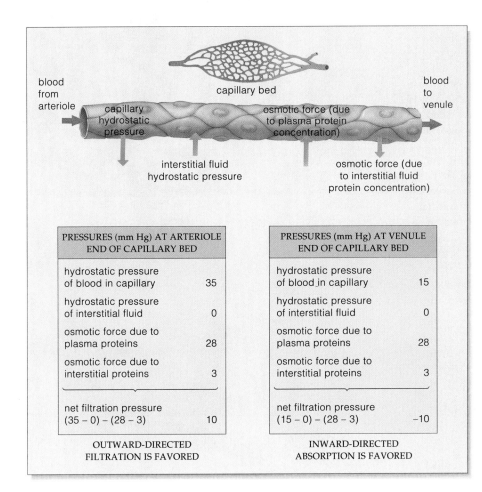

PRESSURES (mm Hg) AT ARTERIOLE END OF CAPILLARY BED	
hydrostatic pressure of blood in capillary	35
hydrostatic pressure of interstitial fluid	0
osmotic force due to plasma proteins	28
osmotic force due to interstitial proteins	3
net filtration pressure (35 − 0) − (28 − 3)	10

OUTWARD-DIRECTED FILTRATION IS FAVORED

PRESSURES (mm Hg) AT VENULE END OF CAPILLARY BED	
hydrostatic pressure of blood in capillary	15
hydrostatic pressure of interstitial fluid	0
osmotic force due to plasma proteins	28
osmotic force due to interstitial proteins	3
net filtration pressure (15 − 0) − (28 − 3)	−10

INWARD-DIRECTED ABSORPTION IS FAVORED

Figure 39.19 Bulk flow in an idealized capillary bed. The fluid movement plays no significant role in diffusion. But it is important in maintaining the distribution of extracellular fluid between the blood and interstitial fluid.

The fluid movements across a capillary wall result from two opposing forces, called ultrafiltration and reabsorption.

At the arteriole end of a capillary, the difference between blood pressure and interstitial fluid pressure leads to filtration. Because of the difference, some plasma (but very few plasma proteins) leaves the capillary by bulk flow through clefts between endothelial cells of the capillary wall. Ultrafiltration is the bulk flow of fluid *out* of the capillary.

Reabsorption is the osmotic movement of some interstitial fluid *into* the capillary. It results from a difference in water concentration between plasma and interstitial fluid. With its dissolved protein components, the plasma has a greater solute concentration, hence a lower water concentration.

Reabsorption near the end of a capillary bed tends to balance ultrafiltration at the beginning. Normally, there is only a small *net* filtration of fluid, which the lymphatic system returns to the blood.

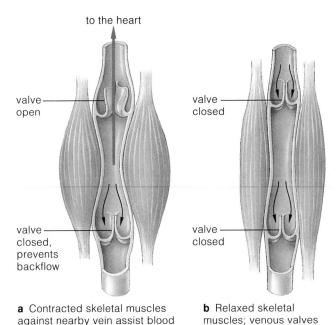

a Contracted skeletal muscles against nearby vein assist blood flow to heart.

b Relaxed skeletal muscles; venous valves shut, no backflow.

Figure 39.20 How the bulging of contracting skeletal muscles helps increase fluid pressure inside a vein. Notice the structure of each valve (membrane flaps) in the vein's lumen.

than an arterial wall. Thus veins can serve as reservoirs for variable volumes of blood. Collectively, the human body's veins can hold up to 50 to 60 percent of the total blood volume.

A vein wall contains some smooth muscle. When blood must circulate faster, as during physical exercise, the smooth muscle contracts. The wall stiffens and the vein bulges less, so that venous pressure rises and drives more blood to the heart. Also, when limbs are moving, skeletal muscles bulge against veins in their vicinity. They help raise the venous pressure and drive blood back to the heart (Figure 39.20). Rapid breathing also contributes to increased venous pressure. Inhaled air pushes down on internal organs and thereby alters the pressure gradient between the heart and veins.

Capillary beds are diffusion zones for exchanges between blood and interstitial fluid. Here also, a slight amount of bulk flow helps maintain the fluid balance between blood and interstitial fluid.

Venules overlap somewhat with capillaries in function.

Veins are highly distensible blood volume reservoirs and help adjust flow volume back to the heart.

39.9

CARDIOVASCULAR DISORDERS

Cardiovascular disorders are the leading cause of death in the United States. They affect at least 40 million people and kill about 1 million each year. The most common are *hypertension*, which is sustained high blood pressure; and *atherosclerosis*, a progressive thickening of the arterial wall and narrowing of the arterial lumen. Both affect blood circulation and so cause most *heart attacks* (damaged or destroyed heart muscle) and *strokes* (brain damage).

In most heart attacks, a "crushing" pain behind the breastbone lasts a half hour or more. Mild to severe pain may radiate into the left arm, shoulder, or neck. Sweating, shortness of breath, erratic heartbeat, nausea, vomiting, and dizziness or fainting may accompany an attack.

RISK FACTORS   Extensive research has correlated the following risk factors with cardiovascular disorders:

1. Smoking (Section 41.7)
2. Genetic predisposition to heart failure
3. High level of cholesterol in the blood
4. High blood pressure
5. Obesity (Section 42.10)
6. Lack of regular exercise
7. Diabetes mellitus (Section 37.7)
8. Age (the older you get, the greater the risk)
9. Gender (until age fifty, males are at much greater risk than females)

The risk factors can be controlled by exercising regularly, eating properly, and not smoking. With too little exercise and too much food, adipose tissue masses increase in the body, more blood capillaries develop to service them, and the heart has to work harder to pump blood through the increasingly divided vascular circuit. In people who smoke, the nicotine in tobacco makes the adrenal glands secrete epinephrine. This hormone, remember, triggers vasoconstriction, which boosts the heart rate and blood pressure. Also, the carbon monoxide in cigarette smoke inhibits the binding of oxygen to hemoglobin, so the heart has to pump harder to get oxygen to cells. Smoking also predisposes people to atherosclerosis even when their blood level of cholesterol is normal.

The next paragraphs describe some of the damage that results from cardiovascular disorders.

HYPERTENSION   Hypertension results from gradual increases in the flow resistance through small arteries. In time, blood pressure remains above 140/90, even when a person is resting. Heredity may be a factor; the disorder tends to run in families. Diet is also a factor. For instance, high salt intake can raise blood pressure in susceptible people and increase the heart's workload. The heart may eventually enlarge and fail to pump blood effectively. High blood pressure also may contribute to a "hardening" of arterial walls that hampers delivery of oxygen to the brain, heart, and other vital organs.

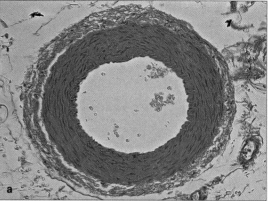

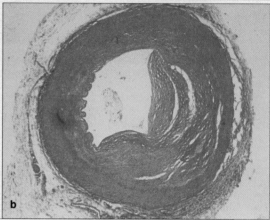

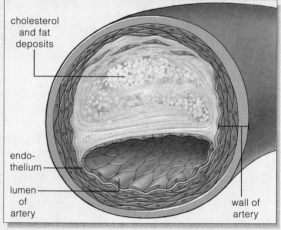

cholesterol and fat deposits

endo-thelium

lumen of artery

wall of artery

c

Figure 39.21   Sections from (**a**) a normal artery and (**b**) one with a narrowed lumen. (**c**) Sketch of an atherosclerotic plaque.

Hypertension is called a "silent killer" because affected individuals may show no outward symptoms. Even when they know their blood pressure is high, some people tend to resist helpful medication, changes in diet, and regular exercise. Of 23 million hypertensive Americans, most do not undergo treatment. About 180,000 die each year.

ATHEROSCLEROSIS   With *arteriosclerosis*, arteries thicken and lose their elasticity. With atherosclerosis, the condition worsens as cholesterol and other lipids build up in the arterial wall and cause the lumen to narrow (Figure 39.21).

Recall, from the introduction to Chapter 16, that the liver produces enough cholesterol to satisfy the body's needs. Food intake increases the cholesterol level. When circulating in blood, cholesterol is bound to proteins as *low-density* lipoproteins, or **LDLs**. These bind to receptors on cells throughout the body. Cells take up LDLs and their cholesterol cargo for use in cell activities. The excess is attached to proteins as *high-density* lipoproteins, or **HDLs**, and transported back to the liver, where it is metabolized.

In some people, for a variety of reasons, not enough LDL is removed from the blood. As the blood level of LDL increases, so does the risk of atherosclerosis. LDLs, with their bound cholesterol, infiltrate the walls of arteries. In those walls, abnormal smooth muscle cells multiply, connective tissue components increase in mass, and cholesterol accumulates in endothelial cells and the clefts between them. On top of the lipids, calcium deposits actually form microscopic slivers of bone. A fibrous net forms over the entire mass. This *atherosclerotic plaque* sticks out into the arterial lumen (Figure 39.21c).

The bony slivers of the plaque shred the endothelium. Platelets gather at the damaged site, and, in the manner described in Section 39.10, they secrete chemicals that initiate clot formation. The condition worsens as fatty globules in the plaque become oxidized. Many globules take on a form that resembles the surface components of common bacteria—including a type that instigates the formation of bonelike calcium deposits in the lungs. As an awful consequence, the call goes out to bacteria-fighting monocytes. Soon an inflammatory response is under way, and certain chemicals released during the fray activate the genes for bone formation. Normally, this is a good thing; it helps wall off invaders and prevents infection from spreading. It is bad news for arteries.

As plaques and clots grow, they can narrow or block an artery. Blood flow to the tissues serviced by the artery may dwindle to a trickle or stop. A clot that stays in place is a *thrombus*. If it becomes dislodged and then travels the bloodstream, it is an *embolus*. Think of coronary arteries, which have narrow diameters. They are highly vulnerable to clogging by a plaque or clot. When they narrow to one-quarter of their former diameter, the outcome ranges from *angina pectoris*, or mild chest pains, to a heart attack.

Physicians can diagnose atherosclerosis in coronary arteries by *stress electrocardiograms*. These are recordings of the electrical activity of the cardiac cycle while a person is exercising on a treadmill. They also can diagnose the condition by *angiography*. In this procedure, a dye that is opaque to x-rays is injected into the bloodstream.

Severe blockage may require surgery. With *coronary bypass surgery*, a section of an artery from the chest is stitched to the aorta and to the coronary artery below the narrowed or blocked region, as shown in Figure 39.22. With *laser angioplasty*, laser beams directed at the plaques vaporize them. With *balloon angioplasty*, a small balloon inflated within a blocked artery breaks up the plaques.

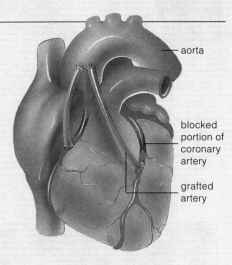

Figure 39.22 Two coronary bypasses (color-coded *green*).

aorta

blocked portion of coronary artery

grafted artery

ARRHYTHMIAS ECGs can be used to detect *arrhythmias*, or irregular heart rhythms (Figure 39.23). Arrhythmias are not always a sign of abnormal conditions. Endurance athletes, for example, may have a below-average resting cardiac rate, or *bradycardia*. As an adaptation to ongoing strenuous exercise, their nervous system has adjusted the cardiac pacemaker's rate of contraction downward. Exercise or stress often causes 100+ heartbeats a minute, or *tachycardia*. *Atrial fibrillation*, an irregular heartbeat, affects more than 10 percent of the elderly and young people with various heart diseases. A coronary occlusion or some other disorder may cause irregular rhythms that rapidly lead to a dangerous condition called *ventricular fibrillation*. In portions of the ventricles, cardiac muscle contracts haphazardly and blood pumping falters. Within seconds, the individual loses consciousness, which might signify impending death. A strong electric shock to the chest may restore normal cardiac function.

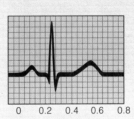

0   0.2   0.4   0.6   0.8
a   time (seconds)

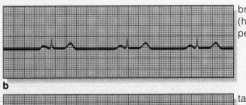

b

bradycardia (here, 46 beats per minute)

tachycardia (here, 136 beats per minute)

c

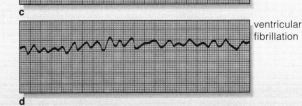

ventricular fibrillation

d

Figure 39.23 (**a**) An ECG of a single, normal human heartbeat. (**b–d**) Three examples of recordings of arrhythmias.

HEMOSTASIS

Small blood vessels are vulnerable to ruptures, cuts, and other damage. **Hemostasis**, a process involving blood vessel spasm, platelet plug formation, and blood coagulation, may repair the damage and stop blood loss.

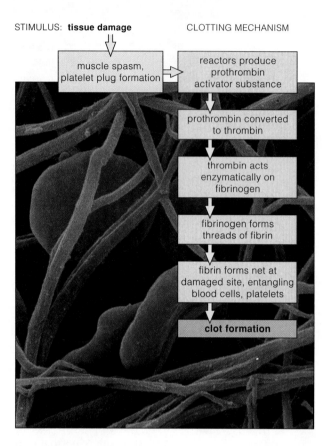

STIMULUS: **tissue damage**   CLOTTING MECHANISM

Figure 39.24 Plasma proteins involved in clot formation. Blood coagulates when damage exposes the collagen fibers in blood vessel walls. Exposure initiates reactions that cause rod-shaped plasma proteins (fibrinogens) to stick together as long, insoluble threads. These adhere to the exposed collagen, forming a net that traps blood cells and platelets. The entire mass is a clot.

As shown in Figure 39.24, smooth muscle in the damaged wall contracts in an automatic response called a spasm. The vasoconstriction temporarily staunches the blood flow. Platelets clump together as a temporary plug in the damaged wall. They also release substances that help prolong the spasm and attract more platelets. Then blood coagulates, or converts to a gel, and forms a clot. Finally, the clot retracts, forming a compact mass, and the breach in the wall is sealed.

The body routinely repairs damage to small blood vessels and prevents blood loss. The repair process includes blood vessel spasm, platelet plug formation, and coagulation.

LYMPHATIC SYSTEM

We conclude this chapter with a brief section on the manner in which the lymphatic system supplements blood circulation. Think of this section as a bridge to the next chapter, on immunity, for the lymphatic system also helps defend the body against injury and attack. The system is composed of drainage vessels, lymphoid organs, and lymphoid tissues. The tissue fluid that has moved into the vessels is called lymph.

Lymph Vascular System

A portion of the lymphatic system called the **lymph vascular system** consists of many tubes that collect and transport water and solutes from interstitial fluid to ducts of the circulatory system. Its main components are **lymph capillaries** and **lymph vessels** (Figure 39.25).

The lymph vascular system serves three functions. First, its vessels are drainage channels for water and plasma proteins that have leaked away from blood at capillary beds and that must be delivered back to the blood circulation. Second, the system also takes up fats that the body has absorbed from the small intestine and delivers them to the blood circulation, in the manner described in Section 42.5. Third, it delivers pathogens, foreign cells and material, and cellular debris from the

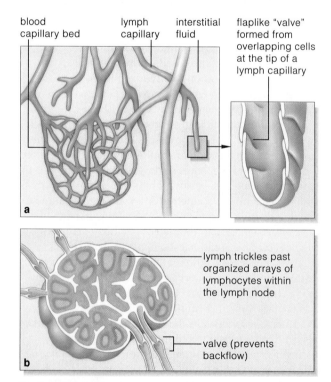

blood capillary bed   lymph capillary   interstitial fluid   flaplike "valve" formed from overlapping cells at the tip of a lymph capillary

lymph trickles past organized arrays of lymphocytes within the lymph node

valve (prevents backflow)

Figure 39.25 (**a**) Diagram of some of the lymph capillaries at the start of the drainage network called the lymph vascular system. (**b**) Cutaway diagram of a single lymph node. Its inner compartments are packed with organized arrays of infection-fighting white blood cells.

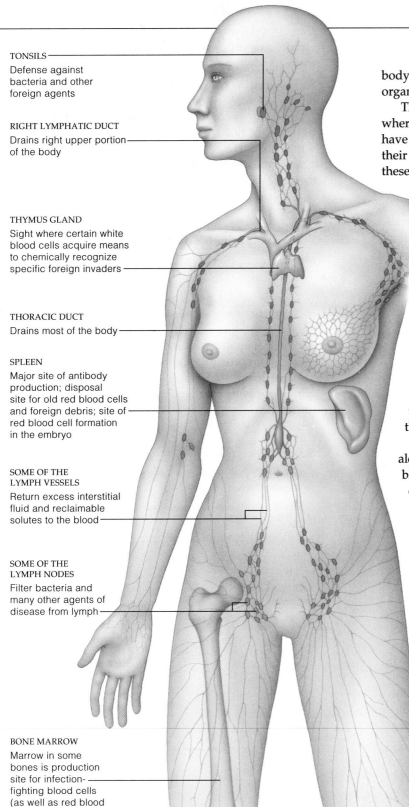

TONSILS
Defense against bacteria and other foreign agents

RIGHT LYMPHATIC DUCT
Drains right upper portion of the body

THYMUS GLAND
Sight where certain white blood cells acquire means to chemically recognize specific foreign invaders

THORACIC DUCT
Drains most of the body

SPLEEN
Major site of antibody production; disposal site for old red blood cells and foreign debris; site of red blood cell formation in the embryo

SOME OF THE LYMPH VESSELS
Return excess interstitial fluid and reclaimable solutes to the blood

SOME OF THE LYMPH NODES
Filter bacteria and many other agents of disease from lymph

BONE MARROW
Marrow in some bones is production site for infection-fighting blood cells (as well as red blood cells and platelets)

Figure 39.26 Components of the human lymphatic system and their functions. The *green* dots represent some of the major lymph nodes. Patches of lymphoid tissue in the small intestine and in the appendix also are part of the lymphatic system.

body's tissues to the lymph vascular system's efficiently organized disposal centers, the lymph nodes.

The lymph vascular system starts at capillary beds, where fluid enters the lymph capillaries. The capillaries have no obvious entrance; water and solutes move into their tips at flaplike "valves." As Figure 39.25a shows, these are areas where endothelial cells overlap. The lymph capillaries merge into lymph vessels, which have a larger diameter. The lymph vessels contain some smooth muscle in their wall as well as valves in their lumen that prevent backflow. They converge into collecting ducts that drain into veins in the lower neck (Figure 39.26).

Lymphoid Organs and Tissues

Other portions of the lymphatic system, called the **lymphoid organs and tissues**, have roles in defending the body against damage and attack. They include lymph nodes, the spleen, and the thymus, as well as tonsils and patches of tissue in the small intestine and appendix.

Lymph nodes are strategically located at intervals along lymph vessels (Figure 39.26). Before entering blood, lymph is filtered as it trickles through at least one node. Masses of lymphocytes take up residence in the nodes after forming in bone marrow. When they recognize an invader, they multiply rapidly and form large armies to destroy it.

The **spleen** is the largest lymphoid organ. It filters pathogens and used-up blood cells from the blood itself. One of its compartments, the red pulp, is a huge reservoir of red blood cells. In human embryos, the red pulp also produces these cells. The other compartment, the white pulp, has masses of lymphocytes associated with blood vessels. If and when a specific invader reaches the spleen during a severe infection, the lymphocytes become mobilized to destroy it, just as they are in lymph nodes.

It is in the **thymus gland** that immature T lymphocytes differentiate in ways that allow them to recognize and respond to specific pathogens. The thymus produces hormones that influence these events. It is central to immunity, the focus of the chapter to follow.

The lymph vascular portion of the lymphatic system returns water and solutes from tissue fluid to blood, and delivers fats and foreign material to lymph nodes. The system's lymph nodes and other lymphoid organs help defend the body against tissue damage and infectious diseases.

SUMMARY

1. The closed circulatory system of humans and other vertebrates consists of a heart (a muscular pump), blood vessels (arteries, arterioles, capillaries, venules, and veins), and blood. The system functions in rapid internal transport of substances to and from cells.

2. Blood helps maintain favorable conditions for cells. This fluid connective tissue consists of plasma, red and white blood cells, and platelets. It delivers oxygen and other substances to the interstitial fluid around cells. It also picks up cell products and wastes from that fluid.

a. Plasma, the liquid portion of blood, is a transport medium for blood cells and platelets. Plasma is a solvent for plasma proteins, simple sugars, lipids, amino acids, mineral ions, vitamins, hormones, and several gases.

b. Red blood cells transport oxygen from the lungs to all body regions. They are packed with hemoglobin, an iron-containing pigment molecule that binds reversibly with oxygen. Red blood cells also transport some carbon dioxide from interstitial fluid to the lungs.

c. Some phagocytic white blood cells cleanse tissues of dead cells, cellular debris, and anything else detected as not belonging to the body. Other white blood cells (lymphocytes) form great armies that destroy specific bacteria, viruses, and other disease agents.

3. The human heart is an incessantly beating double pump. Each half of the heart has two chambers, called an atrium and a ventricle. Blood flows into the atria, then the ventricles, then on into the great arteries. One-way valves enforce the one-way flow.

4. The heart's partition separates blood flow into two circuits, one pulmonary and the other systemic.

a. The pulmonary circuit loops between the heart and lungs. Oxygen-poor blood from the systemic veins enters the *right* atrium, is pumped through pulmonary arteries to both lungs, picks up oxygen, then flows through pulmonary veins to the heart's left atrium.

b. The systemic circuit loops between the heart and all body tissues. Oxygenated blood in the *left* atrium flows into the left ventricle, is pumped into the aorta, then is distributed to capillary beds. There, the blood gives up oxygen and picks up carbon dioxide. Systemic veins return it to the heart's right atrium.

5. Ventricular contraction drives the blood through both circuits. Blood pressure is high in contracting ventricles. It drops successively in arteries, arterioles, capillaries, venules, and veins. It is lowest in relaxed atria.

a. Arteries are rapid-transport vessels and a pressure reservoir (they smooth out pressure changes resulting from heartbeats and thereby smooth out blood flow).

b. Arterioles are sites where the volume of flow through organs can be controlled.

c. Capillary beds are diffusion zones where blood and interstitial fluid exchange substances.

d. Venules overlap capillaries in function. Veins are rapid-transport vessels and a blood volume reservoir that is tapped to adjust flow volume back to the heart.

6. The cardiac conduction system is the basis for the heart's rhythmic, spontaneous contractions.

a. Of all the cardiac muscle cells, 1 percent are not contractile. They are specialized to initiate and distribute action potentials, independently of the nervous system. The nervous system only adjusts the rate and strength of the basic contractions; it does not initiate them.

b. The system's SA node fires action potentials the fastest and is the cardiac pacemaker. Waves of excitation starting here wash over the atria, then down the heart's partition, then up the ventricles. The ventricles contract in a wringing motion that ejects blood from the heart.

7. The lymphatic system has these functions:

a. Its vascular portion takes up water and plasma proteins that seep out of blood capillaries, then returns them to the blood circulation. It transports absorbed fats and delivers pathogens and foreign material to disposal centers. Lymph capillaries and vessels are components.

b. Its lymphoid organs and tissues have production centers for lymphocytes. Some are battlegrounds where organized arrays of lymphocytes fight disease agents.

Review Questions

1. Define the functions of the circulatory system and lymphatic system. Distinguish between blood and interstitial fluid. *39.1*

2. Describe the cellular components of blood. Describe the plasma portion of blood. *39.2*

3. Distinguish between systemic and pulmonary circuits. *39.1*

4. State the functions of arteries, arterioles, capillaries, veins, venules, and lymph vessels. *39.1, 39.5, 39.7, 39.8*

5. Distinguish between the functions of the human heart's atria and ventricles. Then label the heart's components: *39.6*

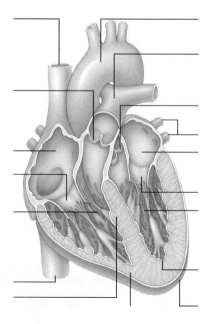

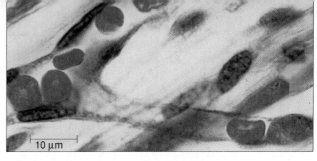

Figure 39.27 Light micrograph of human blood vessels.

Self-Quiz (*Answers in Appendix IV*)

1. Cells directly exchange substances with _____ .
   a. blood vessels   c. interstitial fluid
   b. lymph vessels   d. both a and b

2. Which are *not* components of blood?
   a. plasma
   b. blood cells and platelets
   c. gases and other dissolved substances
   d. all of the above are components of blood

3. The _____ produces red blood cells, which transport _____ and some _____ .
   a. liver; oxygen; mineral ions
   b. liver; oxygen; carbon dioxide
   c. bone marrow; oxygen; hormones
   d. bone marrow; oxygen; carbon dioxide

4. The _____ produces white blood cells, which function in _____ and _____ .
   a. liver; oxygen transport; defense
   b. lymph nodes; oxygen transport; pH stabilization
   c. bone marrow; housekeeping; defense
   d. bone marrow; pH stabilization; defense

5. In the pulmonary circuit, the heart's _____ half pumps blood to lungs, then _____ blood flows to the heart.
   a. right; oxygen-poor   c. right; oxygen-rich
   b. left; oxygen-poor   d. left; oxygen-rich

6. In the systemic circuit, the heart's _____ half pumps _____ blood to all body regions.
   a. right; oxygen-poor   c. right; oxygen-rich
   b. left; oxygen-poor   d. left; oxygen-rich

7. Blood pressure is high in _____ and lowest in _____ .
   a. arteries; veins   c. arteries; ventricles
   b. arteries; relaxed atria   d. arterioles; veins

8. _____ contraction drives blood through the pulmonary circuit and the systemic circuit; blood pressure is highest in contracting _____ .
   a. Atrial; ventricles   c. Ventricular; arteries
   b. Atrial; atria   d. Ventricular; ventricles

9. Which is *not* a function of the lymphatic system?
   a. delivers disease agents to disposal centers
   b. produces lymphocytes
   c. delivers oxygen to cells
   d. returns water and plasma proteins to blood

10. Match the type of blood vessel with its major function.
    ____ arteries   a. diffusion
    ____ arterioles   b. control of blood volume distribution
    ____ capillaries   c. transport, blood volume reservoirs
    ____ venules   d. overlap capillary function
    ____ veins   e. transport, pressure reservoirs

11. Match the components with their most suitable description.
    ____ capillary beds   a. two atria, two ventricles
    ____ lymph vascular system   b. bathes body's living cells
    ____ heart chambers   c. driving force for blood
    ____ blood   d. zones of diffusion
    ____ heart contractions   e. starts at capillary beds
    ____ interstitial fluid   f. fluid connective tissue

Critical Thinking

1. Shirelle, who is using a light microscope to examine a human tissue specimen, sees red blood cells moving single file through thin-walled tubes. She makes a photomicrograph (Figure 39.27). What type of blood vessel has she captured on film?

2. In individuals who have weak or leaky valves in their veins, fluid pressure associated with the backflow of blood causes venous walls below the valves to bulge outward. In time, the walls become stretched and flabbily distorted, a condition called *varicose veins*. Some people are genetically predisposed to develop the condition, but the cumulative mechanical stress associated with prolonged standing, pregnancy, and aging can contribute to it. With chronic varicosity, the legs themselves become swollen. Explain why this might happen. Also explain why veins close to the leg surface are more susceptible to varicosity than those deeper in the leg tissues.

3. Infection by a hemolytic bacterium (*Streptococcus pyogenes*) may trigger an inflammation that ultimately damages valves in the heart. The disease symptoms of *rheumatic fever* follow. Explain how this disease must affect the heart's functioning and what kinds of symptoms would arise as a consequence.

Selected Key Terms

ABO blood typing *39.4*	lymph capillary *39.11*
agglutination *39.4*	lymph node *39.11*
anemia *39.3*	lymph vascular system *39.11*
aorta *39.5*	lymph vessel *39.11*
arteriole *39.5*	lymphatic system *39.1*
artery *39.5*	lymphoid organ, tissue *39.11*
atrium (heart) *39.6*	plasma *39.2*
baroreceptor reflex *39.7*	platelet *39.2*
blood *39.1*	pulmonary circuit *39.1*
blood pressure *39.7*	reabsorption *39.8*
capillary (blood) *39.1*	red blood cell *39.2*
capillary bed *39.1*	Rh blood typing *39.4*
cardiac conduction system *39.6*	spleen *39.11*
cardiac cycle *39.6*	stem cell *39.2*
cardiac pacemaker *39.6*	systemic circuit *39.1*
cell count *39.2*	thymus gland *39.11*
circulatory system *39.1*	transfusion (blood) *39.4*
electrocardiogram *CI*	ultrafiltration *39.8*
endothelium *39.6*	vasoconstriction *39.7*
heart *39.1*	vasodilation *39.7*
hemostasis *39.10*	vein *39.5*
HDL *39.9*	ventricle (heart) *39.6*
interstitial fluid *39.1*	venule *39.5*
LDL *39.9*	white blood cell *39.2*

Readings

Little, R., and W. Little. 1989. *Physiology of the Heart and Circulation.* Fourth edition. Chicago: Year Book Medical.

Robinson, T. F., et al. June 1986. "The Heart as a Suction Pump." *Scientific American* 254(6): 84–91.

Web Site See *http://www.wadsworth.com/biology* for practice quiz questions, hypercontents, BioUpdates, and critical thinking. The Wadsworth Biology Resource Center provides a wealth of information fully organized and integrated by chapter.

40 IMMUNITY

Russian Roulette, Immunological Style

Until about a century ago, smallpox epidemics swept repeatedly through the world's cities. Some outbreaks were so severe that only half of those who were stricken survived. The survivors were left with permanent scars on the face, neck, shoulders, and arms. But they seldom contracted the same disease again; they were said to be "immune" to smallpox.

No one knew what caused smallpox, but the idea of acquiring immunity was dreadfully appealing. In twelfth-century China, people in good health gambled with deliberate infections. They sought out people with mild cases of smallpox (who were only mildly scarred), removed crusts from the scars, ground them up, and inhaled the powder.

By the seventeenth century, Mary Montagu, wife of the English ambassador to Turkey, was championing inoculation. She went so far as to poke bits of smallpox scabs into her children's skin. Others soaked threads in fluid from sores, then poked the threads into skin.

Some individuals who survived the chancy practices acquired immunity to smallpox, but many developed raging infections. As if the odds were not dangerous enough, the crude inoculation procedures also invited the acquisition of several other infectious diseases.

While this immunological version of Russian roulette was going on, Edward Jenner was growing up in the countryside of England. At the time, it was common knowledge that people who contracted cowpox never got smallpox. (Cowpox is a mild disease that can be transmitted from cattle to humans.) No one thought much about this until 1796, when Jenner, by now a physician, injected material from a cowpox sore into the arm of a healthy boy. Six weeks later, after the reaction subsided, Jenner injected some material from *smallpox* sores into the boy (Figure 40.1*a*). He had hypothesized that the earlier injection might provoke immunity to

Figure 40.1 (**a**) Statue honoring Edward Jenner's work to develop an immunization procedure against smallpox, one of the most dreaded diseases in human history. (**b**) False-color scanning electron micrograph of a white blood cell being attacked by HIV (*blue particles*), the virus that causes AIDS. Immunologists are working to develop effective weapons against this modern-day scourge.

smallpox. Fortunately he was right; the boy remained free of smallpox. Jenner had developed an effective immunization procedure against a specific pathogen.

The French mocked Jenner's procedure by calling it **vaccination**. The word literally means "encowment." Much later Louis Pasteur, an influential French chemist, developed similar immunization procedures for other diseases. In acknowledgment of Jenner's pioneering work, Pasteur called his procedures vaccinations, also, and only then did the word become respectable.

By Pasteur's time, improved light microscopes were revealing diverse bacteria, fungal spores, and other previously invisible forms of life. As Pasteur himself discovered, microorganisms abound in ordinary air. Did some cause diseases? Probably. Could they settle into food or drink and make it spoil? He demonstrated that they could and did.

Pasteur also found a way to kill most of the suspect disease agents in food or beverages. As he and others knew, boiling killed the agents. He also knew that you cannot boil wine (or beer or milk, for that matter) and end up with the same beverage. He devised a way to heat such beverages at a temperature low enough not to ruin them but high enough to kill most of the resident microorganisms. We still depend on this antimicrobial method, which was named pasteurization in his honor.

In the late 1870s Robert Koch, a German physician, linked a specific microorganism to a specific disease; namely, anthrax. For one experiment, Koch had injected blood from animals with anthrax into healthy ones. The recipients of the injection ended up with blood that teemed with cells of the bacterium *Bacillus anthracis*—and they developed anthrax. Even more convincing, injections of bacterial cells that were cultured outside the animal body also caused the disease!

Thus, by the beginning of the twentieth century, the promise of understanding the basis of infectious disease and immunity loomed large—and the battles against those diseases were about to begin in earnest. Since that time, spectacular advances in microscopy, biochemistry, and molecular biology have increased our understanding of the body's defenses. We now have greater insight into its responses to tissue damage in general and into its immune responses to specific pathogens or tumor cells. The responses are the focus of this chapter.

KEY CONCEPTS

1. The vertebrate body has physical, chemical, and cellular defenses against pathogenic microorganisms, malignant tumor cells, and other agents that can destroy tissues and even the individual itself.

2. During early stages of tissue invasion and damage, white blood cells and certain proteins dissolved in the plasma portion of blood escape from capillaries and execute a rapid, nonspecific counterattack. We call this a nonspecific inflammatory response. Phagocytic white blood cells ingest the invaders. The plasma proteins promote phagocytosis, and some also destroy invaders directly.

3. If the invasion persists, certain white blood cells make immune responses. Those cells can chemically recognize distinct configurations on molecules that are abnormal or foreign to the body, such as those on bacteria and viruses. If the foreign or abnormal molecule triggers an immune response, it is called an antigen.

4. In one type of immune response, some of the white blood cells produce enormous quantities of antibodies. Antibodies are molecules that bind to a specific antigen and tag it for destruction.

5. In another type of immune response, executioner cells directly destroy body cells that have become abnormal, as by infection or by a tumor-producing process.

You continually cross paths with astoundingly diverse **pathogens**—the viruses, bacteria, fungi, protozoans, parasitic worms, and other agents that cause diseases. Having coevolved with most of them, you and other vertebrates have three lines of defense, so you need not lose sleep over this. Most pathogens cannot breach the body surface. If they do, they face chemical weapons and white blood cells, or leukocytes, that attack anything perceived as foreign. Other white blood cells zero in on specific targets. Table 40.1 lists the three lines of defense.

Surface Barriers to Invasion

Most often, pathogens cannot get past skin or the other linings of body surfaces. Think of skin as a habitat of low moisture, low pH, and thick layers of dead cells. Normally harmless bacteria tolerate these conditions. Few pathogens can compete with dense populations of established types unless conditions change. Repeatedly subject your toes to warm, damp shoes, for example, and you may be inviting certain fungi to penetrate the sodden, weakened tissues and cause *athlete's foot*.

Similarly, resident bacteria on the mucous lining of the gut and vagina help protect you—as when lactate, a fermentation product of *Lactobacillus* populations in the vagina, helps maintain a low pH that most bacteria and fungi cannot tolerate. Barriers in branching, tubular airways leading to your lungs stop airborne pathogens. Air rushing down the tubes flings bacteria against their sticky, mucus-coated walls. In that mucus are protective substances such as **lysozyme**, an enzyme that digests cell walls of bacteria and so invites their death. As a final touch, broomlike cilia in the airways sweep out the trapped and enzymatically whapped pathogens.

The body has additional defenses. Lysozyme and other substances in tears, saliva, and gastric fluid offer protection, as when an outpouring of tears gives the eyes a sterile washing. Urine's low pH and flushing action help bar pathogens from moving into the urinary tract. As another example, diarrhea swiftly flushes irritating pathogens from the gut. Diarrhea must be controlled in children because it causes dehydration, but blocking its action in adults can prolong infection.

Nonspecific and Specific Responses

All animals react to tissue damage. Even simple aquatic invertebrates have phagocytic cells and antimicrobial substances, including lysozymes. But the more complex the animal body, the more complex are the systems that defend it. Think back on the evolution of the circulatory and lymphatic systems in vertebrates that invaded the land (Section 39.1). As circulation of body fluids became more efficient, so did the means for defending the body. Phagocytic cells as well as plasma proteins could move rapidly to tissues under attack, and they could intercept pathogens trickling along the vascular highways.

Vertebrates came to be equipped with sets of plasma proteins. Some proteins promote rapid clot formation after tissue damage; others destroy invading pathogens or target them for phagocytosis. Exquisitely focused responses to specific dangers also evolved.

In short, *internal defenses against a great variety of pathogens are in place even before damage occurs.* Ready and waiting are specialized white blood cells as well as plasma proteins. They take part in a *nonspecific* response to tissue damage in general, not to one pathogen or another. Other white blood cells may recognize unique molecular configurations of a *specific* pathogen. If they do, the resulting "immune" response will run its course whether tissues are damaged or not.

Table 40.1	The Vertebrate Body's Three Lines of Defense Against Pathogens

BARRIERS AT BODY SURFACES (*nonspecific* targets)

Intact skin; mucous membranes at other body surfaces

Infection-fighting chemicals in tears, saliva, etc.

Normally harmless bacterial inhabitants of skin and other body surfaces that can outcompete pathogenic visitors

Flushing effect of tears, saliva, urination, and diarrhea

NONSPECIFIC RESPONSES (*nonspecific* targets)

Inflammation:

1. Fast-acting white blood cells (neutrophils, eosinophils, and basophils)
2. Macrophages (also take part in immune responses)
3. Complement proteins, blood-clotting proteins, and other infection-fighting substances

Organs with pathogen-killing functions (such as lymph nodes)

Some cytotoxic cells (e.g., NK cells) with a range of targets

---

IMMUNE RESPONSES (*specific* targets only)

T cells and B cells; macrophages interact with them

Communication signals (e.g., interleukins) and chemical weapons (e.g., antibodies, perforins)

Intact skin, mucous membranes, antimicrobial secretions, and other barriers at the body surface constitute the first line of defense against invasion and tissue damage.

Inflammation and other internal, nonspecific responses to invasion are the second line of defense.

Immune responses against specific invaders, as executed by armies of white blood cells and their chemical weapons, are the third line of defense.

40.2 COMPLEMENT PROTEINS

A set of plasma proteins has roles in both nonspecific and specific defenses. Collectively, these proteins are called the **complement system**. About twenty kinds of complement proteins circulate in the blood in inactive form. If even a few molecules of one kind are activated, they trigger a huge cascade of reactions. They activate many molecules of another complement protein. Each

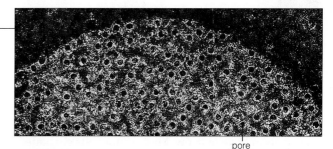

Figure 40.3 Micrograph of a cell surface, showing pores formed by membrane attack complexes.

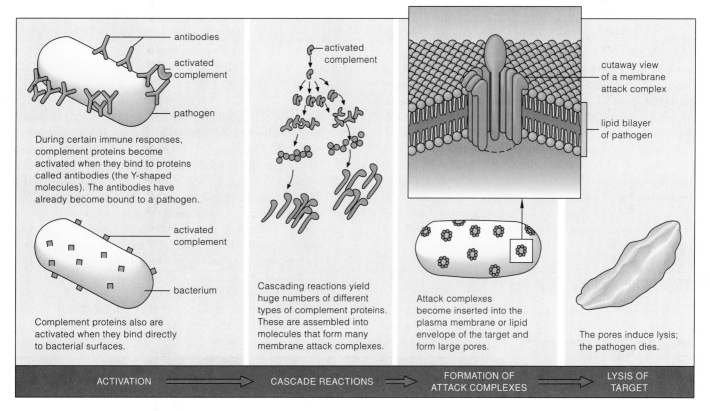

During certain immune responses, complement proteins become activated when they bind to proteins called antibodies (the Y-shaped molecules). The antibodies have already become bound to a pathogen.

Complement proteins also are activated when they bind directly to bacterial surfaces.

Cascading reactions yield huge numbers of different types of complement proteins. These are assembled into molecules that form many membrane attack complexes.

Attack complexes become inserted into the plasma membrane or lipid envelope of the target and form large pores.

The pores induce lysis; the pathogen dies.

ACTIVATION ⟹ CASCADE REACTIONS ⟹ FORMATION OF ATTACK COMPLEXES ⟹ LYSIS OF TARGET

Figure 40.2 Formation of membrane attack complexes. One reaction pathway starts as complement proteins bind to bacterial surfaces; another operates during immune responses to specific invaders. Both pathways produce membrane pore complexes that induce lysis in pathogens.

of these activates many molecules of another kind of protein at the next reaction step, and so on. Deployment of all these molecules has the following effects.

Some of the complement proteins join together to form pore complexes. These molecular structures have an inner channel (Figures 40.2 and 40.3). They become inserted into the plasma membrane of many pathogens, forming pores that induce lysis. **Lysis**, recall, is a gross structural disruption of a plasma membrane or cell wall that leads to cell death. Pore complexes also become inserted into the wall of Gram-negative bacteria. Such cell walls consist of a lipid-rich, outer surface above a peptidoglycan layer. Lysozyme molecules can diffuse through the pores and digest the peptidoglycan.

Some activated proteins promote inflammation, a nonspecific defense response described next. Through their cascades of reactions, they create concentration gradients that attract phagocytic white blood cells to an irritated or damaged tissue. The complement proteins also encourage phagocytes to dine. They can bind to the surface of many types of invaders. When they do so, an invader ends up with a complement "coat." The coat can adhere to phagocytes, and this is rather like putting a basted turkey on a dinner table.

In such ways, the complement proteins target many bacteria, parasitic protistans, and enveloped viruses.

The complement system, a set of twenty or so kinds of plasma proteins circulating in blood, takes part in cascades of reactions that help defend against many bacteria, some parasitic protistans, and enveloped viruses.

The complement system takes part in both nonspecific and specific defenses.

Chapter 40   Immunity   **673**

The Roles of Phagocytes and Their Kin

Certain white blood cells take part in an initial response to tissue damage. White blood cells, recall, arise from stem cells in bone marrow. Many of the cells circulate in blood and lymph. A great many take up stations in the lymph nodes as well as the spleen, liver, kidneys, lungs, and brain. Here you may wish to refer to Section 39.2, which introduces the components of blood, and Section 39.11, which introduces the lymphatic system.

Like SWAT teams, three kinds of white blood cells react swiftly to danger in general but are not adapted for sustained battles. **Neutrophils**, the most abundant kind, phagocytize bacteria. They ingest, kill, and digest bacterial cells to simple molecular bits. **Eosinophils** secrete enzymes that make holes in parasitic worms. **Basophils** secrete histamine and other substances that help keep inflammation going after it starts.

Although slower to act, the white blood cells called **macrophages** are the "big eaters." Figure 40.4 shows one of them. A macrophage engulfs and digests just about any foreign agent. It also helps clean up damaged tissues. Immature macrophages circulating in blood are called monocytes.

The Inflammatory Response

An inflammatory response develops when something damages or kills cells of any given tissue. Infections, punctures, burns, and other insults are the triggers. By a mechanism known as **acute inflammation**, the fast-acting phagocytes and complement proteins, as well as other plasma proteins, escape from the bloodstream at capillary beds in the damaged tissue. There they enter interstitial fluid. Localized signs that acute inflammation is under way include redness, heat, swelling, and pain, as listed in Table 40.2.

Table 40.2	Localized Signs of Inflammation and Their Causes
Redness	Arteriolar vasodilation; increase in blood flow to the affected site
Warmth	Arteriolar vasodilation; more blood, carrying more metabolic heat, arrives at site
Swelling	Chemical signals increase capillary permeability; plasma proteins leak out, disrupt fluid balance across wall of capillaries; localized edema
Pain	Nociceptors (pain receptors) stimulated by increased fluid pressure, local chemical signals

Mast cells, which reside in connective tissues and function like basophils, take part in an inflammatory response. They release **histamine** and other substances into interstitial fluid. Their secretions are local chemical signals that trigger vasodilation of arterioles threading through the damaged tissue. Vasodilation, remember, is an increase in a blood vessel's diameter when smooth muscle in its wall relaxes. When the arterioles become engorged with blood, the affected tissue reddens and becomes warmer, owing to blood-borne metabolic heat.

Released histamine also increases the permeability of the thin-walled capillaries in the tissue. It induces endothelial cells making up the capillary wall to pull apart farther at the narrow clefts between them. Thus

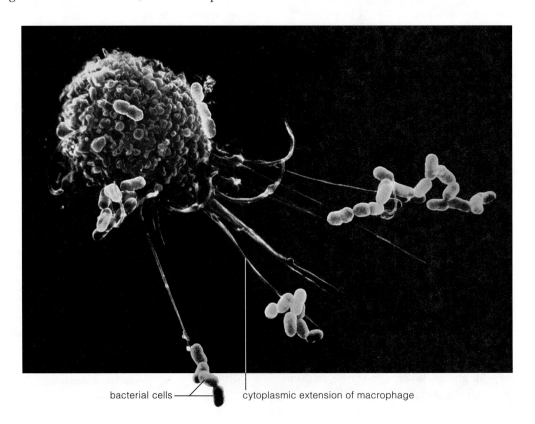

Figure 40.4 False-color scanning electron micrograph of a macrophage (*red*), with some cytoplasmic extensions that made contact with bacterial cells (*green*) in its surroundings. Bacteria are engulfed by this type of phagocytic white blood cell.

bacterial cells — cytoplasmic extension of macrophage

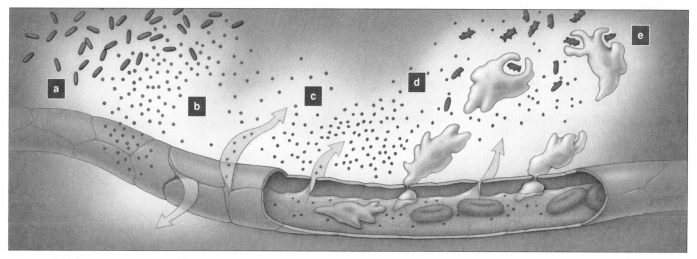

a Bacteria invade a tissue and directly kill cells or release metabolic products that damage tissue.

b Mast cells in tissue release histamine, which then triggers arteriolar vasodilation (hence redness and warmth) as well as increased capillary permeability.

c Fluid and plasma proteins leak out of capillaries; localized edema (tissue swelling) and pain result.

d Complement proteins attack bacteria. Clotting factors wall off inflamed area.

e Neutrophils, macrophages, and other phagocytes engulf invaders and debris. Macrophage secretions attract even more phagocytes, directly kill invaders, and call for fever and for T and B cell proliferation.

Figure 40.5 Acute inflammation in response to a bacterial invasion. The response involves delivering phagocytes and plasma proteins to the tissue. Together, these components of blood inactivate, destroy, or isolate the invaders, remove chemicals and cellular debris, and prepare the tissue for subsequent repair. These are their functions in all inflammatory responses.

the capillaries become "leaky" to plasma proteins that normally do not leave the blood. When some proteins leak out, osmotic pressure increases in the surrounding interstitial fluid. In combination with the higher blood pressure brought about by the increased blood flow to the tissue, ultrafiltration increases and reabsorption decreases across the capillary wall. Localized edema is the outcome of the shift in the fluid balance across the capillary wall. (Here you may wish to review Section 39.8.) The tissue swells with fluid, and its nociceptors give rise to sensations of pain. Typically an individual avoids voluntary movements that might aggravate the pain. This behavior promotes tissue repair.

Within hours of the first physiological responses to the damage, neutrophils are squeezing across capillary walls. They swiftly go to work. Monocytes arrive later, differentiate into macrophages, and engage in sustained action (Figure 40.5). While macrophages are engulfing invaders and debris, they secrete chemical mediators. Mediators called *chemotaxins* attract more phagocytes. An *interleukin* stimulates the formation of B and T cell armies, as described shortly. *Lactoferrin* directly kills bacteria. *Endogenous pyrogen* might trigger the release of prostaglandins, which in turn trigger an increase in the "set point" on the hypothalamic thermostat that controls body temperature. What we call a **fever** is a body temperature that has reached the higher set point.

A fever of about 39°C (100°F) is not a bad thing. It increases body temperature to a level that is "too hot" for the functioning of most pathogens. It also promotes

an increase in a host's defense activities. *Interleukin-1* induces drowsiness, which reduces the body's demands for energy, so more energy can be diverted to the tasks of defense and tissue repair. Macrophages take part in the cleanup and repair operations.

Among the plasma proteins that leak into the tissue are complement proteins and clotting factors of the sort described in Section 39.10. Upon exposure to chemicals secreted by phagocytes and to tissue thromboplastin, fibrin forms and clots develop in the spaces around the inflamed tissue. The clots wall off the inflamed area and typically prevent or delay the spread of invaders and toxic chemicals into the surrounding tissues. After the inflammation subsides, anticlotting factors that had also escaped from the capillaries dissolve the clots.

An inflammatory response develops in a local tissue when cells are damaged or killed, as by infection. It proceeds during both nonspecific and specific defenses of tissues.

Mast cells in damaged or invaded tissues secrete histamine, which causes arterioles to vasodilate and increases capillary permeability to fluid and plasma proteins. The localized vasodilation reddens and warms the tissue. Edema results from the fluid imbalance across the capillary wall. The tissue swelling causes pain.

The response involves phagocytes such as macrophages, which engulf invaders and debris and secrete chemical mediators. It involves plasma proteins, such as complement proteins that target invaders for destruction as well as clotting factors that wall off the inflamed tissue.

Defining Features

Sometimes physical barriers and inflammation are not enough to overwhelm an invader, so an infection may become well established. Then, white blood cells called **B** and **T lymphocytes** form armies that engage in battle.

B and T cells are central to the body's third line of defense—the immune system. We define the **immune system** by two key features. The first is *immunological specificity*, whereby certain kinds of lymphocytes zero in on specific pathogens and eliminate them. The second feature is *immunological memory*, whereby a portion of the T and B cells formed during a first-time confrontation is set aside for a future battle with the same pathogen.

The operating principle for the system is this: *Each kind of cell, virus, or substance bears unique molecular configurations that give it a unique identity.* The unique configurations on an individual's own cells serve as *self* markers. Lymphocytes recognize self markers and will normally ignore them. They also can recognize *nonself* molecular configurations, which are unique to specific foreign agents. When that happens, lymphocytes are stimulated to divide repeatedly, by way of mitosis. And the divisions give rise to huge populations.

As the divisions proceed, subpopulations of the new cells become specialized to respond to the foreign agent in different ways. Some consist of *effector* cells, which are fully differentiated cells that engage and destroy the enemy. Other subpopulations consist of *memory* cells, which enter a resting phase. Instead of engaging in the attack on the specific agent that triggered the initial response, memory cells "remember" it. They will take part in a larger, more rapid response if that same kind of agent shows up again.

Any molecular configuration that triggers formation of lymphocyte armies and is their target is an **antigen**. The most important antigens are certain proteins at the surface of pathogens or tumor cells. As you will see, lymphocytes synthesize receptor molecules that can bind to these configurations. That is how they are able to recognize nonself.

In short, immunological specificity and memory involve three events: *recognition* of antigen, *repeated cell divisions* that form huge populations of lymphocytes, and *differentiation* into subpopulations of effector and memory cells with receptors for one kind of antigen.

Antigen-Presenting Cells—The Triggers for Immune Responses

The plasma membrane of every nucleated cell in the body of every human individual incorporates a variety of proteins. Among these proteins are **MHC markers**, named after the genes that encode the instructions for making them. Certain MHC markers are common at the surface of each nucleated cell in the body. Others are unique to the body's macrophages and lymphocytes.

Think of what happens after a cut allowed bacteria to enter a tissue inside your finger. Lymph vessels pick up some of the inflamed tissue's interstitial fluid and deliver it, along with some invading cells, to the lymph nodes in the vicinity. There, macrophages join the fray. Foreign cells are engulfed and enclosed in vesicles with digestive enzymes that cleave antigen molecules into fragments. The fragments bind to MHC molecules and form **antigen-MHC complexes**. When the vesicles move to the plasma membrane and fuse with it, the complexes are automatically displayed at the macrophage surface.

Any cell displaying processed antigen that is bound with a suitable MHC molecule is an **antigen-presenting cell**. When lymphocytes do encounter such a cell, they take notice (Figure 40.6). *This is the antigen recognition that promotes the cell divisions by which great armies of lymphocytes form.*

Key Players in Immune Responses

The same categories of white blood cells are called into action during each immune response. Figure 40.7 is an overview of how the cells interact. In brief, recognition of antigen-MHC complexes activates **helper T cells**. These produce and secrete substances that induce any responsive T or B lymphocyte to divide and give rise to large populations of effector cells and memory cells. Recognition also activates **cytotoxic T cells**, which can eliminate infected body cells or tumor cells by "touch killing." When they contact a target, they deliver

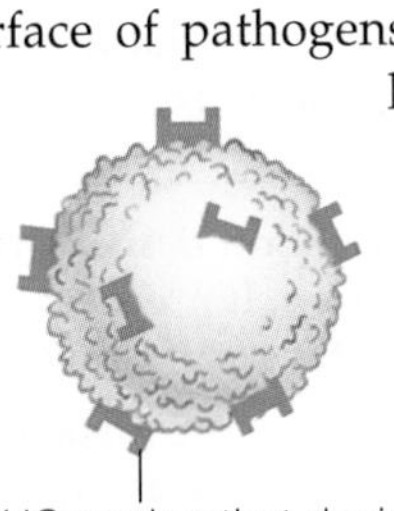

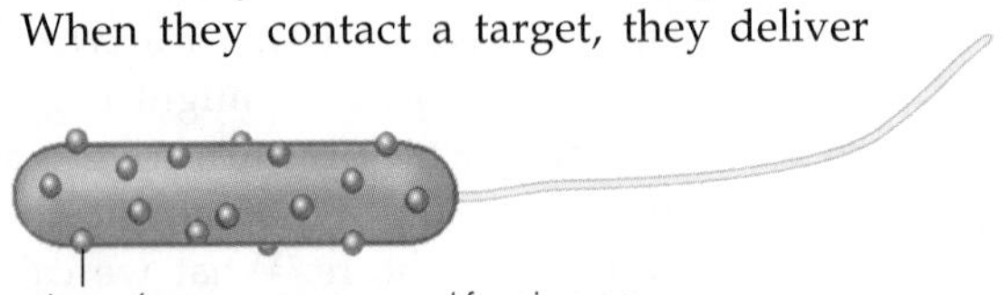

MHC marker that designates "self" (it occurs only at the surface of body's own cells)

T cells and B cells ignore this

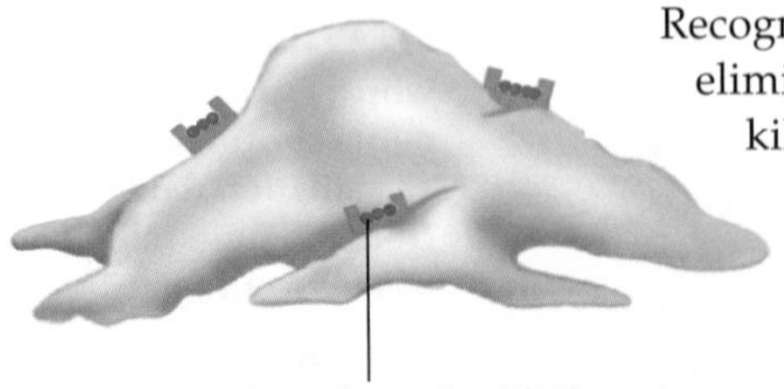

Figure 40.6 Molecular cues that T and B cells either ignore or recognize as a signal to initiate immune responses.

processed antigen, bound to MHC marker, at surface of an antigen-presenting cell

T cells initiate an immune response

antigen (any *unprocessed* foreign or abnormal molecular configuration that lymphocytes recognize as nonself)

B cells initiate an immune response

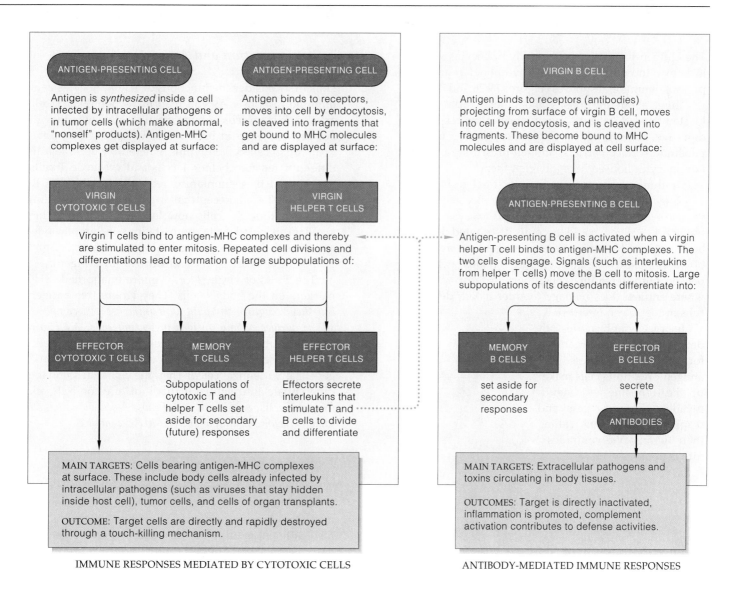

ANTIGEN-PRESENTING CELL

Antigen is *synthesized* inside a cell infected by intracellular pathogens or in tumor cells (which make abnormal, "nonself" products). Antigen-MHC complexes get displayed at surface:

ANTIGEN-PRESENTING CELL

Antigen binds to receptors, moves into cell by endocytosis, is cleaved into fragments that get bound to MHC molecules and are displayed at surface:

VIRGIN B CELL

Antigen binds to receptors (antibodies) projecting from surface of virgin B cell, moves into cell by endocytosis, and is cleaved into fragments. These become bound to MHC molecules and are displayed at cell surface:

VIRGIN CYTOTOXIC T CELLS

VIRGIN HELPER T CELLS

ANTIGEN-PRESENTING B CELL

Virgin T cells bind to antigen-MHC complexes and thereby are stimulated to enter mitosis. Repeated cell divisions and differentiations lead to formation of large subpopulations of:

Antigen-presenting B cell is activated when a virgin helper T cell binds to antigen-MHC complexes. The two cells disengage. Signals (such as interleukins from helper T cells) move the B cell to mitosis. Large subpopulations of its descendants differentiate into:

EFFECTOR CYTOTOXIC T CELLS

MEMORY T CELLS

EFFECTOR HELPER T CELLS

MEMORY B CELLS

EFFECTOR B CELLS

Subpopulations of cytotoxic T and helper T cells set aside for secondary (future) responses

Effectors secrete interleukins that stimulate T and B cells to divide and differentiate

set aside for secondary responses

secrete

ANTIBODIES

MAIN TARGETS: Cells bearing antigen-MHC complexes at surface. These include body cells already infected by intracellular pathogens (such as viruses that stay hidden inside host cell), tumor cells, and cells of organ transplants.

OUTCOME: Target cells are directly and rapidly destroyed through a touch-killing mechanism.

MAIN TARGETS: Extracellular pathogens and toxins circulating in body tissues.

OUTCOMES: Target is directly inactivated, inflammation is promoted, complement activation contributes to defense activities.

IMMUNE RESPONSES MEDIATED BY CYTOTOXIC CELLS

ANTIBODY-MEDIATED IMMUNE RESPONSES

Figure 40.7 Overview of key interactions among B and T lymphocytes during an immune response. Most often, both types of white blood cells are activated when antigen has been detected. An antigen is any large molecule that lymphocytes recognize as not being "self" (normal body molecules). A first-time encounter with antigen elicits a *primary* response. A subsequent encounter with the same type of antigen elicits a *secondary* immune response. A secondary response is larger and more rapid. Memory cells that formed but that were not used during the first battle immediately engage in the second one.

cell-killing chemicals into it. By contrast, B cells make antigen-binding receptor molecules called **antibodies**. When a response is under way, effector B cells secrete staggering numbers of antibody molecules. Only B cells are the basis of *antibody-mediated* responses.

Control of Immune Responses

Antigen provokes an immune response—and removal of antigen stops it. For example, by the time the tide of battle turns, effector cells and their chemical secretions have already killed most of the antigen-bearing agents inside the body. With fewer antigen molecules around

to stimulate the cells, the response declines, then stops. As a final example, inhibitory signals from cells with suppressor roles help shut down immune responses.

Antigens are nonself molecular configurations which, when recognized by certain lymphocytes, trigger immune responses. Helper T cells, cytotoxic T cells, B cells, and their secretions execute these responses.

Following antigen recognition, large T and B cell armies form by repeated mitotic cell divisions. These differentiate into subpopulations of effector cells and memory cells, all of which are sensitized to that one kind of antigen.

The antigen-presenting cells and lymphocytes we have just introduced interact within lymphoid organs that promote immune responses (Figures 39.26 and 40.8).

Think about the tonsils and other lymph nodules located beneath mucous membranes of the respiratory, digestive, and reproductive systems. Just after invaders penetrate surface barriers, antigen-presenting cells and lymphocytes housed here intercept them.

Or think about antigen in interstitial fluid that is entering the lymph vascular system. Because lymph vessels eventually drain into the expressways for blood transport, antigen could become distributed to every body region. However, before antigen can reach the blood, it must trickle through lymph nodes—which are packed with defending cells. Even in those few cases where antigen does manage to enter blood, defending cells in the spleen intercept it.

In the lymph nodes, the defending cells are organized for utmost effectiveness. The antigen-presenting cells make up the front line and engulf invaders. They process and display antigen, thus calling their lymphocyte comrades into action.

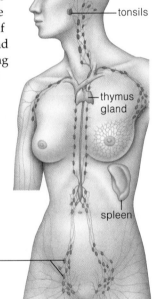

tonsils

thymus gland

spleen

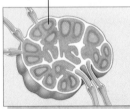

location of antigen-presenting cells and lymphocytes in a lymph node, cross-section

Figure 40.8 Organized arrays of antigen-presenting cells and lymphocytes in lymph nodes.

The cell divisions that produce subpopulations of effector and memory cells proceed in lymph nodes. As lymph drains through, it moves effector activities to the back of the organ and beyond. And all the while, virgin and memory cells circulate through the lymph node, reconnoitering at the front line.

Antigen-presenting cells and lymphocytes intercept and battle pathogens in organized ways within lymphoid organs and tissues, especially the lymph nodes.

T Cell Formation and Activation

Let's first consider the functions of T lymphocytes in an immune response. As you read in Section 39.2, T cells arise from stem cells in bone marrow. However, they do not fully develop in bone marrow. Rather, they travel to an organ called the thymus, where they become fully differentiated into helper T cells and cytotoxic T cells. Specifically, these immature cells acquire their **TCRs** (short for *T-Cell Receptors*) in the thymus. Bristling with receptors, the cells now leave the thymus. They circulate in blood or take up stations in lymph nodes and the spleen, as virgin T cells. In this context, "virgin" means the cells are as yet undisturbed (by antigen).

The TCRs of virgin T cells ignore unadorned MHC markers on the body's cells. They ignore free antigen. *But they recognize and bind with antigen-MHC complexes at the surface of any antigen-presenting cells.* As Figure 40.9 shows, binding induces T cells to divide repeatedly and give rise to large clones. (A clone is a population of genetically identical cells.) Then the clonal descendants differentiate into subpopulations of effector cells and memory cells. *And every one of those descendants has the same TCR for one kind of antigen-MHC complex.*

Functions of Effector T Cells

What actions do subpopulations of effector T cells take? Effector helper T cells secrete interleukins, the chemical mediators that fan repeated mitotic cell divisions and then differentiation of any responsive T and B cells, as described shortly. Effector cytotoxic T cells are killers; they respond to antigen-MHC complexes on body cells that are already infected by intracellular pathogens, such as viruses, and on tumor cells. The complex serves as a "double signal" that tells the killers to attack the cells that bear it (Figure 40.9e).

Effector cytotoxic T cells destroy infected cells with a touch-kill mechanism. They secrete *perforins*, protein molecules that form doughnut-shaped pores in a target cell's plasma membrane. (The pores look similar to the ones shown in Figure 40.3.) These effectors also secrete chemicals that induce cell death by way of **apoptosis**. As described in the introduction to Chapter 15, a target cell is induced to commit suicide. Its cytoplasm dribbles out, its organelles are disrupted, and its DNA becomes fragmented. Having made its lethal hit, the cytotoxic T cell disengages quickly and moves on to new targets.

Cytotoxic T cells also contribute to the rejection of tissue and organ transplants. Parts of MHC markers on donor cells are different enough from the recipient's to be recognized as antigens. But other parts are similar enough to complete the double signal. MHC typing and matching donors to recipients minimize the risk.

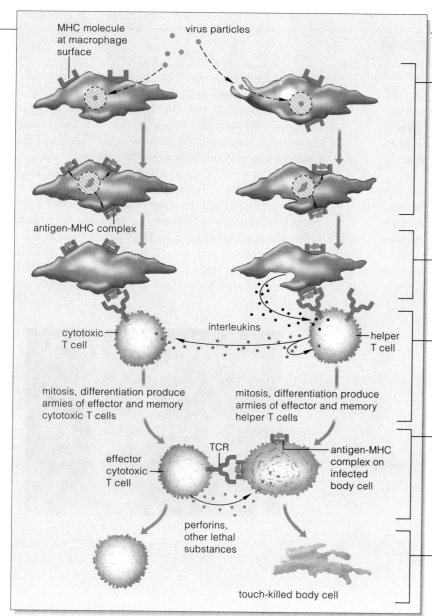

a An intracellular virus penetrates a macrophage and pirates its metabolic machinery. The host cell churns out viral proteins, which are antigenic. These are processed into fragments, bound to MHC molecules, and displayed at the cell surface. And *TCRs of cytotoxic T cells recognize antigens that have been synthesized inside a cell.*

Another macrophage engulfs a particle of the same virus and encloses it in an endocytic vesicle. Digestive enzymes cleave the particle's antigen into fragments. These bind to MHC molecules and are presented at the macrophage surface. *TCRs of helper T cells recognize such engulfed and processed antigens.*

b A responsive T cell binds with the antigen-MHC complexes. Binding stimulates the macrophage to secrete interleukins (*black* dots).

c These communication signals stimulate the helper T cell to secrete different kinds of interleukins (*blue* dots). These new signals stimulate cell divisions and differentiations by which large populations of effector T cells and memory T cells form.

d The same virus penetrated a cell in the lining of the respiratory tract, and antigen that was produced in the cytoplasm has been processed. Antigen is bound to MHC molecules and displayed at the cell's surface.

An effector cytotoxic T cell encounters the target. This type of effector specializes in touch-killing. It releases perforins and toxic substances (*green* dots) onto its target and so programs it for death.

e The effector disengages from the doomed cell and reconnoiters for more targets. Meanwhile, perforins make holes in the cell's plasma membrane. Toxins can move into the cell, disrupt its organelles, and make the DNA disassemble. The infected cell dies.

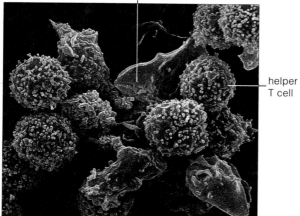

Figure 40.9 Diagram of a T cell-mediated immune response. In this example, the response involves an antigen-MHC complex that activates T cells. The micrograph shows helper T cells associating with an antigen-presenting macrophage.

Regarding the Natural Killer Cells

Other cytotoxic cells, including **natural killer cells** (NK cells), also arise from stem cells in bone marrow. They appear to be lymphocytes, but not T or B cells. Their arousal does not depend on a double signal (that is, an antigen-MHC complex). NK cells reconnoiter for tumor cells and virus-infected cells, then touch-kill them. They possibly recognize odd molecular configurations at the surface of their targets.

T cells arise in bone marrow. Later, in the thymus, they acquire TCRs (receptors for self markers and bound antigen).

Effector helper T cells secrete interleukins that trigger the cell divisions and differentiation into huge armies against specific antigens. Effector cytotoxic cells touch-kill infected cells or tumor cells, even foreign cells of transplants.

B Cells and the Targets of Antibodies

Like T cells, the B cells also arise from stem cells in bone marrow and start down a pathway that will culminate in full differentiation. Along *their* pathway, however, B cells start synthesizing many, many copies of a single kind of antibody molecule.

Although all antibodies are proteins, each kind has binding sites that only match up to a particular antigen. They are more or less Y-shaped, with a tail and with two arms that bear identical antigen receptors. Section 40.9 provides a closer look at these molecules. For now, we can simply think of them as Y-shaped structures, of the sort shown in Figure 40.10.

Each freshly synthesized antibody molecule of a maturing B cell moves to the plasma membrane. Its tail becomes embedded in the membrane's lipid bilayer and the two arms stick out above it. Soon the cell bristles with antigen receptors (the bound antibodies), and it is ready to join the body's defenses as a virgin B cell.

When its antigen receptors lock onto a target, the B cell does something you might not expect. *It becomes an antigen-presenting cell.* First, an endocytic vesicle moves bound antigen into the cell for digestion into fragments. Next, the fragments bind to MHC molecules and are presented at the B cell surface. Now suppose that TCRs of a responsive helper T cell bind to the antigen-MHC complex and that signals are transferred between the

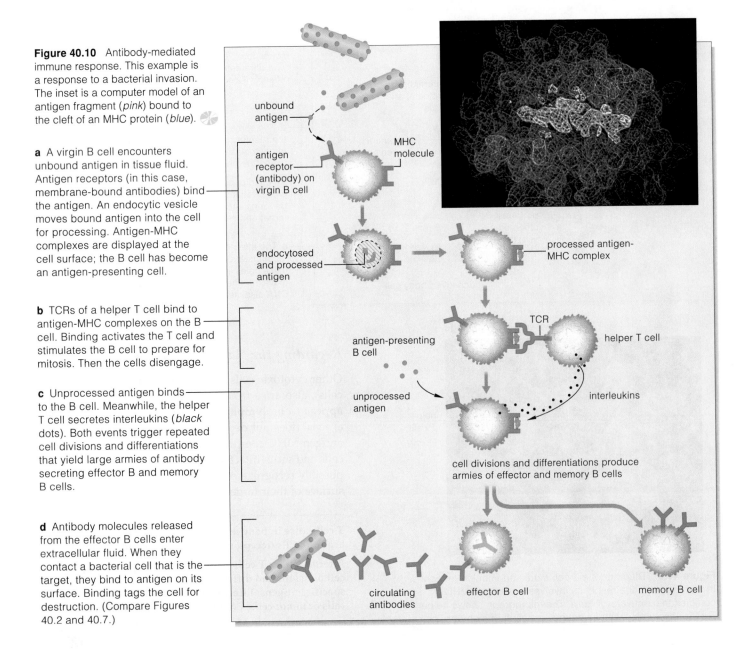

Figure 40.10 Antibody-mediated immune response. This example is a response to a bacterial invasion. The inset is a computer model of an antigen fragment (*pink*) bound to the cleft of an MHC protein (*blue*).

a A virgin B cell encounters unbound antigen in tissue fluid. Antigen receptors (in this case, membrane-bound antibodies) bind the antigen. An endocytic vesicle moves bound antigen into the cell for processing. Antigen-MHC complexes are displayed at the cell surface; the B cell has become an antigen-presenting cell.

b TCRs of a helper T cell bind to antigen-MHC complexes on the B cell. Binding activates the T cell and stimulates the B cell to prepare for mitosis. Then the cells disengage.

c Unprocessed antigen binds to the B cell. Meanwhile, the helper T cell secretes interleukins (*black* dots). Both events trigger repeated cell divisions and differentiations that yield large armies of antibody secreting effector B and memory B cells.

d Antibody molecules released from the effector B cells enter extracellular fluid. When they contact a bacterial cell that is the target, they bind to antigen on its surface. Binding tags the cell for destruction. (Compare Figures 40.2 and 40.7.)

unbound antigen

MHC molecule

antigen receptor (antibody) on virgin B cell

endocytosed and processed antigen

processed antigen-MHC complex

antigen-presenting B cell

TCR

helper T cell

unprocessed antigen

interleukins

cell divisions and differentiations produce armies of effector and memory B cells

circulating antibodies

effector B cell

memory B cell

T and B cell. The cells soon disengage. When the B cell encounters unprocessed antigen, its surface antibodies bind it. The binding, in combination with interleukins secreted from nearby helper T cells, drives the B cell to mitosis. Its clonal descendants differentiate into effector and memory B cells. The effectors (also called plasma cells) produce and secrete huge numbers of antibody molecules. When freely circulating antibody molecules bind antigen, they tag an invader for destruction, as by phagocytes and complement activation.

The main targets of antibody-mediated responses are extracellular pathogens and toxins, which are freely circulating in tissues or body fluids. *Antibodies cannot bind to pathogens or toxins that are hidden in a host cell.*

The Immunoglobulins

During immune responses, B cells produce four classes of antibodies in abundance and lesser quantities of another. Collectively, the five classes of antibodies are called **immunoglobulins**, or **Igs**. They are the protein products of gene shufflings that proceed while B cells mature and while an immune response is under way. The molecules in each class have antigen-binding sites *and* other sites with specialized functions.

IgM antibodies are the first to be secreted during immune responses. They trigger complement cascades, and they bind targets together in clumps—which are more handily eliminated by phagocytes. (Remember the agglutination responses described in Section 39.4?) *IgD* antibodies associate with IgM on virgin B cells, but their function is not yet understood.

IgG antibodies activate complement proteins and neutralize many toxins. These long-lasting antibodies are the only ones to cross the placenta. They can protect the fetus and newborn with the mother's acquired immunities. IgGs also are secreted into the early milk produced by mammary glands, then are absorbed into the suckling newborn's bloodstream.

IgA antibodies enter mucus-coated surfaces of the respiratory, digestive, and reproductive tracts, where they may neutralize infectious agents. Mother's milk delivers them to the mucous lining of a newborn's gut.

IgE triggers inflammation after attacks by parasitic worms and other pathogens. As described later, it also figures in allergies. The tails of IgE antibodies bind to basophils and mast cells, and the antigen receptors face outward. Antigen binding induces basophils and mast cells to release substances that promote inflammation.

Antibodies that are secreted by B cells bind to antigens of extracellular pathogens or toxins and tag them for disposal, as by phagocytes and complement activation.

Carcinomas, sarcomas, leukemia—these chilling words refer to malignant tumors in skin, bone, and other tissues. Such tumors arise when viral attack, irradiation, or chemicals alter genes and cells turn cancerous (Section 15.6). The transformed cells divide repeatedly. Unless they are destroyed or surgically removed, they kill the individual.

Often the surface of a transformed cell bears abnormal proteins and protein fragments bound to MHC. Defense responses to the transformed cells can make the tumor regress but may not be enough to destroy it. Also, some tumors release many copies of the abnormal proteins. If these saturate antigen receptors on the defenders, the tumor may escape detection. If the tumor hides long enough and reaches a critical mass, it may overwhelm the immune system's capacity for an effective response.

Researchers hope to develop procedures to enhance immunological defenses against tumors as well as against certain pathogens. This prospect is called *immunotherapy*.

MONOCLONAL ANTIBODIES A cancer patient might benefit from injections of mass-produced antibodies against tumor-specific antigens. But mature B cells do not live long enough in culture to mass-produce them. Also, being end cells, B cells cannot reproduce. Some time ago, Cesar Milstein and Georges Kohler showed how to make antibody "factories." They injected an antigen into a mouse, which made antibodies against it. They fused antibody-producing B cells from the mouse with cells extracted from B cell tumors. Some of the descendants of the hybrid cells divided nonstop, and they, too, produced the antibody. Clones of such hybrid cells are now being maintained indefinitely. They make identical copies of antibodies in useful amounts. Their products are known as *monoclonal antibodies*.

Several cancer patients have been inoculated with monoclonal antibodies that are expected to home in on the malignant tumors. In some cases, cell-killing chemicals have been artificially attached to such antibodies. At this writing, the success of clinical trials is limited.

MULTIPLYING THE TUMOR KILLERS Lymphocytes often infiltrate tumors. Researchers have removed them from a tumor and exposed them to an interleukin (lymphokine). The result of the deliberate exposure is a population of tumor-infiltrating lymphocytes with enhanced killing abilities. The *LAK* cells (short for lymphokine-activated killers) appear to be somewhat effective when injected back into the patient.

THERAPEUTIC VACCINES *Therapeutic vaccines* might serve as wake-up calls against elusive tumor cells. One idea is to genetically engineer antigen so that it becomes more obvious to killer lymphocytes. For example, Lynn Spitler has been attempting to develop a prostate-specific antigen that is present on all prostate cancers.

Formation of Antigen-Specific Receptors

The variety of antigens in your surroundings is mind-boggling. Collectively, however, antigen receptors of all T and B cell populations in your body show staggering diversity—enough to recognize about a billion different antigens by one estimate. How does the diversity arise?

For the answer, start with the knowledge that all antigen receptors of a single T or B cell are identical—and all are proteins. For example, a Y-shaped antibody molecule consists of four polypeptide chains bonded together (Figure 40.11). Certain parts of each chain, the *variable* regions, fold in ways that produce grooves and bumps with a certain charge distribution. Only antigen that has complementary grooves, bumps, and charge distribution will be able to bind with them.

Receptor diversity begins with rearrangements in the DNA that codes for such variable regions. For example, take a look at Figure 40.12. As a B cell matures, one of many DNA sequences called the V segments becomes joined at random to one of distant DNA sequences called J segments. The DNA intervening between the joined V and J segments loops out and is excised. In this way, the B cell ends up with a unique DNA sequence.

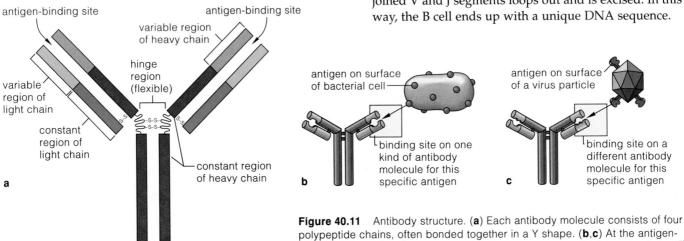

Figure 40.11   Antibody structure. (**a**) Each antibody molecule consists of four polypeptide chains, often bonded together in a Y shape. (**b,c**) At the antigen-binding sites of the molecule, antigen fits into grooves and onto protrusions.

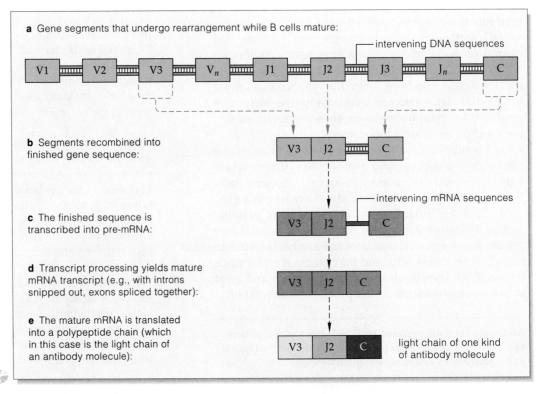

Figure 40.12   Generation of antibody diversity. Antibodies are proteins, and instructions for building proteins are encoded in genes. In the chromosomes with antibody genes, extensive DNA regions contain different versions of segments that code for the variable regions of an antibody molecule.

The different V and J segments shown here are examples. They code for the variable region of a light chain (compare Figure 40.11*a*). As each B cell matures, a recombination event occurs in this region. In this example, any one of the V segments may be joined to any one of the J segments. Afterward, the DNA intervening between them is excised. The new sequence is attached to a C (*Constant*) segment, and this completes a rearranged antibody gene—which also will be present in all of the descendants of the cell.

a Gene segments that undergo rearrangement while B cells mature:

b Segments recombined into finished gene sequence:

c The finished sequence is transcribed into pre-mRNA:

d Transcript processing yields mature mRNA transcript (e.g., with introns snipped out, exons spliced together):

e The mature mRNA is translated into a polypeptide chain (which in this case is the light chain of an antibody molecule):

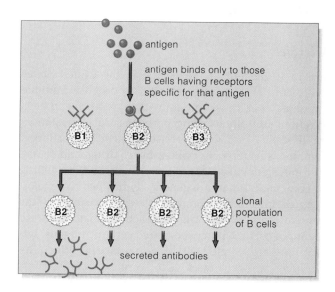

Figure 40.13 Clonal selection of a B cell that produced the specific antibody that can combine with a specific antigen. Only antigen-selected B cells (and T cells) are activated and give rise to a clonal population of immunologically identical cells.

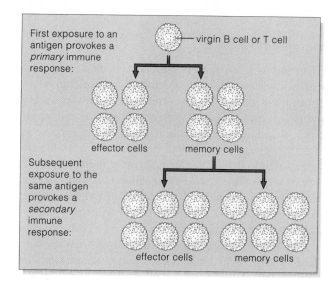

Figure 40.14 Immunological memory. Not all B and T cells are used in a primary immune response to an antigen. A large number continue to circulate as memory cells, which become activated during a secondary immune response.

The same kinds of DNA rearrangements have roles in producing the variable regions of the TCR molecules of maturing T cells.

Some time ago, Mafarlane Burnet developed a *clonal selection* hypothesis that helped point the way to our current view of receptor diversity. He proposed that antigen "chooses" (binds to) one lymphocyte from all the various types in the body, because that lymphocyte has the receptor specific for it. Repeated mitotic cell divisions then give rise to a clone of cells that carry out the response (Figure 40.13).

Immunological Memory

The clonal selection theory explains how an individual can have "immunological memory" of a first encounter with antigen. The term refers to the body's capacity to make a *secondary* immune response to any subsequent encounter with the same type of antigen that provoked the primary response (Figure 40.14).

Memory cells that form during a primary immune response do not engage in that battle. They circulate for years or for decades. Compared to the virgin cells that initiate a primary response, the patrolling battalions consist of far more cells and intercept antigen far sooner. Effector cells form sooner, in greater numbers, so the infection is terminated before the host gets sick. Even greater numbers of memory T and B cells form during a secondary response. Figure 40.15 shows an example of this. In evolutionary terms, the advance preparations against subsequent encounters with a pathogen bestow a survival advantage on the individual.

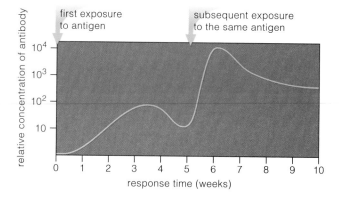

Figure 40.15 Differences in magnitude and duration between a primary and a secondary immune response to the same antigen. In this example, the primary response peaked twenty-four days after it started. The secondary response peaked after only seven days (the span between weeks five and six). Antibody concentration during the secondary response was 100 times greater (10^2 to 10^4).

By recombination of segments, drawn at random from the receptor-encoding regions of DNA, each T cell or B cell receives a gene sequence for one of a billion possible kinds of antigen receptors.

Immunological specificity means the clonal descendants of an antigen-selected cell will react only with the selecting antigen.

Immunological memory refers to the capacity to make a secondary (faster, greater) immune response to a pathogen that caused a primary response in an individual.

Immunization

After reflecting on the chapter introduction, you may already have an understanding that **immunization** refers to various processes that promote increased immunity against specific diseases. With *active* immunization, an antigen-containing preparation known as a **vaccine** is either taken orally or injected into the body, as in Figure 40.16. An initial injection triggers a primary immune response. Later, a subsequent injection (a booster) elicits a secondary response, with the rapid formation of more effector cells and memory cells that can provide long-lasting protection against the disease.

Many vaccines are manufactured from weakened or killed pathogens, as when Sabine polio vaccine is made from weakened polio virus particles. Others are based on inactivated natural toxins, such as the bacterial toxin that causes tetanus. Others are made of harmless viruses, genetically engineered so genes from three or more different viruses are inserted into their DNA or RNA. After vaccination with an engineered virus, the incorporated genes are expressed, antigens are produced, and immunity is established.

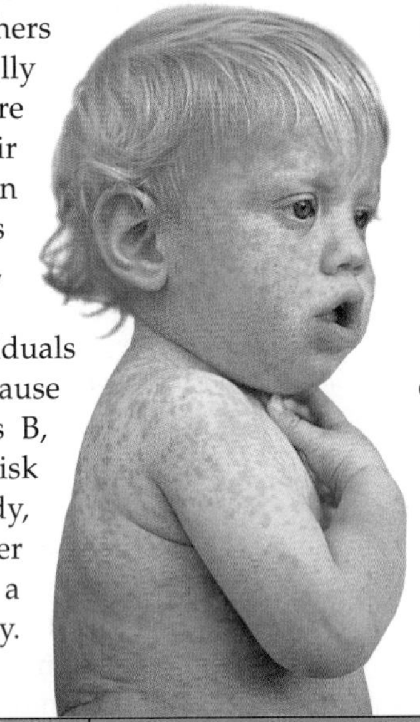

Passive immunization helps individuals already infected with pathogens that cause diphtheria, tetanus, measles, hepatitis B, and some other diseases. A person at risk receives injections of purified antibody, the best source of which is some other individual who already has produced a large amount of the required antibody. The effects do not last long, for the patient's B cells are not producing antibodies. Nevertheless, the injection of antibody molecules often can help counter the immediate attack.

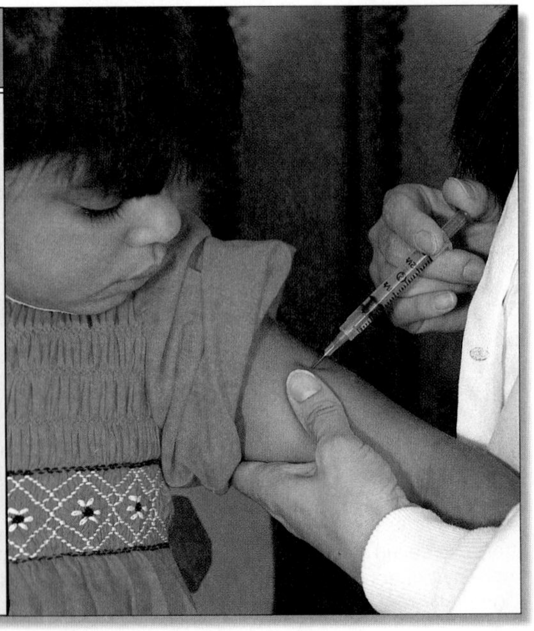

Figure 40.16 From the Centers for Disease Control and Prevention, the 1995 immunization guidelines for children living in the United States. Pediatricians routinely immunize infants and children during office visits. Low-cost or free vaccinations are available at many community clinics and health departments.

Allergies

In 8 to 10 percent of the people in the United States alone, normally harmless substances provoke immune responses that cause inflammation, excessive mucus secretion, and other vexing problems. Such substances are **allergens**, and the response to them is an **allergy**. Common allergens are pollen, many drugs and foods, dust mites, fungal spores, insect venom, and cosmetics.

Some individuals are genetically inclined to develop allergies. Infections, emotional stress, or changes in air temperature also may trigger reactions that otherwise might not occur. Upon exposure to certain antigens, IgE antibodies are secreted and bind to mast cells. When the IgE binds antigen, mast cells secrete prostaglandins, histamine, and other substances that fan inflammation. They stimulate mucus secretion and cause airways to constrict. Stuffed sinuses, labored breathing, a drippy nose, and sneezing are key symptoms of the allergic response in *asthma* and *hay fever* (Figure 40.17).

In a few cases, the inflammatory reactions wash through the body and bring about a life-threatening condition called *anaphylactic shock*. For example, a person who is allergic to wasp or bee venom can die within minutes of one sting. Airways to the lungs constrict massively, and fluid escapes rapidly from dilated, grossly permeable capillaries. Blood pressure plummets and may lead to circulatory collapse.

Antihistamines (anti-inflammatory drugs) often relieve the mild, short-term symptoms of allergies. Over time, a patient might try a desensitization program. First, skin tests identify the offending allergens. Inflammatory responses to some can be

Age	Recommended Vaccines
At birth	Hepatitis B
2 months	Hepatitis B, DPT (diphtheria, whooping cough, tetanus), Hib (*Hemophilus influenzae*)
2–4 months	Hepatitis B
4 months	Polio, DPT, Hib
6 months	DPT, Hib
6–18 months	Hepatitis B, polio
12–15 months	Hib, MMR (measles, mumps, rubella)
12–18 months	DPT
4–6 years	Polio, DPT, MMR (final dose of MMR can be at 11–12 years)
11–12 years	DT (diphtheria, tetanus)

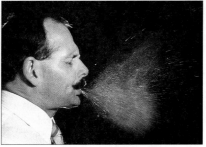

RAGWEED POLLEN . . .   . . . AND SOMETHING IT CAN PROVOKE

Figure 40.17 One of the effects of pollen and other allergens in sensitive people. Allergy sufferers who moved to deserts to escape pollen brought it with them. Half the human population in Tucson, Arizona, is now sensitized to pollen from olive and mulberry trees, planted far and wide in cities and the suburbs.

Figure 40.18 A case of severe combined immunodeficiency, or SCID. Ashanthi DeSilva was born without an immune system. She has a mutated gene for ADA (adenosine deaminase), an enzyme. Without ADA, her cells cannot break down adenosine. As a result, a reaction product accumulates that is toxic to lymphocytes. The disorder's symptoms are caused by infections that cannot be controlled. They include high fever, severe ear and lung infections, diarrhea, and an inability to gain weight.

blocked if the patient can be stimulated to make IgG instead of IgE. Larger doses of specific allergens are administered slowly. Each time, the body makes more circulating IgG and memory cells. The IgG binds with allergen that it encounters and blocks its attachment to IgE, and thereby blocks inflammation.

Autoimmune Disorders

In an **autoimmune response**, the immune system acts against self antigens. Consider *Grave's disorder*, in which the body makes too many thyroid hormone molecules. Feedback mechanisms control the production of the hormones, which affect metabolic rates and the growth and development of many tissues. Affected individuals produce antibodies that bind to receptors on cells that make the hormones. The antibodies do not respond to feedback controls, and they trigger overproduction of thyroid hormones. Disease symptoms typically include elevated metabolic rates, heart fibrillations, excessive sweating, nervousness, and weight loss.

Consider *myasthenia gravis*, a progressive weakening of muscles. Here, antibodies that bind to acetylcholine receptors on skeletal muscle cells cause the disorder.

Finally, consider the chronic inflammation of skeletal joints associated with *rheumatoid arthritis*. Patients are genetically predisposed to this disorder. Macrophages, T cells, and B cells are activated by antigens associated with skeletal joints. Immune responses are made against the body's own collagen molecules and against antibody that has become bound to as-yet-unidentified antigen. Complement activation and inflammation cause more damage in tissues of skeletal joints. So do skewed repair mechanisms. Eventually, the joints fill with synovial membrane cells and become immobilized.

Deficient Immune Responses

When the body has inadequate numbers of functioning lymphocytes, immune responses are not effective. Such

Ashanthi's parents consented to the first federally approved gene therapy for humans. Researchers used genetic engineering methods to splice the ADA gene into the genetic material of a harmless virus. They used the modified virus rather like a hypodermic needle. They allowed it to deliver copies of the "good" gene into Ashanthi's bone marrow cells. Some of the cells incorporated the gene in their DNA and started to synthesize the missing enzyme. At this writing, Ashanthi is now in her teens. Like other ADA-deficient patients who have started treatment, she is doing well.

In other cases, researchers enlist bone marrow stem cells from blood in the umbilical cord of affected newborns. (The cord, which connects the fetus to the placenta during pregnancy, is discarded after childbirth.) They expose the cells to viruses that deliver copies of the ADA gene into them and to factors that stimulate mitotic cell division and growth. The cells are reinserted into the newborns.

severe combined immunodeficiencies (SCIDs) result from heritable disorders as well as from various assaults on the body by outside agents. Deficient or nonexistent immune responses make the person highly vulnerable to infections that are not life threatening to the general population. Figure 40.18 describes one of the heritable disorders. Another is the acquired immunodeficiency syndrome (AIDS). The next section describes how HIV, the virus that causes AIDS, replicates inside certain lymphocytes and destroys the body's capacity to fight infections. We also return to this topic in Section 45.14.

Immunization programs boost immunity to specific diseases.

Certain heritable disorders or attacks by certain pathogens and other agents can result in misdirected, compromised, or nonexistent immunity.

40.11 AIDS—THE IMMUNE SYSTEM COMPROMISED

CHARACTERISTICS OF AIDS   AIDS is a constellation of disorders that follow an infection by a pathogen called the human immunodeficiency virus, or HIV. The virus cripples the immune system, so that the body becomes highly susceptible to usually harmless infections and some otherwise rare forms of cancer.

At this writing, researchers have not developed any vaccine that will work against the known forms of the virus (HIV-1 and HIV-2). Also at this writing, *there is no cure for those already infected.*

By current estimates, more than 1 million Americans are infected. Worldwide, an estimated 22.6 million people are infected; 6 million are already dead. By the turn of the century, the number of infected people may be as high as 60–70 million.

At first an infected person might appear to be in good health, suffering no more than a bout of "the flu." Then he or she starts displaying symptoms that foreshadow AIDS. Typically, symptoms include persistent weight loss, fever, fatigue, bed-drenching night sweats, and many enlarged lymph nodes. In time, diseases resulting from certain opportunistic infections are signs of AIDS. The diseases are rare in the population at large. They include yeast infections of the mouth, esophagus, vagina, and elsewhere, as well as a form of pneumonia caused by *Pneumocystis carinii.* Spots resembling bruises may appear, especially on legs and feet. These are signs of Kaposi's sarcoma, a form of cancer that develops from endothelial cells of blood vessels. The immune system cannot control the infections or cancers, which end up killing the person.

HOW HIV REPLICATES   HIV infects antigen-presenting macrophages and helper T cells. (Immunologists also call helper T cells the CD4 lymphocytes.) HIV is a retrovirus.

Each virus particle has an outermost lipid envelope, which is a bit of plasma membrane that surrounded the particle when it departed from an infected cell. Spiking outward from the envelope are HIV proteins that had become inserted into it. Beneath the lipid envelope, two protein coats—the core proteins—surround two strands of RNA and several copies of reverse transcriptase, an enzyme (compare Sections 16.4 and 22.8).

Once inside a host cell, the viral enzyme uses the RNA as a template for making DNA, which then is inserted into a host chromosome (Figure 40.19). Transcription yields copies of viral RNA. Some transcripts are translated into viral proteins. Others become enclosed in the proteins when new virus particles are put together, and they will function as the hereditary material. The particles are released by budding from the host cell's plasma membrane to start a new round of infection (Figure 40.20). During each round, more macrophages, antigen-presenting cells, and helper T cells are impaired or killed.

The viral genes inserted into the DNA of some host cells remain inactive. They may be activated during a later round of infection.

A TITANIC STRUGGLE BEGINS   Infection marks the onset of a titanic battle between the enemy and the host's immune system. B cells synthesize antibodies in response to HIV antigenic proteins. These are the antibodies that are the basis of diagnostic tests for identifying HIV infection. Armies of helper T cells and cytotoxic T cells also form. However, HIV infects an estimated 2 billion helper T cells and produces 100 million to 1 billion virus particles per day during certain phases of the infection. Every two days, the immune system destroys about half of the virus particles and replaces half of the helper T

Figure 40.19   Life cycle of HIV, one of the retroviruses.

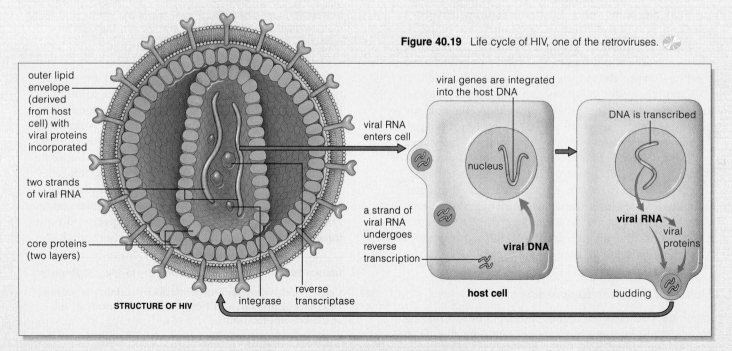

outer lipid envelope (derived from host cell) with viral proteins incorporated

two strands of viral RNA

core proteins (two layers)

integrase   reverse transcriptase

STRUCTURE OF HIV

viral RNA enters cell

a strand of viral RNA undergoes reverse transcription

viral genes are integrated into the host DNA

nucleus

viral DNA

host cell

DNA is transcribed

viral RNA

viral proteins

budding

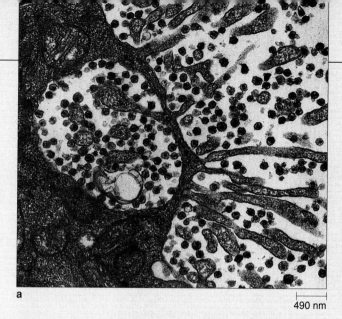

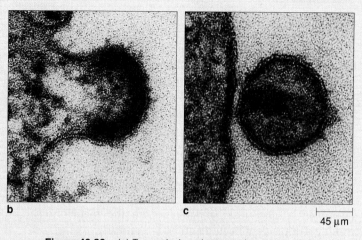

a

490 nm

b

c

45 µm

Figure 40.20 (a) Transmission electron micrograph of HIV particles (black specks) escaping from an infected cell. (b,c) A virus particle budding from the host cell's plasma membrane.

cells lost in the battle. Immense reservoirs of HIV and masses of infected T cells accumulate in lymph nodes. As the battle proceeds, the number of virus particles in the general circulation rises. Gradually, the numbers tilt. The body produces fewer and fewer helper T cells to replace the ones it lost. Although it may take a decade or more, the erosion of the helper T cell count inevitably causes the body to lose its capacity to mount effective immune responses.

Some viruses, including the measles virus, produce far more virus particles in a given day, but the immune system usually wins out. Other viruses, including herpes viruses, can lurk in the body for a lifetime, but the immune system keeps them in check. With HIV, the immune system loses the struggle, then infections and tumors kill the person.

HOW HIV IS TRANSMITTED Like any other virus that infects humans, HIV requires a medium that allows it to leave one host, survive in the environment into which it is released, then enter another host.

HIV is transmitted when some body fluid of an infected person enters the tissues of another person. In the United States, transmission initially occurred most often between males who engaged in homosexual activities, especially anal intercourse. Since then, it has spread among intravenous drug abusers who share blood-contaminated syringes and needles. It also has spread in the heterosexual population, increasingly by vaginal intercourse.

HIV has traveled from infected mothers to offspring during pregnancy, birth, and breast-feeding. Before 1985, contaminated blood supplies accounted for some AIDS cases before health care providers implemented screening. Tissue transplants have caused four infections in 1991. In several developing countries, health care providers have spread HIV by way of contaminated transfusions and reuse of unsterile needles and syringes.

The molecular structure of HIV does not remain stable outside the human body, which is why it must be directly transmitted from one host to another. At this writing, there is no evidence that the virus can be effectively transmitted by way of food, air, water, casual contact, or insect bites. The virus *has* been isolated from human blood, semen,

vaginal secretions, saliva, tears, breast milk, amniotic fluid, urine, and cerebrospinal fluid. It is probably present in other fluids, secretions, and excretions. However, only infected blood, semen, vaginal secretions, and breast milk contain the virus in concentrations that apparently are high enough for successful transmission.

REGARDING PREVENTION AND TREATMENT Developing effective drugs or vaccines against HIV is a formidable challenge. High mutation rates characterize the viral genome, owing to HIV's replication mechanisms and the staggering number of replications in an infected person. Inevitably, natural selection operates in patients who are undergoing drug therapies; it favors drug resistance and fans the evolution of drug-resistant HIV populations. Chemical "cocktails" have had some effect in slowing replication. Among the current drugs of choice are protease inhibitors, AZT (azidothymidine), and ddI (dideoxyinosine). The drugs cannot *cure* infected people, however, for they cannot eliminate the HIV genes that have become incorporated into their DNA.

High mutation rates have another worrisome outcome: They give rise to variations in HIV antigens. The variation makes it difficult for researchers to select effective antigens for vaccines. Another problem is the scarcity of HIV strains with poor cell-killing abilities. Such strains are central to producing antigen that can stimulate the formation of cytotoxic T cell armies—the best protection against HIV. However, over a decade ago, many people in Australia received blood transfusions from a donor whose infection had not been diagnosed. At this writing, neither they nor the donor show any immunodeficiency. And they all carry HIV with a similar defect in the same gene (the nef gene).

In short, *until researchers develop effective vaccines and treatments, checking the spread of HIV depends absolutely on persuading people to avoid or modify social behaviors that put them at risk.* We return to this topic in Section 45.14.

SUMMARY

1. Vertebrates fend off many pathogens with physical and chemical barriers at body surfaces. They also are protected by the nonspecific and specific responses of the white blood cells listed in Table 40.3.

a. Nonspecific responses to irritation or damage of tissues include inflammation and involve organs with phagocytic functions, such as the spleen and liver.

b. Immune responses are made against specific pathogens, foreign cells, or abnormal body cells.

2. Intact skin and mucous membranes that line body surfaces are physical barriers to infection. Glandular secretions (as in tears, saliva, and gastric fluid) are examples of chemical barriers. So are the metabolic products of resident bacteria on body surfaces.

3. An inflammatory response develops in body tissues that have become damaged, as by infection.

a. The response starts with arteriole vasoconstriction that increases blood flow to the tissue, which reddens and becomes warmer as a result. Capillary permeability increases, and local edema causes swelling and pain.

b. Pathogens as well as dead or damaged body cells release the substances that trigger increased permeability of capillaries. White blood cells leave the blood and enter the tissue, where they release a number of chemical mediators and engulf invaders. Plasma proteins also enter the tissue. Complement proteins bind pathogens and induce their lysis, and they attract phagocytes. Blood-clotting proteins wall off the damaged tissue.

4. An immune response has these characteristics:

a. It shows specificity, meaning it is directed against antigen. Each antigen is a molecular configuration that lymphocytes recognize as foreign (nonself).

b. Each response also shows memory, meaning that a subsequent encounter with the same antigen triggers a more rapid, secondary response, of greater magnitude.

c. An immune response normally is not made against the body's own self-marker proteins.

5. Macrophages and other antigen-presenting cells process and bind the fragments of antigen to their own MHC markers. Lymphocytes have receptors that can bind to the displayed antigen-MHC complexes. Binding is the start signal for an immune response.

6. After the recognition of antigen, an immune response proceeds through repeated cell divisions that form clones of B and T lymphocytes, which differentiate into subpopulations of effector and memory cells. Chemical mediators such as interleukins secreted by white blood cells drive the responses. Effector helper T cells, cytotoxic T cells, effector B cells, and antibodies act at once. The memory cells are set aside for secondary responses.

7. T cells arise in bone marrow but continue to develop in the thymus, where they acquire TCRs. These T-cell receptors recognize and bind antigen-MHC complexes on antigen-presenting cells. B cells arise in bone marrow. As they mature, they synthesize antigen receptors (that is, antibodies) that become positioned at their surface.

8. Effector cytotoxic T cells can directly destroy virus-infected cells, tumor cells, and cells of tissue or organ transplants. Effector B cells (plasma cells) produce and secrete great numbers of antibodies that freely circulate.

9. Antibodies are protein molecules, often Y-shaped, each with binding sites for one kind of antigen. Only B cells produce them. When antibody binds to antigen, toxins are neutralized, pathogens are tagged for destruction, or attachment of pathogens to body cells is prevented.

10. In active immunization, vaccines provoke immune responses, with production of effector and memory cells. In passive immunization, injections of purified antibodies help the individual through an infection.

11. Allergic reactions are immune responses to some generally harmless substance. Autoimmune responses are misguided attacks triggered by configurations on the body's own cells. Immunodeficiency is a weakened or nonexistent capacity to mount an immune response.

Table 40.3	Summary of Major White Blood Cells and Their Roles in Defense
Cell Type	Main Characteristics
MACROPHAGE	Phagocyte; acts in nonspecific and specific responses; presents antigen to T cells; cleans up and helps repair tissue damage
NEUTROPHIL	Fast-acting phagocyte; takes part in inflammation, not in sustained responses; most effective against bacteria
EOSINOPHIL	Secretes enzymes that attack certain parasitic worms
BASOPHIL AND MAST CELL	Secrete histamines and other substances that act on small blood vessels, thereby producing inflammation; also contribute to allergic reactions
LYMPHOCYTES:	(All take part in most immune responses; following antigen recognition, all form clonal populations of effector cells and memory cells.)
B cell	Effectors secrete four types of antibodies (IgA, IgE, IgG, and IgM) that protect the host in specialized ways
Helper T cell	Effectors secrete interleukins that stimulate rapid divisions and differentiation of both B cells and T cells
Cytotoxic T cell	Effectors kill infected cells, tumor cells, and foreign cells by a touch-kill mechanism
NATURAL KILLER (NK) CELLS	Cytotoxic cell of undetermined affiliation; kills virus-infected cells and tumor cells by a touch-kill mechanism

Review Questions

1. While jogging barefoot along a seashore, some of your toes accidentally land on a jellyfish. Soon the toes are swollen, red, and warm to the touch. Describe the events that result in these signs of inflammation. *40.3*

2. Distinguish between:
   a. neutrophil and macrophage *40.3*
   b. cytotoxic T cell and natural killer cell *40.4, 40.6*
   c. effector cell and memory cell *40.4*
   d. antigen and antibody *40.4*

3. Describe how a macrophage becomes an antigen-presenting cell. *40.4*

4. Why is a vaccine to control AIDS so elusive? *40.11*

Self-Quiz *(Answers in Appendix IV)*

1. _____ are barriers to pathogens at body surfaces.
   a. Intact skin, mucous membranes   d. Urine flow
   b. Tears, saliva, gastric fluid   e. All of the above
   c. Resident bacteria

2. Macrophages are derived from _____ .
   a. basophils   c. monocytes
   b. neutrophils   d. eosinophils

3. Activated complement functions in defense by _____ .
   a. neutralizing toxins   c. promoting inflammation
   b. enhancing resident   d. forming holes in memory
   bacteria   lymphocyte membranes

4. _____ are certain molecules that lymphocytes recognize as foreign and that elicit an immune response.
   a. Interleukins   d. Antigens
   b. Antibodies   e. Histamines
   c. Immunoglobulins

5. The immunoglobulins _____ increase antimicrobial activity in mucus-coated surfaces of some organ systems.
   a. IgA   b. IgE   c. IgG   d. IgM   e. IgD

6. Antibody-mediated responses work best against _____ .
   a. intracellular pathogens   d. both b and c
   b. extracellular pathogens   e. all of the above
   c. extracellular toxins

7. The most important antigens are _____ .
   a. nucleotides   c. steroids
   b. triglycerides   d. proteins

8. _____ would be a target of an effector cytotoxic T cell.
   a. Extracellular virus particles in blood
   b. A virus-infected body cell or tumor cell
   c. Parasitic flukes in the liver
   d. Bacterial cells in pus
   e. Pollen grains in nasal mucus

9. Development of a secondary immune response is based on populations of _____ .
   a. memory cells   d. effector cytotoxic T cells
   b. circulating antibodies   e. mast cells
   c. effector B cells

10. Match the immunity concepts.
    ____ inflammation   a. neutrophil
    ____ antibody secretion   b. effector B cell
    ____ a phagocyte   c. nonspecific response
    ____ immunological   d. deliberately provoking
    memory   an immune response
    ____ vaccination   e. basis of secondary response
    ____ allergy   f. nonprotective immune
                               response

Critical Thinking

1. As described in the chapter introduction, Edward Jenner lucked out. He performed a potentially harmful experiment on a boy who managed to survive it. What would happen if a would-be Jenner tried to do the same thing today?

2. Rob's bumper sticker reads, "Have you thanked your resident bacteria today?" Explain why he appreciates the bacteria that normally reside on the body's skin and mucous membranes.

3. Researchers are attempting to develop a way to get the immune system to accept foreign tissue as "self." Speculate on some of the clinical applications of such a development.

4. Before each flu season, you get an influenza vaccination. This year you come down with "the flu" anyway. What do you suppose happened? (There are at least three explanations.)

5. Infection by *Ebola* virus results in a hemorrhagic fever with a 90 percent mortality rate (Section 22.9). A patient received a blood serum transfusion from another who survived the disease. Explain why the transfusion might increase chances of survival.

6. Ellen developed *chicken pox* when she was in kindergarten. Later in life, when her children developed chicken pox, she remained healthy even though she was exposed to countless virus particles daily. Explain why.

7. Quickly review Section 33.7 on homeostasis. Then write a short essay on how the immune response contributes to stability in the internal environment.

Selected Key Terms

allergen *40.10*	immune system *40.4*
allergy *40.10*	immunization *40.10*
antibody *40.4*	immunoglobulin (Ig) *40.7*
antigen *40.4*	inflammation, acute *40.3*
antigen-MHC complex *40.4*	lysis *40.2*
antigen-presenting cell *40.4*	lysozyme *40.1*
apoptosis *40.6*	macrophage *40.3*
autoimmune response *40.10*	mast cell *40.3*
B lymphocyte (B cell) *40.4*	MHC marker *40.4*
basophil *40.3*	natural killer (NK) cell *40.6*
complement system *40.2*	neutrophil *40.3*
cytotoxic T cell *40.4*	pathogen *40.1*
eosinophil *40.3*	TCR *40.6*
fever *40.3*	T lymphocyte (T cell) *40.4*
helper T cell	vaccination *CI*
(CD4 lymphocyte) *40.4*	vaccine *40.10*
histamine *40.3*	

Readings

Edelson, R., and J. Fink. June 1985. "The Immunologic Function of Skin." *Scientific American* 252(6): 46–53.

Golub, E., and D. Green. 1991. *Immunology: A Synthesis.* Second edition. Sunderland, Massachusetts: Sinauer.

Janeway, C., Jr. September 1993. "How the Immune System Recognizes Invaders." *Scientific American* 72–79.

Nowak, M., and A. McMichael. August 1995. "How HIV Defeats the Immune System." *Scientific American* 273(2): 58–65.

Tizard, I. 1995. *Immunology: An Introduction.* Fourth edition. Philadelphia: Saunders.

Web Site See *http://www.wadsworth.com/biology* for practice quiz questions, hypercontents, BioUpdates, and critical thinking. The Wadsworth Biology Resource Center provides a wealth of information fully organized and integrated by chapter.

41

RESPIRATION

Conquering Chomolungma

To experienced climbers, Chomolungma may be the ultimate challenge (Figure 41.1). The summit of this Himalayan mountain, also known as Everest, is 9,700 meters (29,128 feet) above sea level. It is the highest place on Earth. Iced-over vertical rock, driving winds, blinding blizzards, and heart-stopping avalanches await the challengers. So does the extreme danger that oxygen-poor air poses to the brain.

Most of us live at low elevations. Of the air we breathe, one molecule in five is oxygen. When we travel 2,400 meters (about 8,000 feet) or more above sea level, the Earth's gravitational pull is weaker, gas molecules spread out more, and the breathing game changes. We face *hypoxia*, or cellular oxygen deficiency. Sensing the deficiency, the brain makes us hyperventilate, or breathe much faster and more deeply than usual. Above 3,300 meters (10,000 feet), hyperventilating can be worrisome. It may lead to significant ion imbalances in cerebrospinal fluid, which can trigger heart palpitations, shortness of breath, headaches, nausea, and vomiting. These are strong clues that our cells are, in a manner of speaking, screaming for oxygen.

Living for months at high elevations helps climbers adapt physiologically to the thinner air. For example, mechanisms kick in that increase the red blood cell count, hence the body's oxygen-carrying capacity. For professional climbers, Chomolungma's base camp is 6,300 meters (19,000 feet) above sea level. At 7,000 meters, oxygen and other gases are extremely diffuse. The oxygen scarcity and low air pressure combine to

Figure 41.1 A climber inching toward the summit of Chomolungma, where oxygen is brutally scarce.

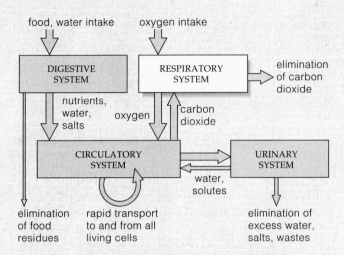

food, water intake   oxygen intake

DIGESTIVE SYSTEM

RESPIRATORY SYSTEM

elimination of carbon dioxide

nutrients, water, salts

oxygen

carbon dioxide

CIRCULATORY SYSTEM

URINARY SYSTEM

water, solutes

elimination of food residues

rapid transport to and from all living cells

elimination of excess water, salts, wastes

Figure 41.2 Interactions between the respiratory system and other organ systems in complex animals.

make blood capillaries leaky. More plasma escapes through the expanded gaps between endothelial cells of the capillary walls. In the brain and lungs, tissues swell with excess fluid. If the edema is not reversed, climbers will become comatose and die. When stricken, they can inhale bottled oxygen. Rescuers can zip them inside airtight bags, then use a device that pumps in oxygen and removes carbon dioxide until the "air" inside the bag more closely approximates the air at 2,400 meters.

Few of us will ever find ourselves near the peak of Chomolungma, pushing our reliance on oxygen to the limits. Here in the lowlands, disease, smoking, and other environmental insults push it in more ordinary ways, although the risks can be just as great.

The point is this: *Each animal has a body plan that is adapted to the oxygen levels of a particular habitat*. One way or another, by a physiological process known as **respiration**, the body plan allows oxygen to move into the internal environment and carbon dioxide to move out. Why is this important? Animals, remember, have great energy demands. Their cells demand a great deal of oxygen—especially for aerobic respiration, the main metabolic pathway that produces considerable energy and carbon dioxide wastes.

This chapter samples a few **respiratory systems**, which function in the exchange of gases between the body and the environment. Together with other organ systems, they also contribute to homeostasis—that is, to maintaining internal operating conditions for all of the body's living cells (Figure 41.2).

KEY CONCEPTS

1. Of all organisms, multicelled animals require the most energy to drive their metabolic activities. At the cellular level, the energy comes mainly from aerobic respiration, an ATP-producing metabolic pathway that requires oxygen and produces carbon dioxide wastes.

2. By a physiological process called respiration, animals move oxygen into their internal environment and give up carbon dioxide to the external environment.

3. Oxygen diffuses into the animal body as a result of a pressure gradient. The pressure of this gas is higher in air than it is in metabolically active tissues, where cells rapidly use oxygen. Carbon dioxide follows its own gradient, in the opposite direction. Its pressure is higher in tissues, where it is a by-product of metabolism, than it is in the air.

4. In most respiratory systems, oxygen and carbon dioxide diffuse across a respiratory surface, such as the thin, moist epithelium of the human lungs. The blood flowing through the body's circulatory system picks up oxygen and gives up carbon dioxide at this respiratory surface.

5. Gas exchange is most efficient when the rate of air flow matches the rate of blood flow. The nervous system brings the rates into balance by controlling the rhythmic pattern and magnitude of breathing.

The Basis of Gas Exchange

A concentration gradient, recall, is a difference in the number of molecules of a substance between two regions. Like all substances, oxygen or carbon dioxide tends to diffuse down its concentration gradient, or show a net outward movement from the region where its molecules are colliding more frequently (Section 5.3). Respiration is based on the tendency of both gases to diffuse down their respective concentration gradients—or, as we say for gases, down **pressure gradients** that occur between the internal and external environments.

The gases do not exert the *same* pressure. Pump air into a flat tire near a beach in San Diego or Miami, and you fill it with about 78 percent nitrogen, 21 percent oxygen, 0.04 percent carbon dioxide, and 0.96 percent other gases. This is true of dry air anywhere at sea level. At sea level, atmospheric pressure is about 760 mm Hg, as measured by a mercury barometer of the sort shown in Figure 41.3. Oxygen exerts only part of that total pressure on the tire wall, and its "partial" pressure is greater than that of carbon dioxide. Said another way, the **partial pressure** of oxygen—its contribution to the total atmospheric pressure—is 760 × 21/100, or about 160 mm Hg. When similarly measured, carbon dioxide's partial pressure is about 0.3 mm Hg.

Gases enter and leave the animal body by crossing a **respiratory surface**. A respiratory surface is a thin layer of epithelium or some other tissue. It must be kept moist at all times, for gaseous molecules cannot diffuse across it unless they are dissolved in fluid. What dictates the number of gas molecules moving across a respiratory surface in a given time? According to **Fick's law**, the more extensive the surface area and the larger the partial pressure gradient, the faster will be the diffusion rate.

Factors That Influence Gas Exchange

SURFACE-TO-VOLUME RATIO All animal body plans promote favorable rates of inward diffusion of oxygen and outward diffusion of carbon dioxide. For example, animals with no respiratory organs are tiny, tubelike, or flattened, and gases diffuse directly across the body surface. These body plans meet a constraint imposed by the surface-to-volume ratio, as described in Section 4.1. To get a sense of the ratio's effects, imagine a flatworm growing in all directions, like an inflating balloon. The worm's surface area does not increase at the same rate as its volume. Once its girth exceeds a single millimeter, the diffusion distance between the body's surface and internal cells will be so great that the flatworm will die.

VENTILATION Large-bodied animals that are highly active have great demands for gas exchange, more than

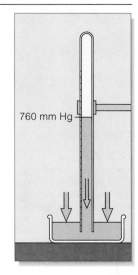

Figure 41.3 Atmospheric pressure as measured with a device called a mercury barometer. Part of the device is a glass tube in which the height of a column of mercury (Hg) can increase or decrease, depending upon air pressure outside the device. At sea level, the mercury rises to about 760 millimeters (29.91 inches) from the base of the tube. At this level, the pressure that the column of mercury exerts inside the tube is equal to the atmospheric pressure on the outside.

760 mm Hg

diffusion alone would be able to satisfy. A variety of adaptations make the exchange rate more efficient. For example, above their gills (respiratory organs), many fishes have tissue flaps that stir the surrounding water by moving back and forth. The stirred water puts more dissolved oxygen closer to the gills, and it carries more carbon dioxide away from them. Among vertebrates, a circulatory system rapidly transports oxygen to cells and carbon dioxide to gills or lungs for disposal. As a final example, you breathe to ventilate your lungs.

TRANSPORT PIGMENTS Rates of gas exchange get a boost with respiratory pigments, mainly **hemoglobin**, that help maintain the steep pressure gradients across a respiratory surface. For example, at the respiratory surfaces inside human lungs, the oxygen concentration is high, and each hemoglobin molecule in blood weakly binds as many as four oxygen molecules. (Here you may wish to refer to Section 3.7.) Then the circulatory system rapidly transports the hemoglobin from the respiratory surface. In oxygen-poor tissues, the oxygen follows its gradient and diffuses out of hemoglobin. By its oxygen-transporting activity, then, hemoglobin helps maintain a pressure gradient that entices oxygen into the lungs.

By a process called respiration, animals take in oxygen for aerobic respiration, an ATP-producing pathway, and remove the pathway's carbon dioxide wastes.

Oxygen and carbon dioxide enter and leave the body by diffusing across a moist respiratory surface. Like other atmospheric gases, they tend to move down their respective pressure gradients. Each gas exerts only part of the total pressure across a respiratory surface.

Gas exchange depends on steep partial pressure gradients between the outside and inside of the animal body. The greater the area of the respiratory surface and the larger the partial pressure gradient, the faster diffusion will proceed.

Flatworms, earthworms, and many other invertebrates are not massive, and their life-styles do not depend on high metabolic rates (Figure 41.4*a*). Demands for gas exchange are simply met by **integumentary exchange**, in which gases diffuse directly across the body's surface covering (integument). This mode of respiration works as long as the surface stays moist; invertebrates that rely solely on it are restricted to aquatic or damp habitats. Integumentary exchange supplements other modes of respiration in amphibians and some other large animals.

Many invertebrates of aquatic habitats have moist, thin-walled respiratory organs called **gills**. Extensively folded gill walls have an increased respiratory surface area that enhances the exchange rates between blood or some other body fluid and the surroundings. Figure 41.4*b* shows the folded gill of a sea hare (*Aplysia*). By supplementing integumentary exchange, the gill helps provide adequate oxygen for this rather large mollusk; some sea hares are 40 centimeters (nearly 16 inches) long.

Invertebrates of dry habitats have small, thick, or hardened integuments that are not well endowed with blood vessels. Although their integuments do conserve

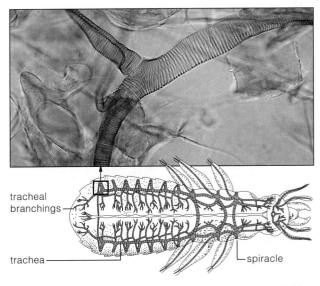

tracheal branchings

trachea

spiracle

Figure 41.5   General plan of insect tracheal systems. Chitin rings reinforce many of the branching tubes of the system.

precious water, these are not good respiratory surfaces. Such animals have *internal* respiratory surfaces. For example, most spiders have book lungs, or respiratory organs with thin, folded walls that resemble book pages (Section 26.16). Like most insects and the millipedes and centipedes, some spiders rely on a system of internal tubes that function in **tracheal respiration**.

Consider the tracheal system of an insect, as in Figure 41.5. Small openings perforate the insect integument. Each opening, called a spiracle, is the start of a tube that branches inside the body. The last branchings dead-end at a fluid-filled tip, where gases diffuse directly into tissues. The tips of the tubes are especially profuse in muscle and other tissues with high oxygen demands.

We find hemoglobin or other respiratory pigments in many invertebrates, although they are rare in insects.

The respiratory pigments increase the capacity of body fluids to transport oxygen, just like they do in vertebrates. In the species with a well-developed head, oxygenated blood tends to circulate first through the head end, then through the rest of the body.

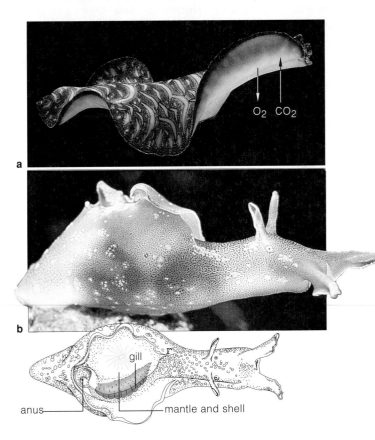

O₂   CO₂

a

b

gill

anus   mantle and shell

Figure 41.4   Invertebrates of aquatic habitats. (**a**) A flatworm, small enough to get along well without an oxygen-transporting circulatory system. Dissolved oxygen in such habitats reaches individual cells by diffusing across the body surface. (**b**) Gill of a sea hare (*Aplysia*), one of the gastropods.

Flatworms and some other invertebrates that live in aquatic or moist habitats and that do not have a massive body rely on integumentary exchange, in which oxygen and carbon dioxide diffuse directly across the body surface.

Most marine invertebrates and many freshwater types have gills of one sort or another. These respiratory organs have moist, thin, and often highly folded walls.

Most insects, millipedes, centipedes, and some spiders use tracheal respiration. Gases flow through open-ended tubes that start at the body surface and end directly in tissues.

Gills of Fishes and Amphibians

Gills are the respiratory organs of many vertebrates. A few kinds of fish larvae and a few amphibians have *external* gills that project into the surrounding water. Adult fishes have a pair of *internal* gills. These are rows of slits or pockets at the back of the mouth that extend to the body surface, as in Figure 41.6a. Whatever their form, all gills have walls of moist, thin, vascularized epithelium.

In fishes, water flows into the mouth and pharynx, then over arrays of filaments in the gills (Figure 41.6b). Blood vessels thread through the respiratory surfaces in each filament. First the water flows past a vessel that is leading to the rest of the body. The blood inside has less oxygen than the water does, so oxygen diffuses into the blood. The same volume of water flows over a vessel leading into the gills. The water already gave up some oxygen, but it still has more than the blood inside the vessel does, so more oxygen diffuses into the filament. Movement of two fluids in opposing directions is called **countercurrent flow**. By this mechanism, a fish extracts about 80 to 90 percent of the dissolved oxygen flowing past. That is more than the fish would get from a one-way flow mechanism, at far less energy cost.

Lungs

Some fishes and all amphibians, birds, and mammals have a pair of **lungs**, or internal respiratory surfaces in the shape of a cavity or sac. Lungs originated in some lineages of fishes more than 450 million years ago, as pouches off the anterior part of the gut wall (Figure 41.7). They evolved rapidly by way of natural selection, probably because lungs afforded fine advantages. They increased the surface area for gas exchange in oxygen-poor habitats, and they worked better than

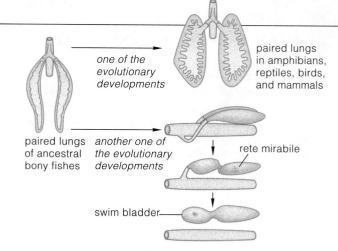

Figure 41.7 Evolution of vertebrate lungs and swim bladders. The esophagus (a tube leading to the stomach) is coded *gold* and respiratory tissues *pink*. Lungs originated as pockets off the anterior part of the gut. They increased the surface area for gas exchange in oxygen-poor habitats. In some fishes, lung sacs became swim bladders. By adjusting the gas volume in these buoyancy devices, a fish can hold its position at different depths.

Trout and other less specialized fishes have a duct between the esophagus and the swim bladder. They surface and gulp air to replenish air in the bladder. Most bony fishes have no such duct. Gases in blood diffuse into their swim bladder, which has a *rete mirabile:* a dense mesh of arteries and veins running in opposing directions. Countercurrent flow through these vessels greatly increases gas concentrations in the bladder. Another region of the bladder promotes reabsorption of gases by body tissues.

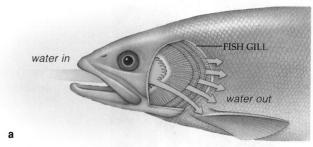

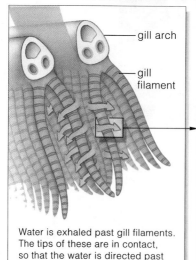

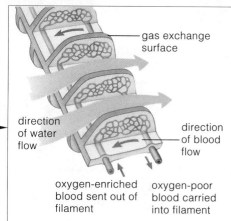

Figure 41.6 Example of a fish gill. (**a**) This is one of a pair of gills. Each gill is located under a bony lid, which has been removed for this sketch. (**b**) Filaments in a fish gill contain vascularized respiratory surfaces. The tips of neighboring filaments touch, so water flowing over them is directed past gas exchange surfaces before being expelled. (**c**) One blood vessel from other body tissues delivers oxygen-poor blood into a filament; another carries oxygenated blood away from it. Blood flowing from one vessel to the other runs counter to the direction of the water flowing over the gas exchange surfaces. Countercurrent flow favors the movement of oxygen (down its partial pressure gradient) into the blood.

Water is exhaled past gill filaments. The tips of these are in contact, so that the water is directed past the gas exchange surface.

A blood vessel carries oxygen-poor blood into each filament. Another carries oxygenated blood out. Blood flowing from one vessel to the other runs counter to the direction of water flowing over gas exchange surfaces.

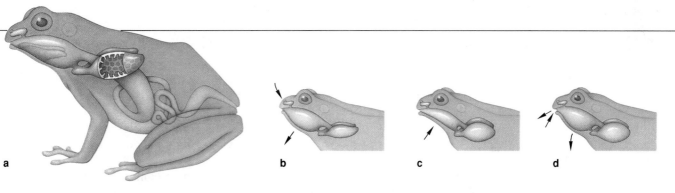

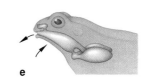

Figure 41.8 (**a**) Respiratory system of frogs, in which the lungs fills by forcing air into them. (**b**) First, the frog lowers the floor of its mouth and inhales air through its two nostrils. (**c**) Then it closes the nostrils, opens the glottis, and elevates the floor of its mouth, so the air has nowhere to go except into the lungs. (**d**) The frog rhythmically ventilates its mouth for a while, which helps move more oxygen into the mouth and more carbon dioxide out of it. (**e**) Finally, the frog contracts muscles in the body wall outside the lungs, the lungs recoil elastically, and air is forced out.

gills could during the move onto dry land. Gills stick together and cannot function at all unless water flows through them and keeps them moist.

Lungfishes of oxygen-poor habitats still have gills. They also use tiny lungs as backups. Amphibians never completed the transition to land. Their skin serves as a respiratory surface for integumentary exchange, in the salamanders especially. Frogs and toads rely more on

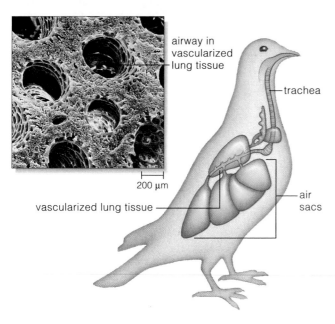

Figure 41.9 Bird respiratory system. Typically, five air sacs are attached to each of two small, inelastic lungs. When a bird inhales, air is drawn into air sacs through tubes, open at both ends, that thread through the vascularized lung tissue. This tissue is the respiratory surface, where gases are exchanged.

When the bird exhales, air is forced out of the sacs, through the small tubes, and out of the trachea. Thus, air is not merely drawn into the bird lungs. Air is continuously drawn through them and across respiratory surfaces. This unique ventilating system supports the high metabolic rates that birds require for flight and other energy-intensive activities.

small lungs for oxygen uptake, but most of the carbon dioxide diffuses across the skin. Frogs also are heavy-duty breathers; they *force* air into the lungs, then empty them by contracting body wall muscles (Figure 41.8).

Frogs can use their lungs for sound production. So can all mammals except whales. Sound originates near the entrance to a larynx, an airway leading to the lungs. Here, part of a mucous membrane folds perpendicularly to the airway. These are **vocal cords**; the gap between them is the **glottis**. Frogs produce sounds by forcing air back and forth through the glottis, between the lungs and paired pouches on the floor of the mouth. Air flow causes the cords to vibrate. Like other vertebrates, frogs produce different sounds by controlling the vibrations.

Paired lungs are the dominant means of respiration in reptiles, birds, and mammals. Breathing moves air by bulk flow into and out of the lungs, where blood capillaries wrap lacily around the respiratory surface. Oxygen and carbon dioxide have steep concentration gradients in the lungs, so they diffuse rapidly across the respiratory surface. Oxygen enters the capillaries and is circulated quickly through the body. In regions where its concentration is low, oxygen diffuses into interstitial fluid, then into cells. Carbon dioxide moves rapidly in the opposite direction and is expelled from the lungs.

This mode of gas exchange is embellished a bit only in the lungs of birds. As described in Figure 41.9, birds are unique in that air not only flows into and out of the lungs, it also flows *through* them. With this exception in mind, we turn next to the human respiratory system, for its operating principles apply to most vertebrates.

A countercurrent flow mechanism in fish gills compensates for low oxygen levels in aquatic habitats. Internal air sacs—lungs—are more efficient in dry land habitats. Amphibians use integumentary exchange, and they also force air into and out of small lungs. Ventilation of paired lungs is the major mode of respiration in reptiles, birds, and mammals.

Functions of the Respiratory System

Obtaining oxygen from the air and removing carbon dioxide from the body are the overriding functions of the human respiratory system. A form of ventilation called breathing alternately moves air into and out of a pair of lungs, each of which contains about 300 million outpouchings. Each outpouching is a tiny air sac called an **alveolus** (plural, alveoli). Controls adjust the rate of breathing so that the inflow and outflow of air match the metabolic demands for gas exchange.

The respiratory system's role in respiration ends at alveoli. From that point on, the circulatory system takes over. As you will see, oxygen and carbon dioxide move by diffusion between the alveoli and the pulmonary capillaries that weave around them. (The Latin *pulmo* means lung.)

However, the respiratory system performs other functions. Breathing also is necessary for speech and other forms of vocalization. As you saw in Section 39.8, it enhances the venous return of blood to the heart. It functions to some extent in eliminating excess heat and water. In addition, control over breathing is vital for adjusting the body's acid-base balance. (Carbon dioxide can be used as a building block for carbonic acid, or H_2CO_3. As you read in Section 2.6, H_2CO_3 accepts or gives up hydrogen ions, H^+, depending on the pH. Fast, deep breathing expels more carbon dioxide, so less carbonic acid forms; hence the body loses acid. Slow, shallow breathing has the opposite effect; carbon dioxide accumulates, more H_2CO_3 forms, and the body gains acid.)

ORAL CAVITY

Supplemental airway when breathing is labored

EPIGLOTTIS

Closes off larynx during swallowing

PLEURAL MEMBRANES

Membranes that separate lungs from other organs; also form a thin, fluid-filled cavity that facilitates breathing

LUNG (ONE OF A PAIR)

Lobed, elastic organ of breathing that enhances gas exchange between the body and the outside air

INTERCOSTAL MUSCLES

Rib cage muscles with roles in breathing

DIAPHRAGM

Muscle sheet between the chest cavity and abdominal cavity with roles in breathing

NASAL CAVITY

Chamber in which air is warmed, moistened, and initially filtered; and in which sounds resonate

PHARYNX (THROAT)

Airway connecting the nasal cavity and mouth with the larynx; enhances speech sounds; connects also with the esophagus that leads to the stomach

LARYNX (VOICE BOX)

Airway where sound is produced and where breathing is blocked during swallowing movements

TRACHEA (WINDPIPE)

Airway connecting the larynx with two bronchi that lead into the lungs

BRONCHIAL TREE

Increasingly branched airways starting with the bronchi and ending at air sacs (alveoli)

ALVEOLI

Thin-walled air sacs where oxygen diffuses into the internal environment and carbon dioxide diffuses out

a

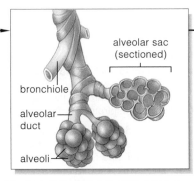

alveolar sac (sectioned)

bronchiole

alveolar duct

alveoli

b Location of alveoli relative to the terminal end of a bronchiole

Figure 41.10 (**a**) Components of the human respiratory system and their functions. Muscles, including the diaphragm, and parts of the axial skeleton have secondary roles in respiration. (**b,c**) Location of the alveoli relative to the lung capillaries. 🌀

The respiratory system also has built-in mechanisms for dealing with airborne foreign cells and substances that have been inhaled with air. Finally, the system removes, inactivates, or otherwise modifies a number of blood-borne substances before they become circulated through the rest of the body.

From Airways to the Lungs

It will take at least 300 million breaths to get you to age seventy-five. You may find yourself going without food for a few hours or days. But stop breathing even for five minutes and normal brain function is over.

Take a deep breath, then look at Figure 41.10 to get an idea of where the air will travel in your respiratory system. Unless you are out of breath and panting, the air has just entered two nasal cavities, not your mouth. There, mucus secretions warm and moisten it. Ciliated epithelium and hairs in the nasal cavities filter dust and particles from air. Also in the nasal cavities are olfactory receptors that function in the sense of smell (Section 36.3). Now the air is poised at the **pharynx**, or throat. The pharynx is the entrance to the **larynx**, an airway with vocal cords. Right now the **epiglottis**, a tissue flap at the start of the larynx, is pointing upward, so the air moves into the **trachea**, or windpipe. When you swallow, the epiglottis points downward and so closes off the entrance to the trachea. At such times, food or fluid being swallowed enters the esophagus, a tube connecting the pharynx with the stomach.

The trachea branches into two airways, one leading into the tissue of each lung. Each airway is a **bronchus** (plural, bronchi). Its epithelial lining has a profusion of cilia and mucus-secreting cells (Figure 41.11). The lining is a barrier to infection. Bacteria and airborne particles stick to the mucus, then cilia sweep debris-laden mucus toward the mouth. Where it goes from there really is up to you, but possibly the sidewalk is not a suitable destination.

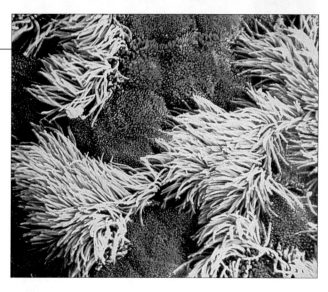

Figure 41.11 False-color scanning electron micrograph of cilia (*gold*) and mucus-secreting cells (*brown*) in a bronchus.

Sites of Gas Exchange in the Lungs

In humans, the lungs are elastic, cone-shaped organs of gas exchange. They are located within the rib cage, to the left and right of the heart and above the **diaphragm**, a muscular partition between the thoracic and abdominal cavity. A thin, pleural membrane lines the outer surface of the lungs and inner surface of the thoracic cavity wall. Visualize the lungs as two baseballs pushed into a partly inflated balloon. The baseballs take up so much space, they press the balloon's opposing sides together. Similarly, the pleural membrane is saclike. The thoracic wall and the lungs press its opposing surfaces together. A thin film of lubricating fluid separates the membrane surfaces and decreases friction between them. During a *pleurisy*, a respiratory ailment, the pleural membrane becomes so inflamed and swollen that friction follows, and breathing can be painful.

Inside each lung, air moves through finer and finer branchings of a "bronchial tree" (Figure 41.10a). These airways are **bronchioles**. Their endings, the *respiratory* bronchioles, bear the cup-shaped alveoli. Most often, alveoli are clustered as larger pouches called alveolar sacs (Figure 41.10b and c). Collectively, alveolar sacs offer a tremendous surface area for gas exchanges with blood. If all the alveolar sacs were stretched out in one layer, they would cover the floor of a racquetball court!

Oxygen uptake and carbon dioxide removal are the major functions of the human respiratory system. In its paired lungs, the circulatory system takes over the remaining tasks of respiration.

The respiratory system also has roles in moving venous blood to the heart, in vocalization, in adjusting the body's acid-base balance, in defense against harmful airborne cells or particles, in removing or modifying a number of blood-borne substances, and in the sense of smell.

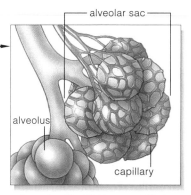

c Location of pulmonary capillaries relative to the alveoli

alveolar sac

alveolus

capillary

The Respiratory Cycle

There is a cyclic pattern to breathing, which ventilates the lungs. Each **respiratory cycle** consists of two actions: *inhalation* (a single breath of air drawn into the airways) and *exhalation* (a single breath out). Inhalation always is an active, energy-requiring action. When someone is breathing quietly, it is brought about by the contraction of the diaphragm and, to a lesser extent, the external intercostal muscles. The outcome is an increase in the thoracic cavity volume. Breathe hard, and the volume increases further because neck muscles contract and so elevate the sternum and first two ribs attached to them.

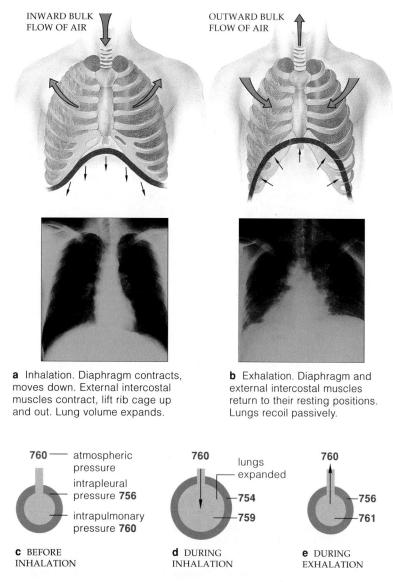

a Inhalation. Diaphragm contracts, moves down. External intercostal muscles contract, lift rib cage up and out. Lung volume expands.

b Exhalation. Diaphragm and external intercostal muscles return to their resting positions. Lungs recoil passively.

760 — atmospheric pressure

intrapleural pressure **756**

intrapulmonary pressure **760**

c BEFORE INHALATION

760 — lungs expanded

— **754**

— **759**

d DURING INHALATION

760

— **756**

— **761**

e DURING EXHALATION

Figure 41.12 (**a,b**) Changes in the size of the thoracic cavity during a respiratory cycle. The *blue* line represents the diaphragm. The x-ray images show how the maximum possible inhalation changes the thoracic cavity volume. (**c–e**) Changes in lung volume and intrapulmonary pressure during a respiratory cycle.

During each respiratory cycle, the thoracic cavity's volume increases, then decreases. *And pressure gradients between air inside and outside the respiratory tract change.* Let's think about the different pressures exerted during a respiratory cycle. The *atmospheric* pressure, 760 mm Hg at sea level, is exerted by the combined weight of all atmospheric gases on all the airways. Before inhalation, *intrapulmonary* pressure (the pressure inside all alveoli) is also 760 mm Hg (Figure 41.12c).

Another pressure gradient helps keep the lungs close to the thoracic cavity wall through the respiratory cycle, even during exhalation, when lungs have a far smaller volume than the thoracic cavity (Figure 41.12b). When the cavity expands, so do the lungs, owing to a pressure gradient that exists across the lung wall. In a person at rest, the *intrapleural* pressure (inside the pleural sac) averages 756 mm Hg, which is less than atmospheric pressure. This pressure is exerted outside the lungs—that is, within the thoracic cavity. While intrapleural pressure is pushing in on the lung's wall, the intrapulmonary pressure is pushing out. The difference in pressure between them (4 mm Hg) is great enough to make the lungs stretch and fill the thoracic cavity.

The cohesiveness of water molecules in the intrapleural fluid (the fluid inside the pleural sac) also helps keep lungs close to the thoracic wall. By analogy, wet two panes of glass and press them together. The panes easily slide back and forth, but they resist being pulled apart. Similarly, intrapleural fluid "glues" the lungs to the wall. Thus, *when the thoracic cavity expands at inhalation, the lungs must expand, also.*

Figure 41.12a shows what happens as you start to inhale. The dome-shaped diaphragm flattens down, and the rib cage is lifted upward and outward. As the thoracic cavity expands, the lungs expand with it. At that time, the air pressure in all alveolar sacs combined is lower than the atmospheric pressure. Fresh air follows the gradient and flows down into the airways, almost to the respiratory bronchioles.

When someone breathes quietly, the second action of the respiratory cycle is passive. The muscles that brought about inhalation relax, and the lungs passively recoil, without further expenditure of energy. The resultant decrease in lung volume compresses the air in the alveolar sacs. Now pressure in the sacs is greater than atmospheric pressure. Air follows the gradient, out from the lungs (Figure 41.12b). Exhalation becomes an active, energy-requiring action only when more air must be expelled rapidly, as

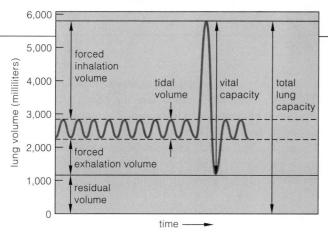

Figure 41.13 Lung volume. During quiet breathing, a tidal volume of air enters and leaves the lungs. Forced inhalation delivers more air to them; forced exhalation releases some air that normally stays in them. A residual volume is trapped in partially filled alveoli even during the strongest exhalation.

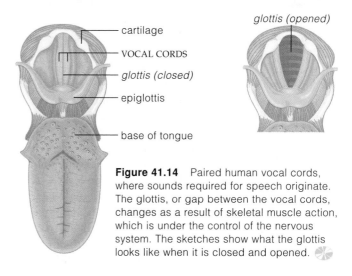

Figure 41.14 Paired human vocal cords, where sounds required for speech originate. The glottis, or gap between the vocal cords, changes as a result of skeletal muscle action, which is under the control of the nervous system. The sketches show what the glottis looks like when it is closed and opened.

during exercise. With active expiration, muscles in the abdominal wall contract, so pressure in the abdomen increases and exerts an upward force on the diaphragm. As the diaphragm is pushed upward, the thoracic cavity volume decreases. Internal intercostal muscles contract and pull the thoracic wall down and inward. The chest wall flattens, and so the thoracic cavity's dimensions decrease further. Lung volume decreases as well when the elastic tissue of the lungs recoils.

Lung Volumes

The lungs can hold up to 5.7 liters or so of air in adult males and 4.2 liters in adult females who are young and in good health. These are average values; a person's age, build, and respiratory health influence the total lung capacity. During quiet breathing, the lungs are far from being fully inflated. In general, they hold 2.7 liters at the end of one inhalation and 2.2 liters at the end of one exhalation. They never do deflate completely. When air flows out and the lung volume is low, the walls of the smallest airways collapse and prevent further air loss.

The volume of air that can move out of the lungs in one breath after maximal inhalation is called the **vital capacity**. Humans rarely use more than half of the total vital capacity, even when they take very deep breaths during strenuous exercise (Figure 41.13). Because the lungs are never empty, gas exchange between alveolar air and blood proceeds even at the end of maximum exhalation. *Hence the concentrations of gases in the blood remain fairly constant throughout the respiratory cycle.*

The volume of air flowing into or out of the lungs in the respiratory cycle, the **tidal volume**, averages 0.5 liter. How much is available for gas exchange? In between breaths, about 0.15 liter stays in the airways; only 0.35 liter of fresh air reaches alveoli. When you breathe, say, ten times a minute, you are supplying your alveoli with (0.35 × 10) or 3.5 liters of fresh air per minute.

Breathing and Sound Production

Near the entrance to the larynx are two paired folds of mucous membrane. The lower pair are the vocal cords (Figure 41.14). During the respiratory cycle, air is forced in and out through the glottis, the gap between them. Air flow makes the cords vibrate, and the vibrations can be controlled in ways that produce different sounds.

Within the folds are thick bands of elastic ligaments connected to various cartilage tissues. When muscles of the larynx contract and relax, the ligaments tighten or slacken, which changes the extent to which the folds are stretched. Under commands from the nervous system, the coordinated action of muscles narrows or widens the glottis. For example, by increasing muscle tension in the vocal cords, a person can decrease the gap between them and make high-pitched sounds or squeaks. The lips, teeth, tongue, and the soft roof over the tongue are enlisted to modify the different sounds into patterns of vocalization, such as speech and song.

When vocal cords are inflamed as an outcome of an infection or irritation, swelling of their mucous lining interferes with their capacity to vibrate. If hoarseness follows, this condition is called *laryngitis*.

Breathing, which ventilates the lungs, has a cyclic pattern. The respiratory cycle consists of inhalation (one breath of air in) and exhalation (one breath of air out).

Inhalation is always an active, energy-requiring process involving contractions mainly of the diaphragm and the external intercostal muscles.

During quiet breathing, exhalation is a passive process. Muscles relax, the thoracic cavity volume decreases, and the lungs recoil elastically. Forceful exhalation is an active process that requires abdominal muscle contraction.

Breathing reverses pressure gradients between the lungs and the air outside the body.

GAS EXCHANGE AND TRANSPORT

Gas Exchange

At each cup-shaped alveolus in the lungs, oxygen and carbon dioxide passively diffuse across the respiratory surface in response to their partial pressure gradients. An inward-directed gradient for oxygen is maintained because inhalations continually replenish oxygen and cells continually use it. An outward-directed gradient for carbon dioxide is maintained because cells continually produce carbon dioxide and exhalation removes it.

Each alveolus in the lungs consists of a single layer of epithelial cells that is surrounded by a thin basement membrane. And each pulmonary capillary consists of a single layer of endothelial cells. Only a thin film of interstitial fluid separates alveoli from the capillaries. The diffusion distances are so small that gases can flow rapidly across the respiratory surface (Figure 41.15).

Oxygen Transport

Oxygen, like carbon dioxide, does not dissolve well in blood, so it cannot be efficiently transported on its own. In all large animals, demands for oxygen are met with the assistance of hemoglobin molecules. These are the respiratory pigments that are packed inside red blood cells. Each hemoglobin molecule, recall, has quaternary structure. As shown earlier in Section 3.7, it actually is an organized, compact array of four polypeptide chains and four iron-containing, nitrogenous groups called **heme groups**. The iron atom of each heme group binds reversibly with oxygen. *Of all the oxygen inhaled into the human body, 98.5 percent of it is bound to the heme groups of hemoglobin.*

Normally, inhaled air that reaches alveoli has plenty of oxygen. The opposite is true of blood flowing past in the pulmonary capillaries. Thus, in the lungs, oxygen tends to diffuse into the plasma portion of blood. Then it diffuses into red blood cells and rapidly binds with hemoglobin. Hemoglobin that has oxygen bound to it is known as **oxyhemoglobin**, or HbO_2.

At any time, the amount of HbO_2 that forms depends on oxygen's partial pressure. The higher the pressure, the greater will be the oxygen concentration. When plenty of oxygen molecules are around, they tend to randomly collide with the heme binding sites at a faster rate. The encounters continue until all four of the binding sites in hemoglobin are saturated.

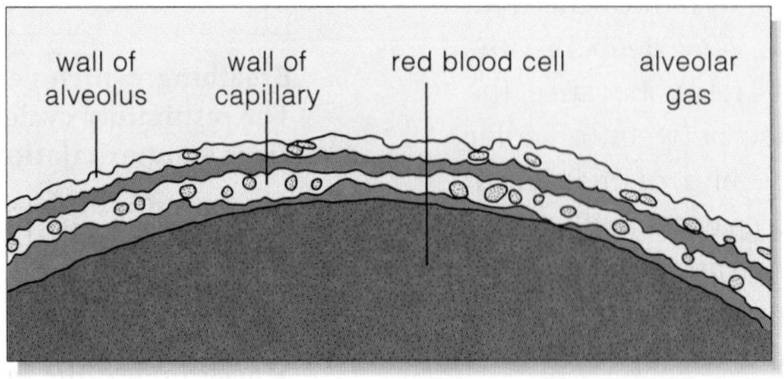

HbO_2 molecules hold onto oxygen rather weakly. They give it up in tissues where the partial pressure of oxygen is lower than in the lungs. They give it up even faster in tissues where the blood is warmer, the pH lower, and the partial pressure of carbon dioxide high. Such conditions exist in contracting muscle and other metabolically whipped-up tissues.

Carbon Dioxide Transport

The systemic portion of the circulatory system delivers carbon dioxide to the lungs. It enters blood capillaries in any tissue where its partial pressure is higher than it is in the blood flowing past.

Three mechanisms transport carbon dioxide from the capillary beds of the systemic circuit to the lungs. About 10 percent of the carbon dioxide simply remains dissolved in the blood. Another 30 percent binds with hemoglobin, forming **carbamino hemoglobin** ($HbCO_2$). Most of it—60 percent—is transported in the form of bicarbonate (HCO_3^-). These bicarbonate molecules form after carbon dioxide combines with water in blood plasma. The resulting carbonic acid dissociates (that is, separates) into bicarbonate and hydrogen ions (H^+):

$$CO_2 + H_2O \rightleftharpoons \underset{\text{CARBONIC ACID}}{H_2CO_3} \rightleftharpoons \underset{\text{BICARBONATE}}{HCO_3^-} + H^+$$

In blood plasma, that reaction converts only 1 of every 1,000 carbon dioxide molecules, which doesn't amount to much. It is a different story in red blood cells, which contain the enzyme **carbonic anhydrase**. The action of that enzyme enhances the reaction rate by 250 times! Most of the carbon dioxide not bound to hemoglobin becomes converted to carbonic acid this way. The enzyme-mediated conversion makes the blood level of carbon dioxide drop rapidly. Thus it helps maintain the gradient that promotes the diffusion of carbon dioxide from interstitial fluid into the bloodstream.

What about the bicarbonate ions that form in the reactions? They tend to move out of the red blood cells and into blood plasma. And what about the hydrogen ions? Hemoglobin acts as a buffer for them and keeps the blood from becoming too acidic. (A buffer, recall, is a molecule that combines with or releases H^+ in response to a shift in cellular pH.)

The reactions are reversed when blood reaches the alveoli, where the partial pressure of

Figure 41.15 What a section through an alveolus and an adjacent lung capillary would look like. Compared to the red blood cell's diameter, the diffusion distance across the capillary wall, the interstitial fluid, and the alveolar wall is extremely small.

(figure labels: wall of alveolus; wall of capillary; red blood cell; alveolar gas)

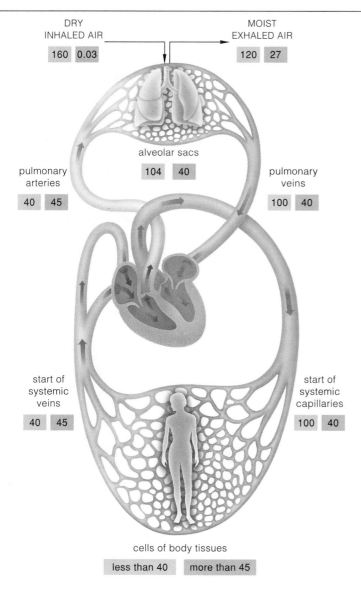

DRY INHALED AIR
160 | 0.03

MOIST EXHALED AIR
120 | 27

alveolar sacs
104 | 40

pulmonary arteries
40 | 45

pulmonary veins
100 | 40

start of systemic veins
40 | 45

start of systemic capillaries
100 | 40

cells of body tissues
less than 40 | more than 45

Figure 41.16 The partial pressure gradients for oxygen (*blue* boxes) and carbon dioxide (*pink* boxes) in the respiratory tract.

carbon dioxide is lower than it is in nearby capillaries. Carbon dioxide and water form here and diffuse into alveolar sacs, then the gas is exhaled from the body.

The partial pressure gradients for oxygen and for carbon dioxide through the human respiratory system are summarized in Figure 41.16.

Matching Air Flow With Blood Flow

Gas exchange is most efficient when the rate of air flow matches the rate of blood flow. The nervous system acts to balance them by controlling the breathing *rhythm* and its *magnitude* (the rate and depth of the tidal volume).

One respiratory center, in the medulla oblongata (part of the brain stem), controls the rhythmic pattern of breathing. There, a group of neurons shows pacemaker activity; it sets the pace for certain neurons that issue

commands to the muscles involved in inhalation. The signals ultimately stimulate the muscles to contract; in their absence, the muscles relax and exhalation occurs. When exhalation must be active and strong, different neurons in the respiratory center issue commands for the activation of the muscles involved in exhalation.

Higher up in the brain stem, in the pons, different centers smooth out the rhythm set by the respiratory pacemakers. The apneustic center prolongs inhalation; the pneumotaxic center curtails it.

Control of the magnitude of breathing depends on the partial pressures of oxygen and carbon dioxide as well as the H^+ concentration in arterial blood. Carbon dioxide levels are the most critical for ongoing adjustments. Chemoreceptors in the brain are highly sensitive to a rise in H^+ in the cerebrospinal fluid that bathes them. (Remember, H^+ can be a by-product of reactions that proceed when the blood has too much carbon dioxide.) The brain responds by commanding the diaphragm and other muscles to alter their activity; it adjusts the rate and depth of breathing. By contrast, aortic bodies (in the aortic arch) and carotid bodies (where the carotid artery branches) notify the brain when the partial pressure of oxygen in arterial blood plummets below 60 mm Hg. Such life-threatening decreases occur at extremely high altitudes and during severe lung diseases.

In some situations, a person can "forget" to breathe. Breathing that is briefly interrupted and then resumes spontaneously is called *apnea*. Especially during REM sleep (Section 35.8), breathing may stop for one or two seconds or minutes—in a few cases as often as 500 times a night. *Sudden infant death syndrome* (SIDS) occurs when a sleeping infant cannot awaken from an apneic episode, perhaps because its respiratory control centers or chemoreceptors (especially carotid bodies) are not yet fully developed. Infants who sleep on their back or sides are at less risk of SIDS than those positioned on their abdomen. The risk is three times as high among the infants of women who smoked cigarettes or were exposed to secondhand smoke during pregnancy.

Driven by its partial pressure gradient, oxygen diffuses from alveolar air spaces, through interstitial fluid, and into lung capillaries. Carbon dioxide, driven by its partial pressure gradient, diffuses in the opposite direction.

Hemoglobin in red blood cells enormously enhances the oxygen-carrying capacity of blood. Most carbon dioxide is transported in blood in the form of bicarbonate, nearly all of which forms by an enzyme action in red blood cells.

Respiratory centers in the brain stem control the rhythmic pattern of breathing and the magnitude of breathing, in ways that maintain appropriate levels of carbon dioxide, oxygen, and hydrogen ions in arterial blood.

41.7 WHEN THE LUNGS BREAK DOWN

In large cities, in certain workplaces, even near a cigarette smoker, airborne particles and certain gases are present in abnormally high concentrations. And they put extra workloads on the respiratory system.

BRONCHITIS   Ciliated, mucous epithelium lines your bronchioles, as Figure 41.11 shows. It is one of the built-in defenses that protect you from respiratory infections. Toxins in cigarette smoke and other airborne pollutants irritate this lining and may lead to *bronchitis*. With this respiratory ailment, excess mucus is secreted when epithelial cells that line the airways become irritated. As mucus accumulates, so do bacteria and other particles stuck in it. Coughing brings up some of the gunk. If the irritation persists, so does coughing. Initial attacks of bronchitis are treatable. However, when the aggravation is allowed to continue, bronchioles become inflamed. Bacteria, chemical agents, or both attack cells of the bronchiole walls. Ciliated cells in the walls are destroyed, and mucus-secreting cells multiply. Fibrous scar tissue forms and in time may narrow or obstruct the airways.

EMPHYSEMA   When bronchitis persists, thick mucus clogs the airways. Inside the lungs, tissue-destroying bacterial enzymes attack the stretchable, thin walls of alveoli. The walls break down, and inelastic fibrous tissue forms around them. Gas exchange proceeds at fewer alveoli, which become enlarged. In time, the fine balance between air flow and blood flow is permanently compromised, for the lungs remain distended and inelastic.

Compare Figure 41.17*a* with *b* to get a sense of what happens. It becomes difficult to run, walk, even exhale. These are symptoms of *emphysema*, a respiratory ailment that affects about 1.3 million people in the United States alone.

A few people are genetically predisposed to the development of emphysema. They don't have a workable gene for the enzyme antitrypsin, which can inhibit bacterial attacks on alveoli. Poor diet and persistent or recurring colds and other respiratory infections also make a person susceptible to emphysema later in life. *But smoking is the major cause of the disease.* Emphysema can develop slowly, over twenty or thirty years. When emphysema is detected too late, the serious damage to lung tissue cannot be repaired.

EFFECTS OF CIGARETTE SMOKE   Worldwide, smoking now kills 3 million people every year. If current patterns do not change, then by the time today's young smokers reach middle age, 10 million may be dying annually because of the habit. That is one person every three seconds. Even now in the United States, one of 50 million cigarette-puffing people dies of emphysema, chronic bronchitis, or heart disease every thirteen seconds. For every eight of them, one nonsmoker dies of ailments brought on by *secondhand smoke*—of prolonged exposure to tobacco smoke in the surrounding air. And how much thought is given to children who breathe secondhand smoke? Parents and others who otherwise care about children are indirectly increasing their vulnerability to allergies and lung ailments. Smoking is the major cause of lung cancer, which is now the leading cause of death among women. Yet every day, 3,000 to 5,000 Americans light a cigarette for the first time. Even children spend a billion dollars a year on cigarettes. The economy shrinks by 22 billion a year because of the direct medical costs of treating smoke-induced respiratory disorders.

How does cigarette smoke lead to lung damage? Noxious particles in the smoke from just one cigarette immobilize cilia in the bronchioles for several hours. The particles also trigger mucus secretions, which in time clog the airways. They can kill infection-fighting macrophages residing in the respiratory tract. What starts as "smoker's cough" can end in bronchitis and emphysema.

Or consider how cigarette smoke contributes to lung cancer. Inside the body, certain compounds in coal tar and in cigarette smoke become converted to highly reactive intermediates. These are the carcinogens; they provoke uncontrolled cell divisions in lung tissues. On the average, 90 of every 100 smokers who develop lung cancer will die from it. If you now smoke, or are thinking about starting or quitting, you may wish to give serious thought to the information in Figure 41.18.

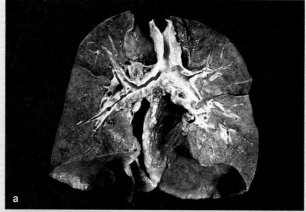
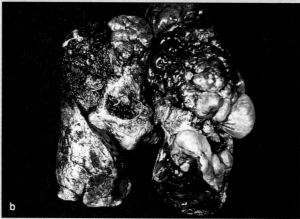

Figure 41.17   (a) Normal appearance of tissues of human lungs. (b) Lungs from someone affected by emphysema.

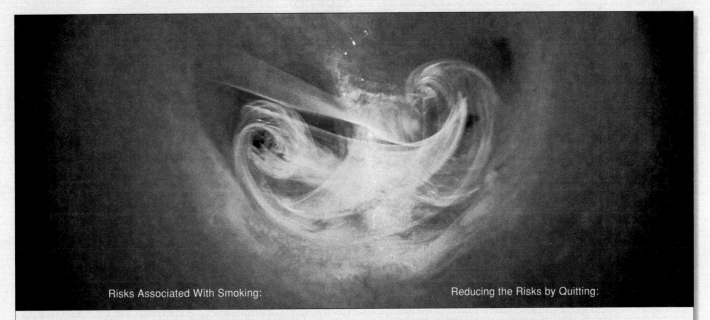

Risks Associated With Smoking: | Reducing the Risks by Quitting:

SHORTENED LIFE EXPECTANCY: Nonsmokers live 8.3 years longer on average than those who smoke two packs daily from the midtwenties on.

Cumulative risk reduction; after 10 to 15 years, life expectancy of ex-smokers approaches that of nonsmokers.

CHRONIC BRONCHITIS, EMPHYSEMA: Smokers have 4–25 times more risk of dying from these diseases than do nonsmokers.

Greater chance of improving lung function and slowing down rate of deterioration.

LUNG CANCER: Cigarette smoking is the major cause of lung cancer.

After 10 to 15 years, risk approaches that of nonsmokers.

CANCER OF MOUTH: 3–10 times greater risk among smokers.

After 10 to 15 years, risk is reduced to that of nonsmokers.

CANCER OF LARYNX: 2.9–17.7 times more frequent among smokers.

After 10 years, risk is reduced to that of nonsmokers.

CANCER OF ESOPHAGUS: 2–9 times greater risk of dying from this.

Risk proportional to amount smoked; quitting should reduce it.

CANCER OF PANCREAS: 2–5 times greater risk of dying from this.

Risk proportional to amount smoked; quitting should reduce it.

CANCER OF BLADDER: 7–10 times greater risk for smokers.

Risk decreases gradually over 7 years to that of nonsmokers.

CORONARY HEART DISEASE: Cigarette smoking is a major contributing factor.

Risk drops sharply after a year; after 10 years, risk reduced to that of nonsmokers.

EFFECTS ON OFFSPRING: Women who smoke during pregnancy have more stillbirths, and weight of liveborns averages less (hence, babies are more vulnerable to disease, death).

When smoking stops before fourth month of pregnancy, risk of stillbirth and lower birthweight eliminated.

IMPAIRED IMMUNE SYSTEM FUNCTION: Increase in allergic responses, destruction of defensive cells (macrophages) in respiratory tract.

Avoidable by not smoking.

BONE HEALING: Evidence suggests that surgically cut or broken bones require up to 30 percent longer to heal in smokers, possibly because smoking depletes the body of vitamin C and reduces the amount of oxygen reaching body tissues. Reduced vitamin C and reduced oxygen interfere with production of collagen fibers, a key component of bone. Research in this area is continuing.

Avoidable by not smoking.

Figure 41.18 From the American Cancer Society, a list of the risks incurred by smoking and the benefits of quitting. The photograph shows a few swirls of cigarette smoke poised at the entrance to the two bronchi that lead into the lungs.

EFFECTS OF MARIJUANA SMOKE In 1994, 17 million Americans smoked *pot*—marijuana (*Cannabis*)—at least once to induce light-headed euphoria. The number is rising, especially in junior high and high schools. About 1.5 million are chronic users. They become enamored of the euphoria, but the sensation does not last long. It fades to apathy, depression, and fatigue. Users keep smoking to keep from feeling "wasted." Besides psychological dependency, chronic use can result in throat irritation, persistent coughing, bronchitis, and emphysema.

Breathing at High Altitudes

From Section 41.6 as well as the introduction to this chapter, you know that the partial pressure of oxygen decreases with increasing altitude. The consequences can be dreadful indeed for the occasional climbers of Chomolungma and other cloud-piercing peaks. Yet, like llamas and other species that evolved at high elevations, millions of people comfortably live out their lives 4,800 meters (16,000 feet) or more above sea level.

Compared to people living at lower elevations, the lungs of permanent residents of high mountains have far more alveoli and blood vessels, which formed while they were growing up. Also, the ventricles of their heart enlarged more, so they pump larger volumes of blood. And their muscle tissue has more mitochondria.

Llamas have an additional advantage. Compared to human hemoglobin, llama hemoglobin contains factors that give it a greater affinity for oxygen (Figure 41.19). It picks up oxygen more efficiently at the lower pressures characteristic of high altitudes.

Does this mean that a healthy person who grew up by the seashore cannot ever pick up roots and move to the mountains? No. By mechanisms of **acclimatization**, an individual often can make long-lasting physiological and behavioral adaptations to a new environment that is markedly different from the one left behind. At high altitudes, gradual changes in the pattern and magnitude of breathing and in cardiac output can supplant the initially acute compensations that are made to cellular oxygen deficiency (hypoxia).

Within a few days, the reduction in oxygen delivery stimulates kidney cells to secrete more **erythropoietin**, a hormone that induces stem cells in bone marrow to proliferate and mature into red blood cells. Each second in an adult human, between 2 million and 3 million red blood cells are being churned out as replacements for the ones continually dying. Stepped-up erythropoietin secretion can increase that astounding pace by six times during extreme stress. Increased numbers of circulating red blood cells increase the oxygen-delivery capacity of blood. When the oxygen level increases sufficiently, the kidneys slow down erythropoietin secretion.

Erythropoietin is the major hormonal mediator of red blood cell production. However, human males have a larger muscle mass and greater demands for oxygen. They depend on testosterone, also. Among other things, this sex hormone stimulates increases in the basic rate of red blood cell production. Normally, a sample of blood from males has a larger percentage of red blood cells than an equivalent sample from females.

The compensatory increase in red blood cells comes at a cost. Having many more cells in the bloodstream increases resistance to blood flow, because the blood is now more viscous, or "thicker." Hence the heart must work harder to pump blood.

Carbon Monoxide Poisoning

High elevations are not the exclusive culprits behind cellular oxygen deficiency. Hypoxia also occurs when the partial pressure of oxygen in arterial blood falls owing to *carbon monoxide poisoning*.

Carbon monoxide, a colorless, odorless gas, is a component of exhaust fumes from cars, trucks, and other vehicles. It is also present in the smoke from tobacco and from burning coal and wood. This gas competes with oxygen for binding sites in hemoglobin. And its binding capacity is at least 200 times greater than that of oxygen.

Even small amounts of carbon monoxide can tie up half of the body's hemoglobin. Thus it can impair oxygen delivery to tissues, sometimes to dangerous degrees.

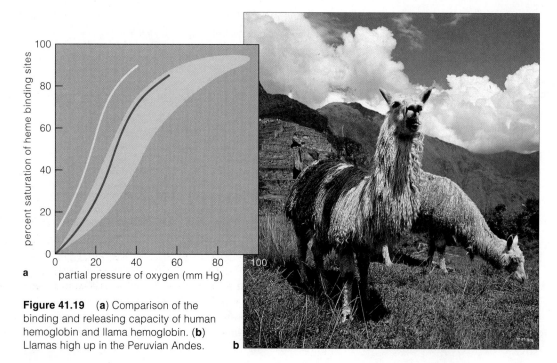

Figure 41.19 (a) Comparison of the binding and releasing capacity of human hemoglobin and llama hemoglobin. (b) Llamas high up in the Peruvian Andes.

a — percent saturation of heme binding sites / partial pressure of oxygen (mm Hg)

Respiration in Diving Mammals

THE PHYSIOLOGY OF DIVING   How do whales, seals, and other air-breathing mammals that routinely dive survive underwater, where they cannot inhale oxygen and exhale carbon dioxide? Apparently such mammals (as well as diving birds) share a common, highly specialized pattern of metabolism during their underwater forays.

Here we are not talking about a quick bob under the water's surface. Weddell seals (*Leptonychotes weddelli*) of Antarctic waters typically remain submerged for twenty to twenty-five minutes, and some do so for more than an hour. They usually limit their dives to 400 meters (1,312 feet) or less, but a few have been observed at depths approaching 600 meters—more than a third of a mile down. As two more examples, one sperm whale (Figure 41.20) submerged itself for eighty-two minutes; another, tracked by sonar, reached a depth of 2,250 meters.

By comparison, humans who are trained to dive without oxygen tanks can fully submerge themselves for about three minutes, at most. Career divers, including the ama of Korea and Japan, do only a bit better.

When diving mammals leave the air behind, they take oxygen with them. Hemoglobin binds much of it. So does **myoglobin**, an oxygen-binding protein that is abundant in skeletal muscles. And the lungs serve as a suitcase filled with oxygen. During short dives, aerobic respiration proceeds as usual. During prolonged dives, however, the oxygen is preferentially distributed to the heart and central nervous system, both of which cannot function without the high energy yields of the aerobic pathway. Basically, blood is directed away from most organs. Blood pumped from the heart travels mainly to the lungs, brain, and back to the heart. Skeletal muscles use their own myoglobin-bound oxygen as well as the hemoglobin-bound oxygen in the dense capillary beds that service them. Later on, they switch to anaerobic pathways, and lactate accumulates in tissues. When the animal surfaces, circulation to the muscles is restored.

Diving mammals also have specialized respiratory adaptations. For example, a sperm whale's respiratory system is adapted to collect and store oxygen. When the whale surfaces, it rapidly blows stale air from its lungs, through a tubelike epiglottis and then a blowhole at the body surface. Before the whale dives again, it breathes rapidly, so that its elastic, extensible lungs fill quickly with large volumes of air. During a dive, special valves close its nostrils; and rings of muscle and of cartilage clamp its bronchioles. The whale's respiratory surface is not that large. However, strategically located valves and local networks of blood vessels (plexuses) store and distribute the blood volume and gases in economical fashion. The heart rate slows, metabolism decreases, and so does the rate of oxygen use and carbon dioxide

Figure 41.20 A sperm whale, adapted for prolonged diving. This aquatic mammal is more than twenty-five meters long, on average. It can dive 2,250 meters below the ocean surface and stay there for well over an hour before coming up for oxygen.

formation. In addition, compared to land mammals, the whale's respiratory center is less sensitive to carbon dioxide levels.

HUMAN DEEP-SEA DIVERS   Professional divers know that water pressure increases greatly with depth. They do not dive in deep water without tanks of compressed air (air under pressure). They also make their ascents from the depths with utmost caution.

Why the caution? Because of the increased pressure at depths, more gaseous nitrogen (N_2) than usual has become dissolved in their body tissues. The outcome is *nitrogen narcosis*, or "raptures of the deep." At depths of 45 meters (150 feet), divers become euphoric and drowsy, as if they were tipsy with alcohol. Some have been known to offer the mouthpiece of their airtank to fishes. At lower depths, the divers become clumsy and weak. Below 105 meters, they can slip into a coma.

Gaseous nitrogen is a lipid-soluble gas, so it readily dissolves in the lipid bilayer of cell membranes. By one hypothesis, when it does this to neurons, it suppresses their capacity to respond to action potentials.

Deep-sea divers also are cautious when they return to the surface. When the pressure decreases, N_2 moves back out of the tissues and into the bloodstream. If an ascent is too rapid, N_2 enters the blood faster than the lungs can dispose of it. Then, bubbles of nitrogen may form in the blood and tissues. Too many bubbles cause pain, especially at joints. Hence the common name, "the bends," for what is otherwise known as *decompression sickness*. If bubbles obstruct the blood flow to the brain, deafness, impaired vision, and paralysis may result.

During brief stays at high elevations, the body makes acute compensatory responses to the thinner air. Over time, it makes adaptive adjustments in breathing and cardiac output.

Diving mammals take along oxygen bound to hemoglobin and myoglobin and in their lungs. They also depend on respiratory adaptations for storing and distributing oxygen.

41.9 RESPIRATION IN LEATHERBACK SEA TURTLES

Late at night, near the shoreline of a Caribbean island, biologist Molly Lutcavage and several colleagues move out on a turtle patrol. Finally a large, dark shape, with flippers flailing, emerges from the pale surf. A female Atlantic leatherback sea turtle (*Dermochelys coriacea*) is returning from the sea to nest in the sand (Figure 41.21).

Sea turtles are members of the reptilian lineage, which extends 300 million years back in time. Green turtles, ridleys, loggerheads, leatherbacks—all are endangered or threatened species. The race is on to gather information about their life histories and physiology that may help pull them back from the brink of extinction.

Leatherbacks are the largest and least understood sea turtles. The adult females may weigh more than 400 kilograms (880 pounds). The adult males may weigh nearly twice that amount. Leatherbacks normally leave the water only to breed and lay eggs. During the rest of their lives, they migrate across vast stretches of the open ocean.

Leatherbacks do something no other reptile on Earth can do. They have the capacity to dive 1,000 meters (3,000 feet) below sea level! Before biologists conducted tracking experiments during the mid-1980s, Weddell seals, fin whales, and some other marine mammals were the only nonhuman deep-sea divers known.

To reach such depths, a turtle would have to swim for nearly forty minutes without taking a breath. How does it dive so deeply, for so long, and still have enough oxygen for aerobic metabolism?

Oxygen reserves in the lungs are not enough for the leatherback's prolonged underwater excursions. Even at depths between 80 and 160 meters, pressure exerted by the surrounding water causes air-filled lungs to collapse. However, like the diving marine mammals described in Section 41.8, the leatherback's skeletal muscle cells have an abundance of the oxygen-binding protein myoglobin. In addition, compared to other diving reptiles, they have more blood per unit of body weight, and their blood contains more red blood cells—hence more hemoglobin (Table 41.1).

Imagine yourself as an observer on Lutcavage's team. Their plan is to draw blood samples from a leatherback so that they can study the oxygen-binding capacity of its hemoglobin. Samples of respiratory gases dissolved in the blood may provide insight into how leatherbacks use oxygen. As the researchers know, a nesting female enters a trancelike state during the twenty or so minutes it takes her to deposit eggs in the sand. During that time, she will not resist being gently handled.

The team gets to work. They fit the head of a nesting female with a helmet that is equipped with an expandable plastic sleeve. The fit is snug but not restrictive, so the turtle is able to breathe normally. With each exhalation, the expired gases flow through a one-way valve into a collecting sac. At the same time, the researchers count the number of breaths required to completely fill the sac. (As a baseline, they also tallied the breathing frequency before placing the helmet over her head.) They store the gas samples according to established procedures and set them aside for laboratory analysis.

Working quickly, a member of the team draws blood from a superficial vein in the turtle's neck, and then packs the sample in ice to preserve it. Once they are back at the laboratory, the sample will be subjected to hemoglobin analysis and a red blood cell count. The oxygen level, carbon dioxide level, and pH also will be measured.

Lutcavage and her coworkers did gather samples of blood and respiratory gases from several turtles. They also analyzed skeletal muscle tissue, taken earlier from a drowned turtle. The data gave insight into how leatherbacks manage their spectacularly deep dives.

As expected, calculations of the oxygen uptake and total tidal volume during breathing revealed that the leatherback cannot inhale and hold enough air in the lungs to sustain aerobic metabolism during a prolonged dive. Other oxygen-supplying mechanisms had to be at

Table 41.1 Physiological Comparison of a Few Diving Reptiles and Mammals					
Characteristic	LEATHERBACK TURTLE	GREEN TURTLE	CROCODILE	KILLER WHALE	WEDDELL SEAL
Maximum diving depth (meters)	Greater than 1,000	Less than 100	Less than 30	260	600
Red blood cell count (percentage of total volume)	39	30	28	44	58
Hemoglobin level (grams/deciliter of blood)	15.6	8.8	8.7	16.0	17–22
Myoglobin level (milligram/gram of muscle tissue)	4.9	—*	—*	—*	44.6
Oxygen-carrying capacity (volume percent)	21	7.5–11.9	12.4	23.7	31.6

* Not known.

work. As further analysis revealed, myoglobin levels are so high in leatherbacks, oxygen remains available for skeletal muscle activity during a dive after the air in the lungs gives out (or when their lungs collapse).

Also, leatherbacks have a notable abundance of red blood cells and a type of hemoglobin with a high affinity for binding oxygen. A leatherback's blood may contain a remarkable 21 percent oxygen by volume. That amount is typical of a human, not a "plodding" turtle.

The oxygen-carrying capacity turns out to be the highest ever recorded for a reptile—and it is close to the capacity of deep-diving mammals. Add to this a streamlined body and massive front flippers, and you have a reptile uniquely adapted for diving and maneuvering at great depths.

Studies of leatherbacks are only now under way. Researchers are starting to investigate newer tracking methods to help them monitor a leatherback's metabolism at sea and during dives. So far, it has proved extremely difficult and expensive to track individual turtles in the open ocean.

Figure 41.21 Two female leatherback sea turtles (*Dermochelys coriacea*) returning to water after laying eggs in the sands of an island in the Caribbean Sea.

1. Aerobic respiration is the main metabolic pathway that provides enough energy for active life-styles. It uses oxygen and produces carbon dioxide. The process by which the animal body as a whole acquires oxygen and disposes of carbon dioxide is called respiration.

2. Air is a mixture of oxygen, carbon dioxide, and other gases, each exerting a partial pressure. Each gas tends to move from areas of higher to lower partial pressure. Respiratory systems make use of this tendency.

3. In respiratory systems, oxygen and carbon dioxide diffuse across a respiratory surface: a moist, thin layer of epithelium that is interposed between the external and internal environments.

4. Modes of respiration differ among animal groups.

   a. Invertebrates with a small body mass depend only on integumentary exchange; oxygen and carbon dioxide simply diffuse across the body surface. This mode also persists in some large animals, including amphibians.

   b. Many marine and some freshwater invertebrates have gills, respiratory organs with moist, thin, and often highly folded walls. Most insects and certain spiders use tracheal respiration. Gases flow through open-ended tubes leading from the body surface directly to tissues. Most spiders have book lungs, with leaflike folds.

   c. Fishes have greater energy demands than small invertebrates. They have gills in which a countercurrent flow mechanism compensates for low oxygen levels in aquatic habitats. Paired lungs are the dominant means of respiration in reptiles, birds, and mammals.

5. The airways of the human respiratory system are the nasal cavities, pharynx, larynx, trachea, bronchi, and bronchioles. Many millions of alveoli at the end of the terminal bronchioles are the main sites of gas exchange.

6. Breathing ventilates the lungs. Each respiratory cycle consists of inhalation (one breath in) and exhalation (one breath out). During inhalation, the chest cavity expands, lung pressure decreases below atmospheric pressure, and air flows into the lungs. The events are reversed during normal exhalation.

7. Driven by its partial pressure gradient, oxygen in the lungs diffuses from alveolar air spaces into pulmonary capillaries. It diffuses into red blood cells and binds weakly with hemoglobin. Hemoglobin gives up oxygen at capillary beds in metabolically active tissues. Oxygen diffuses across the interstitial fluid, then into cells.

8. Driven by its partial pressure gradient, carbon dioxide in tissues diffuses from cells, across interstitial fluid, and into the blood. Most of it reacts with water to form bicarbonate. The reactions are reversed inside the lungs; carbon dioxide diffuses from pulmonary capillaries into the air spaces of alveoli, then it is exhaled.

Figure 41.22 Human female at play in the domain of aquatic animals.

Review Questions

1. Define respiratory surface. Why must the oxygen and carbon dioxide partial pressure gradients across it be steep? *41.1*

2. What is the name of the respiratory pigment in red blood cells? What is the name of a major respiratory pigment in skeletal muscle cells? *41.1, 41.8*

3. A few of your friends who have not taken a biology course ask you what insect lungs look like. What is your answer (assuming your instructor is listening)? *41.2*

4. Briefly describe how countercurrent flow through a fish gill is so efficient at taking up dissolved oxygen from water. *41.3*

5. Does the respiratory system of fishes, amphibians, reptiles, birds, or mammals have air sacs that ventilate the lungs? *41.3*

6. Distinguish between:
   a. aerobic respiration and respiration *CI, 41.1*
   b. pharynx and larynx *41.4*
   c. bronchiole and bronchus *41.4*
   d. pleural sac and alveolar sac *41.4, 41.5*

7. Explain why humans (Figure 41.22) cannot survive on their own, for very long, underwater. *41.8*

8. Define the functions of the human respiratory system. In the diagram below, label its components and the major bones and muscles with which it interacts during breathing. *41.4*

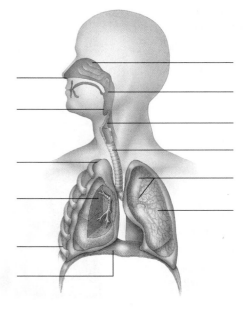

Self-Quiz *(Answers in Appendix IV)*

1. A partial pressure gradient of oxygen exists between _____ .
   a. the atmosphere and the lungs
   b. lungs and metabolically active tissues
   c. air at sea level and air at high altitudes
   d. all of the above

2. The _____ is an airway that connects the nose and mouth with the _____ .
   a. oral cavity; larynx
   b. pharynx; trachea
   c. trachea; pharynx
   d. pharynx; larynx

3. Oxygen in the air diffuses across _____ as it follows its partial pressure gradient into the human body.
   a. pleural sacs
   b. alveolar sacs
   c. a moist respiratory surface
   d. both b and c

4. Each human lung encloses a _____ .
   a. diaphragm
   b. bronchial tree
   c. pleural sac
   d. both b and c

5. In human lungs, gas exchange occurs at the _____ .
   a. two bronchi
   b. pleural sacs
   c. alveolar sacs
   d. both b and c

6. Breathing _____ .
   a. ventilates the lungs
   b. draws air into airways
   c. expels air from airways
   d. reverses pressure gradients
   e. all of the above

7. During quiet breathing, inhalation is _____ and exhalation is _____ .
   a. passive; passive
   b. active; active
   c. passive; active
   d. active; passive

8. After oxygen diffuses into pulmonary capillaries, it diffuses into _____ and binds with _____ .
   a. interstitial fluid; red blood cells
   b. interstitial fluid; carbon dioxide
   c. red blood cells; hemoglobin
   d. red blood cells; carbon dioxide

9. Most carbon dioxide in blood is in the form of _____ .
   a. carbon dioxide
   b. carbon monoxide
   c. carbonic acid
   d. bicarbonate

10. Match the components with their descriptions.
    ____ trachea
    ____ pharynx
    ____ alveolus
    ____ hemoglobin
    ____ bronchus
    ____ bronchiole
    a. airway leading into a lung
    b. throat
    c. fine branch of bronchial tree
    d. windpipe
    e. respiratory pigment
    f. site of gas exchange

Critical Thinking

1. Some cigarette manufacturers conducted public relations campaigns urging their customers to "smoke responsibly." What are some social and biological issues in this controversy? How do these issues apply to the nonsmoking spouse or children of a smoker? To nonsmoking patrons in a restaurant? To the unborn child of a pregnant smoker? In your opinion, what behavior would constitute "responsible smoking"?

2. People sometimes poison themselves with carbon monoxide by building a charcoal fire in an enclosed area. Assuming help arrives in time, what would be the best treatment: (1) placing the victim outdoors in fresh air or (2) rapidly administering pure oxygen? Explain how you arrived at your answer.

3. When you swallow food, muscle contractions force the epiglottis down, to its closed position. This prevents food from going down the trachea and blocking gas flow. Yet each year,

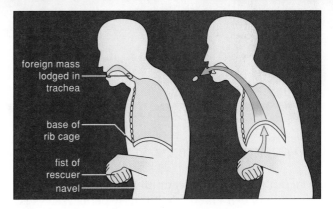

Figure 41.23 Heimlich maneuver to dislodge food stuck in the trachea. Stand behind the victim, make a fist with one hand, then position the fist, thumb-side in, against the victim's abdomen. The fist must be slightly above the navel and well below the rib cage. Now press the fist into the abdomen with a sudden upward thrust. Repeat the thrust several times if needed. The maneuver can be performed on someone who is standing, sitting, or lying down.

several thousand choke to death after food enters the trachea and blocks air flow for as little as four or five minutes. The *Heimlich maneuver*—which is an emergency procedure only— often dislodges food from the esophagus (Figure 41.23). When correctly performed, it elevates the diaphragm forcibly, causing a sharp decrease in the volume of the thoracic cavity and a sudden increase in alveolar pressure. Air forced up the trachea as a result of the increased pressure may be enough to dislodge the obstruction. Once the obstacle is dislodged, a physician must see the person at once. An inexperienced rescuer can inadvertently cause internal injuries or crack a rib.

Reflect on the current social climate in which lawsuits abound. Would you risk performing the Heimlich maneuver to save a relative's life? A stranger's life? Why or why not?

Selected Key Terms

acclimatization *41.8*	larynx *41.4*
alveolus (alveoli) *41.4*	lung *41.3*
bronchiole *41.4*	myoglobin *41.8*
bronchus *41.4*	oxyhemoglobin *41.6*
carbamino hemoglobin *41.6*	partial pressure *41.1*
carbonic anhydrase *41.6*	pharynx *41.4*
countercurrent flow *41.3*	pressure gradient *41.1*
diaphragm *41.4*	respiration *CI*
epiglottis *41.4*	respiratory cycle *41.5*
erythropoietin *41.8*	respiratory surface *41.1*
Fick's law *41.1*	respiratory system *CI*
gill (fish) *41.2*	tidal volume *41.5*
glottis *41.3*	trachea *41.4*
heme group *41.6*	tracheal respiration *41.2*
hemoglobin *41.1*	vital capacity *41.5*
integumentary exchange *41.2*	vocal cord *41.3*

Readings

Hill, R., and G. Wyse. 1992. *Animal Physiology*. Second edition. New York: Harper Collins.

Sherwood, L. 1997. *Human Physiology*. Second edition. Belmont, California: Wadsworth.

Web Site See *http://www.wadsworth.com/biology* for practice quiz questions, hypercontents, BioUpdates, and critical thinking. The Wadsworth Biology Resource Center provides a wealth of information fully organized and integrated by chapter.

DIGESTION AND HUMAN NUTRITION

Lose It—And It Finds Its Way Back

America's fixation on Beautiful People who have nary an ounce of extra fat on their utterly perfect selves is bad enough. After all, we can always rationalize our own extra ounce with the truism that beauty is only skin deep. But the American Medical Association's dire warnings of the Fat Connection to atherosclerosis, heart attacks, strokes, colon cancer, and other ailments is beyond rationalization.

By current standards, the proportion of body fat relative to total tissue mass should be 18 to 24 percent for human females who are less than thirty years old. For males, it should be no more than 12 to 18 percent. An estimated 34 million Americans do not even come close to the standards.

No matter how much we lose by dieting, the lost weight seems to find its way back. This physiological dilemma is an outcome of our evolutionary heritage. Like most other mammals, we have adipose tissue with an abundance of fat-storing cells. Collectively, the cells are an adaptation for survival. They serve as an energy warehouse that can be opened up during times when food is not available. Variations in food intake only influence how empty or full a fat-storing cell will get. Once that cell has formed, it is in the body to stay.

Dieting starts to empty the fat warehouse. The brain interprets this as "starvation" and triggers a metabolic slowdown. The body uses energy far more efficiently—even for the basics, such as breathing and digesting food. That is, it takes less food to do the same things!

As you have probably heard, dieting does no good without *long-term* commitment to physical exercise. Why? When you are dieting, skeletal muscles are also adapting to "starvation," and they burn less energy than before. Jog for four hours or play tennis for eight hours straight, and you may well lose a pound of fat. Meanwhile, your appetite surges, so if and when the dieting stops, "starved" fat cells quickly refill. In 1995, after a decade of careful study, researchers determined that lost weight will be regained unless a person eats less *and* exercises moderately throughout his or her lifetime.

What about gaining weight? Intriguingly, a weight gain triggers an *increase* in metabolism. The body uses 15 to 20 percent more energy than before—until weight drops back to what it was before.

Emotional factors also influence weight gains and losses. Consider *anorexia nervosa*, a potentially fatal eating disorder that is founded on a seriously flawed assessment of body weight. Typically, anorexics have an overwhelming dread of being fat *and* hungry. They starve themselves and often overexercise. Often there are fears about growing up and being sexually mature. Irrational expectations of personal performance are common.

Or consider *bulimia*, an out-of-control, "oxlike appetite." During an hour-long eating binge, a bulimic might ingest more than 50,000 kilocalories' worth of food, then vomit or use laxatives to purge the body

Figure 42.1 Mirror, mirror, on the wall, who is fairest of them all? (And by whose standards?)

of it. Some get into a binge-purge routine as an "easy" way to lose weight. Others suffer severe emotional stress. They may not even like to eat, but at some level the purging relieves them of anger and frustration.

Some bulimics binge-purge once a month. Others do so several times a day. Do they know that repeated purgings can damage the gut? That chronic vomiting brings gastric fluid into the mouth and erodes teeth to stubs? That, at its extreme, bulimia causes the stomach to rupture, and the heart or kidneys to fail?

Are severe eating disorders rare? No. In the United States, an estimated 7 million females and 1 million males are anorexic or bulimic. Most are in their teens and early twenties, but the number also is increasing among preadolescents. Each year, 5 to 6 percent die as a result of complications arising from the disorders.

And with these sobering thoughts in mind, we start a tour of **nutrition**. The word encompasses processes by which an animal ingests and digests food, then absorbs the released nutrients for later conversion to the body's own carbohydrates, lipids, proteins, and nucleic acids.

The nutritional processes proceed at the **digestive system**. This body tube or cavity mechanically and chemically reduces food to particles, then to molecules that are small enough to be absorbed into the internal environment. It also eliminates unabsorbed residues. Other organ systems, especially those shown in Figure 42.2, contribute to the nutritional processes.

KEY CONCEPTS

1. In complex animals, interactions among digestive, circulatory, respiratory, and urinary systems supply the body's cells with raw materials, dispose of wastes, and maintain the volume and composition of extracellular fluid.

2. Most digestive systems have specialized regions for food transport, processing, and storage. Different regions mechanically break apart and chemically break down food, absorb breakdown products, and eliminate the unabsorbed residues.

3. To maintain an acceptable body weight and overall health, energy intake must balance energy output by way of metabolic activity, physical exertion, and so on. Complex carbohydrates provide the most dietary glucose, which typically is the body's main source of immediately usable energy.

4. Nutrition also involves the intake of foods that are good sources of vitamins, minerals, and a number of amino acids and fatty acids that the body itself cannot produce.

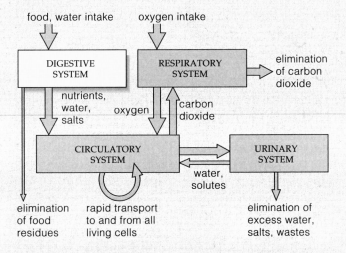

Figure 42.2 Functional links between the digestive, respiratory, circulatory, and urinary systems. These organ systems and others work together to supply cells with raw materials and eliminate wastes.

THE NATURE OF DIGESTIVE SYSTEMS

Incomplete and Complete Systems

Recall, from Chapter 26, that we don't see many organ systems until we get to the flatworms. These are among the invertebrates with an **incomplete digestive system**, a saclike, branching gut cavity that has a single opening at the start of a pharynx (Figure 42.3*a*). After food enters the sac, it is partly digested and circulated to cells even as residues are being sent back out.

Flatworms, cnidarians, some rotifers, and a few other small invertebrates require nothing more complicated than two-way traffic through a saclike gut that has a single opening. But such a system would not work well enough for the other animals. Instead, they depend on a **complete digestive system**. Basically, this is a tube with a mouth (an opening at one end for food intake) and an anus (an opening at the opposite end for eliminating unabsorbed residues). In between its openings, the tube is divided into regions that are specialized for one-way transport, processing, and storage of raw materials.

As an example, think about the complete digestive system of a frog (Figure 42.3*b*). In between the frog's mouth and anus are a pharynx, stomach, and small and large intestines. Functionally connected with the small intestine are a liver, gallbladder, and pancreas, which are organs with accessory roles in digestion. Birds, too, have a complete digestive system, with a few unique regional specializations (Figure 42.3*c*).

Regardless of its complexity, a complete digestive system carries out five overall tasks:

1. **Mechanical processing and motility**. Movements that break up, mix, and propel food material.

2. **Secretion**. Release of digestive enzymes and other substances into the space inside the tube.

3. **Digestion**. Breakdown of food into particles, then into nutrient molecules small enough to be absorbed.

4. **Absorption**. Passage of digested nutrients and fluid across the tube wall and into body fluids.

5. **Elimination**. Expulsion of the undigested and unabsorbed residues from the end of the gut.

In most animals, remember, a digestive system does not act alone. A circulatory system distributes the absorbed nutrients to cells throughout the body. A respiratory system helps cells use the nutrients by supplying them with oxygen for aerobic respiration and removing carbon dioxide wastes. A urinary system counters variations in the kinds and amounts of absorbed nutrients. It helps maintain the composition and volume of the internal environment (Figure 42.2).

Correlations With Feeding Behavior

We can correlate the specializations of any digestive system with feeding behavior. Consider the pigeon in Figure 42.3*c*. A long tube (esophagus) leads from its bill, the *food-gathering* region, to a *food-processing* region. Being compactly centered in the body mass, the food-processing region is an adaptation of an animal that must balance its body during flight. As for other seed-eating birds, a large *crop*, a stretchable storage organ, balloons from the esophagus. It allows the bird to zoom down on food, to eat a lot, and possibly to make a quick getaway before a predator becomes aware of it. Like other birds active during the day, it fills the crop before

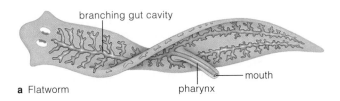

a Flatworm

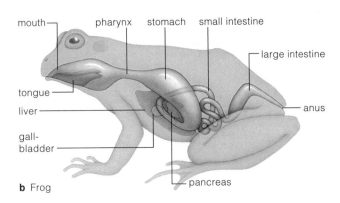

b Frog

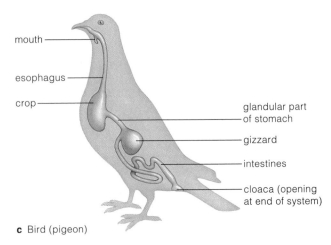

c Bird (pigeon)

Figure 42.3 (**a**) Incomplete digestive system of a flatworm, with two-way traffic of food and undigested material through one opening. (**b,c**) Examples of a complete digestive system, a tube with specialized regions and an opening at each end.

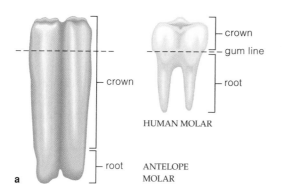

crown — gum line
HUMAN MOLAR

crown
root
ANTELOPE
MOLAR

a

Figure 42.4 Some regional specializations of the complete digestive system of pronghorn antelope (*Antilocapra americana*).

(**a**) Comparison of antelope and human molars. (**b**) Multiple-chambered antelope stomach. The first chamber is a large pouch. The second is smaller, with a honeycombed inner surface (what chefs call tripe). In both chambers, food is mixed with fluid, kneaded, and exposed to the fermentation activities of bacterial and protozoan symbionts. Some of the symbionts degrade cellulose; others synthesize organic compounds, fatty acids, and vitamins. The host uses a portion of these substances. Kneaded food is regurgitated into the mouth, rechewed, then swallowed again. It enters the third chamber, where it is pummeled once more before entering the last stomach chamber.

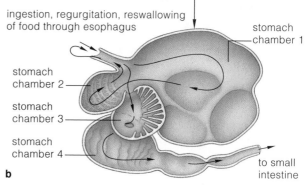

ingestion, regurgitation, reswallowing of food through esophagus

stomach chamber 1

stomach chamber 2

stomach chamber 3

stomach chamber 4

to small intestine

b

the sun goes down. It releases food during the first few hours after dark, thus decreasing the time of overnight fasting. The glandular lining of the first portion of its stomach secretes enzymes and other substances that aid in digestion. The stomach's second portion, a *gizzard*, is a muscular organ that grinds food much as teeth and jaws do. The intestines are proportionally shorter than they are in ducks, ostriches, and other birds that feed on plant parts rich in tough, fibrous cellulose—which requires a much longer processing time.

We also can correlate feeding behavior with the digestive system of pronghorn antelope (*Antilocapra americana*). Fall through winter, on the mountain ridges from central Canada down into northern Mexico, these animals browse on wild sage. Come spring, they move to open grasslands and deserts, there to browse on new growth (Figure 42.4). While they browse, they keep an eye out for coyotes, bobcats, and golden eagles. And can they eat and run! They reach speeds of 95 kilometers per hour and typically leave predators in the dust.

Now think of your own cheek teeth, or molars, each with a flattened crown that acts as a grinding platform. The crown of an antelope's molars dwarfs yours (Figure 42.4*a*). Why the difference? You probably do not rub your mouth against dirt while eating. An antelope does, and abrasive bits of soil enter its mouth along with tough plant material. Its teeth wear down rapidly, and natural selection has favored more crown to wear down.

Antelopes are **ruminants**, a type of hoofed mammal with multiple stomach chambers in which cellulose is slowly broken down. The breakdown gets under way in the first two of four stomach sacs (Figure 42.4*b*). There, bacterial symbionts produce cellulose-digesting enzymes, to the antelope's benefit as well as their own. As enzymes act, the antelope regurgitates and rechews the contents of the first two sacs, then swallows again. (That's what "chewing cud" means.) Pummeling the plant material more than once exposes more surface area to the enzymes and gives them more time to act.

Thus the elaborate stomach of ruminants accepts a steady flow of plant material during lengthy feeding times, then slowly liberates nutrients during times of rest. Predatory or scavenging mammals have different specializations. They gorge on food when they can get it, then may not eat again for some time. Part of their digestive system stores food being gulped down too fast to be digested and absorbed. Other, accessory parts of the digestive system help ensure that there will be an adequate distribution of nutrients between meals.

An incomplete digestive system is a saclike body cavity. Both food and residues enter and leave through the same opening. A complete digestive system is a tube with two openings (mouth and anus) and regional specializations.

Complete digestive systems carry out the following tasks in controlled ways: Movements that break up, mix, and propel food material; secretion of digestive enzymes and other substances; breakdown of food; absorption of nutrients; and expulsion of undigested and unabsorbed residues.

OVERVIEW OF THE HUMAN DIGESTIVE SYSTEM

Let's look at your own body as an example of the kinds of organs in a complete digestive system. The human digestive system is a tube with two openings and many specialized organs (Figure 42.5). If the tube of an adult were fully stretched, it would extend 6.5 to 9 meters (21 to 30 feet). From beginning to end, a mucus-coated epithelium lines all the surfaces facing the lumen. (A *lumen* is the space in a tube.) The thick, moist mucus protects the wall of the tube and enhances diffusion across its inner lining. Substances advance in one direction, from the mouth through the pharynx, esophagus, and gastrointestinal tract (gut). The **gut** starts at the stomach and extends through the small intestine, large intestine (colon), and rectum, to the anus. Accessory organs—the salivary glands, gallbladder, liver, and pancreas—secrete substances into different regions of the tube.

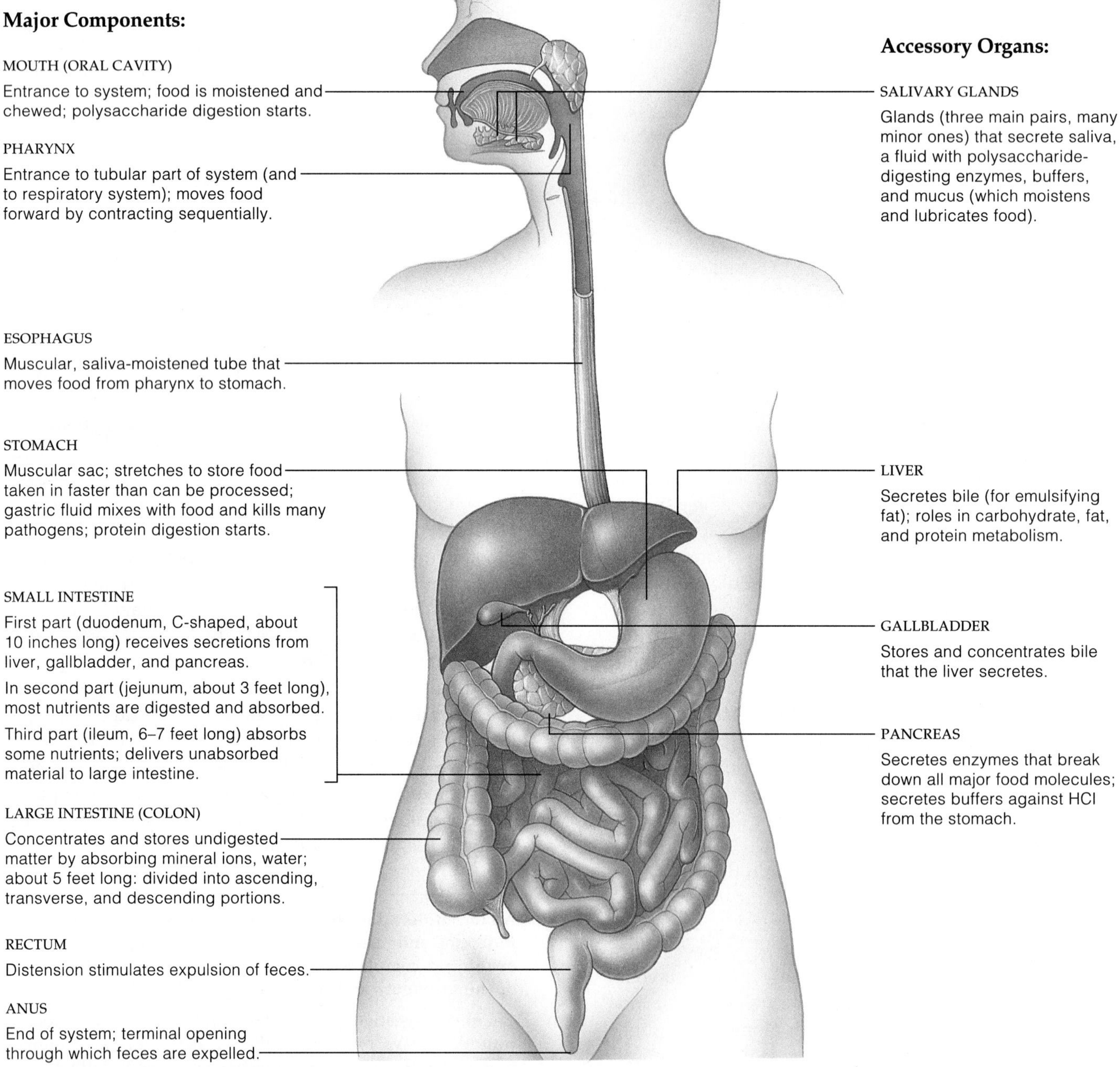

Major Components:

MOUTH (ORAL CAVITY)

Entrance to system; food is moistened and chewed; polysaccharide digestion starts.

PHARYNX

Entrance to tubular part of system (and to respiratory system); moves food forward by contracting sequentially.

ESOPHAGUS

Muscular, saliva-moistened tube that moves food from pharynx to stomach.

STOMACH

Muscular sac; stretches to store food taken in faster than can be processed; gastric fluid mixes with food and kills many pathogens; protein digestion starts.

SMALL INTESTINE

First part (duodenum, C-shaped, about 10 inches long) receives secretions from liver, gallbladder, and pancreas.

In second part (jejunum, about 3 feet long), most nutrients are digested and absorbed.

Third part (ileum, 6–7 feet long) absorbs some nutrients; delivers unabsorbed material to large intestine.

LARGE INTESTINE (COLON)

Concentrates and stores undigested matter by absorbing mineral ions, water; about 5 feet long; divided into ascending, transverse, and descending portions.

RECTUM

Distension stimulates expulsion of feces.

ANUS

End of system; terminal opening through which feces are expelled.

Accessory Organs:

SALIVARY GLANDS

Glands (three main pairs, many minor ones) that secrete saliva, a fluid with polysaccharide-digesting enzymes, buffers, and mucus (which moistens and lubricates food).

LIVER

Secretes bile (for emulsifying fat); roles in carbohydrate, fat, and protein metabolism.

GALLBLADDER

Stores and concentrates bile that the liver secretes.

PANCREAS

Secretes enzymes that break down all major food molecules; secretes buffers against HCl from the stomach.

Figure 42.5 Overview of the component parts of the human digestive system and their specialized functions. Organs with accessory roles in digestion are also listed.

INTO THE MOUTH, DOWN THE TUBE

Food is chewed and polysaccharide breakdown begins in the **oral cavity**, or **mouth**. Thirty-two teeth typically project into an adult's mouth (Figure 42.6). Each **tooth** has an enamel coat (hardened calcium deposits), dentin (a thick, bonelike layer), and an inner pulp (with nerves and blood vessels). It is an engineering marvel, able to withstand years of chemical insults and mechanical stress. Recall, from Section 27.10, that the chisel-shaped incisors shear off chunks of food. Cone-shaped canines tear it. Premolars and molars, with broad crowns and rounded cusps, are good at grinding and crushing food.

Also in the mouth is a **tongue**, an organ consisting of membrane-covered skeletal muscles that function in positioning food in the mouth, swallowing, and speech. On the tongue's surface are circular structures with taste buds embedded in their tissues (Figure 42.7). Each bud contains sensory receptors that can detect chemical

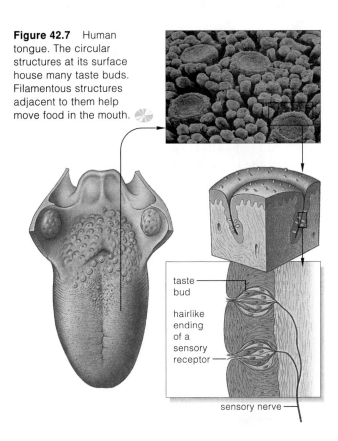

Figure 42.7 Human tongue. The circular structures at its surface house many taste buds. Filamentous structures adjacent to them help move food in the mouth.

taste bud

hairlike ending of a sensory receptor

sensory nerve

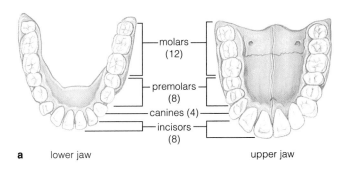

molars (12)

premolars (8)

canines (4)

incisors (8)

a   lower jaw   upper jaw

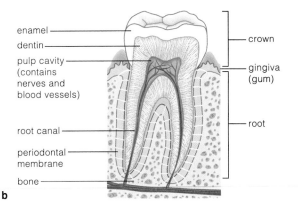

enamel

dentin

pulp cavity (contains nerves and blood vessels)

root canal

periodontal membrane

bone

crown

gingiva (gum)

root

b

Figure 42.6 (**a**) Number and arrangement of human teeth. (**b**) Closer look at a molar's two main regions, the crown and root. Enamel caps the crown. It consists of calcium deposits and is the hardest substance in the body.

Normally harmless bacteria live on and between teeth. Daily flossing, gentle brushing, and avoidance of too many sweets help keep their populations in check. In the absence of such preventive measures, conditions favor bacterial infections. These may result in *caries* (tooth decay), *gingivitis* (inflamed gums), or both. In addition, infections can spread to the periodontal membrane, which anchors teeth to the jawbone. In *periodontal disease*, infection slowly destroys the bone tissue around a tooth.

differences in dissolved substances. The brain uses this information to give rise to our sense of taste.

Chewing mixes food with **saliva**. This fluid contains an enzyme (salivary amylase), a buffer (bicarbonate, or HCO_3^-), mucins, and water. Salivary glands, beneath and in back of the tongue, produce and secrete saliva through ducts to the free surface of the mouth's lining. Salivary amylase breaks down starch. The HCO_3^- helps maintain the mouth's pH when you eat acidic foods. Modified proteins called mucins help form the mucus that binds food into a softened, lubricated ball, or bolus.

When you swallow, contractions of tongue muscles force boluses into the **pharynx**, the tubular entrance to the esophagus *and* the trachea, an airway to the lungs. As food leaves the pharynx, the epiglottis (a flaplike valve) closes off the trachea and prevents breathing. That is why you normally do not choke on food (Figure 41.23). The **esophagus** connects the pharynx with the stomach. Contractions of its muscular wall propel food past a sphincter into the stomach. A **sphincter** is a ring of smooth muscles, the contractions of which can close off a passageway or an opening to the body surface.

By the action of the mouth's teeth and tongue, food gets chewed, mixed with saliva, and bound into soft, lubricated balls that will be propelled through tubes to the stomach. Enzymes in saliva start the digestion of polysaccharides.

DIGESTION IN THE STOMACH AND SMALL INTESTINE

We arrive now at the premier food-processing organs, the stomach and small intestine. Both have layers of smooth muscles, the contractions of which break apart, mix, and directionally move food (Figure 42.8). The lumen of both organs receives digestive enzymes and other secretions with roles in the chemical breakdown of nutrients into fragments, then into molecules small enough to be absorbed. Table 42.1 lists the names and sources of the major digestive enzymes. Carbohydrate breakdown *starts* in the mouth, and protein breakdown *starts* in the stomach. But digestion of nearly all of the carbohydrates, lipids, proteins, and nucleic acids in food is *completed* in the small intestine.

The Stomach

The **stomach**, a muscular, stretchable sac (Figure 42.8*a*), serves three major functions. First, it mixes and stores ingested food. Second, its secretions help dissolve and degrade the food, proteins especially. Third, it helps control passage of food into the small intestine.

Glandular epithelium lines the stomach wall that is exposed to the lumen. Each day, the lining's glandular cells secrete about two liters of hydrochloric acid (HCl), mucus, pepsinogens, and other substances that make up

gastric fluid (stomach fluid). Stomach acidity increases when the HCl dissociates into H^+ and Cl^-, and this helps dissolve food to form a liquid mixture called **chyme**. The acidity can kill many pathogens that are ingested with food. It also can cause *heartburn* when gastric fluid backs up into the esophagus.

Protein digestion starts when the high acidity alters the structure of proteins and exposes the peptide bonds. It also converts pepsinogens to active enzymes (pepsins) that cleave the bonds. Protein fragments accumulate in the stomach's lumen. Meanwhile, glandular cells of the stomach lining secrete gastrin. This hormone stimulates the cells that secrete HCl and pepsinogen.

A *peptic ulcer*, an eroded wall region in the stomach or small intestine, results from insufficient secretion of mucus and buffers (or excess pepsin). Heredity, chronic emotional stress, smoking, and excessive use of alcohol or aspirin are contributing factors. Also, nearly everyone who has intestinal ulcers and 70 percent of those who have stomach ulcers are infected by *Helicobacter pylori*. A full course of antibiotic therapy can cure peptic ulcers in patients who test positive for this bacterium.

The stomach empties by waves of contraction and relaxation, of a type called peristalsis. The waves mix chyme and gain force as they approach a sphincter at the base of the stomach (Figure 42.8*b*). With the arrival of a strong contraction, the sphincter closes, so most of the chyme is squeezed back. A small amount moves into the small intestine. In time, most moves out of the stomach.

The Small Intestine

Figure 42.5 shows three regions of the small intestine: the duodenum, jejunum, and ileum. Each day, the stomach as well as the ducts from three organs, the **pancreas**, **liver**, and **gallbladder**, delivers about 9 liters of fluid to the duodenum. At least 95 percent of the fluid is absorbed across the epithelium that lines the small intestine.

Cells of the intestinal lining and the pancreas secrete digestive enzymes that break down food to monosaccharides (such as glucose), free fatty acids, monoglycerides, amino acids, nucleotides, and nucleotide bases. For instance, like pepsin secreted from cells in the stomach's lining, two pancreatic

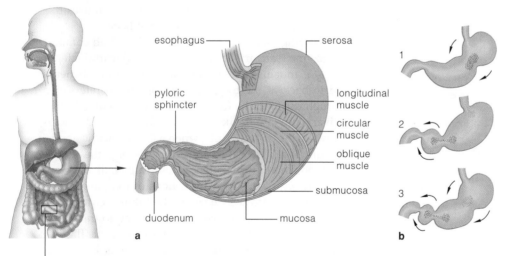

Figure 42.8 (**a**) Stomach structure. (**b**) Peristaltic wave down the stomach. (**c**) Structure of the small intestine of humans. Generally, the gut wall starts with the mucosa (an epithelium and underlying layer of connective tissue) that faces the lumen. The submucosa is a connective tissue layer with blood and lymph vessels and a mesh of nerves that function in local control over digestion. The wall has layers of smooth muscle that differ in orientation. The serosa is an outer layer of connective tissue.

Table 42.1 Major Digestive Enzymes and Their Breakdown Products

Enzyme	Source	Where Active	Substrate	Main Breakdown Products
CARBOHYDRATE DIGESTION:				
Salivary amylase	Salivary glands	Mouth	Polysaccharides	Disaccharides
Pancreatic amylase	Pancreas	Small intestine	Polysaccharides	Disaccharides
Disaccharidases	Intestinal lining	Small intestine	Disaccharides	MONOSACCHARIDES* (e.g., glucose)
PROTEIN DIGESTION:				
Pepsins	Stomach lining	Stomach	Proteins	Protein fragments
Trypsin and chymotrypsin	Pancreas	Small intestine	Proteins	Protein fragments
Carboxypeptidase	Pancreas	Small intestine	Protein fragments	AMINO ACIDS*
Aminopeptidase	Intestinal lining	Small intestine	Protein fragments	AMINO ACIDS*
FAT DIGESTION:				
Lipase	Pancreas	Small intestine	Triglycerides	FREE FATTY ACIDS, MONOGLYCERIDES*
NUCLEIC ACID DIGESTION:				
Pancreatic nucleases	Pancreas	Small intestine	DNA, RNA	NUCLEOTIDES*
Intestinal nucleases	Intestinal lining	Small intestine	Nucleotides	NUCLEOTIDE BASES, MONOSACCHARIDES*

* Breakdown products small enough to be absorbed into the internal environment.

enzymes (trypsin and chymotrypsin) digest proteins to peptides; another cleaves the peptides to free amino acids. The pancreas also secretes bicarbonate, a buffer that helps neutralize HCl arriving from the stomach.

The Role of Bile in Fat Digestion

Fat digestion requires enzyme action. It also requires **bile**, a fluid that the liver secretes continually. The fluid contains bile salts and pigments, cholesterol, and lecithin (one of the phospholipids). When the stomach is empty, a sphincter closes off the main bile duct from the liver. At such times, bile backs up into the saclike gallbladder, where it is stored and concentrated.

By a process called **emulsification**, bile salts speed up fat digestion. Most fats in our diet are triglycerides. Triglyceride molecules are insoluble in water, and they tend to aggregate into large fat globules in the chyme. As movements of the intestinal wall agitate chyme, fat globules break up into small droplets, which become coated with bile salts. Because bile salts carry negative charges, the coated droplets repel each other and remain separated. This suspension of fat droplets, formed by mechanical and chemical action, is the "emulsion."

Compared to fat globules, emulsion droplets give fat-digesting enzymes a much greater surface area to act upon. Thus triglycerides can be broken down much more rapidly to fatty acids and monoglycerides.

Controls Over Digestion

Homeostatic controls, recall, work to counter changes in the internal environment. By contrast, controls over digestion act *before* food is absorbed into the internal environment. The nervous and endocrine systems, and a mesh of nerves in the gut wall, exert the control.

For example, incoming food distends the stomach and stimulates mechanoreceptors in the stomach wall. The resulting signals travel along short reflex pathways to smooth muscles and glands; they also travel longer reflex pathways to the brain. Either way, they stimulate muscles in the gut wall to contract or glandular cells to secrete enzymes, hormones, and other substances into the stomach lumen. Stomach emptying depends on the chyme's volume and composition. For example, a large meal activates more receptors in the stomach wall, so contractions get more forceful and emptying proceeds faster. High acidity or a high fat content in the small intestine calls for hormone secretions that trigger a slowdown in stomach emptying, so food is not moved faster than it can be processed. Fear, depression, and other emotional upsets also trigger such slowdowns.

Gastrointestinal hormones are part of the controls. If chyme contains amino acids and peptides, cells of the stomach lining will secrete the gastrin that stimulates secretion of acid into the stomach. Secretin prods the pancreas to secrete bicarbonate. CCK (cholecystokinin) enhances secretin's action and causes the gallbladder to contract. When the small intestine contains glucose and fat, GIP (glucose insulinotropic peptide) calls for insulin secretion, which stimulates cells to absorb glucose.

Carbohydrate breakdown starts in the mouth, and protein breakdown starts in the stomach.

In the small intestine, most large organic compounds are digested to molecules small enough to be absorbed into the internal environment.

ABSORPTION IN THE SMALL INTESTINE

Structure Speaks Volumes About Function

Unlike the stomach, which can only absorb alcohol and a few other substances, the small intestine is the main site of absorption of the vast majority of nutrients. What makes it so?

Consider Figure 42.9, which shows the structure of the intestinal wall. Focus first on the exuberant folds of the mucosa, which project into the lumen. Look closer and you see amazing numbers of even tinier projections from each one of the large folds. Look closer still and you see that the epithelial cells at the surface of these tiny projections have a brushlike crown of even tinier projections, all exposed to the intestinal lumen.

What is the significance of so much convolution? It has an extremely favorable surface-to-volume ratio, which you read about earlier in Section 4.1. Collectively, the projections from the intestinal mucosa ENORMOUSLY increase the surface area that is available for interacting with chyme and absorbing nutrients from it. Without that immense surface area, absorption would proceed hundreds of times more slowly, which of course would not be enough to sustain human life.

As you can see from Figure 42.9a and b, the absorptive structures on each fold of the intestinal mucosa are **villi** (singular, villus). Although each villus is only about a millimeter long, the mucosa has millions of them. Their very density gives the mucosa a velvety appearance. Inside each villus is an arteriole, a venule, and a lymph vessel, which function in moving substances to and from the general circulation (Figure 42.9e).

The cells making up the epithelial lining of each villus bear **microvilli** (singular, microvillus), which are ultrafine, threadlike projections from their free surface. Each cell has about 1,700 microvilli. Hence its name, "brush border" cell. The cells near the tip of the villus produce some digestive enzymes, as listed in Table 42.1.

Mechanisms of Absorption

Absorption, recall, is the passage of nutrients, water, salts, and vitamins into the internal environment. The vast absorptive surface of the intestinal wall facilitates the process, but so does the action of smooth muscle in the intestinal wall. In **segmentation**, rings of circular muscle contract and relax repeatedly. This creates an oscillating (back and forth) movement that constantly

richly folded mucosa
of small intestine

free surface of mucosa

submucosa

a

b profusion of villi at free surface of the intestinal mucosa

c

one epithelial cell of a villus

d profusion of microvilli at the surface of one epithelial cell | part of the cytoplasm

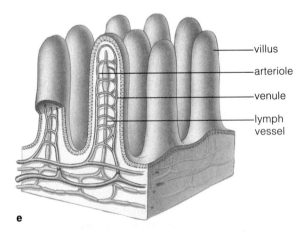

villus
arteriole
venule
lymph vessel

e

Figure 42.9 Surface structures of the mammalian small intestine at increasing magnifications.

(**a,b**) Surface view of the deep, permanent folds of the intestinal mucosa. Each fold bears a profusion of fingerlike absorptive structures called villi. (**c,d**) Individual epithelial cells at the free surface of a villus. Each has a dense crown of microvilli facing the intestinal lumen. (**e**) Blood and lymph vessels in intestinal villi. Monosaccharides and most amino acids that cross the intestinal lining enter blood vessels. Fats enter lymph vessels.

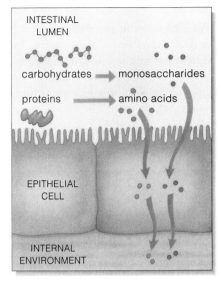
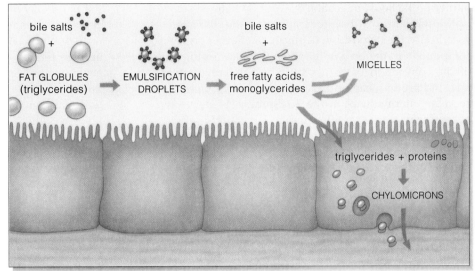

a Digestion of carbohydrates to monosaccharides, and proteins to amino acids, as completed by enzymes secreted from the pancreas and from brush border cells.

b Active transport of monosaccharides and amino acids across the plasma membrane of the cells making up the intestinal lining, then out of the same cells and into the internal environment.

c Emulsification. The movements of the intestinal wall break up fat globules into small, emulsification droplets. Bile salts prevent the globules from re-forming. Pancreatic enzymes digest the droplets to fatty acids and monoglycerides.

d Formation of micelles (as bile salts combine with digestion products and phospholipids). Products can readily slip into and out of the micelles.

e Concentration of monoglycerides and fatty acids in micelles enhances gradients that lead to diffusion of both substances across the lipid bilayer of the plasma membrane of cells making up the lining.

f Reassembly of fat digestion products into triglycerides in cells of the intestinal lining. These become coated with proteins; then they are expelled (by way of exocytosis) into the internal environment.

Figure 42.10 Digestion and absorption in the small intestine.

mixes and forces the contents of the lumen against the wall's absorptive surface:

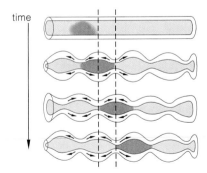

By the time that food is halfway through the small intestine, most of it has been broken apart and digested. Some breakdown products, including monosaccharides and amino acids, are then absorbed in straightforward fashion. So are water and mineral ions, which move across by way of osmosis. Transport proteins in the plasma membrane of brush border cells actively shunt these digestion products across the intestinal lining. By contrast, bile salts assist the absorption of fatty acids and monoglycerides. These products of fat digestion diffuse across the lipid bilayer of the plasma membrane (Figure 42.10).

By a process called **micelle formation**, bile salts combine with products of fat digestion, forming tiny droplets (micelles). Product molecules in the micelles continuously exchange places with product molecules that are dissolved in the chyme. When concentration gradients favor it, the molecules diffuse out of micelles, into epithelial cells. Fatty acids and monoglycerides recombine in the cells, thus forming triglycerides. Then the triglycerides combine with proteins into particles (chylomicrons) that leave the cells, by exocytosis, and enter the internal environment.

Once they have been absorbed, glucose and amino acids enter blood vessels. The triglycerides enter lymph vessels, which eventually drain into blood vessels.

With its richly folded intestinal mucosa, millions of villi, and hundreds of millions of microvilli, the small intestine offers a vast surface area for absorbing nutrients.

Substances pass through the brush border cells that line the free surface of each villus by active transport, osmosis, and diffusion across the lipid bilayer of plasma membranes.

Earlier in the book, in Section 8.6, you considered some of the mechanisms that govern organic metabolism—specifically, the disposition of glucose and other organic compounds in the body as a whole. You saw examples of the conversion pathways by which carbohydrates, fats, and proteins can be broken apart to molecules that serve as intermediates in the ATP-producing pathway of aerobic respiration. Here, Figure 42.11 rounds out the picture by showing all of the major routes by which organic compounds obtained from the diet are shuffled and reshuffled in the body as a whole.

the main energy source. There is no net breakdown of protein in muscle or other tissues during this period.

Between meals, the body taps into the fat stores. In adipose tissue, cells dismantle fats to glycerol and fatty acids and release these into blood. In the liver, cells break apart glycogen and release the glucose units, which also enter the blood. Body cells take up the fatty acids and glucose and use them for ATP production.

Bear in mind, the liver does more than store and interconvert organic compounds. For example, it helps maintain their concentrations in blood. It inactivates

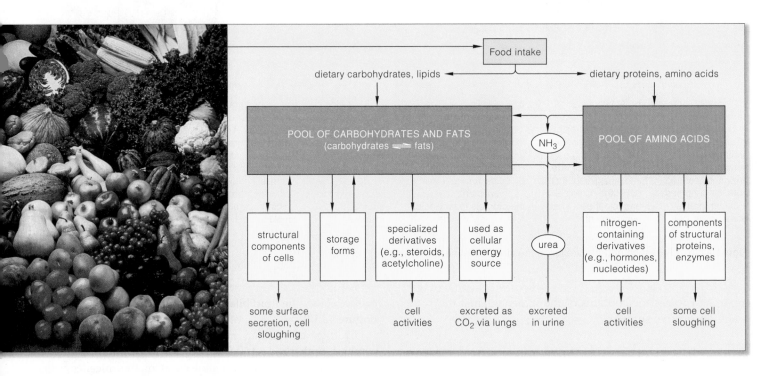

Figure 42.11 Summary of major pathways of organic metabolism. Cells continually synthesize and tear down carbohydrates, fats, and proteins. Urea forms mainly in the liver. See Sections 8.6 and 37.7.

At any time, all living cells in the body are breaking apart most of their own supplies of carbohydrates, lipids, and proteins, then using the breakdown products as energy sources or as building blocks. The nervous system and endocrine system work together to integrate this massive molecular turnover, although a treatment of how they do this would be beyond the scope of this book.

For now, simply consider a few key points about organic metabolism. When you eat, your body builds up pools of materials. Excess amounts of carbohydrates and other organic compounds absorbed from the gut are transformed mostly into fats, which become stored in adipose tissue. Some are converted to glycogen in the liver and in muscle tissue. *While* the compounds are being absorbed and stored, most cells use glucose as

most hormone molecules, which are sent to the kidneys for excretion, in urine. The liver also removes worn-out blood cells from the circulation and inactivates many potentially toxic compounds. For example, ammonia (NH_3), produced during amino acid breakdown, can be toxic in high concentrations. The liver converts NH_3 to urea. This is a less toxic waste product, and it leaves the body by way of the kidneys, in urine.

Just after a meal, the body's cells take up the glucose being absorbed from the gut and use it as a quick energy source. Excess amounts of glucose and other organic compounds become converted mainly to fats that are stored in adipose tissue. Some of the excess is converted to glycogen, which gets stored mainly in the liver and in muscle tissue.

Between meals, the body taps the fat reservoirs as the main energy source. Fats get converted to glycerol and fatty acids, both of which can enter ATP-producing pathways.

THE LARGE INTESTINE

What happens to the material *not* absorbed in the small intestine? It moves into the large intestine, or **colon**. The colon concentrates and stores feces: a mixture of water, undigested and unabsorbed matter, and bacteria. The colon starts at a cup-shaped pouch, the cecum (Figures 42.5 and 42.12). It ascends on the right side of the abdominal cavity, continues across to the other side, then descends and connects with a short tube, the rectum.

Colon Functioning

Material becomes concentrated as water moves across the colon's lining. Cells of the lining actively transport sodium ions out of the lumen. As ion concentrations in the lumen fall, the water concentration increases. Thus water also moves out of the lumen, by way of osmosis. Cells of the lining also secrete mucus, which lubricates the chyme and helps keep it from mechanically damaging the colon wall. Additionally, they secrete bicarbonate, which helps buffer the acidic fermentation products of ingested bacteria that managed to colonize the colon.

Short, longitudinal bands of smooth muscle in the colon wall are gathered at their ends, like a series of full skirts nipped in at elastic waistbands. As they contract and relax, the contents of the lumen move back and forth against the absorptive surface of the colon wall. Nerve plexuses largely control the motion, which is like segmentation in the small intestine but much slower. Whereas the fast transit time through the small intestine does not favor the rapid population growth of ingested bacteria, the colon's slower motion favors it. In general, the colonizers do no harm unless they breach the colon wall and enter the abdominal cavity.

With each new meal, hormonal signals (gastrin) and commands from autonomic nerves induce large portions of the ascending and transverse colon to contract at the same time. Within a few seconds, the lumen's existing contents move as much as three-fourths of the length of the colon and thereby make way for incoming food. The contents are stored in the last portion of the colon until defecation occurs; they distend the rectal wall enough to trigger a reflex action that leads to their expulsion. The nervous system controls the expulsion by stimulating or inhibiting contraction of a muscle sphincter at the anus, the terminal opening of the gut.

The volume of cellulose fiber and other undigested material that cannot be decreased by absorption in the colon is called **bulk**. Bulk contributes to the volume and normal transit time of material through the colon.

Colon Malfunctioning

The frequency of defecation normally ranges from three times a day to once a week. Aging, emotional stress, a

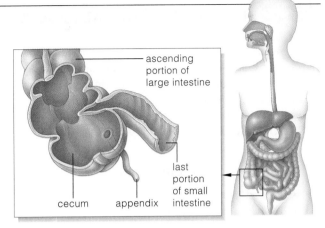

Figure 42.12   Cecum and appendix of the large intestine (colon).

low-bulk diet, injury, or disease can result in delayed defecation, or *constipation*. And the longer the delay, the more water is absorbed, so feces become hard and dry. Besides abdominal discomfort, dull headaches, loss of appetite, and often nausea and depression result.

Also, hard feces may become lodged in the **appendix**, a narrow projection from the cecum (Figure 42.12). The appendix has no known digestive functions. But it is colonized by lymphocytes that provide a line of defense against ingested bacteria. When feces obstruct normal blood flow and mucus secretion to this projection, the stage is set for *appendicitis*, or an inflamed appendix. Unless surgically removed, an inflamed appendix may rupture. Such a rupture will allow bacteria to enter the abdominal cavity and cause life-threatening infections.

The colon also is vulnerable to cancer, which occurs most often among the world's wealthiest and best-fed populations. Too many people skip meals, eat too much and too fast when they do sit down at the table, and generally give their gut erratic workouts. Their diet tends to be rich in sugar, cholesterol, and salt, and low in bulk. Too little bulk extends the transit time of feces through the colon. The longer that irritating, potentially carcinogenic material is in contact with the colon wall, the more damage it can do. Symptoms of *colon cancer* are a change in bowel functioning, rectal bleeding, and blood in feces. Some people appear to be genetically predisposed to develop colon cancer, but a low-fiber diet puts anyone at risk. Intriguingly, colon cancer is almost nonexistent in rural India and Africa, where people cannot afford to eat much more than fiber-rich whole grains. When the same people move to cities in the affluent nations and change their eating habits, the incidence of colon cancer increases.

The large intestine, or colon, functions in the absorption of water and mineral ions from the gut lumen. It also functions in the compaction of undigested residues into feces, for expulsion at the terminal opening of the gut.

You grow and maintain yourself by eating certain foods that are suitable sources of energy and raw materials. Nutritionists measure the energy in **kilocalories**. Each kilocalorie is 1,000 calories of heat energy, or the amount needed to raise the temperature of 1 kilogram of water by 1°C. The raw materials will now be described.

New, Improved Food Pyramids

A few million years ago, our hominid ancestors dined mostly on fresh fruits and other fibrous plant material. From that nutritional beginning, humans in many parts of the world came to prefer low-fiber, high-fat foods. They still do, even when they study **food pyramids**, or charts of well-balanced diets, from elementary school onward. Nutritionists revise the charts on an ongoing basis to reflect additional research. Figure 42.13 shows a recent version. It shows the daily portions of foods that will supply an average-size adult male with 58–60 percent complex carbohydrates, 15-20 percent proteins (less for females), and 20–25 percent fats. Not everyone agrees with the proportions. For example, advocates of *the zone diet* believe that a daily intake of 40 percent complex carbohydrates, 30 percent proteins, and 30 percent fats maintains the body at peak performance levels.

Carbohydrates

Starch and, to a lesser extent, glycogen should be the main carbohydrates in the human diet. These complex carbohydrates are readily broken apart into glucose units, which are the body's main energy source. Starch is abundant in fleshy fruits, cereal grains, and legumes, including beans and peas.

The foods that are rich in complex carbohydrates have an additional advantage—they typically are high in fiber. Section 42.7 highlighted the rather sobering consequences of a low-fiber diet. Epidemiologist Denis Burkitt may have been only half-joking when he put it this way: If you pass a small volume of feces, you have large hospitals.

The carbohydrates called simple sugars do not have the fiber of complex carbohydrates. Neither do they have the vitamins and minerals found in whole foods. Each week, the average American eats as much as 2 pounds of refined sugar (sucrose). You may think this a far-fetched statement, but start taking a closer look at the ingredients listed on the packages of cereal, frozen dinners, soft drinks, and other prepared foods. Many of the sugars are "hidden" as corn syrup, corn sweeteners, dextrose, and so on.

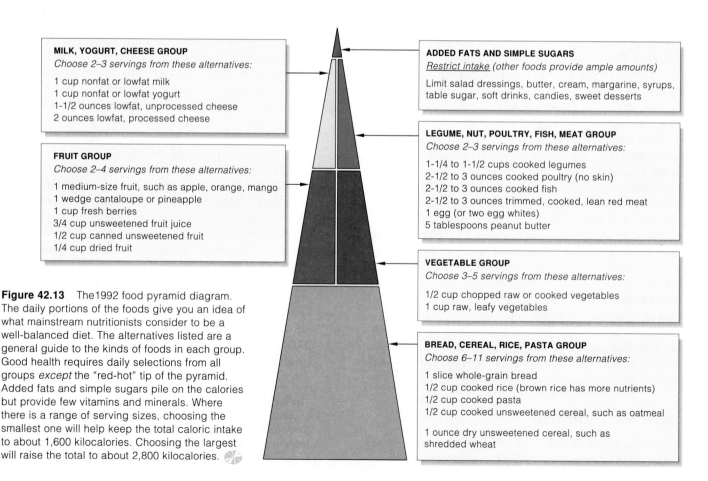

MILK, YOGURT, CHEESE GROUP
Choose 2–3 servings from these alternatives:

1 cup nonfat or lowfat milk
1 cup nonfat or lowfat yogurt
1-1/2 ounces lowfat, unprocessed cheese
2 ounces lowfat, processed cheese

FRUIT GROUP
Choose 2–4 servings from these alternatives:

1 medium-size fruit, such as apple, orange, mango
1 wedge cantaloupe or pineapple
1 cup fresh berries
3/4 cup unsweetened fruit juice
1/2 cup canned unsweetened fruit
1/4 cup dried fruit

ADDED FATS AND SIMPLE SUGARS
Restrict intake (other foods provide ample amounts)

Limit salad dressings, butter, cream, margarine, syrups, table sugar, soft drinks, candies, sweet desserts

LEGUME, NUT, POULTRY, FISH, MEAT GROUP
Choose 2–3 servings from these alternatives:

1-1/4 to 1-1/2 cups cooked legumes
2-1/2 to 3 ounces cooked poultry (no skin)
2-1/2 to 3 ounces cooked fish
2-1/2 to 3 ounces trimmed, cooked, lean red meat
1 egg (or two egg whites)
5 tablespoons peanut butter

VEGETABLE GROUP
Choose 3–5 servings from these alternatives:

1/2 cup chopped raw or cooked vegetables
1 cup raw, leafy vegetables

BREAD, CEREAL, RICE, PASTA GROUP
Choose 6–11 servings from these alternatives:

1 slice whole-grain bread
1/2 cup cooked rice (brown rice has more nutrients)
1/2 cup cooked pasta
1/2 cup cooked unsweetened cereal, such as oatmeal

1 ounce dry unsweetened cereal, such as shredded wheat

Figure 42.13 The 1992 food pyramid diagram. The daily portions of the foods give you an idea of what mainstream nutritionists consider to be a well-balanced diet. The alternatives listed are a general guide to the kinds of foods in each group. Good health requires daily selections from all groups *except* the "red-hot" tip of the pyramid. Added fats and simple sugars pile on the calories but provide few vitamins and minerals. Where there is a range of serving sizes, choosing the smallest one will help keep the total caloric intake to about 1,600 kilocalories. Choosing the largest will raise the total to about 2,800 kilocalories.

No Limiting Amino Acid	Low in Lysine	Low in Methionine, Other Sulfur-Containing Amino Acids	Low in Tryptophan

legumes:
  soybean
  tofu
  soy milk
 	
cereal grains:
  wheat germ
 	
nuts:
 	
milk
cheeses (except
  cream cheese)
yogurt
eggs
meats
	legumes:
	  peanuts
	
	cereal grains:
	  barley
	  buckwheat
	  cornmeal
	  oats
	  rice
	  rye
	  wheat
	
	nuts, seeds:
	  almonds
	  cashews
	  coconut
	  English walnuts
	  hazelnuts
	  pecans
	  pumpkin seeds
	  sunflower seeds
		legumes:
		  beans (dried)
		  black-eyed peas
		  garbanzos
		  lentils
		  lima beans
		  mung beans
		  peanuts
		
		nuts:
		  hazelnuts
		
		fresh vegetables:
		  asparagus
		  broccoli
		  green peas
		  mushrooms
		  parsley
		  potatoes
		  soybeans
		  Swiss chard
			legumes:
			  beans (dried)
			  garbanzos
			  lima beans
			  mung beans
			  peanuts
			
			cereal grains:
			  cornmeal
			
			nuts:
			  almonds
			  English walnuts
			
			fresh vegetables:
			  corn
			  green peas
			  mushrooms
			  Swiss chard

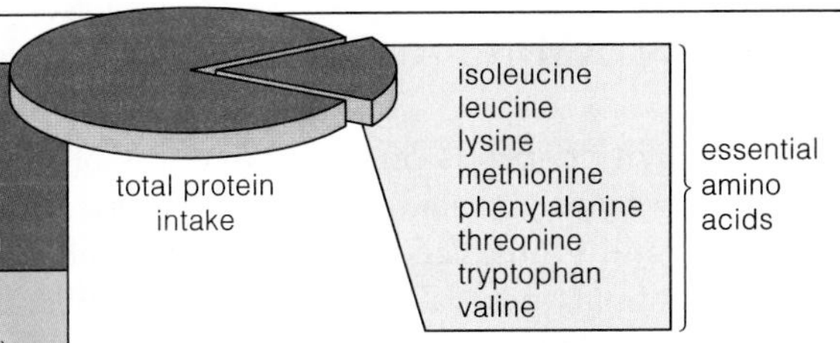

isoleucine
leucine
lysine
methionine
phenylalanine
threonine
tryptophan
valine

} essential amino acids

total protein intake

Figure 42.14 The eight essential amino acids, a small portion of the total protein intake. All must be available at the same time, in certain amounts, if cells are to build their own proteins. Milk and eggs have high amounts of all eight in proportions that humans require; they are among the *complete* proteins.

Nearly all plant proteins are *incomplete*, so vegetarians must be very careful to avoid protein deficiency. For instance, they can combine different foods from the three columns of incomplete proteins at the left. Also, strict vegetarians (who also avoid dairy products and eggs) should be sure to take vitamin B_{12} and vitamin B_2 (riboflavin) supplements. Animal protein is a luxury in most traditional societies. Their cuisines have good combinations of plant proteins, including rice/beans, chili/cornbread, tofu/rice, and lentils/wheat bread.

Lipids

The body cannot function without fats and other lipids. For example, the phospholipid lecithin is a required component of cell membranes; and fats serve as energy reserves, cushion many internal organs, and provide insulation beneath the skin. Dietary fats also help the body store fat-soluble vitamins. However, the body can synthesize most of its own fats, including cholesterol, from proteins and carbohydrates. The ones it cannot make are called **essential fatty acids**, but whole foods can provide plenty of them. Linoleic acid is an example. One teaspoon a day of corn oil, olive oil, or some other polyunsaturated fat in food provides enough of it.

Currently, butter and other fats make up 40 percent of the average diet in the United States. Most of the medical community agrees that the proportion should be 30 percent or less.

Like other animal fats, butter is a saturated fat that tends to raise the level of cholesterol in the blood. As described in earlier chapters, cholesterol is used in the synthesis of bile acids and steroid hormones, and it is a required component of our cell membranes. However, too much cholesterol may interfere seriously with blood circulation through the body (Section 39.9). New on the market are edible but nondigestible oils, such as *sucrose polyester*, for individuals who crave fats enough to put up with the digestive upsets associated with the products.

Proteins

Amino acid components of dietary proteins are needed for the body's own protein-building programs. Of the twenty common types, eight are known as the **essential amino acids**. That is, our body cells cannot synthesize them, so we must obtain them from food. The eight amino acids are methionine (or its metabolic equivalent, cysteine), isoleucine, leucine, lysine, phenylalanine (or tyrosine), threonine, tryptophan, and valine.

Most animal proteins are *complete*, meaning their ratios of amino acids match human nutritional needs. Plant proteins are *incomplete*, meaning they lack one or more essential amino acids. Therefore, to get suitable amounts of amino acids, vegetarians must eat certain combinations of different plants (Figure 42.14).

A measure called **net protein utilization** (**NPU**) is used to compare proteins from different sources. NPU values range from 100 (all essential amino acids occur in ideal proportions) to 0 (one or more amino acids are absent; when eaten alone, the protein is not complete).

Because enzymes and other proteins are vital for the body's structure and function, protein-deficient diets may have severe consequences. Protein deficiency is most damaging to fetuses and infants, for the brain grows and develops rapidly early in life. Unless the mother-to-be eats adequate amounts of protein just before and just after birth, her newborn will be affected by irreversible mental retardation. Even mild protein starvation can retard growth, and it can affect mental and physical performance.

Individuals grow and maintain themselves in good health by eating certain amounts of complex carbohydrates, lipids, and proteins that provide them with adequate raw materials and energy, as measured in kilocalories.

Metabolic activity depends on small amounts of more than a dozen organic substances called **vitamins**. Most plants are able to synthesize all of these substances on their own. Animals generally have lost the ability to do so and must obtain vitamins from food.

At the minimum, human cells require the thirteen different vitamins listed in Table 42.2. Each vitamin has specific metabolic roles. Many reactions require several vitamins, so the absence of one affects the functions of others. Metabolism also depends on certain inorganic substances called **minerals**. For example, most of your cells use calcium and magnesium in many reactions. All cells require iron in electron transport chains. Red blood cells cannot function without iron in hemoglobin, the oxygen-carrying pigment in blood. Neurons simply cannot function without sodium and potassium.

People who are in good health get all the vitamins and minerals they need from a balanced diet of whole

Table 42.2 Vitamins: Sources, Functions, and Effects of Deficiencies or Excesses*

Vitamin	Common Sources	Main Functions	Effects of Chronic Deficiency	Effects of Extreme Excess
FAT-SOLUBLE VITAMINS:				
A	Its precursor comes from beta-carotene in yellow fruits, yellow or green leafy vegetables; also in fortified milk, egg yolk, fish liver	Used in synthesis of visual pigments, bone, teeth; maintains epithelia	Dry, scaly skin; lowered resistance to infections; night blindness; permanent blindness	Malformed fetuses; hair loss; changes in skin; liver and bone damage; bone pain
D	D_3 formed in skin and in fish liver oils, egg yolk, fortified milk; converted to active form elsewhere	Promotes bone growth and mineralization; enhances calcium absorption	Bone deformities (rickets) in children; bone softening in adults	Retarded growth; kidney damage; calcium deposits in soft tissues
E	Whole grains, dark green vegetables, vegetable oils	Counters effects of free radicals; helps maintain cell membranes; blocks breakdown of vitamins A and C in gut	Lysis of red blood cells; nerve damage	Muscle weakness, fatigue, headaches, nausea
K	Enterobacteria form most of it; also in green leafy vegetables, cabbage	Blood clotting; ATP formation via electron transport	Abnormal blood clotting; severe bleeding (hemorrhaging)	Anemia; liver damage and jaundice
WATER-SOLUBLE VITAMINS:				
B_1 (thiamin)	Whole grains, green leafy vegetables, legumes, lean meats, eggs	Connective tissue formation; folate utilization; coenzyme action	Water retention in tissues; tingling sensations; heart changes; poor coordination	None reported from food; possible shock reaction from repeated injections
B_2 (riboflavin)	Whole grains, poultry, fish, egg white, milk	Coenzyme action	Skin lesions	None reported
Niacin	Green leafy vegetables, potatoes, peanuts, poultry, fish, pork, beef	Coenzyme action	Contributes to pellagra (damage to skin, gut, nervous system, etc.)	Skin flushing; possible liver damage
B_6	Spinach, tomatoes, potatoes, meats	Coenzyme in amino acid metabolism	Skin, muscle, and nerve damage; anemia	Impaired coordination; numbness in feet
Pantothenic acid	In many foods (meats, yeast, egg yolk especially)	Coenzyme in glucose metabolism, fatty acid and steroid synthesis	Fatigue, tingling in hands, headaches, nausea	None reported; may cause diarrhea occasionally
Folate (folic acid)	Dark green vegetables, whole grains, yeast, lean meats; enterobacteria produce some folate	Coenzyme in nucleic acid and amino acid metabolism	A type of anemia; inflamed tongue; diarrhea; impaired growth; mental disorders	Masks vitamin B_{12} deficiency
B_{12}	Poultry, fish, red meat, dairy foods (not butter)	Coenzyme in nucleic acid metabolism	A type of anemia; impaired nerve function	None reported
Biotin	Legumes, egg yolk; colon bacteria produce some	Coenzyme in fat, glycogen formation and in amino acid metabolism	Scaly skin (dermatitis), sore tongue, depression, anemia	None reported
C (ascorbic acid)	Fruits and vegetables, especially citrus, berries, cantaloupe, cabbage, broccoli, green pepper	Collagen synthesis; possibly inhibits effects of free radicals; structural role in bone, cartilage, and teeth; role in carbohydrate metabolism	Scurvy, poor wound healing, impaired immunity	Diarrhea, other digestive upsets; may alter results of some diagnostic tests

* The guidelines for appropriate daily intakes are being worked out by the Food and Drug Administration.

Table 42.3   Major Minerals: Sources, Functions, and Effects of Deficiencies or Excesses*

Mineral	Common Sources	Main Functions	Signs of Severe Long-Term Deficiency	Signs of Extreme Excess
Calcium	Dairy products, dark green vegetables, dried legumes	Bone, tooth formation; blood clotting; neural and muscle action	Stunted growth; possibly diminished bone mass (osteoporosis)	Impaired absorption of other minerals; kidney stones in susceptible people
Chloride	Table salt (usually too much in diet)	HCl formation in stomach; contributes to body's acid-base balance; neural action	Muscle cramps; impaired growth; poor appetite	Contributes to high blood pressure in susceptible people
Copper	Nuts, legumes, seafood, drinking water	Used in synthesis of melanin, hemoglobin, and some transport chain components	Anemia, changes in bone and blood vessels	Nausea, liver damage
Fluorine	Fluoridated water, tea, seafood	Bone, tooth maintenance	Tooth decay	Digestive upsets; mottled teeth and deformed skeleton in chronic cases
Iodine	Marine fish, shellfish, iodized salt, dairy products	Thyroid hormone formation	Enlarged thyroid (goiter), with metabolic disorders	Goiter
Iron	Whole grains, green leafy vegetables, legumes, nuts, eggs, lean meat, molasses, dried fruit, shellfish	Formation of hemoglobin and cytochrome (transport chain component)	Iron-deficiency anemia, impaired immune function	Liver damage, shock, heart failure
Magnesium	Whole grains, legumes, nuts, dairy products	Coenzyme role in ATP-ADP cycle; roles in muscle, nerve function	Weak, sore muscles; impaired neural function	Impaired neural function
Phosphorus	Whole grains, poultry, red meat	Component of bone, teeth, nucleic acids, ATP, phospholipids	Muscular weakness; loss of minerals from bone	Impaired absorption of minerals into bone
Potassium	Diet provides ample amounts	Muscle and neural function; roles in protein synthesis and body's acid-base balance	Muscular weakness	Muscular weakness, paralysis, heart failure
Sodium	Table salt; diet provides ample to excessive amounts	Key role in body's acid-base balance; roles in muscle and neural function	Muscle cramps	High blood pressure in susceptible people
Sulfur	Proteins in diet	Component of body proteins	None reported	None likely
Zinc	Whole grains, legumes, nuts, meats, seafood	Component of digestive enzymes; roles in normal growth, wound healing, sperm formation, and taste and smell	Impaired growth, scaly skin, impaired immune function	Nausea, vomiting, diarrhea; impaired immune function and anemia

* The guidelines for appropriate daily intakes are being worked out by the Food and Drug Administration.

foods. Generally speaking, specific vitamin and mineral supplements are necessary only for strict vegetarians, the elderly, and people suffering from a chronic illness or taking medication that affects the use of specific nutrients. For example, in one comprehensive study, supplements of vitamin K helped older women retain calcium and so diminish bone loss (osteoporosis) by 30 percent. Preliminary research suggests two vitamins (C and E) and beta-carotene (the precursor for vitamin A) may inactivate free radicals—those rogue metabolic fragments that may contribute to aging and may impair immune function.

No one should take massive doses of any vitamin or mineral supplement except under medical supervision. As Tables 42.2 and 42.3 indicate, excess amounts of many vitamins and minerals are harmful. For example, large doses of at least two vitamins (A and D) actually can damage the body. Like all fat-soluble vitamins, excess amounts can accumulate in tissues and interfere with normal metabolic function. Similarly, sodium is present in plant and animal tissues and is a component of table salt. Sodium has roles in the body's salt–water balance, muscle activity, and nerve function. However, prolonged, excessive intake of sodium may contribute to high blood pressure.

Severe shortages or self-prescribed, massive excesses of vitamins and minerals can disturb the delicate balances in body function that promote health.

42.10 TANTALIZING ANSWERS TO WEIGHTY QUESTIONS

IN PURSUIT OF THE "IDEAL" WEIGHT To keep your body functioning normally over the long term while simultaneously maintaining an acceptable weight, *caloric intake must be balanced with energy output.*

How many kilocalories should you take in each day to maintain a desired weight? First, multiply that weight (in pounds) by 10 if you are not active physically, by 15 if you are moderately active, and by 20 if you are highly active. Then, depending upon your age, subtract the following amount from the value obtained from the first step:

Age:	25–34	Subtract:	0
	35–44		100
	45–54		200
	55–64		300
	Over 65		400

For example, if you wish to weigh 120 pounds and are highly active, 120 × 20 = 2,400 kilocalories. If you are thirty-five years old, you would take in (2,400 − 100) or 2,300 kilocalories a day. Such calculations provide a rough estimate of caloric intake. Other factors, including height, must also be taken into consideration. An active person 5 feet, 2 inches tall does not require as much energy as an active person who weighs the same but is 6 feet tall.

The question becomes this: *Why* do you wish to weigh a given amount? Are you merely fearful of "being fat"— that is, obese? By definition, **obesity** is an excess of fat in adipose tissues, most often caused by imbalances between caloric intake and energy output. One standard of what constitutes obesity is simply cultural, and it varies from one culture to the next. For example, one female student despaired over her plumpness until she took part in a graduate studies program in Africa. There, great numbers of males considered her to be one of the most desirable females on the planet.

Traditionally, insurance companies have relied on a different standard. They developed charts mainly so that they could identify overweight individuals who might be insurance risks. The medical profession also has used the same charts.

By 1995, researchers were questioning the insurance charts. For example, after a fourteen-year study of nearly 116,000 women, researchers at Harvard University found a strong link between the risk of heart attack and gaining even 11 to 18 pounds in adult life. Women who gained weight after age 18 were by far the group at greatest risk.

In making their correlation, the Harvard researchers took into account age, smoking habits, family history of heart disorders, the use of postmenopausal hormones, and other factors that could influence the risk. *They also found conclusive evidence that exercise is the most crucial factor in controlling weight gain and avoiding heart attacks.*

Generally, thinner people simply live longer. You can use the chart in Figure 42.15 to get an idea of how a 1995 definition of "thin" applies to you.

MY GENES MADE ME DO IT As you have probably noticed, some people have far more trouble keeping off excess weight than others do. Actually, about one of every three Americans is like this. To some extent, age and emotional state have something to do with it. More importantly, *genes* have a lot to do with it.

Figure 42.15 How to estimate the "ideal" weight for an adult. The values given are consistent with a long-term Harvard study into the link between excessive weight and increased risk of heart disorders. Depending upon certain factors (such as having a small, medium, or large frame), the "ideal" may vary by plus or minus 10 percent.

Weight Guidelines for Women:

Starting with an ideal weight of 100 pounds for a woman who is 5 feet tall, add five additional pounds for each additional inch of height. Examples:

Height (feet)	Weight (pounds)
5' 2"	110
5' 3"	115
5' 4"	120
5' 5"	125
5' 6"	130
5' 7"	135
5' 8"	140
5' 9"	146
5' 10"	150
5' 11"	155
6'	160

Weight Guidelines for Men:

Starting with an ideal weight of 106 pounds for a man who is 5 feet tall, add six additional pounds for each additional inch of height. Examples:

Height (feet)	Weight (pounds)
5' 2"	118
5' 3"	124
5' 4"	130
5' 5"	136
5' 6"	142
5' 7"	148
5' 8"	154
5' 9"	160
5' 10"	166
5' 11"	172
6'	178

Researchers have suspected as much for a long time. Consider their studies of identical twins. (*Identical twins are born with identical genes.*) As an outcome of family problems, the twins who were investigated had been separated at birth and had been raised apart, in different households. Even so, at adulthood, the body weights of those separated twins were similar! People, it seemed, were born with a "set point" for body fat. Could it be that whatever set point an individual inherits is the one he or she will be stuck with for life?

Through many studies and experiments, researchers gathered strong evidence that ultimately confirmed this suspicion. It all started in 1950 with an exceptionally obese mouse (Figure 42.16). But it was not until 1995 that molecular geneticists identified one of its genes that has profound influence over the set point for body fat. They named it the *ob* **gene** (guess why). What were the clues that led to its discovery? The nucleotide sequence in the gene region of obese mice differs from that in normal mice. The gene is active in adipose tissue, nowhere else. Its biochemical information translates into a type of protein that cells can release into the bloodstream—that is, the gene might specify a hormonal signal.

Jeffrey Friedman discovered the hormone in 1994 and named it **leptin**. A nearly identical hormone was isolated in humans. Leptin is only one of a number of factors that influence the brain's commands to suppress or whip up appetite, but it's a crucial one. By assessing the blood concentrations of incoming hormonal signals, an appetite control center in the hypothalamus "decides" whether the body has taken in enough food to provide enough fat for the day. If so, commands go out to increase metabolic rates—and to stop eating.

If the *ob* gene is mutated, then its expression alone might be enough to disrupt the control center so that a mouse's appetite skyrockets and metabolic furnaces burn less. By extension, the same disruptions might occur in a human with a similar gene mutation even if he or she fully understands the inherent health risks. During one experiment that supports the hypothesis, researchers injected obese mice with leptin. The mice quickly shed their excess weight.

Will researchers go on to develop therapies for obesity in humans? Maybe, but not for five to ten years. As you saw in Chapter 37, hormonal control in humans is tricky business.

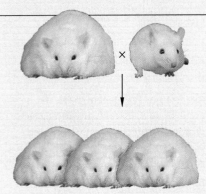

a *1950*. Researchers at the Jackson Laboratories in Maine notice that one of their laboratory mice is extremely obese, with an uncontrollable appetite. Through crossbreeding of this apparent mutant individual with a normal mouse, they produce a strain of obese mice.

b *Late 1960s*. Douglas Coleman of the Jackson Laboratories surgically joins the bloodstreams of an obese mouse and a normal one. The obese mouse now loses weight. Coleman hypothesizes that a factor circulating in blood may be influencing its appetite, but he is not able to isolate it.

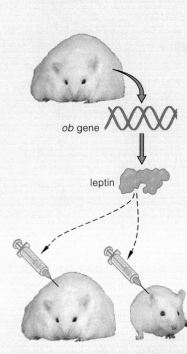

ob gene

leptin

c *1994*. Late in the year, Jeffrey Friedman of Rockefeller University discovers a mutated form of what is now called the *ob* gene in obese mice. Through DNA cloning and gene sequencing, he defines the protein that the mutated gene encodes. The protein, now called leptin, is a hormone that influences the brain's commands to suppress appetite and increase metabolic rates.

d *1995*. Three different research teams develop and use genetically engineered bacteria to produce leptin, which, when injected in obese and normal mice, triggers significant weight loss, apparently without harmful side effects.

Figure 42.16 Chronology of key developments that revealed the identity of a key factor in the genetic basis of body weight.

SUMMARY

1. Nutrition refers to all the processes by which the body takes in, digests, absorbs, and uses food.

2. Mammals have a complete digestive system, basically a tube that has two openings (a mouth and an anus), as well as regional specializations. Protective, mucus-coated epithelium lines all exposed surfaces of the tube and facilitates diffusion across the tube wall.

3. As summarized in Table 42.4, the human digestive system includes a mouth, pharynx, esophagus, stomach, small intestine, large intestine (or colon), rectum, and anus. Salivary glands and the liver, gallbladder, and pancreas have accessory roles in the system's functions.

4. These activities proceed in the digestive system:

   a. Mechanical processing and motility (movements that break up, mix, and propel food material).

   b. Secretion (release of digestive enzymes and other substances from the pancreas, liver, and glandular epithelium into the gut lumen).

   c. Digestion (breakdown of food into particles, then into nutrient molecules small enough to be absorbed).

   d. Absorption (diffusion or transport of digested organic compounds, fluid, and ions from the gut lumen into the internal environment).

   e. Elimination (the expulsion of undigested as well as unabsorbed residues at the end of the system).

5. Starch digestion starts in the mouth, and protein digestion starts in the stomach. Digestion is completed and most nutrients are absorbed in the small intestine. The pancreas secretes the main digestive enzymes. Bile from the liver assists in fat digestion.

6. In absorption, cells of the intestinal lining actively transport glucose and most amino acids out of the gut lumen. Fatty acids and monoglycerides diffuse across the lipid bilayer of these cells. In the cells' cytoplasm they are recombined as triglycerides, then are released, by exocytosis, into interstitial fluid.

7. The nervous system, endocrine system, and nerve plexuses in the gut wall interact to govern activities of the digestive system. Many controls operate in response to the volume and composition of food passing through the stomach and intestines. The controls trigger changes in muscle activity and in the rate at which hormones and enzymes are secreted.

8. Nutritionists advise a daily food intake in certain proportions. For example, for an adult male of average body weight: 58–60 percent complex carbohydrates, 12–15 percent protein, and 20–25 percent fats and other lipids. A well-balanced diet of whole foods normally provides all required vitamins and minerals.

9. To maintain a suitable body weight and overall health, caloric intake must balance energy output.

Table 42.4   Summary of the Digestive System

MOUTH (oral cavity)	Start of digestive system, where food is chewed and moistened; polysaccharide digestion begins here
PHARYNX	Entrance to tubular part of digestive and respiratory systems
ESOPHAGUS	Muscular tube, moistened by saliva, that moves food from pharynx to stomach
STOMACH	Sac where food mixes with gastric fluid and protein digestion begins; stretches to store food taken in faster than can be processed; gastric fluid destroys many microbes
SMALL INTESTINE	The first part (duodenum) receives secretions from the liver, gallbladder, and pancreas Most nutrients are digested and absorbed in the second part (jejunum) Some nutrients are absorbed in the last part (ileum), which delivers unabsorbed material to the colon
COLON (large intestine)	Concentrates and stores undigested matter (by absorbing mineral ions and water)
RECTUM	Distension triggers expulsion of feces
ANUS	Terminal opening of digestive system

Accessory Organs:

SALIVARY GLANDS	Glands (three main pairs, many minor ones) that secrete saliva, a fluid with polysaccharide-digesting enzymes, buffers, and mucus (which moistens and lubricates ingested food)
PANCREAS	Secretes enzymes that digest all major food molecules; secretes buffers against HCl from the stomach
LIVER	Secretes bile (used in fat emulsification); roles in carbohydrate, fat, and protein metabolism
GALLBLADDER	Stores and concentrates bile from the liver

Review Questions

1. Define the five key tasks carried out by a complete digestive system. Then correlate some organs of such a system with the feeding behavior of a particular kind of animal. *42.1*

2. Using the diagram on the next page, list the organs and the accessory organs of the human digestive system. On a separate sheet of paper, list the main functions of each. *42.2; Table 42.4*

3. Name the breakdown products small enough to be absorbed across the intestinal lining, into the internal environment. *42.5*

4. Define segmentation. Does it proceed in the stomach? Does it proceed in the small intestine, colon, or both? *42.5, 42.7*

Self-Quiz (*Answers in Appendix IV*)

1. The _____ maintains the internal environment, supplies cells with raw materials, and disposes of metabolic wastes.
   - a. digestive system
   - b. circulatory system
   - c. respiratory system
   - d. urinary system
   - e. interaction of all of the systems listed

2. Most digestive systems have regions for _____ food.
   - a. transporting
   - b. processing
   - c. storing
   - d. all of the above

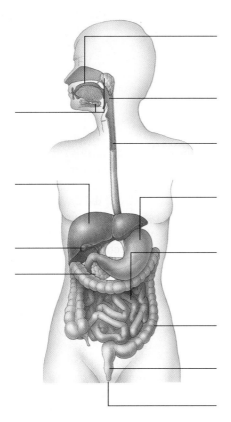

3. Maintaining good health and normal body weight requires that _____ intake be balanced by _____ output.

4. Most of our caloric intake should come from _____ .
   a. complex carbohydrates   c. proteins
   b. simple carbohydrates   d. lipids

5. On its own, the human body cannot produce all of the _____ it requires.
   a. vitamins and minerals   d. a through c
   b. fatty acids   e. a and c
   c. amino acids

6. Secretions from the _____ do *not* assist in digestion and absorption.
   a. salivary glands   c. liver
   b. thymus gland   d. pancreas

7. Digestion is completed and most nutrients are absorbed in the _____ .
   a. mouth   c. small intestine
   b. stomach   d. colon

8. Glucose and most amino acids are absorbed across the gut lining _____ .
   a. by active transport   c. at lymph vessels
   b. by diffusion   d. as fat droplets

9. Bile has roles in _____ digestion and absorption.
   a. carbohydrate   c. protein
   b. fat   d. amino acid

10. Match the organ with its key digestive function(s).
   ____ gallbladder   a. secrete bile and bicarbonate
   ____ stomach   b. digest, absorb most nutrients
   ____ colon   c. store, mix, dissolve food; start protein
   ____ pancreas   breakdown
   ____ salivary   d. store, concentrate bile
        gland   e. concentrate undigested matter
   ____ small   f. secrete substances that moisten food
        intestine   and start polysaccharide breakdown
   ____ liver   g. secrete digestive enzymes, bicarbonate

Critical Thinking

1. A glassful of whole milk contains lactose, proteins, butterfat (mostly triglycerides), vitamins, and minerals. Explain what will happen to each component in your digestive tract.

2. As a person ages, the number of body cells steadily decreases, and energy needs decline. If you were planning an older person's diet, what foods would you emphasize, and why? Which ones would you deemphasize?

3. Using Section 42.10 as a reference, determine your ideal weight and design a well-balanced program of diet and exercise that will help you achieve or maintain that weight.

4. Often, holiday meals are larger than everyday ones and have a high fat content. After stuffing themselves at an early dinner on Thanksgiving Day, Richard and other members of his family feel uncomfortably full for the rest of the afternoon. Based on what you have learned about controls over digestion, propose a biochemical explanation for their discomfort.

5. Your digestive system affords some effective protection against many pathogenic bacteria that can contaminate the kinds of food you eat. Explain some of the ways it can destroy or wash away these microorganisms. (You may wish to refer to Section 40.1, also.)

Selected Key Terms

appendix *42.7*
bile *42.4*
bulk *42.7*
chyme *42.4*
colon *42.7*
complete
   digestive system *42.1*
digestive system *CI*
emulsification *42.4*
esophagus *42.3*
essential amino acid *42.8*
essential fatty acid *42.8*
food pyramid *42.8*
gallbladder *42.4*
gastric fluid *42.4*
gut *42.2*
incomplete
   digestive system *42.1*
kilocalorie *42.8*
leptin *42.10*
liver *42.4*

micelle formation *42.5*
microvillus
   (microvilli) *42.5*
mineral *42.9*
mouth (oral cavity) *42.3*
net protein
   utilization (NPU) *42.8*
nutrition *CI*
ob gene *42.10*
obesity *42.10*
pancreas *42.4*
pharynx *42.3*
ruminant *42.1*
saliva *42.3*
segmentation *42.5*
sphincter *42.3*
stomach *42.4*
tongue *42.3*
tooth *42.3*
villus (villi) *42.5*
vitamin *42.9*

Readings

Blaser, M. J. February 1996. "The Bacteria Behind Ulcers." *Scientific American* (274).

Sherwood, L. 1997. *Human Physiology.* Second edition. Belmont, California: Wadsworth.

Wardlaw, G., P. Insel, and M. Seyler. 1992. *Contemporary Nutrition: Issues and Insights.* St. Louis: Mosby.

Withers, P. 1992. Chapter 18 of *Comparative Animal Physiology.* New York: Saunders/HBJ.

Web Site See *http://www.wadsworth.com/biology* for practice quiz questions, hypercontents, BioUpdates, and critical thinking. The Wadsworth Biology Resource Center provides a wealth of information fully organized and integrated by chapter.

43

THE INTERNAL ENVIRONMENT

Tale of the Desert Rat

Look closely at a fish or some other marine animal, and you will find that the cells of its body are exquisitely adapted to life in a salty fluid. And yet, some lineages of animals that evolved in the seas moved onto dry land about 375 million years ago. They were able to do so partly because they brought some salty fluid along with them, as an *internal* environment for their cells. Even so, it was not a simple transition. On land, those pioneers and their descendants encountered intense sunlight, dry winds, more pronounced swings in temperature, water of dubious salt content, and sometimes no water at all.

How did the pioneers conserve or replace the water and specific salts they lost as a result of their everyday activities? How did they manage to stay comfortably warm when their surroundings became too cold or too hot? They must have done these things, otherwise the volume, composition, and temperature of their internal environment would have spun out of control. In other words, *how did the land-dwelling descendants of marine animals maintain operating conditions inside the body and thus prevent cellular anarchy?*

Any of their existing descendants provides you with some answers. Think of a kangaroo rat living in an isolated desert of New Mexico (Figure 43.1). After a brief rainy season, the sun bakes the sand for months. The only obvious water is imported, sloshing in the canteens of an occasional

researcher or tourist. Yet with nary a sip of free water, this tiny mammal continually counters threats to its internal environment.

The kangaroo rat waits out the daytime heat inside a burrow, then forages in the cool of night for dry seeds and maybe a succulent. It is not sluggish about this. It hops rapidly and far, searching for seeds and fleeing from coyotes and snakes. All that hopping requires ATP energy and water. Seeds, chockful of energy-rich carbohydrates, provide both. Metabolic reactions that release energy from carbohydrates and other organic compounds also yield water. Each day, such "metabolic water" represents a whopping 90 percent of a kangaroo rat's total water intake. By comparison, it is only about 12 percent of the total intake for your body.

Inside its cool burrow, the kangaroo rat conserves and recycles water. As the animal breathes cool air into

	KANGAROO RAT	HUMAN
WATER GAIN (milliliters):		
By ingesting solids	6.0	850
By ingesting liquids	0	1,400
By way of metabolism	54.0	350
	60.0	2,600
WATER LOSS (milliliters):		
In urine	13.5	1,500
In feces	2.6	200
By evaporation	43.9	900
	60.0	2,600

Figure 43.1 Kangaroo rat, a master of water conservation in a New Mexico desert. The chart shows how kangaroo rats and humans gain and lose water. Notice the differences. Also notice that, in both cases, *the losses balance out the gains*—just as they do in all animals. How this happens, and why it absolutely must happen, is the first focus of this chapter.

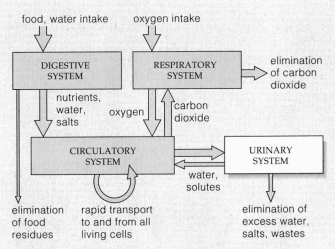

food, water intake   oxygen intake

DIGESTIVE
SYSTEM

RESPIRATORY
SYSTEM

elimination
of carbon
dioxide

nutrients,
water,
salts

oxygen

carbon
dioxide

CIRCULATORY
SYSTEM

URINARY
SYSTEM

water,
solutes

elimination
of food
residues

rapid transport
to and from all
living cells

elimination of
excess water,
salts, wastes

Figure 43.2 Links between the urinary system and other organ systems that contribute to homeostasis, or stability in favorable operating conditions in the animal body.

its warm lungs, water vapor condenses on the epithelial lining inside its nose—so some water can diffuse back inside the body. Also, after a busy night of foraging, the kangaroo rat quickly empties its cheek pouches of seeds, which soak up the small amount of water vapor that escapes by dripping from its nose. When the kangaroo rat eats the dripped-upon seeds, it recycles the water.

A kangaroo rat cannot lose water by perspiring; it has no sweat glands. It can lose water by urinating, but its specialized kidneys do not let it piddle away much. Kidneys filter the blood's water and solutes, including dissolved salts. They adjust *how much water* and *which solutes* return to the blood or leave the body as urine.

Overall, kangaroo rats and all other animals take in enough water and solutes to replace the daily losses (Figure 43.1). How they accomplish their balancing acts will be our initial focus in the chapter. Later, we will take a look at some of the means by which mammals withstand hot, cold, and sometimes unpredictable changes in environmental temperatures on land.

As a starting point, remind yourself of the kinds of fluids inside most animals. **Interstitial fluid** fills the spaces between living cells and other components of tissues. Another fluid, **blood**, transports substances to and from all tissue regions by way of a circulatory system. Taken together, the interstitial fluid and blood are the **extracellular fluid**. In most animals, a well-developed urinary system helps keep the volume and composition of extracellular fluid within tolerable ranges. As you will see, other organ systems, especially those indicated in Figure 43.2, interact with the urinary system in the performance of this homeostatic task.

KEY CONCEPTS

1. Animals are continually gaining and losing water and dissolved substances (solutes). They continually produce metabolic wastes. Even with all the inputs and outputs, the overall volume and composition of the extracellular fluid in the body remain relatively constant.

2. In humans, as in other vertebrates, a urinary system is crucial to balancing the intake and output of water and solutes. This system filters water and solutes from the blood. Then it reclaims both in amounts necessary to maintain extracellular fluid, and it eliminates the rest.

3. Kidneys are blood-filtering organs, and the urinary system of vertebrates has a pair of them. Packed inside each kidney are a great number of tubelike structures called nephrons.

4. At its beginning, each nephron cups around a set of blood capillaries, and it receives water and solutes from them. The nephron returns most of the filtrate to the blood, by giving it up to a second set of capillaries that is intertwined around the nephron's tubular parts.

5. Water and solutes not returned to the blood leave the body as a fluid called urine. During any interval, control mechanisms influence whether the urine is concentrated or dilute. Two hormones, ADH and aldosterone, have key roles in these adjustments.

6. The internal body temperature of animals depends on the balance between heat produced through metabolism, heat absorbed from the environment, and heat lost to the environment.

7. The internal body temperature is maintained within a favorable range through controls over metabolic activity and adaptations in body form and behavior.

The Challenge—Shifts in Extracellular Fluid

Different solid foods and fluids intermittently enter the mammalian gut. Afterward, variable amounts of absorbed water, nutrients, and other substances move into the blood, then into interstitial fluid and on into cells. Such events could easily shift the volume and composition of extracellular fluid beyond tolerable limits. However, the body makes compensatory adjustments that balance out the gains and losses. Within a given time frame, the body takes in as much water and solutes as it gives up.

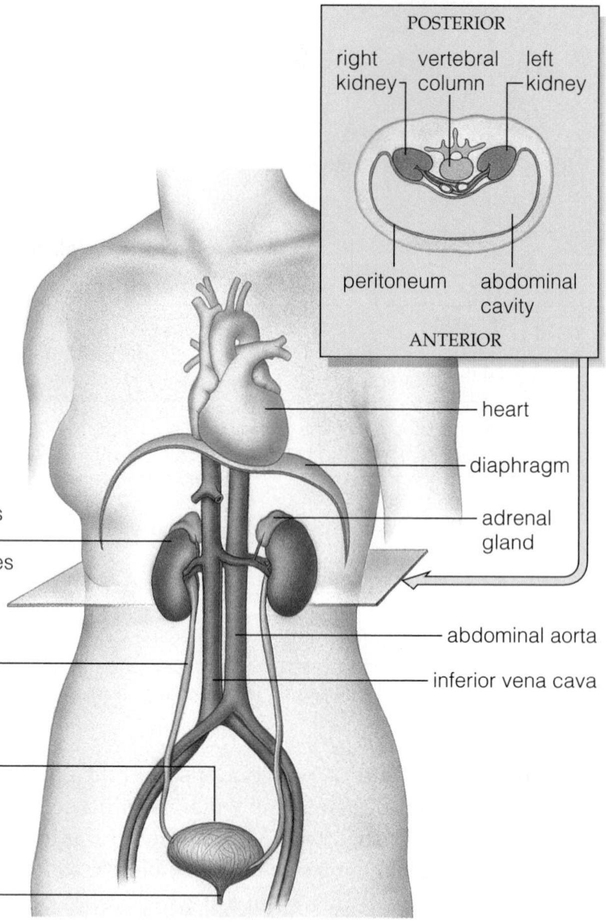

WATER GAINS AND LOSSES To start, think of a human or some other mammal. It *gains* water mainly by two processes:

> *Absorption from gut*
> *Metabolism*

Considerable water is absorbed from solids and liquids in the gut. As you know, water also forms as a normal by-product of many metabolic reactions. In land mammals, how much water enters the gut in the first place depends on a thirst mechanism. When the body loses too much water, such mammals seek out streams, water holes, and so forth. We will consider the thirst mechanism later in the chapter.

KIDNEY (one of a pair)
Constantly filters water and all solutes except proteins from blood; reclaims water and solutes as the body requires and excretes the remainder, as urine

URETER (one of a pair)
Channel for urine flow from a kidney to the urinary bladder

URINARY BLADDER
Stretchable container for temporarily storing urine

URETHRA
Channel for urine flow between the urinary bladder and body surface

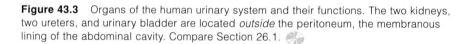

POSTERIOR
right kidney — vertebral column — left kidney
peritoneum — abdominal cavity
ANTERIOR

heart
diaphragm
adrenal gland
abdominal aorta
inferior vena cava

Figure 43.3 Organs of the human urinary system and their functions. The two kidneys, two ureters, and urinary bladder are located *outside* the peritoneum, the membranous lining of the abdominal cavity. Compare Section 26.1.

Normally, the mammalian body *loses* water mainly by way of four physiological processes, as listed here:

> *Urinary excretion*
> *Evaporation from lungs and skin*
> *Sweating, by mammals that sweat*
> *Elimination, in feces*

Urinary excretion affords the most control over water loss. This process eliminates excess water and solutes as urine, a fluid that forms in a urinary system such as that shown in Figure 43.3. Some water also evaporates from respiratory surfaces, and in some species it departs in sweat. A mammal in good health loses very little water from the gut (most is absorbed, not eliminated in feces).

SOLUTE GAINS AND LOSSES There are several ways in which a mammal *gains* solutes, but it does so mainly by these four processes:

> *Absorption from gut*
> *Secretion from cells*
> *Respiration*
> *Metabolism*

For example, nutrients and mineral ions are absorbed from the gut. So are drugs and food additives. Cellular secretions and wastes, including carbon dioxide, enter interstitial fluid, then blood. The respiratory system puts oxygen into blood, and aerobically respiring cells put carbon dioxide into it.

Typically, mammals *lose* solutes in these ways:

> *Urinary excretion*
> *Respiration*
> *Sweating, by some species*

The urine of mammals includes the wastes formed by the breakdown of organic compounds. For example, ammonia forms as amino groups are split from amino acids. Then **urea**, a major waste, forms in the liver when two ammonia molecules join with carbon dioxide. Also in urine are uric acid from nucleic acid breakdown, hemoglobin breakdown products (which give urine much of its color), drugs, and food additives. Besides these solute losses, some mammals also lose mineral ions in sweat. And all mammals lose carbon dioxide, the most abundant waste, by way of respiration.

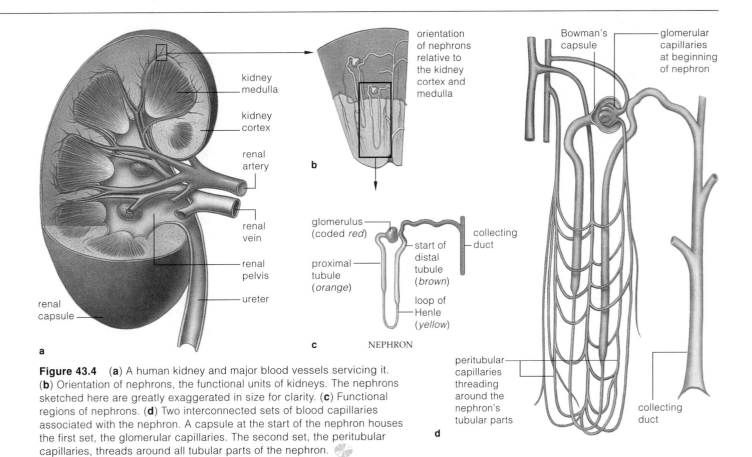

Figure 43.4 (**a**) A human kidney and major blood vessels servicing it. (**b**) Orientation of nephrons, the functional units of kidneys. The nephrons sketched here are greatly exaggerated in size for clarity. (**c**) Functional regions of nephrons. (**d**) Two interconnected sets of blood capillaries associated with the nephron. A capsule at the start of the nephron houses the first set, the glomerular capillaries. The second set, the peritubular capillaries, threads around all tubular parts of the nephron.

Components of the Urinary System

Mammals counter shifts in the composition and volume of extracellular fluid mainly by a **urinary system**. This consists of two kidneys, two ureters, a urinary bladder, and a urethra. The **kidneys** are a pair of bean-shaped organs, about as big as an average fist (Figures 43.3 and 43.4). Each has an outer capsule of connective tissue. Blood capillaries thread through its two inner regions, the cortex and medulla.

Kidneys filter water, mineral ions, organic wastes, and other substances from blood, then they adjust the filtrate's composition and return all but about 1 percent to the blood. The tiny portion of unreclaimed water and solutes is urine. By definition, **urine** is a fluid that rids the body of water and solutes that are in excess of the amounts required to maintain extracellular fluid.

Urine flows from a kidney into a **ureter**, one of two tubular channels to the **urinary bladder**. Urine is briefly stored in that muscular sac before flowing on into the **urethra**, a muscular tube that opens at the body surface. Flow from the bladder, or urination, is a reflex action. When the bladder is filled, smooth muscle present in its balloonlike wall contracts as a sphincter around its neck opens. Thus urine is forced out through the urethra. Skeletal muscle surrounds the urethra. Its contraction, which is under voluntary control, prevents urination.

Nephrons—Functional Units of Kidneys

A human kidney has more than a million **nephrons**, slender tubules packed inside lobes that extend from the cortex down through the medulla. At the nephrons, water and solutes are filtered from the blood, and the amount sent back to blood is adjusted.

Each nephron starts as a **Bowman's capsule**. Here its wall cups around *glomerular* capillaries. The cupped wall region and blood vessels are a blood-filtering unit called a **glomerulus** (Figure 43.4c). Next, the nephron has a **proximal tubule** (closest to the capsule), then a hairpin-shaped **loop of Henle** and **distal tubule** (most distant from the capsule). It ends as a **collecting duct**, which is part of a system that leads into the kidney's central cavity (renal pelvis) and the entrance to a ureter.

Blood does not give up all of its water and solutes. The unfiltered part flows into a second set of capillaries around the nephron's tubular parts. In these *peritubular* capillaries, the blood reclaims water and solutes, then flows into veins and back to the general circulation.

A urinary system counters unwanted shifts in the volume and composition of extracellular fluid. In its paired kidneys, water and solutes are filtered from blood. The body reclaims most of this, but the excess leaves the kidneys as urine.

Urine forms in nephrons by three processes: filtration, tubular reabsorption, and tubular secretion. All three depend on properties of cells of the nephron wall. The cells differ in membrane transport mechanisms and permeability from one part of the nephron to the next.

Blood pressure generated by the heart's contractions drives **filtration**, which proceeds at the glomerulus (Figure 43.5). Pressure "filters" blood by forcing water and all solutes except proteins out of the glomerular capillaries. Then the protein-free filtrate moves from the cupped part of the nephron into the proximal tubule.

Tubular reabsorption proceeds along the nephron's tubular regions. Here, most of the filtrate's water and solutes move out of the nephron's lumen (the space enclosed by its wall). Then they move into neighboring peritubular capillaries (Table 43.1 and Figure 43.6).

Tubular secretion occurs at the tubule wall but in the opposite direction of reabsorption. Solutes from the peritubular capillaries enter cells of the wall, which then secrete them into the nephron's lumen. The main solutes are ions of hydrogen and potassium: H^+ and K^+. The process also prevents some metabolites (such as uric acid) and foreign substances (such as drugs) from accumulating in blood.

Factors Influencing Filtration

Each minute, about 1.5 liters (1-1/2 quarts) of blood flow through an adult's kidneys, and 120 milliliters of water and small solutes are filtered into the nephrons. That's 180 liters of filtrate per day! The high filtration rate is possible mainly because glomerular capillaries are 10 to 100 times more permeable to water and small solutes than other capillaries are. Also, blood pressure is high in glomerular capillaries. Why? Compared with most arterioles in the body, the ones delivering blood to the glomerulus have a wider diameter and less resistance to flow. Thus, hydrostatic pressure generated by the heart does not decline as much in kidneys as elsewhere.

At any time, the blood flow to the kidneys affects the filtration rate. Neural, endocrine, and local controls maintain the flow even when blood pressure changes. For example, when you run a race or dance until dawn, the nervous system diverts an above-normal volume of blood away from kidneys, toward the heart and skeletal muscles. It directs coordinated vasoconstriction and vasodilation in different body regions, so less of the blood flow reaches the kidneys. As one more example, cells in the walls of arterioles

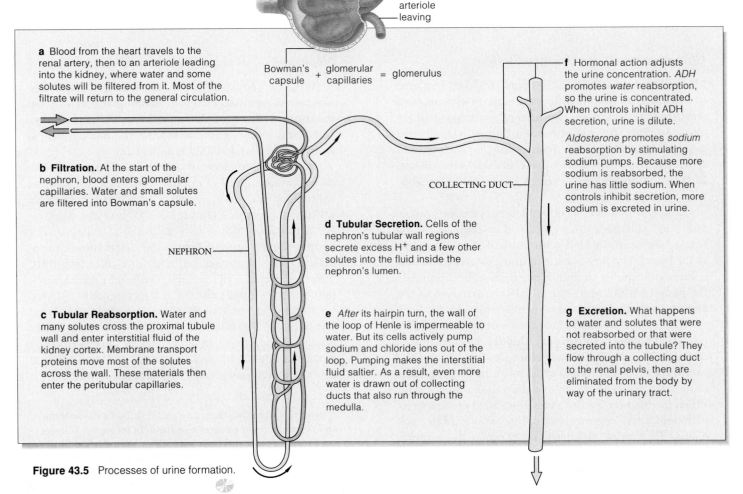

arteriole entering

arteriole leaving

a Blood from the heart travels to the renal artery, then to an arteriole leading into the kidney, where water and some solutes will be filtered from it. Most of the filtrate will return to the general circulation.

Bowman's capsule + glomerular capillaries = glomerulus

f Hormonal action adjusts the urine concentration. *ADH* promotes *water* reabsorption, so the urine is concentrated. When controls inhibit ADH secretion, urine is dilute.

Aldosterone promotes *sodium* reabsorption by stimulating sodium pumps. Because more sodium is reabsorbed, the urine has little sodium. When controls inhibit secretion, more sodium is excreted in urine.

b Filtration. At the start of the nephron, blood enters glomerular capillaries. Water and small solutes are filtered into Bowman's capsule.

COLLECTING DUCT

NEPHRON

d Tubular Secretion. Cells of the nephron's tubular wall regions secrete excess H^+ and a few other solutes into the fluid inside the nephron's lumen.

c Tubular Reabsorption. Water and many solutes cross the proximal tubule wall and enter interstitial fluid of the kidney cortex. Membrane transport proteins move most of the solutes across the wall. These materials then enter the peritubular capillaries.

e *After* its hairpin turn, the wall of the loop of Henle is impermeable to water. But its cells actively pump sodium and chloride ions out of the loop. Pumping makes the interstitial fluid saltier. As a result, even more water is drawn out of collecting ducts that also run through the medulla.

g Excretion. What happens to water and solutes that were not reabsorbed or that were secreted into the tubule? They flow through a collecting duct to the renal pelvis, then are eliminated from the body by way of the urinary tract.

Figure 43.5 Processes of urine formation.

Table 43.1	Average Daily Reabsorption Values for a Few Substances			
	Water (liters)	Glucose (grams)	Sodium (grams)	Urea (grams)
Filtered:	180	180	630	54
Excreted:	1.8	none	3.2	30
Reabsorbed:	99%	100%	99.5%	44%

leading to the glomeruli respond to pressure changes. When blood pressure decreases, they vasodilate, and so the kidneys receive more of the total blood volume. When the pressure rises, they vasoconstrict, so less blood flows in.

Reabsorption of Water and Sodium

REABSORPTION MECHANISM   Kidneys precisely adjust how much water and sodium ions the body excretes or conserves. Suppose you drink too much or too little water or wolf down salty potato chips or lose too much sodium in sweat. Responses start promptly as filtrate enters proximal tubules of nephrons. Cells in the tubule wall actively transport some sodium out of the filtrate, and other ions follow sodium into interstitial fluid. Water also leaves the filtrate, by osmosis. The nephron wall is highly permeable in this region, and about two-thirds of the filtrate's water is reabsorbed here (Figure 43.5c,d).

In the medulla, interstitial fluid is saltiest around the hairpin turn of the loop of Henle. Water moves out of the filtrate, by osmosis, *before* the turn. The fluid left behind gets saltier until it matches the interstitial fluid. The loop wall *after* the turn is impermeable to water. But sodium is pumped out by active transport mechanisms (Figure 43.5e). Interstitial fluid gets saltier and attracts more water out of the filtrate just entering the loop.

Fluid arriving at the distal tubule is dilute. The stage is now set for adjustments, which can lead to urine that is highly dilute, concentrated, or anywhere in between.

HORMONE-INDUCED ADJUSTMENTS   The cells of distal tubules and collecting ducts have hormone receptors, for ADH and aldosterone. When the extracellular fluid volume falls, the hypothalamus triggers the secretion of **ADH**, or antidiuretic hormone. ADH action makes the tubule walls more permeable to water (Figure 43.5f). More water is reabsorbed, so the urine becomes more concentrated. When the body holds excess water, ADH secretion decreases. The walls become less permeable, less water can be reabsorbed, and urine remains dilute.

Aldosterone promotes sodium reabsorption. The extracellular fluid volume decreases when too much sodium is lost. Sensory receptors in the heart and blood vessels detect the decrease and signal glandular cells in

a Sodium ions are pumped out of tubule.

b Pumping is accompanied by the movement of other ions (such as chloride and bicarbonate ions) out of tubule.

c Water follows passively, down the small osmotic gradient that the ion movements have produced.

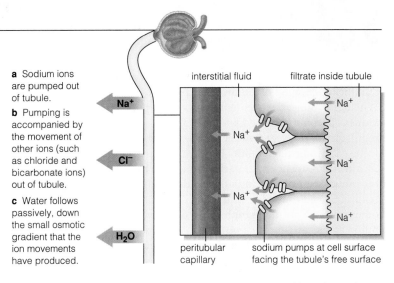

Figure 43.6   Sodium pumping associated with reabsorption of sodium and water in the kidneys.

the walls of the arteriole entering the glomerulus. These cells secrete renin, an enzyme that splits off part of a plasma protein. A second reaction converts the protein fragment to the hormone **angiotensin II**, which acts on aldosterone-secreting cells of the adrenal cortex. This is the outer portion of a gland perched on each kidney (Figure 43.3). Aldosterone stimulates cells of the distal tubules and collecting ducts to reabsorb sodium faster, so less sodium is excreted. Conversely, when the body holds excess sodium, aldosterone secretion is inhibited. Less sodium is reabsorbed, so more is excreted.

THIRST BEHAVIOR   Near the ADH-secreting cells in the hypothalamus and interacting with them is a **thirst center**. When osmoreceptors signal a decrease in blood volume and a rise in blood solute levels, it induces a sensation of thirst that compels the individual to seek water. Also, besides promoting aldosterone secretion, angiotensin II promotes thirst and ADH secretion. Finally, free nerve endings in the mouth initiate thirst behavior when they detect mouth dryness, an early sign of dehydration.

Concentrated or dilute urine forms in kidneys by filtration, tubular reabsorption, and tubular secretion.

Filtration rates depend mainly on heart contractions, which generate high hydrostatic pressure at the glomerulus of the nephron. They also depend on neural, endocrine, and local control of blood flow being directed to the kidneys.

Reabsorption, which can be adjusted by hormonal controls, helps maintain extracellular fluid. The adjustments rid the body of suitable amounts of water and solutes, in the form of dilute or concentrated urine.

ADH promotes water conservation and concentrated urine. When ADH secretion is inhibited, urine is dilute.

Aldosterone promotes sodium conservation. When its secretion is inhibited, more sodium is excreted in urine.

By this point in the chapter, you probably have sensed that the functioning of nephrons is central to good health. Whether by illness or accident, when the nephrons of both kidneys become damaged and no longer perform their regulatory and excretory functions, we call this **renal failure**. Chronic renal failure is irreversible.

Infectious agents that reach the kidneys by way of the bloodstream or through the urethra can cause renal failure. So can ingestion of lead, arsenic, pesticides, and other toxins. Continued high doses of aspirin and some other drugs can do the same thing. Abnormal retention of metabolic wastes, such as the by-products of protein breakdown, can result in *uremic toxicity*. Heart failure, atherosclerosis, hemorrhage, or shock can diminish blood flow and skew the filtration pressure in the kidneys.

Or consider *glomerulonephritis*. In rare instances after a *Streptococcus* throat infection, antibody-antigen complexes become trapped in glomeruli. Unless phagocytes remove them, the complexes continue to activate complement and other agents that bring about widespread inflammation and tissue damage. Or consider how uric acid, calcium salts, and other wastes can settle out of urine and collect in the renal pelvis as *kidney stones*. These hard deposits are usually passed in urine but can become lodged in the ureter or urethra. If they disrupt urine flow, they must be medically or surgically removed to prevent renal failure.

About 13 million people in the United States alone suffer from renal failure. A *kidney dialysis machine* often can restore the proper solute balances. Like the kidney itself, this machine helps maintain the extracellular fluid by selectively removing solutes from blood and adding solutes to it. "Dialysis" refers to an exchange of substances across an artificial membrane that is interposed between two solutions that differ in composition.

In *hemodialysis*, a clinician connects the machine to an artery or to a vein. Then the patient's blood is pumped on through tubes made of a material that is similar to sausage casing or cellophane. The tubes are submerged in a warm saline bath. The mix of salts, glucose, and other substances of the bath sets up the correct concentration gradients with blood. The blood then returns to the body. In *peritoneal dialysis*, fluid of an appropriate composition is introduced into the patient's abdominal cavity, left in place for a specific length of time, then drained out. In this case, the cavity's lining itself, the peritoneum, serves as the membrane for dialysis. For kidney dialysis to have optimum effect, it must be performed three times a week. Each time, the procedure takes about four hours, because the patient's blood must circulate repeatedly in order to improve the solute concentrations in her or his body.

Bear in mind, kidney dialysis is used as a temporary measure in reversible kidney disorders. In chronic cases, the procedure must be used for the rest of the patient's life or until a transplant operation provides her or him with a functional kidney. With treatment and controlled diets, many patients can resume fairly normal activity.

Besides maintaining the volume and composition of extracellular fluid, kidneys help keep it from becoming too acidic or too basic (alkaline). The overall **acid-base balance** is maintained by controlling the concentrations of H^+ and other dissolved ions. To get a sense of the importance of this balance, consider *metabolic acidosis*. This life-threatening condition results when the kidneys cannot secrete enough H^+ to keep pace with all of the H^+ that forms during metabolism.

Buffer systems, respiration, and urinary excretion work in concert to provide control over the acid-base balance. A buffer system, remember, consists of weak acids or bases that help minimize changes in pH by reversibly latching onto and releasing ions (Section 2.6).

Normally, the extracellular pH of the human body should be maintained between 7.37 and 7.43. As you know, acids lower the pH and bases raise it. A variety of acidic and basic substances enter the blood through absorption from the gut and as an outcome of normal metabolism. Typically, cell activities produce an excess of acids, these dissociate into H^+ and other fragments, and pH decreases. The effect is minimized when excess hydrogen ions react with buffer molecules. An example is the *bicarbonate–carbon dioxide* buffer system:

$$H^+ + HCO_3^- \rightleftharpoons H_2CO_3 \rightleftharpoons CO_2 + H_2O$$

BICARBONATE   CARBONIC ACID

In this case, the buffer system neutralizes excess H^+, and the carbon dioxide that forms during the reactions is exhaled from the lungs. Like other buffer systems in the body, however, this one only has a temporary effect. It does not *eliminate* excess H^+. Only the urinary system can do so and thereby restore the buffers.

The same reactions proceed in reverse in cells of the nephron's tubular walls. HCO_3^- formed by the reverse reactions moves into interstitial fluid, then peritubular capillaries. Afterward, it enters the general circulation and buffers excess acid. The H^+ formed in the cells is secreted into the nephron and may join with HCO_3^-. The CO_2 that formed can be returned to blood, then exhaled. The H^+ also may combine with phosphate ions or with ammonia (NH_3), then leave the body in urine.

The kidneys work in concert with buffering systems, which neutralize acids, and with the respiratory system to help keep the extracellular fluid from becoming too acidic or too basic (alkaline).

A bicarbonate–carbon dioxide buffer system temporarily neutralizes excess hydrogen ions. The urinary system alone eliminates excess hydrogen ions and restores these buffers.

The bicarbonate–carbon dioxide buffer system is one of the key mechanisms that help maintain the acid-base balance.

ON FISH, FROGS, AND KANGAROO RATS

Now that you have a general sense of how your body maintains water and solute levels, consider what goes on in some other vertebrates, including that kangaroo rat hopping about at the start of the chapter.

Bony fishes and amphibians of freshwater habitats gain water and lose solutes (Figure 43.7a). Water moves into their internal environment by osmosis; they don't gain water by drinking it. The water diffuses across the thin gill membranes in fishes and across the skin in adult amphibians. Excess water leaves as dilute urine, formed inside a pair of kidneys. In both groups of vertebrates, solute losses are balanced by solutes gained from food and by the inward pumping of sodium by certain cells.

Body fluids of herring, snapper, and other marine fishes are about three times less salty than seawater. Such fishes lose water by osmosis, and they replace it by drinking more. They excrete ingested solutes against concentration gradients (Figure 43.7b). Fish kidneys do not have loops of Henle, so urine cannot ever become saltier than body fluids. Cells in the fish gills actively pump out most of the excess solutes in blood.

Figure 43.7c describes the water-solute balancing act in a salmon. This type of fish spends part of its life cycle in freshwater and another part in seawater.

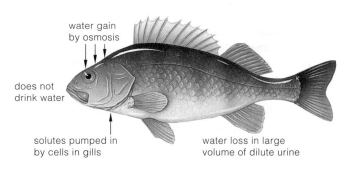

water gain
by osmosis

does not
drink water

solutes pumped in
by cells in gills

water loss in large
volume of dilute urine

a Freshwater bony fish (body fluids far saltier than surroundings)

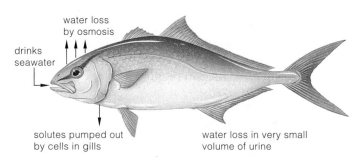

water loss
by osmosis

drinks
seawater

solutes pumped out
by cells in gills

water loss in very small
volume of urine

b Marine bony fish (body fluids less salty than surroundings)

Figure 43.7 (**a**,**b**) Water-solute balance in fishes.

(**c**) Water-solute balance by salmon, a type of fish that lives in saltwater and in freshwater. Salmon hatch in streams and later move downstream to the seas, where they feed and mature. They return to home streams to spawn.

For most salmon, salt tolerance is one outcome of changes in the concentrations of certain hormones. The changes seem to be triggered by increasing daylength in spring. Prolactin, a pituitary hormone, plays a key role in sodium retention in freshwater habitats. We know this because a freshwater fish that has its pituitary gland removed will die from sodium loss— but that fish will live if prolactin is administered to it.

Cortisol, a steroid hormone secreted by the adrenal cortex, is crucial to the development of salt tolerance in salmon. Cortisol secretions correlate with an increase in sodium excretion, in sodium-potassium pumping by cells in the salmon's gills, and in absorption of ions and water in the gut. In young salmon, cortisol secretion increases prior to the seaward movement— and so does salt tolerance.

salmon avoiding grizzly while maintaining solute-water balance

c

And about that kangaroo rat! Proportionally, the loops of Henle of its nephrons are astonishingly long, compared to yours. This means a great deal of sodium is pumped out of the nephron. Therefore, the solute concentration in the interstitial fluid around the loops becomes very high. The osmotic gradient between the fluid in the loops of Henle and the urine is *so* steep that nearly all of the water that does reach the equally long collecting ducts gets reabsorbed. In fact, kangaroo rats give up only a tiny volume of urine, which is three to five times more concentrated than the concentrated urine of humans.

The urinary systems of vertebrates differ in their details, such as the length of the nephron's loop of Henle. They are adapted to balance the body's gains in water and solutes with its losses of water and solutes in particular habitats.

We turn now to another major aspect of the internal environment—its temperature. Start by thinking about the temperature around your body. Is the air too hot? Too cold? If you have been sitting for some time, your cells have not been producing much metabolic heat. If you just finished exercising strenuously, metabolic rates have soared, and so has metabolic heat production.

Despite such differences, and assuming you are in good health, the internal temperature of your body remains much the same. A variety of physiological and behavioral responses to change work to maintain that constancy.

Temperatures Suitable for Life

Enzyme-mediated reactions proceed simultaneously in the millions to trillions of cells of a large-bodied animal. Enzymes of most animals typically function within the range of 0°–40°C (32°–104°F) as Table 43.2 shows. Above 41°C, chemical bonds holding an enzyme molecule in its required shape are disrupted. So is enzyme function. When the temperature decreases by 10 degrees, the rate of enzyme activity commonly plummets by 50 percent or more. Therefore metabolism, and life itself, depends on maintaining the **core temperature** within the range of tolerance of the body's enzymes. "Core" refers to the internal temperature of the animal body, as opposed to temperatures of the tissues near its surface.

Heat Gains and Heat Losses

Each metabolic reaction generates heat. If heat were to accumulate internally, core temperature would steadily increase. However, a warm body tends to lose heat to a cooler environment. The core temperature holds steady when the rate of heat loss balances the rate of metabolic heat production. In

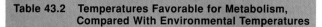

Table 43.2	Temperatures Favorable for Metabolism, Compared With Environmental Temperatures
General range of internal temperatures favorable for metabolism:	0°C to 40°C (32°F to 104°F)
Range of air temperatures above land surfaces:	−70°C to +85°C (−94°F to +185°F)
Range of surface temperatures of open ocean:	−2°C to +30°C (+28.4°F to +86°F)

general, the heat content of a large, complex animal depends on the balance between heat gains and losses, which we can express in the following way:

$$\text{CHANGE IN BODY HEAT} = \text{HEAT PRODUCED} + \text{HEAT GAINED} - \text{HEAT LOST}$$

In this summary equation, the heat gains and losses occur by exchanges at body surfaces, such as the skin. Four processes, called radiation, conduction, convection, and evaporation, drive the exchanges.

With **radiation**, an animal gains heat after exposure to radiant energy (as from sunlight) or to any surface that is warmer than the surface temperature of its body.

With **conduction**, an animal contacting a solid object directly gains heat from it or gives up heat to it in response to a thermal gradient between them. Animals lose heat by resting on objects that are cooler than their own temperature, such as a frozen deck chair (Figure 43.8a). They gain heat when they are in direct contact with warmer objects, such as hot sand (Figure 43.8b).

With **convection**, moving air or water transfers heat. Conduction plays a part in this process; heat moves down a thermal gradient between the body and air or water next to it. Mass transfer also plays a part, with currents carrying heat away from or toward the body.

a

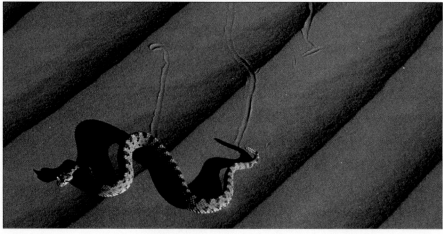

b

Figure 43.8   (**a**) How might this intrepid tourist sitting on a deck chair of a ship off the coast of Antarctica be gaining and losing heat? (**b**) A sidewinder making its signature J-shaped track across hot desert sand at dusk. As near as you can tell, how might this rattlesnake be gaining and losing heat?

When heated, air becomes less dense and moves away from the body. Even when there is no breeze, the body loses heat, for its movements create convection.

With **evaporation**, a liquid converts to gaseous form and heat is lost in the process. As described earlier, in Section 2.5, the liquid's heat content provides energy for the conversion. Evaporation from the body surface has a cooling effect, because the water molecules that are escaping carry away some energy with them.

Evaporation rates depend on humidity and on the rate of air movement. If air next to the body is already saturated with water (that is, when the local relative humidity is 100 percent), water will not evaporate. If air next to the body is hot and dry, evaporation may be the only means of countering the metabolic production of heat and the heat gains from radiation and convection.

Ectotherms, Endotherms, and In-Betweens

Animals can adjust the amount of heat lost or gained through changes in behavior and physiology, although some are better equipped than others to do so. Like most animals, snakes, lizards, and other reptiles have low metabolic rates and poor insulation (Figure 43.8b). They rapidly absorb and gain heat, especially the small species. They protect their core temperature primarily by gaining heat from their environment, not by their metabolic activities. Hence we classify these animals as **ectotherms**, which means "heat from outside."

When outside temperatures change, an ectotherm must alter its behavior. This is *behavioral* temperature regulation. For example, the iguana shown at the start of this unit basks on warm rocks, thus gaining heat by conduction. It keeps reorienting its body to expose the most surface area to the sun's infrared radiation. It loses heat after sunset. Before its metabolic rates decrease, it crawls inside crevices or under rocks, where heat loss is not as great and it is not as vulnerable to predators. As another example, meerkats bask in morning sunlight when they emerge from their burrows (Figure 33.1), and they learn to bask beneath heat lamps after being shipped to zoos in cold climates (Figure 43.9).

Most birds and mammals are **endotherms** ("heat from within"). With high metabolic rates, they can stay active under a wide temperature range. (Compared to a foraging lizard of the same weight, a foraging mouse uses up to thirty times *more* energy.) Core temperatures of these animals depend on a balancing of metabolism, controlled heat loss and conservation, and complex behavior. Adaptations in morphology help conserve or dissipate heat associated with the high metabolic rates. Feathers, fur, fat layers, even clothing reduce heat loss (Figure 43.8a). Some mammals in cold habitats have more massive bodies than close relatives in warmer ones.

Figure 43.9 Meerkats keeping warm on a cold winter night in a zoo in Germany.

For example, the snowshoe hare of Canada has a more compact body, far shorter legs, and far shorter ears than the jackrabbit of the American Southwest. Its body has a greater volume of cells for generating metabolic heat relative to the surface area available for heat loss. And heat dissipation from its legs and ears does not begin to match the losses from a jackrabbit's lengthy extremities.

Certain birds and mammals are **heterotherms**. At some times, their core temperature shifts (as it does in ectotherms). At other times, heat exchange is controlled (as in endotherms). For example, given their small size, hummingbirds have high metabolic rates. They locate and sip nectar only in the day. At night, they may shut down almost entirely. Their metabolic rates plummet, and they may get almost as cool as their surroundings. This way, they conserve precious energy.

In general, ectotherms have the advantage in warm, humid regions, such as the tropics. They need not spend much energy to maintain core temperatures, and more energy can be devoted to reproduction and other tasks. In terms of numbers and diversity, reptiles far exceed mammals in tropical regions. Endotherms have the edge in moderate to cold regions. For example, with their high metabolic rates, snowshoe hares, arctic foxes, and some other endotherms can occupy polar habitats, where you would never find a lizard.

The internal, core temperature of an animal's body is being maintained when heat gains and heat losses are in balance.

Metabolic reactions generate heat inside the body. Radiation, conduction, and convection can move heat down thermal gradients that exist between the body and its surroundings. Evaporative heat loss carries heat away from the body.

Besides being morphologically adapted to their habitats, animals can make behavioral and physiological adjustments to environmental temperatures.

Control centers maintain the core temperature of the mammal's body, and they reside in the hypothalamus (Figure 43.10). The centers continually receive input from peripheral thermoreceptors in the skin and from central thermoreceptors in the body. When the temperature deviates from a set point, the centers integrate complex responses involving skeletal muscles, smooth muscle in the arterioles that service skin, and often sweat glands. (Here you may wish to review Section 33.7.) Negative feedback loops back to the hypothalamus shut off the responses when a suitable temperature is reinstated.

Responses to Cold Stress

Table 43.3 lists the major responses that mammals make to cold stress. (Birds make the same responses.) They are called peripheral vasoconstriction, the pilomotor response, shivering, and nonshivering heat production.

Suppose peripheral thermoreceptors detect a decrease in outside temperature. They notify the hypothalamus, which commands smooth muscle in arterioles in the skin

Table 43.3	Responses to Cold Stress
Core Temperature	Responses
36°–34°C	Shivering response, increase in respiration. Increase in metabolic heat output (about 95°F). Peripheral vasoconstriction routes blood deeper in body. Dizziness and nausea set in.
33°–32°C (about 91°F)	Shivering response stops. Metabolic heat output drops.
31°–30°C (about 86°F)	Capacity for voluntary motion is lost. Eye and tendon reflexes inhibited. Consciousness is lost. Cardiac muscle action becomes irregular.
26°–24°C	Ventricular fibrillation sets in (Section 39.9). Death follows (about 77°F).

to contract. The response is **peripheral vasoconstriction**. The diameters of those arterioles constrict, which limits the blood's convective delivery of heat to body surfaces. When your fingers or toes are chilled, all but 1 percent of blood that otherwise would flow to skin is diverted to other regions of the body.

Also, muscle contractions can make hairs (or feathers) "stand up." This **pilomotor response** creates a layer of still air next to the skin and thereby reduces convective and radiative heat loss. Behavioral changes can further minimize exposed surface areas and reduce heat loss, as when cats curl up or when you hold both arms tightly against the body.

A **shivering response** can be made to prolonged cold. Rhythmic tremors begin as skeletal muscles contract, ten to twenty times a second, in response to commands from the hypothalamus. Shivering increases heat production by several times. It has a high energy cost and is not effective for long.

Prolonged or severe cold exposure also leads to a hormonal response that elevates the rate of metabolism. This **nonshivering heat production** is most notable in *brown* adipose tissue. Animals that hibernate or have become acclimatized to cold have this connective tissue. So do human infants. Adults have little unless they are adapted to cold. The tissue is present in Japanese and Korean women who spend six hours a day diving for shellfish in very cold water.

Failure to defend against cold results in *hypothermia*, a condition in which the core temperature falls below normal. In humans, a drop of only a few degrees affects brain

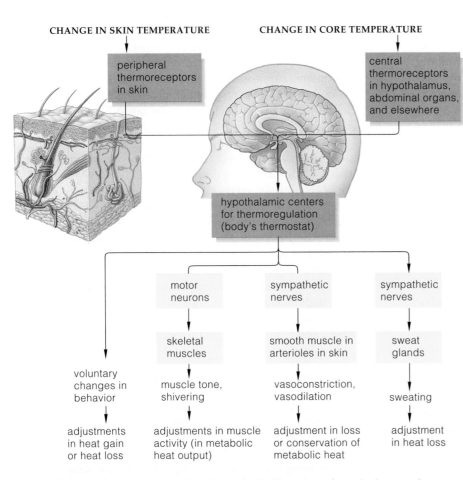

CHANGE IN SKIN TEMPERATURE　　　**CHANGE IN CORE TEMPERATURE**

peripheral thermoreceptors in skin

central thermoreceptors in hypothalamus, abdominal organs, and elsewhere

hypothalamic centers for thermoregulation (body's thermostat)

| motor neurons | sympathetic nerves | sympathetic nerves |

| skeletal muscles | smooth muscle in arterioles in skin | sweat glands |

voluntary changes in behavior

muscle tone, shivering

vasoconstriction, vasodilation

sweating

adjustments in heat gain or heat loss

adjustments in muscle activity (in metabolic heat output)

adjustment in loss or conservation of metabolic heat

adjustment in heat loss

Figure 43.10 Physiological and behavioral adjustments of a typical mammal to changes in outside temperature. This example shows the main pathways for thermoregulation in humans.

Table 43.4	Summary of Mammalian Responses to Changes in Core Temperature	
Stimulus	Main Responses	Outcome
Cold stress	Widespread vasoconstriction in skin; behavioral adjustments (e.g, minimizing surface parts exposed)	Conservation of body heat
	Increased muscle action; shivering; nonshivering heat production	Heat production increases
Heat stress	Widespread vasodilation in skin; behavioral adjustments; in some species, sweating, panting	Dissipation of heat from body
	Decreased muscle action	Heat production decreases

Figure 43.11 Evaporative water loss. Besides losing water at the moist respiratory surface of their lungs, horses, humans, and some other mammals have a large number of sweat glands that move water and solutes through pores to the skin surface.

function and leads to confusion; further cooling leads to coma and death. Many mammals can recover from profound hypothermia. But frozen cells may die unless tissues thaw under close medical supervision. Tissue destruction through localized freezing is called *frostbite*.

Responses to Heat Stress

Peripheral vasodilation and evaporative heat loss are the main responses to heat stress (Table 43.4). With **peripheral vasodilation**, hypothalamic signals cause blood vessels in skin to dilate. More blood flows from deeper body regions to skin, where the excess heat it carries is dissipated. **Evaporative heat loss** occurs at moist respiratory surfaces and across skin. Animals that sweat lose more water by this process (Figure 43.11).

For example, humans and some other mammals have sweat glands, which release water and specific solutes through pores at the skin surface. An average-size human has 2-1/2 million or more sweat glands that can produce 1 to 2 liters of sweat in an hour. For every liter of sweat that evaporates, 600 kilocalories of heat energy are lost. Bear in mind, sweat dripping from skin does not dissipate heat; water in sweat must evaporate for that to happen. On hot, humid days, the evaporation rates typically cannot match the rate of sweat secretion; the air's high water content slows it down.

During strenuous exercise, sweating can balance the high rates of heat production in skeletal muscle. With extreme sweating, as might occur in a marathon race, the body loses an important salt—sodium chloride—as well as water. Such losses disrupt the composition and volume of extracellular fluid, and when great enough, the runner collapses and faints.

What about mammals that sweat little or not at all? Some kinds make behavioral responses, such as licking fur, panting, or resting in shade. "Panting" refers to shallow, very rapid breathing that increases evaporative

water loss from the respiratory tract (compare Figure 33.14). Cooling takes place when the water evaporates from the nasal cavity, mouth, and tongue.

Sometimes peripheral blood flow and evaporative heat loss are not enough to counter heat stress in the body, and *hyperthermia* results. With this condition, the core temperature increases above normal. For humans and other endotherms, increases of only a few degrees above normal can be dangerous.

Fever

A *fever*, recall, is part of the inflammatory response to tissue damage. The hypothalamus resets the body's "thermostat," which dictates what the core temperature is supposed to be (Sections 33.7 and 40.3). Mechanisms that increase metabolic heat production and decrease heat loss are brought into play, but they are carried out to maintain a higher temperature. When they do so at the onset of fever, the person feels chilled. When the fever "breaks," peripheral vasodilation and sweating increase as the body attempts to restore the normal core temperature. Then, the person feels warm. By bringing down a fever, aspirin and other anti-inflammatory drugs may prolong healing time. But, without question, they are necessary when fevers approach dangerous levels.

Mammals counter cold stress by widespread vasoconstriction in skin, behavioral adjustments, increased muscle activity, and shivering and nonshivering heat production. They counter heat stress by widespread peripheral vasodilation in skin and by evaporative heat loss.

Control of Extracellular Fluid

1. For cells inside the animal body, the environment consists of certain types and amounts of substances that are dissolved in water. The extracellular fluid fills tissue spaces and blood vessels. Its volume and composition are maintained only when the animal's daily intake and output of water and solutes are in balance. In mammals, the following processes maintain the balance:

a. Water is gained by absorption from the gut and by metabolism. It is lost by urinary excretion, evaporation from lungs and skin, sweating, and elimination of feces.

b. Solutes are gained by absorption from the gut, secretion, respiration, and metabolism. They are lost by excretion, respiration, and sweating.

c. Losses of water and solutes are controlled mainly by adjusting the volume and composition of urine.

2. The urinary system of vertebrates consists of two kidneys, two ureters, a urinary bladder, and a urethra.

3. Kidneys have many nephrons that filter blood and form urine. Each nephron interacts intimately with two sets of blood capillaries: glomerular and peritubular.

a. The start of a nephron is cup-shaped (Bowman's capsule). It continues as three tubular regions (proximal tubule, loop of Henle, and distal tubule, which empties into a collecting duct).

b. Together, the Bowman's capsule and the set of highly permeable, glomerular capillaries within it are a blood-filtering unit (glomerulus).

c. Blood pressure forces water and small solutes out of the capillaries, into the cup. Most of the filtrate is reabsorbed by the tubules and returned to the blood. A portion is excreted as urine.

4. Urine forms in the nephron by three processes:

a. Filtration of blood at the glomerulus, which puts water and small solutes into the nephron.

b. Reabsorption. Water and solutes to be retained leave the nephron's tubular parts and enter capillaries that thread around them. A small volume of water and solutes remains in the nephron.

c. Secretion. A few substances can leave peritubular capillaries and enter the nephron, for disposal in urine.

5. Urine is made more concentrated or less so by two hormones that act on cells of the wall of distal tubules and collecting ducts. ADH conserves water by enhancing reabsorption across the wall. In its absence, more water is excreted. Aldosterone enhances sodium reabsorption. In its absence, sodium is excreted. Angiotensin II induces aldosterone secretion (sodium conservation) and thirst; it also promotes ADH secretion (water conservation).

6. The urinary system acts in concert with buffers and with the respiratory system to maintain the acid-base balance of extracellular fluid.

Control of Body Temperature

1. Maintaining an animal's core (internal) temperature depends on balancing metabolically produced heat and the heat absorbed from and lost to the environment.

2. Animals exchange heat with their environment by four processes:

a. Radiation. Emission from the body of infrared and other wavelengths. Radiant energy can be absorbed at the body surface, then converted to heat energy.

b. Conduction. Direct transfer of heat energy from one object to another object in contact with it.

c. Convection. Heat transfer by air or water currents; involves conduction and mass transfer of heat-bearing currents away from or toward the animal body.

d. Evaporation. Conversion of liquid to a gas, driven by energy inherent in the heat content of the liquid. Some animals lose heat by evaporative water loss.

3. Core temperatures depend on metabolic rates and on anatomical, behavioral, and physiological adaptations.

a. For ectotherms, the core temperature depends more on heat exchange with the environment than on heat generated by metabolism.

b. For endotherms, the core temperature depends largely on high metabolic rates and precise controls over heat produced and heat lost.

c. For heterotherms, the core temperature fluctuates some of the time, and controls over heat balance come into play at other times.

Review Questions

1. State the function of the urinary system in terms of gains and losses for the internal environment. Name the components of the mammalian urinary system and state their functions. *43.1*

2. Label the component parts of this kidney and nephron. *43.1*

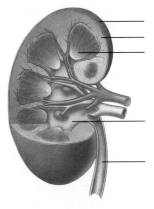 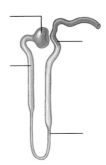

3. Define filtration, tubular reabsorption, and secretion. How does urine formation help maintain the internal environment? *43.2*

4. Which hormones promote (a) water conservation, (b) sodium conservation, and (c) thirst behavior? *43.2*

5. Name and define the physical processes by which animals gain and lose heat. What are the main physiological responses to cold stress and to heat stress in mammals? *43.6, 43.7*

Self-Quiz (Answers in Appendix IV)

1. In mammals, water intake depends on _____ .
   a. absorption from gut   c. a thirst mechanism
   b. metabolism   d. all of the above

2. In mammals, water is lost by way of the _____ .
   a. skin   d. urinary system
   b. respiratory system   e. c and d
   c. digestive system   f. a through d

3. Water and small solutes enter kidneys during _____ .
   a. filtration   c. tubular secretion
   b. tubular reabsorption   d. both a and c

4. Kidneys return water and small solutes to blood by _____ .
   a. filtration   c. tubular secretion
   b. tubular reabsorption   d. both a and b

5. A few substances move out of the peritubular capillaries that thread around tubular parts of the nephron. The substances are moved into the nephron during _____ .
   a. filtration   c. tubular secretion
   b. tubular reabsorption   d. both a and c

6. A nephron's reabsorption mechanism depends on _____ .
   a. osmosis across nephron wall
   b. active transport of sodium across nephron wall
   c. a steep solute concentration gradient
   d. all of the above

7. _____ promotes water conservation.
   a. ADH   c. Low extracellular fluid volume
   b. Aldosterone   d. Both a and c

8. _____ enhances sodium reabsorption.
   a. ADH   c. Low extracellular fluid volume
   b. Aldosterone   d. Both b and c

9. Match the term with the most suitable description.
   ____ glomerulus   a. surrounded by saltiest fluid
   ____ distal tubule   b. extra-long loops of Henle
   ____ loop of Henle   c. involves buffer systems
   ____ acid-base balance   d. blood-filtering unit
   ____ kangaroo rat   e. ADH, aldosterone act here

10. Match the term with the most suitable description.
    ____ ectotherm   a. heat transfer by air or water currents
    ____ endotherm   b. body temperature fluctuates some of
    ____ evaporation   the time, is controlled other times
    ____ heterotherm   c. emission of wavelength energy
    ____ radiation   d. direct heat transfer between one
    ____ conduction   object and another in contact with it
    ____ convection   e. metabolism dictates core temperature
                        f. environment dictates core temperature
                        g. conversion of liquid to gas

Critical Thinking

1. Fatty tissue holds kidneys in place. Extremely rapid weight loss may cause the tissue to shrink and the kidneys to slip from their normal position. If slippage puts a kink in one or both ureters and blocks urine flow, what may happen to the kidneys?

2. Drink one quart of water in an hour. What changes can you expect in your kidney function and in urine composition?

3. In 1912, the ocean liner *Titanic* left Europe on her maiden voyage across the Atlantic to America. In that same year, a chunk of the leading edge of a Greenland glacier broke off and floated out to sea. Late at night, off the Newfoundland coast, the iceberg and the *Titanic* made an ill-fated rendezvous (Figure 43.12). The *Titanic* was said to be unsinkable. Survival drills had been neglected. There were not enough lifeboats to hold even half the 2,200 passengers. The *Titanic* sank in about 2-1/2 hours.

Figure 43.12 Sinking of the *Titanic*, based on eyewitness accounts.

In less than two hours, rescue ships were on the scene, but 1,517 bodies were recovered from a calm sea. All the dead had on life jackets. None had drowned. Probably they died from _____ . If so, how did their blood flow, metabolism, and skeletal muscle action change prior to death?

4. When iguanas have an infection, they rest for a prolonged period in the sun. Propose a hypothesis to explain why.

5. Out on a first date, Jon takes Geraldine's hand in a darkened theater. "*Aha!*" he thinks. "*Cold hands, warm heart!*" What does this tell us about the regulation of core temperature, let alone Jon?

Selected Key Terms

Salt-Water Balance:
acid-base
  balance *43.4*
ADH *43.2*
aldosterone *43.2*
angiotensin II *43.2*
blood *CI*
Bowman's
  capsule *43.1*
collecting duct *43.1*
distal tubule *43.1*
extracellular
  fluid *CI*
filtration *43.2*
glomerulus *43.1*
interstitial fluid *CI*
kidney *43.1*
loop of Henle *43.1*
nephron *43.1*
proximal tubule *43.1*

renal failure *43.3*
shivering
  response *43.7*
thirst center *43.2*
tubular
  reabsorption *43.2*
tubular
  secretion *43.2*
urea *43.1*
ureter *43.1*
urethra *43.1*
urinary
  bladder *43.1*
urinary
  excretion *43.1*
urinary
  system *43.1*
urine *43.1*

Body Temperature:
conduction *43.6*
convection *43.6*
core temperature *43.6*
ectotherm *43.6*
endotherm *43.6*
evaporation *43.6*
evaporative
  heat loss *43.7*
heterotherm *43.6*
nonshivering heat
  production *43.7*
peripheral
  vasoconstriction *43.7*
peripheral
  vasodilation *43.7*
pilomotor
  response *43.7*
radiation *43.6*

Readings

Flieger, K. March 1990. "Kidney Disease: When Those Fabulous Filters Are Foiled." *FDA Consumer* 24: 26–29.

Sherwood, L. 1997. *Human Physiology*. Second edition. Belmont, California: Wadsworth.

Smith, H. 1961. *From Fish to Philosopher*. New York: Doubleday.

Web Site See *http://www.wadsworth.com/biology* for practice quiz questions, hypercontents, BioUpdates, and critical thinking. The Wadsworth Biology Resource Center provides a wealth of information fully organized and integrated by chapter.

PRINCIPLES OF REPRODUCTION AND DEVELOPMENT

From Frog to Frog and Other Mysteries

With a quavering, low-pitched call that only a female of its kind could find seductive, a male frog proclaims the onset of warm spring rains, of ponds, of sex in the night. By August the summer sun will have parched the earth, and his pond dominion will be gone. But tonight is the hour of the frog!

Through the dark, a hormone-primed female moves toward the vocal male. They meet; they dally in the behaviorally prescribed ways of their species. Then he clamps his forelegs above her swollen abdomen and gives her a prolonged squeeze (Figure 44.1a). Out into the water streams a ribbon of hundreds of eggs, which the male blankets with a milky cloud of sperm. Soon afterward, fertilized eggs—**zygotes**—are suspended in the water.

For the leopard frog, *Rana pipiens*, a drama begins to unfold that has been reenacted each spring, with only minor variations, for many millions of years. Within a few hours after fertilization, each zygote divides into two cells, the two divide into four, then the four into eight. In less than twenty hours after fertilization, the

mitotic cell divisions have produced a ball of cells no larger than the zygote. It is an early embryonic stage, of a type known as a blastocyst.

The cells continue to divide, but now they start to interact by way of their surface structures and chemical secretions. At prescribed times, many change shape and migrate to prescribed positions in the embryo. All of those cells inherited the same genetic instructions from the zygote—yet now they are becoming different from one another in appearance and function!

Through their associations, the cells form layers of embryonic tissues and then embryonic organs. A pair of tissue regions at the embryo's surface interact with the tissues beneath them. Together they give rise to a pair of eyes. Within the embryo a heart is forming, and soon it starts an incessant, rhythmic beating. In less than a week, events have transformed the frog embryo into a swimming, algae-eating larva called a tadpole.

Several months pass. Legs form; the tail shortens and disappears. The mouth develops jaws that snap shut on insects and worms. Eventually the transformations lead

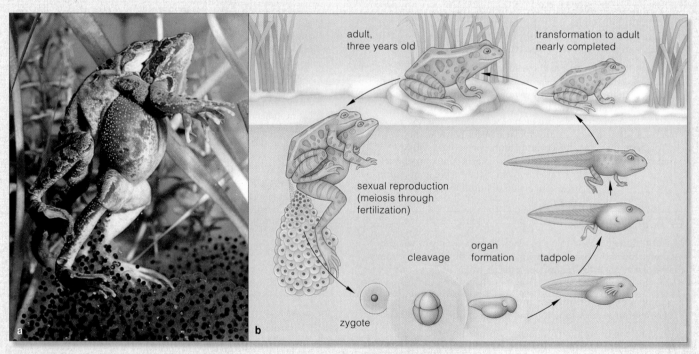

Figure 44.1 Reproduction and development of *Rana pipiens*, the leopard frog. (**a,b**) We zoom in on the life cycle as a male leopard frog clasps a female in a reproductive behavior called amplexus. The female releases her eggs into the water, then the male releases sperm over the eggs. A zygote forms when an egg nucleus and a sperm nucleus fuse at fertilization. (**c**) Frog embryos suspended in the water. (**d**) A tadpole. (**e**) A transitional form between a tadpole and the young adult frog (**f**).

to an adult frog. With luck the frog will avoid predators, disease, and other threats in the months ahead. In time it may even call out quaveringly across a moonlit pond, and the life cycle will turn again.

Many years ago you, too, started a developmental journey when a zygote carved itself up. Three weeks into the journey, your embryonic body had the stamp of "vertebrate" on it. A mere five weeks after that, it was a recognizable human in the making!

With this chapter we turn to one of life's greatest dramas—the development of offspring in the image of sexually reproducing parents. The guiding question is this: *How does the single-celled zygote of a frog, human, or any other complex animal become transformed into all of the specialized cells and structures of the adult form?* Some answers will become apparent as we move through a survey of basic developmental principles.

DEVELOPING EMBRYO

KEY CONCEPTS

1. Sexual reproduction dominates the life cycle of nearly all animals, but separation into sexes has biological costs. It requires the construction and maintenance of specialized reproductive structures. It also requires hormonal control mechanisms and complex forms of behavior attuned to the environment and to potential mates and rivals.

2. Separation into sexes affords a selective advantage. The offspring show variation in traits, which improves the odds that at least some will survive and reproduce despite unexpected challenges from the environment. This reproductive advantage offsets the biological cost of the separation.

3. The life cycle of humans and many other animals proceeds through six stages of embryonic development. The stages are called gamete formation, fertilization, cleavage, gastrulation, organ formation, and growth and tissue specialization.

4. Each stage of embryonic development builds on the tissues and structures that formed during the stage that preceded it.

5. In a developing embryo, the fate of each type of cell depends partly on cleavage, which distributes different regions of the fertilized egg's cytoplasm to different daughter cells. It also depends on interactions among cells of the embryo. These activities are the foundation for cell differentiation and morphogenesis.

6. With cell differentiation, each type of cell selectively uses certain genes and synthesizes proteins not found in other types, and so becomes unique in structure and function.

7. With morphogenesis, tissues and organs change in size, shape, and proportion. They also become organized relative to one another in prescribed patterns, and they do so at prescribed times.

Sexual Versus Asexual Reproduction

In earlier chapters, you read about the cellular basis of **sexual reproduction**. Briefly, by this reproductive mode, meiosis and gamete formation typically occur in two prospective parents. Then, at fertilization, a gamete from one of them fuses with a gamete from the other, thereby forming the zygote, the first cell of a new individual. You also read about **asexual reproduction**, by which a single parent organism produces offspring by various means (but not by gamete formation). Consider now a few structural, behavioral, and ecological aspects of the two reproductive modes.

Think of a scuba diver accidentally kicking a sponge. A tissue fragment breaks away from the sponge body, then grows and develops by mitotic cell divisions and cell differentiation into a new sponge. Or think of one of the flatworms that can undergo transverse fission as it glides through the water. Its body constricts below the midsection. The part behind the constriction grips a substrate and starts a tug-of-war with the part in front of it. After a few hours, it splits off. Both parts go their separate ways, regenerate the missing part, and become a whole worm. Only certain flatworms can do this.

In such cases of *asexual* reproduction, all offspring are genetically the same as the individual parent, or nearly so. Phenotypically they are much the same, also. We can speculate that phenotypic uniformity is useful when each individual's gene-encoded traits are highly adaptive to a limited and more or less consistent set of environmental conditions. Variation introduced into the finely tuned gene package would not do much good, and often it would do harm.

However, most animals live where opportunities, resources, and danger are highly variable. For the most part, such animals reproduce sexually, with female and male parents bestowing different mixes of alleles on the offspring (Section 10.1). The resulting variation in traits improves the odds that some of the offspring, at least, will survive and reproduce even if conditions change in the environment.

Costs and Benefits of Sexual Reproduction

Among animals, separation into sexes is not without cost. Some cells that can serve as gametes must be set aside and nurtured. Housing and delivering gametes near or inside of a prospective mate requires specialized reproductive structures. Often, mating requires special forms of behavior, such as courtship, that can promote fertilization. Mating also requires built-in controls that can synchronize the timing of gamete formation, sexual readiness, even parental behavior in two individuals.

Figure 44.2 *Facing page:* A few examples of where invertebrate and vertebrate embryos develop, how they are nourished, and how (if at all) parents protect them.

(**a**) Snails are *oviparous*, which means that they are egg producers (*ovi-*, egg; *parous*, produce). Parents release the fertilized eggs, which develop on their own, unprotected.

(**b**) Birds also are oviparous. Their fertilized eggs have large reserves of yolk, and they develop and hatch outside the mother's body. Unlike snails, one or both parents expend considerable energy feeding and caring for the young.

Some fishes, lizards, and many snakes are *ovoviviparous*. Their fertilized eggs develop inside the mother; then offspring are born live. Yolk reserves, not the mother's own tissues, sustain the eggs. The copperhead shown in (**c**) is an example. Her offspring are born live inside the relics of egg sacs.

Most mammals are *viviparous*; they produce young that are born live (*vivi-*, alive). (**d,e**) In kangaroos and some other species, embryos are born in unfinished form. The young of this marsupial complete development in a pouch on the mother's ventral surface. Juvenile stages (joeys) continue to draw nourishment from mammary glands inside the mother's pouch. (**f**) By contrast, a human female retains the fertilized egg inside her body. Maternal tissues nourish the developing individual until the time of birth, in the manner described in the next chapter.

Consider the question of *reproductive timing*. How do the mature sperm in one individual become available exactly when the eggs mature in a different individual? Timing depends upon energy outlays for constructing, maintaining, and operating neural as well as hormonal control mechanisms in each parent. Also, parents must produce mature gametes in response to the same cues, such as a seasonal change in daylength, that mark the onset of the most suitable time of reproduction for their species. For example, male and female moose become sexually active only in late summer and early fall. The coordinated timing ensures that their offspring will be born the following spring—when the weather improves and food will be plentiful for many months.

Or consider the challenge of finding and recognizing a potential mate of the same species. Different species invest energy when they synthesize chemical signaling molecules, mating signals such as feathers of certain colors and patterns, and complex sensory receptors that can detect the specific signals being sent. Besides this, males often expend astonishing amounts of energy on courtship routines, as you will read in Chapter 51.

Assuring the survival of offspring is also costly. For example, many invertebrates and bony fishes simply release eggs and motile sperm into the surroundings (Figure 44.1a). The odds for fertilization would not be good if adults produced only *one* sperm or *one* egg each season. Such species invest energy in many gametes, often thousands of them. As another example, nearly all land animals depend on **internal fertilization**, the union

a

b

liveborn snake inside egg sac

c

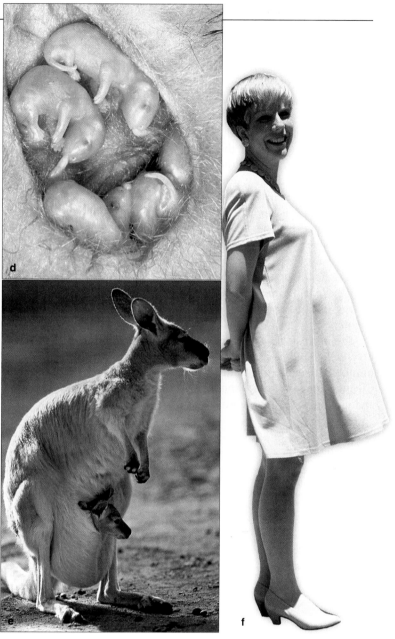

d

e

f

of sperm and egg *inside* a female's body. They invest energy to construct elaborate reproductive organs, such as a penis (by which a male deposits its sperm inside a female) and a uterus (a chamber inside the female where the embryo grows and develops).

Finally, animals set aside energy in forms that *nourish the developing individual* until it is developed enough to feed itself. For instance, nearly all animal eggs contain **yolk**, a protein-rich, lipid-rich substance that nourishes embryonic stages. The eggs of certain species have more yolk than others. Sea urchins make tiny eggs with very little yolk, release large numbers of them, and limit the biochemical investment in each one. Later, each fertilized egg develops into a self-feeding, free-moving larva in less than a day. Sea stars and other predators consume most of the eggs. So for sea urchins, bestowing as little as possible on as many gametes as possible does pay off, in terms of reproductive success.

By contrast, mother birds lay truly yolky eggs. Yolk nourishes the bird embryo through an extended period of development, inside an eggshell that forms after the egg is fertilized. Your mother put tremendous demands on her body to protect and nourish you through nine months of development inside her, starting from the

time you were an egg with almost no yolk. After you implanted yourself in her uterus, physical exchanges with her tissues supported you through the extended pregnancy (Figure 44.2*f*).

As these few examples suggest, animals show great diversity in reproduction and development. However, as you will see in the sections to follow, some patterns are widespread throughout the animal kingdom, and they can serve as a framework for our reading.

Separation into male and female sexes requires special reproductive cells and structures, neural and hormonal control mechanisms, and forms of behavior. A selective advantage—variation in traits among offspring—offsets the biological costs associated with the separation.

STAGES OF DEVELOPMENT—AN OVERVIEW

Embryos are a class of transitional forms on the road from a fertilized egg to an adult. Although they all start out as a single cell, embryos of different species often look different as they grow and develop. For example, you do not look like a frog now, and you did not look like one when you and the frog were early embryos, either. However, despite the differences in appearance, it is possible to identify certain patterns in the way that the embryos of nearly all animal species develop.

Figure 44.3 is an overview of the stages of animal development. During **gamete formation**, the first stage, eggs or sperm develop inside the reproductive organs of one parent's body. **Fertilization**, the second stage, starts when the plasma membrane of a sperm fuses with the plasma membrane of an egg. It is over when the egg nucleus and sperm nucleus fuse, thus forming a zygote.

The third stage, **cleavage**, is a program of mitotic cell divisions that *divide* the volume of egg cytoplasm into a number of smaller, nucleated cells called **blastomeres**. Cleavage only increases the number of cells; it does not change the original volume of egg cytoplasm.

As cleavage draws to a close, the pace of mitotic cell division slackens. The embryo enters **gastrulation**. This fourth stage of animal development is a time of major cellular reorganization. The newly formed cells become arranged into two or three primary tissues, often called germ layers. The cellular descendants of the primary tissues will form all tissues and organs of the adult:

1. **Ectoderm**. This is the *outermost* primary tissue layer, the one that forms first in the embryos of every animal. Ectoderm is the embryonic forerunner of the cell lineages that give rise to tissues of the nervous system and to the integument's outer layer.

2. **Endoderm**. Endoderm is the *innermost* primary tissue layer. It is the embryonic forerunner of the gut's inner lining and of organs derived from the gut.

3. **Mesoderm**. This *intermediate* primary tissue layer is the forerunner of muscle and most of the skeleton; of circulatory, reproductive, and excretory organs; and of connective tissue layers of the gut and integument. Mesoderm originated hundreds of millions of years ago, and it was a pivotal step in the evolution of nearly all large, complex animals.

After they have formed, the primary tissue layers give rise to subpopulations of cells. This marks the onset of **organ formation**. The subpopulations become unique in structure and function, and their descendants give rise to different kinds of tissues and organs.

During the final stage of animal development, **growth and tissue specialization**, organs increase in size and gradually assume their specialized functions. This stage continues into adulthood.

Figure 44.4 shows photographs and diagrams of several stages of embryonic development of a frog.

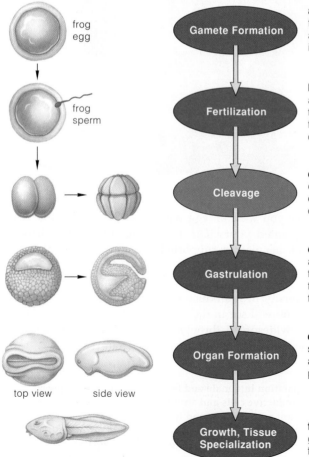

frog egg

frog sperm

top view   side view

Gamete Formation

a Eggs form and mature in female reproductive organs, and sperm form and mature in male reproductive organs.

Fertilization

b A sperm and an egg fuse at their plasma membrane, then the nucleus of one fuses with the nucleus of the other to form the zygote.

Cleavage

c By a series of mitotic cell divisions, different daughter cells receive different regions of the egg cytoplasm.

Gastrulation

d Cell divisions, migrations, and rearrangements produce two or three primary tissues, the forerunners of specialized tissues and organs.

Organ Formation

e Subpopulations of cells are sculpted into specialized organs and tissues in prescribed spatial patterns at prescribed times.

Growth, Tissue Specialization

f Organs increase in size and gradually assume specialized functions.

Figure 44.3 Overview of the stages of animal development. We use a few forms that appear in the frog life cycle as examples.

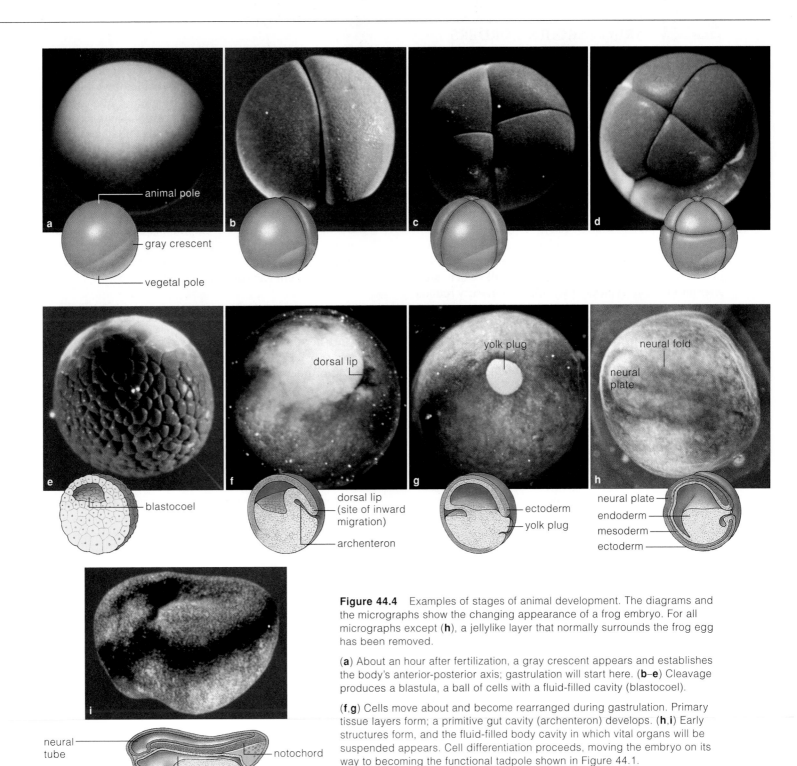

Figure 44.4 Examples of stages of animal development. The diagrams and the micrographs show the changing appearance of a frog embryo. For all micrographs except (**h**), a jellylike layer that normally surrounds the frog egg has been removed.

(**a**) About an hour after fertilization, a gray crescent appears and establishes the body's anterior-posterior axis; gastrulation will start here. (**b-e**) Cleavage produces a blastula, a ball of cells with a fluid-filled cavity (blastocoel).

(**f,g**) Cells move about and become rearranged during gastrulation. Primary tissue layers form; a primitive gut cavity (archenteron) develops. (**h,i**) Early structures form, and the fluid-filled body cavity in which vital organs will be suspended appears. Cell differentiation proceeds, moving the embryo on its way to becoming the functional tadpole shown in Figure 44.1.

Take a moment to study them, for they reinforce a key concept. The structures that form during one stage of development serve as the foundation for the stage that comes after it. Successful development depends on the formation of all of those structures according to normal patterns, in prescribed sequence.

Animal development proceeds from gamete formation and fertilization through cleavage, gastrulation, then organ formation, and finally growth and tissue specialization.

Development cannot proceed properly unless each stage is successfully completed before the next begins.

Information in the Egg Cytoplasm

You probably don't have an arm attached to your nose or toenails growing from your navel. The patterning of body parts for any complex animal, including yourself, is partly mapped out in the cytoplasm of an immature egg, or **oocyte**, even before a sperm enters the picture. A **sperm**, remember, consists of paternal DNA and a bit of equipment that helps the sperm reach and penetrate an egg. Compared to a sperm, an oocyte is much larger and more complex (Section 10.5).

As an oocyte matures, its volume increases. Enzymes, mRNA transcripts, and other factors become stockpiled in different parts of the cytoplasm and will be activated after fertilization. Typically they take part in early rounds of DNA replication and cell division. Also present are tubulin molecules and factors that will govern the angle and timing of their assembly into microtubules for a mitotic spindle. Such factors influence the pattern of cleavage. The cytoplasm also contains yolk, the amount and distribution of which will dictate where cleavage can proceed and how large the blastomeres will be.

We can find evidence of such regionally localized, "maternal messages" as early as fertilization. As a sperm fertilizes a frog egg, for example, it triggers structural reorganization of the egg's cortex, which includes the plasma membrane and the cytoplasm just under it. The cortex has pigment granules concentrated near one pole of the egg and yolk concentrated near the other. Upon fertilization, part of the cortex shifts away from the yolk. A crescent-shaped area of lighter colored cytoplasm is exposed. It is a **gray crescent**, a region of intermediate pigmentation, near the fertilized egg's equator. And it establishes the frog's anterior-posterior body axis.

In itself, the gray crescent is not evidence of regional differences in maternal messages. Such evidence comes from experiments of the sort shown in Figure 44.5b. It also comes from observing embryos in which localized cytoplasmic differences are pronounced enough to be tracked easily during development. For example, if you were to continue tracking the development of fertilized frog eggs, you would see that a gray crescent is always the site where gastrulation normally begins.

Cleavage—The Start of Multicellularity

Once fertilization is over, the zygote enters cleavage. Beneath the plasma membrane, its midsection bears a ring of microfilaments, made of the contractile protein actin. The ring tightens as microfilaments slide past one another and pinch in the cytoplasm. The cell surface above the tightening ring is pulled inward as a **cleavage furrow** (Section 9.5). It is the force of contraction that splits the cytoplasm into blastomeres. A new ring forms from actin subunits in each daughter cell.

Simply by virtue of where they form, blastomeres end up with different maternal messages. This outcome of cleavage, called **cytoplasmic localization**, helps seal

Figure 44.5 Examples of experiments that illustrate how the cytoplasm of a fertilized egg has localized differences that help determine the fate of cells in a developing embryo. The cortex of frog eggs contains granules of dark pigment, concentrated near one pole. At fertilization, a portion of the granule-containing cortex shifts toward the point of sperm entry. The shift exposes lighter colored, yolky cytoplasm in a crescent-shaped gray area:

Normally, the first cleavage puts part of the gray crescent in both of the resulting blastomeres.

(**a**) For one experiment, the first two blastomeres that formed were separated from each other. Each still gave rise to a complete tadpole.

(**b**) For another experiment, a fertilized egg was manipulated so that the cut through the first cleavage plane missed the gray crescent entirely. Only one of the two blastomeres received the gray crescent. It alone developed into a normal tadpole. The blastomere deprived of maternal messages in the gray crescent gave rise to a ball of undifferentiated cells.

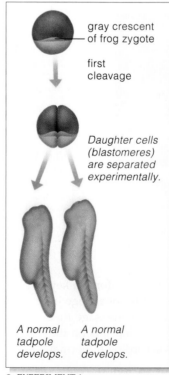

gray crescent of frog zygote

first cleavage

Daughter cells (blastomeres) are separated experimentally.

A normal tadpole develops.   A normal tadpole develops.

a EXPERIMENT 1

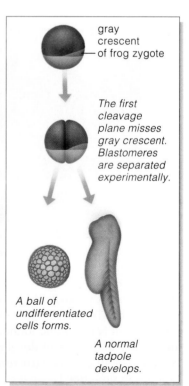

gray crescent of frog zygote

The first cleavage plane misses gray crescent. Blastomeres are separated experimentally.

A ball of undifferentiated cells forms.

A normal tadpole develops.

b EXPERIMENT 2

Table 44.1	Examples of Cleavage Patterns	
Yolk Distribution in the Egg	Cleavage Pattern	Representative Animals
COMPLETE CLEAVAGE		
Even distribution of sparse yolk	Radial	Lancelets, echinoderms
	Spiral	Flatworms, most mollusks, annelids
	Rotational	Mammals
Moderate amount of yolk at one end	Radial	Amphibians
INCOMPLETE CLEAVAGE		
Dense yolk concentrated at one end of egg	At small disk of yolk-free cytoplasm	Fishes, reptiles, birds
Dense yolk all through center	At periphery of yolk	Most insects, other arthropods

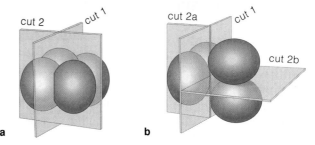

Figure 44.6 Comparison of the early cleavage planes for the egg of (**a**) sea urchins and (**b**) mammals. The cuts are radial in sea urchins and rotational in mammals.

the developmental fate of each cell's descendants. Its cytoplasm alone may have molecules of a protein that can activate, say, the gene coding for a certain hormone. And its descendants alone will make the hormone.

Cleavage Patterns

In the simplest cleavage pattern, complete cuts occur after each nuclear division, and each cleavage plane is perpendicular to the mitotic spindle. In effect, the cuts parcel out a nucleus to each blastomere, which makes all of the blastomeres genetically equivalent.

Blastomeres do not grow in size before they divide again. Cleavage divides the volume of cytoplasm into increasingly smaller cells, often rapidly. A frog zygote becomes 37,000 cells in forty-three hours; a *Drosophila* zygote becomes 50,000 cells in only twelve. The decrease in the ratio of cytoplasmic to nuclear volume influences gene activation. For instance, in frogs, gene transcription starts after the twelfth division. If researchers double the amount of DNA in each blastomere, however, the transition will occur one division cycle ahead of time.

In most animals, the zygote's genes are silent during early cleavage. How fast the cuts proceed and how the blastomeres become arranged are under the control of the proteins and mRNAs stockpiled in the cytoplasm. In mammals, however, certain genes must be activated; cleavage cannot occur without the proteins they specify.

Table 44.1 correlates the amount and distribution of yolk with cleavage patterns of major animal groups. Yolk typically inhibits cleavage. If an egg has little yolk, it may be fully cleaved. If yolk is concentrated at one end, early cuts will be limited to part of the cytoplasm. Such eggs show polarity. The yolk-rich end is called the *vegetal* pole; the *animal* pole is the end closest to the nucleus. A frog egg is like this (Figure 44.4).

Cleavage patterns differ among animals with little yolk in their eggs, so other, heritable factors also must influence the cuts. A sea urchin egg has little yolk and undergoes *radial* cleavage. That is, its cleavage furrows run horizontally and vertically with the animal-vegetal axis (Figure 44.6a). Successive cuts result in a **blastula**, a stage with blastomeres surrounding a fluid-filled cavity called a **blastocoel**. A sea urchin blastula has horizontal rows of blastomeres. Frog eggs also are cleaved radially, but yolk impedes cuts near the vegetal pole. Cleavage is faster and yields more, smaller size blastomeres near the animal pole. The blastocoel forms there, also (Figure 44.4e). Between the time that 16 to 64 blastomeres have formed, any amphibian blastula resembles a mulberry and is called a **morula**, which is Latin for mulberry.

By contrast, eggs of reptiles, birds, and most fishes undergo *incomplete* cleavage. The large volume of yolk restricts those early cuts to a small, caplike region near the animal pole. The result is two flattened layers of cells with a narrow cavity in between.

A mammalian egg undergoes *rotational* cleavage. The first cut passes through the egg's two poles. But then one of the resulting blastomeres is cut the same way and the other is cut horizontally (Figure 44.6b). The blastomeres divide slowly, at different times. And after the third cleavage, they abruptly huddle into a compact ball joined by tight junctions. The sixteen-cell stage is a morula. A blastocoel forms when descendants of the outer cells (trophoblasts) secrete fluid into the morula. As you will see in Chapter 45, trophoblasts will give rise to part of the placenta. The inner cells mass together, and they will give rise to the embryo proper.

The egg cytoplasm contains maternal instructions in the form of regionally distributed enzymes and other proteins, mRNAs, cytoskeletal elements, yolk, and other factors.

Cleavage divides a zygote into blastomeres, each with a localized part of maternal messages inherent in the egg cytoplasm. This outcome is called cytoplasmic localization.

Differences in the amount and distribution of yolk and other, heritable factors give rise to different patterns of cleavage among animal groups.

HOW SPECIALIZED TISSUES AND ORGANS FORM

Nearly all animals have a gut, with tissues and organs that function in digestion and absorption of nutrients. They have surface parts that protect internal parts and detect what is going on outside. In between, most have numerous organs, such as those dealing with structural support, movement, and blood circulation. Their three-layer body plan starts to emerge as cleavage ends and gastrulation begins. The embryo's size increases little, if any, but cells start migrating to new positions to form primary tissue layers—the ectoderm, endoderm, and mesoderm. Figure 44.7 shows how some outer cells of a sea urchin embryo migrate inward and form a lining for a cavity that will become the gut. Figure 44.8 shows the formation of a bird embryo's anterior-posterior axis, the outcome of cell migrations and other rearrangements. In every vertebrate, the anterior-posterior axis defines where a **neural tube**, the forerunner of the brain and spinal cord, will form. All such organs start forming by way of cell differentiation and morphogenesis.

Cell Differentiation

All cells of an embryo descend from the same zygote, so they have the same number and kinds of genes. They all activate the genes for histones, enzymes of glucose metabolism, and other proteins that are absolutely basic to cell survival. However, from gastrulation onward, certain groups of genes are activated in some cells but not others. When a cell selectively activates genes and synthesizes proteins not found in other cell types, we call this process **cell differentiation**. You read about the molecular basis of selective gene expression in Section 15.3. Basically, it results in proteins that are required for distinctive cell structures, products, and functions.

For example, when your eye lenses developed, some cells started synthesizing crystallin, a family of proteins

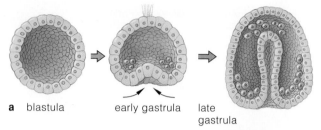

a blastula   early gastrula   late gastrula

Figure 44.7 (**a**) Diagrams of some early developmental stages of a sea urchin (*Lytechinus*). The blastula that resulted from cleavage becomes transformed into a gastrula. (**b**) Scanning electron micrograph of the gastrula's surface cells and region of inward migration.

b   invaginating endoderm

that become incorporated in transparent fibers of the lens. Only those cells could activate the required genes. Long crystallin fibers formed in the cells and forced them to lengthen and flatten. Collectively, those differentiated cells impart unique optical properties to each eye's lens. And those crystallin-producing cells are only one of 150 or so differentiated cell types now present in your body.

As many experiments tell us, nearly all cells become fully differentiated without loss of genetic information. For example, John Gurdon removed the nucleus from unfertilized eggs of the African clawed frog (*Xenopus laevis*). Then he isolated intestinal cells from tadpoles of the same species. He ruptured their plasma membrane, left the nucleus and most of the cytoplasm intact, then inserted the nucleus into the enucleated egg. In some experiments, that nucleus directed the developmental steps leading to a complete frog! The intestinal cell had

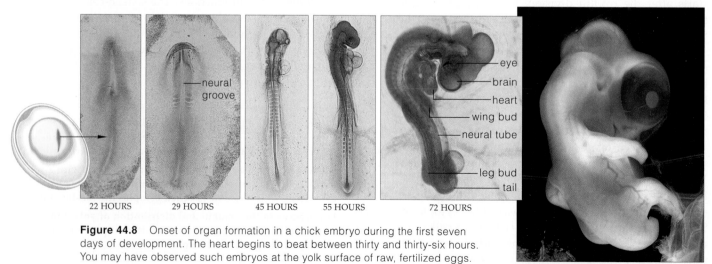

neural groove

eye
brain
heart
wing bud
neural tube

leg bud
tail

22 HOURS   29 HOURS   45 HOURS   55 HOURS   72 HOURS

168 HOURS (SEVEN DAYS OLD)

Figure 44.8 Onset of organ formation in a chick embryo during the first seven days of development. The heart begins to beat between thirty and thirty-six hours. You may have observed such embryos at the yolk surface of raw, fertilized eggs.

the same number and kinds of genes as the zygote, so it had all the genes necessary to form all of the required cell types. Its genes had not been somehow lost when it differentiated into an intestinal cell.

Human cells that normally contribute to only part of an embryo also retain the capacity to produce an entire individual. Consider the spontaneous separation of the two blastomeres of the two-cell human embryo. Their dissociation does not result in two half-embryos. The result is **identical twins**, two normal, fully developed individuals having the same genetic makeup.

Morphogenesis

Morphogenesis refers to a program of orderly changes in an embryo's size, shape, and proportions, the result being specialized tissues and early organs. As part of the program, cells divide, grow, migrate, and change in size. Tissues expand and fold, and cells in some of them die in controlled ways at prescribed locations.

Consider active cell migration. *Cells send out and use pseudopods that move them along prescribed routes.* When they reach their destination, they establish contact with cells already there. For example, forerunners of neurons interconnect this way as a nervous system is forming.

How do cells "know" where to move? They respond to adhesive cues, as when migrating Schwann cells stick to adhesion proteins on the surface of axons but not on blood vessels. And they respond to chemical gradients. Their migrations may be coordinated by the synthesis, release, deposition, and removal of specific chemicals in the extracellular matrix. Adhesive cues also tell the cells when to stop. Cells will migrate to regions of strongest adhesion, but once there, further migration is impeded. Section 23.1 describes such chemotactic behavior among the cells of a slime mold, *Dictyostelium discoideum*.

Besides this, as microtubules lengthen and as rings of microfilaments in cells constrict, *sheets of cells expand and fold inward and outward* (Figure 44.9). The assembly and disassembly of such components of the cytoskeleton are aspects of control mechanisms that bring about the changes. The size, shape, and proportion of body parts emerge through such controlled, localized events. We do not fully understand why some embryonic tissues expand more than others. Apparently, selective controls over the activity of regulatory genes play major roles.

Consider what happens after three primary tissues form in the embryos of amphibians, reptiles, birds, and mammals. At the embryo's midline, ectodermal cells elongate to form a neural plate, the first indication that a region of ectoderm is on its way to becoming nervous tissue. In some cells, microtubules lengthen. Other cells become wedge shaped as a microfilament ring contracts and constricts each cell at one end. Collectively, all the

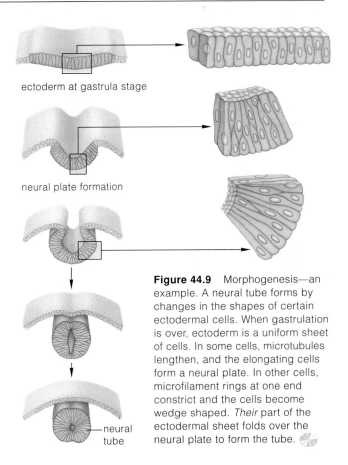

ectoderm at gastrula stage

neural plate formation

neural tube

Figure 44.9 Morphogenesis—an example. A neural tube forms by changes in the shapes of certain ectodermal cells. When gastrulation is over, ectoderm is a uniform sheet of cells. In some cells, microtubules lengthen, and the elongating cells form a neural plate. In other cells, microfilament rings at one end constrict and the cells become wedge shaped. *Their* part of the ectodermal sheet folds over the neural plate to form the tube.

changes in shape cause tissue flaps to fold and meet at the midline, thus forming a neural tube (Figure 44.9).

Finally, *programmed cell death helps sculpt body parts.* That is, cells that function for only limited periods in embryonic tissues can execute themselves. By this form of cell death, called **apoptosis**, molecular signals switch on weapons of self-destruction already stockpiled inside target cells. The Chapter 15 introduction describes the results. For now, think of a hand of a human embryo, which initially looks like a paddle (Section 9.5). Digits of cartilage are forming inside it. Cells in the tissue zones between digits have receptors for certain proteins that form by expression of certain genes. When the receptors bind the proteins, the cells commit mass suicide, and so the digits separate from one another. In some humans, a gene mutation blocks apoptosis and digits stay webbed.

In cell differentiation, a cell selectively uses certain genes and makes certain proteins, not found in other cell types, for distinctive cell structures, products, and functions.

Morphogenesis is a program of orderly changes in body size, shape, and proportions. It results in specialized tissues and organs at prescribed locations, at prescribed times.

Morphogenesis involves cell division, active cell migration, tissue growth and foldings, changes in cell size and shape, and programmed cell death by way of apoptosis.

"Although small children have taboos against stepping on ants because such actions are said to bring on rain, there has never seemed to be a taboo against pulling off the legs or wings of flies. Most children eventually outgrow this behavior. Those who do not either come to a bad end or become biologists."

So wrote Vincent Dethier in *To Know a Fly*, a whimsical yet scientifically sound tribute to flies and their suitability for controlled laboratory experiments. Even before Dethier published his booklet in 1962, plenty of geneticists and developmental biologists had already come to the same conclusion. *Drosophila melanogaster* was and still is the fly of choice. It costs almost nothing to feed or house this tiny insect. It reproduces rapidly in bottles, it has a short life cycle, and disposing of spent bodies is a snap (Section 12.4).

For much of this century, the embryonic development of *Drosophila* has been studied in detail at the anatomical, cytological, biochemical, and genetic levels. The findings tell us much about the embryonic development, even the evolutionary history, of animals in general. They tell us something about our own development, even though fruit fly and mammalian embryos proceed through the stages of development according to different patterns.

SUPERFICIAL CLEAVAGE IN *DROSOPHILA*   For example, mammalian eggs contain relatively little yolk, so cleavage furrows can cut through the entire egg. But the eggs of *Drosophila* (and those of most insects, fishes, reptiles, and birds) contain a large mass of yolk that restricts cleavage to a small portion of egg cytoplasm. The yolk mass of a *Drosophila* egg is centrally located and confines cleavage to the cytoplasm's periphery. This is one example of a *superficial* cleavage pattern. The egg nucleus repeatedly divides, but the cytoplasm does not; cleavage furrows do not form. The outcome is an early embryonic stage called a *syncytial* blastoderm (Figure 44.10).

Compared with a mammalian egg, which undergoes slow cuts every twelve to twenty-four hours, superficial cleavage of a fertilized *Drosophila* egg is rapid. Nuclear division proceeds every eight minutes until 256 nuclei are crowded together in the yolky cytoplasm. This stage of development is a type of *cellular* blastoderm. Each nucleus of the blastoderm has its own small array of microtubules and microfilaments, which will influence the forthcoming patterns of cell formation and elongation.

Most of the nuclei migrate toward the egg periphery. First, plasma membrane wraps around each nucleus that

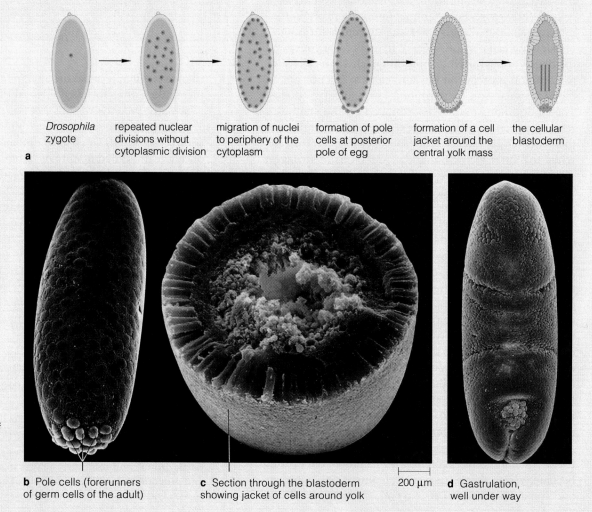

Figure 44.10
Superficial cleavage in the yolk-rich egg of *Drosophila*.

(**a**) A fertilized egg's *nucleus* repeatedly divides. *Cytoplasmic* division is delayed until after the nuclei move to the periphery of the cytoplasm.

(**b**) Pole cells form. (**c,d**) Then other cells form; they are arrayed like a jacket around the central mass of yolk. This early stage of development is a type of blastoderm.

As the blastoderm forms, polar granules in the cytoplasm migrate and become isolated in pole cells that form at one end of the blastoderm. Pole cells are forerunners of germ cells that give rise to eggs or sperm in the adult fly.

a *Drosophila* zygote → repeated nuclear divisions without cytoplasmic division → migration of nuclei to periphery of the cytoplasm → formation of pole cells at posterior pole of egg → formation of a cell jacket around the central yolk mass → the cellular blastoderm

b Pole cells (forerunners of germ cells of the adult)

c Section through the blastoderm showing jacket of cells around yolk

200 µm

d Gastrulation, well under way

reaches the egg's posterior pole axis. The resulting pole cells are the forerunners of germ cells of the adult fly. This physical separation of the future reproductive cells from the rest of the blastoderm is one of the first events of insect development. After the separation occurs, the plasma membrane folds inward and forms a partition around each remaining nucleus and a bit of cytoplasm. Within four hours after fertilization, about 6,000 cells are arranged like a jacket around the yolky mass.

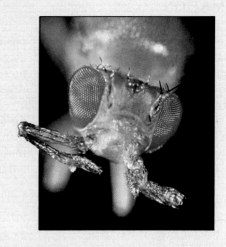

Figure 44.11 Experimental evidence that certain genes are responsible for the specification of body parts. In a normal *Drosophila* larva, a cluster of cells gives rise to antennae on the head. If the larva carries a certain mutated form of the *antennapedia* gene, the adult fly will have legs instead of antennae on its head.

GENES AS MASTER ORGANIZERS   During the late 1940s, Edward Lewis of the California Institute of Technology discovered that the products of certain mutated genes introduce bizarre blips in pattern formation in *Drosophila*. To give one example, as a result of a single gene mutation, a leg grew out of the head region where an antenna should have grown, as shown by the photograph in Figure 44.11.

Look at the *Drosophila* zygote in Figure 44.12a. A **fate map** of its surface is a map that shows where each kind of differentiated cell in the forthcoming adult will originate. The embryo develops with a head and a tail end. In between, a series of body segments will form, each with its own gene-specified identity. In an adult fly, for example, the first segment (part of the thorax) has only legs, the next has legs and wings, the next has legs and balancing devices, and so on.

Researchers have put together a model to explain how polarity in the egg gives rise to segmented polarity of the adult body. First, *maternal effect* genes that specify regulatory proteins are transcribed, translated, or both (Figure 44.12b). The resulting mRNAs and proteins become localized in different parts of the egg cytoplasm. These gene products are activated in the zygote. They activate or suppress *gap* genes that map out broad regions of the body. As the product of one gap gene (hunchback) diffuses through the syncytial blastoderm, it creates a chemical gradient that influences the extent to which other gap genes are expressed. The different concentrations of gap gene products switch on *pair-rule* genes. The products of these genes accumulate in bands (stripes) corresponding to two body segments. And they activate *segment polarity* genes that divide the embryo into segment-size units.

Interactions among the products of the gap, pair-rule, and segment polarity genes control expression of another

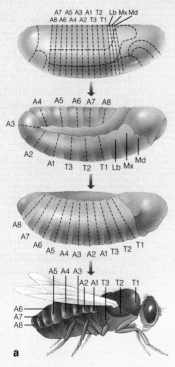

a

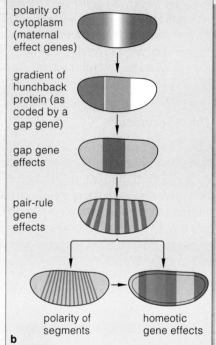

polarity of cytoplasm (maternal effect genes)

gradient of hunchback protein (as coded by a gap gene)

gap gene effects

pair-rule gene effects

polarity of segments

homeotic gene effects

b

Figure 44.12 (**a**) Fate map of a *Drosophila* zygote. The dashed lines indicate regions that normally will develop into different body segments, with specialized structures and appendages. For this type of organism, each segment's fate is sealed at the time of cytoplasmic localization. (**b**) Generalized model of pattern formation in *Drosophila*, as described in the text.

class of genes—*homeotic* genes, which collectively govern the developmental fate of each segment.

When mutated, homeotic genes can transform one body segment into the likeness of another. Thus mutations in the *antennapedia* gene result in abnormal responses to regulatory proteins. In the example cited earlier, a mutated gene product activated the wrong set of homeotic genes in cells that are supposed to produce two antennae on the head. They gave rise to a pair of legs instead.

Researchers do not yet have a complete understanding of the multiple levels of gene interactions that specify body parts in *Drosophila*. But their studies of this tiny organism give us fascinating glimpses into the kinds of regulatory mechanisms that must come into play in other organisms as well. And think about *that* next time you swat at fruit flies, each not much bigger than a rounded pinhead, as they hover over a bowl of pungently ripe fruit.

The developmental fate of an embryonic cell depends partly on which bit of cytoplasm it acquired during cleavage. This is especially true of *Drosophila*, where gene products that were already localized in different parts of the egg cytoplasm give every cell its identity. Sidney Brenner calls this the "European style" of cell determination, in which lineage is everything.

By contrast, the "American style" of cell determination emphasizes the neighborhood. That is, the fate of each cell depends on whom it meets up with in the embryo. The cell retains the capacity to alter its fate, because it contains all of the instructions necessary to give rise to a whole animal. Remember, two blastomeres at the two-cell stage of a human embryo can each give rise to one of two identical twins. However, *when the cell is locked in at a particular place in the embryo, it becomes committed to starting the cell lineages that will form certain body parts in places where we expect those parts to form*. This style of cell determination predominates in echinoderm and vertebrate embryos.

A Theory of Pattern Formation

The sculpting of nondescript clumps of embryonic cells into specialized tissues and organs follows an ordered, spatial pattern. We call this activity **pattern formation**. As you have seen, local differences in the egg cytoplasm influence the formation of functional tissues and organs in all embryos. Later on, cells develop according to their

positions in the embryo. There, the cells interact by way of gene products that are sent, received, and used in prescribed sequences, at prescribed times. Increasingly diverse interactions complete the body plan.

Think about the preceding section, which described the sequence of gene interactions that map out the body plan of *Drosophila*. From studies of gene mutations in this organism and others, researchers have put together a theory of pattern formation. Here are its key points:

1. During development, classes of master genes are activated in orderly sequence, at prescribed times.

2. Regulatory proteins guide interactions among the master genes. The interactions result in the appearance of different gene products that are spatially organized relative to one another in the embryo.

3. Different genes are activated and suppressed in cells along the embryo's anterior-posterior axis and dorsal-ventral axis. Certain protein products of this selective gene expression create chemical gradients in the body of the embryo that help define each cell's identity.

4. **Homeotic genes** are a class of master genes that specify the development of specific body parts.

Embryonic Induction

A change in the developmental fate of an embryonic tissue, as brought about by exposure to a gene product released from an adjacent tissue, is called **embryonic induction**. Experimental modifications of a developing chick wing, as in Figure 44.13, yield examples. The chick wing forms, as a primordial bud, by way of mitotic cell divisions in mesoderm and ectoderm. The mesoderm induces the overlying ectodermal cells to elongate and form a ridge at its apex. This apical ectodermal ridge (AER) forms on the developing limbs of all birds and mammals. Surgically remove the AER before the bud grows fully, and further development ceases.

Rapid cell divisions in mesoderm beneath the AER generate the tissues that elongate the developing limb. As cells exit from this zone, they give rise to the tissues

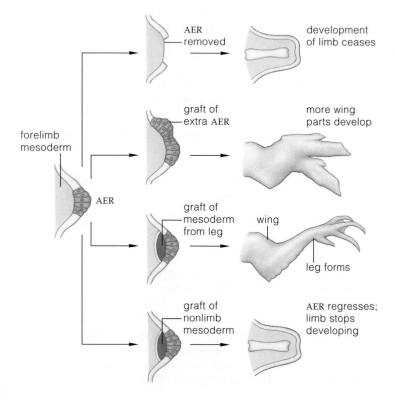

Figure 44.13 Some experiments that reveal an interaction between mesoderm and ectoderm during the formation of a chick wing. This mesoderm induces the overlying ectodermal cells to elongate and form a narrow ridge at its apex. An apical ectodermal ridge (AER) is a population of self-sustaining cells that do not mingle with surrounding cells. Remove AER when a wing bud is developing, and the bud will develop no more. Graft extra AER onto the bud, and the wing that forms has extra parts. Graft leg mesoderm just beneath the AER, and part of a leg (with toes) develops. Graft nonlimb mesoderm beneath the AER, and the AER regresses and wing development stops.

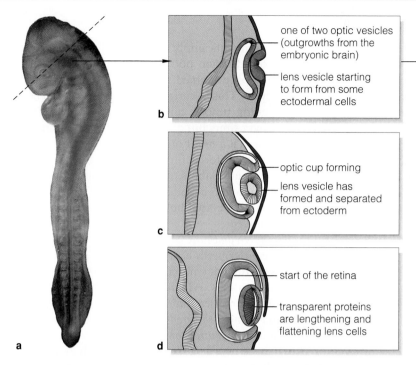

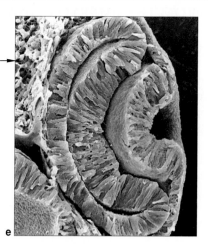

b — one of two optic vesicles (outgrowths from the embryonic brain)
— lens vesicle starting to form from some ectodermal cells

c — optic cup forming
— lens vesicle has formed and separated from ectoderm

d — start of the retina
— transparent proteins are lengthening and flattening lens cells

e

Figure 44.14 Eye formation in a chick embryo. (**a**) Each eye starts forming at an optic vesicle, a fluid-filled pouch beneath ectoderm of the head. (**b**) When ectoderm cells come into contact with cells of the optic vesicle, they are induced to elongate, fold inward, and form a lens vesicle. (**c**) At the same time, the contact induces cells of the optic vesicle to sink inward, forming an optic cup from which most of the eyeball will form. (**d**) The cup's inner layer will produce the retina. (**e**) Scanning electron micrograph of a lens forming in the left optic cup of the embryo's head.

of the limb's outermost parts. Cells pushed out early on give rise to bones closest to the trunk of the body. Cells that depart much later are using their genes differently. They are biochemically different, and they respond to different signals. These cells give rise to the outermost bones and other structures of the limb.

A classic experiment by Hans Spemann, a pioneer in developmental biology, gave us the first stunning example of embryonic induction. In all vertebrates, the retina of the eye originates from the forebrain. The lens, which focuses light onto the retina, originates from the ectoderm (Figure 44.14). Spemann surgically removed retina-forming tissue from the forebrain region of a salamander embryo and inserted it into belly ectoderm. Afterward, a lens formed on the belly! As Spemann suspected, a molecular signal from forerunners of retina cells normally induces ectodermal cells to form a lens.

Today we know that slowly degradable proteins, now called **morphogens**, are among the gene products that form concentration gradients as they diffuse from an inducing tissue into adjoining tissues. The signals they represent are strongest at the start of the gradients, and they weaken with distance. Thus, cells at different positions along the concentration gradients are exposed to different chemical information, which guides selective expression of genes in different parts of the embryo.

Morphogens and other inducers switch on blocks of genes in sequence. Among the targets are homeotic genes, which occur in animals as evolutionarily distant as vertebrates and the roundworm *Caenorhabditis elegans*. While organs are first forming, the gene products interact with regulatory elements to activate or inhibit blocks of genes in similar ways among major animal groups. They direct the global inductive events that map out the body plan, including its major axes. Failure of induction when the organs are forming has global repercussions, as when a heart is constructed at the wrong location. After the basic plan is set, inductions are numerous—but localized. If a lens fails to form, for example, only the eye is affected.

It now appears that the first stages of development are open to evolutionary change, but that early formation of organs is not. This may be why we see no more than a few dozen body plans among all animals. We've known about the physical constraints (such as the surface-to-volume ratio) and architectural constraints (as imposed by the body's axes). Now we know of *phyletic* constraints on change, imposed on each lineage by master genes that operate at the time of early organ formation and govern induction of the basic body plan. Once inductive interactions have produced a structure, it is hard to start over again.

Pattern formation is the orderly, sequential sculpting of embryonic cells into specialized tissues and organs. Cells differentiate partly through cytoplasmic localization and, later, according to their position in the developing embryo.

Inductive interactions among classes of master genes map out the basic body plan and specify where and how specific body parts develop. Products of their selective expression create chemical gradients in the embryo that help seal each cell's developmental fate.

There are physical, architectural, and phyletic constraints on the evolution of animal body plans.

Post-Embryonic Development

What happens after an embryo is released or hatched from the parent? For many species, the new individual is a **juvenile**, a miniaturized version of the adult that simply changes in size and proportion until reaching sexual maturity. For others, **metamorphosis** occurs: the body form changes during the transition from embryo to adult. Metamorphosis requires growth in size, tissue reorganization, and remodeling of body parts, and it is under hormonal control. Section 37.8 gives examples of this transitional time for crustaceans and an insect.

For now, think back on Section 26.19, which showed examples of metamorphosis in insects. Larvae, nymphs, and pupae are examples of the transitional stages. Some insect nymphs are miniature adults. They go through episodes of **molting**, in which they shed a hardened cuticle and grow rapidly before the next cuticle hardens. Some insects show *incomplete* metamorphosis; they undergo gradual, partial change between the first and last molt. Bugs are like this. *Drosophila* is one of the insects that show *complete* metamorphosis. As described in Section 44.5, tissues of immature stages are destroyed and replaced before the adult emerges.

Similarly, frogs and other amphibians show striking and complete metamorphic changes that affect nearly all organs (Figure 44.1). In this case, thyroid hormones called thyroxine and triiodothyronine bring about the transformation of a tadpole into the adult. Increasing levels of these hormones cause some immature organs, such as those of the tail, to degenerate. Yet they bring about further development and differentiation of other organs, including the eye.

Aging and Death

Late in life, the body of all multicellular animals slowly deteriorates. Through processes collectively called **aging**, cells start to break down structurally and functionally, and this leads to structural changes and gradual loss of body functions. The introduction to Chapter 6 provides an example of aging in humans. All animals that show extensive cell differentiation undergo aging.

No one knows what causes aging, although research gives us interesting things to think about. For example, many years ago, Paul Moorhead and Leonard Hayflick cultured normal human embryonic cells. All of the cell lines proceeded to divide about fifty times, then the entire population died off. Hayflick took some cultured cells that were partway through the series of divisions and froze them for several years. Afterward, he thawed the cells and placed them in a culture medium. The cells proceeded to complete the cycle of fifty doublings—whereupon they all died on schedule.

Such experiments imply that normal types of cells have a limited division potential. That is, mitosis may be genetically programmed to decline at a certain time in the life cycle. This is certainly true of neurons. They stop undergoing mitotic cell division after birth, then deteriorate over time. But does the shutdown of mitosis *cause* aging or is it a *result* of aging?

DNA's capacity for self-repair may be gradually lost through an accumulation of gene mutations. As you read in the Chapter 6 introduction, free radicals—rogue molecular fragments that bombard DNA and the other biological molecules—are associated with aging. Also, researchers have just now correlated *Werner's syndrome*, an aging disorder, with a mutation in the gene coding for helicase. This enzyme is vital for DNA replication and repair operations. Whether by free radical attacks or by accumulated replication errors, *structural changes in DNA could endanger the production of functional enzymes and other proteins required for normal life processes.*

Think about how cells depend on smooth exchanges of materials between the cytoplasm and extracellular fluid. Also think about collagen, a structural component of bone and other connective tissues throughout the body. If something shuts down or mutates the collagen-encoding genes, the missing or altered gene product would affect the flow of oxygen, nutrients, hormones, and so forth to and from cells. The repercussions would ripple through the entire body. Finally, think about a deterioration in genes for membrane proteins that serve as self markers. If markers change, do T lymphocytes then perceive the body's own cells as foreign and attack them? If such autoimmune responses increase over time, they would invite the greater vulnerability to disease and stress associated with old age.

For humans, following a low-fat diet and aerobic, low-impact exercise program throughout adulthood is more likely to reduce some of the common effects of aging. Consistent exercise improves cardiovascular and respiratory functioning. It also minimizes bone loss in older adults by helping to maintain bone density.

After embryonic development, different animal groups show variation in the patterns by which they mature into the adult form.

Post-embryonic stages of some species resemble adults in miniature, in that they change little except in size and proportion before the adult form emerges.

Other species undergo metamorphosis. Post-embryonic stages undergo complete or partial changes in growth, tissue reorganization, and tissue remodeling before the adult emerges. Hormones control the changes.

All animals that show extensive cell differentiation also undergo a process of aging, which culminates in death.

44.8 DEATH IN THE OPEN

Everything in the world dies, but we only know about it as a kind of abstraction. If you stand in a meadow, at the edge of a hillside, and look around carefully, almost everything you can catch sight of is in the process of dying, and most things will be dead long before you are. If it were not for the constant renewal and replacement going on before your eyes, the whole place would turn to stone and sand under your feet.

There are some creatures that do not seem to die at all; they simply vanish totally into their own progeny. Single cells do this. The cell becomes two, then four, and so on, and after a while the last trace is gone. It cannot be seen as death; barring mutation, the descendants are simply the first cell, living all over again. . . .

There are said to be a billion billion insects on the Earth at any moment, most of them with very short life expectancies by our standards. Someone estimated that there are 25 million assorted insects hanging in the air over every temperate square mile, in a column extending upward for thousands of feet, drifting through the layers of atmosphere like plankton. They are dying steadily, some by being eaten, some just dropping in their tracks, tons of them around the Earth, disintegrating as they die, invisibly.

Who ever sees dead birds, in anything like the huge numbers stipulated by the certainty of the death of all birds? A dead bird is an incongruity, more startling than an unexpected live bird, sure evidence to the human mind that something has gone wrong. Birds do their dying off somewhere, behind things, under things, never on the wing.

Animals seem to have an instinct for performing death alone, hidden. Even the largest, most conspicuous ones find ways to conceal themselves in time. If an elephant missteps and dies in an open place, the herd will not leave him there; the others will pick him up and carry the body from place to place, finally putting it down in some inexplicably suitable location. When elephants encounter the skeleton of an elephant in the open, they methodically take up each of the bones and distribute them, in a ponderous ceremony, over neighboring acres.

It is a natural marvel. All of the life on earth dies, all of the time, in the same volume as the new life that dazzles us each morning, each spring. All we see of this is the odd stump, the fly struggling on the porch floor of the summer house in October, the fragment on the highway. I have lived all my life with an embarrassment of squirrels in my backyard, they are all over the place, all year long, and I have never seen, anywhere, a dead squirrel.

I suppose that is just as well. If the Earth were otherwise, and all the dying were done in the open, with the dead there to be looked at, we would never have it out of our minds. We can forget about it much of the time, or think of it as an accident to be avoided, somehow. But it does make the process of dying seem more exceptional than it

really is, and harder to engage in at the times when we must ourselves engage.

In our way, we conform as best we can to the rest of nature. The obituary pages tell us of the news that we are dying away, while birth announcements in finer print, off at the side of the page, inform us of our replacements, but we get no grasp from this of the enormity of the scale. There are 4 billion of us on the Earth in this year, 1973, and all 4 billion must be dead, on a schedule, within this lifetime. The vast mortality, involving something over 50 million each year, takes place in relative secrecy. We can only really know of the deaths in our households, among our friends. These, detached in our minds from all the rest, we take to be unnatural events, anomalies, outrages. We speak of our own dead in low voices; struck down, we say, as though visible death can occur only for cause, by disease or violence, avoidably. We send off for flowers, grieve, make ceremonies, scatter bones, unaware of the rest of the 4 billion on the same schedule. All of that immense mass of flesh and bone and consciousness will disappear by absorption into the earth, without recognition by the transient survivors.

Less than half a century from now, our replacements will have more than doubled in numbers. It is hard to see how we can continue to keep the secret, with such multitudes doing the dying. We will have to give up the notion that death is a catastrophe, or detestable, or avoidable, or even strange. We will need to learn more about the cycling of life in the rest of the system, and about our connection in the process. Everything that comes alive seems to be in trade for everything that dies, cell for cell. There might be some comfort in the recognition of synchrony, in the information that we all go down together, in the best of company.

— LEWIS THOMAS, 1973

SUMMARY

1. For animals, sexual reproduction is the dominant reproductive mode. It requires specialized reproductive structures, control mechanisms, and forms of behavior that assist fertilization and support the offspring.

2. Among animals, embryonic development commonly proceeds through six stages:

a. Gamete formation, when oocytes (immature eggs) and sperm form and develop in reproductive organs. Molecular and structural components are stockpiled and become localized in different parts of the oocyte.

b. Fertilization, from the time a sperm penetrates an egg to the fusion of sperm and egg nuclei that results in a zygote (fertilized egg).

c. Cleavage, when mitotic cell divisions transform the zygote into smaller cells, or blastomeres. Cleavage does not increase the original volume of egg cytoplasm; it only increases the number of cells. It regionally divides the yolk, mRNAs, proteins, cytoskeletal elements, and other "maternal messages" among the new blastomeres, an outcome called cytoplasmic localization.

d. Gastrulation, when primary tissue layers (germ layers) form. Cell divisions, cell migrations, and other events lead to the formation of endoderm, ectoderm, and (in most species) mesoderm. All tissues of the adult body develop from primary tissue layers.

e. The onset of organ formation. The different organs start developing by a tightly orchestrated program of cell differentiation and morphogenesis.

f. Growth and tissue specialization, when organs enlarge overall and acquire their specialized chemical and physical properties. The maturation of tissues and organs continues into post-embryonic stages.

3. An embryo cannot develop properly unless each stage of development is successfully completed before the next stage begins.

4. In cell differentiation, a cell selectively uses certain genes and synthesizes proteins not found in other cell types. The outcome is subpopulations of specialized lineages of cells that differ from one another in their structure, biochemistry, and functioning.

5. Starting at the time of gastrulation, morphogenesis is a program by which the embryo changes in size, shape, and proportions. Cell divisions, active cell migrations, changes in cell size and shape, the growth and folding of tissues, and controlled cell death (by way of apoptosis) are involved.

6. Pattern formation is the emergence of the basic body plan and specific body parts in specific regions, in an orderly sequence. Cytoplasmic localization helps seal the fate of cells in this patterning. Among echinoderms and vertebrates, cell determination depends more on interactions among classes of master genes that produce certain products. The products are spatially organized and create chemical gradients that help seal the identity of each cell lineage in the embryo.

7. A change in the developmental fate of an embryonic cell lineage, as brought about by exposure to products released from an adjacent tissue, is called embryonic induction.

8. Following embryonic development, some animals grow directly into the adult form. Other animals show metamorphosis, in which reactivated growth and tissue reorganization produce larvae or some other immature forms before emergence of the sexually mature adult.

9. All animals that show extensive cell differentiation age; they undergo gradual changes in structure and a decline in efficiency.

Review Questions

1. What is the main benefit of sexual reproduction? What are some of its biological costs? *44.1*

2. Define the key events during gamete formation, fertilization, cleavage, gastrulation, and organ formation. At which stage is the frog embryo in the photograph at right? *44.2*

3. Does cleavage increase the volume of cytoplasm, the number of cells, or both, compared to the zygote? *44.2*

4. Define blastula. What is a blastocoel? *44.3*

5. Define cell differentiation and morphogenesis. *44.4*

6. Define cytoplasmic localization. At which stage does it occur? Does it play a larger role in cell determination in insects such as *Drosophila* than in vertebrates? *44.3, 44.5, 44.6*

7. Define pattern formation. Then explain the key points of the theory of pattern formation. *44.6*

Self-Quiz (Answers in Appendix IV)

1. Sexual reproduction among animals is _____ .
   a. biologically costly   c. evolutionarily advantageous
   b. diverse in its details   d. all of the above

2. Development cannot proceed properly unless each stage is successfully completed before the next begins, starting with _____ .
   a. gamete formation   d. gastrulation
   b. fertilization   e. organ formation
   c. cleavage   f. growth and tissue
                         specialization

3. During cleavage, the new blastomeres are allocated different regions of the fertilized egg's cytoplasm. This is called _____ .
   a. cytoplasmic localization   c. cell differentiation
   b. embryonic induction   d. morphogenesis

4. Primary tissue layers first appear _____ .
   a. in the egg cortex   c. in the gastrula
   b. during cleavage   d. in primary organs

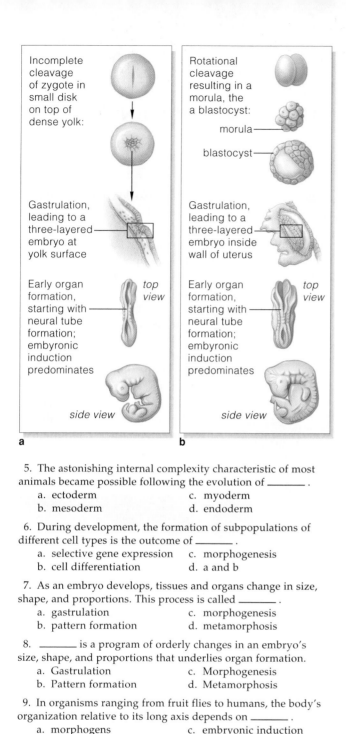

Incomplete cleavage of zygote in small disk on top of dense yolk:

Gastrulation, leading to a three-layered embryo at yolk surface

Early organ formation, starting with neural tube formation; embryronic induction predominates

top view

side view

a

Rotational cleavage resulting in a morula, the a blastocyst:

morula
blastocyst

Gastrulation, leading to a three-layered embryo inside wall of uterus

Early organ formation, starting with neural tube formation; embryronic induction predominates

top view

side view

b

Figure 44.15 *Left*: Comparison of the pattern of embryonic development for (**a**) a chick and (**b**) a human.

Critical Thinking

1. Experimentally, it is possible to divide an amphibian egg so that the gray crescent is wholly within one of the two cells formed. If the two cells are separated from each other, only the cell with the gray crescent will form an embryo with a long axis, notochord, nerve cord, and back musculature. The other cell will form a shapeless mass of immature gut and blood cells. Reflect on Section 44.3. Does cytoplasmic localization or embryonic induction play a greater role in these outcomes?

2. Differences in the amount and distribution of yolk and other factors in the egg cytoplasm give rise to different cleavage patterns among animal groups. Think about the stages shown in Figure 44.15. Then speculate on how some of the differences might contribute to the pattern of incomplete cleavage of fertilized bird eggs (in this example, a *chick* egg) and to the pattern of complete, rotational cleavage of fertilized mammalian eggs (a human egg).

   Also notice this: Despite obvious differences in the early stages of development, chick and human embryos resemble one another once early organ formation is under way. Using the theory of pattern formation, speculate on the role of embryonic induction in bringing about the similarities between these vertebrate lineages.

Selected Key Terms

aging *44.7*
apoptosis *44.4*
asexual reproduction *44.1*
blastocoel *44.3*
blastomere *44.2*
blastula *44.3*
cell differentiation *44.4*
cleavage *44.2*
cleavage furrow *44.3*
cytoplasmic localization *44.3*
ectoderm *44.2*
embryo *44.2*
embryonic induction *44.6*
endoderm *44.2*
fate map *44.5*
fertilization *44.2*
gamete formation *44.2*
gastrulation *44.2*
gray crescent *44.3*

growth, tissue specialization *44.2*
homeotic gene *44.6*
identical twin *44.4*
internal fertilization *44.1*
juvenile *44.7*
mesoderm *44.2*
metamorphosis *44.7*
molting *44.7*
morphogen *44.6*
morphogenesis *44.4*
morula *44.3*
neural tube *44.4*
oocyte *44.3*
organ formation *44.2*
pattern formation *44.6*
sexual reproduction *44.1*
sperm *44.3*
yolk *44.1*
zygote *CI*

Readings

Caldwell, M. November 1992. "How Does a Single Cell Become a Whole Body?" *Discover* 13(11): 86–93.

Gilbert, S. 1994. *Developmental Biology.* Fourth edition. Sunderland, Massachusetts: Sinauer.

McGinnis, W., and M. Kuziora. February 1994. "The Molecular Architects of Body Design." *Scientific American* 270(2): 58–66.

Nusslein-Volhard, C. August 1996. "Gradients That Organize Embryonic Development. *Scientific American*, 54–61.

Raff, R., and T. Kaufman. 1983. *Embryos, Genes, and Evolution.* New York: Macmillan.

Web Site See *http://www.wadsworth.com/biology* for practice quiz questions, hypercontents, BioUpdates, and critical thinking. The Wadsworth Biology Resource Center provides a wealth of information fully organized and integrated by chapter.

5. The astonishing internal complexity characteristic of most animals became possible following the evolution of _____ .
   a. ectoderm   c. myoderm
   b. mesoderm   d. endoderm

6. During development, the formation of subpopulations of different cell types is the outcome of _____ .
   a. selective gene expression   c. morphogenesis
   b. cell differentiation   d. a and b

7. As an embryo develops, tissues and organs change in size, shape, and proportions. This process is called _____ .
   a. gastrulation   c. morphogenesis
   b. pattern formation   d. metamorphosis

8. _____ is a program of orderly changes in an embryo's size, shape, and proportions that underlies organ formation.
   a. Gastrulation   c. Morphogenesis
   b. Pattern formation   d. Metamorphosis

9. In organisms ranging from fruit flies to humans, the body's organization relative to its long axis depends on _____ .
   a. morphogens   c. embryonic induction
   b. homeotic genes   d. all of the above

10. _____ refers to changes in the body form of transitional stages between the embryo and the adult.
    a. Gastrulation   c. Morphogenesis
    b. Pattern formation   d. Metamorphosis

11. Match the development stage with its description.
    ____ cleavage   a. egg and sperm mature in parents
    ____ gametogenesis   b. sperm nucleus, egg nucleus fuse
    ____ organ   c. formation of primary tissue layers
         formation   d. cytoplasmic localization, not
    ____ growth, tissue   gene interactions, dominate in most
         specialization   animals (but not in mammals)
    ____ gastrulation   e. organs, tissues increase in size,
    ____ fertilization   acquire specialized properties
                            f. starts when primary tissue layers
                            split into subpopulations of cells

45

HUMAN REPRODUCTION AND DEVELOPMENT

The Journey Begins

At first nothing appears to be happening to the egg, so recently fertilized in the billowing folds of an oviduct (Figure 45.1a). Before the clock ticks off twenty-four hours, however, a spectacular journey is under way.

Cleavage furrows herald the onset of a programmed series of cuts through the egg cytoplasm. Every twelve to twenty-four hours, a new cut carves up the cytoplasm until, by the fifth day, a fluid-filled ball of thirty-two tiny cells is tumbling along the moist, soft lining of the uterus. It is a type of blastula, an embryonic stage you read about in the preceding chapter. Its surface cells are organized as a thin layer that looks like the shell of a Ping-Pong ball. Inside, some cells huddle together against one part of the surface layer's wall. This is the inner cell mass, the forerunner of an adult body that will someday consist of trillions upon trillions of cells. Yet the whole blastula is smaller than the tip of a pin.

By week's end, molecular signals urge the blastula to anchor itself to the uterine lining and start burrowing into it. Soon afterward, some surface cells of the blastula send out fingerlike cytoplasmic projections. These grow into the domain of maternal blood vessels threading through the lining. The activity establishes functional connections with the mother's tissues that will metabolically support the developing individual through the months ahead.

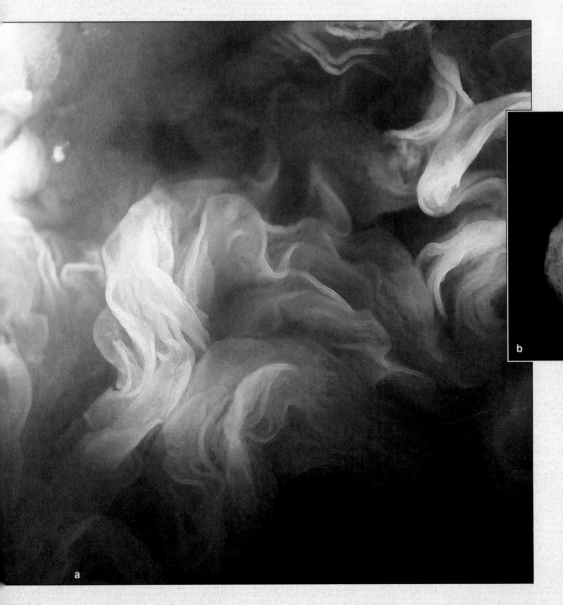

Figure 45.1 (a) Billowing folds on the inner, mucus-coated surface of an oviduct, part of a human female's reproductive system. An unfertilized egg, released earlier from a nearby ovary, was swept into the duct and was on its way to the uterus. A sperm entered the oviduct from the opposite direction. Somewhere in the mucosal folds, that sperm encountered and then penetrated the egg—and a remarkable developmental journey began. (b) A human embryo, as it appears just four weeks after the moment of fertilization.

Now, in an astonishing feat of self-organization, the inner cell mass transforms itself into a three-layered, oval disk. It does so as embryonic cells change in size and shape and migrate to new locations. Gradually, as an outcome of cell differentiation and morphogenesis, a pale, crescent-shaped embryo emerges. Day after day, the sculpting continues. Three weeks after fertilization the neural tube, forerunner of the central nervous system, starts forming. By the fourth week, tissue buds appear; they will develop into upper and lower limbs. A patch of cells in a tubular, primordial heart starts its rhythmic beating. From now on, through birth, adulthood, until the time of death, those cells will beat incessantly as the cardiac pacemaker.

By now, the developing embryo displays bilateral symmetry, cephalization, a tail, and other features that are the hallmarks of a vertebrate (Figure 45.1b). But is it a human? A minnow? A duckling? The answer does not become obvious until eight weeks into the journey, when the embryo is recognizably a human in the making. Later, in a body that is not much bigger than a peanut, complex cell interactions fill in the details of the body plan.

These events parallel your own beginning inside your mother. The story you are about to read is *your* story.

From the preceding chapter, you are acquainted with some principles that govern the development of animals in general. Now you will see how the principles apply to humans, starting with the formation of sperm and eggs in men and women. Gamete formation, remember, is the basis for sexual reproduction. And it is the first stage of animal development.

For both men and women, the reproductive system consists of a pair of gonads (the primary reproductive organs), accessory glands, and ducts. Male gonads are **testes** (singular, testis), and female gonads are **ovaries**. Testes produce sperm, and ovaries produce eggs. Both also secrete sex hormones that influence reproductive functions as well as the development of **secondary sexual traits**. Such traits are features that we associate with maleness and femaleness, although they do not play a direct role in reproduction.

Recall, from Section 12.3, that gonads have the same appearance in all early human embryos. After seven weeks of development, activation of genes on the sex chromosomes as well as hormone secretions trigger their development into testes or ovaries. As you will see, gonads and accessory organs are already formed at birth, but they do not reach their full size and become functional until the individual is a teenager.

KEY CONCEPTS

1. The human reproductive system consists of a pair of primary reproductive organs, or gonads, as well as a number of accessory glands and ducts. Testes are sperm-producing male gonads, and ovaries are egg-producing female gonads.

2. In response to signals from the hypothalamus and the pituitary gland, gonads also release sex hormones that orchestrate reproductive functions and the development of secondary sexual traits.

3. Human males continually produce sperm from puberty onward. The hormones testosterone, LH, and FSH control male reproductive functions.

4. Human females are fertile on a cyclic basis. Each month during their reproductive years, an egg is released from one of a pair of ovaries, and the lining of the uterus is primed for pregnancy. Estrogen, progesterone, FSH, and LH are the hormones that dominate this cyclic activity.

5. Human embryonic development starts with gamete formation and proceeds through fertilization, cleavage, gastrulation, organ formation, and growth and tissue specialization.

REPRODUCTIVE SYSTEM OF HUMAN MALES

Figure 45.2 shows the organs of the male reproductive system and Table 45.1 lists their functions. Again, an adult male has a pair of gonads, of a type called testes. These are *primary* reproductive organs, the equivalent of ovaries in females. They produce sperm and the sex hormones that govern male reproductive functions as well as the development of *secondary* sexual traits. Such traits do not play a direct role in reproduction, but they are still associated in distinct ways with maleness (or femaleness). Examples are the amount and distribution of body fat, hair, and skeletal muscle.

Where Sperm Form

In an embryo that is destined to become male, a pair of testes form on the abdominal cavity wall. Before birth, his testes descend from the abdominal cavity into the scrotum, an outpouching of skin that hangs below the pelvic region. By the time of birth, testes are fully formed miniatures of the adult organs. They start producing sperm at puberty, the time of sexual maturation. Boys typically enter puberty between ages twelve and sixteen.

Figure 45.2a shows the position of the scrotum in an adult male. If sperm cells are to develop as they should, the temperature inside the scrotum must remain a few degrees cooler than the rest of the body's normal core temperature. A control mechanism, which operates by stimulating or inhibiting the contraction of smooth muscles in the scrotum's wall, helps assure that the internal temperature does not stray far from 95°F. When the air just outside the body becomes cold, contractions draw the pouch closer to the warmer body mass. When it is warm outside, muscles relax and thereby lower the pouch.

Packed inside each testis are a great number of small, highly coiled tubes, the **seminiferous tubules**. Sperm formation begins in the tubules, in the manner described in Section 45.2.

Where Semen Forms

Mammalian sperm travel from the testes through a series of ducts leading to the urethra. When they enter one of two long, coiled ducts, the sperm are not quite mature. These first ducts are the epididymides (singular, epididymis). Secretions from glandular cells of the duct wall trigger activities that put the finishing touches on sperm maturation. Until the time that fully mature sperm depart from the body, they are stored in the last stretch of each epididymis.

When the male is sexually aroused, contractions of muscles in the walls of the reproductive organs propel mature sperm into and through a pair of thick-walled tubes called the vas deferentia (singular, vas deferens).

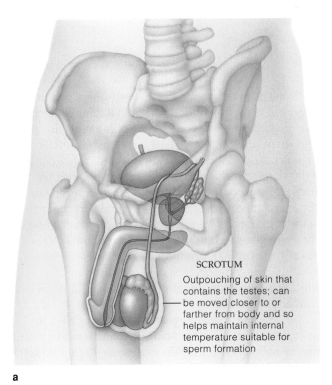

SCROTUM

Outpouching of skin that contains the testes; can be moved closer to or farther from body and so helps maintain internal temperature suitable for sperm formation

a

Figure 45.2 *Above and facing page*: Components of the reproductive system of the human male and their functions.

Table 45.1	Organs and Accessory Glands of the Male Reproductive Tract
REPRODUCTIVE ORGANS:	
Testis (2)	Production of sperm, sex hormones
Epididymis (2)	Sperm maturation site and sperm storage
Vas deferens (2)	Rapid transport of sperm
Ejaculatory duct (2)	Conduction of sperm to penis
Penis	Organ of sexual intercourse
ACCESSORY GLANDS:	
Seminal vesicle (2)	Secretion of large part of semen
Prostate gland	Secretion of part of semen
Bulbourethral gland (2)	Production of lubricating mucus

From there, contractions propel sperm through a pair of ejaculatory ducts, then on through the urethra. This last tube extends through the penis, the male sex organ, and opens at its tip. The urethra, recall, is a duct that also functions in urine excretion.

Glandular secretions become mixed with sperm as they travel to the urethra. The result is **semen**, a thick fluid that eventually is expelled from the penis during sexual activity. Early in the formation of semen, a pair

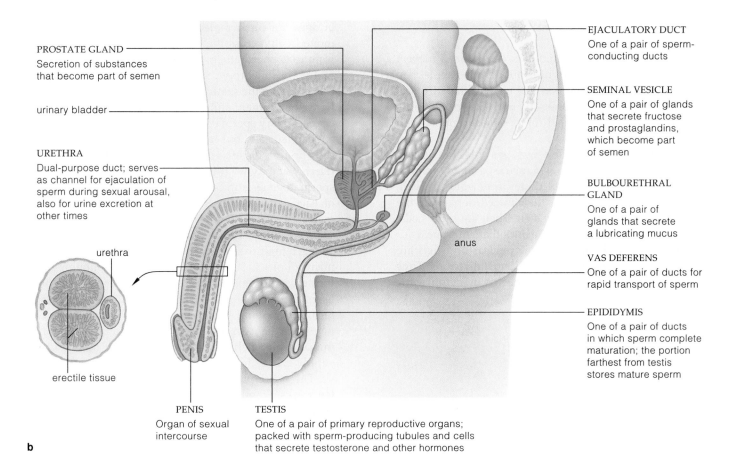

PROSTATE GLAND —
Secretion of substances
that become part of semen

urinary bladder —

URETHRA
Dual-purpose duct; serves —
as channel for ejaculation of
sperm during sexual arousal,
also for urine excretion at
other times

urethra

erectile tissue

PENIS
Organ of sexual
intercourse

TESTIS
One of a pair of primary reproductive organs;
packed with sperm-producing tubules and cells
that secrete testosterone and other hormones

EJACULATORY DUCT
One of a pair of sperm-
conducting ducts

SEMINAL VESICLE
One of a pair of glands
that secrete fructose
and prostaglandins,
which become part
of semen

**BULBOURETHRAL
GLAND**
One of a pair of
glands that secrete
a lubricating mucus

anus

VAS DEFERENS
One of a pair of ducts for
rapid transport of sperm

EPIDIDYMIS
One of a pair of ducts
in which sperm complete
maturation; the portion
farthest from testis
stores mature sperm

b

of seminal vesicles secrete fructose. The sperm use this sugar as an energy source. Seminal vesicles also secrete certain kinds of prostaglandins that can induce muscle contractions. Possibly these signaling molecules take effect during sexual activity. At that time they might induce contractions in the female's reproductive tract and thereby assist sperm movement through it.

The secretions from a prostate gland probably help buffer the acidic conditions that sperm must encounter within the female tract. The vaginal pH is about 3.5–4.0, but the motile capacity of sperm improves at pH 6. Two bulbourethral glands secrete some mucus-rich fluid into the urethra when the male is sexually aroused.

Cancers of the Prostate and Testis

Until recently, cancers of the male reproductive tract did not get much media coverage in the United States, even though *prostate cancer* alone kills 40,000 older men annually. This is surprising, given that the mortality rate for breast cancer, which is highly publicized, is not much higher. You also may be surprised to learn that *testicular cancer* is a frequent cause of death among young men. About 5,000 cases are diagnosed each year in the United States alone.

Both cancers are painless during their early stages. If they are not detected in time, they can spread silently into the lymph nodes of the abdomen, chest, neck, and eventually the lungs. Once the cancer has metastasized, prospects are not good. Testicular cancer, for example, kills as many as 50 percent of those stricken.

Doctors are able to detect prostate cancer in older men through physical examination and blood tests for a rise in prostate-specific antigen. In addition, once a month from high school onward, men should examine each testis after a warm bath or shower, when scrotal muscles are relaxed. The testis should be rolled gently between the thumb and the forefinger to check for any enlargement, hardening, or lump. Such changes might or might not cause noticeable discomfort. But they must be reported so that a physician can order a complete examination. Treatment of testicular cancer has one of the highest success rates, provided the cancer is caught before it starts to spread.

Human males have a pair of testes—primary reproductive organs that produce sperm and sex hormones—as well as accessory glands and ducts. The hormones influence sperm formation and the development of secondary sexual traits.

Sperm Formation

Each testis is about 5 centimeters (less than 2 inches) long. Even so, 125 meters of seminiferous tubules are packed inside it! As many as 300 wedge-shaped lobes, of the sort shown in Figures 45.3 and 45.4, partition its interior. Two or three coiled tubules press against each other inside each lobe.

Just inside each tubule's wall are undifferentiated cells called spermatogonia (singular, spermatogonium). Ongoing cell divisions force them away from the wall, toward the interior. During their forced departure, the cells undergo mitotic cell division. Their daughter cells, primary spermatocytes, are the ones that enter meiosis.

As Figure 45.4a indicates, the nuclear divisions are accompanied by *incomplete* cytoplasmic divisions. Thus the developing cells are interconnected by cytoplasmic

bridges. Ions and molecules required for development move freely across the bridges, so successive divisions from each spermatogonium produce clones of cells that mature in synchrony. **Sertoli cells**, the only other type of cell located in the seminiferous tubules, provide the cells with nourishment and molecular signals.

Secondary spermatocytes form during meiosis I. They are haploid cells, but each one of their chromosomes is in the duplicated state; it consists of two sister chromatids. (Here you may wish to review the general discussion of spermatogenesis in Section 10.5.) The sister chromatids of each chromosome separate from each other during meiosis II. The daughter cells are haploid spermatids. These gradually develop into sperm, the male gametes.

Each mature sperm is a flagellated cell with a core of microtubules in its long tail. An enzyme-containing cap, the acrosome, covers most of the head region. The head is mainly a DNA-packed nucleus (Figure 45.4b). During fertilization, enzymes of the cap itself help the sperm penetrate the extracellular material around an egg. In a midpiece just behind the head, arrays of mitochondria supply energy for the tail's whiplike movements.

The entire process of sperm formation takes nine to ten weeks. From puberty onward, human males produce sperm continuously, so that many millions of cells are in different stages of development on any given day.

Hormonal Controls

Coordinated secretions of LH, FSH, and testosterone govern reproductive function in males. Take a look at Figure 45.3b, which shows the location of **Leydig cells** in tissue between the lobes in testes. Leydig cells secrete **testosterone**, a steroid hormone. Testosterone is central to the growth, form, and functions of the reproductive tract in males. Besides having the main role in sperm formation, it stimulates sexual and aggressive behavior. It promotes development of secondary sexual traits in

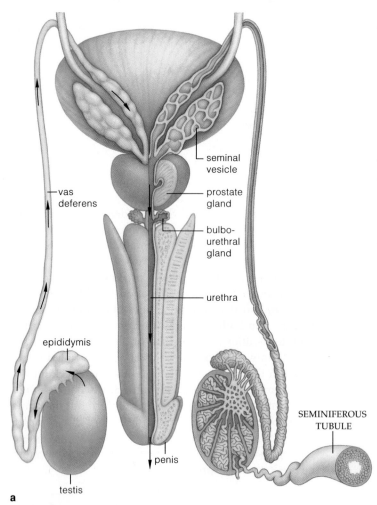

a

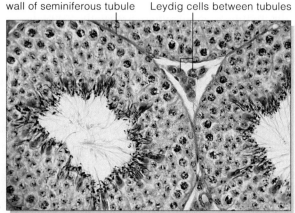

wall of seminiferous tubule   Leydig cells between tubules

b

Figure 45.3 (**a**) Male reproductive tract, posterior view. The arrows show the route that sperm take before ejaculation from a sexually aroused male. (**b**) Light micrograph of cells inside three adjacent seminiferous tubules, cross-section. Leydig cells occupy tissue spaces between the tubules.

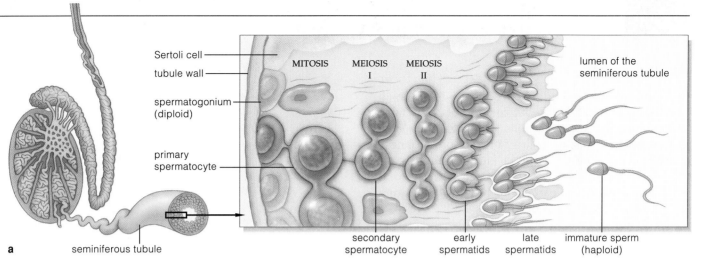

Sertoli cell

tubule wall

spermatogonium (diploid)

primary spermatocyte

MITOSIS

MEIOSIS I

MEIOSIS II

lumen of the seminiferous tubule

secondary spermatocyte

early spermatids

late spermatids

immature sperm (haploid)

a seminiferous tubule

Figure 45.4 (**a**) Sperm formation. The process starts with a spermatogonium (a diploid germ cell). Mitosis, meiosis, as well as incomplete cytoplasmic divisions result in a clone of immature haploid cells that differentiate and develop into mature sperm. (**b**) Structure of a mature sperm from a human male.

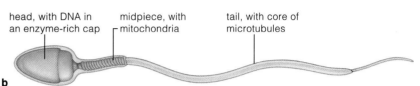

head, with DNA in an enzyme-rich cap

midpiece, with mitochondria

tail, with core of microtubules

b

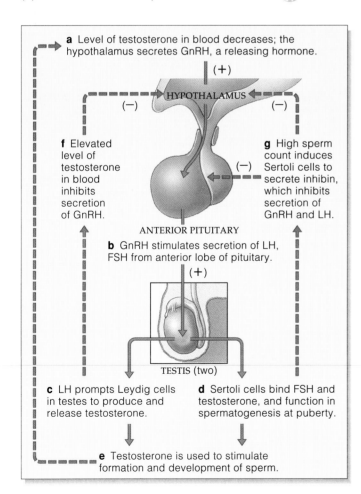

a Level of testosterone in blood decreases; the hypothalamus secretes GnRH, a releasing hormone.

(+)

HYPOTHALAMUS

(−) (−)

f Elevated level of testosterone in blood inhibits secretion of GnRH.

g High sperm count induces Sertoli cells to secrete inhibin, which inhibits secretion of GnRH and LH.

(−)

ANTERIOR PITUITARY

b GnRH stimulates secretion of LH, FSH from anterior lobe of pituitary.

(+)

TESTIS (two)

c LH prompts Leydig cells in testes to produce and release testosterone.

d Sertoli cells bind FSH and testosterone, and function in spermatogenesis at puberty.

e Testosterone is used to stimulate formation and development of sperm.

Figure 45.5 Negative feedback loops to the hypothalamus and to the anterior lobe of the pituitary gland from the testes. Through these loops, excess testosterone production shuts off the mechanisms leading to its production. This helps maintain the testosterone level in amounts required for sperm formation.

males, including an obvious increase in the growth of facial hair and a deepening of the voice at puberty.

LH and **FSH**, remember, are secreted by the anterior lobe of the pituitary gland (Section 37.3). Initially, both hormones were named for their effects in females. (One abbreviation stands for *Luteinizing Hormone*, the other for *Follicle-Stimulating Hormone*.) Later on, researchers discovered that the molecular structure of LH and FSH is identical in both males and females.

The hypothalamus, a part of the forebrain, controls sperm formation by controlling secretion of LH, FSH, and testosterone. As Figure 45.5 shows, in response to low blood levels of testosterone and other factors, the hypothalamus secretes **GnRH**. This releasing hormone stimulates the anterior lobe of the pituitary gland to release LH and FSH, which have targets in the testes. LH prods Leydig cells in testes to secrete testosterone, which helps stimulate the formation and development of sperm. Sertoli cells have receptors for FSH, which is crucial to establishing spermatogenesis at puberty, but researchers don't know whether this hormone has roles in the normal functioning of mature human testes.

A high testosterone level in blood has an inhibitory effect on GnRH release. Also, when the sperm count is high, Sertoli cells release inhibin. This protein hormone acts on the hypothalamus and pituitary to inhibit the release of GnRH and FSH. In this way, feedback loops to the hypothalamus kick in, with a resulting decrease in testosterone secretion and sperm formation.

Sperm formation depends on the hormones LH, FSH, and testosterone. Negative feedback loops from the testes to the hypothalamus and pituitary gland control their secretion.

The Reproductive Organs

We turn now to the reproductive system of a human female. Figure 45.6 shows its components and Table 45.2 summarizes their functions. The female's *primary* reproductive organs, her pair of ovaries, produce eggs and secrete sex hormones. After an immature egg, or **oocyte**, is released from an ovary, it enters the adjacent entrance of an oviduct, part of which is shown in Figure 45.1*a*. A female has a pair of oviducts. Each serves as a channel to the **uterus**, a hollow, pear-shaped organ in which embryos can grow and develop.

The uterine wall is primarily a thick layer of smooth muscle called myometrium. As you will see, the wall's inner lining, the **endometrium**, is central to embryonic development. It consists of connective tissues, glands, and blood vessels. The narrowed portion of the uterus is the cervix. A muscular tube, the vagina, extends from the cervix to the surface of the body. The vagina receives sperm and functions as part of the birth canal.

At the body surface are external genitals (vulva) that include organs for sexual stimulation. Outermost are a pair of fat-padded skin folds, the labia majora. They enclose a smaller pair of skin folds (labia minora) that are highly vascularized but have no fatty tissue. The smaller folds partly enclose the clitoris, a sex organ that is sensitive to stimulation. At the body's surface, the urethra's opening is positioned about midway between the clitoris and the vaginal opening.

Overview of the Menstrual Cycle

Most mammalian females follow an *estrous* cycle. That is, they are fertile and in heat (sexually receptive to males) only at certain times of year. By contrast, female primates, including humans, follow a **menstrual cycle**. Such females are fertile intermittently, on a cyclic basis, and heat is not synchronized with their fertile periods. Said another way, female primates of reproductive age can become pregnant only at certain times of year, but they may be receptive to sex at any time.

During each menstrual cycle, an oocyte matures and escapes from an ovary. Also, the endometrium becomes primed to receive and nourish a forthcoming embryo *if* fertilization takes place. However, if the oocyte does not become fertilized, a total of four to six tablespoons of blood-rich fluid starts flowing out through the vaginal canal. The recurring blood flow is called menstruation. It means "there is no embryo at this time," and it marks the first day of a new cycle. The uterine nest is being sloughed off and is about to be constructed once again.

The events just sketched out proceed through three phases. The cycle starts with a **follicular phase**, a time of menstruation, endometrial breakdown and rebuilding,

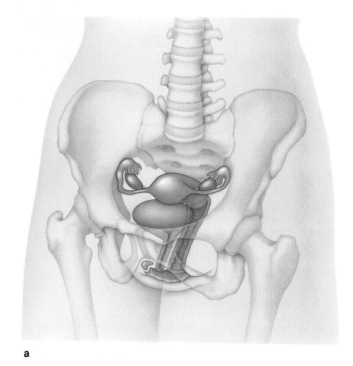

a

Figure 45.6   *Above and facing page*: Components of the human female reproductive system and their functions.

Table 45.2	Female Reproductive Organs
Ovaries	Oocyte production and maturation, sex hormone production
Oviducts	Ducts for conducting oocyte from ovary to uterus; fertilization normally occurs here
Uterus	Chamber in which new individual develops
Cervix	Secretion of mucus that enhances sperm movement into uterus and (after fertilization) reduces embryo's risk of bacterial infection
Vagina	Organ of sexual intercourse; birth canal

and maturation of an oocyte. The next phase, **ovulation**, is restricted to the release of an oocyte from the ovary. Finally, during the **luteal phase** of the menstrual cycle, an endocrine structure (corpus luteum) forms, and the endometrium is primed for pregnancy (Table 45.3).

All three phases are governed by feedback loops to the hypothalamus and pituitary gland from the ovaries. FSH and LH promote cyclic changes in the ovaries. As you will see, the FSH and LH also stimulate the ovaries to secrete sex hormones—**estrogens** and **progesterone**.

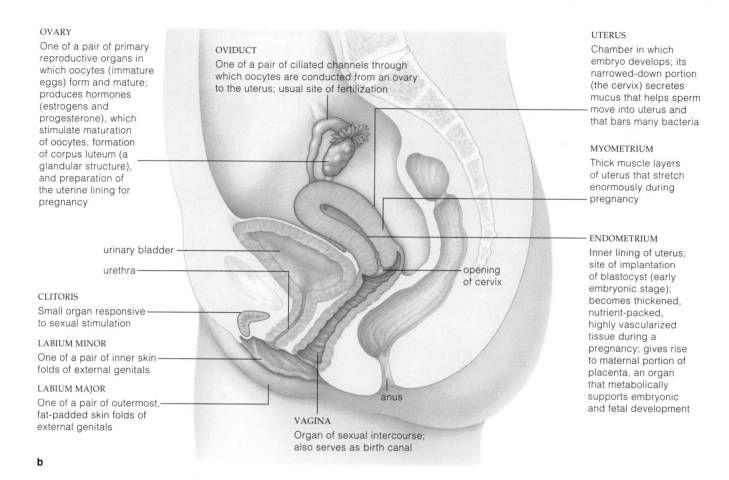

OVARY
One of a pair of primary reproductive organs in which oocytes (immature eggs) form and mature; produces hormones (estrogens and progesterone), which stimulate maturation of oocytes, formation of corpus luteum (a glandular structure), and preparation of the uterine lining for pregnancy

OVIDUCT
One of a pair of ciliated channels through which oocytes are conducted from an ovary to the uterus; usual site of fertilization

UTERUS
Chamber in which embryo develops; its narrowed-down portion (the cervix) secretes mucus that helps sperm move into uterus and that bars many bacteria

MYOMETRIUM
Thick muscle layers of uterus that stretch enormously during pregnancy

urinary bladder

urethra

opening of cervix

ENDOMETRIUM
Inner lining of uterus; site of implantation of blastocyst (early embryonic stage); becomes thickened, nutrient-packed, highly vascularized tissue during a pregnancy; gives rise to maternal portion of placenta, an organ that metabolically supports embryonic and fetal development

CLITORIS
Small organ responsive to sexual stimulation

LABIUM MINOR
One of a pair of inner skin folds of external genitals

LABIUM MAJOR
One of a pair of outermost, fat-padded skin folds of external genitals

anus

VAGINA
Organ of sexual intercourse; also serves as birth canal

b

Table 45.3	Events of the Menstrual Cycle	
Phase	Events	Days of Cycle*
Follicular phase	Menstruation; endometrium breaks down	1–5
	Follicle matures in ovary; endometrium rebuilds	6–13
Ovulation	Oocyte released from ovary	14
Luteal phase	Corpus luteum forms, secretes progesterone; the endometrium thickens and develops	15–28

* Assuming a 28-day cycle.

The secretions of FSH and LH also promote the cyclic changes in the structure of the endometrium.

A human female's menstrual cycles start between ages ten and sixteen. Each cycle lasts for about twenty-eight days, but this is merely the average. It runs longer for some women and shorter for others. The menstrual cycles continue until a woman is in her late forties or early fifties, when her supply of eggs is dwindling and hormonal secretions slow down. This is the onset of *menopause*, the twilight of reproductive capacity.

You may have heard about *endometriosis*, a condition that arises when endometrial tissue abnormally spreads and grows outside the uterus. Endometrial scar tissue may form on ovaries or oviducts and lead to infertility. In the United States, 10 million women may be affected annually. Possibly the condition arises when menstrual flow backs up through the oviducts and spills into the pelvic cavity. Or perhaps some embryonic cells became positioned in the wrong place before birth and were stimulated to grow during puberty, when sex hormones became active. Whatever the case, estrogen still acts on cells in the mislocated tissue. The resulting symptoms include pain during menstruation, sex, or urination.

Ovaries, the female primary reproductive organs, produce oocytes (immature eggs) and sex hormones. Endometrium lines the uterus, a chamber in which embryos develop.

Secretion of the sex hormones estrogen and progesterone is coordinated on a cyclic basis through the reproductive years.

A cycle starts with menstruation, breakdown and rebuilding of the endometrium, and maturation of an oocyte. After release of a mature egg (ovulation), it ends when a corpus luteum forms and the endometrium is primed for pregnancy.

FEMALE REPRODUCTIVE FUNCTION

Cyclic Changes in the Ovary

Take a moment to review Figure 8.9, the generalized picture of meiosis in an oocyte. A normal baby girl has about 2 million *primary* oocytes in her ovaries. By the time she is seven years old, about 300,000 remain; her body resorbed the rest. Primary oocytes already have entered meiosis I, but the nuclear division process was arrested in a genetically programmed way. Meiosis will resume in one oocyte at a time, starting with the first menstrual cycle. Only about 400 to 500 oocytes will be released during her reproductive years.

Figure 45.7 shows a primary oocyte located near the surface of an ovary. A layer of cells, the granulosa cells, surrounds and nourishes it. Together, a primary oocyte and the cell layer around it are a **follicle**. At the start of a menstrual cycle, the hypothalamus is secreting GnRH, which stimulates the anterior pituitary to release FSH and LH (Figure 45.8). In females, these hormones stimulate the follicle to grow. (FSH, remember, is the abbreviation for *Follicle-Stimulating Hormone*.)

The oocyte starts to increase in size, and more layers of cells form around it. Glycoprotein deposits build up between the oocyte and these cell layers. As they do, they widen the space between them. In time the deposits form a noncellular coating, the **zona pellucida**, around the oocyte.

FSH and LH cause cells outside the zona pellucida to secrete estrogens. An estrogen-containing fluid accumulates in the follicle, and estrogen levels in the blood start to rise. About eight to ten hours before its release from the ovary, the oocyte completes meiosis I. Then its cytoplasm divides, forming two cells. One cell, the **secondary oocyte**, ends up with nearly all the cytoplasm. The other cell is the first of three **polar bodies**. The meiotic parceling of chromosomes between the cells gives the secondary oocyte a haploid chromosome number, which is the exact number required for gametes and for sexual reproduction.

About halfway through the cycle, the pituitary gland detects the rising blood level of estrogens. It responds with a brief outpouring of LH. The LH triggers cellular contractions that cause the fluid-engorged follicle to balloon outward, then rupture. The fluid escapes and carries the secondary oocyte with it (Figure 45.7). *Thus the midcycle surge of LH triggers ovulation—the release of a secondary oocyte from the ovary.*

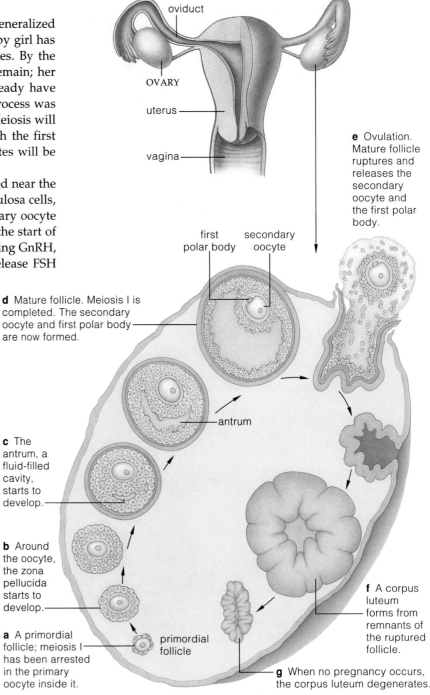

e Ovulation. Mature follicle ruptures and releases the secondary oocyte and the first polar body.

d Mature follicle. Meiosis I is completed. The secondary oocyte and first polar body are now formed.

c The antrum, a fluid-filled cavity, starts to develop.

b Around the oocyte, the zona pellucida starts to develop.

a A primordial follicle; meiosis I has been arrested in the primary oocyte inside it.

primordial follicle

f A corpus luteum forms from remnants of the ruptured follicle.

g When no pregnancy occurs, the corpus luteum degenerates.

Figure 45.7 Cyclic events in a human ovary, cross-section.

A follicle at a given location in the ovary remains there all through the menstrual cycle. It does not move around as in this sketch, which only shows the *order* in which events occur. During the cycle's first phase, the follicle grows and matures. At ovulation, the second phase, the mature follicle ruptures and releases a secondary oocyte. During the third phase, a corpus luteum forms from the follicle's remnants. If pregnancy does not occur, the corpus luteum self-destructs.

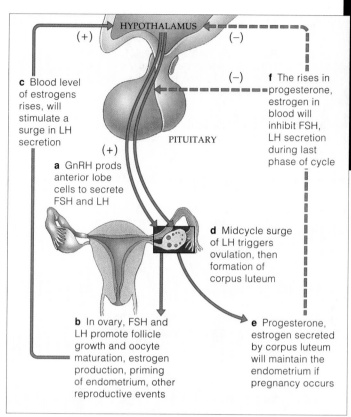

secondary oocyte

surface of ovary

c Blood level of estrogens rises, will stimulate a surge in LH secretion

f The rises in progesterone, estrogen in blood will inhibit FSH, LH secretion during last phase of cycle

HYPOTHALAMUS

(+) (−)

(−)

PITUITARY

(+)

a GnRH prods anterior lobe cells to secrete FSH and LH

d Midcycle surge of LH triggers ovulation, then formation of corpus luteum

b In ovary, FSH and LH promote follicle growth and oocyte maturation, estrogen production, priming of endometrium, other reproductive events

e Progesterone, estrogen secreted by corpus luteum will maintain the endometrium if pregnancy occurs

Figure 45.8 Feedback control of hormonal secretion during a menstrual cycle. A positive feedback loop from an ovary to the hypothalamus causes a surge in LH secretion. The surge triggers ovulation. The light micrograph shows a secondary oocyte being released from an ovary at this time. Afterward, negative feedback loops to the hypothalamus and pituitary inhibit FSH secretion. They prevent another follicle from maturing until the cycle is completed.

Cyclic Changes in the Uterus

The estrogens released during the early phase of the menstrual cycle also help pave the way for pregnancy. They stimulate the growth of the endometrium and its glands. Just before the midcycle surge of LH, cells of the follicle start to secrete some progesterone as well as estrogens. Blood vessels grow rapidly in the thickened endometrium. Then, at the time of ovulation, estrogens act on tissue around the cervical canal, the narrowed-down portion of the uterus that leads to the vagina. The cervix now starts to secrete large quantities of a thin, clear mucus—an ideal medium for sperm travel.

Following ovulation, another structure dominates the cycle. The granulosa cells left behind in the follicle differentiate and form a yellowish glandular structure, the **corpus luteum**. (The name means "yellow body.") The corpus luteum forms as an outcome of the midcycle surge of LH. Hence the name *Luteinizing Hormone*.

The corpus luteum secretes progesterone and some estrogen. Progesterone prepares the reproductive tract for the arrival of a blastocyst, which is the mammalian blastula that forms from a fertilized egg (Section 44.3). For example, this hormone causes cervical mucus to get thick and sticky. The mucus may keep normal bacterial inhabitants of the vagina out of the uterus. Progesterone also maintains the endometrium during a pregnancy.

A corpus luteum persists for about twelve days. All the while, the hypothalamus is calling for minimal FSH secretion and so stops other follicles from developing. If a blastocyst does not burrow into the endometrium, the corpus luteum self-destructs during the last days of the cycle. It does so by secreting certain prostaglandins that apparently disrupt its own functioning.

After this, progesterone and estrogen levels in blood decline rapidly, and so the endometrium starts to break down. Deprived of oxygen and nutrients, the lining's blood vessels constrict and tissues die. Blood escapes as the walls of weakened capillaries start rupturing. The blood and the sloughed endometrial tissues make up a menstrual flow, which continues for three to six days. Then the cycle begins anew, with rising estrogen levels stimulating the repair and growth of the endometrium.

By menopause, the supply of oocytes is dwindling, hormone secretions slow down, and in time menstrual cycles (and fertility) will be over. The eggs that are still to be released late in a woman's life are at some risk of alterations in chromosome number or structure when meiosis resumes. A newborn with Down syndrome, as described in Section 12.9, is one possible outcome.

During a menstrual cycle, FSH and LH stimulate growth of an ovarian follicle (a primary oocyte and the surrounding layer of cells). The first meiotic cell division results in a secondary oocyte and the first polar body.

A midcycle surge of LH triggers ovulation, the release of the secondary oocyte and the polar body from the ovary.

Early on, estrogens call for endometrial repair and growth. Then estrogens and progesterone prepare the endometrium and other parts of the reproductive tract for pregnancy.

By now, you probably have come to the conclusion that the menstrual cycle is not a simple tune on a biological banjo. It is a full-blown hormonal symphony!

Before continuing with your reading, take a moment to review Figure 45.9. It may leave you with a better understanding of the coordinated changes in hormone levels during each menstrual cycle, as correlated with the cyclic changes that they bring about in the ovary and uterus.

Figure 45.9 Changes that occur in the ovary and in the uterus, correlated with changing hormone levels during each turn of the menstrual cycle. *Green* arrows indicate which hormones dominate the cycle's first phase (the time when the follicle matures), then the second phase (when the corpus luteum forms).

(**a**,**b**) FSH and LH secretions bring about the changes in ovarian structure and function. (**c**,**d**) Estrogen and progesterone secretions from the ovary stimulate the changes in the endometrium.

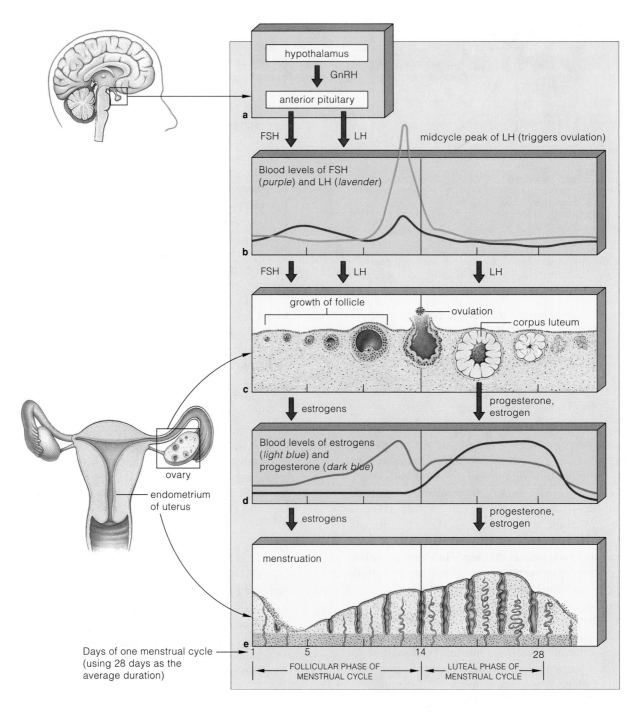

PREGNANCY HAPPENS

Sexual Intercourse

Suppose a secondary oocyte is on its way down an oviduct when a female and male are engaged in sexual intercourse, or **coitus**. The male sex act requires *erection*, whereby the normally limp penis stiffens and lengthens, and *ejaculation*, a forceful expulsion of semen into the urethra and out from the penis. As Figure 45.2 shows, the penis incorporates cylinders of spongy tissue. Many friction-activated sensory receptors occur on the glans penis, the mushroom-shaped tip of one cylinder. In sexually unaroused males, the large blood vessels that lead into the cylinders are vasoconstricted. In aroused males, these vasodilate, so blood flows into the cylinders faster than it flows out. Blood collects in the spongy tissue, and the engorgement stiffens and lengthens the organ, this being helpful for penetration into the vaginal canal.

During coitus, pelvic thrusts stimulate the penis as well as the female's clitoris and vaginal wall. The mechanical stimulation triggers involuntary contractions inside the male reproductive tract. They rapidly force sperm out of each epididymis. They force the contents of seminal vesicles and the prostate gland into the urethra. The resulting mixture, semen, is ejaculated into the vagina. During ejaculation, a sphincter closes off the neck of the bladder and thereby prevents urination.

Emotional intensity, hard breathing, heart pounding, as well as generalized skeletal muscle contractions accompany the rhythmic throbbing of the pelvic muscles. At **orgasm**, the end of the sex act, strong sensations of release, warmth, and relaxation dominate. Similar sensations typify female orgasm. It is a common misconception that a female cannot become pregnant if she doesn't reach orgasm. Don't believe it.

Fertilization

Now sperm are in the vagina. A single ejaculation can put 150 million to 350 million there. If they arrive a few days before or after ovulation or anytime in between, fertilization may be the outcome. Within thirty minutes after ejaculation, muscle contractions move the sperm deeper into the female reproductive tract. Only a few hundred sperm actually reach the upper portion of the oviduct, where fertilization usually takes place.

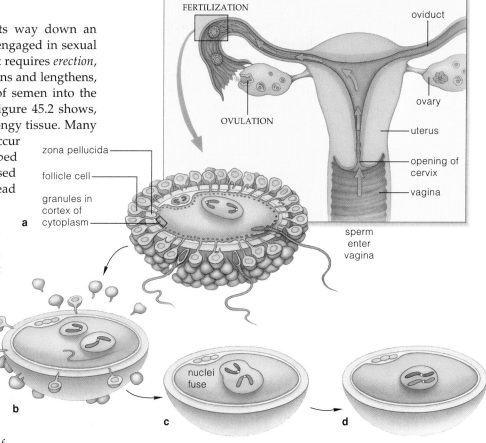

Figure 45.10 Fertilization. (**a**) Many sperm surround a secondary oocyte. Acrosomal enzymes clear a path through the zona pellucida. (**b**) When a sperm does penetrate the secondary oocyte, granules in the egg cortex release substances that make the zona pellucida impenetrable to other sperm. Penetration also stimulates meiosis II of the oocyte's nucleus. (**c**) The sperm's tail degenerates and its nucleus enlarges and fuses with the oocyte nucleus. (**d**) With fusion, fertilization is over. The zygote has formed.

The stunning micrograph that opens Unit II shows living sperm around a secondary oocyte. When sperm contact an oocyte, their cap releases acrosomal enzymes that clear a path through the zona pellucida (Figure 45.10). Although many sperm might get this far, usually only one fuses with the oocyte. Only its nucleus and centrioles do not degenerate in the oocyte's cytoplasm. Penetration induces the secondary oocyte and the first polar body to complete meiosis II. There are now three polar bodies and a mature egg, or **ovum** (plural, ova). As the sperm and egg nuclei fuse, their chromosomes restore the diploid number for a brand new zygote.

The intense physiological events that accompany coitus have one function: to put sperm on a collision course with an egg. Fertilization is over with the fusion of a sperm nucleus and egg nucleus, which results in a diploid zygote.

Pregnancy lasts an average of thirty-eight weeks from the time of fertilization. It takes about two weeks for the blastocyst to form. The time span from the third to the end of the eighth week is the **embryonic** period, when the major organ systems form. When it ends, the new individual has distinctly human features and is called a **fetus**. In the **fetal period**, from the start of the ninth week until birth, organs enlarge and become specialized.

d DAY 5. A blastocoel (a fluid-filled cavity) forms in the morula as a result of secretions from the surface cells. By the thirty-two-cell stage, cells of an inner cell mass are already differentiating. They will give rise to the embryo proper. This embryonic stage is called a blastocyst.

c DAY 4. By 96 hours, there is a ball of sixteen to thirty-two cells that is shaped like a mulberry. It is a morula (after *morum*, Latin for mulberry). Cells of the surface layer will function in implantation and will give rise to a membrane, the chorion.

b DAY 3. After the third cleavage, the cells suddenly huddle together into a compacted ball, which becomes stabilized by numerous tight junctions among the outer cells. Gap junctions form among the interior cells and enhance intercellular communication.

a DAYS 1–2. Cleavage begins within 24 hours after fertilization. The first cleavage furrow extends between the two polar bodies. Subsequent cuts are rotational, so the resulting cells are not symmetrically arranged (compare Section 44.3). Until the eight-cell stage forms, the cells are loosely arranged, with considerable space between them.

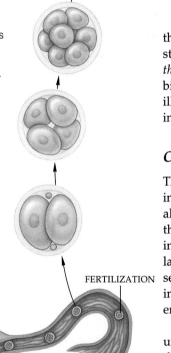

inner cell mass

FERTILIZATION

oviduct   uterus

ovary

endometrium

IMPLANTATION

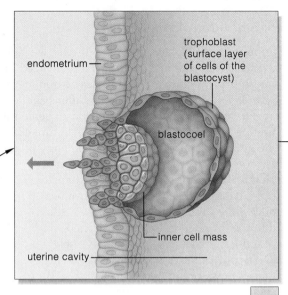

endometrium

trophoblast (surface layer of cells of the blastocyst)

blastocoel

inner cell mass

uterine cavity

e DAYS 6–7. Surface cells of the blastocyst attach to the endometrium and start to burrow into it. Implantation is under way.

actual size

We typically call the first three months of pregnancy the *first* trimester. The *second* trimester extends from the start of the fourth month to the end of the sixth. The *third* trimester extends from the seventh month until birth. Beginning with Figure 45.11, the next series of illustrations show the characteristic features of the new individual at progressive stages of development.

Cleavage and Implantation

Three to four days after fertilization, the zygote is already in the oviduct, and cleavage is under way. Genes are already being expressed; the early divisions depend on their products. At the eight-cell stage, the cells huddle into a compacted ball. By the fifth day, there is a surface layer of cells (the trophoblast), a cavity filled with their secretions (blastocoel), and a cluster of interior cells (the inner cell mass). These are the defining features of the embryonic stage called the **blastocyst** (Figure 45.11*d*).

Six or seven days after fertilization, **implantation** is under way. By this process, the blastocyst adheres to the uterine lining, some of its cells send out projections that invade the mother's tissues, and connections start forming that will metabolically support the developing embryo through the months ahead. While the invasion is proceeding, the inner cell mass develops into two cell

Figure 45.11 From fertilization through implantation. Early on, cleavage produces the blastocyst. Within the blastocyst, the inner cell mass gives rise to the disk-shaped early embryo. Extraembryonic membranes (the amnion, chorion, and yolk sac) start forming.

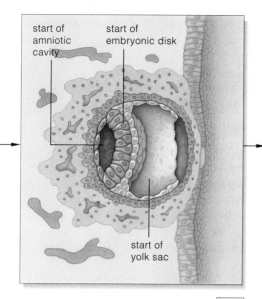

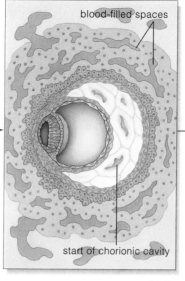

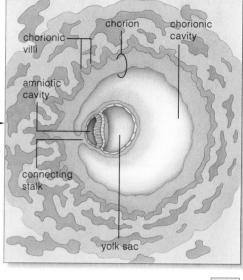

start of amniotic cavity   start of embryonic disk

start of yolk sac

blood-filled spaces

start of chorionic cavity

chorionic villi   chorion   chorionic cavity

amniotic cavity

connecting stalk

yolk sac

f DAYS 10–11. The yolk sac, embryonic disk, and amniotic cavity have started to form from parts of the blastocyst.

actual size

g DAY 12. Blood-filled spaces form in maternal tissue. The chorionic cavity starts to form.

actual size

h DAY 14. A connecting stalk has formed between the embryonic disk and chorion. Chorionic villi, which will be features of a placenta, start to form.

actual size

layers of a flattened, somewhat circular shape. Together, they represent the embryonic disk—and in short order they will give rise to the embryo proper.

Extraembryonic Membranes

As implantation progresses, membranes start to form outside the embryo. First a fluid-filled, *amniotic* cavity opens up between the embryonic disk and part of the blastocyst's surface (Figure 45.11*f*). Then cells migrate around the wall of the cavity and form the **amnion**, a membrane that will enclose the embryo. Fluid inside the cavity will function as a buoyant cradle where the embryo can grow, move freely, and be protected from abrupt temperature changes and mechanical impacts.

While the amnion forms, other cells migrate around the inner wall of the blastocyst's first cavity. They form a lining that becomes the **yolk sac**. This extraembryonic membrane speaks of the evolutionary heritage of land vertebrates (Section 27.7). For most animal species that produce shelled eggs, the sac holds nutritive yolk. In humans, some of the yolk sac becomes a site of blood cell formation, and some will give rise to germ cells, the forerunners of gametes.

Before the blastocyst is fully implanted, spaces open in maternal tissues and fill with the blood seeping in from ruptured capillaries. Inside the blastocyst, another cavity opens around the amnion and yolk sac. Fingerlike projections start to form on the cavity's lining, which is the **chorion**. This new membrane will become part of a spongy, blood-engorged tissue called the placenta.

After the blastocyst is completely implanted, another extraembryonic membrane will form as an outpouching of the yolk sac. This third membrane will become the **allantois**. An allantois has different functions in different animal groups. Among reptiles, birds, and some of the mammals, an allantois serves in respiration and storage of metabolic wastes. In humans, the urinary bladder as well as blood vessels for the placenta form from it.

One more point should be made here. Cells of the blastocyst secrete the hormone **HCG** (*H*uman *C*horionic *G*onadotropin), which stimulates the corpus luteum to keep on secreting progesterone and estrogen. Thus the blastocyst itself prevents menstrual flow and works to avoid being sloughed off until the placenta takes over the task, some eleven weeks later. By the start of the third week, HCG can be detected in the mother's blood or urine. At-home *pregnancy tests* use a treated "dip-stick" that changes color when HCG is present in urine.

A human blastocyst consists of a surface layer of cells around a fluid-filled cavity (blastocoel) and an inner cell mass, which will give rise to the embryo proper.

Six or seven days after fertilization, the blastocyst implants itself in the endometrium. Projections from its surface invade maternal tissues, and connections start to form that in time will metabolically support the developing embryo.

Some parts of the blastocyst give rise to an amnion, yolk sac, chorion, and allantois. These extraembryonic membranes have different roles. But together they are vital for the structural and functional development of the embryo.

By the time a woman has missed her first menstrual period, cleavage is completed and gastrulation is under way. **Gastrulation**, recall, is the stage of extensive cell divisions, migrations, and rearrangements by which the primary tissue layers form (Section 44.4).

By now, the embryonic disk is surrounded by the amnion and chorion, except at the point where a stalk connects it to the chorion's inner wall. Before then, the inner cell mass had behaved *as if* it were still perched on the large ball of yolk that evolved in the reptilian ancestors of mammals. Hence the flattened, two-layered embryonic disk, which also forms in reptiles and birds. Cell divisions and migrations of one layer produced the lining of the yolk sac. The other layer, the epiblast, now starts to generate the embryo proper.

For example, by the eighteenth day, two neural folds have appeared on the embryonic disk (Figure 45.12b). They will merge to form the neural tube. Early in the third week, an allantois, a sausage-shaped outpouching, appears on the yolk sac (Greek *allas*, meaning sausage). In humans, remember, the allantois serves only in the formation of the urinary bladder and of blood vessels for the placenta. Some mesoderm folds into a tube that becomes the notochord. A human notochord is only a structural framework; bony segments of the vertebral column soon form around it. Toward the end of the third week, some mesoderm gives rise to the **somites**. These paired segments are the embryonic source of most bones and skeletal muscles of the head and trunk, as well as the dermis overlying all of those regions. Pharyngeal

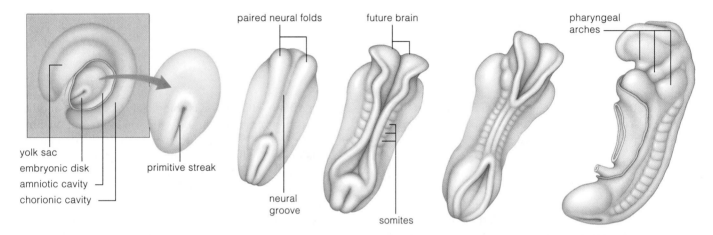

a DAY 15. A faint band appears around a depression along the axis of the embryonic disk. This is the primitive streak, and it marks the onset of gastrulation in vertebrate embryos.

b DAYS 18–23. Cell divisions and migrations, tissue folding, and other events of morphogenesis lead to early organ formation. Neural folds will merge to form the neural tube. Somites (bumps of mesoderm) appear near the embryo's dorsal surface. They will give rise to most of the skeleton's axial portion, skeletal muscles, and much of the dermis.

c DAYS 24–25. By now, some embryonic cells have given rise to pharyngeal arches. These will contribute to the formation of the face, neck, mouth, nasal cavities, larynx, and pharynx.

Figure 45.12 Hallmarks of the embryonic period of humans and other vertebrates. A primitive streak appears, then a notochord. Neural folds appear, then somites and pharyngeal arches. These are dorsal views (of the embryo's back). Compare the diagrams with the photographs in Figure 45.14.

Through cell divisions and migrations, the midline of that layer thickens faintly around a depression at its surface. This "primitive streak" lengthens and thickens further the next day. It marks the onset of gastrulation (Figure 45.12a). It also defines the anterior-posterior axis of the embryo and, in time, its bilateral symmetry. Cells migrating inward at this site give rise to endoderm and mesoderm. Now pattern formation begins, as described in Section 44.6. Interactions among the classes of master genes map out the basic body plan characteristic of all vertebrates. Through embryonic inductions, specialized tissues and organs start forming in orderly sequence in prescribed parts of the embryo.

arches start to form. They will help form the face, neck, mouth, nose, larynx, and pharynx. Small spaces open up in parts of the mesoderm. In time, all of these spaces will interconnect and form the coelomic cavity.

During the third week after fertilization, a time when the woman has missed her first menstrual period, the basic vertebrate body plan emerges in the new individual.

A primitive streak, neural tube, somites, and pharyngeal arches form during the embryonic period of all vertebrates. Formation of the primitive streak establishes the body's anterior-posterior axis and its bilateral symmetry.

ON THE IMPORTANCE OF THE PLACENTA

Even before the onset of the embryonic period, the extraembryonic membranes have been collaborating with the uterus to sustain the embryo's rapid growth. By the third week, tiny fingerlike projections from the chorion have grown profusely into the maternal blood that has pooled in endometrial spaces. The projections, the chorionic villi, enhance the exchange of substances between the mother and the new individual. They are functional components of the **placenta**. This blood-engorged organ consists of both endometrial tissue and extraembryonic membranes. At full term, the placenta will make up a fourth of the inner surface of the uterus (Figure 45.13).

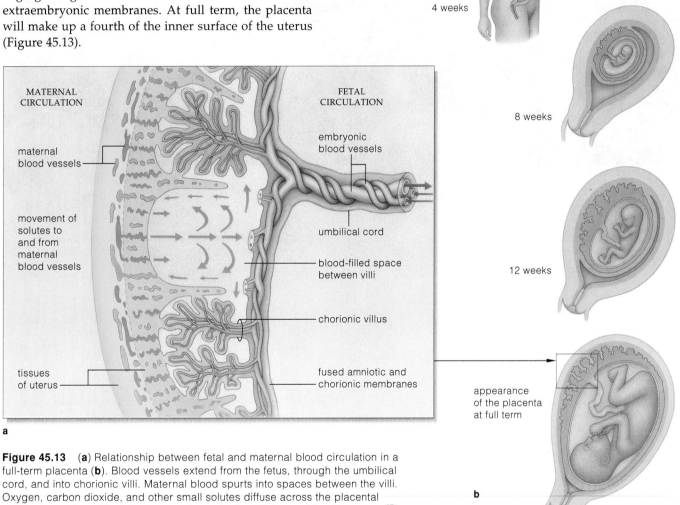

a

Figure 45.13 (**a**) Relationship between fetal and maternal blood circulation in a full-term placenta (**b**). Blood vessels extend from the fetus, through the umbilical cord, and into chorionic villi. Maternal blood spurts into spaces between the villi. Oxygen, carbon dioxide, and other small solutes diffuse across the placental membrane surface. There is no gross intermingling of the two bloodstreams.

The placenta is the body's way of sustaining the new individual while allowing its blood vessels to develop apart from the mother's. Oxygen and nutrients diffuse out of the maternal blood vessels, across the placenta's blood-filled spaces, then into embryonic blood vessels. (These vessels converge in the umbilical cord, the lifeline between the placenta and the new individual.) Carbon dioxide and other wastes diffuse in the other direction. The mother's lungs and kidneys quickly dispose of them.

After the third month, the placenta secretes progesterone and estrogens—and so maintains the uterine lining.

The placenta is a blood-engorged organ of endometrial and extraembryonic membranes. It permits exchanges between the mother and the new individual without intermingling their bloodstreams. Thus it sustains the individual and allows its blood vessels to develop apart from the mother's.

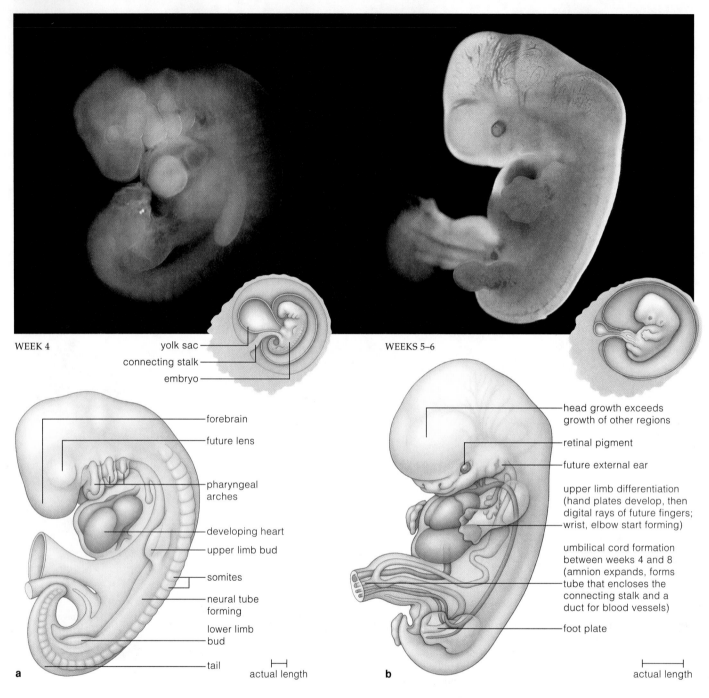

WEEK 4

yolk sac
connecting stalk
embryo

WEEKS 5–6

forebrain

future lens

pharyngeal arches

developing heart

upper limb bud

somites

neural tube forming

lower limb bud

tail

a

actual length

head growth exceeds growth of other regions

retinal pigment

future external ear

upper limb differentiation (hand plates develop, then digital rays of future fingers; wrist, elbow start forming)

umbilical cord formation between weeks 4 and 8 (amnion expands, forms tube that encloses the connecting stalk and a duct for blood vessels)

foot plate

b

actual length

By the end of the fourth week of the embryonic period, the embryo has grown to 500 times its original size. The placenta has been sustaining the growth spurt, but now the pace slows as details of organs fill in. Limbs form, fingers and toes emerge from embryonic paddles, and the umbilical cord forms. The circulatory system becomes intricate. Growth of the all-important head now surpasses that of any other body region (Figure 45.14). The embryonic period ends as the eighth week closes. No longer is

Figure 45.14 (**a**) Human embryo four weeks after fertilization. Like all vertebrates, it has a tail and pharyngeal arches. (**b**) Embryo at five to six weeks. (**c**) Embryo at the boundary between the embryonic and fetal periods. It now has human features. It floats in amniotic fluid. The chorion covers the amniotic sac but has been pulled aside here. (**d**) Fetus at sixteen weeks. Movements begin as nerves make functional connections with forming muscles. Legs kick, arms wave, fingers grasp, the mouth puckers. These reflex actions will be vital skills in the world outside the uterus.

the embryo merely "a vertebrate." By now its features clearly define it as a human fetus.

During the second trimester, the fetus is moving facial muscles. It frowns; it squints. It busily practices

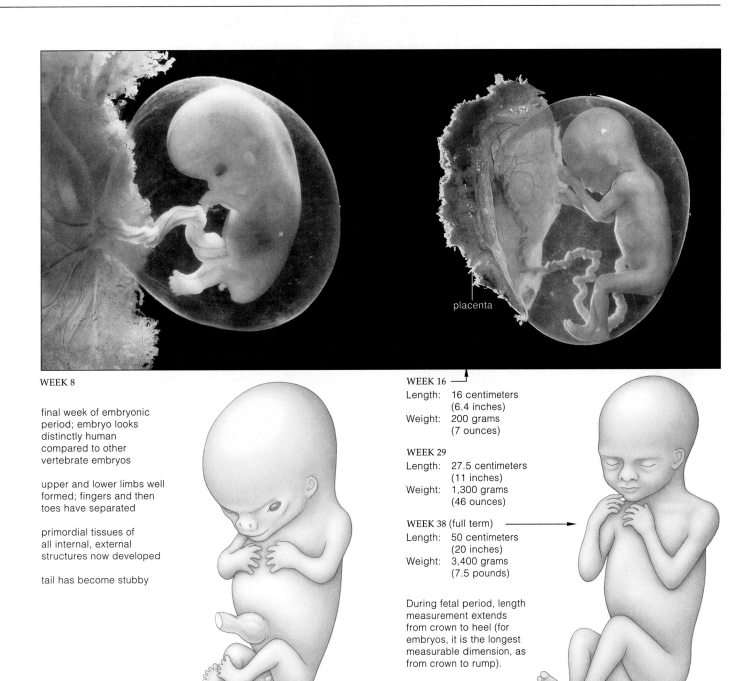

WEEK 8

final week of embryonic
period; embryo looks
distinctly human
compared to other
vertebrate embryos

upper and lower limbs well
formed; fingers and then
toes have separated

primordial tissues of
all internal, external
structures now developed

tail has become stubby

c   actual length

placenta

WEEK 16
Length: 16 centimeters
               (6.4 inches)
Weight: 200 grams
               (7 ounces)

WEEK 29
Length: 27.5 centimeters
               (11 inches)
Weight: 1,300 grams
               (46 ounces)

WEEK 38 (full term)
Length: 50 centimeters
               (20 inches)
Weight: 3,400 grams
               (7.5 pounds)

During fetal period, length
measurement extends
from crown to heel (for
embryos, it is the longest
measurable dimension, as
from crown to rump).

d

the sucking reflex, as shown in the next section. Now
the mother can easily sense movements of the fetal
arms and legs. When the fetus is five months old, she
can hear its heart through a stethoscope positioned on
her abdomen. Soft, fuzzy hair (the lanugo) covers the
fetal body. The skin is wrinkled, reddish, and protected
from abrasion by a thick, cheesy coating. In the sixth
month, delicate eyelids and eyelashes form. During the
seventh month, the eyes open.

A fetus born prematurely before twenty-two weeks
have passed cannot survive. The situation also is grave

for births before twenty-eight weeks, mainly because
lungs have not developed sufficiently. Even with the
best medical care, premature infants have difficulty
breathing and maintaining a normal core temperature.
By the ninth month, however, the rate of survival has
increased to 95 percent.

**During the fetal period, primordial tissues that formed in
the early embryo become sculpted into distinctly human
features.**

45.11 MOTHER AS PROVIDER, PROTECTOR, POTENTIAL THREAT

A woman who decides to become pregnant is committing a large part of her body's resources and functions to the development of a new individual. From fertilization until birth, her future child is absolutely at the mercy of her diet, health habits, and life-style (Figure 45.15).

SOME NUTRITIONAL CONSIDERATIONS How does a pregnant woman best provide nutrients for the embryo, then the fetus? The same balanced diet that is good for her should provide her future child with all of the required carbohydrates, lipids, and proteins. (Chapter 42 is one starting point for an understanding of human nutritional requirements.) The woman's own demands for vitamins and minerals increases during pregnancy. The placenta absorbs enough for her embryo from the bloodstream, except for folate (folic acid). She can reduce the risk that her embryo will develop severe neural tube defects by taking more B-complex vitamins (under supervision by her doctor) before conception and during early pregnancy. Besides this, one study showed that women who smoked while pregnant had depressed blood levels of vitamin C even when their vitamin C intake was identical to that of a

control group. *And so did their fetuses.* Smoking may affect utilization of other nutrients as well.

A pregnant woman also must eat enough so that her body weight increases by between 20 and 25 pounds, on the average. If her weight gain is a great deal lower than that, she is stacking the deck against the fetus. Compared to newborns of normal weight, significantly underweight newborns have more postdelivery complications. They are also at greater risk of having impaired mental functions later in life.

As birth approaches, a fetus makes greater nutritional demands of the mother. Clearly her diet will profoundly influence the remaining developmental events. The brain, like most of the other fetal organs, is especially vulnerable in the weeks just before and after birth, when it undergoes its greatest expansion. By now, all of its neurons have formed. Poor nutrition now will have repercussions on intelligence and other neural functions later in life.

RISK OF INFECTIONS The antibodies circulating in a pregnant woman's blood continually move across the placenta. They protect the new individual from all but

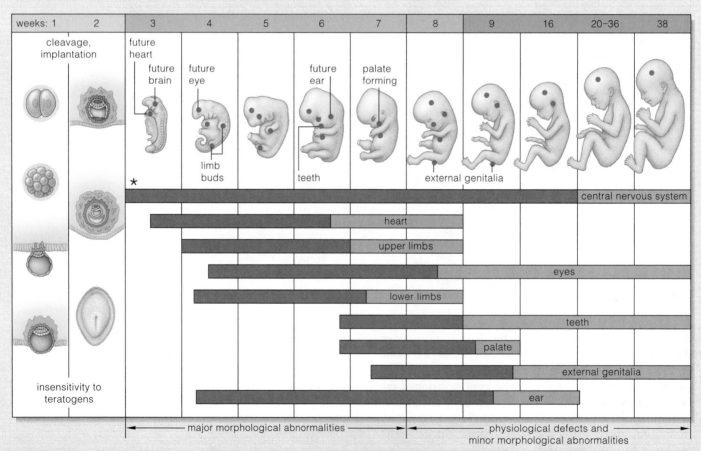

Figure 45.15 Sensitivity to teratogens during pregnancy. *Teratogens* are drugs and any other environmental factors that may induce deformities in the embryo or fetus. They usually have no effect before the onset of organ formation. After that, they can block or abnormally stimulate growth and programmed tissue remodeling and resorption.

★ *dark blue* denotes highly sensitive period; *light blue* denotes less severe sensitivity to teratogens

the most serious bacterial infections. Some virus-induced diseases can be dangerous during the first six weeks after fertilization, which is a critical time of organ formation. Suppose a pregnant woman contracts rubella (German measles) during this critical period. There is a 50 percent chance that some organs of her embryo will not form properly. For example, if she becomes infected when embryonic ears are forming, her newborn may be deaf. If she becomes infected during or after the fourth month of pregnancy, the disease will have no notable effect. However, a woman can avoid this risk entirely, because vaccination before pregnancy can prevent rubella.

EFFECTS OF PRESCRIPTION DRUGS　A pregnant woman absolutely should not take any drugs unless she is under close medical supervision. Think about what happened when the tranquilizer *thalidomide* was being routinely prescribed in Europe. Women who had used thalidomide during the first trimester gave birth to infants who were either missing arms and legs or had grossly deformed ones. As soon as its connection with the deformities was apparent, thalidomide was withdrawn from the market.

However, other tranquilizers, sedatives, and barbiturates are still being prescribed. There is a risk that they may cause similar, although less severe, damage. Even certain *anti-acne drugs* increase the risk of facial and cranial deformities. Or consider two overprescribed antibiotics. One of these, tetracycline, yellows teeth. Streptomycin causes hearing problems and may adversely affect the nervous system.

EFFECTS OF ALCOHOL　As a fetus grows, its physiology becomes increasingly like the mother's. Alcohol passes freely across the placenta. Even one drink has the same effects on the fetus that it has on her. Excessive alcohol intake during pregnancy invites a set of deformities called *fetal alcohol syndrome*, or FAS. Symptoms include reduced brain and head size, mental retardation, facial deformities, poor growth and coordination, and often heart defects (Figure 45.16*b*). In 1992, nearly 4 of every 10,000 newborns were diagnosed with FAS, and the rate is rising. About 60 to 70 percent of newborns of alcoholic women have FAS.

Some researchers suspect that any alcohol at all may be harmful to the fetus. Increasingly, doctors are urging near- or total abstinence during pregnancy.

EFFECTS OF COCAINE　A pregnant woman who uses cocaine, crack especially, disrupts the nervous system of her future child as well as her own. Here you may wish to

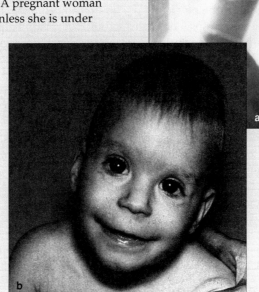

Figure 45.16　(**a**) The fetus at eighteen weeks. (**b**) An infant with fetal alcohol syndrome, or FAS. Obvious symptoms are low and prominent ears, improperly developed cheekbones, as well as an abnormally wide, smooth upper lip. The child can expect growth problems and abnormalities of the nervous system. This disorder currently affects about 1 in 750 newborns in the United States.

read again the start of Chapter 35, which describes the kind of future that offspring of a crack addict face.

EFFECTS OF CIGARETTE SMOKE　Cigarette smoke impairs fetal growth and development. Also, a long-term study at Toronto's Hospital for Sick Children showed that toxic elements in tobacco accumulate even in the fetuses of pregnant nonsmokers who are exposed to *secondhand smoke* at home or work.

Daily smoking during pregnancy leads to underweight newborns even if the woman's weight, nutritional status, and all other relevant variables match those of pregnant nonsmokers. Smoking has other effects. In Great Britain, all infants born the same week were tracked for seven years. Those of smokers were smaller, died of more post-delivery complications, and had twice as many heart abnormalities. At age seven, their average "reading age" was nearly half a year behind children of nonsmokers.

The mechanisms by which smoking affects a fetus are not known. Its demonstrated effects are evidence that the placenta, marvelous structure that it is, cannot prevent all assaults on the fetus that the human mind can dream up.

Giving Birth

On average, pregnancy ends thirty-eight weeks after fertilization, give or take a few weeks. The birth process is called **labor** (or parturition, or delivery). It requires dilation of the cervical canal, so that the fetus can move from the uterus, through the vagina, and out into the world. It also requires very strong uterine contractions as the driving force behind the expulsion.

In the last trimester, mild uterine contractions begin and the cervix softens as its connective tissues loosen. Relaxin, a peptide hormone secreted from the corpus luteum and placenta, may bring about the softening. Relaxin also makes the connections between the pelvic bones loosen up. In the meantime, the fetus "drops," or is shifted downward, usually with its head in contact with the cervix (Figure 45.17a). A *breech birth* is the likely outcome when any part of the fetus other than the head is positioned first near the birth canal.

Rhythmic, usually painless contractions herald the onset of labor. They increase in frequency and intensity over the next two to eighteen hours. What triggers the change in contractility? According to one hypothesis, strong contractions are induced when the concentration of cell receptors for **oxytocin**, another peptide hormone, increases enormously in muscle cells of the uterine wall. Oxytocin, remember, is a hypothalamic hormone that is stored in the posterior pituitary. Some studies indicate that oxytocin may also be produced locally, in the uterus itself, before labor. However, the signal that starts all this activity has not been identified.

Just before birth, the amnion typically ruptures, and "water" (amniotic fluid) gushes from the vagina. Usually contractions expel the fetus in less than an hour after the cervix has fully dilated. Fifteen to thirty minutes later, the contractions cause the placenta to detach from the myometrium; it is expelled as the "afterbirth." The contractions also constrict blood vessels at the placental attachment site and so prevent hemorrhaging. Someone ties and severs the umbilical cord; a few days after it shrivels, the stump will form the newborn's navel. Once the lifeline to the mother has been severed, the newborn embarks on a course of post-embryonic development, starting with the extended period of nurtured existence and learning that is typical of primates.

Many new mothers commonly sink into postpartum depression, or *afterbaby blues*. Apparently corticotropin releasing hormone (CRH) that the placenta produces during pregnancy has a role in this. Its level in blood rises by as much as three times during pregnancy and may influence the timing of labor. Possibly cortisol, a stress hormone, is secreted in response to the elevated CRH levels to help the mother cope with extraordinary stresses that pregnancy and labor place on her body. Once the placenta is expelled, her CRH levels crash to levels that are typical of some clinical depressions. The placental CRH may suppress a feedback loop concerned with the release of CRH from the hypothalamus (Section 37.6). Once the baby has been born, it takes time before the hypothalamus can rebound. The mother's afterbaby blues continue until the hypothalamic secretion of CRH returns to normal.

Nourishing the Newborn

Survival of the newborn depends on an ongoing supply of milk or its nutritional equivalent. Milk production, or **lactation**, occurs in mammary glands in the mother's breasts. Before pregnancy, the breasts consist mainly of adipose tissue and an undeveloped duct system (Figure 45.18a). Their size depends on how much fat they hold, not on milk-producing ability. During pregnancy, they respond to estrogen and progesterone, and a glandular system of milk production develops (Figure 45.18b).

For the first few days after birth, the glands produce a fluid rich in proteins and lactose. Then the anterior

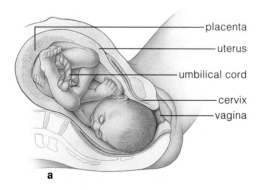

placenta
uterus
umbilical cord
cervix
vagina

a

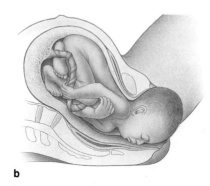

b

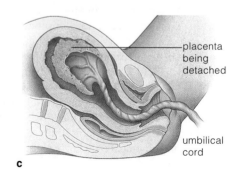

placenta being detached

umbilical cord

c

Figure 45.17 Expulsion of a human fetus during the process of birth. Contractions result in the expulsion of the afterbirth (the placenta, tissue fluid, and blood).

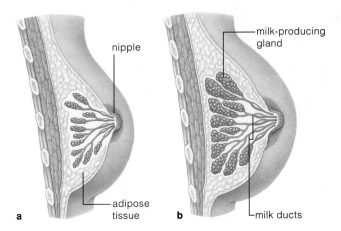

Figure 45.18 (a) Breast of a woman who is not pregnant. (b) Breast of a lactating woman.

pituitary secretes **prolactin**, a hormone that induces the synthesis of the enzymes for milk production (Section 37.3). As a newborn suckles, the pituitary also releases oxytocin. This hormone triggers contractions that force milk into breast tissue ducts and that shrink the uterus back to normal size.

Regarding Breast Cancer

Each year in the United States alone, well over 100,000 women develop *breast cancer*. Obesity, high cholesterol, and high estrogen levels contribute to the cancerous transformation. With early detection and treatment, the chances for cure are excellent. Once a month, about a week after menstruating, a woman should examine her breasts. She can contact her physician or the American Cancer Society (listed in the telephone directory) for a pamphlet on recommended examination procedures.

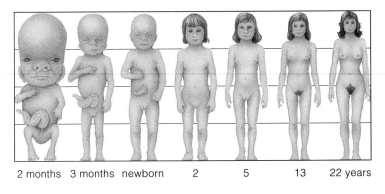

2 months 3 months newborn 2 5 13 22 years

Figure 45.19 Observable, proportional changes in the human body during prenatal and postnatal growth. Changes in overall physical appearance are gradual but quite noticeable until the teenage years. For example, the head becomes proportionally smaller, compared to what it was during the embryonic period. The legs become longer and the trunk becomes shorter.

Table 45.4	Stages of Human Development
PRENATAL PERIOD	
Zygote	Single cell resulting from fusion of sperm nucleus and egg nucleus at fertilization
Morula	Solid ball of cells produced by cleavages
Blastocyst	Ball of cells with surface layer, fluid-filled cavity, and inner cell mass (the mammalian blastula)
Embryo	All developmental stages from two weeks after fertilization until end of eighth week
Fetus	All developmental stages from the ninth week to birth (about thirty-eight weeks after fertilization)
POSTNATAL PERIOD	
Newborn	Individual during the first two weeks after birth
Infant	Individual from two weeks to about fifteen months after birth
Child	Individual from infancy to about ten or twelve years
Pubescent	Individual at puberty, when secondary sexual traits develop; girls between ten and fifteen years, boys between twelve and sixteen years
Adolescent	Individual from puberty until about three or four years later; physical, mental, emotional maturation
Adult	Early adulthood (between eighteen and twenty-five years); bone formation and growth finished. Changes proceed very slowly afterward
Old age	Aging follows late in life

Postnatal Development, Aging, and Death

After birth, a new individual follows a course of further growth and development that leads to the adult, the mature form of the species. Table 45.4 summarizes the *prenatal* (before birth) stages and *postnatal* (after birth) stages. Figure 45.19 is an example of the proportional changes that occur in the body as the life cycle unfolds. Postnatal growth is most rapid between ages thirteen and nineteen. Then, secretions of sex hormones step up and bring about the emergence of secondary sexual traits as well as sexual maturity. Not until adulthood are the bones fully mature. Body tissues normally are maintained in top condition during early adulthood. As the years pass, it becomes more and more difficult to maintain and repair existing tissues, and so the body gradually deteriorates. By processes collectively known as *aging*, the body's cells gradually start to break down. As Section 44.7 indicates, the processes are not fully understood, but they lead to structural changes and gradual loss of bodily functions. All animals that show extensive cell differentiation undergo aging.

The human life cycle flows naturally from the time of birth, growth, and development, to production of the individual's own offspring, and on through aging to the time of death.

Some Ethical Considerations

The transformation of a zygote into an adult of intricate detail raises profound questions. *When does development begin?* As you have seen, major developmental events unfold even before fertilization. *When does life begin?* During her lifetime, a woman can produce as many as 500 eggs, all of which are alive. During one ejaculation, a man can release a quarter of a billion sperm, which are alive. Even before sperm and egg merge by chance and establish the genetic makeup of a new individual, they are as much alive as any other form of life. It is scarcely tenable, then, to say life begins at fertilization. *Life began billions of years ago; and every gamete, every zygote, every sexually mature individual is but a fleeting stage in the continuation of that beginning.*

This greater perspective on life cannot diminish the meaning of conception. It is no small thing to entrust a new individual with the gift of life, wrapped in the unique evolutionary threads of our species and handed down through an immense sweep of time.

Yet how can we reconcile the marvel of individual birth with growing awareness of the astounding birth rate for our whole species? While this book is being written, an average of 10,700 newborns enter the world every single hour. By the time you go to bed tonight, there will be 257,000 more people on Earth than there were last night at that hour. Within a week, the number will reach 1,800,000—about as many people as there are now in the entire state of Massachusetts. *Within one week.* Worldwide, human population growth is rapidly outstripping resources, and each year many millions face the horrors of starvation. Living as we do on one of the most productive continents on Earth, few of us know what it means to give birth to a child, to give it the gift of life, and have no food to keep it alive.

And how can we reconcile the marvel of birth with the confusion that surrounds unwanted pregnancies? Even highly developed countries do not have adequate educational programs concerning fertility. And a great number of people are not inclined to exercise control. Each year in the United States alone, there are 1 million teenage pregnancies. (Many parents actually encourage early boy-girl relationships without thinking through the great risks of premarital intercourse and unplanned pregnancy. Advice is often condensed to a terse "Don't do it. But if you do it, be careful!") And each year, there are 1.6 million abortions among all age groups.

The motivation to engage in sex has been evolving for more than 500 million years. A few centuries of moral and ecological arguments for its suppression have not stopped all that many unwanted pregnancies. Besides this, complex social factors have contributed to a population growth rate that is out of control.

How will we reconcile our biological past and the need for a stabilized cultural present? Whether and how human fertility is to be controlled is one of the most volatile issues of our time. We will return to this issue in the next chapter, in the context of principles that govern the growth and stability of populations. Here, we briefly consider some control options.

Birth Control Options

The most effective method of birth control is complete *abstinence*, no sexual intercourse whatsoever. Data show that it is unrealistic to expect many people to practice it.

A less reliable variation of abstinence is the *rhythm method*. The idea is to avoid intercourse in the woman's fertile period, starting a few days before ovulation and ending a few days after. The woman identifies and tracks her fertile period. She also keeps records of the length of her menstrual cycles, takes her temperature each morning when she wakes up, or both. (The body's core temperature rises by one-half to one degree just before a fertile period.) But ovulation may not be regular, and miscalculations are frequent. Also, sperm deposited in the vagina a few days before ovulation may survive until ovulation. The rhythm method *is* inexpensive; it costs nothing after you buy a thermometer. It does not require fittings and periodic medical checkups. But its practitioners run a risk of pregnancy (Figure 45.20).

Withdrawal, or removing the penis from the vagina before ejaculation, dates back at least to biblical times. But withdrawal requires very strong willpower, and the practice may fail anyway. Fluid released from the penis just before ejaculation may contain some sperm.

Douching (rinsing the vagina with a chemical right after intercourse) is next to useless. Sperm move past the cervix and out of reach of the douche within ninety seconds after ejaculation.

Controlling fertility by surgical intervention is less chancy. In *vasectomy*, a physician makes a tiny incision in a man's scrotum, then severs and ties off each vas deferens. The operation takes only twenty minutes and requires only a local anesthetic. After the operation, sperm cannot leave the testes and cannot be present in semen. So far, there is no firm evidence that vasectomy disrupts hormonal functions or adversely affects sexual activity. Vasectomies can be reversed. But half of those who submit to surgery later develop antibodies against sperm and may not be able to regain fertility.

Females may have a *tubal ligation*. By this surgical intervention, oviducts are cauterized or cut and tied off, usually in a hospital. When the operation is performed correctly, tubal ligation is the most effective means of birth control. A few women have recurring pain in the pelvic area. The operation sometimes can be reversed.

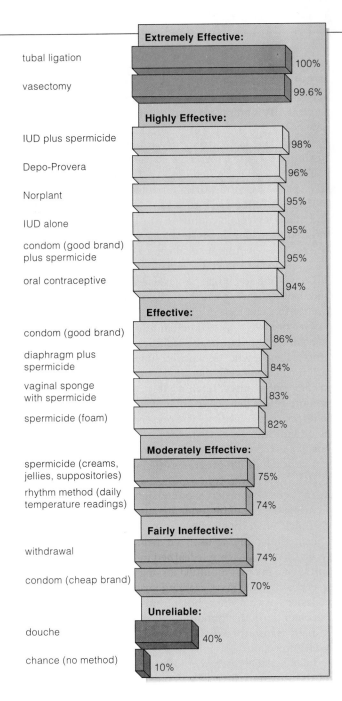

Figure 45.20 Comparison of the effectiveness of methods of contraception in the United States. Percentages reflect the number of unplanned pregnancies per 100 couples who used only that method of birth control for a year. For example, "94% effectiveness" for oral contraceptives (the Pill) means that 6 of every 100 women will still become pregnant, on the average.

Other, less drastic methods of controlling fertility involve physical or chemical barriers to prevent sperm from entering the uterus and moving to the ovarian ducts. *Spermicidal foam* and *spermicidal jelly* are toxic to sperm. They are transferred from an applicator into the vagina just before intercourse. These products are not always reliable unless used with another device, such as a diaphragm or condom.

A *diaphragm* is a flexible, dome-shaped device. It is inserted into the vagina and positioned over the cervix before intercourse. It is relatively effective when fitted initially by a doctor, used with foam or jelly before each sexual contact, inserted correctly each time, and left in place for a prescribed length of time.

Good brands of *condoms*—thin, tight-fitting sheaths worn over the penis during intercourse—are up to 95% effective if used with a spermicide. Only those made of latex provide protection against sexually transmitted diseases (Section 34.19). However, condoms often tear and leak, at which time they become absolutely useless.

The *birth control pill* is an oral contraceptive made of synthetic estrogens and progestins (progesterone-like hormones). Taken daily except for the last five days of a menstrual cycle, it suppresses oocyte maturation and ovulation. Often it corrects erratic menstrual cycles and reduces cramps, but in some women it causes nausea, weight gain, tissue swelling, and headaches. With more than 50 million women taking it, "the Pill" is the most-used fertility control method. It is 94 percent effective. Earlier formulations were linked to blood clots, high blood pressure, and maybe breast cancer. Lower doses of newer formulations might lessen the risk of breast and endometrial cancer. The possibility of a correlation between the Pill and breast cancer is still under study.

Progestin injections or implants inhibit ovulation. A *Depo-Provera* injection works for three months and is 96 percent effective. *Norplant* (six rods implanted under the skin) works for five years and is 95 percent effective. Both may cause sporadic, heavy bleeding, and doctors have some trouble surgically removing Norplant rods.

A pregnancy test doesn't register positive until after implantation. According to one view, a woman is not pregnant until that time. From this viewpoint, RU-486, the *morning-after pill*, intercepts pregnancy. It interferes with hormonal signals that control the events between ovulation and implantation. Three pills, taken within seventy-two hours after intercourse, block fertilization or prevent a blastocyst from burrowing into the uterine lining. RU-486 has side effects (nausea, vomiting, and breast tenderness, for the most part), but not all women experience them. However, RU-486 disrupts complex hormonal interactions and should only be taken under medical supervision. As is true of oral contraceptives, it may trigger elevated blood pressure, blood clots, or breast cancer in some women. At this writing, RU-486 is available in Europe. In the United States, its use is still a subject of controversy.

Whether and how human fertility is to be controlled is a volatile issue. It centers on reconciling our biological past with seriously divergent views about our cultural present.

At some point in their life, at least one of every four people in the United States who engage in sexual intercourse will probably be infected by the pathogens that cause **sexually transmitted diseases** (STDs). STDs have reached epidemic proportions; more than 56 million are already infected. Two-thirds of those are under age twenty-five; one-fourth are teenagers. Women and children are hardest hit. Worse yet, antibiotic-resistant strains of bacteria are on the rise, and some viral diseases simply cannot be cured.

Urban poverty, prostitution, intravenous drug abuse, and sex-for-drugs are fanning the epidemic. Yet STDs also are rampant in high schools and colleges, where students too often think, "It can't happen to me." They reject the idea that no sex—abstinence—is the only safe sex. In one poll of high school students, two-thirds of the respondents said they don't use condoms. More than 40 percent of them have two or more sex partners.

The economics of this health problem are staggering. In 1993, the Centers for Disease Control related to us the annual cost of treating the most prevalent STDs: *Herpes*, $759 million; gonorrhea, $1 billion; chlamydial infection, $2.4 billion; and pelvic inflammatory disease, $4.2 billion. This does not include the accelerating cost of treating patients with AIDS. In many developing countries, AIDS alone threatens to overwhelm health-care delivery systems and to unravel decades of economic progress.

The social consequences are sobering. Mothers bestow a chlamydial infection on one of every twenty newborns in the United States. They bestow type II *Herpes* virus on 1 in 10,000 newborns; one-half of the infected babies die, and one-fourth have severe neural defects. Every year, 1 million women develop pelvic inflammatory disease, and 100,000 to 150,000 become infertile.

AIDS Someone can become infected by HIV, the human immunodeficiency virus, and not even know it. However, the infection marks the start of a titanic battle that the immune system almost certainly will lose (Section 40.11). At first there may be no outward symptoms. Five to ten years later, a set of chronic disorders develops. They are evidence of *AIDS* (acquired immune deficiency syndrome). When the immune system finally does give up, the stage is set for opportunistic infections. Normally harmless, resident bacteria are the first to take advantage of the lowered resistance. Dangerous pathogens also take their toll. In time, the immune-compromised person simply is overwhelmed.

Most commonly, HIV spreads by vaginal, anal, and oral intercourse and by IV drug users. Most of the infections occur by the transfer of blood, semen, urine, or vaginal secretions between people. Cuts or abrasions on the penis, vagina, rectum, and maybe in membranes in the mouth serve as HIV's entrances to the internal environment of a new host.

Today there is no vaccine against HIV and no effective treatment for AIDS. If you get it, you die. *There is no cure.*

HIV was not identified until 1981, but it was present in some parts of Central Africa for at least several decades. In the 1970s and early 1980s, it spread to the United States and other developed countries. Most of those initially infected were male homosexuals. Now a large part of the heterosexual population is infected or at risk. Worldwide, 22.2 million are known to be HIV infected. By early 1992, AIDS was the second leading cause of death among men and the fifth leading cause among women between ages twenty-five and forty-four.

Free or low-cost, confidential testing for HIV exposure is available at public health facilities and at many doctor's offices. People who are worried should know there may be a time lag between their first exposure and the first test to come out positive. It takes a few weeks to six months or more before detectable amounts of antibodies will form in response to infection. (The presence of antibodies in the blood indicates exposure to the virus.) Anyone who tests positive is capable of spreading the virus.

Public education programs attempt to stop the spread of HIV. Most health-care workers advocate safe sex, yet there is great confusion about what "safe" means. Many advocate the use of high-quality latex condoms *and* a spermicide that contains nonoxynol-9 to help block the transmission of the virus. Even then, there is a slight risk. As an unfortunate couple learned in 1997, no one should participate in open-mouthed kissing with someone who tests positive for HIV. Caressing is not risky *if* there are no lesions or cuts where body fluids that harbor the virus can enter the body. Pronounced lesions caused by other sexually transmitted diseases may increase susceptibility to HIV infection.

In sum, AIDS reached epidemic proportions mainly for three reasons. First, it took a while to discover that the virus can travel in semen, blood, and vaginal fluid and that *behavioral* controls can limit its spread. Second, it took time to develop ways to test symptom-free carriers, who infect others. Third, many still don't realize that the medical, economic, and social consequences affect everyone.

GONORRHEA During intercourse, *Neisseria gonorrhoeae* (Figure 45.21) can enter the body at mucous membranes of the urethra, cervix, and anal canal. This bacterium causes *gonorrhea*. An infected female might only notice a slight vaginal discharge or burning

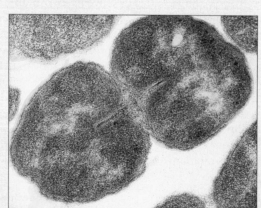

Figure 45.21 Bacterial agents of gonorrhea.

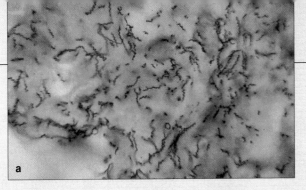
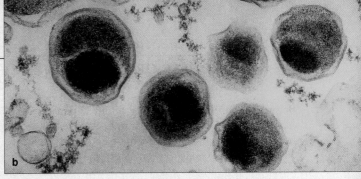

Figure 45.22
The bacterial agents of (**a**) syphilis and (**b**) chlamydia.

sensation while urinating. If the infection spreads to her oviducts, it may induce severe cramps, fever, vomiting, and scar tissue formation, which may cause sterility. Males have more obvious symptoms. Within a week after infection, the penis discharges yellow pus. Urinating is more frequent and may be painful.

This STD is rampant even though prompt treatment quickly cures it. Why? Women experience no troubling symptoms during early stages. Also, infection does not confer immunity to the bacterium, perhaps because there are sixteen or more different strains of it. Contrary to common belief, someone can get gonorrhea over and over again. Also, oral contraceptives, used so widely, promote infection by altering vaginal pH. Populations of resident bacteria decline—and *N. gonorrhoeae* is free to move in.

SYPHILIS *Syphilis*, a dangerous STD, is caused by a spirochete, *Treponema pallidum* (Figure 45.22*a*). Sex with an infected partner puts the motile, spiral bacterium on the surface of genitals or into the cervix, vagina, or oral cavity. It can enter the body through tiny epidermal cuts. One to eight weeks later, new treponemes are twisting about in a flattened, painless chancre (local ulcer). This first chancre is a symptom of the primary stage of syphilis. By then, treponemes are in the blood. The treponemes can cross the placenta of an infected, pregnant woman. The result will be a miscarriage, stillbirth, or syphilitic newborn.

Usually a chancre heals, but in mucous membranes, joints, bones, eyes, spinal cord, and brain, treponemes are multiplying. More chancres and a skin rash develop in this infectious, secondary stage of syphilis. Symptoms subside; immune responses counter the disease in about 25 percent of the cases. Another 25 percent remain infected but symptom-free. In the rest, minor to significant lesions and scars appear in the skin and liver, bones, and other internal organs. Few treponemes occur in this tertiary stage. But the host's immune system is hypersensitive to them. Chronic immune reactions may severely damage the brain and spinal cord and lead to general paralysis.

Probably because the symptoms are so alarming, more people seek early treatment for syphilis than for gonorrhea. Later stages require prolonged treatment.

PELVIC INFLAMMATORY DISEASE *Pelvic inflammatory disease* (PID) is one of the most serious complications of gonorrhea, chlamydial infections, and other STDs. It also can arise when normal bacterial residents of the vagina ascend to the pelvic region. Most commonly, the uterus, oviducts, and ovaries are affected. There are bleeding and vaginal discharge. Pain in the lower abdomen may be as

severe as an acute appendicitis attack. The oviducts may become scarred and invite abnormal pregnancies as well as sterility.

GENITAL HERPES About 25 million Americans have *genital herpes*, caused by the type II *Herpes simplex* virus. Infection requires direct contact with active *Herpes* viruses or sores that contain them. Mucous membranes of the mouth or the genitals are highly susceptible to invasion. Symptoms often are mild or absent. Among infected women, small, painful blisters may appear on the vulva, cervix, urethra, or anal tissues. Among men, blisters form on the penis and anal tissues. Within three weeks, the virus enters latency, and the sores crust over and heal.

This virus is reactivated sporadically. Each time, it causes new, painful sores at or near the original site of infection. Sexual intercourse, menstruation, emotional stress, or other infections can trigger *Herpes* infections. Acyclovir, an antiviral drug, decreases the healing time and often decreases the pain and viral shedding.

GENITAL WARTS More than sixty types of the human papillomaviruses (HPV) are known. A few cause *genital warts*, or benign, bumplike growths. HPV infection of the genitals and anus has become the most prevalent STD in the United States. Type 16 HPV does not usually cause obvious warts but may be linked to precancerous sores and cancers of the cervix, vagina, vulva, penis, and anus. In one Seattle study, 22 percent of female college students who were examined tested positive for the virus.

CHLAMYDIAL INFECTION *Chlamydia trachomatis* (Figure 45.22*b*), a parasitic bacterium, spends part of its life cycle in cells of the genital and urinary tracts. It causes several diseases, including *NGU* (chlamydial nongonococcal urethritis). NGU is far more common than syphilis or gonorrhea. *N. gonorrhoeae* and *C. trachomatis* often are transmitted at the same time. Prompt doses of penicillin cure the gonorrhea—but not NGU, which requires both tetracycline and sulfonamide.

NGU leads to inflammation of the cervix and, in both sexes, the urethra. Infected people may notice a burning sensation while urinating. But when they are symptom-free, they don't seek treatment and complications develop. In males, the prostate becomes swollen and inflamed. In females, the infection may spread into the uterus and the oviducts to cause pelvic inflammatory disease.

Symptom-free infected individuals are unwittingly spreading destructive chlamydial infections through all ethnic groups, among poor and affluent alike.

Because of sterility or infertility, about 15 percent of all couples in the United States cannot conceive. For example, hormonal imbalances in the female body may prevent ovulation, or the male's sperm count may be so low that fertilization would be next to impossible without medical intervention.

In vitro fertilization is conception outside the body (literally "in glass" petri dishes or test tubes). It may occur if the couple can produce normal sperm and oocytes. First the woman receives injections of a hormone that prepares her ovaries for ovulation. Afterward, the preovulatory oocyte is removed from her with a suction device. In the meantime, sperm from the male are placed in a solution that simulates fluid in the oviducts. A few hours after the sperm and suctioned oocyte make contact in a petri dish, fertilization may result. Twelve hours later, the zygote is transferred to a solution that can sustain it through the initial cleavages. Two to four days later, the resulting ball of cells is transferred to the female's uterus.

Each attempt at in vitro fertilization costs about 8,000 dollars, and most attempts aren't successful. In 1994, each "test-tube" baby cost the nation's health-care system about 60,000 to 100,000 dollars, on average. The childless couple may believe no cost is too great. But in an era of increased population growth and shrinking medical coverage, is the cost too great for society to bear? Court battles are being waged over this issue.

At the other extreme is **abortion**, the dislodging and removal of the blastocyst, embryo, or fetus from the uterus. At one time in the United States, abortions were forbidden by law unless the pregnancy endangered the mother's life. Later the Supreme Court ruled that the government does not have the right to forbid abortions during the early stages of pregnancy (typically up to five months). Before this ruling, there were dangerous, traumatic, and often fatal attempts to abort embryos, either by pregnant women themselves or by quacks.

From a clinical standpoint, vacuum suctioning and other methods make abortion painless for the woman, rapid, and free of complications when performed during the first trimester. Abortions performed in the second and third trimesters are extremely controversial unless the mother's life is threatened. Even so, for both medical and humanitarian reasons, the majority of people in this country generally agree that the preferred route to birth control is not through abortion. Rather, it is through sexually responsible behavior that prevents unwanted pregnancy from happening in the first place.

This biology textbook cannot offer you the "right" answer to a moral question because of the reasons given in Section 1.7. It *can* provide you with a serious, detailed description of how a new human individual develops. Your choice of the "right" answer to the question of the morality of abortion will be just that—your choice—and one that can be based on objective insights into the nature of life.

1. Humans have a pair of primary reproductive organs (sperm-producing testes in males and egg-producing ovaries in females), accessory ducts, and glands. Testes and ovaries also produce sex hormones that influence reproductive functions and secondary traits.

   a. The hormones LH, FSH, and testosterone control sperm formation. They are part of feedback loops from the testes to the hypothalamus and anterior lobe of the pituitary gland.

   b. The hormones estrogen, progesterone, FSH, and LH control the maturation and release of oocytes from the ovary, as well as cyclic changes in the endometrium (inner lining of the uterus). Hormonal secretions are part of feedback loops from the ovaries to the hypothalamus and anterior lobe of the pituitary gland.

2. A menstrual cycle is a recurring cycle of intermittent fertility in the reproductive years of female humans and other primates. These events occur during each cycle:

   a. Follicular phase: One of many follicles matures inside an ovary. Each follicle is an oocyte and the cell layer surrounding it. Meanwhile, the endometrium is prepared for a possible pregnancy. It breaks down each cycle if pregnancy does not occur.

   b. Ovulation: A midcycle surge of the LH level in blood triggers ovulation, or the release of a secondary oocyte from the ovary.

   c. Luteal phase: After ovulation a glandular structure, the corpus luteum, develops from the remnants of the follicle. It secretes progesterone and some estrogen that prime the endometrium for fertilization. If fertilization occurs, the corpus luteum is maintained.

3. After fertilization, cleavage produces a blastocyst that becomes implanted in the endometrium. Three primary tissue layers form that will give rise to all organs. They are ectoderm, endoderm, and mesoderm.

4. Four extraembryonic membranes form as embryos of humans (and other vertebrates) develop:

   a. The amnion becomes a fluid-filled sac around the embryo, which it protects from mechanical shocks, abrupt temperature changes, and drying out.

   b. The yolk sac stores nutritive yolk in most shelled eggs, but in humans, part becomes a major site of blood formation and some of its cells give rise to germ cells. (Eventually germ cells give rise to sperm and eggs.)

   c. The chorion is protective membrane around the embryo and the other membranes. It also becomes a major component of the placenta.

   d. In humans, the blood vessels for the placenta arise from the allantois, as does the urinary bladder.

5. The placenta, a blood-engorged organ, is composed of endometrium and extraembryonic membranes. The placenta allows the embryonic blood vessels to develop

independently of the mother's, even while allowing oxygen, nutrients, and wastes to diffuse between them.

6. To some extent, the placenta serves as a protective barrier for the fetus. But it cannot protect the fetus from harmful effects of the mother's nutritional deficiencies, infections, intake of prescription drugs, illegal drugs, alcohol, and cigarette smoking.

7. During pregnancy, both estrogen and progesterone stimulate the growth and development of mammary glands. During labor, strong uterine contractions expel the fetus and the afterbirth. After delivery, nursing stimulates the release of prolactin and oxytocin, which in turn stimulate milk production and release.

8. Whether and how to control human sexuality and fertility are major ethical issues of our time. There is widespread agreement that sexually responsible behavior rather than abortion is the preferred route to birth control.

Review Questions

1. Name the two primary reproductive organs of the human male and where sperm formation starts inside them. Does semen consist of sperm, glandular secretions, or both? 45.1

2. Does sperm formation require mitosis, meiosis, or both? 45.2

3. Study Figure 45.5. Then, on your own, sketch the feedback loops to the hypothalamus and anterior pituitary from the testes that govern sperm formation. Include the names of the releasing hormone, hormones, and cells in the testes involved in these loops. 45.2

4. Name the two primary reproductive organs of the human female. What is the endometrium? 45.3

5. Distinguish between:
   a. oocyte and ovum 45.3, 45.6
   b. ollicular phase and luteal phase of menstrual cycle 45.3
   c. follicle and corpus luteum 45.4
   d. ovulation and implantation 45.3, 45.7

6. What is the menstrual cycle? Name four of the hormones that influence this cycle. Which one triggers ovulation? 45.3, 45.4

7. Study Figure 45.8. Then, on your own, sketch the feedback loops to the hypothalamus and anterior pituitary from the ovaries that govern the menstrual cycle. Include the names of the releasing hormone, hormones, and ovarian structures involved in these loops. 45.4

8. Describe the cyclic changes that occur in the endometrium during the menstrual cycle. What role does the corpus luteum play in the changes? 45.4

9. Distinguish between the embryonic period and fetal period of human development. Then describe the general organization of a human blastocyst. 45.7

10. State the embryonic source of the amnion, yolk sac, chorion, and allantois. Then state the role that each extraembryonic membrane plays in the structure or functioning of the developing individual. 45.7

11. Name some of the early organs that are the hallmark of the embryonic period of humans and other vertebrates. 45.8

12. Describe the placenta's structure and function. Do maternal and fetal bloodstreams grossly intermingle in this organ? 45.9

13. Name the releasing hormone with a key role in labor. Then state the role of estrogen, progesterone, prolactin, and oxytocin in assuring that the newborn will have an ongoing supply of milk. 45.12

14. Label the components of the human male reproductive system and state their functions: 45.1, 45.2

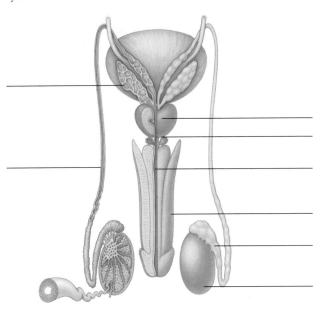

15. Label the components of the human female reproductive system and state their functions: 45.3, 45.4

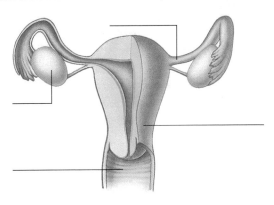

Self-Quiz (Answers in Appendix IV)

1. The _____ and the pituitary gland control secretion of sex hormones from the testes and the ovaries.

2. _____ secretions govern reproductive function in males.
   a. FSH   d. both a and b
   b. LH   e. both b and c
   c. Testosterone   f. all of the above

3. During a menstrual cycle, a midcycle surge of _____ triggers ovulation.
   a. estrogen   b. progesterone   c. LH   d. FSH

4. During a menstrual cycle, _____ and _____ secreted by the corpus luteum prime the uterus for pregnancy.
   a. FSH; LH   c. estrogen; progesterone
   b. FSH; testosterone d. estrogens only

5. In implantation, a _____ burrows into the endometrium.
   a. zygote   c. blastocyst
   b. gastrula d. morula

6. The _____ , a fluid-filled sac, surrounds and protects the embryo from mechanical shocks and keeps it from drying out.
   a. yolk sac   b. allantois   c. amnion   d. chorion

7. At full term, a placenta _____ .
   a. is composed of extraembryonic membranes
   b. directly connects maternal and fetal blood vessels
   c. keeps maternal and fetal blood vessels separated

8. (A) _____ form(s) in all vertebrate embryos.
   a. neural plate   c. pharyngeal arches   e. a through c
   b. somites   d. primitive streak   f. a through d

9. Distinctly human features emerge in the embryo by the end of the _____ week after fertilization.
   a. second   c. fourth   e. eighth
   b. third   d. fifth   f. sixteenth

10. Match the term with the most suitable description.
    ____ seminiferous tubule   a. glandular structure formed from follicle remnants
    ____ allantois   b. most bones, skeletal muscles of head and trunk form from them
    ____ corpus luteum   c. helps form urinary bladder, blood vessels for placenta in humans
    ____ somites   d. blood cell formation site, source of germ cells
    ____ yolk sac   e. sperm form here

Critical Thinking

1. Suppose that, at the time a human zygote is undergoing cleavage, the first two blastomeres that form completely separate from each other. Both blastomeres and their cellular descendants continue to divide on schedule, on the prescribed developmental program. The result is *identical twins*, or two normal, genetically identical individuals. *Nonidentical twins* form when two different secondary oocytes are fertilized at the same time by two different sperm. On the basis of this information, explain why nonidentical twins show considerable genetic variability and identical twins do not.

2. Infection by the rubella virus apparently has an inhibitory effect on mitosis. Serious birth defects result when a woman is infected during the first trimester of pregnancy, but not later. Review the developmental events that unfold during pregnancy and explain why this might be so.

3. Imagine you are an obstetrician advising a woman who has just learned she is pregnant. What instructions would you provide concerning her diet and behavior during pregnancy?

4. In the United States, teenage pregnancies and STD infections are rampant. Suppose the office of the Surgeons General asks you to take part in a task force that will recommend practices that might reduce the incidence of both. What practices might meet with the greatest success? What kind of enthusiasm or resistance might they provoke among the teenagers in your community? Among adults? Explain why.

Selected Key Terms

abortion *45.15*	menstrual cycle *45.3*
allantois *45.7*	oocyte *45.3*
amnion *45.7*	orgasm *45.6*
blastocyst, human *45.7*	ovary *CI*
chorion *45.7*	ovulation *45.3*
coitus *45.6*	ovum *45.6*
corpus luteum *45.4*	oxytocin *45.12*
endometrium *45.3*	placenta *45.9*
estrogen *45.3*	polar body *45.4*
fetus *45.7*	progesterone *45.3*
follicle *45.4*	prolactin *45.12*
follicular phase (menstrual cycle) *45.3*	secondary oocyte *45.4*
	secondary sexual trait *CI*
FSH *45.2*	semen *45.1*
gastrulation *45.8*	seminiferous tubule *45.1*
GnRH *45.2*	Sertoli cell *45.2*
HCG *45.7*	sexually transmitted disease (STD) *45.14*
implantation *45.7*	
in vitro fertilization *45.15*	somite *45.8*
labor (birth process) *45.12*	testis (testes) *CI*
lactation *45.12*	testosterone *45.2*
Leydig cell *45.2*	uterus *45.3*
LH *45.2*	yolk sac *45.7*
luteal phase (menstrual cycle) *45.3*	zona pellucida *45.4*

Readings

Aral, S., and K. Holmes. February 1991. "Sexually Transmitted Diseases in the AIDS Era." *Scientific American* 2(264): 62–69.

Caldwell, M. November 1992. "How Does a Single Cell Become a Whole Body?" *Discover* 13(11): 86–93.

Cohen, J., and I. Stewart. April 1994. "Our Genes Aren't Us." *Discover* 15(4): 78–84. Argument that DNA means nothing in the absence of a developmental context.

Gilbert, S. 1994. *Developmental Biology*. Fourth edition. Sunderland, Massachusetts: Sinauer.

Larsen, W. 1993. *Human Embryology*. New York: Churchill Livingston. Paperback. Recommended for the serious student, but maybe not for the fainthearted.

McGinnis, W., and M. Kuziora. February 1994. "The Molecular Architects of Body Design." *Scientific American* 270(2): 58–66.

Nilsson, L., et al. 1986. *A Child Is Born*. New York: Delacorte Press/Seymour Lawrence.

Sherwood, L. 1997. *Human Physiology*. Fourth edition. Belmont, California: Wadsworth.

Zack, B. July 1981. "Abortion and the Limitations of Science." *Science* 213: 291.

Web Site See *http://www.wadsworth.com/biology* for practice quiz questions, hypercontents, BioUpdates, and critical thinking. The Wadsworth Biology Resource Center provides a wealth of information fully organized and integrated by chapter.

FACING PAGE: *Two organisms—a fox in the shadows cast by a snow-dusted spruce tree. What are the nature and consequences of their interactions with each other, with other organisms, and with their environment? By the end of this last unit, you possibly will see worlds within worlds in such photographs.*

46 POPULATION ECOLOGY

Tales of Nightmare Numbers

Across from Sausalito, California, the steep flanks of Angel Island rise from the waters of San Francisco Bay (Figure 46.1a). The island, set aside as a game reserve, escaped urban development. It did not escape from the descendants of a few deer that well-meaning nature lovers shipped over in the early 1900s. With no natural predators to keep them in check, the few deer became many—far too many for the limited food supply of their isolated habitat. Yet the island attracted a steady stream of picnickers from the mainland. They felt sorry for the malnourished animals and made sure to load the picnic baskets with extra food for them.

The visitors imported so much food that scrawny deer kept on living and reproducing. In time, the herd nibbled away the native grasses, the roots of which had helped slow soil erosion on the steep hillsides. Hungry deer chewed off all the new leaves of seedlings; they killed small trees by stripping the bark and its phloem. The herd was destroying the island habitat.

In desperation, game managers proposed using a few skilled hunters to thin the herd. They were strongly denounced as being cruel. They proposed importing a few coyotes to the island to thin the herd naturally. Animal rights advocates opposed that solution, also.

As a compromise, about 200 of the 300+ deer were captured, loaded onto a boat, and shipped to suitable mainland habitats. A number of them received collars with radio transmitters so that game managers could track them after the release. In less than sixty days, dogs, coyotes, bobcats, hunters, and speeding cars and trucks had killed off most of them. In the end, relocating each surviving deer had cost taxpayers almost 3,000 dollars. The State of California refused to do it again. And no one else, anywhere, volunteered to pick up future tabs.

It is not difficult to define the boundaries of Angel Island or track its inhabitants, so it is easy to draw a lesson from this tale: *A population's growth depends on the resources of its environment. And attempts to "beat nature" by altering the sometimes cruel outcome of limited resources only postpones the inevitable.* Does the same lesson apply to other populations, in other places? Yes, but other considerations may obscure the essence of it.

Figure 46.1 (a) Angel Island, which turned out to be a laboratory for studying population growth. (b) Bathers crowding the banks of the Ganges River in India—a tiny sampling of a human population that has now surpassed 5.8 billion. In this chapter we turn to principles that govern the growth and sustainability of all populations.

Consider this next tale. When 1997 drew to a close, there were more than 5.8 billion people on Earth. About 2 billion already live in poverty, and each year 40 million more join their ranks. Next to China, India is the most populous country, with 950 million inhabitants (Figure 46.1b). By 2025 it may reach 1.38 billion. Forty percent of the people live in rat-infested shantytowns, without adequate food or water. They are forced to wash clothes and dishes in open sewers. Land available to raise their food shrinks by 365 acres a day, on the average. Why? Irrigated soil becomes too salty when it drains poorly and there is not enough water to flush away the salts.

Can wealthier, less densely populated nations help? After all, they use most of the world's resources. Maybe they should learn to get by more efficiently, on less. For example, people might limit their meals to cereal grains and water, give up their private cars, living quarters, air conditioning, televisions, and dishwashers, stop taking vacations, stop laundering so much, close all the malls, restaurants, and theaters at night, and so on.

Maybe wealthier nations also should donate more surplus food than they already donate to less fortunate ones. Then again, would huge donations help, or would they encourage dependency and spur more increases in population size? And what if surpluses run out?

It is a monumental dilemma. At one extreme, the redistribution of resources on a global scale would allow the greatest number of people to survive, but at the lowest comfort level. At the other extreme, foreign aid rationed only to nations that restrict population growth would allow fewer individuals to be born, but their lives would have greater quality.

Currently, the foreign aid program of the United States is based on two premises: (1) that individuals of every nation have an irrevocable right to bear children, even if unrestricted reproduction ruins the environment that must sustain them; and (2) that because human life is precious above all else, the wealthiest nations have an absolute moral obligation to save lives everywhere.

Regardless of the positions that nations take on this issue, ultimately they must come to terms with this fact: *Certain principles govern the growth and sustainability of populations over time.* These principles are the bedrock of **ecology**—the systematic study of how organisms interact with one another and with their physical and chemical environment. Ecological interactions start within and between populations, and they extend on through communities, ecosystems, and the biosphere. They are the focus of this last unit of the book.

KEY CONCEPTS

1. Certain ecological principles govern the growth and sustainability of all populations, including our own.

2. In addition to genetic factors, a population's size, density, distribution, and the number of individuals in various age categories influence its patterns of growth.

3. When the birth rate exceeds the death rate, and when immigration and emigration are in balance, populations may show a pattern of exponential growth.

4. With exponential growth, a population expands by ever increasing increments during successive intervals, because the number of individuals in its reproductive age category (or about to enter it) becomes ever larger.

5. A shortage of any resource that individuals require is a limiting factor on population growth.

6. For all populations, carrying capacity is the maximum number of individuals that can be sustained indefinitely by the resources of a given environment. That number may rise or fall with changes in resource availability.

7. A population may show a pattern of logistic growth. By this pattern, population density—that is, the number of individuals in a specified area at a given point in time—is initially low. Then the population size rapidly increases. It finally levels off as resource scarcity limits its further increase or triggers a decline in numbers.

8. Some populations show fluctuations in numbers that cannot be explained by a single growth model.

9. All populations face limits to growth, because no environment can indefinitely sustain a successively increasing number of individuals. Competition, disease, predation, and other factors control population growth. The controls vary in their relative effects on populations of different species, and they vary over time.

CHARACTERISTICS OF POPULATIONS

By this point in the book you know that a population is a group of individuals of the same species occupying a given area. As described in Section 18.1, its individuals draw from a common gene pool, which is the basis for characteristic ranges of morphological, physiological, and behavioral traits they hold in common. For their investigations, ecologists consider the genetic makeup, modes of reproduction, and reproductive behavior of a population. They also consider the **demographics**, or the vital statistics, of the population, and that will be our focus here. Among those vital statistics are population size, density, distribution, and age structure.

Population size is the number of individuals that share the population's gene pool. **Population density** is the number of individuals in a given area or volume of a habitat, such as the number of guppies per liter of water in a stream. (A *habitat*, recall, is the type of place where a species normally lives. We characterize habitats by physical and chemical features, and by the presence of other species.) **Population distribution** is the general pattern in which the individuals of the population are dispersed through a specified area.

A population also has an **age structure**: the number of individuals in each of several to many age categories. For example, we may divide all of the individuals into *pre-reproductive*, *reproductive*, and *post-reproductive* ages. Those in the first category have a capacity to produce offspring when they mature. Together with actually and potentially reproducing members in the next category, we include them in the population's **reproductive base**.

Let's look first at the relationships that influence the size, structure, and distribution of populations. Later, we will apply the basic principles of population growth to the past, present, and future of the human species.

Regarding Population Density

Populations of a great number of different species often live in the same area, but they generally differ quite a bit in terms of density. You already know this if you have ever counted seashells on a beach or the butterflies and birds in your backyard, or waged a running battle with snails in a small garden. For every species that was well represented in the area, you probably noticed several others that were few and far between.

The *crude* density of each population is no more than the measured number of individuals in a specified area. Such measurements serve as the baseline for tracking changes in population density over time. For a sparsely represented population of trees, snakes, or some other organism, a simple "head count" will do. For dense populations, ecologists commonly make their counts in a small section of the defined area, selected at random, then they use the data to estimate overall density.

Crude density does not tell us how much of the area is being used as living space. As you will see, even the areas that appear to be uniform, such as a long, sandy beach, are more like fine tapestries of light, moisture, temperature, mineral composition, and other variables. Moreover, a tapestry of environmental conditions often changes with the seasons. Thus the population might find one part of the area more suitable for occupancy than others, and it may do so some or all of the time.

Also, more than one species typically occupy the same area. Different species compete with one another for energy, nutrients, living space, and other resources. Many of them interact as predators, parasites, or prey. You will read about species interactions in chapters to come. For now, simply bear in mind that interactions among species influence the density of a population as well as its distribution.

In short, *crude density refers to numbers only, not to how individuals of the population are dispersed through their habitat.* For some or all of the time, environmental conditions and species interactions make parts of the habitat unsuitable or unavailable for occupancy.

Patterns of Dispersion

Theoretically, at least, populations show three patterns of dispersion through a habitat. By these patterns, the individuals are distributed in clumps, nearly uniformly, or randomly (Figure 46.2).

CLUMPED DISPERSION Individuals of most populations form aggregations at specific sites in the habitat. Why is clumping the most common pattern? There are three main reasons.

First, remember that each species is adapted to a limited set of environmental and ecological conditions. And those conditions usually are patchy through the habitat. For example, pasture plants grow luxuriantly in small, scattered patches of soil—where cowpats fell weeks or months before and so enriched the soil with nitrogen. Similarly, different parts of some habitats offer more shade or moisture, better hiding places for prey organisms, or better hunting possibilities for predators.

clumped   nearly uniform   random

Figure 46.2 Diagrams of three generalized patterns of population distribution.

a

b

c

Figure 46.3 (a) A gathering of baboons in a patch of habitat that provides them with water, food, and hiding places. As is the case for many other social species, baboon populations follow the most common pattern of dispersion: clumping. (b) Nearly uniform spacing of creosote bushes near Death Valley, California. (c) Individual of a randomly dispersed population of wolf spiders.

Second, many kinds of animals form social groups, which offer distinct advantages for individual survival and reproduction (Figure 46.3a). Such groups might afford better possibilities for mating. Their individuals might engage in mutual defense against predators and assist one another in raising offspring. Also, with more pairs of eyes to detect danger, individuals can forage more efficiently for food.

Third, many species simply do not have the means to disperse seeds, larvae, or other immature forms of the new generation over large distances. For example, even though sponge larvae can swim, they cannot swim very far from the parent sponge body. They simply settle down on substrates near their parents.

NEARLY UNIFORM DISPERSION Where individuals of a population are more evenly spaced than they would be by chance alone, we call this a regular, nearly uniform pattern of dispersal. An orchard with regularly spaced fruit trees is often cited as an example. Few populations live this way in nature. You may see the pattern when conditions are fairly uniform in the habitat and there is strong competition for resources or territorial behavior. Consider the creosote bush (*Larrea*), a tough-leafed plant adapted to arid conditions in the American Southwest (Figure 46.3b). Large, mature plants deplete the soil all around them of available water. Also, seed-eating ants and rodents forage near them and eat most of the seeds released from mature plants. Thus seeds and seedlings are at a competitive disadvantage, except in areas where established plants weaken or die. For such reasons, the creosote bushes are not clumped together. Neither are they randomly dispersed, for the seeds that rodents and ants overlook and that await opportunity for their day in the sun cannot travel far from the parent plants.

RANDOM DISPERSION Random dispersion occurs only when individuals of a population neither attract nor avoid one another when conditions are fairly uniform through the habitat, and when resources are available all the time. For example, each generation of wolf spiders, which are solitary hunters on forest floors, may well be randomly spaced (Figure 46.3c). But random dispersion is extremely rare. In population studies, this pattern is better used as a theoretical baseline for measuring the degree of clumping or near uniformity.

Some Qualifiers

Our perceptions of a population may be influenced by how wide a view we take. Poke around through a few square meters of the floor of an oak forest, and acorns under an oak tree appear to be randomly spaced. Look at a forest of mixed hardwood trees, and you realize acorns are clustered near parent oaks. Our perceptions also may be influenced by when we do our observing. Population distribution varies over time, as in response to environmental rhythms. Few places yield abundant resources all year long, and many animals often move from one local habitat to another with the seasons. In deciduous tropical forests, many birds and mammals crowd into narrow, gallery forests along watercourses during a dry season. There, trees often provide food and shelter. The density of animal populations becomes much greater than it is during the rainy season.

Each population has its own gene pool and range of traits. It also has a characteristic size, density, distribution pattern, and age structure. Environmental conditions and species interactions influence these characteristics.

How Population Size Changes

Populations are dynamic units of nature. Depending on the species, they may add or lose individuals every half hour or every day, season, or year. Sometimes they glut parts of the habitat with individuals. At other times individuals may be scarce. Populations even can drive themselves or be driven to extinction. Over a specified interval, we can measure such changes in population size in terms of birth rates, death rates, and the number of individuals entering and leaving.

Population size increases as a result of births and **immigration**, or the arrival of new residents from other populations of the species. Population size decreases as a result of deaths and **emigration**, whereby individuals permanently move out of the population.

For many species, population size also changes on a predictable basis as a result of daily or seasonal events called **migrations**. However, migration is a recurring round trip between two areas, so we need not consider its transient effects during this initial consideration of population size.

From Ground Zero to Exponential Growth

For our purposes, assume that immigration is balancing emigration over time, so that we can ignore the effects of both on population size. Doing so allows us to define zero population growth as an interval in which the number of births is balanced by the number of deaths. Population size is stabilized during such an interval of time, with no overall increase or decrease.

Births, deaths, and other variables that may affect population size can be measured in terms of **per capita** rates, or rates per individual. (*Capita* means heads, as in head counts.) Visualize 2,000 mice living in a cornfield. Twenty or so days after their eggs are fertilized, the females produce a litter, then nurse their offspring for a month or so, then get pregnant again. If the female mice collectively give birth to 1,000 mice per month, the birth rate would be 1,000/2,000 = 0.5 per mouse per month. If 200 of the 2,000 die during that interval, the death rate would be 200/2,000 = 0.1 per mouse per month.

If we assume the birth rate and death rate remain constant, we can combine both into a single variable—the **net reproduction per individual per unit time**, or *r* for short. For the mouse population in the cornfield, *r* is 0.5 − 0.1 = 0.4 per mouse per month. This example gives us a way to represent population growth:

$$
\begin{pmatrix} \text{population} \\ \text{growth per} \\ \text{unit time} \end{pmatrix} = \begin{pmatrix} \text{net population} \\ \text{growth rate} \\ \text{per individual} \\ \text{per unit time} \end{pmatrix} \times \begin{pmatrix} \text{number of} \\ \text{individuals} \end{pmatrix}
$$

or, more simply, $G = rN$.

	Net Monthly Increase:	New Population Size:
$G = r \times$ 3,920 =	1,568 =	5,488
$r \times$ 5,488 =	2,195 =	7,683
$r \times$ 7,683 =	3,073 =	10,756
$r \times$ 10,756 =	4,302 =	15,058
$r \times$ 15,058 =	6,023 =	21,081
$r \times$ 21,081 =	8,432 =	29,513
$r \times$ 29,513 =	11,805 =	41,318
$r \times$ 41,318 =	16,527 =	57,845
$r \times$ 57,845 =	23,138 =	80,983
$r \times$ 80,983 =	32,393 =	113,376
$r \times$ 113,376 =	45,350 =	158,726
$r \times$ 158,726 =	63,490 =	222,216
$r \times$ 222,216 =	88,887 =	311,103
$r \times$ 311,103 =	124,441 =	435,544
$r \times$ 435,544 =	174,218 =	609,762
$r \times$ 609,762 =	243,905 =	853,667
$r \times$ 853,677 =	341,467 =	1,195,134

a

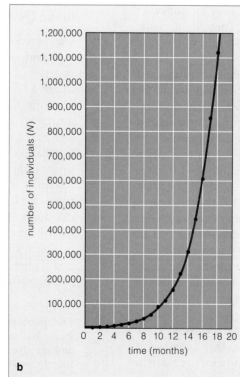

b

c

Figure 46.4 (**a**) Data showing the net monthly increases in a population of field mice living in a cornfield. From start to finish, the numbers listed show a pattern that is typical of exponential growth. (**b**) Graphing the data yields a J-shaped growth curve.

(**c**) Do such growth curves seem far removed from your own experience? Think about the growth of populations of, say, the resident bacteria in your mouth after you present them with a buffet of nutrients in candy or some other sugar-laden food.

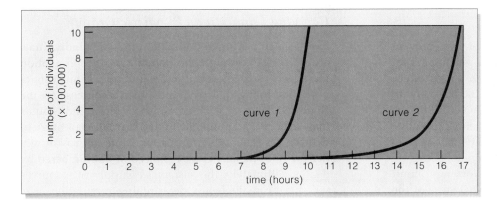

Figure 46.5 Effect of deaths on the rate of increase in two bacterial populations. Growth curve *1* represents a population of bacterial cells that reproduced every half hour. Growth curve *2* represents a different population in which cells divided every half hour, but 25 percent died between divisions.

The graph shows that deaths can slow the rate of increase but cannot in themselves stop exponential growth.

As the next month begins, 2,800 mice are scurrying about. With the net increase of 800 fertile critters, the reproductive base has expanded. If *r* does not change, population size will expand this month also, by a net increase of 0.4 × 2,800 = 1,120 mice. The population is now 3,920. In the months ahead, it just so happens that *r* remains constant, and a growth pattern emerges. As Figure 46.4*a* shows, *in less than two years from the time we first started counting, the number of mice running around in the cornfield increased from 2,000 to more than a million!*

Plot the monthly increases against time, as in Figure 46.4*b*, and you end up with a graph line in the shape of a "J." When population growth over increments of time plots out as a J-shaped curve, you know you are tracking **exponential growth**.

"Exponential" refers to a relationship in which one variable increases much faster than another in a specific mathematical way. In this case, a population's size is a variable that depends on how many individuals make up the reproductive base in successive increments of time. *The larger the reproductive base, the greater will be the expansion in population size during each specified interval.*

Now consider other aspects of exponential growth by supplying a lone bacterium in a culture flask with all the nutrients required for growth. Thirty minutes later, the one cell divides in two. Thirty minutes pass, the two cells divide, and so on every thirty minutes. Assuming no cells die between divisions, the population size will double in each interval—from 1 to 2, then 4, 8, 16, 32, and so on. The length of time it takes for a population to double in size is its **doubling time**.

The larger this population gets, the more cells there are to divide. After 9-1/2 hours (nineteen doublings), it has more than 500,000 bacterial cells. After 10 hours (twenty doublings), it has more than 1 million. Plot the doublings in size against time, and you get the J-shaped curve that is typical of unrestricted, exponential growth. Figure 46.5, curve *1*, shows this.

To examine the effect of deaths on growth rates, start over with one bacterium. Assume that 25 percent

of the cells die in each thirty-minute interval. At this rate, it takes about seventeen hours (not ten) for the population size to reach a million. *But deaths changed only the time scale for population growth.* You still end up with a J-shaped curve—which is curve 2 in Figure 46.5.

Regarding the Biotic Potential

Finally, imagine that a population lives in a place where conditions are ideal. Every one of its individuals has adequate shelter, food, and other vital resources. No predators, pathogens, or pollutants lurk in the habitat. The population may well display its **biotic potential**, which is the maximum rate of increase per individual under ideal conditions.

Each species has a characteristic maximum rate of increase. For many bacteria, it is 100 percent every half hour or so. For humans and other large mammals, it is between 2 and 5 percent per year. But the *actual* rate depends on the age at which each generation starts to reproduce, how often each individual reproduces, and how many offspring are produced. Now think about this. A human female is biologically capable of bearing twenty or more children, yet in each generation, many females do not reproduce at all. The human population has not been displaying its biotic potential. Even so, since the mid-eighteenth century, its growth has been exponential, for reasons that will soon be apparent.

During a specified interval, population size is an outcome of births, deaths, immigration, emigration, and migrations.

With exponential growth, population size expands by ever increasing increments during successive time intervals, because the reproductive base becomes ever larger.

A plot of population size against time has a characteristic J-shaped curve if the population is growing exponentially.

As long as the per capita birth rate remains even slightly above the per capita death rate, a population will grow exponentially.

Limiting Factors

Most of the time, environmental circumstances prevent any population from fulfilling its biotic potential. That is why sea stars, the females of which can produce 2,500,000 eggs each year, do not fill up the oceans with sea stars. That also is why humans will never fill up the planet, even with our capacity for exponential growth.

In natural environments, complex interactions exist within and between populations of different species, so it is not easy to identify all the factors working to limit population growth. To get a sense of what *some* of the factors might be, start again with a lone bacterium in a culture flask, where you can control the variables. First you enrich the culture medium with glucose and other nutrients required for bacterial growth. Then you allow bacterial cells to reproduce for many generations. At first the growth pattern appears to be exponential. Then growth slows, and, after that, population size remains rather stable. But after the stable period, the population size declines rapidly and all of the bacteria die.

What happened? While the population expanded by ever increasing amounts, cells tapped more and more nutrients. The dwindling nutrients signaled the cells to stop dividing (Section 9.3). And when the existing cells exhausted the nutrient supply, they starved to death.

Any essential resource that is in short supply is a **limiting factor** on population growth. Food, minerals of certain types, refuge from predators, living quarters, and a pollution-free environment are examples. The number of such factors can be huge, and their effects can vary. Even so, one factor alone is often enough to put the brakes on population growth at any given time.

Suppose you kept on freshening the supply of all required nutrients for the growing bacterial culture. After an episode of exponential growth, the population would still crash. Like all organisms, bacteria produce metabolic wastes. The wastes produced by such a huge population of bacterial cells would be so great that they would drastically alter living conditions in the culture. By its own activities, the population would end up polluting the experimentally designed habitat and put a stop to its exponential growth.

Carrying Capacity and Logistic Growth

Now visualize a small population in which individuals are dispersed through the habitat. As the population increases in size, more and more individuals must share nutrients, living quarters, and other resources. As the share available to each diminishes, fewer individuals may be born, and more may die by starvation or nutrient deficiencies. Then, the population's rate of growth will decline until the births are balanced or outnumbered by deaths. Ultimately, the *sustainable* supply of resources will be the main factor determining population size. The name **carrying capacity** refers to the maximum number of individuals of a population (or species) that a given environment can sustain indefinitely.

The pattern of **logistic growth** is a fine example of how carrying capacity can affect a population. By this pattern, a population at low density starts growing slowly in size, then grows rapidly, and finally levels off in size once the carrying capacity is reached. We can represent logistic growth in this simplified way:

$$\begin{pmatrix} \text{population} \\ \text{growth per} \\ \text{unit time} \end{pmatrix} = \begin{pmatrix} \text{maximum} \\ \text{net population} \\ \text{growth rate} \\ \text{per individual} \\ \text{per unit time} \end{pmatrix} \times \begin{pmatrix} \text{number} \\ \text{of indi-} \\ \text{viduals} \end{pmatrix} \times \begin{pmatrix} \text{proportion} \\ \text{of resources} \\ \text{not yet used} \end{pmatrix}$$

or $G = r_{max} N [(K - N)/K]$. The K designates carrying capacity. The term enclosed by the brackets is close to 1 when a population is small. It approaches zero when population size is close to carrying capacity.

A plot of logistic growth gives an S-shaped curve (Figure 46.6). Such curves are only an approximation of what goes on in nature. For example, a population that grows too rapidly can overshoot the carrying capacity. The death rate skyrockets and the birth rate plummets. These two responses drive the number of individuals down to the carrying capacity—or lower (Figure 46.7).

Figure 46.6 Idealized S-shaped curve characteristic of logistic growth. After a rapid growth phase (time B to C), growth slows and the curve flattens out as the carrying capacity is reached (time C to D). S-shaped growth curves can show variations, as when changes in the environment bring about a decreased carrying capacity (time D to F). This happened to the human population of Ireland before 1900, when late blight, a disease caused by a water mold, destroyed potato crops that were the mainstay of the diet (Section 23.1).

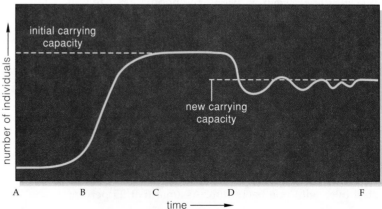

Figure 46.7 Carrying capacity and a reindeer herd. In 1910, four male and twenty-two female reindeer were introduced on St. Matthew Island in the Bering Sea. In less than thirty years, the herd increased to 2,000. Individuals had to compete for dwindling vegetation, and overgrazing destroyed most of it. In 1950, the herd plummeted to eight members. The growth pattern reflects how the population size overshot the carrying capacity, then crashed.

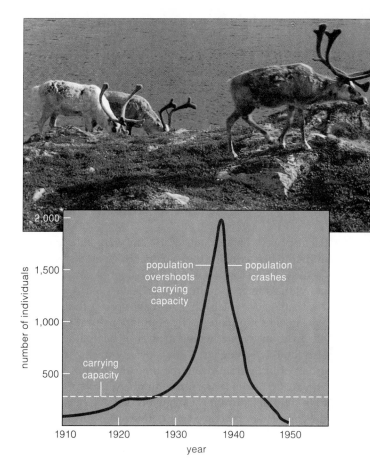

Density-Dependent Controls

The logistic growth equation describes **density-dependent control** of populations. When a population is not large, it may grow rapidly. When it finally bumps up against its carrying capacity, so to speak, it will stop growing, grow very little, or even decline in size. However, other factors besides carrying capacity influence the size of a population. Other natural controls come into play that can put the number of individuals *below* the maximum sustainable level.

Consider this: High density and overcrowding put individuals of the population at greater risk of dying. Under the crowded conditions, predators, parasites, and pathogens interact more intensely with the population and therefore bring about a decline in its density. Once population size declines, density-dependent interactions relax and the population size may increase once again.

Bubonic plague and *pneumonic plague* are examples of density-dependent controls. The bacterium *Yersinia pestis* invades the blood and causes both of these dangerous diseases. *Y. pestis* persists in a large reservoir of rabbits, rats, and certain other small mammals. Hungry fleas transmit it to new hosts. The bacterium multiplies in the flea gut and blocks digestion, the fleas attempt to feed more and more often, and so the disease spreads.

A devastating episode of bubonic plague occurred in the fourteenth century. The plague swept through European cities where humans were crowded together, sanitary conditions were poor, and rats were abundant. By the time the epidemic subsided, urban populations in Europe had declined by 25 million.

Both diseases are still threats. For example, in 1994 in India, they raced through rat-infested cities where garbage and animal carcasses had piled up for months in the streets. Terrified residents fled by the thousands. Some carried the pathogen with them as far away as London. Global panic ensued before concerted efforts to burn the garbage, poison the rats and fleas, and rapidly dispense antibiotics averted a pandemic.

Density-Independent Factors

Sometimes events result in more deaths or fewer births regardless of population density. For example, every year, monarch butterflies migrate from Canada to spend the winter in forested mountains of Mexico. But logging there opened up stands of trees, which normally buffer temperatures. In 1995 the deforestation, in combination with a sudden freeze, killed millions of butterflies. They were **density-independent factors**, meaning their effects came into play independently of population density.

Similarly, heavy applications of pesticides in your backyard may also kill most of the insects, mice, cats, birds, and other animals. And they will do so regardless of how uncrowded or dense their populations are.

Resources in short supply put limits on population growth. Together, all of the limiting factors acting on a population dictate how many individuals can be sustained.

Carrying capacity is the maximum number of individuals of a population that can be sustained indefinitely by the resources in a given environment. The number may rise or fall with changes in resource availability.

The size of a low-density population may increase slowly, go through a rapid growth phase, then level off once the carrying capacity for the population is reached. This is a logistic growth pattern.

Density-dependent controls and density-independent factors can bring about decreases in population size.

So far, we have been considering populations as if all of their individuals are identical during a given time span. For most species, however, different individuals are at different stages of development, so they are interacting in different ways with other organisms and with their environment. In other words, such species have a **life history pattern**, in that their individuals exhibit particular morphological, physiological, and behavioral traits that are adaptive to different conditions at different times in the life cycle. Let's look at a few of the environmental variables that influence age-specific life history patterns.

Life Tables

Each species has a characteristic life span, although few individuals ever reach the maximum age possible. Why? Death looms larger at some ages than at others. Also, individuals tend to reproduce or leave their population at certain ages, which vary from one species to the next.

Age-specific patterns in populations first intrigued life insurance and health insurance companies. They also interest ecologists. Both typically track a **cohort**, a group of individuals from the time of birth until the last one dies. They also track the number of offspring born to individuals during each age interval. A **life table** lists the completed data on a population's age-specific death schedule. Such a schedule is often converted to a more cheery "survivorship" schedule, which lists the number of individuals that reach some specified age (x). The life table in Table 46.1 was constructed for a cohort of 996 phlox plants in Texas. Table 46.2 was constructed for the 1989 human population of the United States.

Often, dividing a population into age classes and then assigning birth rates and mortality risks to each class has practical applications. Unlike a crude census (head count), the data might serve as a basis for informed policy decisions on a variety of issues. Pest management, conservation of endangered species, and social planning for human populations are such issues. For example, birth and death schedules for the northern spotted owl figured in court decisions that halted mechanized logging in old-growth forests, the owl's habitat.

Patterns of Survivorship and Reproduction

Evolutionarily, reproductive success means an individual adds the most offspring to the population during its lifetime. But the typical number of offspring varies among

Table 46.1   Life Table for a Cohort of Annual Plants (*Phlox drummondii*)

Age Interval (days)	Survivorship (number surviving at start of interval)	Number Dying During Interval	Death Rate per Individual During Interval	"Birth" Rate (number of seeds produced per individual) During Interval
0–63	996	328	0.329	0
63–124	668	373	0.558	0
124–184	295	105	0.356	0
184–215	190	14	0.074	0
215–264	176	4	0.023	0
264–278	172	5	0.029	0
278–292	167	8	0.048	0
292–306	159	5	0.031	0.33
306–320	154	7	0.045	3.13
320–334	147	42	0.286	5.42
334–348	105	83	0.790	9.26
348–362	22	22	1.000	4.31
362–	0	0	0	0
		996		

Data from W. J. Leverich and D. A. Levin, *American Naturalist* 1979, 113: 881–903.

Table 46.2   Life Table for the United States Human Population, 1989*

Age Interval (category for individuals between the two ages listed)	Survivorship (number alive at start of age interval, per 100,000 individuals)	Mortality (number dying during the age interval)	Life Expectancy (average lifetime remaining at start of age interval)	Reported Live Births for Total Population
0–1	100,000	896	75.3	
1–5	99,104	192	75.0	
5–10	98,912	117	71.1	
10–15	98,795	132	66.2	11,486
15–20	98,663	429	61.3	506,503
20–25	98,234	551	56.6	1,077,598
25–30	97,683	606	51.9	1,263,098
30–35	97,077	737	47.2	842,395
35–40	96,340	936	42.5	293,878
40–45	95,404	1,220	37.9	44,401
45–50	94,184	1,766	33.4	1,599
50–55	92,418	2,727	28.9	
55–60	89,691	4,334	24.7	
60–65	85,357	6,211	20.8	
65–70	79,146	8,477	17.2	
70–75	70,669	11,470	13.9	
75–80	59,199	14,598	10.9	
80–85	44,601	17,448	8.3	
85 +	27,153	27,153	6.2	Total: 4,040,958

* Compiled by Marion Hansen, based on data from U.S. Bureau of the Census, *Statistical Abstract of the United States*, 1992 (edition 112).

a

Figure 46.8 Three generalized types of survivorship curves. The type I populations show high survivorship until some age and then show high mortality. Type II populations show a fairly constant death rate. Type III populations show low survivorship early in life.

b

c

species. For each species, the number reflects trade-offs in the energy and time allocated to gamete formation, securing mates, parental behavior (if any), and the size of offspring. Section 44.1 gives examples of this. The trade-offs in turn reflect selective pressures associated with species interactions and conditions in the habitat.

For each species, **survivorship curves** are the graph lines that emerge when ecologists plot the age-specific survival of a cohort in a particular habitat. Three types of survivorship curves are common in nature.

Type I curves reflect high survivorship until fairly late in life, then a large increase in deaths. Such curves are typical of elephants and other large mammals that bear only one or a few large offspring at a time and then provide them with extended parental care (Figure 46.8*a*). For example, female elephants give birth to only four or five calves in a lifetime, and they devote several years of parental care to each one.

Type I curves are typical of human populations also, *provided* that people have access to good health care services. Historically, and wherever health care is poor today, infant deaths cause a sharp drop at the start of the curve. Following the drop, the curve then levels off from childhood to early adulthood.

Type II curves reflect a fairly constant death rate at all ages. They are typical of organisms that are just as likely to be killed or to die of disease at any age. This is true of lizards, small mammals, and some songbirds (Figure 46.8*b*).

Type III curves typify populations where the death rate is highest early in life. They are characteristic of species that produce many small offspring, then invest little, if any, parental care in them. Figure 46.8*c* shows how the curve plummets for sea stars. Although sea stars produce mind-boggling numbers of tiny larvae, these must rapidly eat, grow, and finish developing on their own without any support, protection, or guidance from the parents. Corals and other animals that also live in their habitat quickly eat most of them. Plummeting survivorship curves are characteristic of many other kinds of marine invertebrates, most insects, and many fishes, plants, and fungi.

At one time, ecologists thought selection processes favored *either* the early, rapid production of many small offspring *or* the late production of only a few large ones. These two patterns are now known to be extremes, at opposite ends of a range of possible life histories. Also, both patterns—and intermediate ones—may be evident in different populations of the same species, as the next section makes clear.

Tracking a cohort (a group of individuals from birth until the last one dies) reveals patterns of reproduction, death, and migration that typify the populations of a species.

Survivorship curves can reveal differences in age-specific survival among species. In some cases, such differences exist even between populations of the same species.

NATURAL SELECTION AND THE GUPPIES OF TRINIDAD

Several years ago, drenched with sweat and with fish nets in hand, two evolutionary biologists were busily engaged in fieldwork in the mountains of Trinidad, an island in the southern Caribbean Sea. They were after some of the small, live-bearing fishes that were darting about in the shallow freshwater streams (Figure 46.9). The fishes were guppies (*Poecilia reticulata*). And over the course of eleven years the two biologists, David Reznick and John Endler, identified environmental variables that affect guppy life history patterns.

The researchers easily distinguished the male guppies from the females. The males are smaller. Also, they have bright-colored scales in a patterning that functions as a visual signal during mating behavior. The males engage in intricate courtship maneuvers. And once they reach sexual maturity, they stop growing. By contrast, the females are drab colored. Also, they continue to grow in size during the reproductive phase of the life cycle.

In the Trinidad mountains, different guppy populations occupy different streams. In some cases, different guppy populations even occupy different parts of the same stream. There they encounter different kinds of predators. A type of killifish (*Rivulus hartii*) shares some streams with guppies. It is not an especially large fish. It successfully preys on smaller, immature guppies but not on the larger adults. A larger fish, a type of pike-cichlid (*Crenicichla alta*), shares different streams with guppies (Figure 46.10).

It preys on larger, sexually mature guppies and tends not to bother with the small ones.

Using their understanding of natural selection as a starting point, Reznick and Endler hypothesized that predation has profound effects on the life history patterns of guppies. As they had noticed, in streams where the pike-cichlids reign supreme, individual guppies mature faster and their body size is smaller at maturity, compared with the guppies that live and reproduce in killifish-dominated streams. Also, the guppies targeted by pike-cichlids reproduce earlier in life, they produce far more offspring, and they do so more frequently.

However, it was possible that other variables influenced guppy life history patterns in killifish and pike-cichlid streams. To check out this possibility, the researchers carefully shipped two groups of guppies from the two kinds of streams back to a laboratory in the United States. There they allowed them to reproduce for two generations in the absence of predation. They also made sure that the aquarium conditions were identical for the two experimental populations. As it turned out, offspring of the experimental guppy populations showed the same differences as the natural populations. The conclusion? The differences between guppies preyed upon by different predators have a genetic basis. This finding is consistent with the predictions of a theory of life history evolution that is based on such differences in mortality schedules.

Figure 46.9 David Reznick, an evolutionary biologist, contemplating the interactions among guppies (*Poecilia reticulata*) and their predators in a freshwater stream in Trinidad.

GUPPY FROM A KILLIFISH STREAM

GUPPY FROM A PIKE-CICHLID STREAM

PIKE-CICHLID

Figure 46.10 Two guppies and one of the eaters of guppies.

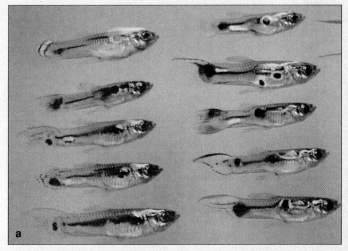
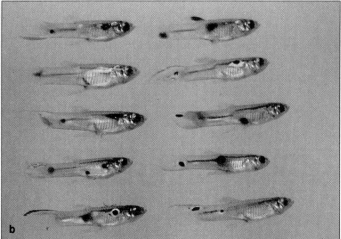

Figure 46.11 Representative guppies that live in streams with killifish (**a**) and streams with pike-cichlids (**b**). Guppies of populations targeted by killifish tend to be larger, less streamlined, and more brightly colored. Guppies of populations targeted by pike-cichlids tend to be smaller, more streamlined, and duller in color patterning. The differences between the groups are accompanied by clear differences in life history patterns.

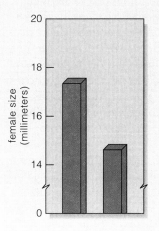
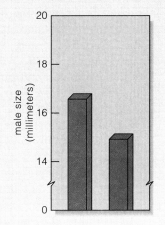
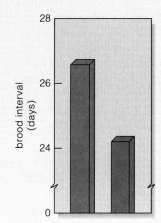
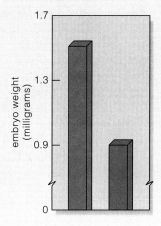

Figure 46.12 Graphs of some of the experimental evidence of natural selection among populations of guppies subjected to different predation pressures. Differences arise in body size and intervals between broods for guppies raised with killifish (*red* bars) or pike-cichlids (*green* bars). Killifish are small and prey on smaller guppies, pike-cichlids are large fish and prey on larger guppies, and so the two predators select for the differences.

Figure 46.11 shows representatives of both kinds of guppy populations.

The researchers looked into the role of predation in the evolution of differences in body size, the length of time between generations, and other aspects of the life history patterns of guppies. They raised many generations of guppies in the laboratory. As predicted, the body size of the guppy lineage that was subjected to predation over time by killifish became larger at maturity (Figure 46.12). The lineage raised with pike-cichlids showed a trend toward earlier maturity.

Finally, when they made their first visit to Trinidad, Reznick and Endler instituted a field experiment by introducing guppies to a site that is upstream from a small waterfall. Before the experiment, the waterfall had been a barrier to dispersal; it had prevented guppies and all predators except the killifish from moving into the upstream site. The researchers chose the introduced guppies from a population that had evolved with pike-cichlids downstream from the waterfall. They designated that part of the stream as the control site.

Eleven years later, researchers revisited the stream. As they discovered, the guppies had evolved. They made comparisons between guppies from the experimental site and the control site. Just as Reznick and Endler had predicted, the body size, frequency of reproduction, and other aspects of guppy life history patterns correlated with the preferences of the neighborhood predator. Later, a laboratory experiment involving two generations of guppies confirmed that the differences were indeed genetic. The researchers concluded that the differences are the product of natural selection.

In 1997, the human population surpassed 5.8 billion. The year before, rates of increase in different nations ranged from below 1 percent to over 3 percent, which averaged out to 1.55 percent annually. Visualize enough people to make another Los Angeles every two weeks or another New York City every four weeks, and you get the picture. Now think about this: Annual additions to the base population mean the *same* percent increase will result in a *larger* absolute increase every year into the future.

Our staggering population growth continues even though as many as 2 billion are already malnourished or starving, without clean drinking water and adequate shelter. It continues when 1.5 billion are already going without the benefits of health care delivery and sewage treatment facilities. And it continues mainly in regions that are already overcrowded. About 5 billion crowd together on 10 percent of the land, and about 3 billion live within 480 kilometers (about 300 miles) of the seas.

Suppose it were possible, by monumental efforts, to double the food supply to keep pace with growth. We would do little more than maintain marginal living conditions for most. The annual deaths from starvation could still be 20 million to 40 million. Even this would come at great cost, for we are drastically modifying the environment that must sustain us. Salted-out cropland, desertification, deforestation, pollution—these are some of the consequences you will read about in Chapter 50, and they do not bode well for our future.

For a while, it would be like the Red Queen's garden in Lewis Carroll's *Through the Looking Glass*, where one is forced to run as fast as one can to remain in the same place. But what happens when our population doubles again? Can you brush this picture aside as being too far in the future to warrant concern? *It is no further removed from you than the sons and daughters of the next generation.*

How We Began Sidestepping Controls

How did we get into this predicament? For most of its history, the human population grew slowly. During the past two centuries, the increases in growth rates became astounding. There are three possible reasons for this:

1. Humans steadily developed the capacity to expand into new habitats and new climate zones.

2. Humans increased the carrying capacity in their existing habitats.

3. Human populations sidestepped limiting factors.

The human population did all three of these things. Consider the first point. Early humans lived mostly in the open grasslands called savannas. They were vegetarians who added scavenged bits of meat to their diet. Starting about 2 million years ago, small bands of hunter-gatherers migrated out of Africa. And by 40,000 years ago, their descendants had spread through much of the world.

Most other species could not have expanded into such a truly broad range of habitats. Humans, with their highly complex brains, could use learning and memory to figure out how to build fires, assemble shelters, make clothing and tools, and plan community hunts. Learned experiences did not die with individuals. They spread quickly from one band to another through language—the capacity for extraordinary cultural communication. Thus, *the human population expanded into diverse new environments in an extremely short time, compared with the long-term geographic dispersal of other kinds of organisms.*

What about the second possibility? About 11,000 years ago, humans began to shift away from hunting and gathering to agriculture. Instead of migrating and following the game herds, they harvested seasonal fruits and grains. They settled down and developed a more dependable basis for life in more favorable settings. A pivotal factor was the domestication of wild grasses, including species ancestral to modern wheats and rice. People now harvested, stored, *and planted* seeds in one place. They domesticated animals and kept them close to home for food and pulling plows. They dug ditches and diverted water to irrigate permanent croplands.

Productivity increased through the new agricultural practices. With larger, more dependable food supplies, population growth rates increased. As towns and cities developed, a social hierarchy emerged that provided a labor base for more intensive agriculture. Much later, food supplies increased further by the use of fertilizers, herbicides, and pesticides. Transportation improved, and so did food distribution. Thus, *even at its simplest, managing food supplies through agriculture increased the carrying capacity for the human population.*

What about the third possibility—did we sidestep limiting factors? Consider what happened as medical practices and sanitary conditions improved. Until about 300 years ago, contagious diseases, malnutrition, and poor hygiene kept death rates high enough to more or less balance birth rates. Contagious diseases, which are density-dependent factors, swept unchecked through crowded settlements and cities that were infested with fleas and rodents. Then came plumbing and methods of sewage treatment. Over time, vaccines, antitoxins, and antibiotics were developed as weapons against many pathogens. The death rates dropped sharply. Births now began to exceed deaths—and rapid population growth was under way.

In the mid-eighteenth century, people discovered how to harness the energy stored in fossil fuels, starting with coal. Within a few decades, large industrialized societies started forming in western Europe and North

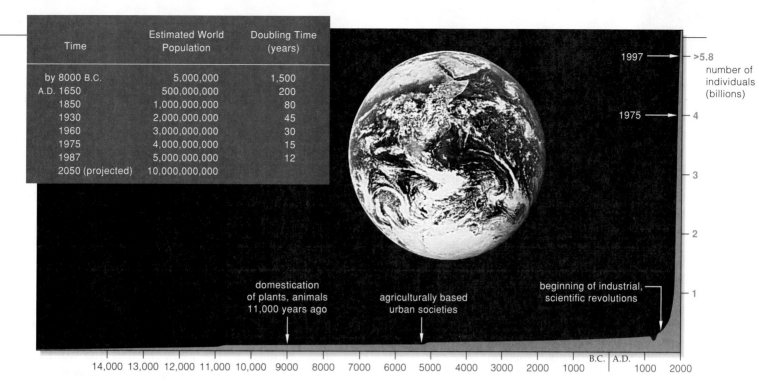

Time	Estimated World Population	Doubling Time (years)
by 8000 B.C.	5,000,000	1,500
A.D. 1650	500,000,000	200
1850	1,000,000,000	80
1930	2,000,000,000	45
1960	3,000,000,000	30
1975	4,000,000,000	15
1987	5,000,000,000	12
2050 (projected)	10,000,000,000	

domestication
of plants, animals
11,000 years ago

agriculturally based
urban societies

beginning of industrial,
scientific revolutions

1997 →►>5.8
number of
individuals
(billions)

1975 →►4

14,000 13,000 12,000 11,000 10,000 9000 8000 7000 6000 5000 4000 3000 2000 1000 | B.C. | A.D. | 1000 2000

Figure 46.13 Growth curve (*red*) for the human population. The vertical axis of the graph represents world population, in billions. (The slight dip between the years 1347 and 1351 is the time when 25 million people died in Europe as a result of bubonic plague.) Agricultural revolutions, industrialization, and improvements in health care have been sustaining the accelerated growth pattern for the past two centuries. The data in the *blue* box tell us how long it took for the human population to double in size at different times in its history.

America. Then efficient technologies developed after World War I. Factories mass-produced cars, tractors, and other affordable goods. Machines replaced many of the farmers needed to produce food, and fewer farmers were able to support a larger population.

Thus, *by bringing many disease agents under control and by tapping into concentrated, existing forms of energy, humans sidestepped factors that had previously limited their population growth.*

Present and Future Growth

Where have the far-flung migrations and the spectacular advances in agriculture, industrialization, and health care taken us? Starting with *Homo habilis,* it took about 2.5 million years for the first populations of humans to reach 1 billion. As you can see from Figure 46.13, it took only 45 years to reach the second billion, 30 years to reach the third, 15 years to reach the fourth—and only 12 years to reach the fifth billion!

From what we know of the principles governing population growth—and unless newer technological breakthroughs increase the carrying capacity—we can expect a dramatic increase in death rates. *Although the*

stupendously accelerated growth of the human population continues, it cannot be sustained indefinitely.

Besides having adverse effects on resource supplies, these skyrocketing numbers invite density-dependent controls. For example, the largest cholera epidemic of this century has been sweeping through southern Asia. Like the six previous epidemics, it began in India. And it may spread through Africa, the Middle East, and then into Mediterranean countries before finally peaking. This epidemic alone may claim 5 million lives. Existing vaccines do not work against the bacterial pathogen, a current strain of *Vibrio cholerae.* People become infected by drinking water or eating food that is contaminated with raw sewage. *V. cholerae* multiplies inside the gut. There it produces a toxin that triggers severe diarrhea and massive fluid loss. Two to seven days later, people who are not treated can die from extreme dehydration.

For centuries, *V. cholerae* has thrived and mutated in India's sewage-enriched rivers (Figure 46.1). In slums of Calcutta alone, millions are forced to bathe in stagnant ponds and polluted waterways. In 1992, cholera struck tens of thousands in that city.

At this writing, a new era of human migration has begun. By some estimates, economic hardship and civil strife have put 50 million people on the move within and between countries. Will their relocations prove peaceable? Where will they find sustainable supplies of food, clean water, and other basic resources?

Through expansion into new habitats, cultural intervention, and technological innovation, the human population has temporarily skirted environmental resistance to growth. Its accelerated growth cannot be sustained indefinitely.

CONTROL THROUGH FAMILY PLANNING

Figure 46.14 shows the 1996 annual rates of increase for populations in different parts of the world. If the annual current average rate of 1.55 percent is maintained, the human population may be *10 billion to 11 billion* by the year 2050. It is mind numbing to think about the resources required to sustain that many people. We will have to increase food production, supplies of drinkable water, energy reserves, and supplies of wood and other materials to meet everyone's basic needs—something we are not even doing now. The gross manipulation of resources is likely to intensify pollution, which almost certainly will adversely affect our water supplies, the atmosphere, and productivity on land and in the seas.

Today there is growing realization that population growth, resource depletion, pollution, and the quality of life are interconnected. As evidence of this, consider that most governments are trying to lower birth rates, as through **family planning programs**. The programs educate individuals about choosing how many children they will have, and when. Their details vary from country to country, but all offer information on the available methods of fertility control, as outlined in Section 45.13. Carefully developed and administered programs may bring about a long-term decline in birth rates.

To arrive at zero population growth, the average "replacement rate" would have to be slightly higher than two children per couple, for some female children die before reaching reproductive age. The replacement rate is about 2.5 children per woman in less developed countries, and 2.1 in the more developed countries. Yet even if each couple on the planet decided to have only two children, the human population would keep on growing for another sixty years! Why? An immense number of existing children will soon be reproducing.

A more useful measure of global trends is the **total fertility rate**. That is the average number of children born to women during their reproductive years, as estimated on the basis of current age-specific birth rates. In 1996, the average rate was 3 children per woman. This is an impressive decline from 1950, when the rate was 6.5. However, it is still far above the replacement level.

Figure 46.15 shows some age structure diagrams for populations growing at different rates. The central part of each diagram includes individuals of reproductive age. The lower part includes children who will move into the reproductive age category over the next fifteen

Figure 46.14 (**a**) The 1996 average annual population growth rate in different regions of the world. (**b**) Population sizes in 1996 (*orange* bars) and projected for the year 2025 (*blue* bars).

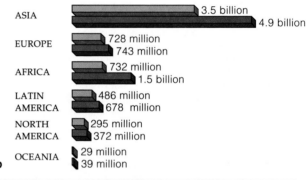

ASIA — 3.5 billion / 4.9 billion
EUROPE — 728 million / 743 million
AFRICA — 732 million / 1.5 billion
LATIN AMERICA — 486 million / 678 million
NORTH AMERICA — 295 million / 372 million
OCEANIA — 29 million / 39 million

b

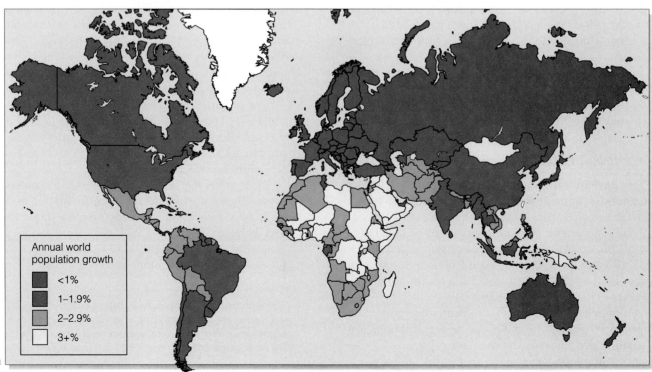

Annual world population growth

<1%
1–1.9%
2–2.9%
3+%

a

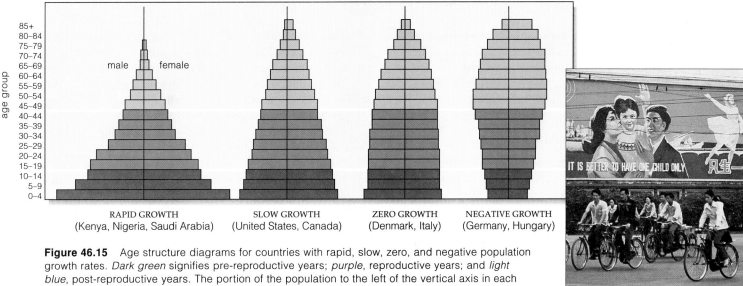

Figure 46.15 Age structure diagrams for countries with rapid, slow, zero, and negative population growth rates. *Dark green* signifies pre-reproductive years; *purple*, reproductive years; and *light blue*, post-reproductive years. The portion of the population to the left of the vertical axis in each diagram represents males, and that to the right, females. Bar widths correspond to the proportion of individuals in each age group. The photograph shows a family planning billboard in China.

years, with 15–44 being the average range for childbearing years. The population of the United States has a narrow base and is an example of slow growth. Figure 46.16 tracks its 78 million *baby-boomers*, a cohort that formed in 1946 after soldiers returned home from World War II. By contrast, age structure diagrams for a rapidly growing population have a broad base.

Currently, *more than one-third of the world population falls in the broad pre-reproductive base.* And the numbers reflect the magnitude of the effort it will take to control world population growth.

One way to slow the birth rate is to bear children during the early thirties, rather than in the mid-teens or early twenties. Delayed reproduction slows the rate of growth and lowers the average number of children in families. In China, for example, the government has established the world's most extensive family planning program. It strongly discourages premarital sex. It calls for pledges to postpone marriage and to limit family size to one child only. Contraceptives, induced abortion, and sterilization are free to married couples; paramedics and mobile units ensure access to these measures in remote rural areas. Couples who pledge to have only one child receive more food, free medical care, better housing, and salary bonuses. Their child will be granted free tuition and preferential treatment when he or she enters the job market. Those who break the pledge forgo benefits and pay higher taxes. Are the measures inhumane? Think about this: Between 1958 and 1962 alone, an estimated

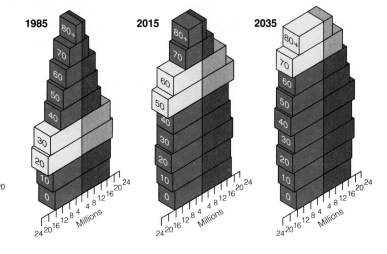

Figure 46.16 Age structure diagrams for the United States population. *Gold* bars track the baby-boom generation.

30 million Chinese died of starvation as a consequence of widespread famines.

Since 1972, China's total fertility rate declined from 5.7 to 1.9. Even so, the population time bomb has not stopped ticking. Its population is now 1.22 billion—and 340 million of its young women are moving into the reproductive age category. By the year 2025, China's projected population will reach 1.5 billion.

Family planning programs on a global scale are designed to help stabilize the size of the human population.

Even if it reaches a level of zero population growth, the human population will continue to grow for sixty years, for its reproductive base already consists of a staggering number of individuals.

Today, there is greater awareness of the relationship between the growth rates and economic development of populations. When individuals are economically secure, they seem to be under less pressure to produce large numbers of children to help them survive.

Demographic Transition Model

We can correlate changes in population growth with changes that typically unfold in four stages of economic development. The four stages are at the heart of the **demographic transition model** (Figure 46.17).

According to this model, living conditions are harsh during a *preindustrial* stage, before widespread use of technology and medical advances. Birth and death rates are both high, so the rate of population growth is low.

Next, during the *transitional* stage, industrialization begins, food production rises, and health care improves. Although death rates drop, the birth rates remain fairly high. As a result, the population grows rapidly. Growth continues at high rates over a long interval. Annual

growth rates tend to be 2.5 to 3 percent, on the average. As living conditions improve and birth rates begin to decline, however, growth starts to level off.

Population growth slows dramatically during the *industrial* stage, when industrialization is in full swing. The slowdown emerges mostly because people move from the countryside to cities—and urban couples have a tendency to control the size of their families. Many couples get caught up in the accumulation of goods. Often they decide that the time and cost of raising more than a few children conflict with that goal. As you will see, this is a sticky issue with respect to population size.

In the *postindustrial* stage, zero population growth is reached. The birth rate falls below the death rate, and population size slowly decreases.

Today, the United States, Canada, Australia, Japan, nations of the former Soviet Union, and most countries of western Europe are in the industrial stage. Their growth rate is slowly decreasing. In Germany, Bulgaria, Hungary, and some other countries, the death rates exceed the birth rates, and the populations are getting smaller.

Mexico and other less developed countries are in the transitional stage. And they do not have enough skilled workers to complete the transition to a fully industrial economy. Fossil fuels and other resources that drive industrialization are being used up there as well as in the industrialized countries. Fuel costs might become prohibitive for countries at the bottom of the economic ladder even before they can enter the industrial stage.

Also, if population growth keeps outpacing economic growth, death rates will increase. Thus many countries may now be stuck in the transitional stage. Some of these countries may return to the harsh conditions of the preceding stage.

Figure 46.17 Diagram of the demographic transition model of changes in the growth characteristics and size of populations, as correlated with changing economic development. The model explains changes that have occurred in western Europe and other industrialized regions.

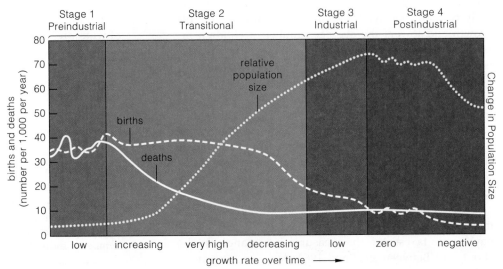

Enormous disparities in economic development are driving forces for immigration and emigration. Think of how in 1995 the United States acquired nearly 900,000 legal immigrants and political refugees. It also acquired at least 300,000 illegal immigrants, mainly from Latin America and Asia. Immigrants accounted for 40 percent of the nation's increase in size that year alone.

Or think of how California's population increased by 12 *million* between 1970 and 1996. With its 2.5 percent growth rate, mainly among legal and illegal newcomers, it may jump by another 16 million by the year 2020. For some time, California has been a symbol of the good life, drawing people from the world over. Yet its schools are overcrowded, funds for social services are shrinking, sewage treatment plants are nearing capacity, and water shortages are severe and frequent. Also, salinization of croplands and air pollution are eroding the agricultural base that is a key factor in California's economic health.

Claiming that population growth affects economic health, many governments restrict immigration. Only the United States, Canada, Australia, and a few other countries do allow large annual increases. Elsewhere, appalling living conditions, harsh government policies, and civil strife promote emigration of far more people than the better-off nations can handle.

A Question of Resource Consumption

This chapter opened with a brief look at the conditions that confront most people in India, with its whopping 16 percent of the human population. By comparison, the United States has merely 4.7 percent. Yet which country is the most "overpopulated"—not in terms of numbers, but rather in terms of resource consumption and instigation of environmental damage?

The highly industrialized United States produces 21 percent of all goods and services. Its people consume fifty times as much as the average person in India. It uses 25 percent of the world's processed minerals and available, nonrenewable sources of energy. It generates at least 25 percent of the global pollution and trash. By contrast, India produces about 1 percent of all goods and services. It uses 3 percent of the available minerals and nonrenewable energy resources. And it generates only about 3 percent of the pollution and trash.

Extrapolating from these numbers, Tyler Miller, Jr., estimated that it would take 12.9 billion impoverished individuals in India to have as much impact on the environment as 258 million Americans.

Differences in population growth among countries correlate with levels of economic development, hence with economic security (or lack of it) of individuals.

For us, as for all species, the biological implications of extremely rapid growth are truly staggering. Yet so are the social implications of what will happen when (and if) the human population declines to the point of zero population growth—and stays there.

For instance, as you read earlier, most individuals of an actively growing population fall in the younger age brackets. Over time, if living conditions ensure constant growth, the age distribution for the population as a whole will guarantee availability of a future work force. This has social implications. Why? *It takes a large work force to support individuals in the older age brackets.* In the United States, older, nonproductive people expect their government to provide them with subsidized medical care, low-cost housing, and other social programs. With improved medicine and hygiene, they live far longer than they did when the nation's social security program was set up. Their cash benefits exceed the contributions they made to the program when *they* were younger.

If the population ever does reach and maintain zero growth over time, a larger proportion of individuals will end up in the older age brackets. Even slow growth poses problems. What happens when baby-boomers retire? Will all those nonproductive members continue to get goods and services if productive ones must carry more and more of the economic burden? This is not an abstract question. Put it to yourself. Exactly how much economic hardship are you willing to bear for the sake of your parents? For your grandparents? How much will your children be willing or able to bear for you?

We have arrived at a major turning point, not only in our biological evolution but in our cultural evolution as well. The decisions awaiting us are among the most difficult we will ever have to make, yet it is clear that they must be made, and soon.

All species face limits to growth. We might think we are different from the rest, for our special ability to undergo rapid cultural evolution has allowed us to postpone the action of most of the factors that limit growth. But the key word here is *postpone*. No amount of cultural intervention can hold back the ultimate check of limited resources and a damaged environment.

We have sidestepped a number of the smaller laws of nature. In doing so, we have become more vulnerable to those laws which cannot be repealed. Today there may be only two options available. Either we make a global effort to limit population growth in accordance with environmental carrying capacity, or we wait until the environment does it for us.

In the final analysis, no amount of cultural intervention can repeal the ultimate laws governing population growth, as imposed by the carrying capacity of the environment.

46.10 SUMMARY

1. A population is a group of individuals of the same species occupying a given area. It has a characteristic size, density, distribution, and age structure as well as characteristic ranges of heritable traits.

2. The growth rate for a population during a specified interval can be determined by calculating the rates of birth, death, immigration, and emigration. To simplify the calculations, we can put aside effects of immigration and emigration, and combine the birth and death rates into a variable r (net reproduction per individual per unit time). Then we can represent population growth (G) as $G = rN$, where N is the number of individuals during the interval specified.

   a. In cases of exponential growth, the population's reproductive base increases and its size expands by ever increasing increments during successive intervals. This trend plots out as a J-shaped growth curve.

   b. As long as the per capita birth rate remains even slightly above the per capita death rate, a population will grow exponentially.

   c. In logistic growth, a low-density population slowly increases in size, goes through a rapid growth phase, then levels off in size once carrying capacity is reached.

3. Carrying capacity is the name ecologists give to the maximum number of individuals in a population that can be sustained indefinitely by the resources available in their environment.

4. The availability of sustainable resources as well as other factors that limit growth dictates population size during a specified interval. The limiting factors vary in their relative effects and vary over time, so population size also changes over time.

5. Limiting factors such as competition for resources, disease, and predation are density-dependent. Density-independent factors, such as weather on the rampage, tend to increase the death rate or decrease the birth rate more or less independently of population density.

6. Patterns of reproduction, death, and migration vary over the life span for a species. Environmental variables also help shape the life history (age-specific) patterns.

7. The human population now exceeds 5.8 billion. Its growth rate varies from below zero in a few developed countries to more than 3 percent per year in some less developed countries. In 1996 the annual growth rate for the entire human population was 1.55 percent.

8. Rapid growth of the human population in the past two centuries occurred through a capacity to expand into new habitats, and because of agricultural, medical, and technological developments that increased the carrying capacity. Ultimately, we must confront the reality of the carrying capacity and limits to our population growth.

Review Questions

1. Define population size, population density, and population distribution. Describe a typical population in terms of several categories for its age structure. 46.1

2. Define exponential growth. Be sure to state what goes on in the age category that underlies its occurrence. 46.2

3. Define carrying capacity, then describe its effect as evidenced by a logistic growth pattern. 46.3

4. Give examples of the limiting factors that come into play when a population of mammals (for example, rabbits or humans) reaches very high density. 46.3, 46.4

5. Define doubling time. At present growth rates, how long will it be before the human population reaches 10 billion? 46.2, 46.6

6. How did earlier human populations expand steadily into new environments? How did they increase the carrying capacity in their habitats? Have they avoided some limiting factors on population growth? Or is the avoidance an illusion? 46.6

Self-Quiz (Answers in Appendix IV)

1. _____ is the study of how organisms interact with one another and with their physical and chemical environment.

2. A _____ is a group of individuals of the same species that occupy a certain area.

3. The rate at which a population grows or declines depends upon the rate of _____ .
   a. births   c. immigration   e. all of the above
   b. deaths   d. emigration

4. Populations grow exponentially when _____ .
   a. birth rate exceeds death rate and neither changes
   b. death rate remains above birth rate
   c. immigration and emigration rates are equal
   d. emigration rates exceed immigration rates
   e. both a and c

5. For a given species, the maximum rate of increase per individual under ideal conditions is the _____ .
   a. biotic potential   c. environmental resistance
   b. carrying capacity   d. density control

6. Resource competition, disease, and predation are _____ controls on population growth rates.
   a. density-independent   c. age-specific
   b. population-sustaining   d. density-dependent

7. Which of the following factors does *not* affect sustainable population size?
   a. predation   c. resources   e. all of the above can
   b. competition   d. pollution   affect population size

8. In 1996, the average annual growth rate for the human population was _____ percent.
   a. 0   c. 1.55   e. 2.7
   b. 1.05   d. 1.6   f. 4.0

9. Match each term with its most suitable description.
   ____ carrying capacity
   ____ exponential growth
   ____ population growth rate
   ____ density-dependent controls

   a. disease, predation
   b. depends on birth rate, death rate, as well as emigration and immigration
   c. the maximum number of individuals sustainable by an environment's resources
   d. population growth plots out as J-shaped curve

Critical Thinking

1. If house cats that have not been neutered or spayed live up to their biotic potential, two can be the start of many kittens—12 the first year, 72 the second year, 429 the third, 2,574 the fourth, 15,416 the fifth, 92,332 the sixth, 553,019 the seventh, 3,312,280 the eighth, and 19,838,741 kittens in the ninth year. Is this a case of logistic growth? Exponential growth? Irresponsible cat owners?

2. A third of the world population is below age fifteen. Describe the effect of this age distribution on the future growth rate of the human population. If you conclude that it will have severe impact, what sorts of humane recommendations would you make to encourage individuals of this age group to limit family size? What are some social, economic, and environmental factors that might keep them from following the recommendations?

3. Write a short essay about a population having one of the age structures shown below. Describe what may happen to younger and older groups when individuals move into new categories.

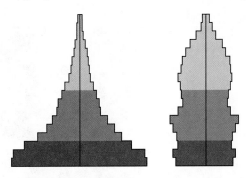

4. Figure 46.18 charts the legal immigration to the United States between 1820 and 1995. (The Immigration Reform and Control Act of 1986 accounted for the most recent dramatic increase; it granted legal status to illegal immigrants who could prove they had lived in the country for years.) During the 1980s and 1990s, an economic downturn fanned resentment against newcomers. Many people now say legal immigration should be restricted to 300,000–450,000 annually and we should crack down on the illegal immigrants. Others argue such a policy would diminish our reputation as a land of opportunity. They also say it would discriminate against legal immigrants during crackdowns on others of the same ethnic background. Do some research, then write an essay on the pros and cons of both positions.

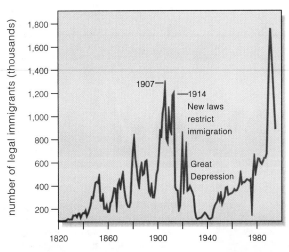

Figure 46.18 Chart of legal immigration to the United States between 1820 and 1995.

5. In his book *Environmental Science*, Miller points out that the unprecedented projected increase in the human population from 5 billion to 10 billion by the year 2050 raises serious questions. Will there be enough food, energy, water, and other resources to sustain twice as many people? Will governments be able to provide adequate education, housing, medical care, and other social services for all of them? Computer models suggest that the answers are no (Figure 46.19). Yet some people claim we can adapt socially and politically to an even more crowded world, assuming harvests improve through technological innovation, every inch of arable land is put under cultivation, and everyone eats only grain. There are no easy answers to the questions. If you have not yet been doing so, start following the arguments in your local newspapers, in magazines, and on television. This will allow you to become an informed participant in a global debate that surely will have impact on your future.

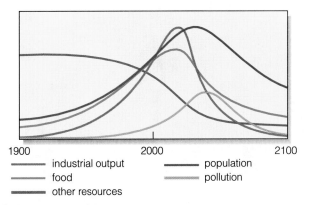

industrial output — population
food — pollution
other resources

Figure 46.19 Computer-based projection of what may happen if the size of the human population continues to skyrocket without dramatic policy changes and technological innovation. The assumptions are that the population has already overshot the carrying capacity and current trends continue unchanged.

Selected Key Terms

age structure *46.1*
biotic potential *46.2*
carrying capacity *46.3*
cohort *46.4*
demographic transition
   model *46.8*
demographics *46.1*
density-dependent control *46.3*
density-independent factor *46.3*
doubling time *46.2*
ecology *CI*
emigration *46.2*
exponential growth *46.2*
family planning program *46.7*
immigration *46.2*

life history pattern *46.4*
life table *46.4*
limiting factor *46.3*
logistic growth *46.3*
migration *46.2*
per capita *46.2*
population density *46.1*
population distribution *46.1*
population size *46.1*
r (net reproduction per
   individual per unit time) *46.2*
reproductive base *46.1*
survivorship curve *46.4*
total fertility rate *46.7*
zero population growth *46.2*

Readings

Cohen, J. E. 1995. *How Many People Can the Earth Support?* New York: Norton. No pat answer to title's question.

Miller, G. T. 1996. *Environmental Science*. Sixth edition. Belmont, California: Wadsworth.

Web Site See *http://www.wadsworth.com/biology* for practice quiz questions, hypercontents, BioUpdates, and critical thinking. The Wadsworth Biology Resource Center provides a wealth of information fully organized and integrated by chapter.

47 COMMUNITY INTERACTIONS

No Pigeon Is An Island

Flying through the rain forests of New Guinea is an extraordinary pigeon with cobalt blue feathers and lacy plumes on its head (Figure 47.1). It is about as big as a turkey, and it flaps so slowly and noisily that its flight sounds like an idling truck. As is true of eight species of smaller pigeons living in the same forest, it perches on branches to eat fruit. How is it possible that *nine* species of large and small fruit-eating pigeons live in the space of the same forest? Wouldn't you think that competition for food would leave one the winner? In fact, in that rain forest, every species lives, grows, and reproduces in a characteristic way, as defined

Figure 47.1 A cobalt blue, turkey-size Victoria crowned pigeon, one of nine species of pigeons living in the same tropical rain forest of New Guinea. Within this habitat, each species has its own niche.

by its relationships with other individuals and with the surroundings.

Big pigeons perch on the sturdiest branches when they feed, and they eat big fruit. Smaller pigeons, with their smaller bills, cannot open big fruit. They eat small fruit hanging from slender branches that are not sturdy enough to support the weight of a turkey-size pigeon. The species of trees in the forest differ with respect to the diameter of their fruit-bearing branches and the size of their fruit. So they attract different pigeons with different characteristics. In such ways, the nine species of pigeons *partition* the fruit supply.

And how do individual trees benefit from enticing the pigeons to dine? Think about it. The seeds inside their fruits have tough coats, which can resist the action of digestive enzymes inside the pigeon gut. During the time it takes for ingested seeds to travel through the gut, the pigeons fly about, so they dispense seed-containing droppings in more than one place. In this way, pigeons tend to disperse seeds some distance away from the parent plants. Later, when seedlings grow, the odds are better that at least some will not have to compete with their parents for sunlight, water, and nutrients. Seeds that drop close to home cannot compete in any significant way with the resource-gathering capacity of the mature trees, which already have extensive, well-developed roots and leafy crowns.

Within the same forest, leaf-eating, fruit-munching, and bud-nipping insects interact with other organisms and the surroundings in certain ways. So do nectar-drinking, flower-pollinating bats, birds, and insects. And so do great numbers of beetles, worms, and other invertebrates that busily extract energy from remains and wastes of other organisms on the forest floor. By their activities, they cycle nutrients back to the trees.

Like humans, then, no pigeon is an island, isolated from the rest of the living world. The nine species of New Guinea pigeons eat fruit of different sizes. They disperse seeds from different sorts of trees. Dispersal influences where new trees will grow and where the decomposers will flourish. Ultimately, tree distribution and decomposition activities influence how the entire forest community is organized.

Directly or indirectly, interactions among coexisting populations organize the community to which they belong. With this chapter, we turn to community interactions that influence all populations over time and in the space of their environment.

KEY CONCEPTS

1. A habitat is the type of place where individuals of a species normally live. A community is an association of all the populations of species that occupy the same habitat.

2. Every species in the community has its own niche, defined as the sum of all activities and relationships in which its individuals engage as they secure and use the resources required for their survival and reproduction.

3. Community structure starts with the adaptive traits that give individuals of each species the capacity to respond to physical and chemical features of the habitat, and to levels and patterns of resource availability over time.

4. Interactions among species influence the structure of a community. They include mutually beneficial interactions, competition, predation, and parasitism.

5. Community structure also depends on the geographic location and size of the habitat, the rates at which the member species arrive and disappear, and the history of physical disturbances to the habitat.

6. The first species to occupy a particular type of habitat are replaced by others, which are replaced by still others, and so on in sequence. This process, known as primary succession, produces a climax community. A climax community is a stable, self-perpetuating array of species in balance with one another and with the environment.

7. Different stages of succession often exist in the same habitat, owing to local differences in the soil and other environmental factors, recurring disturbances such as seasonal fires, and chance events.

FACTORS THAT SHAPE COMMUNITY STRUCTURE

Think of a clownfish darting above a coral reef, a maple tree on a Vermont hillside, or a mole burrowing in your lawn. The type of place where you will normally find a clownfish, maple, or mole is its **habitat**. The habitat of an organism is characterized by physical and chemical features, such as temperature and salinity, and by the array of other species living in it. Directly or indirectly, the populations of all species in a habitat associate with one another as a **community**.

Five factors shape the structure of a community. *First*, interactions between climate and topography help dictate the habitat's temperatures, rainfall, soil types, and other conditions. *Second*, the kinds and amounts of food and other resources that become available through the year influence which species can live there. *Third*, individuals of each species have adaptive traits that allow them to survive and exploit specific resources in the habitat. *Fourth*, species in the habitat interact, as by competition, predation, and mutually helpful activities. *Fifth*, community structure is influenced by the overall pattern and actual history of changing population sizes, the arrival and disappearances of species, and physical disturbances to the habitat.

Together, the five factors help dictate the number of species at different "feeding levels," starting with the producers and continuing through levels of consumers. They influence population sizes. They also help dictate the overall number of species. For example, high solar radiation, warm temperatures, and high humidity in tropical habitats favor growth of many kinds of plants, which support many kinds of animals. Conditions in arctic habitats do not favor great numbers of species.

The chapters to follow deal with energy flow through feeding levels and with geographic factors influencing community structure. Here we begin with interactions among species, using the niche concept as our guide.

The Niche

If the organisms within a community all share the same habitat—that is, if they all have the same "address"—in what respects do they differ? Each kind is distinct in terms of its "profession" within the community—that is, in the sum of activities and relationships in which it engages to secure and use the resources necessary for its survival and reproduction. This is its **niche**.

For each species, the *potential* niche is the one that might prevail in the absence of competition and other factors that could constrain its acquisition and use of resources. However, as you will see, such constraining factors do come into play in all communities. They tend to bring about a more constrained, *realized* niche that shifts in large and small ways over time, as individuals of the species respond to a mosaic of changes.

Categories of Species Interactions

Dozens to hundreds of species interact in diverse ways, even in simple communities. In spite of this diversity, we can identify six categories of interactions that have different effects on population growth (Table 47.1).

Table 47.1 Types of Two-Species Interactions*		
Type of Interaction	Direct Effect on Species 1	Direct Effect on Species 2
Neutral relationship	0	0
Commensalism	+	0
Mutualism	+	+
Interspecific Competition	−	−
Predation	+	−
Parasitism	+	−

* 0 means no direct effect on population growth; + means positive effect; − means negative effect.

A given species has a neutral relationship with most species in its habitat. For example, Canadian lynx and grasses do not affect each other *directly*. Interactions with other species link them only indirectly. The lynx preys on and decreases the number of snowshoe hares that eat plants; and plants fatten the hares, the prey of lynx.

Commensalism directly helps one species but does not affect the other much, if at all. For instance, some birds gain home bases by roosting in trees. The trees get nothing but are not harmed. In **mutualism**, the benefits flow both ways between the interacting species. (Don't think of this as cozy cooperation; the benefits flow from a two-way exploitation.) In **interspecific competition**, disadvantages flow both ways between species. Finally, **predation** and **parasitism** are interactions that directly benefit one species (either the predator or the parasite) and directly hurt the other (the prey or host).

Commensalism, mutualism, and parasitism are all forms of **symbiosis**, which means "living together." For at least part of the life cycle, individuals of one species live near, in, or on individuals of another species.

A habitat is the type of place where individuals of a species normally live. A community consists of all populations that live in the habitat.

Community structure arises from the habitat's physical and chemical features, resource availability over time, adaptive traits of its members, how the members interact, and the history of the habitat and its occupants. Species may have neutral, positive, or negative effects on one another.

A niche is the sum of all activities and relationships in which individuals of a species engage as they secure and use the resources necessary to survive and reproduce.

Mutualistic interactions, in which positive benefits flow both ways, abound in nature. When trees provide New Guinea pigeons with food and when the pigeons help disperse seeds from the trees to new germination sites, they function as mutualists. Many other kinds of plants and animals enter into such interactions. For example, most flowering plants and the insects and birds, bats, and other animals that pollinate them are mutualists. The introduction to Chapter 31 gives vivid examples.

absorptive structures interact in two ways. Either the fungal hyphae penetrate root cells or they form a dense, velvety mat around them. The plant comes to depend on the fungus to absorb enough vital mineral ions from the soil to maintain plant growth. In turn, the fungus withdraws and uses some of the sugar molecules that the plant produces. The plant also affects its partner's reproductive success; when photosynthesis ceases, the fungus stops producing spores.

Figure 47.2 One mutualistic interaction in the high desert of Colorado.

(**a**) Different kinds of flowering plants of the genus *Yucca* are each pollinated exclusively by one species of yucca moth (**b**). This insect cannot complete its life cycle with any other plant.

The adult stage of the moth life cycle coincides with blossoming of yucca flowers. By using her specialized mouthparts, a female moth gathers sticky pollen and rolls it into a ball. Then she wings her way to another flower. She pierces the wall of the flower's ovary, where seeds form and develop, and lays her eggs inside. As she crawls out of the flower, she pushes a ball of pollen onto a pollen-receiving surface.

Pollen grains germinate and grow down through ovarian tissues, carrying sperm to the flower's eggs. The seeds develop after fertilization. Meanwhile, the moth eggs develop to the larval stage. (**c**) When larvae emerge, they eat a few seeds, then gnaw their way out of the ovary. The seeds that moth larvae do not eat give rise to new yucca plants.

a

b

c

Some forms of mutualism are *obligatory*. That is, the individuals of one species cannot grow and reproduce unless they spend their entire life with individuals of the other species, in intimate dependency. This is true of the interaction between yucca plants and yucca moths. Each species of yucca plant is pollinated exclusively by one species of yucca moth. Moth larvae can grow only in yucca plants; they eat only yucca seeds (Figure 47.2).

We also find cases of obligatory mutualism between many fungi and plants. Mycorrhizae, recall, are intimate associations between fungal hyphae and young roots of many plants (Sections 24.4 and 30.2). The two kinds of

And reflect on the apparent endosymbiotic origins of eukaryotes (Section 21.4). Prokaryotic cells engulfed by other prokaryotic cells resisted digestion and lived as guests in their host. They became metabolically interdependent, and the guests evolved into mitochondria, chloroplasts, and other organelles. If those cells had not evolved in such intimate, mutually beneficial ways, you and all other eukaryotic organisms would not even be around today.

In forms of mutualism, each of the participating species reaps benefits from the interaction.

COMPETITIVE INTERACTIONS

Categories of Competition

As you know by now, all organisms do not come into the world with guarantees that they will secure enough energy, nutrients, living space, and other necessities for surviving and reproducing. They typically compete for a share of limited resources. *Intraspecific* competition means individuals of the same population or same species compete with one another. As you probably concluded from the preceding chapter, their interactions can be most fierce. By contrast, *interspecific* competition occurs between populations of different species, and usually it is not as intense. Why? The requirements of two species might be similar, but they never will be as close as they are for individuals of the same species.

Consider two forms of competitive interactions. Sometimes all individuals have equal access to a required resource, but some are better than others at exploiting it. In such cases, competition tends to reduce the supply of a shared, limited resource. (When you and a friend both use straws to share a small milkshake, you might not get nearly as much if your friend uses a jumbo straw.) In other cases, some individuals control access to a resource and partially or wholly prevent others from using it, regardless of its scarcity or abundance. (Thus, even if you shared a ten-gallon milkshake, you still would get less if your friend pinched your straw.)

Competition abounds in nature. For instance, it forces different chipmunk species to use different mountain habitats (Figure 47.3). Reef corals poison and grow on top of other species of reef corals. A strangler fig tree wraps around other trees as a framework for its own growth and finally kills them. From early spring until late summer, a male *broadtailed* hummingbird chases other males and females of its own species away from his blossom-profuse territory in the Rocky Mountains. Come August, however, *rufous* hummingbirds from the Pacific Northwest migrate to their wintering grounds in Mexico. As they pass through the Rockies, rufous males prove to be more aggressive and stronger competitors for the available food. They evict the male broadtails from broadtail territories all along their migratory route.

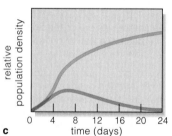

ALPINE CHIPMUNK

LODGEPOLE CHIPMUNK

YELLOW PINE CHIPMUNK

LEAST CHIPMUNK

Figure 47.3 An example of competition in nature. On the eastern slopes of the Sierra Nevada, different chipmunk species occupy different habitats. The alpine habitat is at the highest elevation. Below this are the lodgepole pine, piñon pine, and then the sagebrush habitats. The least chipmunk lives in sagebrush at the base of the mountain range. Its adaptations would allow it to move up into the piñon pine habitat, but the aggressively competitive behavior of yellow pine chipmunks in that habitat won't let it. Dietary preferences keep the yellow pine chipmunk out of the sagebrush habitat.

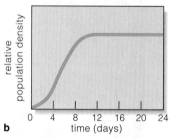

a   time (days)   b   time (days)   c   time (days)

relative population density

Figure 47.4 Results of competitive exclusion between two species of protistans that compete for the same food resource. *Paramecium caudatum* (**a**) and *P. aurelia* (**b**) were grown apart in separate culture tubes and established stable populations. The S-shaped growth curves in the graphs indicate this stability. When the populations were grown together, *P. aurelia* (*red* curve in **c**) drove the other species toward extinction (*blue* curve in **c**). This experiment and others suggest that two species cannot coexist indefinitely in the same habitat *when they require identical resources.* If their requirements do not overlap much, one might influence the population growth rate of the other, but both may still coexist.

Competitive Exclusion

To a greater or lesser extent, any two species differ in their adaptations for getting food or avoiding enemies. Thus one usually competes more effectively for scarce resources. The two are less likely to coexist in the same habitat when they use resources in very similar ways. G. Gause demonstrated this by growing two species of *Paramecium* separately, then together (Figure 47.4*a*–*c*). Both species exploited the same food (bacterial cells) and competed intensely for it. As the experimental results

Figure 47.5 Individuals of two different species of salamanders that coexist in the same habitat: (**a**) *Plethodon glutinosus* complex and (**b**) *P. jordani.*

suggested, two species that require identical resources cannot coexist indefinitely. Many subsequent studies also support this concept, which is now called **competitive exclusion**.

In different experiments, Gause used two other species of *Paramecium* that did not have as much overlap in their requirements. When grown together, one species tended to feed on bacteria suspended in the liquid of the culture tube, and the other on yeast cells at the bottom of the tube. Although the population growth rate was slower for both species, the overlap was not enough for one to exclude the other; the two species continued to coexist.

Field experiments reveal the effects of competition, also. For example, N. Hairston studied salamanders in the Great Smoky Mountains and the Balsam Mountains (Figure 47.5). One kind, *Plethodon glutinosus*, lives at lower elevations than its relative *P. jordani*, but their ranges overlap in some areas. Hairston removed one or the other species from different test plots in the overlap areas. He also left some plots untouched, as controls. After five years, nothing had changed in the control plots; the two species were coexisting. By contrast, the test plots had growing populations. The plots cleared of *P. jordani* had a greater proportion of *P. glutinosus*; and plots cleared of *P. glutinosus* had a greater proportion of *P. jordani*. Hairston concluded that, where populations of the two species are coexisting in nature, competitive interactions suppress the growth rate of both of them.

Resource Partitioning

Think back on the nine species of fruit-eating pigeons in the same forest in New Guinea. They require the same resource: fruit. Yet they overlap only slightly in their use of the resource, because each specializes in fruits of a particular size. All together, they are a good example of **resource partitioning**, whereby competing species coexist by *subdividing* a category of similar resources.

A similar competitive situation arises among three species of annual plants in certain plowed, abandoned fields. Like other plants, all three require sunlight, water, and dissolved mineral ions. Yet each species is adapted to exploiting a different portion of the habitat (Figure 47.6). Drought-tolerant foxtail grasses have a shallow, fibrous root system that quickly absorbs rainwater. They grow where moisture in soil varies from day to day. Mallow plants, with a taproot system, grow in deeper soil that is moist early in the growing season but drier later. Smartweed's taproot system branches in topsoil and in soil below the roots of the other species. It grows where soil is continuously moist.

In sum, species may coexist in the same habitat even if their niches do overlap. Many species compete when experimentally grown together but not in their natural habitats, where they tend to use different foods or other resources. Hairston's salamanders compete yet coexist at suppressed population sizes. New Guinea pigeons coexist through resource partitioning. Birds and people overlap in their need for oxygen, but atmospheric oxygen is so abundant that there is no need to compete for it.

In some competitive situations, all individuals have equal access to a resource that they all require, but some are better than others at exploiting it. In other competitive situations, some individuals control access to a resource.

The more two species in the same habitat differ in their use of resources, the more likely they can coexist.

Two competing species also may coexist by sharing the same resource in different ways or at different times.

Figure 47.6 Resource partitioning among three annual plants in a plowed, abandoned field. All three require water and mineral ions and differ in their adaptations for securing them.

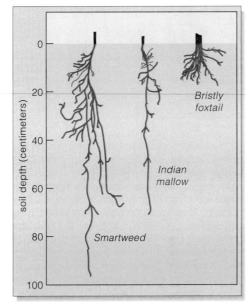

soil depth (centimeters)

0
20
40
60
80
100

Bristly foxtail

Indian mallow

Smartweed

Predation Versus Parasitism

Of all community interactions, predation is so dynamic that it probably is the most riveting of our attention, not to mention the prey's. But what, exactly, is a predator-prey interaction?

Consider the leopard in Figure 47.7 as it closes in on a baboon, one of its preferred foods. A goat pulling up a thistle plant for breakfast, though less dramatic, is also a predator. Its prey is a living organism, killed for food. Can the same be said of a horse grazing upon but not killing a plant? What about a mosquito withdrawing blood from your arm before it flies away? What about ticks or fleas that withdraw blood leisurely before they jump off one host and lay their eggs elsewhere? What about tapeworms, mistletoe, and other organisms that live in or on other species?

For simplicity, let us use two broad definitions for the interactions between various consumers and their victims. **Predators** are animals that feed on other living organisms—their **prey**—but do *not* take up residence on or in them. The prey organism may or may not die from the interaction. **Parasites** take up residence in or on other living organisms—their **hosts**—and feed on specific host tissues for part of the life cycle. The host organisms may or may not die as a result of the interaction.

Dynamics of Predator-Prey Interactions

Many of the adaptations of predators and their victims arose through **coevolution**. The term refers to the joint evolution of two (or more) species that exert selection pressure on each other through their close ecological interaction. Suppose, as an outcome of mutation, a new, heritable means of defense appears in a prey organism and spreads through the population. Some individual predators are more effective than others at countering the new defense. They tend to eat more and have a better chance to survive and reproduce. In time the forms of traits that are most efficient at overcoming the new prey defense increase in frequency in the population. Now the better predators exert selective pressure that favors better defenses among prey—and so on through time.

In any specified interval, the outcome of predator-prey interactions depends partly on the carrying capacity of the prey population. (*Carrying capacity*, remember, is

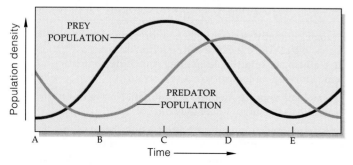

Figure 47.7 Idealized cycling of abundances of predator and prey. (For clarity, this diagram exaggerates the predator density; predators usually are less common than their prey throughout the cycle.) The pattern arises through time lags in predator responses to changes in prey abundance. At time A, prey density is low; predators have more difficulty securing food and their population is declining. In response to the predator decline, prey start increasing. But predators do not start increasing until they start reproducing at time B. Both populations grow until the greater number of predators causes the prey population to decline (time C to E). Predators continue to increase and take out more prey. But the lower prey density leads to starvation among them, and their growth rate slows (starting at time D). At time E, a new cycle starts.

Figure 47.8 Predator-prey interactions between Canadian lynx and snowshoe hares. An apparent correspondence between the abundances of both populations is based on counts of pelts sold by trappers to Hudson's Bay Company over ninety years. The *dashed* line tracks lynx abundances, and the *solid* line, the hare abundances. The chart is a good test of whether you readily accept someone else's conclusions without questioning their scientific basis. (Remember the discussion of scientific methods in Section 1.5?) What other factors might have influenced the cycle? Did weather vary, with more severe winters imposing greater demand for food (required for animals to keep warmer) and higher death rates? Did the lynx have to compete with other predators for prey? Did predators turn to alternative kinds of prey during low points of the hare cycle? Did trapping increase with rising fur prices in Europe, and did it decrease as the pelt supply outstripped the demand?

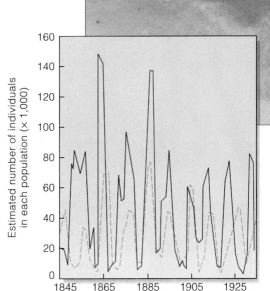

defined as the maximum number of individuals that the resources in a given environment can maintain indefinitely.) The reproductive rates of predator and prey populations also influence the outcome. So do responses of individual predators to increases in prey density.

During times when predation is keeping a prey population from exceeding the carrying capacity, both populations tend to coexist at fairly steady levels. Then, more prey are around, and predators reproduce promptly and eat more of the prey. Population densities fluctuate when predators do not reproduce as fast as the prey, when they can eat only so many prey organisms during a given interval, and when the carrying capacity for the prey population is high.

The graph in Figure 47.7 shows an idealized pattern of the cyclic changes brought about by time lags in a predator's response to changes in prey abundance. In nature, we sometimes observe such a correspondence between changes in predator and prey populations, but other factors also are involved.

For example, Charles Krebs led a long-term study in the Yukon to identify the basis of the ten-year cycle in populations of the snowshoe hare and its predators (Figure 47.8). For eight years, the researchers tracked hare densities in one-square-kilometer experimental plots and control plots. In some plots, electrified fences kept out mammalian predators but not hares. In others, hares were supplied with extra food. In still other plots,

fertilizers enhanced plant growth for the herbivorous hares.

During cyclic peaks and declines, hare densities increased in predator-free plots and in plots having extra food. In plots where both conditions occurred, the densities were thirty-six times greater than in control plots. By itself, increased vegetation (in the fertilized plots) had negligible effect on hare densities. Thus a simple predator-prey model or plant-herbivore model is not enough to explain the results. This particular cycle seems to turn on plants, herbivores, and carnivores—a *three-level* interaction.

Some predator and prey populations coexist at more or less steady levels. Others undergo recurring cycles of abundance and crashes, erratic cycles, or prey extinction.

When predation keeps a prey population from overshooting the carrying capacity, the predator and prey populations may coexist at stable levels. Delays in a predator's response to changes in the abundance of prey contribute to cyclic or irregular fluctuations in population levels.

Predator-prey or plant-herbivore models alone cannot explain the cyclic changes. Other factors, such as changes in the prey's food supply, appear to contribute to the cycles.

THE COEVOLUTIONARY ARMS RACE

Populations of other species are part of the environment of any organism, and the ones that interact as predators and prey exert continual selection pressure on each other. One must defend itself, the other must overcome the defenses. This is the basis of a coevolutionary arms race that has resulted in some truly amazing adaptations.

a

b

c

CAMOUFLAGE   Consider prey species that **camouflage** themselves; they can hide in the open. Such organisms have adaptations in form, patterning, color, and behavior that help them blend with their surroundings and escape detection. Figure 47.9 shows classic examples, including a desert plant (*Lithops*) that resembles a small rock. Only during a brief rainy season does *Lithops* flower. That is when other plants grow profusely (and divert herbivores from *Lithops*), and when free water instead of juicy plant tissues is available to quench an animal's thirst.

WARNING COLORATION   Many prey species taste bad, are highly toxic, or inflict pain on attackers. Often the toxic types have **warning coloration**, or conspicuous patterns and colors that predators learn to recognize as "avoid me" signals. For example, maybe a young, inexperienced bird will spear a yellow-banded wasp or an orange-patterned monarch butterfly—once. It quickly learns to associate the distinctive colors and patterning with a painful sting or with vomiting foul-tasting butterfly toxins.

   Truly dangerous or repugnant species make little or no attempt to conceal themselves. Skunks are like this. So are frogs of the genus *Dendrobates*; they are among the most vivid and most poisonous organisms (Section 33.1).

MIMICRY   Many prey organisms bear close resemblance to dangerous, unpalatable, or hard-to-catch species. Such resemblances are forms of **mimicry**. Figure 47.10 shows how closely some of the tasty but weaponless *mimics* physically resemble their *models*—species that predators have learned to ignore. In "speed" mimicry, a sluggish prey organism closely resembles swift-moving species that predators give up trying to catch.

MOMENT-OF-TRUTH DEFENSES   When luck runs out, survival of prey organisms that are cornered or under attack may turn on a last-ditch trick. Suppose a leopard runs down a tasty baboon. By turning abruptly and displaying formidable canines, the baboon may startle and confuse this predator long enough for a chance at a getaway (Figure 47.7). Other cornered animals release chemicals as disgusting repellants or toxins. Earwigs, skunks, and stink beetles produce awful odors. Several beetles take aim and let loose with noxious sprays.

   Similarly, many plants synthesize predator repellants. Tannins in the foliage and seeds of certain plants taste

Figure 47.9   A few prey organisms demonstrating the fine art of camouflage. (**a**) What bird??? When a predator approaches its nest, the least bittern stretches its neck (which is colored like the surrounding withered reeds), thrusts its beak upward, and sways gently like reeds in the wind. (**b**) An unappetizing bird dropping? No. The body coloration of this caterpillar and the stiff positions it takes help it hide in the open from birds that prey on it. (**c**) Find the plants (*Lithops*) hidden from herbivores owing to their stonelike form, pattern, and coloring.

Figure 47.10 A few examples of mimicry among the insects.

Many predators avoid prey that taste awful, secrete toxins, or inflict painful bites or stings. Commonly, such prey display warning coloration (bright colors, bold markings, or both), and many do not even bother to hide. Many species of prey unrelated to the dangerous or unpalatable ones have evolved striking behavioral and morphological resemblances to them. (**a**) The yellowjacket shown here stings aggressively and is the likely model for nonstinging, edible wasps (**b**) and beetles (**c**) that have a similar appearance. The inedible butterfly in (**d**) is a model for the edible mimic *Dismorphia* (**e**).

a The dangerous model . . .   **b** . . . one of its edible mimics . . .   **c** . . . and another edible mimic

bitter and make the plant tissues hard to digest. Make the mistake of nibbling on the seemingly luscious yellow petals of a buttercup (*Ranunculus*), and you will inflict a chemical burn upon the lining of your mouth.

ADAPTIVE RESPONSES TO PREY As part of the coevolutionary arms race, predators counter prey defenses with their own marvelous adaptations. Among the countering measures are stealth, camouflage, and clever ways of avoiding repellants.

Consider the edible beetles that direct sprays of noxious chemicals at their attackers. Grasshopper mice grab such beetles and plunge the "sprayer" end into the ground, then feast on the unprotected head (Figure 47.11a,b). Chameleons typically hold themselves motionless for extended intervals. Prey may not even "see" them until the amazingly swift chameleon tongue zaps them (Figure 33.10). And it is no accident that stealthy predators blend with backgrounds. Think of snow-white polar bears camouflaged against snow, golden tigers crouched in tall-stalked, golden grasses, and pastel predatory insects lurking in pastel flowers. And hope you never step barefoot on a scorpionfish concealed on the seafloor.

Figure 47.11 Examples of adaptive responses of predators to prey defenses. (**a**) Certain beetles spray noxious chemicals at attackers, which works as a deterrent some of the time. (**b**) At other times, however, grasshopper mice plunge the chemical-spraying tail end of their beetle prey into the ground and feast on the head end. (**c**) Find the scorpionfish, a venomous predator with camouflaging fleshy flaps, multiple colors, and profuse spines. (**d**) Where do pink parts of the flower end and pink parts of the praying mantis begin?

Parasitism, again, is an ecological interaction in which one organism lives on or in a host organism and uses its tissues for nutrients. The parasite benefits; the host does not. This also is true of predation, but you would not expect to come across a predator that characteristically uses another organism as food *and* habitat.

Parasites have pervasive influences in populations. By draining host nutrients, they alter how much energy and how many nutrients the population withdraws from the habitat. Weakened hosts are more vulnerable to predation and less attractive to potential mates. Some infections result in sterility; they may alter the ratio of host males to females. In such ways, parasitic infections lower the birth rate, raise the death rate, and influence intraspecific and interspecific competition.

Kinds of Parasites

Previous chapters introduced you to diverse parasitic species. They include *ecto*parasites (which live on the body surface of a host) as well as *endo*parasites (which live within the host's body). Some live on or in one host for their entire life cycle, and others are free-living some of the time or residents of different hosts at different stages of their life cycle. Many use insects and other arthropods as taxis from one host organism to another.

The **microparasites** among them include a great number of bacteria, viruses, and protistans (especially in the groups called protozoans and sporozoans). Most are microscopically small and reproduce rapidly.

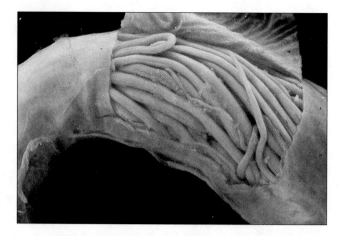

Figure 47.12   Packed inside the small intestine of a host pig, adult roundworms (*Ascaris*), a type of endoparasite.

The **macroparasites** include a variety of invertebrates. Those having worldwide notoriety include numerous flatworms and roundworms (or nematodes), such as the type in Figure 47.12. Also in this category are many arthropods, including fleas, ticks, and mites, and lice that attack birds and mammals. They lay their eggs either directly on the host or in its nests or bedding. As you know from Section 24.4, the fungal kingdom also has its share of macroparasites.

Even the plant kingdom has a few. *Holo*parasitic plants are nonphotosynthetic; they withdraw nutrients and water from young roots of host plants. Members of the broomrape family do this to the roots of oaks and beech trees. *Hemi*parasitic plants retain the capacity for photosynthesis but still withdraw nutrients and water from host plants. Mistletoe is an example; its extensions invade the sapwood of host trees.

Not all parasites feed on host tissues. Some are **social parasites**, which complete their life cycle by manipulating the social behavior of another species. Consider the cuckoo, a social parasite. Adult females lay eggs in the nests of other species. If newly hatched cuckoos eliminate the natural-born offspring, they will end up receiving the undivided attention of their unsuspecting foster parents. Although blind when hatched, they maneuver any egg they contact onto their back and push it out of the nest (Figure 47.13).

Similarly, the North American brown-headed cowbird never builds a nest, incubates its own eggs, or cares for its offspring. It removes an egg from the nest of another bird and lays one as a "replacement." Some birds cannot tell the difference. They hatch the alien egg and raise the hatchling. Typically, a young cowbird aggressively pushes the smaller, remaining rightful occupants out of the nest or demands and gets most of the food.

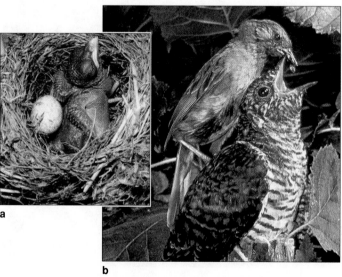

Figure 47.13   Example of a social parasite. (**a**) European cuckoos lay eggs in the nests of other species. In response to an environmental cue, this hatchling is executing a behavior that is innate; it has inherited the knowledge of what to do without having to learn it. Even before its eyes open, it responds to the spherical shape of the host's eggs and shoves them out of the nest. (**b**) The foster parents keep on feeding the usurper.

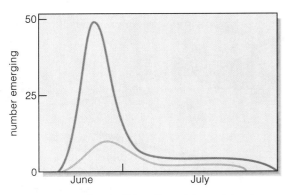

Figure 47.14 Graph of the effect of the activity of a parasitoid wasp on the emergence of adult sawflies from cocoons on the forest floor. The *brown* line represents the number emerging when wasp attacks were experimentally prevented. The *pink* line represents the number emerging after the attacks. Notice that the wasp made most of its attacks on the would-be *early* emergers, from cocoons not buried as deeply in forest litter.

Evolution of Parasitism

Sometimes the slow drain of nutrients during a parasitic infection indirectly causes death, for a host may become so weakened that it dies from secondary infections. In evolutionary terms, however, killing a host is not good for the parasite's reproductive success. An infection of longer duration gives the parasite more time to produce more offspring. Thus natural selection tends to favor parasite and host adaptations that promote some level of mutual tolerance and less-than-fatal effects.

Usually, death results only when a parasite attacks a novel host (which has no coevolved defenses against it) or when too many individual parasites are active at the same time during an infection.

Regarding the Parasitoids

In their own category between predators and parasites are **parasitoids**, insect larvae that always kill what they eat. As parasitoids are growing up, they consume all the soft tissues of their hosts. This sounds a bit horrendous, but fortunately the hosts of parasitoids are not humans but the larvae or pupae of other insect species.

The biologist Peter Price studied the evolutionary effects of a parasitoid wasp that lays eggs on the cocoons of a sawfly species. These sawflies reproduce once a year. They lay their eggs in trees. In due time, fertilized eggs develop into fly larvae, which feed on foliage and grow about as large as a pencil in diameter. Then the larvae drop to the forest floor, where they spin cocoons after they burrow into leaf litter. Some of the larvae burrow more deeply than others.

As Price noticed, the first adult sawflies to emerge did so from cocoons that were the least deeply buried. Later in the season, more sawflies emerged from the more deeply buried cocoons (Figure 47.14). Parasitoid wasps tend to lay eggs on the cocoons closest to the surface of leaf litter. They exert strong selection pressure on the sawfly population, because the deep-burrowing sawfly individuals are more likely to escape detection. There are only so many cocoons near the surface, so the wasp that is able to locate cocoons deeper in the litter will be more competitive in securing food for her larvae.

As Price points out, the host organism stays ahead in this coevolutionary contest. Each time the sawfly larvae burrow deeper, the female wasps have to spend more time searching for them. Fewer wasp eggs are laid—so the wasp population is held in check.

Parasites as Biological Control Agents

As you might well conclude, parasites and parasitoids exert control over the population growth of other insect species. Many kinds are being commercially raised and selectively released as *biological controls*. They are touted as an alternative to chemical pesticides. However, less than 20 percent of existing selections qualify as effective defenses against pests. As suggested by C. Huffaker and C. Kennett, effective control agents have five attributes. Generally, they are well adapted to the host species and their habitat. They are exceptionally good at searching for hosts. Their population growth rate is high relative to that of the host species, and their offspring are mobile enough to ensure adequate dispersal. Finally, the lag time between responses to changes in the numbers of the host population is minimal.

Releasing more than one kind of biological control agent in an area may trigger competition among them and eventually lessen their overall level of effectiveness. Besides this, a shotgun approach to biological control is risky. There is a chance that the parasites or parasitoids may attack nontargeted species. In 1983, for example, F. Howarth reported that populations of butterflies and moths native to the Hawaiian Islands were declining, partly because of the introduction of wasps that were supposed to serve as controls over something else. We will return to the effects of species introductions on native populations later in the chapter.

Like predators and their prey victims, parasites and their hosts are locked into long-term, coevolutionary contests.

Natural selection favors parasitic species that temper their demands in ways that assure an adequate supply of hosts.

Too great a demand quickly kills the hosts and limits the duration of infection, which sets a limit on the number of parasitic offspring that can be produced.

FORCES CONTRIBUTING TO COMMUNITY STABILITY

The Successional Model

How do communities come into being? By the classical model, known as **ecological succession**, a community develops in sequence, from pioneers to an end array of species that remain in equilibrium over some region. A **pioneer species** is an opportunistic colonizer of vacant or vacated habitats, notable for its high dispersal rate and rapid growth. Over time, more competitive species replace the pioneers, then are themselves replaced until the array of species stabilizes under the conditions that prevail in the habitat. This persistent array of species is the **climax community**.

A process of **primary succession** begins as pioneer species colonize a barren habitat. A new volcanic island is such a habitat. So is land that had been covered with ice for thousands of years, then exposed by a retreating glacier (Figure 47.15). Typical pioneer species are small plants with brief life cycles, adapted to grow in exposed areas with intense sunlight, swings in temperature, and nutrient-deficient soil. Each year they produce great numbers of small seeds, which are quickly dispersed.

Once established, the pioneers improve conditions for other species and often set the stage for their own replacement. Many types are mutualists with nitrogen-fixing bacteria and initially outcompete other plants in nitrogen-poor habitats. Then their accumulated wastes and remains add volume to the soil and enrich it with nutrients, which allows other species to take hold. The pioneers also form low-growing mats that shelter seeds of later species yet cannot shade out the seedlings. In time, later successional species crowd out the pioneers, whose seeds travel as fugitives on the wind or water— destined, perhaps, for a new but temporary habitat.

In **secondary succession**, a disturbed area within a community recovers and moves again toward the climax state. The pattern is typical of abandoned fields, burned forests, and storm-battered intertidal zones. It emerges after falling trees open part of an established forest's canopy. Sunlight reaches seeds and seedlings that are already on the forest floor and spurs their growth.

By one hypothesis, the colonizers *facilitate* their own replacement. By another, the earliest colonizers compete against species that could replace them, so the sequence of succession depends on who gets there first.

The Climax-Pattern Model

At one time, some ecologists thought the same general type of community always develops in a given region because of constraints imposed by climate. However, stable communities other than "the climax community" often persist in a region, as when tallgrass prairie extends into the deciduous forests of Indiana.

a

Figure 47.15 (**a**) Primary succession in Alaska's Glacier Bay area. The glacier shown here has been retreating since 1794. (**b**) As a glacier retreats, meltwater tends to leach the newly exposed soil of minerals, including nitrogen. Less than ten years ago, this nutrient-poor soil was still buried under ice. (**c,d**) The first invaders are seeds of mountain avens (*Dryas*) that drift in on the winds. Mountain avens is a pioneer species that benefits from the nitrogen-fixing activities of mutualistic microbes. It grows and spreads rapidly over glacial till.

(**e**) Within twenty years, shrubby deciduous alders, willows, and cottonwood take hold. They are symbionts with nitrogen-fixing microbes. In time, the alders form dense thickets. As the alder thickets mature, the cottonwood as well as hemlock trees grow rapidly. So do a few evergreen spruce trees. (**f**) By eighty years, spruce crowd out the mature alders. (**g**) In areas deglaciated for more than a century, dense forests of Sitka spruce and western hemlock dominate.

By a **climax-pattern model**, a community is adapted to many environmental factors—topography, climate, soil, wind, species interactions, recurring disturbances, chance events, and so on—that vary in their influence over a region. As a result, even in the same region you may see one climax stage extending into another along gradients of influential environmental conditions.

Cyclic, Nondirectional Changes

Many small-scale changes occur over and over again in patches of a habitat. Such recurring changes contribute to the internal dynamics of the community as a whole. Thus, observe a disturbed patch of habitat, and you might conclude that great shifts in species composition are afoot. Observe the community on a larger scale, and you might find that the overall composition includes all of the pioneer species that are colonizing the patch as well as the dominant climax species.

Consider a tropical forest, which develops through phases of colonization by pioneer species, increases in species diversity, and then reaches maturity. Whipping

up if rangers prevent the small fires. Thus the fire-susceptible species can take hold, and the underbrush gradually thickens. The dense underbrush prevents germination of sequoia seeds and fuels hotter fires that can damage the giants. Currently, park rangers set controlled fires, which can rid the habitat of underbrush and promote conditions that are the most favorable for the community's cyclic replacements.

Restoration Ecology

Reflect on the introduction to Chapter 29, which described the regrowth following Mount Saint Helens' violent eruption in 1980, and you already know secondary succession can heal disturbances on a large scale. In terms of a human lifetime, such *natural* restoration of the climax community often takes a long time. Today we also see *active* restoration, or attempts to reestablish biodiversity in abandoned farmlands as well as other large areas disturbed by agriculture, mining, and other human activities. For example, ecologists and volunteers build artificial reefs along coastlines, and repair wetlands and grasslands. In 1972, "lost" species of natural prairie communities were rediscovered in old cemeteries and other neglected bits of land in Illinois. They were transplanted by hand to the same site, which now measures 180 hectares (445 acres). Volunteers maintain this restored patch of prairie, as with controlled burns and hand weeding. They hope to find all of the 150–200 known species of the original community.

sporadically through the successional pattern, however, are heavy winds that cause treefalls. Where trees fall, gaps open in the forest canopy so more light can reach the forest floor. In that local patch, conditions favor the growth of previously suppressed small trees and the germination of pioneers or shade-intolerant species.

Or consider the groves of sequoia trees in the Sierra Nevada in California. Some trees in this type of climax community are giants, more than 4,000 years old. Their persistence depends partly on recurring brush fires that sweep through parts of the forests. Sequoia seeds only germinate in the absence of smaller, shade-tolerant plant species. Too much litter on the forest floor inhibits their germination. Modest fires eliminate trees and shrubs that compete with young sequoias but do not damage the sequoias themselves. Mature sequoias have thick bark that burns poorly and insulates the living phloem cells of the giant trees against modest heat damage.

At one time, fires were prevented in many sequoia groves in national and state parks—not just accidental fires from campsites and discarded cigarettes, but also natural fires sparked by lightning. Litter tends to build

---

Community structure is an outcome of a balance of forces, including predation and competition, operating over time.

A climax community is a stable, self-perpetuating array of species in equilibrium with one another and their habitat.

Similar climax stages can persist along gradients dictated by environmental factors and by species interactions. Also, recurring, small-scale changes help shape many communities.

Natural or deliberate ecological restoration often can repair a damaged climax community, provided that suitable species are available to reinstate the original biodiversity.

The preceding sections might lead you to believe that all communities become stabilized in predictable ways. But this is not always the case. *Community stability is an outcome of forces that have come into uneasy balance.* Resources are sustained, as long as populations do not flirt dangerously with the carrying capacity. Predators and prey coexist, as long as neither wins. Competitors have no sense of fair play. Mutualists are stingy, as when plants produce as little nectar as necessary to attract pollinators, and pollinators take as much nectar as they can for the least effort.

Over the short term, disturbances can hamper the growth of some populations. And long-term changes in climate or some other environmental variable also can have destabilizing effects. If instability becomes great enough, the community may change in ways that will persist even when the disturbance ends or is reversed. If some of its member species happen to be rare or don't compete well with others, they may become extinct.

How Keystone Species Tip the Balance

The uneasy balancing of forces in a community comes into sharp focus with studies of a **keystone species**, a dominant species that can dictate community structure. Robert Paine identified a keystone species by removal experiments in intertidal zones along North America's west coast. Pounding surf, tides, and storms continually disturb these zones, and living space is scarce. In his control plots, Paine included a sea star (*Pisaster*) and its invertebrate prey. The sea star proved to be a keystone species; it controls the abundances of resident mussels (*Mytilus*), limpets, chitons, and assorted barnacles.

After Paine removed all sea stars from experimental plots, mussels took over and crowded out seven other invertebrate species. Mussels are the main prey of sea stars but are the strongest competitors when sea stars are absent. Predation (by sea stars) normally maintains the diversity of prey species by preventing competitive exclusion (by mussels). Remove the sea stars, and the community shrinks from fifteen species to eight.

Or consider periwinkles (*Littorina littorea*), an alga-eating snail of intertidal zones. Jane Lubchenco found that periwinkles can increase *or* decrease the diversity of algal species in different settings. In tidepools, they eat a dominant algal species (*Enteromorpha*). By doing so, they help many other, less competitive algal species survive. By contrast, on rocks exposed only at high tide, *Chondrus* and other red algae predominate. Periwinkles leave these tough, unpalatable species alone—and eat the competitively weak algal species that they pass up in tidepools. Thus, periwinkles increase the diversity of algae in tidepools and yet reduce it on rocks that are recurringly exposed (Figure 47.16).

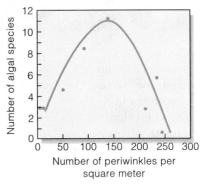

d Algal diversity in tidepools

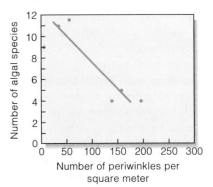

e Algal diversity on rocks that are alternately exposed and submerged

Figure 47.16 Effect of competition and predation on community structure. (**a**) Periwinkles (*Littorina littorea*) influence the number of algal species in different ways in different marine habitats. (**b**) *Chondrus* and (**c**) *Enteromorpha*, two algae in their natural habitat. (**d**) In tidepools, periwinkles graze on the dominant alga (*Enteromorpha*) and so favor less competitive algae that might otherwise be overwhelmed. (**e**) But on rocks that are exposed at low tide, they do not eat the dominant species (*Chondrus* and other red algae), so they decrease algal diversity.

Table 47.2   Detrimental Effects of Some Species Introduced Into the United States

Species Introduced	Origin	Mode of Introduction	Outcome
Water hyacinth	South America	Intentionally introduced (1884)	Clogged waterways; shading out of other vegetation
Dutch elm disease:			
Ophiostoma ulmi (the fungal pathogen)	Europe	Accidentally imported on infected elm timber (1930)	Destruction of millions of elms; great disruption of forest ecology
Bark beetle (carrier of the pathogen)		Accidentally imported on unbarked elm timber (1909)	
Chestnut blight fungus	Asia	Accidentally imported on nursery plants (1900)	Destruction of nearly all eastern American chestnuts; disruption of forest ecology
Argentine fire ant	Argentina	In coffee shipments from Brazil? (1891)	Crop damage; destruction of native ant communities; death of ground-nesting birds
Japanese beetle	Japan	Accidentally imported on irises or azaleas (1911)	Defoliation of more than 250 plant species, including commercially important species such as citrus
Sea lamprey	North Atlantic Ocean	Through Erie Canal (1860s), then through Welland Canal (1921)	Destruction of lake trout and lake whitefish in the Great Lakes
European starling	Europe	Released intentionally in New York City (1890)	Competition with native songbirds; crop damage; transmission of swine diseases; airport runway interference; very noisy and messy in large flocks
House sparrow	England	Released intentionally (1853)	Crop damage; displacement of native songbirds; transmission of some diseases

How Species Introductions Tip the Balance

Finally, consider what can happen when some residents of established communities move out from their home range and successfully take up residence elsewhere. We call such directional movement **geographic dispersal**. It can proceed in three ways.

First, over a number of generations, a population might expand its home range by slowly moving into outlying regions that prove hospitable. Second, some individuals might be rapidly transported across great distances. This is called **jump dispersal**. It often takes an individual across a region where it could not survive on its own, as when an insect travels from the mainland to Maui in a ship's cargo. Third, with imperceptible slowness, a population might be moved away from its home range over geologic time, as by continental drift.

The dispersal and colonization of vacant places can be amazingly rapid as well as successful. Consider one of Amy Schoener's experiments in the Bahamas. She set out plastic sponges on the barren sandy floor of Bimini Lagoon. How fast did aquatic species take up residence in chambers of the artificial hotels? Schoener recorded occupancy by 220 species in thirty days.

Or think about the 4,500 or so nonindigenous species that have become successfully established in the United States following jump dispersal. These are just the ones we know about. Some, including soybeans, rice, wheat, corn, and potatoes, have been put to good use as food resources. Most adversely impact natural communities and farmlands (Table 47.2 and Section 47.9).

If you live in the northeastern United States, you know about imported gypsy moths. They escaped from a research facility near Boston in 1869, and today their descendants defoliate whole forests. You might know about zebra mussels, which probably entered the Great Lakes on a cargo ship's hull. They attach to and block the water intake pipes for cities. The estimated damage exceeds 5 billion dollars. And *Cryphonectria parasitica*, a fungus introduced to North America almost a century ago, killed nearly all American chestnut trees.

If you live in the southeastern United States, you know about the hot sting of fire ants (*Solenopsis invicta*). Nearly a half century ago, the ants were accidentally imported from South America, possibly in the hold of a ship that docked in Mobile, Alabama. Each year, fire ants inflict enough multiple bites to make 70,000 or so people seek medical attention.

And remember those African bees released in South America? As you read in Chapter 8, their Africanized, "killer bee" descendants traveled northward, through Mexico and on into the United States. Swarms have already killed several hundred people in Latin America. At this writing, these aggressive bees have attacked 140 Americans, one of whom died from multiple stings.

Long-term shifts in climate, the rapid introduction and successful establishment of a new species, and many other disturbances can permanently alter community structure.

NILE PERCH AND RABBITS AND KUDZU, OH MY!

When you hear someone bubbling enthusiastically about an **exotic species**, you can safely bet the speaker is not an ecologist. This is a name for a resident of an established community that has moved from its home range and successfully taken up residence elsewhere. It makes no difference whether the importation was deliberate or accidental. Unlike most imports, which cannot take hold outside their home range, an exotic species insinuates itself into the new community. Occasionally the addition is harmless and even has beneficial effects. However, of all the native species that are on rare or endangered lists or have already become extinct, *almost 70 percent owe their precarious existence or demise to displacement by exotic species*. Consider a few of the more spectacular cases.

HELLO VICTORIA, GOOD-BYE CICHLIDS Finding better ways to manage our food supplies is essential, given the astounding growth rate of the human population. Such efforts are well intentioned, but they can have disastrous consequences when ecological principles are not taken into account. For example, several years ago, someone thought it would be a great idea to introduce the Nile perch into Lake Victoria in East Africa. People had been using simple, traditional methods of fishing there for thousands of years (Figure 47.17). Now they were taking too many fish. Soon there would be too few fish to feed the local populations and no excess catches to sell for profit. But Lake Victoria is a very big lake, and the Nile perch is a very big fish (more than 2 meters long). A big fish in a big lake seemed an ideal combination to attract commercial fishermen from the outside world, with their big, elaborate nets—right? Wrong.

Native fishermen had been harvesting native fishes called cichlids, which eat mostly detritus and aquatic plants. The Nile perch eats other fish—including cichlids. Having had no prior evolutionary experience with the new predator, the 200 coexisting species of cichlids that were native to Lake Victoria had no defenses against it.

And so the Nile perch ate its way through the cichlid populations and destroyed the lake's biodiversity. Dozens of cichlid species found nowhere else are extinct. Without the cichlids to clean up the lake bottom, levels of dissolved oxygen plummeted and contributed to frequent fish kills. By 1990, fishermen were catching mostly Nile perch. And now there are signs the Nile perch population is about to crash. By destroying its food source, the Nile perch has undercut its own population growth. It has ceased to be a potentially large, exploitable food source for the people who live around the lake.

As if that weren't enough, the Nile perch is an oily fish. Unlike cichlids, which can be sun dried, the Nile perch must be preserved by smoking—and smoking requires firewood. The local people started cutting down more trees in the local forests, and trees are not rapidly renewable resources. To add insult to injury, the people living near Lake Victoria never liked to eat Nile perch anyway. They prefer the flavor and texture of cichlids.

What is the lesson? A little knowledge and some simple experiments in a contained setting could have prevented the whole mess at Lake Victoria.

THE RABBITS THAT ATE AUSTRALIA During the 1800s, British settlers in Australia just couldn't bond with koalas and kangaroos, so they started to import familiar animals from their homeland. In 1859, in what would be the start of a wholesale disaster, a landowner in northern Australia imported and released two dozen wild European rabbits (*Oryctolagus cuniculus*). Good food and good sport hunting, that was the idea. An ideal rabbit habitat with no natural predators—that was the reality.

Six years later, the landowner had killed 20,000 rabbits and was besieged by 20,000 more. The rabbits displaced livestock, even the kangaroos. And now Australia has 200 million to 300 million rabbits hippityhopping through the southern half of the country. They overgraze perennial grasses in good times and strip bark from shrubs and trees during droughts. You know where they've been; they transform grasslands and shrublands into eroded moonscapes or deserts (Figure 47.18).

Rabbits have been shot and poisoned. Their warrens have been plowed under, fumigated, and dynamited. Even when all-out assaults reduced their population size by 70 percent, the rapidly reproducing imports made a comeback in a year. And did the construction of a 2,000-mile-long fence protect western Australia? No. Rabbits made it to the other side before workers completed the fence.

In 1951, government researchers introduced the myxoma virus by way of mildly infected

Figure 47.17 Native fishermen at the shore of Lake Victoria, East Africa.

Figure 47.18 Part of the fence against 200–300 million rabbits that are destroying Australia's vegetation.

South American rabbits, its normal hosts. The virus causes *myxomatosis*. This disease has mild effects on the South American rabbits that coevolved with the virus, but it nearly always had lethal effects on *O. cuniculus*. Biting insects, mainly mosquitoes and fleas, quickly transmit the virus from host to host. With no coevolved defenses against the virus, the European rabbits died in droves.

As you might expect, however, natural selection has favored the rapid growth of *O. cuniculus* populations that are resistant to the virus.

In 1991, on an uninhabited island in Spencer Gulf, Australian researchers released a population of rabbits that they had injected with a calicivirus. The rabbits died quickly and relatively painlessly from blood clots in their lungs, heart, and kidneys. In 1995, the test virus escaped from the island, maybe on insect vectors. It has been killing between 80 and 95 percent of the adult rabbits in various parts of Australia. At this writing, researchers are questioning whether the calicivirus should be used on a widespread scale, whether it can jump boundaries and infect animals other than rabbits (such as humans), and what the long-term consequences will be.

Meanwhile, the Anti-Rabbit Research Foundation of Australia would very much like to get rid of the cult of the Easter Bunny. Their replacement candidate is the Easter Bilby, the bilby being a marsupial that evolved in Australia, that does not reproduce nearly as rapidly as *O. cuniculus*, and that does not blitz the countryside.

THE PLANT THAT ATE GEORGIA Imported animals and pathogens are not the only sources of ecological disaster. Consider the vine called kudzu (*Pueraria lobata*). Kudzu was deliberately brought over from Japan to the United States, where it faces no serious threats from herbivores, pathogens, or competitor plants.

In temperate regions of Asia, kudzu is a well-behaved legume with a well-developed root system. Somehow, it

Figure 47.19 Kudzu (*Pueraria lobata*) taking over part of Lyman, South Carolina. You can now find kudzu vines from East Texas to Florida and as far north as Pennsylvania.

seemed like a good idea to import it and use it for erosion control on hillsides and near highways in the southeastern United States. However, with nothing to stop it, kudzu's shoots grow one-third of a meter (1 foot) per day. Vines now blanket streambanks, trees, telephone poles, houses, hills, and almost everything else in its path. Attempts to dig them up or burn them do no good. Grazing goats and herbicides can make a dent, but goats are indiscriminate eaters and herbicides can contaminate water supplies. Someone, obviously exasperated with the menace, dubbed kudzu "the plant that ate Georgia" (Figure 47.19). If the global temperature continues to rise, kudzu could reach the Great Lakes by the year 2040.

On the bright side, a Japanese firm is constructing a kudzu farm and processing plant in Alabama. Asians use a starch extract from kudzu in beverages, candy, and herbal medicines. The idea is to export the starch to Asia, where the demand currently exceeds the supply. (And perhaps this gives new meaning to the expression, What comes around, goes around.) Also, kudzu might eventually help reduce the extent of logging operations. At the Georgia Institute of Technology, researchers have reported that kudzu may be useful as an alternative source of paper.

As you might well conclude from your consideration of the preceding discussions, biodiversity differs greatly from one habitat to another. In addition, as many field investigations by ecologists have revealed, biodiversity differs greatly between geographic regions.

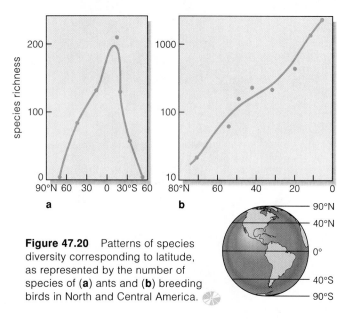

Figure 47.20 Patterns of species diversity corresponding to latitude, as represented by the number of species of (**a**) ants and (**b**) breeding birds in North and Central America.

● MOSSES
● VASCULAR PLANTS

Figure 47.21 Aerial photograph of Surtsey, a volcanic island, at the time of its formation in the sea. Newly formed, isolated islands are natural laboratories for ecologists. The chart indicates the number of colonizing species of mosses and vascular plants recorded on Surtsey between 1965 and 1973.

Mainland and Marine Patterns

The most striking pattern of biodiversity corresponds to distance from the equator. For most groups of plants and animals, *the number of coexisting species on land and in the seas is greatest in the tropics, and it systematically declines from the equator to the poles.* Figure 47.20 shows two examples of this biodiversity pattern. What factors underlie it? Three are paramount.

First, as described in Section 49.1, tropical latitudes intercept more sunlight of consistently greater intensity, and rainfall is heavier, so the growing season is longer. As a result, *resource availability tends to be higher and more reliable in the tropics than elsewhere.* All year long, different tree species in humid tropical forests put out new leaves, flowers, and fruit. Year in and year out, they support many diverse species of herbivores, nectar foragers, and fruit eaters. Specializations cannot evolve to a comparable degree in temperate and arctic regions.

Second, *species diversity might be self-reinforcing.* The diversity of tree species in tropical forests is far greater than in comparable forests at higher latitudes. When more plant species compete and coexist, more herbivore species also evolve and coexist, partly because no single herbivore can overcome the chemical defenses of all of the different plants. Typically, then, more predators and parasites evolve in response to the diversity of prey and hosts. The same is true of diversity on tropical reefs.

Third, evolutionary history tells us this: *The rates of speciation in the tropics have exceeded those of background extinction.* Decreases in biodiversity have occurred, but mainly during mass extinctions. As you read in Section 19.5, global temperatures drop at such times. Species already adapted to cool climates survive; those in the tropics simply have nowhere to go. In addition, millions of species in tropical forests may disappear in the next decade, for reasons described in chapters to follow.

Island Patterns

Islands often serve as terrific laboratories for studying biodiversity. For example, in 1965, a volcanic eruption formed a new island southwest of Iceland. Within six months, bacteria, fungi, seeds, flies, and some seabirds were established on it. One vascular plant appeared after two years, and a moss two years after that (Figure 47.21). As soils improved, the number of plant species increased. All species were colonists from Iceland. None

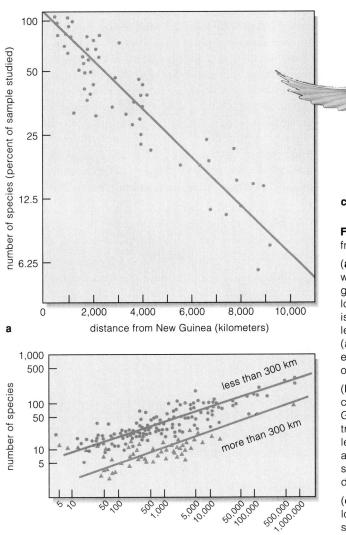

a distance from New Guinea (kilometers)

b area (square kilometers)

c

Figure 47.22 Examples of island biodiversity, correlated with distance from a source of colonizing species and with the island's area.

(**a**) Distance effect. Biodiversity on islands of a specified size declines with increasing distance from the source of colonizing species. Each graph point denotes the number of species of land birds occupying lowland areas of South Pacific islands. The large island of New Guinea is the source of colonists for the smaller islands, each of which is at least 500 kilometers away. To correct for differences in island size (area), the number of bird species on each of the distant islands is expressed as a percentage of the number of bird species on an island of equivalent size close to New Guinea.

(**b**) Area effect. Among islands similarly distant from their source of colonizing species, the larger ones support larger numbers of species. Graph points denote the number of bird species on subtropical and tropical islands. *Solid green* circles and the upper line are for islands less than 300 kilometers from the source of colonists. *Orange* triangles and the lower line denote islands more than 300 kilometers from the source areas. As you can see, this graph incorporates data on the distance effect as well as on the area effect.

(**c**) One travel agent for jump dispersals. Seabirds that island-hop over long distances commonly have a few seeds stuck to their feathers, as shown here. If those seeds successfully germinate in a new island community, they will give rise to a population of new immigrants.

originated on the island, which was named Surtsey. Yet the number of new species will not increase indefinitely. Why not? Studies of community patterns on islands around the world provide us with two models.

First, islands far from a source of potential colonists receive few colonizing species, and the few that arrive are adapted for long-distance dispersal (Figure 47.22). This is the **distance effect**. Second, larger islands tend to support more species than smaller islands at equivalent distances from source areas. This is the **area effect**. The larger islands tend to have more and varied habitats; most have more complex topography and extend farther above sea level. Thus they favor species diversity. Also, being bigger targets, they may intercept more colonists.

Most importantly, extinctions suppress biodiversity on small islands. There, the small populations are far more vulnerable to storms, volcanic eruptions, diseases, and random shifts in birth and death rates. As for any island, the number of species reflects a balance between

immigration rates for new species and extinction rates for established ones. Small islands that are distant from a source of colonists have low immigration rates and high extinction rates, so they support few species once the balance has been struck for their populations.

For most groups of organisms, the number of coexisting species is highest in the tropics and systematically declines from the equator to the poles.

A region's biodiversity depends on many factors, such as its climate, topographical variation, possibilities for dispersal, and evolutionary history—including extinctions and the time available for speciation.

Habitat disturbances tend to work against competitive exclusion and therefore favor increases in biodiversity.

Biodiversity on islands represents a balance between the immigration rates for new species and extinction rates for established species.

SUMMARY

1. A habitat is the type of place where individuals of a given species normally live—that is, their "address." A community consists of all populations of all species that occupy a habitat and that directly or indirectly associate with one another.

2. Each species has its own niche, or "profession," in the community. This is defined as the sum of activities and relationships in which its members engage as they secure and use the resources they require to survive and reproduce.

3. Mutualism, commensalism, competition, predation, and parasitism are species interactions that directly or indirectly link the populations in a community.

4. Two species that require the same limited resource tend to compete, as by using the resource as rapidly or efficiently as possible or by interfering with use of it.

   a. According to the competitive exclusion concept, if two (or more) species require identical resources, they cannot coexist indefinitely.

   b. Species are more likely to coexist if they differ in their use of resources. Also, they may coexist by using a shared resource in different ways or at different times.

5. Some predator and prey populations may coexist at stable levels if predators keep prey from overshooting the carrying capacity. Others show recurring or erratic cycles of abundance and crashes. Factors contributing to such cycles include delays in predator responses to changes in prey density, and changes in the prey's food supply.

6. Predators and their prey coevolve. After some novel, heritable trait appears and gives prey an edge in their contest, selection pressure operates on the population of predators, which also may evolve in response. The same happens when a novel, heritable trait appears in the predator population.

   a. Evolved prey defenses include threat displays, chemical weapons, mimicry, and camouflaging.

   b. Predators overcome prey defenses by adaptive behavior (including stealth), camouflaging, and so on.

7. Parasites and their hosts coevolve in ways that favor resistant hosts and only moderately harmful parasites.

8. By the classical model of ecological succession, a community develops in predictable sequence, from its pioneer species to an end array of species that persists over an entire region.

9. A stable, self-perpetuating array of species that are in equilibrium with one another and with a particular environment is called a climax community.

10. Similar climax stages may persist within the same region, yet show variation as a result of environmental gradients and species interactions.

11. Recurring, small-scale changes are a part of the internal dynamics of communities. Other changes, such as long-term shifts in climate or species introductions, may permanently alter the community configuration.

   a. Community structure represents an uneasy balance of forces, such as predation (as by keystone species) and competition, that have been operating over time.

   b. Community structure can change permanently by the introduction of species that have expanded their geographic range. Dispersals might occur slowly, as when a population gradually expands its home range. Jump dispersals rapidly put individuals into distant habitats. Extremely slow dispersals also have proceeded over evolutionary time, as by continental drift.

12. The number of species in a community depends on the size of a region, the colonization rate, disturbances, and extinction rates. It depends also on the level and pattern of resource availability.

13. Biodiversity tends to be highest in the tropics and to systematically decline toward polar regions, except during times of mass extinction.

Review Questions

1. What is the difference between the habitat and the niche of a species? Why is it difficult to define "the human habitat"? *47.1*

2. Describe competitive exclusion. How might two species that compete for the same resource coexist? *47.3*

3. Define the difference between a predator and a parasite. *47.4*

4. Define primary and secondary succession. *47.7*

5. What is a climax community? How does the climax-pattern model help explain its structure? *47.7*

Self-Quiz *(Answers in Appendix IV)*

1. A habitat _____ .
   a. has distinguishing physical and chemical features
   b. is where individuals of a species normally live
   c. is occupied by various species
   d. both a and b
   e. a through c

2. A two-way flow of benefits in mutualistic interactions between species is an outcome of _____ .
   a. close cooperativeness
   b. two-way exploitation
   c. resource partitioning
   d. competitive coexistence

3. A niche _____ .
   a. is the sum of activities and relationships in a community by which individuals of a species secure and use resources
   b. is unvarying for a given species
   c. shifts in large and small ways
   d. both a and b
   e. both a and c

4. Two species in the same habitat can coexist when they _____ .
   a. differ in their use of resources
   b. share the same resource in different ways
   c. use the same resource at different times
   d. all of the above

5. A predator population and prey population _____ .
   a. always coexist at relatively stable levels
   b. may undergo cyclic or irregular changes in density
   c. cannot coexist indefinitely in the same habitat
   d. both b and c

6. All parasites _____ .
   a. tend to kill their hosts   c. consume host tissues
   b. can kill novel hosts   d. both b and c

7. In _____ , a disturbed site in a community recovers and moves again toward the climax state.
   a. the area effect   c. primary succession
   b. the distance effect   d. secondary succession

8. The number of coexisting species is _____ in the tropics and _____ toward the poles.
   a. highest; declines systematically
   b. lowest; increases systematically
   c. highest; declines sporadically
   d. lowest; increases sporadically

9. Match the terms with the most suitable descriptions.
   ____ jump dispersal   a. opportunistic colonizer of barren or disturbed places
   ____ area effect
   ____ pioneer species   b. dominates community structure
   ____ climax   c. rapid transport over great distance
        community   d. more biodiversity on large islands than small ones at same distance from source areas
   ____ keystone
        species   e. stable, self-perpetuating array of species

Figure 47.24 Water hyacinths choking a waterway in Florida.

Critical Thinking

1. Think of possible examples of competitive exclusion besides the ones used in the chapter, as by considering some of the animals and plants living in your own neighborhood.

2. Flesh flies (Figure 47.23a) have a gray and black body, red eyes, and red tail ends. They are fast fliers, and bird predators soon give up trying to catch them. Many sluggish insects, such as the weevil *Zygops rufitorquis*, resemble flesh flies (Figure 47.23b). This appears to be a case of _____ .

3. The water hyacinth (*Eichhornia crassipes*) is an aquatic plant native to South America. In the 1880s, someone fancied the plant's blue flowers and displayed it at an exposition in New Orleans. Other flower fanciers took home clippings and put them in ponds and streams. Unchecked by natural predators, the fast-growing hyacinths spread through nutrient-rich waters, displaced many native species, and choked rivers and canals (Figure 47.24). They have spread as far west as San Francisco. Do some research to see whether the United States now limits imports and, if so, how effective the restrictions have been.

Selected Key Terms

area effect (on island biodiversity) *47.10*	macroparasite *47.6*
camouflage *47.5*	microparasite *47.6*
climax community *47.7*	mimicry *47.5*
climax-pattern model *47.7*	mutualism *47.1*
coevolution *47.4*	niche *47.1*
commensalism *47.1*	parasite *47.4*
community *47.1*	parasitism *47.1*
competitive exclusion *47.3*	parasitoid *47.6*
distance effect (on island biodiversity) *47.10*	pioneer species *47.7*
	predation *47.1*
ecological succession *47.7*	predator *47.4*
exotic species *47.9*	prey *47.4*
geographic dispersal *47.8*	primary succession *47.7*
habitat *47.1*	resource partitioning *47.3*
host *47.4*	secondary succession *47.7*
interspecific competition *47.1*	social parasite *47.6*
jump dispersal *47.8*	symbiosis *47.1*
keystone species *47.8*	warning coloration *47.5*

Readings

Begon, M., et al. 1990. *Ecology: Individuals, Populations, and Communities.* Second edition. Sunderland, Massachusetts: Sinauer.

Krebs, C. 1994. *Ecology.* Fourth edition. New York: Harper-Collins.

Krebs, C., et al. 25 August 1995. "Impact of Food and Predation on the Snowshoe Hare Cycle." *Science* 269: 1112–1115.

Moore, P. 1987. "What Makes a Forest Rich?" *Nature* 329: 292.

Smith, R. 1996. *Elements of Ecology.* Fifth edition. New York: Harper-Collins.

Worthington, E. B. 1994. "African Lakes Reviewed: Creation and Destruction of Biodiversity." *Environmental Conservation,* 201–204.

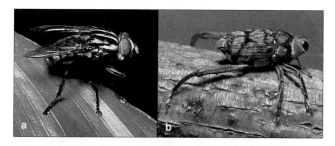

Figure 47.23 (**a**) A flesh fly and (**b**) a weevil (*Zygops*).

Web Site See *http://www.wadsworth.com/biology* for practice quiz questions, hypercontents, BioUpdates, and critical thinking. The Wadsworth Biology Resource Center provides a wealth of information fully organized and integrated by chapter.

48 ECOSYSTEMS

Crêpes for Breakfast, Pancake Ice for Dessert

Think of Antarctica, and you think of ice. Mile-thick slabs of the stuff hide all but a small fraction of a vast continent whipped by fierce winds and kept frozen by murderously low temperatures, on the order of −100°F. Yet, on patches of exposed rocky soil and on nearby islands, mosses and lichens grow. There, during the breeding season, a variety of penguins and seals form great noisy congregations, then reproduce and raise offspring. They cruise offshore or venture out in the open ocean in pursuit of krill, fishes, and squids.

The first explorers of Antarctica called it "The Last Place on Earth." For all of its eerie harshness, however,

Antarctica's unspoiled beauty fired the imagination of those first travelers and others who followed them.

In 1961, thirty-eight nations signed a treaty to set aside Antarctica as a reserve for scientific research. This was the start of scientific outposts—and of the trashing of Antarctica. At first, the researchers discarded a few oil drums and old tires. Then prefabricated villages went up. Onto the ice and into the water went garbage and used equipment. Just offshore, marine life became acquainted with sewage and chemical wastes.

Antarctica also became a destination of cruise ships (Figure 48.1a). Every summer thousands of tourists,

a

Figure 48.1 (a) A boatload of tourists crossing a channel of "pancake ice" off the coast of Antarctica.
(b) On the shore of an island near Antarctica, tourists get close to nature—maybe too close.

fortified by three sumptuous meals a day plus snacks, are ferried from ship to land across channels glistening with pancake ice. They trample the sparse vegetation and bob around the penguins, cameras clicking. We do not know what the long-term impact of tourism will be, but it pales in comparison to other vulnerabilities. Nations looking for new sources of food started licking their chops over Antarctica's krill. Word got out about potentially rich deposits of uranium, oil, and gold. By 1988, treaty nations were poised to authorize digs and drillings. Of course, oil spills or commercial krill harvesting could easily destroy the fragile ecosystem. For example, the tiny, shrimplike krill are all that Adélie penguins eat. They are a key food resource for baleen whales and other marine animals that are, in turn, food for still others.

In 1991, the treaty nations thought about all of this and imposed a fifty-year ban on mineral exploration. Research stations started to bury or incinerate wastes, treat raw sewage, and take other steps to curb pollution. Tour operators promised to supervise tourists. Krill are not being harvested on a massive scale. Yet.

Because Antarctica seems so remote from the rest of the world, it is easier to see how life is interconnected there. We can ask whether harm to one species or one habitat will lead to collapse of the whole, and be fairly sure of the answer. What about places not as sharply defined? Does interconnectedness prevail there, also? Are other places as vulnerable to disturbance or more resilient? These topics will now occupy our attention.

b

KEY CONCEPTS

1. An ecosystem is an association of organisms and their physical environment, interconnected by an ongoing flow of energy and a cycling of materials through it.

2. Every ecosystem is an open system, in that it has inputs and outputs of both energy and nutrients.

3. Over time, energy flows in only one direction through an ecosystem. Most commonly, energy flow begins when photosynthetic autotrophs harness sunlight energy and convert it to forms that they and other organisms of the ecosystem are capable of using. Autotrophs are primary producer organisms for the ecosystem.

4. Energy-rich organic compounds that the primary producers synthesize become incorporated in their body tissues. They are stored forms of energy, and they serve as the foundation for the ecosystem's food webs. Such webs consist of a number of interconnected food chains.

5. Each chain in a food web extends in a straight-line sequence from producers through arrays of consumers, decomposers, and detritivores. And it cross-connects with other chains at different feeding levels.

6. The water, carbon, nitrogen, phosphorus, and other substances required for primary productivity move through biogeochemical cycles that are global in scale. Ions or molecules of the substances move slowly through the environment, then rapidly among organisms of food webs, then back to the environmental reservoir for them.

7. Depending on which part of the environment serves as its largest reservoir, each substance moves through a hydrologic, atmospheric, or sedimentary cycle.

8. Human activities are disrupting the natural cycles of nature and so are endangering ecosystems.

THE NATURE OF ECOSYSTEMS

Overview of the Participants

Diverse natural systems abound on the Earth's surface. In climate, landforms, soil, vegetation, animal life, and other features, deserts differ from hardwood forests, which differ from tundra and prairies. In biodiversity and physical properties, seas differ from reefs, which differ from lakes. *Yet despite the differences, such systems are alike in many aspects of their structure and function.*

With few exceptions, each system runs on energy that plants and other photosynthesizers capture from the sun. Photosynthesizers, recall, are the most common autotrophs (self-feeders). They convert sunlight energy to chemical energy, which they use to construct organic compounds from inorganic raw materials. By securing energy from the environment, autotrophs are **primary producers** for the entire system (Figure 48.2).

All other organisms in the system are heterotrophs, not self-feeders. They extract energy from compounds that primary producers put together. **Consumers** feed on the tissues of other organisms. The consumers called *herbivores* eat plants, *carnivores* eat animals, *omnivores* eat both, and *parasites* extract energy from living hosts that they live in or on. Still other heterotrophs, called **decomposers**, include fungi and bacteria that obtain energy by breaking down the remains or products of organisms. Others, the **detritivores**, obtain nutrients from decomposing particles of organic matter. Crabs and earthworms are examples of detritivores.

Autotrophs secure nutrients as well as energy for the entire system. During growth, they take up water and carbon dioxide from the environment (as sources of oxygen, carbon, and hydrogen) and dissolved minerals (such as nitrogen and phosphorus). Such materials are building blocks for their carbohydrates, lipids, proteins, and nucleic acids. When

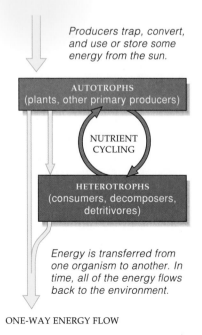

Producers trap, convert, and use or store some energy from the sun.

Energy is transferred from one organism to another. In time, all of the energy flows back to the environment.

ONE-WAY ENERGY FLOW

Figure 48.2 Model of an ecosystem. Energy flows in only one direction, into the ecosystem, through its organisms, and out from it. Nutrients typically are cycled among the arrays of autotrophs and heterotrophs of the ecosystem.

decomposers and detritivores degrade organic matter to small inorganic compounds, they release nutrients to the surroundings. Unless something removes nutrients from the system, as by a creek taking dissolved minerals away from a meadow, autotrophs typically reuse them.

What we have just described in broad outline is an ecosystem. An **ecosystem** is an array of organisms and the physical environment, interacting through a one-way flow of energy and a cycling of materials. It is an open system, unable to sustain itself. Each ecosystem runs on *energy input* (as from the sun) and often *nutrient inputs* (as from a creek that delivers dissolved minerals into a lake). It also has *energy outputs* and *nutrient outputs*. Energy cannot be recycled; in time, most of the energy that autotrophs fix is lost to the environment, mainly as metabolically generated heat. Nutrients are cycled, but some still slip away. Most of this chapter deals with the inputs, internal transfers, and outputs of ecosystems.

Structure of Ecosystems

TROPHIC LEVELS   The organisms of an ecosystem can be classified in terms of their functions in a hierarchy of feeding relationships, called **trophic levels** (from *troph*, meaning nourishment). "Who eats whom?" we can ask. As organism B eats organism A, energy gets transferred to B from A. All organisms at a given trophic level are the same number of transfer steps away from an energy input into the ecosystem. For example, Table 48.1 lists organisms of a forest ecosystem. Being closest to the energy input into this system (sunlight), trees and other primary producers are at the first trophic level. Primary

Table 48.1 Trophic Levels in a Deciduous Forest		
Type of Organism at Trophic Level	Energy Source	Examples of Organisms
FIRST TROPHIC LEVEL		
Primary producers:		
Photoautotrophs	Sunlight	Green algae, trees
Chemoautotrophs	Inorganic substances	Nitrifying bacteria
SECOND TROPHIC LEVEL		
Primary consumers: *Herbivores, decomposers, and detritivores*	Primary producers	Moth larvae, earthworms
THIRD TROPHIC LEVEL		
Secondary consumers: *Primary carnivores*	Primary consumers	Spiders, mice, beetles
FOURTH TROPHIC LEVEL		
Tertiary consumers: *Secondary carnivores and carnivore parasites*	Secondary consumers	Owls, weasels, parasitic flies

consumers, which feed directly on the primary producers, include leaf-eating larvae and other herbivores. They also include detritivores, such as earthworms that feed on organic debris. Such primary consumers are at the second trophic level. At the third trophic level of this forest are primary carnivores (arrays of beetles, spiders, and birds) that prey on the primary consumers. At the fourth level are secondary carnivores, such as owls, that feed on the species of the third trophic level.

By this classification scheme, all of the organisms at a given trophic level interact with the same sets of predators, prey, or both. Remember this, however: Many decomposers, people, and other omnivores feed at several trophic levels, so they must be partitioned among different levels or assigned one of their own.

FOOD WEBS Often a straight-line sequence of who eats whom in an ecosystem is called a **food chain**. We have a hard time finding such simple, isolated cases. Why? Most often, the same species belongs to more than one chain, especially when it is at low trophic levels. Therefore, it is more accurate to visualize food chains as *cross-connecting* with one another, as **food webs**. Figure 48.3 shows key players in a food web in the seas around Antarctica.

A bad-luck story about a fisherman can clarify the difference between a food chain and food web. Suppose a fisherman nets fish that were feeding on algae near the ocean surface. At lunchtime, he cooks some fish. He later loses his footing and falls into the water, where sharks lurk. You may think this is a simple food chain: algae ⟶ fish ⟶ fisherman ⟶ shark. But alternate feeding relationships cut into it. Crustaceans also were grazing on the algae. Small squids and midsize fishes were feeding on the crustaceans. And larger fishes were feeding on smaller ones. Sharks may have been about to feed on those larger and midsize fishes. The fisherman

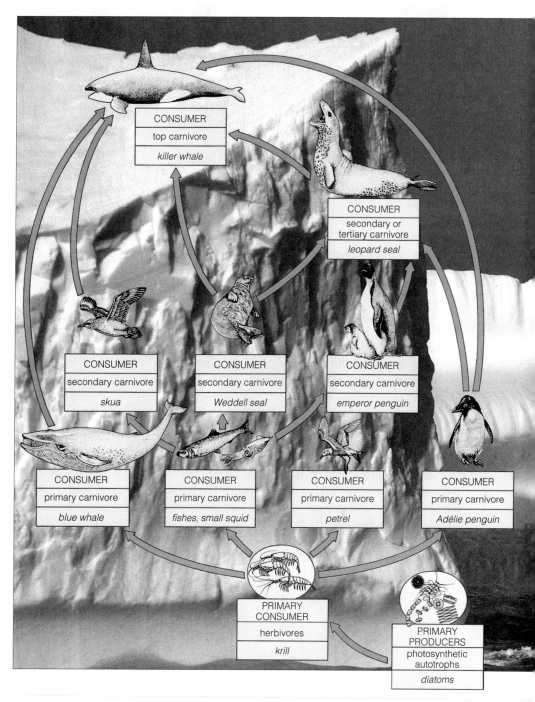

Figure 48.3 Food web in the waters of the Antarctic. Many more participants, including decomposers, are not shown.

ate cooked fish, garlic, and onions and therefore shifted between trophic levels (as carnivore and herbivore). He was even more omnivorous than this, for he sipped wine, a beverage derived from the fermenting activities of the decomposer organisms called yeasts.

An ecosystem consists of producers, consumers, decomposers, and detritivores and the physical environment, connected by a one-way energy flow and a cycling of materials.

A food web is a network of crossing, interlinked food chains involving primary producers, consumers, and decomposers.

Primary Productivity

To get an idea of how energy flow is studied, focus on an ecosystem on land. Such ecosystems typically have multicelled plants as the primary producers. The rate at which the primary producers capture and store a given amount of energy in their tissues during a given time interval is the **primary productivity** for the ecosystem. How much energy actually gets stored depends on how many plants live there. It also depends on the overall balance between photosynthesis (energy trapped) and aerobic respiration (energy used).

Other factors influence the amount of net primary production, its seasonal patterns, and its distribution through a given habitat. They do so in ecosystems in the seas as well as on land (Figure 48.4). For example, the body size and form of primary producers affect how much energy they can trap and store. So do availability of minerals, the temperature range, and the amount of sunlight and rainfall during each growing season. The more harsh the environment, the less new growth on plants—and the lower the productivity.

Major Pathways of Energy Flow

In what direction does energy flow through ecosystems on land? Plants fix only a small part of the energy from the sun. They store half of that in new tissues but lose the rest as metabolic heat. Other organisms tap into the energy stored in plant tissues, remains, or wastes. They, too, lose heat to the environment. *All of these heat losses represent a one-way flow of energy out of the ecosystem.*

Energy from a primary source flows in one direction through two categories of food webs. In **grazing food webs**, the energy flows from photosynthetic organisms to herbivores, then through an array of carnivores. By contrast, in **detrital food webs**, energy flows primarily from photosynthetic organisms through detritivores and decomposers. In most ecosystems, the two kinds of food webs cross-connect. For instance, this happens when a herring gull of a grazing food web opportunistically devours a crab of a detrital food web (Figure 48.5).

The amount of energy moving through food webs differs from one ecosystem to the next and often varies with the seasons. In most cases, however, most of the net primary production passes through the detrital food webs. You may doubt this. After all, when cattle graze heavily on pasture plants, about half the net primary production enters a grazing food web. But cattle do not use all the stored energy. Quantities of undigested plant parts and feces become available for decomposers and detritivores. Or consider a coastal marsh. There, most of the stored energy is not used until parts of marsh grasses die and become available for detrital food webs.

Ecological Pyramids

Often ecologists will represent the trophic structure of an ecosystem in the form of an **ecological pyramid**. In such pyramids, the primary producers form a base for successive tiers of consumers above them.

Some pyramids are based on biomass, or the weight of all the members at each trophic level. For example, for Silver Springs, Florida (a small aquatic ecosystem), a

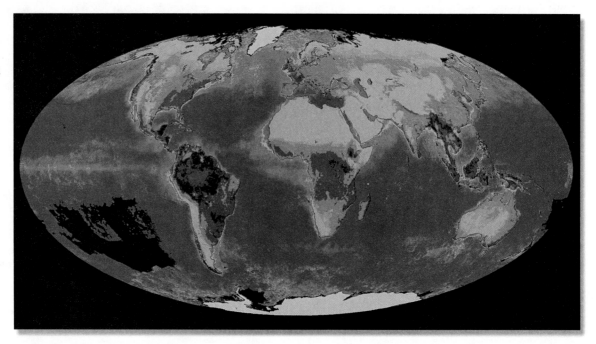

Figure 48.4 A summary of three years of satellite data on the Earth's primary productivity. *Dark green* denotes rain forests and other highly productive regions. *Yellow* denotes deserts (regions of low productivity). The productivity in oceans, ranging from high to low, is color-coded *red* down through *orange*, *yellow*, *green*, and *blue*.

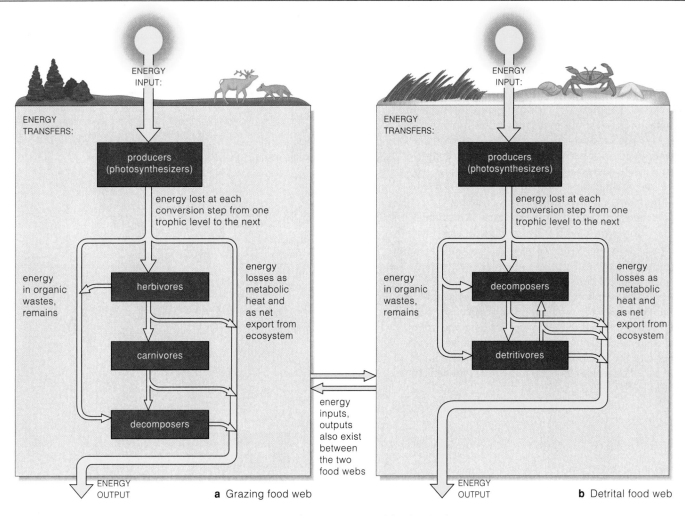

Figure 48.5 One-way flow of energy through two kinds of cross-connected food webs in ecosystems.

pyramid of biomass (measured as grams/square meter during one specified interval) would look like this:

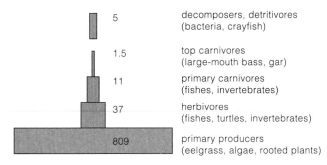

5	decomposers, detritivores (bacteria, crayfish)
1.5	top carnivores (large-mouth bass, gar)
11	primary carnivores (fishes, invertebrates)
37	herbivores (fishes, turtles, invertebrates)
809	primary producers (eelgrass, algae, rooted plants)

This is a typical pyramid of biomass. Some others are "upside-down," with the smallest tier on the bottom. A small pond is like this. Its biomass of producers consists of fast-growing, rapidly reproducing phytoplankton. It supports a much greater biomass of zooplankton in which individuals are bigger, grow slower, and consume less energy per unit of weight.

An **energy pyramid** is a more useful way to depict an ecosystem's trophic structure. Such pyramids show the energy losses at each transfer to a different trophic level in the ecosystem. They have a large energy base at the bottom and are always "right-side up." As you will see from the next section, they provide a better picture of how energy flows in ever diminishing amounts through successive trophic levels of an ecosystem.

Energy flows into food webs of ecosystems from an outside source, usually the sun. Energy leaves ecosystems mainly by losses of metabolic heat, which each organism generates.

Gross primary productivity is an ecosystem's total rate of photosynthesis during a specified interval. The *net* amount is the rate at which primary producers store energy in tissues in excess of their rate of aerobic respiration. Heterotrophic consumption affects the rate of energy storage.

Tissues of living photosynthesizers are the basis of grazing food webs. Remains and wastes of photosynthesizers and consumers are the basis of detrital food webs.

The loss of metabolic heat and the shunting of food energy into organic wastes mean that usable energy flowing through consumer trophic levels declines at each energy transfer.

Imagine you are with ecologists who are bent on gathering data to construct an energy pyramid for a small freshwater spring over the course of one year. You observe them as they measure the energy that each type of individual in the spring takes in, loses as metabolic heat, stores in its body tissues, and loses in waste products. You see that they multiply the energy per individual by population size, then they calculate energy inputs and outputs. In such ways, the ecologists are able to express the flow per unit of water (or land) per unit of time.

The energy pyramid shown in Figure 48.6 summarizes the data from a long-term study of a grazing food web in an aquatic ecosystem—Silver Springs, Florida. The larger diagram in Figure 48.7 shows some of the calculations that ecologists used to construct the pyramid.

Given the metabolic demands of organisms and the amount of energy shunted into their organic wastes, only about 6 to 16 percent of the energy entering one trophic level becomes available for organisms at the next level.

Because the efficiency of the energy transfers is so low, ecosystems in general have no more than four consumer trophic levels.

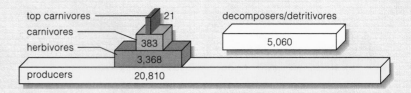

Figure 48.6 Pyramid of energy flow through Silver Springs, Florida, as measured in kilocalories/square meter/year.

Figure 48.7 Breakdown of the annual energy flow through Silver Springs, Florida, as measured in kilocalories/square meter/year.

The primary producers in this small spring are mostly aquatic plants. The carnivores are insects and small fishes; top carnivores are larger fishes. The energy source (sunlight) is available throughout the year. The spring's detritivores and decomposers cycle organic compounds from the other trophic levels.

The producers trapped 1.2 percent of the incoming solar energy, and only a little more than a third of this amount became fixed in new plant biomass (4,245 + 3,368). The producers used more than 63 percent of the fixed energy for their own metabolism.

About 16 percent of the fixed energy was transferred to herbivores, and most of it was used for metabolism or transferred to detritivores and decomposers.

Of the energy that did get transferred to herbivores, only 11.4 percent reached the next trophic level (carnivores). These carnivores used all but about 5.5 percent, which was transferred to top carnivores.

By the end of the specified time interval, all of the 5,060 kilocalories of energy that had been transferred through the system appeared as metabolically generated heat.

Bear in mind, this energy flow diagram is oversimplified, because no community is isolated from others. New individual organisms and substances continually drop from overhead leaves and branches into the springs. Also, organisms and substances are gradually lost by way of a stream that leaves the spring.

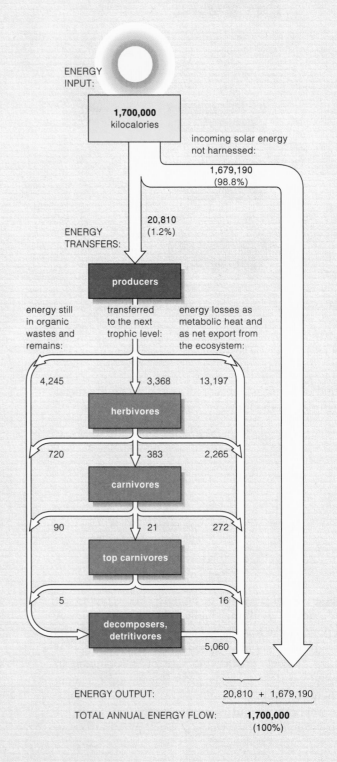

BIOGEOCHEMICAL CYCLES—AN OVERVIEW

Availability of *nutrients* as well as energy profoundly influences the structure of an ecosystem. Photosynthetic producers need carbon, hydrogen, and oxygen, which they get from water and air. They require nitrogen, phosphorus, and other mineral ions. A scarcity of even one of these minerals has widespread adverse effects, for it lowers the ecosystem's primary productivity.

In a **biogeochemical cycle**, ions or molecules of a nutrient are transferred from the environment to organisms, then back to the environment, part of which functions as a vast reservoir for them. The transfers of nutrients into and out of the reservoir are usually less than the exchanges made between and among organisms.

Figure 48.8 is a simple model of the relationship between the geochemical part of the cycles and most ecosystems. This model is based on four factors.

First, mineral elements that producer organisms use for nutrients usually are available in the form of mineral ions, such as ammonium (NH_4^+). *Second*, inputs from the physical environment, together with nutrient cycling activities of all of the decomposers and detritivores, maintain the ecosystem's nutrient reserves. *Third*, the actual amount of a nutrient that is being cycled on through most of the major ecosystems is greater than the amount that is entering and departing per year. *Fourth*, rainfall or snowfall and the gradual weathering of rocks are common sources of inputs to an ecosystem's nutrient reserves. So are the combined effects of metabolic events, such as nitrogen fixation.

Ecosystems on land also have typical outputs from the nutrient reserves. The loss of mineral ions by way of runoff from irrigated cropland is an example.

There are three categories of biogeochemical cycles, based on the part of the environment that contains the greatest portion of the specified ion or molecule. As you will see, in the *hydrologic* cycle, oxygen and hydrogen move in the form of water molecules. This movement is also called the global water cycle. In *atmospheric* cycles,

a large percentage of the nutrient is in the form of an atmospheric gas. For example, this is true of gaseous forms of nitrogen and carbon (mainly carbon dioxide). *Sedimentary* cycles deal with phosphorus and other solid nutrients that do not have gaseous forms. Such nutrients move from land to the seafloor and return to dry land only through geological uplifting, which may

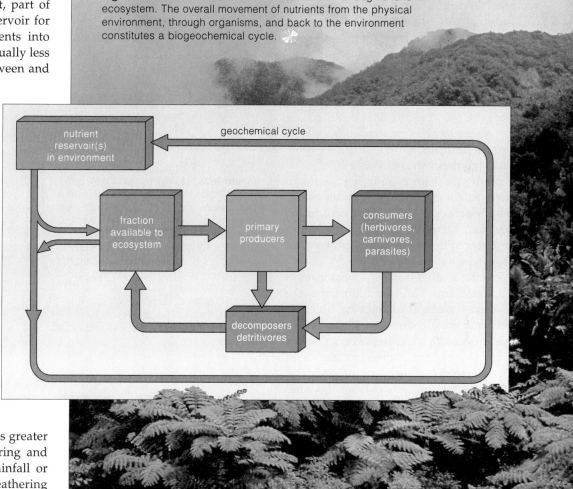

Figure 48.8 Generalized model of nutrient flow through a land ecosystem. The overall movement of nutrients from the physical environment, through organisms, and back to the environment constitutes a biogeochemical cycle.

take millions of years. The Earth's crust is the largest storehouse for sedimentary cycles.

Primary productivity, hence the structure of an ecosystem, depends largely on the availability of nutrients.

In a biogeochemical cycle, ions or molecules of a nutrient move slowly through the environment, then rapidly among organisms, then back to the environmental reservoir for them.

Driven by continual inputs of solar energy, the waters of the Earth move slowly, on a vast scale, from the ocean, into the atmosphere, to the land, and back to the ocean. The water that evaporates into the lower atmosphere remains aloft in the form of vapor, clouds, and crystals of ice. It returns to Earth in the form of precipitation—rain and snow, for the most part. Ocean currents, along with prevailing patterns of winds, have key roles in this global **hydrologic cycle**, which is sketched out in Figure 48.9.

Airborne water molecules remain aloft for ten days or less, on the average. Water reaching the land as rain or snow stays there for 10 to 120 days, with the actual length depending on the season and the location. From there, water evaporates or flows to the ocean, the main reservoir from which it evaporates.

Water in itself is vital for organisms of all ecosystems. But it also functions as a transport medium for moving nutrients into and out of ecosystems. Its key role in the movement of nutrients became clear through long-term studies in watersheds. A **watershed** is a region of any specified size in which all precipitation becomes funneled into a single stream or river. Figure 48.10 shows a simple model of the pathways that water can follow through a watershed.

A watershed may be any size that is of interest to an investigator. To give two very different examples, the Mississippi

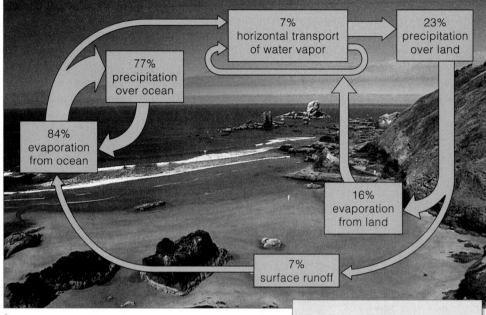

a

Figure 48.9 (**a**) Hydrologic cycle, with percentages of the total evaporation and total precipitation at a given time as well as the percentage of runoff from land to the ocean. The annual net rate of transfer is 37.3 × 10³ cubic kilometers of water to land from the atmosphere. Balancing this is a comparable net loss from the ocean to the atmosphere and a net gain of that amount by the ocean (through runoff).
(**b**) Global water budget. Values reflect the annual movement of water into and out of the atmosphere. ✿

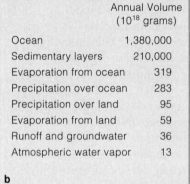

	Annual Volume (10¹⁸ grams)
Ocean	1,380,000
Sedimentary layers	210,000
Evaporation from ocean	319
Precipitation over ocean	283
Precipitation over land	95
Evaporation from land	59
Runoff and groundwater	36
Atmospheric water vapor	13

b

Figure 48.10 Model of the movement of water through a watershed. *Dark blue* indicates inputs to the watershed; *light blue*, distribution within the watershed; and *medium blue*, outputs from it. Transpiration, remember, is the evaporation of water from leaves and other plant parts that are exposed to air. The plants tap water in the soil and in groundwater stores.

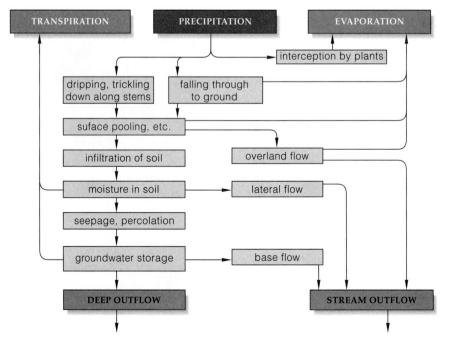

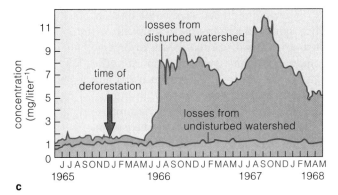

River watershed extends across approximately one-third of the continental United States. By contrast, in New Hampshire, watersheds at Hubbard Brook Valley in the White Mountains measure 14.6 hectares (36 acres), on the average.

Most of the water that enters a watershed seeps into the soil or becomes part of the surface runoff, which enters streams (Figure 48.10). Plants take up the water and dissolved minerals from the soil, then lose water by transpiration.

Measurements of inputs and outputs at watersheds have a number of practical applications. For example, cities that draw on the surface supplies in watersheds can adjust their water usage on the basis of seasonal variations in the volume of water. Watershed studies also revealed the importance of vegetation cover in the movement of nutrients through the ecosystem phase of biogeochemical cycles.

For example, you might think that water draining a watershed would rapidly leach calcium ions and other minerals. Yet in studies of young, undisturbed forests in the Hubbard Brook watersheds, each hectare lost only about 8 kilograms of calcium—and rainfall and the weathering of rocks brought in calcium replacements. Tree roots were also "mining" the soil, so calcium was being stored in a growing biomass of tree tissues.

In some experimental watersheds, nutrient outputs shifted. Researchers stripped the vegetation cover but did not disturb the soil. Then they applied herbicides to the soil for three years to prevent regrowth. After that time, stream outflow from the experimental watersheds carried more than *six times* as much calcium as the stream outflow from undisturbed ones (Figure 48.11c).

Because calcium and other nutrients move so slowly through geochemical cycles, deforestation may disrupt nutrient availability for an entire ecosystem. This is true

Figure 48.11 Studies of disturbances to a forest ecosystem. (**a**) In experimental watersheds in the Hubbard Brook Valley, New Hampshire, researchers studied effects of deforestation. (**b**) All the water that was to drain from an area being studied had to flow over the V-notched concrete structure, where measurements were taken. This area was deforested, then herbicides were applied to prevent regrowth for three years.

(**c**) The arrow marks the time of deforestation. Concentrations of calcium ions and other mineral ions in water that moved through this concrete catchment were compared against those in water passing through a control catchment in an undisturbed part of the same region. As you can see, calcium losses were six times greater from the deforested region.

of forests that cannot regenerate themselves over the short term. Coniferous forests of the Pacific Northwest and elsewhere are like this.

In the hydrologic cycle, water slowly moves on a global scale from the ocean reservoir, through the atmosphere, onto land, then back to the ocean.

In ecosystems on land, plants stabilize the soil and absorb dissolved minerals. By doing so, they minimize the loss of soil nutrients in runoff from land.

In one of the most vital of all atmospheric cycles, carbon moves from vast reservoirs —the ocean and atmosphere—on through organisms in ecosystems, then back to the reservoirs. Figure 48.12*a* shows the global movement, which is known as the **carbon cycle**.

Carbon enters the atmosphere every second. It does so as countless living cells engage in aerobic respiration, as fossil fuels burn, and as volcanoes erupt and release carbon from rocks of the Earth's crust. The world ocean holds most of the carbon, in dissolved form (Figure 48.12*b*). The atmosphere, soils, and plant biomass represent the next largest holding stations for carbon. Most atmospheric carbon is in the form of carbon dioxide (CO_2).

By engaging in carbon dioxide fixation, photosynthetic autotrophs lock up billions of metric tons of carbon atoms in organic compounds every year (Section 7.8). The average time that an ecosystem holds any one carbon atom varies greatly. In tropical rain forests, organic remains decompose rapidly, so not much carbon is tied up in litter on the soil surface. In marshes, bogs, and other anaerobic habitats, decomposers cannot break down organic compounds to smaller bits, so carbon slowly accumulates in compressed organic matter, such as peat.

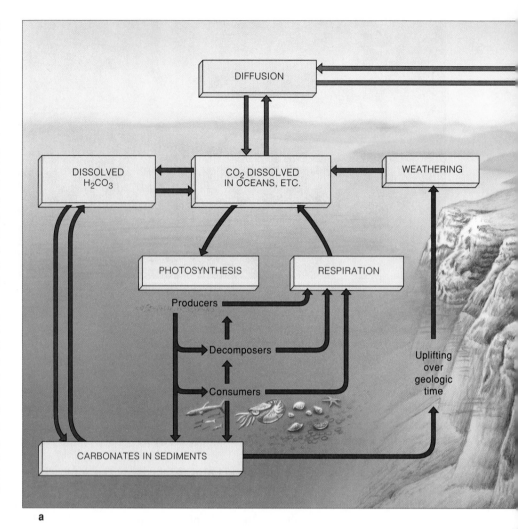

a

Figure 48.12 (**a**) Global carbon cycle. The portion of the diagram on this page illustrates the movement of carbon through typical marine ecosystems. The portion of the diagram on the facing page illustrates the movement of carbon through ecosystems on land.

(**b**) The present-day global carbon budget. The photograph to the far right (**c**) shows a stretch of a Los Angeles freeway system under its self-generated blanket of smog at twilight. Every day, particulate-laden exhaust from gasoline-burning vehicles as well as fossil fuel burning by industries and homes adds carbon and other substances to the atmosphere.

The values in (**b**) are reported in W. Schlesinger, 1991, *Biogeochemistry: An Analysis of Global Change*, New York: Academic Press.

	Amount per Year (10^{15} grams)
GLOBAL CARBON RESERVOIRS AND HOLDING STATIONS:	
Dissolved in ocean	38,000
Present in soil	1,500
Present in atmosphere	720
Plant biomass	560
ANNUAL FLUXES IN THE GLOBAL DISTRIBUTION OF CARBON:	
From atmosphere to plants (carbon fixation)	120
From atmosphere to ocean	107
To atmosphere from ocean	105
To atmosphere from plants	60
To atmosphere from soil	60
To atmosphere from fossil fuel burning	5
To atmosphere from net destruction of plants	2
To ocean from runoff	0.4
Burial in ocean sediments	0.1

b

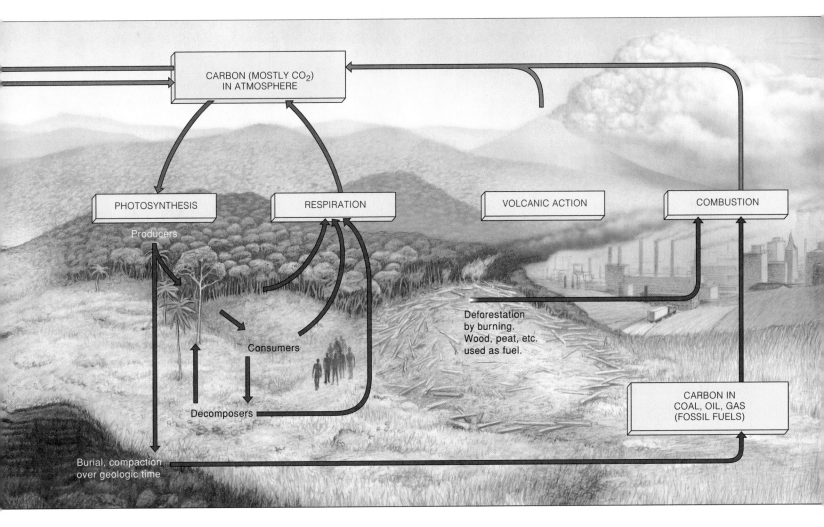

CARBON (MOSTLY CO₂) IN ATMOSPHERE

PHOTOSYNTHESIS

Producers

RESPIRATION

VOLCANIC ACTION

COMBUSTION

Consumers

Deforestation by burning. Wood, peat, etc. used as fuel.

Decomposers

CARBON IN COAL, OIL, GAS (FOSSIL FUELS)

Burial, compaction over geologic time

c

In food webs of ancient aquatic ecosystems, carbon became incorporated into shells and other hard parts. Later, when shelled organisms died, they sank through the water and became buried in sediments. In deeper sediments, carbon remained buried for many millions of years, until geologic forces raised part of the seafloor above the ocean surface (Section 20.2). Also over great time spans, the carbon-containing compounds of many ancient forests were converted to reserves of petroleum, coal, and gas—which we presently tap as fossil fuels.

Now fossil fuel burning and other human activities are putting more carbon into the atmosphere than can be cycled naturally to the ocean reservoir and to other holding stations (Figure 48.12*b*). Because this increase amplifies the greenhouse effect, it may be contributing to global warming. The section to follow describes this effect and some possible outcomes of its modification.

Extensive fossil fuel burning and other human activities may be contributing to imbalances in the global carbon budget.

FROM GREENHOUSE GASES TO A WARMER PLANET?

GREENHOUSE EFFECT   Atmospheric concentrations of gaseous molecules play a profound role in shaping the average temperature near the surface of the Earth. That temperature has enormous effect on the global climate.

Countless molecules of carbon dioxide, water, ozone, methane, nitrous oxide, and chlorofluorocarbons are key players in interactions that dictate the global temperature. Collectively, the gases act somewhat like a pane of glass in a greenhouse—hence their name, the "greenhouse gases." Wavelengths of visible light can pass around them and reach the Earth's surface. However, greenhouse gases impede the escape of longer, infrared wavelengths— that is, heat—from the Earth into space. How? Gaseous molecules can absorb these wavelengths, then reradiate much of the absorbed energy back toward the Earth, as shown in Figure 48.13.

The constant reradiation of heat energy from the greenhouse gases proceeds lockstep with the constant bombardment and absorption of wavelengths from the sun, and so heat builds up in the lower atmosphere. The **greenhouse effect** is the name for this warming action.

GLOBAL WARMING DEFINED   Without the action of greenhouse gases, the Earth's surface would be cold and lifeless. However, there can be too much of a good thing. Largely as an outcome of human activities, greenhouse gases are building to atmospheric levels that are higher than they were in the past. Figure 48.14 documents the increase. As many researchers suspect, greenhouse gases may be contributing to long-term higher temperatures at the Earth's surface, an effect called **global warming**.

What is so alarming about a warmer planet? Suppose the temperature of the lower atmosphere were to rise by only 4°C (7°F). The increase might cause sea levels to rise by about 0.6 meter (2 feet). Why? Temperatures near the ocean surface would increase—and water expands when heated. Also, global warming could make glaciers and the polar ice sheets melt faster. The volume of water released this way alone would flood low coastal regions.

Imagine a long-term rise in sea level, combined with high tides and storm waves. The waterfronts of Vancouver, Seattle, San Diego, New York, Boston, Galveston, Hilo, and all other cities perched along the rim of the world's ocean would be submerged. So would the agricultural lowlands and deltas in India, China, and Bangladesh— where much of the world's rice is grown. Huge tracts of Florida and Louisiana would face saltwater intrusions. And what if global warming disturbed regional patterns of precipitation and temperature? We might expect deserts to expand and the interiors of the great continents to become drier than they already are.

Will the nations that control the world's great grain belts be affected? Will other nations be better or worse off? Imagine the economic and political consequences.

Some predicted effects of climate change are emerging. For example, in 1996, we found that the water 200 meters below the Arctic icecap is 1°C warmer than it was just five years ago. If the rapid warming continues, the icecap will disappear during the next century. The rise in temperature may already be disrupting a vast, deep current in the North Atlantic that affects circulation through the world ocean. Also during the 1990s, record-breaking hurricanes, flooding, and prolonged droughts have adversely affected crop production around the world. Beaches have been diminishing by 0.6 meter or more every year.

EVIDENCE OF AN INTENSIFIED GREENHOUSE EFFECT
During the late 1950s, researchers on the highest peak in the Hawaiian Islands started measuring the atmospheric concentrations of different greenhouse gases. They chose this remote site because it is exceptionally free of local airborne contamination and is representative of average

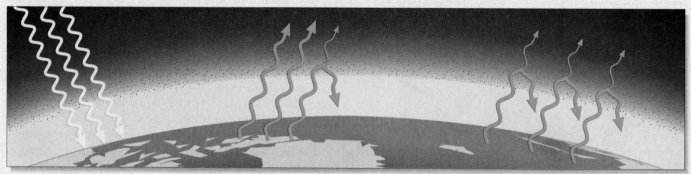

a Rays of sunlight penetrate the lower atmosphere and warm the Earth's surface.

b The Earth's surface radiates heat (infrared wavelengths) to the lower atmosphere. Some heat escapes into space. But greenhouse gases and water vapor absorb some infrared wavelengths and reradiate a portion back toward the Earth.

c As concentrations of greenhouse gases increase in the atmosphere, more heat is trapped near the Earth's surface. The surface temperature of the world ocean rises, more water evaporates into the atmosphere, and the Earth's surface temperature rises.

Figure 48.13   The greenhouse effect, as executed mainly by carbon dioxide, water, ozone, methane, nitrous oxide, and chlorofluorocarbons in the atmosphere.

Figure 48.14 Chart of the shifts in the atmospheric concentrations of carbon dioxide, as correlated with the most recent glaciation and interglacial periods during the past 160,000 years.

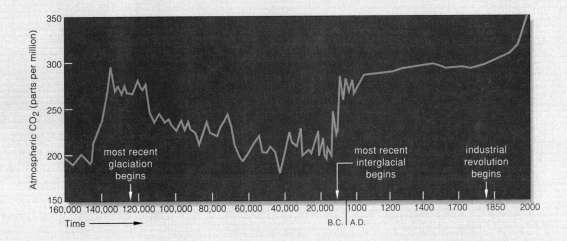

conditions for the Northern Hemisphere. The monitoring activities are still proceeding.

Consider what they found out about carbon dioxide alone. In the Northern Hemisphere, atmospheric levels of carbon dioxide follow the annual cycle of plant growth. They are lower in summer, when photosynthesis rates are highest. They are higher in winter, when photosynthesis slows but aerobic respiration continues. The troughs and peaks around the graph line in Figure 48.15a represent the annual lows and highs. For the first time, scientists had a picture of the integrated effects of the carbon balances for land and water ecosystems of an entire hemisphere. Notice the midline of the troughs and peaks in the cycle. *It steadily increased.* Many scientists take this as evidence of a buildup in carbon dioxide levels that may intensify the greenhouse effect over the next century.

Probably, global burning of fossil fuels is contributing most to the rise in carbon dioxide levels. Deforestation is

also adding to it; when wood burns, its carbon is released. Especially during the past four decades, vast tracts of the world's great forests have been decimated, by deliberate burns and other practices, at astounding rates (Section 50.4). Also, as plant biomass plummets, global absorption of carbon dioxide in photosynthesis may decline.

Will atmospheric levels of greenhouse gases continue to increase until the middle of the twenty-first century? Will global temperature rise by several degrees? If the warming trend is already in motion, we will not be able to reverse it simply by waiting until the last minute to stop fossil fuel burning and deforestation.

There is widespread agreement among scientists that nations must begin preparing for the consequences. For example, we might step up genetic engineering studies to develop drought-resistant and salt-resistant plants. Such plants may prove crucial in regions of saltwater intrusions and climatic change.

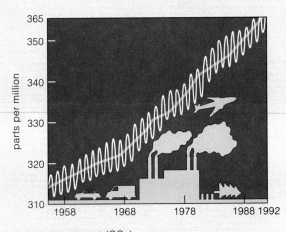

a CARBON DIOXIDE (CO_2)

Of all human activities, fossil fuel burning and deforestation are making the greatest contributions to increases in atmospheric levels of this gas.

Figure 48.15 Recently documented increases in atmospheric concentrations of four greenhouse gases.

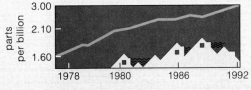

b CHLOROFLUOROCARBONS (CFCs)

Until recent restrictions, these were prevalent in plastic foams, refrigerators, air conditioners, and industrial solvents.

c METHANE (CH_4)

Termite activity and anaerobic bacteria in swamps, landfills, and the stomachs of cattle and other ruminants produce large quantities of methane.

d NITROUS OXIDE (N_2O)

Denitrifying bacteria produce nitrous oxide as a metabolic by-product. The gas also is released in great amounts from fertilizers and animal wastes, as in livestock feedlots.

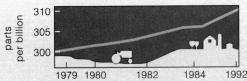

847

Since the time of life's origin, the atmosphere and ocean have contained nitrogen. This component of all proteins and nucleic acids moves in an atmospheric cycle called the **nitrogen cycle**. Gaseous nitrogen (N_2) makes up about 80 percent of the atmosphere, the largest nitrogen reservoir. Triple covalent bonds hold its atoms together ($N\equiv N$). Few organisms have the metabolic means to break them. Only certain bacteria, volcanic action, and lightning convert N_2 into forms that enter food webs.

Of all nutrients required for plant growth, nitrogen often is the scarcest. Today, nearly all of the nitrogen in soils has been put there by nitrogen-fixing organisms. Ecosystems lose it through the activities of bacteria that "unfix" the fixed nitrogen. Land ecosystems lose more through leaching of soils, although this is the basis of nitrogen inputs to aquatic ecosystems such as streams, lakes, and the oceans (Figure 48.16).

Cycling Processes

Let's follow nitrogen atoms through the portion of the nitrogen cycle that proceeds in an ecosystem. The atoms enter the ecosystem and move through it by nitrogen fixation, assimilation and biosynthesis, decomposition, and ammonification.

In **nitrogen fixation**, a few kinds of bacteria convert N_2 to ammonia (NH_3), which quickly dissolves in the cytoplasm, thus forming ammonium (NH_4^+). In aquatic ecosystems, *Anabaena, Nostoc*, and other cyanobacteria are nitrogen fixers. In many land ecosystems, *Rhizobium* and *Azotobacter* fix nitrogen. Collectively, these bacteria fix about 200 million metric tons of nitrogen each year! Plants assimilate and use much of the nitrogen in the biosynthesis of amino acids, proteins, and nucleic acids. Plant tissues are the only nitrogen source for animals, which feed directly or indirectly on plants.

Decomposition and **ammonification** are processes by which bacteria and fungi degrade nitrogenous wastes and remains of organisms. Decomposers use part of the released proteins and amino acids during metabolism. But most of the nitrogen remains in the decay products, in the form of ammonia or ammonium, which plants take up. Nitrifying bacteria also act on the ammonia or ammonium. In **nitrification**, they strip the compounds of electrons, and nitrite (NO_2^-) is the result. Other kinds of bacteria use the nitrite during their metabolism and produce nitrate (NO_3^-), which plants then take up.

Remember, certain plants are better than others at securing nitrogen. For example, clover, beans, peas, and other kinds of legumes are mutualists with nitrogen-fixing bacteria. Section 30.2 describes their interactions. Most plants are mutualists with fungi; together they form mycorrhizae that enhance nitrogen uptake. Sections 24.4 and 30.2 describe the formation and effects of these fungus roots.

Figure 48.16 The nitrogen cycle in an ecosystem on land. The atmosphere is the largest reservoir of nitrogen, which is an element that all organisms require for biosynthesis.

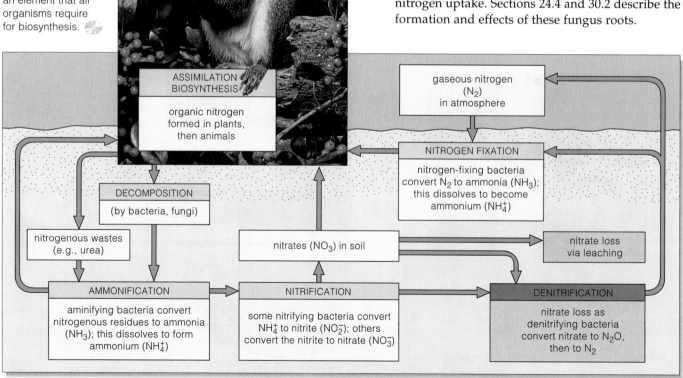

ASSIMILATION BIOSYNTHESIS

organic nitrogen formed in plants, then animals

gaseous nitrogen (N_2) in atmosphere

NITROGEN FIXATION

nitrogen-fixing bacteria convert N_2 to ammonia (NH_3); this dissolves to become ammonium (NH_4^+)

DECOMPOSITION

(by bacteria, fungi)

nitrogenous wastes (e.g., urea)

nitrates (NO_3) in soil

nitrate loss via leaching

AMMONIFICATION

aminifying bacteria convert nitrogenous residues to ammonia (NH_3); this dissolves to form ammonium (NH_4^+)

NITRIFICATION

some nitrifying bacteria convert NH_4^+ to nitrite (NO_2^-); others convert the nitrite to nitrate (NO_3^-)

DENITRIFICATION

nitrate loss as denitrifying bacteria convert nitrate to N_2O, then to N_2

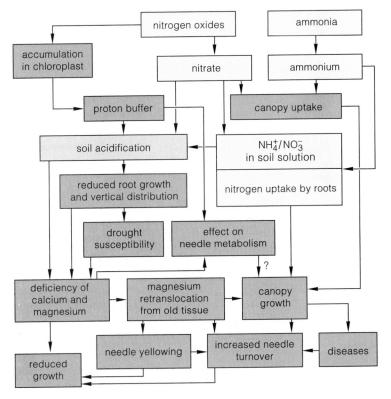

Figure 48.17 How nitrogen oxides, ammonia, ozone, sulfur oxides, and other air pollutants are accelerating the decline of spruce forests. Seedlings growing in acidified soils develop magnesium deficiencies if ammonium levels are high. Needles yellow and drop; photosynthesis suffers. Trees weakened by nutrient imbalances become susceptible to diseases.

Diagram labels: nitrogen oxides; ammonia; accumulation in chloroplast; nitrate; ammonium; proton buffer; canopy uptake; soil acidification; NH_4^+/NO_3^- in soil solution; reduced root growth and vertical distribution; nitrogen uptake by roots; drought susceptibility; effect on needle metabolism; deficiency of calcium and magnesium; magnesium retranslocation from old tissue; canopy growth; needle yellowing; increased needle turnover; diseases; reduced growth

Nitrogen Scarcity

Given the nitrogen cycling processes, you might think that plants have no trouble locating enough nitrogen. However, the ammonium, nitrite, and nitrate that form during the cycle are highly vulnerable to leaching and runoff. With leaching, recall, soil water moves out of an area, which thereby loses the nutrients dissolved in it (Section 30.1).

Besides losses from leaching, some nitrogen also is lost to the air through **denitrification**. By this process, certain bacteria convert nitrate or nitrite to N_2 and a bit of nitrous oxide, or N_2O. Ordinarily, most denitrifying bacteria use the pathway of aerobic respiration. When soil is waterlogged and poorly aerated, however, they switch to anaerobic pathways and use nitrate, nitrite, or nitrous oxide as the final electron acceptor instead of oxygen. (Section 8.5 describes this type of metabolic pathway.) In these reactions, fixed nitrogen is converted to N_2, much of which escapes into the atmosphere.

Also, nitrogen fixation comes at high metabolic cost to the plants that are mutualists with nitrogen fixers. In exchange for nitrogen, plants give up sugars and other photosynthetic products that require large ATP and NADPH investments. Such plants do have the competitive edge in nitrogen-poor soil. However, in nitrogen-rich soil, species that do not have to pay the metabolic price often displace them.

Human Intervention in the Nitrogen Cycle

Humans are altering the cycling of nitrogen in natural ecosystems. Consider how air pollutants contribute to changes in soil. Vehicles, fossil fuel burning power plants, and nitrogenous fertilizers are sources of pollutants, including the oxides of nitrogen. These contribute to soil acidity, which reduces the amounts of magnesium, calcium, and potassium that plants can take up. Figure 48.17 merely hints at the ion interactions in soil water.

And what about those nitrogen fertilizers? To be sure, nitrogen losses from soil are enormous in agricultural regions. With each harvest, nitrogen departs from fields (in the tissues of harvested plants). Soil erosion and leaching remove more. In Europe and North America, farmers traditionally have rotated crops, as when they alternate wheat with legumes. Together with other conservation practices, crop rotation has helped keep the soils stable and productive, sometimes for thousands of years. Today, however, intensive agriculture relies on the application of nitrogen-rich fertilizers. New strains of crop plants are bred for greater capacity to take up fertilizers, and crop yields per hectare have doubled and even quadrupled over the past forty years. Whether pest control and soil management practices can sustain high yields indefinitely remains uncertain, for reasons that will become apparent to you in Chapter 50.

We can't get something for nothing. Production of fertilizers requires huge amounts of energy from fossil fuels—not from free, unending sunlight. Few once believed that fossil fuel supplies might run out, so there was little concern about fertilizer costs. It still is quite common to pour more energy into soil (as in fertilizers) than we get out of it (in the form of food). As long as the human population continues to grow exponentially, farmers will be engaged in a race to grow as much food as they can, as fast as they can, for as many people as possible. Enrichment of soil with nitrogen-containing fertilizers is part of the race, as it is now being run.

The cycling of nitrogen in natural ecosystems depends on the activity of nitrogen-fixing bacteria and on mycorrhizae. Human disruptions to the nitrogen cycle may damage the ecosystems.

A Look at the Phosphorus Cycle

We conclude our look at biogeochemical cycling with one of the sedimentary cycles. In the **phosphorus cycle**, phosphorus moves from the land, to sediments at the bottom of the seas, then back to the land (Figure 48.18). The Earth's crust is the largest reservoir for phosphorus, just as it is for other minerals.

Rock formations on the land contain phosphorus in the form of phosphates, most often. By the natural processes of weathering and soil erosion, phosphates enter streams, then rivers, that transport them to the ocean. There, mainly on the submerged "shelves" of all the continents, phosphorus deposits slowly accumulate with other minerals that form insoluble deposits. Millions of years pass. Where great movements of the crustal plates uplift part of the seafloor, phosphates become exposed on the drained land surfaces. Over time, weathering releases the phosphates from the exposed rocks, and the cycle's geochemical phase begins again.

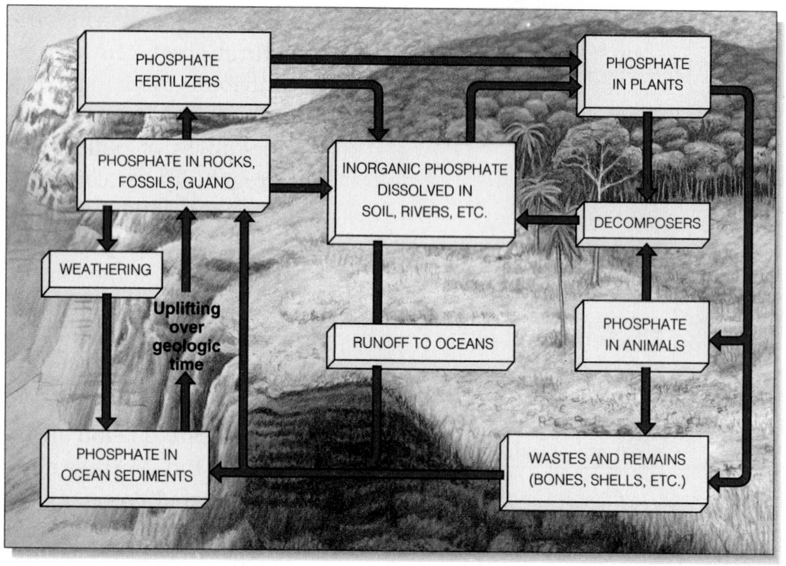

Figure 48.18 The phosphorus cycle.

The ecosystem phase of the cycle is more rapid than the long-term geochemical phase. All organisms require phosphorus for synthesizing phospholipids, NADPH, ATP, nucleic acids, and other compounds. Plants take up dissolved, ionized forms of phosphorus so rapidly and efficiently that they often reduce its concentrations in soil to extremely low levels. Herbivorous animals get phosphates by eating the plants; carnivores obtain it by eating herbivores. Both types of animals excrete phosphates as a waste in urine and feces.

Decomposition in the soil releases phosphates as well. Plants take up the mineral and so help its rapid cycling within the ecosystem.

Eutrophication

Fertilizers use phosphates as key ingredients. Where the fertilizers are heavily applied, phosphorus that becomes concentrated in runoff from fields alters living conditions in lakes and other aquatic ecosystems. Three important nutrients that algae and aquatic plants require for their growth are nitrogen, potassium, and phosphorus. Bacteria fix enough nitrogen. Freshwater ecosystems usually have excess amounts of potassium. But most of the phosphorus is locked in sediments, so it tends to be a limiting factor on plant growth. Therefore, enrich the water with phosphorus-containing fertilizers, and the outcome will be dense algal blooms.

Most mineral elements enter sedimentary cycles. In their dissolved forms, they serve as micronutrients and macronutrients for the growth of primary producers in ecosystems on land as well as in the water provinces. Activities that increase the concentrations of dissolved nutrients can lead to **eutrophication**. The term refers to nutrient enrichment of any aquatic ecosystem. Thus an outpouring of raw sewage into a river or lake, and even logging over the land surrounding it, can increase the nutrient levels and trigger eutrophication. Many field experiments, such as the one shown in Figure 48.19, nicely demonstrate this outcome.

Figure 48.19 Experiment demonstrating lake eutrophication in Ontario, Canada. Researchers stretched a plastic curtain across a channel between two basins of the same lake. They added phosphorus, carbon, and nitrogen to one basin (the basin in the *background*) and carbon and nitrogen to the other (the basin in the *foreground*). Within two months, the phosphorus-enriched basin showed signs of accelerated eutrophication; a dense algal bloom turned the water green.

The Earth's crust is the largest reservoir for phosphorus and for other minerals that move through ecosystems as part of sedimentary cycles.

Ecosystem Modeling

We now come full circle to a premise underlying the story that opened this chapter—that disturbances to one part of an ecosystem can have unexpected effects on other, seemingly unrelated parts.

One approach to predicting unforeseen effects of disturbances is through **ecosystem modeling**. By this method, researchers identify crucial bits of information about different ecosystem components. Then they rely on computer programs and models in order to combine the information. They use the resulting data to predict the outcome of the next disturbance.

For example, an analysis of which species feed on others in the Antarctic food web shown in Figure 48.3 can be turned into a series of equations that describe how many individuals of each population are being consumed. Such equations can be used to predict, say, the impact on the ecosystem if humans overharvest whales or greatly expand the harvesting of krill.

As ecologists attempt to deal with larger and more complex ecosystems, it becomes far more difficult and expensive to run experiments in the field. A modern temptation is to run them instead on the computer. This is a valid exercise *if* the computer model adequately represents the system. The danger is that investigators may not have identified all of the key relationships in the ecosystem and incorporated them accurately into the model. The most crucial fact may be one that we do not yet know, as the following study makes clear.

Biological Magnification— A Case Study

DDT, the first of the synthetic organic pesticides, was first used during World War II. In mosquito-infested, tropical regions of the Pacific, people were vulnerable to a dangerous disease, malaria. DDT helped control the mosquitoes that were transmitting the sporozoan disease agent (a species of *Plasmodium*). In war-ravaged cities of Europe, other people were suffering from the crushing headaches, fevers, and rashes associated with typhus. DDT helped control populations of body lice that were transmitting *Rickettsia rickettsii*, the bacterial agent of this terrible disease. After the war, it seemed like a good idea to use DDT to eliminate many insects that were agricultural pests, forest pests, transmitters of pathogens, or merely nuisances in homes and gardens.

DDT is a relatively stable hydrocarbon compound. It is nearly insoluble in water, so you might think that it would stay put and act only where applied. But winds can carry DDT in vapor form; water can transport fine particles of it. DDT also is highly soluble in fats, so it can accumulate in the tissues of organisms.

Thus, as we now know, DDT can show **biological magnification**. By this occurrence, a nondegradable or slowly degradable substance becomes more and more concentrated in the tissues of organisms at the higher trophic levels of a food web. Most of the DDT from all organisms that a consumer feeds on during its lifetime ends up in its own tissues. Also, in many organisms DDT is converted to modified forms, such as DDE, with different but still harmful effects. Both DDT and the modified compounds disrupt metabolism and are often toxic to many aquatic and terrestrial animals.

After World War II, DDT began to spread through the global environment, infiltrate food webs, and affect organisms in ways that no one had predicted. In cities where DDT was sprayed to control Dutch elm disease, songbirds started dying. In streams flowing through forests where DDT was sprayed to kill larvae of spruce budworms, salmon started dying. In agricultural fields sprayed to control one kind of pest, new kinds of pests moved in. *DDT was indiscriminately killing off natural predators that had been keeping pest populations in check!* It took no great flying leap of the imagination to make the connection. Those organisms were all dying at the same time—and in the same locations where DDT had been sprayed.

Then side effects of biological magnification started showing up in places that were far removed from the areas of DDT application—*and much later in time.* Most devastated were species at the top of food webs, such as bald eagles, peregrine falcons, ospreys, and brown pelicans. It happens that a product of DDT breakdown interferes with important physiological processes. As one of the consequences, birds produced eggs with brittle shells. As a result, many of their chick embryos simply did not make it to hatching time. Some species were at the brink of extinction.

Since the 1970s, DDT has been banned in the United States, except for restricted applications where public health is endangered. Today, many of the species that were hit hardest have partially recovered in numbers. Yet even now, some birds are still laying thin-shelled eggs. They are picking up DDT at their winter ranges in Latin America. As recently as 1990, the California State Department of Health recommended that a fishery off the coast of Los Angeles be closed. DDT from industrial waste discharges that ended twenty years before is still contaminating that ecosystem.

Disturbances to one aspect of an ecosystem typically have unexpected effects on other, seemingly unrelated parts.

Predictions of the consequences, as by ecosystem modeling, must identify all of the key relationships in the ecosystem and incorporate them accurately into the model.

---

1. An ecosystem is an array of producers, consumers, detritivores, and decomposers and their environment. It is an open system, with inputs and outputs of energy and nutrients. There is a one-way flow of energy into and out from an ecosystem and a cycling of materials among its member organisms (Figure 48.20).

a. With few exceptions, sunlight is the initial energy source and photoautotrophs are the ecosystem's primary producers. Photoautotrophs convert sunlight energy to other forms, such as the chemical bond energy of ATP, which is used during the synthesis of organic compounds from simple inorganic substances.

b. Primary producers also assimilate many of the nutrients that are required by all other members of the ecosystem.

2. Primary producers nourish an array of heterotrophs in an ecosystem. They do so either directly or indirectly.

a. Many heterotrophs are consumers. Among these are herbivores that feed on algae and plants, carnivores that feed on animals, parasites that feed on the tissues of living organisms, and omnivores, such as humans, which eat diverse organisms.

b. Other heterotrophs are decomposers. These are mainly fungal and bacterial species that degrade organic remains and wastes. Still others are detritivores, such as crabs and earthworms, that ingest bits of dead or decomposing material.

3. Feeding relationships in ecosystems are structured as trophic levels. Trophic levels are a hierarchy of energy transfers through the ecosystem, sometimes referred to as "Who eats whom."

a. Primary producers make up the first trophic level, herbivores make up the second level, and some array of carnivores make up successively higher levels.

b. Decomposers, humans, and many other kinds of organisms obtain energy from more than one source and cannot be assigned to a single trophic level.

4. Isolated food chains (straight-line sequences of who eats whom in an ecosystem) are rare in nature. They cross-connect with one another, as food webs.

5. The rate at which primary producers capture and store a given amount of energy in a given time interval is the primary productivity.

a. The total rate is the *gross* primary productivity.

b. With respect to primary producers, the rate of energy storage in excess of the rate of aerobic metabolism is the *net* primary productivity.

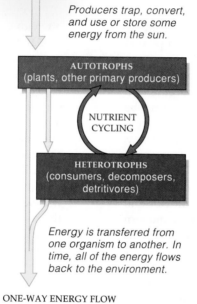

Producers trap, convert, and use or store some energy from the sun.

AUTOTROPHS
(plants, other primary producers)

NUTRIENT CYCLING

HETEROTROPHS
(consumers, decomposers, detritivores)

Energy is transferred from one organism to another. In time, all of the energy flows back to the environment.

ONE-WAY ENERGY FLOW

6. Energy fixed by photosynthesizers passes through grazing food webs and detrital food webs. Both types of food webs commonly are interconnected in the same ecosystem.

a. Both types of food webs lose energy (as heat) as a result of the metabolic activities of all organisms in the ecosystem.

b. In both types of food webs, the amount of useful energy that flows through successive consumer levels declines at each energy transfer. Losses arise through the generation of metabolic heat and through the shunting of food energy into organic wastes.

7. The availability of water and nutrients contributes to primary productivity and so influences the structure of an ecosystem.

8. In a biogeochemical cycle, the ions or molecules of a substance move slowly through the physical environment, then rapidly through organisms, then back to the environmental reservoir for them.

a. Water moves in a hydrologic cycle. In land ecosystems, plants stabilize soil and minimize nutrient losses, as by runoff.

b. In an atmospheric cycle, the largest proportion of some nutrient occurs in gaseous form. Carbon (in carbon dioxide) is an example.

c. In a sedimentary cycle, the nutrient occurs in solid form, and its largest reservoir is the Earth's crust. Phosphorus and other minerals are examples.

9. The carbon cycle is an atmospheric cycle, although the ocean is the largest reservoir (carbon dioxide dissolves easily in water). Photoautotrophs fix billions of metric tons of carbon in carbon compounds each year.

10. Fossil fuel burning and mass conversion of natural ecosystems to cropland or grazing land are contributing to imbalances in the global carbon budget. The increase may be contributing to a global warming trend.

11. Nitrogen availability is often a limiting factor for the total net primary productivity of land ecosystems. Gaseous nitrogen is abundant in the atmosphere, but nitrogen-fixing bacteria must convert it to ammonia and nitrates that primary producers can use. Nitrogen cycling on land is enhanced by mycorrhizae.

12. Phosphorus and most other mineral elements enter sedimentary cycles. When dissolved in water, primary producers can take them up. At high concentrations, phosphorus especially triggers eutrophication (nutrient enrichment of aquatic ecosystems) and fans algal blooms.

13. Disturbance of one aspect of an ecosystem often has unexpected effects on other, seemingly unrelated parts.

14. To predict the consequences of a disruption to an ecosystem, as by computer modeling, researchers must identify all of the ecosystem's key relationships and then incorporate them into the model.

Review Questions

1. Define an ecosystem and its trophic levels. *48.1*

2. Characterize grazing and detrital food webs. *48.2*

3. Define the three types of biogeochemical cycles. *48.4*

4. Define and describe the connections among nitrogen fixation, nitrification, ammonification, and denitrification. *48.8*

Self-Quiz *(Answers in Appendix IV)*

1. Ecosystems have _____ .
   a. energy inputs and outputs
   b. nutrient cycling but not outputs
   c. one trophic level
   d. a and b

2. Trophic levels can be described as _____ .
   a. structured feeding relationships
   b. who eats whom in an ecosystem
   c. a hierarchy of energy transfers
   d. all of the above

3. Primary productivity is affected by _____ .
   a. photosynthesis and respiration by plants
   b. how many plants are neither eaten nor decomposed
   c. rainfall and temperature
   d. all of the above

4. Match the ecosystem terms with the suitable description.
   _____ producers   a. herbivores, carnivores, omnivores
   _____ consumers   b. feed on partly decomposed matter
   _____ decomposers   c. degrade organic remains, wastes
   _____ detritivores d. photoautotrophs

Critical Thinking

1. Imagine and describe an extreme situation in which you would be a participant in a food chain rather than a food web.

2. Marguerite is growing a vegetable garden in Maine. What are the variables that can affect its net primary production?

3. List as many of the agricultural products and manufactured goods as you can identify that you depend upon. Are any implicated in the amplified greenhouse effect?

4. Of all crops grown in the United States, less than 10 percent are completely free from pesticide or fungicide applications. Such *organically grown produce* costs more, spoils faster, and is notably nonuniform in size and appearance. Do you buy such produce? Why or why not?

5. In 1995, the biologists Reed Noss, J. Michael Scott, and Edward LaRoe issued a report on *endangered ecosystems* in the United States. They categorize thirty natural ecosystems as being in danger of disappearing; the once-vast domains have diminished in size by more than 98 percent (Figure 48.21). Agricultural conversion, urban development, and other human activities account for the losses.

   In the next chapter, you will read about the type and extent of the ecosystems in question. The once-largest among them include tallgrass prairie, oak-studded savannas, and eastern deciduous forests, as well as pine forests that once covered much of the coastal plains of the southeastern states. As you can see from Figure 48.21, the most imperiled regions are largely concentrated in the eastern portion of the country. It

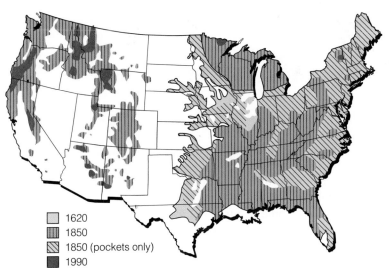

☐ 1620
▦ 1850
▨ 1850 (pockets only)
■ 1990

Figure 48.21 Map showing the extent of deforestation in the United States, starting from the year 1620 through 1990.

shows that more biodiversity has been lost at the ecosystem level than is generally recognized. The finding may have an impact on how (and whether) the government amends its conservation laws, such as the Endangered Species Act. Do we as a nation protect entire ecosystems, not just an endangered species? That is a goal of many conservationists. However, property-rights advocates criticize the goal; they view it as a threat to the long-standing tradition of private ownership of land. They also worry that merely identifying ecosystems in need of protection will endanger property values.

Do some research on a natural habitat that is part of an imperiled ecosystem. Get a sense of its decline in biodiversity, of the kinds of organisms that once flourished there. Then ask yourself: Would you participate in efforts to set aside land for ecological restoration? Or would you consider such efforts too intrusive on individual rights of property ownership?

Selected Key Terms

ammonification *48.8*	food chain *48.1*
biogeochemical cycle *48.4*	food web *48.1*
biological magnification *48.10*	global warming *48.7*
carbon cycle *48.6*	grazing food web *48.2*
consumer *48.1*	greenhouse effect *48.7*
decomposer *48.1*	hydrologic cycle *48.5*
decomposition *48.8*	nitrification *48.8*
denitrification *48.8*	nitrogen cycle *48.8*
detrital food web *48.2*	nitrogen fixation *48.8*
detritivore *48.1*	phosphorus cycle *48.9*
ecological pyramid *48.2*	primary producer *48.1*
ecosystem *48.1*	primary productivity *48.2*
ecosystem modeling *48.10*	trophic level *48.1*
energy pyramid *48.2*	watershed *48.5*
eutrophication *48.9*	

Reading

Krebs, C. 1994. *Ecology*. Fourth edition. New York: Harper-Collins.

Web Site See *http://www.wadsworth.com/biology* for practice quiz questions, hypercontents, BioUpdates, and critical thinking. The Wadsworth Biology Resource Center provides a wealth of information fully organized and integrated by chapter.

49

THE BIOSPHERE

Does a Cactus Grow in Brooklyn?

Suppose you live in the American Southwest but find yourself touring one of the deserts of Africa. There you come across a flowering plant with spines, tiny leaves, and columnlike, fleshy stems—just like some cactus plants back home (Figure 49.1). Or suppose you live in the coastal hills of California and decide to tour the Mediterranean coast, the southern tip of Africa, or even central Chile. There you come across many-branched, tough-leafed woody plants—very much like the many-branched, tough-leafed chaparral plants back home.

In both cases, the plants are separated by enormous geographic and evolutionary distances. Why, then, are they so much alike? The question intrigues you, so you decide to compare their locations on a global map. As you quickly discover, the American and African desert plants grow approximately the same distance from the equator. Chaparral plants and their distant look-alikes grow along the western or southern coasts of continents between latitudes 30° and 40°. As Charles Darwin and other naturalists did long ago, you have just stumbled onto one of many predictable patterns in the world distribution of species.

In part, "accidents of history" put many species in particular places. Go back to Pangea's colossal breakup, more than 100 million years ago. When chunks of that supercontinent began to drift apart, many species had no choice but to go along for the ride. Over evolutionary time, the travelers were dispersed to different isolated locations. And there they proceeded to undergo genetic divergence from the parent populations. Among those

a

b

Figure 49.1 Life in the environment—a case of morphological convergence. (**a**) *Echinocerus*, of the cactus family, grows in deserts of the American Southwest. (**b**) In deserts of southwestern Africa, we find *Euphorbia*, of the spurge family (Euphorbiaceae). The lineages appear to be closely related; but they are geographically and evolutionarily distant. Long ago, plants in both lineages put similar structures (including leaves) to similar uses in similar habitats—and their descendants ended up resembling each other.

changing populations on the fragment that became Australia were the ancestors of eucalyptus trees and platypuses, of wombats and kangaroos.

Species also owe their distribution to topography, climate, and species interactions. With diligence, you might grow a cactus under artificial lights in a heated room in Brooklyn or some other New York City borough. Plant that cactus outside, and it won't last one winter.

This last example reminds us that we humans tinker with the distribution of species. Not all of our tinkering is as harmless as growing a cactus in Brooklyn. Think of our predatory effects on the world's fisheries or the effects of our pesticide battles with insect competitors for food. Earlier chapters provided you with a general picture of predation, competition, and other species interactions. Consider now the physical forces shaping the biosphere itself. This will serve as a foundation for addressing the impact of the human species on the biosphere—the topic of the chapter to follow.

Start with the definition of the **biosphere**—the sum total of all places in which organisms live. It includes the **hydrosphere**, the waters of the Earth, including the ocean, polar ice caps, and other forms of liquid and frozen water. It includes all the soils and sediments of the **lithosphere**, the outer, rocky layer of the Earth. It extends into the lower **atmosphere**, which is made up of gases and airborne particles that envelop the Earth. The air in the upper atmosphere, about 17 kilometers above the Earth's surface, is too thin to support life.

In this vast biosphere you will find ecosystems that range from continent-straddling forests to tiny pools in cup-shaped leaves of plants. As you will see, climate profoundly influences all ecosystems except for a few at hydrothermal vents on the ocean floor.

Climate refers to the average weather conditions, such as temperature, humidity, wind speed, cloud cover, and rainfall, over time. Many factors contribute to climate. The main ones are variations in the amount of incoming solar radiation, the Earth's daily rotation and its path around the sun, the world distribution of continents and oceans, and the elevations of land masses. These factors interact to produce prevailing winds and ocean currents that influence the global patterns of climate. Climate affects the physical and chemical development of sediments and soils.

Together, climate and the composition of sediments and soils affect the growth of primary producers—and, through them, the distribution of entire ecosystems.

KEY CONCEPTS

1. Energy from the sun is the initial energy source for nearly all ecosystems on Earth. In addition, solar energy influences the global distribution of ecosystems. It does so by continually providing heat energy that warms the atmosphere and drives the Earth's weather systems.

2. Interactions among global air circulation patterns, ocean currents, and diverse topographic features result in regional variations in patterns of temperature and rainfall. The patterns influence the composition of soils and sediments, the growth and distribution of primary producers, and, through them, the distribution of ecosystems.

3. A biome is a large, regional unit of land that can be characterized by the climax vegetation of the ecosystems within its boundaries. Deserts and broadleaf forests are examples. Their distribution corresponds roughly with regional variations in climate, topography, and soil type.

4. The water provinces cover more than 71 percent of the Earth's surface. Foremost among them is the world ocean. They also include inland seas, as well as bodies of standing freshwater and moving freshwater.

5. Each freshwater and marine ecosystem has gradients in light availability, temperature, and dissolved gases. The gradients vary daily and seasonally. They affect primary productivity and the composition of species.

AIR CIRCULATION PATTERNS AND REGIONAL CLIMATES

Each winter Pacific Grove, California, is host to great gatherings of tourists and monarch butterflies (Figure 49.2). Similarly, caribou, Canadian geese, sea turtles, and whales undertake vast migrations, moving to and from overwintering grounds. Other kinds of animals stay put, having adaptations in form, physiology, and behavior that allow them to endure seasonal change.

Figure 49.2 Monarch butterflies, migratory insects that each winter gather in the trees of California coastal regions and of Central Mexico. Monarchs travel hundreds of kilometers south to those places, which are cool and humid in winter. If they were to remain in their northern breeding grounds, they would risk being killed by far more severe conditions.

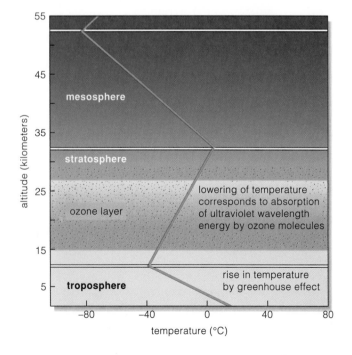

Figure 49.3 Earth's atmosphere. Most of the global air circulation proceeds in the troposphere, where temperature decreases rapidly with altitude. Most of the ultraviolet wavelengths in the sun's rays are absorbed in the upper atmosphere, mainly at the ozone layer. The ozone layer ranges between 17 and 27 kilometers above sea level.

Throughout Canada and the United States, flowering plants form leaves, flower, bear fruit, and drop leaves. In the ocean, uncountable numbers of photosynthetic microorganisms undergo seasonal bursts of primary productivity. Clearly, organisms are exquisitely attuned to climatic conditions, which differ from one region to the next and with the changing seasons. The conditions begin with the incoming rays from the sun.

Of the total amount of solar radiation arriving at the outer atmosphere, only about one-half gets through to the Earth's surface. Atmospheric molecules of ozone (O_3) and oxygen (O_2) absorb most of the wavelengths of ultraviolet radiation. Such wavelengths, remember, are lethal for most organisms. Absorption is greatest in the **ozone layer**, a region between 17 and 27 kilometers above sea level where molecules of ozone are the most concentrated (Figure 49.3). Clouds, bits of dust, and water vapor suspended in the atmosphere absorb some of the other wavelengths or reflect them back into space.

Radiation that manages to penetrate the atmosphere warms the surface of the Earth, which gives up heat by way of radiation or evaporation. Molecules of the lower atmosphere absorb some heat, then reradiate part of it toward the Earth. The effect is somewhat like heat retention in a greenhouse, which lets in the sun's rays while retaining heat that is being lost from the plants and soil inside (Section 48.7). Why is the greenhouse effect important? *Heat energy derived from the sun warms the atmosphere, and it ultimately drives the Earth's weather systems.*

Effects of incoming rays from the sun differ from one latitude to the next. The rays are more concentrated at the equator than at the poles, so air becomes heated more at the equator (Figure 49.4). The global pattern of air circulation starts as warm equatorial air rises and spreads northward and southward. Then the Earth's rotation modifies the circulation into worldwide belts of prevailing east and west winds. Taken together, the differences in solar heating at different latitudes and the modified air circulation patterns define the world's major **temperature zones** (Figure 49.5a).

Global air circulation patterns give rise to differences in rainfall at different latitudes as well. Think about this: Warm air can hold more moisture than cool air. At the equator, air picks up moisture from the seas and rises to cooler altitudes. There it gives up moisture as rain, which supports the luxuriant growth of tropical

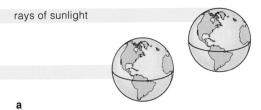

a

Figure 49.4 Global air circulation patterns, as brought about by three interrelated factors. *First,* the sun's rays are less spread out in equatorial regions than in polar regions (**a**). Warm equatorial air rises and spreads northward and southward to produce the initial pattern of air circulation (**b**).

Second, the nonuniform distribution of land masses creates variations in air pressure. Land absorbs and gives up heat faster than the ocean does, so some parcels of air rise (or sink) faster than others. Air pressure decreases where air rises (and increases where air sinks). The differences create winds that disrupt the overall movement from the equator to the poles.

Third, the rotation and overall shape of the Earth introduce easterly and westerly deflections in wind directions. Each time the ball-shaped Earth makes a full rotation, its surface turns faster beneath air masses at the equator and slower beneath those at the poles. Therefore, a rising air mass cannot really move "straight north" or "straight south," but rather is deflected to the east or west. This is the source of prevailing east and west winds (**c**).

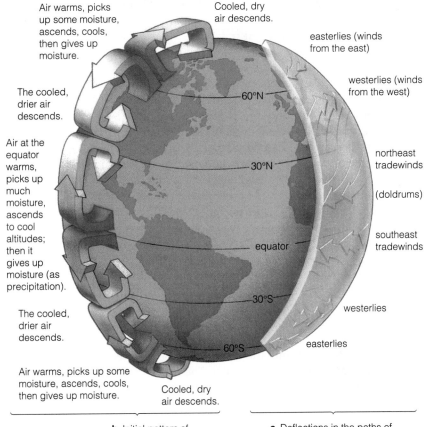

Air warms, picks up some moisture, ascends, cools, then gives up moisture.

Cooled, dry air descends.

easterlies (winds from the east)

westerlies (winds from the west)

The cooled, drier air descends.

Air at the equator warms, picks up much moisture, ascends to cool altitudes; then it gives up moisture (as precipitation).

northeast tradewinds

(doldrums)

southeast tradewinds

60°N

30°N

equator

30°S

60°S

westerlies

easterlies

The cooled, drier air descends.

Air warms, picks up some moisture, ascends, cools, then gives up moisture.

Cooled, dry air descends.

b Initial pattern of air circulation

c Deflections in the paths of air flow near the Earth's surface

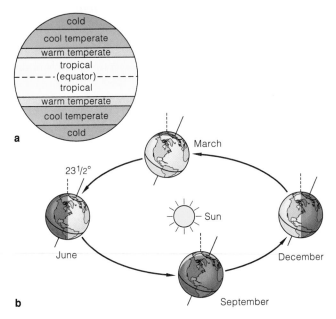

cold
cool temperate
warm temperate
tropical
— — — — (equator) — — — —
tropical
warm temperate
cool temperate
cold

a

March

23 1/2°

Sun

June

December

September

b

Figure 49.5 (**a**) World temperature zones. (**b**) Annual variation in incoming solar radiation. The northern end of the Earth's fixed axis tilts toward the sun in June and away from it in December. As a result, the equator's position relative to the boundary of illumination between day and night varies through the year. Such variations in sunlight intensity and in day length give rise to seasonal variations in temperature in the two hemispheres. The seasonal change becomes more pronounced with distance from the equator. The change is greatest in central regions of continents, where the moderating effects of oceans are minimal.

forests. The air, which is drier now, moves away from the equator. Later, air descends at latitudes of about 30°, and it becomes warmer and drier during this descent. Deserts tend to form at those latitudes. Farther from the equator, air again picks up moisture, ascends to high altitudes, and creates another moist belt at latitudes of about 60°. Then it descends in polar regions, where the low temperatures and almost nonexistent precipitation give rise to the cold, dry polar deserts.

Finally, the amount of solar radiation reaching the surface varies seasonally because of the Earth's rotation around the sun (Figure 49.5*b*). This leads to seasonal changes in daylength, prevailing wind directions, and temperature. And so primary productivity alternately rises and falls on land and in the seas—and migrations shift many kinds of animals to new locations.

Latitudinal differences in the amount of solar radiation reaching the Earth produce global air circulation patterns. The Earth's rotation and its overall shape affect the patterns to produce latitudinal belts of temperature and rainfall.

Also, the Earth's annual rotation around the sun introduces seasonal changes in winds, temperatures, and rainfall.

These factors influence the locations of different ecosystems.

Ocean Currents and Their Effects

The **ocean** is one continuous body of water that covers 71 percent of the Earth's surface. As the uppermost 10 percent circulates in currents, it distributes nutrients in marine ecosystems and affects regional climates. Solar heating and wind friction drive the surface currents.

Latitudinal and seasonal variations in solar heating cause the ocean water to warm and cool on a vast scale. Water increases in volume when heated, and it decreases in volume when cooled. At the equator, the sea level is about 8 centimeters (3 inches) higher than it is at the poles. The sheer volume of water associated with this "slope" is enough to get the surface waters moving in response to gravity.

Surface waters tend to move from the equator to the poles—and they warm air parcels above them during the journey. At midlatitudes, about *10 million billion* calories of heat energy get transferred from warm water to air every second!

The "tug" of mainly the trade winds and westerlies causes a rapid, mass flow of water. The Earth's rotation, the positions of land masses, even the shapes of ocean basins influence the direction and properties of these currents. As Figure 49.6 indicates, the global pattern of circulation is clockwise in the Northern Hemisphere. In

the Southern Hemisphere, it is counterclockwise. Swift, deep, and narrow currents of nutrient-poor waters run parallel with the east coasts of continents. For example, every second, the Gulf Stream moves 55 million cubic meters of warm water northward, along the eastern coast of North America. Slower, shallow, and broad currents paralleling the west coasts of continents move cold water toward the equator. As you will see, these currents have major roles in moving deep, nutrient-rich waters to the surface.

Why do cool, mild, foggy summers prevail along the Pacific Northwest coast? Track the currents shown in Figure 49.6. The California Current moves cold water near the coast as it flows toward the equator. Winds approaching the coast give up heat to the cold waters and thereby become cooler. Why is Washington, D.C., so muggy in summer? The air above the Gulf Stream gains heat and moisture—which winds from the south and east carry to the city. Compared to Ontario in central Canada, why are the winters milder in London and Edinburgh even though they are at the same latitude? The North Atlantic Current picks up warm water from the Gulf Stream, flows past northwestern Europe, and gives up heat energy to prevailing winds.

Regarding Rain Shadows and Monsoons

Mountains, valleys, and other aspects of topography influence regional climates. "Topography" means the physical features of a region, such as its elevation. For example, imagine a warm air mass picking up moisture

Figure 49.6 Surface currents of the world ocean. Warm water moves on a vast scale from the equator toward the poles. The direction of flow depends on winds, the Earth's rotation, gravity, the shape of ocean basins, and the positions of land masses.

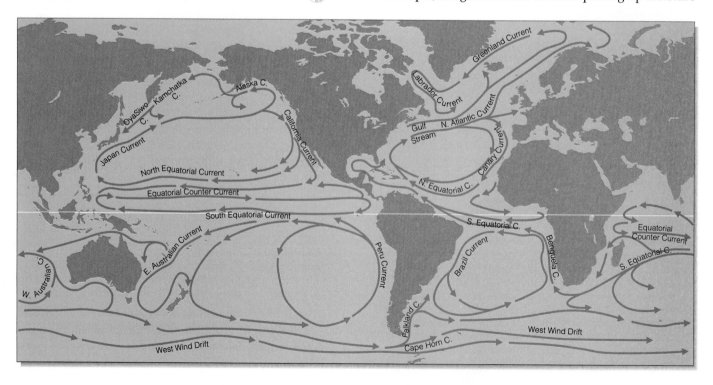

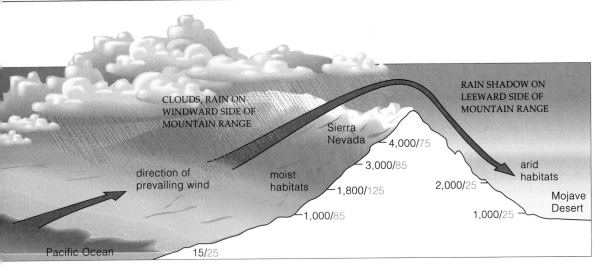

CLOUDS, RAIN ON WINDWARD SIDE OF MOUNTAIN RANGE

RAIN SHADOW ON LEEWARD SIDE OF MOUNTAIN RANGE

Sierra Nevada

— 4,000/75

— 3,000/85

— 1,800/125

— 1,000/85

direction of prevailing wind

moist habitats

2,000/25 —

arid habitats

Mojave Desert

1,000/25 —

Pacific Ocean

15/25

Figure 49.7 Rain shadow effect, a reduction of rainfall on the side of high mountains facing away from prevailing wind. Only plants adapted to arid or semiarid conditions grow in such places.

Blue numbers are average yearly precipitation (in centimeters), measured at different locations on both sides of the mountain range. *Black* numbers signify elevation (in meters).

off California's western coast. Moving inland, it reaches the Sierra Nevada, a mountain range that parallels the coast. The air cools as it ascends to higher altitudes, then loses moisture as rain (Figure 49.7). The vegetation belts at different elevations correspond to changes in air temperature and moisture. Grasslands with species adapted to semiarid conditions prevail at the western base of the range. At higher elevations, more moisture and cool temperatures support deciduous and evergreen forests. Higher still, a subalpine belt supports a few species of evergreen trees that can withstand the rigors of a colder habitat. Above the subalpine belt, only low, nonwoody plants and dwarfed shrubs can grow.

After air flows over the mountain crests and starts its descent, it becomes warmer. Warm air can hold more water, so now it retains moisture and even draws more out of plants and soil. The outcome is a **rain shadow**, an arid or semiarid region of sparse rainfall on the leeward side of high mountains. *Leeward* refers to the direction not facing a wind; *windward* refers to the direction from which the wind is blowing.

The mountain ranges of Europe, the Himalayas of Asia, and the Andes of South America also create great rain shadows. On Hawaii, tropical forests thrive on the windward side of high volcanic peaks, and conditions are arid on the leeward side.

Now consider **monsoons**, patterns of air circulation that affect conditions on the continents lying poleward of warm oceans. There, the land heats up intensely. The resulting low pressures draw in the moisture-laden air that forms above ocean water, as from the Bay of Bengal into India and Bangladesh, or from the Gulf of Mexico into the American Southwest. Those fantastic summer thunderstorms in Tucson are one outcome. Or consider the equatorial zone called the doldrums, where the trade winds converge. There, the direct overhead sun creates intense heating and heavy rain. The zone moves north and south seasonally. It creates the patterns of wet and dry seasons that we see, for example, in East Africa.

Finally, the recurring sea breezes along coastlines are "mini-monsoons." Here again, the heat capacities of land and water differ. In the morning, the temperature of water does not rise as rapidly as it does on land. As warmed air above the land rises, the cooler marine air moves in. After sunset, the land loses heat faster, and land breezes flow in the reverse direction (Figure 49.8).

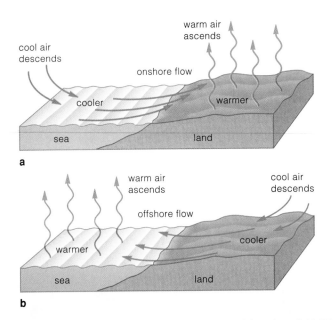

Figure 49.8 Coastal breeze in the afternoon (**a**) and at night (**b**).

Surface ocean currents, in combination with global air circulation patterns, influence regional climates and help distribute nutrients in marine ecosystems.

Air circulation patterns, ocean currents, and landforms interact in ways that influence regional temperatures and moisture levels. Thus they also influence the distribution and dominant features of ecosystems.

Biogeographic Distribution

The circulation patterns in the atmosphere and at the ocean surface, in concert with topography, give rise to regional differences in temperature and moisture. This information helps explain why it is that tundra, forests, grasslands, and deserts form in some places but not others. It also is the focus of **biogeography**—the study of the global distribution of species, each of which is adapted to regional conditions.

For example, the information provides clues to why many evolutionarily distant species look alike. Such species are often results of convergent evolution, which is described in Section 20.4. Briefly, their ancestors were not closely related but lived in similar environments and faced similar pressures. By natural selection, they underwent similar kinds of modifications and ended up looking alike. The ancestors of those cactus and spurge plants shown in Figure 49.1 both evolved in hot, dry deserts where water is scarce. Fleshy plant stems that have thick cuticles conserve precious water. And rows of sharp spines help deter hungy herbivores that might otherwise chew on juicy plant parts.

Biogeographic Realms and Biomes

Long ago W. Sclater, then Alfred Wallace, attached the name **biogeographic realms** to six vast land areas, each with distinguishing plants and animals (Figure 49.9a).

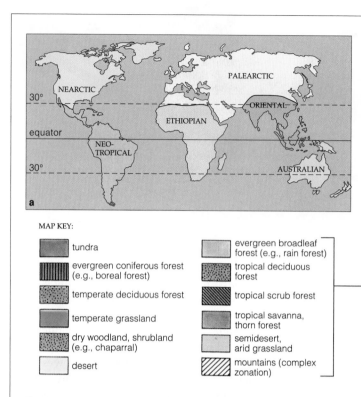

MAP KEY:

- tundra
- evergreen coniferous forest (e.g., boreal forest)
- temperate deciduous forest
- temperate grassland
- dry woodland, shrubland (e.g., chaparral)
- desert
- evergreen broadleaf forest (e.g., rain forest)
- tropical deciduous forest
- tropical scrub forest
- tropical savanna, thorn forest
- semidesert, arid grassland
- mountains (complex zonation)

Figure 49.9 (**a**) Major biogeographic realms. (**b**) Distribution of the world's major biomes. Use the map key above as a guide to this diagram. The overall pattern corresponds roughly with distribution patterns for climate and soil type. Compare this map with the photograph of the Earth's surface preceding Chapter 1.

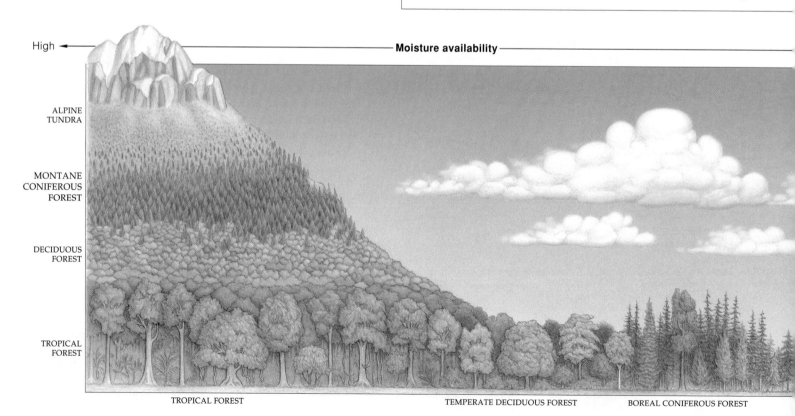

High ◄———————— Moisture availability ————————

ALPINE TUNDRA

MONTANE CONIFEROUS FOREST

DECIDUOUS FOREST

TROPICAL FOREST

TROPICAL FOREST   TEMPERATE DECIDUOUS FOREST   BOREAL CONIFEROUS FOREST

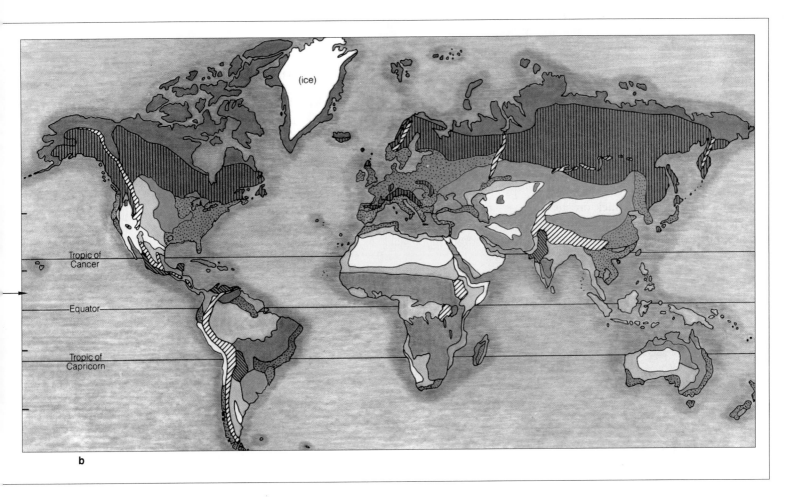

b

Figure 49.10
General changes in plant form along environmental gradients for North America.

This diagram shows gradients in elevation and in the availability of water, both of which have great influence on a biome's primary productivity.

The mean annual air temperature, soil drainage, and other factors also have effects on plant form.

ARCTIC TUNDRA

Low

High

Elevation

Low

The realms retain their distinct identity partly because of climate. They tend to maintain that identity if oceans, mountain ranges, and other major barriers restrict gene flow and so keep their component species isolated.

Each realm may be further divided into biomes. A **biome** is a large region of land that is characterized mainly by the climax vegetation of ecosystems within its boundaries. Take a moment to look over the map of biomes in Figure 49.9b, then see how it correlates with the information in Figure 49.10. As you might deduce, distinctive types of biomes prevail at certain latitudes and elevations. Thus biomes dominated by species of short plants prevail in dry regions, at high elevations, and at high latitudes. Biomes dominated by tall, leafy plant species prevail at tropical and temperate latitudes, and at low elevations with warm temperatures and high rainfall. As you will see next, soils also affect the distribution of biomes.

The distribution and key features of biomes are an outcome of temperatures, soils, and moisture levels (which vary with latitude and elevation), and of evolutionary history.

SOILS OF MAJOR BIOMES

Most land regions have **soils**, which are mixtures of mineral particles and variable amounts of decomposing organic material (humus). As described in Section 30.1, the weathering of hard rocks produces coarse-grained gravel, then sand, silt, and finely grained clay. Water and air infiltrate spaces between the particles. How much the particles are compacted together and the proportions of each kind vary within and between regions.

Soils have a layered structure, or *profile*, that reflects their stage of development. Figure 49.11 shows a few examples. Uppermost is topsoil. It has the most humus and is the most vulnerable to erosion. Topsoil may be less than a centimeter thick on steep slopes and more than a meter thick in grasslands. The lowest layer of soil consists of rocks in varying degrees of weathering.

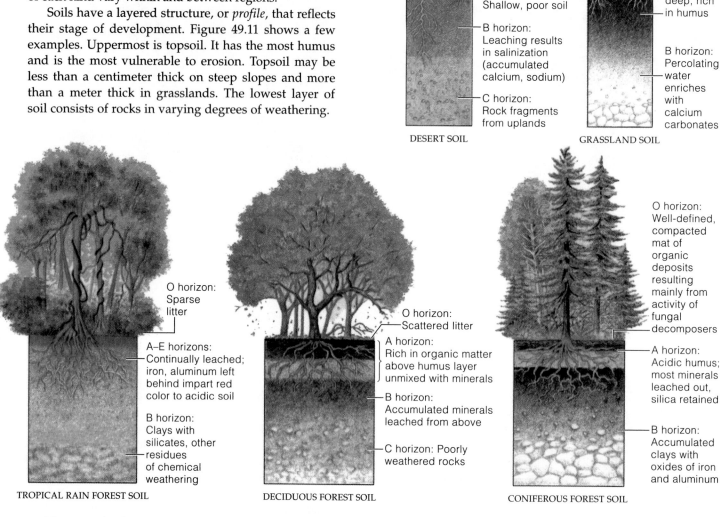

DESERT SOIL

O horizon: Pebbles, little organic matter

A horizon: Shallow, poor soil

B horizon: Leaching results in salinization (accumulated calcium, sodium)

C horizon: Rock fragments from uplands

GRASSLAND SOIL

A horizon: Alkaline, deep, rich in humus

B horizon: Percolating water enriches with calcium carbonates

TROPICAL RAIN FOREST SOIL

O horizon: Sparse litter

A–E horizons: Continually leached; iron, aluminum left behind impart red color to acidic soil

B horizon: Clays with silicates, other residues of chemical weathering

DECIDUOUS FOREST SOIL

O horizon: Scattered litter

A horizon: Rich in organic matter above humus layer unmixed with minerals

B horizon: Accumulated minerals leached from above

C horizon: Poorly weathered rocks

CONIFEROUS FOREST SOIL

O horizon: Well-defined, compacted mat of organic deposits resulting mainly from activity of fungal decomposers

A horizon: Acidic humus; most minerals leached out, silica retained

B horizon: Accumulated clays with oxides of iron and aluminum

Figure 49.11 Soil profiles from a few representative biomes. Compare Section 30.1.

The growth of most plants suffers in poorly aerated, poorly draining soils. Remember, loam topsoils have the best mix of sand, silt, and clay for agriculture. They have enough coarse particles to promote drainage and enough fine particles to retain water-soluble mineral ions that serve as nutrients for plant growth. Gravelly or sandy soils encourage rapid leaching that depletes them of vital mineral ions. Clay soils with fine, closely packed particles are poorly aerated and do not drain well. Few plants can grow in waterlogged clay soils.

As farmers know, soil affects primary productivity. They grow most crops in cleared, former grasslands. Many burn or clear-cut tropical forests for agriculture (Section 50.4). But these biomes have very little topsoil

above poorly draining sublayers. The clearing exposes the topsoil, and heavy rains leach most of the nutrients.

We turn now to the deserts, shrublands, woodlands, grasslands, various forests, and tundra. Bear in mind, none of these major biomes is uniform throughout. Within its borders, local climates, landforms, and other physical features favor patches of distinct communities.

Primary productivity depends on the soil profile and its proportions of sand, silt, clay, gravel, and humus.

DESERTS

Deserts form in lands with less than ten centimeters or so of annual rainfall and high potential for evaporation. Such conditions prevail at latitudes of about 30° north and south. There we find great deserts of the American Southwest, northern Chile, Australia, northern and southern Africa, and Arabia. Farther north are the high deserts of eastern Oregon, and Asia's vast Gobi and the Kyzyl-Kum east of the Caspian Sea. Rain shadows are the main reason why these northern deserts are so arid.

the rains. Deep-rooted species, including mesquite and cottonwood, often grow near the few streambeds that have a permanent underground water supply.

Of all land surfaces, more than one-third is arid or semiarid, without enough rainfall to support crops. The crops growing in California's Imperial Valley and some other deserts require unflagging soil management and extensive irrigation. Without drainage programs, such croplands become unproductive from waterlogging and

Figure 49.12 Warm desert near Tucson, Arizona. The plants include creosote bush, multistemmed ocotillo, columnlike saguaro cacti, and prickly pear cacti with rounded pads.

Deserts do not have lush vegetation. Rain falls in heavy, brief, infrequent pulses that swiftly erode the exposed topsoil. Humidity is so low that the sun's rays easily penetrate the air. They quickly heat the ground surface, which at night radiates heat and cools quickly.

Although arid or semiarid conditions do not favor large, leafy plants, deserts show plenty of biodiversity. In one patch of Arizona's desert, you may come across deep-rooted, evergreen, woody shrubs, such as creosote bush and fleshy-stemmed, shallow-rooted cacti (Figure 49.12). You may see tall saguaro, short prickly pear, and ocotillo, which drops leaves more than once a year, then grows new ones after a rain. The desert perennials and annuals flower profusely, spectacularly, but briefly after

salt buildup. Alarmingly, many parts of the world are now undergoing **desertification**. The term refers to the conversion of vast grasslands and similarly productive biomes to dry wastelands. We take a closer look at this trend in Section 50.6.

Where the potential for evaporation greatly exceeds sparse rainfall, deserts form. This condition prevails at latitudes 30° north and south and in rain shadows.

DRY SHRUBLANDS, DRY WOODLANDS, AND GRASSLANDS

Dry shrublands and **dry woodlands** prevail in western or southern coastal regions of the continents between latitudes 30° and 40°. These semiarid regions get more rain than deserts, but not much more. Most rain falls in mild winters; summers are long, hot, and dry. Often, dominant plants have hardened, tough, evergreen leaves.

We find examples of dry shrublands, which get less than 25 to 60 centimeters of rain per year, in California, South Africa, and Mediterranean regions. Their exotic local names include fynbos and chaparral. California has 2.4 million hectares (6 million acres) of chaparral. In summer, lightning-sparked, wind-driven firestorms can sweep through these biomes (Figure 49.13). Shrubs with highly flammable leaves burn swiftly to the ground. Yet they are highly adapted to episodes of fire and quickly resprout from their root crowns. Trees do not fare as well during the firestorms. The shrubs, which "feed" the fires, have the competitive edge.

Dry woodlands dominate where annual rainfall is about 40 to 100 centimeters. The dominant trees can be tall, but they do not form a dense, continuous canopy. Eucalyptus woodlands of southwestern Australia and oak woodlands of California and Oregon are like this.

Grasslands sweep across much of the interior of continents, in the zones between deserts and temperate forests. Warm temperatures prevail in the summer, and winters are extremely cold. The 25 to 100 centimeters of annual rainfall prevent deserts from forming but are not enough to support forests. Drought-tolerant plants can survive strong winds, light and infrequent rainfall, and rapid evaporation. Dominant animals are grazing and burrowing types. The grazing activities, together with periodic fires, keep shrublands and forests from encroaching on the fringes of many grasslands.

In North America, the main grasslands are *shortgrass* prairie and *tallgrass* prairie. Usually they form where the land is flat or rolling, as in Figure 49.14a. Roots of perennial plants extend profusely through the topsoil. During the 1930s, shortgrass prairie of the American Great Plains was overgrazed and plowed under to grow wheat, which requires more water than this region sometimes receives. Strong winds, prolonged droughts, and unsuitable farming practices turned much of the prairie into a Dust Bowl (Section 50.6). John Steinbeck's *The Grapes of Wrath* and James Michener's *Centennial*, which are two historical novels, speak eloquently of the consequences.

Also in the interior of North America, tallgrass prairie once extended westward from zones of temperate deciduous forests. Diverse legumes and composites, including daisies, thrived in the interior, with its richer topsoil and slightly more frequent rainfall. Farmers converted nearly all of the original tallgrass prairie for agriculture, but some areas are now being restored (Section 47.7).

Between the tropical forests and deserts of Africa, South America, and Australia are broad belts of grasslands with a smattering of shrubs or trees. These are the **savannas** (Figure 49.14c). Rainfall averages 90 to 150 centimeters a year, and prolonged seasonal droughts are common. Where rainfall is low, fast-growing grasses dominate. Acacia and other shrubs grow in regions that get slightly more moisture. Where the rainfall is higher,

Figure 49.13 California chaparral. The dominant plants are multibranched, woody, and no more than a few meters tall. Without periodic fires, they form a nearly impenetrable vegetation cover. The boxed inset shows a firestorm racing through a chaparral-choked canyon above Malibu.

Figure 49.14 (**a**) Rolling shortgrass prairie to the east of the Rocky Mountains. Once, bison were the dominant large herbivore of North American grasslands. At one time, the astoundingly abundant grasses supported 60 million of these hefty ungulates. (Ungulates are hooved, plant-eating mammals.)

(**b**) A rare patch of natural tallgrass prairie in eastern Kansas.

(**c**) The African savanna—warm grasslands with scattered stands of shrubs and trees. More varieties and greater numbers of large ungulates live here than anywhere else. They include migratory wildebeests (shown in this photograph), giraffes, Cape buffalo, zebras, and impalas.

savannas grade into tropical woodlands with tall, coarse grasses, shrubs, and low trees. *Monsoon* grasslands form in southern Asia where heavy rains alternate with a dry season. Dense stands of tall, coarse grasses form, then die back and often burn in the dry season.

Plants of dry shrublands, dry woodlands, and grasslands are supremely adapted to surviving strong winds, grazing animals, and recurring episodes of drought and fire.

In forest biomes, tall trees grow close together and form a fairly continuous canopy over a broad stretch of land. In general, there are three types of trees. Which type prevails in a region depends partly on distance from the equator. The evergreen broadleafs dominate between latitudes 20° north and south. Deciduous broadleafs are common at moist, temperate latitudes where winters are not severe. Evergreen conifers are common at high, colder latitudes and in mountains of temperate zones.

Evergreen broadleaf forests sweep across tropical zones of Africa, the East Indies and Malay Archipelago, southeast Asia, South America, and Central America. Annual rainfall can exceed 200 centimeters and is never less than 130 centimeters. One forest biome, **the tropical rain forest**, depends on regular and heavy rainfall, an annual mean temperature of 25°C, and humidity of 80 percent or more (Figure 49.15). In this highly productive forest, evergreen trees produce new leaves and shed old ones through the year, so these forests produce far more litter than others. Decomposition and mineral cycling are rapid in the hot, humid climate. Soils are weathered, humus-poor, and poor reservoirs of nutrients. We will look more closely at these forests in the next chapter.

Leaving tropical rain forests, we enter regions where temperatures stay mild but rainfall dwindles during

Figure 49.15 Tropical forests, biomes of spectacular biodiversity. Among millions of known species are jaguars and *Rafflesia*, a leafless, foul-smelling plant with a fly-pollinated flower that grows three meters across. Bromeliads, orchids, and other epiphytes grow on trees, obtaining minerals from organic remains of leaves, insects, and other litter that have dissolved in tiny pockets of water.

SPRING

SUMMER

WINTER

AUTUMN

part of the year. This is the start of **deciduous broadleaf forests**. Trees of *tropical* deciduous forests drop some or all of their leaves during a pronounced dry season. The *monsoon* forests of India and southeastern Asia also have such trees. Farther north, in the temperate zone, rainfall is even lower, and temperatures become cold during the winter. *Temperate* deciduous forests prevail here (Figure 49.16). Decomposition is not as rapid as in the humid tropics, and many nutrients are conserved in accumulated litter on the forest floor.

Complex forests of ash, beech, birch, chestnut, elm, and deciduous oaks once stretched across northeastern North America, Europe, and east Asia. They declined

Figure 49.16 Part of a temperate deciduous forest south of Nashville, Tennessee, in spring, summer, autumn, and winter.

drastically when farmers cleared the land. A pathogenic species introduced to North America destroyed nearly all of the chestnuts and many elms (Table 47.8). Now, maple and beech predominate in the Northeast. Farther west, oak-hickory forests prevail, then oak woodlands that eventually grade into tallgrass prairie.

In forest biomes, conditions favor dense stands of tall trees that form a continuous canopy over a broad region.

Conifers (cone-bearing trees) are the primary producers of **coniferous forests**. Most have needle-shaped leaves with a thick cuticle and recessed stomata—adaptations that help the trees conserve water through dry times. They dominate the boreal forests, montane coniferous forests, temperate rain forests, and pine barrens.

Boreal forests stretch across northern Europe, Asia, and North America. (Such a forest also is called *taiga*, meaning "swamp forest.") Most are in glaciated regions with cold lakes and streams (Figure 49.17a). It rains mostly in summer, and evaporation is low in the cool summer air. The cold, dry winters are more severe in

We also find coniferous forests in some temperate lowlands. One such temperate rain forest, paralleling the coast from Alaska into northern California, includes some of the world's tallest trees—Sitka spruce on into the far north and redwoods to the south. Heavy logging destroyed much of this biome. One bright note: In 1994, a major timber company surrendered its rights to log one of the largest intact temperate rain forests outside the tropics. This forest, with trees 800 years old, cloaks the Kitlope Valley of British Columbia.

Sandy, nutrient-poor soil of New Jersey's coastal plain supports **pine barrens**. In these scrub forests,

a

b

eastern parts of these biomes than in the west, where oceanic winds moderate the climate. Spruce and balsam fir dominate the boreal forests of North America. Pines, birches, and aspens take hold in burned or logged areas. Highly acidic bogs dominated by peat mosses, shrubs, and stunted trees prevail in the poorly drained areas. Boreal forests become much less dense to the north, where they grade into arctic tundra.

In the Northern Hemisphere, montane coniferous forests extend southward through the great mountain ranges. Spruce and fir dominate in the north and at higher elevations. They give way to firs and pines in the south and at lower elevations (Figure 49.17b).

Figure 49.17 (a) Spruce-dominated boreal forest. (b) Montane coniferous forest of Yosemite Valley in the Sierra Nevada, California.

grasses and low shrubs grow beneath open stands of pitch pines and oak trees. Pine barrens are adapted to recover quickly from frequent lightning-triggered fires. Managed pine plantations replaced most of the natural pine forests that once dominated the coastal plains of North and South Carolina, Georgia, and Florida.

Coniferous forests prevail in regions in which a cold, dry season alternates with a cool, rainy season.

49.9 TUNDRA

Tundra is derived from *tuntura*, a Finnish word for the great treeless plain between the polar ice cap and belts of boreal forests in Europe, Asia, and North America. Much of this *arctic* tundra is flat, windswept, and wet (Figure 49.18a). Temperatures are cool in summer and below freezing in winter, so not much water evaporates. Little rain or snow falls. Sunlight is nearly continuous during the summer months. Then, short plants grow and flower profusely, and seeds ripen fast.

Snow does not cloak arctic tundra all year long, but summers are too short to thaw much more than surface soil. Just beneath the surface is a perpetually frozen

a

b

Figure 49.18 (**a**) Extensive ponding in the arctic tundra. The rain and snowmelt cannot percolate downward because of the permafrost. (**b**) Short, hardy plants typical of alpine tundra.

layer, the **permafrost**. It is more than 500 meters thick in some places. Because permafrost prevents drainage, the soil above remains waterlogged. Anaerobic conditions and low temperatures limit nutrient cycling. Organic matter decomposes so slowly, it accumulates in soggy masses of organic material. All but about 5 percent of the carbon in the arctic tundra is locked up in peat bogs.

A similar type of biome prevails at high elevations in mountains throughout the world, although there is no permafrost beneath the soil. Figure 49.18b shows an example of this *alpine* tundra. Dominant plants often form low cushions and mats that are able to withstand the buffeting of strong winds. Winter temperatures fall below freezing. Even in the summer, shaded patches of snow persist. The thin, fast-draining soil of an alpine tundra is nutrient-poor, so primary productivity is low.

At high latitudes with short, cool summers and long, very cold winters, we find arctic tundra. In high mountains where seasonal changes vary with latitude but where the climate is too cold to support forests, we find alpine tundra.

The freshwater and saltwater provinces are far more extensive than the Earth's biomes. They include lakes, rivers, ponds, estuaries, and wetlands. They include rocky and sandy shores, coral reefs, parts of the open ocean, even hydrothermal vents on the ocean floor. "Typical" examples are difficult to find. Some ponds can be waded across; Lake Baikal in Siberia is more than 1.7 kilometers deep. All of the aquatic ecosystems have gradients in light penetration, temperature, and dissolved gases, but these differ greatly from one to the next. All we can do here is sample the diversity.

Lake Ecosystems

A **lake** is a body of standing freshwater produced by geologic processes, as when an advancing glacier carves a basin in the Earth. When the glacier retreats, water collects in the exposed basin (Figure 49.19). Gradually over geologic time, erosion, sedimentation, and other events alter the dimensions of the lake, which usually ends up filled or drained completely.

We can subdivide a lake into littoral, limnetic, and profundal zones (Figure 49.20). The littoral extends all around the shore, to a depth at which rooted aquatic plants stop growing. Its waters are shallow and usually well lit. Diversity is greatest here. The limnetic, which is open, sunlit water past the littoral, extends to a depth where photosynthesis is insignificant. Communities of tiny aquatic organisms (plankton) abound in the limnetic zone. The *phyto*plankton include green algae, diatoms, and cyanobacteria. The *zoo*plankton include rotifers, copepods, and many other heterotrophs. The profundal includes all of the open water below the depth at which wavelengths suitable for photosynthesis can penetrate. Detritus sinks through this zone to bottom sediments. Communities of bacterial decomposers that live in and on the sediments release nutrients into the water.

SEASONAL CHANGES IN LAKES In temperate regions, where summers are warm and winters cold, lakes show seasonal changes in density and temperature from the surface to the bottom. A layer of ice forms over many of

Figure 49.19 In the Canadian Rockies, a lake basin formed by glacial action during the last ice age.

them in midwinter. Water near the freezing point is the least dense and accumulates beneath the ice. Water at 4°C is the most dense. It collects in deeper layers that are a bit warmer than the surface layer in midwinter.

In spring, daylength increases and the air is not as cold. As lake ice melts, the surface water slowly warms to 4°C, and temperatures become uniform throughout. Winds blowing over the surface now cause a **spring overturn**. In such overturns, strong vertical movements carry dissolved oxygen from a lake's surface layer to its depths, and nutrients released by decomposition are brought from the bottom sediments to the surface layer.

By midsummer, the surface layer is above 4°C and the lake has a thermocline. This **middle layer** changes abruptly in temperature and prevents vertical mixing. Being warmer and less dense, surface water floats on the thermocline (Figure 49.21). In the water beneath it, decomposers may deplete the dissolved oxygen. Come autumn, the upper layer cools, becomes denser, then sinks. The thermocline vanishes as a result. During this **fall overturn**, water mixes vertically, so that dissolved oxygen moves down and nutrients move up.

Primary productivity corresponds with the seasons. After a spring overturn, the longer daylengths and cycled nutrients support higher rates of photosynthesis. Phytoplankton and rooted aquatic plants quickly take up

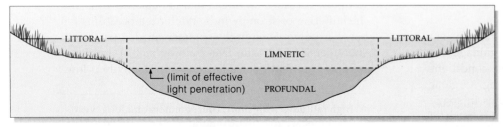

Figure 49.20 Lake zonation. The littoral extends around the lake's edge, from the shore to the depth where aquatic plants stop growing. The profundal is all water below the depth of light penetration. Above the profundal are open, sunlit waters of the limnetic zone.

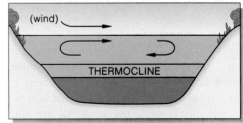

Figure 49.21 Thermal layering in a temperate lake in summer.

Figure 49.22 Stream habitats in North Carolina and Virginia. (a) Leaf detritus in a riffle. (b) Pool. (c) Pool leading into a riffle. (d) Closer look at a riffle. (e) A run, Sinking Creek.

phosphorus, nitrogen, and other nutrients. During a growing season, the thermocline cuts off vertical mixing. Nutrients in the remains of organisms sink to deeper waters, so photosynthesis declines. By late summer, the shortages are limiting photosynthesis. After a fall overturn, nutrient cycling drives another burst of primary productivity. The shorter daylight hours do not sustain a long-lasting burst. Not until spring will primary productivity increase again.

TROPHIC NATURE OF LAKES   The topography, climate, and geologic history of a lake dictate the numbers and kinds of residents, how they are dispersed, and how nutrients are cycled among them. Soils of the lake basin and the surrounding regions contribute to the type and amount of nutrients available to support organisms.

Interplays among the climate, soil, basin shape, and metabolic activities of a lake's residents contribute to conditions ranging from oligotrophy to eutrophy. Water in *oligotrophic* lakes commonly is deep, clear, and nutrient poor. Its primary productivity is not great. The water in *eutrophic* lakes typically is shallow, nutrient rich, and relatively high in primary productivity. These features arise naturally, as sediments gradually accumulate in the lake basin. The lake water becomes less deep and less transparent, compared with oligotrophic lakes. Phytoplankton are the dominant part of the community. More sediments may accumulate. The final successional stage is a filled-in basin, hence no more lake.

Eutrophication is the general name for processes that lead to nutrient enrichment of a lake or any other body of water. As you read earlier, in Section 48.9, human activities also can bring about eutrophication. Consider one case of this. As Val Smith documented, people caused eutrophication of Lake Washington in Seattle—and they also reversed it. From 1941 to 1963, the lake received too much phosphorus-rich sewage. The sewage discharge promoted massive blooms of cyanobacteria, which formed thick mats over the lake each summer and thus made it useless for recreation. With the phosphorus enrichment, nitrogen became the limiting resource in the lake. Then the cyanobacteria— which are superior competitors for nitrogen by their ability to fix N_2—became dominant.

From 1963 to 1968, the sewage discharges were cut back, and finally stopped. By 1975, Lake Washington was almost fully recovered. Lower density populations of diatoms and green algae became dominant.

Stream Ecosystems

The flowing-water ecosystems called **streams** start out as freshwater springs or seeps. They grow and merge as they flow downslope, then often combine to form a river. Between the headwaters and the river's end, we find three kinds of habitats—riffles, pools, and runs. The riffles are shallow, turbulent stretches where water flows swiftly over a rough bottom of sand and rock. Figure 49.22 shows two examples. The pools have deep water flowing slowly over a smooth, sandy, or muddy bottom. The runs are smooth-surfaced but fast-flowing stretches over bedrock or rock and sand.

A stream's average flow volume and temperature depend on rainfall, snowmelt, geography, altitude, and even the shade cast by plants. Its solute concentrations are influenced by the streambed's composition as well as by agricultural, industrial, and urban wastes.

Especially within forests, streams import most of the organic matter that supports food webs. Where trees cast shade, photosynthesis is diminished. Most of the organic matter is litter that enters detrital food webs. Aquatic organisms continually take up and release nutrients as water flows downstream. (Nutrients move upstream only as components of migrating fishes and other animals.) Think of nutrients as spiraling between water and aquatic organisms as the stream flows on its one-way course, usually to a river that flows to the sea.

Ever since cities formed, streams have been sewers for industrial and municipal wastes. The wastes, as well as other pollutants from poorly managed farmlands, choked many streams with sediments and chemically poisoned them. Streams are resilient, however, and they show impressive recovery when pollution is controlled.

Freshwater and saltwater provinces are far more extensive than the Earth's biomes. All of the aquatic ecosystems in these water provinces have characteristic gradients in light penetration, temperature, and dissolved gases.

Beyond land's end are two vast provinces of the open ocean. Together, they cover 71 percent of the Earth's surface. The *benthic* province includes all sediments and rocks of the ocean bottom (Figure 49.23). It starts with continental shelves and extends to deep-sea trenches. The *pelagic* province is the entire volume of ocean water. Its neritic zone is all the water above the continental shelves. Its oceanic zone is the water of the ocean basin.

Walk along the ocean and you may be humbled by its vastness (Figure 49.24). You see an immense expanse extending to the distant horizon, with no mountains or plains or valleys to break it up visually into something less overwhelming. What you do not see are *submerged* mountains and valleys and plains—which are as varied as the ones configuring the dry continents.

Primary Productivity in the Ocean

Within the ocean's upper surface waters, photosynthesis proceeds on a stupendous scale, just as it does on land. The ocean's primary productivity varies seasonally, just as it does on land (Section 7.8 and Figure 49.25). Its vast "pastures" of phytoplankton are the start of food webs that include zooplankton, copepods, shrimplike krill, whales, squids, and fishes. Organic remains and wastes from these marine communities sink through the water. They become the basis of detrital food webs for most communities in the benthic province.

In the 1970s, biologists reevaluated their notions about the ocean's primary productivity. They realized that as much as 70 percent of it may be the contribution of **ultraplankton**. Most of these tiny photosynthetic bacterial cells are only about two micrometers wide (40 millionths of an inch), and some are smaller than that. In tropical and subtropical seas once thought to be nearly devoid of producers, 0.035 gram (one *ounce*) of water holds as many as 3 million of these cells.

Hydrothermal Vents

In 1977, researchers studying the ocean floor near the Galápagos Islands discovered a distinctive ecosystem. In the Galápagos Rift, a volcanically active boundary between two of the Earth's crustal plates, they found communities thriving at **hydrothermal vents**. At this boundary, near-freezing water seeps into fissures and becomes heated to very high temperatures. As the pressure forces heated

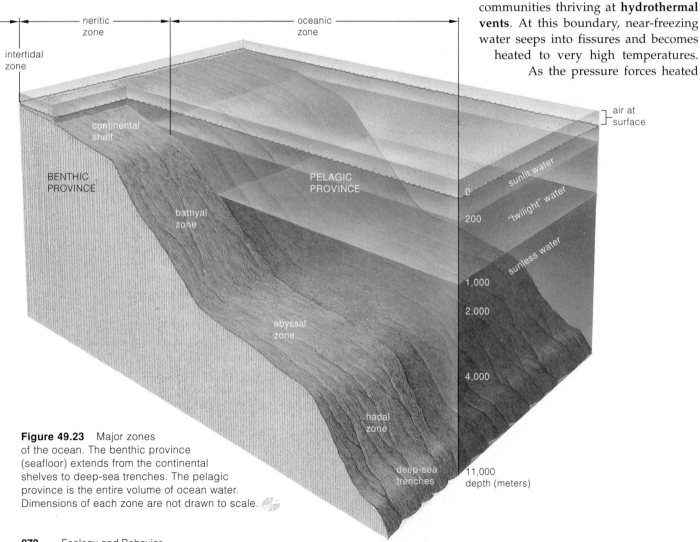

Figure 49.23 Major zones of the ocean. The benthic province (seafloor) extends from the continental shelves to deep-sea trenches. The pelagic province is the entire volume of ocean water. Dimensions of each zone are not drawn to scale.

Figure 49.24 (**a**) From a rocky shore, a view of the open ocean. (**b**) Near a British Columbia coastline, a whale breaching (leaping out of the water). Crustaceans (**c**) and tube worms (**d**) near a hydrothermal vent ecosystem on the ocean floor.

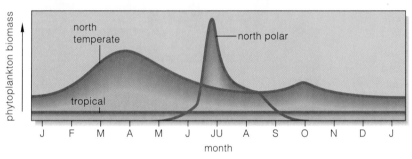

Figure 49.25 Seasonal variations in primary production in the ocean corresponding to latitude. The strongest peak in the *dark-green* graph lines for the north polar and temperate seas corresponds to phytoplankton blooms, brought about by a seasonal increase in daylength. The lower peak corresponds to an increase in availability of nutrients, something like the fall overturn in lakes. In most tropical seas, daylength and nutrient availability do not vary much; neither does primary production.

water upward, minerals are leached from rocks before the water spews out through vents in the seafloor. Iron, zinc, copper sulfides, and sulfates of magnesium and calcium are dissolved in the hydrothermal outpourings. These settle out and form rich mineral deposits. Sulfides in the deposits are energy sources for chemoautotrophic bacteria that are the starting point for hydrothermal vent communities. The bacteria are primary producers for food webs that include tube worms, crustaceans, clams, and fishes (Figure 49.24c,d).

Still more hydrothermal vent ecosystems have been found in the South Pacific near Easter Island; the Gulf of California, about 150 miles south of the tip of Baja California, Mexico; and the Atlantic. In 1990, a team of United States and Russian scientists located one in Lake Baikal, the world's deepest lake. This lake basin seems to be splitting apart (hence the vents) and may mark the beginning of a new world ocean.

Did life originate in such nutrient-rich places on the seafloor? Certainly conditions at the surface of the early Earth were about as inhospitable as you might imagine. At the least, cells on the seafloor would have protection from destructive forms of radiation that bombarded the Earth before the oxygen-rich atmosphere formed. Were cells slowly carted closer to the surface by the uplifting of the seafloor during episodes of crustal crunchings? These are some of the unanswered questions that some evolutionary detectives are now asking.

Although we are most familiar with features of the land, a single great ocean dominates the Earth's surface. Its primary productivity is staggering, and seasonal.

Communities of organisms thrive near hydrothermal vents on the ocean floor, which suggests to some that life itself may not have originated in the shallow waters of the Earth.

CORAL REEFS AND BANKS

The Master Builders

Each wave-resistant formation called a **coral reef** began as the remains of countless organisms accumulated. The hard parts of corals became the structural foundation. Coralline algae added to it; deposits of calcium and magnesium carbonates hardened the cell walls of these red algae, including *Corallina* (Figure 49.26). Secretions from other organisms helped cement things together.

The massive, pocketed spine of an existing reef is home to hundreds of species of corals, and a staggering variety of red algae and other organisms. Section 26.5 and Figure 49.26 merely hint at the wealth of warning colors, spines, tentacles, and stealthy behavior—clues to fierce competition for resources among species that are packed together in limited space.

Fringing reefs, barrier reefs, and atolls are the main reef formations (Figure 49.26). Most of the substantial reefs form in clear, warm waters between latitudes 25° north and south. Farther north or south, solitary corals or small colonies of them have formed **coral banks** in temperate waters, even in cold waters of continental shelves. Among these are smooth, vertical banks in the cold, deep waters near Japan, California, England, and New Zealand. A colonial branched coral, *Lophelia*, has built great banks in the cold waters of Norway's fjords.

The Once and Future Reefs

Colorful dinoflagellates often live as symbionts in the tissues of reef-building corals. In return for protection, they provide a coral polyp with oxygen and recycle its mineral wastes. When stressed, the polyps expel their protistan symbionts. When stressed for more than a few months, they die, and only bleached hard parts remain. Abnormal, widespread bleaching in the Caribbean and tropical Pacific began in the 1980s. So did an increase in sea surface temperature, and this may be a key stress factor. If the damage is one outcome of global warming, as marine biologists Lucy Bunkley-Williams and Ernest Williams suggest, the future looks grim for the reefs.

Human activities are destroying reefs in more direct ways than this. Where raw sewage and other pollutants are being discharged into the nearshore waters around populated islands, including some in the Hawaiian chain, reefs are dead or dying. Where fishermen from Japan, Indonesia, and Kenya move in for profit, reef life is being decimated. No simple nets for these fellows. They drop dynamite or cyanide in the water, so fish hiding among the corals float dead to the surface. They squirt sodium cyanide into the water, and stunned but not dead fish float to the surface. These are the tropical fish in pet stores as well as the rare fish brought live to Japanese restaurants, where they are dispatched and

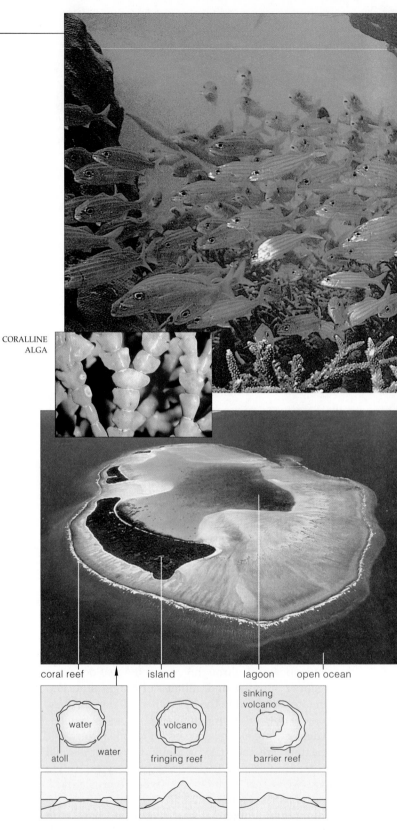

CORALLINE ALGA

coral reef   island   lagoon   open ocean

Figure 49.26 A sampling of the astounding biodiversity of tropical reefs. The above diagrams show three types of coral reefs. *Atolls* are ring-shaped coral reefs that fully or partially enclose a shallow lagoon. *Fringing reefs* form next to the land's edge in regions of limited rainfall, as on the leeward side of tropical islands. *Barrier reefs* form around islands or parallel with the shore of a continent. A calm lagoon forms behind them. Section 26.5 shows one.

LIONFISH

CRAB

MORAY EEL

CHAMBERED NAUTILUS

SEA ANEMONE

DAISY CORAL

PILLAR CORAL

CROWN-OF-THORNS SEA STAR

served up as exorbitantly expensive status symbols. On small, native-owned islands, fishing rights are traded for small sums. Then the reefs are destroyed, so they can no longer sustain the small human populations that once depended on them for their own sustenance.

Coral reefs generally form in clear, warm waters between latitudes 25° north and south. Vertical coral banks form in the temperate or cold waters of continental shelves.

All show spectacular biodiversity, and vulnerability.

LIFE ALONG THE COASTS

Remarkable ecosystems exist in estuaries and intertidal zones along the coasts. Like freshwater ecosystems, they differ in physical and chemical properties, such as light penetration and water temperature, depth, and salinity.

Estuaries

An **estuary** is a partially enclosed coastal region where seawater swirls and mixes with nutrient-rich freshwater from rivers, streams, and runoff from the surrounding land. The confined conditions, slow mixing of water, and tidal action combine to trap the dissolved nutrients. The continually freshened water replenishes nutrients and allows estuaries to support productive ecosystems.

Primary producers are phytoplankton, salt-tolerant plants that can withstand submergence at high tide, and algae living in mud and on plants. Much of the primary production enters detrital food webs in which bacterial and fungal decomposers are the first to feed. Detritus (and bacteria on and in it) feeds nematodes, snails, crabs, and fish. Filter feeders such as clams eat food particles suspended in the slowly moving water. So many larval and juvenile stages of invertebrates and some fishes are present that estuaries are often called marine nurseries. Also, many migratory birds use estuaries as rest stops.

Chesapeake Bay, Mobile Bay, and San Francisco Bay are broad, shallow estuaries. Estuaries in Alaska and British Columbia are narrow and deep; so are Norway's fjords. In Texas and Florida, estuaries lie behind long spits of sand and mud. The New England coast has many estuarine salt marshes (Figure 49.27). In tropical regions, we find mangrove swamps that function much like salt marshes in estuarine ecology.

Today, many estuarine ecosystems are under serious assault from raw sewage, agricultural runoff, industrial, urban, and suburban wastes, and upstream diversion of freshwater for human use. Normal conditions, such as suitable salinity levels, cannot be maintained without inflows of unpolluted freshwater.

The Intertidal Zone

Along rocky and sandy coastlines we find ecosystems of the **intertidal zone**, which is not renowned for its creature comforts. Waves batter its resident organisms. Tides alternately submerge and expose them. The higher they are, the more they may dry out, freeze in winter, or bake in summer, and the less food comes their way. The lower they are, the more they must compete in limited spaces. At low tides, birds, rats, and raccoons move in to feed on them. High tides bring the predatory fishes.

Generalizing about coastlines isn't easy, for waves and tides constantly resculpt them. One feature that both rocky and sandy shores do share is vertical zonation.

Rocky shores often have three zones (Figure 49.28a). The highest zone, the upper littoral, is submerged only during the highest tide of the lunar cycle; it is sparsely populated. The midlittoral (middle zone) is submerged during the highest regular tide and exposed during the lowest. In its tidepools we typically find red, brown, and green algae, hermit crabs, nudibranchs, sea stars, and small fishes (Figure 49.28b). Diversity is greatest in the lowest zone, the lower littoral, which is exposed only during the lowest tide of the lunar cycle. In all three shore zones, rapid erosion prevents detritus from accumulating, so grazing food webs prevail.

Waves and currents continually rearrange stretches of loose sediments, called *sandy* and *muddy* shores. Few large plants grow in these unstable places, so you won't find many grazing food webs. Detrital food webs start with the organic debris imported from offshore or nearby landforms. Vertical zonation is less obvious than along rocky shores. Below the low tide mark in temperate areas are blue crabs and sea cucumbers. Marine worms, crabs, and other invertebrates live between the high and low tide marks. At night, at the high tide mark, beach hoppers and ghost crabs pop out of their burrows, seeking food.

Figure 49.27 Salt marsh of a New England estuary where *Spartina*, a marsh grass, is the main producer. Its microbe-enriched litter feeds consumers in the creeks and sound.

SALT MARSH (estuary)

open ocean   sound   shallow bay   creek   tidal river

Figure 49.28 (a) Vertical zonation in the intertidal zone of a rocky shore in the Pacific Northwest. The difference between high and low tides is about three meters. Elsewhere, it varies from a few centimeters (in the Mediterranean Sea) to more than fifteen meters (in the Bay of Fundy, next to Nova Scotia). (b) Tidepool region typical of the Pacific Northwest.

Upwelling Along Coasts

In the Northern Hemisphere, prevailing winds from the north parallel the west coasts of continents and tug on the ocean surface. Wind friction causes surface waters to begin moving. Under the force of the Earth's rotation, the moving water is deflected westward, away from a coast, and cold, deep, often nutrient-rich water moves in vertically to replace it (Figure 49.29). When cold, deep water moves up this way, we call it **upwelling**. It happens in equatorial currents as well as along the coasts of continents in both hemispheres, and it cools the air above it. For example, those fogbanks that form along the California coast are one outcome of upwelling.

a Wind from the north starts surface ocean water moving. **b** Force of Earth's rotation deflects the moving water westward.

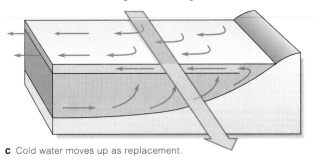

c Cold water moves up as replacement.

Figure 49.29 Upwelling along the west coasts in the Northern Hemisphere. Prevailing winds from the north start surface water moving. The force of Earth's rotation deflects the moving water westward. Cold, deeper water moves up to replace it.

In the Southern Hemisphere, the commercial fishing industries of Peru and Chile depend on wind-induced upwelling. When prevailing coastal winds blow in from the south and southeast, they tug surface water away from shore; and cold, deeper water that the Humboldt Current delivers to the continental shelf moves to the surface. Tremendous amounts of nitrate and phosphate are pulled up, then carried northward by the cold Peru Current. Phytoplankton that depend on these nutrients are the basis of one of the world's richest fisheries.

Every three to seven years, however, warm surface waters of the western equatorial Pacific move eastward. This massive displacement of warm water affects the prevailing wind direction. The eastward flow speeds up so much that it influences the movement of water along the coastlines of Central and South America.

Waters driven toward any coastline will be forced downward and will flow seaward along the continental shelf. Near the coast of Peru, prolonged "downwelling" displaces the cooler waters of the Humboldt Current— and this prevents upwelling. The phenomenon, which local fishermen named **El Niño**, has catastrophic effects on productivity, on birds that feed on anchovetas and other fishes, and on fishing industries. The next section provides a closer look at the El Niño phenomenon.

Major shifts in the circulation patterns of the ocean and atmosphere can have repercussions on the functioning of ecosystems that are global in scope.

49.14   EL NIÑO AND SEESAWS IN THE WORLD'S CLIMATES

We turn, finally, to a story that may help reinforce the unifying concept of this chapter—that conditions in the atmosphere, in the ocean, and on land are interconnected in ways that profoundly influence the world of life. The interconnections are astoundingly complex. The more we learn about them, the more it becomes apparent that we really know very little about them.

Our story begins in the winter of 1982–1983. Record rainfall caused massive mudslides along the California coast. Month after month, storms caused severe flooding along the normally arid and semiarid coasts of Ecuador and Peru. In India, the life-sustaining monsoon rains hardly materialized. Devastating droughts hit natural ecosystems and farms of Australia and the Hawaiian Islands. An ongoing drought in Africa intensified, with a resulting amplification of human starvation and death.

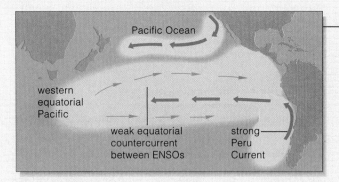

a Westward flow of cold suface water along the equator in between ENSOs.

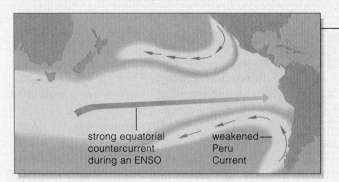

c Massive, eastward dislocation of warm ocean water during an ENSO.

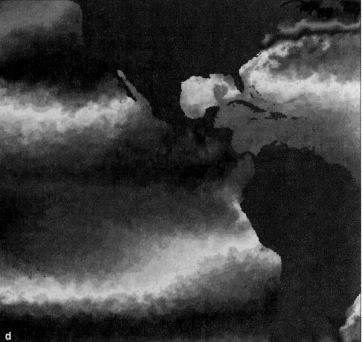

The massive dislocations in the world's patterns of rainfall started with changes in sea surface temperatures and air circulation patterns, which induced drought-related conditions on land. The changes cause a climatic event called the *El Niño Southern Oscillation* (ENSO).

"Southern Oscillation" refers to a recurring seesaw in atmospheric pressure in the western equatorial Pacific. The region is the world's largest reservoir of warm water. More warm air rises here than anywhere else, and heavy

Figure 49.30   (**a**,**b**) Distribution of sea surface temperatures for the Pacific between ENSOs. In the 1988 satellite image in (**b**), warmest water is color-coded *dark red*, and progressively cooler water is *yellow*, then *green*. The western equatorial Pacific had the warmest water. A tongue of relatively cold surface water, which originated with upwelling along South America's coast, flowed westward along the equator from South America. (**c**,**d**) Distribution of sea surface temperatures during an ENSO, as recorded on May 13, 1992. Trade winds weakened and then reversed, leading to a massive, eastward movement of warm water along the equator from the central and eastern Pacific.

rainfall releases much of the heat energy that drives the world's air circulation system. Pulses of heat from the Earth—possibly at clusters of a thousand or more active volcanoes that were recently discovered on the deep ocean floor—may trigger the Southern Oscillation. Whatever the cause, heat moves upward and warms the surface waters.

Normally, the warm reservoir and the heavy rainfall associated with it move westward (Figure 49.30a). Now they move east (Figure 49.30c). Prevailing surface winds in the western equatorial Pacific pick up speed. Stronger winds have a pronounced effect on "dragging" the ocean surface waters to the east. The upper ocean currents are affected to the extent that the westward transport of water slows and the eastward transport increases. *More* warm water in the vast reservoir moves east—and so on in a feedback loop between the ocean and the atmosphere.

As you read earlier, the reversal in the usual westward flow of air and water displaces the cold, deep Humboldt Current. And it stops the upwelling of nutrients along the western coast of South America. Peruvian fishermen call the warm, nutrient-poor current from the east *El Niño*. Usually it reaches their coast around Christmas; hence the name, "the little one," a reference to the baby Jesus.

An ENSO is a recurring event. About 2,000 kilometers west of Peru, the sea level rises. Sea surface temperatures increase, often by 7°C (11°F). Evaporation increases and air pressure falls. Humid updrafts form; they trigger violent storms along coasts and often more rain in inland regions.

The wave of warm water from the 1982 El Niño was 20 centimeters high. It reached South America in two months and spread along the coasts, then slowly moved back across the Pacific. Two more El Niños followed this one, which only now is subsiding in the Bering Sea. Each time, global winds and rainfall patterns changed. So did productivity. For example, for 1970 to 1993, a plot of the corn yields in Zimbabwe matches a plot of the rise and fall of sea surface temperatures in the Pacific.

Researchers use mathematical models to give advance warning of such major episodes of climatic change. At this writing (October 1997), they have done so for the latest ENSO, which is shaping up to be the most powerful in history. Sea surface temperatures in the eastern Pacific are nearly nine degrees above average. Warm water piling up against Peru's coast stretches more than 9,660 kilometers to the west and more than 320 kilometers to the north, which may herald devastating storms in California. An early hurricane has slammed through Acapulco, Mexico. Drought in Australia has ranchers slaughtering herds for lack of water and fodder. El Niño delayed the anticipated monsoons in Southeast Asia, so fires set to clear tracts of rain forests burned out of control for weeks. The warmer, nutrient-poor water does not bode well for the coral reefs off Central America, already hit hard in 1982–1983.

With so much at stake, climatologists plan to use this El Niño to refine ecosystem modeling of the interrelated systems of the ocean, land, and atmosphere.

49.15 SUMMARY

1. The biosphere encompasses only the Earth's waters, lower atmosphere, and uppermost portions of its crust in which organisms live. Energy flows one way through the biosphere, and materials move through it on a grand scale to influence ecosystems everywhere.

2. The distribution of species through the biosphere is an outcome of the Earth's history, topography, climate, and interactions among species.

3. "Climate" refers to the average weather conditions, including temperature, humidity, wind velocity, cloud cover, and rainfall, over time. It results from differences in the amount of solar radiation reaching equatorial and polar regions, the Earth's daily rotation and its annual path around the sun, the distribution of continents and oceans, and the elevation of land masses.

4. Interacting climatic factors produce prevailing winds and ocean currents, which shape the global weather patterns. The weather affects the composition of soil, sedimentation, and water availability in ecosystems—which in turn affect the growth and distribution of their primary producers.

5. The world's land masses are classified as six major biogeographic realms. Each is more or less isolated by oceans, mountain ranges, or desert barriers, which tend to restrict gene flow between other realms. As a result, each tends to maintain its characteristic array of species.

6. A biome (a category of major ecosystems on land) is shaped by regional variations in climate, landforms, and soil composition. Plant species that are adapted to a certain set of conditions dominate each biome. Deserts, dry shrublands, dry woodlands, grasslands, broadleaf forests (such as tropical rain forests), coniferous forests, and tundra are major types.

7. Water provinces cover more than 71 percent of the Earth's surface. They include standing freshwater (such as lakes), running freshwater (such as streams), as well as the world ocean and seas. All aquatic ecosystems show gradients in light penetration, water temperature, salinity, and dissolved gases. These factors vary daily and seasonally. They influence primary productivity.

8. Estuaries, intertidal zones, rocky and sandy shores, tropical reefs, and regions of the open ocean are major marine ecosystems. Photosynthetic activity is greatest in shallow coastal waters and in regions of upwelling. Upwelling is an upward movement of deep, cool ocean water that often carries nutrients to the surface.

9. The interrelatedness of ocean surface temperatures, the atmosphere, and the land is evident in studies of the ENSO (El Niño Southern Oscillation). Abnormally severe droughts and flooding in different regions of the world accompany this recurring event.

Review Questions

1. Define atmosphere, lithosphere, and hydrosphere. *CI*

2. List the major interacting factors that influence climate. *49.1*

3. List some of the ways in which air currents, ocean currents, or both may influence the ecosystem in which you live. *49.1, 49.2*

4. Indicate where the following types of biomes tend to be located around the world. Then describe some of their defining features. *49.3, 49.5–49.9*
   a. deserts
   b. dry shrublands
   c. grasslands
   d. evergreen broadleaf forests
   e. deciduous broadleaf forests
   f. coniferous forests
   g. alpine tundra
   h. arctic tundra

5. Define soils, then explain how the composition of regional soils affects ecosystem distribution. *49.4*

6. Describe the littoral, limnetic, and profundal zones of a large temperate lake in terms of seasonal primary productivity. *49.10*

7. Define the two major provinces of the world ocean. *49.11*

8. What points favor the hypothesis that life may not have originated in the surface waters of the seas? *49.11*

9. Define and characterize the following ecosystems. *49.12, 49.13*
   a. estuary
   b. rocky shore
   c. sandy shore
   d. coral reef

10. Describe an ENSO event and some of its consequences in both hemispheres. *49.13, 49.14*

Self-Quiz (*Answers in Appendix IV*)

1. Solar radiation drives the distribution of weather systems and so influences the distribution of _____ .
   a. every single ecosystem
   b. ecosystems on land only
   c. all ecosystems except those at hydrothermal vents

2. The _____ is a shield against ultraviolet wavelengths from the sun.
   a. upper atmosphere   c. ozone layer
   b. lower atmosphere   d. greenhouse effect

3. Regional variations in the global patterns of rainfall and temperature depend on _____ .
   a. global air circulation   c. topography
   b. ocean currents   d. all of the above

4. A rain shadow is a reduction in rainfall on the _____ of a mountain range.
   a. windward side   c. highest elevation
   b. leeward side   d. lowest elevation

5. Biogeographic realms are _____ .
   a. land and water provinces c. divided into biomes
   b. six major land provinces d. b and c

6. Biome distribution corresponds roughly with regional variations in _____ .
   a. climate   c. topography
   b. soils   d. all of the above

7. _____ are highly adapted to episodes of fire.
   a. Dry shrublands   c. Pine barrens
   b. Grasslands   d. All of the above

8. During _____ , deeper, often nutrient-rich water moves to the surface.
   a. spring overturns   c. upwellings
   b. fall overturns   d. all of the above

9. Match the terms with the most suitable description.
   ____ boreal forest
   ____ permafrost
   ____ chaparral
   ____ desert
   ____ deciduous broadleaf forest
   ____ tropical rain forest

   a. high productivity, poor cycling of mineral nutrients
   b. "swamp forest"
   c. common at moist, temperate latitudes having mild winters
   d. evaporation greatly exceeds sparse, infrequent rainfall
   e. a type of dry shrubland
   f. feature of arctic tundra

10. Match the terms with the most suitable description.
   ____ plankton
   ____ upwelling
   ____ eutrophication
   ____ estuary
   ____ benthic province

   a. deep, cool, often nutrient-rich ocean water moves upward
   b. sediments, rocky formations of the ocean bottom
   c. seawater, freshwater slowly mix
   d. nutrient enrichment of body of water; reduced transparency, phytoplankton blooms
   e. aquatic community of mostly microscopic floating or weakly swimming species

Critical Thinking

1. Kangaroos are furry, pouched mammals. Raccoons are furry placental mammals. Speculate on why Australia but not North America is the original home of kangaroos and why the reverse is true of raccoons. Use your knowledge of geologic history when devising an explanation. (*Compare* Section 27.10.)

2. Reflect on the world distribution of land masses and ocean water, then develop an explanation of why grassland biomes tend to form in the interior of continents, not along their coasts.

3. *Wetlands* are the transitional zones between aquatic and terrestrial ecosystems in which plants are adapted to grow in periodically or permanently saturated soil. They include *marshes*, in which 15 to 300 centimeters of water cover the land. They include *riparian zones* along the fringes of most rivers, such as the shrubby bottomlands of the southern United States, and 14 million hectares of *mangrove swamps*, with diverse species of mangrove trees and shrubs (Figure 49.31). Many species require wetlands to complete at least part of their life cycle. Besides this, different wetlands serve as traps for sediments and waterborne pollutants, buffer zones for controlling erosion and flooding,

Figure 49.31 Mangrove swamp. Mangrove trees root in tidal zones of tropical regions. The salty water would kill most plants. Living cells of mangrove trees accumulate so many organic solutes that they can take up freshwater by osmosis.

a

Figure 49.32 (**a**) Crater Lake, Oregon, a collapsed volcanic cone now filled with water from rains and melted snow. Like the other volcanoes of the Cascade range, it started forming at the dawn of the Cenozoic, when crustal plates underwent major reorganization. (**b**) Characteristics of oligotrophic and eutrophic lakes.

b

OLIGOTROPHIC LAKE	EUTROPHIC LAKE
Deep, steeply banked	Shallow with broad littoral
Large deep-water volume relative to surface-water volume	Small deep-water volume relative to surface-water volume
Highly transparent	Limited transparency
Water blue or green	Water green to yellow or brownish-green
Low nutrient content	High nutrient content
Oxygen abundant through all levels throughout year	Oxygen depleted in deep water during summer
Not much phytoplankton; green algae and diatoms dominant	Abundant masses of phytoplankton; cyanobacteria dominant
Abundant aerobic decomposers favored in profundal zone	Anaerobic decomposers
Low biomass in profundal	High biomass in profundal

and often as carbon dioxide sinks. Wetlands everywhere are being converted for agriculture, home building, and other human activities. Today only 50 percent of the spectacular wetlands called the Florida Everglades remains. California has only 9 percent of its wetlands intact. And Indiana and Missouri have only 10 percent.

Does the great ecological value of wetlands for the nation as a whole outweigh rights of private citizens—who hold title to many of the remaining wetlands in the United States? Should the private owners be forced to transfer title to the government? If so, who should decide the value of a parcel of land *and* pay for it? Or would such seizure of private property violate the United States Constitution?

4. Observe conditions in a lake near your home or a place where you vacation. If lakes aren't part of your life, think about Oregon's Crater Lake instead (Figure 49.32*a*). Using the data in Figure 49.32*b* as a guide, would you conclude Crater Lake is oligotrophic or eutrophic? Will it remain so over time?

5. The United States Forest Service and Geophysical Dynamics Laboratory ran supercomputer programs to predict possible outcomes of a *global warming* trend. (Here you may wish to review Section 48.7.) According to the results, by the year 2030, the United States will experience recurring, devastating forest fires. Severe storms and more frequent rainfall in the western states will accelerate erosion, especially along the coasts and in steep foothills and mountain ranges.

Increases in the deposition of sediments will have adverse effects on rivers, streams, and estuaries. Dry shrublands and woodlands may replace hardwood forests in Minnesota, Iowa, and Wisconsin, and in parts of Missouri, Illinois, Indiana, and Michigan. The breadbasket of the American Midwest will shift north into Canada. Think about where you live now and where you plan to live in the future. How would such changes affect you? Should nations get serious about finding ways to counter global warming? Or do you believe that the scientists who are concerned about this are merely alarmist and that nothing bad will happen? Explain why you have reached your conclusion.

6. Think about the ocean, which covers the majority of the Earth's surface. Its deepest regions are remote from human populations, to say the least. Now think about a modern-day problem—where to dispose of nuclear wastes and other truly hazardous materials. You will be reading about this serious

problem in the next chapter; for now, simply be aware that the United States government has stockpiles of very dangerous wastes and has nowhere to put them. Would it be feasible, let alone ethical, to use the deep ocean as a "burying ground" for them? Why or why not?

Selected Key Terms

atmosphere *CI*
biogeographic realm *49.3*
biogeography *49.3*
biome *49.3*
biosphere *CI*
boreal forest *49.8*
climate *CI*
coniferous forest *49.8*
coral bank *49.12*
coral reef *49.12*
deciduous broadleaf forest *49.7*
desert *49.5*
desertification *49.5*
dry shrubland, woodland *49. 6*
El Niño *49.13*
estuary *49.13*
eutrophication *49.10*
evergreen broadleaf forest *49.7*
fall overturn *49.10*
grassland *49.6*

hydrosphere *CI*
hydrothermal vent *49.11*
intertidal zone *49.13*
lake *49.10*
lithosphere *CI*
monsoon *49.2*
ocean *49.2*
ozone layer *49.1*
permafrost *49.9*
pine barren *49.8*
rain shadow *49.2*
savanna *49.6*
soil *49.4*
spring overturn *49.10*
stream *49.10*
temperature zone *49.1*
tropical rain forest *49.7*
tundra *49.9*
ultraplankton *49.11*
upwelling *49.13*

Readings

Brown, J., and A. Gibson. 1983. *Biogeography*. St. Louis: Mosby.

Garrison, T. 1996. *Oceanography*. Second edition. Belmont, California: Wadsworth.

Smith, R. 1996. *Ecology and Field Biology*. Fifth edition. New York: Harper Collins.

Web Site See *http://www.wadsworth.com/biology* for practice quiz questions, hypercontents, BioUpdates, and critical thinking. The Wadsworth Biology Resource Center provides a wealth of information fully organized and integrated by chapter.

50

HUMAN IMPACT ON THE BIOSPHERE

An Indifference of Mythic Proportions

Of all the concepts introduced in the preceding chapter, the one that should be foremost in your mind is this: *The atmosphere, ocean, and land interact in ways that help dictate living conditions throughout the biosphere.* Driven by energy streaming in continually from the sun, these stupendous interactions give rise to globe-spanning temperatures and circulation patterns upon which life ultimately depends. With this chapter, we turn to a related concept of equal importance. Simply put, *we have become major players in these interactions even before we fully comprehend how they work.*

To gain perspective on what is happening, think about something we take for granted—the air around us. The composition of the present-day atmosphere is a result of geologic and metabolic events—most notably, photosynthesis—that began billions of years ago. The first humans, recall, evolved about 2.5 million years ago. Like us, they breathed oxygen from an atmosphere of ancient origins. Like us, they were protected from the sun's ultraviolet radiation by an ozone shield in the stratosphere. The size of their population was not much

to speak of, and their effect on the biosphere was trivial. About 11,000 years ago, however, agriculture began in earnest, and it laid the foundation for huge increases in population size. A few centuries ago, medical and industrial revolutions expanded that foundation—and human population growth skyrocketed in a mere blip of evolutionary time.

Today we are extracting huge amounts of energy and resources from the environment and giving back monumental amounts of wastes. As we do so, we are destabilizing ecosystems everywhere, even though the magnitude of change might not even be recognized when measured on the scale of a lifetime (Figure 50.1).

In a few developed countries, population growth has more or less stabilized, and the resource use per individual has dropped a bit. But resource utilization levels in those places are already high. In developing countries in Central America, Africa, and elsewhere, population sizes and resource consumption are rapidly growing, even though hundreds of millions of people are already malnourished or starving to death.

1900   1940   1954   1962

Many of the problems sketched out in this chapter are not going to disappear tomorrow. It will take many decades, even centuries, to reverse some trends that are already in motion—and not everyone is ready to make the effort. A few enlightened individuals in Michigan or Alberta or New South Wales can commit themselves to resource conservation and to minimizing pollution. But scattered attempts will not be enough. Individuals of all nations will unite to reverse global trends only when they perceive that the dangers of *not* doing so outweigh the personal benefits of ignoring them.

Does this seem pessimistic? Think of the exhaust fumes released into the air each time you drive a car or truck. Think of oil refineries, food-processing plants, and paper mills that supply you with goods and also release chemical wastes into the nation's waterways. Think of Mexico and other developing countries that produce cheap food by using an unskilled labor force and toxic pesticides. Unregulated pesticide applications poison the people who work the land, the land itself, and sometimes people who buy the exported produce.

Who changes behavior first? We have no answer to the question. We can suggest, however, that a strained biosphere can rapidly impose an answer upon us. As an individual, you might choose to cherish or brood about or ignore any aspect of the world of life. Whatever your choice may be, the bottom line is that you and all other organisms are in this together. Our lives interconnect, to degrees that we are only now starting to comprehend.

Figure 50.1 Tracking human population growth in one small part of the world, based on historical data and satellite imaging. *Red* denotes areas of dense human settlement in and around the San Francisco Bay Area and Sacramento since 1900.

KEY CONCEPTS

1. Human population growth has been skyrocketing ever since the mid-eighteenth century. At present, humans have the population size, the technology, and the cultural inclination to use energy and alter the environment at astonishing rates.

2. Pollutants are substances with which ecosystems have had no prior evolutionary experience, in terms of kinds or amounts, so adaptive mechanisms are not in place that can deal with them. In a more restrictive sense, pollutants are substances that accumulate in amounts that have adverse effects on human health, activities, or survival.

3. Many pollutants, including the kinds that contribute to the formation of smog or acid rain, exert regionally harmful effects. The effects of other pollutants, including chlorofluorocarbons that attack the ozone layer in the stratosphere, are global in scale.

4. Conversion of marginally fertile lands for agriculture, rampant deforestation, and other practices required to meet the demands of the growing human population are leading to loss of soil fertility and desertification. They are also resulting in a decline in the quality and quantity of one of the most crucial of all resources—fresh water.

5. Ultimately, the world of life depends on energy inputs from the sun. That energy drives the complex interactions among the atmosphere, ocean, and land. Our activities are disrupting the interactions in ways that may have severe consequences in the near future.

6. We as a species must come to terms with principles of energy flow and principles of resource utilization that govern all systems of life on Earth.

AIR POLLUTION—PRIME EXAMPLES

Let's start this survey of the impact of human activities by defining pollution, which is central to our problems. **Pollutants** are substances with which ecosystems have had no prior evolutionary experience, in terms of kinds or amounts, so adaptive mechanisms that can deal with them are not in place. From the human perspective, they are substances that accumulate to levels that have adverse effects on our health, activities, or survival.

Air pollutants are prime examples. As listed in Table 50.1, they include carbon dioxide, oxides of nitrogen and sulfur, and chlorofluorocarbons. Also among them are photochemical oxidants, formed as the sun's rays interact with certain chemicals. The United States alone releases 700,000 metric tons of air pollutants *each day*. Whether these remain concentrated at the source or are dispersed during a given interval depends on the local climate and topography, as you will now see.

Smog

During a **thermal inversion**, weather conditions trap a layer of cool, dense air under a layer of warm air. If the trapped air contains pollutants, winds cannot disperse them, and they may accumulate to dangerous levels. Thermal inversions have been key factors in some of the worst air pollution disasters, because they intensify an atmospheric condition called smog (Figure 50.2).

Two types of smog—gray air and brown air—form in major cities. Where winters are cold and wet, **industrial smog** develops as a gray haze over industrialized cities that burn coal and other fossil fuels for manufacturing,

Table 50.1	Major Classes of Air Pollutants
Carbon oxides	Carbon monoxide (CO), carbon dioxide (CO_2)
Sulfur oxides	Sulfur dioxide (SO_2), sulfur trioxide (SO_3)
Nitrogen oxides	Nitric oxide (NO), nitrogen dioxide (NO_2), nitrous oxide (N_2O)
Volatile organic compounds	Methane (CH_4), benzene (C_6H_6), chlorofluorocarbons (CFCs)
Photochemical oxidants	Ozone (O_3), peroxyacyl nitrates (PANs), hydrogen peroxide (H_2O_2)
Suspended particles	Solids (dust, soot, asbestos, lead, etc.) and droplets (sulfuric acid, oils, pesticides, etc.)

heating, and generating electric power. Burning releases airborne pollutants, such as dust, smoke, soot, ashes, asbestos, oil, bits of lead and other heavy metals, and sulfur oxides. If not dispersed by winds and rain, the emissions may reach lethal concentrations. Industrial smog caused 4,000 deaths during the 1952 air pollution disaster in London. Before coal burning was restricted, New York, Pittsburgh, and Chicago also were gray-air cities. Today, most industrial smog forms in the cities of China, India, and other developing countries, as well as cities of the coal-dependent countries of eastern Europe.

In warm climates, **photochemical smog** develops as a brown haze over large cities. It becomes concentrated when the surrounding land forms a natural basin, as it does around Los Angeles and Mexico City (Figure 50.2). The key culprit is nitric oxide. After it is released from vehicles, it reacts with oxygen in air and forms nitrogen dioxide. When exposed to sunlight, nitrogen dioxide reacts with hydrocarbons; photochemical oxidants result. Most of the hydrocarbons are from spilled or partly burned gasoline. Key oxidants are ozone and **PANs** (for peroxyacyl nitrates). Even traces of PANs sting eyes, irritate lungs, and damage crops.

Acid Deposition

Oxides of sulfur and nitrogen are among the worst air pollutants. Coal-burning power plants, metal smelters, and factories emit most sulfur dioxides. Motor vehicles, power plants that burn gas and oil, and nitrogen-rich fertilizers produce nitrogen oxides. In dry weather, fine particles of oxides may briefly stay airborne, then fall to Earth as **dry acid deposition**. When dissolved in atmospheric water, they form weak solutions of sulfuric and nitric acids. Winds may disperse them over great distances. If they fall to Earth in rain and snow, we call this wet acid deposition, or **acid rain**. The pH of normal rainwater is about 5. Acid rain can be 10 to 100 times more acidic, even as potent as lemon juice! The deposited acids eat away at marble buildings, metals, rubber, plastic, nylon

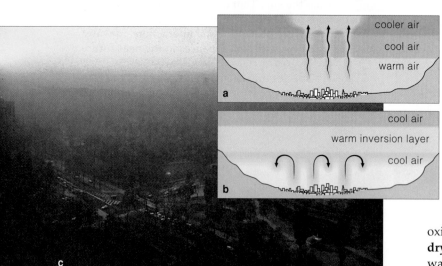

Figure 50.2 (**a**) Normal pattern of air circulation in smog-forming regions. (**b**) Air pollutants trapped under a thermal inversion layer. (**c**) Mexico City on an otherwise bright, sunny morning. Topography, staggering numbers of people and vehicles, and industry combine to make its air among the world's smoggiest. Breathing this city's air is like smoking two packs of cigarettes a day.

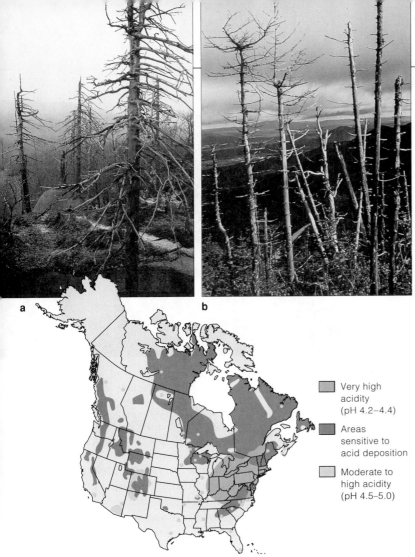

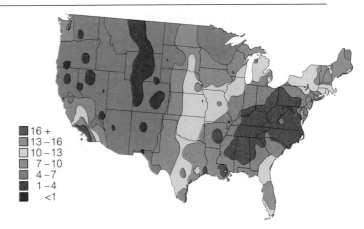

Figure 50.4 Regional differences in the concentrations of fine particles (2.5 μm or less) in micrograms per cubic meter of air, measured in 1992. Data are averages; they understate peak values in individual cities.

16 +
13 – 16
10 – 13
7 – 10
4 – 7
1 – 4
<1

Figure 50.3 Average 1984 acidities of precipitation and soil sensitivities to acid deposition, for regions of North America. Prolonged exposure to air pollutants is contributing to rapid destruction of trees in Germany (**a**), New York's Whiteface Mountains (**b**), and elsewhere. The weakened trees are more vulnerable to drought, disease, and insect attack.

Very high acidity (pH 4.2–4.4)

Areas sensitive to acid deposition

Moderate to high acidity (pH 4.5–5.0)

stockings, and many other materials. They significantly disrupt the physiology of organisms and the chemistry of ecosystems.

Depending on their soil type and vegetation cover, some regions are far more sensitive than others to acid deposition (Figure 50.3). Highly alkaline soil neutralizes acids before they enter streams and lakes of watersheds. Water having a high carbonate content also neutralizes acids. However, in many watersheds throughout much of northern Europe and southeastern Canada, as well as in scattered regions throughout the United States, thin soils overlie solid granite, and these cannot buffer the acids by much.

Rain in much of eastern North America is thirty to forty times more acidic than it was several decades ago. Crop yields are diminishing. Fish populations already have vanished from more than 200 lakes in New York's Adirondack Mountains. By some predictions, fish will disappear from 48,000 lakes in Ontario, Canada, within

the next two decades. Pollution from industrial regions is contributing to the destruction of forest trees and of mycorrhizae that support new growth (Section 24.4).

Researchers confirmed long ago that emissions from power plants, factories, and vehicles are the key sources of air pollutants. In 1995, researchers from the Harvard School of Public Health and Brigham Young University reported this finding from a comprehensive air quality study: Whack a year or so off your life span if you live in cities with the dirtiest air—especially air with fine particles of dust, soot, smoke, or acid droplets. Smaller particles are more easily inhaled and can damage lung tissue. Figure 50.4 summarizes air-quality data gathered in 1992 for the continental United States.

At one time the world's tallest smokestack, in the Canadian province of Ontario, accounted for 1 percent by weight of the annual worldwide emissions of sulfur dioxide. But Canada receives more acid deposition from industrialized regions of the midwestern United States than it sends across its southern border. Most of the air pollutants in Scandinavian countries, the Netherlands, Austria, and Switzerland arise in industrialized regions of western and eastern Europe. *Prevailing winds—hence air pollutants—do not stop at national boundaries.*

Pollutants are substances with which ecosystems have had no prior evolutionary experience, in terms of kinds or amounts, so adaptive mechanisms are not in place to deal with them.

An accumulation of pollutants can harm organisms, as when they reach levels that adversely affect human health.

Smog formation and acid deposition in specific regions are examples of air pollution, although prevailing winds often distribute such pollutants beyond regional boundaries.

Smog forms mainly as a result of fossil fuel burning in urban and industrialized regions. Airborne acidic pollutants drift to Earth as dry particles or as components of acid rain.

Now think about the ozone layer, almost twice as high above sea level as the top of Mount Everest, the highest place on Earth. Each September through mid-October, the layer gets thinner at high latitudes. So pronounced is the seasonal **ozone thinning** that it was once called an "ozone hole." In 1995, ozone thinning above Antarctica spanned an area twice as extensive as Europe. Thinning at high northern latitudes exceeded 10 percent. Sixty years from today, the protective layer may be reduced by 30 percent or more over heavily populated regions of North America, Europe, and Asia.

Why is ozone reduction alarming? It allows more ultraviolet radiation to reach the Earth. The increase may lead to dramatic increases in cancers of the skin, cataracts of the eyes, and weakened immune systems. Ultraviolet radiation also impacts on photosynthesis. Drastic drops in the oxygen-releasing activity of phytoplankton alone could alter the composition of the atmosphere, given their staggering numbers (Section 7.8).

Chlorofluorocarbons (**CFCs**) are the major factors in the ozone reduction. These compounds of chlorine, fluorine, and carbon are odorless and invisible. You find them in refrigerators and air conditioners (as the coolants), solvents, and plastic foams. CFCs slowly escape into air and resist breakdown. When a free CFC molecule absorbs ultraviolet light, it gives up one chlorine atom. If chlorine reacts with ozone, this yields oxygen and chlorine monoxide—which can react with free oxygen to release another chlorine atom. Each released chlorine atom can convert 10,000 ozone molecules or more to oxygen!

The chlorine monoxide levels above Antarctica are 100 to 500 times higher than at mid-latitudes. Why? During the winter, high-altitude ice clouds form there. Winds rotate around the South Pole for most of the winter and isolate the ice clouds from other latitudes (Figure 50.5). This also happens on a lesser scale in the Arctic. Ice provides a surface that promotes the breakdown of chlorine compounds. Chlorine is free to destroy ozone when the air warms in the spring, hence the ozone thinning.

CFCs aren't the only ozone eaters. Methyl bromide, a fungicide, is better at it. It persists only briefly in the atmosphere, but it will account for about 15 percent of the thinning in future years unless production stops.

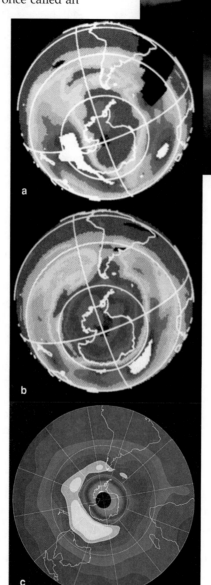

Figure 50.5 Seasonal ozone thinning above Antarctica in (**a**) 1979, (**b**) 1991, and (**c**) 1996. Lowest ozone values are coded *magenta* and *purple*. (**d**) Ice clouds above Antarctica that have a role in the ozone thinning each spring.

A few scientists dismiss the threat of the chemicals that contain chlorine and bromine. But the vast majority of those who have studied ecosystem modeling programs conclude that these chemicals do pose long-term threats—not only to human health and certain crops but also to animal life in general. Substitutes are now available for most applications of CFCs, and others are being developed.

Under international agreement, CFC production has been phased out in the developed countries. By 2010, developing countries will have phased out CFCs. Methyl bromide production will stop by 2010. By recent computer models, the area of thinning should not become any larger. Assuming international goals are met, it still will be about 50 years before the ozone layer is restored to 1985 levels and another 100 to 200 years to total recovery, to pre-1950 levels. Meanwhile, you and the children and grandchildren of future generations will be living with the destructive effects of air pollutants. And if levels of greenhouse gases continue to rise, the stratosphere could cool enough to increase the size and duration of seasonal ozone depletions over the poles.

Air pollution can have global repercussions, as when CFCs and other compounds contribute to a thinning of the ozone layer that shields life from the sun's ultraviolet radiation.

WHERE TO PUT SOLID WASTES, WHERE TO PRODUCE FOOD

Oh Bury Me Not, In Our Own Garbage

In natural ecosystems, one organism's wastes serve as resources for others, so the by-products of existence are cycled through the system. In the developing countries, many resources are scarce. People conserve what they can and discard little. In the United States and some other developed countries, most people use something once, discard it, then buy another. Each year, millions of metric tons of solid wastes are dumped, burned, and buried. Paper products make up half the total volume, which also includes 50 billion nonreturnable cans and bottles. Each week, paper manufactured from 500,000 trees ends up in a Sunday newspaper to Americans. If every reader recycled merely one of ten newspapers, 25 million trees a year could be left standing. Recycling paper would reduce the airborne pollutants released during paper manufacturing by 95 percent and require 30 to 50 percent less energy than making new paper.

A throwaway mentality is unique in the world of life. We are condoning the monumental accumulation of solid wastes in human ecosystems and natural ones. Bury the garbage in landfills? Then what happens when the space around cities runs out? Besides this, landfills "leak" and in time threaten groundwater supplies. Burn the wastes in inefficient incinerators? These spew great volumes of pollutants and ashes into the atmosphere.

Today, recycling is affordable and feasible. At the personal level, individuals can help bring about change by refusing to buy goods that are excessively wrapped, packaged in nondegradable containers, and designed for one-time use. They can engage in curbside recycling, by which they presort recyclable wastes. Such action is more efficient than reliance on huge resource recovery centers. These require so much trash to turn a profit that owners may end up encouraging the throwaway mentality. The centers also produce a toxic ash that must be disposed of in landfills—which eventually leak.

Converting Marginal Lands for Agriculture

The human population already uses nearly 21 percent of the Earth's land surfaces for cropland or for grazing. Another 28 percent is said to be potentially suitable for agriculture, but productivity would be so low that the conversion may not be worth the cost (Figure 50.6).

Asia and several other heavily populated regions now experience recurring, severe food shortages. Yet more than 80 percent of their productive land is already under cultivation. Scientists have made valiant efforts to improve crop production on existing land. Under the banner of the **green revolution**, their research has been directed toward (1) improving the genetic character of crop plants for higher yields and (2) exporting modern

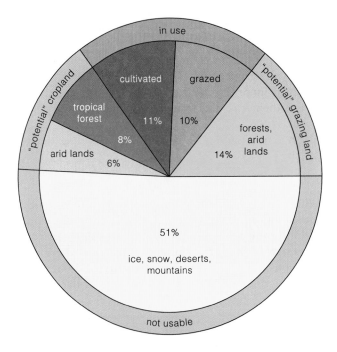

Figure 50.6 Classification of land with respect to its suitability for agriculture. Theoretically, clearing vast tracts of tropical forests and irrigating marginal lands could more than double the world's cropland. Doing so would destroy valuable forest resources, damage the environment, cause severe losses in biodiversity, and possibly cost more than it is worth.

agricultural practices and equipment to the developing countries. Many of these countries rely on *subsistence* agriculture, which runs on energy inputs from sunlight and human labor. They also rely very heavily on *animal-assisted* agriculture, with energy inputs from oxen and other draft animals. By contrast, *mechanized* agriculture requires massive inputs of fertilizers, pesticides, and ample irrigation to sustain high-yield crops. It requires fossil fuel energy to drive farm machines. Crop yields are four times as high. But the modern practices use up 100 times more energy and minerals. Also, there are signs that limiting factors are coming into play to slow down further increases in crop yields.

Pressures for increased food production are greatest in parts of Central and South America, Asia, the Middle East, and Africa where human populations are rapidly expanding into marginal lands. Repercussions extend beyond national boundaries, as you will see next.

Our astounding population growth has impact on the Earth's land. We generate huge amounts of solid wastes but reuse or recycle very little. We also are making the energetically and environmentally costly move of expanding into marginal lands for food production.

DEFORESTATION—CONCERTED ASSAULTS ON FINITE RESOURCES

At one time, tropical forests cloaked regions that were, collectively, twice the size of Europe. For ten thousand years or more, these forests endured in rich complexity, as the homes of an estimated 50 to 90 percent of all land-dwelling species. In less than four decades, we destroyed more than half of these ancient forests, and most of their spectacular arrays of species may be lost forever. With each passing year, we log an additional 38 million acres. That is the equivalent of leveling thirty-four city blocks every minute.

The destruction extends beyond the tropics. Today, highly mechanized logging operations are proceeding in the once-vast temperate forests of the United States, Canada, Europe, Siberia, and elsewhere.

We have a name for the removal of all trees from large tracts of land for logging, agriculture, and grazing operations. It is **deforestation**. Why are we doing this? Paralleling the rapid increases in the size of the human population are rapidly increasing demands for lumber, fuel, and other forest products, as well as for cropland and grazing land. More and more people are competing for diminishing resources. They do this for economic profit, but also because alternative ways of life simply are not available to individuals and families.

The world's great forests have profound influences on ecosystems. Like enormous sponges, the watersheds of forested regions absorb, hold, and then release water gradually. By intervening in the downstream flow of water, they help control soil erosion, flooding, and the accumulation of sediments that can clog rivers, lakes, and reservoirs. When the vegetation cover gets stripped away, the exposed soil becomes vulnerable to leaching of nutrients and to erosion, especially on steep slopes.

Today, deforestation is greatest in Brazil, Indonesia, Colombia, and Mexico. If the clearing and destruction continue at present rates, only Brazil and Zaire will have large tropical forests in the year 2010. By 2035, most of their forests will be gone, also.

Figures 50.7 and 50.8 show close-up and panoramic views of what is now happening in South America's Amazon basin. Section 25.7 includes examples of the effects of rampant deforestation in North America.

In tropical regions, the clearing of forests for agriculture sets the stage for long-term losses in productivity. The irony is that tropical forests are one of the worst places to grow crops or to raise pasture animals. In intact forests, litter does not build up, for the high temperatures and the heavy, frequent rains promote the rapid decomposition of organic remains and wastes. As fast as the decomposers release nutrients, the trees and other plants take them up. Deep, nutrient-rich topsoils simply cannot form.

Long before the advent of highly mechanized logging practices, people were practicing **shifting cultivation** (once referred to as slash-and-burn

Figure 50.7 Countries that are allowing the greatest destruction of tropical forests. *Red* denotes the regions where 2,000 to 14,800 square kilometers are deforested every year. *Orange* denotes "moderate" deforestation, which encompasses areas of 100 to 1,900 square kilometers.

Figure 50.8 The vast Amazon River basin of South America in September 1988. Its features were completely obscured by smoke from fires that had been deliberately set during the dry season to clear tropical forests, pasturelands, and croplands.

Smoke extends to the Andes Mountains near the western horizon, about 650 miles (1,046 kilometers) away. The smoke cover was the largest that astronauts had ever observed. It extended almost 175 million square kilometers (1,044,000 square miles).

The smoke plume near the center of the photograph alone covered an area comparable to the extensive forest fire in Yellowstone National Park in that year. During the El Niño induced drought of 1997, set fires burned out of control and a smoke cover formed again.

Massive deforestation is not confined to equatorial regions. For example, during the past century, 2 million acres of redwood forests along the coast of California have been logged over. Most of the destruction of such temperate forests has been occurring since 1950, owing to the widespread use of chainsaws and tractors and the practice of exporting many of the logs to lumbermills overseas, where wages are low.

agriculture). They cut and burn trees, then till ashes into the soil. The nutrient-rich ashes can sustain crops for one to several seasons. Afterward, cleared plots are abandoned, for heavy leaching leaves the soil infertile. When shifting cultivation is practiced on small, widely scattered plots, a forest ecosystem does not necessarily suffer extensive damage. But soil fertility plummets with increases in population size. Then, larger areas are cleared, and plots are cleared again at shorter intervals.

Now think about the larger picture. Deforestation alters the rates of evaporation, transpiration, runoff, and possibly the regional patterns of rainfall. For example, trees release between 50 and 80 percent of the water vapor above tropical forests. In logged-over regions, annual precipitation declines, and rain rapidly runs off the bare, nutrient-poor soil. As a deforested region gets hotter and drier, soil fertility and moisture decline. In

time, sparse grassland or desertlike conditions might prevail instead of a rich forest biome.

Finally, consider that tropical forests absorb much of the sunlight reaching equatorial regions of the Earth's surface. Deforested land is shinier, so to speak, and it reflects more incoming energy back into space. Also, by their photosynthetic activity, the great numbers of trees in these vast biomes help sustain the global cycling of carbon and oxygen. During extensive tree harvesting or burning, carbon stored in the tree biomass is released to the atmosphere as carbon dioxide. Thus deforestation factors into amplification of the greenhouse effect.

Once-vast forests helped sustain rapid increases in human population growth. Recent and highly mechanized modes of deforestation are rapidly depleting these finite resources.

50.5 YOU AND THE TROPICAL RAIN FOREST

Developing countries in Latin America, Southeast Asia, and Africa have the fastest-growing populations but limited food, fuel, and lumber. Thanks to the relatively recent inventions of chainsaws, tractors, and logging trucks, most of their forests will probably disappear within your lifetime.

Why does it matter? For purely ethical reasons, many condemn the destruction of so much biodiversity. Tropical rain forests have the greatest variety and numbers of insects, and the world's largest ones. They are home to the most bird species and to plants with the largest flowers. Within the forest canopy and understory are monkeys, tapirs, and jaguars in South America and apes, okapi, and leopards in Africa. Massive vines twist around trees. Orchids, mosses, lichens, and other organisms grow on branches, absorbing minerals that rains deliver to them. Entire communities of microbes, insects, spiders, and amphibians live, breed, and die in small pools of water that collect in furled leaves.

For practical reasons, the destruction affects your life. Only a few strains of crop plants and livestock, vulnerable to evolving pathogens, sustain most human populations. Tissue-culture specialists and genetic engineers use genes of forest species to develop new or hybrid strains that can make our food base less vulnerable. Geneticists use them to develop more effective antibiotics and vaccines. Aspirin, the most widely used painkiller, is based on a chemical blueprint of an extract from tropical willow leaves. Many ornamental plants and spices and foods, including cocoa, cinnamon, and coffee, originated in the tropics. So did latex, gums, resins, dyes, waxes, and oils for tires, shoes, toothpaste, ice cream, shampoo, compact discs, condoms, and perfumes. And think about this: Rampant burning of forests is releasing enough air pollutants to change the air you breathe and help heat the planet during your lifetime.

Thus biologists rightly decry the mass extinction, the assaults on species diversity, and the depletion of a major portion of the world's genetic reservoir. Yet something else is going on here. Too many of us become uneasy when we hike or drive through destroyed forests in our own country. Is it because we are losing the comfort of our heritage—a connection with our evolutionary past? Many millions of years ago, our earliest primate ancestors moved into the trees of tropical forests. Through countless generations, their nervous and sensory systems evolved and became highly responsive to information-rich, arboreal worlds.

Does our neural wiring still resonate with rustling leaves, with shafts of light and mosaic shadows? Are we innately attuned to the forests of Eden—or have time and change buried recognition of home?

Figure 50.9 Tropical rain forest in Southeast Asia.

TRADING GRASSLANDS FOR DESERTS

Long-term shifts in climate can convert grasslands to deserts. So can human populations. **Desertification** is the name for the conversion of large tracts of natural grasslands to a more desertlike state. It applies also when conversions of rain-fed or irrigated croplands result in a 10 percent or greater decline in agricultural productivity. Over the past fifty years, 9 million square kilometers worldwide have become desertified. At least

of their water from plants. They also are better water conservers; they lose little in feces, compared to cattle.

In 1978 a biologist, David Holpcraft, began ranching antelopes, zebras, giraffes, ostriches, and other native herbivores. He raised cattle as control groups in order to compare costs and meat yields on the same land. The native herds increased and yielded tasty meat. Range conditions did not deteriorate; they improved. Vexing

Figure 50.10 An awe-inspiring dust storm approaching Prowers County, Colorado, in 1934.

The Great Plains of the American Midwest are dry, windy grasslands that are subjected to severe, recurring droughts. Extensive conversion of these grasslands to agriculture began in the 1870s. Overgrazing left the ground bare across vast tracts. In May 1934, a cloud of topsoil that blew off the land blanketed the entire eastern portion of the United States, giving the Great Plains a dubious new name—the Dust Bowl. About 3.6 million hectares (9 million acres) of cropland were destroyed. Today, without large-scale irrigation and intensive conservation farming, desertlike conditions could prevail.

Figure 50.11 Desertification in the Sahel, a region of West Africa that forms a belt between the hot, dry Sahara Desert and tropical forests. This savanna country is undergoing rapid desertification as a result of overgrazing, overfarming, and prolonged drought.

200,000 square kilometers are still being converted annually. Prolonged droughts accelerate the process, as one did in the Great Plains years ago (Figure 50.10). At present, overgrazing of livestock on marginal lands is the main cause of large-scale desertification.

In Africa, for example, there are too many cattle in the wrong places. Cattle require more water than the region's native wild herbivores, so they move back and forth between grazing areas and watering holes more frequently. As they do this, they trample grasses and compact the soil surface (Figure 50.11). By contrast, gazelles and other native herbivores get most (if not all)

problems remained. African tribes have their own idea of what constitutes "good" meat, and some view cattle as the symbol of wealth in their society.

Without irrigation and conservation practices, grasslands that were converted for agriculture often end up as deserts.

A GLOBAL WATER CRISIS

The Earth has a tremendous supply of water, but most is too salty for human consumption or for agriculture. Imagine all that water in a bathtub. Withdraw all of the fresh, renewable portion (from lakes, rivers, reservoirs, groundwater, and other sources of surface water) and it would barely fill a teaspoon.

Why not consider **desalination**—the removal of salt from seawater? The supply of seawater is essentially unlimited. Desalination processes are available. They either distill seawater or force it through membranes (a method known as reverse osmosis). Costly fuel energy drives these processes, so they may be feasible only in Saudi Arabia and a few other countries with limited population sizes, large energy reserves, and lots of cash. In some situations they may be the only alternative to running out of water, as Santa Barbara and some other California cities nearly did during a prolonged drought. Yet desalination cannot solve the core problem. It may never be cost-effective for large-scale agriculture, and it produces mountains of salts.

Figure 50.12 Irrigated crops in the Sahara Desert, in Algeria.

Consequences of Heavy Irrigation

Large-scale agriculture accounts for nearly two-thirds of the human population's use of freshwater. In many cases, irrigation water from surface sources is piped into vast fields where water-demanding crops could not grow on their own. At its most extreme, irrigation turns some hot deserts into lush gardens, although believing these can be maintained over the long term is rather delusional (Figure 50.12).

Irrigation itself can change the land's suitability for agriculture. Concentrations of mineral salts typically are high in piped-in water. In regions where soil drains poorly, evaporation may cause **salination**—that is, salt buildup in the soil. Salination can stunt the growth of crop plants, eventually kill them, and so decrease yields.

Land that drains poorly also becomes waterlogged. As water accumulates underground, it slowly raises the **water table**, the upper limit at which the ground is fully saturated. When the water table is close to the ground's surface, soil gets saturated with saline water, which can damage plant roots. Properly managing the water-soil system can correct the salination and waterlogging. The economic cost of doing so is high.

Worse yet, water tables are subsiding. The Ogallala aquifer provides irrigation water for 20 percent of all croplands in the United States (Figure 50.13). Today, overdrafts (the amount nature does not replenish) have depleted half of it. Similarly, in the San Joaquin Valley of California, so much groundwater has been removed for irrigation that the water table's surface has subsided by as much as 20 feet in some areas.

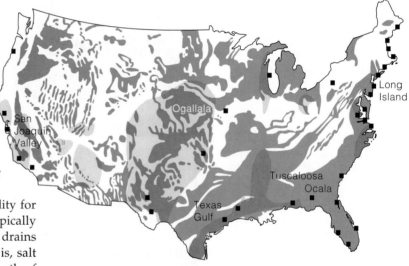

Figure 50.13 Underground aquifers (*blue*) holding 95 percent of all freshwater in the United States. *Gold* indicates where aquifers are being depleted, mainly for agriculture. *Black* boxes indicate aquifers being contaminated by saltwater intrusion.

Water Pollution

Water pollution amplifies the problem of water scarcity. Human sewage, animal wastes, and toxic chemicals make water unfit to drink, even to swim in. Pollutants encourage contamination by pathogens. Agricultural runoff pollutes water with sediments, pesticides, and plant nutrients. Power-generating plants and factories pollute water with chemicals, radioactive materials, and excess heat (thermal pollution).

Pollutants collect in lakes, rivers, and bays before reaching the oceans. Many cities throughout the world dump untreated sewage into their coastal waters. Cities

Figure 50.14 An experimental wastewater treatment facility in Rhode Island. Treatment begins when sewage flows into rows of large water tanks in which water hyacinths, cattails, and other aquatic plants are growing. Decomposers in the tank degrade wastes—which contain nutrients that promote plant growth. Heat from incoming sunlight speeds the decomposition. From these tanks, water flows through an artificial marsh of sand, gravel, and bulrushes that filter out algae and organic wastes. Then it flows into aquarium tanks where zooplankton and snails consume microorganisms suspended in the water—and where zooplankton become food for crayfishes, tilapia, and other fishes that can be sold as bait. After ten days, the now-clear water flows into a second artificial marsh for final filtering and cleansing.

along rivers and harbors maintain shipping channels by dredging the polluted muck and barging it out to sea. They also barge sewage sludge: coarse, settled solids that contain bacteria, viruses, and toxic metals.

In the United States, about 15,000 facilities partially treat liquid wastes from 70 percent of the population and 87,000 industries. The remaining wastes are mostly from suburban and rural populations. These are treated in lagoons or septic tanks or are directly discharged—untreated—into waterways.

There are three levels of **wastewater treatment**. In *primary* treatment, screens and settling tanks remove sludge, which is dried, burned, dumped in landfills, or treated further. Chlorine often is used to kill pathogens. It doesn't kill them all, and it also produces carcinogens whenever it reacts with certain industrial chemicals.

In *secondary* treatment, microbial populations break down organic matter after primary treatment but before chlorination. The wastewater trickles through microbe-containing gravel beds or is aerated in tanks and seeded with microbes. Toxic solutes can poison the microbial helpers. At such times, the facilities are shut down until the microbial populations are reestablished. Primary and secondary treatments do remove most suspended solids and oxygen-demanding wastes, but not all of the nitrogen, phosphorus, and toxic substances, such as heavy metals and pesticides. Usually the water gets chlorinated before being released into the waterways.

Tertiary treatment adequately reduces pollution but is largely experimental and expensive. It is applied to only 5 percent of the nation's wastewater.

In short, most wastewater is not treated adequately. A pattern gets repeated thousands of times along our waterways. Water for drinking is drawn upstream from a city, and wastes from industry and sewage treatment are discharged downstream. It takes no great leap of the imagination to see that water pollution intensifies as rivers flow to the oceans. In Louisiana, waters drained from the central states flow toward the Gulf of Mexico. Its pollution levels are high enough to threaten public health. Water destined for drinking does get treated to remove pathogens, but treatment does not remove toxic wastes dumped by numerous factories upstream.

This rather bleak picture might be numbing to most of us, but not to biologist John Todd. He constructed experimental wastewater treatment facilities in several greenhouses and artificial lagoons (Figure 50.14). When it works properly, the solar-aquatic treatment system produces water fit to drink. Such natural alternatives cannot work for very large urban areas. But they are an attractive alternative for small towns and rural areas.

The Coming Water Wars

If the current rates of population growth and water depletion hold, the amount of freshwater available for everyone on the planet will soon be 55 to 66 percent less than what it was in 1976. Already in the past decade, thirty-three nations have been engaged in conflicts over reductions in water flow, pollution, and silt buildup in major aquifers, rivers, and lakes. The United States and Mexico, Pakistan and India, and Israel and the occupied territories are among the squabblers.

Remember the Persian Gulf War, mainly about oil? Unless we pull off a blue revolution equivalent to the green one, we may be in for upheavals and wars over water rights. Does this sound farfetched? By building dams and irrigation systems at the headwaters of the Tigris and Euphrates rivers, Turkey can, in the view of one of its dam-site managers, stop the water flow into Syria and Iraq for as long as eight months "to regulate their political behavior." Regional, national, and global planning for the future is long overdue.

Water, not oil, may become the most important fluid of the twenty-first century. National, regional, and global policies for water usage and water rights have yet to be developed.

A QUESTION OF ENERGY INPUTS

Paralleling the J-shaped curve of human population growth is a dramatic rise in total and per capita energy consumption. It is due to increased numbers of energy users and to extravagant consumption and waste. For example, in one of the most pleasant of all climates, a major university constructed seven- and eight-story buildings with narrow, sealed windows. The windows can't be opened to catch prevailing ocean breezes. The buildings and windows were not designed or aligned to use sunlight for passive solar heating and breezes for passive cooling. Massive energy-demanding cooling and heating systems were installed.

When you hear talk of abundant energy supplies, bear in mind that there is a huge difference between the *total* and net amounts available. *Net* refers to the amount left over after subtracting the energy that is used to locate, extract, transport, store, and deliver energy to consumers. Some sources, such as direct solar energy, are renewable. And others, such as coal, are not (Figure 50.15).

Fossil Fuels

Fossil fuels are the legacy of forests that disappeared many hundreds of millions of years ago, and as such they are nonrenewable resources (see Section 25.4). The carbon-containing remains of the forests were buried and compressed in sediments, and then transformed into coal, petroleum (oil), and natural gas.

Even with strict conservation, we may use up the known petroleum and natural gas reserves in the next century. As known reserves run out in accessible areas, we explore wilderness areas in Alaska and other fragile environments, such as continental shelves. Net energy declines when the cost of extraction and transportation to and from remote areas increases. The environmental costs of extraction and transportation escalate. The extensive damage and clean-up costs following the 11-million-gallon oil spill from the supertanker *Valdez*, off the coast of Alaska, is a classic example.

What about coal? In theory, known world reserves may meet the energy needs of the human population for at least several centuries. However, coal burning has been the primary source of air pollution. Most of the known coal reserves contain low-quality material with a high sulfur content. Unless the sulfur is removed before or after burning, sulfur dioxides enter the air and add to global acid deposition. Fossil fuel burning also releases carbon dioxide and adds to the greenhouse effect.

Extensive strip mining of coal reserves close to the surface carries its own problems. It reduces the land available for agriculture, grazing, and wildlife. Most strip mines are located in arid and semiarid regions where the absence of sufficient water supplies and poor soils make restoration efforts difficult.

Nuclear Energy

NUCLEAR REACTORS In 1945, as Hiroshima burned, we recoiled in horror from the destructive force of nuclear energy. By the 1950s, however, many were championing nuclear energy as an instrument of progress. A number of energy-poor industrialized nations, including France, now depend heavily upon nuclear power. Yet construction of new nuclear plants has been delayed or even cancelled in most countries. Since 1970 in the United States alone, the plans to build 117 nuclear power plants were shelved. Others under construction were abandoned before completion. A few are being converted, at great cost, to fossil fuel burning. What happened? We started to question the operating cost, efficiency, safety record, and environmental impact of reliance on nuclear power.

By 1990, nuclear power was generating electricity at a cost only slightly above that of coal-burning plants. Putting aside other factors, its electricity-generating costs are now lower. However, by the year 2000, solar energy with natural gas backup also should cost less.

What about safety? Compared to coal-burning plants of the same capacity, a nuclear plant emits less radioactivity and carbon dioxide, and none of the sulfur dioxide. However, there is greater danger in their potential for **meltdown**. When nuclear fuel breaks down (decays), it releases a great deal of

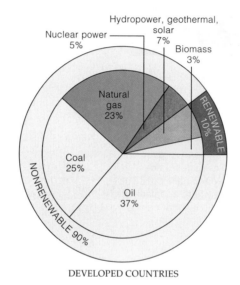

DEVELOPED COUNTRIES

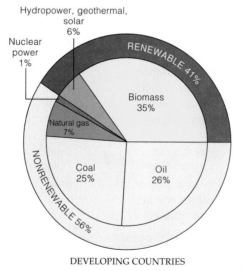

DEVELOPING COUNTRIES

Figure 50.15 Energy consumption in the developed and developing countries, which differ greatly in sources of energy and average per capita energy use. The values indicated do not take into account energy from the sun, which is the foundation for agriculture.

Figure 50.16 Incident at Chernobyl. (**a**) On April 26, 1986, errors in judgment during a routine test procedure resulted in runaway reactions, explosions, and a full core meltdown at the Chernobyl power plant in the Ukraine. Helicopter pilots who were supposed to drop 5,000 tons of lead, sand, clay, and other materials on the blazing core to suppress further release of radioactivity missed the target. Unimpeded and uncovered, nuclear fuel burned for nearly ten days, right on through a six-foot-thick steel and gravel barrier beneath it.

Between 185 and 250 million curies of radioactive matter may have escaped in those ten days alone. Inhaling as little as ten-millionths of a curie of plutonium can cause cancer. Thirty-one people died at once; others died of radiation sickness in the following weeks. Inhabitants of entire villages were relocated; their former homes were bulldozed under. In time, concrete entombed the 180 tons of partially burned nuclear fuel. In 1994, 11,000 square meters of holes in the concrete were still allowing rainwater, air, and other materials to enter or escape.

(**b**) Afterward, the number of people opposed to nuclear plants rose sharply, even in France. You can get a sense of why opposition increased dramatically by studying maps of the global distribution of radioactive fallout within two weeks of the meltdown. The fallout put 300 to 400 million people at risk for leukemia and other serious radiation-induced disorders. Also, throughout Europe, hundreds of millions of dollars were lost when the fallout made crops and livestock unfit for consumption.

a

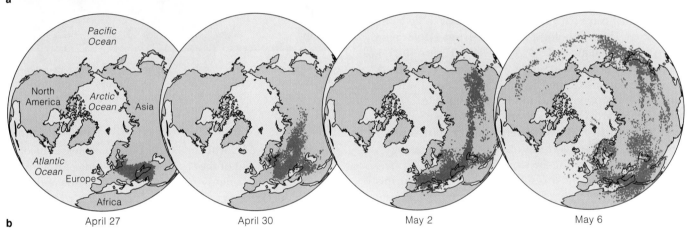

b   April 27   April 30   May 2   May 6

heat. Typically, water circulating over the nuclear fuel absorbs heat and produces steam that drives electricity-generating turbines. If the circulating water system were to develop a leak, water levels might plummet around the fuel, which might then heat past its melting point. Melting fuel on a generator floor would instantly convert the remaining water to steam. Combined with other reactions, steam formation could blow the system apart and release radioactive material. An overheated core could melt through its thick concrete containment slab. Figure 50.16 describes one incident that yielded compelling evidence of the consequences.

NUCLEAR WASTE DISPOSAL Unlike coal, nuclear fuel cannot be burned to harmless ashes. After three years or so, the fuel elements are spent, but they still contain uranium fuel as well as hundreds of new radioisotopes that formed in the reactions. The wastes are extremely radioactive and dangerous. They get extremely hot as they undergo radioactive decay, so they are plunged at once into water-filled pools. The water cools them and

keeps radioactive material from escaping. Even after being stored for several months, the isotopes remaining are lethal. Some must be kept isolated for at least 10,000 years. If a certain isotope of plutonium (^{239}Pu) is not removed, the wastes must be kept isolated for a quarter of a million years! After nearly fifty years of research, scientists still cannot agree on what is the best way to store high-level radioactive wastes. Even if they could, there is no politically acceptable solution. No one wants radioactive wastes anywhere near where they live.

Finally, as if we don't have enough to worry about, following the Soviet Union's breakup, some underpaid workers of a Russian nuclear power plant have been selling fuel elements on the black market. The buyers? Some developing nations that want to produce nuclear weapons—and possibly deliver them into the hands of terrorist organizations.

The nuclear genie is out of the bottle, exploitable by the best and worst elements of the human population.

ALTERNATIVE ENERGY SOURCES

Less than thirty years from now, the projected size of the human population will be such that the demands for fuel will increase by 30 percent—and for electricity by 265 percent. More efficient use and conservation of our existing energy sources alone will not do the trick. We must make the transition to reliance on alternative sources of energy, of the sort described next.

Solar-Hydrogen Energy

Each year, incoming sunlight contains about ten times more energy than that in all of the known fossil fuel reserves. That is 15,000 times as much energy as the human population uses now. Isn't it about time we start to collect it in earnest, in something besides crops?

For example, when exposed to sunlight, electrodes in "photovoltaic cells" produce an electric current that splits water molecules into oxygen and hydrogen gas (H_2)—which can be used directly as fuel or to generate electricity. The technology to tap such **solar-hydrogen energy** has been around since the 1940s (Figure 50.17a). Such energy is stored efficiently for as long as required. It costs less to distribute H_2 than electricity, and water is the only by-product of using it. Space satellites run on it. Here on Earth, fossil fuels are still "cheaper."

Unlike fossil fuels, however, sunlight and seawater are virtually unlimited resources. And the technology's potential to protect the environment is staggering. The environmental scientist G. Tyler Miller, Jr., puts it this way: "If we make the transition to an energy-efficient solar-hydrogen age, we can say good-bye to smog, oil spills, acid rain, and nuclear energy, and perhaps to global warming. The reason is simple. As hydrogen burns in air, it reacts with oxygen gas to produce water vapor—not a bad thing to have coming out of tailpipes, chimneys, and smokestacks." Also, if this technology becomes cost-effective for the developing countries, the great forests now being destroyed for timber and fuel might still be around for future generations.

Recently in the United States, the largest supplier of natural gas and a manufacturer of photovoltaic cells combined forces. They intend to build a solar facility in the Nevada desert that will generate enough energy to supply a city of 100,000 at less cost than for electricity generated by fossil fuels. If the plan succeeds, they may revolutionize the energy industry.

Wind Energy

As you know, solar energy also is converted into the mechanical energy of winds. Where winds that travel faster than 7.5 meters per second prevail, we find cost-effective wind turbines (Figure 50.17b). California gets 1 percent of its electricity from "wind farms." Possibly

Figure 50.17 Harnessing solar energy. (**a**) Electricity-producing photovoltaic cells in panels that collect sunlight energy. (**b**) Wind turbines, which exploit the air circulation patterns that arise from latitudinal variations in the intensity of incoming sunlight.

the winds of North and South Dakota alone can meet all but 20 percent of the current energy needs of the United States. Wind energy has potential for islands and other remote areas far from utility grids. One drawback is that winds don't blow on a regular schedule, so they can't be used as an exclusive or major energy source.

Fusion Power

The sun's gravitational force is enough to compress atomic nuclei to high densities, and its temperatures are high enough to force atomic nuclei to fuse. We call this **fusion power**. Similar conditions do not exist on Earth, but maybe we can mimic them. Researchers confine a certain fuel—a heated gas of two isotopes of hydrogen—in magnetic fields, then hit it with lasers. The fuel implodes, it is compressed to extremely high densities—and energy is released. The more energetic the lasers, the greater the compression, and the more the fuel will burn. The bad news is, although the amount of energy released has been steadily increasing, it will be at least fifty years before fusion reactors are operating, and costs will probably be high. The good news is, that's about the time fossil fuels will start running out.

Sunlight may end up sustaining the energy needs of the human population in more ways than one.

BIOLOGICAL PRINCIPLES AND THE HUMAN IMPERATIVE

Molecules, single cells, tissues, organs, organ systems, multicelled organisms, populations, communities, then ecosystems and the biosphere. These are architectural systems of life, assembled in increasingly complex ways during the past 3.8 billion years. We are latecomers to this immense biological building program. Yet within the relatively short span of 10,000 years, our activities have been changing the very character of the land, ocean, and atmosphere, even the genetic character of species.

It would be presumptuous to think we alone have had profound effects on the world of life. As long ago as the Proterozoic era, photosynthetic organisms irrevocably changed the course of biological evolution by gradually enriching the atmosphere with oxygen. During the past as well as the present, competitive adaptations assured the rise of some groups, whose dominance assured the decline of others. Change is nothing new to this biological building program. What *is* new is the capacity of a species —our own—to comprehend what might be going on.

We now have the population size, the technology, and the cultural inclination to use up energy and modify the environment at frightening rates. Where will rampant, accelerated change lead us? Will feedback controls begin to operate as they do, for example, when population growth exceeds the carrying capacity of the environment? In other words, will negative feedback controls come into play and keep things from getting too far out of hand?

Feedback control will not be enough, for it operates only when deviation already exists. Patterns of resource consumption and growth rates for the human population are founded on an illusion of unlimited resources and a forgiving environment. A prolonged, global shortage of food or the passing of a critical threshold for the global climate can come too fast to be corrected. At some point, deviations may have too great an impact to be reversed.

What about feedforward mechanisms that might serve as early warning systems? For example, when sensory receptors near the surface of skin sense a drop in outside air temperature, each sends messages to the nervous system. That system responds by triggering mechanisms that raise the body's core temperature before the body itself becomes dangerously chilled. If we develop feedforward control mechanisms, maybe we could begin corrective measures before we have altered the environment too significantly.

By themselves, feedforward controls won't work, for they start operating when change is under way. Think of the DEW line—the Distant Early Warning system. It is like a vast sensory receptor, one that can detect the launching of intercontinental ballistic missiles against North America. By the time the system actually detects what it is designed to detect, it may be too late to stop widespread destruction.

It would be naive to assume we can ever reverse who we are at this point in evolutionary time, to de-evolve ourselves culturally and biologically into becoming less complex in the hope of averting disaster. However, there is reason to believe we can avert disaster by using a third kind of control mechanism—a capacity to anticipate events even before they happen. We are not locked into responding only after irreversible change has begun. We have the capacity to anticipate the future—it is the essence of our visions of utopia or of nightmarish hell. *We all have the capacity to adapt to a future that we can partly shape.*

We can, for example, stop trying to "beat nature" and learn to work with it. Individually and collectively, we can work to develop long-term policies at the local, regional, and global levels—policies that take into account biotic and abiotic limits on population growth. Far from being a surrender, this would be one of the most complex and intelligent behaviors of which we are capable.

Having a capacity to adapt and actually using it are not the same thing. We have already put the world of life on dangerous ground because we have not yet mobilized ourselves as a species to work toward self-control.

Our survival depends on predicting possible futures. It depends on preserving, restoring, and even designing and constructing ecosystems that are in harmony with our definition of basic human values and with the biological models available to us. Human values can change; our expectations can and must be adapted to biological reality. *For the principles of energy flow and resource utilization, which govern the survival of all systems of life, do not change.* It is our biological and cultural imperative that we come to terms with these principles, and ask ourselves what our long-term contribution will be to the world of life.

SUMMARY

1. Accompanying the extremely rapid growth of the human population are increases in energy demands and in environmental pollution.

2. Pollutants are substances with which ecosystems have had no prior evolutionary experience (in terms of kinds and amounts) and therefore have no mechanisms for absorbing or cycling them. Many pollutants result from human activities, and they adversely affect the health, activities, or survival of human populations.

3. Smog, a form of air pollution, arises in industrialized and urban regions that rely on fossil fuels. It becomes especially concentrated in land basins having thermal inversions (a trapping of a layer of cool, dense air under a warm air layer). Industrial smog forms in industrial coal-burning regions with cold, wet winters. Large cities with many vehicles in warm climates produce photochemical smog, mainly when sunlight makes nitric oxide (emitted from vehicles) react with hydrocarbons to form photochemical oxidants such as PANs.

4. During dry weather, acidic air pollutants, especially oxides of nitrogen and sulfur, fall to Earth as dry acid deposition. They also dissolve in atmospheric water, then fall to Earth as wet acid deposition, or acid rain.

5. Seasonal thinning of the ozone layer at high latitudes has become pronounced as CFCs (chlorofluorocarbons) and other air pollutants rise to the stratosphere, where they deplete ozone and allow more harmful ultraviolet radiation from the sun to reach the Earth's surface.

6. Human population growth presently depends on the expansion of agriculture, made possible by large-scale irrigation and extensive applications of fertilizers and pesticides. Global freshwater supplies are limited, yet they are being polluted by agricultural runoff (which includes sediments as well as pesticides and fertilizers), industrial wastes, and human sewage.

7. Human populations are damaging land surfaces by:

   a. Passively accepting the dumping, burning, or burial of solid wastes rather than making concerted efforts to recycle or reuse materials and to reduce waste.

   b. Engaging in rampant deforestation (destruction of vast tracts of tropical and temperate forest biomes).

   c. Contributing to desertification (the large-scale conversion of natural grasslands, croplands, or grazing lands to desertlike conditions).

8. Energy in the form of fossil fuels is nonrenewable, dwindling, and environmentally costly to extract and use. Nuclear energy in itself is less polluting, but the costs and risks associated with fuel containment and storing radioactive wastes are enormous. The challenge is to develop affordable alternatives that are based on renewable energy resources, such as solar energy.

Review Questions

1. Define pollution and list some specific examples of water pollutants. *50.1, 50.7*

2. Distinguish among the following conditions: *50.1*
   a. industrial smog
   b. photochemical smog
   c. dry acid deposition
   d. wet acid deposition

3. Define CFCs and describe how they apparently contribute to seasonal thinning of the ozone layer in the stratosphere. *50.2*

4. What percent of the Earth's land masses is under cultivation? What percent is available for new cultivation? *50.3*

5. Define and describe possible consequences of deforestation and of desertification. *50.4, 50.5, 50.6*

6. Which human activity uses the most freshwater? *50.7*

Self-Quiz *(Answers in Appendix IV)*

1. Since the mid-eighteenth century, human population growth has been _____ .
   a. leveling off
   b. growing slowly
   c. accelerating
   d. not much to speak of

2. Pollutants disrupt ecosystems because _____ .
   a. their components differ from those of natural substances
   b. only humans have uses for them
   c. there are no evolved mechanisms to deal with them
   d. their only effect is on ecosystems, not humans

3. During a thermal inversion, weather conditions trap a layer of _____ air under a layer of _____ air.
   a. warm; cool
   b. cool; warm
   c. warm; sooty
   d. cool; sooty

4. _____ is (are) a case of regional air pollution.
   a. Smog
   b. Acid rain
   c. Ozone layer thinning
   d. a and b

5. _____ is (are) a case of air pollution with global effects.
   a. Smog
   b. Acid rain
   c. Ozone layer thinning
   d. b and c

6. Two-thirds of the freshwater used annually goes to _____ .
   a. urban centers
   b. agriculture
   c. treatment facilities
   d. a and c

7. The upper limit at which the ground is fully saturated with water is called _____ .
   a. groundwater
   b. an aquifer
   c. the water table
   d. the salination limit

8. Energy from fossil fuels is _____ ; their extraction and use come at _____ cost to the environment.
   a. renewable; low
   b. nonrenewable; low
   c. renewable; high
   d. nonrenewable; high

9. Nuclear energy normally pollutes _____ than fossil fuels; it poses _____ dangers than fossil fuels.
   a. less; lesser
   b. more; greater
   c. more; lesser
   d. less; greater

10. Match each term with the most suitable description.
    ____ desertification
    ____ deforestation
    ____ green revolution
    ____ solar-hydrogen power

    a. probably one of our best options
    b. soil loss, watershed damage, altered rainfall patterns follow
    c. attempt to improve crop production on existing land
    d. conversion of large tracts of natural grasslands to a more desertlike state

Critical Thinking

1. Investigate where the water for your own city comes from and where it has been. You may find the answer illuminating.

2. Make a list of advantages you personally enjoy as a member of an affluent, industrialized society. List some drawbacks. Do the benefits outweigh the costs? This is not a trick question.

3. List activities you pursue each day that may be contributing to regional or global pollution. Also list ways in which you might help reduce that contribution.

4. Kristen, a recent college graduate, is finding her idealism on a collision course with reality. She strongly believes people who live in the United States are obliged to make the world a level playing field for all human beings, with equality in resources, health, education, economic security, and a pristine environment for all. Yet she also understands that the sheer size of the human population makes this impossible. Kristen recently said she cannot be party to hard choices and actions that go against her ideals and just wants nature to solve the problem for us. Comment on this true story.

5. It has been said that economic wars, more than military wars, will determine the winners and losers among nations in the near future. The economic growth of certain nations, including some in the former Soviet Union and in the Far East, has had devastating impact on the environment. Elsewhere, maintaining standards of environmental protection adds to the cost of goods produced and puts practicing nations at serious disadvantage in this global competition. Should the United States loosen some of its existing environmental laws to help assure its economic survival? Why or why not? Can you think of some pressures that might be imposed on environmentally indifferent nations to encourage them to change their harmful practices?

6. Populations of every species utilize resources and produce wastes. Use Figure 50.18 as a starting point for a brief essay on the accumulation and uses of energy and materials, including wastes, in a human population that is concentrated in a large city. Contrast your description with the flow of energy and the materials cycling that proceed in a natural population of some other organism.

Selected Key Terms

acid rain *50.1*
chlorofluorocarbon (CFC) *50.2*
deforestation *50.4*
desalination *50.7*
desertification *50.6*
dry acid deposition *50.1*
fossil fuel *50.8*
fusion power *50.9*
green revolution *50.3*
industrial smog *50.1*
meltdown *50.8*

ozone thinning *50.2*
PAN (peroxyacyl nitrate) *50.1*
photochemical smog *50.1*
pollutant *50.1*
salination *50.7*
shifting cultivation *50.4*
solar-hydrogen energy *50.9*
thermal inversion *50.1*
wastewater treatment *50.7*
water table *50.7*

Readings

Collins, M. 1990. *The Last Rain Forests*. New York: Oxford University Press.

Frosh, R. September 1995. "The Industrial Ecology of the Twenty-First Century." *Scientific American* 273(3): 178–181.

Gruber, D. 1989. "Biological Monitoring and Water Resources." *Endeavour* 13(3): 135–140.

Miller, G. T., Jr. 1998. *Living in the Environment*. Tenth edition. Belmont, California: Wadsworth. This author consistently puts information from numerous sources into a current, accessible survey of the present and future state of the environment.

Plucknett, D., and D. Winkelmann. September 1995. "Technology for a Sustainable Agriculture." *Scientific American* 273(3): 182–186.

Western, D., and M. Pearl. 1989. *Conservation for the Twenty-First Century*. New York: Oxford University Press.

Wilson, E. 1988. *Biodiversity*. Washington, D.C.: National Academy of Sciences.

Web Site See *http://www.wadsworth.com/biology* for practice quiz questions, hypercontents, BioUpdates, and critical thinking. The Wadsworth Biology Resource Center provides a wealth of information fully organized and integrated by chapter.

Figure 50.18 City as ecosystem.

AN EVOLUTIONARY VIEW OF BEHAVIOR

Deck the Nest With Sprigs of Green Stuff

In 1890, bird fanciers imported more than a hundred starlings (*Sturnus vulgaris*) from Europe and released them in New York City's Central Park. In less than a century, the descendants of the introduced species had expanded their geographic range from coast to coast. They are so good at gleaning food from the agricultural fields of North America that they have outmultiplied native birds. They also have evicted great numbers of them from scarce nesting sites in the cavities of trees.

Once a male and female commandeer a tree cavity, they gather dry grass and twigs and use them to build or rebuild a nest (Figure 51.1). Theirs is no ordinary nest. They *decorate* the nest bowl with sprigs, freshly plucked.

Why do starlings do this? Do the sprigs of greenery camouflage the nest from predators? Not likely. The nests are already concealed, inside the cavities of trees. Do the sprigs function as insulative material that might help keep the forthcoming eggs warm? Actually, the still-moist, green plant parts would promote heat *loss*, not heat conservation. Well, then, do the green sprigs combat parasites? Mites, for example, parasitize birds and infest their nest cavities. In short order, even a few tiny mites can produce thousands of descendants. When present in large numbers, mites can suck enough blood from a nestling to weaken it. They interfere with the nestling's growth, development, and survival.

Larry Clark and Russell Mason decided to test the third hypothesis. As both biologists knew, starlings do not weave just any green plant material into the nests. They choosily favor the leaves of certain plants, such as wild carrot (Figure 51.1). Thus Clark and Mason built a set of experimental nests, some with freshly cut wild carrot leaves and some without. They removed the natural nests that pairs of starlings had constructed and had already started using. Fifty percent of the nesting pairs received replacement nests decorated with sprigs of wild carrot. No sprigs decorated the replacement nests for the other pairs of starlings.

Figure 51.2 shows the results. The number of mites in greenery-free nests was consistently greater than the number in nests decorated with greenery. At the end of one experiment, for example, sprig-free nests teemed with an average of 750,000 mites. The ones with sprigs contained a mere 8,000.

Shoots of wild carrot happen to contain a highly aromatic steroid compound. Almost certainly, the compound repels herbivores and thus helps the plant survive. By coincidence, the compound also prevents mites from becoming sexually mature—and thereby

Figure 51.1 A most excellent fumigator in nature—a European starling (*Sturnus vulgaris*). It combats infestations of mites by decorating previously owned nests with fresh sprigs of wild carrot, shown at the left.

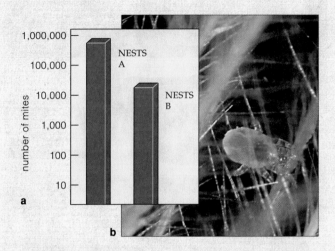

a

b

Figure 51.2 (a) Experimental results that point to the adaptive value of starling nest-decorating behavior. Nests designated *A* were kept free of fresh sprigs of wild carrot and other plants that contain aromatic compounds. The compounds prevent baby mites from developing into adult mites (**b**). Other nests, designated *B*, had fresh sprigs added every seven days. Head counts of mites (*Ornithonyssus sylviarum*) infesting the nests were made at the time the starling chicks left the nest, twenty-one days after the experiment was under way.

prevents mite population explosions inside the nests festooned with wild carrot.

Far from being a trivial behavior, then, "decorating" a nest with aromatic greenery has adaptive value. It fumigates the nest and thereby increases an individual's chance of producing healthy, surviving offspring.

And so starlings lead us into the fascinating world of behavioral research. As you will see, some of the behavioral studies focus on the adaptive value of some behavior to an individual's reproductive success. Other studies focus on the internal mechanisms that enable individuals to behave as they do.

As a starting point, recall that information encoded in genes directs the formation of tissues and organs that make up the animal body, including those of its nervous system. That system detects, processes, and integrates information about stimuli in the internal and external environments, then commands muscles and glands to make suitable responses. Other gene products called hormones contribute to the responses. Because genes specify all the substances required to construct and operate the nervous and endocrine systems, *they are the heritable foundation for the responses that animals make to stimuli.* We call the observable, coordinated responses to stimuli **animal behavior**.

KEY CONCEPTS

1. "Behavior" refers to the coordinated responses that an animal makes to stimuli. A stimulus, remember, is a form of energy that activates a specific type of sensory receptor, either a sensory neuron or a specialized cell adjacent to it.

2. Forms of behavior have a heritable (genetic) basis, as when instructions encoded in certain genes govern the development of the nervous and endocrine systems. These organ systems allow an animal to detect, process, and issue commands for behavioral responses to stimuli.

3. A newly born or hatched animal already has neural wiring and motor systems by which it performs some behaviors without having to learn them by actual experience. Such instinctive behaviors are triggered by sign stimuli—one or two simple, well-defined cues in the environment.

4. The nervous system also has the capacity to process and retain information about specific experiences in the environment, then use the information to vary or change behavioral responses.

5. Like other traits having a genetic basis, behavior has evolved by way of natural selection, which occurs when genetically different individuals in a population have different numbers of surviving offspring. Thus sexual selection and other evolutionary processes have favored behavioral mechanisms that enhance the ability of individuals to pass on their genes to offspring.

6. Evolved modes of communication underlie social behavior. Communication signals, which are encoded in species-specific odors, body coloration and patterning, postures, movements, and other stimuli, hold obvious meaning for both the sender and the receiver of signals.

7. Living in a social group has costs and benefits, as measured by an individual's ability to pass on its genes to offspring. Not every environment favors the evolution of such groups. Under certain circumstances, a solitary life-style allows an individual to leave more descendants than living in a large group would allow.

8. In cases of altruism, an individual helps others in ways that sacrifice personal reproductive success. The evolution of altruism requires circumstances in which individuals can propagate their genes indirectly, by helping their relatives reproduce successfully.

Genes and Behavior

An animal's nervous system, recall, is wired to detect, interpret, and issue commands for response to stimuli—that is, to specific aspects of the external and internal environments. All the steps required to assemble and to operate that system's receptors, nerve pathways, and brain are specified, one way or another, by the animal's genes. If we define an animal's behavior as coordinated responses to stimuli, then its behavior starts with genes.

Stevan Arnold found evidence of the genetic basis of behavior by studying the feeding preferences of coastal and inland populations of a snake species in California. Garter snakes near the coast prefer banana slugs (Figure 51.3a). Snakes living inland prefer tadpoles and small fishes. Offer them a banana slug and they ignore it.

In one set of experiments, Arnold offered captive newborn garter snakes a chunk of slug as the first meal. Offspring of coastal snakes usually ate the chunk. They even flicked their tongue at cotton swabs drenched in essence of slug. (A snake "smells" by tongue-flicking, which draws chemical odors into the mouth.) But the offspring of snakes living inland ignored the swabs, and only rarely did they eat slug meat. Here was a clear difference between captive baby snakes with no prior, direct experience with slugs. Arnold concluded that the snakes had been programmed before birth to accept or reject slugs; they certainly did not learn their feeding preferences through taste trials. Maybe the coastal and inland populations had allelic differences for the genes that affect the formation of odor-detecting mechanisms when a garter snake embryo is developing.

Figure 51.3 (a) Banana slug, food for (b) a grown-up garter snake of coastal California. (c) A newborn garter snake from a coastal population, tongue-flicking at a cotton swab drenched with banana slug tissue fluids.

To test his hypothesis, Arnold crossbred coastal and inland snakes. If the different food preferences between snake populations have a genetic basis, then the hybrid offspring might make an *intermediate* response to slug chunks and odors. The results of the crosses matched the prediction. Compared with typical newborn inland snakes, many baby snakes of mixed parentage tongue-flicked more often at the slug-scented cotton swabs—but less often than typical newborn coastal snakes did.

Hormones and Behavior

Gene products called hormones also contribute to bird song and other kinds of behavior. Think about a male zebra finch, such as the one in Figure 51.4. Each spring he repeats the same song over and over again with such

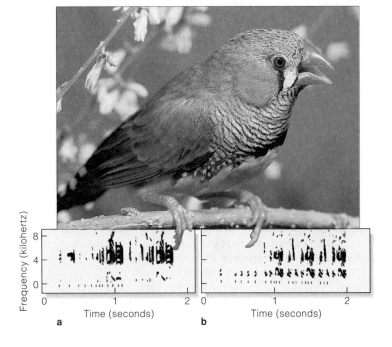

Figure 51.4 Hormones and the territorial song of zebra finches. (a) Sound spectrogram of the male's song. The spectrogram is a visual record of each note's pitch (its frequency, measured in kilohertz). (b) Sound spectrogram of a female zebra finch that was experimentally converted to a singer. When a nestling, she received extra estrogen. At adulthood, the female also received a testosterone implant—and she started singing. Normally, estrogen organizes the song system (parts of the brain) in developing songbird embryos, and testosterone activates the system in adult males.

clarity and such consistency, you might wonder how he does it. His singing is an outcome of seasonal differences in the secretion of melatonin, a hormonal product of the pineal gland (Section 37.8). An increase in the blood level of melatonin suppresses the growth and functions of songbird gonads. Its effect is reversed when photoreceptors located inside the pineal gland absorb sunlight energy and issue signals that lead to a decrease in melatonin secretion.

Winter has fewer hours of daylight than spring, and there is not enough light to inhibit secretion of melatonin. In spring, when daylength increases, the gonads are released from hormonal suppression. Now they increase in size and step up their secretion of estrogen and testosterone. These sex hormones can trigger gender-related differences in bird singing behavior. How? While the embryos of most songbirds develop, estrogen controls formation of a **song system**. The song system consists of several parts of the brain that will govern activity of the muscles of a vocal organ. The regions differ notably in size and structure between male and female birds.

Unlike humans, females of songbird species carry the sex chromosomes XY, and the males carry XX. Certain genes on the Y chromosome prevent the females from making estrogen. But even before a *male* bird hatches, a high blood level of estrogen stimulates development of a masculinized brain. (In his brain tissues, estrogen is converted to testosterone.) Later, his gonads enlarge. As the breeding season begins, his gonads start secreting more testosterone. When it binds to receptors on cells in the song system, it induces metabolic changes that will prepare the male to sing when he is suitably stimulated.

Instinctive Behavior Defined

With their tongue-flicking, body orientation, and strikes at prey, newborn garter snakes offer a fine example of **instinctive behavior**. This means a particular behavior is performed without having been learned by actual experience in the environment. Instead, the nervous system of a newly born or hatched animal is already wired to recognize one or two simple, well-defined cues in the environment, or **sign stimuli**, that can trigger a suitable response. The snake's response is a stereotyped motor program. When the snake recognizes certain sign stimuli, a **fixed action pattern** follows. It is a program of coordinated muscle activity that runs to completion independently of feedback from the environment.

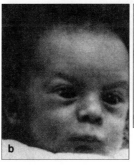

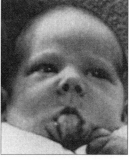

Figure 51.5 Examples of instinctive responses that human offspring make to sign stimuli. (**a**) A close-up face of an adult triggers smiling behavior in very young infants. (**b**) Somewhat older infants instinctively attempt to imitate facial expressions of adults.

As another example, when human infants are two or three weeks old, they tend to smile instinctively when an adult's face comes close to their own (Figure 51.5a). Infants make the same response to an overly simplified stimulus—a flat, face-sized mask with two dark spots where eyes would be on a human face. A mask with one "eye" won't do the trick. As Figure 51.5b suggests, older infants continue to show instinctive behavior.

And remember the cuckoo, a social parasite? Adult females lay their eggs in nests of other species. If newly hatched cuckoos eliminate the natural-born offspring, they will receive the undivided attention of their foster parents (Section 47.6). Although they are blind at birth, when they touch a round object—usually an egg—they execute a fixed action pattern that maneuvers the egg onto their back, then expels it from the nest.

Clearly, then, animals have genetically based means to respond automatically to certain environmental cues. As you will see next, however, they also have the means to process information about specific experiences—then use the information to vary or change their responses. The outcome is what we call learned behavior.

Genes underlie animal behavior—coordinated responses to stimuli. Certain gene products are essential in constructing and operating the nervous system, which governs behavior. Other gene products called hormones also influence the mechanisms required for particular forms of behavior.

Animals start out life neurally wired to recognize simple but important cues in the environment. These sign stimuli trigger suitable responses, such as fixed action patterns.

Let's now consider **learned behavior**, whereby animals process and integrate information gained from specific experiences, then use it to vary or change responses to stimuli. Think of a young toad, with a nervous system that commands it to flip its sticky tongue instinctively at any dark object moving across its field of vision. In the toad world, such objects are usually edible insects. Suppose one dark object is a bumblebee that stings the toad's tongue. The toad's experience leads it to avoid "bumblebee-size black-and-yellow-banded objects that sting." Or think of **imprinting**, a time-dependent form of learning. Imprinting is triggered by exposure to sign stimuli, and it usually occurs during a sensitive period

Figure 51.6 No one can tell these imprinted baby geese that Konrad Lorenz is not Mother Goose. (Refer to Table 51.1.)

when the animal is young. Figure 51.6 gives a classic example. Table 51.1 lists other examples.

Don't fall into the trap of thinking that genes alone specify instinctive behavior and the environment alone governs learned behavior. Behavior arises through gene expression *and* experiences. A bird does have a nervous system prewired to recognize sounds (acoustical cues), but what it *actually hears* influences its response to them. For example, male white-crown sparrows have a genetic capacity to sing—but the ones in different habitats use variations, or dialects, of the species song. As discovered by Peter Marler, male songbirds only acquire the full song by listening to it ten to fifty days after hatching. In this sensitive period, their primed learning mechanism responds to information from the environment.

In addition, the male birds learn parts of a song by picking up cues when other males sing. Marler raised nestlings to maturity in soundproof chambers so they would not hear adult males singing. At maturity, the song of captive males had none of the detailed structure of a typical adult's song. Marler also exposed isolated captives to recordings of white-crown sparrows *and* of song sparrows. At maturity, the captives sang only the white-crown song. They even mimicked the dialects. Evidently, birdsong partly requires a genetically based capacity to learn from specific sounds, or acoustical cues.

This isn't the whole story, however. In still another experiment, Marler allowed young, hand-reared white-crowns to interact with a "social tutor" of a different species, as opposed to listening to the taped songs. The males tended to learn the tutor's song. Their learning mechanisms must be primed to be influenced by social experience as well as by acoustical cues.

Table 51.1 A Few Categories of Learned Behavior

IMPRINTING. This is a time-dependent form of learning involving exposure to sign stimuli, most often early in development. For instance, in response to a moving object and probably to certain sounds, baby geese imprint on their mother and follow her during a short, sensitive period after hatching. They were neurally wired to learn crucial information—the identity of the individual that would protect them in the months ahead. Normally that individual is the mother or father. Imprinting occurs among many animals.

The ethologist Konrad Lorenz must have presented the baby geese in Figure 51.6 with sign stimuli that made them form an attachment to him. As another outcome of imprinting, birds also direct sexual attention to members of the species they had been sexually imprinted on when young.

CLASSICAL CONDITIONING. Ivan Pavlov's classic experiments with dogs are an example of classical conditioning. Dogs will salivate just before eating. Pavlov's dogs were conditioned to salivate, even in the absence of food. They did so in response to the sound of a bell or a flash of light that was initially associated with the presentation of food. In this case, the animals learned to associate an automatic, unconditioned response with a novel stimulus that does not normally trigger the response.

OPERANT CONDITIONING. An animal learns to associate a voluntary activity with its consequences, as when a toad learns to avoid stinging insects after its first attempt to eat them.

HABITUATION. An animal learns by experience *not* to respond to a situation if the response has neither positive nor negative consequences. Thus pigeons and other birds living in cities learn not to flee from people who pose no threat to them.

SPATIAL OR LATENT LEARNING. By inspecting its environment, an animal acquires a mental map of some region, often by learning the position of local landmarks. Thus blue jays can store information about the position of dozens, if not hundreds, of places where they have stashed food.

INSIGHT LEARNING. An animal abruptly solves some problem without trial-and-error attempts at the solution. Chimpanzees often exhibit insight learning in captivity when they suddenly solve a novel problem that their captors devise for them. Some chimps abruptly stacked and stood on several boxes *and* used a stick to reach bananas suspended out of their reach.

In instinctive behavior, animals make complex, stereotyped responses to specific, often simple environmental cues. In learned behavior, responses may vary or change as a result of individual experiences in the environment.

Whether instinctive or learned, behavior develops through interactions between genes and environmental experiences.

THE ADAPTIVE VALUE OF BEHAVIOR

If you accept that genes directly or indirectly govern forms of behavior, then it follows that forms of behavior are subject to evolution by natural selection. **Natural selection**, remember, is a result of differences in survival and reproduction among individuals of a population that differ from one another in heritable traits. Some versions of a trait are better than others at helping the individual survive and reproduce. Thus alleles for those versions increase in a population, and other alleles do not. (Alleles are different molecular forms of the same gene.) In time, such genetic changes lead to increased fitness, or an increase in adaptation to the environment.

By using the theory of evolution by natural selection as our point of departure, we should be able to develop and test possible explanations of why a behavior has persisted. We should be able to discern how those forms bestow reproductive benefits that offset reproductive costs (disadvantages) associated with them. It makes no difference whether we focus on behavior of a solitary animal or of animals in a social group. If the behavior is adaptive, it must promote survival and production of offspring of the *individual.* Keep this thought in mind as you read through the following list of terms, which we will use repeatedly in the remainder of this chapter:

1. **Reproductive success**. The number of surviving offspring that the individual produces.

2. **Adaptive behavior**. Any behavior that promotes the propagation of an individual's genes and tends to occur at increased frequency in successive generations.

3. **Social behavior**. Cooperative, interdependent relationships among individuals of the species.

4. **Selfish behavior**. Within a population, any form of behavior that increases an individual's chance to produce or protect offspring of its own, regardless of the consequences for the group as a whole.

5. **Altruistic behavior**. Within a population, a self-sacrificing behavior. The individual behaves in a way that helps others but diminishes or precludes its own chance of producing offspring.

When biologists speak of selfish or altruistic behavior, they don't mean an individual is consciously aware of what it is doing or of the behavior's reproductive goal. A lion doesn't have to know that eating zebras is good for its reproductive success. Its nervous system simply calls for hunting behavior when the lion is hungry and sees a zebra. The behavior persists in the population because the genes responsible for it are persisting also.

Think of how Norwegian lemmings disperse from their population when density skyrockets and food is scarce. Many die during the exodus. Are they helping the species by committing suicide to get rid of "excess" individuals? Or do they die by starvation, predation, or accidental drowning as they disperse to places where they may reproduce? A wonderfully instructive cartoon by Gary Larson shows lemmings plunging over a cliff above water, presumably in the act of suicide. But one has an inflated inner tube about its waist! If many of the lemmings were altruistic, then only "selfish" lemmings would reproduce. Over time, then, the genes underlying altruism would disappear from the population. More likely, individuals disperse to less crowded places, where they have a better chance to survive and reproduce.

As a more detailed example, in northern forests, ravens scavenge carcasses of deer, elk, or moose, which are few and far between. When one of these large birds comes across a carcass—even in winter, when food is scarce—it often calls out loudly and attracts a crowd of similarly hungry ravens. The calling behavior puzzled Bernd Heinrich, for it seems to go against the caller's interests. Wouldn't a quiet raven eat more, increase its chances of surviving, and leave more descendants than an "unselfish" vocalizer? If the behavior is an outcome of natural selection, then the cost in terms of lost calories and nutrients must be offset by a reproductive benefit for the individual caller. But what could be the benefit?

Maybe a lone bird picking at a carcass is vulnerable to predators that lie in wait for it. If that were so, then other ravens attracted by the calls could help keep an eye out for danger. But ravens are large, agile birds with very few known enemies. Heinrich stealthily watched ravens feeding alone and feeding in pairs at a carcass. He never saw a predator attack any of them.

Then he realized that territory may have something to do with it. A **territory** is an area that one or more individuals defend against competitors. After hauling a cow carcass into a Maine forest, Heinrich observed that solitary or paired ravens don't always vocally advertise food. Maybe those silent ravens were adults that had already staked out a large territory, which happened to include the spot where he put the carcass.

A pair of ravens would gain nothing by attracting others to their territory. But what if that Maine forest had been *subdivided* into territories and was defended by powerful adults? A wandering young bird would be lucky to eat at all in an aggressive pair's territory. But recruiting a gang of other, nonterritorial ravens might overwhelm the resident pair's defensive behavior.

As it turns out, only the wandering ravens advertise carcasses. Basically, their calling behavior is selfish, and adaptive. It gives them a shot at otherwise off-limits food and therefore promotes their reproductive success.

Behavioral biologists generally find it profitable to look for evidence of natural selection of the individual's traits rather than something that benefits the species as a whole.

COMMUNICATION SIGNALS

Competing for food, defending territory, alerting others to danger, advertising sexual readiness, forming bonds with a potential mate, caring for offspring—these are the kinds of intraspecific interactions that are played out among animals. The interactions may be lifelong or as brief as a one-time sexual encounter.

The Nature of Communication Signals

Intraspecific interactions depend on intricate mixes of instinctive and learned behaviors by which individuals send and respond to **communication signals**. These are information-laden cues, encoded in stimuli that hold unambiguous meaning for members of the species. The cues include specific odors, colors, patterning, sounds, postures, and movements.

Communication signals sent by one individual, the **signaler**, induce behavioral changes in others of the species. Responding individuals are the **signal receivers**. The signals evolve or persist when they tend to increase reproductive success of both the sender and the receiver. If one proves disadvantageous to either party, natural selection will favor individuals that either do not send the signal or do not respond to it.

Consider **pheromones**, or chemical signals between individuals of the same species. As you know, chemical odors from food and danger were the most important stimuli that early animals had to deal with. As animals evolved, most began to rely on *signaling* pheromones to bring about rapid responses from a receiver. Some, such as chemical alarm signals, call for aggressive or defensive behaviors. Others, such as the bombykol molecules that female silk moths release, serve as sex attractants (Section 37.1). *Priming* pheromones call

Figure 51.7 A dog soliciting play behavior with a play bow.

for generalized physiological responses. For example, a volatile odor in the urine of certain male mice triggers and enhances estrus in female mice.

Acoustical signals also abound in nature, as when male songbirds sing to secure territory and attract a female. Similarly, tungara frogs issue nighttime calls to females and rival males. The call is a distinctive "whine," followed by an equally distinctive "chuck."

Some signals never vary. For instance, ears laid back against the head of a zebra convey hostility, but ears pointing up convey its absence. Other signals convey the intensity of the signaler's message. A zebra with laid-back ears isn't too riled up when its mouth is open just a bit, but when the mouth is gaping, watch out. The combination is an example of a **composite signal**, which has information encoded in two cues (or more).

Signals can take on different meaning in different contexts. A lion may emit a spine-tingling roar to keep in touch with others of its pride or to threaten a rival. Also, one signal often conveys information about other signals to follow. Dogs and wolves solicit play behavior by means of a play bow, as in Figure 51.7. With this bow, subsequent behavioral patterns that the signal receiver would construe as aggressive, sexual, or exploratory in other contexts will be construed as playful only.

Examples of Communication Displays

The play bow is a **communication display,** a pattern of behavior that serves as a social signal. Another common type is a **threat display**, an unambiguous message that a signaler is prepared to attack a signal receiver. When a rival for a receptive female confronts him, a dominant male baboon rolls his eyes upward and "yawns," thus exposing his formidable canines (Figure 51.8a). Often the

Figure 51.8 (**a**) Exposed canines, a threat display by a male baboon. (**b**) Part of a courtship display, with visual, acoustical, and tactile signals, that often precedes copulation. Here a male albatross spreads his wings, an information-laden cue for the female. See also Section 19.2.

Figure 51.9 Dances of honeybees, examples of tactile displays. (**a**) Bees that visit food sources close to a hive perform a *round* dance on the honeycomb. Worker bees that maintain contact with the forager through the dance will search for food close to the hive.

(**b**) Bees trained to visit feeding stations more than 100 meters from the hive perform a *waggle* dance. During the dance, a bee makes a straight run and waggles its abdomen. The slower the waggles during this part of the dance, the more distant the food. (**c**) As Karl von Frisch discovered, the orientation of the straight run varies, depending on the direction in which food is located. When he put a dish of honey on a direct line between the hive and the sun, foragers that located it returned to the hive and oriented their straight runs right up the honeycomb. When he put the honey at right angles to a line between the hive and the sun, the foragers made their straight runs at 90 degrees to vertical. Thus, a honeybee "recruited" into foraging for food can orient its flight *with respect to the sun and the hive*. By doing so, it will waste less time and energy during its food-gathering expedition.

Waggle dancers also vary the speed of their dance in relation to the distance of the food source from the hive. When a site is 150 meters from a hive, the dance is executed much faster, with more waggles per straight run, compared with a dance concerning a food source that is 500 meters away.

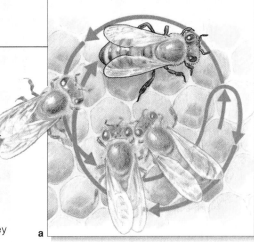
a

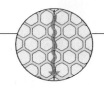

When bee moves straight down comb, recruits fly to source directly away from the sun.

When bee moves to right of vertical, recruits fly at 90° angle to right of the sun.

When bee moves straight up comb, recruits fly straight toward the sun.

b

c

---

rival backs down. In that event, the signaler benefits, for he retains access to the female without putting up a fight. And the signal receiver benefits, because he avoids a serious beating, infected wounds, and possibly death.

Such displays are ritualized, with intended changes in the function of common patterns of behavior. Normal movements may be exaggerated yet simplified; postures may be frozen. Body parts, including feathers, manes, and claws, are often conspicuously enlarged, colored, and patterned. Ritualization is often developed to an amazing degree in **courtship displays** between potential mates. For example, food-enticing behavior is common among many birds. A male bird might emit calls while bowing low, as if to peck the ground for food. When a female comes running, he may spread his wings, as if to focus attention on the ground in front of him (Figure 51.8*b*). Here, one behavior (searching for food) has changed to attract a female, the intent being copulation. As another example, courtship displays of fireflies and some other nocturnal animals incorporate bioluminescent flashes (Section 6.7). A male firefly has a light-generating organ that emits a bright, flashing signal. A few seconds later, a receptive female of his species may answer with a flash. The two will flash back and forth until they meet.

Similarly, in **tactile displays**, a signaler touches the receiver in ritualized ways. After finding a source of pollen or nectar, a foraging honeybee returns to its colony, a hive, and performs a complex dance. It moves about in a circle, jostling in the dark through a crowd of workers. Other bees may follow and maintain physical contact with the dancer. In this way, honeybees acquire information about the general location, distance, and direction of pollen or nectar (Figure 51.9).

Illegitimate Signalers and Receivers

The wrong parties sometimes intercept communication signals. Termites respond with defensive behavior when they catch a whiff of a scent from an invading ant. That scent is meant to identify the ant as a member of an ant colony and elicits cooperative behavior from other ants. Thus the scent has evolved functions, but announcing an invasion of a termite colony is not one of them. When a termite detects the scent and destroys the ant, the termite is an **illegitimate receiver** of a signal *meant for individuals of a different species*.

Or consider **illegitimate signalers**. Certain assassin bugs hook dead and drained bodies of termite prey on their dorsal surface to acquire the odor of their victims. By deceptively signaling that they "belong" to a termite colony, they hunt termite victims more easily. Another illegitimate signaler is the female of certain predatory fireflies. If one observes a flash from a male firefly, she flashes back. If she can lure him into attack range, he becomes her meal—an evolutionary cost of having an otherwise useful response to a come-hither signal.

A communication signal between individuals of the same species is an action that has a net beneficial effect on both the signaler and the receiver. Natural selection tends to favor communication signals that promote reproductive success.

For reasons that we need not explore here, most people find the mating and parenting behaviors of different animals fascinating. How useful is selection theory in helping us interpret such behavior? Let's take a look.

Sexual Selection Theory and Mating Behavior

Competition among members of one sex for access to mates is common. So is choosiness in selecting a mate. Recall, from Section 18.6, that such activities are forms of a microevolutionary process called **sexual selection**. Sexual selection favors traits that give the individual a competitive edge in attracting and holding onto mates.

Typically, male animals produce great numbers of very tiny sperm, and the females produce considerably fewer but larger eggs. Reproductive success for a male generally depends on how many eggs he fertilizes. For a female, success depends largely on how many eggs she produces or how many offspring she can care for. In most cases, the prime factor that influences her sexual preference is the quality of a mate, not the quantity of partners. Hangingflies, sage grouse, and bison (Figures 51.10 through 51.12) are splendid examples of how such females dictate the rules of male competition. They also illustrate how males employ tactics that may help them fertilize as many eggs as possible.

Female hangingflies (*Harpobittacus apicalis*) select the males that offer them superior material goods. And so you might observe males capturing and killing a moth or some other insect. After males do this, they release a sex pheromone that might attract females to the "nuptial

Figure 51.10 A male hangingfly dangling a moth as a nuptial present for a future mate. The females of certain hangingfly species choose sexual partners on the basis of the size of prey that males offer to them.

Figure 51.11 Ornamental feathers of a male sage grouse. These visual signals function in a courtship display—a dance performed at a lek. Males vigorously defend a small patch of this communal mating area. Females (the smaller brown birds) observe the prancing males before choosing the one they will mate with.

gift" (Figure 51.10). A female chooses males that offer the larger, calorie-rich offerings. She permits a male to mate, but only *after* she's been eating a gift for five minutes or so. Then she will accept sperm and store it in a reproductive organ, but only as long as the food holds out. Before twenty minutes are up, she can break off the mating at any point. If the female does so, she might well mate with another male and accept its sperm. And doing so will dilute the reproductive success of her first partner.

In western regions of North and South Dakota, female sage grouse (*Centrocercus urophasianus*) are dispersed among stands of sagebrush. You never find them far from the protective cover of sage during spring courtship and summer nesting. Male sage grouse make no attempt to maintain a large territory. In the breeding season they congregate in a **lek**, a type of communal display ground. There, each male stakes out a few square meters as his territory. Females are attracted to the lek—not to feed or nest, but rather to observe the males. With tail feathers erect and big neck pouches puffed, each male emits booming calls and stamps about, like a wind-up toy, on his display ground (Figure 51.11). Female sage grouse tend to select and mate with one male only. Then they go off to nest by themselves in sagebrush, unassisted by a sexual partner. Many females often choose the same male, so most of the males never mate.

As another example, females of certain species cluster in defendable groups when they are sexually receptive. Where you find such a group, you probably will observe male competition for access to the clusters. Competition for ready-made harems favors combative male lions, sheep, elk, and bison, to name a few animals (Figures 1.6*g* and 51.12).

Figure 51.12   Sexual competition between male bison, which are fighting for access to a cluster of females.

Figure 51.13   Male and female Caspian terns, protecting their chicks. Parental care has costs as well as benefits.

Costs and Benefits of Parenting

What happens after mating, when the offspring arrive? Until the offspring have developed enough to survive on their own, the parents of some species will care for them. For example, adult Caspian terns (Figure 51.13) incubate the eggs, shelter the nestlings and feed them, then accompany and protect them after they start to fly. Such parental behavior comes at a reproductive cost. It drains time and energy that might otherwise be given to improving their own chances of living to reproduce

another time. Yet for many species, parenting improves the likelihood that the current generation of offspring will survive. The benefit of devoting time and energy to immediate reproductive success outweighs the cost of reduced reproductive success at some later time.

Selection theory applies to some aspects of mating behavior, as when sexual selection favors behavioral traits that give the individual a competitive edge in reproductive success. Selection theory applies as well to parental behavior that contributes to reproductive success.

Survey the animal kingdom and you observe a range of social groupings. For some species, individuals spend most of their lives alone or in small family groups. For some other species, individuals live in huge groups of thousands of related individuals. Termite and honeybee societies are like this. Individuals of still other species live in social units composed primarily of nonrelatives. Populations of the human species are like this.

Given such differences, how might we gain insight into the basis for any one social group? To find answers, evolutionary biologists commonly take a **cost-benefit approach**. That is, they attempt to identify the costs and benefits of sociality in terms of the *reproductive success of the individual*, as measured by its contribution to the gene pool of the next generation.

Cooperative Predator Avoidance

First, consider how a group of animals that are acting cooperatively against a predator can reduce the net risk to any one individual. In a flock or herd vulnerable to predation, simply having more pairs of eyes to scan the surrounding area helps individuals detect the predators sooner. Individuals of the group also may join forces to make a counterattack or perform a defensive behavior, as the musk-oxen in Figure 51.14 are doing.

The biologist Birgitta Sillén-Tullberg found tangible evidence of the benefits of sociality. She was studying Australian sawfly caterpillars, which live together in clumps on tree branches. Figure 51.15 shows one of the clumps. When something disturbs the caterpillars, they collectively rear up from the branch and writhe about, all the while regurgitating partially digested food. Their food of choice is eucalyptus leaves. Eucalyptus leaves are heavily impregnated with chemical compounds that are toxic to most animals—including the songbirds that prey on caterpillars.

Sillén-Tullberg hypothesized that individual sawfly caterpillars benefit from the coordinated act of repelling bird predators. She used her hypothesis to predict that the birds are more likely to consume a solitary caterpillar, not a cluster of them.

To test her prediction, Sillén-Tullberg offered young, hand-reared Great Tits (*Parus major*) a chance to feed on sawfly caterpillars, which she offered either one by one or in a group of twenty per offering. She did this for a standard number of presentations. Ten birds that were offered one individual at a time consumed an average of 5.6 caterpillars. But ten birds that were each offered a clump of caterpillars only ate an average of 4.1. As she had predicted for this experiment, the individuals were somewhat safer in a group than on their own.

The Selfish Herd

Simply by their physical position within a group, some individuals form a living shield against predation on others in the group. They all belong to a **selfish herd**, a

Figure 51.15 Social defensive behavior of Australian sawfly caterpillars, which clump together in tree branches. These larvae collectively regurgitate and hold fluid in their mouth (the yellow blobs), where it serves to repel predators that try to grab them. After danger passes, they swallow the fluid, which is toxic to most animals.

Figure 51.14 Social defensive behavior of musk-oxen (*Ovibos moschatus*). In the presence of a perceived threat (usually wolf predators), adults form a circle around their young. They face outward, and the "ring of horns" successfully deters the wolves.

By contrast, some individuals may *help* others survive and reproduce at their own expense. Their helpful behavior may be a cost of belonging to the social group. And it may be that they do not entirely give up their chance at reproductive success. We find some evidence of both possibilities in **dominance hierarchies**, a type of social group in which some of the individuals have adopted subordinate status to others.

In a baboon troop, for example, individuals help one another, but reproductive opportunity is unequal. Upon receiving a threat signal from a dominant member of the troop, subordinate members give up safe sleeping places, choice bits of food, and receptive females. Each recognizes its social status with respect to the others (Figure 51.16). Jane Goodall has studied the dominance hierarchies that also exist among the chimpanzees of Gombe. You read about their social behaviors in Chapter 37.

Figure 51.16 Appeasement behavior between baboons. Notice the assured position of the dominant animal (*left*) and the abject stare and groveling posture of the subordinate one, who is making little conciliatory smacking noises with its lips.

relatively simple society held together by reproductive self-interest, although not consciously so. Investigators have tested the selfish-herd hypothesis for male bluegill sunfishes, which build adjacent nests on the bottom of lakes. Males use their fins to hollow out a depression in lake mud, where females deposit eggs.

If a colony of bluegill males is a selfish herd, then we can predict competition for the "safe" sites—at the center of the colony. Compared to the periphery, eggs laid in nests at the center are less likely to be attacked by snails and largemouth bass. The competition does indeed exist. The largest, most powerful males tend to claim central locations. Other, smaller males assemble around them and bear the brunt of predatory attacks. Even so, they are better off in the group than on their own, fending off a bass singlehandedly, so to speak.

Dominance Hierarchies

As you have seen, individuals of a selfish herd make no personal sacrifice for the others; the personal benefits of living with others simply seem to outweigh the costs.

Why do subordinate adults that don't breed remain at low rungs in the social ladder? As is true for the bluegill sunfish, they probably reap benefits that offset their lowly status. Consider that it simply may not be possible to survive alone, outside the protection afforded by the group. A solitary baboon out in the open surely quickens the pulse of the first leopard that sees it (Section 47.4). Besides, challenging a stronger member may result in injuries that could shorten a life. And it may turn out that self-sacrificing behavior may give a subordinate individual a chance to reproduce if it lives long enough and if predation or weakness in old age removes dominant peers. Some subordinate wolves and baboons do move up the social ladder if dominant members slip down a rung or fall off. Thus, acceptance of subordinate status might pay off in the long term for the patient individual.

We may evaluate the costs and benefits of social life in terms of reproductive success, as measured by the genes that each individual contributes to the next generation.

COSTS OF LIVING IN SOCIAL GROUPS

So far, we have used individual selection theory to help illuminate the advantages of communication signals, territoriality and dominance hierarchies, mating tactics, parenting, and other aspects of living in social groups. Now the question becomes this: If social behavior is so advantageous, *then why are there so few social species among most groups of animals?* One plausible answer is that, in certain environments, the costs to the individual outweigh the benefits of life in a social group.

Most obviously, when more individuals of a species live together in their habitat, the competition for food increases. For example, royal penguins, herring gulls, cliff swallows, and prairie dogs are among the animals

Another cost is the risk of being killed or exploited by others in the group. For example, when presented with the opportunity to do so, breeding pairs of herring gulls will cannibalize the eggs or young chicks of their neighbors in an instant. Or consider lions. These long-lived cats compete for permanent hunting territories, even though they can live for an extended time between kills. When three or more lions hunt cooperatively, they are better at capturing large prey. Yet George Schaller's data revealed that individuals in prides of three or more actually eat *less* well than one or a pair of lions. Also, males intent on taking over a pride show infanticidal behavior; they kill the cubs. For lionesses, then, group

Figure 51.17 Colony of royal penguins on Macquarie Island, between New Zealand and Antarctica.

that live in huge colonies (Figure 51.17). Great numbers of individuals must compete for a fair share of the same pie, so to speak.

In addition, living in social groups can encourage the spread of contagious diseases and parasites. Under crowded living conditions, the individual as well as its offspring is more likely to be weakened by pathogens and parasites that are readily transmitted from host to host in crowded groups. Similarly, plagues spread like wildfire through densely crowded human populations. As described in Section 46.6, this is especially the case in settlements and cities with chronic infestations of rats and fleas, and with inadequate or nonexistent sewage treatment and medical care.

living is costly in terms of food intake and reproductive success. They still stick together. Maybe doing so helps them defend territories against smaller groups of rivals. Also, aggressive males almost always kill the cubs of a single lioness, but occasionally a group of two or more lionesses can save some of the cubs.

Individuals that belong to a social group pay costs in terms of increased competition for food, living quarters, mates, and other limited resources. And they are more vulnerable to contagious diseases and parasitic infections.

Individuals also may risk being exploited or having their offspring killed by others in the social group.

EVOLUTION OF ALTRUISM

A subordinate animal that gives way to a dominant one is acting in its own self-interest. What about altruistic animals? We find them among many vertebrate groups. Consider the wolf pack. Although the females of most mammalian groups make the greatest investment in raising offspring, male wolves also make a contribution by providing their pups with prey, defending a feeding territory, and driving off infanticidal intruders. Unlike herbivores that forage on an individual basis, wolves are carnivores that can share captured prey with others of the social group. Thus a female's reproductive success increases when she monopolizes the benefits that males offer. Usually, a wolf pack contains only one dominant breeding female and male (Figure 51.18). The others are nonbreeding sisters, aunts, brothers, and uncles. They altruistically hunt and bring back food to members that stay inside the den and guard the pups. The nonbreeding females may ovulate and court males, but they fail to reproduce when a dominant female is present.

Altruistic behavior is most extreme in certain insect societies, as when a worker bee plunges its stinger into an invader of the hive. With this act of defending the colony, it also commits suicide. Section 51.9 provides a look at termite and honeybee altruists.

Yet if the altruistic individuals of a social group do not contribute their genes to the next generation, then how is the genetic basis for their altruistic behavior perpetuated over evolutionary time? According to William Hamilton's **theory of indirect selection**, the genes associated with caring for *relatives*—not one's direct descendants—tend to be favored in certain situations. Remember, when a sexually reproducing parent cares for offspring, it isn't helping exact genetic copies of itself. Commonly, such parents are diploid, with pairs of genes. Each gamete they produce, hence each offspring, has only one-half of its genes. If other individuals of the social group have the same ancestors, they share the same genes with parents. Two siblings (brothers, sisters) are as genetically similar as a parent is to one of its offspring. Nephews and nieces are like their uncle in about one-fourth of their genes.

Thus we can think of this form of altruism as an extension of parenting. For example, suppose an uncle helps his niece survive long enough to reproduce. He has made an *indirect* genetic contribution to the next generation, as measured in terms of the genes that he and his relatives share. Altruism costs him; he may lose his own opportunities to reproduce. But if the cost is less than the benefit, altruistic actions will propagate the uncle's genes and favor the spread of his kind of altruism in the species. If an uncle saves two nieces, this is equivalent to saving his own daughter.

Similarly, nonbreeding workers in insect societies indirectly promote their "self-sacrifice" genes through altruistic behavior directed toward relatives. Colonies of honeybees, ants, and termites actually are extended families. The worker force of the family labors on behalf of their siblings, some of which are the future kings and queens. Thus, when a guard bee drives her stinger into a raccoon, she inevitably dies—but her siblings in the hive will perpetuate some of her genes. Sterility and extreme self-sacrifice are rare among social groups of

Figure 51.18 Dominant male of a wolf pack, surrounded by subordinate members of the pack engaged in a ritual display involving lowered heads and drooped tails.

vertebrates. The known exceptions include naked mole-rats. Section 51.10 describes a study of the genetic relationships in a clan of these nearly hairless mammals.

Altruistic behavior occurs among many vertebrate groups such as wolf packs, it reaches its most extreme among some insect societies, such as honeybee and termite colonies.

Altruistic behavior may persist when individuals pass on genes indirectly, by helping relatives survive and reproduce.

By one theory of indirect selection, genes associated with altruistic behavior that is directed toward relatives may be favored in certain situations.

51.9 ABOUT THOSE SELF-SACRIFICING INSECTS

CONSIDER THE TERMITE Picture yourself in a eucalyptus forest in Queensland, Australia. You notice a narrow, brittle tube running along the trunk of a dead tree and decide to chip a few fragments from it. As sunlight pours in through the breach in the tube wall, some small, nearly white insects start banging their heads against the wall before they scurry away.

The head banging is an acoustical signal. It produces vibrations that alert solder termites, which run to the breach and make a defensive stand (Figure 51.19). Each of these small, golden-brown insects has a swollen, eyeless, tapering head with a long, pointed "nose." Disturb a soldier termite and it will shoot thin jets of silvery goo out of its nose! Such silvery strands release volatile compounds that attract still more soldier termites. Together the soldiers battle the danger, which more typically is an invasion by ants.

Insects in that ruptured tunnel belong to a complex termite colony. Whereas the soldier termites protect the colony, the pale ones are workers, which build tunnels that lead to places where they can safely gather fibers of wood. They carry fibers to an underground nest, where termites live and cultivate an edible fungus. Fungal hyphae grow into the wood fibers and absorb nutrients from them. The termites eat some portions of the fungus. They chew the wood into small particles that they can swallow. Unlike most termite species, they don't produce enzymes that can digest the cellulose of wood. Microbial symbionts in their gut do this, and both the microbes and the termites absorb the released nutrients.

Soldier termites and worker termites are sterile; neither can reproduce. They engage in self-sacrificing behavior that contributes to the survival of others. A single queen (with a relatively enormous, egg-producing abdomen) and one or more kings serve as "parents" for the entire society.

CONSIDER THE HONEYBEE The only fertile female in a honeybee colony, or hive, is the queen bee. Figure 51.20*a* shows one, with her court of sterile worker daughters. The daughters feed the queen and relay her pheromones

Figure 51.19 Soldier termites defending their social colony by guarding a break in a foraging tunnel, which worker termites constructed on a dead tree trunk. Members of this caste shoot out strands of glue from a pointed "nose." The glue entangles intruders, such as invading ants.

Figure 51.20 Life in a honeybee colony. (**a**) A court of sterile worker daughters surrounds a queen bee, the only female of the colony that reproduces. (**b**) A stingless drone. (**c**) One of the 30,000 to 50,000 worker bees in the hive. (**d**) Worker bees store honey or pollen in the honeycomb, which also houses new bee generations. Young workers feed the larvae. (**e**) Worker bees transferring food to one another. (**f**) Worker females at the hive entrance serve as guards.

early larval stage   nearly mature pupa

throughout the hive, and these influence the activities of all members. The queen is much larger than the workers, in part because her ovaries are fully developed (unlike those of her sterile daughters).

Only at certain times of year do the stingless drones develop and mature (Figure 51.20*b*). The drones do not work for the colony. Instead, they leave the colony and attempt to mate with queens of other hives—and so perpetuate part of the gene pool elsewhere.

The hive contains between 30,000 and 50,000 worker bees (Figure 51.20*c*). They feed bee larvae, clean and maintain the hive, and construct new honeycomb from wax secretions. The workers store honey or pollen in the honeycomb, which also houses new bee generations from the egg stage, through a series of larval stages, to the emergence of pupae and then the adults (Figure 51.20*d*).

Young workers feed the larvae. Each adult worker lives for about six weeks during the spring and summer. In an overwintering colony, it can survive for about four months.

Workers engage in scent-fanning. Fanned air passes over the bee's exposed scent gland. Pheromones released from the gland help other bees orient to the hive's entrance on their way to or from foraging expeditions. When foraging bees return to the hive after locating a rich source of nectar or pollen, they perform a dance, a tactile display that recruits more workers into taking off for the source (Figure 51.9). In other cooperative behaviors, worker bees transfer food to one another, and worker females at the hive entrance serve as guards (Figure 51.20*e*,*f*). Guard bees will readily sacrifice themselves to repel intruders.

After these glimpses into the lives of self-sacrificing insects, see if you can answer some basic questions. First, by what means do members of a termite or honeybee colony cooperate and benefit the group and themselves? (For example, identify some of the communication signals and the manner in which individuals respond to them.) Second, reflect on the adaptive value of some of their social behaviors. (For example, how might sterility and other extreme forms of self-sacrifice increase individual chances of reproductive success?)

51.10 ABOUT THOSE NAKED MOLE-RATS

We do not mean to be at all judgmental about this, but naked mole-rats (*Heterocephalus glaber*) look rather like bucktoothed sausages with wrinkled, pink skin and a few sprouts of hairs (Figure 51.21). Their behavior is even more fascinating than their looks. Why? Unlike any other known vertebrate, many naked mole-rat individuals live out their lives as nonbreeding helpers in their social group.

These highly social mammals live in arid regions of eastern Africa, in cooperatively excavated burrows. They always live in clans—small social units of anywhere from 25 to 300 individuals. In each clan, you will find only a single reproducing female, and she mates with one to three males. All other members of the clan care for the "queen" and "king" (or kings) and their offspring.

Nonreproducing "diggers" busily excavate extensive subterranean tunnels and special chambers, which serve as living rooms or waste-disposal centers. They locate large tubers growing underground. These they chop up into edible bits, which they then deliver to the queen, her retinue of males, and her offspring.

Besides this, digger mole-rats deliver food to certain other helpers that spend time loafing about, shoulder to shoulder and belly to back, with the reproductive royals. Usually the "loafer" mole-rats are larger than the diggers. And they aren't really loafers at all. They will spring to action when a snake or some other enemy threatens the colony. At great personal risk, they will collectively chase away or attack and kill the predator.

If helpful members fail to have offspring that would perpetuate the genetic basis for helping, then maybe their behavior persists because they are related in some way to offspring of the mole-rats that *do* reproduce. Thus indirect selection theory has yielded a hypothesis about helpers, which in turn can be used to make a prediction:

If helpful behavior is genetically advantageous to some individuals in a naked mole-rat colony (*the hypothesis*), then it follows that the helpers will be related to the reproductive members of the colony that benefit from their altruism (*the prediction*).

To *test* the prediction, we require information about genetic relationships among the members of the naked mole-rat colony. One way to get it would be to know which animals were offspring of which others. However, to construct a family tree of an entire colony would be most laborious and difficult, because there are several breeding males and queens that die and are replaced.

H. Kern Reeve and his colleagues decided on a more practical approach. They relied upon *DNA fingerprinting*, a method of establishing degrees of genetic relatedness among individuals. As described in Section 16.3, this laboratory method starts with the formation of restriction fragments of DNA molecules. Then technicians construct a visual record of different sets of fragments of the DNA from different individuals. Essentially, a set of identical twins would have identical DNA fingerprints (their DNA would have the same base sequence). Individuals having the same father and mother would have similar DNA. Therefore, they would have similar although not identical DNA fingerprints. On the average, DNA fingerprints of genetically unrelated individuals should differ much more than the DNA fingerprints of siblings or other relatives.

When Reeve constructed DNA fingerprints for naked mole-rats, he discovered that all of the individuals from a clan are *very* close relatives, and are very different genetically from members of other clans. The findings suggest that each naked mole-rat clan is highly inbred, a result of generations of brother-sister, mother-son, or father-daughter matings. As you already know from Section 18.8, inbreeding among individuals of a population results in extremely reduced genetic variability.

Therefore, a self-sacrificing naked mole-rat is helping to perpetuate a very high proportion of the forms of genes (alleles) that it carries. As it turns out, the genotypes of helpers and the helped might be as much as 90 percent identical!

Figure 51.21   A peek at a few naked mole-rats (*Heterocephalus glaber*).

AN EVOLUTIONARY VIEW OF HUMAN SOCIAL BEHAVIOR

If we can analyze the evolutionary basis of the behavior of termites, naked mole-rats, and other animals, would it not be rewarding to analyze such a basis of human behavior also? Many people resist the idea. It seems they believe that attempts to identify the adaptive value of a particular human trait are attempts to define its moral or social advantage. Clearly, however, there is a difference between trying to explain something in terms of its evolutionary history and attempting to justify it. "Adaptive" does not mean "morally right." It means valuable with respect to the transmission of an individual's genes.

Consider the case of altruistic behavior that we call **adoption**—the acceptance of offspring of other individuals as one's own. If we wish to gain understanding of this behavior, we might use evolutionary theory to formulate a testable hypothesis about it. But this does not mean we are passing judgment on whether adoption is moral or even socially desirable. These are two entirely separate issues —about which biologists have no more to say than anybody else.

Figure 51.22 Two adult emperor penguins competing to adopt an orphan, which penguins accept as substitute offspring.

Start with a premise that all adaptations have costs and benefits. One cost is that certain adaptive behaviors may be *redirected* under rare or unusual circumstances. In many species, adults that lost offspring will adopt a substitute. Some cardinals have fed goldfish. A whale tried to lift a log out of water as if it were a distressed infant that needed help in reaching the water's surface. Emperor penguins fight to adopt orphans (Figure 51.22).

As John Alcock suggests, such examples constitute a test of a hypothesis about adoption by humans: Human adoptive behavior occurs "by mistake" when otherwise adoptive parental behavior is directed toward unrelated offspring. Suppose it arises through a frustrated desire to bear children (the hypothesis). If so, then we can predict that couples who have lost an only child or who are sterile should be especially prone to adopt strangers.

Now consider this. For most of their evolutionary history, humans have lived in small groups and, later, small villages. They probably had few opportunities or showed little inclination to adopt strangers. They were likely to accept the child of a deceased relative. But

how do adoptive parents gain genetic representation in the next generation? According to theories of natural selection and indirect selection, individuals act in ways that promote genetic self-interest. So we might predict that parents with dependent, care-requiring children of their own are much less likely to adopt substitutes, compared with adults who have no children or who already raised children and are living by themselves. We could test the prediction by piecing together a large enough sampling of information, ideally from diverse existing human cultures, on possible connections between adoption and a childless condition.

Indirect selection also favors adults who focus their parenting behavior on relatives and thereby perpetuate their shared genes in an indirect way. We might even predict that such adults are more likely to be related to the adopted child than we would expect by chance alone. A researcher, Joan Silk, tested the prediction in traditional societies. Her results showed that people will adopt related children far more often than nonrelated ones. Especially in the large, industrialized societies with welfare agencies and other means of adoption assistance, individuals become parents of nonrelated children. Such societies create evolutionary environments. Parenting mechanisms evolved in the past, and it may be that their redirection toward nonrelatives says more about our evolutionary history than it does about the transmission of one's genes.

The point is this: Evolutionary hypotheses about the adaptive value of behavior lend themselves to testing. And through such testing, we can gain understanding about the evolution of human behavior.

It is possible to test evolutionary hypotheses regarding the adaptive value of human behaviors.

There may be greater acceptance of such testing when more people come to understand that adaptive behavior and socially desirable behavior are separate issues.

In biology, "adaptive" means only that some specified trait has proved beneficial in the transmission of the genes of an individual that are responsible for that trait.

SUMMARY

1. Animal behavior (coordinated responses to stimuli) originates with genes that directly or indirectly specify products required for the development and operation of the nervous, endocrine, and skeletal-muscular systems.

2. A behavior performed without having been learned by actual experience is instinctive, a prewired response to one or two simple, well-defined environmental cues (sign stimuli) that trigger a suitable response, such as a fixed action pattern. Individual experiences can lead to variations or changes in responses (learned behavior). Both are outcomes of genetic and environmental inputs.

3. Behavior with a genetic basis is subject to evolution by natural selection. It evolved as a result of individual differences in reproductive success in past generations, when the reproductive benefits of a particular behavior exceeded its reproductive costs (or disadvantages).

4. Discriminating among competing explanations for a particular behavior requires formulating hypotheses to generate predictions, which can be tested by collecting new information and by checking whether test results match the expected observations.

5. Members of the same species often create obstacles to one another's reproductive success, as by selective mate choice and by competition for access to mates.

6. Social groups require cooperative interdependency among individuals of a species. Sociality is promoted by communication signals, cues sent by one individual (a signaler) that can change the behavior of another of the same species (a signal receiver). A signal benefits both the signaler and the receiver.

7. Chemical, visual, acoustical, and tactile signals are components of communication displays.

   a. Pheromones function in chemical communication. Signaling pheromones, such as sex attractants and alarm signals, can induce immediate change in the behavior of a receiver. Priming pheromones elicit a generalized physiological response by the receiver.

   b. Visual signals (observable actions or cues) are key components of courtship displays and threat displays.

   c. Acoustical signals (sounds with precise, species-specific information) include the mating calls of frogs.

   d. Tactile signals are specific forms of physical contact between a signaler and a receiver. Interaction among honeybees during a round dance is an example.

8. Costs and benefits of social life are reflected in the individual's reproductive success, as measured by its genetic contribution to the next generation.

9. Social groups have costs, such as competition for limited resources and greater exposure to contagious diseases and parasites. Benefits outweigh the costs in some cases, as when predation pressure is severe.

10. In self-sacrificing (altruistic) behavior, individuals give up chances to reproduce while helping others of their social group. In most social groups, individuals do not sacrifice reproductive chances to help others.

   a. A dominance hierarchy is a social ranking in which some individuals give way to others of their group. Often, dominant individuals force subordinates to relinquish food or some other resource.

   b. When subordinates give in to dominant members of their group, they may receive compensatory benefits of group living, such as safety from predators. In some species, subordinates may reproduce if they live long enough or if dominant ones slip in the hierarchy or die.

11. By one theory of indirect selection, genes associated with caring for *relatives* (not one's direct descendants but individuals that bear some of the same genes) can be favored in some cases. Indirect selection can lead to extreme altruism, as among workers of some species of social insects and naked mole-rats. Such workers do not usually reproduce. Their altruism helps reproducing relatives survive, so they pass on "by proxy" the genes underlying the development of altruism.

Review Questions

1. Explain how genes and their products, including hormones, influence the mechanisms required for forms of behavior. *51.1*

2. Define these terms: instinctive behavior, sign stimulus, fixed action pattern, and learned behavior. *51.1, 51.2*

3. Contrast altruistic behavior with selfish behavior. Why does either form of behavior persist in a population? *51.3*

4. Describe some characteristics of communication signals. Then give an example of a communication display. *51.4*

5. Describe some feeding behavior or mating behavior in terms of natural (individual) selection. *51.1, 51.5*

6. List some of the benefits and costs of sociality. *51.6, 51.7*

7. Why don't members of a selfish herd live apart if each one is "trying to take advantage" of the others? *51.6*

8. How do sterile termites propagate their genes? *51.8, 51.9*

Self-Quiz (Answers in Appendix IV)

1. "Starlings festoon their nests with sprigs of wild carrot and so minimize nest mites." This statement is _____ .
   a. an untested hypothesis   c. a test of a hypothesis
   b. a prediction   d. a proximate conclusion

2. Genes affect the behavior of individuals by _____ .
   a. influencing the development of nervous systems
   b. affecting the kinds of hormones in individuals
   c. governing the development of muscles and skeletons
   d. all of the above

3. Many kinds of female mammals live in herds. Which of these mating systems might you see in such herds?
   a. a lek mating system
   b. males competing for clusters of receptive females
   c. territorial defense of feeding sites by males
   d. males capturing food to lure females

4. A statement that lemmings dispersing from an overcrowded population bring it in line with available resources _____ .
   a. has been proved to be true
   b. is consistent with Darwinian evolutionary theory
   c. is based on a theory of evolution by group selection
   d. is supported by the finding that most animals behave altruistically during their lives

5. Social behavior evolves because _____ .
   a. social animals are more advanced than solitary ones
   b. under some ecological conditions, the costs of social life are less than its benefits to the group
   c. under some ecological conditions, the costs of social life are less than its benefits to the individual
   d. predator pressures always favor social living

6. The theory of evolution by indirect selection _____ .
   a. focuses on evolution for group benefit
   b. explains that altruism is always adaptive
   c. examines the consequences of the effects of differences among individuals on the reproductive success of relatives
   d. shows how families with many descendants are superior to self-sacrificing families

7. Altruism is helpful behavior that _____ .
   a. cannot evolve
   b. can spread through a species only if it raises the altruist's reproductive success
   c. reduces an individual's reproductive success
   d. always helps spread the altruist's genes

8. The genetic similarity between an uncle and his nephew is _____ .
   a. the same as between a parent and his offspring
   b. greater than between two full siblings
   c. dependent on how many other nephews the uncle has
   d. less than that between a mother and her daughter

9. Match the terms with their most suitable description.

   _____ fixed action pattern   a. time-dependent form of learning requiring exposure to key stimulus
   _____ altruism   b. genes and environmental experience
   _____ basis of instinctive *and* learned behavior   c. instinctive response triggered by a well-defined, often simple stimulus
   _____ imprinting   d. assisting another individual at one's own expense

Critical Thinking

1. You observe mallard ducks in a small pond, then you notice a rooster wading out to the ducks with amorous intent (Figure 51.23). What probably happened to the rooster early in life?

Figure 51.23 A behaviorally confused rooster.

2. A large caterpillar is crawling along a tree branch in a tropical forest. You poke it, and it partly responds by letting go of the branch and puffing up part of its body (Figure 51.24). Propose a mechanism that may underlie its defensive behavior. Also propose how the behavior has adaptive value. How would you test your two hypotheses?

Figure 51.24

3. A hyena scent-marks plants in its territory by releasing specific chemicals from certain glands. What evidence would you need to demonstrate that this action is an evolved communication signal?

4. Suppose you are traveling through Africa and you see a black heron standing in the water with its wings held over its head like an umbrella, as shown in Figure 51.25. You might suppose that, like other herons, it is on the lookout for fish. But most other herons keep their wings down when they are hunting. What is the adaptive value of this behavior? Possibly the black heron is creating shade on the water, and perhaps minnows are being attracted to the shade because it appears to offer shelter. How would you test this hypothesis?

Figure 51.25 A black heron, possibly on a fish-enticing expedition.

5. Develop an adaptive value hypothesis for the observation that male lions kill the offspring of females that they acquire after chasing away the previous pride holders that had mated with those females. How would you test your hypothesis?

6. What are the evolutionary costs and benefits to a male hangingfly of offering a nuptial gift to its potential mate? How might a female's choosiness depend on those costs and benefits?

Selected Key Terms

adaptive behavior *51.3*
adoption *51.11*
altruistic behavior *51.3*
animal behavior *CI*
communication display *51.4*
communication signal *51.4*
composite signal *51.4*
cost-benefit approach
  (to social behavior) *51.6*
courtship display *51.4*
dominance hierarchy *51.6*
fixed action pattern *51.4*
illegitimate receiver *54.1*
illegitimate signaler *51.4*
imprinting *51.2*
indirect selection,
  theory of *51.8*

instinctive behavior *51.1*
learned behavior *51.2*
lek *51.5*
natural selection *51.3*
pheromone *51.4*
reproductive success *51.3*
selfish behavior *51.3*
selfish herd *51.6*
sexual selection *51.5*
sign stimulus *51.1*
signal receiver *51.4*
signaler *51.4*
social behavior *51.3*
song system *51.1*
tactile display *51.4*
territory *51.3*
threat display *51.4*

Readings

Alcock, J. 1993. *Animal Behavior: An Evolutionary Approach*. Fifth edition. Sunderland, Massachusetts: Sinauer.

Dawkins, R. 1989. *The Selfish Gene*. New York: Oxford University Press.

Frisch, K. von. 1961. *The Dancing Bees*. New York: Harcourt Brace Jovanovich. A classic on the behavior of honeybees.

Le Meho, Y. 1977. "Emperor Penguins: A Strategy to Live and Breed in the Cold." *American Scientist* 65(6): 680–693.

Wilson, E. O. 1975. *Sociobiology: The New Synthesis*. Cambridge, Massachusetts: Harvard University Press.

Web Site See *http://www.wadsworth.com/biology* for practice quiz questions, hypercontents, BioUpdates, and critical thinking. The Wadsworth Biology Resource Center provides a wealth of information fully organized and integrated by chapter.

APPENDIX I.   BRIEF CLASSIFICATION SCHEME

The classification scheme that follows is a composite of several that microbiologists, botanists, and zoologists use. The major groupings are agreed upon, more or less. There is not always agreement, however, on what to call a given grouping or where it might fit within the overall hierarchy. There are several reasons for the lack of total consensus.

First, the fossil record varies in its quality and in its completeness. Therefore, the phylogenetic relationship of one group to others is sometimes open to interpretation. Comparative studies at the molecular level are firming up the picture, but this work is still under way.

Second, ever since the time of Linnaeus, classification schemes have been based on the perceived morphological similarities and differences among organisms. Although some original interpretations are now open to question, we are so used to thinking about organisms in certain ways that reclassification often proceeds slowly. Traditionally, for example, birds and reptiles have been considered to be separate classes (Reptilia and Aves). And yet there are now compelling arguments for grouping the lizards and snakes together in one class, and the crocodilians, dinosaurs, and birds in a different class.

Third, researchers in microbiology, mycology, botany, zoology, and the other fields of biological inquiry have inherited a wealth of literature, based on classification schemes that were developed over time in each of those fields. Many see no good reason to give up the established terminology and thereby disrupt access to the past. Until recently, for example, microbiologists and botanists have been using *division*, and zoologists *phylum*, for taxa that are equivalent in the hierarchy of classification. Many still do. Opinions are still polarized with respect to the kingdom Protista, certain members of which could just as easily be grouped in the kingdoms of plants, fungi, or animals. Indeed, the term protozoan is a holdover from an earlier scheme in which amoebas and certain other single-celled organisms were ranked as simple animals.

Given the problems, why do we even bother imposing artificial frameworks on the history of life? We do this for the same reason that a writer might decide to break up the history of civilization into several volumes, a number of chapters, and many paragraphs. Both efforts are attempts to impart obvious structure to what might otherwise be an overwhelming body of knowledge and to enhance the retrieval of information from it.

Finally, bear in mind that we include this classification scheme primarily for your reference purposes. Besides being open to revision, it also is by no means complete. Numerous existing and extinct organisms of the so-called lesser phyla are not represented here. Our strategy is to focus mainly on organisms mentioned in the text. A few examples of organisms also are listed under the entries.

SUPERKINGDOM PROKARYOTA. Prokaryotes. Almost all microscopic species with DNA concentrated in a region of cytoplasm, not inside a membrane-bounded nucleus. All are bacteria, either single cells or simple associations of cells. Autotrophs and heterotrophs (Table 22.2). Reproduce by prokaryotic fission, sometimes by budding and by bacterial conjugation. *Bergey's Manual of Systematic Bacteriology*, the authoritative reference in the field, calls this "a time of taxonomic transition." It groups bacteria mostly by numerical taxonomy (Section 22.6), not on phylogeny. The scheme presented here reflects strong evidence of evolutionary relationships for at least some bacterial groupings.

KINGDOM EUBACTERIA. Gram-negative and gram-positive forms. Peptidoglycan in cell wall. Photosynthetic autotrophs, chemosynthetic autotrophs, and heterotrophs.

PHYLUM GRACILICUTES.   Typical Gram-negative, thin wall. Autotrophs (photosynthetic and chemosynthetic) and heterotrophs. *Anabaena* and other cyanobacteria. *Escherichia, Pseudomonas, Neisseria, Myxococcus.*

PHYLUM FIRMICUTES.   Typical Gram-positive, thick wall. Heterotrophs. *Bacillus, Staphylococcus, Streptococcus, Clostridium, Actinomycetes.*

PHYLUM TENERICUTES.   Gram-negative, wall absent. Heterotrophs (saprobes, pathogens). *Mycoplasma.*

KINGDOM ARCHAEBACTERIA. Methanogens, halophiles, thermophiles. Strict anaerobes, distinct from other bacteria in cell wall, membrane lipids, ribosomes, and RNA sequences. *Methanobacterium, Halobacterium, Sulfolobus.*

SUPERKINGDOM EUKARYOTA. Eukaryotes. Single-celled and multicelled species. Cells start out life with a nucleus (encloses the DNA) and usually other membrane-bound organelles. Chromosomes with numerous proteins attached.

KINGDOM PROTISTA. Diverse single-celled, colonial, and multicelled eukaryotic species, currently most easily classified by what they are *not* (not bacteria, fungi, plants, or animals). Autotrophs, heterotrophs, or both (Table 23.3). Reproduce sexually and asexually (by meiosis, mitosis, or both). Many related evolutionarily to plants, fungi, and possibly animals.

PHYLUM CHYTRIDIOMYCOTA.   Chytrids. Heterotrophs; saprobic decomposers or parasites. *Chytridium.*

PHYLUM OOMYCOTA.   Water molds. Heterotrophs. Decomposers, some parasites. *Saprolegnia, Phytophthora, Plasmopara.*

PHYLUM ACRASIOMYCOTA.   Cellular slime molds. Heterotrophs with free-living, phagocytic amoeboid cells and spore-bearing stages. *Dictyostelium.*

PHYLUM MYXOMYCOTA.   Plasmodial slime molds. Heterotrophs with free-living, phagocytic amoeboid cells and spore-bearing stages. Aggregate into streaming mass of cells that discard plasma membranes. *Physarum.*

PHYLUM SARCODINA.   Amoeboid protozoans. Heterotrophs, free-living or endosymbiotic, some pathogens. Soft-or shelled bodies, locomotion by pseudopods. The rhizopods (naked amoebas, foraminiferans), *Amoeba proteus, Entomoeba.* Also the actinopods (radiolarians, heliozoans).

PHYLUM MASTIGOPHORA.   Animal-like flagellated protozoans. Heterotrophs, free-living, many internal parasites. All with one to several flagella. *Trypanosoma, Trichomonas, Giardia.*

APICOMPLEXA.   Heterotrophs, many parasitic. Complex of rings, tubules, other structures at head end. Most familiar members called sporozoans. *Plasmodium, Toxoplasma.*

PHYLUM CILIOPHORA. Ciliated protozoans. Heterotrophs, predators or symbionts, some parasitic. All have cilia. Free-living, sessile, or motile. *Paramecium,* hypotrichs.

PHYLUM EUGLENOPHYTA. Euglenoids. Mostly heterotrophs, some autotrophs (photosynthetic). Flagellated. *Euglena.*

PHYLUM PYRRHOPHYTA. Dinoflagellates. Photosynthetic, mostly, but some heterotrophs. *Gymnodinium breve.*

PHYLUM CHRYSOPHYTA. Golden algae, yellow-green algae, diatoms. Photosynthetic. Some flagellated, others not. *Mischococcus, Synura, Vaucheria.*

PHYLUM RHODOPHYTA. Red algae. Photosynthetic, some parasitic. Nearly all marine, some freshwater. *Porphyra. Bonnemaisonia, Euchema.*

PHYLUM PHAEOPHYTA. Brown algae. Photosynthetic, nearly all in temperate or marine waters. *Macrocystis, Fucus, Sargassum, Ectocarpus, Postelsia.*

PHYLUM CHLOROPHYTA. Green algae. Mostly photosynthetic, some parasitic. Most freshwater, some marine or terrestrial. *Chlamydomonas, Spirogyra, Ulva, Volvox, Codium, Halimeda.*

KINGDOM FUNGI. Nearly all multicelled eukaryotic species. Heterotrophs; mostly saprobic decomposers, some parasites. Nutrition based on extracellular digestion of organic matter and absorption of nutrients by individual cells. Multicelled species form absorptive mycelium within substrates and structures that produce asexual spores (and sometimes sexual spores).

PHYLUM ZYGOMYCOTA. Zygomycetes. Zygosporangia (zygote inside thick wall) formed by sexual reproduction. Bread molds, related forms. *Rhizopus, Philobolus.*

PHYLUM ASCOMYCOTA. Ascomycetes. Sac fungi. Sac-shaped cells form sexual spores (ascospores). Most yeasts and molds, morels, truffles. *Saccharomycetes, Morchella, Neurospora, Sarcoscypha. Claviceps, Ophiostoma.*

PHYLUM BASIDIOMYCOTA. Basidiomycetes. Club fungi. Most diverse group. Produce basidiospores inside club-shaped structures. Mushrooms, shelf fungi, stinkhorns. *Agaricus, Amanita, Puccinia, Ustilago.*

IMPERFECT FUNGI. Sexual spores absent or undetected. The group has no formal taxonomic status. If better understood, a given species might be grouped with sac fungi or club fungi. *Verticillium, Candida, Microsporum, Histoplasma.*

LICHENS. Mutualistic interactions between fungal species and a cyanobacterium, green alga, or both. *Usnea, Cladonia.*

KINGDOM PLANTAE. Multicelled eukaryotes. Nearly all photosynthetic autotrophs with chlorophylls *a* and *b.* Some parasitic. Nonvascular and vascular species, generally with well-developed root and shoot systems. Nearly all adapted in form and function to survive dry conditions in land habitats; a few in aquatic habitats. Sexual reproduction predominant; also asexual reproduction by vegetative propagation.

PHYLUM RHYNIOPHYTA. Earliest known vascular plants; muddy habitats. Extinct. *Cooksonia, Rhynia.*

PHYLUM PROGYMNOSPERMOPHYTA. Progymnosperms. Ancestral to early seed-bearing plants; extinct. *Archaeopteris.*

PHYLUM PTERIDOSPERMOPHYTA. Seed ferns. Fernlike gymnosperms; extinct. *Medullosa*

PHYLUM CHAROPHYTA. Stoneworts.

PHYLUM BRYOPHYTA. Bryophytes: mosses, liverworts, hornworts. Seedless, nonvascular, haploid dominance. *Marchantia, Polytrichum, Sphagnum.*

PHYLUM PSILOPHYTA. Whisk ferns. Seedless, vascular. No obvious roots, leaves on sporophyte. *Psilotum.*

PHYLUM LYCOPHYTA. Lycophytes, club mosses. Seedless, vascular. Leaves, branching rhizomes, vascularized roots and stems. *Lycopodium, Selaginella.*

PHYLUM SPHENOPHYTA. Horsetails. Seedless, vascular. Some sporophyte stems photosynthetic, others nonphotosynthetic, spore-producing. *Equisetum.*

PHYLUM PTEROPHYTA. Ferns. Largest group of seedless vascular plants (12,000 species), mainly tropical, temperate habitats.

PHYLUM CYCADOPHYTA. Cycads. Type of gymnosperm (vascular, bears "naked" seeds). Tropical, subtropical. Palm-shaped leaves, simple cones on male and female plants. *Zamia.*

PHYLUM GINKGOPHYTA. Ginkgo (maidenhair tree). Type of gymnosperm. Seeds with fleshy outer layer. *Ginkgo.*

PHYLUM GNETOPHYTA. Gnetophytes. Only gymnosperms with vessels in xylem and double fertilization (but endosperm does not form). *Ephedra, Welwitchia.*

PHYLUM CONIFEROPHYTA. Conifers. Most common and familiar gymnosperms. Generally cone-bearing with needle-like or scale-like leaves.
Family Pinaceae. Pines, firs, spruces, hemlock, larches, Douglas firs, true cedars. *Pinus.*
Family Cupressaceae. Junipers, cypresses. *Juniperus.*
Family Taxodiaceae. Bald cypress, redwoods, Sierra bigtree, dawn redwood. *Sequoia.*
Family Taxaceae. Yews.

PHYLUM ANTHOPHYTA. Angiosperms (flowering plants). Largest group of vascular seed-bearing plants. Only organisms that produce flowers, fruits.
Class Dicotyledonae. Dicotyledons (dicots). Some families of several different orders are listed:
Family Nymphaeaceae. Water lilies.
Family Papaveraceae. Poppies.
Family Brassicaceae. Mustards, cabbages, radishes.
Family Malvaceae. Mallows, cotton, okra, hibiscus.
Family Solanaceae. Potatoes, eggplant, petunias.
Family Salicaceae. Willows, poplars.
Family Rosaceae. Roses, apples, almonds, strawberries.
Family Fabaceae. Peas, beans, lupines, mesquite.
Family Cactaceae. Cacti.
Family Euphorbiaceae. Spurges, poinsettia.
Family Cucurbitaceae. Gourds, melons, cucumbers, squashes.
Family Apiaceae. Parsleys, carrots, poison hemlock.
Family Aceraceae. Maples.
Family Asteraceae. Composites. Chrysanthemums, sunflowers, lettuces, dandelions.
Class Monocotyledonae. Monocotyledons (monocots). Some families of several different orders are listed:
Family Liliaceae. Lilies, hyacinths, tulips, onions, garlic.
Family Iridaceae. Irises, gladioli, crocuses.
Family Orchidaceae. Orchids.
Family Arecaceae. Date palms, coconut palms.
Family Cyperaceae. Sedges.
Family Poaceae. Grasses, bamboos, corn, wheat, sugarcane.
Family Bromeliaceae. Bromeliads, pineapples, Spanish moss.

KINGDOM ANIMALIA. Multicelled eukaryotes, nearly all with tissues, organs, and organ systems and with motility during at least part of the life cycle. Heterotrophs; predators (herbivores, carnivores, omnivores), parasites, detritivores. Reproduce sexually (and asexually in many species). Continuous stages of embryonic development.

PHYLUM PLACOZOA. Marine. Simplest known animal. Two cell layers, no mouth, no organs. *Trichoplax.*

PHYLUM MESOZOA. Ciliated, wormlike parasites, about the same level of complexity as *Trichoplax.*

PHYLUM PORIFERA. Sponges. No symmetry, tissues, or organs.

PHYLUM CNIDARIA. Radial symmetry, tissues, nematocysts.
Class Hydrozoa. Hydrozoans. *Hydra, Obelia, Physalia.*
Class Scyphozoa. Jellyfishes. *Aurelia.*
Class Anthozoa. Sea anemones, corals. *Telesto.*

PHYLUM CTENOPHORA. Comb jellies. Modified radial symmetry.

PHYLUM PLATYHELMINTHES. Flatworms. Bilateral, cephalized; simplest animals with organ systems. Saclike gut.
Class Turbellaria. Triclads (planarians), polyclads. *Dugesia.*
Class Trematoda. Flukes. *Schistosoma.*
Class Cestoda. Tapeworms. *Taenia.*

PHYLUM NEMERTEA.   Ribbon worms.

PHYLUM NEMATODA.   Roundworms. *Ascaris, Trichinella.*

PHYLUM ROTIFERA.   Rotifers.

PHYLUM MOLLUSCA.   Mollusks.
  Class Polyplacophora. Chitons.
  Class Gastropoda. Snails (periwinkles, whelks, limpets,
    abalones, cowries, conches, nudibranchs, tree snails, garden
    snails), sea slugs, land slugs.
  Class Bivalvia. Clams, mussels, scallops, cockles, oysters,
    shipworms.
  Class Cephalopoda. Squids, octopuses, cuttlefish, nautiluses.
    Loligo.

PHYLUM BRYOZOA.   Bryozoans (moss animals).

PHYLUM BRACHIOPODA.   Lampshells.

PHYLUM ANNELIDA.   Segmented worms.
  Class Polychaeta. Mostly marine worms.
  Class Oligochaeta. Mostly freshwater and terrestrial worms,
    but many marine. *Lumbricus* (earthworms).
  Class Hirudinea. Leeches.

PHYLUM TARDIGRADA.   Water bears.

PHYLUM ONYCHOPHORA.   Onychophorans. *Peripatus.*

PHYLUM ARTHROPODA.
  Subphylum Trilobita. Trilobites; extinct.
  Subphylum Chelicerata. Chelicerates. Horseshoe crabs,
    spiders, scorpions, ticks, mites.
  Subphylum Crustacea. Shrimps, crayfishes, lobsters, crabs,
    barnacles, copepods, isopods (sowbugs).
  Subphylum Uniramia.
    Superclass Myriapoda. Centipedes, millipedes.
    Superclass Insecta.
      Order Ephemeroptera. Mayflies.
      Order Odonata. Dragonflies, damselflies.
      Order Orthoptera. Grasshoppers, crickets, katydids.
      Order Dermaptera. Earwigs.
      Order Blattodea. Cockroaches.
      Order Mantodea. Mantids.
      Order Isoptera. Termites.
      Order Mallophaga. Biting lice.
      Order Anoplura. Sucking lice.
      Order Homoptera. Cicadas, aphids, leafhoppers,
        spittlebugs.
      Order Hemiptera. Bugs.
      Order Coleoptera. Beetles.
      Order Diptera. Flies.
      Order Mecoptera. Scorpion flies. *Harpobittacus.*
      Order Siphonaptera. Fleas.
      Order Lepidoptera. Butterflies, moths.
      Order Hymenoptera. Wasps, bees, ants.

PHYLUM ECHINODERMATA.   Echinoderms.
  Class Asteroidea. Sea stars.
  Class Ophiuroidea. Brittle stars.
  Class Echinoidea. Sea urchins, heart urchins, sand dollars.
  Class Holothuroidea. Sea cucumbers.
  Class Crinoidea. Feather stars, sea lilies.
  Class Concentricycloidea. Sea daisies.

PHYLUM HEMICHORDATA.   Acorn worms.

PHYLUM CHORDATA.   Chordates.
  Subphylum Urochordata. Tunicates, related forms.
  Subphylum Cephalochordata. Lancelets.
  Subphylum Vertebrata. Vertebrates.
    Class Agnatha. Jawless vertebrates (lampreys, hagfishes).
    Class Placodermi. Jawed, heavily armored fishes; extinct.
    Class Chondrichthyes. Cartilaginous fishes (sharks, rays,
      skates, chimaeras).
    Class Osteichthyes. Bony fishes.
      Subclass Dipnoi. Lungfishes.
      Subclass Crossopterygii. Coelacanths, related forms.
      Subclass Actinopterygii. Ray-finned fishes.
        Order Acipenseriformes. Sturgeons, paddlefishes.
        Order Salmoniformes. Salmon, trout.
        Order Atheriniformes. Killifishes, guppies.

Order Gasterosteiformes. Seahorses.
Order Perciformes. Perches, wrasses, barracudas,
  tunas, freshwater bass, mackerels.
Order Lophiiformes. Angler fishes.
Class Amphibia. Mostly tetrapods; embryo enclosed in
  amnion.
  Order Caudata. Salamanders.
  Order Anura. Frogs, toads.
  Order Apoda. Apodans (caecilians).
Class Reptilia. Skin with scales, embryo enclosed in
  amnion.
  Subclass Anapsida. Turtles, tortoises.
  Subclass Lepidosaura. *Sphenodon*, lizards, snakes.
  Subclass Archosaura. Dinosaurs (extinct), crocodiles,
    alligators.
Class Aves. Birds. (In more recent schemes, dinosaurs,
  crocodilians, and birds are often grouped in the same
  category.)
  Order Struthioniformes. Ostriches.
  Order Sphenisciformes. Penguins.
  Order Procellariiformes. Albatrosses, petrels.
  Order Ciconiiformes. Herons, bitterns, storks,
    flamingoes.
  Order Anseriformes. Swans, geese, ducks.
  Order Falconiformes. Eagles, hawks, vultures, falcons.
  Order Galliformes. Ptarmigan, turkeys, domestic fowl.
  Order Columbiformes. Pigeons, doves.
  Order Strigiformes. Owls.
  Order Apodiformes. Swifts, hummingbirds.
  Order Passeriformes. Sparrows, jays, finches, crows,
    robins, starlings, wrens.
Class Mammalia. Skin with hair; young nourished by
  milk-secreting glands of adult.
  Subclass Prototheria. Egg-laying mammals (duckbilled
    platypus, spiny anteaters).
  Subclass Metatheria. Pouched mammals or marsupials
    (opossums, kangaroos, wombats).
  Subclass Eutheria. Placental mammals.
    Order Insectivora. Tree shrews, moles, hedgehogs.
    Order Scandentia. Insectivorous tree shrews.
    Order Chiroptera. Bats.
    Order Primates.
      Suborder Strepsirhini (prosimians). Lemurs,
        lorises.
      Suborder Haplorhini (tarsioids and anthropoids).
        Infraorder Tarsiiformes. Tarsiers.
        Infraorder Platyrrhini (New World monkeys).
          Family Cebidae. Spider monkeys, howler
            monkeys, capuchin.
        Infraorder Catarrhini (Old World monkeys and
          hominoids).
          Superfamily Cercopithecoidea. Baboons,
            macaques, langurs.
          Superfamily Hominoidea. Apes and humans.
            Family Hylobatidae. Gibbon.
            Family Pongidae. Chimpanzees, gorillas,
              orangutans.
            Family Hominidae. Existing and extinct
              human species (*Homo*) and australopiths.
    Order Carnivora. Carnivores.
      Suborder Feloidea. Cats, civets, mongooses,
        hyenas.
      Suborder Canoidea. Dogs, weasels, skunks, otters,
        raccoons, pandas, bears.
    Order Proboscidea. Elephants; mammoths (extinct).
    Order Sirenia. Sea cows (manatees, dugongs).
    Order Perissodactyla. Odd-toed ungulates (horses,
      tapirs, rhinos).
    Order Artiodactyla. Even-toed ungulates (camels,
      deer, bison, sheep, goats, antelopes, giraffes).
    Order Edentata. Anteaters, tree sloths, armadillos.
    Order Tubulidentata. African aardvarks.
    Order Cetacea. Whales, porpoises.
    Order Rodentia. Most gnawing animals (squirrels,
      rats, mice, guinea pigs, porcupines).

Metric-English Conversions

Length

English		Metric
inch	=	2.54 centimeters
foot	=	0.30 meter
yard	=	0.91 meter
mile (5,280 feet)	=	1.61 kilometer

To convert	multiply by	to obtain
inches	2.54	centimeters
feet	30.00	centimeters
centimeters	0.39	inches
millimeters	0.039	inches

Weight

English		Metric
grain	=	64.80 milligrams
ounce	=	28.35 grams
pound	=	453.60 grams
ton (short) (2,000 pounds)	=	0.91 metric ton

To convert	multiply by	to obtain
ounces	28.3	grams
pounds	453.6	grams
pounds	0.45	kilograms
grams	0.035	ounces
kilograms	2.2	pounds

Volume

English		Metric
cubic inch	=	16.39 cubic centimeters
cubic foot	=	0.03 cubic meter
cubic yard	=	0.765 cubic meters
ounce	=	0.03 liter
pint	=	0.47 liter
quart	=	0.95 liter
gallon	=	3.79 liters

To convert	multiply by	to obtain
fluid ounces	30.00	milliliters
quart	0.95	liters
milliliters	0.03	fluid ounces
liters	1.06	quarts

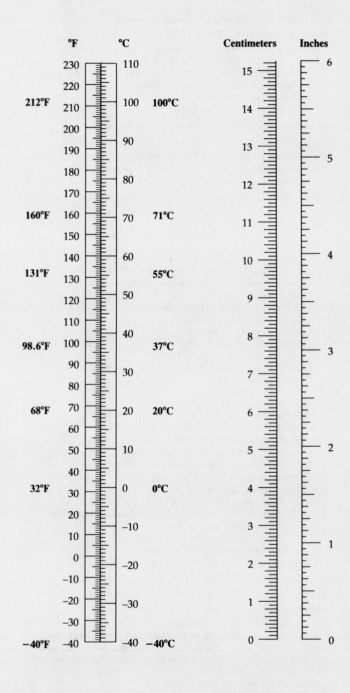

CHAPTER 11

1. a. *AB*
   b. *AB* and *aB*
   c. *Ab* and *ab*
   d. *AB, aB, Ab,* and *ab*

2. a. All offspring will be *AaBB*.

   b. 25% *AABB*; 25% *AaBB*; 25% *AABb*; 25% *AaBb*

   c. 25% *AaBb*; 25% *Aabb*; 25% *aaBb*; 25% *aabb*

   d. 1/16   *AABB*   (6.25%)
      1/8   *AaBB*   (12.5%)
      1/16   *aaBB*   (6.25%)
      1/8   *AABb*   (12.5%)
      1/4   *AaBb*   (25%)
      1/8   *aaBb*   (12.5%)
      1/16   *AAbb*   (6.25%)
      1/8   *Aabb*   (12.5%)
      1/16   *aabb*   (6.25%)

3. Yellow is recessive. Because the first-generation plants have a green phenotype and must be heterozygous, green must be dominant over the recessive yellow.

4. a. *ABC*

   b. *ABc* and *aBc*

   c. *ABC, aBC, ABc,* and *aBc*

   d. *ABC, aBC, AbC, abC, ABc, aBc, Abc,* and *abc*

5. The F_1 plants must all be heterozygous for both genes. When these plants self-pollinate, 1/4 (or 25%) of the F_2 plants will be heterozygous for both genes.

6. a. The mother must be heterozygous for both genes; the father is homozygous recessive for both of the genes. The first child is homozygous recessive for both genes.

   b. The probability that a second child will not be a tongue roller and will have detached earlobes is 1/4 (25%).

7. a. Bill, *Aa Ee Ss* ; and Marie, *AA EE SS* .

   b. The probability that each child will have Bill's phenotype is 100 percent. Because Marie's phenotype is the same as Bill's, the probability that each child will have Marie's phenotype is also 100 percent.

   c. Zero.

8. a. The mother must be heterozygous ($I^A i$). The male with type B blood could have fathered the child if he, too, were heterozygous ($I^B i$).

   b. If the male were heterozygous, then he *could* be the father. However, *any* male who carried the *i* allele could be the father, so one can't say this particular individual absolutely must be. This would include the males with type O blood (*ii*), the heterozygotes with type A blood ($I^A i$), and the heterozygotes with type B blood ($I^B i$).

9. The most direct way to do this would be to allow mice from a true-breeding white-furred strain to mate with mice from a true-breeding brown-furred strain. (True-breeding strains typically are homozygous for alleles influencing the trait being studied.) All F_1 offspring from this cross will be heterozygous for the alleles that influence brown or white fur color. The phenotype of the F_1 offspring should be recorded, then the offspring should be allowed to mate with one another to produce an F_2 population. Assuming that only a single gene locus is involved, several outcomes are possible, and they include the following:

   All F_1 offspring are brown. The F_2 offspring segregate 3 brown: 1 white. *Conclusion:* Segregation is occurring at a single locus where the allele specifying brown is dominant to the allele specifying white.

   All F_1 offspring are white. The F_2 offspring segregate 3 white: 1 brown. *Conclusion:* Segregation is occurring at a single locus where the allele specifying white is dominant to the allele specifying brown.

   All F_2 offspring are tan. The F_2 offspring segregate 1 brown: 2 tan: 1 white. *Conclusion:* Segregation is occurring at a single gene locus where the alleles show incomplete dominance.

10. If the allele that causes the hip disorder is recessive, you *cannot* guarantee that Dandelion's puppies will not carry the allele, inasmuch as Dandelion may be a heterozygous carrier of that allele.

11. Fred should allow his black guinea pig to mate with a white one. This would be a testcross.

    The white one has the genotype *ww*, and the black one might be homozygous dominant (*WW*) or heterozygous (*Ww*). If any offspring from the cross are white, then his black guinea pig is *Ww*. If all offspring are black after Fred allows enough matings to produce a significant number of offspring, then there is a good probability that his guinea pig is *WW*.

    (If ten of the offspring are all black and none is white, then there is about a 99.9 percent chance that his guinea pig is *WW*. The greater the number of offspring, the greater the degree of confidence in the conclusion.)

12. a. $R^1 R^1 \times R^1 R^2$ yields 1/2 red flowers, and 1/2 pink.

    b. $R^1 R^1 \times R^2 R^2$ yields all pink flowers.

    c. $R^1 R^2 \times R^1 R^2$ yields 1/4 red, 1/2 pink, and 1/4 white.

    d. $R^1 R^2 \times R^2 R^2$ yields 1/2 pink and 1/2 white.

13. 9/16 walnut (pea-rose); 3/16 rose; 3/16 pea; 1/16 single.

14. Both parents are heterozygotes ($Hb^A Hb^S$).

    a. 1/4

    b. 1/4

    c. 1/2

15. The mating is M^LM. Genotypes of the offspring will be 1/4 MM, 1/2 M^LM, and 1/4 M^LM^L. Phenotypes will be:

    1/4 homozygous dominant (MM) survivors

    1/2 homozygous (M^LM) survivors

    1/4 lethal (M^LM^L) nonsurvivors

The probability that any one kitten among the surviving progeny will be heterozygous is 2/3.

16. a. Both parents must be heterozygous (Aa). They may produce albino (aa) and unaffected (AA or Aa) children.

    b. Both parents and all their children must be aa.

    c. The albino male must be aa. Because there is an albino child, the female must be heterozygous Aa. If she were AA, they would have no albino children. The one albino child must be aa; the three unaffected children must be Aa. There is a 50 percent chance that any child will be albino. The 3 : 1 ratio is not surprising, given the small number of progeny.

17. a. All offspring of the mating will have an intermediate amount of pigmentation corresponding to genotype $A^1A^2B^1B^2$.

    b. All possible genotypes could occur among offspring of this mating, in these expected proportions:

1/16 $A^1A^1B^1B^1$	(dark red)
1/8 $A^1A^1B^1B^2$	(medium-dark red)
1/16 $A^1A^1B^2B^2$	(medium red)
1/8 $A^1A^2B^1B^1$	(medium-dark red)
1/4 $A^1A^2B^1B^2$	(medium red)
1/8 $A^1A^2B^2B^2$	(light red)
1/16 $A^2A^2B^1B^1$	(medium red)
1/8 $A^2A^2B^1B^2$	(light red)
1/16 $A^2A^2B^2B^2$	(white)

CHAPTER 12

1. a. Human males inherit their X chromosome from their mother.

    b. With respect to an X-linked gene, a human male can produce two kinds of gametes. One kind will have a Y chromosome without the gene. The other kind of gamete will contain an X chromosome—hence it will contain the X-linked gene.

    c. All gametes of a human female who is homozygous for an X-linked gene will have an X chromosome, hence it will contain that gene.

    d. A female who is heterozygous for an X-linked gene will produce two kinds of gametes. One kind will have an X chromosome with one type of allele. The other kind will have an X chromosome with the other allele.

2. a. This gene is only carried on Y chromosomes. So we would not expect females to have hairy pinnae, because females normally do not have Y chromosomes.

    b. Because sons always inherit a Y chromosome from their father and daughters never do, a male having hairy pinnae will always transmit the gene for the trait to his sons and never to his daughters.

3. A 0 percent crossover frequency means that 50 percent of the gametes will be AB and 50 percent will be ab.

4. X rays might have induced a deletion of the dominant allele on one of the fruit fly chromosomes and led to the rare vestigial-wing condition. An alternative explanation is that the irradiation might have induced a mutation in the dominant allele.

5. The likelihood that crossing over will occur between the genes is low if they are very close together, and it is higher if the distance between them is greater.

6. a. Either anaphase I or anaphase II of meiosis.

    b. If a translocation resulted in the attachment of most of human chromosome 21 to the end of a normal chromosome 14, then the new individual would develop Down syndrome. The the cells of that individual would contain the attached chromosome 21 in addition to two normal chromosomes 21. The chromosome number would still be 46.

7. By using c as the symbol for color blindness and C for normal color vision, the cross between these animals can be diagrammed as follows:

$$C(Y) \text{ female} \times Cc \text{ male}$$

In mugwumps, a son receives one sex-linked allele from each of his parents, but a daughter inherits her unpaired sex-linked allele from her father. In this cross, half of the sons will be CC and half Cc, but none will be color blind. Of the daughters, half will be $C(Y)$ and half $c(Y)$. There is a 50 percent chance that a daughter will be color blind. Note: this sex chromosome designation is not the same as it is in humans. But this is the way it is not only for mugwumps, but also for all birds, moths, butterflies, and some other organisms; the Y chromosome is associated with females, not males.

8. To have a daughter who will be affected by childhood muscular dystrophy, the prospective father and the mother must carry the allele for the disorder. Males with the allele have muscular dystrophy and rarely father children.

APPENDIX IV.   ANSWERS TO SELF-QUIZZES

CHAPTER 1
1. cell
2. Metabolism
3. Homeostatsis
4. adaptive
5. mutation
6. d
7. d
8. b
9. c, e, d, b, a

CHAPTER 2
1. b
2. c
3. f
4. e
5. f
6. c, a, b, d

CHAPTER 3
1. d
2. e
3. f
4. b
5. d
6. d
7. d
8. c, e, b, d, a

CHAPTER 4
1. c
2. d
3. d
4. False; many cell types have walls that surround their plasma membrane, which is the outermost *membrane*.
5. c
6. e, d, a, b, c

CHAPTER 5
1. c
2. c
3. a
4. e
5. a
6. centrifuge
7. d
8. b

CHAPTER 6
1. c
2. a
3. a
4. d
5. b
6. e
7. a
8. a
9. d
10. c, g, a, d, e, b, f

CHAPTER 7
1. carbon dioxide, water
2. d
3. c
4. b
5. d
6. c
7. c
8. c, d, e, b, f, a

CHAPTER 8
1. d
2. c
3. b
4. c
5. d
6. c
7. b
8. d
9. b, c, a, d

CHAPTER 9
1. a
2. b
3. b
4. c
5. a
6. d
7. b
8. d, b, c, a

CHAPTER 10
1. d
2. d
3. a
4. b
5. d
6. c
7. c
8. d, a, c, b,

CHAPTER 11
1. a
2. b
3. a
4. c
5. b
6. a
7. d
7. b, d, a, c

CHAPTER 12
1. c
2. c
3. e
4. c
5. d
6. d
7. d
8. c, e, d, b, a

CHAPTER 13
1. c
2. d
3. d
4. c
5. a
6. d
7. d, b, c, a

CHAPTER 14
1. d
2. d
3. b
4. c
5. c
6. a
7. a
8. d
9. c
10. e, c, a, f, d, g , b

CHAPTER 15
1. b
2. b
3. d
4. d
5. d
6. a
7. a
8. d
9. e
10. e
11. b
12. d
13. d, e, b, a, c

CHAPTER 16
1. Plasmids
2. c
3. a
4. b
5. a
6. b
7. b
8. d
9. d, c, f, e, b, a

CHAPTER 17
1. c
2. d
3. d
4. c, d, e, b, a

CHAPTER 18
1. population
2. d
3. c
4. b
5. c
6. c, d, a, b

CHAPTER 19
1. d
2. d
3. c
4. c
5. b
6. c, e, a, d, b

CHAPTER 20
1. a
2. b
3 c
4. d
5. e, b, f, c, a, d

CHAPTER 21
1. c
2. e
3 b
4. d
5. d
6. d
7. b, d, e, a, c

CHAPTER 22
1. c
2. c
3. b
4. c
5. d
6. d
7. e
8. d
9. d, e, b, c, a

CHAPTER 23
1. b
2. d
3. a
4. c
5. a
6. a
7. d
8. b

CHAPTER 24
1. b
2. c
3. b
4. a
5. a
6. e, c, d, a, f, b

CHAPTER 25
1. d
2. b
3. c
4. c
5. b
6. c, e, g, h, f, a, b, d

CHAPTER 26
1. b
2. c
3. a
4. a
5. b
6. c
7. c
8. d, f, c, g, e, i, j, h, a, b

CHAPTER 27
1. d
2. e
3. d
4. b
5. b
6. d
7. d
8. f
9. f
10. d
11. d
12. e
13. d, e, c, b, g, f, a

CHAPTER 28
1. d
2. b
3. c
4. c
5. b
6. c, e, g, h, f, a, b, d

CHAPTER 29
1. a
2. b
3. c
4. a
5. d
6. c
7. c
8. d
9. b, d, e, c, f, a

CHAPTER 30
1. hydrogen bonds
2. e
3. b
4. a
5. c
6. d
7. c
8. d
9. d
10. b
11 b, d, f, c, a, e

CHAPTER 31
1. a
2. pollinators
3. d
4. c
5. b
6. b
7. b
8. a
9. b
10. d
11. c
12. a, c, d, e, b

CHAPTER 32
1. c
2. c
3. e
4. d
5. d
6. a
7. c
8. c
9. a
10. d, e, a, b, c

CHAPTER 33
1. a
2. b
3. b
4. b
5. c
6. d
7. c
8. d
9. c
10. a
11. receptors, integrators, effectors
12. b, a, c, f, g, e, d

CHAPTER 34
1. d
2. b
3. d
4. a
5. d
6. d
7. c
8. b, d,c, a

CHAPTER 35
1. nerve net
2. a
3. false
4. true
5. c
6. b
7. b
8. d
9. a
10. b
11. d
12. e, d, b, c, a

CHAPTER 36
1. stimulus
2. sensation
3. perception
4. d
5. b
6. b
7. e
8. d
9. c
10. d

11. b
12. c
13. e, c, b, d, a

CHAPTER 37
1. f
2. d
3. b
4. d
5. e
6. b
7. b
8. c
9. d, f, c, e, a b

CHAPTER 38
1. integumentary
2. skeletal; muscle
3. smooth; cardiac; skeletal
4. c
5. d
6. d
7. b
8. d
9. f
10. h, f, g, e, a, c, b, i, d

CHAPTER 39
1. c
2. d
3. d
4. c
5. c
6. d
7. b
8. c
9. c
10. e, b, a, d, c
11. d, e, a, f, c, b

CHAPTER 40
1. e
2. c
3. c
4. d
5. a
6. d
7. d
8. b
9. a
10. c, b, a, e, d, f

CHAPTER 41
1. d
2. d
3. d
4. b
5. c
6. e
7. d
8. c

9. d
10. d, b, f, e, a, c

CHAPTER 42
1. e
2. d
3. caloric; energy
4. a
5. d
6. b
7. c
8. a
9. b
10. d, c, e, g, f, b, a

CHAPTER 43
1. d
2. f
3. a
4. b
5. c
6. d
7. d
8. d
9. d, e, a, c, b
10 f, e, g, b, c, d, a

CHAPTER 44
1. d
2. a
3. a
4. c
5. b
6. d
7. c
8. b
9. d
10. d
11. d, a, f, e, c, b

CHAPTER 45
1. hypothalamus
2. f
3. LH
4. c
5. c
6. c
7. c
8. f
9. e
10. e, c, a, b, d

CHAPTER 46
1. Ecology
2. population
3. e
4. e
5. a
6. d
7. e
8. c
9. c, d, b, a

CHAPTER 47
1. e
2. b
3. e
4. d
5. b
6. d
7. d
8. a
9. c, d, a, e, b

CHAPTER 48
1. a
2. d
3. d
4. d, a, c, b

CHAPTER 49
1. c
2. c
3. d
4. b
5. d
6. d
7. d
8. d
9. b, f, e, d, c, a
10. e, a, d, c, b

CHAPTER 50
1. c
2. c
3. b
4. d
5. c
6. b
7. c
8. d
9. d
10. d, b, c, a

CHAPTER 51
1. a
2. d
3. b
4. c
5. c
6. c
7. d
8. d
9. c, d, b, a

ENERGY-
REQUIRING
STEPS OF
GLYCOLYSIS

(two ATP
invested)

ENERGY-
RELEASING
STEPS OF
GLYCOLYSIS

(four ATP
produced)

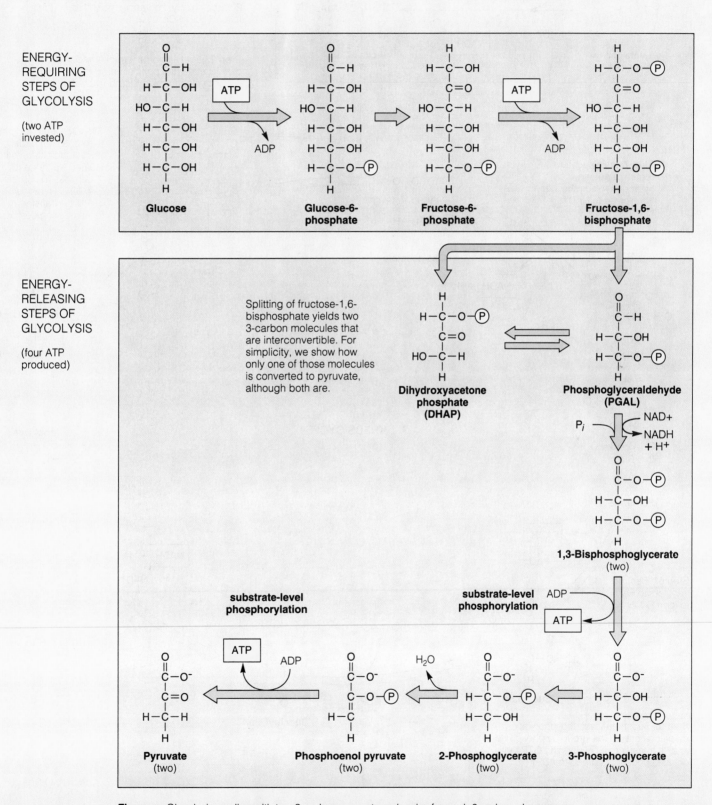

Figure a   Glycolysis, ending with two 3-carbon pyruvate molecules for each 6-carbon glucose entering the reactions. The *net* energy yield is two ATP molecules (two invested, four produced).

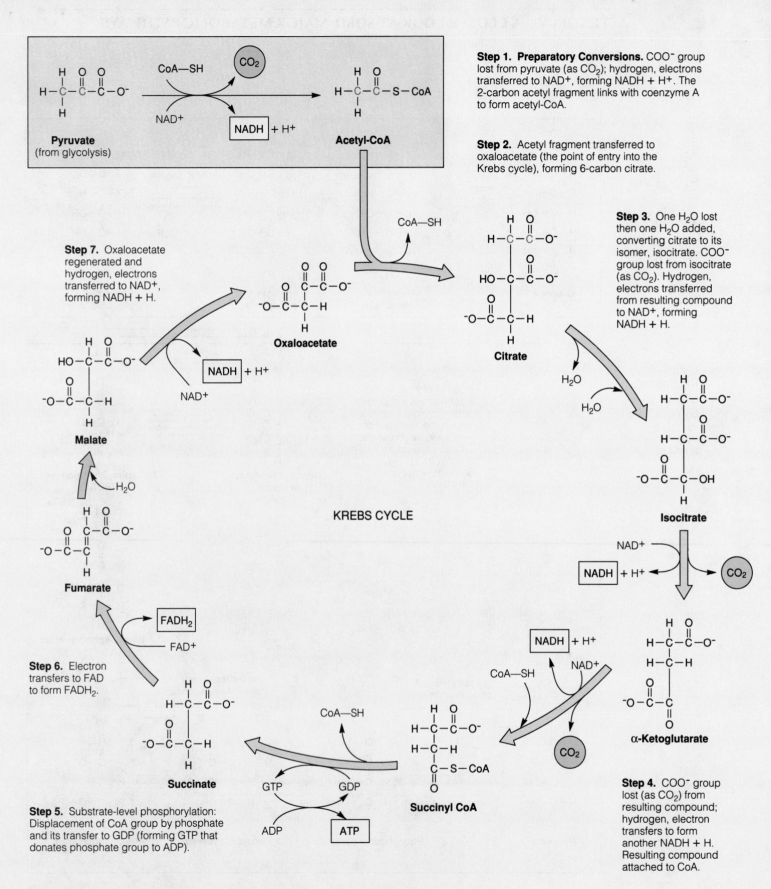

Step 1. Preparatory Conversions. COO^- group lost from pyruvate (as CO_2); hydrogen, electrons transferred to NAD^+, forming $NADH + H^+$. The 2-carbon acetyl fragment links with coenzyme A to form acetyl-CoA.

Step 2. Acetyl fragment transferred to oxaloacetate (the point of entry into the Krebs cycle), forming 6-carbon citrate.

Step 3. One H_2O lost then one H_2O added, converting citrate to its isomer, isocitrate. COO^- group lost from isocitrate (as CO_2). Hydrogen, electrons transferred from resulting compound to NAD^+, forming $NADH + H$.

Step 7. Oxaloacetate regenerated and hydrogen, electrons transferred to NAD^+, forming $NADH + H$.

KREBS CYCLE

Step 6. Electron transfers to FAD to form $FADH_2$.

Step 5. Substrate-level phosphorylation: Displacement of CoA group by phosphate and its transfer to GDP (forming GTP that donates phosphate group to ADP).

Step 4. COO^- group lost (as CO_2) from resulting compound; hydrogen, electron transfers to form another $NADH + H$. Resulting compound attached to CoA.

Pyruvate (from glycolysis)

Acetyl-CoA

Oxaloacetate

Citrate

Isocitrate

Malate

α-Ketoglutarate

Fumarate

Succinyl CoA

Succinate

Figure B Krebs cycle (citric acid cycle). *Red* identifies the carbon atoms entering the cyclic pathway by way of acetyl-CoA.

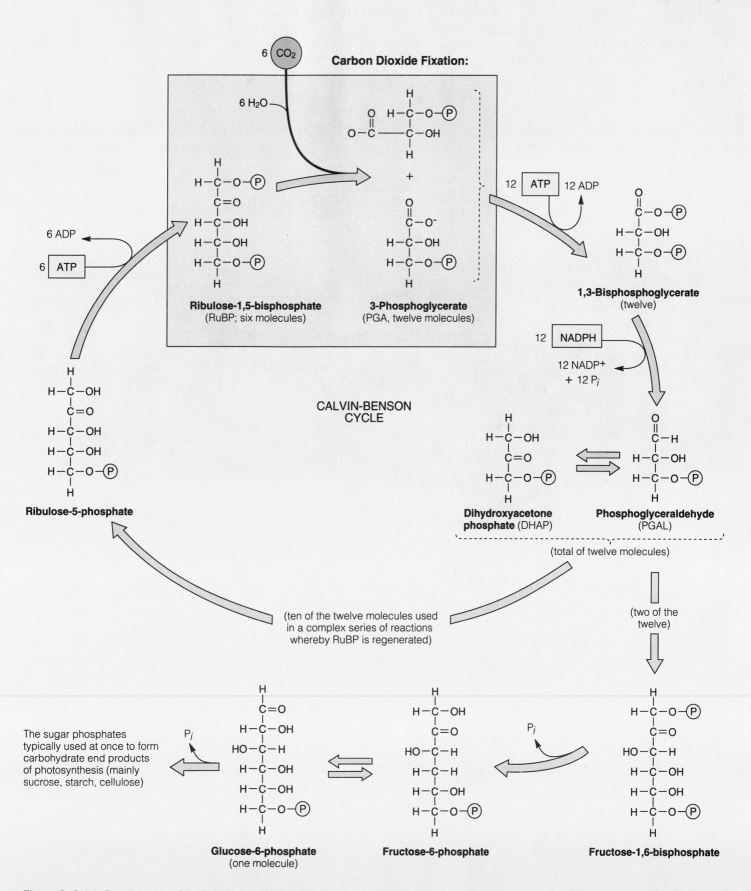

Carbon Dioxide Fixation:

6 CO₂

6 H₂O

CALVIN-BENSON CYCLE

Ribulose-1,5-bisphosphate
(RuBP; six molecules)

3-Phosphoglycerate
(PGA, twelve molecules)

6 ADP

6 ATP

12 ATP   12 ADP

1,3-Bisphosphoglycerate
(twelve)

12 NADPH

12 NADP⁺
+ 12 Pᵢ

Ribulose-5-phosphate

Dihydroxyacetone phosphate (DHAP)

Phosphoglyceraldehyde (PGAL)

(total of twelve molecules)

(ten of the twelve molecules used in a complex series of reactions whereby RuBP is regenerated)

(two of the twelve)

The sugar phosphates typically used at once to form carbohydrate end products of photosynthesis (mainly sucrose, starch, cellulose)

Pᵢ

Glucose-6-phosphate
(one molecule)

Fructose-6-phosphate

Pᵢ

Fructose-1,6-bisphosphate

Figure C Calvin-Benson cycle of the light-independent reactions of photosynthesis.

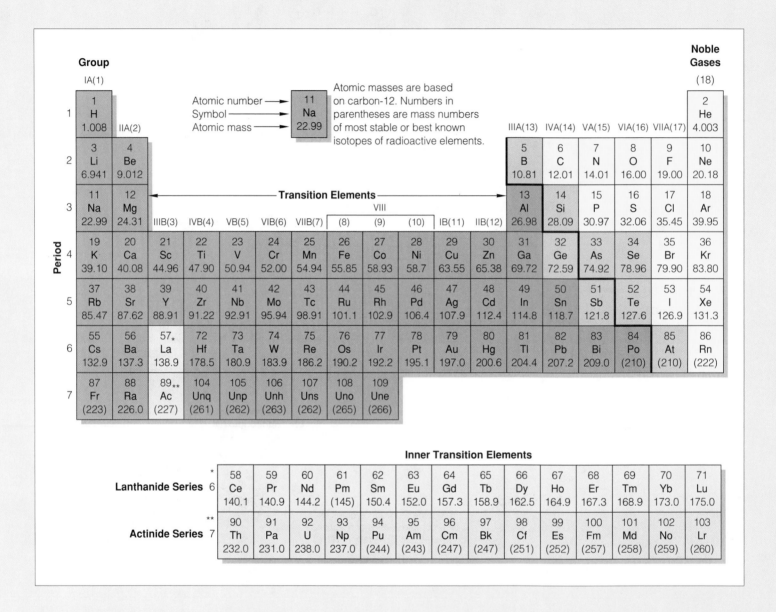

Periodic Table of the Elements

	Group																		Noble Gases

Atomic masses are based on carbon-12. Numbers in parentheses are mass numbers of most stable or best known isotopes of radioactive elements.

Atomic number → 11
Symbol → Na
Atomic mass → 22.99

Transition Elements

Period

Group	IA(1)																		(18)
1	1 H 1.008	IIA(2)																	2 He 4.003
2	3 Li 6.941	4 Be 9.012											IIIA(13) 5 B 10.81	IVA(14) 6 C 12.01	VA(15) 7 N 14.01	VIA(16) 8 O 16.00	VIIA(17) 9 F 19.00	10 Ne 20.18
3	11 Na 22.99	12 Mg 24.31	IIIB(3)	IVB(4)	VB(5)	VIB(6)	VIIB(7)	(8) VIII	(9)	(10)	IB(11)	IIB(12)	13 Al 26.98	14 Si 28.09	15 P 30.97	16 S 32.06	17 Cl 35.45	18 Ar 39.95
4	19 K 39.10	20 Ca 40.08	21 Sc 44.96	22 Ti 47.90	23 V 50.94	24 Cr 52.00	25 Mn 54.94	26 Fe 55.85	27 Co 58.93	28 Ni 58.7	29 Cu 63.55	30 Zn 65.38	31 Ga 69.72	32 Ge 72.59	33 As 74.92	34 Se 78.96	35 Br 79.90	36 Kr 83.80
5	37 Rb 85.47	38 Sr 87.62	39 Y 88.91	40 Zr 91.22	41 Nb 92.91	42 Mo 95.94	43 Tc 98.91	44 Ru 101.1	45 Rh 102.9	46 Pd 106.4	47 Ag 107.9	48 Cd 112.4	49 In 114.8	50 Sn 118.7	51 Sb 121.8	52 Te 127.6	53 I 126.9	54 Xe 131.3
6	55 Cs 132.9	56 Ba 137.3	57* La 138.9	72 Hf 178.5	73 Ta 180.9	74 W 183.9	75 Re 186.2	76 Os 190.2	77 Ir 192.2	78 Pt 195.1	79 Au 197.0	80 Hg 200.6	81 Tl 204.4	82 Pb 207.2	83 Bi 209.0	84 Po (210)	85 At (210)	86 Rn (222)
7	87 Fr (223)	88 Ra 226.0	89** Ac (227)	104 Unq (261)	105 Unp (262)	106 Unh (263)	107 Uns (262)	108 Uno (265)	109 Une (266)									

Inner Transition Elements

Lanthanide Series 6 *	58 Ce 140.1	59 Pr 140.9	60 Nd 144.2	61 Pm (145)	62 Sm 150.4	63 Eu 152.0	64 Gd 157.3	65 Tb 158.9	66 Dy 162.5	67 Ho 164.9	68 Er 167.3	69 Tm 168.9	70 Yb 173.0	71 Lu 175.0
Actinide Series 7 **	90 Th 232.0	91 Pa 231.0	92 U 238.0	93 Np 237.0	94 Pu (244)	95 Am (243)	96 Cm (247)	97 Bk (247)	98 Cf (251)	99 Es (252)	100 Fm (257)	101 Md (258)	102 No (259)	103 Lr (260)

A GLOSSARY OF BIOLOGICAL TERMS

ABO blood typing Method of using two surface proteins (A, B, or both) of red blood cells to characterize an individual's blood. O signifies the absence of both proteins.

abortion The spontaneous expulsion or the dislodging of an embryo or a fetus from the uterus.

abscisic acid (ab-SISS-ik) Plant hormone that promotes stomatal closure, bud dormancy, and seed dormancy.

abscission (ab-SIH-zhun) [L. *abscindere*, to cut off] The dropping of leaves, flowers, fruits, or other plant parts due to hormonal action.

absorption Of most animals, movement of nutrients, fluid, and ions across the gut lining and into the internal environment.

accessory pigment Light-trapping pigment molecule; it contributes to photosynthesis by extending the range of usable wavelengths beyond those absorbed by the chlorophylls.

acid [L. *acidus*, sour] A substance that releases hydrogen ions when dissolved in water.

acid rain The falling to Earth of rain (or snow) that contains sulfur and nitrogen oxides. Also called wet acid deposition (as opposed to dry acid deposition of airborne particles of sulfur and nitrogen oxides).

acoelomate (ay-SEE-luh-mate) Of some of the invertebrates, having no fluid-filled cavity between the gut and body wall.

acoustical signal Sounds that are normally used as a form of communication between animals of the same species.

actin (AK-tin) One of the motor proteins with roles in contraction. Interacts with myosin in muscle cells.

action potential Abrupt, brief reversal in the steady voltage difference across the plasma membrane (the resting membrane potential) of a neuron and other excitable cells.

activation energy The minimum amount of collision energy necessary to drive reactant molecules to an activated state (the transition state) at which a given chemical reaction will proceed spontaneously.

activator A regulatory protein having a role in a positive control system that promotes gene transcription.

active site A crevice in the surface of an enzyme molecule where a specific reaction is catalyzed, or made to proceed far faster than it would spontaneously.

active transport A pumping of one or more specific solutes through the interior of a transport protein that spans the lipid bilayer of a cell membrane. The solute is transported against its concentration gradient. An energy boost, as from ATP, activates the protein.

adaptation [L. *adaptare*, to fit] Of evolution, being adapted (or becoming more adapted) to a given set of environmental conditions. Of a sensory neuron, decreasing frequency of action potentials (or their cessation) even if a stimulus is maintained at constant strength.

adaptive behavior A behavior that promotes propagation of an individual's genes and that tends to increase in frequency in the population over time.

adaptive radiation A burst of divergences from a single lineage that gives rise to many new species, each adapted to an unoccupied or new habitat or to using a novel resource.

adaptive trait Any aspect of form, function, or behavior that helps an individual survive and reproduce under prevailing conditions.

adaptive zone A way of life available for organisms that are physically, ecologically, and evolutionarily equipped to live it, such as "catching insects in the air at night."

adenine (AH-de-neen) A purine; a nitrogen-containing base in certain nucleotides.

adenosine diphosphate (ah-DEN-uh-seen die-FOSS-fate) ADP, an organic compound that can make energy transfers in cells; typically formed by hydrolysis of ATP.

adenosine phosphate Any of a number of relatively small organic compounds, some of which function as chemical messengers within and between cells, and others (such as ATP) that function as energy carriers.

ADH Antidiuretic hormone. Hypothalamic hormone that induces water conservation as required during control of extracellular fluid volume and solute concentrations.

adhesion protein A protein that helps cells of the same type locate one another during development, adhere, and remain in the proper position in the proper tissue.

adipose tissue A type of connective tissue having an abundance of fat-storing cells and blood vessels for transporting fats.

ADP Adenosine diphosphate. A nucleotide coenzyme that accepts unbound phosphate or a phosphate group to become ATP.

ADP/ATP cycle In cells, a mechanism of ATP renewal. During a phosphate-group transfer, ATP reverts to ADP, then it forms again by phosphorylation of ADP.

aerobic respiration (air-OH-bik) [Gk. *aer*, air, + *bios*, life] The main pathway of ATP formation, for which oxygen is the final acceptor of electrons stripped from glucose or another organic compound. It proceeds from glycolysis through the Krebs cycle and electron transport phosphorylation. For each glucose molecule, a typical net yield is 36 ATP.

age structure The number of individuals in each of several or many age categories for a population.

agglutination (ah-glue-tin-AY-shun) In this defensive response, antibodies circulating in blood act against foreign cells (such as transfused red blood cells of the wrong type) and cause them to clump together.

aging A range of processes, including the breakdown of cell structure and function, by which the body gradually deteriorates. All multicelled species that show extensive cell differentiation undergo aging.

AIDS Short for acquired immunodeficiency syndrome. A set of chronic disorders that arises following infection by the human immunodeficiency virus (HIV), which destroys key cells of the immune system.

alcohol An organic compound that has one or more hydroxyl groups (—OH) and readily dissolves in water. Sugars are examples.

alcoholic fermentation Anaerobic pathway of ATP formation. Pyruvate from glycolysis is degraded to acetaldehyde, which accepts electrons from NADH to form ethanol with a net yield of two ATP. NAD$^+$ is regenerated.

aldosterone (al-DOSS-tuh-rohn) Hormone secreted by the adrenal cortex that helps regulate sodium reabsorption.

allantois (ah-LAN-twahz) [Gk. *allas*, sausage] One of four extraembryonic membranes that functions in respiration and in storing the metabolic wastes of embryos of reptiles, birds, and some mammals. In humans, it gives rise to blood vessels for the placenta and to the urinary bladder.

allele (uh-LEEL) At a given gene locus on a chromosome, one of two or more slightly different molecular forms of a gene that arise through mutation and that code for different versions of the same trait.

allele frequency The abundance of each kind of allele in the population as a whole.

allergen Normally harmless substance that provokes inflammation, excessive mucus secretion, and other defense responses.

allergy A response to an allergen.

allopatric speciation [Gk. *allos*, different, L *patria*, native land] Speciation that follows the end of gene flow between populations or subpopulations of a species as a result of their geographic isolation from each other.

allosteric control (AL-oh-STARE-ik) Form of control over a metabolic reaction or pathway that operates by binding a specific substance at a control site on a specific enzyme.

altruism (al-true-ISS-tik) Self-sacrificing behavior; an individual behaves in a way that helps others but decreases its own chances of reproductive success.

alveolus (ahl-VEE-uh-lus), plural **alveoli** [L. *alveus*, small cavity] One of the cupped, thin-walled outpouchings of respiratory bronchioles. A site where oxygen diffuses from air in the lungs to blood, and carbon dioxide diffuses from blood to the lungs.

amino acid (uh-MEE-no) A small organic molecule with a hydrogen atom, an amino group, an acid group, and an R group bonded covalently to a central carbon atom; the subunit of polypeptide chains.

ammonification (uh-moan-ih-fih-KAY-shun) A process by which certain soil bacteria and fungi break down nitrogenous wastes and remains of organisms; part of the nitrogen cycle.

amnion (AM-nee-on) Of land vertebrates, one of four extraembryonic membranes; the boundary layer of a fluid-filled sac that allows the embryo to grow in size, move freely, and be protected from sudden impacts and temperature shifts.

amniote egg An egg, often with a leathery or calcified shell, that has extraembryonic membranes, including the amnion.

amphibian A type of vertebrate somewhere between fishes and reptiles in body plan and reproductive mode; salamanders, frogs and toads, and caecilians are existing groups.

anaerobic pathway (an-uh-ROW-bik) [Gk. *an*, without, + *aer*, air] Metabolic pathway in which a substance other than oxygen serves as the final acceptor of electrons that have been stripped from substrates.

analogous structures (ann-AL-uh-gus) [Gk. *analogos*, similar to one another] Body parts that once differed in evolutionarily distant lineages, then converged in structure and function as the lineages responded to similar environmental pressures.

anaphase (AN-uh-faze) Of anaphase I of meiosis, the stage when each homologous chromosome separates from its partner and both move to opposite spindle poles. Of mitosis and of anaphase II of meiosis, sister chromatids of each chromosome separate and move to opposite poles.

aneuploidy (AN-yoo-ploy-dee) Having one more chromosome or one less relative to the parental chromosome number.

angiosperm (AN-gee-oh-spurm) [Gk. *angeion*, vessel, and *spermia*, seed] A flowering plant.

animal A multicelled heterotroph that feeds on other organisms, that is motile for at least part of the life cycle, that develops by a series of embryonic stages, and that usually has tissues, organs, and organ systems.

annelid A type of invertebrate classified as a segmented worm; an oligochaete (such as an earthworm), leech, or polychaete.

annual A flowering plant that completes its life cycle in one growing season.

anther [Gk. *anthos*, flower] A pollen-bearing part of a stamen.

antibiotic One of many metabolic products of certain microorganisms that can kill their bacterial competitors for nutrients in soil.

antibody [Gk. *anti*, against] One of a diverse array of antigen-binding receptors. Only B cells make antibody molecules and position them at their surface or secrete them.

anticodon A sequence of three nucleotide bases in a tRNA molecule that can base-pair with a codon in an mRNA molecule.

antigen (AN-tih-jen) [Gk. *anti*, against, + *genos*, race, kind] A molecular configuration that white blood cells recognize as foreign and that triggers an immune response. Most antigens are proteins at the surface of pathogens or tumor cells.

antigen-MHC complex Processed antigen fragments bound with a suitable MHC molecule and displayed at the cell surface; basis of antigen recognition that promotes lymphocyte cell divisions.

antigen-presenting cell Any cell displaying antigen-MHC complexes at its surface.

aorta (ay-OR-tah) [Gk. *airein*, to lift, heave] Main artery of systemic circulation; carries oxygenated blood away from the heart to all body regions except the lungs.

apical dominance Inhibitory influence of a terminal bud on growth of lateral buds.

apical meristem (AY-pih-kul MARE-ih-stem) [L. *apex*, top, + Gk. *meristos*, divisible] A mass of self-perpetuating cells responsible for primary growth at root and shoot tips.

apoptosis (APP-oh-TOE-sis) Of multicelled organisms, a form of cell death; molecular signals activate weapons of self-destruction already stockpiled in target cells. It occurs when a cell has completed its prescribed function or becomes altered, as by infection or cancerous transformation.

appendicular skeleton (ap-en-DIK-yoo-lahr) Bones of the limbs, hips, and shoulders.

Archaebacteria A kingdom of prokaryotes; encompasses methanogens, halophiles, and thermophiles, all of which differ from the eubacteria in chemical composition and in cell wall and membrane characteristics.

archipelago An island chain some distance away from a continent.

area effect Larger islands tend to support more species than smaller ones at equivalent distances from sources of colonizing species.

arteriole (ar-TEER-ee-ole) A blood vessel between an artery and capillary; a control point where the blood volume delivered to a given body region can be adjusted.

artery A large-diameter, rapid-transport blood vessel with a thick, muscular wall that smooths out the pulsations in blood pressure caused by heart contractions.

arthropod An invertebrate with a hardened exoskeleton, specialized body segments, and jointed appendages. Spiders, crabs, and insects are examples.

artificial selection Selection of traits among individuals of a population that occurs in an artificial environment, under contrived, manipulated conditions.

asexual reproduction Any of a number of modes of reproduction by which offspring arise from a single parent and inherit the genes of that parent only.

atmosphere A volume of gases, airborne particles, and water vapor that envelops the Earth; 80 percent of its mass is distributed within seventeen miles of the Earth's surface.

atmospheric cycle A biogeochemical cycle in which the atmosphere is the largest reservoir of an element. The carbon cycle and nitrogen cycle are examples.

atom The smallest particle unique to a given element; it has one or more positively charged protons, electrons, and (except for hydrogen), neutrons.

atomic number The number of protons in the nucleus of each atom of an element; the number differs for each element.

ATP Adenosine triphosphate (ah-DEN-uh-seen try-FOSS-fate). A nucleotide of adenine, ribose, and three phosphate groups that acts as an energy carrier. Its phosphate-group transfers drive nearly all energy-requiring metabolic reactions.

australopith (OHSS-trah-low-pith) [L. *australis*, southern, + Gk. *pithekos*, ape] Any of the earliest known hominids; a primate species on or near the evolutionary road that led to modern humans.

autoimmune response Misdirected immune response in which lymphocytes mount an attack against normal body cells.

autonomic nervous system (auto-NOM-ik) All nerves from the central nervous system to the smooth muscle, cardiac muscle, and glands of the viscera (internal organs and structures) of the vertebrate body.

autosome Any of the pairs of chromosomes that are the same in both males and females of the species.

autotroph (AH-toe-trofe) [Gk. *autos*, self, + *trophos*, feeder] Organism that synthesizes its own organic compounds using carbon dioxide (as the carbon source) and energy from the physical environment (such as sunlight energy). *Compare* heterotroph.

auxin (AWK-sin) A plant hormone that influences growth, such as stem elongation.

axial skeleton (AX-ee-uhl) Of a vertebrate skeleton, the skull, backbone, ribs, and breastbone (sternum).

axon A cylindrical extension from the cell body of a neuron, often with finely branched endings, that is specialized for the rapid propagation of action potentials.

B lymphocyte (B cell) The only white blood cell that produces antibodies, then positions them at the cell surface or secretes them as weapons in immune responses.

bacterial conjugation A transfer of plasmid DNA from one bacterial cell to another.

bacterial flagellum Of many bacterial cells, a whiplike motile structure that does not contain a core of microtubules.

bacteriophage (bak-TEER-ee-oh-fahj) [Gk. *baktērion*, small staff, rod, + *phagein*, to eat] Category of viruses that infect bacterial cells.

balancing selection All forms of selection that are maintaining two or more alleles for a trait in a population. When the resulting genetic variation persists, this is a case of balanced polymorphism.

Barr body In cells of female mammals, one of two X chromosomes that was randomly condensed so that its genes are inactivated.

basal body A centriole which, after giving rise to microtubules of a flagellum or cilium, remains attached to its base in the cytoplasm.

base Any substance that accepts hydrogen ions when dissolved in water.

base pair Two nucleotide bases that are located in two adjacent strands of DNA or RNA and that are hydrogen-bonded to each other.

base sequence The particular order in which one nucleotide base follows the next in a strand of DNA or RNA. The order is unique in at least some regions for each species.

basophil Fast-acting white blood cell that secretes histamine and other substances to maintain an inflammatory response.

behavior, animal A response to external and internal stimuli that requires sensory, neural, endocrine, and effector components. Behavior has a genetic basis and can evolve; it also can be modified through learning.

benthic province All sediments and rocky formations of the ocean bottom.

biennial (bi-EN-yul) A flowering plant that lives through two growing seasons.

bilateral symmetry Body plan in which the left and right halves of an animal are mirror-images of each other.

binary fission Of flatworms and some other animals, a mode of asexual reproduction; growth by way of mitotic cell divisions is followed by division of the whole body into two parts of the same or different sizes. Not the same as prokaryotic fission, by which bacteria reproduce.

biogeochemical cycle The movement of an element from the environment to organisms, then back to the environment.

biogeographic realm [Gk. *bios*, life, + *geographein*, to describe the Earth's surface] One of six major land divisions, each with distinguishing plants and animals; it retains its general identity because of climate and geographic barriers to gene flow.

biological clock Internal time-measuring mechanism that helps adjust an organism's daily activities, seasonal activities, or both in response to environmental cues.

biological magnification The increasing concentration of a nondegradable or slowly degradable substance in body tissues as it is passed along food chains.

biological species concept A species is one or more populations of individuals that are interbreeding under natural conditions and producing fertile offspring, and that are reproductively isolated from other such populations. The concept applies only to sexually reproducing species.

bioluminescence A flashing of light that emanates from an organism when excited electrons of luciferins (highly fluorescent substances) return to a lower energy level.

biomass Combined weight of all organisms at a given trophic level in an ecosystem.

biome A broad, vegetational subdivision of a biogeographic realm; shaped by climate, topography, and composition of regional soils.

biosphere [Gk. *bios*, life, + *sphaira*, globe] All regions of the Earth's waters, crust, and atmosphere in which organisms live.

biosynthetic pathway A metabolic pathway by which organic compounds necessary for life are synthesized.

biotic potential Of population growth for a given species, the maximum rate of increase per individual under ideal conditions.

bipedalism Habitually walking on two feet, as by ostriches and humans.

bird The only vertebrate that produces feathers and that has strong resemblances and evolutionary connections to reptiles.

blastocyst (BLASS-tuh-sist) [Gk. *blastos*, sprout, + *kystis*, pouch] Outcome of cleavage of a fertilized mammalian egg; a surface layer of blastomeres around a cavity filled with their secretions (the blastocoel), and an inner cell mass (several blastomeres huddled together against the cavity's inner surface).

blastomere One of the small, nucleated cells produced during the cleavage stage of animal development.

blastula (BLASS-chew-lah) Outcome of one type of cleavage pattern; blastomeres formed by the successive cuts surround a fluid-filled cavity (blastocoel).

blood A fluid connective tissue composed of water, solutes, and formed elements (blood cells and platelets); it carries substances to and from cells and helps maintain an internal environment favorable for cell activities.

blood pressure Fluid pressure, generated by heart contractions, that circulates blood.

blood-brain barrier Mechanism that exerts some control over which solutes enter the cerebrospinal fluid and thus helps protect the brain and spinal cord.

bone Mineral-hardened connective tissue of bones; one of the organs of the vertebrate skeleton that help move the body and its parts, protect of other organs, store minerals, and (in some) produce blood cells.

bottleneck A severe reduction in population size, as brought about by intense selection pressure or a natural calamity.

Bowman's capsule The cup-shaped portion of a nephron that receives water and solutes being filtered from the blood in the kidneys.

brain The most complex integrating center of most nervous systems; it receives, processes, and integrates sensory input and issues coordinated commands for response by muscles and glands.

brain stem The vertebrate nervous tissue that evolved first and that still persists in the hindbrain, midbrain and forebrain.

bronchiole A component of the finely branched bronchial tree inside each lung.

bronchus, plural **bronchi** (BRONG-CUSS, BRONG-kee) [Gk. *bronchos*, windpipe] Tube-like branchings of the trachea that lead into the lungs of most vertebrates.

brown alga A photoautotrophic protistan with a notable abundance of xanthophyll pigments; such algae occupy marine habitats.

bryophyte A nonvascular land plant, such as a moss.

bud An undeveloped shoot of meristematic tissue, primarily; often covered and protected by scales (modified leaves).

buffer system A partnership between a weak acid and the base that forms when it dissolves in water. The two work as a pair to counter slight shifts in pH.

bulk A volume of fiber and other undigested material that absorption processes in the small intestine cannot decrease.

bulk flow In response to a pressure gradient, the movement of more than one kind of molecule in the same direction in the same medium (as in blood, sap, or air).

C4 pathway A pathway of photosynthesis in which carbon dioxide is fixed twice, in two different cell types. Carbon dioxide accumulates in the leaf and helps counter photorespiration. The first compound formed is the 4-carbon oxaloacetate.

Calvin-Benson cycle Cyclic reactions that are the *synthesis* part of the light-independent reactions of photosynthesis. In plants, RuBP or some other compound to which carbon has been affixed undergoes rearrangements; a sugar phosphate forms and the RuBP is regenerated. The cycle runs on ATP and NADPH from the light-dependent reactions.

CAM plant A plant that conserves water by opening stomata only at night, when it fixes carbon dioxide by way of a C4 pathway.

cambium (KAM-bee-um), plural **cambia** One of two types of meristems responsible for secondary growth (increases in stem and root diameter). Vascular cambium gives rise to secondary xylem and secondary phloem; cork cambium gives rise to periderm.

camouflage Adaptations in body color, form, or patterning, and in behavior that function in predator avoidance; they help an organism hide in the open (blend with its surroundings) and escape detection.

cancer A malignant tumor; its cells show gross abnormalities in the plasma membrane and cytoplasm, skewed growth and division, and weakened capacity for adhesion within the parent tissue (leading to metastasis). Unless eradicated, cancer is lethal.

capillary, blood [L. *capillus*, hair] A thin-walled vessel that functions in the exchange of carbon dioxide, oxygen, and some other substances between blood and interstitial fluid, which bathes living cells.

capillary bed A diffusion zone, consisting of a great number of capillaries, where substances are exchanged between blood and interstitial fluid.

carbohydrate [L. *carbo*, charcoal, + *hydro*, water] A molecule that consists of carbon, hydrogen, and oxygen in a 1:2:1 ratio (there are exceptions). All cells use carbohydrates as structural materials, energy reservoirs, and transportable forms of energy. The monosaccharides, oligosaccharides, and polysaccharides are three classes.

carbon cycle A biogeochemical cycle in which carbon moves from the atmosphere (its largest reservoir), through the ocean and organisms, then to the atmosphere.

carbon dioxide fixation Of photosynthesis, the first enzyme-mediated step of the light-independent reactions. Carbon (from CO_2) is affixed to RuBP or another compound for entry into the Calvin-Benson cycle.

carcinogen (kar-SIN-uh-jen) A substance or an agent, such as ultraviolet radiation, that can trigger cancer.

cardiac cycle (KAR-dee-ak) [Gk. *kardia*, heart, + *kyklos*, circle] The sequence of muscle contraction and relaxation for one heartbeat.

cardiac pacemaker Sinoatrial (SA) node; the basis of the normal rate of heartbeat. The self-excitatory cardiac muscle cells that spontaneously generate rhythmic waves of excitation over the heart chambers.

cardiovascular system Of most animals, an organ system of blood, one or more hearts, and blood vessels that functions in the rapid transport of substances to and from cells.

carnivore [L. *caro, carnis*, flesh, + *vovare*, to devour] An animal that eats other animals; a type of heterotroph.

carotenoid (kare-OTT-en-oyds) A light-sensitive, accessory pigment that transfers absorbed energy to chlorophylls. Different types absorb violet and blue wavelengths and transmit red, orange, and yellow.

carpel (KAR-pul) The female reproductive part of a flower; sometimes called a pistil. The lower portion of a single carpel (or of a structure composed of two or more) is an ovary. The upper portion has a stigma (a pollen-capturing surface tissue) and often a style (slender extension of the ovary wall).

carrying capacity The maximum number of individuals in a population (or species) that can be sustained indefinitely by a given environment.

cartilage A type of connective tissue with solid yet pliable intercellular material that resists compression.

Casparian strip A waxy band that is an impermeable barrier between the walls of abutting cells making up the endodermis (and exodermis, if present) inside roots.

cDNA Any DNA molecule copied from a mature mRNA transcript by way of reverse transcription.

cell [L. *cella*, small room] The smallest living unit; an organized unit that can survive and reproduce on its own, given suitable DNA instructions and environmental resources—notably energy and raw materials.

cell count The number of cells of a given type in a microliter of blood.

cell cycle Events by which a cell increases in mass, roughly doubles its number of cytoplasmic components, duplicates its DNA, then undergoes nuclear and cytoplasmic division. It extends from the time a new cell is produced until it completes division.

cell differentiation Developmental process in which different cell populations activate and suppress a fraction of their genes in different ways and so become specialized in composition, structure, and function.

cell junction Of multicelled organisms, a point of contact that links two adjoining cells physically, functionally, or both.

cell plate A disklike structure that forms from remnants of a microtubular spindle when a plant cell divides; it develops into a crosswall that partitions the cytoplasm.

cell theory A theory in biology stating that (1) all organisms are composed of one or more cells, (2) the cell is the smallest unit that retains a capacity for independent life, and (3) all cells arise from preexisting cells.

cell wall A semirigid, permeable structure that helps a cell hold its shape and resist rupturing if internal fluid pressure rises.

central nervous system The brain and spinal cord of vertebrates.

central vacuole A fluid-filled organelle in mature, living plant cells that stores amino acids, sugars, ions, and toxic wastes. As it enlarges, it forces increases in cell surface area that improve nutrient uptake.

centriole (SEN-tree-ohl) A cylinder of triplet microtubules that gives rise to microtubules of cilia and flagella.

centromere (SEN-troh-meer) [Gk. *kentron*, center, + *meros*, a part] A small, constricted region of a chromosome having attachment sites for the microtubules that move the chromosome during nuclear division.

cephalization (sef-ah-lah-ZAY-shun) [Gk. *kephalikos*, head] During the evolution of bilateral animals, the concentration of sensory structures and nerve cells in a head.

cerebellum (ser-ah-BELL-um) [L. diminutive of *cerebrum*, brain] Hindbrain region with reflex centers for maintaining posture and smoothing out limb movements.

cerebral cortex Thin surface layer of the cerebral hemispheres. Some parts receive sensory input, others integrate information and coordinate suitable responses.

cerebrospinal fluid Clear extracellular fluid surrounding and cushioning the brain and spinal cord.

cerebrum (suh-REE-bruhm) Forebrain region that first evolved to integrate olfactory input and select motor responses to it. In mammals, it evolved into the most complex integrating center.

channel protein A transport protein that acts as a channel through which specific ions and other water-soluble substances cross the plasma membrane. Some channels remain open; others are gated, and these open and close in controlled ways.

chemical bond A union between the electron structures of two or more atoms or ions.

chemical synapse (SIN-aps) [Gk. *synapsis*, union] A small cleft between a presynaptic neuron and a postsynaptic cell (another neuron, a muscle cell, or a gland cell) that is bridged by neurotransmitter molecules released from the presynaptic neuron.

chemiosmotic theory (kim-ee-OZ-MOT-ik) An electrochemical gradient across a cell membrane drives ATP formation. Hydrogen ions accumulate in a compartment formed by the membrane. The combined force of the H$^+$ concentration and electric gradients propels ions through transport proteins (ATP synthases) spanning the membrane. By enzyme action at these proteins, ADP and inorganic phosphate combine to form ATP.

chemoreceptor (KEE-moe-ree-sep-tur) A sensory receptor that detects chemical energy (ions or molecules) dissolved in the fluid that bathes it.

chemosynthetic autotroph (KEE-moe-sin-THET-ik) One of a few kinds of bacteria able to synthesize its own organic compounds by using carbon dioxide as the carbon source and certain inorganic substances (such as sulfur) as the energy source.

chlorofluorocarbon (KLORE-oh-FLOOR-oh-car-bun), or **CFC** One of the odorless, invisible compounds of chlorine, fluorine, and carbon, widely used in commercial products, that are contributing to the thinning of the ozone layer above the Earth's surface.

chlorophyll (KLOR-uh-fills) [Gk. *chloros*, green, + *phyllon*, leaf] A light-sensitive pigment that absorbs violet-to-blue and red wavelengths but that transmits green. The main pigments in all but one small group of photoautotrophs. Certain chlorophylls donate electrons to the light-dependent reactions of photosynthesis.

chloroplast (KLOR-uh-plast) An organelle that specializes in photosynthesis in plants and photosynthetic protistans.

chordate An animal having a notochord, a dorsal hollow nerve cord, a pharynx, and gill slits in the pharynx wall for at least part of the life cycle.

chorion (CORE-ee-on) Of placental mammals, one of four extraembryonic membranes; it becomes a key component of the placenta. Absorptive structures (villi) develop at its surface and enhance the rapid exchange of substances between the embryo and mother.

chromatid (CROW-mah-tid) Of a duplicated eukaryotic chromosome, one of two DNA molecules (and associated proteins) that remain attached at their centromere region until separated by mitosis or meiosis; after this, each is a separate chromosome.

chromosome (CROW-moe-some) [Gk. *chroma*, color, + *soma*, body] Of eukaryotes, a DNA molecule with many associated proteins. Of prokaryotes, a DNA molecule without a comparable profusion of proteins.

chromosome number The sum total of chromosomes in cells of a given type. *See* haploidy; diploidy.

cilium (SILL-ee-um), plural **cilia** [L. *cilium*, eyelid] Of eukaryotic cells, a short, hairlike projection with an internal, regular array of microtubules. Cilia can serve as motile or sensory structures or help create currents of fluids. Typically more profuse than flagella.

circadian rhythm (ser-KAYD-ee-un) [L. *circa*, about, + *dies*, day] A cycle of physiological events that is completed every twenty-four hours or so independently of environmental change.

circulatory system An organ system having a muscular pump (heart, most often), blood vessels, and blood; the system transports materials to and from cells and often helps stabilize body temperature and pH.

cladogram [Gk. *clad-*, branch] Evolutionary tree diagram that arranges groups by branch points to show their relative relationships. Groups closer together share a more recent common ancestor than those farther apart.

classification system A way of organizing and retrieving information about species.

cleavage Third stage of animal embryonic development. Mitotic cell divisions divide the volume of egg cytoplasm into a number of smaller, nucleated cells (blastomeres). The number of cells increases, but the original volume of egg cytoplasm does not.

cleavage furrow A ringlike depression that forms during cytoplasmic division of an animal cell and that defines the cleavage plane for the cell. Microfilaments attached to the plasma membrane contract and draw it inward to cut the cell in two.

cleavage reaction A molecule splits into two smaller ones. Hydrolysis is an example.

climate Prevailing weather conditions for an ecosystem, such as temperature, humidity, wind speed, cloud cover, and rainfall.

climax community A self-perpetuating, stable array of species in equilibrium with one another and with their habitat.

climax pattern model Idea that one climax community may extend into another along gradients of environmental conditions, such as variations in climate, topography, and species interactions.

cloaca Of some vertebrates, the last part of a gut that receives feces, urine, and sperm or eggs; of some invertebrates, an excretory, respiratory, or reproductive duct.

cloned DNA Multiple, identical copies of restriction fragments that have been inserted into plasmids or some other cloning vector.

club fungus A fungus with reproductive structures having microscopic, club-shaped cells that produce and bear spores.

cnidarian A radial invertebrate at the tissue level of organization and the only organism to produce nematocysts. Two body forms (medusae and polyps) are common.

coal A nonrenewable source of energy that formed more than 280 million years ago from submerged, undecayed plant remains.

codominance A pair of nonidentical alleles that specify two phenotypes are expressed at the same time in heterozygotes.

codon One of the base triplets in an mRNA molecule, the linear sequence of which corresponds to a linear sequence of amino acids in a polypeptide chain. Of 64 codons, 61 specify different amino acids, and 3 of these also are start signals for translation; 1 serves as a stop signal for translation.

coelom (SEE-lum) [Gk. *koilos*, hollow] A cavity, lined with peritoneum, between the gut and body wall of most animals.

coenzyme A nucleotide; an enzyme helper that accepts electrons and hydrogen atoms stripped from substrates at a reaction site and transfers them elsewhere.

coevolution The joint evolution of two or more closely interacting species; when one species evolves, the change affects selection pressures operating between the two, so the other also evolves.

cofactor A metal ion or coenzyme; it helps an enzyme catalyze a reaction or transfers electrons, atoms, or functional groups from one substrate to another.

cohesion Capacity to resist rupturing when placed under tension (stretched).

cohesion theory of water transport Theory that water moves up through plants due to hydrogen bonding among water molecules confined as narrow columns in xylem. The collective cohesive strength of the bonds allows water to be pulled up in response to transpiration (evaporation from leaves).

collenchyma (coll-ENG-kih-mah) A simple plant tissue that offers flexible support for primary growth, as in lengthening stems.

colon (CO-lun) The large intestine.

commensalism [L. *com*, together, + *mensa*, table] An ecological interaction between species that directly benefits one but does not affect other much, if at all.

communication display A pattern of behavior, often ritualized with intended changes in the function of common behavior patterns, that serves as a social signal.

communication signal A social cue encoded in stimuli that holds unambiguous meaning for individuals of the same species. Specific odors, sounds, coloration and patterning, postures, and movements are examples.

community All populations living in the same habitat. Also, a group of organisms with similar life-styles in a habitat, such as a community of birds.

companion cell A specialized parenchyma cell that helps load organic compounds into conducting cells of phloem.

comparative morphology [Gk. *morph*, form] Study of comparable body parts of adults or embryonic stages of major lineages.

competitive exclusion Theory that species that require identical resources cannot coexist indefinitely.

complement system A set of about twenty proteins circulating in inactive form within vertebrate blood; different kinds induce lysis of pathogens, promote inflammation, and stimulate phagocytes to act during both nonspecific defenses and immune responses.

compound A substance consisting of two or more elements in unvarying proportions.

concentration gradient A difference in the number of molecules or ions of a substance between adjoining regions. Energy inherent in their constant molecular motion makes them collide and career outward from the region of higher to lower concentration. Barring other forces, all substances tend to diffuse down their concentration gradient.

condensation reaction Through covalent bonding, two molecules combine to form a larger molecule, often with the formation of water as a by-product.

cone cell In a vertebrate eye, a photoreceptor that responds to intense light and contributes to sharp daytime vision and color perception.

conifer A pollen- and seed-bearing plant of the dominant group of gymnosperms; mostly evergreen, woody trees and shrubs with needle-like or scale-like leaves.

conjugation, bacterial Of bacteria only, a mechanism by which a donor cell transfers plasmid DNA to a recipient cell.

connective tissue proper A category of animal tissues, all having mostly the same components but in different proportions. They incorporate fibroblasts and other cells, the secretions of which form fibers (mostly of collagen and elastin) and a ground substance of modified polysaccharides.

consumer [L. *consumere*, to take completely] A heterotroph that obtains energy and carbon by feeding on the tissues of other organisms. Herbivores, carnivores, and parasites are examples.

continuous variation Of a population, a more or less continuous range of small differences in a given trait among all of its individuals.

contractile vacuole (kun-TRAK-till VAK-you-ohl) [L. *contractus*, to draw together] Of some protistans, such as a paramecium, an organelle that takes up excess water in the cell body, then contracts; the contractile force is enough to expel the water outside the cell through a pore to its surface.

control group Of an experimental test, a group used to evaluate possible side effects of a test involving an experimental group. Ideally, the control group is identical to the experimental group in all respects except for the variable being studied.

cork cambium A lateral meristem that gives rise to a corky replacement for the epidermis of woody plant parts.

corpus callosum (CORE-pus ka-LOW-sum) A band of axons (200 million in humans) that functionally link two cerebral hemispheres.

corpus luteum (CORE-pus LOO-tee-um) A glandular structure that develops from cells of a ruptured ovarian follicle and secretes progesterone and estrogen.

cortex [L. *cortex*, bark] In general, a rindlike layer such as the kidney or adrenal cortex. In vascular plants, the ground tissue that makes up most of the primary plant body, supports plant parts, and stores food.

cotyledon A seed leaf, which develops as part of the embryo of monocots and dicots; cotyledons provide nourishment for the seedling at the time of germination and initial growth.

courtship display A pattern of ritualized social behavior between potential mates. It may include frozen postures as well as movements that are exaggerated and yet simplified. It may include visual signals, such as body parts that are conspicuously enlarged, distinctively colored or patterned, or some combination of these.

covalent bond (koe-VAY-lunt) [L. *con*, together, + *valere*, to be strong] A sharing of one or more electrons between atoms or groups of atoms. If electrons are shared equally, the bond is nonpolar. If shared unequally, it is polar (slightly positive at one end, slightly negative at the other).

continuous variation Of the individuals of a population, a range of small differences in one or more traits.

cross-bridge formation Of a muscle cell, a reversible interaction between its many actin and myosin filaments that is the basis of contraction.

crossing over During prophase I of meiosis, the breakage and exchange of corresponding segments between nonsister chromatids of a pair of homologous chromosomes; a form of genetic recombination that breaks up old combinations of alleles and puts new ones together in chromosomes.

culture The sum of behavior patterns of a social group, passed between generations by learning and by symbolic behavior, especially language.

cuticle (KEW-tih-kull) A body covering. Of land plants, a transparent cover of waxes and lipid-rich cutin deposited on the outer surface of epidermal cell walls. Of annelids, a thin, flexible surface coat. Of arthropods, a hardened, lightweight cover with protein and chitin components that functions as an exoskeleton.

cyclic AMP (SIK-lik) A nucleotide; cyclic adenosine monophosphate. It functions in intercellular communication, as when it is a second messenger (a cytoplasmic mediator of a cell's response to signaling molecules).

cyclic pathway of ATP formation Ancient photosynthetic pathway occurring at the plasma membrane of some bacteria and at the thylakoid membrane of chloroplasts. A photosystem embedded in the membrane gives up electrons to a transport system, which gives them back to the photosystem. The electron flow sets up concentration and electric gradients across the membrane that drive ATP formation at nearby membrane sites.

cyst Of many microorganisms, a resistant resting stage with thick, tough outer layers that typically forms in response to adverse conditions; of skin, any abnormal, fluid-filled sac without an external opening.

cytochrome (SIGH-toe-krome) [Gk. *kytos*, hollow vessel, + *chrōma*, color] Iron-containing protein molecule; a component of the electron transport systems used in photosynthesis and aerobic respiration.

cytokinesis (SIGH-toe-kih-NEE-sis) [Gk. *kinesis*, motion] Cytoplasmic division; the splitting of a parent cell into daughter cells.

cytokinin (SIGH-tow-KY-nin) Any of the class of plant hormones that stimulate cell division, promote leaf expansion, and retard leaf aging.

cytological marker One or more observable, unusual differences between chromosomes of the same type.

cytomembrane system [Gk. *kytos*, hollow vessel] Organelles functioning as a system to modify, package, and distribute newly formed proteins and lipids. Endoplasmic reticulum, Golgi bodies, lysosomes, and a variety of vesicles are its components.

cytoplasm (SIGH-toe-plaz-um) [Gk. *plassein*, to mold] All cellular parts, particles, and semifluid substances enclosed within the plasma membrane except for the nucleus (or nucleoid, in bacterial cells).

cytoplasmic localization When cleavage divides an animal zygote, each resulting blastomere receives a localized portion of maternal messages in the egg cytoplasm.

cytosine (SIGH-toe-seen) A pyrimidine; one of the nitrogen-containing bases in nucleotides.

cytoskeleton The internal "skeleton" of eukaryotic cells. Its microtubules and other components structurally support the cell and organize and move its internal components. The cytoskeleton also helps free-living cells move through their environment.

cytotoxic T cell A T lymphocyte that uses touch-killing to eliminate infected body cells or tumor cells. When it contacts targets, it delivers cell-killing chemicals into them.

decomposer [partly fr. L. *dis-*, to pieces, + *companere*, arrange] Of ecosystems, a heterotroph that gets energy and carbon by chemically breaking down the remains, products, or wastes of other organisms and helps cycle nutrients back to producers. Certain fungi and bacteria are examples.

deforestation The removal of all trees from a large tract of land, such as the Amazon Basin and the Pacific Northwest.

degradative pathway A metabolic pathway by which organic compounds are broken down in stepwise reactions that lead to products of lower energy.

deletion Loss of a chromosome segment.

denaturation (deh-NAY-chur-AY-shun) Of any molecule, the loss of three-dimensional shape following disruption of hydrogen bonds and other weak bonds.

dendrite (DEN-drite) [Gk. *dendron*, tree] A short, slender extension from the cell body of a neuron; commonly an input zone.

denitrification (DEE-nite-rih-fih-KAY-shun) Conversion of nitrate or nitrite by certain bacteria to gaseous nitrogen (N_2) and a small amount of nitrous oxide (N_2O).

density-dependent control A factor that limits population growth by reducing the birth rate, increasing the rates of death and dispersal, or all of these. Predation, parasitism, disease, and competition for resources are examples.

density-independent factor A factor that tends to cause a population's death rate to increase independently of its density. Storms and floods are examples.

dentition (den-TIH-shun) The type, size, and number of an animal's teeth.

derived trait A novel feature that evolved only once and is shared only by descendants of the ancestral species in which it evolved.

dermal tissue system All the tissues that cover and protect the surfaces of a plant.

dermis The layer of skin underlying the epidermis; consists primarily of dense connective tissue.

desert A biome that typically forms where the potential for evaporation greatly exceeds rainfall and vegetation cover is limited.

desertification (dez-urt-ih-fih-KAY-shun) Conversion of a grassland or an irrigated or rain-fed cropland to a desertlike condition, with a drop in agricultural productivity of 10 percent or more.

detrital food web A network of food chains in which energy flows mainly from plants through arrays of detritivores and decomposers.

detritivore (dih-TRY-tih-vorez) [L. *detritus*; after *deterere*, to wear down] A heterotroph that consumes decomposing particles of organic matter. Earthworms, crabs, and nematodes are examples.

deuterostome (DUE-ter-oh-stome) [Gk. *deuteros*, second, + *stoma*, mouth] A bilateral animal for which the first indentation that forms in the early embryo develops into an anus. An echinoderm or a chordate.

development Of multicelled organisms, the programmed emergence of specialized, morphologically different body parts.

diaphragm (DIE-uh-fram) [Gk. *diaphragma*, to partition] Muscular partition between the thoracic and abdominal cavities; its contraction and relaxation contribute to breathing. Also, a contraceptive device used temporarily to prevent sperm from entering the uterus during sexual intercourse.

dicot (DIE-kot) [Gk. *di*, two, + *kotylēdon*, cup-shaped vessel] A dicotyledon. In general, a flowering plant characterized by seeds having embryos with two cotyledons (seed leaves); net-veined leaves; and floral parts arranged in fours, fives, or multiples of these.

diffusion Net movement of like molecules (or ions) down their concentration gradient. In the absence of other forces, the energy inherent in molecules makes them move constantly and collide at random. Their collisions are most frequent where they are most crowded together; thus they show a net outward movement from regions of higher to lower concentration.

digestive system An internal sac or tube from which ingested food is absorbed into the internal environment.

dihybrid cross An experimental cross in which true-breeding F_1 offspring inherit two gene pairs, each consisting of two nonidentical alleles.

diploidy (DIP-loyd-ee) The presence of two of each type of chromosome (that is, pairs of homologous chromosomes) in the interphase nucleus of somatic cells and germ cells. *Compare* haploidy.

directional selection A mode of natural selection by which the range of variation for some trait shifts in a consistent direction in response to directional change in the environment or to new environmental conditions.

disaccharide (die-SAK-uh-ride) [Gk. *di*, two, + *sakcharon*, sugar] A simple carbohydrate; one of the oligosaccharides consisting of two covalently bonded sugar monomers.

disease Outcome of an infection when the body's defenses cannot be mobilized fast enough; the pathogen's activities interfere with normal body functions.

disruptive selection A mode of natural selection by which forms of a trait at both ends of a range of variation are favored and intermediate forms are selected against.

distal tubule The tubular portion of a nephron farthest from the glomerulus; a region of water and sodium reabsorption.

distance effect Only species adapted for long-distance dispersal are potential colonists of islands far from their home range.

diversity, organismic Sum total of all the variations in form, function, and behavior that have accumulated in different lineages. Variations in traits generally are adaptive to prevailing conditions or were adaptive to conditions that existed in the past.

DNA Deoxyribonucleic acid (dee-OX-ee-RYE-bow-new-CLAY-ik). For all cells and many viruses, a nucleic acid that is the molecule of inheritance. It consists of two nucleotide strands twisted together helically and held together by numerous hydrogen bonds. The nucleotide sequence encodes instructions for synthesizing proteins and, ultimately, new individuals of a particular species.

DNA amplification Any of several methods by which a DNA library is copied again and again to yield multiple, identical copies of DNA fragments (cloned DNA).

DNA-DNA hybridization *See* nucleic acid hybridization.

DNA fingerprint A unique array of RFLPs, inherited in a Mendelian pattern from each parent, that gives each individual a unique identity.

DNA library A collection of DNA fragments produced by restriction enzymes and later incorporated into plasmids.

DNA ligase (LYE-gaze) An enzyme that seals together the new base-pairings during DNA replication; also used by technologists to seal base-pairings between DNA fragments and cut plasmid DNA.

DNA polymerase (poe-LIM-uh-raze) An enzyme that assembles a new strand on a parent DNA strand during replication; also takes part in DNA repair.

DNA probe A short DNA sequence that is synthesized from radioactively labeled nucleotides. Part of the probe is designed to base-pair with part of a gene under study.

DNA repair Following an alteration in the base sequence of a DNA strand, a process that may restore the original sequence, as carried out by DNA polymerases, DNA ligases, and other enzymes.

DNA replication Of cells, the process by which hereditary material is duplicated for distribution to daughter nuclei. Occurs prior to mitosis and meiosis in eukaryotic cells and during prokaryotic fission in bacterial cells.

dominance hierarchy A social organization in which some members of the group have adopted a subordinate status to others.

dominant allele In a diploid cell, an allele that masks the expression of its partner on the homologous chromosome.

dormancy [L. *dormire*, to sleep] A hormone-mediated time of inactivity during which metabolic activities idle. Perennials, seeds, many spores, cysts, and some animals go through dormancy.

double fertilization Of flowering plants only, the fusion of one sperm nucleus with the egg nucleus (to produce a zygote), *and* the fusion of a second sperm nucleus with nuclei of the endosperm mother cell, which gives rise to a nutritive tissue (endosperm).

doubling time The length of time it takes for a population to double in size.

drug addiction Chemical dependence on a drug following habituation and tolerance of it; in time the drug assumes an "essential" biochemical role in the body.

dry shrubland A biome that typically forms where annual rainfall is less than 25 to 60 centimeters; short, multibranched woody shrubs (e.g., chaparral) predominate.

dry woodland A biome that typically forms where annual rainfall is about 40 to 100 centimeters; there may be tall trees, but these do not form a dense canopy.

duplication A repeat of the same linear stretch of an individual's DNA in the same chromosome or in a different one.

ecdysone A hormone with major influence over the development of many insects.

echinoderm A type of invertebrate that has calcified spines, needles, or plates on the body wall. Although radially symmetrical, it has some bilateral features. Sea stars and sea urchins are examples.

ecology [Gk. *oikos*, home, + *logos*, reason] Study of the interactions of organisms with one another and with their physical and chemical environment.

ecosystem [Gk. *oikos*, home] An array of organisms and their physical environment, all of which are interacting through a flow of energy and a cycling of materials.

ecosystem modeling An analytical method, based on computer programs and models, of predicting unforeseen effects of specific disturbances to an ecosystem.

ectoderm [Gk. *ecto*, outside, + *derma*, skin] The first-formed, outermost primary tissue layer of animal embryos; forerunner of cell lineages that give rise to nervous system tissues and the integument's outer layer.

effector Of homeostatic systems, a muscle (or gland) that responds to signals from an integrator, such as the brain, by producing movement (or chemical change) that helps adjust the body to changing conditions.

effector cell A differentiated cell of one of the subpopulations of lymphocytes that form during an immune response; it acts at once to engage and destroy the antigen-bearing agent that triggered the response.

egg A type of mature female gamete; also called an ovum.

El Niño A recurring, massive eastward displacement of warm surface waters of the western equatorial Pacific, which displaces cooler waters off the South American coast. Causes global disruptions in climate.

electromagnetic spectrum The entire range of wavelengths, from the forms of radiant energy less than 10^{-5} nanometer long to radio waves more than 10 kilometers long.

electron A negatively charged unit of matter, with both particulate and wavelike properties, that occupies one of the orbitals around the atomic nucleus. Atoms can gain, lose, or share electrons with other atoms.

electron transfer The donation of one or more electrons stripped from one molecule to another molecule.

electron transport phosphorylation (FOSS-for-ih-LAY-shun) Final stage of aerobic respiration, when electrons from reaction intermediates flow through a membrane transport system that gives them up to oxygen. The flow sets up electrochemical gradients that drive ATP formation at other sites in the membrane.

electron transport system Organized array of enzymes and cofactors, bound in a cell membrane, that accept and donate electrons in series. When it operates, hydrogen ions flow across the membrane, and the flow drives ATP formation and other reactions.

element A substance that cannot be broken down to substances with different properties.

embryo (EM-bree-oh) [Gk. *en*, in, + probably *bryein*, to swell] Of animals generally, a stage formed by cleavage, gastrulation, and other early developmental events. Of seed plants, the young sporophyte, from the first cell divisions after fertilization until germination.

embryonic induction A change in the developmental fate of an embryonic tissue, as brought about by exposure to a gene product released from an adjacent tissue.

embryo sac Common name of the female gametophyte of flowering plants.

emerging pathogen A deadly pathogen, either a newly mutated strain of an existing species or one that evolved long ago and is only now taking great advantage of the increased presence of human hosts.

emulsification Of the chyme in the small intestine, a suspension of droplets of fat coated with bile salts.

end product A substance present at the end of a metabolic pathway.

endangered species A species at the brink of extinction owing to the extremely small size and severely limited genetic diversity of its remaining populations.

endergonic reaction (en-dur-GONE-ik) A chemical reaction having a net gain in energy.

endocrine gland A ductless gland that secretes hormones, which usually enter interstitial fluid and then the bloodstream.

endocrine system System of cells, tissues, and organs, functionally linked to the nervous system, that exerts control by way of its hormones and other chemical secretions.

endocytosis (EN-doe-sigh-TOE-sis) Transport of a substance into a cell by a vesicle, the membrane of which is a patch of plasma membrane that forms around the substance and sinks into the cytoplasm. Phagocytes also engulf prey or pathogens this way.

endoderm [Gk. *endon*, within, + *derma*, skin] The innermost primary tissue layer of animal embryos; gives rise to the inner lining of the gut and organs derived from it.

endodermis A sheetlike wrapping of single cells around the vascular cylinder of a root that functions in controlling the uptake of water and dissolved nutrients.

endometrium (EN-doh-MEET-ree-um) [Gk. *metrios*, of the womb] Innermost lining of the uterus, consisting of connective tissues, glands, and blood vessels.

endoplasmic reticulum or **ER** (EN-doe-PLAZ-mik reh-TIK-yoo-lum) An organelle that begins at the nucleus and curves through the cytoplasm. In rough ER (with many ribosomes on its cytoplasmic side), many new polypeptide chains acquire specialized side chains. Smooth ER (with no attached ribosomes) is a site of lipid synthesis.

endoskeleton [Gk. *endon*, within, + *sklēros*, hard, stiff] An internal framework of bone, cartilage, or both in chordates. Together with skeletal muscle, supports and protects other body parts, helps maintain posture, and moves the body.

endosperm (EN-doe-sperm) Nutritive tissue that surrounds a flowering plant embryo and becomes food for the young seedling.

endospore A resting structure that forms around a copy of the chromosome and part of the cytoplasm of certain bacteria.

endosymbiosis In general, a mutually beneficial interdependence between two species, one of which resides permanently inside the other's body.

energy A capacity to do work.

energy carrier A molecule that delivers energy from one metabolic reaction site to another. ATP is the most common energy carrier in all cells.

energy flow pyramid A pyramid-shaped representation of an ecosystem's trophic structure, illustrating the energy losses at each transfer to a different trophic level.

enhancer A base sequence in DNA that is a binding site for an activator protein.

entropy (EN-trow-pee) A measure of the degree of disorder in a system (how much energy has become so disorganized and dispersed, usually as heat, that it is no longer readily available to do work). Any organized system tends toward entropy without energy inputs to make up for the flow of energy out of it.

enzyme (EN-zime) One of a class of proteins that enormously speed (catalyze) reactions between specific substances, usually at their functional groups.

eosinophil Fast-acting white blood cell; its enzyme secretions digest holes in parasitic worms during an inflammatory response.

epidermis The outermost tissue layer of a multicelled plant and of nearly all animals.

epinephrine (ep-ih-NEF-rin) Hormone of the adrenal medulla; raises blood levels of sugar and fatty acids; increases heart rate and the force of contraction.

epiglottis A flaplike structure at the start of the larynx, the position of which directs the movement of air into the trachea or of food into the esophagus.

epistasis (eh-PISS-tah-sis) An interaction between gene pairs. Two alleles of one gene mask expression of another gene's alleles, so expected phenotypes may not appear.

epithelium (EP-ih-THEE-lee-um) An animal tissue of one or more layers of adhering cells that covers the body's external surfaces and lines its internal cavities and tubes. It has one free surface; the opposite surface rests on a basement membrane between it and an underlying connective tissue. Epidermis is an example.

equilibrium, dynamic [Gk. *aequus*, equal, + *libra*, balance] The point at which a chemical reaction runs forward as fast as in reverse; the concentrations of reactant molecules and product molecules show no net change.

erosion The movement of land under the force of wind, running water, and ice.

erythrocyte (eh-RITH-row-site) [Gk. *erythros*, red, + *kytos*, vessel] Red blood cell.

esophagus (ee-SOF-uh-gus) Tubular portion of a digestive system that receives ingested food and leads to the stomach.

essential amino acid An amino acid that an organism cannot synthesize for itself and must obtain from a food source.

essential fatty acid A fatty acid that an organism cannot synthesize for itself and must obtain from food source.

estrogen (ESS-trow-jen) A sex hormone that helps oocytes mature, induces changes in the uterine lining during the menstrual cycle and pregnancy, and helps maintain secondary sexual traits; also influences bodily growth and development.

estrus (ESS-truss) [Gk. *oistrus*, frenzy] For mammals generally, the cyclic period of a female's sexual receptivity to the male.

estuary (EST-you-ehr-ee) A partly enclosed coastal region where seawater mixes with freshwater and runoff from the surrounding land, as by streams and rivers.

ethylene (ETH-il-een) Plant hormone that stimulates fruit ripening and abscission.

Eubacteria Kingdom of the most common species of bacterial cells.

eukaryotic cell (yoo-CARRY-oh-tic) [Gk. *eu*, good, + *karyon*, kernel] A cell having a "true nucleus" and other distinguishing membrane-bound organelles. *Compare* prokaryotic cell.

eutrophication Nutrient enrichment of a body of water, such as a lake, that typically results in reduced transparency and a phytoplankton-dominated community.

evaporation [L. *e-*, out, + *vapor*, steam] Heat energy converts a substance from the liquid to the gaseous state.

evolution, biological [L. *evolutio*, unrolling] Genetic change in a line of descent over time; brought about by microevolutionary processes (gene mutation, natural selection, genetic drift, and gene flow).

evolutionary systematics The branch of biology that applies evolutionary theory to the task of identifying patterns of diversity over time and in the environment.

evolutionary tree A treelike diagram in which the branches represent separate lines of descent from a common ancestor and branch points represent divergences.

excitatory postsynaptic potential (or EPSP) One of two competing signals at an input zone of a neuron; a graded potential that brings the neuron's plasma membrane closer to threshold.

excretion Any of several processes by which excess water, excess or harmful solutes, or waste materials leave the body by way of a urinary system or certain glands.

exergonic reaction (EX-ur-GONE-ik) A chemical reaction that shows a net loss in energy.

exocrine gland (EK-suh-krin) [Gk. *es*, out of, + *krinein*, to separate] Glandular structure that secretes products, usually through ducts or tubes, to a free epithelial surface.

exocytosis (EK-so-sigh-TOE-sis) Transport of a substance out of a cell by means of a vesicle, the membrane of which fuses with the plasma membrane, so that the vesicle's contents are released outside.

exodermis Layer of cells just inside the root epidermis of most flowering plants; helps control the uptake of water and solutes.

exon Any of the nucleotide sequences of a pre-mRNA molecule that become spliced together to form a mature mRNA transcript and ultimately get translated into protein.

exoskeleton [Gk. *exo*, out, + *skleros*, hard, stiff] An external skeleton, as in arthropods.

experiment A test of potentially falsifiable hypotheses about some aspect of nature. Its premise is that any aspect of the natural world has one or more underlying causes.

exponential growth (EX-po-NEN-shul) A pattern of population growth in which the population size expands by ever increasing increments during successive time intervals because the reproductive base becomes ever larger. The plot of population size against time has a characteristic J-shaped curve.

extinction, background A steady rate of species turnover that characterizes lineages through most of their histories.

extinction, mass An abrupt increase in the rate at which major taxa disappear, with several taxa being affected simultaneously.

extracellular fluid In animals generally, all the fluid not inside cells; includes plasma (the liquid portion of blood) and interstitial fluid (occupying the spaces between cells and tissues).

extracellular matrix A matrix that helps impart shape to many animal tissues; its ground substance contains fibrous proteins and other materials (mostly cell secretions).

FAD Flavin adenine dinucleotide, one of the nucleotide coenzymes that transfers electrons and unbound protons (H^+) from one reaction site to another. At such times it is abbreviated $FADH_2$.

fall overturn The vertical mixing of a body of water in autumn. Its upper layer cools, increases in density, and sinks; dissolved oxygen moves down and nutrients from bottom sediments move up.

family pedigree A chart of the genetic relationship of the individuals in a family through successive generations.

fat A lipid with a glycerol head and one, two, or three fatty acid tails. Tryglycerides (neutral fats) have three. Unsaturated tails have single covalent bonds in their carbon backbone; saturated tails also have one or more double bonds.

fate map A map of the surface of an animal embryo that shows the origin of each kind of differentiated cell in the adult.

fatty acid A molecule with a backbone of up to thirty-six carbon atoms, a carboxyl group (—COOH) at one end, and hydrogen atoms at most or all of the remaining bonding sites.

feedback inhibition Of cells or multicelled organisms, a control mechanism by which the output of a substance changes a specific condition or activity, which then triggers a decrease in further output of the substance or further activity.

fermentation [L. *fermentum*, yeast] A type of anaerobic pathway of ATP formation. It starts with glycolysis, ends with a transfer of electrons back to one of the breakdown products or intermediates, and regenerates NAD^+ required for the reaction. Its has a net yield of two ATP per glucose molecule.

fertilization [L. *fertilis*, to carry, to bear] The fusion of a sperm nucleus with the nucleus of an egg, which thus becomes a zygote.

fever A body temperature higher than a set point in the hypothalamic region that acts as the body's thermostat.

fibrous root system Of most monocots, all the lateral branchings of adventitious roots, which arose earlier from the young stem.

filter feeder An animal that filters food from a current of water that is directed through a body part, such as a sea squirt's pharynx.

filtration Of vertebrates, a process by which blood pressure forces water and solutes from capillaries into interstitial fluid. Filtration occurs in Bowman's capsule of a nephron.

fin Of fishes generally, an appendage that helps propel, stabilize, and guide the body through water.

first law of thermodynamics [Gk. *therme*, heat, + *dynamikos*, powerful] A law of nature stating the total amount of energy in the universe remains constant. Energy cannot be created from nothing and existing energy cannot be destroyed.

fish An aquatic animal of the most ancient, diverse vertebrate lineage, which includes jawless, cartilaginous, and bony fishes.

fitness An increase in adaptation to the environment, as brought about by genetic change.

fixation Of the individuals of a population, only one kind of allele remains at a specified locus; all individuals are homozygous for it.

fixed action pattern Of animals, a program of coordinated, stereotyped muscle activity that runs to completion independently of feedback from the environment.

flagellum (fluh-JELL-um), plural **flagella** [L. whip] Tail-like motile structure of many free-living eukaryotic cells; its core has a 9 + 2 array of microtubules.

flower A reproductive structure that distinguishes angiosperms from other seed plants and often attracts pollinators.

fluid mosaic model All cell membranes consist of a lipid bilayer and proteins. The lipids (phospholipids, mainly) impart basic structure, impermeability to water-soluble molecules, and (through packing variations and movements) fluidity. Diverse proteins spanning the bilayer or attached to one of its surfaces perform most of the membrane functions, including transport, enzyme activity, and reception of molecular signals or substances.

follicle (FOLL-ih-kul) A small sac, pit, or cavity, as around a hair; also a mammalian oocyte with its surrounding layer of cells.

food chain A straight-line sequence of who eats whom in an ecosystem.

food web A network of cross-connecting, interlinked food chains with some number of producers and consumers, as well as decomposers, detritivores, or both.

forebrain Most complex part of a vertebrate brain; it includes the cerebrum (and cerebral cortex), olfactory lobes, and hypothalamus.

forest A biome where tall trees grow together closely enough to form a fairly continuous canopy over a broad region.

fossil Recognizable, physical evidence of an organism that lived in the distant past.

fossil fuel Coal, petroleum, or natural gas; a nonrenewable source of energy formed in sediments by the compression of plant remains over hundreds of millions of years.

fossilization How fossils form. First an organism or traces of it becomes buried in sediments or volcanic ash. Water infiltrates the remains, which become infused with dissolved inorganic compounds. Sediments accumulate and exert pressure above the burial site. Over great spans of time, the pressure and chemical changes transform the remains to stony hardness.

founder effect A form of bottlenecking. By chance, allele frequencies of founders of a new population may not be the same as those of the original population. If there is no gene flow between the two, then natural selection will influence gene frequencies in drastically different ways through its interaction with genetic drift.

free radical A highly reactive, unbound molecular fragment that has the wrong number of electrons.

fruit [L. after *frui*, to enjoy] Of flowering plants, the expanded and ripened ovary of one or more carpels, some with accessory floral structures incorporated.

FSH Follicle-stimulating hormone; one of the hormones produced and secreted by the anterior lobe of the pituitary gland; serves reproductive roles in both sexes.

functional group An atom or group of atoms that is covalently bonded to the carbon backbone of an organic compound and that influences its chemical behavior.

functional-group transfer Donation of a functional group by one molecule to another.

Fungi The kingdom of fungi which, as a group, are major decomposers.

fungus A eukaryotic heterotroph that uses extracellular digestion and absorption; it secretes enzymes that break down an external food source into molecules small enough to be absorbed by its cells. Saprobes feed on nonliving organic matter, parasites feed on living organisms.

gall bladder Organ that stores bile secreted from the liver and that is connected by way of a duct to the small intestine.

gamete (GAM-eet) [Gk. *gametēs*, husband, and *gametē*, wife] A haploid cell, formed by meiosis and cytoplasmic division of a germ cell; required for sexual reproduction. Eggs and sperm are examples.

gamete formation Of animals, the first stage of development, in which sperm or eggs form and mature within reproductive tissues or organs of parents.

gametophyte (gam-EET-oh-fite) [Gk. *phyton*, plant] A haploid, multicelled, gamete-producing body that forms during the life cycle of most plants.

ganglion (GANG-lee-un), plural **ganglia** [Gk. *ganglion*, a swelling] A distinct clustering of cell bodies of neurons in regions other than the brain or spinal cord.

gastrulation (gas-tru-LAY-shun) The fourth stage of animal development; a time of major cellular reorganization when newly formed cells become arranged into two or three primary tissues, or germ layers.

gene [short for German *pangan*, after Gk. *pan*, all + *genes*, to be born] A unit of information about a heritable trait that is passed on from parents to offspring. Each gene has a specific location (locus) on a chromosome.

gene flow A microevolutionary process; the movement of alleles into and out of populations as a result of immigration and emigration.

gene frequency More precisely, allele frequency; the relative abundances of all the different alleles at a given gene locus that are carried by all individuals of a population.

gene locus The particular location of a gene along the length of a chromosome.

gene mutation [L. *mutatus*, a change] A heritable change in a DNA molecule by the deletion, addition, or substitution of one to several bases in its nucleotide sequence.

gene pair Of diploid cells, the two alleles at a particular gene locus (that is, on a pair of homologous chromosomes).

gene pool Sum total of all genotypes in a population.

gene therapy Generally, the transfer of one or more normal genes into an organism to correct or lessen the adverse effects of a genetic disorder.

genetic code [After L. *genesis*, to be born] The correspondence between the nucleotide triplets in DNA (then in RNA) and the specific sequences of amino acids in the resulting polypeptide chains; the basic language of protein synthesis in cells.

genetic disease An illness in which the expression of one or more genes increased the susceptibility of an individual to an infection or weakened its immune response.

genetic disorder An inherited condition that results in mild to severe medical problems.

genetic divergence Build-up of differences in gene pools of two or more populations of a species after a geographic barrier arises and separates them, because gene mutation, natural selection, and genetic drift are free to operate independently in each one.

genetic drift A random change in allele frequencies over the generations brought about by chance alone. The magnitude of its effect on genetic diversity and on the range of phenotypes relates to population size.

genetic engineering Altering the information content of DNA through use of recombinant DNA technology.

genetic equilibrium A state in which a population is not evolving. It occurs only if there is no mutation, if the population is large and isolated from other populations of the species, and if there is no natural selection (all members survive and reproduce equally by random mating).

genetic recombination A nonparental combination of some number of alleles in offspring; an outcome of gene mutation, crossing over, changes in chromosome structure or number, or recombinant DNA technology.

genome All the DNA in a haploid number of chromosomes of a given species.

genotype (JEEN-oh-type) Genetic constitution of an individual; a single gene pair or the sum total of an individual's genes. *Compare* phenotype.

genus, plural **genera** (JEEN-US, JEN-er-ah) [L. *genus*, race, origin] A taxon into which all species with phenotypic similarities and evolutionary relationship are grouped.

geographic dispersal Directional movement in which some residents of an established community leave their home range and take up residence elsewhere; they are considered to be exotic species in the new location.

geologic time scale A time scale for Earth history, the subdivisions of which are based on boundaries marked by episodes of mass extinction and which have been refined by radiometric dating.

germ cell Of animals, a cell of a lineage set aside for sexual reproduction; germ cells give rise to gametes. *Compare* somatic cell.

germ layer Of animal embryos, one of two or three primary tissue layers that form at gastrulation as forerunners of adult tissues. *Compare* ectoderm; endoderm; mesoderm.

germination (jur-min-AY-shun) Of resting spores and seeds, a resumption of activity following a period of arrested development.

gibberellin (JIB-er-ELL-un) A type of plant hormone that promotes stem elongation.

gill A respiratory organ, typically with a moist, thin vascularized layer of epidermis that functions in gas exchange.

gland A secretory cell or structure derived from epithelium and often connected to it.

glomerular capillaries Set of blood capillaries inside Bowman's capsule of the nephron.

glomerulus (glow-MARE-you-luss) [L. *glomus*, ball] First portion of the nephron, where water and solutes are filtered from blood.

glucagon (GLUE-kuh-gone) A hormone that stimulates cells to convert glycogen and amino acids to glucose; secreted by alpha cells of the pancreas when the blood level of glucose decreases.

glyceride (GLISS-er-eyed) One of the fats or oils; a molecule with one, two, or three fatty acid tails attached to a glycerol backbone.

glycerol (GLISS-er-oh) [Gk. *glykys*, sweet, + L. *oleum*, oil] A three-carbon molecule with three hydroxyl groups attached; one of the components of fats and oils.

glycogen (GLY-kuh-jen) A highly branched polysaccharide that is the main storage carbohydrate of animals; cleavage reactions at its many branchings yield an abundance of glucose monomers when required.

glycocalyx A sticky mesh of polysaccharides, polypeptides, or both around the cell wall of many bacteria.

glycolysis (gly-CALL-ih-sis) [Gk. *glykys*, sweet, + *lysis*, loosening or breaking apart] Initial energy-releasing reactions of aerobic and anaerobic pathways by which enzymes break down glucose (or some other organic compound) to pyruvate. It proceeds in the cytoplasm of all cells, it has a net yield of two ATP, and oxygen has no role in it.

glycoprotein A protein having linear or branched oligosaccharides covalently bonded to it. Nearly all surface proteins of animal cells and many proteins circulating in blood are glycoproteins.

gnetophyte Only gymnosperm known to have vessels in its xylem.

Golgi body (GOHL-gee) Organelle of lipid assembly, polypeptide chain modification, and packaging of both in vesicles for export or for transport to locations in the cytoplasm.

gonad (GO-nad) Primary reproductive organ in which animal gametes are produced.

graded potential Of neurons, a local signal that slightly alters the voltage difference across a patch of plasma membrane and that varies in magnitude according to the stimulus. With prolonged or intense stimulation, such signals may spread to a trigger zone of the membrane and initiate an action potential.

granum, plural **grana** In many chloroplasts, any of the stacks of flattened, membranous compartments with chlorophyll and other light-trapping pigments and reaction sites for ATP formation.

grassland A biome with flat or rolling land, 25–100 centimeters of annual rainfall, warm summers, grazing animals, and periodic fires that regenerate dominant species.

gravitropism (GRAV-ih-TROPE-izm) [L. *gravis*, heavy, + Gk. *trepein*, to turn] Tendency of a plant to grow directionally in response to the Earth's gravitational force.

gray matter Inside the brain and spinal cord, the unmyelinated axons, dendrites, and nerve cell bodies and neuroglial cells.

grazing food web A network of food chains in which energy flows from plants to an array of herbivores, then carnivores.

green alga An aquatic protistan with chlorophylls *a* and *b*; early members of its lineage may have given rise to plants.

green revolution In developing countries, the use of improved crop varieties, modern agricultural practices (including massive inputs of fertilizers and pesticides), and equipment to increase crop yields.

greenhouse effect Warming of the lower atmosphere as a result of the presence of increasing levels of greenhouse gases (such as carbon dioxide and methane).

ground meristem (MARE-ih-stem) [Gk. *meristos*, divisible] A primary meristem that produces the ground tissue system, hence the bulk of the plant body.

ground substance Of certain animal tissues, intercellular material made of cell secretions and other noncellular components.

ground tissue system Tissues making up the bulk of a plant body, the most common being parenchyma.

growth Of multicelled organisms, increases in the number, size, and volume of cells.

guanine A nitrogen-containing base; present in one of the four nucleotide building blocks of DNA and RNA.

guard cell Either of two adjacent cells with roles in the movement of carbon dioxide, oxygen, and water vapor across leaf or stem epidermis. An opening (stoma) forms when both swell with water and move apart; it closes when they lose water and collapse against one another.

gut A body region where food is digested and absorbed; of complete digestive systems, the gastrointestinal tract (the portions from the stomach onward).

gymnosperm (JIM-noe-sperm) [Gk. *gymnos*, naked, + *sperma*, seed] A vascular plant that bears seeds at exposed surfaces of reproductive structures, such as cone scales.

habitat [L. *habitare*, to live in] The type of place where an organism normally lives, as characterized by physical and chemical features and by its array of species.

hair cell A mechanoreceptor that may give rise to action potentials when bent or tilted.

half-life The time it takes for half of a given quantity of any radioisotope to decay into a different, less unstable daughter isotope.

halophile A type of archaebacterium that lives in extremely saline habitats.

haploidy (HAP-loyd) The presence of half the parental number of chromosomes in a gamete, as brought about by meiosis; the gamete has one of each pair of homologous chromosomes. *Compare* diploidy.

Hardy-Weinberg rule Allele frequencies will stay the same through the generations if there is no mutation, if the population is infinitely large and is isolated from other populations of the same species, if mating is random, and if all individuals survive and reproduce equally.

HCG Human chorionic gonadotropin. A hormone that helps maintain the lining of the uterus during the menstrual cycle and during the first trimester of pregnancy.

heart Muscular pump that keeps blood circulating through the animal body.

helper T cell A T lymphocyte which, when activated, produces and secretes chemicals that induce responsive T or B lymphocytes to divide and give rise to large populations of effector cells and memory cells.

hemoglobin (HEEM-oh-glow-bin) [Gk. *haima*, blood, + L. *globus*, ball] Iron-containing, oxygen-transporting protein that gives red blood cells their color.

hemostasis (HEE-mow-STAY-sis) [Gk. *haima*, blood, + *stasis*, standing] Stopping of blood loss from a damaged blood vessel through coagulation, blood vessel spasm, platelet plug formation, and other mechanisms.

herbivore [L. *herba*, grass, + *vovare*, to devour] Plant-eating animal.

hermaphrodite An individual with both male and female gonads; two individuals sexually reproduce by the mutual transfer of sperm.

heterocyst (HET-er-oh-sist) A self-modified cyanobacterial cell that makes a nitrogen-fixing enzyme when nitrogen is scarce.

heterotroph (HET-er-oh-trofe) [Gk. *heteros*, other, + *trophos*, feeder] Organism unable to synthesize its own organic compounds; it feeds on autotrophs, other heterotrophs, or organic wastes. *Compare* autotroph.

heterozygous condition (HET-er-oh-ZYE-guss) [Gk. *zygoun*, join together] Of a specified trait, having a pair of nonidentical alleles at a gene locus (on a pair of homologous chromosomes).

higher taxon (plural, **taxa**) One of the ever more inclusive groupings meant to reflect relationships among species. Family, order, class, phylum, and kingdom are examples.

hindbrain One of three divisions of the vertebrate brain; the medulla oblongata, cerebellum, and pons. Has reflex centers for respiration, blood circulation, and other basic functions; also coordinates motor responses and many complex reflexes.

histone Any of a class of proteins that are intimately associated with eukaryotic DNA and largely responsible for the organization of eukaryotic chromosomes.

homeostasis (HOE-me-oh-STAY-sis) [Gk. *homo*, same, + *stasis*, standing] Of animals, a physiological state in which physical and chemical aspects of the internal environment (blood and interstitial fluid) are maintained within ranges suitable for cell activities.

homeotic gene One of a class of master genes that specify the development of specific body part in animals.

hominid [L. *homo*, man] All species on or near the evolutionary road leading to modern humans.

hominoid Apes, humans, and their recent ancestors.

homologous chromosome (huh-MOLL-uh-gus) [Gk. *homologia*, correspondence] Of cells having a diploid chromosome number, one of a pair of chromosomes that are identical in size, shape, and gene sequence, and that interact during meiosis. Nonidentical sex chromosomes in a cell also interact during meiosis and are considered homologues also.

homology Similarity in one or more body parts in different species that is attributable to their descent from a common ancestor.

homologous structures The same body parts, modified in different ways, in different lines of descent from a common ancestor.

homozygous condition (HOE-moe-ZYE-guss) For a specified trait, having a pair of identical alleles at a gene locus (on a pair of homologous chromosomes).

homozygous dominant condition Having a pair of dominant alleles at a gene locus (on a pair of homologous chromosomes).

homozygous recessive condition Having a pair of recessive alleles at a gene locus (on a pair of homologous chromosomes).

hormone [Gk. *hormon*, to stir up, set in motion] Signaling molecule that stimulates or inhibits gene transcription in nonadjacent target cells (any cell having receptors for it).

horsetail A seedless vascular plant with photosynthetic stems that look like horsetails.

human genome project Worldwide basic research project to sequence the estimated 3 billion nucleotides present in the DNA of human chromosomes.

humus Decomposing organic matter in soil.

hydrogen bond A weak interaction between a small, highly electronegative atom of a molecule and a neighboring hydrogen atom that is taking part in a polar covalent bond.

hydrogen ion A free (unbound) proton; a hydrogen atom that has lost its electron and so bears a positive charge (H^+).

hydrologic cycle A biogeochemical cycle, driven by solar energy, in which water moves slowly through the atmosphere, on or through surface layers of land masses, to the ocean, and back again.

hydrolysis (high-DRAWL-ih-sis) [L. *hydro*, water, + Gk. *lysis*, loosening or breaking apart] Cleavage reaction in which covalent bonds break, splitting a molecule into two or more parts. Often H^+ and OH^- (derived from a water molecule) become attached to the exposed bonding sites.

hydrophilic substance [Gk. *philos*, loving] A polar substance that is attracted to the polar water molecule and dissolves easily in water. Sugars are examples.

hydrophobic substance [Gk. *phobos*, dreading] A nonpolar substance that is repelled by the polar water molecule and thus resists being dissolved in water. Oil is an example.

hydrosphere All liquid or frozen water on or near the Earth's surface.

hydrostatic pressure Any volume of fluid that exerts a force directed against a wall, membrane, or another structure enclosing that fluid. The greater its concentration of solutes, the greater will be the pressure that the fluid exerts.

hydrothermal vent ecosystem Ecosystem, near a fissure in the ocean floor, based on chemosynthetic bacteria that use dissolved minerals as their energy source.

hypha (HIGH-fuh), plural **hyphae** [Gk. *hyphe*, web] Of fungi, a filament with chitin-reinforced walls and, often, reinforcing cross-walls; component of the mycelium.

hypodermis A subcutaneous layer having stored fat that helps insulate the body; although not part of skin, it anchors skin and allows it some freedom of movement.

hypothalamus [Gk. *hypo*, under, + *thalamos*, inner chamber or possibly *tholos*, rotunda] Of the forebrain, a major center for homeostatic control of visceral activities (such as salt-water balance, temperature control, and reproduction), related forms of behavior (as in hunger, thirst, and sex), and emotional expression, such as sweating with fear.

hypothesis In science, a possible explanation of a specific phenomenon in nature, one that has the potential to be proved false by experimental tests.

immune response Events by which B and T lymphocytes recognize antigen, undergo cell divisions that form huge populations of lymphocytes, which differentiate into subpopulations of effector and memory cells. The effector cells destroy cells bearing antigen-MHC complexes. Memory cells are not activated until subsequent encounters with the same antigen.

immunization Various processes, including vaccination, that promote increased immunity against specific diseases.

immunoglobulin (Ig) One of five classes of antibodies, each with antigen-binding sites and other sites with specialized functions.

implantation Process by which a blastocyst adheres to the endometrium and establishes connections by which the mother and embryo will exchange substances during pregnancy.

imprinting A time-dependent form of learning that is triggered by exposure to sign stimuli and that usually occurs during a sensitive period when the animal is young.

inbreeding Nonrandom mating among close relatives, which have many identical alleles in common. Inbreeding is a form of genetic drift in a small group of relatives that are preferentially interbreeding.

incomplete dominance One allele of a pair is not fully dominant over its partner, so a heterozygous phenotype in between the two homozygous phenotypes emerges.

independent assortment, theory of By the end of meiosis in a germ cell, each pair of homologous chromosomes—hence the genes that they carry—have been sorted for shipment into gametes independently of how the other pairs were sorted out.

indirect selection A theory in evolutionary biology that self-sacrificing individuals can pass on their genes indirectly by helping relatives survive and reproduce.

induced-fit model A substrate induces change in the shape of an enzyme's active site when bound to it, the result being a more precise molecular fit between the two that promotes reactivity.

infection Invasion and multiplication of a pathogen in host cells or tissues. Disease follows if defenses cannot be mobilized fast enough to prevent the pathogen's activities from interfering with normal functions.

inflammation, acute Important aspect of nonspecific defenses and immune responses when cells of a local tissue are damaged or killed, as by infection; requires action of fast-acting phagocytes and plasma proteins, including complement proteins.

inheritance The transmission, from parents to offspring, of structural and functional patterns that have a genetic basis and are characteristic of their species.

inhibiting hormone A signaling molecule produced and secreted by the hypothalamus that suppresses a particular secretion by the anterior lobe of the pituitary gland.

inhibitor A substance that can bind with an enzyme and interfere with its functioning.

inhibitory postsynaptic potential (IPSP) Of neurons, one of two competing types of graded potentials at an input zone; it tends to drive the resting membrane potential away from threshold.

instinctive behavior A behavior performed without having been learned by experience in the environment. The nervous system of a newly born or hatched animal is prewired to recognize one or two sign stimuli (simple, well-defined cues in the environment) that can trigger a suitable response.

insulin Hormone secreted by beta cells of the pancreas that lowers the glucose level in blood by stimulating cells to take up glucose; also promotes protein and fat synthesis and inhibits protein conversion to glucose.

integration, neural [L. *integrare*, coordinate] Moment-by-moment summation of all the excitatory and inhibitory synapses acting on the neuron; it takes place at each level of synapsing in a nervous system.

integrator Of homeostatic systems, a control point such as a brain where information is pulled together in the selection of responses to stimuli.

integument Of animals, a protective body cover such as skin. Of seed-bearing plants, one or more layers around an ovule that harden, thicken, and form a seed coat.

integumentary exchange (in-teg-you-MEN-tuh-ree) A mode of respiration in which oxygen and carbon dioxide diffuse across a thin, moist, vascularized layer at the body surface of certain animals.

interleukin One of several chemical mediator molecules secreted by helper T cells that fan mitotic cell divisions and differentiation of responsive T and B cells.

intermediate A compound formed between the start and end of a metabolic pathway.

intermediate filament In different types of animal cells, a cytoskeletal element made of particular proteins.

interneuron Any neuron of the vertebrate brain and spinal cord.

internode In vascular plants, the stem region between two successive nodes.

interphase Of the cell cycle, an interval in between nuclear divisions when a cell increases in mass, roughly doubles the number of its cytoplasmic components, and then duplicates its chromosomes (replicates its DNA). The interval differs among species.

interspecific competition Individuals of different species compete with one another for a share of resources in their habitat.

interstitial fluid (IN-ter-STISH-ul) [L. *interstitus*, to stand in the middle of something] That portion of extracellular fluid occupying the spaces between cells and tissues of complex animals.

intertidal zone Generally, part of a rocky or sandy shoreline above the low water mark and below the high water mark; organisms that inhabit it are alternately submerged, then exposed, by tides.

intervertebral disk One of a number of disk-shaped structures containing cartilage that act as shock absorbers and flex points between bony segments of the vertebral column.

intraspecific competition All individuals of a population compete with one another for a share of resources in their habitat.

intron A noncoding portion of a newly formed mRNA molecule.

inversion A linear stretch of DNA within a chromosome that has become oriented in the reverse direction, with no molecular loss.

invertebrate Animal without a backbone.

in vitro fertilization Conception outside the body (literally, "in glass" petri dishes or test tubes).

ion, negatively charged (EYE-on) An atom or a compound that acquired an overall negative charge by gaining one or more electrons.

ion, positively charged An atom or a compound that acquired an overall positive charge by losing one or more electrons.

ionic bond Ions of opposite charge have attracted each other and are staying together.

isotonic condition Equality in the relative concentrations of solutes in two fluids; for two fluids separated by a cell membrane, there is no net osmotic (water) movement across the membrane.

isotope (EYE-so-tope) Of an element, an atom with more or fewer neutrons than the atoms having the most common number.

J-shaped curve The type of curve that emerges when population size is plotted against time; it represents unrestricted, exponential growth.

joint An area of contact or near-contact between bones.

juvenile Of many animals, a miniaturized form between the embryo and adult; it simply changes in size and proportion until it reaches sexual maturity.

karyotype (CARRY-oh-type) For an individual (or a species), a preparation of metaphase chromosomes sorted by length, centromere location, and other defining features.

keratin A tough, water-insoluble protein made by most epidermal cells that becomes concentrated in skin's outermost layers.

keratinization (care-AT-in-iz-AY-shun) Process by which keratin-producing epidermal cells die and accumulate as keratinized bags at the skin surface to form a barrier against dehydration, bacteria, many toxins, and ultraviolet radiation.

keystone species A species that dominates a community and dictates its structure.

key innovation A modified structure or function that allows a lineage to exploit the environment in more efficient or novel ways.

kidney One of a pair of vertebrate organs that filter mineral ions, organic wastes, and other substances from the blood; it controls the amounts returned to blood and thereby helps maintain the volume and solute levels of extracellular fluid.

kilocalorie 1,000 calories of heat energy; the amount of energy required to raise the temperature of 1 kilogram of water by 1°C; unit of measure for the caloric value of foods.

kinase One of a class of enzymes that catalyze phosphate-group transfers.

kinetochore Group of proteins and DNA at a chromosome's centromere where spindle microtubules become attached at mitosis or meiosis. Each chromatid of a duplicated chromosome has its own kinetochore.

Krebs cycle A cyclic pathway that occurs in mitochondria; together with a few preparatory steps, the stage of aerobic respiration in which pyruvate is completely broken down to carbon dioxide and water. Coenzymes accept unbound protons (H^+) and electrons stripped from intermediates and deliver them to the next stage.

lactate fermentation Anaerobic pathway of ATP formation; pyruvate from glycolysis is converted to the three-carbon compound lactate, and NAD^+ is regenerated. The net energy yield is two ATP.

lactation Milk production by hormone-primed mammary glands.

lake A body of standing freshwater with littoral, limnetic, and profundal zones.

large intestine Colon; a gut region that receives unabsorbed food residues from the small intestine and concentrates and stores feces until they are expelled from the body.

larva, plural **larvae** Of many animals, an immature developmental stage between the embryo and adult.

larynx (LARE-inks) A tubular airway to and from the lungs. In humans, it contains vocal cords.

lateral meristem A type of meristem in plants that show secondary growth; either vascular cambium or cork cambium.

lateral root Of taproot systems, a lateral branching from the first, primary root.

leaching The removal of some nutrients in soil as water percolates through it.

leaf For most vascular plants, a structure having chlorophyll-containing tissue that is the major region of photosynthesis.

learned behavior Variation or change in responses to stimuli after an animal has processed and integrated information gained from specific experiences.

lek Of some birds and other animals, a type of communal display ground occupied during times of courtship.

lethal mutation A mutation with drastic effects on phenotype that usually cause the individual's death.

LH Luteinizing hormone. Hormone secreted by the anterior lobe of the pituitary gland, with roles in male and female reproduction.

lichen (LY-kun) A symbiotic interaction between a fungus and a photoautotroph, such as a green alga.

life cycle A recurring pattern of genetically programmed events by which individuals grow, develop, maintain themselves, and reproduce.

life table Tabulation of age-specific patterns of birth and death for a population.

ligament A strap of dense connective tissue that bridges a joint.

light-dependent reactions The first stage of photosynthesis, in which sunlight energy is trapped and converted to the chemical energy of ATP alone (by a cyclic pathway) or ATP and NADPH (by a noncyclic pathway).

light-independent reactions The second stage of photosynthesis, in which ATP makes phosphate-group transfers required to build sugar phosphates. Often NADPH delivers electrons and hydrogen atoms for the synthesis reactions, which also require carbon from carbon dioxide. The sugar phosphates enter other reactions by which starch, cellulose, and other end products of photosynthesis are assembled.

lignification Of land plants, a process by which lignin is deposited in secondary cell walls. Lignin imparts strength and rigidity by anchoring cellulose strands in the walls, stabilizes and protects other components of the wall, and forms a waterproof barrier around the cellulose. A key factor in the evolution of vascular plants.

lignin A complex organic compound that strengthens and waterproofs cell walls in certain tissues of vascular plants.

limbic system Brain centers, located in the middle of the cerebral hemispheres, that interact to govern emotions and influence memory. Distantly related to the olfactory lobes; it still deals with the sense of smell.

limiting factor Any essential resource that, in short supply, limits population growth.

lineage (LIN-ee-age) A line of descent.

linkage group A quantitative measure of the relative positions of the genes along the length of a chromosome.

linkage mapping An investigative approach by which the relative linear positions of a chromosome's genes are deduced by tracking the percentage of recombinants in gametes.

lipid Mainly a hydrocarbon, greasy or oily, that strongly resists dissolving in water but quickly dissolves in nonpolar substances. In nearly all cells, certain lipids are the main reservoirs of stored energy. Other lipids are structural materials (as in membranes) and cell products (such as surface coatings).

lipid bilayer Structural basis of all cell membranes, with two layers of mostly phospholipid molecules. The hydrophobic tails of the molecules are sandwiched in between the hydrophilic heads, which are dissolved in the fluid that bathes them.

lipoprotein A molecule that forms when proteins in blood combine with cholesterol, triglycerides, and phospholipids that were absorbed from the gut, after a meal.

liver A large gland in vertebrates and many invertebrates that stores and interconverts organic compounds and helps maintain their blood concentrations. It additionally inactivates most hormone molecules that have served their functions as well as ammonia and other compounds that can be toxic at high concentrations.

local signaling molecules Secretions from cells of many local tissues that quickly alter chemical conditions in the vicinity, then are degraded before they can travel elsewhere.

logistic population growth (low-JISS-tik) A pattern of population growth in which a low-density population slowly increases in size, then enters a rapid growth phase that levels off once the carrying capacity has been reached.

loop of Henle The hairpin-shaped, tubular region of a nephron that functions in the reabsorption of water and solutes.

lung An internal respiratory surface in the shape of a cavity or sac that can be filled with air; a pair occur in a few fishes and in amphibians, birds, reptiles, and mammals.

lycophyte A seedless vascular plant of mostly wet or shaded habitats; requires free water to complete its life cycle.

lymph (LIMF) [L. *lympha*, water] Tissue fluid that has drained into the vessels of the lymphatic system.

lymph capillary A small-diameter vessel of the lymph vascular system that has no pronounced entrance; tissue fluid moves inward by passing between overlapping endothelial cells at the vessel's tip.

lymph node A lymphoid organ that is a battleground for immune responses; packed, organized arrays of lymphocytes inside cleanse lymph before it reaches the blood.

lymph vascular system [L. *lympha*, water, + *vasculum*, a small vessel] Of the lymphatic system, vessels that take up and transport excess tissue fluid and reclaimable solutes as well as fats absorbed from the digestive tract.

lymphatic system An organ system that supplements the circulatory system. Its vessels deliver fluid and solutes from interstitial fluid to the bloodstream; its lymphoid organs have roles in immunity.

lymphocyte Any of various T and B cells with roles in immune responses that follow recognition of antigen.

lysis [Gk. *lysis*, a loosening] Gross damage to a plasma membrane, cell wall, or both that allows the cytoplasm to leak out and thereby leads to cell death.

lysogenic pathway A common pathway in which a viral replication cycle enters a latent period and viral genes are integrated with a host chromosome. The recombinant molecule is a miniature time bomb that is passed on to all of the host cell's progeny.

lysosome (LYE-so-sohm) The main organelle of intracellular digestion, with enzymes that can break down nearly all polysaccharides, proteins, and nucleic acids, and some lipids.

lysozyme An infection-fighting enzyme in mucous membranes.

lytic pathway A common pathway in which a viral replication cycle proceeds rapidly and ends with lysis of the host cell.

macroevolution The large-scale patterns, trends, and rates of change among families and other more inclusive groups of species.

macrophage A leukocyte (white blood cell) that phagocytizes worn-out cells and tissue debris and anything bearing antigen. It may become an antigen-presenting cell and thus trigger immune responses.

mammal The only vertebrate whose females nourish offspring by milk, produced by and secreted from mammary glands.

mass extinction A sudden rise in rates of extinction above the background level; a catastrophic, global event during which a number of major taxa are wiped out simultaneously.

mass number The total number of protons and neutrons in an atom's nucleus.

mast cell In connective tissues, a basophil-like cell that releases histamine during an inflammatory response.

mechanoreceptor Sensory cell or nearby cell that detects mechanical energy (changes in pressure, position, or acceleration).

medulla oblongata Part of the hindbrain; it has reflex centers for vital tasks, such as respiration and circulation. It coordinates motor responses with complex reflexes, such as coughing. It influences brain regions concerned with sleep and arousal.

medusa (meh-DOO-sah) [Gk. *Medusa*, one of three sisters in Greek mythology having snake-entwined hair] Free-swimming, bell-shaped stage in cnidarian life cycles; oral arms and tentacles extend from the bell.

megaspore A haploid spore that forms in the ovary of seed-bearing plants; one of its cellular descendants develops into an egg.

meiosis (my-OH-sis) [Gk. *meioun*, to diminish] Two-stage nuclear division process that reduces the chromosome number of a germ cell by half, to the haploid number; each daughter nucleus receives one of each type of chromosome. It is the basis of gamete formation and (in plants) spore formation. *Compare* mitosis.

membrane excitability A special membrane property of any cell that can generate action potentials in response to stimulation.

memory The capacity of the brain to store and retrieve information about past sensory experience.

memory cell One of the subpopulations of B or T cells that form during an immune response to antigen but that do not act at once; it enters a resting phase, from which it is released during a secondary immune response.

menopause (MEN-uh-pozz) [L. *mensis*, month, + *pausa*, stop] End of the period of a human female's reproductive potential.

menstrual cycle Cyclic release of a secondary oocyte and priming of the endometrium to receive it, should it become fertilized; the complete cycle averages about twenty-eight days in female humans.

menstruation Cyclic sloughing of the blood-enriched endometrium (lining of the uterus) when pregnancy does not occur.

mesoderm (MEH-zoe-derm) [Gk. *mesos*, middle, + *derm*, skin] Of most animals, intermediate primary tissue layer that forms in the embryo and gives rise to muscle, most of the skeleton; circulatory, reproductive, and excretory organs; and connective tissue layers of the gut and integument.

mesophyll (MEH-zoe-fill) Of vascular plants, a tissue of photosynthetic parenchyma cells with an abundance of air spaces.

messenger RNA (mRNA) A single strand of ribonucleotides transcribed from DNA and translated into a polypeptide chain; the only RNA with protein-building instructions.

metabolic pathway (MEH-tuh-BALL-ik) One of many orderly sequences of enzyme-mediated reactions by which cells maintain, increase, or decrease the concentrations of substances. Different pathways are linear or circular, and one typically interconnects with others.

metabolism (meh-TAB-oh-lizm) [Gk. *meta*, change] All of the controlled, enzyme-mediated chemical reactions by which cells acquire and use energy to synthesize, store, degrade, and eliminate substances in ways that contribute to growth, survival, and reproduction.

metamorphosis (me-tuh-MOR-foe-sis) [Gk. *meta*, change, + *morphe*, form] Major change in body form during the transition from an embryo to the adult owing to hormonally controlled increases in size, reorganization of tissues, and remodeling of body parts.

metaphase Of meiosis I, the stage when all pairs of homologous chromosomes have become positioned midway between the spindle poles. Of mitosis or meiosis II, a stage when each duplicated chromosome has become positioned midway between the spindle poles.

metazoan A multicelled animal.

methanogen A type of archaebacterium that lives in oxygen-free habitats and produces methane gas as a metabolic by-product.

MHC marker Any of a variety of proteins that are self-markers. Some occur on all body cells of an individual; others are unique to the macrophages and lymphocytes.

micelle (my-SELL) Of fat digestion, a small droplet that is assembled from bile salts, fatty acids, and monoglycerides and that assists in fat absorption from the small intestine.

microevolution Change in allele frequencies resulting from mutation, genetic drift, gene flow, and natural selection.

microfilament [Gk. *mikros*, small, + L. *filum*, thread] Thin cytoskeletal element of two twisted polypeptide chains; it has roles in cell movement, especially at the cell surface, and in producing and maintaining cell shapes.

microorganism An organism, usually single-celled, that is far too small to be observed without the aid of a microscope.

microspore Of gymnosperms and flowering plants, a walled haploid spore that develops into a pollen grain.

microtubular spindle Of eukaryotic cells, a temporary bipolar structure composed of organized arrays of microtubules that forms during nuclear division and that moves the chromosomes to prescribed destinations.

microtubule (my-crow-TUBE-yool) A hollow, cylindrical cytoskeletal element, consisting mainly of tubulin, with roles in cell shape, motion, and growth and in the structure of cilia and flagella.

microtubule organizing center (MTOC) In the cytoplasm of eukaryotic cells, a small mass of proteins and other substances. The number, type, and location of MTOCs dictate the particular organization and orientation of microtubules inside the cell.

microvillus (MY-crow-VILL-us) [L. *villus*, shaggy hair] Of many cells, one of a number of very slender cylindrical extensions of the plasma membrane that serve in absorption or secretion.

midbrain Part of the vertebrate brain with coordination centers for reflex responses to visual and auditory input; also swiftly relays sensory signals to forebrain.

migration A recurring pattern of movement between two or more locations in response to environmental rhythms, such as circadian rhythms and seasonal changes in daylength. It requires activation or suppression of internal timing mechanisms that govern physiological and behavioral functions.

mimicry (MIM-ik-ree) In form, behavior, or both, a prey organism (the mimic) closely resembles a dangerous, unpalatable, or hard-to-catch species (its model).

mineral An element or inorganic compound formed by natural geologic processes and required for normal cell functioning.

mitochondrion (MY-toe-KON-dree-on), plural **mitochondria** Of eukaryotic cells, the organelle that specializes in formation of ATP; the second and third stages of aerobic respiration, an oxygen-requiring pathway, occur only in mitochondria.

mitosis (my-TOE-sis) [Gk. *mitos*, thread] Type of nuclear division that maintains the parental chromosome number for daughter cells; for eukaryotes, the basis of growth in size (and often of asexual reproduction).

mixture Two or more elements intermingled in proportions that can and usually do vary.

molar One of the cheek teeth; a tooth with a platform having cusps (surface bumps) that help crush, grind, and shear food.

molecular clock An accumulation of neutral mutations in a lineage that can be measured as a series of predictable "ticks" back through time; a way of calculating the time of origin of one lineage or species relative to others.

molecule A unit of matter in which chemical bonding holds together two or more atoms of the same or different elements.

mollusk An invertebrate with a tissue fold (mantle) draped around a soft, fleshy body; snails, clams, and squids are examples.

molting Shedding of hair, feathers, horns, epidermis, a shell, or cuticle in a process of increases in size or periodic renewal.

Monera In some traditional classification schemes, a kingdom that encompasses both archaebacteria and eubacteria.

monocot (MON-oh-kot) A monocotyledon; a flowering plant with seeds bearing one cotyledon, floral parts generally in threes (or multiples of three), and leaves typically parallel-veined. *Compare* dicot.

monohybrid cross [Gk. *monos*, alone] An experimental cross in which offspring inherit a pair of nonidentical alleles for a single trait being studied, so that they are heterozygous.

monomer Small molecule that is commonly a subunit of polymers, such as the sugar monomers of starch.

monosaccharide (MON-oh-SAK-ah-ride) [Gk. *monos*, alone, single, + *sakharon*, sugar] A simple carbohydrate, with only one sugar monomer. Glucose is an example.

monosomy A chromosome abnormality; the presence of a chromosome that has no homologue in a diploid cell.

morphogenesis (MORE-foe-JEN-ih-sis) [Gk. *morphe*, form, + *genesis*, origin] A program of orderly changes in an animal embryo's size, shape, and proportions, the outcome being specialized tissues and early organs. As part of the program, cells divide, grow, migrate, and change in size. Tissues expand and fold, and cells in some die in controlled ways at prescribed locations.

morphological convergence Lineages only remotely related evolved in response to similar environmental pressures, and they became similar in appearance, functions, or both. Analogous structures are evidence of this macroevolutionary pattern.

morphological divergence Of two or more lineages, change in appearance, functions, or both as they evolve from a shared ancestor. Homologous structures are evidence of this macroevolutionary pattern.

motor neuron A type of neuron that swiftly delivers signals from the brain or spinal cord to muscle cells, gland cells, or both.

multicelled organism An organism with differentiated cells arranged into tissues, organs, and often organ systems.

multiple allele system Three or more different molecular forms of the same gene (alleles) among individuals of a population.

muscle fatigue A decline in tension of a muscle kept in a state of tetanic contraction as a result of continuous, high-frequency stimulation.

muscle tension A mechanical force exerted by a contracting muscle; it resists opposing forces such as gravity and the weight of an object being lifted.

muscle tissue A tissue having cells able to contract under stimulation, then passively lengthen and return to the resting position.

mutagen (MEW-tuh-jen) Any environmental agent that can alter the molecular structure of DNA. Ultraviolet radiation and certain viruses are examples.

mutation [L. *mutatus*, a change, + *-ion*, result or a process or an act] A heritable change in the molecular structure of DNA. Mutations are the source of all alleles and, ultimately, of life's diversity. *See also* lethal mutation; neutral mutation.

mutation frequency The number of times a gene mutation has occurred in a population, as in 1 million gametes from which 500,000 individuals were produced.

mutation rate Of a gene, the probability of it undergoing spontaneous mutation during some specified interval, such as each DNA replication cycle.

mutualism [L. *mutuus*, reciprocal] A two-species, symbiotic ecological interaction that directly benefits both participants.

mycelium (my-SEE-lee-um), plural **mycelia** [Gk. *mykes*, fungus, mushroom, + *helos*, callus] A mesh of tiny, branching filaments (hyphae) that is the food-absorbing part of most fungi.

mycorrhiza (MY-coe-RIZE-uh) "Fungus-root"; a form of mutualism between the hyphae of a fungus and young roots of many plants. The fungus obtains carbohydrates from the plant and in turn releases dissolved mineral ions that plant roots can take up.

myelin sheath Of many sensory and motor neurons, an axonal sheath that enhances the propagation of action potentials; the plasma membranes of Schwann cells are wrapped repeatedly around the axon, and only a small node separates one from the other.

myofibril (MY-oh-FY-brill) One of the many threadlike structures inside a muscle cell; it is divided into many sarcomeres, the basic units of contraction.

myosin (MY-uh-sin) A motor protein with roles in contraction. In muscles, it interacts with actin to bring about contraction.

NAD⁺ Nicotinamide adenine dinucleotide; a nucleotide coenzyme. When carrying electrons and unbound protons (H⁺) between reaction sites, it is abbreviated NADH.

NADP⁺ Nicotinamide adenine dinucleotide phosphate; a phosphorylated nucleotide coenzyme. When carrying electrons and unbound protons (H⁺) between reaction sites, it is abbreviated NADPH₂.

natural killer cell A cytotoxic lymphocyte that reconnoiters for tumor cells and virus-infected cells, then touch-kills them.

natural selection A microevolutionary process; outcome of differences in survival and reproduction among individuals that show variation in heritable traits. Over the generations, it can lead to increased fitness (increased adaptation to the environment).

necrosis (neh-CROW-sis) Of multicelled organisms, the passive death of many cells that results from severe tissue damage.

negative feedback mechanism Homeostatic feedback mechanism in which an activity changes some condition in the internal environment; the altered condition triggers a response that reverses the change.

nematocyst (NEM-ad-uh-sist) [Gk. *nema*, thread, + *kystis*, pouch] A capsule formed only by cnidarians; it houses dischargeable, tubular threads that have prey-piercing barbs and that dispense toxins or sticky substances.

nephridium (neh-FRID-ee-um), plural **nephridia** Of earthworms and some other invertebrates, a system of regulating water and solute levels.

nephron (NEFF-ron) [Gk. *nephros*, kidney] Of the vertebrate kidney, a slender tubule into which water and solutes are filtered from blood, then selectively reabsorbed, and in which urine forms.

nerve Cordlike communication line of the peripheral nervous system, composed of axons of sensory neurons, motor neurons, or both bundled in connective tissue. In the brain and spinal cord, similar cord-like bundles are called nerve tracts.

nerve cord Of many animals, a cordlike communication line of axons of neurons.

nerve impulse *See* action potential.

nerve net A simple nervous system of a diffuse mesh of nerve cells that interact with contractile and often sensory cells in an epithelium.

nervous system An organ system in which nerve cells, such as neurons, are oriented relative to one another in signal-conducting and information-processing pathways. It detects and processes information about changes outside and inside the body and elicits responses from muscle and gland cells.

nervous tissue A type of connective tissue composed of neurons.

net energy Of energy resources available to the human population, the amount of energy that is left over after subtracting the energy used to locate, extract, transport, store, and deliver energy to consumers.

net population growth rate per individual (*r*) Of population growth equations, a single variable combining birth and death rates, which are assumed to remain constant.

neural tube Embryonic forerunner of the brain and spinal cord.

neuroglial cell (NUR-oh-GLEE-uhl) One of the cells that structurally and metabolically support neurons. Neuroglia represent about half the volume of the vertebrate nervous system.

neuromodulator A signaling molecule that can magnify or reduce the effects of a given neurotransmitter on neighboring or distant neurons.

neuromuscular junction A site of chemical synapsing between the axonal endings of a motor neuron and a muscle cell.

neuron (NUR-on) A nerve cell; the basic unit of communication in nervous systems. *See* motor neuron; interneuron; sensory neuron.

neurotransmitter Any of a class of signaling molecules that are secreted from neurons and act on immediately adjacent cells, then are rapidly degraded or recycled.

neutral mutation A mutation with little or no effect on phenotype; natural selection cannot increase or decrease its frequency in a population, for it cannot influence an individual's survival and reproduction. *Compare* molecular clock.

neutron A unit of matter, one or more of which occupies the atomic nucleus and has mass but no electric charge.

neutrophil The most abundant, fast-acting white blood cell; it phagocytizes bacteria during an inflammatory response.

niche (NITCH) [L. *nidas*, nest] The sum total of all activities and relationships in which individuals of a species engage as they secure and use the resources necessary to survive and reproduce.

nitrification (nye-trih-fih-KAY-shun) Of the nitrogen cycle, a process by which certain bacteria strip electrons from ammonia or ammonium in soil. The end product, nitrite (NO_2^-), is broken down to nitrate (NO_3^-) by other bacteria.

nitrogen cycle An atmospheric cycle; a biogeochemical cycle in which the largest nitrogen reservoir is the atmosphere.

nitrogen fixation Process by which a few kinds of bacteria convert gaseous nitrogen (N_2) to ammonia. This dissolves rapidly in their cytoplasm to form ammonium, which can be used in biosynthetic pathways.

nociceptor (NO-SEE-sep-tur) A pain receptor, such as a free nerve ending that detects tissue damage as by burns or distortions.

node A site where one or more leaves are attached to a plant stem.

noncyclic pathway of ATP formation (non-SIK-lik) [L. *non*, not, + Gk. *kylos*, circle] Photosynthetic pathway of ATP formation in which electrons from water molecules flow through two photosystems and two electron transport chains, and end up in NADPH.

nondisjunction The failure of two sister chromatids or of a pair of homologous chromosomes to separate at meiosis or mitosis, so that daughter cells end up with too many or too few chromosomes.

notochord (KNOW-toe-kord) Of chordates, a rod of stiffened tissue (not cartilage or bone) that serves as a supporting structure for the body.

nuclear envelope A double membrane (two lipid bilayers and associated proteins) that is the outermost portion of a cell nucleus.

nucleic acid (new-CLAY-ik) A single-stranded or double-stranded chain of four different kinds of nucleotides joined one after the other at their phosphate groups. They differ in which nucleotide base follows the next in sequence. DNA and RNA are examples.

nucleic acid hybridization Single strands of DNA or RNA from different species are recombined into double-stranded, hybrid molecules, which are then heated to break hydrogen bonds between them. The amount of heat required to separate the strands is a measure of biochemical similarity, which is greatest among closely related species and weakest among distantly related ones.

nucleoid (NEW-KLEE-oid) Of bacteria, the region of the cell in which the bacterial chromosome is physically organized (but not bounded by membrane, as with the nuclear envelope of eukaryotic cells).

nucleolus (new-KLEE-oh-lus) [L. *nucleolus*, a little kernel] In the nucleus of a nondividing cell, a site where the protein and RNA subunits of ribosomes are assembled.

nucleosome (NEW-KLEE-oh-sohm) One of the organizational units of the eukaryotic chromosome; a stretch of DNA looped twice around a "spool" of histone molecules.

nucleotide (NEW-KLEE-oh-tide) A small organic compound with a five-carbon sugar (deoxyribose), nitrogen-containing base, and phosphate group. Nucleotides are the structural units of adenosine phosphates, nucleotide coenzymes, and nucleic acids.

nucleotide coenzyme A protein that assists an enzyme by transporting electrons and hydrogen atoms released at one reaction site to another site.

nucleus (NEW-klee-us) [L. *nucleus*, a kernel] Of atoms, the central core of one or more positively charged protons and (in all but hydrogen) electrically neutral neutrons. In eukaryotic cells, a membranous organelle that physically isolates and organizes DNA out of the way of cytoplasmic machinery.

numerical taxonomy Practice by which traits of an unidentified organism are compared with those of a known group. The greater the total number of traits that it has in common with the known group, the closer is their inferred relatedness.

nutrient An element essential for growth and survival of a given organism; directly or indirectly, it has one or more roles in metabolism that no other element fulfills.

nutrition All of those processes by which food is selectively taken in, digested, absorbed, and later converted to the body's own organic compounds.

obesity An excess of fat in the body's adipose tissues, caused by imbalances between caloric intake and energy output.

oligosaccharide A carbohydrate consisting of a short chain of two or more covalently bonded sugar monomers. The type called a disaccharide has two sugar monomers. *Compare* monosaccharide; polysaccharide.

omnivore [L. *omnis*, all, + *vovare*, to devour] An animal able to obtain energy and carbon from more than one food source rather than being limited to one trophic level.

oncogene (ON-koe-jeen) Any gene having the potential to induce cancerous transformation.

oocyte An immature egg.

oogenesis (oo-oh-JEN-uh-sis) Formation of a female gamete, from a germ cell to a haploid ovum (mature egg).

operator A short base sequence intervening between a promoter and bacterial genes.

operon A promoter-operator sequence that services more than one bacterial gene.

orbital One of a number of volumes of space around the atomic nucleus in which electrons are likely to be at any instant.

organ A structure having definite form and function that consists of more than one tissue.

organ formation A stage of embryonic development in animals in which primary tissue layers split into subpopulations of cells, and different lines of cells become unique in structure and function; basis of growth and tissue specialization.

organ system Two or more organs that interact chemically, physically, or both in performing a common task.

organelle Of cells, an internal, membrane-bounded sac or compartment having one or more specific, specialized metabolic functions.

organic compound A molecule consisting of one or more elements covalently bonded to some number of carbon atoms.

osmoreceptor A sensory receptor that detects changes in water volume (hence solute concentration) in fluid that bathes it.

osmosis (oss-MOE-sis) [Gk. *osmos*, act of pushing] The diffusion of water in response to a water concentration gradient between two regions that are separated from each other by a selectively permeable membrane. The greater the number of molecules and ions dissolved in a solution, the lower its water concentration will be.

osmotic pressure At some point following the development of hydrostatic pressure in a cell, an amount of force that counters the inward diffusion of water and prevents any further increase in fluid volume.

ovary (oh-vuh-ree) A type of primary reproductive organ in which eggs form. In seed-bearing plants, the part of the carpel where eggs develop, fertilization takes place, and seeds mature; at maturity, it and sometimes other plant parts is a fruit.

oviduct (OH-vih-dukt) One of a pair of ducts through which eggs travel from an ovary to the uterus. Formerly called Fallopian tube.

ovulation (AHV-you-LAY-shun) The release of a secondary oocyte from an ovary during a menstrual cycle.

ovule (OHV-youl) [L. *ovum*, egg] Of seed-bearing plants, a female gametophyte with egg cell, surrounding tissue, and one or two protective layers (integuments). After fertilization, it matures into a seed.

ovum (OH-vum) Of vertebrates, the mature female gamete (egg).

oxidation-reduction reaction An electron transfer from one atom or molecule to another. Often hydrogen is also transferred along with an electron or electrons.

ozone thinning A pronounced seasonal thinning of the ozone layer in the lower stratosphere, especially above polar regions.

pancreas (PAN-cree-us) Gland that secretes enzymes and bicarbonate into the small intestine during digestion, and that secretes the hormones insulin and glucagon.

pancreatic islet Any of the 2 million clusters of endocrine cells in the pancreas, including alpha cells, beta cells, and delta cells.

parapatric speciation Neighbor populations become distinct species while maintaining contact along their common border.

parasite [Gk. *para*, alongside, + *sitos*, food] An organism that lives in or on other living organisms (hosts), feeds on specific tissues during part of its life cycle, and usually does not kill its host outright.

parasitism A two-species interaction in which one species lives in or on a host species and uses its tissues for nutrients. The parasite benefits; the host does not.

parasitoid An insect larva that grows and develops inside a host organism (usually another insect), eventually consuming the soft tissues and killing it.

parasympathetic nerve Of the autonomic nervous system, any of the nerves carrying signals that tend to slow down the body's activities and divert energy to basic tasks; such nerves work continually in opposition with sympathetic nerves to make small adjustments in internal organ activity.

parenchyma (par-ENG-kih-mah) A simple tissue that makes up the bulk of the plant body. Its cells take part in photosynthesis, storage, secretion, and other tasks.

parthenogenesis Development of an embryo from an unfertilized egg.

passive transport A transport protein that spans the lipid bilayer of a cell membrane passively allows a solute to diffuse through its interior, down its concentration gradient.

pathogen (PATH-oh-jen) [Gk. *pathos*, suffering, + *-genēs*, origin] A virus, bacterium, fungus, protozoan, parasitic worm, or some other infectious agent that can invade organisms, multiply in them, and cause disease.

pattern formation Of animals, a sculpting of nondescript clumps of embryonic cells into specialized tissues and organs according to an ordered, spatial pattern.

PCR Polymerase chain reaction; a method of enormously amplifying the quantity of DNA fragments cut by restriction enzymes.

peat bog An accumulation of the remains of peat mosses in an compressed, excessively moist, and highly acidic mat.

pedigree A chart of genetic connections among related individuals, constructed according to standardized methods.

pelagic province The full volume of ocean water; it is subdivided into a neritic zone (relatively shallow water over continental shelves) and an oceanic zone (water over ocean basins).

penis A male copulatory organ by which sperm can be deposited inside a female reproductive tract.

peptide hormone A protein hormone or some other water-soluble hormone unable to cross the lipid bilayer of a target cell. It enters by receptor-mediated endocytosis or activates cell surface receptors.

perennial [L. *per-*, throughout, + *annus*, year] A flowering plant that lives for more than two growing seasons.

pericycle (PARE-ih-sigh-kul) [Gk. *peri-*, around, + *kyklos*, circle] Of a root vascular cylinder, one or more layers just inside the endodermis that gives rise to lateral roots and contributes to secondary growth.

periderm Of plants that show extensive secondary growth, a protective covering that replaces epidermis.

peripheral nervous system (per-IF-ur-uhl) [Gk. *peripherein*, to carry around] The nerves leading into and out from a spinal cord and brain and the ganglia along those communication lines.

peristalsis (pare-ih-STAL-sis) Waves of contraction and relaxation of muscles in the wall of a tubular or saclike organ.

peritoneum (pare-ih-tah-NEE-um) A lining of the coelom that also covers and helps maintain the position of internal organs.

peritubular capillary One of the set of blood capillaries around the tubular parts of a nephron; it functions in reabsorption of water and solutes back into the body and in secretion of hydrogen ions and some other substances in the forming urine.

permafrost A permanently frozen, water-impenetrable layer beneath the soil surface in arctic tundra.

peroxisome Enzyme-filled vesicle in which fatty acids and amino acids are digested into hydrogen peroxide and converted to harmless products.

pest resurgence Insecticide resistance arises among a population of pests; the insecticide acts as an agent of directional selection.

PGA Phosphoglycerate (FOSS-foe-GLISS-er-ate) A key intermediate of glycolysis and of the Calvin-Benson cycle.

PGAL Phosphoglyceraldehyde. A key intermediate of glycolysis and of the Calvin-Benson cycle.

pH scale A measure of the concentration of free hydrogen ions in blood, water, and other solutions; pH 0 is the most acidic, 14 the most basic, and 7, neutral.

phagocyte (FAG-uh-sight) [Gk. *phagein*, to eat, + *kytos*, hollow vessel] A cell that obtains nutrients or destroys foreign cells and cellular debris by phagocytosis. The macrophages and amoeboid protozoans are examples.

phagocytosis (FAG-uh-sigh-TOE-sis) [Gk. *phagein*, to eat, + *kytos*, hollow vessel] The engulfment of foreign cells or substances by means of pseudopod formation and then endocytosis.

pharynx (FARE-inks) A muscularized tube by which food enters the gut. Also an organ of gas exchange in some chordates; in land vertebrates, a dual entrance for the tubular part of the digestive tract and windpipe.

phenotype (FEE-no-type) [Gk. *phainein*, to show, + *typos*, image] Observable trait or traits of an individual that arise from the interactions between genes, and between genes and the environment.

pheromone (FARE-oh-moan) [Gk. *phero*, to carry, + *-mone*, as in hormone] A nearly odorless, hormone-like, exocrine gland secretion that is a communication signal between individuals of the same species and that helps integrate social behavior.

phloem (FLOW-um) Of vascular plants, a tissue with living cells that interconnect and form the tubes through which sugars and other dissolved organic compounds are conducted.

phospholipid An organic compound with a glycerol backbone, two fatty acid tails, and a hydrophilic head (a phosphate group and another polar group). Phospholipids are the main structural materials of cell membrane.

phosphorus cycle Movement of phosphorus from rock or soil through organisms, then back to soil.

phosphorylation (FOSS-for-ih-LAY-shun) The attachment of unbound (inorganic) phosphate to a molecule; also the transfer of a phosphate group from one molecule to another, as when ATP phosphorylates glucose.

photoautotroph An organism able to synthesize all organic molecules it requires using carbon dioxide as the carbon source and sunlight as the energy source. All plants, some protistans, and a few bacteria are photoautotrophs.

photolysis (foe-TALL-ih-sis) [Gk. *photos*, light, + *-lysis*, breaking apart] A reaction sequence of the noncyclic pathway of photosynthesis, triggered by photon energy, in which water is split into oxygen, hydrogen, and electrons.

photoperiodism A biological response to a change in the relative length of daylight and darkness.

photoreceptor Light-sensitive sensory cell.

photosynthesis The trapping of energy from the sun and its conversion to chemical energy (ATP, NADPH, or both), followed by synthesis of sugar phosphates that become converted to sucrose, cellulose, starch, and other products. The main pathway by which energy and carbon enter the web of life.

photosystem One of the clusters of light-trapping pigments that are embedded in photosynthetic membranes and that donate electrons to transport systems that are involved in the formation of ATP, NADPH, or both. Examples are photosystems II and I of the thylakoid membrane of chloroplasts.

phototropism [Gk. *photos*, light, + *trope*, turning, direction] An adjustment in the direction and rate of plant growth in response to a source of light.

photovoltaic cell A device that can convert sunlight energy into electricity.

phycobilin (FIE-koe-BY-lins) One of the light-sensitive, accessory pigments that transfer energy they absorb to chlorophylls. The phycobilins are especially abundant in red algae and cyanobacteria.

phylogeny Evolutionary relationships among species, starting with most ancestral forms and including the branches leading to their descendants.

phytochrome Light-sensitive pigment, the activation and inactivation of which triggers plant hormone activities governing leaf expansion, stem branching, stem length and often seed germination and flowering.

phytoplankton (FIE-toe-PLANK-tun) [Gk. *phyton*, plant, + *planktos*, wandering] A community of floating or weakly swimming photoautotrophs in saltwater and freshwater habitats.

pigment A light-absorbing molecule.

pioneer species An opportunistic colonizer of barren or disturbed habitats, notable for its high dispersal rate and rapid growth.

pituitary gland An endocrine gland that interacts with the hypothalamus to control diverse physiological functions, including activity of many other endocrine glands. The posterior lobe stores and secretes hypothalamic hormones; the anterior lobe produces and secretes its own hormones.

placenta (play-SEN-tuh) A blood-engorged organ, consisting of endometrial tissue and extraembryonic membranes, that develops during pregnancy in the female placental mammal. It permits exchanges between the mother and her fetus without intermingling their bloodstreams, thus sustaining the new individual and allowing its blood vessels to develop apart from the mother's.

plankton [Gk. *planktos*, wandering] Any community of floating or weakly swimming organisms, mostly microscopic, living in freshwater and saltwater habitats. *See also* phytoplankton; zooplankton.

plant One of the eukaryotic organisms, nearly all of which are photoautrophic, that have chlorophylls *a* and *b* and that generally have well-developed root and shoot systems.

Plantae The kingdom of plants.

plasma (PLAZ-muh) Liquid component of blood; water and dissolved ions, diverse proteins, sugars, gases, and other substances.

plasma membrane Of cells, the outermost membranous boundary between cytoplasm and external fluid bathing the cell. Its lipid bilayer is the basic structural part of the membrane; diverse proteins embedded in the bilayer or attached to its surfaces carry out most functions, such as transport and reception of extracellular signals.

plasmid Of many bacteria, a small, circular molecule of extra DNA that carries only a few genes and that replicates independently of the bacterial chromosome.

plasmodesma, plural **plasmodesmata** (PLAZ-moe-DEZ-muh) A plant cell junction; a membrane-lined channel across walls of adjacent cells that connects their cytoplasm.

plasmolysis An osmotically induced shrinkage of a cell's cytoplasm.

plate tectonics Theory that slabs, or plates, of the Earth's outer layer (lithosphere) float on a hot, plastic layer of the underlying mantle. All plates are in motion and they raft the continents to new positions over great spans of geologic time.

platelet (PLAYT-let) Any of the fragments of megakaryocytes that release substances necessary for clot formation.

pleiotropy (PLEE-oh-troe-pee) [Gk. *pleon*, more, + *trope*, direction] As a result of expression of alleles at a single gene locus, positive or negative effects on two or more traits. The effects may or may not be manifest at the same time in the individual.

polar body One of four cells that form during the meiotic cell division of an oocyte but that does not become the ovum.

pollen grain [L. *pollen*, fine dust] Depending on the species, the immature or mature, sperm-bearing male gametophyte of gymnosperms and angiosperms.

pollen sac In anthers of flowers, any of the chambers in which pollen grains develop.

pollen tube A tube formed after a pollen grain germinates; grows through carpel tissues, and carries sperm to the ovule.

pollination Of flowering plants, the arrival of a pollen grain on the landing platform (stigma) of a carpel.

pollutant Any substance with which an ecosystem has had no prior evolutionary experience, in terms of kinds or amounts, and that can accumulate to disruptive or harmful levels. Can be naturally occurring or synthetic.

polymer (POH-lih-mur) [Gk. *polus*, many, + *meris*, part] A large molecule with three to millions of subunits.

polymerase (puh-LIM-ur-aze) An enzyme that catalyzes a polymerization reaction. Examples are DNA polymerase (of DNA replication and repair), RNA polymerase (of gene transcription), and the enzyme that catalyzes cellulose synthesis.

polypeptide chain Organic compound with nitrogen atoms arrayed in a regular pattern in its backbone (-N-C-C-N-C-C-). A protein consists of one or more of these chains.

polymorphism (poly-MORE-fizz-um) [Gk. *polus*, many, + *morphe*, form] Of populations, the persistence of two or more distinctive forms of a trait (morphs).

polyp (POH-lip) Vase-shaped, sedentary stage of cnidarian life cycles.

polypeptide chain Three or more amino acids joined by peptide bonds.

polyploidy (POL-ee-PLOYD-ee) Having three or more of each type of chromosome in the interphase nucleus, compared to a diploid chromosome number.

polysaccharide [Gk. *polus*, many, + *sakcharon*, sugar] A straight or branched chain of many covalently linked sugar units, of the same or different kinds. The most common are cellulose, starch, and glycogen.

polysome A number of ribosomes translating the same mRNA molecule one after the other, at the same time, during protein synthesis.

pons Part of the hindbrain; a traffic center for signals passing between the cerebellum and integrating centers of the forebrain.

population Individuals of the same species occupying the same area.

population density A count of individuals of a population occupying a specified area or volume of a habitat.

population distribution The general pattern of dispersion of individuals of a population throughout their habitat.

population size The number of individuals that make up the gene pool of a population.

positive feedback mechanism Homeostatic mechanism by which a chain of events is set in motion that intensifies a change from an original condition; after a limited time, the intensification reverses the change.

predation An ecological interaction in which a predatory species eats (and so directly harms) a prey species.

predator [L. *prehendere*, to grasp, seize] An organism that feeds on other living organisms (its prey), that does not live in or on them, and that may or may not kill them.

prediction A statement about what you can expect to observe in nature if a theory or hypothesis is not false.

pressure flow theory Of vascular plants, a theory that organic compounds move through phloem because of gradients in solute concentrations and pressure between sources (such as photosynthetically active leaves) and sinks (such as growing parts).

primary growth Plant growth originating at root tips and shoot tips and resulting in increases in length.

primary immune response Defensive acts by white blood cells and their secretions, as elicited by a first-time encounter with an antigen; includes both antibody-mediated and cell-mediated activities.

primary productivity, gross Of ecosystems, the rate at which producers capture and store a given amount of energy in their cells and tissues during a specified interval.

primary productivity, net Of ecosystems, the rate of energy storage in the cells and tissues of producers in excess of their rate of aerobic respiration.

primate A mammalian lineage that includes the prosimians, tarsioids, and anthropoids (monkeys, apes, and humans).

primer A short nucleotide sequence that will base-pair with any complementary DNA sequence; later, DNA polymerases recognize them as start tags for replication.

prion A small protein linked with eight rare, fatal degenerative diseases of the nervous system. Prions are altered products of a gene that is present in normal as well as infected individuals.

probability The chance that each outcome of a given event will occur is proportional to the number of ways it can be reached.

procambium (pro-KAM-bee-um) Of vascular plants, a primary meristem that gives rise to primary vascular tissues.

producer Of ecosystems, an autotrophic (self-feeding) organism; it nourishes itself using sources of energy and carbon from its physical environment. Photoautorophs and chemoautotrophs are examples.

progesterone (pro-JESS-tuh-rown) A type of sex hormone secreted by the pair of ovaries in female mammals.

prokaryotic cell (pro-CARRY-oh-tic) [L. *pro*, before, + Gk. *karyon*, kernel] A bacterium; a single-celled organism, usually walled, that does not have the profusion of membrane-bound organelles seen in eukaryotic cells.

prokaryotic fission Division mechanism by which a bacterial cell reproduces.

promoter A short base sequence in DNA that signals the start of a gene; the site where RNA polymerase binds to start transcription.

prophase Of mitosis, the stage when each duplicated chromosome starts to condense, microtubules form a spindle apparatus, and the nuclear envelope starts to break up.

prophase I Of meiosis, the stage when the microtubular spindle starts to form, the nuclear envelope starts to break up, and each duplicated chromosome condenses and pairs with its homologous partner. Nonsister chromatids typically undergo crossing over and genetic recombination.

prophase II Brief first stage of meiosis II when chromosomes start moving toward spindle equator; each has already been separated from its homologue but still consists of two chromatids.

protein Organic compound of one or more chains of amino acids, which peptide bonds join in a linear sequence that is the basis of the protein's three-dimensional structure and chemical behavior.

Protista The kingdom of protistans.

protistan (pro-TISS-tun) [Gk. *prōtistos*, primal, very first] One of the staggeringly diverse eukaryotes, single-celled species of which may resemble the first eukaryotic cells. At present categorized in part by what they are *not* (not bacteria, fungi, plants, or animals).

proto-oncogene A gene sequence similar to an oncogene but that codes for a protein required for normal cell function. When a mutation alters its structure, it may trigger cancerous transformation.

proton Positively charged particle, one or more of which resides in the nucleus of each type of atom; when unbound, a proton is the same as a hydrogen ion (H^+).

protostome (PRO-toe-stome) [Gk. *proto*, first, + *stoma*, mouth] One of the lineage of coelomate, bilateral animals in which the first indentation to form in the embryo develops into a mouth. Includes mollusks, annelids, and anthropods.

protozoan A group of protistans, some predatory, others parasitic, that in certain respects resemble single-celled heterotrophs that presumably gave rise to animals. The amoeboid and animal-like protozoans, as well as sporozoans, are examples.

proximal tubule Of a nephron, the tubular portion into which water and solutes enter after being filtered from the blood at the preceding portion (Bowman's capsule).

pulmonary circuit Blood circulation route leading to and from the lungs.

Punnett-square method A way to predict the probable outcome of a genetic cross in simple diagrammatic form.

purine A type of nucleotide base with a double ring structure. Examples are adenine and guanine.

pyrimidine (phi-RIM-ih-deen) A type of nucleotide bases with a single ring structure. Examples are cytosine and thymine.

pyruvate (PIE-roo-vate) A small organic compound with a backbone of three carbon atoms. Glycolysis produces two molecules of pyruvate as end products.

r A variable used in population growth equations to signify net population growth rate; the assumption is that birth and death rates remain constant and therefore may be combined into this one variable.

radial symmetry Of animals, a body plan having four or more roughly equivalent parts arranged around a central axis.

radiometric dating A method of measuring the proportions of (1) a radioisotope in a mineral rock trapped long ago in a newly formed rock and (2) a daughter isotope that formed from it by radioactive decay in the same rock sample. Used to assign absolute dates to fossil-containing rocks and to a geologic time scale.

radioisotope An unstable atom, with an uneven number of protons and neutrons. It spontaneously emits particles and energy, and over a predictable time span becomes transformed (decays) into a different atom.

rain shadow A reduction in rainfall on the leeward side of high mountains, resulting in arid or semiarid conditions.

reabsorption In kidneys, diffusion or active transport of water and reclaimable solutes from a nephron into peritubular capillaries under the control of ADH and aldosterone. At a capillary bed, osmotic movement of some interstitial fluid into a capillary when there is a difference in the concentration of water between plasma and interstitial fluid.

rearrangement, molecular A conversion of one type of organic compound to another through a juggling of its internal bonds.

receptor, molecular Of cells, a membrane protein that binds a hormone or some other extracellular substance that triggers change in cell activities.

receptor, sensory Of nervous systems, a sensory cell or specialized cell adjacent to it that can detect a particular stimulus.

recessive allele [L. *recedere*, to recede] In heterozygotes, an allele whose expression is fully or partially masked by expression of its partner; only in the homozygous recessive condition is it fully expressed.

recognition protein Of mammalian cell membranes, a glycolipid or glycoprotein that functions as a molecular fingerprint; it identifies a cell as being of a specific type. Self-proteins of the body's own cells are an example.

recombinant technology Procedures by which DNA (genes) from different species may be isolated, cut, spliced together, and the new recombinant molecules multiplied in quantity, as by PCR or by a population of rapidly dividing bacterial cells.

red alga A protistan, most of which are multicelled, aquatic, and photosynthetic, with an abundance of phycobilins.

blood cell An erythrocyte; an oxygen-transporting cell in blood.

red marrow A site of blood cell formation in the spongy tissue of many bones.

reflex [L. *reflectere*, to bend back] A simple, stereotyped movement elicited directly by sensory stimulation.

reflex pathway [L. *reflectere*, to bend back] The simplest route by which stimulation of nerve cells leads to contractile or glandular action. In complex animals, sensory neurons directly stimulate or inhibit motor neurons, without intervention by interneurons.

refractory period Of neurons, the period following an action potential at a given patch of membrane when sodium gates are shut and potassium gates are open, so that the patch is insensitive to stimulation.

regulatory protein A protein involved in a positive or negative control system that operates during transcription or translation or after translation. The protein interacts with DNA, RNA, or a gene product.

releasing hormone A signaling molecule produced by the hypothalamus that enhances or slows down secretions from target cells in the anterior lobe of the pituitary gland.

repressor Regulatory protein that can block gene transcription.

reproduction Of nature, any of a number of processes by which a new generation of cells or multicelled individuals is produced. For eukaryotes, it may be sexual or it may be asexual (as by binary fission, budding, or vegetative propagation). For bacteria only, it occurs by prokaryotic fission.

reproductive isolating mechanism Any heritable feature of body form, functioning, or behavior that prevents interbreeding between one or more genetically divergent populations.

reproductive success The survival and production of offspring by an individual.

reptile A type of carnivorous vertebrate; its ancestors were the first vertebrates to escape dependency on standing water, largely by means of internal fertilization and amniote eggs. Examples are dinosaurs (extinct), crocodiles, lizards, and snakes.

resource partitioning Coexistence among competing species; they share the same resource differently or at different times.

respiration [L. *respirare*, to breathe] Of animals, the exchange of oxygen from the environment for carbon dioxide wastes from cells by way of integumentary exchange, a respiratory system, or both. *Compare* aerobic respiration.

respiratory surface An epithelium or some other surface tissue that functions in gas exchange in animals; it is thin enough to allow oxygen to diffuse into the body and carbon dioxide to diffuse out of it.

resting membrane potential Of a neurons and other excitable cells, a steady voltage difference across the plasma membrane in the absence of outside stimulation.

restoration ecology Attempts to reestablish biodiversity in abandoned farmlands and other large areas altered by agriculture, mining, and other severe disturbances.

restriction enzyme Of bacteria, one of a class of enzymes that can cut apart foreign DNA injected into the cell body, as by viruses; also used in recombinant DNA technology.

reticular formation A mesh of interneurons extending from the upper spinal cord, through the brain stem, and into the cerebral cortex; a low-level pathway through the vertebrate nervous system.

reverse transcriptase A viral enzyme that uses viral RNA as a template to synthesize either DNA or mRNA in a host cell.

reverse transcription A process by which reverse transcriptase assembles DNA on an RNA template. Used by RNA viruses as well as by technologists who use mature mRNA transcripts as the templates for synthesizing cDNA.

RFLP Short for restriction fragment length polymorphism. Such fragments are formed by using restriction enzymes to cut DNA; for each individual, they show slight but unique differences in their banding pattern.

Rh blood typing A laboratory method of characterizing red blood cells on the basis of a protein that serves as a self-marker at their surface; Rh^+ signifies its presence, and Rh^- its absence.

rhizoid A rootlike absorptive structure of some fungi and nonvascular plants.

ribosomal RNA (rRNA) A type of RNA molecule that combines with proteins to form ribosomes, on which the polypeptide chains of proteins are assembled.

ribosome In all cells, the structure at which amino acids are strung together in specified sequence to form the polypeptide chains of proteins. An intact ribosome consists of two subunits, each composed of ribosomal RNA and protein molecules.

RNA Ribonucleic acid. A general category of single-stranded nucleic acids that function in processes by which genetic instructions encoded in DNA are used to build proteins.

rod cell A vertebrate photoreceptor that is sensitive to very dim light and contributes to coarse perception of movement.

root hair Of vascular plants, an extension of a specialized root epidermal cell; root hairs collectively enhance the surface area available for absorbing water and solutes.

root nodule A localized swelling on roots of certain legumes and other plants that contain symbiotic, nitrogen-fixing bacteria.

root A part of a plant that typically grows belowground, absorbs water and dissolved nutrients, helps anchor aboveground parts, and often stores food.

RuBP Ribulose bisphosphate. A compound with a backbone of five carbon atoms; used for carbon fixation by C3 plants and also regenerated in the Calvin-Benson cycle of photosynthesis.

S-shaped curve A curve, obtained when population size is plotted against time, that is characteristic of logistic growth.

salination A salt buildup in soil as a result of evaporation, poor drainage, and heavy irrigation.

salivary gland A gland that secretes saliva, a fluid that initially mixes with food in the mouth and starts the breakdown of starch.

salt A compound that releases ions other than H^+ and OH^- in solution.

saltatory conduction Of myelinated neurons, a rapid form of action potential propagation by a hopping of excitation from one node to another between the jellyrolled membranes of cells making up the myelin sheath.

sampling error A rule of probability: The fewer times that a chance event occurs, the greater will be the variance from the expected outcome of that occurrence. Of experimental groups, large sampling tends to lessen the chance that differences among individuals will distort the results.

saprobe A heterotroph that obtains energy and carbon from nonliving organic matter and so causes its decay. Saprobic fungi are examples.

sarcomere (SAR-koe-meer) Of muscles, the basic unit of contraction; a region of myosin and actin filaments organized in parallel arrays between two Z lines of one of the many myofibrils inside a muscle cell.

sarcoplasmic reticulum (sar-koe-PLAZ-mik reh-TIK-you-lum) A membrane system around a muscle cell's myofibrils that takes up, stores, and releases the calcium ions required for cross-bridge formation in sarcomeres, hence for contraction.

Schwann cell One of a series of neuroglial cells wrapped like jellyrolls around an axon in a nerve and forming a myelin sheath.

sclerenchyma (skler-ENG-kih-mah) A simple plant tissue in which most cells have thick, lignin-impregnated walls. Sclerenchyma supports mature plant parts and commonly protects seeds.

sea-floor spreading Molten rock erupts from great, continuous ridges on the ocean floor, flows laterally in both directions, and hardens to form new crust. As the seafloor spreads out, it forces older crust down into great trenches elsewhere in the seafloor.

second law of thermodynamics A law of nature stating that the spontaneous direction of energy flow is from organized to less organized forms. Overall, the total amount of energy in the universe is spontaneously flowing from forms of higher to lower quality; with each conversion, some energy gets randomly dispersed in a form (usually heat) not as readily available to do work.

second messenger A molecule within a cell that mediates a hormonal signal by triggering a response to it.

secondary immune response An immune response to previously encountered antigen that is more rapid and prolonged than the first owing to the swift participation of memory cells.

secondary sexual trait A trait associated with maleness or femaleness but one with no direct role in reproduction.

secretion A product released across the plasma membrane of a cell that may act singly or as part of glandular tissue.

sedimentary cycle A biogeochemical cycle with no gaseous phase; the element moves from land to the seafloor, then returns only through long-term geological uplifting.

seed Of gymnosperms and angiosperms, a fully mature ovule (contains the plant embryo) with integuments that form the seed coat.

segmentation Of animals, a body plan with a repeating series of units that may or may not be similar to one another in appearance. Also, an oscillating movement created by rings of circular muscle in a tubular wall. In a small intestine, such movement constantly mixes and forces the contents of the lumen against the absorptive wall surface.

segregation, theory of [L. *se-*, apart, + *grex*, herd] A Mendelian theory that diploid organisms inherit pairs of genes for traits (on pairs of homologous chromosomes) and that the two genes segregate during meiosis and end up in separate gametes.

selective gene expression Of a multicelled organism, the activation and suppression of some fraction of the same inherited genes in different populations of cells; it leads to cell differentiation.

selective permeability Of a cell membrane, a capacity to allow some substances but not others to cross it at certain sites, at certain times, owing to its molecular structure.

selfish behavior A behavior by which an individual protects or increases its own chance of producing offspring, regardless of the consequences to the group to which it belongs.

selfish herd A society held together simply by reproductive self-interest.

semen (SEE-mun) [L. *serere*, to sow] Sperm-bearing fluid expelled from a penis during male orgasm.

semiconservative replication [Gk. *hēmi*, half, + L. *conservare*, to keep] How a DNA molecule duplicates itself. A DNA double helix unzips and a complementary strand is assembled on the exposed bases of each strand. Each conserved strand and its new partner are wound together into a double helix. Thus the outcome is two "half-old, half-new" molecules.

senescence (sen-ESS-cents) [L. *senescere*, to grow old] Sum of processes leading to the natural death of an organism or some of its parts.

sensation A conscious awareness of a stimulus. Not the same thing as perception, which is an understanding of the meaning of a sensation.

sensory neuron Any of the nerve cells or cells adjacent to them that detect specific stimuli (such as light energy) and relay signals to the brain and spinal cord.

sensory system Of animals, the "front door" of the nervous system; the components that detect external and internal stimuli and relay information about them into the central nervous system.

sessile animal (SESS-ihl) An animal that remains attached to a substrate during some stage (often the adult) of its life cycle.

sex chromosome Of animals and some plants, one of two types of chromosomes, the combinations of which govern gender. *Compare* autosome.

sexual dimorphism Of a species, having individuals with distinctively male or female phenotypes.

sexual reproduction Production of offspring by way of meiosis, gamete formation, and fertilization.

sexual selection A microevolutionary process; natural selection favors a trait that gives the individual a competitive edge in attracting and holding onto mates, hence in reproductive success.

shifting cultivation The cutting and burning of trees, followed by tilling of ashes into the soil; once called slash-and-burn agriculture.

shell model Model of electron distribution in atoms, in which all orbitals available to electrons occupy a nested series of shells.

shoot system The aboveground parts of a plant, such as stems, leaves, and flowers.

sieve-tube member Of flowering plants, a cellular component of the interconnecting sugar-conducting tubes in phloem.

sign stimulus A simple but important cue in the environment that triggers a suitable response to a stimulus that the nervous system is prewired to recognize.

sink region Of plants, any region using or stockpiling organic compounds for growth and development.

sister chromatid One of two DNA molecules (and associated proteins) of a duplicated chromosome that remain attached at their centromere region until separated at mitosis or meiosis; each ends up as a chromosome in a different daughter nucleus.

skeletal muscle An organ with hundreds to many thousands of muscle cells arrayed in bundles. Connective tissue surrounds the bundles and extends beyond the muscle, in the form of tendons that attach it to bone.

sliding filament model Model of muscle contraction in which each myosin filament in a sarcomere physically slides along and pulls actin filaments toward the center of the sarcomere, which shortens. The sliding requires ATP energy and formation of cross-bridges between the actin and myosin.

small intestine Of vertebrates, the portion of the digestive system where digestion is completed and most nutrients absorbed.

smog, industrial Polluted air, gray colored, that predominates in industrialized cities with cold, wet winters.

smog, photochemical Polluted air, brown and smelly, that occurs in large cities with many gas-burning vehicles and a warm climate.

social behavior Intraspecific interactions that require mixes of the instinctive and learned behaviors by which individuals send and respond to communication signals.

social parasite An animal that exploits the social behavior of another species to gain food, care for young, or some other factor necessary to survive, reproduce, or both.

sodium-potassium pump A membrane transport protein; when activated by ATP, it selectively transports potassium ions across the membrane, against its concentration gradient, and allows the crossing of sodium ions in the opposite direction.

soil A variable mixture of mineral particles and decomposing organic material; air and water occupy spaces between the particles.

solute (SOL-yoot) [L. *solvere*, to loosen] Any substance dissolved in a solution. In water, this means that hydration spheres surround the charged parts of individual ions or molecules and keep them dispersed.

solvent A fluid, such as water, in which one or more substances is dissolved.

somatic cell (so-MAT-ik) [Gk. *somā*, body] Of animals, any body cell that is not a germ cell (which gives rise to gametes).

somatic nervous system Nerves leading from the central nervous system to skeletal muscles.

somite One of many paired segments in a vertebrate embryo that will give rise to most bones, skeletal muscles of the head and trunk, and the overlying dermis.

source region Of plants, a site where organic compounds form by photosynthesis.

speciation (spee-cee-AY-shun) Evolutionary process by which a daughter species forms from a population or subpopulation of the parent species. The process can vary in its details and in the length of time it takes before the required reproductive isolation from each other is complete.

species (SPEE-sheez) [L. *species*, a kind] In general, one kind of organism. Of sexually reproducing organisms only, one or more populations of individuals that are interbreeding under natural conditions and producing fertile offspring, and that are reproductively isolated from other such groups.

sperm [Gk. *sperma*, seed] A type of mature male gamete.

spermatogenesis (sperm-AT-oh-JEN-ih-sis) Process by which mature sperm form from a germ cell in males.

sphere of hydration Through positive or negative interactions, a clustering of water molecules around individual molecules of a substance placed in water. *Compare* solute.

spinal cord The portion of the central nervous system that threads through a canal inside the vertebral column and that affords direct reflex connections between sensory and motor neurons, and that has tracts to and from the brain.

spleen One of the lymphoid organs; it is a filtering station for blood, a reservoir of red blood cells, and a reservoir of macrophages.

spore A single-celled reproductive structure or resistant, resting body; often walled, and often adapted to survive adverse conditions. An asexual spore may directly give rise to a new stage of the life cycle; a sexual spore must unite with another sexual spore to produce a new stage.

sporophyte [Gk. *phyton*, plant] Of plant life cycles, a vegetative body that grows by way of mitosis from a zygote and that produces spore-bearing structures.

spring overturn Of some bodies of water, movement during spring of dissolved oxygen from the surface layer to the depths and movement of nutrients from bottom sediments to the surface.

stabilizing selection A mode of natural selection by which intermediate phenotypes are favored, and extremes at both ends of the range of variation are eliminated.

stamen (STAY-mun) A male reproductive structure in a flower, commonly consisting of a pollen-bearing structure (anther) on a single stalk (filament).

start codon A base triplet in a strand of mRNA that serves as the start signal for the translation stage of protein synthesis.

stem cell A type of self-perpetuating cell that remains unspecialized. Some of its daughter cells also are self-perpetuating; others differentiate into cells of specialized structure and function. Stem cells in bone marrow that give rise to daughter stem cells and to blood cells are an example.

sterol (STAIR-all) A type of lipid with a rigid backbone of four fused carbon rings. Sterols differ in the number, position, and type of functional groups. They occur in eukaryotic cell membranes; cholesterol is the main type in animal tissues.

steroid hormone A lipid-soluble hormone, synthesized from cholesterol, that diffuses directly across the lipid bilayer of a target cell's plasma membrane and binds with a receptor inside the cell.

stigma Of many flowers, a sticky or hairy surface tissue on upper portion of ovary; it captures pollen grains and favor their germination.

stimulus [L. *stimulus*, goad] A specific form of energy, such as mechanical energy, light, and heat, that activates a sensory receptor having the capacity to detect it.

stoma (STOW-muh), plural **stomata** [Gk. *stoma*, mouth] A controllable gap between two guard cells in stems and leaves; any of the tiny passageways across the epidermis by which carbon dioxide moves into a plant and water vapor and oxygen move out.

stomach A muscular, stretchable sac that receives ingested food; of vertebrates, an organ between the esophagus and intestine in which much protein digestion occurs.

stop codon A base triplet in a strand of mRNA that serves as a stop signal during the translation stage of protein synthesis; it blocks further additions of amino acids to a new polypeptide chain.

strain One of two or more organisms with differences that are too minor to classify it as a separate species. An example is a strain of *Escherichia coli* or some other bacterium.

stratification Ancient layers of sedimentary rock resulting from the slow deposition of volanic ash, silt, and other materials, one above the other.

stream A flowing-water ecosystem that starts out as a freshwater spring or seep.

stroma [Gk. *strōma*, bed] The semifluid interior between a thylakoid membrane system and the two outer membranes of a chloroplast; the chloroplast zone where sucrose, starch, cellulose, and other end products of photosynthesis are assembled.

stromatolite A formation of sediments and matted remains of photosynthetic species that lived shallow seas and gradually accumulated in thin, cakelike layers.

substrate A reactant or precursor for a metabolic reaction; a specific molecule or molecules that an enzyme can chemically recognize, bind reversibly to itself, and modify rapidly in a predictable way.

substrate-level phosphorylation A direct, enzyme-mediated transfer of a phosphate group from a substrate of a reaction to a different molecule, as when an intermediate of glycolysis gives up a phosphate group to ADP, thus forming ATP.

succession, primary (suk-SESH-un) [L. *succedere*, to follow after] An ecological pattern by which a community develops in sequence, from the time pioneer species colonize a new, barren habitat to the climax community (an end array of species that remain in equlibrium over the region).

succession, secondary A pattern by which some disturbed area within a community recovers and moves back toward the climax state; typical of abandoned fields, burned forests, and storm-battered intertidal zones.

surface-to-volume ratio A mathematical relationship in which volume increases with the cube of the diameter, but surface area increases only with the square. Of growing cells, the volume of cytoplasm increases faster than the surface area of the plasma membrane that must service the cytoplasm. This constraint generally keeps cells small, elongated, or with membrane foldings.

survivorship curve A plot of age-specific survival of a group of individuals in a given environment, from the time of their birth until the last one dies.

swim bladder An adjustable flotation device that changes in volume as it exchanges gases with blood and that allows many fishes to maintain neutral buoyancy in water.

symbiosis (sim-by-OH-sis) [Gk. *sym*, together, + *bios*, life, mode of life] Literally, living together. For at least part of the life cycle, individuals of one species live near, in, or on individuals of another species. Commensalism, mutualism, and parasitism are examples of symbiotic interactions.

sympathetic nerve Of autonomic nervous systems, one of the nerves dealing mainly with increasing overall body activities at times of heightened awareness, excitement, or danger; also work continually in opposition with parasympathetic nerves to make minor adjustments in internal organ activity.

sympatric speciation [Gk. *sym*, together, + *patria*, native land] Species form within the home range of an existing species, in the absence of a physical barrier. It may happen instantaneously, as by polyploidy.

synaptic integration (sin-AP-tik) Of an individual neuron, the moment-by-moment combining of all excitatory and inhibitory signals arriving at its trigger zone.

syndrome A set of symptoms that may not individually be a telling clue but that collectively characterize a particular genetic disorder or disease.

systemic circuit (sis-TEM-ik) Of a closed circulatory system, a route by which oxygen-enriched blood flows from the lungs to the left half of the heart, through the rest of the body (where it gives up oxygen and takes up carbon dioxide), then back to the right half of the heart.

T lymphocyte One of a class of white blood cells that carry out immune responses. The helper T and cytotoxic T cells are examples.

taproot system A primary root together with its lateral branchings.

target cell Of a signaling molecule such as a hormone, any cell having the type of receptor that can bind with that molecule.

tectum The midbrain's roof. In fishes and amphibians, a center for coordinating most sensory input and initiating motor responses. In most vertebrates (not mammals), a reflex center that swiftly relays sensory input to higher integrative centers in the forebrain.

telophase (TEE-low-faze) Of meiosis I, the stage when one of each pair of homologous chromosomes has arrived at a spindle pole. Of mitosis and of meiosis II, the final stage when chromosomes decondense into threadlike structures and two daughter nuclei form.

temperature A measure of how rapidly the molecules or ions of a substance are moving.

tendon A cord or strap of dense connective tissue that attaches muscle to bones.

territory An area that one or more animals defend against competitors for food, water, mates, or suitable living space.

test A way to determine the accuracy of a prediction, as by conducting experimental or observational tests and by developing models. In science, such predictions must be based on potentially falsifiable hypotheses; that is, they must be tested in the natural world in ways that might disprove them.

testcross For an individual of unknown genotype that shows dominance for a trait, a type of experimental cross that may reveal whether the individual is homozygous dominant or heterozygous.

testis, plural **testes** Male gonad; a primary reproductive organ in which male gametes and sex hormones are produced.

testosterone (tess-TOSS-tuh-rown) Of male vertebrates, a sex hormone with key roles in the development and functioning of the male reproductive system.

tetanus Of muscles, a large contraction in which repeated stimulation of a motor unit causes muscle twitches to mechanically run together. In a disease by the same name, toxins block muscles from relaxing.

thalamus Of the forebrain, a coordinating center for sensory input and a relay station for signals to the cerebrum.

theory, scientific A testable explanation of a broad range of related phenomena, one that has been extensively tested and can be used with a high degree of confidence. A scientific theory remains open to tests, revision, and tentative acceptance or rejection.

thermal inversion The trapping of a layer of dense, cool air beneath a layer of warm air; can cause air pollutants to accumulate to high levels close to the ground.

thermophile A type of archaebacterium of unusually hot aquatic habitats, as in hot springs or near hydrothermal vents.

thermoreceptor A sensory cell or specialized cell by it that detects radiant energy (heat).

thigmotropism (thig-MOE-truh-pizm) [Gk. *thigm*, touch] An orientation of the direction of growth in response to physical contact with a solid object, as when a vine curls around a fencepost.

threshold Of a neuron and other excitable cells, the minimum amount by which the resting membrane potential must change to trigger an action potential.

thylakoid membrane system Of chloroplasts, an internal membrane system commonly folded into flattened channels and disks (grana) that have light-absorbing pigments and enzymes used in the formation of ATP, NADPH, or both during photosynthesis.

thymine A nitrogen-containing base of one of the nucleotides in DNA.

thymus gland A lymphoid organ that has endocrine functions; lymphocytes of the immune system multiply, differentiate, and mature in its tissues, and its hormone secretions affect their functioning.

thyroid gland An endocrine gland, the hormones of which affect overall metabolic rates, growth, and development.

tissue Of multicelled organisms, a group of cells and intercellular substances that function together in the performance of one or more specialized tasks.

tonicity The relative concentrations of solutes in two fluids, such as inside and outside a cell. When solute concentrations are isotonic (equal in both fluids), water shows no net osmotic movement in either direction. When one fluid is hypotonic (has less solutes), the other is hypertonic (has more solutes) and water tends to move into it.

toxin A normal metabolic product of one species with chemical effects that can harm or kill individuals of a different species that encounter it.

trace element Any element that represents less than 0.01 percent of body weight.

tracer A substance that has a radioisotope attached to it so that its pathway or destination in a cell, organism, ecosystem, or some other system can be tracked, as by scintillation counters that detect its emissions.

trachea (TRAY-kee-uh), plural **tracheae** An air-conducting tube of respiratory systems; of land vertebrates, the windpipe, which carries air between the larynx and bronchi.

tracheal respiration Of insects, spiders, and some other animals, a respiratory system consisting of finely branching tracheae that extend from openings in the integument and that dead-end in body tissues.

tracheid (TRAY-kid) Of flowering plants, one of two types of cells in xylem that conduct water and dissolved minerals.

tract A communication line inside the brain and spinal cord; comparable to a nerve of the peripheral nervous system.

transcription [L. *trans*, across, + *scribere*, to write] The first stage of protein synthesis, when an RNA strand is assembled on one of the two strands of a DNA double helix; the base sequence of the resulting transcript is complementary to the DNA template.

transfer RNA (tRNA) An RNA molecule that binds and delivers an amino acid to a ribosome and that pairs with an mRNA codon during the translation stage of protein synthesis.

translation The stage of protein synthesis when the encoded sequence of information in mRNA becomes converted to a sequence of particular amino acids, the outcome being a polypeptide chain; rRNA, tRNA, and mRNA interact to bring this about.

translocation Of cells, a stretch of DNA that physically moved to a different location in the same chromosome or in a different one, with no molecular loss. Of vascular plants, the process by which organic compounds are distributed by way of phloem.

transpiration Evaporative water loss from aboveground plant parts, leaves especially.

transport protein A membrane protein that allows water-soluble substances to move through their interior, which spans the lipid bilayer of a cell membrane.

transposable element A DNA region that moves spontaneously from one location to another in the same DNA molecule or a different one. Often such regions inactivate genes into which they become inserted and cause changes in phenotype.

triglyceride (neutral fat) A lipid having three fatty acid tails attached to a glycerol backbone. Triglycerides are the body's most abundant lipids and richest energy source.

trisomy (TRY-so-mee) Of cells, the presence of three chromosomes of a given type rather than the two characteristic of the parental diploid chromosome number.

trophic level (TROE-fik) [Gk. *trophos*, feeder] All of the organisms in an ecosystem that are the same number of transfer steps away from the energy input into the system.

tropical rain forest A biome where rainfall is regular and heavy, the annual mean temperature is 25°C, humidity is 80 percent or more, and biodiversity is spectacular.

tropism (TROE-pizm) Of plants, a directional growth response to an environmental factor, such as growth toward light.

true breeding Of a sexually reproducing species, a lineage in which the offspring of successive generations are just like the parents in one or more traits being studied.

tumor A tissue mass composed of cells that are dividing at an abnormally high rate.

turgor pressure (TUR-gore) [L. *turgere*, to swell] Internal fluid pressure applied to a cell wall when water moves into the cell by way of osmosis.

ultrafiltration Bulk flow of a small amount of protein-free plasma from a blood capillary when the outward-directed force of blood pressure is greater than the inward-directed osmotic force of interstitial fluid.

uniformity Various theories that mountain building, erosion, and other forces of nature have worked over the Earth's surface in the same repetitive ways through time.

upwelling An upward movement of deep, nutrient-rich water along coasts; it replaces surface waters that move away from shore when the direction of prevailing wind shifts.

uracil (YUR-uh-sill) Nitrogen-containing base of a nucleotide found in RNA molecules; like thymine, it can base-pair with adenine.

ureter A tubular channel for urine flow between the kidney and urinary bladder.

urethra A tubular channel for urine flow between the urinary bladder and an opening at the body surface.

urinary bladder A distensible sac in which urine is stored before being excreted.

urinary excretion A mechanism by which excess water and solutes are removed by way of a urinary system.

urinary system An organ system that adjusts the volume and composition of blood, and so helps maintain extracellular fluid.

urine A fluid formed in kidneys by filtration, reabsorption, and secretion; urine consists of wastes, excess water, and solutes.

uterus (YOU-tur-us) [L. *uterus*, womb] A chamber in which the developing embryo is contained and nurtured during pregnancy.

vaccination An immunization procedure against a specific pathogen.

vaccine An antigen-containing preparation, swallowed or injected, that increases immunity to certain diseases by inducing formation of armies of effector and memory B and T cells.

vagina Part of a reproductive system of mammalian females that receives sperm, forms part of the birth canal, and channels menstrual flow to the exterior.

variable Of an experimental test, a specific aspects of an object or event that may differ over time and among individuals.

vascular bundle Of vascular plants, the arrangement of primary xylem and phloem into multistranded, sheathed cords that extend lengthwise through the ground tissue system.

vascular cambium A lateral meristem that increases stem or root diameter (girth).

vascular cylinder Arrangement of vascular tissues as a central cylinder in roots.

vascular plant A plant with xylem and phloem, and well-developed roots, stems, and leaves.

vascular tissue system Xylem and phloem; conducting tissues that distribute water and solutes through the body of vascular plants.

vein Of the circulatory system, any of the large-diameter vessels that lead back to the heart; of leaves, one of the vascular bundles that thread through photosynthetic tissues.

ventricle (VEN-tri-kuhl) Of the vertebrate heart, one of two chambers from which blood is pumped out. Blood circulation is driven by ventricular contraction.

venule A small blood vessel that accepts blood from capillaries and delivers it to a vein.

vernalization Of flowering plants, stimulation of flowering by exposure to low temperatures.

vertebra, plural **vertebrae** One of a series of hard bones organized, with intervertebral disks, into a backbone.

vertebrate Animal with a backbone.

vesicle (VESS-ih-kul) [L. *vesicula*, little bladder] In the cytoplasm of cells, one of a variety of small membrane-bound sacs that function in the transport, storage, or digestion of substances or in some other activity.

vessel member A cell in xylem; although dead at maturity, its wall becomes part of a water-conducting pipeline.

villus (VIL-us), plural **villi** Any of several fingerlike absorptive structures projecting from the free surface of an epithelium.

viroid An infectious particle consisting only of very short, tightly folded strands or circles of RNA. Viroids may have evolved from introns, which they resemble.

virus A noncellular infectious agent that consists of DNA or RNA and a protein coat; it can replicate only after its genetic material enters a host cell and subverts the host's metabolic machinery.

vision Perception of visual stimuli; requires focusing of light precisely onto a layer of photoreceptive cells that is dense enough to sample details of a light stimulus, followed by image formation in a brain.

visual signal An observable action or cue that functions as a communication signal.

vitamin Any of more than a dozen organic substances that animals require in small amounts for metabolism but that they generally cannot synthesize for themselves.

vocal cord One of the thickened, muscular folds of the larynx that help produce sound waves for speech.

water potential Sum of two opposing forces (osmosis and turgor pressure) that can cause a directional movement of water into or out of a walled cell.

water table The upper limit at which the ground in a specified area has become fully saturated with water.

watershed Any specified region in which all precipitation drains into a single stream or river.

wavelength A wavelike form of energy in motion. The horizontal distance between two crests of every two successive waves.

wax A molecule with long-chain fatty acids packed together and linked to long-chain alcohols or to carbon rings. Waxes have a firm consistency and repel water.

white blood cell One of the eosinophils, neutrophils, macrophages, T and B cells, and other leukocytes which, together with their chemical weapons and mediators, defend the vertebrate body against attacks by pathogens and against tissue damage.

white matter Inside the brain and spinal cord, axons that have glistening white sheaths and that specialize in rapid signal transmission.

wild-type allele For a given gene locus, the allele that occurs normally or with the greatest frequency among individuals of a population.

wing A body part that functions in flight, as among birds, bats, and many insects. A bird wing is a forelimb with feathers, strong muscles, and extremely lightweight bones. An insect wing develops as a lateral fold of the exoskeleton.

X chromosome A sex chromosome with genes that, in humans, causes an embryo to develop into a female, provided that it inherits a pair of these.

X-linked gene Any gene located on an X chromosome.

X-linked recessive inheritance Recessive condition in which the responsible, mutated gene occurs on the X chromosome.

xylem (ZYE-lum) [Gk. *xylon*, wood] Of vascular plants, a tissue that transports water and solutes through the plant body.

Y chromosome A distinctive chromosome present in males or females of many species, but not both. In human males, one is paired with an X chromosome (XY); and in human females, the pairing is XX.

Y-linked gene One of the genes located on a Y chromosome.

yellow marrow A fatty tissue in the cavities of most mature bones that produces red blood cells when blood loss from the body is severe.

yolk sac Of land vertebrates, one of four extraembryonic membranes. In most shelled eggs, it holds nutritive yolk; in humans, part becomes a site of blood cell formation and some of its cells give rise to the forerunners of gametes.

zero population growth A population for which the number of births is balanced by the number of deaths over a specified period, assuming that immigration and emigration also are balanced.

zooplankton A freshwater or marine community of floating or weakly swimming heterotrophs, mostly microscopic, such as rotifers and copepods.

zygote (ZYE-goat) The first cell of a new individual, formed by the fusion of a sperm nucleus with the nucleus of an egg at fertilization; also called a fertilized egg.

CREDITS AND ACKNOWLEDGMENTS

FRONT MATTER **Pages iv-v**, David Macdonald. **vi-vii**, James M. Bell/Photo Researchers. **viii-ix** S. Stammers/SPL/Photo Researchers. **x-xi** © 1990 Arthur M. Greene. **xii-xiii** Gary Head. **xiv-xv** Lennart Nilsson from *Behold Man*, © 1974 Albert Bonniers Forlag and Little, Brown and Company, Boston. **xvi-xvii** John Alcock. **xviii-xix**, Lennart Nilsson from *A Child Is Born*, © 1966, 1977 Dell Publishing Company, Inc. **xx-xxi**, Jim Doran. **xxxii** and **page 1** © 1990 Tom Van Sant/The GeoSphere Project, Santa Monica, California.

CHAPTER 1 **1.1** Frank Kaczmarek. **1.2** Art, American Composition & Graphics (ACG). **1.3, 1.4** Jack deConingh. **1.5** (a) Walt Anderson/ Visuals Unlimited; (b) Gregory Dimijian/ Photo Researchers; (c) Alan Weaving/Ardea London. **1.6** (a) R. Robinson, Visuals Unlimited; (b) Tony Brain/SPL/Photo Researchers; (b) M. Abbey, Visuals Unlimited; (d, f) Edward S. Ross; (e) Dennis Brokaw; (g,h) Pat & Tom Leeson/Photo Researchers. **1.7** J. A. Bishop, L. M. Cook. **1.8** Levi Publishing Company. **1.9** Jack deConingh; art, Raychel Ciemma. **1.11** Gary Head. **1.12** James Carmichael, Jr./NHPA. **Page 19** James M. Bell/Photo Researchers

CHAPTER 2 **2.1** Gary Head for Norman Terry, University of California, Berkeley. **2.2** (a) Lee Newman, University of Washington; (b) Dr. Slavik Dushenkov PhD., Phytotech, Inc. **2.3** Jack Carey. **2.6** (a) Kingsley R. Stern; (b) Chip Clark; (c) Gary Byerly, LSU. **2.9** (a) Hank Morgan/Rainbow; (b) Art, R. Ciemma; (c) Harry T. Chugani, M.D., UCLA School of Medicine. **2.12** Maris and Cramer. **2.14** Photo Researchers. **2.16** (a) Richard Riley/FPG; (b) art, R. Ciemma; (c) copyright © Kennan Ward/The Stock Market. **2.17** H. Eisenbeiss/ Frank Lane Picture Agency. **2.18** Art, R. Ciemma. **2.20** Michael Grecco/Picture Group.

CHAPTER 3 **3.1** Dave Schiefelbein, NASA. **3.4** Tim Davis/Photo Researchers. **3.6** *Above,* Gary Head; *below,* Martin Rogers/FPG. **3.10** David Scharf/Peter Arnold, Inc. **3.12** Clem Haagner/Ardea, London. **3.13** (a) Art, Precision Graphics; (c) Micrograph Lewis L. Lainey; (d) Larry Lefever/Grant Heilman; (e) Kenneth Lorenzen. **3.19** Art, Palay/Beaubois. **3.20** CNRI/ SPL/Photo Researchers; art, Robert Demarest. **3.23** Art, Precision Graphics; A. Lesk/SPL/ Photo Researchers.

CHAPTER 4 **4.1** (a) Corbis-Bettmann. (b) Armed Forces Institute of Pathology; (c) National Library of Medicine; (e) The Francis A. Countway Library of Medicine. **4.2** Art, R. Ciemma. **4.3** George S. Ellmore; art, R. Demarest. **4.5** Driscoll, Youngquist, and Baldeschwieler/Cal Tech/ Science Source/Photo Researchers; art, R. Ciemma. **4.6** Jeremy Pickett-Heaps, School of Botany, University of Melbourne. **4.7** Art, R. Ciemma and ACG. **4.8** M. C. Ledbetter, Brookhaven National Laboratory; art, R. Ciemma and ACG. **4.9** G. L. Decker; art, R. Ciemma and ACG. **4.10** Stephen L. Wolfe. **4.11** (a) *Left,* Don W. Fawcett/Visuals Unlimited; *right,* A. C. Faberge, *Cell and Tissue Research,* 151:403-415, 1974. **4.12** Art, R. Ciemma and ACG. **4.13** (a,b) Micrographs, Don W.Fawcett/ Visuals Unlimited; *below,* art, R. Ciemma. **4.14** Gary W. Grimes; art R. Demarest after a model by J. Kephart. **4.15** Keith R. Porter. **4.16** L. K.

Shumway; *below right* art, Palay/Beaubois. **4.17** J. Victor Small, Gottfried Rinnerthaler. **4.18** Art, Precision Graphics **/ 4.19** (a) C. J. Brokaw; (b) Sidney L. Tamm. **4.20** Art, Precision Graphics after Stephen L. Wolfe, *Molecular and Cellular Biology,* Wadsworth. 1993. **4.21** Art, Precision Graphics after Stephen L. Wolfe, *Molecular and Cellular Biology,* Wadsworth. **4.22** Ron Hoham, Dept. of Biology, Colgate University. **4.23** Art, R. Ciemma; (b) Photo Researchers. **4.24** (a) George S. Ellmore and art, R. Ciemma; (b) Ed Reschke and art, Joel Ito; . **4.25** Art, R. Ciemma. **4.26** (a) Art, R. Ciemma; (b) G. Cohen-Bazire; (c) K. G. Murti/Visuals Unlimited; (d) R. Calentine/Visuals Unlimited; (e) Gary Gaard and Arthur Kelman.

CHAPTER 5 **5.1** (a) Runk/Schoenberger/Grant Heilman; (b) Inigo Everson/Bruce Coleman Ltd. **5.2** (a,b) Art, Precision Graphics; (c) Art, R. Ciemma. **5.3** Art, Raychel Ciemma. **5.4** Charles D. Winters/Photo Researchers; art, Precision Graphics after Stephen L. Wolfe, *Molecular and Cellular Biology,* Wadsworth; **5.5** P. Pinto da Silva, D. Branton, *Journal of Cell Biology,* 45:598, by copyright permission of The Rockefeller University Press; art, Palay/ Beaubois. **5.6** Art, Palay/Beaubois. **5.7** Art, Precision Graphics. **5.9** M. Sheetz, R. Painter, and S. Singer, *Journal of Cell Biology,* 70:193, by copyright permission of The Rockefeller University Press. **5.12** Art, R. Ciemma. **5.15** Art, Leonard Morgan. **5.16** M. M. Perry and A. B. Gilbert. **5.17** M. Abbey/Visuals Unlimited. **5.19** Frieder Sauer/Bruce Coleman Ltd.

CHAPTER 6 **6.1** Gary Head; molecular models, Protein Data Bank. **6.2** Evan Cerasoli. **6.3** *Above* NASA; *below* Manfred Kage/Peter Arnold, Inc. **6.8** Art, Nadine Sokol. **6.10** Art, R. Ciemma; (b,c) art, Precision Graphics. **6.11** (a,b) Thomas A. Steitz; art, Palay/Beaubois. **6.12** Douglas Faulkner/Sally Faulkner Collection. **6.16** Art, R. Ciemma, ACG after B. Alberts et al., *Molecular Biology of the Cell,* Garland Publishing, 1983. **6.17** Art, R. Ciemma and ACG. **6.18** (a) © Oxford Scientific Films/ Animals Animals; (b) © Raymond Mendez/Animals Animals; (c,d) Keith V. Wood. **6.19** (a,b) Photographs, Christopher C. Contag, *Molecular Microbiology,* November 1995, Vol. 18, No. 4, pp. 593-603. "Photonic Detection of Bacterial Pathogens in Living Hosts." Reprinted by permission of Blackwell Science.

CHAPTER 7 **7.1** David R. Frazier/Photo Researchers. **7.2** (a) Hans Reinhard/Bruce Coleman Ltd.; (e,f) David Fisher; art, R. Ciemma and ACG. **7.3** (a) Barker-Blakenship/FPG; (b) Art, Precision Graphics after Govindjee. **7.4** Carolina Biological Supply Company; art, R. Ciemma. **7.5** Art, Precision Graphics after Stephen L. Wolfe, *Molecular and Cellular Biology,* Wadsworth. **7.6** Art, Precision Graphics. **7.7** Larry West/FPG. **7.8** (a) Barker-Blakenship/ FPG; (b) Art, Precision Graphics after Govindjee. **7.9** Art, R. Ciemma. **7.10, 7.11** Art, Precision Graphics. **7.12** E. R. Degginger. **7.13** Art, R. Ciemma and Precision Graphics. **7.14** Art, Precision Graphics. **7.15** (a) Art, R. Ciemma. **7.16** (a) Grant Heilman Inc.; (b) Dick Davis/Photo Researchers; (c) Martin Grosnick/Ardea London. **7.17** NASA. **7.18** Art, R. Ciemma and Precision Graphics. **Page 129** art, R. Ciemma.

CHAPTER 8 **8.1** Stephen Dalton/Photo Researchers. **8.3** (a,b) Gary Head; (c) Janeart/ Image Bank. **8.4** *Right,* art; Palay/Beaubois. **8.5** (a) Keith R. Porter; (b) Art, L. Calver; (c) Art, R. Ciemma. **8.7, 8.8** Art, R. Ciemma. **8.11** (b) Adrian Warren/Ardea London; (c) David M. Phillips/Visuals Unlimited. **8.12** Gary Head. **Page 144** R. Llewellyn/Superstock, Inc. **Page 147** © Lennart Nilsson

CHAPTER 9 **9.1** (*left and right, above*) Chris Huss; (*left inset and right, below*) Tony Dawson. **9.2** C. J. Harrison et al., *Cytogenetics and Cell Genetics* 35:21-27, © 1983 S. Karger A.G., Basel. **9.4** A. S. Bajer, University of Oregon. **9.5** Ed Reschke; art, R. Ciemma. **9.6** (a) *Left,* C. J. Harrison et al., *Cytogenetics and Cell Genetics,* 35:21-27, © 1983 S. Karger A. G., Basel; (b) B. Hamkalo; (c) O. L. Miller, Jr., Steve L. McKnight; (b-d) art, Nadine Sokol. **9.8** B.A. Palevitz and E.H. Newcomb, University of Wisconsin/BPS/Tom Stack & Associates. **9.9** H. Beams, R. G. Kessel, *American Scientist,* 64:279-290, 1976. **9.10** (a-c, e) Lennart Nilsson from *A Child Is Born* © 1966, 1967 Dell Publishing Company, Inc.; (d) L. Nilsson from *Behold Man,* © 1974 by Albert Bonniers Forlag and Little, Brown and Company, Boston.

CHAPTER 10 **10.1** (a) Jane Burton/Bruce Coleman Ltd.; (b) Dan Kline/Visuals Unlimited. **10.2** Art, R. Ciemma. **10.3** CNRI/SPL/Photo Researchers. **10.4** Art, R. Ciemma. **10.5** Art, R. Ciemma and ACG. **10.6** Art, R. Ciemma. **10.9** David M. Phillips/Visuals Unlimited. **10.10** Art, R. Ciemma. **10.12** © Richard Corman/ Outline

CHAPTER 13 **11.1** *Left* Frank Trapper/Sygma; *center* Focus on Sports; *right, above* Fabian/ Sygma; *right, below* Moravian Museum, Brno. **11.2** Jean M. Labat/Ardea London; art, Jennifer Wardrip. **11.5** Art, Hans & Cassady, Inc. **11.7** Art, Hans & Cassady, Inc. **11.8** Art, R. Ciemma. **11.10** William E. Ferguson. **11.12** Micrographs Stanley Flegler/Visuals Unlimited. **11.13** (a,b) Michael Stuckey/Comstock Inc.; (c) Russ Kinne/ Comstock Inc. **11.14** David Hosking. **11.15** Tedd Somes. **11.16** *Top to bottom* Frank Cezus; Frank Cezus; Michael Keller; Ted Beaudin; Stan Sholik/all FPG. **11.17** (a) Dan Fairbanks, Brigham Young University. **11.18** Jane Burton/Bruce Coleman Ltd.; art, D. & V. Hennings. **11.19** (a) William E. Ferguson; (b) Eric Crichton/Bruce Coleman Ltd. **11.20** Evan Cerasoli. **11.21** Leslie Faltheisek/Clacritter Manx. **11.22** Joe McDonald/Visuals Unlimited.

CHAPTER 12 **12.1** Eddie Adams/AP Photo. **12.2** (b) Photograph Omikron/Photo Researchers. **12.4** (a) From Lennart Nilsson, *A Child Is Born,* © 1966, 1977 Dell Publishing Company, Inc.; (b) Redrawn by R. Demarest by permission from page 126 of Michael Cummings, *Human Heredity: Principles and Issues,* Third Edition. © 1994 by West Publishing. All rights reserved; (c) Art, R. Demarest after Patten, Carlson, and others. **12.5** Photograph Carolina Biological Supply Company **/ 12.8** Art, R. Ciemma. **12.10** (b) Photo Dr. Victor A. McKusick; (c) Steve Uzzell. **12.14** Giraudon/Art Resource, NY. **12.16** After Victor A. McKusick, *Human Genetics,* Second edition, © 1969. Reprinted by permission of Prentice-Hall, Inc., Engelwood Cliffs, NJ; photo Corbis-Bettmann. **12.17** (a,b) W. M. Carpenter. **12.18** Art, R. Ciemma.

12.19 (a) Cytogenetics Laboratory, University of California, San Francisco; (b) After Collman, Stoller, *American Journal of Public Health*, 52, 1962; photographs*above* courtesy of Peninsula Association for Retarded Children and Adults, San Mateo Special Olympics, Burlingame, CA; *below* used by permission of Carole Lafrate. **12.21** (a,b) Courtesy G. H. Valentine. **12.23** C. J. Harrison. **12.26** Art, R. Ciemma. **12.27** (a) Fran Heyl Associates © Jacques Cohen, computer enhanced by © Pix Elation; (b) Art, R. Ciemma. **12.29** Bonnie Kamin/Stuart Kenter Associates. **12.31** Carolina Biological Supply Company.

CHAPTER 13 13.1, 13.2 *Above,* A. Lesk/SPL/Photo Researchers; *below,* A. C. Barrington Brown © 1968 J. D. Watson. **13.3** Art, R. Ciemma. **13.4** (b) Lee D. Simon/Science Source/Photo Researchers. **13.6** Biophoto Associates/SPL/Photo Researchers. **13.7** Art, Precision Graphics. **13.10** (a) Ken Greer/Visuals Unlimited; (b) Biophoto Asociates/Science Source/Photo Researchers; (c) James Stevenson/SPL/Photo Researchers; (d) Gary Head.

CHAPTER 14 14.1 *Above,* Kevin Magee/Tom Stack & Associates; *below,* Dennis Hallinan/FPG. **14.3** Gary Head. **14.4** From Stephen L. Wolfe, *Molecular and Cellular Biology*, Wadsworth. **14.5** Art, Palay/Beaubois and Precision Graphics. **14.8** Art, Hans & Cassady, Inc. **14.12** (a) Courtesy of Thomas A. Steitz from *Science*, 246:1135-1142, December 1, 1989. **14.14** Art, R. Ciemma and ACG. **14.15** Dr. John E. Heuser, Washington University School of Medicine, St. Louis, MO; art, Palay/Beaubois. **14.18** *Right* Nik Kleinberg; *left,* Peter Starlinger. **14.19** Art, R. Ciemma.

CHAPTER 15 15.1 (a) Slim Films; (b) Dr. Brian V. Harmon, Queensland University of Technology. From *Methods in Cell Biology*, Volume. 46, 1995, "Anatomical Methods in Cell Death," Academic Press. Reprinted by Permission. **15.2** Brian Matthews, University of Oregon. **15.3** Art, Palay/Beaubois and Hans & Cassady. **15.4** Art, R. Ciemma. **15.5** (a) M. Roth and J. Gall. **15.6** (a) C. J. Harrison et al., *Cytogenetics and Cell Genetics*, 35:21-27, © 1983 S. Karger A. G., Basel; (b) art, Palay/Beaubois. **15.7** (a) Dr. Karen Dyer Montgomery. **15.8** Jack Carey. **15.9** W. Beerman; art, R. Ciemma. **15.10** Art, Precision Graphics; Frank B. Salisbury. **15.11** Art, Betsy Palay/Artemis. **15.12** (a) Art, Betsy Palay/Artemis; (b) Lennart Nilsson © Boehringer Ingelheim International GmbH.

CHAPTER 16 16.1 Lewis L. Lainey. **16.2** (a) Stanley N. Cohen/Science Source/Photo Researchers; (b) Dr. Huntington Potter and Dr. David Dressler. **16.3** Art, R Ciemma. **16.6** Cellmark Diagnostics, Abingdon, U.K. **16.9** Michael Maloney, *San Francisco Chronicle*. **16.10** (a) Stephen L. Wolfe, *Molecular and Cellular Biology*; (b) Keith V. Wood; (c,d) Monsanto Company. **16.11** R. Brinster, R. E. Hammer, School of Veterinary Medicine, University of Pennsylvania. **Page 269** S. Stammers/SPL/Photo Researchers.

CHAPTER 17 17.1 Werner Stoy/Bruce Coleman Ltd; art, Leonard Morgan after R. H. Dott, Jr. and R. L. Batten, *Evolution of the Earth*, Third edition, McGraw-Hill, 1981. Reproduced by permission of McGraw-Hill, Inc. **17.2** (a) Jen & Des Bartlett/Bruce Coleman Ltd.; (b) Kenneth W. Fink/Photo Researchers; (c) Dave Watts/A.N.T. Photo Library. **17.3** Art, R. Ciemma.

17.4 (a)*Left,* Courtesy George P. Darwin, Darwin Museum, Down House; *right,* Heather Angel; (b) Christopher Ralling; (c) Dieter & Mary Plage/Survival Anglia; art, Leonard Morgan. **17.5** (a) Lee Kuhn/FPG; (b) Field Museum of Natural History, Chicago, and the artist, Charles R. Knight (Neg. No. CK21T); (b) © Philip Boyer/Photo Researchers. **17.6** (a) Heather Angel; (b) David Cavagnaro; (c) Dr. P. Evans/Bruce Coleman, Inc.; (d) Alan Root/Bruce Coleman Ltd. **17.7** Down House and The Royal College of Surgeons of England. **17.8** P. Morris/Ardea London; art, R.l Ciemma.

CHAPTER 18 18.1 Elliot Erwitt/Magnum Photos Inc. **18.2** Alan Solem. **18.5** J. A. Bishop and L. M. Cook. **18.7** (a,b) Warren Abrahamson; (c) Forest W. Buchanan/Visuals Unlimited; (d) Kenneth McCrea and Warren Abrahamson. **18.9** (a) Thomas Bates Smith. **18.10** Bruce Beehler. **18.12** *Above,* David Neal Parks; *below,* W. Carter Johnson. **18.13** Jerry Coyne. **18.14** David Cavagnaro. **18.15** Kjell Sandved/Visuals Unlimited. **18.16** D. Avon/Ardea London.

CHAPTER 19 19.1 (a) Gary Head; snail courtesy Larry Reed; (b) adapted from R. K. Selander and D. W. Kaufman, *Evolution*, Vol. 29, No. 3, December 31, 1975. **19.3** (a) Art, R. Ciemma after C.T. Regan and E. Trewavas, 1932. **19.4** (a) Suzanne L. Collins and Joseph T. Collins; (b, c) from *The Amphibians and Reptiles of Missouri* by Tom R. Johnson. Copyright © 1987 by the Conservation Commission of the State of Missouri. Reprinted by permission. **19.5** After F. Ayala and J. Valentine, *Evolving*, Benjamin-Cummings, 1979. **19.6** Alvin E. Staffan/Photo Researchers. **19.7** G. Ziesler/ZEFA. **19.8** After V. Grant, *Organismic Evolution*, W. H. Freeman and Co., 1977. **19.9** Photograph Mike Vasey/San Francisco State University. **19.10** Jen & Des Bartlett/Bruce Coleman Ltd. **19.11** (a) Fred McConnaughey/Photo Researchers; (b) Patrice Geisel/Visuals Unlimited; *right,* Tom Van Sant/The Geosphere Project, Santa Monica, CA. **19.13** Art, R. Ciemma. **19.14** *Left,* Art, Precision Graphics after U. Schliewen et al., *Nature*, 368:629-632, 14 April 1994. Used by permission of Macmillan Magazines, Ltd.; *right* Tom Van Sant/The Geosphere Project, Santa Monica, CA. **19.15** After W. Jensen and F. B. Salisbury, *Botany: An Ecological Approach*, Wadsworth, 1972. **19.16** *Left* © VIREO/Academy of Natural Sciences; *Right,* Robert C. Simpson/Nature Stock. **19.18** After P. Dodson, *Evolution: Process and Product*, Third edition, Prindle, Weber & Schmidt.

CHAPTER 20 20.1 *Left,* Vatican Museums; *right,* Martin Dohrn/SPL/Photo Researchers. **20.2** (a) Donald Baird, Princeton Museum of Natural History; (b) A. Feduccia, *The Age of Birds*, Harvard University Press, 1980; (c) H. P. Banks; (d) Jonathan Blair. **20.4** David Noble/FPG. **20.6** Art, Leonard Morgan. **20.7** *Right,* Martin Land/Photo Researchers; *left,* Maps, Lloyd K. Townsend after A. M. Ziegler, C. R. Scotese, and S. F. Barrett, "Mesozoic and Cenozoic Paleogeographic Maps" and J. Krohn and J. Syndermann, "Paleotides Before the Permian" in F. Brosche and J. SŸndermann (Eds.), *Tidal Friction and the Earth's Rotation II*, Springer-Verlag, 1983. **20.8** (a, b) Gary Head; (c-e) Art, Precision Graphics after E. Guerrant, *Evolution*, 36:699-712, 1982. **20.9** (b) From T. Storer et al., *General Zoology*, Sixth edition, McGraw-Hill, 1979. Reproduced by permission of McGraw-Hill, Inc. **20.10** Art, Victor Royer.

20.11 Art, R. Ciemma. **20.12** *Top,* Douglas P. Wilson/Eric and David Hosking; *center,* Superstock, Inc.; *bottom,* E. R. Degginger. **20.14** *Top,* Kjell B. Sandved/Visuals Unlimited; *center,* Jeffrey Sylvester/FPG; *bottom,* Thomas D. Mangelsen/Images of Nature. **20.15** *Left to right,* Larry Lefever/Grant Heilman Inc.; R.I.M. Campbell/Bruce Coleman Ltd.; Runk and Schoenberger/Grant Heilman Inc.; Bruce Coleman Ltd. **20.17***Clockwise from top left,* Peter Scoones/Seaphot Limited: Planet Earth Pictures; C. B. & D. W. Frith/Bruce Coleman Ltd.; © John Gurche 1989; Roger K. Burnard; Jane Burton/Bruce Coleman Ltd.; W. A. Banaszewski/Visuals Unlimited; Hans Reinhard/ZEFA / **20.18** Art, D. & V. Hennings . **Page 331** © 1990 Arthur M. Greene.

CHAPTER 21 21.1 Jeff Hester and Paul Scowen, Arizona State University, and NASA. **21.2** William K. Hartmann. **21.3** (a) Chesley Bonestell. **21.4** Art, Precision Graphics. **21.5** (a) Sidney W. Fox; (b) W. Hargreaves, D. Deamer. **21.7** Art, Precision Graphics. **21.8** Bill Bachman/ Photo Researchers. **21.9** (a) Stanley W. Awramik; (b-f) Andrew H. Knoll, Harvard University. **21.10** P. L. Walne and J. H. Arnott, *Planta*, 77:325-354, 1967. **21.11** Art, R. Ciemma. **21.12** Robert K. Trench. **21.13** (a) Neville Pledge/South Australian Museum; (b) Chip Clark. **21.14** (a,b,d) Art, R. Ciemma; (c) Patricia G. Gensel. **21.15** *Above,* Megan Rohn courtesy David Dilcher; *below* , Precision Graphics. **21.16** © John Gurche 1989. **21.17** (a) NASA Galileo Imaging Team. **21.18** Art, R. Ciemma. **21.19** (a,b) Paintings © Ely Kish; (c) Field Museum of Natural History, Chicago, and the artist Charles R. Knight (Neg. No. CK8T). **21.20** *Left,* Maps by Lloyd K. Townsend after A. M. Ziegler, C. R. Scotese, and S. F. Barrett, "Mesozoic and Cenozoic Paleogeographic Maps" and J. Krohn, J. Syndermann, "Paleotides Before the Permian" in F. Brosche and J. Syndermann (Eds.), *Tidal Friction and the Earth's Rotation II*, Springer-Verlag, 1983.

CHAPTER 22 22.1 (a-c) Tony Brain and David Parker/SPL/Photo Researchers. **22.2** Lee D. Simon/Photo Researchers. **22.3** Art, R. Ciemma. **22.4** (a) Stanley Flegler/Visuals Unlimited; (b) CNRI/SPL/Photo Researchers. **22.5** L. J. LeBeau, University of Illinois Hospital/BPS. **22.6** L Santo; art, R. Ciemma. **22.9** (a) John D. Cunningham/Visuals Unlimited; (b) Tony Brain/SPL/Photo Researchers; (c) P. W. Johnson and J. McN. Sieburth, University of Rhode Island/BPS. **22.10** T. J. Beveridge, University of Guelph/BPS . **22.11** *Above,* Centers for Disease Control; *below,* Edward S. Ross. **22.12** (a) Richard Blakemore; (b) Hans Reichenbach, Gesellschaft fur Biotechnologische Forschung, Braunschweig, Germany. **22.13** (a) John Clegg/Ardea London; (b) Courtesy Jack Jones, *Archives of Microbiology*, Volume 136, 1983, pp. 254-261. Reprinted by permission of Springer-Verlag; (c) M. Abbey/Visuals Unlimited; (d) G. Shih and R. Kessel/Visuals Unlimited; (e) T. E. Adams/Visuals Unlimited. **22.15** (a) George Musil/Visuals Unlimited; (b) K. G. Murti/Visuals Unlimited; (c,d) Kenneth M. Corbett. **22.16, 22.17** Art, Palay/Beaubois and Precision Graphics. **22.18** R. Regnery, Centers fo Disease Control and Prevention, Atlanta. **22.19** Edmund Zottola.

CHAPTER 23 23.1 (a) Ronald W. Hoham, Dept. of Biology, Colgate University; (b) Steven C. Wilson/Entheos; (c) Edward S. Ross; (d) Gary

W. Grimes and Steven L'Hernault. **23.2** M. S. Fuller, *Zoosporic Fungi in Teaching and Research*, M. S. Fuller and A. Jaworski (eds.), 1987, Southeastern Publishing Company, Athens, GA. **23.3** (a) Heather Angel; (b) W. Merrill. **23.4** (a) Art, Leonard Morgan; (b) M. Claviez, G. Gerish, and R. Guggenheim; (c) London Scientific Films; (d-f) Carolina Biological Supply Company; (g) courtesy Robert R. Kay from R. R. Kay et al. *Development*, 1989 Supplement, pp. 81-90, © The Company of Biologists Ltd. 1989. **23.5** M. Abbey/Visuals Unlimited. **23.6** (a) Dr. Howard Spero; (b) John Clegg/Ardea, London; (c, e) redrawn from V. & J. Pearse and M. & R. Buchsbaum, *Living Invertebrates*, The Boxwood Press, 1987. Used by permission; (d) G. Shih and R. Kessel/Visuals Unlimited; (e) T. E. Adams/ Visuals Unlimited. **23.7** (a) Jerome Paulin/ Visuals Unlimited; (b) David M. Phillips/Visuals Unlimited; (c) John D. Cunningham/Visuals Unlimited. **23.8** Steven L'Hernault; art, Leonard Morgan. **23.9** (a) Frieder Sauer/ Bruce Coleman Ltd.; (b) art, R. Ciemma redrawn from V. & J. Pearse and M. & R. Buchsbaum, *Living Invertebrates*, The Boxwood Press, 1987. Used by permission. (c) Art, R. Ciemma; (d) Sidney L. Tamm; (e) Ralph Buchsbaum, and sketch from V. & J. Pearse and M. & R. Buchsbaum, *Living Invertebrates*, The Boxwood Press, 1987, used by permission. **23.10** Art, R. Ciemma. **23.11** P. L. Walne and J. H. Arnott, *Planta*, 77:325-354, 1967; art, Palay/Beaubois. **23.12** (a) Greta Fryxell, University of Texas, Austin; (b-d) Ronald W. Hoham, Dept. of Biology, Colgate University. **23.13** (a) © S. Berry/Visuals Unlimited; (b) C. C. Lockwood; (c) Florida Department of Environmental Protection, Florida Marine Research Institute, St. Petersburg. **23.14** (a) D. P. Wilson/Eric & David Hosking; (b) Douglas Faulkner/Sally Faulkner Collection. **23.15** Art, R. Ciemma. **23.16** (a) J. R. Waaland, University of Washington/BPS; (b) Lewis Trusty/Animals Animals;*below,* Tom Garrison, *Oceanography: An Invitation to Marine Science*, Wadsworth, 1993. **23.17** (a) Hervé Chaumeton/Agence Nature; (b) © Linda Sims/Visuals Unlimited; (c) Brian Parker/Tom Stack and Associates; (d) Manfred Kage/Peter Arnold, Inc.; (e) Ronald W. Hoham, Colgate University. **23.18** D. J. Patterson/Seaphot Limited: Planet Earth Pictures; art, R. Ciemma. **23.19** Carolina Biological Supply Company.

CHAPTER 24 **24.1** Robert C. Simpson/Nature Stock. **24.2** (a), (b) Robert C. Simpson/Nature Stock. **24.3** Garry T. Cole, University of Texas, Austin/BPS; art, R. Ciemma. **24.4** (a) Jane Burton/Bruce Coleman Ltd.; (b) Thomas J. Duffy. **24.5** Ed Reschke; art, R. Ciemma. **24.6** (a) After T. Rost et al., *Botany*, Wiley, 1979; (b) M. Eichelberger/Visuals Unlimited; (c) Robert C. Simpson/Nature Stock; (d) G. L. Barron, University of Guelph; (e) Garry T. Cole, University of Texas, Austin/BPS. **24.7** N. Allin and G. L. Barron. **24.8** (a) After Raven, Evert, and Eichhorn, *Biology of Plants*, Fourth edition, Worth Publishers, New York, 1986; (b), (c) Gary Head; (d) Mark Mattock/ Planet Earth Pictures; (e) Edward S. Ross. **Page 392** J. D. Cunningham/Visuals Unlimited. **24.9** (a) © 1990 Gary Braasch; (b) F. B. Reeves. **24.10** (a) Dr. P. Marazzi/SPL/Photo Researchers; (b) Eric Crichton/Bruce Coleman Ltd. **24.11** John Hodgin.

CHAPTER 25 **25.1** (a) Pat & Tom Leeson/Photo Researchers; (b, c) Edward S. Ross. **25.2** Art, R. Ciemma after E.O. Dodson and Peter Dodson, *Evolution Process and Product, 3rd Edition*, p. 401, Prindle Weber and Schmidt. **25.4** Craig Wood/Visuals Unlimited; (b) Jane Burton/ Bruce Coleman Ltd.; art, R. Ciemma. **25.5** (a) Fred Bavendam/Peter Arnold; (b) John D. Cunningham/Visuals Unlimited. **25.6** (a,b) Kingsley R. Stern; (c) John D. Cunningham/ Visuals Unlimited. **25.7** (b) Kingsley R. Stern; (c) Ed Reschke/ Peter Arnold; (d) William Ferguson; (e) W. H. Hodge; (f) Kratz/ZEFA. **25.8** Art, R. Ciemma; *inset* photograph A. & E. Bomford/Ardea, London; Lee Casebere. **25.9** Brian Parker/Tom Stack & Associates; art, R. Ciemma; *below,* Field Museum of Natural History, Chicago; (Neg. #7500C). **25.10** (a) Kathleen B. Pigg, Arizona State University; (b) George J. Wilder/Visuals Unlimited. **25.11** (a) Stan Elems/ Visuals Unlimited. **25.12** (a) Jeff Gnass Photography; (b) R. J. Erwin/Photo Researchers; (c) Robert and Linda Mitchell; (d) Doug Sokell/Visuals Unlimited. **25.13** (a) Robert &Linda Mitchell; (b) Ed Reschke. **25.14** (a) Joyce Photographics/Photo Researchers; (b) Runk/Schoenberger/Grant Heilman, Inc.; (c) William Ferguson; (d) Kingsley R. Stern. **25.16** Edward S. Ross; art, R. Ciemma. **25.17** (a) David Hiser, Photographers/ Aspen, Inc.; (b) Terry Livingstone; (c) © 1993 Trygve Steen; (d) © 1994 Robert Glenn Ketchum. **25.18** (a) M.P.L. Fogden/Ardea London; (b) Heather Angel; (c) L. Mellichamp/ Visuals Unlimited; (d) Peter F. Zika/Visuals Unlimited; (e) Art, Jennifer Wardrip. **25.19** Art, R. Ciemma. **25.20** (a,b) © 1989, 1991 Clinton Webb.

CHAPTER 26 **26.1** (a) Courtesy of Department of Library Services, American Museum of Natural History (Neg. # K10273); (c) Lisa Starr, Jack Carey. **26.2** Art, D. & V. Hennings. **26.3** Art, R. Ciemma. **26.4** *Left,* after Laszlo Meszoly in L. Margulis, *Early Life*, Jones and Bartlett. **26.5** Bruce Hall. **26.6** (a-c) R. Ciemma; (d) *left,* Art, R. Ciemma, after Bayer and Owre, *The Free-Living Lower Invertebrates,* © 1968 Macmillan; *right,* Don W. Fawcett/Visuals Unlimited; (e) Marty Snyderman/Planet Earth Pictures. **26.7** Art. R. Ciemma; (c) F. Schensky; (d) Kim Taylor/ Bruce Coleman Ltd. **26.8** R. Ciemma. **26.9** (a) Art, R. Ciemma; (b) Douglas Faulkner/Sally Faulkner Collection; (c) F. S. Westmorland/Tom Stack & Associates. **26.10** Art, Precision Graphics after T. Storer et al., *General Zoology*, Sixth edition, © 1979 McGraw-Hill. **26.11** (a) Christian DellaCorte; (b) Douglas Faulkner/ Photo Researchers; (c) Bill Wood/ Seaphot Limited: Planet Earth Pictures. **26.12** Andrew Mounter/Seaphot Limited: Planet Earth Pictures. **26.13** R. Ciemma; (b) Kathie Atkinson/Oxford Scientific Films. **26.14** Kim Taylor/Bruce Coleman Ltd.; art, R. Ciemma. **26.15** (a) Cath Ellis, University of Hull/SPL/ Photo Researchers; (b) Robert & Linda Mitchell. **26.16** Kjell B. Sandved. **26.17-26.18** R. Ciemma. **27.19** Art, R. Ciemma, micrograph Carolina Biological Supply Company. **26.20** (a) Lorus J. and Margery Milne; (b) Dianora Niccolini. **26.21** Art, R. Ciemma. **Page 430** Art, Palay/Beaubois. **26.22** (a) Gary Head; (b) Alex Kerstitch; (c) Jeff Foott/Tom Stack & Associates; (d) Frank Park/A.N.T. Photo Library; (e) J. Grossauer/ ZEFA. **26.23** (b) Kjell B. Sandved; (e) Hervé Chaumeton/Agence Nature. **26.24** (b) Hervé Chaumeton/Agence Nature. **26.25** (a) Art, R. Ciemma; (b) Bob Cranston; (c) © 1992 Sea World of Califormia, All rights reserved.

26.26 (a) J.A.L. Cooke/Oxford Scientific Films; (b) © Cabisco/Visuals Unlimited; (c) Jon Kenfield/Bruce Coleman Ltd.; (d) *Left,* from Eugene N. Kozloff, *Invertebrates,* copyright © 1990 by Saunders College Publishing. Reproduced by permission of the publisher. *Right,* adapted from Rasmussen, *Ophelia,* Vol. 11 in Eugene N. Kozloff, *Invertebrates,* 1990. **26.27** Art, R. Ciemma. **26.28** Jane Burton/ Bruce Coleman Ltd. **26.29** *Above,* Angelo Giampiccolo/FPG; *below,* Jane Burton/ Bruce Coleman Ltd. **26.30** (a) Redrawn from *Living Invertebrates,* V. & J. Pearse/M. & R. Buchsbaum, The Boxwood Press, 1987. Used by permission; (b) P. J. Bryant, University of California, Irvine/BPS; (c) Ken Lucas/ Seaphot Limited: Planet Earth Pictures; (d) John H. Gerard. **26.31** (a) Franz Lanting/ Bruce Coleman Ltd.; (b) Hervé Chaumeton; (d) Agence Nature; (f) Fred Bavendam/Peter Arnold, Inc. (e), (g) Art, R. Ciemma. **26.32** After David H. Milne, *Marine Life and the Sea,* Wadsworth 1995. **26.33** (a) Z. Leszczynski/ Animals Animals; (b) Steve Martin/Tom Stack & Associates. **26.34** Art, D. & V. Hennings. **26.35** From Georges Pasteur, "Jean Henri Fabre," *Scientific American,* July 1994. Copyright © 1994 by Scientific American, Inc. All rights reserved. **26.36** (a-d) Edward S. Ross; (e) C. P. Hickman, Jr.; (f-i) Edward S. Ross; (j) David Maitland/Seaphot Limited: Planet Earth Pictures. **26.37** (a) Chris Huss/The Wildlife Collection; (b) John Mason/Ardea London; (c) Kjell B. Sandved; (d) Ian Took/Biofotos. **26.38** (a) Art, Lewis Calver; (b-c) Hervé Chaumeton/Agence Nature. **Page 443** Jane Burton/Bruce Coleman Ltd. **26.39** Art, R. Ciemma. **26.40** Walter Deas/Seaphot Limited: Planet Earth Pictures. **26.41** (a) J. Solliday/BPS; (b) Hervé Chaumeton/ Agence Nature.

CHAPTER 27 **27.1** (a) Jean Phillipe Varin/ Jacana/Photo Researchers; (b) Tom McHugh/ Photo Researchers. **27.3** (a) Rick M. Harbo; (b) Peter Parks/Oxford Scientific Films/Animals Animals; (c) redrawn from *Living Invertebrates,* V. & J. Pearse and M. & R. Buchsbaum, The Boxwood Press, 1987. Used by permission. **27.4** (a) Art, R. Ciemma; (b) Runk & Schoenberger/ Grant Heilman, Inc. **27.6** Al Giddings/Images Unlimited; art, R. Ciemma, adapted from A. S. Romer and T. S. Parsons, *The Vertebrate Body*, Sixth edition, Saunders College Publishing, 1986. **27.7** (a, b) After David H. Milne, *Marine Life and the Sea,* Wadsworth, 1995; (c) Heather Angel. **27.8** (a) Erwin Christian/ZEFA; (b) Allan Power/Bruce Coleman Ltd.; (c) Tom McHugh/Photo Researchers. **27.9** (a) Bill Wood/Bruce Coleman Ltd.; (b) R. Ciemma. **27.10** (a) From Tom Garrison, *Oceanography: An Invitation to Marine Science,* Wadsworth, 1993; (b) Robert & Linda Mitchell; (c) Patrice Ceisel/© 1986 John G. Shedd Aquarium; (d) Peter Scoones/Seaphot Limited: Planet Earth Pictures. **27.11** (a) © Marianne Collins; (b, c) art, Laszlo Meszoly and D. & V. Hennings. **27.12** (a) Jerry W. Nagel; (b) Stephen Dalton/ Photo Researchers; (c) John Serraro/Visuals Unlimited; (d) Juan M. Renjifo/Animals Animals. **Page 455** Art, Leonard Morgan adapted from A. S. Romer and T. S. Parsons, *The Vertebrate Body,* Sixth edition, Saunders College Publishing, 1986, and others. **27.13** (a) R. Ciemma; (b) © 1989 D. Braginetz; (c) Zig Leszczynski/Animals Animals. **27.14** D. & V. Hennings. **27.15** (a) Kevin Schafer/Tom Stack & Associates;

(b) D. Kaleth/ Image Bank; (c) Art, R. Ciemma; (d) Andrew Dennis/A.N.T. Photo Library; (e) Stephen Dalton/Photo Researchers; art, R. Ciemma; (f) Heather Angel. **27.16** (a) Gerard Lacz/A.N.T. Photo Library; (b) J.L.G. Grande/Bruce Coleman Ltd.; (c) Rajesh Bedi; (d) Thomas D. Mangelsen/Images of Nature. **27.17** (b) Art, D. & V. Hennings. **27.18** (a) Sandy Roessler/FPG; (b) Leonard Lee Rue III/FPG; (c) Art, R. Ciemma after M. Weiss and A. Mann, *Human Biology and Behavior*, Fifth edition, Harper Collins, 1990. **27.20** (a) D. & V. Blagden/A.N.T. Photo Library; (b) Jack Dermid; (c) Photo Researchers. **27.21** R. Ciemma. **27.22** (a) Clem Haagner/Ardea London; (b) J. Scott Altenbach, University of New Mexico; (c) Douglas Faulkner/Photo Researchers; (d) Christopher Crowley; (e) Leonard Lee Rue III/FPG. **27.23** W. J. Weber/Visuals Unlimited.

CHAPTER 28 **28.1** *Left*, FPG; *right*, Douglas Mazonowicz/Gallery of Prehistoric Art. **28.2** *below*, Roger Burnard. **28.3** (a) Bruce Coleman Ltd.; (b) Tom McHugh/Photo Researchers; (c) Larry Burrows/Aspect Picture Library. **28.4** D. & V. Hennings. **28.5** © Time Inc. 1965/Larry Burrows Collection. **28.8** Art, Precision Graphics after *National Geographic*, February 1997, page 82. **28.9** Art, D. & V. Hennings. **28.10** (a) Dr. Donald Johanson, Institute of Human Origins; (b) Louise M. Robbins; (c,d) Kenneth Garrett/National Geographic Image Collection. **28.11** (a,b) Kenneth Garrett/National Geographic Image Collection; (c) Jean Paul Tibbles. **28.12** Photographs by John Reader ©1981. **28.13** Art, R. Ciemma. **Page 479** Gary Head.

CHAPTER 29 **29.1** (a) Roger Werth; (b) © 1980 Gary Braasch; (c) © 1989 Gary Braasch; (d) Don Johnson/Photo Nats, Inc.. **29.2, 29.4** R. Ciemma. **29.5** James D. Mauseth, *Plant Anatomy*, Benjamin-Cummings, 1988. **29.6** (a-c) Biophoto Associates. **29.7** (a) Jan Robert Factor/Photo Researchers; (b) D. E. Akin and l. L. Risgby, Richard B. Russel Agricultural Research Center, Agricultural Research Service, US, Department of Agriculture, Athens, GA; (c) Kingsley R. Stern. **29.9** George S. Ellmore. **29.10** D. & V. Hennings. **29.11** R. Ciemma. **29.12** (a) Robert & Linda Mitchell; (b) Roland R. Dute; (c) Gary Head. **29.13** D. & V. Hennings; (a) *center*, Ray F. Evert; *right*, James W. Perry; (b) *center*, Carolina Biological Supply Company; *right*, James W. Perry. **29.14** D. & V. Hennings. **29.15** (a) R. Ciemma; (b) C. E. Jeffree et al., *Planta*, 172(1):20-37, 1987. Reprinted by permission of C. E. Jeffree and Springer-Verlag; (c) Jeremy Burgess/SPL/Photo Researchers. **29.16** (a) Heather Angel; (b) Gary Head. **29.17** George Loun/Visuals Unlimited. **29.18** (a) Walter Hodge/Peter Arnold; (b) Dick Davis/Photo Researchers; (c) John Mason/Ardea London; (d) Earl Roberge/Photo Researchers. **29.19** Bob Cerasoli / **29.20** Art, R. Ciemma. **29.21** (a) R. Ciemma; (b) Ripon Microslides, Inc. **29.22** (a) Chuck Brown; (b) Carolina Biological Supply Company. **29.23** Ripon Microslides; sketch after T. Rost et al., *Botany: A Brief Introduction to Plant Biology*, Second edition, © 1984, John Wiley & Sons. **29.24** (a) John Lotter Gurling/Tom Stack and Associates; (b) Art, R. Ciemma, microslide, Biophot. **29.25** R. Ciemma. **29.28** H. A. Core, W. A. Cote, and A. C. Day, *Wood Structure and Identification*, Second edition, Syracuse University Press, 1979. **29.29** (b, c) R. Ciemma. **29.30** Edward S. Ross. **Page 499** Ripon Microslides.

CHAPTER 30 **30.1** (a,b) Robert & Linda Mitchell; (c) John N.A. Lott, *Scanning Electron Microscope Study of Green Plants*, St. Louis: C. V. Mosby, 1976; (d) Robert C. Simpson/Nature Stock. **30.2** (a) David Lavagnaero/Peter Arnold; (b) Dwight R. Kuhn. **30.3** William Ferguson. **30.4** U.S. Department of Agriculture. **30.5** (a) Micrograph Chuck Brown; (b,c) Leonard Morgan. **30.6** Micrograph Jean Paul Revel. **30.7** (b) Mark E. Dudley and Sharon R. Long; (c) Adrian P. Davies/Bruce Coleman Ltd.; (d) NifTAL Project, University of Hawaii, Maui; art, Jennifer Wardrip. **30.8** H. A. Core, W. A. Cote, and A. C. Day, *Wood Structure and Identification*, Second edition, Syracuse University Press, 1979. **30.9** Art, R. Ciemma. **30.10** Frank B. Salisbury. **30.11** George S. Ellmore. **30.12** W. Thomson, *American Journal of Botany*, 57(3):316, 1970. **30.14** T. A. Mansfield. **30.15** (a,b) Jeremy Burgess/SPL/Photo Researchers; John Lawlor/FPG. **30.16** Art, Hans & Cassady, Inc. **30.17** Martin Zimmermann, *Science*, 133:73-79, © AAAS 1961. **30.18** (a-c) Art, Palay/Beaubois. **30.20** James T. Brock.

CHAPTER 31 **31.1** (a) Merlin D. Tuttle, Bat Conservation International; (b) Thomas Eisner, Cornell Univerity. **31.2** Robert A. Tyrrell. **31.3** (a) Gary Head. **31.4** John Shaw/Bruce Coleman Ltd. **31.5** (a,b) David M. Phillips/Visuals Unlimited; (c) David Scharf/Peter Arnold, Inc. **1.6** R. Ciemma. **31.7** (a,b) Patricia Schulz; (c,d) Ray F. Evert; (e,f) Ripon Microslides, Inc. **31.8** (a) Richard H. Gross; (b) Janet Jones; (c) B. J. Miller, Fairfax, VA/BPS. **31.9** (a) Richard H. Gross; (b) Gary Head; (c) ZEFA-Rein. **31.10** (a-c) John Alcock. **31.11** Russell Kaye/ © 1993 The Walt Disney Co. Reprinted with permission of *Discover* Magazine. **31.12** (a) Runk/Schoenberger/Grant Heilman, Inc.; (b) Kingsley R. Stern. **31.13** Gary Head.

CHAPTER 32 **32.1** (a) Michael A. Keller/FPG; (b, c) R. Lyons/Visuals Unlimited. **32.2** Sylvan H. Wittwer/Visuals Unlimited. **32.3** B. Bracegirdle and P. Miles, *An Atlas of Plant Structure*, 1977, Heinemann Educational Books. **32.4** (a,b,d,e) R. Ciemma; (c) Hervé Chaumeton/Agence Nature; (f) Barry L. Runk/Grant Heilman, Inc.; (g) Mauseth. **32.6** (a,b) R. Ciemma; (c) Biophot. **32.7** (a) Kingsley R. Stern. **32.8** (a,b) courtesy of Randy Moore from "How Roots Respond to Gravity," M. L. Evans, R. Moore, and K. Hasenstein, *Scientific American*, December 1986; (c) John Digby and Richard Firn. **32.9** (b) Frank B. Salisbury. **32.10** Gary Head. **32.11** Cary Mitchell. **32.13** Frank B. Salisbury. **32.14** Diana Starr. **32.15** (a) Jan Zeevart; (b) Gary Head. **32.16** N. R. Lersten. **32.17** Larry D. Nooden. **32.18** R. J. Downs. **32.19** Art, R. Ciemma. **32.20** Diana Starr. **32.21** Edward S. Ross; *below*, Gary Head. **32.22** Grant Heilman Inc. **Page 543** © Kevin Schafer.

CHAPTER 33 **33.1** David Macdonald. **33.2** (a) Focus on Sports;*inset*, Manfred Kage/Bruce Coleman Ltd.; art, Palay/Beaubois; (b)*left*, Lennart Nilsson from *Behold Man*, © 1974 by Albert Bonniers Forlag and Little, Brown and Company, Boston; (center) Manfred Kage/Bruce Coleman Ltd.; *right*, Ed Reschke/Peter Arnold, Inc. **33.3** Art, R. Ciemma. **33.4** Gregory Dimijian/Photo Researchers; art, R. Ciemma adapted from C. P. Hickman, Jr., L. S. Roberts, and A. Larson, *Integrated Principles of Zoology*, Ninth edition, Wm. C. Brown, 1995. **33.5** (a-c, e, f) Ed Reschke; (d) Fred Hossler/Visuals Unlimited. **33.6** Roger K. Burnard;*left* art, Joel Ito; *right* art, L. Calver. **33.7, 33.8** Ed Reschke. **33.9** Robert Demarest. **33.10** (a) Lennart Nilsson

from *Behold Man*, © 1974 Albert Bonniers Forlag and Little, Brown and Company, Boston; (b) Kim Taylor/Bruce Coleman Ltd. **33.11** Art, L. Calver. **33.14** Fred Bruemmer

CHAPTER 34 **34.1** Adrian Warren/Ardea London. **Page 559** Art, L. Calver. **34.2** (d) Manfred Kage/Peter Arnold, Inc. / **34.3, 34.5** R. Ciemma. **34.7** Ed Reschke; art, Kevin Somerville. **34.8** (b) Dr. Constantino Sotelo from *International Cell Biology*, page 83, 1977. Used by copyright permission of the Rockefeller University Press. **34.10** R. G. Kessel and R. H. Kardon from *Tissues and Organs: A Text-Atlas of Scanning Electron Microscopy*, W. H. Freeman and Company, 1979. Used by permission of R. G. Kessel; art, Robert Demarest. **34.12** (b) Robert Demarest. **34.13** Painting by Sir Charles Bell, 1809, courtesy of Royal College of Surgeons, Edinburgh.

CHAPTER 35 **35.1** Comstock, Inc. **35.2** Kjell B. Sandved. **35.3, 35.5** (b) R. Ciemma. **35.6** Kevin Somerville. **35.9** (a) Robert Demarest; (b) Manfred Kage/Peter Arnold, Inc. **35.10-35.12** Kevin Somerville. **35.13** (a) Colin Chumbley/Science Source/Photo Researchers; (b) C. Yokochi, J. Rohen, *Photographic Anatomy of the Human Body*, Second edition, Igaku-Shoin Ltd., 1979. **35.14** *Left*, Marcus Raichle, Washington University School of Medicine. **35.15** *Left*, Palay/Beaubois after Penfield and Rasmussen, *The Cerebral Cortex of Man*, copyright © 1950 Macmillan Publishing Company, Inc. Renewed 1978 by Theodore Rasmussen; *right*, Colin Chumbley/Science Source/Photo Researchers. **35.22** (a,b) From Edythe D. London et al., *Archives of General Psychiatry*, 47:567-574 (1990); *right*, Ogden Gigli/Photo Researchers.

CHAPTER 36 **36.1** (a) Eric A. Newman; (b) Merlin D. Tuttle, Bat Conservation International. **36.2** Art, Kevin Somerville. **36.3** From Hensel and Bowman, *Journal of Physiology*, 23:564-568, 1960. **36.4** *Left*, Palay/Beaubois after Penfield and Rasmussen, *The Cerebral Cortex of Man*, copyright © 1950 Macmillan Publishing Company, Inc. Renewed 1978 by Theodore Rasmussen; *right*, Colin Chumbley/Science Source/Photo Researchers. **36.5, 36.6** R. Ciemma. **36.7** Omikron/SPL/Photo Researchers; art, Robert Demarest. **36.9** Edward W. Bower/© 1991 TIB/West; art, Kevin Somerville. **36.10** (b,c) Kevin Somerville . **36.12** Robert Demarest. **36.13** (a,b) Robert E. Preston, courtesy Joseph E. Hawkins, Kresge Hearing Research Institute, University of Michigan Medical School. **36.14** Art, R. Ciemma; (d) Keith Gillett/Tom Stack & Associates. **36.15** E. R. Degginger; sketch after M. Gardiner, *The Biology of Vertebrates*, McGraw-Hill, 1972 . **36.16** G. A. Mazohkin-Porshnykov (1958). Reprinted with permission from *Insect Vision*, © 1969 Plenum Press. **36.17** Chris Newbert. **36.18** Robert Demarest. **36.19** Chase Swift. **36.20, 36.21** Kevin Somerville. **36.22** Lennart Nilsson © Boehringer Ingelheim International GmbH. **36.24** Palay/Beaubois after S. Kuffler and J. Nicholls, *From Neuron to Brain*, Sinauer, 1977. **36.25, 36.26** Gerry Ellis/The Wildlife Collection. **36.27** Douglas Faulkner/Sally Faulkner Collection.

CHAPTER 37 **37.1** Hugo van Lawick. **37.2** Kevin Somerville. **37.5, 37.6** Robert Demarest. **37.7** (a) Mitchell Layton; (b) Syndication International (1986) Ltd.; (c) courtesy of Dr. William H. Daughaday, Washington University School of Medicine, from A. I. Mendelhoff and D. E. Smith, eds., *American Journal of Medicine*, 20:133 (1956). **37.8** Leonard Morgan. **37.9** (a, b)

Art, R. Ciemma; (c) Corbis-Bettmann. **37.11** Biophoto Associates/SPL/Photo Researchers. **37.12** Leonard Morgan. **37.13** (a) John S. Dunning/Ardea London; (b) Evan Cerasoli. **37.14** (a, b) From R. C. Brusca and G. J. Brusca, *Invertebrates*, © 1990 Sinauer Associates. Used by permission; (c) Frans Lanting/Bruce Coleman Ltd.; (d) Robert & Linda Mitchell. **37.15** Roger K. Burnard.

CHAPTER 38 38.1 John Brandenberg/Minden Pictures. **38.2** L. Calver. **38.3** (a) Robert Demarest; (b) CNRI/SPL/Photo Researchers; (c) Gregory Dimijian/Photo Researchers. **38.4** (a) Tom McHugh; (b) M.P.L. Fogden/Bruce Coleman Ltd.; (c) Chaumeton-Lanceau/Agence Nature; sketches from *The Life of Birds*, Fourth edition, by Luis Baptista and Joel Carl Welty, © 1988 by Saunders College Publishing. Reproduced by permission of the publisher. **38.5** Ed Reschke. **38.6** Michael Keller/FPG. **38.7** Linda Pitkin/Planet Earth Pictures. **38.8** D. A. Parry, *Journal of Experimental Biology*, 36:654, 1959. **38.9** (a) Stephen Dalton/Photo Researchers; (b) R. Ciemma. **38.10** D. & V. Hennings. **38.11** (a) R. Ciemma; (b) Ed Reschke, art, Joel Ito. **38.12** K. Kasnot. **38.13** National Osteoporosis Foundation. **38.14** C. Yokochi and J. Rohen, *Photographic Anatomy of the Human Body*, Second edition, Igaku-Shoin Ltd., 1979. **38.15, 38.16** R. Ciemma. **38.17** (a) N.H.P.A./A.N.T. Photo Library; (b) Robert Demarest. **38.18** Kevin Somerville. **38.19** (b) John D. Cunningham; (c) D. W. Fawcett, *The Cell*, Philadelphia: W. B. Saunders Co., 1966; art, Robert Demarest. **38.20** Nadine Sokol. **38.21** Robert Demarest. **38.22** After Stephen L. Wolfe, *Molecular and Cellular Biology*, Wadsworth, 1993. **38.25** Michael Neveux.

CHAPTER 39 39.1 (a) A. D. Waller, *Physiology, The Servant of Medicine*, Hitchcock Lectures, University of London Press, 1910; (b) courtesy The New York Academy of Medicine Library. **39.3** (c) After M. Labarbera and S. Vogel, *American Scientist*, 70:54-60, 1982. **39.4** Precision Graphics. **39.5** Palay/Beaubois. **39.6** CNRI/SPL/Photo Researchers. **39.7** R. Ciemma. **39.8** (a,b) Lester V. Bergman & Associates, Inc.; (c) after F. Ayala and J. Kiger, *Modern Genetics*, © 1980 Benjamin-Cummings. **39.9** Nadine Sokol after Gerard J. Tortora and Nicholas P. Anagnostakos, *Principles of Anatomy and Physiology*, Sixth edition, Copyright © 1990 by Biological Sciences Textbooks, Inc., A & P Textbooks, Inc. and Elia-Sparta, Inc. Reprinted by permission of HarperCollins Publishers. **39.11** Kevin Somerville. **39.12** (a) C. Yokochi and J. Rohen, *Photographic Anatomy of the Human Body*, Second edition, Igaku-Shoin Ltd, 1979; (b, c) R. Ciemma. **39.14** Michael Abbey/ Photo Researchers; art, R. Ciemma. **39.15** R. Ciemma. **39.16** Robert Demarest based on A. Spence, *Basic Human Anatomy*, Benjamin-Cummings, 1982 . **39.18** Sheila Terry/SPL/ Photo Researchers. **39.19** R. Ciemma. **39.20** Kevin Somerville. **39.21** (a) Ed Reschke; (b) F. Sloop and W. Ober/Visuals Unlimited. **39.24** Lennart Nilsson © Boehringer Ingelheim International GmbH. **39.25, 39.26** R. Ciemma. **39.27** Lennart Nilsson from *Behold Man*, © 1974 by Albert Bonniers Forlag and Little, Brown and Company, Boston

CHAPTER 40 40.1 (a) The Granger Collection, New York; (b) Lennart Nilsson © Boehringer Ingelheim International GmbH. **40.2** Nadine Sokol. **40.3** Robert R. Dourmashkin, courtesy

of Clinical Research Centre, Harrow, England. **40.4** Lennart Nilsson © Boehringer Ingelheim International GmbH. **40.5, 40.8** R. Ciemma. **40.9** Morton H. Nielsen and Ole Werdlin, University of Copenhagen; art, R. Ciemma. **40.10** courtesy Don C. Wiley, Harvard University; art, R. Ciemma. **40.11** Hans & Cassady, Inc. **40.13, 40.14** Palay/Beaubois after B. Alberts et al., *Molecular Biology of the Cell*, Garland Publishing Company, 1983. **40.16** *Above*, Lowell Georgia/Science Source/ Photo Researchers; (below) Matt Meadows/ Peter Arnold, Inc. **40.17** (left) David Scharf/ Peter Arnold, Inc.; *right*, Kent Wood/Photo Researchers. **40.18** Ted Thai/Time Magazine. **40.19** After Stephen L. Wolfe, *Molecular B iology of the Cell*, Wadsworth, 1993. **40.20** Z. Salahuddin, National Institutes of Health.

CHAPTER 41 41.1 Galen Rowell/Peter Arnold, Inc. **41.4** (a) Peter Parks/Oxford Scientific Films; (b) Hervé Chaumeton/Agence Nature. **41.5** Ed Reschke . **41.6** R. Ciemma. **41.7** After C. P. Hickman et al., *Integrated Principles of Zoology*, Sixth edition, St. Louis: C. V. Mosby Co., 1979. **41.9** H. R. Duncker, Justus-Liebig University, Giessen, Germany. **41.10** Kevin Somerville . **41.11** CNRI/SPL/Photo Researchers. **41.12** SIU/Visuals Unlimited; art, K. Kosnot. **41.13** From L. G. Mitchell, J. A. Mutchmor, and W. D. Dolphin, *Zoology*, copyright © 1988 by The Benjamin-Cummings Publishing Company; reprinted by permission. **41.14** Modified from A. Spence and E. Mason, *Human Anatomy and Physiology*, Fourth edition, 1992, West Publishing. **41.16** Leonard Morgan. **41.17** O. Auerbach/Visuals Unlimited. **41.18** Lennart Nilsson from *Behold Man*, © 1974 by Albert Bonniers Forlag and Little, Brown and Company, Boston. **41.19** Giorgio Gualco/ Bruce Coleman Ltd. **41.20** From Tom Garrison, *Oceanography: An Invitation to Marine Science*, Wadsworth, 1993. **41.21** Christian Zuber/ Bruce Coleman Ltd. **41.22** Steve Lissau/Rainbow.

CHAPTER 42 42.1 Gary Head. **42.3** R. Ciemma. **42.4** (a) D. Robert Franz/Planet Earth Pictures; (c) adapted from A. Romer and T. Parsons, *The Verterbrate Body*. Sixth edition, Saunders Publishing Company, 1986. **42.5** Art, Kevin Somerville. **42.7** Omikron/SPL/Photo Researchers; art, Robert Demarest. **42.8** After A. Vander et al., *Human Physiology: Mechanisms of Body Function*, Fifth edition, McGraw-Hill, 1990, used by permission; (c) Redrawn from *Human Anatomy and Physiology*, Fourth edition, by A. Spence and E. Mason. Copyright © 1992 by West Publishing Company. All rights reserved. **42.9** (a) R. Ciemma; (b,c)*right*, Lennart Nilsson © Boehringer Ingelheim International GmbH; (c) *left*, Biophoto Associates/SPL/Photo Researchers; art, Robert Demarest. **42.10** R. Ciemma. **Page 719** (in-text art) After A. Vander et al., *Human Physiology: Mechanisms of Body Function*, Fifth edition, McGraw-Hill, 1990. Used by permission. **42.11** Ralph Pleasant/ FPG. **42.15** Gary Head. **42.16** Dr. Douglas Coleman, The Jackson Laboratory.

CHAPTER 43 43.1 *Above*, David Noble/ FPG; *below*, Claude Steelman/Tom Stack & Associates. **43.3, 43.4** Robert Demarest. **43.5** *Above*, Robert Demarest;*below*, Precision Graphics. **43.7** (a, b) From Tom Garrison, *Oceanography: An Invitation to Marine Science*, Wadsworth, 1993; (c) Thomas D. Mangelsen/

Images of Nature. **43.8** (a) Colin Monteath, Hedgehog House, New Zealand; (b) Bob McKeever/Tom Stack & Associates. **43.9** Everett C. Johnson. **43.10** *Left*, Robert Demarest;*right*, Kevin Somerville. **43.11** David Jennings/Image Works / **43.12** Corbis-Bettmann

CHAPTER 44 44.1 (a) Hans Pfletschinger; (b) R. Ciemma; (c-f) John H. Gerard. **44.2** (a) Frieder Sauer/Bruce Coleman Ltd.; (b) Wisniewski/ZEFA; (c) Zig Leszczynski/ Animals Animals; (d) Carolina Biological Supply Company; (e) FredMcKinney/FPG; (f) Gary Head. **44.4** Carolina Biological Supply Company. **44.7** (a) R. Ciemma after V. E. Foe and B. M. Alberts, *Journal of Cell Science*, 61:32, © The Company of Biologists 1983; (b) J. B. Morrill. **44.8** *Left*, Carolina Biological Supply Company;*far right*, Peter Parks/Oxford Scientific Films/Animals Animals. **44.9** R. Ciemma;*right*, after B. Burnside, *Developmental Biology*, 26:416-441, 1971. Used by permission of Academic Press. **44.10** F. R. Turner; art, R. Ciemma. **44.11** Carolina Biological Supply Company. **44.12** (a) Palay/Beaubois after Robert F. Weaver and Philip W. Hedrick, *Genetics*. Copyright © 1989 Wm. C. Brown Publishers; (b) R. Ciemma after Scott Gilbert, *Developmental Biology*, 4/E. Sinauer. **44.13** R. Ciemma after Scott Gilbert, *Developmental Biology*, 4/E. Sinauer. **44.14** (a) Peter Parks/ Oxford Scientific Films/Animals Animals; (b-d) Hans & Cassady, Inc. adapted from W. Freeman and Brian Bracegirdle, *An Atlas of Embryology*, Third edition, Heinemann Educational Books, 1978; (e) S. R. Hilfer and J. W. Yang, *The Anatomical Record*, 197:423-433. 1980. **Page 759** Dennis Green/Bruce Coleman Ltd. **44.15** Palay/Beaubois adapted from R. G. Ham and M. J. Veomett, *Mechanisms of Development*, St. Louis, C. V. Mosby Co., 1980.

CHAPTER 45 45.1 Lennart Nilsson from *A Child Is Born*, © 1966, 1977 Dell Publishing Company, Inc. **45.2** R. Ciemma. **45.3** (a) R. Ciemma; (b) Ed Reschke. **45.4-45.6** R. Ciemma. **45.7***Above*, Robert Demarest;*below*, R. Ciemma. **45.8** Lennart Nilsson from *A Child Is Born*, © 1966, 1977 Dell Publishing Company, Inc. **45.9** Robert Demarest. **45.10-45.13** R. Ciemma. **45.14** Lennart Nilsson, *A Child Is Born*, © 1966, 1977 Dell Publishing Company, Inc.; art, R. Ciemma. **45.15** (a) R. Ciemma modified from Keith L. Moore, *The Developing Human: Clinically Oriented Embryology*, Fourth edition, Philadelphia: W. B. Saunders Co., 1988. **45.16** (a) From Lennart Nilsson, *A Child Is Born*, © 1966, 1977 Dell Publishing Company, Inc.; (b) James W. Hanson, M.D. **45.17** Robert Demarest. **45.18** R. Ciemma. **45.19** R. Ciemma adapted from L. B. Arey, *Developmental Anatomy*, Philadelphia, W. B. Saunders Co., 1965. **45.21** CNRI/SPL/Photo Researchers. **45.22** (a) John D. Cunningham/Visuals Unlimited; (b) David M. Phillips/Visuals Unlimited. **Page 791** Alan and Sandy Carey.

CHAPTER 46 46.1 (a) Gary Head; (b) Antoinette Jongen/FPG. **46.3** (a) Fran Allan/Animals Animals; (b) E. R. Degginger; (c) P. J. Bryant/ BPS. **46.4** (c) Stanley Flegler/Visuals Unlimited. **46.7** E. Vetter/ZEFA. **Page 800** Eric Crichton/ Bruce Coleman Ltd. **46.8** (a) Jonathan Scott/ Planet Earth Pictures; (b) Fred Bavendam/ Peter Arnold, Inc.; (c) Wisniewski/ZEFA. **46.9** Helen Rodd. **46.10, 46.11** John A. Endler. **46.13** NASA. **46.14** After G. T. Miller, Jr.,

Environmental Science, Sixth edition, Wadsworth, 1997. **46.15** United Nations. **46.16** Data from Population Reference Bureau after G. T. Miller, Jr., *Living in the Environment*, Eighth edition, Wadsworth, 1993. **46.17** United Nations. **46.18**, **46.19** After G. T. Miller, Jr., *Environmental Science*, Sixth edition, Wadsworth, 1997.

CHAPTER 47   47.1 *Left*, Donna Hutchins; *right*, Edward S. Ross. **47.2** (a, c) Harlo H. Hadow; (b) Bob and Miriam Francis/Tom Stack & Associates. **47.3** Clara Calhoun/Bruce Coleman Ltd. **47.4** After G. Gause, 1934. **47.5** Stephen G. Tilley. **47.6** After N. Weland and F. Bazazz, *Ecology*, 56:681-188, © 1975 Ecological Society of America. **47.7** John Dominis, Life Magazine, © Time Inc. **47.8** Ed Cesar/Photo Researchers. **47.9** (a) James H. Carmichael; (b) Edward S. Ross; (c) W. M. Laetsch. **47.10** Edward S. Ross. **47.11** (a, b) Thomas Eisner, Cornell University; (c) Douglas Faulkner/Sally Faulkner Collection; (d) Edward S. Ross. **47.12** © C. James Webb, Phototake, NY. **47.13** (a) Eric Hosking; (b) Stephen Dalton/Photo Researchers. **47.14** Data from P. Price and H. Tripp, *Canadian Entymology*, 104:1003-1016, 1972. **47.15** (a-e, g) Roger K. Burnard; (f) E. R. Degginger. **47.16** (a, c) Jane Burton/Bruce Coleman Ltd.; (b) Heather Angel; (d, e) Jane Lubchenco, *American Naturalist*, 112:23-29, © 1978 by The University of Chicago Press. **47.17** R. Slavin/FPG. **47.18** (a) John Carnemolla/Australian Picture Library; (b) Peter Bird/Australian Picture Library / **47.19** Angelina Lax/Photo Researchers / **47.20** After W. Dansgaard et al., *Nature*, 364:218-220, 15 July 1993; D. Raymond et al., *Science*, 259: 926-933, February 1993; W. Post, *American Scientist*, 78:310-326, July-August 1990. **47.21** *Above*, Dr. Harold Simon/Tom Stack & Associates;*below*, after S. Fridriksson, *Evolution of Life on a Volcanic Island*, Butterworth: London, 1975. **47.22** (a,b) After J. M. Diamond, *Proceedings of the National Academy of Sciences*, 69:3199-3201, 1972; (c) David Cavagnaro. **47.23** (a), (b) Edward S. Ross. **47.24** Heather Angel/Biofotos.

CHAPTER 48   48.1 (a,b) Wolfgang Kaehler. **48.3** Sharon R. Chester. **48.4** Gene C. Feldman and Compton J. Tucker/NASA, Goddard Space Flight Center. **48.8** Gerry Ellis/The Wildlife Collection. **48.9** (a) © 1991 Gary Braasch. **48.11** (a) Gene E. Likens from G. E. Likens and F. H. Bormann, *Proceedings First International Congress of Ecology*, pp. 330-335, September 1974, Centre Agric. Publ. Doc. Wagenigen, The Hague, The Netherlands; (b) Gene E. Likens from G. E. Likens et al., *Ecology Monograph*, 40(1):23-47, 1970; (c) after G. E. Likens and F. H. Bormann, "An

Experimental Approach to New England Landscapes" in A. D. Hasler (ed.), *Coupling of Land and Water Systems*, Chapman & Hall, 1975. **48.12** (b) John Lawlor/FPG. **48.14** After W. Dansgaard et al., *Nature*, 364:218-220, 15 July 1993; D. Raymond et al., *Science*, 259: 926-933, February 1993; W. Post, *American Scientist*, 78:310-326, July-August 1990. **48.16** William J. Weber/Visuals Unlimited. **48.17** After E. Schulze, *Science*, 244:776-783, 1989 / **48.19** D. W. Schindler, *Science*, 184:897-899

CHAPTER 49   49.1 (a)*Above*, Edward S. Ross; *below*, David Noble/FPG; (b) *above*, Edward S. Ross; *below*, Richard Coomber/Planet Earth Pictures. **49.2** Edward S. Ross. **49.4** (b) L. Calver. **49.6**, **49.8** From Tom Garrison, *Essentials of Oceanography*, Wadsworth, 1995. **49.9** (b) Victor Royer. **49.10** R. Ciemm. **49.11** .D. & V. Hennings after Whittaker, Bland, and Tilman. **49.12** Harlo H. Hadow. **49.13** *Above*, John D. Cunningham/Visuals Unlimited; *inset*, AP/Wide World Photos; *below*, Jack Wilburn/Animals Animals. **49.14** (a) Kenneth W. Fink/Ardea, London; (b) Ray Wagner/Save the Tall Grass Prairie, Inc.; (c) Jonathan Scott/Planet Earth Pictures. **49.15** *Left*, © 1991 Gary Braasch; (right) Thase Daniel;*below, left*, Adolf Schmidecker/FPG; *right*, Edward S. Ross. **49.16** Thomas E. Hemmerly. **49.17** (a) Dennis Brokaw; (b) Jack Carey. **49.18** (a) Fred Bruemmer; (b) Doug Sokell/Visuals Unlimited. **49.19** D. W. MacManiman. **49.21** Modified after Edward S. Deevy, Jr., *Scientific American*, October 1951. **49.22** (a) David Steinberg; (b-e) E. F. Benfield, Virginia Tech. **49.23** Lloyd K. Townsend. **49.24** (a) Dennis Brokaw; (b) McCutcheon/ZEFA; (c) Robert Hessler; Woods Hole Institution of Oceanography; (d) J. Frederick Grassle, Woods Hole Institution of Oceanography. **49.26** Page 874, *top*, Jim Doran; *center inset*, Sea Studios/Peter Arnold, Inc.; *below*, Douglas Faulkner/Sally Faulkner Collection; page 875, (moray eel) Jeff Rotman; (chambered nautilus) Alex Kerstitch; all others Douglas Faulkner/Sally Faulkner Collection. **49.27** E. R. Degginger; art by D. & V. Hennings. **49.28** (a) Courtesy of J. L. Sumich, *Biology of Marine Life*, Fifth edition, William C. Brown, 1992; (b) © 1991 Gary Braasch. **49.29** Adapted from Tom Garrison, *Essentials of Oceanography*, Wadsworth, 1995. **49.30** (a,b) Illustrations adapted from Tom Garrison, *Oceanography: An Invitation to Marine Science*, Wadsworth, 1993; satelite images courtesy of NOAA/UCAR, Gene Carl Feldman, NASA-GSFC, Otis Brown, University of Miami. **49.31** Glenn M. Oliver/ Visuals Unlimited. **49.32** (a) Jack Carey.

CHAPTER 50   50.1 *Above*, Gary Head; *maps below*, U. S. Geological Survey. **50.2** United Nations. **50.3** (a) USDA Forest Service; (b) Heather Angel; art after G. T. Miller, Jr., *Environmental Science: An Introduction*, Wadsworth, 1986, and the Environmental Protection Agency. **50.4** Center for Air Pollution Impact and Trend Analysis (CAPITA), Washington University, St. Louis, MO. **50.5** (a), (c) computer images, NASA; photograph National Science Foundation. **50.6** Data from G. T. Miller, Jr. / **50.7** *Above*, R. Bieregaard/Photo Researchers; *below*, after G. T. Miller, Jr. *Living in the Environment*, Eighth edition, Wadsworth, 1993. **50.8** NASA. **50.9** Gerry Ellis/The Wildlife Collection. **50.10** Thomas G. Meier/USDA Soil Conservation Service. **50.11** Agency for International Development. **50.12** Dr. Charles Henneghien/ Bruce Coleman Ltd. **50.13** Water Resources Council. **50.14** Ocean Arks International. **50.15** After G. T. Miller, Jr. *Living in the Environment*, Tenth edition, Wadsworth. **50.16** AP/Wide World; art, Precision Graphics after Marvin H. Dickerson, "ARAC: Modeling an Ill Wind" in *Energy and Technology Review*, August 1987. Used by permission of University of California, Lawrence Livermore National Laboratory and U.S. Department of Energy. **50.17** Alex MacLean/Landslides. **Page 897** J. McLoughlin/FPG. **50.18** © 1983 Billy Grimes.

CHAPTER 51   51.1*Left*, John Bova/Photo Researchers; *right*, Robert Maier/Animals Animals. **51.2** (a) from L. Clark, *Parasitology Today*, Vol. 6, No. 11, 1990, Elsevier Trends Journals, Cambridge, U.K.; (b) Jack Clark/ Comstock, Inc. **51.3** (a) Eugene Kozloff; (b, c) Stevan Arnold. **51.4** Hans Reinhard/Bruce Coleman Ltd.; (a, b) from G. Pohl-Apel and R. Sussinka, *Journal for Ornithologie*, 123:211-214. **51.5** (a) Evan Cerasoli; (b) A. N. Meltzoff and M. K. Moore, "Imitation of Facial and Manual Gestures by Human Neonate," *Science*, 198: 75-78. Copyright 1977 by the AAAS. **51.6** Nina Leen in *Animal Behavior*, Life Nature Library. **51.7** Mark Bekoff. **51.8** (a) Edward S. Ross; (b) E. Mickleburgh/Ardea London. **51.9** D. & V. Hennings. **51.10** John Alcock. **51.11** *Left*, David Fritts/Animals Animals; *right*, Ray Richardson/Animals Animals. **51.12** Michael Francis/The Wildlife Collection. **51.13** Frank Lane Agency/Bruce Coleman Inc. **51.14** Fred Bruemmer. **51.15** John Alcock. **51.16** Timothy Ransom. **51.17** A. E. Zuckerman/Tom Stack & Associates. **51.18** Patricia Caulfield. **51.19** John Alcock. **51.20** Kenneth Lorenzen. **51.21** Gregory D. Dimijian/Photo Researchers. **51.22** Yvon Le Maho. **51.23** F. Schutz. **51.24** Lincoln P. Brower.

A

Aardvark, 464t
Abdomen, of crayfish, 438i
Abdominal aorta, 732i
Abdominal cavity, 552i, 732i
ABO blood group, 182, 182i
Abortion, 213, 788
Abscisic acid, 529, 533, 541
Abscission of leaves, flowers, fruits, 538, 538i, 541
Absorption in intestine, 718–719, 718i, 719i, 728
Absorption spectrum of photosynthetic pigments, 117, 117i
Abstention, sexual, 778
Accessory pigments of photosynthesis, 118, 119, 119i, 127
Acclimatization, respiratory, to high altitude, 704
Accommodation, visual, 601, 601i
Acer (maple), 488i, 522, 522i, 538i
Acetabularia, 384i
Acetyl-CoA, 136
Acetylcholine (ACh)
  as neurotransmitter, 564, 567
  drugs interfering with activity of, 585
Acetylcholinesterase, 565
ACh (*See* Acetylcholine)
Achene, 52t
Achondroplasia, 203t, 205, 205t
Achoo syndrome (chronic sneezing), 203t
Acid
  as activator of enzymes, 32
  defined, 32, 34t
  examples, 32
  indoleacetic (auxin), 532
Acid deposition, dry, 884–885, 885i
Acid rain, 33, 33i, 884–885
Acid stomach, 32
Acid-base balance, 736, 742
Acidosis
  of blood, 33
  metabolic, 736
Acne, 630
Acquired immunodeficiency syndrome (*See* AIDS)
Acrasiomycota, 372–373, 373i, 386t
Acromegaly, 616, 616i
Acrosome, of sperm, enzymes, 773
ACTH (*See* Corticotropin)
Actin, 71, 71i, 640–641, 640i, 641i, 642–643, 643i
Actinomycetes, 358
Actinopoda, 374t, 375, 375i, 386t
Action potential, 559, 560, 562, 562i, 563, 563i, 568, 590–591,

591i, 642, 646
Activation energy, defined, 104, 104i
Activator protein, 245
Active transport
  role of ATP, 91, 91i
  role of calcium pump, 91
  role of sodium-potassium pump, 91
  in small intestine, 719i
  of sodium, from nephron, 735, 735i
  through cell membrane, 87, 87i, 91, 91i, 94
Acyclovir, for treatment of genital herpes, 787
Adaptation, sensory, 591
Adaptive behavior, 905
Adaptive radiation
  and adaptive zones, 307
  of mammals, 307, 307i
  of reptiles, 457
Adaptive trait, 10, 10i, 276, 279, 813
Adaptive zone, 307
Addiction, 584
Adductor longus muscle, 639i
Adenine, 50i, 220, 220i
Adenosine deaminase (ADA), deficiency of, 685, 685i
Adenosine diphosphate, conversion to adenosine triphosphate, 102, 102i
Adenosine monophosphate, 102i
Adenosine phosphate, 52t
Adenosine triphosphate (*See* ATP)
Adenovirus, 363t
ADH (*See* Antidiuretic hormone)
Adhesion protein, 83, 83i, 94
Adipose tissue, 548t, 549, 549i, 710, 720, 782–783, 783i
Adolescence, human, 783t
Adoption, of orphans, 917
ADP, conversion to ATP, 102, 102i
Adrenal cortex, 611i, 617t, 618, 618i
Adrenal gland, 555i, 576i, 611i, 618, 618i, 664, 732i
Adrenal medulla, 611i, 617t, 618, 618i
Adrenaline, 617t
Adventitious root, 492
Aerobic respiration
  defined, 132, 132i
  first stage, 134–135, 134i, 135i
  high net energy yield, 132, 133
  links with photosynthesis, 112i
  related to availability of free oxygen, 144
  release of energy by, 112i, 113
  role of glycolysis in, 139, 139i
  second stage, 136–137, 136i, 137i
  summary of energy yield,

138–139, 139i
  third stage, 138–139, 136i, 138i, 139i
  yield of ATP, 139, 139i, 143i
  yield of energy, 139
Aesculus, 488i
Africanized bee, 130–131
Afterbaby blues, 782
Afterbirth, 782
Agar, derived from red algae, 382
Agaricus brunnescens, 391, 391i
Agave (source of fibers), 491
Age spots, 97, 96i
Agent Orange, 533
Agglutination, of blood, 654–655, 654i
Aggregate fruit, 520t
Aggression, anabolic steroid-related, 645
Aggressive behavior, role in sexual selection, 290
Aging, 758–759, 783
Agnatha, 448
Agriculture
  animal-assisted, 887
  and growth of human population, 898
  land suitable for, 887, 887i
  mechanized, 887
  and population size, 882
Agrobacterium tumefaciens, 265, 265i
AIDS, 363t, 366, 670i, 685, 686–687, 686i, 687i, 786
Air circulation patterns, 856–857, 857i, 858–859, 859i
Air pollution 286, 884–886
Aix sponsa (wood duck), secondary sex characters, 39i
Ajellomyces capsulatus, cause of histoplasmosis, 396
Albatross, wandering
  courtship display, 300i, 906
  travels of, 296
Albinism, 185, 185i, 203t
Albumin, in blood, 652i
Alcock, J., 917
Alcohol
  absorption in stomach, 718
  as drug, 584, 585
Alcoholic fermentation, 140–141, 141i, 145
Aldehyde group, formula, 39i
Alder, and symbiotic bacteria, 824
Aldosterone, 617t, 731, 735, 742
Alfalfa, stem, 487i
Algae
  brown, 371, 383, 383i, 386t
  coralline, 874, 874i
  of economic importance, 382, 383
  golden (Chrysophyta), 380–381, 381i, 386t
  green, 371, 384–385, 384i, 385i,

386t, 394, 398, 398i, 414
  red, 371 382, 382i, 386
  single-celled, 380–381, 380i, 381i
  yellow-green, 381, 381i, 386t
Algal bloom, and eutrophication, 850, 850i
Algal diversity, influenced by periwinkle, 826, 826i
Alkalosis, of blood, 33
All-or-nothing event, in transmission of nerve impulse, 563, 563i
Allantois, 775, 776
Allele
  defined, 162, 177, 177i, 194
  fixation of, 292, 292i
  frequency, 283, 284
  multiple, 182
Allergen, 684
Allergy, 684–685, 688
Alligator, 458
Allopatric speciation, 308
Allosteric control of enzyme function, 106–107, 107i
Alpha rhythm, 584i
Alpine tundra, 869, 869i
Altruism, 901, 905, 912, 913, 918
Aluminosilicate, in soil, 502
Alvarez, W. and L., 346
Alveolar duct, 696i
Alveolar sac, 696i, 697, 701i
Alveolus, 696i, 697, 697i, 700, 700i
Alzheimer's disease, 583
Amanita, 391i, 396
Amino acid
  as building block of protein, 4, 46
  defined, 46
  essential, 723, 723i
  examples of, 46i
  formula, 39i, 46i
  in hemoglobin, 229, 229i
  in human diet, 723, 723i
  in sickle-cell hemoglobin, 229, 229i
  spontaneous synthesis of, 335, 336
  structure, 46i
  synthesis, genetic code for, 232, 232i
Aminopeptidase, 717t
Amish, and Ellis-van Creveld syndrome, 293
Ammonia
  conversion by liver to urea, 142, 720
  formation, 732
  from breakdown of protein, 142
Ammonification, by bacteria, fungi, 848, 848i
Ammonites, 310–311i
Amnesia, 583
Amniocentesis, 212–213, 212i
Amnion (amniotic membrane),

196i, 775, 777i, 788
Amniote egg, 456, 456i
Amniotic cavity, 775i, 776i
Amniotic fluid, 782
Amoeba proteus, 72, 93, 93i, 374i
Amoeba, 374–375, 374i
Amoebic dysentery, 374–375
Amphetamine, as stimulant, 584
Amphibian, 448, 455, 455i, 465
  origin of, 454, 454i
  respiratory system, 694–695, 695i
  water-solute balance, 737
Ampulla, of echinoderm tube foot, 443, 443i
Amygdala, of brain, 583, 583i
Amylase
  pancreatic, 717t
  salivary, 715, 717t
Amyloplast, 69
Amylose, bonding of glucose in, 43, 43i
Amyotrophic lateral sclerosis (Lou Gehrig disease), 203t, 207
Anabaena, 358
Anabolic steroid, 645
Anaerobic bacteria, 360
Anaerobic electron transport, in bacteria, 141–142
Anaerobic respiration, 140–141, 141i, 142i
Anaerobic, defined, 132
Anagenesis, defined, 306
Analgesic drug, 585
Analogy, 319
Anaphase
  of meiosis, 163, 163i, 164i, 165i
  of mitosis, 151i, 153, 153i, 155i, 159
Anaphylactic shock, 684
Anapsids, 457
Androgen, 617t
Anemia
  hemolytic, 654
  iron deficiency, 654
  polycythemia, 654
  sickle-cell, resistance to malaria, 290–291, 291i
Anemone, sea, 423, 423i, 572, 573i, 632, 632i, 875i
Aneuploidy, 208, 214
Angel Island, overcrowded deer population, 792
Angiography, 665
Angioplasty, 665
Angiosperm, 344, 344i, 399, 400, 400i, 414t, 481, 517
Angiotensin II, 617t, 735, 742
Angler fish, 298, 298i
Angraecum sesquipedale (orchid), pollination of, 515
Anhidrotic ectodermal dysplasia, 249, 249i
Animal behavior
  adaptive, 905
  altruistic, 901, 905
  defined, 901

genetic basis, 903
heritable, 902–903
and hormones, 902–903
instinctive, 901
learned, 904, 904t
and reproductive success, 905
selfish, 905
social, 901, 905
Animal pole, of frog embryo, 749i
Animalia (kingdom), 8, 9i, 328, 328i, 329i, 417–467
Animal
  general characteristics, 418
  summary of characteristics of cells, 78t
Annelid worm, 418t, 434–435, 434i, 435i, 444
Annual plant, 494
Anomaluridae, 464t
Anorexia nervosa, 710
Antarctic
  disruption of, 834–835, 834i, 835i
  food web, 837, 837i
Anteater, spiny, 464, 464i
Antelope, pronghorn, 713, 713i
Antenna, of crustaceans, 438, 438i
Antennapedia gene, of *Drosophila*, 755, 755i
Anther, 516i
Anthocyanin, 119, 119i, 516
Anthrax, 671
Anthropoid, 469, 470, 472, 477
Anti-anxiety drug, 564
Anti-inflammatory drug, 684–685
Anti-Rb gamma globulin, 655
Antibiotic resistance, 287, 366
Antibiotic
  defined, 11, 287
  produced by fungi, 393
  resistance of microorganisms to, 287
Antibody
  circulating, 680i
  defined, 671, 677
  diversity, 682, 682i
  IgA, 681
  IgD, 681
  IgE, 681
  IgM, 681
  monoclonal, 681
  structure, 682i
  target of, 680
Antibody-mediated response, 680–681, 680i
Anticodon, 232–233, 233i, 234, 234i–235i
Antidiuretic hormone (ADH), 614, 614t, 616, 624, 731, 735, 742
Antigen receptor, diversity, 683, 683i
Antigen
  defined, 671
  primary response to, 677
  processed, 676

recognition, 676, 676i
Antigen-binding site, 682, 682i
Antigen-MHC complex, 678, 679i
Antigen-presenting cell, 676, 676i, 677i, 678i, 680, 688
Antigen-specific receptor, 682–683
Antihistamine, 684–685
Antilocapra americana (pronghorn antelope), 713, 713i
Antimicrobial secretion, 672
Antrum, of ovarian follicle, 770i
Anus, 714i, 728, 728t
Anvil, of middle ear, 596i
Aorta, 656, 656i, 657i, 658i, 732i
Aortic arch, 661
Aphid, 160i, 161, 510, 510i
Apical dominance, 533
Apical ectodermal ridge (AER), 756–757, 756i
Apical meristem, 481, 483, 483i, 486, 486i, 530i
Apicomplexa, 374t, 386t
Aplysia (sea hare), 432i, 693, 693i
Apnea, 701
Apoptosis ((programmed cell death), 242, 678, 753
Appeasement behavior, among baboons, 911, 911i
Appendages, jointed, of arthropods, 436, 436i, 437, 438i, 439i, 440i, 441i
Appendicitis, 721
Appendicular skeleton, 646
Appendix, 721, 721i
Apple scab, caused by fungus, 396, 396i
Apple, fruit of, 521i
Aqueous humor, of eye, 600, 600i, 600t
Aquifer, underground, 892, 892i
Arachnid, 437, 437i
Arceuthobium, 412i
Archaeanthus linnenbergeri, 344i
Archaebacteria (kingdom), 8, 8i, 78t, 328, 328i, 329i, 338, 338–339i, 353, 357, 360, 360i, 361t, 367
Archaeopteryx, 278, 278i, 460
Archean eon, 313, 338, 348–349i, 350
Archenteron, of frog embryo, 749i
Archipelago, speciation in, 303, 303i
Arctic fox, heat balance, 739
Arctostaphylos, 300, 301i
Arenavirus, 363t
Argentine fire ant, 827, 827t
Aristotle, views on diversity, 272
Armadillo, 276, 276i
Armillaria (root fungus), 390, 540
Arnold, S., 902
Arrhythmia, of heart, 665
Arrowhead, variation in leaf form, 298, 298i
Arteriole, 649, 656, 660i, 661, 668
Artery, 657i, 649, 656, 656i, 660i,

668, 733i
Artery, hardening of, 664, 664i
Arthritis, 636, 685
Arthrobotrys dactyloides, capturing nematodes, 393i
Arthropoda, 417, 418t, 436–441, 436i, 437i, 438i, 439i, 440i, 441i, 632, 633i
Artificial selection, 10, 11i, 280–281, 280i
Ascocarp, 393, 393i
Ascomycota, 392–393, 393i, 396
Ascorbic acid, 724t
Ascospore, 392, 393i
Asexual reproduction, in plants, 524–525, 525i, 526
Asinus, 301
Aspergillosis, 396
Aspergillus, 396
Asteroid, catastrophic, 344, 346, 346i
Ateroid impact theory, 346
Asthma, 684
Astigmatism, 604
Atherosclerosis, 664–665, 664i
Atherosclerotic plaque, 45i, 256, 256i
Athlete's foot, caused by fungus, 396, 396i, 672
Atmosphere
  addition of oxygen to, 398
  and climate, 855
  early, 334
  entry of oxygen into, 144, 121
  evolution of, 882
  and ozone layer, 856, 856i
  thermal inversion, 884, 884i
Atmospheric cycle, 852
Atmospheric pressure, of gases, 698
Atoll, 874i
Atom
  bonding activity of, 27
  defined, 6, 21, 22, 34
  models of, 27i
  radioactive isotope of, 23
Atomic nucleus, 22
Atomic number, 22–23, 22t
Atomic weight, 23, 23i
ATP (adenosine triphosphate)
  activator of motor proteins of muscle, 72
  in beating of cilia and flagella, 73i
  composition of, 50, 50i
  cyclic pathway of formation, 120, 126, 132
  defined, 102, 102i
  as energy carrier, 5, 97, 102, 103, 103i, 110
  formation in chloroplasts, 122, 122i, 123
  formation in muscle, 643m, 643i, 646
  formation in photosynthesis, 112i, 113, 114
  function of, 50, 52t

noncyclic pathway of formation, 120–121
probable formation by early forms of life, 338, 339
role in muscle contraction, 103, 103i
transfer of phosphate group to transport protein, 103, 103i
Atrial fibrillation, 665
Atrial natriuretic hormone, 617t
Atrioventricular (AV) node, of heart, 659, 659i
Atrioventricular valve, 658–659, 658i, 659i
Atrium, of heart, 658, 658i
Atropa belladonna, 491
Atropine, 491
Auditory nerve, 596i
Australopith, 473, 473i, 474, 474i
Australopithecus, 473, 473i, 474i, 477, 477i
Autoimmune disorder, 685
Autoimmune response, 685, 688
Autonomic nerve, 586
Autosomal dominant inheritance, disorders, 203t, 204–205, 204i, 205i
Autosomal recessive inheritance, disorders, 203t, 204, 204i
Autosomes, 194
Autotroph, 113, 836, 836i, 836t
Auxin, 532, 533, 533i, 534, 535, 535i, 538, 541
AV node (See Atrioventricular node)
AV valve (See Atrioventricular valve)
Avens, mountain, as pioneer species, 824i
Avery, O., 218
Aves, 448, 460–461, 460i, 461i
Axial skeleton, 646
Axolotl, Mexican, 455
Axon, 560, 560i, 563, 563i, 566i, 567i
Azospirillum, 358
AZT (azidothymidine), in treatment of AIDS, 687

B

B cell
DNA of, 682–683, 682i
effector, 680, 680i
formation stimulated by macrophages, 675, 675i
and immunological memory, 676, 683, 683i
production of antibodies by, 677, 677i
role in immunological memory, 683, 683i
summary of functions, 688, 688t
virgin, 680i
B lymphocyte (See B cell)
Baboon, 906–907, 906i, 911, 911i

Bacillus, 354, 354i, 671
Backbone, of vertebrates, 448
Bacteria (See also Archaebacteria)
abundance in nature, 354i
Actinomycetes, 361t
anaerobic, 141
bioluminescence of, 109, 109i
cell wall, 76, 76i, 366
characteristics, 354–355, 352i, 353i, 354i, 355i
chemoautotrophic, 354, 361t
chemoheterotrophic, 354, 361t
chlamydias, 361t
chromosome of, 258, 258i, 356
denitrifying, 848i, 849
DNA of, 76–77, 76i
endospore-forming, 361t
Eubacteria (See also Eubacteria)
fission, 356–357, 356i, 357i
flagella of, 76, 76i, 77i, 366
genetically engineered, 264
Gram-negative, 361t
Gram-positive, 361t
growth, 356
halophiles, 361t
ice-minus, 264
iron-oxidizing, 361t
metabolic diversity, 76–77, 354
methanogens, 361t
nitrifying, 361t
nitrogen-fixing, in root nodules of plants, 505, 505i, 824
nucleoid, 76, 76i
photoautotrophic, 354, 358, 361t
pathogenic, 264, 358, 359, 361t
photoheterotrophs, 361t
photosynthetic, 354
pilus (pl. pili), 76, 76i
plasma membrane of, 76, 76i
plasmid of, 258, 258i
polysaccharides, 76
replication of DNA, 356, 356i
rickettsias, 361t
of sexually transmitted diseases, 786–787, 786i, 787i
shapes, 354, 354i
sizes, 354
spirochetes, 361t
structure, 76–77, 76i, 77i
sulfate-reducing, 141
sulfur, 361t
summary of characteristics, 367
summary of major groups, 361t
symbiotic, 512
synthesis of insulin and other proteins by, 264
thermophiles, 361t
Bacteriophage
DNA, 218–219, 219i
lysogenic pathway, 364, 364i, 365
lytic pathway, 364, 364i, 365
multiplication cycle, 364, 366, 364i
protein coat, 219, 219i
as research tool, 218–219
structure, 219, 219i

as type of virus, 362, 362i
T4, 219, 219i
in treating bacterial infections, 14, 14–15i
Bacteriorhodopsin, in Archaebacteria, 360
Bahama woodstar (bird), as pollinator, 515i
Baker, H., 314
Balance, sense of, 595
Balanced polymorphism, 290
Balancing selection, 290
Balantidium coli, 379
Banana slug, as prey of garter snakes, 902, 902i
Bark beetle (carrier of Dutch elm disease), 827t
Bark, 496, 496i
Barnacle, 438, 438i, 439
goose, 438i
Baroreceptor reflex, 661
Barr body, 249, 249i
Barr, M., 249
Basal ganglion, of brain, 583i
Base pairing
sequences, 221, 221i
in DNA, 221, 221i, 225
in formation of RNA, 230, 230i
substitution, 236, 236i
Base, 32, 34t
Basement membrane, 546i, 547i
Basidiomycetes, pathogenic and toxic, 396
Basidiomycota, 390, 390i
Basidiospore, 391, 391i
Basilar membrane, of ear, 597i
Bat
echolocation, 588–589, 589i
in flight, 465i
fossil, 312i
as pollinator, 514, 514i
sensory receptors, 588
Bayliss, W, 610
Beadle, G., 228
Beagle, Voyage of, 274–275, 274i, 275i
Bean, germination of seed, 530i–531i
Bean, kidney, 489i
Bear, polar, 821
Bears, biochemical evidence for relationship with pandas, 320–321i
Becquerel, H., 23
Bee, honey, 441i
Beetle
bark (carrier of Dutch elm disease), 827t
bioluminescence of, 109, 109i
dung, 7, 7i
Japanese, 827t
ladybird, 441i
protected by mimicking yellowjacket wasp, 821i
scarab, 441i
spraying toxic secretions, 821,

821i
Behavior
adaptive value, 905
altruistic, 901
appeasement, 911, 911i
defined, 901
heritable basis, 902–903
and hormones, 902–903
instinctive, 901, 903, 903i, 918
mating, 908–909, 908i, 909i
social, 901
traits, 282
Benaron, D., 109
Benthic province, of ocean, 872, 872i
Berry, 520t
Beta-carotene, 117i, 118, 119, 725
Bicarbonate ion, formation in blood, 700
Bicarbonate-carbon dioxide buffer system, 736
Biceps muscle, 638, 638i, 639i
Biennial plant, 494
Big laughing mushroom, 388i
Bilateral symmetry, 444, 444i, 572, 573i
Bilby (Australian marsupial), 829
Bile salts, 719i
Bile, role in fat digestion, 717
Binary fission, 374
Binomial system, for naming organisms, 322–323, 322i, 323i
Biochemistry, providing evidence for evolution, 311, 320–321, 320i, 321i
Biodiversity
on coral reefs, 874, 874i
corresponding to latitude, 830, 830i
on islands, 830–831, 830i, 831i
Biogeochemical cycle, 852, 841, 841i
Biogeographic realm, 860–861, 860i, 879
Biogeography, 272, 860
Biological clock, 536–537, 536i, 537i, 622
Biological control, of pests, 287
Biological magnification of DDT, 851
Biology, methodology of, 3
Bioluminescence, 109, 109i
Biome, 855, 879, 860–863, 860i, 862i
Biosphere, 854–881, 882–899
Biotic potential, 797
Biotin, 724t
Bipedalism, 469, 471, 471i, 473, 474, 477
Bird of paradise, 290, 290i
Bird
body plan, 461i
bones, 461
classification, 448
derivation from reptiles, 457i, 466
digestive system, 712–713, 712i

eggs, 460i
feathers, 460, 461i
flight, 461
fossil, 312i
frigate, 461
migration, 460
oviparity, 746, 747i
parental behavior, 460
pollination by, 514, 515i
respiratory system, 695, 695i
sex chromosomes, 903
song, and sex hormones, 902–903, 902i
song system, 461, 903
Birth control, methods, 784–785, 785i
Birth, human, 782, 782i
Bison, 465, 865i, 908, 909, 909i
Biston betularia, natural selection in, 286, 287i
Bittern, least, and camouflage, 820i
Bivalve, 431, 431i, 433i
Black duck, travels of, 296
Black stem wheat rust, 396t
Black widow spider, 437i
Black-bellied seedcracker, beak of, 289, 289i
Bladder (*See* Urinary bladder)
Bladderwort, 500, 500i
Blastocoel, 749i, 751, 774i
Blastocyst, 762, 771, 774, 774i, 775, 775i, 783i
Blastoderm, 748, 749i, 750, 750i, 751, 754, 754i
Blastula, 749i, 751, 752i
Blight
  chestnut, 827, 827t
  potato, and human population, 798i
Blood
  acidosis of, 33
  agglutination, 654–655, 654i
  alkalosis of, 33
  anemias, 654
  cell count, 653
  clotting, 666, 666i
  composition, 652–653, 652i, 653i
  as connective tissue, 548t, 549, 549i, 556
  disorders, 654
  functions, 649, 650, 652, 731
  hemolytic anemias, 654
  infectious mononucleosis, 654
  iron deficiency anemia, 654
  plasma, 652, 652i
  polycythmias, 654
  red cells, 652, 652i, 668
  types, 655, 654i
  volume, 652
  white cells, 652, 652i, 671, 672, 672t, 688, 688t
Blood plasma, 668
Blood pressure
  arterial, 660–663, 660i, 663i
  capillary, 662, 663i
  measuring, 661i

diastolic, 660, 660i, 661i
and high sodium intake, 725, 725t
measuring, 661i
pressure gradient, 660
pulse, 660, 660i
systolic, 660, 660i, 661i
venous, 660, 662–663, 663i
Blood transfusions, 182
Blood typing, 654–655, 654i, 655i
Blood-brain barrier, 579
Blue-green algae (*See* Cyanobacteria)
Blue-green bacteria (*See* Cyanobacteria)
Bluegill sunfish, behavior, 911
Boa, 459
Body cavity, human, 552, 552i
Body cavity, types of, 419, 419t
Body temperature, 731
Body weight, 726–727, 726i, 727i
Bombykol, of moths, 594, 906
Bond
  chemical, 21, 23, 26
  covalent, 21, 102
  hydrogen, 21
  ionic, 21
Bone
  compact, 549i, 634, 634i
  as connective tissue, 548t, 549, formation, 635, 635i
  functions, 634, 646
  intercellular matrix of, 74, 75i
  marrow, 634i, 635, 667i
  of middle ear, 596, 596i, 597i
  spongy, 634, 634i
  structure, 549i, 634i
  tissue turnover, 635
Bonnemaisonia hamifera, 383i
Bony fish, 452–453, 453i
Booby, blue-footed, 274i
Book lung, of spider, 437, 437i
Boreal forest, 868, 868i
Boron, required by plants, 503t
Borrelia burgdorferi, causative agent of Lyme disease, 359, 359i
Bottleneck, genetic, 292, 294
Bovine spongiform encephalopathy, 366
Bowman's capsule, 733, 733i, 734, 734i, 742
Bradycardia, 665i
Bradyrhizobium, 505i
Brain stem, 578, 579
Brain
  amphibian, 574i
  bird, 574i
  cavities and canals, 579, 579I
  cerebellum, 578, 578i, 580i
  cerebrum, 571, 578i, 579, 580–581, 580i, 581i
  development, human, 578i
  effect of cocaine on, 585i
  embryonic, 752i
  fish, 574i
  forebrain, 571
  hindbrain, 571

human, 575i, 578–585
hypothalamus (*See also* hypothalamus), 571
mammal, 574i
midbrain, 571
olfactory lobe, 574i
PET scan of, 25, 25i
primate, 471
reptile, 456, 574i
shark, 574i
vertebrate, 578–585
Bread mold, in studies of gene mutations, 228, 228i
Breast cancer, 783
Breastbone, of birds, 461i
Breathing
  at high altitude, 704
  human, 698–699, 698i, 699i
  and sound production, 699, 699i
Breech birth, 782
Brenner, S., 266, 756
Brinster, R., 266
Bristlecone pine, 408, 408i
Brittle star, 442, 442i
Bronchial tree, of human lung, 696i
Bronchiole, 696i, 697, 697i
Bronchitis, 702, 703i
Bronchus, 576i, 697, 697i
Brown algae, 383, 383i, 386t
Brown recluse spider, 437i
Brown, R., 54–55
Brown-headed cowbird, as social parasite, 822
Bryophyte, 400, 400i, 401i, 402–403, 402i, 403i, 414t
Bubonic plague, 799
Buckeye, red, 488i
Bud
  breaking dormancy of, 528, 532
  lateral, 486i, 494i
  of plant shoot, 486, 486i
  terminal, 494i
Buffer system, bicarbonate-carbon dioxide, 736
Buffer, defined, 33
Buffon, 273
Bug, 441i
Bulb of Krause, 592, 593i
Bulb, 524t
Bulbourethral gland, 764t, 765i, 766i
Bulimia, 711
Bulk flow, 88
Bulk, dietary, 721
Bull, J., 14
Bult, C., 328
Bunkley-Williams, L., 874
Bunyavirus, 363t
Burgess Shale, 342i, 416i, 417
Burkitt's lymphoma, 253
Burkitt, D., 722
Burnet, M., 683
Buttercup
  root, 493i
  toxicity of, 821

Butterfly
  monarch, 856, 856i
  mouth parts, 440i
  protected by mimicking inedible species, 821i
Byssus, secretion of, 226–227

C

C3 plant, 124–125, 125i, 128
C4 plant, 124–125, 124i, 125i, 509, 512
Cabbage, effect of gibberellin, 528i
Cacao, fruit, 522, 522i
Cactus, 125, 508, 508i, 509
Caecilian, 455, 455i
Caecum, 721i
Caenorhabditis elegans (nematode), 428, 757
Caffeine, as stimulant, 579, 584
Caiman, spectacled, 458i
Calamites, 406, 406i
Calcitonin, 617t
Calcitriol (1,25-hydroxyvitamin D^6), 617t
Calcium pump, in active transport, 91
Calcium
  atomic number and mass number, 22t
  effects of deficiency, 503t
  in human nutrition, 725t
  required by plants, 503t
  role in muscle contraction, 642–643, 642i, 643i, 646
Calicivirus, for controlling rabbits, 829
Calico cat, 249, 249i
California poppy
  effect of gibberellin, 528i
  root system, 492i
Calorie intake, and energy output, 726
Calvin, M., 25
Calvin-Benson cycle, 123, 123i, 124, 124i, 125i
Calyx, 516, 516i, 517i
CAM plant, 125, 125i, 128, 509, 512
Camarhynchus pallidus, 277i
Cambium, 483, 483i, 496, 496i
Cambrian period, 313i, 342, 342i, 348–349i, 416
Camel, 347, 465i
Camouflage, and survival, 820, 820i
cAMP (cyclic adenosine monophosphate), role in transfer of phosphate groups, 50, 102
Camptodactyly, as genetic trait, 186, 203t
Canada goose, 460i
Canada lynx, and snowshoe hare, 819, 819i

Cancer
  caused by viruses, 362
  characteristics, 252–253, 252i, 253i
  colon, 721
  and immunotherapy, 681
  mutations causing, 243
  of prostate gland, 765
  of testis, 765
  risks incurred by smoking, 703i
Candida albicans, 393, 393i
Canidae, 464t
Canine tooth, 462, 462i
Cannabis sativa, 491
Cannibalism, in herring gulls, 912
Capillary, 656, 663
  bed, 650, 651i, 656i, 668
  diffusion through wall, 662
  glomerular, 733i, 742
  of lung, 697i
  peritubular, 733i
  permeability increased by
    histamine, 674
  reabsorption by, 662, 663i
  structure, 660i
  ultrafiltration through, 662, 663i
Capsella, development of seed,
  520, 520i
Capsule, of bacteria, 354i, 355
Carapace, of crustaceans, 438
Carbamino hemoglobin, 700
Carbohydrate
  categories of, 52t
  defined, 42
  examples of, 37, 52t
  in human diet, 722, 722i
  sources of, 722, 722i
Carbon
  atomic number, 22t
  as component of all organisms,
    36–37
  in fossil fuels, 845, 844–845i
  global distribution of, 844–845,
    844i–845i
  mass number, 22t
  proportion in earth's crust, 22i
  proportion in human body, 22i
  proportion in pumpkin, 22i
Carbon-14, 23, 25
Carbon compounds,
  characteristic of living things,
  113
Carbon cycle, 844–845, 844i, 845i,
  852
Carbon dioxide
  as air pollutant, 884, 884t
  in atmosphere, 36, 37, 844–845,
    844–845i, 847, 847i
  diffusion out of animal body, 691
  in freshwater deposits, 844,
    844–845i
  in ocean, 844, 844–845i
  in soil, 844, 844–845i
  as source of carbon for
    autotrophs, 113
  as source of carbon for plant
    photosynthesis, 502–503, 503t
  transport of, in blood, 700

  uptake by plants, 502
Carbon fixation
  defined, 123
  processes of, 124–125
Carbon monoxide
  in cigarette smoke, 664
  poisoning, 704
Carbonic acid, 33, 696
Carbonic anhydrase, 700
Carboniferous period, 313i,
  348–349i, 406, 406i, 407, 407i
Carboxyl group, formula, 39i
Carboxypeptidase, 717t
Carcinogen, causing mutations,
  237
Carcinoma, 224i, 681
Cardiac conduction system, 659,
  668
Cardiac cycle, 658–659, 659i
Cardiac muscle, 550, 550i, 638,
  658, 658i, 659, 659i, 668
Cardiovascular disorders,
  664–665, 664i
Cardiovascular system
  blood pressure, 660–661, 660i,
    661i
  human, 656–657, 656i, 657i
  pulmonary circuit, 656, 656i,
    657
  systemic circuit, 656, 656i, 657
Carnivore, in ecosystem, 836t,
  837, 837i, 839i, 840i, 841i
Carotene, contributing to skin
  color, 630
Carotenoid
  in chloroplasts, 69
  in flowers, 516
  role in absorption of light, 118,
    118i, 119i
Carotid artery, 657i, 661
Carpal bone, 517, 518i, 526, 637i
Carrageenan, derived from red
  algae, 382
Carrot
  effect on mites, 900–901, 900i
  tissue culture propagation,
    524–525, 525i
  used by starling, 900–901, 900i,
    901i
Carrying capacity, of population,
  798, 798i, 799i
Cartilage, 74, 548–549, 548i, 548t,
  549i, 556, 636
Cartilaginous fish, 452, 452i
Cascade reaction, in complement
  system, 673i
Casparian strip, of root, 504, 504i
Caspian tern, parenting, 909, 909i
Castanea (chestnut), 494
Catalase, 97
Cataract, of eye, 605
Catastrophism, 271
Catopithecus, 472
Cave paintings, early, 468,
  468–469i
Cavity, body, human, 552, 552i
CD4 lymphocyte, 686

cDNA, 263, 263i
Cebidae, 464t
Ceboids, 470t
Cell
  of animals, 78t
  of Archaebacteria, 78t
  controlled death, 242–243, 242i,
    243i
  cytoplasm of, 55
  defined, 4, 6, 55
  differentiation of, 246, 760, 783
  division of, 148, 150–153
  DNA of, 56
  of Eubacteria, 78t
  eukaryotic, 60–75, 78t
  of Fungi, 78t
  nucleus of, 55
  organelles of, 55
  origin of term, 54
  origin of, 336–339
  of plants, 78t
  plasma membrane of, 55, 56–57,
    56i
  prokaryotic, 8, 56, 76–77, 76i,
    77i, 78t, 306, 338–339,
    338–339i, 353, 354–361
  of protistans, 78t
  ribosomes, 56
  shape, 57, 59i
  size, 57, 59i
  structure, 56–57, 56i, 57i
  surface-to-volume ratio, 55, 57,
    57i
Cell body, of neuron, 560, 560i,
  567i, 568
Cell cycle, 148–159
Cell junction
  adhesion, 546, 547i
  gap, 546, 547i
  structure and function, 74, 75i
  tight, 546, 547i
  types of, 75, 75i
Cell membrane (plasma
  membrane)
  active transport through, 81,
    87i, 91, 91i, 94
  adhesion proteins, 94
  animal cell, 61, 64i
  of bacteria, 354i, 355
  characteristics, 77
  cycling, 93, 93i
  diffusion through, 81
  fluid mosaic model, 82, 82i, 83,
    83i, 85
  infolding of, in bacteria, 340,
    341i
  lipid bilayer of, 56–57, 56i, 81,
    82, 82i, 83i, 94
  passive transport through, 81,
    87, 87i, 90, 90i, 91, 94
  phospholipid of, 81, 83I
  plant cell, 61i, 62i, 63i
  possible origin, 337
  receptor proteins, 94
  recognition proteins, 94
  selective permeability of, 81, 86,
    86i

  structure and functions, 56–57,
    56i, 80–94
  transport proteins of, 81, 94
Cell movement, structural basis,
  72–74, 72i, 73i, 74i
Cell plate formation in plants,
  156–157, 156i
Cell theory, 55, 77
Cell wall, 61i, 62i
  of bacteria, 354i, 355, 366
  formation in plants, 156–157,
    156i
  inheritance and development of
    shape, 532i
  of plant cells, 74, 74i
  primary wall, 74, 74i
  secondary wall, 74, 74i
Cell-to-cell contact, 546
Cellulose
  bonding of glucose in, 42, 43i
  component of cell wall, 74, 532i
Cenozoic era, 313, 313i, 347, 347i,
  348i–349i, 350, 401i
Centipede, 436i, 439, 439i
Central nervous system, 575, 575i
Central vacuole, of plant cell, 69,
  62i
Centrifuge, 84, 84i
Centriole, 61i, 152, 152i
Centrocercus urophasianus (sage
  grouse), 908, 908i, 909
Centromere, 150, 150i, 152
Cephalization, 573, 573i
Cephalochordata, 448–449, 449i
Cephalopod, 431, 431i, 433, 433i
Cephalothorax, of crayfish, 438i
Cercopithecidae, 464t
Cercopithecoidea, 323
Cerebellum, 578, 578i, 580i
Cerebral cortex, 580, 581, 581i
Cerebral hemisphere, 580, 580i,
  581, 581i
Cerebrospinal fluid, 579, 579i
Cerebrum, 571, 578i, 579,
  580–581, 580i, 581i
Certhidea olivacea, 277i
Cervical nerve, 575i, 576i
Cervix, 768t, 769i, 771, 773i, 782i
Cesium-137, isotope, 21
Cestoda, 427, 427i, 428, 429i
Cestum veneris, 425, 425i
Chaeta, of annelid worms (*See*
  Seta)
Chagas disease, 376
Chameleon, 459
Chambered nautilus, 433, 433i,
  875i
Chameleon, 551, 821
Chanterelle, trumpet, 389i
Chaparral (dry shrubland), 864,
  864i
Chargaff, E., 220
Chase, M., 218, 219
Cheese, fungi in, 393
Cheetah, effect of inbreeding on,
  293, 293i
Chelicera, of arachnids, 437, 437i

Chemical bond, 21, 23, 34
Chemical energy, defined, 110
Chemical equilibrium, 100, 100i
Chemical reaction, 100, 100i, 101, 101i
Chemical synapse, 564–565, 564i–565i
Chemoautotroph, 113, 126, 366, 836t
Chemoheterotroph, 366
Chemoreceptor, 590, 590t, 594, 606
Chernobyl, power plant near, 21i, 895i
Cherry, life cycle, 518i–519i
Chestnut, 494
Chestnut blight, 396, 827, 827t
Chia, F., 432
Chiasma (pl. chiasmata), of chromosomes in meiosis, 166i
Chicken, genetic traits, 185, 185i
Chickenpox virus, 363t
Chimaera, 452, 452i
Chimpanzee, 470t, 477i, 608–609, 904t
Chipmunk, competition between species, 816, 816i
Chironomus, 250, 250i
Chitin, 43, 372, 628
  in cell wall of chytrids, 372
  in cell wall of fungi, 390
Chiton, 431, 431i, 432
Chlamydia trachomatis, 787, 787i
Chlamydial infection, 786, 787, 787i
Chlamydomonas, life cycle, 384i
Chlamydomonas nivalis, 385, 385i
Chloride, in human nutrition, 725t
Chlorine
  atomic number and mass number, 22t
  effects of deficiency, 503t
  monoxide, 886
  required by plants, 503t
Chlorofluorocarbon (CFC), in atmosphere, 884, 846, 846i, 847i, 883, 886, 898
Chlorophyll
  a, 117i, 118, 118i, 119i, 127
  b, 117i, 118, 119i
  of brown algae, 383
  in chloroplasts, 69
  of chrysophytes, 380
  of euglenoids, 380
  of green algae, 384
  porphyrin content, 336, 336i
  of red algae, 382
Chlorophyta, 386
Chloroplast
  ATP formation in, 122, 122i
  defined, 69
  function, 60, 69, 77, 78t
  pigments of, 69
  possible symbiotic origin of, 69, 340, 340, 341i
  of red algae, 382

resemblance to some bacteria, 341
role in photosynthesis, 69
of single-celled algae, 380–381, 380i
structure, 61i, 62i, 69, 69i, 61i, 62i, 114i, 115
thylakoids of, 114i
viewed as symbiotic prokaryotic cells, 815
Cholecalciferol (See Vitamin D)
Cholecystokinin, 717
Cholera, epidemic of, 805
Cholesterol
  abnormalities caused by excess, 45
  and cardiovascular disorders, 665
  in cell membrane, 82, 256
  formula, 45, 82i
  functions, 52t
  precursor of vitamin D, 256
  relationship to atherosclerotic plaques, 256, 256i
  synthesis in liver, 256
Cholesterolemia, 256
Chondrichthyes, 452, 452i
Chondrus (red alga), in tidepool community, 826, 826i
Chordata
  characteristics, 448
  classification, 418t, 446–447
  invertebrate, 448–449
  vertebrate, 450–466
Chorion (See chorionic membrane)
Chorionic
  cavity, 775i, 776i
  gonadotropin, human (HCG), 775
  membrane, 775, 775i, 777i, 788
  villus, 213, 777, 777i
Choroid, of eye, 600, 600i, 600t
Chriacus, 347i
Chromatid, behavior in meiosis, 150, 150i, 152, 153i, 154i, 163–165, 163i, 164i, 165i, 166
Chromatin, 64i, 65, 65t
Chromoplast, defined, 69, function, 69
Chromosome
  autosomes, 194
  of bacteria, 354i, 355
  behavior in meiosis, 163–165, 163i, 164i, 165i
  in cell division, 149, 150–153, 150i, 153i, 154i
  condensed, 65, 65i
  crossing over, 193
  and cytological markers of genes, 199, 199i
  defined, 65
  duplicated, 65, 65i
  homologous, 162, 194
  human, 163i
  proteins of, 65, 223
  sex chromosomes, 194

unduplicated, 65, 65i
Chromosome number
  changes in, 203t, 208–209, 209i
  diploid, 150, 158, 162
  disorders associated with changes, 203t, 208–209, 209i
  haploid, 150, 158, 162
Chromosome structure, changes in
  causing disorders, 203t
  deletion, 210, 210i
  duplication, 210–211, 210i
  inversion, 210, 211, 211i
  translocation, 210, 211, 211i
Chrysanthemum, example of short-day plant, 536, 537i
Chrysaora, 422i
Chrysophyta, 380–381, 380t, 381i, 386t
Chylomicron, 719, 719i
Chyme, in stomach, 716
Chymotrypsin, 717, 717t
Chytrid, 371, 372, 372i, 386t
Chytridiomycota, 371, 372, 372i, 386t
Chytridium confervale, 372i
Cicada, periodical, 300, 300i
Cichlid, African (fish), 304, 304i, 828
Cigarette smoke (See Smoking)
Cilia, 72–73, 72i, 73i, 378, 378i, 379, 379i
Ciliary muscle, of eye, 600, 600i, 601, 601i
Ciliate, 378–379, 378i, 379i, 386
Ciliophora, 374t, 378–379, 378i, 379i, 386t
Circadian rhythm, 536
Circuit, nerve, 566
Circulation, in human body (See also Circulatory system), 552i, 648–669, 777, 777i
Circulatory system (See also Circulation, in human body)
  amphibians, 651, 651i
  of annelids, 435, 435i
  arthropods, 650, 650i
  birds, 651, 651i
  closed, 650, 650i, 651
  earthworm, 650i
  fishes, 651, 651i
  general arrangement in humans, 552i
  general characteristics, 650–651, 650i, 651i
  grasshopper (insect), 650i
  insect, 650, 650i
  links with lymphatic system, 651
  mammals, 651, 651i
  mollusks, 650
  in relation to other systems, 649, 649i
  vertebrates, 651, 651i
Citric acid, produced by fungi, 393
Cladistic taxonomy, 324, 325–327

Cladogenesis, defined, 306
Cladogram, 325, 326–327, 326i–327i
Cladonia rangiferina, 393i
Clam, 431, 431i, 433i
Clark, L., 900
Classification of organisms, schemes for, 323, 329
Clavaria, 388i, 389i
Claviceps purpurea, cause of ergotism, 396
Clavicle, 637i
Clay, in soil, 502
Cleavage
  of animal cells, 157, 156i, 748, 748i
  of frog egg, 744i, 749i
  furrow, 750
  of human egg, 774, 774i, 762
  incomplete, 751t
  patterns of, 751, 751t
  process, 749-750, 750i, 760
  radial, 430, 430i, 751t
  rotational, 751t
  spiral, 430, 430i, 751t
  superficial, in Drosophila, 754, 754i
Climate, 472, 472i, 855, 879
Climax community, 813, 821, 824, 825, 825i, 832
Climax-pattern model, 824
Clitoris, 769i
Clonal selection hypothesis of antigen receptor diversity, 683, 683i
Clone, genetically engineered, 266
Clostridium
  botulinum, 358, 358i, 568
  tetani, 358, 358t, 568
Clotting, of blood, 666, 666i
Clownfish, 423i
Club fungus, 390, 390i, 391i, 395
Club-moss, 343, 343i, 404, 404i
Cnidaria
  body plan, 422–423, 422i, 423i
  classification, 418t
  contractile cells, 423, 572, 573i
  digestive system, 712
  life cycles, 424, 424i, 425i
  nematocysts, 422, 422i, 444
  nervous system, 572, 573i
  planula, 572
Coal
  formation of, 406
  origin and sources, 894, 894i
Cobalt-60, in cancer therapy, 25
Cobra lily, 501i
Cobra, 459
Cocaine, 491, 570–571, 585
Coccidioides immilis, 396
Coccus, 354, 354i, 355i
Coccygeal nerve, 575i
Cochlea, of inner ear, 596, 596i, 597i
Cocklebur, fruit, 522
Codeine, 585
Codium, 384, 384i

Codominance, 182
Codon, 232, 232i, 233, 233i, 234, 234i–235i
Coelacanth, 453i
Coelom
  of annelids, 435, 435i
  false, 444, 444i
  origin of 430, 430i
  as type of body cavity, 419, 419t
Coelomate invertebrates, 430–443
Coenzyme
  activity, inhibition of,107
  defined, 50, 107
Coevolution
  defined, 818
  of camouflage, 820,820i
  of flowering plants and pollinators, 412, 414, 514
  of mimicry, 820, 821i
  of prey and predators, 821, 821i, 832
  of warning coloration, 820
Cofactor, defined, 103
Cohen, S., 620
Cohesion, of molecules, defined, 31, 506, 507, 507i, 512
Cohesion-tension theory of water transport in plants, 506–507, 507i, 512
Cohort, 800
Coiling, in gastropods, 431
Coitus, 773
Colchicine
  inhibitor of microtubule assembly, 70
  use in blocking spindle formation, 195, 195i
Colchicum autumnale, source of colchicine, 70, 195
Cold sore, 364, 631
Cold stress, 740–741, 740t
Cold, common, 362, 363t
Coleoptera, 441i
Coleoptile, 530i, 531i, 532, 533i
Coleus, 486i, 488
Collagen
  fiber, 548, 548i, 549i
  presence in body parts, 49, 52
  three-dimensional structure, 216
Collar cells, of sponges, 420, 421i
Collecting duct, of kidney, 733, 733i, 734i, 742
Collenchyma, of plants, 484, 484i, 498
Colon (See Large intestine)
Colon cancer, 721
Color blindness, as sex-linked trait, 201, 203t, 604
Color, chemical basis of, 118, 118i
Coloration, warning, 820
Comb jelly, 417, 418t, 425, 425i
Comb shape, in chickens, genetic basis of, 185, 185i
Comb, of comb jellies, 425, 425i
Commelina communis, 509i

Commensalism, 814
Communication, animal, 901, 906–907, 906i, 907i, 918
Community
  climax, 813, 821, 824, 825i
  cyclic, nondirectional changes, 824–825
  defined, 6, 814
  instability, 826
  interactions, 812–833
  stability, factors contributing to, 824
  structure, 814, 832
Companion cell, of phloem, 510, 510i, 511, 511i
Comparative
  anatomy, in explaining relationships, 272–273, 273i
  biochemistry, 329
  morphology, 316, 318–319, 318i, 319i, 329
Compartmentalization, in woody plants (responses to attack), 497
Competition
  categories of, 816
  interspecific, 814
  for resources, 285
Competitive
  exclusion, 816–817, 816i, 817i, 832
  interaction, between species, 816–817, 816i, 817i
Complement protein, 673, 673, 673i, 675, 675i
Compound, defined, 27, 34t
Compound eye, insect, 599, 599i
Concentration gradient, 86, 508i, 512
Conclusion, defined, 12, 16
Condensation, 40, 40i, 42i
Conditioning, 904t
Condom, 785, 785i
Conduction, in heat balance, 738
Cone, 408, 409, 407i, 408i, 409i, 410, 410i
Cone cell, of eye, 602, 602i, 603
Cone scale, of conifer, 407i
Conidia, of sac fungi, 393i
Conidiospore, 393i
Conifer, 343, 401i, 407i, 408, 408i, 410–411, 410i, 411i, 414t
Coniferous forest, 868, 868i
Conjugation
  of bacteria, 357, 357i
  of ciilates, 379, 379i
  tube, 357, 357i
  of Spirogyra, 385i
Connective tissue, 556, 548–549, 548i, 548t, 549i
Consciousness, 584
Conservation of mass, law of, 101
Constipation, 721
Consumer, 7, 36, 836, 836i, 836t, 837, 852
Contag, C. and P., 109
Continental drift, 314
Continuous variation, 186–187,

186i, 187i
Contraception, 784–785, 785i
Contractile cell, of cnidarians, 423, 423i
Contractile vacuole, of ciliates, 378–379, 378i
Contraction, sliding filament model, 640–641, 641i
Control agent, biological, attributes of, 823
Control group, in biological research, 13
Controlled cell death, 242–243, 242i, 243i
Convection, 738–739, 742
Convergent evolution, 464, 465
Cooksonia, 312i
Cooperative predator avoidance, 910, 910i
Copepod, 438i, 439
Copernicus, N., 16
Copper
  effects of deficiency, 503t
  in human nutrition, 725t
Copperhead (snake), ovoviviparity, 746, 747i
Coral bank, 874
Coral fungus, 390i
Coral reef, 424–425, 424i, 874–875, 874i, 875i
Coral snake, 459
Corallina (coralline alga), 874
Coralline alga, 874, 874i
Corbin, K., 305
Core temperature, 738, 742
Cork cambium, 496, 496i
Corm, 524t
Corn (Zea mays)
  germination of grain, 530i, 531i
  leaf structure, 124i
  stem, 487i
Cornea
  of human eye, 600, 600i
  of invertebrate eyes, 598, 598i, 599i
  transplant surgery, 605
  of vertebrate eyes, 600, 600i
Corolla, 516, 516i, 517i
Coronary artery, 665, 665i, 657i
Coronavirus, 363t
Corpus
  callosum, 580i, 582, 582i
  luteum, 770i, 771
  striatum, 583i
Cortex (of brain)
  motor, 583i
  prefrontal, 583i
  somatosensory, 592, 592i
  premotor, 583i
Cortex (of kidney), 733i, 735
Cortex (of plants)
  root, 492i, 493, 493i
  stem, 486, 486i, 487i, 495i, 496, 496i
Corticotropin (ACTH), 614, 614t, 618, 624
Corticotropin releasing hormone

(CRH), 782
Cortisol, 617t, 618, 618i, 737i
Cotyledon, of flowering plants, 485, 485i, 520, 520i, 530i, 531i
Courtship display, 906i, 907
Covalent bond, 28–29, 34, 102
Cowbird, brown-headed, as social parasite, 822
Cowpox, 670
Crab, 438, 438i, 439i, 623, 623i, 875i
Crack cocaine, 570–571, 585
Cranial
  bone, 637
  cavity, 552i
  nerves, 575i
Crassulacean acid metabolism (See CAM plant)
Craterellus, 389i
Crayfish, nervous system, 573i
Creatine phosphate
  dephosphorylation, 643, 634i, 646
  phosphorylation, 643, 643i
Creighton, H., 199
Crenicichla lepidota (pike-cichlid), as predator of guppy, 802–803, 802i, 803i
Creosote bush, 795
Cretaceous period, 313i, 348–349i
Cretaceous-Tertiary boundary
  asteroid impacted theory, 346
  and mass extinction, 346, 346i
Creutzfeldt-Jakob disease, 363
CRH (corticotropin releasing hormone), 782
Cri-du-chat syndrome, 203t, 210, 210i
Cricetidae, 464t
Crick, F. D., 216, 216i
Crinoid, 442, 442i
Crocodile, 456i, 457, 458, 458i, 706i
Crop, of bird digestive tract, 712, 712
Cross-bridge, of muscle, 641, 641i, 646
Crossing over, in meiosis, 166, 166i, 193, 194, 199, 199i, 214
Crown, of tooth, 715i
Crown-of-thorns sea star, 875i
Crustacean, 438–439, 438i, 439i, 623, 623i
Cryphonectria parasitica, cause of chestnut blight, 396, 827
Cryptosporidium, 376
Ctenidium, of mollusks, 431
Ctenomyidae, 464t
Ctenophora, 418t, 425, 425i
Cuckoo, 822, 822i, 903
Curie, M., 23
Current, ocean, 858, 858i, 877, 879
Cuticle
  of arthropods, molting of, 623, 623i
  of invertebrates, 628
  of plants, 74, 75i, 400, 485, 485i,

488, 489i, 501, 508, 508i
Cutin, 508
Cuvier, G. L. de, 274
Cyanobacteria, 358, 358i, 361t, 394, 398
Cyanophora paradoxa, 341i
Cycad, 343, 401i, 408, 408i, 414t
Cyclic adenosine monophosphate (*See* cAMP)
Cycling, membrane, 93, 93i
Cyst, of protozoans, 374, 375
Cystic fibrosis, 203t
Cytochrome, 107, 320i
Cytokinin, 529, 532, 533, 538, 541
Cytomembrane system, 66–68, 66i, 67i, 68i
Cytoplasm
  of bacteria, 354i, 355
  as component of cells, 55, 56, 64i
  consitituents of, 77, 78t
  division of, 156–157, 156i
  egg, information in, 750–751, 750i, 751t
  localization of, 750–751
  streaming, role of microfilaments in, 72
Cytosine, 50i, 220, 220i
Cytoskeleton
  components and functions, 60, 61i, 70–71, 77, 78t
  intermediate filaments, 70, 71, 71i
  microfilaments, 70, 70i, 71i
  microtubules, 70, 70i, 71i
Cytotoxic T cell, 676, 679i, 688t

D

2,4–D (2,4–dichlorophenoxyacetic acid), as herbicide, 22, 533
da Vinci, L., speculations about fossils, 310
Daisy coral, 875i
Darlingtonia californica, 501i
Darwin, C., 10, 16, 176, 274–278, 274i, 282, 285, 534, 645, 854
Daucus carota (carrot), tissue culture propagation, 524–525, 525i
Day-neutral plant, 537
Dayflower, 509i
Daylength, and pineal gland, 622
ddI (dideoxyinosine), in treatment of AIDS, 687
DDT, effect on ecosystems, 851
Deamer, D., 337
Death cap mushroom, 391i
Death, 758–759
Deciduous broadleaf forest, 867, 867i
Decomposer, 7, 36, 836, 836i, 836t, 838, 839i, 840i, 841i, 848, 848i, 852
Decomposition, by bacteria, fungi, 848, 848i

Decompression sickness, 705
Deductive logic, 12
Deer, overcrowded population on Angel Island, 792
Deficient immune response, 685, 685i
Deforestation, 411, 411i, 843, 843i, 847, 883, 888–889, 889i, 898
Delbrück, M., 218
Deletion, of base pairs of DNA molecule, 236
Delphinium, 316, 316i
Deltoid muscle, 639i
Demographic transition model, of population growth, 808, 808i
Demographics, 794
Dendrite, of neuron, 560, 560i, 567i, 568
Dendrobates (frog), 547i
Dendrocopus pubescens, 288, 288i
Denitrification, by bacteria, 848i, 849
Density, of population, 794
Density-dependent growth, of population, 799
Density-independent growth, of population, 799
Dentin, of tooth, 715i
Deoxyribonucleic acid (*See* DNA)
Deoxyribose, as five-carbon sugar, 42
Depressant drug, 585
Derived trait, 324
Dermal tissue
  cuticle, 485i
  epidermis, 485i
  periderm, 485
  of plants, 484, 485
Dermaptera, 441i
Dermis, 628, 628i, 630, 631i, 646
Dermochelys coriacea (leatherback turtle), 707i, 706–707, 706i
DeRouin, T., 14
Desalination, 892
Desert, 863, 863i
Desertification, 863, 891, 891i, 898
Dethier, V., 754
Detrital food web, 838, 839i
Detritivore, 836, 836i, 836t, 838, 839i, 840i, 841i, 852
Deuterostomes, 430, 442–443
Deuterostome embryo, 430, 430i
Development
  of animals, 744–761
  human, 773–783
  plant, 528–541
Devonian period, 313i, 343, 343i, 348–349i
Diabetes
  insipidus, 616
  mellitus, 620–621
Dialysis, 735, 737
Diaphragm (contraceptive), 785, 785i
Diaphragm (human body), role in breathing, 696i, 697, 698–699, 698, 732i

Diarrhea, 672
Diastole, of heart, 658, 659i
Diatom, 380, 380t, 381i
Dickensonia, 342i
Dicot, 412, 414t, 485, 485i
Dictyostelium discoideum (slime mold), 373, 373i, 753
Didinium, 371i
Diet, zone, 722
Dieting, 710
Diffusion
  across bilipid layer, 87i
  of carbon dioxide and oxygen in plants, 508
  defined, 86, 87, 86i, 87i
  factors affecting, 86–87
  influence of electric gradient, 87
  influence of pressure gradient, 87
Digestion
  of carbohydrates, 719i, 717t
  controls over, 717
  of disaccharides, 717t
  DNA and RNA, 717t
  of fats, 719i, 717t
  major enzymes, 717i
  nucleic acids, 717t
  nucleotides, 717t
  of polysaccharides, 717t
  of proteins, 719i, 717t
  of triglycerides, 717t
Digestive system, 553i
  accessory organs, 714i
  bird, 712, 712i
  cnidarian, 712
  complete, 712, 713
  flatworm, 712, 712i
  frog, 712, 712i
  functions, 712
  human, 714, 714i, 728, 728t
  incomplete, 712m 712i, 713
  influenced by other systems, 728
  mammal, 728
  relation to other systems, 711, 711i, 712
Digitalis purpurea, 491
Dihybrid cross, 180, 180i, 181i
Dikaryotic mycelium, in mushrooms and other club fungi, 391, 391i
Dimetrodon, Devonian reptile, 343i
Dimorphism, sexual, 290, 290i
Dinoflagellate, 380t, 381, 381i, 386t, 874
Dinosaur, 278, 278i, 344, 345, 346, 456i, 457, 457i, 463
Dionaea muscipula, 500, 500i
Dipeptide, defined, 46
Diphtheria, vaccine, 684, 684i
Diploid phase
  defined, 149, 150, 158
  in plant life cycle, 400–401, 400i
Diptera, 441i
Disaccharidase, 717t
Disaccharide, 42, 52t

Disease
  contagious, 366
  endemic, 366
  and human population, 804, 805
  infectious, 366–367
  lines of defense against, 672, 672t
  pandemic, 366
  sporadic, 366
Dispersal
  area effect, 831, 831i
  geographic, 827
  jump, 827
  of species, 830–831, 830i, 831i
Dispersion, of population, 794–795, 794i
Distal tubule, of nephron, 733i
Distant Early Warning system (DEW), 897
Disulfide bridge, in protein synthesis, 47i, 48
Diversity, basis of, 283
Diving, 705–707, 706i
Division of labor
  in animal body, 545
  in arthropods, 436
Dixon, H., 506
DNA
  amplification, 260, 260i
  of B and T cells, 682–683, 682i
  of bacteria, 354i, 355, 356i
  in bacteriophage, 219, 219i
  base pairing, 221, 221i, 225
  base sequence, 227
  cDNA, formation, 263, 263i
  components, 220–221, 220i, 221i
  composition of, 50–51, 50i, 217
  condensation of, in mitosis, 154
  content, 225
  damaged, 224
  differences between individuals in mole-rat clan, 916
  double helix, 51i, 221, 221i
  as encoder of developmental program, 4
  fingerprint, 262, 262i, 267
  function, 51, 52t
  guiding synthesis of RNA, 227
  heritability of, 4
  hydrogen bonds, 217
  instructions for synthesis of proteins, 227
  library, 259, 259i, 260, 267
  ligase, 222, 223i, 259, 259i
  model of, 51i, 216–217, 216i, 217i
  mutations resulting in aging, 758
  nucleotide strands, 225
  nucleotides of, 217
  polymerase, 222, 223i, 260, 261i
  probe, 263, 263i, 267
  proteins associated with, 223
  rearrangement of, 246
  recombinant, 256–267
  replication and repair, 222–224,

222i, 223i, 225
replication fork, 223, 223i
replication of, 217
role in synthesis of proteins, 4, 149, 151i
sequencing, Sanger method, 261, 261i
similarities and differences among individuals of same species, 262, 262i
structure, 51i, 220–221, 220i, 221i
transmission by sperm and egg, 4
in viruses, 219, 219i, 686, 686i
x-ray diffraction images, 220
DNA-DNA hybridization, in studies on relationships, 321, 320i
Dog, domestication of, 280–281,280i
Dolichohippus, 301
Dolomedes, 17, 17i
Dominance, incomplete, 182, 182i
Dominance hierarchy, 911, 911i, 913, 913i, 918
Dominant species, 826
Dopamine, 564, 585
Dormancy, in plants, 538, 539i, 539, 539i, 541
Dorsal (in anatomy), 419, 552i
Dorsal lip, of frog embryo, 749i
Dosidiscus, 433i
Double fertilization, in flowering plants, 413i, 519, 519i, 526
Double helix, DNA, 221, 221i, 225
Double-blind studies, in genetic research, 209
Doubling time, 797, 797i
Douche method, of birth control, 784, 785i
Douglas fir, 480, 481, 481i
Down syndrome, 203t, 208, 209i
Downy woodpecker, 288, 288i
Dragonfly, 441i
Drew-Baker, K., 382
Drone, of honeybee colony, 915, 915i
Drosophila melanogaster (fruit fly)
development, 754–755, 754i, 754i, 755i
genetics of, 198, 198i, 199-201, 199i, 200i, 201i
Drug addiction, warning signs, 584
Drug resistance, 366
Drugs, abuse of, 584–585
Dry shrubland, 864, 864i
Dryas (mountain avens), role in primary succession, 824, 825i
Duchenne muscular dystrophy, 203t, 206–207
Duck, black, travels of, 296
Duck, wood, secondary sex characters, 39i
Duckweed, 251
Dung beetle, 7, 7i

Dust bowl, 891, 891i
Dutch elm disease, 396, 827t
Dwarfism, pituitary, 616, 616i
Dynein, 71, 73i, 155
Dysentery, amoebic, 374

E

Ear, 595–597, 595i, 596i, 597i, 778i
Eardrum, 596, 596i, 597i
Early Cretaceous, 313i
Earth
    age of, 334
    crust, 334
    history of, summarized, 348, 348i–349i
    mantle of, 334
    tectonic plates, 314–315, 315i
Earthworm, 434, 434i, 435, 435i, 573i, 632, 650i
Earwig, 441i
Ebola virus, 366, 366i
Ecdysiotropin, 623
Ecdysone
    and molting, 623, 623i
    triggering transcription, 250, 250i
ECG (Electrocardiogram), 648, 648i, 665, 665i
Echinocereus (cactus), 854i
Echinoderm, 418t, 442–443, 444, 442i, 443i
Echolocation, 588–589, 589i, 597
Ecological pyramid, 838–839, 839i
Ecology
    defined, 793
    population, 792–811
    restoration, 825
Ecosystem analysis
    at Hubbard Brook Valley, New Hampshire, 843, 843i
    at Silver Springs, Florida, 840, 840i
Ecosystem, 834–853, 879
    defined, 6, 835, 836
    destabilization by humans, 882
    effects of DDT, 851
    energy flow through, 835, 836i, 838–839, 839i, 840, 840i, 841i
    nutrient flow, model, 841i
    human influences, 835
    impact of change, 851
    modeling, 851
    lake, 870–871, 870i
    primary productivity, 836, 836i, 836t, 837, 837i, 838, 838i, 839, 839i, 840i
    stream, 871, 871i
    trophic levels, 836–837, 836i, 836t, 837i
Ectocarpus, 383
Ectoderm, 418, 553, 748, 749i
Ectotherm, 739, 742
Edema, 662, 675, 675i
Ediacaran, 342, 342i
EEG (electroencephalogram), 580,

584, 584i
Eel, moray, 453i, 875i
Effector, 554, 555i, 559, 559i
Egg, formation of, in animals, 168, 169i
Ejaculation, 773
Ejaculatory duct, 764t, 765i
EKG (*See* ECG)
El Niño, 877, 878–879, 878i
Electrocardiogram (*See* ECG)
Electroencephalogram (*See* EEG)
Electromagnetic spectrum, 116, 116i
Electron transport
    phosphorylation, 133, 138, 138i, 139i
Electron transport system, 108, 108i, 120, 120i, 338
Electron transport
    aerobic, 145
    anaerobic, 141–142, 145
Electron
    arrangement in atoms, 26, 26i, 27i
    defined, 21, 22, 34t
    transfer in biological reactions, 108
Element, 21, 22, 23, 34
Elephant seal, northern, 293
Elephantiasis, 428, 429, 429i
Ellis-van Creveld syndrome, 293
Elm disease, Dutch, 827t
Elodea, release of oxygen by, 121i
Embryo
    chick, 752i
    of flowering plants, 518, 519i, 530, 530i
    human, 762i, 763, 774–779, 774i, 775i, 776i, 783t
Embryology, providing evidence for evolution, 316–317, 317i
Embryonic development, 745, 748–757, 748i, 749i, 760
Embryonic disk, 775, 775i, 776, 776i
Embryonic induction, 756–757, 756i, 757i, 760
Embryonic period, human, 774, 778
Emigration, from population, 796
Emperor penguin, adoption of orphans, 917, 917i
Emphysema, 702, 702i
Emulsification, of fats, 717, 719i
Enamel, of tooth, 715i
Endergonic reaction, 101, 101i
Endler, J., 802
Endocarp, of fruit, 521
Endocrine control, 608–625
Endocrine system, 547, 552i, 610–611, 610i, 611i
Endocytosis, 87, 87i, 92, 92i, 93, 93i, 94
Endoderm, 418, 553, 748, 749i, 776
Endodermis, of root, 492i, 493, 493i, 504, 504i
Endogenous pyrogen, 675

Endometriosis, 769
Endometrium 768, 769i, 769t, 771i, 772i, 774i
Endomycorrhiza, 395
Endoplasmic reticulum, 60, 61i, 62i, 63i, 66, 67i
Endorphin, 564
Endoskeleton, 632, 633i, 646
Endosperm, 423, 413i, 518, 519, 519i, 520, 521i, 526, 530i
Endospore
    of *Clostridium botulinum*, 568
    of Eubacteria, 358–359, 358i
Endosymbiosis, theory of origin of organelles, 341, 815
Endothelium, 658, 658i
Endotherm, 739, 742
Energy
    acquisition during photosynthesis, 112i
    chemical, 98
    conservation of, by cells, 108
    consumption by human populations, 894, 894i
    defined, 4, 98
    flow through biological systems, 7, 98–99, 99i, 838–839, 839i, 840, 840i, 852
    heat, 98
    kinetic, 98
    light, 112i, 116, 116i, 117i, 250–251, 251i
    nuclear, 894–895, 895i
    potential, 98
    release of in biological systems, 112i, 130–146
    solar, 855
    sources of, 113
    sources, alternative, 896, 896i
    stepwise release of, 108, 108i
    thermal, 98
    transfer of, 4–5, 102
    wind, 896, 896i
Energy-releasing pathway, 130–145
Englemann, T., 117, 117i
Enhancers, of hormonal influences on genes, 250
Entamoeba histolytica, 374–375
Enterobius vermicularis, 428
Enteromorpha (green alga), as dominant intertidal species, 826, 826i
Enzyme
    active site of, 104, 105i
    activity of in electron transfer system, 108
    as catalyst in biological reaction, 104, 105
    characteristics of, 104, 105
    controls, 106–107
    defined, 103, 110
    effect of pH on, 106, 106i
    effect of temperature on, 106, 106i
    factors influencing activity, 106

formation affected by gene mutation, 228
inhibition of, 107, 107i
restriction, in bacteria, 259, 259i
role of, 40
specificity of, 104
types of reactions mediated by, 40
Eocene epoch, 313i, 347, 347i, 348–349i
Eon, geologic, 313i
Eosinophil, 652i, 653, 653i, 674, 688t
Ephedra, 409, 409i
Epiblast, 775
Epidemic, 366
Epidermal growth factor (EGF), 620
Epidermis, animal
  relationship to other body components, 419i
  of humans and other vertebrates (*See also* Skin), 630, 630i, 646
Epidermis, plant
  of leaves and stems, 484i, 486i, 487i, 488, 489i
  of root, 492, 492i, 493i
  of vertebrate skin, 628, 628i
Epidermophyton floccosum, cause of athlete's foot, 396, 396i
Epididymis, 764, 764t, 765i, 766i
Epiglottis, human, 696i, 697
Epilepsy, 582
Epinephrine (adrenaline), 577, 617t, 618, 664
Epiphyte, 405, 866i
Epistasis, defined, 184
Epithelial tissue, 545, 546–547, 546i, 547i, 556
Epithelium, 423, 546, 546i, 547, 547i
Epoch, geologic, 313i
EPSP (excitatory postsynaptic potential), 565, 565i
Epstein-Barr virus, 363t
Equilibrium, 100, 100i, 595, 606
Equisetum, 404–405, 404i
Equus, 301
Era, geologic, 313i
Ergot, fungus disease of rye, 396
Erosion, soil, 503
Erythroblastosis fetalis, 655
Erythrocyte (*See* Red blood cell)
Erythropoietin, 617t, 704
Escherichia coli
  bacteriophage of, 218, 219i
  DNA, 77
  electron micrograph, 76i
  gene activity related to lactose breakdown, 244–245, 244i, 245i
  importance to humans, 358
  plasmids, 258, 258i
  strains causing diarrhea and other serious disease, 358, 367
Eschscholzia californica, 492i, 528i

Esophagus, 714i, 715, 728
Estrogen, 39i, 617t, 768, 770, 777, 771i, 772i, 777, 782
Estuary, 876, 877
Ethylene, 529, 532, 533, 538, 541
Eubacteria (kingdom) (*See also* Bacteria), 354–359, 354i–359i
Eucalyptus, 412
Euchema, source of carrageenan, 382
Eugenic engineering, 267
Euglenoid, 380, 380i, 380t, 386t
Euglenophyta, 380, 380i, 380t, 386t
Eukaryote, origin, 338, 338–339i, 815
Eukaryotic cell, 55, 56, 60, 61i, 62i, 63i, 64–65, 64i, 65i
Eupenicillium, 393i
Euphorbia, 854
Euplectella, 421i
European starling, 827t
Eurosta solidaginis, 288, 288i
Eutherian, 463
Eutrophication, 850, 850i, 852, 871
Evaporation
  in heat balance, 739, 742
  in hydrologic cycle, 842i
Evaporative heat loss, 741
Evolution, 270–279, 464
  branching, 306
  convergent 465
  cultural, of humans, 475
  defined, 10
  early beliefs, 272
  early scientific theories, 27
  evidence from fossils, 311, 311i, 312–313, 312i, 313i, 329
  evidence from biochemistry, 320–321, 320i, 321i
  evidence from comparative embryology, 316–317, 316i, 317i
  as genetic change through time, 281
  human, 468–478
  plant, 414t
  rates of change, 306
  unbranched, 306
Evolutionary relationships, determining, 324–325, 324–325i
Evolutionary tree, 306, 306i, 307i
Evolutionary history, 457
Excitatory effect, of neurotransmitter, 564
Excitatory postsynaptic potential (EPSP), 565, 565i
Exergonic reaction, 101, 101i, 110
Exhalation, during breathing, 698, 698i, 699, 699i
Exocrine gland, 547, 547i
Exocytosis, 87, 87i, 92, 92i, 94
Exomycorrhiza, 395
Exon, 231, 231i
Exoskeleton, 632, 646
Exoskeleton, of arthropods, 436, 436i

Exotic species, 827, 827t, 828–829, 829t
Experimental design, 12
Experiments, examples of (*See also* Observational tests; Research tools)
  abiotic synthesis of membrane-like lipids, 337
  abiotic synthesis of organic compounds, 333–335, 335i
  abiotic synthesis of protein-like spheres, 337
  aging, 758
  antennapedia gene in *Drosophila*, 755, 755i
  bacterial transformation, 218
  bacteriophage versus antibiotics, 14–15, 14i
  basis of food preferences among garter snakes, 902, 902i
  bee breeding, 130
  cell differentiation, 752
  chromosome mapping, 200
  competitive exclusion, 816–817, 816i
  cooperative predator avoidance among sawfly caterpillars, 910, 910i
  crossing over, 198–199, 198i, 199i
  cytoplasmic localization, 750, 750i
  dihybrid crosses, 180–181
  and discovery of *ob* gene, 727, 727i
  DNA sequencing, 261
  dormancy, 538, 539i
  double-blind study for XYY condition, 209
  effect of nerve growth factor on neuron survival, 620
  effect of parasitoid wasps on sawfly emergence, 823, 823i
  embryonic induction, 757, 757i
  eutrophication, 850, 850i
  exponential growth, 797, 797i
  first smallpox immunization, 670, 689
  flowering response, 537i
  gene transfer, 109, 257, 264–266
  genetic drift, 292
  genetic engineering, 264–266
  gravitropism, 534, 534i
  of HbS and HbA biochemical differences, 228–229
  heart activity in Jimmy the bulldog, 648, 648i
  intense sound and hearing, 597i
  intraspecific competition among salamanders, 817
  logistic growth, 798, 798i
  luciferase gene transfers, 109
  mate recognition, 12–13
  mechanical stress on plant

growth, 535, 535i
  membrane fluidity, 85
  membrane structure, 84–85
  monohybrid cross, 178–179, 178
  natural selection among guppies of Trinidad, 802–803, 802i, 803i
  obesity, 727, 727i
  of one-gene, one-enzyme hypothesis, 228–229, 229i
  pancreatic secretions and discovery of secretin, 610
  pattern formation and apical ectodermal ridge (AER), 756–757, 756i
  pattern formation in *Drosophila*, 755, 755i
  of phototropism, 534–535, 535i
  of potassium in guard cell functioning, 509i
  phytoremediation, 21
  receptive field, cat eye, 603
  reciprocal cross, 198
  of senescence, 538, 538i
  split-brain, 582
  starling nesting behavior, 900–901, 901i
  testcross, 179, 201
  to identify antibiotic resistance genes, 263, 263i
  torsion in mollusks, 432
  tracking steps of photosynthetic pathway, 25
  transcriptional control of phytochrome gene, 251
  vernalization, 539
  watershed, 842–843, 843i
  white-crown sparrow singing behavior, 904
  with auxin, 533i
  with frog reflex pathways, 577
  with keystone species, 826, 826i
  of zebra finch singing behavior, 902–903, 902i
Xenopus development, 752–753
External fertilization, 744, 744i, 746
External oblique muscle, 639i
Extinction, 307
Extracellular fluid, 554, 731, 732, 742
Extraembryonic membrane, 775, 775i
Eye
  color, in humans, genetic basis of, 186, 186i
  disorders, 604–605
  embryonic, 752i
  focusing, 601, 601i
  innervation, 576i
  invertebrate, 598–599, 598i, 599i
  of primates, 470–471
  vertebrate, 599, 600–605

F

Facial bone, 637i
FAD (flavin adenine
    dinucleotide), 50, 107, 133i, 136
Fall overturn, in lakes, 870, 871
False coelom, 444, 444i
Familial hypercholesterolemia,
    203t
Family planning, 806–807
Farsightedness, 604, 605i
Fat
    energy yield from, 142
    in human diet, 722i, 723
    normal proportion to total
        tissue mass, 710
    sources of, 722i, 723
    storage of, 142, 720
Fate map, of *Drosophila* zygote,
    755, 755i
Fatty acid, 44, 44i, 723
Faulty enamel trait, 203t, 207
Feather star, 442, 442i
Feather, 460, 461i, 628, 629i
Feedback
    control of hormonal secretions,
        618–619, 618i
    in hormonal control of of
        menstrual cycle, 619, 619i
    and human modification of
        ecosystems, 897
    inhibition of enzyme activity,
        107, 107i
    negative, 554, 554i, 556, 618,
        618i
    positive, 555, 562, 562i, 618
Feedforward control, 897
Femur, 637i
Fermentation
    alcoholic, 139-140, 141i, 145
    lactate, 140, 140i, 145
    by yeasts, 141, 141i
    yield of energy, 140, 140i
Fern
    in plant classification, 414t
    gametophyte of, 405, 405i
    in geological history, 401i
    life cycle, 405, 405i
    seed, 407, 406i
    sporophyte of, 405, 405i
    whisk, 404, 404i
Fertility, human, control of,
    784–785, 785i
Fertilization
    defined, 168–170, 168i
    double, in flowering plants,
        413i
    of egg by sperm, 147i, 750, 773,
        773i
    in flowering plants, 516, 516i,
        519, 519i
    human, 773, 773i, 774i
    internal, in higher vertebrates,
        456
    as stage in animal reproduction
        and development, 161, 748,
        748i

in vitro, 788
Fertilizer, 20, 849, 850
Fetal alcohol syndrome, 781, 781i
Fetal period, 774
Fetus, human, 778–779, 778i, 779i,
    781i, 783t
Fever, 675, 741
Fiber, of sclerenchyma, 484, 484i
Fibrillation, atrial, 665
Fibrinogen, 652i, 666i
Fibrous root system, 492, 492i,
    493i
Fibula, 637i
Fick's law, 692
Fight-flight response, 577, 618
Filament, 516i, 517
Filovirus, 363t
Filter feeding, 448, 449
Filtration of blood, 734–735, 734i,
    742
Fin, 450
Finch
    African, beak of, 289, 289i
    Galápagos, 276, 277i, 303
    zebra, territorial song, 902,
        902i
Fir, Douglas, 480, 481, 481i
Fire ant, Argentine, 827, 827t
Firefly, illegitimate signalling by,
    907
Fish
    bony, 448, 452–453, 453i, 465
    cartilaginous, 448, 452, 452i,
        465
    early armor-plated types, 342
    jawed, 452–453, 452i, 453i, 465
    jawless, 448, 451, 451i, 465
    kidney of, 737
    lobe-finned, 343, 453, 454, 454i
    ray-finned, 453
    respiratory system of, 694, 694i
    water-solute balance of, 737,
        737i
Fishing spider, 17, 17i
Fission
    of bacteria, 356, 356i, 366
    binary, 374
    multiple, 374
    prokaryotic, 353, 356, 356i, 366
Fitness, in adaptation to
    environment, 285
Fixation, of alleles, 292, 292i
Flagella, of bacteria, 366
Flagellate, 376, 376i, 386
Flagellum, 72–73, 72i, 73i
Flamingo, color of feathers, 119
Flatworm, 417, 418t, 426–427,
    426i, 427i, 572, 573i, 712,
    712i
Flavin adenine dinucleotide (*See*
    FAD)
Flavivirus, 363t
Flavoprotein, possible
    involvement in phototropism,
    534
Flax, 490i
Flea, 441i

Flemming, W., discovery of
    chromosomes, 193
Flight, of birds, 461, 461i
Flowering plant (angiosperms)
    classification, 414t
    flower, 412, 412i, 413i, 514–519
    fruit, 515
    gamete, 515, 516, 516i
    gametophyte, 515, 516, 516i
    in geological history, 401i
    life cycle, 413i
    megaspore, 515, 516, 516i
    in Mesozoic era, 344i
    microspore, 515, 516, 516i
    pollination, 412i, 514–515, 514i,
        515i
    reproduction and development,
        514–527
    seed, 515
    sexual reproduction, 515, 516,
        516i
Flowering, influence of
    gibberellin, 528, 532
Fluid mosaic model, cell
    membrane, 82, 82i, 83, 83i
Fluid pressure, effects on cells, 89,
    89i
Fluid, extracellular, 554, 731, 732,
    742
Fluke, 426–427, 428–429, 429i
Fluorescence, defined, 118
Fluorine, in human nutrition,
    725t
Fly
    fruit, 754–755, 754i, 755i (*See
        also Drosophila melanogaster*)
    Mediterranean fruit, 441i
    gallmaking, 288, 288i
    mouth parts, 440i
Folate (folic acid), 724t
Follicle, ovarian, 770, 770i
Follicle cell, 773i
Follicle-stimulating hormone (*See*
    FSH)
Follicular phase of menstrual
    cycle, 768, 769t, 772i, 788
Fontanel, 636
Food chain, 837, 852
Food pyramid, 722, 722i
Food web, 852, 837, 837i, 838, 839i
Food, of mollusks, 431, 432, 432i,
    433i
Foraminifera, 375, 375i
Forebody, of arachnids, 437
Forebrain, 571, 574i, 578i, 579,
    586
Foregut, of insects, 440
Forelimb, vertebrate,
    morphological divergence, 318,
    318i
Forest
    boreal, 868, 868i
    coniferous, 411, 411i, 868, 868i
    deciduous, broadleaf, 867, 867i
    deciduous, trophic levels,
        836–837, 836t
Formaldehyde, hypothetical

conversion to porphyrin, 336,
    336i
Fossil fuel, 406, 845, 844i, 847,
    852, 894, 894i, 898
Fossil record, 311, 313, 329
Fossil
    examples of, 312i
    early awareness of, 273, 310
    formation, 312, 312i
Fossil fuel, and pollution, 894
Fossilization, 312, 312i
Founder effect, 293, 293i, 294
Fovea, of eye, 600t
Fox, arctic, 465i, 739
Fox, S.. 335, 337
Foxglove, 491
Foxtail grass, and resource
    partitioning, 817, 817i
Fragaria, fruit, 521, 521i
Fragile X syndrome, 203t, 211,
    211i
Franklin, R., 220
Free nerve ending, 592, 593i
Free radical, 96
Freeze-etching, 84, 85i
Freeze-fracturing, 84, 85i
Freshwater province, 870
Friedman, J., 727
Frigate bird, 461
Frilled lizard, 459i
Frisch, K. von, 907i
Frog
    embryonic development, 748,
        748i, 749i
    leap of, 455i, 638i
    leopard, reproduction and
        development, 744–745, 744i,
        745i
    South African clawed, 455
    water-solute balance, 737
Frontal (in anatomy), 552i, 553i
Frontal lobe, 581i
Fructose, formula, 42i
Fruit, 413, 413i, 414, 515, 520,
    520i, 520t, 521, 521i, 522, 522i,
    526
Fruit fly, 754–755, 754i, 755i (*See
    also Drosphila melanogaster*)
    Mediterranean, 441i,
Fruiting bodies, of myxobacteria,
    359, 359i
FSH (Follicle-stimulating
    hormone), 624, 614, 614t, 763,
    767i, 770, 771, 771i
Fucoxanthin, 380, 383
Fungicide, 41
Fungi (kingdom), 8, 8i, 9, 328,
    328i, 388–397
Fungus
    economic importance, 389, 396
    disease-causing, 396, 396t
    toxic, 396, 396t
    coral, 390i
    as decomposers, 388–389
    destructive, 396
    extracellular digestion by, 389,
        390

as heterotrophs, 388–389
imperfect, 393, 393i
life cycles, 390, 391i
major groups, 390–393
mycorrhizae, 505, 505i
nutrition, 390
parasitic, 390
pathogenic, 396
predatory, 393i
purple coral, 388i
root, 540
sac, 390
saprophytic, 390
shelf, 390i
spores of 390i, 391i, 392, 392i
stinkhorn, 390
sulfur shelf, 388i
summary of characteristics of
    cells, 78t
symbiotic association of, 389
symbiotic, 512
yellow coral, 389i
Fusiform initial, of vascular
    cambium, 494, 494i
Fusion power, possibilities, 896

G

GABA (gamma aminobutyric
    acid), 564
Galactosemia, 204, 204i
Galápagos finches, 276, 277i, 303
Galápagos Islands, 275i, 276
Galápagos rift, 872–873
Galilei, G., 16, 54
Gall, produced by insects, 288,
    288i
Gallbladder, 714i, 716, 728, 728t
Gallmaking fly, 288, 288i
Gamete formation, 165i, 168, 168i,
    169i
Gamete, 161, 162, 516, 516i
Gametocyte, of *Plasmodium*, 377,
    377i
Gametophyte
    of bryophytes, 402–403, 402i,
        403i
    of conifers, 410, 410i
    of flowering plants, 413, 413i,
        515, 516, 516i, 518, 518i, 519i,
        526, 413i
    in plant life cycles, 168, 168i,
        400, 400i
Gamma aminobutyric acid
    (GABA), 564
Gamma globulin, anti-Rh, 655
Ganglion, 572, 573i, 576i, 577i
Gap gene, 755, 755i
Gar, long-nose, 453i
Gardenia, effect of auxin, 533i
Gargas caves, 468
Garrod, A., 228, 228i
Garter snake, feeding on slugs,
    902, 902i
Gas exchange (*See* Respiration,
    Respiratory system)

Gassner, G., 539
Gastric fluid, 716
Gastrin, 617t, 721
Gastrocnemius muscle, 639i
Gastrodermis, 422, 423i
Gastropod, 431, 431i, 432, 432i
Gastrula, 752i
Gastrulation, 748, 748i, 754i, 760,
    775, 776
Gause, G., 816
Gel electrophoresis, 228–229,
    229i, 260–261, 261i
Gemmules, of sponges, 421
Gene
    activity, influenced by signaling
        mechanisms, 250–251, 250i,
        251i
    amplification, 246, 267
    antennapedia, 755, 756, 757,
        755i
    control, evidence, 248–249, 248i,
        249i
    cytological marker, 199, 199i
    defined, 162, 175, 177, 189, 194
    expression, 242–244, 246–247
    flow, 283, 291, 291i, 294, 294t,
        299
    gap, 755, 755i
    independent assortment, 181,
        180i, 181, 181i
    linked, 198–199, 198i, 199i, 214
    multiple effects of, 183, 183i
    mutation, 227, 235i, 236–237,
        236i, 294
    pair, 184, 184i, 185i
    pool, defined, 282, 283
    recombination, 199-201 199i,
        200i, 201i
    selective expression, 254
    self-sacrifice, 913
    sex-linked, 198–199, 198i, 199i
    SRY (sex-determining region of
        Y chromosome), 196
    therapy, 109, 257, 267
    transcription, in RNA
        formation, 230, 231i
    transfer, in plants, 265, 265i
Genetic
    code, 232, 232i
    compatibility, between closely
        related species, 301, 301i
    counseling, 212
    disease, 203
    disorder, human (*See also*
        specific disorder), 202–203,
        203t, 212–213, 212i
    drift, 292, 292i, 293, 293i, 294,
        294t
    equilibrium, 283, 294
    recombination, 194
    screening, 212
    traits, morphological, 282
Genetic engineering, 256–268
    of animals, 266, 266i
    of bacteria, 264–265, 264i
    defined, 258
    of plants, 264–265, 265i

risks, 264
Genital herpes, 781, 786, 787
Genitalia
    external, development of, 196,
        197i
    innervation, 576i
Genotype, defined, 177
Genus, as taxonomic category, 8,
    322, 322i
Geographic barrier, in speciation,
    299, 308
Geographic dispersal, 827
Geographic isolation, and
    speciation, 302–303, 302i, 303i
Geologic record, 311
Geologic timetable, and plant
    evolution, 401i
Geomyidae, 464t
Geospiza, 277i
Germ cell, defined, 150, 161
Germination, seed, 530–531, 530i,
    531i, 541
Gey, G. and M., 158
Giant axon, squid, 563, 563i
Giardia lamblia, 376, 376i
Gibberella fukikuroi, 528
Gibberellin, 528–529, 528i, 529i,
    532, 538, 541
Gibbon, 470i, 470t, 477i
Gibbons, R., 73i
Gigantism, pituitary, 616, 616i
Gill slit
    of fishes, 450, 450i
    of chordates, 448
    of tunicates, 448, 449i
Gill
    of fishes and amphibians, 651
    of invertebrates, 693, 693i
    of lower vertebrates, 450
    of sea hare (Aplysia), 693, 693i
Gingiva (gum), of mouth, 715i
Gingivitis, 715
Ginkgo, 343, 401i, 409, 409i, 414t
Giraffe rhinoceros, 347i
Gizzard, of bird digestive tract,
    712i, 713
Glaucoma, 605
Gleditsia, 488i
Global broiling theory, 346
Global warming, 37, 845, 846,
    846i, 847, 847i
Globin molecule, structure of, 48,
    48i
Globulin, in blood, 652i
Glomerular capillaries, 733i, 734,
    734i
Glomerulonephritis, 736
Glomerulus, 733, 733i, 734, 734i,
    742
Glossopteris, 314, 315, 315i
Glottis, 695, 695i, 699, 699i
Glucagon, 142, 617t, 620, 621i
Glucocorticoid, 617t, 618
Glucose, 42i, 115, 123, 123i, 142
Glucose-6-phosphate, 142
Gluteus maximus muscle, 639i
Glycerol, 44, 44i

Glycocalyx, of bacteria, 355, 366
Glycogen, 42, 43i, 142, 643, 720
Glycolipid, cell membrane, 82
Glycolysis, in formation of ATP,
    132, 132i, 133, 133i, 134, 134i,
    135i, 136, 137, 140, 140i, 145,
    145i, 643, 643i, 646
Glycoprotein,
    defined, 49
    role in formation of cell wall, 74
    of virus, 362
Glyptodont, 276, 276i
Gnetophyta, 409, 409i, 414
Gnetum, 409
GnRH, secreted by
    hypothalamus, 767, 770, 771i
Goiter, 619, 619i
Golden algae, 386t
Goldenrod, 288, 288i
Golgi body, 60, 61i, 62i, 63i, 66,
    66i, 67i
Gonad, 617t, 619, 619i, 763
Gonadotropin-releasing
    hormone, 615
Gondwana, 315, 315i, 342, 342i,
    348i
Gonorrhea, 786–787, 786i
Goodall, J., 608, 911
Goose barnacle, 438i
Goose, Canada, 460i
Gopher, 464t
Gorilla, 470t, 471i, 477i, 629i
Graded potential, defined, 562
Gradient, in translocation of
    organic solutes, 511
Grain, 520t
Gram stain, for bacteria, 355, 355i
Grand Canyon, 314i
Granstein cell, of skin, 631
Granum, of chloroplast, 69, 69i,
    114i, 115
Grape, effect of gibberellin, 528i,
    529
Grass, pollen, 517i
Grasshopper
    circulatory system, 650i
    mouth parts, 440i
    nervous system, 573i
Grassland, 864–865, 865i, 891, 891i
Grave's disorder, 685
Graves, J., 302
Gravitropism, 534, 534i, 541
Gray crescent, 749i, 750, 750i, 751i
Gray matter, spinal cord, 577,
    577i
Grazing food web, 838, 839i
Great tit, predator on sawfly
    caterpillars, 910
Green algae, 384–385, 384i, 385i,
    386, 394, 398i, 400i, 414
Green revolution, 887
Green turtle, diving by, 706i
Greenhouse effect, 845, 846–847,
    846i, 847i, 856, 889
Griffith, F., 218
Grouse, sage, sexual selection,
    908, 909, 908i

Growth, animal
  as general stage in
    development, 748, 748i, 760
  influenced by growth hormone
    (somatotropin), 614, 614t
Growth, plant, 494, 494i, 498,
  528–541
Gowth, human population,
  796–799, 796i, 797, 797i, 798i,
  799, 799i
Growth factor, 252, 620
Growth ring, of woody plants,
  497, 497i
Guanine, 50i, 220, 220i
Guard cell, 489i, 508–509, 508i,
  509i, 512
Guilleman, R., 616
Gulf Stream, 858, 858i
Gull, herring, cannibalism, 912
Guppy, natural selection,
  802–803, 802i, 803i
Gurdon, J., 752
Gut, 419, 444, 444i
Gymnodinium breve, 381i
Gymnophilus, 388i
Gymnosperm, 399, 400, 400i,
  408–409, 408i, 409i, 414, 481
Gypsy moth, 827

H

Habitat, 813, 814, 832
Haemanthus, mitosis in, 151i
Hagfish, 451, 451i
Hair, 184, 184i, 626, 628, 628i,
  630
Hair cell, of inner ear, 595, 595i,
  596–597, 597i
Hairston, N., 817
Half-life, of radioisotope, 24, 24i
Halimeda, 385
Halobacterium, 106
Halophile, among
  Archaebacteria, 360, 361i
Hamilton, W., 913
Hammer, of middle ear, 596i
Hamster, response to melatonin
  secretion, 622, 622i
Hand, movements of, 471, 471i
Hangingfly, sexual selection,
  908–909, 908i
Haploid phase, 150, 400–401, 400i
Hardwood tree, 497, 497i
Hardy-Weinberg rule, 284–285,
  284i
Hare, snowshoe, heat balance,
  739
Harpobittacus apicalis
  (hangingfly), sexual selection,
  908–908, 908i
Haversian system, of bone, 634,
  634i
Hawaiian Archipelago,
  speciation in, 303, 303i
Hawkmoth, Madagascar, as
  pollinator, 515

Hay fever, 517, 684, 684i
Hayflick, L., 758
HCG (human chorionic
  gonadotropin), 775
Hearing, 596–597, 596i, 597i, 583i
Heart, 576i, 617t, 650, 650i, 656,
  656i, 658–659, 658i, 659i, 668,
  732i
Heart attack, 664–665, 670
Heartburn, 716
Heartwood, 496, 496i
Heat
  balance, in animals, 738–741
  loss, evaporative, 741
  production, nonshivering, 741
  stress, 741, 741i
Heinrich, B., 905
HeLa cells, 158
Helicobacter pylori, 716
Heliozoa, 375, 375i
Helix aspersa, studies on genetic
  variation, 296–308, 296i
Helper T cell, 676, 679i, 686–687,
  686i, 688t
Heme group, 47i, 48, 700
Hemicellulose, role in formation
  of cell wall, 74
Hemiparasitic plants, 822
Hemiptera, 441i
Hemlock western, 825i
Hemodialysis, 736
Hemoglobin, 104, 229, 229i, 668,
  700, 701, 704i
Hemoglobin molecule, structure
  of, 48i, 49
Hemophilia, 10, 201, 203t, 206,
  206i
Hemophllus influenzae, vaccine,
  684i
Hemostasis, 666
Hemp, Manila (source of fibers),
  491
Henbane, 491
Henle, loop of, 733, 733i, 734i, 735
Henslow, J., 274
Hepadnavirus, 363t
Hepatitis
  A virus, 363t
  B virus, 363t, 684, 684i
Herbicide, 41, 533
Herbivore, 839i, 840i, 841i
Heredity, Weismann's theory of,
  193
Hermaphroditism, defined, 426
  in flatworms, 426
Heroin, 585
Herpes simplex, 363t, 604, 631
Herpes virus, 362, 363i, 363t, 364,
  365, 687, 786, 787
Herring gull, cannibalism, 912
Hershey, A., 218, 219i
Hesperidium, 520t
Heterocephalus glaber (mole rat),
  628i, 916, 916i
Heterocyst, of Cyanobacteria,
  358, 358i
Heterospory, 401

Heterotherm, 739, 742
Heterotroph, 113, 127, 836i, 836t,
  852
Heterozygous, defined, 177, 179
Hindbrain, 571, 574i, 578, 578i,
  586
Hindgut, of insects, 440
Hippocampus, of brain, 583, 583i
Hippocrates, 272
Hippotrigris, 301
Hirsutism, 630
Hirudo medicinalis, 434i
Histamine, 674
Histone, 154, 155i
Histoplasmosis, 396, 604
HIV (Human immunodeficiency
  virus), 362i, 363t, 366, 685,
  686–687, 686i, 687i, 786
Hognose snake, 456i
Holoparasitic plants, 822
Homeostasis, 5, 554–555, 554i,
  555i, 556
Hominid, 469, 470t, 472–473, 473i,
  477
Homo
  erectus, 469, 475, 475i, 476, 477,
    477i, 490
  habilis, 474, 474i, 476, 477, 477i
  sapiens, 469, 475, 475i, 476, 476i,
    477, 477i
Homologous chromosome, 162,
  163, 163i
Homology, defined, 318
Homospory, 401
Homozygous
  dominant, defined, 177, 189
  recessive, defined, 177, 179, 182
Honeybee, 441i
Honeybee, colony of, 907, 907i,
  914–915, 915i
Honeycreepers, Hawaiian,
  speciation, 303, 303i
Hooke, R., early microscopic
  studies by, 54, 54i
Hookworm, 428–429
Hormone
  and behavior, 902–903
  control of molting by, 623, 623i
  discovery of, 610
  effects, 528i, 529i, 532–533, 533i
  and female reproductive
    function, 768–772, 769i, 771i,
    772i
  gastrointestinal, 717
  general properties, 609, 610
  and human reproduction, 788
  influencing gene activity, 250,
    250i
  and male reproductive
    function, 766–767, 767i
  peptide, 612t, 613, 613i
  plant, 528–529, 532–533, 541
  protein, 624
  responses to environmental
    clues, 622–623, 622i, 623i
  sex, and secondary sex
    characters, 783

steroid, 612–613, 612i, 612t, 624
Hornwort, 402, 414
Horse, 301, 301i, 347i
Horseshoe crab, 437, 437i
Horsetail, 343, 401i, 404, 404i, 406,
  406i, 414
  in geologic history, 401i, 402
  giant, 406, 406i
Hot spots (hydrothermal vents),
  872–873, 873i
House sparrow, 827t
Housefly, wing movements, 633i
Howarth, P., 823
Huffaker, C, 823
Huggins, M., 14
Human
  body build, 475, 475i
  cultural evolution, 475
  development, 773–783
  distinctive traits, 474
  early toolmaking, 474, 475i
  embryo, 774–779
  evolution, 468–469
  fetus, 778–779, 778i, 779i, 781i
  genome project, 266
  immunodeficiency virus (See
    HIV)
  population, 804–809, 805i, 810,
    882, 883i
  reflex actions, 778–779
  reproduction and development,
    762–790
  social behavior, 917
Humboldt Current, 877, 879
Humerus, 637i
Hummingbird
  competition between species,
    816
  metabolic rate, 739
  as pollinator, 412i
Humus, in soil, 502
Huntington disorder, 203t,
  204–205, 205i
Husky (dog), 626, 626i
Hutchinson-Gilford progeria
  syndrome, 192
Huxley, T., 460
Hyacinth, water, 827t
Hybrid zone, 305, 305i
Hybrid, defined, 177
Hydra, 422i
Hydrangea macrophylla, effect of
  environment on gene
  expression, 188, 188i
Hydrochloric acid, secretion by
  stomach, 716
Hydrogen bond
  defined, 34
  examples, 29
  formation, 29i
  in ice, 30, 30i
  in water, strength of, 512
  strength of, in water moleules,
    506, 507
Hydrogen ion
  in ATP formation, 122, 122i
  as basis of pH scale, 32

importance of, 32
Hydrogen sulfide, 141, 358
Hydrogen, 22i, 23i, 22t, 26, 360
Hydrologic cycle, 842–843, 842i, 843i, 852
Hydrolysis, defined, 40, 40i
Hydrophilic substance, 30, 34t
Hydrosphere, 855
Hydrostatic pressure, 89
Hydrostatic skeleton, 423, 435
Hydrothermal vent, 872–873, 873i
Hydroxyl group, formula, 39i
Hyla
    chrysoscelis, 298, 299i
    versicolor, 298, 299i
Hylobatid, 470t
Hymenoptera, 441i
Hyocyamus, 491
Hyperopia, 604, 604i
Hypersensitivity, to pollen, 517
Hypertension, 664
Hyperthermia, 741
Hyperthyroidism, 619
Hypertonic solution, defined, 88, 89, 89i, 94
Hypha, 390, 391i, 397, 389
Hypnotic drug, 585
Hypocotyl, 531i
Hypodermis, underlying vertebrate skin, 628
Hypoglycemia, 618
Hypothalamus
    control of emotional states, 583
    control of secretory activity, 618, 618i
    control of sperm formation, 767
    general functions, 571, 578i, 579, 586, 624, 714
    hormones acting on anterior lobe of pituitary, 615, 615i
    inhibitor hormones, 615
    location, 580i, 583i, 610i, 611i
    and production of corticotropin-releasing hormone, 782
    releaser hormones, 615
    role in regulation of menstrual cycle, 619, 619i, 771i, 772i
    role in temperature control, 740, 740i, 741
Hypothermia, 740–741
Hypothesis, in biological research, 12, 16
Hypothyroidism, 619
Hypotonic solution, defined, 88, 89, 89i, 94
Hypotonicity, 508
Hypoxia, 704

I

Ice, hydrogen bonds in, 30
Ice-minus bacterium, 264
Ichthyosaur, 312i, 457i
Ichthyostega, 454i
Imbibition, of water by seeds, 530

Immigration, into population, 796
Immune response, 677i, 672t, 688
Immune system, 676–677
Immunity, 670–689, 683i
Immunization, 684, 684i, 688
Immunodeficiency, 685, 685i, 688
Immunoglobulin, 681
Immunological specificity, 676, 682–683, 682i, 683i
Immunotherapy, and cancer, 681
Implant, contraceptive, 785, 785i
Implantation, of human embryo, 774
Imprinting, 904, 904t
Incineration, as source of pollution, 887
Incisor tooth, 462, 462i
Incomplete dominance, 182, 182i
Independent assortment of genes, 180, 180i, 181, 181i
Indian pipe, 412i
Indirect selection, theory of, 913, 916, 918
Indoleacetic acid (auxin), 532
Indricotherium, 347i
Indriidae, 464t
Induced-fit model of enzyme activity, 104–105, 105i
Induction, embryonic, 756–757, 756i, 757i, 760
Inductive logic, 12
Infancy, human, 783t
Infection, 366–367, 672, 672t
Infectious disease, 366–367
Inferior (in human anatomy), 553i
Inferior vena cava, 732i
Inflammation, 672, 674–675, 674t, 675i
Inflammatory response, 671, 674
Influenza, 362, 363i, 363t
Information flow, 566–567, 568, 566i, 567i
Inhalation, during breathing, 698, 698i, 699, 699i
Inheritance, 182, 182i, 174–189, 902–903
Inhibin, 767
Inhibitory effect, of neurotransmitter, 564
Inhibitory postsynaptic potential (IPSP), 565, 565i
Inner ear, 595, 596–597, 595i, 596i, 597i
Insect
    circulatory system, 650, 650i
    diversity, 440–441, 441i
    early, 343
    economic importance of, 440
    estimates of numbers, 759
    hormonal control of molting, 623, 623i
    life history, 4, 5i
    metamorphosis of, 440, 440i
    mouthparts of, 440, 440i
    pollination by, 514, 515i
    societies of, 913, 914–915

Insecticide, 41
Insertion, of base pairs into DNA molecule, 236, 236i
Insight learning, 904t
Instinctive behavior, 903, 903i, 918
Insulin, 47i, 264, 617t, 620, 621i
Integration, and nervous system, 570–587
Integrator, 554, 555i, 559i
Integument, 627, 629
Integumentary system, 552i, 627, 627i, 628–631, 628i, 629i, 630i, 631i, 646
Intercostal muscles, human, 696i, 699
Intercourse, sexual, 773
Interleukin, 675, 678
Interleukin-1 converting enzyme (ICE), 243
Intermediate filament, structure and function, 70, 71, 71i
Intermediate host, of parasitic worms affecting humans, 427, 428, 428i, 429, 429i
Internal environment, 544, 545, 730–743
Internal fertilization, 746–747
Interneuron, 559, 559i, 567i, 568
Interphase of cell cycle, 15l, 151i, 152i, 153i, 154, 158
Interspecific competition, 814, 816, 816i
Interstitial fluid, 544, 545, 650, 730–743
Intertidal zone
    distribution and diversity of organisms, 876, 877i
    as part of marine environment, 872i
Intervertebral disk, 577i, 636, 637i
Intestine
    innervation, 576i
    large, 714i, 721, 721i, 728, 728t
    small, 617t, 714i, 716–719, 717t, 718i, 719i, 728, 728t
Intrapleural pressure, of gases, 698, 698i
Intrapulmonary pressure, of gases, 698, 698i
Introduced species, in United States, 827t
Intron, 231, 231i
Invertebrate, 416–445, 448–449
Invertebrate eye, 598–599, 598i, 599i
Iodine
    in human nutrition, 725t
    radioisotope used in medical diagnosis, 25, 25i
    required for synthesis of thyroid hormones, 618
Ion, 28, 34t
Ionic bond, 28, 28i
Ionization, by electron transfer, 28, 28i
Ionizing radiation, causing mutations, 237

Insecticide, 41
IPSP (inhibitory postsynaptic potential), 565, 565i
Iridium, on earth, 346   in asteroids, 346
Iris, of eye, 599, 599i, 600, 600i, 600t
Iron
    associated with cytochromes, 107
    in human nutrition, 725t
    required by plants, 503t
Irrigation, consequences of, 892
Isolating mechanism, influencing species formation, 299, 300–301, 300i, 301i, 308, 308t
Isotonic solution, defined, 88, 89i
Isotope, defined, 23, 34t
Itano, H., 228

J

Japanese beetle, 827t
Jaw, evolution of, 450, 450i
Jawless fish, 451, 451i
Jellyfish, 422, 422i, 423, 424, 424i, 572
Jenner, E., 670, 670i
Jet lag, 622
Jet propulsion, by cephalopod mollusks, 433
Joint, 636, 636i
Jordan Valley, catastrophic events in, 270
Jump dispersal, 827
Jumping spider, hydraulic extension of legs, 632, 632i
Junction, cell, 546, 547i
Juniper, 408, 408i
Jurassic period, 313i, 348–349i
Juvenile, defined, 758

K

K-T asteroid impact theory, 346
K-T boundary (*See* Cretaceous-Tertiary boundary)
Kalat, J., 570
Kangaroo, 746, 747i, 464i
Kangaroo rat, 730–731, 730i, 737
Kaposi's sarcoma, 686
Karyotype, 194, 195i
Kelp, 383
Kennett, C., 823
Kentucky bluegrass, as C3 plant, 124
Keratin, 49, 49i, 52t, 226
Keratinocyte, 242, 630
Keratotomy, radial, 605
Ketone group, 39i, 621
Kettlewell, H. B., 286
Key innovations, 307
Keystone species, 826
Khorana, G., 232
Kidney dialysis, 735
Kidney, 576i, 618i, 731, 732, 732i, 742, 736, 617t

Killer bee (Africanized bee), 130–131, 827
Killer whale, diving by, 706i
Killifish, as predator on guppies, 802–803, 802i, 803i
Kilocalorie, defined, 98, 722
Kinesin, 155
Kinetic energy, defined, 110
Kinetochore, 154, 154i, 155, 155i
Kingdom, of organisms
  Animalia, 8, 416–467
  Archaebacteria, 8, 357, 360, 361t
  Eubacteria (See also Bacteria), 8, 357, 358–359, 361t
  Fungi, 8, 388–397
  Plantae, 8, 398–415
  Protista, 8, 370–387
Klinefelter syndrome, 203t, 209
Knoll, A., 340
Koala, 464t
Koch, R., 671
Kohler, G., 681
Koshland, D., 104
Krebs cycle, 133, 133i, 134, 136, 137, 136i, 137i, 139i
Krebs, C., 819
Krebs, H., 136
Kubicek, M., 158
Kudzu, 829, 829i
Kurosawa, E., 528
Kuru disease, 363

L

Labia, 768, 769i
Labor, in human birth, 782
Lactate fermentation, 140, 140i, 145
Lactation, human, 782–783
Lactobacillus, 358, 672
Lactose, 42, 244–245, 244i, 245i
Ladybird beetle, 441i
Lake Baikal, 870
Lake ecosystem, 870–871, 870i
Lamarck, J.-B., 274
Laminaria, 383
Lampbrush chromosomes, transcription in, 248, 248i
Lamprey, 451, 451i, 827t
Lancelet, 448–449, 449i
Landfill, 887
Langerhans cell, 631
Lanugo, of human fetus, 779
Large intestine (colon), 714i, 721, 721i, 728, 728t
Larkspur, evolution of flower types, 316, 316i
Larrea (creosote bush), 795
Larson, G., 905
Larva, 440i, 421, 448, 449i
Laryngitis, 699
Larynx, 576i, 696i, 697
Lascaux cave, 468, 468i–469i
Laser coagulation, in treatment of retinal detachment, 605

Late blight, 372
Late Cretaceous, 313i
Latent learning, 904t
Lateral meristem, 481, 483i
Lateral root, 492, 492i, 493i
Latimeria, 453i
Latissimus dorsi muscle, 639i
Laurentia (supercontinent), 339, 342, 342i, 348i
Law of conservation of mass, 101
Laws of thermodynamics, 110
Leaf
  functions, 489i, 498
  photosynthetic cells, 507i
  primordium, 486, 486i
  structure of, 114i, 124i, 485, 484i, 485i, 489, 489i, 507i
  types, 488, 488i, 489, 489i, 531i
Leakey, M., 474, 475i
Learned behavior, 904, 904t
Learning, 904t
Leech, 434, 434i
Leg bud, embryonic, 752i
Legume (type of fruit), 520t
Lek (display ground), of sage grouse, 909
Lemming, Norwegian, 905
Lemna, 251
Lemur, 464t, 470t
Lens, of eye, 598, 598i, 599, 599i, 600, 600i, 600t, 601, 601t, 757, 757i
Leopard frog, reproduction and development, 744–745, 744i, 745i
Lepidodendron, 406i
Lepidoptera, 441i
Leptin, 727
Leukemia, 363t, 654, 681
Levi-Montalcini, R., 620
Levin, B., 14
Lewis, E., 754
Leydig cell, 766, 766i
LH (luteinizing hormone), 614, 614t, 624, 763, 767i, 770, 771, 771i, 772i
Lichen, 286, 394–395, 394i, 395i, 397
Life history pattern, 800
Life table, 800
Life, chemical origin of, 333–335
Ligament, 627, 636, 636i
Ligase, DNA, 259, 259i
Lignin, 74, 399, 400, 488
Lily, life cycle, 413i
Limb bud, of human embryo, 778, 778i
Limbic system, 578i, 581, 581i
Lindow, S., 264
Lineage, 201, 201i, 311
Linnaeus, C., 322, 323i, 328
Linné, K. von, 322, 323i
Linum usitatissimum, 490i
Lion, cooperative hunting, 912
Lionfish, 875i
Lipase, 717t
Lipid bilayer, of cell membrane,

56–57, 56i, 82, 82i, 83, 83i, 90, 561, 561i
Lipid
  characteristics of, 37
  defined, 44
  examples, 52t
  experimental assembly of, 337, 337i
  functions, 52t
  in human diet, 723
  as reservoirs of energy, 44
Lipoprotein, 49, 256
Lithops, and camouflage, 820, 820i
Lithosphere, 855
Littorina littorea (periwinkle), and diversity of intertidal community, 826, 826i
Live oak, California, 540, 540i
Liver, 576i, 716, 717, 720, 728, 728t, 617t, 714i
Liverwort, 402, 402i, 403i, 414
Lizard, 458–459, 459i
Llama, hemoglobin, 704i
Loam, 502
Lobaria, nitrogen-fixing Cyanobacteria in, 394
Lobe-finned fish, 343
Lobster, 438, 438i
Local potential, defined, 562
Locust, 488i
Logic, 12
Loligo (squid), giant axons, 563i
Long-day plant, 536–537, 537i, 541
Long-nose gar, 453i
Loop of Henle, 733, 733i, 734i, 735, 737
Lophelia, 874
Lorenz, K., 904i, 904t
Loris, 470t
Lou Gehrig disease, 203t, 207
Louse, duck, 441i
LSD (lysergic acid diethylamide), 585
Lubchenco, J., 826
Luciferase, 109
Luciferin, 109
Lucy (australopith), 473, 473i, 474i
Lumbar nerve, 575i
Lung, of human
  changes in volume during breathing, 698–699, 698i
  diseases of, 702–703, 702i, 703i
  exchange of gases in, 700
  innervation, 576i
  location, structure and function, 657i, 696–702, 696i–702i
Lung, of vertebrates other than humans
  of amphibians, 694–695, 695i
  evolution of, 450
  of fishes, 694, 695
Lung, book, of spider, 437, 437i
Lungfish, 695
Luria, S., 218
Lutcavage, M., 706

Luteal phase, of menstrual cycle, 768, 769t, 771, 772i, 788
Luteinizing hormone (See LH)
Lycophyte, 404, 404i, 406, 406i, 407i, 401i, 414t
Lycopodium, 404, 404i
Lyell, C., 274–275
Lyme disease, 359, 359i, 437
Lymph node, 666i, 667, 667i, 672t, 678, 678i
Lymph vessel, 666, 666i, 667, 667i
Lymphatic system, 553i, 651, 666–667, 666i, 667i, 668, 678, 678i
Lymphocyte (See also B cell, T cell), 652i, 653, 653i, 681, 688t
Lymphokine-activated killer (LAK), 681
Lymphoma, 253
Lynx, Canada, and snowshoe hare, 819, 819i
Lyon, M., 249
Lysis, of infective cell, 673
Lysosome, 60i, 63i, 67
Lysozyme, 672, 673
Lystrosaurus, 344
Lytechinus (sea urchin), early development, 752i

M

Macrocystis, 383, 383i
Macroevolution, 310–320
Macroevolutionary patterns of speciation, 309t
Macronucleus, of ciliates, 378i, 379, 379i
Macronutrient, required by plants, 503t
Macroparasites, 822
Macrophage, 652i, 653, 653i, 674, 674i, 675, 675i, 679, 686, 688, 688t
Mad cow disease, 366
Madagascar hawkmoth, as pollinator, 515
Magicicada septendecim, 300, 300i
Magnesium
  atomic number and mass number, 22t
  as component of chlorophyll, 503t
  effects of deficiency, 503t
  in human nutrition, 725t
  in muscle function, 22
  required by plants, 22, 503t
Magnetite, in magnetotactic bacteria, 359, 359i
Magnetotactic bacteria, 359, 359i
Magpie, eggs, 460i
Maiasaura (dinosaur), 456i, 457
Malaria, 377, 377i, 851
Mallophaga, 441i
Mallow, Indian, and resource partitioning, 817, 817i
Malpighian tubule, 437i, 440

Maltose, 42
Malus (apple), fruit, 521i
Mammal
  adaptive radiation of, 307i
  ant-eating, 464t
  aquatic, 464t, 465
  behavioral flexibility, 462
  brain, 462
  carnivorous, 464t
  convergences, 464t
  dentition (teeth), 462, 462i
  distribution, 464t
  diversification in Tertiary
    period, 347
  diving, respiration in, 705, 705i,
    706t
  eutherian, 463
  human impact on, 464–465
  life styles, 464t
  mammary glands, 462, 462i, 782
  marsupial, 463
  monotreme, 463
  origin, 463, 463i
  placental, 463, 464, 465, 465i
  respiratory system, 695–699,
    696i, 697i, 698i, 699i
  seed-eating, 464t
  tuber-eating, 464t
  viviparity, 746, 747i
Mammer, R., 266
Manatee, 465i
Mandible, of crustaceans, 438
Manganese, 503t
Mantle cavity, of mollusks, 432i,
  433, 433i
Mantle, of mollusks, 431, 421i,
  432, 432i, 433, 433i
Manzanita, ecological isolation of
  species, 300, 301i
Maple, 488i, 496, 522, 522i, 538i
Marchantia, 403i
Marchetti, K., 13
Marigold, marsh, nectar guides,
  515i
Marijuana, 491, 703
Mark-release-recapture method,
  in studies on natural selection,
  286
Marler, P., 904
Marsh marigold, nectar guides,
  515i
Marsupial, 347i, 463, 464, 464i,
  746, 747i
Mason, R., 900
Mass number, 22t, 23
Mast cell, 674, 675, 688t
Mastigophora, 374, 374t, 376,
  376i, 386
Maternal effect gene, 755, 755i
Mating behavior, 908–909, 908i,
  909i
Maxillae, of crustaceans, 439
Maxillipeds, of crayfish, 438i
Mayr, E., 298
McClintock, B., 199, 237, 237i
Measles, 363t, 684, 684i, 687
Mechanical stress, effect on

plants, 535, 535i, 541
Mechanoreceptor, 590, 590t, 592,
  606
Medicago, stem, 487i
Medulla, of kidney, 733i
Medulla oblongata, of brain stem,
  576i, 578, 578i, 580i, 661, 701
Medullosa, 406i
Medusa, 422, 422i, 424i, 425i
Meerkat, 544, 544i, 545i, 739, 739i
Megaspore, 407, 410, 410i, 515,
  516, 516i, 518, 519i
Meiosis
  compared with mitosis,
    170–171, 170i, 171i
  in club fungi, 391, 391i
  in conifers, 410i
  defined, 161
  in development of human egg,
    770, 770i
  in flowering plants, 400, 413i,
    518, 518i, 519i, 526
  in formation of sperm, 766, 767i
  of mammalian secondary
    oocyte, 773, 773i
  in oogenesis, 770, 770i
  process, 162–167, 162i, 163i,
    164i–165i, 166i, 167i
  summary, 172, 172i
Meiotic spindle, 164i, 165i
Meissner corpuscle, 592, 593i
Melanin, 184, 184i, 185i, 186, 629i,
  630
Melanocyte, 630
Melanocyte-stimulating
  hormone, 614t
Melanoma, malignant, 224, 224i
Melatonin, 617t, 622
Melosh, H. J., 346
Membrane, basement, 546i, 547i
Membrane, cell (*See* Cell
  membrane)
Membrane cycling, 93, 93i
Membrane fluidity, observing, 85
Memory, 581, 583, 583i
Memory cell, immunological,
  683, 683i
Mendel, G., 174–175, 175i,
  176–177, 176i, 178–182, 178i,
  179i, 180i, 181i
Meninges, 577, 577i
Meningitis, 363t, 367, 577
Menopause, 769
Menstrual cycle, 768–772, 769i,
  769t, 772i, 788
Meristem, 481, 483, 483i, 486i,
  492, 492i, 493i, 498, 530, 530i
Merozoite, of *Plasmodium*, 377,
  377i
Mesocarp, 521
Mesoderm, 418, 553, 748, 749i,
  776
Mesoglea, 423, 422i, 423i
Mesophyll, leaf, 484, 485i, 488,
  489, 489i
Mesozoic era, 313, 313i, 344–345,
  348–349i, 401i, 350

Messenger RNA
  assembly, 230, 230i
  codons, 232–233, 232i, 234,
    234i–235i
  role in protein synthesis, 234,
    234i
  transcription, 230–231, 231i–232i
  translation, 234, 234–235i
Metabolic
  pathway, 103, 103i
  reaction, 110
  water, 730–731, 730i
Metabolism, 5, 96–111, 109i
Metacarpal bone, 637
Metamorphosis
  in arthropods, 436, 440, 440i
  complete, 440i, 758
  incomplete, 440i, 758
  of insects, 440, 440i
  of tunicates, 448, 449i
Metaphase of mitosis, 151i,
  152–153, 153i, 154i, 159
Metaphase, of meiosis, 163, 163i,
  164i, 165i, 167, 167i
Metastasis, defined, 252, 253i
Metatarsal bone, 637i
Methane
  in atmosphere, 846, 846i, 847i
  deposits of, 360
  as electron acceptor for
    methanogenic Archaebacteria,
    360
Methanococcus jannaschii, 328
Methanogenic bacteria, 360
Methyl bromide, 886
MHC molecule, 676, 676i, 677i,
  678
MHK molecule, 680, 680i
Micelle formation, in small
  intestine, 719, 719i
Micrasterias, 370i, 384
Microevolution, 280–284, 309t
Microfilament
  role in morphogenesis, 753, 753i
  structure and function, 61i, 71,
    70i, 71i, 72
Micrometer, 59i
Micronucleus, of ciliates, 378i,
  379, 379i
Micronutrient, required by
  plants, 503t
Microorganism, defined, 352
Microparasite, 822
Microscope, types of, 54, 55, 58,
  58i, 59i
Microscopy, early discoveries,
  54–55, 54i, 55i
Microspore, of plants, 407, 410,
  410i, 413i, 515, 516, 516i, 518,
  518i
Microsporum, 396t
Microtubule organizing center
  (MTOC), 70
Microtubule
  changes in length, 72
  in cilia and flagella, 72, 73I
  in cytoskeleton, 61i

influence on deposition of
    cellulose in cell wall, 532i
  in meiosis, 163
  in mitotic spindle, 151i,
    152–153, 152i, 153i, 155, 155i
  role in cell division, 70
  role in morphogenesis, 753, 753i
  role in shunting of organelles,
    72
  structure and function, 70, 70i,
    71i
Microvillus (pl. microvilli)
  intestinal, 718i
  of respiratory tract, 697i
  of sponge collar cells, 420, 421i
Midbrain, 571, 574i, 576i,
  578–579, 578i, 580i, 586
Middle ear, 596, 596i, 597i
Middle lamella, between plant
  cells, 74, 74i
Midge, 250, 250i
Midgut, of insects, 440
Miescher, J., 216
Migration, 460, 796
Mildew, downy, 372, 372i
Milk, human, 782–783, 783i
Miller, G. T., 896
Miller, S., 334–335, 335i
Miller, T., 809
Millipede, 439, 439i
Milstein, C., 681
Mimicry, 820, 821i
Mineral ion, absorption by plant
  roots, 504–505, 504i, 505i
Mineral, 502, 724, 725, 725t
Mineralocorticoid, 617t
Miocene epoch, 313i, 348–349i,
  469
Mischococcus, 381i
"Missing links," 278
Mistletoe, 412i, 822
Mite, 437, 900–901, 900i, 901i
Mitochondrion
  function, 67, 77, 78t
  in animal cell, 60, 61i, 63i
  cristae of, 68, 68i
  DNA of, 68
  function, 67
  in micrographs, 61i, 62i, 63i, 67i,
    68i
  in plant cell, 62i
  possible origin, 68, 340, 341i
  ribosomes of, 68
  role in aerobic respiration, 132,
    132i
  similarity to bacteria, 341, 341i
  as site of aerobic energy release,
    136, 136i
  in sperm, 767i
  structure, 67, 67i, 68, 68i, 136
  viewed as symbiotic
    prokaryotic cells, 815
Mitosis
  compared with meiosis,
    170–171, 170i, 171i
  defined, 150, 152, 158
  stages of, 151–153, 151i, 152i,

153i, 159–160
Mixture, defined, 27, 34t
Mnemiopsis, 425, 425i
Molar tooth, 462, 462i
Mold, water, 371, 372, 372i, 386
Mole rat, 628i, 916, 916i
Molecular clock, 320
Molecule
　defined, 6, 27, 34t
　random motion of, 100
Mollusk, 417, 418t, 431–433, 431i, 432i, 433i, 444, 650
Molt-inhibiting hormone (MIH), in arthropods, 623i
Molting, of crustaceans, 439
Molybdenum, 503t
Moment-of-truth defense, 820–821, 818i
Monarch butterfly, 856, 856i
Monera, 328, 328i, 329i
Monkey, 464t, 470i, 470t, 477i
Monoclonal antibody, 681
Monocot, 412, 414t, 485, 485i
Monocyte, 652i, 653, 653i, 675
Monohybrid cross, 178–179, 178i, 179i
Monomer, defined, 40
Mononucleosis, infectious, 654
Monophyletic evolutionary lineage, 326
Monosaccharide, 42, 52t
Monotreme, 463, 464, 464i
Monotropa uniflora, 412i
Monsoon, 859
Montagu, M., 670
Moray eel, 453i, 875i
Morchella esculenta, 393i
Morel, 393, 393i
Morgan, T. H., 198–199, 200
Morphogen, 757
Morphogenesis, 745, 753, 753i, 760
Morphological convergence, 319, 319i
Morphological divergence, in evolution, 318–319, 318i
Morphological traits, 282
Morula, 751, 774i, 783t
Mosaic tissue effect, 249, 249i
Mosquito
　as intermediate host for *Wuchereria bancrofti*, 429
　mouth parts, 440i
　in transmission of malaria, 377, 377i, 851
Moss, 402–403, 402i, 403i, 414
Moss, peat, 403, 403i
Moss, reindeer, 394i
Moth
　gypsy, 827
　luna, 441i
　peppered, natural selection in, 286, 287i
Motion sickness, 595
Motor cortex, 581i
Motor neuron, 567i, 568
Motor unit, of nerve and muscle,

644, 646
Moufet, T., 272
Mount Saint Helens, biological succession on, 480–481, 480i–481i, 825
Mountain avens, as pioneer species, 824i
Mouth, 714i, 715, 715i, 728, 728t
Movement, rhythmic, of leaves, 536, 536i
mRNa (*See* Messenger RNA)
MTOC (microtubule organizing center), 70
Mucigel, 492
Mucous membrane, as barrier against infection, 672t
Mucus, secretion by stomach, 716
Muirhead, P., 758
Mule, sterility of, 301
Multicelled organisms, defined, 6
Multiple fission, 374
Multiple fruit, 520i
Multiple sclerosis, 566
Mumps, 363t, 684i
Muscle
　cardiac, 627, 638, 658, 668
　contraction, 642–642, 642i, 643, 643i
　as effector, 555i
　effects of aging, 644
　effects of exercise, 644
　fatigue, 644
　intercostal, 696i
　sense, 592
　skeletal, 640–641, 640i, 641i, 642i
　smooth, 627, 638
　structure and function, 640–644
　system, 552i, 627, 627i
　tension, 644, 644i, 646
　tissue, 550, 550i, 551i, 556
　twitch, 644, 644i, 646
Muscular dystrophy, 203t
Mushroom
　big laughing, 388i
　death cap, 391i
　fly agaric, 391i
　life cycle, 391, 391i
　toxicity, 391i
Musk ox, cooperative predator avoidance, 910, 910i
Mussel
　controlled by sea star predation, 826
　secretion of byssus, 226–227
　zebra, 827
Mutagen, 237
Mutation
　adaptive, in moths, 10, 10i
　advantages, 283
　affecting enzyme formation, 228
　affecting protein synthesis, 236–237, 235i
　beneficial, 283
　defined, 10, 227
　gene, 236–237, 236i, 294
　lethal, 283

neutral, 283
　rate, 237, 265, 283
　resulting in aging, 758
　as source of new alleles, 10, 281, 282, 294
Mutualism (mutualistic symbiosis), 394, 397, 505, 505i, 814, 815, 815i, 824
Myasthenia gravis, 685
Mycelium, 372, 389, 390, 391i, 397
Mycobacterium tuberculosis, 109
Mycobiont, in lichen, 394, 394i
Mycorrhiza, 389, 395, 395i, 397, 400, 501, 505, 512, 815, 848
Myelin sheath, 566, 566i
Myocardium, 658
Myofibril, 627, 640, 640i, 642i, 646
Myoglobin, 705
Myoma virus, for controlling rabbits, 828–829
Myometrium, 769i, 782
Myopia, 604, 604i
Myosin, 71, 103, 103i, 640–641, 640i, 641i
Myrmecophagidae, 464t
Mytilus, controlled by sea star predation, 826
Myxobacteria, 359, 359i
Myxococcus xanthus, 359, 359i
Myxomatosis, controlling rabbits, 829
Myxomycota, 372, 373, 386t

N

NAD+ (nicotinamide adenine dinucleotide), 50, 52t, 107
NADP+ (nicotinamide adenine dinucleotide phosphate)
　as coenzyme in photosynthesis, 107
　conversion to NADPH, 120, 121i
　function, 52t
NADPH, 120, 121, 121i, 123
Names, scientific, 322–323, 322i
Nanometer, defined, 116
Nasal cavity, human, 696i
Natural killer cell (NK cell), 679, 688t
Natural selection, 10, 11, 279, 281, 283, 285, 802–803, 802i, 803i, 905
Nautiloid, 342
Nautilus, chambered, 433, 433, 875i
Neandertal, 475
Nearsightedness, 604, 604i
Nectar, 514, 515, 515i
Negative control system, 254
Negative feedback, 554, 554i, 556, 740
Neisseria gonorrheae, 786–787, 786i
Nematocyst, 422, 422i, 425
Nematoda, 418t, 428–429, 428i, 429i, 393i
Nemertea, 418t, 427, 427i

Nephridium, of annelids, 435, 435i
Nephron, 731, 733, 733i, 734, 734i, 742
Neritic zone, of ocean, 872i
Nerve cord
　of annelids, 435, 435i
　of cephalochordates, 449, 449i
　dorsal, as characteristic of chordates, 448
　of vertebrates, 574, 574i
Nerve ending, free, 592, 593i
Nerve growth factor (NFGH), 620
Nerve net, 423, 423i, 571, 572, 573, 585
Nerve, structure, 566, 566i
Nervous system, human
　autonomic, 571, 576–577, 576i, 586
　bilateral and cephalized, 586
　central, 571
　divisions of, 574
　general arrangement, 552i
　gray matter, 575
　motor, 575i
　parasympathetic, 575i, 576, 576i, 586
　peripheral, 571, 576–577, 576i, 577i
　sensory, 575i
　somatic, 571, 576, 586
　sympathetic, 575i, 576i, 577, 586
　white matter, 575
Nervous system, of invertebrates and lower vertebrates
　invertebrate, 572–573, 573i
　vertebrate, 574–575, 574i, 575i
Net protein utilization (NPU), 723
Neural fold, 776i
Neural groove, 752i, 776i
Neural plate, of frog embryo, 749i
Neural tube, 574, 574i, 578, 578i, 579, 586, 749i, 752, 752i, 762, 776i, 778i
Neuroglia, 551, 571
Neuromodulator, 564, 565i
Neuromuscular junction, 564i
Neuron, 551, 551i, 559–568, 559i, 560i, 567i, 571
Neurospora, 228, 228i, 393
Neurotransmitter, 564, 564i, 565, 565i, 567i, 568, 610
Neutron, 21, 22, 34t
Neutrophil, 652i, 653, 653i, 674, 675, 675i, 688t
Newborn, 782–783, 783t
Newt, 455
Niacin, 724t
Niche, 814, 832
Nicotiana, 491
Nicotinamide adenine dinucleotide (*See* NAD+)
Nicotinamide adenine dinucleotide phosphate (*See* NADP+)
Nicotine, 584, 664

Nightshade, 491
Nile perch, in Lake Victoria, 828
Nirenberg, M., 232
Nitrification, by bacteria, 848, 848i
Nitrobacter, structure, 341i
Nitrogen cycle, 848–849, 848i, 849i, 852
Nitrogen fixation, 394, 848, 848i
Nitrogen oxide, as air pollutants, 884, 884t
Nitrogen
  and decompression sickness, 705
  atomic number, 22t
  in atmosphere, 848, 848i
  mass number, 22t
  proportion in substances, 22i
  required by plants, 20, 503t
  scarcity in some environments, 849
Nitrogen oxides
  in atmosphere, 846, 846i, 847i
  and photochemical smog, 884
Nocireceptor (pain receptor), 674t
Nodule, root, 531i
Nondisjunction, of chromosomes, 208
Nonshivering heat production, 740
Norepinephrine, 564, 617t, 585, 618
Nori, cultivation of, 382
  life cycle of, 382i
Nostoc, 76i
Notochord, 448–449, 449i, 574, 749i, 776
Nuclear energy, 894–895, 895i
Nuclear envelope, 61i, 62, 62i, 63i, 64, 64i, 65i, 65t
Nuclease, 717t
Nucleic acid, 50–51, 52t, 321
Nucleic acid hybridization, 263
Nuclein, 216
Nucleoid, of prokaryotic cells, 76, 76i
Nucleolus, 61i, 62i, 63i, 64–65
Nucleosome, defined, 154, 155i
Nucleotide
  coenzyme, 52
  composition of 50i
  in DNA, 225
  defined, 50, 220
  examples, 52
  function, 52t
  of RNA, 230, 230i
  radioactively labeled, as DNA probe, 263, 263i
Nucleus
  atomic, 22
  first use of term, 55
  as component of cells, 55
  of eukaryotic cells, 60, 64–65, 61i, 62i, 63i, 64i, 65i
  origin of, 340
  origin of term, 55
Nut, 520t

Nutrient, for plant growth, 502–503, 503t
Nutrition, 710–711, 728
Nutritional requirements, 722–723, 722i, 723i
Nymph, of insects, 440, 440i

O

Oak, 488i, 540, 540i
Oat, effect of auxin on seedling, 533i
Ob (obesity) gene, 727
Obelia, 424, 424i
Observational tests, examples of
  of action potential spiking, 563, 565, 565i
  of altruistic naked mole-rats, 916
  of *Archeopteryx* fossils, 278
  with artificial selection, 10–11
  of atmospheric CO2 levels, 37
  of biological controls, 823
  and cell theory, 55
  of continental drift theory, 324
  of denaturation, 49
  of diffusion, 86
  of disruptive selection, 289, 289i
  of energy flow at Silver Springs, 840
  of gene flow, 296–297
  of gene interactions, 184
  of geologic record, 310, 314–315
  of gibberellin's effects, 528, 529i
  of global broiling theory, 346
  of high-frequency stimulation of muscles, 644, 644i
  of imprinting, 904t, 904i
  of incomplete dominance, 182
  of inheritance patterns, 202ff. 185
  of intensified greenhouse effect, 846–847
  of K-T impact theory, 346
  through mark-release-recapture of *Biston betularia*, 286, 287t
  of membrane structure, 84–85, 85i
  for metabolic disorders, 228–229
  of nondisjunction, 208
  of nutrient uptake with and without nitrogen-fixing bacteria, 505, 505i
  of photoperiodism, 536, 536i
  of plant wilting, 508i
  of plate tectonics theory, 314–315
  of raven territorial behavior, 905
  of respiration by leatherback sea turtles, 706–707
  of sea-floor spreading theory, 314
  of signaling mechanisms,

250–251
  of snowshoe hare and Canadian lynx population cycles, 819
  of stabilizing selection, 288, 288i
  of sympatric speciation, 304
  of thigmotropism, 535
  of tonicity, 88–89
  of wavelengths and photosynthetic activity, 117
Occipital lobe, 580, 581i
Ocean, 858, 858i, 872–873, 872i, 873i
Octopus, 431, 431i
Odonata, 441i
Oil gland, of skin, 630, 628i
Oil, derivation from Mesozoic deposits, 345
Okazaki, R., 223
Olfactory
  lobe, 578i, 579, 594
  nerve, 594, 594i
  receptor, 594, 594i
Oligocene epoch, 313i
Oligochaete, 434, 435, 435i
Oligosaccharide, 42, 52t
Ommatidium, 599, 599i
Oncogene, 252
Oocyte, 750, 768, 769t, 770, 770i, 771i, 773, 773i
Oomycota, 372, 386t
Operator, 244, 244i
Operon, lactose, 244, 244i
Ophiostoma ulmi (causative agent of Dutch elm disease), 396, 827t
Opossum, 463, 464i
Opposable movement, of hand, 471, 471i
Optic lobe, 579
Optic nerve, 576i, 580i, 600i, 600t, 602i, 603
Opuntia, stomata of, 508, 508i
Oral cavity, 696i, 715i
Oral contraception, 785, 785i
Orangutan, 470t, 477i
Orbital, 26, 26i, 27i
Orchid, 399i, 515i, 525, 525i
Ordovician period, 313i, 342, 349–349i, 348–349i
Organ, formation of, 748, 748i, 752, 752i, 760
Organ system
  animal, 552–553, 552i, 553i
  defined, 6, 425, 545
  human, 552i, 553i
  integumentary, 552i
  muscular, 552i
  skeletal, 552i
Organelle
  of animal cells, 60, 61i, 63i
  defined, 6
  of eukaryotic cells, 60–75
  origin of, 340–341, 340i, 341i
  of plant cells, 60, 61i, 62i
  viewed as intracellular symbiont, 340, 341i
Organic compound
  bonding in, 38, 38i

carbon backbone of, 38, 38i
  characteristic of living things, 113
  defined, 37
  properties, 38, 52
  spontaneous synthesis of, 334–335, 335i
  use of, by cells, 40
Orgasm, 773
Origin
  of earth, 334
  of life, 333–339, 340–341
Oriole
  Baltimore, 305, 305i
  Bullock's, 305, 305i
Ornithonyssus sylviarum (mite), 901i
Ornithorhynchidae, 464t
Ornithorhynchus anatinus, 446–447, 446i
Orr, A., 304
Orthomyxovirus, 363t
Oryctolagus cuniculus (European rabbit), in Australia, 828–829, 829i
Oryctopodidae, 464t
Osmoreceptor, 590, 590t, 606, 735
Osmosis, 81, 88, 88i, 89, 94, 508i
Osmotic pressure, defined, 89, 89i
Osteichthyes, 448, 452–453, 453i
Osteoarthritis, 636
Osteoblast, 635
Osteocyte, 635
Osteoporosis, 635, 635i
Ostracoderm, 450i, 451, 465
Ostrich, 461
Otolith, of inner ear, 595, 595i
Oval window, 597i
Ovary, 516i, 611i, 617t, 763, 768, 768t, 769i, 770, 465
Overgrazing, 891, 891i
Overturn, spring and fall, in lakes, 870, 871
Ovibos moschatus (musk ox), cooperative predator avoidance, 910, 910i
Oviduct, 768, 768t, 769i, 770i, 774i, 762
Oviparous, 746, 747i
Ovoviviparous, 746, 747i
Ovulation, 768, 769i, 769t, 770, 770i, 771i, 772i, 773i, 788
Ovule
  of flowering plant, 413i, 516i, 517, 518, 519i, 523
  of pine, 407i, 410, 410i
  of seed plants, 407, 407i
Ovum, 773
Oxaloacetate, as entry point in Krebs cyle, 136
Oxidation, 108
Oxygen
  accumulation in atmosphere, 339
  as electron-acceptor, 132, 144, 145
  atomic number, 22t

consumption of, in aerobic respiration, 112i
debt, 643
diffusion into animal body, 691
entry into atmosphere, 121, 144
mass number, 22t
proportion in substances, 22i
release by photolysis, 121, 121i
release of, during photosynthesis, 112i
release of, from chloroplasts, 122, 122i
Oxyhemoglobin, 700
Oxytocin, 555, 614, 614t, 624, 782, 783
Ozone layer, 856, 883, 886, 886i, 898, 882

P

P-32, in studies on bacteriophage, 219, 219i
P680 chlorophyll, 120i, 121, 122i, 128
P700 chlorophyll, 120, 120i, 121, 121i, 128
Pacemaker, heart, 659
Pacinian corpuscle, 592, 593
Pain receptor, 590, 590t, 592, 606, 674t
Pain, 592–593, 593i
Paine, R., 826
Pair-rule gene, 755, 755
Paleocene epoch, 327i, 347, 348i–349i
Paleozoic era, 313, 313i, 342–343, 342i, 343i, 348–349i, 350, 401i
Palmiter, R., 266
Palp, of bivalve mollusk, 433, 433i
Pancreas, 576i, 620–621, 621i, 714i, 716–717, 717i, 728, 728t
Pancreatic islet, 611i, 617t, 620, 621i
Panda, red and giant, 320i
Pandemic, 366
Pangea, 314, 314i, 315i, 343i, 344, 348i, 463, 854
Pantothenic acid, 724
Papillomavirus (HPV), 787
Papovavirus, 363t
Paradisaea raggiana, 290i
Paramecium, 72i, 378–379, 378i, 379i, 816–817, 816i
Paramyxovirus, 363t
Parapatric speciation, 305, 305i, 308
Parapod, of annelid worms, 434, 434i
Parasite, 822–823, 822i, 823i, 832, 903
Parasitism, 814, 818, 822–823, 822i, 823i
Parasitoid, 823
Parasympathetic nervous system, 575i, 576–577, 576i, 586

Parathyroid gland, 611i, 619i, 620
Parathyroid hormone (PTH), 620, 617t
Paratylenchus, 428i
Parenchyma, plant, 484, 484i, 485, 485i, 493i, 496, 498
Parenting, costs and benefits, 909, 909i
Parietal lobe, 581i
Parkinson's disease, 583
Parthenogenesis, in plants, 524, 524t
Partial pressure
of carbon dioxide, 692
of gases in air, 708
of gases in respiratory tract, 700–701, 701i
of gases in lungs and tissues, 708
of oxygen, 692
Parturition (human birth), 782
Parus major (great tit), predator on sawfly caterpillars, 910
Parvovirus, 363t
Passive transport, through cell membrane, 87, 87i, 90. 90i, 91, 94
Pasteur, L., 671
Pasteurization, 671
Patella, 637i
Pathogen, 352–353, 366, 672, 672t
Pattern formation, 756–757, 756i, 757i, 760
Pauling, L., 216, 228
Pavlov, I., 904t
Pea, garden, in studies of inheritance, 174, 176i, 178, 178i
Peacock, 629i, 645
Peat bog, 403, 403i
Pectin, 74, 484, 508
Pectoral girdle, 637i
Pectoralis major muscle, 639i
Pedigree, 202, 202i, 214
Pedipalp, of arachnids, 437, 437i
Pelagic province, of ocean, 872, 872i
Pellicle, of ciliates, 378, 378i
Pelvic
cavity, 552i
girdle, 637i
inflammatory disease (PID), 786, 787
nerve, 576i
Pelvis, renal, 733i
Penguin, 319, 319i, 645, 912, 912i, 917, 917i
Penicillin, 11, 287, 393
Penicillium, 393, 393i
Penis, 764, 764t, 765i, 766i
Pennington, J., 432
Pepo, 520t
Peppered moth, natural selection in, 286, 287i
Pepsin, 716, 717t
Pepsinogen, 716
Peptic ulcer, 716
Peptide bond, 46, 47i

Peptide hormone, 612t, 613, 613i
Peptidoglycan, 354, 673
Peradectes, 347i
Perception, 590, 606
Perch, body plan, 453i
Perch, Nile, in Lake Victoria, 828
Perennial plant, 494
Perforation plate, of xylem vessels, 506i
Perforin, 678, 679i
Pericardium, 658i
Pericarp,, 521, 522
Pericycle, of root, 492i, 493, 493i
Periderm, 496, 496i, 485
Period, geologic, 313i
Periodontal disease, 715
Periodontal membrane, of tooth, 715i
Peripheral nervous system, 575, 575i
Peristalsis, stomach, 716
Peritoneum, 419t, 732i
Peritubular capillaries, 733i
Periwinkle, maintaining diversity of intertidal community, 826, 826i
Permafrost, 869
Permian period, 313i, 348–349i
Peroxisome, 67
Pest resurgence, 287
Pesticide 22, 41, 286–287, 882
Pesticide pollution, 863
PET scan, 25, 25i, 580, 581i, 585i
Petal, 516, 516i, 517i
Petroleum, origin and sources, 894, 894i
PGA (phosphoglycerate), 123, 123i
PGAL (phosphoglyceraldehyde), 123, 123i, 134
pH, 21, 32i, 34, 736
Phaeophyta, 383, 383i, 386t
Phagocyte, 675i
Phagocytosis, 93, 93i, 94, 671, 674, 674i
Phalangeridae, 464t
Phalanges, 637i
Pharyngeal arches, 776i, 778i
Pharynx, 426, 426i, 448, 696i, 697, 714i, 715, 728, 728t
Phaseolus (bean), 489i, 530i–531i
Phenotype, 177, 188, 188i, 212, 288–289, 288i, 289i
Phenylalanine, 100
Phenylketone, 100
Phenylketonuria (PKU), 100, 212
Pheromone, 594, 610, 624, 906, 918, 914–915
Phloem
companion cell, 485, 484i, 487i
distribution of organic solutes by, 400, 498, 501, 510
fibers, 487i
leaf, 489i
primary, 486i
of root, 492i, 493, 493i, 494, 494i, 495i

secondary, 496, 496i
sieve plate, 484i
sieve-tube members, 485, 484i
Phlox drummondii, life table, 800, 800t
Phosphate group, 39i, 102
Phosphoglyceraldehyde (PGAL), 123, 123i, 134
Phosphoglycerate (PGA), 123, 123i
Phospholipid, 45, 45i, 52t, 55, 81, 82, 82i, 83
Phosphorus, 22t, 503t, 725t, 850
Phosphorus cycle, 850, 850i, 852
Phosphorylation, 102, 103, 134, 138, 138i, 139i
Photoautotroph, 113, 126, 126i, 127, 358, 366, 836t, 852
Photobiont, in lichen, 394, 394i
Photochemical oxidants, 884, 884t
Photoheterotroph, 366
Photolysis, 121i, 122, 128
Photon, 116, 127
Photoperiodism, in plants, 536–537, 537i, 541
Photoreceptor, 590, 590t, 598, 598i, 599, 599i, 602–603, 602i, 606
Photorespiration, 124, 128
Photosynthesis
accessory pigments, 127
centered in chloroplasts, 113
cyclic pathway, 128
dependence of organisms on, 112i, 113
formation of glucose in, 123, 123i
general equation, 113, 113i, 115i
light-dependent reactions, 113–114, 114i, 115, 115i, 120–122, 123, 123i
noncyclic pathway, 128
origin of, in Eubacteria, 338, 338–339i
summary of reactants and products, 127i, 128
Photosynthesizer, 836, 852
Photosystem, 120–121, 120i
Phototropism, 534–535, 535i, 541
Phycobilin, 119, 119i, 382
Phycocyanin, 119i
Phycoerythrin, 119i
Phylogeny, defined, 323
Physalia, 425, 425, 425i
Physarum, 371i
Physiological traits, 282
Phytochrome, 251, 251i, 536, 536i, 537, 541
Phytophthora infestans, 372
Phytoplankton, 380, 386, 870, 872, 873i
Phytoremediation, 21
Phytosterol, in cell membranes of plants, 82
Picornavirus, 363t
Pigeon, 812–813, 812i
Pigment, 117, 117i, 118–119, 118i, 119i

Pike-cichlid, as predator of guppy, 802–803, 802i, 803i
Pillar coral, 875i
Pillbug, 438
Pilomotor response, 740
Pilus (pl. pili), of bacteria, 354i, 335, 355i
Pine barren, 868
Pine, 399i, 407i, 408i, 410i
Pineal gland, 578i, 580i, 611i, 617t, 622, 622i
Pineapple, multiple fruit of, 521, 521i
Pine
　life cycle, 410i
　ovule, 407i
　pollen, 407i
　ponderosa, 399i
Pinus longaeva, 408i
Pinworm, 428
Pioneer species, 824, 824i
Pisum sativum, in studies of inheritance, 174, 176i, 178, 178i
Pith, of plants, 486, 486i, 487i, 493i
Pituitary dwarfism, 616, 616i
Pituitary gigantism, 616, 616i
Pituitary gland, 578i, 610, 610i, 611i, 614–615, 614t, 615i, 616, 616i, 618i, 619i, 624, 771i, 772i
Placenta, 464, 465, 465i, 464i, 777, 777i, 778, 779i, 782i, 788–789
Placoderm, 448, 450i, 451, 465
Placozoa, 418t, 420, 420i
Plague
　bubonic, 799
　pneumonic, 799
Planaria, 426, 426i
Plane of symmetry, 552i, 553i
Plankton, 126, 126i, 375, 870, 872, 873i
Plant
　absorption of water by, 504–505, 506–507, 507i
　body, main components, 481–482, 481i, 482i
　CAM, 512
　carnivorous, 500, 500i, 501i
　chlorophyll of, 399
　controlled water loss, 508–509
　development, 528–541
　distribution of organic compounds in, 510–512
　domestication, 490
　early fossil, 398
　evolution, 414
　flowering (See Flowering plant)
　growth, 483, 498, 528–541
　kingdom, 398–415
　nonvascular, 400
　nutrition and transport, 500–513
　physiology, 500–513
　reproduction and development, 514–527, 524t, 525i
　storage products, 510

structure of, 482, 482i
succession, in new habitats, 398
summary of characteristics of cells, 78t
tissues of, 480–499, 482–483, 482i, 483i
translocation of organic compounds, 510, 511
tropisms, 534–535, 534i, 535i
types of, 398–415
vascular, 399, 400, 414
water conservation, 508–509
Plantae (kingdom), 8, 398–415
Planula, 424, 424i, 573, 573i
Plasma membrane
　active transport through, 81, 87i, 91, 91i, 94
　adhesion proteins, 94
　animal cell, 61, 64i
　of bacteria, 354i, 355
　characteristics, 77
　cycling, 93, 93i
　diffusion through, 81
　fluid mosaic model, 82, 82i, 83, 83i, 85
　infolding of, in bacteria, 340, 341i
　lipid bilayer of, 56–57, 56i, 81, 82, 82i, 83i, 94
　passive transport through, 81, 87, 87i, 90, 90i, 91, 94
　phospholipid of, 81, 83I
　plant cell, 61i, 62i, 63i
　possible origin, 337
　receptor proteins, 94
　recognition proteins, 94
　selective permeability of, 81, 86, 86i
　structure and functions, 56–57, 56i, 80–94
　transport proteins of, 81, 94
Plasma protein, 675
Plasma, of blood, 652, 652i, 668
Plasmid
　behavior in bacterial conjugation, 357, 357i
　of bacteria, 258, 258i, 357, 357i
　conjugation, 357, 357i
　genes in, 258, 259, 259i
　and resistance to antibiotics, 258
Plasmodesma, connecting cytoplasm of adjoining plant cells, 74, 74i, 75
Plasmodium, causative agent of malaria, 376, 377, 377i, 851
Plasmolysis, defined, 89, 89i
Plasmopara viticola, 372, 372i
Plate tectonics theory, 314, 329
Platelet, of blood, 652, 652i, 653, 653i, 666, 666i
Platyhelminthes, 418t, 426–427, 426i, 427i
Platypus, duck-billed, 446–447, 446i, 464t
Pleiotropy, 183, 183i, 189
Pleistocene epoch, 347, 348–349i
Plesiosaur, 457i

Plethodon (salamander), competitive exclusion, 817, 817i
Pleural membrane, of lung, 696i, 697
Pliocene epoch, 313i, 348–349i
Plumes, crust-rupturing, 315i
Plutonium, isotope of, 25, 895
Pneumocystis carinii, cause of one type of pneumonia, 396, 396t, 686
Pneumonia, caused by Pneumocystis, 396
Pneumonic plague, 799
Podocarp, 408
Poecilia reticulata (guppy), natural selection, 802–803, 802i, 803i
Polar bear, 821
Polar body, 770, 770i
Pole cells, in Drosophila embryo, 754i
Polio, 363t, 684, 684i
Pollen grain, 414, 518, 518i
Pollen tube, of plants, 410i, 413i, 518, 519, 518i–519i, 526
Pollen, 401, 407i, 410, 410i, 413i, 514, 515, 516, 517, 517i, 523
Pollination, 412, 412i, 413i, 514–515, 514i, 515i, 518, 518i, 526
Pollinator, 412, 412i, 514, 514i
Pollutant, 884, 885, 892–893, 898
Pollution, 509i, 883, 892–893
Polychaete, 434, 434i
Polycythemia, 654
Polydactyly, 203t, 202i
Polymer, defined, 40
Polymerase chain reaction, 260, 260i
Polymorphism, 282, 290
Polyp, 422, 422i, 423i, 424i, 425i
Polypeptide chain, 46, 48, 48i, 229, 229i
Polyploidy, 208, 214, 304, 305i
Polyporus, 388i, 390i
Polysome, 235, 235i
Pome, 520t
Pongid, 470t
Pons, of brain, 578, 578i, 580i, 701
Poplar, 21, 21i, 488i
Poppy, California, effect of gibberellin, 528i
Population
　age structure, 794
　characteristics, 794–795
　defined, 6, 282, 810
　dispersion, 794–795, 794i
　migration, 796
　reproductive base, 794
Population ecology, 792–811
Population growth
　and agriculture, 804, 805i, 882, 898
　average annual, 804, 806i
　carrying capacity of, 793, 810, 818–819
　characteristics, 796–799, 796i, 797i, 798i, 799i

controls, 804
and disease, 804
exponential growth, 810
and family planning, 806–807
human, 804–809, 805i, 806i, 807i, 808, 808i, 810, 883, 898
limiting factors, 793, 804, 810
and predator-prey interaction, 809, 818–819, 818i, 819i
and resource consumption, 809
Populus, 488i, 524, 525i
Pore complex, of bacterial cell wall, 673, 673i
Porifera, 418t, 420–421, 420i, 421i
Porphyra umbilicus, life cycle, 382i
Porphyrin, component of chlorophyll, 336, 336i
Porpoise, 319, 319i
Portuguese man-of-war, 425, 425i
Positive control system, 254
Positive feedback, 555
Post-embryonic development, 758
Post-transcriptional control, 246–247, 247i
Postelsia palmaeformis, 370i, 383i
Posterior (in anatomy), 419, 552i, 553i
Postnatal development, human, 783, 783i, 783t
Postpartum depression, 782
Postzygotic isolating mechanism, 301, 308, 308t
Potassium
　atomic number and mass number, 22t
　in human nutrition, 725t
　required by plants, 503t
　role in functioning of neuron, 561, 561i
　role in controlling turgor of guard cells, 509, 509i
Potato blight, 264, 372, 798i
Potential energy, defined, 110
Potential, of neurons, 562, 563, 563i, 565, 565i, 568
Poultry, genetic traits, 185, 185i
Poxvirus, 363t
Praying mantid, 821i
Precambrian, early adaptive radiation of animals, 339
Precambrian seas, 342
Precipitation, in hydrologic cycle, 842i
Predation, 809, 814, 818–819, 818i, 819i, 910, 910i
Prediction, defined, 12, 16
Prefrontal cortex, 581i, 583i
Pregnancy, 773–781, 781i, 789
Prehensile movement, of hand, 471, 471i
Preimplantation diagnosis, 213
Premolar tooth, 462i
Prenatal development, human, 783, 783i, 783t
　diagnosis of genetic disorders, 212–213, 212i

Pressure flow theory, of translocation of organic compounds, 510–511, 511i
Pressure gradient, of gases, 598, 691, 692
Prey, defenses, 832
Prezygotic isolating mechanism, 300–301, 308, 308t
Price, P., 823
Primary producer, 836, 836i, 836t, 837, 837i, 838, 838i, 839i, 840i, 841i, 852
Primary productivity, 838–839, 838i, 839i, 840i, 841i, 871, 872, 873i
Primary root, 492, 492i, 493i
Primate, 469, 470–473, 470i, 471i, 472i, 473i
Primitive streak, of embryo, 776, 776i
Prion, causative agent of mad cow disease, 363, 366
Procambium, 486i, 495i
Producer, 7, 36
Productivity, primary, 838–839, 838i, 839i, 840i, 841i, 871, 872, 873i
Progeria syndrome, 191i, 192, 192i, 203t
Progesterone, 617t, 768, 771, 771i, 772i, 777, 782
Progymnosperm, in geological history, 401i, 406, 406i
Prokaryotic cell (See Cell, prokaryotic)
Prokaryotic fission, 356, 356i
Prolactin (PRL), 250, 614, 614t, 615, 624, 737i, 783
Promoter, 230, 231i, 244, 245i, 254
Pronghorn antelope, 713, 713i
Prophase, of meiosis, 163, 163i, 164i, 165i, 166–167, 166i
Prophase, of mitosis, 151i, 152, 152i, 153i, 159
Prosimian, 469, 470, 477t
Prostaglandin, 620
Prostate gland, 764t, 765i, 766i
Protein
  adhesion, 83, 83i
  associated with DNA in chromosomes, 223
  of chromosomes, 65
  comparisons of, in studies on relationships, 321
  complete, 723
  controlling cell death, 242–243, 242i, 243i
  conversion of, 142
  defined, 46, 52t
  examples, 52t
  experimental assembly of, from amino acids, 335, 336, 337
  fibrous, 46, 52
  functions, 37, 52t
  globular, 46–47, 52t
  hormone, 624
  in human diet, 723

incomplete, 723
  kinases, 252
  receptor, 83, 83i
  recognition, 83, 83i
  regulatory, 244, 756
  role of ATP, 91, 91i
  role of DNA in synthesis of, 65
  synthesis, 65, 227, 232, 233, 236–237, 236i, 239i
  three-dimensionsal structure of, 48
  transport, 83, 83i, 90–91, 90i, 91i, 103, 103i
Protein synthesis
  affected by mutations, 236–237, 236i
  encoded in DNA and mRNA, 232, 233, 234i–235i
  instructions from DNA, 227
  role of DNA in, 65
  summary, 239i
Proterozoic eon, 313, 313i, 339, 348–349i, 350
Prothrombin, 666i
Protista (kingdom) (See also Algae, Protozoa), 8, 370–387
Proto-oncogene, 252
Protoderm, 486i
Proton, defined, 22, 34t
Protonephridium, 426, 426i, 430, 430i
Protostomes, 430–441
Protostome embryo, 430i
Protozoa, 374–379, 374i, 375i, 376i, 377i, 378i, 379i
  parasitic, 374–375, 376–377, 376i, 377i
Proximal tubule, of nephron, 733, 733i
Prunus, life cycle, 518i–518i
Pseudocoel, 419, 419i
Pseudomonas, 77i, 366, 366i
Pseudopod, 72, 374, 374i, 375
Psilophyta, 404, 404i
Psilophyton, Devonian plant, 343i
Psilotum, 404, 404i
Pterophyta, early, 405, 405i (See also Fern)
Pterosaur, 457i
Ptilodus, 347i
Puberty, human, 622, 783t
Puccinia graminis, 396t
Pueraria lobata (kudzu), 829, 829i
Puffball, 390
Pulmonary
  artery, 657i, 658i, 701i
  circuit, 649, 651, 668
  vein, 656, 656i, 657i, 658i, 701i
Pulp cavity, of tooth, 715i
Pulse, 660, 660i
Punnett-square method, for predicting outcome of genetic cross, 179, 179i, 180
Pupa, of insects, 440, 440i
Pupil, of eye, 600, 600i, 600t, 601
Purine, 220
Purple coral fungus, 388i

Pyramid, ecological, 838–839, 839i, 840i
Pyrenestes ostrinus, beak of, 289, 289i
Pyrimidine, 220
Pyrrhophyta, 380, 381i, 386
Pyruvate
  derived from glucose, 132, 133, 133i
  end product of glycolysis, 134
Python, 459, 588, 588i

Q

Quadriceps femoris muscle, 639i
Quaking aspen, asexual reproduction, 524, 525i
Quaternary period, 313i, 348–349i
Queen, of insect colony, 914, 915, 915i
Quercus agrifolia (California live oak), 488i, 540, 540i

R

Rabbit, 188, 188i, 828
Raccoon, 462i
Radial cleavage, 430, 430i
Radial symmetry, 442, 444, 444i, 572, 573
Radiation, solar, 738, 742, 856, 857, 857i
Radioactive
  decay, 23
  fallout, global distribution of, 895, 895i
  waste, storage, 895
Radioactivity, 23
Radioisotope
  defined, 23, 34t
  half-life of, 24, 24i
  in studies on bacteriophage, 219, 219i
  in therapy, 25, 25i
  use in medical diagnosis, 25
  use in radioactive dating, 24i
Radiolaria, 375, 375i
Radiometric dating, 24, 311, 313i
Radish, effect of gibberellin, 528
Radium-226, in cancer therapy, 25
Radius, 637i
Radula, of mollusks, 431, 432i, 433i
Rafflesia, 866i
Ragweed, 517i
Rain forest, tropical, 866, 866i, 888–890, 888i, 889i, 890i
Rain shadow, 859, 859i
Ramaria, 390i
Rana pipiens (leopard frog), reproduction and development, 744–745, 744i, 745i
Ranunculus (buttercup), toxicity of, 821
Rapid eye movement (REM), during sleep, 584

Rat, kangaroo, 730–731, 730i, 737
Ratfish, 452i
Rattlesnake, 459, 459i
Raven, 905
Ray (cartilaginous fish), 452, 452i
Ray initial, of vascular cambium, 494i, 495
Reabsorption
  though capillary walls, 662
  in nephron, 742
  from tubule of nephron, 734, 734i, 735t, 735
Recent epoch, 313i, 348–349i
Receptacle, of flower, 516, 516i, 517i, 521i
Receptor
  acoustical, 596–596
  for antigen, 682–683
  near body surface, 592, 593i
  as detector of stimuli, 5
  encapsulated, 592
  of hormone, 612, 613
  protein, 83, 83i, 94
  sensory, 554, 559, 559i606
Reciprocal cross, 198, 198i
Recognition protein, 81, 83, 83i
Recombinant DNA technology, 256–267
Recombination
  frequency of, 200–201, 200i
  gene, 194, 199–201, 199i, 200i, 201i
Rectum, 714i, 728, 728t
Rectus abdominis muscle, 639i
Recycling, 887
Red algae, 382, 382i, 386
Red blood cell, 652–653, 652i, 652i
Red snow, 384, 384i
Red tide, caused by dinoflagellates, 381, 381i
Reduction, 108
Redwood, 408, 494i
Reef, coral, 424–425, 424i, 874–875, 874i, 875i
Reeve, H., 916
Reflex, stretch, 567, 567i
Reflex arc, 566–567, 567i
Reflex connections, in spinal cord, 586
Reflex pathway, 572
Reindeer "moss" (reindeer lichen), 394i
Reindeer, carrying capacity of population, 799i
Relaxin, 782
Repressor protein, blocking transcription, 244, 244i, 245i
Reproduction, general
  asexual, 162, 746
  as a characteristic of life, 4
  costs and benefits, 746–747
  defined, 149
  plant
  sexual, 164, 745, 746–747, 760
  success, related to behavior, 905
  timing, 746
  vegetative, 524, 525, 526

Reproduction and life cycles of
plants
  bryophytes, 402–403, 402i, 403i
  conifer, 407–411, 407i, 408i, 409i,
    410i
  fern, 404–405, 405i
  gymnosperm, 407–411, 407i,
    408i, 409i, 410i
  flowering plant, 412–413, 412i,
    413i, 514–525
  haploidy and diploidy, 400–401
  liverwort, 403, 403I
  moss, 402–403, 402i
Reproductive system, human
  female, 768–772, 768i, 768t, 769i,
    770i, 771t, 772i
  male, 764–767, 764i, 764t, 765i,
    766i, 767i
  organs and accessory glands,
    764–772, 764t, 768t
Reptile, 448, 456–459, 456i, 457i,
  458i, 459i, 465, 695
Research tools, examples of,
  amniocentesis, 212, 212i
  aphids as, 510
  bacteriophage as, 14–15,
    218–219
  bell-shaped curves, 187, 187i
  bioluminescence genes, 109
  carbon 14 dating, 25
  centrifuge, 84, 84i
  chorionic villi sampling 213
  colchicine, 70, 155, 195, 208
  computer simulation, 292, 292i,
    851, 879
  cytological markers, 199, 199i
  dentition, 462
  DNA fingerprinting, 262, 262i
  DNA library, 259
  DNA ligase, 259
  DNA probe, 263
  DNA topoisomerase, 154
  double-blind study, 209
  Drosophila, 198, 199, 200
  ECGs, 648, 648i
  ecosystem modeling, 851, 879
  EEGs, 580, 584, 684, 684i
  Escherichia coli, 14–15, 219
  freeze-fracture and freeze-
    etching, 84–85, 85i
  gel electrophoresis, 228–229,
    229i, 260, 261
  gene sequencing analysis, 321,
    328
  genetic code, 292, 292
  Gram staining, 355, 355i
  Hardy-Weinberg rule, 283,
    285–286
  HeLa cells, 158
  human genome project, 266
  karyotyping, 194, 195
  kitchen blender, 219
  laboratory mice, 14–15, 109, 218,
    219, 266
  light microscopes, 58–59
  linkage mapping, 201, 201
  Loligo giant axon, 563i

luciferase, 109
microscopes, 54–55, 58, 59, 278
monohybrid cross, 178, 179
Neurospora crassa, 228
neutral mutations, as molecular
  clocks, 283, 320
nucleic acid hybridization, 263,
  321, 321i, 357
numerical taxonomy, 357
paintbrushes, on feathers, 13
PCR, 260
pedigree analysis, 202–203
PET scanning 25, 25i
pH scale, 32, 32i
pigeons as, 10–11
plasmids, 258–259
phytoremediation, 21
population growth equations,
  796
preimplantation diagnosis, 213
primer, 261
probability, rules of, 179
protein comparisons, 321, 330
Punnett square, 179, 179i,
  204–206
radioisotopes, 24–25, 24i, 25i,
  219
radiometric dating, 24
restriction enzymes, 259
restriction fragments, 259
reverse transcription, for
  cDNA, 263
RFLPs, 262
satellite imaging, 271
scanning electron microscopes,
  58, 59
Southern blot method, 262
taxol, 70
testcross, 179
tissue culture propagation, 264,
  524
tracers, 25
transmission electron
  microscopes, 58, 59
x-ray diffraction imaging,
  220–221
Zea mays, 199, 199i, 237
Residual volume, of lungs, 699i
Resistance to antibiotics, 287
Resource partitioning, 817, 817i
Respiration
  in diving mammals and
    reptiles, 705–707, 705i, 706t
  integumentary, 693, 693i
  in invertebrates, 693, 693i
  gas exchange in, 692
  at high altitude, 704
  summary, 708
Respiration, aerobic
  defined, 5, 132, 132i
  links with photosynthesis,
    112i
  in muscle, 643, 643i, 646
  stages of, 131
  summary of energy yield,
    138–139, 139i, 708
  yield of ATP, 138

Respiration, anaerobic
  defined, 132
  low net energy yield, 131, 132,
    133, 138
  yield of ATP, 131, 140–141, 141i
Respiratory center, in brain, 701
Respiratory cycle, 698–699, 698i,
  699i, 708
Respiratory system, 553i
Respiratory system
  amphibians, 694, 694i
  birds, 695, 695i, 708
  Fick's law, 692
  of fishes, 694, 694i, 708
  gas exchange in, 691, 692–701,
    693i, 694i, 695i, 696i, 697i,
    698i, 699i
  human, 696–699, 696i, 697i,
    698i, 699i
  insects, 693, 693i
  interactions with other systems,
    691i
  of invertebrates, 708
  mammals, 695–699, 696i, 697i,
    698i, 699i, 708
  millipedes, 693
  reptiles, 695, 708
  vertebrates, 694–699
Response, to stimulus, 567i
Resting membrane potential, of
  neurons, 560
Restoration ecology, 825
Restriction enzyme, in bacteria,
  as protection against foreign
  DNA, 259, 259i
Restriction fragment length
  polymorphisms (See RFLP
  analysis)
Rete mirabile, of fish swim
  bladder, 694i
Reticular formation, of brain, 579,
  579i, 585
Retina, of eye, 599, 599i, 600i, 601,
  602–603, 602i, 605
Retrovirus, HIV, 363t, 686
Reverse transcription, for making
  cDNA, 263, 263i
Reznick, D., 802
RFLP analysis (Riff-lips), 262,
  262i
Rhabdovirus, 363t
Rheumatoid arthritis, 636, 685
Rhinovirus, 363t
Rhizobium, 358, 505i
Rhizome, 404, 405i, 524t
Rhizopoda (rhizopods), 374–375,
  374i, 375i, 386t
Rhizopus stolonifer, 392, 392i
Rhodophyta (red algae), 382,
  382i, 386t
Rhodopsin, 602–603
Rhyniophyte, in geological
  history, 401i
Rhythm method, of birth control,
  784, 785i
Rib, 637i
Ribbon worm, 418t, 427, 427i

Ribose, as five-carbon sugar, 42
Ribosomal RNA, function of,
  232–233
Ribosome, 56, 60, 61i, 62i, 63i, 67i,
  354i, 355
Ribulose biphosphate (RuBP),
  123, 123i, 124i
Rickets, 620
Rickettsia rickettsii (causative
  agent of typhus), 851
Riff-lips (RFLP analysis), 262,
  262i
Ringworm, caused by fungus,
  396
Rivulus hartii (killifish), as
  predator of guppy, 802–803,
  802i, 803i
RNA (ribonucleic acid)
  assembly of, 230
  composition, 51
  directing linkage of amino
    acids, 227
  function of, 51, 52t
  guiding synthesis of proteins,
    227
  messenger, 230–231, 232, 232i,
    234, 234i, 235i
  polymerase, 230, 231, 232
  as possible early informational
    molecule, 337
  ribosomal, 232–233
  role in synthesis of proteins, 4
  synthesis, 230–231, 230i, 231i
  three classes of, 230
  transfer, 232–233
  viral, 686, 686i
Robinia, 488i
Rocky Mountain spotted fever,
  437
Rocky shore, diversity of
  organisms, 876, 877i
Rod cell, of eye, 602–603, 602i
Root, of plants
  absorption of water and
    nutrients by, 504–505, 504i,
    505i
  apical meristem, 492i, 530i
  cap, 492i
  Casparian strip, 505, 504i
  cortex, 492i, 493, 493i, 495i
  in desert plants, 493
  dicot, 493, 493i
  endodermis, 492i, 493, 493i, 504,
    504i, 507i
  epidermis, 492, 492i, 493i, 495i
  exodermis, 504, 504i
  fibrous, 492, 492i
  functions, 498
  growth, 483, 483i
  hair, 504, 504i, 507i, 512
  lateral, 492, 492i, 493i
  meristem, 492, 492i, 493
  monocot, 492, 493, 493i
  movement of water through,
    493
  mucigel, 492
  nodule, 505, 505i, 512, 531i

parenchyma, 493i
pericycle, 492i, 493, 493i, 495i
phloem, 492i, 493, 493i, 495, 495i
pith, 493i
primary, 492, 492i, 493i, 530i, 531i
types of, 492, 492i, 531i
uptake of water from soil, 504, 504i
vascular cambium, 495, 495i
vascular cylinder, 492i, 493, 493i, 504, 504i
vascular ray, 495, 495i
vascular tissue, 507i
xylem, 492i, 493, 493i, 495, 495i
Root canal, of tooth, 715i
Rose, pollen, 517i
Rotifera, 417, 418t, 430, 430i
Round window, of ear, 596, 597i
Roundworm, 418t, 428–429, 428i, 429i
Royal penguin, colonies of, 912, 912i
rRNA (See Ribosomal RNA)
Rubisco, 123, 124, 128
RuBP (ribulose biphosphate), 123, 123i
Ruffini ending, 592, 593i
Ruminant, 713, 713i
Runner, in plants, 524t
Rye, 490i

S

S-35, in studies on bacteriophage, 219i
SA node, 659, 659i, 668
Saber-tooth cat, 347i
Sabine polio vaccine, 684
Sac fungi, 390, 392–393, 393i
Saccharomyces, 141, 141i, 393
Saccharum officinarum, 490i
Saccule, of inner ear, 595, 595i
Sacral nerve, 575i, 576i
Sage grouse, sexual selection, 908, 909, 908i
Sage, specializations for pollination, 300, 301i
Sagittal (in anatomy), 552i, 553i
Sagittaria sagittifolia, variation in leaf form, 298, 298i
Saguaro cactus, 514i, 523
Salamander, 455i, 817, 817i
Salination, of irrigated lands, 892
Saliva, 715
Salivary gland, 555i, 576i, 714i, 715, 715i, 728, 728t
Salmon, 148, 148–149i, 737, 737i
Salmonella enteritidis, 109, 366
Salt marsh, 876, 876i
Salt, 33, 34t
Salvia, 300. 301i
Sampling error, 15, 15i, 292
Sand dollar, 442
Sand, in soil, 502
Sanger, F., 47

Saprolegnia, 372i
Sapwood, 496, 496i
Sarcodina, 374–375, 374i, 375i, 386t
Sarcoma, 681
Sarcomere, 627, 640, 641, 640i, 641i, 646
Sarcoplasmic reticulum, 642, 642i
Sarcoscypha coccinea, 393i
Sarcosoma, 388i
Sartorius muscle, 639i
Savanna, 864–865, 865i, 891, 891i
Sawfly, Australian, cooperative predator avoidance, 910. 910i
Scala, of ear, 597i
Scale, of fishes, 452
Scallop, 431i, 433i
Scapula, 637i
Scenedesmus, images produced by different kinds of microscopes, 59i
Schaller, G., 912
Schally, A., 616
Schistosoma japonicum, 428, 428i
Schistosomiasis, 428, 428i
Schleiden, M., 55
Schoener, A., 827
Schopf, T., Jr., 340
Schwann, T., 55
Schwann cells, 566, 566i
Sciatic nerve, 575i
Science, limits of, 16
Sciuridae, 464t
Sclera, of vertebrate eyie, 600, 600t
Sclereid, of sclerenchyma, 484, 484i
Sclerenchyma, of plants, 484, 484i, 487i, 498
Scorpion, 437
Scorpionfish, 821, 821i
Scrotum, 764, 764i
Scutigera coleoptrata, 439
Sea anemone, 423, 423i, 572, 573i, 632, 632i, 875i
Sea biscuit, 442
Sea cucumber, 442, 442i
Sea hare (mollusk), 432i, 693, 693i
Sea horse, 453, 453i
Sea lamprey, 827t
Sea lily, 442
Sea spider, 437
Sea squirt, 448, 449i
Sea star, 442–443, 443i, 826, 875i
Sea urchin, 442, 442i, 572, 572i, 752i
Sea-floor spreading, 314, 315
Seal, 293, 706i
Sebaceous gland (See Oil gland)
Secale, 490i
Second messenger, 613, 613i
Secondary contact, of previously isolated species, 305
Secondary sex characters, 39i
Secretin, 610, 617t, 717
Secretion, in nephron, 734, 734i, 742

Sedimentary cycle, 313, 852
Seed
breaking dormancy of, 528, 532
coat, 519i, 520, 520i, 530, 530i
of conifer, 410, 410i
development, 520, 520i
fern, 406i, 407
formation of, 520i, 521, 523
of flowering plant, 399, 400, 515
germination, 530–531, 530i, 531i
Seedcracker, black-bellied, beak of, 289, 289i
Seeds, evolution of, 401
Segmentation, 419, 434, 434i, 435, 435i, 436, 437
Segmented worms, 417, 418t
Selander, R., 297
Selection
artificial, 10, 11i
balancing, 290
directional, 286, 286i, 294
disruptive, 289, 289i, 294
and extreme phenotypes, 288–289
sexual, 290, 294, 908–909, 908i, 909i
stabilizing, 288, 288i, 294
theory, 908, 909
Selenate, uptake by plants, 20
Selenide, dimethyl, 20
Selenium, as nutrient for plants, 20
Self-replicating systems, possible origins, 336
Selfish behavior, 905
Selfish herd, defined, 910–911
Semen, 764, 773
Semicircular canal, of inner ear, 595, 595i
Semilunar valve, 658–659, 658i, 659i
Seminal vesicle, 764t, 765i, 766i
Seminiferous tubule, 764, 766, 766i, 767i
Senescence, in plants, 538, 541
Sensation, defined, 589, 590, 591, 592–593, 609
Senses, special, 589, 591
Sensory
adaptation, 591
neuron, 568
pathways, 590–591, 591i
reception, 588–607
receptor, 554, 590, 590t, 606, 715, 715i, 606
system, defined, 589
Sepal, 516, 516i, 517i
Septum, of heart, 658
Sequoia sempervirens, 494i, 825
Serotonin, 584, 585
Serratus anterior muscle, 639i
Sertoli cell, 766, 767i
Seta, of annelid worms, 434, 434i, 435
Sewage, discharge into lakes, 871
Sex
attractant, in silk moth, 906

characters, secondary, 763
chromosome, 194, 196–197, 196i, 197i, 903
determination, human, 196–197, 196i, 197i
Sexual
competition, 908–909, 908i, 909i
dimorphism, 290, 290i, 645
intercourse, 773
selection, 290, 908–909, 908i, 909i
Sexual reproduction, in plants (See Reproduction and life cycles of plants.)
Sexually transmitted diseases, 786–787
Shark, 319, 319i, 450i, 452, 452i
Shelf fungus, 390, 390i
Shell, of mollusks, 431, 431i, 432, 432i, 433, 433i
Shepherd's purse, development of seed, 520, 520i
Shifting cultivation (slash-and-burn), 888–889
Shingles virus, 363t
Shivering response, 740, 740i, 740t
Shoemaker-Levy 9, comet, 346
Shoot, of plants, 400, 482, 486i, 530i
Short-day plant, 536–537, 537i, 541
Shrew, tree, 474, 472i
Shrimp, 438
Shrubland, dry, 864, 864i
Siamese cat, effect of environment on gene expression, 188
Sickle-cell anemia, 183, 183i, 203t, 228, 290–291
Sidewinder (snake), 739i
SIDS, 701, 719
Sieve plate, of phloem, 484i
Sieve tube, of phloem, 485, 484i, 487i, 496, 510, 510i, 511i, 512
Signal, communication, 906, 907, 918
Signaling mechanism
influencing physiological reactions, 611–612
influencing gene activity, 250–251, 250i, 25ii
Signaling molecule, 624, 610, 620
Silica, in horsetails, 405
Silk moth, pheromone sex attractant, 906
Silk, J., 917
Sillen-Tullberg, B., 910
Silt, in soil, 502
Silurian period, 313i, 342, 343, 343i, 348–349i
Sink, flow of organic compounds to, 510–511, 511i, 512
Sink region of plants, 510–511, 511i, 512
Sinus gland, of arthropod brain, 623i

Siphonaptera, 441i
Sirrup, of middle ear, 596i, 597i
Sitka spruce, 824i–825i
Skate (cartilaginous fish), 452
Skeletal
  joints, 636–637, 636i
  muscle, 550, 550i, 638–639, 638i, 639i
  structure, functional organization, 640–641, 640i, 641i
  system, 627, 627i 632–637, 636i, 636t, 637i, 636t
Skeleton, 423, 632–636, 632i, 633i, 646
Skewed information flow, 568
Skin
  aging due to tanning, 631, 631i
  as barrier against infection, 672t
  cancer of, 224, 224i, 237
  derivatives of, 628–629, 628i, 629i
  functions, 629–629
  human, 628–629, 628i, 630–631, 630i, 631i
  structure, 628–630, 628i, 629i, 830i
  vertebrate, 628–629, 628i, 629i
  vitamin D production, 629
Skull, 472, 472i, 637i
Slash-and-burn cultivation, 888–889
Sleep, 574, 578i
Sleeping sickness, African, 376, 376i
Sliding-filament model of muscle contraction, 640–641, 641i
Slime mold
  cellular, 371, 372–373, 373i, 386t
  plasmodial, 370i, 371i, 372, 386t
Slipper limpet, reproduction by, 160–161, 160i
Slow-wave sleep pattern, 584i
Slug, banana, as prey of garter snakes, 902, 902i
Slug, sea, 431i, 432i
Small intestine, 617t, 714i, 716–719, 717t, 718i, 719i, 728, 728t
Smallpox, 363t, 670
Smartweed, and resource partitioning, 817, 817i
Smilodon, 347i
Smith, H. W., 14
Smith, S., 517
Smith, T., 289
Smith, V., 871
Smog, 884, 884i, 885, 898
Smoking
  and bronchitis, 702, 703i
  and cancers, 703i
  and cardiovascular disorders, 703i
  effects of, 702–703, 703i
  and emphysema, 702, 703i
  and impaired immune system functioning, 703i

and lung cancer, 702, 703i
and marijuana, 703
and pregnancy, 780
and shortened life-expectancy, 703i
Smooth muscle, 550, 550i, 555i
Smut, of grains, 396
Snail
  as intermediate host for fluke, 428, 428i
  oviparity, 746, 747i
  studies on genetic variation, 296–308, 296i
Snake, 454, 458–459, 456i, 459i
Snowshoe hare, 739, 819, 819i
Social behavior, 905, 910–911, 912, 917, 918
Sodium
  active transport from nephron, 735, 735i
  atomic number and mass number, 22t
  in human nutrition, 725t
  role in functioning of neuron, 561, 561i
Sodium-potassium pump, 91, 561, 561i, 563, 563i
Softwood tree, 497, 497i
Soil
  aluminosilicates of, 502
  erosion of, 503
  minerals, 502
  profile properties, 502, 502i
  salination, 892
  types of, 501, 502, 862
  water, 502
  waterlogging, 892
Solanum esculentum, 482i
Solar energy, harnessing, 896, 896i
Solar-hydrogen energy, 896, 896i
Soldier, of termite colony, 914, 914i
Soldierfish, 453i
Solenopsis invicta (Argentine fire ant), 827, 827t
Solid waste, 887
Solidago altissima, 288
Solute
  absorption in gut, 742
  defined, 31, 34t
  regulation, 732–737
Somatic
  cell, defined, 150
  nerve, 586
  senses, 589
Somatomedin, 617t
Somatostatin, 615, 617t, 620
Somatotropin (STH), 250, 266, 266i, 614, 614t, 615, 616, 616i, 624
Somite, of human embryo, 776, 776i, 778i
Sonchus, 114i, 511, 511i
Song system, of birds, and sex hormones, 461, 903
Soricidae, 464t

Sorus, of fern, 405
Sound production, by frogs, 695
Source region of plant, 510–511, 511i, 512
Southern Oscillation (changes in atmospheric pressure in Pacific Ocean), 879
Sow thistle, 114i, 511, 511i
Spalacidae, 464t
Sparrow, 622, 622i, 827t, 904
Spartina, 20i, 876i
Spatial learning, 904t
Special senses, 589, 591
Speciation, 296–309
  allopatric, 297, 302–303, 302i, 303i, 308
  of animal groups, 830i
  corresponding to latitude, 830, 830i
  decreases during mass extinction, 830, 831
  defined, 297
  in geographically isolated populations, 302–303, 302i, 303i
  on islands, 830–831, 830i
  microevolutionary, 309t
  macroevolutionary, 309t
  models of, 306, 306i
  in ocean basin, 302, 302i
  parapatric, 297, 305, 305i, 308
  patterns of, 306, 309t
  of plant groups, 304, 305i
  by polyploidy, 304, 305i
  rates of, 830
  role of geographic barrier, 299
  sympatric, 297, 304–305, 304i, 305i, 308
Species
  biological concept, 298–299
  dispersal, 830–831, 830i, 831i
  distribution, 855
  diversity patterns, 830–831, 830i, 831i
  genetic divergence, 299, 299i
  geographic isolation, 299
  interactions between, 814, 832t
  introduced, effects of, 827, 827t
  morphological concept, 298
  naming of, 322–323, 322i
  reproductive isolation, 299, 300
  sibling, 300
  as taxonomic category, 8
Spectrin, 71
Speed (amphetamine), 584
Spemann, H., 757
Sperm nucleus, of flowering plant, 518, 519, 518i–519i
Sperm whale, diving by, 705
Sperm
  fertilization of egg, 147i, 750, 750i, 773, 773i
  flagellar movement of, 72i
  formation of, 168, 169i, 764–765
  structure of, 767i
Spermatid, 766, 767i
Spermatocyte, 766, 767i

Spermatogonium, 766, 767i
Spermicidal foam and jelly, for birth control, 785, 785i
Sperry, R., 581
Sphagnum, 40, 403, 403i
Sphenodon, 458, 459i
Sphenophyta, 404–405
Sphincter, 715, 716i
Spicules, of sponges, 420, 421i
Spider monkey, 470i, 470t
Spider, 17, 17i, 437, 437i, 632, 632i
Spider, sea, 437
Spinach, example of long-day plant, 536–537, 537i
Spinal
  canal, 579i
  cavity, 552i
  cord, 575i, 576i, 577, 577i, 578, 578i
Spindle
  in meiosis, 164i, 165i
  microtubules of, 151i, 152–153, 152i, 153i, 154–155, 155i
  mitotic, 151i, 152, 153i, 156i
Spiny anteater, 464, 464i
Spiracle
  of shark, 450i
  of tracheal respiratory system, 693, 693i
Spiral cleavage, 430, 430i
Spirillum, 354, 354i
Spirochetes, 361t
Spirogyra, 117, 117i, 385, 385i
Spleen, 576i, 678, 678i
Spleen, innervation
Split-brain experiments, 582, 582i
Sponge, 417, 418t, 420–421, 420i, 421i, 444
Spongin, 420
Spore, in plants, 169, 390i, 391i, 392, 392i, 400
Sporophyte
  of bryophytes, 402–403, 402i, 403i
  of fungi, 390i, 391i, 392, 392i
  of flowering plant, 400, 400i, 516, 516i, 526, 530–531, 530i–531i
  in life cycles of plants, 168, 168i
Sporozoan, 376–377, 377i, 386t
Sporozoite, 376, 377, 377i
Spotted fever, Rocky Mountain, 437
Spriggina, 342i
Spring overturn, in lakes, 870, 871
Squid, 431, 433, 433i
Squirrel, flying, 464t
SRY gene (sex-determining region gene), 196
Stabilizing selection, 288, 288i
Stamen, 516i, 517
Starch, 42, 123
Starling, E., 610
Starling, European, 827t, 900. 900i
Statolith, in root, 534, 534i
  as modified plastids, 534
Stem cell, of blood, 653, 653i

Stem, of plants (*See also* Phloem, Xylem, Vascular cambium, other specific tissues)
functions, 506–511
growth of, 483, 483I
structure, 482, 482i, 483i, 484–487, 492–497
types of, 487i, 498i
Sternum, 461i, 637i
Steroid hormone, 45, 612–613, 612i, 612t, 624, 645
Sterol, defined, 45
Stethoscope, in measuring blood pressure, 661i
Steward, F., 265, 524
Stigma, of flower, 518, 516i, 518i
Stimulant, 584
Stimulus, 5, 554, 555i, 559, 567i, 589, 606
Stinkhorn fungus, 390
Stoma of leaf, 488, 489i, 501, 507, 507i, 508–509, 508i, 509i, 512, 533
Stomach (*See also* Digestion, Digestive system)
hormones of, 617t
human, 714i, 716, 716i, 726, 728t
innervation, 576
of ruminant mammals, 713, 713i
Storage, of orgnic substances by plants, 510, 511, 511i
Strain, of bacteria, 361
Stratification, of sedimentary deposits, 313
Strawberry, fruit, 521, 521i
Stream ecosystem, 871, 871i
Streptococcus pneumoniae, 218, 218i, 366
Streptomycin, 11, 287
Stress, mechanical, effect on plants, 535, 535i, 541
Stress response, hypothalamus initiated, 618
Stretch reflex, 567, 567i
Strip mining, of coal, 894
Strobilus, of horsetails, 404, 404i
Stroma, of chloroplasts, 114i, 115
Stromatolites, 39, 338, 338i
Strontium-90, isotope, 21
Sturnus vulgaris (starling), introduction to North America, 900, 900i
Style (of pistil), 516i
Subatomic particle, defined, 6
Subcutaneous layer, of vertebrate skin, 628
Subspecies, 305
Substrate, in chemical reaction, defined, 103
Substrate-level phosphorylation, 134, 139i
Succession, 813, 824, 824i
Succulent, as example of CAM plants, 125
Sucrose, 42i
Sudden infant death syndrome (SIDS), 701, 719

Sugar maple, 496, 496
Sugar, blood level of, 5
Sugarcane, 490i
Sulfate-reducing bacteria, 141
Sulfur dioxide, from low-grade coal, 894
Sulfur oxides, as air pollutants, 884, 884t
Sulfur shelf fungus, 388i
Sulfur
atomic number and mass number, 22t
in human nutrition, 725t
required by plants, 503t
Summers, K., 73i
Sun, as source of energy, 110, 883
Sunfish, bluegill, behavior, 911
Sunflower, 21, 21i, 534i
Sunlight (*See* Energy, light)
Sunlight, effect on skin, 631, 631i
Suntan, and aging of skin, 631
Superior (in human anatomy), 553i
Superoxide dismutase, 96
Superplumes, 346
Surface-to-volume ratio, of cells, 57
Suricata suricata (meerkat), 544, 544i, 545i
Surtsey (volcanic island), and colonization by plant species, 830–831, 830i
Survivorship, patterns of, 800–801, 800t, 801i
Suspended particles, as air pollutants, 884, 884t, 885i
Sweat gland, 630, 628i, 740, 740i
Swim bladder, of fishes, 452, 453i, 694i
Symbiosis, 394, 505, 505i, 814
Symmetry
animal, planes of, 552i
bilateral, 418–419, 418i, 444, 444i
human, 553i
radial, 418, 418i, 442, 444, 444i, 572
Sympathetic nervous system, 575i, 576, 576i, 577, 586
Sympatric speciation, 308
Synapse, chemical, 564–565, 564i–565i
Synapsids, 457
Synapsis, of homologous chromosomes in meiosis, 166
Synaptic integration, 565, 565i
Syndrome, defined, 192
Synovial joint, 636
Synura, 381i
Syphilis, 787, 787i
Systematics, biological, 311
Systemic circuit, 651, 668
Systole, of heart, 658, 659i

T

2,4,5-T, 533
T cell

activation, 678
cytotoxic, 676, 677i, 679i, 686, 686, 688i
DNA of, 683
effector, 677i, 678, 683i
formation of, 675, 675i, 678
helper, 676, 677i, 679i, 680, 680i, 686–687, 688, 688t
helper, infected by HIV, 686–687, 680i
memory, 677i, 683, 683i
T lymphocyte (*See* T cell)
T-cell receptor (TCR), of thymus, 678, 680, 680i
T-tubule, of plasma membrane of muscle cells, 642, 642i
Tachycardia, 665, 665i
Tachyglossidae, 464i
Tachyglossus, 464, 464i
Tactile display, of honeybees, 907, 907i
Tadpole, 448, 449i, 744i
Tapeworm, 427, 427i, 428, 429i
Taproot, 492, 492i
Target cell, 624
Tarsal bone, 637i
Tarsier, 470, 470i, 470t
Tarsioid, 469, 470, 470i, 477
Tasmanian wolf, 464t
Taste bud, 594, 594i, 715, 715i
Tatum, E., 228
Taxol, as inhibitor of cell division, 70
Taxonomy
cladistic, 324–325
classical, 324
defined, 311, 322
Taxus brevifolia, source of taxol, 70
Tay-Sachs disorder, 203t, 204
Tectonic plate, 314–315, 315i
Tectorial membrane, 597i
Tectum, of midbrain, 578, 578i
Teeth, of primates, 471, 472, 472i
Teleost, 453
Telophase of mitosis, 151i, 153, 153i, 159
Telophase, of meiosis, 163, 163i, 164i, 165i
Temperature regulation, 739–740
Temperature zones, 856–857, 857i
Temperature, animal body
maintenance of, 738–741
regulation, 740–741, 740i, 741t
Temporal lobe, 580, 581i
Tendon, 627, 638, 638i, 646
Tendril, 535, 535i
Termite, 914, 914i, 907
Tern, Caspian, parenting, 909, 909i
Territoriality, in ravens, 905
Terry, N., 20–21
Tertiary period, 313i, 347, 348–349i
Testcross, in studies of inheritance, 179
Testicular

cancer, 765
feminization syndrome, 203t, 612
Testis
as endocrine gland, 611i
hormones of, 617t
human, 763, 764, 764t, 765i, 766, 766i
secretion of testosterone, 619
Testosterone, 39i, 196, 612, 617t, 619, 645, 763, 766–767, 903, 903i
Tetanus, of muscle, 644, 644i
Tetanus infection, 358, 568, 568i, 684, 684i
Tethys Sea, 343i, 348i
Tetrad scar, of spore of seedless plants, 407, 407i
Thalamus, 578i, 579, 580i, 583i
Thalassoma, 302i
Theobroma cacao, fruit, 522, 522i
Theory of uniformity, 275, 314
Theory, in scientific studies, 13, 16
Thera, volcanic eruption of, 271
Theropod, 278
Therapsid, 463, 457i
Therian, 463
Thermal inversion, in atmosphere, 884, 884i
Thermoanobacter, 106
Thermocline, lake, 870, 870i
Thermodynamics, first law, 98
second law, 99
Thermophile, 360, 361
Thermoreceptor, 588, 588i, 590, 590t, 592, 606
Thiamin, 724t
Thigmotropism, 535, 535i, 541
Thirst center, 735
Thistle, sow, 511, 511i
Thomas, L. , 759
Thoracic cavity, 552i, 698–699, 698i
Thoracic duct, 667i
Thoracic nerve, 575i, 576i
Thorium-234, decay into lead-206, 24
Threat display, of baboon, 906–907, 906i
Threshold level, 562, 562i
Thrombin, 104, 104i, 666i
Thylacinidae, 464t
Thylakoid membrane system, of chloroplasts, 114i, 115, 120, 120i, 122, 122i, 127
Thymine, 50i, 220, 220i
Thymosin, 617t
Thymus gland, 617t, 667, 678
Thyroid gland, 611i, 612t, 613, 617t, 618–619, 619i
Thyroid-stimulating hormone (TSH), 619
Thyrotropin (TSH), 614, 615, 616, 624
Thyroxine, 617t, 618
Tibia, 637i
Tibialis anterior muscle, 639i

Tick, 437
Tidal volume, of lungs, 699, 699i
Tilletia indica, 396t
Tissue, animal
  connective, 545, 548–549, 548i, 549i
  dense, 548–549, 548i
  epithelial, 545, 545i, 546–547, 546i, 547i
  formation of, 752, 752i
  glandular, 547, 547i
  loose, 548, 548i, 548t
  muscle, 545, 550, 550i
  nervous, 545, 551, 551i
  squamous, 546i
  stratified, 546
Tissue, plant
  collenchyma, 484, 484i
  dermal, 484, 485
  epidermis, 484i
  ground, 484
  mesophyll, 484, 485i
  parenchyma, 484, 484i
  sclerenchyma, 484, 484i, 487i
  vascular, 484, 484i
Tissue culture propagation of plants, 524–525, 524t, 525i
Tissue layer, primary, 6, 418
Tissue specializatiion, 748, 748i, 760
Tit, great, predator on sawfly caterpillars, 910
Toad, 455, 455i, 752
Tobacco, 491
Tobin, E., 251
Togavirus, 363t
Tomato, 482i, 535i
Tomato, effect of mechanical stress, 535i
Tongue, 594, 594i, 715, 715i
Tonicity, effects of, 88, 89i, 94
Tonsil, 667i, 678i
Toolmaking, by early humans, 474
Tooth, 713, 713i, 715, 715i
Topsoil, 502
Torsion, in gastropod mollusks, 432, 432i
Tortoise, Galápagos, 458i
Touch, center in brain, 583i
Toxicity, uremic, 736
Toxin, defined, 41
Toxoplasma, 376
Toxoplasmosis, 376
Trace element, 22
Tracer radioisotope (*See* Radioisotope)
Trachea, 693, 693i, 696i, 697
Tracheid, of xylem, 484i, 485, 506
Trachoma, 604–605
Tracts, of nervous system, 575
Transcription, of DNA
  control of, 244–245, 244i, 245i, 246–247, 246i–247i
  in lampbrush chromosomes, 248, 248i
  negative control, 244–245, 245i

positive control, 244, 245
stages in the process, 227, 230i, 231, 231i
Transcription, of RNA, 230–231, 231i
Transfer RNA, function of, 232–233
Transfusion, blood, 182, 654–655, 654i
Transition state, in enzyme activity, 105, 105i
Translation, of messenger RNA, 227, 234
Translocation in plants, 501, 510–511, 511i
Transpiration
  in hydrologic cycle, 842i
  of plants, 501, 506–507, 506i, 507i, 512
Transport protein, 83, 83i, 103
Transport
  active, 87, 87i, 90, 90i, 91
  passive, 90, 90i, 91
  protein-mediated, 91, 91i
Transposable element, causing gene mutation, 236, 237i
Transverse (in anatomy), 552i, 553i
Trapezius muscle, 639i
Tree frog, 298, 299i, 629i
Tree ring, 496–497, 497i
Trematoda, 426–427, 428–429, 429i
Treponema pallidum, 787, 787i
Triassic period, 313i, 344, 348i–349i
Tribolium, 292
Triceps muscle, 638, 638i, 639i
Triceratops, 344
Trichinella spiralis, 429, 429i
Trichocyst, 378, 378i
Trichomonad, 376, 376i
Trichomonas vaginalis, 376, 376i
Trichomoniasis, 376
Trichophyton, 396t
Trichoplax, 420, 420i, 418t, 444
Triglyceride, 44, 44i, 142
Triiodothyronine, 617t
Trilobite, 342, 342i
Trimester, of human pregnancy, 774
Triticale, 490i
Triticum, 304, 305i, 490i
tRNA (Transfer RNA), 232–233
Trophic level, in ecosystem, 836–837, 836t, 837i, 839i, 840i, 841i, 852
Tropical rain forest, 866, 866i, 888–890, 888i, 889i, 890i
Tropism, plant, 534–535, 534i, 535i
Tropomyosin, 643, 643i
Troponin, 643, 643i
Troposphere, 856
Truffle, 393
Trumpet chanterelle, 389i
Trypanosoma, 376, 376i

Trypanosome, 376, 376i
Trypsin, 717, 717t
Tsetse fly, and transmission of trypanosomes, 376, 376i
TSH (*See* Thyrotropin)
Tuatara, 457, 458, 459, 459i
Tubal ligation, for birth control, 785, 785i
Tubastrea, 424i
Tube foot, of echinoderms, 443, 443i
Tuber, 524t
Tuberculosis, microorganism causing, 109
Tubulin, 70, 71i, 152, 154
Tumor
  benign, 252
  breast, 783
  HeLa cell division in, 147, 158
  malignant, 252
Tundra, 869, 869i
Tunicate, 448, 449i
Turbellaria, 426, 426i
Turgor pressure, in cells of plants, 508, 509, 509i
Turner syndrome, 203t, 209
Turtle, 457, 458, 458i, 706–707, 706i, 707i
Twins, identical, 727, 753
Typhus, 851
Tyrosinase, role in production of melanin, 184–185

U

Ulcer, peptic, 716
Ulna, 637i
Ulnar nerve, 575i
Ultrafiltration, through capillary walls, 662
Ultraplankton, 872
Ultrasound, reception of, by bats, 588–589, 589i
Ultraviolet light, protection offered by ozone shield, 882
Ulva, 384i
Underhair, of mammals, 462
Uniformity, theory of, 275, 314
Upwelling, 877, 877i, 878i, 879
Uracil, of RNA, 230, 230i
Uranium-238, 24, 24i
Urea, formation in liver, 720, 720i, 732
Uremic toxicity, 736
Ureter, 732i, 733, 733i
Urethra, 732i, 764, 765i, 766i, 769
Urinary bladder
  location, 732i, 769i
  function, 733,
  innervation, 576i
Urinary excretion, 732
Urinary system, 553i, 731, 732–736, 731i, 732i, 733i, 734i, 735i, 742
Urine, 742, 733, 734–735, 734i
Urochordate, 448, 449i

Ustilago maydis, 396t
Uterus, 576i, 768, 768t, 769i, 770, 773i, 774i, 777i, 782i
Utricle, of inner ear, 595, 595i
Utricularia, 500, 501i

V

Vaccination, 681, 684, 684i
Vacuole
  central, of plant cell, 69, 62i
  contractile, of ciliates, 378–379, 378i
Vagina, 768, 768i, 769i, 773, 773i, 782i
Vagus nerve, 576i
Valium, 564
Valley fever, 396
Valve
  of heart, 658–659, 658i, 659i
  of lymphatic vessels, 666i, 667
  vein, 660i, 662, 663, 663i
Van Leeuwenhoek, A., 54, 54i
Vancomycin, in treatment of genital herpes, 287
Variable
  defined, 13
  identifying, 15
Variation
  continuous, 186–187, 186i, 187i, 282
  qualitatively different, 282
  view of Darwin, 276
Varicella-zoster virus, 363t
Vas deferens, 764, 764t, 766i
Vascular bundle, leaf, 485i, 486, 487i, 489, 489i
Vascular cambium, 494, 494i, 495, 495i, 496, 496i, 507i
Vascular cylinder, of roots, 492i, 493, 493i, 504, 504i
Vascular plants, 399, 400, 404–413
Vascular ray, 495, 495i
Vascular system, lymph, 666
Vascular tissue, of plants, 484–485, 484i, 485i, 498
Vasectomy, for birth control, 785, 785i
Vasoconstriction, 661, 664, 740, 740i, 740t
Vasodilation, 661, 674, 674t, 675i, 740, 740i, 740t, 773
Vaucheria, 381i
Vegetal pole, of frog embryo, 749i
Vegetative propagation, in plants, 524, 524t, 526
Vein, animal, 649, 656, 656i, 657i, 660i, 662–663, 663i
Vein, plant, 489, 489i
Velociraptor (early reptile), 344, 458
Vena cava, 657i, 658i, 732i,
Venter, J., 266
Ventral (in anatomy), 419, 552i
Ventricle
  of brain, 579i
  of heart, 658, 658i

Ventricular fibrillation, 665, 665i
Venturia inaequalis, cause of apple scab, 396i, 396t
Venule, 668
Venus flytrap, 500, 500i
Venus' girdle, 425, 425i
Vernalization, 539
Vertebra, 450, 577, 636, 637i
Vertebrate, 446–467, 651, 651i
Verticillium, cause of plant wilt, 396t
Vesicle, 67, 67i
Vessel member, of xylem 484i, 485, 485i, 506, 506i
Vestibular apparatus, 595, 595i
Vibrio cholerae, epidemic of, 805
Virchow, R., 55
Viroid, 363
Virus
  assembly, 364, 364i
  attachment, 364, 365i
  bacteriophage, 218, 219, 219i, 353i, 362, 362i
  classification, 363t
  complex, 362, 362i
  composition, 219, 219i
  defined, 218, 362, 363
  DNA, 219, 219i, 362, 368
  Ebola, 366, 366i
  effect on cells, 218
  enveloped, 362, 362i
  Epstein-Barr, 654
  helical, 362, 362i
  herpes, 687
  HIV, 362
  latency, 364
  measles, 687
  multiplication cycle, 364, 364i, 365i, 366
  myxoma, for controlling rabbits, 828–829
  penetration, 364, 365i
  plant, 363, 363i
  polyhedral, 362, 362i
  protein coat, 219, 219i
  release, 364, 364i
  replication and synthesis, 364, 364i
  RNA, 362, 368
  of sexually transmitted diseases, 786, 787
  structure, 362, 362i
  summary of characteristics, 368
  types of, 363t
Vision, 470–471, 598–605, 606
Vision, center in brain, 583i
Visual accommodation, 601, 601i
Visual cortex, 581i
Visual field, 582, 582i, 598
Vital capacity, of lungs, 699, 699i
Vitamin
  A, 119, 724t, 725
  B₁ (thiamin), 724t
  B₂ (riboflavin), 724t
  B₆, 724t
  B₁₂, 380, 724t
  C (ascorbic acid), 724t, 725

D, 612t, 613, 620, 631, 724t
E, 724t, 725
K, 358, 724t, 725
and nutrition, 724, 724t, 725, 724t
Vitis (grape), effect of gibberellin, 528i, 529
Vitreous body, of eye, 599i, 600, 600i, 600t
Viviparous, 746, 747i
Vocal cord, frog, 695, 699, 699i
Volvox, 384, 384i
Vomeronasal organ, 594, 610
Vulva, 768

W

Walker, N. M., 14
Wallace, A., 277, 277i
Waller, A., 648, 648i
Walrus, 465i
Wandering albatross, travels of, 296
Warning coloration, 820
Wasp, parasitoid, 823, 823i
Wastewater, treatment, 893, 893i
Water hyacinth, 827t
Water mold, 371, 386t
Water province, 879
Water
  absorption, in gut, 732
  cohesion of, 31
  competition for sources, 893
  conservation in stems and leaves, 508–509, 508i, 509i
  covalent bonds of, 29
  distribution of, in plants, 512
  evaporation of from organisms, 30, 31
  global crisis, 892–893
  hydrogen bonding in ice, 30, 30i
  and hydrologic cycle, 842, 842i
  loss by urinary excretion, 732–735, 732i, 733i, 734i
  metabolic, 730–731, 730i
  molecular structure of, 29
  polarity of, 30, 30i
  pollution, 892–893
  potential, of plants, 508
  production during metabolism, 732
  properties of, 30
  solvent properties of, 31
  source of hydrogen for plant photosynthesis, 502–503, 503t
  temperature-stabilizing effects of, 30
  uptake by plant roots, 502, 504–505, 504i, 505i
Water table, changes in, 892
Water-vascular system, of echinoderms, 443, 443i
Waterlogging, of soils, 892
Watershed, 842i, 843, 843i
Watson, J., 216, 216i, 217i, 221
Wavelength, of energy, 116

correlation with photosynthetic activity, 117i
Wax, 45, 52t
Weddell seal, 706i
Wegener, A., 314
Weight, related to diet, 142
Weismann, A., theory of heredity, 193
Welwitschia mirabilis, 409, 409i
Went, F., 534–535
Wexsler, N., 205
Whale, 465, 705, 706i
Wheat, 490i
Whisk fern, 404, 404i
White blood cell, 652i, 653, 653i, 654, 671, 672, 672t, 688, 688t
White matter, spinal cord, 577, 577i
White-crown sparrow, learning of song, 904
White-throated sparrow, response to melatonin secretion, 622, 622i
Whooping cough, vaccine, 684i
Wilkins, M., 220, 221
Williams, E., 874
Wilting, of plants, 508i
Wind energy, harnessing, 896, 896i
Windpipe (*See* Trachea)
Wing bud, embryonic, 752i
Wing
  of birds, 460, 461i
  of insects, 440, 441i, 633i
Winter blues, related to short daylength, 622
Withdrawal, as method of birth control, 785, 785i
Woese, C., 328
Wolf spider, 437i
Wolf, 913, 913i
Wonder drug, 11
Wood duck, secondary sex characters, 39i
Wood, 497, 497i
Woodland, dry, 864
Woodpecker, downy, 288, 288i
Woody plant, 494–497, 494i, 495i, 496i, 497i
Worker, of insect colony, 914, 914i, 915, 915i
Worm, segmented, 418t
  ribbon, 418t
Wrasse, 302i
Wuchereria bancrofti, 429, 429i

X

X-chromosome, 194, 195i, 196, 196i, 214
X-linked
  anhidrotic ectodermal dysplasia, 203
  inheritance, disorders, 203t, 204i, 206–207, 206i, 207i
X-organ, of arthropod brain, 623i
Xanthophyll, 118

Xenopus laevis (African clawed toad), 455, 752
Xeroderma pigmentosum, 224
Xylem
  cohesion of water molecules in, 506–507, 506i, 512
  conduction of water and dissolved ions, 400, 484–485, 498, 501
  of root, 492i, 493i
  of stem, 484i, 486i, 487i, 492i, 493i.
  of woody plants, 494, 494i, 495, 495i, 496
XYY condition, 203t, 209

Y

Y-chromosome, 194, 195i, 196, 196i, 198i, 214
Y-organ, of arthropod brain, 623i
Yeast, 141, 141i, 393, 393i
Yeast infection, 393, 393i
Yellow coral fungus, 389i
Yellow fever virus, 363t
Yellow-green algae, 381, 381i
Yellowjacket (wasp), and its mimics, 821i
Yersinia pestis, and plague, 799
Yolk, 747, 754, 754i
Yolk sac, 775, 775i, 776, 776i, 788
Yucca moth, 815, 815i
Yucca, pollination of, 815, 815i

Z

Z lines, of muscle, 640–641, 640i, 641i, 642i
Zamia, 408, 408i
Zea mays (corn)
  chromosomes, 198–199, 199i
  genes, 199, 199i
  germination of grain, 530i, 531i
  leaf structure, 124i
  stem, 487i
Zebra, 301, 301i
Zebra, meaning of signals, 906
Zebra finch, territorial song, 902, 902i
Zebra mussel, 827
Zebroid, 301, 301i
Zero population growth, for control of human population, 796, 806, 809
Zinc, 503t, 725t
Zona pellucida, 770, 773i
Zonation
  intertidal, 876, 877i
  of water masses in lakes, 870, 871
Zone diet, 722
Zygomycete, 390, 392, 392i, 395
Zygosporangium, 392, 392i
Zygote
  of flowering plant, 519, 520i
  of frog, 744, 744i
  human, 783t

APPLICATION INDEX

Numbers followed by i refer to illustrations; those followed by t refer to tables.

A

"24-hour flu," 367
ABO blood typing, 182, 655
Abortion, 213, 788
Acclimatization, 704
Achondroplasia, 203t, 205, 205i
Achoo syndrome, 190, 203t
Acid rain (wet acid deposition), 33, 884–885, 885i
Acidosis, 33
Acne, 630
Acromegaly, 616, 616i
ADA gene mutation, 685
Adenoviruses, 363t
Adoption, 917
Aerobic exercise, 644
African sleeping sickness, 376
Afterbaby blues, 782
Age spots, 96i, 97
Agent Orange, 533
Aggressive behavior, 645
Aging, 96–97, 758, 783
  and cardiovascular disorders, 664
  and eye disorders, 605
  and free radicals, 96–97, 725, 758
  and memory loss, 583
  and muscle loss, 644
  and suntans, 631
Agriculture
  land conversion for, 887, 887i
  and leaching, 503, 503i
  mechanized, 887
  and population size, 20, 887
  reliance on fertilizers in, 20–21, 41, 126, 849
  by shifting cultivation, 888–889
  subsistence, 887
AIDS, 111, 362, 363t, 786
  agent of, 686–687
  characteristics of, 686
  chemical cocktails and, 687
  pandemic nature of, 366
  prevention and treatment, 687
  viral agent of, 362
Air pollution
  by acids, 884–885, 885i
  by oxidants, 884, 884t
  by particles, 884t, 885, 885i
  and plant function, 509, 509i
  by volatile organic compounds, 884, 884t, 886
Albinism, 185, 191, 203t
Albinos, 630
Alcohol consumption, 67
  and blood-brain barrier, 579, 585
  and pregnancy, 781
  psychoactive effects of, 585
Algal blooms, 381, 387
Alkalosis, 33
Allergic rhinitis, 517

Allergies
  and antihistamines, 684
  causes of, 517, 684
  to fungi, 396, 396t
  and pesticides, 41
  pollen-induced, 517
Aloe vera, 491
ALS, 207
Alzheimer's disease, 266, 583
Amanita, toxic and hallucinogenic effects of, 391
Amnesia, 583
Amniocentesis, 213
Amoebic dysentery, 374, 375
Amphetamines, 584
Amyotrophic lateral sclerosis, 203t, 207
Anabolic steroids, abuses of, 645
Anaphylactic shock, 684
Anemias, 654
Angiography, 665
Anhydrotic ectodermal dysplasia, 203t, 249, 249i, 557
Anorexia nervosa, 710, 711
Antarctica, human invasion of, 834–835
Anti-acne drugs, 780
Anti-anxiety drugs, 564
Antibiotic resistance, 287, 367
Antibiotic-resistant pathogens, 11, 14–15, 15i
Antibiotics, 11, 14–15, 287, 358
  tracking effects of, 109
Apnea, 701
Apoptosis, and cancer, 242–243
Appendicitis, 721
Apple scab, 396i
Aquarium fishes
  and destruction of reef ecosystems, 874
  water mold attack on, 372
Aquifer depletions, 892, 892i
Argentine fire ants, 827, 827t
Arrhythmia, 665
Arthritis, 636
Asthma, 618
Astigmatism, 604
Atherosclerosis, 664–665
Atherosclerotic plaque, 45i, 256, 664i
Athlete's foot, 396, 396i, 672
Atrial fibrillation, 665
Autoimmune responses, 685
AZT (azidothymidine), 111
  and AIDS, 687

B

Baby boomers, 807, 809
Back pain, 636
Balantidium coli infection, 379
Bald eagle, and DDT, 851
Balloon angioplasty, 665
Basal cell carcinoma, 224, 224i

Benign tumor, 252
Biological controls, of pests, 287, 823
Biological therapy, 14–15
Biosphere, human impact on, 882–885, 897
"Biotech barnyards," 266
Birds, and human predation, 460i
Birth control, 784–785
Bison, extinction of, 465
Blood doping, 654
Blood pressure, measuring, 660, 661i
Blue offspring, 203t, 295
Body building, 645
Body-builder's psychosis, 645
Body-building, 146
Body temperature, 740–741, 740i
  sensation of, 592, 593i
Body weight, 142
Body weight, "ideal," 726, 726i
Botulism, 111, 358, 568
Bradycardia, 665
Breech birth, 782
Bronchitis, 702
Brown algae, commercial products from, 383
Brown pelican, and DDT, 851
Brown rot, of fruit, 396
BSE (bovine spongiform encephalopathy, 367
Bubonic plague, 799
Bulimia, 710–711
Burkitt's lymphoma, 253
Byssus, applications for, 226–227

C

Caffeine, 579, 584, 587
Calicivirus, escape from island "laboratory," 829
Camptodactyly, 186
Cancer
  of bone marrow, 654
  of breast, 783
  causes of, 224, 237, 253
  characteristics of, 243, 252–253
  of colon, 721
  of prostate gland, 765
  in respiratory system, 702, 703i
  of skin, 224, 631
  symptoms of, 224
  of testes, 765
  types of, 224
Carbon monoxide poisoning, 704, 709
Carcinogens, 237, 253, 533
Carcinoma (*See also* Cancer), 681
Cardiovascular disorders, 664–665
  and obesity, 726
Caries, 715i
Carsickness, 595
Cataracts (of lens), 605

Cellular oxygen deficiency, 690
CFCs , and ozone thinning, 886
Chagas disease, 376
Chemical burns, 33
Chernobyl, 21i, 851i
Chestnut tree blight, 396, 396t, 827, 827t
Chicken pox, 689
Childbirth, 555
Chin dimple, 173
Chlamydia, 604
Chlamydial infection, 787
Chocolate, 522i, 584
Choking, 709, 715
Cholera, 95, 287
Chorionic villi sampling, 213
Cigarette smoking, effects of, 701, 702–703, 703i
Cirrhosis, 585
City planning, and watershed studies, 843
Cleft lip, 212
Coal, 406, 406i
Cocaine
  and pregnancy, 781
  effects of, 565, 585
  source of, 491
Coitus, 773
Cold sore, 363t, 364, 631
Cold sweats, 630
Collagen, effects of mutated gene for, 287
Color blindness, 201, 203t, 604
Conifers, uses of, 411
Contraceptives, 784–785, 785t
Corneal transplant surgery, 605
Coronary bypass surgery, 665
Cortisol-like drugs, 618
Crack (cocaine), 570–571
Crack babies, 571–572
Cri-du-chat syndrome, 203t, 210, 210i
Crocodilians, and human predation, 458
Crop plants
  asexual reproduction of, 524
  domestication of, 490, 804
  and dormancy, 539
  examples of, 490–491
  flowering plants as, 412
  fungal attacks on, 396, 528
  genetically engineered, 268
  genetic engineering of, 264–265, 265i
  and hormone applications, 528–529, 529i, 532–533
  and pesticides, 41
  viral infections of, 363
Crop rotation, 849
Crown gall tumor, 265, 265i
Cruetzfeld-Jakob disease, 363, 367
Crustaceans, as food source, 438
Cryptosporidium infection, 376

Culture, 471, 475
CVS, 213
Cyanide poisoning, 111
Cycads, uses of, 408
Cystic fibrosis, 79, 203t, 266

D

Day-care centers, and contagious
    diseases, 366, 367, 607
ddI, and AIDS, 687
DDT, biological magnification of,
    851
Death, 759
Decompression sickness, 705
Deep sea diving, 705, 740
Deforestation
    and endangered ecosystems,
        853, 853i
    and global warming, 847
    in North America, 411, 411i,
        415, 415i
    of tropical forests, 888–889, 888i,
        889i, 890
Dendrobates, and poison darts, 547
Dengue fever, 363t
Depressants, 585
Desalination, 892
Desertification, 863, 891, 891i
Designer (synthetic) organs, 551
Designer dogs, 280–281, 280i
Detached retina, 605
Diabetes, 5
Diabetes insipidus, 616
Diabetes mellitus, 551, 620–621
    Type 1 (juvenile-onset), 625
    Type 2, 621
Diabetics, 264
Diaphragm (contraceptive
    device), 785
Diarrhea, 672, 672t
Digitalis, 491
DNA fingerprinting, 262
Dolly (cloned sheep), 266
Douching, 784
Down syndrome, 203t, 208, 209i,
    213, 215, 771
Downy mildew, 372
DPT vaccine, 684
Drug addiction, 570–571,
    584–585, 584t
Drug evaluations, 109, 362
Drug-resistant pathogens, 367
Dry acid deposition, 884
Duchenne muscular dystrophy,
    203t, 206, 255
Duplications, and human
    evolution, 211
Dust Bowl, 864, 891, 891i
Dutch elm disease, 396t, 827t, 851

E

Earlobe, attached vs. detached,
    174, 190
Easter bilby, 829

Eating disorders, 710–711
Ebola virus, 363t, 366, 366i, 689
Economic development, and
    population growth, 808–809,
    899
Economic wars, and environment,
    899
Edema, 662, 691
EEG patterns, 584, 584i
El Niño, global effects of, 877,
    878–879, 878i, 889i
Elephantiasis, 428, 429, 429i, 662
Ellis-van Creveld syndrome, 203t,
    293
Embolus, 665
Embryonic development, human
    157, 762–763, 774 ff.
Emerging pathogens, 366
Emphysema, 702, 702i
Endangered ecosystems, 853, 853i
Endangered species, 293, 851
    birds, 460i
    mammals, 465
Endangered Species Act, 853
Endometriosis, 769
Endorphins, 564
Energy from photovoltaic cells,
    896, 896i
Epilepsy, 569, 582
Epstein-Barr virus, 363t, 654
Ergotism, 396
Erythroblastosis, 655
Escherichia coli infection, 76i, 367
Eugenic engineering, 267
European rabbits, population
    explosions in Australia, 828
Eutrophication, by human
    activities, 850, 871
Exercise
    aerobic, 644
    and aging, 644, 758
    and bone density, 635
    energy sources for, 140, 643, 643i
    loss of muscle tissue in, 644
    and number of muscle fibers,
        647
    and oxygen debt, 643
    strength training, 643, 644
Exotic species, 828
Experimental wetland, 20–21, 20i
Eye color, 186

F

Familial hypercholesterolemia,
    207t, 256–257
Farsightedness, 604, 605i
Fats, excess, and body weight,
    142, 710–726
Faulty enamel trait, 203t, 207, 207i
Feline infectious peritonitis, 293
Fermentation, commercial uses
    of, 140, 141, 358
Fertilizers, 20–21, 41, 126, 849
    and algal blooms, 387
    and nitrogen cycle, 849

Fetal alcohol syndrome, 781
Fever, 675, 741
Fibers (sclerenchyma), uses of,
    484, 484i, 490i, 491
Fight-flight response, 577, 618
fire ants, 827, 827t
Firestorms, California, 864,
    864i
Fishing, with dynamite and
    cyanide, 874
Flatworm infections, 428
Floral oils, uses of, 491
"Flu shots," 362, 689
Fluke infections, 428, 429i
Food poisoning, 367
Food production, 20, 41, 112–113,
    490, 887
    and genetic engineering,
        264–265
Food pyramids, 722, 722i
Fossil fuels
    alternatives to, 896, 896i
    and fertilizer production, 849
    as finite resources, 894
    formation of, 406
    and global warming, 845
Fragile X syndrome, 203t, 211,
    211i
Frostbite, 741
Fruits, popular, 520t, 521, 521i,
    527, 527i
Fungicides, 41, 886
Fungi
    and air pollution, 394, 395
    effects on human history, 396
    as food spoilers, 392
    in making food products, 393
    as pathogens, 393, 393i, 396,
        396t
Fusion power, 896

G

Galactosemia, 203t, 204
Garden plants
    and apical dominance, 533
    fungal attacks on, 396
    and landscaping, 490
Gene therapy, 109, 257
    ethics of, 266, 267
Genetic disorders, human
    genetic counseling for, 212
    genetic screening for, 212
    phenotypic treatment of, 212
    and prenatal diagnosis,
        212–212, 212i
    types of, 202–203
Genetic engineering, for SCIDs,
    685
Genetically engineered bacteria,
    264
Genetically engineered plants,
    264
Genital herpes, 363t, 787
Giardia I infection, 376
Ginkgo, uses of, 409

Glaucoma, 605
Global warming, 37
    factors suggestive of, 126, 845,
        846–847
    modeling by supercomputer
        programs, 881
Glomerulonephritis, 736
Goiter, 619, 619i
Gonorrhea, 11, 786
Grapes, commercial production
    of, 528i, 529
Grave's disorder, 685
Gypsy moths, and defoliation of
    forests, 827

H

Hair color, 184
Hair thinning, 630
Hairy ears, 215, 215i
Hantaviruses, 363t
Hardwood, commercial use of,
    497i
Hay fever, 517, 684
HDLs, 256–257, 665
Heart attack, 649, 664, 665
    and TPA, 266
Heartburn, 716
Height, 186, 187
Heimlich maneuver, 709
HeLa cells, and cancer research,
    158
Hemodialysis, 736
Hemoglobin production in
    genetically engineered plants,
    265
Hemolytic anemias, 654
Hemophilia, 201, 203t, 206, 206i,
    207i
Hemophilus influenza, 684
Hemorrhagic anemia, 654
Hemorrhagic fevers, 363t, 366,
    689
Henbane, and Hamlet's father,
    491
Hepatitis A, 363t
Hepatitis B, 363t, 684
Herbicides, 41
Herniated disk, 636
Heroin, 585
Herpes infections, 786
Herpes simplex infection, 631
    of the eye, 604
Herpesviruses, 363t
Hib vaccine, 684
High blood pressure, 664
Hirsutism (excessive hairiness),
    630
Histoplasmosis, 396, 604
Hodgkin's disease, 491
Hookworm, 428, 429
Horsetails (Equisetum), as
    potscrubbers, 405
Human birthweight, selection for,
    294i, 295
Human brain, 580–581

Human embryonic development, and blood-brain barrier, 587
Human evolution
    African emergence model, 476
    in African savanna, 468, 469i, 472–476
    in Asia, 475, 476
    in Europe, 468–469, 475, 476
    genes influencing, 211, 317, 321
    migrations during, 347
    multiregional model, 476
    primate origins, 468, 472–473
    trends in, 470–471
Human gene therapy
    ethics of, 266, 267
    examples of, 257
Human genome project, 262, 266
Human history, early interpretations of, 270–271, 310–311
Human nutrition, 142–143, 711, 720, 722–725
    during pregnancy, 780–781
    energy alternatives, 144–145, 146
Human population growth, 793, 804–809, 811
    effect on biosphere, 882–883
    in San Francisco Bay area and Sacramento, 883i
Huntington disorder, 203t, 204–205, 205i
Hutchinson-Gilford progeria syndrome, 192
Hypertension, 664
Hyperthermia, 741
Hyperthydroidism, 619
Hyperventilation, 690
Hypnotics, 585
Hypoglycemia, 618
Hypothermia, 740
Hypothyroidism, 619
Hypoxia, 690, 704

I

Ice-minus bacteria, controversy provoked by, 264, 264i
Identical twins, 727, 753
Immigration to United States, 811, 811i
Immunization, 684
Impetigo, 366
Inbreeding, among Old Order Amish, 293
Infectious diseases, 366–367
Infectious mononucleosis, 654
Inflammation, 674–675, 675i
Insecticides, 41, 491
Insects
    as competitors with humans, 440, 441i
    as crop destroyers, 41, 441, 441i
    as crop pollinators, 130, 441
In-vitro fertilization, 213, 788
Iron-deficiency anemia, 654

Irrigation and drainage programs, 863
Itching, perception of, 592

J

Japanese beetles, and defoliation of plants, 827t
Jet lag, 622

K

Kaposi's sarcoma, 686
Kidney dialysis machine, 736
Kidney stones, 736
"Killer bee" attacks, 130–131, 827
Kissing, and homeostasis, 554
Kitchens, as pathogen quarters, 367
Klinefelter syndrome, 203t, 209
Kudzu, 829, 829i
Kuru, 363

L

Labor, 782
Laboratory-grown epidermis, 551
Lactation, 782–783
LAK cell treatments, 681
Lamprey invasions, effect on commercial fishes, 451, 827t
Language, 471, 580, 581
    and sound production, 695, 699
Laryngitis, 699
Laser angioplasty, 665
Laser coagulation, of retina, 605
Late blight, 372
LDLs 256–257, 665
Leeches, as blood-letting tools, 434i
Leukemia, 654, 681
Lichens, as air pollution monitors, 394
"Lou Gehrig disease," 207
LSD, 585
Lyme disease, 358, 358i, 437

M

Mad cow disease, 367, 369
Madeleine, as a very distant relative of Cambrian animals, 416–417
Malaria
    agents of, 376, 377, 377i
    and DDT spraying, 851
    and human history, 377
    and sickle-cell anemia, 290–291, 291i
Malignant melanoma, 224, 224i
Malignant tumor, 252
Mammalian diversity, human impact on, 464–465
Managed pine plantations, 868

Maple syrup (sap), 496
Marijuana, 491, 585
Measles, 363t
Memory, 581, 583, 583i
Meningitis, 367
Menopause, 769
Metabolic acidosis, 736
Metastasis, 252, 253i
Methyl bromide, and ozone thinning, 886
Mexico City, on a bad air day, 884, 884i
Middle-ear infection, 367, 607
Milk of Magnesia, 33
Minerals, deficiencies and excesses, 724, 724t
Mollusks, as food sources, 431
Monoclonal antibodies, 681
Morning-after pill, 785
Motion sickness, 595
Multiple sclerosis, 566, 569
Mumps, 363t
Muscle fatigue, 644
Muscular dystrophy, 215
Mushrooms,
    as commercial crop, 390
    dangers in hunting for, 390i
Myasthenia gravis, 685
Myobacterium tuberculosis, and lung disease, 109

N

Natural plant toxins, 41
Nearsightedness, 604, 604i
Neem tree extracts, 491
Nicotine, 584
    and blood-brain barrier, 579
Nightshade, and Romeo's demise, 491
Nitrogen cycle, human intervention in, 849, 849i
Nitrogen narcosis, 705
Nori, 382
Norplant, 785
Nuclear energy, 894–895
Nuclear waste disposal, 895

O

Obesity, 664, 726, 727
Olestra (See Sucrose polyester)
Orchids, tissue culture propagation of, 524t, 525, 525i
Orgasm, 773
Ospreys, and DDT, 851
Osteoarthritis, 636
Osteoporosis, 625, 635, 635i
Ozone thinning, 886, 886i

P

Pacemaker, artificial, 25, 649
Pain, perception of, 564, 592–593, 593i

PANs, 884
Papovaviruses, 363t
Parasites, of humans, 822
Parkinson's disease, 583
Passive immunization, 684
Pasteurization, 671
Peat bogs, and ancient human sacrifices, 403
Peat harvesting, 403, 403i
Peat, as poultice, 403i
Pelvic inflammatory disease, 787
Penicillins, 287
Penicillum, 393
Peptic ulcer, 716
Peregrine falcon, and DDT, 851
Periodontal disease, 715i
Periwinkle alkaloids, and cancer, 491
Pest resurgence, 287
Pesticide resistance, 286–287
Pesticides
    effects of, 18, 22, 41
    widespread use of, 41
PET scans, 25, 25i, 381i
Phantom pain, 593
Phytoremediation, 20–21
Pinworms, 428
Pituitary dwarfism, 255, 616, 616i
PKU (phenylketonuria), 100, 203t, 212
Plant lore, 490–491
Plant shoots, uses and abuses of, 490–491
Plastic-producing plants, 265
Pleurisy, 697
Pneumonia, 367, 396, 396t, 686
Polio, 363t
Polio vaccines, 684, 684i
Pollination, of orchards, 130
Polycythemias, 654
Polydactyly, 202i, 203t
Polyploidy, 304
Porphyria, 557
Potato blight, 264
Poxviruses, 363t
Pregnancy, 773, 781
    calcium intake during, 647
    prescription drugs and, 781
Preimplantation diagnosis, 213
Prescription drugs, and pregnancy, 781
Prions, 363, 367
Progeria, 192, 203t, 204
Pseudomonas infections, 358, 367
Puberty, 622

R

Rabies, 363t
Radial keratotomy, 605
Radiation therapy, 25
Radioactive wastes, of nuclear energy, 894–895
Radiometric dating, 24
Rebound effect, after fight-flight response, 577

Recycling programs, 887
Red algae, commercial products from, 382, 382i
Red blood cell disorders, 654
Red-green color blindness, 604
Red tides, 381, 381i, 387
Reefs, destruction by human activities, 874–874
Referred pain, 593, 593i
REM sleep, 584
Renal failure, 736
Reproduction, human, 764–773
Resource consumption by different countries, 809, 899
Restoration ecology, 825
RFLP analysis, 262
RH blood typing, 655
Rheumatic fever, 669
Rheumatoid arthritis, 636–637, 685
Rhythm method, of contraception, 784
Rickets, 620
Ringworm, 396t
Rock concerts, and hearing, 597
Rocky Mountain spotted fever, 437
'Roid rage, 645
Roseola, 363t
Roundworm infections, 428–429
RU-486, 785

S

Salination, of cropland, 892
Salmon, and DDT, 851
Salmonella infection, 367, 369
Sarcoma, 681
SCID, 685, 685i
Scrapie, 363, 367
Seasonal affective disorder (SAD), 625
Secondhand smoke, 781
Sequoia groves, set fires in, 825
Seratonin, and sleep, 584, 585
Severe combined immunodeficiencies, 685, 685i
Sex determination, 267
  human, 196
Sexual intercourse (coitus), 555
"Sexual nose" (vomeronasal organ), 594
Sexually transmitted diseases (STDs), 604, 786–787
Sickle-cell anemia, 203t, 654
  and malaria, 290–291, 291i
  molecular basis of, 228–229, 236
  and parvovirus infection, 363t
  pleiotropic effects in, 183

SIDS, 701
Skin cancer, 224, 224i
  and Granstein cells, 631
  and Langerhans cells, 631
Skin color, 629i, 630
Smallpox, 363t, 670–671
Smell, sense of, 594
Smog
  industrial, 884
  photochemical, 884, 884i
Smoking
  effects of, 702–703, 703t
  "responsible smoking," 709
  and sudden-infant death syndrome, 701
Smut, of grain crops, 396t
Snake bites, 459
Softwood, commercial use of, 497i
Solar-hydrogen energy, 896
Solid wastes, 887
Species introductions
  to Australia, 827–828
  to Hawaii, 823
  to Lake Victoria, 828
  to mainland United States, 827, 827t, 829, 829i
"Speed" (amphetamine), 584
Spelunking, and histoplasmosis, 396
Spermicidal foam and jelly, 785
Spider bites, 437i
Split ends (of hair), 630
Sprain (of ligament or tendon), 636
Squamous cell carcinoma, 224, 224i
Staphylococcus infection, 11
Starvation, and weight gain, 710
Steger's polar crossings, 626–627
Stephen Hawking, and ALS, 207
Stimulants, 584
Strain (of ligament or tendon), 636
Streptococcus infection, 367, 736
Streptomycin, 11
Stress electrocardiogram, 665
Stress response, and cortisol, 618
Stroke (brain damage), 664
Sucrose polyester, 723
Sudden infant death syndrome (SIDS), 701
"Superbugs," 11
"Supermice," 266, 266i
Sweating, 741

T

Tanning, of skin, 631
Tapeworm infection, 428, 429i

Taste, sense of, 594
Taxol, and tumors, 70
Tay-Sachs disorder, 203t, 204
TCE contamination 21, 21i
Tea plants, 490
Teratogen sensitivity, of embryo 780, 780i
Test-tube baby, 788
Testicular feminization syndrome, 203t, 612
Tetanus, 568, 569
Tetany, 33
Thalassemia, 654
Thalidomide, 780
Therapeutic vaccines, 681
Thirst behavior, 557
Thirst, 735
Thyroid gland abnormalities, 25, 619
Tick bites, 437
Titanic, sinking of, 743
Tobacco
  early cultivation of, 491
  and smoking-related deaths, 491
Tongue-rolling trait, 190
Tornado, encounter with, 558–559
Toxoplasmosis, 376
Tracers, as diagnostic tools, 25
Trachoma, 604
Transfusion, blood, 182, 654
Tree farms, 411i
Trichomonad infection, 376
Triticale (hybrid grain), 490i
Tubal ligation, 784
Tuberculosis, 11, 287, 295
Turner syndrome, 203t, 209
Turtles, and human predation, 458, 706
Typhus, 851

U

Ultraviolet radiation, and skin cancer, 224, 224i, 631
Uremic toxicity, 736

V

Vaccination, 671
Vaccines, 377
Valdez oil spill, 894
Valium, 564
Valley fever, 396t
Vancomycin, 287
Varicose veins, 669
Vasectomy, 784

Vegetarian diets, 723, 723i
Ventricular fibrillation, 665
Vertigo, 607
Viroids, 363
Visceral pain, 592
Vitamin C, 53, 725, 725t
Vitamin D, 45, 631, 725t
  deficiency, 620
Vitamin E, 725
Vitamin K, 358, 725
Vitamins, deficiencies and excesses, 725, 725t
Vomiting, 579

W

Wastewater treatment, 21, 893, 893i
Water hyacinth, 827t, 853, 853i
Water pollution
  and agriculture, 892–893
  correlated with human population sizes, 892–893
  and heavy irrigation, 20–21, 892
  of lakes, 871
  by silt, 513, 513i
  of streams, 871
Water wars, 893
Webbed fingers, 753
Weight loss, 710, 726, 743
Weightlifting, 644
Wetlands, conversion of, 880–881
Wheat, genetic origin of, 304, 305i
White blood cell disorders, 654
Whooping cough, 366
Wind energy, 896, 896i
Winter blues, 622, 622i, 625
Withdrawal, as contraceptive method, 784

X

Xeroderma pigmentosum, 224
XYY condition, 203t, 209

Y

"Yeast" (*Candida albicans*) infection, 393, 393i, 686
Yellow fever, 363t

Z

Zebra mussels, 827
Zone diet, 722